ORDINARY AND PARTIAL DIFFERENTIAL EQUATIONS
Theory and Applications

Second Edition

Nita H. Shah
Professor
Department of Mathematics
Gujarat University
Ahmedabad, Gujarat

PHI Learning Private Limited
Delhi-110092
2015

₹ 495.00

ORDINARY AND PARTIAL DIFFERENTIAL EQUATIONS—Theory and Applications
Second Edition
Nita H. Shah

ISBN-978-81-203-5087-8

The export rights of this book are vested solely with the publisher.

Second Printing (Second Edition) ... ... February, 2015

Published by Asoke K. Ghosh, PHI Learning Private Limited, Rimjhim House, 111, Patparganj Industrial Estate, Delhi-110092 and Printed by Mohan Makhijani at Rekha Printers Private Limited, New Delhi-110020.

To the memory of
my sweet mother Damyanti H. Shah
who was the source of light, energy and inspiration to me

Contents

Preface

The Second Edition of the book continues to provide systematic solution techniques for solving ordinary equations, partial differential equations and special functions. An emphasis is given on the core techniques as well as qualitative aspects of the subject. During my academic life of twenty-four years, I observed that students learn through mathematical modelling, solutions and its real life interpretations. Mathematical software plays a vital role in explaining the concepts. The edited chapters are explained with Maple software.

This book is divided into nineteen chapters. In Chapter 1, the concept of differential equations is introduced. Chapter 2 deals with discussion of different methods to solve ordinary differential equations. In Chapter 3, solution procedure for ordinary differential equation of first order and higher degree is explored. Chapters 4 and 5 discuss solution methodology for linear differential equation with constant and variable coefficients, respectively. The exact differential equations are discussed in Chapter 6. Chapter 7 explains solution methodology for linear differential equation of second order. The solution of simultaneous differential equations is discussed in Chapter 8. Chapter 9 is about total differential equations. Chapters 10–12 deal with partial differential equations. The series solution of second order linear differential equations is discussed in Chapter 13. Chapters 14 and 15 deal with Bessel's and Legendre's equations and properties of $J_n(x)$ and $P_n(x)$. In Chapter 16, Laplace transforms and inverse Laplace transforms are discussed. The methods are discussed to solve differential equation using Laplace and inverse Laplace transforms. In Chapter 17, partial differential equations are revisited to solve Laplace equations, wave equations and diffusion equations. In Chapter 18, methods are discussed to solve partial differential equations using Fourier transforms. In Chapter 19, discrete transforms are defined which has wide application in IT industry and communication systems. Each chapter is enriched with solved examples and exercises for the students.

This book is designed for the undergraduate and postgraduate students of Mathematics (B.A., B.Sc. and M.A., M.Sc.), and for the undergraduate students of all branches of Engineering (B.E., B.Tech) and related diploma courses, as part of their course in Engineering Mathematics. For researchers, this book will serve as a ready reference to get geometrical interpretations of the solution of their research problems.

I am sure that the students and the academic community will find this edited book very useful and easy to understand. Any suggestion for the improvement of the book will be highly appreciated and greatly acknowledged.

This book is inspired by my father Shri H.K. Shah and family members without their cooperation the dream of my sweet mother (late) Damyanti H. Shah would have never been possible. I wish to thank my publishers, PHI Learning, for the careful processing of the manuscript, both at the editorial and production level.

Nita H. Shah

CHAPTER 1

Introduction to Differential Equations

1.1 INTRODUCTION

An equation containing an independent variable, dependent variable and differential coefficients of dependent variable with respect to independent variable is called a **differential equation**. For example

$$\frac{dy}{dx} = xy$$

$$\frac{d^2y}{dx^2} = ax$$

$$\frac{d^2y}{dx^2} = \tan x + \sin x$$

$$\frac{dy}{dx} + 4xy = x^3$$

$$\frac{d^2y}{dx^2} - 5\frac{dy}{dx} + 6y = e^x$$

To solve a differential equation, we need to determine an unknown function which satisfies the differential equation. For example, in the equation

$$y'' - 4y - 4y = 3$$

y is regarded as the unknown function.

A function is said to satisfy the differential equation if y is replaced by this function, the differential equation reduces to an identity.

Any function (defined explicitly or implicitly) free of derivatives, satisfying a differential equation is said to be a ***solution*** of the differential equation.
For example

$$y = Ax + \frac{B}{x}$$

is a solution of the differential equation

$$x^2\frac{d^2y}{dx^2}+x\frac{dy}{dx}-y=0$$

because

$$\frac{dy}{dx}=A-\frac{B}{x^2}$$

and

$$\frac{d^2y}{dx^2}=\frac{2B}{x^3}$$

Substituting the values of y, $\frac{dy}{dx}$ and $\frac{d^2y}{dx^2}$ in the LHS of differential equation gives

$$x^2\frac{d^2y}{dx^2}+x\frac{dy}{dx}-y=0$$

Hence, the function $y=Ax+\frac{B}{x}$ is a solution of the given differential equation.

Thus, the solution of a differential equation is a relation between the variables involved which satisfy the differential equation.

A differential equation involving derivatives with respect to a single independent variable is called an **ordinary differential equation**, whereas a differential equation involving partial derivatives with respect to two or more independent variables is called a **partial differential equation**. The examples stated above are ordinary differential equations.

Now,

$$\frac{\partial^2 z}{\partial x^2}+\frac{\partial^2 z}{\partial y^2}+\frac{\partial^2 z}{\partial t^2}=0$$

$$\frac{\partial^2 z}{\partial t^2}=\frac{1}{c^2}\frac{\partial^2 z}{\partial x^2}$$

are examples of partial differential equations.

By *order* of a differential equation, we mean the highest order derivative appearing in the equation. The *degree* of a differential equation is the degree of the highest order derivative when differential coefficients are made free from radicals and fractions. Equivalently, the degree of the differential equation is the power of the highest order derivative occurring in a differential equation when it is expressed in a polynomial in differential coefficients (see Table 1.1). For example

A differential equation is *linear differential equation* if it is expressible in the form

$$A_0\frac{d^ny}{dx^n}+A_1\frac{d^{n-1}y}{dx^{n-1}}+\cdots+A_{n-1}\frac{dy}{dx}+A_ny=B$$

where A_0, A_1, ..., A_n and B are either constants or functions of independent variable x.

Table 1.1

Differential Equation	*Order*	*Degree*
$\left(\frac{dy}{dx}\right)^2 - y\frac{dy}{dx} + \sin x = 0$	1	2
$\frac{d^3y}{dx^3} + \frac{d^2y}{dx^2} + \left(\frac{dy}{dx}\right)^2 = e^x$	3	1
$\left(\frac{dy}{dx}\right)^2 - \frac{1}{dy/dx} = 4$	1	3
$\sqrt{1 + \left(\frac{dy}{dx}\right)^2} = \left(c\frac{d^2y}{dx^2}\right)^{1/3}$	2	2
$x^2\left(\frac{d^2y}{dx^2}\right)^3 + y\left(\frac{dy}{dx}\right)^4 + y^4 = 0$	2	3
$5\frac{d^2y}{dx^2} = \left[1 + \left(\frac{dy}{dx}\right)^2\right]^{3/2}$	2	2
$y = x\frac{dy}{dx} + a\sqrt{1 + \left(\frac{dy}{dx}\right)^2}$	1	2

A differential equation will be non-linear differential equation if

(i) Its degree is more than one.
(ii) Any of the differential coefficient has exponent more than one.
(iii) Exponent of the dependent variable is more than one.
(iv) Product containing dependent variable and its differential coefficients are present.

For example

$$\left(\frac{d^3y}{dx^3}\right)^4 - 6\frac{d^2y}{dx^2} - 4y = 0$$

$$\frac{d^2y}{dx^2} + 2\left(\frac{dy}{dx}\right)^2 + 16y = \sin x$$

$$\frac{dy}{dx} = \frac{x^2 + y^2}{2xy}$$

represents non-linear differential equation.

1.2 APPLICATIONS OF DIFFERENTIAL EQUATIONS

Differential equations are major tools for physical problems arising in applied mathematics. A sampling of examples resulting in differential equations is as follows:

1.2.1 Growth Problem

This problem occurs in various fields like economic growth, book interest, growth of bacteria, decay of radioactive elements in chemistry, etc.

Let $x(t)$ be the quantity of the entity under consideration at any time t is proportional to $x(t)$. This rate of change of quantity can be expressed as differential equation.

$$\frac{dx(t)}{dt} = kx(t) \tag{1.1}$$

where k is a constant of proportionality which is positive if the problem is about growth and is negative if it is about decay.

When we come across physical problems where growth and decay occur simultaneously, the formulation is as given below:

Let $x(t)$ be population at time for only decay (in terms of deaths), the differential equation is

$$\frac{dx(t)}{dt} = -nx(t) \tag{1.2}$$

where n is positive.

When growth and decay are considered simultaneously, similar to Eq. (1.1) and Eq. (1.2), we have the equation

$$\frac{dx(t)}{dt} = (k-n)x(t), \qquad x(t_0) = x_0 \tag{1.3}$$

where x_0 is initial entity at the start of time t_0. At $k = n$, we have $x(t) = x_0$ which suggests that $x(t)$ is constant.

1.2.2 Flow Problem

Flow problem or mixing problem can be described as first order differential equation. Consider the classic problem of mixing in a tank which has two different concentrations of liquid flowing into it. One pipe carries a liquid with a concentration c_1 grams per litre at a volume rate of v_1 litres per second. The second pipe carries a liquid of concentration c_2 at a rate of v_2. The objective is to find the concentration in the tank at any instant of time t, assuming that the mixture is withdrawn at a rate that keeps the volume of the tank constant.

Let $x(t)$ be the quantity of entity. Using indisputable conservation law, we have

$$\frac{dx}{dt} = \text{Rate of inflow} - \text{Rate of outflow}$$

The rate of inflow is $c_1v_1 + c_2v_2$ and that of outflow is $v_1 + v_2$ in order to keep the tank at a constant level. If the volume of the tank is v litres, the concentration of liquid coming out is $\frac{x(t)}{v}$, resulting into the differential equation

$$\frac{dx}{dt} = c_1v_1 + c_2v_2 - \frac{v_1 + v_2}{v}x(t)$$

1.2.3 Electrical Models

The differential equation arises while solving voltage-current problems for electrical circuits. For capacitors and inductors, the rate of change of current at any instant of time can be described by the differential equations

$$L\frac{di(t)}{dt} = V(t) \quad \text{and} \quad C\frac{dV(t)}{dt} = i(t)$$

respectively. The L and C are constants for the device under consideration, called inductance and capacitance respectively.

Let us consider RC-circuit comprising of resistor and capacitor as given in Figure 1.1.

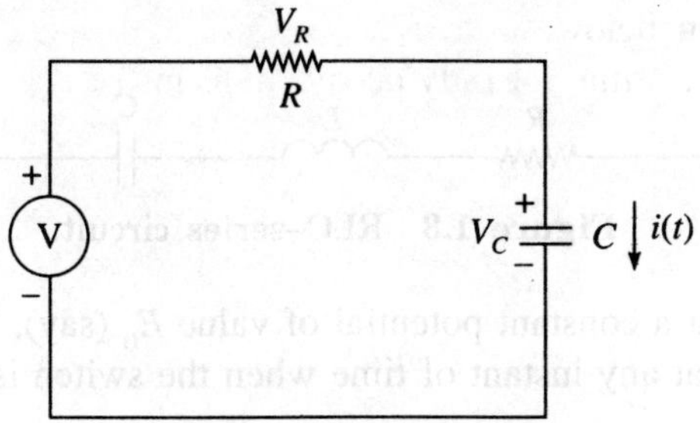

Figure 1.1 RC-circuit.

It is assumed that the current flowing through the capacitor is same as that flowing through the resistors. So,

$$C\frac{dV_C(t)}{dt} = \frac{V(t) - V_C(t)}{R}$$

Let us consider RLC-circuit given in Figure 1.2.

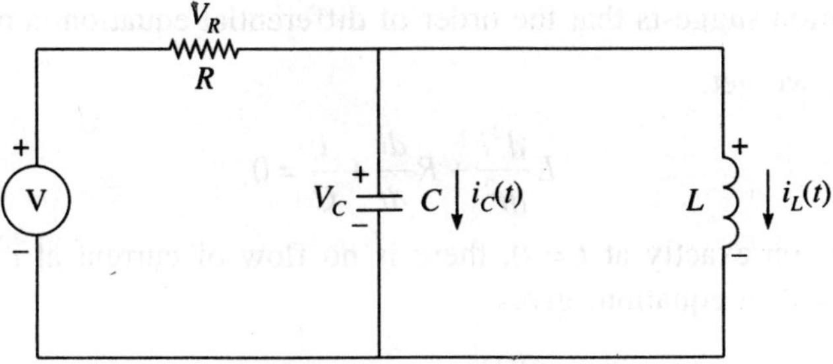

Figure 1.2 RLC-circuit.

In RLC–circuit, the capacitor voltage $V_C(t)$ appears directly across the inductor, and the current flowing into the capacitor is whatever flows through the resistor diminished by the flow down through the inductor. The resulting system is

$$L\frac{di_L(t)}{dt} = V_C(t)$$

$$C\frac{dV_C(t)}{dt} = \frac{V(t) - V_C(t)}{R} - i_L(t)$$

with $i_L(0)$ and $V_C(0)$ known constants.

Let us study an electric circuit which contains in series a capacitor, an inductor and a resistor along with a voltage source as given in Figure 1.3.

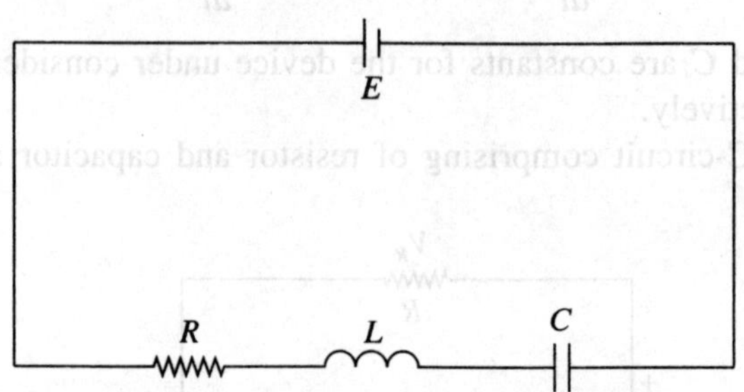

Figure 1.3 RLC–series circuit.

Let E be maintained at a constant potential of value E_0 (say). We want to find the current $i(t)$ passing in the system at any instant of time when the switch is 'on'. We have studied that at any time t, the voltage across the inductor is $L\frac{di}{dt}$, across the resistor is iR and across the capacitor is $\int_0^t \frac{i}{C} ds$, under assumption that switch is 'on' at time $t = 0$. Then, by Kirchhoff's voltage law,

$$L\frac{di(t)}{dt} + Ri + \int_0^t \frac{i}{C} ds = E_0$$

$\int$ sign in the equation suggests that the order of differential equation is not one. Differentiate both sides, w.r.t. t, we get,

$$L\frac{d^2i}{dt^2} + R\frac{di}{dt} + \frac{i}{C} = 0$$

Since, the switch is on exactly at $t = 0$, there is no flow of current at $t = 0$, i.e., $i(0) = 0$.

Substituting $t = 0$ in equation, gives

$$L\left.\frac{di(t)}{dt}\right|_{t=0} + Ri(0) = E_0$$

Since $i(0) = 0$, $Li'(0) = E_0$, the current $i(t)$ is determined by the differential equation

$$L\frac{d^2i}{dt^2} + R\frac{di}{dt} + \frac{i}{C} = 0 \text{ with } i(0) = 0 \text{ and } Li'(0) = E_0$$

Note that the differential equation involves the derivative of second order and two conditions are given to determine a solution. The two conditions are called *initial conditions* and problem is called *initial value problem.*

1.2.4 Pendulum Problem

Let us formulate an equation of motion for a simple pendulum. The friction due to air is neglected. It uses law of conservation of energy. Let a pendulum bob be suspended from a point O and be at rest. Let OA be the vertical position of the pendulum. At time t, P denotes a position of the bob and let OP makes an angle x with OA (see Figure 1.4).

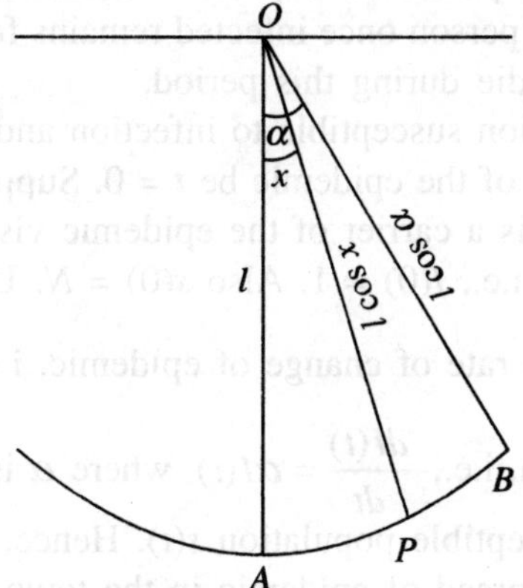

Figure 1.4 Pendulum.

The maximum displacement of the pendulum is denoted by B and the α; the angle that OB makes with OA. Here, α is the maximum displacement. The work done to change the position α is the work required to raise the pendulum bob through a vertical distance.

$$l \cos x - l \cos \alpha$$

where l denotes the length of pendulum.

Since α denotes the angle for maximum displacement, the velocity v, with which the pendulum is swinging, is zero at $x = \alpha$. Using law of conservation of energy;

$$\frac{1}{2}mv^2 = \frac{1}{2}ml^2\left(\frac{dx}{dt}\right)^2 = mg\,(\cos x - \cos \alpha)$$

where m is the mass of the bob of the simple pendulum and g is gravitational force.

Since l and m are never zero, $\frac{1}{2}l\left(\frac{dx}{dt}\right)^2 = g(\cos x - \cos \alpha)$ on differentiating with respect to t, we get

$$l\frac{dx}{dt}\frac{d^2x}{dt^2} = -g \sin x\frac{dx}{dt}.$$

For a swinging pendulum, $\frac{dx}{dt} \neq 0$, hence the equation of motion for pendulum is governed by $\frac{d^2x}{dt^2} + \frac{g}{l}\sin x = 0$.

1.2.5 Spread of Epidemics

Diseases like malaria, common–cold, cholera, tuberculosis, etc. spread by infection. Suppose that the susceptible population of Ahmedabad is N. One person gets infected due to in contact with another susceptible infected person. This process continues to cover the entire susceptible population. Let us assume that a person once infected remains for all the time during the spread of epidemic and none of them die during this period.

Let $s(t)$ denotes the population susceptible to infection and $I(t)$ be the infected population at a time t. Let the starting time of the epidemic be $t = 0$. Suppose one person from out of the population of Ahmedabad who is a carrier of the epidemic visits Delhi, i.e., at $t = 0$, there is one individual who is infected, i.e., $I(0) = 1$. Also $s(0) = N$. Under this assumption, we have relation $s(t) + I(t) = N + 1$. The rate of change of epidemic, i.e., $\frac{dI(t)}{dt}$ is proportional to the infected persons $I(t)$ in the town, i.e., $\frac{dI(t)}{dt} = \alpha I(t)$ where α is *constant of proportionality*. It depends on the size of the susceptible population $s(t)$. Hence, we assume that $\alpha(s(t)) = ks(t)$ where k is constant. Then the spread of epidemic in the town is governed by the differential equation

$$\frac{dI(t)}{dt} = ks(t)I(t) = kI(t)(N + 1 - I(t))$$

Reader can verify that the solution $I(t)$ of the differential equation using $I(0) = 1$ is given by

$$I(t) = \frac{N + l}{l + Ne^{-k(N+l)}}, \; t \geq 0$$

This is the simplest model of epidemics. More variants of epidemic models are available in the literature.

The solutions of the first order differential equations can be visualized as curves in the plane. For example, consider the differential equation $\frac{dy}{dx} = F(y(x), x)$.

This equation specifies the tangent lines to a family of curves in the plane. One can use Maple DEtools to get pictures which generate the direction field associated with a differential equation of first order. For example, consider the differential equation $\frac{dy}{dx} = yx^2$. The Maple code

```
> With (DEtools):
> dfieldplot(diff(y(x), x) = y(x) · x², y(x), x = –5..5, y = –5..5)
```

will generate plot as given in Figure 1.5.

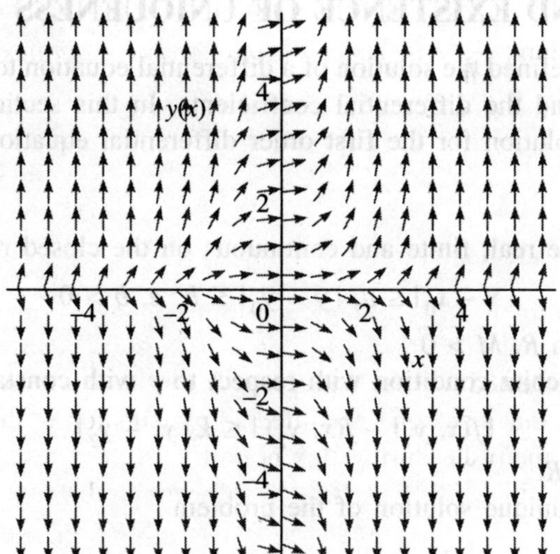

Figure 1.5 Directional field.

It represents the paths of solution trajectories. These arrows resemble motion paths of particles moving with a fluid flow.

EXAMPLE 1.1 Using Maple tools make a plot of the direction field for the first order differential equation

$$\frac{dy}{dx} = y^3 x \exp(-x^2)$$

Solution See Figure 1.6.

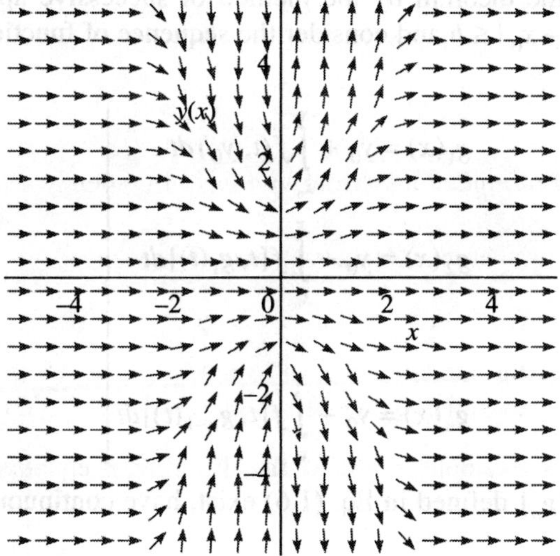

Figure 1.6 Directional field.

It represents the paths of solution trajectories. These arrows resemble motion paths.

1.3 SOLUTION AND EXISTENCE OF UNIQUENESS OF SOLUTION

In Section 1.1, we have defined the solution of a differential equation to be the relation between the variables involved and the differential coefficients. In this section, we try to prove the existence—uniqueness solution for the first order differential equation.

Statement: Let

(i) The function f be real, finite and continuous on the closed rectangle R.

$$|x - x_0| \le a,\ |y - y_0| \le b,\ a, b > 0$$

(ii) $|f(x, y)| \le M$ on R, $M > 0$

(iii) f satisfies a Lipschitz condition with respect to y with constant k in R, i.e.

$$|f(x, y_1) - f(x, y_2)| \le k\,|y_1 - y_2| \tag{1.4}$$

for all (x, y_1), $(x, y_2) \in R$.

Then there exists a unique solution of the problem

$$\frac{dy}{dx} = f(x, y) \text{ with } g(x_0) = y_0 \tag{1.5}$$

on the interval $|x - x_0| \le h$ where $h = \min i\left\{a, \dfrac{b}{M}\right\}$.

i.e. there exists a unique differentiable real function g defined on the interval $|x - x_0| \le h$ which is such that

(i) $\dfrac{dg(x)}{dx} = f[x, g(x)]$ for all x on the given interval

(ii) $g(x_0) = y_0$.

Proof: We will prove the theorem by the method of successive approximations.

Let x be such that $|x - x_0| \le h$ and consider the sequence of functions $g_1, g_2, \ldots, g_n$ defined as follows:

$$\left.\begin{aligned} g_1(x) &= y_0 + \int_{x_0}^{x} f(t, y_0)\,dt \\ g_2(x) &= y_0 + \int_{x_0}^{x} f[t, g_1(t)]\,dt \\ &\vdots \\ g_n(x) &= y_0 + \int_{x_0}^{x} f[t, g_{n-1}(t)]\,dt \end{aligned}\right\} \tag{1.6}$$

Step 1: The functions $\{g_n\}$ defined in Eq. (1.6) exist, have continuous derivatives and satisfy the inequality

$$|g_n(x) - y_0| \le b \tag{1.7}$$

on $|x - x_0| \le h$, and thus, $f[x, g_n(x)]$ is well defined on this interval.

Clearly, g_1 exists and has a continuous derivative on $[x_0, x_0 + h]$.

Also
$$|g_1(x) - y_0| = \left|\int_{x_0}^{x} f(t, y_0)dt\right|$$

$$\le \int_{x_0}^{x} |f(t, y_0)|dt$$
$$\le M|x - x_0|$$
$$\le Mh$$
$$\le b$$

Therefore, $f[x, g_1(x)]$ is defined and continuous on $[x_0, x_0 + h]$.
Assuming g_{n-1} exists, has a continuous derivative and is such that

$$|g_n(x) - y_0| \le b$$

for all x in $[x_0, x_0 + h]$.

Then since $[x, g_{n-1}(x)]$ lies in R, $f[x, g_{n-1}(x)]$ is defined, continuous and satisfies

$$|f[x, g_{n-1}(x)]| \le M \text{ on } [x_0, x_0 + h].$$

Therefore, g_n exists and has continuous derivatives on $[x_0, x_0 + h]$. Also,

$$|g_n(x) - y_0| \le \int_{x_0}^{x} |f[t, g_{n-1}(t)]|dt$$
$$\le M|x - x_0|$$
$$\le b$$

Therefore, $[x, g_n(x)]$ also lies in R, and hence, $f[x, g_n(x)]$ is defined, continuous on $[x_0, x_0 + h]$.

Thus, by mathematical induction, each function g_n in Eq. (1.6) possess the required properties on $[x_0, x_0 + h]$.

Step 2: The functions $\{g_n\}$ are such that

$$|g_n(x) - g_{n-1}(x)| \le \frac{Mk^{n-1}}{n!}h^n \tag{1.8}$$

on $|x - x_0| \le h$.

From Step 1,

$$|g_1(x) - g_0(x)| = |g_1(x) - y_0| \le M|x - x_0|$$

Assume
$$|g_{n-1}(x) - g_{n-2}(x)| \le \frac{Mk^{n-2}}{(n-1)!}(x - x_0)^{n-1} \tag{1.9}$$

on $|x - x_0| \le h$.

Then
$$|g_n(x) - g_{n-1}(x)| = \left|\int_{x_0}^{x} \{f[t, g_{n-1}(t)] - f[t, g_{n-2}(t)]\}dt\right|$$

$$\leq \int_{x_0}^{x} \left| f[t, g_{n-1}(t)] - f[t, g_{n-2}(t)] \right| dt$$

$$\leq k \, |g_{n-1}(t) - g_{n-2}(t)| \qquad \text{[by assumption (iii)]}$$

Using Eq. (1.9)

$$\left| g_n(x) - g_{n-1}(x) \right| \leq \frac{Mk^{n-1}}{(n-1)!} \int_{x_0}^{x} (t - x_0)^{n-1} dt = \frac{Mk^{n-1}}{n!} (x - x_0)^n$$

Therefore, by mathematical induction,

$$\left| g_n(x) - g_{n-1}(x) \right| \leq \frac{Mk^{n-1}}{n!} h^n$$

on $|x - x_0| \leq h$ for all n.

Step 3: As $n \to \infty$, the sequence of functions $\{g_n\}$ converges uniformly to a continuous g on $|x - x_0| \leq h$

Since
$$g_n(x) = y_0 + \sum_{i=1}^{n} [g_i(x) - g_{i-1}(x)]$$

The sequence $\{g_n(x)\}$ is a sequence of partial sums of the infinite series

$$y_0 + \sum_{i=1}^{\infty} [g_i(x) - g_{i-1}(x)] \tag{1.10}$$

Since
$$\sum_{n=1}^{\infty} Mk^{n-1} \frac{h^n}{n!} \to \frac{M}{k} [e^{kh} - 1]$$

By Eq. (1.7) and Weierstrass's M-test, the series given by Eq. (1.10) converges uniformly on $|x - x_0| \leq h$. Hence, the sequence $\{g_n\}$ converges uniformly to g on $|x - x_0| \leq h$. Since each g_n is continuous on $|x - x_0| \leq h$, the limit function g is also continuous on $|x - x_0| \leq h$.

Step 4: The limit function g satisfies the differential equation

$$\frac{dg(x)}{dx} = f(x, y)$$

on $|x - x_0| \leq h$ and is such that $g(x_0) = y_0$.

Since each g_n satisfies Eq. (1.7) on $|x - x_0| \leq h$, $g(x)$ also satisfies (1.7) on $|x - x_0| \leq h$, and hence, $f[x, g(x)]$ is defined on this interval. Therefore, by Lipschitz's condition

$$|f[x, g(x)] - f[x, g_n(x)]| \leq k \, |g(x) - g_n(x)| \tag{1.11}$$

for all x such that $|x - x_0| \leq h$.

Since $\{g_n\}$ is uniformly convergent to g on $|x - x_0| \leq h$, given $\varepsilon > 0$, there exists an integer N depending on ε, alone such that

$$\left| g(x) - g_n(x) \right| < \frac{\varepsilon}{k} \quad \text{for all } n > N$$

and for all x such that $|x - x_0| \leq h$.

From Eq. (1.11)
$|f[x, g(x)] - f[x, g_n(x)]| < \varepsilon$ for all $n > N$ and for all x such that $|x - x_0| \le h$.

Now,
$$\left|\int_{x_0}^{x} \{f[t, g(t)] - f[t, g_n(t)]\}\, dt\right|$$
$$\le \int_{x_0}^{x} |f[t, g(t)] - f[t, g_n(t)]|\, dt$$
$$< \varepsilon\, |x - x_0|$$
$$< \varepsilon\, h$$

for all $n > N$ and on $|x - x_0| \le h$. Therefore,

$$\int_{x_0}^{x} f[t, g(t)]\, dt = \lim_{n\to\infty} \int_{x_0}^{x} f[t, g_n(t)]\, dt$$

Therefore, since

$$g_{n+1}(x) = y_0 + \int_{x_0}^{x} f[t, g_n(t)]dt$$

Taking limit

$$g(x) = y_0 + \int_{x_0}^{x} f[t, g(t)]\, dt \tag{1.12}$$

on $|x - x_0| \le h$.
Differentiating Eq. (1.12)

$$\frac{dg(x)}{dx} = f[x, g(x)]$$

on $|x - x_0| \le h$.
Also from Eq. (1.12),

$$g(x_0) = y_0.$$

Step 5: The function 'g' is the only differentiable on $|x - x_0| \le h$ which satisfies the differential equation

$$\frac{dy}{dx} = f(x, y)$$

with $g(x_0) = y_0$.

If possible let u be another differentiable function defined on $[x_0, x_0 + h]$ such that

$$\left.\begin{aligned} \frac{du(x)}{dx} &= f[x, u(x)] \\ \text{and} \quad u(x_0) &= y_0 \end{aligned}\right\} \tag{1.13}$$

Then
$$|u(x) - y_0| < b \tag{1.14}$$
on
$$(x_0, x_0 + h)$$

Let x_1 be such that $x_0 \le x \le x_1$ and $| u(x_1) - y_0 | = b$.
Let if possible $x_1 < x_0 + h$.

Then $$M_1 = \left|\frac{u(x_1) - y_0}{x_1 - x_0}\right| = \frac{b}{|x_1 - x_0|} > \frac{b}{h} \ge M$$

By Lagrange's mean-value theorem, there exists η with $x_0 < \eta < x_1$ such that

$$M_1 = |u'(\eta)| = |f[\eta, u(\eta)]| \le M$$

Therefore, $x_1 \nless x_0 + h$, and hence, $x_1 \ge x_0 + h$.
Therefore, Eq. (1.14) holds for $x_0 \le x < x_0 + h$, and hence,

$$| u(x) - y_0 | < b \tag{1.15}$$

on $[x_0, x_0 + h]$.

Integrating differential equation in Eq. (1.13)

$$u(x) = \int_{x_0}^{x} f[t, u(t)]\, dt + \text{constant}$$

Since, $$u(x_0) = y_0,\ \text{constant} = y_0.$$

$\therefore$ $$u(x) = y_0 + \int_{x_0}^{x} f[t, u(t)]\, dt \tag{1.16}$$

on $[x_0, x_0 + h]$.

$\therefore$ $$|u(x) - g_1(x)| = \left|\int_{x_0}^{x} \{f[t, u(t)] - f[t, y_0]\}\, dt\right|$$

$$\le \int_{x_0}^{x} \left|f[t, u(t)] - f[t, y_0]\right| dt$$

$$\le k \int_{x_0}^{x} \left|u(t) - y_0\right| dt$$

$$\le bk\left|x - x_0\right|$$

Assume $$|u(x) - g_{n-1}(x)| \le \frac{k^{n-1} b}{(n-1)!}(x - x_0)^{n-1}$$

$$|u(x) - g_n(x)| \le \frac{b(kh)^n}{n!} \tag{1.17}$$

on $[x_0, x_0 + h]$ and for all n.

Now, the series $\sum_{n=1}^{\infty} \frac{b(kh)^n}{n!}$ converges and $\to 0$ as $n \to \infty$.

Therefore, from Eq. (1.17)

$$u(x) = \lim_{n\to\infty} g_n(x) = g(x) \text{ on } [x_0, x_0 + h].$$

Thus, the solution 'g' of the initial value problem is unique on $[x_0, x_0 + h]$.

1.4 GENERAL AND PARTICULAR SOLUTIONS

The solution which contains a number of arbitrary constants equal to the order of the differential equation is called the **general or the complete integral**. The solutions obtained there from, by giving particular values to the constants, are called **particular solution**.

For example, the general solution of $\dfrac{d^2y}{dx^2} + y = 0$ is

$$y = A \sin x + B \cos x$$

because order of the differential equation is 2 and solution contains two arbitrary constants.

The general solution can have various forms, but there will be a relation between the arbitrary constants of one form and those of the other.

1.5 GEOMETRICAL INTERPRETATION OF A DIFFERENTIAL EQUATION OF THE FIRST ORDER AND FIRST DEGREE

Consider, the first order first degree differential equation

$$f\left(x, y, \frac{dy}{dx}\right) = 0 \tag{1.18}$$

We know that when the equation of a curve is given in Cartesian coordinates the tangent of its direction at any point is $\dfrac{dy}{dx}$. For a point (x_1, y_1), there will be a corresponding value of $\dfrac{dy}{dx}$, say m_1, determined by Eq. (1.18). When a point that moves on the curve, on passing through (x_1, y_1), must go in the direction m_1. Suppose it moves from (x_1, y_1) in the direction m_1 for an infinitesimal distance, to a point (x_2, y_2); then, that it moves from (x_2, y_2) in the direction of m_2, the particular direction associated with (x_2, y_2) by the equation, for an infinitesimal distance to a point (x_3, y_3), and so on through successive points. Thus, the point will describe a curve, the coordinates of every point of which, and the direction of the tangent at every point will satisfy the differential equation. The equation of each curve is thus a particular solution of the differential equation; the equation of the system of such curves is the general solution, taken together, gives the locus of the differential equation.

EXAMPLE 1.2 The equation $\dfrac{dy}{dx} = -\dfrac{x}{y}$ indicates that a point moves perpendicular to a line joining it to the origin. It describes a circle with origin at centre.

Solution We have

$$xdx + ydy = 0$$

The general solution is

$$x^2 + y^2 = c^2$$

The circle passing through a point (5, 12) is

$$x^2 + y^2 = 169$$

which is a particular solution.

EXAMPLE 1.3 The equation $\dfrac{dy}{dx} = m$ has the solution $y = mx + c$ which represents locus of all straight lines, infinite in number, with slope m.

Solution Try yourself.

1.6 GEOMETRICAL MEANING OF DIFFERENTIAL EQUATION OF A DEGREE OR AN ORDER HIGHER THAN THE FIRST

If

$$f\left(x, y, \frac{dy}{dx}\right) = 0$$

is of the second degree in $\dfrac{dy}{dx}$, there will be two values of $\dfrac{dy}{dx}$ to each particular point (x_1, y_1). Therefore, the moving point can pass through each point of the plane in either of two directions. Hence, two curves of the system is the locus of the general solution passing through each point.

The general solution

$$\phi(x, y, c) = 0$$

must have two different values of c for each point, and hence, c must appear in the solution in the second degree. It is said that the general solution is a locus of single infinity of curves.

The general solution of a differential equation of the second order

$$f\left(x, y, \frac{dy}{dx}, \frac{d^2y}{dx^2}\right) = 0$$

contains two arbitrary constants, and will, therefore, have for its locus a double infinity of curves.

1.7 FORMATIONS OF DIFFERENTIAL EQUATIONS

EXAMPLE 1.4 Form the differential equation of the family of curves represented by

$$c(y + c)^2 = x^3.$$

Solution Given

$$c(y + c)^2 = x^3 \tag{i}$$

Since only one constant is involved, we shall get a differential equation of first order.

Differentiating Eq. (i) w.r.t. x, we get

$$y + c = \frac{2x}{3}\frac{dy}{dx}$$

$$\Rightarrow \qquad c = \frac{2x}{3}\frac{dy}{dx} - y$$

Substituting this value of c in Eq. (i), we get

$$\left(\frac{2x}{3}\frac{dy}{dx} - y\right)\left(\frac{2x}{3}\frac{dy}{dx}\right)^2 = x^3$$

$$\Rightarrow \qquad 8x\left(\frac{dy}{dx}\right)^3 - 12y\left(\frac{dy}{dx}\right)^2 = 27x$$

is the required differential equation of the family of the curves represented by Eq. (i).

EXAMPLE 1.5 **Form the differential equation representing the family of the curves $y = A \sin (x + B)$, where A and B are constants.**

Solution Given

$$y = A \sin (x + B) \qquad \text{(i)}$$

Since, the given equation contains two arbitrary constants, we shall differentiate it two times and we shall get a differential equation of second order.

Differentiating Eq. (i) w.r.t. x, we get

$$\frac{dy}{dx} - A \cos (x + B) \qquad \text{(ii)}$$

Differentiating Eq. (ii) w.r.t. x, we get

$$\frac{d^2y}{dx^2} = - A \sin (x + B)$$

$$\Rightarrow \qquad \frac{d^2y}{dx^2} = - y$$

$$\Rightarrow \qquad \frac{d^2y}{dx^2} + y = 0$$

is the required differential equation of the given family of the curves.

EXAMPLE 1.6 Form the differential equation corresponding to $y^2 = a(b^2 - x^2)$ by eliminating a and b.

Solution Given

$$y^2 = a(b^2 - x^2) \qquad \text{(i)}$$

Since, there are two arbitrary constants in the given equation, we shall differentiate it twice to get a differential equation of second order.

Differentiating Eq. (i) w.r.t. x, we get

$$2y\frac{dy}{dx} = -2ax$$

$$\Rightarrow \qquad y\frac{dy}{dx} = -ax \qquad \text{(ii)}$$

Differentiating Eq. (ii) w.r.t. x, we get

$$y\frac{d^2y}{dx^2} + \left(\frac{dy}{dx}\right)^2 = -a \qquad \text{(iii)}$$

From Eqs. (ii) and (iii), we get

$$x\left[y\frac{d^2y}{dx^2} + \left(\frac{dy}{dx}\right)^2\right] = y\frac{dy}{dx}$$

which is the required differential equation.

EXAMPLE 1.7 Find the differential equation of all circles touching the x-axis at the origin.

Solution The equation of the family of circles touching x-axis at the origin is

$$(x - 0)^2 + (y - a)^2 = a^2$$

i.e.

$$x^2 + y^2 - 2ay = 0 \qquad \text{(i)}$$

where a is an arbitrary constant.

Equation (i) contains only one arbitrary constant, we differentiate it once w.r.t. x, so that

$$2x + 2y\frac{dy}{dx} - 2a\frac{dy}{dx} = 0$$

$$\Rightarrow \qquad x + y\frac{dy}{dx} - a\frac{dy}{dx} = 0$$

$$\Rightarrow \qquad a = \frac{x + y(dy/dx)}{dy/dx} \qquad \text{(ii)}$$

Putting the value of 'a' from Eq. (ii) in Eq. (i), we get

$$x^2 + y^2 = \frac{2y[x + y\,(dy/dx)]}{dy/dx}$$

$$\Rightarrow \qquad (x^2 - y^2) = 2xy$$

which is the required differential equation.

EXAMPLE 1.8 Obtain the differential equation of all circles of radius r.

Solution The equation of the family of circles of radius r is

$$(x - h)^2 + (y - k)^2 = r^2 \qquad \text{(i)}$$

where h and k are arbitrary constants.

Since Eq. (i) contains two arbitrary constants, we differentiate it two times w.r.t. x and will get the differential equation of second order.
Differentiating Eq. (i) w.r.t. x, we get

$$2(x-h)+2(y-k)\frac{dy}{dx}=0$$

$$\Rightarrow \qquad (x-h)+(y-k)\frac{dy}{dx}=0 \qquad \text{(ii)}$$

Differentiating Eq. (ii) w.r.t. x, we get

$$1+(y-k)\frac{d^2y}{dx^2}+\left(\frac{dy}{dx}\right)^2=0$$

$$\Rightarrow \qquad y-k=-\frac{1+(dy/dx)^2}{d^2y/dx^2}$$

Putting the value of $(y - k)$ in Eq. (ii), we get

$$x-h=\frac{\left[1+\left(\frac{dy}{dx}\right)^2\right]\frac{dy}{dx}}{\frac{d^2y}{dx^2}}$$

Substituting the values of $(x - h)$ and $(y - k)$ in Eq. (i) and on simplification, we get

$$\left[1+\left(\frac{dy}{dx}\right)^2\right]^3=r^2\left(\frac{d^2y}{dx^2}\right)^2$$

which is the required differential equation.

EXAMPLE 1.9 Show that the differential equation that represents all parabolas having their axis of symmetry coincident with the axis of x is $yy_2 + y_1^2 = 0$.

Solution The equation that represents a family of parabolas having their axis of symmetry coincident with the x-axis is

$$y^2 = 4a(x+h) \qquad \text{(i)}$$

where a and h are constants. So we shall differentiate it twice to obtain a second order differential equation.
Differentiating Eq. (i) w.r.t. x, we get

$$2y\frac{dy}{dx}=4a$$

$$\Rightarrow \qquad y\frac{dy}{dx}=2a \qquad \text{(ii)}$$

Differentiating Eq. (ii) w.r.t. x, we get

$$y\frac{d^2y}{dx^2} + \left(\frac{dy}{dx}\right)^2 = 0$$

$$\Rightarrow \qquad yy_2 + y_1^2 = 0 \qquad \text{(iii)}$$

which is the required differential equation.

EXAMPLE 1.10 Show that the differential equation of which $x^2 - y^2 = c(x^2 + y^2)^2$ is the solution of $(x^3 - 3xy^2)\,dx = (y^3 - 3x^2y)\,dy$.

Solution Given

$$x^2 - y^2 = c(x^2 + y^2)^2 \qquad \text{(i)}$$

Differentiating Eq. (i) w.r.t. x, we get

$$2x - 2y\frac{dy}{dx} = 2c(x^2 + y^2)\left[2x + 2y\frac{dy}{dx}\right]$$

$$\Rightarrow \qquad x - y\frac{dy}{dx} = 2c(x^2 + y^2)\left[x + y\frac{dy}{dx}\right] \qquad \text{(ii)}$$

On substituting the value of c from Eq. (i) into Eq. (ii), we get

$$x - y\frac{dy}{dx} = \frac{2(x^2 - y^2)}{x^2 + y^2}(x^2 + y^2)\left[x + y\frac{dy}{dx}\right]$$

$$\Rightarrow \qquad (x^2 + y^2)\left[x - y\frac{dy}{dx}\right] = 2(x^2 - y^2)\left[x + y\frac{dy}{dx}\right]$$

$$\Rightarrow \qquad 3xy^2 - x^3 = \frac{dy}{dx}(3yx^2 - y^3)$$

$$\Rightarrow \qquad (x^3 - 3xy^2)\,dx = (y^3 - 3x^2y)\,dx$$

which is the required differential equation.

EXAMPLE 1.11 Eliminate the constant 'a' from $\sqrt{1 - x^2} + \sqrt{1 - y^2} = a(x - y)$.

Solution Given

$$\sqrt{1 - x^2} + \sqrt{1 - y^2} = a(x - y) \qquad \text{(i)}$$

Put

$$x = \sin\theta \text{ and } y = \sin\phi.$$

$\therefore$ Equation (i) reduce to

$$\cos\theta + \cos\phi = a(\sin\theta - \sin\phi)$$

$$\Rightarrow \qquad 2\cos\left(\frac{\theta + \phi}{2}\right)\cos\left(\frac{\theta - \phi}{2}\right) = 2a\cos\left(\frac{\theta + \phi}{2}\right)\sin\left(\frac{\theta - \phi}{2}\right)$$

$$\Rightarrow \qquad \cot\left(\frac{\theta - \phi}{2}\right) = a$$

$$\Rightarrow \qquad \theta - \phi = 2\cot^{-1} a$$

$$\Rightarrow \qquad \sin^{-1}x - \sin^{-1}y = 2\cot^{-1} a$$

Differentiating w.r.t. x, we get

$$\frac{1}{\sqrt{1-x^2}} - \frac{1}{\sqrt{1-y^2}}\frac{dy}{dx} = 0.$$

EXERCISES

1. Form the differential equation of the family of the curves represented by $y^2 = (x - c)^3$.

$$\left[\textbf{Ans. } 8\left(\frac{dy}{dx}\right)^3 = 27y\right]$$

2. Find the differential equation of the family of curves $y = Ae^{3x} + Be^{-3x}$, where A and B are arbitrary constants. $\left[\textbf{Ans. } \frac{d^2y}{dx^2} - 9y = 0\right]$

3. Find the differential equation of the family of curves $x = A \cos mt + B \sin mt$, where A and B are arbitrary constants. $\left[\textbf{Ans. } \frac{d^2x}{dt^2} + m^2x = 0\right]$

4. Form the differential equation corresponding to $y^2 = a(b - x^2)$ by eliminating a and b.

$$\left[\textbf{Ans. } y\frac{dy}{dx} = x\left[y\frac{d^2y}{dx^2} + \left(\frac{dy}{dx}\right)^2\right]\right]$$

5. Find the differential equation of all the circles which pass through the origin and whose centre lies on x-axis. $\left[\textbf{Ans. } x^2 - y^2 + 2xy\frac{dy}{dx} = 0\right]$

6. Assume that a rain drop evaporates at a rate proportional to its surface area. Form a differential equation involving the rate of change of the radius of the rain drop.

$$\left[\textbf{Ans. } \frac{dr}{dt} = -k\right]$$

7. Find the differential equation of all the parabolas with latus rectum '$4a$' and whose axes are parallel to x-axis. [**Ans.** $2ay_2 + y_1^3 = 0$]

8. Show that the differential equation of which $y = 2(x^2 - 1) + ce^{-x^2}$ is a solution, is $\frac{dy}{dx} + 2xy = 4x^3$.

9. Find the differential equation of all non-vertical lines in a plane. $\left[\textbf{Ans. } \dfrac{d^2y}{dx^2} = 0\right]$

10. Find the differential equation corresponding to $y = e^x(a \cos 2x + b \sin 2x)$.

$$\left[\textbf{Ans. } \frac{d^2y}{dx^2} - 2\frac{dy}{dx} + 5y = 0\right]$$

11. Find the differential equation of all the hyperbolas with coordinate axes as asymptotes.

$$\left[\textbf{Ans. } 2a\frac{d^2y}{dx^2} - 1 = 0\right]$$

12. Find the differential equation of all the ellipses centred at origin.

$$\left[\textbf{Ans. } xy\frac{d^2y}{dx^2} + x\left(\frac{dy}{dx}\right)^2 - y\frac{dy}{dx} = 0\right]$$

13. Form the differential equation of all conics whose axes coincide with the axes of coordinates. $\left[\textbf{Ans. } xy\dfrac{d^2y}{dx^2} + x\left(\dfrac{dy}{dx}\right)^2 - x\dfrac{dy}{dx} = 0\right]$

14. Find the differential equation of all cardioids $r = a(1 - \cos \theta)$.

[**Ans.** $(1 - \cos \theta)\, dr = r \sin \theta\, d\theta$]

15. Find the differential equation of all parabolas with axis parallel to the y-axis.

$$\left[\textbf{Ans. } \frac{d^3y}{dx^3} = 0\right]$$

16. Find the third order differential equation whose solution is the 3-parameter family of curves defined by $x^2 + y^2 + 2ax + 2by + c = 0$.

$$\left[\textbf{Ans. } \left(1+\left(\frac{dy}{dx}\right)^2\right)\frac{d^3y}{dx^3} - 3\frac{dy}{dx}\left(\frac{d^2y}{dx^2}\right)^2 = 0\right]$$

17. Find the differential equation of all the tangent lines to the parabola $y = x^2$.

$$\left[\textbf{Ans. } 4\left(y - x\frac{dy}{dx}\right) + \left(\frac{dy}{dx}\right)^2 = 0\right]$$

18. Find the differential equation whose solution is $\sqrt{1+x} - a\sqrt{1+y} = 1$.

$$\left[\textbf{Ans. } \frac{dy}{dx}(1 + x + \sqrt{1+x}) = 1 + y\right]$$

19. Find the required differential equation for the families of curves $xy = ae^x + be^{-x}$, where a and b are arbitrary constants. $\left[\textbf{Ans. } x\dfrac{d^2y}{dx^2} + 2\dfrac{dy}{dx} = xy\right]$

20. Eliminate c from $\dfrac{x}{c-1}+\dfrac{y}{c+1}=1$. $\left[\textbf{Ans. } \left(y-x\dfrac{dy}{dx}\right)\left(\dfrac{dy}{dx}+1\right)=2\dfrac{dy}{dx}\right]$

21. Find the differential equation of all hyperbolas with coordinate axes as asymptotes.

$\left[\textbf{Ans. } x\dfrac{dy}{dx}+y=0\right]$

22. Find the differential equation of all conics whose axes coincide with the coordinate axes. $\left[\textbf{Ans. } xy\dfrac{d^2y}{dx^2}+x\left(\dfrac{dy}{dx}\right)^2=x\dfrac{dy}{dx}\right]$

23. Find the differential equation of all parabolas with axes parallel to *y*-axis.

$\left[\textbf{Ans. } \dfrac{d^3y}{dx^3}=0\right]$

24. Find the differential equation of all cardioids $r = a(1 - \cos\theta)$.

$\left[\textbf{Ans. } (1-\cos\theta)\dfrac{dr}{d\theta}=r\sin\theta\right]$

25. Find the differential equation of the family of coplanar lines with a given radius '*r*'.

$\left[\textbf{Ans. } \left(\dfrac{d^2y}{dx^2}\right)^2=\left(1+\left(\dfrac{dy}{dx}\right)^2\right)^3\right]$

CHAPTER 2

Differential Equations of First Order and First Degree

2.1 INTRODUCTION

In Chapter 1, we have discussed a method to find a relation between x, y and the derivatives of y w.r.t. x from a given relation between x, y and constants. Here, we will discuss inverse problem, viz. from a given relation between x, y and the derivatives of y, to find a relation between the variables themselves. In this chapter, we discuss methods to find the solution of first order and first degree differential equation. The general form of the first order and first degree differential equation is

$$f\left(x, y, \frac{dy}{dx}\right) = 0 \quad \text{or} \quad \frac{dy}{dx} = f(x, y) \tag{2.1}$$

2.2 VARIABLE SEPARABLE DIFFERENTIAL EQUATIONS

2.2.1 Differential Equation of the Type $\frac{dy}{dx} = f(x)$

Given $$\frac{dy}{dx} = f(x)$$

Then $$dy = f(x)\, dx$$

Integrating both sides, we obtain

$$\int dy = \int f(x)dx$$

$\Rightarrow$ $$y = \int f(x)dx + c$$

where c is constant, is the solution.

EXAMPLE 2.1 Solve $\dfrac{dy}{dx} = \dfrac{2x}{1+x^2}$.

Solution Given

$$\frac{dy}{dx} = \frac{2x}{1+x^2}$$

$$\Rightarrow \qquad dy = \frac{2x}{1+x^2}\, dx$$

Integrating both sides, we get

$$\int dy = \int \frac{2x}{1+x^2}\, dx$$

$$\Rightarrow \qquad y = \log(1 + x^2) + c$$

which is the required solution.

EXAMPLE 2.2 Solve $(e^x + e^{-x})\dfrac{dy}{dx} = (e^x - e^{-x})$.

Solution Given

$$(e^x + e^{-x})\frac{dy}{dx} = (e^x - e^{-x})$$

$$\Rightarrow \qquad dy = \frac{e^x - e^{-x}}{e^x + e^{-x}}\, dx$$

Integrating both sides, we get

$$\int dy = \int \frac{e^x - e^{-x}}{e^x + e^{-x}}\, dx$$

$$\Rightarrow \qquad \int dy = \int \frac{1}{t}\, dt,$$

where $t = e^x + e^{-x}$

$$\Rightarrow \qquad y = \log|t| + c$$

$$\Rightarrow \qquad y = \log|e^x + e^{-x}| + c$$

which is the required solution.

EXAMPLE 2.3 Solve $(x+2)\dfrac{dy}{dx} = x^2 + 4x + 13$.

Solution Given

$$(x+2)\frac{dy}{dx} = x^2 + 4x + 13$$

$$\Rightarrow \qquad dy = \left(\frac{x^2 + 4x + 13}{x+2}\right) dx$$

$$\Rightarrow \qquad dy = \left(x + 2 + \frac{9}{x+2}\right) dx$$

Integrating both sides, we get

$$\int dy = \int\left(x+2+\frac{9}{x+2}\right)dx$$

$$\Rightarrow \qquad y = \frac{x^2}{2}+2x+9\log|x+2|+c$$

which is the required solution.

EXAMPLE 2.4 Solve $\dfrac{dy}{dx} = \dfrac{1}{\sin^4 x+\cos^4 x}$.

Solution Given

$$\frac{dy}{dx} = \frac{1}{\sin^4 x+\cos^4 x}$$

$$\Rightarrow \qquad dy = \frac{1}{\sin^4 x+\cos^4 x}dx$$

Integrating both sides, we get

$$\int dy = \int\frac{1}{\sin^4 x+\cos^4 x}dx$$

$$\Rightarrow \qquad y = \int\frac{\sec^4 x}{1+\tan^4 x}dx$$

$$\Rightarrow \qquad y = \int\frac{\sec^2 x\sec^2 x}{1+\tan^4 x}dx$$

$$\Rightarrow \qquad y = \int\frac{(1+\tan^2 x)\sec^2 x}{1+\tan^4 x}dx$$

$$= \int\frac{1+t^2}{1+t^4}dt, \qquad \text{where } t = \tan^2 x$$

$$= \int\frac{1+\dfrac{1}{t^2}}{t^2+\dfrac{1}{t^2}}dt,$$

dividing numerator and denominator by t^2.

$$= \int\frac{1+\dfrac{1}{t^2}}{\left(t-\dfrac{1}{t}\right)^2+2}dt$$

$$= \int \frac{du}{u^2 + (\sqrt{2})^2}, \text{ where } u = t - \frac{1}{t}$$

$$= \frac{1}{\sqrt{2}} \tan^{-1} \frac{u}{\sqrt{2}} + c$$

$$= \frac{1}{\sqrt{2}} \tan^{-1} \frac{t - \frac{1}{t}}{\sqrt{2}} + c$$

$$= \frac{1}{\sqrt{2}} \tan^{-1} \left(\frac{\tan x - \cot x}{\sqrt{2}} \right) + c$$

which is the required solution.

EXAMPLE 2.5 Solve $\dfrac{dy}{dx} = \dfrac{3e^{2x} + 3e^{4x}}{e^x + e^{-x}}$.

Solution Given

$$\frac{dy}{dx} = \frac{3e^{2x} + 3e^{4x}}{e^x + e^{-x}}$$

$$= \frac{3e^{2x}(1 + e^{2x})}{e^x + \frac{1}{e^x}}$$

$$= \frac{3e^{3x}(1 + e^{2x})}{e^{2x} + 1}$$

$$= 3e^{3x}$$

$$\Rightarrow \quad \frac{dy}{dx} = 3e^{3x}$$

$$\Rightarrow \quad dy = 3e^{3x}\, dx$$

Integrating both sides, we get

$$\int dy = \int 3e^{3x}\, dx$$

$$\Rightarrow \quad y = e^{3x} + c$$

which is the required solution.

EXAMPLE 2.6 Solve $e^{\left(\frac{dy}{dx}\right)} = x + 1$.

Solution Given

$$e^{\left(\frac{dy}{dx}\right)} = x + 1$$

$$\Rightarrow \qquad \frac{dy}{dx} = \log (x + 1)$$

$$\Rightarrow \qquad dy = \log (x + 1)\, dx$$

Integrating both sides, we get

$$\int dy = \int \log (x + 1)\, dx$$

$$\Rightarrow \qquad y = x \log (x + 1) - x + \log (x + 1) + c$$

which is the required solution.

2.2.2 Differential Equation of the Type $\frac{dy}{dx} = f(y)$

Given
$$\frac{dy}{dx} = f(y)$$

Then
$$dx = \frac{1}{f(y)}\, dy$$

Integrating both sides, we obtain

$$\int dx = \int \frac{1}{f(y)}\, dy$$

$$\Rightarrow \qquad x = \int \frac{1}{f(y)}\, dy + c$$

where c is constant, is the solution.

EXAMPLE 2.7 Solve $\frac{dy}{dx} = \frac{1}{y^3 + \cos y}$.

Solution Given

$$\frac{dy}{dx} = \frac{1}{y^3 + \cos y}$$

$$\Rightarrow \qquad dx = (y^3 + \cos y)\, dy$$

Integrating both sides, we obtain

$$\int dx = \int (y^3 + \cos y)\, dy$$

$$\Rightarrow \qquad x = \frac{y^4}{4} + \sin y + c$$

which is the required solution.

EXAMPLE 2.8 Solve $\frac{dy}{dx} - \frac{1 + y^2}{2y} = 0$.

Solution Given

$$\frac{dy}{dx} - \frac{1+y^2}{2y} = 0$$

$$\Rightarrow \quad \frac{dy}{dx} = \frac{1+y^2}{2y}$$

$$\Rightarrow \quad \frac{dy}{dx} = \frac{2y}{1+y^2}$$

$$\Rightarrow \quad dx = \frac{2y}{1+y^2}\,dy$$

Integrating both sides, we obtain

$$x = \log(1+y^2) + c$$

which is the required solution.

EXAMPLE 2.9 Solve $\dfrac{dy}{dx} = \dfrac{1-\cos 2y}{1+\cos 2y}$.

Solution Given

$$\frac{dy}{dx} = \frac{1-\cos 2y}{1+\cos 2y}$$

$$\Rightarrow \quad \frac{dx}{dy} = \frac{1+\cos 2y}{1-\cos 2y}$$

$$\Rightarrow \quad \frac{dx}{dy} = \frac{\cos^2 y}{\sin^2 y}$$

$$\Rightarrow \quad \frac{dx}{dy} = \cot^2 y$$

$$\Rightarrow \quad dx = \cot^2 y\, dy$$

$$\Rightarrow \quad dx = (\operatorname{cosec}^2 y - 1)\, dy$$

Integrating both sides, we obtain

$$x = -\cot y - y + c$$

$$x + \cot y + y = c$$

which is the required solution.

2.2.3 Differential Equation of the Type $\dfrac{dy}{dx} = f(x)\,g(y)$

Given $\quad \dfrac{dy}{dx} = f(x)\, g(y)$

Then $\quad f(x)\, dx = \dfrac{1}{g(y)}\, dy$

Integrating both sides, we obtain

$$\int f(x)\,dx = \int \frac{1}{g(y)}\,dy$$

where c is constant, is the solution of Eq. (2.1).

EXAMPLE 2.10 Solve $(x+a)\dfrac{dy}{dx} = xy$.

Solution Given

$$(x+a)\frac{dy}{dx} = xy$$

$$\Rightarrow \quad (x + a)\,dy = xy\,dx$$

$$\Rightarrow \quad \frac{dy}{y} = \frac{x}{x+a}\,dx$$

$$\Rightarrow \quad \frac{dy}{y} = \frac{x+a-a}{x+a}\,dx$$

$$\Rightarrow \quad \frac{dy}{y} = \left(1 - \frac{1}{x+a}\right)dx$$

Integrating both sides, we obtain

$$\log|y| = x - \log|x+a| + c$$

which is the required solution.

EXAMPLE 2.11 Solve $\cos x\,(a + \cos y)\,dx - \sin y\,(a + \sin x)\,dy = 0$.

Solution Given

$$\cos x\,(a + \cos y)\,dx - \sin y\,(a + \sin x)\,dy = 0$$

$$\Rightarrow \quad \frac{\cos x}{a+\sin x}\,dx - \frac{\sin y}{a+\cos y}\,dy = 0$$

Integrating both sides, we get

$$\Rightarrow \quad \int \frac{\cos x}{a+\sin x}\,dx - \int \frac{\sin y}{a+\cos y}\,dy = \log c$$

$$\Rightarrow \quad \log(a + \sin x) + \log(a + \cos y) = \log c$$

$$\Rightarrow \quad (a + \sin x)(a + \cos y) = c$$

which is the required solution.

EXAMPLE 2.12 Solve $\sec^2 x \tan y\,dx + \sec^2 y \tan x\,dy = 0$.

Solution Given

$$\sec^2 x \tan y\,dx + \sec^2 y \tan x\,dy = 0$$

$$\Rightarrow \quad \frac{\sec^2 x}{\tan x}\,dx + \frac{\sec^2 y}{\tan y}\,dy = 0$$

Integrating both sides, we get

$$\int \frac{\sec^2 x}{\tan x}\,dx + \int \frac{\sec^2 y}{\tan y}\,dy = \log c$$

$$\Rightarrow \quad \log|\tan x| + \log|\tan y| = \log c$$

$$\Rightarrow \quad \tan x \tan y = c$$

which is the required solution.

EXAMPLE 2.13 Solve $(1 + e^{2x})\,dy + (1 + y^2)\,e^x\,dx = 0$, given that when $x = 0$, $y = 1$.

Solution Given

$$(1 + e^{2x})\,dy + (1 + y^2)\,e^x\,dx = 0$$

$$\Rightarrow \quad \frac{1}{1+y^2}\,dy + \frac{e^x}{1+e^{2x}}\,dx = 0$$

Integrating both sides, we get

$$\int \frac{1}{1+y^2}\,dy + \int \frac{e^x}{1+e^{2x}}\,dx = c$$

$$\Rightarrow \quad \int \frac{1}{1+y^2}\,dy + \int \frac{1}{1+t^2}\,dt = c, \quad \text{where } t = e^x$$

$$\Rightarrow \quad \tan^{-1} y + \tan^{-1} t = c$$

$$\Rightarrow \quad \tan^{-1} y + \tan^{-1} e^x = c \qquad \text{(i)}$$

Given that $y = 1$ when $x = 0$.

$$\Rightarrow \quad \tan^{-1} 1 + \tan^{-1} e^0 = c$$

$$\Rightarrow \quad \frac{\pi}{4} + \frac{\pi}{4} = c$$

$$\Rightarrow \quad c = \frac{\pi}{2}$$

Putting $c = \dfrac{\pi}{2}$ in Eq. (i), we get

$$\tan^{-1} y + \tan^{-1} e^x = \frac{\pi}{2}$$

which is the required solution.

EXAMPLE 2.14 Solve $(1 + y^2)(1 + \log x)\,dx + x\,dy = 0$.

Solution Given

$$(1 + y^2)(1 + \log x)\,dy + x dy = 0$$

$$\Rightarrow \quad \frac{(1 + \log x)}{x}\,dx = -\frac{1}{1+y^2}\,dy$$

Integrating both sides, we get

$$\int \frac{(1+\log x)}{x}\,dx = -\int \frac{1}{1+y^2}\,dy$$

$$\Rightarrow \qquad \int t\,dt = -\int \frac{1}{1+y^2}\,dy, \quad \text{where } 1+\log x = t$$

$$\Rightarrow \qquad \frac{t^2}{2} = -\tan^{-1} y = c$$

$$\Rightarrow \qquad \frac{1}{2}(1+\log x)^2 + \tan^{-1} y = c$$

which is the required solution.

EXAMPLE 2.15 Solve $\dfrac{dy}{dx} = 1 + x + y + xy$.

Solution Given

$$\frac{dy}{dx} = 1 + x + y + xy$$

$$\Rightarrow \qquad \frac{dy}{dx} = (1+x)(1+y)$$

$$\Rightarrow \qquad \frac{1}{(1+y)}\,dy = (1+x)\,dx$$

Integrating both sides, we get

$$\int \frac{1}{(1+y)}\,dy = \int (1+x)\,dx$$

$$\Rightarrow \qquad \log|1+y| = x + \frac{x^2}{2} + c$$

which is the required solution.

EXAMPLE 2.16 Solve $y - x\dfrac{dy}{dx} = a\left(y^2 + \dfrac{dy}{dx}\right)$.

Solution Given

$$y - x\frac{dy}{dx} = a\left(y^2 + \frac{dy}{dx}\right)$$

$$\Rightarrow \qquad y - ay^2 = (a+x)\frac{dy}{dx}$$

$$\Rightarrow \qquad \frac{1}{a+x}\,dx = \frac{1}{y-ay^2}\,dy$$

Integrating both sides, we get

$$\int \frac{1}{a+x}\, dx = \int \frac{1}{y-ay^2}\, dy$$

$$\Rightarrow \quad \int \frac{1}{a+x}\, dx = \int \left(\frac{1}{y}+\frac{a}{1-ay}\right) dy$$

$$\Rightarrow \quad \log|a+x| = \log|y| - \log|1-ay| + \log c$$

$$\Rightarrow \quad \log\left|\frac{(a+x)(1-ay)}{y}\right| = \log c$$

$$\Rightarrow \quad \frac{(a+x)(1-ay)}{y} = c$$

$$\Rightarrow \quad (a+x)(1+ay) = cy$$

which is the required solution.

EXAMPLE 2.17 Solve $(x^2 - yx^2)\, dy + (y^2 + x^2y^2)\, dx = 0$.

Solution Given

$$(x^2 - yx^2)\, dy + (y^2 + x^2y^2)\, dx = 0$$

$$\Rightarrow \quad x^2(y-1)\, dy = y^2(1+x^2)\, dx$$

$$\Rightarrow \quad \frac{y-1}{y^2}\, dy = \frac{x^2+1}{x^2}\, dx$$

$$\Rightarrow \quad \left(\frac{1}{y}-\frac{1}{y^2}\right) dy = \left(1+\frac{1}{x^2}\right) dx$$

Integrating both sides, we get

$$\int \left(\frac{1}{y}-\frac{1}{y^2}\right) dy = \int \left(1+\frac{1}{x^2}\right) dx$$

$$\Rightarrow \quad \log|y| + \frac{1}{y} = -\frac{1}{x} + x + c$$

which is the required solution.

EXAMPLE 2.18 Solve $\sin^3 x \dfrac{dx}{dy} = \sin y$.

Solution Given

$$\sin^3 x \frac{dx}{dy} = \sin y$$

$$\Rightarrow \quad \sin^3 x\, dx = \sin y\, dy$$

Integrating both sides, we get

$$\int \sin^3 x\, dx = \int \sin y\, dy$$

$$\Rightarrow \quad \frac{1}{4}\int (3\sin x - \sin 3x)\, dx = \int \sin y\, dy$$

$$\Rightarrow \quad -\frac{3}{4}\cos x + \frac{1}{12}\cos 3x = -\cos y + c$$

$$\Rightarrow \quad \cos y - \frac{3}{4}\cos x + \frac{1}{12}\cos 3x = c$$

which is the required solution.

EXAMPLE 2.19 Solve $\dfrac{dy}{dx} + \sqrt{\dfrac{1-y^2}{1-x^2}} = 0$.

Solution Given

$$\frac{dy}{dx} + \sqrt{\frac{1-y^2}{1-x^2}} = 0$$

$$\Rightarrow \quad \frac{dy}{dx} = -\sqrt{\frac{1-y^2}{1-x^2}}$$

$$\Rightarrow \quad \frac{dy}{\sqrt{1-y^2}} = -\frac{dx}{\sqrt{1-x^2}}$$

Integrating both sides, we get

$$\int \frac{dy}{\sqrt{1-y^2}} = -\int \frac{dx}{\sqrt{1-x^2}}$$

$$\Rightarrow \quad \sin^{-1} y = -\sin^{-1} x + \sin^{-1} c$$

$$\Rightarrow \quad \sin^{-1} x + \sin^{-1} y = \sin^{-1} c$$

$$\Rightarrow \quad \sin^{-1}[x\sqrt{1-y^2} + y\sqrt{1-x^2}] = \sin^{-1} c$$

$$\Rightarrow \quad x\sqrt{1-y^2} + y\sqrt{1-x^2} = c$$

which is the required solution.

EXAMPLE 2.20 Solve $\dfrac{dy}{dx} = e^{x+y} + x^2 e^y$.

Solution Given

$$\frac{dy}{dx} = e^{x+y} + x^2 e^y$$

$$\Rightarrow \qquad \frac{dy}{dx} = e^y(e^x + x^2)$$

$$\Rightarrow \qquad e^{-y}\, dy = (e^x + x^2)\, dx$$

Integrating both sides, we get

$$\int e^{-y}\, dy = \int (e^x + x^2)\, dx$$

$$\Rightarrow \qquad -e^{-y} = e^x + \frac{x^3}{3} + c$$

which is the required solution.

2.2.4 Equation Reducible to Variable Separable Form

The differential equation of the form $\frac{dy}{dx} = f(ax + by + c)$ can be reduced to variable separable by substitution $ax + by + c = v$.

EXAMPLE 2.21 Solve $\sin^{-1}\left(\frac{dy}{dx}\right) = x + y$.

Solution Given

$$\sin^{-1}\left(\frac{dy}{dx}\right) = x + y$$

$$\Rightarrow \qquad \frac{dy}{dx} = \sin(x + y)$$

Let $\qquad x + y = v$

Then $\qquad 1 + \frac{dy}{dx} = \frac{dv}{dx}$

$$\Rightarrow \qquad \frac{dy}{dx} = \frac{dv}{dx} - 1$$

Putting $x + y = v$ and $\Rightarrow \frac{dy}{dx} = \frac{dv}{dx} - 1$ in the given differential equation, we get

$$\frac{dv}{dx} - 1 = \sin v$$

$$\Rightarrow \qquad \frac{dv}{dx} = 1 + \sin v$$

$$\Rightarrow \qquad dx = \frac{dv}{1 + \sin v}$$

Integrating both sides, we get

$$\int dx = \int \frac{dv}{1+\sin v}$$

$$\Rightarrow \qquad \int dx = \int \frac{1-\sin v}{1-\sin^2 v}\,dv$$

$$\int dx = \int \frac{1-\sin v}{\cos^2 v}\,dv$$

$$\Rightarrow \qquad x = \int (\sec^2 v - \tan v \sec v)\,dv$$

$$\Rightarrow \qquad x = \tan v - \sec v + c$$

$$\Rightarrow \qquad x = \tan (x+y) \sec (x+y) + c$$

which is the required solution.

EXAMPLE 2.22 Solve $\dfrac{dy}{dx} = (9x+y+1)^2$.

Solution Given

$$\frac{dy}{dx} = (9x+y+1)^2$$

Let

$$9x + y + 1 = v$$

$$\Rightarrow \qquad 9 + \frac{dy}{dx} = \frac{dv}{dx}$$

$$\Rightarrow \qquad \frac{dy}{dx} = \frac{dv}{dx} - 9$$

Substituting $9x + y + 1 = v$ and $\dfrac{dy}{dx} = \dfrac{dv}{dx} - 9$ in the given differential equation, we get

$$\frac{dv}{dx} - 9 = v^2$$

$$\Rightarrow \qquad \frac{dv}{dx} = v^2 + 9$$

$$\Rightarrow \qquad dx = \frac{dv}{v^2+9}$$

Integrating both sides, we get

$$\int dx = \int \frac{1}{v^2+9}\,dv$$

$$\Rightarrow \qquad x = \frac{1}{3}\tan^{-1}\frac{v}{3} + c$$

$$\Rightarrow \qquad x = \frac{1}{3}\tan^{-1}\left(\frac{9x+y+1}{3}\right)+c$$

which is the required solution.

EXAMPLE 2.23 Solve $(x + y)^2 \dfrac{dy}{dx} = 4.$

Solution Given

$$\frac{dy}{dx} = \frac{4}{(x+y)^2}$$

Let $\qquad x + y = v$

Then $\qquad 1+\dfrac{dy}{dx} = \dfrac{dv}{dx}$

$$\Rightarrow \qquad \frac{dy}{dx} = \frac{dv}{dx}-1$$

Putting $x + y = v$ and $\Rightarrow \dfrac{dy}{dx}=\dfrac{dv}{dx}-1$ in the given differential equation, we get

$$\frac{dv}{dx}-1 = \frac{4}{v^2}$$

$$\Rightarrow \qquad \frac{dv}{dx} = \frac{4}{v^2}+1$$

$$\Rightarrow \qquad \frac{dv}{dx} = \frac{4+v^2}{v^2}$$

$$\Rightarrow \qquad dx = \frac{v^2}{4+v^2}\,dv$$

$$\Rightarrow \qquad dx = \left(1-\frac{4}{4+v^2}\right)dv$$

Integrating both sides, we get

$$\int dx = \int\left(1-\frac{4}{4+v^2}\right)dv$$

$$\Rightarrow \qquad x = v - 2\tan^{-1}\left(\frac{v}{2}\right)+c$$

$$\Rightarrow \qquad x = x+y-2\tan^{-1}\left(\frac{x+y}{2}\right)+c$$

$$\Rightarrow \qquad y-2\tan^{-1}\left(\frac{x+y}{2}\right) = c$$

which is the required solution.

2.2.5 Special Substitution

Some special substitutions reduce differential equation to variable separable method.

Lagrange

Lagrange studied curve minimizing, for example, length using the calculus of variations. One class of equations studied by Lagrange is of the form $y = x f\left(\frac{dy}{dx}\right) + g\left(\frac{dy}{dx}\right)$.

Let $P = \frac{dy}{dx}$. Differentiate given equation with respect to x. This results into linear differential equation as $\frac{dP}{dx} = \frac{P - f(P)}{x f'(P) + g'(P)}$. Alternating the role of the dependent and independent variables, we get a linear differential equation $\frac{dx}{dP} = \frac{-f'(P)}{f(P) - P} x + \frac{g'(P)}{f(P) - P}$

Riccati

The term Riccati equation is associated with quadratic differential equations arising in optimal control theory. The quadratic differential equation is of the form $\frac{dy}{dx} + a(x)y^2 + b(x)y + c(x) = 0$.

Use substitution $y(x) = \frac{w'(x)}{w(x)} \cdot \frac{1}{a(x)}$. Then quadratic differential equation takes the form

$$\frac{w''(x)}{a(x)w(x)} - \frac{w'(x)^2}{a(x)w(x)^2} - \frac{a'(x)w'(x)}{a(x)^2 w(x)} + \frac{w'(x)^2}{a(x)w(x)^2} + \frac{b'(x)w'(x)}{a(x)w(x)} + c(x) = 0$$

Simplification gives second order equation as

$$w''(x) - \left(\frac{a'(x)}{a(x)} - b'(x)\right) w'(x) + a(x)c(x)w(x) = 0$$

2.3 HOMOGENEOUS DIFFERENTIAL EQUATION

A function $f(x, y)$ is called homogeneous function of degree n if

$$f(\lambda x, \lambda y) = \lambda^n f(x, y)$$

For example,

$$f(x, y) = x^3 + y^3 - 3x^2 y$$

is a homogeneous function of degree 3, because

$$\begin{aligned} f(\lambda x, \lambda y) &= \lambda^3 x^3 + \lambda^3 y^3 - 3\lambda^3 x^2 y \\ &= \lambda^3 (x^3 + y^3 - 3x^2 y) \\ &= \lambda^3 f(x, y) \end{aligned}$$

If a first order first degree differential equation is expressible in the form

$$\frac{dy}{dx} = \frac{f(x,y)}{g(x,y)}$$

where $f(x, y)$ and $g(x, y)$ are homogeneous functions of the same degree, then it is called a **homogeneous differential equation**. Such type of equation can be reduced to variable separable form by the substitution as discusses below:

The given differential equation can be written as

$$\frac{dy}{dx} = \frac{x^n f(y/x)}{x^n g(y/x)} = \frac{f(y/x)}{g(y/x)} = F\left(\frac{y}{x}\right)$$

Let $$y = vx.$$

Then $$\frac{dy}{dx} = v + x\frac{dv}{dx}.$$

Substituting these values in $\frac{dy}{dx} = F\left(\frac{y}{x}\right)$, we get

$$v + x\frac{dv}{dx} = F(v)$$

$\Rightarrow$ $$\frac{dv}{F(v) - v} = \frac{dx}{x}$$

On integration, we get

$$\int \frac{dv}{F(v) - v} = \int \frac{dx}{x} + c$$

where c is an arbitrary constant of integration. After integration, replace v by y/x to get the solution of the given differential equation.

EXAMPLE 2.24 Solve $x^2\, dy + y\,(x + y)\, dx = 0$.

Solution Given

$$x^2\, dy + y\,(x + y)\, dx = 0$$

$\Rightarrow$ $$\frac{dy}{dx} = -\left(\frac{y(x+y)}{x^2}\right) \qquad \text{(i)}$$

Since each of the functions $xy + y^2$ and x^2 is a homogeneous function of degree 2, Eq. (i) is a homogeneous differential equation.

Putting $y = vx$ and $\frac{dy}{dx} = v + x\frac{dv}{dx}$ in Eq. (1), we get

$$v + x\frac{dv}{dx} = -\left(\frac{vx^2 + v^2x^2}{x^2}\right)$$

$$\Rightarrow \qquad v + x\frac{dv}{dx} = -(v + v^2)$$

$$\Rightarrow \qquad x\frac{dv}{dx} = -(2v + v^2)$$

$$\Rightarrow \qquad \frac{dv}{v^2 + 2v} = -\frac{dx}{x}$$

Integrating both sides, we get

$$\int \frac{dv}{v^2 + 2v} = -\int \frac{dx}{x}$$

$$\Rightarrow \qquad \int \frac{dv}{v^2 + 2v + 1 - 1} = -\int \frac{dx}{x}$$

$$\Rightarrow \qquad \int \frac{dv}{(v+1)^2 - 1} = -\int \frac{dx}{x}$$

$$\Rightarrow \qquad \frac{1}{2}\log\left(\frac{v+1-1}{v+1+1}\right) = -\log x + \log c$$

$$\Rightarrow \qquad \frac{1}{2}\log\left(\frac{v}{v+2}\right) = -\log x + \log c$$

$$\Rightarrow \qquad \log\left(\frac{v}{v+2}\right) + 2\log x = 2\log c$$

$$\Rightarrow \qquad \log\left(\frac{v}{v+2}\right) + \log x^2 = 2\log c^2$$

$$\Rightarrow \qquad \log\left(\frac{vx^2}{v+2}\right) = \log c^2$$

$$\Rightarrow \qquad \frac{vx^2}{v+2} = k \qquad \text{(say } c^2 = k\text{)}$$

$$\Rightarrow \qquad \frac{\frac{y}{x}x^2}{\frac{y}{x} + 2} = k \qquad (\because \quad v = y/x)$$

$$\Rightarrow \qquad x^2y = k(y + 2x)$$

which is the required solution.

EXAMPLE 2.25 Solve the differential equation $(x + y)\,dy + (x - y)\,dx = 0$.

Solution Given

$$(x + y)\,dy + (x - y)\,dx = 0$$

$$\Rightarrow \qquad \frac{dy}{dx} = \frac{y-x}{x+y} \tag{i}$$

Since each of the functions $x + y$ and $y - x$ is a homogeneous function of degree 1, Eq. (i) is a homogeneous differential equation.

By placing $y = vx$ and $\frac{dy}{dx} = v + x\frac{dv}{dx}$ in Eq. (i), we get

$$v + x\frac{dv}{dx} = \frac{vx-x}{vx+x}$$

$$\Rightarrow \qquad v + x\frac{dv}{dx} = \frac{v-1}{v+1}$$

$$\Rightarrow \qquad x\frac{dv}{dx} = \frac{v-1}{v+1} - v$$

$$\Rightarrow \qquad x\frac{dv}{dx} = -\left(\frac{v^2+1}{v+1}\right)$$

$$\Rightarrow \qquad \frac{v+1}{v^2+1}\,dv = -\frac{dx}{x}$$

Integrating both sides, we get

$$\int \frac{v+1}{v^2+1}\,dv = -\int \frac{dx}{x}$$

$$\Rightarrow \qquad \int \frac{v}{v^2+1}\,dv + \int \frac{1}{v^2+1}\,dv = -\int \frac{dx}{x}$$

$$\Rightarrow \qquad \frac{1}{2}\int \frac{2v}{v^2+1}\,dv + \int \frac{1}{v^2+1}\,dv = -\int \frac{dx}{x}$$

$$\Rightarrow \qquad \frac{1}{2}\log(v^2+1) + \tan^{-1} v = -\log x + c$$

$$\Rightarrow \qquad \log(v^2 + 1) + 2\tan^{-1} v + 2\log x = 2c$$

$$\Rightarrow \qquad \log(v^2 + 1) + 2\tan^{-1} v + \log x^2 = k \qquad (\text{say } 2c = k)$$

$$\Rightarrow \qquad \log[(v^2 + 1)x^2] + 2\tan^{-1} v = k$$

$$\Rightarrow \qquad \log\left[\left(\frac{y^2}{x^2}+1\right)x^2\right]+2\tan^{-1}\frac{y}{x} = k \qquad (\because \quad v = y/x)$$

$$\Rightarrow \qquad \log(x^2+y^2)+2\tan^{-1}\frac{y}{x} = k$$

which is the required solution.

EXAMPLE 2.26 Solve the differential equation $x^2y\,dx - (x^3 + y^3)\,dy = 0$.

Solution Given

$$x^2y\,dx - (x^3 + y^3)\,dy = 0$$

$$\Rightarrow \qquad \frac{dy}{dx} = \frac{x^2y}{x^3+y^3} \qquad \text{(i)}$$

Since each of the functions x^2y and $x^3 + y^3$ is a homogeneous function of degree 3, Eq. (i) is a homogeneous differential equation.

By substituting $y = vx$ and $\frac{dy}{dx} = v + x\frac{dv}{dx}$ in Eq. (i), we get

$$v + x\frac{dv}{dx} = \frac{vx^3}{x^3+v^3x^3}$$

$$\Rightarrow \qquad v + x\frac{dv}{dx} = \frac{v}{1+v^3}$$

$$\Rightarrow \qquad x\frac{dv}{dx} = -\frac{v^4}{1+v^3}$$

$$\Rightarrow \qquad \frac{1+v^3}{v^4}\,dv = -\frac{dx}{x}$$

$$\Rightarrow \qquad \left(\frac{1}{v^4}+\frac{1}{v}\right)dv = -\frac{dx}{x}$$

Integrating both sides, we get

$$\int\left(\frac{1}{v^4}+\frac{1}{v}\right)dv = -\int\frac{dx}{x}$$

$$\Rightarrow \qquad -\frac{1}{3v^3}+\log v = -\log x + c$$

$$\Rightarrow \qquad -\frac{1}{3v^3}+\log v+\log x = c$$

$$\Rightarrow \quad -\frac{1}{3v^3}+\log(vx) = c$$

$$\Rightarrow \quad -\frac{x^3}{3y^3}+\log y = c \qquad (\because \quad v = y/x)$$

which is the required solution.

EXAMPLE 2.27 Solve the differential equation $(x^2 + xy)\,dy = (x^2 + y^2)\,dx$.

Solution Given

$$(x^2 + xy)\,dy = (x^2 + y^2)\,dx$$

$$\Rightarrow \quad \frac{dy}{dx} = \frac{x^2+y^2}{x^2+xy} \tag{i}$$

Since each of the functions $x^2 + y^2$ and $x^2 + xy$ is a homogeneous function of degree 2, Eq. (i) is a homogeneous differential equation.

Putting $y = vx$ and $\frac{dy}{dx} = v + x\frac{dv}{dx}$ in Eq. (i), we get

$$v + x\frac{dv}{dx} = \frac{x^2+v^2x^2}{x^2+vx^2}$$

$$\Rightarrow \quad v + x\frac{dv}{dx} = \frac{1+v^2}{1+v}$$

$$\Rightarrow \quad x\frac{dv}{dx} = \frac{1-v}{1+v}$$

$$\Rightarrow \quad \frac{1+v}{1-v}\,dv = \frac{dx}{x}$$

$$\Rightarrow \quad \frac{2-(1-v)}{1-v}\,dv = \frac{dx}{x}$$

$$\Rightarrow \quad \left(\frac{2}{1-v}-1\right)dv = \frac{dx}{x}$$

Integrating both sides, we get

$$\int\left(\frac{2}{1-v}-1\right)dv = \int\frac{dx}{x}$$

$$\Rightarrow \quad -2\log(1-v) - v = \log x + \log c$$

$$\Rightarrow \quad \log(1-v)^2 + \log x + \log c = -v$$

$$\Rightarrow \quad \log[cx(1-v)^2] = -v$$

$$\Rightarrow \quad cx(1-v)^2 = e^{-v}$$

$$\Rightarrow \qquad cx\left(1-\frac{y}{x}\right)^2 = e^{-y/x} \qquad (\because \quad v = y/x)$$

$$\Rightarrow \qquad c(x-y)^2 = xe^{-y/x}$$

which is the required solution.

EXAMPLE 2.28 Solve the differential equation $(3xy + y^2)\, dx + (x^2 + xy)\, dy = 0$.

Solution Given

$$(3xy + y^2)\, dx + (x^2 + xy)\, dy = 0$$

$$\Rightarrow \qquad \frac{dy}{dx} = -\left(\frac{3xy+y^2}{x^2+xy}\right) \qquad \text{(i)}$$

Since each of the functions $3xy + y^2$ and $x^2 + xy$ is a homogeneous function of degree 2, Eq. (i) is a homogeneous differential equation.

Putting $y = vx$ and $\dfrac{dy}{dx} = v + x\dfrac{dv}{dx}$ in Eq. (i), we get

$$v + x\frac{dv}{dx} = -\left(\frac{3vx^2+v^2x^2}{x^2+vx^2}\right)$$

$$\Rightarrow \qquad v + x\frac{dv}{dx} = -\left(\frac{3v+v^2}{1+v}\right)$$

$$\Rightarrow \qquad x\frac{dv}{dx} = -\left(\frac{4v+2v^2}{1+v}\right)$$

$$\Rightarrow \qquad \frac{1+v}{2v+v^2}\, dv = -2\frac{dx}{x}$$

$$\Rightarrow \qquad \frac{2+2v}{2v+v^2}\, dv = -4\frac{dx}{x}$$

Integrating both sides, we get

$$\int \frac{2+2v}{2v+v^2}\, dv = -4\int \frac{dx}{x}$$

$$\Rightarrow \qquad \log(v^2 + 2v) = -4\log x + \log c$$

$$\Rightarrow \qquad \log(v^2 + 2v) = \log\left(\frac{c}{x^4}\right)$$

$$\Rightarrow \qquad v^2 + 2v = \frac{c}{x^4}$$

$$\Rightarrow \qquad \frac{y^2}{x^2} + 2\frac{y}{x} = \frac{c}{x^4} \qquad (\because \quad v = y/x)$$

$$\Rightarrow \qquad x^2(y^2 + 2xy) = c$$

which is the required solution.

EXAMPLE 2.29 Solve the differential equation $xdy - ydx = \sqrt{x^2+y^2}\,dx$.

Solution Given

$$xdy - ydx = \sqrt{x^2+y^2}\,dx$$

$$\Rightarrow \qquad \frac{dy}{dx} = \frac{\sqrt{x^2+y^2}+y}{x} \qquad \text{(i)}$$

Since each of the functions $\sqrt{x^2+y^2}+y$ and x is a homogeneous function of degree 1, Eq. (i) is a homogeneous differential equation.

Putting $y = vx$ and $\dfrac{dy}{dx} = v + x\dfrac{dv}{dx}$ in Eq. (i), we get

$$v + x\frac{dv}{dx} = \frac{\sqrt{x^2+v^2x^2}+vx}{x}$$

$$\Rightarrow \qquad v + x\frac{dv}{dx} = \sqrt{1+v^2}+v$$

$$\Rightarrow \qquad x\frac{dv}{dx} = \sqrt{1+v^2}$$

$$\Rightarrow \qquad \frac{dv}{\sqrt{1+v^2}} = \frac{dx}{x}$$

Integrating both sides, we get

$$\int\frac{dv}{\sqrt{1+v^2}} = \int\frac{dx}{x}$$

$$\Rightarrow \qquad \log\left(v+\sqrt{1+v^2}\right) = \log x + \log c$$

$$\Rightarrow \qquad v+\sqrt{1+v^2} = cx$$

$$\Rightarrow \qquad \frac{y}{x}+\sqrt{1+\frac{y^2}{x^2}} = cx \qquad (\because \; v = y/x)$$

$$\Rightarrow \qquad y+\sqrt{x^2+y^2} = cx^2$$

which is the required solution.

EXAMPLE 2.30 Solve $y\left[x\cos\left(\frac{y}{x}\right)+y\sin\left(\frac{y}{x}\right)\right]dx - x\left[y\sin\left(\frac{y}{x}\right)-x\cos\left(\frac{y}{x}\right)\right]dy = 0.$

Solution The given differential equation can be written as

$$\frac{dy}{dx}=\frac{y\left[x\cos\left(\frac{y}{x}\right)+y\sin\left(\frac{y}{x}\right)\right]}{x\left[y\sin\left(\frac{y}{x}\right)-x\cos\left(\frac{y}{x}\right)\right]} \tag{i}$$

Putting $y = vx$ and $\frac{dy}{dx}=v+x\frac{dv}{dx}$ in Eq. (i), we get

$$v+x\frac{dv}{dx} = \frac{vx[x\cos(v)+vx\sin(v)]}{x[vx\sin(v)-x\cos(v)]}$$

$$\Rightarrow \quad v+x\frac{dv}{dx} = \frac{v(\cos v+v\sin v)}{v\sin v-\cos v}$$

$$\Rightarrow \quad x\frac{dv}{dx} = \frac{2v\cos v}{v\sin v-\cos v}$$

$$\Rightarrow \quad \frac{v\sin v-\cos v}{2v\cos v}dv = \frac{dx}{x}$$

Integrating both sides, we get

$$\int\frac{v\sin v-\cos v}{2v\cos v}dv = \int\frac{dx}{x}$$

$$\Rightarrow \quad -\log|v\cos v| = 2\log|x| + \log c$$

$$\Rightarrow \quad \log\frac{1}{v\cos v} = \log cx^2$$

$$\Rightarrow \quad \frac{x}{y}\sec\left(\frac{y}{x}\right) = cx^2 \qquad (\because \; v = y/x)$$

$$\Rightarrow \quad xy\cos\left(\frac{x}{y}\right) = k, \quad \text{where } k = 1/c$$

which is the required solution.

EXAMPLE 2.31 Solve $xy\log\left(\frac{x}{y}\right)dx+\left\{y^2-x^2\log\left(\frac{x}{y}\right)\right\}dy=0$.

Solution The given differential equation can be written as

$$\frac{dx}{dy}=\frac{x^2\log\left(\frac{x}{y}\right)-y^2}{xy\log\left(\frac{x}{y}\right)} \tag{i}$$

Putting $x = vy$ and $\frac{dx}{dy} = v + y\frac{dv}{dy}$ in Eq. (i), we get

$$v + y\frac{dv}{dy} = \frac{v^2 y^2 \log v - y^2}{vy^2 \log v}$$

$$\Rightarrow \quad v + y\frac{dv}{dy} = \frac{v^2 \log v - 1}{v \log v}$$

$$\Rightarrow \quad y\frac{dv}{dy} = -\frac{1}{v \log v}$$

$$\Rightarrow \quad v \log v\, dv = -\frac{dy}{y}$$

Integrating both sides, we get

$$\int v \log v\, dv = -\int \frac{dy}{y}$$

$$\Rightarrow \quad \frac{v^2}{2}\left(\log v - \frac{1}{2}\right) = -\log y + c$$

$$\Rightarrow \quad \frac{x^2}{2y^2}\left[\log\left(\frac{x}{y}\right) - \frac{1}{2}\right] + \log y = c \qquad (\because \quad v = x/y)$$

which is the required solution.

2.4 NON-HOMOGENEOUS DIFFERENTIAL EQUATIONS

The general form of non-homogeneous differential equation is

$$\frac{dy}{dx} = \frac{ax + by + c}{dx + ey + f} \tag{2.2}$$

where a, b, c, d, e and f are constants and $\frac{a}{d} \neq \frac{b}{e}$.

Substitute $x = X + h$ and $y = Y + k$ where h and k are the constants to be chosen suitable. With this substitution, Eq. (2.2) reduces to

$$\frac{dY}{dX} = \frac{aX + bY + ah + bk + c}{dX + eY + dh + ek + f} \tag{2.3}$$

Choosing h and k such that

$$ah + bk + c = 0 \tag{2.4}$$

and

$$dh + ek + f = 0 \tag{2.5}$$

Therefore, Eq. (2.3) reduces to

$$\frac{dY}{dX} = \frac{aX + bY}{dX + eY} \tag{2.6}$$

which turns out to be homogeneous differential equation and can be solved by the method discussed in Section 2.3. From the solution of Eq. (2.6), the solution of Eq. (2.2) can be obtained by putting $X = x - h$ and $Y = y - k$.

EXAMPLE 2.32 Solve $(x + 2y - 3)\, dy = (2x - y + 1)\, dx$.

Solution The given equation can be written as

$$\frac{dy}{dx} = \frac{2x - y + 1}{x + 2y - 3} \tag{i}$$

Substitute $x = X + h$ and $y = Y + k$ in Eq. (i) gives

$$\frac{dy}{dx} = \frac{2X - Y + 2h - k + 1}{X + 2Y + h + 2k - 3} \tag{ii}$$

Choosing h and k such that

$$2h - k + 1 = 0$$

and

$$h + 2k - 3 = 0$$

Solving these two equations gives $h = \frac{1}{5}$ and $k = \frac{7}{5}$

Then Eq. (ii) reduces to

$$\frac{dy}{dx} = \frac{2X - Y}{X + 2Y} \tag{iii}$$

Clearly, Eq. (iii) is homogeneous differential equation. Putting $Y = VX$ and $\frac{dY}{dX} = V + X\frac{dV}{dX}$ in Eq. (iii) gives

$$V + X\frac{dV}{dX} = \frac{2X - VX}{X + 2VX}$$

$$\Rightarrow \quad V + X\frac{dV}{dX} = \frac{2 - V}{1 + 2V}$$

$$\Rightarrow \quad X\frac{dV}{dX} = \frac{2 - 2V - 2V^2}{1 + 2V}$$

$$\Rightarrow \quad \frac{1 + 2V}{-1 + V + V^2}\, dV = -2\frac{dX}{X}$$

Integrating both sides, we get

$$\log (V^2 + V - 1) = -2 \log X + \log c$$

$$\Rightarrow \quad \log (V^2 + V - 1) X^2 = \log c$$

$$\Rightarrow \quad (V^2 + V - 1) X^2 = c$$

$$\Rightarrow \quad \left(\frac{Y^2}{X^2} + \frac{Y}{X} - 1\right) X^2 = c \qquad (\because \; V = Y/X)$$

$$\Rightarrow \quad Y^2 + XY - X^2 = c$$

$$\Rightarrow \qquad \left(y-\frac{7}{5}\right)^2+\left(x-\frac{1}{5}\right)\left(y-\frac{7}{5}\right)-\left(x-\frac{1}{5}\right)^2 = c$$

which is the required solution of Eq. (i).

Special case: If in Eq. (2.2) $\frac{a}{d}=\frac{b}{e}$, then we have

$$\frac{dy}{dx}=\frac{k(dx+ey)+c}{dx+ey+f} \tag{2.7}$$

Putting $dx + ey + f = u$ and $d+e\frac{dy}{dx}=\frac{du}{dx}$ in Eq. (2.7), we get

$$\Rightarrow \qquad \frac{1}{e}\left(\frac{du}{dx}-d\right) = \frac{ku+c}{u}$$

$$\Rightarrow \qquad \frac{du}{dx}-d = \frac{keu+ce}{u}$$

$$\Rightarrow \qquad \frac{du}{dx} = \frac{keu+ce+du}{u}$$

$$\Rightarrow \qquad dx = \frac{u}{keu+ce+du}\,du \tag{2.8}$$

On integrating, we get the solution of Eq. (2.8) and then substitute $u = dx + ey + f$ to get the solution of Eq. (2.7).

EXAMPLE 2.33 Solve $(x + 2y - 1)\,dx = (x + 2y + 1)\,dy$.

Solution .The given equation can be written as

$$\frac{dy}{dx}=\frac{x+2y-1}{x+2y+1} \tag{i}$$

Substitute $x + 2y = u$ and $1+2\frac{dy}{dx}=\frac{du}{dx}$ in Eq. (i), we get

$$\frac{1}{2}\left(\frac{du}{dx}-1\right) = \frac{u-1}{u+1}$$

$$\Rightarrow \qquad \frac{du}{dx}-1 = \frac{2u-2}{u+1}$$

$$\Rightarrow \qquad \frac{du}{dx} = \frac{3u-1}{u+1}$$

$$\Rightarrow \qquad dx = \frac{u+1}{3u-1}\,du$$

$$\Rightarrow \qquad dx = \frac{1}{3}\left(1+\frac{4}{3u-1}\right)du$$

$$\Rightarrow \qquad 3dx = \left(1+\frac{4}{3u-1}\right)du$$

Integrating both sides, we get

$$3\int dx = \int\left(1+\frac{4}{3u-1}\right)du$$

$$\Rightarrow \qquad u+\frac{4}{3}\log(3u-1) = 3x + c$$

$$\Rightarrow \qquad x+2y+\frac{4}{3}\log(3x+6y-1) = 3x + c$$

$$\Rightarrow \qquad y+\frac{2}{3}\log(3x+6y-1) = x + c$$

which is the required solution of Eq. (i).

2.5 EXACT DIFFERENTIAL EQUATION

The differential equation

$$Mdx + Ndy = 0 \tag{2.9}$$

is said to be exact if there exists a function $U(x, y)$ such that

$$dU = Mdx + Ndy$$

i.e. if there exists a function $U(x, y)$ such that

$$M = \frac{\partial U}{\partial x} \text{ and } N = \frac{\partial U}{\partial y}.$$

THEOREM 2.1 The necessary and sufficient condition for the ordinary differential equation

$$Mdx + Ndy = 0$$

to be exact is $\dfrac{\partial M}{\partial y} = \dfrac{\partial N}{\partial x}$.

Proof: Let the differential equation be exact

$$Mdx + Ndy = 0.$$

Then by definition, there exists a function $U(x, y)$ such that

$$dU = Mdx + Ndy$$

$$\Rightarrow \qquad Mdx + Ndy = \frac{\partial U}{\partial x}dx + \frac{\partial U}{\partial y}dy$$

$$\Rightarrow \qquad M = \frac{\partial U}{\partial x} \text{ and } N = \frac{\partial U}{\partial y}$$

Assume that M and N both have continuous first order partial derivatives. Then U possesses continuous second order partial derivatives, and hence

$$\frac{\partial^2 U}{\partial x \partial y} = \frac{\partial^2 U}{\partial y \partial x}$$

$$\Rightarrow \qquad \frac{\partial M}{\partial y} = \frac{\partial N}{\partial x}$$

Conversely, let

$$\frac{\partial M}{\partial y} = \frac{\partial N}{\partial x}$$

and
$$\int_{x=\text{constant}} Ndy = V \text{ so that } N = \frac{\partial V}{\partial y}.$$

$$\Rightarrow \qquad \frac{\partial N}{\partial x} = \frac{\partial^2 V}{\partial x \partial y} = \frac{\partial M}{\partial y}$$

$$\Rightarrow \qquad \frac{\partial M}{\partial y} = \frac{\partial}{\partial y}\left(\frac{\partial V}{\partial x}\right)$$

Integrating w.r.t. y, we get

$$M = \frac{\partial V}{\partial x} + \phi(x)$$

where $\phi(x)$ is a function of x.

$$\therefore \qquad Mdx + Ndy = \left(\frac{\partial V}{\partial x} + \phi(x)\right) dx + \frac{\partial V}{\partial y} dy$$

$$= \left(\frac{\partial V}{\partial x} dx + \frac{\partial V}{\partial y} dy\right) + \phi(x)\, dx$$

$$= dV + d\int \phi(x)\, dx$$

$$= d[V + \int \phi(x)\, dx]$$

Therefore, $Mdx + Ndy$ is an exact differential equation, and hence, the equation $Mdx + Ndy = 0$ is exact.

Solution of an exact differential equation

We have seen that to solve an exact differential equation $Mdx + Ndy = 0$, it is necessary to find the function $U(x, y)$.

Integrating,

$$\frac{\partial U}{\partial x} = M$$

partially w.r.t. x, gives

$$U = \int M dx + f(y)$$

$$\therefore \quad \frac{\partial U}{\partial y} = \frac{\partial}{\partial y}\left(\int M dx\right) + \frac{\partial f}{\partial y} \tag{2.10}$$

$$\Rightarrow \quad \frac{\partial f}{\partial y} = \frac{\partial U}{\partial y} - \frac{\partial}{\partial y}\left(\int M dx\right)$$

$$= N - \frac{\partial}{\partial y}\left(\int M dx\right) \tag{2.11}$$

The terms on RHS of Eq. (2.11) is independent of x, because on differentiating w.r.t. x, it gives $\frac{\partial N}{\partial x} - \frac{\partial M}{\partial y}$ which is zero.

Integrating Eq. (2.11) w.r.t. y, gives

$$f(y) = \int \left[N - \frac{\partial}{\partial y}\left(\int M dx\right)\right] dy + c$$

Therefore, from Eq. (2.9),

$$U = \int M dx + \int \left[N - \frac{\partial}{\partial y}\left(\int M dx\right)\right] dy + c$$

which gives the solution of the exact differential equation $Mdx + Ndy = 0$ as

$$\int M dx + \int \left[N - \frac{\partial}{\partial y}\left(\int M dx\right)\right] dy = c \tag{2.12}$$

Thus, to find the solution of an exact differential Eq. (2.10), integrate $\int M dx$ as if y is constant and integrate the terms of N which do not contain x w.r.t. y.

EXAMPLE 2.34 Solve $(x^2 - 4xy - 2y^2)\, dx + (y^2 - 4xy - 2x^2)\, dy = 0$.

Solution We have $M = x^2 - 4xy - 2y^2$ and $N = y^2 - 4xy - 2x^2$.

$$\therefore \quad \frac{\partial M}{\partial y} = -4x - 4y \text{ and } \frac{\partial N}{\partial x} = -4x - 4y$$

$$\therefore \quad \frac{\partial M}{\partial y} = \frac{\partial N}{\partial x}$$

$\therefore$ Given differential equation is an exact.

Now, $$\int_{y=\text{constant}} Mdx = \int_{y=\text{constant}} (x^2 - 4xy - 2y^2)\,dx = \frac{x^3}{3} - 2x^2y - 2xy^2$$

and $$\int Ndy = \int y^2 dy = \frac{y^3}{3} \quad \text{(terms not containing } x\text{)}$$

Therefore, the solution is

$$\frac{x^3}{3} - 2x^2y - 2xy^2 + \frac{y^3}{3} = c.$$

EXAMPLE 2.35 Solve $e^y dx + (1 + xe^y)\, dy = 0$.

Solution We have $M = e^y$ and $N = 1 + xe^y$.

$$\therefore \quad \frac{\partial M}{\partial y} = e^y \text{ and } \frac{\partial N}{\partial x} = e^y$$

$$\therefore \quad \frac{\partial M}{\partial y} = \frac{\partial N}{\partial x}$$

$\therefore$ Given differential equation is an exact.

Now, $$\int_{y=\text{constant}} Mdx = \int_{y=\text{constant}} e^y dx = xe^y$$

and $$\int Ndy = \int 1 dy = y \quad \text{(terms not containing } x\text{)}$$

Therefore, the solution is

$$ex^y + y = c.$$

2.6 INTEGRATING FACTOR (IF)

The differential equation

$$ydx - xdy = 0$$

is not exact, but when multiplied by $\left(\frac{1}{xy}\right)$, it becomes

$$\frac{dx}{x} - \frac{dy}{y} = 0$$

which is exact, and its solution is

$$\log x - \log y = \log c$$

When multiplied by $\frac{1}{x^2}$, it becomes

$$\frac{ydx - xdy}{x^2} = 0$$

which is exact, and its solution is

$$\frac{y}{x} = c$$

which is equivalent to the solution first found. Thus, we have definition:

"A factor, which when multiplied to a non-exact differential equation makes it exact, is known as **integrating factor**".

The number of integrating factors for an equation $Mdx + Ndy = 0$ is infinite. Suppose μ is an integrating factor, then

$$\mu\,(Mdx + Ndy) = dU$$

which gives solution $U = c$.

Multiplication of both sides by any function of U, (say) $f(U)$ gives

$$\mu f(U)\ (Mdx + Ndy) = f(U)\ dU$$

Since $Mdx + Ndy$ is an exact differential $\mu f(U)$ is an integrating factor of the equation

$$Mdx + Ndy = 0$$

and as $f(U)$ is an arbitrary function of U, the number of integrating factors is infinite.

***Rules for finding Integrating Factor* (IF)**

By inspection: Sometimes integrating factor can be obtained by inspection as in the following examples:

EXAMPLE 2.36 Solve $ydx - xdy + \log xdx = 0$.

Solution It is left for the readers to check that the given differential equation is not exact. Dividing by x^2 gives

$$\frac{ydx - xdy}{x^2} + \frac{\log x}{x^2} = 0$$

which can be written as

$$-d\left(\frac{y}{x}\right) + d\left(\int \frac{1}{x^2} \log xdx\right) = 0$$

$$\Rightarrow \qquad d\left(\int \frac{1}{x^2} \log xdx - \frac{y}{x}\right) = 0$$

Therefore, $\frac{1}{x^2}$ is an integrating factor and the solution is

$$\int \frac{1}{x^2} \log x\, dx - \frac{y}{x} = c$$

$$\Rightarrow \qquad c + \frac{y}{x} = \int \frac{1}{x^2} \log xdx$$

$$= -\frac{1}{x} \log x - \frac{1}{x}$$

Therefore, the solution of the given equation is

$$y + \log x + 1 + cx = 0.$$

EXAMPLE 2.37 Solve $xdx + ydy = 2(xdy - ydx)$.

Solution It is left for the readers to check that the given differential equation is not exact. Given equation can be written as

$$d(x^2 + y^2) = 4x^2 d\left(\frac{y}{x}\right)$$

Dividing by $x^2 + y^2$ gives

$$\frac{d(x^2 + y^2)}{x^2 + y^2} = \frac{4x^2 d\left(\frac{y}{x}\right)}{x^2 + y^2}$$

$$\Rightarrow \quad \frac{d(x^2 + y^2)}{x^2 + y^2} = \frac{4d(y/x)}{1 + (y/x)^2}$$

$$\Rightarrow \quad d\left[\log(x^2 + y^2) - 4\tan^{-1}\frac{y}{x}\right] = 0$$

Therefore, $\dfrac{1}{x^2 + y^2}$ is an integrating factor and the solution is

$$\log(x^2 + y^2) - 4\tan^{-1}\frac{y}{x} = c.$$

EXAMPLE 2.38 Solve $(1 + xy)\, ydx\, (1 - xy)\, xdy = 0$.

Solution It is left for the readers to check that the given differential equation is not exact. Given equation can be rearranged as

$$ydx + xdy + xy^2dx - x^2ydy = 0$$

$$\Rightarrow \quad d(xy) + x^2y^2\left(\frac{dx}{x} - \frac{dy}{y}\right) = 0$$

$$\Rightarrow \quad d(xy) + x^2y^2 d\left(\log\frac{x}{y}\right) = 0$$

Dividing by x^2y^2 gives

$$\frac{d(xy)}{x^2y^2} + d\left(\log\frac{x}{y}\right) = 0$$

$$\Rightarrow \quad d\left(\log\frac{x}{y} - \frac{1}{xy}\right) = 0$$

Therefore, $\frac{1}{x^2y^2}$ is an integrating factor and the solution is

$$\log\frac{x}{y} - \frac{1}{xy} = c.$$

EXAMPLE 2.39 Solve $xdy - ydx - \cos\frac{1}{x} = 0.$

Solution It is left for the readers to check that the given differential equation is not exact. Dividing by x^2 gives

$$\frac{xdy - ydx}{x^2} - \frac{1}{x^2}\cos\frac{1}{x} = 0$$

$$\Rightarrow \qquad d\left(\frac{y}{x}\right) + d\left(\sin\frac{1}{x}\right) = 0$$

$$\Rightarrow \qquad d\left(\frac{y}{x} + \sin\frac{1}{x}\right) = 0$$

Therefore, $\frac{1}{x^2}$ is an integrating factor and the solution is

$$\frac{y}{x} + \sin\frac{1}{x} = c.$$

When it is not possible to find out the integration factor by inspection, we will follow below stated rules:

R1: When $Mx + Ny \neq 0$ for a homogeneous Eq. (2.9), $\frac{1}{Mx+Ny}$ is an integrating factor.

Proof:
$$Mdx + Ndy = \frac{1}{2}\left[(Mx+Ny)\left(\frac{dx}{x}+\frac{dy}{y}\right) + (Mx-Ny)\left(\frac{dx}{x}-\frac{dy}{y}\right)\right] \tag{2.13}$$

$$\Rightarrow \qquad \frac{Mdx+Ndy}{Mx+Ny} = \frac{1}{2}\left[\left(\frac{dx}{x}+\frac{dy}{y}\right) + \frac{(Mx-Ny)}{Mx+Ny}\left(\frac{dx}{x}-\frac{dy}{y}\right)\right]$$

Since M and N are of the same degree, we can write $\frac{Mx-Ny}{Mx+Ny} = f\left(\frac{x}{y}\right)$ to be a function of $\frac{x}{y}$.

$$\therefore \qquad \frac{Mdx+Ndy}{Mx+Ny} = \frac{1}{2}\left[\left(\frac{dx}{x}+\frac{dy}{y}\right) + f\left(\frac{x}{y}\right)\left(\frac{dx}{x}-\frac{dy}{y}\right)\right]$$

$$= \frac{1}{2}\left[d\log(xy) + f\left(e^{\log\frac{x}{y}}\right)d\left(\log\frac{x}{y}\right)\right]$$

$$= \frac{1}{2}\left[d\log(xy) + F\left(\log\frac{x}{y}\right)d\left(\log\frac{x}{y}\right)\right]$$

$$= d\left[\frac{1}{2}\log(xy) + \frac{1}{2}\int F\left(\log\frac{x}{y}\right)d\left(\log\frac{x}{y}\right)\right]$$

Therefore, $\dfrac{1}{Mx+Ny}$ is an integrating factor of the Eq. (2.9).

R2: On dividing Eq. (2.13) by $Mx - Ny$, we get

$$\frac{Mdx+Ndy}{Mx-Ny} = F(\log xy)\,d(\log xy) + \frac{1}{2}d\log\left(\frac{x}{y}\right) \tag{2.14}$$

which is an exact differential.

When $Mx + Ny = 0$, then $\dfrac{M}{N} = -\dfrac{y}{x}$. Substitute $\dfrac{M}{N}$ in Eq. (2.9) and integration gives the solution $x = cy$.

When $Mx - Ny = 0$, then $\dfrac{M}{N} = -\dfrac{y}{x}$. Substitute $\dfrac{M}{N}$ in Eq. (2.9) and integration gives the solution $xy = c$.

EXAMPLE 2.40 Solve $y(xy + 2x^2y^2)\,dx + x(xy - x^2y^2)\,dy = 0$.

Solution Here,

$$M = y(xy + 2x^2y^2) \text{ and } N = x(xy - x^2y^2)$$

$$\therefore \quad Mx - Ny = xy(xy + 2x^2y^2) - xy(xy - x^2y^2) = 3x^3y^3 \neq 0$$

$\therefore$ $\dfrac{1}{3x^3y^3}$ is an IF.

Dividing given equation by $3x^3y^3$ gives

$$\frac{1}{3x^3y^3}y(xy+2x^2y^2)dx + \frac{1}{3x^3y^3}x(xy-x^2y^2)dy = 0$$

$$\Rightarrow \quad \frac{dx}{3x^2y} + \frac{dy}{3xy^2} + \frac{2}{3}\frac{dx}{x} - \frac{1}{3}\frac{dy}{y} = 0$$

$$\Rightarrow \quad d\left(-\frac{1}{3}\frac{1}{xy} + \frac{2}{3}\log x - \frac{1}{3}\log y\right) = 0$$

$$\Rightarrow \quad -\frac{1}{xy} + \log\frac{x^2}{y} = c$$

is the required solution.

R3: When $\frac{1}{N}\left(\frac{\partial M}{\partial y}-\frac{\partial N}{\partial x}\right)=c$ is a function of x only (say) $f(x)$, $e^{\int f(x)dx}$ is an integration factor of $Mdx + Ndy = 0$.

Proof: Let $M_1 = Me^{\int f(x)dx}$ and $N_1 = Ne^{\int f(x)dx}$.

$$\therefore \qquad \frac{\partial M_1}{\partial y} = \frac{\partial M}{\partial y}e^{\int f(x)dx}$$

and
$$\frac{\partial N_1}{\partial x} = \frac{\partial N}{\partial x}e^{\int f(x)dx} + Nf(x)\,e^{\int f(x)dx}$$

$$= e^{\int f(x)dx}\left[\frac{\partial N}{\partial x} + Nf(x)\right]$$

$$= e^{\int f(x)dx}\left[\frac{\partial N}{\partial x} + \frac{\partial M}{\partial y} - \frac{\partial N}{\partial x}\right]$$

$$= e^{\int f(x)dx}\frac{\partial M}{\partial y}$$

$$= \frac{\partial M_1}{\partial y}$$

Therefore, the equation $M_1dx + N_1dy = 0$ is an exact.

EXAMPLE 2.41 Solve $(x^2 + y^2)\, dx + dydy = 0$.

Solution Here,

$$M = x^2 + y^2 \text{ and } N = xy.$$

$$\therefore \qquad \frac{\partial M}{\partial y} = 2y \text{ and } \frac{\partial N}{\partial x} = y$$

$$\therefore \qquad \frac{\partial M}{\partial y} - \frac{\partial N}{\partial x} = y$$

$$\therefore \qquad \frac{1}{N}\left(\frac{\partial M}{\partial y}-\frac{\partial N}{\partial x}\right) = \frac{1}{xy}y = \frac{1}{x}$$

which is a function of x only.

$$\therefore \qquad \text{IF} = e^{\int \frac{1}{x}dx} = e^{\log x} = x$$

Multiplying given equation by x, we get

$$x(x^2 + y^2)\, dx + x^2ydy = 0$$

$$\Rightarrow \qquad x^3dx + xy^2dx + x^2ydy = 0$$

$$\Rightarrow \qquad d\left(\frac{x^4}{4}\right)+d\left(\frac{x^2y^2}{2}\right)=0$$

$$\Rightarrow \qquad d\left(\frac{x^4}{4}+\frac{x^2y^2}{2}\right)=0$$

Therefore, the solution is

$$\frac{x^4}{4}+\frac{x^2y^2}{2}=c \text{ or } x^4 + 2x^2y^2 = c$$

R4: When $\frac{1}{M}\left(\frac{\partial N}{\partial x}-\frac{\partial M}{\partial y}\right)$ is a function of y only (say) $f(y)$, $e^{\int f(y)dy}$ is an integration factor of $Mdx + Ndy = 0$.

Proof: Similar to R3.

EXAMPLE 2.42 Solve $(3x^2y^4 + 2xy)\,dx + (2x^3y^3 - x^2)\,dy = 0$.

Solution Here,

$$M = 3x^2y^4 + 2xy \text{ and } N = 2x^3y^3 - x^2$$

$$\therefore \qquad \frac{\partial M}{\partial y}=12x^2y^3+2x \text{ and } \frac{\partial N}{\partial x}=6x^2y^3-2x$$

$$\therefore \qquad \frac{\partial N}{\partial x}-\frac{\partial M}{\partial y} = -6x^2y^3-4x$$

$$\therefore \qquad \frac{1}{M}\left(\frac{\partial N}{\partial x}-\frac{\partial M}{\partial y}\right) = \frac{1}{xy(3xy^3+2)}(-6x^2y^3-4x)$$

$$\therefore \qquad \frac{1}{M}\left(\frac{\partial N}{\partial x}-\frac{\partial M}{\partial y}\right) = -\frac{2}{y}$$

which is a function of y only.

$$\therefore \qquad \text{IF} = e^{-\int \frac{2}{y}dy} = e^{-2\log y} = \frac{1}{y^2}$$

Multiplying given equation by $\frac{1}{y^2}$, we get

$$\frac{1}{y^2}(3x^2y^4+2xy)dx+\frac{1}{y^2}(2x^3y^3-x^2)dy=0$$

$$\Rightarrow \qquad 3x^2y^2dx+2x^3ydy+2\frac{x}{y}dy-\frac{x^2}{y^2}dy=0$$

$$\Rightarrow \qquad x^3y^3 + x^2 = cy$$

which is the required solution.

R5: If the equation is of the form

$$x^{\alpha}y^{\beta}(mydx + nxdy) = 0$$

then its IF is $x^{km-1-\alpha}y^{kn-1-\beta}$ where k can have any value.

Proof: Multiplying the given equation by IF gives

$$x^{km-1}y^{kn-1}(mydx + nxdy) = 0$$

$$\Rightarrow \quad kmx^{km-1}y^{kn}dx + kny^{kn-1}x^{km}dy = 0$$

$$\Rightarrow \quad d(x^{km}y^{kn}) = 0$$

$\therefore$ $x^{km-1-\alpha}y^{kn-1-\beta}$ is an IF.

Remark: An IF of an equation

$$x^{\alpha}y^{\beta}(mydx + nxdy) + x^{\alpha_1}y^{\beta_1}(m_1ydx + n_1xdy) = 0$$

can be taken as $x^{km-1-\alpha}y^{kn-1-\beta}$ for the first term and $x^{k_1m-1-\alpha_1}y^{k_1n-1-\beta_1}$ for the second term, where k and k_1 can have any value, to make respective terms exact differential. These two factors will be identical if

$$km - 1 - \alpha = k_1m - 1 - \alpha_1$$

and

$$km - 1 - \beta = k_1m - 1 - \beta_1$$

which gives values of k and k_1. Then either factor is an IF of the equation.

EXAMPLE 2.43 Solve $x(4ydx + 2xdy) + y^3(3ydx + 5xdy) = 0$.

Solution Here

$$\alpha = 1,\ \beta = 0,\ m = 4,\ n = 2$$

and

$$\alpha_1 = 0,\ \beta_1 = 3,\ m_1 = 3,\ n_1 = 5$$

Then IF of the first term is

$$x^{4k-1-1}y^{2k-1} = x^{4k-2}y^{2k-1}$$

and for the second term is

$$x^{3k_1-1}y^{5k_1-1-3} = x^{3k_1-1}y^{5k_1-4}$$

The two factors will be identical if

$$4k - 2 = 3k_1 - 1$$

and

$$2k - 2 = 5k_1 - 4$$

This gives $k = 1$ and $k_1 = 1$.
Therefore, IF of the given equation is x^2y.
Multiplying the given equation by x^2y gives

$$x^3y(4ydx + 2xdy) + x^2y^4(3ydx + 5xdy) = 0$$

$$\Rightarrow \quad 4x^3y^2dx + 2x^4ydy + 3x^2y^5dx + 5x^3y^4dy = 0$$

$$\Rightarrow \quad d(x^4y^2 + x^3y^5) = 0$$

$\therefore$ $x^4y^2 + x^3y^5 = c$ which is the required solution.

2.7 LINEAR EQUATION

A differential equation is linear if the dependent variable and its derivative appear only in the first degree. The general form of a linear differential equation is

$$\frac{dy}{dx}+Py=Q \tag{2.15}$$

where P and Q are functions of x or constants.
For example

(i) $\frac{dy}{dx}+2y=\cos x$

(ii) $\frac{dy}{dx}+\frac{1}{x}y=e^x$

(iii) $x\frac{dy}{dx}+2y=x^3\sin x$, etc.

are linear differential equations.

When Eq. (2.15) is multiplied by the IF $e^{\int Pdx}$, we get exact differential. That is

$$e^{\int Pdx}\left(\frac{dy}{dx}+Py\right) = Qe^{\int Pdx}$$

$$\Rightarrow \quad \frac{d}{dx}\left(ye^{\int Pdx}\right) = Qe^{\int Pdx}$$

On integrating both sides w.r.t. x, we get the solution of Eq. (2.15) as

$$ye^{\int Pdx} = \int Qe^{\int Pdx}dx+c \tag{2.16}$$

EXAMPLE 2.44 Solve $\frac{dy}{dx}-\frac{y}{x}=2x^2$.

Solution Comparing given equation with Eq. (2.15) gives

$$P = -\frac{1}{x} \text{ and } Q = 2x^2$$

Now,

$$\text{IF} = e^{\int Pdx} = e^{-\int \frac{1}{x}dx} = e^{-\log x} = \frac{1}{x}$$

Substituting IF and Q in Eq. (2.16) gives

$$y\frac{1}{x} = \int 2x^2\frac{1}{x}dx+c$$

$$\Rightarrow \qquad \frac{y}{x} = x^2 + c \text{ or } y = x^3 + cx$$

which is the required solution.

EXAMPLE 2.45 Solve $x \log x \dfrac{dy}{dx} + y = \dfrac{2}{x} \log x$.

Solution Given

$$x \log x \frac{dy}{dx} + y = \frac{2}{x} \log x$$

$$\Rightarrow \qquad \frac{dy}{dx} + \frac{1}{x \log x} y = \frac{2}{x^2}$$

Comparing given equation with Eq. (2.15) gives

$$P = \frac{1}{x \log x} \text{ and } Q = \frac{2}{x^2}$$

Now, $$\text{IF} = e^{\int P dx} = e^{\int \frac{1}{x \log x} dx} = \log x$$

Substituting IF and Q in Eq. (2.16) gives

$$y \log x = \int \frac{2}{x^2} \log x dx + c$$

$$\Rightarrow \qquad y \log x = -\frac{2}{x}(1 + \log x) + c$$

which is the required solution.

EXAMPLE 2.46 Solve $\dfrac{dy}{dx} + y \sec x = \tan x$.

Solution Comparing given equation with Eq. (2.15) gives

$$P = \sec x \text{ and } Q = \tan x$$

Now, $$\text{IF} = e^{\int P dx} = e^{\int \sec x dx} = \sec x + \tan x$$

Substituting IF and Q in Eq, (2.16) gives

$$\begin{aligned} y(\sec x + \tan x) &= \int \tan x(\sec x + \tan x)\, dx + c \\ &= \int (\tan x \sec x + \tan^2 x)\, dx + c \\ &= \int (\tan x \sec x + \sec^2 x - 1)\, dx + c \\ &= \sec x + \tan x - x + c \end{aligned}$$

which is the required solution.

EXAMPLE 2.47 Solve $(x^2+1)\dfrac{dy}{dx}+2xy=\sqrt{x^2+4}$.

Solution Given

$$(x^2+1)\frac{dy}{dx}+2xy = \sqrt{x^2+4}$$

$$\Rightarrow \qquad \frac{dy}{dx}+\frac{2x}{x^2+1}y = \frac{\sqrt{x^2+4}}{x^2+1}$$

Comparing given equation with Eq. (2.15) gives

$$P = \frac{2x}{x^2+1} \text{ and } Q=\frac{\sqrt{x^2+4}}{x^2+1}$$

Now, $$\text{IF} = e^{\int P dx} = e^{\int \frac{2x}{x^2+1}dx} = x^2+1$$

Multiplying given equation by $x^2 + 1$ gives

$$(x^2+1)\frac{dy}{dx} = \sqrt{x^2+4}$$

$$\Rightarrow \qquad \frac{d}{dx}[y(x^2+1)] = \sqrt{x^2+4}$$

Integrating both sides w.r.t. x, gives

$$y(x^2+1) = \int \sqrt{x^2+4}dx+c$$

$$= \frac{x}{2}\sqrt{x^2+4}+\frac{4}{2}\log\left|x+\sqrt{x^2+4}\right|+c$$

$$= \frac{x}{2}\sqrt{x^2+4}+2\log\left|x+\sqrt{x^2+4}\right|+c$$

which is the required solution.

EXAMPLE 2.48 Solve $\dfrac{dy}{dx}-2y=\cos 3x$.

Solution Comparing given equation with Eq. (2.15) gives

$$P = -2 \text{ and } Q = \cos 3x$$

Now, $$\text{IF} = e^{\int P dx} = e^{-\int 2dx} = e^{-2x}$$

Multiplying given equation by e^{-2x} gives

$$e^{-2x}\frac{dy}{dx}-2ye^{-2x} = e^{-2x}\cos 3x$$

$$\Rightarrow \qquad \frac{d}{dx}(ye^{-2x}) = e^{-2x}\cos 3x$$

Integrating both sides w.r.t. x, we get

$$ye^{-2x} = \int e^{-2x}\cos 3x dx + c$$

$$ye^{-2x} = I + c$$

where
$$I = \int e^{-2x}\cos 3x dx$$

$$= \frac{1}{3}e^{-2x}\sin 3x - \int -\frac{2}{3}e^{-2x}\sin 3x dx$$

$$= \frac{1}{3}e^{-2x}\sin 3x + \frac{2}{3}\left[-\frac{1}{3}e^{-2x}\cos 3x - \left(-\frac{2}{3}\right)\int e^{-2x}\cos 3x dx\right]$$

$$= \frac{1}{3}e^{-2x}\sin 3x - \frac{2}{9}e^{-2x}\cos 3x - \frac{4}{9}I$$

$$\Rightarrow \qquad I + \frac{4}{9}I = \frac{e^{-2x}}{9}(3\sin 3x - 2\cos 3x)$$

$$\Rightarrow \qquad \frac{13}{9}I = \frac{e^{-2x}}{9}(3\sin 3x - 2\cos 3x)$$

$$\Rightarrow \qquad I = \frac{e^{-2x}}{13}(3\sin 3x - 2\cos 3x)$$

$$\therefore \qquad ye^{-2x} = \frac{e^{-2x}}{13}(3\sin 3x - 2\cos 3x) + c$$

$$\Rightarrow \qquad y = \frac{1}{13}(3\sin 3x - 2\cos 3x) + ce^{2x}$$

which is the required solution.

EXAMPLE 2.49 Solve $ydx - (x + 2y^2)\, dy = 0$.

Solution Given

$$ydx - (x + 2y^2)\, dy = 0$$

$$\Rightarrow \qquad \frac{dx}{dy} - \frac{1}{y}x = 2y$$

This is linear equation of the form

$$\frac{dx}{dy} + Px = Q$$

where P and Q are functions of y.

We have $P = -\frac{1}{y}$ and $Q = 2y$

Now, $$\text{IF} = e^{\int P dy} = e^{\int -\frac{1}{y} dy} = e^{-\log y} = \frac{1}{y}$$

Multiplying given equation by $\frac{1}{y}$ gives

$$\frac{1}{y}\frac{dx}{dy} - \frac{1}{y^2}x = 2$$

$$\Rightarrow \quad \frac{d}{dx}\left(\frac{x}{y}\right) = 2$$

Integrating both sides w.r.t. y, we get

$$\frac{x}{y} = \int 2dy + c$$

$$\Rightarrow \quad \frac{x}{y} = 2y + c$$

which is the required solution.

EXAMPLE 2.50 Solve $dx + xdy = e^{-y} \sec^2 ydy$.

Solution Given

$$dx + xdy = e^{-y} \sec^2 ydy$$

$$\Rightarrow \quad \frac{dx}{dy} + x = e^{-y} \sec^2 y$$

We have, $P = 1$ and $Q = e^{-y} \sec^2 y$

Now, $$\text{IF} = e^{\int P dy} = e^{\int 1 dy} = e^y$$

Multiplying given equation by e^y gives

$$e^y \frac{dx}{dy} + e^y x = \sec^2 y$$

$$\Rightarrow \quad \frac{d}{dy}(e^y x) = \sec^2 y$$

Integrating both sides w.r.t. y, we get

$$xe^y = \int \sec^2 ydy + c$$

$$xe^y = \tan y + c$$

which is the required solution.

2.8 EQUATION REDUCIBLE TO LINEAR FORM

With proper transformation, some non-linear equation can be transformed to the linear equation. The general form is

$$\frac{dy}{dx} + Py = Qy^n \tag{2.17}$$

$$\Rightarrow \qquad y^{-n}\frac{dy}{dx} + Py^{-n+1} = Q$$

Substitute

$$V = y^{-n+1}.$$

Then

$$\frac{dV}{dx} = (-n+1)y^{-n}\frac{dy}{dx}.$$

$$\Rightarrow \qquad \frac{1}{1-n}\frac{dV}{dx} + PV = Q$$

$$\Rightarrow \qquad \frac{dV}{dx} + (1-n)PV = (1-n)Q$$

which is linear equation. Solve it by the method stated in Section 2.7 and substitute $V = y^{-n+1}$ to get the solution of Eq. (2.17). Equation (2.17) is known as **Bernoulli's Equation**.

EXAMPLE 2.51 Solve $\tan y\dfrac{dy}{dx} + \tan x = \cos y \cos^3 x$.

Solution Given

$$\tan y\frac{dy}{dx} + \tan x = \cos y \cos^3 x$$

$$\sec y \tan y\frac{dy}{dx} + \sec y \tan x = \cos^3 x$$

Put $\sec y = V$, then $\sec y \tan y\dfrac{dy}{dx} = \dfrac{dV}{dx}$

$$\Rightarrow \qquad \frac{dV}{dx} + \tan x V = \cos^3 x \tag{i}$$

$$\therefore \qquad P = \tan x \text{ and } Q = \cos^3 x$$

$$\text{IF} = e^{\int P dx} = e^{\int \tan x dx} = \sec x$$

Multiplying Eq. (i) with sec x gives

$$\sec x\frac{dV}{dx} + \sec x \tan x V = \cos^2 x$$

$$\frac{d}{dx}(V \sec x) = \cos^2 x$$

Integrating both sides w.r.t. x, we get

$$V \sec x = \int \cos^2 x dx + c$$

$$= \frac{1}{2}\int (1+\cos 2x) dx + c$$

$$= \frac{1}{2}\left(x + \frac{\sin 2x}{2}\right) + c$$

$$\Rightarrow \qquad 2 \sec y \sec x = x + \frac{\sin 2x}{2} + c$$

which is the required solution of the given differential equation.

EXAMPLE 2.52 Solve $\frac{dy}{dx} + xy = xy^n$.

Solution Given

$$\frac{dy}{dx} + xy = xy^n$$

We can rewrite given equation as

$$\frac{dy}{dx} + xy^{-n+1} = x \qquad \text{(i)}$$

Substitute $y^{-n+1} = V$, then $(1-n)y^{-n}\frac{dy}{dx} = \frac{dV}{dx}$. Therefore, Eq. (i) becomes

$$\frac{1}{1-n}\frac{dV}{dx} + xV = x \qquad \text{(ii)}$$

$$\frac{dV}{dx} + (1-n)xV = (1-n)x \qquad \text{(iii)}$$

$$\therefore \qquad P = (1-n)\, x \text{ and } Q = (1-n)\, x.$$

$$\text{IF} = e^{\int P dx} = e^{\int (1-n) x dx} = e^{\frac{(1-n)}{2}x^2}$$

Multiplying Eq. (iii) with $e^{\frac{(1-n)}{2}x^2}$ gives

$$\Rightarrow \qquad \frac{d}{dx}\left(Ve^{\frac{(1-n)}{2}x^2}\right) = (1-n)xe^{\frac{(1-n)}{2}x^2}$$

Integrating both sides w.r.t. x, we get

$$Ve^{\frac{(1-n)}{2}x^2} = \int (1-n)xe^{\frac{(1-n)}{2}x^2} dx + c$$

$$Ve^{\frac{(1-n)}{2}x^2} = e^{\frac{(1-n)}{2}x^2} + c$$

$$y^{-n+1}e^{\frac{(1-n)}{2}x^2} = e^{\frac{(1-n)}{2}x^2} + c$$

which is the required solution of the given differential equation.

EXAMPLE 2.53 Show that the equation $\dfrac{dy}{dx} = \dfrac{x^2}{1-y^2}$ is separable and then find an equation for its integral curves.

Solution Rewrite given equation as

$$(1 - y^2)dy = x^2 dx$$

Integrating both sides, we obtain

$$-x^3 + 3y - y^3 = c$$

where c is an arbitrary constant. Observe that the solution is an equation for the integral curves of given differential equation. A directional field and several integral curves are shown in Figures 2.1 and 2.2 respectively.

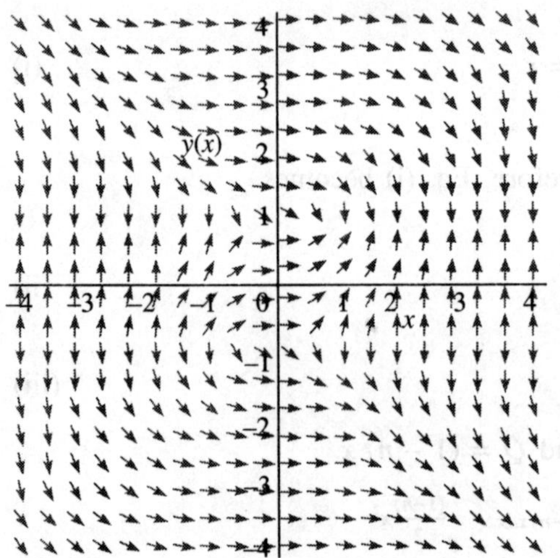

Figure 2.1 Directional field for the differential equation.

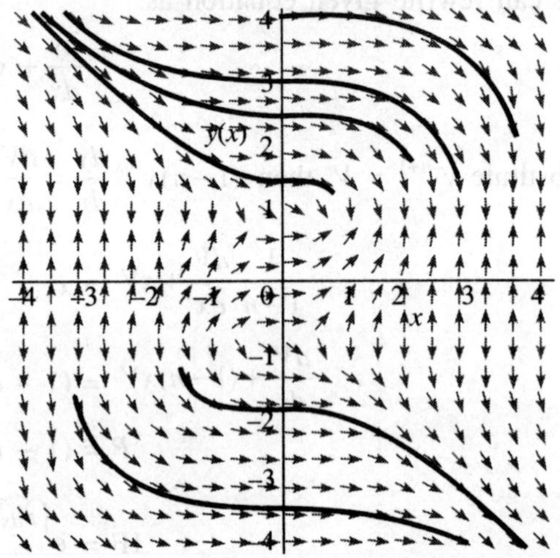

Figure 2.2 Directional field and integral curves for the differential equation.

Maple code is

```
>with(DEtools):
>ode: = D(y)(x) = x^2/(1 - y(x)^2):
>DEplot(ode, y(x), x = -4..4, y = -4..4)
>InitialValues:= [[0, -1], [0, 1.5], [0, -2], [0, 2.5], [0, 3], [0, -3.5], [0, 4]]:
>DEplot(ode, y(x), x = -4..4, y = -4..4, InitialValues, linecolor = black)
```

EXAMPLE 2.54 Solve the initial value problem $\dfrac{dy}{dx} = \dfrac{3x^2 + 4x + 2}{2(y-1)}, y(0) = -1$ and determine the interval in which the solution exists.

Solution The differential equation can be written as

$$2(y - 1)dy = (3x^2 + 4x + 2)dx$$

Integrating both sides, we obtain

$$y^2 - 2y = x^3 + 2x^2 + 2x + c$$

where c is an arbitrary constant. $y(0) = -1$ gives $c = 3$. Hence, the solution of the initial value problem is

$$y = 1 \pm \sqrt{x^3 + 2x^2 + 2x + 4}$$

From two solutions, we obtained only one with "negative" sign that satisfies the given initial condition. So, the required solution is

$$y = 1 - \sqrt{x^3 + 2x^2 + 2x + 4}$$

Now, to determine the interval in which the obtained solution is valid, we must find the interval in which $\sqrt{x^3 + 2x^2 + 2x + 4}$ is positive. The only real zero of this expression is $x = -2$. So, the desired interval is $(-2, \infty)$. The solution of the initial value problem and some other integral curves of the differential equation are shown in Figure 2.3.

Maple code can be generated as in Example 2.53.

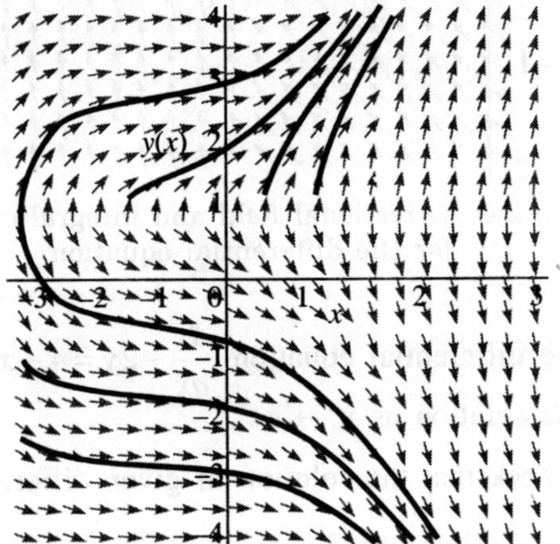

Figure 2.3 Directional field and integral curves for the differential equation.

EXAMPLE 2.55 Solve the differential equation $\dfrac{dy}{dx} + \dfrac{1}{2}y = \dfrac{1}{2}e^{x/3}$. Plot several solutions and find the particular solution whose graph contains the point (0, 1).

Solution Given differential equation is linear. The integrating factor is $e^{x/2}$. Hence, the solution is

$$ye^{x/2} = \frac{3}{5}e^{5x/6} + c \text{ or } y = \frac{3}{5}e^{x/3} + ce^{-x/2}$$

where c is an arbitrary constant. To find the solution passing through (0, 1) gives $c = \frac{2}{5}$. Hence, the solution of the initial value problem is $y = \frac{3}{5}e^{x/3} + \frac{2}{5}e^{-x/2}$. Figure 2.4 represents integral curves described by the given differential equation. The solution passing through (0, 1) is bold curve in the Figure 2.4.

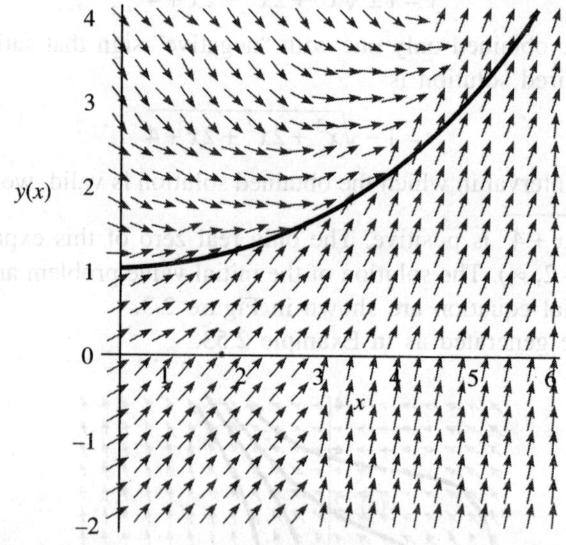

Figure 2.4 Directional field and integral curves for the differential equation.

EXAMPLE 2.56 Solve the differential equation $\frac{dy}{dx} - 2y = 4 - x$. Plot the integral curves. Discuss the behaviour of the solution as $x \to \infty$.

Solution The reader can check that the solution of given differential equation is

$$y = -\frac{7}{4} + \frac{1}{2}x + ce^{2x}$$

where c is an arbitrary constant. The directional field and integral curves are shown in Figure 2.5.

The term ce^{2x} determines the behaviour of the solution for large values of x. If $c \neq 0$, then the solution increases exponentially with the same sign as c itself. This indicates that the solution diverges as $x \to \infty$.

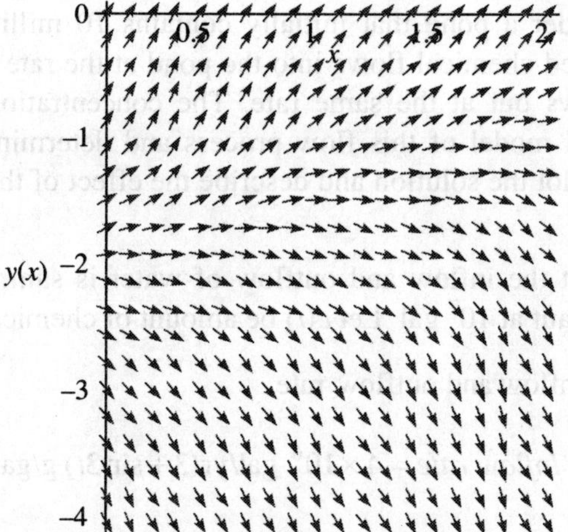

Figure 2.5 Directional field and integral curves for the differential equation.

EXAMPLE 2.57 Solve the differential equation $x\dfrac{dy}{dx}+2y=4x^2, y(1)=2$.

Solution The reader can check that the solution of given differential equation is

$$y=x^2+\frac{c}{x^2}$$

where c is an arbitrary constant. The directional field and integral curves are shown in Figure 2.6.

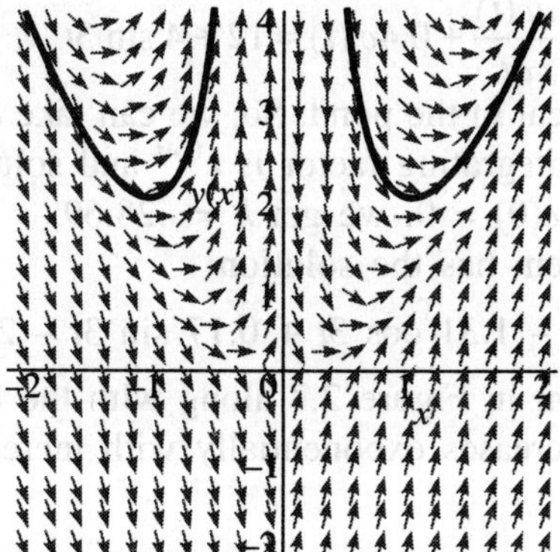

Figure 2.6 Directional field and integral curves for the differential equation.

Using $y(1) = 2$ gives the particular solution as $y=x^2+\dfrac{1}{x^2}, x>0$. This solution is shown by the bold line in Figure 2.6. It is easy to see that the solution is unbounded and is asymptotic to positive y-axis as $x \to 0$ from the right. For $c < 0$, some solutions are asymptotic to the negative y-axis.

EXAMPLE 2.58 Consider a pond that initially contains 10 million gallons of fresh water. Water containing unwanted chemical flows into the pond at the rate of 4 million gal/yr and the mixture in the pond flows out at the same rate. The concentration $a(t) = 3 + \sin 3t$ g/gal. Construct a mathematical model of this flow process and determine the amount of chemical in the pond at any time. Plot the solution and describe the effect of the variation in the incoming concentration.

Solution It is given that the inflow and outflow of water is same. So, the amount of water in the pond remains constant at 10^7 gal. Let $c(t)$ be amount of chemical at any time t is measured in years. Thus, $\frac{dc(t)}{dt} = \text{inflow and outflow rate}$.

$$\textit{Inflow rate} = 4 \times 10^6 \text{ gal/yr}(3 + \sin 3t) \text{ g/gal}$$

The concentration of chemical in pond is $\frac{c(t)}{10^7}$ g/gal.

$$\textit{Outflow rate} = 4 \times 10^6 \text{ gal/yr} \times \frac{c(t)}{10^7} \text{ g/gal} = 0.4\, c(t) \text{ g/yr}$$

Thus, we obtain the differential equation

$$\frac{dc(t)}{dt} = (4 \times 10^6)(3 + \sin 3t) - 0.4c(t)$$

For simplicity, we put $c(t) = \frac{C(t)}{10^7}$, i.e., $C(t) = 10^6 c(t)$. Here, $c(t)$ is measured in millions of grams. With this substitution $\frac{dc(t)}{dt} + 0.4c(t) = 12 + 4 \sin 3t$.

Initially, there is no chemical in the pond. So, we can take $c(0) = 0$. The equation obtained is linear equation. Then, the integrating factor is $e^{0.4t}$ and solution is $c(t) = 30 - 1.31 \cos 3t + 0.17 \sin 3t + ke^{-0.4t}$. Using $c(0) = 0$, we get $k = -28.69$.

So, the initial value problem has the solution

$$c(t) = 30 - 1.31 \cos 3t + 0.17 \sin 3t - 28.69\, e^{-0.4t}$$

A plot of the solution is shown in Figure 2.7 along with the line $c = 30$. It is observed that concentration of chemicals decreases exponentially with increase in time.

EXERCISES

Solve the following differential equations:

1. $\frac{dy}{dx} + 2x = e^{3x}$. $\left[\textbf{Ans. } y + x^2 = \frac{1}{3}e^{3x} + c\right]$

2. $(x+2)\frac{dy}{dx} = x^2 + 3x + 7$. $\left[\textbf{Ans. } y = \frac{x^2}{2} + x + 5 \log|x+2| + c\right]$

3. $\frac{1}{x}\frac{dy}{dx} = \tan^{-1} x.$ $\left[\textbf{Ans. } y = \frac{1}{2}(x^2+1)\tan^{-1} x - \frac{x}{2} + c\right]$

4. $\frac{dy}{dx} = \cos^3 x \sin^2 x + x\sqrt{2x+1}.$

$\left[\textbf{Ans. } y = \frac{1}{3}\sin^3 x - \frac{1}{5}\sin^5 x + \frac{1}{10}(2x+1)^{5/2} - \frac{1}{6}(2x+1)^{3/2} + c\right]$

5. (sin x + cos x) dy + cos x − sin x) dx = 0. [**Ans.** $y + \log|\sin x + \cos x| = c$]

6. $\frac{dy}{dx} - x\sin^2 x = \frac{1}{\log x}.$ $\left[\textbf{Ans. } y = \frac{x^2}{4} - \frac{x \sin 2x}{4} - \frac{\cos 2x}{8} + \log|\log x| + c\right]$

7. $\frac{dy}{dx} = x^5 \tan^{-1}(x^3).$ $\left[\textbf{Ans. } y = \frac{1}{2}(x^6 \tan^{-1} x^3 - x^3 + \tan^{-1} x^3) + c\right]$

8. $e^{\frac{dy}{dx}} = x+1$ with $y(0) = 3$. [**Ans.** $y = (x+1)\log|x+1| - x + 3$]

9. $\sin^4 x \frac{dy}{dx} = \cos x.$ $\left[\textbf{Ans. } y = -\frac{1}{3}\text{cosec}^3 x + c\right]$

10. $\cos x \frac{dy}{dx} - \cos 2x = \cos 3x.$

[**Ans.** $y = \sin 2x - x + 2\sin x + \log|\sec x + \tan x| + c$]

11. $\frac{dy}{dx} = \frac{1}{y^2 + \sin y}.$ $\left[\textbf{Ans. } x = \frac{y^3}{3} - \cos y + c\right]$

12. $\frac{dy}{dx} = \frac{1+y^2}{y^3}.$ $\left[\textbf{Ans. } x = \frac{y^2}{2} - \frac{1}{2}\log(1+y^2) + c\right]$

13. $\frac{dy}{dx} = \sin^2 y.$ [**Ans.** $x + \cot y = c$]

14. $e^x\sqrt{1-y^2}\,dx + \frac{y}{x}dy = 0.$ $\left[\textbf{Ans. } (x-1)e^x + \sqrt{1-y^2} = c\right]$

15. $(1 + y^2)(1 \log x)\,dx + x\,dy = 0.$ $\left[\textbf{Ans. } \frac{1}{2}(1+\log x)^2 + \tan^{-1} y = c\right]$

16. $3e^x \tan y\,dx + (1 - e^x)\sec^2 y\,dy = 0.$ [**Ans.** $c \tan y = (e^x - 1)^3$]

17. $(x-1)\frac{dy}{dx} = 2x^3 y.$ $\left[\textbf{Ans. } \log y = \frac{2}{3}x^3 + x^2 + 2x + 2\log|x-1| + c\right]$

18. $x \cos y dy = (xe^x \log x + e^x)\, dx.$ [**Ans.** $\sin y = e^x \log x + c$]

19. $(e^y + 1) \cos x dx + e^y \sin x dy = 0.$ [**Ans.** $\sin x\, (e^y + 1) = c$]

20. $x \cos^2 y dx = y \cos^2 x dy.$ $\left[\textbf{Ans. } x \tan x - y \tan y = \log\left|\dfrac{\sec x}{\sec y}\right| + c\right]$

21. $\dfrac{dy}{dx} = e^{x+y} + e^y x^3.$ $\left[\textbf{Ans. } e^x + e^{-y} + \dfrac{x^4}{4} = c\right]$

22. $y\sqrt{1+x^2}\,dx + x\sqrt{1+y^2}\,dy = 0.$

$$\left[\textbf{Ans. } \sqrt{1+x^2} + \sqrt{1+y^2} + \frac{1}{2}\log\left|\frac{\sqrt{1+x^2}-1}{\sqrt{1+x^2}+1}\right| + \frac{1}{2}\log\left|\frac{\sqrt{1+y^2}-1}{\sqrt{1+y^2}+1}\right| = c\right]$$

23. $\sqrt{1+x^2}\,dy + \sqrt{1+y^2}\,dx = 0.$ $\left[\textbf{Ans. } (y+\sqrt{1+y^2})(x+\sqrt{1+x^2}) = c\right]$

24. $\sqrt{1+x^2+y^2+x^2y^2} + xy\dfrac{dy}{dx} = 0.$ $\left[\textbf{Ans. } \sqrt{1+x^2} + \sqrt{1+y^2} + \dfrac{1}{2}\log\left|\dfrac{\sqrt{1+x^2}-1}{\sqrt{1+x^2}+1}\right| = c\right]$

25. $\dfrac{dy}{dx} = \dfrac{e^x(\sin^2 x + \sin 2x)}{y(2\log y + 1)}.$ [**Ans.** $y^2 \log y = e^x \sin 2x + c$]

26. $\dfrac{dy}{dx} = \dfrac{x(2\log x + 1)}{\sin y + y\cos y}.$ [**Ans.** $y \sin y = x^3 \log x + c$]

27. $\tan y dx + \sec^2 y \tan x dy = 0.$ [**Ans.** $\sin x \tan y = c$]

28. $\tan y \dfrac{dy}{dx} = \sin(x+y) + \sin(x-y).$ [**Ans.** $2 \cos x + \sec y = c$]

29. $\sin^{-1}\left(\dfrac{dy}{dx}\right) = x + y.$ [**Ans.** $x = \tan(x + y) - \sec(x + y) + c$]

30. $\cos(x + y)\, xy = dx.$ [**Ans.** $y + \cot(x + y) - \operatorname{cosec}(x + y) = c$]

31. $\dfrac{dy}{dx} = \cos(x+y) + \sin(x+y).$ $\left[\textbf{Ans. } y + x^2 = \dfrac{1}{3}e^{3x} + c\right]$

32. $(x+y+1)\dfrac{dy}{dx} = 1.$ [**Ans.** $x = ce^{3y} - y - 2$]

33. $(x + y)\, (dx - dy) = dx + dy.$ $\left[\textbf{Ans. } \dfrac{1}{2}(y-x) + \dfrac{1}{2}\log|x+y| = c\right]$

34. $(x^2 - y^2)\, dx + 2xydy = 0$, given that $y(1) = 1.$ [**Ans.** $x^2 + y^2 = 2x$]

35. $x^2 y dx - (x^3 + y^3)\, dy = 0.$ $\left[\textbf{Ans. } \log y - \dfrac{x^3}{3y^3} = c\right]$

36. $x\dfrac{dy}{dx} = y - x\tan\left(\dfrac{y}{x}\right).$ $\left[\textbf{Ans. } x\sin\left(\dfrac{y}{x}\right) = c\right]$

37. $x^2 dy + y(x + y)\, dx = 0.$ $[\textbf{Ans. } x^2 y = c(y + 2x)]$

38. $x^2\dfrac{dy}{dx} = x^2 - 2y^2 + xy.$ $\left[\textbf{Ans. } \dfrac{x + \sqrt{2}y}{x - \sqrt{2}y} = (cx)^{2\sqrt{2}}\right]$

39. $\dfrac{dy}{dx} = \dfrac{y}{x} - \sqrt{\dfrac{y^2}{x^2} - 1}.$ $\left[\textbf{Ans. } y + \sqrt{y^2 - x^2} = c\right]$

40. $3x^2 dy = (3xy + y^2)\, dx.$ $\left[\textbf{Ans. } -\dfrac{3x}{y} = \log|x| + c\right]$

41. $\dfrac{dy}{dx} = \dfrac{y}{x} + \sin\left(\dfrac{y}{x}\right).$ $\left[\textbf{Ans. } \tan\left(\dfrac{y}{2x}\right) = cx\right]$

42. $y^2 dx + (x^2 - xy + y^2)\, dy = 0.$ $[\textbf{Ans. } y = ce^{\tan-1(y/x)}]$

43. $x\dfrac{dy}{dx} = y - x\cos^2\left(\dfrac{y}{x}\right).$ $\left[\textbf{Ans. } \tan\left(\dfrac{y}{x}\right) = \log\left(\dfrac{c}{x}\right)\right]$

44. $xy\log\left(\dfrac{x}{y}\right)dx + \left[y^2 - x^2\log\left(\dfrac{x}{y}\right)\right]dy = 0.$ $\left[\textbf{Ans. } \dfrac{x^2}{y^2}\left[\log\left(\dfrac{x}{y}\right) - \dfrac{1}{2}\right] + \log y^2 = c\right]$

45. $(1 + e^{x/y})dx + 1 + e^{x/y}\left(1 - \dfrac{x}{y}\right)dy = 0.$ $[\textbf{Ans. } x + ye^{x/y} = c]$

46. $x\cos\left(\dfrac{y}{x}\right)(ydx + xdy) = y\sin\left(\dfrac{y}{x}\right)(xdy - ydx).$ $\left[\textbf{Ans. } \sec\left(\dfrac{y}{x}\right) = cxy\right]$

47. $(x^2 + 3xy + y^2)\, dx - x^2 dy = 0.$ $\left[\textbf{Ans. } \dfrac{x}{x + y} + \log x = c\right]$

48. $(6x - 2y - 7)\, dx = (2x + 3y - 6)\, dy.$ $\left[\textbf{Ans. } 3y^2 + 4xy - 6x^2 + 14x - 12y - \dfrac{9}{2} = c\right]$

49. $(x - y - 1)\, dx + (x + 4y - 1)\, dy = 0.$ $\left[\textbf{Ans. } \log(4y^2 + (x-1)^2) + \tan^{-1}\dfrac{2y}{x - 1} = c\right]$

50. $(2x - y + 1)dx + (2y - x - 1)dy = 0.$ $[\textbf{Ans. } x^2 - xy + y^2 + x - y = c]$

51. $(4x+6y+5)\,dy=(2x+3y+4)\,dx.$ $\left[\textbf{Ans. } 3x-6y+\frac{9}{7}\log(14x+21y+22)=c\right]$

52. $(3y-7x-3)\,dx+(7y-3x-7)\,dy=0.$ $[\textbf{Ans. } (x+y-1)^5(x-y+1)^2=c]$

53. $(x+y-2)\,dx+(x-y+4)\,dy=0.$ $[\textbf{Ans. } (x+1)^2+2(x+1)(y-3)-(y-3)^2=c]$

54. $(2x-5y+3)\,dx-(2x+4y-6)\,dy=0.$ $[\textbf{Ans. } (2x+y-3)^2(x-4y+3)=c]$

55. $2x(ye^{x^2}-1)\,dx+e^{x^2}\,dy=0.$ $[\textbf{Ans. } ye^{x^2}-x^2=c]$

56. $e^y dx+(1+xe^y)dy=0.$ $[\textbf{Ans. } xe^y+y=c]$

57. $(x^4-2x^2y+y^4)\,dx-(2x^2y-4xy^3+\sin y)\,dy=0.$ $\left[\textbf{Ans. } \frac{x^5}{5}-x^2y^2+xy^4+\cos y=c\right]$

58. $(x^2-4xy-2y^2)\,dx+(y^2-4xy-2x^2)\,dy=0.$ $[\textbf{Ans. } x^3-6x^2y+6xy^2+y^3=c]$

59. $y(2xy+e^x)\,dx-e^x dy=0.$ $\left[\textbf{Ans. } x^2+\frac{e^x}{y}=c\right]$

60. $y(1+xy)\,dx+x(1-xy)\,dy=0.$ $\left[\textbf{Ans. } \log\left(\frac{x}{y}\right)-\frac{1}{xy}=c\right]$

61. $(xy-2y^2)\,dx-(x^2-2xy)\,dy=0.$ $\left[\textbf{Ans. } \frac{x}{y}+\log\left(\frac{y^2}{x^2}\right)=c\right]$

62. $xdy+ydx=2x^2ydx.$ $[\textbf{Ans. } xy=ce^{x^2}]$

63. $(x^2+y^2+x)\,dx+xydy=0.$ $[\textbf{Ans. } 4x^3+3x^4+6x^2y^2=c]$

64. $(2xy^4e^y+2xy^3+y)\,dx+(x^2y^4e^y-x^2y^2-3x)\,dy=0.$ $\left[\textbf{Ans. } x^2e^y+\frac{x^2}{y}+\frac{x}{y^3}=c\right]$

65. $y^2dx+(x^2-xy-y^2)\,dy=0.$ $[\textbf{Ans. } (x-y)y^2=c(x+y)]$

66. $y(2xy+1)\,dx+x(1+2xy-x^3y^3)\,dy=0.$ $\left[\textbf{Ans. } y=ce^{-(3xy+1)/3x^3y^3}\right]$

67. $(2x^3y^2+4x^2y+2xy^2+xy^4+2y)\,dx+2(y^3+x^2y+x)\,dy=0.$

$[\textbf{Ans. } (2x^2y^2+4xy+y^4)e^{x^2}=c]$

68. $(y^4+2y)\,dx+(xy^3+2y^4-4x)\,dy=0.$ $\left[\textbf{Ans. } xy+y^2+\frac{2x}{y^2}=c\right]$

69. $(y^3 - 2x^2 y)\,dx + (2xy^2 - x^3)\,dy = 0.$ [**Ans.** $x^2y^2(y^2 - x^2) = c^2$]

70. $y^3(2ydx + xdy) - (5ydx + 7xdy) = 0.$ [**Ans.** $2y^3 - 5 = cx^{10/3}y^{14/3}$]

71. $x^2y^3(2ydx + xdy) - (5ydx + 7xdy) = 0.$ [**Ans.** $x^{4/3}y^{2/3} + 2x^{-5/3}y^{-7/3} = c$]

72. $x(4ydx + 2xdy) + y^3(3ydx + 5xdy) = 0.$ [**Ans.** $x^4y^2 + x^3y^5 = c$]

73. $4\dfrac{dy}{dx} + 8y = 5e^{-3x}.$ $\left[\textbf{Ans. } ye^{2x} = -\dfrac{5}{4}e^{-x} + c\right]$

74. $x\dfrac{dy}{dx} + y = x\log x.$ [**Ans.** $4xy = 2x^2\log x - x^2 + c$]

75. $x\dfrac{dy}{dx} + y = xe^{-x}.$ [**Ans.** $xy = -e^{-x}(x+1) + c$]

76. $\dfrac{dy}{dx} + y = \sin x.$ $\left[\textbf{Ans. } y = ce^{-x} + \dfrac{1}{2}(\sin x - \cos x)\right]$

77. $\dfrac{dy}{dx} = y\tan x - 2\sin x.$ [**Ans.** $2y\cos x = \cos 2x + c$]

78. $\dfrac{dy}{dx} + 2x = e^{3x}.$ $\left[\textbf{Ans. } y + x^2 = \dfrac{1}{3}e^{3x} + c\right]$

79. $\dfrac{dy}{dx} + y\tan x = x^2\cos^2 x.$ [**Ans.** $y\sec x = x^2\sin x + 2x\cos x - 2\sin x + c$]

80. $xdy = (2y + 2x^4 + x^2)dx.$ [**Ans.** $y = x^4 + x^2\log x + cx^2$]

81. $dx + xdy = e^{-y}\sec^2 y\,dy.$ [**Ans.** $xe^y = \tan y + c$]

82. $\sin x\dfrac{dy}{dx} + y\cos x = 2\sin^2 x\cos x.$ $\left[\textbf{Ans. } y\sin x = \dfrac{2}{3}\sin^3 x + c\right]$

83. $x\dfrac{dy}{dx} + 2y = x\cos x.$ [**Ans.** $x^2y = x^2\sin x + 2x\cos x - 2\sin x + c$]

84. $(x + 2y^2)\dfrac{dy}{dx} = y$, given $y(2) = 1$. [**Ans.** $x = 2y^2$]

85. $y^2\dfrac{dx}{dy} + x - \dfrac{1}{y} = 0.$ $\left[\textbf{Ans. } x = 1 + \dfrac{1}{y} + ce^{1/y}\right]$

86. $\dfrac{dy}{dx} + \dfrac{1}{x}\sin 2y = x^2\cos^2 y.$ $\left[\textbf{Ans. } \tan y = \dfrac{x^3}{5} + cx^{-2}\right]$

87. $\dfrac{1}{y}\dfrac{dy}{dx}+\dfrac{x}{1-x^2}=xy^{-1/2}.$ $\left[\textbf{Ans. } \sqrt{y}=c(1-x^2)^{1/4}-\dfrac{1}{3}(1-x^2)\right]$

88. $\dfrac{dy}{dx}+y\sec x=\tan x.$ $[\textbf{Ans. } (y-1)(\sec x+\tan x)+x=c]$

89. $\dfrac{dy}{dx}+y\sec^2 x=\tan x\sec^2 x.$ $[\textbf{Ans. } y=\tan x-1+ce^{-\tan x}]$

90. $(x+1)\dfrac{dy}{dx}-ny=e^x(x+1)^{(n+1)}.$ $[\textbf{Ans. } y=(e^x+c)(x+1)^n]$

91. $\tan x\cos y\dfrac{dy}{dx}+\sin y+e^{\sin x}=0.$ $[\textbf{Ans. } \sin x\sin y+e^{\sin x}=c]$

92. $(4r^2s-6)dr+r^3ds=0.$ $\left[\textbf{Ans. } s=\dfrac{3}{r^2}+\dfrac{c}{r^4}\right]$

93. $x\dfrac{dy}{dx}-y=(x-1)e^x.$ $\left[\textbf{Ans. } \dfrac{y}{x}=c+\dfrac{e^x}{x}\right]$

94. $(1-x^2)\dfrac{dy}{dx}-xy=1.$ $[\textbf{Ans. } y\sqrt{1-x^2}=c+\sin^{-1}x]$

95. $x\sin\theta d\theta+(x^3-2x^2\cos\theta+\cos\theta)dx=0.$ $[\textbf{Ans. } 2\cos\theta=x(1+c\exp(-x^2))]$

96. $x\cos x\dfrac{dy}{dx}+y(x\sin x+\cos x)=1.$ $[\textbf{Ans. } xy\sec x=c+\tan x]$

97. $x\dfrac{dy}{dx}+y=x^2y^2\cos x.$ $\left[\textbf{Ans. } \dfrac{1}{xy}+x\sin x+\cos x=c\right]$

98. $\dfrac{dy}{dx}+y\cos x=y^n\sin 2x.$ $\left[\textbf{Ans. } \dfrac{1}{y^{n-1}}=ce^{(n-1)\sin x}+2\sin x+\dfrac{2}{n-1}\right]$

99. $(x^2y^2+xy)dy=dx.$ $\left[\textbf{Ans. } \dfrac{1}{x}=2-y^2+c\exp\left(-\dfrac{1}{2}y^2\right)\right]$

100. $x\dfrac{dy}{dx}+y=y^2\log x.$ $\left[\textbf{Ans. } \dfrac{1}{xy}=c+\dfrac{1}{x}(1+\log x)\right]$

101. $\dfrac{dy}{dx}+2xy=y^2e^{x^2}\log x.$ $\left[\textbf{Ans. } \dfrac{1}{y}=ce^{x^2}-x(\log x-1)e^{x^2}\right]$

102. $\dfrac{dy}{dx}-2y\tan x=y^2\tan^2 x.$ $\left[\textbf{Ans. } \dfrac{\sec^2 x}{y}+\dfrac{\tan^2 x}{3}=c\right]$

103. For the following problems

(i) Draw a directional field.

(ii) Comment on the solutions for large x.

(iii) Find the general solution and use it to determine the behaviour of the solution as $x \to \infty$.

(a) $\dfrac{dy}{dx} - 2y = x^2 e^{2x}$ $\left[\textbf{Ans. } y = ce^{2x} + \dfrac{x^3}{3}e^{2x};\ y \to \infty \text{ as } x \to \infty\right]$

(b) $\dfrac{dy}{dx} + y = xe^{-x} + 1$ $\left[\textbf{Ans. } y = ce^{-x} + 1 + \dfrac{x^2}{2}e^{-x};\ y \to 1 \text{ as } x \to \infty\right]$

(c) $\dfrac{dy}{dx} + \dfrac{1}{x}y = 3\cos 2x,\ x > 0$

$\left[\textbf{Ans. } y = \dfrac{c}{x} + \dfrac{3\cos 2x}{4x} + \dfrac{3\sin 2x}{2};\ y \text{ is asymptotics to } \dfrac{3\sin 2x}{2} \text{ as } x \to \infty\right]$

(d) $\dfrac{dy}{dx} + 2xy = 2xe^{-x^2}$ $[\textbf{Ans. } y = x^2 e^{-x^2} + ce^{-x^2};\ y \to 0 \text{ as } x \to \infty]$

(e) $(1 + x^2)\dfrac{dy}{dx} + 4xy = (1 + x^2)^{-2}$ $\left[\textbf{Ans. } y = \dfrac{\tan^{-1} x + c}{(1 + x^2)^2};\ y \to 0 \text{ as } x \to \infty\right]$

104. Find the value of y_0 for which the solution of the initial value problem $\dfrac{dy}{dx} - y = a + 3\sin x,\ y(0) = y_0$ remains finite as $x \to \infty$. $\left[\textbf{Ans. } y_0 = -\dfrac{5}{2}\right]$

105. Invest a principle P_0 at an annual rate of return r compounded continuously.

(a) Find the time t required for P_0 to double in value as a function of r. $\left[\textbf{Ans. } t = \dfrac{\ln 2}{r}\right]$

(b) Determine t if $r = 8\%$. [**Ans.** $t = 8.66$ years]

(c) Find the return rate that must be kept to double P_0 in 8 years. [**Ans.** $r = 8.66\%$]

106. Nita borrows ₹ 8000 to buy a scooter. The bank charges interest at an annual rate of 10%. Assuming that interest is compounded continuously and Nita makes payments continuously at a constant annual rate r. Determine the payment rate r that is required to pay-off the loan in 3 years. Also calculate how much interest she will have to pay during the 3 years period. [**Ans.** $r = 30.86\%$, interest paid = ₹ 1259.92]

107. Nita can afford to spend no more than ₹ 800/month as EMI. Suppose that the interest rate is 9% and 20 years is EMI time. Assume that interest is compounded continuously and that the payments are also made continuously.

(a) Determine the maximum amount that she can afford to borrow.
[Ans. ₹ 89,034.79]

(b) Determine the total interest paid during the term of EMI.
[Ans. ₹ 1,02,965.21]

108. The American Chemist W.F. Libby (Nobel prize winner in chemistry in 1960) developed radiocarbon dating, an important tool in archeological research. This is used to determine the age of certain wood and plant remains, hence of animal or human bones. Radiocarbon dating is based on the fact that some wood or plant remains contain residual amounts of carbon-14, a radioactive isotope of carbon. This isotope is accumulated during the lifetime of plant and begins to decay as its death. Since the half-life of carbon-14 is long (approximately 5730 years), McGraw-Hill Encyclopedia of Science and Technology, 8th edition, New York, 1997, Vol. 5, p. (48), significant amount of carbon-14 remains after many thousands of years. If $Q(t)$ is the amount of carbon-14 at time t and Q_0 is the original amount, then the ratio $\dfrac{Q(t)}{Q_0}$ can be computed.

(a) $\dfrac{dQ(t)}{dt} = -rQ(t)$ is the required differential equation. Determine the decay rate r for carbon-14.
[Ans. $r = 0.00012097$/year]

(b) Find $Q(t)$ with initial condition $Q(0) = Q_0$.
[Ans. $Q(t) = Q_0 \exp(-rt)$; where r is as given in (a)]

(c) Suppose that certain remains are found in which the current residual amount of carbon-14 is 20% of the original amount. Determine the age of these remains.
[Ans. 13,305 years]

109. Heat transfer from a body to its surroundings by radiation, based on Stefan–Boltzmann law, is described by the differential equation $\dfrac{dx(t)}{dt} = -\alpha(x(t)^4 - T^4)$ where $x(t)$ is the absolute temperature of the body at time t, T is the absolute temperature of surroundings, and α is a constant depending on the physical parameters of the body.

However, if $x(t) >> T$, then the differential equation reduces to $\dfrac{dx(t)}{dt} = -\alpha x(t)^4$.

Suppose that a body with initial temperature 2000 K is surrounded by a medium with temperature 300 K and $\alpha = 2.0 \times 10^{-12}$ k^{-3}/s.

(a) Find body temperature at any time t.
$\left[\textbf{Ans. } x(t) = \dfrac{2000}{(1+0.048t)^{1/3}}\right]$

(b) Plot a graph of $x(t)$ versus t.

(c) Find the time t_0 at which $x(t_0) = 600$.
[Ans. $t_0 = 750.77$ seconds]

110. Consider an insulated box with internal temperature $x(t)$. According to Newton's law of cooling $x(t)$ satisfies the differential equation $\dfrac{dx(t)}{dt} = -k(x(t) - T(t))$ where $T(t)$ is the ambient temperature. Let $T(t) = T_0 + T_1 \cos wt$.

(a) Solve differential equation.

$$\left[\textbf{Ans. } x(t) = ce^{-kt} + T_0 + \frac{kT_1}{k^2 + w^2}(k \cos wt + w \sin wt)\right]$$

Note: The first term in the solution $\to 0$ as $t \to \infty$. This is called the transient part. The remainder of the solution is called the steady state, $S(t)$ (say).

(b) Let $T_0 = 60\ °\text{F}, T_1 = 15\ °\text{F}, k = 0.2/h$ and $w = \dfrac{\pi}{12}$. Draw graphs of $S(t)$ and $T(t)$ versus t on the same axis. From the graph, estimate amplitude R of the oscillating part of $S(t)$. Also calculate the time lag L between corresponding maxima of $S(t)$ and $T(t)$. [**Ans.** R = 9.11 °F, L = 3.51 hours]

111. A rocket sled having an initial speed of 160 km/h is slowed by a channel of water. Assume that, during the braking process, the acceleration a is given by $a(v) = -\mu v^2$, where v is the velocity and μ is a constant.

(a) Using the relation $\dfrac{dv}{dt} = v\dfrac{dv}{dx}$, write the differential equation in terms of v and x.

$$\left[\textbf{Ans. } \frac{dv}{dx} = -\mu v\right]$$

(b) If it requires a distance of 2000 m to slow the sled to 15 km/h, determine the value of μ. [**Ans.** μ = 6.2492/km]

112. A body of mass m is projected vertically upward with an initial velocity v_0 in a medium offering a resistance kv, where k is a constant. Assume that the gravitational force of the earth is constant.

(a) Find the velocity $v(t)$ of the body at any time.

$$\left[\textbf{Ans. } v(t) = -\frac{mg}{k} + \left[v_0 + \frac{mg}{k}\right]\exp\left(-\frac{kt}{m}\right)\right]$$

(b) Calculate $v(t)$ as $k \to 0$. [**Ans.** $v(t) = v_0 - gt$]

Note: This agrees with the velocity of a mass m projected upward with an initial velocity v_0 in vaccum.

113. By means of substitution $y = vx$, solve the equation $\dfrac{dy}{dx} = \dfrac{y^2}{x^2} + \dfrac{y}{x} + 1$.

[**Ans.** $\tan^{-1} v = \ln |cx|$]

114. Find the general solution of $\dfrac{dy}{dx} = \ln x$. Find the particular solution satisfying $y(1) = 1$.

[**Ans.** $y = x\ln|x| - x + c,\ y = x\ln|x| - x + 2$]

115. Find the general solution of $\dfrac{dy}{dx} = x(y-2)$. Find the particular solution satisfying $y(0) = 5$.

[**Ans.** $y = 2 + ce^{t^2/2},\ y = 2 + 3e^{t^2/2}$]

116. RL-circuit with ramp input: The differential equation governing current $i(t)$ at any time t in RL-circuit when a voltage $u(t)$ is applied, is given by $iR + L\dfrac{di}{dt} = t, t \geq 0$ and $i(0) = 0$. Solve this equation to find $i(t)$. $\left[\textbf{Ans. } i(t) = \dfrac{t}{R} + \dfrac{L}{R^2}(e^{-Rt/L} - 1)\right]$

117. Find the general solution of $\dfrac{dy}{dx} = 2y + 4x$. What is the particular solution which satisfies $y(1) = 2$? [**Ans.** $y = -2x - 1 + ce^{2t}, y = -2t - 1 + 5e^{2(t-1)}$]

118. Solve $\dfrac{dy}{dx} = x - yx, \quad y(0) = 0.$ $\left[\textbf{Ans. } y = 1 - e^{-\frac{x^2}{2}}\right]$

119. Use an integrating factor to obtain the general solution of $iR + L\dfrac{di}{dt} = \sin wt$, where R, L and w are constants. $\left[\textbf{Ans. } i(t) = \dfrac{L}{R^2 + L^2w^2}\left[\dfrac{R}{L}\sin wt - w\cos wt\right] + ce^{-\frac{Rt}{L}}\right]$

120. Find the general solution of $\dfrac{dy}{dx} + 2xy = x$. Find the particular solution satisfying the condition $y(0) = -1$. $\left[\textbf{Ans. } y = \dfrac{1}{2}(1 - 3e^{-t^2})\right]$

Chapter 3

Differential Equations of First Order and of Higher Degree

3.1 INTRODUCTION

In this chapter, we shall discuss the methods to solve the differential equations of first order and of higher degree. Write $\frac{dy}{dx}$ as p. The most general form of the equation is

$$p^n + Q_1 p^{n-1} + Q_2 p^{n-2} + \cdots + Q_n = 0 \tag{3.1}$$

where Q_1, Q_2, ..., Q_n are functions of x and y. Two cases may arise:

(i) When Eq. (3.1) can be factorized into linear factors;
(ii) When Eq. (3.1) can not be factorized into linear factors.

3.2 DIFFERENTIAL EQUATIONS WHICH CAN BE FACTORIZED INTO LINEAR FACTORS

We can write Eq. (3.1) as

$$(p - R_1)(p - R_2) \dots (p - R_n) = 0 \tag{3.2}$$

To obtain the solution of Eq. (3.2) equate each of the factors in Eq. (3.2) to zero. Solve the equations by the methods stated in Chapter 2.

Let the solutions be

$$\left.\begin{aligned} f_1(x, y, c_1) &= 0 \\ f_2(x, y, c_2) &= 0 \\ &\vdots \\ f_n(x, y, c_n) &= 0 \end{aligned}\right\} \tag{3.3}$$

where each of the constants c_k can take infinite number of values. To get the general solution take $c_1 = c_2 = \cdots = c_n = c$. Then the n solutions are

$$\left.\begin{aligned} f_1(x, y, c) &= 0 \\ f_2(x, y, c) &= 0 \\ &\vdots \\ f_n(x, y, c) &= 0 \end{aligned}\right\} \tag{3.4}$$

Therefore, the general solution of Eq. (3.1) is

$$f_1(x, y, c)\, f_2(x, y, c) \ldots f_n(x, y, c) = 0 \tag{3.5}$$

EXAMPLE 3.1 Solve $p^2 - 4p + 3 = 0$.

Solution Given

$$p^2 - 4p + 3 = 0$$

$$\Rightarrow \qquad (p - 3)\,(p - 1) = 0$$

$$p - 3 = 0 \text{ gives } \frac{dy}{dx} = 3 \Rightarrow y = 3x + c$$

and $$p - 1 = 0 \text{ gives } \frac{dy}{dx} = 1 \Rightarrow y = x + c$$

Therefore, the general solution is

$$(y - 3x - c)\,(y - x - c) = 0.$$

EXAMPLE 3.2 Solve $p - \dfrac{1}{p} = \dfrac{x}{y} - \dfrac{y}{x}$.

Solution Given

$$p^2 + p\left(\frac{x}{y} - \frac{y}{x}\right) - 1 = 0$$

$$\Rightarrow \qquad \left(p + \frac{y}{x}\right)\left(p - \frac{x}{y}\right) = 0$$

Then $$p + \frac{y}{x} = 0 \tag{i}$$

and $$p - \frac{x}{y} = 0 \tag{ii}$$

From Eq. (i), $$\frac{dy}{dx} + \frac{y}{x} = 0$$

$$\Rightarrow \qquad xdy = ydx = 0$$

$$\Rightarrow \qquad d(xy) = 0$$

Integrating, $$xy = c.$$

From Eq. (ii), $$\frac{dy}{dx} - \frac{x}{y} = 0$$

$\Rightarrow$ $$ydy = xdx = 0$$

$\Rightarrow$ $$\frac{1}{2}d(y^2 - x^2) = 0$$

Integrating, $$y^2 - x^2 = c$$

Therefore, the general solution is

$$(xy - c)(y^2 - x^2 - c) = 0.$$

EXAMPLE 3.3 Solve $p^3 + 2xp^2 - y^2p^2 - 2xy^2p = 0$.

Solution Given

$$p^3 + 2xp^2 - y^2p^2 - 2xy^2p = 0$$

$\Rightarrow$ $$p(p + 2x)(p - y^2) = 0$$

$\Rightarrow$ $$p = 0$$

$\Rightarrow$ $$y = c$$

$$p + 2x = 0$$

$\Rightarrow$ $$\frac{dy}{dx} + 2x = 0$$

$\Rightarrow$ $$dy + 2xdx = 0$$

$\Rightarrow$ $$d(y + 2x^2) = 0$$

Integrating, $$y + 2x^2 = c$$

and $$p - y^2 = 0$$

$\Rightarrow$ $$\frac{dy}{dx} - y^2 = 0$$

$\Rightarrow$ $$\frac{dy}{y^2} - dx = 0$$

$\Rightarrow$ $$d\left(-\frac{1}{y} - x\right) = 0$$

Integrating, $$-\frac{1}{y} - x = c$$

$\Rightarrow$ $$cy = -xy - 1$$

Therefore, the general solution is

$$(y - c)(y + x^2 - c)(xy + cy + 1) = 0.$$

EXAMPLE 3.4 Solve $4y^2p^2 + 2pxy(3x+1) + 3x^3 = 0.$

Solution Given

$$4y^2p^2 + 2pxy(3x+1) + 3x^3 = 0$$

$\Rightarrow$ $$(2yp + 3x^2)(2yp + x) = 0$$

Now, $$2yp + x = 0$$

$\Rightarrow$ $$2y\frac{dy}{dx} + 3x^2 = 0$$

$\Rightarrow$ $$2ydy + 3x^2dx = 0$$

Integrating, $$y^2 + x^3 = c$$

and $$2yp + x = 0$$

$\Rightarrow$ $$2y\frac{dy}{dx} + x = 0$$

$\Rightarrow$ $$2ydy + xdx = 0$$

Integrating, $$y^2 + \frac{x^2}{2} = c$$

Therefore, the general solution is

$$(x^3 + y^2 - c)\left(\frac{x^2}{2} + y^2 - c\right) = 0.$$

EXAMPLE 3.5 Solve $p^2 + p\left(x + y - \frac{2y}{x}\right) + xy + \frac{y^2}{x^2} - y - \frac{y^2}{x} = 0.$

Solution Given

$$p^2 + p\left(x + y - \frac{2y}{x}\right) + xy + \frac{y^2}{x^2} - y - \frac{y^2}{x} = 0$$

$\Rightarrow$ $$\left(p - \frac{y}{x} + x\right)\left(p - \frac{y}{x} + y\right) = 0$$

Now, $$p - \frac{y}{x} + x = 0$$

$\Rightarrow$ $$\frac{dy}{dx} - \frac{1}{x}y = -x$$

which is linear equation with IF $= \frac{1}{x}$.

$\therefore$ Solution of this equation is

$$y\frac{1}{x} = -x + c$$

and $$p-\frac{y}{x}+y=0$$

$$\Rightarrow \qquad \frac{dy}{dx}-y\left(\frac{1}{x}-1\right)=0$$

$$\Rightarrow \qquad \frac{dy}{y}=\left(\frac{1}{x}-1\right)dx$$

Integrating, $$\log y = \log x - x + c$$

$$\Rightarrow \qquad y = cxe^{-x}$$

Therefore, the general solution is

$$(x^2 + y - cx)\,(y - cxe^{-x}) = 0.$$

3.3 DIFFERENTIAL EQUATIONS WHICH CANNOT BE FACTORIZED INTO LINEAR FACTORS

Consider the equation of the form

$$f(x, y, p) = 0 \tag{3.6}$$

It may take following forms:

(i) Solvable for y
(ii) Solvable for x
(iii) Solvable for p
(iv) Either x or y is absent
(v) Homogeneous in x and y
(vi) First degree in x and y

Solution methodology for equations of type (iii) is discussed in Section 3.2.

3.4 DIFFERENTIAL EQUATION SOLVABLE FOR *y*

Let the solution of Eq. (3.6) be

$$y = F(x, p) \tag{3.7}$$

Differentiate Eq. (3.7) w.r.t. x.

$$\Rightarrow \qquad p=\phi\left(x, p, \frac{dp}{dx}\right) \tag{3.8}$$

Equation (3.8) is an equation in two variables x and p and may be solved. Let its solution be

$$\phi(x, p, c) = 0 \tag{3.9}$$

Eliminate p from Eqs. (3.7) and (3.9), we shall get a solution.

Note: If it is not possible to eliminate p from Eqs. (3.7) and (3.9), the values of x and y in terms of p as a parameter can be found and these together give a solution.

EXAMPLE 3.6 Solve $y = 2px + p^n$.

Solution Given

$$y = 2px + p^n \tag{i}$$

Differentiating Eq. (i) w.r.t. x gives

$$\frac{dy}{dx} = 2p + 2x\frac{dp}{dx} + np^{n-1}\frac{dp}{dx}$$

$$\Rightarrow \quad p = 2p + 2x\frac{dp}{dx} + np^{n-1}\frac{dp}{dx}$$

$$\Rightarrow \quad p\frac{dx}{dp} + 2x = -np^{n-1}$$

$$\Rightarrow \quad \frac{dx}{dp} + \frac{2}{p}x = -np^{n-2} \tag{ii}$$

This is linear equation in x and p.

$$\therefore \quad \text{IF} = e^{\int \frac{2}{p}dp} = e^{\log p^2} = p^2$$

$\therefore$ Solution of Eq. (ii) is

$$xp^2 = \int -np^{n-2}p^2 dp + c$$

$$\Rightarrow \quad xp^2 = -n\int p^n dp + c$$

$$\Rightarrow \quad xp^2 = -\frac{np^{n+1}}{n+1} + c$$

$$\Rightarrow \quad x = cp^{-2} - \frac{np^{n-1}}{n+1} \tag{iii}$$

Substitute this value of x in Eq, (i), we get

$$y = \frac{2c}{p} + \frac{1-n}{1+n}p^n \tag{iv}$$

The Eqs. (iii) and (iv) taken together with parameter p, gives general solution of given equation.

EXAMPLE 3.7 Solve $y - 2px = \tan^{-1}(xp^2)$.

Solution Given

$$y - 2px = \tan^{-1}(xp^2) \tag{i}$$

Differentiating Eq. (i) w.r.t. x gives

$$\frac{dy}{dx} = 2p + 2x\frac{dp}{dx} + \frac{p^2 + 2xp\frac{dp}{dx}}{1 + x^2p^4}$$

$$\Rightarrow \qquad p = 2p + 2x\frac{dp}{dx} + \frac{p}{1+x^2p^4}\left(p + 2x\frac{dp}{dx}\right)$$

$$\Rightarrow \qquad p + 2x\frac{dp}{dx} + \frac{p}{1+x^2p^4}\left(p + 2x\frac{dp}{dx}\right) = 0$$

$$\Rightarrow \qquad \left(p + 2x\frac{dp}{dx}\right)\left(1 + \frac{p}{1+x^2p^4}\right) = 0$$

Then $\qquad p + 2x\dfrac{dp}{dx} = 0$

$$\Rightarrow \qquad 2\frac{dp}{p} + \frac{dx}{x} = 0$$

Integrating, $\qquad 2 \log p + \log x = \log c$

$$\Rightarrow \qquad xp^2 = c$$

$$\Rightarrow \qquad p = \sqrt{\frac{c}{x}} \qquad \text{(ii)}$$

Eliminating p from Eqs. (i) and (ii), gives

$$y = 2x\sqrt{\frac{c}{x}} + \tan^{-1} c$$

$$\Rightarrow \qquad y = 2\sqrt{cx} + \tan^{-1} c$$

which is the general solution of given equation.

EXAMPLE 3.8 Solve $xp^2 - yp - y = 0$.

Solution Given

$$xp^2 - yp - y = 0 \qquad \text{(i)}$$

$$\Rightarrow \qquad y = \frac{xp^2}{p+1} \qquad \text{(ii)}$$

Differentiating Eq. (ii) w.r.t. x gives

$$p = \frac{p^2}{p+1} + x\left[\frac{2p}{p+1} - \frac{p^2}{(p+1)^2}\right]\frac{dp}{dx}$$

$$\Rightarrow \qquad \frac{p}{p+1} = \frac{x(2p+p^2)}{(p+1)^2}\frac{dp}{dx}$$

$$\Rightarrow \qquad p(p+1) - xp(p+2)\frac{dp}{dx} = 0$$

$$\Rightarrow \qquad p\left[(p+1) - x(p+2)\frac{dp}{dx}\right] = 0$$

Then either $p = 0$ or

$$\Rightarrow \quad (p+1) - x(p+2)\frac{dp}{dx} = 0$$

$$\Rightarrow \quad \frac{dx}{x} = \frac{p+2}{p+1}dp$$

$$\Rightarrow \quad \frac{dx}{x} = \left(1+\frac{1}{p+1}\right)dp$$

Integrating, $\log x = p + \log [(p+1)c]$

$$\Rightarrow \quad x = c(p+1)e^p \tag{iii}$$

Putting value of x in Eq. (ii) gives

$$y = cp^2e^p \tag{iv}$$

Equations (iii) and (iv) together give the general solution.

EXAMPLE 3.9 Solve $x - yp = ap^2$.

Solution Given

$$x - yp = ap^2 \tag{i}$$

$$y = \frac{x-ap^2}{p} \tag{ii}$$

Differentiating Eq. (ii) w.r.t. x and simplifying it gives

$$(ap^2+x)\frac{dp}{dx} = p(1-p^2)$$

$$\Rightarrow \quad \frac{dx}{dp} - \frac{1}{p(1-p^2)}x = \frac{ap}{1-p^2}$$

which is linear equation whose solution is

$$x = \frac{p}{\sqrt{1-p^2}}(c - a\sin^{-1} p)$$

Substituting value of x in Eq. (ii) gives

$$y = -ap + \frac{1}{\sqrt{1-p^2}}(c + a\sin^{-1} p)$$

which is the general solution of given equation in parameter p.

3.5 DIFFERENTIAL EQUATION SOLVABLE FOR x

Here, the differential equation is of the form

$$x = F(y, p) \tag{3.10}$$

Differentiating Eq. (3.10) w.r.t. y, let we get

$$\frac{1}{p} = \phi\left(y, p, \frac{dp}{dy}\right) \tag{3.11}$$

Let the solution of Eq. (3.11) be

$$\varphi(y, p, c) = 0 \tag{3.12}$$

The elimination of p between Eqs. (3.10) and (3.12) gives the general solution of the Eq. (3.10).

EXAMPLE 3.10 Solve $y = 2xp + y^2p^3$.

Solution Given

$$y = 2xp + y^2p^3 \tag{i}$$

$$\Rightarrow \qquad x = \frac{y - y^2p^3}{2p} \tag{ii}$$

Differentiating Eq. (ii) w.r.t. y gives

$$\frac{dx}{dy} = \frac{1}{2}\frac{p\left(1 - 2yp^3 - y^3p^2\frac{dp}{dy}\right) - (y - y^2p^3)\frac{dp}{dy}}{p^2}$$

$$\Rightarrow \qquad \frac{2}{p} = \frac{1}{p^2}\left(p - 2yp^4 - y^3p^3\frac{dp}{dy}\right) - (y - y^2p^3)\frac{dp}{dy}$$

$$\Rightarrow \qquad p + 2yp^4 + 2y^2p^3\frac{dp}{dy} + y\frac{dp}{dy} = 0$$

$$\Rightarrow \qquad p(1 + 2yp^3) + y(1 + 2yp^3)\frac{dp}{dy} = 0$$

$$\Rightarrow \qquad \left(p + y\frac{dp}{dy}\right)(1 + 2py^3) = 0$$

Then

$$p + y\frac{dp}{dy} = 0$$

$$\Rightarrow \qquad \frac{d}{dy}(py) = 0$$

Integrating,

$$py = c$$

$$\Rightarrow \qquad p = \frac{c}{y}$$

Substituting $p = \frac{c}{y}$ in Eq. (i) gives

$$y^2 = 2cx + c^3$$

which is the required solution.

EXAMPLE 3.11 Solve $p^3 - 2xyp + 4y^2 = 0$.

Solution Given equation can be written as

$$x = \frac{p^2}{2y} + \frac{2y}{p} \tag{i}$$

Differentiating Eq. (i) w.r.t. y gives

$$\frac{1}{p} = -\frac{p^2}{2y^2} + \frac{2}{p} + \frac{dp}{dy}\left(\frac{p}{y} - \frac{2y}{p^2}\right)$$

$$\Rightarrow \quad \left(\frac{1}{p} - \frac{p^2}{2y^2}\right) + \left(\frac{p}{y} - \frac{2y}{p^2}\right)\frac{dp}{dy} = 0$$

$$\Rightarrow \quad \left(\frac{1}{p} - \frac{p^2}{2y^2}\right) - \frac{2y}{p}\left(\frac{1}{p} - \frac{p^2}{2y^2}\right)\frac{dp}{dy} = 0$$

$$\Rightarrow \quad \left(1 - \frac{2y}{p}\frac{dp}{dy}\right)\left(\frac{1}{p} - \frac{p^2}{2y^2}\right) = 0$$

Now, $$1 - \frac{2y}{p}\frac{dp}{dy} = 0$$

$$\Rightarrow \quad 2\frac{dp}{p} = \frac{dy}{y}$$

Integrating, $$p^2 = cy$$

$$\Rightarrow \quad p = \sqrt{cy}$$

Substitute value of p in given equation, we have

$$16y = c(2x - c)^2$$

which is the required solution.

EXAMPLE 3.12 Solve $x = y + a \log p$.

Solution Given

$$x = y + a \log p \tag{i}$$

Differentiating Eq. (i) w.r.t. y gives

$$\frac{dx}{dy} = 1 + \frac{a}{p}\frac{dp}{dy}$$

$$\Rightarrow \quad \frac{1}{p} = 1 + \frac{a}{p}\frac{dp}{dy}$$

$$\Rightarrow \quad 1 - p = a\frac{dp}{dy}$$

$$\Rightarrow \qquad dy = \frac{adp}{1-p}$$

Integrating, $y = c - a\log(1-p)$

Substituting y in given equation gives

$$x = c + a\log\left(\frac{p}{1-p}\right)$$

which is the solution in parameter p.

3.6 DIFFERENTIAL EQUATION IN WHICH EITHER *x* OR *y* IS ABSENT

(a) Suppose Eq. (3.4) does not contain x. Then given equation will be of the form

$$f(y, p) = 0 \tag{3.13}$$

If it is solvable for p, we get

$$p = F(y)$$

$$\Rightarrow \qquad \frac{dy}{dx} = F(y)$$

which can be integrated by method of variable separation discussed in Chapter 2. If Eq. (3.13) is solvable for y, then proceed as Section 3.4.

(b) Suppose Eq. (3.4) does not contain y. Then given equation will be of the form

$$f(x, p) = 0 \tag{3.14}$$

If it is solvable for p, we get

$$p = F(x)$$

$$\Rightarrow \qquad \frac{dy}{dx} = F(x)$$

which can be integrated by method of variable separation discussed in Chapter 2.

If Eq. (3.14) is solvable for x, then proceed as Section 3.5.

EXAMPLE 3.13 Solve $y = mp + mp^2$, where m and n are constants.

Solution Given $y = mp + mp^2$ (i)

Differentiating w.r.t. x, gives

$$\frac{dy}{dx} = (m + 2np)\frac{dp}{dx}$$

$$\Rightarrow \qquad p = (m + 2np)\frac{dp}{dx}$$

$$\Rightarrow \qquad dx = \left(\frac{m}{p} + 2n\right)dp$$

On integrating, $x = m\log p + 2np + c$ (ii)

Equation (ii) together with Eq. (i) gives the general solution of the given differential equation.

EXAMPLE 3.14 Solve $x^2 = 4(1 + p^2)$.

Solution Solving given equation for x gives

$$x = 2\sqrt{1+p^2} \tag{i}$$

Differentiating w.r.t. y, gives

$$\frac{1}{p} = \frac{2p}{\sqrt{1+p^2}}\frac{dp}{dy}$$

$$dy = 2\frac{p^2}{\sqrt{1+p^2}}dp$$

$$= 2\left(\sqrt{1+p^2} - \frac{1}{\sqrt{1+p^2}}\right)dp$$

On integrating, $$y = 2\left[\frac{p\sqrt{1+p^2}}{2} + \frac{1}{2}\log(p+\sqrt{1+p^2}) - \log(p+\sqrt{1+p^2})\right] + c$$

$\Rightarrow$ $$y = 2\left[\frac{p\sqrt{1+p^2}}{2} - \frac{1}{2}\log(p+\sqrt{1+p^2})\right] + c$$

$\Rightarrow$ $$y = p\sqrt{1+p^2} - \log(p+\sqrt{1+p^2}) + c$$

This with given equation gives the general solution.

3.7 DIFFERENTIAL EQUATION HOMOGENEOUS IN x AND y

Here, the general form of the equation is

$$f\left(p, \frac{y}{x}\right) = 0$$

(a) If it is possible to solve for p, then proceed as Chapter 2 to find the solution of the given differential equation.

(b) If it is solvable for $\frac{y}{x}$ then obtain $y = xf(p)$. Find solution using method discussed in Section 3.4.

3.8 DIFFERENTIAL EQUATION OF FIRST DEGREE IN x AND y

Let the Eq. (3.4) is of the first degree in x and y. If it is solvable for x and y both, we get the form

$$y = xf(p) + g(p) \tag{3.15}$$

which can be solved by the method discussed in Section 3.4.

If in Eq. (3.15),

$$f(p) = p$$

Then we get

$$y = px + f(p) \tag{3.16}$$

which is known as **Clairaut's equation**.
Differentiating Eq. (3.16) w.r.t. x gives

$$p = p + [x + f'(p)]\frac{dp}{dx}$$

where $$f'(p) = \frac{df}{dp}.$$

$$\Rightarrow \qquad [x + f'(p)]\frac{dp}{dx} = 0$$

Then either

$$x + f'(p) = 0 \tag{3.17}$$

or

$$\frac{dp}{dx} = 0 \Rightarrow p = c$$

Substituting $p = c$ in Eq. (3.16) gives

$$y = cx + f(c)$$

which is the required solution of the Eq. (3.16).

Eliminating p between Eqs. (3.16) and (3.17), a relation between x and y containing no arbitrary constant can be obtained. However, this will not be the general solution.

EXAMPLE 3.15 Solve $p = \sin(y - xp)$.

Solution Given

$$p = \sin(y - xp)$$

$$\Rightarrow \qquad y = px + \sin^{-1} p$$

which is the Clairaut's equation. Therefore, the solution is

$$y = cx + \sin^{-1} c.$$

EXAMPLE 3.16 Solve $y = (1 + p)\,x + p^2$.

Solution Given

$$y = (1 + p)\,x + p^2 \tag{i}$$

Differentiating w.r.t. x gives

$$\frac{dy}{dx} = 1 + p + (x + 2p)\frac{dp}{dx}$$

$$\Rightarrow \qquad p = 1 + p + (x + 2p)\frac{dp}{dx}$$

$$\Rightarrow \qquad \frac{dx}{dp} + x = -2p$$

which is linear equation whose solution is given by

$$x = 2(1 - p) + ce^{-p} \tag{ii}$$

Therefore, Eq. (i) becomes

$$y = (1+p)[2(1-p) + ce^{-p}] + p^2 = 2(1-p^2) + c(1+p)e^{-p} + p^2 \quad \text{(iii)}$$

Therefore, the solution of given differential equation is the relation between Eqs. (ii) and (iii).

EXAMPLE 3.17 Solve $y = 2px + y^2p^3$.

Solution Given

$$y = 2px + y^2p^3$$

Put

$$y^2 = V.$$

Then

$$\frac{dy}{dx} = \frac{1}{2\sqrt{V}}\frac{dV}{dx} = \frac{1}{2\sqrt{V}}P$$

where

$$P = \frac{dV}{dx}.$$

$\Rightarrow$

$$p = \frac{1}{2\sqrt{V}}P$$

Then, given equation reduces to

$$\sqrt{V} = 2x\frac{1}{2\sqrt{V}}P + V\frac{1}{8V\sqrt{V}}P^3$$

$\Rightarrow$

$$p = xP + \frac{1}{8}P^3$$

which is the Clairaut's equation. Therefore, the solution is

$$V = cx + \frac{1}{8}c^3$$

Hence, the solution of given differential equation is

$$y^2 = cx + \frac{1}{8}c^3.$$

EXAMPLE 3.18 Solve $y = 2px + y^{n-1}p^n$.

Solution Given

$$y = 2px + y^2p^3$$

Put

$$y^2 = V.$$

Then

$$\frac{dy}{dx} = \frac{1}{2\sqrt{V}}\frac{dV}{dx} = \frac{1}{2\sqrt{V}}P$$

where

$$P = \frac{dV}{dx}.$$

$\Rightarrow$

$$p = \frac{1}{2\sqrt{V}}P$$

Then, given equation reduces to

$$\sqrt{V} = 2x\frac{1}{2\sqrt{V}}P + V^{\frac{n-1}{2}}\frac{1}{\sqrt{V}}\left(\frac{P}{2}\right)^n$$

$\Rightarrow$
$$V = xP + \left(\frac{P}{2}\right)^n$$

which is the Clairaut's equation. Therefore, the solution is

$$V = cx + \left(\frac{c}{2}\right)^n$$

Hence, the solution of given differential equation is

$$y^2 = cx + \left(\frac{c}{2}\right)^n.$$

EXAMPLE 3.19 Solve $xyp^2 - (x^2 + y^2 - 1)p + xy = 0$.

Solution Given

$$xyp^2 - (x^2 + y^2 - 1)p + xy = 0$$

Put $x^2 = U$ and $y^2 = V$.

Then
$$p = \sqrt{\frac{U}{V}}P$$

where
$$P = \frac{dV}{dU}.$$

Therefore, given equation reduces to

$$UP - V + \frac{P}{P-1} = 0$$

$\Rightarrow$
$$V = UP + \frac{P}{P-1}$$

which is the Clairaut's equation and its solution is

$$V = Uc + \frac{c}{c-1}.$$

Therefore, the solution of the given equation is

$$y^2 = cx^2 + \frac{c}{c-1}.$$

EXAMPLE 3.20 Solve $(xp - y)^2 = a(1 + p^2)(x^2 + y^2)^{3/2}$.

Solution Put $x = r\cos\theta$ and $y = r\sin\theta$.

Then $r^2 = x^2 + y^2$ and $\theta = \tan^{-1}\frac{y}{x}$.

$$\therefore \qquad \frac{d\theta}{dx} = \frac{1}{1+\dfrac{y^2}{x^2}}\,\frac{xp-y}{x^2} = \frac{xp-y}{x^2+y^2}$$

$$\therefore \qquad r^2\frac{d\theta}{dx} = xp-y$$

Also,
$$2r\frac{dr}{dx} = 2x+2px$$

$$\Rightarrow \qquad r\frac{dr}{dx} = x+px$$

$$\therefore \qquad \frac{1}{r}\frac{dr}{d\theta} = \frac{x+yp}{xp-y}$$

$$\therefore \qquad 1+\left(\frac{1}{r}\frac{dr}{d\theta}\right)^2 = \frac{(1+p^2)(x^2+y^2)}{(xp-y)^2}$$

$$\Rightarrow \qquad \frac{(xp-y)^2}{1+p^2} = \frac{r^4}{r^2+\left(\dfrac{dr}{d\theta}\right)^2}$$

$$\Rightarrow \qquad \frac{r^4}{r^2+\left(\dfrac{dr}{d\theta}\right)^2} = ar^3 \quad \text{(from given equation)}$$

$$\Rightarrow \qquad \frac{r}{a} = r^2+\left(\frac{dr}{d\theta}\right)^2$$

$$\Rightarrow \qquad \frac{dr}{d\theta} = \sqrt{\frac{r}{a}(1-ar)}$$

$$\Rightarrow \qquad d\theta = \sqrt{a}\,\frac{dr}{\sqrt{r(1-ar)}}$$

Put
$$r = \frac{1}{a}\sin^2\alpha.$$

Then
$$d\theta = \frac{\sqrt{a}\,\dfrac{1}{a}\,2\sin\alpha\cos\alpha\, d\alpha}{\sqrt{\dfrac{1}{a}\sin^2\alpha\cos^2\alpha}} = 2d\alpha$$

On integration,
$$2\alpha = \theta + c$$

$$\Rightarrow \qquad 2\sin^{-1}\sqrt{ar} = \theta + c$$

Therefore, the solution of the given differential equation is

$$2\sin^{-1}[\sqrt{a}(x^2+y^2)^{1/4}] = \tan^{-1}\frac{y}{x}+c.$$

EXAMPLE 3.21 Solve $8y = px(12 - 9p^2)$.

Solution On differentiating the given equation with respect to x, we get

$$8p = p(12 - 9p^2) + x(12 - 27p^2)\frac{dp}{dx}$$

which simplifies to $\frac{dx}{x} + 3\frac{dp}{p} = 0.$

Integrating equation gives $xp^3 = c$ or $p = \left(\frac{c}{x}\right)^{1/3}$.

Therefore, the solution is $(8y + 9c)^3 = 1728cx^2$.

EXAMPLE 3.22 Solve $(py + nx)^2 = (y^2 + nx^2)(1 + p^2)$.

Solution Substitute $y = vx$. Then $p = xP + v$ where $P = \frac{dv}{dx}$. With this substitution, given equation reduces to $x\frac{dv}{dx} = \pm\sqrt{\frac{n-1}{n}}\sqrt{(n + v^2)}$

Separating variables gives

$$\frac{dv}{\sqrt{(n + v^2)}} = \pm\sqrt{\frac{n-1}{n}}\frac{dx}{x}$$

On integrating, we get

$$\log(v + \sqrt{n + v^2}) + c = \pm\sqrt{\frac{n-1}{n}}\log x \Rightarrow v + \sqrt{n + v^2} = cx^{\pm\sqrt{\frac{n-1}{n}}}$$

Therefore, the solution of given equation is $y + \sqrt{nx^2 + y^2} = cx^{1\pm\sqrt{\frac{n-1}{n}}}$.

EXAMPLE 3.23 Solve $(xp - y)^2 = (x^2 - y^2)\sin^{-1}\frac{y}{x}$.

Solution Substitute $y = vx$. Then $p = xP + v$ where $P = \frac{dv}{dx}$. With this substitution, given equation reduces to $x^2P^2 = (1 - v^2)\sin^{-1} v \Rightarrow \frac{dv}{\sqrt{1 - v^2}\sqrt{\sin^{-1} v}} = \pm\frac{dx}{x}$.

On integration, we have

$$2\sqrt{\sin^{-1} v} = \pm\log x + c \Rightarrow x = ce^{\pm\sqrt{\sin^{-1}(y/x)}}$$

which is the solution of the given equation.

EXAMPLE 3.24 Solve $3p^2y^2 - 2xyp + 4y^2 - x^2 = 0$.

Solution The given equation can be rewritten as $(x - 3py)^2 - 4(x^2 - 3y^2) = 0$.

Substitute $x^2 - 3y^2 = v^2$. Then $x - 3yp = vP$ where $P = \dfrac{dv}{dx}$. With this substitution, given equation reduces to $P^2 = 4$ or $P = \pm 2$.

Taking positive sign, integration results $v = c + 2x \Rightarrow v^2 = (c + 2x)^2$. Therefore, the solution of given equation is $3x^2 + 3y^2 + 4cx + c^2 = 0$.

EXERCISES

Solve the following differential equations:

1. $xp^2 - (y - x)\,p - y = 0$. $\left[\textbf{Ans. } (y + x - c)\left(\dfrac{y}{x} - c\right) = 0\right]$

2. $xyp^2 + (x^2 + xy + y^2)p + x^2 + xy = 0$. $[\textbf{Ans. } (xy + x^2 - c)\,(x^2 + y^2 - c) = 0]$

3. $p^2 + 2py\cot x - y^2 = 0$. $[\textbf{Ans. } [y\,(1 + \cos x) - c]\,[y(1 - \cos x) - c] = 0]$

4. $p^3 - (x^2 + xy + y^2)\,p + xy(x + y) = 0$.

$[\textbf{Ans. } (2y + x^2 - c)\,(y - ce^x)\,(x + y - 1 - ce^{-x}) = 0]$

5. $p^2 + (y + x)\,p + xy = 0$. $[\textbf{Ans. } (2y + x^2 - c)\,(x + \log y - c) = 0]$

6. $xp^2 + (y - 1 - x^2)\,p - x(y - 1) = 0$. $[\textbf{Ans. } (2y - x^2 + c)\,(xy - x + c) = 0]$

7. $x^2p^2 - 2xyp + y^2 = x^2y^2 + x^4$. $[\textbf{Ans. } [y - x\sin h\,(x + c)]\,[y - x\sin h\,(x - c)] = 0]$

8. $xp^2 - (y - x)p - y = 0$. $\left[\textbf{Ans. } (y + x - c)\left(\dfrac{y}{x} - c\right) = 0\right]$

9. $xy^2(p^2 + 2) = 2py^3 + x^3$. $[\textbf{Ans. } (x^2 - y^2 - c)(x^2 - y^2 + cx^4) = 0]$

10. $\left(p^2 - \dfrac{1}{4x^2}\right)\left(p - \sqrt{\dfrac{y}{x}}\right) = 0$.

$\left[\textbf{Ans. } \left(y - \sin^{-1}\dfrac{x}{2} - c\right)\left(y - \cos^{-1}\dfrac{x}{2} - c\right)(\sqrt{y} - \sqrt{x} - c) = 0\right]$

11. $p^3 + p = e^y$. $\left[\textbf{Ans. } x = -\dfrac{1}{p} + 2\tan^{-1} p + c \text{ and } y = \log p + \log\,(1 + p^2)\right]$

12. $x - yp = 3p^2$.

$\left[\textbf{Ans. } x = -\dfrac{p}{\sqrt{1 - p^2}}(c + 3\sin^{-1} p) \text{ and } y = -3p + \dfrac{1}{\sqrt{1 - p^2}}(c + 3\sin^{-1} p)\right]$

13. $y = x + a\tan^{-1} p$. $\left[\textbf{Ans. } x + c = \frac{a}{2}[(p-1)\log(1+p^2)^{-1/2} - \tan^{-1} p] \text{ and } y = x + a\tan^{-1} p\right]$

14. $y = p + xp^2$. [**Ans.** $(p-1)^2 x = a - p + \log p$ and $(p-1)^2 y = p^2(a - p + \log p) + p$]

15. $y = p\sin p + \cos p$. [**Ans.** $x = c + \sin p$ and $y = p\sin p + \cos p$]

16. $y = x(1+p) + p^2$. [**Ans.** $x = 2(1-p) + e^{-p}$ and $y = 2 - p^2 + c(1+p)e^{-p}$]

17. $y = 2px + p^3$. $\left[\textbf{Ans. } x = \frac{c}{p^2} - p^2 \text{ and } y = \frac{2c}{p} - p^3\right]$

18. $y = 1 + xp^3$. [**Ans.** $x^{2/3} = (y-1)^{2/3} + c$]

19. $x^2 + p^2 x = yp$. $\left[\textbf{Ans. } x = c\sqrt{p} - \frac{1}{3}p^2 \text{ and } y = \frac{x^2}{p} + px\right]$

20. $y = 2px + xp^2$. [**Ans.** $y = 2c\sqrt{x} + c^2$]

21. $p^3 - 2xyp + 4y^2 = 0$. [**Ans.** $16y = c(2x - c)^2$]

22. $y = 2px + 4p^2y^2$. [**Ans.** $y^3 = 2cx + 4c^2$]

23. $y^2 \log y = xyp + p^2$. [**Ans.** $\log y = cx + c^3$]

24. $x = 4p(1+p)^2$. [**Ans.** $y = 2p^2 + 3p + c$ and $x = 4p(1+p)^2$]

25. $p^3 - 4xyp + 8y^2 = 0$. [**Ans.** $y = c(x-c)^2$]

26. $yp^2 - 2xp + y = 0$. [**Ans.** $2cx = y^2 + c^2$]

27. $y^2 = a^2(1+p^2)$. [**Ans.** $x + c = a\log(p + \sqrt{1+p^2})$ and $y = a\sqrt{1+p^2}$]

28. $y^2(1-p^2) = b$. [**Ans.** $y^2 - b = (x+c)^2$]

29. $y = b + \dfrac{1}{\sqrt{1+p^2}}$. [**Ans.** $(x+c)^2 + (y-b)^2 = 1$]

30. $y^2 + xyp - x^2p^2 = 0$. [**Ans.** $y^2 + cyx^{2/(\sqrt{5}-1)} = c^2x^{4/(\sqrt{5}-1)}$]

31. $y = 3px + 6y^2p^2$. [**Ans.** $y^2 = 3cx + c^2$]

32. $(px - y)(py + x) = a^2 p$. $\left[\textbf{Ans. } y^2 = cx^2 - \frac{ca^2}{c+1}\right]$

33. $x^2(y - px) = yp^2$. [**Ans.** $y^2 = cx^2 + c^2$]

34. $p = \log(px - y)$. [**Ans.** $y = cx - e^c$]

35. $y = 2px + f(xp^2)$. $\left[\textbf{Ans. } y = c\sqrt{x} + f\left(\frac{1}{4}c^2\right)\right]$

36. $y = 2px + x^2p^4$. $\left[\textbf{Ans. } y = c\sqrt{x} + \frac{c^4}{16}\right]$

37. $p^2 - 9p + 18 = 0$. [**Ans.** $(y - 6x + c)(y - 3x + c) = 0$]

38. $x^2p^2 - 2xyp + 2y^2 - x^2 = 0$. [**Ans.** $y = \sin(x \log ax)$]

39. $(xp - y)^2 = p^2 - \frac{2y}{x}p + 1$. $\left[\textbf{Ans. } \sinh^{-1}\frac{y}{x} + \sin^{-1}\frac{1}{x} = c\right]$

40. $e^{3x}(p-1) + p^3e^{2y} = 0$. [**Ans.** $e^y = ce^x + c^3$]

41. $p^2 - 4p + 3 = 0$. [**Ans.** $(y - 3x - c)(y - x - c) = 0$]

42. $x^2p^2 + xyp - 2y^2 - 0$. $\left[\textbf{Ans. } (y - cx)\left(y - \frac{c}{x^2}\right) = 0\right]$

43. $x + yp^2 = p(1 + xy)$. [**Ans.** $(2y - x^2 - c)(2x - y^2 - c) = 0$]

44. $y^2 + xyp - x^2p^2 = 0$. $\left[\textbf{Ans. } (y - cx^k)\left(y - \frac{c}{x^k}\right) = 0; k = \frac{1+\sqrt{5}}{2}\right]$

45. $xyp^2 + p(3x^2 - 2y^2) - 6xy = 0$. [**Ans.** $(y - cx^2)y^2 + 3x^2 - c = 0$]

46. $p(p + y) = x(x + y)$ [**Ans.** $(2y - x^2 - c)(y + x - 1 - ce^{-x}) = 0$]

47. $x(x+1)p^2 + (x^2 + x - 2y(x+1))p + y(y - x) = 0$. [**Ans.** $(y - c(x+1))(y + x\log cx) = 0$]

48. $p^3 - (x^2 + xy + y^2)p^2 + xy(x^2 + xy + y^2)p - x^3y^3 = 0$.

[**Ans.** $(xy + cy + 1)(x^2 - 3y + c)(cy + e^{x^2/2}) = 0$]

49. $\left(p^2 - \frac{1}{a^2 - x^2}\right)\left(p - \sqrt{\frac{y}{x}}\right) = 0$.

$\left[\textbf{Ans. } \left(y - \sin^{-1}\frac{x}{a} - c\right)\left(y - \cos^{-1}\frac{x}{a} - c\right)\left(\sqrt{y} - \sqrt{x} - c\right) = 0\right]$

50. $4y = x^2 + p^2$. $\left[\textbf{Ans. } \log(p - x) = \frac{x}{p - x} + c\right]$

51. $y = p\tan p + \log(\cos p)$. [**Ans.** $x = \tan p + c$, with given relation]

52. $p^2(p+m) = a(mx+y)$.

$$\left[\textbf{Ans. } ax + c = \frac{3}{2}p^2 - mp + m^2 \log(m+p), \text{ with given relation}\right]$$

53. $y = 2p + \sqrt{1+p^2}$.

$$\left[\textbf{Ans. } x = 2\log p + \log(p + \sqrt{1+p^2}), \text{ with given relation}\right]$$

54. $xp^2 - 2yp + ax = 0$.

$$\left[\textbf{Ans. } 2y = cx^2 + \frac{a}{c}\right]$$

55. $y = xp^2 + p^3$.

$$\left[\textbf{Ans. } x = \frac{1}{(1-p)^2}\left\{\frac{3}{2}p^2 - p^3 + c\right\}, \text{ with given relation}\right]$$

56. $y = 2xp + p^2$.

$$[\textbf{Ans. } (xy+c)^2 = 4(x^2+y)(y^2-cx)]$$

57. $16x^2 + 2p^2y - p^3x = 0$.

$$[\textbf{Ans. } 2 - c^2y - c^3x^2 = 0]$$

58. $y + px = p^2x^4$.

$$\left[\textbf{Ans. } y = \frac{c}{x} + c^2\right]$$

59. $y - 2px + xp^2 = 0$.

$$[\textbf{Ans. } y = 2c\sqrt{x} - c^2]$$

60. $x^2 - yp + p^2x = 0$.

$$\left[\textbf{Ans. } x = c\sqrt{p} - \frac{1}{3}p^2, \text{ with given relation}\right]$$

61. $x - y - p^2 = 0$.

$$[\textbf{Ans. } y = c - \{p^2 + 2p + 2\log(p-1)\}, \text{ with given relation}]$$

62. $x + py - p^3 = 0$.

$$\left[\textbf{Ans. } y = \frac{c}{\sqrt{1+p^2}} + p^2 - 2, \text{ with given relation}\right]$$

63. $y - x - p^3 = 0$.

$$\left[\textbf{Ans. } x = \frac{3}{2}p^2 + 3p + 3\log(p-1) + c, \text{ with given relation}\right]$$

64. $y = 2px + p^2y$.

$$[\textbf{Ans. } y^2 = 2cx + c^2]$$

65. $y = 2px + p^3y^2$.

$$\left[\textbf{Ans. } x = \frac{y^2}{2c} - \frac{c^2}{2}\right]$$

66. $xp^2 - 2yp + 4x = 0$.

$$[\textbf{Ans. } cy = x^2 + c^2]$$

67. $x(1+p^2) = 1$.

$$\left[\textbf{Ans. } y = c + \sqrt{x(1-x)} - \tan^{-1}\sqrt{\frac{1-x}{x}}; \text{ with given relation}\right]$$

68. $p^2 - 2px + 1 = 0.$ $\left[\textbf{Ans. } x = \frac{1}{2}\left(p + \frac{1}{p}\right), y = \frac{1}{4}p^2 - \frac{1}{2}\log p + c\right]$

69. $p = (a - x)\sqrt{1 + p^2}.$ [**Ans.** $(x - a)^2 + (y + c)^2 = 1$]

70. $y = xp + \frac{a}{p}.$ $\left[\textbf{Ans. } y = cx + \frac{a}{c}\right]$

71. $(y - px)^2 = 1 + p^2.$ [**Ans.** $(y - cx)^2 = 1 + c^2$]

72. $y = p(x - a) + \frac{b}{p}.$ $\left[\textbf{Ans. } y = c(x - a) + \frac{b}{c}\right]$

73. $yp^2 - 2xp + y = 0.$ $\left[\textit{Hint: } y^2 = v, \textbf{ Ans. } y^2 = cx - \frac{c^2}{4}\right]$

74. $\cos^2 yp^2 + \sin x \cos x \cos yp - \sin y \cos^2 x = 0.$

[*Hint:* $u = \sin x, v = \sin y,$ **Ans.** $\sin y = c \sin x + c^2$]

CHAPTER 4

Linear Differential Equations with Constant Coefficients

4.1 INTRODUCTION

In this chapter, we shall consider differential equations of order higher than one. In these differential equations the dependent variable and its derivatives appear only in the first degree, and are not multiplied together, their coefficients are constants or the functions of x. The general form of the equation is

$$\frac{d^n y}{dx^n} + P_1 \frac{d^{n-1} y}{dx^{n-1}} + P_2 \frac{d^{n-2} y}{dx^{n-2}} + \cdots + P_n y = X \tag{4.1}$$

where P_1, P_2, …, P_n and X are constants or functions of X.

We shall use the following two results:

R1: If $y = f(x)$ is the general solution of

$$\frac{d^n y}{dx^n} + P_1 \frac{d^{n-1} y}{dx^{n-1}} + P_2 \frac{d^{n-2} y}{dx^{n-2}} + \cdots + P_n y = 0 \tag{4.2}$$

and $y = F(x)$ is a general solution of Eq. (4.1) then

$$y = f(x) + F(x)$$

is a general solution of Eq. (4.1).

R2: If $y = y_k$, $k = 1, 2, \ldots, n$ are the solutions of Eq. (4.2) then

$$y = c_1 y_1 + c_2 y_2 + \cdots + c_n y_n$$

where c_1, c_2, …, c_n are arbitrary constants; is also solution of Eq. (4.2).

R1 describes the method of finding the general solution of Eq. (4.2). First find the general solution of the Eq. (4.2) and call it $y = f_1(c_1, c_2, \ldots, c_n, x)$. Then find the general solution of Eq. (4.1) which does not contain any arbitrary constants. Call it $y = f_2(x)$. Then $y = f_1 + f_2$ is the general solution of the Eq. (4.1). f_1 is known as the **complementary function** (denoted by CF) and f_2 is known as the **particular integral** (denoted by PI).

In rest of the chapter, D stands for $\frac{d}{dx}$. With this notation, the Eqs. (4.2) and (4.1) can be written as

$$(D^n + P_1 D^{n-1} + \cdots + P_n)\, y = 0 \tag{4.3}$$

and

$$(D^n + P_1 D^{n-1} + \cdots + P_n)\, y = X \tag{4.4}$$

4.2 SYMBOLIC OPERATOR

We have

$$(D - m_1)y = Dy - m_1 y = \frac{dy}{dx} - m_1 y.$$

$(D - m_1)(D - m_2)y$ is defined as the expression obtained on operating $(D - m_1)$ on $\left(\frac{dy}{dx} - m_2 y\right)$.

i.e.

$$\frac{d^2y}{dx^2} - (m_1 + m_2)\frac{dy}{dx} + m_1 m_2 y$$

where m_2 is constant. The operation $(D - m_2)(D - m_1)y$ is equivalent to above expression where m_1 is constant. That is, if m_1 and m_2 are both constants, the expressions $(D - m_1)(D - m_2)y$ and $(D - m_2)(D - m_1)y$ are equivalent. Thus, the expression is independent of the order of the operational factors.

4.3 METHOD FOR FINDING CF

Consider the Eq. (4.3) and let it be equivalent to

$$(D - m_1)(D - m_2) \cdots (D - m_n)y = 0 \tag{4.5}$$

Then

$$\left.\begin{array}{l}(D - m_1)y = 0 \\ (D - m_2)y = 0 \\ \vdots \\ (D - m_n)y = 0\end{array}\right\}$$

is also solution of Eq. (4.5).

Consider the equation

$$(D - m_k)y = 0$$

$$\Rightarrow \qquad \frac{dy}{dx} = m_k y$$

$$\Rightarrow \qquad \frac{dy}{dx} = m_k dx$$

which on integration gives

$$y = c_k e^{m_k x}, \; k = 1, 2, \ldots, n$$

Then
$$y = \sum_{k=1}^{n} c_k e^{m_k x} \tag{4.6}$$

is a solution of Eq. (4.5). If $m_1, m_2, \ldots, m_n$ are distinct, the functions $e^{m_1 x}, e^{m_2 x}, \ldots, e^{m_n x}$ are linearly independent, and hence, Eq. (4.6) is the general solution of Eq. (4.5). $m_1, m_2, \ldots, m_n$ are the roots of the equation

$$m^n + P_1 m^{n-1} + \cdots + P_n = 0 \tag{4.7}$$

The Eq. (4.7) is known as the Auxiliary Equation (AE). Three cases may arise depending on the values of the roots.

Case 1: If all roots are distinct then Eq. (4.6) is the CF of given differential equation.

Case 2: If roots are repeated then proceed as discussed below:

Let m_1 and m_2 be two equal roots of the auxiliary equation. Then Eq. (4.6) becomes

$$y = (c_1 + c_2)\, e^{m_1 x} + c_3 e^{m_3 x} + \cdots + c_r e^{m_n x}$$

Putting $c_1 + c_2 = c$, the obtained solution will have only $(n - 1)$ arbitrary constants and so it is not the general solution.

We have

$$(D - m_1)^2\, y = 0$$

$$\Rightarrow \quad (D - m_1)\, V = 0 \text{ where } V = (D - m_1)y$$

$$\Rightarrow \quad (D - m_1)y = c_1 e^{m_1 x}$$

$$\Rightarrow \quad \frac{dy}{dx} - m_1 y = c_1 e^{m_1 x}$$

which is linear equation whose solution is

$$y = (c_1 x + c_2) c_1 e^{m_1 x}.$$

Similarly, if m_1, m_2 and m_3 are equal roots of the auxiliary equation then arguing as above, the corresponding part of CF is

$$y = (c_1 x^2 + c_2 x + c_3) e^{m_1 x}.$$

In general, if the auxiliary equation has r equal roots then the corresponding part of the CF is

$$y = (c_1 x^{r-1} + c_2 x^{r-2} + \cdots + c_r) e^{m_1 x}.$$

Case 3: The auxiliary equation has complex roots. Let a pair of complex roots of the auxiliary equation be $m_1 = a + ib$ and $m_2 = a - ib$. The corresponding part of the CF is

$$c_1 e^{(a+ib)x} + c_2 e^{(a-ib)x} = e^{ax}(c_1 e^{ibx} + c_2 e^{-ibx}) = e^{ax}\left[(c_1 + c_2)\cos bx + i(c_1 - c_2)\sin bx\right]$$

$$= e^{ax}(A \cos bx + B \sin bx)$$

where $A = c_1 + c_2$ and $B = i(c_1 - c_2)$ are arbitrary constants.

If the complex roots m_1 and m_2 occur twice, the corresponding part of CF is

$$(c_1 + c_2 x)\, e^{(a+ib)x} + (c_3 + c_4 x)\, e^{(a-ib)x} = e^{ax}\left[(A_1 + A_2 x)\cos bx + (B_1 + B_2 x)\sin bx\right]$$

EXAMPLE 4.1 Solve $\dfrac{d^2y}{dx^2} + \dfrac{dy}{dx} - 2y = 0.$

Solution Given equation in symbolic form is

$$(D^2 - D - 2)y = 0$$

The auxiliary equation is

$$m^2 - m - 2 = 0$$

$$\Rightarrow \quad (m + 2)(m - 1) = 0$$

$$\therefore \quad m = -2, 1.$$

Hence, the CF is

$$y = c_1 e^{-2x} + c_2 e^x.$$

EXAMPLE 4.2 Solve $(D^3 - D^2 - 12D)y = 0.$

Solution Given

$$(D^3 - D^2 - 12D)y = 0$$

$$\Rightarrow \quad D(D^2 - D - 12)y = 0$$

Therefore, the auxiliary equation is

$$m(m^2 - m - 12) = 0$$

$$\Rightarrow \quad m(m - 4)(m + 3) = 0$$

$$\Rightarrow \quad m = 0, 4, -3$$

Hence, the CF is

$$y = c_1 + c_2 e^{4x} + c_3 e^{-3x}.$$

EXAMPLE 4.3 Solve $\dfrac{d^2x}{dt^2} - 6\dfrac{dx}{dt} + 9x = 0.$

Solution Given equation in symbolic form is

$$(D^2 - 6D + 9)x = 0$$

The auxiliary equation is

$$m^2 - 6m + 9 = 0$$

$$\Rightarrow \quad (m - 3)^2 = 0$$

$$\therefore \quad m = 3, 3.$$

Hence, the CF is

$$x = (c_1 + c_2 t)\, e^{3t}.$$

EXAMPLE 4.4 Solve $\dfrac{d^4y}{dx^4} - \dfrac{d^3y}{dx^3} - 9\dfrac{d^2y}{dx^2} - 11\dfrac{dy}{dx} - 4y = 0.$

Solution Given equation in symbolic form is

$$(D^4 - D^3 - 9D^2 - 11D - 4)y = 0$$

The auxiliary equation is

$$m^4 - m^3 - 9m^2 - 11m - 4 = 0$$

$$\Rightarrow \quad (m + 1)^3 (m - 4) = 0$$

$$\therefore \quad m = -1, -1, -1, 4.$$

Hence, the CF is

$$y = (c_1x^2 + c_2x + c_3) e^{-x} + c_4e^{4x}.$$

EXAMPLE 4.5 Solve $\dfrac{d^3y}{dx^3} + \dfrac{d^2y}{dx^2} + 4\dfrac{dy}{dx} + 4y = 0.$

Solution Given equation in symbolic form is

$$(D^3 + D^2 + 4D + 4) y = 0$$

The auxiliary equation is

$$m^3 + m^2 + 4m + 4 = 0$$

$$\Rightarrow \quad (m + 1) (m^2 + 4) = 0$$

$$\therefore \quad m = -1, 2i, -2i.$$

Hence, the CF is

$$y = c_1e^{-x} + e^{0x}(c_2 \cos 2x + c_3 \sin 2x)$$

$$y = c_1e^{-x} + c_2 \cos 2x + c_3 \sin 2x.$$

EXAMPLE 4.6 Solve $\dfrac{d^4y}{dx^4} + 4y = 0.$

Solution Given equation in symbolic form is

$$(D^4 + 4)y = 0$$

The auxiliary equation is

$$m^4 + 4 = 0$$

$$m^4 + 4m^2 + 4 - 4m^2 = 0$$

$$\Rightarrow \quad (m^2 + 2)^2 - (2m)^2 = 0$$

$$\Rightarrow \quad (m^2 + 2m + 2) (m^2 - 2m + 2) = 0$$

$$\therefore \quad m = -1 \pm i, 1 \pm i.$$

Hence, the CF is

$$y = e^{-x} (c_1 \cos x + c_2 \sin x) + e^x (c_3 \cos x + c_4 \sin x).$$

4.4 SYMBOLIC OPERATOR $\dfrac{1}{f(D)}$

$\dfrac{1}{f(D)}X$ is that operator of x, not containing arbitrary constants which when operated by $f(D)$ gives X, i.e.

$$f(D)\left(\frac{1}{f(D)}X\right) = X.$$

Thus, $\dfrac{1}{f(D)}X$ satisfies the equation $f(D)y - X$ and is its particular integral. Clearly, $f(D)$ and $\dfrac{1}{f(D)}$ are inverse operators. Thus,

$$\frac{1}{D}X = \int X dx.$$

Let
$$f(D) = (D - m_1)(D - m_2) \ldots (D - m_n).$$

Then
$$\frac{1}{f(D)}X = \frac{1}{(D-m_1)(D-m_2)\ldots(D-m_n)}X$$
$$= \frac{1}{(D-m_1)}\frac{1}{(D-m_2)}\ldots\frac{1}{(D-m_n)}X$$

which means that the resulting of operating $\dfrac{1}{(D-m_n)}$ on X is to be operated up on by $\dfrac{1}{(D-m_{n-1})}$ and so on.

Consider
$$f(D) = (D - \alpha)(D - \beta).$$

Then
$$\frac{1}{f(D)}X = \frac{1}{(D-\alpha)(D-\beta)}X.$$

Now, $\dfrac{1}{\alpha-\beta}\left(\dfrac{1}{D-\alpha} - \dfrac{1}{D-\beta}\right)X$. Operate this function with $(D - \alpha)(D - \beta)$.

$$\therefore \quad \frac{1}{\alpha-\beta}(D-\alpha)(D-\beta)\left(\frac{1}{D-\alpha} - \frac{1}{D-\beta}\right)X$$
$$= \frac{1}{\alpha-\beta}\left[(D-\alpha)(D-\beta)\frac{1}{D-\alpha}X - (D-\alpha)(D-\beta)\frac{1}{D-\beta}X\right]$$
$$= \frac{1}{\alpha-\beta}\left[(D-\beta)(D-\alpha)\frac{1}{D-\alpha}X - (D-\alpha)(D-\beta)\frac{1}{D-\beta}X\right]$$
$$= \frac{1}{\alpha-\beta}[(D-\beta)X - (D-\alpha)X]$$
$$= \frac{1}{\alpha-\beta}[(D-\beta) - (D-\alpha)]X$$
$$= \frac{1}{\alpha-\beta}(\alpha-\beta)X = X$$

Therefore,
$$\frac{1}{f(D)} = \frac{1}{(D-\alpha)(D-\beta)} = \frac{1}{\alpha-\beta}\left[\frac{1}{(D-\alpha)} - \frac{1}{(D-\beta)}\right].$$

4.5 METHODS OF FINDING PI

Let us solve $$\frac{1}{D-\alpha}X.$$

Let $$\frac{1}{D-\alpha}X = U$$

$$\Rightarrow \quad (D-\alpha)\,U = X$$

$$\Rightarrow \quad \frac{dU}{dx} - \alpha U = X$$

which is linear equation whose solution is

$$Ue^{-\alpha x} = \int Xe^{-\alpha x}dx + \text{constant.}$$

Since, U should be free of arbitrary constants, $U = e^{\alpha x}\int Xe^{-\alpha x}\,dx.$

Now consider, $$\frac{1}{f(D)}X$$

We need to discuss following two cases:

Case 1: $\frac{1}{f(D)}$ may be resolved into factors.

Let $$\frac{1}{f(D)} = \frac{1}{(D-m_1)(D-m_2)\cdots(D-m_n)}.$$

Then
$$\frac{1}{f(D)}X = \frac{1}{(D-m_1)(D-m_2)\cdots(D-m_n)}X$$
$$= \frac{1}{(D-m_1)}\frac{1}{(D-m_2)}\cdots\frac{1}{(D-m_{n-1})}e^{m_n x}\int Xe^{-m_n x}dx$$
$$= \frac{1}{(D-m_1)}\frac{1}{(D-m_2)}\cdots\frac{1}{(D-m_{n-2})}e^{m_{n-1}x}\int Xe^{-m_{n-1}x}dx\,\{e^{m_n x}\int Xe^{-m_n x}dx\}$$
$$= \frac{1}{(D-m_1)}\frac{1}{(D-m_2)}\cdots\frac{1}{(D-m_{n-2})}e^{m_{n-1}x}\int e^{(m_n-m_{n-1})x}\int Xe^{-m_n x}\,(dx)^2$$

Continuing, we get

$$\frac{1}{f(D)}X = e^{m_1 x}\int e^{(m_2-m_1)x}\int e^{(m_3-m_2)x}\cdots\int e^{(m_n-m_{n-1})x}\int Xe^{-m_n x}\,(dx)^n.$$

Case 2: The operator $\frac{1}{f(D)}$ can be resolved into partial fractions.

Let
$$\frac{1}{f(D)}=\frac{A_1}{(D-m_1)}+\frac{A_2}{(D-m_2)}+\cdots+\frac{A_n}{(D-m_n)}$$
$$\frac{1}{f(D)}=\frac{A_1}{(D-m_1)}+\frac{A_2}{(D-m_2)}+\cdots+\frac{A_n}{(D-m_n)}$$

Then
$$\frac{1}{f(D)}X=A_1\frac{1}{(D-m_1)}X+A_2\frac{1}{(D-m_2)}X+\cdots+A_n\frac{1}{(D-m_n)}X$$
$$=A_1e^{m_1x}\int Xe^{-m_1x}\,dx+A_1e^{m_2x}\int Xe^{-m_2x}dx+\cdots+A_1e^{m_nx}\int Xe^{-m_nx}\,dx.$$

EXAMPLE 4.7 Solve $\frac{d^2y}{dx^2}-3\frac{dy}{dx}+2y=e^{3x}$.

Solution Given equation in symbolic form is
$$(D^2-3D+2)y=e^{3x}$$
Its auxiliary equation is
$$m^2-3m+2=0$$
$$\Rightarrow \qquad m=1,2$$
Therefore, CF is
$$y=c_1e^x+c^2e^{3x}.$$
and
$$\text{PI}=\frac{1}{D-2}\frac{1}{D-1}e^{3x}$$
$$=\frac{1}{D-2}e^x\int e^{3x}e^{-x}\,dx$$
$$=\frac{1}{D-2}e^x\int e^{2x}\,dx$$
$$=\frac{1}{D-2}e^x\frac{e^{2x}}{2}$$
$$=\frac{1}{2}\frac{1}{D-2}e^{3x}$$
$$=\frac{1}{2}e^{2x}\int e^{3x}e^{-2x}\,dx$$
$$=\frac{1}{2}e^{2x}e^x=\frac{1}{2}e^{3x}$$

Therefore, the general solution of given differential equation is
$$y=c_1e^x+c_2e^{3x}+\frac{1}{2}e^{3x}.$$

EXAMPLE 4.8 Solve $\frac{d^2y}{dx^2}+y=\sec^2 x$.

Solution Given equation in symbolic form is

$$(D^2 + 1)y = \sec^2 x$$

Its auxiliary equation is

$$m^2 + 1 = 0$$

$\Rightarrow$

$$m = \pm i$$

Therefore, CF is

$$y = c_1 \cos x + c^2 \sin x$$

and

$$\text{PI} = \frac{1}{D^2+1}\sec^2 x$$

$$= \frac{1}{(D+i)(D-i)}\sec^2 x$$

$$= \frac{1}{2i}\left(\frac{1}{D-i} - \frac{1}{D+i}\right)\sec^2 x$$

$$= \frac{1}{2i}\left[e^{ix}\int e^{-ix}\sec^2 x dx - e^{-ix}\int e^{ix}\sec^2 x dx\right]$$

$$= \frac{1}{2i}\left[e^{ix}\int \frac{\cos x - i\sin x}{\cos^2 x}dx - e^{-ix}\int \frac{\cos x + i\sin x}{\cos^2 x}dx\right]$$

$$= \frac{1}{2i}\left[e^{ix}\int (\sec x - i\sec x\tan x)\,dx - e^{-ix}\int (\sec x + i\sec x\tan x)\,dx\right]$$

$$= \frac{1}{2i}\left[(e^{ix} - e^{-ix})\int \sec x dx - i\,(e^{ix} + e^{-ix})\int \sec x \tan x\, dx\right]$$

$$= \frac{1}{2i}\left[2i\sin x\log|\sec x + \tan x| - 2i\cos x\sec x\right]$$

$$= \sin x\log|\sec x + \tan x| - 1$$

$\therefore$ The general solution is $y = c_1 \cos x + c_2 \sin x + \sin x \log|\sec x + \tan x| - 1$.

EXAMPLE 4.9 Solve $\dfrac{d^2y}{dx^2} + n^2 y = \operatorname{cosec} nx$.

Solution Given equation in symbolic form is

$$(D^2 + n^2)y = \operatorname{cosec} nx$$

Its auxiliary equation is

$$m^2 + n^2 = 0$$

$\Rightarrow$

$$m = \pm ni$$

Therefore, CF is

$$y = c_1 \cos nx + c_2 \sin nx.$$

and

$$\text{PI} = \frac{1}{D^2+n^2}\operatorname{cosec} nx$$

$$= \frac{1}{(D+ni)(D-ni)}\operatorname{cosec} nx$$

$$= \frac{1}{2ni}\left[\frac{1}{D-ni} - \frac{1}{D+ni}\right] \text{cosec } nx$$

$$= \frac{1}{2ni}\left[e^{inx}\int e^{-inx}\text{cosec } nx dx - e^{-inx}\int e^{inx}\text{cosec } nx dx\right]$$

$$= \frac{1}{2ni}\left[e^{inx}\int (\cos nx - i\sin nx) \text{ cosec } nx dx - e^{-ix}\int (\cos nx + i\sin nx) \text{ cosec } nx dx\right]$$

$$= \frac{1}{2ni}\left[e^{inx}\int (\cot nx - i)\, dx - e^{-inx}\int (\cot nx + i)\, dx\right]$$

$$= \frac{1}{2ni}\left[(e^{inx} - e^{-inx})\int \cot nx dx - i(e^{inx} + e^{-inx})\int dx\right]$$

$$= \frac{1}{2ni}\left[2i\sin nx \frac{\log|\sin nx|}{n} - 2ix\cos nx\right]$$

$$= \frac{1}{n^2}\sin nx \log|\sin nx| - \frac{1}{n}x\cos nx$$

$\therefore$ The general solution is $y = c_1\cos nx + c_2\sin nx + \frac{1}{n^2}\sin nx\log|\sin nx| - \frac{1}{n}x\cos nx.$

4.6 PI WHEN $X = e^{ax}$, WHERE 'a' IS ANY CONSTANT

We know that

$$De^{ax} = ae^{ax}$$

$$D^2e^{ax} = a^2e^{ax}$$

and so on.

Therefore,

$$f(D)e^{ax} = f(a)e^{ax} \tag{4.8}$$

Two cases may arise:

Case 1: Let $f(a) \neq 0$.

Operating Eq. (4.8) by $\frac{1}{f(D)}$, we get

$$\frac{1}{f(D)}f(D)e^{ax} = \frac{1}{f(D)}f(a)e^{ax}$$

$$\Rightarrow \quad e^{ax} = f(a)\frac{1}{f(D)}e^{ax}$$

$$\Rightarrow \quad \frac{1}{f(D)}e^{ax} = \frac{1}{f(a)}e^{ax}$$

Case 2: Let $f(a) = 0$.

Then $(D - a)$ is at least one factor of $f(D)$.

Let $$f(D) = (D - a)^k\, g(D)$$

where $$g(a) \neq 0.$$

Then
$$\frac{1}{f(D)}e^{ax} = \frac{1}{(D-a)^k}\frac{1}{g(D)}e^{ax}$$
$$= \frac{1}{(D-a)^k}\frac{1}{g(a)}e^{ax}$$
$$= \frac{1}{g(a)}\frac{1}{(D-a)^k}e^{ax}$$
$$= \frac{1}{g(a)}\frac{1}{(D-a)^{k-1}}e^{ax}\int e^{ax}e^{-ax}dx$$
$$= \frac{1}{g(a)}\frac{1}{(D-a)^{k-1}}xe^{ax}$$
$$= \frac{1}{g(a)}\frac{1}{(D-a)^{k-2}}e^{ax}\int xe^{ax}e^{-ax}dx$$
$$= \frac{1}{g(a)}\frac{1}{(D-a)^{k-2}}\frac{x^2}{2!}e^{ax}$$
$$\vdots$$
$$= \frac{1}{g(a)}\frac{x^k}{k!}e^{ax}$$

Since $f(D) = (D - a)^k\, g(D)$

$f^k(D) = k!\, g(D)$ + terms containing powers of $(D - a)$.

$f^k(D) = k!\, g(D)$

$$\therefore \quad \frac{1}{f(D)}e^{ax} = \frac{1}{g(a)}\frac{x^k}{k!}e^{ax}$$

$$\therefore \quad \frac{1}{f(D)}e^{ax} = \frac{x^k e^{ax}}{f^k(a)}.$$

EXAMPLE 4.10 Solve $(D^2 + 4D + 3)y = e^{-3x}$.

Solution The auxiliary equation of given equation is
$$m^2 + 4m + 3 = 0$$
$$\Rightarrow \quad m = -1, -3$$

Therefore, $\quad CF = c_1e^{-x} + c_2e^{-3x}$

and
$$PI = \frac{1}{D^2 + 4D + 3}e^{-3x}$$
$$= x\frac{1}{2D+4}e^{-3x}$$
$$= -\frac{1}{2}xe^{-3x}$$

Therefore, the general solution of given differential equation is

$$y = c_1 e^{-x} + c_2 e^{-3x} - \frac{1}{2} x e^{-3x}.$$

EXAMPLE 4.11 Solve $(D^3 - 5D^2 + 7D - 3)y = e^{2x} \sinh x$.

Solution The auxiliary equation of given equation is

$$m^3 - 5m^2 + 7m - 3 = 0$$

$$\Rightarrow \quad (m - 1)^2 (m - 3) = 0$$

$$\Rightarrow \quad m = 1, 1, 3$$

Therefore, $\quad \text{CF} = c_1 e^x + x c_2 e^x + e^{3x}$

and

$$\begin{aligned}
\text{PI} &= \frac{1}{D^3 - 5D^2 + 7D - 3} e^{2x} \sinh x \\
&= \frac{1}{D^3 - 5D^2 + 7D - 3} e^{2x} \left(\frac{e^x - e^{-x}}{2} \right) \\
&= \frac{1}{2} \frac{1}{D^3 - 5D^2 + 7D - 3} e^{3x} - \frac{1}{2} \frac{1}{D^3 - 5D^2 + 7D - 3} e^x \\
&= \frac{1}{2} x \frac{1}{3D^2 - 10D + 7} e^{3x} - \frac{1}{2} x \frac{1}{3D^2 - 10D + 7} e^x \\
&= \frac{1}{8} x e^{3x} - \frac{1}{2} x^2 \frac{1}{6D - 10} e^x \\
&= \frac{1}{8} x e^{3x} + \frac{1}{8} x^2 e^x
\end{aligned}$$

$$\therefore \quad y = c_1 e^x + x c_2 e^x + e^{3x} + \frac{1}{8} x e^{3x} + \frac{1}{8} x^2 e^x$$

is the general solution of given differential equation.

4.7 PI WHEN X = cos ax OR sin ax, WHERE 'a' IS ANY CONSTANT

Let $\quad X = \sin ax.$

Then

$$\begin{aligned}
DX &= a \cos ax \\
D^2 X &= (-a^2) \sin ax \\
D^3 X &= (-a^3) \cos ax \\
D^4 X &= (-a^2)^2 \sin ax
\end{aligned}$$

Therefore, in general, $\quad (D^2)^n X = (-a^2)^n \sin ax.$

Thus, if $f(D^2)$ is a rational integral function of D^2, then

$$g(D^2) \sin ax = g(-a^2) \sin ax \tag{4.9}$$

Two cases may arise:

Case 1: Let $g(-a^2) \neq 0$.

Operating Eq. (4.9) by $\dfrac{1}{g(D^2)}$, we get

$$\frac{1}{g(D^2)} g(D^2) \sin ax = \frac{1}{g(D^2)} g(-a^2) \sin ax$$

$$\Rightarrow \qquad \sin ax = g(-a^2) \frac{1}{g(D^2)} \sin ax$$

$$\Rightarrow \qquad \frac{1}{g(D^2)} \sin ax = \frac{1}{g(-a^2)} \sin ax \tag{4.10}$$

Case 2: Let $g(-a^2) = 0$.
Then $(D^2 + a^2)$ is at least one factor of $g(D^2)$.
Let $g(D^2) = (D^2 + a^2)^k\, h(D^2)$ where $h(-a^2) \neq 0$.

Then
$$\frac{1}{g(D^2)} \sin ax = \frac{1}{(D^2 + a^2)^k} \frac{1}{h(D^2)} \sin ax$$
$$= \frac{1}{h(-a^2)} \frac{1}{(D^2 + a^2)^k} \sin ax$$

Let
$$\frac{1}{(D^2 + a^2)} \sin ax = U.$$

Therefore,
$$U = \frac{1}{2ai} \left(\frac{1}{D - ai} - \frac{1}{D + ai} \right) \sin ax$$
$$= \frac{1}{2ai} \left(e^{iax} \int e^{-iax} \sin ax dx - e^{-iax} \int e^{iax} \sin ax dx\right)$$
$$= \frac{1}{2ai} \left[e^{iax} \int e^{-iax} \left(\frac{e^{iax} - e^{-iax}}{2i} \right) dx - e^{-iax} \int e^{iax} \left(\frac{e^{iax} - e^{-iax}}{2i} \right) dx \right]$$
$$= -\frac{1}{4a} \left[e^{iax} \int (1 - e^{-2iax})\, dx - e^{-iax} \int (e^{2iax} - 1)\, dx\right]$$
$$= -\frac{1}{4a} \left[e^{iax} \left(x + \frac{e^{-2iax}}{2ai} \right) - e^{-iax} \left(\frac{e^{2iax}}{2ai} - x \right) \right]$$
$$= -\frac{1}{4a} \left[x(e^{iax} + e^{-iax}) + \frac{1}{a} \left(\frac{e^{-iax} - e^{iax}}{2i} \right) \right]$$
$$= -\frac{1}{4a} \left(2x \cos ax - \frac{1}{a} \sin ax \right)$$

$$= \frac{1}{4a^2}\sin ax - \frac{x}{2a}\cos ax \tag{4.11}$$

$$\therefore \quad \frac{1}{g(D^2)}\sin ax = \frac{1}{h(-a^2)}\left(\frac{1}{4a^2}\sin ax - \frac{x}{2a}\cos ax\right) \tag{4.12}$$

Since $(D^2 + a^2)$ is at least one factor of $g(D^2)$ and CF $= c_1 \cos ax + c_2 \sin ax$, the first term of PI can be combined with CF

$$\therefore \quad \frac{1}{g(D^2)}\sin ax = -\frac{1}{h(-a^2)}\frac{x}{2a}\cos ax.$$

Similarly, if $X = \cos ax$, we have

$$\therefore \quad \frac{1}{g(D^2)}\cos ax = \frac{1}{g(-a^2)}\cos ax, \; g(-a^2) \neq 0 \tag{4.13}$$

and

$$\frac{1}{D^2 + a^2}\cos ax = \frac{x}{2a}\sin ax \tag{4.14}$$

EXAMPLE 4.12 Solve $(D^2 - 5D + 6)y = \cos 3x$.

Solution The auxiliary equation of given equation is

$$m^2 - 5m + 6 = 0$$

$$\Rightarrow \quad (m - 3)(m - 2) = 0$$

$$\Rightarrow \quad m = 2, 3$$

Therefore,

$$\text{CF} = c_1 e^{2x} + c_2 e^{3x}$$

and

$$\text{PI} = \frac{1}{D^2 - 5D + 6}\cos 3x$$

$$= \frac{1}{(-3^2) - 5D + 6}\cos 3x$$

$$= \frac{1}{-5D - 3}\cos 3x$$

$$= -(5D - 3)\frac{1}{25D^2 - 9}\cos 3x$$

$$= \frac{1}{234}(5D - 3)\cos 3x$$

$$= \frac{1}{234}(-15\sin 3x - 3\cos 3x)$$

$$= -\frac{1}{234}(15\sin 3x + 3\cos 3x)$$

$$\therefore \qquad y = c_1e^{2x} + c_2e^{3x} - \frac{1}{234}(15\sin 3x + 3\cos 3x)$$

is the general solution of given differential equation.

EXAMPLE 4.13 Find PI of $(D^3 + 1)y = \cos(2x - 1)$.

Solution

$$\text{PI} = \frac{1}{D^3+1}\cos(2x-1)$$

$$= \frac{1}{D^3+1}\cos(2x-1)$$

$$= \frac{1}{D(-4)+1}\cos(2x-1)$$

$$= \frac{1}{1-4D}\cos(2x-1)$$

$$= (1+4D)\frac{1}{1-16D^2}\cos(2x-1)$$

$$= (1+4D)\frac{1}{1-16(-4)}\cos(2x-1)$$

$$= \frac{1}{65}(1+4D)\cos(2x-1)$$

$$= \frac{1}{65}[\cos(2x-1) - 8\sin(2x-1)].$$

4.8 PI WHEN $X = x^m$, m IS A POSITIVE INTEGER

Here
$$\text{PI} = \frac{1}{f(D)}x^m = [f(D)]^{-1}x^m.$$

Expand $[f(D)]^{-1} x^m$ in ascending powers of D up to the terms in D^m and operate on x^m term by term. Since the $(m + 1)$th and higher derivatives of x^m are zero, we need not consider terms of D^{m+1} and so on.

EXAMPLE 4.14 Find PI of $\dfrac{d^2y}{dx^2} + \dfrac{dy}{dx} = x^2 + 2x + 4.$

Solution The symbolic form of given equation is

$$(D^2 + D)y = x^2 + 2x + 4$$

Therefore,
$$\text{PI} = \frac{1}{D^2+D}(x^2+2x+4)$$

$$= \frac{1}{D(1+D)}(x^2+2x+4)$$

$$= \frac{1}{D}(1+D)^{-1}(x^2+2x+4)$$
$$= \frac{1}{D}(1-D+D^2-\cdots)(x^2+2x+4)$$
$$= \frac{1}{D}[x^2+2x+4-(2x+2)+2]$$
$$= \frac{1}{D}(x^2+4)$$
$$= \int(x^2+4)\,dx$$
$$= \frac{x^3}{3}+4x.$$

EXAMPLE 4.15 Find solution of $(D^4+m^2D^2+n^2D^2+m^2n^2)y = \cos\left(\frac{m+n}{2}\right)x\cos\left(\frac{m-n}{2}\right)x.$

Solution The symbolic form of given equation is

$$(D^2+m^2)(D^2+n^2)y = \frac{1}{2}(\cos mx+\cos nx)$$

Reader can check that

$$\text{CF} = (c_1\cos mx + c_2\sin mx)(c_3\cos nx + c_4\sin nx)$$

Now,

$$\text{PI} = \frac{1}{2}\frac{1}{(D^2+m^2)(D^2+n^2)}\cos mx + \frac{1}{2}\frac{1}{(D^2+m^2)(D^2+n^2)}\cos nx$$
$$= \frac{1}{2}\frac{1}{(n^2-m^2)(D^2+m^2)}\cos mx + \frac{1}{2}\frac{1}{(m^2-n^2)(D^2+n^2)}\cos nx$$
$$= \frac{1}{2}\frac{1}{(n^2-m^2)}\left\{\frac{x}{2m}\sin mx - \frac{x}{2n}\sin nx\right\}$$
$$= \frac{1}{4}\frac{x}{(n^2-m^2)}\left\{\frac{1}{m}\sin mx - \frac{1}{n}\sin nx\right\}$$

Therefore, the solution of the given differential equation is

$$y = (c_1\cos mx + c_2\sin mx)(c_3\cos nx + c_4\sin nx) + \frac{1}{4}\frac{x}{(n^2-m^2)}\left\{\frac{1}{m}\sin mx - \frac{1}{n}\sin nx\right\}$$

EXAMPLE 4.16 Find solution of $(D^2+2n\cos\alpha D+n^2)y = L\cos nx$ given that $y=0, \frac{dy}{dx}=0$ when $x = 0$.

Solution AE is $m^2+2mn\cos\alpha+n^2=0$. The roots of AE are $m=-n\cos\alpha\pm in\sin\alpha$. So,

$\text{CF}=e^{-n\cos\alpha x}\{c_1\cos(n\sin\alpha x)+c_2\sin(n\sin\alpha x)\}$.

$$\text{PI}=L\frac{1}{D^2+2n\cos\alpha D+n^2}\cos nx$$

$$=L\frac{1}{-n^2+2n\cos\alpha D+n^2}\cos nx$$

$$=L\frac{1}{2n\cos\alpha}\frac{1}{D}\cos nx$$

$$=L\frac{1}{2n\cos\alpha}\frac{\sin nx}{n}=\frac{L\sin nx}{2n^2\cos\alpha}$$

Therefore, the general solution of the given equation is

$$y=e^{-n\cos\alpha x}\{c_1\cos(n\sin\alpha x)+c_2\sin(n\sin\alpha x)\}+\frac{L\sin nx}{2n^2\cos\alpha}$$

With given conditions: when x = 0, y = 0 and $\frac{dy}{dx}=0$, we have

$$c_1=0 \text{ and } c_2=-\frac{L}{2n^2\sin\alpha\cos\alpha}=-\frac{L}{n^2\sin 2\alpha}$$

Therefore, the particular solution of the given differential equation is

$$y=e^{-n\cos\alpha x}\left\{\frac{-L\sin(n\sin\alpha x)}{n^2\sin 2\alpha}\right\}+\frac{L\sin nx}{2n^2\cos\alpha}$$

4.9 PI WHEN $X = e^{ax}V$, WHERE a IS A CONSTANT AND V IS A FUNCTION OF x

We have

$$D(e^{ax}u)=e^{ax}Du+ae^{ax}u=e^{ax}(D+a)u$$

$$D^2(e^{ax}u)=e^{ax}(D+a)^2u$$

In general, $$D^k(e^{ax}u) = e^{ax}(D+a)^k u.$$

$\therefore$ $$f(D)(e^{ax}u) = e^{ax}f(D+a)u.$$

Operating both sides by $\frac{1}{f(D)}$ gives

$$e^{ax}V=\frac{1}{f(D)}[e^{ax}f(D+a)u]$$

Put $$f(D + a)u = V.$$

$$\Rightarrow \qquad u = \frac{1}{f(D+a)}V$$

$$\therefore \qquad \frac{1}{f(D)}(e^{ax}V) = e^{ax}\frac{1}{f(D+a)}V. \tag{4.15}$$

EXAMPLE 4.17 Find PI of $(D^2 - 2D + 4)y = e^x \cos x$.

Solution

$$\text{PI} = \frac{1}{D^2 - 2D + 4}(e^x \cos x)$$

$$= e^x \frac{1}{(D+1)^2 - 2(D+1) + 4}\cos x, \text{ replacing } D \text{ by } D + 1$$

$$= e^x \frac{1}{D^2 + 3}\cos x$$

$$= e^x \frac{1}{-1+3}\cos x$$

$$= \frac{1}{2}e^x \cos x.$$

EXAMPLE 4.18 Find PI of $(D^2 + 2)y = x^2 e^{3x}$.

Solution

$$\text{PI} = \frac{1}{D^2 + 2}x^2 e^{3x}$$

$$= e^{3x}\frac{1}{(D+3)^2 + 2}x^2, \text{ replacing } D \text{ by } D + 3$$

$$= e^{3x}\frac{1}{D^2 + 6D + 11}x^2$$

$$= \frac{1}{11}e^{3x}\left(1 + \frac{6D + D^2}{11}\right)^{-1} x^2$$

$$= \frac{1}{11}e^{3x}\left(1 - \frac{6D + D^2}{11} + \frac{36}{121}D^2 + \cdots\right)x^2$$

$$= \frac{1}{11}e^{3x}\left(1 - \frac{6D}{11} + \frac{35}{121}D^2 + \cdots\right)x^2$$

$$= \frac{1}{11}e^{3x}\left(x^2 - \frac{12}{11}x + \frac{70}{121}\right).$$

EXAMPLE 4.19 Find PI of $(D^4 + D^2 + 1)y = ax^2 + be^{-x} \sin 2x$.

Solution

$$\text{PI} = \frac{1}{D^4 + D^2 + 1}(ax^2 + be^{-x} \sin 2x)$$

$$= a\frac{1}{D^4 + D^2 + 1}x^2 + b\frac{1}{D^4 + D^2 + 1}e^{-x} \sin 2x$$

$$= a[1 + (D^2 + D^4)]^{-1} x^2 + be^{-x} \frac{1}{(D-1)^4 + (D-1)^2 + 1} \sin 2x$$

$$= a[1 - D^2 + \cdots]x^2 + be^{-x} \frac{1}{D^4 - 4D^3 + 7D^2 - 6D + 3} \sin 2x$$

$$= a(x^2 - 2) + be^{-x} \frac{1}{16 + 16D - 28 - 6D + 3} \sin 2x$$

$$= a(x^2 - 2) + be^{-x} \frac{1}{10D - 9} \sin 2x$$

$$= a(x^2 - 2) + be^{-x}(10D + 9)\frac{1}{100D^2 - 81} \sin 2x$$

$$= a(x^2 - 2) - \frac{1}{481}be^{-x}(10D + 9) \sin 2x$$

$$= a(x^2 - 2) - \frac{1}{481}be^{-x}(20 \cos 2x + 9 \sin 2x).$$

4.10 PI WHEN $X = xV$, WHERE V IS A FUNCTION OF x

By Leibnitz's theorem

$$D^n(xV) = xD^nV + nD^{n-1}V$$

$$= xD^nV + \frac{d}{dD}(D^n)V, \text{ where } n \text{ is a positive integer.}$$

$$f(D)(xV) = (D^n + Q_1D^{n-1} + \cdots + Q_n)(xV)$$

$$= \left(xD^nV + \frac{d}{dD}(D^n)V\right) + Q_1\left(xD^{n-1}V + \frac{d}{dD}(D^{n-1})V\right) + \cdots + Q_n(xV)$$

$$= x(D^n + Q_1D^{n-1} + \cdots + Q_n)V + \left[\frac{d}{dD}(D^n + Q_1D^{n-1} + \cdots + Q_n)\right]V$$

$$= xf(D)V + \frac{d}{dD}[f(D)]V$$

$$= xf(D)V + f'(D)V \tag{4.16}$$

where

$$f'(D) = \frac{d}{dD}[f(D)].$$

Let $$f(D)V = U.$$

$$\Rightarrow \quad V = \frac{1}{f(D)}U$$

Therefore, Eq. (4.16) becomes

$$f(D)\left(x\frac{1}{f(D)}U\right) = xU + f'(D)\frac{1}{f(D)}U$$

Operating on both sides by multiplying $\frac{1}{f(D)}$, we get

$$\frac{1}{f(D)}f(D)\left(x\frac{1}{f(D)}U\right) = \frac{1}{f(D)}xU + \frac{1}{f(D)}f'(D)\frac{1}{f(D)}U$$

$$\Rightarrow \quad \frac{1}{f(D)}(xU) = x\frac{1}{f(D)}U - \frac{1}{f(D)}f'(D)\frac{1}{f(D)}U$$

$$\Rightarrow \quad x\frac{1}{f(D)}U = \frac{1}{f(D)}(xU) + \frac{1}{f(D)}f'(D)\frac{1}{f(D)}U$$

Replacing U by V gives

$$\frac{1}{f(D)}(xV) = x\frac{1}{f(D)}V - \frac{1}{f(D)}f'(D)\frac{1}{f(D)}V$$

$$\Rightarrow \quad \frac{1}{f(D)}(xV) = \left[x - \frac{1}{f(D)}f'(D)\right]\frac{1}{f(D)}V.$$

EXAMPLE 4.20 Find PI of $(D^4 + 2D^2 + 1)y = x^2 \sin x$.

Solution

$$\text{PI} = \frac{1}{(D^2+1)^2}(x^2 \sin x)$$

$$= \text{Imaginary part of } \frac{1}{(D^2+1)^2}(x^2 e^{ix})$$

$$= \text{Imaginary part of } e^{ix}\frac{1}{[(D+i)^2+1]^2}x^2$$

$$= \text{Imaginary part of } e^{ix}\frac{1}{(D^2+2iD)^2}x^2$$

$$= \text{Imaginary part of } e^{ix}\frac{1}{-4D^2}\left(1+\frac{D}{2i}\right)^{-2}x^2$$

$$= \text{Imaginary part of } \frac{1}{-4}\left[e^{ix}\frac{1}{D^2}\left(1-\frac{2D}{2i}+\frac{3D^2}{4i^2}\right)x^2\right]$$

$$= \text{Imaginary part of } \frac{1}{-4}\left[e^{ix}\frac{1}{D^2}\left(x^2 + 2ix - \frac{3}{2}\right)\right]$$

$$= \text{Imaginary part of } \frac{1}{-4}\left[e^{ix}\left(\frac{x^4}{12} + \frac{ix^3}{3} - \frac{3x^2}{4}\right)\right]$$

$$= \text{Imaginary part of } \frac{1}{-4}\left[(\cos x + i\sin x)\left(\frac{x^4}{12} + \frac{ix^3}{3} - \frac{3x^2}{4}\right)\right]$$

$$= \frac{1}{-4}\left[\frac{x^3\cos x}{3} + \sin x\left(\frac{x^4}{12} - \frac{3x^2}{4}\right)\right].$$

EXAMPLE 4.21 Find PI of $(D^2 - 2D + 1)y = x^{ex} \cos x$.

Solution

$$\text{PI} = \frac{1}{D^2 - 2D + 1}(xe^x \cos x)$$

$$= \frac{1}{(D-1)^2}(xe^x \cos x)$$

$$= e^x \frac{1}{D^2}(x \cos x)$$

$$= e^x \left[x - \frac{1}{D^2}2D\right](-\cos x)$$

$$= -xe^x \cos x - 2\frac{1}{D^2}\sin x$$

$$= -xe^x \cos x + 2\sin x.$$

EXAMPLE 4.22 Find solution of $D(D+1)^2 = e^{2x} + x(x+1)$.

Solution The AE is $m(m + 1)^2 = 0$. The roots are $m = 0, -1, -1$.

Therefore, $\text{CF} = c_1 + (c_2 + c_3 x)e^{-x}$

$$\text{PI} = \frac{1}{D(D+1)^2}e^{2x} + \frac{1}{D(D+1)^2}(x^2 + x)$$

$$= \frac{1}{18}e^{2x} + \frac{1}{D}(1 + 2D + D^2)^{-1}(x^2 + x)$$

$$= \frac{1}{18}e^{2x} + \frac{1}{D}(1 + 2D + D^2)^{-1}(x^2 + x)$$

$$= \frac{1}{18}e^{2x} + \frac{1}{D}(1-(2D+D^2)+4D^2)^{-1}(x^2+x)$$

$$= \frac{1}{18}e^{2x} + \frac{1}{D}(1-2D+3D^2+\cdots)(x^2+x)$$

$$= \frac{1}{18}e^{2x} + \frac{1}{D}((x^2+x)-2(2x+1)+6)$$

$$= \frac{1}{18}e^{2x} + \frac{1}{D}(x^2-3x+4)$$

$$= \frac{1}{18}e^{2x} + \frac{x^3}{3} - \frac{3x^2}{3} + 4x$$

$$= \frac{1}{18}e^{2x} + \frac{x}{6}(2x^2-9x+24)$$

Therefore, the general solution of the given differential equation is

$$y = c_1 + (c_2 + c_3 x)e^{-x} + \frac{1}{18}e^{2x} + \frac{x}{6}(2x^2-9x+24)$$

EXERCISES

Solve the following differential equations:

1. $(D^2+3aD-4a^2)y=0.$ **[Ans.** $y = c_1 e^{-4ax} + c_2 e^{ax}$]
2. $(D^3-3D^2+3D+1)y=0.$ **[Ans.** $y = (c_1 + c_2 x + c_3 x^2)e^x$]
3. $(D^4+8D^2+16)y=0.$ **[Ans.** $y = (c_1 + c_2 x)\cos 2x + (c_3 + c_4 x)\sin 2x$]
4. $(D^2-2aD+a^2+b^2)y=0.$ **[Ans.** $y = e^{ax}(c_1 \cos bx + c_2 \sin bx)$]
5. $(D^4-6D^3+12D^2-8D)y=0.$ **[Ans.** $y = c_1 + c_2 e^{2x} + c_3 x e^{2x} + c_4 x^2 e^{2x}$]
6. $(D^4+4)y=0.$ **[Ans.** $y = (c_1 e^x + c_2 e^{-x})\cos x + (c_3 e^x + c_4 e^{-x})\sin x$]
7. $(D^3-2D^2-4D+8)y=0.$ **[Ans.** $y = c_1 e^{-2x} + (c_2 + c_3 x)e^{2x}$]
8. $(D^2-1)y=2+5x.$ **[Ans.** $y = c_1 e^x + c_2 e^{-x} - 2 - 5x$]
9. $(D^2-5D+6)y=e^{4x}.$ $\left[\textbf{Ans. } y = c_1 e^{3x} + c_2 e^{2x} + \frac{1}{2}e^{4x}\right]$
10. $(D^2-3D+2)y=\sin(e^{-x}).$ **[Ans.** $y = c_1 e^x + c_2 e^{2x} - e^{2x}\sin(e^{-x})$]
11. $(D^2+1)y=\operatorname{cosec}^2 x.$ **[Ans.** $y = c_1 \cos x + c_2 \sin x - 1 - \cos x \log|\operatorname{cosec} x - \cot x|$]
12. $(D^2+4D+4)y=2\sinh 2x.$ $\left[\textbf{Ans. } y = (c_1 + c_2 x)e^{-2x} + \frac{1}{16}e^{2x} - \frac{1}{2}x^3 e^{-2x}\right]$

13. $(D^2+1)y=(e^x+1)^2$.

$$\left[\textbf{Ans. } y=c_1e^{-x}+e^{x/2}\left(c_2\cos\frac{\sqrt{3}}{2}x+c_3\sin\frac{\sqrt{3}}{2}x\right)+\frac{1}{9}e^{2x}+e^x+1\right]$$

14. $(D^4+D^3-6D^2)y=3e^{2x}$.

$$\left[\textbf{Ans. } y=c_1+c_2x+c_3e^{-3x}+c_4e^{2x}+\frac{3}{20}xe^{2x}\right]$$

15. $(2D+1)^2y=e^{-x/2}$.

$$\left[\textbf{Ans. } y=(c_1+c_2x)e^{-x/2}+\frac{x^2}{2}e^{-x/2}\right]$$

16. $\dfrac{d^2y}{dx^2}=y+e^x+e^{-x}$.

$$\left[\textbf{Ans. } y=c_1e^x+c_2e^{-x}+\frac{x}{2}(e^x-e^{-x})\right]$$

17. $(D^2-2D+5)y=10\sin x$.

$$\left[\textbf{Ans. } y=e^x(c_1\cos 2x+c_2\sin 2x)+\cos x+2\sin x\right]$$

18. $(D^2+4)y=e^x+\sin 2x$.

$$\left[\textbf{Ans. } y=c_1\cos 2x+c_2\sin 2x+\frac{1}{5}e^x-\frac{x\cos 2x}{4}\right]$$

19. $(D^3+1)y=\sin 3x-\cos^2\dfrac{x}{2}$.

$$\left[\begin{aligned}&\textbf{Ans. } y=c_1e^{-x}+e^{x/2}\left(c_2\cos\frac{\sqrt{3}}{2}x+c_3\sin\frac{\sqrt{3}}{2}x\right)\\&\quad+\frac{\sin 3x+27\cos 3x}{730}-\frac{1}{2}+\frac{\sin x-\cos x}{4}\end{aligned}\right]$$

20. $(D^2-4)y=\cos^2 x$.

$$\left[\textbf{Ans. } y=c_1e^{2x}+c_2e^{-2x}-\frac{1}{8}-\frac{1}{16}\cos 2x\right]$$

21. $(D^2+4)y=\sin 3x+e^x+x^2$.

$$\left[\textbf{Ans. } y=c_1\cos 2x+c_2\sin 2x+\frac{1}{5}(e^x-\sin x)+\frac{1}{8}(2x^2-1)\right]$$

22. $(D^3-13D+12)y=x$.

$$\left[\textbf{Ans. } y=c_1e^x+c_2e^{3x}+c_3e^{-4x}+\frac{1}{12}\left(x+\frac{13}{12}\right)\right]$$

23. $(D^4+D^3-6D^2)y=x^2+3e^{2x}+4\sin x$.

$$\left[\textbf{Ans. } y=c_1+c_2x+c_3e^{-3x}+c_4e^{2x}-\frac{1}{6}\left(\frac{x^4}{12}+\frac{x^3}{18}+\frac{7x^2}{36}\right)+\frac{3xe^{2x}}{20}+\frac{2}{25}(7\sin x+\cos x)\right]$$

24. $(D^2-4D+3)y=2x^2$.

$$\left[\textbf{Ans. } y=c_1e^x+c_2e^{3x}+\frac{2}{3}\left(x^2+\frac{8}{3}x+\frac{26}{9}\right)\right]$$

25. $(D^2-4)y=x^2$.

$$\left[\textbf{Ans. } y=c_1e^{2x}+c_2e^{-2x}-\frac{1}{4}\left(x^2+\frac{1}{2}\right)\right]$$

Find particular integral of the following differential equations:

26. $(D^2+3D-4)y = xe^{-2x}$. $\left[\textbf{Ans.}\ \text{PI} = -\frac{1}{6}e^{-2x}\left(x-\frac{1}{6}\right)\right]$

27. $(D^2+3D+2)y = e^{2x}\sin x$. $\left[\textbf{Ans.}\ \text{PI} = \frac{1}{170}e^{2x}(11\sin x - 7\cos x)\right]$

28. $(D^2+2)y = x^2e^{3x} + e^x\cos 2x$.

$$\left[\textbf{Ans.}\ \text{PI} = \frac{1}{121}e^{3x}\left(11x^2-12x+\frac{50}{11}\right)+\frac{1}{17}e^x(4\sin 2x - \cos 2x)\right]$$

29. $(D^3-D^2+3D+5)y = e^x\cos 2x$. $\left[\textbf{Ans.}\ \text{PI} = \frac{x}{16}e^x(\sin 2x - \cos 2x)\right]$

30. $(D^2-5D+6)y = x^3e^x$. $\left[\textbf{Ans.}\ \text{PI} = \frac{e^x}{2}\left(x^3+\frac{9}{2}x^2+\frac{21}{2}x+\frac{45}{4}\right)\right]$

31. $(D^4-1)y = e^x\cos x$. $\left[\textbf{Ans.}\ \text{PI} = -\frac{1}{5}e^x\cos x\right]$

32. $(D^2+4)y = x\sin x$. $\left[\textbf{Ans.}\ \text{PI} = \frac{1}{3}x\sin x - \frac{2}{9}\cos x\right]$

33. $(D^2-4D+4)y = 8x^2e^{2x}\sin 2x$. $[\textbf{Ans.}\ \text{PI} = e^{2x}(3\sin 2x - 2x^2\sin 2x - 4x\cos 2x)]$

34. $(D^4-1)y = x^2\sin x$. $\left[\textbf{Ans.}\ \text{PI} = \frac{1}{4}\left[\left(\frac{x^3}{3}-\frac{5x}{4}\right)\cos x - \left(\frac{3x^2}{2}-\frac{x}{2}\right)\sin x\right]\right]$

35. $(D^2+1)y = 3\cos^2 x + 2\sin^3 x$. $\left[\textbf{Ans.}\ \text{PI} = \frac{3}{2}-\frac{1}{2}\cos 2x - \frac{3}{4}x\cos x + \frac{1}{16}\sin 3x\right]$

Miscellaneous exercises:

36. Solve $D(D^2-D-12)y=0$. $[\textbf{Ans.}\ y = c_1 + c_2e^{4x} + c_3e^{-3x}]$

37. Solve $(D^4+2n^2D^2+n^4)y=0$. $[\textbf{Ans.}\ y = (c_1+c_2x)\cos nx + (c_3+c_4x)\sin nx]$

38. Solve $(D^3+2D^2-5D-6)y=0$. $[\textbf{Ans.}\ y = c_1e^{2x} + c_2e^{-x} + c_3e^{-3x}]$

39. Solve $D(D^3-6D^2+12D-8)y=0$. $[\textbf{Ans.}\ y = c_1 + (c_2+c_3x+c_4x^2)e^{2x}]$

40. Solve $(D^4+D^3+2D^2-D+3)y=0$.

$$\left[\begin{aligned}\textbf{Ans.}\ y &= e^{-x}[c_1\cos(\sqrt{2}x)+c_2\sin(\sqrt{2}x)]\\ &+e^{x/2}\left[c_3\cos\left(\frac{\sqrt{3}}{2}x\right)+c_4\sin\left(\frac{\sqrt{3}}{2}x\right)\right]\end{aligned}\right]$$

41. Solve $(D^3 - 2D^2 - 4D + 8)y = 0$. [**Ans.** $y = c_1 e^{-2x} + (c_2 + c_3 x)e^{2x}$]

42. Solve $(D^2 - 6D + 9)y = x^2 e^{3x}$. $\left[\textbf{Ans. } y = (c_1 + c_2 x)e^{3x} + \frac{x^4}{12}e^{3x}\right]$

43. Solve $(D^2 + 1)y = \text{cosec } x$. [**Ans.** $y = c_1 \cos x + c_2 \sin x + \sin x \log \sin x - x \cos x$]

44. Solve $(D^2 + n^2)y = \cot nx$.

$\left[\textbf{Ans. } y = c_1 \cos nx + c_2 \sin nx + \frac{1}{n^2} \sin nx(\text{cosec } nx - \cot nx)\right]$

45. Solve $(D^2 - 9D + 18)y = e^{e^{-3x}}$. $\left[\textbf{Ans. } y = c_1 e^{3x} + c_2 e^{6x} + \frac{1}{9}e^{e^{-3x}}e^{6x}\right]$

46. Solve $(D^2 - 1)y = (1 + e^{-x})^{-2}$. [**Ans.** $y = c_1 e^x + c_2 e^{-x} - 1 + e^{-x}\log(1 + e^x)$]

47. Solve $(D^2 + 4D + 3)y = e^{-3x}$. $\left[\textbf{Ans. } y = c_1 e^{-x} + c_2 e^{-3x} + \frac{1}{2}xe^{-3x}\right]$

48. Solve $(D^2 - 1)^2 y = e^x$. $\left[\textbf{Ans. } y = (c_1 + c_2 x)e^x + (c_3 + c_4 x)e^{-x} + \frac{1}{8}x^2 e^x\right]$

49. Solve $(D - 2)^n y = e^{2x}$. $\left[\textbf{Ans. } y = (c_1 + c_2 x + \cdots + c_n x^{n-1})e^{2x} + \frac{x^n}{n!}e^{2x}\right]$

50. Solve $(D^2 - 4D + 3)y = 2e^{3x}$. [**Ans.** $y = c_1 e^x + c_2 e^{3x} + xe^{3x}$]

51. Solve $(D^2 - 6D + 8)y = (e^{2x} - 1)^3$. $\left[\textbf{Ans. } y = c_1 e^{4x} + c_2 e^{2x} + \frac{x}{2}e^{4x} + xe^{2x} + \frac{1}{8}\right]$

52. Solve $(D^3 + 1)y = 3 + e^{-x} + 5e^{2x}$.

$$\left[\begin{aligned} &\textbf{Ans. } y = c_1 e^{-x} + e^{x/2}\left[c_2 \cos\left(\frac{\sqrt{3}}{2}x\right) + c_3 \sin\left(\frac{\sqrt{3}}{2}x\right)\right] \\ &\qquad + 3 + \frac{1}{3}xe^{-x} + \frac{5}{9}e^{2x} \end{aligned}\right]$$

53. Solve $(D^3 - 6D^2 + 11D - 6)y = e^{2x}$. [**Ans.** $y = c_1 e^x + c_2 e^{2x} + c_3 e^{3x} - xe^{2x}$]

54. Solve $(D^3 - 7D + 6)y = e^{2x}$. $\left[\textbf{Ans. } y = c_1 e^x + c_2 e^{2x} + c_3 e^{-3x} - \frac{x}{5}e^{2x}\right]$

55. Solve $(D^2 - 2D + 1)y = e^{2x}$. [**Ans.** $y = (c_1 + c_2 x)e^x + e^{2x}$]

56. Solve $(D^2+4D-12)y=e^{3x}$. $\left[\textbf{Ans. } y=c_1e^{2x}+c_2e^{-6x}+\frac{1}{9}e^{3x}\right]$

57. Solve $(D^4-1)y=\sin x$. $\left[\textbf{Ans. } y=c_1\cos x+c_2\sin x+c_3e^{x}+c_4e^{-x}+\frac{x}{4}\cos x\right]$

58. Solve $(D^2+D-2)y=\sin x$. $\left[\textbf{Ans. } y=c_1e^{x}+c_2e^{-2x}-\frac{1}{10}(\cos x+3\sin x)\right]$

59. Solve $(D^2-6D+9)y=\sin 2x$. $\left[\textbf{Ans. } y=(c_1+c_2x)e^{2x}+\frac{1}{8}\cos 2x\right]$

60. Solve $(D^2+16)y=\sin 2x$. $\left[\textbf{Ans. } y=c_1\cos 4x+c_2\sin 4x-\frac{x}{4}\cos 2x\right]$

61. Solve $(D^2+a^2)Dy=\sin ax$. $\left[\textbf{Ans. } y=c_1+c_2\cos ax+c_3\sin ax-\frac{x}{2a^2}\sin ax\right]$

62. Solve $(D^2+9)y=e^{x}-\cos 2x$. $\left[\textbf{Ans. } y=c_1\cos 3x+c_2\sin 3x+\frac{1}{10}e^{x}-\frac{1}{5}\cos 2x\right]$

63. Solve $(D^2-D-2)y=e^{x}+\sin 2x$.

$$\left[\textbf{Ans. } y=c_1e^{2x}+c_2e^{-x}-\frac{1}{2}e^{x}+\frac{1}{40}\{2\cos 2x-6\sin 2x\}\right]$$

64. Solve $(D^2+D+1)y=\sin 2x$.

$$\left[\begin{aligned}&\textbf{Ans. } y=e^{-x/2}\left[c_1\cos\left(\frac{\sqrt{3}}{2}x\right)+c_2\sin\left(\frac{\sqrt{3}}{2}x\right)\right]\\&-\frac{1}{13}(2\cos 2x+3\sin 2x)\end{aligned}\right]$$

65. Solve $(4D^2+16D-9)y=4e^{x/2}+3\sin\frac{x}{4}$.

$$\left[\begin{aligned}&\textbf{Ans. } y=c_1e^{x/2}+c_2e^{-9x/2}\\&+\frac{x}{5}e^{x/2}-\frac{12}{1625}\left\{16\cos\frac{x}{4}+37\sin\frac{x}{4}\right\}\end{aligned}\right]$$

66. Solve $(D^3+1)y=\cos 2x$.

$$\left[\begin{aligned}&\textbf{Ans. } y=c_1e^{-x}+e^{x/2}\left[c_2\cos\left(\frac{\sqrt{3}}{2}x\right)+c_3\sin\left(\frac{\sqrt{3}}{2}x\right)\right]\\&+\frac{1}{65}(\cos 2x-8\sin 2x)\end{aligned}\right]$$

67. Solve $(D^3 - 4D^2 + D - 4)y = \cos x + e^x$.

$$\left[\textbf{Ans. } y = c_1 e^{4x} + c_2 \cos x + c_3 \sin x - \frac{x}{34}(\cos x + 4\sin x)\right]$$

68. Solve $(D^3 - 2D^2 + 9D - 18)y = 6\cos 3x + 48\sin 3x + 3e^{2x}$.

$$\left[\textbf{Ans. } y = c_1 e^{2x} + c_2 \cos 3x + c_3 \sin 3x + \frac{3}{13} x e^{2x} + x(\cos 3x - 2\sin 3x)\right]$$

69. Solve $(D^4 + 10D^2 + 9)y = \cos(2x + 3)$.

$$\left[\textbf{Ans. } y = c_1 \cos x + c_2 \sin x + c_3 \cos 3x + c_4 \sin 3x - \frac{1}{15}\cos(2x + 3)\right]$$

70. Solve $(D^2 - 2D^2 + 1)y = \cos x$.

$$\left[\textbf{Ans. } y = (c_1 + c_2 x)e^x + (c_3 + c_4 x)e^{-x} + \frac{1}{4}\cos x\right]$$

71. Solve $(D^2 + D + 1)y = \sin 2x$.

$$\left[\textbf{Ans. } y = e^{-x/2}\left[c_1 \cos\left(\frac{\sqrt{3}}{2}x\right) + c_2 \sin\left(\frac{\sqrt{3}}{2}x\right)\right] - \frac{1}{13}(2\cos 2x + 3\sin 2x)\right]$$

72. Solve $(D^2 + D - 2)y = x + \sin x$.

$$\left[\textbf{Ans. } y = ae^x + be^{-2x} - \frac{1}{4}(2x + 1) - \frac{1}{10}(\cos x + 3\sin x)\right]$$

73. Solve $(D^2 - 2D + 3)y = \cos x + x^2$.

$$\left[\textbf{Ans. } y = e^x(a\cos\sqrt{2}x + b\sin\sqrt{2}x) + \frac{1}{4}(\cos x - \sin x) + \frac{1}{27}(9x^2 + 12x + 2)\right]$$

74. Solve $(D^2 + D + 1)y = \cos 2x + x^2$.

$$\left[\textbf{Ans. } y = e^{-\frac{x}{2}}\left\{c_1 \cos\frac{\sqrt{3}}{2}x + c_2 \sin\frac{\sqrt{3}}{2}x\right\} + x^2 - 2x + \frac{1}{13}(2\sin 2x - 3\cos 2x)\right]$$

75. Solve $(D^2 - 3D + 2)y = e^{2x} + x^2$.

$$\left[\textbf{Ans. } y = c_1 e^x + c_2 e^{2x} + xe^{2x} + \frac{1}{2}\left(x^2 + 3x + \frac{7}{2}\right)\right]$$

76. Solve $(D^3 + 2D^2 + 4D + 8)y = x^2$.

$$\left[\textbf{Ans. } y = c_1 e^{-2x} + c_2 \cos 2x + c_3 \sin 2x + \frac{1}{8}(x^2 - x)\right]$$

77. Solve $(D^4 - 16)y = x^4$. $\left[\textbf{Ans. } y = c_1 e^{2x} + c_2 e^{-2x} + c_3 \cos 2x + c_4 \sin 2x - \dfrac{x^4}{16} - \dfrac{24}{256}\right]$

78. Solve $(D^4 + D^2 + 16)y = 16x^2 + 256$.

$$\left[\textbf{Ans. } y = \left(c_1 e^{\frac{\sqrt{7}}{2}x} + c_2 e^{-\frac{\sqrt{7}}{2}x}\right)\cos\frac{3}{2}x + \left(c_3 e^{\frac{\sqrt{7}}{2}x} + c_4 e^{-\frac{\sqrt{7}}{2}x}\right)\sin\frac{3}{2}x + x^2 + \frac{127}{8}\right]$$

79. Solve $(D^2 + 5D + 6)y = e^{-2x}\sec^2 x(1 + 2\tan nx)$. $[\textbf{Ans. } y = c_1 e^{-3x} + c_2 e^{-2x} + e^{-2x}\tan x]$

80. Solve $(D-2)^2 y = x^{-2}e^{2x}$. $[\textbf{Ans. } y = (c_1 + c_2 x)e^{2x} - e^{-2x}\log x]$

81. Solve $(D^2 + 4)y = e^x x^4$.

$$\left[\textbf{Ans. } y = c_1 \cos 2x + c_2 \sin 2x + \frac{e^x}{5}\left[x^4 - \frac{4x^2(2x+3)}{5} + \frac{24(2x^2+4x+1)}{25} - \frac{96(2x+3)}{125} + \frac{384}{625}\right]\right]$$

82. Solve $(D^2 + 4)y = e^x \cos 2x$. $\left[\textbf{Ans. } y = c_1 \cos 2x + c_2 \sin 2x + \dfrac{e^x}{17}(\cos 2x + 4\sin 2x)\right]$

83. Solve $(D+2)^2 y = x^2 e^{-2x}$. $\left[\textbf{Ans. } y = (c_1 + c_2 x)e^{-2x} + \dfrac{x^4}{12}e^{-2x}\right]$

84. Solve $(D^2 + 1)y = xe^{2x}$. $\left[\textbf{Ans. } y = c_1 \cos x + c_2 \sin x + \dfrac{1}{5}\left(x - \dfrac{4}{5}\right)e^{2x}\right]$

85. Solve $(D^2 + 1)y = \cos x + e^x \sin x$.

$$\left[\textit{Ans. } y = c_1 \cos x + c_2 \sin x + \frac{x}{2}\sin x - \frac{e^x}{5}(2\cos x - \sin x)\right]$$

86. Solve $(D^3 - 1)y = xe^x + \cos^2 x$.

$$\left[\textbf{Ans. } y = c_1 e^x + e^{-\frac{x}{2}}\left(c_2 \cos\frac{\sqrt{3}}{2}x + c_3 \sin\frac{\sqrt{3}}{2}x\right) + \frac{e^x}{3}\left(\frac{x^2}{2} - x\right) - \frac{1}{2} - \frac{1}{130}(8\sin 2x + \cos 2x)\right]$$

87. Solve $(D^3 + 1)y = e^{2x}\sin x$.

$$\left[\textbf{Ans. } y = c_1 e^{-x} + e^{\frac{x}{2}}\left(c_2 \cos\frac{\sqrt{3}}{2}x + c_3 \sin\frac{\sqrt{3}}{2}x\right) - \frac{1}{130}e^{2x}(11\cos x - 3\sin x)\right]$$

88. Solve $(D^3-3D^2+3D-1)y=(x+1)e^x$.

$$\left[\textbf{Ans. } y=(c_1+c_2x+c_3x^2)e^x+e^x\left(\frac{x^4}{24}+\frac{x^3}{6}\right)\right]$$

89. Solve $(D^3-7D-6)y=e^{2x}(1+x)$. $\left[\textbf{Ans. } y=c_1e^{-x}+c_2e^{-2x}+c_3e^{3x}-\frac{e^{2x}}{12}\left(x+\frac{17}{12}\right)\right]$

90. Solve $(D^4-1)y=e^x\cos x$. $\left[\textbf{Ans. } y=c_1e^x+c_2e^{-x}+c_3\cos x+c_4\sin x-\frac{1}{5}e^x\cos x\right]$

91. Solve $(D^2-1)y=x\sin x+(x^2+1)e^x$.

$$\left[\textbf{Ans. } y=c_1e^x+c_2e^{-x}-\frac{1}{2}(x\sin x+\cos x)+\frac{1}{12}xe^x(2x^2-3x+9)\right]$$

92. Solve $(D^2+9)y=x\cos x$. $\left[\textbf{Ans. } y=c_1\cos 3x+c_2\sin 3x+\frac{1}{8}x\cos x+\frac{1}{32}\sin x\right]$

93. Solve $(D-1)^2y=x\sin x$. $\left[\textbf{Ans. } y=(c_1+c_2x)e^x+\frac{1}{2}x\cos x+\frac{1}{2}(\cos x-\sin x)\right]$

94. Solve $(D^2+2)y=x^2e^{3x}+e^x\cos 2x$.

$$\left[\textbf{Ans. } y=c_1\cos\sqrt{2}x+c_2\sin\sqrt{2}x+\frac{e^{3x}}{11}\left(x^2-\frac{12}{11}x+\frac{50}{121}\right)-\frac{e^x}{17}(\cos 2x-4\sin 2x)\right]$$

95. Solve $(D^2-5D+6)y=x\sin 3x$.

$$\left[\textbf{Ans. } y=c_1e^{2x}+c_2e^{3x}-\frac{1}{234}\left\{x(3\sin 3x-15\cos 3x)+\frac{1620\sin 3x-846\cos 3x}{234}\right\}\right]$$

96. Solve $(D^3+3D^2-4)y=x\cos x$.

$$\left[\textbf{Ans. } y=c_1e^x+(c_2+c_3x)e^{-2x}+\frac{1}{50}\left\{\left(\frac{6}{5}-7x\right)\cos x+\left(\frac{33}{5}-x\right)\sin x\right\}\right]$$

97. Solve $(D^4-1)y=x\sin x$.

$$\left[\textbf{Ans. } y=c_1e^{-x}+c_2e^x+c_3\cos x+c_4\sin x+\frac{1}{8}(x^2\cos x-3x\sin x)\right]$$

98. Solve $(D^4+D^2+1)y=x\cos x$.

$$\left[\begin{array}{l}\textbf{Ans. } y=e^{\frac{x}{2}}\left(c_1\cos\frac{\sqrt{3}}{2}x+c_2\sin\frac{\sqrt{3}}{2}x\right)+e^{-\frac{x}{2}}\left(c_3\cos\frac{\sqrt{3}}{2}x+c_4\sin\frac{\sqrt{3}}{2}x\right)\\ \qquad +x\cos x-4\sin x+2\cos x\end{array}\right]$$

99. Solve $(D^2 - 4D + 3)y = (1 + e^{-x})^{-1}$.

$$\left[\textbf{Ans. } y = c_1 e^x + c_2 e^{3x} + \frac{1}{2}e^{3x} + \frac{1}{2}(e^x - e^{3x})\log(1 + e^{-x})\right]$$

100. Solve $(D^2 - 1)y = x^2 \sin 3x$. $\left[\textbf{Ans. } y = c_1 e^x + c_2 e^{-x} - \frac{25x^2 - 13}{250}\sin 3x - \frac{3}{25}x\cos 3x\right]$

101. Solve $(D^2 + 1)y = 3\cos^2 x + 2\sin^3 x$.

$$\left[\textbf{Ans. } y = c_1 \cos x + c_2 \sin x + \frac{3}{2} - \frac{1}{2}\cos 2x - \frac{3}{4}x\cos x + \frac{1}{16}\sin 3x\right]$$

102. Solve $(D - 1)^2(D^2 + 1)^2 y = e^x + \sin^2 \frac{x}{2}$.

$$\left[\textbf{Ans. } y = (c_1 + c_2 x)e^x + (c_3 + c_4 x)\cos x + (c_5 + c_6 x)\sin x + \frac{1}{8}x^2 e^x - \frac{1}{32}x^2 \sin x + \frac{1}{2}\right]$$

103. Solve $(D^3 - 2D^2 - 5D + 6)y = (e^{2x} + 3)^2$.

$$\left[\textbf{Ans. } y = c_1 e^x + c_2 e^{3x} + c_3 e^{-2x} + \frac{e^{4x}}{18} - \frac{3}{2}e^{2x} + \frac{3}{2}\right]$$

104. Solve $(D^3 - 2D + 4)y = x^4 + 3x^2 - 5x + 2$.

$$\left[\textbf{Ans. } y = c_1 e^{-2x} + e^x(c_2 \cos x + c_3 \sin x) + \frac{1}{4}x^4 + \frac{1}{2}x^3 + \frac{3}{2}x^2 - \frac{5}{4}x - \frac{7}{8}\right]$$

105. Solve $(D^3 + 1)y = e^{2x}\sin x + e^{\frac{x}{2}}\sin\left(\frac{\sqrt{3}}{2}x\right)$.

$$\left[\begin{aligned} \textbf{Ans. } y = c_1 e^{-x} + e^{\frac{x}{2}}\left\{c_2 \cos\frac{\sqrt{3}}{2}x + c_3 \sin\frac{\sqrt{3}}{2}x\right\} + \frac{e^{2x}}{130}(3\sin x - 11\cos x) \\ - \frac{1}{6}xe^{\frac{x}{2}}\left(\sin\frac{\sqrt{3}}{2}x + \sqrt{3}\cos\frac{\sqrt{3}}{2}x\right) \end{aligned}\right]$$

106. Solve $D^2(D^2 + 2D - 3)y = x^2 + 3e^{2x} + 4\sin x$.

$$\left[\textbf{Ans. } y = c_1 + c_2 x + c_3 e^x + c_4 e^{-3x} + \frac{x^2}{108}(3x^2 + 8x + 28)\right]$$

Homogeneous Linear Differential Equations with Variable Coefficients

5.1 INTRODUCTION

In this chapter, we will consider linear differential equations in which the coefficients are the functions of x. We shall only discuss the homogeneous linear equation. The general form of homogeneous linear equation of order n is

$$x^n \frac{d^n y}{dx^n} + Q_1 x^{n-1} \frac{d^{n-1} y}{dx^{n-1}} + \cdots + Q_n y = X \tag{5.1}$$

where Q_1, Q_2, ..., Q_n are constants and X is a constant or a function of x.

5.2 SOLUTION METHODOLOGY

Equation (5.1) can be reduced to an Eq. (4.1) with constant coefficients by changing the independent variable by the substitution

$$z = \log x$$

$$\Rightarrow \qquad x = e^z$$

Then $$\frac{dy}{dx} = \frac{dy}{dz}\frac{dz}{dx} = \frac{1}{x}\frac{dy}{dz}.$$

$$\Rightarrow \qquad x\frac{dy}{dx} = \frac{dy}{dz} = Dy,$$

where $$D = \frac{d}{dz}.$$

Now, $$\frac{d^2y}{dx^2} = -\frac{1}{x^2}\frac{dy}{dz} + \frac{1}{x}\frac{d^2y}{dz^2}\frac{dz}{dx} = \frac{1}{x^2}\left(\frac{d^2y}{dz^2} - \frac{dy}{dz}\right)$$

$$\Rightarrow \qquad x^2\frac{d^2y}{dx^2} = \frac{d^2y}{dz^2} - \frac{dy}{dz} = D(D-1)y$$

Continuing in this way, we get

$$x^n \frac{d^n y}{dx^n} = D(D-1)\cdots[D-(n-1)]y$$

Substitute these values in Eq. (5.1), we get

$$D(D-1)\cdots[D-(n-1)]y + Q_1 D(D-1)\cdots[D-(n-2)]y + \cdots + Q_{n-2}D(D-1)y$$
$$+Q_{n-1}Dy + Q_n y = Z \qquad (5.2)$$

where Z is a function obtained from X after transformation.

Equation (5.2) is an equation with constant coefficients. Solve this equation by the methods discussed in Chapter 4.

EXAMPLE 5.1 Solve $x^2 \dfrac{d^2y}{dx^2} + 8x\dfrac{dy}{dx} + 8y = \log x$.

Solution Substitute $z = \log x$.

$$\Rightarrow \qquad x\frac{dy}{dx} = Dy$$

where
$$D = \frac{d}{dz}$$

and
$$x^2 \frac{d^2y}{dx^2} = D(D-1)y$$

in the given equation. Therefore,

$$[D(D-1) - 8D + 8]y = Z$$

$$\Rightarrow \qquad (D^2 - 9D + 8)y = Z \qquad \text{(i)}$$

Its auxiliary equation is

$$m^2 - 9m + 8 = 0 \Rightarrow m = 1, 8.$$

Therefore, CF of Eq. (i) is $c_1e^z + c_2e^{8z}$

$$\begin{aligned}
\text{PI} &= \frac{1}{D^2 - 9D + 8} Z \\
&= \frac{1}{8} \frac{1}{1 + \dfrac{D^2 - 9D}{8}} Z \\
&= \frac{1}{8}\left(1 + \frac{D^2 - 9D}{8}\right)^{-1} Z \\
&= \frac{1}{8}\left(1 - \frac{D^2 - 9D}{8}\right) Z \\
&= \frac{1}{8}\left(Z + \frac{9}{8}\right)
\end{aligned}$$

Therefore, solution of Eq. (i) is $y = c_1 e^z + c_2 e^{8z} + \frac{1}{8}\left(Z + \frac{9}{8}\right)$. Hence, the solution of the given equation is

$$y = c_1 x + c_2 x^8 + \frac{1}{8}\left(\log x + \frac{9}{8}\right).$$

EXAMPLE 5.2 Solve $x^2 \frac{d^2 y}{dx^2} - 3x\frac{dy}{dx} + y = \frac{\log x \sin(\log x) + 1}{x}$.

Solution Compare the given equation with Eq. (5.2), we have

$$Q_1 = -3, Q_2 = 1 \text{ and } z = \frac{\log x \sin(\log x) + 1}{x}$$

The corresponding linear differential equation with constant coefficients is

$$[D(D-1) - 3D + 1]y = e^{-z}(z \sin z + 1)$$

or

$$[D^2 - 4D + 1]y = e^{-z}(z\sin z + 1)$$

The AE is $m^2 - 4m + 1 = 0$ which gives $m = 2 \pm \sqrt{3}$.

Therefore,
$$\text{CF} = e^{2z}\{c_1 e^{\sqrt{3}z} + c_2 e^{-\sqrt{3}z}\}$$

$$\begin{aligned}
\text{PI} &= \frac{1}{D^2 - 4D + 1}(e^{-z} z \sin z) + \frac{1}{D^2 - 4D + 1} e^{-z} \\
&= e^{-z}\frac{1}{(D-1)^2 - 4(D-1) + 1}(z\sin z) + \frac{1}{6}e^{-z} \\
&= e^{-z}\frac{1}{D^2 - 6D + 6}(z\sin z) + \frac{1}{6}e^{-z}
\end{aligned}$$

Let us compute

$$\begin{aligned}
\frac{1}{D^2 - 6D + 6}(z\sin z) &= \left\{z - \frac{1}{D^2 - 6D + 6}(2D - 6)\right\}\frac{1}{D^2 - 6D + 6}(\sin z) \\
&= \left\{z - \frac{1}{D^2 - 6D + 6}(2D - 6)\right\}\frac{1}{5 - 6D}(\sin z) \\
&= \left\{z - \frac{1}{D^2 - 6D + 6}(2D - 6)\right\}(5 + 6D)\frac{1}{25 - 36D^2}(\sin z) \\
&= \frac{1}{61}\left\{z - \frac{1}{D^2 - 6D + 6}(2D - 6)\right\}(5\sin z + 6\cos z)
\end{aligned}$$

$$=\frac{1}{61}\left\{z(5\sin z+6\cos z)-\frac{2}{D^2-6D+6}(D-3)(5\sin z+6\cos z)\right\}$$

$$=\frac{1}{61}\left\{z(5\sin z+6\cos z)+\frac{2}{D^2-6D+6}(21\sin z+13\cos z)\right\}$$

$$=\frac{1}{61}\left\{z(5\sin z+6\cos z)+\frac{2}{5-6D}(21\sin z+13\cos z)\right\}$$

$$=\frac{1}{61}\left\{z(5\sin z+6\cos z)+2(5+6D)\frac{1}{25-36D^2}(21\sin z+13\cos z)\right\}$$

$$=\frac{1}{61}\left\{z(5\sin z+6\cos z)+\frac{2}{61}(5+6D)(21\sin z+13\cos z)\right\}$$

$$=\frac{1}{61}\left\{z(5\sin z+6\cos z)+\frac{2}{61}(27\sin z+191\cos z)\right\}$$

Therefore, the solution of the transformed equation is

$$y=e^{2z}\{c_1e^{\sqrt{3}z}+c_2e^{-\sqrt{3}z}\}+\left\{\frac{z}{61}(5\sin z+6\cos z)+\frac{2}{3721}(27\sin z+191\cos z)+\frac{1}{6}\right\}e^{-z}$$

So, the solution of given differential equation is

$$y=x^2\{c_1x^{\sqrt{3}}+c_2x^{-\sqrt{3}}\}$$
$$+\left[\frac{\log x}{61}\{5\sin(\log x)+6\cos(\log x)\}+\frac{2}{3721}\{27\sin(\log x)+191\cos(\log x)\}+\frac{1}{6}\right]\frac{1}{x}$$

5.3 METHOD OF FINDING CF

The CF is the solution of Eq. (5.1) obtained when $X = 0$.
Put $y = x^m$, m is a positive integer. Then

$$\frac{d^n y}{dx^n}=m(m-1)\cdots(m-n+1)x^{m-n}.$$

Therefore, Eq. (5.1) with $X = 0$ reduces to

$$[m(m-1)\cdots(m-n+1)+Q_1m(m-1)\cdots(m-n+2)+\cdots+Q_n]x^m=0.$$

$$\therefore \quad m(m-1)\cdots(m-n+1)+Q_1m(m-1)\cdots(m-n+2)+\cdots+Q_n=0 \tag{5.3}$$

Clearly, $y = x^m$ satisfies Eq. (5.1). Equation (5.3) is nth degree in m. Let its roots be m_1, m_2, ..., m_n. Three cases may arise:

Case 1: All roots m_1, m_2, ..., m_n are different.

Since $y=x^{m_1}, y=x^{m_2}, ..., y=x^{m_n}$ are n-independent solutions of Eq. (5.1) with $X = 0$, the CF is $y=c_1x^{m_1}+c_2x^{m_2}+\cdots+c_nx^{m_n}$, where c's are arbitrary constants.

EXAMPLE 5.3 Solve $x^3 \frac{d^3y}{dx^3} + 3x^2 \frac{d^2y}{dx^2} + x\frac{dy}{dx} + y = 0.$

Solution Substitute $y = x^m$ in the given equation.

$$\therefore \quad [m(m-1)(m-2) + 3m(m-1) + m + 1]\, x^m = 0$$

Therefore, $y = x^m$ satisfies given differential equation if

$$m(m-1)(m-2) + 3m(m-1) + m + 1 = 0$$

$$\Rightarrow \quad m^3 + 1 = 0$$

$$\Rightarrow \quad m = -1, \frac{1 \pm \sqrt{3}i}{2}$$

$$\therefore \quad y = c_1 e^{-x} + e^{x/2}\left(c_2 \cos\frac{\sqrt{3}}{2}x + c_3 \sin\frac{\sqrt{3}}{2}x\right)$$

is the required solution of given differential equation.

Case 2: All n-roots $m_1, m_2, \ldots, m_n$ are not different.

Arguing as in Chapter 4, if Eq. (5.3) has a root m_1 repeated k-times, the CF of equation is

$$\text{CF} = e^{m_1 z}(c_1 + c_2 z + \cdots + c_k z^{k-1})$$

$$= x^{m_1}[c_1 + c_2 \log x + \cdots + c_k (\log x)^{k-1}].$$

EXAMPLE 5.4 Solve $x^3 \frac{d^3y}{dx^3} - x^2 \frac{d^2y}{dx^2} - 6x\frac{dy}{dx} + 18y = 0.$

Solution Substitute $y = x^m$ in the given equation.

$$\therefore \quad [m(m-1)(m-2) - m(m-1) - 6m + 18]x^m = 0$$

Therefore, $y = x^m$ satisfies given differential equation if

$$m(m-1)(m-2) - m(m-1) - 6m + 18 = 0$$

$$\Rightarrow \quad m^3 - 4m^2 - 3m + 18 = 0$$

$$\Rightarrow \quad (m+2)(m-3)^2 = 0$$

$$\Rightarrow \quad m = -2, 3, 3$$

Therefore, CF is $$y = \frac{c_1}{x^2} + x^3(c_2 + c_3 \log x).$$

Case 3: If Eq. (5.3) has a pair of complex roots (say) $a \pm ib$ then corresponding CF is

$$\text{CF} = e^{az}(c_1 \cos bz + c_2 \sin bz)$$

$$= x^a [c_1 \cos (b \log x) + c_2 \sin (b \log x)].$$

EXAMPLE 5.5 Solve $x^2\dfrac{d^2y}{dx^2}+x\dfrac{dy}{dx}+9y=0.$

Solution Substitute $y = x^m$ in the given equation.

$$\therefore \qquad [m(m-1)+m+9]\,x^m = 0$$

Therefore, $y = x^m$ satisfies given differential equation if

$$m(m-1)+m+9=0$$

$$\Rightarrow \qquad m^2+9=0$$

$$\Rightarrow \qquad m=\pm 3i$$

Therefore, $\qquad \text{CF} = c_1 \cos(3\log x) + c_2 \sin(3\log x).$

5.4 OPERATOR θ

In Section 5.2, we have proved that

$$x^n\frac{d^ny}{dx^n}=D(D-1)\cdots[D-(n-1)]y,\ \text{where}\ D=x\frac{d}{dx}$$

$$x^n\frac{d^ny}{dx^n}=x\frac{d}{dx}\left(x\frac{d}{dx}-1\right)\cdots\left[x\frac{d}{dx}-(n-1)\right]y$$

$$=\theta(\theta-1)(\theta-2)\cdots[\theta-(n-1)]y,\ \text{where}\ D=x\frac{d}{dx}$$

Therefore, Eq. (5.1) can be written as

$$\{\theta(\theta-1)\cdots[\theta-(n-1)]+Q_1\theta(\theta-1)\cdots[\theta-(n-2)]+\cdots+Q_n\}\,y=X \tag{5.4}$$

Denote this equation as

$$f(\theta)y=X$$

Then Eq. (5.3) in m becomes $f(m) = 0$.
Hence, to find CF of Eq. (5.4), one can solve

$$f(m)=0 \text{ or } f(\theta)=0.$$

The function $\dfrac{1}{f(\theta)}X$ is defined to be that function which when operated by $f(\theta)$ gives X.

5.5 METHOD OF FINDING PI

The PI of Eq. (5.4) is the function $\dfrac{1}{f(\theta)}X$. It can be obtained in following two ways.

1. The operator $\dfrac{1}{f(\theta)}$ is expressible in factor form as $\dfrac{1}{\theta-\alpha_1}\dfrac{1}{\theta-\alpha_2}\cdots\dfrac{1}{\theta-\alpha_n}$.

Therefore, $$\frac{1}{f(\theta)}X = \frac{1}{\theta-\alpha_1}\frac{1}{\theta-\alpha_2}\cdots\frac{1}{\theta-\alpha_n}X.$$

Start operating from right to left to find PI.

2. The operator $\frac{1}{f(\theta)}$ can be resolved into partial fractions:

Let $$\frac{1}{f(\theta)} = \frac{A_1}{\theta-\alpha_1}+\frac{A_2}{\theta-\alpha_2}+\cdots+\frac{A_n}{\theta-\alpha_n}$$

$\therefore$ $$\frac{1}{f(\theta)}X = \left[\frac{A_1}{\theta-\alpha_1}+\frac{A_2}{\theta-\alpha_2}+\cdots+\frac{A_n}{\theta-\alpha_n}\right]X \text{ will give PI.}$$

To find the value of $\frac{1}{\theta-\alpha}X$.

Let $$\frac{1}{\theta-\alpha}X = V$$

$\Rightarrow$ $$x\frac{dV}{dx} - \alpha V = X$$

$\Rightarrow$ $$\frac{dV}{dx} - \frac{\alpha}{x}V = \frac{1}{x}X$$

This is linear equation whose solution is

$$\frac{V}{x^{\alpha}} = \int \frac{X}{x^{\alpha+1}}dx$$

$\Rightarrow$ $$V = x^{\alpha}\int \frac{X}{x^{\alpha+1}}dx$$

$\Rightarrow$ $$\frac{1}{\theta-\alpha}X = x^{\alpha}\int \frac{X}{x^{\alpha+1}}dx.$$

Therefore, PI in point (1) is

$$x^{\alpha_1}\int x^{\alpha_2-\alpha_1-1}\int x^{\alpha_3-\alpha_2-1}\cdots\int x^{-\alpha_n-1}\,(dx)^n$$

and in point (2) is

$\Rightarrow$ $$A_1x^{\alpha_1}\int\frac{X}{x^{\alpha_1+1}}dx + A_2x^{\alpha_2}\int\frac{X}{x^{\alpha_2+1}}dx+\cdots+A_nx^{\alpha_n}\int\frac{X}{x^{\alpha_n+1}}dx$$

EXAMPLE 5.6 Solve $x^2\frac{d^2y}{dx^2}+2x\frac{dy}{dx}-12y=(x+1)^2$.

Solution Here $$f(\theta) = \theta(\theta-1)+2\theta-12$$
$$= \theta^2+\theta-12$$
$$= (\theta+4)(\theta-3)$$

Therefore, the roots of the equation $f(\theta) = 0$ are -4, 3.

$$\text{PI} = \frac{1}{(\theta-3)(\theta+4)}(x+1)^2$$

$$= \frac{1}{(\theta-3)(\theta+4)}(x^2+1+2x)$$

$$= \frac{1}{7}\left[\frac{1}{(\theta-3)} - \frac{1}{(\theta+4)}\right](x^2+1+2x)$$

$$= \frac{1}{7}x^3\int x^{-4}(x^2+1+2x)\,dx - \frac{1}{7}x^{-4}\int x^3(x^2+1+2x)\,dx$$

$$= \frac{1}{7}x^3\int\left(\frac{1}{x^2}+\frac{1}{x^4}+\frac{2}{x^3}\right)dx - \frac{1}{7}x^{-4}\int(x^5+x^3+2x^4)\,dx$$

$$= \frac{1}{7}x^3\left(-\frac{1}{x}-\frac{1}{3x^3}-\frac{1}{2x^2}\right) - \frac{1}{7}x^{-4}\left(\frac{x^6}{6}+\frac{x^4}{4}+\frac{2x^5}{5}\right)$$

$$= \frac{1}{7}\left(-x^2-\frac{1}{3}-\frac{x}{2}-\frac{x^2}{6}-\frac{1}{4}-\frac{2x}{5}\right)$$

$$= \frac{1}{7}\left(-\frac{7}{12}-\frac{9x}{10}-\frac{7x^2}{6}\right)$$

$$= -\frac{1}{7}\left(\frac{7}{12}+\frac{9x}{10}+\frac{7x^2}{6}\right).$$

5.6 SOLUTION OF $\frac{1}{f(\theta)}x^m$, m IS POSITIVE INTEGER

We know that

$$\theta x^m = x\frac{d}{dx}x^m = mx^m$$

$$\theta^2 x^m = \theta[\theta(x^m)] = \theta(mx^m) = m^2x^m$$

In general, $\theta^n x^m = m^n x^m$

$\therefore$

$$f(\theta)x^m = f(m)x^m \tag{5.5}$$

Let $f(m) \neq 0$ and operate with $\frac{1}{f(\theta)}$ on both sides of Eq. (5.5).

$$\frac{1}{f(\theta)}f(\theta)x^m = \frac{1}{f(\theta)}[f(m)x^m]$$

$$\Rightarrow \qquad x^m = f(m)\frac{1}{f(\theta)}x^m$$

$$\Rightarrow \qquad \frac{1}{f(\theta)}x^m = \frac{1}{f(m)}x^m.$$

If $f(m) = 0$ then $(\theta - m)$ is at least one factor of $f(\theta)$.

Let $\qquad f(\theta) = (\theta - m)^k F(\theta)$ where $f(m) \neq 0$.

Then
$$\frac{1}{f(\theta)}x^m = \frac{1}{(\theta - m)^k}\frac{1}{f(\theta)}x^m$$

$$= \frac{1}{F(m)}\frac{1}{(\theta - m)^k}x^m$$

$$= \frac{1}{F(m)}\frac{1}{(\theta - m)^{k-1}}\left(\frac{1}{\theta - m}x^m\right)$$

$$= \frac{1}{F(m)}\frac{1}{(\theta - m)^{k-1}}x^m\int x^{-m-1}x^m dx$$

$$= \frac{1}{F(m)}\frac{1}{(\theta - m)^{k-1}}(x^m \log x)$$

$$= \frac{1}{F(m)}\frac{1}{(\theta - m)^{k-2}}x^m\int x^{-m-1}x^m \log x dx$$

$$= \frac{1}{F(m)}\frac{1}{(\theta - m)^{k-2}}x^m\int \frac{\log x}{x}dx$$

$$= \frac{1}{F(m)}\frac{1}{(\theta - m)^{k-2}}\frac{1}{2}x^m(\log x)^2$$

$$= \frac{1}{F(m)}\frac{1}{k!}x^m(\log x)^k$$

Now, $\qquad f(\theta) = (\theta - m)^k F(\theta).$

$\therefore \qquad f^k(\theta) = k!\ F(\theta)$ + terms containing $(\theta - m)$ and its higher powers.

$$\therefore \qquad f^k(m) = k!\ F(m)$$

$$\therefore \qquad \frac{1}{f(\theta)}x^m = \frac{x^m(\log x)^k}{f^k(m)}.$$

EXAMPLE 5.7 Find PI of $x^3\dfrac{d^3y}{dx^3} + 2x^2\dfrac{d^2y}{dx^2} + 2y = 8\left(x + \dfrac{1}{x}\right)$.

Solution We have

$$f(\theta) = \theta(\theta - 1)(\theta - 2) + 2\theta(\theta - 1) + 2$$
$$= \theta^3 - \theta^2 + 2$$

Therefore,
$$\text{PI} = \frac{1}{\theta^3 - \theta^2 + 2} 8\left(x + \frac{1}{x}\right)$$
$$= 8\left[\frac{1}{\theta^3 - \theta^2 + 2}\left(x + \frac{1}{x}\right)\right]$$
$$= 8\left(\frac{x}{2} + \frac{1}{5}\frac{\log x}{x}\right).$$

5.7 EQUATIONS REDUCIBLE TO HOMOGENEOUS LINEAR DIFFERENTIAL EQUATIONS

Consider the equation

$$(a+bx)^n \frac{d^n y}{dx^n} + Q_1(a+bx)^{n-1}\frac{d^{n-1}y}{dx^{n-1}} + \cdots + Q_n y = X \tag{5.6}$$

where a, b, Q_1, Q_2, ..., Q_n are all constants and X is a function of x. Eq. (5.6) can be reduced to homogeneous linear equation, by substituting

$$z = a + bx$$

With this substitution

$$\frac{dy}{dx} = b\frac{dy}{dz}$$
$$\frac{d^2y}{dx^2} = b^2\frac{d^2y}{dz^2}$$
$$\vdots$$
$$\frac{d^n y}{dx^n} = b^n\frac{d^n y}{dz^n}$$

Then Eq. (5.6) reduces to

$$b^n z^n \frac{d^n y}{dz^n} + Q_1 b^{n-1} z^{n-1}\frac{d^{n-1}y}{dz^{n-1}} + \cdots + Q_n y = Z \tag{5.7}$$

where Z is the transform of X with the substitution $z = a + bx$. Equation (5.7) is a linear homogeneous equation and can be solved by the methods discussed in Sections 5.3 and 5.5.

EXAMPLE 5.8 Find CF of $(2x+5)^2\frac{d^2y}{dx^2} - 6(2x+5)\frac{dy}{dx} + 8y = 0.$

Solution Let $2x + 5 = z$. Then given equation reduces to

$$4z^2\frac{d^2y}{dz^2} - 12z\frac{dy}{dz} + 8y = 0$$

$$\Rightarrow \qquad z^2\frac{d^2y}{dz^2} - 3z\frac{dy}{dz} + 2y = 0$$

Here $$f(\theta) = \theta(\theta-1) - 3\theta + 2 = \theta^2 - 4\theta + 2$$

The roots of $f(\theta) = 0$ are $\theta = 2 \pm \sqrt{2}$.

Therefore, CF $= c_1 z^{2+\sqrt{2}} + c_2 z^{2-\sqrt{2}}$, where $z = 2x + 5$.

EXAMPLE 5.9 Solve $(3x+2)^2 \dfrac{d^2y}{dx^2} + 3(3x+2)\dfrac{dy}{dx} - 36y = 3x^2 + 4x + 1$.

Solution Put $z = 3x + 2$

Then given equation reduces to

$$9z^2 \frac{d^2y}{dx^2} + 9z\frac{dy}{dx} - 36y = \frac{1}{3}(z^2 - 1) \quad \text{or} \quad z^2 \frac{d^2y}{dx^2} + z\frac{dy}{dx} - 4y = \frac{1}{27}(z^2 - 1)$$

Here, $$f(\theta) = \theta(\theta - 1 + \theta - 4) = \theta^2 - 4 = (\theta + 2)(\theta - 2)$$

Then roots of $f(\theta) = 0$ are $\theta = \pm 2$.

$\therefore$ CF of the transformed equation is

$$\text{CF} = c_1 z^2 + \frac{c_2}{z^2}$$

Now $$\text{PI} = \frac{1}{27}\frac{1}{(\theta^2 - 4)}(z^2 - 1) = \frac{1}{27}\left\{\frac{z^2 \log z}{4} + \frac{1}{4}\right\}$$

$\therefore$ Solution of the transformed equation is

$$y = C_1 z^2 + \frac{C_2}{z^2} + \frac{1}{27}\left\{\frac{z^2 \log z}{4} + \frac{1}{4}\right\}$$

$\therefore$ The solution of given differential equation is

$$y = c_1(3x+2)^2 + \frac{c_2}{(3x+2)^2} + \frac{1}{108}\{(3x+2)^2 \log(3x+2) + 1\}$$

EXAMPLE 5.10 Solve $x^2 \dfrac{d^2y}{dx^2} + 3x\dfrac{dy}{dx} + y = \dfrac{1}{(1-x)^2}$.

Solution Here, $f(\theta) = \theta(\theta - 1) + 3\theta + 1 = \theta^2 + 2\theta + 1 = (\theta + 1)^2$

$\therefore$ The roots of $f(\theta) = 0$ are $\theta = -1, -1$

$\therefore$ $$\text{CF} = (c_1 + c_2 \log x)\frac{1}{x}$$

Now $$\text{PI} = \frac{1}{(\theta+1)^2}\frac{1}{(1-x)^2}$$

$$= \frac{1}{\theta+1}\frac{1}{x}\int\frac{1}{(1-x)^2}dx$$

$$= \frac{1}{\theta+1}\frac{1}{x}\frac{1}{1-x}$$

$$= x^{-1}\int\frac{1}{x(1-x)}dx$$

$$= \frac{1}{x}\int\left\{\frac{1}{1-x}+\frac{1}{x}\right\}dx$$

$$= \frac{1}{x}\log\frac{x}{1-x}$$

$\therefore$ The solution of given differential equation is

$$y = (c_1 + c_2 \log x)\frac{1}{x} + \frac{1}{x}\log\frac{x}{1-x}$$

EXAMPLE 5.11 Solve $x^2\dfrac{d^2y}{dx^2} - (2m-1)x\dfrac{dy}{dx} + (m^2+n^2)y = n^2x^m \log x.$

Solution Put CF $= z = \log x$

Let $$\frac{d}{dz} = D' \text{ and } \frac{d^2}{dz^2} = D'^2$$

With these substitution and assumptions, the given equation can be rewritten as

$$\{D'(D'-1) - (2m-1)D' + m^2 + n^2\}y = n^2e^{mz}z$$

$$\Rightarrow \quad \{D'^2 - 2mD' + m^2 + n^2\}y = n^2e^{mz}z$$

Then AE is

$$M^2 - 2mM + m^2 + n^2 = 0 \;\Rightarrow\; M = m \pm in$$

$\therefore$ CF of the transformed equation is

$$\text{CF} = e^{mz}(c_1 \cos nz + c_2 \sin nz)$$

Now $$\text{PI} = \frac{1}{D'^2 - 2mD' + m^2 + n^2}n^2e^{mz}z$$

$$= n^2\frac{1}{(D'-m)^2 + n^2}ze^{mz}$$

$$= n^2 e^{mz} \frac{1}{D'^2 + n^2} z$$

$$= n^2 e^{mz} \frac{1}{n^2}\left(1 + \frac{D'^2}{n^2}\right)^{-1} z$$

$$= e^{mz}\left\{1 - \frac{D'^2}{n^2} + \cdots\right\} z$$

$$= ze^{mz}$$

∴ Solution of the transformed equation is

$$y = e^{mz}\{c_1 \cos nz + c_2 \sin nz\} + ze^{mz}$$

∴ Solution of the given differential equation is

$$y = x^m\{c_1 \cos(n \log x) + c_2 \sin(n \log x)\} + x^m \log x$$

EXERCISES

Solve the following differential equations:

1. $x^2 \frac{d^2y}{dx^2} - x\frac{dy}{dx} + y = \log x.$ [**Ans.** $y = (c_1 + c_2 \log x)x + 2 + \log x$]

2. $(x^2D^2 - 3xD + 4)y = 2x^2.$ [**Ans.** $y = (c_1 + c_2 \log x)x^2 + x^2(\log x)^2$]

3. $(x^3D^3 + 2xD - 2)y = x^2 \log x + 3x.$

[**Ans.** $y = c_1x + x[c_2 \cos(\log x) + c_3 \sin(\log x)] + \frac{1}{2}x^2(\log x - 2) + 3x \log x$]

4. $[(x+1)^2 D^2 + (x+1)D]y = (2x+3)(2x+4).$

[**Ans.** $y = c_1 + c_2 \log(x+1) + [\log(x+1)]^2 + x^2 + 8x$]

5. $(x^3D^3 - x^2D^2 + 2xD - 2)y = x^3 + 3x.$

[**Ans.** $y = (c_1 + c_2 \log x)x + c_3x^2 + \frac{1}{4}x^3 - \frac{3}{2}x(\log x)^2$]

6. $[(1+x)^2 D^2 + (1+x)D + 1]y = 4\cos[\log(x+1)].$

[**Ans.** $y = c_1 \cos[\log(x+1)] + c_2 \sin[\log(x+1)] + 2[\log(x+1)]\sin[\log(x+1)]$]

7. $(x^2D^2 + xD - 1)y = x^3.$ [**Ans.** $y = c_1x + \frac{c_2}{x} + \frac{1}{8}x^3$]

8. $(x^2D^2+7xD+5)y=x^5.$ $\left[\textbf{Ans. } \text{PI}=\frac{1}{60}x^5\right]$

9. $(D^2-\frac{6}{x^2})y=x\log x.$ $\left[\textbf{Ans. } \text{PI}=\frac{x^3}{5}\left(\frac{(\log x)^2}{2}-\frac{\log x}{5}\right)\right]$

10. $[(1+2x)^2D^2-6(1+2x)D+16]\,y=8(1+2x)^2.$ $[\textbf{Ans. } \text{PI}=(1+2x)^2\,[\log(1+2x)]^2\,]$

11. $[(x+1)^2D^2-3(x+1)D+4]\,y=x^2.$ $\left[\textbf{Ans. } \text{PI}=\frac{1}{2}(x+1)^2\,[\log(x+1)]^2-2(x+1)+\frac{1}{4}\right]$

12. $(x^2D^2+4xD+2)y=e^x.$ $\left[\textbf{Ans. } =\frac{e^x}{x^2}\right]$

13. $xD(xD+3)y=\frac{1}{x}.$ $\left[\textbf{Ans. } y=c_1+\frac{c_2}{x^2}-\frac{1}{x}\right]$

14. $(x^2D^2+8xD+13)y=\log x.$

$$\left[\textbf{Ans. } y=x^{-7/2}\left\{c_1\cos\left(\frac{\sqrt{3}}{2}\log x\right)+c_2\sin\left(\frac{\sqrt{3}}{2}\log x\right)\right\}+\frac{1}{13}\left(\log x-\frac{7}{13}\right)\right]$$

15. $(x^2D^2+xD-1)y=x^4.$ $\left[\textbf{Ans. } y=c_1x+\frac{c_2}{x}+\frac{1}{15}x^4\right]$

16. $(x^2D^2-xD+2)y=x\log x.$ $[\textbf{Ans. } y=x\{(c_1\cos(\log x)+c_2\sin(\log x))+x\log x\}]$

17. $(x^2D^2-2xD+2)y=x+x^2\log x+x^3.$

$$\left[\textbf{Ans. } y=c_1x^2+c_2x-x\log x+\frac{x^2}{6}\left(\log x-\frac{1}{6}\right)+\frac{x^3}{2}\right]$$

18. $(x^2D^2-2xD+2)y=(\log x)^2-\log x^2.$

$$\left[\textbf{Ans. } y=c_1x+c_2x^2+\frac{1}{2}((\log x)^2+\log x)+\frac{1}{4}\right]$$

19. $(x^2D^2-2)y=x^2+\frac{1}{x}.$ $\left[\textbf{Ans. } y=c_1x^2+\frac{c_2}{x}+\frac{1}{3}\log\left(x^2-\frac{1}{x}\right)\right]$

20. $(x^2D^2+4xD+2)y=e^x.$ $\left[\textbf{Ans. } y=\frac{c_1}{x}+\frac{c_2}{x^2}+\frac{e^x}{x^2}\right]$

21. $xD^2(xD+1)y=1.$ $\left[\textbf{Ans. } y=c_1+(c_2+c_3\log x)x+\frac{x(\log x)^2}{2}\right]$

22. $((x+4)^2D^2-4(x+4)D+6)y=x.$ $\left[\textbf{Ans. } y=c_1t^2+c_2t^3+\frac{1}{6}(3x+8);\ t=x+4\right]$

23. $((x+1)^2D^2-3(x+1)D+4)y=x^2.$

$$\left[\textbf{Ans. } y=(c_1+c_2\log t)t^2+\frac{t^2(\log t)^2}{2}-2t+\frac{1}{4};\ t=x+1\right]$$

24. $((x+3)^2D^2-4(x+3)D+6)y=x.$ $\left[\textbf{Ans. } y=c_1t^2+c_2t^3+\frac{t}{2}-\frac{1}{2};\ t=x+3\right]$

25. $((3x+2)^2D^2+5(3x+2)D-3)y=x^2+x+1.$

$$\left[\textbf{Ans. } y=\frac{c_1}{t}+c_2t^{1/3}+\frac{1}{27}\left\{\frac{t^2}{15}-\frac{t\log t}{8}-7\right\};\ t=3x+2\right]$$

26. $(x^2D^2-3xD+3)y=x^2(2x-5).$ $[\textbf{Ans. } y=c_1x+c_2x^3+x^3\log x+5x^2]$

27. $(x^2D^2-3xD+4)y=2x^2+\log x.$

$$\left[\textbf{Ans. } y=(c_1+c_2\log x)x^2+x^2(\log x)^2+\frac{1}{4}(1+\log x)\right]$$

28. $(x^3D^3-4x^2D^2+5xD-2)y=x^3.$

$$\left[\textbf{Ans. } y=c_1x^2+x^{5/2}(c_2x^a+c_3x^{-a})-\frac{x^3}{5};\ a=\frac{\sqrt{21}}{2}\right]$$

29. $(x^3D^3+2x^2D^2+3xD-3)y=x(x+1).$

$$\left[\textbf{Ans. } y=c_1x+c_2\cos(\sqrt{3}\log x)+c_3\sin(\sqrt{3}\log x)+\frac{x^2}{7}+\frac{x\log x}{4}\right]$$

30. $(x^3D^3+3x^2D^2+xD+1)y=x\log x.$

$$\left[\textbf{Ans. } y=\frac{c_1}{x}+\sqrt{x}\left(c_2\cos\left(\frac{\sqrt{3}}{2}\log x\right)+c_3\sin\left(\frac{\sqrt{3}}{2}\log x\right)\right)+\frac{x}{2}\left(\log x-\frac{3}{2}\right)\right]$$

31. $(x^3D^3-4x^2D^2+5xD-2)y=x^3.$

$$\left[\textbf{Ans. } y=c_1x^2+c_2x^a+c_3x^b-\frac{x^3}{5};\ a=\frac{5+\sqrt{21}}{2},\ b=\frac{5-\sqrt{21}}{2}\right]$$

32. $(x^3D^3+2x^2D^2+2)y=\frac{1}{x}.$ $\left[\textbf{Ans. } y=\frac{c_1}{x}+x(c_2\cos(\log x)+c_3\sin(\log x))+\frac{\log x}{5x}\right]$

33. $(x^4D^3+2x^3D^2-x^2D+x)y=1.$ $\left[\textbf{Ans. } y=(c_1+c_2\log x)x+\frac{c_3}{x}+\frac{\log x}{4x}\right]$

34. $(16(x+1)^4D^4+96(x+1)^3D^3+104(x+1)^2D^2+8(x+1)D+1)y=(x+1)(x+3).$

$$\left[\textbf{Ans. } y=(c_1+c_2\log t)t^{1/2}+(c_3+c_4\log t)t^{-1/2}+\frac{(t^3+50t)}{225};\ t=x+1\right]$$

CHAPTER 6 Exact Differential Equations and Equations of Higher Order

6.1 INTRODUCTION

A differential equation

$$f\left(\frac{d^n y}{dx^n}, \frac{d^{n-1} y}{dx^{n-1}}, \ldots, \frac{dy}{dx}, y\right) = X \tag{6.1}$$

is said to be exact when it can be derived by simple differentiation from an equation of the next lower order

$$f\left(\frac{d^{n-1} y}{dx^{n-1}}, \ldots, \frac{dy}{dx}, y\right) = \int X dx + c$$

We have studied exact differential equations of first order in Chapter 2.
The equation

$$xy\frac{d^2 y}{dx^2} + x\left(\frac{dy}{dx}\right)^2 + y\left(\frac{dy}{dx}\right) = 0 \tag{6.2}$$

can be obtained from

$$xy\frac{dy}{dx} = \text{Constant}$$

by one differentiation. Therefore, Eq. (6.2) is exact differential equation.

6.2 CONDITION FOR THE EXACTNESS OF THE LINEAR DIFFERENTIAL EQUATION

Consider the differential equation

$$Q_0 \frac{d^n y}{dx^n} + Q_1 \frac{d^{n-1} y}{dx^{n-1}} + \cdots + Q_n y = g(x) \tag{6.3}$$

where Q_0, Q_1, …, Q_n are functions of x only.

Let the successive derivatives of Q's be denoted by $Q^{(1)}, Q^{(2)}, \ldots, Q^{(n)}$. Clearly, the first term of Eq. (6.1) can be obtained from $Q_0 \dfrac{d^{n-1}y}{dx^{n-1}}$ by direct differentiation.

Let the first integral of Eq. (6.3) be

$$Q_0 \frac{d^{n-1}y}{dx^{n-1}} + R_1 \frac{d^{n-2}y}{dx^{n-2}} + \cdots + R_{n-1}y = \int g(x)\,dx \tag{6.4}$$

So that L.H.S. of Eq. (6.3)

$$= \frac{d}{dx}\left[Q_0 \frac{d^{n-1}y}{dx^{n-1}} + R_1 \frac{d^{n-2}y}{dx^{n-2}} + \cdots + R_{n-1}y \right]$$

$$= Q_0 \frac{d^n y}{dx^n} + (Q_0^{(1)} + R_1)\frac{d^{n-1}y}{dx^{n-1}} + (R_1^{(1)} + R_2)\frac{d^{n-2}y}{dx^{n-2}} + \cdots + (R_{n-2}^{(1)} + R_{n-1})\frac{dy}{dx} + R^{(1)}{}_{n-1}y$$

Equating the coefficients

$$Q_1 = Q_0^{(1)} + R_1$$

$$Q_2 = R_1^{(1)} + R_2$$

$$\vdots$$

$$Q_{n-1} = R_{n-2}^{(1)} + R_{n-1}$$

$$Q_n = R_{n-1}^{(1)}$$

$$\therefore \quad Q_n = R_{n-1}^{(1)}$$

$$= Q_{n-1}^{(1)} - R_{n-2}^{(2)}$$

$$= Q_{n-1}^{(1)} - (Q_{n-2}^{(2)} - R_{n-3}^{(3)})$$

$$= Q_{n-1}^{(1)} + (-1)Q_{n-2}^{(2)} + (-1)^2 R_{n-3}^{(3)}$$

$$\vdots$$

$$= Q_{n-1}^{(1)} + (-1)Q_{n-2}^{(2)} + (-1)^2 Q_{n-3}^{(3)} + \cdots + (-1)^{n-1} Q_0^{(n)}$$

$$\Rightarrow \quad Q_n + (-1)Q_{n-1}^{(1)} + (-1)^2 Q_{n-2}^{(2)} + (-1)^3 Q_{n-3}^{(3)} + \cdots + (-1)^n Q_0^{(n)} = 0 \tag{6.5}$$

Equation (6.5) gives the condition for the Eq. (6.3) to be exact. When this condition is satisfied, Eq. (6.4) is the first integral of Eq. (6.3).

EXAMPLE 6.1 Solve $x\dfrac{d^3y}{dx^3} + (x^2 - 3)\dfrac{d^2y}{dx^2} + 4x\dfrac{dy}{dx} + 2y = 0.$

Solution We have $Q_0 = x,\; Q_1 = (x^2 - 3),\; Q_2 = 4x,\; Q_3 = 2.$

Then $Q_3 - Q_2{}^{(1)} + Q_1{}^{(2)} - Q_0{}^{(3)} = 2 - 4 + 2 = 0.$

Therefore, given equation is exact.
Integration gives

$$x\frac{d^2y}{dx^2}+(x^2-4)\frac{dy}{dx}+2xy=c_1$$

Verify that this is also exact.
Again integrating,

$$x\frac{dy}{dx}+(x^2-5)y=c_1x+c_2$$

which is linear equation of the first order and can be solved by the method discussed in Chapter 2.

EXAMPLE 6.2 Solve $(x^3-4x)\dfrac{d^3y}{dx^3}+(9x^2-12)\dfrac{d^2y}{dx^2}+18x\dfrac{dy}{dx}+6y=6.$

Solution Given

$$(x^3-4x)\frac{d^3y}{dx^3}+(9x^2-12)\frac{d^2y}{dx^2}+18x\frac{dy}{dx}+6y=6 \tag{i}$$

Here $Q_0=x^3-4x,\ Q_1=9x^2-12,\ Q_2=18x,\ Q_3=6.$

Then $Q_3-Q_2^{(1)}+Q_1^{(2)}-Q_0^{(3)}=6-18+18-6=0.$

Therefore, given equation is exact.

Let

$$(x^3-4x)\frac{d^2y}{dx^2}+R_1\frac{dy}{dx}+R_2y=\int 6dx+c_1 \tag{ii}$$

be the first integral of Eq. (i). Then Eq. (ii) is identical with the equation

$$\frac{d}{dx}\left[(x^3-4x)\frac{d^2y}{dx^2}+R_1\frac{dy}{dx}+R_2y\right]=6$$

$$\Rightarrow \quad (x^3-4x)\frac{d^3y}{dx^3}+(3x^2-4+R_1)\frac{d^2y}{dx^2}+(R_1^{(1)}+R_2)\frac{dy}{dx}+R_2^{(1)}y=6$$

Comparing the coefficients,

$$3x^2-4+R_1=9x^2-12 \quad\Rightarrow\quad R_1=6x^2-8$$

and $R_1^{(1)}+R_2=18x \quad\Rightarrow\quad R_2=18x-12x=6x$

Therefore, the first integral is

$$(x^3-4x)\frac{d^2y}{dx^2}+(6x^2-8)\frac{dy}{dx}+6xy=6x+c_1 \tag{iii}$$

For Eq. (iii)

$$Q_0=x^3-4x,\ Q_1=6x^2-8,\ Q_2=6x.$$

$$\therefore \quad Q_2-Q_1^{(1)}+Q_0^{(2)}=6x-12x+6x=0.$$

Therefore, Eq. (iii) is exact.

Let its first integral be

$$(x^3-4x)\frac{dy}{dx}+Ry=3x^2+c_1x+c_2$$

Then Eq. (iii) is identical with the equation

$$\frac{d}{dx}\left[(x^3-4x)\frac{dy}{dx}+Ry\right]=6x+c_1$$

$$\Rightarrow \qquad (x^3-4x)\frac{d^2y}{dx^2}+(3x^2-4+R)\frac{dy}{dx}+R^{(1)}y=6x+c_1$$

Comparing the coefficients,

$$3x^2 - 4 + R = 6x^2 - 8 \;\Rightarrow\; R = 3x^2 - 4.$$

Therefore, the integral of Eq. (iii) is

$$(x^3-4x)\frac{dy}{dx}+(3x^2-4)y=3x^2+c_1x+c_2 \tag{iv}$$

For Eq. (iv), $\quad Q_0 = x^3 - 4x,\ Q_1 = 3x^2 - 4.$

$\therefore \quad Q_1 - Q_0^{(1)} = 3x^2 - 4 - (3x^2 - 4) = 0.$

Therefore, Eq. (iv) is exact and

$$(x^3-4x)y=x^3+c_1\frac{x^2}{2}+c_2x+c_3.$$

is its integral.

Therefore, solution of given equation is

$$(x^3-4x)y=x^3+c_1\frac{x^2}{2}+c_2x+c_3$$

EXAMPLE 6.3 Solve $x^5\frac{d^2y}{dx^2}+3x^3\frac{dy}{dx}+(3-6x)x^2y=x^4+2x-5.$

Solution Given

$$x^5\frac{d^2y}{dx^2}+3x^3\frac{dy}{dx}+(3-6x)x^2y=x^4+2x-5 \tag{i}$$

Here $\quad Q_0 = x^5,\ Q_1 = 3x^3,\ Q_2 = (3 - 6x)\,x^2.$

Then $\quad Q_2 - Q_1^{(1)} + Q_0^{(2)} = x^2(3 - 6x) - 9x^2 + 20x^3 \neq 0.$

Therefore, given equation is not exact. Multiply Eq. (i) by x^m.

$$x^{m+5}\frac{d^2y}{dx^2}+3x^{m+3}\frac{dy}{dx}+(3x^{m+2}-6x^{m+3})y=x^{m+4}+2x^{m+1}-5x^m \tag{ii}$$

If Eq. (ii) is exact then

$$3x^{m+2} - 6x^{m+3} - 3(m+3)x^{m+2} + (m+5)(m+4)x^{m+3} = 0$$

$$\Rightarrow \quad (m+7)(m+2)x^{m+3} - 3(m+2)x^{m+2} = 0$$

$$\Rightarrow \quad (m+2)[(m+7)x^{m+3} - 3x^{m+2}] = 0$$

$$\Rightarrow \quad m = -2$$

Substituting in Eq. (ii) gives

$$x^3\frac{d^2y}{dx^2} + 3x\frac{dy}{dx} + (3-6x)y = x^2 + \frac{2}{x} - \frac{5}{x^2} \tag{iii}$$

Let its first integral be

$$x^3\frac{dy}{dx} + Ry = \frac{x^3}{3} + 2\log x + \frac{5}{x} + c_1$$

Then Eq. (iii) is identical with

$$\frac{d}{dx}\left[x^3\frac{dy}{dx} + Ry\right] = x^2 + \frac{2}{x} - \frac{5}{x^2}$$

$$\Rightarrow \quad x^3\frac{d^2y}{dx^2} + (3x^2 + R)\frac{dy}{dx} + Q^{(1)}y = x^2 + \frac{2}{x} - \frac{5}{x^2}$$

Comparing the two equations, we get

$$3x^2 + R = 3x \quad \Rightarrow \quad R = 3x - 3x^2$$

Therefore, the first integral of Eq. (iii) is

$$x^3\frac{dy}{dx} + (3x - 3x^2)y = \frac{x^3}{3} + 2\log x + \frac{5}{x} + c_1$$

$$\Rightarrow \quad \frac{dy}{dx} + \frac{3}{x^2}(1-x)y = \frac{1}{3} + 2\frac{\log x}{x^3} + \frac{5}{x^4} + \frac{c_1}{x^3}$$

which is linear equation with IF $= \dfrac{1}{x^3}e^{-3/x}$. Therefore, solution of given equation is

$$\frac{y}{x^3}\,\text{IF} = \frac{1}{3}\int\frac{1}{x^3}e^{-3/x}dx + 2\int\frac{1}{x^6}e^{-3/x}dx + 5\int\frac{1}{x^7}e^{-3/x}dx + c_1\int\frac{1}{x^6}e^{-3/x}dx.$$

EXAMPLE 6.4 Solve $x\dfrac{d^3y}{dx^3} + (x^2 + x + 3)\dfrac{d^2y}{dx^2} + (4x+2)\dfrac{dy}{dx} + 2y = 0$.

Solution Given

$$x\frac{d^3y}{dx^3} + (x^2 + x + 3)\frac{d^2y}{dx^2} + (4x+2)\frac{dy}{dx} + 2y = 0 \tag{i}$$

Here $Q_0 = x, Q_1 = x^2 + x + 3, Q_2 = 4x + 2, Q_3 = 2.$

$$\therefore \qquad Q_3 - Q_2^{(1)} + Q_1^{(2)} - Q_0^{(3)} = 2 - 4 + 2 - 0 = 0$$

$\therefore$ Given equation is exact.

Let
$$x\frac{d^2y}{dx^2} + P_1\frac{dy}{dx} + P_2 y = \text{Constant (say) } c_1$$

be the first integral of the given equation. Then given equation is identical with

$$\frac{d}{dx}\left\{x\frac{d^2y}{dx^2} + P_1\frac{dy}{dx} + P_2 y\right\} = 0$$

$$\Rightarrow \qquad x\frac{d^3y}{dx^3} + (1+P_1)\frac{d^2y}{dx^2} + \left(\frac{dP_1}{dx} + P_2\right)\frac{dy}{dx} + \frac{dP_2}{dx}y = 0$$

Comparing the two equations, we get

$$1 + P_1 = x^2 + x + 3 \;\Rightarrow\; P_1 = x^2 + x + 2$$

and
$$\frac{dP_1}{dx} + P_2 = 4x + 2$$

$$\Rightarrow \qquad 2x + 1 + P_2 = 4x + 2 \;\Rightarrow\; P_2 = 2x + 1$$

Therefore, the first integral is

$$x\frac{d^2y}{dx^2} + (x^2 + x + 2)\frac{dy}{dx} + (2x+1)y = c_2 \qquad \text{(ii)}$$

Then
$$Q_0 = x, Q_1 = x^2 + x + 2, Q_2 = 2x + 1$$

$$\therefore \qquad Q_2 + Q_1^{(1)} + Q_0^{(2)} = 2x + 1 - (2x+1) + 0 = 0$$

Let the first integral of Eq. (ii) be

$$x\frac{dy}{dx} + Py = c_1 x + c_2$$

where c_2 is constant.

Then Eq. (ii) is identical with the equation $\dfrac{d}{dx}\left(x\dfrac{dy}{dx} + Py\right) = c_1.$

$$\Rightarrow \qquad x\frac{d^2y}{dx^2} + (1+P)\frac{dy}{dx} + \frac{dP}{dx}y = c_1$$

Comparing the two equations, we obtain

$$1 + P = x^2 + x + 2 \;\Rightarrow\; P = x^2 + x + 1$$

Therefore, the first integral of Eq. (ii) is

$$x\frac{dy}{dx}+(x^2+x+1)y=c_1x+c_2 \tag{iii}$$

Equation (iii) is second integral of given Eq. (i). Equation (iii) is not exact but it is linear. Hence,

$$\text{IF}=e^{\int\frac{x^2+x+1}{x}dx}=e^{\frac{x^2}{2}+x+\log x}=xe^{x+\frac{x^2}{2}}$$

Therefore, the general solution of given equation is

$$y=\frac{1}{x}e^{-\left(x+\frac{x^2}{2}\right)}\int(c_1x+c_2)e^{x+\frac{x^2}{2}}dx+\frac{c_3}{x}e^{-\left(x+\frac{x^2}{2}\right)}$$

EXAMPLE 6.5 Solve $\sqrt{x}\frac{d^2y}{dx^2}+2x\frac{dy}{dx}+3y=x$.

Solution Given

$$\sqrt{x}\frac{d^2y}{dx^2}+2x\frac{dy}{dx}+3y=x \tag{i}$$

Here $Q_0=\sqrt{x}, Q_1=2x, Q_2=3$

$$\therefore \qquad Q_2-\frac{dQ_1}{dx}+\frac{d^2Q_0}{dx^2}=3-2-\frac{1}{4}x^{-\frac{3}{2}}\neq 0$$

Hence, given equation is not exact. We proceed as follows to compute IF.

Multiply Eq. (i) by x^m

$$x^{m+\frac{1}{2}}\frac{d^2y}{dx^2}+2x^{m+1}\frac{dy}{dx}+3x^my=x^{m+1} \tag{ii}$$

For Eq. (ii) to be exact, we have

$$3x^m-2(m+1)x^m+\left(m+\frac{1}{2}\right)\left(m-\frac{1}{2}\right)x^{m-\frac{3}{2}}=0$$

$$\Rightarrow \qquad x^m(1-2m)\left[1-\frac{(2m+1)}{4}x^{-\frac{3}{2}}\right]=0$$

$$\Rightarrow \qquad m=\frac{1}{2}$$

$\therefore$ Equation (ii) is exact for $m = 2$ and $\sqrt{x}$ is an I.F. of Eq. (i).

For $m=\frac{1}{2}$, Eq. (ii) is

$$x\frac{d^2y}{dx^2}+2x^{3/2}\frac{dy}{dx}+3\sqrt{x}y=x^{3/2} \tag{iii}$$

Let the first integral of Eq. (iii) be

$$x\frac{dy}{dx}+Py=\int x^{3/2}dx+c_1=\frac{2}{5}x^{5/2}+c_1$$

Then Eq. (iii) is identical with

$$\frac{d}{dx}\left(x\frac{dy}{dx}+Py\right)=x^{3/2} \quad \text{or} \quad x\frac{d^2y}{dx^2}+(1+P)\frac{dy}{dx}+\frac{dP}{dx}y=x^{3/2}$$

Comparing the two equations

$$1+P=2x^{3/2} \quad \Rightarrow \quad P=2x^{3/2}-1$$

Therefore, the first integral of Eq. (iii) is

$$\Rightarrow \qquad \frac{dy}{dx}+\left(2x^{1/2}-\frac{1}{x}\right)y=\frac{2}{5}x^{3/2}+\frac{c_1}{x}$$

which is linear.

$$\therefore \qquad \text{I.F.}=e^{\int\left(2x^{1/2}-\frac{1}{x}\right)dx}=e^{\frac{4}{3}x^{3/2}-\log x}=\frac{1}{x}e^{\frac{4}{3}x^{3/2}}$$

Reader can check that

$$y=\frac{1}{5}x+c_1xe^A\int x^{-2}e^{1/A}dx+c_2e^A; \quad \text{where } A=e^{-\frac{4}{3}x^{3/2}}$$

6.3 SOLUTION OF NON-LINEAR EQUATION WHICH ARE EXACT

We illustrate the procedure to solve non-linear equation by an example.

EXAMPLE 6.6 Solve the equation $\left(x^2+2y^2\frac{dy}{dx}\right)\frac{d^2y}{dx^2}+2y\left(\frac{dy}{dx}\right)^3+3x\frac{dy}{dx}+y=0$.

Solution First we need to show that given equation is exact.

Under the assumption that this equation is exact, rewrite equation as

$$dU=(y+3xp+2yp^3)dx+(x^2+2y^2p)dp; \quad \text{where } p=\frac{dy}{dx}$$

Let U_1 be U if p is the only variable, i.e., $U_1=U(p)$

$$\Rightarrow \qquad U_1=x^2p+y^2p^2$$

Let the restriction be neglected, so that

$$dU_1=(2xp+2yp^3)dx+(x^2+2y^2p)dp$$

$$\therefore \qquad dU-dU_1=(y+xp)dx=d(xy)$$

Integration gives

$$U - U_1 = xy + c_1 \quad \text{or} \quad U = U_1 + xy + c_1 = px^2 + y^2p^2 + xy + c_1$$

Therefore, the first integral is

$$x^2\frac{dy}{dx} + y^2\left(\frac{dy}{dx}\right)^2 + xy = c.$$

6.3.1 Computational Steps

Consider the exact differential equation of nth order. Since, it can be derived from one of order $(n-1)$ by direct differentiation, it will contain $\dfrac{d^ny}{dx^n}$ only in its first degree. If it is not so, the equation is not exact. Then

1. Let the equation be written in the form $U = 0$.
2. Integrate Udx assuming that $\dfrac{d^{n-1}y}{dx^{n-1}}$ is the only variable and $\dfrac{d^ny}{dx^n}$ is its derivative. Call this expression to be U_1.
3. Obtain dU_1 relaxing the restriction that $\dfrac{d^{n-1}y}{dx^{n-1}}$ is the only variable.
4. Find $Udx - dU_1$.
5. Repeat steps (1) to (4) till $Udx - dU_1 - dU_2 - \cdots = 0$.
6. Then the first integral of the given equation is $U_1 + U_2 + \cdots =$ constant

EXAMPLE 6.7 Compute the first integral of the equation

$$x^2\frac{d^3y}{dx^3} + x\frac{d^2y}{dx^2} + (2xy-1)\frac{dy}{dx} + y^2 + 2x = 0.$$

Solution Here $U = x^2\dfrac{d^3y}{dx^3} + x\dfrac{d^2y}{dx^2} + (2xy-1)\dfrac{dy}{dx} + y^2 + 2x$

Then $\quad U_1 = x^2y_2 \Rightarrow dU_1 = x^2y_3 + 2xy_2dx$

$\therefore \quad Udx - dU_1 = \{-xy_2 + (2xy-1)y_1 + y^2 + 2x\}dx$

Consider, $v_1 = -xy_2 + (2xy-1)y_1 + y^2 + 2x$

$\therefore \quad U_2 = -xy_1 \Rightarrow dU_2 = -xdy_1 - y_1dx$

$\therefore \quad V_1dx - dU_2 = (2xyy_1 + y^2 + 2x)dx = d(xy^2 + x^2)$

$\therefore \quad V_1dx = dU_2 + d(xy^2 + x^2)$

$\therefore \quad Udx = dU_1 + V_1dx = dU_1 + dU_2 + d(xy^2 + x^2) = d(U_1 + U_2 + xy^2 + x^2)$

Therefore, the first integral is

$$U_1 + U_2 + xy^2 + x^2 = c$$

$$\Rightarrow \qquad x^2 \frac{d^2 y}{dx^2} - x\frac{dy}{dx} + xy^2 + x^2 = c$$

6.4 DIFFERENTIAL EQUATIONS OF THE FORM $\frac{d^n y}{dx^n} = f(x)$

Given
$$\frac{d^n y}{dx^n} = f(x) \tag{6.6}$$

Equation (6.6) can be solved by repeated integration.

EXAMPLE 6.8 Solve $\frac{d^3 y}{dx^3} = x - \sin x$.

Solution On integration, we get

$$\frac{d^2 y}{dx^2} = \frac{x^2}{2} + \cos x + c_1$$

Again integrating,

$$\frac{dy}{dx} = \frac{x^3}{6} + \sin x + c_1 x + c_2$$

which on integration gives

$$y = \frac{x^4}{24} - \cos x + c_1 \frac{x^2}{2} + c_2 x + c_3$$

as the solution of the given equation.

6.5 DIFFERENTIAL EQUATIONS OF THE FORM $\frac{d^2 y}{dx^2} = f(y)$

We want to solve
$$\frac{d^2 y}{dx^2} = f(y) \tag{6.7}$$

Multiply both sides by $2\frac{dy}{dx}$.

$$\therefore \qquad 2\frac{dy}{dx}\frac{d^2 y}{dx^2} = f(y)\, 2\frac{dy}{dx}.$$

On integrating,

$$\left(\frac{dy}{dx}\right)^2 = 2\int f(y)\frac{dy}{dx}\, dx + c$$

$$\Rightarrow \qquad \left(\frac{dy}{dx}\right)^2 = 2\int f(y)dy + c$$

$$\Rightarrow \qquad \frac{dy}{dx} = \sqrt{2\int f(y)dy + c}$$

$$\Rightarrow \qquad dx = \frac{dy}{\sqrt{2\int f(y)dy + c}}$$

Integrating again,

$$x + c_1 = \int \frac{dy}{\sqrt{2\int f(y)dy + c}}$$

is the required solution of Eq. (6.7).

EXAMPLE 6.9 Solve $\frac{d^2y}{dx^2} + \text{cosec}^2 y \cot y = 0$ given that $y = \frac{\pi}{2}$, $\frac{dy}{dx} = 1$ when $x = 0$.

Solution Multiply given equation by $2\frac{dy}{dx}$.

$$2\frac{dy}{dx}\frac{d^2y}{dx^2} + 2\,\text{cosec}^2\, y \cot y \frac{dy}{dx} = 0$$

On integrating,

$$\left(\frac{dy}{dx}\right)^2 - \text{cosec}^2 y = c.$$

When $y = \frac{\pi}{2}, \frac{dy}{dx} = 1$. Then $c = 0$.

$$\therefore \qquad \left(\frac{dy}{dx}\right)^2 = \text{cosec}^2 y$$

$$\Rightarrow \qquad \frac{dy}{dx} = \text{cosec}\, y$$

$$\Rightarrow \qquad dx = \sin y dy$$

Integrating,

$$x + c = -\cos y$$

When $x = 0$, $y = \frac{\pi}{2}$. Then $c = 0$.

Therefore, $x + \cos y = 0$ is the required solution.

EXAMPLE 6.10 Find the equation of the family of curves which touches the line $y = 1$ and are the integral curves of the differential equation $\frac{d^2y}{dx^2} - y = \frac{1}{y^3}$.

Solution Given

$$\frac{d^2y}{dx^2} = y + \frac{1}{y^3}.$$

Multiply given equation by $2\frac{dy}{dx}$ and integrate.

$$\therefore \qquad \left(\frac{dy}{dx}\right)^2 = y^2 - \frac{1}{y^2} + c.$$

Since the line $y = 1$ touch the curves, $\frac{dy}{dx} = 0$. Then $c = 0$.

$$\therefore \qquad \left(\frac{dy}{dx}\right)^2 = \frac{y^4 - 1}{y^2}$$

$$\therefore \qquad dx = \frac{y}{\sqrt{y^4 - 1}}dy$$

Integration gives

$$x + c = \frac{1}{2}\log\left(y^2 + \sqrt{y^4 - 1}\right)$$

where c is constant.

$$\Rightarrow \qquad ae^{2x} = y^2 + \sqrt{y^4 - 1}$$

where a is constant.

$$\Rightarrow \qquad y^4 - 1 = (ae^{2x} - y^2)^2$$

$$\Rightarrow \qquad 2y^2 = ae^{2x} + \frac{1}{ae^{2x}}$$

is the required equation of the family of the curves.

EXAMPLE 6.11 Solve $\frac{d^2y}{dx^2} + n^2y = 0$.

Solution The equation can be written as $\frac{d^2y}{dx^2} = -n^2y$.

Multiplying by $2\frac{dy}{dx}$

$$2\frac{dy}{dx}\frac{d^2y}{dx^2} = -2n^2y\frac{dy}{dx}$$

By integrating, we obtain

$$\left\{\frac{dy}{dx}\right\}^2 = -n^2y^2 + c^2n^2 \text{ (say)}$$

$$\Rightarrow \qquad \frac{dy}{dx} = n\sqrt{(c^2 - y^2)}$$

$$\Rightarrow \qquad \frac{dy}{\sqrt{(c^2 - y^2)}} = ndx$$

By integrating, we obtain

$$\sin^{-1}\frac{y}{c} = nx + m \quad \Rightarrow \quad y = c\sin(nx + m)$$

is the required solution.

6.6 DIFFERENTIAL EQUATIONS THAT DO NOT CONTAIN *y* EXPLICITLY

The general form of the equation is

$$f\left(\frac{d^n y}{dx^n}, \frac{d^{n-1} y}{dx^{n-1}}, \ldots, \frac{dy}{dx}, x\right) = 0 \tag{6.8}$$

In order to solve Eq. (6.8), take the derivative of the lowest order appearing in the differential equation as dependent variable and thereby reducing the order. Writing $p = \frac{dy}{dx}$, $p^{(1)} = \frac{d^2 y}{dx^2}$, ..., $p^{(n)} = \frac{d^n y}{dx^n}$, the Eq. (6.8) reduces to

$$f(p^{(n-1)}, p^{(n-2)}, \ldots, p^{(1)}, p, x) = 0 \tag{6.9}$$

which is an equation of order one less than that of the given Eq. (6.8). Solve Eq. (6.9) for p. Let the solution be

$$p = F(x).$$

Then
$$\frac{dy}{dx} = F(x).$$

$\therefore \; y = \int F(x)dx + c$ is the solution.

EXAMPLE 6.12 Solve $(1 - x^2)\frac{d^2 y}{dx^2} - x\frac{dy}{dx} = 2.$

Solution Substitute $\frac{dy}{dx} = p.$

Then given equation reduces to

$$(1 - x^2)\frac{dp}{dx} - xp = 2$$

$$\Rightarrow \qquad \frac{dp}{dx} - \frac{x}{1-x^2}p = \frac{2}{1-x^2}.$$

This is linear equation. Its IF is

$$e^{-\int \frac{x}{1-x^2}dx} = e^{\frac{1}{2}\log(1-x^2)} = \sqrt{1-x^2}$$

$$\therefore \qquad p\sqrt{1-x^2} = 2\int \frac{1}{\sqrt{1-x^2}}dx + c = 2\sin^{-1}x + c$$

where c is constant.

$$\therefore \qquad p = \frac{2\sin^{-1}x}{\sqrt{1-x^2}} + \frac{c}{\sqrt{1-x^2}}$$

Integrating,

$$y = \int \frac{2\sin^{-1}x}{\sqrt{1-x^2}}dx + \int \frac{c}{\sqrt{1-x^2}}dx + c_1$$

$$= (\sin^{-1}x)^2 + c\sin^{-1}x + c_1$$

where c_1 is constant.

EXAMPLE 6.13 Solve $2x\dfrac{d^3y}{dx^3}\dfrac{d^2y}{dx^2} = \left(\dfrac{d^2y}{dx^2}\right)^2 - 4.$

Solution Substitute $\dfrac{d^2y}{dx^2} = p.$

Then given equation reduces to

$$2x\frac{dp}{dx}p = p^2 - 4$$

$$\Rightarrow \qquad \frac{2pdp}{p^2-4} = \frac{dx}{x}$$

Integrating,

$$\log(p^2 - 4) = \log x + \log c,$$

where c is a constant.

$$\Rightarrow \qquad p^2 = cx + 4$$

$$\Rightarrow \qquad \frac{d^2y}{dx^2} = \sqrt{cx+4}$$

Integrating

$$\frac{dy}{dx} = \frac{2}{3c}(cx+4)^{3/2} + c_1$$

Again integrating,

$$y = \frac{4}{15c^2}(cx+4)^{5/2} + c_1 x + c_2$$

where c_1 and c_2 are arbitrary constants.

EXAMPLE 6.14 Solve $(1-x^2)\dfrac{d^2y}{dx^2} - x\dfrac{dy}{dx} = a$; where a is constant.

Solution Put $\dfrac{dy}{dx} = p$. Equation reduces to

$$(1-x^2)\frac{dp}{dx} - xp = a$$

$$\Rightarrow \qquad \frac{dp}{dx} - \frac{x}{1-x^2}p = \frac{a}{1-x^2}$$

The obtained form is linear. Readers can check that the solution is

$$p = \frac{2\sin^{-1}x}{\sqrt{1-x^2}} + \frac{c}{\sqrt{1-x^2}}$$

Again integrating gives the general solution as

$$y = (\sin^{-1}x)^2 + c\sin^{-1}x + d$$

where d is constant.

6.7 DIFFERENTIAL EQUATIONS WHICH DO NOT CONTAIN *x* EXPLICITLY

The general form of the equation is

$$f\left(\frac{d^n y}{dx^n}, \frac{d^{n-1}y}{dx^{n-1}}, \ldots, \frac{dy}{dx}, y\right) = 0 \tag{6.10}$$

Put

$$p = \frac{dy}{dx}.$$

Then

$$\frac{d^2y}{dx^2} = \frac{dp}{dx} = \frac{dp}{dy}\frac{dy}{dx} = p\frac{dp}{dy} = pp_1 \text{ where } p_1 = \frac{dp}{dy}.$$

Similarly, $\dfrac{d^3y}{dx^3} = pp_1^2 + p^2 p_2$ and so on. Then Eq. (6.10) takes the form

$$f(p_{n-1}, p_{n-2}, \ldots, p_1, p, y) = 0 \tag{6.11}$$

which may be solvable for p. Let the solution be

$$p = F(y).$$

Then $x = \int \frac{dy}{F(y)} + c$ is the required solution.

EXAMPLE 6.15 Solve $y(1-\log y)\frac{d^2y}{dx^2} + (1+\log y)\left(\frac{dy}{dx}\right)^2 = 0.$

Solution Substitute $\frac{dy}{dx} = p$. Then $\frac{d^2y}{dx^2} = pp_1$ where $p_1 = \frac{dp}{dy}$.
Then given equation reduces to

$$y(1 - \log y)\, pp_1 + (1 + \log y)\, p^2 = 0$$

$$\Rightarrow \qquad \frac{dp}{p} = \frac{1+\log y}{y(\log y - 1)}dy.$$

Integrating,
$$\log p + c = \int \frac{1+\log y}{y(\log y - 1)}dy$$

Put
$$\log y = u$$

$$\therefore \qquad \log p + c = \int \frac{1+u}{u-1}du$$

$$\Rightarrow \qquad \log p + c = \int \left(1 + \frac{2}{u-1}\right)du$$

$$\Rightarrow \qquad \log p + c = u + 2\log(u-1)$$

$$\Rightarrow \qquad \log p + c = \log y + 2\log(\log y - 1)$$

Taking $c = -\log k$, we get

$$\frac{p}{k} = y(\log y - 1)^2$$

$$\Rightarrow \qquad kdx = \frac{dy}{y(\log y - 1)^2}$$

Integrating,

$$kx + c_1 = -\frac{1}{\log y - 1}$$

$\Rightarrow \quad \log y - 1 + \frac{1}{kx + c_1} = 0$ is the required solution.

EXAMPLE 6.16 Solve $y_1^2 - yy_2 = n\sqrt{y_1^2 + a^2y_2^2}$.

Solution Substitute $y_1 = p$. Then $y^2 = pp_1$ where $p_1 = \frac{dp}{dy}$.

Therefore, given equation reduces to

$$p^2 - ypp_1 = n\sqrt{p^2 + a^2p^2p_1^2}$$

$$\Rightarrow \qquad p - yp_1 = n\sqrt{1 + a^2p_1^2}$$

which is Clairaut's form in p_1 and y.
Therefore, its solution is

$$p = cy + n\sqrt{1 + a^2c^2}$$

where c is constant.

$$\Rightarrow \qquad \frac{dy}{cy + n\sqrt{1 + a^2c^2}} = dx$$

Integrating.

$$\frac{1}{c}\log\,(cy + n\sqrt{1 + a^2c^2}) = x + c_1$$

$$\Rightarrow \qquad \log\,(cy + n\sqrt{1 + a^2c^2}) = cx + \log d$$

$\Rightarrow \quad cy + n\sqrt{1 + a^2c^2} = de^{cx}$ is the solution of the given differential equation.

EXAMPLE 6.17 Solve $\dfrac{d^2y}{dx^2} + \dfrac{1}{y}\left(\dfrac{dy}{dx}\right)^2 + y^2 = 0.$

Solution Put $\dfrac{dy}{dx} = p$. Then $\dfrac{d^2y}{dx^2} = p\dfrac{dp}{dx}$.
Given equation reduces to

$$p\frac{dp}{dx} + \frac{p^2}{y} = -y^2$$

Put $p^2 = v$. Then equation reduces to

$$\frac{dv}{dx} + \frac{2}{y}v = -2y^2$$

which is linear and I.F. $= y^2$

$$\therefore \qquad vy^2 = -\frac{2}{5}y^5 + c_1$$

$$\Rightarrow \qquad p^2 = v = -\frac{2}{5}y^3 + \frac{c_1}{y^2}$$

$$\Rightarrow \qquad \frac{dy}{dx} = \frac{1}{y}\sqrt{c_1 - \frac{2}{5}y^5}$$

$$\Rightarrow \qquad dx = \frac{ydy}{\sqrt{c_1 - \frac{2}{5}y^5}}$$

Integrating $x + c_2 = \int \frac{ydy}{\sqrt{c_r - \frac{2}{5}y^5}}$ is the general solution.

EXAMPLE 6.18 Solve $y\frac{d^2y}{dx^2} - \left(\frac{dy}{dx}\right)^2 = y^2 \log y.$

Solution Put $p = \frac{dy}{dx}$. Then $\frac{d^2y}{dx^2} = p\frac{dp}{dx}$.

With these substitutions, the given equation reduces to $yp\frac{dp}{dx} - p^2 = y^2 \log y.$

Put $p^2 = v$.

$$\therefore \qquad y\frac{dv}{dx} - 2v = 2y^2 \log y$$

$$\therefore \qquad \frac{dv}{dx} - \frac{2}{y}v = 2y \log y$$

which is linear with I.F. $= \frac{1}{y^2}$.

Therefore, the solution is

$$v \cdot \frac{1}{y^2} = (\log y)^2 + a$$

$$\therefore \qquad p^2 = y^2(\log y)^2 + ay^2$$

Separating variables, $dx = \frac{dy}{y\sqrt{a + (\log y)^2}}$

Integrating

$$x + \log b = \log\left\{\log y + \sqrt{a + (\log y)^2}\right\} \Rightarrow \log y + \sqrt{a + (\log y)^2} = be^x$$

$$\Rightarrow \qquad \log y = \frac{1}{2}\left\{be^x - \frac{a^2}{b}e^{-x}\right\}$$

6.8 DIFFERENTIAL EQUATIONS IN WHICH *y* APPEARS IN ONLY TWO DERIVATIVES WHOSE ORDERS DIFFER BY TWO

The general form of the equation is

$$f\left(\frac{d^n y}{dx^n}, \frac{d^{n-2} y}{dx^{n-2}}, x\right) = 0 \tag{6.12}$$

Substitute $\dfrac{d^{n-2}y}{dx^{n-2}} = p$ in Eq. (6.12).

$$\therefore \qquad f(p^{(2)}, p, x) = 0 \tag{6.13}$$

which may be solvable for p. Let its solution be

$$p = F(x) \quad \Rightarrow \frac{d^{n-2}y}{dx^{n-2}} = F(x).$$

Then y can be obtained by successive integrations.

EXAMPLE 6.19 Solve $\dfrac{d^4 y}{dx^4} + a^2 \dfrac{d^2 y}{dx^2} = 0.$

Solution Substitute $\dfrac{d^2 y}{dx^2} = p$

Therefore, given equation reduces to

$$p^{(2)} + a^2 p = 0.$$

$$\Rightarrow \qquad (D^2 + a^2)\, p = 0 \quad \text{where } D = \frac{d}{dx}.$$

Its auxiliary equation is

$$m^2 + a^2 = 0 \quad \Rightarrow \quad m = \pm\, ai.$$

$$\therefore \qquad p = c_1 \cos ax + c_2 \sin ax$$

$$\Rightarrow \qquad \frac{d^2 y}{dx^2} = c_1 \cos ax + c_2 \sin ax$$

Integrating,

$$\frac{dy}{dx} = \frac{c_1}{a} \sin ax - \frac{c_2}{a} \cos ax + c_3$$

Again integrating,

$y = -\dfrac{c_1}{a^2} \cos ax - \dfrac{c_2}{a^2} \sin ax + c_3 x + c_4$ is the required solution of the given equation.

EXAMPLE 6.20 Solve $8\frac{d^3y}{dx^3}=\frac{dy}{dx}\left[1+\left(\frac{dy}{dx}\right)^2\right]^2$.

Solution Substitute $\frac{dy}{dx}=p$.

Therefore, given equation reduces to

$$8p^{(2)} = p(1 + p^2)^2.$$

Multiplying by $2p^{(1)}$ gives

$$8(2p^{(1)}p^{(2)} = 2p(1 + p^2)^2p^{(1)}$$

Integrating,

$$8(p^{(1)})^2 = \frac{1}{3}(1+p^2)^3+\frac{c_1}{3}$$

$\Rightarrow$
$$dx=\frac{\sqrt{3}8^{3/2}\,dp}{\sqrt{c_1+(1+p^2)^3}}$$

Integrating,

$$x+c_2=\sqrt{3}8^{3/2}\int\frac{dp}{\sqrt{c_1+(1+p^2)^3}} \tag{i}$$

Also
$$\frac{dy}{dx}=p$$

$\Rightarrow$
$$dy=pdx=\sqrt{3}8^{3/2}\frac{pdp}{\sqrt{c_1+(1+p^2)^3}}$$

Integrating,

$$y+c_3=\sqrt{3}8^{3/2}\int\frac{pdp}{\sqrt{c_1+(1+p^2)^3}} \tag{ii}$$

Equations (i) and (ii) gives the solution.

EXAMPLE 6.21 Solve $x^2\frac{d^4y}{dx^4}=a\frac{d^2y}{dx^2}$.

Solution Put $p=\frac{d^2y}{dx^2}$

Equation can be written as

$$x^2\frac{d^2p}{dx^2}-ap=0$$

Put $v = \log x$ and let $\frac{d}{dv}=D$.

Then, we have

$$[D(D-1)-a]p=0 \;\Rightarrow\; (D^2-D-a)p=0$$

Auxiliary equation is

$$m^2 - m - a = 0 \;\Rightarrow\; m = \frac{1 \pm \sqrt{1+4a}}{2}$$

$$p = c_1 e^{a_1 v} + c_2 e^{a_2 v} = c_1 x^{a_1} + c_2 x^{a_2}$$

where $a_1 = \dfrac{1+\sqrt{1+4a}}{2}, a_2 = \dfrac{1-\sqrt{1+4a}}{2}$.

$\therefore$
$$\frac{d^2 y}{dx^2} = c_1 x^{a_1} + c_2 x^{a_2}$$

Integrating twice, we get general solution as

$$y = \frac{c_1 x^{a_1+2}}{(a_1+1)(a_1+2)} + \frac{c_2 x^{a_2+2}}{(a_2+1)(a_2+2)} + c_3 x + c_4$$

6.9 DIFFERENTIAL EQUATIONS IN WHICH *y* APPEARS IN ONLY TWO DERIVATIVES WHOSE ORDERS DIFFER BY ONE

The general form of the equation is

$$f\left(\frac{d^n y}{dx^n}, \frac{d^{n-1} y}{dx^{n-1}}, x\right) = 0 \tag{6.14}$$

Substitute $p = \dfrac{d^{n-1}y}{dx^{n-1}}$ in Eq. (6.14).

$\therefore$
$$f(p^{(1)}, p, x) = 0 \tag{6.15}$$

which may be solvable for p. Let its solution be

$$p = F(x) \;\Rightarrow\; \frac{d^{n-1}y}{dx^{n-1}} = F(x).$$

Then y can be obtained by successive integrations.

EXAMPLE 6.22 Solve $\dfrac{d^2 y}{dx^2} + \dfrac{dy}{dx} = e^x$.

Solution Substitute $\dfrac{dy}{dx} = p$

Therefore, given equation reduces to

$$p^{(1)} + p = e^x$$

which is linear equation in p and x. Its integrating factor is $e^{\int 1 dx} = e^x$. Hence, solution is

$$pe^x = \int e^x e^x dx + c$$

$$\Rightarrow \quad pe^x = \frac{e^{2x}}{2} + c$$

$$\Rightarrow \quad p = \frac{e^x}{2} + ce^{-x}$$

$$\Rightarrow \quad \frac{dy}{dx} = \frac{e^x}{2} + ce^{-x}$$

$$\Rightarrow \quad dy = \left(\frac{e^x}{2} + ce^{-x}\right) dx$$

Integrating,

$$y = \frac{e^x}{2} - ce^{-x} + c_1$$

which is the required solution of the given equation.

EXAMPLE 6.23 Solve $2\dfrac{d^2y}{dx^2} = \sqrt{1 + \left(\dfrac{dy}{dx}\right)^2}$.

Solution Substitute $\dfrac{dy}{dx} = p$

Therefore, given equation reduces to

$$2p^{(1)} = \sqrt{1 + p^2}$$

$$\Rightarrow \quad \frac{2dp}{\sqrt{1 + p^2}} = dx$$

Integrating,

$$2\log(p + \sqrt{1 + p^2}) = x + c$$

$$\therefore \quad p + \sqrt{1 + p^2} = ce^{x/2} \tag{i}$$

Also $$(1 + p^2) - p^2 = 1 \tag{ii}$$

Dividing Eq. (ii) by Eq. (i), we get

$$\sqrt{1 + p^2} - p = \frac{1}{c} e^{-x/2} \tag{iii}$$

Subtracting Eq. (iii) and Eq. (i) gives

$$2p = ce^{x/2} - \frac{1}{c} e^{-x/2}$$

$$\Rightarrow \qquad 2\frac{dy}{dx} = ce^{x/2} - \frac{1}{c}e^{-x/2}$$

Integrating,

$$2y = 2ce^{x/2} + \frac{2}{c}e^{-x/2} + c_1$$

$$\Rightarrow \qquad y = ce^{x/2} + \frac{1}{c}e^{-x/2} + c_2$$

where c and c_2 are arbitrary constants.

EXERCISES

Solve the following differential equations:

1. $x\frac{d^2y}{dx^2} + 2x\frac{dy}{dx} + 2y = 0.$ $\left[\textbf{Ans. } y = xe^{-2x}\left(c_1\int \frac{e^{2x}}{x^2}dx + c_2\right)\right]$

2. $\sin x\frac{d^2y}{dx^2} - \cos x\frac{dy}{dx} + 2\sin xy = 0.$ $\left[\textbf{Ans. } y = c_1 \sin^2 x + c_2 \cos x - c_2 \sin^2 x \log \tan \frac{x}{2}\right]$

3. $(1+x+x^2)\frac{d^3y}{dx^3} + (3+6x)\frac{d^2y}{dx^2} + 6\frac{dy}{dx} = 0.$ $[\textbf{Ans. } (1+x+x^2)\,y = c_1x^2 + c_2x + c_3]$

4. $\frac{d^2y}{dx^2} = x^2 \sin x.$ $[\textbf{Ans. } y = c_1x + c_2 - x^2 \sin x - 4x\cos x + 6\sin x]$

5. $\sqrt{(16-x^2)^3}\,\frac{d^2y}{dx^2} = x.$ $\left[\textbf{Ans. } y = c_1x + c_2 + \sin^{-1}\frac{x}{4}\right]$

6. $x^2\frac{d^4y}{dx^4} + 1 = 0.$ $\left[\textbf{Ans. } y = c_1x^3 + c_2x^2 + c_3x + c_4\frac{x^2 \log x}{2} - \frac{3}{4}x^2\right]$

7. $\frac{d^2y}{dx^2} = \frac{1}{\sqrt{ay}}$, where a is arbitrary constant.

   $\left[\textbf{Ans. } 3x = 2a^{1/4}(\sqrt{y} - 2c_1)\sqrt{(\sqrt{y} + c_1)} + c_2\right]$

8. $\frac{d^2y}{dx^2} = e^{2y}$, given that $y = 0$, $\frac{dy}{dx} = 1$ when $x = 0$. $[\textbf{Ans. } x = 1 - e^{-y}]$

9. $2x\frac{d^3y}{dx^3}\frac{d^2y}{dx^2} = \left(\frac{d^2y}{dx^2}\right)^2 - a^2.$ $\left[\textbf{Ans. } y = \frac{4}{15c^2}(a^2 + c_1x)^{5/2} + c_2x + c_3\right]$

10. $\dfrac{d^2y}{dx^2}+\dfrac{1}{y}\left(\dfrac{dy}{dx}\right)^2+y^2=0.$ $\left[\textbf{Ans. } x+b=\int\dfrac{ydy}{\sqrt{a-\dfrac{2}{5}y^5}}\right]$

11. $y\dfrac{d^2y}{dx^2}-\left(\dfrac{dy}{dx}\right)^2=y^2\log y.$ [**Ans.** $\log y=c_1e^x+c_2e^{-x}$]

12. $\dfrac{d^2y}{dx^2}+2\dfrac{dy}{dx}+4\left(\dfrac{dy}{dx}\right)^2=0.$ [**Ans.** $4y=\log(c_2-c_1e^{-2x})$]

13. $y\dfrac{d^2y}{dx^2}=\left(\dfrac{dy}{dx}\right)^2\left(1-\cos y\dfrac{dy}{dx}+y\sin y\dfrac{dy}{dx}\right).$ [**Ans.** $x=c_1+c_2\log y+\sin y$]

14. $\dfrac{d^5y}{dx^5}+m^2\dfrac{d^3y}{dx^3}=e^{ax}.$

$\left[\textbf{Ans. } y=c_1\cos mx+c_2\sin mx+c_3x^2+c_4x+c_5+\dfrac{e^{ax}}{a^3(m^2+a^2)}\right]$

15. $2\dfrac{d^2y}{dx^2}\dfrac{d^3y}{dx^3}=1.$ [**Ans.** $15y=4(x+c_1)^{5/2}+c_2x+c_3$]

16. $\dfrac{d^2y}{dx^2}=4+k^2\left(\dfrac{dy}{dx}\right)^2.$ $\left[\textbf{Ans. } y=c_1+\dfrac{1}{k^2}\log\sec 2k(x-c)\right]$

17. $\dfrac{d^2y}{dx^2}+2e^x\left(y+\dfrac{dy}{dx}\right)=x^2.$ $\left[\textbf{Ans. } ye^{2e^x}=\dfrac{1}{3}\int x^3e^{2e^x}dx+c_1\int e^{2e^x}dx+c_2\right]$

18. $\dfrac{d^2y}{dx^2}+2\tan x\dfrac{dy}{dx}+3y=\tan^2x\sec x.$

$\left[\textbf{Ans. } y\sec^3x=c_1+c_2\int\sec^4x\,dx+\int\sec^4x(\tan x-x)dx\right]$

19. $\dfrac{d^3y}{dx^3}+\cos x\dfrac{d^2y}{dx^2}-2\sin x\dfrac{dy}{dx}-y\cos x=\sin 2x.$

$\left[\textbf{Ans. } y=e^{-\sin x}\int e^{\sin x}(c_1x+c_2)dx+c_3e^{-\sin x}+\dfrac{1-\sin x}{2}\right]$

20. $x\dfrac{d^3y}{dx^3}-(2x^2-1)\dfrac{d^2y}{dx^2}-8x\dfrac{dy}{dx}-4y=e^x.$

$\left[\textbf{Ans. } y=xe^{x^2}\int\dfrac{e^{x-x^2}}{x^2}dx+c_1xe^{x^2}\int\dfrac{e^{-x^2}}{x}dx+c_2xe^{x^2}\int\dfrac{e^{-x^2}}{x^2}dx+c_3xe^{x^2}\right]$

21. Find the first integral of $x^3\dfrac{d^3y}{dx^3}+4x^2\dfrac{d^2y}{dx^2}+x(x^2+2)\dfrac{dy}{dx}+3x^2y=2x.$

$$\left[\textbf{Ans. } x^2\left[x\frac{d^2y}{dx^2}+\frac{dy}{dx}+xy\right]=x^2+c\right]$$

22. $\dfrac{dy}{dx}\left(\dfrac{d^2y}{dx^2}-yx^2\right)=xy^2.$ $\left[\textbf{Ans. } \dfrac{dy}{dx}=\sqrt{x^2y^2+c^2}\right]$

23. $\dfrac{dy}{dx}\left(\dfrac{d^2y}{dx^2}-yx^2\right)=xy^2.$ $\left[\textbf{Ans. } \dfrac{dy}{dx}=\sqrt{x^2y^2+c^2}\right]$

24. $\left(y^2+2x^2\dfrac{dy}{dx}\right)\dfrac{d^2y}{dx^2}+2(x+y)\left(\dfrac{dy}{dx}\right)^2+x\dfrac{dy}{dx}+y=0.$ $\left[\textbf{Ans. } \dfrac{dy}{dx}\left(y^2+x^2\dfrac{dy}{dx}\right)+xy=c\right]$

25. $\dfrac{d^3y}{dx^3}=x-\sin x.$ $\left[\textbf{Ans. } y=c_1x^2+c_2x+c_3-\cos x+\dfrac{x^4}{24}\right]$

26. $\dfrac{d^3y}{dx^3}=xe^x.$ [**Ans.** $y=e^x(x-3)+c_1x^2+c_2x+c_3$]

27. $\dfrac{d^2y}{dx^2}=-y.$ [**Ans.** $y=c\sin(x+c)$]

28. Solve $\dfrac{d^2y}{dx^2}+\operatorname{cosec}^2 y\cot y=0$ given that $y=\dfrac{\pi}{2}, \dfrac{dy}{dx}=1$ when $x=0.$

[**Ans.** $x+\cos y=0$]

29. $\left(\dfrac{d^2y}{dx^2}\right)^2=1+\left(\dfrac{dy}{dx}\right)^2.$ [**Ans.** $y=c_1+\cosh(x+a)$]

30. $x\dfrac{d^2y}{dx^2}-\dfrac{dy}{dx}=-\log x-\dfrac{2}{x}.$ [**Ans.** $y=c_1x^2+c_2+(x+1)\log x$]

31. $4x\dfrac{d^2y}{dx^2}-(1-2x)\dfrac{dy}{dx}=e^{-\frac{x}{2}}.$ $\left[\textbf{Ans. } y=c_1+c_2\int x^{\frac{1}{4}}e^{-\frac{x}{2}}dx+2e^{-\frac{x}{2}}\right]$

32. $(1+x^2)\dfrac{d^2y}{dx^2}+\left[1+\left(\dfrac{dy}{dx}\right)^2\right]=0.$ [**Ans.** $y=ax+(1+a^2)\log(x-a)+b$]

33. $(1+x^2)\dfrac{d^2y}{dx^2}+2x\dfrac{dy}{dx}-\dfrac{2}{x^3}=0.$ $\left[\textbf{Ans. } y=a+b\tan^{-1}x+\dfrac{1}{x}\right]$

34. $\dfrac{d^2y}{dx^2}+\dfrac{dy}{dx}\left(1+\left(\dfrac{dy}{dx}\right)^2\right)=0.$ [**Ans.** $y+\sin^{-1}(ae^{-x})=c$]

35. $\left(\dfrac{d^3y}{dx^3}\right)^2(1+x)-\dfrac{d^2y}{dx^2}=0.$ **[Ans.** $y=a+bx+cx^3+18c^2x^2$ **]**

36. $\dfrac{d^2y}{dx^2}+\left(\dfrac{dy}{dx}\right)^2=1.$ **[Ans.** $y^2=x^2+c_1x+c_2$ **]**

37. $y\dfrac{d^2y}{dx^2}-\dfrac{dy}{dx}\left(2+\dfrac{dy}{dx}\right)=0.$ **[Ans.** $c_1y+2=c_2e^{c_1x}$ **]**

38. $2y\dfrac{d^2y}{dx^2}-3\left(\dfrac{dy}{dx}\right)^2-4y^2=0.$ **[Ans.** $y=c_1\sec^2(x+a)$ **]**

39. $y\dfrac{d^2y}{dx^2}-\left(\dfrac{dy}{dx}\right)^2\left[1-\cos y\dfrac{dy}{dx}+y\sin y\dfrac{dy}{dx}\right]=0.$ **[Ans.** $x=c_1+c_2\log y+\sin y$ **]**

40. $2(2a-y)\dfrac{d^2y}{dx^2}=1+\left(\dfrac{dy}{dx}\right)^2.$ $\left[\textbf{Ans. } x+c=\int\sqrt{\dfrac{2a-y}{b-2a+y}}\,dy\right]$

41. $4\dfrac{d^4y}{dx^4}-\dfrac{d^2y}{dx^2}=0.$ $\left[\textbf{Ans. } y=c_1e^{\frac{x}{2}}+c_2e^{-\frac{x}{2}}+c_3x+c_4\right]$

42. $x^2\dfrac{d^4y}{dx^4}+4\dfrac{d^2y}{dx^2}=0.$ $\left[\textbf{Ans. } y=\begin{cases} c_1+c_2x+x^{\frac{5}{2}}\left(c_3x^{\frac{1}{2}\sqrt{1-4a^2}}+c_4x^{-\frac{1}{2}\sqrt{1-4a^2}}\right), & a<\dfrac{1}{2}\\[2ex] c_1+c_2x+c_3x^{\frac{5}{2}}\cos\left(\dfrac{\sqrt{4a^2-1}}{2}\log\dfrac{x}{c_4}\right), & a>\dfrac{1}{2}\end{cases}\right]$

43. $a\dfrac{d^2y}{dx^2}-\dfrac{dy}{dx}=0.$ **[Ans.** $a\log(y+b)=x+c$ **]**

44. $-4\dfrac{d^2y}{dx^2}+\left[1+\left(\dfrac{dy}{dx}\right)^2\right]^{\frac{3}{2}}=0.$ **[Ans.** $(x+a)^2+(y-b)^2=16$ **]**

45. $x\dfrac{d^2y}{dx^2}-\dfrac{dy}{dx}=0.$ **[Ans.** $y=c_1x^2+c_2$ **]**

46. $\dfrac{d^2y}{dx^2}\cdot\dfrac{dy}{dx}=x.$ $\left[\textbf{Ans. } y=\dfrac{1}{2}\left[x\sqrt{x^2+c^2}+c^2\log\left(x+\sqrt{x^2+c^2}\right)+d\right]\right]$

47. $3\dfrac{d^3y}{dx^3}=\dfrac{d^2y}{dx^2}.$ $\left[\textbf{Ans. } y=c_1e^{\frac{x}{3}}+c_2x+c_3\right]$

48. $\dfrac{d^3y}{dx^3}\cdot\dfrac{d^2y}{dx^2}=\dfrac{1}{2}.$ $\left[\textbf{Ans. } 15y=4(x+c_1)^{\frac{5}{2}}+c_2x+c_3\right]$

CHAPTER 7

Linear Differential Equations of Second Order

7.1 INTRODUCTION

The second order differential equations appear in physics. We will study methods to solve these differential equations in the following chapters. The general form of the linear differential equation of second order is

$$\frac{d^2y}{dx^2}+P\frac{dy}{dx}+Qy=R \tag{7.1}$$

where P, Q, R are functions of x only. For example,

$$\frac{d^2y}{dx^2}+e^x\frac{dy}{dx}-xy=x^2\cos x$$

is a linear differential equation of second order. The equation which is not linear is called non-linear. If $R = 0$ in Eq. (7.1) then it is called homogeneous.

The general solution of Eq. (7.1) contains two constants of integration which can be obtain by the condition $y(x_0) = y_0$ and $\dfrac{dy(x_0)}{dx} = y_0^*$. The problem of solving Eq. (7.1) subject to above stated condition is called **initial-value problem**.

For the initial-value problem

$$\left.\begin{aligned}&\frac{d^2y}{dx^2}+P\frac{dy}{dx}+Qy=R\\&\text{subject to}\\&y(x_0)=y_0;\quad \frac{dy(x_0)}{dx}=y_0^*\end{aligned}\right\} \tag{7.2}$$

if P, Q, R are continuous in (α, β) and also $x_0 \in (\alpha, \beta)$, there exists a unique solution of Eq. (7.2) valid for $\alpha < x < \beta$.

EXAMPLE 7.1 Find the longest interval in which the solution of the initial-value problem

$$x(x-4)\frac{d^2y}{dx^2}+3x\frac{dy}{dx}+4y=0$$

subject to $$y(3)=0,\quad \frac{dy(3)}{dx}=-1$$

is certain to have a unique solution.

Solution Given equation can be written as

$$\frac{d^2y}{dx^2}+\frac{3}{(x-4)}\frac{dy}{dx}+\frac{4}{x(x-4)}y=0$$

Here, $$P=\frac{3}{(x-4)},Q=\frac{4}{x(x-4)} \text{ and } R=0.$$

Clearly, $x = 0$ and $x = 4$ are points of discontinuity. Therefore, the longest open interval containing $x = 3$, in which both P and Q are continuous is $0 < x < 4$. This is the longest interval in which solution exists by existence—uniqueness theorem.

If the coefficient P and Q are constant, Eq. (7.1) can be solved by the methods stated in Chapter 4. Here, we will concentrate to solve the Eq. (7.1) when P and Q are functions of x.

Substitute $y = Vz$, where V is a function of x. Then

$$\frac{dy}{dx}=Vz^{(1)}+V^{(1)}z$$

and $$\frac{d^2y}{dx^2}=Vz^{(2)}+2V^{(1)}z^{(1)}+V^{(2)}z$$

Therefore, Eq. (7.1) becomes

$$[Vz^{(2)}+2V^{(1)}z^{(1)}+V^{(2)}z]+P[Vz^{(1)}+V^{(1)}z]+QVz=R$$

$$\Rightarrow \quad V[z^{(2)}+Pz^{(1)}+Qz]+V^{(1)}[2z^{(1)}+Pz]+V^{(2)}z=R \tag{7.3}$$

7.2 SOLUTION METHODOLOGY

7.2.1 When an Integral Included in the CF is Known

Let $y = z$ be an integral included in the CF. Then

$$z^{(2)} + Pz^{(1)} + Qz = 0$$

Then Eq. (7.3) reduces to

$$V^{(2)}z + V^{(1)}[2z^{(1)} + Pz] = R \tag{7.4}$$

Put $p = V^{(1)}$ in Eq. (7.4). Therefore,

$$p^{(1)}z + p[2z^{(1)} + Pz] = R$$

$$\Rightarrow \qquad p^{(1)} + p\left[P + \frac{2z^{(1)}}{z}\right] = \frac{R}{z} \tag{7.5}$$

which is linear.

$$\therefore \qquad \text{IF} = e^{\int\left[P+\frac{2z^{(1)}}{z}\right]dx} = e^{\int Pdx + 2\log z} = z^2 e^{\int Pdx}.$$

Therefore, the solution of Eq. (7.5) is

$$pz^2 e^{\int Pdx} = \int Rze^{\int Pdx} dx + c_1$$

$$\Rightarrow \qquad V^{(1)} = p = \frac{1}{z^2} e^{-\int Pdx} \int Rze^{\int Pdx} dx + \frac{c_1}{z^2} e^{-\int Pdx}$$

Integrating,

$$V = \int \frac{1}{z^2} e^{-\int Pdx} \int Rze^{\int Pdx} (dx)^2 + c_1 \int \frac{1}{z^2} e^{-\int Pdx} dx + c_2 \tag{7.6}$$

where c_1 and c_2 are arbitrary constants. Therefore, the general solution of Eq. (7.1) is given by

$$y = Vz$$

where V is given by Eq. (7.6).

Hint for substitutions:

1. If $1 + P + Q = 0$, take $z = e^x$.
2. If $P + Qx = 0$, take $z = x$.
3. If $1 - P + Q = 0$, take $z = e^{-x}$.
4. If $2 + 2Px + Qx^2 = 0$, take $z = x^2$.
5. If $1 + \frac{P}{a} + \frac{Q}{a^2} = 0$, take $z = e^{ax}$.

EXAMPLE 7.2 Solve $xy^{(2)} - (x - 2)\, y^{(1)} - 2y = x^3$.

Solution Rewrite given equation as

$$y^{(2)} - \frac{(x-2)}{x} y^{(1)} - \frac{2}{x} y = x^2 \tag{i}$$

Here $\qquad P = -\frac{(x-2)}{x}, Q = -\frac{2}{x}$ and $R = x^2$.

$$\therefore \qquad 1 + P + Q = 0$$

$\therefore\quad y = e^x$ is an integral included in CF. Put $y = Ve^x$.

Then $\qquad y^{(1)} = (V + V^{(1)})\, e^x$

and $\qquad y^{(2)} = (V + 2V^{(1)} + V^{(2)})e^x$

Substituting in Eq. (i) gives

$$(V+2V^{(1)}+V^{(2)})e^x-\left(1-\frac{2}{x}\right)(V+V^{(1)})e^x-\frac{2}{x}Ve^x=x^2$$

$$(V+2V^{(1)}+V^{(2)})e^x-\left(1-\frac{2}{x}\right)(V+V^{(1)})e^x-\frac{2}{x}Ve^x=x^2$$

$$\Rightarrow \qquad V^{(2)}+\left(1+\frac{2}{x}\right)V^{(1)}=x^2e^{-x} \qquad \text{(ii)}$$

Put $V^{(1)} = p$ in Eq. (ii).

$$\therefore \qquad p^{(1)}+\left(1+\frac{2}{x}\right)p=x^2e^{-x}$$

which is linear.

$$\text{IF} = e^{x+2\log x}=e^{x+\log x^2}=x^2e^x.$$

$$\therefore \qquad px^2e^x = \int x^2e^{-x}x^2e^x\,dx+c_1$$

$$\therefore \qquad px^2e^x = \frac{x^5}{5}+c_1$$

$$\Rightarrow \qquad p=\frac{x^3e^{-x}}{5}+\frac{c_1e^{-x}}{x^2}$$

$$\Rightarrow \qquad V^{(1)} = \frac{x^3e^{-x}}{5}+\frac{c_1e^{-x}}{x^2}$$

Integrating,

$$V=\int\frac{x^3e^{-x}}{5}\,dx+\int\frac{c_1e^{-x}}{x^2}\,dx+c_2$$

$$=-\frac{1}{5}e^{-x}(x^3+3x^2+6x+6)+c_1\int\frac{e^{-x}}{x^2}\,dx+c_2$$

$$\therefore \qquad y=Ve^x=-\frac{1}{5}(x^3+3x^2+6x+6)+c_1e^x\int\frac{e^{-x}}{x^2}\,dx+c_2e^x$$

is the solution of the given differential equation.

EXAMPLE 7.3 Solve $xy^{(1)} - y = (x-1)(y^{(2)} - x + 1)$.

Solution Rewrite given equation as

$$y^{(2)}-\frac{x}{x-1}y^{(1)}+\frac{1}{x-1}y=x-1 \qquad \text{(i)}$$

Here $\qquad P=\dfrac{x}{x-1}, Q=\dfrac{1}{x-1}$, and $R = x-1$.

$$\therefore \qquad P + Qx = 0$$

$\therefore$ $y = x$ is an integral included in CF. Put $y = Vx$.

Then $$y^{(1)} = V + xV^{(1)}$$

and $$y^{(2)} = 2V^{(1)} + xV^{(2)}$$

Substituting in Eq. (i) gives

$$xV^{(2)} + 2V^{(1)} - \frac{x}{x-1}(V + xV^{(1)}) + \frac{1}{x-1}Vx = x - 1$$

$$\Rightarrow \qquad xV^{(2)} + \left(2 - \frac{x^2}{x-1}\right)V^{(1)} = x - 1$$

$$\Rightarrow \qquad V^{(2)} + \left(\frac{2}{x} - \frac{x}{x-1}\right)V^{(1)} = \frac{x-1}{x} \qquad \text{(ii)}$$

Put $V^{(1)} = p$ in Eq. (ii).

$$\therefore \qquad p^{(1)} + \left(\frac{2}{x} - \frac{1}{x-1} - 1\right)p = 1 - \frac{1}{x} \qquad \text{(iii)}$$

which is linear equation.

$$\therefore \qquad \text{IF} = e^{\int\left(\frac{2}{x} - \frac{1}{x-1} - 1\right)dx} = \frac{x^2}{x-1}e^{-x}.$$

Therefore, the solution of Eq. (iii) is

$$\frac{x^2}{x-1}e^{-x}p = \int xe^{-x}\,dx + c_1$$

$$\Rightarrow \qquad \frac{x^2}{x-1}e^{-x}p = c_1 - xe^{-x} - e^{-x}$$

$$\Rightarrow \qquad p = c_1\left(\frac{x-1}{x^2}\right)e^{x} - \frac{(x+1)(x-1)}{x^2}$$

$$\Rightarrow \qquad V^{(1)} = c_1\left(\frac{1}{x} - \frac{1}{x^2}\right)e^{x} - \left(1 - \frac{1}{x^2}\right)$$

Integrating,

$$V = c_2 + c_1\int\left(\frac{1}{x} - \frac{1}{x^2}\right)e^{x}dx - \left(x + \frac{1}{x}\right)$$

$$\Rightarrow \qquad V = c_2 + \frac{c_1}{x}e^{x} - \left(x + \frac{1}{x}\right)$$

$\therefore$ $y = c_2x + c_1e^{x} - (x^2 + 1)$ is the solution of Eq. (i).

EXAMPLE 7.4 Solve $(x \sin x + \cos x)\, y^{(2)} - x \cos x y^{(1)} + y \cos x = x$.

Solution Rewrite given equation as

$$y^{(2)} - \frac{x\cos x}{x\sin x + \cos x} y^{(1)} + \frac{\cos x}{x\sin x + \cos x} y = \frac{x}{x\sin x + \cos x} \qquad \text{(i)}$$

Here $\quad P = -\dfrac{x\cos x}{x\sin x + \cos x}, \quad Q = \dfrac{\cos x}{x\sin x + \cos x}, \quad$ and $\quad R = \dfrac{x}{x\sin x + \cos x}$.

$$\therefore \qquad P + Qx = 0$$

$\therefore \quad y = x$ is an integral included in CF. Put $y = Vx$.

Then
$$y^{(1)} = V + xV^{(1)}$$

and
$$y^{(2)} = 2V^{(1)} + xV^{(2)}.$$

Substituting in Eq. (i) gives

$$x(x\sin x + \cos x)V^{(2)} + (2x\sin x + 2\cos x - x^2\cos x)V^{(1)} = x \qquad \text{(ii)}$$

Put $V^{(1)} = p$ in Eq. (ii).

$$\therefore \qquad p^{(1)} + \left(\frac{2}{x} - \frac{x\cos x}{x\sin x + \cos x}\right)p = \frac{1}{x\sin x + \cos x} \qquad \text{(iii)}$$

which is linear equation.

$$\therefore \qquad \text{IF} = e^{\int\left(\frac{2}{x} - \frac{x\cos x}{x\sin x + \cos x}\right)dx} = \frac{x^2}{x\sin x + \cos x}$$

Therefore, the solution of Eq. (iii) is

$$p\frac{x^2}{x\sin x + \cos x} = c_1 + \int \frac{x^2}{(x\sin x + \cos x)^2}\,dx$$

$$= c_1 + \int \frac{x}{(x\sin x + \cos x)^2}\,\frac{x}{\cos x}\,dx$$

$$= c_1 + \left[-\frac{1}{x\sin x + \cos x}\,x\sec x + \int \frac{(\sec x + x\sec x\tan x)}{x\sin x + \cos x}\,dx\right]$$

$$= c_1 - \frac{x\sec x}{x\sin x + \cos x} + \int \frac{\sec^2 x(x\sin x + \cos x)}{x\sin x + \cos x}\,dx$$

$$= c_1 - \frac{x\sec x}{x\sin x + \cos x} + \tan x$$

$$\therefore \qquad p = c_1\left[\frac{\sin x}{x} + \frac{\cos x}{x^2}\right] - \frac{\sec x}{x} + \tan x\left[\frac{\sin x}{x} + \frac{\cos x}{x^2}\right]$$

$$\Rightarrow \qquad V^{(1)} = c_1\left[\frac{\sin x}{x} + \frac{\cos x}{x^2}\right] - \frac{\sec x}{x}(1 - \sin^2 x) + \frac{\sin x}{x^2}$$

$$= c_1\left[\frac{\sin x}{x} + \frac{\cos x}{x^2}\right] - \frac{\cos x}{x} + \frac{\sin x}{x^2}$$

Integrating,

$$V = c_1\left[-\frac{\cos x}{x}\right] - \frac{\sin x}{x} + c_2$$

$$\therefore \qquad y = Vx = -c_1 \cos x - \sin x + c^2 x$$

where c_1 and c_2 are arbitrary constants is the required solution of given equation.

EXAMPLE 7.5 Solve $x^2 y^{(2)} + xy - y = 0$ given that $y = x + \frac{1}{x}$ is one integral.

Solution Divide given equation by x^2.

So we get
$$y^{(2)} + \frac{1}{x} y^{(1)} - \frac{1}{x^2} y = 0 \qquad \text{(i)}$$

Put
$$y = v\left(x + \frac{1}{x}\right).$$

Then $y^{(1)} = v\left(1 - \frac{1}{x^2}\right) + v^{(1)}\left(x + \frac{1}{x}\right)$ and $y^{(2)} = \frac{2v}{x^3} + 2v^{(1)}\left(1 - \frac{1}{x^2}\right) + v^{(2)}\left(x + \frac{1}{x}\right)$.

Substituting values in Eq. (i) gives $\left(x + \frac{1}{x}\right) v^{(2)} + \left(3 - \frac{1}{x^2}\right) v^{(1)} = 0 \qquad \text{(ii)}$

Put $p = v^{(1)}$. Then Eq. (ii) reduces to $\left(x + \frac{1}{x}\right) p^{(1)} + \left(3 - \frac{1}{x^2}\right) p = 0$

$$\Rightarrow \qquad \frac{dp}{p} = \frac{3x^2 - 1}{x(x^2 + 1)} dx$$

or
$$\frac{dp}{p} = \left\{\frac{1}{x} - \frac{4x}{x^2 + 1}\right\} dx$$

Integration gives

$$\log p = \log x - 2\log (x^2 + 1) + \log c_1$$

$$\Rightarrow \qquad v^{(1)} = p = \frac{c_1 x}{(x^2 + 1)^2}$$

Integration gives

$$v = -\frac{c_1}{2(x^2 + 1)} + c_2$$

Therefore, the general solution of given equation is

$$y = \left\{c_2 - \frac{c_1}{2(x^2 + 1)}\right\}\left(x + \frac{1}{x}\right) = c_2\left(x + \frac{1}{x}\right) + \frac{c_3}{x}$$

EXAMPLE 7.6 Solve $(1-x^2)y^{(2)}+xy^{(1)}-y=x(1-x^2)^{\frac{3}{2}}$

Solution The equation can be rewritten as

$$y^{(2)}+\frac{x}{(1-x^2)}y^{(1)}-\frac{1}{(1-x^2)}y=x(1-x^2)^{\frac{1}{2}} \quad \text{(i)}$$

Then $$P=\frac{x}{(1-x^2)},\ Q=-\frac{1}{(1-x^2)} \text{ and } R=x(1-x^2)^{\frac{1}{2}}$$

$\therefore$ $$P+Qx=0$$

This suggests that $y = x$ is an integral in C.F. Put $y = vx$. Then

$$y^{(1)} = v + xv^{(1)} \text{ and } y^{(2)} = 2v^{(1)} + xv^{(2)}$$

With these substitutions in given equation, we get

$$x(1-x^2)v^{(2)}+(2-x^2)v^{(1)}=x(1-x^2)^{\frac{3}{2}} \quad \text{(ii)}$$

Dividing by $x(1 - x^2)$ results

$$v^{(2)}+\frac{(2-x^2)}{x(1-x^2)}v^{(1)}=\sqrt{(1-x^2)}$$

$\Rightarrow$ $$v^{(2)}+\left[\frac{2}{x}+\frac{x}{(1-x^2)}\right]v^{(1)}=\sqrt{(1-x^2)} \quad \text{(iii)}$$

Equation (iii) is linear in $v^{(1)}$ and I.F. $=\dfrac{x^2}{\sqrt{1-x^2}}$.

$\therefore$ Solution of Eq. (iii) is

$$v^{(1)}\frac{x^2}{\sqrt{1-x^2}}=\frac{x^3}{3}+c_1$$

$\Rightarrow$ $$v^{(1)}=\frac{x}{3}\sqrt{1-x^2}+c_1\frac{1}{x^2}\sqrt{1-x^2}$$

Integrating, $$v=\frac{1}{3}\int x\sqrt{1-x^2}\,dx+c_1\int\frac{1}{x^2}\sqrt{1-x^2}\,dx+c_2=-\frac{1}{9}(1-x^2)^{\frac{3}{2}}+c_1I+c_2$$

where $I=\displaystyle\int\frac{\sqrt{1-x^2}}{x^2}dx$. Put $x = \sin\theta$. Then

$$I=\int\frac{\cos^2\theta}{\sin^2\theta}d\theta=\int\cot^2\theta\,d\theta=-\cot\theta-\theta=-\frac{\sqrt{1-x^2}}{x}-\sin^{-1}x$$

$$\therefore \qquad v = c_2 - c_1\left(\frac{\sqrt{1-x^2}}{x} + \sin^{-1} x\right) - \frac{1}{9}(1-x^2)^{\frac{3}{2}}$$

Therefore, the general solution of given differential equation is

$$y = vx = c_2 x - c_1\left(\sqrt{1-x^2} + x\sin^{-1} x\right) - \frac{x}{9}(1-x^2)^{\frac{3}{2}}$$

EXAMPLE 7.7 Solve $xy^{(2)} - y^{(1)} + (1-x)y = x^2 e^{-x}$.

Solution Rewrite equation as

$$y^{(2)} - \frac{1}{x}y^{(1)} + \left(\frac{1}{x} - 1\right)y = xe^{-x} \qquad \text{(i)}$$

Evidently $y = e^x$ is an integral included in C.F.
Put $y = ve^x$. Then

$$y^{(1)} = (v + v^{(1)})e^x \text{ and } y^{(2)} = (v + 2v^{(1)} + v^{(2)})e^x.$$

Substituting these values in Eq. (i) and by simplifying we get

$$v^{(2)} + \left(2 - \frac{1}{x}\right)v^{(1)} = xe^{-2x} \qquad \text{(ii)}$$

Put $v^{(1)} = p$. Therefore, Eq. (ii) becomes

$$p^{(1)} + \left(2 - \frac{1}{x}\right)p = xe^{-2x}$$

which is linear and I.F. $= \dfrac{1}{x}e^{2x}$.

$$\therefore \qquad p \cdot \frac{1}{x}e^{2x} = x + c_1$$

or

$$v^{(1)} = p = x^2 e^{-2x} + c_1 x e^{-2x}$$

Integrating,

$$v = c_2 + c_1(2x+1)e^{-2x} - \frac{1}{4}(2x^2 + 2x + 1)e^{-2x}$$

$\therefore$ Solution of given differential equation is

$$y = ve^x = c_2 e^x + c_1(2x+1)e^{-x} - \frac{1}{4}(2x^2 + 2x + 1)e^{-x}$$

7.2.2 Changing the Dependent Variable

In Eq. (7.3), let

$$P_1 = P + \frac{2z^{(1)}}{z} \qquad (7.7)$$

and $$Q_1 = \frac{1}{z}(z^{(2)} + Pz^{(1)} + Qz) \tag{7.8}$$

$$\therefore \quad V^{(2)} + P_1V^{(1)} + Q_1V = \frac{R}{Z} = R_1 \quad \text{(say)} \tag{7.9}$$

Any desired value can be assigned to P_1 or Q_1 by the proper choice of z. The choice of z should be made so that $Q_1 = 0$. Equation (7.7) implies

$$\frac{2z^{(1)}}{z} = P_1 - P.$$

Integrating,

$$\log z = \frac{1}{2}\int (P_1 - P)\,dx \Rightarrow z = e^{\frac{1}{2}\int (P_1 - P)dx} \tag{7.10}$$

Thus, by assigning any arbitrary value to P_1, z can be obtained from Eq. (7.10). In particular if $P_1 = 0$, then

$$z = e^{\frac{1}{2}\int -P\,dx}$$

$$\therefore \quad z^{(1)} = \left(-\frac{P}{2}\right)e^{\frac{1}{2}\int -P\,dx}$$

and $$z^{(2)} = \left(\frac{P^2}{4} - \frac{P^{(1)}}{2}\right)e^{\frac{1}{2}\int -P\,dx}$$

Therefore, from Eq. (7.8)

$$Q_1 = \frac{1}{z}\left[\left(\frac{P^2}{4} - \frac{P^{(1)}}{2}\right)e^{\frac{1}{2}\int -P\,dx} - \frac{P^2}{2}e^{\frac{1}{2}\int -P\,dx} + Qe^{\frac{1}{2}\int -P\,dx}\right]$$

$$\Rightarrow \quad Q_1 = Q - \frac{1}{2}P^{(1)} - \frac{P^2}{4} \tag{7.11}$$

Thus, with the value of z given by Eq. (7.10), Eq. (7.9) reduces to

$$V^{(2)} + Q_1V = R_1 \tag{7.12}$$

Equation (7.12) is known as the **normal form** of the Eq. (7.1) and can be integrable. Transformation of Eq. (7.1) to the form of Eq. (7.12) is known as **removal of the first derivative**.

EXAMPLE 7.8 Solve $\sin^2 xy^{(2)} + 2 \sin x \cos xy^{(1)} + \sin^2 xy = 0$.

Solution The given equation is equivalent to

$$y^{(2)} + 2 \tan xy^{(1)} + y = 0.$$

Here, $$P = 2 \tan x,\ Q = 1 \text{ and } R = 0.$$

$$\therefore \quad Q_1 = Q - \frac{1}{2}P^{(1)} - \frac{P^2}{4} = 1 - \sec^2 x + \tan^2 x = 0$$

and $$z = e^{\frac{1}{2}\int -P\,dx} = e^{\frac{1}{2}\int -2\tan x\,dx} = \cos x$$

Therefore, the normal form of the given equation is

$$V^{(2)} = 0$$

$$\Rightarrow \quad V^{(1)} = c_1$$

$$\Rightarrow \quad V = c_1 x + c_2$$

$$\therefore \quad y = Vz = (c_1 x + c_2)\cos x$$

is the required solution of the given equation.

EXAMPLE 7.9 Solve $\dfrac{d}{dx}[\cos^2 x\, y^{(1)}] + \cos^2 x\, y = 0.$

Solution The equation is simplified to

$$\cos^2 x\, y^{(2)} - 2\sin x\cos x\, y^{(1)} + \cos^{(2)} x\, y = 0$$

$$\Rightarrow \quad y^{(2)} - 2\tan x\, y^{(1)} + y = 0 \qquad \text{(i)}$$

Here, $P = -2\tan x, Q = 1, R = 0$. Therefore,

$$Q_1 = Q - \frac{1}{2}P^{(1)} - \frac{P^2}{4} = 1 + \sec^2 x - \tan^2 x = 2$$

and $$z = e^{-\frac{1}{2}\int P dx} = e^{\int \tan x dx} = \sec x$$

Therefore, normal form of Eq. (i) is

$$v^{(2)} + 2v = 0$$

$$\Rightarrow \quad (D^2 + 2)v = 0$$

$$\therefore \quad v = c_1 \cos\sqrt{2}x + c_2 \sin\sqrt{2}x$$

The solution of given differential equation is

$$y = vz = (c_1 \cos\sqrt{2}x + c_2 \sin\sqrt{2}x)\sec x$$

7.2.3 Changing the Independent Variable

Change the independent variable from x to z by the transformation $z = f(x)$. Then

$$y^{(1)} = \frac{dy}{dz}\frac{dz}{dx} = y_1 z^{(1)} \quad \text{where } y_1 = \frac{dy}{dz}$$

and $$y^{(2)} = y_2(z^{(1)})^2 + y_1 z^{(2)}$$

Substitute these values in Eq. (7.1) gives

$$y_2 + \left[\frac{z^{(2)} + Pz^{(1)}}{(z^{(1)})^2}\right] y_1 + \frac{Q}{(z^{(1)})^2} y = \frac{R}{(z^{(1)})^2}$$

$$\Rightarrow \qquad y^2 + P'y_1 + Q'y = R' \tag{7.13}$$

where

$$\left.\begin{aligned} P' &= \frac{z^{(2)} + Pz^{(1)}}{(z^{(1)})^2} \\ Q' &= \frac{Q}{(z^{(1)})^2} \\ R' &= \frac{R}{(z^{(1)})^2} \end{aligned}\right\} \tag{7.14}$$

P', Q' and R' are functions of x, but can be transformed to z.

(i) Let $Q' = a^2$, a is constant. Then

$$Q = a^2(z^{(1)})^2 \quad \Rightarrow \quad adz = \sqrt{Q}\, dx.$$

Integrating,

$$az = \int \sqrt{Q} dx.$$

(ii) If z is so chosen that $P' = 0$.

$$\Rightarrow \qquad z^{(2)} + Pz^{(1)} = 0.$$

Integrating,

$$z^{(1)} = e^{-\int P\,dx}.$$

Again integrating,

$$z = \int e^{-\int P\,dx} dx.$$

If with this value of z, Q' becomes constant or constant divided by z^2, Eq. (7.13) can be integrated.

EXAMPLE 7.10 Solve $x^4y^{(2)} + x^3y^{(1)} + 4y = 0$.

Solution The given equation can be written as

$$y^{(2)} + \frac{1}{x} y^{(1)} + \frac{4}{x^4} y = 0.$$

Here, $$P = \frac{1}{x}, Q = \frac{4}{x^4} \text{ and } R = 0.$$

Choose z such that $P' = 0$. Then

$$z = \int e^{-\int Pdx} dx$$

$$= \int e^{-\int \frac{1}{x}dx} dx = \int e^{-\log x} dx = \frac{1}{x}.$$

Now, $$Q' = \frac{Q}{(z^{(1)})^2} = \frac{4/x^4}{(-1/x^2)^2} = 4 \text{ and } R' = 0.$$

Therefore, given equation reduces to

$$y^2 + 4y = 0$$

$$\Rightarrow \quad (D^2 + 4)y = 0$$

$$\Rightarrow \quad y = a\cos 2z + b\sin 2z$$

$$\Rightarrow \quad y = a\cos\left(\frac{2}{x}\right) + b\sin\left(\frac{2}{x}\right)$$

where a and b are arbitrary constants, is the required solution.

EXAMPLE 7.11 Solve $y^{(2)} + (3\sin x - \cot x)\, y^{(1)} + 2y\sin^2 x = e^{-\cos x}\sin^2 x$.

Solution Here, $P = 3\sin x - \cot x$, $Q = 2\sin^2 x$ and $R = e^{-\cos x}\sin^2 x$. Choose z such that $P' = 0$ and so

$$z^{(1)} = e^{-\int P dx}$$

$$= e^{-\int (3\sin x - \cot x)\, dx}$$

$$= e^{3\cos x + \log\sin x} = \sin x e^{3\cos x}$$

$$\therefore \quad z = \int \sin x e^{3\cos x} dx = -\frac{1}{3} e^{3\cos x}$$

$$\therefore \quad Q' = \frac{Q}{(z^{(1)})^2} = \frac{2\sin^2 x}{(\sin x e^{3\cos x})^2} = \frac{2}{9z^2}$$

and $$R' = \frac{R}{(z^{(1)})^2} = \frac{e^{-\cos x}\sin^2 x}{(\sin x e^{3\cos x})^2} = e^{-7\cos x} = -\frac{1}{3^{7/3}} z^{-7/3}$$

Therefore, given equation reduces to

$$y_2 + \frac{2}{9z^2} y = -\frac{1}{3^{7/3}} z^{-7/3}$$

$$\Rightarrow \quad (9z^2 D^2 + 2)y = -\frac{1}{3^{7/3}} z^{-7/3}$$

Put $$X = \log z.$$

$$\Rightarrow \quad [9D(D-1) + 2]\, y = -\frac{1}{3^{1/3}} e^{-X/3} \text{ where } D = \frac{d}{dX}$$

$$\Rightarrow \quad (9D^2 - 9D + 2)y = -\frac{1}{3^{1/3}} e^{-X/3}$$

Therefore, the auxiliary equation is

$$9m^2 - 9m + 2 = 0 \quad \Rightarrow \quad m = \frac{2}{3}, \frac{1}{3}$$

$$\therefore \qquad \text{CF} = c_1 e^{2X/3} + c_2 e^{X/3}$$

$$= c_1 z^{2/3} + c_2 z^{1/3}$$

$$= c_1 \left(-\frac{1}{3} e^{3\cos x}\right)^{2/3} + c_2 \left(-\frac{1}{3} e^{3\cos x}\right)^{1/3}$$

$$= c_3 e^{2\cos x} + c_4 e^{\cos x}$$

and

$$\text{PI} = -\frac{1}{3^{1/3}} \frac{1}{9D^2 - 9D + 2} e^{-X/3}$$

$$= -\frac{1}{6 \cdot 3^{1/3}} e^{-X/3}$$

$$= -\frac{1}{6 \cdot 3^{1/3}} z^{-1/3}$$

$$= -\frac{1}{6 \cdot 3^{1/3}} \left(-\frac{1}{3}\right)^{-1/3} e^{-\cos x}$$

$$= -\frac{1}{6} e^{-\cos x}$$

Therefore, y = CF + PI is the required solution of the given differential equation.

EXAMPLE 7.12 Solve $x^6 y^{(2)} + 3x^5 y^{(1)} + 9y = \dfrac{1}{x^2}$.

Solution Dividing by x^6, we get $y^{(2)} + \dfrac{3}{x} y^{(1)} + \dfrac{9}{x^6} y = \dfrac{1}{x^8}$

Here

$$P = \frac{3}{x}, Q = \frac{9}{x^6} \text{ and } R = \frac{1}{x^8}$$

We choose z such that $P' = 0$. Then $z = \int e^{-\int P dx} = \int \dfrac{1}{x^3} dx = -\dfrac{1}{2x^2}$.

$$\therefore \qquad Q' = \frac{Q}{\left(z^{(1)}\right)^2} = 9 \text{ and } R' = \frac{R}{\left(z^{(1)}\right)^2} = \frac{1}{x^2} = -2z$$

Therefore, equation reduces to $y^{(2)} + 9y^2 = -2z$

$$\Rightarrow \qquad (D^2 + 9)y = -2z, \quad \text{where } D = \frac{d}{dz}$$

$$\therefore \qquad \text{C.F.} = c_1 \cos 3z + c_2 \sin 3z \cdot$$

$$\text{P.I.} = \frac{1}{D^2+9}(-2z)$$

$$= -\frac{2}{9}\left\{1+\frac{D^2}{9}\right\}^{-1} z = -\frac{2}{9}\left\{1-\frac{D^2}{9}+\cdots\right\} z = -\frac{2z}{9}$$

Hence, $y = c_1 \cos 3z + c_2 \sin 3z - \dfrac{2z}{9}$.

The solution of given differential equation is

$$y = c_1 \cos\left(\frac{3}{2x^2}\right) - c_2 \sin\left(\frac{3}{2x^2}\right) + \frac{1}{9x^2}.$$

EXAMPLE 7.13 Solve $y^{(2)} + \tan x\, y^{(1)} + 81 y \sec^2 x = 0$.

Solution Here $P = \tan x, Q = 81 \sec^2 x$ and $R = 0$

We need to choose z such that $P' = 0$. Then

$$z = \int e^{-\int P dx} dx = \int e^{-\int \tan x dx} dx = \int \sec x dx = \log\left(\tan\left(\frac{\pi}{4}+\frac{x}{2}\right)\right)$$

$\therefore \quad Q' = 81$ and $R' = 0$

$\therefore \quad$ Equation reduces to $y^{(2)} + 81y = 0 \quad \Rightarrow \quad (D^2+81)y = 0$

$$\therefore \qquad \text{C.F.} = c_1 \cos 9z + c_2 \sin 9z = c_1 \cos\left(\log\left(\tan\left(\frac{\pi}{4}+\frac{x}{2}\right)\right)\right) + c_2 \sin\left(\log\left(\tan\left(\frac{\pi}{4}+\frac{x}{2}\right)\right)\right)$$

7.2.4 By Factorization of the Operator

Suppose that Eq. (7.1) can be written in the form

$$f(D)y = R \tag{7.15}$$

and $f(D)$ can be factorized in the form $f_2(D)\, f_1(D)$ such that when $f_1(D)$ operates on y and $f_2(D)$ operates upon the resultant of this operation is giving same result when $f(D)$ operates on y, i.e.

$$f(D)y = f_2(D)\,[f_1(D)y]$$

Note: The factors $f_1(D)$ and $f_2(D)$, in general, are not commutative as they contain variable.

EXAMPLE 7.14 Solve $xy^{(2)} - (1+x)y^{(1)} + y = x^2$.

Solution Given

$$xy^{(2)} - (1+x)\, y^{(1)} + y = x^2 \tag{i}$$

The symbolic form is

$$[xD^2 - (1 + x)D + 1]y = x^2$$

$$\Rightarrow \quad (xD - 1)(D - 1)y = x^2 \qquad \text{(ii)}$$

Put $$(D - 1)y = U$$

$$\therefore \quad (xD - 1)U = x^2$$

$$\Rightarrow \quad xU^{(1)} - U = x^2$$

$$\Rightarrow \quad U^{(1)} - \frac{1}{x}U = x$$

which is linear equation. Its integrating factor is

$$\text{IF} = e^{-\int \frac{1}{x}dx} = \frac{1}{x}$$

$$\therefore \quad U\frac{1}{x} = x + c_1$$

$$\Rightarrow \quad U = x^2 + c_1 x$$

$$\Rightarrow \quad (D - 1)y = x^2 + c_1 x$$

$$\Rightarrow \quad y^{(1)} - y = x^2 + c_1 x$$

which is linear equation. Its integrating factor is e^{-x}.

$$\therefore \quad y = e^x \int (x^2 + c_1 x)\, e^{-x} dx + c_2 e^x$$

$$= -e^x e^{-x}[x^2 + 2x + 2 + c_1(x+1)] + c_2 e^x$$

$$= -[x^2 + 2x + 2 + c_1(x+1)] + c_2 e^x$$

is the solution of given differential equation.

EXAMPLE 7.15 Solve $[3x^3D^2 - 2xD + 4] = 0$.

Solution Given

$$[3x^3D^2 - 2xD + 4 + 4]y = 0 \qquad \text{(i)}$$

$$\Rightarrow \quad (xD - 2)(3x^2D + 2)y = 0 \qquad \text{(ii)}$$

Put $$(3x^2D + 2)y = U$$

$$\therefore \quad (xD - 2)U = 0$$

$$\Rightarrow \quad xU^{(1)} - 2U = 0$$

$$\Rightarrow \quad x\frac{dU}{dx} - 2U = 0$$

$$\Rightarrow \quad 2\frac{dx}{x} - \frac{dU}{U} = 0$$

Integrating,

$$U = c_1 x^2$$

$\therefore$
$$(3x^2 D + 2)y = c_1 x^2$$

$\Rightarrow$
$$3x^2 y^{(1)} + 2y = c_1 x^2$$

$\Rightarrow$
$$y^{(1)} + \frac{2}{3x^2} y = \frac{c_1}{3}$$

which is linear equation.

$\therefore$
$$\text{IF} = e^{\frac{2}{3}\int \frac{dx}{x^2}} = e^{-\frac{2}{3x}}$$

$\therefore$
$$ye^{-\frac{2}{3x}} = \frac{c_1}{3}\int e^{-\frac{2}{3x}}\, dx + c_2$$

$\Rightarrow$
$$y = \frac{c_1}{3} e^{\frac{2}{3x}} \int e^{-\frac{2}{3x}}\, dx + c_1 e^{\frac{2}{3x}}$$

is the required solution of the given equation.

EXAMPLE 7.16 Solve $xy^{(2)} + (x - 2)y^{(1)} - 2y = x^3$.

Solution Rewriting the given equation in D-form results to

$$[xD^2 + (x-2)D - 2]y = x^3$$

$\Rightarrow$
$$(xD - 2)(D + 1)y = x^3 \tag{i}$$

Put $(D + 1)y = U$.

$\therefore$ Equation (i) reduces to $(xD - 2)U = x^3$

$\Rightarrow$
$$xU^{(1)} - 2U = x^3$$

$\Rightarrow$ $U^{(1)} - \frac{2}{x}U = x^2$, which is linear and I.F. $= \frac{1}{x^2}$.

$\therefore$ U is given by

$$U = x^3 + c_1 x^2$$

$\Rightarrow$
$$(D + 1)y = U = x^3 + c_1 x^2$$

$\Rightarrow$
$$y^{(1)} + y = x^3 + c_1 x^2$$

which is linear and I.F. $= e^x$.

$\therefore$ The solution of given differential equation is given by

$$ye^x = \int e^x x^3 dx + c_1 \int x^2 e^x dx + c_2 = e^x x^3 + (c_1 - 3)(x^2 - 2x + 2)e^x + c_2$$

$\Rightarrow$
$$y = x^3 + (c_1 - 3)(x^2 - 2x + 2) + c_2 e^{-x}$$

EXAMPLE 7.17 Solve $xy^{(2)} + (1+x)y^{(1)} + y = e^x$.

Solution Put $U = \frac{d}{dx}$. Then we get $\left[xD^2 + (1+x)D + 1\right]y = e^x$.

$$\Rightarrow \qquad (xD+1)(D+1)y = e^x \qquad \text{(i)}$$

Put $(D+1)y = U$.

Therefore, Eq. (i) reduces to $(xD + 1)U = e^x$

$$\Rightarrow \qquad xU^{(1)} + U = e^x$$

or
$$U^{(1)} + \frac{1}{x}U = e^x$$

which is linear with I.F. = x.

$\therefore$ Solution is
$$Ux = e^x + c_1$$

$$\Rightarrow \qquad U = \frac{1}{x}e^x + \frac{c_1}{x}$$

$$\Rightarrow \qquad (D+1)y = \frac{1}{x}e^x + \frac{c_1}{x}$$

$$\Rightarrow \qquad y^{(1)} + y = \frac{1}{x}e^x + \frac{c_1}{x}$$

which is linear and I.F. = e^x.

$\therefore$ Solution is

$$ye^x = c_2 + c_1\int\frac{1}{x}e^x\,dx + \int\frac{1}{x}e^{2x}\,dx$$

$$\Rightarrow \qquad y = c_2e^{-x} + c_1e^{-x}\int\frac{1}{x}e^x\,dx + e^{-x}\int\frac{1}{x}e^{2x}\,dx$$

7.2.5 Variation of Parameters

Let U and V be linearly independent solutions corresponding to homogeneous equation

$$y^{(2)} + Py^{(1)} + Qy = 0 \qquad (7.16)$$

Then
$$y = c_1U + c_2V \qquad (7.17)$$

where c_1 and c_2 are arbitrary constants, is the CF of Eq. (7.16). The method is to replace c_1 and c_2 in Eq. (7.1) by respective functions 'f' and 'g' which are to be determined so that the resulting function

$$f \cdot U + g \cdot V$$

satisfies Eq. (7.1).

Let $$y = f \cdot U + g \cdot V \tag{7.18}$$

Then $$y^{(1)} = f^{(1)} \cdot U + f \cdot U^{(1)} + g^{(1)} \cdot V + g \cdot V^{(1)}$$

Since there are two functions f and g, but only one condition on them, the second condition can be taken as zero unless it does not rule out the first one.

$\therefore$ $$f^{(1)} \cdot U + g^{(1)} \cdot V = 0 \tag{7.19}$$

Then $$y^{(1)} = f \cdot U^{(1)} + g \cdot V^{(1)}$$

$\Rightarrow$ $$y^{(2)} = f \cdot U^{(2)} + f^{(1)} \cdot U^{(1)} + g^{(1)} \cdot V^{(1)} + g \cdot V^{(2)}$$

Substitute y, $y^{(1)}$ and $y^{(2)}$ in Eq. (7.1).

$\therefore$ $$f[U^{(2)} + PU^{(1)} + QU] + g[V^{(2)} + PV^{(1)} + QV] + f^{(1)}U^{(1)} + g^{(1)}V^{(1)} = R$$

Since U and V satisfy Eq. (7.16), we have

$$f^{(1)}U^{(1)} + g^{(1)}V^{(1)} = R \tag{7.20}$$

From Eqs. (7.19) and (7.20),

$$\frac{f^{(1)}}{-RV} = \frac{g^{(1)}}{-RU} = \frac{1}{UV^{(1)} - VU^{(1)}}$$

Therefore, $$\left.\begin{aligned} f^{(1)} &= \frac{-RV}{UV^{(1)} - VU^{(1)}} \\ g^{(1)} &= \frac{RU}{UV^{(1)} - VU^{(1)}} \end{aligned}\right\} \tag{7.21}$$

The denominator in Eq. (7.21) is not zero as U and V are linearly independent. Integrating Eq. (7.21), we have f and g, and hence, the solution of Eq. (7.1) can be obtained.

EXAMPLE 7.18 Solve $y^{(2)} + 4y = \text{cosec } 2x$.

Solution Given

$$y^{(2)} + 4y = \text{cosec } 2x \tag{i}$$

Here, $$U = \sin 2x \text{ and } V = \cos 2x.$$

Let $$y = f \sin 2x + g \cos 2x \tag{ii}$$
be the solution. Then

$$y^{(1)} = f^{(1)} \sin 2x + 2f \cos 2x + g^{(1)} \cos 2x - 2g \sin 2x$$

Take f and g such that

$$f^{(1)} \sin 2x + g^{(1)} \cos 2x = 0 \tag{iii}$$

Then $$y^{(1)} = 2f \cos 2x - 2g \sin 2x$$

$\therefore$ $$y^{(2)} = 2[f^{(1)} \cos 2x - g^{(1)} \sin 2x] - 4[f \sin 2x + g \cos 2x]$$

Substitute the values of y and $y^{(2)}$ in Eq. (i).

$$\therefore \quad 2[f^{(1)}\cos 2x - g^{(1)}\sin 2x] - 4[f\sin 2x + g\cos 2x] + 4f\sin 2x + 4g\cos 2x = \operatorname{cosec} 2x$$

$$\Rightarrow \quad 2\cos 2x f^{(1)} - 2\sin 2x g^{(1)} = \operatorname{cosec} 2x \qquad \text{(iv)}$$

Solving Eqs. (iii) and (iv), we get

$$f^{(1)} = \frac{1}{2}\cot 2x \text{ and } g^{(1)} = \frac{1}{2}.$$

Integrating,

$$f = \frac{1}{4}\log(\sin 2x) + c_1 \text{ and } g = \frac{x}{2} + c_2$$

Therefore, Eq. (ii) becomes

$$y = c_1\sin 2x + c_2\cos 2x + \frac{1}{4}\sin 2x\log(\sin 2x) + \frac{1}{2}x\cos 2x.$$

EXAMPLE 7.19 Solve $x^2y^{(2)} + xy^{(1)} - y = x^2e^x$.

Solution Rewrite given equation as

$$y^{(2)} + \frac{1}{x}y^{(1)} - \frac{1}{x^2}y = e^x \qquad \text{(i)}$$

Here,
$$U = x \text{ and } V = \frac{1}{x}.$$

Let
$$y = fx + g\frac{1}{x} \qquad \text{(ii)}$$

be the solution. Then

$$y^{(1)} = f^{(1)}x + f + g^{(1)}\frac{1}{x} - g\frac{1}{x^2}$$

Take f and g such that

$$f^{(1)}x + g^{(1)}\frac{1}{x} = 0 \qquad \text{(iii)}$$

Then
$$y^{(1)} = f - g\frac{1}{x^2}$$

and
$$y^{(2)} = f^{(1)} - g^{(1)}\frac{1}{x^2} + g\frac{2}{x^3}$$

Substitute the values of y, $y^{(1)}$ and $y^{(2)}$ in Eq. (i). We get

$$\left(f^{(1)} - g^{(1)}\frac{1}{x^2} + g\frac{2}{x^3}\right) + \frac{1}{x}\left(f - g\frac{1}{x^2}\right) - \frac{1}{x^2}\left(fx + g\frac{1}{x}\right) = e^x$$

$$\Rightarrow \quad f^{(1)} - g^{(1)}\frac{1}{x^2} = e^x \qquad \text{(iv)}$$

Solving Eqs. (iii) and (iv) gives

$$f^{(1)} = \frac{1}{2}e^x \text{ and } g^{(1)} = -\frac{1}{2}x^2e^x$$

Integrating, $$f = \frac{1}{2}e^x + c_1 \text{ and } g = -\frac{1}{2}e^x(x^2 - 2x + 2) + c_2$$

Substituting in Eq. (ii) gives

$$y = c_1x + \frac{c_2}{x} + e^x - \frac{1}{x}e^x$$

which is the required solution of the given equation.

EXAMPLE 7.20 Solve $x^2y^{(2)} - 2x(1+x)y^{(1)} + 2(1+x)y = x^3$.
The integrals in the C.F. are $y = x$ and $y = xe^{2x}$.

Solution Given

$$x^2y^{(2)} - 2x(1+x)y^{(1)} + 2(1+x)y = x^3 \tag{i}$$

Here $U = x, V = xe^{2x}$.

Let $$y = fx + gxe^{2x} \tag{ii}$$

be the solution.

Then $$y^{(1)} = f + f^{(1)}x + g^{(1)}xe^{2x} + g(2x+1)e^{2x}$$

Take f and g such that $f^{(1)}x + g^{(1)}xe^{2x} = 0$

$\Rightarrow$ $$f^{(1)} + g^{(1)}e^{2x} = 0 \tag{iii}$$

Then $$y^{(1)} = f + g(2x+1)e^{2x}$$

and $$y^{(2)} = f^{(1)} + g^{(1)}(2x+1)e^{2x} + 4g(2x+1)e^{2x}$$

Substituting y, $y^{(1)}$ and $y^{(2)}$ in Eq. (i) and by simplifying, we get

$$f^{(1)} + g^{(1)}(2x+1)e^{2x} = x \tag{iv}$$

Solving (iii) and (iv) gives

$$f^{(1)} = -\frac{1}{2} \text{ and } g^{(1)} = \frac{1}{2}e^{-2x}$$

Integrating,

$$f = -\frac{1}{2}x + c_1 \text{ and } g = -\frac{1}{4}e^{-2x} + c_2$$

Substituting in Eq. (ii) gives

$$y = c_1x + c_2xe^{2x} - \frac{x^2}{2} - \frac{1}{4}x.$$

EXAMPLE 7.21 Solve $y^{(2)} + 4y = \sec 2x$.

Solution Given

$$y^{(2)} + 4y = \sec 2x \tag{i}$$

Here $U = \cos 2x, V = \sin 2x$

Let $y = f\cos 2x + g\sin 2x$ (ii)

be the solution.

Then $y^{(1)} = f^{(1)} \cos 2x - 2f \sin 2x + g^{(1)} \sin 2x + 2g\cos 2x$

Take f and g such that

$$f^{(1)} \cos 2x + g^{(1)} \sin 2x = 0 \tag{iii}$$

Then $y^{(1)} = 2g\cos 2x - 2f \sin 2x$

and $y^{(2)} = 2\left[g^{(1)} \cos 2x - f^{(1)} \sin 2x\right] - 4[g \sin 2x + f \cos 2x]$

Substituting y, $y^{(1)}$ and $y^{(2)}$ in Eq. (i) and by simplifying, we get

$$g^{(1)}(2\cos 2x) - f^{(1)}(2\sin 2x) = \sec 2x \tag{iv}$$

Solving (iii) and (iv) gives

$$f^{(1)} = -\frac{1}{2}\tan 2x \text{ and } g^{(1)} = \frac{1}{2}$$

Integrating

$$f = \frac{1}{4}\log(\cos 2x) + c_1 \text{ and } g = \frac{1}{2}x + c_2$$

Therefore solution of Eq. (ii) is

$$y = c_1 \cos 2x + c_2 \sin 2x + \frac{\cos 2x}{4}\log(\cos 2x) + \frac{x}{2}\sin 2x$$

7.2.6 Method of Variation of Parameters for Linear Equations of Order Higher than Two

In Section 7.2.5, we have discussed to solve the linear equations of order two. In this section we extend the method to linear equations of order higher than two. Here, the solution is obtained by changing the arbitrary constants of the C.F. So, this method is known as method of variation of parameters. Let us discuss this method for an equation of order three.

Consider

$$y^{(3)} + Py^{(2)} + Qy^{(1)} + Ry = S \tag{7.22}$$

Let $y = U$, $y = V$ and $y = W$ be linearly independent solutions of the corresponding homogeneous equation.

$$y^{(3)} + Py^{(2)} + Qy^{(1)} + Ry = 0 \tag{7.23}$$

Let $$y = fU + gV + hW \tag{7.24}$$

be the solution of Eq. (7.22), where f, g and h are functions of x to be determined.

Then $$y^{(1)} = \left(f^{(1)}U + g^{(1)}V + h^{(1)}W\right) + \left(fU^{(1)} + gV^{(1)} + hW^{(1)}\right)$$

Let $$f^{(1)}U + g^{(1)}V + h^{(1)}W = 0 \tag{7.25}$$

Then $$y^{(1)} = fU^{(1)} + gV^{(1)} + hW^{(1)}$$

and $$y^{(2)} = f^{(1)}U^{(1)} + g^{(1)}V^{(1)} + h^{(1)}W^{(1)} + fU^{(2)} + gV^{(2)} + hW^{(2)}$$

As we need to determine three unknown functions, we assume one more condition. Let it be

$$f^{(1)}U^{(1)} + g^{(1)}V^{(1)} + h^{(1)}W^{(1)} = 0 \tag{7.26}$$

Then $$y^{(2)} = fU^{(2)} + gV^{(2)} + hW^{(2)}$$

$\therefore$ $$y^{(3)} = f^{(1)}U^{(2)} + g^{(1)}V^{(2)} + h^{(1)}W^{(2)} + fU^{(3)} + gV^{(3)} + hW^{(3)}$$

Substituting values of y, $y^{(1)}$, $y^{(2)}$ and $y^{(3)}$ in Eq. (7.22), we get

$$f^{(1)}U^{(2)} + g^{(1)}V^{(2)} + h^{(1)}W^{(2)} = S \tag{7.27}$$

Solve Eqs. (7.25)–(7.27) for $f^{(1)}$, $g^{(1)}$ and $h^{(1)}$. On integrating, we get three functions f, g and h. Hence, the solution of the given equation can be computed.

Note: Since U, V and W are linearly independent, the determinant of coefficient of three equations is non-zero.

EXAMPLE 7.22 Solve $y^{(3)} - 6y^{(2)} + 11y^{(1)} - 6 = e^x$.

Solution Given

$$y^{(3)} - 6y^{(2)} + 11y^{(1)} - 6 = e^x \tag{i}$$

The corresponding homogeneous equation is

$$y^{(3)} - 6y^{(2)} + 11y^{(1)} - 6 = 0$$

or $$(D^3 - 6D^2 + 11D - 6)y = 0$$

or $$(D-1)(D-2)(D-3)y = 0$$

$\therefore$ $$U = e^x, V = e^{2x}, W = e^{3x}$$

Let $$y = fe^x + ge^{2x} + he^{3x} \tag{ii}$$

be the solution.

Then $\qquad y^{(1)} = f^{(1)}e^x + g^{(1)}e^{2x} + h^{(1)}e^{3x} + fe^x + 2ge^{2x} + 3he^{3x}$

Let $\qquad f^{(1)}e^x + g^{(1)}e^{2x} + h^{(1)}e^{3x} = 0 \qquad$ (iii)

Then $\qquad y^{(1)} = fe^x + 2ge^{2x} + 3he^{3x}$

$\therefore \qquad y^{(2)} = f^{(1)}e^x + 2g^{(1)}e^{2x} + 3h^{(1)}e^{3x} + fe^x + 4ge^{2x} + 9he^{3x}$

Let $\qquad f^{(1)}e^x + 2g^{(1)}e^{2x} + 3h^{(1)}e^{3x} = 0 \qquad$ (iv)

Then $\qquad y^{(2)} = fe^x + 4ge^{2x} + 9he^{3x}$

$\therefore \qquad y^{(3)} = f^{(1)}e^x + 4g^{(1)}e^{2x} + 9h^{(1)}e^{3x} + fe^x + 8ge^{2x} + 27he^{3x}$

Substituting values of y, $y^{(1)}$, $y^{(2)}$ and $y^{(3)}$ in Eq. (i), we get

$$f^{(1)}e^x + 4g^{(1)}e^{2x} + 9h^{(1)}e^{3x} = e^{2x} \qquad \text{(v)}$$

Solving Eqs. (iii)–(v), we get

$$\frac{f^{(1)}}{e^{5x}} = \frac{g^{(1)}}{-2e^{4x}} = \frac{h^{(1)}}{e^{3x}} = a \text{ (say)}$$

$$\Rightarrow \qquad f^{(1)} = ae^{5x},\ g^{(1)} = -2ae^{4x},\ h^{(1)} = ae^{3x}$$

Substituting these in Eq. (v) gives $a = \frac{1}{2}e^{-4x}$

$$\therefore \qquad f^{(1)} = \frac{1}{2}e^x,\ g^{(1)} = -1,\ h^{(1)} = -\frac{1}{2}e^{-x}$$

Integrating,

$$f = \frac{1}{2}e^x + c_1,\ g = -x + c_2,\ h = -\frac{1}{2}e^{-x} + c_3$$

$\therefore \quad y = c_1e^x + c_2e^{2x} + c_3e^{3x} - xe^{2x} \quad$ [constants are rearranged]

7.3 METHOD OF UNDETERMINED COEFFICIENTS

This method is applicable to compute P.I. when R.H.S. of the equation consists of the sum of terms each of which possesses a finite number of essentially different derivatives.

We discuss this method in following example:

EXAMPLE 7.23 Solve $y^{(2)} + y^{(1)} - 2y = -2e^{-x} - 5\cos x$.

Solution Given

$$y^{(2)} + y^{(1)} - 2y = -2e^{-x} - 5\cos x \qquad \text{(i)}$$

Readers can check that

$$\text{C.F.} = c_1e^{-2x} + c_2e^x$$

R.H.S. contains the terms e^{-x} and $\cos x$. Their possible derivatives are e^{-x}, $\cos x$, $\sin x$ except for constant multipliers.

Let a, b, c be constants so that $y = \text{P.I.} = ae^{-x} + b\cos x + c\sin x$

Now $$y^{(1)} = -ae^{-x} - b\sin x + c\cos x$$

and $$y^{(2)} = ae^{-x} - b\cos x - c\sin x$$

Substituting y, $y^{(1)}$ and $y^{(2)}$ in Eq. (i) and by simplifying, we get

$$-2ae^{-x} + (c-3b)\cos x - (b+3c)\sin x = -2e^{-x} - 5\cos x$$

Comparing coefficients results to $a = 1$, $c - 3b = -5$ and $b + 3c = 0$

$$\therefore \quad a = 1,\ b = \frac{3}{2},\ c = -\frac{1}{2}$$

$$\therefore \quad \text{P.I.} = e^{-x} + \frac{3}{2}\cos x - \frac{1}{2}\sin x$$

$\therefore$ The solution of given differential Eq. (i) is

$$y = c_1 e^{-2x} + c_2 e^{x} + e^{-x} + \frac{3}{2}\cos x - \frac{1}{2}\sin x$$

EXERCISES

Solve the following differential equations:

1. $xy^{(2)} - (2x-1)y^{(1)} + (x-1)\,y = 0.$ **[Ans.** $y = (c_1 + c_2 \log x)\,e^x$**]**

2. $xy^{(2)} - 2x(x+1)\,y^{(1)} + 2(x+1)\,y = x^2.$ $\left[\textbf{Ans.}\ y = \frac{x}{2}(c_1 e^{2x} + c_2 - x)\right]$

3. $\sin^2 x y^{(2)} - 2y = 0.$ **[Ans.** $y = c_1\,(1 - x\cot x) + c_2 \cot x$**]**

4. $x^2(1+x)\,y^{(2)} - x(x^2+4x+2)\,y^{(1)} + (x^2+4x+2)\,y = -x^3\,(x+1)^2.$

$\left[\textbf{Ans.}\ y = c_1 x^2 e^x + c_2 x + \frac{x^2}{2}(x+2)\right]$

5. $(3-x)\,y^{(2)} - (9-4x)\,y^{(1)} + (6-3x)\,y = 0.$

[Ans. $y = c_1 e^x + c_2 e^{3x}\,(4x^3 - 42x^2 + 150x - 183)$**]**

6. $y^{(2)} + (1 - \cot x)\,y^{(1)} - y\cot x = \sin^2 x.$

$\left[\textbf{Ans.}\ y = c_1\,(\sin x - \cos x) + c_2 e^{-x} - \frac{1}{10}(\sin 2x - 2\cos 2x)\right]$

7. $xy^{(2)} - (2x-1)\,y^{(1)} + (x-1)\,y = 0.$ **[Ans.** $y = (c_1 + c_2 \log x)\,e^x$**]**

8. $x^2 y^{(2)} - 2x(x+1)\, y^{(1)} + (x^2 + 2x + 2)\, y = 0.$ [**Ans.** $y = (c_1 x^2 + c_2 x)\, e^x$]

9. $y^{(2)} - 2\tan x y^{(1)} + 5y = e^x \sec x.$

$$\left[\textbf{Ans. } y = (c_1 \cos\sqrt{6}x + c_2 \sin\sqrt{6}x)\sec x + \frac{1}{7} e^x \sec x\right]$$

10. $x\dfrac{d}{dx}\left(x\dfrac{dy}{dx} - y\right) - 2xy^{(1)} + (x^2 + 2)\, y = 0.$ [**Ans.** $y = x\,(c_1 \cos x + c_2 \sin x)$]

11. $y^{(2)} + 2n \cot nx y^{(1)} + (m^2 - n^2)\, y = 0.$ [**Ans.** $y \sin nx = c_1 \cos mx + c_2 \sin mx$]

12. $(y^{(2)} + y)\cot x + 2(y^{(1)} + y\tan x) = 0.$ [**Ans.** $y = (c_1 + c_2 x)\cos x$]

13. $x^2 y^{(2)} - 2xy^{(1)} + 2(x^2 + 1)\, y = 0.$ [**Ans.** $y = x(c_1 \cos\sqrt{2}x + c_2 \sin\sqrt{2}x)$]

14. $y^{(2)} + \tan x y^{(1)} + y\cos^2 x = 0.$ [**Ans.** $y = c_1 \cos(\sin x) + c_2 \sin(\sin x)$]

15. $x^4 y^{(2)} + 2x^3 y^{(1)} + 4y = 0.$ $\left[\textbf{Ans. } y = c_1 \cos\left(\dfrac{2}{x}\right) + c_2 \sin\left(\dfrac{2}{x}\right)\right]$

16. $(x^3 - x)\, y^{(2)} + y^{(1)} + 8x^3 y = 0.$ $\left[\textbf{Ans. } y = c_1 \sin(2\sqrt{x^2 - 1}) + c_2 \cos(2\sqrt{x^2 - 1})\right]$

17. $xy^{(2)} - y^{(1)} + 4x^3 y = 8x^3 \sin x^2.$ [**Ans.** $y = c_1 \cos x^2 + c_2 \sin x^2 - x^2 \cos x^2$]

18. $xy^{(2)} - (1+x)\, y^{(1)} + y = x^2.$ [**Ans.** $y = c_2 e^x - x^2 - (2 + c_1)(x+1)$]

19. $(e^x - 2)\, y^{(2)} + e^x y^{(1)} + 2y = 0.$ [**Ans.** $y = c_1 + 2c_1 e^{-x} \log(e^x - 2) + c_2 e^{-x}$]

20. $y^{(2)} - 2xy^{(1)} + y = xe^x \log x,\ x > 0.$ $\left[\textbf{Ans. } y = c_1 e^x + c_2 x e^x - \dfrac{5}{36} x^3 e^x + \dfrac{1}{6} x^3 e^x \log x\right]$

21. $y^{(2)} + y = \dfrac{1}{1 + \sin x}.$

$$\left[\textbf{Ans. } y = c_1 \cos x + c_2 \sin x + \sin x \log(1 + \sin x) - x\cos x - \frac{\cos^2 x}{1 + \sin x}\right]$$

22. $(1 + e^x)\, y^{(2)} - (1 + e^x)\, y = 2e^x.$

[**Ans.** $y = c_1 e^x + c_2 e^{-x} - (e^x - e^{-x}) \log(e^x + 1) + xe^x - 1$]

23. $(x+2)\, y^{(2)} - (2x+5)\, y^{(1)} + 2y = (x+1)\, e^x.$ [**Ans.** $y = c_1 e^{2x} + c_2\,(2x+5) - e^x$]

24. $(x-1)\, y^{(2)} + (x-1)^2 = xy^{(1)} - y.$ [**Ans.** $y = c_1 e^x + c_2 x + x^2 + x + 1$]

25. $y^{(2)} + 2y^{(1)} + y = x - e^x.$ $\left[\textbf{Ans. } y = (c_1 + c_2 x)\, e^{-x} + x - 2 - \dfrac{e^x}{4}\right]$

26. $y^{(2)} - 4y^{(1)} + 4y = e^x \sin x.$ $\left[\textbf{Ans. } y = (c_1 + c_2 x) e^{2x} + \dfrac{1}{2} e^x \cos x\right]$

27. $y^{(2)} - 2y^{(1)} + y = x^2 e^x$. $\left[\textbf{Ans. } y = (c_1 + c_2 x)\, e^x + \dfrac{x^4 e^x}{12}\right]$

28. $y^{(2)} - x^2 y^{(1)} + xy = x$. $\left[\textbf{Ans. } y = c_1 x + c_2 x \int x^{-2} e^{x^3/3} dx + 1\right]$

29. $y^{(2)} - \cot x y^{(1)} - (1 - \cot x) y = e^x \sin x$.

$\left[\textbf{Ans. } y = c_1 e^x + c_2 e^{-x} (\cos x + 2 \sin x) - \dfrac{1}{2} e^x \cos x\right]$

30. $xy^{(2)} + (1 - x) y^{(1)} - y = e^x$. $\left[\textbf{Ans. } y = e^x \log x + c_1 e^x \int x^{-1} e^{-x} dx + c_2 e^x\right]$

31. $x^2 y^{(2)} - 2x(1 + x) y^{(1)} + 2(1 + x) y = x^2$. $\left[\textbf{Ans. } 2y = x(c_1 e^{2x} + c_2 - x)\right]$

32. $x^2 y^{(2)} + xy^{(1)} - y = x^2 e^x$. $\left[\textbf{Ans. } y = c_1 x + \dfrac{c_2}{x} + x \int e^x \dfrac{(x^2 - 2x + 2)}{x^3} dx\right]$

33. $x^3 y^{(2)} + 2xy^{(1)} - 2y = 1$. $\left[\textbf{Ans. } y = c_1 x + c_2 x e^{2/x} - \dfrac{1}{2}\right]$

34. $(x + 2) y^{(2)} - (4x + 9) y^{(1)} + (3x + 7) y = 0$. $\left[\textbf{Ans. } y = c_1 e^x + c_2 e^{3x} (2x + 3)\right]$

35. $(x - 1) y^{(2)} - xy^{(1)} + y = (x - 1)^2$. $\left[\textbf{Ans. } y = c_1 e^x + c_2 x - (x^2 + x + 1)\right]$

36. $x(1 + x^2) y^{(2)} - 2x^2 y^{(1)} + 2xy = 1 - x^2$. $\left[\textbf{Ans. } y = c_1 (x^2 - 1) + c_2 x + x \log x\right]$

37. $y^{(2)} - 2 \tan x y^{(1)} + 5y = e^x \sec x$. $\left[\textbf{Ans. } y = (c_1 \cos \sqrt{6} x + c_2 \sin \sqrt{6} x) \sec x + \dfrac{1}{7} e^x \sec x\right]$

38. $y^{(2)} - 4xy^{(1)} + (4x^2 - 1) y + 3e^{x^2} = 0$. $\left[\textbf{Ans. } y = e^{x^2} (c_1 \cos x + c_2 \sin x + \dfrac{3}{2} x \cos x\right]$

39. $y^{(2)} - 2bxy^{(1)} + b^2 x^2 y = x$. $\left[\textbf{Ans. } y = c e^{bx^2/2} \sin(x\sqrt{b} + a) + e^{bx^2/2} \left\{\dfrac{1}{D^2 + b} x e^{-bx^2/2}\right\}\right]$

40. $y^{(2)} + 4xy^{(1)} + 4x^2 y = 2x^3 e^{-x^2}$. $\left[\textbf{Ans. } y = (c_1 e^{\sqrt{2}x} + c_2 e^{-\sqrt{2}x} - x^3 - 3x) e^{-x^2}\right]$

41. $x^2 y^{(2)} - 2x(x + 1) y^{(1)} + (x^2 + 2x + 2) y = 0$. $\left[\textbf{Ans. } y = e^x (c_1 x^2 + c_2 x)\right]$

42. $4x^2 y^{(2)} + 4x^5 y^{(1)} + (x^8 + 6x^4 + 4) y = 0$. $\left[\textbf{Ans. } y = c_1 e^{-x^4/8} \sqrt{x} \sin\left(\dfrac{\sqrt{3}}{2} \log x + c_2\right)\right]$

43. $4x^2 y^{(2)} - 4x^{3/2} y^{(1)} + (x + \sqrt{x} - 8) y = 0$. $\left[\textbf{Ans. } y = \left(c_1 x^2 + \dfrac{c_2}{x}\right) e^{\sqrt{x}}\right]$

44. $xy^{(2)} - y^{(1)} - 4x^3 y = x^5$. $\left[\textbf{Ans. } y = c_1 e^{x^2} + c_2 e^{-x^2} - \dfrac{x^2}{4}\right]$

45. $xy^{(2)} - y^{(1)} + 4x^3y = 8x^3 \sin x^2.$ $\left[\textbf{Ans. } y = c_1 \cos x^2 + c_2 \sin x^2 - x^2 \cos x^2\right]$

46. $x^4y^{(2)} + 2x^3y^{(1)} + 9y = 0.$ $\left[\textbf{Ans. } y = c_1 \cos\left(\frac{3}{x}\right) + c_2 \sin\left(\frac{3}{x}\right)\right]$

47. $x(x^2 - 1)y^{(2)} + y^{(1)} + 64x^3y = 0.$ $\left[\textbf{Ans. } y = c_1 \cos\left(4\sqrt{x^2-1}\right) + c_2 \sin\left(4\sqrt{x^2-1}\right)\right]$

48. $(1 + x)^2 y^{(2)} + (1 + x)\, y^{(1)} + y = 25 \cos \log(1 + x).$

$\left[\textbf{Ans. } y = c_1 \cos\{\log(1 + x)\} + c_2 \sin\{\log(1 + x)\} + 5 \log(1 + x) \sin\{\log(1 + x)\}\right]$

49. $(1 + x^2)^2 y^{(2)} + 2x(1 + x^2)\, y^{(1)} + y = 0.$ $\left[\textbf{Ans. } y = c_1 \cos(2 \tan^{-1} x) + c_2 \sin(2 \tan^{-1} x)\right]$

50. $\cos x y^{(2)} + \sin x y^{(1)} - 2\cos^3 x y = 2\cos^5 x.$ $\left[\textbf{Ans. } y = c_1 e^{\sqrt{2}\sin x} + c_2 e^{-\sqrt{2}\sin x} + \sin^2 x\right]$

51. $3x^2y^{(2)} + 2(1 - 3x^2)\, y^{(1)} - 4y = 0.$ $\left[\textbf{Ans. } y = c_1 e^{2x} + c_2 e^{2x} \int e^{\frac{2}{3x} - 2x}\, dx\right]$

52. $y^{(2)} + y = \operatorname{cosec} x.$ $\left[\textbf{Ans. } y = (c_1 - x)\cos x + (c_2 + \log \sin x)\sin x\right]$

53. $y^{(2)} - y = \dfrac{2e^x}{1 + e^x}.$ $\left[\textbf{Ans. } y = c_1 e^x + c_2 e^{-x} - (e^x - e^{-x}) \log(e^x + 1) + xe^x - 1\right]$

54. $y^{(2)} + 9y = \sec^2 3x.$ $\left[\textbf{Ans. } y = c_1 \cos 3x + c_2 \sin 3x + \frac{\sin 3x}{9}(\log(\sec 3x + \tan 3x)) - \frac{1}{4}\right]$

55. $(x - 1)y^{(2)} - xy^{(1)} + y + (x - 1)^2 = 0.$ $\left[\textbf{Ans. } y = c_1 e^x + c_2 x + x^2 + x + 1\right]$

56. $y^{(2)} + 2y^{(1)} + 2y = 1 + x^2.$ $\left[\textbf{Ans. } y = e^{-x}(c_1 \cos x + c_2 \sin x) + 1 - x + \frac{x^2}{2}\right]$

57. $y^{(2)} - 3y^{(1)} + 2y = e^x + 2e^{2x} + \sin x.$

$\left[\textbf{Ans. } y = c_1 e^x + c_2 e^{2x} - xe^x + 2xe^{2x} + \frac{1}{10}(3\cos x + \sin x)\right]$

58. $y^{(2)} - 4y^{(1)} + 4y = e^x \sin x.$ $\left[\textbf{Ans. } y = (c_1 + c_2 x)e^{2x} + \frac{1}{2}e^x \cos x\right]$

59. $4y^{(2)} + 4y^{(1)} + y = e^x + 2\cos 2x.$

$\left[\textbf{Ans. } y = c_1 e^{-x/2} + c_2 xe^{-x/2} + \frac{e^x}{9} - \frac{30}{289}\cos 2x + \frac{16}{289}\sin 2x\right]$

60. $y^{(2)} - 2y^{(1)} + y = x^2e^x.$ $\left[\textbf{Ans. } y = (c_1 + c_2 x)e^x + \frac{x^4}{12}e^x\right]$

CHAPTER 8

Simultaneous Linear Differential Equations

8.1 INTRODUCTION

Sometimes, in mathematical physics, we come across differential equations containing one independent variable and two or more dependent variables. Such equations are called **simultaneous differential equations**. The method of solution of such equation is by elimination or substitution, analogous to that of solving two linear simultaneous equations in algebra. The number of equations is same as the number of dependent variables.

8.2 SIMULTANEOUS LINEAR DIFFERENTIAL EQUATIONS WITH CONSTANT COEFFICIENTS

We will illustrate by the following examples:

(a) Method of elimination

EXAMPLE 8.1 Solve the following simultaneous equations:

$$\frac{dx}{dt} = 7x - y, \frac{dy}{dt} = 2x + 5y.$$

Solution Put $D = \frac{d}{dt}$. The given equations in symbolic form are

$$(D - 7)\, x + y = 0 \tag{i}$$

$$-2x + (D - 5)\, y = 0 \tag{ii}$$

To eliminate y, operating on both sides of Eq. (i) by $(D - 5)$ and then subtracting, we get

$$(D - 5)\,(D - 7)\, x + 2x = 0$$

$$\Rightarrow \quad [(D - 5)\,(D - 7) + 2]\, x = 0$$

$$\Rightarrow \quad (D^2 - 12D + 37)\, x = 0$$

Its auxiliary equation is

$$m^2 - 12m + 37 = 0$$

$$\Rightarrow \qquad m = 6 \pm i$$

$$\therefore \qquad CF = e^{6t}(c_1 \cos t + c_2 \sin t)$$

$$\Rightarrow \qquad x = e^{6t}(c_1 \cos t + c_2 \sin t)$$

From Eq. (i),

$$y = 7x - \frac{dx}{dt}$$

$$= 7e^{6t}(c_1 \cos t + c_2 \sin t) - 6e^{6t}(c_1 \cos t + c_2 \sin t) - e^{6t}(-c_1 \sin t + c_2 \cos t)$$

$$\therefore \qquad y = e^{6t}[(c_1 - c_2) \cos t + (c_1 + c_2) \sin t]$$

Hence,

$$x = e^{6t}(c_1 \cos t + c_2 \sin t)$$

$$y = e^{6t}[(c_1 - c_2) \cos t + (c_1 + c_2) \sin t]$$

EXAMPLE 8.2 Solve following simultaneous equations:

$$\frac{dx}{dt} + 2\frac{dy}{dt} - 2x + 2y = 3e^t$$

$$3\frac{dx}{dt} + \frac{dy}{dt} + 2x + y = 4e^{2t}.$$

Solution Given

$$\frac{dx}{dt} + 2\frac{dy}{dt} - 2x + 2y = 3e^t \qquad \text{(i)}$$

$$3\frac{dx}{dt} + \frac{dy}{dt} + 2x + y = 4e^{2t} \qquad \text{(ii)}$$

Multiplying Eq. (ii) by 2 and then subtracting Eq. (i) from Eq. (ii), we get

$$5\frac{dx}{dt} + 6x = 8e^{2t} - 3e^t$$

$$\Rightarrow \qquad \frac{dx}{dt} + \frac{6}{5}x = \frac{8}{5}e^{2t} - \frac{3}{5}e^t$$

which is linear differential equation in x and t. Its

$$IF = e^{\int \frac{6}{5} dt} = e^{\frac{6t}{5}}$$

Therefore, the solution is

$$xe^{\frac{6t}{5}} = \int \left(\frac{8}{5}e^{2t} - \frac{3}{5}e^t \right) e^{\frac{6t}{5}} \, dt + c_1$$

$$\Rightarrow \qquad x = c_1 e^{-\frac{6t}{5}} + \frac{1}{2}e^{2t} - \frac{3}{11}e^t$$

Now, multiplying Eq. (i) by 3 and then subtracting Eq. (ii) from Eq. (i), we get

$$5\frac{dy}{dt}+5y=8x+9e^{t}-4e^{2t}$$

$$=8c_1e^{-\frac{6t}{5}}+4e^{2t}-\frac{24}{11}e^{t}+9e^{t}-4e^{2t}$$

$$=8c_1e^{-\frac{6t}{5}}+\frac{75}{11}e^{t}$$

$\therefore$
$$\frac{dy}{dt}+y=\frac{8c_1}{5}e^{-\frac{6t}{5}}+\frac{15}{11}e^{t}$$

which is also linear differential equation in y and t. Its

$$\text{IF}=e^{\int 1dt}=e^{t}$$

Therefore, the solution is

$$ye^{t}=\int\left(\frac{8c_1}{5}e^{-\frac{6t}{5}}+\frac{15}{11}e^{t}\right)e^{t}\,dt+c_2$$

$$=-8c_1e^{\frac{t}{5}}+\frac{15}{22}e^{2t}+c_2$$

$$y=c_2e^{-t}+\frac{15}{22}e^{t}-8c_1e^{-\frac{4t}{5}}$$

Hence,
$$x=c_1e^{-\frac{6t}{5}}+\frac{1}{2}e^{2t}-\frac{3}{11}e^{t}$$

$$y=c_2e^{-t}+\frac{15}{22}e^{t}-8c_1e^{-\frac{4t}{5}}.$$

EXAMPLE 8.3 Solve $\frac{dx}{dt}=ax+by+c$ and $\frac{dy}{dt}=a'x+b'y+c'$.

Solution Writing D for $\frac{d}{dt}$, the given equations can be written as

$$(D-a)x-by=c \tag{i}$$

$$-a'x+(D-b')y=c' \tag{ii}$$

Eliminating y, we get

$$[(D-b')(D-a)-a'b]x=bc'-b'c$$

$\Rightarrow$
$$[D^2-(a+b')D+ab'-a'b]x=bc'-b'c$$

Its auxiliary equation is

$$m^2 - (a + b')m + ab' - a'b = 0$$

$\therefore$
$$m = \frac{(a + b') \pm \sqrt{(a - b')^2 + 4a'b}}{2} = \lambda, \mu \text{ (say)}$$

where
$$\lambda = \frac{(a + b') \pm \sqrt{(a - b')^2 + 4a'b}}{2}$$

and
$$\mu = \frac{(a + b') \pm \sqrt{(a - b')^2 + 4a'b}}{2}$$

Therefore,
$$\text{C.F.} = c_1 e^{\lambda t} + c_2 e^{\mu t}$$

$$\text{PI} = (bc' - b'c)\frac{1}{(D^2 - (a + b')D + ab' - a'b)} \cdot 1$$

$$= (bc' - b'c)\frac{1}{(ab' - a'b)}, \text{ if } ab' - a'b \neq 0$$

Therefore,
$$x = c_1 e^{\lambda t} + c_2 e^{\mu t} + \frac{bc' - b'c}{ab' - a'b}$$

and
$$\frac{dx}{dt} = c_1 \lambda e^{\lambda t} + c_2 \mu e^{\mu t}$$

From Eq. (ii),
$$y = -\frac{1}{b}\left[c + ax - \frac{dx}{dt}\right] = -\frac{1}{b}[c + ax - c_1 \lambda e^{\lambda t} - c_2 \mu e^{\mu t}]$$

$\therefore$
$$y = -\frac{1}{b}\left[c + e^{\lambda t}c_1(a - \lambda) + e^{\mu t}c_2(a - \mu) + \frac{a(bc' - b'c)}{ab' - a'b}\right]$$

EXAMPLE 8.4 Solve $\dfrac{d^2x}{dt^2} - 3x - 4y = 0$ and $\dfrac{d^2y}{dt^2} + x + y = 0$.

Solution The given equations are equivalent to

$$(D^2 - 3)x - 4y = 0 \tag{i}$$

$$x + (D^2 + 1)y = 0 \tag{ii}$$

Eliminating y gives $(D^2 - 1)^2 x = 0$

Therefore,
$$x = (c_1 + c_2 t)e^t + (c_3 + c_4 t)e^{-t}$$

Then
$$\frac{d^2x}{dt^2} = \{c_1 + c_2(t + 2)\}e^t + \{c_3 + c_4(t - 2)\}e^{-t}$$

So
$$y = \frac{1}{4}[e^t\{-2c_1 - 2c_2(t-1)\} + e^{-t}\{-2c_2 - 2c_4(t+1)\}]$$

$$= \frac{1}{2}[(c_2 - c_1 - c_2 t)e^t - (c_3 + c_4 + c_4 t)e^{-t}]$$

EXAMPLE 8.5 **Currents i_1 and i_2 in two coupled circuit are given by**

$$L\frac{di_1}{dt} + Ri_1 + R(i_1 - i_2) = E$$

$$L\frac{di_2}{dt} + Ri_2 - R(i_1 - i_2) = 0$$

where L, R, E are constants find i_1 and i_2 in terms of t given that $i_1 = i_2 = 0$ at $t = 0$.

Solution Writing D for $\frac{d}{dt}$, the given equations become

$$(LD + 2R)\, i_1 - Ri_2 = E \quad \text{(i)}$$

$$(LD + 2R)\, i_2 - Ri_2 = 0 \quad \text{(ii)}$$

From Eq. (ii), we have

$$(LD + 2R)\, i_2 = Ri_1 \quad \text{(iii)}$$

To eliminate, i_2, operating on both sides of Eq. (i) by $(LD + 2R)$, we get

$$(LD + 2R)^2\, i_1 - R(LD + 2R)\, i_2 = 2RE$$

$$\Rightarrow \quad (LD + 2R)^2\, i_1 - R(Ri_1) = 2RE$$

$$\Rightarrow \quad [(LD + 2R)^2 - R^2]\, i_1 = 2RE$$

$$\Rightarrow \quad [L^2D^2 + 4RLD + 3R^2]\, i_1 = 2RE$$

Its auxiliary equation is

$$L^2m^2 + 4RLm + 3R^2 = 0$$

$$\Rightarrow \quad m = -\frac{R}{L}, \frac{-3R}{L}$$

$$\therefore \quad \text{CF} = c_1 e^{-\frac{Rt}{L}} + c_2 e^{-\frac{3Rt}{L}}$$

$$\text{PI} = \frac{1}{L^2D^2 + 4RLD + 3R^2} 2REe^{0t}$$

$$= 2RE \frac{1}{0 + 0 + 3R^2} e^{0t}$$

$$= \frac{2E}{3R}$$

Therefore, general solution for i_1 will be

$$i_1 = c_1 e^{-\frac{Rt}{L}} + c_2 e^{-\frac{3Rt}{L}} + \frac{2E}{3R} \quad \text{(iv)}$$

Now, $$(LD + 2R)\, i_1 = L\frac{di_1}{dt} + 2Ri_1$$

$$= L\left[-\frac{c_1 R}{L} e^{-\frac{Rt}{L}} - c_2 \frac{3R}{L} e^{-\frac{3Rt}{L}}\right] + 2R\left[c_1 e^{-\frac{Rt}{L}} + c_2 e^{-\frac{3Rt}{L}} + \frac{2E}{3R}\right]$$

$\therefore$ $$(LD + 2R)\, i_1 = c_1\, Re^{-\frac{Rt}{L}} - c_2\, Re^{-\frac{3Rt}{L}} + \frac{4E}{3} \quad \text{(v)}$$

From Eq. (i), we have

$$i_2 = \frac{1}{R}[(LD + 2R)\, i_1 - E] \quad \text{(vi)}$$

$$i_2 = c_1 e^{-\frac{Rt}{L}} - c_2 e^{-\frac{3Rt}{L}} + \frac{E}{3R} \quad \text{(vii)}$$

Initially, when $t = 0$, $i_1 = 0$ and $i_2 = 0$. Hence, from Eq. (iv) and Eq. (vii), we have

$$c_1 + c_2 + \frac{2E}{3R} = 0$$

and $$c_1 - c_2 + \frac{E}{3R} = 0$$

$\Rightarrow$ $$c_1 = -\frac{E}{2R} \text{ and } c_2 = -\frac{E}{6R}$$

Putting these values of c_1 and c_2 in i_1 and i_2, we obtain

$$i_1 = \frac{E}{R}\left[\frac{2}{3} - \frac{1}{2}e^{-\frac{Rt}{L}} - \frac{1}{6}e^{-\frac{3Rt}{L}}\right]$$

$$i_2 = \frac{E}{R}\left[\frac{1}{3} - \frac{1}{2}e^{-\frac{Rt}{L}} + \frac{1}{6}e^{-\frac{3Rt}{L}}\right]$$

(b) Method of differentiation: If two equations containing x, y, x_1 and y_1 are given, two more equations containing x, y and y_1 derivatives up to second order are obtained on differentiating the given equations w.r.t. 't'. From these four equations, an equation containing x (or y) and its derivatives only can be obtained by eliminating y (or x) and its derivatives. The obtained equation may be solvable for x (or y). Then y (or x) is obtained by substituting the value of x (or y) in only to the given equations and solution can be obtained.

EXAMPLE 8.6 Solve the following simultaneous equations:

$$\frac{dx}{dt} + y = \sin t, \frac{dy}{dt} + x = \cos t$$

given that when $t = 0$, $x = 1$ and $y = 0$.

Solution Writing D for $\frac{d}{dt}$, the given equations become

$$Dx + y = \sin t \tag{i}$$

$$Dy + x = \cos t \tag{ii}$$

Differentiating Eq. (i) w.r.t. t, we get

$$D^2x + Dy = \cos t \tag{iii}$$

Subtracting Eq. (ii) from Eq. (iii), we get

$$D^2x - x = 0$$

$$\Rightarrow \qquad (D^2 - 1)\, x = 0$$

Its auxiliary equation is

$$m^2 - 1 = 0$$

$$\Rightarrow \qquad m = \pm 1$$

CF is

$$x = c_1e^t + c_2e^{-t} \tag{iv}$$

and

$$\frac{dx}{dt} = c_1e^t - c_2e^{-t} \tag{v}$$

Now, substituting value of $\frac{dx}{dt}$ in Eq. (i), we get

$$y = \sin t - c_1e^t + c_2e^{-t} \tag{vi}$$

and

$$x = c_1e^t + c_2e^{-t}$$

Given at $t = 0$, $x = 1$ and $y = 0$

$$\therefore \qquad c_1 + c_2 = 1$$

and

$$c_2 - c_1 = 0$$

$$\Rightarrow \qquad c_1 = c_1 = \frac{1}{2}$$

Hence,

$$x = \frac{1}{2}(e^t + e^{-t})$$

and

$$y = \sin t - \frac{1}{2}(e^t - e^{-t})$$

$$\therefore \qquad x = \cosh t$$

$$y = \sin t - \sinh t.$$

EXAMPLE 8.7 Solve $x_1 + y_1 + 2x - y = 3(t^2 - e^{-t})$

$$2x_1 - y_1 - x - y = 3(2t - e^{-t}).$$

Solution Given

$$x_1 + y_1 + 2x - y = 3(t^2 - e^{-t}) \tag{i}$$

$$2x_1 - y_1 - x - y = 3(2t - e^{-t}) \tag{ii}$$

Differentiating Eqs. (i) and (ii) w.r.t. t, we get

$$x_2 + y_2 + 2x_1 - y_1 = 3(2t^2 - e^{-t}) \tag{iii}$$

and

$$2x_2 + y_2 + 2_1 - y_1 = 3(t^2 - e^{-t}) \tag{iv}$$

Adding Eqs. (iii) and (iv) gives

$$x_1 + 3x_2 = 6(1 + t + e^{-t}) \tag{v}$$

Subtracting Eqs. (iii) and (iv) gives

$$x_1 - 2y_1 - e^x = 3(2t - t_2) \tag{vi}$$

Subtracting Eqs. (v) and (vi),

$$3x_2 + 3x = 6 + 3t^2 + 6e^{-t}$$

$$\Rightarrow \quad (D^2 + 1)\, x = 2 + t^2 + 2e^{-t} \tag{vii}$$

Therefore, the auxiliary equation is

$$m^2 + 1 = 0$$

$$\Rightarrow \quad m = \pm i$$

$$\therefore \quad \text{CF} = c_1 \cos t + c_2 \sin t$$

and

$$\text{PI} = \frac{1}{D^2 + 1}(2 + t^2 + 2e^{-t})$$

$$= 2 + e^{-t} + (1 + D^2)^{-1}\, t^2$$

$$= 2 + e^{-t} + (1 - D^2 + \cdots)\, t^2$$

$$= 2 + e^{-t} + t^2 - 2$$

$$= t^2 + e^{-t}$$

$$\therefore \quad x = c_1 \cos t + c_2 \sin t + t^2 + e^{-t}$$

$$\therefore \quad x_1 = -c_1 \sin t + c_2 \cos t + 2t - e^{-t}$$

Now, from Eq. (vi),

$$2y_1 = x_1 - 3x - 6t + 3t^2$$

$$= (-c_1 \sin t + c_2 \cos t + 2t - e^{-t}) - 3(c_1 \cos t + c_2 \sin t + t^2 + e^{-t}) - 6t + 3t^2$$

$$\therefore \quad 2y_1 = (c_2 - 3c_1) \cos t - (3c_2 + c_1) \sin t - 4t - 4e^{-t}$$

Integrating, $\quad 2y = (c_2 - 3c_1) \sin t + (3c_2 + c_1) \cos t - 2t_2 + 4e^{-t}$

$$\Rightarrow \quad y = \frac{1}{2}(c_2 - 3c_1) \sin t + \frac{1}{2}(c_1 + 3c_2) \cos t - t^2 + 2e^{-t}.$$

EXAMPLE 8.8 Solve $\dfrac{d^2x}{dt^2}+n^2y=0$ and $\dfrac{d^2y}{dt^2}+n^2x=0.$

Solution Given equations can be written as

$$D^2x + n^2y = 0 \tag{i}$$

and

$$-n^2x + D^2y = 0 \tag{ii}$$

where $D=\dfrac{d}{dt}$. Eliminating y gives $(D^4 + n^4)x = 0$.

Then auxiliary equation is $m^4 + n^4 = 0$.

$$\Rightarrow \qquad m=(-1)^{1/4}n=n\left\{\cos\frac{2k\pi+\pi}{4}+i\sin\frac{2k\pi+\pi}{4}\right\},\ k=0,1,2,3$$

$$\Rightarrow \qquad m=n\left(\frac{1\pm i}{\sqrt{2}}\right),\ n\left(\frac{-1\pm i}{\sqrt{2}}\right)$$

$$\therefore \qquad x=e^{a}\{c_1\cos a+c_2\sin a\}+e^{-a}\{c_3\cos a+c_4\sin a\},\ a=\frac{nt}{\sqrt{2}}$$

Then,

$$\frac{d^2x}{dt^2}=\frac{n^2}{2}e^{a}\{2c_2\cos a-2c_1\sin a\}+\frac{n^2}{2}e^{-a}\{-2c_4\cos a+2c_3\sin a\}$$

From Eq. (ii),

$$y=-\frac{1}{n^2}x=-\frac{1}{2}e^{a}\{2c_2\cos a-2c_1\sin a\}-\frac{1}{2}e^{-a}\{-2c_4\cos a+2c_3\sin a\}.$$

8.3 SIMULTANEOUS LINEAR DIFFERENTIAL EQUATIONS WITH VARIABLE COEFFICIENTS

Given m-simultaneous equation, we want to obtain the relations between m dependent variables. When the equations of the system are linear they will be reduced in symmetrical form. The method of solution of such equations will be discussed in Section 8.4.

8.4 SOLUTION METHODOLOGY

8.4.1 In Symmetrical Form

We will discuss the solution procedure for three variables however, it can be generalized for m-variables.

Let the equation be in the form

$$\frac{dx}{P}=\frac{dy}{Q}=\frac{dz}{R} \tag{8.1}$$

where P, Q and R are functions of x, y and z.

By equating two out of three members in Eq. (8.1), it may be possible to get an equation in only two variables. The solution of this equation gives one of the relations in the complete integral of Eq. (8.1). Repeat the same method to get another relation with any other two members. The two relations so obtained together give the complete solution of Eq. (8.1).

Sometimes it is possible to find multipliers *l*, *m*, *n* and *L*, *M*, *N* such that at least one of the equations

$$\frac{dx}{P} = \frac{dy}{Q} = \frac{dz}{R} = \frac{ldx + mdy + ndz}{lP + mQ + nR} = \frac{Ldx + Mdy + Ndz}{LP + MQ + NR}$$

can be integrated.

Any combination of the methods stated above will help in obtaining two independent relations between the variables each containing an arbitrary constant.

EXAMPLE 8.9 Solve $\frac{dx}{yz} = \frac{dy}{zx} = \frac{dz}{xy}$.

Solution Consider first two ratios.

$$\therefore \quad \frac{dx}{y} = \frac{dy}{x}$$

$$\Rightarrow \quad xdx = ydy$$

$$\Rightarrow \quad \frac{1}{2}d(x^2 - y^2) = 0$$

Integrating,
$$x^2 - y^2 = c_1 \quad \text{(i)}$$

Now, consider last two ratios.

$$\therefore \quad \frac{dy}{z} = \frac{dz}{y}$$

$$\Rightarrow \quad ydy = zdz$$

$$\Rightarrow \quad \frac{1}{2}d(y^2 - z^2) = 0$$

Integrating,
$$y^2 - z^2 = c_2 \quad \text{(ii)}$$

Equations (i) and (ii) constitute the complete solution.

EXAMPLE 8.10 Solve $\frac{dx}{bz - cy} = \frac{dy}{cx - az} = \frac{dz}{ay - bx}$

Solution Each ratio $= \frac{xdx + ydy + zdz}{0} = \frac{adx + bdy + cdz}{0}$

$$\therefore \quad xdx + ydy + zdz = 0$$

and
$$adx + bdy + cdz = 0$$

Integrating,
$$x^2 + y^2 + z^2 = c_1 \quad \text{(i)}$$

and
$$ax + by + cz = c_2 \quad \text{(ii)}$$

Equations (i) and (ii) constitute the complete solution.

EXAMPLE 8.11 Solve $\dfrac{dx}{\sin(x+y)} = \dfrac{dy}{\cos(x+y)} = \dfrac{dz}{z}$.

Solution Taking first two ratios

$$\frac{dy}{dx} = \cot(x+y) \tag{i}$$

Put
$$x + y = V$$

$\therefore$
$$\frac{dy}{dx} = \frac{dV}{dx} - 1$$

Putting this in Eq. (i), we get

$$\frac{dV}{dx} = 1 + \cot V$$

$\Rightarrow$
$$dx = \frac{dV}{1+\cot V}$$

$\Rightarrow$
$$dx = \frac{\sin V}{\sin V + \cos V}\, dV$$

$$= \frac{1}{2}\left[\frac{(\cos V + \sin V + \sin V - \cos V)}{\sin V + \cos V}\right]$$

Integrating,
$$2x + c_1 = \int \left(1 + \frac{\sin V - \cos V}{\cos V + \sin V}\right) dV$$

$$2x + c_1 = V - \log(\sin V + \cos V)$$

$$= x + y - \log[\sin(x+y) + \cos(x+y)]$$

$\therefore$
$$x - y + c_1 = -\log[\sin(x+y) + \cos(x+y)]$$

$\Rightarrow$
$$c_1 = y - x - \log[\sin(x+y) + \cos(x+y)] \tag{ii}$$

Also
$$\frac{dz}{z} = \frac{dx}{\sin(x+y)} = \frac{dy}{\cos(x+y)} = \frac{d(x+y)}{\sin(x+y) + \cos(x+y)}$$

Put
$$x + y = U$$

$\therefore$
$$\frac{dz}{z} = \frac{dU}{\sin U + \cos U}$$

Integrating,
$$\log z + c_2 = \int \frac{dU}{\sin U + \cos U}$$

$$= \frac{1}{\sqrt{2}} \int \operatorname{cos\,sec}\left(U + \frac{\pi}{4}\right) dU$$

$$= \frac{1}{\sqrt{2}} \log \tan\left(\frac{\pi}{8} + \frac{U}{2}\right)$$

$$\therefore \qquad z^{\sqrt{2}} \cot\left(\frac{\pi}{8} + \frac{x+y}{2}\right) = c_2 \qquad \text{(iii)}$$

Equations (ii) and (iii) constitute the complete solution.

EXAMPLE 8.12 Solve $-\dfrac{dx}{x(x+y)} = \dfrac{dy}{y(x+y)} = \dfrac{dz}{(x-y)(2x+2y+z)}$.

Solution Taking first two ratios

$$-\frac{dx}{x} = \frac{dy}{y}$$

Integrating, $$xy = c_1 \qquad \text{(i)}$$

Also $$\frac{dz}{(x-y)(2x+2y+z)} = -\frac{d(x+y)}{(x+y)(x-y)}$$

$$\Rightarrow \qquad \frac{dz}{2(x+y)+z} = -\frac{d(x+y)}{x+y}$$

Put $$x + y = U$$

$$\Rightarrow \qquad \frac{dz}{dU} = \frac{2U+z}{-U}$$

$$\Rightarrow \qquad \frac{dz}{dU} = -2 - \frac{z}{U}$$

$$\Rightarrow \qquad \frac{dz}{dU} + \frac{z}{U} = -2$$

which is linear equation in z and U.

$\therefore$ Its $$\text{IF} = e^{\int \frac{1}{U} dU} = e^{\log U} = U$$

$\therefore$ Solution is

$$zU = -U_2 + c_2$$

$$\Rightarrow \qquad z(x+y) + (x+y)^2 = c_2 \qquad \text{(ii)}$$

Equations (i) and (ii) constitute the complete solution.

EXAMPLE 8.13 Solve $\dfrac{dx}{y^2 + yz + z^2} = \dfrac{dy}{z^2 + zx + x^2} = \dfrac{dz}{x^2 + xy + y^2}$.

Solution Each ratio $= \dfrac{d(y-x)}{(x-y)(x+y+z)} = \dfrac{d(y-z)}{(z-y)(x+y+z)}$

$$\Rightarrow \qquad \frac{d(y-x)}{y-x} = \frac{d(y-z)}{y-z}$$

Integrating, $$\frac{y-x}{y-z} = c_1 \qquad \text{(i)}$$

Again $$\text{Each ratio} = \frac{xdx + ydy + zdz}{(x+y+z)(xy+yz+zx)}$$

$$\text{Each ratio} = \frac{(ydx + xdy) + (zdx + xdz) + (zdy + ydz)}{2(x+y+z)(x^2+y^2+z^2)}$$

$$= \frac{d(xy+yz+zx)}{2(x+y+z)(x^2+y^2+z^2)}$$

$\therefore$ $$\frac{xdx + ydy + zdz}{(x+y+z)(xy+yz+zx)} = \frac{d(xy+yz+zx)}{2(x+y+z)(x^2+y^2+z^2)}$$

$\Rightarrow$ $$2(x^2+y^2+z^2)\frac{d(x^2+y^2+z^2)}{2} = (xy+yz+zx)\,d(xy+yz+zx)$$

Integrating, $$(x^2 + y^2 + z^2)^2 = (xy + yz + zx)^2 + c_2 \qquad \text{(ii)}$$

Equations (i) and (ii) constitute the complete solution.

EXAMPLE 8.14 Solve $\dfrac{dx}{z-y} = \dfrac{dy}{x-z} = \dfrac{dz}{y-x}$.

Solution $$\text{Each ratio} = \frac{dx + dy + dz}{0}$$

$\Rightarrow$ $$d(x + y + z) = 0$$

Integrating, $$x + y + z = c_1 \qquad \text{(i)}$$

Again $$\text{Each ratio} = \frac{xdx + ydy + zdz}{0}$$

$\Rightarrow$ $$\frac{1}{2}d(x^2+y^2+z^2) = 0$$

Integrating, $$x^2 + y^2 + z^2 = c_2 \qquad \text{(ii)}$$

Equations (i) and (ii) constitute the complete solution.

EXAMPLE 8.15 Solve $\dfrac{dx}{y+z} = \dfrac{dy}{z+x} = \dfrac{dz}{x+y}$.

Solution Each ratio $= \dfrac{d(x-y)}{-(x-y)} = \dfrac{d(y-z)}{-(y-z)} = \dfrac{d(x+y+z)}{2(x+y+z)}$.

Taking first two ratios

$$\frac{d(x-y)}{-(x-y)} = \frac{d(y-z)}{-(y-z)}$$

Integrating, $$\frac{x-y}{y-z} = c_1 \qquad \text{(i)}$$

Also
$$\frac{d(x-y)}{-(x-y)}=\frac{d(x+y+z)}{2(x+y+z)}$$

Integrating, $$(x+y+z)(x-y)^2=c_2 \qquad \text{(ii)}$$

Equations (i) and (ii) constitute the complete solution.

EXAMPLE 8.16 Solve $\dfrac{dx}{mz-ny}=\dfrac{dy}{nx-lz}=\dfrac{dz}{ly-mx}$.

Solution Each ratio $=\dfrac{xdx+ydy+zdz}{0}=\dfrac{ldx+mdy+ndz}{0}$.

Therefore, $$xdx+ydy+zdz=0$$

and $$ldx+mdy+ndz=0$$

Integrating, $$x^2+y^2+z^2=c_1 \qquad \text{(i)}$$

and $$lx+my+nz=c_2 \qquad \text{(ii)}$$

Equations (i) and (ii) constitute the complete solution.

EXAMPLE 8.17 Solve $\dfrac{dx}{y^3x-2x^4}=\dfrac{dy}{2y^4-x^3y}=\dfrac{dz}{9z(x^3-y^3)}$.

Solution Each ratio $=\dfrac{\dfrac{dx}{x}+\dfrac{dy}{y}+\dfrac{dz}{3z}}{0}$.

$\therefore$ $$\frac{dx}{x}+\frac{dy}{y}+\frac{dz}{3z}=0$$

Integrating, $$x^3y^3z=c_1 \qquad \text{(i)}$$

The first two ratios gives $y(2y^3-x^3)dx+x(2x^3-y^3)dy=0$. Dividing by x^3y^3 results into

$$\left(\frac{2y}{x^3}dx-\frac{1}{y^2}dx\right)+\left(\frac{2x}{y^3}dy-\frac{1}{x^2}dy\right)=0 \quad\Rightarrow\quad -d\left(\frac{y}{x^2}+\frac{x}{y^2}\right)=0$$

Integrating, $$\frac{y}{x^2}+\frac{x}{y^2}=c_2 \qquad \text{(ii)}$$

Equations (i) and (ii) constitute the complete solution.

EXAMPLE 8.18 Solve $\dfrac{dx}{y(x+y)+az}=\dfrac{dy}{x(x+y)-az}=\dfrac{dz}{z(x+y)}$.

Solution Taking first two ratios

$$\text{Each ratio}=\frac{dx+dy}{(x+y)^2}$$

Then $$\frac{dx + dy}{(x + y)^2} = \frac{dz}{z(x + y)} \Rightarrow \frac{dx + dy}{x + y} = \frac{dz}{z}$$

Integrating, $$\frac{x + y}{z} = c_1 \qquad \text{(i)}$$

Also $$\text{Each ratio} = \frac{xdx - ydy}{az(x + y)}$$

$\therefore$ $$\frac{xdx - ydy}{az(x + y)} = \frac{dz}{z(x + y)}$$

$\Rightarrow$ $$xdx - ydy - adz = 0$$

Integrating, $$x^2 - y^2 - az = c_2 \qquad \text{(ii)}$$

Equations (i) and (ii) constitute the complete solution.

8.4.2 By Introducing a New Variable

Consider the equations

$$\frac{dx}{P} = \frac{dy}{Q} = \frac{dz}{R}$$

where

$$P = ax + by + cz + d$$
$$Q = a'x + b'y + c'z + d'$$
$$R = a''x + b''y + c''z + d''$$

Introduce the new variable 't' such that

$$\frac{dx}{P} = \frac{dy}{Q} = \frac{dz}{R} = \frac{dt}{t}$$

Then $$\frac{dt}{t} = \frac{ldx + mdy + ndz}{lP + mQ + nR}$$

$$= \frac{ldx + mdy + ndz}{(al + a'm + a''n)\,x + (bl + b'm + b''n)\,y + (cl + c'm + c''n)\,z + (dl + d'm + d''n)}$$

where l, m, n are suitable multipliers.

Choose l, m, n such that

$$\left.\begin{aligned} al + a'm + a''n &= \lambda l \\ bl + b'm + b''n &= \lambda m \\ cl + c'm + c''n &= \lambda n \end{aligned}\right\}$$

$\Rightarrow$ $$\left.\begin{aligned} (a - \lambda)l + a'm + a''n &= 0 \\ bl + (b' - \lambda)m + b''n &= 0 \\ cl + c'm + (c'' - \lambda)n &= 0 \end{aligned}\right\} \qquad (8.2)$$

Eliminating l, m, n from Eq. (8.2),

$$\begin{vmatrix} a-\lambda & a' & a'' \\ b & b'-\lambda & b'' \\ c & c' & c''-\lambda \end{vmatrix} = 0 \tag{8.3}$$

Solve the cubic equation obtained in Eq. (8.2) for λ. Let roots be λ_1, λ_2, λ_3. Let corresponding to $\lambda = \lambda_1$, the values of l, m, n be l_1, m_1, n_1.

Then $$\frac{dt}{t} = \frac{l_1 dx + m_1 dy + n_1 dz}{\lambda_1(l_1 x + m_1 y + n_1 z) + r_1}$$

where $$r_1 = l_1 d + m_1 d' + n_1 d''$$

Integrating, $$c_1 t = (l_1 x + m_1 y + n_1 z + r_1 \lambda_1^{-1})^{\lambda_1^{-1}}$$

Similarly, $$c_2 t = (l_2 x + m_2 y + n_2 z + r_2 \lambda_2^{-1})^{\lambda_2^{-1}}$$

and $$c_3 t = (l_3 x + m_3 y + n_3 z + r_3 \lambda_3^{-1})^{\lambda_3^{-1}}$$

Eliminating t, the complete solution is

$$(l_1 x + m_1 y + n_1 z + r_1 \lambda_1^{-1})^{\lambda_1^{-1}} = A(l_2 x + m_2 y + n_2 z + r_2 \lambda_2^{-1})^{\lambda_2^{-1}} = B(l_3 x + m_3 y + n_3 z + r_3 \lambda_3^{-1})^{\lambda_3^{-1}}.$$

EXAMPLE 8.19 Solve $\dfrac{dx}{x+y-z} = \dfrac{dy}{y+z-x} = \dfrac{dz}{z+x-y}$.

Solution Introduce new variable t such that

$$\frac{dt}{t} = \frac{dx}{x+y-z} = \frac{dy}{y+z-x} = \frac{dz}{z+x-y} = \frac{l dx + m dy + n dz}{x(l-m+n) + y(l+m-n) + z(-l+m+n)}$$

Choose l, m, n such that

$$\left.\begin{aligned} l-m+n &= \lambda l \\ l+m-n &= \lambda m \\ -l+m+n &= \lambda n \end{aligned}\right\} \Rightarrow \left.\begin{aligned} l(1-\lambda)-m+n &= 0 \\ l+m(1-\lambda)-n &= 0 \\ -l+m+n(1-\lambda) &= 0 \end{aligned}\right\} \tag{i}$$

Eliminating l, m, n from Eq. (i)

$$\begin{vmatrix} 1-\lambda & -1 & 1 \\ 1 & 1-\lambda & -1 \\ -1 & 1 & 1-\lambda \end{vmatrix} = 0$$

$$\Rightarrow \qquad \lambda^3 - 3\lambda^2 + 6\lambda - 4 = 0$$

$$\Rightarrow \qquad (\lambda-1)(\lambda^2 - 2\lambda + 4) = 0$$

Then $\lambda = 1, 1 \pm i\sqrt{3} = 1, -2w, -2w^2$, where $w^3 = 1$.

For
$$\lambda = 1, l = m = n = 1$$

$\therefore$
$$\frac{dt}{t} = \frac{d(x+y+z)}{x+y+z}$$

Integrating,
$$\log t = \log(x+y+z) + A \qquad \text{(ii)}$$

For
$$\lambda = -2w$$
$$l(1+2w) - m + n = 0$$
$$l + m(1+2w) - n = 0$$

$\therefore$
$$\frac{l}{-2w} = \frac{m}{2+2w} = \frac{n}{2+4w+4w^2}$$

$\Rightarrow$
$$\frac{l}{-2w} = \frac{m}{2(1+w)} = \frac{n}{2(1+2w+2w^2)}$$

$\Rightarrow$
$$\frac{l}{-2w} = \frac{m}{-2w^2} = \frac{n}{-2}$$

$\Rightarrow$
$$\frac{l}{w} = \frac{m}{w^2} = \frac{n}{1}$$

$\therefore$
$$\frac{dt}{t} = \frac{d(wx+w^2y+z)}{-2w(wx+w^2y+z)}$$

$\Rightarrow$
$$\frac{dt}{t} = \frac{1}{-2w}\frac{d(x+wy+w^2z)}{(x+wy+w^2z)}$$

Integrating,
$$\log t = -\frac{1}{2}w^2 \log(x+wy+w^2z) + B \qquad \text{(iii)}$$

Similarly, for $\lambda = -2w^2$,
$$\log t = -\frac{1}{2}w \log(x+w^2y+wz) + C \qquad \text{(iv)}$$

From Eqs. (ii)–(iv),
$$2\log t + w^2 \log(x+wy+w^2z) = a$$

and
$$2\log t + w \log(x+w^2y+wz) = b$$
where a and b are constants.

EXERCISES

Solve the following simultaneous equations:

1. $\dfrac{dx}{dt}+2x+3y=0, 3x+\dfrac{dy}{dt}+2y=2e^{2t}$.

$$\left[\textbf{Ans. } x=c_1e^t+c_2e^{-5t}-\frac{6}{7}e^{2t},\ y=-c_1e^t+c_2e^{-5t}+\frac{8}{7}e^{2t}\right]$$

2. $\dfrac{dx}{dt}+5x-2y=t, \dfrac{dy}{dt}+2x+y=0$; given that $x=y=0$ at $t=0$.

$$\left[\textbf{Ans. } x=-\frac{1}{27}(1+6t)\,e^{-3t}+\frac{1}{27}(1+3t),\ y=-\frac{2}{27}(2+3t)\,e^{-3t}+\frac{9}{27}(2-3t)\right]$$

3. $\dfrac{dx}{dt}+4x+3y=t, \dfrac{dy}{dt}+2x+5y=e^t$.

$$\left[\textbf{Ans. } x=c_1e^{-2t}+c_2e^{-7t}-\frac{1}{8}e^t+\frac{5}{14}t-\frac{31}{196}, y=-\frac{2}{3}c_1e^{-2t}+c_2e^{-7t}+\frac{5}{24}e^t-\frac{1}{7}t+\frac{9}{98}\right]$$

4. $\dfrac{dx}{dt}-7x+y=0, \dfrac{dy}{dt}-2x-5y=0$.

$$[\textbf{Ans. } x=e^{6t}[c_1\cos t+c_2\sin t],\ y=e^{6t}[(c_1-c_2)\cos t+(c_1+c_2)\sin t]$$

5. $5\dfrac{dy}{dt}-2\dfrac{dx}{dt}+4y-x=e^{-t}, \dfrac{dy}{dt}+8y-3x=5e^{-t}$.

$$[\textbf{Ans. } x=3c_1e^t+2c_2e^{-2t}+3e^{-t},\ y=c_1e^t+c_2e^{-2t}+2e^{-t}]$$

6. $4\dfrac{dx}{dt}+9\dfrac{dy}{dt}+11x+31y=e^t, 3\dfrac{dx}{dt}+7\dfrac{dy}{dt}+8x+24y=e^{2t}$.

$$\left[\textbf{Ans. } x=(c_1+c_2t)\,e^{-4t}+\frac{31}{25}e^t-\frac{49}{36}e^{2t}, y=-c_1e^{-4t}-c_2\,(1+t)\,e^{-4t}-\frac{11}{25}e^t+\frac{19}{36}e^{2t}\right]$$

7. $\dfrac{d^2x}{dt^2}+4x+y=te^t, \dfrac{d^2y}{dt^2}+y-2x=\sin^2 t$.

$$\left[\begin{array}{l}\textbf{Ans. } x=c_1\cos\sqrt{3}t+c_2\sin\sqrt{3}t+c_3\cos\sqrt{2}t+c_4\sin\sqrt{2}t+\frac{1}{4}\cos 2t-\frac{1}{12}+\frac{e^t}{6}\left(t-\frac{1}{6}\right)\\ y=-[c_1\cos\sqrt{3}t+c_2\sin\sqrt{3}t+2c_3\cos\sqrt{2}t+2c_4\sin\sqrt{2}t]+te^t-\frac{5}{6}e^t\left(t-\frac{1}{6}\right)-\frac{1}{3}(e^t-1)\end{array}\right]$$

8. $\dfrac{dx}{dt}+wy=0, \dfrac{dy}{dt}-wx=0$. $[\textbf{Ans. } x=a\cos wt+b\sin wt,\ y=a\sin wt-b\cos wt]$

9. $\dfrac{dx}{dt} - 3x + 4y = e^{-2t}, \dfrac{dy}{dt} - x + 2y = 3e^{-2t}$, given that $x = 12$, $y = 7$ for $t = 0$.

$$\left[\textbf{Ans. } x = 6e^{2t} + 9e^{-t} - 3e^{-2t}, y = \frac{3}{2}e^{2t} + 9e^{-t} - \frac{7}{2}e^{-2t}\right]$$

10. $\dfrac{d^2x}{dt^2} + \dfrac{dx}{dt} + \dfrac{dy}{dt} - y = 0, \dfrac{d^2y}{dt^2} - \dfrac{dy}{dt} - \dfrac{dx}{dt} + x = 0.$

[**Ans.** $x = c_1e^t + c_2 \cos t + c_3 \sin t, y = c_4e^t + c_2 \sin t - c_3 \cos t$]

11. $\dfrac{dx}{y^3x - 2x^4} = \dfrac{dy}{2y^4 - x^3y} = \dfrac{dz}{9z(x^3 - y^3)}.$ $\left[\textbf{Ans. } x^3y^3z = c_1, \dfrac{y}{x^2} + \dfrac{x}{y^2} = c_2\right]$

12. $\dfrac{dx}{xy} = \dfrac{dy}{y^2} = \dfrac{dz}{zxy - 2x^2}.$ $\left[\textbf{Ans. } x = yc_1, e^{-x}\left(z - 2\dfrac{x}{y}\right) = c_2\right]$

13. $\dfrac{dx}{y^2} = \dfrac{dy}{x^2} = \dfrac{dz}{x^2y^2z^2}.$ $\left[\textbf{Ans. } x^3 - y^3 = c_1, x^3 + \dfrac{3}{z} = c_2\right]$

14. $\dfrac{adx}{(b-c)\,yz} = \dfrac{bdy}{(c-a)\,zx} = \dfrac{cdz}{(a-b)\,xy}.$

[**Ans.** $ax^2 + by^2 + cz^2 = c_1, a^2x^2 + b^2y^2 + c^2z^2 = c_2$]

15. $\dfrac{dx}{x^2(y^3 - z^3)} = \dfrac{dy}{y^2(z^3 - x^3)} = \dfrac{dz}{z^2(x^3 - y^3)}.$ $\left[\textbf{Ans. } \dfrac{1}{x} + \dfrac{1}{y} + \dfrac{1}{z} = c_1, x^2 + y^2 + z^2 = c_2\right]$

16. $\dfrac{yzdx}{y - z} = \dfrac{zxdy}{z - x} = \dfrac{xydz}{x - y}.$ [**Ans.** $x + y + z = c_1, xyz = c_2$]

17. $\dfrac{dx}{z(x+y)} = \dfrac{dy}{z(x-y)} = \dfrac{dz}{x^2 + y^2}.$ [**Ans.** $x^2 - 2xy - y^2 = c_1, 2xy - z^2 = c_2$]

18. $\dfrac{dx}{x + y - xy^2} = \dfrac{dy}{x^2y - x - y} = \dfrac{dz}{z(y^2 - x^2)}.$

[**Ans.** $z(1 - xy) = c_1, x^2 + y^2 - 2\log(1 - xy) = c_2$]

19. $\dfrac{dx}{1} = \dfrac{dy}{2} = \dfrac{dz}{5z + \tan(y - 2x)}.$ [**Ans.** $y - 2x = c_1, 5z + \tan(y - 2x) = c_2e^{5x}$]

20. $t\dfrac{d^2x}{dt^2} + 2\dfrac{dx}{dt} + tx = 0, \dfrac{dy}{dt} + \dfrac{2}{t}y - \dfrac{dx}{dt} = 0.$

$$\left[\textbf{Ans. } x = \frac{1}{t}(c_1 \cos t + c_2 \sin t), y = \frac{c_3}{t^2} + \frac{1}{t}(c_1 \cos t + c_2 \sin t) - \frac{2}{t^2}(c_1 \sin t - c_2 \cos t)\right]$$

21. $\dfrac{dx}{xz(z^2+xy)}=\dfrac{dy}{-yz(z^2+xy)}=\dfrac{dz}{x^4}.$ **[Ans.** $xy=c_1, x^4-z^4-2xyz^2=c_2$**]**

22. $\dfrac{dx}{xy^3-2x^4}=\dfrac{dy}{2y^4-x^3y}=\dfrac{dz}{2z(x^3-y^3)}.$ **[Ans.** $x^2y^2z^3=c_1, (x^3+y^3)z^3=c_2$**]**

23. $\dfrac{dx}{x+y}=\dfrac{dy}{x+y}=\dfrac{dz}{-(x+y+2z)}.$ **[Ans.** $x-y=c_1, xy+yz+zx=c_2$**]**

24. $\dfrac{dx}{x^2+a^2}=\dfrac{dy}{xy-az}=\dfrac{dz}{xz+ay}.$ **[Ans.** $xz+ay=c_1(az-xy), a(y^2+z^2)=c_2(az-xy)$**]**

25. $\dfrac{dx}{x}=\dfrac{dy}{y}=\dfrac{dz}{xyz^2(x^2-y^2)}.$ $\left[\textbf{Ans. } x^4+\dfrac{4x^3}{yz(x^2-y^2)}=c_1, \dfrac{y}{x}=c_2\right]$

26. $\dfrac{dx}{dt}=4x-2y, \dfrac{dy}{dt}=3x-y$ $\left[\textbf{Ans. } x=\dfrac{2}{3}c_1e^t+c_2e^{2t}, y=c_1e^t+c_2e^{2t}\right]$

27. $\dfrac{dx}{dt}=x+2y, \dfrac{dy}{dt}=y$ **[Ans.** $x=c_2e^t+2c_1te^{2t}, y=c_1e^t$**]**

28. $\dfrac{dx}{dt}=7x-y, \dfrac{dy}{dt}=2x+5y$

[Ans. $x=e^{6t}(c_1\cos t+c_2\sin t), y=e^{6t}\{(c_1-c_2)\cos t+(c_1+c_2)\sin t\}$**]**

29. $\dfrac{dx}{dt}+wy=0, \dfrac{dy}{dt}-wx=0$ **[Ans.** $x=c_1\cos wt+c_2\sin wt, y=c_1\sin wt-c_2\cos wt$**]**

30. $\dfrac{dx}{dt}+4x+3y=t, \dfrac{dy}{dt}+2x+5y=e^t$

$\left[\textbf{Ans. } x=c_1e^{-2t}+c_2e^{-7t}+\dfrac{5}{14}t-\dfrac{1}{8}e^t-\dfrac{31}{196}, y=-\dfrac{2}{3}c_1e^{-2t}+c_2e^{-7t}-\dfrac{1}{7}t+\dfrac{5}{24}e^t+\dfrac{9}{98}\right]$

31. $\dfrac{dx}{dt}+2x+3y=0, \dfrac{dy}{dt}+3x+2y=2e^{2t}$

$\left[\textbf{Ans. } x=c_1e^t+c_2e^{-5t}-\dfrac{6}{7}e^{2t}, y=-c_1e^t+c_2e^{-5t}+\dfrac{8}{7}e^{2t}\right]$

32. $\dfrac{dx}{dt}+2x-3y=t, \dfrac{dy}{dt}-3x+2y=e^{2t}$

$\left[\textbf{Ans. } x=c_1e^t+c_2e^{-5t}-\dfrac{2}{5}t+\dfrac{3}{7}e^{2t}+\dfrac{3}{25}, y=c_1e^t+\dfrac{c_2}{3}e^{-5t}-\dfrac{3}{5}t+\dfrac{4}{7}e^{2t}-\dfrac{4}{75}\right]$

33. $\dfrac{dx}{dt}+5x+y=e^t, \dfrac{dy}{dt}+3y-x=e^{2t}$

$$\left[\textbf{Ans. } x=(c_1+c_2t)e^{-4t}+\frac{4}{25}e^t-\frac{1}{36}e^{2t}, y=-(c_1+c_2+c_2t)e^{-4t}+\frac{1}{25}e^t+\frac{7}{36}e^{2t}\right]$$

34. $\dfrac{dx}{dt}+\dfrac{dy}{dt}+2x+y=0, \dfrac{dy}{dt}+5x+3y=0$

$$\left[\textbf{Ans. } x=c_1\cos t+c_2\sin t, y=\frac{1}{2}(c_2-3c_1)\cos t-\frac{1}{2}(c_1-3c_2)\sin t\right]$$

35. $\dfrac{dx}{dt}+\dfrac{dy}{dt}-y=2t+1, 2\dfrac{dx}{dt}+2\dfrac{dy}{dt}+y=t$ $\left[\textbf{Ans. } x=\dfrac{t^2}{2}+\dfrac{4}{3}t+c_1, y=-t-\dfrac{2}{3}\right]$

36. $4\dfrac{dx}{dt}+9\dfrac{dy}{dt}+2x+31y=e^t, 3\dfrac{dx}{dt}+7\dfrac{dy}{dt}+x+24y=3$

$$\left[\textbf{Ans. } x=(c_1\cos t+c_2\sin t)e^{-4t}+\frac{31}{26}e^t-\frac{93}{17}, y=[(c_2-c_1)\sin t-(c_1+c_2)\cos t]e^{-4t}-\frac{2}{13}e^t+\frac{6}{17}\right]$$

37. $\dfrac{dx}{dt}-3x+4y=e^{-2t}, \dfrac{dy}{dt}+2y-x=3e^{-2t}$, given that $x = 12$, $y = 7$ for $t = 0$.

$$\left[\textbf{Ans. } x=6e^{2t}+9e^{-t}-3e^{-2t}, y=\frac{3}{2}e^{2t}+9e^{-t}-\frac{7}{2}e^{-2t}\right]$$

38. $\dfrac{d^2x}{dt^2}-3x-4y+3=0, \dfrac{d^2y}{dt^2}+x+y+5=0$

$$\left[\begin{aligned}&\textbf{Ans. } x=(c_1+c_2t)e^t+(c_3+c_4t)e^{-t}-23,\\ &y=\left(\frac{1}{2}(c_2-c_1)-\frac{c_2}{2}t\right)e^t+\left(-\frac{1}{2}(c_3+c_4)-\frac{c_4}{2}t\right)e^{-t}+18\end{aligned}\right]$$

39. $\dfrac{dx}{yz}=\dfrac{dy}{zx}=\dfrac{dz}{xy}$ $[\textbf{Ans. } x^2-y^2=c_1, y^2-z^2=c_2]$

40. $\dfrac{dx}{y^2}=\dfrac{dy}{y}=\dfrac{dz}{z}$ $\left[\textbf{Ans. } 2x^2-y^2=c_1, \dfrac{y}{z}=c_2\right]$

41. $\dfrac{xdx}{y^2z}=\dfrac{dy}{zx}=\dfrac{dz}{y^2}$ $[\textbf{Ans. } x^3-y^3=c_1, x^2-z^2=c_2]$

42. $\dfrac{dx}{x(y-z)}=\dfrac{dy}{y(z-x)}=\dfrac{dz}{z(x-y)}$ $[\textbf{Ans. } x+y+z=c_1, xyz=c_2]$

43. $\dfrac{dx}{x(y^2-z^2)}=\dfrac{dy}{y(z^2-x^2)}=\dfrac{dz}{z(x^2-y^2)}$ $[\textbf{Ans. } x^2+y^2+z^2=c_1, xyz=c_2]$

44. $\dfrac{dx}{x^2 - yz} = \dfrac{dy}{y^2 - zx} = \dfrac{dz}{z^2 - xy}$ $\left[\textbf{Ans. } \dfrac{x-y}{y-z} = c_1, xy + yz + zx = c_2\right]$

45. $\dfrac{ldx}{mn(y-z)} = \dfrac{mdy}{nl(z-x)} = \dfrac{ndz}{lm(x-y)}$

$[\textbf{Ans. } l^2x^2 + m^2y^2 + n^2z^2 = c_1, l^2x + m^2y + n^2z = c_2]$

46. $\dfrac{dx}{x(y^2 - z^2)} = \dfrac{dy}{-y(z^2 + x^2)} = \dfrac{dz}{z(x^2 + y^2)}$ $\left[\textbf{Ans. } x^2 + y^2 + z^2 = c_1, \dfrac{yz}{x} = c_2\right]$

47. $\dfrac{dx}{y^2 + z^2} = \dfrac{dy}{-xy} = \dfrac{dz}{-xz}$ $\left[\textbf{Ans. } x^2 + y^2 + z^2 = c_1, \dfrac{y}{z} = c_2\right]$

48. $\dfrac{dx}{y^2 + z^2 - x^2} = \dfrac{dy}{-2xy} = \dfrac{dz}{-2xz}$ $\left[\textbf{Ans. } \dfrac{x^2 + y^2 + z^2}{z} = c_1, \dfrac{y}{z} = c_2\right]$

49. $\dfrac{dx}{1+y} = \dfrac{dy}{1+x} = \dfrac{dz}{z}$ $[\textbf{Ans. } z(x-y) = c_1, 2 + x + y = c_2 z]$

50. $\dfrac{dx}{x^2} = \dfrac{dy}{y^2} = \dfrac{dz}{4xy}$ $\left[\textbf{Ans. } \dfrac{1}{y} - \dfrac{1}{x} = c_1, \log\dfrac{x}{y} - \dfrac{z}{4}\left(\dfrac{1}{y} - \dfrac{1}{x}\right) = c_2\right]$

51. $\dfrac{dx}{xz - y} = \dfrac{dy}{yz - x} = \dfrac{dz}{1 - z^2}$ $[\textbf{Ans. } (x+y)(z+1) = c_1, (x-y)(z-1) = c_2]$

52. $\dfrac{dx}{y - xz} = \dfrac{dy}{yz + x} = \dfrac{dz}{x^2 + y^2}$ $[\textbf{Ans. } xy - z = c_1, x^2 - y^2 + z^2 = c_2]$

CHAPTER 9

Total Differential Equations

9.1 INTRODUCTION

In this chapter, we will consider single equation having one independent variable and more than one dependent variable. In other words, in total differential equations the differential coefficients of several dependent variables with respect to the single independent variable will appear.

Consider
$$f(x, y, z) = c \tag{9.1}$$
where x, y and z are variables and c is a constant. Then

$$df = 0$$

$$\Rightarrow \quad \frac{\partial f}{\partial x}dx + \frac{\partial f}{\partial y}dy + \frac{\partial f}{\partial z}dz = 0 \tag{9.2}$$

If $\frac{\partial f}{\partial x}, \frac{\partial f}{\partial y}, \frac{\partial f}{\partial z}$ have any common factor then eliminate it and the general form of the total differential equation in three variables can be written as

$$Pdx + Qdy + Rdz = 0 \tag{9.3}$$

where P, Q and R are function of x, y and z. The Eq. (9.3) is known as **Pfaffian differential equation** in three variables.

9.2 CONDITION OF INTEGRABILITY

Consider the Eq. (9.3)

$$Pdx + Qdy + Rdz = 0$$

Let Eq. (9.3) be integrable and its integral be Eq. (9.1)

$$f(x, y, z) = c_2$$

so that the expression

$$df = \frac{\partial f}{\partial x}dx + \frac{\partial f}{\partial y}dy + \frac{\partial f}{\partial z}dz$$

is equal to or multiplier of

$$Pdx + Qdy + Rdz$$

Thus,

$$\frac{\partial f}{\partial x} = \lambda P$$

$$\frac{\partial f}{\partial y} = \lambda Q$$

$$\frac{\partial f}{\partial z} = \lambda R$$

where λ is some function.

$$\therefore \qquad \frac{\partial}{\partial y}(\lambda P) = \frac{\partial}{\partial x}(\lambda Q)$$

$$\Rightarrow \qquad \lambda\left(\frac{\partial P}{\partial y} - \frac{\partial Q}{\partial x}\right) = Q\frac{\partial \lambda}{\partial x} - P\frac{\partial \lambda}{\partial y} \tag{9.4}$$

Similarly,

$$\lambda\left(\frac{\partial Q}{\partial z} - \frac{\partial R}{\partial y}\right) = R\frac{\partial \lambda}{\partial y} - Q\frac{\partial \lambda}{\partial z} \tag{9.5}$$

and

$$\lambda\left(\frac{\partial R}{\partial x} - \frac{\partial P}{\partial z}\right) = P\frac{\partial \lambda}{\partial z} - R\frac{\partial \lambda}{\partial x} \tag{9.6}$$

Multiplying Eqs. (9.4), (9.5) and (9.6) by R, P and Q, respectively and adding

$$\lambda\left[R\left(\frac{\partial P}{\partial y} - \frac{\partial Q}{\partial x}\right) + P\left(\frac{\partial Q}{\partial z} - \frac{\partial R}{\partial y}\right) + Q\left(\frac{\partial R}{\partial x} - \frac{\partial P}{\partial z}\right)\right] = 0$$

$$\Rightarrow \qquad P\left(\frac{\partial Q}{\partial z} - \frac{\partial R}{\partial y}\right) + Q\left(\frac{\partial R}{\partial x} - \frac{\partial P}{\partial z}\right) + R\left(\frac{\partial P}{\partial y} - \frac{\partial Q}{\partial x}\right) = 0 \tag{9.7}$$

which is the necessary condition for the Eq. (9.3) to possess an integral of the form given in Eq. (9.1).

Note: If $P_1 = \mu P$, $Q_1 = \mu Q$, $R_1 = \mu R$ where μ is any quantity then

$$\sum P_1\left(\frac{\partial Q_1}{\partial z} - \frac{\partial R_1}{\partial y}\right) = \sum \mu P\left(\frac{\partial(\mu Q)}{\partial z} - \frac{\partial(\mu R)}{\partial y}\right)$$

$$= \sum \mu^2 P\left(\frac{\partial Q}{\partial z} - \frac{\partial R}{\partial y}\right) + \sum \mu\left(PQ\frac{\partial \mu}{\partial z} - PR\frac{\partial \mu}{\partial y}\right)$$

$$= \mu^2 \sum P\left(\frac{\partial Q}{\partial z} - \frac{\partial R}{\partial y}\right)$$

Thus, P_1, Q_1, R_1 satisfy Eq. (9.7) if P, Q, R.

Next, we establish sufficient condition.

Let the coefficients P, Q, R in Eq. (9.3) satisfy the condition given in Eq. (9.7).

Let $$Pdx + Qdy$$
be an exact.

$$\therefore \quad Pdx + Qdy = dV$$

$$= \frac{\partial V}{\partial x}dx + \frac{\partial V}{\partial y}dy$$

$$\therefore \quad P = \frac{\partial V}{\partial x}$$

and $$Q = \frac{\partial V}{\partial y}$$

Substituting these values of P and Q in Eq. (9.6), we get

$$\frac{\partial V}{\partial x}\left(\frac{\partial^2 V}{\partial z \partial y} - \frac{\partial R}{\partial y}\right) + \frac{\partial V}{\partial y}\left(\frac{\partial R}{\partial x} - \frac{\partial^2 V}{\partial z \partial x}\right) + R\left(\frac{\partial^2 V}{\partial y \partial x} - \frac{\partial^2 V}{\partial x \partial y}\right) = 0$$

$$\Rightarrow \quad \frac{\partial V}{\partial x}\left(\frac{\partial^2 V}{\partial z \partial y} - \frac{\partial R}{\partial y}\right) + \frac{\partial V}{\partial y}\left(\frac{\partial R}{\partial x} - \frac{\partial^2 V}{\partial z \partial x}\right) = 0$$

$$\Rightarrow \quad \begin{vmatrix} \dfrac{\partial V}{\partial x} & \dfrac{\partial}{\partial x}\left(\dfrac{\partial V}{\partial z} - R\right) \\ \dfrac{\partial V}{\partial y} & \dfrac{\partial}{\partial y}\left(\dfrac{\partial V}{\partial z} - R\right) \end{vmatrix} = 0$$

which implies that a relation independent of x and y exists between V and $\frac{\partial V}{\partial z} - R = R_1$.

Now, $$\frac{\partial V}{\partial z} - R = R_1$$

$$\Rightarrow \quad R = \frac{\partial V}{\partial z} - R_1$$

$$\therefore \quad Pdx + Qdy + Rdz = \frac{\partial V}{\partial x}dx + \frac{\partial V}{\partial y}dy + \left(\frac{\partial V}{\partial z} - R_1\right)dz$$

$$= dV - R_1 dz$$

$\therefore$ Equation (9.3) reduces to

$$dV - R_1 dz = 0 \tag{9.8}$$

Equation (9.8) has only two variables V and z and may be integrated. Let its integral be

$$g(V, z) = \text{Constant} \tag{9.9}$$

Then $$\frac{\partial g}{\partial V} dV + \frac{\partial g}{\partial z} dz = 0$$

is identical with Eq. (9.8). Comparing

$$\frac{\frac{\partial g}{\partial V}}{1} = \frac{\frac{\partial g}{\partial z}}{-R_1} = \lambda \quad \text{(say)}$$

$\Rightarrow$ $$\frac{\partial g}{\partial V} = \lambda$$

and $$\frac{\partial g}{\partial z} = -\lambda R_1$$

$\therefore$ $$\lambda[Pdx + Qdy + Rdz] = \lambda[dV - R_1 dz]$$
$$= \frac{\partial g}{\partial V} dV + \frac{\partial g}{\partial z} az$$
$$= dg$$

Equation (9.3) is satisfied by the Eq. (9.8). Replacing V by its value in terms of x, y and z, Eq. (9.3) has a primitive of the form

$$f(x, y, z) = c$$

Remark: Let $X = (P, Q, R)$

Then $$\text{curl } X = \left(\frac{\partial R}{\partial y} - \frac{\partial Q}{\partial z}, \frac{\partial P}{\partial z} - \frac{\partial R}{\partial x}, \frac{\partial Q}{\partial x} - \frac{\partial P}{\partial Y}\right)$$

The condition of integrability as shown in Eq. (9.6) is

$$X \cdot \text{curl } X = 0$$

Note: If $Pdx + Qdy$ is not exact, an integrating factor μ can be found for it so that

$$\mu Pdx + \mu Qdy + \mu Rdz = 0$$

can be considered.

9.3 SOLUTION METHODOLOGY TO OBTAIN THE PRIMITIVE

Under the assumption that the condition given by Eq. (9.6) for inegrability is satisfied, which guarantees the existence of primitive Eq. (9.1), follow the method to find the primitive as discussed below:

(i) Assume, $$z = \text{Constant}$$

Then $$dz = 0$$

Equation (9.3) reduces to

$$Pdx + Qdy = 0 \tag{9.10}$$

Let integral of Eq. (9.10) be

$$f(x, y) = V = \text{Constant} \tag{9.11}$$

where f may contain z.

(ii) From Eqs. (9.11) and (9.11)

$$\lambda P = \frac{\partial V}{\partial x}$$

and
$$\lambda Q = \frac{\partial V}{\partial y}$$

λ can be obtained from either of these equations.

(iii) $\lambda P dx + \lambda Q dy + \lambda R dz$

$$= \left(\frac{\partial V}{\partial x}dx + \frac{\partial V}{\partial y}dy + \frac{\partial V}{\partial z}dz\right) + \left(\lambda R - \frac{\partial V}{\partial z}\right)\partial z$$

$$= dV + Sdz$$

where $S = \lambda R - \dfrac{\partial V}{\partial z}$; a function of z and V.

(iv) Solve the equation

$$dV + Sdz = 0$$

and in its integral replace V by its value in x, y and z. The obtained relation will be the primitive of Eq. (9.3).

EXAMPLE 9.1 Solve $(y + z)dx + dy + dz = 0$.

Solution Here $P = y + z$; $Q = 1$; $R = 1$

$\therefore$
$$P\left(\frac{\partial Q}{\partial z} - \frac{\partial R}{\partial y}\right) + Q\left(\frac{\partial R}{\partial x} - \frac{\partial P}{\partial z}\right) + R\left(\frac{\partial P}{\partial y} - \frac{\partial Q}{\partial x}\right)$$

$$= (y + z)(0 - 0) + 1(0 - 1) + (1 - 0)$$

$$= -1 + 1 = 0$$

Therefore, condition of integrability is satisfied. Assuming x = Constant, equation reduces to

$$dy + dz = 0$$

$\Rightarrow$
$$d(y + z) = 0$$

Integrating,
$$y + z = \text{Constant}$$

Take
$$V = y + z$$

$\therefore$
$$\lambda Q = \frac{\partial V}{\partial y} = 1$$

$\therefore$
$$\lambda = 1$$

$\therefore$
$$S = \lambda P - \frac{\partial V}{\partial x}$$

$$= y + z$$

$$= V$$

The given equation reduces to

$$dV + Vdx = 0$$

$$\Rightarrow \quad \frac{dV}{V} + dx = 0$$

Integrating,

$$\log V + x = \log c$$

$$\Rightarrow \quad Ve^x = c$$

$$\Rightarrow \quad (y + z)\, e^x = c$$

is the primitive of the given equation.

EXAMPLE 9.2 Solve $yzdx + 2zxdy - 3xydz = 0$.

Solution Here $P = yz$; $Q = 2zx$; $R = -3xy$

$$\therefore \quad P\left(\frac{\partial Q}{\partial z} - \frac{\partial R}{\partial y}\right) + Q\left(\frac{\partial R}{\partial x} - \frac{\partial P}{\partial z}\right) + R\left(\frac{\partial P}{\partial y} - \frac{\partial Q}{\partial x}\right)$$

$$= yz\,(2x + 3x) + 2zx\,(-3y - y) - 3xy(z - 2z)$$

$$= 5xyz - 8xyz + 3xyz$$

$$= 0$$

Therefore, the condition of integrability is satisfied. Assuming x = Constant, equation reduces to

$$2zxdy - 3xydz = 0$$

$$\Rightarrow \quad 2zdy - 3ydz = 0$$

$$\Rightarrow \quad 2\frac{dy}{y} - 3\frac{dz}{z} = 0$$

Integrating,

$$\Rightarrow \quad \log y^2 - \log z^3 = \text{Constant}$$

$$\Rightarrow \quad \frac{y^2}{z^3} = \text{Constant}$$

$$\therefore \quad V = \frac{y^2}{z^3}$$

$$\lambda Q = \frac{\partial V}{\partial y}$$

$$\Rightarrow \quad \lambda(2zx) = \frac{2y}{z^3}$$

$$\Rightarrow \quad \lambda = \frac{y}{xz^4}$$

$$\therefore \qquad S = \lambda P - \frac{\partial V}{\partial x}$$

$$S = \frac{y}{xz^4} yz - 0$$

$$S = \frac{y^2}{xz^3}$$

$$= \frac{V}{x}$$

The given equation reduces to

$$dV + \frac{V}{x} dx = 0$$

$$\Rightarrow \qquad \frac{dV}{V} + \frac{dx}{x} = 0$$

Integrating, $\qquad Vx = c$

$\therefore$ The primitive is

$$\frac{xy^2}{z^3} = c$$

EXAMPLE 9.3 Solve $(2xy^2 + 2xy + 2xz^2 + 1)dx + dy + 2zdz = 0$.

Solution Here $P = 2xy^2 + 2xy + 2xz^2 + 1; Q = 1; R = 2z$

$$\therefore \qquad P\left(\frac{\partial Q}{\partial z} - \frac{\partial R}{\partial y}\right) + Q\left(\frac{\partial R}{\partial x} - \frac{\partial P}{\partial z}\right) + R\left(\frac{\partial P}{\partial y} - \frac{\partial Q}{\partial x}\right)$$

$$= (2xy^2 + 2xy + 2xz^2 + 1)(0) + (-4xz) + 2z(2x) = 0$$

Therefore, the condition of integrability is satisfied. Assuming x = constant, equation reduces to

$$dy + 2zdz = 0$$

Integrating,

$$\Rightarrow \qquad y + z^2 = \text{Constant}$$

$$\therefore \qquad V = y + z^2$$

$$\lambda Q = \frac{\partial V}{\partial y} = 1$$

$$\therefore \qquad S = \lambda P - \frac{\partial V}{\partial x}$$

$$S = 2xy^2 + 2xy + 2xz^2 + 1$$

$$S = 2xy^2 + 2x(y + z^2) + 1 = 2xy^2 + 2xV + 1$$

The given equation reduces to

$$dV + (2xy^2 + 2xV + 1) = 0$$

$$\Rightarrow \qquad \frac{dV}{dx} + 2xV = -(2x^2 + 1) = 0$$

which is linear equation and its If = e^{x2}. The solution of this linear equation is

$$(V + x)e^{x^2} = c$$

$\therefore$ The primitive is

$$(y + x + z^2)e^{x^2} = c$$

EXAMPLE 9.4 Solve $yz(1+4xz)dx - xz(1+2xz)dy - xydz = 0.$

Solution Here $P = yz(1+4xz); Q = -xz(1+2xz); R = -xy$. It is left for the reader to check that

$$P\left(\frac{\partial Q}{\partial z} - \frac{\partial R}{\partial y}\right) + Q\left(\frac{\partial R}{\partial x} - \frac{\partial P}{\partial z}\right) + R\left(\frac{\partial P}{\partial y} - \frac{\partial Q}{\partial x}\right) = 0$$

Therefore, the condition of integrability is satisfied. Assuming x = Constant, equation reduces to

$$-xz(1+2xz)dy - xydz = 0$$

$$\Rightarrow \qquad \frac{dy}{y} + \frac{dz}{z(1+2xz)} = 0$$

$$\Rightarrow \qquad \frac{dy}{y} + \left\{\frac{1}{z} - \frac{2x}{1+2xz}\right\}dz = 0$$

Integrating,

$$\log y + \log z - \log(1+2xz) = c$$

$$\Rightarrow \qquad \frac{yz}{1+2xz} = \text{constant}$$

$$\therefore \qquad V = \frac{yz}{1+2xz}$$

$$\therefore \qquad \lambda Q = \frac{\partial V}{\partial y} = \frac{z}{1+2xz}$$

$$\Rightarrow \qquad -xz(1+2xz)\lambda = \frac{z}{1+2xz}$$

$$\therefore \qquad \lambda = \frac{-1}{x(1+2xz)^2}$$

$$\therefore \qquad S = \lambda P - \frac{\partial V}{\partial x} = -\frac{yz}{1+2xz} = -\frac{V}{x}$$

The given equation reduces to

$$dV + \frac{V}{x}dx = 0$$

$\Rightarrow$
$$\frac{dV}{V} - \frac{dx}{x} = 0$$

Integrating,

$$\frac{V}{x} = c \ \Rightarrow \ V = cx$$

$\therefore$ The primitive is

$$\frac{yz}{1+2xz} = cx$$

9.4 SOLUTION BY INSPECTION

When the condition of integrability is satisfied, by rearranging the given equation, one can find primitive.

EXAMPLE 9.5 Solve $(y + z)dx + (z + x)dx + (x + y)dz = 0$.

Solution Here

$$P = y + z,$$
$$Q = z + x$$
$$R = x + y$$

$\therefore$
$$\sum P\left(\frac{\partial Q}{\partial z} - \frac{\partial R}{\partial y}\right) = (y+z)(1-1) + (z+x)(1-1) + (x+y)(1-1)$$

$\therefore$ Given equation is integrable.
Equation can be written as

$$(xdy + ydx) + (zdy + ydz) + (xdz + zdx) = 0$$

$\Rightarrow$
$$d(xy + yz + zx) = 0$$

Integrating,
$$xy + yz + zx = \text{Constant.}$$

EXAMPLE 9.6 Solve $(ydx + xdy)(a - z) + xydz = 0$.

Solution Reader can check integrability condition. Here

$$P = (a - z)y$$
$$Q = (a - z)x$$
$$R = xy$$

Given equation can be written as

$$\frac{ydx + xdy}{xy} + \frac{dz}{a-z} = 0$$

$\Rightarrow$
$$\frac{d(xy)}{xy} - \frac{d(a-z)}{a-z} = 0$$

Integrating, $$\frac{xy}{a-z} = c$$

$\therefore$ $$xy = c(a - z)$$

is the required primitive.

EXAMPLE 9.7 Solve $yz \log zdx - zx \log zdy + xydz = 0$.

Solution Here

$$P = yz \log x$$

$$Q = zx \log z$$

$$R = xy$$

The reader can check that given equation is integrable.
Rewriting equation

$$\frac{dx}{x} - \frac{dy}{y} + \frac{dz}{z \log z} = 0$$

Integrating, $$\log x - \log y + \log (\log z) = \log c$$

$$x \log z = cy$$

is the required primitive.

EXAMPLE 9.8 Solve $(y+z)(x^2y^2z^2 - 1)\,dx + x^2y^2z^2(xy+xz)\left(\frac{dy}{y} + \frac{dz}{z}\right) + x(dy+dz) = 0.$

Solution Dividing whole equation by $x(y + z)$ gives

$$\left(xy^2z^2 - \frac{1}{x}\right)dx + x^2y^2z^2\left(\frac{dy}{y} + \frac{dz}{z}\right) + \frac{d(y+z)}{y+z} = 0$$

$\Rightarrow$ $$(xy^2z^2dx + x^2yz^2dy + x^2y^2zdz) - \frac{dx}{x} + \frac{d(y+z)}{y+z} = 0$$

$\Rightarrow$ $$d\left(\frac{1}{2}x^2y^2z^2 - \log x + \log (y+z)\right) = 0$$

Integrating, $$\frac{1}{2}x^2y^2z^2 - \log x + \log (y+z) = c$$

which is required primitive.

EXAMPLE 9.9 Solve $(z^3 + zx^2 + 2yx^2 - y^2z)\,dy + (y^3 + yx^2 + 2zx^2 - z^2y)\,dz = 2x(y^2 + z^2 + yz)\,dx.$

Solution Given equation can be rewritten as

$$(z^3dy - y^2zdy + y^3dz - z^2ydz) + x^2[(z+2y)\,dy + (y+2z)\,dz] = 2x(y^2 + z^2 + yz)\,dx$$

$\Rightarrow$ $$(z^2 - y^2)(zdy - ydz) = (y^2 + z^2 + yz)\,d(x^2) - x^2d(y^2 + z^2 + yz)$$

$$\Rightarrow \quad z^2(z^2 - y^2)d\left(\frac{y}{z}\right) = d\left(\frac{x^2}{y^2 + z^2 + yz}\right)(y^2 + z^2 + yz)^2$$

$$\Rightarrow \quad \frac{z^2(z^2 - y^2)}{(y^2 + z^2 + yz)^2} d\left(\frac{y}{z}\right) = d\left(\frac{x^2}{y^2 + z^2 + yz}\right)$$

Put
$$\frac{y}{z} = U$$

Then
$$\frac{1 - U^2}{(U^2 + U + 1)^2} dU = d\left(\frac{x^2}{y^2 + z^2 + yz}\right)$$

Integrating,
$$\frac{x^2}{y^2 + z^2 + yz} = \int \frac{1 - U^2}{(U^2 + U + 1)^2} dU + c.$$

EXAMPLE 9.10 Solve $z(1 - z^2)\,dx + z dy - (x + y + xz^2)\,dz = 0.$

Solution Equation can be written as

$$z(dx + dy) - z^2(zdx + xdz) - (x + y)\,dz = 0$$

$$\Rightarrow \quad zd(x + y) - z^2 d(xz) - (x + y)\,dz = 0$$

$$\Rightarrow \quad \frac{zd(x + y) - (x + y)dz}{z^2} - d(xz) = 0$$

$$\Rightarrow \quad d\left(\frac{x + y}{z}\right) - d(xz) = 0$$

$$\Rightarrow \quad d\left(\frac{x + y}{z} - xz\right) = 0$$

Integrating,
$$\frac{x + y}{z} - xz = c$$

is the required primitive.

EXAMPLE 9.11 Solve

$$(2y^2 + 4az^2x^2)xdx + \left\{3y + 2x^2 + \frac{1}{\sqrt{y^2 + z^2}}\right\}ydy + \left\{4z^2 + 2ax^4 + \frac{1}{\sqrt{y^2 + z^2}}\right\}zdz = 0.$$

Solution Equation can be written as

$$(2y^2xdx + 2x^2ydy) + (4az^2x^3dx + 2ax^4zdz) + 3y^2dy + 4z^3dz + \frac{1}{\sqrt{y^2 + z^2}}(ydy + zdz) = 0$$

$$\Rightarrow \quad d(x^2y^2 + ax^4z^2 + y^3 + z^4 + \sqrt{y^2 + z^2}) = 0$$

Integrating,

$$x^2y^2 + ax^4z^2 + y^3 + z^4 + \sqrt{y^2 + z^2} = c$$

is the required primitive.

EXAMPLE 9.12 Solve $(y+z)(x^2y^2z^2 - 1)dx + x^2y^2z^2(xy + xz)\left(\frac{dy}{y} + \frac{dz}{z}\right) + x(dy + dz) = 0.$

Solution Dividing whole equation by $x(y + z)$ gives

$$\left(xy^2z^2 - \frac{1}{x}\right)dx + x^2y^2z^2\left(\frac{dy}{y} + \frac{dz}{z}\right) + \frac{d(y+z)}{y+z} = 0$$

$$\Rightarrow \qquad (xy^2z^2dx + x^2yz^2dy + x^2y^2zdz) - \frac{dx}{x} + \frac{d(y+z)}{y+z} = 0$$

$$\Rightarrow \qquad d\left(\frac{1}{2}x^2y^2z^2 - \log x + \log(y+z)\right) = 0$$

Integrating,

$$\frac{1}{2}x^2y^2z^2 - \log x + \log(y+z) = c$$

is the required primitive.

EXAMPLE 9.13 Solve $(3x^2 + 2xy - y^2 + z)dx + (x^2 - 2xy - 3y^2 + z)dy + (x + y)dz = 0.$

Solution Rewrite equation as

$$(3x^2dx - 3y^2dy) + (2xydx + x^2dy) - (y^2dx + 2xydy) + zd(x+y) + (x+y)dz = 0$$

$$\Rightarrow \qquad d\{x^3 - y^3 + x^2y - xy^2 + z(x+y)\} = 0$$

Integrating,

$$x^3 - y^3 + x^2y - xy^2 + z(x+y) = c$$

is the required primitive.

EXAMPLE 9.14 Solve $(x^2z - y^3)dx + 3xy^2dy + x^3dz = 0.$

Solution Rewrite equation as

$$x^2(zdx + xdz) - y^3dx + 3xy^2dy = 0$$

$$\Rightarrow \qquad d(zx) - \frac{y^3}{x^2}dx + \frac{3y^2}{x}dy = 0$$

$$\Rightarrow \qquad d\left(zx + \frac{y^3}{x}\right) = 0$$

Integrating,

$$zx + \frac{y^3}{x} = c$$

is the required primitive.

9.5 HOMOGENEOUS EQUATIONS

Let in the equation

$$Pdx + Qdy + Rdz = 0 \tag{9.12}$$

coefficients P, Q and R are homogeneous functions of the same degree in x, y and z. Then one of the variables can be separated from the other two by the substitution

$$x = Uz \text{ and } y = Vz$$

$\therefore$ If

$$P = z^n f_1\left(\frac{x}{z}, \frac{y}{z}\right)$$

$$Q = z^n f_2\left(\frac{x}{z}, \frac{y}{z}\right)$$

and

$$R = z^n f_3\left(\frac{x}{z}, \frac{y}{z}\right)$$

then Eq. (9.12) will reduce to

$$z^{n+1}[f_1(U,V)dU + f_2(U,V)dV] + z^n[Uf_1(U,V) + Vf_2(U,V) + f_3(U,V)]dz = 0 \tag{9.13}$$

Case 1: If the coefficient of dz in Eq. (9.13) is zero, then

$$f_1(U,V)\,dU + f_2(U,V)\,dV = 0 \tag{9.14}$$

can be solved for U and V. The solution of Eq. (9.12) can be obtained by replacing U and V by $\frac{x}{z}$ and $\frac{y}{z}$, respectively in the solution of Eq. (9.14).

Case 2: If the coefficient of in Eq. (9.13) is not zero, then

$$\frac{f_1(U,V)\,dU + f_2(U,V)\,dV}{Uf_1(U,V) + Vf_2(U,V) + f_3(U,V)} + \frac{dz}{z} = 0 \tag{9.15}$$

Clearly, second term is perfect differential if the condition of integrability is satisfied the first term will also be a perfect differential.

Hence, Eq. (9.14) can be integrated which is required primitive of Eq. (9.11) where $U = \frac{x}{z}$ and $V = \frac{y}{z}$.

Note: Since Eq. (9.12) becomes exact on dividing by the factor

$$z^{n+1}\left[Uf_1(U,V)+Vf_2(U,V)+f_3(U,V)\right]=Px+Qy+Rz$$

The expression

$$\frac{1}{Px+Qy+Rz}$$

(Provided $Px + Qy + Rz \neq 0$) is an integrating factor of the Eq. (9.11).

EXAMPLE 9.15 Solve $(y^2 + yz + z^2)\,dx + (x^2 + xz + z^2)\,dy + (x^2 + xy + y^2)\,dz = 0$.

Solution Given $(y^2 + yz + z^2)\,dx + (x^2 + xz + z^2)\,dy + (x^2 + xy + y^2)\,dz = 0$ (i)

It is left for the reader to check integrability of Eq. (i). The given equation is homogeneous.

Put $x = Uz$

and $y = Vz$

$\therefore$ $dx = zdU + Udz$

and $dy = zdU + Vdz$

with these substitutions, Eq. (i) can be written as

$$z^3[(V^2+V+1)\,dU+(U^2+U+1)\,dV]+z^2(U+V+1)(UV+U+V)\,dz=0$$

$$\Rightarrow\quad \frac{(V^2+V+1)\,dU+(U^2+U+1)\,dV}{(U+V+1)(UV+U+V)}+\frac{dz}{z}=0 \tag{ii}$$

Let $p = (U + V + 1)(UV + U + V)$

$$dp=(dU+dV)(UV+U+V)+dU(V+1)(U+V+1)+dV(U+1)(U+V+1)$$
$$=2(UV+U+V)(dU+dV)+(V^2+V+1)\,dU+(U^2+U+1)\,dV$$

$$\Rightarrow\quad (V^2 + V + 1)\,dU + (U^2 + U + 1)\,dV = dp - 2(UV + U + V)(dU + dV)$$

$\therefore$ Equation (ii) reduces to

$$\frac{dp}{p}-\frac{2(dU+dV)}{U+V+1}+\frac{dz}{z}=0$$

$$\Rightarrow\quad d\left[\log\left(\frac{zp}{(U+V+1)^2}\right)\right]=0$$

Integrating,

$$\frac{pz}{(U+V+1)^2}=c$$

$$\Rightarrow\quad \frac{(U+V+1)+(UV+U+V)z}{(U+V+1)^2}=c$$

$$\Rightarrow \quad \frac{(UV+U+V)z}{(U+V+1)} = c$$

$$\Rightarrow \quad (xy + yz + zx) = c(x + y + z).$$

EXAMPLE 9.16 Solve $(yz + z^2)\, dx - xzdy + dydz = 0.$

Solution It is left for the reader to check the integrability of the given equation. The equation being homogeneous, put

$$x = Uz$$

and

$$y = Vz$$

So that

$\therefore$
$$dx = zdU + Udz$$

and

$$dy = zdV + Vdz$$

$\therefore$ Given equation is equivalent to

$$z(V + 1)\, dU - zUdV + U(V + 1)\, dz = 0$$

$$\Rightarrow \quad \frac{dU}{U} - \frac{dV}{V+1} + \frac{dz}{z} = 0$$

$$\Rightarrow \quad d\left(\log \frac{Uz}{V+1}\right) = 0$$

Integrating,

$$\log \frac{Uz}{V+1} = \log c$$

$$\Rightarrow \quad Uz = c(V + 1)$$

$$\Rightarrow \quad xz = c(y + z).$$

EXAMPLE 9.17 Solve $yz^2(x^2 - yz)\, dx + zx^2\, (y^2 - zx)\, dy + xy^2\, (z^2 - xy)\, dz = 0.$

Solution It is left for the reader to check the integrability condition. The equation being homogeneous, put

$$x = Uz$$

and

$$y = Vz$$

$\therefore$
$$dx = zdU + Udz$$

and

$$dy = zdV + Vdz$$

with these substitutions, given equation reduces to

$$V(U^2 - V)\, dU + U^2(V^2 - U)\, dV = 0$$

$$\Rightarrow \quad \left(\frac{1}{V} - \frac{1}{U^2}\right) dU + \left(1 - \frac{U}{V^2}\right) dV = 0$$

$$\Rightarrow \quad \left(\frac{dU}{V} - \frac{U}{V^2}\, dV\right) - \frac{1}{U^2}\, dU + dV = 0$$

$$\Rightarrow \qquad d\left(\frac{U}{V}+\frac{1}{U}+V\right)=0$$

Integrating,
$$\frac{U}{V}+\frac{1}{U}+V=c$$

$$\Rightarrow \qquad \frac{x}{y}+\frac{z}{x}+\frac{y}{z}=c.$$

EXAMPLE 9.18 Solve $(y^2z-y^3+x^2y)dx-(x^2z+x^3-xy^2)dy+(x^2y-xy^2)dz=0.$

Solution It is left for the reader to check the integrability condition. The equation being homogeneous, put

$$x = Uz$$

and
$$y = Vz$$

$\therefore$
$$dx = zdU + Udz$$

and
$$dy = zdV + Vdz$$

with these substitutions, given equation reduces to

$$(V^2-V^3+U^3V)dU+(-V^2-U^3+UV^2)dV=0$$

$$\Rightarrow \qquad (U^2-V^2)(VdU-UdV)+V^2dU-U^2dV=0$$

Dividing by U^2V^2, we get

$$\Rightarrow \qquad \left(1-\frac{V^2}{U^2}\right)d\left(\frac{U}{V}\right)+\frac{1}{U^2}dU-\frac{1}{V^2}dV=0$$

$$\Rightarrow \qquad d\left(\frac{U}{V}+\frac{V}{U}-\frac{1}{U}+\frac{1}{V}\right)=0$$

Integrating,
$$\frac{U}{V}+\frac{V}{U}-\frac{1}{U}+\frac{1}{V}=c$$

$$\Rightarrow \qquad \frac{x}{y}+\frac{y}{x}-\frac{z}{x}+\frac{z}{y}=c$$

is the required primitive.

EXAMPLE 9.19 Solve $y(y+z)dx+z(x+z)dy+y(y-x)dz=0.$

Solution It is left for the reader to check the integrability condition. The equation being homogeneous, put

$$x = Uz$$

and
$$y = Vz$$

$\therefore$ $$dx = zdU + Udz$$

and $$dy = zdV + Vdz$$

with these substitutions, given equation reduces to

$$z^3[(V^2 + V)dU + (U+1)dV] + z^2[(V^2 + V)(U+1)]dz = 0$$

Dividing by $z^3(V^2 + V)(U + 1)$, we get

$$\Rightarrow \quad \frac{dU}{U+1} + \frac{dV}{V(V+1)} + \frac{dz}{z} = 0$$

$$\Rightarrow \quad \frac{dU}{U+1} + \left\{\frac{1}{V} - \frac{1}{(V+1)}\right\} dV + \frac{dz}{z} = 0$$

$$\Rightarrow \quad d\left(\log \frac{zV(U+1)}{V+1}\right) = 0$$

Integrating,

$$\frac{zV(U+1)}{V+1} = c$$

$$\Rightarrow \quad \frac{(x+z)y}{y+z} = c$$

is the required primitive.

9.6 NON-INTEGRABLE SINGLE DIFFERENTIAL EQUATIONS

In this section, we will discuss the method when the coefficients P, Q, R of the Eq. (9.3),

$$Pdx + Qdy + Rdz = 0$$

does not satisfy the condition of integrability.

Consider an arbitrary relation

$$f(x, y, z) = 0 \tag{9.16}$$

between x, y and z.

On differentiating Eq. (9.16),

$$= \frac{\partial f}{\partial x} dx + \frac{\partial f}{\partial y} dy + \frac{\partial f}{\partial z} dz = 0 \tag{9.17}$$

when f in Eq. (9.16) is known, Eqs. (9.16) and (9.17) determine one of the variables and its differential in terms of the other two variables and their differentials. Thus, one of the variables and its differential can be eliminated between Eqs. (9.3), (9.16) and (9.17). Eliminating z and dz between these three equations, we get

$$Mdx + Ndy = 0 \tag{9.18}$$

where M and N are functions of x and y. Integrate Eq. (9.18). The integral containing an arbitrary constant together with the relation given in Eq. (9.16) constitute a solution of the differential equation.

EXAMPLE 9.20 Solve $2zdx + dy + ydz = 0$ consistent with $x + y + z = 0$.

Solution It is left for the reader to check that the given equation

$$2zdx + dy + ydz = 0 \tag{i}$$

is not integeable.

Also given $$x + y + z = 0 \tag{ii}$$

$\therefore$ $$z = -x - y$$

and $$dz = -dx - dy$$

Substituting the values of z and dz in Eq. (i) gives

$$-2(x + y)\,dx + dy - y(dx + dy) = 0$$

Integrating, $$(2x + y + 2)^2 = c(2x + y + 1) \tag{iii}$$

Eqs. (ii) and (iii) constitute the solution.

9.7 EQUATIONS CONTAINING MORE THAN THREE VARIABLES

Consider the equation

$$Pdx + Qdy + Rdz + Tdt = 0 \tag{9.19}$$

which contains four variables x, y, z and t. Let Eq. (9.19) has a complete integral of the form

$$f(x, y, z, t) = 0 \tag{9.20}$$

Then $$df = 0$$

$\Rightarrow$ $$\frac{\partial f}{\partial x}dx + \frac{\partial f}{\partial y}dy + \frac{\partial f}{\partial z}dz + \frac{\partial f}{\partial t}dt = 0 \tag{9.21}$$

Comparing Eqs. (9.19) and (9.21)

$$\left.\begin{aligned} \lambda P &= \frac{\partial f}{\partial x} \\ \lambda Q &= \frac{\partial f}{\partial y} \\ \lambda R &= \frac{\partial f}{\partial z} \\ \text{and} \quad \lambda T &= \frac{\partial f}{\partial t} \end{aligned}\right\} \tag{9.22}$$

From the first two relations of Eq. (9.22)

$$\frac{\partial}{\partial y}(\lambda P) = \frac{\partial^2 f}{\partial y \partial x} = \frac{\partial}{\partial x}(\lambda Q)$$

$$\Rightarrow \qquad \lambda\left(\frac{\partial P}{\partial y}-\frac{\partial Q}{\partial x}\right)=Q\frac{\partial \lambda}{\partial x}-P\frac{\partial \lambda}{\partial y} \tag{9.23}$$

Similarly,

$$\lambda\left(\frac{\partial Q}{\partial z}-\frac{\partial R}{\partial y}\right)=R\frac{\partial \lambda}{\partial y}-Q\frac{\partial \lambda}{\partial z} \tag{9.24}$$

$$\lambda\left(\frac{\partial R}{\partial t}-\frac{\partial T}{\partial z}\right)=T\frac{\partial \lambda}{\partial z}-R\frac{\partial \lambda}{\partial t} \tag{9.25}$$

$$\lambda\left(\frac{\partial R}{\partial x}-\frac{\partial P}{\partial z}\right)=P\frac{\partial \lambda}{\partial z}-R\frac{\partial \lambda}{\partial x} \tag{9.26}$$

$$\lambda\left(\frac{\partial T}{\partial x}-\frac{\partial P}{\partial t}\right)=P\frac{\partial \lambda}{\partial t}-T\frac{\partial \lambda}{\partial x} \tag{9.27}$$

$$\lambda\left(\frac{\partial Q}{\partial t}-\frac{\partial T}{\partial y}\right)=T\frac{\partial \lambda}{\partial y}-Q\frac{\partial \lambda}{\partial t} \tag{9.28}$$

Multiplying Eqs. (9.23), (9.24) and (9.26) by R, P and Q, respectively and adding

$$P\left(\frac{\partial Q}{\partial z}-\frac{\partial R}{\partial y}\right)+Q\left(\frac{\partial R}{\partial x}-\frac{\partial P}{\partial z}\right)+R\left(\frac{\partial P}{\partial y}-\frac{\partial Q}{\partial x}\right)=0 \tag{9.29}$$

Similarly,

$$P\left(\frac{\partial Q}{\partial t}-\frac{\partial T}{\partial y}\right)+Q\left(\frac{\partial T}{\partial x}-\frac{\partial P}{\partial t}\right)+T\left(\frac{\partial P}{\partial y}-\frac{\partial Q}{\partial x}\right)=0 \tag{9.30}$$

$$T\left(\frac{\partial Q}{\partial z}-\frac{\partial R}{\partial y}\right)+Q\left(\frac{\partial R}{\partial t}-\frac{\partial T}{\partial z}\right)+R\left(\frac{\partial T}{\partial y}-\frac{\partial Q}{\partial t}\right)=0 \tag{9.31}$$

$$P\left(\frac{\partial T}{\partial z}-\frac{\partial R}{\partial t}\right)+T\left(\frac{\partial R}{\partial x}-\frac{\partial P}{\partial z}\right)+R\left(\frac{\partial P}{\partial t}-\frac{\partial T}{\partial x}\right)=0 \tag{9.32}$$

The relations between Eq. (9.29) to Eq. (9.32) are the necessary conditions for Eq. (9.19) to have a primitive of the form given in Eq. (9.20). All these conditions are not independent as any one of these can be derived from the remaining three.

Note: In general, for n-variables, the number of independent equations is $\frac{1}{2}(n-1)(n-2)$.

9.8 SOLUTION METHODOLOGY

After establishing integrability, follow the method as stated below to find the solution.

Treat any two variables constant and integrate the function. Replace the arbitrary constant by an arbitrary function of all those variables which were treated as constant. Differentiate the equation so obtained with regard to all the variables and compare it with the given equation to obtain a relation free of the two variables (which were not taken constant) and their differentials taking the help of the integral relation if required. The new relation will be an equation involving $(n - 1)$ differentials instead of n. Repeat the procedure to this new relation to decrease the number of differentials, Continue till the solution is obtained.

EXAMPLE 9.21 Solve $z(y + z)\,dx + z(t - x)\,dy + y(x - t)\,dz + y(y + z)\,dt = 0$.

Solution It is left for the reader to check the integrability of given equation. We have

$$z(y + z)\,dx + z(t - x)\,dy + y(x - t)\,dz + y(y + z)\,dt = 0 \quad \text{(i)}$$

Put $\quad y = \text{Constant}$

and $\quad z = \text{Constant}$

So that $\quad dy = 0 = dz$

$\therefore$ Equation (i) reduces to

$$z(y + z)\,dx + y(y + z)\,dt = 0$$

$$zdx + ydt = 0$$

Integrating, $\quad xz + yt = \text{Constant} = f \quad$ (ii)

where f is a function of y and z.

Differentiating Eq. (ii) w.r.t. all the variables,

$$xdz + zdx + ydt + tdy + df \quad \text{(iii)}$$

Comparing Eqs. (i) and (iii)

$$df = dy + dz$$

Integrating, $\quad f = c(y + z)$

$\therefore \quad xz + yt = c(y + z)$

is the primitive.

EXAMPLE 9.22 Solve $y \cos tdx + x \cos tdy + xy \cos tdz + xy \sin tdt = 0$.

Solution Given

$$y \cos tdx + x \cos tdy + xy \cos tdz + xy \sin tdt = 0 \quad \text{(i)}$$

Put $\quad z = \text{Constant}$

and $\quad t = \text{Constant}$

So that $\quad dz = 0 = dt$

$\therefore$ Equation (i) reduces to

$$y \cos tdx + x \cos tdy = 0$$

$\Rightarrow \quad \cos t\,(ydx + xdy) = 0$

$\Rightarrow \quad ydx + xdy = 0$

$\Rightarrow \quad d(xy) = 0$

Integrating, $\quad xy = \text{Constant} = f \quad$ (ii)

where f is a function of z and t.

Differentiating Eq. (ii) w.r.t. all the variables.

$$ydx + xdy = df \quad \text{(iii)}$$

Eliminating x, y, dx and dy from Eqs. (i) to (iii)

$$\cos t df + f \cos t dz + f \sin t dt = 0$$

$$\frac{df}{f} + dz + \sin t dt = 0$$

$$\Rightarrow \quad d[\log f + z - \log \cos t] = 0$$

Integrating,

$$\log f + z - \log \cos t = 0$$

$$\Rightarrow \quad \frac{f}{\cos t} e^z = 0$$

$$\Rightarrow \quad xy = ce^z \cos t$$

is the primitive.

9.9 EQUATIONS OF HIGHER DEGREE

Consider the equation of the form

$$Pdx^2 + Qdy^2 + Rdz^2 + 2P'dydz + 2Q'dzdx + 2R'dxdy = 0 \tag{9.33}$$

where P, Q, R, P', Q', and R' are functions of x, y and z.

Equation (9.33) can be written as

$$\frac{1}{R}[(Rdz + P'dy + Q'dx)^2 - \{(Q'^2 - RP)dx^2 + 2(P'Q' - RR')\,dxdy + (P'^2 - RQ)dy^2\}] = 0$$

In order to resolve Eq. (9.32) into two factors, the expression

$$(Q'^2 - RR)\,dx^2 + 2(P'Q' - RR')\,dxdy + (P'^2 - RQ)\,dy^2$$

must be a perfect square, which requires

$$(P'Q' - RR')^2 = (Q'^2 - RP)\,(P^2 - RQ)$$

$$\Rightarrow \quad R(PQR + 2P'Q'R' - PP'^2 - QQ'^2 - RR'^2) = 0$$

Since $R \neq 0$, the condition reduces to

$$PQR + 2P'Q'R' - PP'^2 - QQ'^2 - RR'^2 = 0 \tag{9.34}$$

when condition given in Eq. (9.34) is satisfied, the Eq. (9.33) may be replaced by two equations each of the form

$$Pdx + Qdy + Rdz = 0$$

The solution of either of these will be a particular solution of Eq. (9.33) and the two general solution taken together will be the general solution.

If the solution of the linear equations is of the forms

$$f_1(x, y, z) = c_1$$

and

$$f_2(x, y, z) = c_2$$

then the general solution will be

$$[f_1(x, y, z) - c_1]\,[f_2(x, y, z) - c_2] = 0 \tag{9.35}$$

EXAMPLE 9.23 Solve

$$xyz(dx^2 + dy^2 + dz^2) + x(y^2 + z^2)\,dydz + y(z^2 + x^2)\,dzdx + z(x^2 + y^2)\,dxdy = 0.$$

Solution Given

$$xyz(dx^2 + dy^2 + dz^2) + x(y^2 + z^2)\,dydz + y(z^2 + x^2)\,dzdx + z(x^2 + y^2)\,dxdy = 0 \quad \text{(i)}$$

Here
$$P = xyz + Q = R$$

$$P' = \frac{1}{2}x(y^2 + z^2)$$

$$Q' = \frac{1}{2}y(z^2 + x^2)$$

and
$$R' = \frac{1}{2}z(x^2 + y^2)$$

Now,
$$PQR + 2P'Q'R' - PP'^2 - QQ'^2 - RR'^2$$

$$= x^3y^3z^3 + \frac{1}{4}xyz(y^2 + z^2)(z^2 + x^2)(x^2 + y^2) - \frac{1}{2}x^2yz(y^2 + z^2) - \frac{1}{2}xy^2z(z^2 + x^2) - \frac{1}{2}xyz^2(x^2 + y^2)$$

$= x^3y^3z^3 - x^3y^3z^3$

$= 0$

∴ Condition given in Eq. (9.33) is satisfied. The equation can be written as

$$(xdx + ydy + zdz)\,(yzdx + zxdy + xydz) = 0$$

∴ Two linear equations are

$$xdx + ydy + zdz = 0$$

and
$$yzdx + zxdy + dydz = 0$$

Integrating,

and
$$\left.\begin{aligned} x^2 + y^2 + z^2 &= c_1 \\ xyz &= c_2 \end{aligned}\right\} \quad \text{(ii)}$$

where c_1 and c_2 are constants. Equation (ii) constitutes the complete solution.

EXERCISES

Solve the following differential equations:

1. $zy(1 + 4xz)dx - xz(1 + 2xz)dy - xydz = 0.$ **[Ans.** $yz = cx(1 + 2xz)$]

2. $(3x^2 + 2xy - y^2 + z)dx + (x^2 - 2xy - 3y^2 + z)dy + (x + y)dz = 0.$

   [Ans. $x^3 - y^3 + x^2y - xy^2 + z(x + y) = c$]

3. $zydx - zxdy - y^2dz = 0.$ **[Ans.** $z = ce^{x/y}$]

4. $yz \log z dx - zx \log z dy + xy dz = 0.$ **[Ans.** $y = cx \log z$ **]**

5. $a^2y^2z^2dx + b^2z^2x^2dy + c^2x^2y^2dz = 0.$ $\left[\textbf{Ans.}\ \dfrac{a^2}{x} + \dfrac{b^2}{y} + \dfrac{c^2}{z} = k\right]$

6. $(y^2 + z^2 + x^2)dx - 2xydy - 2xzdz = 0.$ $\left[\textbf{Ans.}\ \dfrac{y^2 + z^2}{x} - x = c\right]$

7. $(yz + xyz)dx + (zx + xyz)dy + (xy + xyz)dz = 0.$ **[Ans.** $\log(xyz) + x + y + z = c$ **]**

8. $(2y^2 + 4az^2x^2)xdx + \{3y + 2x^2 + \sqrt{y^2 + z^2}\}\, ydy + \{4z^2 + 2ax^4 + \sqrt{y^2 + z^2}\}\, zdz = 0.$

$\left[\textbf{Ans.}\ x^2y^2 + 2a^4x^2 + y^3 + z^4 + \sqrt{y^2 + z^2} = c\right]$

9. $(y + a)^2\, dx + zdy - (y + a)\, dz = 0.$ **[Ans.** $z = (x + b)(y + a)$**]**

10. $z(z + y^2)dx + z(z + x^2)dy - xy(x + y)dz = 0.$ **[Ans.** $x(y^2 + z) = z(x + y)(1 - y)$**]**

11. $(ay - bz)dx + (cz - ax)dy + (bx - cy)\, dz = 0.$ **[Ans.** $ax - cz = c_1(ay - bz)$**]**

12. $y^2z(x \cos x - \sin x)dx + x^2z(y \cos y - \sin y)dy + xy(y \sin x + x \sin y + xy \cos z)dz = 0.$

$\left[\textbf{Ans.}\ y\left(\dfrac{\sin x}{x} + \dfrac{\sin y}{y}\right) + \sin z = c\right]$

13. $(2x + yz)\, ydx - x^2dy + (x + 2z)y^2dz = 0.$ $\left[\textbf{Ans.}\ \dfrac{x^2}{y} + xz + z^2 = c\right]$

14. $(1 + y)(2 + 2x + y + z)dx - x(1 + z + x)dy + (1 + y)(1 + x + y)dz = 0.$

[Ans. $(1 + x + y)(1 + z + x) = c(1 + y)$**]**

15. $\dfrac{xdx + ydy + zdz}{\sqrt{x^2 + y^2 + z^2}} + \dfrac{zdx - xdz}{z^2 + x^2} + 3ax^2dx + 2bydy + cdz = 0.$

$\left[\textbf{Ans.}\ \sqrt{x^2 + y^2 + z^2} + \tan^{-1}\dfrac{x}{z} + ax^3 + by^2 + cz = k\right]$

16. $z^2dx + (z^2 - 2yz)dy + (2y^2 - yz - xz)dz = 0.$ $\left[\textbf{Ans.}\ \dfrac{y^2}{z^2} - \dfrac{y}{z} - \dfrac{x}{z} = c\right]$

17. $(y^2 + z^2 + 2xy + 2xz)dx + (x^2 + z^2 + 2xy + 2yz)dy + (x^2 + y^2 + 2xz + 2yz)dz = 0.$

[Ans. $x(y^2 + z^2) + y(x^2 + z^2) + z(x^2 + y^2) = c$**]**

18. $(2xz - yz)dx + (2yz - xz)dy - (x^2 - xy + y^2)dz = 0.$ **[Ans.** $x^2 - xy + y^2 = cz$**]**

19. $\dfrac{y+z-2x}{(y-x)(z-x)}dx+\dfrac{z+x-2y}{(z-y)(x-y)}dy+\dfrac{x+y-2z}{(x-z)(y-z)}dz=0.$

[**Ans.** $(x-y)(y-z)(z-x)=c$]

20. $(x^2y-y^3-y^2z)dx+(xy^2-x^2z-x^3)dx+xy(x+y)dz=0.$ $\left[\textbf{Ans. } \dfrac{y+z}{x}+\dfrac{z+x}{y}=c\right]$

21. $2(2y^2+yz-z^2)dx+x(4y+z)dy+x(y-2z)dz=0.$ [**Ans.** $x^2(2y^2+yz-z^2)=c$]

22. $ydx+xdy-(x+y+2z)dz=0$ consistent with $x+y+2z=0.$ [**Ans.** $xy=c$]

23. $dz=4ydx+3dy.$ $\left[\textbf{Ans. } z=4\int f(x)dx+3f(x)+c,\ y=f(x)\right]$

24. $(x^3-yzt)dx+(y^3-xzt)dy+(z^3-xyt)dz+(t^3-xyz)dt=0.$

[**Ans.** $x^4+y^4+z^4+t^4-4xyzt=c$]

25. $(y+z+t)dx+(z+t+x)dy+(t+x+y)dz+(x+y+z)dt=0.$

[**Ans.** $xy+xz+xt+yz+yt+zt=c$]

26. $(2x+y^2+2xt-z)dx+2xydy-xdz+x^2dt=0.$ [**Ans.** $x^2+xy^2-xz+x^2t=c$]

27. $dxdydz=0.$ [**Ans.** $x=c_1, y=c_2, z=c_3$]

28. $(ye^x-e^z)dx+(e^x+ze^y)dy-(ye^x+ze^y-e^y)dz=0.$ [**Ans.** $ye^x+ze^y-xe^z=ce^z$]

29. $yzdx+(x^2yz-x)dy+(x^2z-xy)dz=0.$ $\left[\textbf{Ans. } y^2+z^2-\dfrac{2yz}{x}=c\right]$

30. $x(y^2-z^2)dx+y(z^2-x^2)dy+z(x^2-y^2)dz=0.$ [**Ans.** $(y^2-z^2)=c(z^2-x^2)$]

31. $y(y+z)dx-xzdy+xydz=0.$ [**Ans.** $xz=c(y+z)$]

32. $yz(y+z)dx+xz(x+z)dy+xy(x+y)dz=0.$ [**Ans.** $xyz=c(x+y+z)$]

33. $2(2y^2+yz-z^2)dx+x(4y+z)dy+x(y-2z)dz=0.$ [**Ans.** $x^2(2y^2+yz-z^2)=c$]

34. Solve $ydx+xdy-(x+y+2z)dz=0$ consistent with $x+y+2z=0.$

[**Ans.** $xz=c(y+z)$]

Partial Differential Equations of First Order

10.1 INTRODUCTION

Let f be a function of two independent variables x and y, and z be a dependent variable, i.e.

$$z = f(x, y)$$

Then the partial derivatives $\dfrac{\partial z}{\partial x}$ and $\dfrac{\partial z}{\partial y}$ will be denoted by p and q, respectively. In general, if

$$z = f(x_1, x_2, \ldots, x_n)$$

then the partial derivatives $\dfrac{\partial z}{\partial x_1}, \dfrac{\partial z}{\partial x_2}, \ldots, \dfrac{\partial z}{\partial x_n}$ will be denoted by $p_1, p_2, \ldots, p_n$, respectively.

Thus, a partial differential equation is a relation between the independent variables, the dependent variable and its partial derivatives.

e.g. (i) $px + qy + xy = z^2$

(ii) $r - 4s + \sin xt = p + q$

are partial differential equations.

The order of the partial differential equation is the order of the highest order partial derivative. The equation is of first order when all the partial derivatives occurring in it are of the order one and is of the second order when the partial derivatives are of order two.

In this chapter, we will discuss P.D.E. of first order only.

10.2 TYPES OF INTEGRALS AND THEIR GEOMETRICAL INTERPRETATIONS

Complete Integral

Consider a relation

$$f(x, y, z, a, b) = 0 \tag{10.1}$$

where a and b are arbitrary constants.

Differentiate Eq. (10.1) partially w.r.t. x and y, we get

$$\frac{\partial f}{\partial x}+\frac{\partial f}{\partial z}p=0 \tag{10.2}$$

and

$$\frac{\partial f}{\partial y}+\frac{\partial f}{\partial z}q=0 \tag{10.3}$$

Eliminating constants a and b from Eqs. (10.1) and (10.3) gives a relation

$$F(x, y, z, p, q) = 0 \tag{10.4}$$

which is partial differential equation of first order.

Conversely, the relation shown in Eq. (10.1) is a solution of Eq. (10.4) then it can not have more than two arbitrary constants because if it has, these cannot be eliminated between Eqs. (10.1) to (10.3) and to eliminate these more equations are needed which can be obtained by differentiating Eqs. (10.2) and (10.3). Clearly, the resulting equation will be of higher order than one.

The relation given in Eq. (10.1) is known as *complete integral* of Eq. (10.4). Thus, *complete integral* of an equation is a relation between the variables as many arbitrary constants as there are independent values.

If particular values are given to the arbitrary constants, the solution so obtained is known as a *particular integral.*

A complete integral, being a relation between x, y and z represents a surface. Since it contains two arbitrary parameters, it belongs to a double infinite system of surfaces or to a single infinite system of family of surfaces.

EXAMPLE 10.1 Eliminate 'a' and 'b' from $z = ax^3 + by^3$.

Solution Let

$$f = z - ax^3 - by^3 \tag{i}$$

Differentiating Eq. (i) partially w.r.t. x and y

$$p - 3ax^2 = 0 \tag{ii}$$

and

$$q - 3by^2 = 0 \tag{iii}$$

Multiply Eq. (ii) by x and Eq. (iii) by y.

$$xp - 3ax^3 = 0$$

and

$$xq - 3by^3 = 0$$

Adding,

$$xp + yq - 3(ax^3 + by^3) = 0$$

$\Rightarrow$

$$xp + yq = 3z.$$

EXAMPLE 10.2 Eliminate 'a' and 'b' from $2z = (ax + y)^2 + b$.

Solution Let

$$f = 2z - (ax + y)^2 + b \tag{i}$$

Differentiating Eq. (i) partially w.r.t. x and y gives

$$2p - 2a(ax + y) = 0$$

and

$$2q - 2(ax + y) = 0$$

$\Rightarrow$ $$p - a(ax + y) = 0 \tag{ii}$$

and $$q - (ax + y) = 0 \tag{iii}$$

Multiply Eq. (ii) by x and Eq. (iii) by y.

$\therefore$ $$xp - ax(ax + y) = 0$$

$$yq - y(ax + y) = 0$$

Adding, $$xp + yq = (ax + y)^2$$

$\Rightarrow$ $$xp + yq = q^2.$$

EXAMPLE 10.3 Eliminate arbitrary constants a, b, and c from $ax^2 + by^2 + cz^2 = 1$.

Solution Let $$f = ax^2 + by^2 + cz^2 - 1 \tag{i}$$

$\therefore$ $$\frac{\partial f}{\partial x} = 2ax + 2czp = 0 \tag{ii}$$

$$\frac{\partial f}{\partial y} = 2by + 2czq = 0 \tag{iii}$$

$$\frac{\partial^2 f}{\partial x^2} = 2a + 2cp^2 + 2czr = 0 \tag{iv}$$

and $$\frac{\partial^2 f}{\partial y^2} = 2b + 2cq^2 + 2czt = 0 \tag{v}$$

Multiply (iv) by x and subtract from (ii)

$$2c[xp^2 + xzr - pr] = 0$$

$\Rightarrow$ $$pz = xp^2 + xzr \tag{vi}$$

Similarly from (iii) and (v) one can get

$$qz = yq^2 + yzt \tag{vii}$$

Singular Integral

If 'a' and 'b' are assumed to be functions of the independent variables those do not change the forms of p and q, the differential equation obtained by the elimination of the functions will be same as in the case when 'a' and 'b' were arbitrary constants.

Differentiating Eq. (10.1) partially w.r.t. x and y, we get

$$\frac{\partial f}{\partial x} + \frac{\partial f}{\partial z} p + \frac{\partial f}{\partial a} \cdot \frac{\partial a}{\partial x} + \frac{\partial f}{\partial b} \cdot \frac{\partial b}{\partial x} = 0 \tag{10.5}$$

and $$\frac{\partial f}{\partial y} + \frac{\partial f}{\partial z} q + \frac{\partial f}{\partial a} \cdot \frac{\partial a}{\partial y} + \frac{\partial f}{\partial b} \cdot \frac{\partial b}{\partial y} = 0 \tag{10.6}$$

Equations (10.5) and (10.6) to be similar to Eqs. (10.2) and (10.3), we have

$$\frac{\partial f}{\partial a}\cdot\frac{\partial a}{\partial x}+\frac{\partial f}{\partial b}\cdot\frac{\partial b}{\partial x}=0 \tag{10.7}$$

and

$$\frac{\partial f}{\partial a}\cdot\frac{\partial a}{\partial y}+\frac{\partial f}{\partial b}\cdot\frac{\partial b}{\partial y}=0 \tag{10.8}$$

Define

$$R=\begin{vmatrix}\dfrac{\partial a}{\partial x} & \dfrac{\partial b}{\partial x}\\ \dfrac{\partial a}{\partial y} & \dfrac{\partial b}{\partial y}\end{vmatrix}$$

On solving Eqs. (10.7) and (10.8), we get

$$\left.\begin{aligned}R\frac{\partial f}{\partial a}&=0\\ \text{and}\qquad R\frac{\partial f}{\partial y}&=0\end{aligned}\right\} \tag{10.9}$$

when $R \neq 0$, we get

$$\left.\begin{aligned}\frac{\partial f}{\partial a}&=0\\ \text{and}\qquad \frac{\partial f}{\partial b}&=0\end{aligned}\right\} \tag{10.10}$$

We can determine the values of 'a' and 'b' from two equations of (10.10). Then the relation shown by Eq. (10.1) is a solution with change in the quantities 'a' and 'b' when the values are substituted in Eq. (10.1) a solution of Eq. (10.4) not having any arbitrary constant, is obtained. In general this solution will be different from any solution obtained from Eq. (10.1) by giving a particular constant values to 'a' and 'b'. Thus, the elimination of 'a' and 'b' between Eqs. (10.1) and (10.10) gives a new solution which is known as *singular integral*. Thus. singular integral is a relation between the variables having no arbitrary constants.

General Integral

The Eq. (10.9) is also satisfied if

$$R = 0$$

i.e. 'a' and 'b' are functionally related.

Let

$$b = \phi(a) \tag{10.11}$$

where ϕ is an arbitrary function. Multiplying Eqs. (10.7) and (10.8) by dx and dy, respectively and adding

$$\frac{\partial f}{\partial a}\,da+\frac{\partial f}{\partial b}\,db=0 \tag{10.12}$$

From Eq. (10.11)

$$db = \frac{\partial \phi}{\partial a} da$$

∴ Equation (10.12) becomes

$$\frac{\partial f}{\partial a} + \frac{\partial f}{\partial b} \cdot \frac{\partial \phi}{\partial a} = 0 \tag{10.13}$$

From Eq. (10.13), the value of 'a' involving the arbitrary function ϕ may be obtained, and hence, 'b' can be obtained. Substitute these values in Eq. (10.1). We will get a new solution which will be different from above discussed two integrals. This solution is known as *general integral*.

A general integral gives the system of families of surfaces as a representative family and finding its envelope. The Eqs. (10.11) and (10.13) represent a curve drawn on the surface of the family whose parameter is 'a' while the equation obtained by eliminating 'a' between them is the envelope of the family.

The Singular Integral represents the general envelope of all the surfaces included in the complete integral.

10.3 SINGULAR INTEGRAL OF P.D.E. BY INSPECTION

Consider a partial differential Eq. (10.1)

$$f(x, y, z, p, q) = 0$$

whose complete integral is

$$F(x, y, z, a, b) = 0 \tag{10.14}$$

The singular integral is given by Eqs. (10.14) and (10.10).

The values of z, p, q derived from Eq. (10.14) when substituted in Eq. (10.1) will give integral equation. Do this substitution in Eq. (10.1) so that p and q gets replaced by functions of x, y, z, a and b. Then the singular integral is given by Eq. (10.1) and the equations obtained on differentiating Eq. (10.1) partially w.r.t. a and b. The equations are

$$\frac{\partial f}{\partial p} \cdot \frac{\partial p}{\partial a} + \frac{\partial f}{\partial q} \cdot \frac{\partial q}{\partial a} = 0 \tag{10.15}$$

and

$$\frac{\partial f}{\partial p} \cdot \frac{\partial p}{\partial b} + \frac{\partial f}{\partial q} \cdot \frac{\partial q}{\partial b} = 0 \tag{10.16}$$

If $\frac{\partial f}{\partial p} \neq 0$ and $\frac{\partial f}{\partial q} \neq 0$, Eqs. (10.15) and (10.16) hold if

$$\frac{\partial p}{\partial a} \cdot \frac{\partial q}{\partial b} - \frac{\partial p}{\partial b} \cdot \frac{\partial q}{\partial a} = 0$$

which gives relation between p and q and it does not contain a and b explicitly. Let this functional relation be

$$\phi(p, q) = 0 \tag{10.17}$$

If both the constants a and b occur in p and q (not necessary), the Eq. (10.17) would mean that one of them is a function of the other and the equations using them gives the general integral.

Equations (10.15) and (10.16) are also satisfied if

$$\frac{\partial f}{\partial p} = 0 \tag{10.18}$$

$$\frac{\partial f}{\partial q} = 0 \tag{10.19}$$

Eliminating p and q between Eqs. (10.1), (10.18) and (10.19) will give a relation between x, y, z independent of any arbitrary constants. The solution obtained here will be singular integral.

EXAMPLE 10.4 Eliminate the function 'f' from $z = e^{ax} f(x + y)$.

Solution We have

$$z = e^{ax} f(x + y) \tag{i}$$

Differentiating Eq. (i) partially w.r.t. x and y, respectively.

$$p = ae^{ax} f(x + y) + e^{ax} f'(x + y) \tag{ii}$$

and

$$q = e^{ax} f'(x + y) \tag{iii}$$

Subtracting Eqs. (ii) and (iii) gives

$$p - q = ae^{ax} f(x + y)$$

$$\Rightarrow \quad p - q = az.$$

EXAMPLE 10.5 Eliminate function 'f' from

$$z = xy + f(x^2 + y^2).$$

Solution We have

$$z = xy + f(x^2 + y^2) \tag{i}$$

Differentiate Eq. (i) partially w.r.t. x and y, respectively.

$$\therefore \quad p = y + 2xf'(x^2 + y^2) \tag{ii}$$

and

$$q = x + 2yf'(x^2 + y^2) \tag{iii}$$

Multiply Eq. (ii) by x and Eq. (iii) by y and subtract. We get

$$px - qy = 0.$$

EXAMPLE 10.6 Eliminate arbitrary functions 'f' and 'g' from $z = f(xy) + g\left(\frac{x}{y}\right)$.

Solution We have

$$z = f(xy) + g\left(\frac{x}{y}\right) \tag{i}$$

Differentiate Eq. (i) partially w.r.t. x and y, respectively.

$$\therefore \quad p = yf'(xy) + \frac{1}{y}g'\left(\frac{x}{y}\right) \tag{ii}$$

and $$q = xf'(xy) - \frac{x}{y^2} g'\left(\frac{x}{y}\right) \tag{iii}$$

Again differentiate Eq. (ii) w.r.t. x

$$\therefore \quad r = y^2 f''(xy) + \frac{1}{y^2} g''\left(\frac{x}{y}\right) \tag{iv}$$

and differentiate Eq. (iii) w.r.t. y

$$\therefore \quad t = x^2 f''(xy) + \frac{2x}{y^3} g'\left(\frac{x}{y}\right) + \frac{x^2}{y^4} g''\left(\frac{x}{y}\right) \tag{v}$$

Multiply Eq. (iv) by x^2 and Eq. (v) by y^2 and subtract.

$$\therefore \quad x^2 r - r^2 t = -\frac{2x}{y} g'$$

$$= -px + qy$$

$$\therefore \quad x^2 r - y^2 t + px - qy = 0.$$

10.4 LINEAR PARTIAL DIFFERENTIAL EQUATION OF ORDER ONE

We have seen that a partial differential equation of order one is said to be linear if it is of first degree in p and q. e.g.

$$xp + yq = xy$$

$$x^2 p + y^2 q = z^2$$

are linear as these are of first degree in p and q.

The equation

$$p + e^x = 2z^2$$

$$x^2 p^2 + y^2 q^2 = z^2$$

are non-linear equations.

The general form of linear partial differential equation of first order is

$$Pp + Qq = R \tag{10.20}$$

where P, Q and R are functions of x, y and z. The Eq. (10.20) is known as *Lagrange's equation.*

In next section, we will discuss the method to solve Eq. (10.20).

10.5 LAGRANGE'S METHOD

We will explore the method in the following theorem.

THEOREM 10.1 The general solution of the linear partial differential equation

$$Pp + Qq = R$$

is

$$f(U, V) = 0$$

where f is an arbitrary function and $U(x, y, z) = c_1$ and $V(x, y, z) = c_2$ form two independent solutions of the equations

$$\frac{dx}{P} = \frac{dy}{Q} = \frac{dz}{R}.$$

Proof: Taking total differential of the equation $U(x, y, z) = c_1$, we get

$$\frac{\partial U}{\partial x}dx + \frac{\partial U}{\partial y}dy + \frac{\partial U}{\partial z}dz = 0$$

Since $U(x, y, z) = c_1$ is a solution of the equation

$$\frac{dx}{P} = \frac{dy}{Q} = \frac{dz}{R}$$

we have,
$$P\frac{\partial U}{\partial x} + Q\frac{\partial U}{\partial y} + R\frac{\partial U}{\partial z} = 0 \tag{10.21}$$

Similarly, $V(x, y, z) = c_2$ is a solution of the equation

$$\frac{dx}{P} = \frac{dy}{Q} = \frac{dz}{R}$$

we have,
$$P\frac{\partial V}{\partial x} + Q\frac{\partial V}{\partial y} + R\frac{\partial V}{\partial z} = 0 \tag{10.22}$$

Solving Eqs. (10.21) and (10.22) for P, Q and R, we get

$$\frac{P}{\dfrac{\partial U}{\partial y}\dfrac{\partial V}{\partial z} - \dfrac{\partial U}{\partial z}\dfrac{\partial V}{\partial y}} = \frac{Q}{\dfrac{\partial U}{\partial z}\dfrac{\partial V}{\partial x} - \dfrac{\partial U}{\partial x}\dfrac{\partial V}{\partial z}} = \frac{R}{\dfrac{\partial U}{\partial x}\dfrac{\partial V}{\partial y} - \dfrac{\partial U}{\partial y}\dfrac{\partial V}{\partial x}}$$

$$\frac{P}{\dfrac{\partial(U, V)}{\partial(y, z)}} = \frac{Q}{\dfrac{\partial(U, V)}{\partial(z, x)}} = \frac{R}{\dfrac{\partial(U, V)}{\partial(x, y)}} \tag{10.23}$$

The relation $f(U, V) = 0$ leads to the partial differential equation

$$\frac{\partial(U, V)}{\partial(y, z)}p + \frac{\partial(U, V)}{\partial(z, x)}q = \frac{\partial(U, V)}{\partial(x, y)} \tag{10.24}$$

Substituting from Eq. (10.23) into Eq. (10.24), we get the equation

$$Pp + Qq = R \tag{10.25}$$

Hence, we have seen that $f(U, V)$ is a solution of the Eq. (10.25) if only if $U(x, y, z) = c_1$ and $V(x, y, z) = c_2$ are the solutions of

$$\frac{dx}{P} = \frac{dy}{Q} = \frac{dz}{R}.$$

EXAMPLE 10.7 Solve $\dfrac{y^2 z}{x}\dfrac{\partial z}{\partial x} + x^2 z \dfrac{\partial z}{\partial y} = y^2.$

Solution We have

$$\frac{y^2 z}{x} p + x^2 z q = y^2$$

The subsidiary equations are

$$\frac{dx}{y^2 z} = \frac{dy}{x^2 z} = \frac{dz}{y^2}$$

Taking the first two ratios, we have

$$\frac{dx}{y^2} = \frac{dy}{x^2}$$

$\Rightarrow$ $$x^2 dx = y^2 dy$$

Integrating, $$x^3 - y^3 = c_1 \quad \text{(i)}$$

Now, take first and third ratios.

$\therefore$ $$xdx = zdz$$

Integrating, $$x^2 - z^2 = c^2 \quad \text{(ii)}$$

From Eqs. (i) and (ii), the general solution is

$$f(x^3 - y^3, x^2 - z^2) = 0.$$

EXAMPLE 10.8 Find the general solution of the differential equation $x^2 p + y^2 q = (x + y) z$.

Solution The subsidiary equations are

$$\frac{dx}{x^2} = \frac{dy}{y^2} = \frac{dz}{(x+y)z} \quad \text{(i)}$$

Taking first two ratios, we have

$$\frac{dx}{x^2} = \frac{dy}{y^2}$$

Integrating, $$\frac{1}{y} - \frac{1}{x} = c_1 \quad \text{(ii)}$$

$$\text{Each ratio } = \frac{dx - dy}{x^2 - y^2} = \frac{dz}{(x+y)z}$$

$\Rightarrow$ $$\frac{dx - dy}{x - y} = \frac{dz}{z}$$

$\Rightarrow$ $$\frac{d(x-y)}{x-y} = \frac{dz}{z}$$

Integrating, $$\frac{x-y}{z} = c_2 \tag{iii}$$

From Eqs. (ii) and (iii), the general solution is

$$f\left(\frac{1}{y}-\frac{1}{x}, \frac{x-y}{z}\right)=0.$$

EXAMPLE 10.9 Solve $(y^2 + z^2)\,p - xyq - xz = 0$.

Solution The subsidiary equations are

$$\frac{dx}{y^2+z^2}=\frac{dy}{-xy}=\frac{dz}{-xz}$$

Taking the last two ratios, we get

$$\frac{dy}{y}=\frac{dz}{z} \tag{i}$$

Integrating, $$\frac{y}{z}=c_1 \tag{ii}$$

$$\text{Each ratio } = \frac{xdx+ydy+zdz}{0}$$

$$\therefore \qquad xdx + ydy + zdz = 0$$

Integrating, $$x^2 + y^2 + z^2 = c^2 \tag{iii}$$

Hence, the general solution of given equation is

$$f\left(\frac{y}{z}, x^2+y^2+z^2\right)=0.$$

EXAMPLE 10.10 Solve $y^2p - xyq = x(z - 2y)$.

Solution The subsidiary equations are

$$\frac{dx}{y^2}=\frac{dy}{-xy}=\frac{dz}{x(z-2y)} \tag{i}$$

Taking first two ratios, we get

$$xdx + ydy = c_1$$

Integrating, $$x^2 + y^2 = c_1 \tag{ii}$$

Taking second and third ratios, we get

$$\frac{dy}{y}=\frac{dz}{2y-z}$$

$$\Rightarrow \qquad 2ydy - (zdy + ydz) = 0$$

$$\Rightarrow \qquad 2ydy - d(zy) = 0$$

Integrating, we have

$$y^2 - yz = c \tag{iii}$$

From Eqs. (ii) and (iii), the general solution of given equation is

$$f(x^2 + y^2, y^2 - yz) = 0.$$

EXAMPLE 10.11 Solve $(y + z)\, p + (z + x) = x + y$.

Solution The subsidiary equations are

$$\frac{dx}{y+z} = \frac{dy}{z+x} = \frac{dz}{x+y} \tag{i}$$

$$\text{Each ratio } = \frac{dx+dy+dz}{2(x+y+z)} = \frac{dx-dy}{y-x} = \frac{dy-dz}{z-y}$$

Taking the first two ratios, we get

$$\frac{1}{2}\frac{d(x+y+z)}{x+y+z} = -\frac{d(x-y)}{x-y}$$

Integrating, $$\frac{1}{2}\log(x+y+z) = -\log(x-y) + \log c_1$$

$\Rightarrow$ $$\log(x+y+z) + 2\log(x+y) = \log c_1$$

$\Rightarrow$ $$(x+y+z)(x+y)^2 = c_1 \tag{ii}$$

Taking last two ratios, we get

$$-\frac{d(x-y)}{x-y} = -\frac{d(y-z)}{y-z}$$

Integrating, $$\log(x-y) = \log(y-z) + \log c_2$$

$\Rightarrow$ $$\frac{x-y}{y-z} = c_2 \tag{iii}$$

From Eqs. (ii) and (iii), the general solution of the equation is

$$f\left[(x+y+z)(x+y)^2, \frac{x-y}{y-z}\right] = 0.$$

EXAMPLE 10.12 Solve $(z^2 - 2yz - y^2)\, p + (xy + zx)\, q = xy - zx$.

Solution The subsidiary equations are

$$\frac{dx}{z^2-2yz-y^2} = \frac{dy}{xy+zx} = \frac{dz}{xy-zx} \tag{i}$$

$$\text{Each ratio } = \frac{xdx + ydy + zdz}{0}$$

$$\therefore \qquad xdx + ydy + zdz = 0$$

Integrating, $$x^2 + y^2 + z^2 = c_1 \qquad \text{(ii)}$$

Taking the last two ratios, we have

$$\frac{dy}{y+z} = \frac{dy}{y-z}$$

$$\Rightarrow \qquad ydy - (zdy + ydz) - zdz = 0$$

$$\Rightarrow \qquad ydy - d(yz) - zdz = 0$$

Integrating, $$\frac{y^2}{2} - yz - \frac{z^2}{2} = c_2$$

$$\Rightarrow \qquad y^2 - 2yz - z^2 = c_2 \qquad \text{(iii)}$$

From Eqs. (ii) and (iii), the general solution is

$$f(x^2 + y^2 + z^2,\ y^2 - 2yz - z^2) = 0.$$

EXAMPLE 10.13 Solve $x(y^3 - 2x^3)\,p + y(2y^3 - x^3)\,q = 9z(x^3 - y^3)$.

Solution The subsidiary equations are

$$\frac{dx}{x(y^3 - 2x^3)} = \frac{dy}{y(2y^3 - x^3)} = \frac{dz}{z(x^3 - y^3)} \qquad \text{(i)}$$

$$\text{Each ratio} = \frac{\dfrac{dx}{x} + \dfrac{dy}{y} + \dfrac{dz}{3z}}{0}$$

$$\therefore \qquad \frac{1}{x}\,dx + \frac{1}{y}\,dy + \frac{1}{3z}\,dz = 0$$

Integrating, $$\log x + \log y + \frac{1}{3}\log z = \log c_1$$

$$\Rightarrow \qquad xyz^{1/3} = c_1 \qquad \text{(ii)}$$

Taking the first two ratios, we get

$$y(2y^3 - x^3)\,dx + x(2x^3 - y^3)\,dy = 0$$

Divide by x^3y^3

$$\therefore \qquad \left(\frac{2y}{x^3} - \frac{1}{y^2}\right)dx + \left(\frac{2x}{y^3} - \frac{1}{x^2}\right)dy = 0 \qquad \text{(iii)}$$

Here, $$M = \frac{2y}{x^3} - \frac{1}{y^2} \quad \text{and} \quad N = \frac{2x}{y^3} - \frac{1}{x^2}$$

Check that $$\frac{\partial M}{\partial y} = \frac{\partial N}{\partial x}$$

$\therefore$ Equation (iii) is an exact equation.

Therefore, solution of Eq. (ii) is

$$\int M dx = \int \left(\frac{2y}{x^3} - \frac{1}{y^2} \right) dx = -\frac{y}{x^2} - \frac{x}{y^2}$$

$$\therefore \qquad \frac{x}{y^2} + \frac{y}{x^2} = c_2 \tag{iv}$$

From Eqs. (ii) and (iv), the general solution is

$$f\left(xyz^{1/3}, \frac{x}{y^2} + \frac{y}{x^2} \right) = 0.$$

EXAMPLE 10.14 Solve $\cos(x + y)\, p + \sin(x + y)\, q = z$.

Solution The subsidiary equations are

$$\frac{dx}{\cos(x+y)} = \frac{dy}{\sin(x+y)} = \frac{dz}{z} \tag{i}$$

$$\text{Each ratio} = \frac{dx + dy}{\cos(x+y) + \sin(x+y)}$$

$$= \frac{d(x+y)}{\cos(x+y) + \sin(x+y)}$$

Put

$$u = x + y$$

$$= \frac{du}{\cos u + \sin u}$$

$$= \frac{1}{\sqrt{2}} \cdot \frac{du}{\cos\left(\frac{\pi}{4} - u \right)}$$

$$\Rightarrow \qquad \sqrt{2}\, \frac{dz}{z} = \frac{du}{\cos\left(\frac{\pi}{4} - u \right)}$$

Integrating,

$$\sqrt{2} \log z = -\log \tan\left(\frac{\pi}{4} + \frac{\pi}{8} - \frac{u}{2} \right) + \log c_1$$

$$\Rightarrow \qquad z^{\sqrt{2}} \tan\left(\frac{3\pi}{8} - \frac{x+y}{2} \right) = c_1 \tag{ii}$$

$$\text{Again each ratio} = \frac{dx - dy}{\cos(x+y) - \sin(x+y)}$$

$$= \frac{d(x-y)}{\cos(x+y) - \sin(x+y)}$$

$$\therefore \qquad \frac{d(x-y)}{\cos(x+y)-\sin(x+y)} = \frac{d(x+y)}{\cos(x+y)+\sin(x+y)}$$

$$\therefore \qquad d(x-y) = \frac{\cos u - \sin u}{\cos u + \sin u}\, du$$

Integrating, $\quad x - y = \log(\cos u + \sin u) + c_2$

$$\therefore \qquad e^{y-x}[\cos(x+y) + \sin(x+y)] = c_2 \qquad \text{(iii)}$$

From Eqs. (ii) and (iii), the general solution is

$$f\left\{z^{\sqrt{2}} \tan\left(\frac{3\pi}{8} - \frac{x+y}{2}\right), e^{y-x}[\cos(x+y) + \sin(x+y)]\right\} = 0.$$

EXAMPLE 10.15 Solve $(x + y - z)(p - q) + a(px - qy + x - y) = 0$.

Solution We have

$$(x + y - z)(p - q) + a(px - qy + x - y) = 0 \qquad \text{(i)}$$

Put $\quad U = x + y$

and $\quad V = x - y$

$$\therefore \qquad p = \frac{\partial z}{\partial x} = \frac{\partial z}{\partial U}\cdot\frac{\partial U}{\partial x} + \frac{\partial z}{\partial V}\cdot\frac{\partial V}{\partial x}$$

$$= \frac{\partial z}{\partial U} + \frac{\partial z}{\partial V}$$

and

$$q = \frac{\partial z}{\partial y} = \frac{\partial z}{\partial U}\cdot\frac{\partial U}{\partial y} + \frac{\partial z}{\partial V}\cdot\frac{\partial V}{\partial y}$$

$$= \frac{\partial z}{\partial U} - \frac{\partial z}{\partial V}$$

$\therefore$ Equation (i) reduces to

$$2(U - z)\frac{\partial z}{\partial V} + a\left(U\frac{\partial z}{\partial V} + V\frac{\partial z}{\partial U} + V\right) = 0$$

$$\Rightarrow \qquad (2U - 2z + aU)\frac{\partial z}{\partial V} + aV\frac{\partial z}{\partial U} = -aV \qquad \text{(ii)}$$

Subsidiary equations are

$$\frac{dV}{2U - 2z + aU} = \frac{dU}{aV} = \frac{dz}{-aV} \qquad \text{(iii)}$$

Considering last two ratios, we have

$$dU = -dz$$

Integrating, $\quad U + z = b \qquad$ (iv)

Taking first two ratios, we have

$$\frac{dV}{2U-2z+aU}=\frac{dU}{aV}$$

$$\Rightarrow \qquad \frac{dV}{4U-2b+aU}=\frac{dU}{aV}$$

$$\Rightarrow \qquad aVdV = (4U - 2b + aU)\, dU$$

Integrating,

$$\frac{1}{2}aV^2-2U^2+2bU-\frac{aU^2}{2}=c_1$$

$$\Rightarrow \qquad \frac{1}{2}aV^2-2U^2+2U\,(U+z)-\frac{aU^2}{2}=c_1$$

$$\Rightarrow \qquad \frac{1}{2}aV^2+2Uz-\frac{aU^2}{2}=c_1 \qquad \text{(v)}$$

From Eqs. (iv) and (v), the general solution is

$$f\left(U+z, \frac{1}{2}aV^2+2Uz-\frac{aU^2}{2}\right)=0$$

where

$$U = x + y$$
$$V = x - y.$$

EXAMPLE 10.16 Solve $(x^2-y^2-z^2)p+2xyq=2xz.$

Solution We have subsidiary equations as

$$\frac{dx}{x^2-y^2-z^2}=\frac{dy}{2xy}=\frac{dz}{2xz}$$

Consider last two ratios

$$\frac{dy}{2xy}=\frac{dz}{2xz}$$

Integrating this we get, $y = c_1z$.

Again, $\dfrac{dx}{x^2-y^2-z^2}=\dfrac{dz}{2xz}$ gives $2x\dfrac{dx}{dz}-\dfrac{x^2}{z}=-(1+{c_1}^2)z.$

Let $x^2 = t$. Then

$$2x\frac{dx}{dz}=\frac{dt}{dz}$$

With these substitutions, equation reduces to

$$\frac{dt}{dz}-\frac{1}{z}=-(1+{c_1}^2)z$$

which is linear and its IF $= \dfrac{1}{z}$. So solution of this differential equation is

$$t\frac{1}{z} = -(1+c_1^2)x + c_2 \Rightarrow \frac{x^2}{z} + \left(1 + \frac{y^2}{z^2}\right)z = c_2$$

$$\Rightarrow \qquad \frac{x^2+y^2+z^2}{z} = c_2$$

Therefore, the solution is $f\left(\frac{y}{z}, \frac{x^2+y^2+z^2}{z}\right) = 0$.

EXAMPLE 10.17 Solve $(p-q)(e^{x+y}+xyz)+(yq-xp)(z+xye^{-(x+y)})+(x+y)(1-z^2)=0$.

Solution Rewrite equation in standard form as

$$\{e^{x+y}+xyz-xz-x^2ye^{-(x+y)}\}p+\{yz+y^2xe^{-(x+y)}-e^{x+y}-xyz\}q=(y-x)(1-z^2)$$

We have subsidiary equations as

$$\frac{dx}{e^{x+y}+xyz-xz-x^2ye^{-(x+y)}} = \frac{dy}{yz+y^2xe^{-(x+y)}-e^{x+y}-xyz} = \frac{dz}{(y-x)(1-z^2)}$$

$$\Rightarrow \qquad \frac{d(x+y)}{(y-x)(z+xye^{-(x+y)})} = \frac{d(xy)}{(y-x)(xyz+e^{(x+y)})} = \frac{dz}{(y-x)(1-z^2)}$$

$$\Rightarrow \qquad \frac{d(x+y)}{(z+xye^{-(x+y)})} = \frac{d(xy)}{(xyz+e^{(x+y)})} = \frac{dz}{(1-z^2)}$$

$$\Rightarrow \qquad \frac{e^{(x+y)}d(x+y)}{(ze^{(x+y)}+xy)} = \frac{d(xy)}{(xyz+e^{(x+y)})} = \frac{dz}{(1-z^2)}$$

$$= \frac{d(e^{(x+y)})}{(ze^{(x+y)}+xy)}$$

$$\therefore \qquad \frac{dz}{(1-z^2)} = \frac{zd(xy)-d(e^{(x+y)})}{xy(z^2-1)}$$

$$\Rightarrow \qquad xydz + zd(xy) - d(e^{(x+y)}) = 0$$

$$\Rightarrow \qquad d\{xyz - e^{(x+y)}\} = 0$$

Integrating this we get, $xyz - e^{(x+y)} = c_1$.

Again, $\frac{dz}{(1-z^2)} = \frac{d(xy)}{2xyz-c_1} = \frac{d(u)}{2uz-c_1}$, where $u = xy$.

Therefore, $\frac{du}{dz} - 2\frac{uz}{1-z^2} = -\frac{c_1}{1-z^2}$ which is a linear equation and IF = (z^2-1). So solution of this differential equation is

$$u(z^2-1)=c_1z+c_2 \Rightarrow xy(z^2-1)=(xyz-e^{(x+y)})z+c_2$$
$$\Rightarrow ze^{(x+y)}-xy=c_2$$

Therefore, the solution is $f(xyz-e^{(x+y)}, ze^{(x+y)}-xy)=0$.

EXAMPLE 10.18 Solve $(x+y-z)(p-q)+a(px-qy+x-y)=0$.

Solution Let $u = x + y$ and $v = x - y$.

Then $p=\dfrac{\partial z}{\partial u}+\dfrac{\partial z}{\partial v}$ and $q=\dfrac{\partial z}{\partial u}-\dfrac{\partial z}{\partial v}$.

With these substitutions, given equation reduces to

$$2(u-z)\frac{\partial z}{\partial v}+a\left(u\frac{\partial z}{\partial v}+v\frac{\partial z}{\partial u}+v\right)=0$$

$$\Rightarrow \qquad (2u-2z+au)\frac{\partial z}{\partial v}+av\frac{\partial z}{\partial u}=-av$$

We have subsidiary equations as

$$\frac{du}{2u-2z+au}=\frac{du}{av}=\frac{dz}{-av}$$

Consider last two ratios

$$du = -dz$$

Integrating, we get $u + z = c_1$.
Consider first two ratio

$$\frac{du}{2u-2z+au}=\frac{du}{av}$$

$$\Rightarrow \qquad \frac{du}{4u-2b+au}=\frac{du}{av}$$

$$\Rightarrow \qquad avdv=(4u-2b+au)du$$

Integrating,

$$\frac{1}{2}av^2+2uz-\frac{au^2}{2}=c_2$$

Therefore, the solution is $f\left(u+z, \dfrac{1}{2}av^2+2uz-\dfrac{au^2}{2}\right)=0$.

EXAMPLE 10.19 Solve $x_2x_3zp_1+x_3x_1zp_2+x_1x_2zp_3=x_1x_2x_3$.

Solution Subsidiary equations are

$$\frac{dx_1}{x_2x_3z}=\frac{dx_2}{x_3x_1z}=\frac{dx_3}{x_1x_2z}=\frac{dz}{x_1x_2x_3}$$

The first two ratios results to $x_1 dx_1 - z dz = 0$. Integrating, $x_1^2 - z^2 = c_1$.

Similarly, second and third ratios will give $x_2^2 - z^2 = c_2$ and last two ratios will give $x_3^2 - z^2 = c_3$. Therefore, the solution is $f(x_1^2 - z^2, x_2^2 - z^2, x_3^2 - z^2) = 0$.

10.6 NON-LINEAR PARTIAL DIFFERENTIAL EQUATIONS OF ORDER ONE

The most general partial differential equation of order one is

$$F(x, y, z, p, q) = 0 \tag{10.26}$$

We will discuss Charpit's method to solve Eq. (10.26) in the next section.

10.7 CHARPIT'S METHOD

Introduce a partial differential equation of order one

$$f(x, y, z, p, q, a) = 0 \tag{10.27}$$

Containing the arbitrary constant 'a' and satisfying the conditions.

1. Equations (10.26) and (10.27) can be solved to obtain

$$p = p(x, y, z, a) \text{ and } q = q(x, y, z, a)$$

2. The equation

$$dz = p(x, y, z, a)\, dx + q(x, y, z, a)\, dy \tag{10.28}$$

is integrable.

When a function 'f' satisfying the conditions (1) and (2) is found, the solution of Eq. (10.27) containing two arbitrary constants (including 'a') will give the solution of Eq. (10.26). The conditions (1) will be satisfied if

$$J = \frac{\partial(F, f)}{\partial(p, q)} = \begin{vmatrix} \dfrac{\partial F}{\partial p} & \dfrac{\partial f}{\partial p} \\ \dfrac{\partial F}{\partial q} & \dfrac{\partial f}{\partial q} \end{vmatrix} \neq 0 \tag{10.29}$$

Assuming Eq. (10.29) is satisfied, Eq. (10.28) exists and the condition (2) will be satisfied if

$$p\left(\frac{\partial q}{\partial z}\right) + q\left(-\frac{\partial p}{\partial z}\right) - \left(\frac{\partial p}{\partial y} - \frac{\partial q}{\partial x}\right) = 0$$

$$\Rightarrow \qquad p\frac{\partial q}{\partial z} + \frac{\partial q}{\partial x} = q\frac{\partial p}{\partial z} + \frac{\partial p}{\partial y} \tag{10.30}$$

Substituting the values of p and q as functions of x, y and z in Eqs. (10.26) and (10.27) and differentiating with respect to x, we have

$$\frac{\partial F}{\partial x}+\frac{\partial F}{\partial p}\cdot\frac{\partial p}{\partial x}+\frac{\partial F}{\partial q}\cdot\frac{\partial q}{\partial x}=0$$

and
$$\frac{\partial f}{\partial x}+\frac{\partial f}{\partial p}\cdot\frac{\partial p}{\partial x}+\frac{\partial f}{\partial q}\cdot\frac{\partial q}{\partial x}=0$$

Therefore,
$$\left[\frac{\partial F}{\partial p}\frac{\partial f}{\partial q}-\frac{\partial F}{\partial q}\frac{\partial f}{\partial p}\right]\frac{\partial q}{\partial x}=\frac{\partial F}{\partial x}\frac{\partial f}{\partial p}-\frac{\partial F}{\partial p}\frac{\partial f}{\partial x}$$

$$\frac{\partial q}{\partial x}=\frac{1}{J}\left[\frac{\partial F}{\partial x}\frac{\partial f}{\partial p}-\frac{\partial F}{\partial p}\frac{\partial f}{\partial x}\right]$$

Similarly,
$$\left.\begin{aligned}\frac{\partial p}{\partial y}&=\frac{1}{J}\left[\frac{\partial F}{\partial q}\frac{\partial f}{\partial y}-\frac{\partial F}{\partial y}\frac{\partial f}{\partial q}\right]\\ \frac{\partial p}{\partial z}&=\frac{1}{J}\left[\frac{\partial F}{\partial q}\frac{\partial f}{\partial z}-\frac{\partial F}{\partial z}\frac{\partial f}{\partial q}\right]\\ \frac{\partial q}{\partial z}&=\frac{1}{J}\left[\frac{\partial F}{\partial z}\frac{\partial f}{\partial p}-\frac{\partial F}{\partial p}\frac{\partial f}{\partial z}\right]\end{aligned}\right\}\tag{10.31}$$

and

Substituting Eq. (10.31) into Eq. (10.30)

$$\left(-\frac{\partial F}{\partial p}\right)\frac{\partial f}{\partial x}+\left(-\frac{\partial F}{\partial q}\right)\frac{\partial f}{\partial y}+\left(-p\frac{\partial F}{\partial p}-q\frac{\partial F}{\partial q}\right)\frac{\partial f}{\partial z}+\left(p\frac{\partial F}{\partial z}+\frac{\partial F}{\partial x}\right)\frac{\partial f}{\partial p}+\left(q\frac{\partial F}{\partial z}+\frac{\partial F}{\partial y}\right)\frac{\partial f}{\partial q}=0 \tag{10.32}$$

The Eq. (10.32) being linear in variables x, y, z, p and q has its subsidiary equations

$$\frac{dx}{-\frac{\partial F}{\partial p}}=\frac{dy}{-\frac{\partial F}{\partial q}}=\frac{dz}{-p\frac{\partial F}{\partial p}-q\frac{\partial F}{\partial q}}=\frac{dp}{\frac{\partial F}{\partial x}+x\frac{\partial F}{\partial z}}=\frac{dq}{\frac{\partial F}{\partial y}+q\frac{\partial F}{\partial z}} \tag{10.33}$$

Any integrals of Eq. (10.33) satisfy Eq. (10.32). If such an integral has p or q, it can be taken for Eq. (10.27).

Once the Eq. (10.27) is obtain, the problem reduces to solve Eqs. (10.26) and (10.27) for p and q and then integrate Eq. (10.28).

Note: J is defined as Jacobian.

EXAMPLE 10.20 Find the complete integral of the equation $p^2 + q^2 - 2px - 2qy + 1 = 0$.

Solution Given
$$p^2 + q^2 - 2px - 2qy + 1 = 0 \tag{i}$$

Subsidiary equations are

$$\frac{dx}{-(2p-2x)}=\frac{dy}{-(2q-2y)}=\frac{dp}{-2p}=\frac{dq}{-2q} \tag{ii}$$

Considering last two ratios

$$\frac{dp}{p} = \frac{dq}{q}$$

Integrating, $$p = aq \qquad \text{(iii)}$$

where 'a' is an arbitrary constant. Substituting $p = aq$ in Eq. (i) gives

$$q^2(1 + a^2) - 2q(ax + y) + 1 = 0$$

$\Rightarrow$ $$q = (ax + y) + \sqrt{(ax + y)^2 - (1 + a^2)}$$

$\therefore$ $$dz = pdx + qdy \text{ gives}$$

$$dz = aqdx + qdy$$

$$= q(adx + dy)$$

$$= d(ax + y)\,[(ax + y) + \sqrt{(ax + y)^2 - (1 + a^2)}]$$

Integrating,

$$z + b = \frac{1}{2}(ax + y)^2 + \frac{1}{2}(ax + y)\sqrt{(ax + y)^2 - (1 + a^2)} - \frac{(a^2 + 1)}{2}\log[(ax + y)\sqrt{(ax + y)^2 - (1 + a^2)}]$$

where b is an arbitrary constant.

EXAMPLE 10.21 Find the complete integral of the equation $p^2 + q^2 - 2pq \tanh 2y = \text{sech}^2 2y$.

Solution Given

$$p^2 + q^2 - 2pq \tanh 2y = \text{sech}^2 2y \qquad \text{(i)}$$

Subsidiary equations are

$$\frac{dx}{-(2p - 2q\tanh 2y)} = \frac{dy}{-(2q - 2p\tanh 2y)} = \frac{dp}{0} = \frac{dq}{-4pq\,\text{sech}^2 2y + 4\,\text{sech}^2 2y \tanh 2y}$$

$\therefore$ $$dp = 0$$

$\Rightarrow$ $$p = a \text{ (Constant)}$$

Therefore, from Eq. (i)

$$q^2 - 2a \tanh 2yq + a^2 - \text{sech}^2 2y = 0$$

$\therefore$ $$q = a\tanh 2y + \sqrt{a^2 \tanh^2 2y - a^2 + \text{sech}^2 2y}$$

$$= a\tanh 2y + \sqrt{1 - a^2}\,\text{sech}\, 2y$$

$\therefore$ $$dz = pdx + qdy$$

gives $$dz = adx + [a\tanh 2y + \sqrt{1 - a^2}\,\text{sech}\, 2y]\, dy$$

$$= d\left(ax + \frac{a}{2}\log\cosh 2y\right) + \sqrt{1-a^2}\ \text{sech}\ 2ydy$$

Integrating,
$$z + b = ax + \frac{a}{2}\log\cosh 2y + \sqrt{1-a^2}\int \frac{2e^{2y}}{e^{2y}+e^{-2y}}$$

$$= ax + \frac{a}{2}\log\cosh 2y + \sqrt{1-a^2}\int \frac{2e^{2y}dy}{1+e^{4y}}$$

$$= ax + \frac{a}{2}\log\cosh 2y + \sqrt{1-a^2}\ (\tan^{-1} e^{2y}).$$

EXAMPLE 10.22 Find complete integral of the equation $xp + 3yq = 2(z - x^2q^2)$.

Solution Given
$$xp + 3yq = 2(z - x^2q^2) \qquad \text{(i)}$$
Subsidiary equations are

$$\frac{dx}{x} = \frac{dy}{-3y - 4x^2q} = \frac{dp}{p - 2p + 4xq^2} = \frac{dq}{3q - 2q}$$

$$\therefore \qquad \frac{dx}{x} = \frac{dq}{q}$$

$$\Rightarrow \qquad qx = a \text{ (Constant)}$$

$$\Rightarrow \qquad q = \frac{a}{x}$$

Substituting in Eq. (i) gives

$$p = \frac{2(z-a^2)}{x} - \frac{3ya}{x^2}$$

$$\therefore \qquad dz = pdx + qdy$$

$$= \left[\frac{2(z-a^2)}{x} - \frac{3ya}{x^2}\right]dx + \frac{a}{x}dy$$

Multiplying by x^2

$$x^2dz = 2x(z - a^2)\ dx - 3aydx + axdy$$

$$\Rightarrow \qquad x^4 d\left(\frac{z-a^2}{x^2}\right) = -3aydx + axdy$$

$$\Rightarrow \qquad d\left(\frac{z-a^2}{x^2}\right) = \frac{a}{x^3}dy - \frac{3ay}{x^4}dx$$

$$\Rightarrow \qquad d\left(\frac{z-a^2}{x^2}\right) = d\left(\frac{ay}{x^3}\right)$$

Integrating, $$\frac{z-a^2}{x^2} = \frac{ay}{x^3}+b$$

$\Rightarrow$ $$z = a\left(a+\frac{y}{x}\right)+bx^2$$

where a and b are arbitrary constants.

EXAMPLE 10.23 Find complete integral of the equation $p^2+q^2-2px-2qy+2xy=0$.

Solution Subsidiary equations are

$$\frac{dx}{-2(p-x)}=\frac{dy}{-2(q-y)}=\frac{dp}{2(y-p)}=\frac{dq}{2(x-q)}$$

Then $dx+dy=dp+dq$.

Integrating,

$$p+q=x+y+c_1$$

Therefore, $$(p-x)+(q-y)=c_1 \tag{i}$$

Given equation can be rewritten as

$$(p-x)^2+(q-y)^2=(x-y)^2$$

Now, $$\{(p-x)+(q-y)\}^2+\{(p-x)-(q-y)\}^2=2\{(p-x)^2+(q-y)^2\}$$

$\therefore$ $$(p-x)+(q-y)=\sqrt{2(x-y)^2-c_1{}^2} \tag{ii}$$

Solve Eqs. (i) and (ii) for p and q. We get,

$$p=\frac{c_1}{2}+x+\frac{1}{2}\sqrt{2(x-y)^2-c_1{}^2}$$

and $$q=\frac{c_1}{2}+y-\frac{1}{2}\sqrt{2(x-y)^2-c_1{}^2}$$

$\therefore$ $dz = pdx + qdy$ gives

$$dz=\left\{\frac{c_1}{2}+x+\frac{1}{2}\sqrt{2(x-y)^2-c_1{}^2}\right\}dx+\left\{\frac{c_1}{2}+y-\frac{1}{2}\sqrt{2(x-y)^2-c_1{}^2}\right\}dy$$

$$=\frac{1}{2}d(x^2+y^2)+\frac{a}{2}d(x+y)+\frac{1}{2}\sqrt{2(x-y)^2-c_1{}^2}\,d(x-y)$$

Integrating,

$$z+c_2=\frac{x^2+y^2}{2}+\frac{a}{2}(x+y)+\frac{1}{2}\int\sqrt{(2u^2-c_1^2)}du, \quad u=x-y$$

$$z+c_2=\frac{x^2+y^2}{2}+\frac{a}{2}(x+y)+\frac{1}{\sqrt{2}}\left\{\frac{u\sqrt{u^2-\frac{c_1^2}{2}}}{2}-\frac{c_1^2}{4}\log\left(u+\sqrt{u^2-\frac{c_1^2}{2}}\right)\right\}$$

$$z+c_2=\frac{x^2+y^2}{2}+\frac{a}{2}(x+y)+\frac{(x-y)\sqrt{2(x-y)^2-c_1^2}}{4}-\frac{a^2}{4\sqrt{2}}\log\left(x-y+\sqrt{(x-y)^2-\frac{c_1^2}{2}}\right)$$

EXAMPLE 10.24 Find complete integral of the equation $p^2+q^2-2px-2qy+1=0$.

Solution Given
$$p^2+q^2-2px-2qy+1=0 \tag{i}$$
Subsidiary equations are
$$\frac{dx}{-2(p-x)}=\frac{dy}{-2(q-y)}=\frac{dp}{-2p}=\frac{dq}{-2q}$$
Consider last two ratios
$$\frac{dp}{p}=\frac{dq}{q}$$
Integration gives
$$pq=c_1 \tag{ii}$$
Substitute value of p from Eq. (ii) in Eq. (i) and solve for q. We get,

$$q=(c_1x+y)+\sqrt{(c_1x+y)^2-(1+c_1^2)}$$

$$\therefore \quad dz=pdx+qdy=q(c_1dx+dy)$$
$$=d(c_1dx+dy)\left\{(c_1x+y)+\sqrt{(c_1x+y)^2-(1+c_1^2)}\right\}$$

Integrating,

$$z+c_2=\frac{1}{2}(c_1x+y)^2+\frac{(c_1x+y)+\sqrt{(c_1x+y)^2-(1+c_1^2)}}{2}$$
$$+\frac{(1+c_1^2)}{2}\log\left\{(c_1x+y)+\sqrt{(c_1x+y)^2-(1+c_1^2)}\right\}$$

is the required complete integral.

EXAMPLE 10.25 Find complete integral of the equation $xp+3yq=2(z-x^2q^2)$.

Solution Given
$$xp+3yq=2(z-x^2q^2) \tag{i}$$
Subsidiary equations are
$$\frac{dx}{-x}=\frac{dy}{-3y-4x^2q}=\frac{dp}{-p+4xq^2}=\frac{dq}{q}$$

Then $\dfrac{dx}{-x} = \dfrac{dq}{q}$ gives $qx = c_1$. Substitute this in Eq. (i). We get,

$$p = \frac{2(z - c_1^2)}{x} - \frac{3yc_1}{x^2}$$

$\therefore \quad dz = pdx + qdy$ gives

$$dz = \left\{\frac{2(z - c_1^2)}{x} - \frac{3yc_1}{x^2}\right\}dx + \frac{c_1}{x}dy$$

Simplifying this we get,

$$d\left(\frac{z - c_1^2}{x^2}\right) = d\left(\frac{ay}{x^3}\right)$$

Integrating, $\dfrac{z - c_1^2}{x^2} = \dfrac{ay}{x^3} + c_2$ which is complete integral of given equation.

10.8 COMPATIBLE SYSTEMS OF FIRST ORDER EQUATIONS

Consider, first order partial differential equations

$$f(x, y, z, p, q) = 0 \tag{10.34}$$

and

$$g(x, y, z, p, q) = 0 \tag{10.35}$$

These two equations are said to be *compatible* if every integral of one is also an integral of the order.

If

$$J = \frac{\partial(f, g)}{\partial(p, q)} \neq 0$$

Eqs. (10.34) and (10.35) can be solved to obtain

$$p = p(x, y, z), \; q = q(x, y, z)$$

Then Eqs. (10.34) and (10.35) will be compatible if

$$dz = pdx + qdy$$

is integrable.

By Eq. (10.26), given equations are compatible if

$$\left[\frac{\partial f}{\partial x}\frac{\partial g}{\partial p} - \frac{\partial g}{\partial x}\frac{\partial f}{\partial p}\right] + \left[\frac{\partial f}{\partial y}\frac{\partial g}{\partial q} - \frac{\partial f}{\partial q}\frac{\partial g}{\partial y}\right] + p\left[\frac{\partial f}{\partial z}\frac{\partial g}{\partial p} - \frac{\partial g}{\partial z}\frac{\partial f}{\partial p}\right] + q\left[\frac{\partial f}{\partial z}\frac{\partial g}{\partial q} - \frac{\partial f}{\partial q}\frac{\partial g}{\partial z}\right] = 0$$

$$\Rightarrow \quad \frac{\partial(f, g)}{\partial(x, p)} + p\frac{\partial(f, g)}{\partial(z, p)} + \frac{\partial(f, g)}{\partial(y, q)} + q\frac{\partial(f, g)}{\partial(z, q)} = 0$$

$$\Rightarrow \quad [f, g] = 0$$

where

$$[f, g] = \frac{\partial(f, g)}{\partial(x, p)} + p\frac{\partial(f, g)}{\partial(z, p)} + \frac{\partial(f, g)}{\partial(y, q)} + q\frac{\partial(f, g)}{\partial(z, q)} \tag{10.36}$$

EXAMPLE 10.26 Show that equations $xp = yq$, $z(xp + yq) = 2xy$ are compatible and solve them.

Solution Let

$$f = xp - yq = 0$$
$$g = z(xp + yq) - 2xy = 0$$

Then

$$\frac{\partial(f, g)}{\partial(x, p)} = \frac{\partial f}{\partial x}\frac{\partial g}{\partial p} - \frac{\partial f}{\partial p}\frac{\partial g}{\partial x}$$
$$= p(zx) - x(zp - 2y)$$
$$= 2xy$$

$$\frac{\partial(f, g)}{\partial(z, p)} = -px^2 - xyq$$

$$\frac{\partial(f, g)}{\partial(y, q)} = -2xy$$

and

$$\frac{\partial(f, g)}{\partial(z, q)} = xyp - qy^2$$

Then $[f, g] = 2xy - p^2x^2 - xypq - 2xy + xypq - q^2y^2 = 0$ (using f)

$\therefore$ Equations are compatible.

Solving given equations

$$p = \frac{y}{z}$$

$$q = \frac{x}{z}$$

$\therefore$
$$dz = pdx + qdy$$

gives
$$dz = \frac{y}{z}dx + \frac{x}{z}dy$$

$\Rightarrow$
$$zdz = ydx + xdy$$

$\Rightarrow$
$$zdz = d(xy)$$

Integrating, $z^2 = 2xy + c$

where c is arbitrary constant.

EXAMPLE 10.27 Show that the equation $z = px + qy$ is compatible with any equation $f(x, y, z, p, q) = 0$ which is homogeneous in x, y and z.

Solution Given f is homogeneous in x, y and z.

$\therefore$
$$f = z^n\phi\left(\frac{x}{z}, \frac{y}{z}, p, q\right)$$

Put $$u = \frac{x}{z}, v = \frac{y}{z}$$

$\therefore$ Given equation reduces to

$$g = px + qy - z = 0 \tag{i}$$

and $$h = \phi(u, v, p, q) = 0 \tag{ii}$$

$$\frac{\partial(g, h)}{\partial(x, p)} = \frac{\partial g}{\partial x}\frac{\partial h}{\partial p} - \frac{\partial g}{\partial p}\frac{\partial h}{\partial x}$$
$$= p\frac{\partial \phi}{\partial p} - x\frac{\partial \phi}{\partial x}$$
$$= p\frac{\partial \phi}{\partial p} - x\frac{\partial \phi}{\partial u}\cdot\frac{\partial u}{\partial x}$$
$$= p\frac{\partial \phi}{\partial p} - \frac{x}{z}\frac{\partial \phi}{\partial u}$$
$$= p\frac{\partial \phi}{\partial p} - u\frac{\partial \phi}{\partial u}$$

Similarly, $$\frac{\partial(g, h)}{\partial(z, p)} = \frac{\partial g}{\partial z}\frac{\partial h}{\partial p} - \frac{\partial g}{\partial p}\frac{\partial h}{\partial z}$$
$$= -\frac{\partial \phi}{\partial p} - x\frac{\partial \phi}{\partial z}$$

$$\frac{\partial(g, h)}{\partial(y, q)} = \frac{\partial g}{\partial y}\frac{\partial h}{\partial q} - \frac{\partial g}{\partial q}\frac{\partial h}{\partial y}$$
$$= q\frac{\partial \phi}{\partial q} - y\frac{\partial \phi}{\partial v}\frac{\partial v}{\partial y}$$
$$= q\frac{\partial \phi}{\partial q} - v\frac{\partial \phi}{\partial y}$$

and $$\frac{\partial(g, h)}{\partial(z, q)} = \frac{\partial g}{\partial z}\frac{\partial h}{\partial q} - \frac{\partial g}{\partial q}\frac{\partial h}{\partial z}$$
$$= -\frac{\partial \phi}{\partial q} - y\frac{\partial \phi}{\partial z}$$

$\therefore$ $$[f, g] = p\frac{\partial \phi}{\partial p} - u\frac{\partial \phi}{\partial u} + p\left(-\frac{\partial \phi}{\partial p} - x\frac{\partial \phi}{\partial z}\right) + q\frac{\partial \phi}{\partial q} - v\frac{d\phi}{dv} + q\left(-\frac{\partial \phi}{\partial q} - y\frac{\partial \phi}{\partial z}\right)$$
$$= -u\frac{\partial \phi}{\partial u} - v\frac{\partial \phi}{\partial v} - (px + qy)\frac{\partial \phi}{\partial z}$$
$$= -u\frac{\partial \phi}{\partial u} - v\frac{\partial \phi}{\partial v} - z\frac{\partial \phi}{\partial z}$$
$$= -u\frac{\partial \phi}{\partial u} - v\frac{\partial \phi}{\partial v} - z\left[\frac{\partial \phi}{\partial u}\frac{\partial u}{\partial z} + \frac{\partial \phi}{\partial v}\frac{\partial v}{\partial z}\right]$$

$$= -u\frac{\partial\phi}{\partial u} - v\frac{\partial\phi}{\partial v} - z\left[-\frac{x}{z^2}\frac{\partial\phi}{\partial u} - \frac{y}{z^2}\frac{\partial\phi}{\partial v}\right]$$

$$= -u\frac{\partial\phi}{\partial u} - v\frac{\partial\phi}{\partial v} + u\frac{\partial\phi}{\partial u} + v\frac{\partial\phi}{\partial v} = 0$$

$\therefore$ Given equations are compatible.

EXAMPLE 10.28 If $u_1 = \dfrac{\partial u}{\partial x}, u_2 = \dfrac{\partial u}{\partial y}, u_3 = \dfrac{\partial u}{\partial z}$, show that the equations

$$f(x, y, z, u_1, u_2, u_3) = 0,\ g(x, y, z, u_1, u_2, u_3) = 0$$

are compatible if

$$\frac{\partial(f,g)}{\partial(x,u_1)} + \frac{\partial(f,g)}{\partial(y,u_2)} + \frac{\partial(f,g)}{\partial(z,u_3)} = 0.$$

Solution Differentiating given equations partially w.r.t. x gives

$$\frac{\partial f}{\partial x} + \frac{\partial f}{\partial u_1}\frac{\partial u_1}{\partial x} + \frac{\partial f}{\partial u_2}\frac{\partial u_2}{\partial x} + \frac{\partial f}{\partial u_3}\frac{\partial u_3}{\partial x} = 0$$

and

$$\frac{\partial g}{\partial x} + \frac{\partial g}{\partial u_1}\frac{\partial u_1}{\partial x} + \frac{\partial g}{\partial u_2}\frac{\partial u_2}{\partial x} + \frac{\partial g}{\partial u_3}\frac{\partial u_3}{\partial x} = 0$$

Eliminating $\dfrac{\partial u_1}{\partial x}$ gives

$$\left[\frac{\partial f}{\partial x}\frac{\partial g}{\partial u_1} - \frac{\partial g}{\partial x}\frac{\partial f}{\partial u_1}\right] + \left[\frac{\partial f}{\partial u_2}\frac{\partial g}{\partial u_1} - \frac{\partial f}{\partial u_1}\frac{\partial g}{\partial u_2}\right]\frac{\partial u_2}{\partial x} + \left[\frac{\partial f}{\partial u_3}\frac{\partial g}{\partial u_1} - \frac{\partial f}{\partial u_1}\frac{\partial g}{\partial u_3}\right]\frac{\partial u_3}{\partial x} = 0 \quad \text{(i)}$$

Similarly, differentiating given equations partially w.r.t. y, z and eliminating $\dfrac{\partial u_2}{\partial y}, \dfrac{\partial u_3}{\partial z}$, respectively, gives

$$\left[\frac{\partial f}{\partial y}\frac{\partial g}{\partial u_2} - \frac{\partial f}{\partial u_2}\frac{\partial g}{\partial y}\right] + \left[\frac{\partial f}{\partial u_1}\frac{\partial g}{\partial u_2} - \frac{\partial f}{\partial u_2}\frac{\partial g}{\partial u_1}\right]\frac{\partial u_1}{\partial y} + \left[\frac{\partial f}{\partial u_3}\frac{\partial g}{\partial u_2} - \frac{\partial f}{\partial u_2}\frac{\partial g}{\partial u_3}\right]\frac{\partial u_3}{\partial y} = 0 \quad \text{(ii)}$$

and

$$\left[\frac{\partial f}{\partial z}\frac{\partial g}{\partial u_3} - \frac{\partial f}{\partial u_3}\frac{\partial g}{\partial z}\right] + \left[\frac{\partial f}{\partial u_1}\frac{\partial g}{\partial u_3} - \frac{\partial f}{\partial u_3}\frac{\partial g}{\partial u_1}\right]\frac{\partial u_1}{\partial z} + \left[\frac{\partial f}{\partial u_2}\frac{\partial g}{\partial u_3} - \frac{\partial f}{\partial u_3}\frac{\partial g}{\partial u_2}\right]\frac{\partial u_2}{\partial z} = 0 \quad \text{(iii)}$$

Adding Eqs. (i), (ii) and (iii) gives

$$\left[\frac{\partial f}{\partial x}\frac{\partial g}{\partial u_1} - \frac{\partial f}{\partial u_1}\frac{\partial g}{\partial x}\right] + \left[\frac{\partial f}{\partial y}\frac{\partial g}{\partial u_2} - \frac{\partial f}{\partial u_2}\frac{\partial g}{\partial y}\right] + \left[\frac{\partial f}{\partial z}\frac{\partial g}{\partial u_3} - \frac{\partial f}{\partial u_3}\frac{\partial g}{\partial z}\right] = 0$$

$$\Rightarrow \quad \frac{\partial(f,g)}{\partial(x,u_1)} + \frac{\partial(f,g)}{\partial(y,u_2)} + \frac{\partial(f,g)}{\partial(z,u_3)} = 0.$$

10.9 JACOBI'S METHOD

Let the partial differential equation be

$$F(x, y, z, p, q) = 0 \tag{10.37}$$

Let $u(x, y, z) = 0$ (10.38)

be a relation in x, y and z.

Differentiating Eq. (10.38) w.r.t. x gives

$$u_1 + pu_3 = 0 \quad \Rightarrow \quad p = -\frac{u_1}{u_3}$$

Similarly, $q = -\dfrac{u_2}{u_3}$

Therefore, Eq. (10.37) becomes

$$F(x, y, z, u_1, u_2, u_3) = 0 \tag{10.39}$$

which does not contain new dependent variable u.

Any function 'g' compatible with Eq. (10.39) is given by

$$\frac{\partial(f, g)}{\partial(x, u_1)} + \frac{\partial(f, g)}{\partial(y, u_2)} + \frac{\partial(f, g)}{\partial(z, u_3)} = 0$$

$$\Rightarrow \quad \left[\frac{\partial f}{\partial x}\frac{\partial g}{\partial u_1} - \frac{\partial f}{\partial u_1}\frac{\partial g}{\partial x}\right] + \left[\frac{\partial f}{\partial y}\frac{\partial g}{\partial u_2} - \frac{\partial f}{\partial u_2}\frac{\partial g}{\partial y}\right] + \left[\frac{\partial f}{\partial z}\frac{\partial g}{\partial u_3} - \frac{\partial f}{\partial u_3}\frac{\partial g}{\partial z}\right] = 0$$

$$\Rightarrow \quad \left(-\frac{\partial f}{\partial u_1}\right)\frac{\partial g}{\partial x} + \left(-\frac{\partial f}{\partial u_2}\right)\frac{\partial g}{\partial y} + \left(-\frac{\partial f}{\partial u_3}\right)\frac{\partial g}{\partial z} + \frac{\partial f}{\partial x}\frac{\partial g}{\partial u_1} + \frac{\partial f}{\partial y}\frac{\partial g}{\partial u_2} + \frac{\partial f}{\partial z}\frac{\partial g}{\partial u_3} = 0 \tag{10.40}$$

Subsidiary equations of Eq. (10.40) are

$$\frac{dx}{-\dfrac{\partial f}{\partial u_1}} = \frac{dy}{-\dfrac{\partial f}{\partial u_2}} = \frac{dz}{-\dfrac{\partial f}{\partial u_3}} = \frac{\partial u_1}{\dfrac{\partial f}{\partial x}} = \frac{\partial u_2}{\dfrac{\partial f}{\partial y}} = \frac{\partial u_3}{\dfrac{\partial f}{\partial z}} \tag{10.41}$$

Any integral of Eq. (10.41) satisfies Eq. (10.40).

Jacobi's method consists in finding two integrals, as

$$\left.\begin{aligned} g(x, y, z, u_1, u_2, u_3, a) &= 0 \\ \text{and} \quad h(x, y, z, u_1, u_2, u_3, b) &= 0 \end{aligned}\right\} \tag{10.42}$$

involving two arbitrary constants a and b from Eq. (10.41). These integrals should be such that Eqs. (10.39) and (10.42) can be solved for u_1, u_2 and u_3. Substitute these values in

$$du = u_1dx + u_2dy + u_3dz \tag{10.43}$$

and integrate it. Get integral of the form

$$u = \phi(x, y, z, a, b, c) \tag{10.44}$$

Using Eq. (10.40), we get

$$\phi(x, y, z, a, b, c) = 0 \qquad (10.45)$$

Equation (10.45) satisfies the given Eq. (10.37). It contains three arbitrary constants, whereas complete integral of given equation can contain only two arbitrary constants. So complete integral is obtained from Eq. (10.45) by making a choice of third constant.

EXAMPLE 10.29 Solve $p^2x + q^2y = z$.

Solution Substitute $p = \dfrac{-u_1}{u_3}$ and $q = \dfrac{-u_2}{u_3}$ in given equation

$$\therefore \qquad xu_1^2 + yu_2^2 - zu_3^2 = 0 \qquad \text{(i)}$$

Subsidiary equations are

$$\frac{dx}{-2xu_1} = \frac{dy}{-2yu_2} = \frac{dz}{2zu_3} = \frac{du_1}{u_1^2} = \frac{du_2}{u_2^2} = \frac{du}{-u_3^2}$$

$$\Rightarrow \qquad \frac{dx}{x} + 2\frac{du_1}{u_1} = 0$$

and

$$\frac{dy}{y} + 2\frac{du_2}{u_2} = 0$$

Integrating these equations give

$$xu_1^2 = a$$

and

$$yu_2^2 = b$$

where a and b are arbitrary constants.

$$\therefore \qquad u_1 = \sqrt{\frac{a}{x}} \text{ and } u_2 = \sqrt{\frac{b}{y}}$$

$\therefore$ Equation (i) gives

$$u_3 = \sqrt{\frac{a+b}{z}}$$

$$\therefore \qquad du = u_1dx + u_2dy + u_3dz$$

$$= \sqrt{\frac{a}{x}}dx + \sqrt{\frac{b}{y}}\,dy + \sqrt{\frac{a+b}{z}}\,dz$$

Integrating,

$$u = 2\sqrt{ax} + 2\sqrt{by} + 2\sqrt{(a+b)z} + c$$

where c is arbitrary constant.

EXAMPLE 10.30 Solve $x_3^2 p_1^2 p_2^2 p_3^2 + p_1^2 p_2^2 - p_3^2 = 0.$

Solution Subsidiary equations are

$$\frac{dx_1}{-(2x_3^2 p_1 p_2^2 p_3^2 + 2p_1 p_2)} = \frac{dx_2}{-(2x_3^2 p_1^2 p_2 p_3^2 + 2p_1^2 p_2)} = \frac{dx_3}{-(2x_3^2 p_1^2 p_2^2 p_3 - 2p_3)}$$

$$= \frac{dp_1}{0} = \frac{dp_2}{0} = \frac{dp_3}{2x_3 p_1^2 p_2^2 p_3^2}$$

Consider ratios,

$$dp_1 = 0$$

and

$$dp_2 = 0$$

∴ Two independent integrals are

$$p_1 = c_1 \text{ (Constant)}$$

and

$$p_2 = c_2 \text{ (Constant)}$$

∴ From given equation

$$p_3 = \pm \frac{c_1 c_2}{\sqrt{1 - (c_1 c_2 x_3)^2}}$$

∴

$$dz = p_1 dx_1 + p_2 dx_2 + p_3 dx_3$$

$$= c_1 dx_1 + c_2 dx_2 \pm \frac{c_1 c_2}{\sqrt{1 - (c_1 c_2 x_3)^2}} dx_3$$

Integrating, $z = c_1 x_1 + c_2 x_2 + \sin^{-1}(c_1 c_2 x_3) + c_3$

where c_3 is also constant.

EXAMPLE 10.31 Solve $p_3(p_1 + p_2)\, x_3 + x_1 + x_2 = 0.$

Solution Subsidiary equations are

$$\frac{dx_1}{-p_3 x_3} = \frac{dx_2}{-p_3 x_3} = \frac{dx_3}{-x_3(p_1 + p_2)} = \frac{dp_1}{1} = \frac{dp_2}{1} = \frac{dp_3}{p_3(p_1 + p_2)}$$

The third and sixth ratios, we get

$$\frac{dp_3}{p_3} = \frac{dx_3}{-x_3}$$

Integrating, $p_3 x_3 = c_1$ (Constant)

Fourth and fifth ratios give

$$dp_1 = dp_2$$

Integrating, $$p_3 x_3 = c_1 \text{ (Constant)} \qquad \text{(i)}$$

∴

$$p_3 = \frac{c_1}{x_3}$$

∴ Given equation gives

$$p_1 + p_2 = -\frac{x_1 + x_2}{c_1} \quad \text{(ii)}$$

Solving Eqs. (i) and (ii)

$$p_1 = \frac{c_2}{2} - \frac{x_1 + x_2}{2c_1}$$

and

$$p_2 = -\frac{c_2}{2} - \frac{x_1 + x_2}{2c_1}$$

∴

$$dz = p_1 dx_1 + p_2 dx_2 + p_3 dx_3$$

$$= \left(\frac{c_2}{2} - \frac{x_1 + x_2}{2c_1}\right) dx_1 + \left(-\frac{c_2}{2} - \frac{x_1 + x_2}{2c_1}\right) dx_2 + \frac{c_1}{x_3} dx_3$$

$$= \frac{c_2}{2} d(x_1 - x_2) - \frac{x_1 + x_2}{2c_1} d(x_1 + x_2) + \frac{c_1}{x_3} dx_3$$

Integrating,

$$z = \frac{c_2}{2}(x_1 - x_2) - \frac{(x_1 + x_2)^2}{4c_1} + c_1 \log x_3 + c_3$$

where c_3 is also constant.

EXAMPLE 10.32 Solve $pq = xz$.

Solution Given

$$pq = xz \quad \text{(i)}$$

Substitute $\frac{z - c_1^2}{x^2} = \frac{ay}{x^3} + c_2$ in the given equation

∴

$$u_1 u_2 - x u_3^2 = 0 \quad \text{(ii)}$$

Subsidiary equations are

$$\frac{dx}{-u_2} = \frac{dy}{-u_1} = \frac{dz}{2xzu_3} = \frac{du_1}{-zu_3^2} = \frac{du_2}{0} = \frac{du}{-xu_3^2}$$

⇒

$$du_2 = 0 \text{ and } \frac{dz}{2z} = \frac{du_1}{-u_3}$$

Integrating these equations give

$$u_2 = a$$

and

$$zu_3^2 = b$$

where a and b are arbitrary constants.

∴

$$u_2 = a \text{ and } u_3 = \sqrt{\frac{b}{z}}$$

$\therefore$ Equation (ii) gives

$$u_1 = \frac{b}{a} x$$

$$\therefore \quad du = u_1 dx + u_2 dy + u_3 dz$$

$$du = \frac{b}{a} x dx + a dy + \sqrt{\frac{b}{z}}\, dz$$

Integrating,
$$u = \frac{b}{2a} x^2 + ay + 2\sqrt{bz} + c$$

where c is arbitrary constant. Hence, complete integral is $\frac{b}{2a} x^2 + ay + 2\sqrt{bz} + c = 0.$

EXAMPLE 10.33 Solve $p_1 x_1 + p_{12} x_2 = p_3^2$.

Solution Subsidiary equations are

$$\frac{dx_1}{-x_1} = \frac{dx_2}{-x_2} = \frac{dx_3}{2p_3} = \frac{dp_1}{p_1} = \frac{dp_2}{p_2} = \frac{dp_3}{0}$$

Consider ratios

$$\frac{dx_1}{-x_1} = \frac{dp_1}{p_1}$$

and
$$\frac{dx_2}{-x_2} = \frac{dp_2}{p_2}$$

So we have $p_1 x_1 = a$ and $p_2 x_2 = b$. Then $p_1 = \frac{a_1}{x_1}$, $p_2 = \frac{a_2}{x_2}$ and $p_3 = \pm\sqrt{a_1 + a_2}$.

$$\therefore \quad dz = p_1 dx_1 + p_2 dx_2 + p_3 dx_3 = \frac{a_1}{x_1} dx_1 + \frac{a_2}{x_2} dx_2 \pm \sqrt{a_1 + a_2}\, dx_3$$

Integrating,

$$z = a_1 \log x_1 + a_2 \log x_2 \pm \left(\sqrt{a_1 + a_2}\right) x_3 + b$$

is the required solution.

10.10 STANDARD FORMS

In this section, we will discuss few standard forms of Eq. (10.57) which can be integrated easily.

10.10.1 Form I

Consider, the equation of the form

$$F(p, q) = 0 \tag{10.46}$$

Subsidiary equations for Eq. (10.46) are

$$\frac{dp}{0} = \frac{dq}{0}$$

$$\therefore \quad dp = 0$$

Integrating,

$$p = a \text{ (Constant)} \tag{10.47}$$

Substituting it in Eq. (10.46)

$$F(a, q) = 0$$

gives

$$q = b \text{ (Constant)} \tag{10.48}$$

Thus, 'b' is such that $F(a, b) = 0$.

$$\therefore \quad dz = pdx + qdy \tag{10.49}$$

gives

$$dz = adx + bdy$$

Integrating,

$$z = ax + by + c \tag{10.50}$$

where c is an arbitrary constant. Equations (10.50) and (10.49) give complete integral of the Eq. (10.46).

General Integral

Let on solving Eq. (10.49) for 'b' gives $b = \phi(a)$.

$\therefore$ Equation (10.50) gives

$$z = ax + \phi(a)\, y + c$$

To obtain general integral, put $c = \psi(a)$ where ψ is an arbitrary function. The general integral is obtained by eliminating 'a' between the equations

$$z = ax + \phi(a)y + \psi(a)$$

and

$$0 = x + \phi'(a)y + \psi'(a)$$

Singular Integral

Singular integral (if it exists) is determined by the equations

$$z = ax + \phi(a)y + c$$

$$0 = x + \phi'(a)y$$

$$0 = 1$$

The last equation suggests that singular integral does not exist.

EXAMPLE 10.34 Solve $p^2 + q^2 = 4$.

Solution Given

$$F(p, q) = p^2 + q^2 - 4 \tag{i}$$

Put

$$p = a \text{ (Constant)}$$

$$\therefore \quad q = \sqrt{4 - a^2} \tag{ii}$$

$$\therefore \qquad dz = pdx + qdy$$

$$= adx + \sqrt{4-a^2}\,dy$$

Integrating, $$z = ax + \sqrt{4-a^2}\,y + b$$

where b is a constant.

EXAMPLE 10.35 Find complete integral of the equation $(1 - x^2)\, yp^2 + x^2q = 0$.

Solution Given $$(1 - x^2)\, yp^2 + x^2q = 0$$

It can be rewritten as

$$\left(\frac{\sqrt{1-x^2}}{x}\,p\right)^2 + \frac{a}{y} = 0$$

Put $$X = \sqrt{1-x^2}$$

and $$Y = \frac{y^2}{2}$$

Then $$p = \frac{\partial z}{\partial x}$$

$$= \frac{\partial z}{\partial X}\cdot\frac{\partial X}{\partial x}$$

$$= \frac{-x}{\sqrt{1-x^2}}\,\frac{\partial z}{\partial X}$$

and $$q = \frac{\partial z}{\partial y}$$

$$= \frac{\partial z}{\partial Y}\cdot\frac{\partial Y}{\partial y}$$

$$= y\frac{\partial z}{\partial Y}$$

Therefore, Eq. (i) reduces to

$$\left(\frac{\partial z}{\partial X}\right)^2 + \frac{\partial z}{\partial Y} = 0$$

$\therefore$ Complete integral is

$$z = aX + (-a^2)Y + c$$

$$= a\sqrt{1-x^2} - \frac{a^2}{2}\,y^2 + c.$$

EXAMPLE 10.36 Find the complete integral of the equation $(y - x)(qy - px) = (p - q)^2$.

Solution Given

$$(y - x)(qy - px) = (p - q)^2 \tag{i}$$

Put

$$y + x = U$$

and

$$xy = V$$

Then

$$p = \frac{\partial z}{\partial x}$$

$$= \frac{\partial z}{\partial U}\frac{\partial U}{\partial x} + \frac{\partial z}{\partial V}\frac{\partial V}{\partial x}$$

$$= \frac{\partial z}{\partial U} + y\frac{\partial z}{\partial V}$$

and

$$q = \frac{\partial z}{\partial y}$$

$$= \frac{\partial z}{\partial U}\frac{\partial U}{\partial y} + \frac{\partial z}{\partial V}\frac{\partial V}{\partial y}$$

$$= \frac{\partial z}{\partial U} + x\frac{\partial z}{\partial V}$$

Substitute these values in Eq. (i), gives

$$\frac{\partial z}{\partial U} = \left(\frac{\partial z}{\partial V}\right)^2$$

∴ Complete integral is

$$z = aU + \sqrt{a}V + c$$

$$= a(x + y) + \sqrt{a}xy + c.$$

10.10.2 Form II

Consider the partial differential equation which does not contain independent variable, i.e.

$$F(z, p, q) = 0 \tag{10.51}$$

Subsidiary equations for Eq. (10.51) are

$$\frac{dp}{p\frac{\partial F}{\partial z}} = \frac{dq}{q\frac{\partial F}{\partial z}}$$

⇒

$$\frac{dp}{p} = \frac{dq}{q}$$

Integrating,

$$p = aq \tag{10.52}$$

where 'a' is a constant.

Substituting in Eq. (10.51) gives

$$F(z, aq, q) = 0 \tag{10.53}$$

The equation

$$\begin{aligned} dz &= pdx + qdy \\ &= q(adx + dy) \\ &= qd(ax + y) \end{aligned}$$

$\therefore$
$$\frac{dz}{dX} = q$$

where
$$X = ax + y$$

Substituting in Eq. (10.53) gives

$$F\left(z, a\frac{dz}{dX}, \frac{dz}{dX}\right) = 0 \tag{10.54}$$

Equation (10.54) is an ordinary differential equation of first order which can be solved for z as a function of X. Complete integral is then obtained by putting $X = y + ax$. General and singular integrals are obtained by the method discussed in Section 10.10.1.

EXAMPLE 10.37 Find complete integral of the equation $p^2 = z^2(1 - pq)$.

Solution Given
$$p^2 = z^2(1 - pq) \tag{i}$$

Replace p by $a\dfrac{dz}{dX}$ and q by $\dfrac{dz}{dX}$ in Eq. (i), we get

$$a^2\left(\frac{\partial z}{\partial X}\right)^2 = z^2\left[1 - a\left(\frac{\partial z}{\partial X}\right)^2\right]$$

$\Rightarrow$
$$a(a + z^2)\left(\frac{dz}{dX}\right)^2 = z^2$$

$\Rightarrow$
$$dX = \sqrt{a}\,\frac{\sqrt{a + z^2}}{z}\,dz$$

$$= \sqrt{a}\,\frac{(a + z^2)}{z\sqrt{a + z^2}}\,dz$$

$\Rightarrow$
$$\frac{dX}{\sqrt{a}} = \frac{z}{\sqrt{a + z^2}}\,dz + \frac{a}{z\sqrt{z^2 + a}}\,dz$$

Integrating,
$$\frac{X}{\sqrt{a}} + b = \sqrt{a + z^2} + \int \frac{dz}{z\sqrt{a + z^2}}$$

Put $$a + z^2 = u^2$$

$$dz = \frac{udu}{z}$$

$$\therefore \quad \frac{X}{\sqrt{a}} + b = \sqrt{a+z^2} + a\int \frac{du}{u^2 - a}$$

$$= \sqrt{a+z^2} + \frac{\sqrt{a}}{2}\log\left(\frac{u-\sqrt{a}}{u+\sqrt{a}}\right)$$

$$= \sqrt{a+z^2} + \frac{\sqrt{a}}{2}\log\left(\frac{\sqrt{a+z^2}-\sqrt{a}}{\sqrt{a+z^2}+\sqrt{a}}\right)$$

$$\therefore \quad \frac{y}{\sqrt{a}} + \sqrt{a}x + b = \sqrt{a+z^2} + \frac{\sqrt{a}}{2}\log\frac{\sqrt{a+z^2}-\sqrt{a}}{\sqrt{a+z^2}+\sqrt{a}}$$

is complete integral where a and b are constants.

EXAMPLE 10.38 Find the complete integral of the equation $q^2y^2 = z(z - px)$.

Solution Given $$q^2y^2 = z(z - px) \tag{i}$$

Put $$U = \log x$$

and $$V = \log y$$

Then
$$p = \frac{\partial z}{\partial x}$$

$$= \frac{\partial z}{\partial U}\cdot\frac{\partial U}{\partial x}$$

$$= \frac{1}{x}\frac{\partial z}{\partial U}$$

and
$$q = \frac{\partial z}{\partial y}$$

$$= \frac{\partial z}{\partial V}\cdot\frac{\partial V}{\partial y}$$

$$= \frac{1}{y}\frac{\partial z}{\partial V}$$

Substituting in Eq. (i) gives

$$\left(\frac{\partial z}{\partial V}\right)^2 = z\left(z - \frac{\partial z}{\partial U}\right)$$

Let $$Q = \frac{\partial z}{\partial V}$$

and $$P = \frac{\partial z}{\partial U} \tag{ii}$$

$\therefore$ $$Q_2 = z(z - P)$$

Replacing 'P' by $a\dfrac{dz}{dx}$ and 'Q' by $\dfrac{dz}{dX}$ in Eq. (ii) gives

$$\left(\frac{dz}{dX}\right)^2 = z\left(z - a\frac{dz}{dX}\right)$$

$\Rightarrow$ $$\left(\frac{dz}{dX}\right)^2 + az\left(\frac{dz}{dX}\right) - z^2 = 0$$

$\therefore$ $$\frac{dz}{dX} = \frac{z}{2}(-a + \sqrt{a^2 + 4})$$

$\Rightarrow$ $$\frac{2dz}{z} = (-a + \sqrt{a^2 + 4})\, dX$$

Integrating, $$2 \log z = (-a + \sqrt{a^2 + 4})X + b$$
$$= (-a + \sqrt{a^2 + 4})\,(V + aU) + b$$
$$= (-a + \sqrt{a^2 + 4}) \log y + a(-a + \sqrt{a^2 + 4}) \log X + b$$

is the complete integral.

10.10.3 Form III

Consider the form

$$f_1(x, p) = f_2(y, q) \tag{10.55}$$

Subsidiary equations for Eq. (10.55) are

$$\frac{dx}{-\frac{\partial f_1}{\partial p}} = \frac{dy}{\frac{\partial f_2}{\partial p}} = \frac{dp}{\frac{\partial f_1}{\partial x}} = \frac{dq}{-\frac{\partial f_2}{\partial y}}$$

$\therefore$ $$\frac{\partial f_1}{\partial p} dp + \frac{\partial f_1}{\partial x} dx = 0$$

$\Rightarrow$ $$df_1 = 0$$

Integrating, $$f_1 = a \text{ (Constant)}$$

Therefore, Eq. (10.55) gives

$$f_1(x, p) = f_2(y, q) = a \tag{10.56}$$

Let Eq. (10.56) on solving gives

$$p = F_1(a, x) \text{ and } q = F_2(a, y)$$

Therefore,

$$dz = pdx + qdy$$
$$= F_1(a, x)\, dx + F_2(a, y)\, dy$$

Integrating, $$z = \int F_1(a, x)\, dx + \int F_2(a, y)\, dy + b \tag{10.57}$$

where b is an arbitrary constant.

Equation (10.57) is complete integral of Eq. (10.56).

EXAMPLE 10.39 Find the complete integral of the equation $\frac{p^2}{x} - \frac{q^2}{y} = \frac{1}{z}\left(\frac{1}{x} + \frac{1}{y}\right)$.

Solution Given $$\frac{p^2}{x} - \frac{q^2}{y} = \frac{1}{z}\left(\frac{1}{x} + \frac{1}{y}\right) \tag{i}$$

Put $$Z = \frac{2}{3} z^{3/2}$$

Then $$\frac{\partial Z}{\partial x} = \frac{\partial Z}{\partial z} \cdot \frac{\partial z}{\partial x} = \sqrt{z}\, p$$

and $$\frac{\partial Z}{\partial y} = \frac{\partial Z}{\partial z} \cdot \frac{\partial z}{\partial y} = \sqrt{z}\, q$$

Substituting in Eq. (i) gives

$$\frac{P^2}{x} - \frac{Q^2}{y} = \frac{1}{x} + \frac{1}{y} \tag{ii}$$

where $$P = \frac{\partial Z}{\partial x} \quad \text{and} \quad Q = \frac{\partial Z}{\partial y}$$

$$\Rightarrow \qquad \frac{P^2 - 1}{x} = \frac{Q^2 + 1}{y}$$

Put $$\frac{P^2 - 1}{x} = \frac{Q^2 + 1}{y} = a$$

$$\therefore \qquad P = \sqrt{1 + ax}$$

and $$Q = \sqrt{ay - 1}$$

Therefore, the equation

$$dz = Pdx + Qdy$$

$$= \sqrt{ax+1}\,dx + \sqrt{ay-1}\,dy$$

Integrating, $$z = \frac{2}{3a}(1+ax)^{3/2} + \frac{2}{3a}(ay-1)^{3/2} + b$$

$$az^{3/2} = (1 + ax)^{3/2} + (ay - 1)^{3/2} + b$$

where a and b are arbitrary constants.

EXAMPLE 10.40 Find complete integral of $\sqrt{p}+\sqrt{q}=2x$.

Solution Given $$\sqrt{p}+\sqrt{q}=2x \tag{i}$$

$\Rightarrow$ $$\sqrt{p}-2x=-\sqrt{q}=a\text{ (Constant)}$$

$\Rightarrow$ $$p = (a + 2x)^2 \text{ and } q = a^2$$

Therefore, the equation

$$dz = pdx + qdy$$

gives $$dz = (a + 2x)^2\, dx + a^2 dy$$

Integrating, $$z = \frac{1}{6}(a+2x)^3 + a^2 y + b$$

is the complete integral where a and b are arbitrary constants.

10.10.4 Form IV

The general form is

$$z = px + qy + f(p, q) \tag{10.58}$$

Subsidiary equations for Eq. (10.58) are

$$\frac{dp}{-p+p} = \frac{dq}{-q+q}$$

$\Rightarrow$ $$\frac{dp}{0} = \frac{dq}{0}$$

$\Rightarrow$ $$dp = 0 = dq$$

Integrating, $$p = a\text{ (Constant)}$$

and $$q = b\text{ (Constant)}$$

Substitute these values in Eq. (10.58).

$\therefore$ $$z = ax + by + f(a, b) \tag{10.59}$$

Also $$dz = pdx + qdy$$

gives $$dz = adx + bdy$$

Integrating, $$z = ax + by + c \tag{10.60}$$

Comparing Eqs. (10.59) and (10.60), Eq. (10.59) can be taken as complete integral of the Eq. (10.58).

To obtain general integral

Put $b = \phi(a)$ in Eq. (10.59)

where ϕ is an arbitrary function.

$\therefore$ $$z = ax + \phi(a)\, y + \phi[a, \phi(a)] \tag{10.61}$$

Eliminate 'a' from Eq. (10.61) and

$$0 = x + \phi'(a)\, y + f'(a) \tag{10.62}$$

to obtain general integral.

Singular integral is obtained by eliminating 'a' and 'b' from Eq. (10.59) and

$$0 = x + \frac{\partial f}{\partial a} \tag{10.63}$$

$$0 = y + \frac{\partial f}{\partial b} \tag{10.64}$$

EXAMPLE 10.41 Solve $z = px + qy + \sqrt{\alpha p^2 + \beta q^2 + \gamma}$.

Solution Complete integral of equation is

$$z = ax + by + \sqrt{\alpha a^2 + \beta b^2 + \gamma} \tag{i}$$

Let us compute singular integral.

Differentiate Eq. (i) w.r.t. 'a' and 'b', respectively.

$$0 = x + by + \frac{\alpha a}{\sqrt{\alpha a^2 + \beta b^2 + \gamma}} \tag{ii}$$

and $$0 = ax + y + \frac{\beta b}{\sqrt{\alpha a^2 + \beta b^2 + \gamma}} \tag{iii}$$

Eliminate 'a' and 'b' from Eqs. (i), (ii) and (iii) gives singular integral as

$$\frac{x^2}{\alpha} + \frac{y^2}{\beta} + \frac{z^2}{\gamma} = 1.$$

EXAMPLE 10.42 Find complete integral of $pqz = p^2(xq + p^2) + q^2(yp + q^2)$.

Solution Rewriting given equation gives

$$pqz = p^2qx + q^2py + p^4 + q^4$$

$\Rightarrow$ $$z = px + qy + \frac{p^3}{q} + \frac{q^3}{p}$$

∴ Complete integral is

$$z = ax + by + \frac{a^3}{b} + \frac{b^3}{a}$$

where a and b are arbitrary constants.

EXAMPLE 10.43 Find complete integral of $z = px + qy + p^2 + q^2$.

Solution Complete integral is

$$z = ax + by + a^2 + b^2 \tag{i}$$

where a and b are arbitrary constants. To obtain general integral put $b = f(a)$ in Eq. (i). We get

$$z = ax + f(a)y + a^2 + \{f(a)\}^2 \tag{ii}$$

General integral is obtained by eliminating 'a' from (ii) and

$$0 = x + f'(a)y + 2a + 2f(a)f'(a) \tag{iii}$$

Singular integral is obtained by eliminating 'a' and 'b' between Eq. (i) and

$$x + 2a = 0 \text{ and } y + 2b = 0.$$

Then, singular integral is $z = -\dfrac{x^2 + y^2}{4}$. Reader can check that it satisfies given equation.

EXAMPLE 10.44 Find complete integral of $z = px + qy - 2\sqrt{pq}$.

Solution Complete integral is

$$z = ax + by - 2\sqrt{ab} \tag{i}$$

where a and b are arbitrary constants. To obtain general integral put $b = f(a)$ in (i). We get

$$z = ax + f(a)y + a^2 - 2\sqrt{af(a)} \tag{ii}$$

General integral is obtained by eliminating 'a' from (ii) and

$$0 = x + f'(a)y - \frac{1}{\sqrt{af(a)}}\{f(a) + af'(a)\} \tag{iii}$$

Singular integral is obtained by eliminating 'a' and 'b' between Eq. (i) and

$$x - \sqrt{\frac{b}{a}} = 0 \text{ and } y - \sqrt{\frac{a}{b}} = a$$

Then, singular integral is $(x - z)(y - z) = 1$. Reader can check that it satisfies given equation.

EXERCISES

Eliminate the arbitrary constants *a*, *b* and *c* from the following relations:

1. $z = ax^2 + bxy + cy^2$. **[Ans.** $x^2 r + 2xys + y^2 t = 2z$ **]**

2. $z = (x-a)^2 + (y-b)^2$. **[Ans.** $4z = p^2 + q^2$ **]**

3. $-z + \frac{x^2}{y} = \log\left(\frac{ax^2}{y^2} + \frac{b}{y}\right)$. $\left[\textbf{Ans. } xp + 2yq = 2\left(z - \frac{x^2}{y}\right)^2\right]$

4. $ax^2 + by^2 + cz^2 - 1 = 0$. **[Ans.** $pq + zs = 0$ **]**

5. $z^2 = ax^3 + by^3 + ab$. **[Ans.** $9x^2y^2z = 6px^3y^2 + 6qx^2y^3 + 4zpq$ **]**

Eliminate arbitrary functions from the following relations:

6. $z = e^{4x} f(x+y)$. **[Ans.** $p - q = 4z$ **]**

7. $z = f(\sqrt{x^2 + y^2})$. **[Ans.** $yp - xq = 0$ **]**

8. $z = xy + f(x^2 + y^2)$. **[Ans.** $py - qx = y^2 - x^2$ **]**

9. $z = f(xy) + g\left(\frac{x}{y}\right)$. **[Ans.** $rx^2 - ty^2 + px - qy = 0$ **]**

10. $lx + my + nz = f(x^2 + y^2 + z^2)$. **[Ans.** $(m + nq)x - (x-y)yp - (x-y)xq = 0$ **]**

Solve the following differential equations:

11. $xzp + yzq = xy$. $\left[\textbf{Ans. } f\left(z^2 - xy, \frac{y}{x}\right) = 0\right]$

12. $x^2 p + y^2 q = x + y$. $\left[\textbf{Ans. } \frac{1}{x} - \frac{1}{y} = c_1, x - y = c_2 e^z\right]$

13. $z(xp - yq) = y^2 - x^2$. **[Ans.** $x^2 + y^2 + z^2 = f(xy)$**]**

14. $(y+z)p + (z+x)q = x + y$. $\left[\textbf{Ans. } f\left[(x-y)^2(x+y+z), \frac{x-y}{y-z}\right] = 0\right]$

15. $\frac{y^2 z}{x} p + xzq = y^2$. **[Ans.** $x^2 - z^2 = f(x^3 - y^3)$**]**

16. $(y^2 + z^2)p - xyq + xz = 0$. **[Ans.** $f(x^2y^2 + y^4, yz) = 0$ **]**

17. $z = xp + yq + \sqrt{x^2 + y^2 + z^2}$. $\left[\textbf{Ans. } z + \sqrt{x^2 + y^2 + z^2} = f\left(\frac{y}{x}\right)\right]$

18. $z(z^2 - xy)(px - qy) = x^4.$ [**Ans.** $f[xy, (z^2 - xy)^2 - x^4] = 0$]

19. $p \tan x + q \tan y = \tan z.$ $\left[\textbf{Ans. } f\left(\frac{\sin x}{\sin y}, \frac{\sin x}{\sin z}\right) = 0\right]$

20. $p(y^3 x - 2x^4) + q(2y^4 - x^3 y) = 2z(x^3 - y^3).$ [**Ans.** $(x^3 + y^3)z^3 = f(x^2y^2z^3)$]

21. $p_1 + p_2 + p_3 (1 + \sqrt{z + x_1 + x_2 + x_3}) + 3 = 0.$

$\left[\textbf{Ans. } f(z + 3x_1, z + 3x_2, z + 6\sqrt{z + x_1 + x_2 + x_3}) = 0\right]$

22. $(x_3 - x_2) p_1 + x_1 p_2 - x_3 p_3 = x_2 (x_1 + x_3) - x_2^{\,2}.$

[**Ans.** $f(z - x_1 x_2, x_1 + x_2 + x_3, x_2 x_3) = 0$]

23. $p^2 + px + q = z.$ [**Ans.** $z = ax + a^2 + be^y$]

24. $p^2 x + q^2 y = z.$ $\left[\textbf{Ans. } \sqrt{z} = a\sqrt{x} + \sqrt{1 - a^2}\,\sqrt{y} + b\right]$

25. $z = px + qy + p^4 + q^4.$ [**Ans.** $z = ax + by + a^4 + b^4$]

26. $p = (qy + z)^2.$ [**Ans.** $yz = ax + 2\sqrt{ay} + b$]

27. $pq + x(2y + 1)p + y(y + 1)q - (2y + 1)z = 0.$ [**Ans.** $z = ax + b(a + y + y^2)$]

28. $16p^2z^2 + 9q^2z^2 + 4z^2 - 4 = 0.$ $\left[\textbf{Ans. } \sqrt{16a^2 + 9}\,\sqrt{1 - z^2} + 2(ax + y) = 0\right]$

29. $4(y + zq) = q(xp + yq^2).$ $\left[\textbf{Ans. } z = ax^2 + by^2 - \frac{1}{4b}\right]$

30. $(p^2 + q^2)\,x = pz.$ [**Ans.** $z^2 = a^2x^2 + (ay + b)^2$]

31. $p_1^{\,3} + p_2^{\,2} + p_3 = 1$ by Jacobi's method. [**Ans.** $z = c_1 x + c_2 x + (1 - c_1^{\,3} - c_2^{\,2})x_3 + c_3$]

32. $p^2 + q^2 = npq.$ [**Ans.** $z = ax + by + c$, where $a^2 + b^2 = nab$]

33. $q = e^{-p/a}.$ [**Ans.** $z = mx + ye^{-m/a} + n$]

34. $p^3 + q^3 = 27z.$ $\left[\textbf{Ans. } \frac{1}{2}z^{2/3} = ax + (1 - a^3)^{1/3} y + c\right]$

35. $x^2p^2 = yq^2.$ [**Ans.** $(z - b - a \log x)^2 = 4a^2 y$]

36. $yp + x^2q^2 = 2x^2 y.$ [**Ans.** $(3z - ax^3 - b)^2 = 4(2 - a)y^3$]

37. $z = p^2 - q^2.$ $\left[\textbf{Ans. } 2\sqrt{z} = \sqrt{1 + a^2}\,x + ay + b\right]$

38. $p^2 y(1 + x^2) = qx^2.$ $\left[\textbf{Ans. } z = a\sqrt{1 + x^2} + \frac{a^2y^2}{2} + b\right]$

39. $(1-y^2)xq^2 - y^2 p = 0.$ **[Ans.** $(2z - ax^2 - b)^2 = a(1-y^2)$]

40. $zy^2 p = x(y^2 + z^2 q^2).$ $\left[\textbf{Ans. } z^2 = ax^2 + \sqrt{a-1}y^2 + c\right]$

41. $z = p^n q^n.$ $\left[\textbf{Ans. } z^{2n-1} = (Ax + By + c)^{2n} \text{ where } AB = \left(\frac{2n-1}{2n}\right)^2\right]$

42. $p(q^2+1) + (3-z)q = 0.$ **[Ans.** $4(Az - 3A - 1) = (x + Ay + B)^2$]

43. $z^2(1 + p^2 + q^2) = c^2.$ **[Ans.** $(1+a^2)(c^2 - z^2) = (y + ax + b)^2$]

44. $16z^2p^2 + 25z^2q^2 + 9z^2 - 81 = 0.$ **[Ans.** $(16a^2 + 25)(9 - z^2) = 9(y + ax + b)^2$]

45. $zpq = p + q.$ $\left[\textbf{Ans. } z^2 = 2(a+1)\left(x + \frac{y}{a}\right) + b\right]$

46. $p - q - x + y = 0.$ **[Ans.** $2z = (x+a)^2 + (y+a)^2 + b$]

47. $z(p^2 - q^2) = x - y.$ **[Ans.** $z^{3/2} = (y+a)^{3/2} + (y+a)^{3/2} + b$]

48. $q = xp + p^2.$ $\left[\textbf{Ans. } z = -\frac{x^2}{4} \pm \left[\frac{x\sqrt{x^2+4a}}{4} + a\log(x + \sqrt{x^2+4a})\right] + ay + b\right]$

49. $p^2q^2 + x^2y^2 = x^2q^2(x^2+y^2).$ $\left[\textbf{Ans. } z = \frac{1}{3}(x^2+a^2)^{3/2} + (y^2-a^2)^{3/2} + b\right]$

50. $z = px + qy - 2\sqrt{pq}.$ **[Ans.** Singular integral: $(x-z)(cy-z) = 1$]

51. $z = xp + yq - \log pq.$ **[Ans.** $z = ax + by - \log ab$, Singular integral: $z = 2 + \log xy$]

52. $py + qx + pq = 0.$ **[Ans.** $2z = ax^2 + by^2 + c$]

53. $xp - yq = xqf(z - px - qy).$ **[Ans.** $(z-a)^2 = b\{2xy + x^2 f(a)\}$]

54. $z^2 = pqxy.$ **[Ans.** $z = bx^a y^{1/a}$]

55. $px^2 + z = y^2q^2.$ $\left[\textbf{Ans. } z = \left(a + \frac{1}{2}\log y\right)^2 + be^{1/x}\right]$

56. Find the integral surface of the equation $(y-z)[2xyp + (x^2-y^2)q] + z(x^2-y^2) = 0$ through the curve $x = t^2, y = 0, z = t^3.$ **[Ans.** $x^3 - 3xy^2 = z^2 - 2yz$]

57. Find the integral surface of the equation $x(y^2+z)p - y(x^2+z)q = (x^2-y^2)z$ through the straight line $x + y = 0, z = 1.$ **[Ans.** $x^2 + y^2 + 2xyz - 2z + 2 = 0$]

58. Find the equation of the integral surface of the differential equation $2y(z-3)p+(2x-z)q=y(2x-3)$ which passes through the circle $x^2+y^2-2x=0$, $z=0$.
[**Ans.** Try yourself.]

59. Find the integral surface of the equation $(x-y)y^2p+(y-x)x^2q=(x^2+y^2)z$ through the curve $xz=a^3$, $y=0$. [**Ans.** $z^3(x^3+y^3)^2=a^9(x-y)^3$]

60. Find the integral surface of $px^2+qy^2+z^2=0$ which passes through the hyperbola $xy=x+y$, $z=1$. $\left[\textbf{Ans. } \dfrac{1}{x}+\dfrac{1}{y}+\dfrac{2}{z}=3\right]$

61. Solve $x^2p+y^2q+z^2=0$. $\left[\textbf{Ans. } f\left(\dfrac{1}{y}-\dfrac{1}{x},\dfrac{1}{y}+\dfrac{1}{z}\right)=0\right]$

62. Solve $x^2p+y^2q=(x+y)z$. $\left[\textbf{Ans. } f\left(\dfrac{xy}{z},\dfrac{x-y}{z}\right)=0\right]$

63. Solve $p+q=\dfrac{1}{x+y}$. $\left[\textbf{Ans. } f\left(y-x,\dfrac{x+y}{e^{2z}}\right)=0\right]$

64. Solve $\dfrac{y-z}{yz}p+\dfrac{z-x}{zx}q=\dfrac{x-y}{xy}$. [**Ans.** $x+y+z=f(xyz)$]

65. Solve $(z-y)p+(x-z)q=y-x$. [**Ans.** $xy+yz+zx=f(x+y+z)$]

66. Solve $z(p+q)=z^2+(x-y)^2$. [**Ans.** $f(x-y,\log(z^2+(x-y)^2-2x)=0$]

67. Solve $xyp+y(2x-y)q=2xz$. $\left[\textbf{Ans. } f\left(xy-x^2,\dfrac{z}{xy}\right)=0\right]$

68. Solve $\cos(x+y)p+\sin(x+t)q=z+\dfrac{1}{z}$.

$$\left[\textbf{Ans. } (x^2+1)^{\frac{1}{\sqrt{2}}}\tan\left(\frac{3\pi}{8}-\frac{x+y}{2}\right)=f(e^{y-x}\{\cos(x+y)+\sin(x+y)\})\right]$$

69. Solve $x_1p_1+x_2p_2+x_3p_3=x_1x_2x_3$. $\left[\textbf{Ans. } x_1x_2x_3-3z=f\left(\dfrac{x_2}{x_1},\dfrac{x_1}{x_3}\right)\right]$

70. Solve $pq=xz$. $\left[\textbf{Ans. } z=\left(\dfrac{x^2}{2}+c_1\right)(y+c_2)\right]$

71. Solve $pq=z$. $\left[\textbf{Ans. } z=\left(\dfrac{x^2}{2}+c_1\right)(y+c_2)\right]$

72. Solve $px + qy = z\sqrt{1 + pq}$

$$\left[\textbf{Ans. } \log z + \frac{u^2}{2} + \frac{u}{2}\sqrt{u^2 - 1} - \frac{1}{2}\log\left[u + \sqrt{u^2 - 1}\right] = c_2,\ u = \frac{c_1 x + y}{\sqrt{c_1 z}}\right]$$

73. Solve $p(1 + q^2) = q(z - a)$ [**Ans.** $4\{c(z - a) - 1\}z = (x + by + c)^2$]

74. Solve $p + q = \dfrac{1}{(z - px - qy)}$ [**Ans.** $(c_1 + c_2)(z - c_1 x - c_2 y) = 1$]

75. Solve $pq = 4$ $\left[\textbf{Ans. } z = ax + \dfrac{4}{a}y + b\right]$

76. Solve $x^2p^2 + y^2q^2 = z^2$ $\left[\textbf{Ans. } z = cx^a y^{\sqrt{1-a^2}}\right]$

77. Solve $x^2p^2 + y^2q^2 = z$ [**Ans.** $4z = \{a\log x + \sqrt{1 - a^2}\log y + b\}^2$]

78. Solve $p^3 = qz$ [**Ans.** $4a^3z = (y + ax + b)^2$]

79. Solve $p(1 + q) = qz$ $\left[\textbf{Ans. } z - a = be^x + \dfrac{y}{a}\right]$

80. Solve $\sqrt{p} + \sqrt{q} = 2x$ $\left[\textbf{Ans. } z = \dfrac{1}{6}(2x - a)^3 + a^2y + b\right]$

81. Solve $p^2 + q^2 = x + y$ $\left[\textbf{Ans. } \dfrac{3}{2}z = (x + a)^{3/2} + (y + a)^{3/2} + b\right]$

82. Solve $z = px + qy + 3(pq)^{1/3}$ [**Ans.** $z = ax + by + 3(ab)^{1/3}$]

83. Solve $2(z + px + qy) = yp^2$ $\left[\textbf{Ans. } yz = \dfrac{ax}{y} - \dfrac{a^2}{4y^2} + b\right]$

CHAPTER 11

Linear Partial Differential Equations with Constant Coefficients

11.1 INTRODUCTION

In this chapter, we will consider linear differential equations with constant coefficients. The differential equation has two independent variables. The general form of an equation is

$$f\left(\frac{\partial}{\partial x}, \frac{\partial}{\partial y}\right) z = V(x, y) \tag{11.1}$$

where f is a national integral algebraic function with constant coefficients and V is any function of the independent variables x and y.

As discussed in Chapter 4, the complete integral of the Eq. (11.1) consists of two parts:

(i) The most general integral of

$$f\left(\frac{\partial}{\partial x}, \frac{\partial}{\partial y}\right) z = 0 \tag{11.2}$$

which is known as Complementary Function (CF).

(ii) Any particular solution of Eq. (11.1), which is known as Particular Integral (PI).

For simplicity, $\dfrac{\partial}{\partial x}$ and $\dfrac{\partial}{\partial y}$ will be denoted by D and D', respectively.

11.2 HOMOGENEOUS LINEAR PDE WITH CONSTANT COEFFICIENTS

Let $$f(D, D') = A_0 D^n + A_1 D^{n-1} D' + A_2 D^{n-2} D'^2 + \cdots + A_n D'^n \tag{11.3}$$

where $A_0, A_1, \ldots, A_n$ are constants.

Then Eq. (11.1) is known as homogeneous equation and takes the form

$$(A_0 D^n + A_1 D^{n-1} D' + A_2 D^{n-2} D'^2 + \cdots + A_n D'^n)\, z = V(x, y) \tag{11.4}$$

11.3 COMPUTATION OF COMPLEMENTARY FUNCTION

Consider an equation

$$(A_0 D^n + A_1 D^{n-1} D' + A_2 D^{n-2} D'^2 + \cdots + A_n D'^n) z = 0 \tag{11.5}$$

Let
$$z = \phi(y + mx) \tag{11.6}$$
be a solution of Eq. (11.5).

Now,
$$D^r z = m^r \phi^{(r)} (y + mx)$$
$$D'^s z = \phi^{(s)} (y + mx)$$
and
$$D^r D'^s z = m^r \phi^{(r+s)} (y + mx)$$

Therefore, substituting Eq. (11.6) in Eq. (11.5)

$$(A_0 m^n + A_1 m^{n-1} + A_2 m^{n-2} + \cdots + A_n)\, \phi^{(n)} (y + mx) = 0$$

which will be satisfied if

$$A_0 m^n + A_1 m^{n-1} + A_2 m^{n-2} + \cdots + A_n = 0 \tag{11.7}$$

Equation (11.7) is known as the Auxiliary Equation (AE).
Let $m_1, m_2, \ldots, m_n$ be the roots of the Eq. (11.7).
Three cases may arise:

Case 1: Let $m_1, m_2, \ldots, m_n$ be distinct.
Then part of CF corresponding to $m = m_1$ is

$$z = \phi_1(y + m_1 x)$$

where ϕ_1 is an arbitrary function.
Part of CF corresponding to $m = m_2$ is

$$z = \phi_2(y + m_2 x)$$

where ϕ_2 is an arbitrary function and so on.
Equation (11.5) being linear, the sum of the solutions is also a solution. Hence,

$$\text{CF} = \phi_1(y + m_1 x) + \phi_2(y + m_2 x) + \cdots + \phi_n(y + m_n x) \tag{11.8}$$

Note:

(i) Equation (11.7) is obtained by putting $D = m$ and $D' = 1$ in Eq. (11.3).

(ii) If and $A_0 = 0$ and $A_n \neq 0$, obtain an auxiliary equation by $D = 1$ and $D' = m$. In this case

$$\text{CF} = \phi_1(x + m_1 y) + \phi_2(x + m_2 y) + \cdots + f_n(x + m_n y) \tag{11.9}$$

(iii) If $A_0 = 0 = A_n$, auxiliary equation will be of degree $(n - 1)$. Here, if auxiliary equation is obtained by Note (i), the nth term of CF is of the form $\phi(x)$ and if auxiliary equation is obtained by Note (ii), the nth term of CF is of the form $\phi(y)$.

EXAMPLE 11.1 Solve $(D^3 - 6D^2 D' + 11 DD'^2 - 6D'^3)\, z = 0$.

Solution Auxiliary equation is

$$m^3 - 6m^2 + 11m - 6 = 0$$

$$\Rightarrow \quad (m - 1)(m - 2)(m - 3) = 0$$

$$\Rightarrow \quad m = 1, 2, 3$$

$$\therefore \quad \text{CF} = \phi_1(y + x) + \phi_2(y + 2x) + \phi_3(y + 3x)$$

EXAMPLE 11.2 Solve $(D^2D' - 3DD'^2 + 2D'^3)\, z = 0.$

Solution Write $D' = m$ and $D = 1$

Auxiliary equation is

$$m - 3m^2 + 2m^3 = 0$$

$$\Rightarrow \quad m(2m^2 - 3m + 1) = 0$$

$$\Rightarrow \quad m(m - 1)\,(2m - 1) = 0$$

$$\Rightarrow \quad m = 0,\ 1,\ \frac{1}{2}$$

$$\therefore \quad \text{CF} = \phi_1(x) + \phi_2(x + y) + \phi_3\left(x + \frac{1}{2}y\right).$$

EXAMPLE 11.3 Solve $(D^2D' - 9DD'^2)\, z = 0.$

Solution According to Note (i), the auxiliary equation is

$$m^2 - 9m = 0$$

$$\Rightarrow \quad m(m - 9) = 0$$

$$\Rightarrow \quad m = 0,\ 9$$

$$\therefore \quad \text{CF} = \phi_1(x) + \phi_2(y) + \phi_3(y + 9x)$$

Case 2: Complex roots

Let a pair of complex roots of the Eq. (11.7) be

$$u + iv$$

Then corresponding part of CF is

$$z = \phi_1(y + ux + ivx) + \phi_2(y + 4x - ivx)$$

$$= \phi_1(X + iY) + \phi_2(X - iY)$$

where $\quad X = y + ux$ and $Y = vx$

Let $\quad \psi_1 = \phi_1 + \phi_2$

and $\quad i\psi_2 = \phi_1 - \phi_2$

Then $\quad \phi_1 = \frac{1}{2}(\psi_1 + i\psi_2)$

and $\quad \phi_2 = \frac{1}{2}(\psi_1 - i\psi_2)$

$$\therefore \quad z = \frac{1}{2}\psi_1(X + iY) + \frac{1}{2}i\psi_2(X + iY) + \frac{1}{2}\psi_1(X - iY) - \frac{1}{2}i\psi_2\,(X - iY)$$

$$= \frac{1}{2}[\psi_1(X + iY) + \psi_2(X - iY)] + \frac{1}{2}i[\psi_1(X + iY) - \psi_2(X - iY)].$$

EXAMPLE 11.4 Solve $(D^4 - 16D'^4)\,z = 0$.

Solution Auxiliary equation is

$$m^4 - 16 = 0$$

$$\Rightarrow \quad (m - 2)(m + 2)(m^2 + 4) = 0$$

$$\Rightarrow \quad m = 2, -2, \pm 2i$$

$$\therefore \quad \text{CF} = \phi_1(y + 2x) + \phi_2(y - 2x) + \phi_3(y + 2ix) + \phi_4(y - 2ix)$$

Case 3: Repeated roots

Let the root m of Eq. (11.7) repeats twice.

Consider the equation

$$(D - mD')(D - mD')\,z = 0 \tag{11.10}$$

Put

$$(D - mD')\,z = U \tag{11.11}$$

Then Eq. (11.10) reduces to

$$(D - mD')\,U = 0$$

$$\Rightarrow \quad \frac{\partial U}{\partial x} - m\frac{\partial U}{\partial y} = 0 \tag{11.12}$$

Equation (11.12) being linear has its subsidiary equations

$$\frac{dx}{1} = \frac{dy}{-m} = \frac{dU}{0} \tag{11.13}$$

Two independent integrals of Eq. (11.13) are

$$y + mx = \text{Constant}$$

and

$$U = \text{Constant}$$

$$\therefore \quad U = \phi(y + mx)$$

is a solution of Eq. (11.12) where ϕ is an arbitrary function.

$\therefore$ Equation (11.11) gives

$$\frac{\partial z}{\partial x} - m\frac{\partial z}{\partial y} = \phi(y + mx) \tag{11.14}$$

which has its subsidiary equations

$$\frac{dx}{1} = \frac{dy}{-m} = \frac{dz}{\phi(y + mx)} \tag{11.15}$$

Two independent integrals of Eq. (11.15) are

$$y + mx = \text{Constant}$$

and

$$z = x\phi(y + mx) + \text{Constant}$$

$$\therefore \quad z = x\phi(y + mx) + y(y + mx) \tag{11.16}$$

is the solution of Eq. (11.10) where ψ is an arbitrary function.

In general, if the root 'm' is repeated r times, the corresponding part of CF is

$$z = x^{r-1}\phi_1(y + mx) + x^{r-2}\phi_2(y + mx) + \cdots + \phi_r(y + mx) \tag{11.17}$$

where $\phi_1, \phi_2, \ldots, \phi_r$ are arbitrary functions.

Note: If auxiliary equation is obtained by changing 'D' to '1' and 'D'' to 'm' the part of CF corresponding to r times repeated root 'm' is

$$z = y^{r-1}\phi_1(x + mx) + y^{r-2}\phi_2(x + mx) + \cdots + \phi_r(x + mx) \tag{11.18}$$

EXAMPLE 11.5 Solve $(D^3D'^2 + D^2D'^3)\,z = 0$.

Solution Changing D as m and D' as 1, the auxiliary equation is

$$m^3 + m^2 = 0$$

$$\Rightarrow \quad m^2(m + 1) = 0$$

$$\Rightarrow \quad m = 0, 0, -1$$

$$\therefore \quad z = \phi_1(y) + x\phi_2(y) + \phi_3(y - x).$$

EXAMPLE 11.6 Solve $(D^4D' - 3D^3D'^2 - 5D^2D'^3 + 3DD'^4 + 4D'^5)\,z = 0$.

Solution Changing D as 1 and D' as m, the auxiliary equation is

$$m - 3m^2 - 5m^3 + 3m^4 + 4m^5 = 0$$

$$\Rightarrow \quad m(4m^4 + 3m^3 - 5m^2 - 3m + 1) = 0$$

$$\Rightarrow \quad m = 0, -1, -1, +1, \frac{1}{4}$$

$$\therefore \quad z = \phi_1(x) + \phi_2(x - y) + y\phi_3(x - y) + \phi_4(x + y) + \phi_5\left(x + \frac{1}{4}y\right)$$

where $\phi_1, \phi_2, \phi_3, \phi_4$ and ϕ_5 are arbitrary functions.

11.4 COMPUTATION OF PARTICULAR INTEGRAL (PI)

Under symbolic notations, Eq. (11.1) becomes

$$f(D, D')\,z = V(x, y) \tag{11.19}$$

The operator $\dfrac{1}{f(D, D')}$ is defined by an identity

$$f(D, D')\frac{1}{f(D, D')}V(x, y) = V(x, y)$$

Then PI is given by

$$\frac{1}{f(D, D')}V(x, y)$$

Let $m_1, m_2, ..., m_n$ be roots of Eq. (11.7). Then Eq. (11.3) can be written as

$$f(D, D') = A_0 (D - m_1)(D - m_2) \cdots (D - m_n) \tag{11.20}$$

$$\therefore \quad \text{PI} = \frac{1}{A_0(D - m_1)(D - m_2)\dots(D - m_n)} V(x, y) \tag{11.21}$$

EXAMPLE 11.7 Solve $(D^2 + 4D'^2) z = x$.

Solution Auxiliary equation is

$$m^2 - 4 = 0$$

$$\Rightarrow \quad (m - 2)(m + 2) = 0$$

$$\Rightarrow \quad m = 2, -2$$

$$\therefore \quad \text{CF} = \phi_1(y + 2x) + \phi_2(y - 2x)$$

where ϕ_1 and ϕ_2 are arbitrary functions.

$$\text{PI} = \frac{1}{D^2 - 4D'^2} x$$

$$= \frac{1}{D^2}\left(1 - 4\frac{D'^2}{D^2}\right)^{-1} x$$

$$= \frac{1}{D^2}\left(1 + 4\frac{D'^2}{D^2} + \cdots\right) x$$

$$= \frac{1}{D^2} x$$

$$= \frac{x^3}{6}$$

or

$$\text{PI} = \frac{1}{D^2 - 4D'^2} x$$

$$= \frac{1}{-4D'^2}\left(1 - \frac{D^2}{4D'^2}\right)^{-1} x$$

$$= -\frac{1}{4D'^2}\left(1 + \frac{D^2}{4D'^2} + \cdots\right) x$$

$$= -\frac{1}{4}\frac{1}{D'^2} x$$

$$= -\frac{1}{4}\frac{xy^2}{2}$$

$$= -\frac{1}{8} xy^2$$

Therefore, complete integral is

$$z = \phi_1(y + 2x) + \phi_2(y \;\; 2x) \;\; \frac{x^3}{6}$$

or

$$z = \phi_1(y+2x) + \phi_2(y-2x) - \frac{1}{8}xy^2.$$

EXAMPLE 11.8 Find PI of equation $(D^2 - D'^2)\, z = x - y$.

Solution

$$\text{PI} = \frac{1}{D^2 - D'^2}(x-y)$$

$$= \frac{1}{D^2}\left(1 - \frac{D'^2}{D^2}\right)^{-1}(x-y)$$

$$= \frac{1}{D^2}\left(1 + \frac{D'^2}{D^2}\right)(x-y)$$

$$= \frac{1}{D^2}(x-y)$$

$$= \frac{x^3}{6} - \frac{yx^2}{2}.$$

EXAMPLE 11.9 Find PI of equation $(D^2 - 2DD' + 15D'^2)\, z = 12xy$.

Solution

$$\text{PI} = \frac{1}{D^2 - 2DD' - 15D'^2}12xy$$

$$= \frac{1}{D^2}\left[1 - \left(\frac{2D'}{D} + \frac{15D'^2}{D^2}\right)\right]^{-1} 12xy$$

$$= 12\frac{1}{D^2}\left[1 + \frac{2D'}{D} + \frac{15D'^2}{D^2}\right]xy$$

$$= 12\frac{1}{D^2}\left[xy + 2\frac{x^2}{2}\right]$$

$$= 12\frac{1}{D^2}[xy + x^2]$$

$$= 12\left[\frac{1}{6}x^3y + \frac{x^4}{12}\right]$$

$$= 2x^3y + x^4.$$

Note: If $V(x, y)$ is rational integral algebraic function, Eq. (11.21) can be evaluated by expanding the symbolic function $\dfrac{1}{f(D, D')}$ in an infinite series of ascending powers of D or D'.

EXAMPLE 11.10 Find PI of equation $(D^2 + 3DD' + 2D'^2)z = x + y$.

Solution The particular integral is given by

$$\text{PI} = \frac{1}{(D^2 + 3DD' + 2D'^2)} x + y$$

$$= \frac{1}{2D'^2}\left\{1 + \left(\frac{D^2}{2D'^2} + \frac{3}{2}\frac{D}{D'}\right)\right\}^{-1} (x + y)$$

$$= \frac{1}{2D'^2}\left\{1 - \frac{3}{2}\frac{D}{D'}\right\}(x + y)$$

$$= \frac{1}{2D'^2}\left(x + y - \frac{3}{2}y\right)$$

$$= \frac{1}{2D'^2}\left(x - \frac{y}{2}\right)$$

$$= \frac{xy^2}{4} - \frac{y^3}{24}$$

Note: If $V(x, y)$ is rational integral algebraic function, Eq. (11.21) can be evaluated by expanding the symbolic function $\dfrac{1}{f(D, D')}$ in an infinite series of ascending powers of D or D'.

11.4.1 $V(x, y)$ is a function of $ax + by$

Let
$$V(x, y) = \phi(ax + by)$$

In the following theorem, we will discuss the method to find PI when $V(x, y) = \phi(ax + by)$.

THEOREM 11.1 If $f(D, D')$ is of degree n, then $\dfrac{1}{f(D, D')}\phi^{(n)}(ax + by) = \dfrac{1}{f(a, b)}\phi(ax + by)$ provided $f(a, b) \neq 0$.

Proof: By direct differentiation

$$D^r\phi(ax + by) = a^r\phi^{(r)}(ax + by)$$

$$D'^s\phi(ax + by) = b^s\phi^{(s)}(ax + by)$$

and
$$D^rD'^s\phi(ax + by) = a^rb^s\phi^{(r+s)}(ax + by)$$

$\therefore$
$$f(D, D')\,\phi(ax + by) = f(a, b)\,\phi^{(n)}(ax + by)$$

Operating upon both sides by $\dfrac{1}{f(D, D')}$

$$\phi(ax + by) = f(a, b)\frac{1}{f(D, D')}\phi^{(n)}(ax + by)$$

Since $\qquad f(a, b) \neq 0$

$$\frac{1}{f(a,b)}\phi(ax+by) = \frac{1}{f(D,D')}\phi^{(n)}(ax+by)$$

$$\Rightarrow \qquad \frac{1}{f(D,D')}\phi^{(n)}(ax+by) = \frac{1}{f(a,b)}\phi(ax+by)$$

When $\qquad f(a, b) = 0$

Let $\qquad f(D, D') = (bD - aD')^r\, g(D, D')$

where $g(D, D')$ is of degree $n - r$ and $g(a, b) \neq 0$. Then

$$\frac{1}{f(D,D')}\phi^{(n)}(ax+by)$$

$$= \frac{1}{(bD-aD')^r}\frac{1}{g(D,D')}\phi^{(n)}(ax+by)$$

$$= \frac{1}{(bD-aD')^r}\frac{1}{g(a,b)}\phi^{(r)}(ax+by)$$

$$= \frac{1}{g(a,b)}\frac{1}{(bD-aD')^r}\phi^{(r)}(ax+by)$$

Put $$\frac{1}{(bD-aD')}\phi^{(r)}(ax+by) = U \tag{11.22}$$

So that $$b\frac{\partial U}{\partial x} - a\frac{\partial U}{\partial y} = \phi^{(r)}(ax+by) \tag{11.23}$$

which has its subsidiary equations

$$\frac{dx}{b} = \frac{dy}{-a} = \frac{dU}{\phi^{(r)}(ax+by)} \tag{11.24}$$

Two independent integrals of Eq. (11.24) are

$$ax + by = \text{Constant}$$

and $$U = \frac{x}{b}\phi^{(r)}(ax+by)$$

$$\therefore \qquad \frac{1}{(bD-aD')^2}\phi^{(r)}(ax+by)$$

$$= \frac{1}{bD-aD'}U$$

$$= \frac{1}{bD-aD'}\frac{x}{b}\phi^{(r)}(ax+by)$$

$$= \frac{x^2}{2!b^2}\phi^{(r)}(ax+by)$$

In general, $$\frac{1}{(bD - aD')^r}\phi^{(r)}(ax+by) = \frac{x^r}{r!}\frac{1}{b^r}\phi^{(r)}(ax+by)$$

Substituting in Eq. (11.22) gives

$$\frac{1}{f(D, D')}\phi^{(n)}(ax+by) = \frac{1}{g(a,b)}\frac{x^r}{r!}\frac{1}{b^r}\phi^{(r)}(ax+by) \tag{11.25}$$

EXAMPLE 11.11 Find PI of equation $(D^2 + D'^2)\, z = \sin mx \cos ny$.

Solution

$$\text{PI} = \frac{1}{D^2 + D'^2}\sin mx \cos ny$$

$$= \frac{1}{2}\frac{1}{D^2 + D'^2}[\sin(mx+ny) + \sin(mx-ny)]$$

$$= -\frac{1}{2}\frac{1}{m^2 + n^2}[\sin(mx+ny) + \sin(mx-ny)]$$

$$= -\frac{1}{m^2+n^2}\sin mx \cos ny.$$

EXAMPLE 11.12 Solve $(D^2 + DD' - 2D'^2)\, z = \sqrt{2x+y}$.

Solution Auxiliary equation is

$$m^2 + m - 2 = 0$$

$$\Rightarrow \quad (m+2)(m-1) = 0$$

$$\Rightarrow \quad m = -2, 1$$

$$\therefore \quad \text{CF} = \phi_1(y - 2x) + \phi_2(y + x)$$

$$\text{PI} = \frac{1}{D^2 + DD' - 2D'^2}\sqrt{2x+y}$$

$$= \frac{1}{4}\int\!\!\int \sqrt{U}\,(dU)^2$$

where $U = 2x + y$

$$= \frac{1}{15}(2x+y)^{5/2}$$

$$\therefore \quad z = \phi_1(y-2x) + \phi_2(y+x) + \frac{1}{15}(2x+y)^{5/2}.$$

EXAMPLE 11.13 Find PI of $(D^2 - 3DD' + 2D'^2)\, z = e^{2x-y} + e^{x+y} + \cos(x + 2y)$.

Solution

$$\text{PI} = \frac{1}{D^2 - 3DD' + 2D'^2}[e^{2x-y} + e^{x+y} + \cos(x+2y)]$$

$$= \frac{1}{(D-2D')(D-D')}[e^{2x-y} + e^{x+y} + \cos(x+2y)]$$

$$= \frac{1}{12}e^{2x-y} - \frac{1}{D-D'}e^{x+y} - \frac{1}{3}\cos(x+2y)$$

$$= \frac{1}{12}e^{2x-y} - xe^{x+y} - \frac{1}{3}\cos(x+2y).$$

11.4.2 *V(x, y)* Contains Either sin (*ax* + *by*) or cos (*ax* + *by*)

In this case PI can be obtained by the method of undetermined coefficients.

EXAMPLE 11.14 Solve $(D^2 + DD' - 6D'^2)\, z = y \cos x$.

Solution Auxiliary equation is

$$m^2 + m - 6 = 0$$

$$\Rightarrow \quad (m+3)(m-2) = 0$$

$$\Rightarrow \quad m = -3, 2$$

$$\therefore \quad \text{CF} = \phi_1(y - 3x) = \phi_2(y + 2x)$$

To find PI, let

$$z = Ay \cos x + By \sin x + C \cos x + D \sin x \qquad \text{(i)}$$

Then

$$D^2 z = -Ay \cos x - By \sin x - C \cos x - D \sin x$$

$$DD'z = -A \sin x + B \cos x$$

$$D'^2 z = 0$$

Substituting in given equation, we get

$$-Ay \cos x - By \sin x - (A + D) \sin x + (B - C) \cos x = y \cos x$$

Comparing coefficients gives

$$A = -1$$

$$B = 0$$

$$C = 0$$

and

$$D = 0$$

Substituting in Eq. (i) gives

$$z = -y \cos x + \sin x$$

Hence,

$$\text{PI} = -y \cos x + \sin x$$

$$\therefore \quad z = \phi_1(y - 3x) + \phi_2(y + 2x) - y \cos x + \sin x.$$

EXAMPLE 11.15 Find PI of equation $(D^3 + D^2D' - DD'^2 - D'^3)\, z = e^y \cos 2x$.

Solution Let

$$z = Ae^y \cos 2x + Be^y \sin 2x \tag{i}$$

where A and B are constants to be determined such that Eq. (i) satisfies given equation.

We have

$$D^3z = 8Ae^y \sin 2x - 8Be^y \cos 2x$$

$$D^2D'z = -4Ae^y \cos 2x - 4Be^y \sin 2x$$

$$DD'^2z = -2Ae^y \sin 2x + 2Be^y \cos 2x$$

and

$$D'3z = Ae^y \cos 2x + Be^y \sin 2x$$

Substituting in given equation gives

$$(10A - 5B)\, e^y \sin 2x + (-5A - 10B)\, e^y \cos 2x = e^y \cos 2x$$

Comparing coefficients,

$$10A - 5B = 0$$

$$5A + 10B = 1$$

$\therefore$

$$A = -\frac{1}{25}$$

and

$$B = -\frac{2}{25}$$

$\therefore$

$$\text{PI} = -\frac{1}{25} e^y \cos 2x - \frac{2}{25} e^y \sin 2x.$$

11.5 GENERAL METHOD FOR FINDING PI

If $V(x, y)$ is any function, Eq. (11.21) can be evaluated by solving n-equations of first order.

$$\left.\begin{aligned} U_1 &= \frac{1}{D - m_n D'} V(x, y) \\ U_2 &= \frac{1}{D - m_{n-1} D'} U_1 \\ &\vdots \\ U_n &= \frac{1}{D - m_1 D'} U_{n-1} \end{aligned}\right\} \tag{11.26}$$

Each of the Eq. (11.26) is of the form

$$p - mq = g(x, y) \tag{11.27}$$

which is linear equation and has subsidiary equations

$$\frac{dx}{1} = \frac{dy}{-m} = \frac{dU}{g(x, y)} \tag{11.28}$$

Two independent integrals of Eq. (11.28) are

$$y + mx = a \text{ (Constant)} \tag{11.29}$$

and

$$U = \int g(x, a - mx)\, dx \tag{11.30}$$

After integrating Eq. (11.30) replace a by $y + mx$. The resulting function will be a solution of Eq. (11.27).

Similarly, we can obtain $U_1, U_2, \ldots, U_n$ of Eq. (11.26), and hence, PI can be found.

EXAMPLE 11.16 Solve $(D^2 - DD' - 2D'^2)\, z = (y - 1)\, e^x$.

Solution Auxiliary equation is

$$m^2 - m - 2 = 0$$

$$\Rightarrow \quad (m - 2)(m + 1) = 0$$

$$\Rightarrow \quad m = -1, 2$$

$$\therefore \quad \text{CF} = \phi_1(y - x) + \phi_2(y + 2x)$$

$$\text{PI} = \frac{1}{(D - 2D')(D + D')}(y - 1)\, e^x$$

$$= \frac{1}{D + D'}\left[\int_{a=y+2x} (a - 2x - 1)\, e^x dx\right]$$

$$= \frac{1}{D + D'}[(a - 2x - 1)\, e^x + 2e^x]_{a=y+2x}$$

$$= \frac{1}{D + D'}(y + 1)\, e^x$$

$$= \int_{a=y-x} (a + x + 1)\, e^x dx$$

$$= [(a + x + 1)\, e^x - e^x]_{a=y-x}$$

$$= ye^x.$$

EXAMPLE 11.17 Solve $(D^2 - D'^2)\, z = \tan^3 x \tan y - \tan x \tan^3 y$.

Solution Auxiliary equation is

$$m^2 - 1 = 0$$

$$\Rightarrow \quad (m - 1)(m + 1) = 0$$

$$\Rightarrow \quad m = 1, -1$$

$$\therefore \quad \text{CF} = \phi_1(y + x) + \phi_2(y - x)$$

$$\text{PI} = \frac{1}{(D - D')(D + D')}(\tan^3 x \tan y - \tan x \tan^3 y)$$

$$= \frac{1}{(D - D')(D + D')} \tan x \tan y\, (\tan^2 x - \tan^2 y)$$

$$= \frac{1}{(D-D')(D+D')} \tan x \tan y (\sec^2 x - \sec^2 y)$$

$$= \frac{1}{D-D'} \left\{ \int_{a=y-x} \tan x \tan (a+x) [\sec^2 x - \sec^2 (a+x)]\, dx \right\}$$

$$= \frac{1}{D-D'} \begin{bmatrix} \frac{1}{2}\tan^2 x \tan (a+x) - \frac{1}{2}\int \tan^2 x \sec^2 (a+x)\, dx \\ -\frac{1}{2}\tan x \tan^2 (a+x) + \frac{1}{2}\int \sec^2 x \tan^2 (a+x)\, dx \end{bmatrix}_{a=y-x}$$

$$= \frac{1}{D-D'} \left[\frac{1}{2}\tan^2 x \tan(a+x) - \frac{1}{2}\tan x \tan^2 (a+x) + \frac{1}{2}\int \{\sec^2 (a+x) - \sec^2 x\}\, dx \right]_{a=y-x}$$

$$= \frac{1}{D-D'} \left[\frac{1}{2}\tan^2 x \tan(a+x) - \frac{1}{2}\tan x \tan^2 (a+x) + \frac{1}{2}\tan(a+x) - \frac{1}{2}\tan x \right]_{a=y-x}$$

$$= \frac{1}{D-D'} \left[\frac{1}{2}\tan y \sec^2 x - \frac{1}{2}\tan x \sec^2 y \right]$$

$$= \frac{1}{2} \int_{a=y+x} \left[\tan (a-x) \sec^2 x - \tan x \sec^2 (a-x) \right] dx$$

$$= \frac{1}{2} [\tan x \tan (a-x) + \int \tan x \sec^2 (a-x)\, dx - \int \tan x \sec^2 (a-x)\, dx]_{a=y+x}$$

$$= \frac{1}{2} \tan x \tan y$$

$$\therefore \qquad z = \phi_1 (y+x) + \phi_2 (y-x) + \frac{1}{2} \tan x \tan y.$$

EXAMPLE 11.18 Solve $(D^2 - 2DD' - 15D'^2)\, z = 12xy$.

Solution Auxiliary equation is

$$m^2 - 2m - 15 = 0$$

$$\Rightarrow \qquad (m-5)(m+3) = 0$$

$$\Rightarrow \qquad m = -3, 5$$

$$\therefore \qquad \text{CF} = \phi_1(y - 3x) + \phi_2(y + 5x)$$

$$\text{PI} = \frac{1}{(D-5D')(D+3D')} 12xy$$

$$= \frac{1}{D-5D'} \int_{a=y-3x} 12x(a+5x)\, dx$$

$$= \frac{1}{D-5D'} [6ax^2 + 12x^3]_{a=y-3x}$$

$$= \frac{1}{D-5D'}[6yx^2 - 6x^3]$$

$$= \int_{a=y+5x} [6(a-5x)x^2 - 6x^3]\,dx$$

$$= \int_{a=y+5x} [6ax^2 - 36x^3]\,dx$$

$$= 2ax^3 - 9x^4\Big|_{a=y+5x}$$

$$= 2yx^3 + x^4$$

$\therefore$ $$z = \phi_1(y-3x) + \phi_2(y+5x) + 2x^3y + x^4.$$

EXAMPLE 11.19 Solve $(D^2 - DD' - 2D'^2)z = (2x^2 + xy - y^2)\sin xy - \cos xy$.

Solution Auxiliary equation is

$$m^2 - m - 2 = 0$$

$\Rightarrow$ $$(m-2)(m+1) = 0$$

$\Rightarrow$ $$m = -1, 2$$

$\therefore$ $$\text{CF} = \phi_1(y-x) + \phi_2(y+2x)$$

Now, $$\text{PI} = \frac{1}{(D-2D')(D+D')}\{(2x^2 + xy - y^2)\sin xy - \cos xy\}$$

$$= \frac{1}{(D-2D')(D+D')}\{(2x-y)(x+y)\sin xy - \cos xy\}$$

$$= \frac{1}{(D-2D')}\left[\int \{(x-a)(2x+a)\sin(a+x) - \cos(a+x)\}dx\right];\ a = y-x$$

$$= \frac{1}{(D-2D')}\left[\int \{(x-a)(2x+a)\sin(ax+x^2) - \cos(ax+x^2)\}dx\right];\ a = y-x$$

$$= \frac{1}{(D-2D')}(y-2x)\cos xy$$

$$= \int (a-4x)\cos(ax-2x^2)dx;\ a = y+2x$$

$$= \sin xy$$

$\therefore$ Complete solution is $\phi_1(y-x) + \phi_2(y+2x) + \sin xy$.

11.6 NON-HOMOGENEOUS LINEAR PDE WITH CONSTANT COEFFICIENTS

If all the terms of LHS of Eq. (11.1) are not of the same degree, equation is said to be **non-homogeneous PDE**. The equation is said to be reducible if the symbolic function $f(D, D')$ can be factorized and each of the factor is of first degree in D and D' otherwise it is called irreducible. For example

$$f(D, D')\,z = (D^3 - 3DD'^2 - 2D'^3)\,z = e^y\,(2x + 3)$$

is reducible because

$$(D^3 - 3DD'^2 - 2D'^3)\,z = (D + D')\,(D - D')\,(D - 2D')\,z = e^y\,(2x + 3)$$

whereas $f(D, D')\,z = D^3D' + DD'^2)\,z = DD'\,(D^2 + D')\,z = 12xy$ is irreducible.

11.7 REDUCIBLE NON-HOMOGENEOUS PDE

Let $$f(D, D') = (a_1D + b_1D' + c_1)\,(a_2D + b_2D' + c_2) \cdots (a_nD + b_nD' + c_n) \tag{11.31}$$

where a's, b's and c's are constants.

Then Eq. (11.2) takes the form

$$(a_1D + b_1D' + c_1)\,(a_2D + b_2D' + c_2) \cdots (a_nD + b_nD' + c_n)\,z = 0 \tag{11.32}$$

Any solution of the equation

$$(a_kD + b_kD' + c_k)\,z = 0 \tag{11.33}$$

is a solution of the Eq. (11.32). Subsidiary equations of Eq. (11.33) are

$$\frac{dx}{a_k} = \frac{dy}{b_k} = \frac{dz}{-c_k z} \tag{11.34}$$

Two independent integrals of Eq. (11.34) are

$$b_kx - a_ky = \text{Constant}$$

and $$z = \text{Constant } e^{-\frac{c_k}{a_k}x} \text{ if } a_k \neq 0$$

or $$z = \text{Constant } e^{-\frac{c_k}{b_k}y} \text{ if } b_k \neq 0$$

Therefore, general solution of Eq. (11.33) is

$$z = e^{-\frac{c_k}{a_k}x}\,\phi_k\,(b_kx - a_ky) \text{ if } a_k \neq 0 \tag{11.35}$$

or $$z = e^{-\frac{c_k}{b_k}y}\,\psi_k\,(b_kx - a_ky) \text{ if } b_k \neq 0 \tag{11.36}$$

where ϕ_k and ψ_k are arbitrary functions.

If no two factors of Eq. (11.31) are linearly independent, the general solution of Eq. (11.32) is the sum of the general solution of the equation of type like Eq. (11.33) corresponding to each factor in Eq. (11.31).

EXAMPLE 11.20 Solve $(D^2 - D'^2 - 2D + 2D')\,z = 0$.

Solution Given $$(D^2 - D'^2 - 2D + 2D')\,z = 0$$

$$\Rightarrow \quad (D - D')(D + D' - 2)z = 0$$

$$\therefore \quad \text{CF} = \phi_1(y + x) + e^{2x}\phi_2(x - y)$$

or
$$\text{CF} = \psi_1(y + x) + e^{2y}\psi_2(x - y).$$

11.8 SOLUTION METHODOLOGY FOR REPEATED FACTORS

Let $(aD + bD' + c)$ be root repeated twice of the Eq. (11.31). Consider the equation

$$(aD + bD' + c)(aD + bD' + c)z = 0 \tag{11.37}$$

Put
$$(aD + bD' + c)z = U \tag{11.38}$$

Then, Eq. (11.37) becomes

$$(aD + bD' + c)U = 0 \tag{11.39}$$

As in Section 11.7, general solution of Eq. (11.39) is

$$U = e^{-\frac{c}{a}x}\phi(bx - ay) \text{ if } a \neq 0 \tag{11.40}$$

or
$$U = e^{-\frac{c}{b}y}\psi(bx - ay), \text{ if } b \neq 0 \tag{11.41}$$

Substituting Eq. (11.40) in Eq. (11.37) gives

$$(aD + bD' + c)z = e^{-\frac{c}{a}x}\phi(bx - ay) \tag{11.42}$$

Subsidiary equations of Eq. (11.42) are

$$\frac{dx}{a} = \frac{dy}{b} = \frac{dz}{e^{-\frac{c}{a}x}\phi(bx - ay) - cz} \tag{11.43}$$

Two independent integrals of Eq. (11.43) are

$$bx - ay = \lambda \text{ (Constant)} \tag{11.44}$$

and
$$\frac{dz}{dx} + \frac{c}{a}z = \frac{1}{a}e^{-\frac{c}{a}x}\phi(bx - ay)$$

$$\Rightarrow \quad \frac{dz}{dx} + \frac{c}{a}z = \frac{1}{a}e^{-\frac{c}{a}x}\phi(\lambda)$$

which is linear equation. Its solution is

$$ze^{\frac{c}{a}x} = \frac{1}{a}x\phi(\lambda) + \text{Constant}$$

$$\Rightarrow \quad ze^{\frac{c}{a}x} = \frac{1}{a}x\phi(bx - ay) + \text{Constant}$$

Therefore, general solution of Eq. (11.42) is

$$z = \frac{x}{a} e^{-\frac{c}{a}} \phi(bx - ay) + e^{-\frac{c}{a}} \phi(bx - ay)$$

$$= e^{-\frac{c}{a}x} [\phi_1(bx - ay) + x\phi_2(bx - ay)] \qquad (11.45)$$

where ϕ_1 and ϕ_2 are arbitrary functions.
Similarly from Eq. (11.41), we get

$$z = e^{-\frac{c}{b}y} [\psi_1(bx - ay) + y\psi_2(bx - ay)] \qquad (11.46)$$

where ψ_1 and ψ_2 are arbitrary functions.
In general for r-times repeated factor $(aD + bD' + c)$, the general solution is

$$z = e^{-\frac{c}{a}x} \sum_{i=1}^{r} x^{i-1} \phi_i(bx - ay) \text{ if } a \neq 0$$

or

$$z = e^{-\frac{c}{b}y} \sum_{i=1}^{r} y^{i-1} \psi_i(bx - ay) \text{ if } b \neq 0$$

where ϕ_i and ψ_i, $i = 1, 2, ..., r$ are arbitrary functions.

11.9 IRREDUCIBLE NON-HOMOGENEOUS PDE'S

Consider the equation

$$f(D, D')\, z = 0 \qquad (11.47)$$

Put

$$z = ce^{ax+by} \qquad (11.48)$$

where a, b and c are constants.
Now,

$$D^r z = ca^r e^{ax+by}$$

$$D^r D'^s z = ca^r b^s e^{ax+by}$$

and

$$D'^s z = cb^s e^{ax+by}$$

Substituting these in Eq. (11.47) give

$$cf(a, b)\, e^{ax+by} = z$$

which will hold if

$$f(a, b) = 0 \qquad (11.49)$$

and c is arbitrary constant.
For any chosen value of 'a' (or b), Eq. (11.49) gives one or more value of 'b' (or a). Thus, there exists infinitely many pairs of (a_i, b_i) satisfying Eq. (11.49).

i.e.

$$z = \sum_{i=1}^{\infty} c_i e^{a_i x + b_i y} \qquad (11.50)$$

where

$$f(a_i, b_i) = 0 \qquad (11.51)$$

is a solution of the Eq. (11.47). If

$$f(D, D') = (D + hD' + k)\, g(D, D') \tag{11.52}$$

any pair (a, b) is such that

$$a + hb + k = 0 \tag{11.53}$$

satisfies Eq. (11.49). Such pairs are infinite in number.

By Eq. (11.53)

$$a = -(hb + k)$$

Thus,

$$z = \sum_{i=1}^{\infty} c_i e^{-(hb_i + k)x + b_i y}$$

$$= e^{-kx} \sum_{i=1}^{\infty} c_i e^{b_i (y - hk)} \tag{11.54}$$

is a part of CF corresponding to linear factor $(D + hD' + k)$ in Eq. (11.52). One can write Eq. (11.54) as

$$e^{-kx}\, \phi(y - hk)$$

where ϕ is an arbitrary function.

EXAMPLE 11.21 Solve $(2D^2 - D'^2 + D)\, z = 0$.

Solution The CF is given by

$$z = \sum_{i=1}^{\infty} c_i e^{a_i x + b_i y}$$

where a_i and b_i are related by

$$2a_i^2 - b_i^2 + a_i = 0$$

$$\Rightarrow \qquad b_i^2 = a_i(2a_i + 1)$$

$$\Rightarrow \qquad b_i = \sqrt{a_i(2a_i + 1)}$$

$$\therefore \qquad \text{CF} = z = \sum_{i=1}^{\infty} c_i e^{a_i x + \sqrt{a_i(2a_i + 1)}\, y}.$$

EXAMPLE 11.22 Solve $(D^2 + DD' + D + D' + 1)\, z = 0$.

Solution The CF is given by

$$z = \sum_{i=1}^{\infty} c_i e^{a_i x + b_i y}$$

where a_i and b_i are related by

$$a_i^2 + a_i b_i + a_i + b_i + 1 = 0.$$

11.10 PARTICULAR INTEGRAL

Consider the equation

$$f(D, D')\, z = V(x, y) \tag{11.55}$$

Here $f(D, D')$ is a non-homogeneous function of D and D'.

$$\text{PI} = \frac{1}{f(D, D')} V(x, y) \tag{11.56}$$

11.10.1 When $V(x, y) = e^{ax+by}$, Where 'a' and 'b' are constants

We will state the method to evaluate Eq. (11.56) in the following theorem.

THEOREM 11.2 If $f(a, b) \neq 0$, then

$$\frac{1}{f(D, D')} e^{ax+by} = \frac{1}{f(a, b)} e^{ax+by}$$

Proof: By differentiation

$$D^r e^{ax+by} = a^r e^{ax+by}$$

$$D'^s e^{ax+by} = b^s e^{ax+by}$$

$$D^r D'^s e^{ax+by} = a^r b^s e^{ax+by}$$

$\therefore$

$$f(D, D')\, e^{ax+by} = f(a, b)\, e^{ax+by}$$

Operating upon by $f(D, D')$, we get

$$e^{ax+by} = f(a, b) \frac{1}{f(D, D')} e^{ax+by}$$

$\Rightarrow$

$$\frac{1}{f(D, D')} e^{ax+by} = \frac{1}{f(a, b)} e^{ax+by}$$

EXAMPLE 11.23 Solve $(D^3 - 3DD' + D + 1)\, z = e^{2x+3y}$.

Solution Given

$$(D^3 - 3DD' + D + 1)\, z = e^{2x+3y} \tag{i}$$

For

$$\text{CF} = (D^3 - 3DD' + D + 1)\, z = 0$$

Then,

$$\text{CF} = z = \sum_{i=1}^{n} c_i e^{a_i x + b_i y}$$

where

$$a_i^3 - 3a_i b_i + a_i + 1 = 0$$

For

$$\text{PI} = \frac{1}{D^3 - 3DD' + D + 1} e^{2x+3y}$$

$$= \frac{1}{2^3 - 3 \cdot 2 \cdot 3 + 2 + 1} e^{2x+3y}$$

$$= -\frac{1}{7} e^{2x+3y}$$

Therefore, general solution of Eq. (i) is

$$z = \sum_{i=1}^{n} c_i e^{a_i x + b_i y} - \frac{1}{7} e^{2x+3y}$$

where $a_i^3 - 3a_i b_i + a_i + 1 = 0.$

EXAMPLE 11.24 Solve $(D^2 - D')\, z = e^{2x+y}$.

Solution Given $(D^2 - D')\, z = e^{2x+y}$ (i)

For $\text{CF} = (D^2 - D')\, z = 0$

Then, $a_i^2 - b_i = 0$

$\Rightarrow$ $b_i = a_i^2$

$\therefore$ $\text{CF} = z = \sum_{i=1}^{\infty} c_i e^{a_i x + a_i^2 y}$

For

$$\text{PI} = \frac{1}{D^2 - D'} e^{2x+y}$$

$$= \frac{1}{2^2 - 1} e^{2x+y}$$

$$= \frac{1}{3} e^{2x+y}$$

$\therefore$ General solution of given Eq. (i) is

$$z = \sum_{i=1}^{\infty} c_i e^{a_i x + a_i^2 y} + \frac{1}{3} e^{2x+y}.$$

EXAMPLE 11.25 Solve $(D^2 - D'^2 - 3D + 3D')z = e^{x-2y}$.

Solution Given equation is equivalent to $(D - D')(D + D' - 3)z = e^{x-2y}$.

$$\text{CF} = \phi_1(y + x) + e^{3x}\phi_2(y - x)$$

and $\text{PI} = \dfrac{1}{(D - D')(D + D' - 3)} e^{x-2y} = -\dfrac{1}{12} e^{x-2y}$

Therefore solution is

$$z = \phi_1(y + x) + e^{3x}\phi_2(y - x) - \frac{1}{12} e^{x-2y}$$

11.10.2 When *V*(*x*, *y*) = sin (*ax* + *by*) or cos (*ax* + *by*)

We know that

$$D^2 \sin(ax + by) = (-a^2) \sin(ax + by)$$

$$DD' \sin (ax + by) = (-ab) \sin (ax + by)$$

and
$$D'^2 \sin (ax + by) = (-b^2) \sin (ax + by)$$

In Eq. (11.56) replace D^2 by $(-a^2)$, DD' by $(-ab)$ and D'^2 by $(-b^2)$.
Same procedure will be followed by $V(x, y) = \cos (ax + by)$.

EXAMPLE 11.26 Solve $(D + D')(D + D' - 2) z = \sin (3x + 4y)$.

Solution $\text{CF} = \phi_1(y - x) + e^{2x}\phi_2(x - y)$.

$$\text{PI} = \frac{1}{(D + D')(D + D' - 2)} \sin (3x + 4y)$$

$$= \frac{1}{D^2 + 2DD' + D'^2 - 2D - 2D'} \sin (3x + 4y)$$

$$= \frac{1}{(-3^2) + 2(-3 \cdot 4) + (-4^2) - 2(D + D')} \sin (3x + 4y)$$

$$= -\frac{1}{49 + 2D + 2D'} \sin (3x + 4y)$$

$$= -\frac{(49 - 2D - 2D')}{2401 - 4(D^2 + D'^2 + 2DD')} \sin (3x + 4y)$$

$$= -\frac{(49 - 2D - 2D')}{2401 - 4(-9 - 16 - 24)} \sin (3x + 4y)$$

$$= -\frac{(2D + 2D' - 49)}{2205} \sin (3x + 4y)$$

$$= -\frac{1}{2205} [6 \cos (3x + 4y) + 8 \cos (3x + 4y) - 49 \sin (3x + 4y)]$$

$$= \frac{1}{2205} [14 \cos (3x + 4y) - 49 \sin (3x + 4y)]$$

$\therefore$ General solution is

$$z = \phi_1 (y - x) + e^{2x} \phi_2 (x - y) + \frac{1}{2205} [14 \cos (3x + 4y) - 49 \sin (3x + 4y)].$$

EXAMPLE 11.27 Find PI of equation $(D^2 - D') z = A \cos (lx + my)$ where A, l, m are constants.

Solution

$$\text{PI} = \frac{1}{D^2 - D'} A \cos (lx + my)$$

$$= \frac{1}{-l^2 - D'} A \cos (lx + my)$$

$$= -\frac{1}{l^2 + D'} A \cos (lx + my)$$

$$= -\frac{(l^2 - D')}{l^4 - D'^2} A \cos (lx + my)$$

$$= -\frac{(D' - l^2)}{l^4 + m^2} A \cos(lx + my)$$

$$= \frac{A}{l^4 + m^2}[-m \sin(lx + my) - l^2 \cos(lx + my)]$$

$$= -\frac{A}{l^4 + m^2}[m \sin(lx + my) + l^2 \cos(lx + my)].$$

11.10.3 When $V(x, y) = x^m y^n$, Where m and n are Positive Integers

In this case Eq. (11.56) can be evaluated by expanding the symbolic function $\frac{1}{f(D, D')}$ in an infinite series of ascending powers of D and D'.

EXAMPLE 11.28 Find PI of the equation $(DD' + D - D' - 1)\, z = x^2 y$.

Solution

$$\text{PI} = \frac{1}{(D'+1)(D-1)} x^2 y$$

$$= -(1 + D')^{-1}(1 - D)^{-1} x^2 y$$

$$= -(1 - D' + D'^2 - \cdots)(1 + D + D^2 + D^3 + \cdots) x^2 y$$

$$= -(1 + D + D^2 - D' - DD' - D^2 D' + \cdots) x^2 y$$

$$= -(x^2 y + 2xy + 2y - x^2 - 2x - 2).$$

EXAMPLE 11.29 Find PI of the equation

$$D(D + D' - 1)(D + 3D' - 2)\, z = x^2 - 6xy + 3y^2.$$

Solution

$$\text{PI} = \frac{1}{D(D + D' - 1)(D + 3D' - 2)}(x^2 - 6xy + 3y^2)$$

$$= \frac{1}{2D}[1 - (D + D')]^{-1}\left[1 - \frac{D + 3D'}{2}\right]^{-1}(x^2 - 6xy + 3y^2)$$

$$= \frac{1}{2D}[1 + (D + D') + (D + D')^2 + \cdots]\left[1 + \frac{D + 3D'}{2} + \left(\frac{D + 3D'}{2}\right)^2 + \cdots\right](x^2 - 6xy + 3y^2)$$

$$= \frac{1}{2D}\left[1 + \frac{3}{2}D + \frac{5}{2}D' + \frac{7}{4}D^2 + \frac{19}{4}D'^2 + \frac{22}{4}DD' + \cdots\right](x^2 - 6xy + 3y^2)$$

$$= \frac{1}{2D}\left[(x^2 - 6xy + 3y^2) + \frac{3}{2}(2x - 6y) + \frac{5}{2}(-6x + 6y) + \frac{7}{4}(2) + \frac{19}{4}\cdot 6 + \frac{22}{4}(-6)\right]$$

$$= \frac{1}{2D}[x^2 - 6xy + 3y^2 + 3x - 9y - 15x + 15y - 1]$$

$$= \frac{1}{2D}[x^2 - 6xy + 3y^2 - 12x - 6y - 1]$$

$$= \frac{1}{2}\left[\frac{x^3}{3} - 3x^2y + 3xy^2 - 6x^2 - 6xy - x\right].$$

EXAMPLE 11.30 Find PI for the equation

$$(D^2 - DD' - 2D'^2 + 2D + 2D')\, z = e^{3x+4y} + \sin(x + 2y) + xy.$$

Solution Given

$$(D^2 - DD' - 2D'^2 + 2D + 2D')\, z = e^{3x+4y} + \sin(x + 2y) + xy$$

$$\Rightarrow \qquad (D + D')(D - 2D' + 2)\, z = e^{3x+4y} + \sin(x + 2y) + xy$$

$$\text{PI} = \frac{1}{(D + D')(D - 2D' + 2)}[e^{3x+4y} + \sin(x + 2y) + xy]$$

PI corresponding to e^{3x+4y}

$$= \frac{1}{(D + D')(D - 3D' + 2)} e^{3x+4y}$$

$$= \frac{1}{(3 + 4)(3 - 12 + 2)} e^{3x+4y}$$

$$= -\frac{1}{49} e^{3x+4y}$$

PI corresponding to $\sin(x + 2y)$

$$= \frac{1}{D^2 - DD' - 2D'^2 + 2D + 2D'} \sin(x + 2y)$$

$$= \frac{1}{(-1) - (-2) - 2(-4) + 2D + 2D'} \sin(x + 2y)$$

$$= \frac{1}{9 + 2D + 2D'} \sin(x + 2y)$$

$$= \frac{1}{9 + 2(D + D')} \sin(x + 2y)$$

$$= \frac{9 - 2(D - D')}{81 - 4(D + D')^2} \sin(x + 2y)$$

$$= \frac{9 - 2(D - D')}{81 - 4(D^2 + 2DD' + D'^2)} \sin(x + 2y)$$

$$= \frac{9 - 2(D - D')}{81 - 4(-1 - 4 - 4)} \sin(x + 2y)$$

$$= \frac{1}{45}[9\sin(x + 2y) - 2\cos(x + 2y) + 4\cos(x + 2y)]$$

$$= \frac{1}{45}[9\sin(x + 2y) + 2\cos(x + 2y)]$$

PI corresponding to xy

$$= \frac{1}{(D+D')(D-2D'+2)} xy$$

$$= \frac{1}{2D}\left(1+\frac{D'}{D}\right)^{-1}\left(1+\frac{D-2D'}{2}\right)^{-1} xy$$

$$= \frac{1}{2D}\left(1-\frac{D'}{D}+\frac{D'^2}{D^2}-\frac{D'^3}{D^2}+\cdots\right)\left(1-\frac{D-2D'}{2}+\frac{(D-2D')^2}{4}\cdots\right)xy$$

$$= \frac{1}{2D}\left(1-\frac{D'}{D}+\frac{D'^2}{D^2}-\frac{D'^3}{D^2}+\cdots\right)\left(xy-\frac{y}{2}+x-1\right)$$

$$= \frac{1}{2D}\left(xy-\frac{y}{2}+x-1-\frac{x^3}{2}+\frac{x}{2}\right)$$

$$= \frac{1}{2D}\left(xy-\frac{y}{2}+\frac{3}{2}x-\frac{x^2}{2}-1\right)$$

$$= \frac{1}{2}\left(\frac{x^2y}{2}-\frac{yx}{2}+\frac{3}{4}x^2-\frac{x^3}{6}-x\right)$$

$$= \frac{x}{24}(6xy-6y+9x-2x^2-12)$$

PI of given equation is

$$= -\frac{1}{49}e^{3x+4y}+\frac{1}{45}[9\sin(x+2y)+2\cos(x+2y)]+\frac{x}{24}(6xy-6y+9x-2x^2-12).$$

11.10.4 When $V(x, y) = e^{ax+by}\phi(x, y)$, Where a and b are Arbitrary Constants

We will explain procedure to find out PI in the following theorem:

THEOREM 11.3 Let $\phi(x, y)$ be any function, then

$$\frac{1}{f(D, D')}e^{ax+by}\phi(x, y) = e^{ax+by}\frac{1}{f(D+a, D'+b)}\phi(x, y).$$

Proof: By Leibnitz's theorem for successive differentiation

$$D^r[e^{ax+by}\phi(x, y)] = e^{ax+by}(D+a)^r\phi(x, y)$$

$$D'^s[e^{ax+by}\phi(x, y)] = e^{ax+by}(D'+b)^s\phi(x, y)$$

and

$$D^rD'^s[e^{ax+by}\phi(x, y)] = e^{ax+by}(D+a)^r(D'+b)^s\phi(x, y)$$

$\therefore \qquad f(D, D')[e^{ax+by}\phi(x, y)] = e^{ax+by} f(D + a, D' + b)\phi(x, y) \qquad (11.57)$

Put $\qquad f(D + a, D' + b)\,\phi(x, y) = \psi(x, y)$

Also since $\phi(x, y)$ is any function, so will be $\psi(x, y)$.

$$\therefore \qquad \phi(x, y) = \frac{1}{f(D + a, D' + b)} \psi(x, y)$$

Substituting in Eq. (11.57)

$$f(D, D')\left[e^{ax+by} \frac{1}{f(D + a, D' + b)} \psi(x, y)\right] = e^{ax+by}\psi(x, y)$$

Operating upon by $\dfrac{1}{f(D, D')}$

$$e^{ax+by} \frac{1}{f(D + a, D' + b)} \psi(x, y) = \frac{1}{f(D, D')}[e^{ax+by}\psi(x, y)]$$

Replacing by $\psi(x, y)$ by $\phi(x, y)$, we get

$$\frac{1}{f(D, D')}[e^{ax+by}\phi(x, y)] = e^{ax+by} \frac{1}{f(D + a, D' + b)} \phi(x, y).$$

EXAMPLE 11.31 Find PI of the equation $(3D^2 - 2D'^2 + D + 1)\, z = 4e^{x+y} \cos(x + y)$.

Solution

$$\text{PI} = 4 \frac{1}{3D^2 - 2D'^2 + D + 1} e^{x+y} \cos(x + y)$$

$$= 4e^{x+y} \frac{1}{3(D + 1)^2 - 2(D' + 1)^2 + D + 1 + 1} \cos(x + y)$$

$$= 4e^{x+y} \frac{1}{3D^2 + 7D - 2D'^2 - 4D' + 1} \cos(x + y)$$

$$= 4e^{x+y} \frac{1}{3(-1^2) + 7D - 2(-1^2) - 4D' + 1} \cos(x + y)$$

$$= 4e^{x+y} \frac{1}{7D - 4D'} \cos(x + y)$$

$$= 4e^{x+y} (7D + 4D') \frac{1}{49D^2 - 16D'^2} \cos(x + y)$$

$$= 4e^{x+y} (7D + 4D') \frac{1}{-49 + 16} \cos(x + y)$$

$$= -\frac{4}{33} e^{x+y} [-11 \sin(x + y)]$$

$$= \frac{4}{3} e^{x+y} \sin(x + y).$$

EXAMPLE 11.32 Solve $(D - 3D' - 2)^2 z = 2e^{2x} \tan(y + 3x)$.

Solution Reader can check that

$$\text{CF} = e^{2x}[\phi_1(3x + y) + x\phi_2(3x + y)]$$

$$\text{PI} = 2\frac{1}{(D - 3D' - 2)^2} e^{2x} \tan(y + 3x)$$

$$= 2e^{2x} \frac{1}{(D - 3D')^2} \tan(y + 3x)$$

$$= 2e^{2x} \frac{1}{D - 3D'} \int_{a=y+3x} \tan a\, dx$$

$$= 2e^{2x} \frac{1}{D - 3D'} [x \tan(y + 3x)]$$

$$= 2e^{2x} \int_{a=y+3x} x \tan a\, dx$$

$$= x^2 e^{2x} \tan(y + 3x)$$

$\therefore$ The general solution of given equation is

$$z = e^{2x}[\phi_1(3x + y) + x\phi_2(3x + y)] + x^2 e^{2x} \tan(y + 3x)$$

where ϕ_1 and ϕ_2 are arbitrary functions.

11.11 SOLUTION OF $f(xD, yD') = V(x, y)$

Consider an equation

$$f(xD, yD') = V(x, y) \tag{11.58}$$

where $$f(xD, yD') = \sum_{r,s} c_{rs} x^r y^s D^r D'^s$$

and c_{rs} are constants.

Equation (11.58) can be reduced to a linear partial differential equation with constant coefficients by the substitution

$$u = \log x,\ v = \log y \tag{11.59}$$

we have

$$xD = x\frac{\partial}{\partial x}$$

$$= x\frac{\partial}{\partial u} \cdot \frac{\partial u}{\partial x}$$

$$= \frac{\partial}{\partial u} = d \quad \text{(say)}$$

$$x^2 D^2 = x^2 D\left(\frac{1}{x}\frac{\partial}{\partial u}\right)$$

$$= x^2\left(-\frac{1}{x^2}\frac{\partial}{\partial u} + \frac{1}{x^2}\frac{\partial^2}{\partial u^2}\right)$$

$$= \frac{\partial^2}{\partial u^2} - \frac{\partial}{\partial u}$$

$$= d^2 - d$$

$$= d(d-1)$$

In general, $x^r D^r = d(d-1)(d-2)\cdots[d-(r-1)]$

Similarly, $y^s D'^s = d'(d'-1)(d'-2)\cdots[d'-9\,s-1)]$

$\therefore$ $f(xD, yD') = \sum_{r,s} c_{rs} d(d-1)\cdots[d-(r-1)]\, d'(d'-1)\cdots[d'-(s-1)]$

$= g(d, d')$ (say)

where the coefficients of $g(d, d')$ are constants.

Thus, by substitution of Eq. (11.59) in Eq. (11.58) gives

$$g(d, d')\, z = V(e^u, e^v) = U(u, v) \tag{11.60}$$

Equation (11.60) can be solved by the methods discussed in Sections 11.3 and 11.4.

EXAMPLE 11.33 Solve $(x^2D^2 - y^2D'^2 + xD - yD')\, z = 0$.

Solution Given

$$(x^2D^2 - y^2D'^2 + xD - yD')\, z = 0 \tag{i}$$

Put $u = \log x$

and $v = \log y$

$\therefore$ Equation (i) reduces to

$$[d(d-1) - d'(d'-1) - d' + d]\, z = 0$$

$\Rightarrow$ $(d^2 - d'^2)\, z = 0$

Auxiliary equation is

$$m^2 - 1 = 0$$

$\Rightarrow$ $m = 1, -1$

$\therefore$ $\text{CF} = z = \phi_1(v + u) + \phi_2(v - u)$

$$= \phi_1(\log xy) + \phi_2\left(\log\frac{y}{x}\right)$$

$$= \psi_1(xy) + \psi_2\left(\frac{y}{x}\right).$$

EXAMPLE 11.34 Solve $(x^2D^2 + 2xyDD' + y^2D'^2)\, z = (x^2 + y^2)^{5/2}$.

Solution Put $u = \log x$

and $v = \log y$

in given equation.

$\therefore \quad [d(d-1)+2dd'+d'(d'-1)]\,z=(e^{2u}+e^{2v})^{5/2}$ (i)

$\therefore \quad [(d+d')^2-(d+d')]\,z=(e^{2u}+e^{2v})^{5/2}$

$\Rightarrow \quad (d+d')(d+d'-1)\,z=(e^{2u}+e^{2v})^{5/2}$

$\therefore$

$$\text{CF}=\phi_1(u-v)+e^u\phi_2(u-v)$$

$$=\phi_1\left(\log\frac{x}{y}\right)+x\phi_2\left(\log\frac{x}{y}\right)$$

$$=\psi_1\left(\frac{x}{y}\right)+x\psi_2\left(\frac{x}{y}\right)$$

$$\text{PI}=\frac{1}{(d+d')(d+d'-1)}(e^{2u}+e^{2v})^{5/2}$$

Put

$$z=\frac{1}{d+d'-1}(e^{2u}+e^{2v})^{5/2}$$

$\Rightarrow$

$$\frac{\partial z}{\partial u}+\frac{\partial z}{\partial v}=z+(e^{2u}+e^{2v})^{5/2} \qquad \text{(ii)}$$

Subsidiary equations of Eq. (ii) are

$$\frac{du}{1}=\frac{dv}{1}=\frac{dz}{z+(e^{2u}+e^{2v})^{5/2}} \qquad \text{(iii)}$$

Two independent integrals of Eq. (iii) are

$$u-v=a \text{ (Constant)} \qquad \text{(iv)}$$

and

$$\frac{\partial z}{\partial v}-z=(e^{2u}+e^{2v})^{5/2}$$

$$=e^{sv}(e^{2a}+1)^{5/2} \qquad \text{(v)}$$

which is being linear, its solution is

$$ze^{-v}=\frac{e^{4v}}{4}(e^{2a}+1)^{5/2}$$

$$=\frac{e^{5v}}{4}(e^{2a}+1)^{5/2}$$

$$=\frac{(e^{2u}+e^{2v})^{5/2}}{4}$$

$$\text{PI} = \frac{1}{d+d'}\left[\frac{(e^{2u}+e^{2v})^{5/2}}{4}\right]$$

$$= \frac{1}{4}\int_{a=v-u} (e^{2u}+e^{2a+2u})^{5/2}\, du$$

$$= \frac{1}{4}[(e^{2a}+1)^{5/2}\int e^{5u}\, du]_{a=v-u}$$

$$= \frac{1}{20}[e^{5u}\,(e^{2a}+1)^{5/2}]_{a=v-u}$$

$$= \frac{1}{20}(e^{2u}+e^{2v})^{5/2}$$

$$= \frac{1}{20}(x^2+y^2)^{5/2}$$

$$\therefore \qquad z = \psi_1\left(\frac{x}{y}\right) + x\psi_2\left(\frac{x}{y}\right) + \frac{1}{20}(x^2+y^2)^{5/2}.$$

EXAMPLE 11.35 Solve $(x^2D^2 + 2xyDD' + y^2D'^2)\, z + nz = n(xD + yD')\, z + x^2 + y^2 + x^3$.

Solution Put $\qquad u = \log x$

and $\qquad v = \log y$

$\therefore$ Given equation reduces to

$$[d(d-1) + 2dd' + d'(d'-1)]\, z - n(d+d')\, z + nz = e^{2u} + e^{2v} + e^{3u}$$

$$\Rightarrow \qquad [(d+d')^2 - (d+d')]\, z - n(d+d')\, z + nz = e^{2u} + e^{2v} + e^{3u}$$

$$\Rightarrow \qquad [(d+d')\,(d+d'-1) - n(d+d') + n]\, z = e^{2u} + e^{2v} + e^{3u}$$

$$\Rightarrow \qquad [(d+d')^2 - (n+1)\,(d+d') + n]\, z = e^{2u} + e^{2v} + e^{3u}$$

$$\Rightarrow \qquad (d+d'-n)\,(d+d'-1)\, z = e^{2u} + e^{2v} + e^{3u} \qquad \text{(i)}$$

$$\therefore \qquad \text{CF} = e^{nu}\phi_1\,(u-v) + e^{u}\phi_2(u-v)$$

$$= x^n\psi_1\left(\frac{x}{y}\right) + x\psi_2\left(\frac{x}{y}\right)$$

$$\text{PI} = \frac{1}{(d+d'-n)\,(d+d'-1)}(e^{2u} + e^{2v} + e^{3u})$$

$$= \frac{1}{2-n}\, e^{2u} + \frac{1}{2-n}\, e^{2v} + \frac{1}{(3-n)2}\, e^{3u}$$

$$= -\frac{x^2+y^2}{n-2} - \frac{1}{2}\cdot\frac{1}{n-3}\, x^3$$

$$\therefore \qquad z = x^n \psi_1\left(\frac{x}{y}\right) + x\psi_2\left(\frac{x}{y}\right) - \frac{x^2+y^2}{n-2} - \frac{1}{2}\cdot\frac{1}{n-3}x^3.$$

EXAMPLE 11.36 Solve $(x^2D^2 - xyDD' - 2y^2D'^2 + xD - 2yD')\, z = \log \dfrac{y}{x}$.

Solution Put $u = \log x$

and $v = \log y$

in the given equation

$$\therefore \qquad [d(d-1) - dd' - 2d'(d'-1) + d - 2d']\, z = v - u$$

$$\Rightarrow \qquad (d^2 - dd' - 2d'^2)\, z = v - u$$

$$\Rightarrow \qquad (d - 2d')(D + d')\, z = v - u$$

$$\therefore \qquad \text{CF} = \phi_1(2u + v) + \phi_2(u - v)$$

$$= \psi_1(x^2y) + \psi_2\left(\frac{x}{y}\right)$$

$$\text{PI} = \frac{1}{(d-2d')(d+d')}(v-u)$$

$$= \frac{1}{d-2d'}\frac{1}{d}\left(1+\frac{d'}{d}\right)^{-1}(v-u)$$

$$= \frac{1}{d-2d'}\frac{1}{d}\left(1-\frac{d'}{d}+\cdots\right)(v-u)$$

$$= \frac{1}{d-2d'}\frac{1}{d}(v-u-u)$$

$$= \frac{1}{d-2d'}\frac{1}{d}(v-2u)$$

$$= \frac{1}{d-2d'}(uv-u^2)$$

$$= \frac{1}{d}\left(1-\frac{2d'}{d}\right)^{2}(uv-u^2)$$

$$= \frac{1}{d}\left(1+\frac{2d'}{d}+\frac{4d'^2}{d^2}+\cdots\right)(uv-u^2)$$

$$= \frac{1}{d}(uv-u^2+u^2)$$

$$= \frac{u^2v}{2}$$

$$= \frac{1}{2}(\log x)^2 \log y$$

$$\therefore \qquad z = \psi_1(x^2 y) + \psi_2\left(\frac{x}{y}\right) + \frac{1}{2}(\log x)^2 \log y.$$

EXAMPLE 11.37 Find the surface passing through the two lines $z = x = 0$ and $z - 1 = x - y = 0$ and satisfying the differential equation $(D^2 - 4DD' + 4D'^2)\, z = 0$.

Solution The auxiliary equation is

$$m^2 - 4m + 4 = 0$$

$$\Rightarrow \qquad m = 2, 2$$

$$\therefore \qquad z = \phi_1(y + 2x) + x\phi_2\,(y + 2x) \qquad \text{(i)}$$

Given that surface of Eq. (i) will pass through the lines

$$z = x = 0 \text{ and } z - 1 = x - y = 0$$

$$\therefore \qquad 0 = \phi_1(y)$$

and

$$1 = \phi_1(3x) + x\phi_2(3x)$$

$$\therefore \qquad x\phi_2(3x) = 1$$

$$\Rightarrow \qquad \phi_2(3x) = \frac{1}{x}$$

$$\Rightarrow \qquad \phi_2(x) = \frac{3}{x}$$

$\therefore$ Equation (i) becomes

$$z = \frac{3x}{y + 2x}$$

or

$$z(y + 2x) = 3x$$

is the required surface.

EXAMPLE 11.38 Find a surface satisfying $(D^2 - DD')\, z = 0$ and touching the elliptical paraboloid $z = 9x^2 + y^2$ along its section by the plane $y = 3x + 1$.

Solution The auxiliary equation is

$$m^2 - m = 0$$

$$\Rightarrow \qquad m = 0, 1$$

$$\therefore \qquad z = \phi_1(y) + \phi_2(y + x) \qquad \text{(i)}$$

Since Eq. (i) touches the curve given by

$$z = 9x^2 + y^2 \text{ and } y = 3x + 1$$

The values of p and q obtained from $z = 9x^2 + y^2$ and Eq. (i) must be equal for any point on $y = 3x + 1$.

$$\therefore \qquad \phi_2'\,(y + x) = 18x$$

$$\Rightarrow \qquad \phi_2'\,(4x + 1) = 18x \qquad \text{(ii)}$$

and $\phi_1'(y) + \phi_2'(y + x) = 2y$ for $y = 3x + 1$

$\Rightarrow$ $$\phi_1'(3x + 1) + \phi_2'(4x + 1) = 6x + 2 \quad \text{(iii)}$$

From Eq. (ii)

$$\phi_2'(x) = \frac{9}{2}(x - 1)$$

Integrating, $$\phi_2(x) = \frac{9}{2}\left(\frac{x^2}{2} - x\right) + \text{Constant}$$

From Eqs. (ii) and (iii)

$$\phi_1'(3x + 1) = 2 - 12x$$

$\therefore$ $$\phi_1'(x) = 6 - 4x$$

Integrating, $$f_1(x) = 6x - x^2 + \text{Constant}$$

$\therefore$ $$z = 6y - y^2 + \frac{9}{2}[(y - x)^2 - (y - x)] + \text{Constant}$$

$\Rightarrow$ $$z = 6y - y^2 + \frac{9}{2}y^2 - 9xy + \frac{9x^2}{2} - \frac{9y}{2} + \frac{9x}{2} + \text{Constant}$$

$\Rightarrow$ $$z = \frac{3y}{2} + \frac{7}{2}y^2 - 9xy + \frac{9x^2}{2} + \frac{9x}{2} + \text{Constant} \quad \text{(iv)}$$

Eliminating y and z between equations

$$z = 9x^2 + y^2,\ y = 3x + 1 \text{ and}$$

Equation (iv) gives

$$9x^2 + (3x + 1)^2 = \frac{9x^2}{2} + \frac{7}{2}(3x + 1)^2 - 9x(3x + 1) + \frac{9x}{2} + \frac{3}{2}(3x + 1) + \text{Constant}$$

$\Rightarrow$ $$\text{Constant} = -\frac{1}{2}$$

Therefore, the equation of the required surface is

$$9x^2 + 7y^2 - 18xy + 9x + 3y - 2z - 1 = 0.$$

EXERCISES

Solve the following equations:

1. $(D^3 + DD'^2 - 10D'^3)z = 0.$ **[Ans.** $z = \phi_1(y + 2x) + \phi_2(y - x + 2ix) + \phi_3(y - x - 2ix)$]
2. $(D^3 - 3D^2D' + 3DD'^2 - D'^3)z = 0.$ **[Ans.** $\text{CF} = \phi_1(y + x) + x\phi_2(y + x) + x^2\phi_3(y + x)$]
3. $(2D^2 - 5DD' + 2D'^2)z = 24(y - x).$ **[Ans.** $z = \phi_1(y + 2x) + \phi_2(2y + x) + 6x^2y + 3x^3$]

4. $(D^2 - 6DD' + 9D'^2)\, z = 12x^2 + 16xy.$

$$\left[\textbf{Ans. } z = \phi_1(y+3x) + x\phi_2(y+3x) + 5x^4 + \frac{8}{3}x^3 y\right]$$

5. $(D^2 D' + 2DD'^2 + D'^3)\, z = \dfrac{1}{x^2}.$ $\left[\textbf{Ans. } z = \phi_1(x) + \phi_2(x+y) + y\phi_3(x+y) + \dfrac{1}{2x}y\right]$

6. $(D^2 - DD' - 2D'^2)\, z = x - y.$ $\left[\textbf{Ans. } z = \phi_1(y+2x) + \phi_2(y-x) + \dfrac{1}{6}x(x-y)^2\right]$

7. $(D - 3D')^2 (D + 3D')\, z = e^{3x+y}.$

$$\left[\textbf{Ans. } z = \phi_1(y+3x) + x\phi_2(y+3x) + \phi_3(y-3x) + \frac{x^2}{12} e^{3x+1}\right]$$

8. $(4D^2 - 4DD' + D'^2)\, z = 16 \log(x+2y).$

$$[\textbf{Ans. } z = \phi_1(x+2y) + y\phi_2(x+2y) + 2x^2 \log(x+2y)]$$

9. $(D^3 - 7DD'^2 - 6D'^3)\, z = \cos(x-y) + x^2 + xy^2 + y^3.$

$$\left[\begin{aligned}\textbf{Ans. } z = \phi_1(y-x) + \phi_2(y-2x) + \phi_3(y+3x) + \frac{1}{4}x\cos(x-y)\\ + \frac{5}{72}x^6 + \frac{1}{60}x^5 + \frac{7}{20}x^5 y + \frac{1}{24}x^4 y^3 + \frac{1}{6}x^3 y^3\end{aligned}\right]$$

10. $(D^3 - 7DD'^2 - 6D'^3)\, z = \sin(x+2y) + e^{3x+y}.$

$$\left[\textbf{Ans. } z = \phi_1(y-x) + \phi_2(y-2x) + \phi_3(y+3x) - \frac{1}{75}\cos(x+2y) + \frac{1}{20}xe^{3x+y}\right]$$

11. $(D^2 + 5DD' + 5D'^2)\, z = x \sin(3x - 2y).$

$$\left[\begin{aligned}\textbf{Ans. } z = \phi_1(y+\alpha x) + \phi_2(y-\alpha x) + x\sin(3x-2y)\\ + 4\cos(3x-2y) \text{ where } \alpha = \frac{1}{2}(-5+\sqrt{5})\end{aligned}\right]$$

12. $(D^3 + 2D^2 D' - DD'^2 - 2D'^3)\, z = (y+2)\, e^x.$

$$[\textbf{Ans. } z = \phi_1(y+x) + \phi_2(y-x) + \phi_3(y-2x) + ye^x]$$

13. $(D^3 - 3DD'^2 - 2D'^3)\, z = \cos(x+2y) - e^y(2x+3).$

$$\left[\textbf{Ans. } z = \phi_1(y-x) + x\phi_2(y-x) + \phi_3(y+2x) + \frac{1}{27}\sin(x+2y) + xe^y\right]$$

14. $(DD' + D'^2 - 3D')\, z = 0.$ $[\textbf{Ans. } z = \psi_1(x) + e^{3y}\psi_2(x-y)]$

15. $(D - 2D' + 5)(D^2 + D' + 3)\, z = 0.$ $\left[\textbf{Ans. } z = e^{-5x}\phi(y+2x) + \displaystyle\sum_{i=1}^{\infty} c_i e^{a_i x - (a_i^2+3)y}\right]$

16. $(D - D' - 1)(D + D' - 2)\, z = e^{2x-y}$. $\left[\textbf{Ans. } z = e^x \phi_1(x + y) + e^{2y}\phi_2(x - y) - \frac{1}{2}e^{2x-y}\right]$

17. $(3DD' - 2D'^2 - D')\, z = \sin(2x + 3y)$.

$$\left[\textbf{Ans. } z = \phi_1(x) + e^{x/3}\phi_2(2x + 3y) + \frac{1}{3}\cos(2x + 3y)\right]$$

18. $(D^3 - DD'^2 - D^2 + DD')\, z = \dfrac{x+2}{x^3}$.

$$[\textbf{Ans. } z = \phi_1(y) + \phi_2(x + y) + e^x\phi_3(y - x) + \log x]$$

19. $(D - D'^2)\, z = \cos(x - 3y)$.

$$\left[\textbf{Ans. } z = \sum_{i=1}^{\infty} c_i e^{b_i^2 x + b_i y} + \frac{1}{82}\{9\cos(x - 3y) + \sin(x - 3y)\}\right]$$

20. $(D + D' - 1)(D + D' - 3)(D + D')\, z = e^{x+y}\sin(2x + y)$.

$$\left[\textbf{Ans. } z = e^x\phi_1(x - y) + e^{3x}\phi_2(x - y) + \phi_3(x - y) + \frac{1}{130}e^{x+y}\{3\cos(2x + y) - 2\sin(2x + y)\}\right]$$

21. $(D^2 + DD' + D' - 1)\, z = 4\sin hx$. $[\textbf{Ans. } z = e^{-x}\phi_1(y) + e^x\phi_2(x - y) + xe^x + xe^{-x}]$

22. $(x^2D^2 - 2xyDD' + y^2D'^2 - xD + 3yD')\, z = \dfrac{8y}{x}$. $\left[\textbf{Ans. } z = \psi_1(xy) + x^2\psi_2(xy) + \dfrac{y}{x}\right]$

23. $(x^2D^2 - y^2D'^2 + xD - yD')\, z = \log x$. $\left[\textbf{Ans. } z = \phi_1(xy) + \phi_2\left(\dfrac{y}{x}\right) + \dfrac{1}{6}(\log x)^3\right]$

24. $(x^2D^2 - 4y^2D'^2 - 4yD' - 1)\, z = x^2y^2\log y$.

$$\left[\textbf{Ans. } z = \sum_{i=1}^{\infty} c_i x_i^{a_i} y_i^{b_i} - \frac{1}{225}x^2y^2(15\log y - 16)\; a_i^2 - 4b_i^2 - a_i - 1 = 0\right]$$

25. Find a surface satisfying $(D^2 - 2DD' + D'^2)\, z = 6$ and touching the hyperbolic paraboloid $z = xy$ along its section by the plane $y = x$. $[\textbf{Ans. } z = x^2 - xy + y^2]$

26. A surface is drawn satisfying $(D^2 + D'^2)\, z = 0$ and touching $x^2 + y^2 = 1$ along its section by $z = 0$. Prove that the required surface is $y^2(x^2 + y^2 - 1) = z^2(x^2 + y^2)$.

27. $(D^3D'^2 + D^2D'^3)\, z = 0$. $[\textbf{Ans. } z = \phi_1(y) + x\phi_2(y) + \phi_3(y - x) + \phi_4(x) + y\phi_5(x)]$

28. $(D^2 - D'^2)z = x - y$. $\left[\textbf{Ans. } z = \phi_1(y + x) + \phi_2(y - x) + \dfrac{x^3}{6} - \dfrac{x^2y}{2}\right]$

29. $(D^2 + 2DD' + D'^2)z = 12xy$. $[\textbf{Ans. } z = \phi_1(y - x) + x\phi_2(y - x) + 2x^3y - x^4]$

30. $(D^2 + (a+b)DD' + abD'^2)z = xy.$

$$\left[\textbf{Ans. } z = \phi_1(y-ax) + \phi_2(y-bx) + \frac{x^3y}{6} - \frac{1}{24}(a+b)x^4\right]$$

31. $(D^3 - 7DD'^2 - 6D'^3)z = x^2 + xy^2 + y^3.$

$$\left[\textbf{Ans. } z = \phi_1(y-x) + \phi_2(y-2x) + \phi_3(y+3x) + \frac{5}{72}x^6 + \frac{x^5}{60}(1+21y) + \frac{1}{24}x^4y^2 + \frac{1}{6}x^3y^3\right]$$

32. $(D^3 - 4D^2D' + 5DD'^2 - 2D'^3)z = e^{y+2x} + \sqrt{x+y}.$

$$\left[\textbf{Ans. } z = \phi_1(y+x) + x\phi_2(y+x) + \phi_3(y+2x) + xe^{y+2x} - \frac{1}{3}x^2(x+y)^{3/2}\right]$$

33. $(D^3 - 3DD'^2 - 2D'^3)z = \cos(x+2y).$

$$\left[\textbf{Ans. } z = \phi_1(y-x) + x\phi_2(y-x) + \phi_3(y+2x) + \frac{1}{27}\sin(x+2y)\right]$$

34. $(D^2 - DD' - 2D)z = \cos(3x+4y).$

$$\left[\textbf{Ans. } z = \phi_1(y) + e^{2x}\phi_2(y+x) + \frac{1}{15}\{\cos(3x+4y) - 2\sin(3x+4y)\}\right]$$

35. $(x^2D^2 - y^2D'^2)z = xy.$ $\left[\textbf{Ans. } z = \phi_1(xy) + x\phi_2\left(\frac{x}{y}\right) + xy\log x\right]$

CHAPTER 12 Partial Differential Equations of Order Two with Variable Coefficients

12.1 INTRODUCTION

In this chapter, we discuss methods to solve the partial differential equations of order two with variable coefficients. We shall consider only two independent variables and one dependent variable. Denote second order partial derivatives by

$$r = \frac{\partial^2 z}{\partial x^2},\ s = \frac{\partial^2 z}{\partial x \partial y} \text{ and } t = \frac{\partial^2 z}{\partial y^2}$$

and as taken in Chapter 10,

$$p = \frac{\partial z}{\partial x},\ q = \frac{\partial z}{\partial y}$$

Thus, the general form of equation of order two is

$$f(x, y, z, p, q, r, s, t) = 0 \tag{12.1a}$$

or

$$Rr + Ss + Tt + Pp + Qq + Zz = F \tag{12.1b}$$

where R, S, T, P, Q, Z and F may be functions of x and y and not all R, S and T are zero.

12.2 SOLUTION OF EQUATION (12.1)

12.2.1 Type I

In this type, we have equations of the form

$$\left.\begin{aligned} r &= \frac{\partial^2 z}{\partial x^2} = \frac{F}{R} \\ s &= \frac{\partial^2 z}{\partial x \partial y} = \frac{F}{S} \\ \text{and} \quad t &= \frac{\partial^2 z}{\partial y^2} = \frac{F}{T} \end{aligned}\right\} \tag{12.2}$$

Equations (12.2) are reducible equations with constant coefficients and can be solved directly.

EXAMPLE 12.1 Solve $t = \cos(xy)$.

Solution Given $t = \cos(xy)$

$$\Rightarrow \qquad \frac{\partial^2 z}{\partial y^2} = \cos(xy)$$

Integrating w.r.t. y gives

$$\frac{\partial z}{\partial y} = \frac{1}{x}\sin(xy) + \phi_1(x)$$

where ϕ_1 is arbitrary function. Again integrating w.r.t. y gives

$$z = -\frac{1}{x^2}\cos(xy) + y\phi_1(x) + \phi_2(x)$$

where ϕ_2 is an arbitrary function.

EXAMPLE 12.2 Solve r = x^2e^y.

Solution We have

$$\frac{\partial^2 z}{\partial x^2} = x^2 e^y$$

Integrating w.r.t. x

$$\frac{\partial z}{\partial x} = \frac{x^3}{3}e^y + \phi_1(y)$$

where ϕ_1 is an arbitrary function.
Again integrating w.r.t. x

$$z = \frac{x^4}{12}e^y + x\phi_1(y) + \phi_2(y)$$

where ϕ_2 is an arbitrary function.

EXAMPLE 12.3 Solve $s = x^2 - y^2$.

Solution We have

$$\frac{\partial^2 z}{\partial x \partial y} = x^2 - y^2$$

Integrating w.r.t. x

$$\frac{\partial z}{\partial y} = \frac{x^3}{3} - y^2 x + \phi_1(y)$$

where ϕ_1 is an arbitrary function.
Integrating w.r.t. y

$$z = \frac{x^3 y}{3} - \frac{y^3 x}{3} + \psi_1(y) + \psi_2(x)$$

where ψ_1 and ψ_2 are arbitrary functions.

12.2.2 Type II

In type II, we will try to solve equations of the form

$$\left.\begin{aligned} Rr + Pp &= F(x, y)\\ Ss + Pp &= F(x, y)\\ Ss + Qq &= F(x, y)\\ Tt + Qq &= F(x, y)\end{aligned}\right\} \tag{12.3}$$

These equations can be written as

$$\left.\begin{aligned} R\frac{\partial p}{\partial x} + Pp &= F\\ S\frac{\partial p}{\partial y} + Pp &= F\\ S\frac{\partial q}{\partial x} + Qq &= F\\ \text{and}\qquad T\frac{\partial q}{\partial y} + Qq &= F\end{aligned}\right\} \tag{12.4}$$

which are linear differential equations of order one in which p (or q) is the dependent variable.

EXAMPLE 12.4 Solve $xr + p = 9x^2y^2$.

Solution Rewriting equation in the form of Eq. (12.4)

$$\frac{\partial p}{\partial x} + \frac{p}{x} = 9xy^2$$

Integrating factor is

$$\text{IF} = e^{\int \frac{1}{x}dx} = e^{\log x} = x$$

$\therefore$ Solution is

$$px = 3x^3y^2 + \phi_1(y)$$

where ϕ_1 is an arbitrary function

$$\Rightarrow \qquad \frac{\partial z}{\partial x} = 3x^3y^2 + \frac{1}{x}\phi_1(y)$$

Integrating w.r.t. x gives

$$z = x^3y^2 + \log x\phi_1(y) + \phi_2(y)$$

where ϕ_2 is an arbitrary function.

EXAMPLE 12.5 Solve $yt - q = xy$.

Solution Rewriting given equation in the form

$$\frac{\partial q}{\partial y} - \frac{1}{y}q = x$$

$$\text{IF} = e^{-\int \frac{1}{y} dy} = e^{-\log y} = \frac{1}{y}$$

$\therefore$ Solution of equation is

$$\frac{1}{y} q = \int \frac{x}{y} dy + \phi_1(x)$$

$\Rightarrow$ $$q = xy \log y + y\phi_1(x)$$

where ϕ_1 is an arbitrary function.

$\Rightarrow$ $$\frac{\partial z}{\partial y} = xy \log y + y\phi_1(x)$$

Integrating, $$z = \frac{1}{2} xy^2 \log y - \frac{xy^2}{4} + \frac{y^2}{2} \phi_1(x) + \phi_2(x)$$

where ϕ_2 is an arbitrary function.

EXAMPLE 12.6 Solve $ys + p = \cos(x + y) - y \sin(x + y)$.

Solution Rewriting given equation

$$\frac{\partial p}{\partial y} + \frac{p}{y} = \frac{1}{y} \cos(x + y) - \sin(x + y)$$

$$\text{IF} = e^{\int \frac{1}{y} dy} = y$$

$\therefore$ $$py = y \cos(x + y) + \phi_1(x)$$

where ϕ_1 is an arbitrary function.

$\therefore$ $$y \frac{\partial z}{\partial x} = y \cos(x + y) + \phi_1(x)$$

Integrating w.r.t. x gives

$$yz = y \sin(x + y) + \psi_1(x) + \psi_2(y)$$

where ψ_1 and ψ_2 are arbitrary functions.

12.2.3 Type III

Here equations are of the type

$$\left.\begin{aligned} Rr + Ss + Pp &= F \\ Ss + Tt + Qq &= F \end{aligned}\right\} \tag{12.5}$$

$\Rightarrow$ $$\left.\begin{aligned} R\frac{\partial p}{\partial x} + S\frac{\partial p}{\partial y} &= F - Pp \\ S\frac{\partial q}{\partial x} + T\frac{\partial q}{\partial y} &= F - Qq \end{aligned}\right\} \tag{12.6}$$

Equations (12.6) are linear with p (or q) as dependent variable and x and y as independent variables.

EXAMPLE 12.7 Solve $r + \frac{y}{x}s = 5xy^2$.

Solution Rewrite given equation in the form

$$\frac{\partial p}{\partial x} + \frac{y}{x}\frac{\partial p}{\partial y} = 5xy^2 \tag{i}$$

Lagrange's subsidiary equations of Eq. (i) are

$$\frac{dx}{1} = \frac{dy}{\frac{y}{x}} = \frac{dp}{5xy^2} \tag{ii}$$

Considering first two ratios

$$\frac{dx}{x} = \frac{dy}{y}$$

Integrating, $\frac{y}{x} = c$ (Constant)
Considering first and third ratios of Eq. (ii),

$$dp = 5xy^2dx$$

$$\Rightarrow \quad dp = 5c^2x^3dx$$

Integrating,

$$p = \frac{5}{4}c^2x^4 + \text{Constant}$$

$$= \frac{5}{4}x^2y^2 + \text{Constant}$$

$\therefore$ General solution of Eq. (i) is

$$p = \frac{5}{4}x^2y^2 + \phi\left(\frac{y}{x}\right)$$

$$\Rightarrow \quad \frac{\partial z}{\partial x} = \frac{5}{4}x^2y^2 + \phi\left(\frac{y}{x}\right)$$

Integrating,

$$z = \frac{5}{12}x^3y^2 + \int\phi\left(\frac{y}{x}\right)dx + \phi_1(y)$$

$$= \frac{5}{12}x^3y^2 + y\int\frac{1}{-\frac{y^2}{x^2}}\phi\left(\frac{y}{x}\right)d\left(\frac{y}{x}\right) + \phi_1(y)$$

$$= \frac{5}{12}x^3y^2 + y\phi_2\left(\frac{y}{x}\right) + \phi_1(y)$$

where ϕ, ϕ_1 and ϕ_2 and are arbitrary functions.

EXAMPLE 12.8 Solve $sy - 2xr - 2p = 6xy$.

Solution Given

$$y\frac{\partial p}{\partial y} - 2x\frac{\partial p}{\partial x} = 6xy + 2p \tag{i}$$

Lagrange's subsidiary equations are

$$\frac{dy}{y} = \frac{dx}{-2x} = \frac{dp}{6xy + 2p} \quad \text{(ii)}$$

Taking first two ratios

$$2\frac{dy}{y} + \frac{dx}{x} = 0$$

Integrating, $xy^2 = c$ (Constant)
Each ratio in Eq. (ii)

$$= \frac{(2yp + 2xy^2)\, dy - 2y^3 dx - y^2\, dp}{0}$$

$$\Rightarrow \quad (2yp + 2xy^2)\, dy - 2y^3 dx - y^2 dp = 0$$

$$\Rightarrow \quad 2y(p + 2x)\, dy - y^2(dp + 2xdy + 2yd) = 0$$

$$\Rightarrow \quad 2y(p + 2x)\, dy - y^2 d(p + 2xy) = 0$$

$$\Rightarrow \quad \frac{2dy}{y} - \frac{d(p+2xy)}{p+2xy} = 0$$

Integrating,

$$\frac{y^2}{p+2xy} = \text{Constant}$$

$$\Rightarrow \quad p + 2xy = y^2 \text{ Constant}$$

$\therefore$ General solution of Eq. (ii) is

$$p + 2xy = y^2\phi(xy^2)$$

$$\Rightarrow \quad \frac{\partial z}{\partial x} = y^2\phi(xy^2) - 2xy$$

where ϕ is an arbitrary function.

Integrating,

$$z = \int y^2 \phi(xy^2)\,dx - x^2 y + \phi_2(y)$$

$$= \phi_1(xy^2) + \phi_2(y) - x^2 y$$

where ϕ_1 and ϕ_2 are arbitrary functions.

12.2.4 Type IV

Here equations are of the form

$$\left.\begin{aligned} Rr + Pp + Zz = F \\ Tt + Qq + Zz = F \end{aligned}\right\} \quad (12.7)$$

These are linear ordinary differential equations with x (or y) as independent variable and z as dependent variable.

EXAMPLE 12.9 Solve $xs + q - xp - z = (3 - y)(1 + \log x)$.

Solution Given
$$\frac{\partial^2 z}{\partial x \partial y} + \frac{1}{x}\frac{\partial z}{\partial y} - \left(\frac{\partial z}{\partial x} + \frac{z}{x}\right) = \frac{3-y}{x}(1 + \log x)$$

$$\Rightarrow \quad \frac{\partial}{\partial y}\left(\frac{\partial z}{\partial x} + \frac{1}{x}z\right) - \left(\frac{\partial z}{\partial x} + \frac{1}{x}z\right) = \frac{3-y}{x}(1 + \log x)$$

Put
$$V = \frac{\partial z}{\partial x} + \frac{1}{x}z$$

$$\Rightarrow \quad \frac{\partial V}{\partial y} - V = \frac{3-y}{x}(1 + \log x)$$

$$\text{IF} = e^{-y}$$

$$\therefore \quad Ve^{-y} = \frac{(y-2)e^{-y}}{x}(1 + \log x) + \phi_1(x)$$

$$\Rightarrow \quad V = \frac{(y-2)}{x}(1 + \log x) + e^y \phi_1(x)$$

$$\Rightarrow \quad \frac{\partial z}{\partial x} + \frac{1}{x}z = \frac{(y-2)}{x}(1 + \log x) + e^y \phi_1(x)$$

$$\text{IF} = e^{\int \frac{1}{x} dx} = x$$

$$\Rightarrow \quad zx = x(y-2)\log x + e^y \psi_1(x) + \psi_2(y)$$

where ψ_1 and ψ_2 are arbitrary functions.

12.3 NON-LINEAR PARTIAL DIFFERENTIAL EQUATIONS OF ORDER TWO

Only in some cases Eq. (12.1) can be integrated. Monge gave a method which is applicable to a wide class (again not all) of such equations. It consists of one or two integrals of the form

$$U = \psi(V) \tag{12.8}$$

where U and V are known functions of x, y, z, p and q and ψ is an arbitrary function. Eq. (12.8) is called an intermediate integral of Eq. (12.1).

Note that every Eq. (12.1) does not possess an intermediate integral. The general partial differential equation having intermediate integral of the form Eq. (12.8) is

$$Rr + Ss + Tt + U(rt - s^2) = V \tag{12.9}$$

where R, S, T, U and V are functions of x, y, z, p and q.

12.4 MONGE'S METHOD TO SOLVE $Rr + Ss + Tt = V$

Consider the equation

$$Rr + Ss + Tt = V \tag{12.10}$$

Since z is a function of x and y

$$dz = \frac{\partial z}{\partial x}dx + \frac{\partial z}{\partial y}dy = pdx + qdy \tag{12.11}$$

$$dp = \frac{\partial p}{\partial x}dx + \frac{\partial p}{\partial y}dy = rdx + sdy \tag{12.12}$$

and

$$dq = \frac{\partial q}{\partial x}dx + \frac{\partial q}{\partial y}dy = sdx + tdy \tag{12.13}$$

Then Eq. (12.12) gives $r = \dfrac{dp - sdy}{dx}$

and Eq. (12.13) gives $t = \dfrac{dq - sdx}{dy}$

Substituting in Eq. (12.10) gives

$$R\left(\frac{dp - sdy}{dx}\right) + Ss + T\left(\frac{dq - sdx}{dy}\right) = V$$

$$\Rightarrow \quad (Rdpdy + Tdqdx - Vdxdy) - s\,[R(dy)^2 - Sdxdy + T(dx)^2] = 0 \tag{12.14}$$

The equations

$$Rdpdy + Tdqdx - Vdxdy = 0 \tag{12.15}$$

and

$$R(dy)^2 - Sdxdy + T(dx)^2 = 0 \tag{12.16}$$

are known as Monge's equations.

Any relation between x, y, z, p and q satisfying Eqs. (12.15) and (12.16) will satisfy Eq. (12.14) and hence Eq. (12.10). Equation (12.16) being quadratic in $\dfrac{dy}{dx}$ can be factorized into two equations.

Case 1: When left hand side of Eq. (12.16) is a perfect square.

Let $\quad R(dy)^2 - Sdxdy + T(dx)^2 = (Ady + Bdx)^2$

Then the system of equations to be solved is

$$\left.\begin{aligned} Rdpdy + Tdqdx - Vdxdy = 0 \\ Ady + Bdx = 0 \end{aligned}\right\} \tag{12.17}$$

If $\quad U = U(x, y, z, p, q) = \text{Constant}$

and $\quad V = V(x, y, z, p, q) = \text{Constant}$

satisfy Eq. (12.17) then

$$U = \psi(V) \tag{12.18}$$

is an intermediate integral of Eq. (12.10) where ψ is an arbitrary function.

Case 2: When the left hand side of Eq. (12.16) is not a perfect square.

Let $$R(dy)^2 - Sdxdy + T(dx)^2 = (A_1dy + B_1dx)(A_2dy + B_2dx)$$

where $$A_1B_2 - A_2B_1 \neq 0$$

Then the system of equations to be solved is

$$\left.\begin{aligned} A_1dy + B_1dx = 0 \\ Rdpdy + Tdqdx - Vdxdy = 0 \end{aligned}\right\} \tag{12.19}$$

and

$$\left.\begin{aligned} A_2dy + B_2dx = 0 \\ Rdpdy + Tdqdx - Vdxdy = 0 \end{aligned}\right\} \tag{12.20}$$

If only one system is integrable then proceed as case I to find complete integral. If both the systems are integrable, there will be two intermediate integrals. Solve these integrals for p and q and substitute these values in Eq. (12.11) and integrate the resultant equation to compute complete integral.

EXAMPLE 12.10 Solve $r = 4t$.

Solution Monge's equations are

$$dpdy - 4dqdx = 0 \tag{i}$$

and $$(dy)^2 - 4(dx)^2 = 0 \tag{ii}$$

Equation (ii) can be factorized as

$$(dy - 2dx)(dy + 2dx) = 0$$

∴ Two systems are

$$\left.\begin{aligned} dpdy - 4dqdx = 0 \\ dy - 2dx = 0 \end{aligned}\right\} \tag{iii}$$

and

$$\left.\begin{aligned} dpdy - 4dqdx = 0 \\ dy + 2dx = 0 \end{aligned}\right\} \tag{iv}$$

Integrating second equation of Eq. (iii) gives

$$y - ax = \text{Constant}$$

Eliminating $\frac{dy}{dx}$ between the equations of Eq. (iii) gives

$$dp - 2dq = 0$$

$$\Rightarrow \qquad p - 2q = \text{Constant}$$

∴ Intermediate integral corresponding to Eq. (iii) is

$$p - 2q = \phi_1(y - 2x) \tag{v}$$

Similarly Eq. (iv) gives

$$p + 2q = \phi_2(y + 2x) \tag{vi}$$

Solving Eqs. (v) and (vi) for p and q gives

$$p = \frac{1}{2}[\phi_1(y-2x) + \phi_2(y+2x)]$$

and
$$q = \frac{1}{4}[\phi_2(y+2x) - \phi_1(y-2x)]$$

Substituting in $dz = pdx + qdy$ gives

$$dz = \frac{3}{4}\phi_2(y+2x)\, d(y+2x) + \frac{1}{4}\phi_1(y-2x)\, d(y-2x)$$

Integrating,
$$z = \psi_2(y + 2x) + \psi_1(y - 2x)$$

where ϕ_1, ϕ_2, ψ_1 and ψ_2 are arbitrary functions.

EXAMPLE 12.11 Solve $q(1 + q)\, r - (p + q + 2pq)\, s + p(1 + p)\, t = 0$.

Solution Given
$$q(1 + q)\, r - (p + q + 2pq)\, s + p(1 + p)\, t = 0 \tag{i}$$

Monge's equations are

$$q(1 + q)\, dpdy + p(1 + p)\, dqdx = 0 \tag{ii}$$

and
$$q(1 + q)\, (dy)^2 + (p + q + 2pq)\, dxdy + p(1 + p)\, (dx)^2 = 0 \tag{iii}$$

$\Rightarrow$
$$\{(1 + q)\, dy + (1 + p)\, dx\}\, \{pdx + qdy\} = 0$$

$\therefore$ Two systems are

$$\left.\begin{aligned} q(1+q)\, dpdy + p(1+p)\, dqdx = 0 \\ (1+q)\, dy + (1+p)\, dx = 0 \end{aligned}\right\} \tag{iv}$$

and
$$\left.\begin{aligned} q(1+q)\, dpdy + p\,(1+p)\, dqdx = 0 \\ pdx + qdy = 0 \end{aligned}\right\} \tag{v}$$

Second equation of Eq. (iv) is

$$dy + dx + (pdx + qdy) = 0$$

$\Rightarrow$
$$dx + dy + dz = 0$$

Integrating,
$$x + y + z = \text{Constant}$$

Eliminating $\dfrac{dy}{dx}$ between the equations of Eq. (iv)

$\Rightarrow$
$$qdp - pdq = 0$$

$\Rightarrow$
$$\frac{dp}{p} - \frac{dq}{q} = 0$$

On integrating

$$\frac{p}{q} = \text{Constant}$$

Therefore, the immediate integral corresponding to Eq. (iv) is

$$p = q\phi_1(x + y + z) \tag{vi}$$

Second equation of Eq. (v) gives $dz = 0$

$\Rightarrow$ $$z = \text{Constant}$$

Eliminating $\frac{dy}{dx}$ between the equations of Eq. (v)

$\Rightarrow$ $$(1 + q)\, dp - (1 + p)\, dq = 0$$

$\Rightarrow$ $$\frac{dp}{1+p} - \frac{dq}{1+q} = 0$$

which on integrating gives

$$\frac{1+p}{1+q} = \text{Constant}$$

$\therefore$ The intermediate integral corresponding to Eq. (v) is

$$1 + p = (1 + q)\, \phi_2(z) \tag{vii}$$

Solving Eqs. (vi) and (vii) gives

$$p = \frac{\phi_1(x+y+z)(\phi_2(z)-1)}{\phi_1(x+y+z) - \phi_2(z)}$$

and $$q = \frac{\phi_2(z) - 1}{\phi_1(x+y+z) - \phi_2(z)}$$

$\therefore$ $$dz = pdx + qdy$$

gives $$dz = \frac{\phi_2(z) - 1}{\phi_1(x+y+z) - \phi_2(z)} [\phi_1(x+y+z)\, dx + dy]$$

$\Rightarrow$ $$[\phi_1(x + y + z) - \phi_2(z)]\, dz = [\phi_2(z) - 1]\, [\phi_1(x + y + z)\, dx + dy]$$

$\Rightarrow$ $$[\phi_1(x + y + z) - 1]\, dz = [\phi_2(z) - 1]\, d\,(x + y + z) + [\phi_2(z) - 1]\, [\phi_1(x + y + z) - 1]\, dx$$

$\Rightarrow$ $$\frac{dz}{\phi_2(z) - 1} = \frac{d(x+y+z) + dx}{\phi_1(x+y+z) - 1}$$

Integrating, $$\psi_2(z) = \psi_1(x + y + z) + x$$

where ϕ_1, ϕ_2, ψ_1 and ψ_2 are arbitrary functions.

EXAMPLE 12.12 Solve $x^2r + 2xys + y^2t = 0$.

Solution Monge's equations are

$$x^2dpdy + y^2dqdx = 0 \tag{i}$$

and $$x^2(dy)^2 - 2xydxdy + y^2(dx)^2 = 0 \tag{ii}$$

$\Rightarrow$ $$(xdy - ydx)^2 = 0$$

$\therefore$ The system to be integrated is

$$\left.\begin{aligned} x^2 dpdy + y^2 dqdx = 0 \\ xdy - ydx = 0 \end{aligned}\right\} \tag{iii}$$

Integrating second equation of Eq. (iii) gives $\frac{y}{x} = \text{Constant}$

Eliminating $\frac{dy}{dx}$ between equations of Eq. (iii)

$$\Rightarrow \qquad xdp + ydq = 0$$

$$\Rightarrow \qquad d(xp + yq) - (pdx + qdy) = 0$$

$$\Rightarrow \qquad d\,(xp + yq - z) = 0$$

Integrating, $\qquad xp + yq - z = \text{Constant}$

∴ The intermediate integral corresponding to Eq. (iii) is

$$xp + yq - z = \phi\left(\frac{y}{x}\right) \tag{iv}$$

Lagrange's subsidiary equations of Eq. (iv) are

$$\frac{dx}{x} = \frac{dy}{y} = \frac{dz}{z + \phi\left(\frac{y}{x}\right)} \tag{v}$$

Then two independent integrals of Eq. (v) are

$$\frac{y}{x} = \text{Constant}$$

and

$$z + \phi\left(\frac{y}{x}\right) = x \cdot \text{Constant}$$

Therefore, general solution of Eq. (iv) is

$$z + \phi\left(\frac{y}{x}\right) = x\psi\left(\frac{y}{x}\right)$$

where ϕ and ψ are arbitrary functions.

EXAMPLE 12.13 Solve $(x - y)\,(xr - xs - ys - yt) = (x + y)\,(p - q)$.

Solution Monge's equations are

$$x(x - y)\,dpdy + y(x - y)\,dqdx - (x + y)\,(p - q)\,dxdy = 0 \tag{i}$$

and

$$x(x - y)\,dy^2 + (x - y)\,(x + y)\,dxdy + y(x - y)\,dx^2 = 0 \tag{ii}$$

$$\Rightarrow \qquad (xdy + ydx)\,(dy + dx) = 0$$

∴ Two systems are

$$\left.\begin{aligned} x(x - y)dpdy + y(x - y)dqdx - (x + y)\,(p - q)\,dxdy = 0 \\ xdy + ydx = 0 \end{aligned}\right\} \tag{iii}$$

and

$$\left.\begin{aligned} x(x - y)\,dpdy + y(x - y)\,dqdx - (x + y)\,(p - q)\,dxdy = 0 \\ dy + dx = 0 \end{aligned}\right\} \tag{iv}$$

Integrating second equation of Eq. (iii) gives

$$xy = \text{Constant}$$

Putting $xdy + ydx = 0$ in first equation of Eq. (iii) simplifies to

$$(x - y)\, dp - (x - y)\, dq = (p - q)\,(dx - dy)$$

$$\Rightarrow \qquad \frac{d(p-q)}{p-q} = \frac{d(x-y)}{x-y}$$

which on integration gives

$$\frac{p-q}{x-y} = \text{Constant}$$

$\therefore$ The intermediate integral corresponding to Eq. (iii) is

$$p - q = (x - y)\,\phi(xy) \tag{v}$$

Lagrange's subsidiary equations of Eq. (v) are

$$\frac{dx}{1} = \frac{dy}{-1} = \frac{dz}{(x-y)\,\phi(xy)} \tag{vi}$$

Considering first two ratios

$$dx + dy = 0$$

$$\Rightarrow \qquad x + y = c \text{ (Constant)}$$

Consider first and third ratios

$$dz = (2x - c)\,\phi(cx - x^2)\, dx$$

Integrating,

$$z = \int (2x - c)\,\phi(cx - x^2)dx + \text{Constant}$$

$$= \int -\phi(cx - x^2)\, d(cx - x^2) + \text{Constant}$$

$$= \psi_1(cx - x^2) + \text{Constant}$$

$$= \psi_1(xy) + \text{Constant}$$

$\therefore$ General solution is

$$z = \psi_1(xy) + \psi_2(x + y)$$

where ϕ_1, ψ_1 and ψ_2 are arbitrary functions.

EXAMPLE 12.14 Solve $(1 + q)^2 r - 2(1 + p + q + pq)\, s + (1 + p)^2 t = 0$.

Solution Monge's equations are

$$(1 + q)^2\, dpdy + (1 + p)^2\, dqdx = 0 \tag{i}$$

and

$$(1 + q)^2\, dy^2 + 2(1 + p + q + pq)\, dxdy + (q + p)^2\, dx^2 = 0 \tag{ii}$$

$$\Rightarrow \qquad [(1 + q)\, dy + (1 + p)\, dx]^2 = 0$$

$\therefore$ The system to be solved is

$$\left.\begin{aligned}(1+q)^2 dpdy + (1+p)^2 dqdx = 0\\ (1+q)\,dy + (1+p)\,dx = 0\end{aligned}\right\} \tag{iii}$$

Second equation of Eq. (iii) is

$$dx + dy + pdx + qdy = 0$$

$$\Rightarrow \qquad dx + dy + dz = 0$$

Integrating, $x + y + z = \text{Constant}$

Equation (iii) gives

$$(1 + q)\, dp - (1 + p)\, dq = 0$$

$\Rightarrow$ $$\frac{dp}{1+p} - \frac{dq}{1+q} = 0$$

Integrating, $$\frac{1+p}{1+q} = \text{Constant}$$

Therefore, the intermediate integral of Eq. (iii) is

$$1 + p = (1 + q)\, \phi(x + y + z) \quad \text{(iv)}$$

Lagrange's subsidiary equations of Eq. (iv) are

$$\frac{dx}{1} = \frac{dy}{-\phi(x+y+z)} = \frac{dz}{\phi(x+y+z) - 1}$$

$$\text{Each ratio} = \frac{dx + dy + dz}{0}$$

$\Rightarrow$ $$dx + dy + dz = 0$$

Integrating, $x + y + z = c$ (Constant)

Considering first two ratios

$$dy = -\phi(c)\, dx$$

Integrating, $$y = -x\phi(c) + \text{Constant}$$

$$= -x\phi(x + y + z) + \text{Constant}$$

$\therefore$ General solution of Eq. (iv) is

$$y = \psi(x + y + z) - x\phi(x + y + z)$$

where ψ and ϕ are arbitrary functions.

EXAMPLE 12.15 Solve $2x^2r - 5xys + 2y^2t + 2(px + qy) = 0$.

Solution Monge's equations are

$$2x^2dpdy + 2y^2dqdx + 2(px + qy)\, dxdy = 0 \quad \text{(i)}$$

and $$2x^2dy^2 + 5xydxdy + 2y^2dx^2 = 0 \quad \text{(ii)}$$

$\Rightarrow$ $$(2xdy + ydx)(xdy + 2ydx) = 0$$

$\therefore$ Two systems to be solved are

$$\left.\begin{aligned} 2x^2dpdy + 2y^2dqdx + 2(px + qy)dxdy = 0 \\ 2xdy + ydx = 0 \end{aligned}\right\} \quad \text{(iii)}$$

and $$\left.\begin{aligned} 2x^2dpdy + 2y^2dqdx + 2(px + qy)dxdy = 0 \\ xdy + 2ydx = 0 \end{aligned}\right\} \quad \text{(iv)}$$

Solution of second equation of Eq. (iii) is

$$y^2x = \text{Constant}$$

Using second equation, first equation of Eq. (iii) is equivalent to

$$xdp - 2ydq + pdx - 2qdy = 0$$

$$\Rightarrow \quad d(xp - 2yq) = 0$$

Integrating, $\quad xp - 2yq = \text{Constant}$

$\therefore$ The intermediate integral corresponding to Eq. (iii) is

$$xp - 2yq = \phi_1(xy^2) \qquad \text{(v)}$$

Solution of second equation of Eq. (iv) is

$$x^2y = \text{Constant}$$

Using second equation, first equation of Eq. (iv) is equivalent to

$$2xdp - ydq + 2pdx - qdy = 0$$

$$\Rightarrow \quad d(2xp - yq) = 0$$

Integrating, $\quad 2xp - yq = \text{Constant}$

$\therefore$ The intermediate integral of Eq. (iv) is

$$2xp - yq = \phi_1\,(yx^2) \qquad \text{(vi)}$$

Solving Eqs. (v) and (vi), p and q give

$$p = \frac{1}{3x}[2\phi_2(yx^2) - \phi_1(xy^2)]$$

and

$$q = \frac{1}{3y}[\phi_2(yx^2) - 2\phi_1(xy^2)]$$

Substituting in $dz = pdx + qdy$

$$\Rightarrow \quad 3dz = \phi_2(yx^2)\left[\frac{2dx}{x} + \frac{dy}{y}\right] - \phi_1(xy^2)\left[\frac{dx}{x} + \frac{2dy}{y}\right]$$

$$= \phi_2(yx^2)\, d(\log x^2y) - \phi_1(xy^2)\, d(\log xy^2)$$

Integrating, $\quad 3z = 3\psi_2(yx^2) + 3\psi_1(xy^2)$

$$\Rightarrow \quad z = \psi_2(yx^2) + \psi_1(xy^2)$$

is the general solution of the given partial differential equation where ϕ_1, ϕ_2, ψ_1 and ψ_2 are arbitrary functions.

EXAMPLE 12.16 Solve $(x - y)\,[x(r - s) - y(s - t)] = (x + y)\,(p - q)$.

Solution Monge's equations are

$$xdpdy + ydqdx - \frac{(x+y)}{x-y}(p-q)\,dxdy = 0 \qquad \text{(i)}$$

and $$xdy^2 + (x + y)\, dydx + ydx^2 = 0 \qquad \text{(ii)}$$

$\Rightarrow$ $$(xdy + ydx)\,(dx + dy) = 0$$

$\Rightarrow$ $$xy = c \text{ and } x + y = c$$

Then Eq. (i) can be written as

$$dp - dq - \frac{dx - dy}{x - y}(p - q) = 0$$

$\Rightarrow$ $$\frac{d(p-q)}{p-q} - \frac{d(x-y)}{x-y} = 0$$

$\Rightarrow$ $$p - q = (x - y)\,\phi(xy)$$

Hence, we get intermediate integral. Its subsidiary equations are

$$\frac{dx}{1} = \frac{dy}{-1} = \frac{dz}{(x - y)\,\phi(xy)}$$

The first two ratios give $x + y = a$.
Using this, we have

$$dz = -\phi(ax - x^2)\,(a - 2x)\,dx$$

Integrating, $$z = \phi_1(ax - x^2) + b$$

Writing $b = \phi_2(a)$ and $a = x + y$, we get

$$z = \phi_1(xy) + \phi_2(x + y)$$

which is the general solution of given differential equation, where ϕ, ϕ_1 and ϕ_2 are arbitrary functions.

EXAMPLE 12.17 Solve $y^2r - 2ys + t = p + 6y$.

Solution Monge's equations are

$$y^2dpdy + dqdx + (p + 6y)\,dydx = 0 \qquad \text{(i)}$$

and $$y^2dy^2 + 2ydydx + dx^2 = 0 \qquad \text{(ii)}$$

$\Rightarrow$ $$(ydy + dx)^2 = 0$$

$\Rightarrow$ $$ydy + dx = 0$$

Integrating, $$y^2 + 2x = c$$

Putting $ydy = -dx$ in Eq. (i),

$$(ydp - dq) + (p + 6y)\,dy = 0$$

$\Rightarrow$ $$(ydp + pdy) - dq + 6ydy = 0$$

Integrating, $$py - q + 3y^2 = c$$

$\therefore$ The intermediate integral is

$$py - q + 3y^2 = \phi(y^2 + 2x)$$

The Lagrange's subsidiary equations are

$$\frac{dx}{y} = \frac{dy}{-1} = \frac{dz}{\phi(y^2 + 2x) - 3y^2}$$

From the first two ratios,

$$y^2 + 2x = a$$

Second and third ratios give

$$dz + [\phi(y^2 + 2x) - 3y^2]\, dy = 0$$

Integrating,

$$z + y\phi(y^2 + 2x) - y^3 = b$$

Hence, the general solution is

$$z = y^3 - y\phi(y^2 + 2x) + \psi(y^2 + 2x)$$

where ϕ and ψ are arbitrary functions.

12.5 MONGE'S METHOD TO SOLVE $Rr + Ss + Tt + U(rt - s^2) = V$

Consider

$$Rr + Ss + Tt + U(rt - s^2) = V \tag{12.21}$$

with

$$U \neq 0$$

Substitute

$$r = \frac{dp - sdy}{dx}$$

and

$$t = \frac{dq - Udx}{dy}$$

in Eq. (12.21).

$$\Rightarrow \quad Rdpdy + Tdqdx + Udpdq - Vdxdy - s[R(dy)^2 - Sdxdy + T(dx)^2 + Udqdy + Udpdx] = 0 \tag{12.22}$$

The equations

$$Rdpdy + Tdqdx + Udpdq - Vdxdy = 0 \tag{12.23}$$

and

$$R(dy)^2 - Sdxdy + T(dx)^2 + Udqdy + Udpdx = 0 \tag{12.24}$$

are called Monge's equations.

Multiply Eq. (12.24) by λ and add to the Eq. (12.23), we get

$$Rdpdy + Tdqdx + Udpdq - Vdxdy + \lambda[R(dy)^2 - Sdxdy + T(dx)^2 + Udqdy + Udpdx] = 0 \tag{12.25}$$

Let Eq. (12.25) be equivalent to

$$(ady + bdx + cdp) + \alpha dy + \beta dx + \gamma dq) = 0$$

$$\Rightarrow \quad a\alpha(dy)^2 + (a\beta + b\alpha)\, dxdy + b\beta(dx)^2 + c\alpha dpdy + c\beta dpdx + c\gamma dpdq + \gamma adydq + b\gamma dxdq = 0 \tag{12.26}$$

Comparing coefficients in Eqs. (12.25) and (12.26),

$$\alpha a = \lambda R,\ a\beta + b\alpha = -V - s\lambda,\ b\beta = \lambda T,\ c\alpha = R,\ c\beta = \lambda U,\ c\gamma = U,\ \beta a = \lambda U \text{ and } b\gamma = T$$

The first relation will be satisfied by taking

$$a = \lambda \text{ and } \alpha = R$$

$$\Rightarrow \quad c = 1,\ \gamma = U,\ b = \frac{T}{U} \text{ and } \beta = \lambda U$$

Substituting values in second relation, we get

$$\lambda^2 U + \frac{TR}{U} = -V - s\lambda \tag{12.27}$$

$$\Rightarrow \qquad \lambda^2 U^2 + sU\lambda + TR + UV = 0$$

Let λ_1 and λ_2 be the roots of Eq. (12.27)
Let $\lambda_1 \neq \lambda_2$. Thus, Eq. (12.25) can be factorized as

$$(\lambda_1 Udy + Tdx + Udp)\,(Rdy + \lambda_2 Udx + Udq) = 0 \tag{12.28}$$

$$(\lambda_2 Udy + Tdx + Udp)\,(Rdy + \lambda_2 Udx + Udq) = 0 \tag{12.29}$$

There are four systems to be considered.
The system

$$\lambda_1 Udy + Tdx + Udp = 0$$

$$\lambda_2 Udy + Tdx + Udp = 0$$

$$\Rightarrow \qquad (\lambda_1 - \lambda_2)\, Udy = 0$$

But $\lambda_1 \neq \lambda_2$. Therefore, $Udy = 0$ which will not give any solution.
Similarly, the system

$$Rdy + \lambda_1 Udx + Udq = 0$$

$$Rdy + \lambda_2 Udx + Udq = 0$$

$\Rightarrow \quad Udx = 0$ may not be useful.
Thus, the only two systems are to be considered, viz

$$\left.\begin{aligned} \lambda_1 Udy + Tdx + Udp &= 0 \\ Rdy + \lambda_2 Udx + Udq &= 0 \end{aligned}\right\} \tag{12.30}$$

and

$$\left.\begin{aligned} \lambda_2 Udy + Tdx + Udp &= 0 \\ Rdy + \lambda_1 Udx + Udq &= 0 \end{aligned}\right\} \tag{12.31}$$

Each system, if integrable, will give an intermediate integral of Eq. (12.21). Then solve intermediate integrals by the methods discussed in Section 12.4.

EXAMPLE 12.18 Solve $s^2 - rt = 4$.

Solution Comparing with Eq. (12.21) gives $R = 0 = S = T$, $U = -1$ and $V = 4$.
$\therefore$ Equation (12.27) gives $\lambda = \pm 2$.
$\therefore$ Two system to be considered as

$$\left.\begin{aligned} -2dy - dp &= 0 \\ 2dx - dq &= 0 \end{aligned}\right\} \tag{i}$$

and

$$\left.\begin{aligned} 2dy - dp &= 0 \\ -2dx - dq &= 0 \end{aligned}\right\} \tag{ii}$$

Equations (i) on integration gives

$$p + 2y = \text{Constant}$$

and

$$q - 2x = \text{Constant}$$

$\therefore$ The intermediate integral corresponding to Eq. (i) is

$$p + 2y = \phi_1(q - 2x) \qquad \text{(iii)}$$

where ϕ_1 is an arbitrary function.
Similarly, system Eq. (ii) gives

$$p - 2y = \phi_2(q + 2x) \qquad \text{(iv)}$$

From Eqs. (iii) and (iv), we get

$$p = \frac{1}{2}[\phi_1(q-2x) + \phi_2(q+2x)] \qquad \text{(v)}$$

and

$$y = \frac{1}{4}[\phi_1(q-2x) - \phi_2(q-2x)] \qquad \text{(vi)}$$

Now, $dz = pdx + qdy$ on integration gives

$$z = \int pdx + \int qdy$$

$$= \int pdx + qy - \int ydq$$

$$\Rightarrow \qquad z - yq = \frac{1}{2}\int[\phi_1(q-2x) + \phi_2(q+2x)]\,dx - \frac{1}{4}\int[\phi_1(q-2x) + \phi_2(q+2x)]\,dq$$

$$= \frac{1}{2}\int \phi_1(q-2x)\frac{d(2x-q)}{2} + \frac{1}{2}\int \phi_2\,(q+2x)\frac{d(q+2x)}{2}$$

$$= -\frac{1}{4}\int \phi_1(q-2x)\,d(q-2x) + \frac{1}{4}\int \phi_2(q+2x)\,d(q+2x)$$

$$= \psi_1(q - 2x) + \psi^2\,(q + 2x) \qquad \text{(vii)}$$

where

$$-4\psi_1'(q - 2x) = \phi_1(q - 2x)$$

and

$$4\psi_2'(1 + 2x) = \phi_2(q + 2x)$$

Substituting in Eq. (vi)

$$\Rightarrow \qquad y = -\psi_1'(q - 2x) - \psi_2'(q + 2x) \qquad \text{(viii)}$$

Eliminate q from Eqs. (vii) and (viii) to obtain the final integral.

EXAMPLE 12.19 Solve $rt - s^2 - s(\sin x + \sin y) = \sin x \sin y$.

Solution Here $R = 0 = T$, $S = -(\sin x + \sin y)$, $U = 1$ and $V = \sin x \sin y$.

$\therefore$ Equation (12.27) gives $\lambda = \sin x \sin y$

$\therefore$ Two systems are

$$\left.\begin{array}{l}\sin xdy + dp = 0\\ \sin ydx + dq = 0\end{array}\right\} \qquad \text{(i)}$$

and $$\left.\begin{aligned} \sin y dy + dp = 0 \\ \sin x dx + dq = 0 \end{aligned}\right\} \quad \text{(ii)}$$

Clearly, Eq. (i) is not integrable. Integration of Eq. (ii) gives

$$p - \cos y = \text{Constant}$$

and $$q - \cos x = \text{Constant}$$

∴ The intermediate integral corresponding to Eq. (iii) is

$$p - \cos y = \phi(q - \cos x) \quad \text{(iii)}$$

Applying Charpit's method to Eq. (iii) gives

$$\frac{dx}{-1} = \frac{dy}{\phi'(q - \cos x)} = \frac{dp}{-\sin x\phi'(q - \cos x)} = \frac{dq}{\sin y} \quad \text{(iv)}$$

Equation (iv) can not be integrated. So to obtain an integral of most general form assume the arbitrary function 'ϕ' to be linear.

Let $$p - \cos y = a(q - \cos x) + b \quad \text{(v)}$$
where a and b are constants.
Lagrange's subsidiary equation of Eq. (v) are

$$\frac{dx}{1} = \frac{dy}{-a} = \frac{dz}{\cos y - a\cos x + b} \quad \text{(vi)}$$

Considering first two ratios and integrating

$$y + ax = \text{Constant} = \lambda \text{ (say)}$$

Considering first and third ratios

$$dz = [\cos(\lambda - ax) - a\cos x + b]\,dx$$

Integrating, $$z = -\frac{1}{a}\sin(\lambda - ax) - a\sin x + bx + \text{Constant}$$

⇒ $$az + \sin y + a^2 \sin x - abx = a\,\text{Constant}$$

∴ Most general integral solution is

$$az + \sin y + a^2 \sin x + abx + a\phi(y + ax)$$

where ϕ is an arbitrary function.

EXAMPLE 12.20 Solve $xqr + (p + q)s + ypt + (xy - 1)(rt - s^2) + pq = 0$.

Solution $R = xq$, $S = (p + q)$, $T = yp$, $U = xy - 1$ and $V = -pq$.

∴ Equation (12.27) gives $\lambda = \dfrac{-p}{xy - 1}, \dfrac{-q}{xy - 1}$

∴ Two systems to be solved are

$$\left.\begin{aligned} -pdy + ypdx + (xy - 1)dp = 0 \\ xqdy - qdx + (xy - 1)dq = 0 \end{aligned}\right\} \quad \text{(i)}$$

and
$$\left.\begin{aligned}-qdy + ypdx + (xy-1)\,dp = 0\\ xqdy - pdx + (xy-1)\,dq = 0\end{aligned}\right\} \qquad \text{(ii)}$$

System shown in Eq. (ii) is not integrable. Multiply second equation of Eq. (ii) by y and add to first.

$\Rightarrow$ $$(xy - 1)\,qdy + (xy - 1)\,dp + (xy - 1)\,ydq = 0$$

$\Rightarrow$ $$qdy + dp + ydq = 0$$

$\Rightarrow$ $$d(yq + p) = 0$$

Integrating, $$yq + p = \text{Constant}$$

Now, multiply first equation of Eq. (ii) by x and add to second

$\Rightarrow$ $$pdx + xdp + dq = 0$$

$\Rightarrow$ $$d(px + q) = 0$$

Integrating, $$px + q = \text{Constant}$$

The intermediate integral corresponding to Eq. (ii) is
$$px + q = \phi(yq + p)$$

To obtain general solution, consider
$$px + q = a\,(yq + p) + b \qquad \text{(iii)}$$

where a and b are constants.

Lagrange's subsidiary equations of Eq. (iii) are
$$\frac{dx}{x-a} = \frac{dy}{1-ay} = \frac{dz}{b}$$

The first two ratios on integration give
$$(x - a)^a\,(1 - ay) = \text{Constant}$$

Considering first and third ratios and integrating
$$z = b\log(x - a) + \text{Constant}$$

$\therefore$ Most general integral solution is
$$z = b\log(x - a) + \psi[(x - a)^a\,(1 - ay)]$$

where ϕ and ψ are arbitrary functions.

EXERCISES

1. Solve $(b + cq)^2 r - 2(b + cq)(a + cp)s + (a + cp)^2 t = 0$.

   [Ans. $y + x\phi(ax + by + cz) = \psi(ax + by + cz)$]

2. Solve $x^2 r + 2xys + y^2 t = 0$. $\left[\textbf{Ans.}\ z + \phi\left(\frac{y}{x}\right) = x\psi\left(\frac{y}{x}\right)\right]$

3. Solve $(x - z)[xq^2 r - q(x + z + 2px)s + (z + px + pz + p^2 x)r] = (1 + p)q^2(x + z)$.

   [Ans. $y = \psi_1(x + z) + \psi_2(xz)$]

4. Solve $t - r\sec^4 y = 2q\tan y$. [**Ans.** $z = \psi_1(x - \tan y) + \psi_2(x + \tan y)$]

5. Solve $(1 + pq + q^2)r + (q^2 - p^2)s - (1 + pq + p^2)t = 0$.

[**Ans.** $\sqrt{(2 + u^2)}\, z = \psi_2\,(x - y) + \int u(u^2 + 2)^{-1/2}\phi_2'(u)\,du$ where $u = x + y$]

6. Solve $x^2r - y^2t - xp + yq = xy$.

$$\left[\textbf{Ans. } z = \frac{xy}{4}[(\log x)^2 - (\log y)^2] + xy\psi_1\left(\frac{y}{x}\right) + \psi_2(xy)\right]$$

7. Solve $(x - y)(x^2r - 2xys + y^2t) - 2xy(p - q) = 0$. [**Ans.** $z = (x + y)\phi(xy) + \psi(xy)$]

8. Solve $e^{2y}(r - p) = e^{2x}(t - q)$. [**Ans.** $z = \psi_1(e^x + e^y) + \psi_2(e^x - e^y)$]

9. Solve $z(qs - pt) = pq^2$. [**Ans.** $y = \phi(z) + 2\psi(x)$]

10. Solve $r + (a + b)s + abt = xy$.

$$\left[\textbf{Ans. } z = \frac{1}{b - a}[\psi_1(y - ax) + \psi_2(y - bx)] + \frac{1}{6}x^2y - (a + b)\frac{x^4}{24}\right]$$

11. Solve $(r - s)y + (s - t)x + q - p = 0$. [**Ans.** $z = \psi_1(x + y) + \psi_2(x^2 - y^2)$]

12. Solve $r - t\cos^2 x + p\tan x = 0$. [**Ans.** $z = \psi_1(y + \sin x) + \psi_2(y - \sin x)$]

13. Solve $(q + 1)s = (p + 1)t$. [**Ans.** $z = \phi_1(x) + \phi_2(x + y + z)$]

14. Solve $xy(t - r) + (x^2 - y^2)(s - 2) = py - qx$. $\left[\textbf{Ans. } z = \phi_1\left(\frac{y}{x}\right) + \phi_2(x^2 + y^2) + xy\right]$

15. Solve $q^2r - 2pqs + p^2t = pq^2$. [**Ans.** $y = \phi_1(z) + e^x\phi_2(z)$]

16. Solve $qr + (p + x)s + yt + q - y(s^2 - rt) = 0$.

$$\left[\textbf{Ans. } z = \lambda x - \frac{1}{2}x^2 + \psi_1(\lambda) + \psi_2\left(\frac{\lambda}{y}\right)\right]$$

17. Solve $2s + (rt - s^2) = 1$. [**Ans.** $z = xy - \phi(a)x - ay + \psi(a)$ and $0 = \phi'(a)x - y + \psi'(a)$]

18. Solve $yr - ps + t + y(rt - s^2) + 1 = 0$.

$$\left[\textbf{Ans. } 6a^2z = 2y^3 - 3a^2y^2 + 6axy + 6by + \phi\left(ax + \frac{1}{2}y^2\right)\right]$$

19. Solve $qxr + (x + y)s + pyt + xy(rt - s^2) = 1 - pq$.

[**Ans.** $mz = n\log x - y - m^2x + \phi(x^my)$]

20. Solve $r + 4kt = 4s$.

[**Ans.** $z = \phi_1[y + 2(1 - m)x] + \phi_2[y + 2(1 + m)x]$, where $m^2 = 1 - k$]

CHAPTER 13

Power Series Methods

13.1 INTRODUCTION

In mathematics and mathematical physics, we come across the differential equations whose solution cannot be expressed in terms of polynomials, rational functions, exponentials, trigonometric functions, etc. For such differential equation, it is possible to get the solution in terms of the power series. The solution in the form of an infinite series is called 'integration in series.' In this chapter, we discuss methods of obtaining solutions in the form of infinite series.

13.2 SERIES SOLUTION NEAR AN ORDINARY POINT

A function f defined on an interval containing the point $x = x_0$ is said to be analytic at x_0 if its Taylor series

$$\sum_{n=0}^{\infty} \frac{f^n(x_0)}{n!}(x - x_0)^n$$

exists and converges to $f(x)$ for all x in open interval containing x_0. All polynomial functions, exponential functions, trigonometric functions (whenever defined) are analytic. A rational function $\dfrac{f(x)}{g(x)}$ is analytic except at those values of x for which $g(x) = 0$.

The point $x = x_0$ is an *ordinary point* of the differential equation

$$y'' + P(x)\, y' + Q(x)\, y = 0 \tag{13.1}$$

if the functions $P(x)$ and $Q(x)$ are analytic at $x = x_0$. If either or both of these functions are not analytic at $x = x_0$, then $x = x_0$ is called a **singular point** of the Eq. (13.1).

Let $x = x_0$ is an ordinary point of the Eq. (13.1). Let its powers series solution be

$$y = \sum_{n=0}^{\infty} a_n (x - x_0)^n \tag{13.2}$$

and this power series converges in the interval $|x - x_0| < 0$ (where $R > 0$ called **radius of convergence**) about x_0. Then

$$y' = \sum_{n=1}^{\infty} na_n (x - x_0)^{n-1}$$

and
$$y'' = \sum_{n=2}^{\infty} n(n-1) a_n (x - x_0)^{n-2}$$

substituting the values of y, y' and y'' in Eq. (13.1), the coefficients a's are obtained. With these values Eq. (13.2) is the desired solution of the differential Eq. (13.1).

EXAMPLE 13.1 Solve $(x^2 - 1) y'' + xy' - xy = 0$ in powers of x (i.e. about $x_0 = 0$).

Solution We want to solve

$$(x^2 - 1) y'' + xy' - xy = 0 \tag{i}$$

Let power series solution of Eq. (i) about $x_0 = 0$ be

$$y = \sum_{n=0}^{\infty} a_n x^n \tag{ii}$$

Differentiating twice successively,

$$y' = \sum_{n=1}^{\infty} na_n x^{n-1} \tag{iii}$$

and
$$y'' = \sum_{n=2}^{\infty} n(n-1) a_n x^{n-2} \tag{iv}$$

Substituting in Eq. (i), we get

$$(x^2 - 1) \sum_{n=2}^{\infty} n(n-1)\, a_n x^{n-2} + x \sum_{n=1}^{\infty} na_n x^{n-1} - x \sum_{n=0}^{\infty} a_n x^n = 0$$

$$\Rightarrow \quad \sum_{n=2}^{\infty} n(n-1)\, a_n x^n - \sum_{n=2}^{\infty} n(n-1)\, a_n x^{n-2} + \sum_{n=1}^{\infty} na_n x^n - \sum_{n=0}^{\infty} a_n x^{n+1} = 0$$

$$\Rightarrow \quad \sum_{n=2}^{\infty} n(n-1)\, a_n x^n - \sum_{n=0}^{\infty} (n+2)(n+1)\, a_{n+2} x^n + \sum_{n=1}^{\infty} na_n x^n - \sum_{n=1}^{\infty} a_{n-1} x^n = 0$$

$$\Rightarrow \quad -2a_2 - (6a_3 + a_1 - a_0)\, x + \sum_{n=2}^{\infty} [n(n-1)\, a_n - (n+2)(n+1)\, a_{n+2} + na_n - a_{n-1}]\, x^n = 0$$

$$\Rightarrow \quad -2a_2 = 0$$

$$\Rightarrow \quad a_2 = 0 \tag{v}$$

$$-(6a_3 + a_1 - a_0) = 0$$

$$\Rightarrow \quad a_3 = -\frac{1}{6}(a_1 - a_0) \tag{vi}$$

and $$n(n-1)\,a_n - (n+2)\,(n+1)\,a_{n+2} + na_n - a_{n-1} = 0$$

or $$-(n+2)\,(n+1)\,a_{n+2} + n^2a_n - a_{n-1} = 0,\ n \geq 2 \tag{vii}$$

For $n = 2$, Eq. (vii) gives

$$-12a_4 + 4a_2 - a_1 = 0$$

$$a_4 = -\frac{1}{12}a_1$$

For $n = 3$, Eq. (vii) gives

$$-20a_5 + 9a_3 - a_2 = 0$$

$$\Rightarrow \quad -20a_5 - \frac{9}{6}(a_1 - a_0) = 0$$

$$\Rightarrow \quad a_5 = -\frac{3}{40}(a_1 - a_0)$$

and so on.

$$\therefore \quad y = a_0\left(1 + \frac{1}{6}x^3 + \frac{3}{40}x^5 + \cdots\right) + a_1\left(x - \frac{1}{6}x^3 - \frac{1}{12}x^4 - \cdots\right).$$

EXAMPLE 13.2 Solve $y'' - 2x^2y' + 4xy = x^2 + 2x + 4$ about $x_0 = 0$.

Solution Let series solution be

$$y = \sum_{n=0}^{\infty} a_n x^n \tag{i}$$

Differentiating Eq. (i) twice successively,

$$y' = \sum_{n=1}^{\infty} n a_n x^{n-1} \tag{ii}$$

and $$y'' = \sum_{n=2}^{\infty} n(n-1)a_n x^{n-2} \tag{iii}$$

Substituting values of y, y' and y'' in given differential equation, we get

$$\sum_{n=2}^{\infty} n(n-1)\,a_n x^{n-2} - 2x^2\sum_{n=1}^{\infty} na_n x^{n-1} + 4x\sum_{n=0}^{\infty} a_n x^n = x^2 + 2x + 4$$

$$\Rightarrow \quad \sum_{n=0}^{\infty}(n+2)\,(n+1)\,a_{n+2}x^n - 2\sum_{n=2}^{\infty}(n-1)\,a_{n-1}x^n + 4\sum_{n=1}^{\infty} a_{n-1}x^n = x^2 + 2x + 4$$

$$\Rightarrow \quad (2a_2 - 4) + (6a_3 + 4a_0 - 2)x + (12a_4 + 2a_1 - 1)\,x^2$$

$$+ \sum_{n=3}^{\infty}[(n+2)\,(n+1)\,a_{n+2} - 2(n-1)\,a_{n-1} + 4a_{n-1}]x^n = 0$$

Then $$2a_2 - 4 = 0$$

$$\Rightarrow \quad a_2 = 2 \tag{iv}$$

$$6a_3 + 4a_2 - 2 = 0 \tag{v}$$

$$12a_4 + 2a_1 - 1 = 0 \tag{vi}$$

and $$(n + 2)(n + 1)a_{n+2} - 2(n - 1)a_{n-1} + 4a_{n-1} = 0, n \geq 3$$

$\Rightarrow$ $$(n + 2)(n + 1)a_{n+2} = 2(n - 3)a_{n-1}, n \geq 3 \tag{vii}$$

From Eqs. (v) and (vi), we get

$$a_3 = \frac{1}{3} - \frac{2}{3}a_0$$

and $$a_4 = \frac{1}{12} - \frac{1}{6}a_1$$

Put $n = 3, 4, 5, 6, \ldots$, in Eq. (vii)

$\Rightarrow$ $$a_5 = 0$$

$$30a_6 = 2a_3 \quad \Rightarrow \quad a_6 = \frac{1}{45} - \frac{2}{45}a_0$$

$$42a_7 = 4a_4 \quad \Rightarrow \quad a_7 = \frac{1}{126} - \frac{1}{63}a_1$$

$$56a_8 = 6a_5 \quad \Rightarrow \quad a_8 = 0$$

and so on.

Therefore, series solution of given differential equation is

$$y = a_0\left(1 - \frac{2}{3}x^3 - \frac{2}{45}x^6 - \cdots\right) + a_1\left(x - \frac{1}{6}x^4 - \frac{1}{63}x^7 - \cdots\right) + 2x^2 + \frac{1}{3}x^3 + \frac{1}{12}x^4 + \frac{1}{45}x^6 + \frac{1}{126}x^7 + \cdots$$

EXAMPLE 13.3 Find the power series solution in powers of $(x - 1)$ of the initial value problem

$$xy'' + y' - 2y = 0 \quad \text{with} \quad y(1) = 1, y'(1) = 2 \tag{i}$$

Solution Clearly $x = 1$ is an ordinary point of the given differential equation. Let power series solution of Eq. (i) be

$$y = \sum_{n=0}^{\infty} a_n (x - 1)^n \tag{ii}$$

Differentiating twice successively, we get

$$y' = \sum_{n=1}^{\infty} n a_n (x - 1)^{n-1} \tag{iii}$$

and $$y'' = \sum_{n=2}^{\infty} n(n - 1)a_n (x - 1)^{n-2} \tag{iv}$$

Substituting y, y' and y'' in Eq. (i) gives

$$x\sum_{n=2}^{\infty} n(n - 1)a_n (x - 1)^{n-2} + \sum_{n=1}^{\infty} n a_n (x - 1)^{n-1} - 2\sum_{n=0}^{\infty} a_n (x - 1)^n = 0$$

$$\Rightarrow \qquad [(x-1)+1]\sum_{n=2}^{\infty} n(n-1)a_n(x-1)^{n-2} + \sum_{n=1}^{\infty} na_n(x-1)^{n-1} - 2\sum_{n=0}^{\infty} a_n(x-1)^n = 0$$

$$\Rightarrow \qquad \sum_{n=2}^{\infty} n(n-1)a_n(x-1)^{n-1} + \sum_{n=2}^{\infty} n(n-1)a_n(x-1)^{n-2} + \sum_{n=1}^{\infty} na_n(x-1)^{n-1} - 2\sum_{n=0}^{\infty} a_n(x-1)^n = 0$$

$$\Rightarrow \sum_{n=1}^{\infty}(n+1)na_{n+1}(x-1)^n + \sum_{n=0}^{\infty}(n+2)(n+1)a_{n+2}(x-1)^n + \sum_{n=0}^{\infty}(n+1)a_{n+1}(x-1)^n - 2\sum_{n=0}^{\infty}a_n(x-1)^n = 0$$

$$\Rightarrow \qquad (2a_2 + a_1 - 2a_0) + \sum_{n=1}^{\infty}[(n+1)na_{n+1} + (n+2)(n+1)a_{n+2} + (n+1)a_{n+1} - 2a_n](x-1)^n = 0$$

$$\therefore \qquad 2a_2 + a_1 - 2a_0 = 0 \;\Rightarrow\; a_2 = a_0 - \frac{a_1}{2} \tag{v}$$

and
$$(n+1)(n+2)\,a_{n+2} + (n+1)^2\,a_{n+1} - 2a_n = 0,\; n \ge 1 \tag{vi}$$

Put $x = 1$ in Eqs. (ii) and (iii)

$\therefore$
$$a_0 = 1$$

and
$$a_1 = 2$$

Then Eq. (v) gives
$$a_2 = 0$$

Put $n = 1, 2, 3, \ldots$, in Eq. (vi).

$$6a_3 + 4a_2 - 2a_1 = 0 \;\Rightarrow\; a_3 = \frac{2}{3}$$

$$12a_4 + 9a_3 - 2a_2 = 0 \;\Rightarrow\; a_4 = -\frac{1}{2}$$

and so on.

$\therefore$ Series solution in powers of $(x - 1)$ is

$$y = 1 + 2(x-1) + \frac{2}{3}(x-1)^3 - \frac{1}{2}(x-1)^4 + \cdots$$

13.3 SERIES SOLUTIONS NEAR REGULAR SINGULAR POINTS

Consider the Eq. (13.1)

$$y'' + P(x)\,y' + Q(x)\,y = 0$$

and assume that at least one of the function $P(x)$ or $Q(x)$ is not analytic at $x = x_0$. In Section (13.2), we called such point to be singular point. Define new functions as

$$P_1(x) = (x - x_0)\,P(x)$$

and
$$Q_1(x) = (x - x_0)^2\,Q(x)$$

If $P_1(x)$ and $Q_1(x)$ both are analytic at $x = x_0$, then x_0 is called a **regular singular point** of Eq. (13.1) otherwise irregular.

For example, consider the equation

$$x^2y'' - xy' + (x^2 + 2)\, y = 0$$

$$\Rightarrow \qquad y'' - \frac{1}{x}y' + \frac{x^2+2}{x^2}y = 0$$

Then $P(x) = -\dfrac{1}{x}$ and $Q(x) = \dfrac{x^2+2}{x^2}$

$\therefore$ $P_1(x) = xP(x) = -1$ and $Q_1(x) = x^2Q(x) = (x^2 + 2)$

Since both $P(x)$ and $Q(x)$ are not analytic at $x = 0$, $x = 0$ is a singular point of given differential equation. Moreover, both the functions $P_1(x)$ and $Q_1(x)$ are analytic at $x = 0$, $x = 0$ is the regular singular point of given differential equation.
e.g. consider the equation

$$(x - 1)^2 (x - 3)^2 y'' + 2(x - 3) y' + (x + 6) y = 0$$

Here $P(x) = \dfrac{2}{(x-1)^2(x-3)}$ and $Q(x) = \dfrac{x+6}{(x-1)^2(x-3)^2}$

Clearly, $P(x)$ and $Q(x)$ are not analytic at $x = 1$ and $x = 3$, and hence, both these points are singular points of given differential equation.

For $x = 1$

$$P_1(x) = \frac{2}{(x-1)\,(x-3)} \text{ and } Q_1(x) = \frac{x+6}{(x-3)^2}$$

$P_1(x)$ is not analytic at $x = 1$ so $x = 1$ is an irregular singular point.

For $x = 3$

$$P_2(x) = \frac{2}{(x-1)^2} \text{ and } Q_2(x) = \frac{x+6}{(x-1)^2}$$

Since both the functions $P_2(x)$ and $Q_2(x)$ are analytic at $x = 3$, $x = 3$ is regular singular point. Next we discuss the method (called method of Frobenius) to find series solution of the given differential equation when $x = x_0$ is a regular singular point. If $x = x_0$ is regular singular point of Eq. (13.1) the equation has solution of the form

$$(x - x_0)^r \sum_{n=0}^{\infty} a_n (x - x_0)^n \tag{13.3}$$

where r is a definite real constant which may be determined and the solution remains valid in some deleted neighbourhood of x_0.
Equation (13.3) can be written as

$$y = \sum_{n=0}^{\infty} a_n (x - x_0)^{n+r} \quad \text{where } a_0 \neq 0 \tag{13.4}$$

Differentiating twice successively, we get

$$y' = \sum_{n=0}^{\infty} (n+r) a_n (x - x_0)^{n+r-1}$$

and
$$y'' = \sum_{n=0}^{\infty} (n+r)(n+r-1) a_n (x - x_0)^{n+r-2}$$

Substituting y, y' and y'' in Eq. (13.3) and comparing coefficients, a quadratic equation in r is obtained. This equation is known as the **indicial equation** of Eq. (13.1). The two roots r_1 and r_2 ($r_1 \geq r_2$) of the indicial equation are called the exponents of Eq. (13.1) and are the only possible values for the constant r in Eq. (13.3). There are four cases depending on the values of r_1 and r_2.

Case 1: $r_1 - r_2 \neq 0$ or a positive integer.

Case 2: $r_1 - r_2 = 0$.

Case 3: $r_1 - r_2$ a positive integer making a coefficient of y infinite.

Case 4: $r_1 - r_2$, a positive integer making a coefficient of y indeterminate.

Case 1: $r_1 - r_2 \neq 0$ or a positive integer.

EXAMPLE 13.4 Find the solution near $x = 0$ of the differential equation

$$9x(1 - x)\, y'' + 12y' + 4y = 0 \tag{i}$$

Solution Check that $x = 0$ is regular singular point of Eq. (i). Let series solution of Eq. (i) be

$$y = \sum_{n=0}^{\infty} a_n x^{n+r}, \; a_0 \neq 0 \tag{ii}$$

Differentiating twice in succession, we get

$$y' = \sum_{n=0}^{\infty} (n+r) a_n x^{n+r-1}$$

$$y'' = \sum_{n=0}^{\infty} (n+r)(n+r-1) a_n x^{n+r-2}$$

Substituting y, y' and y'' in Eq. (i) gives

$$9x(1-x)\sum_{n=0}^{\infty} (n+r)(n+r-1)a_n x^{n+r-2} - 12\sum_{n=0}^{\infty} (n+r)a_n x^{n+r-1} + 4\sum_{n=0}^{\infty} a_n x^{n+r} = 0$$

$$\Rightarrow \quad 9\sum_{n=0}^{\infty} (n+r)(n+r-1)a_n x^{n+r-1} - 9\sum_{n=0}^{\infty} (n+r)(n+r-1)a_n x^{n+r}$$

$$-12\sum_{n=0}^{\infty} (n+r)a_n x^{n+r-1} + 4\sum_{n=0}^{\infty} a_n x^{n+r} = 0$$

$$\Rightarrow \quad 3\sum_{n=0}^{\infty}(n+r)(3n+3r-7)a_n x^{n+r-1} - \sum_{n=0}^{\infty}[9(n+r)^2 - 9(n+r) - 4]\, a_n x^{n+r} = 0$$

$$\Rightarrow \quad 3\sum_{n=0}^{\infty}(n+r)(3n+3r-7)\, a_n x^{n+r-1} - \sum_{n=0}^{\infty}(3n+3r-4)(3n+3r+1)\, a_n x^{n+r} = 0$$

$$\Rightarrow \quad 3\sum_{n=-1}^{\infty}(n+r+1)(3n+3r-4)a_{n+1} x^{n+r} - \sum_{n=0}^{\infty}(3n+3r-4)(3n+3r+1)a_n x^{n+r} = 0$$

$$\Rightarrow \quad 3r(3r-7)a_0 x^{r-1} + \sum_{n=0}^{\infty}[3(n+r+1)(3n+3r-4)a_{n+1} - (3n+3r-4)(3n+3r+1)a_n]\, x^{n+r} = 0$$

Therefore indicial equation is

$$3r(3r-7) = 0 \Rightarrow r = 0, \frac{7}{3}$$

and
$$3(n+r+1)(3n+3r-4)a_{n+1} - (3n+3r-4)(3n+3r+1)a_n = 0,\ n \ge 0$$

$$\Rightarrow \quad a_{n+1} = \frac{(3n+3r+1)}{3(n+r+1)} a_n,\ n \ge 0$$

For
$$r = 0$$

$$a_{n+1} = \frac{3n+1}{3(n+1)} a_n$$

Putting $n = 0, 1, 2, \ldots$, gives

$$a_1 = \frac{1}{3} a_0$$

$$a_2 = \frac{4}{6} a_1 = \frac{1 \cdot 4}{3 \cdot 6} a_0$$

$$a_3 = \frac{7}{9} a_2 = \frac{1 \cdot 4 \cdot 7}{3 \cdot 6 \cdot 9} a_0$$

and so on. Therefore, the series solution corresponding to $r = 0$ is

$$y_1 = a_0 \left(1 + \frac{1}{3}x + \frac{1 \cdot 4}{3 \cdot 6}x^2 + \frac{1 \cdot 4 \cdot 7}{3 \cdot 6 \cdot 9}x^3 + \cdots\right)$$

For
$$r = \frac{7}{3}$$

$$a_{n+1} = \frac{3n+8}{3n+10}$$

Putting $n = 0, 1, 2, \ldots$, we get

$$a_1 = \frac{8}{10} a_0$$

$$a_2 = \frac{11}{13}a_1 = \frac{8 \cdot 11}{10 \cdot 13}a_0$$

$$a_3 = \frac{14}{16}a_2 = \frac{8 \cdot 11 \cdot 14}{10 \cdot 13 \cdot 16}a_0$$

and so on. Therefore, the series solution corresponding to $r = \frac{7}{3}$ is

$$y_2 = a_0 x^{7/3}\left(1 + \frac{8}{10}x + \frac{8 \cdot 11}{10 \cdot 13}x^2 + \frac{8 \cdot 11 \cdot 14}{10 \cdot 13 \cdot 16}x^3 + \cdots\right)$$

Clearly, the solutions y_1 and y_2 are linearly independent. So the general solution of given differential equation is

$$y = c_1 y_1 + c_2 y_2$$

where c_1 and c_2 are arbitrary constants.

Case 2: $r_1 - r_2 = 0$

EXAMPLE 13.5 **Find series solution near $x = 0$ of the differential equation**

$$x(1 - x)\, y'' + (1 - x)\, y' + y = 0 \tag{i}$$

Solution Clearly, $x = 0$ is a regular singular point of given differential Eq. (i). Let the series solution of Eq. (i) be

$$y = \sum_{n=0}^{\infty} a_n x^{n+r}, \; a_0 \neq 0 \tag{ii}$$

Differentiating twice successively, we get

$$y' = \sum_{n=0}^{\infty} (n + r) a_n x^{n+r-1}$$

and

$$y'' = \sum_{n=0}^{\infty} (n + r)(n + r - 1) a_n x^{n+r-2}$$

Substituting y, y' and y'' in Eq. (i) gives

$$x(1 - x)\sum_{n=0}^{\infty} (n + r)(n + r - 1) a_n x^{n+r-2} + (1 - x)\sum_{n=0}^{\infty} (n + r) a_n x^{n+r-1} + \sum_{n=0}^{\infty} a_n x^{n+r} = 0$$

$$\Rightarrow \quad \sum_{n=0}^{\infty} (n + r)(n + r - 1) a_n x^{n+r-1} - \sum_{n=0}^{\infty} (n + r)(n + r - 1) a_n x^{n+r} + \sum_{n=0}^{\infty} (n + r) a_n x^{n+r-1}$$

$$- \sum_{n=0}^{\infty} (n + r) a_n x^{n+r} + \sum_{n=0}^{\infty} a_n x^{n+r} = 0$$

$$\Rightarrow \quad \sum_{n=0}^{\infty}(n+r)^2 a_n x^{n+r-1} - \sum_{n=0}^{\infty}[(n+r)(n+r-1)+(n+r)-1]a_n x^{n+r} = 0$$

$$\Rightarrow \quad \sum_{n=-1}^{\infty}(n+r+1)^2 a_{n+1} x^{n+r} - \sum_{n=0}^{\infty}[(n+r)^2-1]a_n x^{n+r} = 0$$

$$\Rightarrow \quad r^2 a_0 x^{r-1} + \sum_{n=0}^{\infty}[(n+r+1)^2 a_{n+1} - [(n+r)^2-1]a_n]\, x^{n+r} = 0$$

Therefore, indicial equation is

$$r^2 = 0 \quad \Rightarrow \quad r = 0$$

and

$$a_{n+1} = \frac{(n+r)^2-1}{(n+r+1)^2} a_n$$

or

$$a_{n+1} = \frac{n+r-1}{n+r+1} a_n$$

Putting $n = 0, 1, 2, \ldots,$ gives

$$a_1 = \frac{r-1}{r+1} a_0$$

$$a_2 = \frac{r}{r+2} a_1 = \frac{(r-1)r}{(r+1)(r+2)} a_0$$

$$a_3 = \frac{r+1}{r+3} a_2 = \frac{(r-1)\,r(r+1)}{(r+1)(r+2)(r+3)} a_0$$

and so on.

$$\therefore \quad y = a_0 x^r \left[1 + \frac{r-1}{r+1}x + \frac{(r-1)r}{(r+1)(r+2)}x^2 + \frac{(r-1)\,r(r+1)}{(r+1)(r+2)(r+3)}x^3 + \cdots\right] \qquad \text{(iii)}$$

is a solution for $r = 0$.

With y given by Eq. (iii), left hand member of Eq. (i) becomes $r^2 a_0 x^{r-1}$. So that

$$x(1-x)\, y'' + (1-x)\, y' + y = r^2 a_0 x^{r-1}$$

$$\therefore \quad \frac{\partial}{\partial r}[x(1-x)y'' + (1-x)y' + y] = 2r a_0 x^{r-1} + r^2 a_0 x^{r-1} \log x$$

$$\Rightarrow \quad \frac{\partial}{\partial r}\left[x(1-x)\frac{d^2}{dx^2} + (1-x)\frac{d}{dx} + 1\right] y = 2r a_0 x^{r-1} + r^2 a_0 x^{r-1} \log x$$

$$\Rightarrow \quad \left[x(1-x)\frac{d^2}{dx^2} + (1-x)\frac{d}{dx} + 1\right]\frac{\partial y}{\partial r} = 2r a_0 x^{r-1} + r^2 a_0 x^{r-1} \log x$$

Therefore, for $r = 0$, $\frac{\partial y}{\partial r}$ is a second solution of given differential Eq. (i). From Eq. (iii)

$$\frac{\partial y}{\partial r} = a_0 x^r \log x \left[1 + \frac{r-1}{r+1}x + \frac{(r-1)r}{(r+1)(r+2)}x^2 + \cdots\right] + a_0 x^r \left[\frac{2}{(r+1)^2}x + \cdots\right] \quad \text{(iv)}$$

Putting $r = 0$ in Eqs. (iii) and (iv) gives

$$y_1 = a_0[1 - x]$$

and

$$y_2 = a_0 \log x\, [1 - x] + a_0(2x + \cdots)$$

$$= y_1 \log x + a_0\, (2x + \cdots)$$

Since, y_1 and y_2 are linearly independent, the general solution is

$$y = c_1 y_1 + c_2 y_2$$

where c_1 and c_2 are arbitrary constants.

Case 3: $r_1 - r_2$, a positive integer making a coefficient of y infinite.

EXAMPLE 13.6 Find the general solution near $x = 0$ of the differential equation

$$x(1 + x + x^2)\, y'' + 3x^2 y' - 2y = 0 \quad \text{(i)}$$

Solution Clearly, $x = 0$ is regular singular point of differential Eq. (i). Let the series solution of Eq. (i) be

$$y = \sum_{n=0}^{\infty} a_n x^{n+r},\ a_0 \neq 0 \quad \text{(ii)}$$

Then

$$y' = \sum_{n=0}^{\infty} (n+r)\, a_n x^{n+r-1}$$

and

$$y'' = \sum_{n=0}^{\infty} (n+r)(n+r-1)\, a_n x^{n+r-2}$$

Substituting y, y' and y'' in Eq. (i) gives

$$x(1+x+x^2)\sum_{n=0}^{\infty}(n+r)(n+r-1)\,a_n x^{n+r-2} + 3x^2\sum_{n=0}^{\infty}(n+r)a_n x^{n+r-1} - 2\sum_{n=0}^{\infty} a_n x^{n+r} = 0$$

$$\Rightarrow \sum_{n=-1}^{1}(n+r+1)(n+r)a_{n+1}x^{n+r} + \sum_{n=1}^{\infty}(n+r-1)(n+r+1)a_{n-1}x^{n+r}$$

$$+ \sum_{n=0}^{\infty}[(n+r)(n+r-1) - 2]a_n x^{n+r} = 0$$

$$\Rightarrow \quad r(r-1)a_0 x^{r-1} + \{(r+1)ra_1 + [r(r-1) - 2]a_0\}x^r + \sum_{n=1}^{\infty}\{(n+r+1)(n+r)a_{n+1}$$

$$+ (n+r-1)(n+r+1)a_{n-1} + [(n+r)(n+r-1) - 2]a_n\}\, x^{n+r} = 0$$

Therefore, indicial equation is

$$r(r-1)=0 \Rightarrow r=0, 1$$

$$a_1=-\frac{r(r-1)-2}{r(r+1)}a_0$$

and $\quad (n+r+1)(n+r)a_{n+1}+(n+r-1)(n+r+1)a_{n-1}+[(n+r)(n+r-1)-2]a_n=0$

Putting $n = 1, 2, \ldots$, gives

$$a_2=-\frac{(3r-2)}{r(r+1)}a_0$$

$$a_3=\frac{r^3+3r^2-5r-2}{r(r+1)(r+2)}a_0$$

and so on. Therefore, the series solution is

$$y=a_0x^r\left[1-\frac{r(r-1)-2}{r(r+1)}x-\frac{3r-2}{r(r+1)}x^2+\cdots\right]$$

The coefficient of y becomes infinite for $r = 0$. To overcome this difficulty, replace a_0 by kr. Then

$$y=kr\left[r-\frac{r^2-r-2}{r+1}x-\frac{3r-2}{r+1}x^2\cdots\right] \tag{iii}$$

Substituting y in left hand side of Eq. (i) gives

$$x(1+x+x^2)y''+3x^2y'-2y=kr^2(r-1)x^{r-1}$$

Now, proceed as Case 2 to find other solution. It is given by $\frac{\partial y}{\partial r}$. Check that for $r = 0$,

$$y_2=y_1\log x+k[1-x-5x^2-\cdots]$$

Putting $r = 1$ in Eq. (iii) gives

$$y=kx\left[1+x-\frac{1}{2}x^2-\frac{1}{2}x^3\cdots\right]$$

$$=k\left[x+x^2-\frac{1}{2}x^3-\frac{1}{2}x^4\cdots\right]$$

Thus, we have $2y_3 = y_1$. So out of three solutions only two are linearly independent. Hence, general solution is

$$y=c_1y_1+c_2y_2$$

where c_1 and c_2 are arbitrary constants.

Case 4: $r_1 - r_2$, a positive integer making a coefficient of y indeterminate.

EXAMPLE 13.7 **Find the general solution near $x = 0$ of the differential equation**

$$x^2y''+x(1+x)y'+(x-9)y=0 \tag{i}$$

Solution Check that $x = 0$ is regular singular point of the differential Eq. (i). Let the series solution of Eq. (i) be

$$y = \sum_{n=0}^{\infty} a_n x^{n+r}, n \neq 0 \tag{ii}$$

Then
$$y' = \sum_{n=0}^{\infty} (n+r) a_n x^{n+r-1}$$

and
$$y'' = \sum_{n=0}^{\infty} (n+r)(n+r-1) a_n x^{n+r-2}$$

Substituting y, y' and y'' in Eq. (i) after simplification, we get

$$(r^2 - 9) a_0 x^r + \sum_{n=1}^{\infty} \{[(n+r)^2 - 9] a_n + (n+r) a_{n-1}\} x^{n+r} = 0$$

Therefore, indicial equation is

$$r^2 - 9 = 0 \Rightarrow r \pm 3$$

and
$$a_n = \frac{n+r}{9 - (n+r)^2} a_{n-1}, n \geq 1$$

Putting, $n = 1, 2, \ldots$, gives

$$a_1 = \frac{(r+1)}{9 - (r+1)^2} a_0$$

$$a_2 = \frac{(r+1)(r+2)}{[9 - (r+1)^2][9 - (r+2)^2]} a_0$$

and so on.
For $r = 3$,

$$y_1 = a_0 x^3 \left[1 - \frac{4}{7} x + \frac{4 \cdot 5}{7 \cdot 16} x^2 - \cdots\right]$$

and for $r = -3$, the coefficients of x^3, x^4, ... becomes zero and the coefficient of remaining terms indeterminate. Thus, for $r = -3$,

$$y_2 = a_0 x^{-3} \left[1 - \frac{2}{5} x + \frac{2 \cdot 1}{5 \cdot 8} x^2\right]$$

Therefore, the general solution is

$$y = c_1 y_1 + c_2 y_2$$

where c_1 and c_2 are arbitrary constants.

Observations

1. If the indicial equation has two an equal roots differing by a quantity which is not an integer, two independent solutions are obtained by substituting these values of r in the series for y.

2. If the indicial equation has two equal roots, two independent solutions are obtained by substituting the value of r in y and $\dfrac{\partial y}{\partial r}$.
3. If the indicial equation has two roots r_1 and r_2 (with $r_1 > r_2$) differing by an integer and if some of the coefficients of y become infinite for $r = r_2$, modify y by replacing a_0 by $k(r = r_2)$. Two independent solutions are then obtained by substituting $r = r_2$, in y and $\dfrac{\partial y}{\partial r}$.
4. If the indicial equation has two roots r_1 and r_2 (with $r_1 > r_2$) differing by an integer and if some of the coefficients of y become indeterminant when $r = r_2$, the solution is given by putting $r = r_2$ in y which then contains two arbitrary constants.

EXAMPLE 13.8 Find the series solution of

$$x^3y'' + x(1 - x)\,y' + y = 0 \text{ near } x = \infty \tag{i}$$

Solution Put

$$x = \frac{1}{z}$$

Then

$$y' = -z^2\frac{dy}{dz}$$

and

$$y'' = z^4\frac{d^2y}{dz^2} + 2z^3\frac{dy}{dz}$$

Therefore, Eq. (i) becomes

$$z\frac{d^2y}{dz^2} + (3 - z)\frac{dy}{dz} + y = 0 \tag{ii}$$

Clearly $z = 0$ is regular singular point of Eq. (ii).
Let the series solution of Eq. (ii) be

$$y = \sum_{n=0}^{\infty} a_n z^{n+r}$$

Then

$$\frac{dy}{dz} = \sum_{n=0}^{\infty} (n + r)\,a_n z^{n+r-1}$$

$$\frac{d^2y}{dz^2} = \sum_{n=0}^{\infty} (n + r)\,(n + r - 1)\,a_n z^{n+r-2}$$

Substituting $y, \dfrac{dy}{dz}$ and $\dfrac{d^2y}{dz^2}$ and simplifying, we get

$$[r(r-1) + 3r]\,a_0 z^{n-1} + \sum_{n=0}^{\infty} \{[(n + r + 1)\,(n + r) + 3(n + r + 1)]a_{n+1} - (n + r + 1)\,a_n\}\,z^{n+r} = 0$$

The indicial equation is

$$r(r-1) + 3r = 0 \Rightarrow r = 0, -2$$

and
$$a_{n+1} = \frac{n+r-1}{(n+r+1)(n+r+3)} a_n, \; n \geq 0$$

Putting $n = 0, 1, 2, 3, \ldots$, we get

$$a_1 = \frac{r-1}{(r+1)(r+3)} a_0$$

$$a_2 = \frac{r(r-1)}{(r+1)(r+2)(r+3)(r+4)} a_0$$

$$a_3 = \frac{r(r-1)}{(r+2)(r+3)^2(r+4)(r+5)} a_0$$

and so on.

$$\therefore y = a_0 z^r \left[1 + \frac{(r-1)}{(r+1)(r+3)} z + \frac{r(r-1)}{(r+1)(r+2)(r+3)(r+4)} z^2 + \frac{r(r-1)}{(r+2)(r+3)^2(r+4)(r+5)} z^3 + \cdots \right]$$

For $r = -2$, coefficients of z^2, z^3, ... become infinite. So replace a_0 by $k(r+2)$. Therefore,

$$y = kz^r \left[(r+2) + \frac{(r-1)(r+2)}{(r+1)(r+3)} z + \frac{r(r-1)}{(r+1)(r+3)(r+4)} z^2 + \frac{r(r-1)}{(r+3)^2(r+4)(r+5)} z^3 + \cdots \right]$$

As mentioned in observation (4), the second solution will be given by $\frac{\partial y}{\partial r}$ at r = – 2. One can check that for $r = -2$.

$$y_1 = kz^{-2}(-3z^2 + z^3) = k(-3 + z) = k\left(-3 + \frac{1}{x}\right)$$

and
$$\frac{\partial y}{\partial r} = ky_1 \log z + kz^{-2}\left[1 + 3z + 4z^2 - \frac{11}{3} z^3 + \cdots\right]$$

$$= ky_1 \log \frac{1}{x} + k\left[x^2 + 3x + 4 - \frac{11}{3x} + \frac{1}{8x^2} + \cdots\right]$$

$$= y_2$$

Therefore, the general solution is

$$y = c_1 y_1 + c_2 y_2$$

where c_1 and c_2 are arbitrary constants.

EXERCISES

Find the series solution of the following differential equations:

1. $y'' + xy' + x^2 y = 0$ about $x_0 = 0$.

$$\left[\textbf{Ans. } y = a_0\left(1 - \frac{1}{12}x^4 + \cdots\right) + a_1\left(x - \frac{1}{6}x^3 - \frac{1}{40}x^5 \cdots\right)\right]$$

2. $(1 - x^2)\, y'' + 2xy' - y = 0$ about $x_0 = 0$.

$$\left[\textbf{Ans. } y = a_0\left(1 - \frac{1}{2}x^2 - \frac{1}{24}x^4 \cdots\right) + a_1\left(x - \frac{1}{6}x^3 - \frac{1}{120}x^5 \cdots\right)\right]$$

3. $(2 + x^2)\, y'' + xy' - (1 + x)\, y = 0$ about $x_0 = 0$.

$$\left[\textbf{Ans. } y = a_0\left(1 + \frac{1}{4}x^2 + \frac{x^3}{12} \cdots\right) + a_1\left(x + \frac{1}{24}x^4 \cdots\right)\right]$$

4. $y'' + (x - 3)\, y' + y = 0$ about $x_0 = 2$.

$$\left[\textbf{Ans. } y = a_0\left[1 - \frac{1}{2}(x-2)^2 - \frac{1}{6}(x-2)^3 + \cdots\right] + a_1\left[(x-2) + \frac{1}{2}(x-2)^2 - \frac{1}{6}(x-2)^3 \cdots\right]\right]$$

5. $(x^2 - 1)\, y'' + 3xy' + xy = 0$ with $y(0) = 2$, $y'(0) = 3$.

$$\left[\textbf{Ans. } y = 2 + 3x + \frac{11}{6}x^3 + \frac{1}{4}x^4 + \cdots\right]$$

6. $(2x + x^2)\, y'' - y' - 6xy = 0$ about $x_0 = 0$.

$$\left[\textbf{Ans. } y = c_1\left(1 + 3x^2 - \frac{3}{5}x^4 - \frac{1}{15}x^6 + \cdots\right) + c_2 x^{3/2}\left(1 + \frac{3}{8}x^2 - \frac{2\cdot 3}{8\cdot 16}x^4 + \cdots\right)\right]$$

7. $2x^2 y'' + xy' + (x^2 - 3)\, y = 0$ about $x_0 = 0$.

$$\left[\textbf{Ans. } y = c_1 x^{3/2}\left(1 - \frac{1}{18}x^2 + \frac{1}{936}x^4 \cdots\right) + c_2 x^{-1}\left(1 + \frac{3}{2}x^2 - \frac{1}{24}x^4 \cdots\right)\right]$$

8. $2x(1 - x)\, y'' + (1 - x)\, y' + 3y = 0$ about $x_0 = 0$.

$$\left[\textbf{Ans. } y = c_1\left(1 - 3x + \frac{3x^2}{1\cdot 3} + \frac{3x^3}{3\cdot 5} + \cdots\right) + c_2 x^{1/2}(1 - x)\right]$$

9. $4x^2(1 - x^2)\, y'' + 8x^3 y' - y = 0$ about $x_0 = 0$.

$$\left[\begin{aligned} &\textbf{Ans. } y = \sqrt{x}(c_1 + c_2 \log x)\left[1 + \frac{1\cdot 3}{4^2}x^2 + \frac{1\cdot 3\cdot 5\cdot 7}{4^2\cdot 8^2}x^4 + \cdots\right] \\ &+ 2c_2\sqrt{x}\left[\frac{1\cdot 3}{4^2}\left(1 + \frac{1}{3} - \frac{1}{2}\right)x^2 + \frac{1\cdot 3\cdot 5\cdot 7}{4^2\cdot 8^2}\left(1 + \frac{1}{3} - \frac{1}{2} + \frac{1}{5} + \frac{1}{7} - \frac{1}{4}\right)x^4 + \cdots\right]\end{aligned}\right]$$

10. $x(1-x)\,y'' + (1-5x)\,y' - 4y = 0$ about $x_0 = 0$.

$$\left[\textbf{Ans. } y = c_1(1^2 + 2^2 x + 3^2 x^2 + \cdots) + c_2 \log x\,(1^2 + 2^2 x + 3^2 x^2 + \cdots) - 2c_2(1\cdot 2x + 2\cdot 3x^2 + 3\cdot 4x^3 \cdots)\right]$$

11. $2xy'' + y' - y = x + 1$ about $x_0 = 0$.

$$\left[\textbf{Ans. } y = c_1\left(1 + x + \frac{x^2}{6} + \frac{x^3}{90} + \cdots\right) + c_2\sqrt{x}\left(1 + \frac{x}{3} + \frac{x^2}{30} + \cdots\right) + \frac{1}{6}x^2\left(1 + \frac{x}{15} + \frac{x^2}{420} + \cdots\right) - 1\right]$$

12. $x^2y'' - 3xy' + (3 + 4x)\,y = 0$ about $x_0 = 0$.

$$\left[\textbf{Ans. } y = (c_1 + c_2 \log x)\left[-8x^3 + \frac{32}{3}x^4 + \cdots\right] + c_2(x + 4x^2 + 4x^3 \cdots)\right]$$

13. $xy'' + x(x+4)\,y' + 4y = 0$ about $x_0 = 0$.

$$\left[\textbf{Ans. } y = (c_1 + c_2 \log x)\,(-4x + 16x^2 + \cdots) + c_2(1 - 16x^3 \cdots)\right]$$

14. $8x^3y'' + 2x^2y' + y = 0,\ x > 0$.

$$\left[\textbf{Ans. } y = c_1\left(1 - \frac{1}{14x} + \frac{1}{616x^2} \cdots\right) + c_2x^{3/4}\left(1 - \frac{1}{2x} + \frac{1}{40x^3} \cdots\right)\right]$$

CHAPTER 14

Bessel's Equation and Bessel Functions

14.1 BESSEL'S EQUATION AND ITS SOLUTION

The differential equation of the form

$$x^2y'' + xy' + (x^2 - n^2)\,y = 0 \tag{14.1}$$

is called **Bessel's equation** of order n, n being a non-negative constant. We discuss solution of Eq. (14.1) by method of Frobenius.

Let the series solution of Eq. (14.1) be

$$y = \sum_{m=0}^{\infty} c_m x^{k+m},\ c_0 \neq 0 \tag{14.2}$$

Then
$$y' = \sum_{m=0}^{\infty} c_m (k+m)\, x^{k+m-1}$$

and
$$y'' = \sum_{m=0}^{\infty} c_m\,(k+m)\,(k+m-1)\,x^{k+m-2}$$

Substituting y, y' and y'' in (14.1) gives

$$x^2 \sum_{m=0}^{\infty} c_m\,(k+m)\,(k+m-1)\,x^{k+m-2} + x\sum_{m=0}^{\infty} c_m\,(k+m)\,x^{k+m-1} + (x^2-n^2)\sum_{m=0}^{\infty} c_m x^{k+m} = 0$$

$$\Rightarrow \quad \sum_{m=0}^{\infty} c_m\,[(k+m)\,(k+m-1)+(k+m)-n^2\,]\,x^{k+m} + \sum_{m=0}^{\infty} c_m x^{k+m+2} = 0$$

$$\Rightarrow \quad \sum_{m=0}^{\infty} c_m (k+m+n)\,(k+m-n)\,x^{k+m} + \sum_{m=0}^{\infty} c_m x^{k+m+2} = 0 \tag{14.3}$$

Equating to zero the smallest power of x, namely x^k, Eq. (14.3) gives the indicial equation

$$c_0\,(k+n)\,(k-n) = 0$$

But $c_0 \neq 0$. Therefore, $(k + n)(k - n) = 0$
So the roots of the indicial equation are n and $-n$.
Next equating to zero the coefficient of x^{k+1} gives $c_1(k + 1 + n)(k + 1 - n) = 0$, so that $c_1 = 0$ for $k = n$ and $-n$.
Finally, equating to zero the coefficient of x^{k+m} gives

$$c_m(k + m + n)(k + m - n) + c_{m-2} = 0$$

$$\Rightarrow \qquad c_m = \frac{c_{m-2}}{(k+m+n)(k+m-n)} \tag{14.4}$$

Putting $m = 3, 5, 7, \ldots$ in Eq. (14.4) and using $c_1 = 0$, we have

$$c_m = 0 \text{ for } m = 1, 3, 5, \ldots \tag{14.5}$$

Putting $m = 2, 4, 6, \ldots$ in Eq. (14.4), we get

$$c_2 = \frac{1}{(n+k+2)(n-k-2)} c_0$$

$$c_4 = \frac{1}{(n+k+4)(n-k-4)} c_2 = \frac{1}{(n+k+4)(n-k-4)(n+k+2)(n-k-2)} c_0$$

and so on. Putting these values in Eq. (14.2), we get

$$y = c_0 x^k \left[1 + \frac{x^2}{(n+k+2)(n-k-2)} + \frac{x^4}{(n+k+2)(n-k-2)(n+k+4)(n-k-4)} + \cdots\right]$$

Putting $k = n$ and $-n$ and replacing c_0 by a and b in the above equation gives

$$y = ax^n \left[1 - \frac{x^2}{4(1+n)} + \frac{x^4}{4 \cdot 8(1+n)(2+n)} - \cdots\right] \tag{14.6}$$

and

$$y = bx^{-n} \left[1 - \frac{x^2}{4(1-n)} + \frac{x^4}{4 \cdot 8(1-n)(2-n)} - \cdots\right] \tag{14.7}$$

The solutions of Eqs. (14.6) and (14.7) are general solutions of the Bessel's equation. The particular solution of Eq. (14.1) is obtained from Eq. (14.6) by taking the arbitrary constant $a = \frac{1}{2^n \sqrt{n+1}}$, is called the **Bessel function of the first kind of order *n***. It will be denoted by $J_n(x)$. Thus, we have

$$J_n(x) = \frac{x^n}{2^n \sqrt{n+1}} \left[1 - \frac{x^2}{4(n+1)} + \frac{x^4}{4 \cdot 8(n+1)(n+2)} + \cdots\right] \tag{14.8}$$

or

$$J_n(x) = \sum_{r=0}^{\infty} (-1)^r \frac{1}{r!\overline{)(n+r+1)}} \left(\frac{x}{2}\right)^{2r+n} \tag{14.9}$$

Putting $b = \dfrac{1}{2^n \sqrt{n+1}}$ in Eq. (14.7), will give

$$J_{-n}(x) = \sum_{r=0}^{\infty} (-1)^r \frac{1}{r!\,\Gamma(-n+r+1)} \left(\frac{x}{2}\right)^{2r-n} \tag{14.10}$$

Let n be non-integer. We know that $\Gamma m = \infty$, if m is zero or a negative integer and $\sqrt{m}$ is finite otherwise. Since n is not an integer and r is always an integer, the term $\Gamma(-n+r+1)$ in Eq. (14.10) is always finite and non-zero. For $2r < n$, Eq. (14.10) shows that $J_{-n}(x)$ contains negative powers of x. On the other hand, Eq. (14.9) shows that $J_n(x)$ does not contain negative powers of x at all. Therefore, at $x = 0$, $J_n(x)$ is finite while $J_{-n}(x)$ is infinite, and so one cannot be expressed as a integral multiple of the other. This establishes that $J_n(x)$ and $J_{-n}(x)$ are two linearly independent solutions of Eq. (14.1) when n is not an integer. Hence, we can write the general solution of Bessel's Eq. (14.1) when n is not an integer as

$$y = AJ_n(x) + BJ_{-n}(x)$$

where A and B are arbitrary constants.

14.2 BESSEL'S FUNCTION OF THE FIRST KIND OF ORDER *n*

Bessel's function of the first kind and of order n is denoted by $J_n(x)$ as defined in Eq. (14.9) where n is any non-negative integer.

Remark: When n is an integer, $\Gamma(n+r+1) = (n+r)!$ and so Eq. (14.9) can be rewritten as

$$J_n(x) = \sum_{r=0}^{\infty} (-1)^r \frac{1}{r!(n+r)!} \left(\frac{x}{2}\right)^{2r+n}$$

For $n = 0$ and $n = 1$, we have

$$J_0(x) = 1 - \frac{x^2}{2^2} + \frac{x^4}{2^2 \cdot 4^2} - \frac{x^6}{2^2 \cdot 4^2 \cdot 6^2} + \cdots$$

and $$J_1(x) = \frac{x}{2} - \frac{x^3}{2^2 \cdot 3} + \frac{x^5}{2^2 \cdot 4^2 \cdot 6} - \cdots$$

respectively.

Note: We shall denote $J_n(x)$ by J_n and $\dfrac{dJ_n(x)}{dx}$ as J_n', etc.

14.3 NEUMANN FUNCTION OF ORDER *n*

THEOREM 14.1 Show that

(i) When n is positive integer, $J_{-n}(x) = (-1)^n J_n(x)$

(ii) For any integer, $J_{-n}(x) = (-1)^n J_n(x)$

Proof:

(i) Let n be a positive integer. We know that

$$J_{-n}(x)=\sum_{r=0}^{\infty}(-1)^r\frac{1}{r!\,\Gamma(-n+r+1)}\left(\frac{x}{2}\right)^{2r-n}$$

Since n is positive integer, $\Gamma(-n+r+1)$ is infinite and so $\frac{1}{\Gamma(-n+r+1)}$ is zero. for $r = 0, 1, 2, \ldots, (n-1)$. So we can take sum from $r = n$ to infinity.

$$\Rightarrow \qquad J_{-n}(x)=\sum_{r=n}^{\infty}(-1)^r\frac{1}{r!\,\Gamma(-n+r+1)}\left(\frac{x}{2}\right)^{2r-n}$$

Changing the variable of summation to $m = r - n$ so that $r = m + n$ and so $m = 0$ when $r = n$ and $m = \infty$.
When $r = \infty$.

$$\therefore \qquad J_{-n}(x)=\sum_{m=0}^{\infty}(-1)^{m+n}\frac{1}{(m+n)!\,\Gamma(m+1)}\left(\frac{x}{2}\right)^{2(m+n)-n}$$

$$=\sum_{m=0}^{\infty}(-1)^{m+n}\frac{1}{m!\,\Gamma(m+n+1)}\left(\frac{x}{2}\right)^{2m+n}$$

$$=(-1)^n\sum_{r=0}^{\infty}(-1)^r\frac{1}{r!\,\Gamma(n+r+1)}\left(\frac{x}{2}\right)^{2r+n}$$

$$=(-1)^n J_n(x)$$

from Eq. (14.9)

Thus, for $n > 0$, $J_{-n}(x) = (-1)^n J_n(x)$.

(ii) Let $n < 0$. Let P be a positive integer such that $n = -P$. Since $P > 0$, from (i), we have

$$J_{-P}(x) = (-1)^P J_P(x)$$

$$\Rightarrow \qquad J_P(x) = (-1)^{-P} J_{-P}(x)$$

But $P = -n$. Hence, we get

$$J_{-n}(x) = (-1)^n J_n(x).$$

Remark 1: $J_{-n}(x) = (-1)^n J_n(x)$ for any integer n.

Remark 2: When n is an integer, $J_{-n}(x)$ is constant multiple of $J_n(x)$. Hence, $J_n(x)$ and $J_{-n}(x)$ are not two linearly independent solutions of Bessel's Eq. (14.1). So $y = AJ_n(x) + BJ_{-n}(x)$ is not the general solution of Bessel's equation when n is not an integer. We established in Section 14.1 that $y = AJ_n(x) + BJ_{-n}(x)$ is the general solution of Bessel's equation when n is not an integer. Next, we try to discuss the general solution of Bessel's equation, when n is an integer.

THEOREM 14.2 The two independent solutions of Bessel's equation may be taken to be $J_n(x)$ and

$$Y_n(x) = \frac{\cos n\pi J_n(x) - J_{-n}(x)}{\sin n\pi} \tag{14.11}$$

for all values of n.

Proof:

Case 1: Let n be not an integer. Then $\sin n\pi \neq 0$. Then (14.11) proves that $Y_n(x)$ is a linear combination of $J_n(x)$ and $J_{-n}(x)$ which contradicts the remark in Section 14.1 that $J_n(x)$ and $J_{-n}(x)$ are linearly independent solution when n is not an integer. $J_n(x)$ and a linear combination of $J_n(x)$ and $J_{-n}(x)$ will also be linearly independent solution. Thus, we find that $J_n(x)$ and $Y_n(x)$ are two independent solution of Bessel's equation.

Case 2: Let n be an integer. Then $\cos n\pi = (-1)^n$, $\sin n\pi = 0$ and $J_{-n}(x) = (-1)^n J_n(x)$. Using these values in Eq. (14.11), we have $Y_n(x)$ in the form 0/0 and so $Y_n(x)$ is undefined. To define $Y_n(x)$, consider

$$Y_n(x) = \lim_{\nu\to n} Y_\nu(x)$$

$$= \lim_{\nu\to n} \frac{\cos \nu\pi J_\nu(x) - J_{-\nu}(x)}{\sin \nu\pi}$$

$$= \frac{\left\{\dfrac{\partial}{\partial \nu}[\cos \nu\pi J_\nu(x) - J_{-\nu}(x)]\right\}_{\nu=n}}{\left[\dfrac{\partial}{\partial \nu}\sin \nu\pi\right]_{\nu=n}}$$

(by *L*' Hospital's rule)

$$= \frac{\left[-\pi \sin \nu\pi J_\nu(x) + \cos \nu\pi \dfrac{\partial}{\partial \nu} J_\nu(x) - \dfrac{\partial}{\partial \nu} J_{-\nu}(x)\right]_{\nu=n}}{[\pi \cos \nu\pi]_{\nu=n}}$$

$$= \frac{\cos n\pi \left[\dfrac{\partial}{\partial \nu} J_\nu(x)\right]_{\nu=n} - \left[\dfrac{\partial}{\partial \nu} J_{-\nu}(x)\right]_{\nu=n}}{\pi \cos n\pi}$$

$$= \frac{(-1)^n \left[\dfrac{\partial}{\partial \nu} J_\nu(x)\right]_{\nu=n} - (-1)^{2n}\left[\dfrac{\partial}{\partial \nu} J_{-\nu}(x)\right]_{\nu=n}}{\pi(-1)^n}$$

$$= \frac{1}{\pi}\left[\frac{\partial}{\partial \nu} J_\nu(x) - (-1)^n \frac{\partial}{\partial \nu} J_{-\nu}(x)\right]_{\nu=n} \tag{14.12}$$

We need to establish

(i) $Y_n(x)$ is a solution of Bessel's equation.

(ii) $Y_n(x)$ is a solution independent of $J_n(x)$.

Proof of (i): Since $J_v(x)$ and $J_{-v}(x)$ are solutions of Bessel's equation of order v, we have

$$x^2 \frac{d^2 J_v}{dx^2} + x \frac{dJ_v}{dx} + (x^2 - v^2) J_v = 0 \tag{14.13}$$

and

$$x^2 \frac{d^2 J_{-v}}{dx^2} + x \frac{dJ_{-v}}{dx} + (x^2 - v^2) J_{-v} = 0 \tag{14.14}$$

Differentiating Eqs. (14.13) and (14.14) w.r.t. v gives

$$x^2 \frac{d^2}{dx^2}\left(\frac{\partial J_v}{\partial v}\right) + x \frac{d}{dx}\left(\frac{\partial J_v}{\partial v}\right) + (x^2 - v^2)\frac{\partial J_v}{\partial v} - 2vJ_v = 0 \tag{14.15}$$

and

$$x^2 \frac{d^2}{dx^2}\left(\frac{\partial J_{-v}}{\partial v}\right) + x \frac{d}{dx}\left(\frac{\partial J_{-v}}{\partial v}\right) + (x^2 - v^2)\frac{\partial J_{-v}}{\partial v} - 2vJ_{-v} = 0 \tag{14.16}$$

Multiplying Eq. (14.16) by $(-1)^v$ and subtracting from Eq. (14.15) gives

$$x^2 \frac{d^2}{dx^2}\left[\frac{\partial J_v}{\partial v} - (-1)^v \frac{\partial J_{-v}}{\partial v}\right] + x \frac{d}{dx}\left[\frac{\partial J_v}{\partial v} - (-1)^v \frac{\partial J_{-v}}{\partial v}\right]$$

$$+ (x^2 - v^2)\left[\frac{\partial J_v}{\partial v} - (-1)^v \frac{\partial J_{-v}}{\partial v}\right] - 2v[J_v - (-1)^v J_{-v}] = 0 \tag{14.17}$$

Putting $v = n$ in Eq. (14.17) and using Eq. (14.11), we get

$$x^2 \frac{d^2}{dx^2}[\pi Y_n(x)] + x \frac{d}{dx}[\pi Y_n(x)] + (x^2 - n^2)\,\pi Y_n(x) - 2n[J_n(x) - (-1)^n J_{-n}(x)] = 0$$

Using $J_n(x) = (-1)^n J_{-n}(x)$ for an integer n, we have above equation as

$$x^2 Y_n'' + xY_n' + (x^2 - n^2)\, Y_n = 0 \tag{14.18}$$

This proves that Y_n is a solution of Bessel's equation of order n.
Proof of (ii): For integral value of n, $Y_n(x)$ is

$$Y_n(x) = \frac{2}{\pi}\left\{\log\frac{x}{2} + \gamma - \frac{1}{2}\sum_{r=0}^{n}\frac{1}{r}\right\} J_n(x) - \frac{1}{\pi}\sum_{m=0}^{\infty}(-1)^n \frac{1}{m!\,(n+m)!}\left(\frac{x}{2}\right)^{n+2m} \sum_{r=1}^{m}\left[\frac{1}{r} + \frac{1}{r+n}\right]$$

$$- \frac{1}{\pi}\sum_{m=0}^{n-1}\frac{(n-m-1)!}{m!}\left(\frac{x}{2}\right)^{-n+2m} \tag{14.19}$$

where γ is Euler's constant.
From Eq. (14.19), we observe that $Y_n(x)$ is infinite when $x = 0$, whereas $J_n(x)$ is infinite when $x = 0$. So $Y_n(x)$ given in Eq. (14.11) and $J_n(x)$ are two independent solutions of Bessel's equation of order n.

Remark 1: The general solution of Bessel's equation when n is an integer is given by

$$y = AJ_n(x) + BY_n(x)$$

$Y_n(x)$ is known as Bessel's function of order n of the second kind or the Neumann function of order n and is denoted by $N_n(x)$.

Remark 2: Equation reducible to Bessel's equation

$$x^2y' + xy'' + (\lambda^2x^2 - n^2)\, y = 0 \tag{14.20}$$

Substitute $z = \lambda x$. So that

$$\frac{dy}{dx} = \lambda\frac{dy}{dz} \text{ and } \frac{d^2y}{dx^2} = \lambda\frac{d^2y}{dz^2}$$

Then Eq. (14.20) becomes

$$z^2\frac{d^2y}{dz^2} + z\frac{dy}{dz} + (z^2 - n^2)\, y = 0 \tag{14.21}$$

which is Bessel's equation of order n. From above remark, the general solution of Eq. (14.21) is

$$y = AJ_n(z) + BY_n(z)$$

or

$$y = AJ_n(\lambda x) + BY_n(\lambda x) \tag{14.22}$$

The Eq. (14.21) is the *modified Bessel's equation.*

14.4 BESSEL'S FUNCTION OF THE SECOND KIND OF ORDER *n*

This is denoted by $Y_n(x)$ and defined as

$$Y_n(x) = \frac{J_n(x)\cos n\pi - J_{-n}(x)}{\sin n\pi}, \qquad n \neq \text{integer}$$

and

$$Y_n(x) = \lim_{\nu \to n} \frac{J_\nu(x)\cos \nu\pi - J_{-\nu}(x)}{\sin \nu\pi}, \; n \text{ is an integer}$$

14.5 RECURRENCE FORMULAE FOR $J_n(x)$

R1:

$$\frac{d}{dx}[x^n J_n(x)] = x^n J_{n-1}(x)$$

Proof:

$$\frac{d}{dx}[x^n J_n(x)]$$

$$= \frac{d}{dx}\left[x^n \sum_{r=0}^{\infty} (-1)^r \frac{1}{r!\,\Gamma(n+r+1)}\left(\frac{x}{2}\right)^{2r+n}\right]$$

$$= \sum_{r=0}^{\infty} (-1)^r \frac{1}{r!\,\Gamma(n+r+1)} \cdot \frac{1}{2^{2r+n}} \frac{d}{dx} x^{2r+2n}$$

$$= \sum_{r=0}^{\infty} \frac{(-1)^r (2r+2n)}{r!\,\overline{)(n+r+1)}} \frac{1}{2^{2r+n}} \cdot x^{2r+2n-1}$$

$$= \sum_{r=0}^{\infty} \frac{(-1)^r\, 2(r+n)}{r!(n+r)\,\overline{)(n+r)}} \cdot \frac{x^n \cdot x^{2r+n-1}}{2^{2r+n}} \qquad (\because \overline{)(n+1)} = n\overline{)n})$$

$$= x^n \sum_{r=0}^{\infty} \frac{(-1)^r}{r!\,\overline{)(n-1+r+1)}} \left(\frac{x}{2}\right)^{2r+n-1}$$

$$= x^n J_{n-1}(x) \qquad \text{[by the definition of } J_n(x)\text{]}$$

R2:
$$\frac{d}{dx}[x^{-n} J_n(x)] = -\, x^{-n} J_{n+1}(x)$$

Proof: Using the definition of $J_n(x)$

$$\frac{d}{dx}[x^{-n} J_n(x)]$$

$$= \frac{d}{dx}\left[x^{-n} \sum_{r=0}^{\infty} (-1)^r \frac{1}{r!\,\overline{)(n+r+1)}} \left(\frac{x}{2}\right)^{2r+n} \right]$$

$$= \sum_{r=0}^{\infty} \frac{(-1)^r}{r!\,\overline{)(n+r+1)}} \cdot \frac{1}{2^{2r+n}} \frac{d}{dx} x^{2r}$$

$$= \sum_{r=0}^{\infty} \frac{(-1)^r}{r(r-1)!\,\overline{)(n+r+1)}} \cdot \frac{2r}{2^{2r+n}} \cdot x^{2r-1}$$

$$= \sum_{r=0}^{\infty} \frac{(-1)^r}{(r-1)!\,\overline{)(n+r+1)}} \cdot \frac{1}{2^{2r+n-1}} \cdot x^{2r-1}$$

Since $(r-1)! = \infty$ when $r = 0$, the term corresponding to $r = 0$ vanishes.

$$= \sum_{r=1}^{\infty} \frac{(-1)^r}{(r-1)!\,\overline{)(n+r+1)}} \cdot \frac{1}{2^{2r+n-1}} \cdot x^{2r-1}$$

$$= \sum_{m=0}^{\infty} \frac{(-1)^{m+1}}{m!\,\overline{)(n+m+2)}} \cdot \frac{1}{2^{2r+2+n-1}} \cdot x^{2m+2-1} \qquad \text{(Put } m = r-1\text{)}$$

$$= \sum_{m=0}^{\infty} \frac{(-1)^{m+1}}{m!\,\overline{)(n+m+2)}} \cdot \frac{x^n \cdot x^{-n}}{2^{2r+2+n-1}} \cdot x^{2m+2-1}$$

$$= -\, x^{-n} \sum_{m=0}^{\infty} \frac{(-1)^m}{m!\,\overline{)(n+m+2)}} \left(\frac{x}{2}\right)^{2m+n+1}$$

$$= -x^{-n}\sum_{r=0}^{\infty}\frac{(-1)^r}{r!\,\overline{)(n+1+r+1)}}\left(\frac{x}{2}\right)^{n+1+2r}$$

$$= -x_{-n}\, J_{n+1}(x) \qquad \text{[by definition of } J_n(x)\text{]}$$

R3:

$$J_n'(x) = J_{n-1}(x) - \frac{n}{x}J_n(x)$$

Proof: From (R1)

$$\frac{d}{dx}[x^n J_n(x)] = x^n J_{n-1}(x)$$

$$\Rightarrow \qquad nx^{n-1}J_n(x) + x^n J_n'(x) = x^n J_{n-1}(x)$$

Dividing both sides by x^{n-1} and rearranging terms, we get

$$J_n'(x) = J_{n-1}(x) - \frac{n}{x}J_n(x)$$

R4:

$$J_n'(x) = \frac{n}{x}J_n(x) - J_{n+1}(x)$$

Proof: From (R2)

$$\frac{d}{dx}[x^{-n}\, J_n(x)] = -\, x^{-n}J_{n+1}(x)$$

$$\Rightarrow \qquad -nx^{-n-1}J_n(x) + x^{-n}J_n(x) = -\, x^{-n}J_{n+1}(x)$$

Dividing both sides by x^{-n} and rearranging terms, we get

$$J_n'(x) = \frac{n}{x}J_n(x) - J_{n+1}(x)$$

R5:

$$2J_n'(x) = J_{n-1}(x) - J_{n+1}(x)$$

Proof: Adding (R3) and (R4), we get the result.

R6:

$$J_{n-1}(x) + J_{n+1}(x) = \frac{2n}{x}J_n(x)$$

Proof: Subtracting (R2) from (R1), we get the result.

14.6 GENERATING FUNCTION FOR THE BESSEL'S FUNCTION $J_n(x)$

Prove that

$$\exp\left\{\frac{1}{2}x\left(z - \frac{1}{z}\right)\right\} = \sum_{n=-\infty}^{\infty} z^n J_n(x)$$

Proof: We have

$$\exp\left\{\frac{1}{2}x\left(z - \frac{1}{z}\right)\right\} = e^{(x/2)z - (x/2)z^{-1}}$$

$$= e^{(x/2)z}\cdot e^{(x/2)z^{-1}}$$

$$= \left[1 + \left(\frac{x}{2}\right)z + \left(\frac{x}{2}\right)^2 \frac{z^2}{2!} + \cdots + \left(\frac{x}{2}\right)^n \frac{z^n}{n!} + \left(\frac{x}{2}\right)^{n+1} \cdot \frac{z^{n+1}}{(n+1)!} + \cdots\right]$$

$$\times \left[1 - \left(\frac{x}{2}\right)z^{-1} + \left(\frac{x}{2}\right)^2 \frac{z^{-2}}{2!} + \cdots + \left(\frac{x}{2}\right)^n \frac{z^{-n}}{n!} + \left(\frac{x}{2}\right)^{n+1} \cdot \frac{z^{-(n+1)}}{(n+1)!} + \cdots\right] \quad (14.23)$$

The coefficient of z^n on multiplying is

$$= \left(\frac{x}{2}\right)^n \cdot \frac{1}{n!} - \left(\frac{x}{2}\right)^{n+2} \cdot \frac{1}{(n+1)!} + \left(\frac{x}{2}\right)^{n+4} \cdot \frac{1}{(n+2)!2!} - \cdots$$

$$= \sum_{r=0}^{\infty} \frac{(-1)^r}{r!(n+r)!} \left(\frac{x}{2}\right)^{n+2r}$$

$$= \sum_{r=0}^{\infty} \frac{(-1)^r}{r!\,\Gamma(n+r+1)} \cdot \left(\frac{x}{2}\right)^{n+2r}$$

The coefficient of z^{-n} on multiplying is

$$= \left[\left(\frac{x}{2}\right)^n \cdot \frac{(-1)^n}{n!} + \left(\frac{x}{2}\right)^{n+1} \cdot \frac{(-1)^{n+1}}{(n+1)!} \cdot \frac{x}{2} + \left(\frac{x}{2}\right)^{n+2} \frac{(-1)^{n+2}}{(n+2)!\,2!} \left(\frac{x}{2}\right)^2 + \cdots\right]$$

$$= (-1)^n \left[\left(\frac{x}{2}\right)^n \cdot \frac{1}{n!} - \left(\frac{x}{2}\right)^{n+2} \cdot \frac{1}{(n+1)!} + \left(\frac{x}{2}\right)^{n+4} \cdot \frac{1}{(n+2)!\,2!} + \cdots\right]$$

$$= (-1)^n J_n(x)$$

Thus, the coefficient of z^{-n} is $(-1)^n J_n(x)$.

$\therefore$ $J_n(x) = (-1)^n$ the coefficient of z^{-n}.

The coefficient of z^0 is

$$= 1 - \left(\frac{x}{2}\right)^2 + \left(\frac{x}{2}\right)^4 \left(\frac{1}{2!}\right)^2 - \left(\frac{x}{2}\right)^6 \left(\frac{1}{3!}\right)^2 + \cdots$$

$$= 1 - \frac{x^2}{2^2} + \frac{x^4}{2^2 \cdot 4^4} - \cdots = J_0(x)$$

Thus, the coefficients of z^0, $(z - z^{-1})$, $(z^2 + z^{-2})$, ..., $[z_n + (-1)^n z^{-n}]$, ... are J_0, J_1, J_2, ..., J_n, ...

Hence, Eq. (14.23) gives

$$\exp\left\{\frac{1}{2}x\left(z - \frac{1}{z}\right)\right\} = J_0(x) + (z - z^{-1})\,J_1(x) + (z^2 + z^{-2})\,J_2(x) + \cdots + (z^n + (-1)^n z^{-n})\,J_n(x) + \cdots$$

$$= \sum_{n=-\infty}^{\infty} z^n J_n(x) \qquad [\text{as } J_{-n}(-1)^n J_n]$$

14.7 ORTHOGONALITY OF BESSEL'S FUNCTIONS

If λ_j and λ_k are roots of the equation $J_n(\lambda a) = 0$ then

$$\int_0^a x\, J_n(\lambda_j x) J_n(\lambda_k x)\, dx = \begin{cases} 0, & \text{if } j \neq k \\ \dfrac{a^2}{2} J^2{}_{n+1}(\lambda_j a), & \text{if } j = k \end{cases}$$

Let $I \neq j$. This means λ_j and λ_k are different roots of equation $J_n(\lambda a) = 0$.

$$\therefore \qquad J_n(\lambda_j a) = 0 \text{ and } J_n(\lambda_k a) = 0 \tag{14.24}$$

Define $$u(x) = J_n(\lambda_j x) \text{ and } v(x) = J_n(\lambda_k x) \tag{14.25}$$

Then u and v are Bessel's functions satisfying the modified Bessel's Eq. (14.20). Therefore,

$$x^2 u'' + xu' + (\lambda_j^2 x^2 - n^2)\, u = 0 \tag{14.26}$$

and $$x^2 v'' + xv' + (\lambda_k^2 x^2 - n^2)\, v = 0 \tag{14.27}$$

Multiplying Eq. (14.26) by v and Eq. (14.27) by u and then subtracting, we get

$$x^2(vu'' - uv'') + x(vu' - uv') + x^2(\lambda^2{}_j - \lambda^2{}_k)\, uv = 0$$

$$\Rightarrow \qquad x(vu'' - uv'') + (vu' - uv') = x(\lambda^2{}_k - \lambda^2{}_j)\, uv$$

$$\Rightarrow \qquad x\frac{d}{dx}(vu' - uv') + (vu' - uv') = x(\lambda^2{}_k - \lambda^2{}_j)\, uv$$

$$\Rightarrow \qquad \frac{d}{dx}[x(vu' - uv')] = x(\lambda^2{}_k - \lambda^2{}_j)\, uv$$

Integrating both sides w.r.t. x over $[0, a]$, we get

$$(\lambda^2{}_k - \lambda^2{}_j)\int_0^a xuv\, dx = [x(vu' - uv')]_0^a \tag{14.28}$$

Using Eqs. (14.25) and (14.28), we get

$$(\lambda^2{}_k - \lambda^2{}_j)\int_0^a xJ_n(\lambda_j x)\, J_n(\lambda_k x)\, dx$$

$$= \{x\,[J_n(\lambda_k x)\, J_n'(\lambda_j x) - J_n(\lambda_j x)\, J_n'(\lambda_k x)]\}_0^a$$

$$= a[J_n(\lambda_k a)\, J_n'(\lambda_j a) - J_n(\lambda_j a)\, J_n'(\lambda_k a)]$$

$$= 0 \qquad [\because \text{ of Eq. (14.24)}]$$

Since $\lambda_j \neq \lambda_k$,

$$\int_0^a xJ_n(\lambda_j x)\, J_n(\lambda_k x)\, dx = 0 \qquad j \neq k \tag{14.29}$$

Now, consider $j = k$. Here roots are equal.
Multiply Eq. (14.26) by $2u'$. Then

$$2x^2u''u' + 2xu'^2 + 2(\lambda^2{}_jx^2 - n^2)\,uu' = 0$$

$$\Rightarrow \quad \frac{d}{dx}[x^2u'^2 - n^2u^2 + \lambda^2{}_jx^2u^2] - 2\lambda^2{}_jxu^2 = 0$$

$$\Rightarrow \quad 2\lambda^2{}_jxu^2 = \frac{d}{dx}[x^2u'^2 - n^2u^2 + \lambda^2{}_jx^2u^2]$$

Integrating both sides w.r.t. x over $[0, a]$, we have

$$2\lambda^2{}_j\int_0^a xu^2dx = (x^2u'^2 - n^2u^2 + \lambda^2{}_jx^2u^2)_0^a$$

Using $J_n(0) = 0$, Eqs. (14.24) and (14.25), we get

$$2\lambda^2{}_j\int_0^a xJ^2{}_n(\lambda_jx)\,dx = a^2[\{J_n'(\lambda_jx)\}^2]_0^a \tag{14.30}$$

From recurrence relation (R4), we have

$$J_n'(x) = \frac{n}{x}J_n(x) - J_{n+1}(x)$$

$$\therefore \quad \frac{d}{d(\lambda_jx)}J_n(\lambda_jx) = \frac{n}{\lambda_jx}J_n(\lambda_jx) - J_{n+1}(\lambda_jx)$$

$$\Rightarrow \quad \frac{1}{\lambda_j}\frac{d}{dx}J_n(\lambda_jx) = \frac{n}{\lambda_jx}J_n(\lambda_jx) - J_{n+1}(\lambda_jx)$$

$$\Rightarrow \quad J_n'(\lambda_jx) = \frac{n}{x}J_n(\lambda_jx) - \lambda_jJ_{n+1}(\lambda_jx)$$

$$\Rightarrow \quad \{[J_n'(\lambda_jx)]^2\}_{x=a} = \left\{\left[\frac{n}{x}J_n(\lambda_jx) - \lambda_jJ_{n+1}(\lambda_jx)\right]^2\right\}_{x=a}$$

$$= \lambda_j^2J_{n+1}^2\,[\lambda_j(a)]$$

Using this value in Eq. (14.30), we get

$$\int_0^a xJ^2{}_n(\lambda_jx)\,dx = \frac{a^2}{2}J^2{}_{n+1}(\lambda_ja).$$

14.8 EXAMPLES

EXAMPLE 14.1 Prove that

(i) $J_{-1/2}(x) = \sqrt{\dfrac{2}{\pi x}}\cos x$

(ii) $J_{1/2}(x) = \sqrt{\frac{2}{\pi x}} \sin x$

(iii) $[J_{1/2}(x)]^2 + [J_{-1/2}(x)]^2 = \frac{2}{\pi x}$

Proof: By definition of $J_n(x)$

$$J_n(x) = \frac{x^n}{2^n \overline{)(n+1)}} \left[1 - \frac{x^2}{2 \cdot 2(n+1)} + \frac{x^4}{2 \cdot 4 \cdot 2^2 (n+1)(n+2)} + \cdots\right]$$

Put $n = -1/2$ and simplify, we get

$$J_{-1/2}(x) = \frac{x^{-1/2}}{2^{-1/2} \overline{)\left(\frac{1}{2}\right)}} \left[1 - \frac{x^2}{1 \cdot 2} + \frac{x^4}{1 \cdot 2 \cdot 3 \cdot 4} - \cdots\right]$$

$$= \sqrt{\frac{2}{\pi x}} \left[1 - \frac{x^2}{2!} + \frac{x^4}{4!} - \cdots\right] \qquad \left(\text{note that } \overline{)\left(\frac{1}{2}\right)} = \sqrt{\pi}\right)$$

$$= \sqrt{\frac{2}{\pi x}} \cos x$$

Next put $n = \frac{1}{2}$. Then

$$J_{1/2}(x) = \frac{x^{1/2}}{2^{1/2} \overline{)\left(1 + \frac{1}{2}\right)}} \left[1 - \frac{x^2}{1 \cdot 2 \cdot 3} + \frac{x^4}{1 \cdot 2 \cdot 3 \cdot 4 \cdot 5} - \cdots\right]$$

$$= \left(\frac{x}{2}\right)^{1/2} \cdot \frac{1}{\overline{)\frac{3}{2}}} \cdot \frac{1}{x} \left[x - \frac{x^3}{1 \cdot 2 \cdot 3} + \frac{x^5}{1 \cdot 2 \cdot 3 \cdot 4 \cdot 5} - \cdots\right]$$

$$= \left(\frac{x}{2}\right)^{1/2} \cdot \frac{1}{\frac{1}{2} \overline{)\frac{1}{2}}} \cdot \frac{1}{x} \left[x - \frac{x^3}{3!} + \frac{x^5}{5!} - \cdots\right]$$

$$= \left(\frac{2}{\pi x}\right)^{1/2} \left[x - \frac{x^3}{3!} + \frac{x^5}{5!} - \cdots\right]$$

$$= \left(\frac{2}{\pi x}\right)^{1/2} \sin x$$

Squaring and adding, above two, we get

$$[J_{1/2}(x)]^2 + [J_{-1/2}(x)]^2 = \frac{2}{\pi x}(\sin^2 x + \cos^2 x) = \frac{2}{\pi x}.$$

EXAMPLE 14.2 Prove that if $n > m > -1$, then

$$J_n(x) = \frac{2\left(\frac{x}{2}\right)^{n-m}}{\Gamma(n-m)} \int_0^1 (1-t^2)^{n-m-1} t^{m+1} J_m(xt)\,dt.$$

Proof: Let

$$I = \int_0^1 (1-t^2)^{n-m-1} t^{m+1} J_m(xt)\,dt \qquad \text{(i)}$$

Using definition of $J_m(xt)$, we get

$$I = \int_0^1 (1-t^2)^{n-m-1} t^{m+1} \sum_{r=0}^{\infty} (-1)^r \frac{1}{r!\,\Gamma(m+r+1)} \left(\frac{xt}{2}\right)^{2r+m} dt$$

$$= \sum_{r=0}^{\infty} \frac{(-1)^r \left(\frac{x}{2}\right)^{2r+m}}{r!\,\Gamma(m+r+1)} \int_0^1 (1-t^2)^{n-m-1} t^{2m+2r+1}\,dt$$

$$= \sum_{r=0}^{\infty} \frac{(-1)^r \left(\frac{x}{2}\right)^{2r+m}}{r!\,\Gamma(m+r+1)} \int_0^1 (1-t^2)^{n-m-1} (t^2)^{m+r}\, t\,dt$$

(Put $t^2 = z$ so $2t\,dt = dz$)

$$= \sum_{r=0}^{\infty} \frac{(-1)^r \left(\frac{x}{2}\right)^{2r+m}}{r!\,\Gamma(m+r+1)} \cdot \frac{1}{2} \int_0^1 (1-z)^{n-m-1} z^{m+r+1-1}\,dz$$

$$= \sum_{r=0}^{\infty} \frac{(-1)^r \left(\frac{x}{2}\right)^{2r+m}}{r!\,\Gamma(m+r+1)} \cdot \frac{1}{2} \frac{\Gamma(n-m)\,\Gamma(m+r+1)}{\Gamma(n+r+1)}$$

(provided $n - m > 0$ and $m + r + 1 > 0$, i.e. $n > m > -1$)

$$= \frac{\Gamma(n-m)}{2} \left(\frac{x}{2}\right)^{m-n} \sum_{r=0}^{\infty} (-1)^r \frac{1}{r!\,\Gamma(n+r+1)} \left(\frac{x}{2}\right)^{n+2r}$$

$$= \frac{\Gamma(n-m)}{2} \left(\frac{x}{2}\right)^{m-n} J_n(x)$$

$$\therefore \qquad J_n(x) = \frac{2\left(\frac{x}{2}\right)^{m-n}}{\overline{)n-m}} I$$

$$\therefore \qquad J_n(x) = \frac{2\left(\frac{x}{2}\right)^{m-n}}{\overline{)n-m}} \int_0^1 (1-t^2)^{n-m-1} t^{m+1} J_m(xt)\, dt.$$

EXAMPLE 14.3 Prove that

(i) $J_n J'_{-n} - J'_n J_{-n} = -\dfrac{2 \sin n\pi}{x\pi}$

(ii) $\dfrac{d}{dx}\left(\dfrac{J_{-n}}{J_n}\right) = -\dfrac{2 \sin n\pi}{\pi x J_n^{\,2}}$.

Proof:

(i) We know that J_n and J_{-n} are solutions of Bessel's equation

$$y'' + \left(\frac{1}{x}\right) y' + \left(1 - \frac{n^2}{x^2}\right) y = 0$$

$$\therefore \qquad J''_n + \left(\frac{1}{x}\right) J'_n + \left(1 - \frac{n^2}{x^2}\right) J_n = 0 \tag{14.31}$$

and

$$J''_{-n} + \left(\frac{1}{x}\right) J'_{-n} + \left(1 - \frac{n^2}{x^2}\right) J_{-n} = 0 \tag{14.32}$$

Multiply Eq. (14.31) by J_{-n} and Eq. (14.32) by J_n and subtract. The result is

$$J''_n J_{-n} - J''_{-n} J_n + \left(\frac{1}{x}\right)(J'_n J_{-n} - J'_{-n} J_n) = 0 \tag{14.33}$$

Let

$$J'_n J_{-n} - J'_{-n} J_n = v \tag{14.34}$$

Differenting Eq. (14.34) w.r.t. x, gives

$$J_n'' J_{-n} + J_n' J_{-n}' - (J_{-n}'' J_n + J_{-n}' J_n') = v'$$

$\Rightarrow$

$$J_n'' J_{-n} - J_{-n}'' J_n = v'$$

Then Eq. (14.33) becomes

$$v' + \frac{1}{x}\, dv = 0$$

Integrating, $\qquad \log v + \log x = \log c \;$ or $\; v = \dfrac{c}{k}$

$$\therefore \qquad J_n{}'J_{-n} - J'_{-n}J_n = \frac{c}{x} \tag{14.35}$$

Now,
$$J_n(x) = \frac{1}{2^n\Gamma(n+1)}\left[x^n - \frac{x^{n+2}}{4(n+1)} + \frac{x^{n+4}}{4\cdot 8(n+1)(n+2)} - \cdots\right]$$

and
$$J_{-n}(x) = \frac{1}{2^{-n}\Gamma(-n+1)}\left[x^{-n} - \frac{x^{-n+2}}{4(-n+1)} + \frac{x^{-n+4}}{4\cdot 8(-n+1)(-n+2)} - \cdots\right]$$

Using the above values of J_n and J_{-n}, Eq. (14.35) gives

$$\frac{1}{2^n\Gamma(n+1)}\left[nx^{n-1} - \frac{(n+2)x^{n+1}}{4(n+1)} + \frac{(n+4)x^{n+3}}{4\cdot 8(n+1)(n+2)} - \cdots\right]$$

$$\times\frac{1}{2^{-n}\Gamma(-n+1)}\left[x^{-n} - \frac{x^{-n+2}}{4(-n+1)} + \frac{x^{-n+4}}{4\cdot 8(-n+1)(-n+2)} - \cdots\right]$$

$$-\frac{1}{2^{-n}\Gamma(-n+1)}\left[-nx^{-n-1} - \frac{(-n+2)x^{-n+1}}{4(-n+1)} + \frac{(-n+4)x^{-n+3}}{4\cdot 8(-n+1)(-n+2)} - \cdots\right]$$

$$\times\frac{1}{2^n\Gamma(n+1)}\left[x^n - \frac{x^{n+2}}{4(n+1)} + \frac{x^{n+4}}{4\cdot 8(n+1)(n+2)} - \cdots\right] = \frac{c}{x}$$

Now, comparing the coefficients of $1/x$ from both sides, we get

$$c = \frac{n}{2^n\Gamma(n+1)\,2^{-n}\Gamma(-n+1)} + \frac{n}{2^{-n}\Gamma(-n+1)\,2^n\Gamma(n+1)}$$

$$\Rightarrow \qquad c = \frac{2n}{n\Gamma(n)\,\Gamma(1-n)} \qquad [\because\ \Gamma(n+1) = n\Gamma(n)]$$

$$= \frac{2}{\pi/\sin n\pi} = \frac{2\sin n\pi}{\pi} \qquad \left[\because\ \Gamma(n)\,\Gamma(1-n) = \frac{\pi}{\sin n\pi}\right]$$

Hence,
$$J_nJ'_{-n} - J'_nJ_{-n} = -\frac{2\sin n\pi}{\pi x} \tag{14.36}$$

(ii) Dividing both sides of Eq. (14.36) by J_n^2 gives

$$\frac{J_nJ'_{-n} - J'_nJ_{-n}}{J_n^2} = -\frac{2\sin n\pi}{\pi x J_n^2}$$

$$\Rightarrow \qquad \frac{d}{dx}\left(\frac{J_{-n}}{J_n}\right) = -\frac{2\sin n\pi}{\pi x J_n^2}.$$

EXAMPLE 14.4 Prove that

$$\int_0^x t[J_n(t)]^2 dt = \frac{1}{2}x^2[J_n^2(x) - J_{n-1}(x)J_{n+1}(x)].$$

Proof:

$$\frac{d}{dt} = \left\{\frac{t^2}{2}[J_n^2(t) - J_{n-1}(t)J_{n+1}(t)]\right\}$$

$$= t[J_n^2(t) - J_{n-1}(t)J_{n+1}(t)] + \frac{1}{2}t^2\left[2J_n(t)\,J_n'(t) - J_{n-1}'(t)\,J_{n+1}(t) - J_{n-1}(t)\,J_{n+1}'(t)\right]$$

$$= t[J_n^2(t) - J_{n-1}(t)J_{n+1}(t)] + \frac{1}{2}t^2\left[2J_n(t)\cdot\frac{1}{2}[J_{n-1}(t) - J_{n+1}(t)]\right.$$

$$\left. - J_{n+1}(t)\left(\frac{n-1}{t}J_{n-1}(t) - J_n(t)\right) - J_{n-1}(t)\left(J_n(t) - \frac{n+1}{t}J_{n+1}(t)\right)\right]$$

(Using R3, R4, R5)

$$= tJ_n^{\,2}(t).$$

$$\therefore \qquad tJ_n^2(t) = \frac{d}{dt}\left\{\frac{t^2}{2}[J_n^2(t) - J_{n-1}(t)J_{n+1}(t)]\right\}$$

Integrating above equation w.r.t. t from 0 to x gives

$$\therefore \qquad \int_0^x tJ_n^2(t)\,dt = \left[\frac{t^2}{2}[J_n^2(t) - J_{n-1}(t)J_{n+1}(t)]\right]_0^x$$

$$= \frac{x^2}{2}[J_n^{\,2}(x) - J_{n-1}(x)J_{n+1}(x)] \qquad [\because\ J_n(0) = J_{n+1}(0) = J_{n-1}(0) = 0]$$

EXAMPLE 14.5 Prove that

(i) $J_{-3/2}(x) = \sqrt{\dfrac{2}{\pi x}}\left(-\dfrac{\cos x}{x} - \sin x\right)$

(ii) $J_{3/2}(x) = \sqrt{\dfrac{2}{\pi x}}\left(\dfrac{\sin x}{x} - \cos x\right).$

Proof: In Example 14.1, we have proved that

$$J_{-1/2}(x) = \sqrt{\frac{2}{\pi x}}\cos x$$

and

$$J_{1/2}(x) = \sqrt{\frac{2}{\pi x}}\sin x$$

In recurrence relation R6, put $n = -1/2$

$$\therefore \quad J_{-3/2}(x) + J_{1/2}(x) = -\left(\frac{2}{2x}\right) J_{-1/2}(x)$$

$$\Rightarrow \quad J_{-3/2}(x) = J_{1/2}(x) - \frac{1}{x} J_{-1/2}(x)$$

$$= \sqrt{\frac{2}{\pi x}} \sin x - \frac{1}{x}\sqrt{\frac{2}{\pi x}} \cos x$$

$$= \sqrt{\frac{2}{\pi x}}\left(-\frac{\cos x}{x} - \sin x\right)$$

Again, in recurrence relation R6, put $n = 1/2$.

$$\therefore \quad J_{-1/2}(x) + J_{3/2}(x) = \left(\frac{2}{2x}\right) J_{1/2}(x)$$

$$\Rightarrow \quad J_{3/2}(x) = \frac{1}{x} J_{1/2}(x) - J_{-1/2}(x)$$

$$= \frac{1}{x}\sqrt{\frac{2}{\pi x}} \sin x - \sqrt{\frac{2}{\pi x}} \cos x$$

$$= \sqrt{\frac{2}{\pi x}}\left(\frac{\sin x}{x} - \cos x\right).$$

EXAMPLE 14.6 Prove that

(i) $\frac{1}{2} xJ_n = (n+1) J_{n+1} - (n+3) J_{n+3} + (n+5) J_{n+5} - \cdots$

(ii) $J_{n-1} = \frac{2}{x}[nJ_n - (n+2) J_{n+2} + (n+4) J_{n+4} - \cdots].$

Proof: (i) Recurrence relation R6 is

$$2nJ_n = x(J_{n-1} + J_{n+1})$$

Replacing n by $n + 1$ in the above relation, we get

$$2(n+1) J_{n+1} = x(J_n + J_{n+2})$$

$$\Rightarrow \quad \frac{1}{2} xJ_n = (n+1) J_{n+1} - \frac{1}{2} xJ_{n+2} \tag{i}$$

Replacing n by $n + 2$ in Eq. (i) gives

$$\frac{1}{2} xJ_{n+2} = (n+3) J_{n+3} - \frac{1}{2} xJ_{n+4} \tag{ii}$$

Putting the values of $\frac{1}{2}xJ_{n+2}$ from Eq. (i) in Eq. (ii) gives

$$\frac{1}{2}xJ_n = (n+1)J_{n+1} - (n+3)J_{n+3} + \frac{1}{2}xJ_{n+4} \quad \text{(iii)}$$

Replacing n by $n + 4$ in Eq. (i) gives

$$\frac{1}{2}xJ_{n+4} = (n+5)J_{n+5} - \frac{1}{2}xJ_{n+6} \quad \text{(iv)}$$

Putting the value of $\frac{1}{2}xJ_{n+4}$ from Eq. (iv) in Eq. (iii) gives

$$\frac{1}{2}xJ_n = (n+1)J_{n+1} - (n+3)J_{n+3} + (n+5)J_{n+5} - \frac{1}{2}xJ_{n+6}$$

Continuing this way and noting that $J_n \to 0$ as $n \to \infty$, we get

$$\frac{1}{2}xJ_n = (n+1)J_{n+1} - (n+3)J_{n+3} + (n+5)J_{n+5} - \cdots \quad \text{(v)}$$

(ii) Put $n = -1$ in Eq. (v) and multiply both sides by $\frac{2}{x}$ gives

$$J_{n-1} = \frac{2}{x}[nJ_n - (n+2)J_{n+2} + (n+4)J_{n+4} - \cdots].$$

EXAMPLE 14.7 Prove that

$$\frac{d}{dx}(J_n^2 + J_{n+1}^2) = 2\left(\frac{n}{x}J_n^2 - \frac{n+1}{2}J_{n+1}^2\right).$$

Proof: The recurrence relation R3 and R4 are

$$J_n' = -\frac{n}{x}J_n + J_{n-1} \quad \text{(i)}$$

and

$$J_n' = \left(\frac{n}{x}\right)J_n - J_{n+1} \quad \text{(ii)}$$

Replacing n by $n + 1$ in Eq. (i), we get

$$J_{n+1}' = -\frac{n+1}{x}J_{n+1} + J_n \quad \text{(iii)}$$

Now,

$$\frac{d}{dx}(J_n^2 + J_{n+1}^2) = 2J_nJ_n' + 2J_{n+1}J_{n+1}'$$

$$= 2J_n\left(\frac{n}{x} - J_{n+1}\right) + 2J_{n+1}\left(-\frac{n+1}{x}J_{n+1} + J_n\right)$$

$$= 2\left(\frac{n}{x}J_n^2 - \frac{n+1}{x}J_{n+1}^2\right).$$

EXAMPLE 14.8 Prove that

(i) $J_0^2 + 2(J_1^2 + J_2^2 + \cdots) = 1$

(ii) $|J_0(x)| \le 1$

(iii) $|J_n(x)| \le 2^{-1/2}, n \ge 1.$

Proof: (i) From Example 14.7, we have

$$\frac{d}{dx}(J_n^2 + J_{n+1}^2) = 2\left(\frac{n}{x}J_n^2 - \frac{n+1}{x}J^2{}_{n+1}\right) \tag{i}$$

Putting n = 0, 1, 2, ..., we get

$$\frac{d}{dx}(J_0^2 + J_1^2) = 2\left(0 - \frac{1}{x}J_1{}^2\right)$$

$$\frac{d}{dx}(J_1^2 + J_2^2) = 2\left(\frac{1}{x}J_1^2 - \frac{2}{x}J_2^2\right)$$

$$\frac{d}{dx}(J_2^2 + J_3^2) = 2\left(\frac{2}{x}J_2{}^2 - \frac{3}{x}J_3{}^2\right)$$

$$\vdots \qquad\qquad \vdots$$

Adding these column wise and noting that $J_n \to 0$ as $n \to \infty$, we get

$$\frac{d}{dx}[J_0^2 + 2(J_1^2 + J_2^2 + \cdots)] = 0$$

On integration we have

$$J_0^2(x) + 2[J_1^2(x) + J_2^2(x) + \cdots] = c \tag{ii}$$

Put x = 0 and using $J_0(0)$ = 1, $J_n(0)$ = 0 for $n \ge 1$, we get c = 1. Hence, Eq. (ii) becomes

$$J_0^2 + 2\,[J_1^2 + J_2^2 + \cdots] = 1 \tag{iii}$$

(ii) From Eq. (iii), we have

$$J_0^2 = 1 - 2[J_1^2 + J_2^2 + \cdots + J_n^2] \tag{iv}$$

Since J_1^2, J_2^2, ... are all positive or zero, Eq. (iv) gives

$$J_0^2 \le 1, \text{ i.e. } |J_0(x)| \le 1$$

(iii) Solving Eq. (iv) for J_n^2, we have

$$J_n^2 = \frac{1}{2}(1 - J_0^2) - [J_1{}^2 + J_2{}^2 + \cdots + J_{n+1}^2 + J_{n+2}^2 + \cdots] \tag{v}$$

Since J_0^2, J_1^2, J_2^2, ... are all positive or zero, Eq. (v) gives

$$J_n^2 \le \frac{1}{2} \Rightarrow |J_n(x)| \le 2^{-1/2}, n \ge 1.$$

EXAMPLE 14.9 Prove that

(i) $\dfrac{d}{dx}[xJ_n(x)J_{n+1}(x)] = J_n^2(x) - J_{n+1}^2(x)$

(ii) $x = 2J_0J_1 + 6J_1J_2 + \cdots + 2(2n+1)J_nJ_{n+1} + \cdots$

Proof: (i) $\dfrac{d}{dx}[xJ_nJ_{n+1}] = J_nJ_{n+1} + (xJ_n'J_{n+1} + J_nJ_{n+1}')$

$$= J_nJ_{n+1} + J_{n+1}(xJ_n') + J_n(xJ_{n+1}') \qquad \text{(i)}$$

From recurrence relation R3 and R4, we have

$$xJ_n' = nJ_n - xJ_{n+1} \qquad \text{(ii)}$$

and

$$xJ_n' = -nJ_n + xJ_{n-1} \qquad \text{(iii)}$$

Replacing n by $n + 1$ in Eq. (iii), we have

$$xJ'_{n+1} = (-n+1)\,J_{n+1} + xJ_n \qquad \text{(iv)}$$

Putting the values xJ_n' and xJ'_{n+1} from Eqs. (ii) and (iv) in Eq. (i), we get

$$\frac{d}{dx}[xJ_nJ_{n+1}] = J_nJ_{n+1} + J_{n+1}(nJ_n - xJ_{n+1}) + J_n[-(n+1)J_{n+1} + xJ_n]$$

$$= x(J_n^{\ 2} - J^2_{\ n+1})$$

(ii) From Eq. (i), we have

$$\frac{d}{dx}(xJ_nJ_{n+1}) = x(J_n^{\ 2} - J^2_{\ n+1})$$

Putting $n = 0, 1, 2, \ldots$, we get

$$\frac{d}{dx}(xJ_0J_1) = x(J_0^2 - J_1^2) \qquad \text{(i)}$$

$$\frac{d}{dx}(xJ_1J_2) = x(J_1^2 - J_2^2) \qquad \text{(ii)}$$

$$\frac{d}{dx}(xJ_2J_3) = x(J_2^2 - J_3^2) \qquad \text{(iii)}$$

$$\vdots \qquad \vdots$$

Multiply Eq. (i) by 1, (ii) by 3, (iii) by 5, … and add them. Then

$$\frac{d}{dx}[x(J_0J_1 + 3J_1J_2 + 5J_2J_3 + \cdots)]$$

$$= x[(J_0^2 - J_1^2) + 3(J_1^2 - J_2^2) + 5(J_2^2 - J_3^2) + \cdots]$$

$$= x[J_0^2 + 2(J_1^2 + J_2^2 + \cdots)]$$

$$= x \times 1 = x \qquad \text{(by Example 14.8)}$$

Integrating both sides w.r.t. x, we have

$$x(J_0J_1 + 3J_1J_2 + 5J_2J_3 + \cdots) = \frac{x^2}{2} + c \qquad \text{(iv)}$$

For $x = 0$, $c = 0$. Hence,

$$2J_0J_1 + 6J_1J_2 + \cdots = x^2$$

or

$$x = 2J_0J_1 + 6J_1J_2 + \cdots$$

EXAMPLE 14.10 Show that $J_n(x) = 0$ has no repeated roots at $x = 0$.

Proof: If possible, suppose $J_n(x) = 0$ has repeated roots; then at least two roots must be equal (say α). Thus, α is a double root of $J_n(x) = 0$. Then from the theory of equations, we have

$$J_n(\alpha) = 0 \text{ and } J_n'(\alpha) = 0 \qquad \text{(i)}$$

Recurrence relations R3 and R4 are

$$J_{n-1}(x) = \frac{n}{x}J_n(x) + J_n'(x) \qquad \text{(ii)}$$

and

$$J_{n+1}(x) = \frac{n}{x}J_n(x) - J_n'(x) \qquad \text{(iii)}$$

Replacing x by α in Eqs. (ii) and (iii) and using Eq. (i), we have

$$J_{n+1}(\alpha) = 0 \text{ and } J_{n-1}(\alpha) = 0 \text{ except when } x = 0.$$

Since two different power series have district sum functions, so $J_{n+1}(\alpha) = 0$ and $J_{n-1}(\alpha) = 0$ is absurd. Hence, $J_n(x) = 0$ has no repeated roots except at $x = 0$.

EXAMPLE 14.11 If $n > -1$, show that $\int_0^x x^{n+1}J_n(x)\,dx = x^{n+1}J_{n+1}(x)$.

Proof: From recurrence relation R1, we have

$$\frac{d}{dx}[x^n J_n(x)] = x^n J_{n-1}(x)$$

Put $n = n + 1$

$$\therefore \quad \frac{d}{dx}[x^{n+1}J_{n+1}(x)] = x^{n+1}J_n(x)$$

Integrate both sides w.r.t. x between the limits 0 and x.

$$\therefore \quad [x^{n+1}J_{n+1}(x)]_0^x = \int_0^x x^{n+1}J_n(x)\,dx$$

$$\Rightarrow \quad \int_0^x x^{n+1}J_n(x)\,dx = x^{n+1}J_{n+1}(x).$$

EXAMPLE 14.12 Show that

$$\text{(i)}\quad \int_0^x x^{-n} J_{n+1}(x)\,dx = \frac{1}{2^n\,\Gamma(n+1)} - x^{-n} J_n,\ n > 1$$

$$\text{(ii)}\quad \int_0^\infty x^{-n} J_{n+1}(x)\,dx = \frac{1}{2^n\,\Gamma(n+1)},\ n > -\frac{1}{2}$$

Proof: (i) From recurrence relation R2, we have

$$\frac{d}{dx}[x^{-n} J_n(x)] = -\,x^{-n} J_{n+1}(x) \qquad \text{(i)}$$

Integrating both sides w.r.t. x between the limits 0 to x. Then

$$[x^{-n} J_n(x)]_0^x = -\int_0^x x^{-n} J_{n+1}(x)\,dx$$

$$\therefore \qquad x^{-n} J_n(x) - \lim_{x\to 0} x^{-n} J_n(x) = -\int_0^x x^{-n} J_{n+1}(x)\,dx \qquad \text{(ii)}$$

But

$$\lim_{x\to 0} x^{-n} J_n(x) = \lim_{x\to 0} \frac{1}{x^n}\cdot\frac{x^n}{2^n\,\Gamma(n+1)}\left[1 - \frac{x^2}{2\cdot 2(n+1)} + \cdots\right]$$

$$= \frac{1}{2^n\,\Gamma(n+1)}$$

$\therefore$ Equation (ii) may be written as

$$\int_0^x x^{-n} J_{n+1}(x)\,dx = \frac{1}{2^n\,\Gamma(n+1)} - x^{-n} J_n(x)$$

(ii) Integrating Eq. (i) w.r.t. x, from 0 to ∞ gives

$$[x^{-n} J_n(x)]_0^\infty = -\int_0^\infty x^{-n} J_{n+1}(x)\,dx$$

$$\therefore \qquad \lim_{x\to\infty} \frac{J_n(x)}{x^n} - \lim_{n\to 0} \frac{J_n(x)}{x^n} = -\int_0^\infty x^{-n} J_{n+1}(x)\,dx \qquad \text{(iii)}$$

As in part (i),

$$\lim_{n\to\infty} \frac{J_{n+1}(x)}{x^n}\ \frac{1}{2^n\,\Gamma(n+1)} \qquad \text{(iv)}$$

We know that for large values of x, the $J_n(x)$'s approximate value is

$$J_n(x) = \sqrt{\left(\frac{2}{\pi x}\right)}\cos\left\{x - \left(n + \frac{1}{2}\right)\frac{\pi}{2}\right\},\ n > -\frac{1}{2}$$

$$\therefore \qquad \lim_{x\to\infty} \frac{J_n(x)}{x} = 0$$

Using Eqs. (iv) and (v) in Eq. (iii), we get

$$\int_0^\infty x^{-n} J_{n+1}(x)\, dx = \frac{1}{2^n \Gamma(n+1)}, \; n > -\frac{1}{2}.$$

EXAMPLE 14.13 Evaluate $\int J_3(x)\, dx$ in terms of J_0 and J_1.

Solution From R2, we have

$$-x^n J_{n+1} = -\frac{d}{dx}[x^{-n} J_n(x)]$$

Integration gives

$$\int x^{-n} J_{n+1} dx = -x^{-n} J_n \tag{i}$$

Now,

$$\begin{aligned}\int J_3(x)\, dx &= \int (x^2)(x^{-2} J_3)\, dx \\ &= x^2(-x^{-2} J_2) - \int 2x(-x^{-2} J_2)\, dx \\ &= -J_2 + 2\int x^{-1} J_2\, dx \\ &= -J_2 - 2x^{-1} J_1 + c \end{aligned} \tag{ii}$$

From recurrence relation R6, we have

$$\frac{2n}{x} J_n = J_{n-1} + J_{n+1}$$

Put $n = 1$. Then

$$J_2 = \frac{2J_1}{x} - J_0$$

$\therefore$ Equation (ii) becomes

$$\begin{aligned}\int J_3(x)\, dx &= -\frac{2J_1}{x} + J_0 - \frac{2J_1}{x} + c \\ &= J_0 - \frac{4J_1}{x} + c.\end{aligned}$$

EXAMPLE 14.14 Using generating function, prove the following:

(i) $\cos(x \sin\phi) = J_0 + 2\cos 2\phi J_2 + 2\cos 4\phi J_4 + \cdots$

(ii) $\sin(x \sin\phi) = 2\sin\phi J_1 + 2\sin 3\phi J_3 + \cdots$

(iii) $\cos(x\cos\phi) = J_0 - 2\cos 2\phi J_2 + 2\cos 4\phi J_4 - \cdots$

(iv) $\sin(x\cos\phi) = 2\cos\phi J_1 - 2\cos 3\phi J_3 + 2\cos 5\phi J_5 - \cdots$

(v) $\cos x = J_0 - 2J_2 + 2J_4 - \cdots$

(vi) $\sin x = 2J_1 - 2J_3 + 2J_5 - \cdots$

Proof: We know that

$$\exp\left\{\frac{x}{2}\left(z-\frac{1}{z}\right)\right\} = J_0 + (z - z^{-1})\, J_1 + (z^2 + z^{-2})\, J_2 + (z^3 - z^{-3})\, J_3 + \cdots \qquad \text{(i)}$$

Let $z = e^{i\phi}$. Then $z^n = e^{in\phi}$ and $z^{-n} = e^{-in\phi}$. Then Eq. (i) becomes

$$\exp\left\{\frac{x}{2}(e^{i\phi} - e^{-i\phi})\right\} = J_0 + (e^{i\phi} - e^{-i\phi})\, J_1 + (e^{2i\phi} - e^{-2i\phi})\, J_2 + (e^{3i\phi} - e^{-3i\phi})\, J_3 + \cdots \qquad \text{(ii)}$$

We know that

$$\cos n\phi = \frac{1}{2}(e^{in\phi} + e^{-in\phi}) \text{ and } \sin n\phi = \frac{1}{2i}(e^{in\phi} - e^{-in\phi})$$

Therefore, Eq. (ii) can be written as

$$\exp(ix \sin\phi) = J_0 + 2i \sin\phi J_1 + 2\cos 2\phi J_2 + 2i \sin 3\phi J_3 + \cdots$$

$$\Rightarrow \quad \cos(x\sin\phi) + i\sin(x\sin\phi) = (J_0 + 2\cos 2\phi J_2 + \cdots) + 2i(\sin\phi J_1 + \sin 3\phi J_3 + \cdots) \qquad \text{(iii)}$$

Comparing real and imaginary parts, proves (i) and (ii). Replacing ϕ by $\frac{\pi}{2} - \phi$ in (i) and (ii) proves (iii) and (iv). Substituting $\phi = 0$ in (iii) and (iv) gives (v) and (vi).

EXAMPLE 14.15 Show that

(i) $J_n(x) = \dfrac{1}{\pi}\displaystyle\int_0^{\pi} \cos(n\phi - x\sin\phi)\, d\phi$, when n is a positive integer.

(ii) $J_n(x) = \dfrac{1}{\pi}\displaystyle\int_0^{\pi} \cos(n\phi - x\sin\phi)\, d\phi$, when n is any integer.

(iii) $J_0(x) = \dfrac{1}{\pi}\displaystyle\int_0^{\pi} \cos(x\sin\phi)\, d\phi = \dfrac{1}{\pi}\displaystyle\int_0^{\pi} \cos(x\cos\phi)\, d\phi$

(iv) Deduce that $J_0(x) = 1 - \dfrac{x^2}{2^2} + \dfrac{x^4}{2^2 \cdot 4^2} + \cdots = \displaystyle\sum_{r=0}^{\infty} \frac{(-1)^r x^{2r}}{(2^r r!)^2}$.

Proof: (i) We know that

$$\int_0^{\pi} \cos m\phi \cos n\phi\, d\phi = \int_0^{\pi} \sin m\phi \sin n\phi\, d\phi = \begin{cases} \pi/2, & m = n \\ 0, & m \neq n \end{cases} \qquad \text{(i)}$$

$$\cos(x\sin\phi) = J_0 + 2J_2 \cos 2\phi + 2J_4 \cos 4\phi + \cdots \qquad \text{(ii)}$$

and

$$\sin(x\sin\phi) = 2J_1 \sin\phi + 2J_3 \sin 3\phi + 2J_5 \sin 5\phi + \cdots \qquad \text{(iii)}$$

Multiply both sides of Eq. (ii) by $\cos n\phi$ and then integrate w.r.t. ϕ from 0 to π and using Eq. (i), we get

$$\int_0^{\pi} \cos(x\sin\phi)\cos n\phi\, d\phi = \begin{cases} 0, & n \text{ is odd} \\ \pi J_n, & n \text{ is even} \end{cases} \qquad \text{(iv)}$$

Similarly, $$\int_0^{\pi} \sin(x\sin\phi)\sin n\phi\, d\phi = \begin{cases} \pi J_n, & n \text{ is odd} \\ 0, & n \text{ is even} \end{cases} \qquad \text{(v)}$$

Let n be odd. Adding Eq. (iv) and (v) gives

$$\int_0^{\pi} [\cos(x\sin\phi)\cos n\phi + \sin(x\sin\phi)\sin n\phi]\, d\phi = \pi J_n$$

$$\Rightarrow \quad \int_0^{\pi} \cos(n\phi - x\sin\phi)\, d\phi = \pi J_n$$

$$\Rightarrow \quad J_n(x) = \frac{1}{\pi}\int_0^{\pi} \cos(n\phi - x\sin\phi)\, d\phi \qquad \text{(vi)}$$

Next, let n be even. Verify that you get Eq. (vi).

(ii) Let n be any integer. For positive integer we have proved in (i) that

$$J_n(x) = \frac{1}{\pi}\int_0^{\pi} \cos(n\phi - x\sin\phi)\, d\phi$$

Now, let n be a negative integer and $n = -m$ where m is positive integer. We must prove that

$$J_{-m}(x) = \frac{1}{\pi}\int_0^{\pi} \cos(-m\phi - x\sin\phi)\, d\phi$$

Let $\phi = \pi - \theta$ so that $d\phi = -d\theta$

$$\therefore \quad \frac{1}{\pi}\int_0^{\pi} \cos(-m\phi - x\sin\phi)\, d\phi$$

$$= \frac{1}{\pi}\int_{\pi}^{0} \cos[-m(\pi-\theta) - x\sin(\pi-\theta)]\,(-d\theta)$$

$$= \frac{1}{\pi}\int_0^{\pi} \cos[(m\theta - x\sin\theta) - m\pi]\, d\theta$$

$$= \frac{1}{\pi}\int_0^{\pi} [\cos(m\theta - x\sin\theta)\cos m\pi + \sin(m\theta - x\sin\theta)\sin m\pi]\, d\theta$$

$$= \frac{1}{\pi}\int_0^{\pi} (-1)^m \cos(m\theta - x\sin\theta)\, d\theta \qquad [\because \cos m\pi = (-1)^m \ \sin m\pi = 0]$$

$$= (-1)^m \cdot \frac{1}{\pi}\int_0^{\pi} \cos(m\phi - x\sin\phi)\, d\phi$$

$$= (-1)^m J_m(x)$$

$$= J_{-m}(x)$$

Hence, $J_n(x) = \frac{1}{\pi}\int_0^{\pi} \cos(n\phi - x\sin\phi)\, d\phi$ for any integer n.

(iii) Integrating Eq. (ii) w.r.t. ϕ between the limits 0 to π and using

$$\int_0^{\pi} \cos p\phi d\phi = 0 \tag{vii}$$

if p is an even integer, we have

$$\int_0^{\pi} \cos(x\sin\phi)\, d\phi = J_0(x)\int_0^{\pi} d\phi + 0 + 0 + \cdots$$

$$= \pi J_0(x)$$

$$\therefore \qquad J_0(x) \frac{1}{\pi}\int_0^{\pi} \cos(x\sin\phi)\, d\phi \tag{viii}$$

Replacing ϕ by $\frac{\pi}{2} - \phi$, and simplifying, we get

$$\cos(x\cos\phi) = J_0 - 2J_2\cos 2\phi + 2J_4\cos 4\phi - \cdots$$

Integrating w.r.t. ϕ between the limits 0 to π and using Eq. (vii), we get

$$J_0(x) = \frac{1}{\pi}\int_0^{\pi} \cos(x\cos\phi)\, d\phi \tag{ix}$$

Eq. (iv) From Eq. (ii), we have

$$J_0(x) \frac{1}{\pi}\int_0^{\pi}\left(1 - \frac{x^2\cos^2\phi}{2!} + \frac{x^4\cos^4\phi}{4!} - \cdots\right) d\phi \tag{x}$$

But
$$\int_0^{\pi} \cos^{2n}\phi d\phi = \frac{1\cdot 3\cdot 5\cdots(2n-1)}{2\cdot 4\cdot 6\cdots(2n)}\pi \tag{xi}$$

Using Eq. (xi), Eq. (x) becomes

$$J_0(x) = \frac{1}{\pi}\left[\pi - \frac{x^2}{2!}\cdot\frac{1}{2}\pi + \frac{x^4}{4!}\cdot\frac{1\cdot 3}{2\cdot 4}\pi - \cdots\right]$$

$$= \left[1 - \frac{x^2}{2^2} + \frac{x^4}{2^2\cdot 4^2} - \cdots\right]$$

$$= \sum_{r=0}^{\infty} \frac{(-1)^r x^{2r}}{(2^r\cdot r!)^2}.$$

Note: (i) and (ii) are called **Bessel's integral**.

EXAMPLE 14.16 Prove that

(i) $$J_n(x) = \frac{1}{\sqrt{\pi}\,\Gamma\left(n+\frac{1}{2}\right)}\left(\frac{x}{2}\right)^n \int_0^{\pi} \cos(x \sin\phi) \cos^{2n}\phi\, d\phi$$

(ii) $$J_n(x) = \frac{1}{\sqrt{\pi}\,\Gamma\left(n+\frac{1}{2}\right)}\left(\frac{x}{2}\right)^n \int_0^{\pi} \cos(x \cos\phi) \sin^{2n}\phi\, d\phi.$$

Proof: (i) Let
$$I = \int_0^{\pi} \cos(x \sin\phi) \cos^{2n}\phi\, d\phi \tag{i}$$

$\therefore$
$$I = 2\int_0^{\pi/2} \cos(x \sin\phi) \cos^{2n}\phi\, d\phi$$

$$= 2\int_0^{\pi/2} \cos^{2n}\phi\left[1 - \frac{x^2 \sin^2\phi}{2!} + \frac{x^4 \sin^4\phi}{4!} - \cdots\right] d\phi \tag{ii}$$

We know that

$$\int_0^{\pi/2} \cos^p\phi \cos^q\phi\, d\phi = \frac{\Gamma\left(\frac{p+1}{2}\right)\Gamma\left(\frac{q+1}{2}\right)}{2\,\Gamma\left(\frac{p+q+2}{2}\right)} \tag{iii}$$

Using Eq. (iii), Eq. (ii) can be written as

$$I = \left[\frac{\Gamma\left(n+\frac{1}{2}\right)\Gamma\left(\frac{3}{2}\right)}{2\,\Gamma(n+1)} - \frac{x^2}{2!}\,\frac{\Gamma\left(n+\frac{1}{2}\right)\Gamma\frac{3}{2}}{2\,\Gamma(n+2)} + \frac{x^4}{4!}\,\frac{\Gamma\left(n+\frac{1}{2}\right)\Gamma\left(\frac{5}{2}\right)}{2\,\Gamma(n+3)} - \cdots\right]$$

$$= \Gamma\left(n+\frac{1}{2}\right)\left[\frac{\sqrt{\pi}}{\Gamma(n+1)} - \frac{x^2}{2!}\cdot\frac{\frac{1}{2}\sqrt{\pi}}{(n+1)\,\Gamma(n+1)} + \frac{x^4}{4!}\cdot\frac{\frac{3}{2}\cdot\frac{1}{2}\cdot\sqrt{\pi}}{(n+2)(n+1)\,\Gamma(n+1)} - \cdots\right]$$

$$= \frac{\Gamma\left(n+\frac{1}{2}\right)\sqrt{\pi}}{\Gamma(n+1)}\left[1 - \frac{x^2}{4(n+1)} + \frac{x^4}{4\cdot 8\,(n+1)(n+2)} - \cdots\right] \tag{iv}$$

We know that

$$J_n(x) = \frac{x^n}{2^n \Gamma(n+1)}\left[1 - \frac{x^2}{4(n+1)} + \frac{x^4}{4\cdot 8\,(n+1)\,(n+2)} - \cdots\right] \qquad \text{(v)}$$

Multiplying both sides of Eq. (iv) by $\dfrac{x^n}{2^n \Gamma\left(n+\frac{1}{2}\right)\sqrt{\pi}}$, we get

$$\frac{x^n}{2^n \Gamma\left(n+\frac{1}{2}\right)\sqrt{\pi}} I = \frac{x^n}{2^n \Gamma(n+1)}\left[1 - \frac{x^2}{4(n+1)} + \frac{x^4}{4\cdot 8(n+1)\,(n+2)} - \cdots\right] \qquad \text{(vi)}$$

Using Eqs. (i) and (v), Eq. (vi) becomes

$$J_n(x) = \frac{1}{\sqrt{\pi}\,\Gamma\left(n+\frac{1}{2}\right)}\left(\frac{x}{2}\right)^n \int_0^{\pi} \cos\,(x \sin \phi) \cos^{2n} \phi\, d\phi$$

(ii) Try yourself.

EXAMPLE 14.17 Prove that, if n is a positive integer

$$J_n(x+y) = \sum_{r=0}^{n} J_r(x) J_{n-r}(y) + \sum_{r=1}^{\infty} (-1)^r \{J_r(x) J_{n+r}(y) + J_{n+r}(x) J_r(y)\}.$$

Proof: Since $J_n(x) = (-1)^n J_{-n}(x)$ and $J_n(x)$ is the coefficient of z^n and $\left(-\frac{1}{z}\right)^n$ in the expansion of $e^{\frac{x}{2}\left(z-\frac{1}{z}\right)}$ we have $e^{\frac{x}{2}\left(z-\frac{1}{z}\right)} = \sum_{m=-\infty}^{\infty} J_m(x) z^m$.

Transform $x \to x + y$.

$$e^{\frac{1}{2}(x+y)\left(z-\frac{1}{z}\right)} = \sum_{m=-\infty}^{\infty} J_m(x+y) z^m$$

$$\therefore \qquad \sum_{m=-\infty}^{\infty} J_m(x+y) z^m = e^{\frac{1}{2}x\left(z-\frac{1}{z}\right)} e^{\frac{1}{2}y\left(z-\frac{1}{z}\right)}$$

$$\Rightarrow \qquad \sum_{m=-\infty}^{\infty} J_m(x+y) z^m = \left\{\sum_{r=-\infty}^{\infty} J_r(x) z^r\right\}\left\{\sum_{s=-\infty}^{\infty} J_s(y) z^s\right\} = \sum_{r=-\infty}^{\infty} \sum_{s=-\infty}^{\infty} J_r(x) J_s(y) z^{r+s}$$

Equating coefficients of z^n

$$J_n(x+y) = \sum_{r=-\infty}^{\infty} J_r(x) J_{n-r}(y)$$

$$= \sum_{r=-\infty}^{-1} J_r(x) J_{n-r}(y) + \sum_{r=0}^{n} J_r(x) J_{n-r}(y) + \sum_{r=n}^{\infty} J_r(x) J_{n-r}(y)$$

$$= \sum_{r=-\infty}^{-1} (-1)^r J_{-r}(x) J_{n-r}(y) + \sum_{r=0}^{n} J_r(x) J_{n-r}(y) + \sum_{r=1}^{\infty} J_{n+r}(x) J_{-r}(y)$$

$$= \sum_{r=1}^{\infty} (-1)^r J_r(x) J_{n+r}(y) + \sum_{r=0}^{n} J_r(x) J_{n-r}(y) + \sum_{r=1}^{\infty} J_{n+r}(x) J_{-r}(y)$$

$$= \sum_{r=0}^{n} J_r(x) J_{n-r}(y) + \sum_{r=1}^{\infty} (-1)^r \{J_r(x) J_{n+r}(y) + J_{n+r}(x) J_r(y)\}$$

EXERCISES

1. Show that $y = x^{-n/2} J_n(2\sqrt{x})$ that satisfies the equation $xy'' + (n+1)y' + y = 0$.

2. If n is a positive integer, show that

$$J_n(x) = (-2)^n x^n \frac{d^n}{d(x^2)^n} \{J_0(x)\}.$$

3. Prove that if n is a positive integer

$$J_n(x+y) = \sum_{r=0}^{n} J_r(x) J_{n-r}(y) + \sum_{r=1}^{\infty} (-1)^r [J_r(x) J_{n+r}(y) + J_{n+r}(x) J_r(y)]$$

4. Show that $\int_0^y \frac{x \sin ax}{\sqrt{y^2 - x^2}} dx = \frac{\pi y}{2} J_1.$

5. Prove that $\int_0^1 \frac{u J_0(xu)}{\sqrt{1-u^2}} du = \frac{\sin x}{x}.$

6. Show that

(i) $x \sin x = 2(2^2 J_2 - 4^2 J_4 + 6^2 J_6 - \cdots)$

(ii) $x \cos x = 2(J_1 - 3^2 J_3 + 5^2 J_5 - \cdots).$

7. If $a > 0$, prove that $\int_0^\infty e^{-ax} J_0(bx)\, dx = \frac{1}{\sqrt{a^2 + b^2}}.$

8. Verify that $J_0(x) = \dfrac{1}{\pi}\displaystyle\int_0^{\pi} \cos(x \sin \theta)\, d\phi$ satisfies Bessel's equation in which $n = 0$.

9. Prove that $J_n(x) = \dfrac{x^n}{2^{n-1}\,\Gamma\left(n+\dfrac{1}{2}\right)\sqrt{\pi}}\displaystyle\int_0^1 (1-t^2)^{n-1/2} \cos xt\, dt.$

10. Prove that $J_0(x) = \dfrac{2}{x}\displaystyle\int_0^1 \dfrac{\cos xt}{\sqrt{1-t^2}}\, dt.$

11. Show that $\displaystyle\int_0^{\infty} \dfrac{J_n(x)}{x}\, dx = \dfrac{1}{n}.$

CHAPTER 15

Legendre's Equation and Its Polynomials

15.1 LEGENDRE'S EQUATION AND ITS SOLUTION

The differential equation of the form

$$(1 - x^2)\, y'' - 2xy' + n(n + 1)\, y = 0 \tag{15.1}$$

is called **Legendre's equation**, where n is a positive integer.

Let the series solution of Eq. (15.1) be

$$y = \sum_{m=0}^{\infty} c_m x^{k-m},\ c_0 \neq 0 \tag{15.2}$$

Then
$$y' = \sum_{m=0}^{\infty} c_m (k-m)\, x^{k-m-1}$$

and
$$y'' = \sum_{m=0}^{\infty} c_m (k-m)\,(k-m-1)\, x^{k-m-2}$$

Substitute values of y, y' and y'' in Eq. (15.1), we have

$$(1-x^2)\sum_{m=0}^{\infty} c_m (k-m)(k-m-1)\, x^{k-m-2} - 2x \sum_{m=0}^{\infty} c_m (k-m) x^{k-m-1} + n(n+1)\sum_{m=0}^{\infty} c_m x^{k-m} = 0$$

$$\Rightarrow \quad \sum_{m=0}^{\infty} c_m (k-m)(k-m-1) x^{k-m-2} - \sum_{m=0}^{\infty} c_m [(k-m)(k+m-1) + 2(k+m) - n(n+1)] x^{k-m} = 0$$

$$\Rightarrow \sum_{m=0}^{\infty} c_m (k-m)(k-m-n)\, x^{k-m-2} - \sum_{m=0}^{\infty} c_m (k-m-n)(k-m+n+1) x^{k-m} = 0 \tag{15.3}$$

To get the indicial equation, we equate to zero the coefficient of the highest power of x (namely, x^k) in Eq. (15.3) and obtain

$$c_0(k - n)\,(k + n + 1) = 0$$

$$\Rightarrow \qquad (k - n)(k + n + 1) = 0, \; c_0 \neq 0 \tag{15.4}$$

Hence, the roots of Eq. (15.3) are $k = n, -(n + 1)$. They are unequal and differ by an integer. Next, lower power of x is $k - 1$. So we equate to zero the coefficient of x^{k-1} in Eq. (15.3) and obtain

$$c_1(k - 1 - n)(k + n) = 0 \tag{15.5}$$

For $k = n$ and $-(n + 1)$, neither $(k - 1 - n)$ nor $(k + n)$ is zero, so from Eq. (15.5), $c_1 = 0$. When coefficient of x^{k-m} in Eq. (15.3) is equated to zero, we get

$$c_{m-2}(k - m + 2)(k - m + 1) - c_m(k - m - n)(k - m + n + 1) = 0$$

$$\Rightarrow \qquad c_m = \frac{(k-m+2)(k-m+1)}{(k-m-n)(k-m+n+1)} c_{m-2} \tag{15.6}$$

Putting $m = 3, 5, 7, \ldots$ in Eq. (15.6) and using $c_1 = 0$, we have

$$c_1 = c_3 = c_5 = \cdots = 0 \tag{15.7}$$

which holds true for both $k = n$ and $k = -(n + 1)$.

Next we compute $c_2, c_4, c_6, \ldots$, etc. There are two cases.

Case 1: When $k = n$. Then Eq. (15.6) becomes

$$c_m = -\frac{(n-m+2)(n-m+1)}{m(2n-m+1)} c_{m-2} \tag{15.8}$$

Putting $m = 2, 4, 6, \ldots$ in Eq. (15.8), we get

$$c_2 = -\frac{n(n-1)}{2(2n-1)} c_0$$

$$c_4 = -\frac{(n-2)(n-3)}{4(2n-3)} c_2 = \frac{n(n-1)(n-2)(n-3)}{2 \cdot 4 \cdot (2n-1)(2n-3)} c_0$$

and so on.

For $k = n$, Eq. (15.2) is

$$y = c_0 x^n + c_1 x^{n-1} + c_2 x^{n-2} + c_3 x^{n-3} + c_4 x^{n-4} + \cdots \tag{15.9}$$

Label c_0 as a.

$$\therefore \qquad y = a\left[x^n - \frac{n(n-1)}{2(2n-1)} x^{n-2} + \frac{n(n-1)(n-2)(n-3)}{2 \cdot 4 \cdot (2n-1)(2n-3)} x^{n-4} + \cdots\right] \tag{15.10}$$

Case 2: When $k = -(n + 1)$. Then Eq. (15.6) gives

$$c_m = \frac{(n+m-1)(n+m)}{m(2n+m-1)} c_{m-2} \tag{15.11}$$

Putting $m = 2, 4, 6, \ldots$ in Eq. (15.11), we have

$$c_2 = -\frac{(n+1)(n+2)}{2(2n+3)} c_0$$

$$c_4 = -\frac{(n+3)(n+4)}{4(2n+5)} c_2 = \frac{(n+1)(n+2)(n+3)(n+4)}{2 \cdot 4 \cdot (2n+3)(2n+5)} c_0$$

and so on. For $k = -(n + 1)$, Eq. (15.2) is

$$y = c_0 x^{-n-1} + c_1 x^{-n-2} + c_2 x^{-n-3} + c_3 x^{-n-4} + c_4 x^{-n-5} + \cdots \tag{15.12}$$

Using Eq. (15.17) and c_2, c_4, ... in Eq. (15.12) and writing $c_0 = b$, we get

$$y = b\left[x^{-n-1} + \frac{(n+1)(n+2)}{2(2n+3)} x^{-n-3} + \frac{(n+1)(n+2)(n+3)(n+4)}{2 \cdot 4(2n+3)(2n+5)} x^{-n-5} + \cdots\right] \tag{15.13}$$

Thus, two independent solutions of Eq. (15.1) are given by Eqs. (15.10) and (15.13). If we take $a = \frac{1 \cdot 3 \cdot 5 \cdots (2n-1)}{n!}$, the solution of Eq. (15.10) is denoted by $P_n(x)$ and is called **Legendre's function of the first kind** or **Legendre's polynomial of degree** ***n***. If we take $b = \frac{n!}{1.3.5 \cdots (2n+1)}$, the solution of Eq. (15.13) is denoted by $Q_n(x)$ and is called **Legendre's function of second kind**. Thus, $P_n(x)$ and $Q_n(x)$ are two linearly independent solutions of Eq. (15.1). Hence, the general solution of Eq. (15.1) is

$$y = AP_n(x) + BQ_n(x) \tag{15.14}$$

where A and B are arbitrary constants.

Remark: Legendre's polynomial of degree n is defined by

$$P_n(x) = \frac{1 \cdot 3 \cdot 5 \cdots (2n-1)}{n!}\left[x^n - \frac{n(n-1)}{2(2n-1)} x^{n-2} + \frac{n(n-1)(n-2)(n-3)}{2 \cdot 4(2n-1)(2n-4)} x^{n-4} + \cdots\right] \tag{15.15}$$

The general term of polynomial $P_n(x)$ is

$$\frac{1 \cdot 3 \cdot 5 \cdots (2n-1)}{n!}(-1)^r \frac{n(n-1) \cdots (n-2r+1)}{2 \cdot 4 \cdots 2r(2n-1)(2n-3) \cdots (2n-2r+1)} x^{n-2r}$$

$$= \frac{2n!}{2^n \cdot n!}(-1)^r \frac{n!}{(n-2r)!} \frac{1}{2^r \cdot r!} \frac{2^n (2n-2r)! n!}{(2n)!(n-r)!} x^{n-2r}$$

$$= (-1)^r \frac{(2n-2r)!}{2^n r!(n-r)!(n-2r)!} x^{n-2r} \tag{15.16}$$

Since Eq. (15.15) is a polynomial of degree n, r must be chosen so that $n - 2r \geq 0$, i.e. $r \leq n/2$. Thus, if n is even, r varies from 0 to $n/2$, while if n is odd, r varies from 0 to $(n - 1)/2$, i.e. for the complete polynomial Eq. (15.15) r varies from to $[n/2]$, where

$$[n/2] = \begin{cases} n/2, & \text{if } n \text{ is even} \\ (n-1)/2, & \text{if } n \text{ is odd} \end{cases}$$

Hence, the Legendre polynomial of degree n is given by

$$p_n(x) = \sum_{r=0}^{[n/2]} (-1)^r \frac{(2n-2r)!}{2^n r!(n-r)!(n-2r)!} x^{n-2r} \tag{15.17}$$

15.2 LEGENDRE'S FUNCTION OF THE FIRST KIND OR LEGENDRE POLYNOMIAL OF DEGREE *n*

Equation (15.15) [or (15.17)] is called **Legendre's function of the first kind** or **Legendre polynomial of degree *n***. The first four Legendre polynomials are

$$P_0(x) = 1$$

$$P_1(x) = x$$

$$P_2(x) = \frac{1}{2}(3x^2 - 1)$$

$$P_3(x) = \frac{1}{2}(5x^3 - 3x)$$

and so on.

15.3 LEGENDRE'S FUNCTION OF THE SECOND KIND

As seen in Section (15.1), Legendre's function of the second kind is denoted by $Q_n(x)$ and is defined as

$$Q_n(x) = \frac{n!}{1\cdot 3\cdots(2n+1)}\left[x^{-(n+1)} + \frac{(n+1)(n+2)}{2(2n+3)}x^{-(n+3)} + \frac{(n+1)(n+2)(n+3)(n+4)}{2\cdot 4(2n+3)(2n+5)}x^{-(n+5)} + \cdots\right]$$

15.4 GENERATING FUNCTION FOR LEGENDRE POLYNOMIALS

THEOREM 15.1 $(1 - 2xz + z^2)^{-1/2} = \sum_{n=0}^{\infty} z^n p_n(x), |x| \le 1, |z| < 1.$

Proof: Given $|x| \le 1$, and $|z| < 1$

$\therefore$

$$(1 - 2xz + z^2)^{-1/2}$$

$$= [1 - z(2x - z)]^{-1/2}$$

$$= 1 + \frac{1}{2}z(2x - z) + \frac{1\cdot 3}{2\cdot 4}z^2(2x-z)^2 + \cdots + \frac{1\cdot 3\cdots(2n-3)}{2\cdot 4\cdots(2n-2)}z^{n-1}(2x-z)^{n-1}$$

$$+ \frac{1\cdot 3\cdots(2n-1)}{2\cdot 4\cdots(2n)}z^n(2x-z)^n + \cdots \tag{15.18}$$

Now, the coefficient of z^n in $\frac{1\cdot 3\cdots(2n-1)}{2\cdot 4\cdots(2n)}z^n(2x-z)^n$

$$= \frac{1\cdot 3\cdots(2n-1)}{2\cdot 4\cdots(2n)}(2x)^n$$

$$= \frac{1\cdot 3\cdots(2n-1)}{2^n\cdot n!}2^n\cdot x^n$$

$$= \frac{1.3\cdots(2n-1)}{n!}x^n \tag{15.19}$$

Again the coefficient of z^n in $\dfrac{1\cdot 3\cdots(2n-3)}{2\cdot 4\cdots(2n-2)}z^{n-1}(2x-z)^{n-1}$

$$= \frac{1\cdot 3\cdots(2n-3)}{2^{n-1}\,1\cdot 2\cdot 3\cdots(n-1)}[-(n-1)(2x)]^{n-2}$$

$$= -\frac{1\cdot 3\cdots(2n-3)}{2^{n-1}\,1\cdot 2\cdot 3\cdots(n-1)}\cdot\frac{2n-1}{n}\cdot\frac{n}{2n-1}\cdot(n-1)\,2^{n-2}\,x^{n-2}$$

$$= -\frac{1.3\cdots(2n-1)}{n!}\frac{n(n-1)}{2(2n-1)}x^{n-2} \tag{15.20}$$

and so on. Using Eqs. (15.19) and (15.20), we see that the coefficient of z^n in the expansion of Eq. (15.18) is given by

$$\frac{1.3.5\cdots(2n-1)}{n!}\left[x^n - \frac{n(n-1)}{2(2n-1)}x^{n-2} + \frac{n(n-1)(n-2)(n-3)}{2\cdot 4(2n-1)(2n-3)}x^{n-4} - \cdots\right]$$

i.e. $P_n(x)$, a Legendre polynomial.

We find that $P_1(x)$, $P_2(x)$, ... are the coefficients of z, z^2, ... in the expansion of $(1 - 2xz + z^2)^{-1/2}$.

Thus, $$(1-2xz+z^2)^{-1/2} = 1 + zP_1(x) + z^2P_2(x) + \cdots + z^nP_n(x) + \cdots = \sum_{n=0}^{\infty} z^n p_n(x).$$

Note: $(1 - 2xz + z^2)$ $|x| \le 1$, and $|z| < 1$ is called the **generating function for the Legendre polynomial** $P_n(x)$.

15.5 LAPLACE'S FIRST INTEGRAL FOR $P_n(x)$

When n is positive integer, Laplace's first integral for $P_n(x)$ is defined as

$$P_n(x) = \frac{1}{\pi}\int_0^{\pi}[x \pm \sqrt{x^2-1}\cos\phi]^n\,d\phi.$$

Proof: We know that

$$\int_0^{\pi}\frac{d\phi}{a \pm b\cos\phi} = \frac{\pi}{\sqrt{a^2-b^2}},\ \text{when } a^2 > b^2 \tag{15.21}$$

Let $$a = 1 - zx \text{ and } b = z\sqrt{x^2 - 1}$$

$\therefore$ $$a^2 - b^2 = 1 - 2zx + z^2$$

Using these values of a, b and $(a^2 - b^2)$, Eq. (15.21) can be written as

$$\pi(1 - 2zx + z^2)^{-1/2} = \int_0^{\pi} [1 - zx \pm z\sqrt{(x^2 - 1)} \cos\phi]^{-1}\, d\phi$$

$\Rightarrow$ $$\pi\sum_{n=0}^{\infty} z^n P_n(x) = \int_0^{\pi} (1 - zt)^{-1}\, d\phi, \text{ where } t = x \pm \sqrt{(x^2 - 1)} \cos\phi$$

$$= \int_0^{\pi} (1 + zt + z^2t^2 + \cdots)\, d\phi$$

$$= \int_0^{\pi} \sum_{n=0}^{\infty} (zt)^n\, d\phi$$

$$= \sum_{n=0}^{\infty} z^n \int_0^{\pi} t^n\, d\phi$$

$\therefore$ $$\pi\sum_{n=0}^{\infty} z^n P_n(x) = \sum_{n=0}^{\infty} z^n \int_0^{\pi} [x \pm \sqrt{(x^2 - 1)} \cos\phi]^n\, d\phi$$

Equating coefficients of z^n gives

$$\pi p_n(x) = \int_0^{\pi} [x \pm \sqrt{(x^2 - 1)} \cos\phi]^n\, d\phi$$

$\Rightarrow$ $$P_n(x) = \frac{1}{\pi}\int_0^{\pi} [x \pm \sqrt{(x^2 - 1)} \cos\phi]^n\, d\phi \qquad (15.22)$$

Deductions:

(i) $P_n(\cos\theta) = \dfrac{1}{\pi}\displaystyle\int_0^{\pi} (\cos\theta + i\sin\theta\cos\phi)^n\, d\phi.$

Proof: Let $x = \cos\theta$ in Eq. (15.22). Then we have

$$\sqrt{x^2 - 1} = \sqrt{\cos^2\theta - 1} = \sqrt{(-1)\sin^2\theta} = i\sin\theta$$

Hence, Eq. (15.22) gives

$$P_n(\cos\theta) = \frac{1}{\pi}\int_0^{\pi} (\cos\theta + i\sin\theta\cos\phi)^n\, d\phi$$

(ii) $P_1(x) = \dfrac{1}{\pi}\int_0^{\pi} [x + \sqrt{(x^2-1)}\cos\phi]\, d\phi.$

Proof: Put $n = 1$ in Eq. (15.22) using positive sign.

15.6 LAPLACE'S SECOND INTEGRAL FOR $P_n(x)$

When n is positive integer,

$$p_n(x) = \frac{1}{\pi}\int_0^{\pi} \frac{1}{[x \pm \sqrt{(x^2-1)}\cos\phi]^{n+1}}\, d\phi$$

is called **Laplace's second integral for $P_n(x)$.**

Proof: Recall Eq. (15.21). Let $a = zx - 1$ and $b = z\sqrt{x^2-1}$. Then $a^2 - b^2 = 1 - 2zx + z^2$. Using these values, Eq. (15.21) is

$$\pi(1 - 2zx + z^2)^{-1/2} = \int_0^{\pi} [-1 + zx \pm z\sqrt{(x^2-1)}\cos\phi]^{-1}\, d\phi$$

$$\Rightarrow \quad \frac{\pi}{z}\left(1 - 2x\frac{1}{z} + \frac{1}{z^2}\right)^{-1/2} = \int_0^{\pi}\left[-1 + z\{x \pm \sqrt{(x^2-1)}\cos\phi\}\right]^{-1} d\phi \tag{15.23}$$

Let
$$t = x \pm \sqrt{(x^2-1)}\cos\phi \tag{15.24}$$

We know that
$$(1 - 2zx + z^2)^{-1/2} = \sum_{n=0}^{\infty} z^n P_n(x) \tag{15.25}$$

Replacing z by $\dfrac{1}{z}$ in Eq. (15.25), we have

$$\left(1 - 2x\cdot\frac{1}{z} + \frac{1}{z^2}\right) = \sum_{n=0}^{\infty} \frac{1}{z^n} P_n(x) \tag{15.26}$$

Using Eqs. (15.24) and (15.26), we have

$$\frac{\pi}{z}\sum_{n=0}^{\infty} \frac{1}{z^n} p_n(x) = \int_0^{\pi} (-1 + zt)^{-1}\, d\phi$$

$$= \int_0^{\pi} (zt)^{-1}\left(1 - \frac{1}{zt}\right)^{-1} d\phi$$

$$= \int_0^{\pi} \frac{1}{zt}\sum_{n=0}^{\infty}\left(\frac{1}{zt}\right)^n d\phi$$

$$= \sum_{n=0}^{\infty} \frac{1}{z^{n+1}}\int_0^{\pi} \frac{d\phi}{t^{n+1}}$$

$$\therefore \qquad \sum_{n=0}^{\infty} \frac{\pi}{z^{n+1}} P_n(x) = \sum_{n=0}^{\infty} \frac{1}{z^{n+1}} \int_0^{\pi} \frac{d\phi}{[x \pm \sqrt{(x^2-1)}\cos\phi]^{n+1}}$$

Equating the coefficients of $\frac{1}{z^{n+1}}$ gives

$$P_n(x) = \frac{1}{\pi} \int_0^{\pi} \frac{d\phi}{[x \pm \sqrt{(x^2-1)}\cos\phi]^{n+1}} \tag{15.27}$$

Replacing n by $-(n+1)$ in Eq. (15.27), we have

$$P_{-(n+1)}(x) = \frac{1}{\pi} \int_0^{\pi} \frac{d\phi}{[x \pm \sqrt{(x^2-1)}\cos\phi]^{-n}}$$

$$= \frac{1}{\pi} \int_0^{\pi} [x \pm \sqrt{(x^2-1)}\cos\phi]^n \, d\phi$$

$= P_n(x)$, by article given Eq. (15.5)

Thus, $\qquad P_n(x) = P_{-(n+1)}(x).$

15.7 ORTHOGONAL PROPERTIES OF LEGENDRE POLYNOMIALS

THEOREM 15.2 $\displaystyle\int_{-1}^{1} P_m(x)\, P_n(x)\, dx = \begin{cases} 2, & \text{if } m \neq n \\ \dfrac{2}{2n+1}, & \text{if } m = n \end{cases}$

Proof: Let $m \neq n$. Since $P_m(x)$ and $P_n(x)$ satisfy Legendre's equation, we have

$$(1-x^2)\,P_m'' - 2xP_m' + m(m+1)\,P_m = 0 \tag{15.28}$$

$$(1-x^2)\,P_n'' - 2xP_n' + n(n+1)\,P_n = 0 \tag{15.29}$$

Multiply Eq. (15.28) by P_n and Eq. (15.29) by P_m and then subtract. The result is

$$(1-x^2)(P_nP_m'' - P_mP_n'') - 2x(P_nP_m' - P_mP_n') + [m(m+1) - n(n+1)]P_mP_n = 0$$

$$\Rightarrow \qquad (1-x^2)\frac{d}{dx}(P_nP_m' - P_mP_n') - 2x(P_nP_m' - P_mP_n') = [n^2 - m^2 + n - m]\,P_mP_n$$

$$\Rightarrow \qquad \frac{d}{dx}[(1-x^2)(P_nP_m' - P_mP_n')] = (n-m)(n+m+1)\,P_mP_n$$

Integrating both sides w.r.t. x from -1 to 1, we get

$$(n-m)(n+m+1)\int_{-1}^{1} P_m(x)\,P_n(x)\,dx = [(1-x^2)(P_nP_m' - P_mP_n')]_{x=-1}^{1}$$

$$\Rightarrow \qquad \int_{-1}^{1} P_m(x)P_n(x)\,dx = 0 \tag{15.30}$$

Next, consider $m = n$. Then we need to prove

$$\int_{-1}^{1} [P_n(x)]^2\, dx = \frac{2}{2n+1}$$

We consider generating function

$$(1 - 2xz + z^2)^{-1/2} = \sum_{n=0}^{\infty} z^n P_n(x)$$

and
$$(1 - 2xz + z^2)^{-1/2} = \sum_{m=0}^{\infty} z^m P_m(x)$$

Multiplying the corresponding sides of the above two equation gives

$$(1 - 2xz + z^2)^{-1} = \sum_{m=0}^{\infty} \sum_{n=0}^{\infty} P_m(x)\, P_n(x) z^{m+n}$$

Integrating both sides w.r.t. x between -1 to 1 gives

$$\int_{-1}^{1} (1 - 2xz + z^2)^{-1}\, dx = \sum_{m=0}^{\infty} \sum_{n=0}^{\infty} \left\{ \int_{-1}^{1} P_m(x)\, P_n(x)\, dx \right\} z^{m+n} \tag{15.31}$$

Using Eq. (15.30), Eq. (15.31) reduces to

$$\sum_{n=0}^{\infty} \left[\int_{-1}^{1} [P_n(x)]^2\, dx \right] z^{2n} = \int_{-1}^{1} \frac{dx}{1 - 2xz + z^2}$$

$$= \left[\frac{\log(1 - 2xz + z^2)}{-2z} \right]_{-1}^{1}$$

$$= -\frac{1}{2z} \left[\log(1 - z)^2 - \log(1 + z)^2 \right]$$

$$= -\frac{1}{2z} [2 \log(1 - z) - 2 \log(1 + z)]$$

$$= \frac{1}{z} [\log(1 + z) - \log(1 - z)]$$

$$= \frac{1}{z} \left[\left(z - \frac{z^2}{2} + \frac{z^3}{3} - \cdots \right) - \left(-z - \frac{z^2}{2} - \frac{z^3}{3} - \cdots \right) \right]$$

$$= \frac{2}{z} \left(z + \frac{z^3}{3} + \frac{z^5}{5} + \cdots \right)$$

$$= \frac{2}{z} \sum_{n=0}^{\infty} \frac{z^{2n+1}}{2n+1}$$

$$= \sum_{n=0}^{\infty} \frac{2}{2n+1} z^{2n}$$

Equating coefficients of z^{2n} gives

$$\int_{-1}^{1} [P_n(x)]^2 \, dx = \frac{2}{2n+1} \tag{15.32}$$

15.8 LEGENDRE SERIES

THEOREM 15.3 (a) If $f(x)$ is a polynomial of degree n, then $f(x) = \sum_{r=0}^{n} c_r P_r(x)$

where
$$c_r = \left(r + \frac{1}{2}\right) \int_{-1}^{1} f(x) \, P_r(x) \, dx$$

(b) If $f(x)$ is even (or odd), only those c_r with even (or odd) suffixes are non-zero.

Proof: (a) Since $f(x)$ is a polynomial of degree n, we write

$$f(x) = a_n x^n + a_{n-1} x^{n-1} + \cdots + a_1 x + a_0 \tag{15.33}$$

Again, we know that $P_n(x)$ is a polynomial of degree n of the form

$$P_n(x) = k_n x^n + k_{n-1} x^{n-1} + \cdots + k_1 x + k_0 \tag{15.34}$$

Consider $f(x) - \left(\frac{a_n}{k_n}\right) P_n(x)$. Two cases may aries.

Case 1: $f(x) - \left(\frac{a_n}{k_n}\right) P_n(x) = 0$ so that $f(x) - \left(\frac{a_n}{k_n}\right) P_n(x)$ which proves the result in (a).

Case 2: $f(x) - \left(\frac{a_n}{k_n}\right) P_n(x) = g_{n-1}(x)$, $g_{n-1}(x)$ being a polynomial of degree $n - 1$. Taking

$c_n = \frac{a_n}{k_n}$

We may write

$$f(x) = c_n P_n(x) + g_{n-1}(x) \tag{15.35}$$

Taking $g_{n-1}(x)$ in place of $f(x)$ and arguing as above, we have

$$g_{n-1}(x) = c_{n-1} P_{n-1}(x) + g_{n-2}(x)$$

Then Eq. (15.35) may be rewritten as

$$f(x) = c_n P_n(x) + c_{n-1} P_{n-1}(x) + g_{n-2}(x)$$

Continue the steps till we get

$$f(x) = c_n P_n(x) + c_{n-1} P_{n-1}(x) + \cdots + c_1 P_1(x) + c_0 P_0(x)$$

Note that $P_0(x) = 1$.

$$\therefore \qquad f(x) = \sum_{r=0}^{n} c_r P_r(x) \qquad (15.36)$$

Since $\sum_{r=0}^{n} c_r P_r(x) = \sum_{r=0}^{n} c_s P_s(x)$, Eq. (15.36) gives

$$f(x) = \sum_{s=0}^{n} c_s P_s(x) \qquad (15.37)$$

Multiply both sides of Eq. (15.37) by $P_r(x)$ and then integrating w.r.t. x from –1 to 1. We get

$$\int_{-1}^{1} f(x)\, P_r(x)\, dx = \sum_{s=0}^{n} \left\{ c_s \int_{-1}^{1} P_s(x)\, P_r(x)\, dx \right\}$$

$$= c_r \cdot \frac{2}{2r+1}, \text{ when } r = s$$

$$\therefore \qquad c_r = \left(r + \frac{1}{2}\right) \int_{-1}^{1} f(x)\, P_r(x)\, dx \qquad (15.38)$$

(b) Try yourself.

Remark: Suppose $f(x)$ is continuous and has continuous derivatives in [–1, 1] then it is easy to prove that series converges uniformly in [–1, 1] and the series expansion is unique.

15.9 RECURRENCE FORMULAE

R1: $\qquad nP_n = (2n - 1)\, xP_{n-1} - (n - 1)\, P_{n-2},\ n \geq 2$

Proof: From the generating function, we have

$$(1 - 2xz + z^2)^{-1/2} = \sum_{n=0}^{\infty} z^n P_n(x)$$

Differentiating both sides w.r.t. z gives

$$-\frac{1}{2}(1 - 2xz + z^2)^{-3/2}(-2x + 2z) = \sum_{n=0}^{\infty} nz^{n-1} P_n(x)$$

Multiply both sides by $(1 - 2xz + z^2)$

$$\Rightarrow \qquad (x - z)\,(1 - 2xz + z^2)^{-1/2} = (1 - 2xz + z^2) \sum_{n=0}^{\infty} nz^{n-1} P_n(x)$$

$$\Rightarrow \qquad (x - z) \sum_{n=0}^{\infty} z^n P_n(x) = (1 - 2xz + z^2) \sum_{n=0}^{\infty} nz^{n-1} P_n(x)$$

$$\Rightarrow \qquad x\sum_{n=0}^{\infty} z^n P_n - \sum_{n=0}^{\infty} z^{n+1} P_n = \sum_{n=0}^{\infty} nz^{n-1} P_n - 2x\sum_{n=0}^{\infty} nz^n P_n + \sum_{n=0}^{\infty} nz^{n+1} P_n$$

Equating coefficients of z^n gives

$$xP_n - P_{n-1} = (n+1)P_{n+1} - 2xnP_n + (n-1)P_{n-1}$$

$$\Rightarrow \quad (n+1)P_{n+1} = (2n+1)xP_n - nP_{n-1}$$

Replacing n by $(n - 1)$ gives

$$nP_n = (2n - 1)\, xP_{n-1} - (n - 1)\, P_{n-1}$$

R2: $$nP_n = xP'_n - P'_{n-1}$$

Proof: We know that

$$(1-2xz+z^2)^{-1/2} = \sum_{n=0}^{\infty} z^n P_n(x) \tag{15.39}$$

Differentiating w.r.t. z gives

$$-\frac{1}{2}(1-2xz+z^2)^{-3/2}(-2x+2z) = \sum_{n=0}^{\infty} nz^{n-1}P_n(x)$$

$$\Rightarrow \quad (x-z)(1-2xz+z^2)^{-3/2} = \sum_{n=0}^{\infty} nz^{n-1}P_n(x) \tag{15.40}$$

Again differentiate Eq. (15.39) w.r.t. x, gives

$$z(1-2xz+z^2)^{-3/2} = \sum_{n=0}^{\infty} z^n P'_n$$

$$\Rightarrow \quad z(x-z)(1-2xz+z^2)^{-3/2} = (x-z)\sum_{n=0}^{\infty} z^n P'_n$$

$$\Rightarrow \quad z\sum_{n=0}^{\infty} nz^{n-1}P_n(x) = (x-z)\sum_{n=0}^{\infty} z^n P'_n \qquad \text{[Using (15.40)]}$$

$$\Rightarrow \quad \sum_{n=0}^{\infty} nz^n P_n = x\sum_{n=0}^{\infty} z^n P'_n - \sum_{n=0}^{\infty} z^{n+1}P'_n$$

Equating coefficient of z^n on both sides gives

$$nP_n = xP'_n - P'_{n-1}$$

R3: $$(2n + 1)\, P_n = P'_{n+1} - P'_{n-1}$$

Proof: From R1, we have

$$(2n + 1)\, xP_n = (n + 1)\, P_{n+1} + nP_{n-1}$$

Differentiating w.r.t. x gives

$$(2n+1)xP'_n + (2n+1)P_n = (n+1)P'_{n+1} + nP'_{n-1}$$

$$\Rightarrow \quad (2n+1)(nP_n + P'_{n-1}) + (2n+1)P_n = (n+1)P'_{n+1} + nP'_{n-1} \qquad \text{(Using R2)}$$

$$\Rightarrow \quad (2n+1)(n+1)P_n = (n+1)P'_{n+1} - (n+1)P'_{n-1}$$

$$\Rightarrow \quad (2n+1)P_n = P'_{n+1} - P'_{n-1}$$

R4: $$(n + 1)\, P_n = P'_{n+1} - xP'_n$$

Proof: From R3 – R2, one can prove the result.

R5: $$(1 - x^2)\, P'_n = n(Pn_{-1} - xP_n)$$

Proof: R2 is $nP_n = xP'_n - P'_{n-1}$ and
R4 is $(n + 1)\, P_n = P'_{n+1} - xP'_n$
Replace n by $(n - 1)$ in R4, we get

$$nP_n = P'_n - xP'_{n-1} \tag{15.41}$$

Multiply both sides of R2 by x. Therefore,

$$xnP_n = x^2P'_n - xP'_{n-1} \tag{15.42}$$

Subtracting Eqs. (15.41) and (15.42) gives

$$n(P_{n-1} - xP_n) = (1 - x^2)\, P'_n$$

R6: $$(1 - x^2)\, P'_n = (n + 1)\, (xP_n - P_{n+1})$$

Proof: Rewrite R1 as

$$(n + 1)\, (xP_n - P_{n+1}) = n(P_{n-1} - xP_n)$$

R5 is $$(1 - x^2)\, P'_n = n(P_{n-1} - xP_n)$$

Hence, $$(1 - x^2)\, P_n = (n + 1)\, (xP_n - P_{n+1}).$$

15.10 RODRIGUE'S FORMULA

It is given by

$$P_n(x) = \frac{1}{2^n\, n!}\frac{d^n}{dx^n}(x^2 - 1)^n$$

Proof: In Section 15.1 [Eq. (15.17)], we have Legendre's polynomial as

$$P_n(x) = \sum_{r=0}^{[n/2]} (-1)^r \frac{(2n - 2r)!\, x^{n-2r}}{2^n\, r!(n - r)!\,(n - 2r)!}$$

where $$\left[\frac{n}{2}\right] = \begin{cases} \dfrac{1}{2}n, & \text{if } n \text{ is even} \\ \dfrac{1}{2}(n-1), & \text{if } n \text{ is odd} \end{cases}$$

By binomial theorem,

$$(x^2 - 1)^n = \sum_{r=0}^{n} c(n, r)\, (-1)^r\, (x^2)^{n-r}$$

$$= \sum_{r=0}^{n} c(n, r)\, (-1)^r\, x^{2n-2r}$$

$$\frac{1}{2^n\, n!}\frac{d^n}{dx^n}(x^2-1)^n = \frac{1}{2^n\, n!}\sum_{r=0}^{n} c(n,r)\,(-1)^r \frac{d^n}{dx^n}x^{2n-2r} \tag{15.43}$$

But, $\dfrac{d^n}{dx^n}x^m = 0$, if $m < n$, $\dfrac{d^n}{dx^n}x^m = \dfrac{m!}{(m-n)!}x^{m-n}$ if $m \ge n$

$$\Rightarrow \qquad \frac{d^n}{dx^n}x^{2n-2r} = 0 \text{ if } 2n-2r < n, \text{ i.e. } r > \frac{n}{2}$$

Hence, we see that we must replace $\sum\limits_{r=0}^{n}$ by $\sum\limits_{r=0}^{n/2}$ if n is even and by $\sum\limits_{r=0}^{(n-1)/2}$ if n is odd.

i.e. we must replace $\sum\limits_{r=0}^{n}$ by $\sum\limits_{r=0}^{n/2}$ in Eq. (15.43).

$$\Rightarrow \qquad \frac{1}{2^n\cdot n!}\frac{d^n}{dx^n}(x^2-1)^n = \frac{1}{2^n\cdot n!}\sum_{r=0}^{[n/2]} c(n,r)\,(-1)^r \frac{d^n}{dx^n}x^{2n-2r}$$

$$= \frac{1}{2^n\cdot n!}\sum_{r=0}^{[n/2]} c(n,r)\,(-1)^r \frac{(2n-2r)!}{(2n-2r-n)!}x^{2n-2r-n}$$

$$= \sum_{r=0}^{[n/2]} \frac{1}{2^n\cdot n!}\cdot\frac{n!(-1)^r}{r!(n-r)!}\,\frac{(2n-2r)!}{(n-2r)!}x^{n-2r}$$

$$= P_n(x)$$

$$\therefore \qquad P_n(x) = \frac{1}{2n\cdot n!}\frac{d^n}{dx^n}(x^2-1)^n.$$

15.11 EXAMPLES

EXAMPLE 15.1 Prove that

$$1 + \frac{1}{2}P_1(\cos\theta) + \frac{1}{3}P_2(\cos\theta) + \cdots = \log\frac{1+\sin\dfrac{\theta}{2}}{\sin\dfrac{\theta}{2}}.$$

Proof: From the generating function, we have

$$\sum_{n=0}^{\infty} z^n P_n(x) = (1-2xz+z^2)^{-1/2} \tag{i}$$

Integrating Eq. (i) w.r.t. z from 0 to 1, we get

$$\sum_{n=0}^{\infty}\int_0^1 z^n P_n(x)\,dz = \int_0^1 \frac{dz}{\sqrt{1-2xz+z^2}} \tag{ii}$$

Put $$x = \cos\theta$$

$$\Rightarrow \quad \sum_{n=0}^{\infty} P_n \cos\theta \left[\frac{z^{n+1}}{n+1}\right]_0^1 = \int_0^1 \frac{dz}{\sqrt{1 - 2z\cos\theta + z^2}}$$

$$\Rightarrow \quad \sum_{n=0}^{\infty} \frac{P_n \cos\theta}{n+1} = \int_0^1 \frac{dz}{\sqrt{(z - \cos\theta)^2 + \sin^2\theta}}$$

$$= \left[\log(z - \cos\theta) + \sqrt{(z - \cos\theta)^2 + \sin^2\theta}\right]_0^1$$

$$= \log\left[1 - \cos\theta + \sqrt{(1 - \cos\theta)^2 + \sin^2\theta}\right] - \log(1 - \cos\theta)$$

$$= \log\left[(1 - \cos\theta) + \sqrt{2(1 - \cos\theta)}\right] - \log(1 - \cos\theta)$$

$$= \log\frac{(1 - \cos\theta) + \sqrt{2}\sqrt{(1 - \cos\theta)}}{1 - \cos\theta}$$

$$= \log\frac{\sqrt{1 - \cos\theta} + \sqrt{2}}{\sqrt{1 - \cos\theta}}$$

$$= \log\frac{\sqrt{2\sin^2\frac{\theta}{2}} + \sqrt{2}}{\sqrt{2\sin^2\frac{\theta}{2}}}$$

$$= \log\frac{1 + \sin\frac{\theta}{2}}{\sin\frac{\theta}{2}}$$

$$\therefore \quad P_0(\cos\theta) + \frac{1}{2}P_1(\cos\theta) + \frac{1}{3}P_2(\cos\theta) + \cdots = \log\frac{1 + \sin\frac{\theta}{2}}{\sin\frac{\theta}{2}}$$

$$\Rightarrow \quad 1 + \frac{1}{2}P_1(\cos\theta) + \frac{1}{3}P_2(\cos\theta) + \cdots = \log\frac{1 + \sin\frac{\theta}{2}}{\sin\frac{\theta}{2}}.$$

EXAMPLE 15.2 Prove that

$$P_n\left(-\frac{1}{2}\right) = P_0\left(-\frac{1}{2}\right)P_{2n}\left(\frac{1}{2}\right) + P_1\left(-\frac{1}{2}\right)P_{2n-1}\left(\frac{1}{2}\right) + \cdots + P_{2n}\left(-\frac{1}{2}\right)P_0\left(\frac{1}{2}\right).$$

Proof: We have generating function

$$(1 - 2xz + z^2)^{-1/2} = \sum_{n=0}^{\infty} z^n P_n(x) \qquad \text{(i)}$$

Replacing x by $\left(\frac{1}{2}\right)$ and $\left(-\frac{1}{2}\right)$ successively in Eq. (i) gives

$$(1-2z+z^2)^{-1/2} = \sum_{n=0}^{\infty} z^n P_n\left(\frac{1}{2}\right) \tag{ii}$$

and
$$(1+2z+z^2)^{-1/2} = \sum_{n=0}^{\infty} z^n P_n\left(-\frac{1}{2}\right) \tag{iii}$$

Next replacing z by z^2 in Eq. (iii), we have

$$(1+z^2+z^4)^{-1/2} = \sum_{n=0}^{\infty} z^{2n} P_n\left(-\frac{1}{2}\right) \tag{iv}$$

But
$$1 + z^2 + z^4 = (1 + z + z)\ (1 - z + z^2)$$

$$\Rightarrow \quad (1 + z^2 + z^4)^{-1/2} = (1 + z + z^2)^{-1/2}\ (1 - z + z^2)^{-1/2}$$

$$\Rightarrow \quad \sum_{n=0}^{\infty} z^{2n} P_n\left(-\frac{1}{2}\right) = \sum_{n=0}^{\infty} z^n P_n\left(-\frac{1}{2}\right) + \sum_{n=0}^{\infty} z^n P_n\left(\frac{1}{2}\right)$$

Equating the coefficients of z^{2n} from both sides of the above equation, obtain the result.

EXAMPLE 15.3 Prove $(1-z^2)\,(1-2xz+z^2)^{3/2} = \sum_{n=0}^{\infty} (2n+1) z^n P_n(x)$.

Proof: We have
$$(1-2xz+z^2)^{-1/2} = \sum_{n=0}^{\infty} z^n P_n \tag{i}$$

Differentiate both sides w.r.t. z, we get

$$-\frac{1}{2}(1-2xz+z^2)^{-3/2}(-2x+2z) = \sum_{n=0}^{\infty} nz^{n-1} P_n$$

$$\Rightarrow \quad (x-z)\,(1-2xz+z^2)^{-3/2} = \sum_{n=0}^{\infty} nz^{n-1} P_n$$

Multiply both sides by $2z$, we have

$$2z(x-z)\,(1-2xz+z^2)^{-3/2} = 2\sum_{n=0}^{\infty} nz^n P_n \tag{ii}$$

Adding Eqs. (i) and (iii) gives

$$\sum_{n=0}^{\infty} (2n+1) z^n P_n = \frac{1}{(1-2xz+z^2)^{1/2}} + \frac{2z(x-z)}{(1-2xz+z^2)^{3/2}}$$

$$= \frac{1-z^2}{(1-2xz+z^2)^{3/2}}.$$

EXAMPLE 15.4 Prove that $(1 - 2xz + z^2)^{-1/2}$ is a solution of the equation

$$z\frac{\partial^2 (zv)}{\partial z^2} + \frac{\partial}{\partial x}\left\{(1-x^2)\frac{\partial v}{\partial x}\right\} = 0.$$

Proof:

$$v = (1-2xz+z^2)^{-1/2} = \sum_{n=0}^{\infty} z^n P_n \tag{i}$$

$$\Rightarrow \quad zv = (1-2xz+z^2)^{-1/2} = \sum_{n=0}^{\infty} z^{n+1} P_n$$

$$\Rightarrow \quad \frac{\partial^2 (zv)}{\partial z^2} = \frac{\partial^2}{\partial z^2}\left[\sum_{n=0}^{\infty} z^{n+1} P_n\right] = \sum_{n=0}^{\infty} n(n+1) z^{n-1} P_n$$

$$\Rightarrow \quad z\frac{\partial^2 (zv)}{\partial z^2} = \sum_{n=0}^{\infty} n(n+1)\, z^n P_n \tag{ii}$$

From Eq. (i),

$$\frac{\partial v}{\partial x} = \sum_{n=0}^{\infty} \frac{\partial}{\partial x}(z^n P_n) = \sum_{n=0}^{\infty} z_n P_n' \tag{iii}$$

$$\frac{\partial^2 v}{\partial x^2} = \sum_{n=0}^{\infty} z_n P_n'' \tag{iv}$$

Now,

$$\frac{\partial}{\partial x}\left\{(1-x^2)\frac{\partial v}{\partial x}\right\} = (1-x^2)\frac{\partial^2 v}{\partial x^2} - 2x\frac{\partial v}{\partial x}$$

$$\Rightarrow \quad \frac{\partial}{\partial x}\left\{(1-x^2)\frac{\partial v}{\partial x}\right\} = (1-x^2)\sum_{n=0}^{\infty} z^n P_n'' - 2x\sum_{n=0}^{\infty} z^n P_n' \tag{v}$$

Adding Eqs. (ii) and (v), we get

$$z\frac{\partial^2 (zv)}{\partial v^2} + \frac{\partial}{\partial x}\left\{(1-x^2)\frac{\partial v}{\partial x}\right\}$$

$$= \sum_{n=0}^{\infty} n(n+1)\, z^n p_n + \sum_{n=0}^{\infty} (1-x^2)\, z^n p_n'' - 2x\sum_{n=0}^{\infty} z^n p_n'$$

$$= \sum_{n=0}^{\infty} z^n [(1-x^2)\, P_n'' - 2xP_n' + n(n+1)\, P_n]$$

$= 0$, as P_n is a solution of Legendre's equation.

EXAMPLE 15.5 Show that

$$P_n(\cos\theta) = \frac{1\cdot 3\cdot 5\cdots(2n-1)}{2^{n-1} n!}\left[\cos n\theta + \frac{1\cdot n}{1(2n-1)}\cos(n-2)\,\theta + \frac{1\cdot 3\cdot n(n-1)}{1\cdot 2(2n-1)(2n-3)}\cos(n-4)\,\theta + \cdots\right].$$

Proof: We have $\sum_{n=0}^{\infty} z^n P_n(x) = (1-2xz+z^2)^{-1/2}$

Therefore,

$$\sum_{n=0}^{\infty} z^n P_n(\cos\theta)$$

$$= (1-2z\cos\theta+z^2)^{-1/2}$$

$$= [1-z(e^{i\theta}+e^{-i\theta})+z^2]^{-1/2}$$

$$= [(1-ze^{i\theta})(1-ze^{-i\theta})]^{-1/2}$$

$$= (1-ze^{i\theta})^{-1/2}(1-ze^{-i\theta})^{-1/2}\left[1+\frac{1}{2}ze^{i\theta}+\frac{1\cdot 3}{2\cdot 4}z^2e^{2i\theta}+\cdots+\frac{1\cdot 3\cdot 5\cdots(2n-1)}{2\cdot 4\cdot 6\cdots(2n)}z^n e^{in\theta}+\cdots\right]$$

$$\times\left[1+\frac{1}{2}ze^{-i\theta}+\frac{1\cdot 3}{2\cdot 4}z^2e^{-2i\theta}+\cdots+\frac{1\cdot 3\cdot 5\cdots(2n-1)}{2\cdot 4\cdot 6\cdots(2n)}z^n e^{-in\theta}+\cdots\right]$$

Equating the coefficients of z^n from both sides, we get

$$P_n(\cos\theta) = \frac{1\cdot 3\cdot 5\cdots(2n-1)}{2\cdot 4\cdot 6\cdots(2n)}(e^{in\theta}+e^{-in\theta})$$

$$+\frac{1\cdot 3\cdot 5\cdots(2n-3)}{2\cdot 4\cdot 6\cdots(2n-2)}\cdot\frac{1}{2}(e^{i(n-2)\theta}+e^{-i(n-2)\theta})$$

$$+\frac{1\cdot 3\cdot 5\cdots(2n-5)}{2\cdot 4\cdot 6\cdots(2n-4)}\cdot\frac{1\cdot 3}{2\cdot 4}(e^{i(n-4)\theta}+e^{-i(n-4)\theta})$$

$$= \frac{1\cdot 3\cdot 5\cdots(2n-1)}{2\cdot 4\cdot 6\cdots(2n)}\left[2\cos n\theta+\frac{2n}{2n-1}\cdot 2\cos(n-2)\theta+\frac{2n(2n-2)}{(2n-1)(2n-3)}\frac{1\cdot 3}{2\cdot 4}2\cos(n-4)\theta+\cdots\right]$$

$$= \frac{1\cdot 3\cdot 5\cdots(2n-1)}{2^{n-1}\cdot n!}\left[\cos n\theta+\frac{1\cdot n}{1\cdot(2n-1)}\cos(n-2)\theta+\frac{1\cdot 3n(n-1)}{1\cdot 2(2n-1)(2n-3)}\cos(n-4)\theta+\cdots\right].$$

EXAMPLE 15.6 **If $f(x)$ is a polynomial of degree $n < l$, prove that $\int_{-1}^{1} f(x)P_l(x)\,dx = 0$.**

Proof: In Section 4.8, we have discussed that if $f(x)$ is a polynomial of degree n then

$$f(x) = \sum_{r=0}^{n} c_r P_r(x)$$

Multiply both sides by $P_l(x)$ and then integrate over $[-1, 1]$, we have

$$\int_{-1}^{1} f(x)\,P_l(x)\,dx = \sum_{r=0}^{n} c_r \int_{-1}^{1} P_r(x)\,P_l(x)\,dx \qquad \text{(i)}$$

Since $n < l$, for $r = 0, 1, 2, \ldots, n$ we find that $r \neq l$. But we know that

$$\int_{-1}^{1} P_r(x) P_l(x) = 0, \quad \text{if } r \neq l$$

$\therefore$ Equation (i) gives $\int_{-1}^{1} f(x)\, P_l(x)\, dx = 0.$

EXAMPLE 15.7 Find series of Legendres polynomials for x^2.

Solution Since x^2 is a polynomial of degree two, from Legendre series, we have

$$x^2 = \sum_{r=0}^{2} c_r P_r(x) = c_0 P_0(x) + c_1 P_1(x) + c_2 P_2(x) \quad \text{(i)}$$

where
$$c_r = \left(r + \frac{1}{2}\right) \int_{-1}^{1} x^2 P_r(x)\, dx \quad \text{(ii)}$$

we know that
$$P_0(x) = 1,\ P_1(x) = x,\ P_2(x) = \frac{1}{2}(3x^2 - 1) \quad \text{(iii)}$$

Putting $r = 0, 1, 2$ successively in Eq. (ii) and using Eq. (iii), we have

$$c_0 = \frac{1}{2} \int_{-1}^{1} x^2\, dx = \frac{1}{3}$$

$$c_1 = \frac{3}{2} \int_{-1}^{1} x^3\, dx = 0$$

$$c_2 = \frac{1}{2} \cdot \frac{5}{2} \int_{-1}^{1} x^2 (3x^2 - 1)\, dx = \frac{2}{3}$$

Hence Eq. (i) becomes

$$x^2 = \frac{1}{3} P_0(x) + \frac{2}{3} P_2(x).$$

EXAMPLE 15.8 Prove that $\int_{x}^{1} P_n(x)\, dx = \frac{1}{2n+1} [P_{n-1}(x) - P_{n+1}(x)]$.

Proof: The recurrence relation R3 is

$$P_n(x) = \frac{1}{2n+1} [P'_{n+1}(x) - P'_{n-1}(x)]$$

Integrate both sides w.r.t. x between limits x to 1, we get

$$\int_{x}^{1} P_n(x)\, dx = \frac{1}{2n+1} [P_{n+1}(x) - P_{n-1}(x)]_x^1$$

$$= \frac{1}{2n+1}[P_{n+1}(1) - P_{n-1}(1) - P_{n+1}(x) + P_{n-1}(x)]$$

$$= \frac{1}{2n+1}[P_{n-1}(x) - P_{n+1}(x)] \qquad [\because \; P_{n+1}(1) = 1, P_{n-1}(1) = 1]$$

EXAMPLE 15.9 Prove that $(2n+1)(x^2-1)P_n' = n(n+1)(P_{n+1} - P_{n-1})$.

Note: This result is known as Beltramics result.

Proof: The recurrence relations R5 and R6 are

$$(1-x^2)P_n' = n(P_{n-1} - xP_n) \tag{i}$$

and

$$(1-x^2)P_n' = (n+1)(xP_n - P_{n+1}) \tag{ii}$$

Multiply Eq. (i) by $(n + 1)$ and Eq. (ii) by n and add them. We get

$$(n+1)(1-x^2)P_n' + n(1-x^2)P_n' = n(n+1)P_{n-1} - n(n+1)P_{n+1}$$

$$\Rightarrow \quad (2n+1)(1-x^2)P_n' = n(n+1)(P_{n-1} - P_{n+1})$$

$$\Rightarrow \quad (2n+1)(x^2-1)P_n' = n(n+1)(P_{n+1} - P_{n-1}).$$

EXAMPLE 15.10 Prove that

(i) $$\sum_{k=0}^{m}(2k+1)P_k(x)P_k(y) = \frac{m+1}{x-y}[P_{m+1}(x)P_m(y) - P_m(x)P_{m+1}(y)]$$

(ii) $$\sum_{k=0}^{m}(2k+1)P_m(x) = \frac{m+1}{x-y}[P_{m+1}(x) - P_m(x)].$$

Note: The (i) is called Christoffel's summation formula.

Proof: From recurrence relation R1, we have

$$(2k+1)xP_k(x) = (k+1)P_{k+1}(x) + kP_{k-1}(x) \tag{i}$$

and

$$(2k+1)yP_k(y) = (k+1)P_{k+1}(y) + kP_{k-1}(y) \tag{ii}$$

Multiplying Eq. (i) by $P_k(y)$ and Eq. (ii) by $P_k(x)$ and then subtracting, we get

$$(2k+1)(x-y)P_k(x)P_k(y) = (k+1)[P_{k+1}(x)P_k(y) - P_{k+1}(y)P_k(x)]$$
$$-k[P_{k-1}(x)P_k(y) - P_{k-1}(y)P_k(x)] \tag{iii}$$

Putting $k = 0, 1, 2, \ldots, m$ in Eq. (iii) and adding the resulting equations, we have

$$(x-y)\sum_{k=0}^{m}(2k+1)P_k(x)P_k(y) = (m+1)[P_{m+1}(x)P_m(y) - P_{m+1}(y)P_m(x)]$$

$$\sum_{k=0}^{m}(2k+1)P_k(x)P_k(y) = \frac{(m+1)}{(x-y)}[P_{m+1}(x)P_m(y) - P_{m+1}(y)P_m(x)]$$

EXAMPLE 15.11 Prove that

$$P_n' = (2n-1)\,P_{n-1} + (2n-5)\,P_{n-3} + (2n-9)\,P_{n-5} + \cdots$$

The last terms of the series being $3P_1$ or P_0 according as n is even or odd.

Proof: In recurrence relation R3 replace n by $n - 1$.
Therefore, we have

$$P_n' = (2n-1)\,P_{n-1} + P_{n-2}' \tag{i}$$

Case 1: Let n be even. Replacing n by $n, n-2, n-4, \ldots, 4, 2$ successively in Eq. (i) and using $P_0 = 1$ and $P_0' = 0$, we get

$$\begin{aligned} P_n' &= (2n-1)\,P_{n-1} + P_{n-2}' \\ P_{n-2}' &= (2n-5)\,P_{n-3} + P_{n-4}' \\ P_{n-4}' &= (2n-9)\,P_{n-5} + P_{n-6}' \\ &\vdots \\ P_4' &= 7P_3 + P_2' \\ P_2' &= 3P_1 + P_0' \end{aligned}$$

Adding these and simplifying, we get

$$P_n' = (2n-1)\,P_{n-1} + (2n-5)\,P_{n-3} + \cdots + 3P_1 \tag{ii}$$

Case 2: Let n be odd. Replace n by $n, n-2, n-4, \ldots, 5, 3$ successively in Eq. (i) and using $P_1(x) = x$, $P_0 = L$, $P_1' = 1$ and $P_0 = 1$, we get

$$\begin{aligned} P_n' &= (2n-1)\,P_{n-1} + P_{n-2}' \\ P_{n-2}' &= (2n-5)\,P_{n-3} + P_{n-4}' \\ P_{n-4}' &= (2n-9)\,P_{n-5} + P_{n-6}' \\ &\vdots \\ P_5' &= 9P_4 + P_3' \\ P_3' &= 5P_2 + P_1' \end{aligned} \qquad [\because p_1' = P_0]$$

Adding these and simplifying, we get

$$P_n' = (2n-1)\,P_{n-1} + (2n-5)\,P_{n-3} + \cdots + 5P_2 + P_0 \tag{iii}$$

Combining Eqs. (ii) and (iii), we have

$$P_n' = \sum_{r=0}^{[(n-1)/2]} (2n-4r-1)\,P_{n-2r-1}(x).$$

EXAMPLE 15.12 Prove that $\displaystyle\int_{-1}^{1} xP_n(x)\,P_{n-1}(x)\,dx = \frac{2n}{4n^2-1}.$

Proof: From recurrence relation R1, we have

$$xP_n(x) = \frac{n+1}{2n+1} P_{n+1}(x) + \frac{n}{2n+1} P_{n-1}(x) \tag{i}$$

Multiply both sides by $P_{n-1}(x)$ and then integrate w.r.t. x over the interval [–1, 1], we get

$$\int_{-1}^{1} xP_n(x)\, P_{n-1}(x)\, dx = \frac{n+1}{2n+1} \int_{-1}^{1} P_{n+1}(x)\, P_{n-1}(x)\, dx + \frac{n}{2n+1} \int_{-1}^{1} [P_{n-1}(x)]^2\, dx \tag{ii}$$

We know that

$$\int_{-1}^{1} P_m(x)\, P_n(x)\, dx = \begin{cases} 0, & \text{if } m \neq n \\ \dfrac{2}{(2n+1)}, & \text{if } m = n \end{cases} \tag{iii}$$

Using Eq. (iii) in Eq. (ii), we get

$$\int_{-1}^{1} xP_n(x)\, P_{n-1}(x)\, dx = 0 + \frac{n}{2n+1} \cdot \frac{2}{2(n-1)+1}$$

$$= \frac{2n}{(2n+1)(2n-1)} = \frac{2n}{4n^2-1}.$$

EXAMPLE 15.13 Show that $\displaystyle\int_{-1}^{1} x^2 P_{n+1}(x)\, P_{n-1}(x)\, dx = \frac{2n(n+1)}{(2n-1)(2n+1)(2n+3)}$.

Deduce the value of $\displaystyle\int_{0}^{1} x^2 P_{n+1}(x)\, P_{n-1}(x)\, dx$.

Proof: From recurrence relation R1, we have

$$(2n-1)\, xP_{n-1} = nP_n + (n-1)\, P_{n-2} \tag{i}$$

Replace n by $n + 2$ in Eq. (i), we have

$$(2n+3)\, xP_{n+1} = (n+2)\, P_{n+2} + (n-1)\, P \tag{ii}$$

Multiplying two equations give

$$(2n-1)(2n+3)\, x^2 P_{n-1}\, P_{n+1} = n(n+1) P_n^2 + n(n+2) P_{n+2} P_n + (n-1)(n+2)\, P_{n-2} P_{n+1}$$
$$+ (n-1)(n+1)\, P_{n-2} P_n$$

Integrating w.r.t. x over [–1, 1] and using orthogonality of the Legendre polynomials, we have

$$(2n-1)(2n+3) \int_{-1}^{1} x^2 P_{n-1}(x) P_{n+1}(x)\, dx = n(n+1) \frac{2}{2n+1}$$

$$\Rightarrow \qquad \int_{-1}^{1} x^2 P_{n-1}(x)\, P_{n+1}(x)\, dx = \frac{2n(n+1)}{(2n-1)(2n+1)(2n+3)}.$$

EXAMPLE 15.14 Prove that $xP'_n = nP_n + (2n-3)P_{n-2} + (2n-7)P_{n-4} + \cdots$, and hence, or otherwise prove that

(i) $\displaystyle\int_{-1}^{1} xP_nP'_n\,dx = \frac{2n}{2n+1}$

(ii) $\displaystyle\int_{-1}^{1} xP_nP'_m\,dx = \text{either } 0 \text{ or } 2 \text{ or } \frac{2n}{2n+1}$.

Proof: From recurrence relation R2, we have

$$xP'_n - P'_{n-1} = nP_n \tag{i}$$

Again, from recurrence relation R3, we have

$$P'_{n+1} - P'_{n-1} = (2n+1)P_n \tag{ii}$$

Replacing n by $n-2$, $n-4$, $n-6$, ... successively in Eq. (i), we get

$$\left.\begin{aligned} P'_{n-1} - P'_{n-3} &= (2n-3)\,P_{n-2} \\ P'_{n-3} - P'_{n-5} &= (2n-7)\,P_{n-4} \\ &\vdots \end{aligned}\right\} \tag{iii}$$

Adding Eqs. (i) and (iii) and simplifying, we get

$$xP'_n = nP_n + (2n-3)\,P_{n-2} + (2n-7)\,P_{n-4} + \cdots \tag{iv}$$

Multiply both sides of Eq. (iv) by P_n, we get

$$xP_nP'_n = nP_n^2 + (2n-3)\,P_{n-2}P_n + (2n-7)\,P_{n-4}P_n + \cdots$$

and integrate w.r.t. x over $[-1, 1]$. Therefore,

$$\int_{-1}^{1} xP_nP'_n\,dx = n\,\frac{2}{2n+1} + 0 + 0 + \cdots = \frac{2n}{2n+1}$$

(iii) Replace n by m in Eq. (iv) and multiply both sides by P_n, we get

$$xP_nP'_m = mP_mP_n + (2m-3)\,P_{m-2}P_n + (2m-7)\,P_{m-4}P_n + \cdots \tag{v}$$

We have the following three cases:

Case 1: When n is different from m, $m-2$, $m-4$, ... and so on. Then we have

$$\int_{-1}^{1} xP_nP'_m\,dx = 0$$

Case 2: When $n = m$, then we have

$$\int_{-1}^{1} xP_nP'_n\,dx = \frac{2n}{2n+1} \qquad \text{(as proved above)}$$

Case 3: Let $n = m - 2$. Then $n \neq m$, $n \neq m - 4$, ... and so on. So

$$\int_{-1}^{1} xP_n P_m' \, dx = 0 + (2m-3)\frac{2}{2n+1}$$

$$= [2(n+2)-3]\frac{2}{2n+1}$$

$$= (2n+1)\frac{2}{2n+1} = 2$$

Similarly, one can prove that

$$\int_{-1}^{1} xP_n P_m' \, dx = 2, \text{ when } n = m - 4 \text{ or } n = m - 6, \text{ etc.}$$

Thus, $$\int_{-1}^{1} xP_n P_m' \, dx = 0 \text{ or } 2 \text{ or } \frac{2n}{2n+1}.$$

EXAMPLE 15.15 Prove that $\int_{-1}^{1} (P_n')^2 \, dx = n(n+1)$.

Proof: From Christoffel's expansion, we get

$$P_n' = (2n-1)P_{n-1} + (2n-5)P_{n-3} + (2n-9)P_{n-5} + \cdots \quad \text{(i)}$$

The last term on RHS of Eq. (i) is $3P_1$ or P_0 according as n is even or odd.

$$\therefore \quad (P_n')^2 = (2n-1)^2 P^2{}_{n-1} + (2n-5)^2 P_{n-3}^2 + \cdots + (3P_1)^2 + 2\sum(2n-1)(2n-5)P_{n-1}P_{n-3} \quad \text{(ii)}$$

$$\text{or} \quad (P_n')^2 = (2n-1)^2 P^2{}_{n-1} + (2n-5)^2 P_{n-3}^2 + \cdots + P_0^2 + 2\sum(2n-1)(2n-5)P_{n-1}P_{n-3} \quad \text{(iii)}$$

We consider two cases:

Case 1: When n is even. Integrating both sides of (2A) w.r.t. x from -1 to 1, we get

$$\int_{-1}^{1} (P_n')^2 \, dx = (2n-1)^2 \frac{2}{2(n-1)+1} + (2n-5)^2 \frac{2}{2(n-3)+1} + \cdots + 3^2 \cdot \frac{2}{2 \cdot 1 + 1}$$

$$= 2[(2n-1) + 92n-5) + \cdots + 3]$$

$$= 2 \cdot \frac{\frac{n}{2}}{2}(2n-1+3) = n(n+1)$$

Case 2: When n is odd. Integrate (ii) w.r.t. x over $[-1, 1]$, we get

$$\int_{-1}^{1} (P_n')^2 \, dx = (2n-1)^2 \cdot \frac{2}{2(n-1)+1} + (2n-5)^2 \cdot \frac{2}{2(n-3)+1} + \cdots + \frac{2}{2 \cdot 0 + 1}$$

$$= 2[(2n-1)+(2n-5)+\cdots+1]$$

$$= 2\frac{\frac{(n+1)}{2}}{2}(2n-1+1) = n(n+1).$$

EXAMPLE 15.16 Show that all the roofs of $P_n(x) = 0$ are real and lie between -1 and 1.

Proof: Consider $f(x) = (x^2 - 1)^n = (x - 1)^n (x + 1)^n$ (i)

From Eq. (i), we observe that $f(x)$ vanishes for $x = 1$, and $x = -1$, hence, by Rolle's theorem, $f'(x)$ must vanish at least for one $\alpha \in (-1, 1)$.
Now, from Eq. (i), we have

$$f'(x) = n(x-1)^{n-1}(x+1)^n + n(x+1)^{n-1}(x-1)^n \tag{ii}$$

Equation (ii) proves that $f'(x)$ vanishes at $x = 1$ and $x = -1$. But we have argued that $f'(x)$ vanishes at $x = \alpha \in (-1, 1)$. Hence, by applying Rolle's theorem to $f'(x)$ two times, we conclude that $f''(x)$ must vanish at $x = \beta \in (-1, 1)$ and also at $x = \gamma \in (0, 1)$. Proceeding this way, we conclude that $f^n(x) = 0$ must have n real roots lying between $(-1, 1)$. Rodrigue's formula gives

$$P_n(x) = \frac{1}{2^n n!}\frac{dn}{dx^n}(x^2-1)^n = \frac{1}{2^n \cdot n!} f^{(n)}(x) = 0$$

$$\Rightarrow \qquad P_n(x) = 0$$

Hence, $P_n(x) = 0$ have n real roots between -1 and 1.

EXAMPLE 15.17 Prove that all the roots of $P_n(x)$ are distinct.

Proof: If possible, let the roots of $P_n(x) = 0$ be not distinct. Then at least two roots must be equal. Let α be the repeated root. Then from the theory of equations we have

$$P_n(\alpha) = 0 \text{ and } P_n'(\alpha) = 0 \tag{i}$$

Since $P_n(x)$ satisfies Legendre's equation, we get

$$(1-x^2)P_n'' - 2xP_n' + n(n+1)P_n = 0 \tag{ii}$$

Differentiating r-times and using Leibnitz theorem, Eq. (ii) gives

$$(1-x^2)\frac{d^{r+2}}{dx^{r+2}}P_n(x) + c(r,1)(-2x)\frac{d^{r+1}}{dx^{r+1}}P_n(x) + c(r,2)(-2)\frac{d^r}{dx^r}P_n(x)$$

$$-2\left[\frac{d^{r+1}}{dx^{r+1}}P_n(x) + c(r,1)\cdot\frac{d^r}{dx^r}P_n(x)\right] + n(n+1)\frac{d^r}{dx^r}P_n(x) = 0$$

$$\Rightarrow (1-x^2)\frac{d^{r+2}}{dx^{r+2}}P_n(x) - 2x[1+c(r,1)]\frac{d^{r+1}}{dx^{r+1}}P_n(x) - [2c(r,2)+2c(r,1)+n(n+1)]\frac{d^r}{dx^r}P_n(x) = 0 \tag{iii}$$

Putting $r = 0$ and $x = \alpha$ in Eq. (iii) and using Eq. (i), we get

$$(1-\alpha^2)P_n''(\alpha) - 0 - 0 = 0$$

$$\Rightarrow \qquad P_n''(\alpha) = 0 \tag{iv}$$

Putting $r = 1$, and $x = \alpha$ in Eq. (iii) and using Eqs. (i) and (iv), we get

$$P_n'''(\alpha) = 0 \tag{v}$$

Similarly, $$P_n^{(n)}(\alpha) = 0$$

But $$P_n(x) = \frac{1\cdot 3\cdots(2n-1)}{n!}\left[x^n - \frac{n(n-1)}{2(2n-1)}x^{n-2} + \cdots\right]$$

$$\therefore \quad \frac{d^n}{dx^n}P_n(x) = \frac{1\cdot 3\cdots(2n-1)}{n!}n! \tag{vi}$$

Hence, $$\left[\frac{d^n}{dx^n}P_n(x)\right]_{x=\alpha} \neq 0 \tag{vii}$$

Since Eqs. (vi) and (vii) are contradictory results, our supposition that the roots of $P_n(x) = 0$ are not distinct is absurd. Hence, all the roots of $P_n(x) = 0$ must be distinct.

EXAMPLE 15.18 Prove that

(i) $\int_{-1}^{1} P_n(x)\,dx = 2$, if $n = 0$

(ii) $\int_{-1}^{1} P_n(x)\,dx = 0$, if $n \geq 1$

Proof: (i) Try yourself

(ii) Using Rodrigue's formula, we have

$$\int_{-1}^{1} P_n(x)\,dx$$

$$= \frac{1}{2^n\cdot n!}\int_{-1}^{1} D^n(x^2-1)^n\,dx, \text{ where } D^n = \frac{d^n}{dx^n}$$

$$= \frac{1}{2^n n!}\{D^{n-1}[(x^2-1)^n]\}_{-1}^{1}$$

$$= \frac{1}{2^n n!}\{D^{n-1}[(x-1)^n\,(x+1)^n]\}_{-1}^{1}$$

$$= \frac{1}{2^n\cdot n!}[(x+1)^n D^{n-1}(x-1)^n + c(n-1,1)D^{n-2}(x-1)^n D(x+1)^n + \cdots + (x-1)^n D^{n-1}(x+1)^n]$$

using Leibnitz's theorem

$$= \frac{1}{2^n\cdot n!}[n!(x-1)\,(x+1)^n + \cdots + n!(x+1)\,(x-1)^n]_{-1}^{1}$$

$$= 0. \qquad \left[\because\ D^n(ax+b)^m = a^n\frac{m!}{(m-n)!}(ax+b)^{m-n}\right]$$

EXAMPLE 15.19 If $m > n - 1$ and n is a positive integer, prove that

$$\int_0^1 x^m P_n(x)\,dx = \frac{m(m-1)(m-2)\cdots(m-n+2)}{(m+n+1)(m+n-1)\cdots(m-n+1)}.$$

Proof: Using Rodrigue's formula, we have

$$\int_0^1 x^m P_n(x)\,dx$$

$$= \frac{1}{2^n \cdot n!}\int_0^1 x^m D^n (x^2-1)^n\,dx$$

$$= \frac{1}{2^n n!}\left\{[x^m D^{n-1}(x^2-1)^n]_0^1 - m\int_0^1 x^{m-1} D^{n-1}(x^2-1)^n\,dx\right\} \tag{i}$$

$$= \frac{(-1)m}{2^n \cdot n!}\int_0^1 x^{m-1} D^{n-1}(x^2-1)^n\,dx$$

$$= \frac{(-1)^2 m(m-1)}{2^n \cdot n!}\int_0^1 x^{m-2} D^{n-2}(x^2-1)^n\,dx$$

$$= \frac{(-1)^n m(m-1)\cdots(m-n+1)}{2^n \cdot n!}\int_0^1 x^{m-n}(x^2-1)^n\,dx$$

$$= \frac{(-1)^n m(m-1)\cdots(m-n+1)}{2^n \cdot n!}\int_0^1 x^{m-n}(-1)^n(1-x^2)^n\,dx$$

$$= \frac{m(m-1)\cdots(m-n+1)}{2^n \cdot n!}\int_0^1 x^{m-n}(1-x^2)^n\,dx$$

Substitute $x^2 = t$, so that $dx = \dfrac{dt}{2t^{1/2}}$

$$= \frac{m(m-1)\cdots(m-n+1)}{2^{n+1} \cdot n!}\int_0^1 (t^{1/2})^{(m-n-1)}(1-t)^n\,dt$$

$$= \frac{m(m-1)\cdots(m-n+1)}{2^{n+1} \cdot n!}\int_0^1 t^{1/2(m-n+1)-1}(1-t)^{(n+1)-1}\,dt$$

$$= \frac{m(m-1)\cdots(m-n+1)}{2^{n+1} \cdot n!}\,\frac{\Gamma\left(\dfrac{m-n+1}{2}\right)\Gamma(n+1)}{\Gamma\left(\dfrac{m-n+1}{2}+n+1\right)}$$

$$= \frac{m(m-1)\cdots(m-n+1)}{2^{n+1}\cdot n!}\frac{\Gamma\left(\frac{m-n+1}{2}\right)n!}{\Gamma\left(\frac{m+n+3}{2}\right)}$$

$$= \frac{m(m-1)\cdots(m-n+1)\Gamma\left(\frac{m-n+1}{2}\right)}{2^{n+1}\cdot\frac{m+n+1}{2}\cdot\frac{m+n-1}{2}\cdots\frac{m-n+1}{2}\Gamma\left(\frac{m-n+1}{2}\right)}$$

$$= \frac{m(m-1)\cdots(m-n+1)}{(m+n+1)(m+n-1)\cdots(m-n+1)}$$

EXERCISES

1. Prove the following:

(i) $P_n(1) = 1$

(ii) $P_n(-1) = (-1)^n$

(iii) $P_n'(1) = \frac{1}{2}n(n+1)$

(iv) $P_n'(-1) = (-1)^{n-1}\frac{n(n+1)}{2}$

(v) $P_{2n}(0) = (-1)^n \frac{2n!}{2^{2n}(n!)^2}$

(vi) $P_{2n+1}(0) = 0.$

2. Prove that $\dfrac{1+z}{z\sqrt{1-2xz+z^2}} - \dfrac{1}{z} = \sum_{n=0}^{\infty}(P_n + P_{n+1})\, z^n.$

3. Prove that $|P_n(\cos\theta) \le 1|$ when θ is real.

4. Expand $f(x) = x^4 + 2x^3 + 5x^2 - x - 2$ in terms of Legendre polynomials.

$$\left[\textbf{Ans. } f(x) = \frac{8}{35}P_4 + \frac{4}{5}P_3 + \frac{82}{21}P_2 + \frac{1}{5}P_1 - \frac{2}{15}P_0\right]$$

5. If $P_n(x)$ is a Legendre polynomial of degree n and α is such that $P_n(\alpha) = 0$ then show that $P_{n-1}(\alpha)$ and $P_{n+1}(\alpha)$ are of opposite signs.

6. Prove that $\sum_{r=0}^{n}(2r+1)P_r = P_{n+1}' + P_n'.$

7. Prove that $P'_{2n+1} = (2n+1)P_{2n} + 2nxP_{2n-1} + (2n-1)x^2P_{2n-2} + \cdots + 2x^{n-1}P_1 + x^{2n}$.

8. Prove that $\int_0^1 P_n(x)\,dx = \dfrac{(-1)^{(n-1)/2}\,(n-1)!}{2^n\left(\dfrac{n}{2}+1\right)!\left(\dfrac{n}{2}-1\right)!}$ when n is even.

9. Prove that $P_0^2 + 3P_1^2 + 5P_2^2 + \cdots + (2n+1)P_n^2 = (n+1)^2[P_nP'_{n+1} - P_{n+1}P'_n]$.

10. Prove that $P'_n - P'_{n-1} = (2n-1)P_{n-1}$.

11. Prove $\int_{-1}^{1}(1-x^2)P'_nP'_m\,dx = \begin{cases} 0, & \text{if } n \neq m \\ \dfrac{2n(n+1)}{2n+1}, & \text{if } n = m \end{cases}$.

12. Prove that $\int_{-1}^{1} x^2P_n^2\,dx = \dfrac{1}{8(2n-1)} + \dfrac{3}{4(2n+1)} + \dfrac{1}{8(2n+3)}$.

13. If $x > 1$, show that $P_n(x) < P_{n+1}(x)$.

14. If $m < n$, show that $\int_0^m x^mP_n(x)\,dx = 0$.

15. Prove that $\int_{-1}^{1} x^nP_n(x)\,dx = \dfrac{2^{n+1}(n!)^2}{(2n+1)!}$.

16. Show that $\int_{-1}^{1} x^4P_6(x)\,dx = 0$.

17. Use the Rodrigue's formula for Legendre polynomial $P_n(x)$ to prove that

$$\int_{-1}^{1} f(x)\,P_n(x)\,dx = \frac{(-1)^n}{2^n \cdot n!}\int_{-1}^{1} f^n(x)\,(x^2-1)^n\,dx$$

where $f(x)$ is a continuous function in $(-1, 1)$ and $f^n(x)$ denotes the nth derivative of f. Hence, deduce that

$$\int_{-1}^{1} x^mP_n(x)\,dx = \begin{cases} 0, & \text{if } m < n \\ \dfrac{2^{n+1}(n!)^2}{(2n+1)!}, & \text{if } m = n \end{cases}.$$

CHAPTER 16

Laplace Transforms and Its Applications

16.1 INTRODUCTION

Laplace transforms have wide applications in mathematical physics. It provides an easy and effective tool for the solution of problems arising in engineering. The advantage of the Laplace transforms is obtaining the solution of initial value problems without finding the general solution and arbitrary constants.

16.2 LAPLACE TRANSFORM OF SOME ELEMENTARY FUNCTIONS

Let $f(t)$ be a function of t defined for all $t > 0$. Then the Laplace transform of $f(t)$, denoted by $L\{f(t)\}$, is defined as

$$L\{f(t)\} = \int_0^\infty e^{-st} f(t)\, dt \tag{16.1}$$

provided that the integral exists. s is a parameter which may be real or complex.

Note: Laplace transform $L\{f(t)\}$ is also denoted by $F(s)$.

1. $L(1) = \dfrac{1}{s}, s > 0.$

Proof: $L(1) = \int_0^\infty e^{-st} 1\, dt = \left[-\dfrac{e^{-st}}{s}\right]_0^\infty = \dfrac{1}{s}$

2. $L(t^n) = \dfrac{n!}{s^{n+1}}$, $n = 0, 1, 2, 3, \ldots$.

Proof: $L(t^n) = \int_0^\infty e^{-st} t^n\, dt$

$$= \int_0^\infty e^{-y} \left(\frac{y}{s}\right)^n \frac{dy}{s}, \text{ putting } st = y$$

$$= \frac{1}{s^{n+1}} \int_0^{\infty} e^{-y} y^n \, dy$$

$$= \frac{\Gamma(n+1)}{s^{n+1}}, \; n > -1 \text{ and } s > 0$$

$$= \frac{n!}{s^{n+1}}$$

3. $L(e^{at}) = \dfrac{1}{s-a}, \; s > a.$

Proof: Try yourself.

4. $L(\sin at) = \dfrac{a}{s^2 + a^2}, \; s > 0.$

Proof: $L(\sin at) = \displaystyle\int_0^{\infty} e^{-st} \sin at \, dt$

$$= \left[\frac{e^{-st}}{s^2 + a^2} (-s \sin at - a \cos at) \right]_0^{\infty}$$

$$= \frac{a}{s^2 + a^2}.$$

5. $L(\cos at) = \dfrac{s}{s^2 + a^2}, \; s > 0.$

Proof: Try yourself.

6. $L(\sinh at) = \dfrac{a}{s^2 + a^2}, \; s > |a|.$

Proof: $L(\sinh at) = \displaystyle\int_0^{\infty} e^{-st} \sinh at \, dt$

$$= \int_0^{\infty} e^{-st} \left(\frac{e^{at} - e^{-at}}{2} \right) dt$$

$$= \frac{1}{2} \left[\int_0^{\infty} e^{-(s-a)t} dt - \int_0^{\infty} e^{-(s+a)t} dt \right]$$

$$= \frac{1}{2} \left[\frac{1}{s-a} - \frac{1}{s+a} \right]$$

$$= \frac{a}{s^2 - a^2}, \; s > |a|.$$

7. $L(\cosh at) = \dfrac{s}{s^2 - a^2}$, $s > |a|$.

Proof: Try yourself.

The table below shows Laplace transforms of various elementary functions.

$f(t)$	$F(s)$	*Condition*
1	$\dfrac{1}{s}$	$s > 0$
T	$\dfrac{1}{s^2}$	$s > 0$
t^n	$\dfrac{n!}{s^{n+1}}$	$s > 0$
e^{at}	$\dfrac{1}{s-a}$	$s > a$
$\sin at$	$\dfrac{a}{s^2+a^2}$	$s > 0$
$\cos at$	$\dfrac{s}{s^2+a^2}$	$s > 0$
$\sinh at$	$\dfrac{a}{s^2-a^2}$	$s > \lvert a \rvert$
$\cosh at$	$\dfrac{s}{s^2-a^2}$	$s > \lvert a \rvert$
$J_o(at)$	$\dfrac{1}{\sqrt{s^2+a^2}}$	
$J_n(at)$	$\dfrac{(\sqrt{s^2+a^2}-s)^n}{a^n\sqrt{s^2+a^2}}$	

16.3 PROPERTIES OF LAPLACE TRANSFORMS

1. **Linearity Property:** If a and b are any constants and f and g are functions of t, $t > 0$, then

$$L\{af(t) + bg(t)\} = aL\{f(t)\} + bL\{g(t)\}.$$

Proof:

$$L\{af(t) + bg(t)\} = \int_0^\infty e^{-st}[af(t) + bg(t)]\, dt$$

$$= a\int_0^\infty e^{-st} f(t)\, dt + b\int_0^\infty e^{-st} g(t)\, dt$$

$$= aL\{f(t)\} + bL\{g(t)\}$$

This result can be generalized.

2. **First Shifting Property:** If $L\{f(t)\} = F(s)$, then $L\{e^{at} f(t)\} = F(s - a)$.

Proof: By definition of Laplace transform, we have

$$L\{e^{at} f(t)\} = \int_0^\infty e^{-st} e^{at} f(t)\, dt$$

$$= \int_0^\infty e^{-(s-a)t} f(t)\, dt$$

$$= F(s - a)$$

Using First shifting property, following formulae can be obtained, for $s > a$.

1. $L(e^{at} t^n) = \dfrac{n!}{(s-a)^{n+1}}$.

2. $L(e^{at} \sin bt) = \dfrac{b}{(s-a)^2 + b^2}$.

3. $L(e^{at} \cos bt) = \dfrac{s-a}{(s-a)^2 + b^2}$.

4. $L(e^{at} \sinh bt) = \dfrac{b}{(s-a)^2 - b^2}$.

5. $L(e^{at} \cosh bt) = \dfrac{s-a}{(s-a)^2 - b^2}$.

EXAMPLE 16.1 Find the Laplace transform of $f(t) = (3e^{2t} - 4)^2$.

Solution

$$L\{f(t)\} = L\{(3e^{2t} - r)\}^2$$

$$= L\{9e^{4t} - 24e^{2t} + 16\}$$

$$= \frac{9}{s-4} - \frac{24}{s-2} + \frac{16}{s}, s > 4.$$

EXAMPLE 16.2 Find the Laplace transform of

(i) $\sin 3t \cos t$ and (ii) $\sin^2 4t$.

Solution

(i) $\sin 3t \cos t = \dfrac{1}{2}(\sin 4t + \sin 2t)$

$\therefore$

$$L\{\sin 3t \cos t\} = \frac{1}{2} L\{\sin 4t + \sin 2t\}$$

$$= \frac{1}{2}[L\{\sin 4t\} + L\{\sin 2t\}]$$

$$= \frac{1}{2}\left[\frac{4}{s^2 + 16} + \frac{2}{s^2 + 4}\right]$$

(ii) $\sin^2 4t = \frac{1}{2}(1 - \cos 8t)$

$$\therefore \quad L\{\sin^2 4t\} = \frac{1}{2}\{L(1 - \cos 8t)\}$$

$$= \frac{1}{2}[L\{1\} - L\{\cos 8t\}]$$

$$= \frac{1}{2}\left[\frac{1}{s} - \frac{s}{s^2 + 64}\right]$$

EXAMPLE 16.3 Find the Laplace transform of e^{-4t} (sin 5t + 3 cos 2t).

Solution

$$L\{e^{-4t}(\sin 5t + 3\cos 2t)\}$$

$$= L\{e^{-4t}\sin 5t\} + 3L\{e^{-4t}\cos 2t\}$$

$$= \frac{4}{(s+4)^2 + 5^2} + \frac{3(s+4)}{(s+4)^2 + 5^2}$$

$$= \frac{35 + 16}{(s+4)^2 + 5^2}.$$

EXAMPLE 16.4 Given that $L\left\{\frac{\sin t}{t}\right\} = \tan^{-1}\left(\frac{1}{s}\right)$, find $L\left\{\frac{\sin 2t}{t}\right\}$.

Solution

$$L\left\{\frac{\sin 2t}{t}\right\} = L\left\{2\frac{\sin 2t}{2t}\right\}$$

$$= 2 \cdot L\left\{\frac{\sin 2t}{2t}\right\}$$

$$= 2 \cdot \frac{1}{2}\tan^{-1}\frac{2}{s}$$

$$= \tan^{-1}\frac{2}{s}.$$

EXAMPLE 16.5 Find the Laplace transform of $f(t)$ defined as $f(t) = \begin{cases} t/a, & 0 < t < a \\ 1, & t > a \end{cases}$.

Solution

$$L\{f(t)\} = \int_0^a e^{-st} \cdot \frac{t}{a}\,dt + \int_a^\infty e^{-st} \cdot 1\,dt$$

$$= \frac{1}{a}\left[t \cdot \frac{e^{-st}}{-s}\Bigg|_0^a - \int_0^a 1 \cdot \frac{e^{-st}}{-s}\,dt\right] + \left|\frac{e^{-st}}{-s}\right|_a^\infty$$

$$= \frac{1}{a}\left[\frac{ae^{-as}}{-s} - \frac{e^{-as}}{s^2} + \frac{1}{s^2}\right] + \frac{e^{-as}}{s}$$

$$= \frac{1 - e^{-as}}{as^2}.$$

EXAMPLE 16.6 Find the Laplace transform of $F(t) = \begin{cases} \sin\left(t - \dfrac{\pi}{4}\right), t > \dfrac{\pi}{4} \\ 0 \qquad\qquad ,t < \dfrac{\pi}{4} \end{cases}.$

Solution $L\{f(t)\} = e^{-\frac{\pi}{4}s} L\{\sin t\} = e^{-\frac{\pi}{4}s} \dfrac{1}{s^2 + 1}.$

16.4 LAPLACE TRANSFORMS OF DERIVATIVES

1. If $f'(t)$ is continuous and $L\{f(t)\} = F(s)$, then $L\{f'(t)\} = sF(s) - f(0)$.

Proof:

$$L\{f'(t)\} = \int_0^\infty e^{-st} f'(t)\, dt$$

$$= \left|e^{-st} f(t)\right]_0^\infty - \int_0^\infty (-s)\, e^{-st} f(t)\, dt \qquad \text{integrate by parts.}$$

Now,

$$\lim_{t\to\infty} e^{-st} f(t) = 0.$$

$\therefore$

$$L\{f'(t)\} = -f(0) + s\int_0^\infty e^{-st} f(t)\, dt$$

$$= -f(0) + sF(s)$$

$$= sF(s) - f(0).$$

2. If $f'(t)$ and its first $(n - 1)$ derivatives are continuous, then

$$L\{f^n(t)\} = s^n F(s) - s^{n-1} f(0) - s^{n-2} f'(0) \cdots - f^{n-1}(0).$$

Proof: Using the general rule of integration, we have

$$L\{f^n(t)\} = \int_0^\infty e^{-st} f^n(t)\, dt$$

$$= [e^{-st} f^{n-1}(t) - (-s)e^{-st} f^{n-2}(t) + \cdots + (-1)^{n-1}(-s)^{n-1} e^{-st} f(t)]_0^\infty + (-1)^n(-s)^n \int_0^\infty e^{-st} f(t)\, dt$$

$$= -f^{n-1}(0) - sf^{n-2}(0) - \cdots - s^{n-1}f(0) + s^n \int_0^\infty e^{-st} f(t)\,dt$$

$$= s^n F((s) - s^{n-1}f(0) - \cdots - sf^{n-2}(0) - f^{n-1}(0).$$

EXAMPLE 16.7 Assuming $L\{sin\ at\} = \dfrac{a}{s^2 + a^2}$, find $L\{\cos at\}$.

Solution $f(t) = \sin at$

$\therefore$ $$f'(t) = a \cos at$$

Now, $$L\{f'(t)\} = sF(s) - f(0)$$

$\Rightarrow$ $$aL\{\cos at\} = s \cdot \frac{a}{s^2 + a^2} - 0$$

$$= \frac{as}{s^2 + a^2}$$

$\therefore$ $$L\{\cos at\} = \frac{s}{s^2 + a^2}.$$

EXAMPLE 16.8 Show that $L\{1\} = \dfrac{1}{s}$, $L(t) = \dfrac{1}{s^2}$, ..., $L\{t^n\} = \dfrac{n!}{s^{n+1}}$.

Solution We know that

$$L\{f'(t)\} = sF(s) - f(0)$$

Put $$f(t) = 1, \text{ then } f'(t) = 0$$

$\therefore$ $$L\{0\} = sL\{1\} - 1$$

$\Rightarrow$ $$0 = sL\{1\} - 1$$

$\Rightarrow$ $$L\{1\} = \frac{1}{s}$$

Now, take $$f(t) = t$$

$$L\{1\} = sL\{t\} - 0$$

$$\frac{1}{s} = sL\{t\} - 0$$

$\Rightarrow$ $$L(t) = \frac{1}{s^2}$$

Now, take $$f(t) = t^2$$

Then $$L\{2t\} = sL\{t^2\} - 0$$

$$\frac{2}{s^2} = sL\{t^2\}$$

$$\Rightarrow \qquad L\{t^2\} = \frac{2}{s^3}$$

Continuing this way, take $\qquad f(t) = t^n$

$$L\{nt^{n-1}\} = sL(t^n) - 0$$

$$\Rightarrow \qquad n \cdot \frac{(n-1)!}{s^n} = sL\{t^n\}$$

$$\Rightarrow \qquad L\{t^n\} = \frac{n!}{s^{n+1}}.$$

EXAMPLE 16.9 Find $L\{\cos 2t\}$.

Solution Let $f(t) = \cos 2t$. Then $f'(t) = -2\sin 2t$, $f''(t) = -4\cos 2t$.

Now, $\qquad L\{f''(t)\} = s^2 L\{f(t)\} - sf(0) - f'(0)$

$$\Rightarrow \qquad -4L\{\cos 2t\} = s^2 L\{\cos 2t\} - s$$

$$L\{\cos 2t\} = \frac{s}{s^2+4}.$$

16.5 LAPLACE TRANSFORM OF INTEGRALS

THEOREM 16.1 If $L\{f(t)\} = F(s)$, then $L\left\{\int_0^t f(u)\,du\right\} = \frac{F(s)}{s}$.

Proof: Let $\qquad g(t) = \int_0^t f(u)\,du$

Then $\qquad g'(t) = f(t)$

Now, $\qquad L\{g'(t)\} = sL\{g(t)\} - g(0)$

$$\Rightarrow \qquad L\{f(t)\} = sL\left\{\int_0^t f(u)\,du\right\}$$

$$\Rightarrow \qquad L\left\{\int_0^t f(u)\,du\right\} = \frac{F(s)}{s}.$$

EXAMPLE 16.10 Find $L\left\{\int_0^t \frac{\sin u}{u}\,du\right\}$.

Solution We know that $L\left\{\frac{\sin t}{t}\right\} = \tan^{-1}\frac{1}{s}$

$$\therefore \quad L\left\{\int_0^t \frac{\sin u}{u}\,du\right\} = \frac{1}{s}\tan^{-1}\frac{1}{s}.$$

THEOREM 16.2 If $L\{f(t)\} = F(s)$, then $L\{t^n f(t)\} = (-1)^n F^n(s)$.

Proof: We have

$$F(s) = \int_0^\infty e^{-st} f(t)\,dt$$

$$\Rightarrow \quad F'(s) = \int_0^\infty -te^{-st} f(t)\,dt = -\int_0^\infty e^{-st}[tf(t)]\,dt$$

Again differentiating gives

$$F''(s) = \int_0^\infty e^{-st}[t^2 f(t)]\,dt$$

Similarly,

$$F^n(s) = (-1)^n \int_0^\infty e^{-st}[t^n f(t)]\,dt$$

$$= (-1)^n L[t^n f(t)]$$

$$\therefore \quad L\{t^n f(t)\} = (-1)^n F^n(s).$$

EXAMPLE 16.11 Evaluate $L\{t^2 \cos 2t\}$.

Solution

$$L\{\cos 2t\} = \frac{s}{s^2+4}.$$

$$\therefore \quad L\{t^2 \cos 2t\} = \frac{d^2}{ds^2}\left\{\frac{s}{s^2+4}\right\}$$

$$= \frac{d}{ds}\left\{\frac{4-s^2}{(s^2+4)^2}\right\}$$

$$= \frac{2s^3-24s}{(s^2+4)^3}.$$

THEOREM 16.3 If $L\{f(t)\} = F(s)$, then $L\left\{\frac{f(t)}{t}\right\} = \int_s^\infty F(u)\,du$, provided $\lim_{t\to 0}\frac{f(t)}{t}$ exists.

Proof: Let

$$\frac{f(t)}{t} = g(t).$$

$\Rightarrow$ $$f(t) = tg(t)$$

$\therefore$ $$L\{f(t)\} = L\{tg(t)\}$$

$$= -\frac{d}{ds}L\{g(t)\}$$

$\Rightarrow$ $$F(s) = -\frac{d}{ds}L\{g(t)\}$$

Integrating both sides, we get

$$L\{g(t)\} = -\int_\infty^s F(u)\,du$$

$\Rightarrow$ $$L\left\{\frac{f(t)}{t}\right\} = \int_s^\infty F(u)\,du.$$

EXAMPLE 16.12 Prove that $L\left\{\frac{\sin t}{t}\right\} = \tan^{-1}\frac{1}{s}$.

Proof: We know that

$$L\{\sin t\} = \frac{1}{1+s^2}$$

and $$\lim_{t\to 0}\frac{\sin t}{t} = 1$$

$\therefore$ $$L\left\{\frac{\sin t}{t}\right\} = \int_s^\infty \frac{1}{1+u^2}\,du$$

$$= \tan^{-1} u\Big|_s^\infty$$

$$= \frac{\pi}{2} - \tan^{-1} s = \tan^{-1}\frac{1}{s}.$$

EXAMPLE 16.13 Prove that $\int_0^\infty \frac{\sin t}{t} = \frac{\pi}{2}$.

Proof: We know that

$$L\{\sin t\} = \frac{1}{1+s^2}$$

$$\therefore \quad \int_0^\infty \frac{\sin t}{t} = \int_0^\infty \frac{du}{u^2+1} = \tan^{-1} u \Big|_0^\infty = \frac{\pi}{2}$$

EXAMPLE 16.14 Find the Laplace transform of $\dfrac{1-e^t}{t}$.

Solution Since $$L\{1 - e^t\} = L\{1\} - L\{e^t\}$$

$$= \frac{1}{s} - \frac{1}{s-1}$$

$$\therefore \quad L\left\{\frac{1-e^t}{t}\right\} = \int_s^\infty \left(\frac{1}{u} - \frac{1}{u-1}\right) du$$

$$= [\log u - \log (u-1)]_s^\infty$$

$$= \log\left(\frac{s-1}{s}\right).$$

THEOREM 16.4 Let $f(t)$ be a periodic function with period p (> 0), then

$$L\{f(t)\} = \frac{1}{1-e^{-sp}} \int_0^p e^{-su} f(u)\, du.$$

Proof: $$L\{f(t)\} = \int_0^\infty e^{-st} f(t)\, dt$$

$$= \int_0^p e^{-st} f(t)\, dt + \int_p^{2p} e^{-st} f(t)\, dt + \cdots + \int_{np}^{(n+1)p} e^{-st} f(t)\, dt + \cdots$$

Consider $$I_n = \int_{np}^{(n+1)p} e^{-st} f(t)\, dt$$

Put $$t = u + np$$

Then $$I_n = \int_o^p e^{-s(u+np)} f(u+np)\, dp$$

Since, $f(t)$ is periodic with period p,

$$f(u) = f(u + np)$$

$$\therefore \quad I_n = e^{-snp} \int_0^p e^{-su} f(u)\, du$$

$$\therefore \quad L\{f(t)\} = (1 + e^{-sp} + e^{-2sp} + \cdots) \int_0^p e^{-su} f(u)\, du = \frac{1}{1-e^{-sp}} \int_0^p e^{-su} f(u)\, du.$$

THEOREM 16.5 **(Initial Value Theorem):** Let $L\{f(t)\} = F(s)$, then $\lim_{t\to 0} f(t) = \lim_{s\to\infty} sF(s)$ provided the limits exist.

Proof: We have proved that

$$L\{f'(t)\} = sF(s) - f(0+)$$

Assuming $f'(t)$ to be piece-wise continuous and of exponential form

$$\lim_{s\to\infty} \int_0^\infty e^{-st} f'(t) = 0$$

$$\Rightarrow \qquad \lim_{s\to\infty} L\{f'(t)\} = 0$$

$$\therefore \qquad 0 = \lim_{s\to\infty} sF(s) - f(0+)$$

$$\Rightarrow \qquad f(0+) = \lim_{s\to\infty} sF(s)$$

$$\Rightarrow \qquad \lim_{t\to 0} f(t) = \lim_{s\to\infty} sF(s).$$

THEOREM 16.6 **(Final Value Theorem):** If $L\{f(t)\} = F(s)$, then $\lim_{t\to\infty} f(t) = \lim_{s\to 0} sF(s)$ provided the limits exist.

Proof: We have

$$L\{f'(t)\} = sF(s) - f(0+)$$

Taking limit as $s \to 0$

$$\begin{aligned}
\lim_{s\to 0} sF(s) - f(0+) &= \lim_{s\to 0} L\{f'(t)\} \\
&= \lim_{s\to 0} \int_0^\infty e^{-st} f'(t)\, dt \\
&= \int_0^\infty f'(t)\, dt \\
&= \lim_{a\to\infty} \int_0^a f'(t)\, dt \\
&= \lim_{a\to\infty} \{f(a) - f(0+)\} \\
&= \lim_{a\to\infty} f(a) - f(0+)
\end{aligned}$$

$$\Rightarrow \qquad \lim_{s\to 0} sF(s) = \lim_{t\to\infty} f(t).$$

EXAMPLE 16.15 Verify Theorems 16.5 and 16.6 for the function e^{-2t}.

Solution Given

$$f(t) = e^{-2t}$$

$$\therefore \qquad F(s) = L\{f(t)\} = \frac{1}{s+2}$$

$$\therefore \qquad \lim_{s\to\infty} sF(s) = \lim_{s\to\infty} \frac{s}{s+2} = 1$$

Also $$\lim_{t\to 0} f(t) = \lim_{t\to 0} e^{-2t} = 1$$

$$\therefore \qquad \lim_{s\to\infty} sF(s) = \lim_{t\to 0} f(t)$$

This verifies Theorem 16.5.

Also $$\lim_{t\to\infty} f(t) = \lim_{t\to\infty} e^{-2t} = 0$$

and $$\lim_{s\to 0} sF(s) = \lim_{s\to 0} \frac{s}{s+2} = 0$$

$$\therefore \qquad \lim_{t\to\infty} f(t) = \lim_{s\to 0} sF(s)$$

This verifies Theorem 16.6.

EXAMPLE 16.16 Show that $L\{t^n\} = \dfrac{\Gamma(n+1)}{s^{n+1}}, n > -1, s > 0.$

Solution We know that $L\{t^n\} = \int_0^\infty e^{-st} t^n dt$. Put $st = u$. Then

$$L\{t^n\} = \frac{1}{s^{n+1}} \int_0^\infty e^{-u} u^n du = \frac{\Gamma(n+1)}{s^{n+1}}$$

EXAMPLE 16.17 Find $L\{J_0(t)\}$ and deduce that $\int_0^\infty J_0(t)\, dt = 1$.

Solution In Chapter 14, we have derived that

$$J_0(t) = \sum_{n=0}^{\infty} \frac{(-1)^n}{n!\,\Gamma(n+1)} \left(\frac{t}{2}\right)^{2n}$$

$$= 1 - \frac{t^2}{2^2} + \frac{t^4}{2^2 \cdot 4^2} - \cdots$$

$$\therefore \qquad L\{J_0(t)\} = \frac{1}{s} - \frac{1}{2^2} \cdot \frac{2!}{s^3} + \frac{1}{2^2 \cdot 4^2} \cdot \frac{4!}{s^5}$$

$$= \frac{1}{s}\left\{1 - \frac{1}{2} \cdot \frac{1}{s^2} + \frac{1 \cdot 3}{2 \cdot 4} \cdot \frac{1}{s^4} - \cdots\right\}$$

$$= \frac{1}{s}\left\{\left(1+\frac{1}{s^2}\right)^{-1/2}\right\}$$

$$= \frac{1}{\sqrt{s^2+1}}$$

$$\therefore \qquad \int_0^\infty e^{-st} J_0(t)\, dt = \frac{1}{\sqrt{s^2+1}}$$

Taking limit $s \to 0+$, we get

$$\int_0^\infty J_0(t)\, dt = 1.$$

EXAMPLE 16.18 Find $L\{J_1(t)\}$.

Solution We know that $J_0'(t) = -J_1(t)$.

$$\therefore \qquad L\{J_1(t)\} = -L\{J_0'(t)\}$$

$$= -1[sL\{J_0(t)\} - J_0(0)]$$

$$= -1\left\{s \cdot \frac{1}{\sqrt{s^2+1}} - 1\right\}$$

$$= 1 - \frac{s}{\sqrt{s^2+1}}.$$

EXAMPLE 16.19 Find $L\{\sin\sqrt{t}\}$. Deduce the value of $L\left\{\frac{\cos\sqrt{t}}{\sqrt{t}}\right\}$.

Solution The sine series is given by

$$\sin\sqrt{t} = t^{1/2} - \frac{t^{3/2}}{3!} + \frac{t^{5/2}}{5!} - \cdots$$

$$\therefore \qquad L\{\sin\sqrt{t}\} = L\left\{t^{1/2} - \frac{t^{3/2}}{3!} + \frac{t^{5/2}}{5!} - \cdots\right\}$$

$$= \frac{\Gamma(3/2)}{s^{3/2}} - \frac{\Gamma(5/2)}{3!\,s^{5/2}} + \frac{\Gamma(7/2)}{5!\,s^{7/2}} - \cdots$$

$$= \frac{\sqrt{\pi}}{2s^{3/2}}\left\{1 - \frac{1}{2^2 s} + \frac{\left(-\frac{1}{2^2 s}\right)^2}{2!} \cdots\right\}$$

$$= \frac{\sqrt{\pi}}{2s^{3/2}} e^{-\frac{1}{2^2 s}}$$

$$= \frac{\sqrt{\pi}}{2s^{3/2}} e^{-\frac{1}{4s}}$$

Now, $L\{f'(t)\} = sL(f(t)\} - F(0)$

Put

$$f(t) = \sin \sqrt{t}$$

$$\therefore \qquad f'(t) = \frac{\cos \sqrt{t}}{2\sqrt{t}}$$

$$\therefore \qquad \frac{1}{2} L\left\{\frac{\cos \sqrt{t}}{2\sqrt{t}}\right\} = \frac{s\sqrt{\pi}}{2s^{3/2}} e^{-1/4s}$$

$$\Rightarrow \qquad L\left\{\frac{\cos \sqrt{t}}{2\sqrt{t}}\right\} = \sqrt{\frac{\pi}{s}}\, e^{-1/4s}.$$

EXAMPLE 16.20 Show that $L\{\log t\} = \dfrac{\Gamma'(1) - \log s}{s}$.

Solution We have seen in Example 16.16 that

$$\int_0^{\infty} e^{-st} t^n dt = \frac{\Gamma(n+1)}{s^{n+1}}$$

Differentiate w.r.t. n

$$\int_0^{\infty} e^{-st} t^n \log t dt = \frac{\Gamma'(n+1) - \Gamma(n+1)\log s}{s^{n+1}}$$

$$\therefore \qquad \lim_{n\to 0} \int_0^{\infty} e^{-st} t^n \log t dt = \frac{\Gamma'(1) - \log s}{s}$$

$$\Rightarrow \qquad \int_0^{\infty} e^{-st} \log t dt = \frac{\Gamma'(1) - \log s}{s}$$

$$\Rightarrow \qquad L\{\log t\} = \frac{\Gamma'(1) - \log s}{s}.$$

16.6 INVERSE LAPLACE TRANSFORMS

If $L\{f(t)\} = F(s)$, then

$$L\left\{\int_0^t f(u)\, du\right\} = \frac{1}{s} F(s)$$

or $$L^{-1}\left\{\frac{1}{s}F(s)\right\} = \int_0^t f(u)\,du.$$

The following table gives inverse Laplace transform of elementary functions.

$F(s)$	$L^{-1}\{F(s)\} = f(t)$
$\frac{1}{s}$	1
$\frac{1}{s^2}$	t
$\frac{1}{s^{n+1}}, n = 0, 1, 2, \ldots$	$\frac{t^n}{n!}$
$\frac{1}{s-a}$	e^{at}
$\frac{1}{s^2+a^2}$	$\frac{\sin at}{a}$
$\frac{s}{s^2+a^2}$	$\cos at$
$\frac{1}{s^2-a^2}$	$\frac{\sinh at}{a}$
$\frac{s}{s^2-a^2}$	$\cosh at$
$\frac{1}{\sqrt{s^2+a^2}}$	$J_0(at)$

EXAMPLE 16.21 Find the inverse Laplace transform of $\frac{1}{s(s^2+16)}$.

Solution
$$L^{-1}\left\{\frac{1}{(s^2+16)}\right\} = \frac{1}{4}\sin 4t$$

$$\therefore \quad L^{-1}\left\{\frac{1}{s(s^2+16)}\right\} = \int_0^t \frac{1}{4}\sin 4t\,dt$$

$$= \frac{1}{16}[-\cos 4t]_0^t$$

$$= \frac{1-\cos 4t}{16}.$$

16.7 PROPERTIES OF INVERSE LAPLACE TRANSFORMS

P1: Linearity property

If $L^{-1}\{F(s)\} = f(t)$ and $L^{-1}\{G(s)\} = g(t)$, then $L^{-1}\{aF(s) + bG(s)\} = af(t) + bg(t)$ where a and b are any constants.

P2: First translation property

If $L^{-1}\{F(s)\} = f(t)$, then $L^{-1}\{F(s-a)\} = e^{at} f(t)$.

P3: Second translation property

If $L^{-1}\{F(s)\} = f(t)$, then $L^{-1}\{e^{-as}F(s)\} = \begin{cases} f(t-a), & t > a \\ 0, & t < a \end{cases}$.

P4: Change of scale property

If $L^{-1}\{F(s)\} = f(t)$, then $L^{-1}\{F(ks)\} \dfrac{1}{k} f\left(\dfrac{t}{k}\right)$.

P5: Inverse Laplace transform of derivatives

If $L^{-1}\{F(s)\} = f(t)$, then $L^{-1}\{F^{n}(s)\} = (-1)^n t^n f(t)$.

P6: Inverse Laplace transform of integrals

If $L^{-1}\{F(s)\} = f(t)$, then $L^{-1}\left\{\displaystyle\int_s^{\infty} F(u)\, du\right\} = \dfrac{f(t)}{t}$.

P7: Multiplication by t^n

If $L^{-1}\{F(s)\} = f(t)$, and $f(0) = 0$, then $L^{-1}\{sf(s)\} = f'(t)$. If $f(0) \neq 0$, then

$$L^{-1}\{sF(s) - f(0)\} = f'(t)$$

or

$$L^{-1}\{sF(s)\} = f'(t) + f(0)\delta(t)$$

where $\delta(t)$ is called Dirac delta function or unit impulse function. It is defined as the limiting function of $f_\varepsilon(t)$ as $\varepsilon \to 0$ where

$$f_\varepsilon(t) = \begin{cases} \dfrac{1}{\Sigma}, & 0 \le t < \varepsilon \\ 0, & t > \varepsilon \end{cases}$$

P8: Division by s

If $L^{-1}\{F(s)\} = f(t)$, then $L^{-1}\left\{\dfrac{F(s)}{s}\right\} = \displaystyle\int_0^t f(u)\, du$.

P9: Convolution property

If $L^{-1}\{F(s)\} = f(t)$ and $L^{-1}\{G(s)\} = g(t)$, then $L^{-1}\{F(s)\,G(s)\} = \int_0^t f(u)g(t-u)\,du = f * g$

$f * g$ is called convolution of f and g.

EXAMPLE 16.22 Find $L^{-1}\left\{\dfrac{1}{s^2 - 4s + 20}\right\}$.

Solution

$$L^{-1}\left\{\frac{1}{s^2 - 4s + 20}\right\} = L^{-1}\left\{\frac{1}{(s-2)^2 + 16}\right\}$$

$$= e^{2t}L^{-1}\left\{\frac{1}{s^2 + 16}\right\} \qquad \text{(using P2)}$$

$$= \frac{1}{4}e^{2t}\sin 4t.$$

EXAMPLE 16.23 Find $L^{-1}\left\{\dfrac{1}{\sqrt{3s+4}}\right\}$.

Solution

$$L^{-1}\left\{\frac{1}{\sqrt{3s+4}}\right\} = \frac{1}{\sqrt{3}}L^{-1}\left\{\frac{1}{\left(s+\frac{4}{3}\right)^{1/2}}\right\}$$

$$= \frac{1}{\sqrt{3}}e^{-4t/3}\,L^{-1}\left\{\frac{1}{s^{1/2}}\right\}$$

$$= \frac{1}{\sqrt{3}}e^{-4t/3}\cdot\frac{t^{-1/2}}{\Gamma(1/2)}$$

$$= \frac{1}{\sqrt{3\pi}}e^{-4t/3}t^{-1/2}.$$

EXAMPLE 16.24 Find $L^{-1}\left\{\dfrac{s}{(s^2+9)^2}\right\}$.

Solution By P5

$$L^{-1}\left\{\frac{s}{(s^2+9)^2}\right\} = -tL^{-1}\left\{\int\frac{s}{(s^2+9)^2}\,ds\right\}$$

$$= -tL^{-1}\left\{\frac{-1}{3(s^2+9)}\right\}$$

$$= \frac{t}{3} L^{-1}\left\{\frac{1}{s^2+9}\right\}$$

$$= \frac{t}{9}\sin 3t.$$

EXAMPLE 16.25 Find $L^{-1}\left\{\log\left(1+\frac{1}{s^2}\right)\right\}$.

Solution Let $F(s) = \log\left(1+\frac{1}{s^2}\right)$

$$= \log(1+s^2) - 2\log s$$

$$\therefore \quad F'(s) = \frac{2s}{1+s^2} - \frac{2}{s}$$

$$\therefore \quad L^{-1}\{F'(s)\} = 2\cos t - 2 = -2(1-\cos t)$$

$$\Rightarrow \quad -tL^{-1}\{F(s)\} = -2(1-\cos t)$$

$$\Rightarrow \quad L^{-1}\{F(s)\} = \frac{2(1-\cos t)}{t}.$$

EXAMPLE 16.26 Given that $L^{-1}\left\{\frac{s}{(s^2+1)^2}\right\} = \frac{1}{2}t\sin t$, find $L^{-1}\left\{\frac{1}{(s^2+1)^2}\right\}$.

Solution

$$L^{-1}\left\{\frac{1}{(s^2+1)^2}\right\} = L^{-1}\left\{\frac{1}{s}\cdot\frac{s}{(s^2+1)^2}\right\}$$

$$= \frac{1}{2}\int_0^t u\sin u\,du \qquad \text{(by P8)}$$

$$= \frac{1}{2}(\sin t - t\cos t).$$

EXAMPLE 16.27 Evaluate $L^{-1}\left\{\frac{1}{(s+1)(s^2+1)}\right\}$ by using convolution property.

Solution We know that

$$L^{-1}\left\{\frac{1}{s+1}\right\} = e^{-t} \text{ and } L^{-1}\left\{\frac{1}{s^2+1}\right\} = \sin t$$

$$\therefore \quad L^{-1}\left\{\frac{1}{(s+1)(s^2+1)}\right\} = \int_0^t e^{-u}\sin(t-u)\,du$$

$$= \frac{1}{2}(\sin t - \cos t + e^{-t}).$$

EXAMPLE 16.28 Evaluate $L^{-1}\left\{\dfrac{e^{-2s}}{(s-2)^4}\right\}$.

Solution $L^{-1}\left\{\dfrac{1}{(s-2)^4}\right\} = e^{2t}L^{-1}\left\{\dfrac{1}{(s-2)^4}\right\} = \dfrac{1}{6}e^{2t}t^3.$

Using second translation property,

$$L^{-1}\left\{\frac{e^{-2s}}{(s-2)^4}\right\} = \frac{1}{6}e^{2(t-2)}(t-2)^3 H(t-2).$$

16.8 SOLUTION OF ORDINARY DIFFERENTIAL EQUATIONS WITH CONSTANT COEFFICIENTS

In this section, we will discuss the Laplace transform method to solve ordinary differential equations.

EXAMPLE 16.29 Solve $y''' + 2y'' - y' - 2y = 0$ given $y(0) = y'(0) = 0$ and $y''(0) = 6$ by method of Laplace transform.

Solution Taking the Laplace transforms of both sides of given differential equation, we get

$$s^3\bar{y} - s^2y(0) - sy'(0) - y''(0) + 2[s^2\bar{y} - sy(0) - y'(0)] - [s\bar{y} - y(0)] - 2\bar{y} = 0$$

Using given conditions, it reduces to

$$(s^3 + 2s^2 - s - 2)\bar{y} = 6$$

$$\therefore \quad \bar{y} = \frac{6}{(s-1)(s+1)(s+2)}$$

$$= \frac{1}{s-1} - \frac{3}{s+1} + \frac{2}{s+2}$$

Taking inverse Laplace transform, we get

$$y = L^{-1}\left\{\frac{1}{s-1}\right\} - 3L^{-1}\left\{\frac{1}{s+1}\right\} + 2L^{-1}\left\{\frac{1}{s+2}\right\}$$

$\Rightarrow$ $y = e^t - 3e^{-t} + 2e^{-2t}$ is the desired solution.

EXAMPLE 16.30 Solve $y'' + y = \cos 2t$ given that $y(0) = 1, y'(0) = -2$.

Solution Taking Laplace transform of both sides of given differential equation gives

$$s^2\bar{y} - sy(0) - y'(0) + \bar{y} = \frac{s}{s^2 + 4}$$

Using initial conditions, we get

$$(s^2 + 1)\bar{y} = s - 2 + \frac{s}{s^2 + 4}$$

$$\therefore \qquad \bar{y} = \frac{s}{s^2 + 1} - \frac{2}{s^2 + 1} + \frac{s}{(s^2 + 1)(s^2 + 4)}$$

$$= \frac{s}{s^2 + 1} - \frac{2}{s^2 + 1} + \frac{1}{3}\left[\frac{s}{(s^2 + 1)} - \frac{s}{(s^2 + 4)}\right]$$

$$= \frac{4}{3}\frac{s}{s^2 + 1} - \frac{2}{s^2 + 1} - \frac{1}{3}\cdot\frac{s}{(s^2 + 4)}$$

Taking inverse Laplace transforms, the solution of given differential equation is

$$y = \frac{4}{3}L^{-1}\left\{\frac{s}{s^2 + 1}\right\} - 2L^{-1}\left\{\frac{1}{s^2 + 1}\right\} - \frac{1}{3}L^{-1}\left\{\frac{s}{(s^2 + 4)}\right\}$$

$$= \frac{4}{3}\cos t - 2\sin t - \frac{1}{3}\cos 2t.$$

EXAMPLE 16.31 Solve $y'' + 9y = t$, given that $y(0) = c$, $y(\pi/2) = 0$.

Solution Taking Laplace transform of both sides of given differential equation, we get

$$s^2\bar{y} - sy(0) - y'(0) + 9\bar{y} = \frac{1}{s^2}$$

Let $y'(0) = c$ and using initial conditions,

$$(s^2 + 9)\bar{y} = c + \frac{1}{s^2}$$

$$\Rightarrow \qquad \bar{y} = \frac{c}{s^2 + 9} + \frac{1}{s^2(s^2 + 9)}$$

$$= \frac{c}{s^2 + 9} + \frac{1}{9}\left(\frac{1}{s^2} - \frac{1}{s^2 + 9}\right)$$

$$= \left(c - \frac{1}{9}\right)\frac{1}{s^2 + 9} + \frac{1}{9}\cdot\frac{1}{s^2}$$

Taking inverse Laplace transforms, we get

$$y=\left(c-\frac{1}{9}\right)L^{-1}\left\{\frac{1}{s^2+9}\right\}+\frac{1}{9}L^{-1}\left\{\frac{1}{s^2}\right\}$$

$$=\left(c-\frac{1}{9}\right)\frac{\sin 3t}{3}+\frac{1}{9}t$$

To determine c, use $y(\pi/2)=0$

$$\Rightarrow \qquad 0=-\frac{1}{3}\left(c-\frac{1}{9}\right)+\frac{\pi}{18}$$

$$\Rightarrow \qquad c=\frac{\pi}{6}+\frac{1}{9}$$

$\therefore$ The solution of the given differential equation is $y=\dfrac{\pi}{18}\sin 3t+\dfrac{1}{9}t.$

EXAMPLE 16.32 Solve $y'''+3y''+3y'+y=t^2e^{-t}$, given that $y(0)=1, y'(0)=0, y''(0)=-2$.

Solution Taking Laplace transform of both sides of given differential equation, we get

$$\{s^3\bar{y}-s^2y(0)-s\,y'(0)-y''(0)\}+3\{s^2\bar{y}-sy(0)-y'(0)\}+3\{s\bar{y}-y(0)\}+\bar{y}=\frac{2}{(s+1)^3}$$

Using $y(0)=1, y'(0)=0, y''(0)=-2$, we get

$$\{s^3+3s^2+3s+1\}\bar{y}=s^2+3s+1+\frac{2}{(s+1)^3}$$

$$\Rightarrow \qquad \bar{y}=\frac{s^2+3s+1}{(s+1)^3}+\frac{2}{(s+1)^6}$$

$$=\frac{(s+1)^2+(s+1)-1}{(s+1)^3}+\frac{2}{(s+1)^6}$$

$$=\frac{1}{s+1}+\frac{1}{(s+1)^2}-\frac{1}{(s+1)^3}+\frac{2}{(s+1)^6}$$

$$\therefore \qquad y=L^{-1}\left(\frac{1}{s+1}+\frac{1}{(s+1)^2}-\frac{1}{(s+1)^3}+\frac{2}{(s+1)^6}\right)$$

$$=e^{-t}\left(1+t-\frac{t^2}{2}+\frac{t^5}{60}\right)$$

16.9 SOLUTION OF ORDINARY DIFFERENTIAL EQUATIONS WITH VARIABLE COEFFICIENTS

We will illustrate the method in following examples.

EXAMPLE 16.33 Solve $y'' - ty' + y = 1$, given that $y(0) = 0$ and $y'(0) = 2$.

Solution Taking Laplace transform of both sides of given differential equation, we get

$$\{s^2\bar{y} - sy(0) - y'(0)\} + \frac{d}{ds}(L\{y'\}) + \bar{y} = \frac{1}{s}$$

$$\Rightarrow \quad s^2\bar{y} - s - 2 + \frac{d}{ds}(s\bar{y} - y(0)) + \bar{y} = \frac{1}{s}$$

$$\Rightarrow \quad s^2\bar{y} - s - 2 + \bar{y} + s\frac{d\bar{y}}{ds} + \bar{y} = \frac{1}{s}$$

$$\Rightarrow \quad s\frac{d\bar{y}}{ds} + (s^2 + 2)\,\bar{y} = s + 2 + \frac{1}{s}$$

$$\Rightarrow \quad \frac{d\bar{y}}{ds} + \left(s + \frac{2}{s}\right)\bar{y} = 1 + \frac{2}{s} + \frac{1}{s^2}$$

This is linear differential equation.

$$\text{IF} = e^{\int (s+2/s)\,ds} = s^2 e^{s^2/2}$$

$$\therefore \quad \bar{y}(s^2 e^{s2/2}) = \int (s^2 + 2s + 1)e^{s2/2}\,ds + c$$

$$= \int s^2 e^{s2/2}\,ds + 2\int se^{s2/2}\,ds + \int e^{s^2}\,ds + c$$

$$= (s + 2)e^{s2/2} + c$$

$$\therefore \quad \bar{y} = \frac{s+2}{s^2} + \frac{c}{s^2}e^{-s2/2}$$

$$= \frac{1}{s} + \frac{2}{s^2} + \frac{c}{s^2}\left(1 - \frac{s^2}{2} + \frac{s^4}{8} - \cdots\right)$$

$$= \frac{1}{s} + \frac{2}{s^2} + c\left(\frac{1}{s^2} - \frac{1}{2} + \frac{s^2}{8} - \cdots\right)$$

$$= \frac{1}{s} + \frac{c+2}{s^2} - c\left(\frac{1}{2} - \frac{s^2}{8} - \cdots\right)$$

$$\therefore \quad y = L^{-1}\left\{\frac{1}{s} + \frac{c+2}{s^2} - c\left(\frac{1}{2} - \frac{s^2}{8} - \cdots\right)\right\}$$

$$= 1 + (c + 2)t \qquad [\because L^{-1}(s^k) = 0,\ k = 0, 1, \ldots]$$

Now, $y'(0) = 2 \Rightarrow c = 0.$

$\therefore$ The solution of given differential equation is $y = 1 + 2t$.

EXAMPLE 16.34 Solve $ty'' + y' + ty = 0$ with $y(0) = 2$, $y'(0) = 0$.

Solution Taking Laplace transform of both sides of given differential equation gives

$$-\frac{d}{ds}\{s^2\bar{y} - sy(0) - y'(0)\} + s\bar{y} - y(0) - \frac{dy}{ds} = 0$$

$$\Rightarrow \qquad (s^2+1)\frac{d\bar{y}}{ds} + s\bar{y} = 0$$

$$\Rightarrow \qquad \frac{d\bar{y}}{\bar{y}} + \frac{sds}{s^2+1} = 0$$

On integrating,

$$\Rightarrow \qquad \log \bar{y} + \frac{1}{2}\log(s^2+1) = c$$

$$\Rightarrow \qquad \bar{y}\sqrt{s^2+1} = c$$

$$\Rightarrow \qquad \bar{y} = \frac{c}{\sqrt{s^2+1}}$$

$$\therefore \qquad y = L^{-1}\left\{\frac{c}{\sqrt{s^2+1}}\right\} = cJ_0(t)$$

Given $y(0) = 2.$

$\Rightarrow$ $cJ_o(0) = 2 \Rightarrow c = 2$ $\qquad [\because J_o(0) = 1]$

$\therefore$ $y = 2J_0(t)$ is the required solution of the given differential equation.

16.10 SIMULTANEOUS ORDINARY DIFFERENTIAL EQUATIONS

In this section, we discuss method to solve simultaneous ordinary differential equations using Laplace transforms.

EXAMPLE 16.35 Solve $x' + 5x + 2y = t$,

$$y' + 2x + y = 0, \quad x(0) = 0, \; y(0) = 0$$

Here 'desh' represents denotes differentiation w.r.t. t.

Solution Taking the Laplace transforms of the given equations, we get

$$[s\bar{x} - x(0)] + 5\bar{x} - 2\bar{y} = \frac{1}{s^2}$$

$$\Rightarrow \qquad (s+5)\,\overline{x} - 2\overline{y} = \frac{1}{s^2} \qquad \text{(i)}$$

and
$$[5\overline{y} - y(0)] + 2\overline{x} + \overline{y} = 0$$

$$\Rightarrow \qquad 2\overline{x} + (s+1)\,\overline{y} = 0 \qquad \text{(ii)}$$

Solving Eqs. (i) and (ii) for $\overline{x}$, we get

$$\overline{x} = \begin{vmatrix} 1/s^2 & -2 \\ 0 & s+1 \end{vmatrix} \div \begin{vmatrix} s+5 & -2 \\ 2 & s+1 \end{vmatrix}$$

$$= \frac{s+1}{s^2(s+3)^2} = \frac{1}{27s} + \frac{1}{9s^2} - \frac{1}{27(s+3)} - \frac{2}{9(s+3)^2}$$

Substituting $\overline{x}$ in Eq. (ii) gives

$$\overline{y} = -\frac{2}{s^2(s+3)^2} = \frac{4}{27s} - \frac{2}{9s^2} - \frac{4}{27(s+3)} - \frac{2}{9(s+3)^2}$$

Taking inverse Laplace transform, we get

$$x = \frac{1}{27} + \frac{t}{9} - \frac{1}{27}e^{-3t} - \frac{2}{9}te^{-3t}$$

and
$$y = \frac{4}{27} - \frac{2t}{9} - \frac{4}{27}e^{-3t} - \frac{2}{9}te^{-3t}.$$

EXAMPLE 16.36 Solve $x'' + 3x - 2y = 0$

$$x'' + y'' - 3x + 5y = 0$$

with
$$x(0) = 0,\ x'(0) = 0,\ y(0) = 0,\ y'(0) = 0.$$

Solution Taking Laplace transforms of both the equations, we get

$$[s^2\overline{x} - sx(0) - x'(0)] + 3\overline{x} - 2\overline{y} = 0$$

$$\Rightarrow \qquad (s^2+3)\,\overline{x} - 2\overline{y} = 3 \qquad \text{(i)}$$

and
$$[s^2\overline{x} - sx(0) - x'(0)] + [s^2\overline{y} - sy(0) - y'(0)] - 3\overline{x} + 5\overline{y} = 0$$

$$\Rightarrow \qquad (s^2-3)\,\overline{x} + (s^2+5)\,\overline{y} = 5 \qquad \text{(ii)}$$

Solving Eqs. (i) and (ii) for $\overline{x}$ and $\overline{y}$, gives

$$\overline{x} = \frac{3s^2+25}{(s^2+1)(s^2+9)} = \frac{11}{4}\cdot\frac{1}{s^2+1} + \frac{1}{4}\cdot\frac{1}{s^2+9}$$

and
$$\overline{y} = \frac{2s^2+24}{(s^2+1)(s^2+9)} = \frac{11}{4}\cdot\frac{1}{s^2+1} - \frac{3}{4}\cdot\frac{1}{s^2+9}$$

Taking inverse Laplace transform, we get

$$x = \frac{11}{4}\sin t + \frac{1}{12}\sin 3t$$

and

$$y = \frac{11}{4}\sin t - \frac{1}{4}\sin 3t.$$

16.11 SOLUTION OF PARTIAL DIFFERENTIAL EQUATIONS

THEOREM 16.7 Let $u(x, t)$ be a function defined for $t > 0$ and $x \in [a, b]$. Show that

(i) $L\left(\frac{\partial u}{\partial t}\right) = sU - u(x, 0)$

where $U = U(x, s) = L\{u(x, t)\}$

(ii) $L\left(\frac{\partial u}{\partial x}\right) = \frac{dU}{dx}$

(iii) $L\left(\frac{\partial^2 u}{\partial t^2}\right) = s^2U - su(x, 0) - u_t(x, 0)$

where $u_t(x, 0) = \frac{\partial u}{\partial t}$ at $t = 0$

(iv) $L\left(\frac{\partial^2 u}{\partial x^2}\right) = \frac{d^2U}{dx^2}$.

Proof: (i)
$$L\left(\frac{\partial u}{\partial t}\right) = \int_0^\infty e^{-st}\frac{\partial u}{\partial t}\,dt$$
$$= \lim_{k\to\infty}\int_0^k e^{-st}\frac{\partial u}{\partial t}\,dt$$
$$= \lim_{k\to\infty}\left\{e^{-st}\,u(x, t)\Big|_0^k + s\int_0^k e^{-st}u(x, t)\,dt\right\}$$
$$= s\int_0^\infty e^{-st}u(x, t)\,dt - u(x, 0)$$
$$= sU - u(x, 0)$$

(ii)
$$L\left(\frac{\partial u}{\partial x}\right) = \int_0^\infty e^{-st}\frac{\partial u}{\partial x}\,dt$$
$$= \frac{d}{dx}\int_0^\infty e^{-st}u(x, t)\,dt$$
$$= \frac{dU}{dx}$$

(iii) Put $v = \frac{\partial u}{\partial t}$

$$\therefore \quad L\left(\frac{\partial^2 u}{\partial t^2}\right) = L\left(\frac{\partial v}{\partial t}\right)$$

$$= sL\{v\} - v(x, 0)$$

$$= sL\left(\frac{\partial u}{\partial t}\right) - v(x, 0)$$

$$= s\{sU - u(x, 0)\} - u_t(x, 0)$$

$$= s^2U - su(x, 0) - u_t(x, 0)$$

(iv) Put $v = \frac{\partial u}{\partial x}$

$$\therefore \quad L\left(\frac{\partial^2 u}{\partial x^2}\right) = L\left(\frac{\partial v}{\partial x}\right)$$

$$= \frac{d}{dx}(L\{v\})$$

$$= \frac{d}{dx} L\left\{\frac{\partial u}{\partial x}\right\}$$

$$= \frac{d}{dx}\left(\frac{dU}{dx}\right)$$

$$= \frac{d^2U}{dx^2}.$$

EXAMPLE 16.37 Solve $\frac{\partial u}{\partial t} = 3\frac{\partial^2 u}{\partial x^2}$ given that $u(0, t) = 0, u(5, t) = 0, u(x, 0) = \sin \pi x$.

Solution Taking Laplace transform of the partial differential equation, we get

$$sU - u(x, 0) = 3\frac{d^2U}{dx^2}$$

$$\Rightarrow \quad \frac{d^2U}{dx^2} - \frac{s}{3}U = -\frac{1}{3}u(x, 0)$$

$$\Rightarrow \quad \frac{d^2U}{dx^2} - \frac{s}{3}U = -\frac{1}{3}\sin \pi x$$

$$\therefore \quad \text{CF} = Ae^{\sqrt{\frac{s}{3}}x} + Be^{-\sqrt{\frac{s}{3}}x}$$

$$\text{PI} = \frac{1}{D^2 - \frac{s}{3}}\left(-\frac{1}{3}\sin \pi x\right)$$

$$= \frac{1}{3}\cdot\frac{1}{\pi^2 + \frac{s}{3}}\sin \pi x$$

$\therefore$
$$U = Ae^{\sqrt{\frac{s}{3}}x} + Be^{-\sqrt{\frac{s}{3}}x} + \frac{1}{3}\cdot\frac{1}{\pi^2 + \frac{s}{3}}\sin \pi x \qquad \text{(i)}$$

Taking the Laplace transform of conditions which involve t,

$$L\{u(0, t)\} = U(0, s) = 0$$
$$L\{u(5, t)\} = U(5, s) = 0$$

Then Eq. (i) becomes

$$A + B = 0$$

and
$$Ae^{5\sqrt{\frac{s}{3}}} + Be^{-5\sqrt{\frac{s}{3}}} = 0$$

$\Rightarrow$
$$A = 0 = B$$

$\therefore$ Equation (i) becomes

$$U = \frac{1}{s + 3\pi^2}\sin \pi x$$

Taking inverse Laplace transform, we get

$$u(x, t) = e^{-3\pi^2 s}\sin \pi x.$$

EXAMPLE 16.38 Solve $\dfrac{\partial^2 u}{\partial t^2} = \dfrac{\partial^2 u}{\partial x^2}, x > 0, t > 0$ given that $u(0, t) = 10 \sin 2t$, $u(x, 0) = 0$, $u_t(x, 0) = 0$ and $\lim_{x\to\infty} u(x, t) = 0$.

Solution Taking the Laplace transform of the given partial differential equation gives

$$s^2U - su(x, 0) - u_t(x, 0) = \frac{d^2U}{dx^2}$$

$\Rightarrow$
$$\frac{d^2U}{dx^2} - s^2U = 0$$

$\Rightarrow$
$$U = Ae^{sx} + Be^{-sx} \qquad \text{(i)}$$

Taking the Laplace transform of given conditions which involve t

$$U(0, s) = 10 \cdot \frac{2}{s^2 + 4} = \frac{20}{s^2 + 4}$$

and

$$\lim_{x \to \infty} u(x, t) = 0.$$

$\therefore$ Equation (i) gives $A + B = \dfrac{20}{s^2 + 4}$ and $A = 0$

$$\Rightarrow \qquad B = \frac{20}{s^2 + 4}$$

$\therefore$ Equation (i) becomes

$$U = \frac{20}{s^2 + 4} e^{-sx}$$

Taking inverse Laplace transform, we get

$$u(x, t) = \begin{cases} 10 \sin 2(t - x), & t > x \\ 0, & t < x \end{cases}.$$

EXERCISES

1. Find the Laplace transforms of

(a) $e^{2t} + 4t^3 - 2 \sin 3t + 3 \cos 3t$ $\quad \left[\textbf{Ans. } \dfrac{1}{s-2} + \dfrac{24}{s^4} + \dfrac{3(s-2)}{s^2+9}\right]$

(b) $\cosh at - \cos at$ $\quad \left[\textbf{Ans. } \dfrac{2a^4 s}{s^4 - a^4}\right]$

(c) $\cos^3 2t$ $\quad \left[\textbf{Ans. } \dfrac{s(s^2+28)}{(s^2+4)(s^2+36)}\right]$

(d) $t - \sinh 2t$ $\quad \left[\textbf{Ans. } \dfrac{4+s^2}{s^2(4-s^2)}\right]$

(e) $e^{-t} \cos^2 t$ $\quad \left[\textbf{Ans. } \dfrac{s^2+2s+3}{(s+1)(s^2+2s+5)}\right]$

(f) $t^3 e^{-3t}$ $\quad \left[\textbf{Ans. } \dfrac{6}{(s+3)^4}\right]$

(g) $te^{-4t} \sin 3t$ $\quad \left[\textbf{Ans. } \dfrac{6(s+4)}{(s^2+8s+25)^2}\right]$

(h) $f(t) = \begin{cases} t, & 0 < t < 4 \\ 5, & t > 4 \end{cases}$ $\left[\textbf{Ans. } \dfrac{1}{s^2} + e^{-4s}\left(\dfrac{1}{s} - \dfrac{1}{s^2}\right)\right]$

(i) $f(t) = \begin{cases} \sin t, & 0 < t < \pi \\ 0, & \pi < t < 2\pi \end{cases}$. $\left[\textbf{Ans. } \dfrac{1 + e^{-\pi s}}{s^2 + 1}\right]$

2. Find the inverse Laplace transform of

(a) $\dfrac{3(s^2 - 2)^2}{2s^5}$ $\left[\textbf{Ans. } \dfrac{3}{2} - 3t^2 + \dfrac{1}{2}t^4\right]$

(b) $\dfrac{s^2}{s^2 + 4s + 8}$ $[\textbf{Ans. } -4e^{-2t} \cos 2t]$

(c) $\dfrac{s^2 + s - 2}{s(s+3)(s-2)}$ $\left[\textbf{Ans. } \dfrac{1}{3} + \dfrac{4}{15}e^{3t} + \dfrac{2}{5}e^{-2t}\right]$

(d) $\dfrac{s}{(s^2 - 1)^2}$ $\left[\textbf{Ans. } \dfrac{1}{2}t \sinh t\right]$

(e) $\dfrac{1}{s(s+4)}$ $[\textbf{Ans. } \cos 2t]$

(f) $\dfrac{s}{(s+1)^2(s^2+1)}$ $\left[\textbf{Ans. } \dfrac{1}{2}(\sin t - te^{-t})\right]$

(g) $\dfrac{s}{(s+1)(s^2 + 2s + 2)}$ $[\textbf{Ans. } e^{-t}(1 - \cos t)]$

(h) $\dfrac{1}{s^2(s^2 + a^2)}$ $\left[\textbf{Ans. } \dfrac{1}{a^2}(at - \sin at)\right]$

(i) $\dfrac{s}{(s+a)^2}$ $[\textbf{Ans. } (1 - at)e^{-at}]$

(j) $\log\left(\dfrac{s+1}{s-1}\right)$. $\left[\textbf{Ans. } \dfrac{2 \sinh t}{t}\right]$

3. Find Laplace transforms of the following:

(a) $t \sin^2 t$ $\left[\textbf{Ans. } \dfrac{2(3s^2 + 4)}{s^2(s^2 + 4)^2}\right]$

(b) $t^2 \cos at$ $\left[\textbf{Ans. } \dfrac{2s^2 - 6a^2 s}{(s^2 + a^2)^3}\right]$

(c) $t \sinh at$ $\left[\textbf{Ans. } \dfrac{2as}{(s^2 - a^2)^2}\right]$

(d) $\frac{1}{t}(\cos 2t - \cos 3t)$ $\left[\textbf{Ans. } \frac{1}{2}\log\left(\frac{s^2+9}{s^2+4}\right)\right]$

(e) $\frac{1}{t^2}(1-\cos t)$. $\left[\textbf{Ans. } \cot^{-1} s - \frac{s}{2}\log\left(1+\frac{1}{s^2}\right)\right]$

4. Using convolution theorem, evaluate $L^{-1}\left\{\frac{s}{(s^2+a^2)^2}\right\}$. $\left[\textbf{Ans. } \frac{t-\sin at}{2a}\right]$

5. Evaluate $L^{-1}\left\{\frac{s^2}{(s^2+a^2)(s^2+b^2)}\right\}$. $\left[\textbf{Ans. } \frac{1}{a^2-b^2}(a\sin at - b\sin bt)\right]$

6. Solve $y_2 + w^2 y = 0$ with $y(0) = A$ and $y'(0) = B$. $\left[\textbf{Ans. } y = A\cos wx + \frac{B}{w}\sin wx\right]$

7. Solve $\frac{dx}{dt} + x = \sin wt,\ x(0) = 2$. $\left[\textbf{Ans. } x = \left(2 + \frac{w}{w^2+1}\right)e^{-t} + \frac{\sin wt - w\cos wt}{1+w^2}\right]$

8. $y'' + 4y' + 3y = e^{-t},\ y(0) = y'(0) = 1$. $\left[\textbf{Ans. } y = \frac{7}{4}e^{-t} - \frac{3}{4}e^{-3t} - \frac{1}{2}te^{-t}\right]$

9. $(D^2 - 1)x = a\cosh t,\ x(0) = x'(0) = 0$. $\left[\textbf{Ans. } x = \frac{at}{2}\sinh t\right]$

10. $y'' + y' - 2y = t,\ y(0) = 1,\ y'(0) = 0$. $\left[\textbf{Ans. } y = e^t + \frac{1}{4}e^{-2t} - \frac{t}{2} - \frac{1}{4}\right]$

11. $y'' + 2y' + 2y = \frac{17}{2}\sin 5t,\ y = 2$ and $y' = -4$ at $t = 0$.

$\left[\textbf{Ans. } y = e^{-t}\left(4\cos 2t - \frac{1}{2}\sin 2t\right) - 2\cos 2t + \frac{1}{2}\sin 2t\right]$

12. Solve $y''' + 2y'' - y' - 2y = 0$, where $y = 1,\ y' = 2,\ y'' = 2$ at $t = 0$.

$\left[\textbf{Ans. } y = \frac{1}{3}(5e^t + e^{-2t}) - e^{-t}\right]$

13. Solve $\frac{dx}{dt} - y = e^t,\ \frac{dy}{dt} + x = \sin t$, given that $x(0) = 1,\ y(0) = 0$.

$\left[\textbf{Ans. } x = \frac{1}{2}(e^t + \cos t + 2\sin t - t\cos t),\ y = \frac{1}{2}(t\sin t - e^t + \cos t - \sin t)\right]$

14. Solve $\frac{3dx}{dt} + \frac{dy}{dt} + 2x = 1,\ \frac{dx}{dt} + \frac{4dy}{dt} + 3y = 0$ given $x = 3,\ y = 0$ when $t = 0$.

$\left[\textbf{Ans. } x = \frac{1}{10}(5 - 2e^{-t} - 3e^{-6t/11}),\ y = \frac{1}{5}(e^{-t} - e^{-6t/11})\right]$

15. Solve $ty'' + 2y' + ty = 0$, given that $y(0) = 1, y'(\pi) = 0$. $\left[\textbf{Ans. } y = \frac{\sin t}{t}\right]$

16. Solve $y'' + ty' - y = 0$, given that $y(0) = 0, y'(0) = 2$. [**Ans.** $y = 2t$]

17. Solve $\frac{\partial u}{\partial t} = \frac{\partial^2 u}{\partial x^2}$, $u(x, 0) = \sin 2\pi x, u(1, t) = 0, 0 < x < 1$. [**Ans.** $u(x, t) = e^{-4\pi^2 t} \sin 2\pi x$]

18. Solve $\frac{\partial u}{\partial t} = 3\frac{\partial^2 u}{\partial x^2}$, $u_x(0, t) = 0, u\left(\frac{\pi}{2}, t\right) = 0, u(x, 0) = \cos 3x - 2\cos 9x$.

[**Ans.** $u(x, t) = e^{-27t} \cos 3x - 2e^{-243t} \cos 9x$]

19. Solve $\frac{\partial^2 u}{\partial t^2} = 9\frac{\partial^2 u}{\partial x^2}$ given that

$$u(0, t) = u(2, t) = 0, u_t(x, 0) = 0 \text{ and } u(x, 0) = 10 \sin 2\pi x - 20 \sin 5\pi x.$$

[**Ans.** $u(x, t) = 10 \cos 6\pi t \sin 2\pi x - 20 \cos 15\pi t \sin 5\pi x$]

CHAPTER 17 Applications to Partial Differential Equations of Order Two

17.1 INTRODUCTION

In mathematical physics, we come across the partial differential equations. Sometimes, it is difficult to find the general solution of partial differential equations containing arbitrary constants which are difficult to evaluate so as to satisfy the given boundary conditions. In this chapter, we discuss method to overcome this difficulty.

17.2 VARIABLE SEPARATION METHOD

In this method, we disintegrate the given partial differential equation into a product of functions each of which contain only one of the variables.

EXAMPLE 17.1 Solve $\frac{\partial^2 z}{\partial x^2} - 4\frac{\partial z}{\partial x} + \frac{\partial z}{\partial y} = 0$ by the method of separation of variables.

Solution Let the solution of given partial differential equation be

$$z = X(x)\ Y(y) \tag{i}$$

where X is a function of x only and Y is that of y only. Substituting this value of z in the given equation, we get

$$X''Y - 4X'Y - XY' = 0$$

where $$X' = \frac{dX}{dx},\ Y' = \frac{dY}{dy} \text{ and } X'' = \frac{d^2X}{dx^2}$$

Separating the variables, we get

$$\frac{X'' - 4X'}{X} = -\frac{Y'}{Y} \tag{ii}$$

Since x and y are independent variables, Eq. (ii) is valid if each side is equal to the same constant, k (say).

$$\Rightarrow \qquad \frac{X'' - 4X'}{X} = k$$

$$\Rightarrow \qquad X'' - 4X' - kX = 0 \tag{iii}$$

and $$Y' + ky = 0 \tag{iv}$$

Equations (iii) and (iv) are ordinary differential equations.

The auxiliary equation for Eq. (iii) is

$$m^2 - 4m - k = 0$$

$$\Rightarrow \qquad m = 2 \pm \sqrt{4+k}$$

$\therefore$ Solution of Eq. (iii) is

$$X = c_1 e^{(2+\sqrt{4+k})x} + c_2 e^{(2-\sqrt{4+k})x}$$

and solution of Eq. (iv) is $Y = c_3 e^{-ky}$.

Substituting these values of x and y in Eq. (i) gives $z = (c_1 e^{(2+\sqrt{4+k})x} + c_2 e^{(2-\sqrt{4+k})x}) c_3 e^{-ky}$ is the required solution of given partial differential equation.

17.3 LAPLACE'S EQUATION AND ITS SOLUTION

The Laplace's equation is

$$\nabla^2 V = 0 \tag{17.1}$$

∇^2 is called Laplacian operator. This elliptic equation occurs in mathematical physics. V is called a potential function.

17.3.1 In Cartesian Coordinates

We know that

$$\nabla^2 = \frac{\partial^2}{\partial x^2} + \frac{\partial^2}{\partial y^2} + \frac{\partial^2}{\partial z^2}$$

$\therefore$ Equation (17.1) takes form

$$\frac{\partial^2 V}{\partial x^2} + \frac{\partial^2 V}{\partial y^2} + \frac{\partial^2 V}{\partial z^2} = 0 \tag{17.2}$$

Put $V = X(x)Y(y)Z(z)$. Then Eq. (17.2) becomes

$$\frac{1}{X}\frac{d^2 x}{dx^2} + \frac{1}{Y}\frac{d^2 y}{dy^2} + \frac{1}{Z}\frac{d^2 z}{dz^2} = 0$$

This is possible only when each term on LHS is constant.

Let
$$\left.\begin{aligned} \frac{1}{X}\frac{d^2X}{dx^2} &= -\ell^2 \\ \frac{1}{Y}\frac{d^2Y}{dy^2} &= -m^2 \\ \frac{1}{Z}\frac{d^2Z}{dz^2} &= -n^2 \end{aligned}\right\} \tag{17.3}$$

Then
$$l^2 + m^2 + n^2 = 0 \tag{17.4}$$

The first equation of Eq. (17.3) give

$$(D^2 + l^2)X = 0, D = \frac{d}{x}$$

$\Rightarrow$
$$X = A_1 e^{i\ell x} + A_2 e^{-i\ell x}$$

Similarly,
$$Y = B_1 e^{imx} + B_2 e^{-imx}$$

and
$$Z = C_1 e^{inx} + C_2 e^{-inx}$$

$\therefore$ The solution of Eq. (17.2) is

$$V = (A_1 e^{i\ell x} + A_2 e^{-i\ell x})(B_1 e^{imx} + B_2 e^{-imx})(C_1 e^{inx} + C_2 e^{-inx}) \tag{17.5}$$

where A's, B's and C's are all arbitrary constants.

EXAMPLE 17.2 Find the function $V(x, y, z)$ satisfying $\nabla^2 V = 0$ in the region $0 \le x \le a$, $0 \le y \le b$, $0 \le z \le c$ and satisfying the conditions.

(i) $V = 0$ on $x = 0$, $x = a$, $y = 0$, $y = b$, $z = 0$,

(ii) $V = f(x, y)$ for $z = c$, $0 \le x \le a$, $0 \le y \le b$.

Solution We have seen that solution of $\nabla^2 V = 0$ in Cartesian coordinates is of the form $\exp\{\pm ilx \pm imy \pm inz\}$ where $l^2 + m^2 + n^2 = 0$ and l, m, n all may not be real. Since solution is to satisfy condition (i), l, m must be taken as real and n to be complex.
Let $n = i\gamma$. Then $l^2 + m^2 = \gamma^2$ and solution becomes

$\Rightarrow$
$$(A_1 \cos lx + A_2 \sin lx)(B_1 \cos my + B_2 \sin my)(C_1 \cosh \gamma z + C_2 \sinh \gamma z)$$

It is given that $V = 0$ for $x = 0$, $y = 0$ and $z = 0$.

$\Rightarrow$
$$A_1 = 0, B_1 = 0, C_1 = 0$$

$\therefore$ Solution takes the form

$$V = \sum_{l,m} A_{lm} \sin lx \sin my \sinh \gamma z$$

where A_{lm} is constant.
Now, at $x = a$, $V = 0$ for all y and z.

$\therefore$
$$\sin la = 0$$

$\Rightarrow$ $$la = p\pi \text{ or } l = \frac{p\pi}{a}$$

where p is an integer.

Similarly, $$m = \frac{q\pi}{b}$$

where q is an integer.

$\therefore$ $$\gamma^2 = \pi^2\left[\frac{p^2}{q^2} + \frac{q^2}{b^2}\right] = \gamma_{pq} \quad \text{(say)}$$

$\therefore$ $$V = \sum_{p,q} A_{pq} \sin\frac{p\pi x}{a} \sin\frac{q\pi y}{b} \sinh(\gamma_{pq} z) \tag{i}$$

By Fourier series, $f(x, y)$ can be written as

$$f(x, y) = \sum_p \sum_q f_{pq} \sin\frac{p\pi x}{a} \sin\frac{q\pi y}{b}$$

where $$f_{pq} = \frac{4}{ab}\int_0^a\int_0^b f(x, y) = \sin\frac{p\pi x}{a} \sin\frac{q\pi y}{b}$$

Using condition (ii), $V = f(x, y)$ for $z = c$ gives

$$A_{pq} \sin(\gamma_{pq} c) = f_{pq}$$

$\Rightarrow$ $$A_{pq} = f_{pq} \operatorname{cosech}(\gamma_{pq} c)$$

Therefore, Eq. (i)

$$V = \sum_{p,q} f_{pq} \operatorname{cosech}(\gamma_{pq} c) \sin\frac{p\pi x}{a} \sin\frac{q\pi y}{b} \sinh(\gamma_{pq} z)$$

is the required solution.

17.3.2 In Spherical Polar Coordinates

In spherical polar coordinate, r, θ and ϕ Eq. (17.2) takes the form

$$\frac{\partial^2 v}{\partial r^2} + \frac{2}{r}\frac{\partial v}{\partial r} + \frac{1}{r^2}\frac{\partial^2 v}{\partial \theta^2} + \frac{\cot\theta}{r}\frac{\partial v}{\partial \theta} + \frac{1}{r^2 \sin^2\theta}\frac{\partial^2 v}{\partial \phi^2} = 0 \tag{17.6}$$

Put $$V = R(r)\,\Theta(\theta)\,\Theta(\phi) \text{ in Eq. (17.6)} \tag{17.7}$$

$\therefore$ $$\frac{1}{R}\left[\frac{d^2R}{dr^2} + \frac{2}{r}\frac{dR}{dr}\right] + \frac{1}{\Theta}\left[\frac{1}{r^2}\frac{d^2\Theta}{d\theta^2} + \frac{\cot\theta}{r^2}\frac{d\Theta}{d\theta}\right] + \frac{1}{\Phi r^2 \sin^2\theta}\frac{d^2\Phi}{d\phi^2} = 0$$

$\Rightarrow$ $$\frac{r^2\sin^2\theta}{R}\left[\frac{d^2R}{dr^2} + \frac{2}{r}\frac{dR}{dr}\right] + \frac{\sin^2\theta}{\theta}\left[\frac{d^2\Theta}{d\theta^2} + \cot\theta\frac{d\Theta}{d\theta}\right] + \frac{1}{\phi}\frac{d^2\Phi}{d\phi^2} = 0$$

This is possible only when

$$\frac{1}{\Phi}\frac{d^2\Phi}{d\phi^2} = -m^2 \tag{17.8}$$

and

$$\frac{r^2 \sin^2\theta}{R}\left[\frac{d^2R}{dr^2} + \frac{2}{r}\frac{dR}{dr}\right] + \frac{\sin^2\theta}{\Theta}\left[\frac{d^2\Theta}{d\theta^2} + \cot\theta\frac{d\Theta}{d\theta}\right] = m^2 \tag{17.9}$$

Solution of Eq. (17.8) is

$$\Phi = c_1 e^{im\phi} + c_2 e^{-im\phi}$$

where c_1, c_2 are arbitrary constants and Eq. (17.9) becomes

$$\frac{1}{R}\left[r^2\frac{d^2R}{dr^2} + 2r\frac{dR}{dr}\right] + \frac{1}{\Theta}\left[\frac{d^2\Theta}{d\theta^2} + \cot\theta\frac{d\Theta}{d\theta} - \frac{m^2}{\sin^2\theta}\Theta\right] = 0$$

Here, again each term should be constant.

Put

$$\frac{1}{R}\left[r^2\frac{d^2R}{dr^2} + 2r\frac{dR}{dr}\right] = n(n+1) \tag{17.10}$$

Then

$$\frac{1}{\Theta}\left[\frac{d^2\Theta}{d\theta^2} + \cot\theta\frac{d\Theta}{d\theta} - \frac{m^2}{\sin^2\theta}\Theta\right] = -n(n+1) \tag{17.11}$$

Equation (17.10) gives

$$r^2\frac{d^2R}{dr^2} + 2r\frac{dR}{dr} - n(n+1)R = 0$$

Take $u = \log r$

$\Rightarrow$ $[D(D-1) + 2D - n(n+1)]\,R = 0, \; D = \dfrac{d}{du}$

$\Rightarrow$ $[D^2 + D - n(n+1)]\,R = 0$

$\Rightarrow$ $[D+(n+1)]\,(D-n)\,R = 0$

$\therefore$ $R = A_n e^{nu} + B_n e^{-(n+1)u} = A_n r^n + \dfrac{B_n}{r^{n+1}}$

where A_n, B_n are arbitrary constants.
Equation (17.11) implies

$$\frac{d^2\Theta}{d\theta^2} + \cot\theta\frac{d\Theta}{d\theta} + \left[n(n+1) - \frac{m^2}{\sin^2\theta}\right]\Theta = 0 \tag{17.12}$$

Put $\cos\theta = v$

$\therefore$ $\dfrac{d}{d\theta} = \dfrac{d}{dv}\cdot\dfrac{dv}{d\theta} = -\sin\theta\dfrac{d}{dv}$

and
$$\frac{d^2}{d\theta^2} = -\cos\theta\frac{d}{dv} - \sin\theta\frac{d^2}{dv^2}\cdot\frac{dv}{d\theta}$$
$$= \sin^2\theta\frac{d^2}{dv^2} - \cos\theta\frac{d}{dv}$$
$$= (1-v^2)\frac{d^2}{dv^2} - v\frac{d}{dv}$$

$\therefore$ Equation (17.12) reduces to

$$(1-v^2)\frac{d^2\Theta}{dv^2} - 2v\frac{d\Theta}{dv} + \left[n(n+1) - \frac{m^2}{1-v^2}\right]\Theta = 0 \tag{17.13}$$

which is Legendre's associated equation.
Therefore, solution of Eq. (17.6) is

$$V = \left[A_n r^n + \frac{B_n}{r^{n+1}}\right]\Theta\,[c_1 e^{im\theta} + c_2 e^{-im\theta}] \tag{17.14}$$

where Θ is a solution of Eq. (17.13).

Remark: If $m = 0$, Eq. (17.13) becomes Legendre's equation whose solution is

$$\Theta = c_n p_n(\cos\theta) + D_n q_n(\cos\theta)$$

where c_n, D_n are constants.

It is known that for $\cos\theta = \pm 1$, Q_n ($\cos\theta$) is infinite, so if Θ is to remain finite, $\theta = 0$ (i.e. $\cos\theta = 1$), D_n must be taken identically zero.
Then $\Theta = c_n P_n(\cos\theta)$ and Eq. (17.14) can be written in the form

$$V = \left[A_n r^n + \frac{B_n}{r^{n+1}}\right]P_n(\cos\theta)$$

and hence,
$$V = \sum_n \left[A_n r^n + \frac{B_n}{r^{n+1}}\right]p_n(\cos\theta).$$

17.3.3 In Cylindrical Coordinates

In cylindrical coordinates, Eq. (17.2) takes the form

$$\frac{\partial^2 V}{\partial\rho^2} + \frac{1}{\rho}\frac{\partial V}{\partial\rho} + \frac{1}{\rho^2}\frac{\partial^2 V}{\partial\phi^2} + \frac{\partial^2 V}{\partial z^2} = 0 \tag{17.15}$$

Put $V = R(\rho)\,\Theta\,(\phi)\,Z(z)$. Then Eq. (17.15) reduces to

$$\frac{1}{R}\left[\frac{d^2R}{d\rho^2} + \frac{1}{\rho}\frac{dR}{d\rho}\right] + \frac{1}{\rho^2}\cdot\frac{1}{\Phi}\frac{d^2\Phi}{d\phi^2} + \frac{1}{z}\frac{d^2Z}{dz^2} = 0$$

This is possible only when

$$\frac{1}{Z}\frac{d^2Z}{dz^2} = m^2 \tag{17.16}$$

and
$$\frac{1}{R}\left[\frac{d^2R}{d\rho^2}+\frac{1}{\rho}\frac{dR}{d\rho}\right]+\frac{1}{\rho^2}\cdot\frac{1}{\Phi}\frac{d^2\Phi}{d\phi^2} = -m^2 \tag{17.17}$$

where m is constant.
The solution of Eq. (17.16) is $Z = c_1e^{mz} + c_2e^{-mz}$, c_1, c_2 are constants.
Equation (17.17) gives

$$\frac{\rho^2}{R}\left[\frac{d^2R}{d\rho^2}+\frac{1}{\rho}\frac{dR}{d\rho}\right]+m^2\rho^2+\frac{1}{\Phi}\frac{d^2\Phi}{d\phi^2} = 0$$

Here again each term should be constant.

Put
$$\frac{1}{\Phi}\frac{d^2\Phi}{d\phi^2} = -n^2 \tag{17.18}$$

Then
$$\frac{\rho^2}{R}\left[\frac{d^2R}{d\rho^2}+\frac{1}{\rho}\frac{dR}{d\rho}\right]+m^2\rho^2 = n^2$$

$\Rightarrow$
$$\frac{d^2R}{d\rho^2}+\frac{1}{\rho}\frac{dR}{d\rho}+\left(m^2-\frac{n^2}{\rho^2}\right)R = 0 \tag{17.19}$$

The solution of Eq. (17.18) is

$$\Phi = D_1e^{in\phi} + D_2e^{-in\phi}$$

where D_1, D_2 are constants.
Substitute $m\rho = t$ in Eq. (17.19), we get

$$t^2\frac{d^2R}{dt^2}+t\frac{dR}{dt}+(t^2-n^2)R = 0$$

which is Bessel's equation and its general solution is

$$R = A_{mn}J_n(t) + B_{mn}J_{-n}(t) = A_{mn}J_n(m\rho) + B_{mn}J_{-n}(m\rho)$$

where A_{mn}, B_{mn} are constants.
Function $J_{-n}(m\rho) \to \infty$ for $\rho = 0$. So if V is to remain finite for $\rho = 0$, B_{mn} must be zero, and hence,

$$R = A_{mn}J_n(m\rho).$$

$\therefore$ The solution of Eq. (17.15) is

$$V = A_{mn}J_n(m\rho)\, e^{\pm mz \pm in\phi}$$

$\therefore$ General solution is

$$V = \sum_m\sum_n A_{mn}J_n(m\rho)\, e^{\pm mz\pm in\phi}$$

Remark 1: For problems, with symmetry about the z-axis, n can be taken to be zero to obtain solution in the form

$$V = \sum_m A_m J_0(m\rho) e^{\pm mz}$$

Remark 2: If V is symmetrical about z-axis and tends to zero as $\rho \to 0$ and $z \to \infty$, solution is given by

$$V = \sum_m A_m J_0(m\rho) e^{-mz}$$

EXAMPLE 17.3 Find the potential function $V(\rho, z)$ in the region $0 \le \rho \le 1$, $z \le 0$, satisfying the conditions

(i) $V \to 0$ as $z \to \infty$
(ii) $V = 0$ on $\rho = 1$
(iii) $V = f(\rho)$ on $z = 0$ for $0 \le \rho \le 1$.

Solution By remark Eq. (i),

$$V = \sum_m A_m J_0(m\rho) e^{\pm mz}$$

Since $V \to 0$ as $z \to \infty$, solution is

$$V = \sum_m A_m J_0(m\rho) e^{-mz} \tag{i}$$

This will satisfy (ii) if $J_0(m) = 0$.
$\Rightarrow$ m is a roof of $J_0(x) = 0$.
We know that if $f(\rho) = \sum \lambda_m J_0(m\rho)$

then
$$\lambda_m = \frac{2}{\{J_1(m)\}^2} \int_0^1 \rho f(\rho)\, J_0(m\rho)\, d\rho$$

$\therefore$ V of Eq. (i) will satisfy condition (iii) if $A_m = \lambda_m$.

17.4 SOLUTION OF WAVE EQUATION

The wave equation is

$$\nabla^2 V = \frac{1}{c^2} \frac{\partial^2 V}{\partial t^2} \tag{17.20}$$

Put $V = ST(t)$ where S is not dependent on t.
$\therefore$ Equation (17.20) can be written as

$$\frac{1}{S} \nabla^2 S = \frac{1}{c^2 T} \frac{d^2 T}{dt^2}$$

Put
$$\frac{1}{c^2 T} \frac{d^2 T}{dt^2} = -k^2 \tag{17.21}$$

Then
$$\frac{1}{S}\nabla^2 S = -k^2 \tag{17.22}$$

where k is constant.
The solution of Eq. (17.21) is
$$T = c_1 e^{ickt} + c_2 e^{-ickt}$$

c_1, c_2 are constants.
Equation (17.22) implies
$$(\nabla^2 + k^2)\, S = 0 \tag{17.23}$$

Equation (17.23) is known as space form of the wave equation or *Helmboltz's equation.*

17.4.1 In Cartesian Coordinates

Equation (17.23) is of the form
$$\frac{\partial^2 S}{\partial x^2} + \frac{\partial^2 S}{\partial y^2} + \frac{\partial^2 S}{\partial z^2} + k^2 S = 0 \tag{17.24}$$

Put $S = X(x)\, Y(y)\, Z(z)$ in Eq. (17.24), we get
$$\frac{1}{X}\frac{d^2 X}{dx^2} + \frac{1}{Y}\frac{d^2 Y}{dy^2} + \frac{1}{Z}\frac{d^2 Z}{dz^2} + k^2 = 0$$

Put
$$\frac{1}{X}\frac{d^2 X}{dx^2} = -k^2$$
$$\frac{1}{Y}\frac{d^2 Y}{dy^2} = -k^2$$
$$\frac{1}{Z}\frac{d^2 Z}{dz^2} = -k^2$$

Then
$$l^2 + m^2 + n^2 = k^2$$

We have
$$X = A_1 e^{ikx} + A_2 e^{-ikx}$$
$$Y = B_1 e^{iky} + B_2 e^{-iky}$$
$$Z = D_1 e^{ikz} + D_2 e^{-ikz}$$

$\therefore$ Solution of wave Eq. (17.20) is
$$V = (A_1 e^{ikx} + A_2 e^{-ikx})\,(B_1 e^{iky} + B_2 e^{-iky})\,(D_1 e^{ikz} + D_2 e^{-ikz})\,(c_1 e^{ickt} + c_2 e^{-ickt}).$$

17.4.2 In Spherical Polar Coordinates

Here the Eq. (17.23) takes the form
$$\frac{\partial^2 S}{\partial r^2} + \frac{2}{r}\frac{\partial S}{\partial r} + \frac{1}{r^2}\frac{\partial^2 S}{\partial \theta^2} + \frac{\cot\theta}{r^2}\frac{\partial S}{\partial \theta} + \frac{1}{r^2 \sin^2\theta}\frac{\partial^2 S}{\partial \phi^2} + k^2 S = 0 \tag{17.25}$$

Put $\quad S = R(r)\,\Theta(\theta)\,\Theta(\phi)$ in Eq. (17.25), we get

$$\frac{1}{R}\left[\frac{d^2R}{dr^2}+\frac{2}{r}\frac{dR}{dr}\right]+\frac{1}{\Theta}\left[\frac{1}{r^2}\frac{d^2\Theta}{d\theta^2}+\frac{\cot\theta}{r^2}\frac{d\Theta}{d\theta}\right]+\frac{1}{r^2\sin^2\theta}\cdot\frac{1}{\Phi}\frac{d^2\Phi}{d\phi^2}+k^2=0$$

$$\Rightarrow\quad \frac{r^2\sin^2\theta}{R}\left[\frac{d^2R}{dr^2}+\frac{2}{r}\frac{dR}{dr}\right]+\frac{\sin^2\theta}{\Theta}\left[\frac{d^2\Theta}{d\theta^2}+\cot\theta\frac{d\Theta}{d\theta}\right]+k^2r^2\sin^2\theta+\frac{1}{\Phi}\frac{d^2\Phi}{d\phi^2}=0$$

Put
$$\frac{1}{\Phi}\frac{d^2\phi}{d\phi^2}=-m^2$$

$$\Rightarrow\quad \Phi=D_1e^{im\phi}+D_2e^{-im\phi}$$

where D_1 and D_2 are constants.

Then
$$\frac{r^2\sin^2\theta}{R}\left[\frac{d^2R}{dr^2}+\frac{2}{r}\frac{dR}{dr}\right]+\frac{\sin^2\theta}{\Theta}\left[\frac{d^2\Theta}{d\theta^2}+\cot\theta\frac{d\Theta}{d\theta}\right]+k^2r^2\sin^2\theta=m^2$$

$$\Rightarrow\quad \left\{\frac{r^2}{R}\left[\frac{d^2R}{dr^2}+\frac{2}{r}\frac{dR}{dr}\right]+k^2r^2\right\}+\left\{\frac{1}{\Theta}\left[\frac{d^2\Theta}{d\theta^2}+\cot\theta\frac{d\Theta}{d\theta}\right]-\frac{m^2}{\sin^2\theta}\right\}=0$$

Again each bracket must be constant.

Put
$$\frac{r^2}{R}\left[\frac{d^2R}{dr^2}+\frac{2}{r}\frac{dR}{dr}\right]+k^2r^2=n(n+1)\tag{17.26}$$

and
$$\frac{1}{\Theta}\left[\frac{d^2\Theta}{d\theta^2}+\cot\theta\frac{d\Theta}{d\theta}\right]-\frac{m^2}{\sin^2\theta}=-n(n+1)\tag{17.27}$$

where n is constant.
In Eq. (17.26), put $R = r^{-1/2}R_1$. Then

$$\frac{dR}{dr}=-\frac{1}{2}r^{-3/2}R_1+r^{-1/2}\frac{dR_1}{dr}$$

$$\frac{d^2R}{dr^2}=\frac{3}{4}r^{-5/2}R_1-r^{-3/2}\frac{dR_1}{dr}+r^{-1/2}\frac{d^2R_1}{dr^2}$$

Therefore, Eq. (17.26) becomes

$$r^2\frac{d^2R_1}{dr^2}+r\frac{dR_1}{dr}+R_1\left\{k^2r^2-\left(n+\frac{1}{2}\right)^2\right\}=0$$

Put $kr = r_1$. Therefore, equation reduces to

$$r_1^2\frac{d^2R_1}{dr_1^2}+r_1\frac{dR_1}{dr_1}+R_1\left\{r_1^2-\left(n+\frac{1}{2}\right)^2\right\}=0$$

which is Bessel's equation of order $(n + 1/2)$.

$$R_1 = A_1 J_{n+1/2}(r_1) + A_2 J_{-(n+1/2)}(r_1)$$

and hence,
$$R = r^{-1/2}[A_1 J_{n+1/2}(kr) + A_2 J_{-(n+1/2)}(kr)]$$

where A_1, A_2 are constants.

In Eq. (17.27), put $\cos\theta = v$. Arguing as in Section 17.3.2, we get Legendre's associated equation as:

$$(1 - v^2)\frac{d^2\Theta}{dv^2} - 2v\frac{d\Theta}{dv} + \left\{n(n+1) - \frac{m^2}{1 - v^2}\right\}\Theta = 0$$

$$\therefore \qquad \Theta = B_1 P_n^m(v) + B_2 Q_n^m(v)$$

$$= B_1 P_n^m(\cos\theta) + B_2 Q_n^m(\cos\theta)$$

$\therefore$ General solution of Eq. (17.25) is

$$V = r^{-1/2}[A_1 J_{n+1/2}(kr) + A_2 J_{-(n+1/2)}(kr)][B_1 P_n^m(\cos\theta) + B_2 Q_n^m(\cos\theta)]$$
$$[D_1 e^{im\phi} + D_2 e^{-im\phi}][C_1 e^{ickt} + C_2 e^{-ickt}]$$

Remark 1: If V is to remain finite for $\theta = 0$, then B_2 must be zero. In this case

$$V = r^{-1/2}[A_{mn} J_{n+1/2}(kr) + B_{mn} J_{-(n+1/2)}(kr)]P_n^m(\cos\theta)e^{\pm im\phi \pm ickt} \tag{17.28}$$

If further, V is to be such that

$$V(r,\ \theta + \pi,\ \phi) = V(r,\ \theta,\ \phi)$$

and
$$V(r,\ \theta + \pi,\ \phi) = V(r,\ \theta,\ \phi)$$

We must have
$$P_n^m[\cos(\pi + \theta)] = \pm P_n^m(\cos\theta)$$

$$\Rightarrow \qquad P_n^m(-\cos\theta) = \pm P_n^m(\cos\theta)$$

and
$$e^{im(2\pi+\phi)} = e^{\pm im\phi}$$

$$\Rightarrow \qquad e^{\pm m2\pi i} = 1$$

which are possible only when m, n are integers.

$\therefore$ The solution of Eq. (17.28) can be written as:

$$V = r^{-1/2} J_{\pm(n+1/2)}(kr)P_n^m(\cos\theta)e^{\pm im\phi \pm ickt} \tag{17.29}$$

The function $J_{\pm(n+1/2)}$ are called *spherical Bessel function.*

Remark 2: If there is axil symmetry as z-axis, V will be independent of ϕ, and hence, Eq. (17.29) gives

$$V = r^{-1/2} J_{\pm(n+1/2)}(kr)\, P_n(\cos\theta)e^{\pm ickt}$$

Remark 3: In the case of spherical symmetry, V will be independent of θ and ϕ. Hence, solution of wave equation is

$$V = r^{-1/2} J_{\pm 1/2}(kr)\, e^{\pm ickt}$$

EXAMPLE 17.4 Show that if there is spherical symmetry, the solution of wave equation can be obtained in the form

$$V = \frac{1}{r}[f(r - ct) + g(r + ct)]$$

where f, g are arbitrary functions.

Solution Here V is independent of θ and ϕ so wave equation reduces to

$$\frac{\partial^2 V}{\partial r^2} + \frac{2}{r}\frac{\partial V}{\partial r} = \frac{1}{c^2}\frac{\partial^2 V}{\partial t^2} \tag{i}$$

Put
$$V = \frac{n}{r}$$

Then
$$\frac{\partial V}{\partial r} = -\frac{1}{r^2} n + \frac{1}{r}\frac{\partial n}{\partial r}$$

and
$$\frac{\partial^2 V}{\partial r^2} = \frac{2}{r^3} n - \frac{2}{r^2}\frac{\partial n}{\partial r} + \frac{1}{r}\frac{\partial^2 n}{\partial r^2}$$

$\therefore$ Equation (i) reduces to

$$\frac{\partial^2 n}{\partial r^2} = \frac{1}{c^2}\frac{\partial^2 n}{\partial t^2}$$

$\Rightarrow$
$$(D_t^2 - c^2 D_r^2)\, n = 0, \quad D_t \equiv \frac{\partial}{\partial t},\ D_r \equiv \frac{\partial}{\partial r}$$

$\Rightarrow$
$$n = f(r - ct) + g(r + ct)$$

$\Rightarrow$
$$V = \frac{1}{r}[f(r - ct) + g(r + ct)]$$

where f, g are arbitrary functions.

17.4.3 In Cylindrical Coordinates

In cylindrical coordinates (ρ, ϕ, z), the Eq. (17.23) is

$$\frac{\partial^2 S}{\partial \rho^2} + \frac{1}{\rho}\frac{\partial S}{\partial \rho} + \frac{1}{\rho^2}\frac{\partial^2 S}{\partial \phi^2} + \frac{\partial^2 S}{\partial z^2} + k^2 S = 0 \tag{17.30}$$

Put $S = R(\rho)\, \Theta\, (\phi) Z(z)$ in Eq. (17.30) and simplicity.

$\therefore$
$$\frac{1}{R}\left[\frac{d^2 R}{d\rho^2} + \frac{1}{\rho}\frac{dR}{d\rho}\right] + \frac{1}{\rho^2}\cdot\frac{1}{\Phi}\frac{d^2 \Phi}{d\phi^2} + \frac{1}{Z}\frac{d^2 Z}{dz^2} + k^2 = 0$$

Put
$$\frac{1}{Z}\frac{d^2Z}{dz^2} = -\gamma^2 \tag{17.31}$$

Then
$$\frac{1}{R}\left[\frac{d^2R}{d\rho^2}+\frac{1}{\rho}\frac{dR}{d\rho}\right]+\frac{1}{\rho^2}\cdot\frac{1}{\Phi}\frac{d^2\Phi}{d\phi^2}+k^2=\gamma^2$$

$\Rightarrow$
$$\frac{\rho^2}{R}\left[\frac{d^2R}{d\rho^2}+\frac{1}{\rho}\frac{dR}{d\rho}\right]+\frac{1}{\rho^2}\cdot\frac{1}{\Phi}\frac{d^2\Phi}{d\phi^2}+(k^2-\gamma^2)\rho^2=0$$

Put
$$\frac{1}{\Phi}\frac{d^2\Phi}{d\phi^2}=-m^2 \tag{17.32}$$

and
$$\frac{\rho^2}{R}\left[\frac{d^2R}{d\rho^2}+\frac{1}{\rho}\frac{dR}{d\rho}\right]+\rho^2(k^2-\gamma^2)=m^2$$

$\Rightarrow$
$$\rho^2\frac{d^2R}{d\rho^2}+\rho\frac{dR}{d\rho}+(\rho^2w^2-m^2)R=0 \tag{17.33}$$

where $w^2 = k^2 - \gamma^2$.

Equation (17.31) gives $Z = e^{\pm i\gamma z}$

and Eq. (17.23) gives $\Phi = e^{\pm imk\phi}$

The solution of Eq. (17.33) is $R = J_{\pm m}(\rho w)$.

$\therefore$
$$V = J_{\pm m}(\rho w)e^{\pm i\gamma z\pm im\phi\pm ickt}$$

Note: Sign + or – are to be chosen depending upon the physical conditions on V.

17.5 SOLUTION OF DIFFUSION EQUATION

The diffusion equation is
$$\nabla^2V=\frac{1}{k}\frac{\partial V}{\partial t} \tag{17.34}$$

where k is constant, V is temperature and t is time variable.

Put $V = ST(t)$, S is independent of t.

$\therefore$ Equation (17.34) reduces to
$$\frac{1}{S}\nabla^2S=\frac{1}{k}\cdot\frac{1}{T}\frac{dT}{dt}$$

Put
$$\frac{1}{k}\cdot\frac{1}{T}\frac{dT}{dt}=-\lambda^2 \tag{17.35}$$

where λ is real constant as $\dfrac{dT}{dt}<0$

and
$$\frac{1}{S}\nabla^2S=-\lambda^2$$

$\Rightarrow$
$$(\nabla^2+\lambda^2)s=0 \tag{17.36}$$

The solution of Eq. (17.35) is

$$T = Ae^{-\lambda 2kt}.$$

17.5.1 In Cartesian Coordinates

Here Eq. (17.36) is of the form

$$\frac{\partial^2 S}{\partial x^2}+\frac{\partial^2 S}{\partial y^2}+\frac{\partial^2 S}{\partial z^2}+\lambda^2 S=0 \tag{17.37}$$

Also V, and hence, S decreases with x, y, z.
Put $S = X(x)\ Y(y)\ Z(z)$. Then Eq. (17.37) reduces to

$$\frac{1}{X}\frac{d^2X}{dx^2}+\frac{1}{Y}\frac{d^2Y}{dy^2}+\frac{1}{Z}\frac{d^2Z}{dz^2}+\lambda^2=0$$

Put

$$\frac{1}{X}\frac{d^2X}{dx^2}=-l^2$$

$$\frac{1}{Y}\frac{d^2Y}{dy^2}=-m^2$$

$$\frac{1}{Z}\frac{d^2Z}{dz^2}=-n^2$$

where l, m, n are real constants and $l^2 + m^2 + n^2 = \lambda^2$. The solutions are

$$X = A_1 \cos lx + A_2 \sin lx$$

$$Y = B_1 \cos my + B_2 \sin my$$

$$Z = C_1 \cos nz + C_2 \sin nz$$

$\therefore \quad S = (A_1 \cos lx + A_2 \sin lx)\ (B_1 \cos my + B_2 \sin my)\ (C_1 \cos nz + C_2 \sin nz)$

and hence, the general solution of diffusion equation is

$$V = Ae^{-\lambda^2 kt} S.$$

17.5.2 In Spherical Polar Coordinates

Here Eq. (17.36) takes the form

$$\frac{\partial^2 S}{\partial r^2}+\frac{2}{r}\frac{\partial S}{\partial r}+\frac{1}{r^2}\frac{\partial^2 S}{\partial \theta^2}+\frac{\cot\theta}{r^2}\frac{\partial S}{\partial \theta}+\frac{1}{r^2\sin^2\theta}\frac{\partial^2 S}{\partial \phi^2}+\lambda^2 S=0 \tag{17.38}$$

Put $S = R(r)\ \Theta\ (\theta)\ \Phi\ (\phi)$ and arguing as in Section 17.4.2, we get

$$\frac{r^2\sin^2\theta}{R}\left[\frac{d^2R}{dr^2}+\frac{2}{r}\frac{dR}{dr}\right]+\frac{\sin^2\theta}{\Theta}\left[\frac{d^2\Theta}{d\theta^2}+\cot\theta\frac{d\Theta}{d\theta}\right]+\lambda^2 r^2\sin^2\theta+\frac{1}{\Phi}\frac{d^2\Phi}{d\phi^2}=0$$

Put $$\frac{1}{\Phi}\frac{d^2\Phi}{d\phi^2} = -m^2$$

$\Rightarrow$ $$\Phi = c_1 e^{im\phi} + c_2 e^{-im\phi}$$

Then $$\frac{r^2 \sin^2\theta}{R}\left[\frac{d^2R}{dr^2} + \frac{2}{r}\frac{dR}{dr}\right] + \frac{\sin^2\theta}{\Theta}\left[\frac{d^2\Theta}{d\theta^2} + \cot\theta\frac{d\Theta}{d\theta}\right] + \lambda^2 r^2 \sin^2\theta = m^2$$

$\Rightarrow$ $$\left\{\frac{1}{R}\left[r^2\frac{d^2R}{dr^2} + 2r\frac{dR}{dr}\right] + \lambda^2 r^2\right\} + \left\{\frac{1}{\Theta}\left[\frac{d^2\Theta}{d\theta^2} + \cot\theta\frac{d\Theta}{d\theta}\right] - \frac{m^2}{\sin^2\theta}\right\} = 0$$

Each bracket must be constant.

Put $$\frac{1}{R}\left[r^2\frac{d^2R}{dr^2} + 2r\frac{dR}{dr}\right] + \lambda^2 r^2 = n(n+1) \tag{17.39}$$

Then $$\frac{1}{\Theta}\left[\frac{d^2\Theta}{d\theta^2} + \cot\theta\frac{d\Theta}{d\theta}\right] - \frac{m^2}{\sin^2\theta} = -n(n+1) \tag{17.40}$$

In Eq. (17.40), put $\cos\theta = v$, we get Legendre's associated equation

$$(1-v^2)\frac{d^2\Theta}{dv^2} - 2v\frac{d\Theta}{dv} + \left\{n(n+1) - \frac{m^2}{1-v^2}\right\}\Theta = 0$$

$\therefore$ $$\Theta = A_1 P_n^m(\cos\theta) + A_2 Q_n^m(\cos\theta)$$

Equation (17.39) is

$$r^2\frac{d^2R}{dr^2} + 2r\frac{dR}{dr} + [\lambda^2 r^2 - n(n+1)]R = 0$$

$\therefore$ $$R = r^{-1/2}[B_1 J_{n+1/2}(\lambda r) + B_2 J_{-(n+1/2)}(\lambda r)]$$

$$\therefore\ S = r^{-1/2}[B_1 J_{n+1/2}(\lambda r) + B_2 J_{-(n+1/2)}(\lambda r)][A_1 P_n^m(\cos\theta) + A_2 Q_n^m(\cos\theta)][C_1 e^{im\phi} + C_2 e^{-im\phi}]$$

$\therefore$ General solution of diffusion equation in spherical polar coordinates is

$$V = Ae^{-\lambda^2 kt}S.$$

17.5.3 In Cylindrical Coordinates

Here, Eq. (17.36) takes the form

$$\frac{\partial^2 S}{\partial\rho^2} + \frac{1}{\rho}\frac{\partial S}{\partial\rho} + \frac{1}{\rho^2}\frac{\partial^2 S}{\partial\phi^2} + \frac{\partial^2 S}{\partial z^2} + \lambda^2 S = 0 \tag{17.41}$$

Put $S = R(\rho)\ \Theta\ (\phi)Z(z)$ in Eq. (17.41) and simplify, we get

$$\frac{1}{R}\left[\frac{d^2R}{d\rho^2}+\frac{1}{\rho}\frac{dR}{d\rho}\right]+\frac{1}{\rho^2}\frac{1}{\Phi}\frac{d^2\Phi}{d\phi^2}+\frac{1}{Z}\frac{d^2Z}{dz^2}+\lambda^2=0$$

Put $$\frac{1}{Z}\frac{d^2Z}{dz^2}=-\mu^2$$

$\Rightarrow$ $$Z = B_1e^{i\mu z}+B_2e^{-i\mu z}$$

Then $$\frac{1}{R}\left[\frac{d^2R}{d\rho^2}+\frac{1}{\rho}\frac{dR}{d\rho}\right]+\frac{1}{\rho^2}\cdot\frac{1}{\Phi}\frac{d^2\Phi}{d\phi^2}+\lambda^2-\mu^2=0$$

$\Rightarrow$ $$\frac{1}{R}\left[\rho^2\frac{d^2R}{d\rho^2}+\rho\frac{dR}{d\rho}\right]+(\lambda^2-\mu^2)\rho^2+\frac{1}{\Phi}\frac{d^2\Phi}{d\phi^2}=0$$

Again each term must be constant.

Put $$\frac{1}{\Phi}\frac{d^2\Phi}{d\phi^2}=-\gamma^2$$

$\Rightarrow$ $$\Phi = C_1e^{i\gamma\phi}+C_2e^{-i\gamma\phi}$$

and $$\frac{1}{R}\left[\rho^2\frac{d^2R}{d\rho^2}+\rho\frac{dR}{d\rho}\right]+(\lambda^2-\mu^2)\rho^2=\gamma^2$$

$\Rightarrow$ $$\rho^2\frac{d^2R}{d\rho^2}+\rho\frac{dR}{d\rho}+[(\lambda^2-\mu^2)\rho^2-\gamma^2]R=0$$

$\Rightarrow$ $$R = A_1J_\gamma(\rho\sqrt{\lambda^2-\mu^2})+A_2J_{-\gamma}(\rho\sqrt{\lambda^2-\mu^2})$$

$\therefore$ The solution of Eq. (17.41) is

$$S = [A_1J_\gamma(\rho\sqrt{\lambda^2-\mu^2})+A_2J_{-\gamma}(\rho\sqrt{\lambda^2-\mu^2})][C_1e^{i\gamma\phi}+C_2e^{-i\gamma\phi}][B_1e^{i\mu z}+B_2e^{-i\mu z}]$$

$\therefore$ The general solution of diffusion equation in cylindrical coordinates is

$$V = Ae^{-\lambda^2kt}S.$$

EXERCISES

1. Find the deflection $v(x, y, t)$ of the square membrane with $a = b = 1$ and $c = 1$, if the initial velocity is zero and the initial deflection is $f(x, y) = A(\sin \pi x \sin 2\pi y$.

[**Ans.** $v(x, y, t) = A\sin\pi x\sin 2\pi y\cos(\sqrt{5}\pi t)$]

2. A tightly stretched unit square membrane starts vibrating from rest and its initial displacement is $k \sin 2\pi x \sin \pi y$. Show that the deflection at any instant is $k \sin 2\pi x \sin \pi y \cos (\sqrt{5}\pi t)$.

3. Find the potential in the interior of a sphere of unit radius when the potential on the surface is $f(\theta) = \cos^2 \theta$.
$$\left[\textbf{Ans. } v(r, \theta) = \frac{1}{2} + r^2\left(\cos^2 \theta - \frac{1}{3}\right)\right]$$

4. The potential on the surface of a unit sphere is $f(\theta) = \cos^2 \theta$. Show that the potential at all points of the space is given by

$$u(r, \theta) = 2r^2\left(\cos^2 \theta - \frac{1}{3}\right), r < 1$$

and
$$u(r, \theta) = 2r^{-3}\left(\cos^2 \theta - \frac{1}{3}\right) - r^{-1/3}, r < 1.$$

5. For an infinite cylinder, temperature T is a function of ρ and t only. Determine $T(\rho, t)$ for $0 \le \rho \le a$ subject to the condition.

(i) $T(\rho, 0) = f(\rho)$

(ii) Surface $\rho = a$ is maintained at zero temperature.

$$\left[\textbf{Ans. } T = \sum_{n=1}^{\infty} A_n e^{-\lambda_n^2 kt} J_0(\lambda_n \rho)\right]$$

Chapter 18 Fourier Transforms and Its Applications to Partial Differential Equations

18.1 FOURIER TRANSFORM

Let f be a function. The Fourier transform of f is defined by

$$\Im[f(s)] = \frac{1}{\sqrt{2\pi}} \int_{-\infty}^{\infty} f(t)e^{ist}\, dt \tag{18.1}$$

where s is real. It is denoted by $\hat{f}(s)$. The above definition is stated under the assumption that integral in Eq. (18.1) converges.

Remark 1: If f is continuous and piecewise continuously differentiable such that $\int_{-\infty}^{\infty} f(t)\, e^{ist}\, dt$ converges for each s and $\lim_{t\to\pm\infty} f(t) = 0$, then $\Im[f'] =$ is $\Im[f]$.

$$\int_{-a}^{b} f'(t)e^{ist} dt = \left| f(t)e^{ist} \right|_{-a}^{b} - is\int_{-a}^{b} f(t)e^{ist} dt$$

Let $a \to \infty$ and $b \to \infty$.

$$\int_{-\infty}^{\infty} f'(t)e^{ist} dt = -is\int_{-a}^{b} f(t)e^{ist} dt$$

$\Rightarrow$

$$\Im[f'] = -\ is\ \Im[f].$$

Remark 2: Let f be real.

Then $$\hat{f}(s) = \frac{1}{\sqrt{2\pi}}\int_{-\infty}^{\infty} f(t)e^{-ist} dt = \overline{\left(\frac{1}{\sqrt{2\pi}}\int_{-\infty}^{\infty} f(t)e^{ist} dt\right)} = \overline{\hat{f}(s)}$$

i.e. real part of $\hat{f}$ is an even function of s and the imaginary part is an odd function of s.

Remark 3: We know that Fourier series for a function f defined on $(-c, c)$ is

$$f(x)=\sum_{-\infty}^{\infty} a_n e^{-in\pi x/c}$$

where
$$a_n=\frac{1}{2c}\int_{-c}^{c} f(t)e^{in\pi t/c}\,dt$$

$$\therefore \quad f(x)=\sum_{\infty}^{\infty}\frac{1}{2c}\int_{-c}^{c} f(t)e^{in\pi t/c}\,dt e^{-in\pi x/c}$$

$$=\frac{1}{2\pi}\sum_{\infty}^{\infty}\frac{\pi}{c}\int_{-c}^{c} f(t)e^{in\pi(t-x)/c}\,dt$$

Let
$$c\to\infty.$$

The set of numbers $\frac{n\pi}{c}$ becomes more and more dense on the real axis with $n = 0, \pm 1, \pm 2, \ldots$

Hence, we can put $\frac{n\pi}{c} = s$ and $\frac{\pi}{c}$ as ds.
Clearly $-\infty < s < \infty$.

$$f(x)=\frac{1}{2\pi}\int_{-\infty}^{\infty} ds\int_{-\infty}^{\infty} f(t)e^{is(t-x)}\,dt$$

$$=\frac{1}{2\pi}\int_{-\infty}^{\infty}\int_{-\infty}^{\infty} f(t)e^{is(t-x)}\,dt\,ds$$

$$=\frac{1}{2\pi}\int_{\infty}^{\infty} e^{-isx}\,ds\int_{\infty}^{\infty} f(t)e^{ist}\,dt \tag{18.2}$$

The integral Eq. (18.2) is called **Fourier complex integral**.

Remark 4: By definition of Fourier transform

$$\hat{f}(s)=\frac{1}{\sqrt{2\pi}}\int_{-\infty}^{\infty} f(t)e^{ist}\,dt$$

Then Eq. (18.2) implies

$$f(t)=\frac{1}{\sqrt{2\pi}}\int_{-\infty}^{\infty}\hat{f}(s)e^{-ist}\,ds \tag{18.3}$$

Equation (18.3) gives method of finding $f(t)$ from its Fourier transform $\hat{f}(s)$. This is known as inversion theorem and $f(x)$ is called inverse Fourier transform of $\hat{f}(s)$.

Remark 5: It is often convenient to define the Fourier transform in the form

$$\hat{f}(s) = \int_{-\infty}^{\infty} f(t)e^{ist}\,dt \tag{18.4}$$

Then
$$f(t) = \frac{1}{2\pi}\int_{-\infty}^{\infty} \hat{f}(s)e^{-ist}\,ds \tag{18.5}$$

The factor $\dfrac{1}{\sqrt{2\pi}}$ is being omitted in defining transform. This is useful however it neglects the partial symmetry between the function $f(t)$ and its transform $\hat{f}(s)$.

EXAMPLE 18.1 Find the Fourier transform of $f(x) = \begin{cases} 1, \text{for } |x| < 1 \\ 0, \text{for } |x| > 1 \end{cases}$. Hence evaluate $\displaystyle\int_0^{\infty} \frac{\sin x}{x}\,dx$.

Solution The Fourier transform of $f(x)$

$$\hat{f}(s) = \int_{\infty}^{\infty} f(x)e^{isx}\,dx$$

$$= \int_{-1}^{1} 1e^{isx}\,dx$$

$$= \left|\frac{e^{isx}}{is}\right|_{-1}^{1}$$

$$= \frac{e^{is} - e^{-is}}{is} = 2, \frac{\sin s}{s} \cdot s \neq 0$$

For
$$s = 0,\ \hat{f}(s) = 2$$

Now, from the inversion formula,

$$f(x) = \frac{1}{2\pi}\int_{\infty}^{\infty} \hat{f}(s)e^{-isx}\,ds$$

$$\Rightarrow \qquad \frac{1}{2\pi}\int_{\infty}^{\infty} \frac{2\sin s}{s}e^{-isx}\,ds = \begin{cases} 1, & \text{for } |x| < 1 \\ 0, & \text{for } |x| > 1 \end{cases}$$

Putting $x = 0$, we get $\displaystyle\int_{\infty}^{\infty} \frac{\sin s}{s}\,ds = \pi$

Clearly, $\dfrac{\sin s}{s}$ is an even function.

$$\therefore \qquad \int_{o}^{\infty} \frac{\sin s}{s}\,ds = \frac{\pi}{2}.$$

EXAMPLE 18.2 Find the Fourier transform of $e^{-|x|}$.

Solution Let $f(x) = e^{-|x|}$.
Then Fourier transform is

$$\hat{f}(s) = \frac{1}{\sqrt{2\pi}} \int_{-\infty}^{\infty} e^{-|x|} e^{isx}\, dx$$

$$= \frac{1}{\sqrt{2\pi}} \left[\int_{-\infty}^{0} e^{x} e^{isx}\, dx + \int_{0}^{\infty} e^{-x} e^{isx}\, dx \right]$$

$$= \frac{1}{\sqrt{2\pi}} \left[\int_{0}^{\infty} e^{-x} e^{-isx}\, dx + \int_{0}^{\infty} e^{-x} e^{isx}\, dx \right]$$

$$= \frac{1}{\sqrt{2\pi}} \left[\int_{0}^{\infty} e^{-x(1+is)}\, dx + \int_{0}^{\infty} e^{x(is-1)}\, dx \right]$$

$$= \frac{1}{\sqrt{2\pi}} \left[-\frac{e^{-x(1+is)}}{1+is} \right]_0^{\infty} + \frac{1}{\sqrt{2\pi}} \left[\frac{e^{x(is-1)}}{is-1} \right]_0^{\infty}$$

$$= \frac{1}{\sqrt{2\pi}} \left[\frac{1}{1+is} - \frac{1}{is-1} \right]$$

$$= \frac{1}{\sqrt{2\pi}} \left[\frac{1-is}{1+s^2} + \frac{1+is}{1+s^2} \right]$$

$$= \sqrt{\frac{2}{\pi}} \cdot \frac{1}{1+s^2}.$$

EXAMPLE 18.3 Show that $\Im[f(ax-b)] = \frac{1}{|a|} e^{isb/a}\, \hat{f}\left(\frac{s}{a}\right)$.

Solution By definition

$$\Im[f(ax-b)] = \frac{1}{\sqrt{2\pi}} \int_{-\infty}^{\infty} f(ax-b) e^{isx}\, dx$$

Put $\qquad y = ax - b.$

Then $\qquad \Im[f(y)] = \frac{1}{\sqrt{2\pi}} \left(\pm\frac{1}{a} \right) \int_{-\infty}^{\infty} e^{is\left(\frac{y+b}{a}\right)} f(y)\, dy$

Depending upon $a > 0$ or $a < 0$.

$$= \frac{1}{|a|} e^{isb/a} \frac{1}{\sqrt{2\pi}} \int_{-\infty}^{\infty} f(y)\, e^{isy/a}\, dy$$

$$= \frac{1}{|a|} e^{isb/a}\, \hat{f}\left(\frac{s}{a}\right).$$

EXAMPLE 18.4 Find the Fourier transform of $f(x) = \int_{-\infty}^{x} e^{-(x-y)} g(y)dy$.

Solution

$$\hat{f}(s) = \frac{1}{\sqrt{2\pi}} \int_{-\infty}^{\infty} f(x)\, e^{isx} dx$$

$$= \frac{1}{\sqrt{2\pi}} \int_{-\infty}^{\infty} e^{isx} \int_{-\infty}^{x} e^{-(x-y)} g(y)\, dy$$

$$= \frac{1}{\sqrt{2\pi}} \int_{y=-\infty}^{\infty} e^{y} g(y)\, dy \int_{x=y}^{\infty} e^{-(1-is)x} dx$$

$$= \frac{1}{\sqrt{2\pi}} \int_{y=-\infty}^{\infty} g(y) \frac{e^{isy}}{1-is} dy$$

$$= \frac{\hat{g}(s)}{1-is}.$$

18.2 FOURIER SINE AND COSINE TRANSFORMS

If $f(t)$ is an odd function then $f(-t) = -f(t)$ and $\int_{-\infty}^{\infty} f(t) \cos wt\, dt = 0$

$$\therefore \quad \hat{f}(s) = \frac{1}{\sqrt{2\pi}} \int_{-\infty}^{\infty} f(t)\, e^{ist} dt$$

$$= \frac{1}{\sqrt{2\pi}} \int_{-\infty}^{\infty} f(t) (\cos st + i \sin st)\, dt$$

$$= \frac{2i}{\sqrt{2\pi}} \int_{0}^{\infty} f(t) \sin st\, dt$$

If $f(x)$ is defined on $(0, \infty)$, its sine transform is defined as

$$\hat{f}_s(s) = \Im_s[f] = \sqrt{\frac{2}{\pi}} \int_{0}^{\infty} f(t) \sin st\, dt \tag{18.6}$$

If $f(x)$ is an odd function then $\hat{f}(s) = i\hat{f}_s(s)$.
Therefore, by inversion theorem,

$$f(t) = \frac{1}{\sqrt{2\pi}} \int_{-\infty}^{\infty} (e^{-ist})\, i\hat{f}_s(s)\, ds$$

Similarly, Fourier cosine transform is defined as

$$\hat{f}_c(s) = \sqrt{\frac{2}{\pi}} \int_{0}^{\infty} f(t) \cos st\, dt \tag{18.7}$$

For a function $f(x)$ defined on $(0, \infty)$. If $f(x)$ is extended on $(-\infty, \infty)$ as an even function then

$$\hat{f}(s) = \hat{f}_c(s).$$

Remark 1: $\hat{f}_s(s)$, sine transform is an odd function of s and $f(x) = \Im_s[\Im_s(f)]$.

Remark 2: $\hat{f}_c(s)$, cosine transform is an even function of s and $f(x) = \Im_c[\Im_c(f)]$.

Remark 3: If $f(t)$ and $g(t)$ be two functions, then

$$\int_0^\infty \hat{f}_c(s)\hat{g}_c(s)\cos(st)\,ds$$

$$= \sqrt{\frac{2}{\pi}}\int_0^\infty \hat{f}_c(s)\cos ts\,ds\int_0^\infty g(y)\cos(sy)\,dy$$

$$= \sqrt{\frac{2}{\pi}}\int_0^\infty g(y)\,dy\int_0^\infty \hat{f}_c(s)\cos(ts)\cos(sy)\,ds$$

$$= \frac{1}{\sqrt{2\pi}}\int_0^\infty g(y)\,dy\int_0^\infty \hat{f}_c(s)\{\cos(t-y)s + \cos(t+y)s\}\,ds$$

$$= \frac{1}{2}\int_0^\infty g(y)\{f(t-y) + f(t+y)\}\,dy \tag{18.8}$$

Similarly,

$$\int_0^\infty \hat{f}_s(s)\,\hat{g}_c(s)\sin(ts)\,ds = \frac{1}{2}\int_0^\infty f(y)\{g(t-y) - g(t+y)\}\,dy \tag{18.9}$$

$$\int_0^\infty \hat{f}_c(s)\,\hat{g}_s(s)\sin(ts)\,ds = \frac{1}{2}\int_0^\infty g(y)\{f(t-y) - f(t+y)\}\,dy \tag{18.10}$$

and

$$\int_0^\infty \hat{f}_s(s)\,\hat{g}_s(s)\cos(ts)\,ds = \frac{1}{2}\int_0^\infty g(y)\{f(t-y) + f(t+y)\}\,dy \tag{18.11}$$

Taking $t = 0$ in Eqs. (18.8) and (18.11) gives

$$\int_0^\infty \hat{f}_c(s)\,\hat{g}_c(s)\,ds = \int_0^\infty f(y)\,g(y)\,dy \tag{18.12}$$

$$\int_0^\infty \hat{f}_s(s)\,\hat{g}_s(s)\,ds = \int_0^\infty f(y)\,g(y)\,dy \tag{18.13}$$

Also,

$$\int_{-\infty}^\infty \hat{f}(s)\,\hat{g}(s)\,ds = \int_{-\infty}^\infty \hat{f}(s)\,ds\frac{1}{\sqrt{2\pi}}\int_{-\infty}^\infty g(y)\,e^{isy}\,dy$$

$$= \int_{-\infty}^{\infty} g(y)\, dy \frac{1}{\sqrt{2\pi}} \int_{-\infty}^{\infty} \hat{f}(s)\, e^{isy}\, ds$$

$$= \int_{-\infty}^{\infty} g(y)\, f(y)\, dy \tag{18.14}$$

When f is identical with g, Eqs. (18.12) and (18.13) reduce to

$$\int_{0}^{\infty} \{\hat{f}_c(s)\}^2\, ds = \int_{0}^{\infty} \{f(y)\}^2\, dy \tag{18.15}$$

and

$$\int_{0}^{\infty} \{\hat{f}_s(s)\}^2\, ds = \int_{0}^{\infty} \{f(y)\}^2\, dy \tag{18.16}$$

The Eqs. (18.12) to (18.16) are known as Parseval's relations for Fourier transforms.

EXAMPLE 18.5 Find the Fourier sine transform of $\dfrac{e^{-\lambda x}}{x}$, $x > 0$.

Solution The sine transform of function $f(x) = \dfrac{e^{-\lambda x}}{x}$ is

$$\hat{f}_s(s) = \sqrt{\frac{2}{\pi}} \int_{0}^{\infty} f(x) \sin sx\, dx$$

$$= \sqrt{\frac{2}{\pi}} \int_{0}^{\infty} \frac{e^{-\lambda x}}{x} \sin sx\, dx$$

Differentiating w.r.t. s gives

$$[\hat{f}_s(s)]' = \sqrt{\frac{2}{\pi}} \int_{0}^{\infty} \frac{e^{-\lambda x}\, x \cos sx}{x}\, dx$$

$$= \sqrt{\frac{2}{\pi}} \int_{0}^{\infty} e^{-\lambda x} \cos sx\, dx$$

$$= \sqrt{\frac{2}{\pi}} \frac{\lambda}{\lambda^2 + s^2}$$

Integrating w.r.t. s gives

$$\hat{f}_s(s) = \sqrt{\frac{2}{\pi}} \tan^{-1}\left(\frac{s}{\lambda}\right) + \text{constant}$$

For $s = 0$, $\hat{f}_s(s) = 0$. Therefore, constant $= 0$.

$$\therefore \qquad \hat{f}_s(s) = \sqrt{\frac{2}{\pi}} \tan^{-1}\left(\frac{s}{\lambda}\right).$$

EXAMPLE 18.6 Find Fourier cosine transform of $x^n e^{-ax}$.

Solution
$$\hat{f}_c(s) = \sqrt{\frac{2}{\pi}} \int_0^\infty x^n e^{-ax} \cos sx \, dx$$

We know that

$$\int_0^\infty e^{-ax} \cos sx \, dx = \frac{a}{a^2 + s^2} = \frac{1}{2}\left[\frac{1}{a - is} + \frac{1}{a + is}\right]$$

Differentiating n-times w.r.t. 'a' gives

$$\int_0^\infty (-1)^n x^n e^{-ax} \cos sx \, dx = \frac{1}{2}\left[\frac{(-1)^n n!}{(a - is)^{n+1}} + \frac{(-1)^n n!}{(a + is)^{n+1}}\right]$$

$$\Rightarrow \qquad \int_0^\infty x^n e^{-ax} \cos sx \, dx = \frac{n!}{2}\left[\frac{(a + is)^{n+1} + (a - is)^{n+1}}{(a^2 + s^2)^{n+1}}\right]$$

Put $a = r \cos \theta$ and $s = r \sin \theta$ in RHS.

$$\therefore \qquad \hat{f}_c(s) = \sqrt{\frac{2}{\pi}} \cdot \frac{n!}{r^{n+1}} \cos [(n + 1)\theta].$$

EXAMPLE 18.7 Prove that $\Im_c[f''] = -\sqrt{\frac{2}{\pi}} f'(0) - s^2 \Im_s[f]$.

Solution By definition

$$\Im_c[f''] = \sqrt{\frac{2}{\pi}} \int_0^\infty f''(x) \cos sx \, dx$$

$$\therefore \qquad \sqrt{\frac{\pi}{2}} \Im_c[f''] = [f'(x) \cos sx]_0^\infty + s \int_0^\infty f'(x) \sin sx \, dx$$

$$= - f'(0) + s\left[\left|f(x) \sin sx\right|_0^\infty - s \int_0^\infty f(x) \cos sx \, dx\right]$$

$$= - f'[0] - s^2 \Im_c[0] \sqrt{\frac{\pi}{2}}$$

Assuming $f, f'' \to 0$ as $x \to 0$.

$$\therefore \qquad \Im_c[f''] = -\sqrt{\frac{2}{\pi}} f'(0) - s^2 \Im_c[f].$$

18.3 CONVOLUTION PRODUCT

Let $f(x)$ and $g(x)$ be two functions. The convolution product of f and g is denoted by $f * g$ and defined as

$$f * g(t) = \frac{1}{\sqrt{2\pi}} \int_{-\infty}^{\infty} f(t-x)\, g(x)\, dx \tag{18.17}$$

The convolution product is also known as **Faltung** of the functions f and g over the interval $(\infty, -\infty)$.

THEOREM 18.1 Fourier transform of a product of two products is the convolution product of their Fourier transforms.

Proof: We need to prove that $\Im[f \cdot g] = \hat{f} * \hat{g}(s)$.

$$\Im[f \cdot g] = \frac{1}{\sqrt{2\pi}} \int_{-\infty}^{\infty} e^{isx} f(x)\, g(x)\, dx$$

By inversion theorem

$$f(x) = \frac{1}{\sqrt{2\pi}} \int_{-\infty}^{\infty} e^{-i\lambda x} \hat{f}(\lambda)\, d\lambda$$

$$\therefore \qquad \Im[f \cdot g] = \frac{1}{\sqrt{2\pi}} \int_{-\infty}^{\infty} e^{isx} g(x) \left\{ \frac{1}{\sqrt{2\pi}} \int_{-\infty}^{\infty} e^{-i\lambda x} \hat{f}(\lambda)\, d\lambda \right\} dx$$

$$= \frac{1}{\sqrt{2\pi}} \int_{-\infty}^{\infty} \hat{f}(\lambda) \left\{ \frac{1}{\sqrt{2\pi}} \int_{-\infty}^{\infty} e^{i(s-\lambda)x} g(x)\, dx \right\} d\lambda$$

$$= \frac{1}{\sqrt{2\pi}} \int_{-\infty}^{\infty} \hat{f}(\lambda)\, \hat{g}(s-\lambda)\, d\lambda$$

$$= \hat{f} * \hat{g}(s).$$

THEOREM 18.2 **(Convolution theorem):** Fourier transform of convolution product of two functions is the product of their Fourier transforms.

Proof: We need to prove $\Im[f * g] = (\hat{f} \cdot \hat{g})(s)$
By definition

$$f * g(x) = \frac{1}{\sqrt{2\pi}} \int_{-\infty}^{\infty} f(x-t)\, g(t)\, dt$$

$$\Im[f * g] = \frac{1}{\sqrt{2\pi}} \int_{-\infty}^{\infty} e^{isx} \left\{ \frac{1}{\sqrt{2\pi}} \int_{-\infty}^{\infty} f(x-t)\, g(t)\, dt \right\} dx$$

$$= \frac{1}{2\pi} \int_{-\infty}^{\infty} e^{is(x-t)} f(x-t)\, e^{ist} g(t)\, dt\, dx$$

$$= \frac{1}{\sqrt{2\pi}} \int_{-\infty}^{\infty} e^{ist} g(t) \left\{ \frac{1}{\sqrt{2\pi}} \int_{-\infty}^{\infty} e^{is(x-t)} f(x-t)\, dx \right\} dt$$

$$= \frac{1}{\sqrt{2\pi}} \int_{-\infty}^{\infty} e^{ist} g(t)\, dt \frac{1}{\sqrt{2\pi}} \int_{-\infty}^{\infty} e^{is(x-t)} f(x-t)\, dx$$

$$= \hat{g}(s)\, \hat{f}(s)$$

$$= (\hat{f} \cdot \hat{g})\,(s).$$

18.4 FINITE FOURIER TRANSFORM

Consider the function $f(x)$ in interval $(0, c)$ such that

(i) $f(x)$ has only a finite number of maxima and minima in C, and

(ii) $f(x)$ has only a finite number of finite discontinuities in $(0, c)$ and no infinite discontinuity.

Such a function is said to satisty Dirichlet's condition in the interval $(0, c)$.
For such a function $f(x)$, the finite Fourier cosine transform is defined by

$$\hat{f}_c(n) = \int_0^c f(x) \cos \frac{n\pi x}{c}\, dx \tag{18.18}$$

Then at each point of continuity in $(0, c)$, function $f(x)$ is given by

$$f(x) = \frac{1}{c} \hat{f}_c(0) + \frac{2}{c} \sum_{n=1}^{\infty} \hat{f}_c(n) \cos \frac{n\pi x}{c} \tag{18.19}$$

Similarly, finite Fourier sine transform is defined by

$$\hat{f}_s(n) = \int_0^c f(x) \sin \frac{n\pi x}{c}\, dx \tag{18.20}$$

and then each point of continuity in $(0, c)$, $f(x)$ is given by

$$f(x) = \frac{2}{c} \sum_{n=1}^{\infty} \hat{f}_s(n) \sin \frac{n\pi x}{c} \tag{18.21}$$

Remark: The finite Fourier sine transform is used for problems with boundary conditions of heat distribution on two parallel boundaries while the finite cosine transform is used for problems in which the velocities normal to two parallel boundaries.

18.5 APPLICATIONS TO PARTIAL DIFFERENTIAL EDUCATIONS

EXAMPLE 18.8 Determine the distribution of temperature in the semi-infinite medium $x \geq 0$, when the end $x = 0$ is maintained at zero temperature and the initial distribution of temperature is $f(x)$.

Solution Let $v(x, t)$ be the temperature at any point x and at any time t. We have to solve the heat-flow equation

$$\frac{\partial v}{\partial t}=c^2\frac{\partial^2 v}{\partial x^2}, x>0, t>0 \quad \text{(i)}$$

subject to the initial condition

$$v(x, 0) = f(x) \quad \text{(ii)}$$

and the boundary condition

$$v(0, t) = 0 \quad \text{(iii)}$$

Taking Fourier sine transform of Eq, (i), we get

$$\frac{dv_s}{dt}=c^2[sv(0,t)-s^2\hat{v}_s]$$

$$\Rightarrow \quad \frac{dv_s}{dt}+c^2s^2\hat{v}_s=0 \quad \text{(iv)}$$

Also the Fourier sine transform of Eq. (ii) is

$$\hat{v}_s = \hat{f}_s(s) \text{ at } t = 0 \quad \text{(v)}$$

Solving Eq. (iv) and using Eq. (v), we get

$$\hat{v}_s = \hat{f}_s(s)e^{-c^2s^2t}$$

Hence, taking inverse Fourier sine transform, we get

$$v(x,t)=\frac{2}{\pi}\int_0^\infty \hat{f}_s(s)\, e^{-c^2s^2t} \sin xs\, ds.$$

EXAMPLE 18.9 An infinitely long string having one end at $x = 0$, is initially at rest along the x-axis. The end $x = 0$ is given a transverse displacement $f(t)$, $t > 0$. Find the displacement of any point of the string at any time.

Solution Let $y(x, t)$ be the transverse displacement of any point x of the string at any time t. We have to solve the wave equation

$$\frac{\partial^2 y}{\partial t^2}=c^2\frac{\partial^2 y}{\partial x^2}, x>0, t>0 \quad \text{(i)}$$

subject to the condition $y(x, 0) = 0$, $(x, 0) = 0$, $y(0, t) = f(t)$ and the displacement $y(x, t)$ is bounded.

The Laplace transform of Eq. (i) is

$$s^2y-s.y(x,0)-\frac{\partial y(x,0)}{\partial t}=c^2\frac{d^2\overline{y}}{dx^2}$$

Using $y(x, 0) = 0$ and $y_t(x, 0) = 0$, we get

$$\frac{d^2\overline{y}}{dx^2}=\left(\frac{s}{c}\right)^2\overline{y} \quad \text{(ii)}$$

Taking Laplace transforms of Eqs. (iii) and (iv) give

$$y(0, s) = (s) \text{ at } x = 0 \quad \text{(iii)}$$

and $\bar{y}(x, s)$ is bounded. (iv)

The solution of Eq. (ii) is

$$\bar{y}(x, s) = c_1 e^{sx/c} + c_2 e^{-sx/c}$$

Equation (iv) can be satisfied if $c_1 = 0$ and using Eq. (iii), we get $c_2 = \bar{f}(s)$.

$$\therefore \quad \bar{y}(x, s) = \bar{f}(s)\, e^{-sx/c}$$

Using complex inversion formula, we obtain

$$y = \frac{1}{2\pi i} \int_{a-i\infty}^{a+i\infty} e^{(t-x/c)^2} \bar{f}(s)\, ds$$

$$= f(t - x/c).$$

EXAMPLE 18.10 If the initial temperature of an infinite bar is given by $T(x) = \begin{cases} T_0, & \text{for } |x| < a \\ 0, & \text{for } |x| > a \end{cases}$.

Determine the temperature at any point x and at any time t.

Solution We need to solve

$$\frac{\partial T}{\partial t} = c^2 \frac{\partial^2 T}{\partial t^2} \; (t > 0) \quad \text{(i)}$$

subject to the initial condition

$$T(x, 0) = \begin{cases} T_0, & \text{for } |x| < a \\ 0, & \text{for } |x| > a \end{cases} \quad \text{(ii)}$$

Taking the Fourier transform of Eq. (i) gives

$$\frac{d\hat{T}}{dt} = -c^2 s^2 \hat{T} \quad \text{(iii)}$$

and that of Eq. (ii) is

$$\hat{T}(s, 0) = \int_{-\infty}^{\infty} T(x, 0)\, e^{isx}\, dx$$

$$= \int_{-a}^{a} T_0 e^{isx}\, dx$$

$$= T_0 \left[\frac{e^{isa} - e^{-isa}}{is} \right]$$

$$= 2T_0 \frac{\sin as}{s} \quad \text{(iv)}$$

Solving Eq. (iii) and using Eq. (iv), we get

$$\hat{T} = 2T_0 \frac{\sin as}{s} e^{-c^2 s^2 t}$$

Hence, taking its inverse Fourier transform, we get

$$T(x, t) = \frac{1}{2\pi} \int_{-\infty}^{\infty} 2T_o \frac{\sin as}{s} e^{-c^2 s^2 t} e^{-ixs}\, ds$$

$$= \frac{T_0}{\pi} \int_{-\infty}^{\infty} \frac{\sin as}{s} e^{-c^2 s^2 t} (\cos xs - i \sin xs)\, ds$$

$$= \frac{2T_0}{\pi} \int_0^{\infty} \frac{\sin as}{s} \cos xs e^{-c^2 s^2 t}\, ds$$

$$= \frac{T_0}{\pi} \int_0^{\infty} e^{-c^2 s^2 t} \left[\frac{\sin (a+x)s + \sin (a-x)s}{s} \right] ds$$

Put $\gamma^2 = c^2 s^2 t$

$$= \frac{T_0}{\pi} \int_0^{\infty} e^{-\gamma^2} \left[\sin \frac{(a+x)\gamma}{c\sqrt{t}} + \sin \frac{(a-x)\gamma}{c\sqrt{t}} \right] \frac{d\gamma}{\gamma}$$

$$= \frac{T_o}{2} \left[erf \frac{(a+xc)}{2c\sqrt{t}} + erf \frac{(a-x)}{2c\sqrt{t}} \right].$$

EXERCISES

1. Show that $\Im[e^{icx} f(x)] = \hat{f}(s+c)$.

2. Find the Fourier transform of $f(x) = e^{-ax^2}$, where a is constant. $\left[\textbf{Ans.}\ \frac{1}{\sqrt{2a}} e^{-s^2/4a}\right]$

3. Find the Fourier transform of $f(x) = \begin{cases} \frac{1}{n}, & \text{for } |x| \le n \\ 0, & \text{otherwise} \end{cases}$ and evaluate limit of this transform as $n \to 0$. $\left[\textbf{Ans.}\ \sqrt{\frac{2}{\pi}} \cdot \frac{\sin sn}{sn}, \sqrt{\frac{2}{\pi}}\right]$

4. Show that $\Im\left[\frac{1}{\sqrt{2t}} e^{-x^2/4t}\right] = e^{-w^2 t}$.

5. Show that $\Im_s[f'] = \sqrt{\frac{2}{\pi}}\, wf(0) - w^2 \Im_s[f]$.

6. Find the Fourier cosine transform of $f(x) = e^{-x} + 2e^{-2x}$.

$$\left[\textbf{Ans.}\ \sqrt{\frac{2}{\pi}}\left(\frac{1}{1+s^2}+\frac{4}{4+s^2}\right)\right]$$

7. Find the Fourier sine transform of xe^{-x}.

$$\left[\textbf{Ans.}\ \sqrt{\frac{2}{\pi}}\cdot\frac{2s}{(1+s^2)^2}\right]$$

8. Find the Fourier cosine transform of the function $f(x)=\begin{cases}1, 0<x<a\\ 0,\ x\geq a\end{cases}$.

$$\left[\textbf{Ans.}\ \sqrt{\frac{2}{\pi}}\frac{\sin sa}{s}\right]$$

9. Find the inverse transform of $\dfrac{1}{(s^2+1)^2}$ using convolution theorem.

$$\left[\textbf{Ans.}\ \frac{1}{2}\sqrt{\frac{\pi}{2}}e^{-|x|}(1+|x|)\right]$$

10. Find finite Fourier sine transform of $f(x) = x,\ 0 < x < 1$.

$$\left[\textbf{Ans.}\ \frac{(-1)^{n+1}}{n\pi}\right]$$

11. Find the finite Fourier cosine transform of $\sin ax$.

$$\left[\textbf{Ans.}\ \begin{cases}0, & \text{if } n \text{ is even}\\ \dfrac{2a}{a^2-n^2}, & \text{if } n \text{ is odd}\end{cases}\right]$$

12. Determine the temperature at any point x at any instant of time t in a solid bar bounded by planes $x = 0$ and $x = \pi$. It is given that ends of the bar are kept at zero temperature always while initial temperature of point x is $2x$.

$$\left[\textbf{Ans.}\ T(x,t)=4\sum_{n=1}^{\infty}\frac{(-1)^{n+1}}{n}e^{-n^2kt}\sin(nx)\right]$$

13. Solve $\dfrac{\partial U}{\partial t}=\dfrac{\partial^2 U}{\partial x^2}, 0<x<\infty, t>0$ subject to $U(0, t) = 0$, $U(x, 0) = f(x)$, $U(x, t)$ is bounded.

$$\left[\textbf{Ans.}\ \sqrt{\frac{2}{\pi}}\int_0^{\infty}\hat{f}_s(w)e^{-w^2t}\sin wx\,dw\right]$$

14. Solve $\dfrac{\partial U}{\partial t}=\dfrac{\partial^2 U}{\partial x^2}, t>0$ given that $U(x, 0) = e^{-x^2}$.

$$[\textbf{Ans.}\ (1+4t)^{-1/2}e^{-x^2/1+4t}]$$

15. Solve $\dfrac{\partial U}{\partial t}+\dfrac{\partial^2 U}{\partial x^2}+tU=0, x>0, t>0$ given that $U(x, 0) = e^{-x}$, $U_x(0, t) = 0$ and $U(x, t) \to 0$ uniformly in t as $x \to \infty$.

$$\left[\textbf{Ans.}\ \frac{2}{\pi}e^{-t^2/2}\int_0^{\infty}\frac{e^{-w^2t}\cos wx}{1+w^2}dw\right]$$

CHAPTER 19

Z-Transforms and Its Applications

Previously, we studied Fourier transforms and Laplace transforms. These transforms are defined through integrals and so they are also known as *integral transforms*. These transforms deal with functions of continuous variables.

In this chapter, we discuss Z-transforms which deal with functions of discrete variables. The Z-transforms have its applications in communication systems, control systems, analysis of signals and linear time invariant (LTI) systems.

19.1 Z-TRANSFORM

Consider $\langle u_n \rangle = \langle u_0, u_1, ..., u_n, ... \rangle$ a sequence of real numbers. Suppose that

$$u_0 + u_1 \frac{1}{z} + u_2 \frac{1}{z^2} + \cdots + u_n \frac{1}{z^n} + \cdots, \qquad z \in \mathbb{C},$$

set of complex numbers be a series generated by the sequence $\langle u_n \rangle$, which is convergent for all z in a domain D. Then a cmplex-valued function $\bar{u}(z)$ is called the Z-transform of $\langle u_n \rangle$ which is defined as:

$$\bar{u}(z) = \mathrm{Z}\{\langle u_n \rangle\} = \sum_{n=0}^{\infty} u_n z^{-n} \tag{19.1}$$

Note: 1. The transform defined in Eq. (19.1) is known as one-sided or right-handed or unilateral Z-transform.

2. When n takes all integer values, Z-transform is said to be a two-sided or bilateral.

19.2 Z-TRANSFORM OF SOME SPECIAL SEQUENCE

19.2.1 Unit Constant Sequence

Given $\qquad u_n = 1 \quad \forall n \in \mathbb{N}$

$$\therefore \qquad \langle u_n \rangle = \langle 1, 1, 1, ... \rangle$$

Now, $Z\{\langle u_n \rangle\} = Z\{\langle 1 \rangle\} = \sum_{n=0}^{\infty} u_n z^{-n} = \sum_{n=0}^{\infty} z^{-n} = \dfrac{z}{z-1}$ (Infinite geometric series which converges absolutely for $|z| > 1$).

19.2.2 Alternating Unit Constant Sequence

Given $\qquad u_n = (-1)^n \quad \forall n \in \mathbb{N}$

$$\therefore \qquad \langle u_n \rangle = \langle 1, -1, 1, -1, ... \rangle$$

Now, $$Z\{\langle u_n \rangle\} = Z\{\langle (-1)^n \rangle\} = \sum_{n=0}^{\infty} u_n z^{-n} = \sum_{n=0}^{\infty} (-1)^n z^{-n} = \frac{z}{z+1}$$

19.2.3 Geometric Sequence

Given $\qquad u_n = a^n \quad \forall n \in \mathbb{N}, a \in \mathbb{R}(\mathbb{C})$

$$\therefore \qquad \langle u_n \rangle = \langle 1, a, a^2, ... a^n, ... \rangle$$

Now, $$Z\{\langle u_n \rangle\} = Z\{\langle a^n \rangle\} = \sum_{n=0}^{\infty} u_n z^{-n} = \sum_{n=0}^{\infty} a^n z^{-n} = \frac{z}{z-a}$$

(Infinite geometric series which converges absolutely for $|z| > |a|$).

19.2.4 Natural Number Sequence

Given $\qquad u_n = n \quad \forall n \in \mathbb{N}$

$$\therefore \qquad \langle u_n \rangle = \langle 0, 1, 2, ..., n, ... \rangle$$

Now,

$$Z\{\langle u_n \rangle\} = Z\{\langle n \rangle\} = \sum_{n=0}^{\infty} u_n z^{-n} = \sum_{n=0}^{\infty} n z^{-n} = 0 + 1\frac{1}{z} + 2\frac{1}{z^2} + \cdots + n\frac{1}{z^n} + \cdots$$

$$= \frac{1}{z}\left(1 + \frac{2}{z} + \frac{3}{z^2} + \cdots + \frac{n}{z^{n-1}} + \cdots\right)$$

$$= \frac{1}{z}\left(1 - \frac{1}{z}\right)^{-2} = \frac{z}{(z-1)^2}$$

19.2.5 Power–Reciprocal Factorial Sequence

Given $$u_n = \frac{a^n}{n!} e^{-a} \quad \forall n \in \mathbb{N}, a \in \mathbb{R}$$

$\therefore$ $$\langle u_n \rangle = \left\langle e^{-a}, \frac{a}{1!}e^{-a}, \frac{a^2}{2!}e^{-a}, \ldots, \frac{a^n}{n!}e^{-a}, \ldots \right\rangle$$

Now,

$$Z\{\langle u_n \rangle\} = Z\left\{\left\langle \frac{a^n}{n!} e^{-a} \right\rangle\right\} = \sum_{n=0}^{\infty} \frac{1}{n!} e^{-a} \frac{a^n}{z^n}$$

$$= e^{-a}\left(+ \frac{1}{1!}\frac{a}{z} + \frac{1}{2!}\frac{a^2}{z^2} + \cdots + \frac{1}{n!}\frac{a^n}{z^n} + \cdots \right) = e^{-a} e^{(a/z)} = e^{a(1/z-1)}$$

Note: For $a = 1$, we get $Z\left\{\left\langle \frac{1}{n!} \right\rangle\right\} = e^{1/z}$.

19.2.6 Binomial Coefficient Sequence

Given $$u_n = {}^nC_k, 0 \le k \le n, n \in \mathbb{N}$$

$\therefore$ $$\langle u_n \rangle = \left\langle {}^nC_0, {}^nC_1, {}^nC_2, \ldots, {}^nC_n \right\rangle$$

Now, $$Z\{\langle u_n \rangle\} = Z\{\langle {}^nC_k \rangle\} = \sum_{k=0}^{n} {}^nC_k \frac{1}{z^k} = {}^nC_0 1 + {}^nC_1 \frac{1}{z} + {}^nC_2 \frac{1}{z^2} + \cdots + {}^nC_n \frac{1}{z^n} = \left(1 + \frac{1}{z}\right)^n$$

19.3 RECURSIVE FORMULA FOR THE SEQUENCE OF A POWER OF NATURAL NUMBER

Given $$u_n = n^p, n, p \in \mathbb{N}$$

$\therefore$ $$\langle u_n \rangle = \left\langle 0, 1^p, 2^p, \ldots, n^p, \ldots \right\rangle$$

Prove that $Z\{\langle n^p \rangle\} = z\frac{d}{dz}\left[Z\{\langle n^{p-1} \rangle\}\right]$ or $-zZ'\{\langle n^{p-1} \rangle\}$.

Proof: We have

$$Z'\{\langle n^p \rangle\} = \frac{d}{dz} Z\{\langle n^p \rangle\} = \frac{d}{dz}\left(\sum_{n=0}^{\infty} n^p z^{-n} \right) = \sum_{n=0}^{\infty} n^p (-n) z^{-n-1} = -\frac{1}{z}\sum_{n=0}^{\infty} n^{p+1} z^{-n}$$

Replace p by $p-1$ in above argument.

$$Z'\{\langle n^{p-1}\rangle\} = \frac{d}{dz}Z\{\langle n^{p-1}\rangle\} = -\frac{1}{z}\sum_{n=0}^{\infty} n^p z^{-n} = -\frac{1}{z}Z\{\langle n^p\rangle\}$$

Hence, we have proved.

Note: 1. If we put $p=1$, we get $Z\{\langle u_n\rangle\} = Z\{\langle n\rangle\} = \frac{z}{(z-1)^2}$ (as discussed in Section 19.2.4).

2. If we put $p=2$, we get $Z\{\langle u_n\rangle\} = Z\{\langle n^2\rangle\} = -z\frac{d}{dz}\left[\frac{z}{(z-1)^2}\right] = \frac{z^2+z}{(z-1)^3}$.

In Table 19.1, we exhibit Z-transforms of standard functions.

Table 19.1 Z-transforms of some standard functions

Sequence $\langle u_n\rangle$	*Z-transform* $\bar{u}(z)$
$\langle 1\rangle$	$\frac{z}{z-1}$
$\langle n\rangle$	$\frac{z}{(z-1)^2}$
$\langle n^2\rangle$	$\frac{z^2+z}{(z-1)^2}$
$\langle n^p\rangle, p\in\mathbb{N}$	$-z\frac{d}{dz}\left[Z\{\langle n^{p-1}\rangle\}\right]$
$\langle a^n\rangle$	$\frac{z}{z-a}$
$\langle nu_n\rangle$	$-z\frac{d}{dz}(\bar{u}(z))$
$\langle a^n u_n\rangle$	$\bar{u}\left(\frac{z}{a}\right)$
$\langle H_n\rangle$, unit step function	$\frac{z}{z-1}$
$\langle \delta_n\rangle$, unit impulse sequence	1
$\langle u_{n+1}\rangle$	$z(\bar{u}(z)-u_0)$
$\langle u_{n+2}\rangle$	$z^2\left(\bar{u}(z)-u_0-\frac{u_1}{z}\right)$
$\langle u_{n+3}\rangle$	$z^3\left(\bar{u}(z)-u_0-\frac{u_1}{z}-\frac{u_2}{z^2}\right)$
$\langle u_{n-k}\rangle$	$z^{-k}\bar{u}(z)$
u_0	$\lim_{z\to\infty}(z-1)\bar{u}(z)$
$\lim_{n\to\infty} u_n$	$\lim_{z\to 1}(z-1)\bar{u}(z)$
$\langle \cos nt\rangle$	$\frac{z(z-\cos t)}{z^2-2z\cos t+1}$
$\langle \sin nt\rangle$	$\frac{z\sin t}{z^2-2z\cos t+1}$

19.4 PROPERTIES OF Z-TRANSFORM

P1: Z-transform preserves sum, i.e.,

$$Z\{\langle u_n + v_n\rangle\} = Z\{\langle u_n\rangle + \langle v_n\rangle\} = Z\{\langle u_n\rangle\} + Z\{\langle v_n\rangle\}.$$

Proof: $Z\{\langle u_n + v_n\rangle\} = \sum_{n=0}^{\infty}(u_n + v_n)z^{-n} = \sum_{n=0}^{\infty} u_n z^{-n} + \sum_{n=0}^{\infty} v_n z^{-n} = Z\{\langle u_n\rangle\} + Z\{\langle v_n\rangle\}.$

P2: Z-transform preserves scalar multiplication, i.e., $Z\{\langle ku_n\rangle\} = kZ\{\langle u_n\rangle\}$, k is any scalar.

Proof: $Z\{k\langle u_n\rangle\} = Z\{\langle ku_n\rangle\} = \sum_{n=0}^{\infty}(ku_n)z^{-n} = k\sum_{n=0}^{\infty} u_n z^{-n} = kZ\{\langle u_n\rangle\}.$

P3: (Damping rule) If $Z\{\langle u_n\rangle\} = \bar{u}(z)$ then $Z\{\langle a^{-n}u_n\rangle\} = \bar{u}(az)$, k is any scalar.

Proof: $Z\{\langle a^{-n}u_n\rangle\} = \sum_{n=0}^{\infty} a^{-n}u_n z^{-n} = \sum_{n=0}^{\infty} u_n (az)^{-n} = \bar{u}(az)$.

Note: 1. $Z\{\langle a^n u_n\rangle\} = \bar{u}\left(\dfrac{z}{a}\right).$

2. $Z\{\langle na^n\rangle\} = \dfrac{az}{(z-a)^2}.$

Hint: In $Z\{\langle n\rangle\} = \dfrac{z}{(z-1)^2}$ replace z by $\dfrac{z}{a}$.

P4: Prove that $Z\{\langle \cos nt\rangle\} = \dfrac{z(z-\cos t)}{z^2 - 2z\cos t + 1}$ and $Z\{\langle \sin nt\rangle\} = \dfrac{z\sin t}{z^2 - 2z\cos t + 1}$.

Proof: We have $Z\{\langle 1\rangle\} = \dfrac{z}{z-1}$.

By the damping rule, $Z\{\langle a^n \cdot 1\rangle\} = \dfrac{z/a}{z/a - 1} = \dfrac{z}{z-a}$.

$$\therefore \quad Z\{\langle e^{-int}\rangle\} = Z\{\langle (e^{-it})^n \cdot 1\rangle\} = \frac{z}{z - e^{-it}} = \frac{z}{z - e^{-it}}\,\frac{z + e^{-it}}{z + e^{-it}} = \frac{z[z - (\cos t + i\sin t)]}{z^2 - 2z\cos t + 1}$$

$$\therefore \quad Z\{\langle \cos nt - i\sin nt\rangle\} = \frac{z[z - (\cos t + i\sin t)]}{z^2 - 2z\cos t + 1}$$

$$\Rightarrow \quad Z\{\langle \cos nt\rangle\} - iZ\{\langle \sin nt\rangle\} = \frac{z(z-\cos t)}{z^2 - 2z\cos t + 1} - \frac{iz\sin t}{z^2 - 2z\cos t + 1}$$

Comparing real and imaginary parts, we get

$$Z\{\langle \cos nt\rangle\} = \frac{z(z-\cos t)}{z^2 - 2z\cos t + 1} \text{ and } Z\{\langle \sin nt\rangle\} = \frac{z\sin t}{z^2 - 2z\cos t + 1}$$

Note:

1. Replace t by 'it', we get

$$Z\{\langle \cosh nt \rangle\} = \frac{z(z-\cosh t)}{z^2 - 2z\cosh t + 1} \text{ and } Z\{\langle \sinh nt \rangle\} = \frac{z\sinh t}{z^2 - 2z\cosh t + 1}$$

2. $$Z\left\{\left\langle a^n \cos nt \right\rangle\right\} = \frac{z(z - a\cos t)}{z^2 - 2az\cos t + a^2} \text{ and } Z\left\{\left\langle a^n \sin nt \right\rangle\right\} = \frac{az\sin t}{z^2 - 2az\cos t + a^2}.$$

Hint: Apply damping rule on replacing z by z/a.

19.5 SHIFTING PROPERTIES

THEOREM 19.1 First Shifting Theorem

If $Z\{\langle u_n \rangle\} = \bar{u}(z)$ then $Z\{\langle u_{n-k} \rangle\} = z^{-k}\bar{u}(z)$, for $k > 0$.

[Interpretation: u_n are shifted to the right.]

Proof: By definition

$$\begin{aligned} Z\{\langle u_{n-k} \rangle\} &= \sum_{n=0}^{\infty} u_{n-k} z^{-n} \\ &= u_0 z^{-k} + u_1 z^{-(k+1)} + \cdots \qquad [\because u_n = 0, \text{ for } n < 0] \\ &= z^{-k} \sum_{n=0}^{\infty} u_n z^{-n} \\ &= z^{-k}\bar{u}(z) \end{aligned}$$

THEOREM 19.2 Second Shifting Theorem

If $Z\{\langle u_n \rangle\} = \bar{u}(z)$ then $Z\{\langle u_{n+k} \rangle\} = z^k\left[\bar{u}(z) - \left(u_0 + \frac{u_1}{z} + \cdots + \frac{u_{k-1}}{z^{k-1}}\right)\right]$.

[Interpretation: u_n are shifted to the left.]

Proof: By definition

$$\begin{aligned} Z\{\langle u_{n+k} \rangle\} &= \sum_{n=0}^{\infty} u_{n+k} z^{-n} \\ &= \sum_{n=0}^{\infty} u_{n+k} z^{-(n+k)} \\ &= z^k \left[u_k z^{-k} + u_{1+k} z^{-(1+k)} + \cdots\right] \end{aligned}$$

Adding and subtracting $\left(u_0 + u_1 z^{-1} + \cdots + u_{k-1} z^{-(k-1)}\right)$

$$\begin{aligned} &= z^k \left[\sum_{n=0}^{\infty} u_n z^{-n} - \sum_{n=0}^{k-1} u_n z^{-n}\right] \\ &= z^k \left[\bar{u}(z) - \left(u_0 + u_1 z^{-1} + \cdots + u_{k-1} z^{-(k-1)}\right)\right] \end{aligned}$$

In particular,

For $k = 1$, we get $Z\{\langle u_{n+1}\rangle\} = z[\bar{u}(z) - u_0]$.

For $k = 2$, we get $Z\{\langle u_{n+2}\rangle\} = z^2\left[\bar{u}(z) - u_0 - \dfrac{u_1}{z}\right]$.

For $k = 3$, we get $Z\{\langle u_{n+3}\rangle\} = z^3\left[\bar{u}(z) - u_0 - \dfrac{u_1}{z} - \dfrac{u_2}{z^2}\right]$.

THEOREM 19.3 Multiplication by n.

If $Z\{\langle u_n\rangle\} = \bar{u}(z)$ then $Z\{\langle nu_n\rangle\} = -z\dfrac{d}{dz}\bar{u}(z)$.

Proof: By definition

$$
\begin{aligned}
Z\{\langle nu_n\rangle\} &= \sum_{n=0}^{\infty} nu_n z^{-n} \\
&= (-z)\sum_{n=0}^{\infty} u_n(-n)z^{-n-1} \\
&= (-z)\sum_{n=0}^{\infty} \frac{d}{dz}(u_n z^{-n}) \\
&= (-z)\frac{d}{dz}\sum_{n=0}^{\infty}(u_n z^{-n}) \\
&= -z\frac{d}{dz}\bar{u}(z)
\end{aligned}
$$

Sum and differentiation are interchangeable under the assumption that the series is uniformly convergent. In general,

$$Z\{\langle n^p u_n\rangle\} = (-z)^p \frac{d^p}{dz^p}\bar{u}(z),\ p = 0, 1, 2, \ldots$$

THEOREM 19.4 Division by n.

If $Z\{\langle u_n\rangle\} = \bar{u}(z)$ then $Z\left\{\left\langle \dfrac{u_n}{n}\right\rangle\right\} = -\displaystyle\int_0^z \frac{1}{t}\bar{u}(t)dt$.

Proof: By definition

$$
\begin{aligned}
Z\left\{\left\langle \frac{u_n}{n}\right\rangle\right\} &= \sum_{n=0}^{\infty} \frac{u_n}{n} z^{-n} \\
&= -\sum_{n=0}^{\infty} u_n\left(\frac{z^{-n}}{-n}\right)
\end{aligned}
$$

$$= -\sum_{n=0}^{\infty} u_n \int_0^z t^{-n-1} dt$$

$$= -\int_0^z \sum_{n=0}^{\infty} u_n t^{-n-1} dt$$

$$= -\int_0^z t^{-1} \left(\sum_{n=0}^{\infty} u_n t^{-n} \right) dt$$

$$= -\int_0^z t^{-1} \bar{u}(t) dt$$

Sum and integration are interchangeable under the assumption that the series is uniformly convergent.

THEOREM 19.5 Initial Value Theorem

If $Z\{\langle u_n \rangle\} = \bar{u}(z)$ then $u_0 = \lim_{z \to \infty} \bar{u}(z)$.

Proof: $\bar{u}(z) = Z\{\langle u_n \rangle\} = \sum_{n=0}^{\infty} u_n z^{-n} = u_0 + u_1 \frac{1}{z} + u_2 \frac{1}{z^2} + \cdots$

$$\therefore \qquad \lim_{z \to \infty} \bar{u}(z) = u_0$$

Also, $$\lim_{z \to \infty} z[\bar{u}(z) - u_0] = \lim_{z \to \infty} \left[u_1 + u_2 \frac{1}{z} + u_3 \frac{1}{z^2} + \cdots \right] = u_1$$

and $$\lim_{z \to \infty} z^2 \left[\bar{u}(z) - u_0 - u_1 \frac{1}{z} \right] = u_2$$

THEOREM 19.6 Final Value Theorem

If $Z\{\langle u_n \rangle\} = \bar{u}(z)$ then $\lim_{n \to \infty} u_n = \lim_{z \to 1} (z-1)\bar{u}(z)$.

Proof: By linearity property

$$Z\{\langle u_{n+1} - u_n \rangle\} = Z\{\langle u_{n+1} \rangle\} - Z\{\langle u_n \rangle\} = z[\bar{u}(z) - u_0] - \bar{u}(z) = (z-1)\bar{u}(z) - u_0 z$$

$$\Rightarrow \qquad (z-1)\bar{u}(z) - u_0 z = \sum_{n=0}^{\infty} (u_{n+1} - u_n) z^{-n}$$

$$\Rightarrow \qquad \lim_{z \to 1} [(z-1)\bar{u}(z) - u_0 z] = \lim_{n \to \infty} \sum_{n=0}^{\infty} (u_{n+1} - u_n)$$

Hence, $\lim_{z \to 1} (z-1)\bar{u}(z) = \lim_{n \to \infty} u_{n+1}$.

19.6 INVERSE Z-TRANSFORM

Let the infinite series on the RHS of Eq. (19.1) converges to $\bar{u}(z)$, the mapping is said to be invertible and known as the inverse Z-transform; denoted by $Z^{-1}\{\bar{u}(z)\}$ and defined as $Z^{-1}\{\bar{u}(z)\} = \langle u_n \rangle$.

THEOREM 19.7 Convolution Theorem

If $Z^{-1}\{\bar{u}(z)\} = u_n$ and $Z^{-1}\{\bar{v}(z)\} = v_n$, then $Z^{-1}\{\bar{u}(z)\bar{v}(z)\} = u_n * v_n = \sum_{m=0}^{n} u_m v_{n-m}$, where * is convolution operation.

Proof: By definition,

$$\bar{u}(z)\bar{v}(z) = \left(\sum_{n=0}^{\infty} u_n z^{-n}\right)\left(\sum_{n=0}^{\infty} v_n z^{-n}\right)$$

$$= u_0 v_0 + (u_0 v_1 + u_1 v_0) z^{-1} + (u_0 v_2 + u_1 v_1 + u_2 v_0) z^{-2} + \cdots (u_0 v_n + u_1 v_{n-1} + \cdots + u_n v_0) z^{-n} + \cdots$$

$$= \sum_{n=0}^{\infty} (u_0 v_n + u_1 v_{n-1} + \cdots + u_n v_0) z^{-n}$$

$$\therefore \qquad Z\{\langle u_0 v_n + u_1 v_{n-1} + \cdots + u_n v_0 \rangle\} = Z\left\{\sum_{m=0}^{n} u_m v_{n-m}\right\}$$

Taking inverse Z-transform of both sides, we get

$$Z^{-1}\{\bar{u}(z)\bar{v}(z)\} = Z\left\{\sum_{m=0}^{n} u_m v_{n-m}\right\}.$$

19.7 METHODS TO COMPUTE INVERSE Z-TRANSFORM

19.7.1 Method 1: Using Table 19.1 of Z-transform

EXAMPLE 19.1 Find $Z^{-1}\left(\frac{z}{z-1}\right)$.

Solution From Table 19.1, $Z\{\langle 1 \rangle\} = \frac{z}{z-1}$.

Therefore, the inverse Z-transform of $\frac{z}{z-1}$ is the unit constant sequence $\langle 1 \rangle$.

EXAMPLE 19.2 Find $Z^{-1}\left(\frac{5z}{(z-1)^2}+3z^{1/z}+7z\log\left(z-\frac{1}{z}\right)+\frac{1}{3}\frac{z(z-\cos\theta)}{z^2-2z\cos\theta+1}\right)$.

Solution By linearity property,

$$Z^{-1}\left(\frac{5z}{(z-1)^2}+3z^{1/z}+7\log\left(z-\frac{1}{z}\right)+\frac{1}{3}\frac{z(z-\cos\theta)}{z^2-2z\cos\theta+1}\right)$$

$$=5Z^{-1}\left(\frac{z}{(z-1)^2}\right)+3Z^{-1}\left(z^{1/z}\right)+7Z^{-1}\left(\log\left(z-\frac{1}{z}\right)\right)+\frac{1}{3}Z^{-1}\left(\frac{z(z-\cos\theta)}{z^2-2z\cos\theta+1}\right)$$

$$=5\langle n\rangle+3\left\langle\frac{1}{n!}\right\rangle-7\left\langle\frac{1}{n+1}\right\rangle+\frac{1}{3}\langle\cos n\theta\rangle$$

Therefore, $\langle u_n\rangle=\left\langle 5n+\frac{3}{n!}-\frac{7}{n+1}+\frac{1}{3}\cos n\theta\right\rangle$ is the required inverse Z-transform of given function.

19.7.2 Method 2: Using Partial Fraction

If $\bar{u}(z)$ is a rational function of z then convert $\frac{\bar{u}(z)}{z}$ into partial fractions. Multiply both sides by z and write inverse transform of each fraction. By linearity property, their sum is the inverse Z-transform of the given function.

EXAMPLE 19.3 Find $Z^{-1}\left(\frac{z+1}{z^2-3z+2}\right)$.

Solution Let $\bar{u}(z)=\frac{z+1}{z^2-3z+2}=\frac{z+1}{(z-1)(z-2)}$.

Write
$$\frac{\bar{u}(z)}{z}=\frac{z+1}{z(z-1)(z-2)}$$

So,
$$\frac{\bar{u}(z)}{z}=\frac{A}{z}+\frac{B}{(z-1)}+\frac{C}{(z-2)} \tag{i}$$

where A, B, C are constants to be determined. Readers can check that $[A, B, C]=\left[\frac{1}{2},-2,\frac{3}{2}\right]$.

Multiply both sides of Eq. (i) by z.

$$\bar{u}(z)=\frac{1}{2}-2\frac{z}{(z-1)}+\frac{3}{2}\frac{z}{(z-2)}$$

Take inverse Z-transform.

$$\therefore \qquad \langle u_n \rangle = \frac{1}{2}\langle \delta_n \rangle - 2\langle 1^n \rangle + \frac{3}{2}\langle 2^n \rangle = \left\langle \frac{1}{2}\delta_n - 2^n + 3\cdot 2^{n-1} \right\rangle$$

where δ_n is the unit impulse function.

EXAMPLE 19.4 Find $Z^{-1}\left(\dfrac{2z^2+7z}{(3z-1)(3z+1)}\right)$.

Solution Let $\bar{u}(z) = \dfrac{2z^2+7z}{(3z-1)(3z+1)}$.

$$\therefore \qquad \frac{\bar{u}(z)}{z} = \frac{2z+7}{(3z-1)(3z+1)}$$

So,

$$\frac{\bar{u}(z)}{z} = \frac{A}{3z-1} + \frac{B}{3z+1} \qquad \text{(i)}$$

where A and B are constants to be determined. Readers can check that $[A, B] = \left[\dfrac{23}{6}, -\dfrac{19}{6}\right]$.

Multiply both sides of Eq. (i) by z.

$$\therefore \qquad \bar{u}(z) = \frac{23}{6}\frac{z}{3z-1} - \frac{19}{6}\frac{z}{3z+1}$$

$$= \frac{23}{6}\frac{z}{z-(1/3)} - \frac{19}{6}\frac{z}{z+(1/3)}$$

$$\therefore \qquad Z^{-1}[\bar{u}(z)] = \frac{23}{6}Z^{-1}\left[\frac{z}{z-(1/3)}\right] - \frac{19}{6}Z^{-1}\left[\frac{z}{z+(1/3)}\right]$$

$$= \frac{23}{6}\left\langle \left(\frac{1}{3}\right)^n \right\rangle - \frac{19}{6}\left\langle \left(-\frac{1}{3}\right)^n \right\rangle$$

$$= \frac{1}{6}\left\{23\langle 3^{-n}\rangle - 19\langle -3^n \rangle\right\}$$

$$= \frac{1}{6}\left\{(23 - 19(-1)^n)\langle -3^n \rangle\right\}$$

19.7.3 Method 3: Long Division Method

If $\bar{u}(z)$ is a rational function of z and denominator can be factorized or not factorized then use long division method.

EXAMPLE 19.5 Find $Z^{-1}\left(\dfrac{10z}{z^2-3z+2}\right)$.

Solution Let $\bar{u}(z)=\dfrac{10z}{z^2-3z+2}$. Multiply the numerator and denominator by z^{-2}. We get,

$$\bar{u}(z)=\frac{10z^{-1}}{1-3z^{-1}+2z^{-2}}$$

Perform long division method

$$\begin{aligned}\bar{u}(z)&=10z^{-1}+30z^{-2}+70z^{-3}+150z^{-4}+\cdots\\&=10[z^{-1}+3z^{-2}+7z^{-3}+15z^{-4}+\cdots]\\&=10[1+z^{-1}+3z^{-2}+7z^{-3}+15z^{-4}+\cdots-1]\\&=10[(2^0-1)+(2^1-1)z^{-1}+(2^2-1)z^{-2}+(2^3-1)z^{-3}+(2^4-1)z^{-4}+\cdots-1]\\&=10\sum_{n=0}^{\infty}(2^n-1)z^{-n}\end{aligned}$$

$$\therefore \qquad \langle u_n\rangle=10\langle 2^n-1\rangle,\ n=0,1,2,\ldots$$

19.7.4 Method 4: Residues Method

Using the theory of complex variables, the inverse Z-transform if a function $\bar{u}(z)$ is given by

$$\langle u_n\rangle=\frac{1}{2\pi i}\int_c \bar{u}(z)z^{n-1}dz$$

where c is a closed contour in the region of convergence of $\bar{u}(z)$ containing origin and all the isolated singularities of the integrand. Then use residue theorem.

EXAMPLE 19.6 Find $Z^{-1}\left(\dfrac{z}{z^2+3z+2}\right)$.

Solution Let $\bar{u}(z)=\dfrac{z}{z^2+3z+2}=\dfrac{z}{(z+1)(z+2)}$.

The inverse Z-transform $\bar{u}(z)$ is

$$\langle u_n\rangle=\frac{1}{2\pi i}\int_c \bar{u}(z)z^{n-1}dz = \text{sum of residues } \sum r_k.$$

$$\langle u_n\rangle=\frac{1}{2\pi i}\int_c \bar{u}(z)z^{n-1}dz=\frac{1}{2\pi i}\int_c \frac{z}{(z+1)(z+2)}z^{n-1}dz$$

$$=\frac{1}{2\pi i}\int_c \frac{z^n}{(z+1)(z+2)}dz$$

where c is a simple closed curve containing the singularities at $z = -1$ and $z = -2$. These two singularities are simple pole.

Now, $$r_1 = \underset{z=-1}{\text{Res}}\,(\bar{u}(z)z^{n-1}) = \lim_{z\to -1}(z+1)\frac{z^n}{(z+1)(z+2)} = (-1)^n$$

and $$r_2 = \underset{z=-2}{\text{Res}}(\bar{u}(z)z^{n-1}) = \lim_{z\to -2}(z+2)\frac{z^n}{(z+1)(z+2)} = -(-2)^n$$

$\therefore$ $$\langle u_n\rangle = \text{sum of residues} = \langle(-1)^n\rangle + \langle -(-2)^n\rangle = \langle(-1)^n(1-2^n)\rangle$$

EXAMPLE 19.7 Find $Z^{-1}\left(\dfrac{z^2+z}{(z-1)(z^2+1)}\right)$.

Solution Let $\bar{u}(z) = \dfrac{z^2+z}{(z-1)(z^2+1)}$.

The inverse Z-transform $\bar{u}(z)$ is

$$\langle u_n\rangle = \frac{1}{2\pi i}\int_c \bar{u}(z)z^{n-1}dz = \text{sum of residues } \sum r_k.$$

$$\langle u_n\rangle = \frac{1}{2\pi i}\int_c \bar{u}(z)z^{n-1}dz = \frac{1}{2\pi i}\int_c \frac{z^2+z}{(z-1)(z^2+1)}z^{n-1}dz = \frac{1}{2\pi i}\int_c \frac{z+1}{(z-1)(z^2+1)}z^n dz$$

where c is a simple closed curve containing the singularities at $z = 1, \pm i$. These singularities are simple pole.

Now, $$r_1 = \underset{z=1}{\text{Res}}(\bar{u}(z)z^{n-1}) = \lim_{z\to 1}(z-1)\frac{z+1}{(z-1)(z^2+1)}z^n = 1$$

$$r_2 = \underset{z=i}{\text{Res}}(\bar{u}(z)z^{n-1}) = \lim_{z\to i}(z-i)\frac{z+1}{(z-1)(z+i)(z-i)}z^n = -\frac{i^n}{2}$$

and $$r_3 = \underset{z=-i}{\text{Res}}(\bar{u}(z)z^{n-1}) = \lim_{z\to -i}(z+i)\frac{z+1}{(z-1)(z+i)(z-i)}z^n = -\frac{(-i)^n}{2}$$

$\therefore$ $$\langle u_n\rangle = \text{sum of residues} = \langle 1\rangle + \left\langle -\frac{i^n}{2}\right\rangle + \left\langle -\frac{(-i)^n}{2}\right\rangle$$

$$= \langle 1\rangle - \frac{1}{2}\langle i^n + (-i)^n\rangle$$

$$= \langle 1\rangle - \frac{1}{2}2\left\langle \cos\frac{n\pi}{2}\right\rangle = \left\langle 1 - \cos\frac{n\pi}{2}\right\rangle$$

19.7.5 Method 5: Power Series Method

Expand $\bar{u}(z)$ in powers of $1/z$ using standard formulae and then identify the sequence $\langle u_n \rangle$ which is required inverse Z-transform of $\bar{u}(z)$.

EXAMPLE 19.8 Find $Z^{-1}\left(\log\frac{z}{z+1}\right)$ using power series method.

Solution Let $\bar{u}(z) = \log\left(\frac{z}{z+1}\right) = -\log\left(\frac{z+1}{z}\right) = -\log\left(1+\frac{1}{z}\right)$

$$= -\frac{1}{z} + \frac{1}{2}\frac{1}{z^2} - \frac{1}{3}\frac{1}{z^3} + \cdots + \frac{(-1)^n}{n}\frac{1}{z^n} + \cdots$$

$$\therefore \qquad \langle u_n \rangle = \begin{cases} 0, & \text{for } n = 0 \\ \frac{(-1)^n}{n}, & \text{for } n \neq 0 \end{cases}$$

EXAMPLE 19.9 Find $Z^{-1}\left(\frac{z^2}{z^2+1}\right)$ using power series method.

Solution Let $\bar{u}(z) = \frac{z^2}{z^2+1} = \frac{1}{1+\frac{1}{z^2}} = \left(1+\frac{1}{z^2}\right)^{-1} = 1 - \frac{1}{z^2} + \frac{1}{z^4} + \cdots = \sum_{n=0}^{\infty} (-1)^{n/2} z^{-n}$, n is even.

Then $\qquad \langle u_0, u_1, u_2, u_3, \ldots \rangle = \langle 1, 0, 1, 0, \ldots \rangle = \left\langle \cos\frac{n\pi}{2} \right\rangle$

$$\Rightarrow \qquad \bar{u}(z) = \langle u_n \rangle = \left\langle \cos\frac{n\pi}{2} \right\rangle, n = 0, 1, 2, \ldots$$

19.7.6 Method 6: Convolution Method

When the product of two transforms is given, one can apply convolution theorem to find inverse Z-transform.

i.e., If $Z^{-1}[\bar{u}(z)] = \langle u_n \rangle$ and $Z^{-1}[\bar{v}(z)] = \langle v_n \rangle$ then $Z^{-1}[\bar{u}(z)\bar{v}(z)] = \sum_{m=0}^{n} u_m v_{n-m} = \langle u_n \rangle * \langle v_n \rangle$.

EXAMPLE 19.10 Find $Z^{-1}\left(\frac{z^2}{(z-a)(z-b)}\right)$ using convolution theorem.

Solution Given $\bar{u}(z) = \frac{z}{z-a}$. Then $\langle u_n \rangle = Z^{-1}[\bar{u}(z)] = Z^{-1}\left[\frac{z}{z-a}\right] = \langle a^n \rangle$

and $\bar{v}(z) = \frac{z}{z-b}$. Then $\langle v_n \rangle = Z^{-1}[\bar{v}(z)] = Z^{-1}\left[\frac{z}{z-b}\right] = \langle b^n \rangle$. Then

$$Z^{-1}\left(\frac{z^2}{(z-a)(z-b)}\right)=Z^{-1}\left(\frac{z}{z-a}\cdot\frac{z}{z-b}\right)$$

$$=\langle a^n\rangle * \langle b^n\rangle$$

$$=\sum_{m=0}^{n} a^m b^{n-m}$$

$$=b^n\sum_{m=0}^{n}\left(\frac{a}{b}\right)^m$$

$$=b^n\left[\frac{(a/b)^{n+1}-1}{(a/b)-1}\right]$$

$$=\frac{a^{n+1}-b^{n+1}}{a-b}$$

19.7.7 Difference Equation

We will see applications of Z-transforms to solve difference equations. An equation involving differences of an unknown function at one or more values of the argument is known as *difference equation*. For example,

$$\Delta y_{n+1}+y_n=e^x$$

$$\Rightarrow \qquad y_{n+2}-y_{n+1}+y_n=e^x$$

Using $E = 1 + \Delta$, we have $(E^2 - E + 1)y_n = e^x$.

The solution of difference equation is sum of complementary solution and particular solution as we studied in Chapter 2.

For computing complementary solution, we proceed as follows:

Write $y_{n+k}=E^k y_n, k=0,1,2,\ldots,n$ in linear difference equation

$$y_{n+m}+a_1y_{n+m-1}+a_2y_{n+m-2}+\cdots+a_my_n=0$$

where $a_1, a_2, \ldots, a_n$ are constants.

$$\Rightarrow \qquad \phi(E)y_n=(E^m+a_1E^{m-1}+a_2E^{m-2}+\cdots+a_m\cdot 1)y_n=0 \qquad (19.2)$$

Let $y_n = \lambda^n$ be the solution of Eq. (19.2).

$$\Rightarrow \qquad E^k y_n=\lambda^{n+k}, k=0,1,2,\ldots,m$$

and we have

$$(\lambda^m+a_1\lambda^{m-1}+a_2\lambda^{m-2}+\cdots+a_m\cdot 1)\lambda^n=0$$

$$\lambda^m+a_1\lambda^{m-1}+a_2\lambda^{m-2}+\cdots+a_m\cdot 1=0 \qquad (19.3)$$

Equation (19.3) is called auxiliary equation of the difference equation. It has *m*-roots (say) $\lambda_1, \lambda_2, \ldots, \lambda_m$. Use Table 19.2 to compute complementary function.

Table 19.2 Complementary solution

Root of AE	*Nature of roots*	*Complementary solution*
$\lambda_1, \lambda_2, ..., \lambda_m$	Real and distinct	$c_1\lambda_1^n + c_2\lambda_2^n + \cdots + c_m\lambda_m^n$
$\lambda_1, \lambda_2, ..., \lambda_m$	Real and two roots are equal	$(c_1 + c_2 n)\lambda_1^n + c_3\lambda_3^n + \cdots + c_m\lambda_m^n$
$\lambda_1, \lambda_2, ..., \lambda_m$	Real and k roots are equal	$(c_1 + c_2 n + \cdots + c_k n^{k-1})\lambda_1^n + c_{k+1}\lambda_{k+1}^n + \cdots + c_m\lambda_m^n$
$\alpha \pm i\beta$	A pair of complex roots	$r^n(c_1 \cos n\theta + c_2 \sin n\theta), r = \sqrt{\alpha^2 + \beta^2}, \theta = \tan^{-1}\dfrac{\beta}{\alpha}$

Next, we need to work out particular integral. Operate given difference equation by the inverse operator $\dfrac{1}{\phi(E)}$. Then we get $\text{PI} = \dfrac{1}{\phi(E)} f(n)$.

1. If $f(n) = a^n$ and $\phi(a) \neq 0$ then $\text{PI} = \dfrac{1}{\phi(E)} a^n = \dfrac{1}{\phi(a)} a^n$.

2. If $\phi(E) = (E - a)\psi(a)$ and $\psi(a) \neq 0$ then $\text{PI} = \dfrac{1}{\psi(E)(E - a)} a^n = \dfrac{1}{\psi(a)} n a^{n-1}$.

3. If $\phi(E) = (E - a)^2 \psi(a)$ and $\psi(a) \neq 0$ then $\text{PI} = \dfrac{1}{\psi(E)(E - a)^2} a^n = \dfrac{1}{\psi(a)} n(n-1)a^{n-2}$.

4. If $f(n) = \sin \alpha x$ or $\cos \alpha x$

   We know that
$$\sin \alpha n = \frac{e^{i\alpha n} - e^{-i\alpha n}}{2i} = \frac{1}{2i}[(e^{i\alpha})^n - (e^{-i\alpha})^n]$$
   and
$$\cos \alpha n = \frac{e^{i\alpha n} + e^{-i\alpha n}}{2} = \frac{1}{2}[(e^{i\alpha})^n + (e^{-i\alpha})^n]$$
   Apply (1) by replacing '$i\alpha$' with 'α'.

5. If $f(n) = n^p$ then $\text{PI} = \dfrac{1}{\phi(E)} n^p = \dfrac{1}{\phi(1+\Delta)} n^p = \phi^{-1}(1+\Delta)n^p$.

   Expand $\phi^{-1}(1 + \Delta)$ in increasing powers of Δ up to Δ^p and express n^p in the factorial form and apply transform on each term of the expansion.

EXAMPLE 19.11 Form difference equation for $y^n = a \cdot 1^n + b \cdot 2^n$.

Solution
$$y^n = a \cdot 1^n + b \cdot 2^n \tag{i}$$

Apply shift operator E twice.
$$Ey_n = y_{n+1} = 1 \cdot a \cdot 1^n + 2 \cdot b \cdot 2^n \tag{ii}$$
$$E^2 y_n = y_{n+2} = 1 \cdot a \cdot 1^n + 4 \cdot b \cdot 2^n \tag{iii}$$

Eliminating a and b from Eqs. (i)–(iii), we get

$$\begin{vmatrix} y_n & 1 & 1 \\ y_{n+1} & 1 & 2 \\ y_{n+2} & 1 & 4 \end{vmatrix} = 0 \;\Rightarrow\; 2y_n - 3y_{n+1} + y_{n+2} = 0 \text{ or } y_{n+2} - 3y_{n+1} + 2y_n = 0$$

EXAMPLE 19.12 Form difference equation for Fibonacci sequence and solve it.

Solution We know that Fibonacci sequence is $\langle F_n \rangle = \langle 0, 1, 1, 2, 3, 5, 8, \ldots \rangle$. Let y_n be the nth term of the sequence. Then $y_n = y_{n-1} + y_{n-2}, n > 0$.

$$\Rightarrow \qquad y_{n+2} = y_{n+1} + y_n \text{ or } y_{n+2} - y_{n+1} - y_n = 0. \qquad \text{(i)}$$

This is the required difference equation. Eq. (i) can be written as

$$(E^2 - E - 1)y_n = 0 \quad \Rightarrow \quad m^2 - m - 1 = 0$$

which gives $m = \dfrac{1 \pm \sqrt{5}}{2}$.

$$\therefore \qquad \text{CF} = c_1\left(\frac{1+\sqrt{5}}{2}\right)^n + c_2\left(\frac{1-\sqrt{5}}{2}\right)^n, n > 0$$

where c_1 and c_2 are constants to be determined.

$$\text{For } n = 1, y_1 = 0 \quad \Rightarrow \quad c_1\left(\frac{1+\sqrt{5}}{2}\right) + c_2\left(\frac{1-\sqrt{5}}{2}\right) = 0$$

$$\text{For } n = 2, y_2 = 1 \quad \Rightarrow \quad c_1\left(\frac{1+\sqrt{5}}{2}\right)^2 + c_2\left(\frac{1-\sqrt{5}}{2}\right)^2 = 1$$

Solving these two equations, we get $c_1 = \dfrac{5-\sqrt{5}}{10}$ and $c_2 = \dfrac{5+\sqrt{5}}{10}$.

Hence, the solution is $y_n = \dfrac{5-\sqrt{5}}{10}\left(\dfrac{1+\sqrt{5}}{2}\right)^n + \dfrac{5+\sqrt{5}}{10}\left(\dfrac{1-\sqrt{5}}{2}\right)^n$.

EXAMPLE 19.13 Solve $u_{n+2} - 4u_{n+1} + 4u_n = 2^n$.

Solution The corresponding operator form is

$$(E^2 - 4E + 4)u_n = (E-2)^2 u_n = 2^n$$

AE is $m^2 - 4m + 4 = 0 \Rightarrow m = 2, 2$.

Therefore, $\qquad \text{CF} = (c_1 + c_2 n)2^n$.

$$\text{PI} = \frac{1}{(E-2)^2}2^n = \frac{n(n-1)}{2!}2^{n-2} = n(n-1)2^{n-3}$$

$\therefore$ The solution is $u_n = (c_1 + c_2 n)2^n + n(n-1)2^{n-3}$.

EXERCISES

1. Find the Z-transform of the unit impulse sequence $\langle \delta_n \rangle$ where $\delta_n = \begin{cases} 1, & n = 0 \\ 0, & n \neq 0 \end{cases}$.

[Ans. 1**]**

2. Find the Z-transform of $u_n = 2^{-n}, -3 \le n \le 3$.

$\left[\textbf{Ans.}\ 8z^3 + 4z^2 + 2z + 1 + \frac{1}{2}\frac{1}{z} + \frac{1}{4}\frac{1}{z^2} + \frac{1}{8}\frac{1}{z^3}\right]$

3. Find the Z-transform of $u_n = \left(\frac{3}{4}\right)^n, n \in \mathbb{N}$.

$\left[\textbf{Ans.}\ \frac{4z}{4z-3}\right]$

4. Find the Z-transform of $\langle u_n \rangle$ where $u_0 = 4, u_1 = -5, u_2 = 0, u_3 = 0$ and $u_4 = 2$.

$\left[\textbf{Ans.}\ 4 - \frac{5}{z} + \frac{2}{z^4}\right]$

5. Find the Z-transform of $\langle u_{n+2} \rangle$ if $Z\{\langle u_n \rangle\} = \frac{z}{z+1} + \frac{z}{z^2+1}$.

$\left[\textbf{Ans.}\ \frac{z(z^2 - z + 2)}{(z+1)(z^2+1)}\right]$

6. Find the Z-transform of $Z\{\langle na^n \rangle\}$.

$\left[\textbf{Ans.}\ \frac{az}{(z-a)^2}\right]$

7. Find $Z^{-1}\left[\frac{8z^2}{(2z-1)(4z-1)}\right]$.

$\left[\textbf{Ans.}\ \left\langle \frac{1}{2^{n-1}} - \frac{1}{4^n} \right\rangle\right]$

8. Find $Z^{-1}\left[\frac{2(z^2 - z)}{(z^2+1)^2}\right]$.

$\left[\textbf{Ans.}\ \left\langle 2n \sin\frac{n\pi}{2} \right\rangle\right]$

9. Find $Z^{-1}\left[\frac{z^2}{(z-1)(z-3)}\right]$.

$\left[\textbf{Ans.}\ \frac{1}{2}\left\langle 3^{n+1} - 1 \right\rangle\right]$

10. Solve $y_{n+2} - 5y_{n+1} - 6y_n = 2^n$.

$\left[\textbf{Ans.}\ y_n = c_1 6^n + c_2(-1)^n - \frac{1}{3}2^{n-2}\right]$

11. Solve $u_{n+2} - 3u_{n+1} + 2u_n = 4^n$ given that $u_0 = 0$ and $u_1 = 1$.

$\left[\textbf{Ans.}\ u_n = -\frac{2}{3} + 2^{n-1} + \frac{1}{3}2^{n-1}\right]$

12. Solve $y_{n+3} - 7y_{n+2} + 14y_{n+1} - 8y_n = 0$.

$[\textbf{Ans.}\ y_n = c_1 + c_2 2^n + c_3 2^{2n}]$

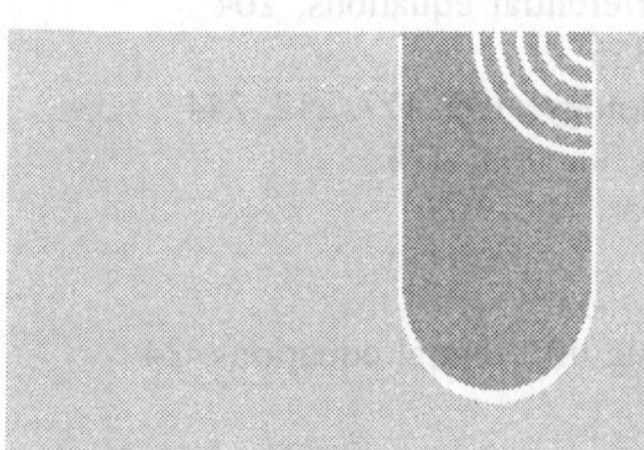

Index

Gynecological Algorithms in Clinical Practice

Editor

Alok Sharma

MD (OBG) DHA MICOG FICOG

Consultant Obstetrician and Gynecologist

Department of Obstetrics and Gynecology

Maternal and Child Health Center

Civil Hospital

Mandi, Himachal Pradesh, India

Forewords

Jaydeep Tank

Sunita Tandulwadkar

Hrishikesh D Pai

JAYPEE BROTHERS MEDICAL PUBLISHERS

The Health Sciences Publisher

New Delhi | London

Jaypee Brothers Medical Publishers (P) Ltd

Headquarters
Jaypee Brothers Medical Publishers (P) Ltd
EMCA House, 23/23-B
Ansari Road, Daryaganj
New Delhi 110 002, India
Landline: +91-11-23272143, +91-11-23272703
+91-11-23282021, +91-11-23245672
Email: jaypee@jaypeebrothers.com

Corporate Office
Jaypee Brothers Medical Publishers (P) Ltd
4838/24, Ansari Road, Daryaganj
New Delhi 110 002, India
Phone: +91-11-43574357
Fax: +91-11-43574314
Email: jaypee@jaypeebrothers.com

Overseas Office
JP Medical Ltd
83 Victoria Street, London
SW1H 0HW (UK)
Phone: +44 20 3170 8910
Fax: +44 (0)20 3008 6180
Email: info@jpmedpub.com

Website: www.jaypeebrothers.com
Website: www.jaypeedigital.com

Inquiries for bulk sales may be solicited at: jaypee@jaypeebrothers.com

Gynecological Algorithms in Clinical Practice

First Edition: **2025**

ISBN: 978-93-5696-516-4

Printed at: Sterling Graphics Pvt. Ltd. India.

Dedication

To the challenges that made me stronger
To my parents Sh Hans Raj Sharma and Smt Dhanwanti Devi Sharma, who planted the seed of knowledge in my mind and nurtured it
To my wife Dr Pratibha Sharma, who has been my most profound inspiration
For my daughter Hiranya, who fills my heart with joy each and every day

Contributors

Aakriti Batra MBBS MS (Obs & Gyne)
Consultant
Department of Obstetrics and Gynecology
Vardhman Mahavir Medical College and Safdarjung Hospital
New Delhi, India

Aarthi S Jayraj MBBS MD MCh
Senior Resident
Department of Obstetrics and Gynecology
All India Institute of Medical Sciences
New Delhi, India

Aastha Chauhan MBBS MS
Senior Resident
Department of Obstetrics and Gynecology
Pandit Bhagwat Dayal Sharma Postgraduate Institute of Medical Sciences
Rohtak, Haryana, India

Achla Batra MBBS DGO DNB
Professor and Consultant
Department of Obstetrics and Gynecology
Vardhman Mahavir Medical College and Safdarjung Hospital
New Delhi, India
President, Association of Obstetricians Gynecologists of Delhi (AOGD)

Alpesh Gandhi MBBS DGO FRCOG FICOG PGDHM
Senior Consultant Critical Care in Obstetrics and High-risk, Pregnancy Expert
Arihant Women's Hospital
Ahmedabad, Gujarat, India
Past President, FOGSI

Amrita S Jaipuriar MBBS MS
Consultant
Department of Obstetrics and Gynecology
Jaipuriar Medical and Research Centre
Gorakhpur, Uttar Pradesh, India

Ananta Kanwar MBBS MS DNB
Consultant
Department of Obstetrics and Gynecology
Clio Mother and Child Institute
Ludhiana, Punjab, India

Ananya Aggarwal MBBS Student
SGT Medical College, Hospital, and Research Institute
Gurugram, Haryana, India

Anil E MD
Consultant, Obstetrics and Gynecology
Department of Telemedicine
Postgraduate Institute Medical Education and Research
Chandigarh, India

Anju Soni MBBS MS
Consultant
Soni Hospital
Jaipur, Rajasthan, India
President, Indian Menopause Society (IMS) (2024–2025)

Ankita Srivastava MBBS MS DNB MRCOG (UK)
Associate Consultant
Department of Obstetrics and Gynecology
Sir Ganga Ram Hospital
New Delhi, India

Anupama Bahadur MBBS MD DNB MNAMS
Professor
Department of Obstetrics and Gynecology
All India Institute of Medical Sciences
Rishikesh, Uttarakhand, India

Arihant Arora MBBS MD
Resident
Department of Obstetrics and Gynecology
Government Medical College
Jammu, Jammu and Kashmir, India

Arun Arora MBBS Dip Sports Medicine MD (Obs & Gyne) FICOG
Head
Department of Obstetrics and Gynecology
Government Medical College
Udhampur, Jammu and Kashmir, India

Arunav Arora MBBS MD
Resident
Department of Obstetrics and Gynecology
Government Medical College
Jammu, Jammu and Kashmir, India

Asmita Kaundal MBBS MD
Assistant Professor
Department of Obstetrics and Gynecology
All India Institute of Medical College
Bilaspur, Himachal Pradesh, India

Avneet Kaur MBBS MS
Senior Resident
Department of Gynecologic Oncology and Surgical Oncology
Max Institute of Cancer Care, Max Hospital
New Delhi, India

Ayush C Srivastava MBBS MD
Pursuing Fellowship in Surgical Oncology
Rosewell Park Cancer Center
Buffalo, New York, USA

Bhupinder Deep Kaur MBBS MD
Senior Residents
Department of Obstetrics and Gynecology
Government Medical College
Patiala, Punjab, India

Bindiya Gupta MD MAMS FICOG
Professor
Department of Obstetrics and Gynecology
University College of Medical Sciences and
Guru Teg Bahadur Hospital
New Delhi, India
Commonwealth Fellow, Gyne Oncology, Birmingham, UK
IFCPC IARC Certified Colposcopist
ISCCP Colposcopy Trainer
Member Oncology Committee, AOGD and FOGSI
Joint Secretary, Asia-Oceania Research Organization
in Genital Infection and Neoplasia (AOGIN), India
Editorial Board Member, ISCCP
Editor, AOGD

Chandrika Anand MBBS MD DNB (OBG)
Director
PCOS centre
Saraswati Gynoworld
Bengaluru, Karnataka, India

David Simson MBBS DNB (Radiation Oncology)
Consultant and Radiation Oncologist
Rajiv Gandhi Cancer Institute and Research Center
New Delhi, India

Dilpreet Kaur Pandher MBBS MD
Professor
Department of Obstetrics and Gynecology
Government Medical College and Hospital
Chandigarh, India

Disha Dave MS
Assistant Professor
Department of Obstetrics and Gynecology
Mahatma Gandhi Institute of Medical Sciences
Wardha, Maharashtra, India

Divya Dwivedi MBBS MS FICOG
Associate Professor
Department of Obstetrics and Gynecology
Ganesh Shankar Vidyarthi Memorial Medical College
Kanpur, Uttar Pradesh, India

Divya Mahindru MBBS MS
Centre Head—Gynecologist and IVF Specialist
Indira IVF Fertility and IVF Centre
Jalandhar, Punjab, India

Ezzat Khalda MBBS MD
Associate Professor
Department of Radiodiagnosis
Sikkim Manipal Institute of Medical Sciences
Gangtok, Sikkim, India

Gayatri Suresh MBBS MD
Senior Resident
Department of Obstetrics and Gynecology
All India Institute of Medical College
Rishikesh, Uttarakhand, India

Geetika Gupta Syal MBBS MD
Associate Professor
Department of Obstetrics and Gynecology
Kamla Nehru State Hospital for Mother and Child
Shimla, Himachal Pradesh, India

Geetu Mahajan MBBS MD
Associate Professor
Department of Obstetrics and Gynecology
Government Medical College
Udhampur, Jammu and Kashmir, India

Girija Shankar Mohanty MD
Assistant Professor
Department of Obstetrics and Gynecology
Postgraduate Institute of Medical Education and Research
Chandigarh, India

Girija Wagh MD FICOG Dip Endoscopy FICS
Professor and Head
Department of Obstetrics and Gynecology
Bharati Vidyapeeth University Medical College
Pune, Maharashtra, India
Director, Primrose Wellness and Care, Pune
Consultant, Apollo Spectra Hospitals, Pune
Vice President, FOGSI

Hafizur Rahman MBBS DGO DNB
Additional Professor and Head
Department of Obstetrics and Gynecology
All India Institute of Medical Sciences
Kalyani, West Bengal, India

Janki Munjal Pandya
MS (Obs & Gyne) FICOG Diploma in Minimal Invasive Surgery
Assistant Professor
AMC MET Medical College
Sheth Lallubhai Gordhandas Municipal General Hospital
Ahmedabad, Gujarat, India

Japleen Kaur MD DNB Fellow (REI) MNAMS MRCOG (UK)
Assistant Professor
Department of Obstetrics and Gynecology
Dr BR Ambedkar Institute of Medical Sciences
Mohali, Punjab, India

Jasmeen Kaur MBBS MS
Senior Consultant
Department of Obstetrics and Gynecology
Fortis Bloom IVF Centre, Fortis Hospital
Mohali, Punjab, India

Jyotsna Gupta
MBBS DNB (Obs & Gyne) PGDHHM MBBS FICMCH MNAMS (Obs & Gyne)
Consultant
Dr Jyotsna's Clinic
New Delhi, India

K Aparna Sharma MBBS MD
Professor
Department of Obstetrics and Gynecology
All India Institute of Medical Sciences
New Delhi, India

Kalpana Koli MS DNB
Deaprtment of Obstetrics and Gynecology
HBT Medical College and Dr RN Cooper Municipal Hospital
Mumbai, Maharashtra, India

Kanan Kotdawala MBBS MD
Consultant Gynecologist
Kotdawala Maternity Nursing Home
Ahmedabad, Gujarat, India

Kanica Batra Modi MBBS DNB-MS Fellow IARC IFCPC WHO Fellow (ESGO) KEM Germany, Fellow (IGCS) Gynecology Oncology Royal Hospital for Women, Sydney
Senior Consultant
Department of Gynecologic Oncology and Surgical Oncology
Max Institute of Cancer Care, Max Hospital
New Delhi, India

Kanika Chopra MBBS MS
Consultant
Department of Obstetrics and Gynecology
Sri Ganga Ram Hospital
New Delhi, India

Karishma Thariani MBBS MS (Obs & Gyne) DNB (Obs & Gyne)
Consultant Urogynecologist
Center for Urogynecology and Pelvic Health
New Delhi, India

Kavita Mandrelle Bhatti MBBS MD
Professor and Head
Department of Obstetrics and Gynecology
Christian Medical College and Hospital
Ludhiana, Punjab, India

KD Kamal MBBS MD DM (Endocrinology)
Consultant Endocrinology
Indian Naval Hospital Ship
Mumbai, Maharashtra, India

Kirti Saini MBBS
PG Student
Department of Obstetrics and Gynecology
Pandit Bhagwat Dayal Sharma Postgraduate Institute of Medical Sciences
Rohtak, Haryana, India

Madhva Prasad S MS DNB PGDMLS PGDCR
Associate Professor
Department of Obstetrics and Gynecology
Vydehi Institute of Medical Sciences and Research Centre
Bengaluru, Karnataka, India

Mala Srivastava MBBS DNB DGO FICOG
Senior Consultant
Department of Obstetrics and Gynecology
Sir Ganga Ram Hospital
New Delhi, India

Manish Subhash Mahajan MBBS DGO DNB (Obs & Gyne)
Fellowship in Cosmetic Gynaecology Dip Cosmetic Gynecology
Fellowship in Stem Cells and Regenerative Medicine
Obstetrician and Gynecologist, Cosmetic Gynecologist,
Madalasaa Aesthetic Gynecology
and Feminine Rejuvenation Centre
Thane, Maharashtra, India

Meenakshi B Chauhan MBBS MD
Senior Professor and Unit Head
Department of Obstetrics and Gynecology
Pandit Bhagwat Dayal Sharma Postgraduate Institute of Medical Sciences
Rohtak, Haryana, India

Megha Agarwal MBBS MS
Private Practitioner
The Gynae Clinic
New Delhi, India

Megha Mittal MBBS MS DNB ICOG Fellowship Gyne-oncology MICC
Consultant
Department of Gynecologic Oncology and Surgical Oncology
Max Institute of Cancer Care, Max Hospital
New Delhi, India

Minakshi Rohilla MD
Professor
Department of Obstetrics and Gynecology
Postgraduate Institute of Medical Education and Research
Chandigarh, India

Monika Gupta MD DNB MNAMS ACME FICOG FICMCH
Fellow Urogynaecology & Pelvic Reconstructive Surgery (UK)
Professor
Department of Obstetrics and Gynecology
In-Charge, Urogynecology Specialty Clinic
Vardhman Mahavir Medical College and Safdarjung Hospital
New Delhi, India

Monika Thakur MBBS MD
Assistant Professor
Dr Radhakrishnan Government Medical College
Hamirpur, Himachal Pradesh, India

Munjal Jayeshkumar Pandya MS (Obs & Gyne) FICOG
Diploma in Minimal Invasive Surgery
Assistant Professor
AMC MET Medical College
Sheth Lallubhai Gordhandas Municipal General Hospital
Ahmedabad, Gujarat, India

Namita Batra MBBS MS
Senior Resident
Department of Obstetrics and Gynecology
University College of Medical Sciences
and Guru Teg Bahadur Hospital
New Delhi, India

Neema Tufchi MSc PhD (Biotechnology)
Research Scientist
ICMR Project at Institute of Obstetrics and Gynecology
Sir Ganga Ram Hospital
New Delhi, India

Nidhi Jindal MBBS MD (PGIMER) DNB FMAS
Assistant Professor
Department of Obstetrics and Gynecology
Indira Gandhi Medical College
Shimla, Himachal Pradesh, India

Nilesh Balkawade MBBS MS DNB (Obs & Gyne) MNAMS FIAGE (Fellow, Indian Association of Gynaec Endoscopists) Fellowship in Reproductive Medicine, Dip in Gyn Endoscopy (Kiel, Germany)
Clinical Head and Fertility Specialist
Oasis Fertility
Pune, Maharashtra, India
Chairperson, Quiz Committee, FOGSI (2021–2025)
Past Clinical Secretary, POGS (2021–2022)
Member, Infertility Committee, AMOGS (2020–2022)

Nisha Malik MBBS MD DNB MRCOG
Assistant Professor
Department of Obstetrics and Gynecology
All India Institute of Medical Sciences
Bilaspur, Himachal Pradesh, India

Parneet Kaur MBBS MD DNB
Professor and Head
Department Obstetrics and Gynecology
Government Medical College
Patiala, Punjab, India

Parul Kotdawala MBBS MD
Consultant Gynecologist
Kotdawala Women's Clinic
Ahmedabad, Gujarat, India

Pavika Lal MBBS MD (AIIMS) DNB FICOG
Associate Professor
Department of Obstetrics and Gynecology
Ganesh Shankar Vidyarthi Memorial Medical College
Kanpur, Uttar Pradesh, India

Pikee Saxena MBBS MD FICOG MNAMS FMAS FIMCH PGCC (Hospital Management) PGDCR (Clinical Research)
Director Professor
Department of Obstetrics and Gynecology
Lady Hardinge Medical College
New Delhi, India

Pooja Mehta MBBS MD
Head
Department of Obstetrics and Gynecology
Fortis Bloom IVF Centre, Fortis Hospital
Mohali, Punjab, India

Poonam Shivkumar MS
Director Professor
Department of Obstetrics and Gynecology
Unit Chief and Medical Superintendent
Mahatma Gandhi Institute of Medical Sciences
Wardha, Maharashtra, India

Priya Ganesh Kumar MBBS MD DGO DFP FICOG
Medical Director
Sai Niwas Chain of Preventive Oncology Centers
PAN, India

Radhika Gupta MBBS MS
Senior Resident
Department of Obstetrics and Gynecology
Government Medical College and Hospital
Chandigarh, India

Rajesh Kumar Verma MBBS MD (Anesthesia)
Associate Professor
Department of Anesthesiologist
Indira Gandhi Medical College
Shimla, Himachal Pradesh, India

Rajlaxmi Mundhra MBBS MD DNB MNAMS
Associate Professor
Department of Obstetrics and Gynecology
All India Institute of Medical College
Rishikesh, Uttarakhand, India

Rajni Singh MBBS MS
Senior Resident
Department of Obstetrics and Gynecology
All India Institute of Medical College
Rishikesh, Uttarakhand, India

Rakhi Singh MBBS DGO FICOG DPE DART
Senior Fertility Consultant
Department of Obstetrics and Gynecology
Abalone Clinic and IVF Center
Noida, Uttar Pradesh, India

Raman Singla MBBS FAIS FMAS
Head and Consultant
Department of Obstetrics and Gynecology
Singla Mediclinic and Radiance Hospital
Mohali, Punjab, India

Reena Wani MD FRCOG FICOG DNBE FCPS DGO DFP
Professor and Head
Department of Obstetrics and Gynecology
HBT Medical College and Dr RN Cooper Municipal Hospital
Mumbai, Maharashtra, India

Reeti Mehra MBBS MD
Professor
Department of Obstetrics and Gynecology
Government Medical College and Hospital
Chandigarh, India

Renu Sharma MBBS MD
Consultant
Department of Obstetrics and Gynecology
Shri Mata Vaisho Devi Narayana Superspecialty Hospital
Jammu, Jammu & Kashmir, India

Richa Vatsa MBBS MD MRCOG
Associate Professor
Department of Obstetrics and Gynecology
All India Institute of Medical Sciences
New Delhi, India

Rimmy Singla MBBS DGO MRCOG (UK)
Dip Reproductive Med. (Germany)
Head and Consultant
Singla Mediclinic and Radiance Hospital
Mohali, Punjab, India

Rohini Rao MBBS MD (Obs & Gyne)
Associate Professor
Department of Obstetrics and Gynecology
Indira Gandhi Medical College and Hospital
Shimla, Himachal Pradesh, India

Roopa Malik MBBS MS
Professor
Department of Obstetrics and Gynecology
Pandit Bhagwat Dayal Sharma Postgraduate Institute of
Medical Sciences
Rohtak, Haryana, India

Ruchika Garg MBBS MD
Professor
Department of Obstetrics and Gynecology
SN Medical College
Agra, Uttar Pradesh, India

Sachin Paprikar MBBS DGO DNB FMAS F Gynaec Endoscopy
Obstetrician Gynecologist and Gyne
Laparoscopic Surgeon, Consultant and HOD
Department of Obstetrician and Gynecologist
District Hospital Bijapur
Bijapur, Chhattisgarh, India

Sadhana Gupta MBBS MS
Senior Consultant
Department of Obstetrics and Gynecology
Jeevan Jyoti Hospital and Medical Research Centre
Gorakhpur, Uttar Pradesh, India
Co-chair, SAFOG NCD Committee
Vice President, FOGSI 2016

Sajjan Rajpurohit MBBS MD DNB
Senior Director
Department of Gynecologic Oncology and Surgical Oncology
Max Institute of Cancer Care, Max Hospital
New Delhi, India

Sarabjeet Singh MBBS DGO PhD Fellowship ART (Fellowship Advanced Reproductive Medicine)
Medical Director
Sarab Healthcare and Dove Art Bank, Jalandhar
Consultant
Reproductive Medicine and Endoscopy
Virk Fertility Services
Jalandhar, Punjab, India

Saritha Shamsunder MBBS MD FRCOG
Consultant
Department of Obstetrics and Gynecology
Vardhman Mahavir Medical College and Safdarjung Hospital
New Delhi, India

Savita Rani Singhal MBBS MD
Senior Professor
Department of Obstetrics and Gynecology
Pandit Bhagwat Dayal Sharma Postgraduate Institute of Medical Sciences
Rohtak, Haryana, India

Sayanti Paul MBBS MS MRCOG
Assistant Professor
Department of Obstetrics and Gynecology
All India Institute of Medical Sciences
Kalyani, West Bengal, India

Seema Sharma MBBS DGO MD FICOG FRCOG (London, UK)
Clinical Director and Senior Consultant
Department of Obstetrics and Gynecology
Apollo Cardle and Children's Hospital
Moti Nagar, New Delhi, Delhi

Seema Singhal MS MCh (Gyne & Onco) FACS FICOG FCLS MNAMS
Additional Professor
Department of Obstetrics and Gynecology
All India Institute of Medical Sciences
New Delhi, India

Shail Kaur MBBS MS (Obs & Gyne)
Associate Professor
Punjab Institute of Medical Sciences
Jalandhar, Punjab, India
Former Registrar, PGI Chandigarh

Shalini Sharma MBBS MD
Senior Resident
Department of Obstetrics and Gynecology
Kamla Nehru State Hospital for Mother and Child
Shimla, Himachal Pradesh, India

Sharda Patra MBBS MD
Associate Professor
Department of Obstetrics and Gynecology
Lady Hardinge Medical College and Hospitals
New Delhi, India

Shikha Rani MS FMAS
Assistant Professor
Department of Obstetrics and Gynecology
Government Medical College and Hospital
Chandigarh, India

Smiti Nanda MBBS MD
Retired Professor and Head
Department Obstetrics and Gynecology
Postgraduate Institute of Medical Sciences
Rohtak, Haryana, India

Sneha Bhuyar MBBS MD FICMCH FICOG
Consultant
Sukhkarta Hospital
Yavatmal, Maharashtra, India

Sonalica Suresh MBBS DGO DNB
Consultant
Department of Obstetrics and Gynecology
NMC Specialty Hospital
Dubai, UAE

Soni Bharti MBBS MD DNB
Senior Resident
Department of Obstetrics and Gynecology
All India Institute of Medical Sciences
New Delhi, India

Soniya Dhiman MBBS MD
Assistant Professor
Department of Obstetrics and Gynecology
All India Institute of Medical Sciences
New Delhi, India

Soumitra Barik DMRT DNB
Consultant
Department of Radiation Oncology
Bagchi Sri Shankara Cancer Centre and Research Institute
Bhubaneswar, Odisha, India

Subash Mallya
MBBS DGO DNB MNAMS Diploma in Endoscopy (Germany)
Consultant
Department of Obstetrics and Gynecology
AKG Memorial Cooperative Hospital
Kannur, Kerala, India

Suparna Grover MBBS MD
Professor
Department of Obstetrics and Gynecology
Government Medical College
Amritsar, Punjab, India

Supriya Kumari MD
Resident Doctor
Department of Obstetrics and Gynecology
Postgraduate Institute Medical Education and Research
Chandigarh, India

Surekha Tayade MBBS MD (Obs & Gyne) DNB (Obs & Gyne) MNAMS FICOG FAIMER Fellow
Professor
Department of Obstetrics and Gynecology
Mahatma Gandhi Institute of Medical Sciences
Wardha, Maharashtra, India

Sushila Saini MBBS MS (Obs & Gyne) DEGS (Kiel University Germany) FMAS (USA) FICMCH
Managing Director
Jaipur Doorbeen Hospital
Jaipur, Rajasthan, India

Suvarna Satish Khadilkar MBBS MD DGO FICOG CIMP PG Diploma in Endocrinology
Consultant Endocrinologist and Gynecologist
Professor and Head
Department of Obstetrics and Gynecology
Bombay Hospital Institute of Medical Sciences
and Medical Research Centre
Mumbai, Maharashtra, India
Treasurer, FOGSI (2018–2021)
Editor-in-Chief, Journal of Obs/Gyne of India (2018–2021)
Deputy Secretary General, FOGSI (2022–2023)

Swarupa Mitra MBBS MD (Radiation Oncology)
Director (Radiation Oncology)
Fortis Hospital
Gurugram, Haryana, India

Sweety Soni MBBS MD MRCOG (UK) DRANZCOG (Australia)
Consultant
Department of Obstetrics and Gynecology
Saket Hospital
Jaipur, Rajasthan, India

Tanu Priya MBBS MD
Senior Resident
Department of Obstetrics and Gynecology
Dr Rajender Parshad Medical College
Kangra, Himachal Pradesh, India

Tapasya Dhar MBBS MD
Professor
Department of Obstetrics and Gynecology
Christian Medical College and Hospital
Ludhiana, Punjab, India

Urvashi Miglani MBBS MD FICOG FMCH
Consultant
Sai Niwas Chain of Preventive Oncology Centers
PAN India

Vandana Mohan Fellow Diploma Clinical ART
Fellow Diploma Clinical ART
Southend Fertility and IVF Centre
New Delhi, India

Vandana R Bhuriya MBBS MD
Professor
Department of Obstetrics and Gynecology
Pandit Bhagwat Dayal Sharma Postgraduate Institute
of Medical Sciences
Rohtak, Haryana, India

Venus Bansal MBBS MS (Obs & Gyne) FICOG FICMCH
Senior Consultant
Department of Obstetrics and Gynecology
Clio Mother and Child Institute
Ludhiana, Punjab, India

Versha Dhama MBBS MS
Senior Resident
Department of Obstetrics and Gynecology
Vardhman Mahavir Medical College and Safdarjung Hospital
New Delhi, India

Vinay K Dogra MBBS MD (Medicine) DM (Endocrinology)
Assistant Professor
Department of Endocrinology
Dr Rajendra Prasad Government Medical College
Kangra, Himachal Pradesh, India

Vivek Kaushal MBBS MD (Anesthesia)
Senior Resident
Department of Anesthesia
Indira Gandhi Medical College
Shimla, Himachal Pradesh, India

Yashodhara Pradeep MBBS MD
Professor and Head (Retd)
Department of Obstetrics and Gynecology
Dr Ram Manohar Lohia Institute of Medical Sciences
Lucknow, Uttar Pradesh, India

Yogita Dogra MBBS MD (Obs & Gyne) DM (Reproductive Medicine) FICOG
Senior Consultant and Director
Arriva IVF Superspecialty Centre
Shimla, Himachal Pradesh, India

Tapasya Dhar MBBS MD
Professor
Department of Obstetrics and Gynecology
Christian Medical College and Hospital
[illegible] Punjab, India

[illegible]
Pandit Bhagwat Dayal Sharma Postgraduate Institute of Medical Sciences
Rohtak, Haryana, India

Venus Bansal MBBS MD [illegible]
Senior Consultant
Department of Obstetrics and Gynecology
Clio Mother and Child Institute
Ludhiana, Punjab, India

Vertika Dhama MBBS MS
Senior Resident
Department of Obstetrics and Gynecology
Vardhman Mahavir Medical College and Safdarjung Hospital
New Delhi, India

Vinay K Dogra MBBS MD (Medicine) DM (Endocrinology)
Assistant Professor
Department of Endocrinology
Dr Rajendra Prasad Government Medical College
Kangra, Himachal Pradesh, India

Vivek Kaushal MBBS MD [illegible]
Senior Resident
Department of Anesthesia
Indira Gandhi Medical College
Shimla, Himachal Pradesh, India

Yashodhara Pradeep MBBS MD
Professor and Head (Retd)
Department of Obstetrics and Gynecology
Dr Ram Manohar Lohia Institute of Medical Sciences
Lucknow, Uttar Pradesh, India

[illegible]
Shimla, Himachal Pradesh, India

Suparna Grover MBBS MD
Professor
Department of Obstetrics and Gynecology
[illegible]
[illegible] India

[illegible]

Sushila Saini MBBS MS (Obst & Gynae) [illegible]
FMAS [illegible]
Managing Director
[illegible]
Jaipur, Rajasthan, India

Suvarna Satish Khadilkar MBBS MD DGO DFP FICOG CIMP PG [illegible]
[illegible]
Consultant Endocrinologist and Gynecologist
Professor and Head
Department of Obstetrics and Gynecology
Bombay Hospital Institute of Medical Sciences
and Medical Research Centre
Mumbai, Maharashtra, India
Treasurer, FOGSI (2016–2021)
Editor-in-Chief, Journal of Obstetrics and Gynecology of India (2015–2021)
Deputy Secretary General, FOGSI (2012–2013)

Swarupa Mitra MBBS MD (Radiation Oncology)
Director (Radiation Oncology)
Fortis Hospital
Gurugram, Haryana, India

Sweety Soni MBBS MD [illegible]
Consultant
Department of Obstetrics and Gynecology
Soni Hospital
Jaipur, Rajasthan, India

Tanvi Rajput MBBS [illegible]
[illegible]
[illegible] Medical College
[illegible] Pradesh, India

Foreword

I am very pleased to write the foreword for this new book, *Gynecological Algorithms in Clinical Practice*.

The book covers a wide spectrum of gynecological topics in a very succinct manner. The line diagrams, tables, flowcharts, and color photographs make it easy to understand the condition and its management. The book itself is styled in such a manner that makes it easy to read, understand and follow. Further recommendations for reading make it possible to gain further understanding if needed. The book will be useful for trainees as well as practicing clinicians.

I commend the efforts of all the contributors to this book, which provides us with an evidence-based approach to the management of gynecological problems in practice. I am very happy to congratulate Dr Alok Sharma for his efforts and this contribution to women's health literature.

Warm regards,

Jaydeep Tank
MD DNB DGO FCPS FICOG FRCOG (Hon. Causa)
Obstetrician, Gynecologist, and IVF Consultant
Ashwini Maternity and Surgical Hospital
Mumbai, Maharashtra, India
President, FOGSI (2024–2025)

Foreword

It gives me great pleasure to introduce this comprehensive guide on *Gynecological Algorithms in Clinical Practice*, edited by the esteemed Dr Alok Sharma. This book is an invaluable resource for gynecologists, addressing the full spectrum of gynecological issues from adolescence to menopause.

Dr Alok Sharma and his contributors have meticulously covered a broad range of topics, from infertility and gynecologic oncology to the latest advancements in our field. The inclusion of sections on stress urinary incontinence and pelvic organ prolapse is particularly noteworthy, as these are common yet often challenging conditions to manage.

In addition, the book delves into emerging areas of interest such as cosmetic gynecology, violence against women, and sexual assault. These chapters are crucial in providing comprehensive care and support to our patients. The attention to detail extends to critical aspects like blood transfusion safety, ensuring that all practices adhere to the highest standards of care.

This book's blend of foundational knowledge and cutting-edge advancements makes it a practical tool for daily practice. I am confident that it will serve as an essential reference for both seasoned practitioners and those new to the field.

I commend Dr Sharma and his team for their dedication to advancing gynecological practice and for providing such a valuable resource to our community.

Sincerely,

Sunita Tandulwadkar
MD (Obstetrics and Gynecology) FICS (Gyne-Endoscopy) FICOG
Gynecological Endoscopist and ART IVF Specialist
Chief and Medical Director
Department of Obstetrics and Gynecology, Ruby Hall Clinic
Pune, Maharashtra, India
Chief, Ruby Hall IVF and Endoscopy Center, Pune
Director, Solo Clinic, Center of Excellence Infertility and Endoscopy
Founder and Medical Advisor
Solo Stem Cells, Stem Cells Research and Application Center, Pune
Co-Founder, Solo Research Foundation—Sponsor a Birth
President, IAGE (2019–2020)
Chairperson, Maharashtra Chapter ISAR (2018–2020)

Foreword

Dear Colleagues, Warm greetings!!

It gives me immense pleasure to present to you this book on *Gynecological Algorithms in Clinical Practice*. This book is a crisp compilation of chapters on all clinically relevant topics related to the field of Gynecology by none other than stalwarts in the field of Obstetrics and Gynecology.

It is estimated that approximately 1.2 billion women worldwide live in less-resourced locations and many of these women have little or no access to good-quality gynecological care. For such women there is, therefore, a significant need for evidence-based, cost-effective medical care. This carefully structured textbook seeks to support various levels of health personnel (e.g., junior gynecologists, general doctors, midwives, clinical officers, nurses at MCH clinics, etc.) in less-resourced locations by providing evidence-based information about low-tech diagnostic and treatment options for common gynecological conditions such Puberty and its Aberrations, Breast in Gynecology Neuroendocrinology, Male and Female Infertility, Obstetrical Complications, Gyne-oncology with Radiations Therapy and Chemotherapy, Laparoscopy and Robotics and many more related topics...... FOGSI and its parent bodies FIGO, AOFOG, and SAFOG have always played a vital role in spreading the knowledge, both amongst doctors and patients. Last year, as FOGSI President, my social responsibility theme was defined by Change (Badlaav in Hindi) which included the three objectives of Integration of thought and action (Ekikaran), Equality of treatment irrespective of economic status (Samanata) and Technology to achieve these objectives (Takniki). I continue to propound this philosophy for the next two years at the international level as the Trustee of FIGO for the Asia Oceania region. I am inspired to do this under the visionary leadership of our FIGO President, Anne Kihara and FIGO President Elect, Frank Louwen. In fact, FIGO has initiated a new Committee on Breast Health from this year, underlying the significance of breast wellbeing in the modern woman's life. This book is a step towards our collective goal of improving women's health in the World, by providing updated information about the relevant topics in Women's Health. It will be a ready reckoner for both the students and clinicians to update their knowledge on evidence-based management. I appreciate the efforts of all the authors for their efforts to write, collate, edit, and publish this book. I sincerely hope that this book will benefit and empower you and your patients. Wish you all happy reading.

Sincerely,

Hrishikesh D Pai
MD FRCOG (UK-HON) FCPS FICOG MSc (USA)
Founder and Medical Director
Bloom IVF Group
Consultant Gynecologist
Lilavati Hospital IVF Center
Mumbai, Maharashtra, India
Professor of Reproductive Medicine
DY Patil University, Navi Mumbai
Past President FOGSI (2022–2023)
Trustee, FIGO Asia Oceania (2023–2025)

Preface

I have welcomed, with enthusiasm, the opportunity to edit this very first edition of the textbook *Gynecological Algorithms in Clinical Practice*. I have tried to provide a text to help undergraduates, postgraduates, and practitioners to clarify their doubts and to organize some of the intricacies regarding the subject. I have assembled a group of contributors, nationally with an exceptionally broad range of backgrounds and similar interests. All the contributors have emphasized the latest clinical management approaches with deep understanding. I hope that the reader will enjoy being stimulated by the fascination and intellectual stimulation that comes from the study of gynecology. I wish to thank all the contributors who have submitted chapters for the reader's pleasure, to impart their knowledge and to hopefully see this translated into high-quality clinical practice in a ready reckoner form. I would like to thank all the contributors whose contributions span the breadth of the field of Gynecology, and we hope that these individual chapters will stimulate the reader to a greater understanding and thereafter a healthy appetite for more knowledge. They have all contributed to the same goal, that is, to create a source of information that will help in the care of women from birth to menopause, including puberty, reproductive age group and perimenopause in algorithm form, to have a quick look in your outpatient department, wards and operation theater. I have tried to include problems of puberty, principles of radiotherapy and chemotherapy in gynecological cancers, basics of infertility, laparoscopy and hysteroscopy surgeries in gynecology.

It is my level best effort to cover all the relevant aspects pertaining to these phases of women's life where gynecologists may need guidance in decision making. I hope that the aggregate of my efforts has resulted in a text that will be a worthy resource for those who want to care for women.

Alok Sharma

Acknowledgments

Giving thanks is a pleasant job, but it is, nonetheless, difficult when one sincerely tries to put such ideas into words. The following humble words of expression and gratitude cannot really convey the deep feelings of my heart. In any attempt to create and produce a textbook of obstetrics and gynecology, one must be fortunate enough to have the assistance and support of many talented professionals, both within and outside the obstetrics and gynecology department. To begin with, I am indebted to all my authors for their generous contributions to this book; all of them, despite their busy schedules, provided recent, up-to-date, and evidence-based chapters on various aspects in the field of gynecology.

I have selected contributors from different parts of India.

It has been a pleasure to work with the dedicated professionals from M/s Jaypee Brothers Medical Publishers (P) Ltd, New Delhi, India. This publisher has been gracious in offering support without any interference whatsoever, and the team has also ensured that the quality of work is superb.

I am deeply indebted to Ms Chetna Malhotra (Senior Director—Professional Publishing, Marketing and Business Development) for her unconditional support; she has brought her considerable intelligence, energetic work ethic, and creativity to this edition. Her dedication for creating the best textbook possible equaled my efforts to produce an appropriate style for the textbook.

This textbook would have not seen the light of the day without the untiring efforts of Priyansh Saxena (Development Editor) and Manpreet Kaur (Development Editor), who skillfully kept my project on track through an array of potential hurdles.

I acknowledge my respected parents, Smt Dhanwanti Devi Sharma and Hans Raj Sharma, who have laid the foundation stone of literacy in me and given me the courage to face the challenges of the world by inculcating good attributes in me. I thank my wife, Dr Pratibha Sharma, for her tolerance and understanding of the time I spent away from her during my own career and in editing this book. Her constant encouragement, moral support, and love have been a source of inspiration to me for completing this work. I thank my brother, Dr Amit Sharma, for guiding me at every step of the way in shaping the book. Finally, but certainly not last, I thank my lovely daughter, Hiranya, for her immense patience.

With love and warm regards

Jai Hind

Alok Sharma

Contents

Section 8: Gynecology Oncology

Section 9: Endoscopy/Laparoscopy

Section 10: Surgery for Obstetrical Complications

Section 11: Miscellaneous

Section

1 General Gynecology

Chapter 1

History Taking and Examination

Shail Kaur, Dilpreet Kaur Pandher

Interaction with any doctor is an anxiety-generating event, especially so in case of sensitive matters such as gynecology which therefore demands:

- Highest standards of ethical and professional behavior
- Strong rapport (make her as comfortable as possible) and empathic listening
- Respect for privacy
- A thorough assessment of underlying medical problems
- Observation of non-verbal cues.

INTRODUCTION

A comprehensive history is a keystone to proper treatment with examination and investigations being essential pillars to reach a tentative diagnosis. The general outline would be in the following format:

- Age
- Chief complaints
- History of presenting complaints
- Menstrual history
- Obstetric history
- Contraceptive history
- Sexual history
- Past medical/surgical
- Family history
- Social/occupational history
- History of recreational drug/substance abuse/smoking.

AGE

Disorders may be related to the age of the patient with more emphasis on the menstrual relationship:[1]

- *Childhood/prepubertal age*
  - Foreign body
  - Vaginitis
  - Tumors of ovary/vagina
- *Adolescence/peripubertal age*
  - Menstrual irregularity
  - Bleeding disorders
  - Uterovaginal anomalies
  - Polycystic ovary syndrome (PCOS)
  - Precocious/delayed puberty
  - Germ cell tumors
- *Reproductive age group*
  - Pregnancy and its complications
  - Menstrual disorders
  - Infections
  - Benign lesions of the genital tract
- *Peri-menopausal age group*
  - Menstrual irregularity
  - Menopausal symptoms
  - Benign tumors of genital tract
  - Premalignant/malignant lesions
- *Post-menopausal age*
  - Menopausal symptoms
  - Malignancies of genital tract.

History of Complaints

It is essential to remain empathic and non-judgmental while eliciting the history of gynecological complaints as many times the patients are reluctant in expressing the issue due to embarrassment or they find it to be socially inappropriate. It is essential to provide a relaxed and private setting in order for the patient to fully disclose the issues they are facing.

It is better to start with open-ended questions and then to allow the patient to describe in her own words the complaints she faces. Expressing attention to what she is saying and observing non-verbal clues are imperative:

- Type and site of symptoms
- *Timing*

- Onset
- Duration
- *Cyclical:* Are they associated with menstruation?
- Intermittent or continuous

- Evolution of symptoms
- Aggravating/relieving factors if any
- Previous treatment history
- Relation to menstruation/micturition/defecation.

VAGINAL DISCHARGE

Vaginal discharge is most often a normal and regular occurrence. However, a change in the character of discharge can indicate *infection.* Inquire about color, consistency, amount, and smell.

Vaginal discharge is a prominent symptom of vaginitis but may be difficult to distinguish from normal vaginal discharge. Characteristics of normal vaginal discharge vary by the hormonal status of the patient.

- In reproductive-aged females, normal vaginal discharge consists of 1–4 mL of fluid (per 24 hours), which is white or transparent, thick or thin, and mostly odorless. The pH is typically 4.0–4.5 as measured with pH paper. This physiologic discharge is formed by mucoid endocervical secretions in combination with sloughing epithelial cells, normal vaginal flora, and vaginal transudate.
- *Minimal additional symptoms:* Although normal discharge may be yellowish, slightly malodorous, and accompanied by mild irritative symptoms, it is not accompanied by pruritus, pain, burning or significant irritation, erythema, local erosions, or cervical or vaginal friability. The absence of these signs and symptoms helps to distinguish normal vaginal discharge from discharge related to a pathological process, such as vaginitis or cervicitis and are therefore to be inquired during history.
  - *Premenarche:* Prior to estrogenization, physiologic vaginal discharge ranges from absent or scant. If present, fluid can be clear, white, or mucoid. The vaginal pH is alkaline.
  - *Postmenopause:* The hypoestrogenic state of menopause leads to a reduction of vaginal discharge and increase in vaginal pH (typically >4.5).

ABNORMAL DISCHARGE

Disruption of the normal ecosystem can lead to conditions favorable for development of vaginitis with vaginal discharge. Some of these potentially disruptive factors include sexually transmitted diseases, antibiotics, foreign body, estrogen level, use of hygienic products, pregnancy, sexual activity, and contraceptive choice.[2]

History of change in the volume, color, or odor of vaginal discharge suggest pathology. Ask for:

- Pruritus
- Burning
- Irritation
- Erythema
- Dyspareunia
- Spotting
- Dysuria

ABNORMAL GENITAL TRACT BLEEDING

Abnormalities in Frequency[3]

- *Frequent:* Frequent menstrual bleeding refers to periods that start at intervals <24 days.
- *Infrequent:* Infrequent menstrual bleeding refers to periods that start at intervals >38 days.
- *Absent:* Absence of menses is either primary (absence of menarche by age 15 years) or secondary (absence of spontaneous menstrual bleeding for 6 months in a patient who previously had menstrual bleeding) amenorrhea; these are discussed in detail separately
- *Irregular bleeding:* The definition of irregular bleeding depends on patient age:
  - *18–25 years:* Cycle length variance >9 days
  - *26–41 years:* Cycle length variance >7 days
  - *42–45 years:* Cycle length variance >9 days

  For patients <18 or >45 years, the >9-day definition is also applied, although the evidence defining normal in these groups is less clear.
- *Prolonged menstrual bleeding* is defined as menstrual bleeding consistently lasting >8 days; this is often, but not always, associated with heavy menstrual bleeding (HMB).

  There is no consensus on the lower limit of normal for the duration of menstrual bleeding.

Abnormalities in Volume

- *Heavy:* For clinical purposes, HMB is defined as a volume that interferes with the patient's physical, social, emotional, and/or material quality of life. It is based on the patient's perception of increased daily or total monthly volume of menstrual blood flow, regardless of the duration, frequency, or regularity.
- *Light:* Light menstrual bleeding is uncommon and rarely related to pathology, although it may be a

presenting symptom of cervical stenosis or intrauterine synechiae. Individuals who express concern about light periods may perceive a heavy, red bleed as a sign of good health.

For research purposes, <5 mL is considered "low volume", a metric that can only be assessed quantitatively with methods like the alkaline hematin assay.

- *Intermenstrual bleeding* refers to abnormal uterine bleeding (AUB) that occurs between well-defined cyclical menses. The distinction between bleeding and spotting is based on the patient's need for menstrual product use. Intermenstrual bleeding can also be difficult to distinguish from irregular and/or very frequent menses; thus, care must be taken before applying this term to patients with these other abnormalities.

  Intermenstrual bleeding can be cyclical or acyclical.
- *Cyclical midcycle intermenstrual bleeding:* A small amount of bleeding arising from the endometrium around midcycle occurs in approximately 9% of all reproductive-age females. This is thought to be associated with the midcycle drop in circulating estradiol levels that occurs just after ovulation.
- *Acyclical intermenstrual bleeding:* Intermenstrual bleeding that is not cyclical or predictable is typically associated with nonmalignant lesions, such as chronic cervicitis/endometritis or polyps of the cervix or endometrium or intracavitary uterine fibroids; postcoital bleeding is a frequent symptom. Less commonly, such bleeding can be indicative of pathologic process, such as cervical or endometrial cancer.

Menstrual History and Bleeding Pattern

The following questions can be used to help elicit the patient's menstrual history and bleeding pattern:

- What was the first day of the last menstrual period and several previous menstrual periods?
- For how many days does bleeding continue? How many days of full bleeding and how many days of light bleeding or brown staining does this include?
- Does bleeding occur between menstrual periods?
- If bleeding is irregular, how many bleeding episodes have there been in the past 6–12 months? What is the average time from the first day of one bleeding episode to the next?
- Is the bleeding associated with dysmenorrhea or dyspareunia (i.e., suggestive of endometriosis and/or adenomyosis)?
- How heavy is the bleeding?
- *Is the patient certain that the bleeding is from the vagina?*
  - Does the patient see the blood in the toilet only during or after either urination or defecation?
  - Does the patient see the bleeding only when wiping with toilet tissue? If so, has the patient tried to separately dab the urethra, vagina, and anus with toilet tissue to check the source of the bleeding?
  - If the patient uses a pad, on what part of the pad is blood visible? If the patient uses a tampon, is bleeding visible while a tampon is in the vagina?
- Were there precipitating factors, such as trauma, intercourse, or a procedure? Bleeding related to trauma or intercourse suggests a vaginal or cervical source of bleeding (e.g., cervical dysplasia, cervicitis, vulvovaginal atrophy, and cervical polyp), while bleeding after a procedure [e.g., pregnancy termination, intrauterine device (IUD) placement] may suggest a uterine source of bleeding.
- *Sexual history:* A sexual history helps to determine whether the patient might be pregnant, as pregnancy is a common cause of uterine bleeding; however, a pregnancy test is generally also performed.

A sexual history may also help determine the patient's risk for sexually transmitted infections (e.g., *Chlamydia trachomatis, Neisseria gonorrhoeae,* trichomonas, herpes simplex), which can cause cervicitis and present with cervical bleeding. Patients at risk for sexually transmitted infections should be asked if they have lower abdominal pain, fever, and/or vaginal discharge, all which suggest pelvic infection [e.g., pelvic inflammatory disease (PID), endometritis].

- *Initial evaluation:* The primary goal in the diagnostic evaluation of postmenopausal patients with uterine bleeding is to exclude malignancy since increasing age is a risk factor for endometrial carcinoma **(Table 1)**. A secondary goal is to exclude other intraabdominal processes (e.g., diverticular disease, infection, and other malignancies) and nonuterine sources of bleeding.

History in patients presenting with post-menopausal bleeding: Important information that can be elicited from the patient's history includes the following:

TABLE 1: Parameters for evaluation of menstrual cycle.

Parameter	Normal	Abnormal
Frequency	≥24 and ≤38 days	• Absent (no bleeding)—amenorrhea • Frequent (<24 days) • Infrequent (>38 days)
Duration	≤8 days	Prolonged (>8 days)
Regularity	Regular—shortest to longest cycle variation: ≤7–9 days	Irregular—shortest to longest cycle variation: ≥10 days
Flow volume (patient determined)	Patient considers normal	• Patient considers light • Patient considers heavy
Intermenstrual bleeding (bleeding between cyclically regular onset of menses)	None	Random cyclic (predictable): • Early cycle • Mid cycle • Late cycle
Unscheduled bleeding on progestin ± estrogen gonadal steroids (contraceptive pills, rings, patches, IUDs, or injections)	• Not applicable for patients not on gonadal steroid medication • None (for patients on gonadal steroid medication)	Present
Intermenstrual bleeding (bleeding between cyclically regular onset of menses)	None	Random cyclic (predictable): • Early cycle • Mid cycle • Late cycle

(IUDs: intrauterine devices)

- When did the bleeding start?
- Were there precipitating factors, such as trauma, intercourse, or intimate partner violence?
- What is the nature of the bleeding (e.g., duration, quantity)?
- Are there any associated symptoms such as pain, fever, or changes in bladder or bowel function?
- Does the patient take any medications (e.g., hormone replacement therapy, anticoagulants), or herbal or dietary supplements?
- Does the patient have a history of obesity or diabetes mellitus?
- Is there a family history of breast, colon, and endometrial cancer?

The answers to these questions may help direct the clinician toward one of the major categories of uterine bleeding. As an example, obesity, diabetes mellitus, and use of tamoxifen are risk factors for endometrial cancer; vaginal dryness and soreness with dyspareunia and bleeding after intercourse suggest atrophy. By contrast, a history of trauma, sexual assault, or intimate partner violence, which is prevalent in postmenopausal and vulnerable (e.g., nursing home) populations, may help direct the clinician toward a nonuterine source of bleeding.

PELVIC PAIN

Use the acronym *SOCRATES* to elicit the key features of any abdominal or pelvic pain:

- **S**ite
- **O**nset
- **C**haracter
- **R**adiation
- **A**ssociations
- **T**ime course
- **E**xacerbating and relieving factors
- **S**everity

Pelvic pain that has lasted <3 months is considered as acute pelvic pain.

Chronic pelvic pain is non-cyclic pain perceived in pelvic area that has persisted for >3 months and is unrelated to pregnancy.

Pain limited to menstrual phase is defined as dysmenorrhea:

- If underlying anatomic cause is not identifiable it is termed primary dysmenorrhea.
- If underlying anatomic cause is identifiable it is termed secondary dysmenorrhea.

FEMALE SEXUAL PAIN

Pain that occurs just before, during, or after sexual intercourse or other sexual activities that involve the clitoris, vulva, vagina, and/or perineum is called female sexual pain (FSP).

- Terms that are commonly used when discussing FSP include:[4]
  - *Apareunia:* Inability to achieve vaginal insertion
  - *Dyspareunia:* Recurrent or persistent pain with sexual intercourse that causes distress.
  - *Superficial dyspareunia:* Dyspareunia limited to the vulva or vaginal entrance
  - *Deep dyspareunia:* Dyspareunia associated with deep penetration that manifests in the deeper parts of the vagina or lower pelvis
  - *Primary dyspareunia:* Pain with vaginal insertion has occurred since onset of coitarche
  - *Secondary dyspareunia:* Pain begins after some time of pain-free sexual activity.
  - *Genito-pelvic pain/penetration disorder (GPPPD):* An all-encompassing term for FSP as defined by the American Psychiatric Association.

Gynecological disorders often have urinary or bowel manifestations due to proximity:

- Urinary symptoms—prolapse, myomas and pelvic masses, endometriosis
- Constipation with enteroceles
- Diarrhea and tenesmus in pelvic abscess, endometriosis.

INFERTILITY

- Age at marriage
- Duration of marriage
- Whether staying together regularly
- Incidents of separation-due to occupational/societal activities, marital breakdown.

MASS PER ABDOMEN

Symptoms in the history may suggest a cause for the pelvic mass:

- *Vaginal bleeding and pelvic pain with a positive pregnancy test:* Ectopic pregnancy, another obstetric complication, or, rarely, gestational trophoblastic disease
- *Dysmenorrhea:* Endometriosis, adenomyosis, or uterine fibroids
- *Fever and pelvic pain:* Tubo-ovarian abscess
- *Postmenopausal bleeding:* Increased suspicion of uterine cancer
- *Hormonal effects, such as precocious puberty, virilization, or abnormal uterine bleeding:* A masculinizing or feminizing ovarian tumor.

Size of Mass

- Duration
- First noticed at which site
- Any change in size
- Association with pain/fever
- Presence of any other lumps
- Any loss of weight
- Any secondary change
- Anorexia
- Features of pressure effects (bowel/bladder).

MENSTRUAL HISTORY

- Age at menarche
- Cycle length
- Cycle duration
- *Blood loss volume:*
  - Number of pads/tampons used per day
  - Whether pads are fully soaked
  - Passage of clots

A menstrual history is a key part of any gynecological assessment. It has the following components:[3]

- Frequency—average 28 days
  - <24 days frequent, >38 days infrequent
- Duration—average 5 days
  - >8 days prolonged, <4.5 days shortened
- Volume—average 40 mL menstrual blood loss over course of menses
  - >80 mL heavy (features of anemia), <5 mL light
  - Women may describe "flooding" and clots passed
- Date of last menstrual period (LMP)
  - If late, consider performing a pregnancy test.

Abnormal vaginal bleeding exists in three main forms:

- Intermenstrual (between menstrual periods)—causes include infection, malignancy, fibroids, endometriosis, pregnancy, and hormonal contraception.

- Post-coital (after sexual intercourse)—causes include cervical ectropion, infection, vaginitis, and malignancy.
- Post-menopausal (after the menopause)—causes include malignancy, vaginal atrophy, and use of hormone replacement therapy.

If postmenopausal, record the date of the menopause and any episode of bleeding after that.

OBSTETRIC HISTORY

For pregnancies[3] not carried beyond 24 weeks, inquire about:

- Gestation—miscarriages can be classified into early pregnancy (12 weeks or less) or second trimester (13–24 weeks).
- Miscarriages—outcome (spontaneous, medical management, surgical management—evacuation of retained products of conception).
- Terminations—method of management—medical or surgical.
- Identified causes of miscarriage/stillbirth—e.g., abnormal parental karyotype, fetal anomaly.

For ectopic pregnancies, ask about:

- Site of the ectopic
- Management—expectant [monitoring of serum human chorionic gonadotropin (hCG) levels], medical (methotrexate injection), surgical (laparoscopy or laparotomy; salpingectomy (removal of tube) or -otomy (cutting of tube and suctioning of trophoblastic tissue)

For each previous pregnancy carried beyond 24 weeks, inquire about the following:

- Gestation—previous preterm labor is a risk factor for subsequent preterm labor
- Mode of delivery—spontaneous vaginal, assisted vaginal or cesarean
- Gender
- Birth weight—a previous small for gestational age (SGA) baby increases the risk of a subsequent one
- Complications—e.g., preeclampsia, gestational hypertension, gestational diabetes, obstetric anal sphincter injury (3rd, 4th degree tears), postpartum hemorrhage
- Assisted reproductive therapies (ART)—e.g., ovulation induction with clomiphene, in vitro fertilization (IVF).

PAST HISTORY

- Cervical smear—ascertain the date of the last smear, its result, and any treatment arising
- Surgical history—particularly any pelvic or abdominal surgery
- Previous gynecological problems
- Previous sexually transmitted infection.

History of obstetric or gynecologic surgery—a history of obstetric or gynecologic surgery may suggest an underlying disorder that can be associated with AUB.

- A prior cesarean birth can result in a cesarean scar defect (particularly if a patient has had multiple cesarean births).
- Past myomectomy suggests the possibility of recurrent or persistent uterine fibroids.
- Past excisional cervical procedures performed for diagnosis or management of cervical intraepithelial neoplasia suggest the possibility of recurrent or persistent cervical neoplasia.
- Rarely, uterine surgery can result in an arteriovenous malformation (also referred to as enhanced myometrial vascularity).

Many prescribed (and over the counter) medications can have an effect on gynecological health. Inquire about the use of the following:

- *Contraception:*
    - Type and brand name
    - Correct use
    - Previous contraception history
- *Hormone replacement therapy:*
    - Duration of use
    - Cyclical or continuous
    - Combined or estrogen-only
    - Method of delivery
    - Recent antibiotic use—some antibiotics are associated with vaginal candidiasis.
- Any other medications—including over the counter medications
- Known allergies

A *family history* of any of the following is particularly relevant in a gynecological history:

- Breast/ovarian cancer/endometrial cancer—can be familial [e.g., breast cancer (BRCA) 1/2 gene].
- Diabetes—associated with some reproductive abnormalities.
- Bleeding disorders—can be associated with heavy menstrual bleeding.

The *social/personal history* is an important part of a gynecological history. In many cases, the patient's lifestyle can influence disease and/or symptomatology:

- Weight:
  - Rapid weight loss/gain can cause oligo/amenorrhea.
  - Obesity can cause menstrual changes and increase the risk of endometrial cancer.
- Occupation—some industrial exposures are risk factors for gynecological disease.
- Smoking and alcohol intake
- Diet and exercise

Gynecological Examination

- Introduce yourself to the patient.
- Wash your hands.
- Obtain verbal consent after carefully explaining to the patient what the examination involves and why it is necessary.

For example: *"I will be passing a speculum, which is a plastic/metal instrument, through the vagina to visualize the neck of the womb".*

- Reassure them that this should not be painful, but you will stop immediately if it becomes too uncomfortable.
- Request a chaperone in c/o male doctors especially.
- The patient ought to have an empty bladder, as this can make the examination less uncomfortable.
- Ask the patient to remove all clothing from the waist down and any sanitary protection.
- Cover with sheet when appropriate.
- Prepare your equipment—gloves, lubricant, speculum (for example, Cusco's speculum) +/– smear, and swabs.

General Examination

- Pallor is a common finding in women of reproductive age group but me pronounced in patients with a history of heavy menstrual bleeding.
- Lymphadenopathy—chronic infections, tuberculosis, and metastasis of malignancy
- Edema
  - Non-pitting—hypothyroidism, filariasis
  - Unilateral pitting—malignant growth involving the lymphatics
  - Bilateral pitting—large abdominal tumors
- Hirsutism in patients with PCOS, virilizing tumors, danazol (Ferriman Gallwey system)
- Breast secretion (galactorrhea) in hyperprolactinemia.

Abdominal Examination

- *Inspection*
  - Visible mass
  - Distension
  - Eversion of the umbilicus with raised intra-abdominal pressure
  - Scars of any previous surgery
  - Striae gravidarum and linea nigra
  - Movement of abdominal wall:
    - Separate from underlying tumors causing an apparent change with respiration.
    - Restricted movement suprapubically in cases associated with pelvic peritonitis.
- *Palpation*
  - Stand on the right side:
    - Use the ulnar border of hand.
    - Start from hypochondrium and progress downward.
      - In patients with pain or tenderness it is better to start from a non-painful area.
  - Palpate the abdomen for masses and tenderness gently:
    - Hepato-splenomegaly
    - Abdominal mass
      - Site (which quadrant)
      - Extent
      - Size
        - Pelvic mass described in terms of size of gravid uterus in weeks, e.g., 16 weeks size
      - Margins
      - Lower border palpable?
        - Lower margin cannot be felt suggests pelvic origin
      - Consistency
      - Mobility
        - Pelvic masses usually have side to side mobility but up-down mobility may be seen in pedunculated masses
    - Smooth, firm suggestive of myomas
    - Cystic, fluctuant suggestive of ovarian mass
    - Hard, irregular with restricted mobility suggestive of a malignancy
    - Percussion
      - Dullness in flanks which shifts on turning the patient to the lateral position aka shifting dullness is seen with ascites and hemoperitoneum

    - Central dullness which does not shift, and resonant flanks are seen with large ovarian cysts
  - Auscultation
    - Absent bowel sounds with peritonitis
    - Increased sounds with intestinal obstruction
    - Bruit over highly vascular tumors
- Palpate the groin for inguinal lymphadenopathy.

EXTERNAL EXAMINATION

The patient should be laid on their back, with legs bent at the hip (feet toward their buttocks) and asked to flop their knees apart.

- Put on a pair of gloves.
- *Inspect the external genitalia for:*
  - Deficiency associated with childbirth
  - Abnormal secondary sexual characteristics—hair distribution, clitoromegaly
  - Skin abnormalities—lesions, warts, erythema
  - Discharge—color, consistency
  - Bleeding
  - Swellings of the vulva—tumors, cysts (sebaceous, Bartholin's)
- Ask the patient to cough or strain to observe any incontinence or prolapse.
- Palpate the labia majora with the index finger and thumb for any swellings.

SPECULUM EXAMINATION (CUSCO)

- Lubricate the speculum and warn the patient.
- Part the labia using your left hand.
- *Gently insert the speculum with your right hand:*
  - Fully insert the speculum with the screw facing sideways and the blades vertical.
  - Rotate 90° during insertion so the screw faces upward and the blades become horizontal.
- Slowly open the blades and use light to inspect the cervix.
  - Tighten the screw to hold open the speculum so you can use your right hand for swabs or pipelle biopsy if necessary.
- *Look for:*
  - Abnormal discharge
  - Erosions
  - Ulcerations
  - Growths
  - Inflammation
  - Bleeding
  - Polyps
  - Ectropion
- At this point swabs/endometrial biopsy should be taken if required.
- To remove the speculum, undo the screw to allow the blades to close (leave open slightly to not pinch the vaginal walls), rotating back 90° and gently remove.

A Sim's speculum can be used to assess prolapse:

- Insert the blade of the speculum along the posterior wall of the vagina to hold it back.
- Ask the woman to cough while looking for uterine descent and cystocele.
- Repeat while holding back the anterior wall, looking for rectocele/enterocoele.

Bimanual pelvic examination (abdominopelvic examination):

- Index and middle finger of gloved dominant hand introduced into vagina after separating the labia with the nondominant hand.
- Nondominant hand is then placed on the abdomen.
- *Palpation*
  - Cervix-hypertrophy, mass, tenderness on motion
  - Fingers shifted to posterior fornix and uterus lifted upward and palpated between the two hands
    - Anteverted/retroverted/midposition
    - Size-normal/enlarged
    - Shape
      - Uniformly enlarged
      - Adenomyosis
      - Endometrial polyps
      - Pyo/hematometra
      - Pregnancy
      - CA (cancer) endometrium
    - Irregularly enlarged with myomas
    - Consistency
      - Soft-pregnancy, pyo/hemato-metra
      - Firm-adenomyosis, myoma
      - Hard-malignancy, calcified myomas
    - Mobility-restricted in PID, endometriosis, malignancy
    - Tenderness
  - Palpation of adnexa
    - Palpable/not palpable
    - If palpable,
      - Size (normal/enlarged in terms of weeks gravid size)
      - Shape
      - Side/bilateral
      - Consistency

    - Mobility
    - Tenderness
    - Attachment to uterus (presence of groove between mass and uterus)
- Posterior fornix
  - Soft bulge-fluid/pus/blood in pouch of Douglas
  - Nodules-endometriosis, metastasis
  - Fixed ovarian masses
- Rectal-vaginal examination (middle finger in rectum and index finger in vagina)
  - Recto-vaginal septum
    - Induration in endometriosis/PID
    - Tender nodules
    - Parametrial infiltration

REFERENCES

1. Seshadri L. Gynaecological History and Physical Examination. Essentials of Gynaecology. Walters Kluwer; 2017. pp. 56-75.
2. Motchell H. Vaginal discharge-causes, diagnosis and treatment: BMJ. 2004;328(7451):1306-8.
3. Munro MG, Critchley HOD, Fraser IS, FIGO Menstrual Disorders Committee. The two FIGO systems for Normal and abnormal uterine bleeding symptoms and classification of causes of abnormal uterine bleeding in the reproductive years: 2018 revision. Int J Gynaecol Obstet. 2018;143(3):393-408.
4. Sorensen J, Bautista KE, Lamvu G, Feranec J. Evaluation and treatment of female sexual pain: a clinical review. Cureus. 2018;10(3):e2379.

Chapter 2

Ultrasound in Gynecology

Jyotsna Gupta, Sarabjeet Singh

INTRODUCTION

The invention and the first clinical application of an ultrasound were done in *1956 in Glasgow* by *Ian Donald—the obstetrician* and *Tim Brown—the engineer.*[1] In the last six decades of its existence, field of ultrasonography (USG) has grown in leaps and bounds. The trust and accuracy of the ultrasound has established itself as a first diagnostic modality in almost all specialties of the medical science so that it is simply impossible to imagine a medical science without ultrasound. The role of ultrasound in obstetrics and gynecology is immense and well-known and it has served as a savior in various issues related to the field thereby improving on the success rates of the treatment and decreasing morbidity and mortality massively.

BASICS OF THE ULTRASOUND

The basic origin of ultrasound or ultrasonic waves is from *the sound waves.* The human ear can hear the sound waves between the range of *20 and 20,000 Hz* (Hz = hertz or wave cycle per second). Any wave <20 Hz *(infrasonic)* or >20,000 Hz *(ultrasonic or ultrasound)* cannot be heard by human ear. Typically an ultrasound uses a frequency of 2–10 MHz (megahertz: million hertz: way above the threshold of human ear).[2] The ultrasound waves are typically generated from the piezoelectric crystals [or nowadays by synthetic lead zirconate titanate (PZT)] lying within the ultrasound probes or transducers. These waves cannot travel in the vacuum and need a medium to pass therefore a watery gel is applied so as to facilitate the movement of these waves from the transducer to the body. The ultrasonic waves which penetrate into the body are ultimately reflected back by the body tissues. The change in frequency of the returning waves is decided by the consistency of the tissues. These waves are interpreted finally by the machine to make an impression in the form of an image. The tissues like bone or air which are strong reflectors of the waves are interpreted as hyperechogenic images whereas tissues with poor reflection of the waves like fluid or soft tissues will form a hypoechoic image. It must be emphasized over here that ultrasound waves travel longitudinally and hence for a good image the angle of the transducer should be kept perpendicular to the image of the interest. Various modes used in ultrasound are described in **Table 1** and ultrasound echogenicity seen in various tissues is described in **Table 2.**

TABLE 1: Various modes used in ultrasound **(Figs. 1A to D).**[3-5]

Mode	*Clinical application*
A-mode (amplitude mode)	• Historical and basic mode where a graph showing X and Y axis representing depth of the tissue and amplitude of the returning beam, respectively • No longer used. Earlier used in obstetrics and gynecology to measure biparietal diameter (BPD)
B-mode (brightness mode)	• Real-time imaging mode where the change in frequency of the returning waves is decided by the consistency of the tissues • Most important mode used widely in obstetrics and gynecology and other specialty ultrasounds
M-mode (motion mode)	Specifically used in tissues in motion like fetal heart to document the viability and structural and rhythm issues like congenital heart disease or arrhythmias
Color and spectral (pulsed) Doppler mode	• Uses the Doppler effect where a relative change in frequency shift is perceived by the observer and interpreted to understand the flow of the blood to the concerned tissue • Used in conjunction with real-time B-mode in calculating blood flows in uterine artery, umbilical, middle cerebral and ductus venosus arteries, paraovarian, and endometrial flows and many more

Ultrasound has long been debated about the *safety of its radiation particularly in obstetrics*. Numerous studies have been conducted based on the *thermal effects* which raise the tissue temperature *[measured as thermal index (TI)]* and mechanical effects causing tissue cavitation [measured as mechanical index (MI)].[9,10] The concern is especially raised in first trimester when the fetus is developing. However, the general consensus is that ultrasound is a safe modality and no significant effect of its radiation is seen unless the time of exposure is *>60 minutes*. It is advised to all the sonologist to follow as low as reasonably achievable (*ALARA*) *principal* and keep both TI and MI values below 1 for all obstetrical ultrasounds. The risk of damage is *lowest in B mode* and *higher with color Doppler and 3D/4D* imaging with highest exposure in spectral Doppler. These advanced features should be applied only when needed and should preferably be avoided in first trimester.[11,12]

TABLE 2: Echogenic nature of various body tissues.[6-8]

Tissue	*Echogenicity*	*Tissue*	*Echogenicity*
Bone	Hyper-echogenic	Brain	Slight echogenic
Stomach	Anechoic	Intestines	Echogenic
Lung	Echogenic	Liver	Slight echogenic
Kidneys	Echogenic	Bladder	Anechoic
Placenta	Echogenic	Amniotic fluid	Anechoic
Fibroids	Mostly hypoechoic but can be isoechoic or hyperechoic, especially with calcifications or degeneration		

ULTRASOUND IN GYNECOLOGY: THE ROUTES AND TECHNIQUE (FIGS. 2A TO D)[13]

Routes

- Transvaginal is the most preferred route for gynecological scans.
- Transabdominal route is second preferred modality.
- Transperineal and transrectal routes can be chosen according to special indications.

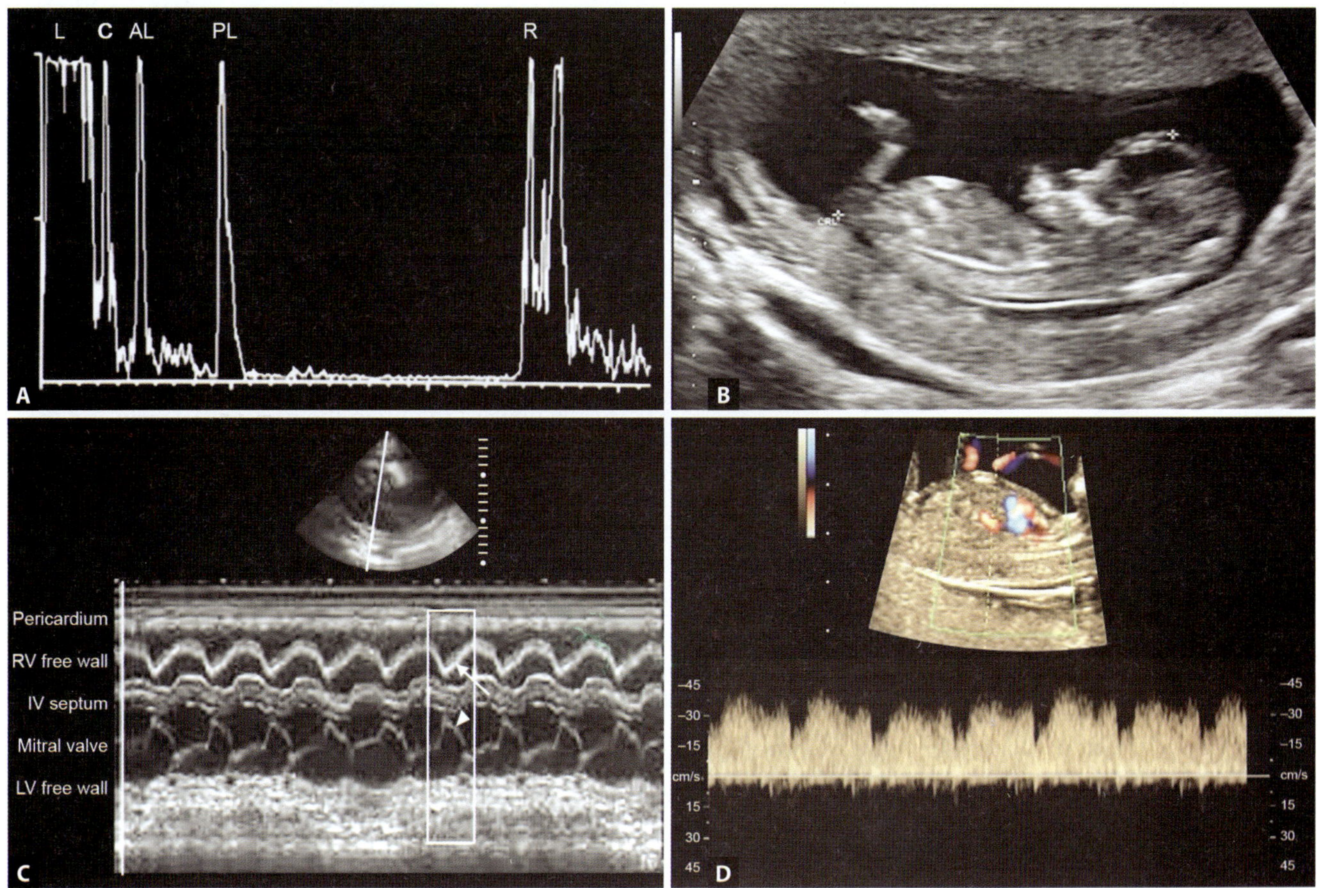

Figs. 1A to D: (A) A-mode; (B) B-mode; (C) M-mode; (D) Color and spectral mode.

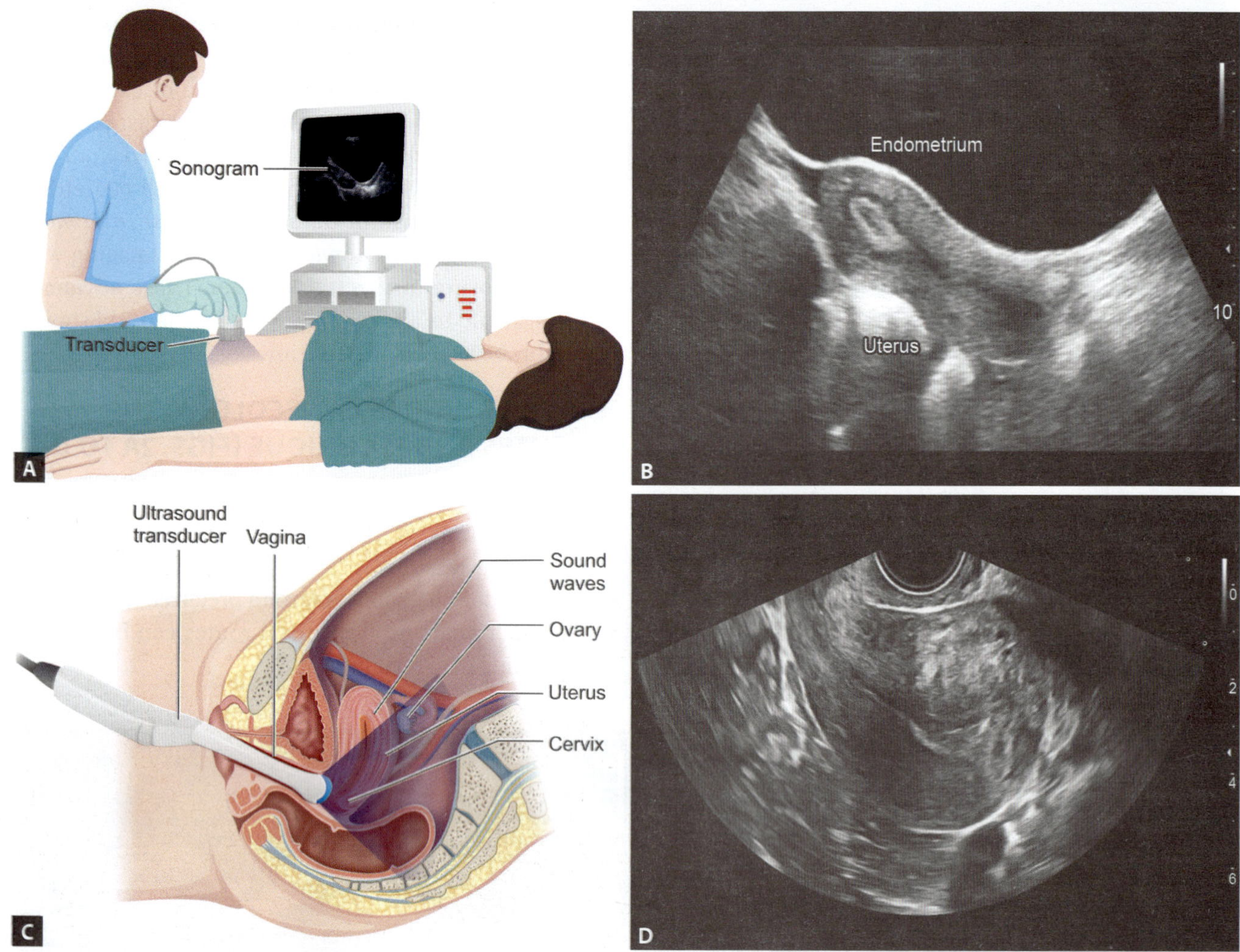

Figs. 2A to D: (A and B) Transabdominal scan; (C and D) Transvaginal scan.

Patient Position

Transvaginal scan requires the patient to lie down in dorsal lithotomy position with empty bladder and perineum at the edge of the table.

Transabdominal scan requires patient to lie down supine with full bladder to displace the bowel.

Technique

In the transvaginal route, the probe is positioned in the upper fornix and midsagittal view of the uterus is obtained showing cervix to fundus along with entire endometrium. Then the probe is moved right and left to look for respective ovaries and adnexa.

Transvaginal probe can be maneuvered in right, left, up, and down positions only.

In transabdominal route also the probe is positioned vertically upward to obtain the midsagittal view and then rocked right and left accordingly.

ULTRASOUND OVARY, ENDOMETRIUM, AND FALLOPIAN TUBE—FROM FERTILITY PERSPECTIVE

Assessment of reproductive function of women is an important part of fertility evaluation in couples suffering from subfertility and pelvis ultrasound (especially transvaginal route) is typically used in this regard.

The ultrasound evaluation of women for fertility workup includes assessment of ovarian reserve, development of follicle(s) in natural or stimulated cycles, rupture and formation of adequate corpus luteum (CL), and development of endometrium in various phases of menstrual cycle in synchronous with developing follicle along with associated gynecological pathologies like myomas.

ASSESSMENT OF OVARIES—FROM FERTILITY PROSPECTIVE (OVARIAN RESERVE AND FOLLICLE MONITORING)

Ovarian reserve testing (ORT) is a subjective evaluation of the ovary with an aim to determine the "quality" and "quantity" of the oocytes that it possesses. It does not necessary means the reproductive potential of the women.

It is important to understand that a recruitment of follicles begin from the late luteal phase of the previous cycle (from preantral follicles which cannot be seen on ultrasound). As the menstruation begins, they become antral follicles and can be seen on ultrasound (2–8 mm in diameter). It is the antral follicle count (AFC) which forms the basis of any assisted reproductive technology (ART). According to Bologna criteria any AFC <5–7 is considered a poor reserve. Any AFC >12 is recognized as polycystic ovary according to Rotterdam criteria.[14,15]

Antral follicle count in combination with anti-Müllerian hormone (AMH), day 2 follicle-stimulating hormone (FSH) and inhibin B can even be more predictive in ORT.[16]

As the cycle advances, all the recruited follicles start growing under the effect of rising FSH from pituitary. It is generally after 5–7 days. One of these takes dominance [the dominant follicle (DF)] which becomes apparently bigger in size and the other follicles regress. The growth of DF leads to increase in estrogen which is secreted by its granulosa cells. An increase in estradiol (E2) negatively affects FSH and leads to luteinizing hormone (LH) surge which ruptures the follicle to release the oocyte (ovulation) and the collapsed follicle becomes CL cyst. The smooth development of follicle with documentation of ovulation is an important reassuring step in fertility evaluation. Few ultrasound features which help predicting ovulation are:[17]

- Sudden collapse of growing follicle with appearance of free fluid in the pouch of Douglas.
- The crenated margins of collapsed follicle with increased vascularity and echogenicity.
- The "progestinization" of endometrium (discussed earlier) with increased blood flow in spiral and uterine artery **(Figs. 3A to D)**.

ASSESSMENT OF ENDOMETRIUM—FROM FERTILITY PERSPECTIVE[18,19]

Endometrial thickness (ET) tracking can guide a sonologist for escalating estrogen levels. It should be measured in long axis or sagittal plane in a way that the complete endometrial lining till endocervical canal can be visualized and thickest part measured. Normally, three phases of endometrium can be seen:

1. *Early menstrual proliferative phase:* At the beginning of menstruation, endometrium is seen as a thin, bright echogenic stipe (<4 mm) which correspond to an E2 Levels of <50 pg/mL and represent only the basal layer. A thick endometrium on day 2/3 indicates either a polyp or persistent CL or an inadequate down regulated cycle.

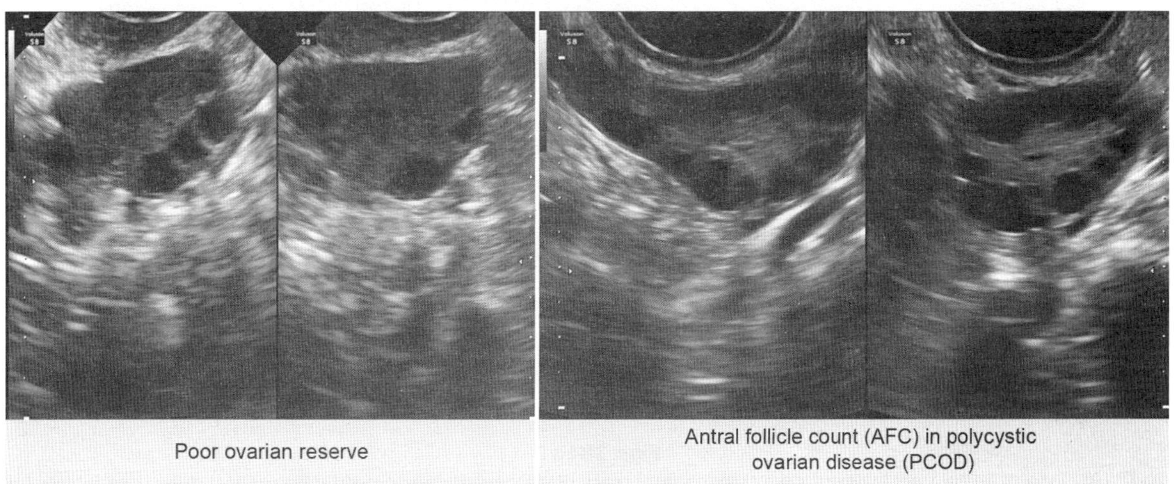

Fig. 3A

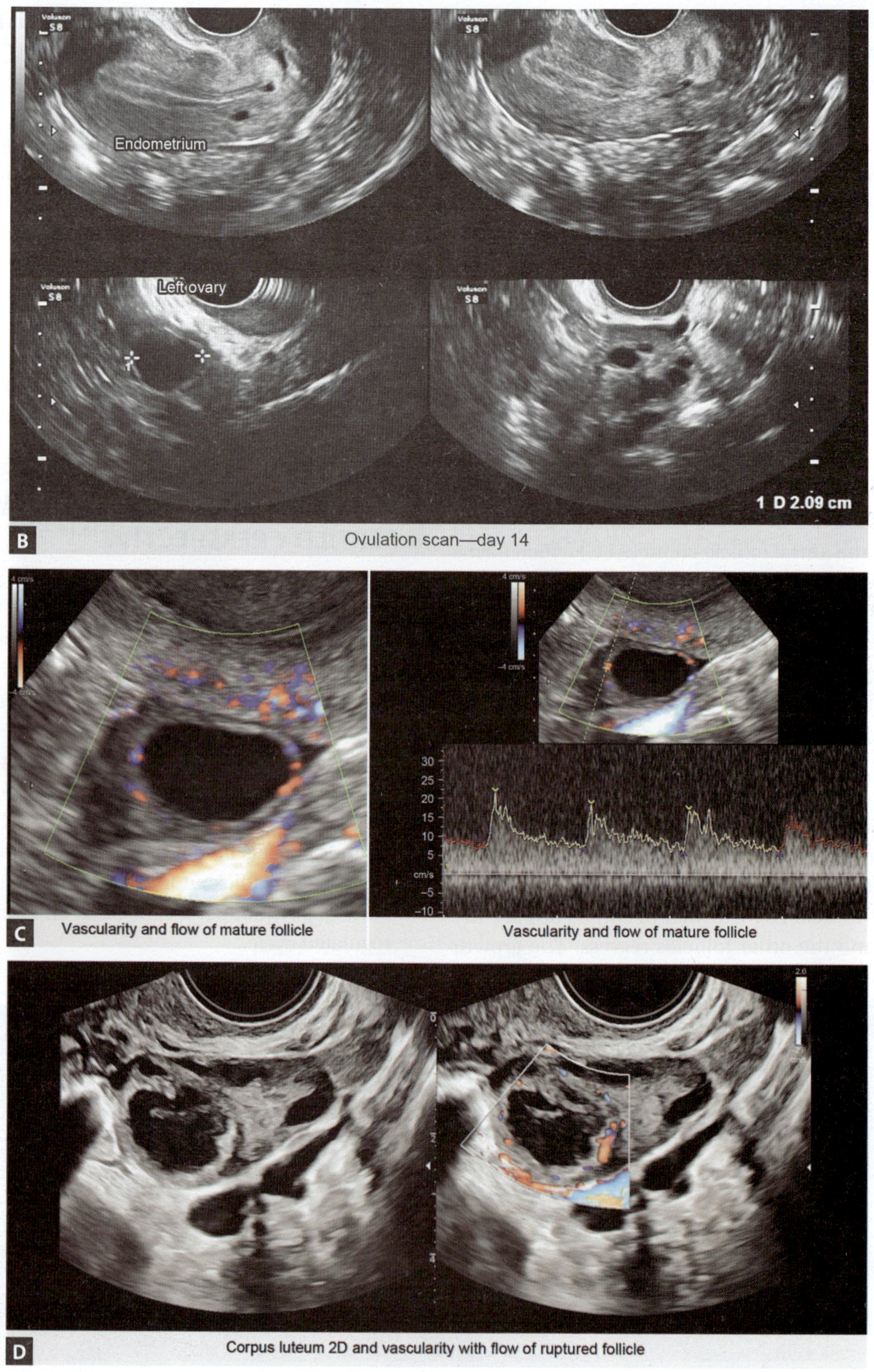

Figs. 3B to D

Figs. 3A to D: (A) Depicts antral follicle count on a transvaginal scan; (B) Shows the endometrial lining and dominant follicle on day 14 of cycle; (C) Shows the mature follicle with vascularity on color flow and pulse Doppler; and (D) Shows a corpus luteum in 2D and color flow mode.

2. *Late proliferative phase:* As the cycle advances, the endometrium starts developing; acquiring a characteristic three-layer pattern (trilaminar appearance). This represents the outer echogenic basal layer, middle hypoechogenic functional layer, and inner most echogenic interphase.
3. *Secretory phase:* This represents the progestational endometrium where the whole endometrium becomes uniformly isoechogenic like basal layer. It indicates the secretory activity in the endometrial glands. The endometrium gets compacted with decreasing thickness than seen in preovulatory stage.

TABLE 3: Applebaum scoring of endometrial vascularity.

Zone I	Blood vessels reaching till endometrial-myometrial junction
Zone II	Blood vessels penetrating into endometrium till outer hyperechoic region
Zone III	Blood vessels penetrating into endometrium till hypoechoic zone
Zone IV	Blood vessels penetrating into endometrial cavity

Zone III and zone IV vascularity is considered an "Ideal" and should cover a 5 mm^2 area. The resistance index (RI) of endometrial vessels should be <0.50. The absence of above features could have a detrimental effect on implantation of the embryo or can increase a chance of miscarriage

Endometrial Doppler

Endometrial blood flow evaluation has been proposed as a mechanism to know the endometrial receptivity and is used as a tool to decide the adequacy of maturity of the endometrium especially during the start of luteal phase in embryo transfer in an in vitro fertilization (IVF) cycle or deciding the day of human chorionic gonadotropin (hCG) trigger in intrauterine insemination (IUI) cycles in synchronous with the mature follicle [Applebaum classification **(Table 3)**] **(Fig. 4)**.

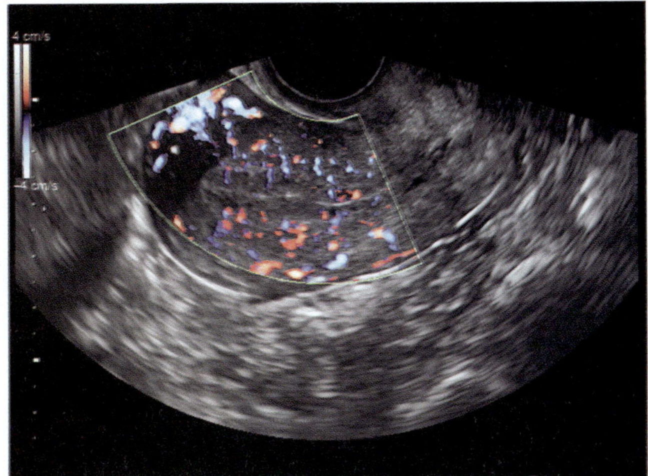

Fig. 4: Endometrium.

Uterine Artery Doppler

Similarly, assessment of *uterine artery blood flow* can also indicate the implantation potential of the endometrium with few studies emphasizing that a high uterine artery pulsatility index (PI) > 3.2, implantation is low and embryo transfer or an IUI should be withheld.

ASSESSMENT OF FALLOPIAN TUBES—FROM FERTILITY PERSPECTIVE

Fallopian tubes are generally not visualized on routine USG; however, certain pathologies can make them apparent on ultrasound such as thickened tubes in pelvic inflammatory disease (PID), fluid filled tubes in hydrosalpinx or pyosalpinx, torsion, and ectopic pregnancy.

Hydrosalpinx can be recognized by the following USG features:[20,21]

- Thin walled (in acute cases) or thick walled (in chronic cases)
- Elongated, C-shaped, S-shaped fluid filled structures with typical folds giving a "cogwheel appearance"
- Indentation of side walls *(waist sign)*
- Separated from uterus and ovary
- *Bead on string sign:* In cases of *incomplete septa.*

Ultrasound-based Methods to Detect Fallopian Tube Patency (SSG/HyCoSy)

Although hysterosalpingography (HSG) has been conventionally used as a primary mode of detecting tubal patency, its use has been questioned due to exposure of iodinated X-ray contrast media and ionizing radiation. Alternatively, ultrasound-based procedures like sonosalpingography (SSG) or hysterosalpingo-contrast sonography (HyCoSy) gained acceptance as the saline-based solutions used in these procedures were safe and inexpensive. Moreover, the performance of these techniques was evaluated to be equivalent to HSG and since these procedures are done in real time USG, a better diagnosis of uterine pathologies such as endometrial polyps and adhesions can be made **(Figs. 5A and B)**.[22,23]

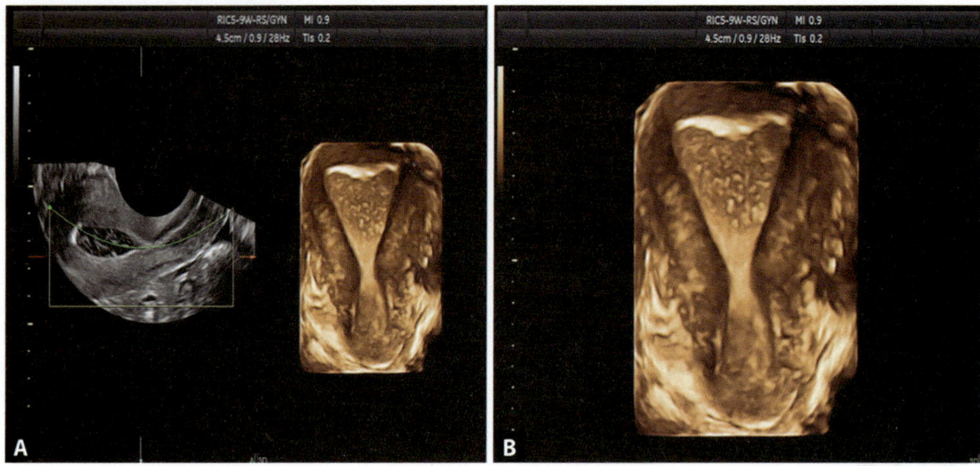

Figs. 5A and B: 3D rendering of distended endometrial cavity after saline instillation.

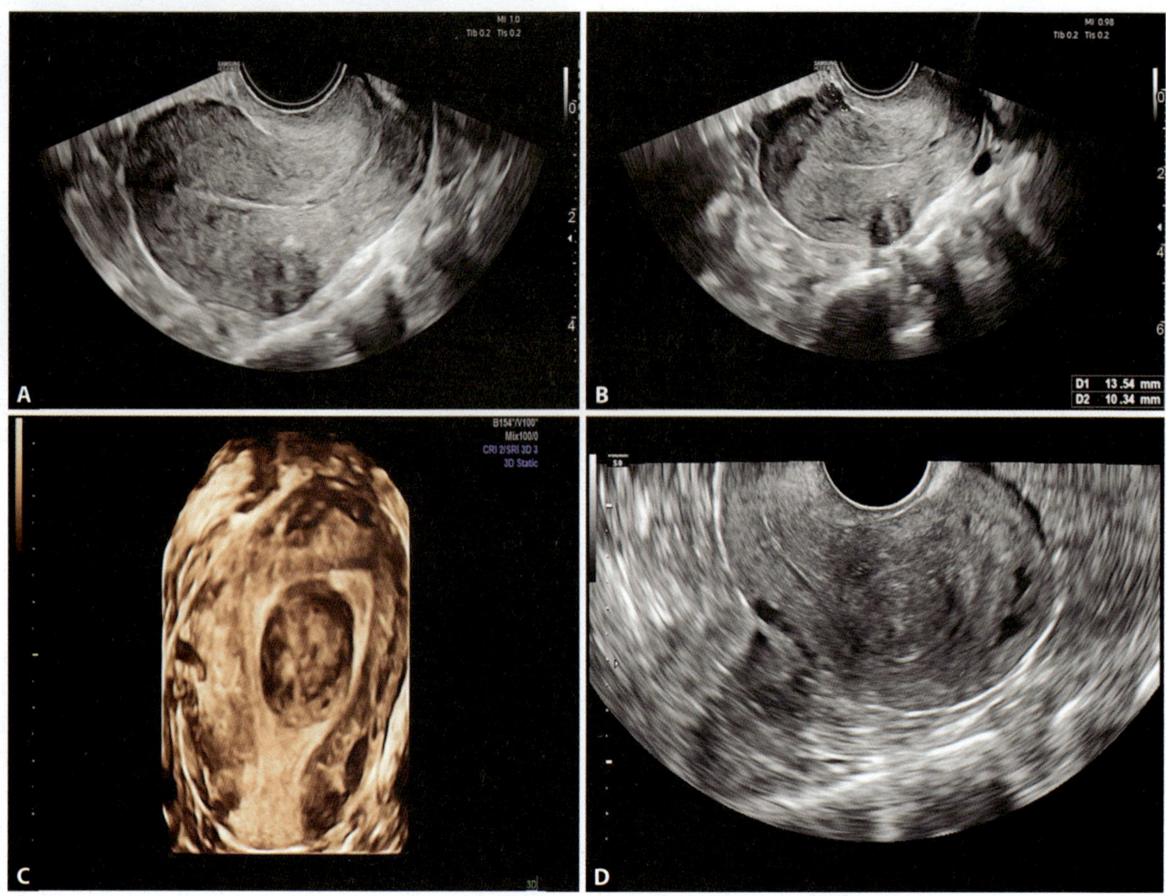

Figs. 6A to D: Fibroids . (A) Intramural myoma; (B) Intramural and subserosal myoma; (C) Submucosal myoma in 3D rendering mode; (D) 2D view of submucosal myoma.

ULTRASOUND IN GYNECOLOGY

Fibroids (Myomas)[24]

Fibroids typically are isolated or multiple.

Ultrasound Appearance (Figs. 6A to D)

- Well-defined solid lesions with whorled appearance.
- They generally are *isoechogenic* (appear same as myometrium) but can be hypo-hyper echogenic in some cases.

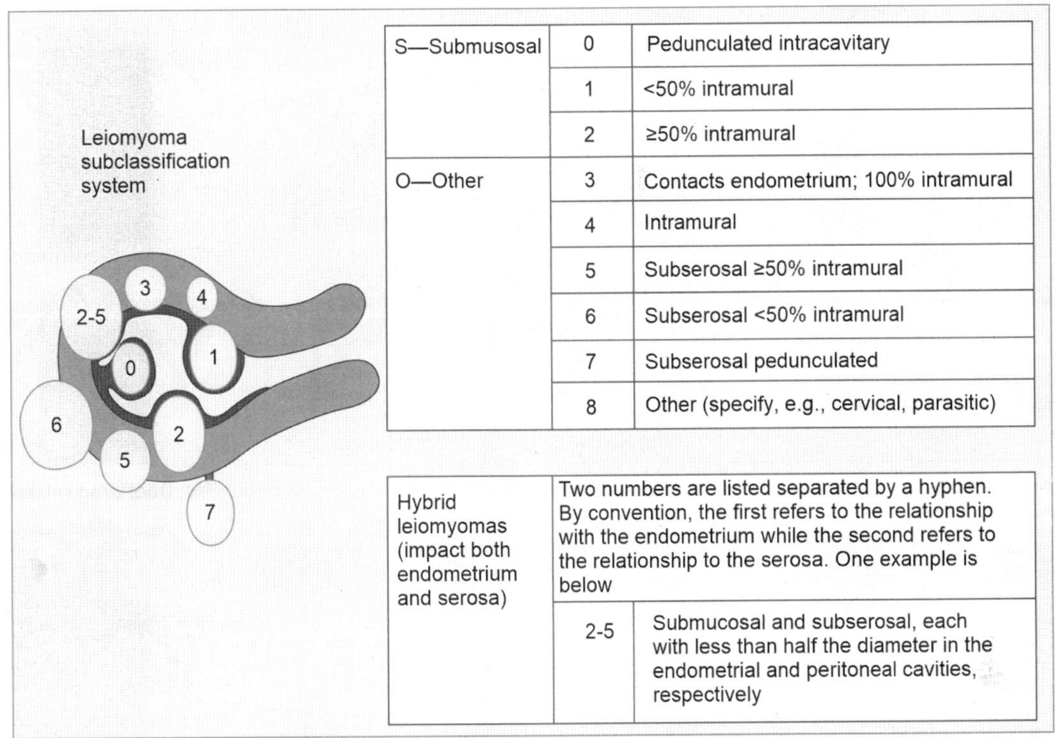

S—Submusosal	0	Pedunculated intracavitary
	1	<50% intramural
	2	≥50% intramural
O—Other	3	Contacts endometrium; 100% intramural
	4	Intramural
	5	Subserosal ≥50% intramural
	6	Subserosal <50% intramural
	7	Subserosal pedunculated
	8	Other (specify, e.g., cervical, parasitic)
Hybrid leiomyomas (impact both endometrium and serosa)	Two numbers are listed separated by a hyphen. By convention, the first refers to the relationship with the endometrium while the second refers to the relationship to the serosa. One example is below	
	2-5	Submucosal and subserosal, each with less than half the diameter in the endometrial and peritoneal cavities, respectively

Fig. 7: Tabular depiction of various types of myomas and their location (subclassification of leiomyomas).

- A *posterior acoustic shadowing* is noted in most of them more so in myomas with calcifications.
- An *alteration of uterine contour* can be seen depending on the location and number of myomas. Areas of cystic changes can be noticed in fibroids undergoing degeneration.
- *Doppler application* in fibroids show a *circumferential vascular flow* can be seen which is absent in necrotic fibroids or fibroids with torsion.
- An ultrasonologist should try to map all the fibroids seen and report should be issued mentioning number, location, size, and proximity to uterine cavity (see later).
- Submucosal fibroids can sometimes be difficult to differentiate from polyps. Sonosalpingography can typically be used in these cases which seem to carry a better specificity than transvaginal sonography (TVS) in this regard **(Fig. 7)**.[22,23]

ADENOMYOSIS[25]

It is of two types, i.e., focal adenomyoma and diffuse adenomyosis.

Ultrasonographic Features

Appearance of adenomyoma is less distinct than the myomas (fibroid) as it lacks the capsule. The uterus appears bulky, globular, with characteristic subendometrial echogenic linear striations or lines extending into the myometrium *(rain shower sign or venetian blind sign)*.[25] The endometrial-myometrial junction is irregular and less distinct.

Small anechoic cysts (1-5 mm) are typically seen scattered in the myometrial or subendothelial regions.

The typical appearance of multiple heterogeneous areas with subendothelial nodules along with the linear striations give an appearance of liver cirrhosis pattern on ultrasound and hence adenomyosis is also referred to as "cirrhosis of the uterus".[25]

Focal adenomyosis is local or small area of myometrial thickening with irregular margins heterogeneous appearance and cystic spaces.

Color Doppler shows irregular vessel coursing all through the substance with no definite pattern.[26]

Müllerian Duct Abnormalities

Müllerian duct anomalies (MDAs) are congenital developmental abnormities of the Müllerian ducts **(Fig. 8)**.

Developmental failure of one or both the Müllerian ducts (categories I and II) (hypoplasia/agenesis and unicornuate uterus).

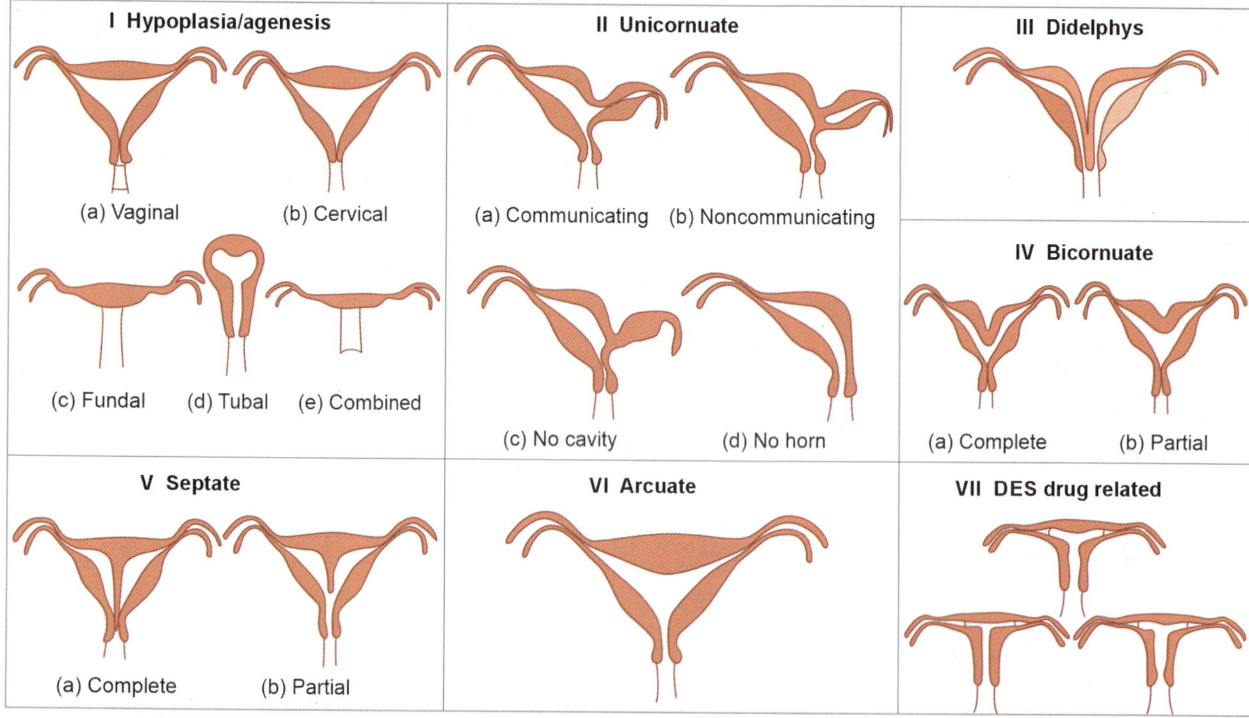

Fig. 8: Classification of various müllerian anomnalies.

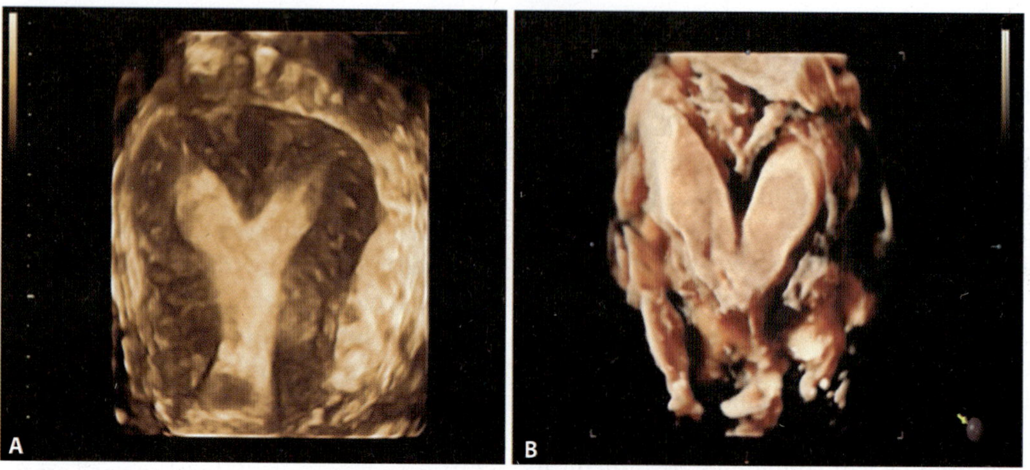

Figs. 9A and B: (A) Septate uterus; (B) Bicornuate uterus.

Fusion issues with the ducts (categories III and IV) (didelphys and bicornuate uterus).

Failure of resorption of the septum (categories V and VI) (septate and arcuate uterus) **(Figs. 9A and B)**.[27]

Müllerian duct anomalies are common although the exact prevalence of these is unknown (varied from 1 to 5%). Majority of MDAs are asymptomatic but a higher association is seen in cases of subfertility, recurrent miscarriages, dysmenorrhea, and endometriosis. They also seem to have a role in troubled pregnancies leading to preterm, premature rupture of membrane (PPROM), and malpresentations.[28]

Hysterosalpingography is generally the first imaging modality where suspected diagnosis of an MDA is made coincidently while checking tubal patency. The sensitivity and specificity of HSG for MDA is 44% and 96%, respectively. HSG might not differentiate between categories of MDA, especially bicornuate and septate uterus. Similarly, 2D USG has a limitation of visualizing fundus completely and hence is not a good choice for diagnosing MDA.

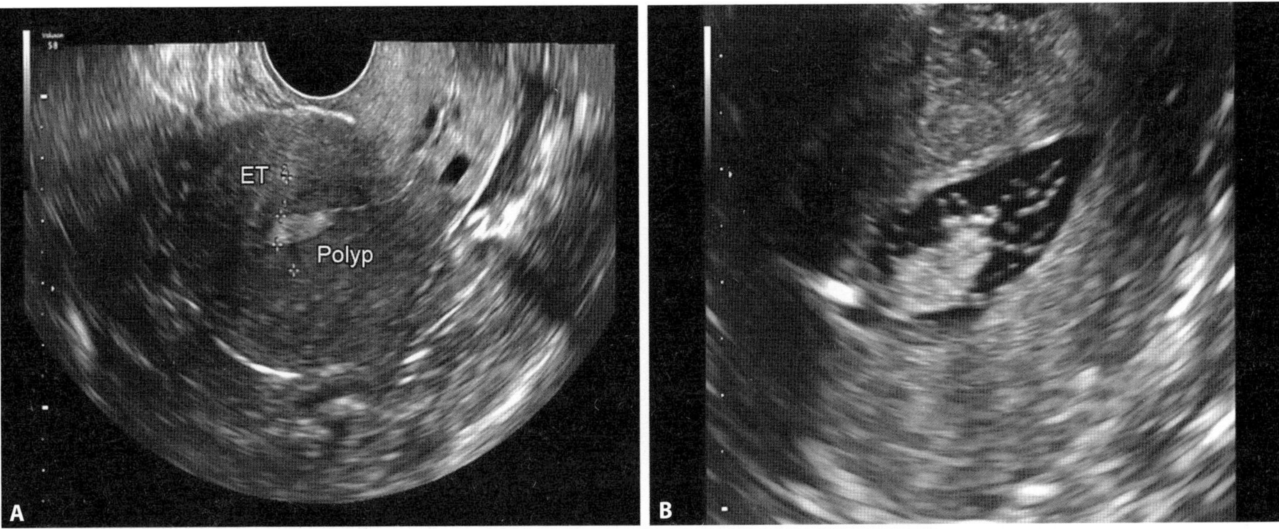

Figs. 10A and B: (A) Endometrial (ET) polyp on 2D; (B) Endometrial polyp after Saline sonosalpingography.

The 3D/4D USG is particularly helpful in diagnosing and categorizing MDA with a great accuracy. The sensitivity, specificity, positive predictive value (PPV), and negative predictive value (NPV) of 3D/4D in this regard is reported to be almost 100%. 3D/4D USG for MDAs should be done in luteal phase as endometrium is hyperechoic and easily identified **(Figs. 9A and B)**.[29]

Endometrial Pathologies

Endometrium is one of the most dynamic tissues of the human body that changes every day with the period of menstruation.

Endometrial Polyps

Single or multiple echogenic lesions in the endometrium which can easily be identified with following features:[30,31]

- *Interrupted mucosa sign:* The polyp disrupts the normal cavity contour
- *The bright edge sign:* Echogenic lines seen on the border of the polyp
- *Pedicle artery sign:* A feeding vessel may be seen extending from the base. The stalk or the base may be thin or broad based.
- Sometimes multiple small polyps are identified at the margins (polypoidal endometrium).

Polyps may be seen with cystic spaces representing *"proteinaceous content"* inside or may be surrounded by *endometrial fluid* **(Figs. 10A and B)**.

Uterine Synechiae/Asherman's Syndrome

The endometrium appears interrupted with unclear endometrial-myometrial junction. *Multiple echogenic foci* can be seen scattered with areas of fluid collections that do not vanish in further visits. Endometrium remains thin despite adequate hormone therapy. A *hyperechogenic lesion with posterior shadow* can be noticed in cases of calcifications. In severe cases the *narrowing of uterine cavity* can be seen.[32]

Postmenopausal Bleeding and Endometrial Hyperplasia

In premenopausal women, thickness of the endometrium varies with the days of the menstrual cycle however any ET >15 mm is considered upper limit of secretary endometrium. Any ET <8 mm carries a meagre chance of any abnormality. *In Postmenopausal women, any ET > 5 mm* should be further evaluated especially with symptoms like postmenopausal bleeding (PMB).[33,34]

On ultrasound, it is difficult to differentiate between endometrial hyperplasia from endometrial carcinoma unless the lesion/mass is large enough to involve myometrium or cervix and is vascular on Doppler. Following features may help identify both the lesions: The endometrium is *asymmetrically thickened with irregular surface. Cystic areas* may be noticed within the endometrium. *A fluid can be noticed* along with a polypoidal mass that might be seen invading myometrium.[34]

OVARIAN PATHOLOGIES

Ovarian Cysts

Ovarian cysts are defined as *fluid containing structures of >30 mm* in diameter. Ovarian cysts are common among females of all ages.

It is estimated that up to 10% of premenopausal women worldwide have undergone surgery for the ovarian cysts during their lifetime.

Almost all the cysts in *premenopausal women are benign* but the risk increases with increasing age (from 1 in 1,000 to 3 in 1,000 at the age of 50 years). The associated fear of malignancy in every cyst diagnosed generates a considerable anxiety among the treating physician and the concerned women; hence it becomes crucial to evaluate the nature of the ovarian cyst.[35]

A pelvic ultrasound along with biochemical markers is the standard mode of evaluation of ovarian cysts in females.

Transvaginal ultrasound is preferred over transabdominal route due to its better sensitivity.

Transabdominal route can particularly be useful in cases of large cysts which cannot be completely mapped by TVS or in patients which deny TVS for their personal reasons. Addition of color flow Doppler and 3D/4D image can further enhance the findings.

While doing pelvic ultrasound of ovarian cysts, the morphological assessment of its ultrasound features is the most important step as it helps to differentiate a benign from a malignant cyst.

Simple cyst is generally recognized as a *round or oval shaped structure* which has a *thin wall* and carries *no septations or solid areas.* A characteristic *posterior wall enhancement* with *anechoic fluid* can be seen. A simple cyst diagnosed by earlier mentioned criteria is *95–99% established to be a benign cyst.*

Complex Cyst

Irregular or *dense, solid nodules with septations or papillary projections* is highly suspicious and is termed as complex cyst.

Two most common ultrasound methods used to evaluate ovarian cysts are *risk of malignancy index (RMI)* and the *International Ovarian Tumor Analysis Group (IOTA group)* **(Box 1 and Table 4)**. Four biochemical markers which are used along with pelvic ultrasound are *CA-125, alpha-fetoprotein (AFP), lactate dehydrogenase (LDH), and hCG.* CA-125 is unreliable in premenopausal women as it can be seen raised in endometriosis, pelvic infections, fibroids and adenomyosis hence can avoided in case of clear cysts however in complex masses a minimum of LDH, AFP and hCG should be done to rule out germ cell masses **(Table 5 and Fig. 11)**.[36,37]

BOX 1: Risk of malignancy index.

Risk of malignancy index (*RMI*) = *Menopausal status* × *CA-125* value (*IU/mL*) × *ultrasound features*

Menopausal status: *1 = premenopausal, 3 = postmenopausal*

(Postmenopausal *is the one with an amenorrhea of >1 year or the one with an age of >50 years with hysterectomy)*

Ultrasound features: (Remember by pneumonic) = *ABCDE*

A = Ascites

B = Bilateral lesions

C = Cysts multilocular

D = Dense (solid) areas in the cysts

E = Extravasation (malignant)

CA-125 value can be from zero to in thousands IU/mL

The sensitivity and specificity of RMI is 78% and 87%, respectively

TABLE 4: International Ovarian Tumor Analysis Group (IOTA group).

Using data derived from IOTA Group, ultrasound features (rules) were defined to categorize the ovarian masses as benign and malignant. The sensitivity and specificity of these rules is better than risk of malignancy index (RMI) (95% and 91%, respectively)

Benign masses	*Malignant masses*
Unilocular cysts	Irregular solid tumor
Presence of solid components with largest <7 mm	Ascites
Acoustic shadowing	At least four papillary structures
Smooth multilocular tumor with a largest diameter of <100 mm	Irregular multilocular solid tumor with a largest diameter of >100 mm
No blood flow	Very strong blood flow

Mature Teratomas

Dermoid cysts or mature teratomas are a category of germ cell tumors primarily seen in the premenopausal females. They do not secrete any hormones and pelvic ultrasound is the principal modality to diagnose them. Dermoid cysts are typically benign cysts that contain fat, fluid, and solid areas and these features help to diagnose the dermoid while doing USG. Following notable features are seen on ultrasound:[38]

TABLE 5: Differential diagnosis of adnexal masses.		
Benign nonovarian masses	***Benign ovarian masses***	***Malignant masses***
• Hydrosalpinx • Paraovarian cysts/fimbrial cysts • Adhesion pockets • Tubo-ovarian mass/abscess • Ectopic pregnancy/chronic ectopic • Pelvic kidney • Peritoneal cysts • Appendicular masses • Diverticular abscess • Wandering fibroid	• Functional cysts • Endometrioma • Corpus luteal cysts • Dermoids/mature teratomas • Serous cystadenoma • Mucinous cystadenoma	• Germ cell tumors • Epithelial carcinomas • Sex chord tumors • Secondaries from breast and gastrointestinal tract (GIT) and other organs

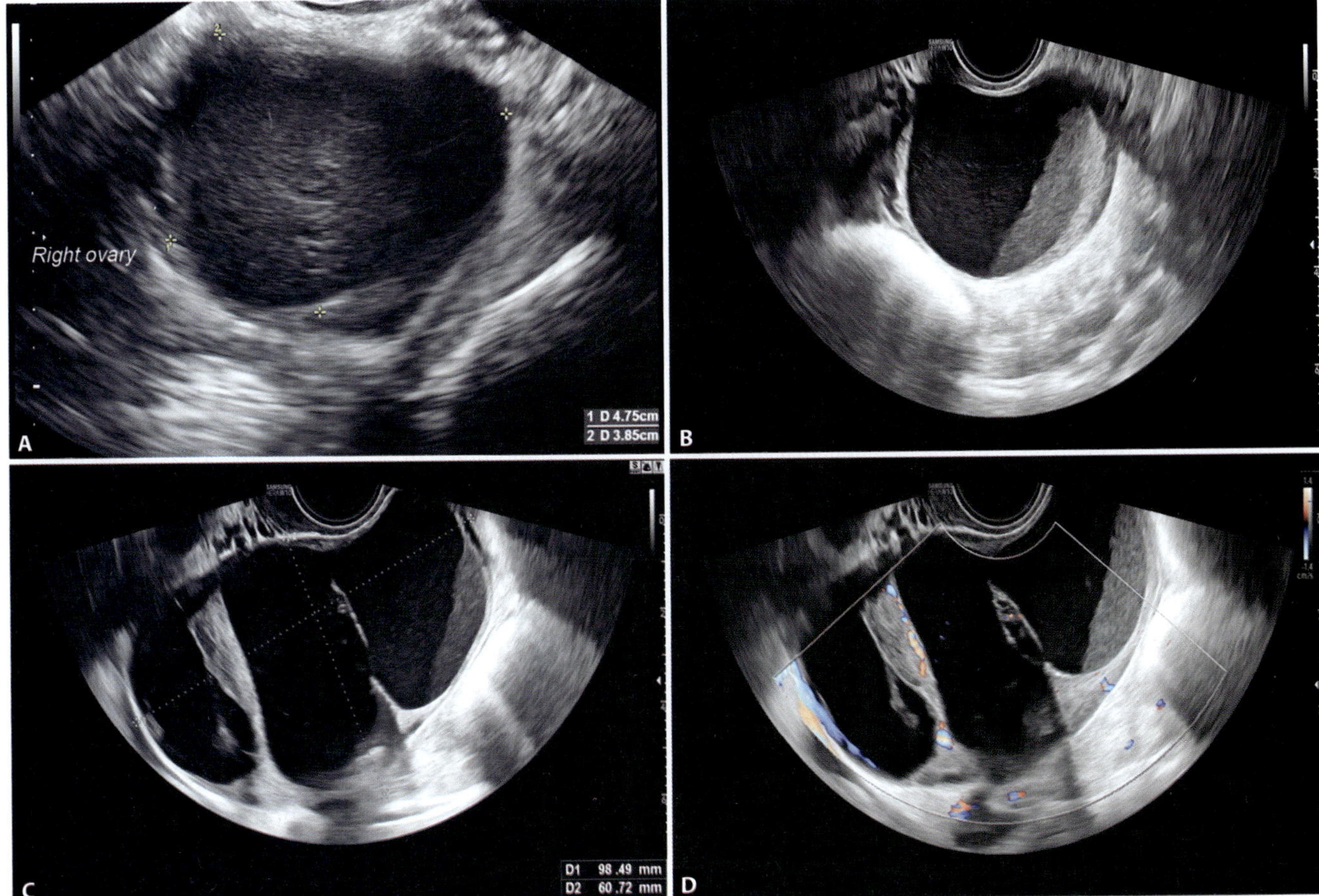

Figs. 11A to D: (A) Endometrioma; (B) Complex hemorrhagic cyst with retracted clot; (C) Complex multiseptated cyst 2D; (D) Complex multiseptated cyst on color Doppler application.

- *Dermoid plug sign:* It is an echogenic mass seen on USG which represent a teeth, bone, or bunch of hairs. This the most common feature noted in this regard.
- *Dermoid mesh sign:* The echogenic foci created by small hairs floating in the cyst fluid give an appearance of multiple small hypogenic lines or dots.
- *Tip of an iceberg sign:* This is a mass created by mixture of matted hairs and sebum giving an echogenic shadow whose base cannot be visualized.
- *Fat-free level:* Produced due to layering of serous fluid and sebum.

Box 1 and **Table 4** shows most prevelant and accepted method of evaluating malignancies and providing a risk score of being malignant based on ultrasound findings and CA125.

SUMMARY

The contribution of ultrasound in obstetrics and gynecology is enormous and has contributed in almost every part of the field. After history and clinical examination a handy ultrasound evaluation in gynecological clinic is a great blessing in case management. In fertility science and gynecological disorders, its utility ranges from tracking ovulation to monitoring controlled ovarian hyperstimulation (COH) in IVF and assessment of endometrial-myometrial-ovarian issues related to oncology.

REFERENCES

1. Campbell S. A short history of sonography in obstetrics and gynaecology. Facts Views Vis Obgyn. 2013;5(3):213-29.
2. Benson CB. Ultrasound in Obstetrics and Gynecology: A Practical Approach. Stuttgart: Thieme; 2000.
3. Sassaroli E, Crake C, Scorza A, Kim DS, Park MA. Image quality evaluation of ultrasound imaging systems: advanced B-modes. J Appl Clin Med Phys. 2019;20(3):115-24.
4. Rathbun KM, Zweig AS. A Realistic and Inexpensive Ultrasound Phantom for Teaching M-Mode Measurement of Fetal Heart Rate. J Emerg Med. 2021;61(6):744-8.
5. Maulik D, Heitmann E, Maulik D. New horizons in Doppler ultrasound technology: relevance for obstetrical applications. Clin Obstet Gynecol. 2010;53(4):831-41.
6. Indrielle-Kelly T, Frühauf F, Burgetová A, Fanta M, Fischerová D. Diagnosis of endometriosis 2nd part—Ultrasound diagnosis of endometriosis (adenomyosis, endometriomas, adhesions) in the community. Ceska Gynekol. 2019;84(4):260-8.
7. Abinader R, Warsof SL. Benefits and Pitfalls of Ultrasound in Obstetrics and Gynecology. Obstet Gynecol Clin North Am. 2019;46(2):367-78.
8. Recker F, Weber E, Strizek B, Gembruch U, Westerway SC, Dietrich CF. Point-of-care ultrasound in obstetrics and gynecology. Arch Gynecol Obstet. 2021;303(4):871-6.
9. Nowicki A. Safety of ultrasonic examinations; thermal and mechanical indices. Med Ultrason. 2020;22(2):203-10.
10. Fatahi Asl J, Farzanegan Z, Tahmasbi M, Birgani SM, Malekzade M, Yazdaninejad H. Evaluation of the Scan Duration and Mechanical and Thermal Indices Applied for the Diagnostic Ultrasound Examinations. J Ultrasound Med. 2021;40(9):1839-50.
11. Schiska A. Teaching Radiography Students the ALARA Principle. Radiol Technol. 2021;93(2):228-31.
12. Frane N, Bitterman A. Radiation Safety and Protection. Treasure Island (FL): StatPearls Publishing; 2022.
13. Buscarini E, Lutz H, Mirk P. Manual of Diagnostic Ultrasound, 2nd edition. Geneva, Switzerland: World Health Organization; 2013.
14. Younis JS, Ben-Ami M, Ben-Shlomo I. The Bologna criteria for poor ovarian response: a contemporary critical appraisal. J Ovarian Res. 2015;8:76.
15. Gleicher N, Darmon S, Patrizio P, Barad DH. Reconsidering the Polycystic Ovary Syndrome (PCOS). Biomedicines. 2022;10(7):1505.
16. Khan HL, Bhatti S, Suhail S, Gul R, Awais A, Hamayun H, et al. Antral follicle count (AFC) and serum anti-Müllerian hormone (AMH) are the predictors of natural fecundability have similar trends irrespective of fertility status and menstrual characteristics among fertile and infertile women below the age of 40 years. Reprod Biol Endocrinol. 2019;17(1):20.
17. Kondagari L, Kahn J, Singh M. Sonography Gynecology Infertility Assessment, Protocols, and Interpretation. Treasure Island (FL): StatPearls Publishing; 2022.
18. Khan MS, Shaikh A, Ratnani R. Ultrasonography and Doppler Study to Predict Uterine Receptivity in Infertile Patients Undergoing Embryo Transfer. J Obstet Gynaecol India. 2016;66(Suppl 1):377-82.
19. Check JH. The importance of sonographic endometrial parameters in influencing success following embryo transfer in the modern era and therapeutic options–part 1: the importance of late proliferative phase endometrial thickness. Clin Exp Obstet Gynecol. 2011;38(3):197-200.
20. Della Grotta LM, Dyer RB, Holbert BL. The "cogwheel" sign of hydrosalpinx. Abdom Radiol. 2019;44:3486-7.
21. Rezvani M, Shaaban AM. Fallopian tube disease in the nonpregnant patient. Radiographics. 2011;31(2):527-48.
22. Acholonu UC, Silberzweig J, Stein DE, Keltz M. Hysterosalpingography versus sonohysterography for intrauterine abnormalities. JSLS. 2011;15(4):471-4.
23. Lim SL, Jung JJ, Yu SL, Rajesh H. A comparison of hysterosalpingo-foam sonography (HyFoSy) and hysterosalpingo-contrast sonography with saline medium (HyCoSy) in the assessment of tubal patency. Eur J Obstet Gynecol Reprod Biol. 2015;195:168-72.
24. Qi H, Zhou C, Huang Z, Yang N, Wu Q. Value of Transabdominal Combined Transvaginal Color Doppler Ultrasonography in the Distinguish between Uterine Adenomyoma and Uterine Fibroids. Biomed Res Int. 2022;2022:9599571.
25. Cunningham RK, Horrow MM, Smith RJ, Springer J. Adenomyosis: A Sonographic Diagnosis. Radiographics. 2018;38(5):1576-89.
26. Van den Bosch T, Van Schoubroeck D. Ultrasound diagnosis of endometriosis and adenomyosis: State of the art. Best Pract Res Clin Obstet Gynaecol. 2018;51:16-24.

27. Pfeifer SM, Attaran M, Goldstein J, Lindheim SR, Petrozza JC, Rackow BW, et al. ASRM müllerian anomalies classification 2021. Fertil Steril. 2021;116(5):1238-52.
28. Hiersch L, Yeoshoua E, Miremberg H, Krissi H, Aviram A, Yogev Y, et al. The association between Müllerian anomalies and short-term pregnancy outcome. J Matern Fetal Neonatal Med. 2016;29(16):2573-8.
29. Passos IMPE, Britto RL. Diagnosis and treatment of müllerian malformations. Taiwan J Obstet Gynecol. 2020;59(2):183-8.
30. Caspi B, Appelman Z, Goldchmit R, Ashkenazi M, Haruvy Y, Hagay Z. The bright edge of the endometrial polyp. Ultrasound Obstet Gynecol. 2000;15(4):327-30.
31. Kamaya A, Yu PC, Lloyd CR, Chen BH, Desser TS, Maturen KE. Sonographic Evaluation for Endometrial Polyps: The Interrupted Mucosa Sign. J Ultrasound Med. 2016;35(11):2381-7.
32. Amin TN, Saridogan E, Jurkovic D. Ultrasound and intrauterine adhesions: a novel structured approach to diagnosis and management. Ultrasound Obstet Gynecol. 2015;46(2):131-9.
33. Nalaboff KM, Pellerito JS, Ben-Levi E. Imaging the endometrium: disease and normal variants. Radiographics. 2001;21(6):1409-24.
34. Van Den Bosch T, Verbakel JY, Valentin L, Wynants L, De Cock B, Pascual MA, et al. Typical ultrasound features of various endometrial pathologies described using International Endometrial Tumor Analysis (IETA) terminology in women with abnormal uterine bleeding. Ultrasound Obstet Gynecol. 2021;57(1):164-72.
35. Farghaly SA. Current diagnosis and management of ovarian cysts. Clin Exp Obstet Gynecol. 2014;41(6): 609-12.
36. American College of Obstetricians and Gynecologists. ACOG Practice Bulletin. Management of adnexal masses. Obstet Gynecol. 2007;110(1):201-14.
37. Ngu SF, Chai YK, Choi KM, Leung TW, Li J, Kwok GST, et al. Diagnostic Performance of Risk of Malignancy Algorithm (ROMA), Risk of Malignancy Index (RMI) and Expert Ultrasound Assessment in a Pelvic Mass Classified as Inconclusive by International Ovarian Tumour Analysis (IOTA) Simple Rules. Cancers (Basel). 2022; 14(3):810.
38. Outwater EK, Siegelman ES, Hunt JL. Ovarian teratomas: tumor types and imaging characteristics. Radiographics. 2001;21(2):475-90.

Chapter

3

Puberty and its Aberrations

Pavika Lal, Divya Dwivedi

INTRODUCTION

Puberty is defined as sequential coordinated physiological events that occur in a predefined order in an individual to attain reproductive/sexual maturity. Most notable and significant changes are development of secondary sexual characters, attainment of peak height, and achievement of fertility along with cognitive and psychosocial maturation.

Factors affecting the age at onset and progression of pubertal development[1]

• Genetic	• Geographical location
• Nutrition	• Psychological state
• General health	• Exposure to light

- Children residing in urban areas, nearer to equatorial regions and at lower altitudes experience puberty earlier as compared to children living in rural areas far from equator and at higher altitudes.
- Blind girls experience puberty earlier than sighted girls.[2]
- 17% of the body fat must be achieved for the initiation of menarche and body fat must be 22% to have regular periods beyond the age of 16 years.[3]

ENDOCRINOLOGY OF NORMAL PUBERTY

Rhythmic regular pulsatile secretion of hypothalamic gonadotropin-releasing hormone (GnRH) by the hypothalamus (pulse generator) which in turn regulates secretion of gonadotropins and sex steroids by anterior pituitary and gonads respectively is the most characteristic hormonal change occurring at the time of puberty.

Fetal life and infancy:
Hypothalamic pulse generator activity remains suppressed.

Childhood and early adolescence:
- *Juvenile pause:* Although hypothalamic gonadotropin-releasing hormone (GnRH) pulse generator is active, but the frequency and amplitude of GnRH secretion is irregular and very low.[4]
- *Gonadostat theory:* The typical "diphasic" pattern of gonadotropin secretion from infancy to puberty results primarily from changing levels of central inhibition of pulsatile GnRH secretion and, to a lesser extent, from high sensitivity to low levels of gonadal steroids feedback.
- *Puberty:* During puberty, luteinizing hormone (LH) amplitude increases 10 folds as compared to follicle-stimulating hormone (FSH) amplitude which increase two folds, later pulse frequency increases but the increase in LH is more significant than FSH due to influence of increasing estradiol levels at both hypothalamus and pituitary levels.[5,6]
- *Early puberty:* Negative effect of low levels of sex steroid decreases in early puberty.[6]
- *Late puberty:* LH maturation of stimulatory feedback response to estrogens is responsible of LH peak resulting in ovulation during late puberty.

Sequence of pubertal development in males and females:
In male: Increased testicular volume → pubarche → height spurt

First recognized sign of male puberty is pigmented pubic hair although it occurs 6 months after initiation of genital growth.
In female: Accelerated growth → thelarche → pubarche → menarche

**Girls reach peak height velocity earlier during the pubertal development as compared to boys owing to which boys and girls have an average of 28 cm and 25 cm respectively during this period.*
**Gonadotropins, insulin like growth factor-1, and sex steroid play a major role in growth spurt during pubertal development.*

Central and Peripheral Mechanisms of Puberty

A coordinated central nervous system programming is essential for puberty, i.e., it inhibits puberty until an appropriate time. Various features which govern and control the "Neuro endocrine switch" for GnRH pulse generation, which is on in early infancy, switched off during childhood and then switches on during puberty have been identified. There are both inhibitory and excitatory signals which regulate the activity of hypothalamic pituitary ovarian axis that mark the onset and timing of pubertal development.[7]

Excitatory signals:	*Inhibitory signals:*
• *Glutamate* signaling may play a role in resurgence of pulsatile GnRH secretion by hypothalamus via N methyl D aspartate receptor.[8] • *Kisspeptins: They are neuropeptides that have emerged as critical* component which *controls the activity of* GnRH between infancy and puberty and act as the primary transducer of both internal and external environmental stimulus.[9]	• *Gamma-aminobutyric acid (GABA)* is produced in specialized neurons in hypothalamus that restricts/prevents GnRH pulse generator activity during childhood. • Release of GABA decreases at onset of puberty leading to increase in pulsatile GnRH secretion.[10] • *Neuropeptide Y:* Hypothalamic peptide which is involved in control of food intake, behavior, and reproductive function. • Both GABA and neuropeptide Y act as neurological break for GnRH pulse.[11]
Leptin: Produced by adipocytes and serum concentration of leptin is strongly associated with body fat as well as its content. It plays permissive role in onset of puberty in which metabolic signals are communicated to higher centers at the onset of puberty.[12,13]	*Other candidate metabolic signals:* • Insulin • Ghrelin (endogenous ligand of growth hormone) • Ghrelin-like peptide • Free fatty acids[14,15]

Tanner staging—breast

Stage	Features
Stage 1	• Prepubertal stage • No palpable breast tissue, with areola <2 cm in diameter • Nipples may be inverted, flat, or raised
Stage 2	• Breast budding occurs • Visible palpable breast tissue • Areola begins to enlarge • Nipple develops to varying degrees
Stage 3	• Further growth and elevation of entire breast • Nipple is generally at or above the midplane of breast tissue when viewed from side
Stage 4	Projection of areola and papilla above the general breast contour
Stage 5	Mature breast in both contour and proportion

Tanner stage—pubic hairs

Stage	Features
Stage 1	No sexually stimulated pubic hair present
Stage 2	First appearance of coarse, long, crinkly pubic hair along the labia majora.
Stage 3	Coarse, curly hair extends up to the mons pubis
Stage 4	Adult hair in thickness and texture, typically does not extend to inner aspect of thighs
Stage 5	Pubic hair extends onto the thighs

*Tanner staging is *Not* used for axillary hair development.

PRECOCIOUS PUBERTY

Require thorough evaluation to determine the cause in following cases:

- <6 years of age with either breast or pubic hair development
- <8 years of age with both breast and pubic hair development
- In girls who are <8 years of age with only isolated thelarche, pubarche, or adrenarche.

Between 6 and 8 years clinician should individualize regarding the psychosocial consequences of early puberty development in an individual and in her parents and then decide accordingly the level of extent of evaluation.[16]

Precocious puberty it is of two types:

- Central precocious puberty/GnRH dependent/true precocious puberty
- Peripheral precocious puberty/GnRH independent/pseudo puberty

Central (true) precocious puberty:
- Constitutional (idiopathic) precocious puberty
- Hypothalamic neoplasms (most commonly hamartomas)
- Congenital malformations
- Infiltrative processes (Langerhans cell histiocytosis)
- After irradiation
- Trauma
- Infection

Precocious puberty of peripheral origin (precocious pseudo puberty):
- McCune–Albright syndrome
- Congenital adrenal hyperplasia
  - 21 hydroxylase deficiency
  - 11 beta hydroxylase deficiency
  - 3 beta hydroxysteroid dehydrogenase deficiency

- Iatrogenic ingestion/absorption of estrogens or androgens
- Primary hypothyroidism
- Gonadotropin secreting neoplasms
- Gonadal neoplasms
- Adrenal neoplasms
- Polycystic ovarian syndrome
- Non-classic forms of congenital adrenal hyperplasia
- Idiopathic hirsutism

Evaluation of precocious pubertal development

↓

Points in history which should be considered in evaluation:
- When was the 1st physical change observed
- Any similar pattern of physical change in other siblings
- History of pubertal development in parents
- Previous history of neurological disease or central nervous system (CNS) trauma
- History of exposure to exogenous sexual steroids
- History of headache, seizure, and abdominal pain

↓

Examination
- Height, weight, calculation of growth velocity (centimeter/year)
- Fundoscopy may be required in patients suggestive of increased intracranial tension.
- Visual field assessment may be required in cases of suprasellar lesions.
- Tanner staging of breast and pubic hair should be determined.

Bone age assessment should be done if examination reveals early signs of precocious puberty.

Incomplete Precocious Puberty

Isolated premature thelarche: Isolated breast development in girls <8 years which can be unilateral or bilateral without nipple or areolar development and not accompanied by any other signs of sexual maturity. It may be due to increased sensitivity of breast to low levels of circulating estradiol levels or increase estradiol secretion by follicular cysts.

Reassurance is required in majority of cases and further pubertal development is normal. Ultrasonography (USG) measurement of uterine volume may be useful test to differentiate between premature thelarche and early menarche.

Exaggerated Thelarche

It represents a spectrum between premature thelarche and precocious puberty as such individuals show increased growth velocity and advanced bone age.[17,18]

Isolated premature adrenarche: It is caused by increased sensitivity to low levels of androgens and must be distinguished from late onset congenital adrenal hyperplasia. There is an increased risk of polycystic ovary syndrome (PCOS), hyperinsulinemia, acanthosis nigricans, and dyslipidemia in adolescence and adult life. Counselling regarding diet and lifestyle modifications are required in individuals especially who are obese. Follow-up should be considered to detect early development of PCOS.

Isolated premature menarche: It usually occurs at the age of 1–9 years with self-limiting episode followed by normal course of pubertal development. It always becomes important to rule out history of foreign body, child abuse, and very rarely vaginal rhabdomyosarcoma.

Management of precocious puberty:

Aims of treatment:
- Stop or slow the development until normal pubertal age is attained.
- To maximize adult height
- To decrease the risk of psychological and mental consequences associated with easily sexual maturation

Treatment depends upon the underlying cause, i.e., whether the precocious puberty is gonadotropin dependent or independent.

Gonadotropin-dependent Precocious Puberty

When to start treatment?
- When sexual maturity progresses to the next stage within 3–6 months.
- When growth velocity is accelerated to >6 cm/year.
- Bone age is advanced by 1 year or more.
- When predicted adult height is below the target range or is decreasing on serial determinations.[19]

Gonadotropin-releasing hormone agonist treatment causes a brief initial "flare" of gonadotropin release, followed by pituitary desensitization and down-regulation.

The choice among the available GnRH agonist formulations depends mostly on physician preference and availability, depot preparations being preferred because of their improved compliance.

- Buserelin 6.3 mg every 2 months
- Goserelin 3.6 mg every month or 9.8 mg every 3 months
- Histrelin 50-mg implant every year
- Leuprolide 3.75–7.5 mg monthly or 11.25 mg every 3 months
- Triptorelin 3.0–3.75 mg monthly or 11.25 mg every 3 months

- The adequacy of GnRH agonist treatment can be monitored simply by measuring the serum LH concentration 30–60 minutes after each repeated injection of the agonist; the LH level should be <3.0 IU/L, consistent with prepubertal norms after acute GnRH agonist stimulation.[20]
- Gonadotropin-releasing hormone agonist treatment should be monitored at 3–6-month intervals with serial physical examinations along with evaluation of bone age.[21]
- Prompt reactivation of the pituitary-gonadal axis and pubertal development, in a pattern similar to that in normal adolescents, generally follows the discontinuation of treatment.[22]

Gonadotropin-independent Precocious Puberty

- The treatment of gonadotropin-independent precocious puberty is aimed at the underlying pathology.
- Girls with functional tumors involving the ovaries or adrenals are treated surgically, human chorionic gonadotropin (hCG)-secreting tumors may also require adjunctive radiation or chemotherapy, depending on the type and location of the tumor.
- In girls with McCune-Albright syndrome, treatment can be aimed at blocking aromatization and estrogen production.[23-26]
- Basal and GnRH stimulation with gonadotropins level differentiates between central and precocious puberty.

If no associated cause is ascertained the decision of treatment is based mainly on rate of pubertal development and on estimated adult height.

Delayed Puberty

- *Definition:* Delayed or interrupted puberty exists in girls who fail to develop any secondary sexual characteristics by age 13, have not had menarche by age 15 (95th centile is 14.5 years), or have not attained menarche 5 or more years since the onset of pubertal development.
- Asynchronous pubertal development is characterized by pubertal development that deviates from the normal pattern of puberty.
- *Evaluation*: A detailed careful history and examination forms an important clue to the cause of delayed puberty and helps to narrow down the differential diagnosis.
- *Most important cause in both sexes is constitutional delay of growth and development which has a genetic component, but* it is always important to differentiate congenital GnRH deficiency from constitutional delay.

Persistent hypogonadism >18 years of age suggests congenital GnRH deficiency.

Causes of delayed puberty[27]*:*
- *Anatomic abnormalities of the genital tract*
  - Müllerian dysgenesis [Mayer–Rokitansky–Küster–Hauser (MRKH) syndrome]
  - Distal genital tract obstruction
    - Imperforate hymen
    - Transverse vaginal septum
- *Hypergonadotropic (FSH >30 mIU/mL) hypogonadism (gonadal failure)*
  - Gonadal dysgenesis with stigmata of Turner syndrome
  - Pure gonadal dysgenesis (46XX, 46XY)
- *Hypogonadotropic (LH and FSH <10 mIU/mL) hypogonadism*
  - Kallmann syndrome
  - Prader–Willi syndrome
  - Laurence–Moon–Bardet–Biedl syndrome
- *Neoplasms of hypothalamic—pituitary area*
  - Craniopharyngiomas
  - Pituitary adenomas
  - Others
- *Infiltrative process (Langerhans cell type histiocytosis)*
- Post chemotherapy (especially alkylating agents), post radiotherapy of CNS
- *Severe chronic illness with malnutrition*
- *Anorexia nervosa and related disorders*
- *Antidopaminergic and gonadotropin-releasing inhibiting drugs* (especially psychotropic agents, opiates)
- *Primary hypothyroidism*

Important points to be elicited in history:
- History of pubertal development that has started and then stopped or not yet initiated
- History of previous growth pattern should be taken into account (e.g., important gonadotropins, childhood development)
- History of dietary habits (to rule out eating disorders)
- History of previous and present medical/chronic illness (chronic liver disease, chronic kidney disease, hypothyroidism, inflammatory bowel disease/Crohn's disease, chemotherapy especially alkylating agents, history of galactorrhea)

- History of drugs that might delay or retard the pubertal development
- History of pubertal milestones in elder siblings or parents

Physical examination includes:

- Height—compared with norms for that age, also bone age should be checked and then monitoring should be done for at least 6 months.
- Weight
- Arm span
- Tanner staging of breast and pubic hair
- Stigmata of Turner syndrome

Investigations required for evaluation of delayed puberty:

- Serum FSH, serum LH, serum Estradiol, serum prolactin, serum TSH
- Karyotype—main role in cases with increase FSH concentration with delayed pattern
- Bone age—X-ray of nondominant hand
- Magnetic resonance imaging (MRI) brain/computed tomography (CT) head if there is suspicious of any intracranial lesion/signs and symptoms of increased intracranial tension—headache, vomiting
- Done in patients with chronic illness—complete blood count (CBC), erythrocyte sedimentation rate (ESR), liver function test (LFT), renal function test (RFT)

Management of Delayed Puberty[28]

- Counselling, reassurance and psychological support is the mainstay of treatment as constitutional delay is the most important in both sexes.
- Patients with severe pubertal delay, hormone therapy should be considered especially in females with age >12 years without signs of sexual maturation to allay anxiety and depression with an aim to cause age-appropriate sexual maturation.
- Increase in growth spurt along with increase in bone mineral density without epiphyseal closure should also be considered while prescribing the hormonal treatment.
- Follow-up and serial growth monitoring are must in all cases.
- During the treatment bone age must be monitored at a 6 monthly interval.
- Correct the specific cause:
  - Levothyroxine in cases of hypothyroidism
  - Dopamine agonist treatment in cases of hyperprolactinemia
  - Excision of craniopharyngioma or any other central lesion may be required.
  - Growth hormone is indicated in patients with growth hormone deficiency.

Long-term hormonal treatment is required in patients with isolated GnRH deficiency:

- To maintain sex hormones levels in normal physiological range
- To induce ovulation with exposure to gonadotropins when patient comes for fertility treatment.
- 0.25 - 0.5 mg oral estradiol is started initially, gradually increasing at intervals of 3–6 months according to response with an aim to achieve complete sexual maturation over a time span of 2–3 years.
- Progestin is not added with estrogen unless there is substantial breast development because early institution of progestins may adversely affect the breast contour.
- Progestin treatment can be started once patient has attained menarche or after 12–24 months of estrogen treatment.

Hormonal treatment can be discontinued for varying intervals (1–3 months) once the patient has attained complete sexual maturation along with resumption of spontaneous menses which usually happens in cases of constitutional delay.

CONCLUSION

Puberty is a highly coordinated complex process with interplay of central and peripheral signaling mechanisms that require in depth understanding by the clinicians. Any deviation from the normal ensures a detailed workup via thorough history and examination and investigations which would help in establishing a proper diagnosis. A multidisciplinary approach may becomes essential with involvement of pediatric endocrinologist for institution of adequate treatment and long follow up.

REFERENCES

1. Tanner JM. Growth at Adolescence. 2nd edition. Oxford, UK: Blackwell Scientific Publications; 1962.
2. Flynn-Evans EE, Stevens RG, Tabandeh H, Schernhammer ES, Lockley SW. Effect of light perception on menarche in blind women. Ophthalmic Epidemiol. 2008;16:243-8.
3. Baker ER. Body weight and the initiation of puberty. Clin Obstet Gynecol. 1985;28:573-9.
4. Yazigi RA, Quintero CH, Salameh WA. Prolactin disorders. Fertil Steril. 1997;67:215.
5. Lloyd RV, Chandler WF, Kovacs K, Ryan N. Ectopic pituitary adenomas with normal anterior pituitary glands. Am J Surg Pathol. 1986;10:546,
6. Hoffman WH, Gala RR, Kovacs K, Subramanian MG. Ectopic prolactin secretion from a gonadoblastoma. Cancer. 1987;60:2690.

7. Palmer PE, Bogojavlensky S, Bhan AK, Scully RE, Prolactinoma in wall of ovarian dermoid cyst with hyperprolactinemia. Obstet Gynecol. 1990;75:540.
8. Testa G, Vegetti W, Motta T, Alagna F, Bianchedi D, Carlucci C, et al. Two-year treatment with oral contraceptives in hyperprolactinemic patients. Contraception. 1998;58:69.
9. Bentzen JG, Forman JL, Pinborg A, Lidegaard Ø, Larsen EC, Friis-Hansen L, et al. Ovarian reserve parameters: a comparison between users and non-users of hormonal contraception. Reprod Biomed Online. 2012;25(6):612.
10. Zanettini R, Antonini A, Gatto G, Gentile R, Tesei S, Pezzoli G. Valvular heart disease and the use of dopamine agonists for Parkinson's disease. N Engl J Med. 2007;356:39,
11. Ginsburg J, Hardiman P, Thomas M. Vaginal bromocriptine–clinical and biochemical effects. Gynecol Endocrinol. 1992;6:119.
12. Bedecarrats GY, Kaiser UB. Mutations in the human gonadotropin-releasing hormone receptor: insights into receptor biology and function. Semin Reprod Med. 2007;25:368.
13. Layman LC, McDonough PG, Cohen DP, Maddox M, Tho SP, Reindollar RH. Familial gonadotropin releasing hormone resistance and hypogonadotropic hypogonadism in a family with multiple affected individuals. Fertil Steril. 2001;75:1148.
14. Reinhold C, Hricak H, Forstner R, Ascher SM, Bret PM, Meyer WR, et al. Primary amenorrhea: evaluation with MR imaging. Radiology. 1997;203:383.
15. Polasek PM, Erickson LD, Stanhope CR, Transverse vaginal septum associated with tubal atresia. Mayo Clin Proc. 1995;70:965.
16. DiVall SA, Radovick S, Pubertal development and menarche. Ann N Y Acad Sci. 2008;1135:19.
17. Codner E, Roman R, Premature thelarche from phenotype to genotype. Pediatr Endocrinol Rev. 2008;5:760.
18. Garibaldi LR, Aceto T Jr, Weber C. The pattern of gonadotropin and estradiol secretion in exaggerated thelarche. Acta Endocrinol. 1993;128:345.
19. Carel JC, Leger J, Clinical practice. Precocious puberty. N Engl J Med. 2008;358:2366.
20. Bhatia S, Neely EK, Wilson DM. Serum luteinizing hormone rises within minutes after depot leuprolide injection: implications for monitoring therapy. Pediatrics. 2002;109:E30.
21. Carel JC, Eugster EA, Rogol A, Ghizzoni L, Palmert MR, Antoniazzi F, et al. Consensus statement on the use of gonadotropin-releasing hormone analogs in children. Pediatrics. 2009;123:e752,
22. Jay N, Mansfield MJ, Blizzard RM, Crowley WF Jr, Schoenfeld D, Rhubin L, et al. Ovulation and menstrual function of adolescent girls with central precocious puberty after therapy with gonadotropin releasing hormone agonists. J Clin Endocrinol Metab. 1992;75:890.
23. Nunez SB, Calis K, Cutler GB Jr, Jones J, Feuillan PP. Lack of efficacy of fadrozole in treating precocious puberty in girls with the McCune-Albright syndrome. J Clin Endocrinol Metab. 2003;88:5730.
24. Feuillan P, Calis K, Hill S, Shawker T, Robey PG, Collins MT. Letrozole treatment of precocious puberty in girls with the McCune-Albright syndrome: a pilot study. J Clin Endocrinol Metab. 2007;92:2100.
25. Shulman DI, Francis GL, Palmert MR, Eugster EA. Use of aromatase inhibitors in children and adolescents with disorders of growth and adolescent development. Pediatrics. 2008;121:e975.
26. Mieszczak J, Lowe ES, Plourde P, Eugster EA. The aromatase inhibitor anastrozole is ineffective in the treatment of precocious puberty in girls with McCune-Albright syndrome. J Clin Endocrinol Metab. 2008;93:2751,
27. Berek JS. Berek and Novak's Gynecology. Lippincott Williams and Wilkins; 2019. Chapter 8: Puberty. pp. 132-64.
28. Taylor HS, Pal L, Sell E. Speroff's Clinical Gynecologic Endocrinology and Infertility. Wolters Kluwer; 2019. Chapter 9: Normal and abnormal growth and pubertal development. pp. 311-35.

4

Chapter

Premenstrual Syndrome, Dysmenorrhea, and Puberty Menorrhagia

Reena Wani, Sachin Paprikar, Kalpana Koli

PREMENSTRUAL SYNDROME[1-3]

Definition

Premenstrual syndrome (PMS) encompasses a vast array of psychological symptoms such as depression, anxiety, irritability, loss of confidence, mood swings, and physical symptoms such as bloatedness and mastalgia. Symptoms must be present during the luteal phase, resolve with the onset of menstruation, followed by a symptom-free week.

Etiopathogenesis

- Serotonin receptors sensitivity to progesterone and progestogens.
- Reduced levels of allopregnanolone, a progesterone metabolite which modulates gamma-aminobutyric acid (GABA) levels.

Diagnosis

- Self-screening with symptom diaries over at least two consecutive menstrual cycles, before any form of treatment is initiated.
- Symptoms should be recorded prospectively, as retrospective recall of symptoms is unreliable.
- Many patient-rated questionnaires available. However, the Daily Record of Severity of Problems (DRSP) remains the most widely used and is simple for patients to use.

Management

- Women with severe PMS may benefit from being managed by a multidisciplinary team comprising a general practitioner, a gynecologist, a mental health professional (psychiatrist, clinical psychologist, or counselor), and a dietician.
- *Lifestyle modifications:*
  - Exercise
  - Adequate sleep
  - Quit smoking
  - Choose appropriate clothing
  - Calcium, magnesium supplementation
  - Yoga
  - Meditation
- Cognitive behavior therapy
- Vitamin B_6
- *Hormonal:*
  - Combined new generation pill [combined oral contraceptives (COCs)], continuous better than cyclically.
  - First-line: Drospirenone containing COCs
  - Estradiol patches—100 µg + micronized progesterone—100 mg or 200 mg (days 17–28) orally or vaginally or levonorgestrel intrauterine device (LNG-IUD)
  - Micronized progesterone should be considered as first-line progestogens rather than newer progestogens
- *Nonhormonal:*
  - First-line pharmaceutical management in severe PMS: Low dose selective serotonin reuptake inhibitors (SSRIs).
    - Continuously or in luteal phase (days 15–28)
    - Citalopram/escitalopram 10 mg
  - Gonadotropin-releasing hormone (GnRH) analogs: Highly effective in treating severe PMS.
    - When using GnRH analogs for >6 months, add-back hormone therapy should be used.
    - GnRH analogs + add-back hormone replacement therapy (HRT) (continuous combined estrogen + progesterone).

- Example, estradiol patches or estradiol gel combined with micronized progesterone
- *Surgical treatment:*
  - Hysterectomy and bilateral oophorectomy
  - Indications:
    - Failure of medical management
    - Long-term GnRH analog treatment is required
    - Other gynecological conditions requiring surgery.

PUBERTY MENORRHAGIA[2,4-6]

Puberty menorrhagia is defined as excessive bleeding in amount (>80 mL) or duration (>7 days) between menarche and 19 years of age.

Pathophysiology

The majority of cases of puberty menorrhagia are caused by anovulatory cycles caused by hypothalamic-pituitary-ovarian axis immaturity. Because ovulation has not occurred, the endometrium continues to be stimulated by unopposed estrogen, resulting in excessive and prolonged bleeding during menstruation (*see* **Box 1**).

Treatment

Acute Management

- Adolescents who are hemodynamically unstable or actively bleeding heavily should be hospitalized for management.

BOX 1: Causes of puberty menorrhagia.

Immaturity of hypothalamic–pituitary–ovarian axis
Anovulatory cycles
Polycystic ovary syndrome
Organic diseases
- Coagulation defects:
  - Von Willebrand disease
  - Platelet function defects
  - Thrombocytopenia
  - Clotting factor deficiencies
- Estrogen producing tumors
- Arteriovenous malformations of vessels supplying the uterus
- Pelvic tuberculosis
- Endocrine:
  - Diabetes mellitus
  - Thyroid disease
- Renal and hepatic disease

Local causes
- Genital trauma
- Foreign body

- *Volume expansion* with crystalloid.
- *Hormonal therapy:*
  - Available options:
    - Intravenous conjugated estrogen every 4–6 hours.
    - Monophasic combined oral contraceptive pills (OCPs) every 6–8 hours until cessation of bleeding.
    - Progesterone only therapy if estrogens are contraindicated or not tolerated. Oral norethindrone acetate 5–10 mg 6 hourly or medroxyprogesterone 10–20 mg every 6–12 hourly.
- *Hemostatic agents* such as tranexamic acid, aminocaproic acid, and desmopressin.
- *Transfusion:* Hemodynamically unstable patients or those with severe symptoms of anemia.
- *Surgical options:* When there is a lack of response to medical therapy, if the patient is clinically unstable despite initial measures—intrauterine balloon tamponade and suction evacuation or suction curettage.

Long-term Management (see Box 2)

- *First-line therapy:* COC
  - Monophasic pills with 30–50 µg of ethinylestradiol and second-generation progesterone.
  - Regimen: One pill twice or thrice a day until the bleeding stops, then one pill a day for at least 21 days; treatment is continued for 3–6 months.
- *Second-line therapy:* Progestin therapy, if estrogen-containing therapy is not tolerated is contraindicated. Available options:
  - Micronized progesterone (200 mg/day)
  - Medroxyprogesterone (10 mg/day)
  - Norethindrone acetate (2.5–5 mg/day)
  - Depot medroxyprogesterone acetate (DMPA)
  - Levonorgestrel-releasing intrauterine device (less preferred)
- *Treatment for iron deficiency anemia*: Oral iron supplementation along with dietary counseling to increase iron intake.

DYSMENORRHEA[6-10]

Greek, dys—difficult/painful/abnormal; meno—month; and rrhea—flow.

Types

Dysmenorrhea is classified as primary or secondary *on the basis of associated anatomic pathology.*

BOX 2: Evaluation of cases.

Severity classification

Mild: Longer menses (>7 days) or shorter cycles (<3 weeks) for 2 months in succession, with slightly or moderately increased bleeding, a usually normal (≥12 g/dL) or mildly decreased hemoglobin (Hb) (10–12 g/dL)

Moderate: Moderately prolonged or frequent (every 1–3 weeks) menses, with moderate to heavy bleeding and a hemoglobin level of ≥10 g/dL

Severe: Heavy bleeding with a hemoglobin level of <10 g/dL

Evaluation

History:
- Detailed menstrual history
- Sexual history
- History of prolonged bleeding after surgery or tooth extraction
- Epistaxis
- Bleeding gums
- Easy bruising
- Family history of coagulation disorders

Physical examination:
- Hemodynamic stability
- Dermatologic signs of anemia and bleeding disorders: Pallor, presence of bruises, and petechiae
- Abdominal examination: Tenderness, distention, hepatosplenomegaly, or masses
- Vaginal examination is rarely indicated in an adolescent
- Evidence of injuries which may point toward sexual assault

Laboratory investigations:
- Urine pregnancy test
- Complete blood count
- Peripheral blood smear
- Blood group
- Prothrombin time
- Activated partial thromboplastin time
- Fibrinogen
- Von Willebrand panel: Plasma for Von Willebrand factor (vWF) antigen, functional tests for vWF, factor VIII activity

Pelvic ultrasound:
- If bleeding accompanied by palpable mass
- Pregnancy and pregnancy related conditions are suspected
- Patients who do not respond to initial management

- *Primary dysmenorrhea* is defined as menstrual pain in the absence of pelvic pathology.
- *Secondary dysmenorrhea* is defined as painful menses associated with underlying pathology.

Clinical Varieties

- *Spasmodic dysmenorrhea:* Cramping pain, generally most pronounced on the first and second day.

BOX 3: Causes of secondary dysmenorrhea.

- *Common:*
  - Endometriosis
  - Chronic pelvic inflammatory disease
  - Adenomyosis
  - Intrauterine polyps
  - Submucosal fibroids
  - Intrauterine contraceptive devices syndrome
- *Less common:*
  - Congenital uterine abnormalities
  - Cervical stenosis
  - Asherman syndrome
  - Uterine retroversion
  - Pelvic congestion
  - Ovarian cysts

- *Congestive dysmenorrhea:* Increasing pelvic discomfort and pain a few days before menses begin, with relief in symptoms after the onset of menses.
- *Membranous dysmenorrhea:* Special group in which the endometrium is shed as a cast during menses accompanied by painful uterine cramps.

Pathophysiology

- Pathophysiologically, prostaglandins are implicated in dysmenorrhea.
- As the progesterone levels begin to decline in the late luteal phase, lytic enzymatic actions begin. These trigger various pathways [including the cyclooxygenase (COX) pathway] resulting in the release of prostaglandins as menstruation begins.
- Higher levels of prostaglandins in dysmenorrheic females result in higher uterine tone, high amplitude contractions, and ischemia causing dysmenorrhea.
- Depending on the type of concomitant pelvic disease, prostaglandins and anatomical mechanisms are involved in secondary dysmenorrhea.

Diagnosis (Table 1)

- Diagnosis of primary dysmenorrhea is based on history and presence of a normal pelvic examination.
- Women suspected of suffering from secondary dysmenorrhea may require review of a pain diary and tests to confirm the clinical diagnosis, the extent and type of underlying pathology need to be carried out. An ultrasound examination or laparoscopy or hysteroscopy or both may be required.

TABLE 1: How to distinguish between primary and secondary dysmenorrhea.

	Primary dysmenorrhea	*Secondary dysmenorrhea*
Onset	Shortly after menarche	During reproductive age group, 30s or 40s
Duration	8–72 hours during menses	Prior to onset of menses and throughout menstrual cycle
Nature	Cramping pelvic pain, with or without nausea and vomiting, which commences with the start of menstrual flow; the pain may radiate to the lower back or upper legs	Variable number of days; noncyclical and cyclical episodes
Coexistent gynecological symptoms	Usually, no other gynecological symptoms	Coexistent gynecological symptoms, e.g., heavy menstrual bleeding (HMB), dyspareunia, vaginal discharge, intermenstrual bleeding, postcoital bleeding, and chronic pelvic pain
Clinical examination	Normal pelvis	Abnormal pelvic examination; findings based on underlying disorder
Pelvic ultrasound	Normal pelvic ultrasound	Adenomyosis, uterine fibroids, and ovarian endometriosis

Treatment

- Dysmenorrhea treatments aim to relieve pain and symptoms by affecting the physiological mechanisms behind menstrual cramps (such as prostaglandin production) or symptom relief.
- For women with menstrual cramps and no other associated findings or symptoms, no additional evaluation is initially required; can be treated empirically.
- Analgesics like paracetamol, piroxicam or antispasmodics like hyoscine compounds, camylofin, drotaverine or nonsteroidal anti-inflammatory drugs (NSAIDs) like mefenamic acid provide pain relief.
- Steroid hormone contraception inhibit ovulation, decrease endometrial proliferation and in turn lower endometrial prostaglandin levels.
- Cyclic COC pills can be used; extended or continuous administration regimen in women with pain not controlled by the traditional cyclic pill schedule.
- Progestin-only contraceptives such as the levonorgestrel-releasing intrauterine system (LNG-IUS), depot medroxyprogesterone acetate injection, and progestin-releasing implanted rods can also be used.
- Oral vitamins E, fish oil, low-fat diet, and Chinese herbal medicine improve dysmenorrhea.
- Nonpharmacologic pain management, in particular acupuncture or transcutaneous electrical nerve stimulation (TENS) may also be useful.
- Surgery is indicated if medical measures fail to provide relief and in women with secondary dysmenorrhea to treat the underlying pelvic pathology. Surgical interventions may be based on severity of symptoms, patient's age, desire for childbearing, and menstrual functions.

REFERENCES

1. AOGD Secretaria. (2020). AOGD Bulletin. [online] Available from http://aogd.org/AOGD-Bulletin-January-2020.pdf. [Last accessed December, 2023].
2. Management of Premenstrual Syndrome: Green-top Guideline No. 48. BJOG. 2017;124(3):e73–105.
3. Barthwal NG. (2021). Pre-Menstrual Syndrome: The Emotional Roller-Coaster. [online] Available from https://www.fogsi.org/wp-content/uploads/committee-2020-activities/vol-22-endocrinology-committee-newsletter.pdf. [Last accessed December, 2023].
4. ACOG. (2019). Screening and Management of Bleeding Disorders in Adolescents with Heavy Menstrual Bleeding. [online] Available from https://www.acog.org/en/clinical/clinical-guidance/committee-opinion/articles/2019/09/screening-and-management-of-bleeding-disorders-in-adolescents-with-heavy-menstrual-bleeding. [Last accessed December, 2023].
5. The Royal Children's Hospital Melbourne. (2020). Adolescent Gynaecology—Heavy Menstrual Bleeding. [online] Available from https://www.rch.org.au/clinicalguide/guideline_index/Adolescent_Gynaecology_Menorrhagia/. [Last accessed December, 2023].

6. Mavrelos D, Saridogan E. Treatment options for primary and secondary dysmenorrhoea. Prescriber. 2017;28(11):18-25.
7. Proctor M, Farquhar C. Diagnosis and management of dysmenorrhoea. BMJ. 2006;332(7550):1134–8.
8. Wallace S, Keightley A, Gie C. Dysmenorrhoea. Obstet Gynaecol. 2010;12(3):149-54.
9. FOGSI. Adolescent Health Committee FOGSI: Adolescence. [online] Available from https://www.fogsi.org/wp-content/uploads/fogsi-focus/Adolescence-Issue-11.pdf. [Last accessed December, 2023].
10. ACOG. (2018). Dysmenorrhea and Endometriosis in the Adolescent. [online] Available from https://www.acog.org/en/clinical/clinical-guidance/committee-opinion/articles/2018/12/dysmenorrhea-and-endometriosis-in-the-adolescent. [Last accessed December, 2023].

Chapter 5

Leiomyoma

Anupama Bahadur, Rajlaxmi Mundhra, Gayatri Suresh

INTRODUCTION

Leiomyomas are benign and monoclonal tumors of the uterine smooth muscles. They are the most common benign tumors in women of reproductive age group with a prevalence of 70–80% by 50 years of age.[1] While most women with fibroid uterus remain asymptomatic, 25–30% have symptoms affecting quality of life.[2] Fibroid uterus constitutes the most common indication for performing hysterectomy accounting for 30–40% of cases.

RISK FACTORS (FIG. 1)[3]

- *Race:* Incidence of fibroids is particularly higher in the African race compared to Caucasians. Fibroids are thrice more commonly found in African American women in comparison to the white race. This is probably a result of racial differences in the serum follicle-stimulating hormone (FSH) levels.
- *Age:* Advancing age is a significant factor for the development of uterine fibroids. For example, incidence of fibroids in African American women increase from 60% in the age group of 35–49 years to >80% in the age group of >50 years.
- *Obesity:* Obesity has a direct role in creating hormone imbalance. Studies have demonstrated that increased body mass index (BMI) proportionally increases one's risk for development of fibroids.
- *Parity:* Increased parity has a protective role against development of fibroids.
- *Hypertension:* Women with hypertension have a five times higher risk of suffering from fibroids.

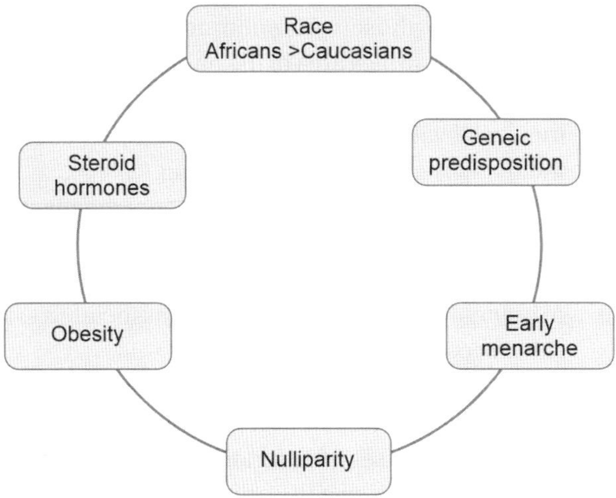

Fig. 1: Risk factors of leiomyomas.

PATHOGENESIS (FLOWCHART 1)[4]

The various risk factors for development of fibroids result in chronic inflammation and alteration in local hormonal milieu which in turn results in mutations leading to fibroid development. Further steroid hormones play an important role in the growth of fibroids. Progesterone helps in fibroid growth whereas estrogen is implicated in inducing progesterone receptors on the fibroid.

Types of Fibroids (Fig. 2)

Classification of fibroids is based on location of fibroid within the uterus.

Flowchart 1: Pathogenesis of fibroids.[4]

Tumorigenesis by mutation

Estradiol induces progesterone receptor on fibroid ↔ Progesterone facilitates fibroid growth

Cell proliferation | Extracellular matrix development | Decreased apoptosis

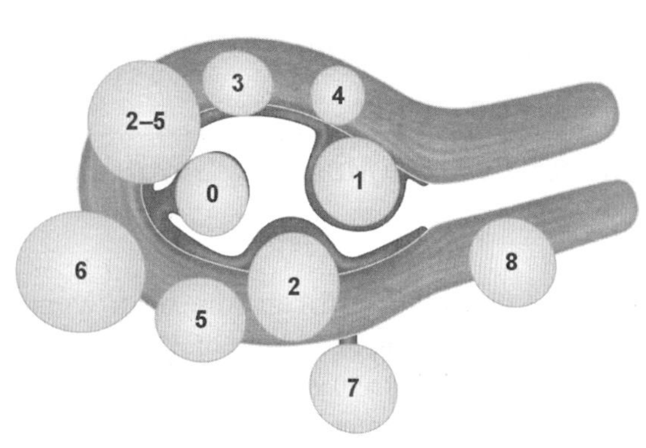

SM-Submucous	0	Pedunculated intracavitary
	1	<50% intramural
	2	≥50% intramural
	3	Contacts endometrium; 100% intramural
O-Other	4	Intramural
	5	Subserous ≥50% intramural
	6	Subserous <50% intramural
	7	Subserous pedunculated
	8	Other (specify e.g., cervical, parasitic)

Hybrid	Two numbers are listed separated by a hyphen. By convention, the first refers to the relationship with the endometrium while the second refers to the relationship to the serosa. One example is below	
(contact both the endometrium and the serosal layer)	2–5	Submucous and subserous, each with less than half the diameter in the endometrial and peritoneal cavities, respectively.

Fig. 2: The International Federation of Gynaecology and Obstetrics (FIGO) 2018 classification.

The International Federation of Gynaecology and Obstetrics 2018 Classification[5]

Changes in 2018:

- Inclusion of type 3 as submucous leiomyoma
- Distinction between following leiomyoma subtypes:
  - *Types 0 and 1:* Stalk diameter ≤10% of mean leiomyoma diameter in type 0
  - *Type 6 and 7:* Stalk diameter ≤10% of mean leiomyoma diameter in type 7
  - *Type 2 and 3:* Based upon the lowest filling pressure that allows visualization of the endometrial cavity
  - *Type 4 and 5:* Based upon observation of distortion of the serosa by ultrasonography or MRI.

The European Society for Gynaecological Endoscopy (ESGE) classification of submucous myomas [determine by saline infusion sonography (SIS) or hysteroscopy] is given in **Table 1**.

TABLE 1: Classification of submucous myomas.

Type 0	Entirely within endometrial cavity No myometrial extension (pedunculated)
Type I	<50% myometrial extension (sessile) <90-degree angle of myoma surface to uterine wall
Type II	≥50% myometrial extension (sessile) ≥90-degree angle of myoma surface to uterine wall

CLINICAL FEATURES

- *Asymptomatic (50–60%):*[3] Almost half the patients with fibroids remain asymptomatic with fibroids being diagnosed incidentally.
- *Symptomatic:* The most common symptoms associated with fibroids include **(Table 2)**:
  - *Heavy menstrual bleeding (HMB):* Heavy bleeding and prolonged cycles are associated with fibroids, particularly those with submucosal component. Increase in surface area of the endometrium, endometrial proliferation, altered uterine contractility, increased vascularity around the fibroid is some of the reasons causing heavy bleeding.
  - *Dysmenorrhea:* Spasmodic dysmenorrhea is typically associated with submucosal and intramural fibroids.
  - *Intermenstrual bleeding:* May be associated with submucosal fibroids.
  - *Pressure symptoms:* The bulk of the tumor may result in symptoms such as increased urinary frequency in cases of anterior uterine wall fibroids causing pressure over the urinary bladder. Similarly large fibroids, particularly posterior wall fibroids

TABLE 2: Common symptoms of fibroids.[3]

Heavy menstrual bleeding (HMB)	*Pressure symptoms*	*Reproductive dysfunction*
• Most common symptom • Heavy or prolonged menses-typical pattern • Associated with dysmenorrhea	• Pelvic pain/pressure • Urinary tract or bowel complaints • Anterior wall fibroid-increased urinary frequency • Posterior wall-urinary retention • Large fibroids—>6 cm or uterus >18 weeks—risk of Hemolytic disease of the newborn (HDN) • Venous compression	• Infertility • Recurrent pregnancy loss • Pregnancy complications

TABLE 3: Fibroids presenting as emergency.[3]

Torsion	*Prolapse*	*Degeneration*
• *Presentation:* Acute pain • *Risk factors:* Large subserosal fibroid and long thin stalk • *Diagnosis:* Ultrasonography (USG) Doppler • *Management:* Surgical	• *Presentation:* Mass protruding through vagina • *Risk factors:* Pedunculated type 0 and 1 fibroid and postmedical management (GnRH agonist) • *Diagnosis:* Clinical • *Management:* Surgical	• *Types:* – Fatty – Cystic – Calcific – Sarcomatous – Hyaline – Red (common in pregnancy) • *Presentation:* Pelvic pain and low grade fever • *Investigation:* MRI • *Management:* Medical

can cause compression over the ureters leading to hydroureteronephrosis. Large fibroids can even result in venous compression and resultant lower limb edema.

- *Reproductive dysfunction:*
  - Recurrent pregnancy loss—submucosal fibroids and large intramural fibroids that distort uterine cavity may result repeated abortions.
  - Fibroids may also result in infertility. Alteration in the local hormonal environment, increased presence of inflammatory cytokines could be contributory factors in infertility. Mechanical factors such as cornual fibroids disturbing tubal function may also contribute to infertility.
  - In pregnancy, fibroids may lead to complications such as preterm labor, malpresentation, abortions, antepartum and postpartum hemorrhage, red degeneration of fibroid, and dysfunctional labor.

- Fibroids presenting as emergency **(Table 3)**

DIAGNOSIS[3,6]

See **Flowchart 2**.

MANAGEMENT

See **Flowchart 3**.

Medical Management

Nonhormonal

- *Nonsteroidal anti-inflammatory drugs (NSAIDs):*
  - Not effective for treatment of menorrhagia in women with fibroids
  - May decrease fibroid related dysmenorrhea
  - They act by decreasing the levels of prostaglandins which are implicated in causing dysmenorrhea
- *Antifibrinolytic agents:*
  - Tranexamic acid (500 mg to 1 g TDS) used as a first-line nonhormonal therapy for HMB due to fibroids. The drug prevents conversion of plasminogen to plasmin thereby inhibiting fibrinolysis and decreasing bleeding. They result in up to 40% decrease in bleeding during menstruation
  - *Hormonal:*
    - *Oral contraceptive pills (OCPs):*[7]
      - Low dose OCPs are not known to cause fibroid growth and hence can be safely used

Flowchart 2: Approach to the diagnosis of fibroids.[3,6]

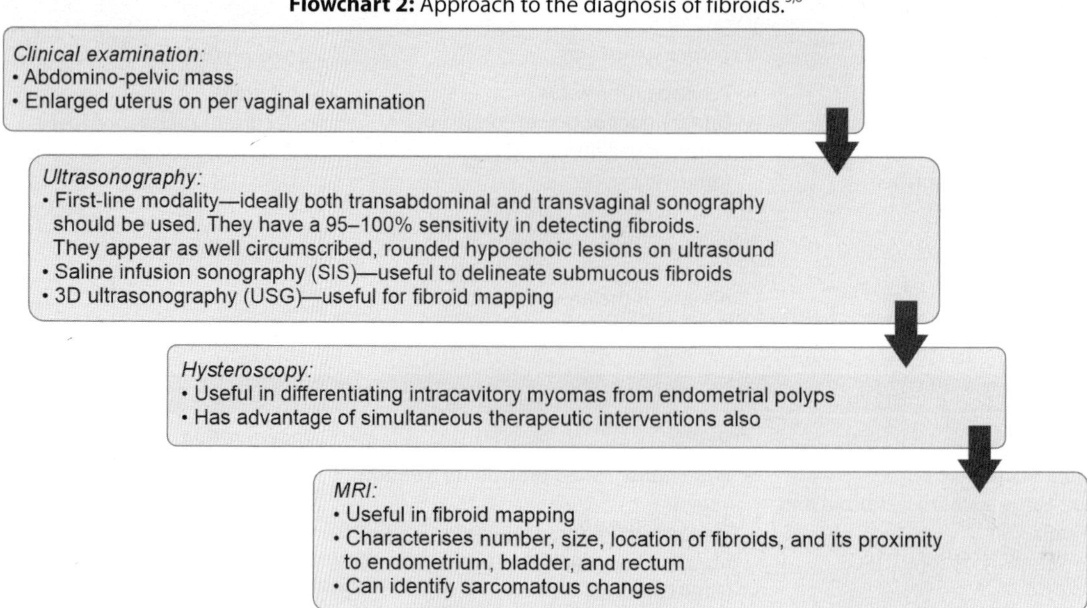

Flowchart 3: Management options for fibroid uterus.[6]

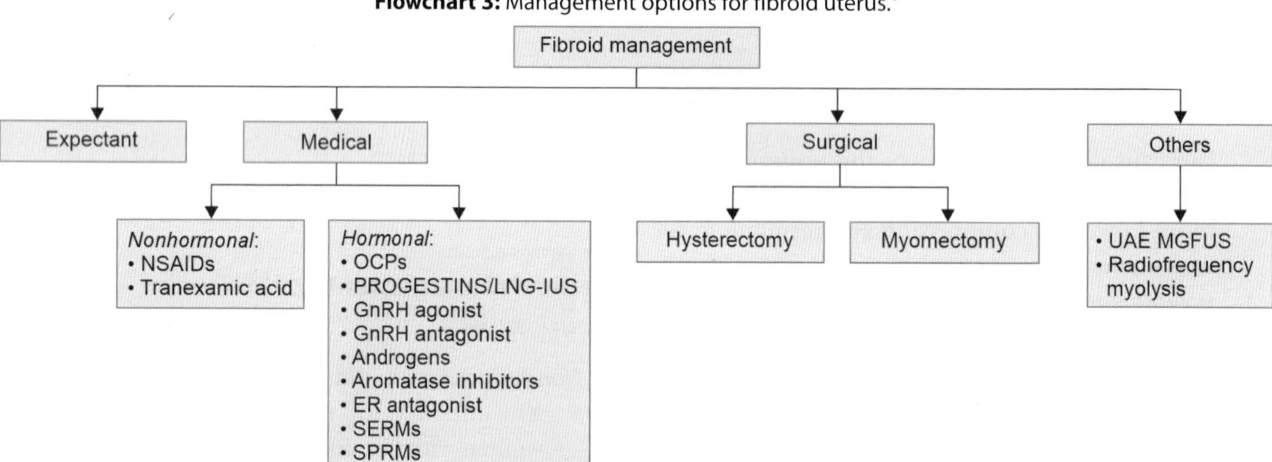

(ER: estrogen receptor; LNG-IUS: levonorgestrel-releasing intrauterine system; NSAIDs: nonsteroidal anti-inflammatory drugs; OCPs: oral contraceptive pills; SERMs: selective estrogen receptor modulators; SPRMs: selective progesterone receptor modulators)

- They reduce menstrual blood loss and suppress new fibroid development. They also provide additional benefits of providing contraception, decreasing anemia, and also dysmenorrhea
- *Progestins/levonorgestrel-releasing intrauterine system (LNG-IUS):*[8,9]
  - They cause downregulation of estrogen and progesterone receptors on fibroids and cause their shrinkage
  - Additionally, they cause endometrial atrophy and decreases bleeding. They however may not be appropriate in cases of submucosal fibroids or large intramural fibroids causing cavity distortion where expulsion rates of the device are high
- *GnRH agonist:*[10,11]
  - GnRH agonist can result in shrinkage of fibroids by up to 50% in 3 months

- They act by downregulating and desensitizing GnRH receptors

  Uses:

  i. Preoperatively to reduce fibroid volume
  ii. Decrease bleeding and resultant anemia

  Disadvantages:

  i. Initial flare of symptoms may be encountered
  ii. Use restricted to 3–6 months (risk of menopausal symptoms)
  iii. Prolonged use needs add back therapy
  iv. Withdrawal of drug leads to regrowth of fibroids

- *GnRH antagonist:*
  - They are competitive blockers and have action similar to GnRH agonists
  - *Advantage:* Fast onset of action, no risk of flare up
  - *Disadvantage:* Long-acting preparations are not available; thus, daily injections needed
- *Androgens (danazol):*[12]
  - Competes with estrogen and progesterone receptors
  - Decrease gonadotropin secretion and reduces ovarian steroid production
  - Associated with 20–25% reduction in volume
  - *Disadvantage:* Androgenic side effects
- *Aromatase inhibitors:*[13]
  - Inhibits conversion of androgens to estrogen
  - Under evaluation for use in HMB
- *Estrogen receptor antagonist (fulvestrant):*[14]
  - Degrades estrogen receptors and suppresses their action
- Selective estrogen receptor modulators (SERM):
  - Nonsteroidal agents with tissue-specific agonist/antagonist action on estrogen receptor
  - Raloxifene (60 mg/day dosage)—useful adjunct to GnRH agonist to shrink fibroids
- *Selective progesterone receptor modulators:*
  - Mifepristone (RU-486)
    - Pure progesterone antagonist
    - Dosage: 2.5 mg daily for 3–6 months
    - Causes fibroid shrinkage
  - Ulipristal acetate
    - Dosage: 5–10 mg daily for 3 months
    - Causes fibroid size reduction by 12–20%
    - MHRA safety alert against ulipristal issued in 2018 in view of transaminitis associated with drug use

Surgical Management

- *Myomectomy:*[15]
  - Indicated in women with symptomatic fibroids, not responding to medical management or fibroids causing pressure symptoms or reproductive dysfunction
  - Preferred surgical intervention in young women desirous of future fertility
  - *Approaches:* Laparotomy, hysteroscopy, laparoscopy, and robot-assisted laparoscopy
  - *Disadvantages:*
    - Increased blood loss
    - Recurrence risks—15%; 10% patients eventually require hysterectomy in 5–10 years
- *Hysterectomy:*
  - Option in women who have completed childbearing
  - Preferred treatment in postmenopausal women with enlarging fibroids
  - *Approaches:* Abdominal/vaginal/minimally invasive

Other Management Options

- *Uterine artery embolization (UAE):*[16]
  - Minimally invasive technique
  - May cause fibroid size reduction by around 50%; however, risk of reintervention is around 30%
  - Contraindicated in PID, contrast allergy, renal failure, pregnancy, connective and tissue disorder
  - *Complications*: Postembolization syndrome, puncture site thrombosis, uterine artery dissection, and sepsis
- *Magnetic Resonance-guided focused Ultrasound Surgery (MRgFUS):*
  - *Indications:* Few fibroids and dimension <10 cm
  - *Contraindications:* Degenerated fibroids, multiple fibroids (≥5), fibroids close to bladder/bowel, and subserosal fibroid that can detach
  - *Complications:* Damage to adjacent structures (bowel/bladder), abdominal burns, deep vein thrombosis (DVT), and time consuming
- *Transcervical radiofrequency ablation:*[17]
  - Causes coagulative necrosis of fibroid
  - Outpatient procedure
  - Fibroid size reduction by 60–70%
  - Reintervention rates: 12–15%

Flowchart 4: Leiomyoma management algorithm.[6]

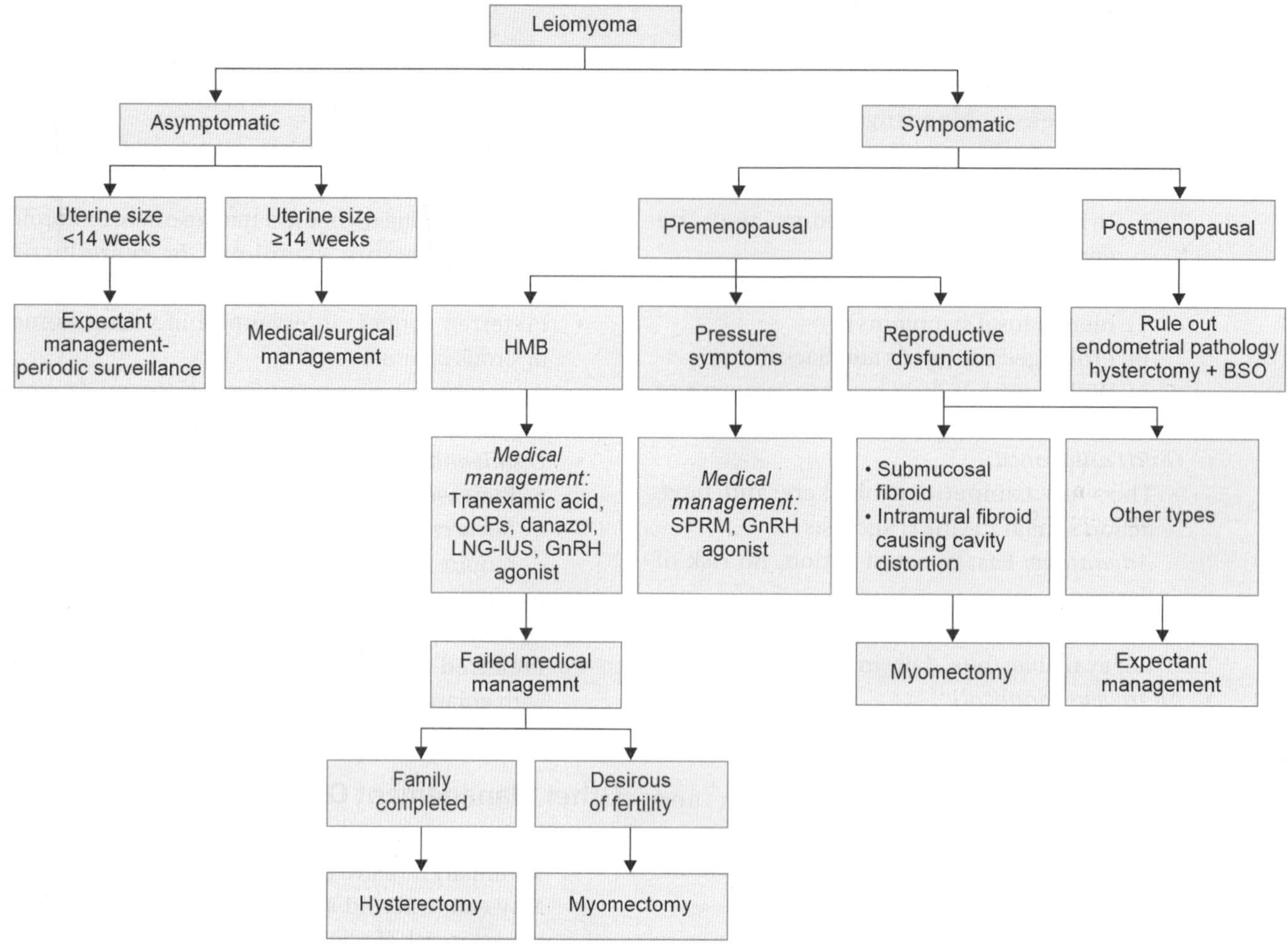

(HMB: heavy menstrual bleeding; LNG-IUS: levonorgestrel-releasing intrauterine system; OCPs; oral contraceptive pills; SPRM: selective progesterone receptor modulator)

REFERENCES

1. Baird DD, Dunson DB, Hill MC, Cousins D, Schectman JM. High cumulative incidence of uterine leiomyoma in black and white women: ultrasound evidence. Am J Obstet Gynecol. 2003;188(1):100-7.
2. Okolo S. Incidence, aetiology and epidemiology of uterine fibroids. Best Pract Res Clin Obstet Gynaecol. 2008;22(4):571-88.
3. Donnez J, Dolmans MM. Uterine fibroid management: from the present to the future. Hum Reprod Update. 2016;22(6):665-86.
4. Bulun SE. Uterine fibroids. N Engl J Med. 2013;369(14):1344-55.
5. Munro MG, Critchley HOD, Fraser IS, FIGO Menstrual Disorders Committee. The two FIGO systems for normal and abnormal uterine bleeding symptoms and classification of causes of abnormal uterine bleeding in the reproductive years: 2018 revisions. Int J Gynaecol Obstet. 2018;143(3):393-408.
6. Khan AT, Shehmar M, Gupta JK. Uterine fibroids: current perspectives. Int J Womens Health. 2014;6:95-114.
7. Sayed GH, Zakherah MS, El-Nashar SA, Shaaban MM. A randomized clinical trial of a levonorgestrel-releasing intrauterine system and a low-dose combined oral contraceptive for fibroid-related menorrhagia. Int J Gynaecol Obstet. 2011;112(2):126-30.
8. Maruo T, Ohara N, Yoshida S, Nakabayashi K, Sasaki H, Xu Q, et al. Translational research with progesterone receptor modulator motivated by the use of levonorgestrel-releasing intrauterine system. Contraception. 2010;82(5):435-41.
9. Englund K, Blanck A, Gustavsson I, Lundkvist U, Sjöblom P, Norgren A, et al. Sex steroid receptors in human

myometrium and fibroids: changes during the menstrual cycle and gonadotropin-releasing hormone treatment. J Clin Endocrinol Metab. 1998;83(11):4092-6.

10. Carr BR, Marshburn PB, Weatherall PT, Bradshaw KD, Breslau NA, Byrd W, et al. An evaluation of the effect of gonadotropin-releasing hormone analogs and medroxyprogesterone acetate on uterine leiomyomata volume by magnetic resonance imaging: a prospective, randomized, double blind, placebo-controlled, crossover trial. J Clin Endocrinol Metab. 1993;76(5):1217-23.
11. Friedman AJ, Daly M, Juneau-Norcross M, Rein MS, Fine C, Gleason R, et al. A prospective, randomized trial of gonadotropin-releasing hormone agonist plus estrogen-progestin or progestin "add-back" regimens for women with leiomyomata uteri. J Clin Endocrinol Metab. 1993;76(6):1439-45.
12. ACOG Committee on Practice Bulletins-Gynecology. ACOG practice bulletin. Surgical alternatives to hysterectomy in the management of leiomyomas. Number 16, May 2000 (replaces educational bulletin number 192, May 1994). Int J Gynaecol Obstet. 2001;73(3):285-93.
13. Song H, Lu D, Navaratnam K, Shi G. Aromatase inhibitors for uterine fibroids. Cochrane Database Syst Rev. 2013;(10):CD009505.
14. Wakeling AE, Dukes M, Bowler J. A potent specific pure antiestrogen with clinical potential. Cancer Res. 1991;51(15):3867-73.
15. García CR. Management of the symptomatic fibroid in women older than 40 years of age. Hysterectomy or myomectomy? Obstet Gynecol Clin North Am. 1993;20(2):337-48.
16. Ukybassova T, Terzic M, Dotlic J, Imankulova B, Terzic S, Shauyen F, et al. Evaluation of Uterine Artery Embolization on Myoma Shrinkage: Results from a Large Cohort Analysis. Gynecol Minim Invasive Ther. 2019;8(4):165-71.
17. Miller CE, Osman KM. Transcervical Radiofrequency Ablation of Symptomatic Uterine Fibroids: 2-Year Results of the SONATA Pivotal Trial. J Gynecol Surg. 2019;35(6):345-9.

Chapter 6

Endometriosis and Adenomyosis

Kavita Mandrelle Bhatti

ENDOMETRIOSIS

Endometriosis is a chronic inflammatory disease with presence of endometrium-like tissue outside the uterine cavity. It is defined as a disease characterized by the presence of endometrial glands and/or stroma outside the endometrium and myometrium.

The disease is estrogen-dependent and occurs in women of reproductive age.

Prevalence

The exact prevalence of endometriosis is unknown. Incidence ranges from 2 to 10% within the general female population.

In women with infertility the estimated prevalence is up to 50% and 90% in those with chronic pelvic pain.[1]

Endometriosis can involve a number of sites. Some of the common and less common sites are described in **Table 1**.

Etiological Factors for Endometriosis

- Delayed childbearing and low parity
- Genital outflow obstruction and developmental disorders of the genital tract
- Environmental toxins and family history of endometriosis.

TABLE 1: Sites of endometriosis.

Common anatomic sites	*Less common sites*
Pelvic peritoneum	Sigmoid colon
Ovaries	Appendix
Pouch of Douglas	Fallopian tube
Uterosacral ligaments	Umbilicus
Rectovaginal septum	Episiotomy
	Caesarean and surgical scars
	Extra-abdominal sites Ureter, kidneys, lungs, and pleura

Pathogenesis (Figs. 1A and B)

Endometrial tissue at ectopic sites is responsive to hormones which create a proinflammatory milieu due to the local production of cytokines, prostaglandins, interleukins, vascular endothelial growth factors, and neurogenic growth factors. The inflammation leads to abnormal peritoneal environment that causes adhesions.

There are various theories to explain the occurrence of endometriosis at different sites:

- *Sampson's retrograde menstruation theory:* Retrograde flow of menstrual blood seen in cases of genital tract obstruction and Müllerian developmental abnormalities.
- *Coelomic metaplasia theory:* Menstrual blood may cause chronic irritation of the peritoneum and coelomic metaplasia.
- *Direct implantation theory:* The implantation of endometrial or decidual tissue at scar tissue sites following surgeries may explain occurrence of scar endometriosis, at surgical sites after caesarean section, episiotomy, and myomectomy.
- *Genetic basis and familial disposition:* It is commonly seen to occur in first-degree relatives and have a family predisposition. Ethnic groups like Asian women are also predisposed to occurrence of endometriosis.
- *Lymphatic and vascular theory:* Vascular and lymphatic spread can explain the occurrence of endometriosis at distant sites.
- *Environmental toxins* such as dioxins can cause somatic mutations of cells.

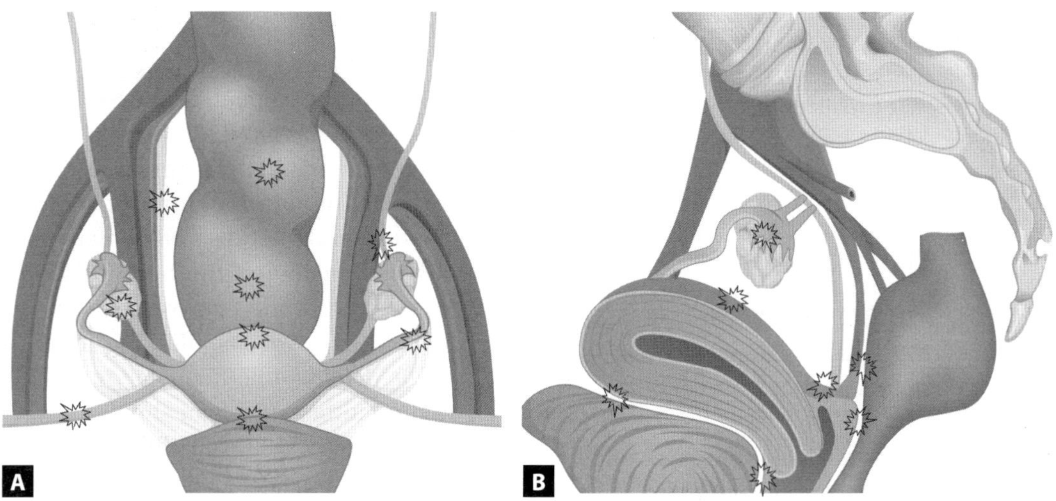

Figs. 1A and B: Drawing of the female pelvis in the (A) ventral and (B) lateral view illustrating the primary locations of endometriotic lesions.[2]

A single theory cannot explain the occurrence of endometriosis at different sites.

Pathology

Active endometrial tissue at the ectopic sites shed blood which acts as an irritant and leads to adhesions and puckering at the sites of the deposits. Encysted blood in the ovarian tissue enlarge to form the chocolate cyst. A pseudocyst is formed by an accumulation of menstrual blood which becomes chocolate in color and resembles "anchovy sauce". The adhesions cause distortion of the tubo-ovarian anatomic relation.

The fallopian tubes remain patent in pelvic endometriosis.

The endometriotic deposits on the peritoneum may appear typically as brownish black spots known as "powder burns" or as red flame shaped, yellow gelatinous patches, white puckered fibrotic areas, or as peritoneal defects.

Microscopically, the cyst lining is cuboidal and have hemosiderin laden large phagocytic cells also called pseudoxanthoma cells. Histopathologically, the endometriotic deposits have both endometrial glands and endometrial stroma.

Pelvic/abdominal endometriosis is clinically subdivided into:

- Superficial (peritoneal/serosal) lesions
- Ovarian endometriotic cysts (endometrioma)
- Deep endometriosis (defined as >5 mm below the serosal/peritoneal surface).[3]

Clinical Features

Cyclical and noncyclical signs and symptoms:

- Dysmenorrhea
- Deep dyspareunia
- Dysuria
- Dyschezia
- *Abdominopelvic pain:* Pain is unspecific and severity may not correlate with the extent of disease
- Heavy menstrual bleeding
- Painful rectal bleeding or hematuria
- Shoulder tip pain
- Catamenial pneumothorax
- Cyclical cough/hemoptysis/chest pain
- Cyclical scar swelling and pain
- Infertility
- Previous diagnosis of ovarian cyst
- Symptoms of irritable bowel syndrome or pelvic inflammatory disease.

Patients reporting multiple symptoms have increased chances of endometriosis ***(Box 1)****.*

Clinical Examination

Inspection and palpation of the abdomen—there may be no specific findings. If endometriotic cysts are large they may be felt as firm, tender, and fixed masses.

Speculum examination of vagina and cervix show bluish nodules in the posterior fornix which are tender on touch.

Bimanual and rectovaginal palpation may reveal an immobile/fixed uterus and adnexal mass (bilateral or

BOX 1: Causes for infertility in endometriosis.

Causes for infertility in endometriosis:
- Distorted pelvic anatomy due to adhesions
- Anovulation or oligo-ovulation
- Poor oocyte quality
- Impaired fertilization
- Luteal phase defect
- Poor endometrial receptivity
- Implantation failure
- Peritoneal factors—increased macrophages, growth factors, and cytokines

unilateral) or tenderness in the vaginal fornices. A tender nodule may be palpable in the rectovaginal septum.

Investigations

- *Biomarkers:*
  - No biological markers currently exist that reliably can rule in and rule out endometriosis. CA-125 can be considered as a screening marker for symptomatic patients.
  - *CA-125:*
    - Specificity of 93% [95% confidence interval (CI) 89–95%]
    - Sensitivity of only 52% (95% CI 38–66%)
  - It is inexpensive and widely available.
  - A negative result does not rule out the disease
  - It is considered that even a positive test is not clinically relevant, and may cause anxiety in the patient, and possible overtreatment. Cancer antigen CA-125 may be moderately raised in severe endometriosis. It does not have a diagnostic value since it is also raised in other conditions. In endometriosis, CA-125 may be useful to monitor therapeutic response and to detect recurrence of the disease.
  - As such, CA-125 testing is not considered relevant in the diagnosis of endometriosis.
- *Imaging techniques:*
  - *Transvaginal ultrasonography* is the initial imaging of choice. The chocolate cysts appear as adnexal masses with homogenous internal echoes also known as "ground glass appearance". The cysts may be unilateral or bilateral. Transvaginal and transrectal ultrasonography are useful for identifying lesions of the rectovaginal septum and uterosacral ligaments.
  - *Renal ultrasonography* should be done in cases of severe endometriosis to rule out ureteric involvement and consequent hydroureteronephrosis.
  - *Magnetic resonance imaging:* T1- and T2-weighted magnetic resonance imaging is superior to transvaginal and computed tomography to diagnose endometriosis particularly small nodules. MRI is considered the best imaging technique for mapping endometriosis as it provides a more reliable map of deep infiltrating endometriosis.[4]

    For pelvic (superficial) endometriosis: Transvaginal ultrasound scan showed good specificity (95%; 95% CI 89–100%), but poor sensitivity (65%; 95% CI 27–100%). MRI showed both poor specificity and sensitivity (72% and 79%, respectively).

    For ovarian endometriotic cysts (endometrioma): Transvaginal ultrasound showed good mean specificity and sensitivity with reasonable confidence intervals and heterogeneity (specificity 96%, 95% CI 92–99%; sensitivity 93%, 95% CI 87–99%).

    For MRI, mean specificity and sensitivity were similar to transvaginal ultrasound scan (91% and 95%), respectively.

    Deep endometriosis: Deep endometriosis can involve many areas in the pelvis such as visceral organs (e.g., bowel and bladder), the pelvic wall and its retroperitoneal structures (ureters, nerves, blood vessels, etc.). For transvaginal ultrasound overall specificity and sensitivity estimates have been reported as 94% and 79%, respectively, whereas sensitivity may be slightly improved with 3D ultrasound (87%).

    An MRI has an overall mean specificity of 77% and a mean sensitivity of 94% in diagnosing deep endometriosis of the pelvis.[5]

    It is recommended to use imaging (US or MRI) in the diagnostic work-up for endometriosis, but one needs to be aware that a negative finding does not exclude endometriosis, particularly superficial peritoneal disease.

    In patients with negative imaging results or where empirical treatment was unsuccessful, one can consider offering laparoscopy for the diagnosis and treatment of suspected endometriosis.
- Laparoscopy has been considered as the "gold standard" for the diagnosis of endometriosis with the

typical appearance of "powder burn" or "match stick" appearance as well as gelatinous lesions and peritoneal defects. *Biopsy and histopathological confirmation* of the lesions will show endometrial glands and stroma with hemosiderin laden macrophages. A negative histology does not entirely rule out the disease.

- *Cystoscopy and intravenous urography* will help in diagnosis of bladder and ureteric involvement by deep infiltrating endometriosis.
- *Rectosigmoidoscopy* can be done where rectal involvement is suspected.
- *Staging* is done according to the extent of the disease seen on laparoscopy.

The staging of endometriosis is based on laparoscopic findings in accordance with the Revised American Fertility Society Scoring System **(Fig. 2)**.

Differential diagnosis of endometriosis is given in **Box 2**.

Complications of endometriosis include:

- Infertility
- Risk of ovarian cancer (endometrioid type)
- Inflammation with scar tissue and adhesion formation
- Intestinal and bladder complications
- Rupture and infection of endometriotic cyst

Treatment options are based on following factors:

- Age of the patient
- Severity of symptoms
- Location and extent of the disease
- Desire for fertility
- Previous treatment and surgery.

Expectant treatment: Expectant management may be followed in women who are young, and unmarried with minimal or mild disease and are asymptomatic. Women who are approaching menopause can also be managed and observed expectantly.

Medical Management (Box 3)

Nonsteroidal Anti-inflammatory Drugs

Ibuprofen, naproxen, and mefenamic acid inhibit the release of prostaglandins.

Hormonal Therapy

Combined oral contraceptive pills given in a continuous fashion may lead to a pseudopregnancy state and help in regression of the disease.

Danazol and gonadotropin-releasing hormone (GnRH) agonists cause a "pseudomenopause" or a "medical oophorectomy" state.

Progestins: Progestins such as dydrogesterone, medroxyprogesterone acetate, norethisterone, depot medroxyprogesterone acetate (DMPA), and norethisterone acetate are some of the progestins which can be given for pain relief and dysmenorrhea.

Dienogest is a synthetic progesterone-19 nortestosterone that causes anovulation. The dose is 2 mg daily and is effective for relief of pain in endometriosis.

Aromatase inhibitors: Letrozole and anastrozole are the common aromatase inhibitors used in endometriosis for pain relief.

The GnRH agonists: The commonly used GnRH agonists are leuprolide acetate 3,75 mg every four weekly, goserelin 3.6 mg subcutaneously every 4 weeks, triptorelin 3.75 mg 4 weekly and nafarelin 200 mg twice daily as a nasal spray.

Side effects of GnRH agonists are related to the hypoestrogenic state and include hot flushes, irritability, headache, mood swings, insomnia, vagina dryness, and decreased libido. It decreases bone mineral density particularly at the spine and hip bones. It should not be given to young girls and for >6 months. Add back therapy can be given in the form of low dose combined estrogen—0.625 mg of conjugated equine estrogen and medroxyprogesterone acetate 2.5 mg daily. Low dose combined oral contraceptives may be used effectively as add back therapy.

The GnRH antagonists such as cetrorelix 3 mg subcutaneously for 6–8 weeks is as effective as GnRH agonists. Elagolix a nonapeptide is equally effective.

Progesterone antagonists such as mifepristone (RU 486) and ulipristal acetate have also been used for pain relief in endometriosis.

Depot medroxyprogesterone acetate in a dose of 150 mg intramuscularly monthly can cause amenorrhea and suppression of endometriosis. Side effects include decreased bone mineral density and prolonged amenorrhea even after stopping the treatment.

Androgens such as danazol which is an isoxazole derivative of 17alpha-ethynylestradiol testosterone suppresses gonadotropins and leads to anovulation. It causes hypoestrogenic and hyperandrogenic state that

AMERICAN SOCIETY FOR REPRODUCTIVE MEDICINE
REVISED CLASSIFICATION OF ENDOMETRIOSIS

Patient's Name____________________ Date____________________

Stage I (Minimal) - 1-5
Stage II (Mild) - 6-15
Stage III (Moderate) - 16-40
Stage IV (Severe) - >40
Total __________

Laparoscopy________ Laparotomy________ Photography________
Recommended treatment ____________________

Prognosis____________________

	ENDOMETRIOSIS	< 1 cm	1-3 cm	> 3 cm
PERITONEUM	Superficial	1	2	4
	Deep	2	4	6
OVARY	R Superficial	1	2	4
	Deep	4	16	20
	L Superficial	1	2	4
	Deep	4	16	20
	POSTERIOR CULDESAC OBLITERATION	Partial		Complete
		4		40
	ADHESIONS	< 1/3 Enclosure	1/3-2/3 Enclosure	> 2/3 Enclosure
OVARY	R Filmy	1	2	4
	Dense	4	8	16
	L Filmy	1	2	4
	Dense	4	8	16
TUBE	R Filmy	1	2	4
	Dense	4'	8'	16
	L Filmy	1	2	4
	Dense	4'	8'	16

*If the fimbriated end of the fallopian tube is completely enclosed, change the point assignment to 16.
Denote appearance of superficial implant types as red [(R), red, red-pink, flamelike, vesicular blobs, clear vesicles], white [(w), opacifications, peritoneal defects, yellow-brown], or black [(B) black, hemosiderin deposits, blue]. Denote percent of total described as R___%, W___% and B___%. Total should equal 100%.

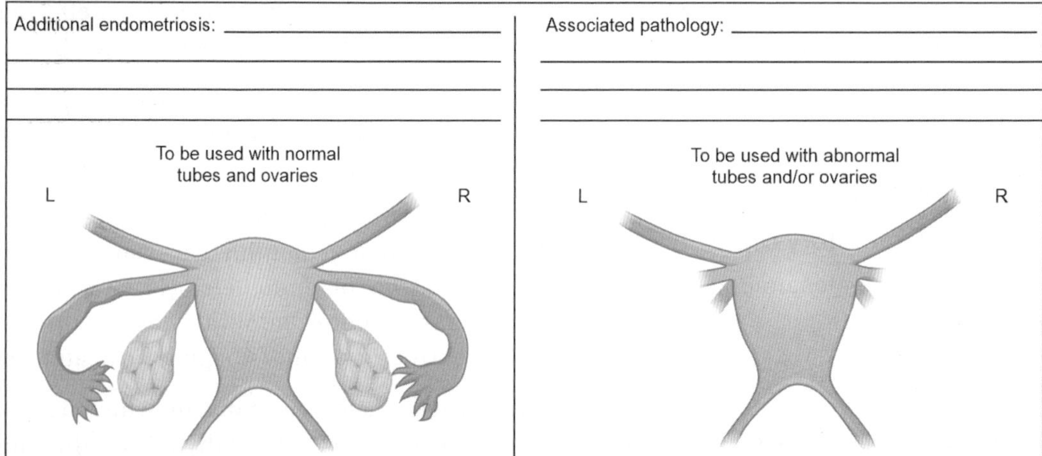

Fig. 2: The American Society for Reproductive Medicine revised classification of endometriosis.[6]

BOX 2: Differential diagnosis.

- Pelvic inflammatory disease
- Ovarian cysts
- Irritable bowel syndrome

BOX 3: Medications used in endometriosis.[7]

Hormonal combined oral contraceptives:

- Progesterone containing contraceptives
  - Oral or injectable
  - Implant
  - Levonorgestrel containing intrauterine system (LNG-IUS)
- Selective progesterone receptor modulators
  - Mifepristone
  - Ulipristal acetate
  - Onapristone
- Gonadotropin-releasing hormone agonists
  - Leuprolide acetate
  - Nafarelin
  - Goserelin
- Gonadotropin-releasing hormone antagonists
  - Cetrorelix

Nonhormonal:

- Nonsteroidal anti-inflammatory drugs (NSAIDs)
- Aromatase inhibitors
- Danazol

causes atrophy of the ectopic endometrial tissue. Dose may vary from 200 mg three times a day to four times a day. Side effects are androgenic such as breast atrophy, hoarseness of voice, acne, and hirsutism.

Gestrinone in a dose of 3–10 mg weekly in divided doses reduces the pain in endometriosis.

Levonorgestrel–intrauterine system (LNG-IUS) are effective in pelvic pain, dysmenorrhea, dyspareunia, and menstrual abnormalities associated with endometriosis, particularly rectovaginal deep infiltrative lesions.

Surgery in Endometriosis for Pain Relief

Surgery may be either conservative with retention of ovarian tissue or can be definitive.

Indications of surgery include:

- Endometriomas larger than 3 cm
- Endometriosis not responding to medical treatment or disease progression
- Severe endometriosis or deep infiltrative disease.

Conservative surgery is reserved for young women desirous of pregnancy and in minimal to mild disease. Laparoscopic adhesiolysis, cauterization of the ectopic endometrial implants, laser vaporization, resection of rectovaginal nodules, chocolate cyst aspiration, cystectomy, and laparoscopic uterosacral nerve ablation.

Laparoscopy for endometriosis is for diagnosis and treatment. It should include removal of visible endometriotic lesions. Ovarian endometriomas can be removed laparoscopically. Ovarian cystectomy is preferred over cyst aspiration and cyst wall ablation.

Deep infiltrative endometriosis may also need surgical resection but the complication rates are high. Deep infiltrative lesions may require superficial shaving and removal or segmental resection. Injury to ureter and bowel are common. Also, there may be recurrence of the disease.

Definitive surgery-hysterectomy with bilateral salpingo-oophorectomy is reserved for women who have severe disease and have completed the family. Those not responding to conservative surgery or with recurrence of disease can be offered definitive surgery after careful consideration of the age, parity, and future fertility prospects. Hysterectomy may be the last option in multiparous women where other medical options have failed.

Definitive surgery includes hysterectomy with bilateral salpingo-oophorectomy. Resection of ectopic endometrial implants may be required along with hysterectomy and bilateral salpingo-oophorectomy.

Alternative medicine: Studies are required to confirm the role of acupuncture and Chinese herbal medicine for the treatment of endometriosis.

Future therapies are mentioned in **Box 4**.

BOX 4: Future therapies.[7]

- Antiangiogenesis factors
- Statins
- Tumor necrosis factor (TNF)-α blockers
- Peroxisome proliferator activated-receptor gamma ligand (PPAR-γ)
- Pentoxifylline

Management of Infertility Related to Endometriosis

Expectant management: Spontaneous conception may occur in women with minimal or mild disease.

Laparoscopic ablation and fulguration of endometriotic lesions have doubtful benefit for future fertility in a woman with mild to moderate endometriosis.

Intrauterine insemination with superovulation is an option for achieving pregnancy in cases with mild to moderate endometriosis.

Controlled ovarian hyperstimulation (COH) and in vitro fertilization (IVF) improves pregnancy rate in moderate to severe endometriosis and is also effective in cases where other associated factors causing infertility are present such as tubal factor or male factor.

ADENOMYOSIS UTERI

It is the presence of endometrial glands and stroma within the myometrial fibers. When the endometrial glands extend >2.5 mm beyond the endometrial-myometrial interface it is called adenomyosis.

Adenomyosis is described as a disease of premenopausal, multiparous women in their thirties to forties. Recent data suggests a prevalence of 20–35%.

Risk Factors for Adenomyosis

- Increased body mass index
- Advanced age
- Multiparity
- Repeated uterine curettage—especially postpartum or postabortal, cesarean section, and myomectomy.
- Oral contraceptive pill and tamoxifen use
- Endometriosis and leiomyoma.

Pathogenesis

Adenomyosis may be diffuse or confined at one area when it is focal adenomyosis. The focal areas can appear as a leiomyoma but lack a capsule and cannot be enucleated. In diffuse adenomyosis there is globular enlargement of the uterus. On cut section the endometrium within the myometrium appears as hemosiderin laden brown deposits. There is hypertrophy and hyperplasia of the myometrium. The endometrium which consists of the endometrial glands and stroma invade the endometrial-myometrial interface and come to lie within the myometrium. Genetic and immunological factors may also be involved in the pathogenesis of the condition.

Other conditions such as leiomyoma, endometriosis, endometrial hyperplasia, and rarely endometrial carcinoma may be associated with adenomyosis.

Clinical Features

- Abnormal uterine bleeding—heavy menstrual loss and intermenstrual bleeding
- Congestive dysmenorrhea.
- Dyspareunia due to enlarged and tender uterus.
- Infertility and miscarriages due to defective retrograde myometrial contractions, dyspareunia, interference in sperm transport, reduced endometrial receptivity, and defective implantation.
- On examination the uterus is diffusely and uniformly enlarged and rarely exceeds 14 weeks pregnant uterus size. The uterus can be tender on examination and may be associated with features of leiomyoma and pelvic endometriosis.

Investigations

Imaging (Table 2)

Transvaginal ultrasound is preferred. Findings show a diffuse enlargement of uterus due to myohyperplasia. Focal adenomyosis can be at the anterior myometrium or more commonly in the posterior wall. Absence of a capsule differentiates adenomyosis from fibroid. Ultrasound also shows cystic spaces and loss of a distinct end myometrial junction, a heterogeneous appearance

TABLE 2: Direct and indirect imaging features of adenomyosis.[8]

Imaging feature	*Ultrasound description*	*MR description*
Direct features	Tiny myometrial cysts	Tiny myometrial cysts
	Hyperechoic nodules or striations	Myometrial foci of high signal intensity on T1-weighted images
	Poor definition of the endometrial-myometrial interface	
Indirect features	Diffuse myometrial heterogeneity associated thin hypoechoic linear striations within a heterogeneous myometrium	Junctional zone thickening Abnormal myometrial signal intensity
	Diffuse asymmetric or symmetric widening of the myometrial walls	Large, regular, asymmetric uterus without leiomyomas

of the myometrium which is typically known as "salt and pepper" appearance.

Magnetic resonance imaging is useful to confirm the diagnosis of adenomyosis. It shows poor identification of the junctional zone (JZ) and a thickness of >12 mm of the JZ. Areas of high signal intensity within area of low signal intensity. There is asymmetrical enlargement of uterine wall with heterogeneous appearance of myometrial echotexture.

Management

Medical management: Tranexamic acid and mefenamic acid controls bleeding and pain in adenomyosis.

Hormonal treatment in the form of continuous use of combined oral contraceptive pills, progestins and GnRH analogs may help in subsiding the symptoms as long as the treatment is continued.

Danazol as oral tablets or as intrauterine device controls heavy menstrual bleeding and dysmenorrhea associated with adenomyosis.

Aromatase inhibitor and mifepristone (RU486) have also been used for treatment of adenomyosis with limited success.

Levonorgestrel-intrauterine system has been found to be effective for control of dysmenorrhea and heavy menstrual bleeding associated with adenomyosis.

Surgical Treatment of Adenomyosis[8]

- *Uterine artery embolization*—has been tried for relief of symptoms
- *Endometrial resection or ablation* in women not desirous of future pregnancy
- *Adenomyomectomy* is the localized excision of focal adenomyosis
- *Hysterectomy* for women who have completed their family.

REFERENCES

1. Chamié LP, Blasbalg R, Pereira RM, Warmbrand G, Serafini PC. Findings of pelvic endometriosis at transvaginal US, MR imaging, and laparoscopy. Radiographics. 2011;31(4):E77-100.
2. Foti PV, Farina R, Palmucci S, Vizzini IAA, Libertini N, Coronella M, et al. Endometriosis: clinical features, MR imaging findings and pathologic correlation. Insights Imaging. 2018;9(2):149-72.
3. Koninckx PR, Meuleman C, Demeyere S, Lesaffre E, Cornillie FJ. Suggestive evidence that pelvic endometriosis is a progressive disease, whereas deeply infiltrating endometriosis is associated with pelvic pain. Fertil Steril. 1991;55:759-65.
4. Bazot M, Lafont C, Rouzier R, Roseau G, Thomassin-Naggara I, Daraï E. Diagnostic accuracy of physical examination, transvaginal sonography, rectal endoscopic sonography, and magnetic resonance imaging to diagnose deep infiltrating endometriosis. Fertil Steril. 2009;92(6): 1825-33.
5. Nisenblat V, Bossuyt PM, Farquhar C, Johnson N, Hull ML. Imaging modalities for the non-invasive diagnosis of endometriosis. Cochrane Database Syst Rev. 2016;2(2):CD009591.
6. Revised American Society for Reproductive Medicine classification of endometriosis: 1996. Fertil Steril. 1997;67:817-21.
7. Rafique S, Decherney AH. Medical Management of Endometriosis. Clin Obstet Gynecol. 2017;60(3):485-96.
8. Dessouky R, Gamil SA, Nada MG, Mousa R, Libda Y. Management of uterine adenomyosis: current trends and uterine artery embolization as a potential alternative to hysterectomy. Insights Imaging. 2019;10(1):48.

Chapter 7

Pelvic Inflammatory Disease

Nidhi Jindal

INTRODUCTION

Pelvic inflammatory disease (PID) is defined as inflammation of upper genital tract which comprises endometrium, fallopian tubes, and other adnexal structures including broad ligament and pouch of Douglas (POD). Most common factor associated is ascending infection from lower genital tract but definitely there are other risk factors associated also. The incidence varies between 0.28 and 1.67% globally.[1] However, in India it affects around 24–32% of women and contributory to 3–10% of gynecological admissions.[2]

ETIOLOGY

- *Neisseria gonorrhoeae (NG)* and *Chlamydia trachomatis (CT)* are the main causative agents of PID. These are sexually transmitted organisms. *Chlamydia* is more common causative agent and causes around 10% of infections of lower genital tract in women of reproductive age group.[3]
- Commensals of vagina such as *Gardnerella vaginalis, Haemophilus influenza, Streptococcus agalactiae,* and anaerobes are other likely causes of PID.
- Genital tuberculosis is also another common reason found in India and other developing countries.

RISK FACTORS

- Early start of sexual activity
- Multiple partners
- Immunocompromised conditions such as human immunodeficiency virus (HIV) and diabetes
- Intrauterine contraceptive devices (IUCD)
- Previous history of PID
- Poor genital hygiene

PATHOGENESIS

See **Flowchart 1**.

Flowchart 1: Pathogenesis.

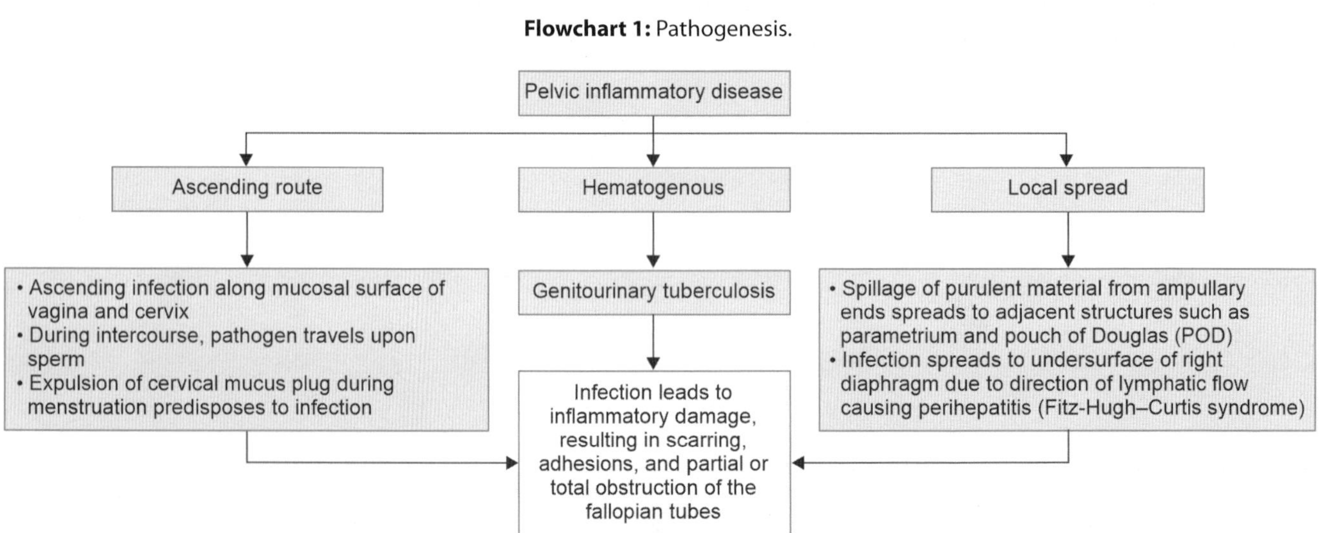

CLINICAL FEATURES

Symptoms	*Signs*
• Lower abdominal pain which is typically bilateral • Deep dyspareunia • Recent onset dysmenorrhea • Abnormal vaginal bleeding, including postcoital, intermenstrual, and menorrhagia • Abnormal vaginal or cervical discharge which is often purulent	• Lower abdominal tenderness which is usually bilateral • Adnexal tenderness on bimanual vaginal examination • Cervical motion tenderness on bimanual vaginal examination • Fever (>38°C)

About 60% of cases are subclinical, 36% have mild-to-moderate disease, and 4% have severe disease. Clinical features are severe in tubo-ovarian (TO) abscess and pelvic peritonitis.

DIFFERENTIAL DIAGNOSIS

- Ectopic pregnancy
- Acute appendicitis, diverticulitis, and cholecystitis
- Endometriosis
- Ovarian cyst torsion or rupture
- Urinary tract infection
- Functional pain (*see* **Flowchart 2**)

MANAGEMENT OF PELVIC INFLAMMATORY DISEASE

See **Flowchart 3**.

COMPLICATIONS OF PELVIC INFLAMMATORY DISEASE

- TO abscess
- Pelvic peritonitis
- Long-term sequelae such as chronic pelvic pain, ectopic pregnancy, tubal factor infertility, and implantation failure in assisted reproductive technique (ART)

INDICATIONS OF HOSPITAL ADMISSION

- Uncertain diagnosis
- Pregnancy with PID
- Failure to respond to outpatient management
- Inability to tolerate outpatient treatment
- Pelvic abscess or TO mass on imaging
- Failure to improve after 72 hours of oral therapy
- Immune-deficient state like HIV with low CD4 counts

TREATMENT REGIMENS

Outpatient regimen:
- Oral Rx should be used to treat mild to moderate disease
- If no response in 72 hours, then needs reevaluation and parenteral treatment
- Clinical improvement is shown by defervescence, decrease in direct and rebound tenderness, decrease in uterine, adnexal, and cervical motion tenderness.
- Perform a follow-up examination 48–72 hours after prescribing to ensure clinical improvement

The National AIDS Control Organization (NACO) guidelines for Indian population (Kit-6 yellow kit):
- Tablet cefixime 400 × 1 tablet
- Tablet metronidazole 400 mg × (1 BD for 14 days)
- Tablet doxycycline 100 mg (1 BD for 14 days)

The Centers for Disease Control and Prevention (CDC) guidelines:[4]
- Ceftriaxone 250 mg IM in a single dose PLUS
  Doxycycline 100 mg orally twice a day for 14 days WITH or WITHOUT
  Metronidazole 500 mg orally twice a day for 14 days
  OR

Flowchart 2: Investigations.

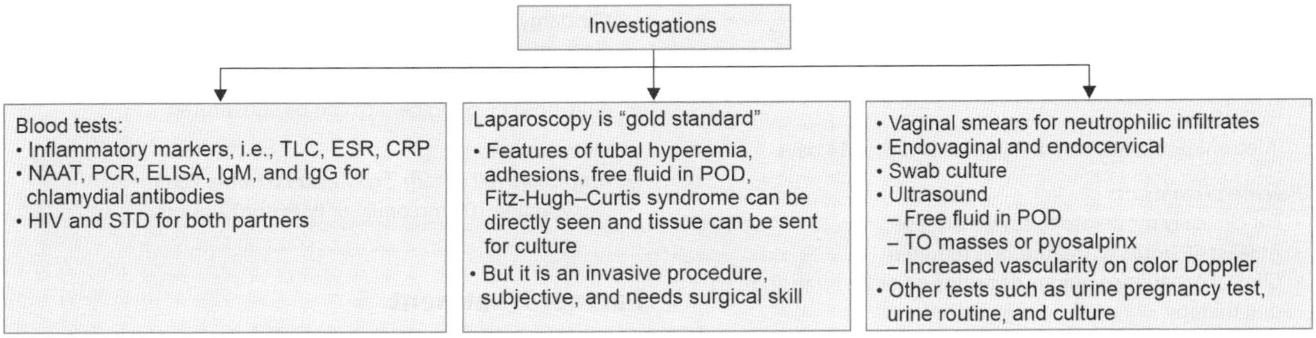

Investigations

Blood tests:
- Inflammatory markers, i.e., TLC, ESR, CRP
- NAAT, PCR, ELISA, IgM, and IgG for chlamydial antibodies
- HIV and STD for both partners

Laparoscopy is "gold standard"
- Features of tubal hyperemia, adhesions, free fluid in POD, Fitz-Hugh–Curtis syndrome can be directly seen and tissue can be sent for culture
- But it is an invasive procedure, subjective, and needs surgical skill

- Vaginal smears for neutrophilic infiltrates
- Endovaginal and endocervical
- Swab culture
- Ultrasound
  - Free fluid in POD
  - TO masses or pyosalpinx
  - Increased vascularity on color Doppler
- Other tests such as urine pregnancy test, urine routine, and culture

(CRP: C-reactive protein; ELISA: enzyme-linked immunosorbent assay; ESR: erythrocyte sedimentation rate; HIV: human immunodeficiency virus; IgG: Immunoglobulin G; IgM: immunoglobulin M; NAAT: nucleic acid amplification testing; PCR: polymerase chain reaction; POD: pouch of Douglas; STD: sexually transmitted disease; TLC: total leukocyte count; TO: tubo-ovarian)

Flowchart 3: Diagnosis.

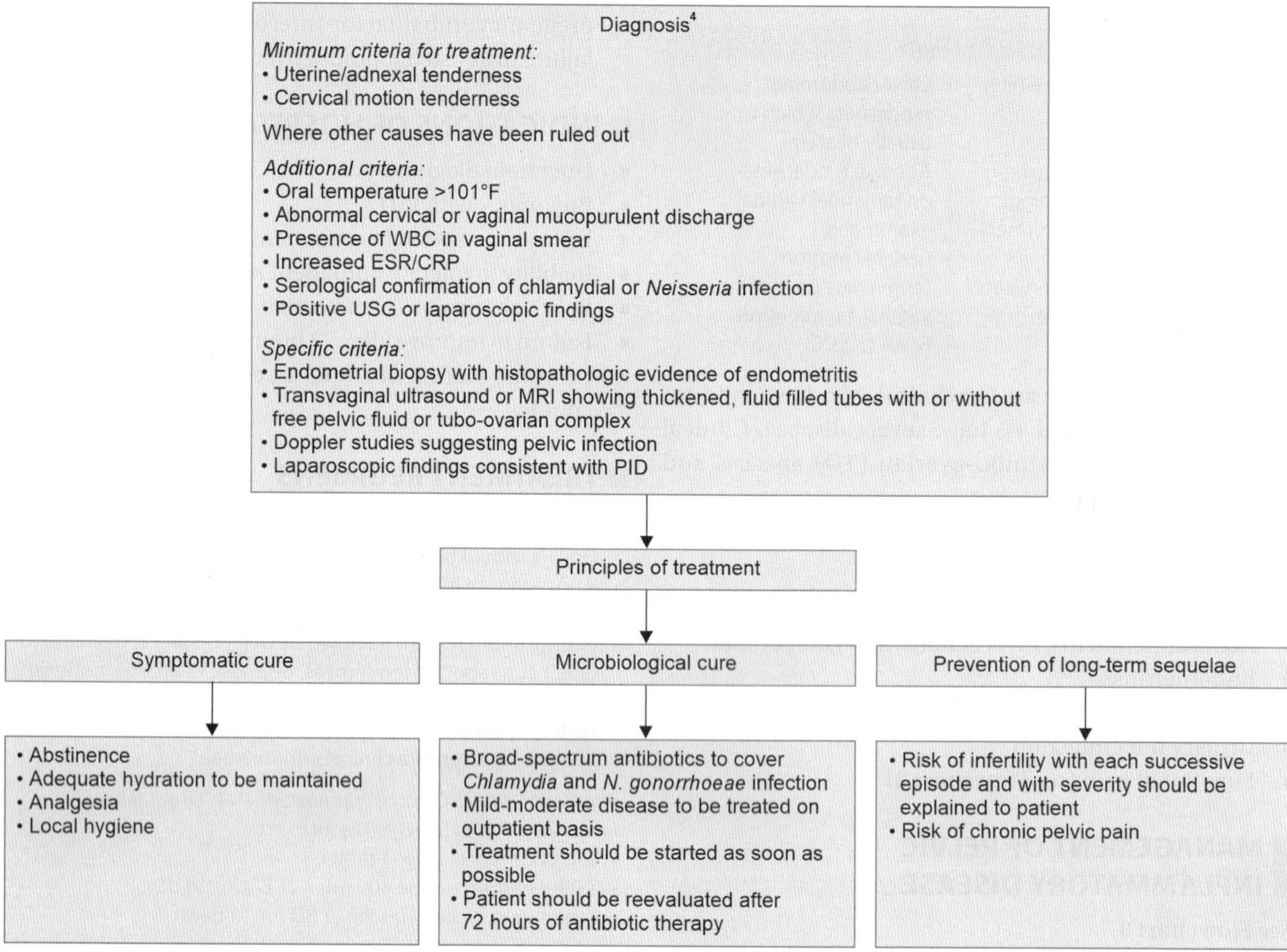

(CRP: C-reactive protein; ESR: erythrocyte sedimentation rate; PID: pelvic inflammatory disease; USG: ultrasonography; WBC: white blood cell)

- Cefoxitin 2 g IM in a single dose and probenecid, 1 g orally administered concurrently in a single dose PLUS
  Doxycycline 100 mg orally twice a day for 14 days WITH or WITHOUT
  Metronidazole 500 mg orally twice a day for 14 days
  OR
- Other parenteral third-generation cephalosporin (e.g., ceftizoxime or cefotaxime) PLUS
  Doxycycline 100 mg orally twice a day for 14 days WITH or WITHOUT
  Metronidazole 500 mg orally twice a day for 14 days

Parenteral regimens:
- For clinically moderate to severe disease
- At least 48 hours of treatment is required
- Clinical experience guides decisions regarding transition to oral therapy
- Usually within 24 hours of clinical improvement can be switched to oral

Regimen A:
- Cefotetan 2 g IV every 12 hours OR
- Cefoxitin 2 g IV every 6 hours PLUS
- Doxycycline 100 mg orally or IV every 12 hours

Regimen B:
- Clindamycin 900 mg IV every 8 hours PLUS
- Gentamicin: Loading dose IV or IM (2 mg per kg), followed by a maintenance dose (1.5 mg per kg) every 8 hours; a single daily dose (3–5 mg per kg) can be substituted

Alternative regimen:
- Ampicillin/sulbactam 3 g IV every 6 hours PLUS
- Doxycycline 100 mg orally or IV every 12 hours

Partner Treatment

- Male sex partners of women with PID should be examined and treated if they have had sexual contact with the patient during the 60 days preceding the onset of symptoms in the patient.

- They should be treated with regimens effective against NG and CT irrespective of the pathogens isolated from infected women. According to the NACO guidelines, Kit-6 is given to male partner too.
- Sexual abstinence during treatment.
- Educate about correct and consistent use of condoms.
- Schedule return visit after 7 days.

Treatment of Pelvic Inflammatory Disease in Special Cases

- *Adolescent PID:*
  - Knowledge regarding menstrual hygiene, sex education, and contraception should be imparted to adolescents through "adolescent friendly clinics"
  - Confidentiality should be maintained while treating this age group
  - Treatment regimens remain same with dose adjustments
- *PID with pregnancy:*
  - High risk of maternal septicemia and preterm labor
  - Patient should be hospitalized and given parenteral therapy
  - Treatment regimens should be adjusted according to pregnancy safety of drugs, e.g., doxycycline is contraindicated
- *PID with IUCD in situ:*
  - Risk of PID is there during first 3 weeks of IUCD insertion
  - No need to remove in mild disease
  - Should be removed in recurrent PID or severe disease where symptoms are not improving after 72 hours of therapy
- *PID with HIV:*
  - More chances of having TO abscess, especially in patients with low CD4 counts
  - Generally respond well like other patients.

REFERENCES

1. Shinde SA, Shinde US, Aher GS. Pelvic Inflammatory Disease (PID): A Cross Sectional Prospective Study at a Tertiary Care Centre. Int J Clin Biomed Res. 2018;4(3):61-4.
2. Pachori R, Kulkarni N. Studies on Incidence of Pelvic Inflammatory Disease and associated Clinical Consequences in Reproductive Women. World J Pharma Pharma Sci. 2016;5(3):1329-37.
3. Kreisel K, Torrone E, Bernstein K, Hong J, Gorwitz R. Prevalence of pelvic inflammatory disease in sexually experienced women of reproductive age—United States, 2013–2014. MMWR Morb Mortal Wkly Rep. 2017;66:80-3.
4. Workowski KA, Berman S; Centers for Disease Control and Prevention (CDC). Sexually transmitted diseases treatment guidelines, 2010. MMWR Recomm Rep. 2010;59(RR-12):1-110.

Chapter 8

Endometrial Tuberculosis

Poonam Shivkumar, Disha Dave

INTRODUCTION

Endometrial tuberculosis (TB) is one of the forms of genital TB primarily affecting young women in the reproductive age group of 15–40 years of age. Endometrial TB in females is by no means uncommon, particularly in communities where pulmonary or other forms of extragenital TB are common. TB can affect any organ in the body, can exist as silent killer without any symptoms, and can recur. Endometrial TB is caused by *Mycobacterium tuberculosis (MTB)* (rarely *Mycobacterium bovis* and/or atypical mycobacteria) generally secondary to TB of the lungs or other organs with infection reaching through hematogenous, lymphatic route or direct spread from abdominal TB. In female genital tuberculosis (FGTB), main pelvic organ affected are fallopian tubes in 90% women, whereas uterine endometrium is affected in 70% and ovaries in about 25% women.

Tuberculosis was recognized as a clinical entity as far back as 1000 BC. However, it was not until 1744 that Morgagni,[1] following a postmortem examination of a 20-year-old woman who died of TB and whose uterus and fallopian tubes were found to be filled with caseous material, described the first case of genital TB. The term *tuberculosis* was first used in 1834 and tubercle bacilli was discovered by Koch in 1882.

Tuberculosis is considered the most important communicable disease across the world. Since the beginning of the 20th century, the incidence of TB in general and genital TB more particularly has been declining in developed countries. However, TB remains a major health burden in many developing countries like India, and in these areas, genital TB is responsible for a major proportion of females presenting with infertility.[2] TB affects almost 50% of the population in developing countries. An estimated 30 million persons have active TB, and 7–10 million people die each year of TB.[3,4]

A review of the literature reveals that the highest incidence of TB is still in India, followed by Scandinavia and Scotland. It is estimated that almost 50% of the population of India has TB and that one person dies every minute from TB. The true incidence of endometrial TB is not known given that, owing to its subtle presentation, many cases remain undiagnosed. It has been estimated that approximately 5% of females presenting to subfertility clinics have genital TB.[5] Tripathy and Tripathy stated that genital TB is mostly a secondary manifestation of primary TB, the most common primary site being the lungs.[6] They reported that the genital tract is vulnerable to this disease after puberty, and most cases occur during the childbearing period. Endometrial TB is typically understood as a disease of young women, with 80–90% of cases diagnosed in patients 20–40 years old, often during workup for infertility.[5]

EPIDEMIOLOGY OF ENDOMETRIAL TUBERCULOSIS

Genitourinary TB is accounts for 27.1% cases of extrapulmonary tuberculosis (EPTB) with genital TB being seen in 9% cases.[7] The exact incidence of FGTB is not known due to underreporting of cases, asymptomatic cases, nonspecific symptomatology, and the lack of reliable diagnostics with high sensitivity.[8-11] The reported incidence is varies in different nations being 1% in infertility clinics of the USA,[5] 1% in the Scandinavian countries,[12] 4–8% in Pakistan,[13,14] 15–21.1% in South Africa,[15] and 1–19% in different areas of India.[16,17] High incidence (26%) has been reported from tertiary centers due to referred cases,[18] as also (up to 48%) in women registering for assisted reproduction.[19]

Human immunodeficiency virus (HIV) infection has increased the incidence of EPTB and FGTB both in India and Africa[20,21] due to poor immunity. The mean age of developing FGTB is 40 years in developed nations.[9,12] However, the disease presents in younger age group (20–30 years) in Asia due to early marriage and childbearing in them.[8-11]

India is regarded as hub of tuberculosis infection (TBI) globally, with nearly 35–40 crores Indian population having TBI, of which 26 lakhs people (18–36 lakhs) are estimated to develop TB disease annually. This may change after the ongoing national TB prevalence survey data becomes available.

In a series conducted in India, 800 women with pelvic inflammatory disease were assessed; 6% of 48 cases were attributed to TB.[22] Tripathy and Tripathy studied patients with a history of infertility, menstrual irregularity, and lower abdominal pain from July 1971 to February 1983.[16] In 165 patients, endometrial TB was diagnosed histopathologically, with approximately 13 new cases each year.

Khilnani et al. reported that TB is one of the most common diseases in India.[23] In cases of infertility, the incidence of genital TB is 17.4%. In women who died of pulmonary TB, postmortem studies revealed genital TB in 8%. Careful histologic examination by serial section of fallopian tubes resulted in an increased incidence of tubal TB from 7.7 to 20% in autopsy series.

In surgically removed adnexa, the frequency of genital TB, as reported in the examination of operative specimens, varies from 2 to 20%. The care with which excised fallopian tubes are examined histologically in different laboratories is an important and confounding factor in determining the incidence of tuberculous salpingitis or endometritis.

MODE OF TRANSMISSION

Endometrial TB is almost always secondary to TB elsewhere in the body—usually pulmonary and rarely renal, gastrointestinal, bone, or joint; occasionally it is part of a generalized miliary disease process. The mode of spread is usually hematogenous or lymphatic and occasionally occurs by way of direct contiguity with an intra-abdominal or peritoneal focus.[5,24] The focus in the lung often heals, and the lesion may lie dormant in the genital tract for years, only to reactivate at a later age. It can be sexually-transmitted through infected semen of male partner having active genitourinary TB.[12]

Hematogenous Spread

Tubercle bacilli can also reach the bloodstream and thus the genital tract from extrapulmonary and chronic pulmonary lesions. The fallopian tube is the most favorable nidus for tubercle bacilli, with the earliest lesion seen in the mucosa. The tendency of the tubercle bacillus to affect bilateral organs results in both tubes being involved in the tuberculous process. Initially there is uniform pelvic involvement of both tubes, with subsequent dissemination to other genital organs and the peritoneum. Tuberculous peritonitis is commonly associated with genital tract involvement and may also be associated with rupture of a caseous abdominal lymph node or, rarely, with spread from an intestinal focus.[25]

Lymphatic Spread

A less common mode of infection is lymphatic spread which occurs when the primary lesion is in the abdominal cavity. In some countries where people drink raw milk (unpasteurized), infection spreads by way of the alimentary tract and caused by the bovine tubercle bacillus is still reported. Gavaller et al. reported that, 33% of cases of FGTB were caused by bovine bacillus, which was spread to the fallopian tubes by way of the lymphatics.[26]

Direct Spread

Direct spread to the genital tract organs from tuberculous abdominal viscera, such as the bladder, rectum, appendix, and intestines, has been reported. Some experts believe that this spread is along the peritoneal surface. However, peritoneal involvement can also be the result of spillage of infected material from the fallopian tubes; thus, the primary process is not always clear. It can also occur when adhesions are present between the bladder or intestine and fallopian tubes or perforation of a tuberculous ulcer can result in direct spread to the genital organs.

There are reports of primary cervical and vulvar disease in which sexual partners may be the source of infection. This type of disease may also occur in a woman who has TB of another organ and who excretes tubercle bacilli in her stool, urine, or sputum. When these excretions come into contact with the external genitalia, TB of the vulva or vagina may result, particularly if the skin is abraded or broken. Sutherland reported that of 128 women with proven TB, 5 (3.9%) of their husbands had active genitourinary TB; however, 3 of the 5 wives of these men also had evidence of TB outside the genital tract.[27] Lattimer et al. confirmed the presence of MTB in the semen of men with genitourinary TB in 1954.[28]

RISK FACTORS

Risk factors for the development of TB include immunocompromised states such as HIV and diabetes, chronic liver and renal diseases, malnutrition, and alcoholism. Poor housing, homelessness, pneumoconiosis, genetics, vitamin deficiency, immunosuppressive drug intake, and renal transplantation are also been reported as the risk factors for the development of TB. A person on renal dialysis for the treatment of chronic renal failure also has more chances of developing TB than normal population.[29,30]

Smoking is an independent risk factor causing approximately a twofold increase not only in active TB disease but also in latent TB infection and mortality.[31] Few of the modifiable socioeconomic risk factors, including indoor air pollution, living in a poor housing condition with a low number of windows per room, and socioeconomic position of the household, can be powerful predictors of TBI and disease.[32] Moreover, people who have had one episode of TB in the past have more chances of developing TB again than compared who had not to have an episode of TB and thus further exacerbating the vicious cycle of poverty and TB.[33]

PATHOGENESIS

The endometrium is affected in 50–80% (average 70%) patients through hematogenous, lymphatic, or contagious route. Initially, there are no visible lesions, later ulcers are formed and in advanced stages, distortion of the uterine cavity occurs due to adhesions. The lesions can be focal or multiple ulcers, necrosis, or hemorrhagic areas. Destruction of the endometrium may occur with resulting Asherman's syndrome manifesting as secondary amenorrhea and infertility.[34] On microscopy, typical picture of caseative epithelioid granulomas, epithelioid cells, and specialized Langerhans giant cells may not form due to the shedding of the endometrium during menstruation. The presence of a focal collection of lymphocytes on endometrial biopsies may also indicate endometrial TB.[17]

PATHOLOGY

Grossly, the size and shape of the uterus may appear normal. The tuberculous process is generally localized to the endometrium, mostly affecting fundus, and decreases toward the cervix. The myometrium is not usually involved. In premenopausal patients, much of the infected tissue is shed during the menstruation, only to have the endometrium reinfected from the tubes with each cycle.

In genital TB, there is a high incidence of involvement of the endometrium. Schaefer reported an incidence of 50–60%;[5] Onuigbo reported an incidence of 60%,[35] and Nogales-Ortiz et al. reported an incidence of 79%,[36] whereas Sutherland estimated 90% involvement of the endometrium in genital TB.[27]

In a large series of 1,436 cases, Nogales-Ortiz et al. reported 79% endometrial involvement using extensive endometrial sampling.[36] Grossly, the endometrium appears normal in most cases, probably because of the cyclic menstrual shedding. However, when extensive involvement of the endometrium occurs, there may be ulcerative, granular, or fungating lesions present, or the endometrial cavity may be obliterated with intrauterine adhesions. Sometimes, the macroscopic appearance may resemble carcinoma, and TB has been suggested microscopically.

In 2.5% of cases of tuberculous endometritis, Nogales-Ortiz et al.,[36] Schenker and Margalioth,[37] and Hasselgren and Bolin[38] noted total destruction of the endometrium with resulting amenorrhea secondary to end-organ failure and major predisposition to pyometra when the internal os becomes occluded.

MICROSCOPIC APPEARANCE

On histologic examination, the picture is one of infrequent, usually isolated, small tubercles scattered irregularly through the endometrium. In most cases, the lesions are extremely scanty, and hence careful search through all the sections of the endometrium removed at curettage may reveal only one or two foci of TB **(Fig. 1)**.

Tuberculosis of the endometrium resembles TB in other tissues, but the advanced stages—caseation, fibrosis, and calcification—rarely are seen during the reproductive period of the regular cyclical shedding of the endometrium. The classic lesion in tuberculous endometritis is the non-caseating granuloma, composed of epithelial cells, Langhans giant cells, and lymphocytes. These granulomata are located throughout the endometrium but occur in greater density in the more superficial layers. They occasionally perforate into gland lumina, causing an acute inflammatory reaction and giving the appearance of microabscesses. Endometrial glands adjacent to granulomata may not reveal a secretory response or may become compressed, resulting in a pseudoadenomatous appearance **(Fig. 2)**.[36,39,40]

Most often, the endometrial lesion consists of circumscribed lesions of endothelioid cells surrounded

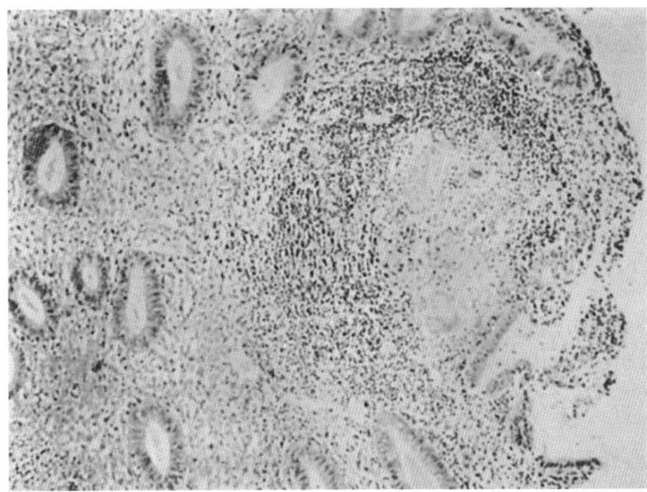

Fig. 1: Tuberculous endometritis. Photomicrograph of a single tuberculous granuloma is seen on the left, consisting of central epithelioid cells, with a Langerhans-type giant cell surrounded by a cuff of lymphocytes. No central caseation is present. The surrounding endometrium appears completely normal; the glands are proliferative, and there is no infiltrate in the stroma, seen on the right (×100).

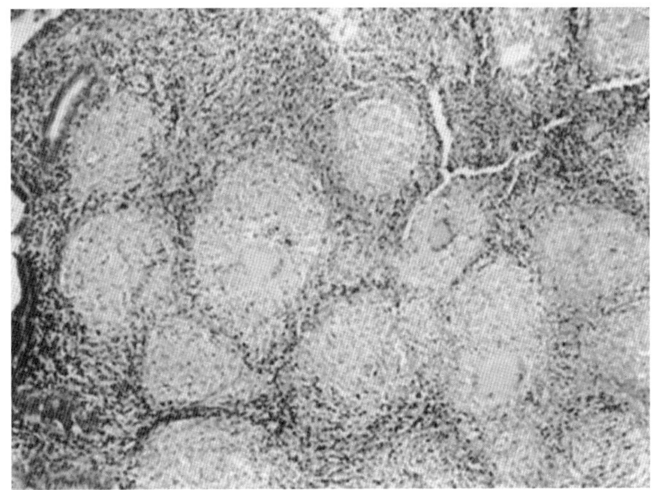

Fig. 2: Tuberculous endometrium. A much more florid picture of numerous granulomata with dense lymphocyte infiltrate in the surrounding stroma is seen in this photomicrograph (×100).

by a zone of lymphocytes and plasma cells. These inflammatory cells may be present in the stroma without focal lesions. However, the giant cells are not always indicative of tuberculous infection. Endometrial tuberculous lesions are frequently focal and immature because they tend to be shed monthly except in postmenopausal women or women with amenorrhea. Sutherland stated that the endometrium is reinfected on a regular basis from the tubes or from infections of the basalis by organisms in menstrual blood after sloughing of the superficial endometrium.[41] The granulomatous lesions are usually best recognized on cycle days 24–26 or within 12 hours of the onset of menses.[39]

TABLE 1: Symptoms and signs in female genital tuberculosis.

Puberty menorrhagia	Vague or definite abdominal or pelvic lump
Heavy menstrual bleeding in early phase	Ascites
Postmenopausal bleeding	Doughy feel of abdomen
Oligomenorrhea Hypomenorrhea	Soft tender enlarged uterus (pyometra)
Primary and secondary amenorrhea	Tenderness and induration in fornices
Dysmenorrhea	Lump in adnexa
Infertility	Fullness and tenderness in pouch of Douglas
Abdominal or pelvic mass	Unusual finding
Abdominal and pelvic pain	Solid lesions in external genitalia
Acute abdomen	Ulcer on external genitalia
Vaginal discharge	Genital fistula
Douglas ulcer or growth	
Urinary incontinence or fecal incontinence	

CLINICAL FEATURES (TABLE 1)

Symptoms and signs in female genital tuberculosis given in **Table 1**.

DIFFERENTIAL DIAGNOSIS

- Endometriosis
- Pelvic inflammatory disease (acute and chronic)
- Abnormal uterine bleeding
- Endometrial hyperplasia
- Endometrial polyp
- Endometrial cancer.

DIAGNOSIS (TABLE 2)

The possibility of endometrial TB should always be considered, especially when patient is from an endemic area, a patient having family history of other exposure to TB, or a patient with some proven extragenital manifestations of the disease.

TABLE 2: Diagnosis of endometrial TB.

Test	*Patients*	*Comments*
X-ray of chest	All	All patients presenting with symptoms consistent with TB should have a chest X-ray to look for evidence of previous or active pulmonary TB
HIV test	All	EPTB is associated with HIV infection. All patients should be offered integrated counseling and testing
Pregnancy test	All of childbearing age	To rule out pregnancy as possible cause of symptoms, and to ensure further testing is safe and appropriate
Pelvic ultrasound	All	Part of initial assessment of most patients presenting with gynecological symptoms
Hysterosalpingogram	Selected	May be done as part of the investigation of infertility, but many women with genital TB will have a normal HSG
CT pelvis or MRI pelvis	Selected	To further characterize lesions and plan surgical intervention in selected patients. Disadvantage of CT is exposure to ionizing radiation, which is particularly a concern in women of childbearing age
FDG-PET CT	Selected	Although not widely available, PET scans may give more information about the presence and activity of tubercular tubo-ovarian mass lesions. Further evidence about the diagnostic accuracy of PET CT for detecting and monitoring the progression of FGTB is needed
Endometrial aspirate	Selected	Where facilities exist, endometrial aspirate can be obtained and sent for (1) staining and microscopy for AFB and (2) culture and drug susceptibility testing. Sensitivity is low, and negative results cannot rule out FGTB
Laparoscopy	Selected	Laparoscopy with biopsy of lesions is required when: • Other less invasive tests are inconclusive • Malignancy is also suspected • As part of infertility investigations when less invasive tests are inconclusive Laparoscopy offers the dual advantage of pelvic organ visualization and specimen collection from otherwise inaccessible sites. Specimens should be subject to (1) staining and microscopy, (2) culture and drug susceptibility testing, and (3) histopathology

(AFB: acid-fast bacilli; EPTB: extrapulmonary tuberculosis; FDG: fluorodeoxyglucose; FGTB: female genital tuberculosis; HIV: human immunodeficiency virus; HSG: hysterosalpingography; PET: positron emission tomography; TB: tuberculosis)

Infertility for which no obvious cause can be found, chronic pelvic inflammatory disease resistant to standard antibiotic therapy, or adnexal disease with ascites in virgin females should alert the doctor to look genital TB.

A negative chest radiograph does not rule out the diagnosis because most pulmonary lesions are dormant by the time the genital tract becomes involved in the disease process.

Routine laboratory studies are of little value; most patients have a normal white blood cell count with differential, although there is a tendency to lymphocytosis. Anemia is sometimes seen, whereas microscopic examination of urine may show hematuria or abacteriuric pyuria if there is concomitant urinary tract involvement.[42]

Investigations to confirm the diagnosis of endometrial TB are listed in **Table 2**.

- Complete blood count
- Chest radiographs
- Tuberculin test
- Menstrual blood for culture
- Endometrial curettage
- Histologic examination
- Culture for MTB
- Peritoneal fluid for culture
- Peritoneal biopsy for culture and histology
- Hysterosalpingography
- Ultrasonography
- Cervical cytology
- Endoscopy
- Laparoscopy
- Hysteroscopy
- Cystoscopy.

Other Serologic Tests

The diagnosis of TB is based on the identification of MTB or others of the MTB complex—*M. bovis, M. africanum,* and *M. microti*—in culture. Isolation of mycobacteria from clinical specimens in pure cultures pose a great challenge because of the longer period of cultivation required for most of them. Cultivation of MTB requires special culture media. The most popular are Löwenstein-Jensen or other egg-based solid media, Middlebrook 7H10/7H11 agar, Middlebrook 7H9 broth, and Middlebrook 7H12 broth that are manufactured as radiolabeled BACTEC (Becton-Dickinson Diagnostic Instrument Systems, Sparks, MD) medium.[43]

The egg-based media are the most commonly used for the isolation of MTB, although it takes 3–8 weeks to achieve visible growth. Growth on agar-based media can be detected within 3 weeks for most MTB isolates. The BACTEC 7H12 broth can be used for rapid detection of growth, often within 1 week for most isolates. Specimens are usually collected in sterile containers with liquid in the case of tissue (7H9 broth or saline) and transported to the laboratory in dry ice.

Endometrial Curettage

Endometrial biopsy is the first diagnostic step, and although its sensitivity suffers from sampling errors, it can be very helpful if granulomata are found or, less commonly, if smears or cultures are positive for MTB.[12,44]

A definitive diagnosis of TB requires isolation of tubercle bacilli, although many authorities accept histologic confirmation of granulomata as the gold standard diagnostic modality.[45,46] Although the newer culture techniques have increased the growth rate of the organism, it still takes 40 days or more for 75% of all positive cultures to show growth.[6] An acid-fast bacilli (AFB) stain is rapid, simple, and easy technique which could be used initially to make the diagnosis while waiting for culture results to be available; however, it requires 10 organisms/mL for a positive test.

The histologic examination of endometrial tissues removed by biopsy or curettage, especially from the cornual area, affords a rapid method of diagnosing endometrial TB in at least 50% of cases. AFB stain and culture and rarely, guinea pig inoculation, yield better microbiologic assessment. The golden time for sampling is at the end of the menstrual cycle or within 12 hours after the onset of menstrual flow to allow the endometrial granulomata to develop. One negative biopsy or one negative curettage does not exclude the diagnosis of genital TB. Czernobilsky stated that serial sections needed to be studied because the lesions are frequently patchy.[39] Falk and associates stated that a positive endometrial culture for TB was found only in 25% of cases of tuberculous endometritis.[12] Dilation and curettage may increase the yield on endometrial specimens merely by increasing the amount of tissue available for histologic evaluation and cultures.

The demonstration of tuberculous endometritis may be assumed to indicate tuberculous salpingitis in practically 100% of cases. Negative endometrial evaluation may be the result of sampling error, or the disease may be in other genital organs without an associated tuberculous endometritis. Sometimes, the disease is diagnosed or suspected only at exploratory laparotomy with subsequent positive culture for tubercle bacilli or histologic confirmation. AFB culture or stain of peritoneal fluid shows variable results, but the sensitivity can be increased by peritoneal biopsy for culture, stain, and histologic evaluation. Laparoscopy with directed biopsies of suspicious areas may be helpful if less invasive methods fail to provide the needed diagnostic information.

Tissue has been obtained with percutaneous needle techniques, peritoneoscopy, culdoscopy, and laparoscopy, but all carry a significant risk of bowel injury secondary to a closed approach in the presence of dense adhesive disease. A small open biopsy can obtain tissue more safely.[25,41,47]

The diagnosis of FGTB should be made based on any one of:

- Laparoscopic appearance typical for FGTB
- Gynecological specimen positive for AFBs on microscopy or positive for MTB on culture
- Gynecological specimen with findings consistent with FGTB on histopathological examination.

Microbiological Tests

Endometrial biopsy, curettage, or aspirate performed in the luteal phase (day 21) are the most important tests for the diagnosis of endometrial TB.[8-11] All efforts should be made to achieve microbiological confirmation. Demonstration of typical caseous granuloma with or without Langerhans giant cells on histopathology is diagnostic of genital TB. Multiple biopsies may increase pick up rate.[48] The WHO approved rapid diagnostic tests (WRDTs) are cartridge-based nucleic acid amplification test (CBNAAT/GeneXpert

MTB/RIF) and liquid culture by mycobacterial growth indicator tube (MGIT) with phenotypic drug sensitivity testing (DST) or genotypic DST by line probe assay (LPA) are done on endometrial sample sent in saline.[49-51] Polymerase chain reaction (PCR) is a fast, sensitive, and specific test to diagnose mycobacterial DNA in endometrial sample by targeting 65 kDa and the *mpt64* gene.[52,53] It has a high sensitivity and can detect 1–10 bacilli per mL.[52,53] However, the false-positive rate of PCR is high. Many researchers have advised against PCR for the rapid detection of TB59. Rana et al. observed better results of reverse transcriptase-PCR in the detection of FGTB. In our experience, there is high false positivity of PCR and it is recommended not to start antitubercular treatment (ATT) only on positive PCR unless there is evidence of FGTB clinically on examination, or on radiology or endoscopy.[8,11,54]

A positive CBNAAT result is a useful confirmation, but a negative test does not always rule out TB. The Xpert MTB/RIF is an automated test which can detect purified DNA of MTB on the endometrial sample. It can also detect rifampicin resistance and can give report within 2 hours. Sharma et al.[55] observed GeneXpert to have a 33–50% sensitivity, 100% specificity for the diagnosis of FGTB. We also observed 35% sensitivity and 100% specificity in the detection of FGTB with GeneXpert.[56]

Loop-mediated isothermal amplification (LAMP) assay has been used on endometrial biopsy. Sethi et al.[57] observed high sensitivity of 66.18% for LAMP in contrast to 29.4% for AFB smear, 10.29% for AFB culture, 8.82% for histopathology, and 95% for PCR. They also observed a high concordance of 63% between PCR and LAMP.

Hysteroscopy

It is endoscopic visualization of endometrial cavity and may be normal in the absence of endometrial TB (as the uterus is involved in only about 70% of cases of TB) and in early disease. Usually, there is a pale looking cavity, the presence of tubercles, small white caseous nodules, varying grades of intrauterine synechiae **(Fig. 3)** and Asherman's syndrome.[58] Kumar and Kumar observed white caseous nodules as stars shining against the blue background of methylene blue dye in their study of endometrial TB on hysteroscopy at higher magnification. As there is a high-risk of complications and difficulties in performing hysteroscopy in FGTB cases due to the small shrunken cavity, the procedure needs to be performed by expert gynecologists preferably under laparoscopic guidance.[59]

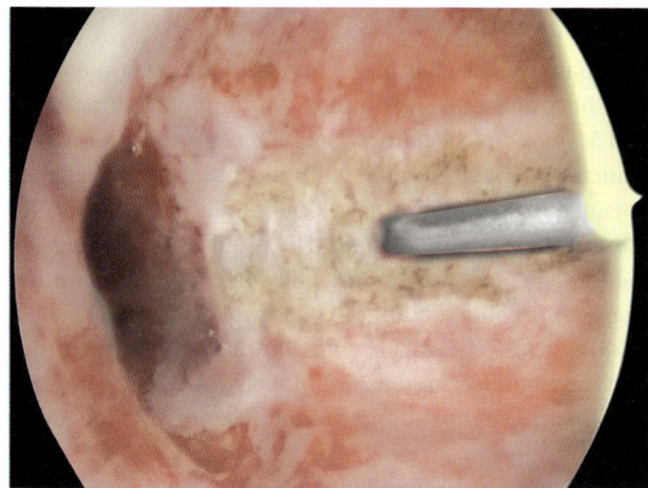

Fig. 3: Hysteroscopy showing grade II adhesions in female genital tuberculosis being resected.

POLYMERASE CHAIN REACTION

Polymerase chain reaction-based tests (either commercially available or in-house assays) are increasingly used to diagnose endometrial TB. A literature review prepared for this guideline found that estimates of sensitivity and specificity varied widely across reports in the literature, and the Technical Advisory Subcommittee for FGTB raised concerns about their experience of high rates of false positives when these tests are applied to peritoneal/gynecological specimens. There were no data at the time of publication looking at the use of Xpert MTB/RIF for the diagnosis of FGTB. The guideline group decided that a recommendation was not possible at this time regarding the use of PCR-based tests in FGTB, and noted that high quality diagnostic test accuracy studies are needed to address this question.

NUCLEIC ACID AMPLIFICATION TECHNIQUES FOR TUBERCULOSIS

During the 1990s, several nucleic acid amplification (NAA) techniques evolved that dramatically changed the way in which we can detect and identify MTB. Unlike the signal amplification technique, which relies on large numbers of organisms for detection and, thus, is useful only for cultured isolates, NAA techniques allow for detection of MTB from samples containing relatively few MTB bacilli. For this reason, NAA techniques can be utilized to identify MTB directly from clinical specimens, avoiding the most time-consuming aspect of MTB identification, the time required to culture the isolate.

In the mid-1990s, the United States Food and Drug Administration (FDA) approved two direct NAA tests for identification of MTB complex in respiratory specimens that were smear-positive for AFB: The Gen-Probe *M. tuberculosis* Direct (MTD) test and the Roche Amplicor MTB test.[60] The Amplicor test is a PCR-based test that amplifies MTB DNA, whereas the Gen-Probe MTD test is a transcription-mediated amplification test that amplifies MTB ribosomal ribonucleic acid (rRNA). The Gen-Probe MTD test takes only 3 hours to perform.

When used for AFB smear-positive respiratory samples, the Gen-Probe MTD test is quite reliable, with high sensitivity (83–100%), positive predictive value (94–100%), and negative predictive value (96–100%) reported in the literature.[61-63] This high level of accuracy allows for early identification of AFB as either MTB or nontuberculous mycobacteria (NTM), and treatment/perfection control decisions can be based on this information.

When used in AFB smear-negative respiratory samples, early studies suggested diminished reliability of this test with sensitivity of 50% and positive predictive value of 3–50%, although specificity remained high.[60] More recent data with a new, enhanced MTD test (MTD2) suggest that even in AFB smear-negative respiratory samples, sensitivity and specificity approach 100%.[63] Many studies suggest that the Gen-Probe MTD test is also useful in assessing extrapulmonary specimens with high sensitivity (84–100%), specificity (98–100%), and positive predictive value (90–100%).[61-63]

These new, rapid, highly sensitive molecular techniques do not solve all the problems detecting MTB. Like any test, NAA assays are not perfect and should be used only in conjunction with traditional culture isolation methods to maximize sensitivity of the laboratory diagnosis of TB. Current NAA technology allows for the detection of MTB in patient specimens within hours compared with the 14–28 days required for culture. The potential clinical benefits of this early detection, for both an individual and at public health level, are unimaginable.

TREATMENT

Aims of treatment:

- To achieve TB cure
- To prevent the long-term sequelae
- To restore normal anatomy if has been distorted.

Three basic principles have emerged in the years following the advent of chemotherapy for TB:

1. The regimen for treatment must contain multiple drugs to which the organism is susceptible.
2. The drugs should be taken regularly.
3. The drug therapy should continue for a sufficient period of time.

An algorithm has been developed for FGTB in INDEX-TB guidelines, an initiative of Central TB Division, Ministry of Health and Family Welfare, Government of India and the WHO in 2016 for the diagnosis and management of FGTB.[64] Another Indian team is also working to develop an algorithm suitable for the detection and management of FGTB.[65] It is mandatory to notify every case diagnosed and treated under NIKSHAY, a web-based notification system maintained by the Ministry of Health and Family Welfare, Government of India (*http://nikshay.in*).

All new cases, whether microbiologically confirmed or clinically diagnosed and drug sensitive previously treated patients (nonresponders, failures, recurrent TB, and lost to follow-up for 1 month after receiving 1 month of ATT) are given combination therapy. Medical therapy with antitubercular drugs for 6–9 months is effective for these FGTB cases.[66] In a randomized controlled trial, 6 months ATT was found to be equally effective as 9 months therapy.[67]

Directly Observed Treatment, Short-course Strategy

The treatment of FGTB is shown in **Flowchart 1**. Directly observed treatment, short-course (DOTS) is favored and recommended by the WHO.[68] Daily treatment is given under direct supervision of a treatment supporter. Four drugs isoniazid (H), rifampicin (R), pyrazinamide (Z), and ethambutol (E) are given for 2 months (HRZE), followed by H, R, and E (HRE) daily for 4 months **(Flowchart 1)**. Thrice-weekly regimen is no longer used.

Non-DOTS Treatment

Self-administered therapy for women not willing to take medicines from the DOTS center can buy combination kits containing four drugs (HRZE) for 2 months followed by kit containing three drugs (HRE) for 4 months.

Treatment of HIV-TB and DRTB

If the patient is both HIV and TB positive, both need to be treated. ATT is started to all patients with drug-sensitive or drug-resistant TB (DRTB) and HIV irrespective of CD4 cell count, as early as possible, usually by 2 weeks of initiation

Flowchart 1: Treatment regimens for female genital tuberculosis (FGTB). Moxifloxacin to be given, if resistance to levofloxacin. Recently, four-drug fixed dose combination (FDC) HRZE (75, 150, 400, and 275) has been introduced.[68,69]

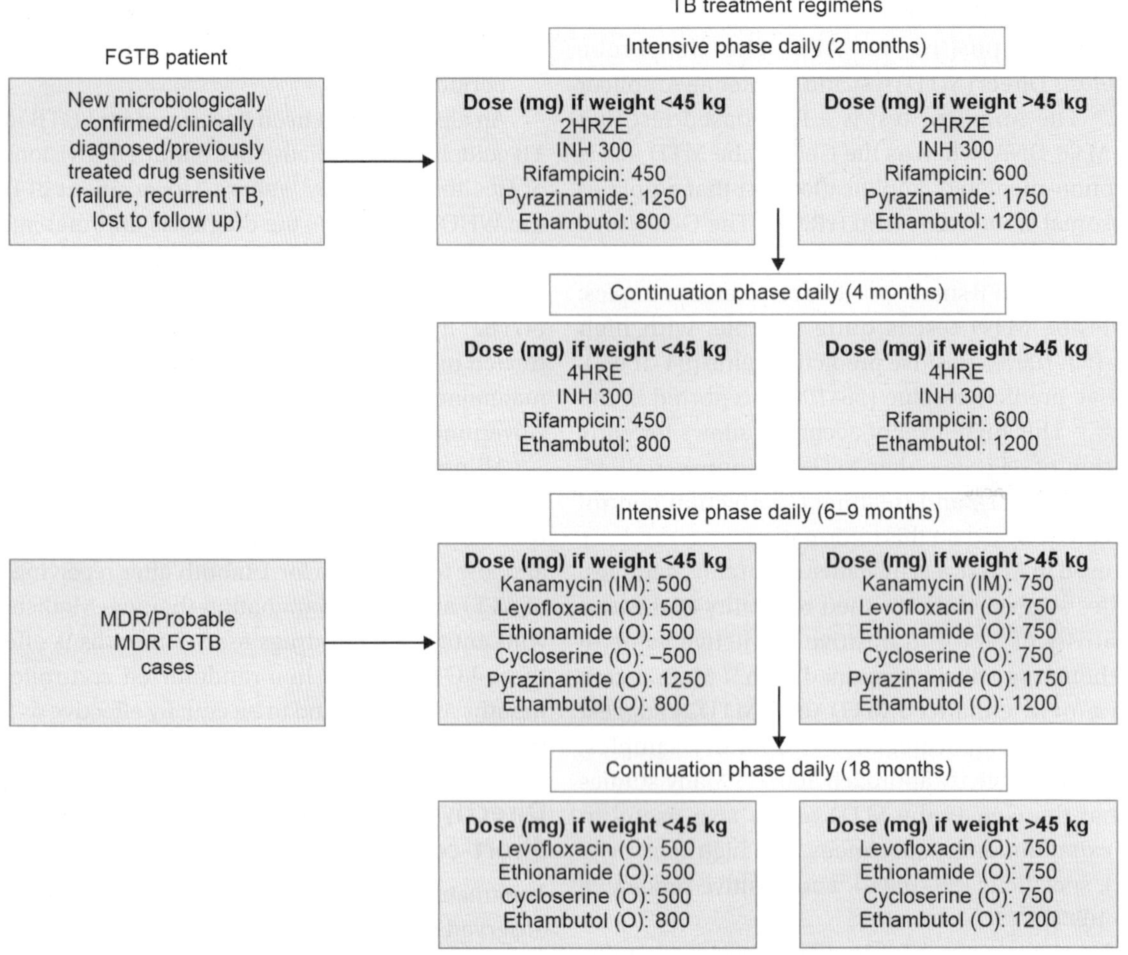

(HRZE: isoniazid, rifampicin, pyrazinamide, and ethambutol; INH: isoniazid; MDR: multidrug-resistant; IM: intramuscular; O: oral; TB: tuberculosis).

of ATT. In drug-sensitive TB patient on the second-line ART, rifampicin is replaced by rifabutin (150 mg daily).[68,69] Any patient who fails to respond clinically and/or radiologically despite 2 months of supervised adequate ATT is a presumptive DRTB case and should be offered WRDTs to confirm drug sensitivity. The diagnosis of DRTB is based on the results of WRDT. Drug-resistant [RR/MDR/XDR (extensively DR)]. FGTB is treated with reserve drugs for 18–24 months like for pulmonary MDR.

Surgical Treatment

Schaefer[5] had advised surgical treatment by removal of the uterus, both tubes and ovaries for persistent disease, tubercles, pyosalpinx, tubo-ovarian mass, nonhealing ulcers in multidrug-resistant TB (MDR TB) despite medical treatment. Besides, surgical treatment in TB is difficult and hazardous with more complications as observed by us in hysteroscopy and laparoscopy, laparotomy, and vaginal hysterectomy performed in women with genital TB. However, fertility preserving conservative surgery in the form of drainage of persistent pelvic or tubo-ovarian abscess despite medical treatment can be performed as recommended by the American Thoracic Society.[70]

Repeat Laparoscopy and Hysteroscopy after Antitubercular Treatment

Repeat laparoscopy and hysteroscopy can be performed after completion of a full course of ATT for the disappearance of disease. This also helps in prognostication for treatment of infertility. It was observed that after ATT, minor

TABLE 3: Adverse effects of antitubercular drugs.

First-line drugs	*Side effects*
Isoniazid	Peripheral neuropathy, seizures, erythema and skin rash, hepatitis, and lethargy
Rifampicin	Gastrointestinal irritation, hepatitis, skin reaction, flu-like syndrome, anemia, and thrombocytopenic purpura
Pyrazinamide	Hepatitis, gastrointestinal irritation (nausea and vomiting), and hyperuricemia
Ethambutol	Optic neuritis
Injectable streptomycin, kanamycin, and amikacin	Ototoxicity, renal toxicity, vertigo, and electrolyte imbalance
Quinolones	Gastrointestinal irritation, abdominal pain, nausea vomiting, hepatitis, seizures, dizziness, skin rash, joint pains, and photosensitivity QTc prolongation
Ethionamide	Gastrointestinal irritation, abdominal pain, nausea, vomiting, hepatitis, hypothyroidism and goiter, hallucination and depression, and neuropathy
Cycloserine	Neurological effects (dizziness, seizures, headache, tremors, and insomnia), hypersensitivity, suicidal tendency, and depression
Para-amino salicylic acid	Gastrointestinal irritation, abdominal pain, nausea, vomiting, skin rash, hypothyroidism, and hypokalemia

TB lesions disappeared, but major adhesions, frozen pelvis, blocked tubes (advanced disease) persisted with poor prognosis for fertility.[43,71] Bahadur et al. observed significant improvement in grades I and II adhesions only on repeat hysteroscopy after completion of ATT in FGTB patients, but major adhesions (grade III onward) persisted. We observed the persistence of advanced Asherman's syndrome and adhesions on repeat hysteroscopy in FGTB patients.[72] Tuboplasty should be avoided in case of FGTB as it does not improve the prognosis for fertility but there is risk of flare-up of TB and further ectopic gestation after ATT.[10,11]

Adverse Effects of Drugs

First-line drugs are safe and may have minor side effects but rarely serious side effects like hepatitis. However, reserve drugs are toxic and are to be given for longer duration. Various adverse effects of ATT are shown in **Table 3**.

RECENT ADVANCES AND FUTURE DIRECTION

Research is being conducted to develop newer drugs. Recently injection-free regimen using bedaquiline and delamanid have been recommended by the WHO in their rapid advise and update 2018[73] and the WHO Consolidated Guidelines on Drug-Resistant Tuberculosis Treatment (2019)[74] against resistant strains of MTB and to shorten therapy period. Stem cells, nanotechnology, and colostrum are being tested to regenerate fallopian tubes, endometrium, and ovaries to treat permanent damage.[11] The WHO rapid communication[73] has stressed upon injection free regimens for all types of TB, including MDR and rifampicin-resistant (RR) TB.[74] It has also advised discontinuation of category II regimen. For new or all previously treated patients if rifampicin or isoniazid sensitive after ensuring strict clinical, radiological, and bacteriological follow-up (if available) (2HRZE/4HRE) is given.[68,69] For isoniazid-resistant TB patients the uniphasic regimen (6LfxRZE) is given. For MDR, pre-XDR, and XDR-TB patients the recommended regimen is bedaquiline, levofloxacin (or moxifloxacin), linezolid, cycloserine, and clofazimine. Streptomycin, amikacin instead of kanamycin and capreomycin[73-75] are given if injectable drugs are to be used.

REFERENCES

1. Morgagni JB. De Sedibus et Causis meorboram. Epistola. 1744;38:34.
2. Muir DG, Belsey MA. Pelvic inflammatory disease and its consequences in the developing world. Am J Obstet Gynecol. 1980;138:913-28.
3. Ellner JJ. Tuberculosis. In: Kelley A (Ed). Textbook of Internal Medicine. Philadelphia, JB Lippincott & Co, 1989. p. 1569.
4. Dhillon SS, Gosewehr JA, Julian TM, Huey J. Genital tuberculosis: case report and literature review. Wis Med J. 1990;89:14-7.

5. Schaefer G. Female genital tuberculosis. Clin Obstet Gynecol. 1976;19:223-39.
6. Tripathy SN, Tripathy SN. Endometrial tuberculosis. J Indian Med Assoc. 1987;85:136-40.
7. Golden MP, Vikram HR. Extrapulmonary tuberculosis: An overview. Am Fam Physician. 2005;72:1761-8.
8. Kumar S, Sharma JB. Female genital tuberculosis. In: Sharma SK, Mohan A (Eds). Tuberculosis, 3rd edition. Delhi: Jaypee Brothers Medical Publishers (P) Ltd.; 2015. pp. 362-71.
9. Neonakis IK, Spandidos DA, Petinaki E. Female genital tuberculosis: a review. Scand J Infect Dis. 2011;43: 564-72.
10. Sharma JB. Tuberculosis and obstetric and gynecological practice. In: Studd J, Tan SL, Chervenak FA (Eds). Progress in Obstetrics and Gynaecology. Philadelphia: Elsevier; 2008. pp. 395-427.
11. Sharma JB. Current diagnosis and management of female genital tuberculosis. J Obstet Gynaecol India. 2015;65:362-71.
12. Falk V, Ludviksson K, Agren G. Genital tuberculosis in women. Analysis of 187 newly diagnosed cases from 47 Swedish hospitals during the ten-year period 1968 to 1977. Am J Obstet Gynecol. 1980;138:974-7.
13. Khan SM. Incidence of genital tuberculosis in infertile women. J Pak Med Assoc. 1985;35:280-1.
14. Shahzad S. Investigation of the prevalence of female genital tract tuberculosis and its relation to female infertility: An observational analytical study. Iran J Reprod Med. 2012;10:581-8.
15. Margolis K, Wranz PA, Kruger TF, Joubert JJ, Odendaal HJ. Genital tuberculosis at Tygerberg hospital—prevalence, clinical presentation and diagnosis. S Afr Med J. 1992;81:12-5.
16. Tripathy SN, Tripathy SN. Infertility and pregnancy outcome in female genital tuberculosis. Int J Gynaecol Obstet. 2002;76:159-63.
17. Bazaz-Malik G, Maheshwari B, Lal N. Tuberculous endometritis: A clinicopathological study of 1000 cases. Br J Obstet Gynaecol. 1983;90:84-6.
18. Gupta N, Sharma JB, Mittal S, Singh N, Misra R, Kukreja M, et al. Genital tuberculosis in Indian infertility patients. Int J Gynaecol Obstet. 2007;97:135-8.
19. Singh N, Sumana G, Mittal S. Genital tuberculosis: A leading cause for infertility in women seeking assisted conception in North India. Arch Gynecol Obstet. 2008;278:325-7.
20. Duggal S, Duggal N, Hans C, Mahajan RK. Female genital TB and HIV co-infection. Indian J Med Microbiol. 2009;27:361-3.
21. Ali AA, Abdallah TM. Clinical presentation and epidemiology of female genital tuberculosis in Eastern Sudan. Int J Gynaecol Obstet. 2012;118:236-8.
22. Shah HN, Patel S, Nagpol S. Pelvic inflammation (A study of 800 cases). J Obstet Gynecol India. 1978;28:429-35.
23. Khilnani PH, Pandit AA, Krishna UR. Cytology as an aid in the diagnosis of genital tuberculosis. J Post Grad Med. 1988;34:100-2.
24. Siegler AM, Kontopoulous V. Female genital tuberculosis and the role of hysterosalpingography. Semin Roentgenol. 1979;14:295-304.
25. DesPrez RM, Goodwin RA. *Mycobacterium tuberculosis*. In: Mandell GL, Douglas RG, Bennett JE (Eds). Principles and Practices of Infectious Diseases, 2nd edition. New York: John Wiley & Sons; 1985. p. 1383.
26. Gavaller J, Suranyi S, Berencsi G. Neue Gesichttpunkte in der Klinik der Genital tuberkulose. Zentralbl Gynakol. 1956;78:496.
27. Sutherland AM. Postmenopausal tuberculosis of the female genital tract. Obstet Gynecol. 1982;59:54S-7S.
28. Lattimer JK, Colmore HP, Sanger G, Robertson DH, McLellan FC. Transmission of genital tuberculosis from husband and wife via the semen. Am Rev Tuberc. 1954; 69:618-24.
29. Muneer A, Macrae B, Krishnamoorthy S, Zumla A. Urogenital tuberculosis—epidemiology, pathogenesis and clinical features. Nat Rev Urol. 2019;16(10):573-98.
30. Furin J, Cox H, Pai M. Tuberculosis. Lancet. 2019; 393(10181):1642-56.
31. Loddenkemper R, Brönnecke M, Castell S, Diel R. Tuberculosis and Tobacco Smoking. Pneumologie. 2016; 70(1):17-22.
32. Saunders MJ, Wingfield T, Tovar MA, Baldwin MR, Datta S, Zevallos K, et al. A score to predict and stratify risk of tuberculosis in adult contacts of tuberculosis index cases: a prospective derivation and external validation cohort study. Lancet Infect Dis. 2017;17(11):1190-9.
33. Albuquerque MFPM, Coimbra I d'Arc Batista J, Maruza M, Ximenes RAA, Lacerda HR. Empirical treatment for TB in HIV: lessons from a cohort study of people living with HIV treated in Recife, Brazil. BMC Public Health. 2014;14:289.
34. Sharma JB, Roy KK, Pushparaj M, Gupta N, Jain SK, Malhotra N, et al. Genital tuberculosis: An important cause of Asherman's syndrome in India. Arch Gynecol Obstet. 2008;277:37-41.
35. Onuigbo WI. Genital tuberculosis and reproductive function. J Reprod Med. 1978;21:249-50.
36. Nogales-Ortiz F, Tarancion I, Nogales FF Jr. The pathology of female genital tuberculosis: A 31-year study of 1436 cases. Obstet Gynecol. 1979;53:422-8.
37. Schenker JG, Margalioth EJ. Intrauterine adhesions: an updated appraisal. Fertil Steril. 1982;37:593610.
38. Hasselgren PO, Bolin T. Postmenopausal tuberculous pyometra. Acta Obstet Gynecol Scand. 1977;56:23-5.
39. Czernobilsky B. Endometritis and infertility. Fertil Steril. 1978;30:119-30.
40. Macintosh OC, Saxon RD. Endometritis due to Mycobacterium tuberculosis. Can Med Assoc J. 1985; 133:667-8.

41. Sutherland AM. Tuberculosis: Gynaecological tuberculosis. Br J Hosp Med. 1979;22:569-76.
42. Burnie JC. The age of onset of genital tuberculosis in women. J Obstet Gynaecol Br Emp. 1956;63:96-9.
43. Heifets L. Mycobacteriology laboratory. Clin Chest Med. 1997;18:35-53.
44. Chattopadhyay SK, Sengupta BS, Edress YB, Al-Meshari AA. The pattern of female genital tuberculosis in Riyadh, Saudi Arabia. Br J Obstet Gynaecol. 1986;93:367-71.
45. Francis WJ. Female genital tuberculosis. A review of 135 cases. J Obstet Gynaecol Br Commonw. 1964;71:418-28.
46. Daly JW, Monif GRG. Mycobacteria. In: Monif GRG (Ed). Infectious Diseases in Obstetrics and Gynaecology, 2nd edition. Philadelphia: Harper & Row; 1982. p. 301.
47. Simon HB, Weinstein AJ, Pasternak MS, Swartz MN, Kunz LJ. Genitourinary tuberculosis. Clinical features in a general hospital population. Am J Med. 1977;63:410-20.
48. Mondal SK. Histopathologic analysis of female genital tuberculosis: A fifteen-year retrospective study of 110 cases in Eastern India. Turk J Pathol. 2013;29:41-5.
49. Thangappah RBP, Paramasivan CN, Narayanan S. Evaluating PCR, culture and histopathology in the diagnosis of female genital tuberculosis. Indian J Med Res. 2011;134:40-6.
50. Goel G, Khatuja R, Radhakrishnan G, Agarwal R, Agarwal S, Kaur I, et al. Role of newer methods of diagnosing genital tuberculosis in infertile women. Indian J Pathol Microbiol. 2013;56:155-7.
51. Katoch VM. Newer diagnostic techniques for tuberculosis. Indian J Med Res. 2004;120:418-28.
52. Bhanu NV, Singh UB, Chakraborty M, Suresh N, Arora J, Rana T, et al. Improved diagnostic value of PCR in the diagnosis of female genital tuberculosis leading to infertility. J Med Microbiol. 2005;54:927-31.
53. Abdissa S, Abebe T, Ameni G, Teklu S, Bekuretsion Y, Abebe M, et al. Endometrial tuberculosis among patients undergoing endometrial biopsy at Tikur Anbesa specialized hospital, Addis Ababa, Ethiopia. BMC Infect Dis. 2018;18:304.
54. Arora R, Sharma JB. Female genital tuberculosis—a diagnostic and therapeutic challenge. Indian J Tuberc. 2014;61:98-102.
55. Sharma SK, Kohli M, Chaubey J, Yadav RN, Sharma R, Singh BK, et al. Evaluation of Xpert MTB/RIF assay performance in diagnosing extrapulmonary tuberculosis among adults in a tertiary care centre in India. Eur Respir J. 2014;44: 1090-3.
56. Sharma JB, Kriplani A, Dharmendra S, Chaubey J, Kumar S, Sharma SK, et al. Role of geneXpert in diagnosis of female genital tuberculosis: A preliminary report. Eur J Obstet Gynecol Reprod Biol. 2016;207:237-8.
57. Sethi S, Dhaliwal L, Dey P, Kaur H, Yadav R, Sethi S, et al. Loop-mediated isothermal amplification assay for detection of *Mycobacterium tuberculosis* complex in infertile women. Indian J Med Microbiol. 2016;34:322-7.
58. Sharma JB, Roy KK, Pushparaj M, Kumar S. Hysteroscopic findings in women with primary and secondary infertility due to genital tuberculosis. Int J Gynaecol Obstet. 2009;104:49-52.
59. Sharma JB, Roy KK, Pushparaj M, Karmakar D, Kumar S, Singh N, et al. Increased difficulties and complications encountered during hysteroscopy in women with genital tuberculosis. J Minim Invasive Gynecol. 2011;18:660-5.
60. Rapid diagnostic tests for tuberculosis: what is the appropriate use? American Thoracic Society Workshop. Am J Respir Crit Care Med. 1997;155:1804-14.
61. Gamboa F, Fernandez G, Padilla E, Manterola JM, Lonca J, Cardona PJ, et al. Comparative evaluation of initial and new versions of the Gen-Probe Amplified Mycobacterium Tuberculosis direct test for direct detection of Mycobacterium tuberculosis in respiratory and nonrespiratory specimens. J Clin Microbiol. 1998;36: 684-9.
62. Vlaspolder F, Singer P, Roggeveen C. Diagnostic value of an Amplification Method (Gen-Probe) compared with that of culture for diagnosis of tuberculosis. J Clin Microbiol. 1999;33:2699-703.
63. Chedore P, Jamieson FB. Routine use of the Gen-probe MTD2 Amplification Test for detection of Mycobacterium tuberculosis in clinical specimens in a large public mycobacteriology laboratory. Diagn Microbiol Infect Dis. 1999;35:185-91.
64. Sharma SK, Ryan H, Khaparde S, Sachdeva KS, Singh AD, Mohan A, et al. INDEX-TB guidelines: Guidelines on extrapulmonary tuberculosis for India. Indian J Med Res. 2017;145(4):448-63.
65. Ramakant B. (2014). Indian Scientists Developing Diagnostic Algorithm for Female Genital TB. [online] Available from https://www.thehansindia.com/posts/index/2014-09-21/Indianscientists-developing-diagnostic-algorithm-for-female-genital-TB-108475. [Last accessed December, 2023].
66. Arora R, Rajaram P, Oumachigui A, Arora VK. Prospective analysis of short course chemotherapy in female genital tuberculosis. Int J Gynaecol Obstet. 1992;38:311-4.
67. Sharma JB, Singh N, Dharmendra S, Singh UB, Vanamail P, Kumar S, et al. Six months versus nine months anti-tuberculous therapy for female genital tuberculosis: a randomized controlled trial. Eur J Obstet Gynecol Reprod Biol. 2016;203:264-73.
68. World Health Organization. WHO global tuberculosis report 2018. Geneva: World Health Organization; 2018.
69. Ministry of Health and Family Welfare. India TB Report: Revised National Tuberculosis Control Programme: Annual status report. New Delhi: Ministry of Health and Family Welfare; 2018.
70. Lee J, Warner L, Khaleghian R. Sonographic features of tuberculous endometritis. J Clin Ultrasound. 1983;11: 331-3.

71. Sutherland AM. Surgical treatment of tuberculosis of the female genital tract. Br J Obstet Gynaecol. 1980;87: 610-2.
72. Barter JF, Addison WA, Rosenberg ER, Livengood CH 3rd. Calcific pelvic lymphadenopathy presenting as a postmenopausal adnexal mass: a case report. J Reprod Med. 1984;29:209-10.
73. World Health Organization. Rapid Communication: Key Changes to Treatment of Multidrug-and Rifampicin-Resistant Tuberculosis (MDR/RR-TB) (WHO/CDS/TB/2018.18). Geneva: World Health Organization; 2018.
74. World Health Organization. WHO Consolidated Guidelines on Drug-resistant Tuberculosis Treatment. Geneva: World Health Organization; 2019.
75. Tripathy SN, Tripathy SN. Tuberculosis: Manual for Obstetrics and Gynaecologists, 1st edition. New Delhi: Jaypee Brothers Medical Publisher (P) Ltd.; 2015. pp. 155-275.

Chapter 9

Contraception and Sterilization

Shikha Rani, Japleen Kaur, Anil E

INTRODUCTION

India's population (2022) is 1.42 billion. As per the National Family Health Survey-5 (NFHS-5) current use of any family planning method is 66.7% and unmet need for the family planning method is 9.4%.[1] As per the NFHS-5, user of different family planning methods is given in **Table 1**.

TABLE 1: Use of different family planning methods as per NFHS-5.

Family planning method	*Percentage of user*
Any contraceptive method	66.7
Any modern contraceptive	56.5
Female sterilization	37.9
Male sterilization	0.3
IUD/PPIUD	2.1
Pill	5.1
Condom	9.5
Injectables	0.6

(IUD: intrauterine device; NFHS-5: National Family Health Survey-5; PPIUD: postpartum intrauterine device)

FERTILITY AWARENESS METHODS

Refer to **Table 2**.

Advantages of All Fertility Awareness Methods

- It utilizes fertility awareness to either prevent or achieve pregnancy.
- No side effects.

TABLE 2: Fertility awareness of methods.[2]

Method	*Abstinence period*	*Method specific advantage/ disadvantage*	*Failure rate—perfect use (%)*	*Failure rate—typical use (%)*
Standard days	Avoid unprotected intercourse during cycle days 8–19	• Easiest to teach • Fewer days of alternative contraception or abstinence	05	12
Billings and Creighton or cervical mucus method	On all days of cervical secretions, and until 4 days after the last day secretions	Difficult to teach and follow	2.9	15.5
Two days method	On all days of cervical secretions, and until first after the last day secretions	Difficult to follow	04	14
Symptothermal	Combines basal body temperature and cervical mucus method	Difficult to measure body temperature every day in the morning and track of cervical secretion	<1	2
Basal body temperature	On all days before the rise in basal body temperature	• Tells that ovulation has already occurred • Not reliable if woman is having fever		6.6
Withdrawal (coitus interruptus)	Withdrawal of the penis from the vagina during the act of ejaculation	Require lot of control by male partner to withdraw just before ejaculation	04	20

Disadvantages

- Not very effective
- Not suitable for women with irregular periods, perimenopausal women, women with abnormal uterine bleeding (AUB), and during vaginal or cervical infection
- Does not protect against reproductive tract infections (RTIs).

Lactation Amenorrhea Method

Women must be exclusively breastfeeding, at least every 4 hours in the day and every 6 hours in the night. Reliable for the first 6 months after delivery if menstruation has not restarted and effectiveness is up to 92–100%.

Birth Control App (Natural Cycles)

Natural cycles is a birth control app which is cleared by the Food and Drug Administration (FDA). This app calculates the fertile period of a women by her menstrual cycle and basal body temperature. Temperature is measured as a first thing in the morning with the help of a basal body thermometer. It is approved for premenopausal women ≥18 years. The FDA allows marketing of first direct-to-consumer app for contraceptive use to prevent pregnancy.

Effectiveness: Perfect use (98%) and typical use (93%).[3]

Contraindications for Birth Control App

- Women using ovulation inhibition birth control methods.
- Women suffering from medical conditions where pregnancy can pose significant risk to mother and fetus. As in these cases women should be using more effective contraceptive methods.
- Women who want protection against RTIs.

BARRIER CONTRACEPTION

Refer to **Table 3**.

Advantages

- To use only at the time of sexual intercourse.
- Can be used in all medical conditions.

Nonhormonal Contraceptive Vaginal Gel

Phexxi is a nonhormonal contraceptive gel which contains lactic acid, citric acid, and potassium bitartrate 1.8, 1, and 0.4%, respectively. It is not spermicidal but a vaginal pH modulator. It is off white to tan in color and is odorless and tasteless.

How to Use?

Action starts immediately upon vaginal application of one dose of Phexxi (5 g) and last for 1 hour, which provides

TABLE 3: Different barrier contraceptive methods.[4]

Contraceptive	*To be used*	*Failure rate—perfect use (%)*	*Failure rate—typical use (%)*
Male condom	At time of coitus	2	17.4
Female condom	At time of coitus	5	27.0
Diaphragm	Applied at time of coitus must remain in place for 6 hours after sex, but not >24 hours total. A tablespoon of spermicidal agent is applied at the dome of diaphragm at the time of insertion. For any additional act of coitus spermicidal agent should be applied on the vagina	6	16
Today Sponge (nonoxynol-9)	To be put in immediately or up to 24 hours before sexual intercourse and should be left in place for at least 6 hours after that	20	32
Spermicidal agents (foam, gel, suppository, and film)	• 10–15 minutes before act of coitus and reinsert at every act of coitus • Effective for 1 hour	18	29
Cervical cap	Applied at time of coitus must remain in place for 6 hours after sex, but not >48 hours total. Spermicidal agent is applied at the dome at the time of insertion. No additional dose of spermicidal gel is required for any additional act of intercourse[5]	9	20

protection for one act of intercourse. For any more act of intercourse beyond or before 1 hour, requires repeat application.[7] It comes in a box with pack of 12 prefilled foil pouches. Each pouch has 5 g of Phexxi. After intercourse, the gel gets mixed with vaginal secretions and semen, slowly comes out of the vagina.

Mode of Action

It makes the vaginal pH acidic which causes the reduction in the sperm motility and also makes a viscous barrier.[6]

Contraindications

- Avoid if history of recurrent urinary tract infection (UTI) or urinary tract abnormalities
- Avoid use with vaginal rings
- Pregnancy

Adverse Effects

- Vulvovaginal burning sensation, pruritus, or vaginal discharge
- Genital discomfort in both partners
- Urinary tract symptoms (dysuria, cystitis, and rarely pyelonephritis)
- Discontinue and seek medical advice if any of the above symptoms are severe or prolonged and in case of local hypersensitivity reaction.

Effectiveness

- *Perfect use:* 93%
- *Typical use:* 86%

Advantages

Can be used by all females except the above contraindications.

Disadvantages

Does not protect against human immunodeficiency virus (HIV) or any RTIs.

COMBINED HORMONAL CONTRACEPTIVE METHODS

Combined hormonal contraceptive (CHC) contains estrogen paired with progesterone in different formulations. Mostly CHC are comprised of ethinylestradiol (EE) in a dosage between 20 and 35 µg (low dose) to prevent health hazards.[8] The progesterone in CHC belongs to different generations and form many formulations.

- *First:* Norethisterone enanthate (NET-EN)
- *Second:* Levonorgestrel (LNG)
- *Third:* Desogestrel, gestodene, and norgestimate
- *Newer/other:* Drospirenone (DRSP), dienogest, and nomegestrol acetate.[9]

Main Mechanism of Action

They act by inhibiting ovulation, they act on hypothalamic-pituitary-ovarian axis and inhibit luteinizing hormone (LH) and follicle-stimulating hormone (FSH). Progesterone causes cervical change and this also helps in contraceptive effect.

Combined Oral Contraceptive

Initiation of Combined Oral Contraceptive

Combined oral contraceptive (COC) must be taken daily at the same time and leads to decreased efficacy if taken >24 hours apart. It is usually taken on the first day of menses and it is a better strategy because of its maximum efficacy but can be taken till day 5 of menstrual cycle. The other methods are to start on any day or to start on Sundays in which additional birth control is advised for 7 days.[10] It can be started after 6 weeks postpartum and till day 5 postabortion/suction and evacuation or ectopic pregnancy.[11]

Missed Pill

If a woman misses one pill, advise to take the missed pill as soon as possible and no backup method is necessary. If a woman misses two pills during first/second week, advise to take two pills daily for 2 days then finish pack and advised back up method for 7 days. If two pills are missed during the third week, then start new pack and backup method to be used for 7 days. If she misses three pills, then to start new pack immediately and to have a backup method for 7 days.[12]

Return to fertility: Almost 99% of women return to fertility in 90 days.

Efficacy: Refer to **Table 3**.

Contraindications

- Absolute contraindications [medical eligibility criteria 4 (MEC 4)]:
  - Thrombophlebitis, thromboembolic disorders, and family history of vascular diseases/coronary artery diseases (CADs)

- Markedly impaired liver function
- History of CAD/cardiovascular disease (CVD)
- Migraine with aura
- Diabetes mellitus with vascular disease
- Suspected or diagnosed case of breast cancer
- Pregnancy
- Unevaluated vaginal bleeding
- Women >35 years and smoke
- Severe hypercholesterolemia/hypertriglyceridemia
- Uncontrolled hypertension (HTN)

- *Relative contraindications (MEC 2 and 3)*:
  - Women <35 years of age who have migraine without aura
  - Controlled HTN
  - Uterine fibroids
  - Gestational diabetes mellitus
  - Elective surgery
  - Seizure disorders
  - Obstructive jaundice in pregnancy
  - Sickle cell disease
  - Gall bladder disease
  - Mitral valve prolapse
  - Systemic lupus erythematosus
  - Smoking
  - Hepatic disease.[4]

Noncontraceptive Uses of Combined Oral Contraceptive

- AUB
- Dysmenorrhea
- Ovulation pain
- Acne and hirsutism
- Endometriosis
- Hormone therapy for hypothalamic amenorrhea
- Prevention of menstrual porphyria.

Transdermal Combined Contraceptive

Transdermal patches are the most convenient noninvasive method of contraceptive.

Advantages over Oral Combined Contraceptives

- Minimal change in plasma concentration of estrogen in the body.
- Lesser side effects
- Better compliance
- Patch have to be applied weekly once. They can be applied on dry skin of buttocks, lower abdomen, upper outer arm and upper body and to be applied within 5 days of menstruation. Three weekly patches are followed by one patch free week. However, they have shown a slight increase in venous thromboembolism when compared to oral combined pills.[13] EVRA is used in UK and it contains 6 mg norelgestromin and 600 µg EE. Each 20 cm^2 transdermal patch releases 203 µg of norelgestromin and 33.9 µg of EE in 24 hours.[14]

Combined Hormonal Vaginal Rings

It is a flexible light weight vaginal ring. NuvaRing is a lightweight flexible ring with a diameter of 54 mm and 4 mm cross section which releases 120 µg etonogestrel and 150 µg EE per day, respectively. A ring is placed for 3 weeks followed by ring free week. Vaginal ring can be self-placed over cervix. Contraceptive benefit was not significantly superior to COC though prolonged and frequent bleeding was more in COC group.[15]

NONHORMONAL ORAL CONTRACEPTIVE

Centchroman (Ormeloxifene) (Fig. 1)

- Selective estrogen receptor modulator.
- Nonsteroidal and nonhormonal contraceptive method. It is available in government supply with the name of Chaya.
- Administer twice a week on fixed days for the first 3 months, thereafter once a week.[16]

Missed Pill

- Take a pill as soon as possible after it is missed.
- Missed by 1–2 days but ≤7 days, continue normal schedule with back-up method till the next periods.
- Missed pill ≥7 days, start all over again like a new user.

Fig. 1: Centchroman tablet.

Efficacy

Perfect use failure rate one to two pregnancy per 100 women, failure rate with typical use is not available.

Advantages

- Safe for lactating mother.
- Delayed period is the only side effect, if delay is ≥15 days then pregnancy test is to be done.

PROGESTERONE-ONLY CONTRACEPTIVES

It is available in the form of pills, injectables, implants, and intrauterine devices. Pills, injectables, and implants are covered in this section whereas progesterone-containing intrauterine system is covered under the "intrauterine devices".

Progesterone-only Pills

It is especially useful in cases where estrogen is contraindicated or needs to be avoided. In women such as lactating women, women smokers over the age of 35 years, women with serious medical conditions (diabetes with vascular disease, severe systemic lupus erythematosus, HTN, CVD), and previous episodes of thrombosis.

Main mechanism of action: Inhibition of ovulation and less important cervical mucus thickening and implantation.

Side effects of progesterone-only pills (POPs): Headache, breast tenderness, decreased libido, irregular bleeding, and amenorrhea.

Return to fertility: Immediate.

Efficacy: Refer to **Table 3**.

Contraindications (MEC 4)

- Known or suspected breast cancer
- Undiagnosed AUB
- Liver disorders such as benign or malignant liver tumor, severe cirrhosis, or acute liver disease.
- Medical eligibility criteria can be referred to, for defining safety of use in various medical or surgical conditions.

Formulations and Prescription

- *Desogestrel containing pill*: To be taken continuously, without a pill free period, intake delay of up to 12 hours does not affect its efficacy, e.g., Cerazette (desogestrel 75 μg).[17]

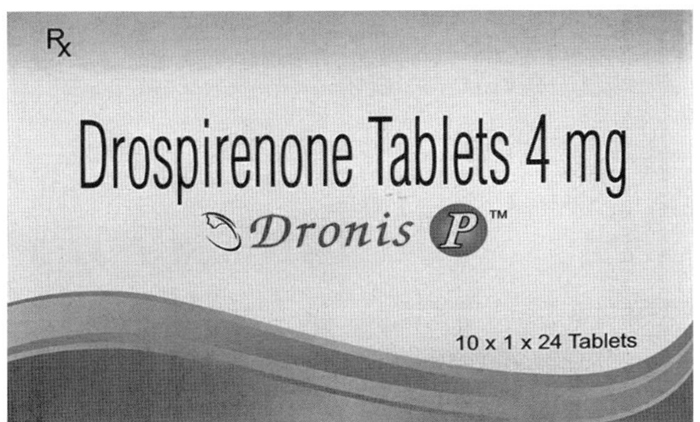

Fig. 2: Drospirenone tablet.

- *Drospirenone containing pill* **(*Fig. 2*)**: It is prescribed in 24 days on and 4 days off pattern and is associated with less unscheduled bleeding than the combined pills. It has a 24-hour missed pill safety window.[18] Its additional antimineralocorticoid and antiandrogenic action has a role for noncontraceptive benefits such as weight loss, blood pressure, and acne. It has very low risk of adverse cardiovascular, venous, and arterial thromboembolic events. Some contraindications are renal impairment, adrenal insufficiency, presence or history of cervical cancer, liver tumors, benign or malignant, or hepatic impairment, and undiagnosed AUB, e.g., Dronis P (drospirenone 4 mg).[19]

Progesterone-only Injections

Use: As with POPs, except not recommended in previous history of thrombosis.

Mechanism of action and safety: Same as POP. Additionally bone mineral density, especially in adolescents may be adversely affected, with longer use (>2 years).[20]

Return to fertility: Delayed up to months and unpredictable.

Efficacy: Refer to **Table 3**.

Formulations and Prescription

- *Depot medroxyprogesterone acetate (DMPA):* 150 mg intramuscular every 3 months, e.g., Antara (available free of cost in government supply) **(Fig. 3)**. A subcutaneous preparation is DMPA 105 mg.
- *Norethisterone enanthate*: 200 mg intramuscular every 2 months. Although less widely studied than DMPA, it causes relatively lesser menstrual disturbance, e.g., Noristerat.

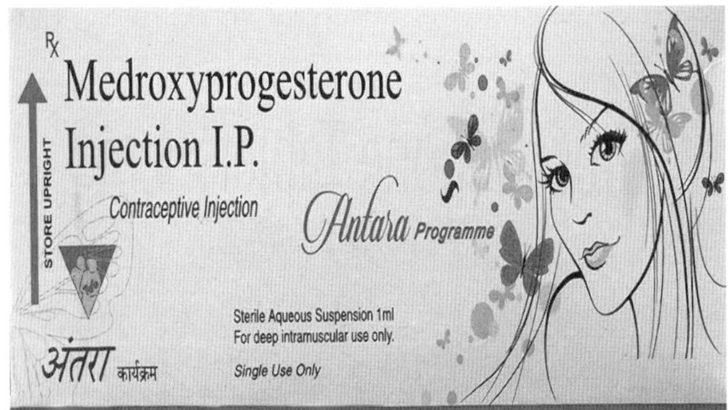

Fig. 3: Medroxyprogesterone injection.

TABLE 4: Percentage of women experiencing unintended pregnancy within the first year of use with typical use and perfect use of progesterone-only contraceptives.[21]

Type of progesterone-only contraceptive	*Failure rate—perfect use (%)*	*Failure rate—typical use (%)*
Progestogen-only pills	0.3	9
Progestogen-only injections	0.2	6
Progestogen-only implant	0.05	0.05
Levonorgestrel-releasing intrauterine system	0.2	0.2
Combined hormonal contraception	0.3	9

Starting Progesterone-only Contraception (Table 4)

No additional contraception is necessary if progesterone-only contraception (for which the woman is medically eligible) is initiated immediately or within 5 days of menstruation or abortion (or 21 days of childbirth). However, if progesterone-only contraception is started 5 days or more after abortion (or 21 days of childbirth), a woman is at risk of another pregnancy and thus additional contraceptive precautions (e.g., barrier methods/abstinence) are required up to 7 days for injectables and implants, whereas only 2 days for POPs.

LONG-ACTING REVERSIBLE CONTRACEPTION

In the US the use of long-acting reversible contraception (LARC) has gone up to 18% recently.[22,23] These include

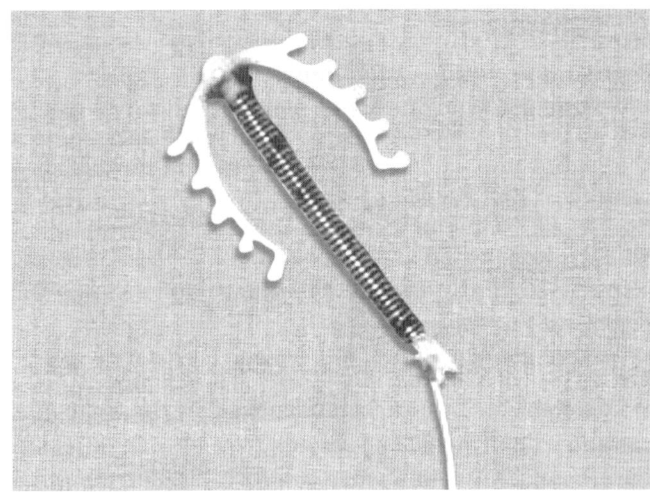

Fig. 4: Copper intrauterine device (IUD).

nonhormonal IUD, hormonal IUD, implants, and progesterone-only injections. Progesterone-only injections has been covered in progesterone-only contraceptive.

Advantages

- Most effective reversible contraceptive methods
- Single administration.
- Do not need continuous effort on part of the patient for prolonged use
- Efficacy comparable to sterilization procedure
- High satisfaction and continuation rates.

Nonhormonal Intrauterine Devices

Types

- *Unmedicated IUD:* These include Lippes loop, and Saf-T-coil. Lippes loop is still widely used in the world. USA has stopped manufacturing these in view of high removal and expulsion rate. These are not available in India too.
- *Copper IUD:* These have been approved for use up to 10 years after insertion. They can be inserted immediately after delivery [postpartum intrauterine contraceptive device (PPIUCD)] and after first or second trimester abortion with a failure rate of 0.8 per 100 women in a year.[24] Kelly's forceps was used for immediate or within 48 hours postpartum insertion. Also, interval insertion can be done. The types are CuT380A, Multiload 375, CuSAFE-300 IUD, and Nova T. CuT380A and Multiload 375 are available free of cost **(Fig. 4)**.

Clinical Considerations

- They are appropriate for nulliparous women and adolescents and can be advised (MEC 2).[25]
- They can be inserted immediate postpartum (vaginal delivery and cesarean section) and post first trimester (MEC 1) and second trimester abortion (MEC 2).
- Copper containing IUDs have no effect on breastfeeding.[26]
- Equally effective in obese women.
- Routine gynecological procedures such as endometrial biopsy, colposcopy, cervical excision, and ablation procedures can be done with IUD in situ.
- They can be also used as emergency contraception and method of choice in obese women.

Contraindications for the Use of IUDS (MEC 4)

- Pregnancy
- Active sexually transmitted infections (STIs) and active genital tuberculosis at the time of insertion
- Distorted uterine cavity due to leiomyomas or congenital abnormality
- Any septic abortion or uterine infection in last 3 months
- Any suspected or confirmed uterine malignancies
- Any unevaluated AUB.[27]

Hormone-releasing Intrauterine Device

Levonorgestrel containing IUD have different compositions, i.e., 13.5 mg (Skyla), 52 mg (Liletta), 19.5 mg (Kyleena), and 52 mg (Mirena) **(Table 5)**. The most commonly used is 52 mg LNG-IUD which releases 20 µg per day. The duration of usage for Skyla, Liletta, Kyleena, and Mirena are 3, 5, 8, and 6 years, respectively as FDA approved, while Mirena and Liletta can be used up to 8 years as per evidence. It mainly acts by bringing endometrial changes and altering the frequency of bleeding. Pregnancy is prevented by making cervical mucus thick which makes it not possible for sperm penetration. Routine gynecological procedures such as endometrial biopsy, colposcopy, cervical excision, and ablation procedures can be done with IUD in situ.

TABLE 5: Types of levonorgestrel-releasing intrauterine system (LNG-IUS).[28]

Type of LNG-IUS	LNG (mg)	Failure rate—perfect use (%)	Failure rate—typical use (%)
Skyla	13.5	0.3	0.4
Kyleena	19.5	0.2	0.2
Liletta and Mirena	52	0.1	0.1

Noncontraceptive Benefits with Levonorgestrel IUD

- Treatment of primary dysmenorrhea
- Improvement of symptoms, i.e., pain and decrease in uterine volume in adenomyosis
- Decrease in myoma size and helps in reducing volume of uterus and bleeding in fibroids
- Used in the management of endometriosis like in decreasing pain
- Protection against endometrial cancer and hyperplasia
- Helps in preventing pelvic inflammatory disease
- Used in patients with hemostatic disorders to decrease menstrual blood flow.

Complications of Intrauterine Device

- Displacement or accidental removal of IUD after insertion
- Problems with cervical dilatation and bradycardia/vasovagal symptoms
- Rare chances of unintended pregnancy (0.2–0.6%)
- Remote risk of uterine perforation while insertion of IUD
- AUB patterns (mostly in copper containing IUDs).[29]

Implants

This mode of LARC usage has been rapidly increasing, especially among adolescents and constitutes 4% of all contraceptive users. Organon (etonogestrel implant) which is manufactured in US is a radiopaque single flexible rod which contains 68 mg etonogestrel and is 4 cm long and 2 mm in diameter. It causes inhibition of ovulation by preventing LH surge.[30] It can be used for up to 5 years in the nonobese population. The major drawback is light and unpredictable bleeding pattern. Types of implants are detailed in **Table 6**.

TABLE 6: Types of progesterone-only implants.

Type of device	Number of rods	Approved validity (years)	Hormonal content
Norplant	6	5	216 mg levonorgestrel
Jadelle	2	3	75 mg levonorgestrel
Norplant 2	2	3	75 mg levonorgestrel
Implanon	1	3	68 mg etonogestrel

Contraindications of Implants (MEC 4)

- Breast cancer
- Decompensated liver disease
- Hepatocellular adenoma
- Malignant liver tumors.

Timing and Site of Placement

It can be placed anytime 4 weeks postpregnancy; it can be placed immediately after medical abortion/uterine aspiration in first trimester abortion and can be placed after giving mifepristone in abortions.[31] It is inserted in upper arm over triceps at about 8–10 cm proximal to the medial epicondyle. Insertion as well as removal can be done under local anesthesia.

Emergency Contraception

It protects against pregnancy in unprotected sexual intercourse. Both copper T and progesterone pills are available. Although copper T is the most effective form of emergency contraception which can be inserted up to 120 hours of unprotected intercourse. In case it is unacceptable, or eligibility criteria are not met, progesterone-only oral emergency contraception can be offered **(Table 6)**. It is of two type:

1. *Levonorgestrel emergency contraceptive (LNG-EC)*: Recommended usage is single dose, 1.5 mg, up to 72 hours of unprotected sexual intercourse. It acts by inhibiting LH surge and ovulation. It may not be effective if the estimated date of ovulation is within 5 days of the unprotected intercourse or if ovulation has already occurred. Double dosage of 3 mg is recommended for obese women and those on antiepileptic drugs. It is safe to use during breastfeeding. Regular hormonal contraception can be started immediately, i.e., quick start. If periods do not happen within 3 weeks of using LNG-EC, urine pregnancy test is to be done. Brand names—I-Pill, E-pill (government supply).[6] It is available in government supply **(Fig. 5)**.
2. *Ulipristal acetate emergency contraceptive (UPA-EC)*: Recommended usage is single dose, 30 mg, up to 120 hours of unprotected intercourse. It is effective even if taken within 5 days prior to the estimated date of ovulation. Same dosage for obese women. It could be less effective in women who has recently taken a progesterone or is on enzyme inducing drugs. It is contraindicated in severe asthma, managed with glucocorticoids. It is to be avoided in breastfeeding women. If needs to be taken, breast milk must be discarded for the next 7 days. Regular hormonal contraception can be started only after 5 days of use of UPA-EC. Brand name—ellaOne, not available in government supply. More expensive compared to LNG-EC.[32]

Fig. 5: Levonorgestrel tablets I.P.

Efficacy

Overall pregnancy rate with copper T is <0.1%. It is 1–2% with oral EC, however the failure rate is two to three times more with LNG-EC as compared to UPA-EC.

MALE STERILIZATION

Vasectomy is safer, simpler, less expensive, and just as effective as female sterilization. The first vasectomy program on a national scale was launched in 1954 in India. Currently, the two most common surgical techniques for approaching the vas during vasectomy are:

1. *Incisional method:* Scalpel to make one or two incisions (each 1–2 cm in length) to deliver the vas.
2. *No-scalpel method (NSV):* No-scalpel technique uses a sharp, pointed, and forceps-like instrument to puncture the scrotum.

Eligibility of Male Sterilization Procedure

- *Male sterilization:* Age ≤60 years.
- Clients should be ever married.
- The couple should have at least one child above 1 year of age unless the sterilization is medically indicated.
- Clients or their spouses must not have undergone sterilization in the past (not applicable in the cases of failure of previous sterilization).
- Clients must be in a sound state of mind, to understand the full implications of sterilization.

- Mentally ill clients must be certified by a psychiatrist, and a statement should be given by the legal guardian/ spouse regarding the soundness of the client's state of mind.[33]

Key Points During Counseling

- Intended to provide lifelong, permanent, and very effective protection against pregnancy.
- Involves a safe and simple surgical procedure.
- 3 or sometimes more months delay in taking effect.
- The couple must use condoms or another contraceptive method for 3 or more months, till semen analysis shows azoospermia after the vasectomy.
- Does not affect male sexual performance.
- Reversal can be possible but involves a complex surgery and success cannot be guaranteed.

Timing of surgical procedure: Can be done at any convenient time on healthy and eligible clients.

Who can perform: Any trained MBBS.

Failure rate: 1–4 per 5,000 vasectomies.

FEMALE STERILIZATION

Eligibility of Female Sterilization Procedure

Female age ≥22 years and ≤49 years, rest eligibility criteria are same as of male sterilization. **Table 7** shows the timing of female sterilization. **Table 8** shows the technique of female sterilization.

Absolute Contraindication (MEC 4)

- Hemodynamically unstable
- Any active pelvic infection.

Key Points During Counseling

- Permanent procedure.
- A surgical procedure that has the possibility of complications, including failure, requiring further management.
- No effect on the client's strength or her ability to perform normal day-to-day functions.
- Does not protect against RTIs, STIs, or HIV/acquired immunodeficiency syndrome (AIDS).
- A reversal of this surgery is possible, but it involves a major surgery whose success cannot be guaranteed.
- It does not affect sexual pleasure, ability, or performance. Also, conditions such as menstrual irregularities and chronic pelvic inflammatory disease are not caused by tubal ligation.

TABLE 7: Timing of female sterilization.

Interval	Within 7 days of the menstrual period (in the follicular phase of the menstrual cycle)
Postpartum	After 24 hours up to 7 days of delivery
Switching from other contraceptive method: • COCs • IUD	• Anytime if she has not missed any pills and finish the ongoing pill pack to maintain menstrual regularity • Concurrently with the removal of IUD
MTP: • Medical • Surgical	• After next menstrual period • Concurrently or within 7 days post MTP
Following spontaneous abortion	Can be performed, provided the client fulfils the medical eligibility criteria

(COCs: combined oral contraceptives; IUD: intrauterine device; MTP: medical termination of pregnancy)

TABLE 8: Techniques of female sterilization.

Approach to fallopian tubes	*Method of sterilization*
Minilap (isthmic part of tube)	• Pomeroy (2.5 cm of tube is removed after making a tubal loop) • Parkland (same as Pomeroy but tubal loop is not formed) • Irvin (proximal tubal end is buried in myometrium) • Uchida (proximal tubal end is buried in mesosalpinx) • Kroener (fimbriectomy)
Laparoscopy (isthmic part of tube)	• Electrocauterization with monopolar or bipolar • Silastic rings (Falope or Yoon ring) • Hulka–Clemens clip or Filshie clip
Transcervical (interstitial part of tube)	• Essure metal coil device with polyester fibers inserted with hysteroscope • Electrocautery • Cryosurgery • Laser • Quinacrine pellets
Vaginal	Rarely used due to the high risk of infection and pelvic abscess formation

Who can Perform Sterilization?

- *Minilap*: Trained MBBS doctor.
- *Laparoscopic*: DGO, MD (Obstetrics and Gynecology) MS (Surgery) who are trained in laparoscopic sterilization.

Essure and Adiana

Use backup contraception for 3 months after the procedure, then a hysterosalpingogram is done to confirm occlusion. If tubes are not occluded repeat hysterosalpingogram after 3 months more till then continue with backup contraception.[34] Due to financial concerns and declining sales both Essure and Adiana had been withdrawn from the US market.

Risk reducing salpingectomy is a new surgical approach for females who desire sterilization and reduction in risk of subsequent cancer ovary.[35]

Type of Anesthesia

Mostly local anesthesia with sedation, however spinal or general anesthesia can be given if required.

Complications

- *Bleeding from mesosalpinx*: Control by laparotomy/laparoscopy
- *Uterine perforation*: Nonbleeding—observation, bleeding—repair immediately
- Any abdominal organ injury (bladder/bowel/vascular) to be repaired immediately

Follow-up

- *First:* By any healthcare provider within 48 hours.
- *Second:* On day 7 at healthcare facility for stitch removal.
- *Third:* After 1 month or first menstrual period whichever is earlier at the healthcare facility. Also, a certificate of sterilization is issued by a medical officer.

Failure Rate

Different sterilization methods have different efficacy. Failure rates ranges from 0.75 to 3.65% depending on method used.[3]

As per the WHO tubal ligation failure rate is 0.5%.[2]

REFERENCES

1. International Institute for Population Sciences (IIPS) and ICF. 2021. National Family Health Survey (NFHS-5), 2019-21: India. Mumbai: IIPS; 2021.
2. WHO. (2023). Family Planning and Contraception Fact Sheets. [online] Available from www.who.int/news-room/fact-sheets/detail/family-planning-contraception. [Last accessed December, 2023].
3. Pearson JT, Chelstowska M, Rowland SP, Benhar E, Kopp-Kallner H, Berglund Scherwitzl E, et al. Contraceptive Effectiveness of an FDA-Cleared Birth Control App: Results from the Natural Cycles U.S. Cohort. J Womens Health (Larchmt). 2021;30:782-8.
4. Taylor HS, Pal L, Seli E. Speroff's Clinical Gynecologic Endocrinology and Infertility, 9th edition. Connecticut: Wolter Kluwer; 2019.
5. ACOG. (2022). Barrier Method of Birth Control: Spermicide, Condom, Sponge, Diaphragm, and Cervical Cap. [online] Available from www.acog.org/womens-health/faqs/barrier-methods-of-birth-control-spermicide-condom-sponge-diaphragm-and-cervical-cap. [Last accessed December, 2023].
6. Bayer LL, Jensen JT. ACIDFORM: a review of the evidence. Contraception. 2014;90:11-8.
7. Amaral E, Perdigão A, Souza MH, Mauck C, Waller D, Zaneveld L, et al. Postcoital testing after the use of a bio-adhesive acid buffering gel (ACIDFORM) and a 2% nonoxynol-9 product. Contraception. 2004;70:492-7.
8. Dragoman MV. The combined oral contraceptive pill—recent developments, risks and benefits. Best Pract Res Clin Obstet Gynaecol. 2014;28:825-34.
9. Sitruk-Ware LR, Mishell DR. Progestins and Antiprogestins in Clinical Practice. New York: Marcel Dekker; 2000.
10. Curtis KM, Jatlaoui TC, Tepper NK, Zapata LB, Horton LG, Jamieson DJ, et al. U.S. Selected Practice Recommendations for Contraceptive Use, 2016. MMWR Recomm Rep. 2016;29:65:1-66.
11. Faculty of Sexual and Reproductive Healthcare (FSRH). FSRH Clinical Guideline: Contraception After Pregnancy (January 2017, amended October 2020). England: Faculty of Sexual and Reproductive Healthcare; 2020.
12. Letterie GS, Chow GE. Effect of "missed" pills on oral contraceptive effectiveness. Obstet Gynecol. 1992;79:979-82.
13. Jick SS, Kaye JA, Russman S, Jick H. Risk of nonfatal venous thromboembolism in women using a contraceptive transdermal patch and oral contraceptives containing norgestimate and 35 μg of ethinylestradiol. Contraception. 2006;73(3):223-8.
14. Janssen-Cilag Ltd electronic MC (eMC). (2017). Summary of Product Characteristics: Evra transdermal patch. [online] Available from https://www.medicines.org.uk/emc/medicine/12124/SPC/Evra++transdermal+patch/. [Last accessed December, 2023].
15. Lopez LM, Grimes DA, Gallo MF, Stockton LL, Schulz KF, et al. Skin patch and vaginal ring versus combined oral contraceptives for contraception. Cochrane Database Syst Rev. 2013; 2013(4):CD003552.

16. Ministry of Health and Family Welfare. Reference Manual for Oral Contraceptive Pills. New Delhi: Family Planning Division, Ministry of Health and Family Welfare; 2016.
17. Korver T, Klipping C, Heger-Mahn D, Duijkers I, van Osta G, Dieben T. Maintenance of ovulation inhibition with the 75-microg desogestrel-only contraceptive pill (Cerazette) after scheduled 12-h delays in tablet intake. Contraception. 2005;71:8-13.
18. Archer DF, Ahrendt HJ, Drouin D. Drospirenone-only oral contraceptive: results from a multicenter noncomparative trial of efficacy, safety and tolerability. Contraception. 2015;92:439-44.
19. Palacios S, Regidor PA, Colli E, Skouby SO, Apter D, Roemer T, et al. Oestrogen-free oral contraception with a 4 mg drospirenone-only pill: new data and a review of the literature. Eur J Contracept Reprod Health Care. 2020;25:221-7.
20. Berenson AB, Rahman M, Breitkopf CR, Bi LX. Effects of depot medroxyprogesterone acetate and 20-microgram oral contraceptives on bone mineral density. Obstet Gynecol. 2008;112:788-99.
21. Trussell J. Contraceptive efficacy. In: Hatcher R, Trussell J, Nelson A (Eds). Contraceptive Technology, 20th edition. New York, NY: Ardent Media; 2011.
22. Kavanaugh ML, Pliskin E. Use of contraception among reproductive-aged women in the United States, 2014 and 2016. F S Rep. 2020;1:83-93.
23. Bansode OM, Sarao MS, Cooper DB. Contraception. Treasure Island (FL): StatPearls Publishing; 2023.
24. Trussell J. Contraceptive failure in the United States. Contraception. 2011;83:397-404.
25. Kavanaugh ML, Jerman J, Finer LB. Changes in use of long-acting reversible contraceptive methods among U.S. women, 2009–2012. Obstet Gynecol. 2015;126:917-27.
26. Shaamash AH, Sayed GH, Hussien MM, Shaaban MM. A comparative study of the levonorgestrel-releasing intrauterine system Mirena versus the Copper T380A intrauterine device during lactation: breast-feeding performance, infant growth and infant development. Contraception. 2005;72:346-51.
27. Curtis KM, Tepper NK, Jatlaoui TC, Berry-Bibee E, Horton LG, Zapata LB, et al. U.S. Medical Eligibility Criteria for Contraceptive Use, 2016. MMWR Recomm Rep. 2016;65:1-103.
28. Baker CC, Crenin MD. Long-Acting Reversible Contraception. Obstet Gynecol. 2022;140(5):883-97.
29. Farmer M, Webb A. Intrauterine device insertion-related complications: can they be predicted? J Fam Plann Reprod Health Care. 2003;29:227-31.
30. Bennink HJ. The pharmacokinetics and pharmacodynamics of Implanon, a single-rod etonogestrel contraceptive implant. Eur J Contracept Reprod Health Care. 2000;5:12-20.
31. Zheng W, Tang Y, Wang C, Niu X, Qian Z, Huang L. Three-year observational study of immediate post-abortion insertion versus menstrual insertion of etonogestrel contraceptive implant. BMC Womens Health. 2021;21:436.
32. Chaplin S. Update of the FSRH guideline on emergency contraception. Prescriber. 2017;28:27-31.
33. Ministry of Health and Family Welfare. Reference Manual for Male Sterilization. New Delhi: Family Planning Division, Ministry of Health and Family Welfare; 2013.
34. Cooper JM, Carignan CS, Cher D, Kerin JF; Selective Tubal Occlusion Procedure 2000 Investigators Group. Microinsert nonincisional hysteroscopic sterilization. Obstet Gynecol. 2003;102:59-67.
35. Creinin MD, Zite N. Female tubal sterilization: the time has come to routinely consider removal. Obstet Gynecol. 2014;124:596-9.

Chapter 10

Breast in Gynecology

Sneha Bhuyar

INTRODUCTION

The breast is composed of glands, fat, and fibrous tissue. In females, the breasts, under the influence of hormones are constantly going through changes from the time of their development through puberty, menstrual cycles, pregnancy, and lactation until menopause. Being the primary physician for women, gynecologists must have a thorough knowledge of the breast and the various breast symptoms for which a woman seeks treatments.

Breast symptoms such as lumps, mastalgia, or nipple discharges are common of which >90% of symptoms are benign.[1] More often than not breast malignancies may mimic benign disorders and thus importance should be given to early detection and diagnosis of breast diseases by both the patient and doctors, so as to avoid unnecessary surgery and to alleviate patient's fear of carcinoma.

Benign breast diseases (BBDs) are nonmalignant lesions of the breast with lesions consisting of heterogeneous conditions, from nonspecific pain to lumps. BBDs have become a major women's health issue due to the rising global prevalence and awareness about breast cancers.

Breast cancer: Incidence of malignant breast diseases has superseded cervical cancer to become the most common cancer among women in India. As per the Globocan 2020, in India, breast cancer accounts for 13.5% of all cancer cases and 10.6% of all deaths.[2] It has now surpassed lung cancer as the leading cause of global cancer incidence in 2020, with an estimated 2.3 million new cases.[3]

CLASSIFICATION

Classification of Benign Breast Diseases[4]

It has been controversial due to lack of a clarity in the clinical and histopathological separation between the physiological and pathological changes occurring in breast.

- *College of American Pathologists (CAP) classification*: Histological classification based on findings of biopsy
  - Nonproliferative lesions:
    - No increased risk
    - Cysts
    - Mild hyperplasia of the usual type
    - Epithelial-related calcifications
    - Fibroadenoma
    - Papillary apocrine change
  - Proliferative lesions without atypia = 1.5–2.0
    - Sclerosing adenosis
    - Radial and complexing sclerosing lesions
    - Moderate and florid hyperplasia of the usual type
    - Intraductal papillomas
  - Atypical proliferative lesions = 4.0–5.0
    - Atypical lobular hyperplasia
    - Atypical ductal hyperplasia
- *Aberrations of normal development and involution (ANDI):* It encompasses the physiological changes, growth, development, and involution of the breast. However, some authors believe that the ANDI classification lacks clarity in distinguishing between normal physiological and pathologic changes.
- Clinical classification **(Box 1)**.

BOX 1: Clinical classification of benign breast diseases (BBDs).

- Breast pain
- Palpable lump
- Nodularity
- Physiological swelling and tenderness
- Nipple discharge
- Infection or inflammation

Classification of Breast Carcinoma[5]

See **Flowchart 1**.

RISK FACTORS (TABLE 1)[6]

See **Table 1**.

PATHOPHYSIOLOGY OF BREAST CANCER

See **Flowchart 2**.

CLINICAL PRESENTATION[7-12]

Benign Breast Diseases

Breast Pain

Also known as mastalgia or mastodynia.

Cyclic breast pain: Usually occurs due to benign mammary dysplasia in the late luteal phase of the menstrual cycle, in association with the premenstrual syndrome or in the second half of the cycle, and resolves at the onset of menses. Common in women in third or fourth decade of lives.

Flowchart 1: Classification of breast carcinoma.

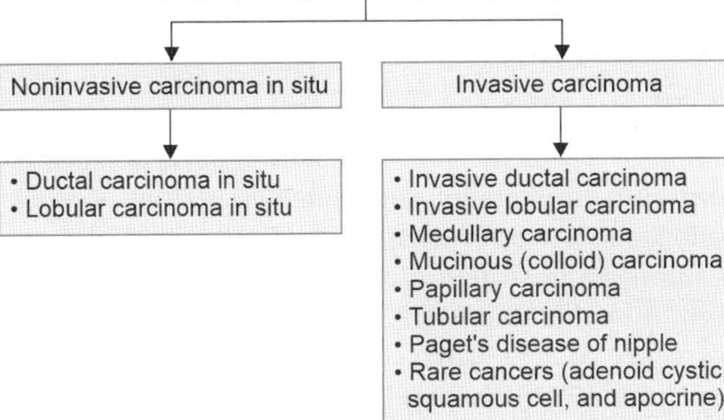

Noncyclic breast pain: It is unrelated to the menstrual cycle and usually occurs in postmenopausal women. Can present as dragging or burning-like pain, which may be intermittent or constant and usually unilateral. Acute enlargement of cysts and periductal mastitis may cause severe and localized pain of sudden onset.

Nonbreast pain: Extramammary pain arising from the chest wall or radiating pain may be mistakenly attributed to the breast pain. Pain is localized to a limited area and characterized as burning or knife like. Several distinct subtypes can be distinguished including localized or diffuse lateral chest wall pain, radicular pain from cervical arthritis, Tietze's syndrome, costochondritis, herpes zoster infection, or rib fracture.

Fibroadenoma (Breast Mouse)

It is one of the most common benign tumors of breast and accounts for 50% of all breast biopsies and occurs in women of 20–35 years of age and is rare after menopause. It arises from hyperplasia of a single lobule and is surrounded by a well-circumscribed capsule. They are firm, rubbery masses with a well-circumscribed border. Usually, fibroadenomas form as a single tumor. However, in up to 15% of patients, multiple tumors are present, and 10% will be bilateral. Fibroadenomas grow rapidly under

TABLE 1: Factors influencing breast diseases.

Low risk	*Moderate risk*	*High risk*	*Protective*
Alcohol consumption	Dense breast	Carrier of BRCA1 and BRCA2 genetic mutations	Consumption of foods rich in carotenoids such as melons, carrots, sweet potatoes, beans, and nuts
Obesity	Age >40 years	Family history of first-degree relative with breast cancer	Breastfeeding
Age >30 years for first pregnancy	Nulliparity	Past history of breast cancer	Oral contraceptive use has a protective effect on the risk of fibrocystic breast disease
Oral contraceptives	Early menarche <12 years	Benign breast disease with atypia	
Hormone replacement therapy	Late menopause >55 years	Lobular carcinoma in situ and ductal carcinoma in situ	
Smoking	Benign breast disease with proliferation without atypia	Chest irradiation	

Flowchart 2: Pathophysiology of breast cancer.

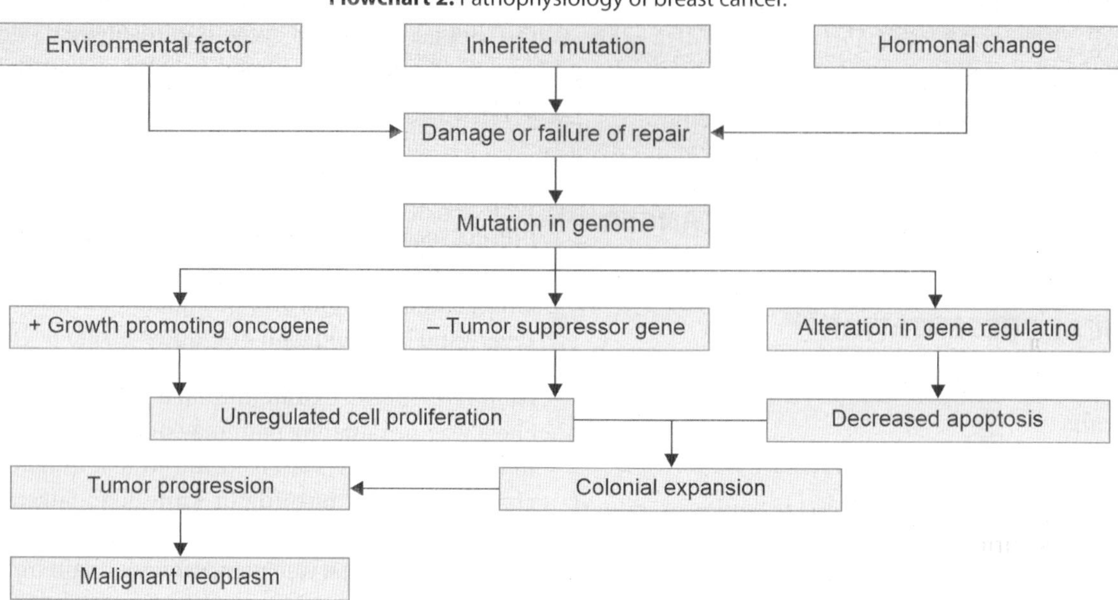

hormonal influence and during pregnancy and swelling may enlarge during pregnancy and lactation. Those larger than 5 cm are termed giant fibroadenomas. They carry little to no increased risk of breast cancer. However, on rare occasions, intraductal or invasive cancer can develop in a fibroadenoma, just as it can in any other part of the breast. In addition, multiple fibroadenomas are associated with some rare cancer syndromes, such as Maffucci syndrome, Cowden syndrome, and Carney complex. High-risk fibroadenomas are in patients with a family history of malignancy, larger than 5 cm and those persisting after 35 years of age, those harboring apocrine metaplasia, epithelial calcifications, and sclerosing adenosis immediate excision.

Fibrocystic Disease of Breast

It is one of the most common lesions of breast referring to histological picture of fibrosis, cyst formation, and epithelial hyperplasia and is observed clinically in up to 50% and histologically in 90% of women. Commonly in women of 35–55 years of age, peaks in the fourth decade when hormonal function is at its peak, and sharply diminishes after menopause. Exact pathogenesis is not clear, hormonal imbalance, particularly estrogen predominance over progesterone, seems to play an important role in its development. The lesions comprise both cysts and solid lesions, including adenosis. Most women are asymptomatic but smooth mobile compressible lump with pain and tenderness or sometimes nipple discharge may be present.

Cysts

Cystic disease is caused by dilation of ducts and acini to form cysts, proliferation, and metaplasia of their epithelial lining, and multiplication of ducts and acini (adenosis) resulting in obstruction of the terminal ductal lobular unit. It occurs in one-third of women between 35 and 50 years of age. Ultrasonography and fine-needle aspiration (FNA) cytology is required to diagnose cysts as clinical examination or mammography cannot distinguish cysts from solid lumps. Cysts are not associated with an increased risk of carcinoma development and management is routine follow-up of the patient, without further therapy.

Nipple Discharge

Nipple discharge can be physiologic and pathologic. Characteristics of physiologic discharge include non-spontaneous, multiple duct, bilateral, and nonbloody. Pathologic discharge is characterized as spontaneous, serous, or bloody, usually unilateral and usually single duct. A clear or milky color represents hormonal influence. Green black, it states a blocked milk duct. If the discharge is bloody, it can be related to an injury, infection or tumor either benign or malignant. Treatment depends on the underlying cause.

Papillomas

Occur in major ducts and consist of epithelial proliferation on a fibrovascular stalk. They are typically located within a few centimeters of the nipple and grow within the duct, it can result in obstruction. Can be asymptomatic or manifest as serous or serosanguinous nipple discharge.

Duct Ectasia

Duct ectasia predominantly affects the ducts in the retroareolar region and is defined as a nonspecific dilation of one or more ducts, typically larger than 2 mm in diameter. It may present as a palpable lump or nipple discharge.

Mastitis

Acute mastitis: Also known as puerperal or lactation mastitis, usually occurs during the first 3 months postpartum and is cellulitis of the interlobular connective tissue within the mammary gland, which can result in abscess formation and septicemia. Improper nursing techniques can lead to milk stasis and cracks on the nipple, which may facilitate entrance of microorganisms through the skin.

Granulomatous mastitis: Granulomatous reactions resulting from an infectious etiology, foreign material, or systemic autoimmune diseases such as sarcoidosis and Wegener's granulomatosis can involve the breast. Identification of the etiology requires microbiologic and immunologic testing in addition to histopathologic evaluation.

Lipoma

These are common and are made up of mostly fat tissue. Can be felt as a lump and treatment is by surgical removal.

Less common conditions:

- *Amazia*: Congenital absence of the breast, may occur on one or both sides.
- *Polymazia*: Accessory breasts have been recorded in the axilla, groin, buttock, and thigh.
- Hematoma
- Traumatic fat necrosis
- Tuberculosis of breast
- *Mondor's disease*: It presents as thrombophlebitis of superficial veins of breast and anterior chest wall, usually subsides; however, treatment is by restricted arm movements and analgesics
- *Galactocele*: Presents as a solitary subareolar cyst dating from lactation
- Calcification
- Gynecomastia
- *Phyllodes tumors*: Benign tumors, previously known as serocystic disease of Brodie, usually occur in women over the age of 40 years. Presents as a large mobile tumor with an uneven surface.

Signs and Symptoms of Breast Cancer

- Lump or thickened tissue in breast or armpit
- Pain in breast or armpit without any cyclic correlation
- Change in size or contour of breast
- Redness, pitting of skin, and ulcers
- Change in appearance of areola or nipple
- Inverted or sunken nipple
- Nipple discharge or bleeding
- Fixation of the mass to the chest wall or to the overlying skin
- Matted or fixed axillary lymph nodes.

DIAGNOSIS

History and Physical Examination

Triple assessment, which includes clinical examination, imaging, and histopathological examination is now considered a gold standard approach to the diagnosis of breast lump (*see* **Flowchart 3**).

History

See **Box 2**.

Physical Examination

- Palpate the four breast quadrants while patient is sitting and lying down

Flowchart 3: Triple assessment.

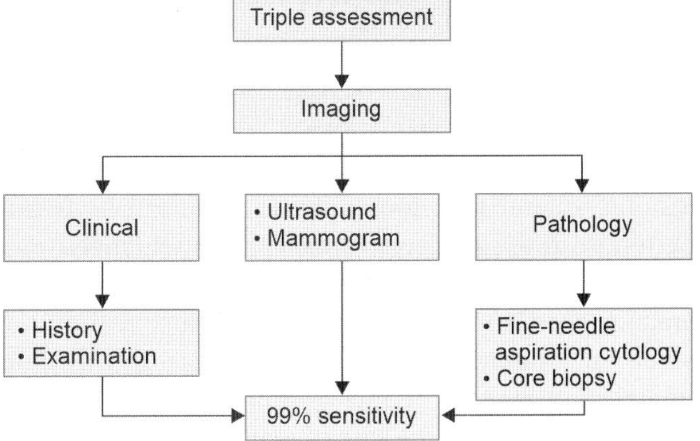

BOX 2: History taking.

- *Personal characteristics:*
  - Age at menarche/menopause
  - Age at first childbearing and parity
  - Lactation status and history of breastfeeding
- *Breast lump characteristics:*
  - Duration of mass
  - Changes in size
  - Change relative to menstrual cycle
  - Pain, swelling, redness, fever, and discharge
- *Family and past history:*
  - Personal history of breast cancer or previous breast masses
  - History of breast disease in family
  - Relative's age at onset
  - Medical and surgical history
  - Recent breast trauma, biopsy, or surgery
  - Recent radiation therapy or chemotherapy
- *Diet and medications:*
  - Current medications
  - History of hormone therapy
- *Social history:*
  - Smoking and alcohol
  - Radiation and chemical exposure

- Observe symmetry of the breasts
- Identify palpable lumps and examine the axillary nodes
- Examine overlying skin, areola, and axilla
- Examine nipple for any discharge and determine the color and consistency of the discharge
- Seek to elicit chest-wall pain and examine costochondral junctions
- Examine lateral chest wall in lying down position with patient turned on her side (at 90°), to move breast away from the chest wall, compare pain elicited by squeezing breast with pain elicited by palpation of chest wall.

DIAGNOSTIC TOOLS[13,14]

Imaging

Mammography

Mammography is useful for evaluation of palpable lesions. Digital mammography is preferred as it can penetrate through dense breast tissue which is seen especially in younger women. It is recommended that women aged 40 years and older should have annual screening mammography preferably using full-field digital mammography and computer-aided detection (CAD). It improves the effectiveness of perception of small nonpalpable malignancies that have the optimum prognosis. Mammograms are reported in a format following the guidelines of Breast Imaging-Reporting and Data System (BI-RADS).

Ultrasound

Ultrasound is used for the initial evaluation of a palpable mass in women under the age of 40 years. Ultrasound is useful in differentiating a fibroadenoma from a cyst. It has a 98% accuracy to diagnose a cyst better than a mammogram. For some cases, ultrasound is done in conjunction with mammography to fully evaluate the lump.

Magnetic Resonance Imaging

It is more sensitive than digital mammography but high risk of false positives. A routine annual MRI is recommended for women whose lifetime risk of breast cancer is >20%.

Invasive

Fine-needle Aspiration

Fine-needle aspiration of a suspicious palpable lump is a well-established diagnostic technique and a routine outpatient procedure. A fine needle is inserted with a small handheld syringe holder creating negative pressure within the syringe at suspicious sites. In the case of cysts, it can serve both as a diagnostic as well as a therapeutic modality by drainage of the cyst.

With an adequate cell sample, an FNA cytologic diagnosis of a palpable solid breast mass can be obtained in as many as 90% of cases.

Breast Biopsy

Histological tissue diagnosis of palpable or radiological abnormalities can be achieved by core biopsy using automated vacuum-assisted devices, under stereotactic, ultrasound, or MRI guidance. They are used to evaluate microcalcifications and therapeutically remove the probable benign lesion. With the advent of percutaneous tissue sampling techniques, the traditional open surgical biopsy or lumpectomy is rarely required nowadays in order to establish the histological diagnosis.

MANAGEMENT[7,15,16]

A detailed history and physical examination systematically evaluate the entire breast and chest wall and focuses on

Flowchart 4: Breast lump.

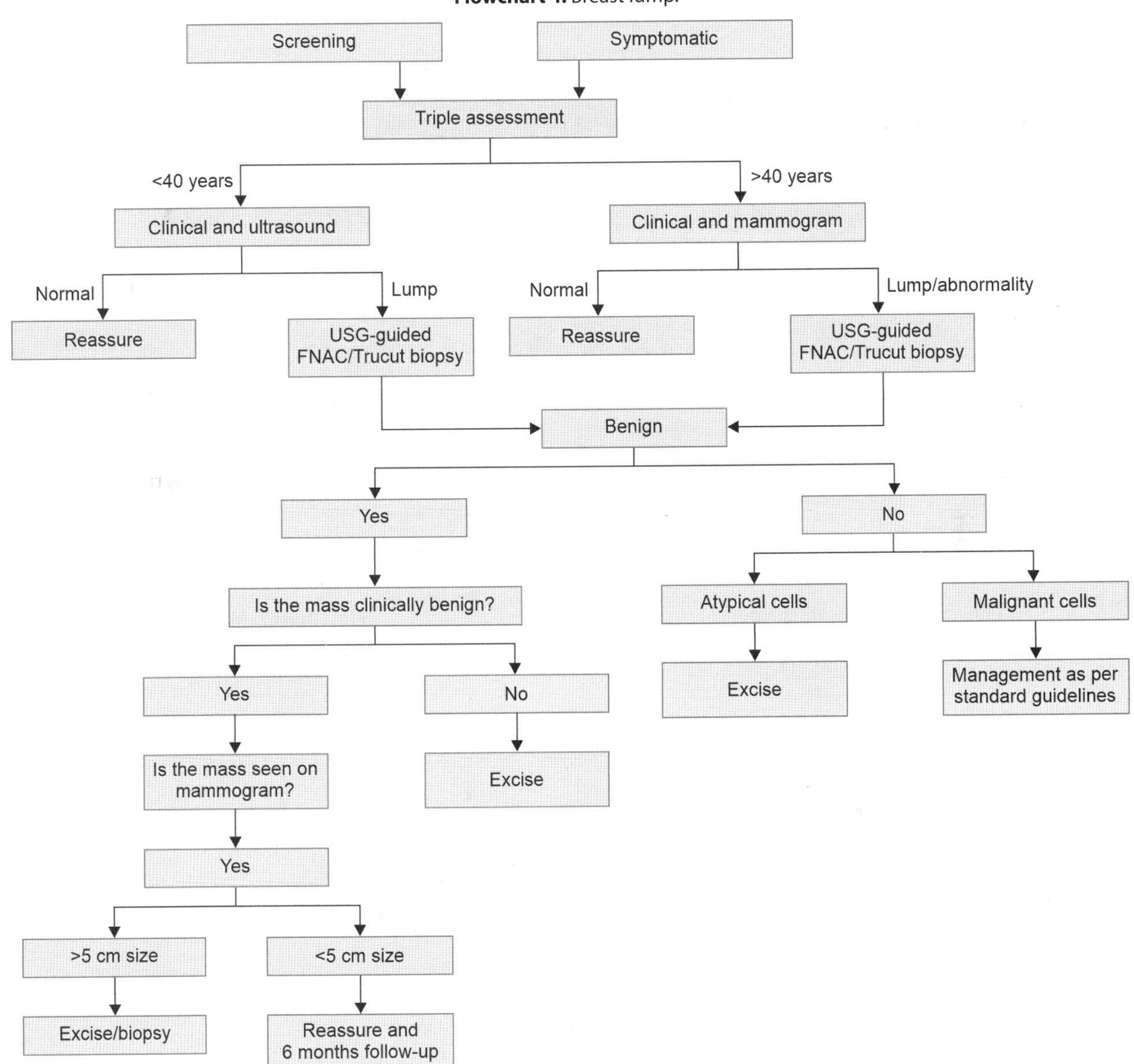

(FNAC: fine-needle aspiration cytology)

areas involving the patient's symptoms. Triple assessment, including clinical examination, imaging, and histopathological examination is the gold standard approach to the diagnosis of a breast lump.

Breast Lump

See **Flowchart 4**.

Breast Cyst

See **Flowchart 5**.

Nipple Discharge

See **Flowchart 6**.

MASTITIS

See **Flowchart 7**.

Breast Cancer

Treatment depends on the:

- Stage
- Type

Flowchart 5: Breast cyst.

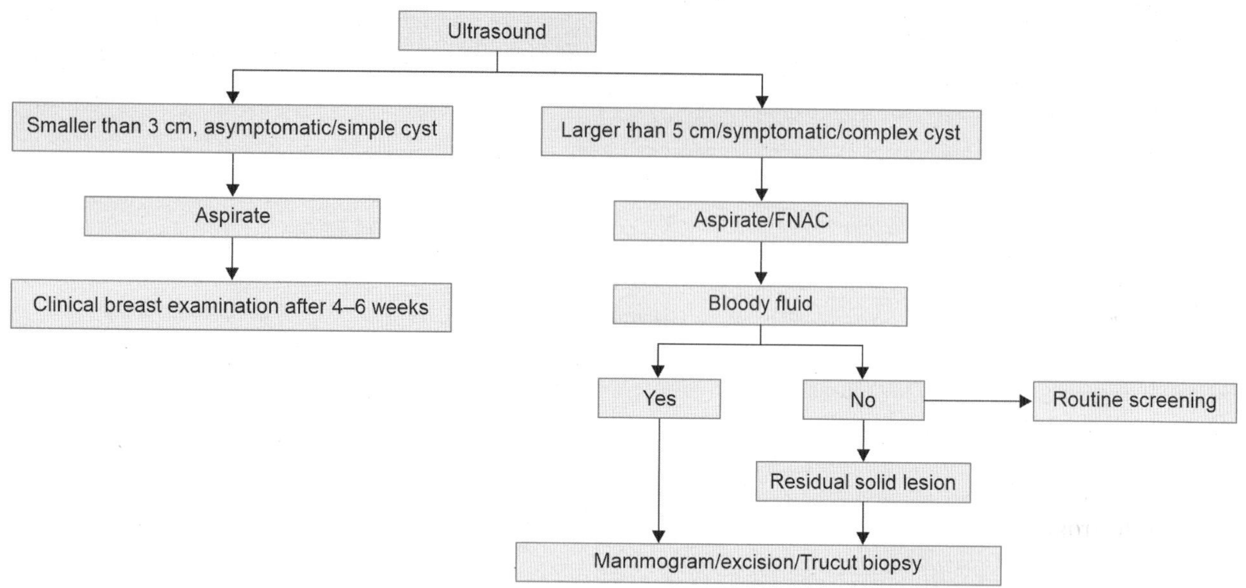

(FNAC: fine-needle aspiration cytology)

Flowchart 6: Nipple discharge.

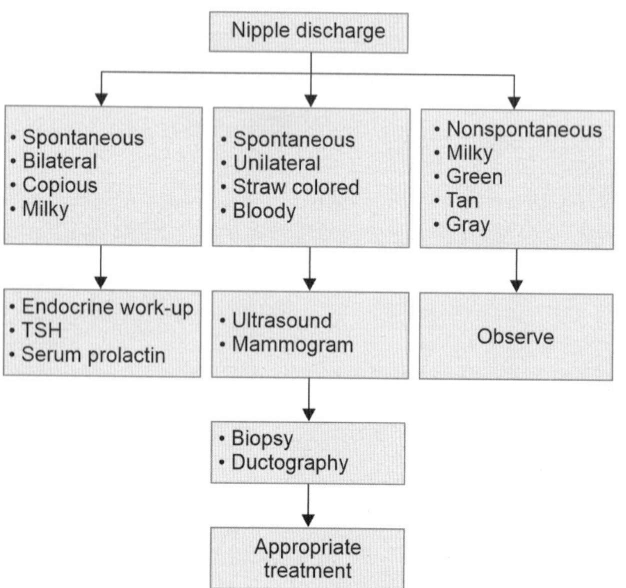

(TSH: thyroid-stimulating hormone)

Flowchart 7: Mastitis.

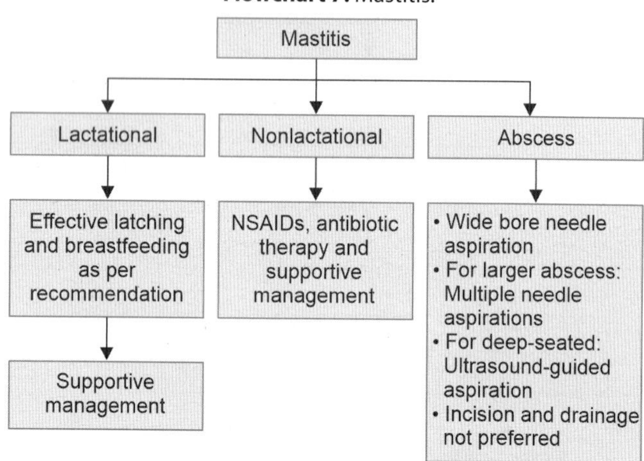

(NSAIDs: nonsteroidal anti-inflammatory drugs)

- Sensitivity to hormones
- General condition of the patient

Modalities for treatment:

- *Surgical management:*
  - *Conservative breast surgery:* For treatment of early stage cancers as favors cosmesis and gives good oncological control. Lumpectomy involves removal of lump along with 1 cm margin of healthy tissue. A quadrantectomy involves removal of a sector of breast with tumor free margins of 2 cm.
  - *Mastectomy:* It involves removal of breast lobules, ducts, fatty tissue, nipple, areola, and skin.
  - *Radical mastectomy:* It involves removal of all the earlier discussed along with pectoralis major muscle and axillary lymph nodes.
  - *Modified radical mastectomy:* It involves mastectomy with en block removal of the axillary nodes.

- *Breast reconstructive surgery:* Oncoplasty can be done to recreate the breast to prevent the disfigurement after a mastectomy.
- *Radiation:* It is a treatment with high energy rays indicated after a conservative surgery, if the margins are not tumor free, lump >5 cm, or stage 4 carcinoma.
- *Chemotherapy:* Cytotoxic drugs are used as adjuvants if there is a high risk of recurrence or spread. Neoadjuvant chemotherapy can be administered preoperatively to decrease the size of the tumor. It can also be given for palliation.
- *Hormone blocking agents:* Agents such as tamoxifen, aromatase inhibitors, ovarian suppression, and luteinizing hormone-releasing hormone (LHRH) agonists are used to prevent recurrence in hormone sensitive [estrogen receptor (ER) and progesterone receptor (PR) positive] breast cancers.
- Biological targeted drug therapy*:* Newer drugs such as Herceptin and pertuzumab can alter the behaviors of malignant cells and are effective in HER2 positive patients (*See* **Flowchart 8**).

PREVENTION[17]

As primary physician for women it is pertinent to create awareness through seminars, screening camps, and teaching women self-breast examination (SBE).

Primary Screening

It should be clinical breast examination (CBE) done by a physician in resource-poor setting. In settings where mammography is available, it should be used as a screening modality.

Mammography

It is the most widely used screening modality, with evidence of benefit for women aged 40–74 years. The National Cancer Grid: Resource Stratified Guidelines 2019/ Preventive Oncology and Primary Care recommends: Frequency of screening: 2 yearly

- *Women aged 40–49 years:* CBE + mammography depending on risk assessment and advise of the physician.
- *Women aged >50–75 years:* CBE + mammography

Flowchart 8: Biological targeted drug therapy.

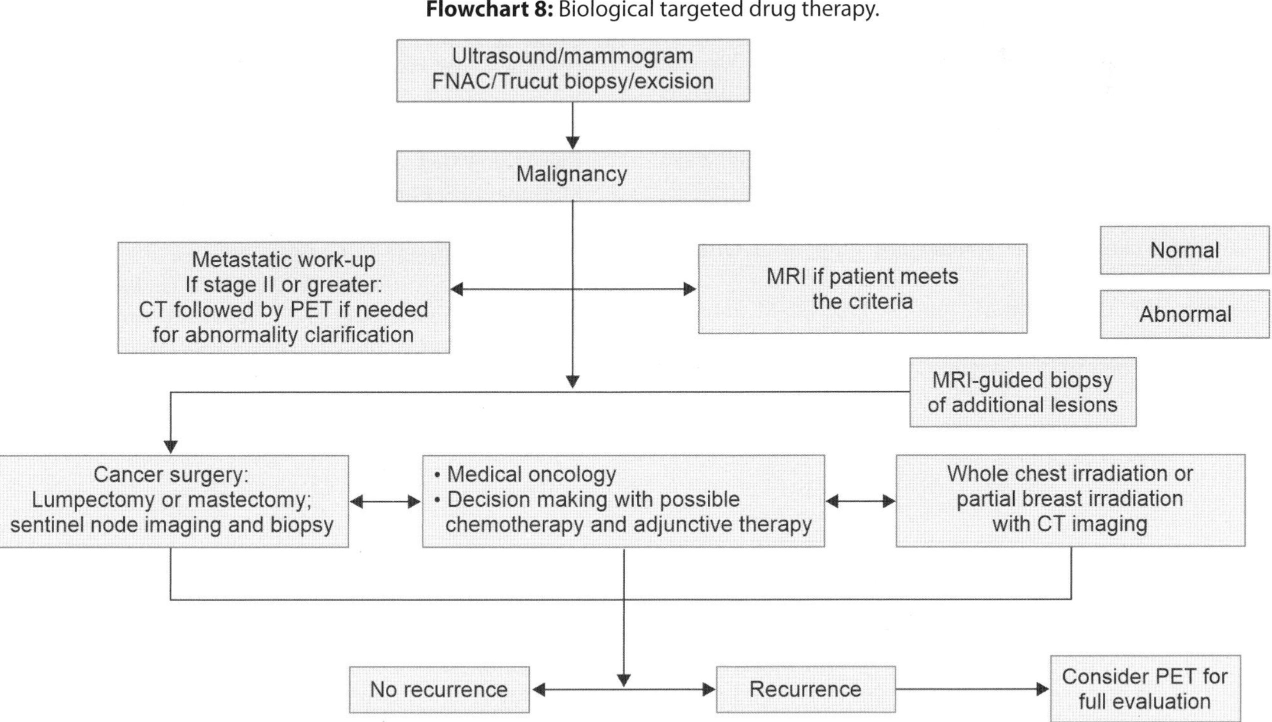

(FNAC: fine-needle aspiration cytology; PET: positron emission tomography)

Self Breast Examination

Screening by breast self-examination is a valuable tool where women can detect and report any abnormal changes in the breast. Awareness should be raised and the correct technique should be taught to women early so that it becomes a normal part of self-care. Breast cancer screening was found to reduce the risk of mortality by 20%. Despite the presence of various screening methods, majority of breast cancer (48%) cases are detected by women themselves, stressing the importance of SBE. SBE is considered to be a simple, inexpensive, quick, noninvasive, and nonhazardous intervention. This could be a useful measure for early identification of breast cancer in resource-poor countries where accessibility to better screening methods is less.

REFERENCES

1. Goehring C, Morabia A. Epidemiology of benign breast disease, with special attention to histologic types. Epidemiol Rev. 1997;19(2):310-27.
2. International Agency for Research on Cancer. (2021). India Source: Globocan 2020. [online] Available from https://gco.iarc.fr/today/data/factsheets/populations/900-world-fact-sheets.pdf. [Last accessed December, 2023].
3. Sung H, Ferlay J, Siegel RL, Laversanne M, Soerjomataram I, Jemal A, et al. Global Cancer Statistics 2020: GLOBOCAN Estimates of Incidence and Mortality Worldwide for 36 Cancers in 185 Countries. CA Cancer J Clin. 2021;71:209-49.
4. Chilakala A, Navya KCN. Clinical study of benign breast diseases based on aberrations in normal development and involution classification and development. J Evolution Med Dent Sci. 2019;8(31):2467-74.
5. Weigelt B, Geyer FC, Reis-Filho JS. Histological types of breast cancer: how special are they? Mol Oncol. 2010;4(3):192-208.
6. Johansson A, Christakou AE, Iftimi A, Eriksson M, Tapia J, Skoog L, et al. Characterization of Benign Breast Diseases and Association With Age, Hormonal Factors, and Family History of Breast Cancer Among Women in Sweden. JAMA Netw Open. 2021;4(6):e2114716.
7. Stachs A, Stubert J, Reimer T, Hartmann S. Benign Breast Disease in Women. Dtsch Arztebl Int. 2019; 116(33-34):565-74.
8. Maldonado-Martinez HA, Perez-Sanchez VM, Sanchez CHC, Juarez-Sanchez P, Meneses-Garcia A. (2014). Benign breast diseases: Current pathological and molecular concepts. [online] Available from https://www.researchgate.net/publication/292915563_Benign_breast_diseases_Current_pathological_and_molecular_concepts. [Last accessed December, 2023].
9. Navneet K, Nitin A, Panwar P, Mishra K. Clinicopathologic profile of benign breast conditions in Indian women. Prospective study based on aberrations of normal development and involution classification. World J Surg. 2012;36:2252-8.
10. Kumar N, Monika K. Benign breast diseases in tertiary center in North Bihar: A clinico-pathological study. Int J Sci Stud. 2016;4(2):56-9.
11. Malik R, Bhardwaj VK. Breast lesions in young females. A 20 year study for significance of early recognition. Indian J Pathol Microbiol. 2003;46(4):559-62.
12. Katsura C, Ogunmwonyi I, Kankam HK, Saha S. Breast cancer: presentation, investigation and management. Br J Hosp Med (Lond). 2022;83(2):1-7.
13. Takkar N, Kochhar S, Garg P, Pandey AK, Dalal UR, Handa U. Screening methods(clinical breast examination and mammography) to detect breast cancer in women aged 40–49 years. J Mid-life Health. 2017;8:2-10.
14. Guray M, Sahin AA. Benign Breast Diseases: Classification, Diagnosis, and Management. Oncologist. 2006;11:435-49.
15. Vaidyanathan L, Barnard K, Elnicki DM. Benign breast disease: when to treat, when to reassure, when to refer. Cleve Clin J Med. 2002;69(5):425-32.
16. McDonald ES, Clark AS, Tchou J, Zhang P, Freedman GM. Clinical Diagnosis and Management of Breast Cancer. J Nucl Med. 2016;57 Suppl 1:9S-16S.
17. Pashayan N, Antoniou AC, Ivanus U, Esserman LJ, Easton DF, French D, et al. Personalized early detection and prevention of breast cancer: ENVISION consensus statement. Nat Rev Clin Oncol. 2020;17(11):687-705.

Chapter 11

Abnormal Uterine Bleeding

Girija Wagh

INTRODUCTION

Abnormal uterine bleeding (AUB) is a term that refers to menstrual bleeding of abnormal quantity, duration, or schedule.[1] It is a common gynecologic complaint and generally means abnormal menstruation, accounting for one-third of outpatient visits to gynecologists. AUB can be caused by a wide variety of local and systemic diseases or related to medications. The most common etiologies in nonpregnant women are structural uterine pathology (e.g., fibroids, endometrial polyps, adenomyosis, or neoplasia). Dysfunctional uterine bleeding (DUB) or dysfunctional bleeding is a terminology typically restricted to AUB present in the absence of structural pelvic pathology such as ovulatory dysfunction, disorders of hemostasis, secondary to thyroid dysfunction etc.; and therefore is a functional abnormality.[2] DUB as a terminology is deemed not to be used to avoid confusion. AUB is a terminology which includes both the structural as well as functional causes of AUB. Likewise, it includes not only heavy menstrual bleeding (HMB) but also intermenstrual bleeding and oligomenorrhea.

BRIEF BACKGROUND

Abnormal uterine bleeding is known to affect 9–14% women globally while 17.9% is the Indian prevalence.[3] AUB is also described as menorrhagia, metrorrhagia, polymenorrhagia, DUB, and HMB. The terminologies such as menorrhagia, metrorrhagia, polymenorrhagia, and DUB are best avoided due to their unspecified descriptions.

Before the diagnosis of what is abnormal the normal menstrual pattern **(Table 1)** should be understood with its normal and abnormal variations and documented with detailed history for better approach of investigation as well as treatment.

DEFINITION OF ABNORMAL UTERINE BLEEDING

Abnormal menstrual bleeding considerably affecting the quality of the life of the woman and the one outside the limits is described in **Table 1** and it is defined as AUB. Clinical presentation can be in the form of acute AUB, chronic AUB or acute AUB superimposed on chronic AUB unrelated to pregnancy **(Table 2)**. Reduced bleeding or amenorrhea also is categorized as AUB.

Since 2011 a new system of nomenclature is introduced by The International Federation of Gynecology and Obstetrics (the FIGO) known as PALM-COEIN[3] **(Table 3)** system of classification which is based on the etiopathogenesis. This is an easy-to-follow classification and can be documented and the management can be planned rationally. The FIGO system facilitates both basic science and clinical investigation, as well as the practical, rational, and consistent application of medical and surgical therapy for affected women. In 2018 "Intermenstrual bleeding" was added in the classification by the FIGO and irregular bleeding is defined as the one which lies outside the 75th percentile.[1]

CAUSES OF ABNORMAL UTERINE BLEEDING

The FIGO classification system reiterated in 2018 depicts the etiology of AUB categorically by the *"FIGO AUB PALM-COEIN Classification System."*

HISTORY AND INITIAL EXAMINATION

Women presenting with abnormal bleeding should be carefully evaluated with detailed history, associated symptoms, imaging, and supportive laboratory investigation. Three important questions to be answered

TABLE 1: The International Federation of Gynecology and Obstetrics (FIGO) System—normal and abnormal limits of menstrual bleeding plus the 2018 additions.

Clinical dimensions of menstruation and the cycle	*Descriptive term*	*Normal limits*
Frequency of menses (days)	• Frequent • Normal • Infrequent	<24 24–38 >38
Regularity of menses: Cycle-to-cycle variation over 12 months, days *(depends on age) *18–25 and 42–45 age:* >9 days; 26–41 age: >7 days	• Absent • Regular • Irregular	No bleeding Variations ± 2–20 Variation >20
Duration of flow (days)	• Prolonged • Normal • Shortened	>8 4.5–8 <4.5
Volume of monthly blood loss (mL)**	• Heavy • Normal • Light	>80 20–80 <20
Intermenstrual bleeding: Bleeding between cyclically regular cycles	• Early cycle • Mid cycle • Late cycle	• Predictable • Unpredictable
Unscheduled bleeding: Hormonal preparations	Iatrogenic bleeding	

*Irregularity interpretation should take into context the age of the woman.
**Menstrual blood flow volume is determined on perception of the patient and by various methods such as the menstrual hygiene products used, pictographs and presence of clots, flooding or hemogram.

TABLE 2: Clinical classification of abnormal uterine bleeding.

Acute abnormal uterine bleeding (AUB)	A considerable bout of bleeding mandating immediate treatment and evaluation
Chronic AUB	Moderate bleeding over
Acute AUB superimposed on chronic AUB	

TABLE 3: The International Federation of Gynecology and Obstetrics (FIGO) abnormal uterine bleeding (AUB) PALM-COEIN classification.

PALM		*COEIN*	
Structural causes		*Non-structural causes*	
P	Polyp	*C*	Coagulopathy
A	Adenomyosis	*O*	Ovulatory disorders
L	Leiomyoma	*E*	Endometrial factors
M	Malignancy and hyperplasia	*I*	Iatrogenic
		N	Not classified

are also the gross issues under which one can categorize the patients:

- Is the woman pregnant? (urine pregnancy test or blood levels of β-hCG (β-human chorionic gonadotropin) help to rule out)
- Is the bleeding from the uterus? **(Flowchart 1)**
- Is the bleeding due to coagulopathy? **(Table 3**—coagulopathy screen)
- Are there any extrauterine sources of bleeding? **(Table 4)**
- Pregnancy is to be ruled out and suspected in all women despite their marital status and confident history as they may not understand the importance of such a history. Tactful and assuring questioning usually yields the correct answers which can be confirmed by urine or serum beta human chorionic gonadotropin (β-hCG) testing and ultrasound. Pregnancy of unknown location ectopic gestation, miscarriages, recent miscarriage, or hCG secreting tumors of ovary occur rarely but can lead to considerable diagnostic dilemma. Previous pregnancy events can be responsible to cause a scar ectopic, isthmocele, and arteriovenous (A-V) malformations.
- Is the bleeding occurring from the uterus? Bleeding per vaginal can be either from the vaginal, urethra, rectum cervix, or other sites and thorough gynecological evaluation can help establish this along with imaging or endoscopy as aids. The extrauterine causes of bleeding are enlisted in **Table 4**.

History and clinical examinations should be able to determine the bleeding pattern, associated symptoms,

Flowchart 1: Algorithm—overview of clinical approach.

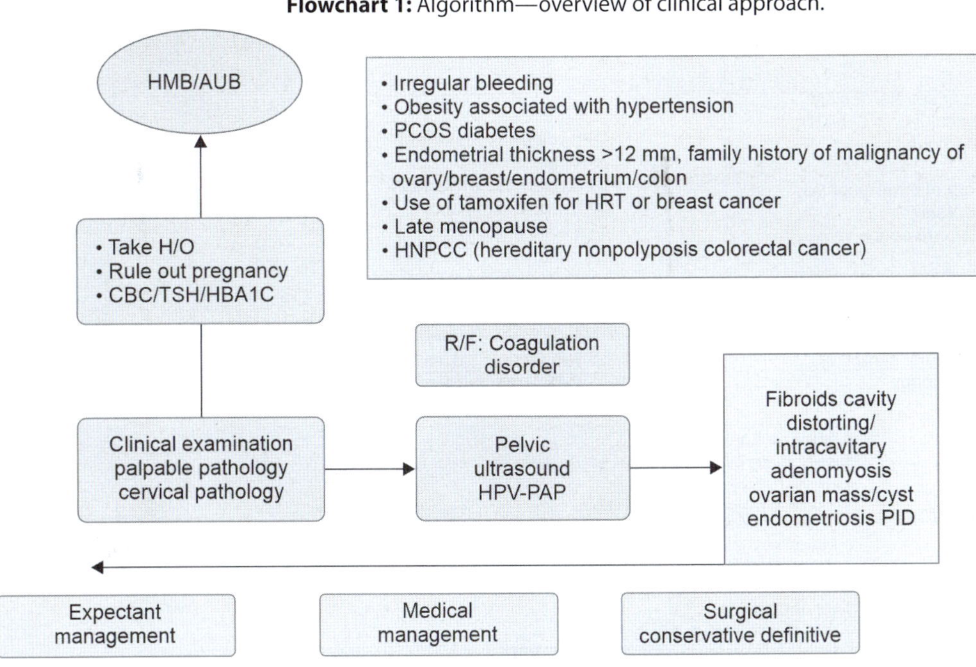

(CBC: complete blood count; HBAC: glycosylated hemoglobin; HPV-PAP: human papillomavirus-Papanicolaou test; HRT: hormone replacement therapy; PCOS: polycystic ovary syndrome; PID: pelvic Inflammatory disease; TSH: thyroid stimulating hormone)

TABLE 4: Extrauterine sources of bleeding presenting as abnormal uterine bleeding (AUB).

Sources of bleeding other than the uterus	*Cervical polyp on per speculum examination in 22-year-old nulliparous women presenting with AUB*
• Vagina • Cervix • Fallopian tube • Ovary • Urethra/bladder/hematuria • Lower gastrointestinal tract (GIT)	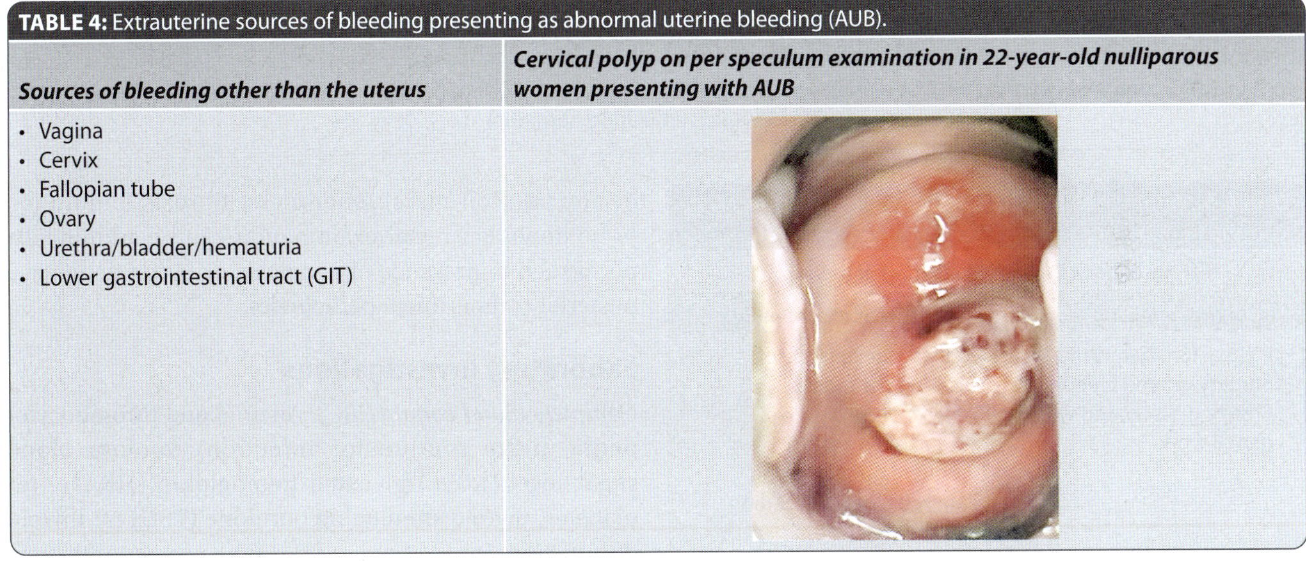

plausible cause and help in reaching a probable diagnosis to guide the further investigations and treatment options. A systematic clinical approach from history, clinical examination, laboratory tests, imaging, and additional diagnostic approaches should be undertaken **(Flowchart 2)**.

Coagulopathy can be suspected as the cause of AUB if any of the following is present **(Table 5)**.

CLINICAL EXAMINATION

General examination for assessment of obesity [body mass index (BMI) >30 kg/M^2], pallor which depicts anemia, hypertension (BP >140/90 mm Hg) should be done. Presence of tachycardia, fever, lymphadenopathy, and jaundice can be signs of infection. Systemic examination should include clinical breast examination, thyroid examination, auscultation of the lungs and the heart for any

Flowchart 2: Algorithm—clinical evaluation in abnormal uterine bleeding.

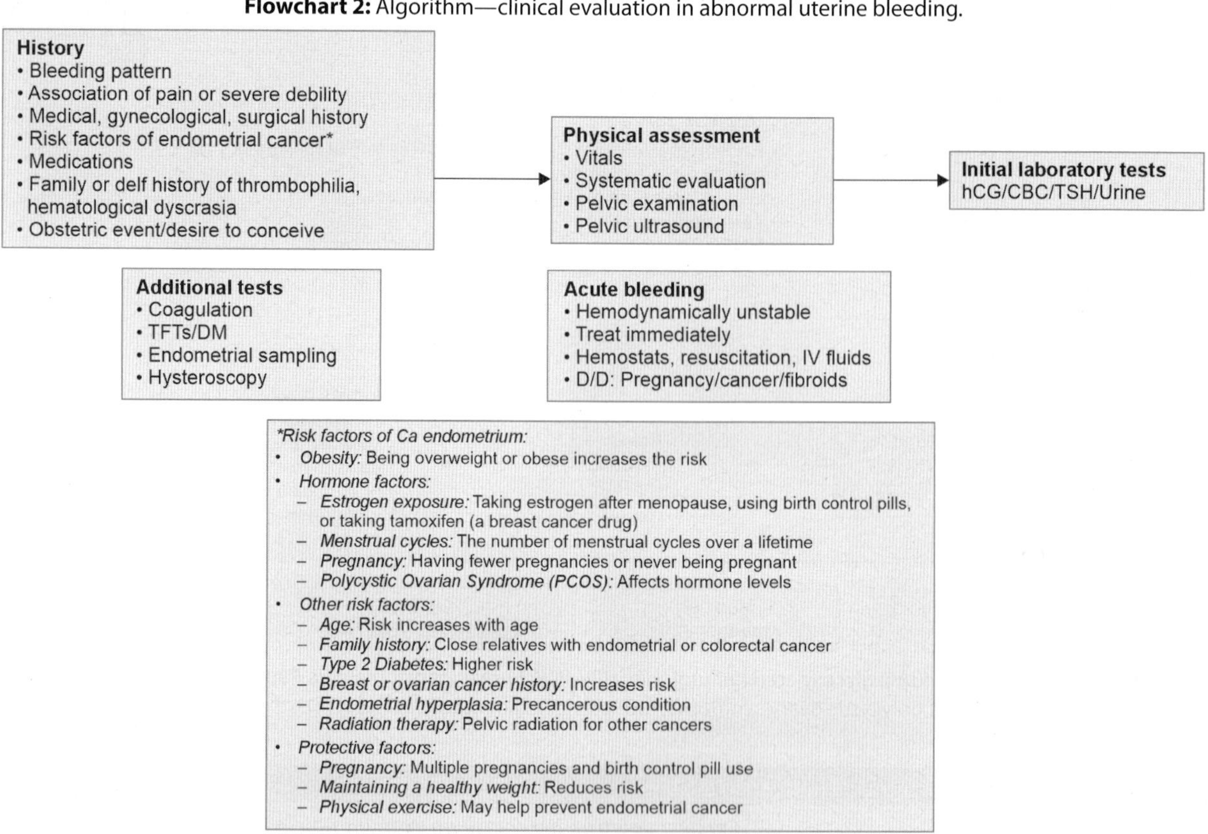

(CBC: complete blood count; DM: diabetes mellitus; hCG: human chorionic gonadotropin; TFTs: thyroid function tests; TSH: thyroid stimulating hormone)

TABLE 5: Parameters for coagulopathy (checklist)—can be assessed from history.

History of heavy bleeding starting at menarche	
One of the following	***At least two of the following***
• Postpartum hemorrhage • Surgery-related bleeding • Bleeding associated with dental work	• At least one episode of bruising per month • At least one episode of epistaxis per month • Frequent gum bleeding • Family history of bleeding symptoms

abnormalities to be done. Per abdominal examination for any herniation, ascites, presence of lump, and divaricated rectus muscle should be done.

Per-speculum examination to examine the cervix, presence of cervicitis, pelvic organ prolapse, polyp, mass or bleeding should be done and VIA/Pap test can be done. Per-vaginal bimanual examination to ascertain the size, position, consistency, contour of the uterus with presence of any adnexal mass, tenderness, fibrosis, induration to be done. Per rectal examination can be additionally performed in presence of neoplasia or better demarcate posterior compartment pathologies.

Laboratory Investigations

Complete blood count (CBC): Anemia and thrombocytopenia, urine routine for infection, random blood sugar level (RBSL)/glycated hemoglobin (HBA1c) for diabetes, thyroid stimulating hormone (TSH) for thyroid dysfunction.

In case of adolescent or reproductive age woman desiring a pregnancy, hormonal tests can be done. In case of suspicion of coagulopathy appropriate tests to be done.

Imaging with Ultrasonography

Transvaginal sonography gives better assessment of the internal genital abnormalities and with evolution of the machines most structural abnormalities can be diagnosed

TABLE 6: Common symptoms and imaging in abnormal uterine bleeding (AUB) etiologies.

Etiology	*Symptoms*	*Signs*	*Imaging*
Polyp	Prolonged uncontrolled bleeding, Inter-menstrual bleeding	Pallor, infertility uterus is usually normal in size, if big, cervical os may be patulous, polyp extruding outside	• Endometrium looks thickened • *SIS:* Echogenic smooth intracavitary mass outlined by fluid
Adenomyosis	HMB with marked dysmenorrhea	Uterus enlarged up to 12 weeks uniformly enlarged, globular. May or may not be tender	Globular uterine enlargement up to 18 weeks gravid uterus. Thickening of uterine wall, which may be asymmetrical in focal disease. Obscuring of endo-myometrial junction; multiple hypo-echoic halo zones of ≥12 mm thickness; heterogeneous texture of endometrium
Leiomyoma	• *Submucous:* More prolonged uncontrolled bleeding, intramural variable amount of HMB • *Subserous:* May be asymptomatic	Uterus irregularly enlarged	Firm well-defined, solid masses with a whorled appearance; similar echogenicity to myometrium, occasionally hypoechoic alteration of the normal uterine contour of uterus 3D-USG—for exact location in selected patients MRI-fibroid mapping when indicated. Pressure changes—hydronephrosis etc.
Malignancy	Postmenopausal bleeding, Irregular bleeding pattern at perimenopause	Normal to mildly enlarged uterus mobility may be restricted	US-thickened endometrium >12 mm at premenopausal age; endometrial carcinoma-thickened endometrium, irregular endometrial lining, and loss of endo-myometrial junction
Coagulopathy	Puberty menorrhagia, heavy bleeding at menarche, history suggestive of bleeding diathesis, and family history	Uterus normal size pallor petechiae, tendency to bruise easily	Normal scan might have hemorrhagic cysts in ovary
Ovulatory disorders	Irregular cycles, amenorrhea, bloating, hirsutism, and weight gain	Anovulation polycystic ovary syndrome oligomenorrhea; signs of insulin resistance, amenorrhea followed by HMB	Uterus normal size polycystic ovaries on ultrasound-thickened endometrium
Endometrial vaginal discharge cervical erosion	Inter-menstrual spotting prolonged spotting	Vaginal discharge Cervical erosion	Uterus normal size fluid in endometrial cavity
Iatrogenic	History of hormonal medication intake Copper T use	No abnormality	Uterus normal size Copper T in situ
Not classified	HMB	Refer to PALM-COEIN	Arteriovenous malformations

(HMB: heavy menstrual bleeding; SIS: saline infusion sonography)

and treatments planned. Uterine size, myometrial architecture, presence of leiomyomas (fibroids) with their size, location, and endometrial indentation can be studied. Endometrial thickness and presence of intracavitary pathologies such as polyp or submucous fibroids can be assessed. Additional assessment can be done by performing a saline sonography (SIS) if necessary **(Table 6)**. USG also helps in evaluating adnexal masses.

Endometrial Sampling or Curettage

Recommendations[4]

- All postmenopausal women with uterine bleeding (ET >5 mm is taken as a cut-off for D and C)

TABLE 7: Treatment of abnormal uterine bleeding (AUB) based on etiology.

Etiology	*Treatment*
Polyp	• Hysteroscopic surgical removal • Multiple polyps or polypoidal endometrium • LNG-IUS (if fertility is not desired) can be combined with surgical removal
Adenomyosis	• OCP, NSAIDs, progestogens, SERMs LNG-IUS, GnRH agonists with add back therapy (with combined oral contracepptives) • Hysterectomy is severe dysmenorrhea and no relief with above therapy
Leiomyoma	• Only symptomatic may be treated • *Medical therapy:* Tranexamic acid, NSAIDs, ulipristal acetate, progestogens • *Surgical:* Myomectomy, hysteroscopic submucous myomectomy, hysterectomy if all the other modalities have failed or not accepted
Malignancy	• Atypical endometrial hyperplasia—surgical treatment continued. Fertility not desired—hysterectomy • Hyperplasia without atypia LNG-IUS followed by oral progestins or PRMs • *Malignancy:* Ovarian tumor in older woman, fibro-myosarcoma suspected or endometrial carcinoma hysterectomy would be the part of the radical approach to cancer. Treatment has to be individualized especially in younger women
COEIN	LNG-IUS or tranexamic acid, NSAIDs, followed by OCPs or cyclic oral progestins; medical or surgical treatment failed or contraindicated: GnRH agonists with add-back hormone therapy when steroidal and other options unsuitable: SERMs (e.g., ormeloxifene). Thermal balloon ablation or TCRE can be considered

(LNG-IUS: levonorgestrel intrauterine system; NSAID: non-steroidal anti-inflammatory drugs, e.g., mefenamic acid; OCP: oral contraceptive pills; PRM: progesterone receptor modulator, e.g., ulipristal acetate; SERM: selective estrogen receptor modulators; TCRE: transcervical resection of the endometrium)

- In women >40 years with thickened endometrium, suspicion of endometrial pathology or polyp, and all the factors underlined below.
- In women <40 years who have high risk factors for carcinoma endometrium such as irregular bleeding, obesity associated with hypertension, polycystic ovary syndrome (PCOS), diabetes, endometrial thickness >12 mm, family history of malignancy of ovary/breast/endometrium/colon, use of tamoxifen for hormone replacement therapy (HRT) or breast cancer, late menopause, HNPCC (hereditary nonpolyposis colorectal cancer).
- Abnormal uterine bleeding unresponsive to medical treatment.

Endometrial aspiration should be the preferred procedure for obtaining endometrial sample for histopathology.

If endometrium is thick on imaging, but where holoprosencephaly (HPE) is inadequate or atrophic, hysteroscopy should be performed to rule out polyps.

Dilatation and curettage should not always be the procedure of choice for endometrial assessment.

TREATMENT OF ABNORMAL UTERINE BLEEDING

Abnormal uterine bleeding treatment is specifically planned once the diagnosis is arrived at. It is easy to categorize treatment based on the FIGO etiological classification. As far as possible treatment should be optimized by the use of medications such as hemostatic medicines or hormonal therapy. Conservative surgical approaches can be used especially in the younger women. Hysterectomy should be chosen as the last resort for benign conditions. Age of the woman, desire for infertility, and affection of the quality of life are to be considered in planning treatment. Treatment overview specified for the etiology is tabulated in **Table 7**.

Medical therapy in AUB should be appropriately used to confirm its effectivity. the medications should be prescribed with appropriate doses **(Table 8)** and instructions. Compliance of the patient should be ensured by planned follow up, and supportive medications such as antacids etc.

HYSTERECTOMY: PLACE IN TREATMENT OF ABNORMAL UTERINE BLEEDING

Hysterectomy to be done only when indicated and when medical or conservative surgical management has failed to relieve symptoms. It is done in the following conditions **(Table 9)**.

CONCLUSION

The prognosis and outcome of AUB usually is favorable and depends on the etiology. The primary objective of evaluating

TABLE 8: Medical management of abnormal uterine bleeding (AUB)—ready reckoner.

Drug	*Doses*	*Duration/schedule*
Non-steroidal anti-inflammatory drugs (NSAID) • Mefenamic acid • Naproxen	 • Tablets 250–500 mg tid in acute bleeding and pain • Tablets 250–500 mg tid in acute bleeding and pain	 • 4–5 days • 4–5 days
• *Tranexamic acid* (antifibrinolytic drug) • Injection	• Two tablets three times daily as long as needed for up to 4 days. If very HMB, dosage may be increased. A total dose of 4 g daily (8 tablets) should not be exceeded, • 0.5 g (1 ampoule of 5 mL) to 1 g (1 ampoule of 10 mL or two ampoules of 5 mL), tranexamic acid by slow intravenous injection (= 1 mL/minute) two to three times daily till needed in acute heavy menstrual bleeding (HMB).	
Progesterone • Medroxyprogesterone acetate/ norethisterone • Injections depot medroxyprogesterone acetate (DMPA) LNG-IUS • Levonorgestrel intrauterine system • Dienogest*	 • 10 mg tablet tid/qid in acute bleeding • Maintenance dose 10 mg bid • 250 mg IM/3 months • 104 mg SC/3 months • 52 mg of LNG covered by a rate-controlling membrane which serves to regulate the rate of hormonal release: Initially, 20 µg and after 5 years 11 µg of levonorgestrel is released every 24 hours from this polymer cylinder. • 2 mg daily	 • For 4–5 days • For 21 days • 6 months–2 years • 5–7 years • 6 months to several years
SERM Ormeloxifene	60 mg tablet	Twice a week for 3 months, followed by once a week for another 3 months
COCs	Formulations should be individualized as per the need	6 months–5 years
SPRMs Ulipristal acetate (UPA)	5–10 mg daily	12 weeks

*Dienogest is primarily used for treatment of endometriosis and long-term use can be used effectively for pain relief of adenomyosis. It can be used till menopause.

(COCs: combined oral contraceptives; LNG-IUS: levonorgestrel-releasing intrauterine system; SERM: selective estrogen receptor modulators; SPRMs: selective progesterone receptor modulators)

TABLE 9: Indications of hysterectomy in abnormal uterine bleeding (AUB).

Fibroid uterus	Pressure symptoms or severe bleeding not responding to treatment
Benign ovarian tumor	Some tumors may cause AUB and in older women necessitate hysterectomy
AUB	No response to medical/ conservative treatment
Postmenopausal bleeding	After thorough evaluation
Adenomyosis/endometriosis	After medicated treatment has failed

and treating AUB is to ensure a good quality of life and screen for and rule out malignancy. While doing this the woman's desires for fertility, associated confounders such as medical or metabolic disorders are correlated. Treatments for AUB are predominantly medical line further divided as hormonal and nonhormonal. Surgical approach as well can be conservative or radical and usually surgical approach is to be offered only if indicated. Non-hormonal treatment with anti-fibrinolytic and non-steroidal anti-inflammatory medications are quite efficacious and help in alleviating the pain and blood loss to the tune of 50%. Combination pills of estrogen and progesterone are rampantly prescribed and sometimes show relief as well but there is no randomized data supporting these indications as well as the efficacy. Hormones, especially progesterone, can be delivered in various forms such as uterine systems or depot injections. Various receptor modulators as well have found to be beneficial. Conservative surgical techniques have demonstrated to be as efficacious as the intrauterine systems, e.g., is the comparison between the LNG-IUS and endometrial ablation. Hysterectomy should be based reserved for those women where the alternative therapeutics do not work and the assessment of her pathology and its risk factors.

REFERENCES

1. Munro MG, Critchley HOD, Fraser IS, FIGO Menstrual Disorders Committee. The two FIGO systems for normal and abnormal uterine bleeding symptoms and classification of causes of abnormal uterine bleeding in the reproductive years: 2018 revisions. Int J Gynaecol Obstet. 2018;143(3):393-408. Erratum in: Int J Gynaecol Obstet. 2019;144(2):237.
2. Munro MG, Critchley HO, Fraser IS; FIGO Menstrual Disorders Working Group. The FIGO classification of causes of abnormal uterine bleeding in the reproductive years. Fertil Steril. 2011;95(7):2204-8, 2208.e1-3.
3. Sharma A, Dogra Y. Trends of AUB in tertiary centre of Shimla hills. J Midlife Health. 2013;4(1):67-68.
4. Goldstein RB, Bree RL, Benson CB, Benacerraf BR, Bloss JD, Carlos R, et al. Evaluation of the woman with postmenopausal bleeding: Society of Radiologists in Ultrasound-Sponsored Consensus Conference statement. J Ultrasound Med. 2001;20(10):1025-36.

Chapter 12

Congenital Malformation of Female Genital Organs

Mala Srivastava, Ankita Srivastava, Neema Tufchi

INTRODUCTION

Congenital malformations of female genital organs are present in different ways in different individuals.

VESTIBULAR ANUS

Vestibular anus is diagnosed in newborns. Anal opening is evidenced by anal pit. The anal opening is situated either close to the posterior end of the vestibule. It needs surgery for repositioning.

HYMEN ABNORMALITY (FIGS. 1 TO 4)

There are many abnormalities of hymen.

The girls with imperforate hymen when starts menstruating, they develop hematocolpos. These girls present with primary amenorrhea and pain abdomen. Sometimes the hematocolpos may give rise to mass abdomen due to collection of menstrual blood into the vagina. The diagnosis is by ultrasonography. The treatment is simple by giving cruciate incision over the bulging membrane of imperforate hymen.

VAGINAL MALDEVELOPMENTS (FIGS. 5 TO 8)

Common variations of vaginal maldevelopments are as follows:

- Agenesis of vagina
- Failure of vertical fusion
- Failure of lateral fusion
- Disorders of lateral fusion

Surgical procedures are required to correct these conditions.

- Cervical Abnormalities
- Single hemicervix
- Septate cervix

Uterine Anomalies[1-3]

Utero-vaginal agenesis, a rare congenital anomaly, was first documented in the 16th century by Columbo et al. (1600). This condition, characterized by the absence of the uterus and upper part of the vagina, has prevalence in the general population ranging from 0.1% to 3.5%, as noted by Byrene et al.[4] Among fertile women, the

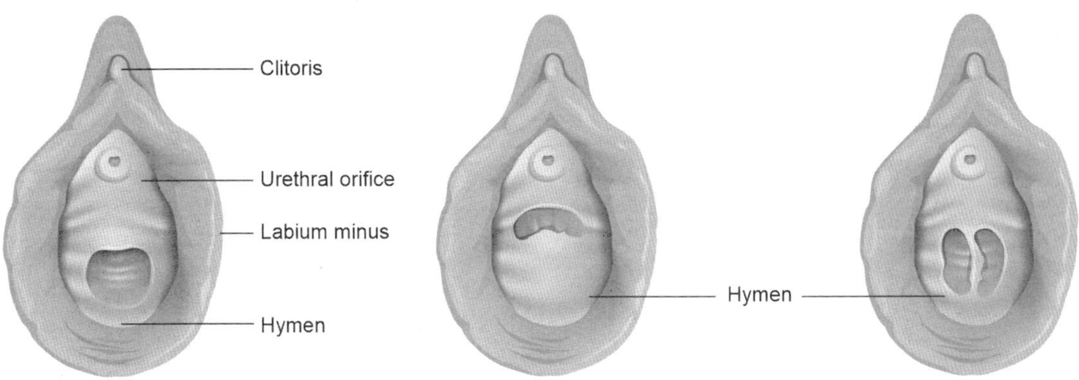

Fig. 1: Normal hymen.

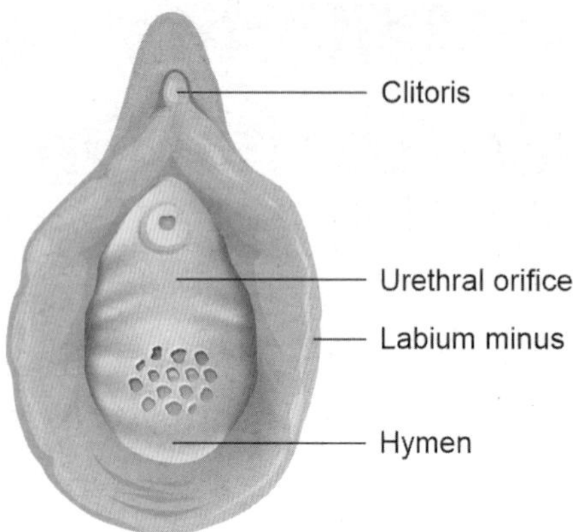

Fig. 2: Cribriform hymen.

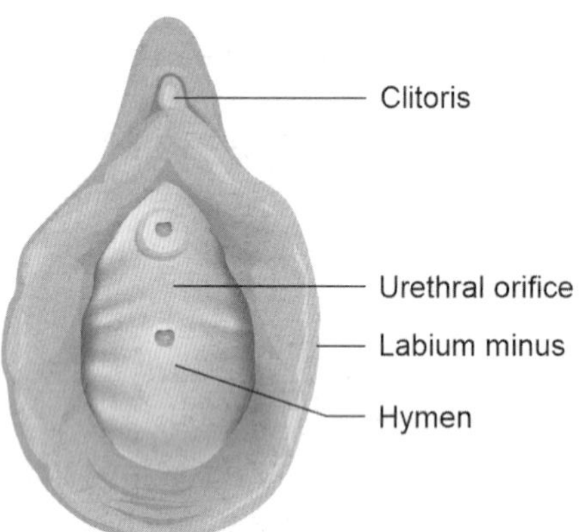

Fig. 3: Pinpoint hymen.

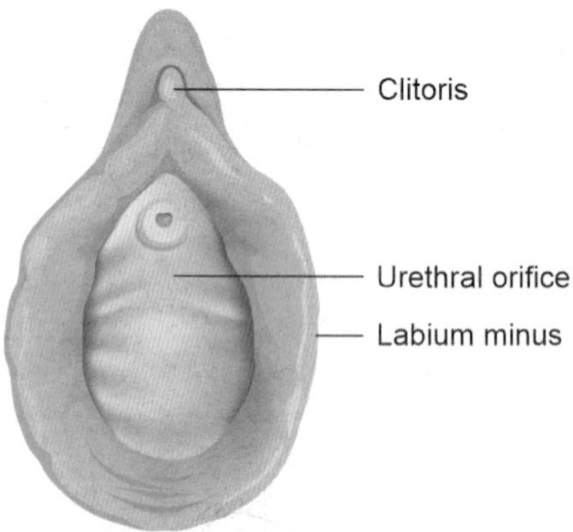

Fig. 4: Imperforate hymen.

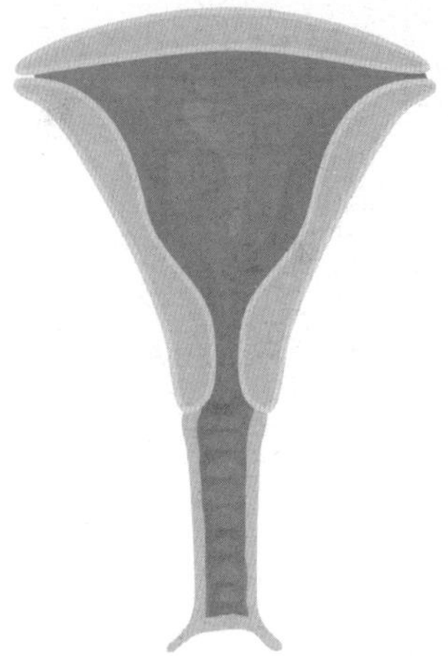

Fig. 5: Agenesis of vagina.

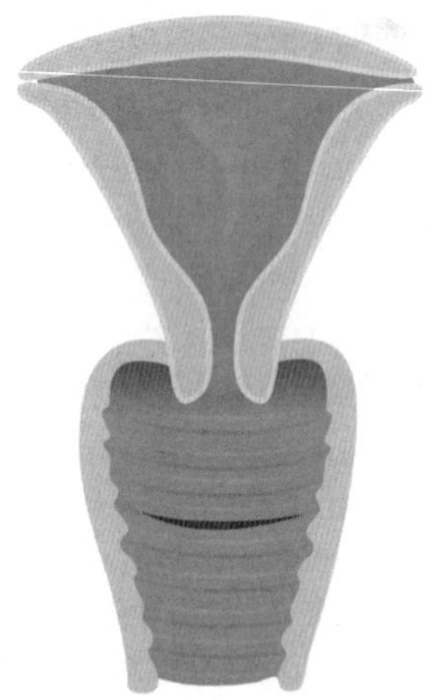

Fig. 6: Failure of vertical fusion.

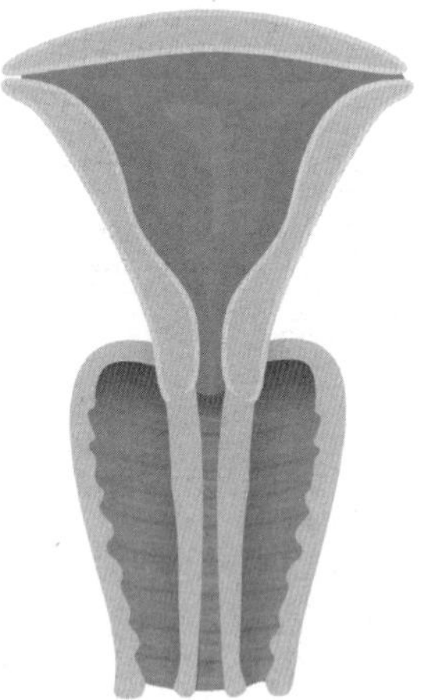

Fig. 7: Failure of lateral fusion.

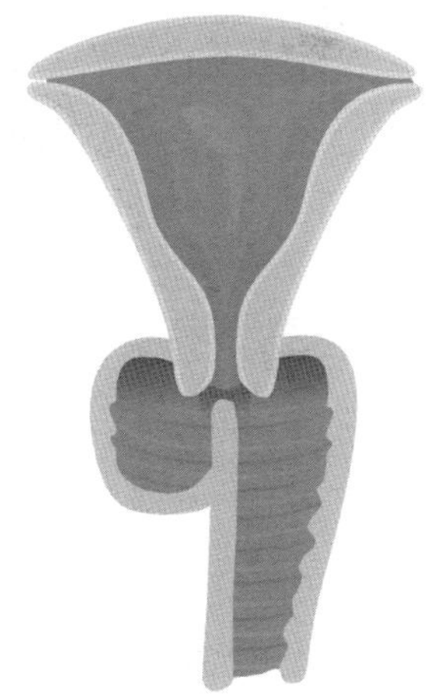
Fig. 8: Disorders of lateral fusion.

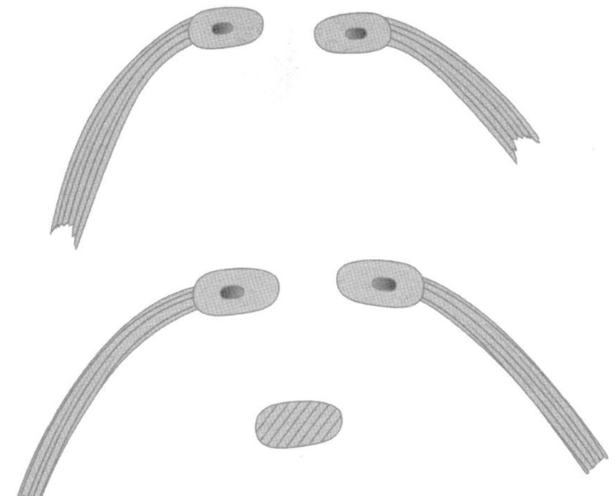
Fig. 9: Müllerian agenesis/Mayer–Rokitansky–Küster–Hauser syndrome complete and partial.

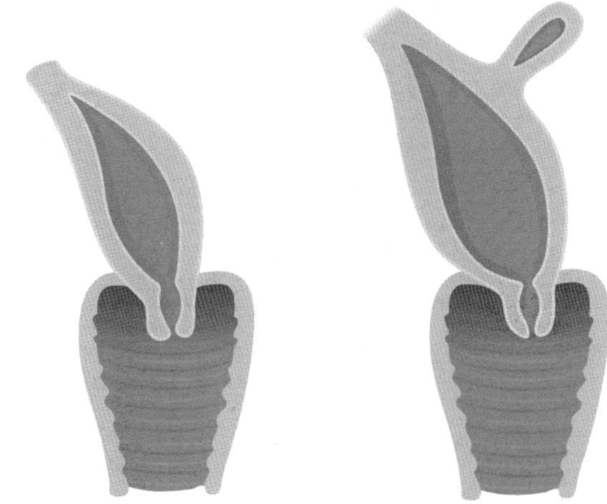
Fig. 10: Unicornuate uterus with or without a rudimentary horn.

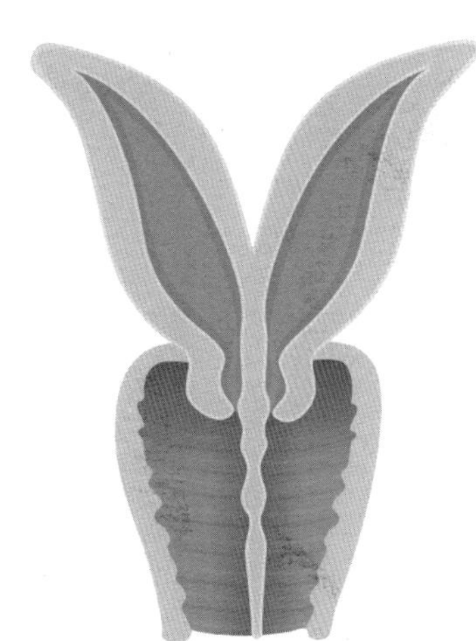
Fig. 11: Uterus didelphys.

incidence is slightly higher at 4.3%, while infertile women show a prevalence of 3.6%. In the sterile population, the incidence drops to 2.4%. Notably, Grimbizis et al.[5] reported that recurrent aborters exhibit a significantly higher prevalence, ranging from 5 to 13%, indicating a strong correlation between utero-vaginal agenesis and reproductive challenges.

Genetics of Müllerian Anomalies

- Sporadic
- Familial
- Multifactorial
- Autosomal dominant
- Autosomal recessive
- X linked
- Variants of GALT (Galactose 1 phosphate uridyl transferase enzyme defect)

Genes Associated: HOXA 9, 13 & WNT 4.

Seven major categories of uterine anomalies (AFS/ASRM):

1. *Class I*: Müllerian agenesis/hypoplasia **(Fig. 9)**
2. *Class II*: Unicornuate uterus with or without a rudimentary horn **(Fig. 10)**
3. *Class III*: Uterus didelphys **(Fig. 11)**
4. *Class IV*: Bicornuate uterus **(Fig. 12)**
5. *Class V*: Septate uterus **(Fig. 13)**
6. *Class VI*: Arcuate uterus **(Fig. 14)**
7. *Class VII*: Diethylstilbestrol-related anomalies **(Fig. 15)**

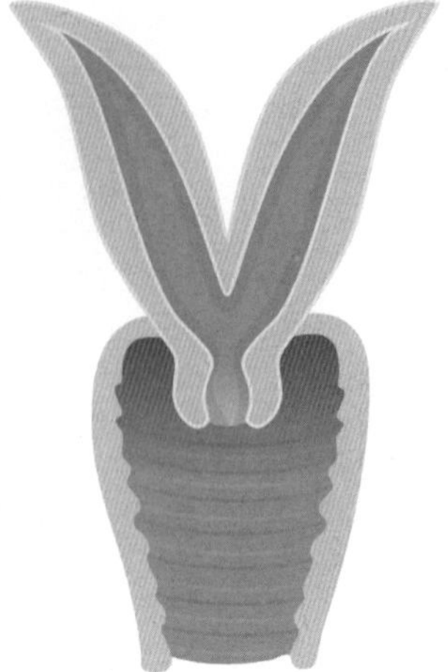

Fig. 12: Bicornuate uterus.

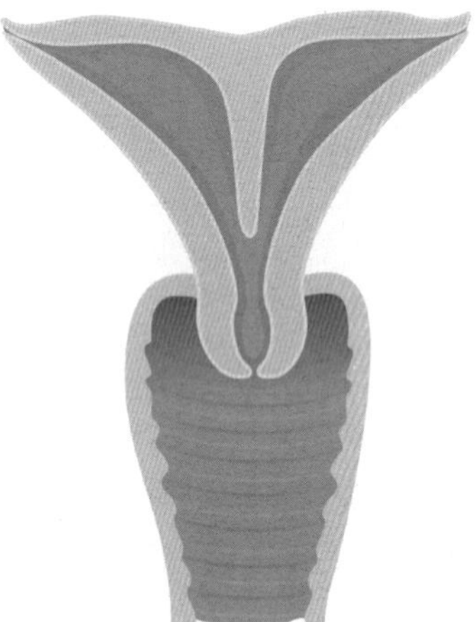

Fig. 13: Septate uterus.

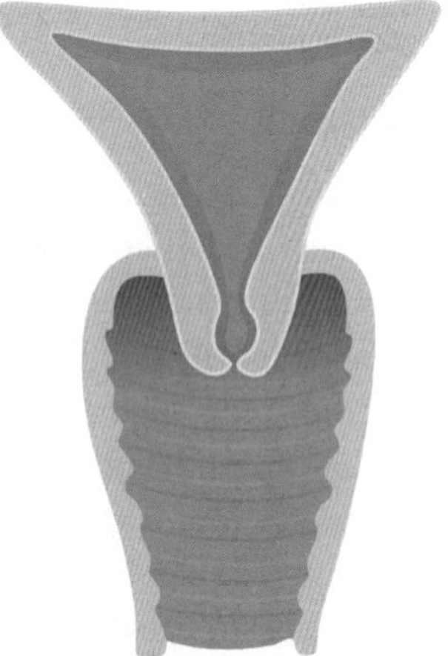

Fig. 14: Arcuate uterus.

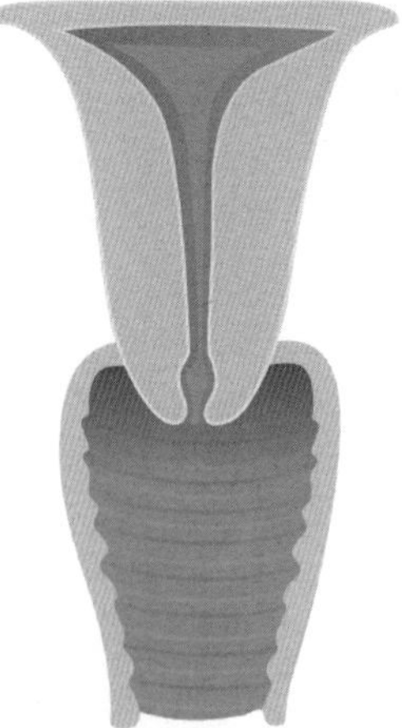

Fig. 15: Diethylstilbestrol-related anomalies.

American Fertility Society Classification of Müllerian Anomalies (Fig. 16).[6]
I. Segmental müllerian hypoplasia or agenesis
A. Vaginal
B. Cervical
C. Uterine fundus
D. Tubal
E. Combined anomalies
II. Unicornuate uterus
A. Communicating rudimentary horn
B. Noncommunicating cavity
C. No endometrial horn
D. No rudimentary horn
III. Uterine didelphys
IV. Bicornuate uterus
A. Complete (division to internal os)
B. Partial
V. Septate uterus
A. Complete (septum to internal os)
B. Partial
VI. Arcuate
VII. Diethylstilbestrol related

Incidence of Müllerian Anomalies According to AFS

- Arcuate uterus 32.8%
- Septate uterus 33.6%
- Bicornuate uterus 20.0%
- DES exposed uterus 0.8%
- Unicornuate }
- Uterine didelphys } 33%

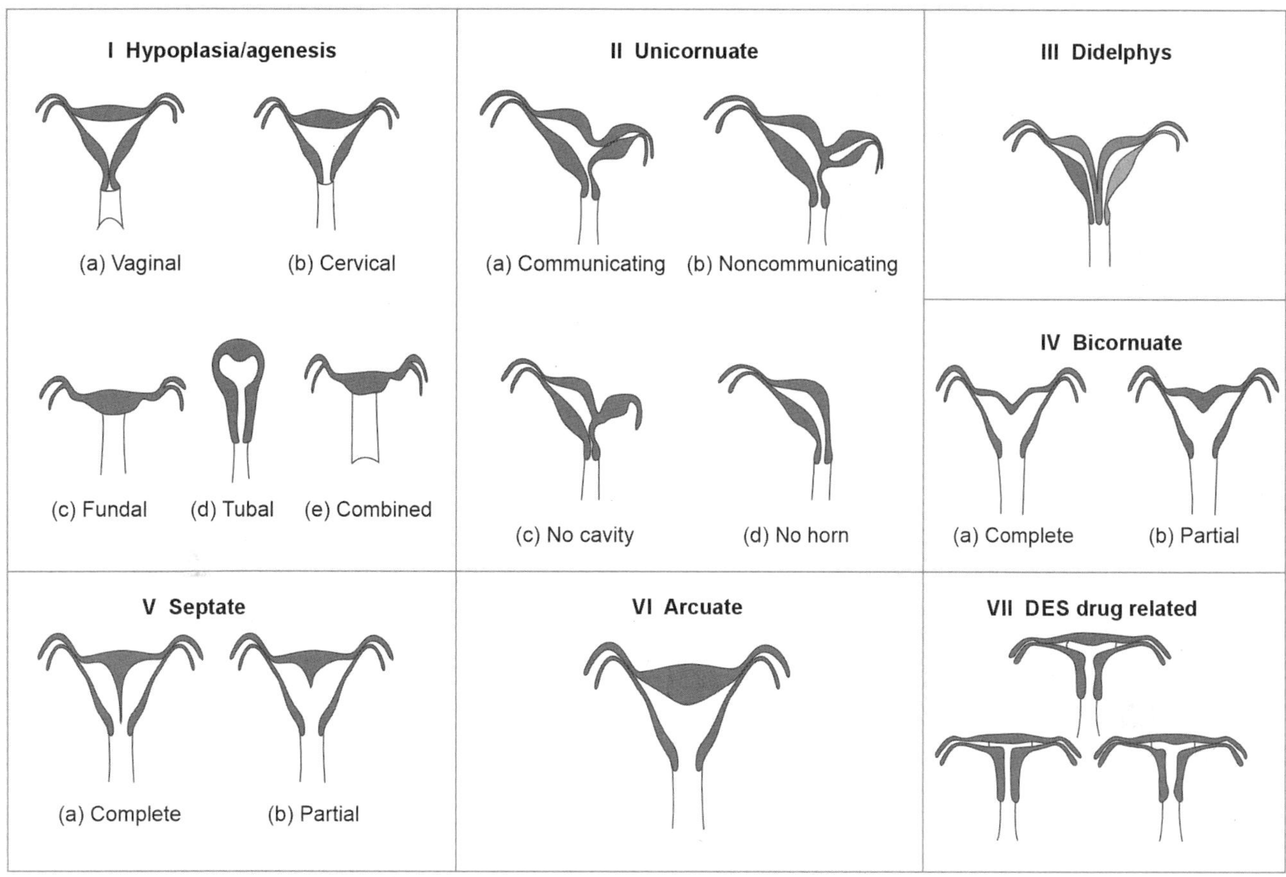

Fig. 16: I Hypoplasia/agenesis; II Unicornuate; III Didelphus; IV Bicornuate; V Septate; VI Arcuate; VII DES drug related.

Effect of Müllerian Anomalies upon Reproduction

- Infertility
- Endometriosis
- Ectopic pregnancy
- Recurrent Pregnancy Loss
- Prematurity, IUGR, fetal malposition
- Uterine dysfunction
- Uterine rupture
- Increased perinatal morbidity and mortality

The Diagnosis of Müllerian Anomalies is Based on

- Clinical features
- Hystero salphingogram
- Sonosalphingogram
- MRI – 100% accuracy
- Hystero laparoscopy
- Laparotomy or LSCS

Class I: Rokitansty Syndrome Presents as

- Primary amenorrhea
- Feminine patients
- Short vagina

Investigations Required to Diagnose the Condition

- Karyotyping
- USG/MRI
- Hormone assay
- IVP (associated vertebral anomalies can be detected) and renal sonography
- Diagnostic Laparoscopy is not routinely done.

The Treatment Includes

- Vaginal Reconstruction
    - Vagino plasty: Mac Indoes Vaginoplasty; Williams vulvovaginoplasty, Vecchietti procedure

- Fertility - by surrogacy
- Psychological support

Unicornuate Uterus (Class II)

Women with a unicornuate uterus have an increased incidence of infertility, endometriosis, and dysmenorrhea. Implantation in the normal-sized hemiuterus is associated with increased incidence of:

- Spontaneous abortion
- Preterm delivery
- Iintrauterine fetal demise
- Unilateral failure of development of MDA
- Incidence: 2.5-13%

Clinical Features Includes

- Haematometra
- Endometriosis
- Preterm labour—43%
- IUGR
- Mal presentation
- Ectopic—4.3%
- Pregnancy in accessory horn—2%
- Rupture uterus.

DIAGNOSIS AND SURGICAL MANAGEMENT

- HSG—non communicating horn cannot be diagnosed
- USG—3D or High Resolution
- MRI—banana shaped uterus
- Laparoscopy - indicated for excision of rudimentary horn which has endometrium
- IVU or renal sonography
  Cervical encirclage is mandatory if patient conceives

Uterine Didelphys (Class III)

This anomaly is distinguished from bicornuate and septate uteri by the presence of complete nonfusion of the cervix and hemiuterine cavity. Except for ectopic and rudimentary horn pregnancies, problems associated with uterine didelphys are similar but less frequent than those seen with unicornuate uterus.

- Failure of midline fusion of MD either completely or partially
- Incidence: 11%
- Types: Total Septum, Partial Septum, Transverse Septum

Clinical Features Includes

- Asymptomatic
- Failure of tampons to obstruct menstrual flow
- Hematometrocolpos
- Hematometra
- Hematosalpinx
- Endometriosis
- Other associated anomalies: bladder exstrophy, congenital VVF, cervical agenesis

DIAGNOSIS AND SURGICAL MANAGEMENT

- Clinical features
- USG
- MRI-2 widely separated uterine horns, 2 cervices are typical identified. Intercornual angle >60 degree
- Laparoscopy
- IVP

Bicornuate and Septate Uteri (Classes IV and V)

Marked increase in miscarriages that is likely due to the abundant muscle tissue in the septum. Pregnancy losses in the first 20 weeks were reported by Buttram and Gibbons.[7]

- 70% for bicornuate
- 88% for septate uteri

There also is an increased incidence of preterm delivery, abnormal fetal lie, and cesarean delivery.

BICORNUATE UTERUS

A bicornuate uterus, occurring in approximately 20% of cases, arises from the incomplete fusion of the Müllerian ducts (MD) at the level of the uterine fundus. This congenital uterine anomaly can present in two forms: complete and incomplete. In the complete form, known as bicornuate bicollis, there are two cervices, each corresponding to a separate uterine cavity. In the incomplete form, termed bicornuate unicollis, there is a single cervix leading to two partially fused uterine cavities. This malformation can impact reproductive outcomes, often necessitating specialized medical care for affected individuals.

DIAGNOSIS

The bicornuate uterus should be differentiated from septate uterus by:

- HSG
- USG which shows 2 endometrial cavities with a deep dimple in the fundus.
- MRI - Ideal
- Intercornual distance is >105 degrees

- Myometrial tissue is seen in bicornuate uterus vs septum in septate uterus with angle of <75 degree
- Laparoscopy

Septate uterus occurs due to:
- Incomplete resorption of medial septum
- The incidence is: 33.6%
- The types can be: Complete and Incomplete
- DD: Uterus didelphys
- Renal tract anomalies are rare

Clinical Features Includes

- Dyspareunia
- Dysmenorrhoea
- Primary or secondary infertility
- Poor reproductive performance

Reproductive Outcome in Septate Uterus are

- Spontaneous abortion 33–75%
- Live birth 62%
- Term deliveries 51%
- Preterm labour 10%
- Ectopic 2%

Metroplasty increases the incidence of live birth to 82%.

Class VI

Arcuate Uterus

This malformation is only a mild deviation from the normally developed uterus.

Diagnosis and Treatment of Arcuate Uterus

- HSG
- MRI
- IVP and renal ultrasound
- Hysteroscopy
- Resection indicated in poor performers

Diethylstilbestrol-induced Reproductive Tract Abnormalities

- Development of rare vaginal clear cell adenocarcinoma.
- Increased risk of developing
  - Cervical intraepithelial neoplasia
  - Small-cell cervical carcinoma
  - Vaginal adenosis,
  - Non-neoplastic structural abnormalities

Structural Abnormalities which are due to Diethylstilbestrol-Induced Reproductive Tract Abnormalities are:
- Transverse septa
- Circumferential ridges involving the vagina and cervix
- Cervical collars
- Smaller uterine cavities
- Shortened upper uterine segments
- T-shaped and irregular
- Oviduct abnormalities

Their incidences of miscarriage, ectopic pregnancy, and preterm delivery are also increased, especially in women with structural abnormalities.

MANAGEMENT OF T SHAPED UTERUS

- Lateral metroplasty
- Encerclage is mandatory in the event of pregnancy

Abnormalities of Fallopian Tubes

- Anomalies of the Ovaries
- Wolffian Remnant Abnormalities
- Parovarian Cyst

Other Abnormalities[1-3]

- Labia minora
- Labia majora
- Clitoral abnormalities
- Perineum

CONCLUSION

Müllerian duct anomalies (MDAs) are not as uncommon as previously thought and can present at various stages of life, manifesting as primary amenorrhea, infertility, recurrent abortion, or preterm labor. Accurate diagnosis of MDAs is often facilitated by MRI, which provides detailed imaging crucial for identifying these anomalies. Diagnostic laparoscopy and hysteroscopy (DHL) are indicated primarily when intervention is necessary. Corrective surgery for MDAs has been shown to significantly improve pregnancy outcomes, offering hope for many women experiencing reproductive challenges due to these congenital anomalies.

REFERENCES

1. Kachhawa G, Kriplani A. Management of Reproductive Tract Anomalies. J Obstet Gynaecol India. 2017;67(3): 162-7.

2. Grimbizis GF, Campo R, Tarlatzis BC, Gordts S. Female Genital Tract Congenital Malformations: Classification, Diagnosis and Management. New York: Springer; 2015.
3. Sardo ADS, Gloria C, Angioni S, Luigi N. Management of congenital uterine and vaginal anomalies. In: Connor ME, Clark TJ (Eds). Diagnostic and Operative Hysteroscopy. Cambridge: Cambridge University Press; 2020. pp. 167-79.
4. Byrne J, Nussbaum-Blask A, Taylor WS, Rubin A, Hill M, O'Donnell R, Shulman S. Prevalence of Müllerian duct anomalies detected at ultrasound. Am J Med Genet. 2000;94(1): 9-12.
5. Grimbizis GF, Campo R; On behalf of the Scientific Committee of the Congenital Uterine Malformations (CONUTA) common ESHRE/ESGE working group: Stephan Gordts, Sara Brucker, Marco Gergolet, Vasilios Tanos, T.-C. Li, Carlo De Angelis, Attilio Di Spiezio Sardo. Clinical approach for the classification of congenital uterine malformations. Gynecol Surg. 2012;9(2):119-29.
6. Pfeifer SM, Attaran M, Goldstein J, Lindheim SR, Petrozza JC, Rackow BW, Siegelman E, Troiano R, Winter T, Zuckerman A, Ramaiah SD. ASRM müllerian anomalies classification 2021. Fertil Steril. 2021;116(5): 1238-52.
7. Buttram VC Jr, Gibbons WE. Müllerian anomalies: a proposed classification. (An analysis of 144 cases). Fertil Steril. 1979;32(1):40-6.

Chapter 13

Sexuality and Female Sexual Dysfunction

Seema Sharma

INTRODUCTION

Sexual problems are reported by approximately 40% of females worldwide, and approximately 12% (one in every eight females) have a sexual dysfunction associated with personal or interpersonal distress.[1]

In this chapter we will broadly discuss sexual dysfunction in its various aspects:

- Definition and diagnostic criteria
- Etiology and risk factors
- Evaluation
- Management
- Sexuality in cancer survivors
- Sexuality in the elderly population
- Sexuality in palliative care.

DEFINITION

Female sexual dysfunction (FSD) means a recurrent or persistent sexual issue causing significant personal distress to a woman, lasting for at least 6 months and present in three out of four sexual encounters.

She may complaint of lack of sexual desire, impaired arousal, inability to achieve orgasm, or pain with sexual activity, few or all of them at the same time.[2]

Female sexual dysfunction may be global (present all the time) or partner specific or situation specific.

ETIOLOGY

Sexual function is strongly affected by relationship and sociocultural factors, medical conditions, and psychological issues.

A history of physical or sexual abuse is strongly associated with FSD.

- Depression, anxiety, a few medical conditions, including urinary or fecal incontinence, chronic pain conditions, cardiovascular and pulmonary disease, substance use disorders, renal failure, and cancer.
- Medications, specifically certain antidepressants (selective serotonin reuptake inhibitors), are associated with FSD. Other agents, including beta blockers, antipsychotics, and aromatase inhibitors may contribute to sexual problems.
- The effects of age and menopause on FSD vary considerably among patients. In general, sexual problems increase with aging, but distressing sexual problems peak in midlife.
- The lack of estrogen associated with menopause may cause vulvovaginal changes, hot flashes, and night sweats; resulting dyspareunia and fatigue contribute to sexual problems at midlife.
- Hormonal contraception, including estrogen-progestin contraceptives, is unlikely or only very mildly adversely affects female sexual function.

Relationship issues, fatigue, stress, menopause, psychiatric issues and medications, pregnancy and childbirth, pelvic organ prolapse, urinary incontinence, endometriosis, fibroids, genitourinary syndrome of menopause (GSM), endocrine disorders, hypertension, hormonal contraceptives, and substance abuse are commonly considered in the etiology of sexual dysfunction.[3]

EVALUATION

For proper evaluation of a couple a through medical and sexual history including childhood abuse for both the partners should be taken. It is important to be open ended and nonjudgmental while talking to her.

Sexual history of sexually dysfunction woman is given in **Box 1**.

BOX 1: Sexual history of sexually dysfunction woman.

- *Chief complaint:*
  - Since when it is present (new/chronic/intermittent)
  - Is it severe enough to cause interpersonal distress/ difficulty?
  - It is generalized or present with specific activity/or partner
- *Other useful questions:*
  - What is your main concern (low desire and difficulty with orgasm/pain)?
  - When were you last satisfied with your sex life? Describe the frequency of sexual activity and your level of desire, arousal, and response at that time versus now
  - When did you start to notice a change that was bothersome to you? Can you associate any incident at that time that could have attributed to your problem (childbirth/depression/cancer diagnosis/menopause, etc.)
  - Could stress, fatigue, or limited privacy contributing to your problem?
  - Have you and your partner tried anything to improve your sex life?
  - Describe your partner in terms of health, relationship, sexual interest, etc.
  - Describe your relationship with your partner, time spent together, vacation, etc. Is your sexual problem causing stress in your relationship?
  - Do you and your partner talk about sexuality?
  - Have you ever met with a counselor or sex therapist to discuss sexual problem?

Physical examination should focus of relevant symptoms first and then the local examination to elicit sexual dysfunction, especially if pain is the issue.

Laboratory evaluation of hormones such as estradiol, follicle-stimulating hormone (FSH), and luteinizing hormone (LH) in women are not recommended routinely. Measurements of androgens, especially testosterone, are also not recommended due to lack of reliable data. Most of the diagnosis is quite clear by history and examination.

CLINICAL APPROACH

- Sexual function issues should be addressed whenever a woman approaches a gynecologist for any health issue. Simple asking her a question on "does she have any sexual health concern" may be enough. In a busy clinic, sometimes it is wise to reschedule another visit rather than give her half-hearted advise.
- Complete her evaluation and diagnosis before treatment.
- Assess her and her partner's treatment goals in realistic terms.
- Counsel the couple yourself or send to your sexual health counselor.
- Treat associated conditions and use multidisciplinary approach.

MANAGEMENT

Management strategies have to be individualized to suit the couple and there are a few basic principles that should be adopted before putting her on any medication.

- *Counseling* should be targeted on a couple's requirement. Couples who have marriage issues and breakdown in communication can benefit from couples therapy and those who have poor knowledge about the anatomy and other aspects may benefit from sex therapy. Women having poor body image and low self-esteem can benefit from psychotherapy.
- *Lifestyle changes* such as weight loss, exercise, and keeping euglycemic status help to improve libido and quality of sexual intimacy.
- *Improving body image* by education and counseling.
- *Treating pelvic floor dysfunction* by pelvic physical or magnetic therapy. Pelvic organ prolapse and correction of early incontinence should be applied early in the treatment of sexual dysfunction.
- *Specific treatment therapies* can be targeted toward specific problems.

The Diagnostic and Statistical Manual of Mental Disorders, Fifth Edition (DSM-V) has now classified FSD into three major headings; the interest and desire disorders, pain disorders, and orgasmic dysfunctions. We will take specific treatment for each and try to specify here.

Sexual Interest/Arousal Disorders

Hormones Therapy

Androgens:

- Testosterone is the most common androgen used to treat postmenopausal FSD. Testosterone production rate in postmenopausal women is only 10% of men. No specific preparation is available exclusively for use in women so far. Data on its effectiveness is not yet available. These preparations are not Food and Drug Administration (FDA) approved yet for use in women. Testosterone should only be prescribed to peri- or postmenopausal women with low libido associated with marital distress who otherwise are healthy, in a stable relationship, and having no contra indications to testosterone therapy. Ideally low dose vaginal estrogens, relationship interventions, counseling, etc. should have been tried before the trial of testosterone. *Testosterone* is available as two main formulations; testosterone and methyltestosterone. It can be used as oral preparations, transdermal patch, injectables, and topical gel formulations each with its unique effects and side effects. Few randomized trials using transdermal patch delivering 300 μg/day testosterone per day for 6 months in postmenopausal women have shown benefit in many aspects of sexuality like improved desire, satisfaction, and orgasm ability.[4-7]

- *Dehydroepiandrosterone (DHEA)* is another androgen which may help to improve libido in women, especially with adrenal insufficiency. It is available as oral and vaginal insert preparation. Oral DHEA is ineffective for FSD. Vaginal insert as 1% gel is effective to improve desire, libido, and satisfaction and may also reduce pain associated with intercourse. Androgen treatment should be avoided in women having liver diseases, cardiovascular diseases, endometrial hyperplasia, or breast and endometrial cancer. This therapy can produce side effects such as hirsutism, acne, voice change, dyslipidemia, and clitoromegaly. Mostly trial of treatment should not go beyond 6 months and it should be stopped if there is no perceived benefit by the women.[8]

Estrogens: Estrogens can be tried in a perimenopausal or postmenopausal women within the window of opportunity if her sexual problems are associated with hot flashes, night sweats, sleep disruption, etc. Vaginal estrogen creams may help to reduce the GSM, improve lubrication, reduce pain, and improve her sexual experience. The cream comes in different formulations of conjugated estrogens (Premarin), estradiol, and estriol (Evalon) cream. The starting dose is usually, 1 g intravaginal application every day followed by 0.5 g twice a week till symptoms are alleviated. Vasomotor symptoms can be corrected by use of oral or estrogen patches.

Tibolone

- It is a synthetic steroid with estrogenic, progestogenic, and androgenic properties.
- In randomized trials, tibolone is more effective than estrogen and progestin combination for treatment of FSD in postmenopausal women.
- 2.5 mg per day dose to be swallowed as whole with water at same time irrespective of meal status.
- It is not yet FDA approved due to concerns about breast and endometrial cancer.

Serotonergic or Dopaminergic Agents

Flibanserin:

- Flibanserin is a centrally acting serotonin receptor agonist/antagonist that causes transient increase in dopamine and decreases serotonin in the brain.
- It is also called as the Pink pill or the female Viagra.
- Approved by the FDA for treatment of FSD in premenopausal women.
- It is given orally at 100 mg once a day and may increase frequency of sexually satisfying sexual experiences.
- Common side effects are dizziness, hypotension, and daytime sleepiness. Alcohol should be avoided simultaneously as it can cause interactions and hypersensitivity reactions.

Bupropion:

- It is a centrally acting, FDA approved drug to treat depression and helps in smoking cessation.
- It is a safer option than flibanserin as long-term usage data is available.
- 300 mg/d, given in the morning by mouth in premenopausal, nondepressed women increases sexual pleasure, arousal, and orgasm.

Buspirone: It is an anxiolytic which can be used in selective serotonin reuptake inhibitor (SSRI) associated sexual dysfunction.

Others: Apomorphine has been found to be useful to improve sexual function in premenopausal women.

Bremelanotide

- It is a melanocortin receptor agonist used in the treatment of reduced libido and desire in premenopausal women.
- FDA approved it for use in 2019 and has to be taken as a self-administered subcutaneous injection, 1.75 mg at least 45 minutes before sexual activity.
- Common side effects include nausea, flushing, vomiting, headache, and possible teratogenicity. Women on bremelanotide should use an effective contraception.

Phosphodiesterase Inhibitors

- Sildenafil use in women with sexual dysfunction has given inconsistent results.

- Though widely used to treat male sexual dysfunction, in various studies 10–100 mg used per day over 12 weeks has improved desire and arousal in some women which has been equivalent to placebo response.
- 50–100 mg per day dose is useful to improve arousal in women having SSRI's-induced sexual dysfunction.
- Not many studies have been performed with longer acting PDE-5 inhibitors such as Tadalafil and Vardenafil.
- Common side effects are headache, flushing, and nausea.
- Use of PDE-5 inhibitors by the male partner may improve satisfaction in both partners when needed.

Orgasmic Disorders

- A large percentage of women report not achieving orgasm at all.
- Treatment of a female orgasmic disorder mainly is by psychosocial interventions and the use of vibrators and devices which act by increasing the clitoral blood flow.

Sexual Pain

Genito-pelvic pain and penetration disorders are managed depending on the cause of pain. Three main etiologies of sexual pain in women are:

- GSM
- Provoked pelvic floor hypertonicity (vaginismus)
- Vulvar pain syndromes like vulvodynia.

Management options for sexual pain associated with GSM include vaginal lubricants and moisturizers, low-dose vaginal estrogen, vaginal DHEA, and ospemifene, vaginal laser, or radiofrequency devices.

Vaginismus can be treated with pelvic floor therapy, use of vaginal dilators, biofeedback, and desensitization techniques.

Recently botulinum toxin has been used to relax a particular muscle group in vaginismus.

SEXUAL DYSFUNCTION IN CANCER SURVIVORS

Sexual dysfunction is quite common but is a less discussed entity in cancer survivors due to mortality fears. Women suffering from cancers of genitourinary systems and breast are more likely to be affected.

Emotions such as love, intimacy, body image concerns, and fear of mortality rank higher than physical intercourse in cancer survivors. Clinicians dealing with such patients should assess each patient for her sexual functionality and offer counseling and management options if needed.

Her sexuality can get affected not only by primary cancer therapies such as surgery, chemotherapy, and radiation and endocrine therapy but also financial and emotional distress of having such a disease.

A large number of adolescents who are cancer survivors would have sexual dysfunction, reduced fertility, and premature menopause.

Management in cancer survivors needs management of specific symptoms such as dyspareunia, vaginal stenosis, decreased libido, and arousal. A multidisciplinary team should be included in the management of FSD.

SEXUAL DYSFUNCTION IN OLDER ADULTS

Women usually associate declining libido, sexual responsiveness, and dryness with postmenopausal changes.

Similarly, due to declining testosterone and associated physical changes due to aging, men also report a decrease in sexual desire, erections, and frequency of ejaculation.

The most important factor determining sexual activity in this age group is physical health and quality of interpersonal relationship.

Management of FSD in older population includes reassurance, sex therapy, and medications to treat comorbidities. Specific condition treatment remains the same as in younger population.

SEXUALITY IN PALLIATIVE CARE

Sexual dysfunction is extremely common in terminally ill patients and is highly dependent on the specific condition they are suffering from.

Most caregivers do not enquire about their sexual issues despite patients wanting to discuss and correct it.

The PLISSIT model (permission, limited information, specific suggestions, and intensive therapy) is useful to follow for disseminating information and management.

Management in general should be directed at treating the underlying cause and unless contraindicated same management principles should be followed as for their peers in same age.

REFERENCES

1. Shifren JL, Monz BU, Russo PA, Segreti A, Johannes CB. Sexual problems and distress in United States women: prevalence and correlates. Obstet Gynecol. 2008;112:970-8.

2. American Psychiatric Association. Sexual dysfunctions. In: Diagnostic and Statistical Manual of Mental Disorders, 5th edition. Arlington: American Psychiatric Association; 2013.
3. Wierman ME, Aritt W, Basson R, Davis SR, Miller KK, Murad MH, et al. Androgen therapy in women: a reappraisal: an endocrine Society clinical practice guideline. J Clin Endocrinol Metab. 2014;99:3489-510.
4. Huang G, Basaria S, Travison TG, Ho MH, Davda M, Mazer NA, et al. Testosterone dose-response relationship in hysterectomized women with or without oophorectomy: effects on sexual function, body composition, muscle performance and physical function in a randomized trial. Menopause. 2014;21:612-23.
5. Davis SR, Baber R, Panay N, Bitzer J, Perez SC, Islam RM, et al. Global consensus position statement on the use of testosterone therapy for women. Climacteric. 2019;22:429-34.
6. Somboopom W, Davis S, Seif MW, Bell R. Testosterone for peri- and postmenopausal health. Cochrane Database Syst Rev. 2005:(4):CD004509.
7. Davis SR, Vander Mooren MJ, Van Lunsen RH, Lopes P, Ribot C, Rees M, et al. Efficacy and safety of a testosterone patch for the treatment of hypoactive sexual desire disorder in surgically menopause women: a randomized, placebo controlled trial. Menopause. 2006;13:387-96.
8. Vierhapper H, Nowotny P, Mair H, Waldhausal W. Production of dihydrotestosterone in healthy men and women and in men with male pattern baldness: determination by stable isotope/dilution and mass spectrometry. J Clin Endocrinol Metab. 2001;86:5762-4.

Chapter

14

HIV in Women: Newer Challenges

Anju Soni, Sweety Soni

INTRODUCTION

Human immunodeficiency virus (HIV) is the longest pandemic that has hovered over mankind for almost four decades, since its identification in 1981. It spread very rapidly, challenging the healthcare as well as socioeconomic structure. The breakthrough in combined antiretroviral therapy (cART) tremendously reduced HIV-associated morbidity and mortality. It has helped women live longer and healthier along with HIV. Due to increased longevity, unique problems related to women's health have increased in this population.

BURDEN OF HIV INFECTION

As per WHO **(Table 1)**, in 2021, 38.4 million of the world population and more than 19.7 million women were living with HIV. Around 1.5 million were newly infected and women and girls accounted for around 49% of all new infections.[1,2]

PATHOGENESIS

The overall prevalence of HIV in India is 0.22%, with an estimate of 0.20% in adult females and 0.24% in adult males. India has estimated 2.4 million people living with HIV and 63,000 new cases in 2021. Almost 4 out of 10 people living with HIV in India are women.

HIV is a small ds-RNA retrovirus with two strains HIV-I and HIV-II. It encodes an enzyme reverse transcriptase. Their main target is CD4 helper lymphocytes. HIV causes progressive immunosuppression by gradually affecting T-lymphocytes, B-lymphocytes, monocytes, macrophages,

TABLE 1: Summary of the global HIV epidemic, 2021.

People living with HIV in 2021		*People acquiring HIV in 2021*	*People dying from HIV-related causes in 2021*
Total	38.4 million (33.9–43.8 million)	1.5 million (1.1–2.0 million)	650,000 (510,000–860,000)
Adults	36.7 million (32.3–41.9 million)	1.3 million (990,000–1.8 million)	560,000 (430,000–740,000)
Women (15+ years)	19.7 million (17.6–22.4 million)	640,000 (480,000–870,000)	240,000 (180,000–320,000)
Men	16.9 million (14.6–19.7 million)	680,000 (500,000–920,000)	320,000 (250,000–430,000)
Children (<15 years)	1.7 million (1.3–2.1 million)	160,000 (110,000–230,000)	98,000 (67,000–140,000)

Source: UNAIDS/WHO estimates

etc. It is a multistage disease: (1) Acute symptoms (1 week to 3 months), (2) Immune response to the virus, (3) Clinical latency (about 10 years, can be further increased with ART), and (4) Acquired immunodeficiency syndrome (AIDS) and related complex.[3]

TRANSMISSION

HIV gets transmitted from an infected person via contact with blood and body fluids which can be by parenteral routes (blood transfusion, needle sharing, needle stick injury) or by sexual routes. Penile vaginal intercourse is the most common mode of transmission.[4] Women are twice as likely to get infected as compared to men. Vertical transmission of HIV is definitely demonstrated in the antenatal, intrapartum, and postpartum (breastfeeding) phase.

ISSUES RELATED TO HIV IN WOMEN

Gynecological Health Issues

HIV-positive has higher prevalence of gynecological problems and comprises an important component of ongoing HIV care in women.

- *Sexually transmitted diseases (STDs):* HIV-positive women are more likely to get infected with STDs such as syphilis, genital herpes, chancroid, and pelvic inflammatory disease (PID). Not only is the presentation atypical, these conditions tend to be more severe and difficult to treat as compared to HIV-negative women. Pelvic inflammatory disease presents more commonly with tubo-ovarian abscess and need for surgical intervention.[5,6]
- *Yeast infections:* Vaginal candidiasis tends to be recurrent and more difficult to treat. Oral thrush is also more common in HIV affected women (than men).[5]
- *Bacterial vaginosis:* It is more prevalent in HIV-positive women.[5]
- *Human papilloma virus (HPV) related problems:* Women with HIV are at higher risk for developing cervical intraepithelial neoplasia (CIN) lesions, invasive cervical malignancy, vulvar intraepithelial neoplasia (VIN), and vulvar neoplasia. HPV 16 and 18 are more common in HIV-positive women. Due to lower general immunity, they are less likely to clear these HPV infections. It is important that these women should be counseled and screened appropriately. Regular PAP smear test and HPV vaccine are strongly recommended.[5]
- *Menstrual cycle disorders:* Menstrual dysfunction like amenorrhea is common. Most women experience anovulatory cycles. Weight loss, mental stress, drug abuse, premature ovarian failure, and HIV-related liver or renal disease are some of the contributing factors. Heavy menstrual bleeding or inter menstrual bleeding is also commonly experienced due to thrombocytopenia associated with HIV infection and indinavir based ART regimes.[6] Decrease in premenstrual syndrome is common due to increase in anovulatory cycles.
- *Fertility-related issues:* Women who are HIV-positive are more likely to be subfertile due to STD/PID related tubal blockage and menstrual irregularity. Use of other illicit drugs, weight loss, stress, and premature ovarian failure (POF) leading to anovulation is also commonly seen.

General Health Issues

HIV-positive patients are generally more likely to get infection due to lower immunity. Women are more prone to develop Bacterial Pneumonia **(Box 1)**.

- *Cardiac issue:* Women with HIV are three times more likely to have a heart attack than HIV-negative women. Heart disease is a leading cause of death in HIV-positive women, even if they are on regular and proper treatment.[5]
- *Osteoporosis:* Bone loss occurs faster and earlier in women living with HIV. Osteoporosis in HIV-positive individuals is multifactorial and occurs due to low body weight, poor nutrition, low vitamin D levels, physical inactivity, and side effects of drugs like tenofovir.[5,7]

Age-related Issues

- *Menopause:* With improvement in treatment options and availability of services, more HIV-positive women live longer and go through phase of menopause.
  i. *Impact of menopause on HIV infection:* Approximately 5% of new HIV diagnosis in the USA occurred in adults greater than 55 years of age. HIV in elderly women is diagnosed later in the course of disease. This adversely affects the function of the immune system, treatment response and ultimately the prognosis and survival. Many older individuals are sexually active and tend to use less barrier protection as there is no fear of getting pregnant. Changes in vaginal tissue like atrophic changes cause easy bruising and predispose women to

BOX 1: Issues related to HIV in women.

- *Gynecological health issues:*
  - Sexually transmitted diseases (STDs)
  - Yeast infections
  - Bacterial vaginosis
  - Human papilloma virus-related problems
  - Menstrual cycle disorders
  - Fertility-related issues
- *General health issues:*
  - Cardiac issue
  - Osteoporosis
- *Age-related issues:*
  - In menopause:
    - Impact of menopause on HIV infection
    - Impact of HIV infection on menopause
  - In adolescence
- Contraception and HIV
- HIV and pregnancy
- HIV treatment adverse effects in women
- HIV and intimate partner violence (IPV)

acquire HIV infection. More recent studies have generally yielded promising results, finding that menopausal status itself does not influence response to antiretroviral treatment.[8]

ii. *Impact of HIV infection on menopause:* Average age at menopause is earlier in HIV-positive women.[9] They tend to have more severe vasomotor symptoms. Early menopause is associated with increased incidents of osteoporosis, cardiovascular and neurological morbidity. Many HIV-positive women, going through menopausal transition, experience various mental health issues like depression, panic attacks, and sleep disturbances. There is also a link between depression and menopausal symptoms. And the two have been associated with negative attitude toward menopause, reduced adherence to therapy leading to reduced quality of life. Therefore, it is important for clinicians to discuss aging and menopause with their HIV-positive patients, so that they can be better prepared for any changes or symptoms they might develop.[8]

- *Adolescence:* The HIV-positive adolescent population acquires infection by two means—perinatal transmission or sexual transmission. CDC reports that youth of 13–24 years of age accounted for 21% of new HIV cases in 2018. Some problems are unique to this group given the physiological (puberty and rapid growth), developmental, and psychosocial characteristics. They tend to indulge in high-risk sexual behavior leading to unwanted pregnancy and alcohol and illicit drug use. They are also more likely to hide their seropositive status and not avail healthcare facilities. Medication adherence is also a challenge in this age group. Considering these factors, it is crucial for the clinician to optimize the effective treatment and implement adherence support strategy.[5]

Contraception and HIV

Women should be cautioned that some ART medicines can interact with oral contraceptive pills (OCPs) and depot preparations. Double protection is always recommended.[5] Regular use of condoms is the best prevention strategy against HIV transmission.

HIV and Pregnancy

The most common route of HIV infection in the pediatric age group is mother to child transmission (MTCT) during antenatal, intrapartum, and breastfeeding phase. Appropriate measures to eliminate MTCT (EMTCT) can reduce the risk of transmission effectively from 30 to 1%.

The National EMTCT program is based on a four pronged strategy. This includes primary HIV prevention in women of childbearing age; prevention of unwanted pregnancies in HIV-positive women; prevention of parent to child transmission (PPTCT) in HIV-positive pregnant women; and provision of ongoing care and support of women and their families. The aim is to sensitize and engage the private sector and integrate the EMTCT services into general health services. The government, along with National AIDS Control Organisation (NACO), Federation of Obstetricians and Gynaecologists Society of India (FOGSI), and other bodies, is determined to achieve UNAIDS laid goal of elimination of HIV by 2030.

Prevention of parent to child transmission (PPTCT): Current WHO guidelines stated that:

- All HIV-positive pregnant and breastfeeding women to start on ART irrespective of CD4 count, WHO stage, and duration of gestation
- Different regime for HIV 1 infection and HIV 2 and/or mixed infection
- ART should be lifelong and triple drug—TLD (Tenofovir + Lamivudine + Dolutegravir)
- Other commonly used regimens: TLE (Tenofovir + Lamivudine + Efavirenz)—for HIV-I, TLL (Tenofovir + Lamivudine + Lopinavir) and TLR (Tenofovir +

TABLE 2: Dosage, schedule, and side effects of common antiretroviral drugs used in pregnancy.

Name of ARV drug	*Dose schedule*	*Side effects*
Tenofovir (TDF)	300 mg OD	Nephrotoxicity, hypophosphatemia
Lamivudine (3TC)	300 mg OD	Rarely pancreatitis
Dolutegravir (DTG)	50 mg OD	Insomnia, glucose intolerance, weight gain
Lopinavir/ritonavir (LPV/r)	400/100 mg BID	Gastrointestinal disturbance, glucose intolerance, lipodystrophy, dyslipidemia

(BID: twice a day; OD: once a day)

Lamivudine + Ritonavir) (for HIV-I and -II) for women who are not willing to take dolutegravir **(Table 2)**.

HIV Treatment Adverse Effects in Women

Most of the ART data available is based on clinical trials where subjects were usually men. Although treatment response is similar in both, adverse effects may vary. The possible reason being that standard dosing is based on research in men. Side effects in women are perhaps due to smaller physique and lower weight.[5]

- *Rashes:* Women are more likely to experience allergic rash to ART than men.
- *Liver problems:* Especially drugs like nevirapine
- *Body shape changes/image:* Women tend to gain more fat around their breasts and waist when on ART.

HIV and Intimate Partner Violence

Intimate partner violence (IPV) can be both—a risk factor and consequence to HIV. IPV includes physical and sexual violence, threats of violence, and psychological aggression. Exposure to IPV can increase women's risk for HIV infection through forced sex with an infected partner, limited or compromised negotiation of safer sex practices and increased sexual risk-taking behaviors.[10] Across a number of studies, the rate of IPV, childhood sexual abuse, and childhood physical abuse were much higher in HIV-positive women.[11] HIV-positive women may experience abuse that is more frequent and more severe.[12,13] IPV can hinder HIV testing, treatment initiation, and adherence. Clinicians should screen for IPV in suspected cases in a nonjudgemental manner and provide support and referral to mental health workers and IPV advocates.[14]

HIV IN WOMEN VERSUS MEN[15]

- Women tend to be diagnosed with HIV at a later stage when compared to male counterparts.
- Women have lower CD4 counts than men (with similar viral loads).
- Women have lower viral load at the time of diagnosis.
- Women are most often diagnosed during pregnancy, planning pregnancy or when hospitalized with some acute infection.
- Fewer women avail treatment and/or there is delay in starting appropriate treatment.
- Not able to access the healthcare system due to lack of awareness and lack of financial or social support.
- Women have higher rates of herpes infection and oral thrush than men.

PRINCIPLES OF TREATMENT OF HIV IN WOMEN

There is no permanent cure for HIV. The main objective of possible treatment options is to reduce the mortality and morbidity caused by the virus and associated conditions, increasing survival, improving the quality of life, and preventing HIV transmission.[16] Women with HIV, not only have to deal with only medical conditions but often must cope with various social, economic, and psychological issues as well. All these problems are best managed through a patient centered, multidisciplinary approach. Every patient should have a complete medical history, physical examination, and laboratory evaluation.

General Measures

- Counseling regarding epidemiology of disease and risks of transmission
- Counseling regarding prevention of disease
- Healthy diet and proper nutrition
- Taking care of mental health issues
- Education regarding safe sexual practices and regular use of condoms.

Specific Treatment

Combination ART is the recommended mode of treatment. The aim of treatment is to effectively suppress the viral load, preserve (or improve) immune function, and reduce the risk of opportunistic infections **(Box 2)**. Since 2016,

BOX 2: Goals of antiretroviral therapy (ART)*.

- *Clinical goal:*
  - To prolong life and improve quality of life
- *Immunological goal:*
  - Quantitative (CD4 within normal range)
  - Qualitative (pathogen specific immune response)
- *Virological goal:*
  - Greatest possible reduction in viral load as long as possible to halt disease progression and to prevent or delay resistance
- *Therapeutic goal:*
  - Rational sequencing of drugs to achieve previous three goals while:
    - Maintaining future therapeutic options
    - Minimizing drug toxicities and side effects
    - Maximizing treatment adherence
- *Epidemiological goal:*
  - Reduce HIV transmission

*National AIDS Control Organization. National Guidelines for HIV Care and Treatment, 2021. New Delhi: NACO, Ministry of Health and Family Welfare, Government of India; 2021.

WHO recommended *Treat All*: that all people living with HIV be provided with lifelong ART therefore decision to commence treatment is not influenced by the CD4 cell count, viral load, and general condition of the patient.[17] The best result is obtained with strict adherence to first-line therapy.

PREVENTION CHALLENGES[5]

- Social barriers such as racism, discrimination, and HIV stigma have a major impact on health and wellbeing. These factors prevent women from seeking and receiving quality healthcare like HIV testing, treatment, and other prevention services.
- Lack of knowledge of HIV status
- Lack of knowledge of HIV preexposure prophylaxis
- Lack of awareness regarding partner's high-risk behavior like injectable drug abuse and multiple partners
- *Mental health:* Women with HIV who experience depression, anxiety, and post-traumatic stress disorder are less likely to seek HIV care or medicine adherence.

PREVENTION STRATEGY

- Protecting women's rights
- Providing safe reporting and protection in cases of IPV
- Improving education among girl children
- Economic empowerment of women
- Creating awareness and promotion of condom use
- Equitable and accessible health facilities
- Combat the social stigma associated with HIV and IPV
- Provision of preexposure and postexposure prophylaxis for HIV.

CONCLUSION

Women are more vulnerable with respect to the causes and consequences of HIV infection. It is a complex phenomenon involving biological and socioeconomic inequities. The nature of the HIV pandemic has changed over the last few years. In earlier stages of pandemic, the main focus was sexual and reproductive health of women and main aim was to prevent vertical transmission.[8] With the arrival of extremely effective ART and combined effort of all sectors, there has been an overall decline in HIV and AIDS infection; and a tremendous improvement in life expectancy and quality of life of women living with HIV. As a consequence, age-related comorbidities including menopause represent an emerging and more complex aspect of HIV care.[8] A holistic approach to management, in conjunction with recommended guidelines and considering epidemiological changes is pivotal to successful long-term healthcare outcomes in women living with HIV.

REFERENCES

1. UNAIDS. Global HIV & AIDS statistics — Fact sheet. [online] Available from https://www.unaids.org/en/resources/fact-sheet. [Last accessed March, 2024]
2. WHO. HIV-AIDS. [online] Available from https://www.who.int/data/gho/data/themes/hiv-aids. [Last accessed March, 2024]
3. Sharma JB. Sexually Transmitted diseases and RTIs (Reproductive Tract Infections): Text Book of Gynaecology, 1st edition. Sirmour (HP): Avichal Publishing Company; 2018. pp. 358-77.
4. Kaul R, Liu CM, Park DE, Galiwango RM, Tobian AAR, Prodger JL. The Penis, the Vagina and HIV Risk: Key Differences (Aside from the Obvious). Viruses. 2022;14(6):1164-79.
5. HIVinfo.NIH.gov. (2023). HIV and Women. [online] Available from https://hivinfo.nih.gov/understanding-hiv/fact-sheets/hiv-and-women#:~:text=The%20most%20common%20way%20that,is%20riskier%20than%20insertive%20sex. [Last accessed March, 2024]
6. Cejtin HE. Gynecologic issues in the HIV-infected woman. Infect Dis Clin North Am. 2008;22(4):709-39.
7. Harris VW, Brown TT. Bone Loss in the HIV-Infected Patient: Evidence, Clinical Implications, and Treatment Strategies. J Infect Dis. 2012;205(Suppl 3):S391-8.

8. Andany N, Kennedy VL, Aden M, Loutfy M. Perspectives on menopause and women with HIV. Int J Womens Health. 2016;8:1-22.
9. Yalamanchi S, Dobs A, Greenblatt RM. Gonadal function and reproductive health in women with human immunodeficiency virus infection. Endocrinol Metab Clin North Am. 2014;43(3):731-41.
10. Maman S, Campbell J, Sweat MD, Gielen AC. The intersections of HIV and violence: directions for future research and interventions. Soc Sci Med. 2000;50(4):459-78.
11. Machtinger EL, Wilson TC, Haberer JE, Weiss DS. Psychological trauma and PTSD in HIV-positive women: a meta-analysis. AIDS Behav. 2012;16(8):2091-100.
12. Gielen AC, Ghandour RM, Burke JG, Mahoney P, McDonnell KA, O'Campo P. HIV/AIDS and intimate partner violence: intersecting women's health issues in the United States. Trauma Violence Abuse. 2007;8(2):178-98.
13. Wyatt GE, Myers HF, Williams JK, Kitchen CR, Loeb T, Carmona JV, et al. Does a history of trauma contribute to HIV risk for women of color? Implications for prevention and policy. Am J Public Health. 2002;92(4):660-5.
14. Sullivan T. The Intersection of Intimate Partner Violence and HIV: Detection, Disclosure, Discussion, and Implications for Treatment Adherence. Top Antivir Med. 2019;27(2):84-7.
15. The Well Project. (2023). Women and HIV. [online] Available from https://www.thewellproject.org/hiv-information/women-and-hiv. [Last accessed March, 2024]
16. WHO. Global HIV Programme. [online] Available from https://www.who.int/teams/global-hiv-hepatitis-and-stis-programmes/hiv/treatment. [Last accessed March, 2024]
17. WHO. (2023). HIV and AIDS. [online] Available from https://www.who.int/news-room/fact-sheets/detail/hiv-aids. [Last accessed March, 2024]

Chapter 15

Gynecological Examination in Children and Adolescents

Savita Rani Singhal, Roopa Malik

INTRODUCTION

Pediatric gynecological examination is very simple; however, it is usually unknown without specific experience in pediatric gynecology.

Prerequisites for examination in children are:

- A thorough knowledge of anatomy and physiology of urogenital system of neonate and children, which is slightly different from adults.[1,2]
- A very gentle examination technique
- Good communication skill
- Trusting patient–physician relationship

PRINCIPLES OF EXAMINATION

- Confidence of child must be gained by thorough explanation of procedure to both the child, if child can understand as well as parents. In case of older children, she herself can explain the reason for her visit. However, when child is not able to explain, or the accompanying adult tries to explain, one should make sure that the child is active participant of talk and nods her head to approve what is being told by the adult.
- All necessary equipment required must be available at the time of examination:
  - A height-adjustable examination table
  - Bright light
  - Handheld mirrors so that patient can observe the examination
  - A pair of Latex free non-powdered nitrile gloves
    For vaginal examination, various speculums of varying sizes should be available, preferably:
  - Small Pederson or Huffman specula for prepubertal patients
  - Otoscope
  - Nasal speculum may also be used alternatively
- A quick examination must be done to reduce the time for which the patient is exposed.
- Examination which exposes the child to minimum invasive procedures should be done.
- In adolescent age group, one should be more careful about psychological problems like puberty, sexuality, and issues related to future fertility.

INDICATIONS OF EXAMINATION

- *Gynecological complaints* like vulvar rash, labial adhesions, pruritus vulvae, vulvar or lower abdominal pain, vaginal discharge or bleeding suspicious of underlying tumor, infection, trauma and suspected congenital malformations.
- *Endocrinal reasons* like abnormal pubertal course, virilization, abnormal menstrual cycles and disordered sexual development.
- *Suspected or confirmed sexual abuse.*

POSITIONS FOR EXAMINATION

Patient positioning is the most important factor of a successful gynecological examination. More than one position may be required to obtain adequate visualization of genitals in some patients. Examination position is also dependent on age of the patient.

- *Baby in lap:* In children under 2 years, mother may be asked to help in examination of child while keeping the child in her lap.
- *Frog-leg position:* For young prepubertal patients, frog-leg position **(Fig. 1)** is most frequently used position. It allows patient to directly visualize the examiner while examination is being performed.
- *Dorsal lithotomy position:* For older children, a leg brace and a dorsal lithotomy position can help better examination.

- *Knee chest position:* This position is particularly important especially in cases presenting with vaginal discharge or suspected case of foreign body.[3]
- *Examination under anesthesia:* In spite of best efforts, sometimes child may not cooperate and optimal genital examination may not be possible. In such cases depending on severity of complaints and clinical consequences of the pathology, decision for examination under anesthesia may be undertaken.

GYNECOLOGICAL EXAMINATION IN CHILDREN

For better visualization of clitoris and lower genital tract, hymen and lower vagina **(Figs. 2A to C)**:

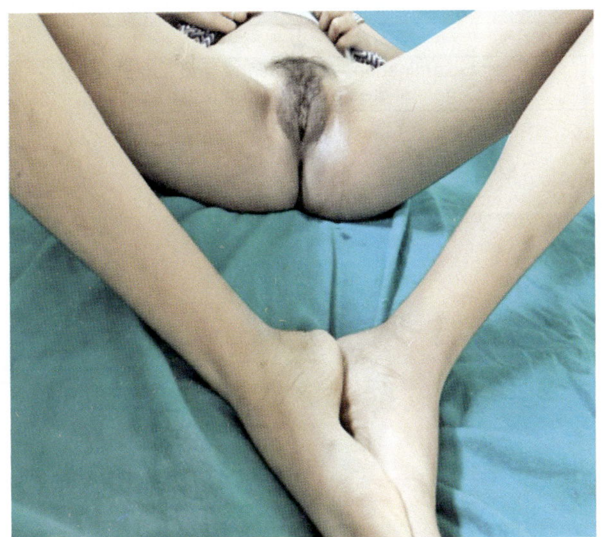

Fig. 1: Vulval examination: Frog-leg position.

- Clitoral hood should be pulled back to assess size of clitoris.
- Labia majora are separated by pulling in backwards and lateral direction **(Fig. 3)**.
- Labia majora may be pulled anteriorly **(Fig. 4)**.
- Child may be asked to cough or to take deep breath.
- Direct lateral pulling of labia majora should be avoided as it may cause stretching of the posterior fourchette and may split it, thus provoking a withdrawal reaction from the child which may further pose difficulty in examination.

GYNECOLOGICAL EXAMINATION OF NEWBORN

- One should not forget that examination in newborn is conducted as per **Flowchart 1**.
- One should be aware that in newborns, the genitalia are under effect of maternal estrogens due to which following findings may be observed **(Fig. 5)**:
  - Vulva may be swollen, pink, and may be covered with mucoid discharge due to hyperestrogenism.
  - Labia minora are more prominent and may be seen extending beyond the labia majora.
  - Clitoris may be slightly enlarged.
  - Hymen may be seen pouting out of the introitus.
  - Mother may complain of vaginal bleeding, which may be present for initial 1–2 weeks of age of neonate.
- Signs of hyperestrogenism are seen usually for first two weeks and normalize within 6 weeks[4] **(Table 1)**.

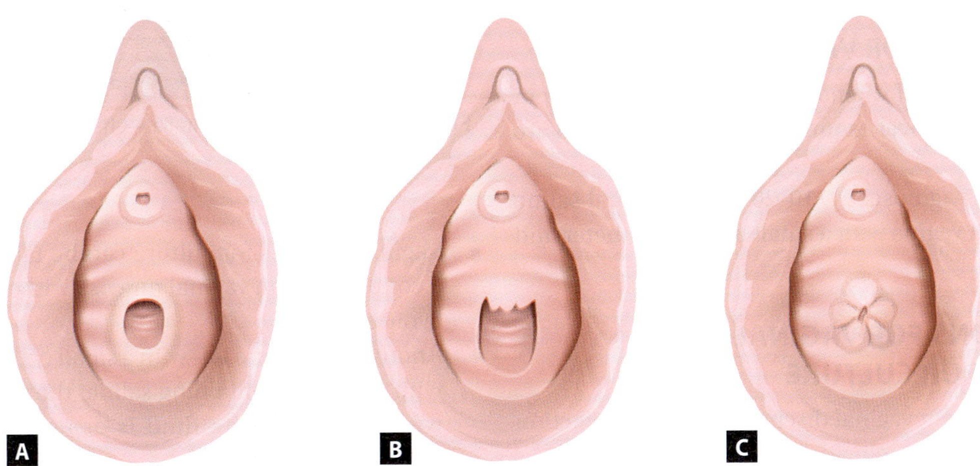

Figs. 2A to C: Configuration of hymen in prepubertal girls. (A) Annular hymen; (B) Crescentic hymen; (C) Redundant hymen.

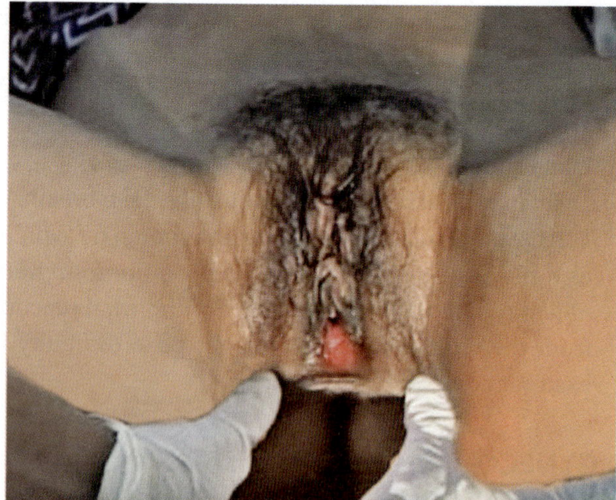

Fig. 3: Vulval examination: Pulling labia majora backwards and laterally.

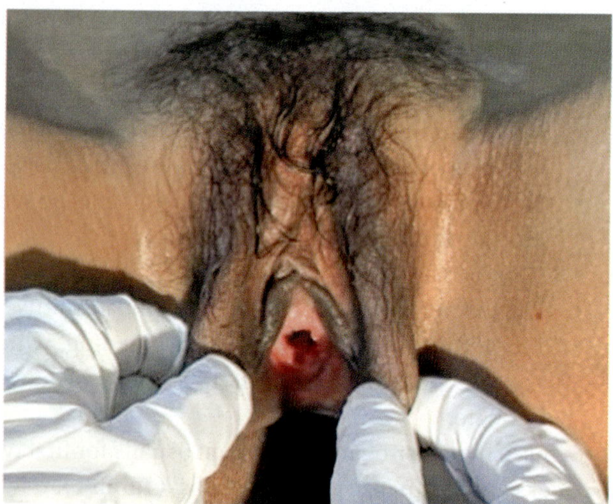

Fig. 4: Vulval examination: Pulling labia majora anteriorly.

INDICATIONS FOR VAGINOSCOPY

- Strong suspicion of vaginal tumors or foreign body in vagina, when clinical examination and ultrasound are inconclusive.
- While doing vaginoscopy, 10–20 mL of normal saline can be put in vagina with a small urethral catheter and a syringe.

INDICATIONS FOR VAGINAL SAMPLES FOR CULTURE

- When we suspect vaginal discharge is due to infection and not due to physiological leukorrhea.

Flowchart 1: Systematic gynecological examination of children/adolescents.

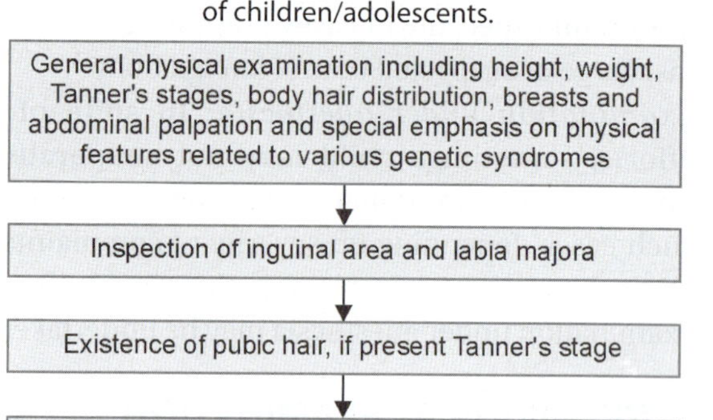

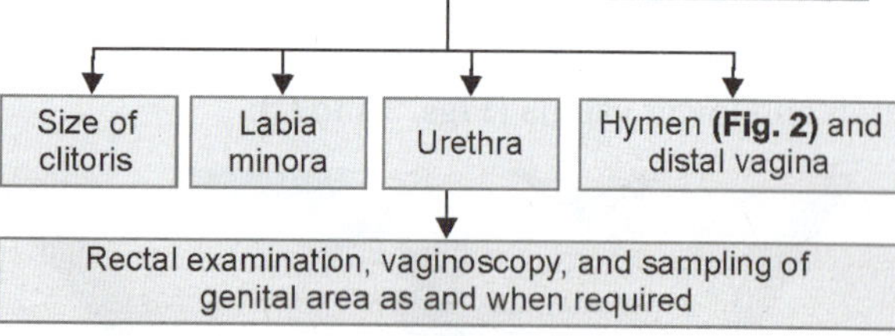

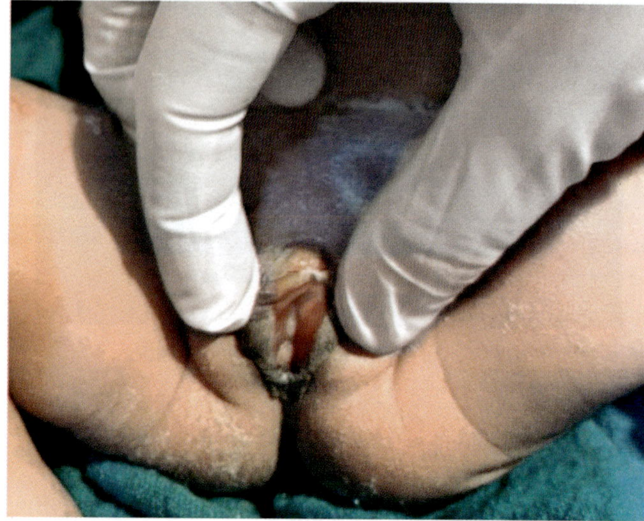

Fig. 5: Newborn female genitalia—effect of maternal estrogens.

- Vaginal samples for culture can be taken by touching a cotton swab on vaginal mucosa without rubbing it to avoid any pain.
- Another technique is—with a small catheter connected with a syringe containing saline; the vagina is flushed with saline and the fluid is aspirated and sent for culture and sensitivity.

TABLE 1: Anatomy of genital system in newborns, prepubertal, and adolescent girls.

Site	*Newborn*	*Childhood*	*Teenage*
• Labia majora	• Edema may be present	• Small and average	• Thickens and feels different from surrounding tissue
• Labia minora	• May be thick and protruded toward labia majora	• Thin and short, rudimentary extension of clitoral sheath extending to 1/3rd of the length of labia majora	• Larger but not fully developed until puberty[5]
• Clitoris	• Clitoral sheath may be thicker	• Mean clitoral length 3 mm, width 3 mm	• Mean clitoral width 2–4 mm, longitudinal diameter of 5 mm
• Hymen	• May be thicker, fimbriated, and protruded from introitus	• Annular, atrophic crescentic form, thin with visible blood vessels • 4–7 cm long atrophic walls, red and thin lining	• Nitrogenized and thickened
• Vagina	• Thickening of the vaginal mucosa, with white discharge, length 3–6 cm	• Flushed with vault with small opening in middle	• 5–9 cm long, moist, pale pink with rugosities
• Cervix	• Prominent, fundus/cervix ratio ½	• Length includes cervix 3–3.5 cm, thickness 1 cm, endometrium thin and may not be visible	• Adult form, usually pale pink and has a prominent ectopic appearance 4–5 cm long
• Uterus	• Mean length 3–4 cm, thickness 1.5 cm, endometrium echogenic	• Volume 0.3–0.7 cm^3	
• Ovaries	• Volume 1.0–1.2 cm^3		• Volume 2–3 cm^3

CONCLUSION

Gynecological examination of children and adolescents should be done after building a trustworthy relationship with her care giver and herself, in a conducive environment and availability of all necessary equipments, with utmost gentleness, in presence of clear indications of examination and every step of examination having therapeutic intent.

REFERENCES

1. Cheikhelard A, Chaktoura Z, Thibaud E. Gynecologic clinical examination of the child and adolescent. Endocr Dev. 2012;22:1-10.
2. Sultan C. Pediatric and Adolescent Gynecology: Evidence-based Clinical Practice, Volume 22. Basel, Switzerland: Karger Medical and Scientific Publishers; 2012.
3. Emans SJ, Goldstein DP. The gynecologic examination of the prepubertal child with vulvovaginitis: the use of the knee-chest position. Paediatrics. 1980;65:758-60.
4. Rusda M, Bacin BEP, Rambe AYM. (2022). Gynecological Examination in Children and Adolescents. [online] Available from https://www.cabidigitallibrary.org/doi/pdf/10.5555/20220209562. [Last accessed March, 2024].
5. Ventegodt S. Every contact with the patient should be therapeutic. J Paediatr Adolesc Gynecol. 2007;20:323-4.

Section

2

Reproductive and Clinical Endocrinology

Chapter 16

Neuroendocrinology: Steroidogenesis, Metabolism, and Mechanism of Action

Vandana Mohan, Pikee Saxena

INTRODUCTION

Hormones, a substance produced by glands with internal secretion, which serve to carry signals through the blood to the target organs, were originally described by Starling in 1905. Recent advancements in the understanding of human physiology has shown that a single definition cannot be used to accurately define the molecules.[1] The world-renowned Ganong's Review of Medical Physiology defines hormones as the soluble messengers of the endocrine system.[2] According to the Encyclopaedia Britannica hormones are organic substances secreted by plants and animals which function in the regulation of physiological activities and in maintaining the homeostasis.[3] The classic endocrine hormones travel via bloodstream to reach the distant sites, but the cellular communication, such as paracrine, autocrine, and intracrine, is also vital for local sites.

All steroid hormones have similar basic structure with relatively minor chemical differences which leads to striking differences in their biochemical activity. The sex steroids are subdivided into three main groups according to the number of carbon atoms.

1. 21-carbon series—corticoids and progestins; based on the pregnane nucleus
2. 19-carbon series—androgens; based on the androstane nucleus
3. 18-carbon series—estrogens; based on the estrane nucleus.

During steroidogenesis, the number of carbon atoms in cholesterol or any other steroid molecule can be reduced but can never be increased. **Flowchart 1** shows the steroid series.

Flowchart 1: The steroid series.

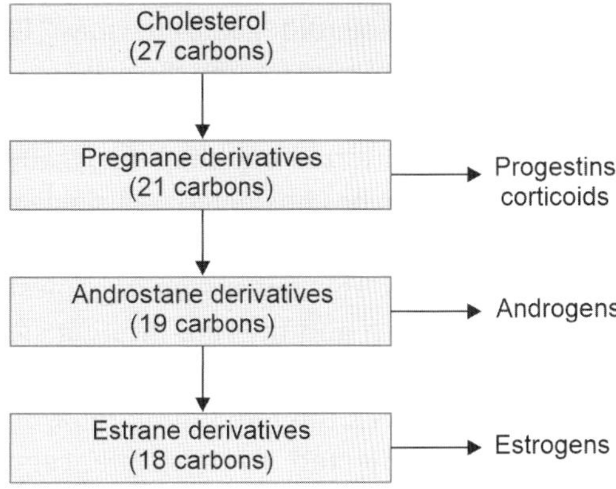

STEROIDOGENESIS

Cholesterol is the common precursor of all the steroid hormones and can be synthesized in all the steroid producing organs, except the placenta, from the precursor, acetate. However, the major source of cholesterol for steroidogenesis is the blood cholesterol, as the in situ synthesis from acetate is not enough to meet the demands of cholesterol required for steroidogenesis. Cholesterol circulating in the bloodstream enters the cells via the cell membrane receptor for low-density lipoprotein (LDL) which is the bloodstream carrier for cholesterol.

The tropic hormones secreted from the anterior pituitary bind to the cell surface receptor of the G protein system, activate the adenylate cyclase, and increase the intracellular concentration of cyclic adenosine monophosphate (AMP). Cyclic AMP activity leads to the transcription of the genes that encodes the steroidogenic enzymes and the accessory proteins. The trophic hormones also stimulate the endocytosis of LDL cholesterol esters into the cell via the clathrin-coated pits. Cyclic AMP also stimulates the hydrolysis of cholesteryl esters and the transport of free cholesterol to the mitochondria. The transfer of cholesterol from the outer to the inner mitochondrial membrane, which is the rate limiting step in steroidogenesis, is mediated by the protein activation stimulated by the tropic hormones. Various proteins are suggested to act as the regulators of acute intracellular cholesterol transfer: Steroidogenic acute regulator (StAR), steroidogenesis activator polypeptides (SAP), and peripheral benzodiazepine receptor (PBR).[4] StAR mediates the transport of cholesterol into the mitochondria for steroidogenesis in the adrenal gland and the gonads but not in the placenta and brain. Congenital lipoid adrenal hyperplasia, an autosomal recessive disorder due to mutation in the StAR gene that results in premature insertion of a stop codon, is characterized by failure of adrenal and gonadal steroidogenesis. Mutations of the StAR gene, located on the chromosome 8p11.2, are the only inherited disorder of steroidogenesis not caused by a defect in one of the steroidogenic enzymes.[5] **Figure 1** illustrates the cholesterol transfer into the cell.

Steroidogenesis includes all the processes by which cholesterol is finally converted into the biologically active steroid hormones. It can be considered as a single process that is duplicated in each endocrine gland with some cell-type-specific variations. The conversion of cholesterol to pregnenolone is the first and the rate-limiting step, and is mediated by a single enzyme, P450scc. To regulate the quantity of steroidogenesis, this rate-limiting enzymatically complex step is subjected to multiple regulatory mechanisms. Whereas, the qualitative regulation, that is the determination of the type of steroid to be produced, is mediated by many other enzymes and cofactors.

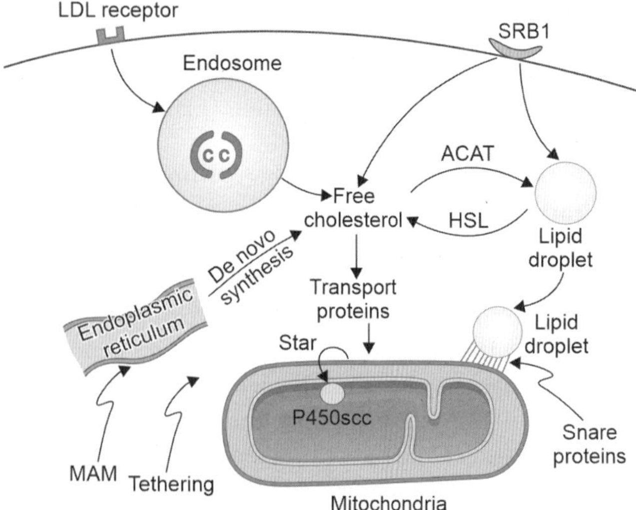

Fig. 1: Cholesterol transfer into the cell.
(ACAT: acyl-coenzyme A: cholesterol acyltransferase; HSL: hormone-sensitive lipase; LDL: low-density lipoprotein; MAM: mitochondria-associated ER membrane; SRB1: scavenger receptor B type 1)

Steroidogenic enzymes can be divided into two groups: (1) Cytochrome P450 enzymes and (2) hydroxysteroid dehydrogenases. Cytochrome P450 is a generic term used for a group of oxidative enzymes, all of which contain 500 amino acids and a single heme group. They are called P450 (pigment 450) because in their reduced state, all of them absorb light of 450 nm wavelength. The human genome contains genes for 57 cytochrome P450 enzymes: 7 in the mitochondria termed as "type 1" and the rest 50 in the endoplasmic reticulum termed as "type 2".[6] **Table 1** shows the P450 enzymes associated with steroidogenesis.

Once inside the mitochondria, cholesterol is converted to pregnenolone by P450scc. Ovarian steroid synthesis can proceed further by one of the two pathways:

1. Δ^5-3β-hydroxysteroids pathway—17-hydroxy-pregnenolone and dehydroepiandrosterone (DHEA)
2. Δ^4-3-ketone pathway—progesterone and 17α-hydroxyprogesterone.

The P450c17 enzyme is bound to smooth endoplasmic reticulum and mediates four reactions involved in converting pregnenolone and progesterone to their 17-hydroxylated products. This enzyme has two different functions, 17-hydroxylase and 17,20-lyase, resulting from the post-translational influencing factors. In ovarian theca cells, testicular Leydig cells and the adrenal reticulum, both the activities are expressed, directing the steroidogenic pathway toward DHEA. Whereas in the corpus luteum, the principal pathway is via the progesterone.

Both 17α-hydroxyprogesterone and DHEA are the immediate precursors of the C-19 series of androgens, forming androstenedione and later testosterone. Both the C-19 steroids (androstenedione and testosterone) can be converted into C-18 steroid estrogens (estrone and estradiol) by microsomal aromatization. The P450arom,

TABLE 1: The P450 enzymes associated with steroidogenesis.

Enzyme	*Cellular location*	*Reactions*
P450scc	Mitochondria	Cholesterol side chain cleavage
P450c11	Mitochondria	• 11-hydroxylase • 18-hydroxylase • 19-methyloxidase
P450c17	Endoplasmic reticulum	• 17-hydroxylase • 17,20-lyase
P450c21	Endoplasmic reticulum	21-hydroxylase
P450arom	Endoplasmic reticulum	Aromatase

found the in endoplasmic reticulum and encoded by *CYP19A1* gene, mediates this aromatization. Endometrial cancer, which results from excessive estrogenic stimulation of the endometrium, has been linked to a specific haplotype of genetic polymorphism in CYP19.[7] Transcription of the aromatase is regulated by several tissue-specific promoter sites that respond to cytokines, cyclic nucleotides, gonadotropins, glucocorticoids, and growth factors. In the ovary, the expression of the aromatase is regulated by cyclic AMP and gonadotropins, in the adipose tissue by the prostaglandin (PGE2) and is unregulated in the placenta and adipose tissue. Aromatase inhibitors (e.g., letrozole and anastrozole), specific inhibitors of P450arom, block the estrogen production and thereby are useful in the treatment of abnormal uterine bleeding and breast cancer.

The 17β-hydroxysteroid dehydrogenase enzyme converts estrone to estradiol, androstenedione to testosterone, and DHEA to androstenediol, and vice versa. Eight different isozymes have been identified.[8] Type 1 is active in the placenta and the granulosa cells and converts estrone to estradiol. The type 2 and 4 are found in several tissues and convert testosterone and estradiol into androstenedione and estrone, respectively. Type 3 and 5 are located in the testis and reduce androstenedione into testosterone. Type 6 is found only in the rodents, while type 7 and 8 are widespread but have limited activity. This cell-specific presence of isozymes helps in regulating the local concentration of estrogens and androgens. **Flowchart 2** illustrates the ovarian steroids biosynthesis pathways.

Flowchart 2: The ovarian steroids biosynthesis pathways.

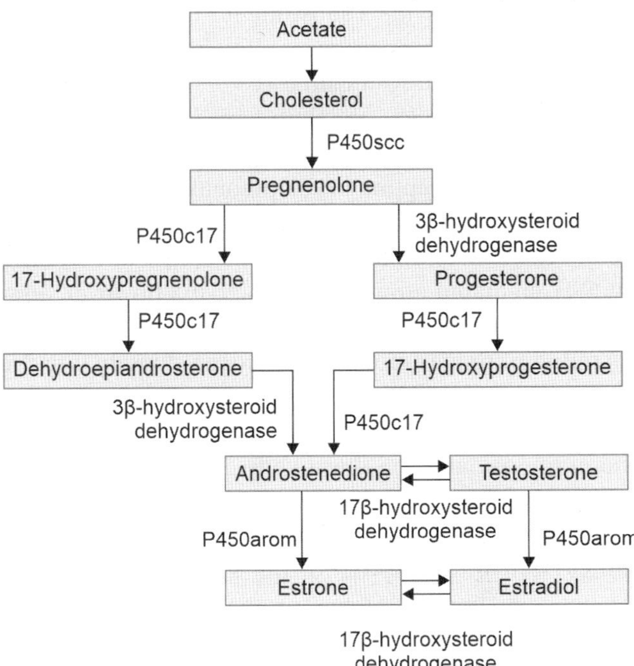

THE TWO-CELL TWO-GONADOTROPIN HYPOTHESIS

In the early stages of an ovarian follicle, the theca cells express only the luteinizing hormone (LH) receptors whereas the granulosa cells express only the follicle-stimulating hormone (FSH) receptors. LH acts on the theca cells to produce androgens which diffuse into the adjacent granulosa cells, where they undergo aromatization into estrogen under the influence of FSH.[9] As the follicle grows, FSH induces the appearance of LH receptors on the granulosa cells. **Figure 2** depicts the two-cell two-gonadotropin hypothesis. All these actions are modulated by the autocrine and the paracrine factors secreted by the theca and the granulosa cells. Insulin-like growth factor (IGF), which is secreted by the theca cells, enhances the LH-mediated stimulation of the androgen production in the theca cells as well as the FSH-mediated aromatization

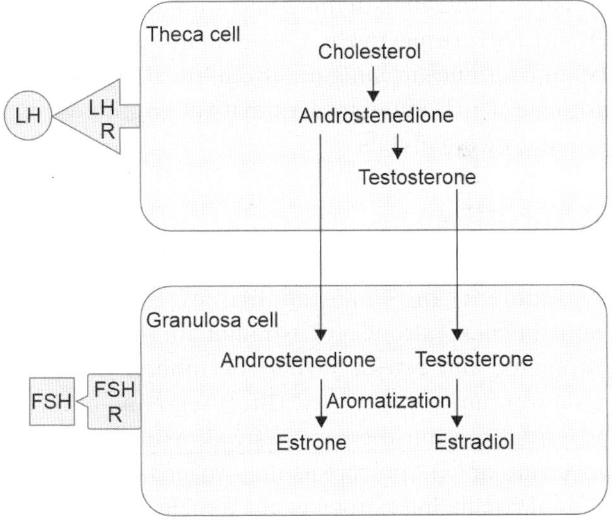

Fig. 2: The two-cell two-gonadotropin hypothesis.
(FSH: follicle-stimulating hormone; LH: luteinizing hormone)

in the granulosa cells. Also, FSH upregulates the activity of its own receptor gene by a cyclic AMP-mediated mechanism which is inhibited by several inhibitory agents, such as epidermal growth factor, fibroblast growth factor, and GnRH-like protein. Inhibin and activin are produced in the granulosa cell in response to FSH. In an autocrine fashion, activin enhances the action of FSH, especially the production of FSH receptor. Inhibin, on the other hand, enhances stimulation of androgen synthesis in theca cells by LH so as to provide a substrate for the aromatization to estrogen in the granulosa cell, whereas the activin suppresses the androgen synthesis.

After the ovulation, the dominance of the luteinized granulosa layer depends on the induction of an adequate number of LH receptors in the preovulatory phase, and therefore, is dependent on the adequate action of FSH. Before the ovulation, the granulosa cell is characterized by FSH-mediated aromatization activity; however, after the ovulation, granulosa cell is known for LH-mediated secretion of progesterone and estrogen directly into the bloodstream.

Although the androgen aromatase is present in both the granulosa and the theca cells, its activity is higher in the granulosa cell and, therefore, the granulosa cell is the main biosynthetic source of estrogen in the growing follicle. On the other hand, the granulosa cells lack P450c17, and hence the rate of aromatization in granulosa cell is directly related to and dependent on the androgen substrate produced by the theca cells. Thus, in the pre-ovulatory phase, the secretion of estrogen is the result of combined stimulation of theca and granulosa cells by the LH and FSH, respectively. Post-ovulation, the two-cell system continues to function; theca luteal cells produce androgen which undergoes aromatization into estrogen by granulosa luteal cells.

ESTROGEN

Estrogens are steroid hormones which control the development and maintenance of female sexual characteristics. Estrone and estradiol are produced from androgens, and estriol is their peripheral metabolite with lower biological activity. Direct secretion of estrone and estradiol from the ovary as well as the peripheral conversion of C-19 precursors in estrogens contributes to the total circulating estrogens in a female. The glandular estrogen synthesis occurs in the granulosa and theca cells of the ovaries, as well as in the corpus luteum whereas, the extra-glandular synthesis occurs in fat cells and bone by aromatization. **Figure 3** shows the various sources of estrogens. The rate of production of estradiol is 100–300 μg/day.[10]

PROGESTERONE

Progesterone is secreted from ovaries, placenta, and adrenals. The total production of progesterone in a female in the preovulatory phase is less than 1 mg/day and increases to 20–30 mg/day in the luteal phase.[10] About 10–20% of metabolized progesterone is excreted as pregnanediol whereas pregnanetriol is the chief urinary metabolite of 17α-hydroxyprogesterone. **Figure 4** shows the metabolism of progesterone. In adrenogenital syndrome, a syndrome of virilizing adrenal hyperplasia, an enzyme defect results in the inability to synthesize glucocorticoids and causes an accumulation of 17α-hydroxyprogesterone along with an increased excretion of pregnanetriol.

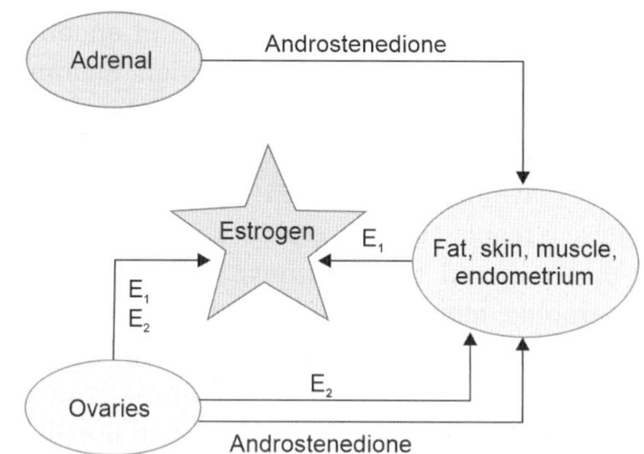

Fig. 3: Various sources of estrogens.

Fig. 4: The fate of progesterone.

ANDROGENS

Although androgens are typically considered as male hormones, they play an important physiological roles in females also. The main androgens produced by the ovary are DHEA and androstenedione along with very small amounts of testosterone. These are produced mainly by the stromal tissue derived from theca cells. At the time of ovulation, there is a rise in the circulating levels of androgens due to mid-cycle accumulation of the stromal tissue. The adrenal sex steroids represent the intermediate byproducts in the synthesis of glucocorticoids and mineralocorticoids.

Androstenedione is equally produced by the adrenal and the ovary. 50% of the DHEA is produced by the adrenal, while ovary and peripheral tissue contribute 25%

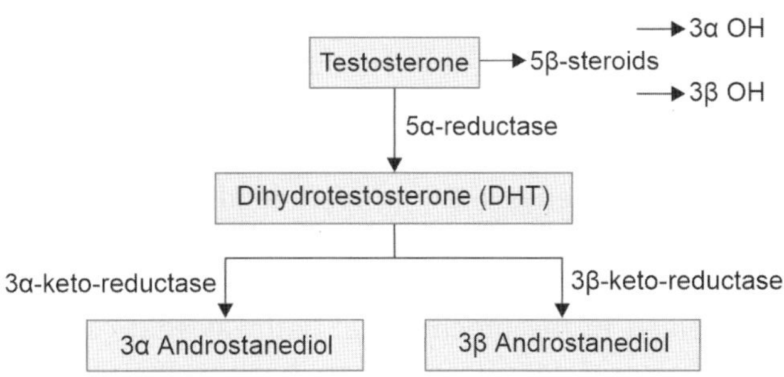

Fig. 5: The testosterone metabolism.

each. About 50% of the testosterone is derived from the peripheral conversion androstenedione, and 25% each is directly produced by the ovary and adrenal. The rate of testosterone production is 0.2–0.3 mg/day in normal female.[11] The androgenic effects of androgens are mainly dependent on the unbound fraction which can freely move into the target cell. The major androgens are excreted as 17-ketosteroids in urine.

Androgens can function in any of the following three ways:

1. By intracellular conversion of testosterone to 5α-derivative, dihydrotestosterone (DHT)—intracrine activity
2. By testosterone itself—endocrine activity
3. By intracellular conversion of testosterone to estradiol (aromatization)—intracrine activity.

The conversion of testosterone to DHT is an irreversible reaction and is carried by the enzyme 5α-reductase. DHT is the main androgenic hormone in a variety of target tissues. It is formed from testosterone in the males and mainly from androstenedione in the skin in the females. The 5α-reductase enzyme exists in two forms: type 1 in skin and type 2 in reproductive tissues.[12] The blood DHT levels are very low as it is mainly metabolized intracellularly. During masculine differentiation, the development of wolffian duct structures (epididymis, vas deferens, seminal vesicle) is dependent on testosterone, while the development of urogenital sinus and tubercle into the male external genitalia, urethra, and prostrate requires DHT.[13] Muscle development requires testosterone, while hair follicles need DHT. **Figure 5** shows the testosterone metabolism.

MECHANISM OF ACTION

Hormones activate target cells by either of the two ways: (1) the lipid soluble hormones diffuse through the plasma membrane and bind a receptor protein within the cytoplasm, (2) whereas the water soluble hormones bind a receptor protein in the cell membrane of the target cell. A small amount of hormone can produce a biologic response because of its high affinity, specificity, and activity of the receptors, along with the large concentration of receptors in the cell. The receptors can be divided into following categories:

- *Intracellular receptors:* Receptors for estrogen and thyroid hormones
- *G Protein receptors:* Receptors for tropic hormones and prostaglandins
- *Ion gate channels:* Acetylcholine receptor
- *Receptors with intrinsic enzyme activity:* Receptors for insulin and growth factors (tyrosine kinase) and receptors for activin and inhibin (serine kinase)
- *The system of internalization:* Receptors for LDL, prolactin, and growth hormone.

The mechanism of action of sex steroid hormones includes:

- *Steroid hormone diffusion across the cell membrane*—concentration of free hormone in bloodstream is an important factor. Estrogen, progestins, and androgens are transferred across the nuclear membrane to bind to their receptors within the nucleus.[14]
- *Steroid hormone binding to a receptor protein*—transformation or activation of the receptor occurs.
- *Interaction of a hormone-receptor complex with nuclear DNA*—at specific DNA sites (hormone-responsive elements) located upstream of the gene.
- *Synthesis of messenger RNA (mRNA)*
- *Transport of the mRNA to the ribosomes*
- *Protein synthesis in the cytoplasm resulting in specific cellular activity.*

Biologic activity is maintained only while the nuclear site is occupied with the hormone-receptor complex. Dose and duration of exposure to a hormone are equally important for it.

THE HYPOTHALAMIC-PITUITARY-OVARIAN AXIS AND HORMONAL FEEDBACK MECHANISM

The hypothalamic-pituitary-ovarian (HPO) axis plays a key role in the regulation of hormonal synthesis by the ovaries. The hypothalamus is connected to the pituitary gland via a portal vascular system that allows transport of hypothalamic releasing factors from the brain to the pituitary. The hypothalamus, being the coordinating center, provides precise signals via pulsatile release of GnRH to the gonadotrophs, which in turn secrete LH and

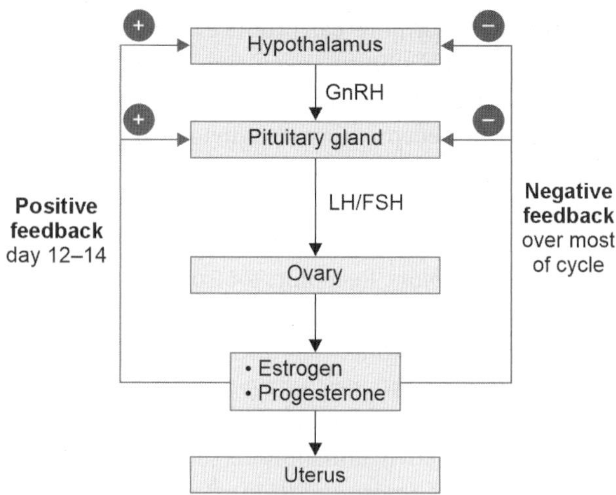

Fig. 6: The hypothalamic–pituitary–ovarian axis and the feedback mechanism.
(FSH: follicle-stimulating hormone; LH: luteinizing hormone)

FSH. LH and FSH act on ovaries to help in development of follicles. The developed follicles secrete steroid hormones (estradiol and progesterone) and peptide hormones (inhibin, activin, and follistatin), all collectively control secretion of gonadotropins. Estradiol and progesterone, depending on concentration, have either positive or negative feedback and can alter the frequency and/or amplitudes of pulses at the level of both the hypothalamus and pituitary. **Figure 6** shows the hypothalamic–pituitary–ovarian axis and the feedback mechanism.

CLINICAL IMPLICATIONS

- Disorders in the initial steps of steroidogenesis, i.e., internalization of LDL particles and subsequent intracellular processing of cholesterol, include adrenoleukodystrophy, Wolman disease, and Niemann–Pick type C disease.[15,16] These may present as adrenal insufficiency.
- Severe mutations in StAR gene, regulating the delivery of cholesterol to the inner mitochondrial membrane, cause classic congenital lipoid adrenal hyperplasia (CAH).[17] It is characterized by adrenal insufficiency and 46,XY disorders of sexual development (DSD).
- *Polycystic ovarian syndrome (PCOS):* The follicles of polycystic ovary lack a sufficient number of mature granulosa cells to produce enough estrogens. On the other hand, there is hypertrophy of stromal and thecal tissue leading to high ovarian androgen production and thus anovulation. There is disordered gonadotropin secretion with elevated mean LH, low or normal FSH, and a persistently rapid frequency of GnRH pulse secretion. This leads to unregulated follicular growth resulting in the absence of dominant follicle formation that is necessary for normal ovulation.
- *Premature ovarian insufficiency (POI):* Factors involved in steroidogenesis, the process leading to estrogen production, can cause POI. POI-causing pathogenic variants can occur in StAR gene. Enzymes directly involved in the biochemical pathway that converts cholesterol to estrogen can harbor POI-causing variants. Variants in the CYP17A1 and CYP19A1 genes, for example, encoding for pregnenolone/progesterone hydroxylation and estrogen aromatization, respectively, can cause POI.[18,19]
- With an increased incidence of female reproductive disorders, there has been an increasing concern for the potential reproductive impact of exogenous factors, particularly of environmental pollutants with endocrine disrupting properties. Studies have reported that ovarian steroid hormone production is an important target for the action of endocrine disrupting chemicals (EDCs) like pesticides, phthalate and phenol derivatives, and halogenated aryl hydrocarbons.[20]

CONCLUSION

Cholesterol is the common precursor of all the steroid hormones. Steroidogenesis includes all the processes by which cholesterol is finally converted into the biologically active steroid hormones. The process involves various hormones, transport proteins, enzymes, and cofactors. Most steroidogenic enzymes are either forms of cytochrome P450 or are hydroxysteroid dehydrogenases.

Major sites of steroid synthesis include the adrenals, ovaries, testes, and placenta. The sex steroids are subdivided into three main groups-21 carbon series (corticoids and progestins), 19 carbon series (androgens) and 18 carbon series (estrogens).

The pituitary gonadotropins, luteinizing hormone and follicle-stimulating hormone, regulate the ovarian production of sex steroids. According to the two-cell–two-gonadotropin theory, LH stimulates thecal cells to produce androgens, and FSH stimulates granulosa cells to produce estrogens from androgens.

The hypothalamic-pituitary-ovarian axis and the hormonal feedback mechanism (positive or negative) maintains the menstrual and reproductive systems of the female. Any deviation in the process of steroidogenesis, gene mutations, disorders of HPO axis and harmful exogenous factors (EDCs) have significant clinical implications.

REFERENCES

1. Starka L, Duskova M. What is a hormone? Physiol Res. 2020;69(Suppl 2):S183-5.
2. Barret KE, Barman SM, Yuan J, Brooks HL. Ganong's Review of Medical Physiology, 26th edition. New York: McGraw-Hill Education; 2019.
3. Barrington EJW; Encyclopedia Britannica. (2020). "Hormone". [online] Available from https://www.britannica.com/science/hormone. [Last accessed March, 2024]
4. Selvaraj V, Stocco DM, Clark BJ. Current knowledge on the acute regulation of steroidogenesis. Biol Reprod. 2018;99(1):13-26.
5. Bizzarri C, Pisaneschi E, Mucciolo M, Pedicelli S, Galeazzi D, Novelli A, et al. Lipoid congenital adrenal hyperplasia by steroidogenic acute regulatory protein (STAR) gene mutation in an Italian infant: an uncommon cause of adrenal insufficiency. Ital J Pediatr. 2017;43(1):57.
6. Miller WL. Steroidogenic electron-transfer factors and their diseases. Ann Pediatr Endocrinol Metab. 2021;26(3):138-48.
7. Berstein LM, Imyanitov EN, Suspitsin EN, Grigoriev MY, Sokolov EP, Togo A, et al. CYP19 gene polymorphism in endometrial cancer patients. J Cancer Res Clin Oncol. 2001;127(2):135-8.
8. Labrie F, Luu-The V, Lin S-X, Simard J, Labrie C. Role of 17b-hydroxysteroid dehydrogenases in sex steroid formation in peripheral intracrine tissues. Trends Endocrinol Metab. 2000;11:421.
9. Hillier SG, Whitelaw PF, Smyth CD. Follicular oestrogen synthesis: the 'two-cell, two-gonadotrophin' model revisited. Mol Cell Endocrinol. 1994;100(1-2):51-4.
10. Taylor HS, Pal L, Seli E. Chapter 1: Hormone biosynthesis, metabolism, and mechanism of action. Speroff's clinical gynaecologic endocrinology and infertility, 9th edition. Philadelphia: Lippincott Williams and Wilkins; 2019. pp. 3-57.
11. Vierhapper H, Nowotny P, Waldhäusl W. Determination of testosterone production rates in men and women using stable isotope/dilution and mass spectrometry. J Clin Endocrinol Metab. 1997;82(5):1492-6.
12. Marks LS. 5alpha-reductase: history and clinical importance. Rev Urol. 2004;6 Suppl 9 (Suppl 9):S11-21.
13. Lucas-Herald AK, Bashamboo A. Gonadal development. Endocr Dev. 2014;27:1-16.
14. Mauvais-Jarvis F, Lange CA, Levin ER. Membrane-Initiated Estrogen, Androgen, and Progesterone Receptor Signaling in Health and Disease. Endocr Rev. 2022;43(4):720-42.
15. Engelen M, Kemp S, de Visser M, van Geel BM, Wanders RJ, Aubourg P, et al. X-linked adrenoleukodystrophy (X-ALD): clinical presentation and guidelines for diagnosis, follow-up and management. Orphanet J Rare Dis. 2012;7:51.
16. Xu M, Liu K, Swaroop M, Porter FD, Sidhu R, Firnkes S, et al. δ-Tocopherol reduces lipid accumulation in Niemann-Pick type C1 and Wolman cholesterol storage disorders. J Biol Chem. 2012;287(47):39349-60.
17. Kim CJ. Congenital lipoid adrenal hyperplasia. Ann Pediatr Endocrinol Metab. 2014;19(4):179-83.
18. Fortuno C, Labarta E. Genetics of primary ovarian insufficiency: a review. J Assist Reprod Genet. 2014;31(12):1573-85.
19. Kim S, Pyun JA, Cha DH, Ko JJ, Kwack K. Epistasis between FSHR and CYP19A1 polymorphisms is associated with premature ovarian failure. Fertil Steril. 2011;95:2585-8.
20. Mlynarcikova A, Fickova M, Scsukova S. Impact of endocrine disruptors on ovarian steroidogenesis. Endocrine Regulations. 2014;48(4):201-24.

17

Chapter

Uterus, Endometrium, and Regulation of Menstruation

Renu Sharma

INTRODUCTION

The uterus is a dynamic female reproductive organ that develops from the Mullerian system during the embryonic stage. Throughout the reproductive years, the endometrium, which is its deepest layer, undergoes cyclical modifications and becomes receptive to the implantation of fertilized ovum. Menstruation happens when anticipated fertilization is not achieved. Understanding the development of the uterus, its vasculature, and the menstrual cycle and how it is regulated forms the basis of many gynaecological disease processes.[1]

A FEW INTERESTING HISTORICAL POINTS

- Soranus of Ephesus was the first one to delineate the appearance of the uterus (98–138 AD).
- Gabriele Falloppio depicted the ovaries and tubes precisely in 1561 and the tubes are known after him as fallopian tubes.
- Andreas Vesalius discovered the endometrial cavity inside the uterus in the 16th century.

DEVELOPMENT OF MULLERIAN SYSTEM

- By the 5th week of intrauterine life, the genital ridge develops on the medial side of the mesonephros, which will form future gonads.
- Till 6–7 weeks, the genital ridge remains undifferentiated and has the potential to develop into either testis or ovaries (bipotential/ambisexual).
- After 7 weeks, gonadal differentiation starts depending on the chromosomal pattern.[2]
- In an 8 weeks embryo, two duct systems are present lateral to the genital ridge:
  1. Wolffian system (mesonephric duct system)
  2. Mullerian system (paramesonephric duct system).

Flowchart 1: Development of Wolffian system.

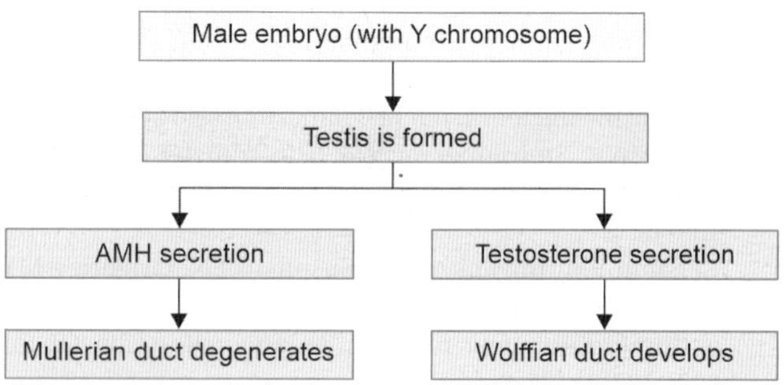

Only one of the duct systems further grows in an embryo depending on the presence or absence of the Y chromosome (SRY gene on its short arm) and the type of gonad.

- *If the Y chromosome is present,* testis is formed. Testis secretes anti-Mullerian hormone (AMH) and testosterone **(Flowchart 1)**
- AMH is a transforming growth factor β (TGF-β) family of glycoproteins secreted by Sertoli cells of the testis in early fetal life. It leads to regression of the Mullerian system (paramesonephric duct system).
- Testosterone is secreted by the Leydig cells of the testis and is responsible for the development of the Wolffian system (mesonephric duct system).
- *If Y chromosome is absent,* ovaries are formed, which start secreting estrogen, leading to the development of the Mullerian system (paramesonephric ducts) **(Flowchart 2)** In the presence of estrogen and absence of AMH, paramesonephric ducts lead to the formation of the uterus, cervix, upper one-third of the vagina, and the tubes.
- Mullerian ducts from two sides join in the midline in the caudal part to form the uterovaginal canal by the 9–10th week **(Fig. 1)**.

Flowchart 2: Development of Mullerian system.

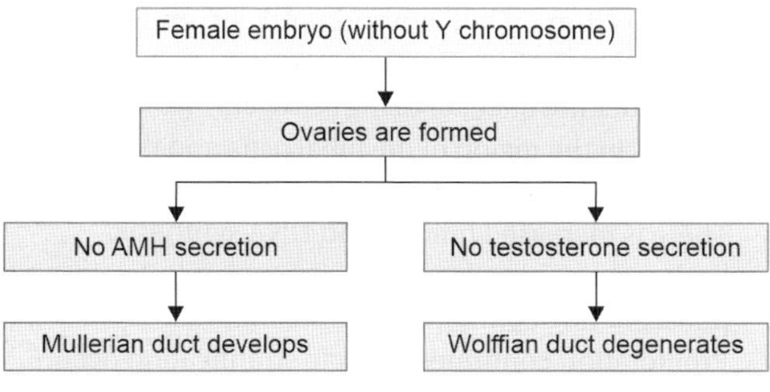

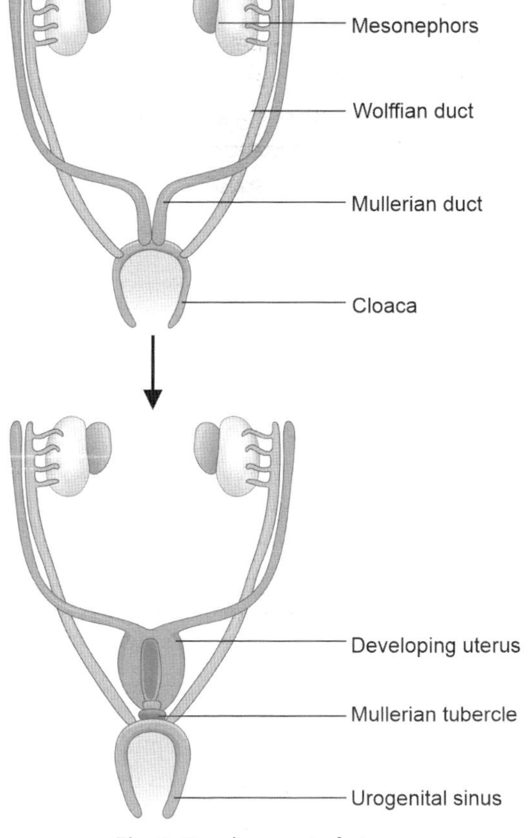

Fig. 1: Development of uterus.

- The unfused cranial parts of these ducts form the two fallopian tubes.
- Endometrium is differentiated by 20 weeks
- Canalization of the uterus, cervix, and upper part of the vagina is completed by 20–22 weeks.

VASCULATURE OF THE UTERUS

- The primary blood supply of the uterus is from the uterine artery **(Flowchart 3 and Fig. 2)**.
- Uterine artery is a branch of the anterior division of the internal iliac artery.
- Uterine artery divides into ascending branch and a descending vaginal branch before ending.
- Arcuate arteries arise from the ascending uterine artery and run parallel to the endometrial cavity.[3]
- Radial arteries arise from arcuate arteries, which run perpendicular to the endometrial cavity.
- The radial artery gives rise to the small basal artery, which supplies the basal layer of endometrium and the spiral artery, which supplies the functional layer up to the superficial endometrium.

Flowchart 3: Main blood supply (arterial) of the Uterus.

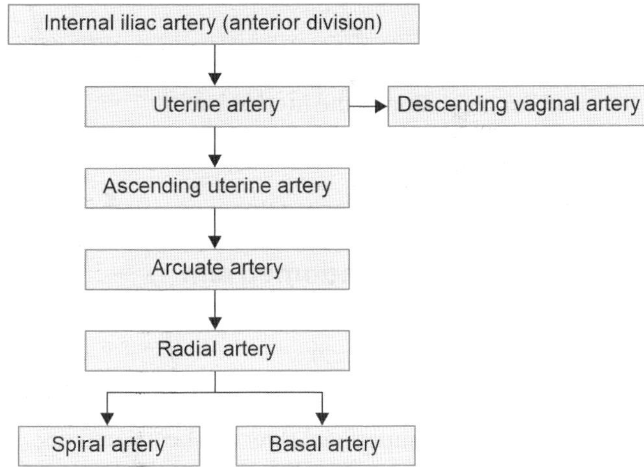

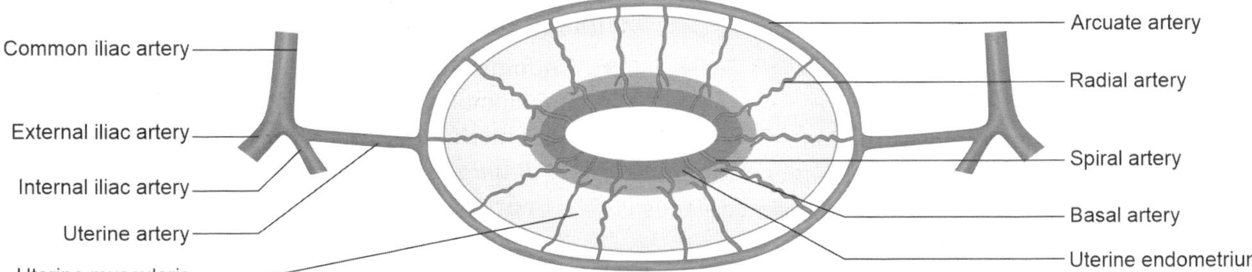

Fig. 2: Schematic representation of uterine vasculature.

PHYSIOLOGY OF MENSTRUATION

Menstruation is the cyclical shedding of the superficial functional layer of the endometrium. It is the external manifestation of the three closely synchronized cycles, which are:

1. Uterine cycle (endometrial cycle)
2. Ovarian cycle
3. Endocrine cycle (HPO axis)

Histological Changes in the Endometrium during the Menstrual Cycle

Out of the three layers of the uterus (serosa, myometrium, and endometrium), the endometrium is the most dynamic layer. It undergoes cyclical changes throughout the reproductive life of a female.

The endometrium has two parts[4]:

1. *Basal layer:*
   - It is the deeper one-third part of endometrium, which is in relation to the myometrium.
   - It is nonresponsive to cyclical hormonal changes.
   - It is not shed during the menstrual cycle.
   - Regeneration of superficial endometrium occurs from this layer.
2. *Functional layer:*
   - It is the superficial two-third part of the endometrium in relation to the uterine cavity.
   - It undergoes cyclical changes in response to hormones.
   - It is shed during menstruation.

Different Phases of Endometrium during Menstruation

Menstrual Phase (Day 0–2)

- Menstrual endometrium is composed of the basal layer only as the superficial layer has been shed off.
- It also has some fragmented glands, vessels, leukocytic and RBC infiltrates.
- Its thickness is just 0.5 mm.

Regeneration Phase (Day 3–5)

- It starts within 48 hours of menstruation.
- Regeneration occurs from the stem cells present around the blood vessels and the cells of the fragmented glands in the basal layer.
- Regeneration starts as a healing response to injury at first and continues due to increasing estrogen levels.

Proliferative Phase (Day 5–14)

- Proliferation of all components of endometrium (epithelium, glands, stroma) occurs.
- It is under the effect of rising estrogen.
- It starts in the region of the tubal ostia and the isthmus.
- Reepithelization is completed by day 5-6.
- There is increased mitosis and DNA synthesis.
- Endometrial glands are narrow and tubular in early phase. Lining epithelium is low columnar.
- Proliferation continues, pseudo stratification occurs.
- Adjacent glands get linked to each other.
- Spiral vessels are unbranched, not coiled. They form a capillary network under the epithelium.
- Stroma becomes loose and expanded. There is diffuse distribution of lymphocytes and macrophages.
- The thickness of the endometrium increases to 3.5-6 mm.

Secretory Phase (Day 15–21)

- It is under the effect of mainly progesterone after ovulation.
- There is a decrease in DNA synthesis and mitosis.
- Epithelial proliferation stops 3 days after ovulation.
- Endometrial glands grow in length and become highly tortuous.
- Intracytoplasmic glycogen vacuole appears in the glandular cells on days 17-18 (first histological sign of ovulation). The vacuole is subnucleolar initially and then moves toward the cell surface.
- Glycogen vacuole releases its secretions in the lumen of the glands by day 21-22 and glands give the sawtoothed appearance.
- The endometrial secretions are formed from active secretion of glycoproteins and peptides, transudation from plasma and immunoglobulins from circulation.
- Spiral vessels become highly coiled.
- Thickness of endometrium is up to 6 mm.

Implantation Phase (Day 21–25)

- Endometrium prepares itself so that it is ready for implantation if fertilization occurs in this cycle.
- Endometrium develops three different zones:
  1. *Basal zone (lower 1/4):* This layer does not undergo any change.
  2. *Spongy layer (middle 2/4):* It has dilated tortuous glands, tightly coiled vessels, and loose edematous stroma.

3. *Compact layer (upper 1/4):* Stromal cells become large, polyhedral and abut on each other to form a compact layer beneath the epithelium. These are pre-decidual cells.

- Spiral vessels and capillary network are engorged.
- Marked infiltration of stroma occurs with leukocytes, including macrophages and natural killer cells.

Phase of Endometrial Breakdown

- In the absence of fertilization, the corpus luteum degenerates.
- Levels of progesterone and estrogens wane off, resulting in the endometrial breakdown, breach in superficial vessels and capillaries, matrix degeneration, and interstitial hemorrhage.
- The thickness of endometrium is around 1 mm.

Causes are:

- Ischemia: Shrinkage of the endometrium occurs and blood flow in spiral arteries decreases. Venous drainage is hampered. Prolonged vasoconstriction causes ischemia and tissue necrosis.
- Intracellular release of lysosomal lytic enzymes into stromal, epithelial, and endothelial cells is there. It leads to autodigestion of cells. Prostaglandins are released which cause tissue necrosis, extravasation of blood, and apoptosis.
- Metalloproteinases (MMPs) are released which degrade the extracellular matrix.
- Increase in tumor necrosis factor-α (TNF-α) induces apoptosis and damages vascular endothelium.

CESSATION OF MENSTRUAL BLEEDING

The following mechanisms play a role:

- Formation of thrombin platelet plugs in superficial vessels.[5]
- Stromal cells secrete tissue factor (TF) and plasminogen activator and inhibitor. The balance between these factors determines the amount of blood loss. TF stimulates coagulation. PAI-1 limits fibrinolysis and proteolytic activity.
- Constriction of spiral arteries by myofibroblasts surrounding them.
- Healing occurs with the start of reepithelization.

REGULATION OF MENSTRUATION

The menstrual cycle requires a coordinated interplay of endocrine, autocrine, and paracrine factors at three essential levels:

1. Hypothalamus
2. Pituitary
3. Ovary

The hypothalamus serves as the master neuroendocrine gland, which is connected to the pituitary below and higher cortical centers above.

Gonadotropin-releasing hormone (GnRH) from hypothalamus acts on the pituitary gland in the sella turcica.

Ovaries, in turn, are controlled by hormones from the pituitary.

The endometrium is directly affected by steroid hormones (estrogen and progesterone) secreted from the ovaries.

The proliferative phase of the endometrium corresponds to the follicular phase (preovulatory) of the ovary and is under the control of estrogen.

The secretory phase of endometrium corresponds to the luteal phase (postovulatory) and is mainly under the influence of progesterone.

Hence, menstruation requires an intact and well-coordinated hypothalamus–pituitary–ovarian uterine axis (HPO axis) so that these cyclical changes continue for around 400 cycles in the reproductive life of a woman **(Fig. 3)**.

There exist positive and negative feedback loops at different levels of the HPO axis for coordination.

The arcuate nucleus of the hypothalamus starts secreting GnRH at puberty in a pulsatile fashion.

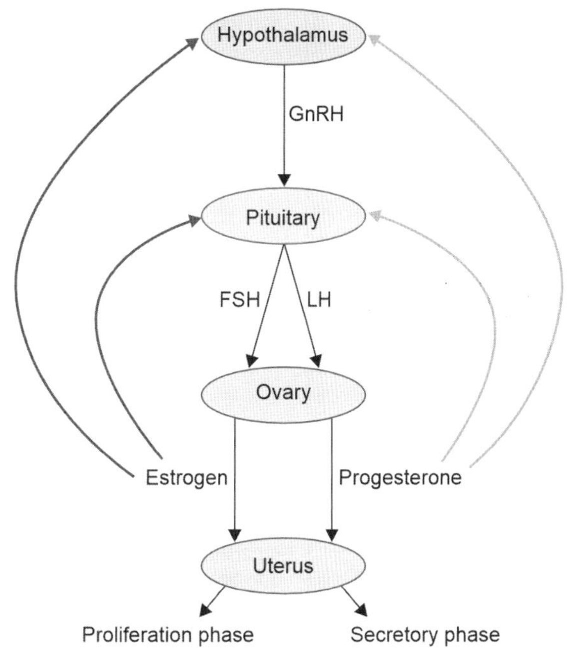

Fig. 3: Hypothalamus–pituitary–ovarian axis.

GnRH is a decapeptide (10 amino acids) with a short half-life of 2–4 minutes.

GnRH reaches the anterior pituitary (adenohypophysis) through the portal system of vessels.

GnRH is the single hormone that controls the release of FSH and LH from the anterior pituitary.

The pulses of GnRH vary in different phases of the menstrual cycle, leading to either FSH or LH secretion.[6] The pulses of high frequency and low amplitude (1 in every 1 hour) cause release of FSH and pulses of low frequency and high amplitude (1 in every 3 hours) cause release of LH.

Control of GnRH secretion is by different mechanisms:

- Autoregulation: Ultrashort feedback loop
- Short feedback loop: Gonadotropins (LH and FSH) from the anterior pituitary exert negative feedback on GnRH secretion
- Long feedback loop: Sex steroid hormones from the ovary
- Neurotransmitters and neuromodulators
- Peptides

Anterior Pituitary (Adenohypophysis)

The pituitary gland, located in the sella turcica at the base of the skull, has two parts—anterior and posterior.

Gonadotropins (FSH and LH) are secreted by the anterior pituitary's basophil cells (β cells).

The FSH acts on the ovaries and is responsible for the recruitment of a cohort of follicles for the next cycle, the growth of primordial and dominant follicle in the follicular phase (preovulatory phase).

The FSH stimulates the release of estrogen from the ovary. Estrogen is synthesized by the granulosa cells of the developing follicles by aromatization of the androgens. This estrogen acts on the endometrium and leads to regenerative and proliferative changes **(Fig. 4)**.

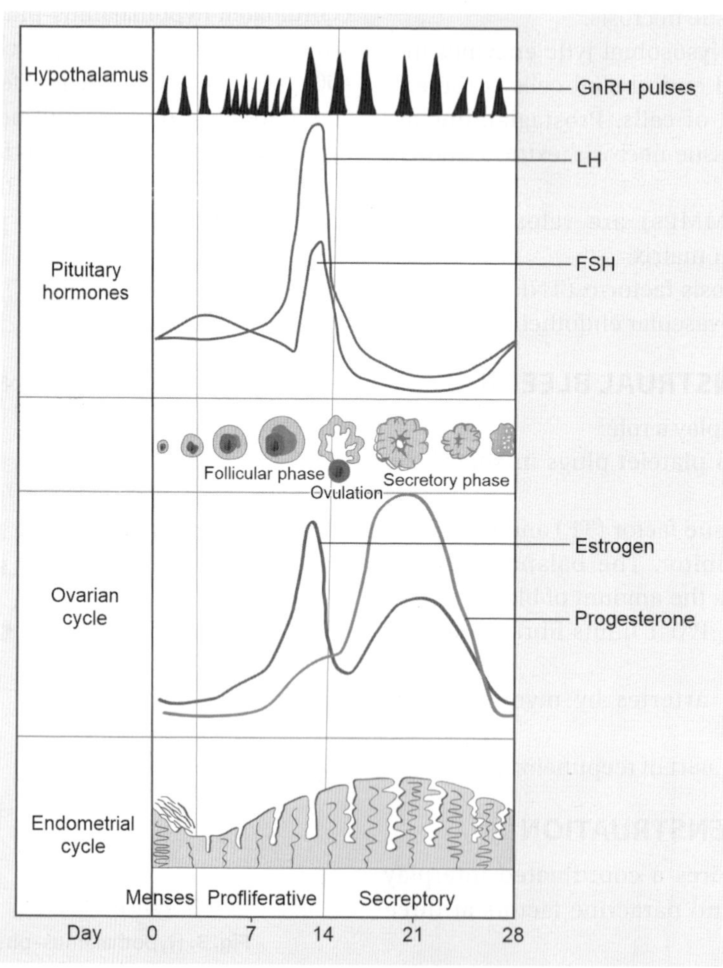

Fig. 4: Hormonal patterns at different phases of menstrual cycle.

The increasing estrogen and inhibin from ovary levels exert a negative feedback effect on the hypothalamus and pituitary and FSH level falls.

On the other hand, LH levels start rising due to the positive feedback effect of estrogen.

Peak levels of estrogens occur 48 hours before ovulation and LH surge starts 24–36 hours before ovulation.

The LH causes full maturation of Graafian follicle and oocyte.

It causes luteinization of granulosa cells which secrete progesterone.

A small preovulatory rise of progesterone causes a small rise in FSH due to positive feedback.

The LH surge with FSH rise leads to ovulation.

Corpus luteum is formed and its luteinized theca cells secrete mainly progesterone and some estrogen.

The endometrium undergoes secretory changes under the effect of progesterone.

If fertilization does not happen, the corpus luteum matures by day 22 and then degenerates.

Levels of progesterone and estrogen decline. Withdrawal of hormones results in breakdown of the superficial layer of endometrium and menstruation starts.

It exerts a positive feedback effect on the hypothalamus and pituitary and levels of FSH start rising and a new cascade of events starts.

CONCLUSION

The precisely coordinated endometrial cycle with the pituitary, ovarian, and hypothalamic axis is a prerequisite for a regular menstrual cycle. Disturbances at any of these levels lead to reproductive dysfunction and menstrual disorders.

REFERENCES

1. Taylor HS, Pal L, Seli E (Eds). The Uterus, Endometrial Physiology, and Menstruation Normal and abnormal sexual development. In Speroff's Clinical Gynecologic Endocrinology and Infertility, 9th edition. Lippincott Williams & Wilkins; 2020. pp. 593-617.
2. Sadler TW. Langman's Medical Embryology, 14th ed., Philadelphia: Wolters Kluwer; 2019. pp. 286-91.
3. Farrer-Brown G, Beilby JOW, Tarbit MH. The blood supply of the uterus: 1. Arterial Vasculature. BJOG Int J Obstet Gynaecol. 1970;77(8):673-81.
4. Olive DL, Palter SF, Moreno GLG. Reproductive Physiology. Berek & Novak's gynecology, 16th edition. Philadelphia: Wolters Kluwer; 2020: 282-314.
5. McCartney CR, Marshall JC. Neuroendocrinology of reproduction. In: Strauss JF, Barbieri. Yen & Jaffe's reproductive Endocrinology. Philadephia: Elsevier; 2019, 8th edition, 1, 1-28, e8.
6. Christiaens GCML, Sixma JJ, Haspels AA. Hemostasis in menstrual endometrium: a review. Obstet Gynecol Surv. 1982;37:281.

Chapter 18

Disorders Related to Menstruation

Rajlaxmi Mundhra, Anupama Bahadur

INTRODUCTION

Menstrual disorders are one of the commonest gynaecological complaints for women. Issues related to these continues from menarche to menopause in the form of missed periods or amenorrhea, abnormal uterine bleeding (menorrhagia, oligomenorrhea, polymenorrhea, hypo menorrhea), dysmenorrhea and premenstrual syndrome.

AMENORRHEA

Amenorrhea means absence of menstruation.

- Normal menstruation requires co-ordination in four main compartments:[1]

1. *Central nervous system and hypothalamus:* The gonadotropin-releasing hormone (GnRH) must be released in a pulsatile fashion from hypothalamus.
2. *Anterior pituitary:* GnRH stimulates anterior pituitary to secrete follicle-stimulating hormone (FSH) and luteinizing hormone (LH).
3. *Ovary:* FSH and LH promote ovarian follicular development and ovulation.
4. *Uterus/outflow tract:* A normal ovarian follicle secretes estrogen and after ovulation, it gets converted into corpus luteum and also secretes progesterone. These two hormones act on endometrium and if pregnancy does not occur, their withdrawal results in bleeding.

Defect in any of these compartments (hypothalamus, pituitary, ovary, uterus, and outflow tract) results in abnormal or absent menstruation.

- *Classification of amenorrhea* ***(Flowchart 1)****:*
- *Causes of amenorrhea*[2] ***(Fig. 1)****:*

Evaluating a case of primary amenorrhea[2-4] *(rule out pregnancy first):*

Most important step is to look for the presence and absence of uterus.

First step in diagnostic workup is to evaluate for secondary sexual characters and get FSH and LH done to

Flowchart 1: Classification of amenorrhea.

- Types
  - Physiological
    - Pre puberty
    - Pregnancy
    - Menopause
  - Pathological
    - • *Primary:* No menses by 13 years of age in the absence of secondary sexual characteristics
      • No menses by 15 years of age in presence of secondary sexual characteristics
    - • *Secondary:* No menses for at least 6 months or for a duration equivalent to at least three previous menstrual cycles

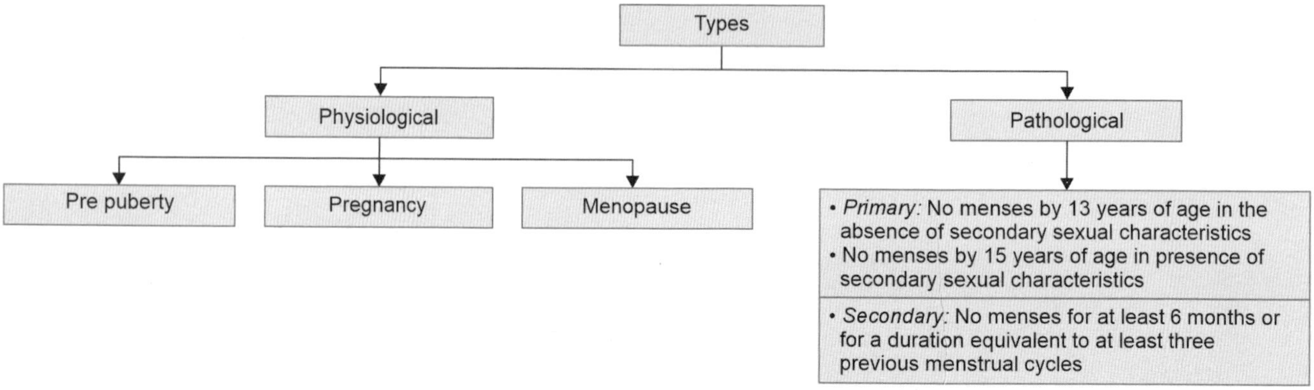

distinguish hyper- and hypogonadotropic hypogonadism **(Box 1; Flowchart 2)**.

Evaluating a case of primary amenorrhea due to outflow tract obstruction ***(Flowchart 3 and Box 2)****:*

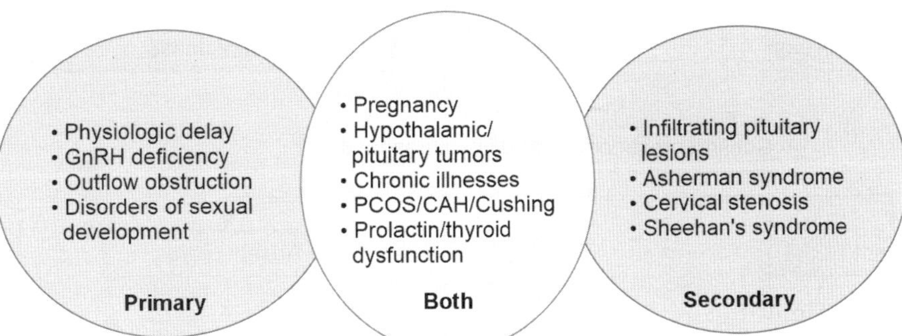

Fig. 1: Causes of amenorrhea.

BOX 1: Causes of hypo- and hypergonadotropic hypogonadism.

- *Hypogonadotropic hypogonadism:* Delayed puberty, isolated GnRH deficiency, Kallmann's syndrome, GnRH deficiency due to weight loss, excessive exercises, anorexia nervosa, chronic illnesses, and stress
- *Hypergonadotropic hypogonadism:* Gonadal dysgenesis (Turner syndrome), premature ovarian failure (autoimmune/chemotherapy and radiotherapy induced/galactosemia/genetic/mumps/enzyme deficiency due to 17-hydroxylase deficiency syndrome)

Flowchart 2: Basic diagnostic work up of a case of primary amenorrhoea.

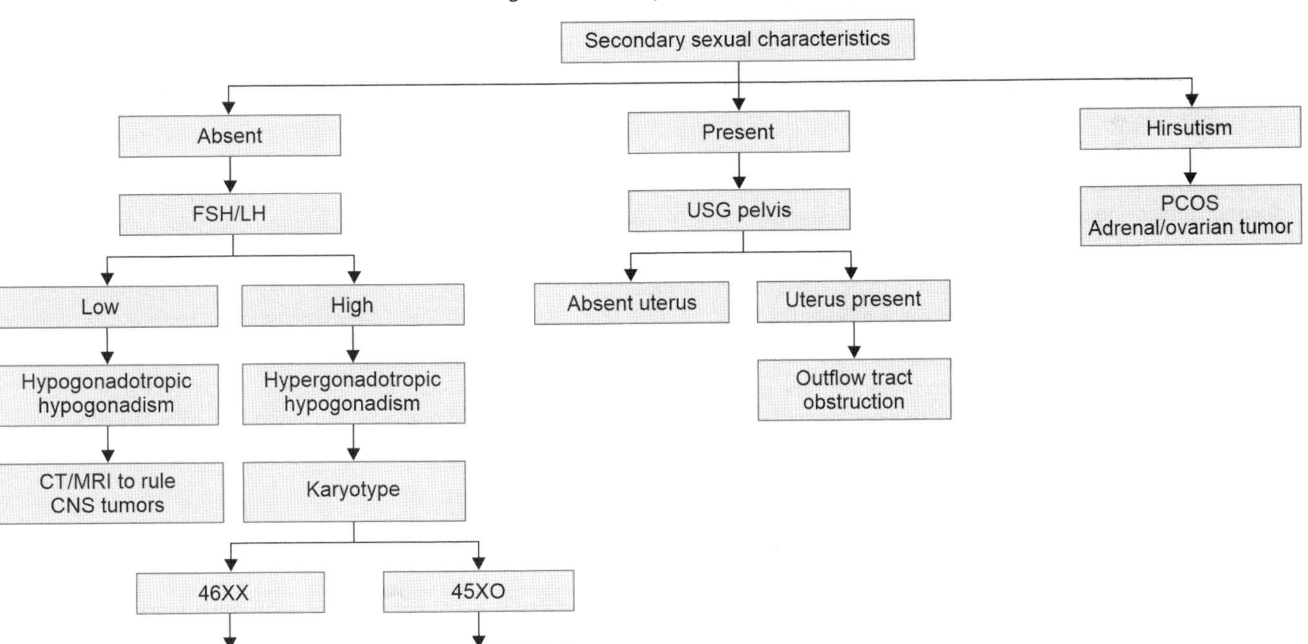

Note: If FSH is high and karyotype is normal, rule out 17α-hydroxylase deficiency as it may be a life-threatening condition. Symptoms of 17α-hydroxylase deficiency include hypertension, hypokalemia, and abnormal sexual development.[2]

(FSH: follicle-stimulating hormone; LH: luteinizing hormone; PCOS: polycystic ovarian syndrome; USG: ultrasonography)

Flowchart 3: Evaluating a case of primary amenorrhoea due to outflow tract obstruction.

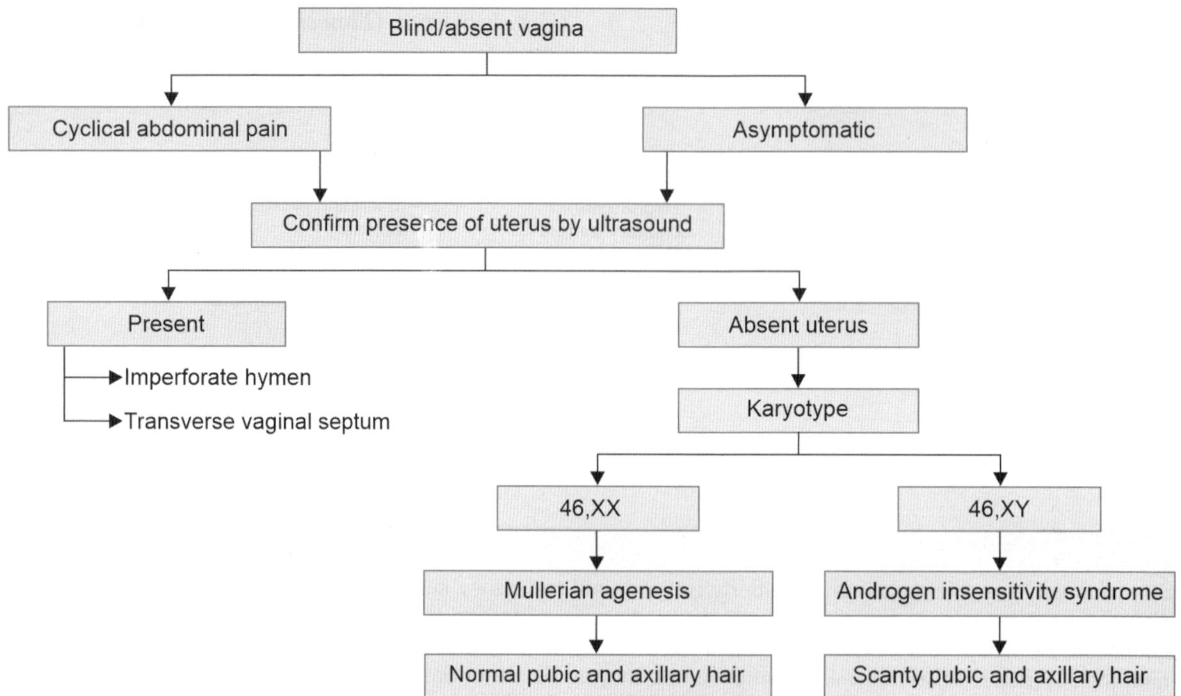

Note: Imperforate hymen has tense bluish membrane at introitus which becomes prominent with Valsalva maneuver. No such effect seen in transverse vaginal septum.

BOX 2: Treatment of Primary amenorrhea.

Focus on underlying cause and treat accordingly. Counseling forms a major step

Developmental anomalies:
- Imperforate hymen—cruciate incision to open the vaginal orifice
- Transverse septum—surgical removal is required

Hypothalamic pituitary ovarian axis defects:[2]
- Cyclical estrogen and progestogen therapy for the initiation, maturation, and maintenance of secondary sexual characteristics. Start with 0.3–0.625 mg per day of oral conjugated estrogens or 0.5–1 mg per day of oral estradiol along with MPA at a dosage of 5–10 mg for 12–14 days per month
- Craniopharyngiomas need surgery
- Hyperprolactinemia and prolactinomas require treatment with dopamine agonists
- Hormone replacement therapy/GnRH for Kallmann syndrome or other etiologies for hypothalamic amenorrhea

Chromosomal abnormalities:[5]
- XY gonads need to be removed due to increased risk of gonadoblastoma and subsequent gonadal germ cell tumors; and cyclical estrogen and progestogen therapy should be started
- Those without uterus should be counseled that though pregnancy is not possible but sexual function can be achieved through vaginoplasty done prior to marriage

Enzyme deficiency:
17α-hydroxylase deficiency—corticosteroid replacement and estrogen

Evaluating a case of secondary amenorrhea ***(Flowchart 4 and Box 3)****:*

Evaluating a case of secondary amenorrhea with abnormal TSH and prolactin ***(Flowchart 5; Boxes 4 and 5)****:*

ABNORMAL UTERINE BLEEDING (BOXES 6 TO 9)

Abnormal uterine bleeding (AUB) is defined as any bleeding from genital tract having deviation from

Flowchart 4: Evaluating a case of secondary amenorrhoea.

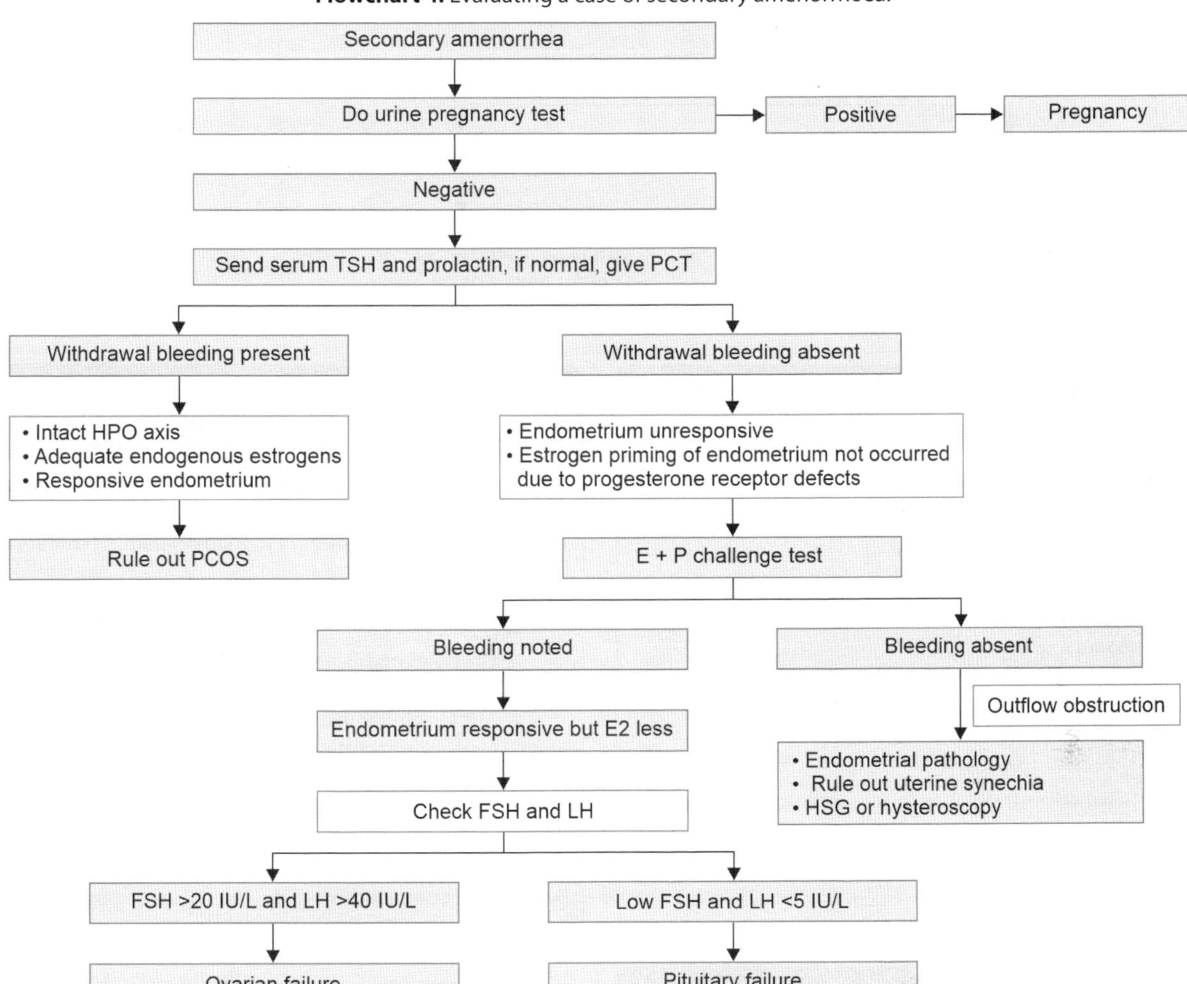

(FSH: follicle-stimulating hormone; LH: luteinizing hormone; PCOS: polycystic ovarian syndrome; PCT: progesterone challenge test; E + P: estrogen and progesterone challenge test)

BOX 3: Common regimen of PCT and E + P challenge test.

Progesterone challenge test:
- Medroxyprogesterone acetate (MPA)—10 mg once daily orally for 7–10 days
- Norethisterone 5 mg orally once daily for 7–10 days

Estrogen and progesterone challenge test:
- Conjugated equine estrogen/premarin 1.25 mg orally once daily for 21 days with MPA 10 mg orally added in 3rd week

(PCT: progesterone challenge test; E + P: estrogen and progesterone challenge test)

Flowchart 5: Evaluating a case of secondary amenorrhoea with abnormal TSH and prolactin.

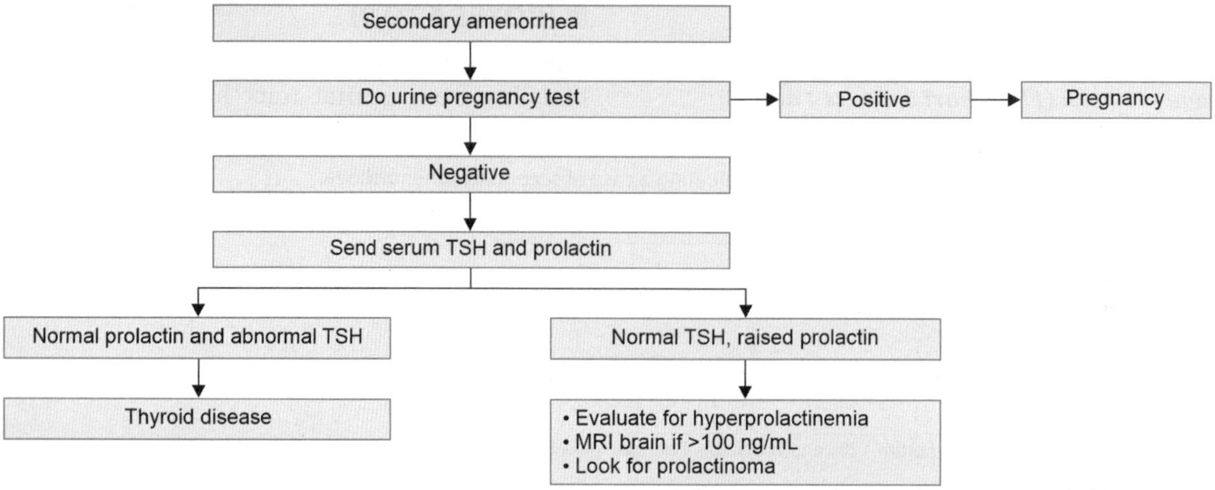

Secondary amenorrhea
↓
Do urine pregnancy test → Positive → Pregnancy
↓
Negative
↓
Send serum TSH and prolactin
↓
- Normal prolactin and abnormal TSH → Thyroid disease
- Normal TSH, raised prolactin →
  - Evaluate for hyperprolactinemia
  - MRI brain if >100 ng/mL
  - Look for prolactinoma

BOX 4: Causes of hyperprolactinemia.

Physiological:
- Pregnancy
- Lactation
- Breast stimulation
- Stress

Drugs:
- Antihypertensive
- Antidepressants
- Antipsychotics
- Oral contraceptive pills
- Dopamine receptor antagonists

Pathological:
- Hypothyroidism
- Prolactinoma

BOX 6: Normal menstrual cycle.

- *Cycle length:* 28 days
  - <21 days: Polymenorrhea
  - >35 days: Oligomenorrhea
- *Volume:* 30–80 mL
  - >80 mL: Heavy
- *Duration:* 2–8 days
  - >8 days: Prolonged
  - <2 days: Shortened

normal cyclicity, frequency, duration and flow. The term dysfunctional uterine bleeding (DUB) is no longer used.

BOX 5: *Treatment of secondary amenorrhea:* Depends on correction of underlying pathology. Counseling is the most important part with focus on fertility concerns of the women.

- *Hypothalamic pituitary dysfunction:* Lifestyle modifications, weight optimization, hormone replacement, calcium/vitamin D supplementation, surgery for hypothalamic/pituitary tumors
- *Ovarian causes:*
  - *PCOS:*
    - Lifestyle modification
    - Weight reduction
    - Metformin for insulin resistance
    - Oral contraceptive pills for cycle regularization
    - Ovulation induction for fertility concerns
    - Antiandrogens for hirsutism
  - *Premature ovarian failure:*
    - Evaluate for the cause
    - Oral contraceptive pills
    - Calcium/vitamin D supplementation
  - *Uterine causes:*
    - Tuberculosis—antitubercular therapy
    - *Asherman syndrome:* Hysteroscopic adhesiolysis
    - Cervical stenosis*:* Surgical dilatation
  - *Correct thyroid diseases and hyperprolactinemia*

So, the terminology becomes AUB-P, AUB-A, AUB-L, AUB-M, AUB-C, AUB-O, AUB-E, AUB-I, AUB-N to denote the various causes from PALM-COEIN classification **Table 1**.

Evaluation of AUB in adolescents ***(Flowchart 6):***

Evaluation of AUB in reproductive and perimenopausal age group ***(Flowchart 7):***

BOX 7: Causes of AUB in adolescents.

Irregular bleeding (timing and pattern of flow is unpredictable):
- Anovulatory due to immature HPO axis: Such pattern is common in first 2 years following menses
- PCOS: Look for other features (acanthosis nigricans/acne/hirsutism)
- Thyroid disorders
- Hyperprolactinemia

Regular menses with heavy flow:
- Bleeding disorders (Von Willebrand disease/leukemia/thrombocytopenia)
  Suspect it when:
  - Heavy menstrual bleeding since menarche
  - Family history of bleeding disorder
  - Personal history of bruising, epistaxis, gum bleeding
  - Family history of bleeding disorder
  - Bleeding necessitating blood transfusion
  - Not controlled on conventional treatment
- Drug intake—anticoagulant therapy
- Structural causes (endometrial polyp, fibroid)

Intermenstrual bleeding:
- Hormonal contraception
- Ectropion
- Sexually transmitted diseases

BOX 8: Management of acute uterine bleeding.[1]

Hemodynamically stable:
- High-dose OCPs (containing 30–35 µg ethinyl estradiol):
  1 pill TDS × 48–72 hours
  2 pills OD × 1 week
  1 pill OD × 1 week
  Stop
  Consider restarting at one pill daily

Hemodynamically unstable:
- Intravenous estrogen therapy (conjugated equine estrogen 25 mg every 4 hourly until bleeding stops usually 24 hours)
- Bleeding stops in 70% cases within 4–8 hours
- Start 2.5 mg conjugated estrogen or 2 mg micronized estradiol every 6 hourly followed by once daily dose or start OCP

(OCP: oral contraceptive pill)

BOX 9: Management of AUB.

Medical:
- Hormonal therapy (OCPs/cyclical or continuous progestin therapy/LNG-IUS)
- *Prostaglandin synthetase inhibitors:* Mefenamic acid
- *Antifibrinolytic therapy:* Tranexamic acid

Surgical: Choice of preference depends on severity of bleeding, patient preferences, and desire for future fertility
- Endometrial ablation/resection
- Conservative management of leiomyoma (hysteroscopy guided/laparoscopic or robotic myomectomy/uterine artery embolization/magnetic resonance-guided focused ultrasonography ablation)
- Hysterectomy

TABLE 1: Classification of AUB.

FIGO PALM-COEIN system:[6]	
PALM (structural causes):	*COEIN (functional causes):*
• *P:* Polyp	• *C:* Coagulopathy
• *A:* Adenomyosis	• *O:* Ovulatory dysfunction
• *L:* Leiomyoma	• *E:* Endometrial
• *M:* Malignancy + hyperplasia	• *I:* Iatrogenic
	• *N:* Not yet classified

DYSMENORRHEA (BOX 10)

Dysmenorrhea means painful menstruation **(Flowchart 8)**.

Primary dysmenorrhea: It occurs in ovulatory cycle. Following ovulation, progesterone levels rise and

Flowchart 6: Evaluation of AUB in adolescents.

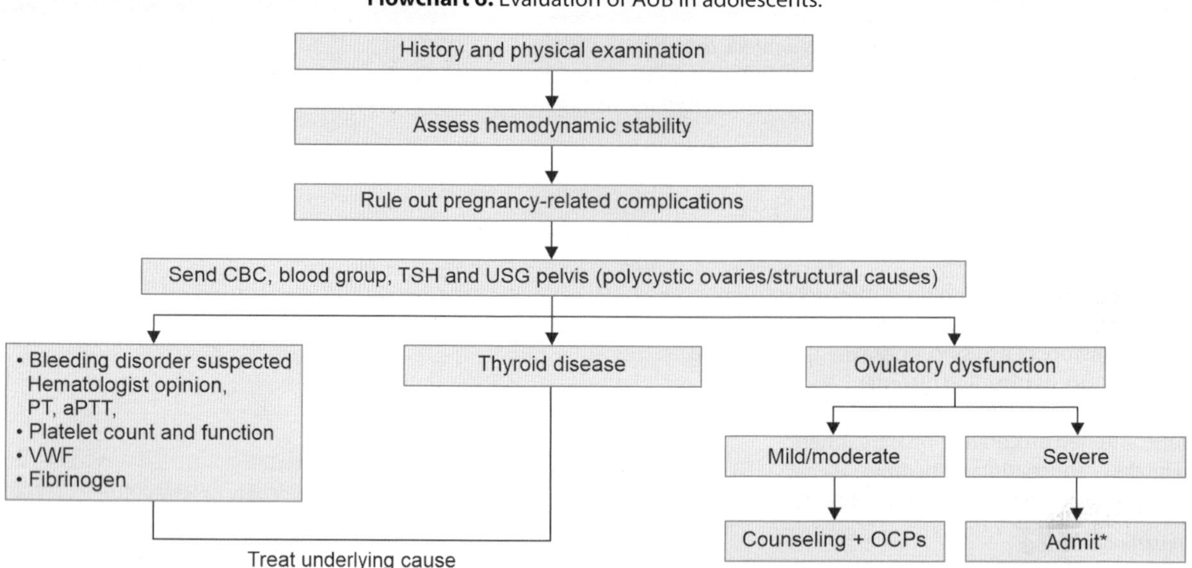

(aPTT: activated partial thromboplastin time; CBC: complete blood count; PT: prothrombin time; TSH: thyroid-stimulating hormone; USG: ultrasonography; VWF: von Willebrand factor)

Flowchart 7: Evaluation of AUB in reproductive and perimenopausal age group.

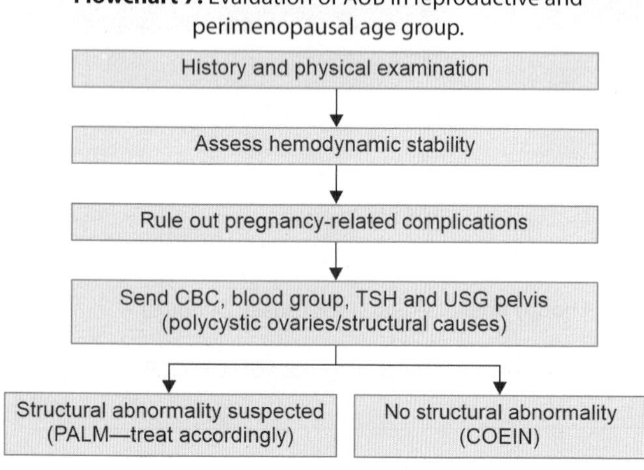

Note: Endometrial aspiration should be done in women at risk for endometrial pathology, including polyps, hyperplasia, or carcinoma.

Flowchart 8: Classification of dysmenorrhea.

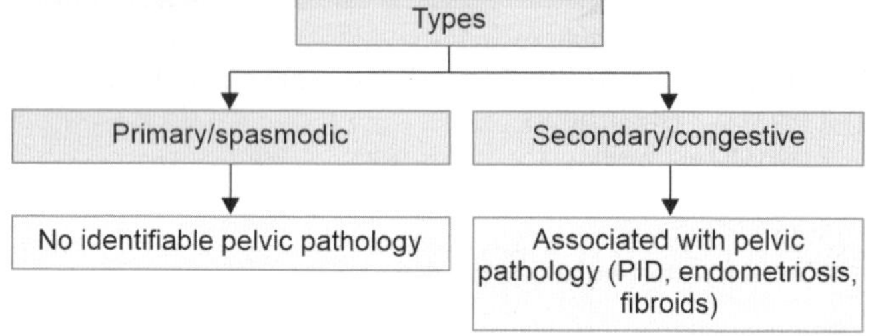

increase PGF2α in endometrium with subsequent rise in myometrial contractions, vasoconstriction, and uterine hypoxia. It manifests as cramping pains, more intense on day 1 and day 2 of cycles. The pain is suprapubic in site with radiation to back and thighs. It usually disappears a few years later especially following vaginal delivery.

Secondary/congestive dysmenorrhea: Pain usually starts a few days prior to menses and increases with onset of menses and gets relieves thereafter. Treatment consists of identifying and managing the underlying pathology.

PREMENSTRUAL SYNDROME[3,4] (BOXES 11 TO 13)

The premenstrual syndrome (PMS) is defined by presence of both physical and behavioral (including affective) symptoms occurring repetitively during the luteal phase of menstrual cycle and interferes with her quality of life.

Premenstrual dysphoric disorder is the most severe form wherein the women suffers from at least one affective symptom such as irritability, depression, and mood swings.

Diagnosis: There is no specific biochemical abnormalities associated with this disorder. Detailed menstrual history and drug history needs to be taken. Physical examination may not show any finding. It is a diagnosis of exclusion.

BOX 10: Treatment.

- Lifestyle modifications to alleviate stress and anxiety
- Nonsteroidal anti-inflammatory drugs: Prostaglandin synthetase inhibitors
  - Mefenamic acid—500 mg TDS
  - Ibuprofen—400 mg TDS
  - Naproxen—500 mg BD
- Oral contraceptive pills: Drug of choice for those desiring contraception
- *Progestins:* LNG-IUS for those desiring contraception is also an option. Depot medroxyprogesterone acetate injection at 3 months interval causes amenorrhea and relieves pain.
- *Surgery:* Laparoscopic uterine nerve ablation (LUNA) and laparoscopic presacral neurectomy can be tried but not routinely recommended.

BOX 11: Symptoms of PMS.

Somatic:
- Breast tenderness, bloating, weight gain, altered bowel habits, and headache

Affective:
- Mood swings
- Irritability
- Depression
- Crying spells
- Anxiety
- Increased appetite
- Insomnia

(PMS: premenstrual syndrome)

BOX 12: Differential diagnosis.

- Mood and anxiety disorders
- Menopausal transition
- Thyroid disorders
- Substance abuse

BOX 13: Treatment.

Counseling and supportive therapy:
- *Lifestyle modifications:*
  - Adequate rest
  - Stress reduction
  - Avoid caffeine
  - Smoking cessation
  - Regular exercise
  - Intake of nutritious diet, consisting of complex carbohydrates
- *Spironolactone in fluid retention*
- *OCPs*
- *Selective serotonin reuptake inhibitors (SSRIs):* Fluoxetine, sertraline, and paroxetine are FDA approved
- *GnRH agonists*

(GnRH: gonadotropin-releasing hormone; OCP: oral contraceptive pill)

REFERENCES

1. Taylor HS, Pal L, Seli E. Speroff's Clinical Gynecologic Endocrinology and Infertility, 9th edition. Philadelphia: Wolters Kluwer; 2020.
2. Baker VL, Beall SA. Amenorrhoea. In: Berek Jonathan S (Ed). Berek & Novak's Gynecology, 16th edition. Philadelphia: Wolters Kluwer; 2020. pp. 2034-88.
3. Konar H. DC Dutta Textbook of Gynaecology, 5th edition. Kolkatta: New Central Book Agency; 2009.
4. Sheshadri L. Essentials of Gynaecology, 2nd edition. Gurugram: Wolters Kluwer; 2017.
5. Jorgensen PB, Kjartansdóttir KR, Fedder J. Care of women with XY karyotype: a clinical practice guideline. Fertil Steril. 2010;94:105-15.
6. Munro MG, Critchley HO, Broder MS, Fraser IS; FIGO Working Group on Menstrual Disorders. FIGO classification system (PALM-COEIN) for causes of abnormal uterine bleeding in nongravid women of reproductive age. Int J Gynaecol Obstet. 2011;113(1):3-13.

Chapter 19

Physiology of Conception

Munjal Jayeshkumar Pandya, Janki Munjal Pandya

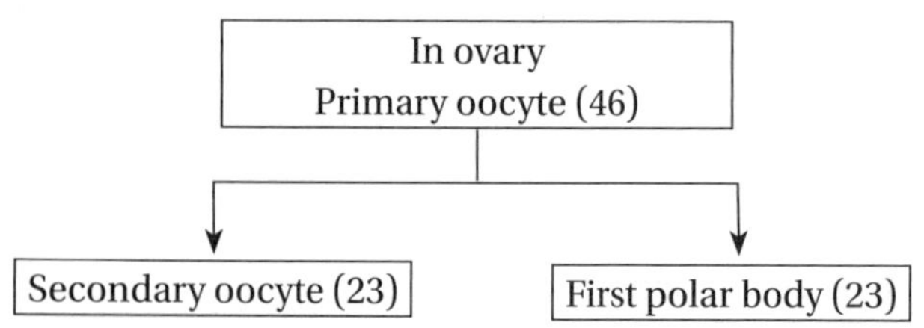

- *Ovulation:*

At the time of ovulation

↓

Secondary oocyte gets released in abdominal cavity

↓

Picked up by one of the Fallopian tubes

- *Intercourse:*

Semen enters vagina

↓

Sperms travel through vagina and uterus, reaching ampullary part of fallopian tube (Prostaglandins of semen and oxytocin from female lead to contractions of uterine musculature along with increased motility of Fallopian tube, facilitating sperm travel)

↓

Sperm capacitation (increases sperm motility) occurs, cholesterol membrane is lost and cellular membrane becomes permeable to calcium, making sperm capable for flagellar reaction[1]

↓

Fertilization

Sperm penetration through granulosa cell layer (corona radiata) and glycoprotein membrane (zona pellucida) of ovum by zona reaction

Acrosome reaction: Hyaluronidase and proteolytic enzymes in acrosome penetrate through these layers

Zona reaction: After sperm entry through zona pellucida, plasma membrane properties change, thus preventing entry of other sperms through zona pellucida

↓

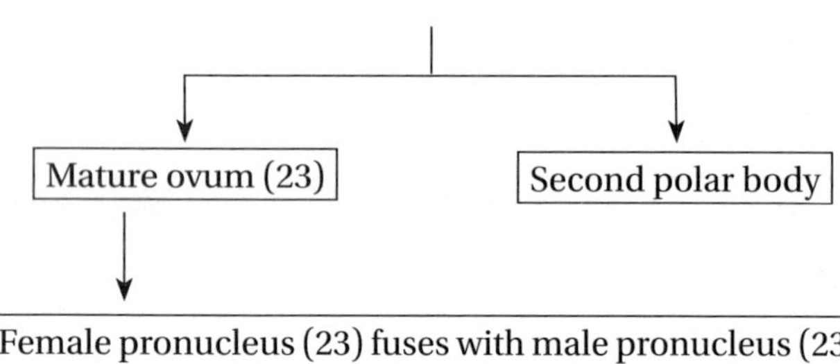

↓

Female pronucleus (23) fuses with male pronucleus (23), along with fusion of cell membrane, forming *Zygote* (46)

↓

Zygote divides into two blastomeres, which further get divided into 16 cell stage, called "Morula"

↓

Day 4 (after fertilization): Continued cell division, along with fluid accumulation between cells, leading to formation of "early blastocyst" (58 cell stage), which consists of inner cell mass (5 cells), which will form embryo and trophectoderm (53 cells), giving rise to trophoblasts

↓

"Embryo hatching": Proteases of endometrial glands separate blastocyst from zona pellucida
Blastocyst secretes hormones and cytokines, increasing endometrial receptivity
Nutrition of embryo comes from uterine secretions

↓

Day 6/7: Implantation **(Fig. 1)**
Apposition: Blastocyst gets placed fundal region of posterior uterine wall
Adhesion: Blastocyst gets adhered to uterine wall
Invasion: Penetration of blastocyst into one-third of myometrium and uterine vasculature
(Corpus luteum releases estrogen and progesterone, mediating this process)

↓

Day 8: Trophoblasts' differentiation into syncytiotrophoblasts (multinucleated cells forming outer layer of blastocyst) and cytotrophoblasts (mononuclear cells forming inner layer of blastocyst)

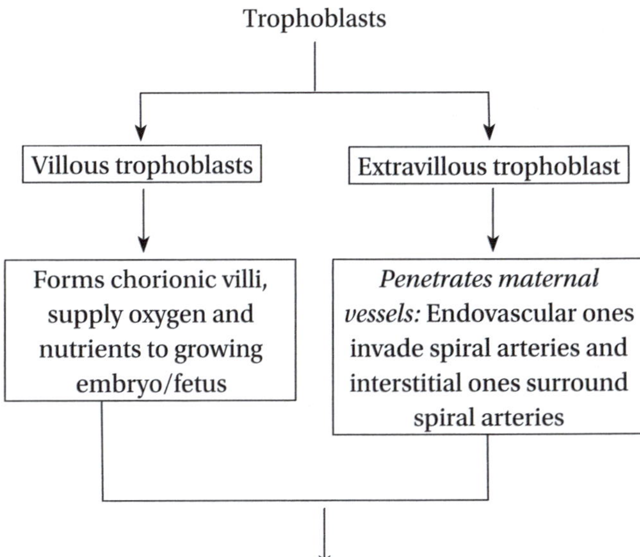

↓

Day 10: Complete implantation of blastocyst into endometrium (now called as *"Decidua"*)
Inner cell mass forms ectoderm and endoderm

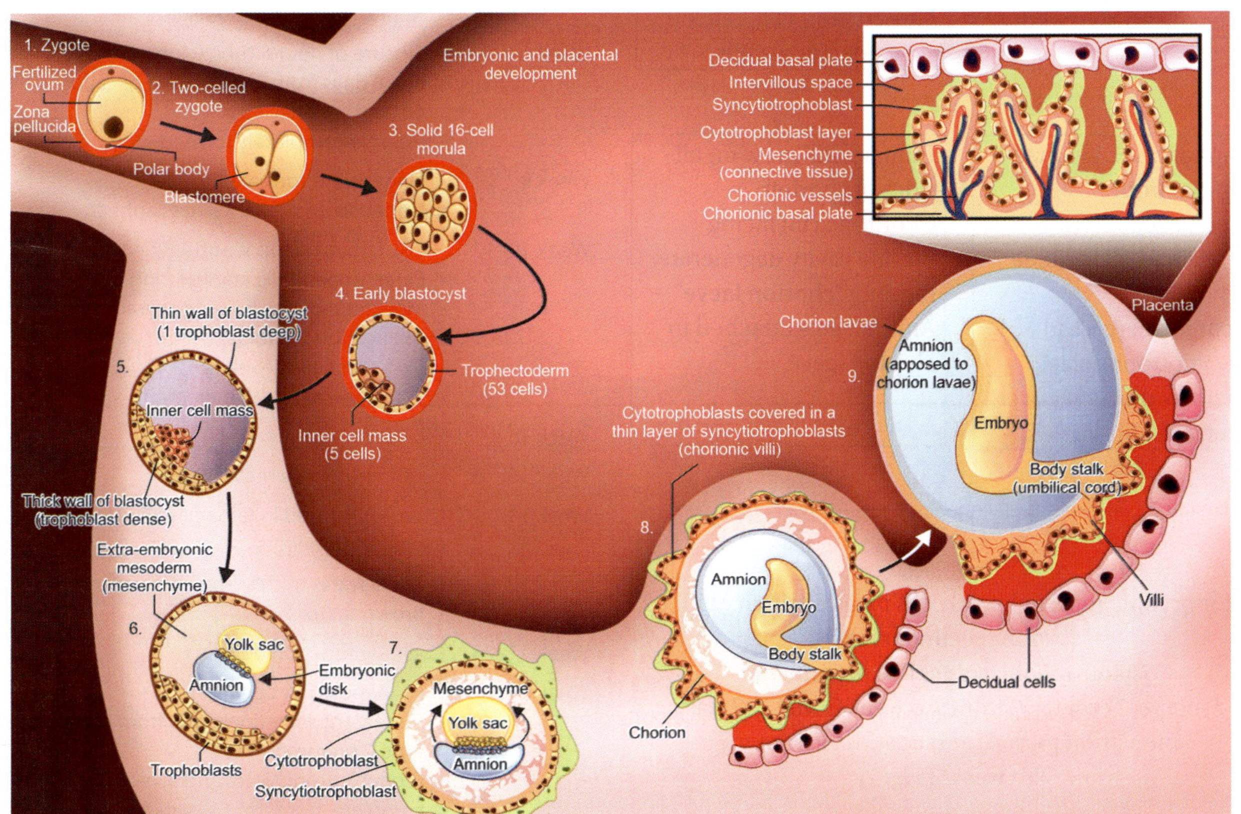

Fig. 1: Implantation.

Mesenchymal cells are formed from mesoderm, filling cavity of blastocyst
Ectoderm forms amniotic epithelium, forming amniotic sac

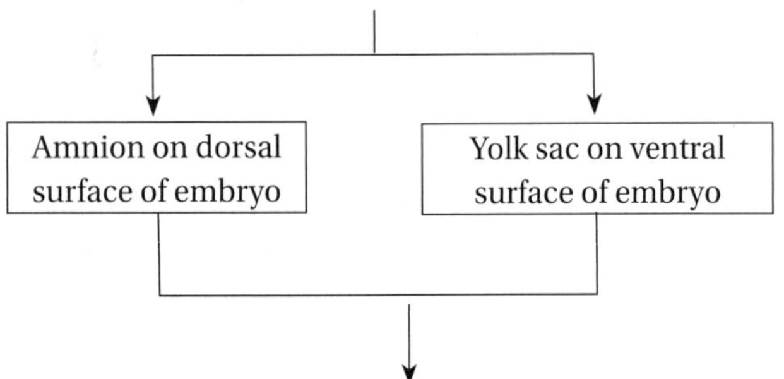

Dorsal part of yolk sac forms gut
Allantois is formed, connected with hindgut
Body stalk is formed from mesenchymal cells, allantois gets connected into it
Yolk sac: Hematopoiesis
Amnion obliterates yolk sac, covering body stalk and filles uterine space

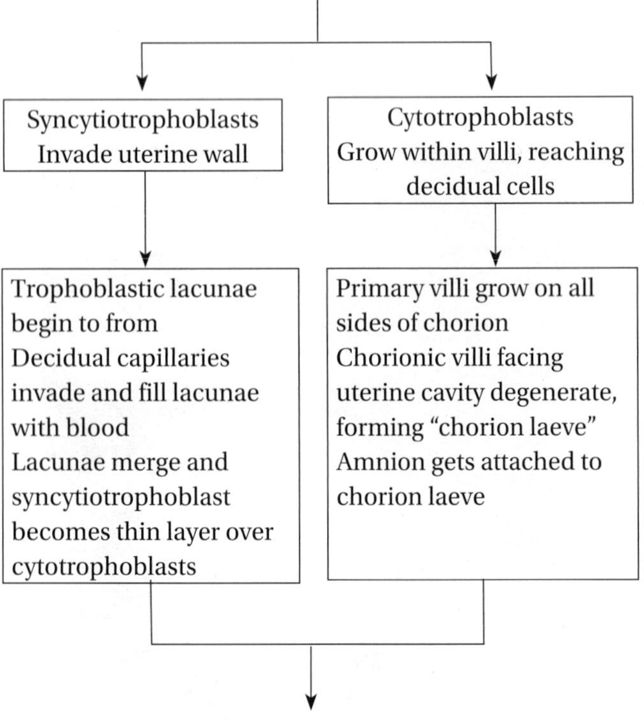

Embryonic mesenchymal cells invade primary villi, forming secondary villi
Cytotrophoblasts get less prominent
Embryo: Term used from 3rd week from LMP (last menstrual period) and continues through 8th week[2] **(Fig. 2)**

↓

Angiogenesis of mesenchymal cells, forming chorionic vessels, thus, forming tertiary villi
Fetoplacental circulation is established due to connection between chorionic vessels and embryonic vessels **(Figs. 3A to C)**
Anchoring villi are formed by extension of trophoblasts, connecting decidual plate and chorionic plate

↓

Weeks 8–10: Membranes are formed
Amnion with chorion, which in turn with decidua get attached, as embryo is almost occupying whole of uterine cavity
Amnion: Gives tensile strength to the membranes, produces amniotic fluid, which is derived from maternal plasma, and then from skin and urine of fetus
Amniotic fluid: Provides space of growth and movement of fetus

Organogenesis (Figs. 4A to E)	
Week 3	• Formation of fetal blood vessels
Week 4	• Cardiovascular system is formed • Fetoplacental circulation is established • Beginning of partition of primitive heart • Neural plate formation starts
Week 6	• Closure of neural tube (closure of cranial part by 38 days from LMP, closure of caudal part by 40 days) • Heart is completely formed
Week 8	• Fingers and toes are formed • Upper lip and external ears appear
Week 10–12	• CRL (crown rump length) is 5–6 cm • External genitalia start getting differentiated • Most fetal bones have appearance of ossification centers • Skin and nails develop • Spontaneous fetal movements start
Week 16	CRL 12 cm
Week 16–18	Beginning of eye movements
Week 18	Uterus is formed in female fetus, vaginal canalization begins
Week 20	• Testis starts to descend in male fetus • Brown fat forms
Week 22–25	Cochlear function develops
Week 24	• Fat deposition begins • Surfactant secretion starts from secretory type II pneumocytes • Bronchi, bronchioles and alveoli formation complete; but terminal sacs are not formed yet

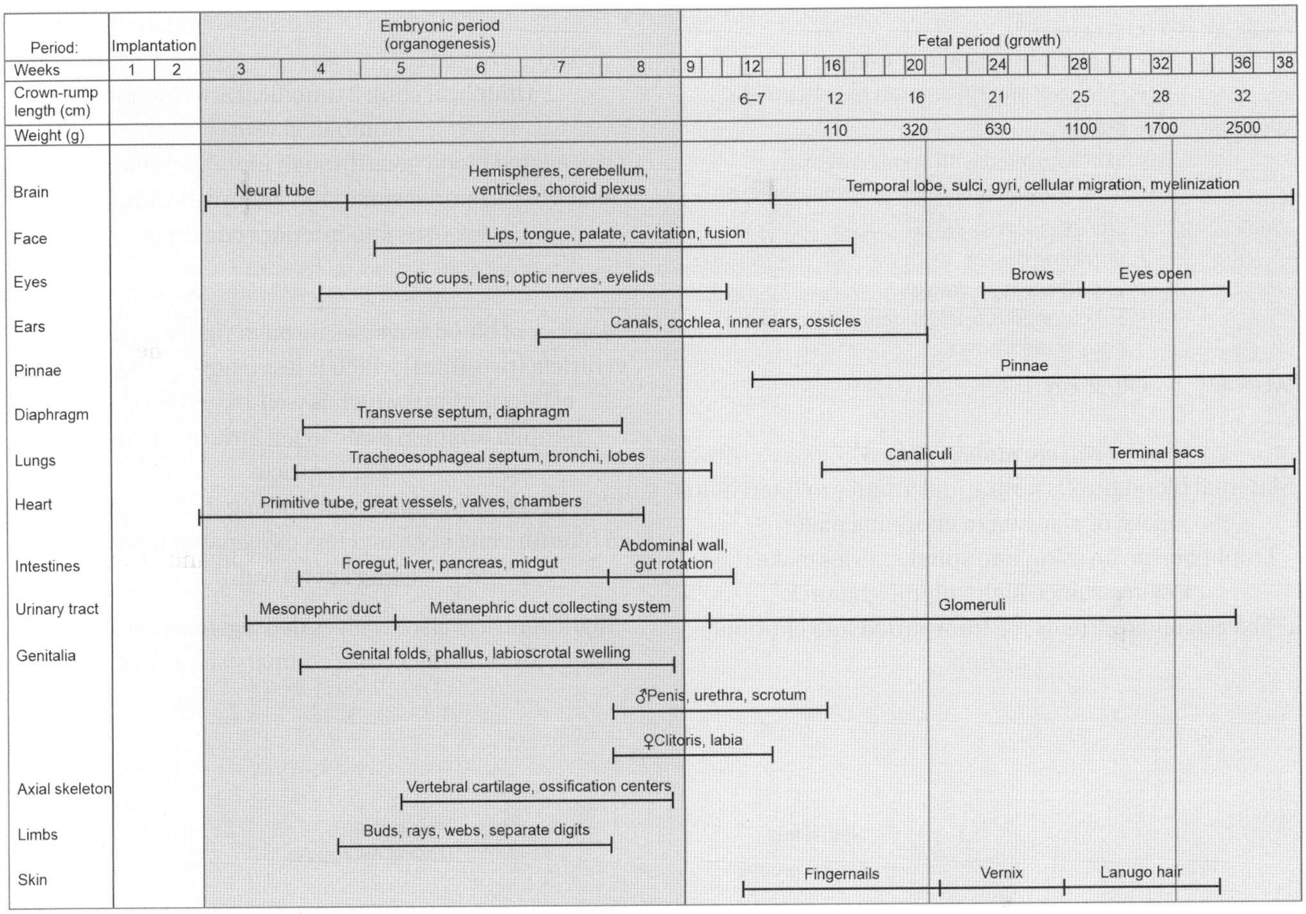

Period:
Implantation
Embryonic period (organogenesis)
Fetal period (growth)
Weeks
1 2 3 4 5 6 7 8 9 12 16 20 24 28 32 36 38
Crown-rump length (cm)
6–7 12 16 21 25 28 32
Weight (g)
110 320 630 1100 1700 2500
Brain
Neural tube
Hemispheres, cerebellum, ventricles, choroid plexus
Temporal lobe, sulci, gyri, cellular migration, myelinization
Face
Lips, tongue, palate, cavitation, fusion
Eyes
Optic cups, lens, optic nerves, eyelids
Brows
Eyes open
Ears
Canals, cochlea, inner ears, ossicles
Pinnae
Pinnae
Diaphragm
Transverse septum, diaphragm
Lungs
Tracheoesophageal septum, bronchi, lobes
Canaliculi
Terminal sacs
Heart
Primitive tube, great vessels, valves, chambers
Intestines
Foregut, liver, pancreas, midgut
Abdominal wall, gut rotation
Urinary tract
Mesonephric duct
Metanephric duct collecting system
Glomeruli
Genitalia
Genital folds, phallus, labioscrotal swelling
♂Penis, urethra, scrotum
♀Clitoris, labia
Axial skeleton
Vertebral cartilage, ossification centers
Limbs
Buds, rays, webs, separate digits
Skin
Fingernails
Vernix
Lanugo hair

Fig. 2: Development of embryo/fetus according to gestational age.

Yolk sac
Embryo
Amnion
Chorion
Allantois
Body stalk
Developing villi
A
Developing neural groove
Yolk sac
Amnion
Allantois
Chorion
Body stalk
B
C

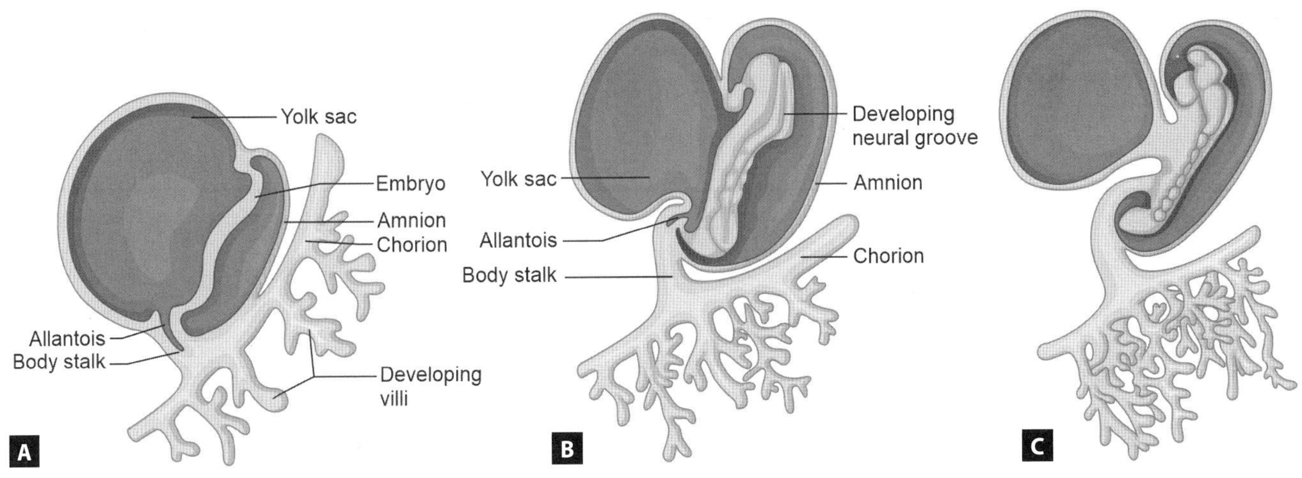

Figs. 3A to C: Early human embryos.

Week 26	• Eyes open • Neural pain system is developed • Fetal liver and spleen start hematopoiesis
Week 28	• CRL 25 cm • Skin is covered with vernix caseosa • *Bone marrow:* Major site of hematopoiesis
Week 32–36	• CRL 28 cm, at 32 weeks • Wrinkling of skin • CRL 32 cm at 36 weeks • Subcutaneous fetal fat deposition making body more rotund
Week 40	CRL 36 cm

Fetal circulation **(*Fig. 5*)**

Maternal blood travels from spiral arteries to intervillous space of placenta to chorionic plate

Trophoblastic invasion into spiral arteries make them low-resistance high-volume channels

Oxygen and nutrients are transported into chorionic vessels

↓

Umbilical cord: 2 umbilical arteries and 1 umbilical vein

Oxygenated blood goes through umbilical vein, partly to fetal liver via portal sinus, and majority through ductus venosus into inferior vena cava

↓

Majority of blood traverses through right atrium to right ventricle (two-third) which is less oxygenated, and some of it (one-third) which is well-oxygenated, goes through foramen ovale to left atrium and then to left ventricle

(Well-oxygenated blood course along dorsomedial aspect of inferior vena cava, and less oxygenated blood courses along lateral wall)

Right ventricle also receives deoxygenated blood from superior vena cava, from coronary sinus, making the

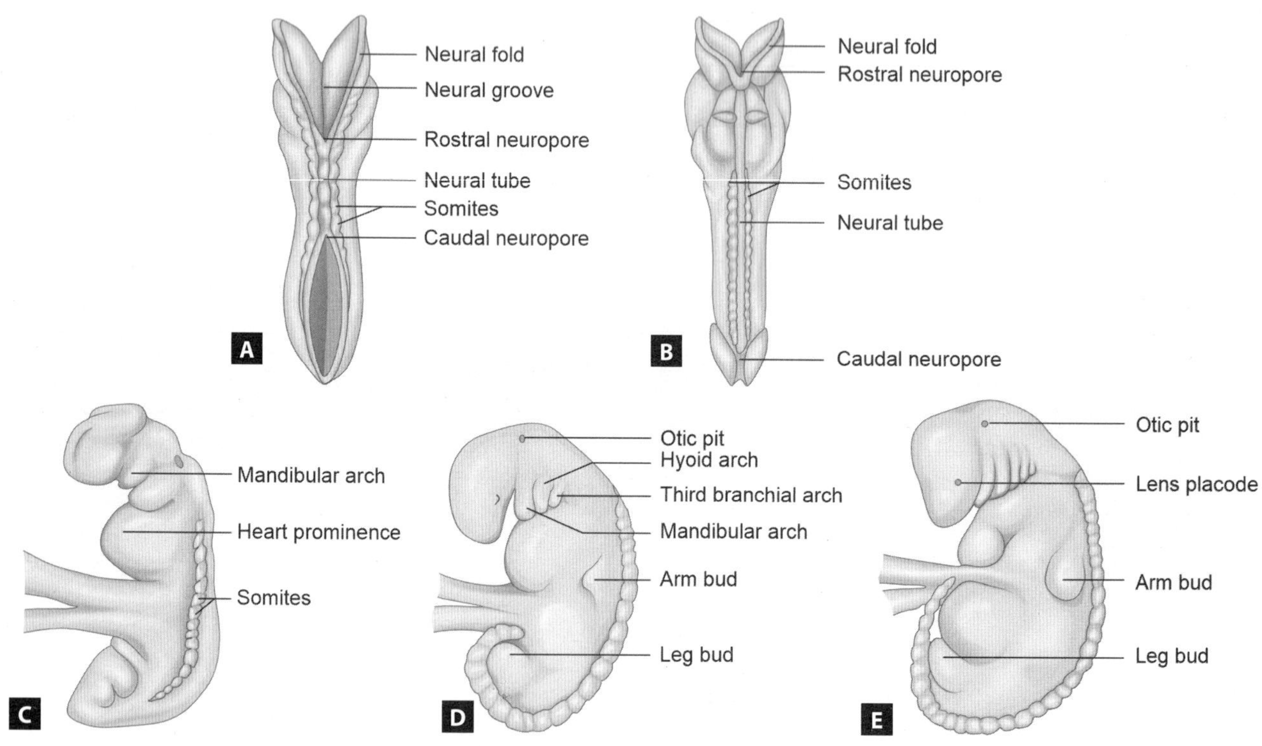

Figs. 4A to E: Embryo (3–4 weeks).

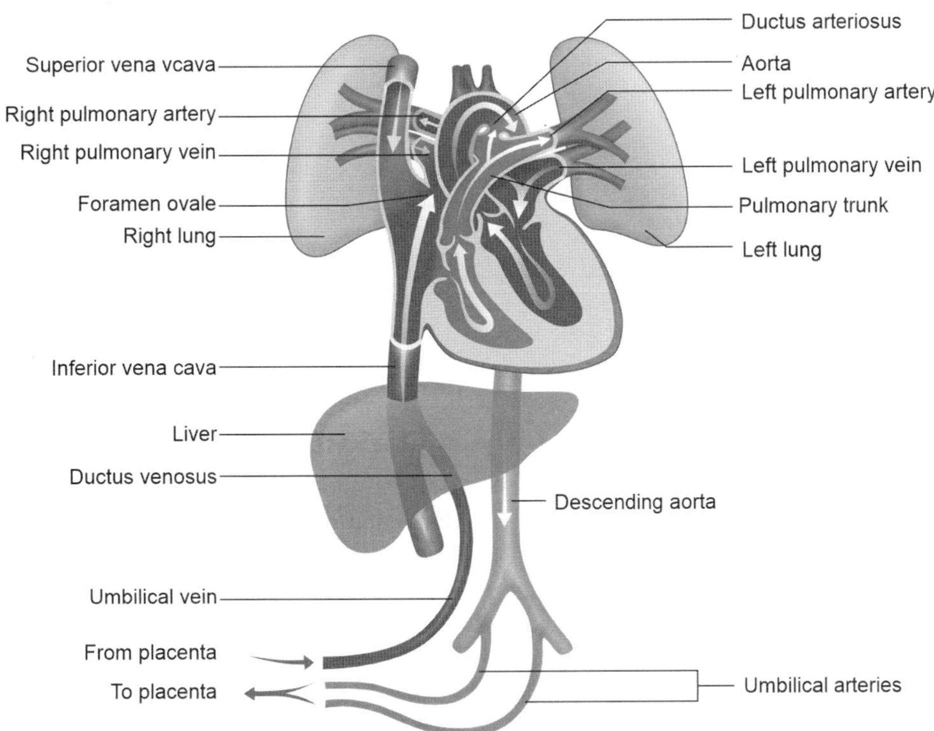

Fig. 5: Fetal circulation.

blood 15–20% less saturated with oxygen, as compared to blood in left ventricle

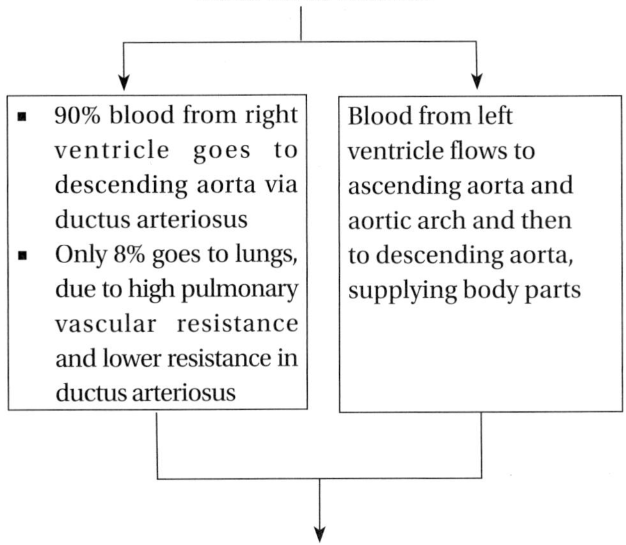

One-third of blood passing through ductus arteriosus is passed to the body

Remaining right ventricular output goes back to placenta through 2 hypogastric arteries, which are continued as umbilical arteries in umbilical cord

↓

The blood gets oxygenated in placenta and recirculated through umbilical vein

REFERENCES

1. Peltz S, Srinivasan R, Chaudhry S. Conception and Pregnancy. [online] Available from https://www.pathophys.org/conception-and-pregnancy/. [Last accessed March, 2024]
2. Cunningham FG, Leveno KJ, Dashe JS, Hoffman BL, Spong CY, Casey BM. Embryogenesis and Fetal Development. Williams Obstetrics, 26th edition. New York: McGraw Hill Medical; 2022. Chapter 7. pp. 330-80.

Chapter 20

Amenorrhea

K Aparna Sharma, Asmita Kaundal

INTRODUCTION

Complete absence of menstrual cycles in a women of reproductive age is termed as amenorrhea and is one of the commonest gynecological condition with which women present to the gynecological outpatient department. Most common cause of amenorrhea is physiological and occurs due to pregnancy, breastfeeding or menopause. Amenorrhea which is not physiological needs to be investigated. Amenorrhea is not a disease but can be an important symptom of an underlying disease. While investigating for amenorrhea history, examination and workup should be directed to rule out following important causes: outflow tract abnormalities, ovarian insufficiency/failure, hypothalamic-pituitary disorders and other endocrinological disease. Sometimes amenorrhea can be induced due to exogenous hormones and medications. Treating underlying etiology can lead to successful resumption of menstrual cycles in many women. A multidisciplinary approach is required to diagnose and treat women with amenorrhea. Since normal cyclical menstruation is considered as an indicator of normal reproductive health. Women experiencing any disturbance in menstrual cycles may be distressed and should be provided emotional and psychological support.

DEFINITION

Absence or cessation of menstruation is called amenorrhea.[1] It can be physiological in case like pregnancy, breastfeeding, or menopause. Amenorrhea can also be induced while using hormonal contraceptive or exogenous androgens. Amenorrhea, which is neither physiological nor induced, needs to be investigated. The prevalence of amenorrhea is 3–4% after excluding amenorrhea due to pregnancy, lactation, or menopause.[2,3]

CLASSIFICATION

Depending upon the onset of amenorrhea, it can be divided into primary and secondary amenorrhea. If a girl has never attained menarche it is primary and if the cessation of cycle occur after menarche it is called secondary amenorrhea. Secondary amenorrhea is seen more commonly than primary amenorrhea **(Tables 1 and 2)**.

CAUSES

Amenorrhea can be due to various causes but the most common causes seen in the routine practice are polycystic ovarian syndrome, hypothalamic amenorrhea, hyperprolactinemia, and ovarian failure[7-11] **(Table 3)**.

TABLE 1: Classification of amenorrhea.

Primary amenorrhea	• Lifelong absence of mensturation[4] • Absence of menarche by the age of 16 years in a girl with normal secondary sexual characteristics or absence of menarche by the age of 14 years in a girl with absent secondary sexual characteristic is termed as primary amenorrhea[5]
Secondary amenorrhea	Absence of menstrual cycle for 3 months in a women with previously normal cycles or for 9 month in women with oligomenorrhea is called secondary amenorrhea[5]

TABLE 2: As per WHO, amenorrhea can be divided into three categories.[6]

Type I	• No evidence of endogenous estrogen production • Normal or low FSH level • Normal prolactin • No lesion in hypothalamic–pituitary region
Type II	• Evidence of estrogen production • Normal prolactin and FSH
Type III	Elevated FSH indicating gonadal failure

(FSH: follicle-stimulating hormone)

TABLE 3: Causes of amenorrhea.

S. no.	*Condition*
1.	*Normogonadotropic:*
	• *Congenital:* Androgen insensitivity syndrome, Mullerian agenesis (Mayer–Rokitansky–Kuster–Hauser syndrome), endometrial hypoplasia, or aplasia • *Hyperandrogenic anovulation:* Acromegaly, androgen-secreting tumor (ovarian or adrenal), Cushing's disease, exogenous androgens, non-classic congenital adrenal hyperplasia, polycystic ovary syndrome, thyroid disease • *Outflow tract obstruction:* Asherman's syndrome, cervical stenosis, cervical agenesis, vaginal agenesis, imperforate hymen, transverse vaginal septum • *Other:* Pregnancy, thyroid disease
2.	*Hypogonadotropic hypogonadism:*
	• *Anorexia or bulimia nervosa* • *Central nervous system tumor* • *Constitutional delay of growth and puberty* • *Chronic illness:* Chronic liver disease, chronic renal insufficiency, diabetes • *Immunodeficiency* • *Inflammatory bowel disease* • *Thyroid disease* • *Severe depression or psychological stressors* • *Cranial radiation* • *Excessive exercise* • *Excessive weight loss or malnutrition* • *Hypothalamic or pituitary destruction* • *Kallmann syndrome* • *Sheehan's syndrome*
3.	*Hypergonadotropic hypogonadism:*
	• *Gonadal dysgenesis:* Turner's syndrome • *Postmenopausal ovarian failure* • *Premature ovarian failure:* Autoimmune, chemotherapy, galactosemia, genetic, 17-α hydroxylase deficiency syndrome, idiopathic, mumps, and pelvis radiation
4.	*Hyperprolactinemia:*
	• *Prolactin ≤100 ng per mL (100 µg per L)* • *Altered metabolism:* Liver failure, renal failure • *Ectopic production:* Bronchogenic (carcinoma), gonadoblastoma, hypopharynx, ovarian dermoid cyst, renal cell carcinoma, and teratoma • *Breastfeeding* • *Breast stimulation* • Hypothyroidism • *Medications:* Oral contraceptive pills, antipsychotics, antidepressants, antihypertensive, histamine H2 receptors blockers, opiates, and cocaine • *Prolactin >100 ng per mL* • *Empty sella syndrome* • *Pituitary adenoma*

EVALUATION AND DIAGNOSIS

A detailed history and thorough physical examination is the first step in the evaluation of amenorrhea.

History

A detailed history should include the details about onset of amenorrhea and duration to know whether it is primary or secondary amenorrhea. In case of primary amenorrhea history of any cyclical abdominal pain, breast tenderness should be asked to rule out possibility of outflow tract obstruction. In case of secondary amenorrhea age of menarche, previous cycle length, regularity/pattern of the cycle should be enquired. History related to fertility, i.e., previous pregnancies (if any), any problem in conception and treatment required. History of significant weight

TABLE 4: Salient features in history and examination and their association with type of amenorrhea.

Findings	***Associations***
Menarche and menstrual history	Primary versus secondary amenorrhea
Sexual activity	Pregnancy
Excessive exercise, weight loss, illicit drug use, current or previous chronic illness, previous central nervous system chemotherapy or radiotherapy	Hypothalamic amenorrhea
Previous pelvic irradiation, vasomotor symptoms like hot flushes, night sweats	Premature ovarian failure
Anosmia	Kallmann syndrome
Illicit drug use	Multiple associations, hyperprolactinemia
Cyclic abdominal pain, breast changes	Outflow tract obstruction or Mullerian agenesis
History of dilatation and curettage	Asherman's syndrome
Galactorrhea, headache, and visual disturbances	Pituitary tumor
Hirsutism, acne	Polycystic ovarian syndrome, hyperandrogenism, Cushing's syndrome, adrenal or ovarian tumors, congenital adrenal hyperplasia
Heat/cold intolerance, palpitation, skin changes, diarrhea or constipation	Thyroid disease
History of postpartum hemorrhage in previous pregnancy	Sheehan's syndrome
Family history: Menarche and menstrual history, puberty history	Constitutional delay of growth and puberty
Physical examination:	
Growth chart, BMI	Constitutional delay of growth and puberty
BMI: High *BMI:* Low	PCOS Hypothalamic amenorrhea
Male pattern baldness, increased facial hairs and acne	Hyperandrogenism, congenital adrenal hyperplasia, Cushing syndrome, PCOS, ovarian or adrenal tumor
Webbed neck, short stature, widely spaced nipples	Turner's syndrome
Rudimentary or absent uterus; pubic hair	Mullerian agenesis
Central obesity, blue/purple striae, buffalo hump	Cushing's disease
Thyroid examination	Thyroid disease
Imperforate hymen, transverse vaginal septum	Outflow tract obstruction
Virilization, clitoral hypertrophy	Androgen secreting tumor, congenital adrenal hyperplasia, 5α-reductase deficiency
Undescended testes, external genital appearance	Androgen insensitivity syndrome
Acanthosis nigricans, skin tags	Hyperinsulinemia (PCOS)
Thin vaginal mucosa, reddened vaginal mucosa	Decreased endogenous estrogen

(PCOS: polycystic ovary syndrome)

changes (gain/loss), eating habits, psychological health, stressors, medication, past and present medical or surgical history should be asked. Associated symptoms like night sweats, hot flushes to rule out vasomotor symptoms, family history of early menopause should also be asked. History suggestive of galactorrhea, headaches and visual field defects can be associated with hypothalamic or pituitary disorders[12] **(Table 4)**.

Physical Examination

A general physical examination including height, weight and BMI of the patient, her general condition and wellbeing should be carried out. Look for presence or absence of secondary sexual characteristics. Look for evidence of hyperandrogenism (acne, hirsutism, male pattern baldness, etc.). Any feature suggestive

TABLE 5: Investigations for amenorrhea.

Urine pregnancy test	***Rule out pregnancy***
FSH and LH	• *High:* Premature ovarian failure • *Low:* Functional hypothalamic amenorrhea • *Normal:* PCOS, intrauterine adhesions
TSH	Hypo-/hyperthyroidism
Prolactin	Raised in pituitary adenoma, medication, hypothyroidism other neoplasms
Testosterone free and total and dehydroepiandrosterone sulfate (DHEAS)	*High:* Hyperandrogenism, PCOS, ovarian and adrenal tumor, Cushing syndrome, congenital adrenal hyperplasia
17-hydroxyprogesterone levels (08:00 AM)	Raised in late onset congenital adrenal hyperplasia
Estradiol	*Low:* Poor endogenous estrogen secretion (poor ovarian reserve)
Complete blood count and metabolic panel	Abnormal in chronic disease, liver enzymes can be raised in hypothalamic amenorrhea
AMH	• *High:* Functional hypothalamic amenorrhea, PCOS • *Low:* Primary ovarian insufficiency
Karyotype	*Abnormal:* Turner's and other chromosomal disorders
Ultrasound/MRI pelvis	Anatomy of reproductive tract, PCOS, androgen secreting tumors of ovary
MRI brain	Tumor (micro-/macroadenoma)
MRI adrenal glands	Androgen secreting tumors (adrenal)
Dual energy X-ray absorptiometry (DEXA)	Fracture risk evaluation

(AMH: anti-Müllerian hormone; FSH: follicle-stimulating hormone; LH: luteinizing hormone; MRI: magnetic resonance imaging; PCOS: polycystic ovary syndrome; TSH: thyroid stimulating hormone)

of dysmorphism will be associated with congenital disorders. Look for the vaginal length to rule out outflow tract obstruction.[12]

Investigations

First step in the evaluation of amenorrhea is to rule out pregnancy. Once pregnancy is excluded check for serum FSH, LH, TSH and prolactin to rule out other endocrinological causes of amenorrhea.[4,7,13-16] Serum free and total testosterone and DHEAS (dehydroepiandrosterone sulfate) should be ordered in case of signs suggestive of hyperandrogenism are present. In patients with short stature karyotyping can be offered to rule out Turner's syndrome.[17-20]

Ultrasound pelvis will be helpful in evaluation of reproductive tract anatomy. In case there are features suggestive of any pituitary or hypothalamic lesions, MRI brain should be advised. A progesterone challenge test with 10 mg medroxyprogesterone acetate daily for 10 days or norethindrone 5 mg per oral daily for 10 days can be done to check for estrogen exposure.[20] If patient does not bleed after progesterone challenge, estrogen plus progesterone challenge should be given with conjugated equine estrogen, 1.25 mg, per oral for 21 days or estradiol 2 mg, per oral for 21 days followed by medroxyprogesterone acetate 10 mg once a day for 10 days starting from day 16 to 25 of the cycle[21] **(Table 5)**.

MANAGEMENT OF PRIMARY AMENORRHEA (FLOWCHART 1)

Evaluation of primary amenorrhea should be done in context with the presence and absence of secondary sexual characteristics. If breast development is normal with normal pubic hairs distribution an ultrasound pelvis or MRI should be performed to know the presence of normal uterus as Mullerian agenesis either congenital absence of vagina or abnormal development of uterus usually a rudimentary uterus is associated with at least 15% of the case of primary amenorrhea.[22,23] If the uterus is present and is normal, amenorrhea could be because of outflow tract obstruction (imperforate hymen or transverse vaginal septum). Patient generally presents with cyclical

Flowchart 1: Evaluation and diagnosis of primary amenorrhea.

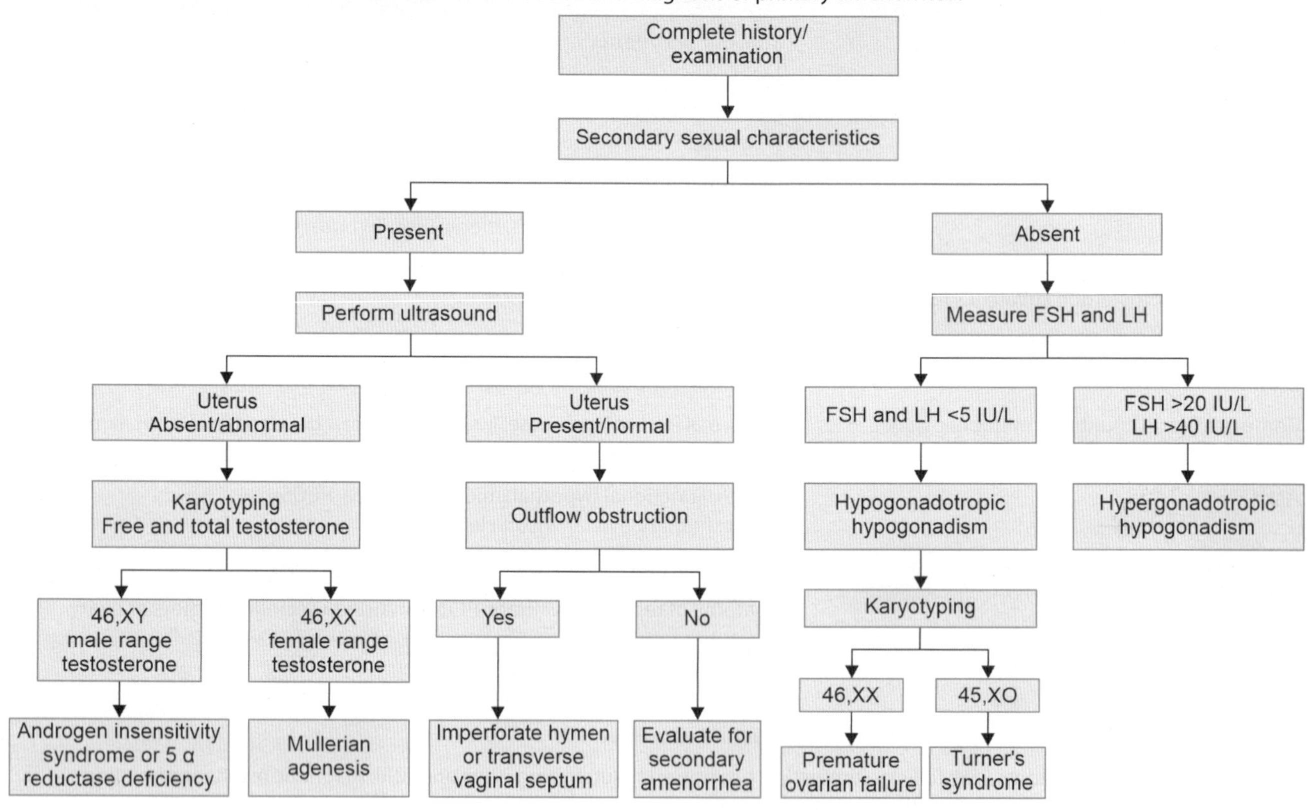

(FSH: follicle-stimulating hormone; LH: luteinizing hormone)

abdominal pain due to distention of the uterus because of accumulation of blood.

In cases where breast development is normal but pubic hair development is sparse or absent most common cause for primary amenorrhea is androgen insensitivity syndrome. A karyotyping should be done to confirm the diagnosis and on examination if undescended testes are present gonadectomy should be offered because of high chances of malignancy.

If the secondary sexual characteristics are absent, serum FSH and LH levels should be measured. If the LH and FSH levels are low (<5 IU/L), a diagnosis of hypogonadotropic hypogonadism is made. Most common cause of hypogonadotropic hypogonadism in cases of primary amenorrhea is constitutional delay in growth or puberty. Family history of menarche should be asked as most of the cases are familial and watchful expectancy is advocated. Kallmann syndrome can also be associated be hypogonadotropic hypogonadism and along with amenorrhea patient will also complain of anosmia.

Hypergonadotropic hypogonadism, i.e., FSH (>20 IU/L) and LH (>40 IU/L) is suggestive of gonadal dysgenesis or premature ovarian failure. Most common cause of premature ovarian failure is Turner's syndrome and karyotyping (45, XO) will be helpful in diagnosis. Patients with Turner's syndrome will also have webbed neck, widely spaced nipples, short stature, streak ovaries on ultrasound along with amenorrhea.

MANAGEMENT OF SECONDARY AMENORRHEA (FLOWCHART 2)

Management of amenorrhea will depend upon the underlying cause. Most of the times treating the underlying pathology will help in return of menstrual function.

Hypothyroidism: Women with hypothyroidism can present with oligomenorrhea, hypermenorrhea, or amenorrhea. Treatment of hypothyroidism restores the menstrual cycle but it may take several months.

Hyperprolactinemia: Galactorrhea, headache, visual field, amenorrhea can be related to high level of hormone

Flowchart 2: Evaluation and diagnosis of secondary amenorrhea.

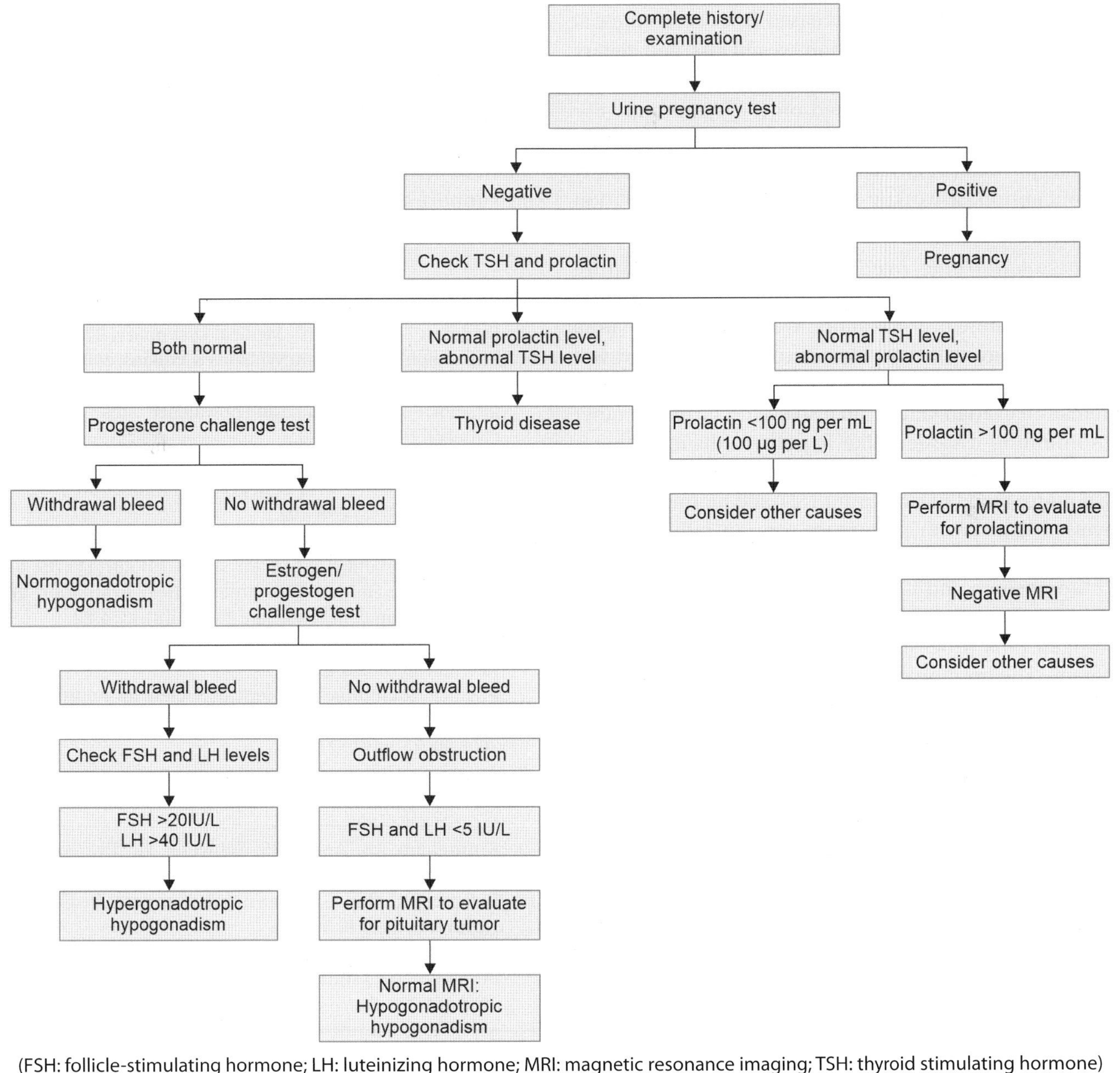

(FSH: follicle-stimulating hormone; LH: luteinizing hormone; MRI: magnetic resonance imaging; TSH: thyroid stimulating hormone)

prolactin. Hyperprolactinemia can be associated with prolactinomas, microadenomas, or macroadenomas. Medications like antipsychotic drugs, combined hormonal contraceptives, anti-histamines, and antiemetics are also associated with raised prolactin levels. Microadenomas less than 10 mm needs follow-up for progression. Dopamine agonists like bromocriptine and cabergoline will be helpful in relieving the symptoms related to raised prolactin levels. Macroadenoma will require surgical removal.

Normogonadotropic amenorrhea: Two most common causes of normogonadotropic amenorrhea are outflow tract obstruction and chronic anovulation. Outflow tract obstruction could be due to Asherman's syndrome, cervical stenosis, cervical fibroid obstructing the outflow tract. Ultrasound, hysteroscopy, and hysterosalpingography can be used to confirm the diagnosis.

Polycystic ovarian disease is one of the leading cause of chronic anovulation. PCOS is a complex metabolic syndrome associated with obesity, insulin resistance, hyperandrogenism, and subfertility. It is also associated with long-term increased risk of endometrial carcinoma. Lifestyle modification, weight loss, and dietary management are the first-line management of PCOS. Further management is based on the symptoms. For regulation of menstrual cycles, combined oral hormonal contraceptive can be given. If the patient is having difficulty in conception, a workup for infertility and ovulation induction can be tried.

Hypergonadotropic hypogonadism: Ovarian failure leading to amenorrhea if occurs prior to 40 years of age is called premature ovarian failure. Women with premature ovarian failure will have hypoestrogenic state and will be at increased risk of heart problems and osteoporosis. Premature ovarian failure can also occur because of other autoimmune disorders like hypothyroidism, diabetes mellitus, and Addison's disease. Hence, screening should be done for blood sugars, serum TSH, and morning cortisol levels. Women with premature ovarian failure before 30 years of age should undergo karyotyping to look for Y chromosome and if present need for gonadectomy.[22] Women with premature ovarian failure should receive estrogen plus progesterone as hormonal replacement therapy. The hormonal replacement therapy helps in the pubertal growth and prevention of osteoporosis. The goal of hormone replacement therapy in adolescent is to mimic the physiological pubertal growth. In such cases, gradually increasing dose of estrogens is given to stimulate breast development. Progestins are avoided till the development of breast mound and areola.

Hypogonadotropic hypogonadism: It usually occurs when there is disruption of normal functioning of hypothalamic–pituitary–ovarian axis. Excessive weight loss, eating disorders, and stress can be related to hypothalamic amenorrhea.[9,24-26] Athletes involved in vigorous physical training exhibit symptoms of amenorrhea, eating disorders, and osteoporosis. The triad is also known as female athlete triad. Increasing calories intake or decreasing physical exercise may help. Screening for eating disorders in women with excessive weight loss is recommended. Treatment with oral contraceptive pills and hormonal replacement therapy can be given.[27] Weight-bearing exercises can be suggested. Calcium and vitamin D supplements should be given.

CONCLUSION

Amenorrhea is an important symptom of various underlying pathologies. When not physiological, it can be challenging to reach the diagnosis. A multidisciplinary approach involving gynecologists, physician, endocrinologist and radiologist is required for through evaluation and successful management of women with amenorrhea. Care should be taken to maintain privacy and confidentiality. Any psychological issues or distress related to the disease or otherwise should be addressed and support should be provided to the women by an experienced health care provider.

REFERENCES

1. Stedman TL. Stedman's Medical Dictionary, 27th edition. Philadelphia: Lippincott Williams & Wilkins; 2000. p. 56.
2. Pettersson F, Fries H, Nillius SJ. Epidemiology of secondary amenorrhea. I. Incidence and prevalence rates. Am J Obstet Gynecol. 1973;117:80-6.
3. Bachmann G, Kemmann E. Prevalence of oligomenorrhea and amenorrhea in a college population. Am J Obstet Gynecol. 1982;144:98-102.
4. Klein DA, Poth MA. Amenorrhea: an approach to diagnosis and management. Am Fam Physician. 2013;87(11):781-8.
5. Hunter TM, Heiman DL. Amenorrhea: Evaluation and treatment. Am Fam Physician. 2006;73(8):1374-82.
6. Insler V. Gonadotopin therapy: new trends and insights. Int J Fertil. 1988;33:85-97.
7. Practice Committee of American Society for Reproductive Medicine. Current evaluation of amenorrhea. Fertil Steril. 2008;90(5 Suppl):S219-25.
8. American College of Obstetricians and Gynecologists. Amenorrhea (ACOG Technical Bulletin 128). Washington, D.C.: ACOG; 1989.
9. Kiningham RB, Apgar BS, Schwenk TL. Evaluation of amenorrhea. Am Fam Physician. 1996;53:1185-94.
10. Pletcher JR, Slap GB. Menstrual disorders. Pediatr Clin North Am. 1999;46:505-18.
11. Reindollar RH, Byrd JR, McDonough PG. Delayed sexual development: a study of 252 patients. Am J Obstet Gynecol. 1981;140:371-80.
12. Klein DA, Paradise SL, Reeder RM. Amenorrhea: A systematic approach to diagnosis and management. Am Fam Physician. 2019;100(1):39-48.
13. Gordon CM, Ackerman KE, Berga SL, Kaplan JR, Mastorakos G, Misra M, et al. Functional hypothalamic amenorrhea: an Endocrine Society clinical practice guideline. J Clin Endocrinol Metab. 2017;102(5):1413-39.
14. Nelson LM. Clinical practice. Primary ovarian insufficiency. N Engl J Med. 2009;360(6):606-14.
15. American College of Obstetricians and Gynecologists. Committee opinion no. 605: primary ovarian insufficiency in adolescents and young women. Obstet Gynecol. 2014;124(1):193-7.
16. American College of Obstetricians and Gynecologists' Committee on Practice Bulletins—Gynecology. ACOG

Practice Bulletin no. 194: polycystic ovary syndrome. Obstet Gynecol. 2018;131(6):e157-71.
17. Goodman NF, Cobin RH, Futterweit W, Glueck JS, Legro RS, Carmina E. American Association of Clinical Endocrinologists, American College of Endocrinology, and Androgen Excess and PCOS Society disease state clinical review: guide to the best practices in the evaluation and treatment of polycystic ovary syndrome—part 1. Endocr Pract. 2015;21(11):1291-300.
18. Martin KA, Anderson RR, Chang RJ, Ehrmann DA, Lobo RA, Murad MH, et al. Evaluation and treatment of hirsutism in premenopausal women: an Endocrine Society clinical practice guideline. J Clin Endocrinol Metab. 2018;103(4):1233-57.
19. Nelson SM. Biomarkers of ovarian response: current and future applications. Fertil Steril. 2013;99(4):963-9.
20. Tosi F, Fiers T, Kaufman JM, Dall'Alda M, Moretta R, Giagulli VA, et al. Implications of androgen assay accuracy in the phenotyping of women with polycystic ovary syndrome. J Clin Endocrinol Metab. 2016;101(2):610-8.
21. Warren MP, Biller BM, Shangold MM. A new clinical option for hormone replacement therapy in women with secondary amenorrhea: effects of cyclic administration of progesterone from the sustained-release vaginal gel Crinone (4% and 8%) on endometrial morphologic features and withdrawal bleeding. Am J Obstet Gynecol. 1999; 180(pt 1):42-8.
22. Nakamura S, Douchi T, Oki T, Ijuin H, Yamamoto S, Nagata Y. Relationship between sonographic endometrial thickness and progestin-induced withdrawal bleeding. Obstet Gynecol. 1996;87(5 pt 1):722-5.
23. Folch M, Pigem I, Konje JC. Müllerian agenesis: etiology, diagnosis, and management. Obstet Gynecol Surv. 2000;55:644-9.
24. Kalantaridou S, Davis SR, Nelson LM. Premature ovarian failure. Endocrinol Metab Clin North Am. 1998;27:989-1006.
25. Speroff L, Fritz MA. Amenorrhea. In: Clinical gynecologic endocrinology and infertility, 7th edition. Philadelphia, Pa.: Lippincott Williams & Wilkins; 2005. pp. 401-64.
26. Speroff L, Glass RH, Kase NG. Normal and abnormal sexual development. In: Clinical gynecologic endocrinology and infertility, 6th edition. Baltimore, Md.: Lippincott Williams & Wilkins; 1999. pp. 339-79.
27. Hergenroeder AC, Smith EO, Shypailo R, Jones LA, Klish WJ, Ellis K. Bone mineral changes in young women with hypothalamic amenorrhea treated with oral contraceptives, medroxyprogesterone, or placebo over 12 months. Am J Obstet Gynecol. 1997;176:1017-25.

Chapter

21 Chronic Anovulation in Polycystic Ovary Syndrome

Rakhi Singh, Chandrika Anand

INTRODUCTION

In reproductive age group polycystic ovary syndrome (PCOS) is amongst the most common endocrine disorders. The most common causes are oligo-anovulation and hyperandrogenism, acne, and obesity. These have substantial psychosocial, body image, and economic consequences. Three important diagnostic characteristics of polycystic ovary syndrome have been proposed, these include hyperandrogenism, chronic anovulation, and polycystic ovaries on ultrasonography. According to Rotterdam definition there are four phenotypes of polycystic ovary syndrome which are enlisted in **Table 1**.

There are multiple features of polycystic ovary syndrome; amongst them these three primary characteristics must be evaluated to confirm and established the type of phenotypes in women as summarized in **Table 1**.[1] PCOS is associated with anovulatory PCOS is ~80% of cases of infertility. The importance of gonadotropins luteinizing hormone (LH) and follicle-stimulating hormone (FSH) is well established in regard to timely appearance, concentrations, and sequence. In the normal cycle, the final maturation of the follicle corresponds because the timing of the gonadotropin surge which is controlled by the level of gonadal feedback signals, and it reflects the degree of dominant follicle growth and functional maturity of the oocyte.[2]

CHRONIC ANOVULATION

Anovulation is one of the most common gynecological problems encountered in women of reproductive age. In most of the cases the patients complain of irregular menstruation or amenorrhea. These patients have optimum estrogen, but their cycles are anovulatory with progesterone insufficiency. A study conducted in Danish women showed that the prevalence of women having irregular menstruation is 10.7–28.5%.[3] A woman who has longer intervals between their menstrual cycles for >35 days, or amenorrhea for at least 6 months are facing chronic anovulation. Various diseases and conditions can cause this symptom which includes polycystic ovary syndrome (PCOS), Cushing's syndrome, hypo or hyperthyroidism, hyperprolactinemia, hypothalamic and pituitary disorders, late-onset congenital adrenal hyperplasia (CAH), adrenal, and ovarian tumors.[4-6]

Below are the key causes of chronic anovulation:[7]

- Polycystic ovarian syndrome
- Prolactin disorders

TABLE 1: Phenotypes for polycystic ovary syndrome (PCOS) based on 2003 Rotterdam criteria.

	Severe PCOS	*Hyperandrogenism and chronic anovulation*	*Ovulatory PCOS*	*Mild PCOS*
Periods	Irregular	Irregular	Normal	Irregular
Ovaries on ultrasonography	Polycystic	Normal	Polycystic	Polycystic
Androgen concentrations	High	High	High	Mildly raised
Insulin concentrations	Increased	Increased	Increased	Normal
Risks	Potential long-term	Potential long-term	Unknown	Unknown
Prevalence in affected women	61%	7%	16%	16%

- Hyperprolactinemia
    - Pituitary microadenoma
    - Pituitary macroadenoma
- Idiopathic chronic anovulation
- Thyroid disorders
    - Subclinical hypothyroidism
    - Hypothyroidism
    - Hyperthyroidism
    - Hashimoto thyroiditis
- Congenital adrenal hyperplasia
    - Non-classical
    - Heterozygous
- Adrenal tumor
- Cushing's syndrome
- Indefinite diagnosis

CHRONIC ANOVULATION IN POLYCYSTIC OVARY SYNDROME

Adequate evidences are available for dysregulation of hypothalamic gonadotropin secretion PCOS women which may lead to the mechanism of anovulation.[8] Comprehensive analysis in the pattern of gonadotropin release has highlighted the high levels and frequency of LH release in women with anovulatory cycles which indicates a high frequency of pulsatile release—gonadotropin releasing hormone (GnRH). Anovulatory cycles are characterized by the absence of progesterone-mediated negative feedback control of gonadotropins.[9,10] During the luteal phase in normal ovulatory cycles, LH pulse frequency is slow and mean LH levels are lowest at this phase of the cycle. To add another level of difficulty, studies suggest that progesterone-mediated negative feedback on LH is affected in PCOS women.[11,12] In particular the role of gamma-aminobutyric acid has to be further explored in regard to anti-Müllerian hormone (AMH).[13] The possibility of hypothalamic involvement in the genesis of anovulatory PCOS has implicated a central role for AMH in the genesis of abnormal GnRH activity thus increases LH and androgens. AMH levels, primarily produced by GCs of large preantral and small antral follicles, cause increase in PCOS women and its serum levels reflect the number of antral follicle count in ovary seen by ultrasonography (USG) viewed **(Fig. 1)**.[14]

The patient clinically presenting with galactorrhea can be easily diagnosed and does not require extensive investigation to identify the root cause of anovulation. However, patients who have no clear clinical signs might need more extensive investigations to recognize the causes of anovulation by doing investigations to assessment of thyroid profile, prolactin, FSH, LH and estradiol levels, transabdominal sonography (TAS) USG to also see for early

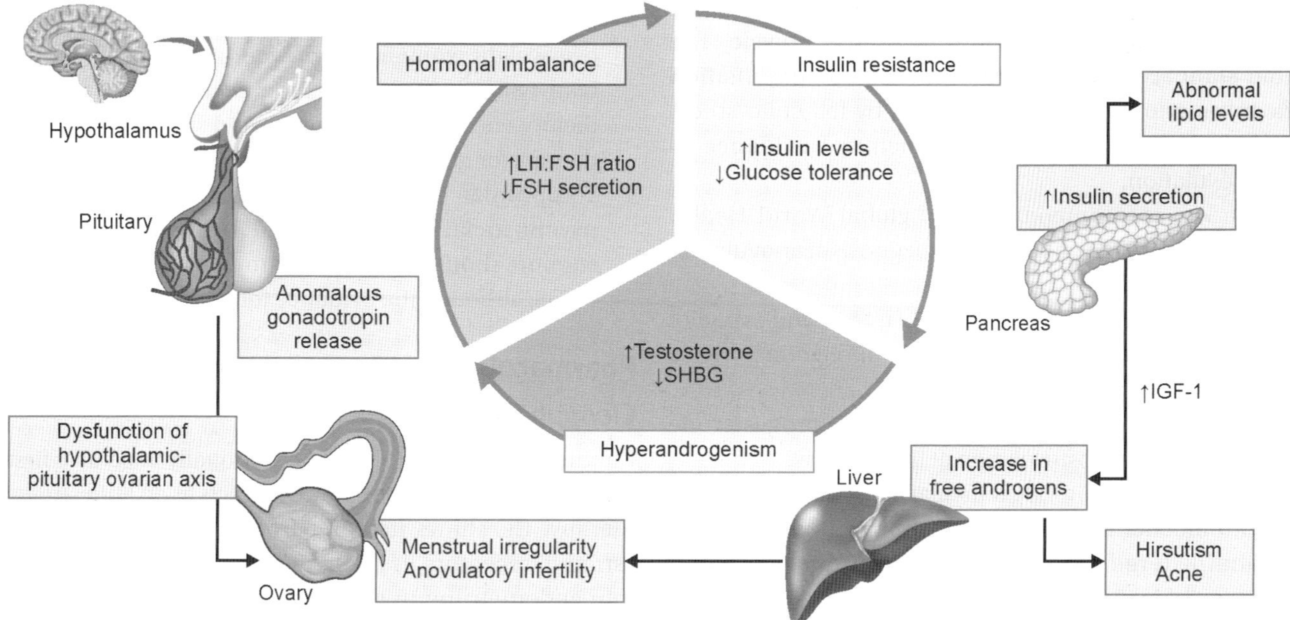

Fig. 1: Diagnosing chronic anovulation (pathophysiology of PCOS).
(FSH: follicle-stimulating hormone; IGF: insulin-like growth factor; LH: luteinizing hormone; PCOS: polycystic ovary syndrome; SHBG: sex hormone binding globulin)

Flowchart 1: Laboratory investigations for women who had chronic anovulation without obvious clinical manifestation.

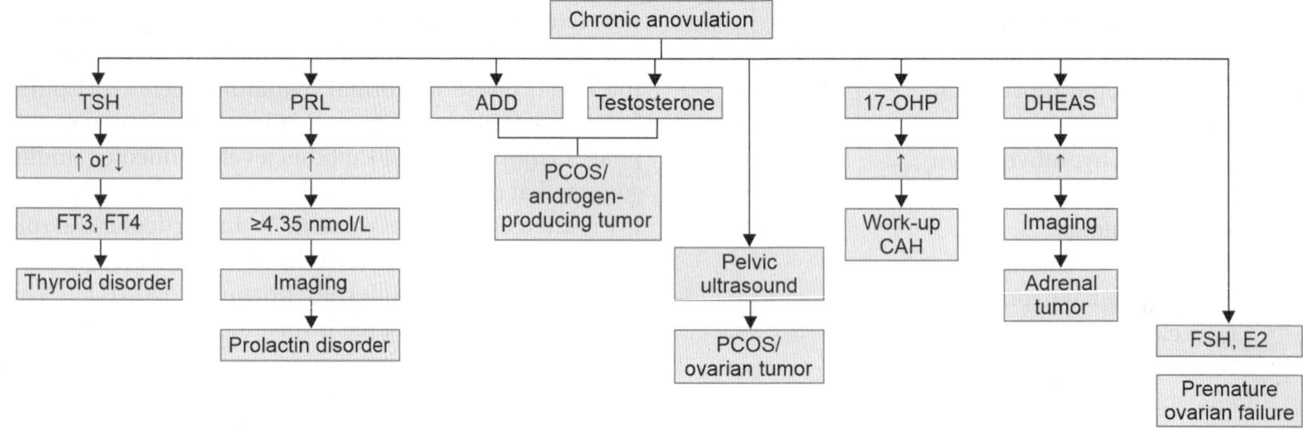

(17-OHP: 17-hydroxyprogesterone; ADD: androstenedione; CAH: congenital adrenal hyperplasia; DHEAS: dehydroepiandrosterone sulfate; E2: estradiol; FSH: follicle-stimulating hormone; FT3: free triiodothyronine; FT4: free thyroxine; PCOS: polycystic ovary syndrome; PRL: prolactin; TSH: thyroid-stimulating hormone)

ovarian insufficiency, dexamethasone suppression test for Cushing's assessment, testosterone, androstenedione, dehydroepiandrosterone sulfate (DHEAS), and 17-hydroxyprogesterone (17-OHP) **(Flowchart 1)**.[7]

Diagnosing Chronic Anovulation in Polycystic Ovary Syndrome

If the cycles are irregular cycles (>35 or <21 days) for >2 years post-menarche oligo-ovulation or anovulation. Ovulation is evaluated by measurement of serum progesterone during the mid-luteal stage of cycle. The "gold standard" investigation would be a combination of confirmation of ovulation on USG by follicular study, endometrial thickness, and biochemical evidence to assess ovulation.

The key recommendations of global World Health Organization (WHO) guidance for diagnosis of anovulation are as follows:[15]

- To check on day 2–5 of cycle for LH, FSH, and estradiol levels or randomly in a woman with oligomenorrhea or amenorrhea.
- To document ovulation and check for progesterone level on 7-day postovulation at luteal phase.
- Ultrasound monitoring follicular study can be coupled with biochemical assessment of ovulation should be administered to provide more information about timing of ovulation and ovarian function.
- Higher AMH may be a biochemical marker for polycystic ovaries, although normative ranges are also seen in PCOS and therefore this is a weak recommendation.

Management of Chronic Anovulation in Polycystic Ovary Syndrome

The key recommendations of global WHO guidance for management of anovulation are as follows:[15]

- Lifestyle management (single or combined approaches of diet, exercise, and/or behavioral interventions) for weight loss or prevention of weight gain, or for general health benefits, should be recommended in all women with PCOS.
- Lifestyle management targeting weight loss in overweight women and prevention of weight gain in lean women should include exercise and reduced dietary energy (caloric) intake and should be first-line therapy for all women with PCOS.
- To optimize adherence with lifestyle interventions, psychosocial factors should be considered, and support should be provided to infertile women with PCOS, although the evidence for this is weak.

Pharmacological Management

Clomiphene Citrate

Clomiphene citrate (CC) was traditionally used as a first-line therapy in anovulatory PCOS. It acts as an antiestrogen therapy. CC shows its antiestrogen action by blocking estradiol receptors in the hypothalamus, thus leading to thin endometrium in the stimulated cycles. The change in GnRH pulse frequency, FSH hormone is released from anterior pituitary thus follicular development occurs. The LH hormone if it continuously increases during the follicular stage, may be an obstacle to the success of the treatment. A meta-analysis conducted has confirmed that CC is effective in increasing pregnancy rates compared with placebo as first-line therapy. CC starting dose is 50 mg/day for 5 days.

Aromatase Inhibitors

Aromatase inhibitors is now considered as the gold standard for ovulation induction in the anovulatory cycles. As it acts by preventing estrogen biosynthesis from androgens by hypothalamic/pituitary feedback and increases in FSH secretion. Theoretically, these agents do not have adverse effects of CC, as they do not have effect on estrogen receptors centrally or on endometrium. Moreover, by conserving ovarian/pituitary feedback, which reduces the risk of multiple follicle development compared with CC.

Metformin

Metformin can be used alone to improve ovulation and also pregnancy rate in PCOS women with anovulatory cycles and are suffering from infertility with no other infertility factors. Metformin can be given in combination with CC to improve fertility outcomes in women with PCOS, who are CC-resistant.

Gonadotropin Therapy

Gonadotropins are used in the treatment of anovulatory cycles either after the use of oral ovulogens or as an adjunct to induce ovulation in the CC resistance cases or the failed ovulation. Gonadotropins are usually given in low doses to avoid the chances of overstimulation instead of the standard the step up or the step-down protocol.

CONCLUSION

- Anovulation is a common gynecological problem among women of reproductive age.
- Patients commonly present with irregular menstruation or amenorrhea.
- PCOS is most common cause attributed to anovulation followed by Thyroid disorders and Hyperprolactinemia.
- Infertility is most common complication arising due to Chronic anovulation having high level of LH hormone secretion.
- In LOD the number of punctures in one ovary should be maximum of 4.

REFERENCES

1. Norman RJ, Dewailly D, Legro RS, Hickey TE. Polycystic ovary syndrome. Lancet. 2007;370(9588):685-97.
2. Kase N. Chronic anovulation and the polycystic ovary syndrome. In: Deligdisch L, Kase N, Cohen C (Eds). Altchek's Diagnosis and Management of Ovarian Disorders. Cambridge: Cambridge University Press; 2013. pp. 151-88.
3. Hahn KA, Wise LA, Riis AH, Mikkelsen EM, Rothman KJ, Banholzer K, et al. Correlates of menstrual cycle characteristics among nulliparous Danish women. Clin Epidemiol. 2013;5:311-9.
4. Hamilton-Fairley D, Taylor A. Anovulation. BMJ. 2003; 327:546-9.
5. Li RH, Ng EH. Management of anovulatory infertility. Best Pract Res Clin Obstet Gynaecol. 2012;12:757-68.
6. Practice Committee of American Society for Reproductive Medicine. Current evaluation of amenorrhea. Fertil Steril. 2008;90(5 Suppl):S219-25.
7. Chandeying P, Pantasri T. Prevalence of conditions causing chronic anovulation and the proposed algorithm for anovulation evaluation. 2015;41(7):1074-9.
8. Coutinho EA, Kauffman AS. The role of the brain in the pathogenesis and physiology of polycystic ovary syndrome (PCOS). Med Sci (Basel). 2019;7(8):84.
9. Baird DT, Corker CS, Davidson DW, Hunter WM, Michie EA, Van Look PF. Pituitary-ovarian relationships in polycystic ovary syndrome. J Clin Endocrinol Metab. 1977;45:798-801.
10. Polson DW, Mason HD, Franks S. Bromocriptine treatment of women with clomiphene-resistant polycystic ovary syndrome. Clin Endocrinol. 1987;26:197-203.
11. Pastor CL, Griffin-Korf ML, Aloi JA, Evans WS, Marshall JC. Polycystic ovary syndrome: evidence for reduced sensitivity of the gonadotropin-releasing hormone pulse generator to inhibition by estradiol and progesterone. J Clin Endocrinol Metab. 1998;83:582-90.
12. Eagleson CA, Gingrich MB, Pastor CL, Arora TK, Burt CM, Evans WS, et al. Polycystic ovarian syndrome: evidence that flutamide restores sensitivity of the gonadotropin-releasing hormone pulse generator to inhibition by estradiol and progesterone. J Clin Endocrinol Metab. 2000;85:4047-52.
13. Sullivan SD, Moenter SM. Prenatal androgens alter GABAergic drive to gonadotropin-releasing hormone neurons: implications for a common fertility disorder. Proc Natl Acad Sci USA. 2004;101:7129-34.
14. Teede H, Misso M, Tassone EC, Dewailly D, Ng EH, Azziz R, et al. Anti-mullerian hormone in PCOS: a review informing international guidelines. Trends Endocrinol Metabol. 2019;30:467-78.
15. Balen AH, Morley LC, Misso M, Franks S, Legro RS, Wijeyaratne CN, et al. The management of anovulatory infertility in women with polycystic ovary syndrome: an analysis of the evidence to support the development of global WHO guidance. Hum Reprod Update. 2016;22(6): 687-708.

Chapter

22 Hirsutism

Monika Thakur

INTRODUCTION

The presence of dark, coarse, terminal hairs distributed in a male pattern is called as hirsutism. The central and facial parts of the body show the increased growth of sexual hair which are androgen dependent and bother the patient and need a concerned and sympathetic response. The incidence of hirsutism is between 5 and 10% among reproductive age women.[1]

It is generally distinguished from *hypertrichosis* in which there is generalized increase in vellus hair. It is associated with, few systemic illnesses (e.g., dermatomyositis, malnutrition, anorexia nervosa, hypothyroidism), with use of certain drugs (e.g., penicillamine, cyclosporin, minoxidil, diazoxide, phenytoin), or malignancy (paraneoplastic syndrome).

Virilization is an extreme form of androgen excess and refers to a condition of hirsutism associated with presence of other male characteristics such as, hoarse voice, temporal balding, enlargement of clitoris along with features of defeminization such as amenorrhea and breast atrophy. The most common causes of virilization are androgen-producing tumors of ovary or adrenal and congenital adrenal hyperplasia.

ENDOCRINOLOGY

Hirsutism is interplay between circulating androgen levels and the stimulation of hair follicles to androgen as per the sensitivity. In women, the major circulating androgens are dehydroepiandrosterone (DHEA), dehydroepiandrosterone sulfate (DHEA-S), androstenedione, testosterone, and Dihydrotestosterone (DHT).[4] DHEA, DHEA-S, and androstenedione are considered pre-hormones because they require conversion to testosterone to exert androgenic effects and have little or no intrinsic androgenic activity.

TABLE 1: Distinct phases of hair growth & Different types of hair.

Distinct phases of hair growth[2]	*Types of hair*[3]
Telogen (quiescent phase)	Vellus (fine, soft, short, and unpigmented)
Anagen (growth phase)	Terminal (long, coarse, and pigmented)
Catagen (involution phase)	Lanugo (vellus hair that covers the body of infants)

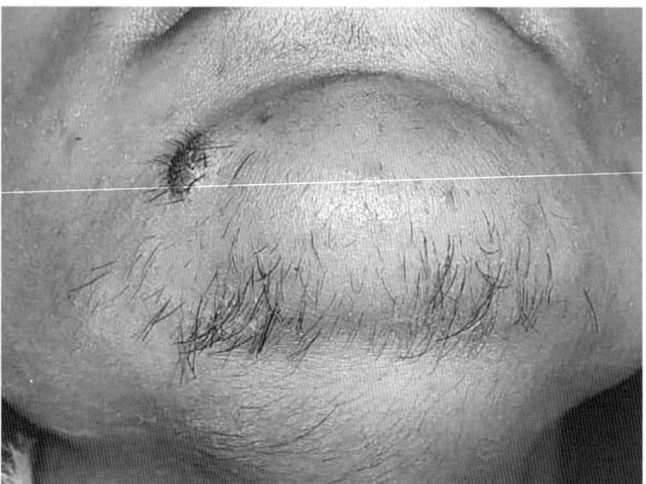

Fig. 1: Excessive hair growth over chin.

The various sources of androgens in a female are depicted in **Table 2**.

A total of 80% of testosterone in the circulation are bound to sex hormone-binding globulin (SHBG), another 19% is loosely bound to albumin leaving only 1% of fraction unbound or free. The specific assays available to measure the level of free testosterone are costly and rarely needed. The presence of hirsutism is a sign of androgen excess and in a woman who has hirsutism, but total testosterone levels

TABLE 2: Sources of androgens in a female.

Androgen	Source	Daily production	Serum concentration
DHEA-S	Adrenal gland (exclusively)	3.5–20 mg/day[5]	100–350 µg/dL
DHEA	• Adrenal gland—50% • Ovaries—20% • Peripheral conversion of DHEA-S—30%	6–8 mg/day[6]	1–10 ng/mL
Androstenedione	• Adrenal gland—50% • Ovaries—50%	1.4–6.2 mg/day[7]	0.5–2.0 ng/mL[8]
Testosterone	• Adrenals—25% • Ovaries—25% • Peripheral conversion of androstenedione—50%	0.1–0.4 mg/day	• Total: 20–80 ng/dL • Free testosterone (calculated) 0.6–6.8 pg/mL Sex hormone-binding globulin (SHBG) 18–114 nmol/L

are normal itself implies decreased SHBG and increased free testosterone.

Dihydrotestosterone is produced by intracellular conversion of testosterone (via 5α-reductase), only in the periphery. Therefore, serum DHT levels are very low and does not correlate with the of 5α-reductase activity.[9] Instead, 3α-androstanediol glucuronide (3α-AG), can be used as a marker of androgen metabolism in periphery. The pathophysiology of idiopathic hirsutism correlates with increased activity of 5α-reductase at the peripheral level and levels of 3α-AG in the serum are raised almost in all hirsute women even when serum androgen levels are normal which indirectly states that serum 3α-AG levels correlate highly with levels of 5α-reductase activity in genital skin. However, serum testing for 3α-AG is rarely done because the results will have no impact either on the diagnosis or on the treatment of hirsutism.

PATHOPHYSIOLOGY OF HIRSUTISM

- *Raised androgen levels* have a big role in determining the distribution and type of hair. The conversion of testosterone to the active form DHT happens by the enzyme 5α-reductase within a hair follicle. Both DHT and testosterone convert soft, short vellus hair to coarse terminal hair but DHT is found to be more effective than testosterone. This way hairs get converted to terminal hairs only in androgen-sensitive areas and this conversion is irreversible. Due to this the areas which are affected most commonly with increased number of hairs are chin, sideburns, upper lip, lower abdomen (linea alba), and chest.
- *Racial and ethnic factors:* Although, there is no difference among the number of hair follicles per unit between men and women, however, there is difference among propensity of number of hairs in different races and as per ethnicity.

  Mediterranean descent > Northern Europeans > Asians
- *Genetic factors:* In addition, the tendency which runs in families for development of hirsutism is strong and the basis lies in genetic differences of 5α-reductase activity and also target tissue sensitivity to androgens.

CAUSES OF HIRSUTISM

Hirsutism is the result of excess androgen production either from the adrenals or ovaries or excess stimulation of the end organs, i.e., hair follicles. Hirsutism is the early manifestation before other features of virilization become visible. Whereas virilism is almost always associated with hirsutism (except at birth) but hirsutism may not be associated with virilism.

- *Ovarian*
  - Polycystic ovarian syndrome (PCOS)
  - Masculinizing tumors of ovary e.g., Sertoli-Leydig cell tumor, Hilus cell tumor, Lipoid cell tumor
  - Luteoma of pregnancy
  - Hyperthecosis
- *Adrenal*
  - Congenital adrenal hyperplasia (or late onset)
  - Adrenal tumor
  - Cushing's syndrome
- *Exogenous*
  - Androgens
  - Anabolic steroids
  - Phenothiazines
  - Synthetic progestogens
  - Danazol
  - Metoclopramide

- Methyldopa
- Reserpine

- *Pituitary Tumor*
  - Acromegaly (excess growth hormone)
  - Cushing's disease [excess adrenocorticotropic hormone (ACTH)]
- *Others*
  - HAIR-AN syndrome (hyperandrogenic insulin resistant-acanthosis nigricans)
  - Obesity
- *Idiopathic*

Polycystic ovarian syndrome is the most common cause of androgen excess in women and is responsible for 70–80% of cases of hirsutism in adolescent women. The second most common cause is idiopathic hirsutism.

Congenital adrenal hyperplasia (non-classical): The deficiency of 21-hydroxylase enzyme causes increased production of androgens from adrenal is proposed to be most common cause of congenital adrenal hyperplasia (CAH).

Pathophysiology

↓ Cortisol production → compensatory rise in pituitary ACTH secretion → rise in level of steroid hormones proximal to enzyme block → alternatively gets converted to androgens.

Routine testing of serum 17OH-progesterone is not recommended, and testing is only done selectively in cases where hirsutism is of early onset, family history of congenital adrenal hyperplasia and in cases where women belong to high-risk ethnic groups (Mediterranean, Hispanic or Ashkenazi Jews).

The schematic representation of diagnosis of CAH is depicted in **Flowchart 1**.

Cushing's disease: The screening for Cushing syndrome is indicated when hirsutism is associated by feature suggestive of hypercortisolism, such as central obesity, moon face, buffalo hump, muscle weakness, severe fatigue, easy bruising or purple striae, hyperpigmentation, menstrual disorders, diabetes, and cognitive impairment.

Causes of Cushing syndrome:

- Exogenous administration of glucocorticoids (most common cause)
- Pituitary secreting adenoma (ACTH-secreting corticotroph adenoma)
- Cortisol secreting adrenal adenoma
- Ectopic production of ACTH (by bronchial carcinoid or other rare tumors).

The methods of screening available are given in **Table 3**.

The schematic representation of diagnosis of Cushing syndrome is given in **Flowchart 2**.

CLINICAL FEATURES

The history taking has to be taken very sensitively and tactfully which should also include questions regarding the

Flowchart 1: Schematic representation of diagnosis of congenital adrenal hyperplasia (CAH).

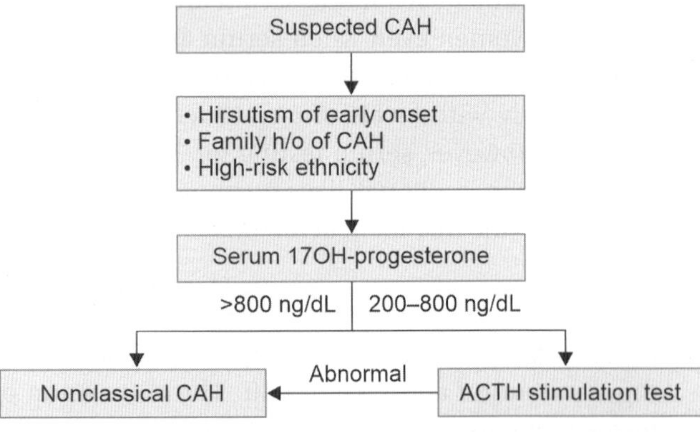

TABLE 3: Methods of screening.

Test	*Procedure*	*Cut-off values*	*Precautions*
24-hour urinary free cortisol excretion	Measured twice in order to avoid fluctuations	>3 times the upper normal limit is clearly abnormal	Avoid excessive fluid intake and use of products containing corticosteroids
Late-night salivary cortisol level	11:00 PM-midnight salivary cortisol level measured	Free cortisol in blood is in equilibrium with cortisol in saliva	Avoid chewing tobacco and avoid smoking because it falsely raises salivary cortisol levels
Overnight dexamethasone suppression test	1 mg of dexamethasone given 11:00 PM to midnight. Serum cortisol measured at 8:00 AM in morning	Serum cortisol <1.8 µg/dL is normal	False positive in patients on OCP's

Flowchart 2: Schematic representation of diagnosis of cushing syndrome.

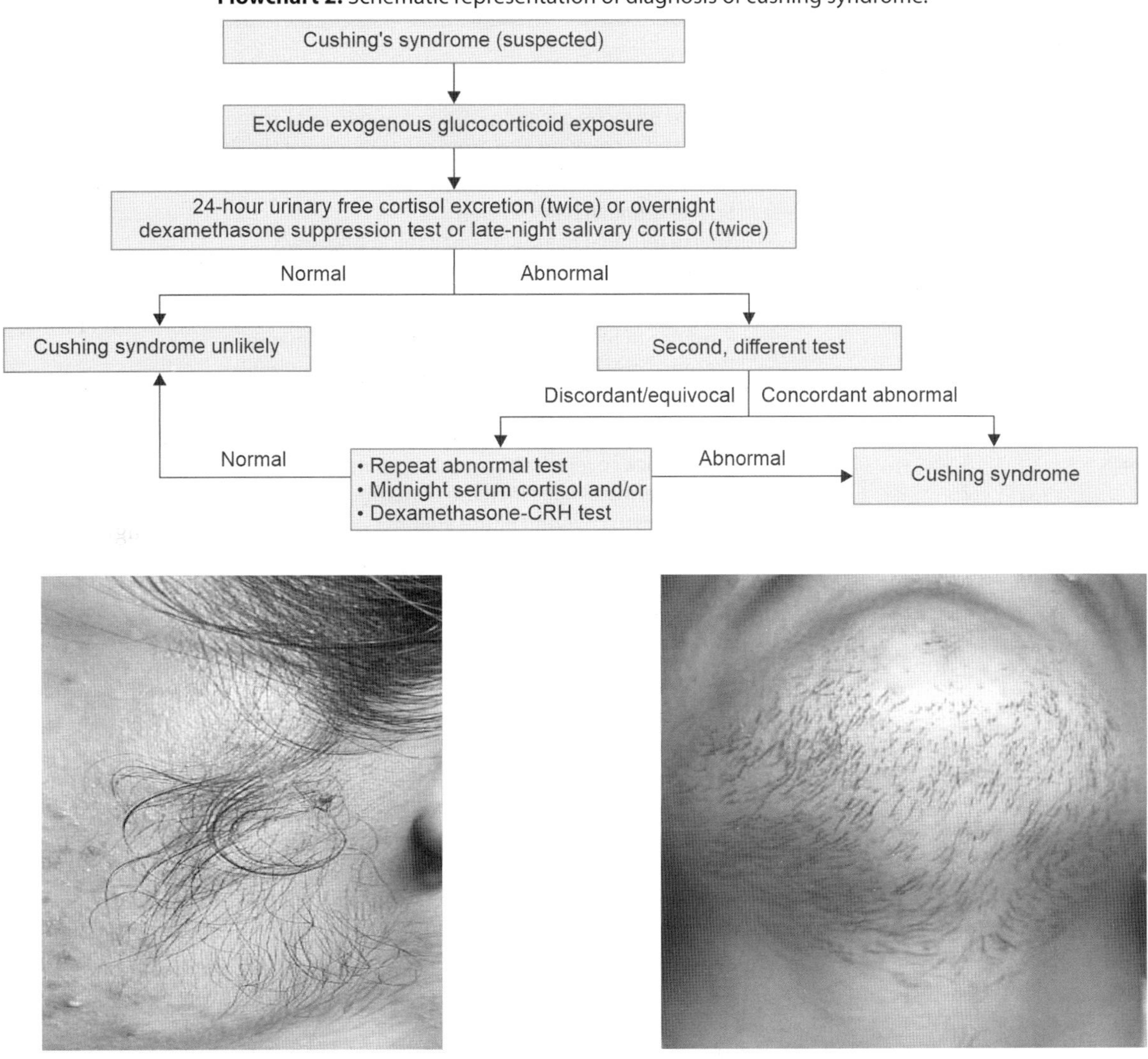

Fig. 2: Excessive facial hair

Fig. 3: Hirsutism in an adolescent female.

frequency of hair removal method and the method adopted by the patient for hair removal. The quantification of extent of hirsutism is usually subjective and most doctors classify it as mild, moderate, or severe arbitrarily. The assessment based on certain scoring system is always helpful especially to formulate a baseline from which results of therapy can be tested.

Ferriman-Gallwey (FG) scoring system: This is the scoring system in which extent of hair growth is scaled in nine androgen sensitive body areas on a scale of 0–4 (upper lip, chin, chest, upper back, lower back, upper abdomen, lower abdomen, arm, forearm, thigh, and leg). Increasing numeric scores correspond to greater hair density within a given area. This scoring system was developed in 1961[10] and later modified in 1981.[11] Hair growth over the forearm and lower leg was noted to be less sensitive and subsequent modifications of the Ferriman-Gallwey method have excluded scoring these areas and hence nine body areas are used in the modified method. A total score of >8 is defined as hirsutism. A score of 8–15 indicates mild hirsutism and >15 indicate moderate or severe hirsutism, respectively.

Limitations of FG scoring system **(Table 4)** are:

- Does not include the sideburn, perianal area, and buttocks.
- Does not tell the extent to which hirsutism might impact woman's well-being.
- Substantial hirsutism might be confined to 1 or 2 areas without exceeding the cut off value score of 8.

TABLE 4: Modified Ferriman–Gallwey scoring system.

Site	*Score*	
Upper lip	1 2 3 4	Small number of terminal hairs over upper lip and outer lip border Thin moustache covering <50% of upper lip or at the outer border Moustache covering 50% from outer margin of the lip or 50% the lip height Moustache covering most of the upper lip and crossing the midline lip
Chin	1 2 3 4	Sparse terminal hairs on chin Sparse terminal hairs with small thickened areas Light growth covering the whole chin Heavy growth covering the whole chin
Chest	1 2 3 4	Presence of terminal hairs in midline or circumareolar Presence of terminal hairs in midline of circumareolar Terminal hairs covering 75% of chest area Terminal hairs covering the whole area
Upper back	1 2 3 4	Upper back covered with sparse terminal hairs Spread terminal hairs increase in number Light growth covering entire area Heavy growth covering entire area
Lower back	1 2 3 4	<4 cm wide coverage of sacral area with hair Increase inside coverage Terminal hair occupying 75% of area of lower back Heavy growth covering the whole area
Upper abdomen	1 2 3 4	Presence of terminal hairs in midline which are scattered The terminal hairs increased in number but are still midline Terminal hairs covering 50% of area of upper abdomen Terminal hairs covering the whole area
Lower abdomen	1 2 3 4	Small number of scattered midline terminal hairs the length of linea alba Midline concentration of terminal hair the length of linea alba A midline thickened band of terminal hair <½ width of pubic hair at base An inverted V-shaped coverage ½ width of pubic hair at base
Upper arms	1 2 3 4	<25% area of upper arm covered with terminal hairs which are scattered The number increases but coverage is incomplete Light growth covering the whole area Heavy growth covering the whole area
Thigh	1 2 3 4	<25% area of thigh covered with terminal hairs which are scattered The number increases but coverage is incomplete Light growth covering the whole area Heavy growth covering the whole area

INVESTIGATIONS

History

The important points to be taken in history of a hirsute women are:

- *Menstrual history:* Age at menarche, the regularity of menses and information regarding any previous pregnancies and method of contraception.
- *Rate of progression of hirsutism:* The rapidity with which the symptoms ensue and speed of progression helps in identifying the cause of hirsutism. The breast shrinking, amenorrhea, and change in voice indicate the possibility of androgen secreting ovarian tumor.
- *Medical history:* History of drug intake should be sought. Infertility if present points toward anovulation and possible presence of PCOD.
- *Family history.*

Examination

- Body mass index (BMI)
- Degree of hirsutism should be noted, (Ferriman-Gallwey Scoring)

- Any signs of virilization. Acne, seborrhea, temporal balding, deepening of the voice, increased muscle mass, breast atrophy, and clitoromegaly (clitoral length >10 mm[12])
- Acanthosis nigricans indicate insulin resistance, and thin skin, striae, or bruising are signs of hypercortisolism.
- Abdominal and pelvic examination should be carried out to rule out androgen-secreting tumor.

Hormonal study: The testing for raised androgen levels are done for women with Ferriman–Gallwey score >8 or when the hirsutism is sudden in onset and rapidly progressive along with features of defeminization (2008 Endocrine Society Clinical Practice Guidelines). As per the guidelines the testing is not indicated for mild hirsutism, e.g., PCOS, because the chances of a medical disorder which will change the management, or the outcome is very low.[13]

Thyroid function tests and serum prolactin should be done in women with menstrual irregularities. This includes measurement of serum testosterone and DHEA-S. The serum levels of testosterone >150 ng/dL identify all women with androgen-producing tumor.[14-19] In adrenal tumor, DHEAS levels are raised >700–800 µg/dL. It gives *a better estimate than 24-hour urine estimation of 17-ketosteroid.*

Preovulatory LH, FSH levels will need to be estimated.

In PCOD, LH level exceeds 10 U/L; testosterone >2.5 nmol/L, and SHBG <30 nmol/L.

Serum 17α-OH progesterone concentration. Levels >800 ng/dL are virtually diagnostic for congenital adrenal hyperplasia and concentration between 200 and 800 ng/dL should be confirmed by ACTH stimulation test.

Ultrasound scan (transvaginal): It may help to detect an ovarian tumor and adrenal tumor when serum total testosterone concentration (≥150 ng/dL).

Computed tomography (CT) scan and magnetic resonance imaging (MRI) are needed in cases where pituitary or adrenal tumor is suspected. Laparoscopic visualization of pelvic organs is sometimes needed to establish the diagnosis.

The routine use of contraceptive steroids, dexamethasone suppression test, or use of gonadotropin-releasing hormone (GnRH) agonist to differentiate adrenal and ovarian origin is not recommended.[20-22]

Insulin resistance: As the sheer number of different measures of insulin resistance illustrates, there is currently no uniformly accepted, validated, simple test for measuring insulin resistance in clinical practice. A baseline 75 g glucose tolerance test (OGTT) is carried out in all women with polycystic ovaries because 35% have impaired glucose tolerance (glucose 140–199 mg/dL) and 10% have diabetes (glucose ≥200 mg/dL).

The schematic representation of various investigations used for diagnosis of hirsutism is given in **Flowchart 3**.

Flowchart 3: Schematic representation of various investigations used for diagnosis of hirsutism.

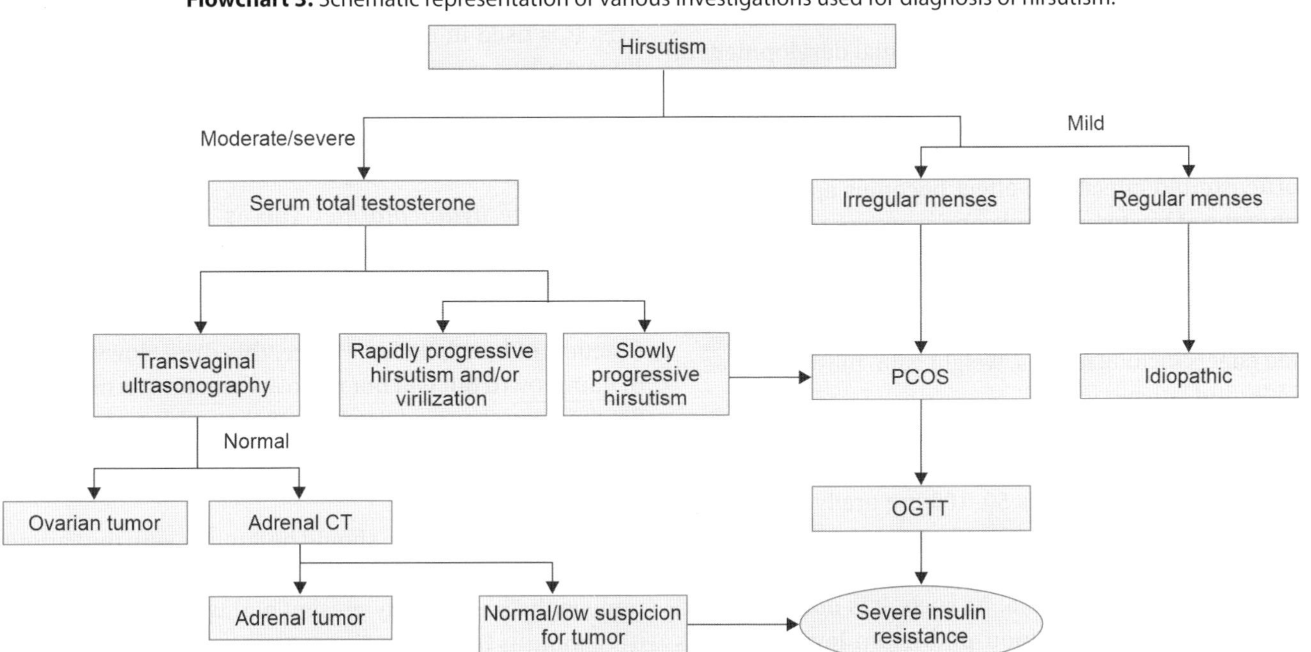

(OGTT: glucose tolerance test; PCOS: polycystic ovarian syndrome)

MANAGEMENT

Aims of Treatment

- Decreasing the level of androgen to halt the further conversion of vellus hairs to terminal hairs
- Increasing the binding
- Blocking the action of androgens.

It takes almost 6 months for medical treatment to showcase results, which equals half-life of a hair follicle growth cycle.

Weight reduction is associated with improved insulin levels and lowers down androgen levels also. Weight loss increases SHBG levels thereby reducing the free testosterone and its action on hair follicles in periphery.

Removal of the source which means surgical removal of adrenal or ovarian tumor if any. Cushing's disease can be treated by adrenalectomy, radiation to the pituitary, or removal of ACTH producing tumor by transsphenoidal surgery. In iatrogenic cases, the offending drug should be stopped.

Medical therapy: Suppression of androgens with combined oral OCPs, not containing androgenic progestogen such as norethisterone and levonorgestrel, will suppress ovarian androgens. COCs are effective in establishing regular menses and are first-line option in women who are actually candidates for COC.

Anti-androgens

Antiandrogens are the most effective for treatment of hirsutism but always used along with oral contraceptive pills because they might affect the sexual development of a male fetus if patients fall pregnant during the treatment.

Cyproterone acetate is a potent progestogen, it inhibits DHT binding to its receptor at the periphery and has a weak corticosteroid effect. The effect becomes apparent after 4 months of treatment. Oral contraceptives regularize the cycle, and prevent pregnancy. Estrogen present in the pills avoids menopausal symptoms and also raises the serum hormone binding globulin, which binds the androgen. Side effects include nausea, weight gain, mastalgia, loss of libido, and fatigue.

Spironolactone is a competitive inhibitor of androgen binding to the androgen receptor and also directly inhibits 5α-reductase. The dose is 50–100 mg orally twice daily. The drug effectively treats hirsutism but carries important risks. The important side effects are intermenstrual bleeding and hyperkalemia. As such, it is not prescribed with agents that raise blood potassium levels, such as potassium supplements, angiotensin-converting enzyme inhibitors, nonsteroidal anti-inflammatory drugs such as indomethacin, or other potassium-sparing diuretics.

Flutamide is another antiandrogen marketed for the treatment of prostate cancer. It is uncommonly used for hirsutism due to its potential hepatotoxicity. As an antiandrogen, teratogenicity again is a concern.

Finasteride is a 5α-reductase inhibitor, which blocks conversion of testosterone to DHT. The studies show finasteride to be effective modestly for hirsutism when used in the dose of 5-mg. Side effects are few, although decreased libido has been noted. As with other antiandrogens, the risk of male fetal teratogenicity is present, and effective contraception must be used concurrently.

Insulin-sensitizing drugs: It is always thought that this group of drugs might help to improve insulin resistance, which is associated with PCOS, one of the common causes for hirsutism. However, recent meta-analysis including nine placebo-controlled trials conclude that insulin-sensitizing drugs have no benefits for the treatment of hirsutism.[23-25]

The mainstay of medical treatment for hirsutism had been, and remains, estrogen-progestin contraceptives, with the addition of an antiandrogen after approximately 6 months if the desired cosmetic result has not yet been achieved.

Gonadotropin-releasing hormone agonists are used to suppress selectively ovarian steroid production by inhibiting LH secretion from the pituitary. This is used only when other antiandrogens have failed to give a response.

Dexamethasone acts by suppressing pituitary-adrenal axis. It is used in adrenal or mixed adrenal and ovarian hyperandrogenism. The dose varies from 0.25 to 1 mg daily at bedtime. The aim of treatment is to suppress androstenedione and bring 17-hydroxyprogesterone to normal range.

Eflornithine hydrochloride: This is an FDA-approved drug which acts by irreversibly inhibiting the enzyme ornithine decarboxylase and is applied locally twice daily. The enzyme is essential for the division and function of hair follicles. Hence, this drug slows the rate of hair growth but will not remove the hairs permanently. So, the women are advised to use the routine methods of hair removal concurrently. Side effects are erythema and local burning.

Hair removal: Physician should be familiar with the various methods of hair removal since medical treatment will not remove the hair already present and will take a period of 6–12 months before any improvement becomes apparent clinically.

TABLE 5: Comparison of various drugs management of hirsutism.

	Formulations/doses	*Mechanism of action*	*Side effects/remarks*
Estrogen-progestin contraceptives (60–100% women report improvement)	• Ethinyl estradiol (20–30 μg) + Drospirenone (3 mg) • Transdermal contraceptive patch (delivering 20 μg ethinyl estradiol and 150 μg norelgestromin daily) • Vaginal contraceptive ring (releasing 15 μg ethinyl estradiol and 120 μg etonogestrel daily)	• Suppresses pituitary LH secretion • Increases hepatic SHBG (estrogen induced) • Decreases adrenal DHEA-S secretion • Inhibits 5α-reductase activity in skin	
Progestin only	Medroxyprogesterone acetate (150 mg intramuscularly every 3 months, or 10–20 mg orally daily)	• Suppresses gonadotropins secretion (LH suppression) • Increases clearance of testosterone due to induction of hepatic enzyme activity	
Antiandrogens			
Cyproterone acetate	2 mg cyproterone acetate and 35 μg ethinyl estradiol	• Inhibits gonadotropin secretion • Competitive androgen receptor antagonist and • Inhibits enzyme involved in androgen synthesis	Fatigue, edema, loss of libido, weight gain, nausea, headache, and mastalgia
Spironolactone (aldosterone antagonist)	50–100 mg twice daily	• Competing with DHT for binding to androgen receptor • Inhibits ovarian and adrenal androgen synthesis	Fatigue, increased diuresis during early days of treatment, irregular bleeding, occasionally hyperkalemia
Flutamide	250–750 mg daily	Inhibits hair growth directly	Higher cost and potential for causing severe hepatotoxicity make it an unattractive therapeutic choice
Finasteride	5 mg daily	5α-reductase inhibitor (thus blocks the conversion of testosterone to DHT)	Decreased libido, potential risk of male fetal teratogenicity
Insulin-sensitizing drugs			
Metformin and thiazolidinediones (rosiglitazone and pioglitazone)	-	Decreases circulating insulin and androgen levels in women with PCOS	Guidelines issued by the endocrine society suggest against their use for the treatment of hirsutism
Gonadotropin-releasing hormone agonists	Leuprolide, nafarelin, goserelin	Decreases serum androgen levels drastically	Not recommended for routine use, primarily because they induce a severe hypoestrogenism, and more costly along with inconvenient to use
Glucocorticoids	Dexamethasone 0.25–1 mg daily at bedtime	• Suppresses endogenous ACTH secretion • Used in long term management of women with classical congenital adrenal hyperplasia	Less effective than oral contraceptives or antiandrogens for the treatment of hirsutism

Contd…

Contd...

	Formulations/doses	*Mechanism of action*	*Side effects/remarks*
Eflornithine hydrochloride	13.9% cream topically applied twice daily	Inhibitor of ornithine decarboxylase, an enzyme essential for hair growth	• Best suited for patients with mild facial hirsutism, such as occurs after menopause. • Side effects—local burning, erythema
Temporary hair removal		Removal of hair by plucking, waxing, shaving, or use of depilatory agents	Results achieved are temporary.
Permanent hair removal		Electrolysis and photoepilation (laser and pulsed light therapies)	• Electrolysis is painful and can cause scarring. • Laser techniques are less painful but are costly and can cause dyspigmentation.

Flowchart 4: Schematic representation of a case of hirsutism in adolescent/adult female.

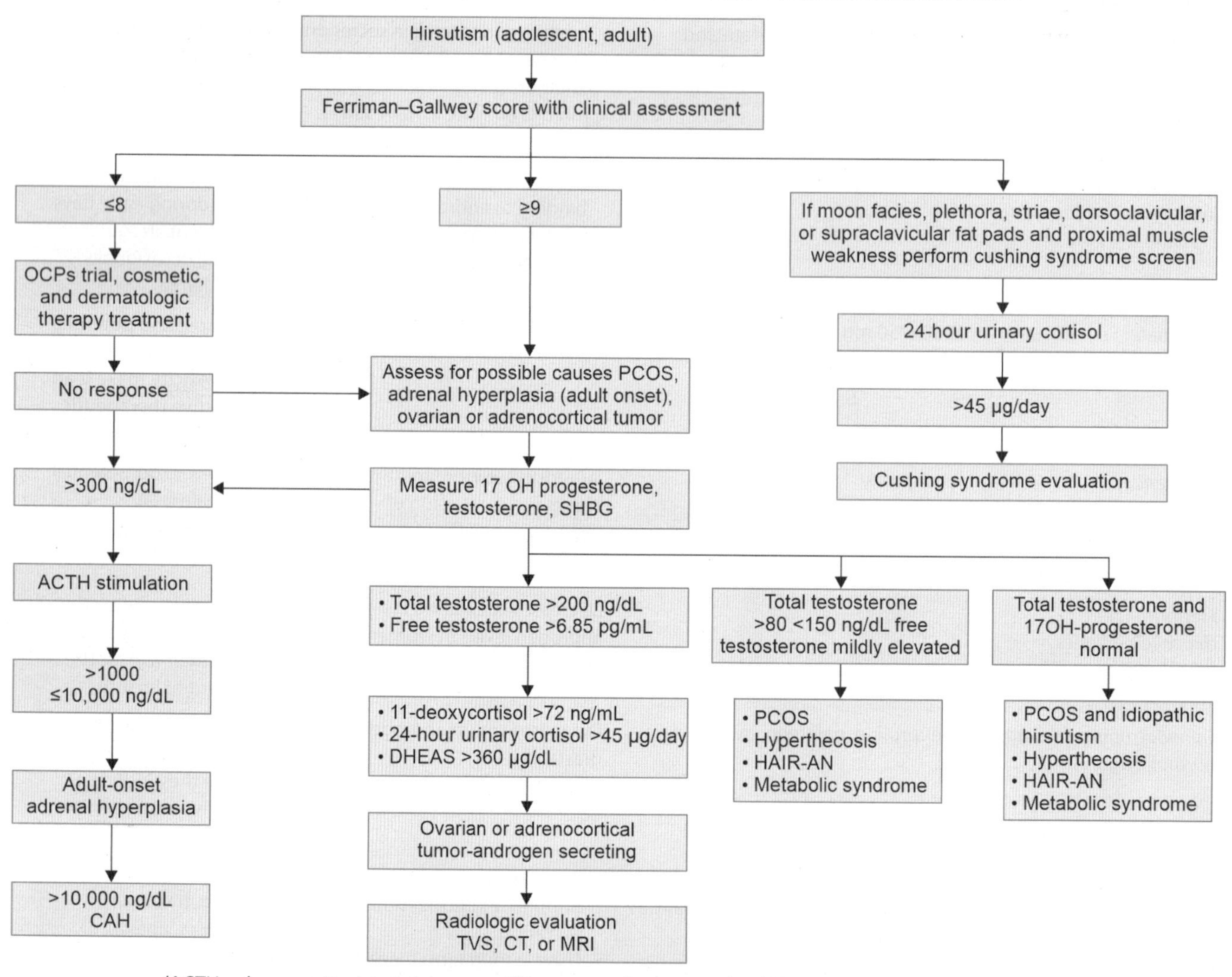

(ACTH: adrenocorticotropic hormone; CT: computed tomography; MRI: magnetic resonance imaging; PCOS: polycystic ovarian syndrome; TVS: transvaginal sonography)

Depilation includes shaving, or use of other topical depilatories. Shaving is the most common method used and does not increase the hirsutism but will increase the density of hair follicle.

Epilation includes methods such as threading, waxing, plucking, laser treatment, or electrolysis. Electrolysis and laser treatment are the permanent methods of hair removal.

Electrolysis needs repetitive sessions span over weeks to months and is executed by a trained professional in which electric current is passed to destroy the individual hair follicle. This can result in scarring and can be painful.

Alternatively, laser therapy can be employed in which only target tissue absorbs the light of selective wavelength, and the surrounding tissues receive minimal thermal damage. Usually performed for light-skinned women.

Advantages
- Cover wider surface area
- Fewer sessions needed
- Less pain.

But this procedure is costly and can result in dyspigmentation.

Permanent hair removal using electrolysis or photoepilation therapies (laser, pulsed light), when necessary, is best postponed until hormonal suppression has achieved its maximum benefits.

The comparison of various drugs management of hirsutism is given in **Table 5**.

The schematic representation of a case of hirsutism in adolescent/adult female is given in **Flowchart 4**.

REFERENCES

1. Mcknight E. The prevalence of "hirsutism" in young women. Lancet. 1964; 1:410-3.
2. Stenn KS, Paus R. Controls of hair follicle cycling. Physiol Rev. 2001:81(1);449-94.
3. Freinkel RK, Woodley DT (Eds). Hair. In: The Biology of the Skin. 1st edition. CRC Press; 2001.
4. Mihailidis J, Dermesropian R, Taxel P, Luthra P, Grant-Kels JM. Endocrine evaluation of hirsutism. Int J Womens Dermatol. 2017;3(1 Suppl):S6-S10.
5. Longcope C. Adrenal and gonadal androgen secretion in normal females. J Clin Endocrinol Metab. 1986;15:213-28.
6. Parker CR Jr. Androgens throughout the life of women. In: Azziz R, Nestler JE, Dewailly D (Eds). Androgen Excess Disorders in Women. Polycystic Ovary Syndrome and Other Disorders. 2nd edition. Totowa, NJ: Human Press; 2006. pp. 35-47.
7. Endoh A, Kristiansen SB, Casson RP, Buster JE, Hornsby PJ. The zona reticularis is the site of biosynthesis of dehydroepiandrosterone and dehydroepiandrosterone sulfate in the adult human adrenal cortex, resulting from its low expression of 3β-hydroxysteroid dehydrogenase. J Clin Endocrinol Metab. 1991;78:C113-8.
8. Hanings RV, Flood CA, Hackett RJ, Loughlin JS, McClure NRJ. Metabolic clearance rate of dehydroepiandrosterone sulfate, its metabolism to testosterone and its intrafollicular metabolism to dehydroepiandrosterone, androstenedione, testosterone and dihydrotestosterone in vivo. J Clin Endocrinol Metab. 1991;72:1088-95.
9. Serafini P, Ablan F, Lobo RA. 5 alfa-reductase activity in the genital skin of hirsute women. J Clin Endocrinol Metab. 1985;60(2):349-55.
10. Ferriman D, Gallwey JD. Clinical assessment of body hair growth in women. J Clin Endocrinol Metab. 1961;21:1440-7.
11. Hatch R, Rosenfield RL, Kim MH, Tredway D. Hirsutism: Implications, etiology, and management. Am J Obstet Gynecol. 1981;140:815-30.
12. Remeithe SA, Wherrett DK. Disorders of sex development in Fanaroff and Martin's neonatal-perinatal medicine. Elsevier Saunders. 2016:2;1524-6.
13. Martin KA, Chang RJ, Ehrmann DA, Ibanez L, Lobo RA, Rosenfield RL, et al. Evaluation and treatment of hirsutism in postmenopausal women: an endocrine society clinical practice guideline. J Clin Endocrinol Metab. 2008; 93:1105-20.
14. Ehrmann DA, Barnes RB, Rosenfield RL, Cavaghan MK, Imperial J. Prevalence of impaired glucose tolerance and diabetes in women with polycystic ovary syndrome. Diabetes Care. 1999;22:141-6.
15. Lobo RA. Hirsutism in polycystic ovary syndrome: current concepts. Clin Obstet Gynecol. 1991;34:817-26.
16. Hoffman DI, Klove K, Lobo RA. The prevalence and significance of elevated dehydroepiandrosterone sulfate levels in anovulatory women. Fertil Steril. 1984;42:76-81.
17. Deslypere JP, Verdonck L, Vermeulen A. Fat tissue: a steroid reservoir and site of steroid metabolism. J Clin Endocrinol Metab. 1985;61:564-70.
18. Thomson RL, Buckley JD, Lim SS, Noakes M, Clifton PM, Norman RJ, et al. Lifestyle management improves quality of life and depression in overweight and obese women with polycystic ovary syndrome. Fertil Steril. 2010;94:1812-6.
19. Kerchner A, Lester W, Stuart SP, Dokras A. Risk of depression and other mental health disorder in women with polycystic ovary syndrome: a longitudinal study. Fertil Steril. 2009:91:207-12.
20. Mathur R, Alexander CJ, Yano J, Trivax B, Azziz R. Use of metformin in polycystic ovary syndrome. Am J Obstet Gynecol. 2008;199:596-609.

21. Creanga AA, Bradley HM, McCormick C, Witkop CT. Use of metformin in polycystic ovary syndrome: a meta-analysis. Obstet Gynecol. 2008;111:959-68.
22. Legro RS, Barnhart HX, Schlaff WD, Carr BR, Diamond MP, Carson SA, et al. Clomiphene, metformin, or both for infertility in the polycystic ovary syndrome. N Engl J Med. 2007;356:551-66.
23. Moran LJ, Pasquali R, Teede HJ, Hoeger KM, Norman RJ. Treatment of obesity in polycystic ovary syndrome: a position statement of the Androgen excess and Polycystic Ovary Syndrome Society. Fertil Steril. 2009;92: 1966-82.
24. Glueck CJ, Phillips H, Cameron D, Sieve-Smith L, Wang P. Continuing metformin throughout pregnancy in women with polycystic ovary syndrome appears to safely reduce first-trimester spontaneous abortion: a pilot study. Fertil Steril. 2001;75:46-52.
25. Tang T, Glanville J, Orsi N, Barth JH, Balen AH. The use of metformin for women with PCOS undergoing IVF treatment. Hum Reprod. 2006;21(6):1416-25.

Chapter

23

Role of Thyroid in Gynecology

Surekha Tayade

INTRODUCTION

Menarche, pubertal growth and development, menstrual cycle, fertility, fetal development, postpartum period, reproductive years, and postmenopausal years are all significantly influenced by thyroid hormones.[1] Thyroid disorder is more likely to affect women than men. Thyroid problems in women can result in a wide range of reproductive illnesses, from infertility to aberrant sexual development. Infertility can be impacted by thyroid dysfunction in a number of ways, including anovulatory cycles, luteal phase defects, excessive prolactin (PRL) levels, and sex hormone imbalances. Thus, cyclic menstruation, fertility, and pregnancy require adequate thyroid function.[2]

PREVALENCE

The prevalence of hypothyroidism in pregnancy is around 2.5% according to the Western literature.[3] The prevalence of Grave's disease is around 0.1–0.4% and that of thyroid autoimmunity (TAI) is around 5–10%.[4] According to one of the studies conducted on Indian population, the prevalence of hypothyroidism, Graves' disease, gestational transient thyrotoxicosis, and TAI was 4.8%, 0.6%, 6.4%, and 12.4%, respectively.[5]

PHYSIOLOGY OF HYPOTHALAMIC-PITUITARY-THYROID AXIS

The integration of hormonal processes at the thyroid, hypothalamus, and pituitary levels is necessary for normal thyroid function. Thyrotropin, also known as thyroid-stimulating hormone (TSH), is a physiological indicator of thyroid hormones (THs) effects. It is secreted by the anterior pituitary's thyrotrophic cells and controls the thyroid's production of T3 and T4. In addition, the hypothalamus secretes thyrotropin-releasing hormone (TRH), which controls pituitary TSH secretion. The hypothalamic-pituitary-thyroid (HPT) axis is made up of TSH, TRH, and THs. In general, increased blood levels of THs prevent the release of TRH and TSH, whereas decreased serum TH levels have the reverse effect.[6]

Two tyrosyl residues, joined by an ether bond, make up the building blocks of T3 and T4, which have three and four iodine residues, respectively. Iodide entrance into the thyroid follicle is necessary for the manufacture of these hormones by the thyroid, which is reliant on the action of two transmembrane glycoproteins found in the thyroid, sodium-iodide symporter (NIS), and pendrin **(Fig. 1)**. Iodide is oxidized by thyroperoxidase (TPO) after entering the thyroid follicle and is then combined with Tg to produce

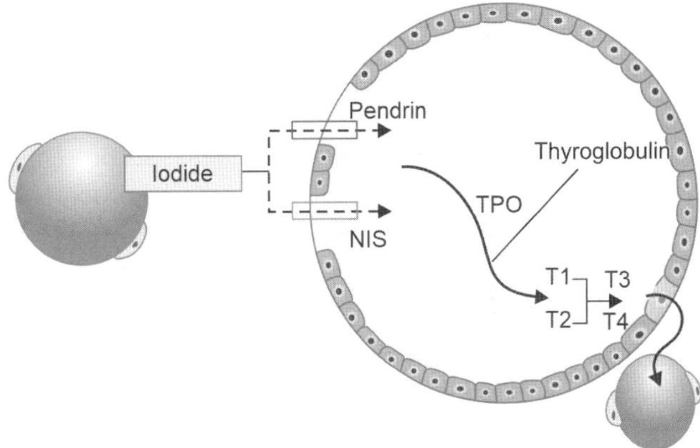

Fig. 1: Biosynthesis of thyroid hormones by the thyroid. Through two transmembrane glycoproteins, sodium-iodide symporter (NIS) and pendrin, iodide travels across the basolateral plasma membrane of thyrocytes and enters the thyroid follicle. Iodide enters the thyroid follicle, where it is oxidized by the enzyme thyroperoxidase (TPO) and combined with thyroglobulin to produce monoiodothyronine (T1) and diiodothyronine (T2), which are then converted into T3 and T4. The expression of the transcription factor Pax8 is necessary for the expression of NIS, pendrin, TPO, and thyroglobulin.

monoiodothyronine (T1) and diiodothyronine (T2), which are then converted into T3 and T4. The expression of the transcription factor Pax8, which is essential for the growth and efficient functioning of the thyroid, determines the expression of NIS, pendrin, TPO, and Tg.[7]

TRANSPORT AND BIOAVAILABILITY OF THYROID HORMONES

T4 is the primary hormone secreted by the thyroid and is thought to be a precursor of T3 or a prohormone. T3 is the biologically active hormone. T3 is about four times as powerful as T4, but its plasma half-life and circulating concentration are substantially shorter. Deiodinases (D1, D2, and D3) deiodinate T4 in peripheral tissues, such as the liver, resulting in the generation of T3 and/or reverse T3 (rT3). Reverse T3 does not appear to have any genomic effects, whereas T3 acts by binding to four distinct nuclear receptors, each of the receptors expresses differently depending on the target tissue.[8]

Most of the circulating T3 is produced by deiodinase type 1 (D1), whereas D2 regulates the production of intracellular T3. T4 can likewise be rendered inactive by D1 by becoming rT3. THs are inactivated by a third deiodinase, D3, which transforms T4 and T3 into rT3 and T2, respectively, in tissues. The bioavailability of THs is regulated by sulfation in addition to the action of the deiodinases, and 80% of the T4 generated by the thyroid is converted to inactive sulfated biological molecules such T4S, T3S, and rT3S.[9]

Less than 0.1% of the total quantity of circulating TH (T3 or T4) is in its free state, unattached to plasma proteins, and can enter cells by certain carrier-mediated processes. Thyroxine-binding globulin (TBG), transthyretin (TTR), and albumin are three distinct transporter proteins that T3 and T4 bind to reversibly when released into the bloodstream. T3 and T4 can be carried by any of the three proteins, while T4 has a stronger affinity for the three proteins. A minor portion of THs can also be bound to by lipoproteins. TBG, on the other hand, serves as the main carrier protein in humans due to its higher affinity for THs.[10]

Diffusion or carrier-mediated transport utilizing membrane transporters such as MCT8, MCT10, and Oatp1a2 allow T3 and T4 to enter the target cell. In the target cell, THs carry out their function directly by activating their nuclear receptors, promoting, or inhibiting the expression of transcription genes that are reliant on the recruitment of coactivators, such as the steroid receptor coactivator (SRC) and/or the dimerization of the retinoic acid X receptor (RXR) **(Fig. 2)**. In addition to nuclear receptors, THs can also operate indirectly by interacting with the membrane protein v3 integrin, which then triggers the signaling pathways mitogen activated protein kinase (MAPK) and ERK1/2 (extracellular signal-regulated kinase) to control the transcription and phosphorylation of its nuclear receptors.[11]

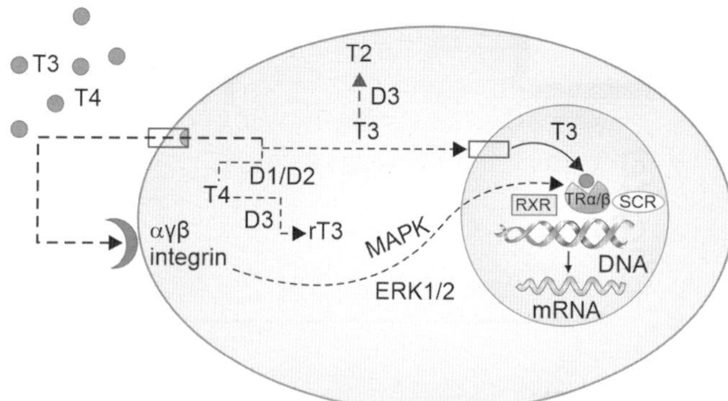

Fig. 2: Mechanism of action of thyroid hormones on target cells. The majority of T4s, or THs, enter the target cell through diffusion or carrier-mediated transport. Type 1 (D1) and type 2 deiodinases in the target cell convert T4 to T3 (D2). By converting T4 and T3 into rT3 and T2, respectively, deiodinase type 3 (D3) is in charge of inactivating THs. T3 reaches the cell nucleus and activates nuclear receptors in the deoxyribonucleic acid (DNA), promoting or inhibiting the activation of transcriptional genes that depend on the dimerization of the retinoic acid X receptor (RXR) and/or the recruitment of coactivators, such as the steroid receptor coactivator (SRC). In addition to nuclear receptors, THs can also function by attaching to the cell membrane protein known as v3 integrin, which then triggers a signaling cascade involving MAPK and ERK1/2 to control nuclear receptor transcription and phosphorylation.

All cells' metabolism of carbohydrates, proteins, and lipids is regulated by THs after activating their receptors, and they also control cell differentiation and proliferation. As a result, adjustments in TH plasma levels may have negative effects on the reproductive system as well as other organs and organ systems.

RISK FACTORS OF THYROID DISORDERS

Old age, female gender, goiter, and poorly controlled diabetes in type 2 diabetes mellitus patient found to be risk factors for thyroid dysfunction. Known risk factors for developing thyroid disease during pregnancy, including a family history of autoimmune thyroid disease, presence of a goiter, signs and symptoms of thyroid disease, known thyroid dysfunction, history of type 1 diabetes mellitus or other autoimmune diseases, prior neck irradiation and previous miscarriages or preterm deliveries.[12]

TYPES OF THYROID DISORDERS

- *Hypothyroidism*
  - Primary—TSH increases, T3 and T4 decrease.
  - Secondary or central—pituitary secretion of TSH is decreased so TSH, T3, and T4 are decreased.
- *Hyperthyroidism*
  - Primary—TSH decreases, T3 and T4 increase.
  - Secondary—TSH, T3, and T4 increase.

ETIOLOGY

Hypothyroidism

- Goiter
- Surgery or radioactive iodine to correct a thyroid problem
- Radiation treatment to the thyroid, neck, or chest
- Family history of thyroid disease
- Pregnancy
- Turner syndrome
- Celiac disease
- Sjögren syndrome
- Pernicious anemia
- Diabetes
- Rheumatoid arthritis
- Lupus

Hyperthyroidism

- Graves' disease
- Overactive thyroid nodules
- Inflammation of the thyroid gland, called thyroiditis
- Too much iodine
- A noncancerous tumor of the pituitary gland
- Family history of thyroid disease
- Pernicious anemia
- Diabetes
- Primary adrenal insufficiency
- Use nicotine products.
- Pregnancy

CLINICAL MANIFESTATION (TABLE 1)

Although the hormonal and other biochemical anomalies in these two conditions are different, both hyper and hypothyroidism can cause irregular menstruation and infertility. Through modifications in sex hormone levels, gonadotropin release, and perhaps ovarian function, thyroid disease affects both menstrual flow and fertility. In women, irregular bleeding patterns, perinatal difficulties, decreased fertility, and early menopause are common illnesses linked to thyroid dysfunction. Women with hyperthyroidism frequently experience oligomenorrhea and polymenorrhea, while hypothyroidism is more frequently associated with menorrhagia, polymenorrhea, and infrequently amenorrhea. Thyroid disease frequently results in anovulation. Women with thyroid disorders are more likely to experience prenatal complications such as miscarriage, pregnancy-induced hypertension (PIH), gestational diabetes mellitus (GDM), stillbirth, and prematurity as well as postnatal complications such as congenital abnormalities, postnatal depression, and abnormal post-partum bleeding.[13] Infertility and other gynecological diseases can be brought on by undiagnosed and untreated thyroid dysfunction.

TABLE 1: Clinical manifestations of thyroid disorders.

Hyperthyroidism	*Hypothyroidism*
Weight loss	Weight gain
Anxiety	Depression
Increased sweating	Dry skin
Tremulousness	lethargy
Diarrhea	Voice change
Palpitation	Changes in menses
Muscular weakness	Muscle cramps
Heat intolerance	Cold intolerance

EFFECT OF THYROID HORMONES ON OTHER HORMONES

Sex Steroid

Sex hormone-binding globulin (SHBG) is the transporter protein that carries sex steroids (testosterone, dihydrotestosterone, and estradiol) through the blood. THs influence the production of this transporter protein by modifying the production of hepatic SHBG via hepatocyte nuclear factor-4 (HNF4). Low levels of SHBG in the serum under hypothyroid conditions lead to lower levels of total circulating steroid levels and higher levels of free fraction. In contrast, increased SHBG in hyperthyroid circumstances raises overall circulating steroid levels, with normal or decreased free fraction. Both hypo- and hyperthyroidism slow down the metabolic clearance of sexual hormones. Thyroid hyperfunction is linked to higher plasma levels of estrogen, androstenedione, and testosterone as a result of increased androstenedione and testosterone synthesis, decreased clearance of 17-estradiol, and increased metabolism of androstenedione to estrone and testosterone to estradiol.[14] Sex steroids are also affected by THs in terms of their synthesis as well as their transport and elimination rates.

Hypothyroidism has also been demonstrated to induce a delay in sexual maturity or precocious puberty and galactorrhea. Hyperthyroidism in females before puberty causes delayed menstruation and an increased prevalence of oligomenorrhea or amenorrhea.

Prolactin

Since lactotropic cells, which are similar to thyrotropic cells and express membrane receptors for this releasing hormone, the PRL rise in hypothyroidism is related to TRH activation. So, following L-thyroxine therapy, hyperprolactinemia is frequently corrected when euthyroidism is restored. Additionally, a rise in pituitary vasoactive intestinal peptide, which influences PRL secretion by functioning as a paracrine or autocrine regulator, is also a contributing factor to the hypothyroidism-induced increase in PRL. The presence of hypogonadotropic anovulation, amenorrhea, and impaired fertility are also linked to hyperprolactinemia in women, just as they are with hypothyroidism. Dopaminergic agonist therapy can reduce PRL secretion and improve fertility. These reproductive changes result from the inhibition of the pulsatile secretion of LH caused by excess PRL that inhibits the activity of gonadotropin-releasing hormone (GnRH) neurons.[15]

EFFECT OF THYROID HORMONE ON OVARIAN MORPHOPHYSIOLOGY

Folliculogenesis and Ovulation

The proper gonad development, oocyte maturation, granulosa cell proliferation and differentiation, and the interaction of multiple hormones and growth factors that coordinate cyclic ovary alterations throughout folliculogenesis are all necessary for female fertility. Thus, elements of autocrine, endocrine, and/or paracrine origin work directly or indirectly in follicular cells to influence their differentiation, either for follicular growth or atresia, at each stage of follicular development. T3 and T4 are two of these elements that have been found in the follicular fluid of human ovarian follicles. T3 and T4 act directly on ovarian tissue, as shown by the expression of TH receptors in oocytes, granulosa, ovarian stromal, and cumulus cells. In granulosa cells, T3 can increase proliferation while lowering apoptosis. The interaction between T3 and gonadotropic hormones also prevents theca cells from producing too many androgens and induces aromatization, which results in granulosa cells producing more estrogen.[16]

Luteogenesis

The amount of progesterone produced by the corpus luteum determines the length of the female menstrual cycle and the gestation period. Due to a reduction in the production of 20 alpha-hydroxysteroid dehydrogenase (HSD), an enzyme that converts progesterone in the corpus luteum to the inactive form 20 alpha-hydroxyprogesterone, women with hypothyroidism may experience a prolonged luteal phase. Along with high levels of PRL, the rise of progesterone promotes lower basal gonadotropin (LH and FSH) levels by negatively affecting the hypothalamus' and pituitary's ability to secrete gonadotropins. The decrease in estradiol levels associated with hypothyroidism may be brought on by decreased ovarian granulosa cell responsiveness to FSH or suppression of FSH release brought on by increased progesterone levels. Thus, the rise in luteal LH receptors brought on by hyperprolactinemia may have had an impact on estradiol levels. Since there are more receptors, LH has a higher impact in the corpus luteum, boosting the production of androstenedione, and estradiol from progesterone.[17]

Premature luteolysis brought on by hyperthyroidism might cause premature labor. Pregnant women with hyperthyroidism exhibit increased apoptosis and corpus luteum expression of cyclooxygenase-2 (COX-2), prostaglandin F2 alpha (PGF2), and hydrosteroid dehydrogenase (HSD), a decrease in luteotropic factors such prostaglandin E2 (PGE2) and estrogen receptor (ER), and higher luteal expression of vascular endothelial growth factor (VEGF) and fetal liver kinase 1 (FLK1). Additionally, there is a rise in the proliferation of pericytes and endothelial cells. All of these findings show that thyroid dysfunctions impact luteolysis and luteal vascularization in pregnant women.

THYROID DYSFUNCTION AND OVARIAN CYSTS

The development of large ovarian cysts is facilitated by the presence of human chorionic gonadotropin, and the occurrence of ovarian cysts in women with severe hypothyroidism may be linked to changes in circulating LH concentrations and preovulatory secretion of LH and FSH, particularly during gestation. Since they typically do not exhibit the clinical indications of ovarian

TABLE 2: Effect of thyroid disorders on ovaries.

Hyperthyroidism	*Hypothyroidism*
Increases folliculogenesis	Decreases folliculogenesis
Decreases follicular atresia	Increases follicular atresia
Increases ovulation	Decreases ovulation
Accelerates luteolysis	Retards luteolysis
Increases luteal angiogenesis	Decreases luteal angiogenesis

TABLE 3: Effect of thyroid disorders on uterus and uterine tubes.

Hyperthyroidism	*Hypothyroidism*
Increases uterine weight	Decreases uterine weight
Increases endometrial thickness	Decreases endometrial thickness
Increases uterine vascularization	Decreases decidualization
Increases ampulla epithelium	Decreases infundibulum epithelium

hyperstimulation, such as abdominal distension, hemoconcentration, ascites, or pleural effusion, cases of hypothyroidism producing ovarian hyperstimulation are underdiagnosed in women, particularly in nonpregnant women. Despite this finding, since TSH and FSH have a structural relationship, it has been hypothesized that the development of ovarian cysts in hypothyroidism may be linked to high levels of TSH that can activate FSH receptors in the ovary. Another theory is connected to the previously described hyperprolactinemia in hypothyroidism, which affects LH secretion by inhibiting GnRH. Given that mutations in these receptors were found in pregnant women with spontaneous ovarian hyperstimulation syndrome, it is also thought that mutations in the FSH receptor enhance the activation brought on by human chorionic gonadotropin (hCG) or TSH **(Table 2)**.

EFFECT OF THYROID HORMONES ON THE UTERUS AND UTERINE TUBE

Through their intracellular receptors, thyroid hormones influence the uterus and uterine tube, controlling how receptive these organs are to estrogen. While the expression of deiodinases decreases in the secretory phase and is inversely related to the increase in progesterone, the expression of T3 and T4 receptors increases in the middle of the secretory phase in the uterine epithelium. Therefore, it is conceivable that variations in T3 and T4 serum levels have an impact on uterine and uterine tube morphophysiology by impairing the proper activation of their receptors throughout the menstrual cycle as well as by affecting plasma sex steroid concentrations and the trophic action of these hormones on the genital tract.

Similar to alterations seen in the uterus, TH deficiency in the uterine tube affects the villus height of the infundibulum as well as the quantity and size of villus-lining cells, dramatically lowering the segment's epithelial height. All of these changes to the uterus and uterine tubes have the potential to impair embryo fertilization, differentiation, nutrition, and implantation, which explains the embryonic death and lower implantation rate seen in hypothyroid patients.

In contrast to hypothyroidism, there is an increase in the height of the uterine tube's ampulla's epithelium in hyperthyroidism. Additionally, hyperthyroidism increases the uterine tube's secretory activity and the endometrium and myometrium's thickness, which results in a thicker uterine wall.[18]

EFFECT OF THYROID HORMONES ON THE UTERUS DURING GESTATION

Leukemia inhibitory factor, a crucial component of the process of decidualization and embryo implantation, has also been reported to increase in blood levels with hypothyroidism, and TSH has been demonstrated to stimulate the production of this factor in cultures of endometrial stromal cells. This result supports the idea that TSH and THs both play crucial roles in the implantation and decidualization processes.

For the embryo to implant and survive as well as to anchor and regulate fetal-placental development, the endometrium must become decidualized. Hypothyroidism affects decidualization during implantation and that women with hypothyroidism have decreased decidual cell expression of IL-4 and IL-10. Triiodothyronine also stimulates the in vitro production of inflammatory cytokines and angiogenic factors by human decidual cells, and this response is gestational period dependent[19] **(Table 3)**.

EFFECT OF THYROID HORMONES ON PLACENTAL MORPHOPHYSIOLOGY

Maternal THs play an important role in trophoblastic cells' proliferation, differentiation, survival, invasion, and endocrine functions throughout pregnancy, especially with regard to the placenta. The primary cause of this participation in trophoblast cell activity is the direct action of THs on certain nuclear receptors found in the syncytiotrophoblast and villous cytotrophoblast of the human villous placenta. The effects of maternal thyroid dysfunction in women include abortion, preterm delivery, hypertension, fetal death, and mental impairments in offspring.

According to several research, hypothyroidism in women impacts placental and/or fetal weight in relation to fetal-placental development. Although thyroid dysfunctions have an impact on fetal-placental development, their impact on a woman's ability to get pregnant will depend on when the endocrine dysfunction first appears in connection to conception as well as how severe the hypo- or hyperthyroidism is. After conception, moderate-to-severe maternal hypothyroidism irreversibly slows placental weight gain and fetal growth.

It is critical to emphasize that, in contrast to subclinical hypo- and hyperthyroidism, clinical hypo- and hyperthyroidism in pregnant women necessitate therapy. There is currently little evidence supporting the efficacy of treating subclinical maternal hypothyroidism, and there is disagreement among scientists as to whether medication is even necessary. However, if the cause of the gestational subclinical hypothyroidism is an autoimmune thyroid condition, therapy may be advantageous. Still a rare condition, maternal subclinical hyperthyroidism does not need to be treated while a woman is pregnant[20] **(Table 4)**.

TABLE 4: Effect of thyroid disorders on placenta and fetus.

Hyperthyroidism	*Hypothyroidism*
Increases offspring premature birth	Decreases placental and fetal weight
Increases trophoblast proliferation	Decreases trophoblast proliferation
Increases placental lactogen	Increases trophoblast apoptosis
Increases Nitric oxide synthase (NOS), Interferon gamma (IFN-γ), Microphage Migratory Inhibitory factor (MIF)	Decreases NOS2, INFγ
Decreases TNF, VEGF, FLK1	Decreases VEGF, PIGF, Matrix metalloproteinase (MMPs), Toll like receptor 4, IL10
	Compromises trophoblastic differentiation
	Affects placental morphogenesis
	Compromises placental vascularization
	Abortion preeclampsia

THYROID DYSFUNCTION AND ABNORMAL UTERINE BLEEDING

An irregular menstrual cycle is linked to thyroid problems in women. Many physiological processes in females, including pubertal growth and development, menarche, menstrual cycles, fertility, and menopause, are initiated by thyroid hormones. The alteration of TSH response, rise in prolactin levels, alteration of luteinizing hormone (LH) response, alteration of peripheral androgen to estrogen conversion, alteration of SHBG, and alteration of coagulation pathways, in addition to effect on lipid profile, may be the mechanisms by which thyroid disorders are associated with abnormal uterine bleeding (AUB).

In women of reproductive age, hypothyroidism frequently contributes to a variety of problems, including infertility, menstrual irregularities, and sexual dysfunction. For women with hypothyroidism, irregular periods and abnormal blood flow are the earliest signs of menstrual disorders. Women with subclinical hypothyroidism frequently appear with undetected menorrhagia. If hyperthyroidism develops before puberty, it delays the start of menstruation, but in women of reproductive age, it is linked to oligomenorrhea and amenorrhea.[21]

THYROID DYSFUNCTION AND INFERTILITY

Thyroid hormone receptors are present in a variety of cells, including oocytes, cumulus cells, and granulose cells as is well known. Both the local impact of thyroid hormones on these receptors and the inhibition of pulsatile GnRH release from the hypothalamus in hypothyroidism result in the failure of healthy ovulation and the absence of high-quality oocytes. Additionally, corpus luteum insufficiency brought on by LH release disruption and hyperprolactinemia brought on by excessive TRH release may both have detrimental effects on the regular menstrual cycle. Although high TSH may lower the sex hormone binding protein (SHBP), hence boosting free testosterone and estradiol levels, total serum levels of these hormones are often low in hypothyroidism. Additionally, a decrease in the metabolism's ability to remove androstenedione and estrone may lead to an increase in the aromatization of peripheral estrogens. Infertility, menorrhagia, and polymenorrhea are among the symptoms of hypothyroidism that may be brought on by these mechanisms as well as a decrease in hemostasis components without consideration to these mechanisms.

The most prevalent autoimmune illnesses in women of reproductive age are autoimmune thyroid diseases (AITD), which are characterized by the development of antibodies against specific thyroid gland structures such thyroglobulin (TG), TPO, and thyroid microsomal (TM). Autoimmunity that is organ-specific or non-specific is stimulated by an overactive immune response, which is typically 5–10 times more prevalent in females. Organ-specific ATA that occurs against TG causes thyroid follicular tissue destruction, while ATA that occurs against TPO prevents tyrosine from being iodinated. Non-specific antibodies develop when T-cells are stimulated to attack particular thyroid chemicals, which then influences B cells. A dysfunctional immune response alters the profile of endometrial T-cells, which hinders effective implantation while also causing the release of cytokines such interferon-, IL-4, and IL-10.

Hormonal alterations that affect the reproductive system may occur in hyperthyroid women. According to Joshi et al.,[22] hyperthyroidism causes an increase in androstenedione and testosterone synthesis, which in turn causes estrone and estradiol levels to rise. Higher plasma estrogen levels are caused by both this mechanism and a decrease in the metabolic clearance of estrogen in hyperthyroid women. A few authors have also claimed to have found elevated serum levels of LH, FSH, and SHBG. Infertility is around 5–8% more common in women with hyperthyroidism.[22]

DIAGNOSIS

Methods of diagnosis for hyperthyroidism include:

- Analysis of medical history and symptoms
- Physical examination
- An ultrasound or nuclear medicine scan of your thyroid to see if it has nodules, or to determine whether it is inflamed or overactive.
- Blood tests to measure levels of thyroid hormones and related substances, including thyroid stimulating hormone (TSH), triiodothyronine (T3), thyroxine (T4), and thyroid peroxidase antibody (TPO).

Diagnosis of hypothyroidism:
The blood level of T4 will be low, while the TSH level will be high.

TREATMENT

Hypothyroidism

Levothyroxine is recommended as the preparation of choice for the treatment of hypothyroidism due to its efficacy in resolving the symptoms of hypothyroidism, long-term experience of its benefits, favorable side effect profile, ease of administration, good intestinal absorption, long serum half-life, and low cost.

Hyperthyroidism

- *Antithyroid drugs:* These medications interfere with the production of thyroid hormones. Methimazole (brand name Tapazole) is the most commonly prescribed hyperthyroidism medication, whereas propylthiouracil (also known as PTU) is preferred during the first trimester of pregnancy.
- *Radioactive iodine (RAI) treatment:* RAI therapy, also called radioiodine, is recommended for those who are particularly vulnerable to antithyroid medication and surgery.
- *Surgical removal of the thyroid gland or nodule:* Also known as a thyroidectomy, surgery to remove the thyroid (either one part of the gland or the entire gland) can be an effective and permanent treatment for hyperthyroidism.

REFERENCES

1. Ajmani NS, Sarbhai V, Yadav N, Paul M, Ahmad A, Ajmani AK. Role of thyroid dysfunction in patients with menstrual disorders in tertiary care center of walled city of Delhi. J Obstet Gynaecol India. 2016;66(2):115-9.
2. Orazulike NC, Odum EP. Evaluation of thyroid function in infertile female patients in Port Harcourt, Nigeria. Trop J Obstet Gynaecol. 2018;35(1):38-43.
3. LeBeau SO, Mandel SJ. Thyroid disorders during pregnancy. Endocrinol Metab Clin North Am. 2006;35(1):117-36, vii.
4. Hollowell JG, Staehling NW, Flanders WD, Hannon WH, Gunter EW, Spencer CA, et al. Serum TSH, T(4), and thyroid antibodies in the United States population (1988 to 1994): National Health and Nutrition Examination Survey (NHANES III). J Clin Endocrinol Metab. 2002;87(2):489-99.
5. Nambiar V, Jagtap VS, Sarathi V, Lila AR, Kamalanathan S, Bandgar TR, et al. Prevalence and impact of thyroid disorders on maternal outcome in Asian-Indian pregnant women. J Thyroid Res. 2011;2011:e429097.
6. Choksi NY, Jahnke GD, St Hilaire C, Shelby M. Role of thyroid hormones in human and laboratory animal reproductive health. Birth Defects Res B Dev Reprod Toxicol. 2003;68(6):479-91.
7. Di Palma T, Nitsch R, Mascia A, Nitsch L, Di Lauro R, Zannini M. The paired domain-containing factor Pax8 and the homeodomain-containing factor TTF-1 directly interact and synergistically activate transcription. J Biol Chem. 2003;278(5):3395-402.

8. Harvey CB, Williams GR. Mechanism of thyroid hormone action. Thyroid. 2002;12(6):441-6.
9. Wu SY, Huang WS, Polk D, Chen WL, Reviczky A, Williams J, et al. The development of a radioimmunoassay for reverse triiodothyronine sulfate in human serum and amniotic fluid. J Clin Endocrinol Metab. 1993;76(6):1625-30.
10. GALE. (2018). Thyroid hormones and female reproduction - Document - Gale Academic OneFile. [online] Available from: https://go.gale.com/ps/i.do?id=GALE%7CA586025987&sid=googleScholar&v=2.1&it=r&linkaccess=abs&issn=00063363&p=AONE&sw=w&userGroupName=anon%7Ea53c83c3 [Last accessed March, 2024].
11. Cheng SY, Leonard JL, Davis PJ. Molecular aspects of thyroid hormone actions. Endocr Rev. 2010;31(2):139-70.
12. American Thyroid Association. (2021). Thyroid and pregnancy. [online] Available from: https://www.thyroid.org/patient-thyroid-information/ct-for-patients/vol-3-issue-11/vol-3-issue-11-p-9-10/ [Last accessed March, 2024].
13. Koutras DA. Disturbances of menstruation in thyroid disease. Ann N Y Acad Sci. 1997;816:280-4.
14. Dittrich R, Beckmann MW, Oppelt PG, Hoffmann I, Lotz L, Kuwert T, et al. Thyroid hormone receptors and reproduction. J Reprod Immunol. 2011;90(1):58-66.
15. Grattan DR, Jasoni CL, Liu X, Anderson GM, Herbison AE. Prolactin regulation of gonadotropin-releasing hormone neurons to suppress luteinizing hormone secretion in mice. Endocrinology. 2007;148(9):4344-51.
16. Gregoraszczuk EL, Skalka M. Thyroid hormone as a regulator of basal and human chorionic gonadotrophin-stimulated steroidogenesis by cultured porcine theca and granulosa cells isolated at different stages of the follicular phase. Reprod Fertil Dev. 1996;8(6):961-7.
17. Frasor J, Gibori G. Prolactin regulation of estrogen receptor expression. Trends Endocrinol Metab. 2003;14(3):118-23.
18. Oliveira TS, Nunes VA, Nascimento EF, Serakides R. Histomorphometry and histochemistry of the uterine tubes and uterus of puberal and prepuberal rats induced for hyperthyroidism. Arq Bras Med Vet Zootec. 2005;57:448-56.
19. Vasilopoulou E, Loubière LS, Lash GE, Ohizua O, McCabe CJ, Franklyn JA, et al. Triiodothyronine regulates angiogenic growth factor and cytokine secretion by isolated human decidual cells in a cell-type specific and gestational age-dependent manner. Hum Reprod. 2014;29(6):1161-72.
20. Mintziori G, Kita M, Duntas L, Goulis DG. Consequences of hyperthyroidism in male and female fertility: pathophysiology and current management. J Endocrinol Invest. 2016;39(8):849-53.
21. Mansourian AR. Female reproduction physiology adversely manipulated by thyroid disorders: a review of literature. Pak J Biol Sci. 2013;16(3):112-20.
22. Joshi JV, Bhandarkar SD, Chadha M, Balaiah D, Shah R. Menstrual irregularities and lactation failure may precede thyroid dysfunction or goitre. J Postgrad Med. 1993;39(3):137-41.

Chapter 24

Gynecological Disorders and Adrenals

Tanu Priya, Vinay K Dogra, KD Kamal

INTRODUCTION

The suprarenal glands or the adrenal glands are paired organs with triangle shape present on superior aspect of bilateral kidneys. Adrenal glands produce glucocorticoids (cortisol), mineralocorticoid (aldosterone), androgen precursors and a small amount of androgens (testosterone and 11-oxygenated 19-carbon androgens/precursors) from its cortex (90% of total adrenal gland volume) and catecholamines (norepinephrine and epinephrine) from central portion of the adrenal gland called adrenal medulla (10% of total adrenal gland volume). These hormones play a vital role in the body's fight or flight response. The suprarenal gland also secretes several essential hormones that play a significant role in regulating the body's immune system, metabolism, and salt and water balance. Adrenal androgens seem to play a role in the adult woman and in both sexes before puberty. This chapter reviews adrenal glands' anatomy, physiology, biosynthesis, regulation and biological actions of adrenal cortex hormones and pathological aspects of these in females.

ADRENAL GLAND ANATOMY

The adult adrenal glands are pyramidal-shaped retroperitoneal structure. Each gland is approximately 2 cm × 5 cm × 1 cm thick and weighs 4 g each lying above the upper poles of the kidneys. Each gland is composed of outer adrenal cortex and the inner medulla, contained within a single capsule. Its arterial blood supply from branches of the phrenic arteries, the aorta, and the renal arteries, which give rise to the superior, middle, and inferior adrenal arteries, respectively. Arterial blood enters from the outer cortex, flows through fenestrated capillaries between the cords of cells, and drains into venules in the medulla. The right adrenal vein directly enters the inferior vena cava and the left one drains into the left renal vein.

ADRENAL STEROIDS AND STEROIDOGENESIS

The adrenal cortex is divided into three histologic and functional zones—the outer zone is called zona glomerulosa (ZG), the intermediate zone is called zona fasciculate (ZF), and the inner zone is called zona reticularis (ZR) **(Fig. 1)**. Each one is characterized by the expression of specific steroidogenic enzymes, which result in the production of different steroid hormones. ZG contributes 15% of cortical volume and secrets

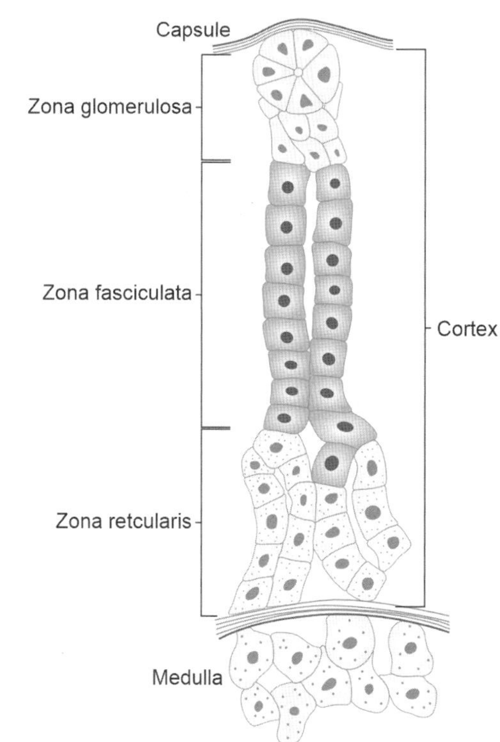

Fig. 1: Schematic diagram of the structure of the human adrenal cortex, depicting the outer zona glomerulosa and inner zona fasciculata and zona reticularis.[2]

aldosterone. ZF is the thickest part of the adrenal cortex, contributing about 75% of the cortex, produces cortisol and corticosterone as well as small amounts of androgens and estrogens. ZR surrounds the medulla and produces the adrenal androgens; dehydroepiandrosterone (DHEA), dehydroepiandrosterone sulphate (DHEAS), androstenedione, 11β-hydroxyandrostenedione, and small amounts estrogens.[1]

REGULATION OF ADRENAL STEROIDOGENESIS

Under the influence of adrenocorticotropic hormone (ACTH) glucocorticoids are secreted in relatively high amounts (cortisol, 10–20 mg/day) from the ZF **(Fig. 2)**; angiotensin II controls mineralocorticoids which are secreted in low amounts (aldosterone, 100–150 μg/day) from the ZG. Most abundant steroids secreted from the adult adrenal gland (>20 mg/day) are adrenal androgen precursors (DHEA, DHEAS, androstenedione, 11β-hydroxyandrostenedione) **(Flowchart 1)**.[2,3]

DISEASES OF ADRENAL GLANDS

Different pathological conditions of adrenal as congenital adrenal hyperplasia (CAH), Cushing syndrome (CS) and androgen secreting adrenal tumors can present in females with different gynecological disorders **(Table 1)**. Polycystic ovary syndrome (PCOS) is primarily an ovary-related disorder of chronic anovulation. New evidence suggests that the adrenal steroid 11-hydroxyandrostenedione may be converted to 11-ketotesosterone and 11-keto-dihydrotestosterone in peripheral or target tissues. These latter steroids may act as potent androgens[4] and may have a role in clinical features of PCOS. There is also increased adrenal output of DHEAS, which (after peripheral conversion to DHEA that is further converted to androstenedione) contributes indirectly to extraovarian

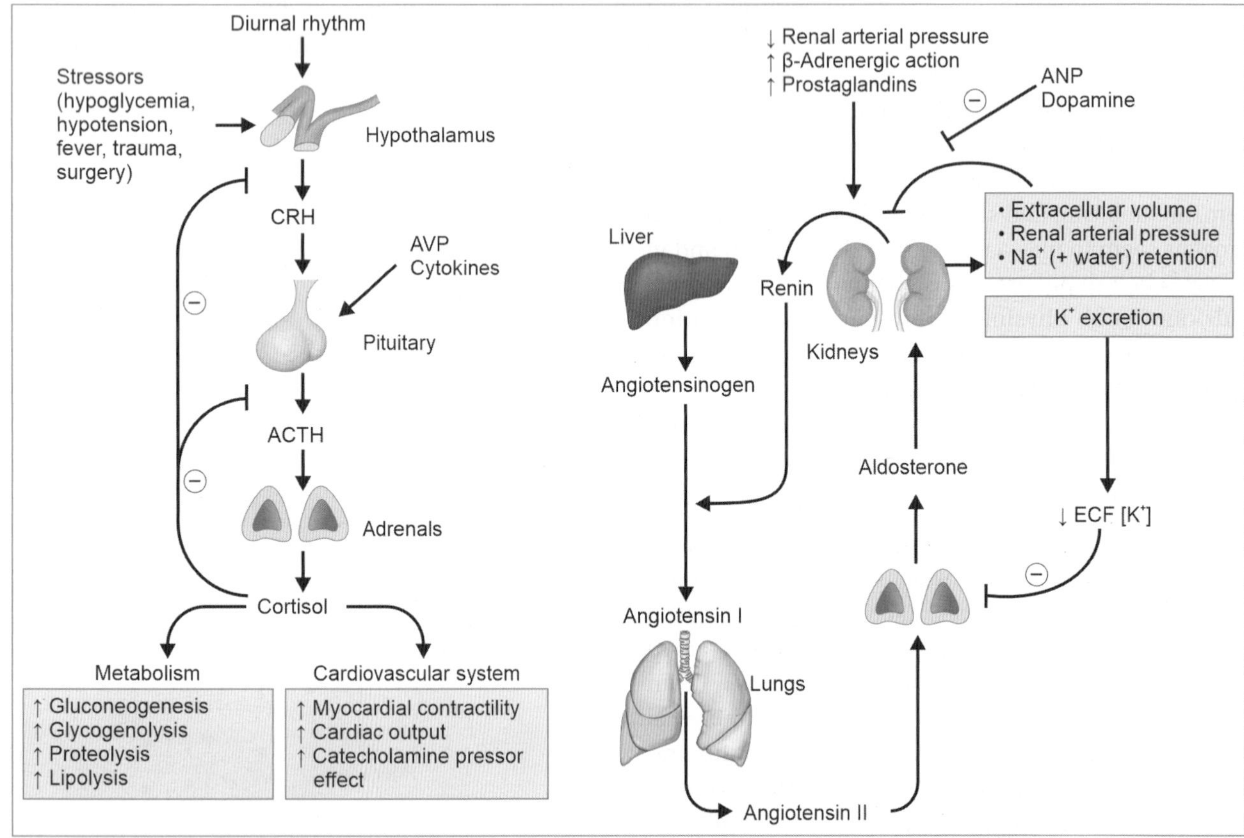

Fig. 2: Normal negative feedback regulation of cortisol (A) and aldosterone secretion (B). Hypothalamic-pituitary-adrenal axis (A). (ACTH: adrenocorticotropic hormone; CRH: corticotropin-releasing hormone; AVP: arginine vasopressin; ECF: extracellular fraction; ANP: atrial natriuretic peptide)

Flowchart 1: Adrenal steroidogenesis.

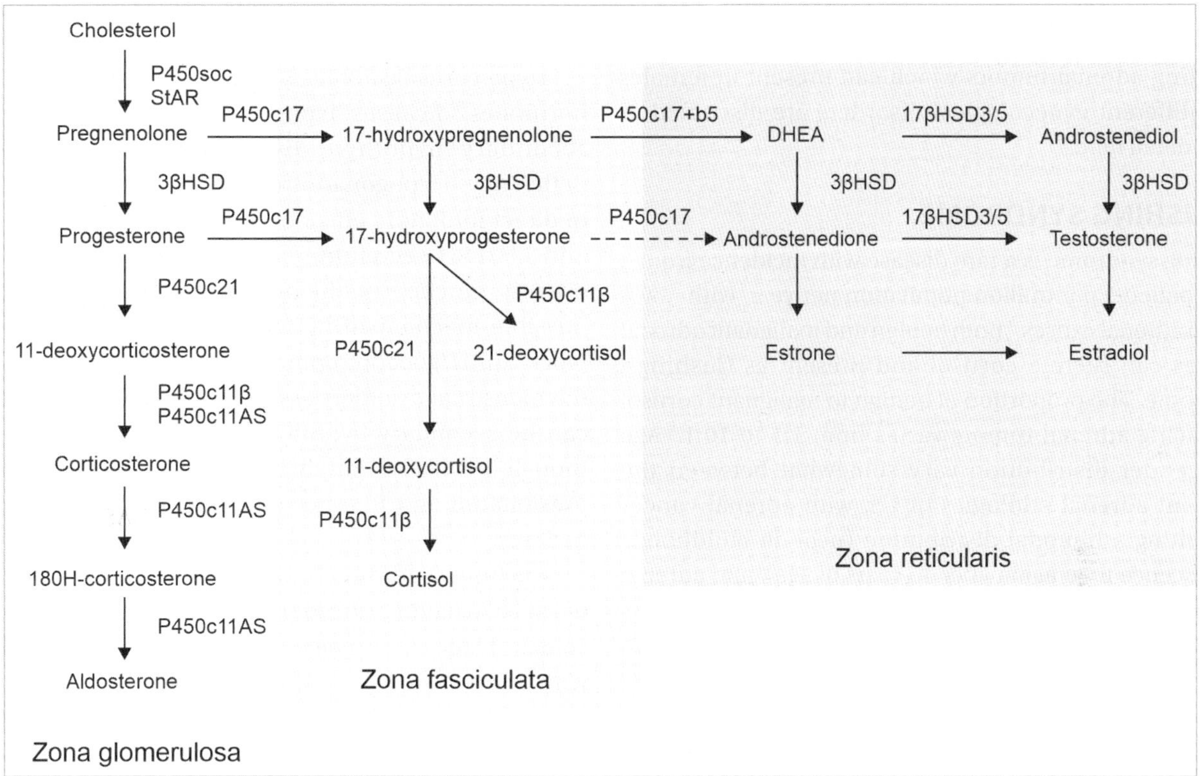

Cholesterol
↓ P450soc StAR
Pregnenolone —P450c17→ 17-hydroxypregnenolone —P450c17+b5→ DHEA —17βHSD3/5→ Androstenediol
↓ 3βHSD ↓ 3βHSD ↓ 3βHSD ↓ 3βHSD
Progesterone —P450c17→ 17-hydroxyprogesterone - - P450c17 - -→ Androstenedione —17βHSD3/5→ Testosterone
17-hydroxyprogesterone —P450c11β→ 21-deoxycortisol
↓ P450c21 ↓ P450c21 ↓ ↓
11-deoxycorticosterone 11-deoxycortisol Estrone ——→ Estradiol
↓ P450c11β P450c11AS ↓ P450c11β
Corticosterone Cortisol
↓ P450c11AS
18OH-corticosterone
↓ P450c11AS
Aldosterone

Zona glomerulosa | Zona fasciculata | Zona reticularis

(DHEA: dehydroepiandrosterone; P450: cytochrome; SCC: side chain cleavage; StAR: steroidogenic acute regulatory protein; 3β-HSD: 3β-hydroxysteroid dehydrogenase; P450c21, 21-hydroxylase; P450c11β; 11β-hydroxylase; P450c11AS, aldosterone synthase; P450c17; 17α hydroxylase; P450c17+b5, 17/20-lyase; 17β-HSD: 17β-hydroxysteroid dehydrogenase)

TABLE 1: Diseases of adrenal glands.

Adrenocortical diseases	***Adrenal medullary disease***
Glucocorticoid excess • Cushing syndrome	Pheochromocytoma
Glucocorticoid resistance	
Glucocorticoid deficiency • Primary hypoadrenalism • Secondary hypoadrenalism • Post chronic corticosteroid replacement therapy	
Congenital adrenal hyperplasia (CAH) Deficiencies of 21-hydroxylase, 3βHSD, 17α-hydroxylase, 11β-hydroxylase, P450 oxidoreductase, P450 side-chain cleavage, and StAR	
Mineralocorticoid excess or AME/11βHSD deficiency	
Mineralocorticoid deficiency • Defects in aldosterone synthesis • Defects in aldosterone action • Hyporeninemic hypoaldosteronism	
Adrenal Incidentalomas, adenomas, and carcinomas	

[AME: apparent mineralocorticoid excess; HSD: Hydroxysteroid dehydrogenase; StAR: steroidogenic acute regulatory (protein)] testosterone formation in PCOS patients. The topic of PCOS is discussed in detail elsewhere in this book.

Congenital adrenal hyperplasia, CS, and androgen secreting adrenal tumors which can present in females with different gynecological disorders are discussed in more detail here.

CUSHING SYNDROME

Cushing syndrome is a rare disease with incidence from 2 to 10 patients per million population per year, with <50% due to adrenal causes.[5] Both benign and malignant adrenal masses can secrete cortisol and present as Cushing's syndrome. The proportion of benign to malignant cortisol producing adrenal tumors varies from 1:1 to 10:1.[5] Age and gender distribution vary somewhat between the different adrenal etiologies of CS, with adrenal cancer presenting a bimodal distribution (peak in childhood/adolescence then again after 45 years of age), and adrenal adenoma occurring mostly in young adults (25–45 years of age). Primary pigmented nodular adrenal dysplasia (PPNAD) mostly in infancy and adolescence; ACTH-independent macronodular adrenocortical hyperplasia (AIMAH) mostly in adults older than 40 years of age and the McCune-Albright syndrome (MAS) are few genetic diseases with adrenal CS.

Most notably in adrenocortical cancer and rarely in adenomas in addition to cortisol hypersecretion, adrenal androgens may also be produced in excess. Androgen secretion is due to inefficient steroidogenesis, as enzymes involved in the latter steps of cortisol synthesis (e.g., 11β-hydroxylase or 21β-hydroxylase) are unable to transform their substrate overload. Precursors accumulate and are diverted to alternative biosynthetic pathways which leads to accumulation of DHEA, DHEAS, androstenedione, and estrogen.

Gynecological Features of Cushing's Syndrome

Gynecological clinical features of CS are hirsutism; acne over face, chest, and back; temporal balding; menstrual alterations as oligomenorrhea or amenorrhea; secondary polycystic ovary syndrome; loss of libido, infertility.

Menstrual abnormalities are seen in about 80% of patients with pituitary-dependent CS at diagnosis, with amenorrhea being the most common presentation.[6]

Hirsutism, a common feature of ACTH-dependent Cushing's syndrome, is usually absent in benign adrenal tumors that secrete only cortisol but, conversely, accompanies malignant transformation because of co-secretion of different androgens.[2]

Hypercortisolism causes suppression of other pituitary hormones so causes hypogonadotropic hypogonadism, secondary hypothyroidism, and growth hormone deficiency. Hypogonadotropic hypogonadism is result of direct inhibitory effect of cortisol on GnRH pulsatility and LH/FSH secretion.[7] GnRH neurons are known to express glucocorticoid receptors,[8] thus indicating a possible mechanism by which hypercortisolemia can block GnRH/gonadotropin release and be responsible for the menstrual disorders associated with CS. Other causes can be secondary hypothyroidism, hyperandrogenism though less likely.[6] Rarely infertility may be the primary presentation in CS[9] especially in mild or subclinical CS patients.

Cushing syndrome in pregnancy is fortunately rare, because ovulatory disorders and consequently infertility constitute the clinical picture in 75% of untreated patients with CS.

In 60% of cases of pregnancy with CS are due to ACTH-independent causes (48% adenoma and 10% carcinoma). Rarely true pregnancy-induced CS are seen which regresses after delivery.[10] The onset of adrenal-dependent CS may relate to the aberrant expression of LH receptors on the tumor, cross-reacting with human chorionic gonadotropin (hCG).

Diagnosis of Cushing Syndrome

High clinical index of suspicion needed for diagnosis of CS. Widespread indiscriminate biochemical screening in obese, hypertensive, infertile and diabetic populations is not recommended. One of four highly sensitive screening tests are recommended—urinary free cortisol, at least two measurements (greater than the normal range for the assay); late-night salivary cortisol, at least two measurement (>145 ng/dL or >4 nmol/L); overnight dexamethasone (>1.8 μg/dL or >50 nmol/L); or the 2-mg/48-hour dexamethasone suppression test (>1.8 μg/dL or >50 nmol/L). Abnormal test results in a patient with clinically suspected CS needs confirmation with one of the additional tests (**Flowchart 2**); if both test results are abnormal, patients should be investigated for the cause of the CS. The clinician's next step after confirming CS is to determine the cause biochemically and on imaging detail of which is beyond the scope of this book.[11]

Management

In CS with primary adrenal disease, removal of adrenal tumor is the rational approach to cure hypercortisolism.

Flowchart 2: Algorithm for testing patients with suspected cushing syndrome.

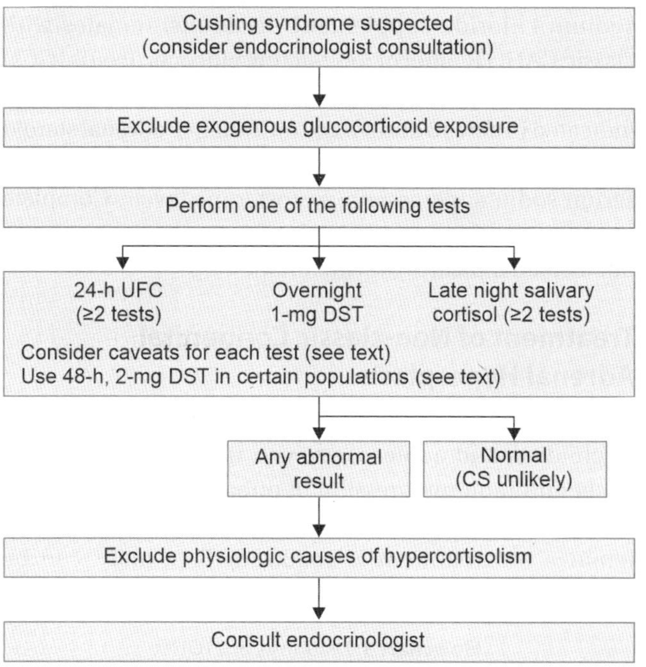

(UFC: urinary free cortisol; DST: dexamethasone suppression test; CS: Cushing syndrome)

Inoperable patients may be treated via medical therapy by agents—ketoconazole, metyrapone and mitotane with some success. After treatment of underlying pathology gynecological features of CS usually reverses.

CONGENITAL ADRENAL HYPERPLASIA

Congenital adrenal hyperplasia includes a group of autosomal recessive disorders caused by defects in one of the steroidogenic enzymes involved in cortisol biosynthesis or in the electron-providing factor, POR. Between 90 and 95% of cases of CAH are caused by 21-hydroxylase (CYP21A2) deficiency[12] so we will discuss here this most common cause only. It happens due to defective conversion of 17 hydroxy-progesterone (17OHP) to 11-deoxycortisol which results in reduced cortisol biosynthesis in turn resulting in reduced negative feedback drive and increased ACTH secretion. Resulting in excess production of adrenal androgens. So clinical features in CAH are due to glucocorticoid and mineralocorticoid deficiency, and due to androgen excess. Depending upon their residual enzymatic activity and age of presentation CAH is divided into classic and non-classic types. Classic type is further divided into types, classic salt wasting type, and simple virilizing type **(Table 2)**. Data regarding CAH is scarce from India but in western societies, the incidence of classic 21-hydroxylase deficiency varies from one in 10,000 to one in 15,000 live births. Non-classic 21-hydroxylase

TABLE 2: Types of 21-hydroxylase deficiency.

Phenotype	*Classic salt wasting*	*Simple virilizing*	*Non-classic*
Age at diagnosis	Newborn to 6 months	*Female:* Newborn to 2 years *Male:* 2–4 years	Child to adult
Genitalia	*Female:* Ambiguous *Male:* Normal	*Female:* Ambiguous *Male:* Normal	*Female:* Virilized *Male:* Normal
Incidence	1:20,000	1:60,000	1:1,000
Hormones			
Aldosterone	Reduced	Normal	Normal
Renin	Increased	Normal or increased	Normal
Cortisol	Reduced	Reduced	Normal
17OHP	>50,000 ng/dL	10,000–50,000 ng/dL	1,000–10,000 ng/dL (ACTH stimulation)
Testosterone	Increased	Increased	Variable, increased
Growth	–2 to –3 SD	–1 to – 2 SD	Probably normal
21-Hydroxylase activity (% of wild type)	0	1–5	5–20

(17OHP: 17-Hydroxyprogesterone; ACTH: adrenocorticotropic hormone; SD: standard deviation)

deficiency is more common, with an incidence of about one in 500–1,000 live births.[2]

Simple Virilizing Form

In the simple virilizing form of 21-hydroxylase deficiency, there is in utero virilization of an affected female fetus due to enhanced ACTH drive to adrenal androgen secretion. The sexual ambiguity at birth and even inappropriate sex assignment occur which depends on the severity of clitoral enlargement, labial fusion, and development of a urogenital sinus.

Salt-wasting Form

In classic 21-hydroxylase deficiency, 75% patients of both sexes have the salt-wasting form and also have aldosterone deficiency. Neonates commonly present after the first 2 weeks of life with a salt-wasting crisis and hypotension if not diagnosed from screening. The neonates also have features of simple virilizing form. The clinical signs and symptoms of salt wasting include poor feeding, vomiting, failure to thrive, lethargy, and sepsis-like symptoms.

NON-CLASSIC CONGENITAL ADRENAL HYPERPLASIA

Patients with non-classic congenital adrenal hyperplasia (NCCAH) have 5–20% residual 21 hydroxylase activity. These patients present in childhood or early adulthood with premature pubarche or with a phenotype that may masquerade as PCOS.[13] The recognized secondary cause of PCOS in 12% of patients is non classic 21-hydroxylase deficiency.[14] Various presenting features are hirsutism, primary or secondary amenorrhea, or anovulatory infertility, androgenic alopecia, and acne.[15]

Diagnosis

In suspected females an early morning (before 8 AM) baseline serum 17OHP measurement should be done as screening method. In females with borderline 17OHP levels, 17OHP is measured after a cosyntropin (ACTH 1-24) stimulation test (0.25 mg given intravenously and 17OHP measured at baseline and after 60 minutes). Genetic analysis is done, if cosyntropin stimulation test is also equivocal, or it cannot be accurately performed (in patient taking glucocorticoids), or for the purpose of genetic counselling. Randomly measured 17OHP levels can be normal in NCCAH hence. So, 17OHP levels should be screened early in the morning (before 8 AM). For menstruating females, steroid measurements should be obtained in the follicular phase **(Flowchart 3)**.[16]

Management

Classic CAH in children is treated with maintenance therapy of hydrocortisone (HC), fludrocortisone, and sodium chloride supplement. In adults, females with classic CAH, HC and/or long-acting glucocorticoids (GCs) plus mineralocorticoids (MCs), can be used as clinically indicated **(Tables 3 and 4)**. Optimization of adrenal steroid treatment is done by serum 17OHP, androstenedione, serum sodium, potassium, testosterone levels. Complete suppression of serum 17OHP level is not a treatment goal but instead indicates overtreatment.

Treatment of Non-classic Congenital Adrenal Hyperplasia

Glucocorticoid treatment of NCCAH is recommended in children and adolescents with inappropriately early onset and rapid progression of pubarche or bone age or with overt virilization. In asymptomatic nonpregnant females no treatment is required. Treatment can be

Flowchart 3: Diagnosis of 21OHD.

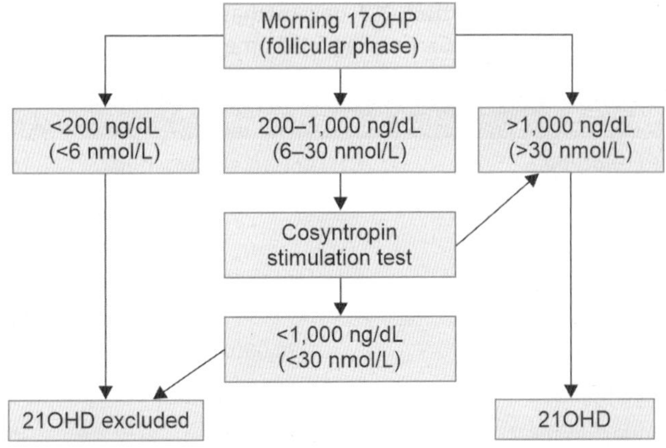

(17OHP: 17 hydroxy-progesterone; 21OHD: 21-hydroxylase deficiency)

TABLE 3: Maintenance therapy in pediatric patients with congenital adrenal hyperplasia (CAH).

Drugs	*Total daily dose*	*Daily distribution*
GCs: HC tablets	10–15 mg/m²·d	3 times·d
MCs: Fludrocortisone tablets	0.05–0.2 mg/d	1–2 times/d·d
Sodium chloride supplements	1–2 g/d (17–34 mEq/d) in infancy	Divided into several feedings

TABLE 4: Maintenance therapy in adult patients with congenital adrenal hyperplasia (CAH).

Type of long-acting corticosteroid	*Suggested dose (mg/d)*	*Daily doses*
HC	15–25	2–3
Prednisone	5–7.5	2
Prednisolone[a]	4–6	2
Methylprednisolone	4–6	2
Dex[a]	0.25–0.5	1
Fludrocortisone	0.05–0.2	1–2

[a]Suspension or elixir may permit improved dose titration for these drugs.

TABLE 5: Stress dose of steroids in congenital adrenal hyperplasia (CAH).

Patients' age	*Initial parenteral HC dose*
Infants and preschool children	25 mg
School-age children	50 mg
Adults	100 mg

discontinued when adult height is attained, or other symptoms resolve. GC treatment is suggested in females with hyperandrogenism or infertility. Within 3 months of GC treatment irregular menses and acne usually reversed, whereas hirsutism remission is more difficult with GC monotherapy. An oral contraceptive with or without anti-androgens is likely the best approach for treating hirsutism in women with NCCAH.

Stress Dosing

An increase in the GC dosage is recommended for situations such as febrile illness, gastroenteritis with dehydration, major surgery accompanied by general anesthesia, and major trauma and continued till the stress is over. In NCCAH HC stress dose is suggested only if a patient has a suboptimal (14–18 mg/dL, 400–500 nmol/L) cortisol response to cosyntropin or iatrogenic adrenal suppression **(Table 5)**.

Successive IV HC may be administered as one-quarter of the initial parenteral HC dose (above) given every 6 hourly.

PREGNANCY AND CONGENITAL ADRENAL HYPERPLASIA

Counseling is needed for partners of patients with CAH regarding the possibility of CAH in offspring and should be offered genetic testing.

The fetus will have 25% chance of having CAH and a 50% chance of being female; thus, there is a 12.5% chance that the fetus will be female and have CAH. As there are no universal protocols regarding prenatal treatment for CAH, so there should be an institute-based protocol regarding management of such cases. Treatment must be started by 6–7 weeks of gestation as the period during which fetal genitalia may become virilized begins 6 weeks after conception. All pregnancies at risk for CAH would need to be treated as chorionic villus biopsy (CVS) done at 12 weeks of gestation for confirming the affected fetus. Treatment can be discontinued after CAH is ruled out on genetic analysis on chorionic villus sampling. It aims to reduce female genital virilization and associated social stigma, the need for reconstructive surgery, and the emotional distress associated with the birth of a child with atypical sexual development, this warrants prenatal treatment with dexamethasone.[17] Prenatal treatment is to be stopped when chorionic villus biopsy shows male fetus as this form of CAH does not disrupt the development of male genitalia. In contrast to HC and prednisolone, dexamethasone is not inactivated by placental 11-beta hydroxysteroid dehydrogenase 2 (11bHSD2) enzyme and readily reaches the fetus so it is the drug of choice in pregnancy.

In NCCAH the rate of miscarriages is about 6.5% for pregnancies obtained with glucocorticoid treatment versus 26.3% without. So glucocorticoid treatment should be considered in NCCAH women wanting pregnancy.[18] No treatment is required if women with NCCAH becomes pregnant. However infertile or have history of prior miscarriage in women with NCCAH, treatment with a GC is recommended as it does not traverse the placenta. While managing a pregnant NCCAH patient risk and benefit should be assessed carefully as use of GC during pregnancy is also associated with maternal and fetal adverse effects as increased prevalence of gestational diabetes (20% vs. 10%), hypertension, increased maternal weight gain, and teratogenic effects in fetus.[19]

VIRILIZING TUMORS OF ADRENALS

Most of adrenal masses are incidentally discovered. In up to 4% of patients imaged for non-adrenal disease, an adrenal mass is uncovered.[20] Clinically, two issues are associated with adrenal masses whether the lesion is functional (i.e., secreting hormones) and whether it is malignant. Adrenal tumors can be adrenocortical adenomas, myelolipomas, pheochromocytoma, hamartomas, ganglioneuroma, granulomatous infiltrations of the adrenal gland, metastasis or adrenocortical carcinoma, adrenal cyst.[21] Adrenal cortical tumors can secrete glucocorticoids, aldosterone or androgens. So, androgen secreting tumors can be benign (adenoma) or malignant (adrenocortical carcinoma). Virilizing adrenal tumors in females should be suspected if marked hirsutism, deepening of the voice, oligomenorrhoea/amenorrhea or clitoromegaly is found. Sometimes these tumors can be large enough to be manually palpable.

It has bimodal age incidence, with most cases presenting at ages 40–50 years of age.

Imaging techniques, such as ultrasonography, computed tomography (CT), magnetic resonance imaging (MRI), or sometimes adrenal venous sampling, are used for distinguishing between an ovarian and an adrenal tumor as a cause of virilization.[22] Smooth, homogeneous adenomas on an enhanced adrenal CT scan with a Hounsfield Unit (HU) score (a marker of radiodensity) <10 HU or signal dropout on out-of-phase MRI sequences are invariably benign. Malignancy is suspected in irregular, inhomogeneous adenomas with a score >20 HU.[23]

Treatment is surgery in initial stages and chemotherapy (doxorubicin, etoposide, and mitotane), and/or radiotherapy and/or immunotherapy in later stages.

CONCLUSION

The suprarenal glands or the adrenal glands are triangular-shaped paired organs on the superior aspect of bilateral kidneys. Adrenals play a significant role in the endocrine system by secreting various important hormones. Adrenals are an important source of androgens in females. Various clinical conditions related to adrenals are associated with either deficiency or excess of different adrenal hormones. Because of the increase in awareness of clinical conditions related to adrenals and the availability of required lab investigations now clinical conditions of adrenals are becoming more prevalent, especially in young females the NCCAH which previously being wrongly treated as PCOS.

REFERENCES

1. Hall JE, Hall ME. Guyton and Hall Textbook of Medical Physiology. 14th edition. Elsevier. 2020.
2. Melmed S, Koenig R, Rosen CJ, Auchus RJ, Goldfine AB. Williams Textbook of Endocrinology. 14th edition. Elsevier. 2019.
3. Miller WL. The hypothalamic-pituitary-adrenal axis: a brief history. Horm Res Paediatr. 2018;89(4):212-23.
4. Turcu AF, Auchus RJ. Clinical significance of 11-oxygenated androgens. Curr Opin Endocrinol Diabetes Obes. 2017; 24(3):252-259.
5. Lindholm J, Juul S, Jørgensen JO, Astrup J, Bjerre P, Feldt-Rasmussen U, et al. Incidence and late prognosis of Cushing's syndrome: a population-based study. J Clin Endocrinol Metab. 2001;86(1):117-23.
6. Lado-Abeal J, Rodriguez-Arnao J, Newell-Price JD, Perry LA, Grossman AB, Besser GM, et al. Menstrual abnormalities in women with Cushing's disease are correlated with hypercortisolemia rather than raised circulating androgen levels. J Clin Endocrinol Metab. 1998;83(9):3083-8.
7. Dubey AK, Plant TM. A suppression of gonadotropin secretion by cortisol in castrated male rhesus monkeys (Macaca mulatta) mediated by the interruption of hypothalamic gonadotropin-releasing hormone release. Biol Reprod. 1985;33(2):423-31.
8. Ahima RS, Harlan RE. Glucocorticoid receptors in LHRH neurons. Neuroendocrinology. 1992;56(6):845-50.
9. Valea A, Morar A, Dumitru DP, Carsote M, Ghemigian A, Dumitrache C. Infertility as the onset of Cushing's disease: is pasireotide a treatment option? ARS Medica Tomitana. 2015;21(3).
10. Wallace C, Toth EL, Lewanczuk RZ, Siminoski K. Pregnancy-induced Cushing's syndrome in multiple pregnancies. J Clin Endocrinol Metab. 1996;81(1):15-21.
11. Nieman LK, Biller BM, Findling JW, Newell-Price J, Savage MO, Stewart PM, et al. The diagnosis of Cushing's syndrome: an Endocrine Society Clinical Practice Guideline. Clin Endocrinol Metab. 2008;93(5):1526-40.
12. White PC, Speiser PW. Congenital adrenal hyperplasia due to 21-hydroxylase deficiency. Endocr Rev. 2000;21(3):245-91.
13. New MI. Extensive clinical experience: nonclassical 21-hydroxylase deficiency. J Clin Endocrinol Metab. 2006; 91(11):4205-14.
14. Azziz R, Sanchez LA, Knochenhauer ES, Moran C, Lazenby J, Stephens KC, et al. Androgen excess in women: experience with over 1000 consecutive patients. J Clin Endocrinol Metab. 2004;89(2):453-62.
15. Azziz R, Dewailly D, Owerbach D. Clinical review 56: Nonclassic adrenal hyperplasia: current concepts. J Clin Endocrinol Metab. 1994;78(4):810-5.
16. Speiser PW, Arlt W, Auchus RJ, Baskin LS, Conway GS, Merke DP, et al. Congenital adrenal hyperplasia due to steroid 21-hydroxylase deficiency: An Endocrine Society Clinical Practice Guideline. J Clin Endocrinol Metab. 2018;103(11):4043-88.
17. Meyer-Bahlburg HFL, Reyes-Portillo JA, Khuri J, Ehrhardt AA, New MI. Syndrome-Related stigma in the general social environment as reported by women with classical congenital adrenal hyperplasia. Arch Sex Behav. 2017;46(2):341-51.
18. Bidet M, Bellanné-Chantelot C, Galand-Portier MB, Golmard JL, Tardy V, Morel Y, et al. Fertility in women with nonclassical congenital adrenal hyperplasia due

to 21-hydroxylase deficiency. J Clin Endocrinol Metab. 2010;95(3):1182-90.

19. US Preventive Services Task Force. Screening for Gestational Diabetes: US Preventive Services Task Force Recommendation Statement. JAMA. 2021;326(6):531-8.
20. Kloos RT, Gross MD, Francis IR, Korobkin M, Shapiro B. Incidentally discovered adrenal masses. Endocr Rev. 1995;16(4):460-84.
21. Diab DL, Faiman C, Siperstein AE, Zhou M, Zimmerman RS. Virilizing adrenal ganglioneuroma in a woman with subclinical Cushing syndrome. Endocr Pract. 2008; 14(5):584-7.
22. Gaudio AD, Gaudio GAD. Virilizing adrenocortical tumors in adult women. Report of 10 patients, 2 of whom each had a tumor secreting only testosterone. Cancer. 1993;72(6):1997-2003.
23. Johnson PT, Horton KM, Fishman EK. Adrenal imaging with multidetector CT: evidence-based protocol optimization and interpretative practice. Radiographics. 2009;29(5):1319-31.

Chapter

25 Hyperprolactinemia

Richa Vatsa

INTRODUCTION

Prolactin (PRL) hormone:

- Polypeptide of 199 amino acids
- 40% structural similarity with growth hormone and placental lactogen
- Encoded by single gene on chromosome 6

PATHOPHYSIOLOGY[1]

Various hormones affect PRL secretion acting either on hypothalamic dopamine secretion or directly on pituitary lactotrophs **(Table 1 and Flowchart 1)**.

Effect of dopamine on prolactin:[2]

- Binds to D2 receptor on lactotrophs
- Suppresses PRL gene transcription and PRL secretion **(Table 1)**
- Inhibits lactotroph growth and development

TABLE 1: Mediators of prolactin (PRL) secretion.

Increase	***Decrease***
Estrogen	Serotonin
Endogenous opioids	Prolactin
Thyrotropin-releasing hormone (TRH)	Neuropeptide Y

Functions of prolactin in mammals:

- Lactogenesis
- Marked elevation after sexual arousal persisting for at least 1 hour might be responsible for suppression of sexuality immediately after arousal and orgasm[3]

Definition of hyperprolactinemia:

Serum PRL level >20 ng/mL in men and postmenopausal women or >25 ng/mL in premenopausal women.

The causes of hyperprolactinemia are described in **Table 2** and **Flowchrt 2** describes the clininal features of hyperprolactinemia in different age groups.

EVALUATION OF HYPERPROLACTINEMIA

History Taking

Chief complaints:

- Milky discharge from breast and inability to conceive
- Hypothyroidism—excessive weight gain, cold intolerance, lethargy, and fatiguability
- Hypogonadism—decreased libido, bone pain, and hot flashes
- Symptoms of chiasma compression—headache and visual symptoms

Flowchart 1: Control of prolactin secrection.

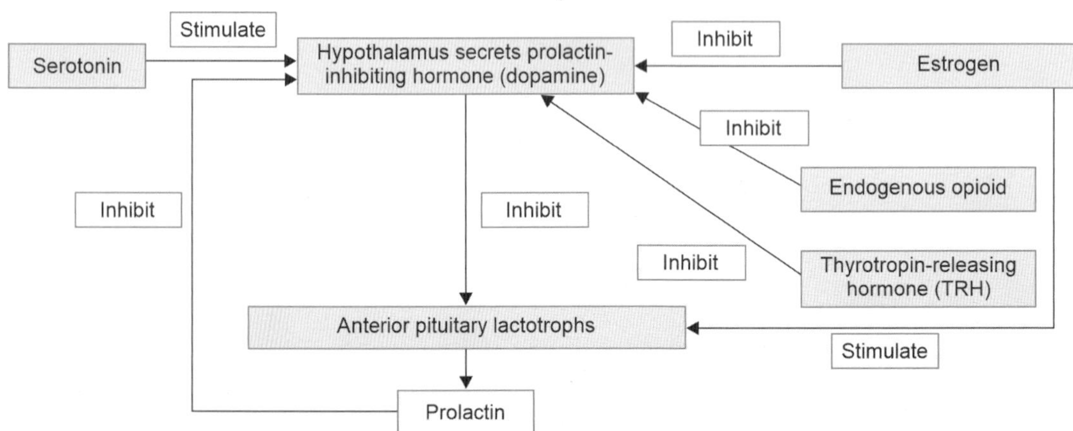

TABLE 2: Causes of hyperprolactinemia.[4-7]

Physiological: • Pregnancy and breast feeding • Suckling, nipple stimulation • Stress, hypoglycemia, myocardial infarction, surgery, and venipuncture stress • Sleep • Coitus • Exercise • Meals • Neonatal period	
Pathological: *Hypothalamic-pituitary stalk damage (disrupt delivery of dopamine):* • Tumors craniopharyngioma • Suprasellar mass extension • Meningioma • Dysgerminoma • Rathke cyst • Infiltrations—histiocytosis disease • Irradiation—decreased hypothalamic dopamine secretion • Trauma—pituitary stalk section, sellar surgery, and head trauma	
Pituitary cause: • Prolactinoma • Macroadenoma (compressive) • Plurihormonal adenoma • Lymphocytic hypophysitis • Parasellar mass • Idiopathic	
Systemic disorders: • *Hypothyroidism:* TRH stimulates prolactin secretion • *Chronic renal disease:* Decrease prolactin clearance • *Cirrhosis:* Hyperestrogenic state	
Macroprolactinemia: • *Chest wall disorders:* Activation of sensory pathway stimulating PRL secretion in the mechanism same as nipple stimulation, neural reflex pathway, and neurogenic hyperprolactinemia – Chest wall trauma – Herpes zoster (affecting dermatome of breast) – Chest wall surgery	
• *Ectopic PRL secretion:* – Bronchogenic carcinoma – Renal cell carcinoma – Gonadoblastoma – Non-Hodgkin lymphoma – Benign or malignant ovarian teratoma – Colorectal adenocarcinoma – Aberrantly located pituitary tissues	
Pharmacological cause: • Most frequent cause of non-tumoral hyperprolactinemia • Either lower dopamine levels or inhibit dopamine action	
Drug name	***Level of prolactin elevation***
Antipsychotic drugs:	
Fluphenazine	High
Haloperidol	High
Risperidone	High

Contd...

Contd...

Drug name	*Level of prolactin elevation*
Paliperidone	High
Perphenazine	Moderate
Thiothixene	Moderate
Trifluoperazine	Moderate
Quetiapine	None or low
Aripiprazole	None or low
Antidepressants:	
Clomipramine	High
Amitriptyline	Low
Nortriptyline	None
Antiemetics:	
Metoclopramide	High
Domperidone	High
Prochlorperazine	Low
Antihypertensives:	
Verapamil	Low
Methyldopa	Moderate
H2 Antihistaminics:	
Promethazine	High
Cimetidine	High

Flowchart 2: Hyperprolactinemia—clinical manifestation.

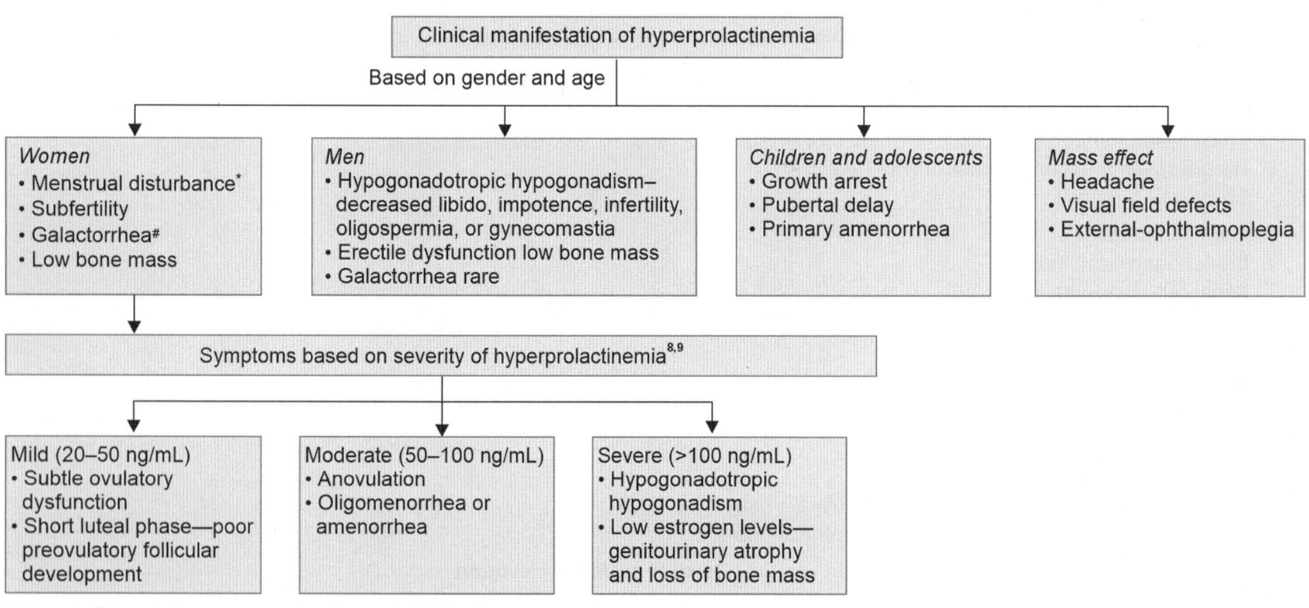

*Cause of secondary amenorrhea in up to 30% of women.

#Only one-third women with hyperprolactinemia have galactorrhea, the possible hypothesis:

- Milk secretion requires estrogen and hyperprolactinemia may result in profound hypogonadotropic hypogonadism, resulting in low estrogen levels.
- Presence of biologically inactive big PRL and macroprolactin.

- Ask history—history of coitus, nipple stimulation, and exercise.

Menstrual history: Delayed prolonged cycles, reduced flow, oligomenorrhea, or amenorrhea. **Flowchart 3** describes the mechanism of menstrual disturbance in hyperprolactinemia.

Obstetrics history: Subfertility

Past history:
- History of intake of oral medication—antipsychotics
- History of renal disease—dialysis
- History of chest wall injury, herpes zoster.

Examination

General examination:
- Body mass index (BMI), skin for acne and hair growth—indicative of polycystic ovary syndrome (PCOS), hypothyroidism, and precocious puberty
- Signs of chest wall injury.

Flowchart 3: Mechanism of menstrual disturbance in hyperprolactinemia.

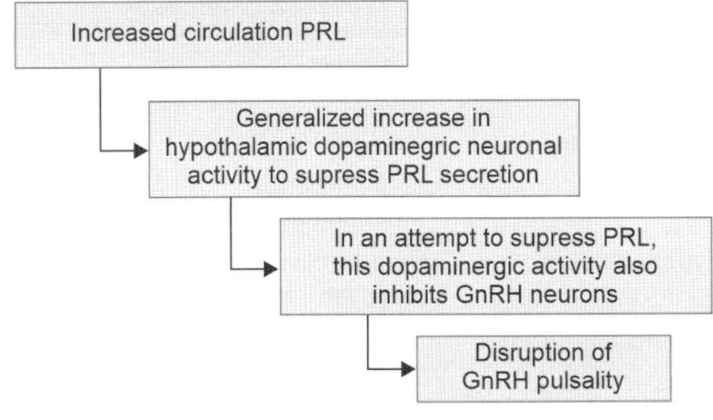

Gynecological examination:
- Breast development—Tanner stage
- Assess mammary glands for galactorrhea—single duct/multiple ducts, unilateral/bilateral

Ophthalmologic examination—visual field testing (bitemporal visual field loss)

Any clinical sign linked to certain medical disorder: Café-au-lait spots, asymmetrical bone growth—McCune-Albright syndrome.

Diagnostic Workup

- Rule out pregnancy—Urine pregnancy test
- Sample collection—morning hours, fasting sample, and avoid excessive stress.

Table 3 describes the various types of prolacting found in blood.

Imaging:

When to order imaging:
- Raised PRL levels above recommended levels in drug induced hyperprolactinemia
- All symptomatic hyperprolactinemic—after ruling out physiological cause
- Dynamic contrast magnetic resonance imaging (MRI) (gadolinium enhanced) to look for:
  - Size of mass lesion
  - Extent of tumor
  - Suprasellar extension
  - Compression of optic chiasma
  - Invasion of cavernous sinus
- Magnetic resonance imaging without gadolinium—patient with renal impairment, and pregnancy.

Different types of PRL detachable in blood are described in **Table 3**.

TABLE 3: Different types of prolactin detectable in blood.[10-15]

	Monomeric	*Big prolactin*	*Macroprolactin*
Important points	Predominant circulating form (80–95%)		Noncovalent or interchain disulfide bonding between separate molecules of prolactin
Molecular structure	Monomer 23 kDa	Dimers or trimers 50–60 kDa	Aggregation of smaller PRL molecules bound together with immunoglobulins >100 kDa
Biological activity	More active than larger forms	Little bioactivity, accounts for 10–25% of hyperprolactinemia reported by commercial assays[16]	
Clearance	Fast	More slowly	
Symptoms	Present	Minimal or no menstrual disturbance or galactorrhea	
Workup			If suspected, pretreat the serum with polyethylene glycol to precipitate macroprolactin before performing assay

Macroprolactinemia: When majority of PRL is macroprolactin in a patient of hyperprolactinemia, the condition is known as macroprolactinemia.[15]

Pitfalls in diagnosis are described in **Table 4**.

TREATMENT

Medical Management (Table 5, Flowcharts 4 to 6)

Dopamine agonist: Bromocriptine, cabergoline, and quinagolide. Comparison of various dopamine agnists have been described in **Table 6**.

Effect of treatment:

- Resumption of ovulation and menstrual cycle—in several weeks
- Stoppage of milk secretion—might take weeks to months

Necessity of treatment of hyperprolactinemia:

- For ovulation induction and troublesome galactorrhea: Dopamine agonist
- Hyperprolactinemia itself does not pose a health risk
- Oligomenorrhea without hypogonadism—cyclical progestin withdrawal
- With hypogonadism—cyclical or combine estrogen and progesterone.

The management of hyperprolactinemic infertility is described in **Table 7**.

TABLE 4: Pitfalls in diagnosis.

False negative: • Hook effect—very high prolactin concentration in serum saturates both capture and signal antibodies, preventing them from forming a sandwich in immunoradiometric and chemiluminescent assays • Sample subjected to 1:100 dilutions prior to assay
False positive: Macroprolactins: • Physiologically inactive forms of prolactin, bound to IgG • Pretreatment with polyethylene glycol (PEG)
Heterophile antibodies: • RA autoantibodies • PRL assay interference in absence of PRL • Heterophile antibodies blocking tubes
Radiological pitfall: Pituitary incidentaloma—found in 10% sellar MRI in normal individual

PROLACTINOMA

- Benign tumor arising from lactotroph cells
- Functional lactotroph adenomas accounts for about 40% of all clinically recognized adenomas
- Can occur as a part of MEN 1 syndrome
- Diagnosis of prolactinoma is likely with PRL >200 ng/mL
- Prevalence—10/1,00,000 in men and 30/1,00,000 in women
- Prevalence in children and adolescents—0.1/10,00,000

TABLE 5: Modality of treatment depends on etiology.

Etiology	*Treatment*
Idiopathic hyperprolactinemia	• Dopamine agonist—lowest possible dose • If normal PRL level for 2 years—discontinue on trial basis
Hypothyroidism	Thyroid replacement therapy, confirm normal PRL after treatment
Hypothalamic disease	• Removal of cause—first choice • Dopamine agonists—second choice

Flowchart 4: Management of hyperprolactinemia.

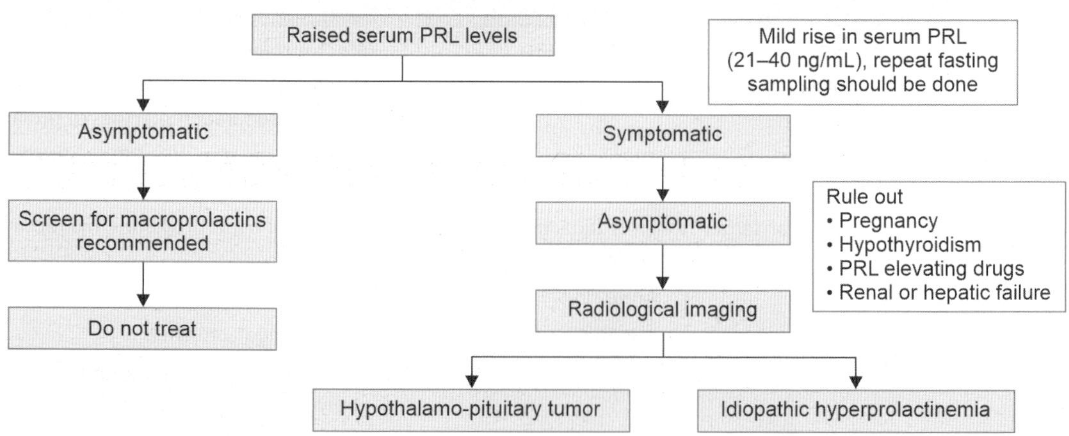

Flowchart 5: Follow up on medical management.

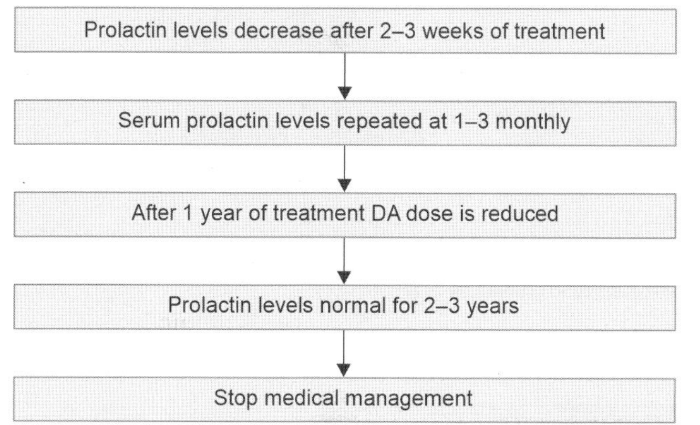

Flowchart 6: Drug-induced hyperprolactinemia.

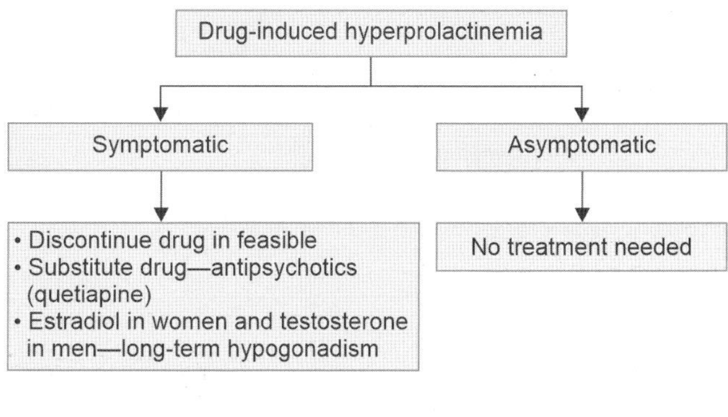

TABLE 6: Comparison of bromocriptine and cabergoline.[17-25]

	Cabergoline	***Bromocriptine***
Type of dopamine agonist	• Ergot derivative • Selective type 2 dopamine receptor agonist	• Ergot derivative • Nonselective dopamine receptor agonist
Brand name	Cabgolin	B crip
Duration of action	7–14 days	8–12 hours
Half life	65 hours	3.3 hours
Side effects	Milder and less frequent nausea and orthostatic hypotension	Nausea, headache, dizziness, orthostatic hypotension
Advantages	• Higher efficacy • Superior in reducing prolactin secretion (95%) • Greater reduction in tumor volume (90%) • Works in bromocriptine resistant cases (90%) • Weekly or twice weekly dose	• Safe in fetus • Inexpensive • Restores ovulation (80–90% cases) • Reduces prolactinoma size (90% cases)
Disadvantages	Possible risk of fibrotic valvular heart disease	• Tolerance; resistance • Multiple daily dose
Dosage		
Starting dose	0.25–0.5 mg/week, bedtime, snack	0.625–1.25 mg/daily, single night dose (HS)
Dose regimen	Total dose 3.5 mg/week or less	• 2–3 divided daily doses if total doses >2.5 mg daily (improves effectiveness and tolerance) • Daily dose rarely exceeds 10 mg
Dose titration	4–6 weekly Frequent titration—large tumor with mass effect	4 weekly interval
Dose route	Oral or vaginal	Oral or vaginal

TABLE 7: Management of hyperprolactinemic infertility.

- Dopamine agonist—treatment of choice
- Bromocriptine theoretically better first choice however available data suggest cabergoline also safe in early pregnancy
- Prolactin levels normalizes—60–85%
- Cyclic menses restored—70–90%
- Ovulatory cycles return in— 50–75%
- Start with low dose and increase gradually (bromocriptine 1.25–2.5 mg/day; cabergoline 0.25 mg twice a week)
- Dose increased until PRL level—normal or near normal and ovulatory function is restored
- If ovulation does not occur despite normal PRL—cyclical treatment with letrozole follwed by gonadotropins therapy can be added

Flowchart 7: Treatment of prolatinoma based on size and symptom.

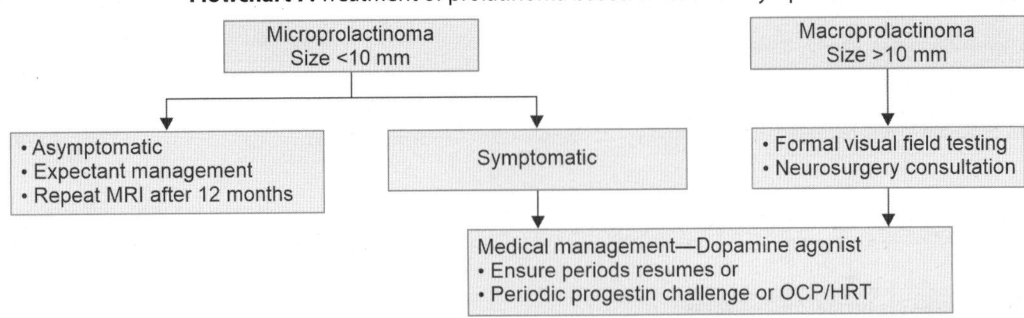

Flowchart 8: Follow up after medical management.

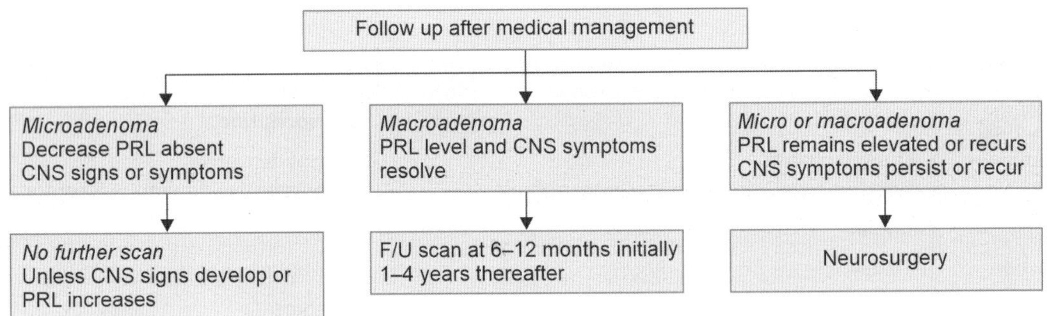

- About 10% of PRL secreting adenomas also secrets GH
- Tumor classified arbitrarily by their size measured by CT or MRI imaging.
- Treatment of prolactinoma based on size and follow up of patients on medical managmenet is given in **Flowchart 7 and 8** respectively.

Management of Prolactinoma in Pregnancy and in Postpartum Period (Tables 8 and 9 and Flowchart 9)

- Begins before conception—pre-conceptional counseling regarding potential risk to mother and the fetus
- Multidisciplinary approach—endocrinologist, neurosurgeon, psychiatrist, and gynecologist
- Maternal concern—adenoma growth
- Fetal concern—exposure to dopamine agonist
- Patient with macroadenoma—advised decrease in size before conception, whether prolactinoma elevates optic chiasma or not
- Clinically significant tumor growth occurs in 2.4% women with microadenomas and 21% with macroadenomas
- Cabergolin preferred to bromocriptine—high therapeutic/adverse effect ratio
- No increase in adverse pregnancy outcome or fetal malformation compared to general population.

GALACTORRHEA

- Secretion of milky fluid from one or both breasts in women who have either never been pregnant or have stopped nursing for over 12 months.
- Post-partum galactorrhea associated with amenorrhea—self-limiting condition
- Treatment usually involves managing the underlying disorder.

Galactorrhea without Hyperprolactinemia

- Serum PRL level—often normal
- Not due to any ongoing disease
- Diagnosis: Ensure that breast secretion is clear or milky.
- Measure serum PRL concentration → if elevated, the cause should be sought.
- If PRL is not elevated → no further test is required.

Treatment

- Does not need to be treated.
- If bothersome, treat with low dose of dopamine agonist → 0.25 mg cabergoline twice a week.

TABLE 8: Postnatal follow up.

- Breastfeeding—not contraindicated in prolactinoma
- Dopamine agonist not used—impair lactation
- Women with visual field impairment, patient should not breast feeding—treated with dopamine agonist
- Re-evaluation with serum prolactin measurement 3 months after delivery—in women who do not breast feed, after breast feeding is over

TABLE 9: Other treatment modalities.

Surgery for prolactinoma:
- Successful outcome post-surgery—75% microprolactinoma, 34–38% macroprolactinoma

Indication
- Dopamine resistance or intolerance
- Presence of invasive macroadenoma with compromised vision—not improving after medical management
- Apoplexy with severe neurological signs
- Patient preference

Radiation therapy:
- Limited role
- Macroadenoma not responding to DA agonist or surgery
- Stereotactic technique—deliver photons (gamma knife, cyber-knife) or protons selectively to target
- Complications—hypopituitarism—lifelong hormone supplements, neuropathies, stroke, and secondary tumor formations

Flowchart 9: Management of patients of prolactinoma in pregnancy.

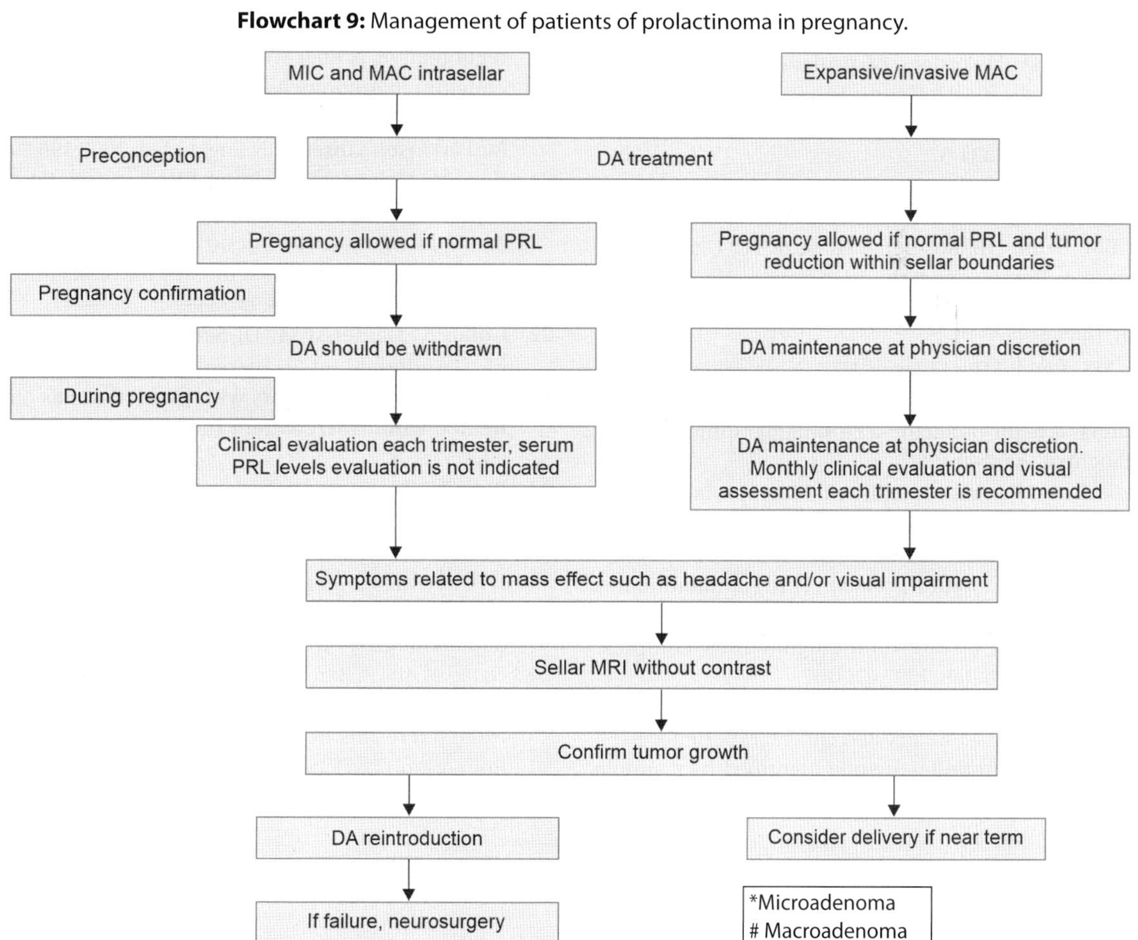

REFERENCES

1. Speroff L. Neuroendocrinology. In: Taylor HS, Pal L, Seli E (Eds). Speroff's Clinical Gynecologic Endocrinology and Infertility, 9th edition. India: Wolters Kluwer; 2019. p. 107.
2. Speroff L. The Breast. In: Taylor HS, Pal L, Seli E (Eds). Speroff's Clinical Gynecologic Endocrinology and Infertility, 9th edition. India: Wolters Kluwer; 2019. pp. 541-2.
3. Exton MS, Kruger TH, Koch M, Paulson E, Knapp W, Hartmann U, et al., Coitus-induced orgasm stimulates prolactin secretion in healthy subjects. Psychoneuroendocrinology 2001;26:287.
4. Vilar L, Abucham J, Albuquerque JL, Araujo LA, Azevedo MF, Boguszewski CL, et al. Controversial issues in the management of hyper-prolactinemia and prolactinomas—An overview by the Neuroendocrinology Department of the Brazilian Society of Endocrinology and Metabolism. Arch Endocrinol Metab. 2018;62(2):236-63.
5. Vilar L, Fleseriu M, Bronstein MD. Challenges and pitfalls in the diagnosis of hyperprolactinemia. Arq Bras Endocrinol Metabol. 2014;58(1):9-22.
6. Melmed S, Casanueva FF, Hoffman AR, Kleinberg DL, Montori VM, Schlechte JA, et al. Diagnosis and treatment of hyperprolactinemia: an Endocrine Society clinical practice guideline. J Clin Endocrinol Metab. 2011;96(2):273-88.
7. Bronstein MD. Disorders of prolactin secretion and prolactinomas. In: DeGroot LJ, Jameson JL (Eds). Endocrinology. 6th edition. Philadelphia: Saunders/Elsevier; 2010. pp. 333-57.
8. Seppala M, Ranta T, Hirvonen E. Hyperprolactinaemia and luteal insufficiency. Lancet. 1976;1:229.
9. Biller BM, Baum HB, Rosenthal DI, Saxe VC, Charpie PM, Klibanski A. Progressive trabecular osteopenia in women with hyperprolactinemic amenorrhea, J Clin Endocrinol Metab. 1992;75:692.
10. Elllis MJ, Reed MR, Livesey IM. Cross-reactivities of macroprolactin and big-prolactin in three commercial immunoassays for prolactin: a chromatographic analysis. Clin Biochem. 2007;40:1285.
11. Fahie-Wilson MN, John R, Ellis AR. Macroprolactin; high molecular mass forms of circulating prolactin, Ann Clin Biochem. 2005;42:175.
12. Hattori N, Inagaki C. Anti-prolactin (PRI) autoantibodies cause asymptomatic hyperprolactinemia; bioassay and clearance studies of PRL-immunoglobulinG complex. J Clin Endocrinol Metab. 1997;82:3107.
13. Leslie H, Courtney CH, Bell PM, Hadden DR, McCance DR, Ellis PK, et al. Laboratory and clinical experience in 55 patients with macroprolactinemia identified by a simple polyethylene glycol precipitation method, I Clin Endocrinol Metab. 2001;86:2743.
14. Kasum M, Orešković S, Čehić E, Šunj M, Lila A, Ejubovicć E. Laboratory and clinical significance of macroprolactinemia in women with hyperprolactinemia. Taiwan J Obstet Gynecol. 2017;56(6):719-24.
15. Kasum M, Oreskovic S, Zec I, Jezek D, Tomic V, Gall V, et al. Macroprolactinemia: new insights in hyperprolactinemia. Biochem Med (Zagreb). 2012;22(2):171-9.
16. Smilh TR, Suliman AM, Fahie-Wilson MN, McKenna TI. Gross variability in the detection of prolactin in sera containing big prolactin (macroprolactin) by commercial immunoassays. Clin Endocrinol Metab. 2002; 87:5410.
17. Webster J, Piscitelli C, Polli A, Ferrari Ch, Ismail I, Scanion ME. A comparison of cabergoline and bromocriptine in the treatment of hyperprolactinemic amenorrhea: Cabergoline Comparative Study Group. N Engl j Med. 1994;331:904.
18. Verhelst J, Abs R, Maiter D, Van Den Bruel A, Vandeweghe M, Yelleniers B, et al. Cabergoline in the treatment of hyperprolactinemia: a study in 455 patients. Clin Endocrinol Metab. 1999;84:2518.
19. Schade R, Andersohn F, Suissa S, Haverkamp W, Garbe E. Dopamine agonists and the risk of cardiac-valve regurgitation, N Engl Med. 2007;356:29.
20. Zanettini R, Antonini A, Gatto G, Gentile R, Tesei S, Pezzoli G. Valvular heart disease and the use of dopamine agonists for Parkinson's disease. N Engl J Med. 2007;356:39.
21. Kars M, Delgado V, Holman ER, Feelders RA, Smit JW, Romijn JA, et al. Aortic valve calcification and mild tricuspid regurgitation but no clinical heart disease after 8 years of dopamine agonist therapy for prolactinoma. J Clin Endocrinol Metab. 2008;93:3348.
22. Colao A, Galderisi M, Di Sarno A, Pardo M, Gaccione M, D'Andrea M, et al. Increased prevalence of tricuspid regurgitation in patients with prolactinomas chronically treated with cabergoline. J Clin Endocrinol Metab. 2008; 93:3777.
23. Roth BL. Drugs and valvular heart disease. N Engl J Med. 2007;356:6.
24. Motta I, de Vincentiis S, Marchini M, Colombo N, D'Alberton A. Vaginal cabergoline in the treatment of hyperprolactinemic patients intolerant to oral dopaminergics. Fertil Steril. 1996;65:440.
25. Galao A, Abs R, Barena DG, Chanson P, Paulus W, Kleinberg Di. Pregnancy outcomes following cabergoline treatment: extended results from a 12-year observational study. Clin Endocrinol (Oxf). 208;68:66.

Section

3

Infertility

Chapter 26

Basic Management of Infertile Couples

Hafizur Rahman, Sayanti Paul, Ezzat Khalda

INTRODUCTION

Almost 15% of the world's population suffers from infertility or subfertility. Though both the words "subfertility" and "infertility" are used interchangeably in clinical practice, majority of these couples conceive either spontaneously or after treatment and only 4% of the couples remain absolutely childless. With the increasing age of the mothers and decreasing trend of fecundity in modern era, the problem of infertility is rising. Timely supportive management of infertility has significant impact on improvement of the physical and psychological health of the suffering couples.

DEFINITION AND SUBTYPES

Infertility is defined as failure to conceive spontaneously after 12 or more months of regular unprotected intercourse.[1] When woman is more than 35 years old, a shorter waiting time of 6 months has been recommended to define infertility.[1,2] It is further classified into primary and secondary subtypes as follows:

Subtype	*Criteria*
Primary infertility	Couple have never conceived
Secondary infertility	Couple conceived previously irrespective of the outcome

ETIOLOGICAL FACTORS

Infertility may be due to various reasons and more than one factor may be present in a single scenario. Identification of the causative factors is of utmost importance. It helps to formulate right treatment, presume success of the treatment, and address associated medical conditions which have long-term health consequences. Common causative factors of infertility and their approximate prevalence are given in **Flowchart 1**.[3]

Ovulatory Dysfunction

Ovulatory dysfunctions are classified by World Health Organizations (WHO) into three groups as per the underlying defect in the hypothalamic pituitary ovarian axis (HPO).

Group I Ovulatory Dysfunction

Hypothalamic pituitary failure or hypogonadotropic hypogonadism. It comprises 10% of the ovulatory disorders. It is characterized by low gonadotropin, normal estrogen, and normal prolactin level. Various causes of Group I ovulatory dysfunction are as given in **Table 1**.

Group II Ovulatory Dysfunction

It is a type of hypothalamic pituitary ovarian dysfunction. It is also known as normogonadotropic normoestrogenic anovulation. It comprises 85% of the ovulatory dysfunction. Polycystic ovarian syndrome (PCOS) and hyperprolactinemic amenorrhea are the most common form of Group II ovulatory dysfunction. Rotterdam criteria are used for diagnosis of PCOS in adult patients.

Flowchart 1: Common causative factors of infertility and their approximate prevalence.

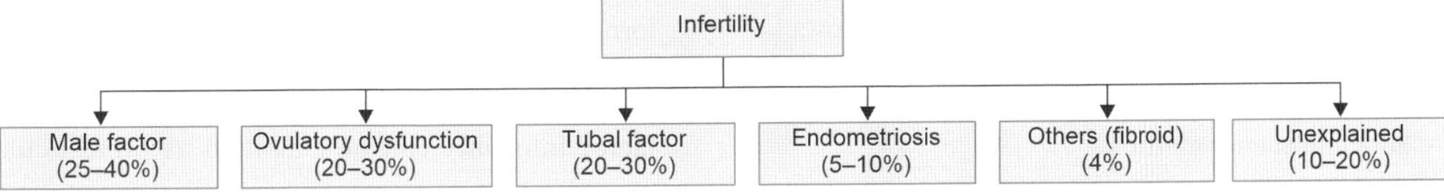

TABLE 1: Causes of Group I ovulatory dysfunction.

Hypothalamic	• Low body weight • Stress or excessive exercise • Craniopharyngioma or other hypothalamic tumors • Kallmann's syndrome • Idiopathic
Pituitary	• Sheehan's syndrome • Prolactinoma or other functional tumors of pituitary • Non functional pituitary lesions • Brain irradiation • Pituitary surgery

TABLE 2: Causes of Group III ovulatory dysfunction.

Causes of ovarian failure	• Chromosomal (Turner syndrome, 45, XO) • Gonadal dysgenesis • Androgen insensitivity syndrome (46, XY) • Chemotherapy • Radiotherapy • Autoimmune • Oophorectomy • Idiopathic

Group III Ovulatory Dysfunction

It is also known as hypergonadotropic hypogonadism and characterized by high gonadotropin and low estrogen levels due to ovarian failure. It comprises only 4–5% of ovulatory disorders. There may be different causes of ovarian failure **(Table 2)**.

Serum follicle-stimulating hormone (FSH) measurement is the most useful test for diagnosing type of ovulatory disorder. However, to differentiate between Group I and Group II ovulatory dysfunction one need to consider history examination hormonal and imaging evaluation.

Unexplained Infertility

By definition unexplained infertility is one where the basic infertility investigations are normal (evidence of ovulation, tubal patency, and normal semen analysis). Basically, it is a diagnosis of exclusion, and its prevalence depends on the thoroughness of investigations. On an average, it affects almost 10–20% of the infertile couples. The causes underlying unexplained infertility might be as follows:[3]

- Subtle abnormalities in oocyte and sperm function
- Defective endometrial receptivity
- Subclinical endometriosis
- Nutritional factors like deficiency of zinc, magnesium, and selenium
- Celiac disease
- Immunological factors, etc.

TABLE 3: Causes of uterine and tubal factor infertility.

Uterine factor infertility	• Fibroids particularly cavity distorting ones • Uterine polyps • Intrauterine adhesions or Asherman's syndrome (may occur after dilatation and curettage, surgical treatment of miscarriage or overzealous hysteroscopic resection) • Congenital anomalies of female genital tract like uterine septum, vaginal atresia, etc.
Tubal factor infertility	• Tubal blockage (proximal or distal) • Peritubal adhesions • Hydrosalpinx

TABLE 4: Terminologies of semen abnormality.

Terminology	*Description*
Oligozoospermia	Sperm count <15 million/mL
Asthenozoospermia	Sperm motility <32% (progressive)
Teratozoospermia	Morphologically normal sperm percentage <5%
Azoospermia	Absence of sperm in the ejaculate
Aspermia	No ejaculate
Necrozoospermia	Only dead sperms are present in semen
Globozoospermia	Morphologically abnormal sperm where acrosome is absent

Uterine and Tubal Factor Infertility

Any factor which affects transport of sperms to the tubes, or which inhibits implantation of embryo may lead to infertility. Similarly, any factor which affects normal anatomy or physiology of fallopian tubes can cause infertility. Common causes of uterine and tubal factor infertility are described in **Table 3**.

Male Factor Infertility

- *A number of terminologies* are used to describe the specific type of seminal abnormality. Oligoteratozoospermia (OAT) is the most common form of seminal abnormality. Other commonly used terminologies are described in **Table 4.**
- *Causes of male factor infertility*: The causes of male factor infertility can be classified as pretesticular,

TABLE 5: Causes of male factor infertility.

Pretesticular (hypothalamic–pituitary) causes	Congenital	• Kallmann's syndrome • Genetic syndromes (Laurence-Moon-Biedel, Prader-Willi syndrome, etc.) • Hemochromatosis • Pituitary/hypothalamic tumors
	Acquired	• Trauma • Surgery • Irradiation • Infiltrative disorders (tuberculosis, sarcoidosis) • Drugs/ hormonal • Systemic causes
Testicular	Congenital	• Klinefelter syndrome • Cryptorchidism • Y-chromosome deletions • 5 Androgen insensitivity • α reductase deficiency
	Acquired	• Varicocele • Infection • Drugs, toxins • Smoking • Hyperthermia • Antisperm antibodies • Systemic illness
Post-testicular	Sperm tansport problems at various levels	• Epididymis • Vas deference • Seminal vesicle and prostate • Defective ejaculation

testicular, and post-testicular causes as described in **Table 5**. But in significant proportion of cases, the cause remains unknown or unexplained.

MANAGEMENT OF THE INFERTILE COUPLE

When a couple attends clinic with history of failure to conceive, a holistic approach should be taken by the treating gynecologist to identify the cause of infertility, to counsel regarding pre-pregnancy care, to screen for relevant medical and genetic conditions, to give necessary information regarding the available options for management, and finally make a shared decision with the couple regarding further treatment **(Box 1)**.

BOX 1: Management of the infertile couple.

- Preconceptional counseling
- Infertility workup for diagnosis
- Discussion regarding chances of natural conception and available treatment options
- Shared decision making

When to start evaluation?

Whenever a couple presents with the complaints of subfertility, irrespective of the duration of subfertility, preconception counseling and evaluation should be done. Basic evaluation should start if the couple fulfils the definition of infertility or at high risk of infertility (12 or more months of unwanted nonconception). Expedited evaluation without waiting for 12 months should be done under the conditions:

- Female partner's age >35 years and failed attempt of conception for 6 months
- Female partner's age >40 years
- Women suffering from irregular menstrual cycles, oligomenorrhea, or amenorrhea
- Female partner suffering from uterine tubal or peritoneal disease and stage III/IV endometriosis which affects fertility
- Male partner with known or at high risk of having male subfertility

TABLE 6: Preconceptional counseling and evaluation of infertility patients.

Conditions	*Advice*
Pre-existing medical condition	• Stabilize the health condition and achieve good control • Explain the risk on pregnancy and the risk of pregnancy on the condition • Multidisciplinary approach should be taken for complicated cases • Due consideration should be given to change to safer alternative drugs if available
Weight	Check BMI and advice appropriate weight loss or gain whenever applicable
Smoking and other addictions	Both partners should stop smoking and using other substances which might affect fertility or the fetus
Regular intercourse	Ideal frequency of sexual intercourse should be once in every 2/3 days
Virology and STI screening	• Screening for Rubella IgG should be done • If not immune against rubella, immunization should be offered • Screen for HIV, hepatitis B, and hepatitis C should be done for high-risk group or if assisted conception is planned • Screen for thalassemia carrier status of both the partners in regions where it is prevalent
Folic acid supplementation to prevent neural tube defects	• Recommended dose of folic acid is 0.4 mg daily for low-risk women • Daily dose has to be increased to 5 mg daily in presence of high-risk factors (previous baby with NTD, diabetic, on anti-epileptic, etc.)

(BMI: body mass index; NTD: neural tube defect; STI: sexually transmitted infection)

- Sexual dysfunction
- Known genetic or environmental conditions causing poor ovarian reserve
- History of chemotherapy, radiotherapy, fragile X mutation carrier, etc.

The above list is not conclusive, and it considers only heterosexual couples. In any other condition when the gynecologist suspects that the couple is at high risk of infertility or when dealing with same sex couples where natural method of conception is impossible, infertility evaluation should start immediately.

TABLE 7: Basic infertility workup.

Evaluation of female partner	• History • Physical examination • Evaluation for ovulatory dysfunction • Evaluation of tubal factor • Evaluation of uterine factor
Evaluation of male partner	• History • Physical examination when clinically indicated • Semen analysis

Preconceptional Counseling and Evaluation

As the couples presenting with subfertility are actually contemplating pregnancy, preconceptional counseling forms an integral part of the infertility management. It aims to optimize the health condition of the mother by timely identification of modifiable risk factors, immunization, and education **(Table 6)**. Information regarding the fertile period and ideal frequency of intercourse can be given to improve fertility.

Basic Infertility Workup

Basic infertility workup aims to find out any abnormality of semen, ovulation, and anatomical integrity of female pelvis. Evaluation starts with detailed history and clinical examination and ends with relevant investigations to identify any abnormalities in the above-mentioned domains. The components of basic infertility evaluation are tabulated in **Table 7**.

Workup for Female Partner

- *History:* Infertility evaluation starts with a careful and detailed history taking **(Table 8)**. The specific points that need to be addressed during history taking is tabulated below. It has been adopted from the committee opinion by American Society of Reproductive Medicine.[2]
- *Physical examination:* Physical examination should be targeted to identify underlying or associated factors causing infertility. Though speculum examination should be done for all, routine bimanual examination is not needed. It has to be done depending on the

TABLE 8: Detailed history of female partner.

Points	***Routine assessment***	***Additional consideration***
Fertility history	• Current conception attempts • Length of time of unprotected intercourse • Coital frequency • Use of ovulation monitoring • Partner status and are they contributing sperm or oocytes to the patient's reproductive efforts • Presence of sexual dysfunction, including: decreased libido, dyspareunia, vaginismus • History of previous conception attempts • Prior periods of intercourse without contraception or with low efficacy contraception • Any prior fertility evaluation or treatment	• Patient may incorrectly identify attempts at pregnancy as only conscientious efforts for conception, rather than periods of active sexual activity without contraception • Coital frequency may change over time • If using urine LH kits, assess whether patient has been successful in detecting ovulatory surges • If using a fertility tracking app, discuss its limitations in accurately predicting the fertile window
Gynecologic history	• Menstrual history: – Age at menarche – Cycle length (range), duration, and amount of bleeding – Presence of intermenstrual bleeding – Presence of dysmenorrhea – Presence of molimina • General gynecologic history: – Cervical screening history including related treatments – Contraceptive use including type and duration – Sexually transmitted infections and/or pelvic inflammatory disease – Dyspareunia or chronic pelvic pain – History of abnormal cervical screening (pap smear—human papillomavirus testing)	• If menses onset <8 years of age or >14 years of age, was evaluation performed and were menses ever achieved spontaneously?[4] • If menstrual interval is <21 days or >35 days or there is a significant variation in range, perform a review of systems including: – Thyroid symptoms – Hirsutism – Visual field defects – Galactorrhea – Stressors – Dietary and exercise habits – Vasomotor symptoms – If abnormal menstrual bleeding, were any investigations performed and was a diagnosis made? – Have any surgical cervical excision procedures been performed?
Obstetric history	• Total number of pregnancies and outcomes, including: – Biochemical miscarriage – Clinical miscarriage – Pregnancy of unknown location – Terminations – Ectopic pregnancy – Stillbirth – Live birth – Conceived with current vs. prior partner(s) – Details of any fertility treatment required • Obstetrical complications, including: – Gestational diabetes – Hypertensive disorders – Preterm delivery – Placental disease – Intrauterine growth restriction – Congenital disease or birth defects in offspring	If outcome other than live birth, inquire about related evaluations

Contd...

Contd...

Points	*Routine assessment*	*Additional consideration*
Medical history	• Past medical and surgical history: – Medical disorders with particular attention to endocrine, autoimmune, genetic, psychiatric, or malignant disorders – Endocrine history should include evaluation of the thyroid, and the presence of galactorrhea and hirsutism – Prior hospitalizations – Surgical procedures • Medications and allergies: – Use of gonadotoxic medications or radiotherapy – Current medications including any supplements – Known drug allergies and type of reaction	If diagnosed with an endocrine disease, what is the status of the disease, including medications and last hormonal testing?
Family history	• Any family members with known history of: – Inherited disorders – Endocrinopathies – Birth defects – Developmental delay – Infertility – Early menopause (<40 years of age) – Multiple spontaneous abortions – Heritable cancer syndromes	• If known or suspected history of inherited disorder, construct family pedigree and assess whether patient had carrier testing. Consider referral to a genetic counselor • If family history of developmental delay, assess whether the individual was evaluated for fragile X syndrome • If family history of infertility, assess whether there was a known associated diagnosis • If family history of early menopause, assess whether there was a known autoimmune or genetic cause
Social history	• Occupation and potential exposure to toxic agents • Use of tobacco, alcohol, or recreational drugs • History of psychological, physical, and/or sexual trauma • Gender identity • Race and ethnicity • Diet and exercise habits	

availability of resources like ultrasound and when tactile examination gives additional information like vaginismus or deeply infiltrating endometriosis. The relevant examinations are tabulated in **Table 9**.[2]

- *Ovulatory dysfunction:* Ovulatory dysfunction accounts for approximately 40% of the overall infertility. Ovulatory function can be assessed in the following ways **(Table 10)**.
- *Ovarian reserve:* Ovarian reserve refers to the remaining pool of ovarian follicles available for fertilization. Physiologically ovarian reserve diminishes with age. So, female partner's age should be the initial predictor of her chance of natural conception. There may be other factors (e.g., chemotherapy, radiation, and genetic mutations) underlying poor ovarian reserve at an early age. Diminished ovarian reserve is associated with poor response to ovarian stimulation and successful treatment outcome. Ovarian reserve can be assessed through serum tests or transvaginal ultrasound. The tests commonly performed are charted in **Table 11**.

  If the female partner has unexplained ovarian insufficiency before the age of 40 fragile X carrier screening should be done to look for FMR1 mutation.
- *Tubal factor:* Tubal patency can be assessed in a number of ways. Commonly used methods are discussed in **Table 12**.[1]
- *Uterine factor:* Many uterine pathologies are associated with infertility, specially when they affect the normal anatomy or functioning of endometrial cavity.

TABLE 9: Physical examination of female with infertility.

Examination	*Indication*	*Evaluate for*	*Additional consideration*
Skin	• Polymenorrhea • Oligomenorrhea • Amenorrhea • Signs or symptoms of androgen excess (e.g., hirsutism, acne, scalp hair loss)	• Hirsutism • Acne • Androgenic alopecia • Acanthosis nigricans	Evaluate for evidence of biochemical androgen excess (hyperandrogenemia); and similar/mimicking disorders including thyroid dysfunction, hyperprolactinemia, and 21-hydroxylase deficient nonclassic adrenal hyperplasia
Thyroid examination	• Abnormal thyroid function tests • Goiter	Thyroid texture and size and the presence of nodularity, tenderness, or cervical adenopathy	• Refer for a thyroid ultrasound if the patient reports rapid growth of the thyroid or if the examination identifies nodularity, asymmetry, or tenderness • Referral to a specialist • Fine-needle aspiration may also be indicated based on examination and ultrasound findings
Breast examination	• Breast pain • Breast mass • Nipple discharge	• Palpable tenderness • Masses • Skin changes • Expressed or spontaneous nipple discharge • Nearly all breast abnormalities without a known cause should be imaged	• Refer for breast ultrasound • Ultrasonography is the preferred initial modality in women <30 years of age and diagnostic mammography is preferred in women over 30 years of age
Speculum examination	• Dyspareunia • Postcoital spotting	• Vaginal and cervical abnormalities • Lesions • Cervical polyps	
Bimanual pelvic examination	Not routinely indicated for infertility evaluation	• Tenderness • Rectovaginal masses or nodularity • Uterine masses • Ovarian masses	• Bimanual pelvic examination will rarely add clinical information to the infertility evaluation that cannot be assessed with pelvic ultrasound • Perform as an adjunct to ultrasound when a tactile examination may add additional useful information to the evaluation. The consideration of a bimanual examination may be influenced by the availability of resources for affordable ultrasonography

It may be found in 16% of the women with infertility.[2] Endometrial polyps, submucous fibroids, intrauterine adhesions, and Mullerian anomalies are some of the commonly found uterine factors of infertility. Following investigations can be done to assess uterine abnormalities **(Table 13)**.

Tests That Should Not be Done Routinely[1,2]

The following tests are not done routinely unless suspected from history or examination **(Table 14)**.

Workup for Male Partner

- *History:* Following points need to be considered—
  - Frequency, timing and in difficulty in intercourse
  - History of erectile or ejaculatory dysfunction
  - Prior fertility history
  - Urologic history
  - Medical and surgical history (including surgery for cryptorchidism)
  - Current medications including any supplements

TABLE 10: Assessment of ovulation.

Methods	*Comments*
History	• Regular menstrual cycle at 21–35 days interval is highly suggestive of ovulation in 99.5% of non-hirsute women[2] • Presence of premenstrual symptoms suggests ovulatory cycles • In presence of hirsutism even with regular cycles ovulation rate decreases to 60% and needs further investigations to confirm ovulation • Clinically anovulation is established from the history of oligomenorrhea or amenorrhea. Further investigations should be aimed to identify the cause
Midluteal progesterone	• Should ideally be obtained 1 week before the expected date of next menstruation • Ovulation is diagnosed if midluteal progesterone level is >3 ng/mL
Ovulation predictor kits	• It identifies midcycle luteinizing hormone (LH) surge which precedes ovulation by 36–48 hours • Midday or evening urine specimens are better to identify the peak LH level • False positive result may be found in PCOS patients due to tonically elevated LH levels
Transvaginal ultrasonography (TVS)	• When performed in luteal phase it can identify the evidence of ovulation (e.g., crenated follicle, fluid in pouch of Douglas) and luteinization • Additionally helps to identify other pelvic pathologies
Basal body temperature	• Biphasic basal body temperature pattern is found in ovulatory cycles • Patient needs to record temperature throughout the cycle • Due to lack of accuracy, it is not recommended routinely
Hormonal assays: Specific hormonal assays to be ordered to identify the cause of anovulation. Tests should be individualized depending on the clinical scenario	• Serum thyroid stimulating hormone (TSH): Routinely done to rule out thyroid disorder. If TSH is abnormal, serum FT4 and thyroid antibodies should be tested • Serum prolactin—should be tested in presence of galactorrhea, oligomenorrhea, or amenorrhea • Serum follicle stimulating hormone (FSH) and estradiol—it is tested to diagnose hypothalamic amenorrhea • Total testosterone, sex hormone binding globulin, free androgen index—tested in cases of suspected hyperandrogenism, oligomenorrhea, or polycystic ovarian syndrome (PCOS)

TABLE 11: Tests for ovarian reserve.

Tests	*Comments*
Serum anti-Mullerian hormone (AMH)	• Secreted by the granulosa cells of small antral follicles • Can be measured on any day of menstrual cycle as serum level remains stable throughout the cycle • AMH <1 ng/mL[2]/≤5.4 pmol/L[5] is associated with poor ovarian reserve
Basal FSH and estradiol	• Assessed on cycle day 2–5 • FSH >10 IU/L[2]/>8.9 IU/L[5] is associated with poor ovarian reserve • Serum estradiol level <60–80 pg/mL aids the diagnosis of poor ovarian reserve
Antral follicle count (AFC)	• TVS is done to count antral follicles in both ovaries measuring 2–10 mm • Low antral follicle count <5 follicles[2,5] • AFC may be elevated in PCOS • AFC may be depressed in hypothalamic amenorrhea or when on hormonal contraception

(FSH: follicle-stimulating hormone; TVS: transvaginal ultrasonography)

TABLE 12: Assessment of tubal factors.

Test	*Comments*
Hysterosalpingography (HSG)	• Radiopaque contrast dye is instilled through cervix and pelvic X-rays are taken to visualize uterus, tubes, and spillage from fimbrial end • Proximal and distal tubal block, salpingitis isthmica nodosa, tubal dilatation can be visualized • Positive and negative predictive value for assessment of tubal patency is 38% and 94% respectively • Due to low specificity, further confirmatory tests are needed to be done when tubal block is diagnosed in HSG
Sonohysterography	Uterus and adnexa are visualized in ultrasound after instillation of fluid through a transcervical catheter.
Hysterosalpingo-contrast sonography (HyCoSy)	• Uterus and tubes are visualized in USG after instilling contrast agent containing air bubbles inside the uterine cavity • It is a newer method and is an extension of sonohysterography • Contrast agents like perflutren lipid microspheres and sulfur hexafluoride lipid-type A microsphere, agitated saline have been used • No contrast agent has USFDA approval till now • To diagnose tubal patency sensitivity is between 76 and 96%, specificity is between 67 and 100% • For both sonohysterography and hysteron-contrast-sonography, findings are more dependent on the observer than HSG
Laparoscopy with chromopertubation	• It is the most definitive method to diagnose abnormalities of tubal factor where scopes are directly inserted in the abdomen to visualize spillage of dye from the fimbrial end of tubes • It is an invasive test and needs anesthesia • It is considered only when some block is suspected in HSG or when laparoscopy is needed for evaluation of other associated pelvic pathology (history of tubal surgery, endometriosis, etc.)
Salpingography and falloposcopy	• Transcervical tubal cannulation done under USG guidance for selective salpingography. It may allow flushing of tubal lumen • Falloposcope is used to directly visualize the tubal lumen

TABLE 13: Uterine factor evaluation.

Investigation	*Comments*
Ultrasonography (USG)	• It is most widely available and most commonly used modality to look for uterine and adnexal pathology • It can be done through transabdominal (TAS) or transvaginal route (TVS) • TVS is more sensitive to identify uterine pathologies
Saline infusion sonography (SIS) or sonohysterography	• It is a relatively newer and better method of delineating uterine cavity by infusion of saline inside the cavity • It has very high positive predictive value (>90%) for diagnosis of intrauterine pathologies
Hysterosalpingography	• Identifies size and shape of the uterine cavity • Any space-occupying lesions of uterine cavity appears as filling gap in the uterine cavity • It has low sensitivity (50%) for diagnosis of endometrial polyps or submucous fibroids
Hysteroscopy	• Most definitive method of diagnosis and treatment of cavitary abnormalities • It needs operative setup and anesthesia
MRI/3D USG	• These investigations are only advised for better evaluation of pelvic anatomy when some abnormality is suspected in first-line investigations • It should not be considered for basic workup

TABLE 14: Special tests for infertility.

Unless specifically indicated these tests should not form an part of basic infertility workup	• Laparoscopy for unexplained infertility • Advanced sperm function testing (e.g., DNA fragmentation) • Postcoital testing • Thrombophilia testing • Immunologic testing • Karyotyping • Endometrial biopsy • Serum prolactin • Serum progesterone • Serum estradiol • Serum FSH and LH

(FSH: follicle-stimulating hormone; LH: luteinizing hormone)

TABLE 15: Normal semen parameters (WHO reference value 2021).

Factor	*Lower reference value*
Semen volume	1.4 mL or more (1.3–1.5)
Total sperm count per ejaculate	39 million or more (35–40)
Total motility (progressive and nonprogressive)	42% or more
Progressive motility	30% or more
Vitality	54% or more live spermatozoa
Sperm morphology (normal forms)	4% or more (strict criteria, amended in 2013)

  - Exogenous steroid use (anabolic steroids, testosterone)
  - History related to sexual practices and history of sexually transmitted infections
  - Exposure to trauma to gonads or toxin exposure
  - Family history
- *Semen analysis:* Semen analysis is one of the routine first-line investigations for infertility workup. Even in presence of female factor infertility, semen analysis has to be performed to rule out male factor infertility. Ideally samples are obtained at the laboratory by masturbation after 3–5 days of abstinence. If samples are not collected on site, immediate transportation at room or body temperature should be done within 1 hour. If first semen analysis report is abnormal repeat confirmatory test should be done after 3 months to allow time for the current batch of spermatozoa be replaced by a fresh batch.[5] But if the diagnosed abnormalities are of severe nature (e.g., azoospermia) then immediate repeat test should be ordered to confirm. Any abnormality of semen analysis warrants further evaluation by the andrologist or urologist specially trained in male reproductive endocrinology. Semen parameters are assessed using WHO reference value 2021 and is tabulated in **Table 15**.[6]

Screening for antisperm antibodies should not be offered as its treatment has failed to improve fertility outcome.[5]

On account of severe oligozoospermia or azoospermia, the male partner should be further evaluated by hormonal assays like serum FSH, LH, testosterone, and estradiol to identify the cause of infertility. Patients having reduced libido and low testosterone should have their prolactin level tested.

Discussion Regarding Natural Conception Chance

The management of infertility should include a detailed discussion about chances of natural conception if the couple opts for expectant management. It is also important for the specialist to choose the potential candidates for infertility treatment. Spontaneous chance of conception is 15–20% after first month of trying and subsequently it falls to 5% per cycle at the end of 12 months. Cumulative conception rate is 60% in first 6 months, 85% after 12 months, and 95% in 2 years.[7,8] Age of female and male partner is the most important prognostic factor affecting chance of natural conception. Increased infertility is mainly due to decline in the quantity and quality of the available gametes and only 50% women in their 40s are expected to conceive.[7] Chance of natural conception after 1 year for women aged 19–26 years is 92% but for women aged 35–39 years it declines to 80%.[5] Various other factors may affect the probability of natural conception and described in **Table 16**.

There are certain conditions associated with very poor chance of conception. The presence of any such factors should be sought for and discussed with the couple when present.

Some of these factors are:
- Sperm count <2 million/mL or azoospermia
- Bilateral tubal block with grossly damaged tubes on laparoscopy

- Total absence of natural period
- Inability to have natural intercourse.

Discussion of Available Treatment Options and Shared Decision Making

After initial workup, the infertile couples should be informed regarding available treatment options for the specific type of infertility and a shared decision should be made with the couple regarding further management plan. With the advent of assisted reproductive technologies (ART), in vitro fertilization (IVF) has become one of the important management options in all cases of infertility either as primary or secondary treatment. While making informed decision due respect should be given to the financial, social, cultural, and religious values and individual preference of the couple should be adjusted against the risks involved in the treatment. *The treatment options of different conditions are summarized here.*

TABLE 16: Prognostic factors for infertility.

Factors	*Good prognostic marker*	*Poor prognostic marker*
Woman's age	35 year or less	40 year or more
Previous pregnancy	Yes	None
Duration of subfertility	<3 years	<3 years
Cause of subfertility	Unexplained	Other

MANAGEMENT ALGORITHM OF OVULATORY DYSFUNCTION

According to the type of ovulatory dysfunction, the management protocols will be different and are charted in **Table 17**.[5]

Ovulation Induction

Ovulation induction (OI) is a method where ovaries are stimulated to induce ovulation in anovulatory women and is the mainstay of treatment in Group II ovulatory dysfunction. Various methods both medical and surgical have been used for ovulation induction.

- *Clomiphene citrate:* It is a selective estrogen receptor modifier (SERM) which blocks the estrogen receptor in the anterior pituitary and its negative feedback on FSH secretion. Thus, it increases FSH secretion.

  Dose: 50–150 mg/day for 5 consecutive days starting from day 2/3 of cycle.

  It achieves ovulation in 70–85% of patients. The cumulative conception rate is 67% after 6 months of use. Common risks involved are the risk of multiple pregnancy (11%) and very small risk of ovarian hyperstimulation syndrome (OHSS).

  If ovulation is not achieved after maximum dose of clomiphene, it is considered as clomiphene resistance and other drugs should be used.
- *Letrozole:* It is an aromatase inhibitor and widely used for OI. It has higher live birth rate than clomiphene. 2.5–7.5 mg/day dose is used for OI. Risk of multiple

TABLE 17: Treatment of ovulatory dysfunction.

Treatment of Group I ovulatory dysfunction	• Lifestyle modifications • Treatment of the underlying cause • Administration of gonadotropin including both FSH and LH can induce ovulation in 95% cases • Pulsatile administration of gonadotropin-releasing hormone or pump is an effective treatment option for ovulatory dysfunction of hypothalamic origin. But it is not easily available and associated with high cost of therapy	• by increasing body weight if BMI <19 • By controlling exercise if they are involved in high physical exercise • Brain lesions • Prolactinoma
Treatment of Group II ovulatory dysfunction	• Lifestyle modification • Ovulation induction	• Reduction of even 5–10% body weight in obese patients can achieve ovulation • Using oral drugs • Insulin sentitizers like metformin • Ovarian drilling for PCOS • Gonadotropin injections

(FSH: follicle-stimulating hormone; LH: luteinizing hormone; PCOS: polycystic ovarian syndrome) pregnancy and OHSS are much lesser compared to clomiphene citrate.

- *Metformin:* Metformin is most widely used insulin sensitizer used for OI in PCOS. It reduces serum insulin and androgen concentration and induces ovulation in 40–50% of PCOS patients. When used alone, the risk of multiple pregnancy and OHSS can be avoided. It can be used alone or along with other oral drugs to improve the conception rate.

  Regarding the dose there is no uniform consensus. Mostly researched dose is 1,500 mg in divided dose and maximum 2,500 mg daily have been used when BMI is very high.
- *Gonadotropins:* Injectable gonadotropins are highly effective agents for OI and successful ovulation occurs in >95% cases. Many preparations are available like human menopausal gonadotropins, recombinant FSH, etc. In PCOS patients, preparations having only FSH activity are preferred.

  It has very high risk of multiple pregnancy and OHSS. So, it is considered as second-line treatment when patient is resistant to clomiphene or letrozole. Close folliculometry should be done to diagnose multifollicular development and aggressive cancellation of the cycle should be done if there is development of >2 dominant follicle (DF)
- *Ovarian drilling:* It is done laparoscopically where diathermy is used to make holes in the ovary with an aim to destroy ovarian stroma in PCOS patients.

  Ovarian surface cysts are punctured with monopolar current cautery (20–30 W) up to a depth of 2–4 mm. Usually 4–8 punctures are made in each ovary. It can achieve ovulation in 80% cases if done adequately.

  However, it carries the risk of ovarian adhesion, diminished ovarian reserve if healthy tissue is damaged and inherent risk of anesthesia and laparoscopy. So, it is considered as second line of treatment.
- Additional points to remember while performing ovulation induction:
  - Ovulation inducing agents should be used in the lowest effective dose and duration.
  - For women taking clomiphene or letrozole, folliculometry should be done at least in the first cycle to ensure the adequacy of the dose with minimum risk of multiple pregnancy.
  - While using gonadotropins risk of multiple pregnancy and ovarian hyperstimulation should be discussed priorly.
  - Serial folliculometry should always be an integral part of the treatment while using gonadotropins.
  - Up to date information about the risks of ovarian hyperstimulation should be given to the couple.
  - Though no direct link has proven till now between ovulation inducing agents and ovarian cancer, there is a weak association between these agents and development of borderline ovarian tumor (BOT) in later life.[9]

Ovarian Hyperstimulation Syndrome

Ovarian hyperstimulation syndrome (OHSS) is a systemic disorder triggered by the release of vasoactive substances from the stimulated ovaries. It is a complication of ovarian stimulation for ovulation induction. Though risk of OHSS is minimal with oral ovulation induction agents, maximum risk is found while using injectable gonadotropins particularly in high dose. Patients of PCOS are more at risk of developing OHSS.

Symptoms of OHSS are due to third space loss of fluid following increased capillary permeability. Clinically, it is evident in the form of abdominal pain, bloating, nausea, vomiting, extreme thirst, anuria or oliguria and respiratory distress. Depending on the severity, it is divided into four categories—mild, moderate, severe, and critical.

Risk factors of OHSS are young age, PCOS, prior OHSS, high-dose gonadotropin injection, exposure to HCG injection, very high serum estradiol level, conception of multiple pregnancy, etc.

MANAGEMENT ALGORITHM OF TUBAL AND UTERINE FACTOR INFERTILITY (TABLE 18)

Additional comments:

- Proper selection of cases is very important before proceeding for tubal surgery to get optimum results
- Ectopic pregnancy is an inherent risk of any tubal surgery
- IVF is the first-line treatment for moderate severe or bilateral tubal disease. But in the presence of hydrosalpinx prior salpingectomy should be considered.

MANAGEMENT ALGORITHM OF MALE FACTOR INFERTILITY (TABLE 19)

Additional considerations:

- Men with idiopathic semen abnormalities should not be treated with antiestrogen, gonadotropins,

TABLE 18: Management protocol of *tubal and uterine factor* infertility.

Cause of infertility	*Management protocol*
Mild tubal disease	Surgical correction is more effective than no treatment and should be considered if surgical expertise is available[5]
Proximal tubal blockage	Consider tubal catheterization or hysteroscopic cannulation as these procedures have shown to improve pregnancy rate[5,10]
Hydrosalpinx	Laparoscopic salpingectomy should be offered to women with hydrosalpinx before proceeding for IVF as it improves chance of live birth[5]
Severe tubal disease	IVF should be considered as first-line treatment
Female sterilization	Tubal reanastomosis is the first choice if the woman had previous history of sterilization
Intrauterine adhesions	Hysteroscopic adhesiolysis under laparoscopic or USG guidance should be done
Endometriosis	• Minimal to mild endometriosis—consider management in line of unexplained infertility • Moderate to severe—surgical management should be considered to improve pregnancy rate • For ovarian endometrioma consider laparoscopic surgical treatment • In vitro fertilization (IVF) should be considered when other methods fail
Endometrial polyp	Recent Cochrane review failed to show any improvement of pregnancy rate prior to fertility treatment
Uterine fibroid	• Myomectomy for submucous or cavity distorting intramural fibroids is associated with higher pregnancy rate, but live birth rate is not increased. One has to judge the risk of major surgical procedures with the expected benefit. Minimally invasive surgeries are preferred when expertise is available[5,11] • Role of uterine artery embolization, magnetic resonance guided laser coagulative necrosis and high-intensity focused ultrasound is being evaluated[5]

TABLE 19: Management of male factor infertility.

Type of male factor infertility	*Treatment option*
Hypogonadotropic hypogonadism	• Gonadotropins can improve fertility in patients with hypogonadotropic hypogonadism and should be tried first where feasible • If it fails consider sperm retrieval and intracytoplasmic sperm injection (ICSI) • Consider donor insemination if no sperm is available
Hyperprolactinemia	• Any underlying cause or history of drug intake if present should be reversed first • Cabergoline or bromocriptine is used to treat hyperprolactinemia • If serum testosterone level fails to normalize within 6 months of control of prolactin level, human gonadotropin (hCG) injections or pulsatile GnRH can be tried
Obstructive azoospermia (OA)	• Correction of epididymal blockage to restore duct patency should be done if facility and expertise of surgery is available • For irreversible obstructive azoospermia surgical sperm retrieval from epididymis and testis followed by IVF/ICSI can be done alternatively
Nonobstructive azoospermia (NOA) and severe deficits in semen quality	• If sperm retrieval possible then consider ICSI • Donor insemination is considered when no sperm can be retrieved
Mild male factor infertility	• Treatment is done in the line of unexplained infertility • If initial treatment fails consider IVF with husband's sperm
Management of ejaculatory failure	• Medical management of anejaculation and retrograde ejaculation should be tried first when possible • Sperm retrieval by penile electrovibration or sperm recovery from urine followed by IUI, IVF, ICSI, or GIFT if medical management fails • Surgical sperm retrieval and IVF/ICSI when all method fails

(GnRH: gonadotropin-releasing hormone; GIFT: gamete intrafallopian transfer; IUI: intrauterine insemination; IVF: in vitro fertilization) androgens bromocriptine or kinin enhancing drugs as they have not been found to be effective.[5]

- Men with leukocytes in semen should not be treated with antibiotic unless specific infection has been identified.[5]
- Surgical correction of varicocele should not be done as a part of infertility treatment as it does not increase pregnancy rates.[5]
- Intrauterine insemination with husband's semen is considered only when there is mild form of abnormality of semen or when there is any sociocultural or religious objection to IVF. Donor insemination is an alternative to many couples who are not willing for IVF or when ART procedures have failed.
- Type of ART for treatment of male factor infertility depends on the quality and quantity of available spermatozoa. For mild to moderate severity often IVF is feasible. For severe oligo-astheno-teratozoospermia, ICSI is the mode of fertilization.
- Intracytoplasmic sperm injection (ICSI) is a method where a spermatozoon is directly injected into the cytoplasm of oocyte to achieve fertilization. It is particularly helpful for severe OAT syndrome when semen parameters are grossly compromised.
- Different methods of surgical sperm retrieval are available as described in **Table 20**.

MANAGEMENT ALGORITHM OF UNEXPLAINED INFERTILITY

For all the couples with unexplained infertility, mild male factor infertility, and mild endometriosis, same management protocol is followed. After diagnosis of unexplained infertility, almost 15% couples will conceive in 1 year and 35% in 2 years without any treatment. So NICE recommends expectant management for 2 years which includes up to 1 year before starting infertility investigations when woman's age is less than 36 years.[5] Various treatment options are tabulated in **Table 21** and

TABLE 20: Sperm retrieval techniques and procedures.

Name of sperm retrieval technique	*Description*
Microsurgical epididymal sperm aspiration (MESA)	Sperm is aspirated microsurgical from epididymis
Percutaneous epididymal sperm aspiration (PESA)	Sperm is aspirated from the epididymis using a needle without need of surgical exploration
Percutaneous testicular sperm aspiration	Fine-needle aspiration of sperm is done from the testes
Open testicular sperm extraction (TESE)	Biopsy is taken from the testes followed by extraction of the sperm from the specimen. Depending on the incision, it might be conventional testicular sperm extraction (cTESE) or microdissection testicular sperm extraction (mTESE)

TABLE 21: Management protocol for unexplained infertility.

Management option	*Remarks*
Expectant management	Considering better chance of conception these couples are advised for regular unprotected sexual intercourse for a total of 2 years which may include up to 1 year before the fertility investigations
Ovarian stimulation	Ovulation induction with oral agents alone should not be offered routinely as it does not increase the chance of conception or live birth
Intrauterine insemination	• Routinely intrauterine insemination (IUI) with or without ovarian stimulation should not be offered when regular intercourse is possible, unless there is social, cultural, or religious objection to IVF • IUI should be considered as an alternative to vaginal intercourse if: – Vaginal intercourse is not possible due to physical or psychosexual illness – People in same sex relation – When sperm washing is needed (male partner HIV positive) • In the above context consider IVF after 12 cycles of failed intrauterine insemination
IVF	Consider IVF if patient fails to conceive after 2 years of regular unprotected intercourse or after 12 cycles of inseminations

(IVF: in vitro fertilization)

the recommendations have been endorsed from NICE guideline.[5]

Many researchers have challenged the NICE recommendation against IUI and ovarian stimulation for unexplained infertility. Farquhar et al. in their paper concluded that IUI with ovarian stimulation is a safe and effective treatment of unexplained infertility.[4] Even in practice while treating patients in resource-poor settings, it is often considered as a cost effective treatment option before proceeding for IVF.

ASSISTED REPRODUCTIVE TREATMENTS

It encompasses all procedures where gametes (sperm and eggs) are directly handled in vitro. It includes intrauterine insemination (IUI), in vitro fertilization (IVF), and intracytoplasmic sperm injection (ICSI).

Intrauterine Insemination

Intrauterine insemination is a process of insemination of sperm artificially inside the uterine cavity at appropriate time. It may or may not be associated with ovarian stimulation. Its clinical application is based on the assumption that placing processed sperm in high number in the uterine cavity makes them more accessible and improves chances of fertilization. Before doing IUI patency of fallopian tubes and ovulation should be ensured. IUI can be done using husband's or donor sperm. There may be a number of indications for IUI as described below:

- Unexplained infertility
- Minimal/mild endometriosis
- Mild male factor infertility
- Cervical factor infertility
- Physical or psychological disability to have vaginal intercourse
- As part of donor insemination

Risks of IUI:

- Flaring of pelvic infection
- Increased risk of ectopic pregnancy
- When associated with ovarian stimulation risk of multiple pregnancy and OHSS.

IVF and ICSI

In vitro fertilization is a procedure where gametes are fertilized in vitro, embryo is developed in the culture media from the fertilized ovum and then one or more embryos are transferred to uterine cavity for further development. ICSI involved direct injection of sperm into the cytoplasm of ovum. The common indications of IVF and/or ICSI are described below:

- Tubal damage/disease
- Male factor infertility when other options fail. ICSI is considered for severe male factor infertility
- Endometriosis moderate to severe degree
- Unexplained infertility
- Ovulatory disorders when other treatment fails
- Premature ovarian failure when ovum was preserved or using donor ovum
- In surrogacy
- When preimplantation genetic screening (PGS) or preimplantation genetic diagnosis (PGD) is indicated.

CONCLUSION

To conclude basic management of infertile couples starts with preconceptional counselling and diagnostic workup of both partners to find out the cause of subfertility. Couple should be provided with detailed information regarding available treatment options and a realistic chance of conception. Finally any management should be mutually agreed by the couple and the treating physician.

REFERENCES

1. ACOG. Infertility Workup for the Women's Health Specialist: ACOG Committee Opinion, Number 781. Obstet Gynecol. 2019;133(6):e377-84.
2. Penzias A, Azziz R, Bendikson K, Cedars M, Falcone T, Hansen K, et al. Fertility evaluation of infertile women: a committee opinion. Fertil Steril. 2021;116(5):1255-65.
3. RCOG. Subfertility Available from: https://elearning.rcog.org.uk//tutorials/core-knowledge/subfertility. [Last accessed March, 2024]
4. Farquhar CM, Liu E, Armstrong S, Arroll N, Lensen S, Brown J. Intrauterine insemination with ovarian stimulation versus expectant management for unexplained infertility (TUI): a pragmatic, open-label, randomised, controlled, two-centre trial. Lancet (London, England). 2018;391(10119):441-50.
5. National Collaborating Centre for Women's and Children's Health (UK). (2013). Fertility: Assessment and Treatment for People with Fertility Problems. Available from: https://pubmed.ncbi.nlm.nih.gov/25340218/. [Last accessed March, 2024]
6. Boitrelle F, Shah R, Saleh R, Henkel R, Kandil H, Chung E, et al. The Sixth Edition of the WHO Manual for Human Semen Analysis: A Critical Review and SWOT Analysis. Life (Basel). 2021;11(12):1368.
7. Dunson DB, Baird DD, Colombo B. Increased infertility with age in men and women. Obstet Gynecol. 2004;103(1):51-6.

8. Te Velde ER, Eijkemans R, Habbema HDF. Variation in couple fecundity and time to pregnancy, an essential concept in human reproduction. Lancet. 2000;355(9219):1928-9.
9. Rizzuto I, Behrens RF, Smith LA. Risk of ovarian cancer in women treated with ovarian stimulating drugs for infertility. Cochrane Database Syst Rev. 2013;2013(8):CD008215.
10. Thurmond AS, Machan LS, Maubon AJ, Rouanet JP, Hovsepian DM, Van Moore A, et al. A review of selective salpingography and fallopian tube catheterization. Radiographics. 2000;20(6):1759-68.
11. Pritts EA. Fibroids and infertility: a systematic review of the evidence. Obstet Gynecol Surv. 2001;56(8):483-91.

Chapter 27

Female Infertility

Yogita Dogra

INTRODUCTION

Infertility is typically defined as the inability to achieve pregnancy after 1 year of unprotected intercourse.[1] Infertility can be either due to male factor or female factor or both. Infertility can be primary or secondary. In the case of primary infertility, the couple has never been able to conceive a child, whereas in the case of secondary infertility, the couple has previously conceived a child but is now experiencing problems conceiving again, after either carrying the pregnancy to term or suffering a miscarriage.[2]

In today's society, delaying pregnancy is a prevalent choice among women. In recent years, there has been an increase in the number of women in their late 30s and 40s who are attempting pregnancy and having children. You may be unaware that your fertility begins to decline significantly in your mid-30s, and that this decline accelerates in your late 30s. Some women experience a decline in fecundity as early as their late 20s and early 30s. Fertility decreases with age because fewer eggs remain in the ovaries, and the quality of the residual eggs is lower than it was when you were younger.[1]

CAUSES OF FEMALE INFERTILITY[3]

The causes of female infertility are shown in **Figure 1**.

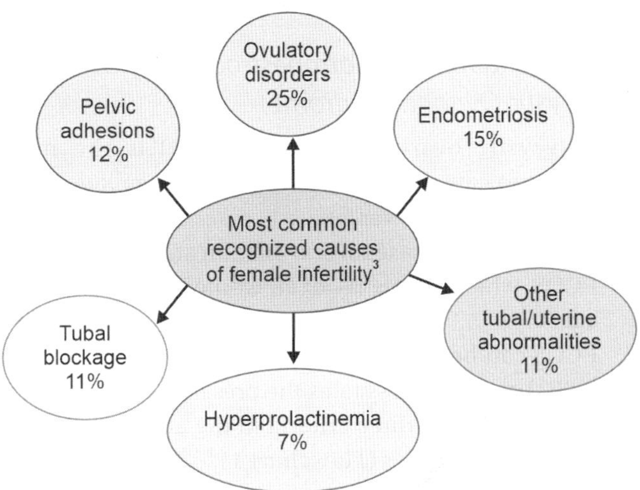

Fig. 1: Causes of female infertility.

Ovulatory Disorders

25% of the known causes of female infertility are attributed to ovulatory disorders. Anovulation or oligo-ovulation leads to infertility because no oocytes are released each month and subsequently there is no possibility of fertilization and pregnancy in the absence of an oocyte.[3]

To aid in treatment and classification, the World Health Organization (WHO) has subdivided ovulatory disorders into four categories as shown in **Figure 2**.[3]

Ovulatory disorders (WHO classification)

1. Hypogonadotropic hypogonadal anovulation (hypothalamic amenorrhea)	2. Normogonadotropic normoestrogenic anovulation [PCOS (polycystic ovarian syndrome)]
3. Hypergonadotropic hypoestrogenic anovulation (premature ovarian failure)	4. Hyperprolactinemic anovulation (pituitary adenoma)

Fig. 2: Categories of ovulatory disorders.

Hypogonadotropic Hypogonadal Anovulation (Hypothalamic Amenorrhea)

- A decrease in hypothalamic gonadotropin-releasing hormone (GnRH) secretion is linked with eating disorders and excessive exercise, leading to hypothalamic amenorrhea or functional hypothalamic amenorrhea (FHA).[4]

- Reduced caloric intake, weight reduction, or strenuous activity can all increase cortisol levels, which in turn suppresses GnRH.[5]
- When GnRH pulsatility is low or nonexistent, the anterior pituitary gland produces less of the gonadotropins follicle-stimulating hormone (FSH) and luteinizing hormone (LH).
- Low estrogen levels and irregular follicular growth are the results of these two deficits.
- There will be a wide range of FSH and LH levels, from normal to low, but the hormone ratio will be similar to that of a prepubescent girl's.[6]

Normogonadotropic Normoestrogenic Anovulation (Polycystic Ovarian Syndrome)

- Polycystic ovary syndrome (PCOS) is the most common cause of anovulation in normogonadotropic normoestrogenic women.
- Up to 85% of all cases of anovulation are attributable to PCOS, which affects 8% of all women of reproductive age.[7]
- PCOS can be diagnosed using the Rotterdam criteria (2003), which necessitates the presence of at least two of the following three criteria in the absence of other pathological causes **(Flowchart 1)**.[8]
- PCOS can also be diagnosed using 2006 Androgen Excess Society (AES) Guidelines (AE-PCOS 2006 criteria) **(Flowchart 2)**.[9]
- PCOS-related infertility is believed to be caused by a dysfunction in the maturation of follicles, which results in anovulation.

Flowchart 1: Rotterdam criteria (2003).

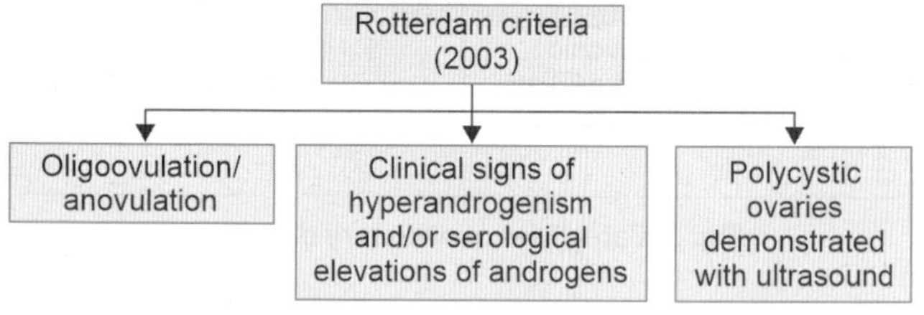

Flowchart 2: AE-PCOS (Androgen Excess and Polycystic Ovary Syndrome) 2006 criteria.

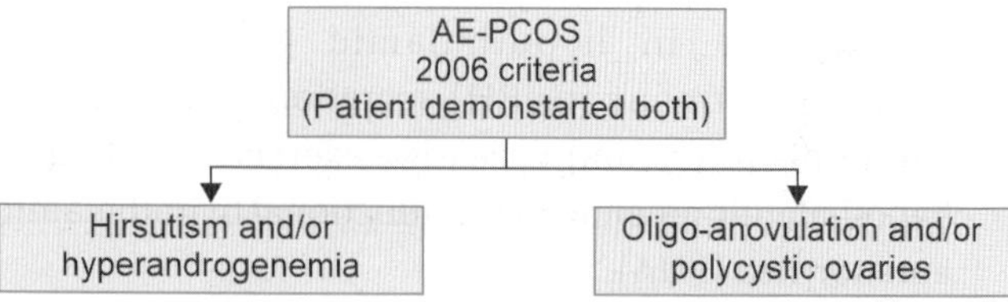

- Both the FSH and estrogen levels will be within acceptable laboratory ranges.
- In terms of LH, it can be either normal or increased.
- Classical descriptions of PCOS and infertility point to aberrant pulsatility of GnRH, although the pathophysiology behind these conditions is poorly understood. Anti-Mullerian hormone (AMH) levels tend to rise in tandem with the prevalence of ovarian cysts and follicular arrest.[10]

Hypergonadotropic Hypoestrogenic Anovulation (Premature Ovarian Failure)

- Hypergonadotropic hypoestrogenic anovulation is a subtype of premature ovarian insufficiency and age-related ovarian resistance in females.
- As previously stated, a woman's age affects her fertility due to a well-studied phenomenon involving a steady decline in the quality and quantity of her oocytes.
- At 20 weeks of gestation, the female fetus has approximately 6 million follicles. There are approximately one million follicles in a neonate. The number of follicles falls to 300,000 at the advent of puberty.[11]
- In a woman's mid-30s, the rate of follicle loss begins to accelerate. Follicle loss persists throughout a woman's existence.[12]
- External factors are also linked to diminished follicular number. Cigarette smoking is the most studied and well-known habit. Both fecundability and follicular quantity are negatively correlated with cigarette consumption. Early menopause (age <40 years) is also associated with cigarette smoking; the prevalence of early menopause among ever-smokers increases by >30%.[13]
- Additionally, ovarian health is essential to overall fecundity. Meiotic nondisjunction, which results in aneuploidy, is linked to the deterioration of oocyte quality throughout a woman's existence. This is believed to be a result of accumulated injury and age-related alterations in the granulosa cells. As women age, the number of meiotic nondisjunction events and corresponding aneuploid or chromosomally aberrant oocytes and embryos increases significantly.[14]
- Primary ovarian insufficiency (POI) is defined as hypergonadotropic hypogonadism before 40 years of age. This disease is marked by a lack of folliculogenesis, a reduction in estrogen, oocyte loss, and infertility.[15]

- Turner syndrome, monosomy of the sex chromosomes leading to a 45X karyotype, is the most prevalent cause of POI.[16]

Hyperprolactinemic Anovulation (Pituitary Adenoma)

- Prolactin suppresses hypothalamic GnRH secretion, resulting in a low LH, which causes anovulation, oligomenorrhea, or amenorrhea.
- 20–50 ng/mL of prolactin serum causes insufficient progesterone release from the corpus luteum, thereby shortening the luteal phase.
- Contradictory, prolactinemia is described as the cause of luteal phase defects that result in infertility. Due to abnormal feedback on the hypothalamic-pituitary-ovarian axis, prolactin levels between 50 and 100 ng/mL cause amenorrhea or oligomenorrhea. A concentration >100 ng/mL is most commonly associated with pituitary adenomas and overt hypogonadism and amenorrhea.[17]
- As stated previously, the WHO recognizes prolactinemia as a primary cause of female infertility. However, the ASRM (American Society for Reproductive Medicine) published guidelines stated that prolactin is not required as part of the initial workup.[18]

Endometriosis

Endometriosis is defined as the presence of endometrial tissue outside of the uterus. The diagnosis is centered on the identification of endometrial glands and/or stroma outside of the uterus via histology. Endometriosis is most commonly found in the pelvis, but it can extend to the entire abdomen, affecting 10–15% of women of reproductive age. 40–50% of women with endometriosis will experience infertility.[19]

- The ASRM classifies endometriosis into four stages, with stage I being minimal and stage IV being severe. Endometriosis is known to induce infertility, but it is believed that the pathophysiology varies depending on the stage.[20]
- For stages I and II, infertility is believed to be linked to inflammation characterized by an increase in the production of prostaglandins and cytokines, macrophages, and natural killer cells.[21] Inflammation impairs ovarian and tubal function, resulting in abnormal follicular development, fertilization, and implantation.[22]
- Stages III and IV are associated with pelvic adhesions and/or masses that distort pelvic anatomy; consequently, tubal motility, oocyte release, and sperm motility will be impaired. Also, it is hypothesized that advanced endometriosis impairs folliculogenesis, which reduces the potential for fertilization.[23]

Pelvic/Tubal Adhesions

Along with uterine and tubal abnormalities, pelvic and tubal adhesions are largely responsible for female infertility. Infectious processes within the abdomen are the primary cause of pelvic/tubal adhesions.

- Pelvic inflammatory disease (PID) is the most prevalent infectious processes associated with infertility.
- *Chlamydia trachomatis* is the microorganism that poses the greatest risk of infertility in relation to PID. One in four women with tubal factor infertility will have chlamydia-specific antibodies, which are inversely proportional to the likelihood of pregnancy.[24]
- The likelihood of infertility is affected by both the number and severity of PID episodes. A study revealed that the pregnancy rate following PID was 89% following one episode, 77% following two episodes, and 46% following three episodes.[25]
- The live birth rates for mild, moderate, and severe PID severity were 90%, 82%, and 57%, respectively.[26]

Hydrosalpinges is a tubal abnormality induced by acute and chronic inflammation that compromises the fallopian's structural integrity. This damage results in tubal obstruction, which impedes the flow of physiologic fluid through the fallopian tube and causes fluid accumulation. It is believed that hydrosalpinges impair fertility through the retrograde passage of toxins and prostaglandins into the endometrium, thereby reducing endometrial receptivity and creating an unfriendly environment for implantation.[27] Hydrosalpinx has been shown to reduce the likelihood of pregnancy by 50% in patients undergoing in vitro fertilization (IVF).[28]

Uterine Causes

Either space-occupying lesions or reduced endometrial receptivity are associated with uterine causes of infertility.

- In a meta-analysis of uterine leiomyomas (fibroids), only submucosal or intracavitary fibroids were found to impair implantation and pregnancy rates relative to other infertile controls.[29]

- Although uncommon, congenital uterine abnormalities (CUA) are also associated with infertility. Uterine septa are the most prevalent and are also associated with recurrent pregnancy loss. CUA-related infertility is believed to account for approximately 8% of female infertility; however, 25% of women with late first-trimester or second-trimester miscarriages have CUAs.[30]

UNEXPLAINED INFERTILITY[1]

- There is no readily identifiable cause of infertility in approximately 10% of couples attempting to conceive for whom all of the above tests are normal.
- In a much higher proportion of couples, only minor abnormalities that should not be severe enough to cause infertility are identified. In these instances, infertility is considered "unexplained".
- Egg quality, fertilization, genetics, tubal function, or sperm function may be difficult to diagnose and/or treat in couples with unexplained infertility.
- Couples with unexplained infertility have used fertility medications and intrauterine insemination (IUI) with limited success.
- IVF has been shown to be the most effective treatment for unexplained infertility if no pregnancy is achieved within three to six treatment cycles.

EVALUATION OF FEMALE INFERTILITY

When analyzing female infertility, it is critical to keep in mind that a male infertility examination is also necessary and ought to begin at the same time as the evaluation of female infertility.

History[3]

The most important aspects of the chronicles of the infertile woman are mentioned in **Box 1**.

BOX 1: History taking.

- Duration of infertility
- Obstetrical history
- Menstrual history
- Medical, surgical, and gynecological history
- Sexual history (premature ejaculation, couples overall coital frequency and timing)
- Social and lifestyle history (smoking, drinking, drug usage)
- Pregnancy loss, family history, genetic history, etc.

Physical Examination[3]

The following should be covered in a thorough physical examination:

- Vital signs and body mass index (BMI)
- Thyroid evaluation
- Breast examination for galactorrhea
- *Signs of androgen excess:* Dermatological and external genitalia examination
- The appearance of abnormal vaginal or cervical anatomy
- Pelvic masses or tenderness
- Uterine enlargement or irregularity
- Transvaginal ultrasonography.

Evaluation[3]

The evaluation of female infertility is shown in **Fig. 3**.

Assessment of Ovarian Function and Reserve

Evaluation of ovarian function is as straightforward as analyzing the menstrual cycle history.

- Women who have regular, predictable menstrual cycles and menstrual discharge as well as molimina (bloating, fatigue, breast tenderness) are most likely ovulating.
- At-home LH predictor devices detect a midcycle LH surge, which is indirect evidence of ovulation and aids in determining the fertile window.[31]
- At day 21 of a cycle, progesterone serum level, or more precisely a midluteal phase progesterone level, can also detect ovulation.
- A progesterone assay value >3 ng/mL is evidence of ovulation and should be obtained approximately 1 week prior to menstruation.[32]
- Daily ultrasounds for monitoring the growth and demise of a follicle are a more invasive, more accurate, but probably superfluous method for determining ovulation **(Flowchart 3)**.[33]
- The theory behind day 3 FSH and estradiol is that women with a healthy ovarian reserve produce enough

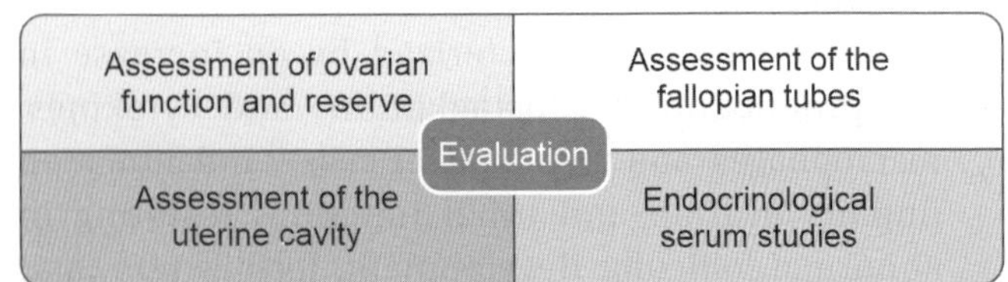

Fig. 3: Evaluation of female infertility.

Flowchart 3: Common assays available to evaluate ovarian reserve.

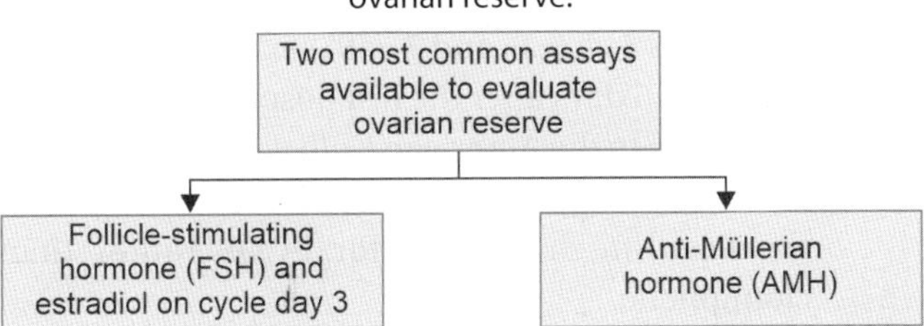

ovarian hormones from small follicles to keep FSH levels low. It is crucial to identify women with a low follicle count that generates insufficient hormones, resulting in a lack of inhibition and an elevated FSH. Due to insufficient ovarian reserve, FSH levels >20 IU/mL indicate a poor prognosis for spontaneous ovulation. Normal ovarian reserve is indicated by FSH levels <10 IU/mL. Less than 80 pg/mL on day 3 is regarded normal and indicative of adequate ovarian reserve. Greater than 80 pg/mL resulted in reduced pregnancy rates, and >100 pg/mL resulted in no pregnancy.[34]

- AMH is a hormone expressed by preantral and antral follicles that can be measured at any time during a woman's menstrual cycle as a marker of ovarian function.[35] Throughout a woman's natural reproductive life, AMH levels will decrease until they are undetectable at menopause.[36] AMH levels appear to be an excellent indicator of exogenous gonadotropin response:[37]
    - <0.5 ng/mL predicts difficulty getting >3 follicles to grow.
    - <1.0 ng/mL shows limited egg supply that may require more aggressive ovulation induction protocols.
    - 1.0–3.5 ng/mL shows normal values.
    - >3.5 shows ample supply and may require mild induction to prevent ovarian hyperstimulation syndrome (OHSS).
- Antral follicle counts, determined by transvaginal ultrasound in the early follicular phase and measuring the number of follicles <9 mm in the ovaries, are also an accurate measure of ovarian reserve and a predictor of ovarian response to stimulation.[38]

Assessment of the Fallopian Tubes

- Laparoscopy with chromopertubation represents the gold standard for evaluating tubal patency.
- Laparoscopy is a first-line diagnostic test for suspected pelvic adhesions, endometriosis, and other pelvic pathologies. However, the hysterosalpingogram (HSG) is more commonly used for the first-line evaluation of tubal patency and abnormalities due to its high specificity and low invasiveness.[39]
- The HSG has weak predictability for intrauterine and tubal adhesions.[40] With oil-soluble media, another added benefit of the HSG is an increase in pregnancy and live birth rates. After HSG, pregnancy and live birth rates increased relative to controls, according to a meta-analysis [odds ratio (OR): 2.98, 95% confidence interval (CI): 1.05–6.37].[41]

Assessment of the Uterine Cavity

- The gold standard for evaluating the uterine cavity is hysteroscopy, which permits direct visualization of intrauterine pathology and permits immediate surgical correction.
- Despite the fact that hysteroscopy is considered the gold standard, saline infusion sonography (SIS) is a less invasive alternative.
- The SIS is an adequate screening instrument prior to infertility treatment, with or without 3D model rendering, as it is highly sensitive and specific for all intrauterine abnormalities.[42]

MANAGEMENT OF FEMALE INFERTILITY

Lifestyle Changes

- Extremes of BMI are frequently associated with infertility and ovulatory dysfunction in women.[43]
- Women with a BMI of <17 kg/m^2 and a history of intensive exercise regimens or eating disorders are prone to develop hypogonadotropic hypogonadism, which is characterized by pituitary gonadotropin secretions that have ceased.[44]
- One study demonstrated the significance of behavioral modification in ovulation induction. 87% of the women who received individualized therapy to treat energy deficiencies or behavioral issues regained normal ovarian function.[45]
- Women with anovulation and a BMI >27 kg/m^2 can enhance ovulation through weight loss alone. Multiple studies have demonstrated that a 10% weight loss can restore normal ovulation in 50–100% of women in <1 year.[46]
- Obese women who received counseling and interventions for weight loss prior to infertility treatment did not have higher pregnancy or live birth

Flowchart 4: Medications for ovarian stimulation.

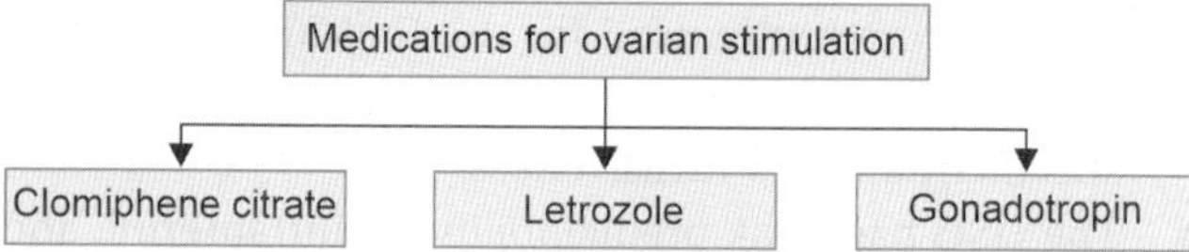

rates than obese women who received infertility treatment without weight loss interventions. Therefore, there is no requirement for a particular BMI to initiate fertility treatment.[47]

Controlled Ovarian Hyperstimulation

The medications for ovarian stimulation are shown in **Flowchart 4**.

- Clomiphene citrate (CC) is the first-line treatment for infertility of unknown cause and the medication most providers use.
  - Clomiphene is a selective estrogen receptor modulator (SERM) with estrogen antagonist and agonist effects that ultimately enhance the release of gonadotropin from the anterior pituitary.
  - Clomiphene is effective for treating WHO class 2 anovulation, but ineffective for treating WHO classes 1 and 3.
  - Starting on day 2, 3, 4, or 5 of the cycle, 50 mg of clomiphene is administered daily for 5 days. The couple is encouraged to engage in sexual activity every other day for 1 week, commencing 5 days after the last pill has been taken. However, when clomiphene is combined with IUI, pregnancy rates may increase. Between cycle days 2 and 5, there is little difference in ovulation, pregnancy, and live birth regardless of when the medication is administered.[48]
- Letrozole is another commonly used oral medication for ovulation induction.
  - Letrozole is an aromatase inhibitor that inhibits the conversion of androstenedione and testosterone to estrone and estradiol, thereby preventing the production of estrogen.[49]
  - Letrozole is approved by the Food and Drug Administration (FDA) for the protracted adjuvant treatment of breast cancer, but its use to induce ovulation is considered off-label.
  - However, an abundance of scientific literature and numerous committee opinions support the safety and efficacy of its use in inducing ovulation.
  - Similar to clomiphene, letrozole is administered at a commencing dose of 2.5, 5, or 7.5 mg/day on cycle days 3, 4, 5, 6, and 7 with sexual activity every other day 5 days after the last dose.
  - According to the American College of Obstetricians and Gynecologists, letrozole should be regarded the first-line treatment for PCOS rather than clomiphene.[50]

Letrozole's advantages over clomiphene include:

- Higher rate of monofollicular development and a corresponding decrease in twin gestations
- Shorter half-life
- No antiestrogenic effects on the endometrium and central nervous system
- Lower estradiol levels, which is a benefit for women with breast cancer undergoing IVF
- Gonadotropin therapy is a more intensive medical treatment for anovulatory disorders classified as WHO class 1, 2, or 3.
  - As a second-line treatment option, gonadotropins are advantageous for women who have tried numerous cycles of clomiphene without success. One study demonstrated a higher live birth rate with gonadotropins than with continuous clomiphene use.[51]
  - It is essential to emphasize the need for close monitoring when using more invasive and intensive gonadotropins.
  - In order to evaluate mature follicles, transvaginal ultrasounds are used to monitor follicular development every 2–3 days during the late follicular phase.
  - A mature follicle has a diameter greater than 18 mm and an estradiol concentration greater than 200 pg/mL.
  - Once a mature follicle has been identified, recombinant human chorionic gonadotropin (hCG) 250 mg subcutaneous injection or 10,000 U of urinary-derived hCG is administered to induce ovulation.[52]
  - After the trigger injection, an IUI is performed 24–36 hours later. IUIs are compatible with all ovarian induction agents, and IUI with medication is recommended to increase pregnancy rates.

Tubal and Pelvic Adhesions

- First-line treatment for bilateral tubal factor infertility is IVF.

- There are poorer pregnancy outcomes and an increased risk of ectopic pregnancy following tubal corrective surgery.
- To increase the pregnancy rate of IVF, women with severe tubal disease, including hydrosalpinx, are encouraged to undergo bilateral salpingectomy.[53]
- Fimbrioplasty is an option for women with moderate distal tubal disease to enable multiple pregnancies without IVF.
- One small study revealed that the pregnancy rate for benign tubal disease was equivalent to IVF, but the risk of ectopic pregnancy was 15%, compared to 0.7% for IVF.[54]
- Patients who have undergone bilateral salpingectomy or tubal ligation for contraception in the past are a significant tubal factor population.
- The likelihood of pregnancy following tubal reanastomosis depends on the patient's age, the type of ligation performed, and the extent of available tubal tissue. Younger women who had a ring or clip with >4 cm of tubal length are the best candidates and have pregnancy rates comparable to IVF.[55] However, pregnancy is substantially slower to achieve after tubal surgery compared to IVF.

Uterine Abnormalities

- Insufficient evidence exists regarding the effect of leiomyomas on infertility and live birth rates.
- The location of fibroids is the most essential factor to consider. Fibroids that invade the endometrium and distort the uterine cavity inhibit implantation and increase the risk of miscarriage.[56]
- The pregnancy rates of women with submucosal or submucosal-intramural fibroids that distort the uterine cavity have been shown to be lower. Pregnancy and live birth rates increase following the removal of fibroids.[57]
- Surgical hysteroscopy is the treatment of choice for removing the most harmful fibroids.
- Other uterine pathologies, such as uterine synechiae and septa, are primarily associated with recurrent pregnancy loss, but they can also cause infertility.
- Women with both synechiae and septa who undergo hysteroscopy experience a significant decrease in miscarriages.[58]
- Infertility has also been associated with asymptomatic lesions. According to one study, polypectomy performed on asymptomatic infertile women prior to IUI increased pregnancy rates from 28 to 63%.[59]

COMPLICATIONS OF INFERTILITY TREATMENT

Multiple gestations, ectopic pregnancy, and OHSS are the three primary complications associated with infertility treatments.

Multiple Gestations

- Since their inception, multiple births have posed a problem for artificial reproductive technologies.
- Compared to gonadotropins, oral ovarian induction agents, clomiphene and letrozole, carry a lower risk of multiple pregnancies, with percentages of twins, triplets, and quadruplets of 7%, 0.5%, and 0.3%, respectively. Gonadotropins increase the likelihood of multiple pregnancies, including triplets, by 13%.[60]

Ectopic Pregnancy

Another risk associated with the treatment of infertile patients is ectopic pregnancy, which requires extensive counseling. Patients with infertility experience a two- to threefold increase in ectopic pregnancies. This is believed to be linked to a high incidence of tubal factor infertility.[61] The highest risk of ectopic pregnancy is associated with tubal surgery performed to treat tubal factor infertility. Estimates of ectopic pregnancy rates following tubal reconstructive surgery range from 9% to 30%.[54] Fresh embryo transfer carries a higher risk of ectopic pregnancy than frozen embryo transfer, but the overall rate of ectopic pregnancies with IVF is approximately 1.3%.[62] In a large study comparing ovulation induction agents, clomiphene, letrozole, and gonadotropins were associated with 4%, 6%, and 8% of ectopic pregnancies, respectively.[63]

Ovarian Hyperstimulation Syndrome[64]

- This is an iatrogenic complication of controlled ovarian hyperstimulation that causes a wide variety of signs and symptoms, including abdominal distention, nausea, vomiting, enlarged ovaries, third-spacing of fluids, renal failure, and venous thrombosis, acute respiratory distress syndrome, electrolyte abnormalities, cardiac arrhythmias, and sepsis.
- If severe OHSS is not treated and monitored, the enumerated complications can result in death.

- The underlying pathophysiological characteristic is an increase in capillary permeability, which results in a shift of fluid into the third compartment.
- Patients with >20 mature follicles who also receive an hCG trigger injection are at the greatest risk for developing OHSS.

CONCLUSION

- Women should see their physicians for a referral to an infertility subspecialist if they are unable to conceive after 1 year of unprotected timed intercourse or 6 months of unprotected timed intercourse if they are older than 35.
- It is essential to convey that infertility can be caused by both male and female factors, or a combination of both.
- It is essential to remember this, as the majority of infertile couples seek care from the female partner's health care provider, potentially overlooking the male's contribution.
- Alleviate the patient's physical, emotional, social, and interpersonal stressors associated with infertility is to conduct an immediate and comprehensive investigation of both partners.
- The medications and procedures available to treat female infertility have been thoroughly researched and their risk profiles are well-known.
- The use of clomiphene or letrozole with timed intercourse alone can correct a known cause of anovulation, but referral to a subspecialist should not be delayed.
- When counseling infertile couples, primary care providers must set realistic expectations regarding the likelihood of pregnancy and the possibility of complications.
- The patient must understand the possibility of multiple pregnancies, ectopic pregnancy, and OHSS.
- The patient must also comprehend the likelihood of pregnancy per treatment cycle for each modality.
- Even with IVF, inability to conceive is a possibility that must be thoroughly understood prior to devoting significant resources to infertility treatment.

REFERENCES

1. American Society for Reproductive Medicine (2012). Infertility: An Overview. A Guide for Patients. [online]. Available from https://www.fertilityanswers.com/wp-content/uploads/2016/04/infertility-an-overview-booklet.pdf [Last accessed March, 2024].
2. Olooto WE, Amballi AA, Banjo TA. A review of female infertility; important etiological factors and management. J Microbiol Biotech Res. 2012;2(3):379-85.
3. Walker MH, Tobler KJ. Female infertility. In: StatPearls [Internet]. Treasure Island (FL): StatPearls Publishing; 2024.
4. Santoro N, Filicori M, Crowley WF Jr. Hypogonadotropic disorders in men and women: diagnosis and therapy with pulsatile gonadotropin-releasing hormone. Endocr Rev. 1986;7(1):11-23.
5. Ackerman KE, Patel KT, Guereca G, Pierce L, Herzog DB, Misra M. Cortisol secretory parameters in young exercisers in relation to LH secretion and bone parameters. Clin Endocrinol (Oxf). 2013;78(1):114-9.
6. Perkins RB, Hall JE, Martin KA. Neuroendocrine abnormalities in hypothalamic amenorrhea: spectrum, stability, and response to neurotransmitter modulation. J Clin Endocrinol Metab. 1999;84(6):1905-11.
7. Hull MG. Epidemiology of infertility and polycystic ovarian disease: endocrinological and demographic studies. Gynecol Endocrinol. 1987;1(3):235-45.
8. Rotterdam ESHRE/ASRM-Sponsored PCOS consensus workshop group. Revised 2003 consensus on diagnostic criteria and long-term health risks related to polycystic ovary syndrome (PCOS). Hum Reprod. 2004;19(1):41-7.
9. Azziz R, Carmina E, Dewailly D, Diamanti-Kandarakis E, Escobar-Morreale HF, Futterweit W, et al. Task Force on the Phenotype of the Polycystic Ovary Syndrome of The Androgen Excess and PCOS Society. The Androgen Excess and PCOS Society criteria for the polycystic ovary syndrome: the complete task force report. Fertil Steril. 2009;91(2):456-88.
10. Fauser BC, Van Heusden AM. Manipulation of human ovarian function: physiological concepts and clinical consequences. Endocr Rev. 1997;18(1):71-106.
11. Baker TG. Radiosensitivity of mammalian oocytes with particular reference to the human female. Am J Obstet Gynecol. 1971;110(5):746-61.
12. Richardson SJ, Senikas V, Nelson JF. Follicular depletion during the menopausal transition: evidence for accelerated loss and ultimate exhaustion. J Clin Endocrinol Metab. 1987;65(6):1231-7.
13. Westhoff C, Murphy P, Heller D. Predictors of ovarian follicle number. Fertil Steril. 2000;74(4):624-8.
14. Broekmans FJ, Soules MR, Fauser BC. Ovarian aging: mechanisms and clinical consequences. Endocr Rev. 2009;30(5):465-93.
15. Nelson LM, Anasti JN, Kimzey LM, Defensor RA, Lipetz KJ, White BJ, et al. Development of luteinized graafian follicles in patients with karyotypically normal spontaneous premature ovarian failure. J Clin Endocrinol Metab. 1994;79(5):1470-5.

16. Nelson LM. Clinical practice. Primary ovarian insufficiency. N Engl J Med. 2009;360(6):606-14.
17. Seppälä M, Ranta T, Hirvonen E. Hyperprolactinaemia and luteal insufficiency. Lancet. 1976;1(7953):229-30.
18. The American College of Obstetricians and Gynecologists. Infertility Workup for the Women's Health Specialist: ACOG Committee Opinion, Number 781. Obstet Gynecol. 2019;133(6):e377-e384.
19. Prescott J, Farland LV, Tobias DK, Gaskins AJ, Spiegelman D, Chavarro JE, et al. A prospective cohort study of endometriosis and subsequent risk of infertility. Hum Reprod. 2016;31(7):1475-82.
20. Macer ML, Taylor HS. Endometriosis and infertility: a review of the pathogenesis and treatment of endometriosis-associated infertility. Obstet Gynecol Clin North Am. 2012;39(4):535-49.
21. Bulun SE. Endometriosis. N Engl J Med. 200;360(3):268-79.
22. Gupta S, Goldberg JM, Aziz N, Goldberg E, Krajcir N, Agarwal A. Pathogenic mechanisms in endometriosis-associated infertility. Fertil Steril. 2008;90(2):247-57.
23. Toya M, Saito H, Ohta N, Saito T, Kaneko T, Hiroi M. Moderate and severe endometriosis is associated with alterations in the cell cycle of granulosa cells in patients undergoing in vitro fertilization and embryo transfer. Fertil Steril. 2000;73(2):344-50.
24. Weström L, Joesoef R, Reynolds G, Hagdu A, Thompson SE. Pelvic inflammatory disease and fertility. A cohort study of 1,844 women with laparoscopically verified disease and 657 control women with normal laparoscopic results. Sex Transm Dis. 1992;19(4):185-92.
25. Weström L. Incidence, prevalence, and trends of acute pelvic inflammatory disease and its consequences in industrialized countries. Am J Obstet Gynecol. 1980;138(7 Pt 2):880-92.
26. Lepine LA, Hillis SD, Marchbanks PA, Joesoef MR, Peterson HB, Westrom L. Severity of pelvic inflammatory disease as a predictor of the probability of live birth. Am J Obstet Gynecol. 1998;178(5):977-81.
27. Meyer WR, Castelbaum AJ, Somkuti S, Sagoskin AW, Doyle M, Harris JE, et al. Hydrosalpinges adversely affect markers of endometrial receptivity. Hum Reprod. 1997;12(7):1393-8.
28. Van Voorhis BJ, Mejia RB, Schlaff WD, Hurst BS. Is removal of hydrosalpinges prior to in vitro fertilization the standard of care? Fertil Steril. 2019;111(4):652-6.
29. Pritts EA. Fibroids and infertility: a systematic review of the evidence. Obstet Gynecol Surv. 2001;56(8):483-91.
30. Chan YY, Jayaprakasan K, Zamora J, Thornton JG, Raine-Fenning N, Coomarasamy A. The prevalence of congenital uterine anomalies in unselected and high-risk populations: a systematic review. Hum Reprod Update. 2011;17(6):761-71.
31. Practice Committee of American Society for Reproductive Medicine in collaboration with Society for Reproductive Endocrinology and Infertility. Optimizing natural fertility. Fertil Steril. 2008;90(5 Suppl):S1-6.
32. Wathen NC, Perry L, Lilford RJ, Chard T. Interpretation of single progesterone measurement in diagnosis of anovulation and defective luteal phase: observations on analysis of the normal range. Br Med J (Clin Res Ed). 1984;288(6410):7-9.
33. Ecochard R, Boehringer H, Rabilloud M, Marret H. Chronological aspects of ultrasonic, hormonal, and other indirect indices of ovulation. BJOG. 2001;108(8):822-9.
34. Smotrich DB, Widra EA, Gindoff PR, Levy MJ, Hall JL, Stillman RJ. Prognostic value of day 3 estradiol on in vitro fertilization outcome. Fertil Steril. 1995;64(6):1136-40.
35. Dewailly D, Andersen CY, Balen A, Broekmans F, Dilaver N, Fanchin R, et al. The physiology and clinical utility of anti-Mullerian hormone in women. Hum Reprod Update. 2014;20(3):370-85.
36. de Vet A, Laven JS, de Jong FH, Themmen AP, Fauser BC. Antimüllerian hormone serum levels: a putative marker for ovarian aging. Fertil Steril. 2002;77(2):357-62.
37. Toner JP, Seifer DB. Why we may abandon basal follicle-stimulating hormone testing: a sea change in determining ovarian reserve using antimüllerian hormone. Fertil Steril. 2013;99(7):1825-30.
38. Vrontikis A, Chang PL, Kovacs P, Lindheim SR. Antral follice counts (AFC) predict ovarian response and pregnancy outcomes in oocyte donation cycles. J Assist Reprod Genet. 2010;27(7):383-9.
39. Mol BW, Swart P, Bossuyt PM, van der Veen F. Is hysterosalpingography an important tool in predicting fertility outcome? Fertil Steril. 1997;67(4):663-9.
40. Papaioannou S, Bourdrez P, Varma R, Afnan M, Mol BW, Coomarasamy A. Tubal evaluation in the investigation of subfertility: a structured comparison of tests. BJOG. 2004;111(12):1313-21.
41. Luttjeboer F, Harada T, Hughes E, Johnson N, Lilford R, Mol BW. Tubal flushing for subfertility. Cochrane Database Syst Rev. 2007;(3):CD003718.
42. Seshadri S, El-Toukhy T, Douiri A, Jayaprakasan K, Khalaf Y. Diagnostic accuracy of saline infusion sonography in the evaluation of uterine cavity abnormalities prior to assisted reproductive techniques: a systematic review and meta-analyses. Hum Reprod Update. 2015;21(2):262-74.
43. Frisch RE. The right weight: body fat, menarche and ovulation. Baillieres Clin Obstet Gynaecol. 1990;4(3):419-39.
44. Yen SS, Rebar R, Vandenberg G, Ehara Y, Siler T. Pituitary gonadotrophin responsiveness to synthetic LRF in subjects with normal and abnormal hypothalamic-pituitary-gonadal axis. J Reprod Fertil Suppl. 1973;20(0):137-61.
45. Berga SL, Marcus MD, Loucks TL, Hlastala S, Ringham R, Krohn MA. Recovery of ovarian activity in women with functional hypothalamic amenorrhea who were treated with cognitive behavior therapy. Fertil Steril. 2003;80(4):976-81.

46. Crosignani PG, Colombo M, Vegetti W, Somigliana E, Gessati A, Ragni G. Overweight and obese anovulatory patients with polycystic ovaries: parallel improvements in anthropometric indices, ovarian physiology and fertility rate induced by diet. Hum Reprod. 2003;18(9):1928-32.
47. Mutsaerts MA, van Oers AM, Groen H, Burggraaff JM, Kuchenbecker WK, Perquin DA, et al. Randomized Trial of a Lifestyle Program in Obese Infertile Women. N Engl J Med. 2016;374(20):1942-53.
48. Wu CH, Winkel CA. The effect of therapy initiation day on clomiphene citrate therapy. Fertil Steril. 1989;52(4):564-8.
49. Cole PA, Robinson CH. Mechanism and inhibition of cytochrome P-450 aromatase. J Med Chem. 1990; 33(11):2933-42.
50. ACOG Committee Opinion No. 738: Aromatase Inhibitors in Gynecologic Practice. Obstet Gynecol. 2018;131(6):1.
51. Weiss NS, Nahuis MJ, Bordewijk E, Oosterhuis JE, Smeenk JM, Hoek A, et al. Gonadotrophins versus clomifene citrate with or without intrauterine insemination in women with normogonadotropic anovulation and clomifene failure (M-OVIN): a randomised, two-by-two factorial trial. Lancet. 2018;391(10122):758-65.
52. Ludwig M, Doody KJ, Doody KM. Use of recombinant human chorionic gonadotropin in ovulation induction. Fertil Steril. 2003;79(5):1051-9.
53. Johnson NP, Mak W, Sowter MC. Surgical treatment for tubal disease in women due to undergo in vitro fertilisation. Cochrane Database Syst Rev. 2004;(3):CD002125.
54. Audebert A, Pouly JL, Bonifacie B, Yazbeck C. Laparoscopic surgery for distal tubal occlusions: lessons learned from a historical series of 434 cases. Fertil Steril. 2014;102(4): 1203-8.
55. Boeckxstaens A, Devroey P, Collins J, Tournaye H. Getting pregnant after tubal sterilization: surgical reversal or IVF? Hum Reprod. 2007;22(10):2660-4.
56. Pritts EA, Parker WH, Olive DL. Fibroids and infertility: an updated systematic review of the evidence. Fertil Steril. 2009;91(4):1215-23.
57. Casini ML, Rossi F, Agostini R, Unfer V. Effects of the position of fibroids on fertility. Gynecol Endocrinol. 2006;22(2):106-9.
58. March CM, Israel R. Gestational outcome following hysteroscopic lysis of adhesions. Fertil Steril. 1981; 36(4):455-9.
59. Pérez-Medina T, Bajo-Arenas J, Salazar F, Redondo T, Sanfrutos L, Alvarez P, et al. Endometrial polyps and their implication in the pregnancy rates of patients undergoing intrauterine insemination: a prospective, randomized study. Hum Reprod. 2005;20(6):1632-5.
60. Hansen KR, He AL, Styer AK, Wild RA, Butts S, Engmann L, et al, Eunice Kennedy Shriver National Institute of Child Health and Human Development Reproductive Medicine Network. Predictors of pregnancy and live-birth in couples with unexplained infertility after ovarian stimulation-intrauterine insemination. Fertil Steril. 2016;105(6):1575-83.e2.
61. Li C, Zhao WH, Zhu Q, Cao SJ, Ping H, Xi X, et al. Risk factors for ectopic pregnancy: a multi-center case-control study. BMC Pregnancy Childbirth. 2015;15:187.
62. Londra L, Moreau C, Strobino D, Garcia J, Zacur H, Zhao Y. Ectopic pregnancy after in vitro fertilization: differences between fresh and frozen-thawed cycles. Fertil Steril. 2015;104(1):110-8.
63. Diamond MP, Legro RS, Coutifaris C, Alvero R, Robinson RD, Casson P, et al., NICHD Reproductive Medicine Network. Letrozole, Gonadotropin, or Clomiphene for Unexplained Infertility. N Engl J Med. 2015;373(13):1230-40.
64. Golan A, Ron-el R, Herman A, Soffer Y, Weinraub Z, Caspi E. Ovarian hyperstimulation syndrome: an update review. Obstet Gynecol Surv. 1989;44(6):430-40.

Chapter 28

Male Factor Infertility

Arun Arora, Geetu Mahajan, Arunav Arora, Arihant Arora

INTRODUCTION

Infertility is described as inability to conceive despite regular intercourse without using contraception for 1 year.

Both male and female partners are equally responsible for causation of infertility, accounting for approximately 40% each. It is always advisable to investigate both partners simultaneously **(Figs. 1A and B)**.

EVALUATION OF MALE INFERTILITY

The mainstay of evaluation of male infertility is semen analysis. If the results of the first semen analysis are abnormal, a repeat confirmatory test should be offered ideally after 1 month or earlier, if there is a gross abnormality in semen report.

ABNORMAL SEMEN ANALYSIS

The results from semen analysis (SA) should be used to guide management of the patient. In general, results are of greatest clinical significance when multiple SA abnormalities are present **(Table 1)**.

Complete male evaluation with history, physical examination, and hormonal evaluation including follicle-stimulating hormone (FSH) and testosterone for infertile men with impaired libido, erectile dysfunction, oligozoospermia or azoospermia, atrophic testes, or evidence of hormonal abnormality on physical evaluation should be done **(Flowchart 1)**.

A special situation of obese males with oligospermia needs following protocol **(Flowchart 2)**.

Azoospermic men should be clinically evaluated to differentiate genital tract obstruction from impaired sperm

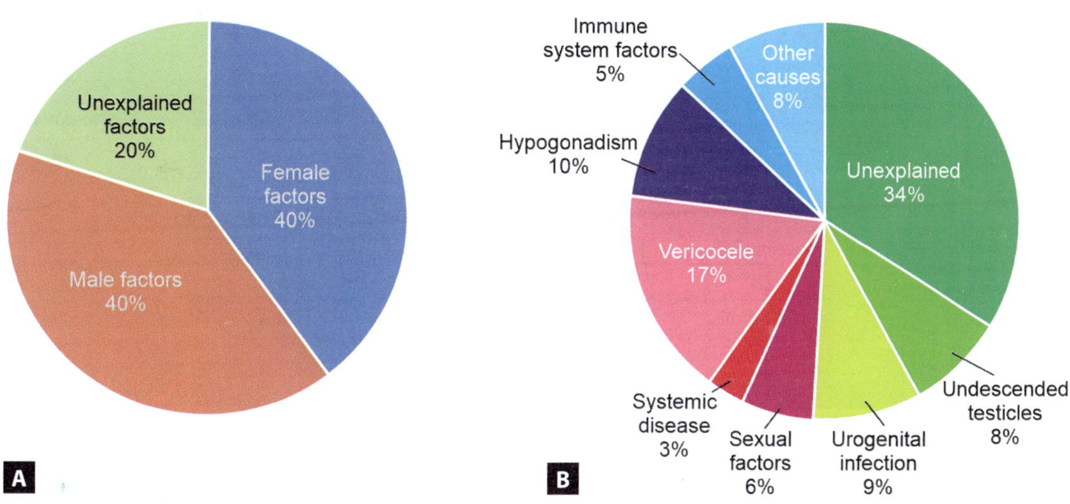

Figs. 1A and B: Causes of infertility; (A) General causes of infertility; (B) Causes of male infertility.

TABLE 1: Semen analysis.

Semen analysis parameter	*Normal values*
Sperm count number of sperm in ejaculate	15 million/mL or more
Total motility proportion of sperm that can swim	40% or more
Progressive motility proportion of sperm swimming in one direction	32% or more
Morphology size and shape of sperm	4% or more normal forms
Vitality proportion of sperm that are alive	58% or more live sperm
Semen volume amount of semen produced	1.5 mL or more
White blood cells can indicate an infection	<1 million/mL

Flowchart 1: Male factor evaluation.

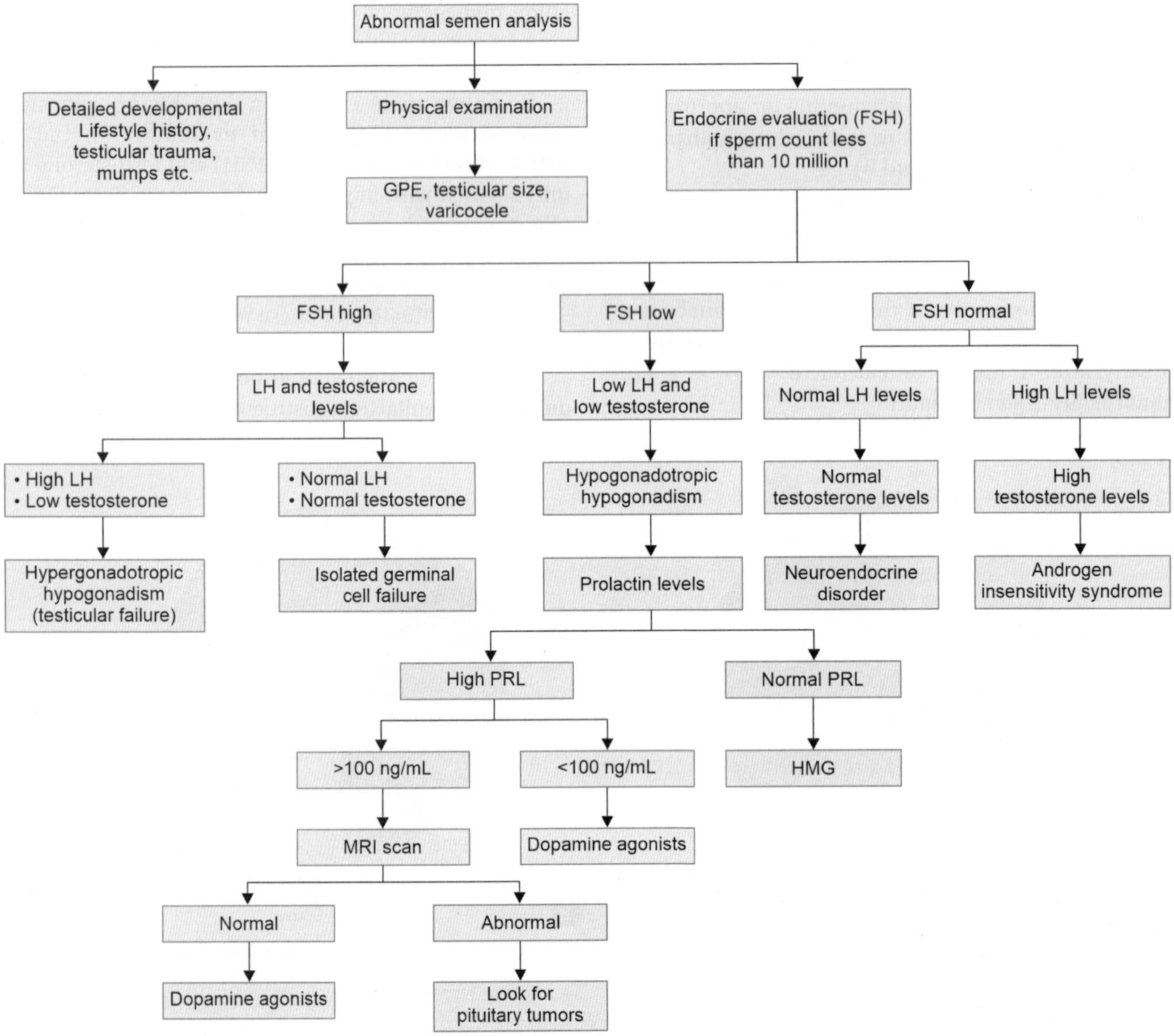

Abnormal semen analysis
- Detailed developmental Lifestyle history, testicular trauma, mumps etc.
- Physical examination → GPE, testicular size, varicocele
- Endocrine evaluation (FSH) if sperm count less than 10 million
  - FSH high → LH and testosterone levels
    - • High LH • Low testosterone → Hypergonadotropic hypogonadism (testicular failure)
    - • Normal LH • Normal testosterone → Isolated germinal cell failure
  - FSH low → Low LH and low testosterone → Hypogonadotropic hypogonadism → Prolactin levels
    - High PRL
      - >100 ng/mL → MRI scan
        - Normal → Dopamine agonists
        - Abnormal → Look for pituitary tumors
      - <100 ng/mL → Dopamine agonists
    - Normal PRL → HMG
  - FSH normal
    - Normal LH levels → Normal testosterone levels → Neuroendocrine disorder
    - High LH levels → High testosterone levels → Androgen insensitivity syndrome

Flowchart 2: Evaluation of oligospermia with obesity.

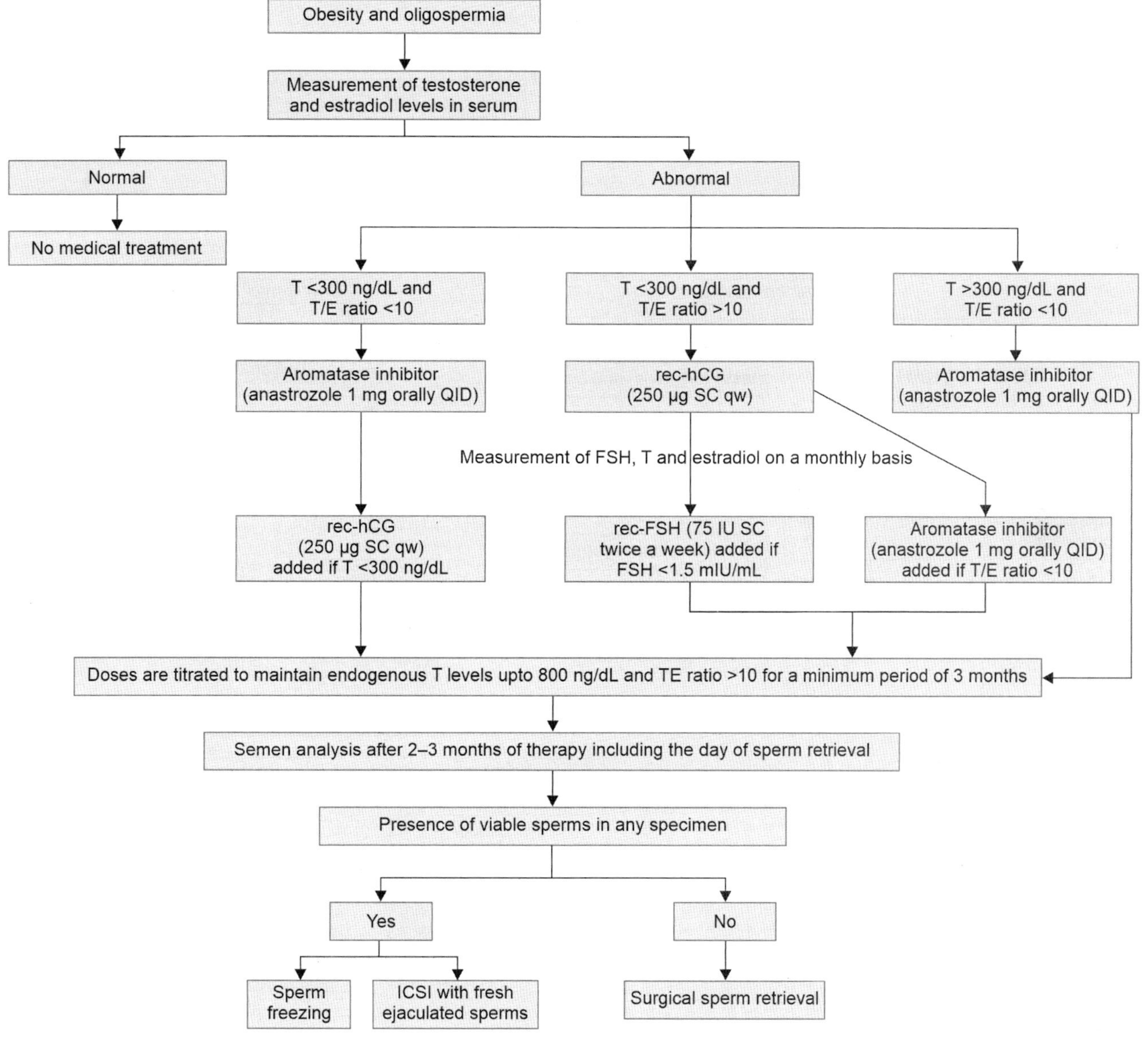

production initially based on semen volume, physical exam, and FSH levels **(Flowchart 3)**.

PEARLS

Testing and Male Infertility

- Sperm deoxyribonucleic acid (DNA) fragmentation analysis is not recommended in the initial evaluation of infertile couple.
- Men with increased round cells on SA should be evaluated further to differentiate white blood cells (WBCs) (pyospermia) from germ cells.
- Patients with pyospermia should be evaluated for the presence of infection.
- Antisperm antibody (ASA) testing should not be done in the initial evaluation of male infertility.
- Karyotype and Y-chromosome microdeletion analysis should be recommended for men with primary infertility and azoospermia or severe oligospermia (<5 million sperm/mL) with elevated FSH or testicular atrophy or a presumed diagnosis of impaired sperm production as a cause of azoospermia.
- Cystic fibrosis transmembrane conductance regulator (CFTR) mutation carrier testing should be advised

Flowchart 3: Evaluation of azoospermia.

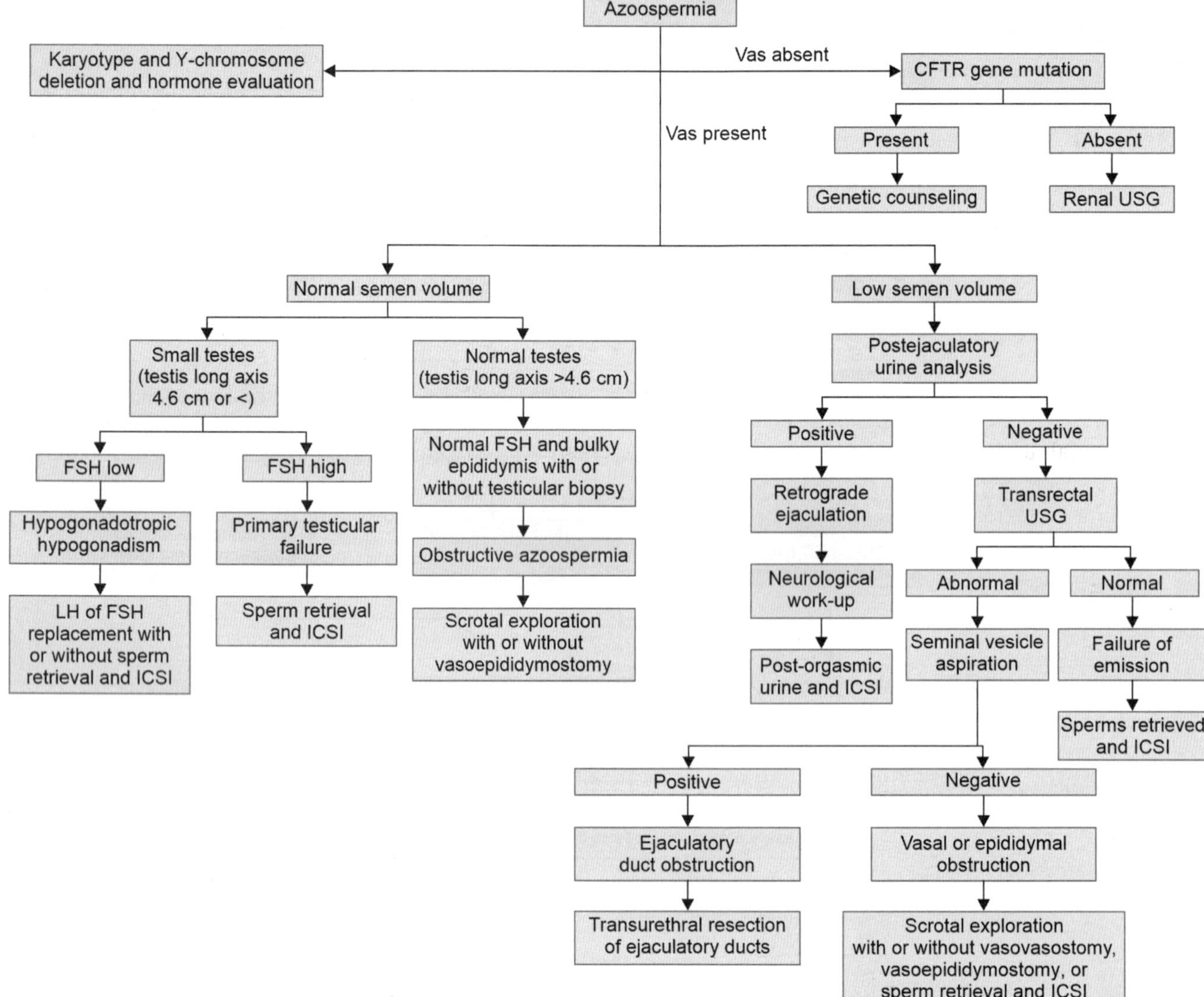

in men with vasal agenesis or idiopathic obstructive azoospermia. For men who are positive for CFTR mutation, genetic evaluation of the female partner should be recommended.

- Diagnostic testicular biopsy should not be routinely performed to differentiate between obstructive and nonobstructive azoospermia (NOA).

Varicocele and Male Infertility

- Surgical varicocelectomy should be considered in men attempting to conceive, who have palpable varicocele, infertility, and abnormal semen parameters, except for azoospermic men.
- Clinicians should not recommend varicocelectomy for men with nonpalpable varicocele detected solely by imaging.
- For men with clinical varicocele and NOA, couples should be informed of the absence of definitive evidence supporting varicocele repair prior to assisted reproductive technology (ART).
- ASA testing should not be done in the initial evaluation of male infertility.

Surgical Sperm Retrieval and Male Infertility

- For men with NOA undergoing sperm retrieval, microdissection testicular sperm extraction (TESE) should be performed.

- In men undergoing surgical sperm retrieval, either fresh or cryopreserved sperm may be used for intracytoplasmic sperm injection (ICSI).
- In men with azoospermia due to obstruction undergoing surgical sperm retrieval, sperm may be extracted from either the testis or the epididymis.
- For men with aspermia, surgical sperm extraction or induced ejaculation, including sympathomimetic, vibratory stimulations, and electroejaculation, may be performed depending on the patient's condition and clinician's experience.

Imaging and Male Infertility

- Scrotal ultrasound should not be routinely performed in the initial evaluation of the infertile male.
- Transrectal ultrasonography (TRUS) should not be performed as part of the initial evaluation. It is only recommended in men with SA suggestive of ejaculatory duct obstruction (i.e., acidic, azoospermia, semen volume <1.5 mL, with normal serum T, palpable vas deferens).
- Abdominal imaging for the sole indication of an isolated small or moderate right varicocele.
- Patients with vasal agenesis should be recommended for renal sonography to evaluate for renal abnormalities.

Gonadotoxic Therapies and Male Infertility

- Clinicians should discuss the effects of gonadotoxic therapies and other cancer treatments on sperm production with patients prior to commencement of therapy.
- Clinicians should inform patients undergoing chemotherapy and/or radiation therapy to avoid pregnancy for at least 12 months after completion of treatment.
- Clinicians should encourage men to bank sperm, preferably multiple specimens, when possible, prior to commencement of gonadotoxic therapy or other cancer treatment that may affect fertility in men.
- Clinicians should inform patients that a SA performed after gonadotoxic therapies should be done at least 12 months (and preferably 24 months) after treatment completion.
- Clinicians should inform patients undergoing a retroperitoneal lymph node dissection of the risk of aspermia.
- Clinicians should obtain a post-orgasmic urinalysis for men with aspermia after retroperitoneal lymph node dissection who are interested in fertility.
- Clinicians should inform men seeking paternity who are persistently azoospermic after gonadotoxic therapies that TESE is a treatment option.

Medical Therapy and Male Infertility

- Testosterone monotherapy should not be prescribed.
- There are limited benefits of use of selective estrogen receptor modulators (SERMs) in idiopathic infertility relative to results of ART.
- Benefits of supplements (e.g., antioxidants, vitamins) are of questionable clinical utility in treating male infertility.
- For men with idiopathic infertility, treatment with FSH analogue with the aim of improving sperm concentration may be considered.
- There is limited data supporting pharmacologic manipulation with SERMs, aromatase inhibitors (AIs), and gonadotropins prior to surgical intervention.

CONCLUSION

In conclusion, 40% of all causes of infertility can be attributed to male factors. Therefore, the male and female partners should always be assessed together. An accurate semen analysis is the cornerstone of investigations. The report on semen analysis provides guidance on the next course of action. Most patients are now able to father their children thanks to advancements in diagnostic modalities like genetic testing and treatments like assisted reproductive technologies.

SUGGESTED READING

1. Behre HM, Bergmann M, Simoni M, et al. Primary testicular failure. [updated 2015 Aug 30]. South Dartmouth, MA: MDText.com, Inc., 2000.
2. Bernie AM, Mata DA, Ramasamy R, et al. Comparison of microdissection testicular sperm extraction, conventional testicular sperm extraction, and testicular sperm aspiration for nonobstructive azoospermia: a systematic re view and meta-analysis. Fertil Steril. 2015;104:1099.
3. Chillon M, Casals T, Mercier B, et al. Mutations in the cystic fibrosis gene in patients with congenital absence of the vas deferens. N Engl J Med. 1995;332:1475.
4. Finkelstein JS, Whitcomb RW, O'Dea LS, et al. Sex steroid control of gonadotropin secretion in the human male. I. Effects of testosterone administration in normal and gonadotropin-releasing hormone-deficient men. J Clin Endocrinol Metab. 1991;73:609.
5. Guzick DS, Overstreet JW. Factor-Litvak P, et al. Sperm morphology, motility, and concentration in fertile and infertile men. N Engl J Med. 2001;345:1388.

6. Jarow JP. Transrectal ultrasonography of infertile men. Fertil Steril. 1993;60:1035.
7. Kolettis PN, Sandlow JI. Clinical and genetic features of patients with congenital unilateral absence of the vas deferens. Urology. 2002;60:1073.
8. Mak V, ZielenskiJ, Tsui LC, et al. Proportion of cystic fibrosis gene mutations not detected by routine testing in men with obstructive azoospermia. Jama. 1999;281:2217.
9. Melmed S, Casanueva FF, Hoffman AR, et al. Diagnosis and treatment of hyperprolactinemia: an endocrine society clinical practice guideline. J Clin Endocrinol Metab. 2011;96:273.
10. Sandro C Esteves. Clinical management of infertile men with nonobstructive azoospermia. Asian J Androl. 2015;17:459-70.
11. Yu J, Chen Z, Ni Y, et al. CFTR mutations in men with congenital bilateral absence of the vas deferens (CBAVD): a systemic review and meta-analysis. Hum Reprod. 2012;27:25.

Chapter

29

Tubal Factor in Infertility

Nilesh Balkawade

INTRODUCTION

Tubal factor in infertility accounts for almost 30% cases of female infertility. This chapter will try to throw light on all these aspects **(Box 1)**.

BOX 1: Introduction.

- Incidence: 25–35% cases of female infertility[1]
- Around 50% of the cases are due to salpingitis (PID)[2]
- IVF techniques have taken over upper hand in treatment over the surgical modalities
- "Infertility surgery is dead, only the obituary remains"[3]

(IVF: in vitro fertilization; PID: pelvic inflammatory disease)

TUBAL ANATOMY AND PHYSIOLOGY

Figures 1 and 2 depict the tubal anatomy and physiology.

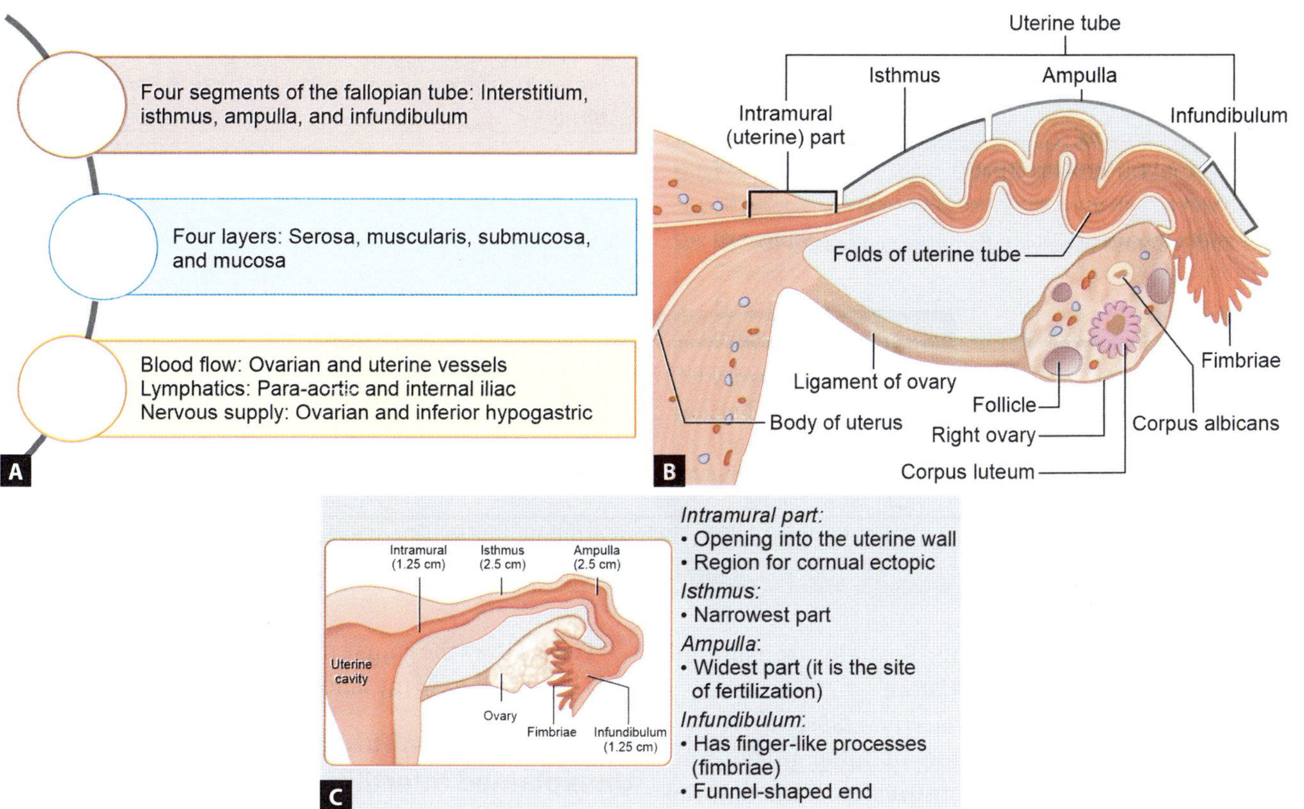

Figs. 1A to C: (A and B) Parts of fallopian tube and (C) fallopian tube affection in pelvic inflammatory disease (PID).

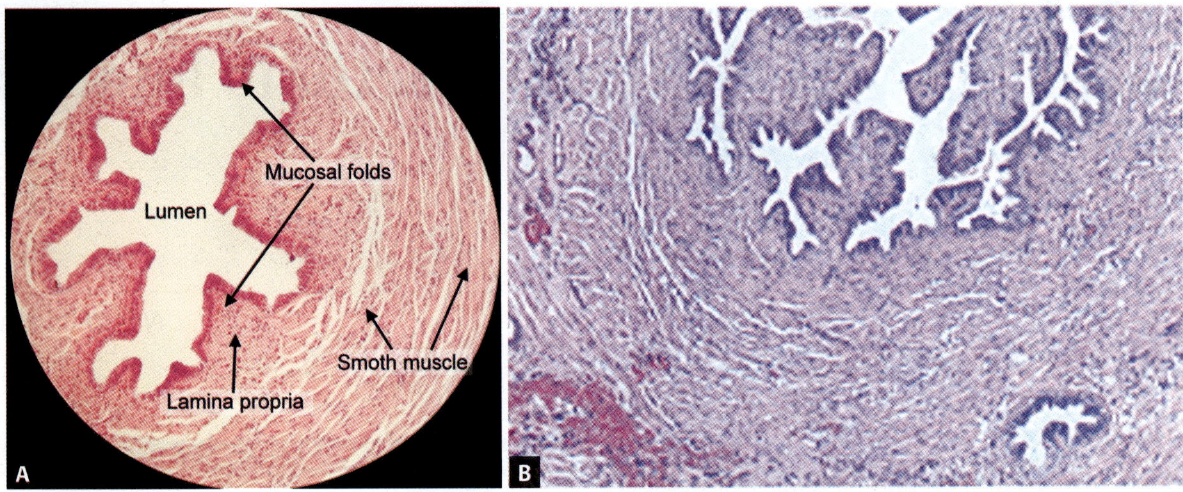

Figs. 2A and B: Fallopian tube histopathology: (A) Normal tube and (B) chronic inflammation.

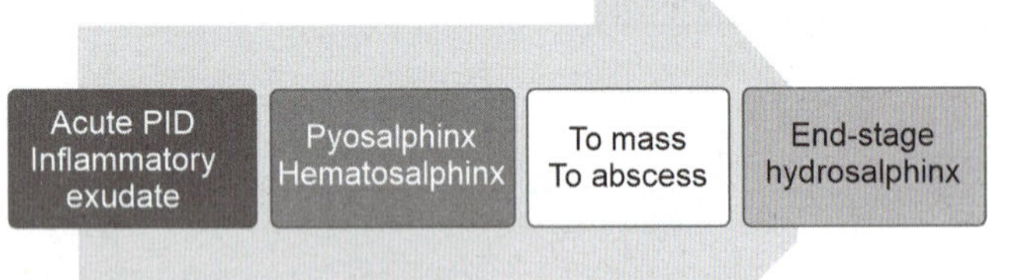

Fig. 3: Course of pelvic inflammatory disease (PID).[6]

BOX 2: Functions of fallopian tube.

- Catching the oocyte
- Movement and capacitation of sperms
- Movement of oocyte/zygote
  Peristaltic contraction (muscularis layer) + ciliary activity(mucosa) + fluid (secretory cells in the oviduct), move oocyte/zygote toward the uterus
- Protection of the embryo against heat, stress, and immune response

BOX 3: Role of fallopian tubes in unexplained fertility.

Proximal tube: Entry of spermatozoa through tubal ostia

Middle part:
- Transport of gametes
- Sperm capacitation
- Fertilization failure
- Protection against heat, stress, and immune response

Distal tube: Oocyte capture by fimbria

- Mostly ascending infection: Sexually transmitted *Chlamydia, Neisseria gonorrhoeae*[4]
- *Hematogenous route: Mycobacterium tuberculosis*
- *Acute inflammation:* Fusion of tubal plicae and obliteration of the ostium; obstruction of the fimbrial end of fallopian tube and, less commonly, of the intramural or isthmic portion may lead to infertility.[5]
- *Chronic inflammation:* Fibrotic and serosal adhesion

Fig. 3 shows the course of pelvic inflammatory disease. The functions of the fallopian tube are shown in **Box 2**.

Role of Fallopian Tubes in Unexplained Infertility[7]

The role of fallopian tubes and their function in unexplained infertility are shown in **Box 3** and **Figure 4**, respectively.

Figures 5A and B depict normal and abnormal tubal mucosa in unexplained infertility.

CAUSES OF TUBAL FACTOR INFERTILITY

Boxes 4 to 7 mention the various causes leading to tubal factor infertility.

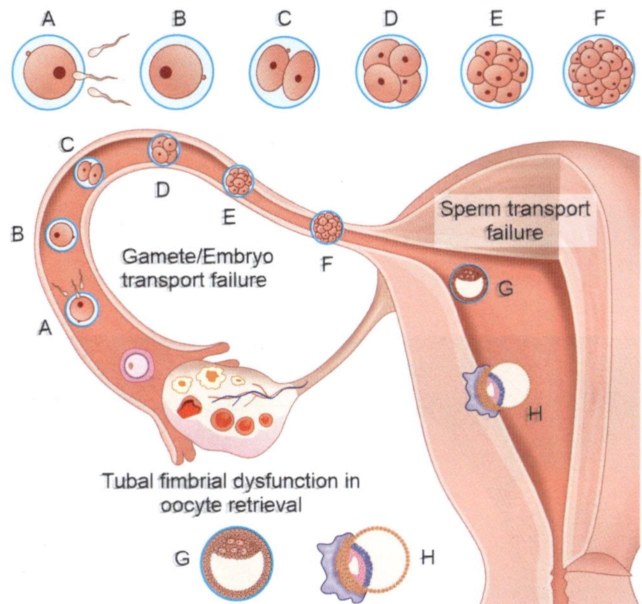

Fig. 4: Fallopian tube function in unexplained infertility.

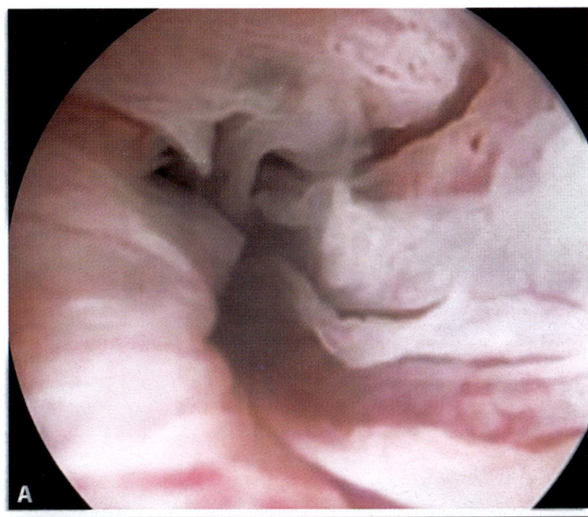

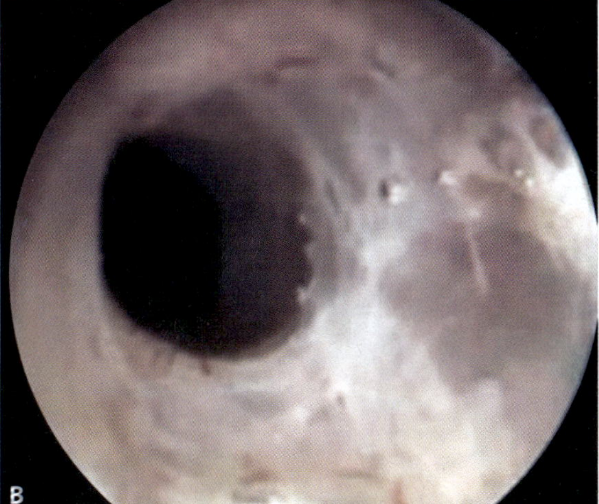

Figs. 5A and B: Normal and abnormal tubal mucosa in unexplained infertility.

BOX 4: Proximal tubal disease.[8]

Reversible:
- Muscular spasm
- Stromal edema
- Amorphous debris
- Mucosal agglutination and viscous secretions
- Cornual polyps
- Parasite infection

Irreversible:
- Chronic salpingitis
- Endometriosis
- Salpingitis isthmica nodosa
- Intrauterine synechiae

BOX 5: Distal and midtubal disease.[8]

Less severe:
- Distal tubal stenodilatations
- Fimbrial adhesions
- Tubal kinking
- Fixity from adhesions with preserved tubal patency

Severe:
- Fimbrial end involvement with hydrosalpinx **(Figs. 6 to 10)**
- Fimbrial agglutination
- Midtubal disease
- Steno-occlusions
- Bulbous termination from scarring and fibrosis

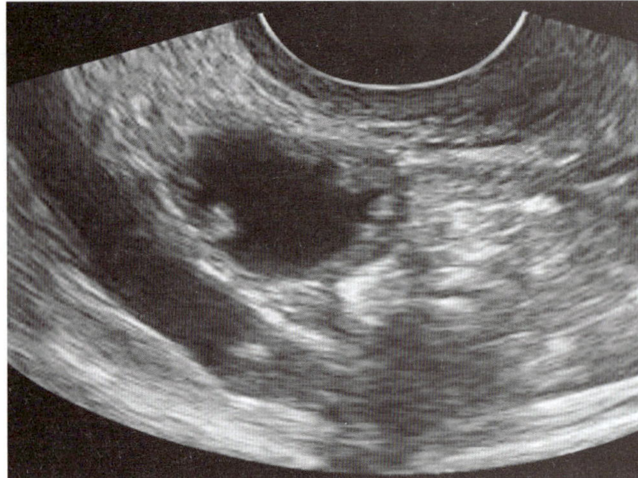

Fig. 6: Ultrasound signs in hydrosalpinx: Cogwheel sign.[9]

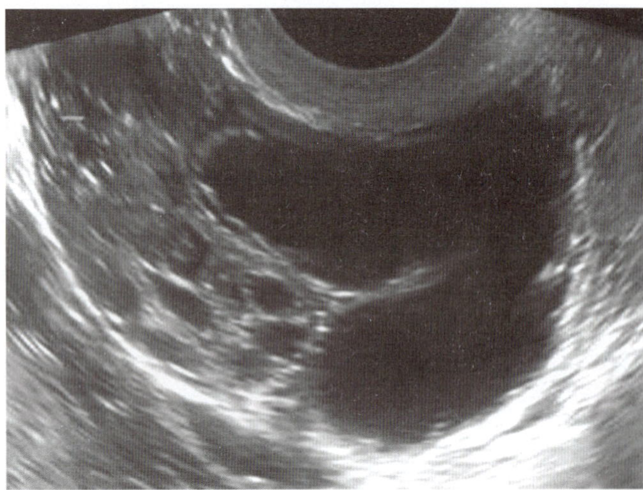

Fig. 9: "C" or "S"-shaped cystic lesion.

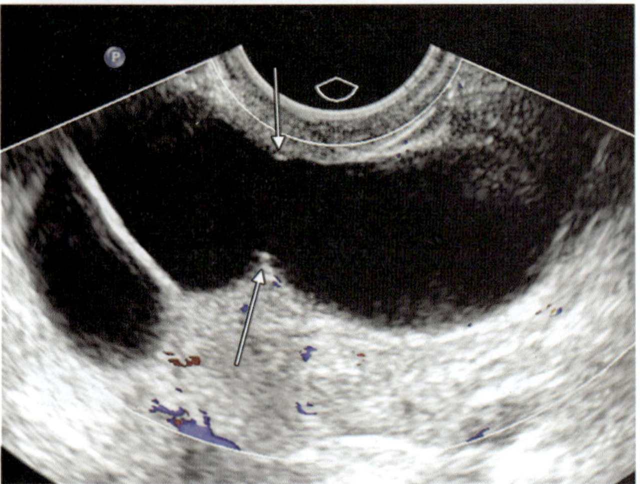

Fig. 7: Ultrasound signs in hydrosalpinx: Waist sign.[9]

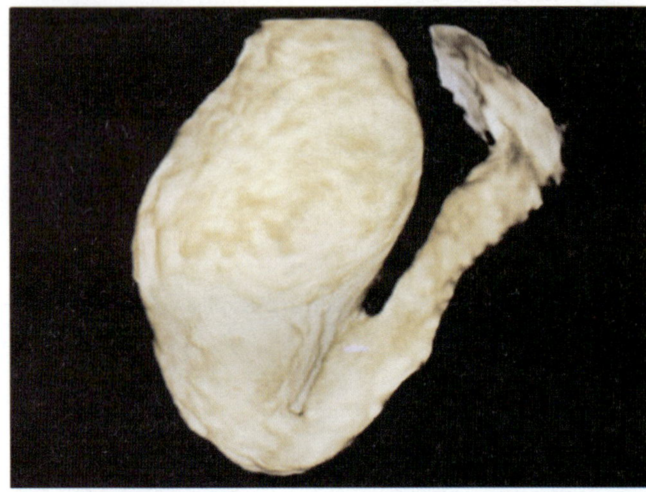

Fig. 10: 3D ultrasonography (USG) of hydrosalpinx.

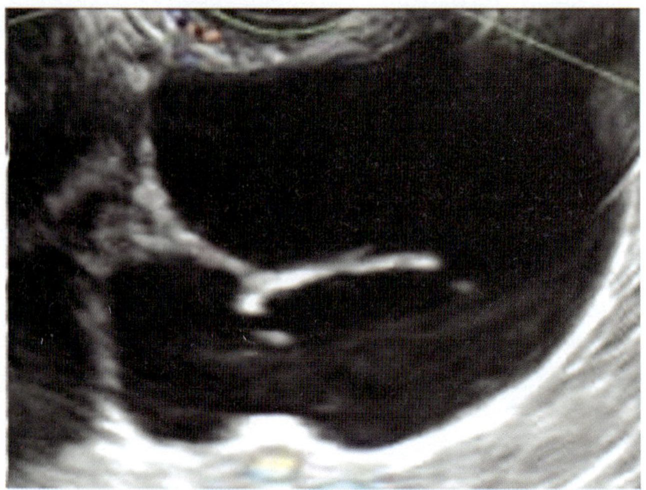

Fig. 8: Incomplete septa.

BOX 6: Tuberculous involvement of the tube.[8]

Mild:
Damage to the tubal lining

Severe:
- Tubal scarring
- Rigidity
- Fibrosis
- Steno-occlusions
- Dilatations
- Hydrosalpinx
- Peritubal and pelvic adhesions

BOX 7: Organisms causing salpingitis.[8]

- *Acute salpingitis* **(Fig. 11)**
- *Escherichia coli*
- *Staphylococcus aureus*
- *Chronic salpingitis* **(Fig. 12)**
- *Neisseria gonorrhea*
- *Chlamydia trachomatis*
- *Mycobacterium tuberculosis*

Figure 13 shows the Hull and Rutherford classification of tubal disease.[13]

Classification by Zou et al. for pregnancy prognosis of tubal factor is given in **Table 1**.

DIAGNOSIS OF TUBAL FACTOR INFERTILITY

Figure 14 illustrates the diagnosis of tubal factor infertility.

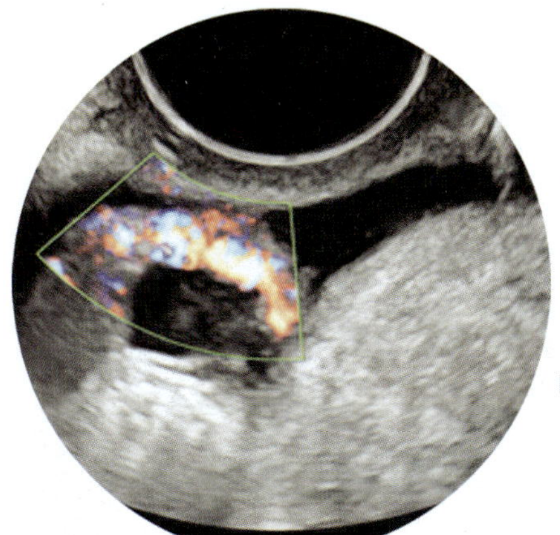

Fig. 11: Acute salpingitis.[10,11]

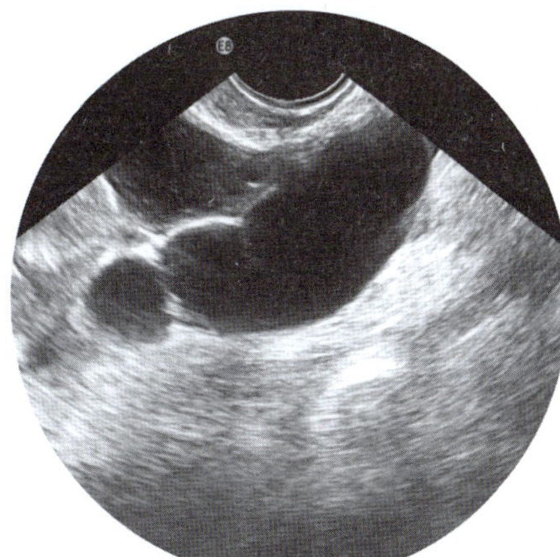

Fig. 12: Chronic salpingitis.[12]

Minor:
- Proximal occlusion without tubal fibrosis
- Distal occlusion without tubal distension
- Healthy mucosal appearance at hysterosalpingography (HSG), salpingoscopy
- Flimsy peritubal/ovarian adhesions
- Pregnancy rate = 84.7%
- Ectopic rate = 7.1%

Grade I

Intermediate:
- Unilateral severe tubal damage
- Limited dense adhesions of tubes and ovaries
- Pregnancy rate = 72%
- Ectopic rate = 23.3%

Grade II

Severe:
- Bilateral severe tubal damage
- Extensive tubal fibrosis
- Tubal distension >1.5 cm
- Abnormal mucosal appearance
- Bipolar occlusion
- Extensive dense adhesion
- Pregnancy rate = 28%
- Ectopic rate = 57.7%

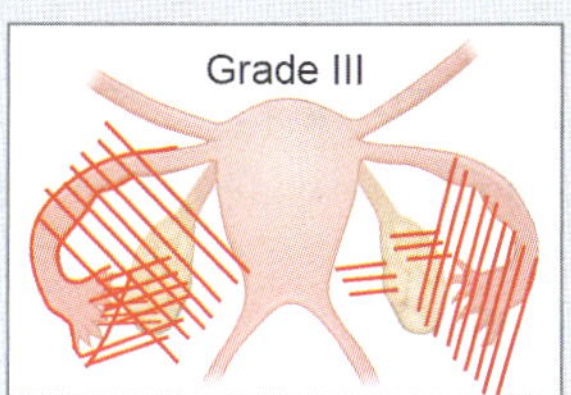

Fig. 13: Hull and Rutherford classification of tubal disease.

TABLE 1: Zou et al. classification for pregnancy prognosis of tubal factor.[14]

Nature of adhesion	*Adhesion range*	*Peritubal*	*Partial ovarian*	*Completely wrapped/extensive*
Patency	Membranous Dense Patent 0	1 2 Passable 4	2 4 Subpassable 6	4 8 Completely blocked 8
Morphology	Soft 0	Partially narrowed/twisted 4	Stiffness/hydrosalpinx 8	
Fimbrial structure	Complete 0	Partially destroyed 4	Destroyed 8	
Mild: 0–7; moderate: 8–15; severe: >16				

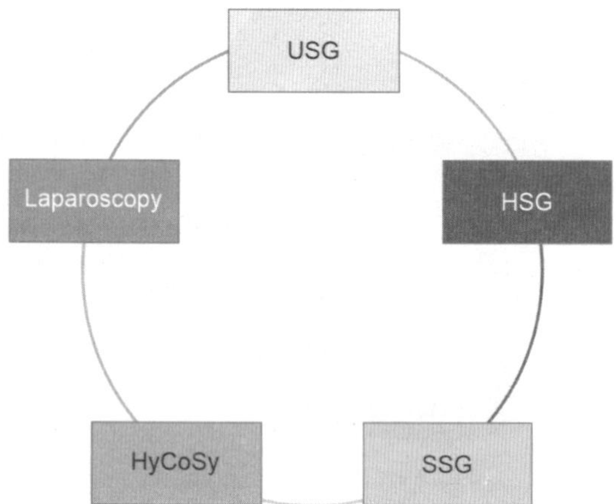

Fig. 14: Diagnosis of tubal factor infertility. (HSG: hysterosalpingography; HyCoSy: hysterosalpingo-contrast sonography; SSG: sonosalpingography; USG: ultrasonography)

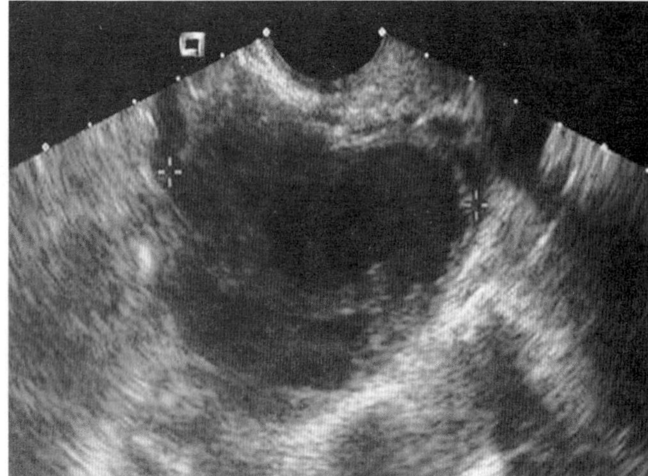

Fig. 15: Tubo-ovarian abscess (cystic lesion with thick walls).

Ultrasonography

Pelvic inflammatory Disease

Pelvic inflammatory disease (PID) includes endometritis, salpingitis, salpingo-oophoritis, tubo-ovarian abscess **(Fig. 15)** and pelvic peritonitis.

It is a result of the extension of the inflammatory process to the ovary resulting in tubo-ovarian (TO) mass. Extent further can cause break in the walls of tubes and ovaries, causing TO abscess.[12]

Tubal patency tests are given in **Table 2**.

TABLE 2: Sonographic markers for tubal inflammation by Timor-Tritsch et al.[12]

	Acute (%)	*Chronic (%)*
Thickened tube wall	100	3
Incomplete septa		93
Cogwheel sign	86	3
Beads on string appearance	0	57
Tubo-ovarian (TO) complex	36	2

Hysterosalpingography

Figures 16 to 19 show hysterosalpingography.

Saline Sonography[15]

Sonosalpingography (SSG) uses transvaginal sonography (TVS) for tubal patency by visualizing turbulence near the fimbrial end when saline or mixture of air and saline is injected through a Foley's catheter placed within the uterus.

Types:

- Negative contrast SSG (saline and air mixture)
- Positive contrast SSG (dye)

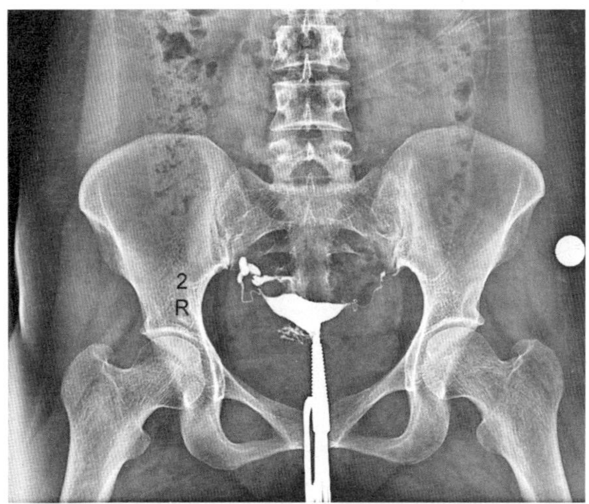

Fig. 16: Hysterosalpingography (HSG)—*small terminal hydrosalpinx* is noted on right side with no free peritoneal spill noted on right side. Left free spillage of contrast is suggestive of tubal patency.
Courtesy: Dr Sandeep Pansare, Radiologist, Pune.

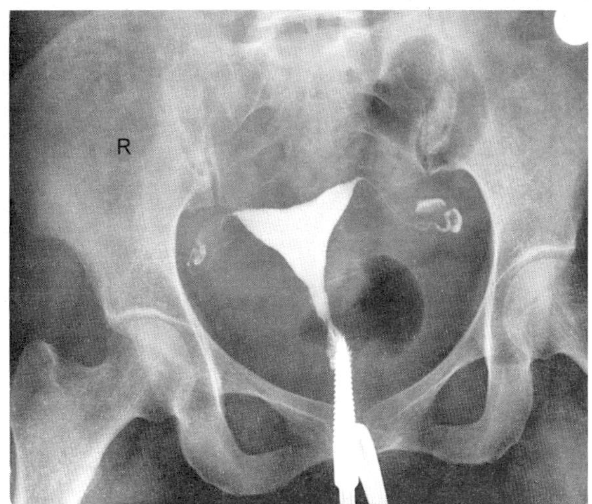

Fig. 17: Hysterosalpingography (HSG)—bilateral distal tubal block with terminal dilatation.
Courtesy: Dr Sandeep Pansare, Radiologist, Pune.

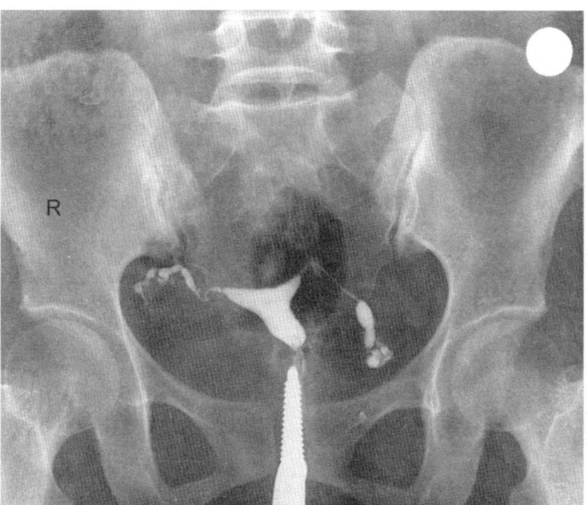

Fig. 18: Hysterosalpingography (HSG)—bilateral small terminal hydrosalpinges.
Courtesy: Dr Sandeep Pansare, Radiologist, Pune.

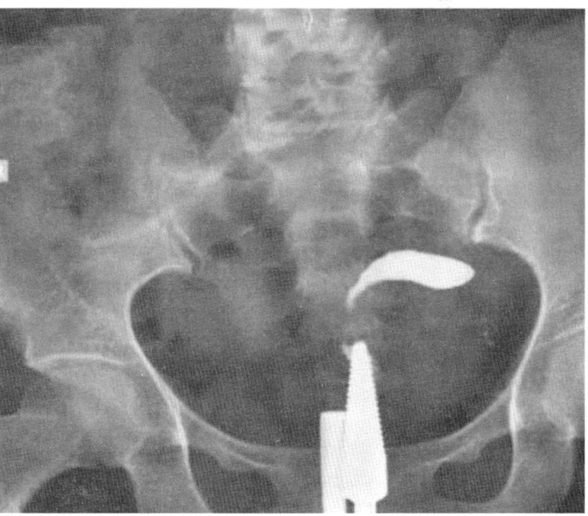

Fig. 19: Hysterosalpingography (HSG)—single tube with Unicornuate uterus.
Courtesy: Dr Sandeep Pansare, Radiologist, Pune.

- Popularly known as the "Sion test", popularized by Dr Gautam Allahabadia[15]
- It has the advantage of detecting endometrial pathologies more effectively than routine TVS.[16]

Hystero-contrast-salpingography

Hysterosalpingo-contrast sonography (HyCoSy) can be effectively used as an alternative to HSG and laparoscopy as it avoids radiation exposure, discomfort, or even abdominal pain in HSG and the invasiveness of laparoscopy. At the same time, the accuracy of the tubal patency is not compromised **(Figs. 20 and 21)**.[17,18]

Laparoscopic Tubal Patency

- A direct way of visualizing and true picture available.
- Positive image instead of negative image or black and white images.
- "See and treat" possible; in the case of proximal tubal (cornual) blocks, tubal cannulation, and treatment is possible in the same sitting.
- Fimbrioplasty can be done for terminal block.[19]

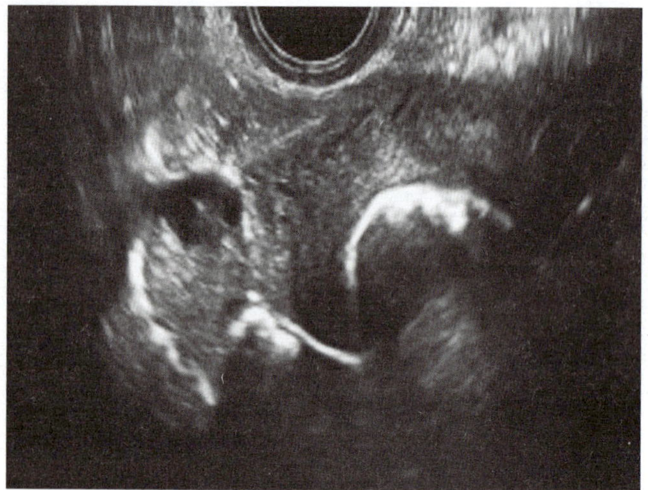

Fig. 20: Hysterosalpingo-contrast sonography (HyCoSy).

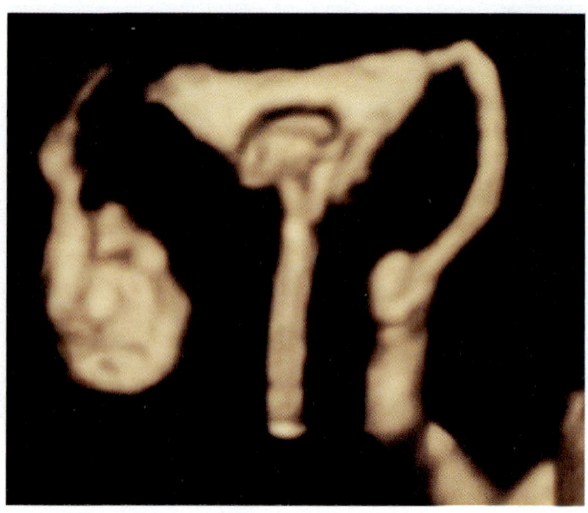

Fig. 21: 3D hysterosalpingo-contrast sonography (HyCoSy).

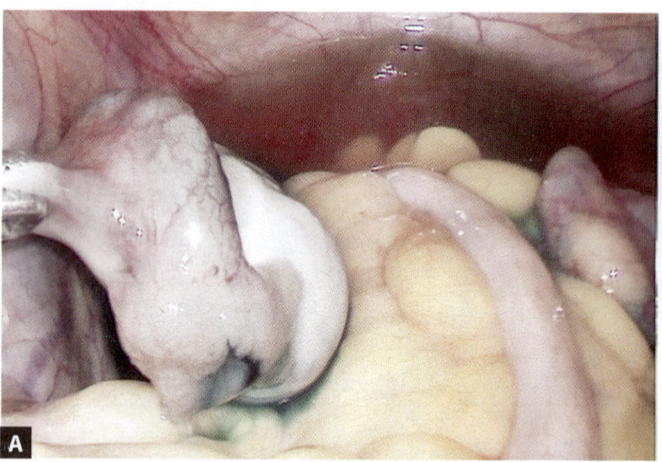

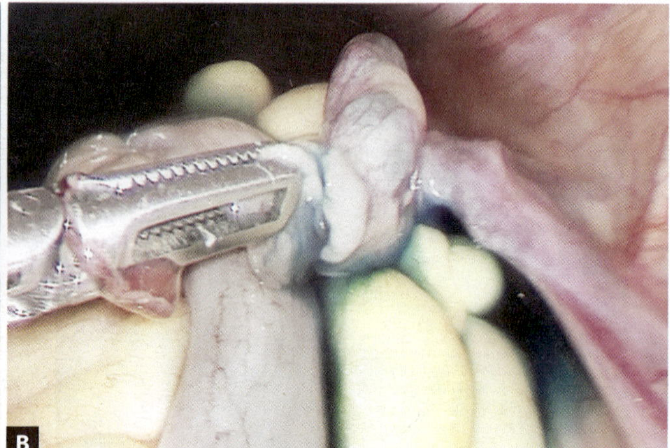

Figs. 22A and B: Tubal Patency test by laparoscopy with bilateral patent tubes.

MANAGEMENT OF TUBAL FACTOR INFERTILITY (FIGS. 22 TO 27)

Surgeries in the Era of ART

- Hysteroscopic tubal cannulation
- Tubal recanalization
- Tubotubal anastomosis
- Fimbrioplasty

(Reconstructive)

- Laparoscopic tubal delinking
- Laparoscopic salpingectomy

(Destructive)

Factors to be considered when planning surgery are mentioned in **Box 8.**

ASRM COMMITTEE OPINION

Box 9 shows the American Society for Reproductive Medicine (ASRM) committee opinion.

The results of tubal surgery in the era of assisted reproductive technology (ART) are shown in **Table 3**.

IN VITRO FERTILIZATION

In vitro fertilization is considered as an important treatment for tubal disease, especially in cases of:

- Widespread tubal disease
- Hydrosalpinx/pyosalpinx with tubal block

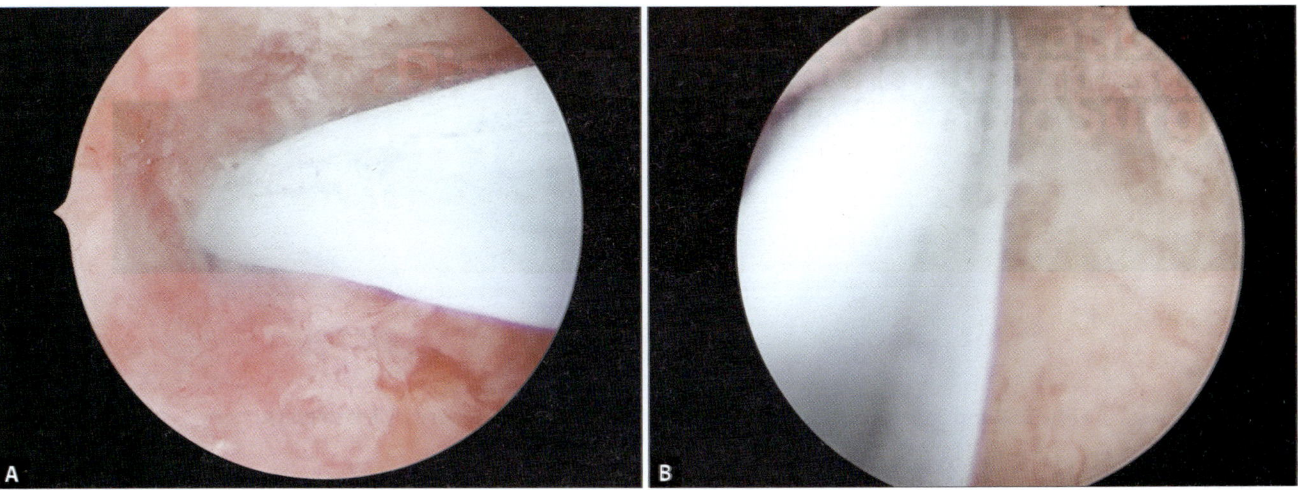

Figs. 23A and B: Bilateral (B/L) tubal cannulation for B/L cornual block.

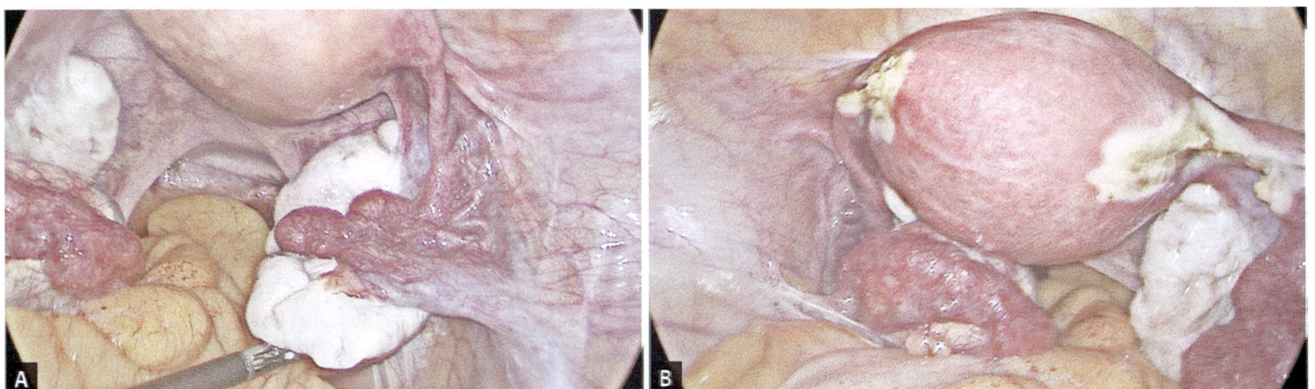

Figs. 24A and B: (A) Bilateral (B/L) terminal hydrosalpinx; (B) B/L tubal delinking.

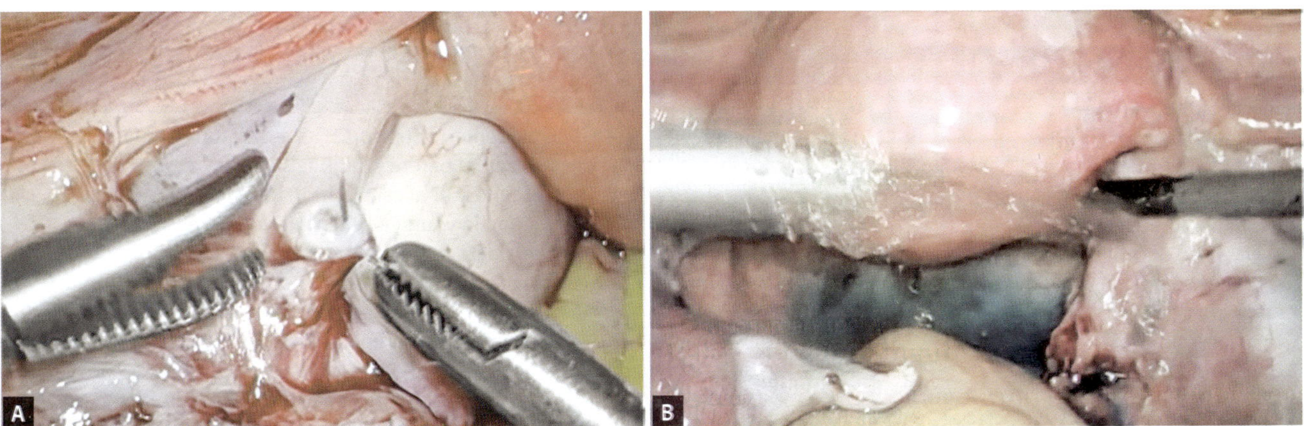

Figs. 25A and B: Tubal recanalization.
Courtesy: Dr Rajesh Modi, Akola.

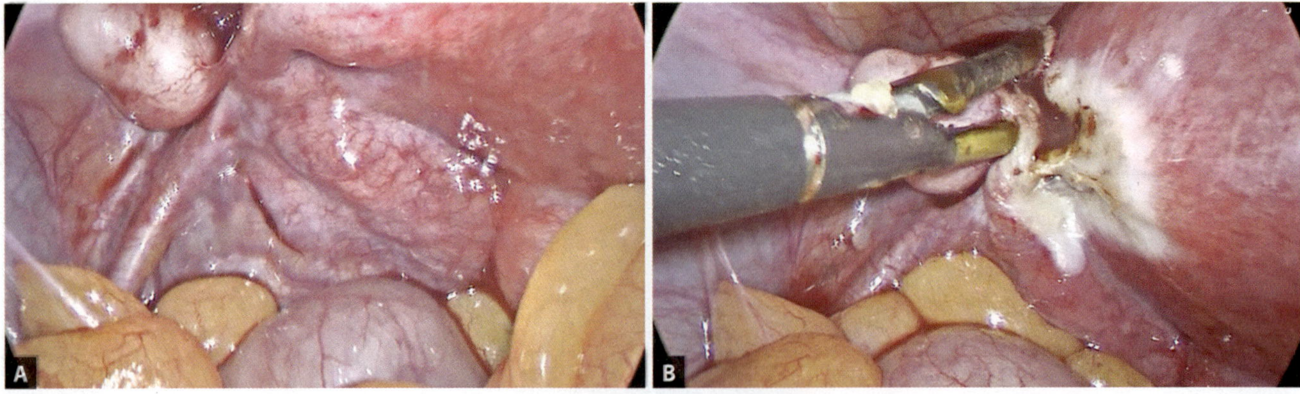

Figs. 26A and B: Left fallopian tube hydrosalpinx with adhesions dealt by left salpingectomy.

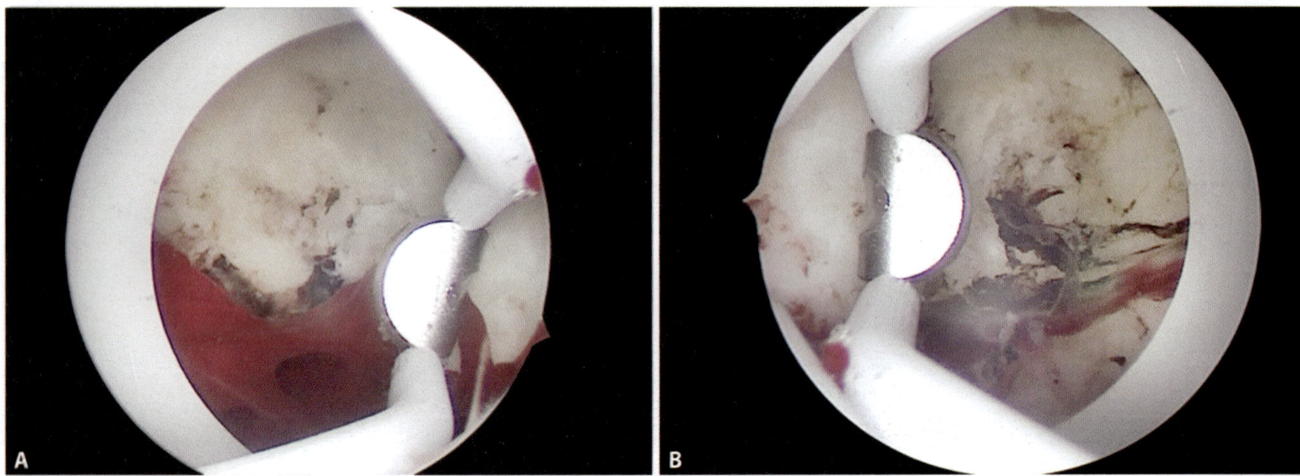

Figs. 27A and B: Bilateral (B/L) hydrosalpinx with pelvic adhesions taken care by ostial fulguration.

BOX 8: Factors to be considered when planning surgery.[19]

- Age of the woman
- Ovarian reserve
- Husband semen analysis
- Number of children desired
- Site and extent of tubal disease
- Presence of other infertility factors
- Risk of ectopic pregnancy and other complications
- Experience of the surgeon
- Success rates of the IVF program
- Cost and patient preference

(IVF: in vitro fertilization)

BOX 9: American Society for Reproductive Medicine (ASRM) committee opinion.[19]

- Good evidence to support HSG as the standard first-line test to assess tubal patency; high false positive
- Recommends tubal cannulation for proximal tubal obstruction in young women with no other significant infertility factors
- Laparoscopic fimbrioplasty or neosalpingostomy for the treatment of mild hydrosalpinges—FAIR EVIDENCE
- No adequate trials comparing pregnancy rates with tubal surgery versus IVF. IVF has a higher per cycle pregnancy rate
- Tubal anastomosis for reversal of tubal sterilization has a significantly higher cumulative pregnancy rate than IVF, and it is more cost-efficient, even in women 40 years of age or older

(HSG: hysterosalpingography; IVF: in vitro fertilization)

TABLE 3: Results of tubal surgery in the era of assisted reproductive technology (ART).[20]

Surgery	*Indication*	*Term pregnancy rate [mean (range)]*	*Ectopic pregnancy rate [mean (range)]*
Salpingectomy	Hydrosalpinx prior to IVF (limited data for natural conception)	25% LBR (IVF)	1–2%
Resection anastomosis	SIN	44% (11–57%)	7% (0–12%)
Salpingo-ovariolysis	Peri-adnexal adhesions	50% (17–64%)	5% (0–16%)
Fimbrioplasty	Fimbrial phimosis	50% (15–60%)	7% (2–23%)
Salpingostomy	Hydrosalpinx with >50% normal mucosa or bilateral or well-informed patient	30% (0–42%)	9% (7–24%)
Reversal of sterilization	Tubal sterilization	52% (33–86%)	5% (0–14%)

(ART: assisted reproductive technology; IVF: in vitro fertilization; LBR: live birth rate; SIN: salpingitis isthmica nodosa)

Flowchart 1: Management protocol for tubal factor infertility.

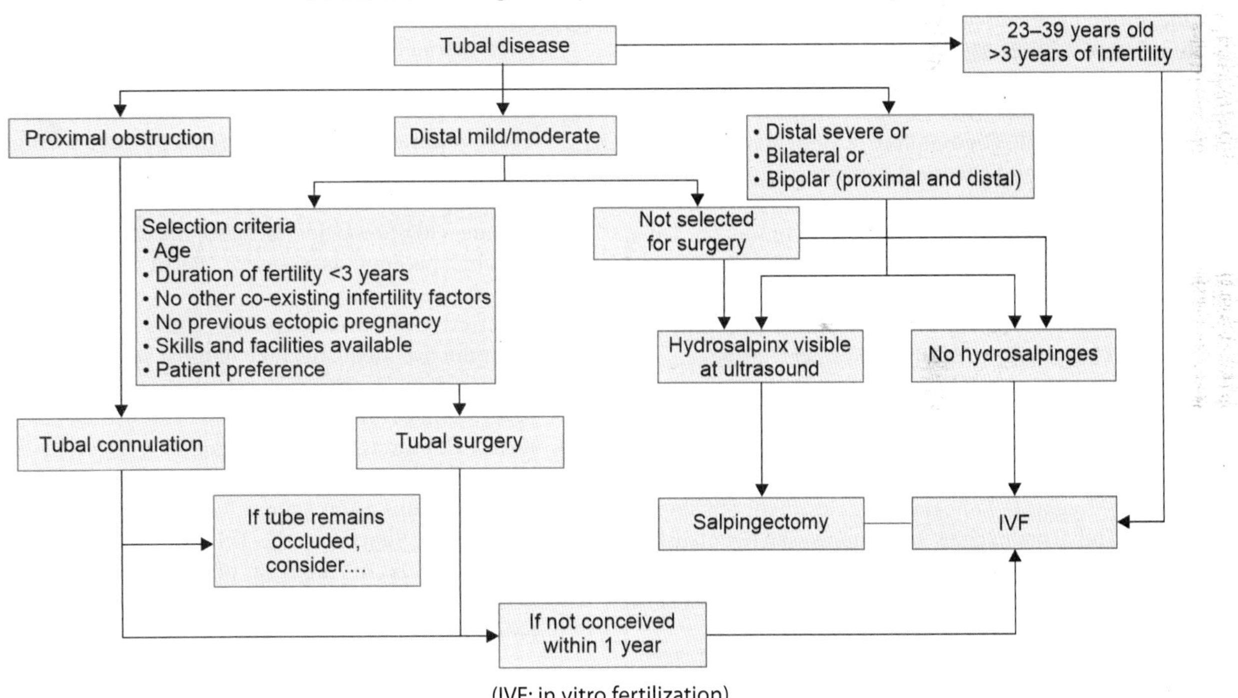

(IVF: in vitro fertilization)

- Distal tubal occlusions
- Failed tubal recanalization
- Small amount of residual tube post tubectomy.

The presence of hydrosalpinx is associated with decreased pregnancy rate and increased rates of ectopic pregnancy and miscarriage. Management of hydrosalpinx, regardless of the type of treatment, leads to an increased chance of pregnancy **(Table 3)**.

Salpingectomy, tubal occlusion, and hydrosalpinx aspiration are considered better forms of treatment before undertaking embryo transfer.[21]

CONCLUSION

At the end, we concluded the chapter in **Flowchart 1**.

ACKNOWLEDGMENTS

I extend gratitude to all my team for their support in my endeavor and contributing in my surgeries. I thank Dr Rajesh Modi and Dr Sandeep Pansare for their picture contributions for this chapter.

REFERENCES

1. Kawwass JF, Crawford S, Kissin DM, Session DR, Boulet S, Jamieson DJ. 2013. Tubal factor infertility and perinatal risk after assisted reproductive technology. Obstet Gynecol. 2013;121:1263-71.
2. Honoré GM, Holden AE, Schenken RS. Pathophysiology and management of proximal tubal blockage. Fertil Steril. 1999;71:785-95.
3. Feinberg EC, Levens ED, Decherney AH. Infertility surgery is dead: only the obituary remains? Fertil Steril. 2018;89:232-6.
4. Thor AD, Young RH, Clement PB. Pathology of the fallopian tube, broad ligament, peritoneum, and pelvic soft tissues. Hum Pathol. 1991;22:856-67.
5. Fortier KJ, Haney AF. The pathologic spectrum of uterotubal junction obstruction. Obstet Gynecol. 1985;65:93-8.
6. David A, Garcia CR, Czernobilsky B. Human hydrosalpinx. Histologic study and chemical composition of fluid. Am J Obstet Gynecol. 1969;105:400-11.
7. Ikemoto Y, Kuroda K, Kuribayashi Y, Inoue M. Tubal function abnormalities with tubal patency in unexplained infertility. In: Kuroda K, Brosens J, Quenby S, Takeda S (Eds). Treatment Strategy for Unexplained Infertility and Recurrent Miscarriage. Singapore: Springer; 2018.
8. Patil M. Assessing tubal damage. J Hum Reprod Sci. 2009;2(1):2-11.
9. Della Grotta LM, Dyer RB, Holbert BL. The "cogwheel" sign of hydrosalpinx. Abdom Radiol (NY). 2019;44:3486-7.
10. Panchal S, Nagori C. Sonographic assessment of fallopian tubes and tubal pathologies. In: Jayaprakasan K, Polanski L, Ojha K (Eds). Gynaecological Ultrasound Scanning: Tips and Tricks. Cambridge: Cambridge University Press; 2020. pp. 127-44.
11. Kupesic S, Kurjak A, Pasalic L, Benic S, Ilijas M. The value of transvaginal color Doppler in the assessment of pelvic inflammatory disease. Ultrasound Med Biol. 1995;21(6):733-8.
12. Timor-Tritsch IE, Lerner JP, Monteagudo A, Murphy KE, Heller DS. Transvaginal sonographic markers of tubal inflammatory disease. Ultrasound Obstet Gynecol. 1998;12(1):56-66.
13. Rutherford AJ, Jenkins JM. Hull and Rutherford classification of infertility. Hum Fertil (Camb). 2002;5 (1 Suppl):S41-5.
14. Zou SE, Jin Y, Ko YL, Zhu J. A new classification system for pregnancy prognosis of tubal factor infertility. Int J Clin Exp Med. 2014;7(5):1410-6. eCollection 2014.
15. Allahbadia GN, Nalawade YV, Patkar VD, Niyogi GM, Shah PK. The Sion test. Aust N Z J Obstet Gynaecol. 1992;32(1):67-70.
16. Sanin-Ramirez D, Carriles I, Graupera B, Ajossa S, Neri M, Rodriguez I, et al. Two-dimensional transvaginal sonography vs saline contrast sonohysterography for diagnosing endometrial polyps: systematic review and meta-analysis. Ultrasound Obstet Gynecol. 2020;56(4):506-15.
17. Deichert U, Schleif R, van De Sandt M, Juhnke I, Transvaginal hysterosalpingo-contrast-sonography (Hy-CoSy) compared with conventional tubal diagnostics. Hum Reprod. 1989;4(4):418-24.
18. Saunders RD, Shwayder JM, Nakajima ST. Current methods of tubal patency assessment. Fertil Steril. 2011;95(7): 2171-9.
19. Practice Committee of the American Society for Reproductive Medicine. Role of tubal surgery in the era of assisted reproductive technology: a committee opinion. Fertil Steril. 2015;103(6):e37-43.
20. Posaci C, Camus M, Osmanagaoglu K, Devroey P. Tubal surgery in the era of assisted reproductive technology: clinical options. Hum Reprod. 1999;14 Suppl 1:120-36.
21. Capmas P, Suarthana E, Tulandi T. Management of Hydrosalpinx in the Era of Assisted Reproductive Technology: A Systematic Review and Meta-analysis. J Minim Invasive Gynecol. 2021;28(3):418-41.

Chapter 30

Ovulation Induction Protocols

Pooja Mehta, Jasmeen Kaur

INTRODUCTION

This chapter aims to impart an indepth knowledge about different ovulation induction protocols and appropriate patient selection. But before that an overview of the physiology of ovulation is a must.

PHYSIOLOGY OF OVULATION: AN OVERVIEW

Ovulation is a very complex process involving interplay of many hormones, its clear understanding is a must before we go ahead with understanding its abnormalities and their treatment. At birth, each ovary contains 1–2 million primordial follicles. These primordial follicles are arrested in Prophase I of meiosis until the onset of puberty. At the onset of puberty, the gonadotropic hormones began to induce the maturation of the primordial follicle allowing for completion of Meiosis I forming a secondary follicle. The secondary follicle begins Meiosis II, but this phase will not be completed unless that follicle is fertilized.

In a natural cycle, both follicle-stimulating hormones (FSH) start increasing in the late luteal phase of the previous cycle and increase steadily till FSH threshold is reached and a cohort of follicles have been recruited. These follicles secrete estrogen which has a negative feedback on hypothalamus thereby decreasing the secretion of FSH and luteinizing hormone (LH). When the dominant follicle is mature enough and the estrogen level is at its peak, it leads to stimulates the pituitary leading to the LH surge. Under the effect of LH surge, the oocyte completes meiosis with the expulsion of the first polar body and ovulation occurs. This is followed by rise and maintenance of progesterone level for the next 14 days by the corpus luteum. In an event of no ongoing pregnancy and the nonrescue of the corpus luteum by human chorionic gonadotropin (hCG), it withers, and the cycle starts again.

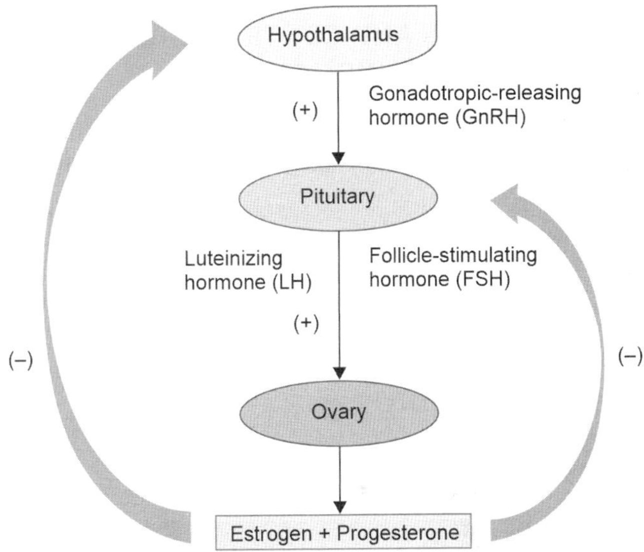

Fig. 1: The hypothalamic–pituitary–ovarian axis of the female reproductive system. (+) represents positive feedback and (–) represents negative feedback.[1]

ANOVULATION

Failure to ovulate or anovulation which occurs due to hypothalamic-pituitary-ovarian (HPO) axis dysfunction has been classified by World Health Organization (WHO)[2] as under and has been dealt in details in previous chapters **(Fig. 1)**:

- *WHO group 1:* Hypogonadotropic hypogonadism (5–10%) has low levels of LH, FSH, and estrogen as well
- *WHO group 2:* Normogonadotropic normoestrogenic anovulation (75–85%) has normal FSH and estrogen but a high or normal LH (PCOS)
- *WHO group 3:* Hypergonadotropic anovulation (10–20%) has high FSH levels but low estrogen (premature ovarian failure)
- *WHO group 4:* Hyperprolactinemic anovulation (5–10%) has high prolactin with low or normal FHS and low estrogen levels.

The treatment for infertility is subdivided into nonpharmacological and pharmacological.

Nonpharmacological Treatment

- Weight loss
- Dietary modification
- Exercise and yoga
- Behavioral modification

Pharmacological Treatment

- Gonadotropins for controlled ovarian stimulation
- Adjuvant therapy

INDICATIONS AND CONTRAINDICATIONS OF OVULATION INDUCTION[3]

Indications

- Patients of anovulation like polycystic ovary syndrome (PCOS)
- Unexplained infertility
- Superovulation in intrauterine insemination (IUI)
- Thin endometrium
- Luteal phase defects
- COH (controlled ovarian hyperstimulation) as in in vitro fertilization (IVF)
- Poor responders
- Hyperprolactinemia

Contraindications

- Liver diseases
- Medical illness like hypertension, DM (theses should be controlled before starting stimulation)
- Large ovarian cysts (this suggests the significance of day 2 ultrasound before starting stimulation)
- Premature ovarian failure
- Genetic abnormalities like Turner syndrome, Down syndrome, etc.

OBJECTIVE OF OVULATION INDUCTION

It depends upon the modality of treatment, i.e., a natural trial, IUI, or IVF.

In IUI or natural trial, the main aim is to limit the number of dominant follicles to a minimum 1 to maximum 3, i.e., leading follicle of >18 mm.

In IVF procedure a controlled ovarian stimulation is done to achieve a follicle growth of 5-20 follicles without adversely affecting the health of the patient.

Intrauterine Insemination (Table 1)

Intrauterine insemination is a simple, noninvasive, and less expensive, so is widely used as a first line of treatment for patients with infertility. Success of the procedure depends on the patient selection. IUI gives good results

TABLE 1: Drugs and protocols for IUI.

Drugs	*Mode of action*	*Treatment*	*Success rate*	
Clomiphene citrate[6,7]	Antiestrogenic	• Start from day 2 to day 5 of periods for 5 consecutive days • Starting dose of 50 mg/day and can be increased to a maximum 150 mg/day • In cases of nonresponse extended therapy can be given for up to 7–10 days	• Ovulation rate 67% • Cycle fecundity 15–22% • Further increases by 9.5% with IUI • CC resistance	Can be used for maximum 6 consecutive months or 12 months in a lifetime[8]
Letrozole[9]	Aromatase inhibitor	• Start from day 3 to day 7 of periods for 5 consecutive days • In a single dose of 2.5 to 5 mg/day • Single dose of 20 mg on day 3 can also be given	• Ovulation rate 87% • Cycle fecundity 30%[10] • Further increases by 9.5% with IUI	
Gonadotropins (FSH)* along with CC		5 days CC followed by 75–150 IU/day from day 6 is added till dominant follicle develops		
Gonadotropin		• Step-up protocol • Step-up low-dose protocol • Step-down protocol • Sequential step-up step-down regime • Low-dose gonadotropins		

*As CC increases LH, FSH is preferred than HMG.

(CC: clomiphene citrate; FSH: follicle-stimulating hormone; HMG: human menopausal gonadotropin; IUI: intrauterine insemination; LH: luteinizing hormone)

in patients <35 years of age with functional tubes and infertility due to a cervical factor, anovulation, moderate male factor, unexplained factors, immunological factor, and ejaculatory disorders with clinical pregnancy rates per cycle ranging from 10 to 20%. It, however, has limited use in patients with endometriosis, severe male factor infertility, tubal factor infertility, and advanced maternal age ≥35 years.[4]

For IUI, strict patient selection criteria and individualized stimulation protocols are very important as these not only improve the success rate but also decrease the cycle cancellation rate either due to ovarian hyperstimulation syndrome (OHSS) or inadequate response.

Intrauterine insemination with or without controlled ovarian stimulation (COS) is done with the aim of obtaining 1–3 dominant follicles (>18 mm). Oral ovulogens like clomiphene citrate or letrozole are used either alone or in combination with gonadotropin to achieve this target.[5]

Strict patient selection criteria and individualized stimulation protocols not only improve the success rate but also decrease the cycle cancellation rate due to OHSS or inadequate response.

Three to six IUI cycles may be offered before considering IVF. However, patients with advanced maternal age, severe male factor infertility, tubal pathology, or severe endometriosis will benefit from a direct referral to IVF/intracytoplasmic sperm injection (ICSI).[4]

Clomiphene resistance: Clomiphene resistance defined as failure to ovulate after receiving 150 mg of CC daily for 5 days per cycle, for at least three cycles, is common and occurs in approximately 15–40% in women with PCOS.

Clomiphene failure: It is defined as failure to conceive despite successful ovulation for six cycles.

In Vitro Fertilization

Indication

- Tubal diseases (9%)
- Ovulatory (7%)
- Moderate to severe endometriosis (5%)
- Poor ovarian reserve (10%)
- Uterine causes (1%)
- Advanced maternal age
- Male factor (18%)
- Unexplained infertility (12%)
- Multiple factors (30%)
- Preimplantation genetic testing (PGT)

Contraindications of IVF

There are no absolute contraindications to the procedure of IVF. However, it should not be performed in women who have a significant risk of morbidity and mortality of pregnancy if IVF was successful.

Steps in IVF

- Initial evaluation of the couple
- Suppression of natural hormones
- Controlled ovarian hyperstimulation (COH)
- Oocyte retrieval and sperm collection
- In vitro fertilization of oocytes
- Embryo transfer

Drugs and Protocols for COS

Basic principle is administration of gonadotropins to stimulate multiple oocytes and administration of either gonadotropin-releasing hormone (GnRH) antagonist or agonist to prevent premature ovulation.

Individualization of the stimulation protocol by keeping in mind the age, ovarian reserve (AMH, FSH, and LH), and previous response to COS is the first step toward a successful IVF.

Gonadotropins: Follicular-stimulating hormone (FSH), luteinizing hormone (LH), or human menopausal gonadotropin (FSH + LH) preparations are available in urinary, purified, highly purified, recombinant formulations.

Human menopausal gonadotropin: It was first obtained from the urine of postmenopausal women. It exhibits both LH and FSH activity in equal proportions. It is available in both urinary and highly purified preparations.

- HMG 150 IU = 75 IU FSH + 75 IU LH
- HMG-HP (highly purified) 150 IU = 75 IU FSH + 75 IU LH. With <5% urinary protein
- Urinary FSH (uFSH) = 75 IU FSH + 1 IU LH
- FSH-HP (highly purified) = 75 IU FSH + <0.1 IU LH + >5% urinary protein
- rFSH (recombinant) = 75 IU FSH with no LH activity
- rFSh + rLH

INDUCTION PROTOCOLS

Minimal Stimulation Protocol

Minimal stimulation protocol utilizes clomiphene citrate in conjunction with human menopausal gonadotropin (HMG), which is more effective compared to administering

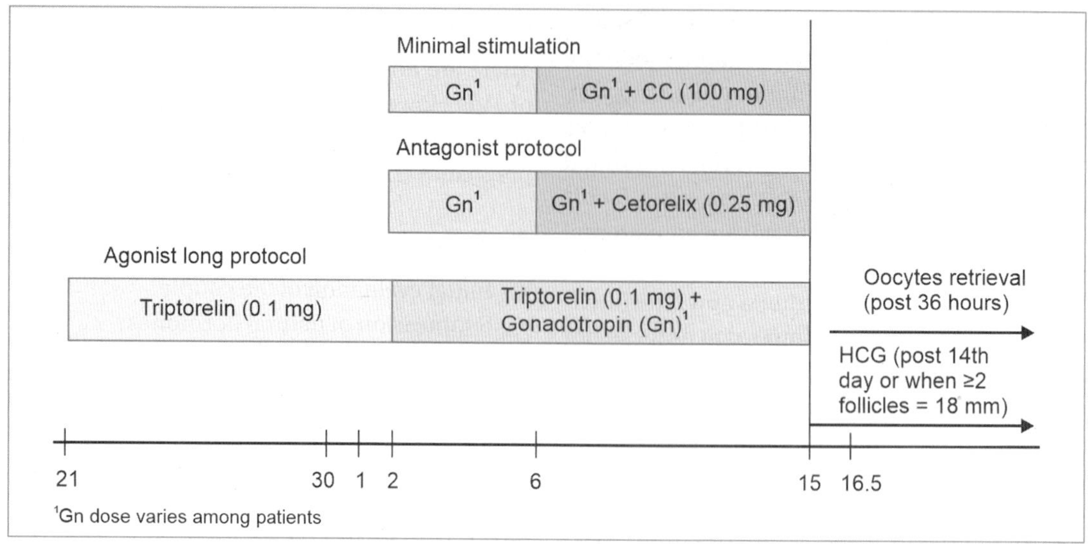

Fig. 2: Stimulation procedures for agonist long, antagonist and minimal stimulation protocols for IVF. (CC: clomiphene citrate; Gn: gonadotropin mixtures; IVF: in vitro fertilization)[31]

HMG alone (46% vs. 25.9%).[11-13] In this treatment protocol **(Fig. 2)**, administration of CC occurs on the 6th day, or earlier depending on LH level rise, and continues until HCG administration. This is followed by the retrieval of mature oocyte and IVF.

Letrozole, an aromatase inhibitor, is used alternative to clomiphene for minimal stimulation protocol in some clomiphene resistant patients.

Long Agonist Protocol

The GnRH long agonist protocol starts with administration of 0.1 mg GnRH agonist (e.g., triptorelin) from cycle-day 21 onwards followed by administration of gonadotropin at 150–225 IU daily starting on cycle-day 2 after a routine baseline scan. The adjustment of gonadotropin dose is based on follicular development which is assessed by regular ultrasound follow-up. Continual administration of GnRH agonist and gonadotropin lasts until follicles reached from 16 to 18 mm in size and are ready for hCG administration.

Physiology: Initial downregulation under effect of agonist followed by ovarian stimulation with gonadotropins

Benefit: Better cycle control

Drawback: Long duration, multiple injections, and higher cost because of more gonadotropin requirement

Ideal patients: Normo responders

Short Agonist Protocol

In this protocol, GnRH is administered along with gonadotropin injection from day 1 or 2 of cycle after a routine baseline scan. Both are administered continuously till adequate follicle size is achieved.

Benefit: Initial flare effect under effect of agonist leading to better follicular recruitment.

Drawback: Lesser pregnancy rate than long protocol[14] as there can high LH level from the previous cycle

Ideal patients: Poor responders but not used nowadays

Microdose Agonist Flare Protocol

This protocol aims to minimize the ovarian suppression under the effect of GnRH-agonist as seen in long agonist protocols but uses the initial up-regulatory effect of the agonist on follicular recruitment, especially useful in patient of POR. It gives similar results as an antagonist protocol.

Antagonist Protocol

For the GnRH antagonist (Cetrorelix, Ganirelix) protocol administration of gonadotropin (150–225 IU) is started after a baseline day 2 or 3 scan. Gonadotropin dosage is adjusted according to ovarian response judged by subsequent ultrasounds. Depending upon when the antagonist is added, the protocol is further divided into:

- *Fixed:* The antagonist is started from day 6 of the cycle
- *Flexible:* The antagonist is added once the leading follicle is >14 mm.

Antagonist and gonadotropins are continued till the day of trigger.

Physiology: There is initial good follicular development under effect of gonadotropins before the antagonists are started leading to immediate blockage of the receptors.

Benefit: Shorter duration, lesser number of injection, and lesser side effects. Lower incidence of OHSS.

Ideal patients: Normo responders, poor responders

In all the protocols, there is routine monitoring of patients with transvaginal sonography (TVS) to assess follicular size. Once the follicle reaches 16–18 mm, ovulation trigger is given and after 34–36 hours of the trigger injection the mature oocytes are retrieved by ovum pick-up.

IVF PROTOCOL SELECTION

For protocol selection, the patients are categorized into three groups depending upon their response to the stimulation. Many screening tests such as ovarian reserve, CC challenge, GnRH, GnRH agonist, measurement of anti-Müllerian hormone (AMH), and antral follicular count (AFC) along with maternal age aid in this classification.[15,16]

- High responders*:* Antagonist protocol
- Normo responders*:* Long agonist protocol or antagonist protocol
- *Poor responders:*[15,17]

  Poor responders are patients with a low number of oocytes (<4) or no ovarian response with FSH >300 IU.[18] Poor response has been shown to be associated with advanced maternal age that affects oocyte quality and follicle numbers. This has also been observed in some young patients, but the causes are unclear.[15,16]

The diagnostic criteria for poor ovarian responders was laid down by the Bologna the va criteria in 2011. According to this classification, the women were divided into four categories based on qualitative and quantitative parameters:[1]

1. Age (≥40 years)
2. Antral follicle count (AFC <5–7)
3. Anti-Müllerian hormone (AMH <0.5–1.1 ng/mL).
4. Ovarian response to previous stimulation (≤3 oocytes with a conventional stimulation)

IDEAL RESPONSE EXPECTED AND CYCLE CANCELLATION

- Less than 5 oocyte retrieval is suboptimal response.
- More than 20 oocyte retrieval is hyperstimulation and increases chances of OHSS.

Cycle Monitoring

All patients are closely monitored for follicle number, size, rate of growth of follicles, perifollicular flow, synchronous growth in bilateral ovaries, endometrial lining, and presence of any free fluid in the POD.

Cycle conversion to IUI is offered to the patient in cases of poor ovarian response (<3 follicles of >12 mm).

TRIGGER

Ovulation trigger is a crucial step in an ART cycle. The agent used mimics the natural mid cycle LH surge leading to final maturation of the oocytes.

Human Chorionic Gonadotropin

Human chorionic gonadotropin was the first trigger used due to its molecular similarity to LH, but it is associated with high risk for OHSS due to prolonged LH activity.[19] Both different uhCG and rhCG have been used as trigger and change in the pregnancy rate has been observed. Different doses for both uhCG (2,000–10,000 IU) and rhCG (250 ug and 500 ug) have been described in the literature but do not seem to affect the success rate. The probability of OHSS was reduced when lower doses of hCG were administered but this did not reach statistical significance.[18]

Gonadotropin-releasing Hormone Agonist

The GnRH agonist is used as trigger in GnRH antagonist stimulation cycles or in cycles where no agonist has been used for stimulation. A meta-analysis including three RCTs (275 women) reported a significant difference in clinical pregnancy rate in favor of hCG,[20] neither has a significant difference in OHSS rate been reported between GnRH agonist and hCG use.[21]

Commonly used GnRH agonists are leuprorelin, buserelin, nafarelin, and triptorelin.

Dual Trigger

Combination of GnRHa and hCG together constitutes a dual trigger. The meta-analysis found a significant higher pregnancy rate with dual trigger as compared to hCG trigger though there was no difference in the number of oocytes retrieved.[22]

Side Effects and Complications

- Discomfort, rash, or pain at injection site
- Allergic reaction
- Mood swings
- Bloating
- Pain or heaviness in lower abdomen
- Risk of multiple pregnancy
- Ovarian hyperstimulation syndrome (OHSS).

ADJUVANTS

Adjuvant therapy is the treatment given either before or along with the ovarian stimulation to enhance the ovarian response. It has a significant role to play in patients with poor ovarian response (POR).

According to ESHRE, Bologna criteria in order to define the poor response in IVF, at least two of the following three features must be present:

1. Advanced maternal age or any other risk factor for POR
2. A previous POR
3. An abnormal ovarian reserve test (ORT).

Two episodes of POR after maximal stimulation are sufficient to define a patient as poor responder in the absence of advanced maternal age or abnormal ORT.

These patients not only exhibit a poor response but also need higher doses of gonadotropins. The adjuvant therapies recommended not only help in eliciting a better response, but also help in reducing the dose of gonadotrophins required, offering both economical and health benefits.

Growth Hormone

Low-dose growth hormone (GH) supplementation significantly increased ovarian response, the number of oocyte retrieved, clinical pregnancy rate, and the ongoing pregnancy rate.

Growth hormone acts on the ovary by increasing the production of insulin-like growth factor-1 (IGF-1), which in turn increases the response of granulosa cells to FSH and LH.

Beneficial effect of GH is also seen on the probability of clinical pregnancy rate (CPR) and live birth rate (LBR) in women with decreased ovarian reserve (DOR) and poor responders. But comparing different studies has brought us to the conclusion that more studies need to be conducted about the efficacy and safety of GH to treat a woman with a poor response, till then we should use it cautiously and judiciously.

The only drawback is that it increases the cost of treatment significantly.

Growth hormone is given daily starting from day of stimulation as a subcutaneous injection of 4–8 IU until trigger.

Clomiphene Citrate

Clomiphene citrate is an antiestrogenic, which acts by increasing GnRH production and in turn increases endogenous gonadotropin production, aids in more follicular recruitment and development.

The drawback that comes to mind is its antiestrogenic effect on endometrial lining, which can be bypassed if go ahead with freeze all cycle, or decision for transfer in same cycle can be taken after a thorough evaluation of the endometrial lining.

It usually is administered at a dose of 50–150 mg per day starting from day of stimulation for 5 days.

Letrozole

It acts by increasing androgen production by preventing aromatization to estrogen, leading to sensitization of the follicles to FSH leading to decreased FSH requirement.

It is given in the dose of 2.5 mg from day 3 to day 7.

Estrogen Priming

Estrogen priming is done in the luteal phase of the previous cycle in antagonist protocol. It is given as a 0.1 mg transdermal patch or oral preparation of 4 mg estradiol valerate from day 21 of previous cycle to till 3 days after menstruation or up to day of trigger.

L-arginine

L-arginine enhances periovulatory dilatation under the effect of nitric oxide. It is given in a daily dose of 16 mg.

Pyridostigmine

Pyridostigmine or growth hormone releasing factor is given from day 21 of the previous cycle till the day of trigger, in the daily dose of 120 mg/day. It increases the endogenous growth hormone by increasing acetylcholine.

Exogenous Testosterone

- *Dehydroepiandrosterone:* Dehydroepiandrosterone or DHEA is produced majorly from reticular zone in adrenals and in small amount from theca cells in the ovary. It increases the fertility outcome by increasing the production if IGF-I as well as androgen production. Androgens as we know has a role to play in both recruitment and growth of follicles, preventing follicular atresia.[23,24]

  Dehydroepiandrosterone is given as 75 mg in daily or divided doses for at least 1–4 months to get an optimum response.
- *Transdermal testosterone:* It acts by increasing expression of insulin-like growth factor-1 (IGF-1) and in turn increases response to gonadotropins. It improves the number and quality of follicles as well as increases the cumulus oocyte complex.

  It can be used in both agonist and antagonist cycles. In antagonist cycles 12.5 mg of transdermal testosterone gel or spray is applied on the external aspect of the thigh for 15–21 days before starting the stimulation in the previous cycle.

  The benefit of the transdermal route is that it allows a steady delivery of testosterone in a more physiological way than the oral administration.

Glucocorticoids

It acts by stimulating the secretion of growth hormone and IGF-I.

Dexamethasone 1 mg is given daily in long agonist protocol.

OVARIAN HYPERSTIMULATION SYNDROME

Ovarian hyperstimulation syndrome is an iatrogenic, potentially life-threatening complication of ovarian stimulation, characterized by ovarian enlargement and fluid extravasation from intravascular compartment to the third spaces (mainly thoracic and abdominal cavities).

Milder OHSS is common and characterized by weight gain, abdominal discomfort, and enlarged ovaries; whereas, severe OHSS is accompanied by severe ascites, pleural effusion, electrolyte imbalance, and hypovolemia with oliguria.[25,26]

Depending on the time of onset of symptoms, OHSS is classified as early and late.

- *Early OHSS:* The symptoms begin within 3–7 days after hCG administration, and resolve within 10–14 days in absence of pregnancy. Exogenous hCG is the cause for this type of OHSS.
- *Late OHSS:* The symptoms here develop nearly 10 days after hCG injection administration, as the cause here is the endogenous hCG production.

Pathophysiology

Beta-hCG and its analogs, estrogens, estradiol, prolactin, histamine, and prostaglandins have all been implicated in development of OHSS. But now, it is increasingly better understood that the vasoactive substances such as interleukins, tumor necrosis factor (TNF)-alfa, endothelin-1, and vascular endothelial growth factor (VEGF) secreted by ovaries have been implicated in increasing vascular permeability.

Risk Factors[27]

Primary risk factors:

- Young age <30 years
- Higher dose of gonadotropins
- Polycystic ovary syndrome
- Previous history of OHSS
- Lower body weight

Secondary risk factors:

- Rapidly rising serum estradiol levels
- Use of hCG as opposed to progesterone for luteal phase support after IVF
- Large number of oocytes (>20)

Clinical Presentation

Mild manifestation of OHSS is relatively common and includes abdominal pain, nausea, vomiting, diarrhea, and abdominal distension (sometimes as much as 25 cm). Advancement of illness is recognized when symptoms persist, worsen, or pain associated with ascites, oliguria, hemoconcentration, respiratory difficulty (dyspnea), hypercoagulable state, acute renal failure, and hypovolemia.

Classification[28,29]

Mild OHSS:

- *Grade I:* Abdominal distension and discomfort
- *Grade II:* Grade I disease plus nausea, vomiting and/or diarrhea plus ovarian enlargement from 5–12 cm.

Moderate OHSS:

- *Grade III*: Features of mild OHSS plus ultrasonography evidence of ascites.

Severe OHSS:

- *Grade IV:* Features of moderate OHSS plus clinical evidence of ascites and/or hydrothorax and breathing difficulties.
- *Grade V:* All of the above plus change in blood volume, increased blood viscosity due to hemoconcentration, coagulation abnormalities, and diminished renal perfusion and function.

Treatment of OHSS[30]

Conservative treatment with intravenous fluids and close monitoring are usually done in mild and moderate OHSS. In severe OHSS, proper evaluation and investigations are done to evaluate hematocrit, electrolytes, kidney and liver function. Intensive care admission is indicated in cases with severe respiratory distress or major electrolyte imbalance with elevated serum creatinine. Crystalloids in the form of IV saline and colloids as albumin or hydroxyethyl starch are commonly used to expand intravascular volume, and aspiration of ascetic fluid if required.

CONCLUSION

Individualization of every case of infertility is a must and a proper categorization into high, poor or normo - responders needs to be done before starting a patient on a customized ovulation induction protocol for best results.

REFERENCES

1. Sun J, Zhang Q, Wang Z, Yan B. Effects of nanotoxicity on female reproductivity and fetal development in animal models. Int J Mol Sci. 2013;14(5):9319-37.
2. Dhont M. WHO-classification of anovulation: background, evidence and problems. Int Congr Ser. 2005;1279:3-9.
3. Sharma M, Balasundaram P. (2023). Ovulation Induction Techniques. Available from: https://www.ncbi.nlm.nih.gov/books/NBK574564/. [Last accessed March, 2024]
4. Allahbadia GN. Intrauterine Insemination: Fundamentals Revisited. J Obstet Gynaecol India. 2017;67(6):385-92.
5. Liu J, Li T-C, Wang J, Wang W, Hou Z, Liu J. The impact of ovarian stimulation on the outcome of intrauterine insemination treatment: an analysis of 8893 cycles. BJOG. 2016;123(S3):70-5.
6. Garcia J, Jones GS, Wentz AC. The use of clomiphene citrate. Fertil Steril. 1977;28;707-17.
7. Practice Committee of the American Society for Reproductive Medicine. Use of clomiphene citrate in women. Fertil Steril. 2004;82;S90-6.
8. Rust LA, Israel R, Mishell DR Jr. An individualized graduated therapeutic regimen for clomiphene citrate. Am J Obstet Gynecol. 1974;120(6):785-90.
9. American College of Obstetricians and Gynecologists. ACOG Committee opinion No 738: Aromatase inhibitor in gynaecologic practice. Obstet Gynecol. 2018;131:e194-9.
10. Mitwally MF, Casper RF. Aromatase inhibition improves ovarian response to follicle-stimulating hormone in poor responders. Fertil Steril. 2002;77(4):776-80.
11. Zhang J, Chang L, Sone Y, Silber S. Minimal ovarian stimulation (mini-IVF) for IVF utilizing vitrification and cryopreserved embryo transfer. Reprod Biomed Online. 2010;21(4):485-95.
12. Ibrahim AE. The Minimal Stimulation Protocol for ICSI: An Alternative Protocol for Ovarian Stimulation. N Y Sci J. 2014;7:19-23.
13. Ziadeh SM, Zakaria MR, Abu-Hieja A. Pregnancy rates using CC/hMG or hMG alone. J Obstet Gynaecol Res. 1997;23:97-101.
14. Oehninger S. Poor responders in in vitro fertilization (IVF) therapy: the challenge continues. Facts Views Vis Obgyn. 2011;3:101-8.
15. Davis OK. IVF stimulation: protocols for poor responders. Methods Mol Biol. 2014;1154:329-41.
16. Oudendijk JF, Yarde F, Eijkemans MJ, Broekmans FJ, Broer SL. The poor responder in IVF: is the prognosis always poor?: a systematic review. Hum Reprod Update. 2012;18:1-11.
17. Data S. Gonadotrophin releasing hormone agonist protocols for pituitary desensitization in in vitro fertilization and gamete intrafollicular cycles. Cochrane Database Syst Rev.2002;(2):CD001299.
18. Fauser BC, De Jong D, Olivenes F, Wramsby H, Tay C, Itskovitz-Eldor J, et al. Endocrine profiles after triggering of final oocyte maturation with GnRH agonist after cotreatment with the GnRH antagonist ganirelix during ovarian hyperstimulation for in vitro fertilization. J Clin Endocrinol Metab. 2002;87:709-15.
19. Humaidan P, Westergaard LG, Mikkelsen AL, Fukuda M, Yding Andersen C. Levels of the epidermal growth factor-like peptide amphiregulin in follicular fluid reflect the mode of triggering ovulation: a comparison between gonadotrophin-releasing hormone agonist and urinary human chorionic gonadotrophin. Fertil Steril. 2011;95:2034-8.
20. Youssef MA, Abou-Setta AM, Lam WS. Recombinant versus urinary human chorionic gonadotrophin for final oocyte maturation triggering in IVF and ICSI cycles. Cochrane Database Syst Rev. 2016;(4):CD003719.
21. Griesinger G, Diedrich K, Devroey P, Kolibianakis EM. GnRH agonist for triggering final oocyte maturation in the

GnRH antagonist ovarian hyperstimulation protocol: a systematic review and meta-analysis. Hum Reprod Update. 2006;12:159-68.
22. Ding N, Liu X, Jian Q, Liang Z, Wang F. Dual trigger of final oocyte maturation with a combination of GnRH agonist and hCG versus a hCG alone trigger in GnRH antagonist cycle for in vitro fertilization: A Systematic Review and Meta-analysis. Eur J Obstet Gynecol Reprod Biol. 2017;218:92-8.
23. Magamage MPS, Zengyo M, Moniruzzaman M, Miyano T. Testosterone induces activation of porcine primordial follicles in vitro. Reprod Med Biol. 2011;10(1):21-30.
24. Sen A, Prizant H, Light A, Biswas A, Hayes E, Lee HJ, et al. Androgens regulate ovarian follicular development by increasing follicle stimulating hormone receptor and microRNA-125b expression. Proc Natl Acad Sci U S A. 2014;111(8):3008-13.
25. Kronenberg H, Williams RH. Williams Textbook of Endocrinology, 11th edition. Philadelphia: Saunders; 2008.
26. Papanikolaou EG, Tournaye H, Verpoest W, Camus M, Vernaeve V, Van Steirteghem A, et al. Early and late ovarian hyperstimulation syndrome: early pregnancy outcome and profile. Hum Reprod. 2005;20(3):636-41.
27. The Practice Committee of the American Society for reproductive Medicine. Ovarian hyperstimulation syndrome. Fertil Steril. 2008;90:S188-93.
28. Delvigne A, Rozenberg S. Review of clinical course and treatment of ovarian hyperstimulation syndrome (OHSS). Hum Reprod Update. 2003;9:77-96.
29. Golan A, Ron-El R, Herman A, Soffer Y, Weinraab Z, Caspi E. Ovarian hyperstimulation syndrome: an update review. Obstet Gynecol Surv. 1989;6:430-40.
30. Aboulghar M. Treatment of Ovarian Hyperstimulation Syndrome. Semin Reprod Med. 2010;28(6):532-9.
31. Shrestha D, La X, Feng HL. Comparison of different stimulation protocols used in in vitro fertilization: a review. Ann Transl Med. 2015;3(10):137.

Chapter 31

Assisted Reproductive Technology

Sarabjeet Singh, Jyotsna Gupta

INTRODUCTION

Assisted Reproductive Technology (ART) encompasses various medical procedures designed to aid couples in achieving pregnancy when natural conception is challenging. Techniques such as intrauterine insemination (IUI), In vitro fertilization (IVF) and Intracytoplasmic sperm injection (ICSI) are common ART methods. These technologies are utilized when infertility issues arise due to factors like blocked fallopian tubes, low sperm count, or ovulation disorders. ART procedures typically involve the extraction and manipulation of Gametes (eggs and sperms) outside the human body (In vitro) and developing embryos to transfer them back in the in the uterus to facilitate conception. Ethical considerations and success rates vary, making ART a complex field at the intersection of medical science, ethics, and reproductive rights.

DEFINITION

Assisted reproductive technology (ART) is defined as a fertility treatment where either eggs or embryos are handled outside the human body.[1] Primarily, ART included fertility treatments like in vitro fertilization (IVF) or intracytoplasmic sperm injection (ICSI). In these treatments, ovaries are stimulated with hormones (gonadotropin injections) to generate multiple oocytes through a process termed as controlled ovarian hyperstimulation (COH). The growing follicles are monitored with an ultrasound. Generally, after 9–12 days of injections, follicles become mature, after which these follicles are punctured and aspirated under the transvaginal ultrasound guidance to retrieve oocytes, a procedure known as oocyte retrieval (OCR) or oocyte pick up (OPU). These oocytes are then either inseminated with washed spermatozoa or injected using micromanipulators to develop embryos. Former procedure of inseminating oocyte with washed sperms is termed as in vitro fertilization (conventional IVF) whereas injecting spermatozoa into oocyte is a more advanced procedure called as intracytoplasmic sperm injection. The developing embryos are assessed on daily basis for the growth and development. These embryos are then transferred back into the women's uterus through a catheter either on day 2/3 (cleavage stage) or day 5 (blastocyst) stage. Apparently, any fertility treatments lesser than IVF/ICSI are not included in the definition of ART like an intrauterine insemination (IUI) or ovulation induction (OI) with an aim of natural intercourse,[1] but worldwide 35 countries approve intrauterine insemination too as an ART procedure.[2] This chapter focuses on the various ART treatments as per above definition and includes IUI. Since ART science itself is so vast, compiling the complete ART in a single chapter is difficult, the author would brief down the various ART treatments available in the modern era.

Infertility (subfertility) is defined as a disease when a couple who is trying to conceive fails to do so within 1 year of a regular and unprotected intercourse. It is worthy to note here that a single woman who wants a pregnancy fails to achieve it with six cycles of IUI would also be termed as subfertile in the recent definition laid down by American Society for Reproductive Medicine (ASRM).[3] It is important to understand that fertility is a science of probabilities and it is measured by chance of conception per menstrual cycle (fecundity rate). Hence, it becomes important to convey to the couples that 84% would conceive within 1 year and another 8–10% would conceive in next year.[4] A fertility evaluation should be offered to the couples after 1 year and initial evaluation should include a semen analysis, documentation of ovulation and fallopian tubes assessment.[5] In cases where apparently no cause is found (unexplained) or cases where mild endometriosis or mild male factor exists, National Institute for Health and Clinical Excellence (NICE) guidelines advise to wait for another 1 or 2 years.[6] An earlier evaluation (within 6 months) should be offered in couples with an advanced age of the women

Flowchart 1: Evaluation of infertile couple.

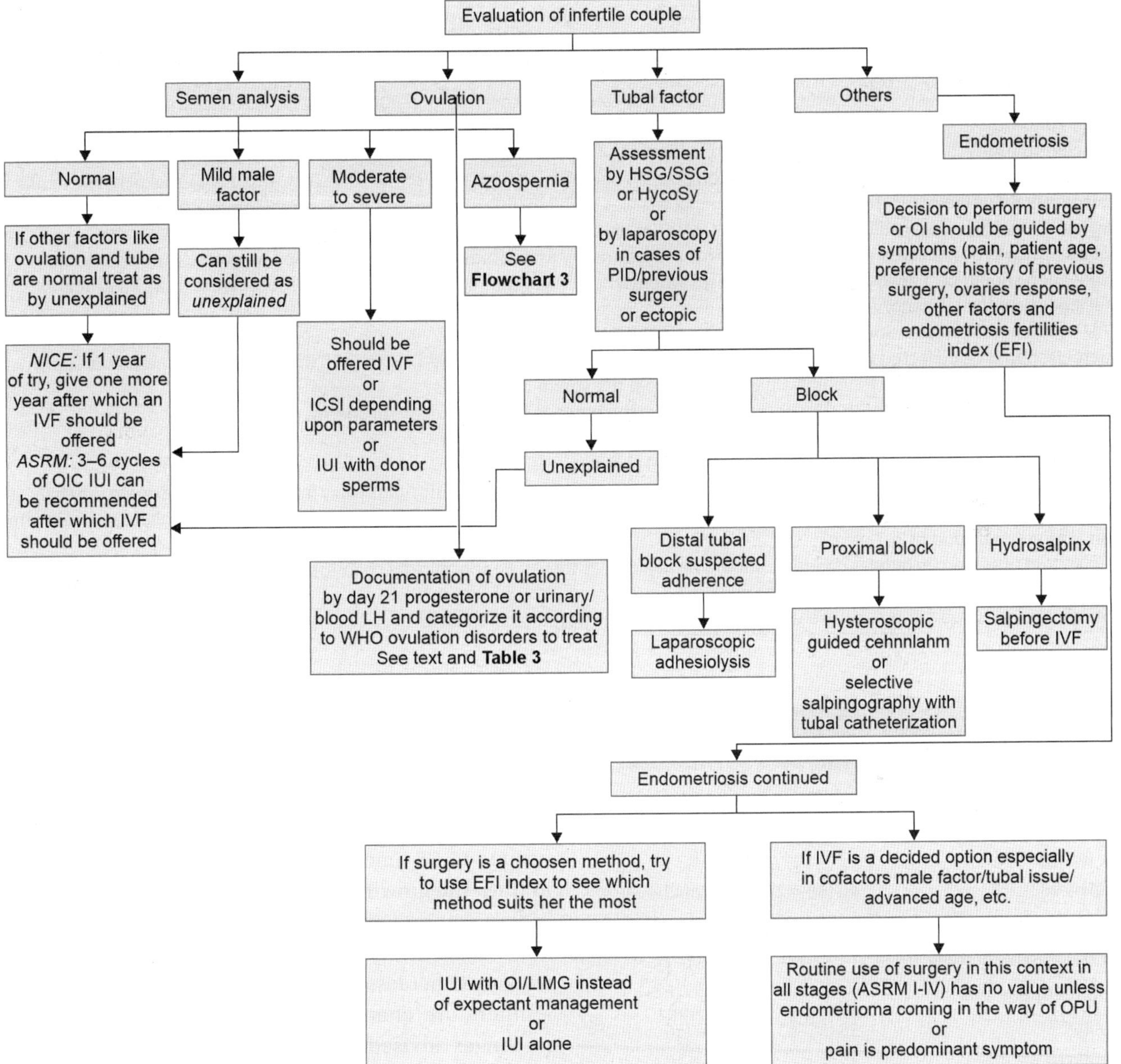

(ASRM: American Society for Reproductive Medicine; HSG: hysterosalpingography; HyCoSy: hystero-contrast-salpingography; ICSI: intracytoplasmic sperm injection; IUI: intrauterine insemination; IVF: in vitro fertilization; LH: luteinizing hormone; HMG: human menopausal gonadotropins; NICE: National Institute for Health and Clinical Excellence; OI: ovulation induction; OS with IUI: ovarian stimulation with intrauterine insemination; OPU: oocyte pick up; SSG: sonosalpingography WHO: World Health Organization)

(>35 years), women with a history of ovarian/pelvic surgery, radio-/chemotherapy, or clinical symptoms suggesting of some underlying fertility pathology like dysmenorrhea or oligo-/hypomenorrhea[7] **(Flowchart 1)**.

Subfertility affects around 12–15% of the heterosexual couples with a global estimation of 180 million individuals suffering from it.[8] The main causes of subfertility are described in **Table 1**.

TABLE 1: Causes of subfertility.

Cause	%
Ovulation disorders	25%
Fallopian tube disorders	20%
Male subfertility	30%
Uterine or peritoneal disorders	10%
Unexplained subfertility	25%
In 40% of the cases a pathology exists in both male and female	

BOX 1: Summary of semen collection and analysis.

Semen of a person varies with time. If a semen analysis is found to be normal, a repeat analysis is not needed unless some new development is suspected. In an abnormal sample, however, a repeat analysis should be done after a gap of 2–4 weeks to conclude the final result (a normal spermatogenic cycle is 3 months approximately)

A normal semen volume is 1.5 mL which is a mixture of prostate (major contributor—90%) and seminal vesicles and accessory ducts (minor contributor—10%). A low semen volume is seen in obstruction of the ejaculatory duct, congenital bilateral absence of the vas deferens (CBAVD), collection problems (loss of a fraction of the ejaculate), partial retrograde ejaculation or androgen deficiency. A high semen volume may reflect active exudation in cases of active inflammation of the accessory organs. Normal pH of semen is 7.2 which is a mixture of prostate (acidic) and seminal vesicles (alkaline) fluids. It should be measured before 1 hour as CO_2 may get washed. A low pH and low volume may indicate CBAVD or obstruction

- Semen should be collected in a private room near the laboratory as it avoids fluctuations in temperature and time and home collection should be allowed only in special cases where a person cannot masturbate in the clinic or there is lack of adequate facilities near the laboratory
- Always give clear written and spoken instructions concerning the collection
- Semen should be collected via masturbation. Coitus withdrawal is not a preferred method as there is fear of losing initial sperm rich fractions and exposure to vaginal fluids may hamper the sperm parameters
- The glass or plastic wide mouth containers should have the batch tested for toxicity for sperms
- Temperature should be maintained while carrying and throughout assessment
- Minimum of 2 days and a maximum of 7 days of sexual abstinence is desired
- Ask for the completeness of the sample as initial semen with sperm-rich prostatic fluids carry major share of spermatozoa than later fractions (dominated by seminal vesicular fluid). If the sample is incomplete, a second sample should be collected, again after an abstinence period of 2–7 days
- Collection of semen by regular condoms should be avoided as they have spermicidal chemicals. Only special nontoxic condoms designed for semen collection should be used
- Sterile methods and safeguards should be used as semen is a potential biohazard

TABLE 2: Semen parameters as per WHO.

	WHO, 2010	*WHO, 2021*
Semen volume (mL)	1.5 (1.4–1.7)	1.4 (1.3–1.5)
Total sperm number (10^6 per ejaculate)	39 (33–46)	39 (35–40)
Total motility (%)	40 (38–42)	42 (40–43)
Progressive motility (%)	32 (31–34)	30 (29–31)
Nonprogressive motility (%)	1	1 (1–1)
Immotile sperm (%)	22	20 (19–20)
Vitality (%)	58 (55–63)	54 (50–56)
Normal forms (%)	4 (3–4)	4 (3.9–4)

SEMEN ANALYSIS

Recently WHO has updated its much awaited "Laboratory Manual for the Examination and Processing of Human Semen".[9] Although debated by many experts, it has brought a few changes in various semen parameters and omitted few tests compared to its previous version.[10] Interested readers can freely download and read this manual in detail from the official WHO website.[9] The summary points for semen analysis are tabulated in **Box 1** and comparison of semen parameters in WHO 2010 and 2021 Manual is provided in **Table 2**.

DOCUMENTATION OF THE OVULATION

Although a regular menstrual cycle usually indicates that ovulation is happening correctly, a urinary luteinizing hormone (LH) kit or day 21 progesterone (in 28-day cycle) (value >16–28 nmol/L) is recommended.[11] An ultrasound to track the growth and rupture can be more handful in Indian setting where this modality is readily available and can provide additional information about the fertility workup. Other less reliable modalities such as basal body temperature changes or cervical mucus changes are no longer advised.[12] Based on the pattern of ovulation, WHO has classified these women into three categories[13] **(Table 3)**.

EVALUATION OF THE FALLOPIAN TUBES

Various modalities used for diagnosing the patency of fallopian tubes are: Hysterosalpingography (HSG), hystero-contrast-salpingography (HyCoSy), sonosalpingography (SSG), or the laparoscopy with chromopertubation which is considered as a gold standard but is more invasive and expansive.[14] It is advised that HSG (or HyCoSy/SSG, if facility is available) should be recommended as an initial tool for fallopian tubes assessment whereas laparoscopy should be the chosen modalities in cases

TABLE 3: WHO classification of ovulation disorders.

Group 1	Hypothalamic pituitary failure (Hypothalamic amenorrhea or hypogonadotropic hypogonadism) • *Prevalence:* 10% • *Diagnostic points:* Low gonadotropins (FSH and LH) and low estrogen • *Treatment:* Medical management with gonadotropins with LH or pulsatile GnRH and/or weight gain
Group 2	Hypothalamic pituitary dysfunction • *Prevalence:* 85% • Predominantly involves PCOS • *Treatment:* – *First line:* Lifestyle modification and/or ovulation induction with oral ovulogens and/or insulin sensitizers – *Second line:* Gonadotropins with IUI or laparoscopic drilling – *Third line:* IVF
Group 3	Ovarian failure • *Prevalence:* 5% • *Diagnostic points:* High gonadotropins (FSH and LH) and low estrogen • *Treatment:* IVF donor egg program

(FSH: follicle-stimulating hormone; GnRH: gonadotropin-releasing hormone; IUI: intrauterine insemination; IVF: in vitro fertilization; LH: luteinizing hormone; PCOS: polycystic ovarian syndrome)

where comorbidities like pelvic inflammatory disease, previous pelvic surgery, previous ectopic pregnancy, or endometriosis are suspected.[15]

INTRAUTERINE INSEMINATION

Intrauterine insemination is an ART procedure where sperms are washed, prepared, and inseminated into the uterine cavity through a IUI catheter around the time of release of oocyte (follicle rupture). IUI is a simple procedure that can be done in a stimulated or unstimulated cycle. Sperms are inseminated into uterus using various sperm preparation techniques (see below). The sperm preparation separates the debris, prostaglandins, infectious agents, antigenic proteins, dead spermatozoa, and leukocytes from the main fraction thereby reducing the chance of infections, painful uterine cramps, and incidence of collapse after IUI procedure and improves the concentration of motile spermatozoa.[16] Dead and immotile sperms generate 10–20% more reactive oxygen species (ROS) which decrease the fertilizing capacity of spermatozoa.[17] The clinical pregnancy rates in IUI range from 10 to 20% per cycle.[18] A patent fallopian tube(s) and a moderate to good semen parameters are a prerequisite for an IUI treatment **(Box 2)**. Following are the proposed indications of IUI:[19,20]

BOX 2: Learning points for intrauterine insemination (IUI).

- A clinical pregnancy rate per cycle of IUI performed ranges from 10 to 20%. Almost 50% of the females less than age 40 years would conceive with 6 cycles of IUI and another 25% would conceive with further 6 cycles of IUI (cumulative pregnancy rates of 75%). Important prognostic markers for IUI success are: Advanced age of the female (>35 years), duration of infertility, protocol used for ovulation induction (better chances are expected with gonadotropins than oral agents but at the cost of multiple pregnancy), number of cycles performed, timing of insemination, and male semen parameters.[23]
- Delaying IUI after rupture of follicle has detrimental effect on its success rate hence timing of IUI is important with fresh washed sperms should be inseminated 4–6 hours after ovulation and thawed donor sperms should be inseminated as close to ovulation as possible. A delay of more than 12 hours decreases the success rates and increases chances of chromosomal abnormalities.[24]
- IUI using fresh sperms is associated with higher chances of conception. As per Indian ART law, donor insemination is to be done using frozen quarantined semen only (ART Laws have been revised, subject to an amendment—Please visit official website of ministry of health, government of India.[25]
- In those scenarios where women are ovulating normally, unstimulated natural cycle IUI (without using ovulation drugs) should be offered as it provides equal chances of conception without raising risk of multiple pregnancy. Examples of such scenarios are: donor insemination, persons where IUI is being done for physical (hypospadias, etc.) or psychosexual disorders (erectile dysfunction) or HIV positive husband.[19]
- NICE guidelines state that women with unexplained subfertility should not be offered IUI with or without ovulation drugs, instead another 1 year of actively natural try can be more rewarding and is highly recommended. Those couples who do not become pregnant even after two years of active try. A direct IVF should be offered. This can be an issue in Indian scenario where IVF is self-sponsored event and still marriages are happening at early age, hence advising Indian couples for an IVF after 2 years of marriage in unexplained infertility can be tricky. IUI for 3–6 cycles seems a reasonable option. This also has been recommended by ASRM.[26]
- Women with mild endometriosis and couples with mild male subfertility, IUI can be offered as a treatment option especially in Indian scenario although NICE advises not to.
- A minimum recommended threshold of semen parameters to achieve a desired conception rate is unknown but accepted minimum values below which the success rates of an IUI are reduced are: Prewash total motile count $>10 \times 10^6$, post wash total motile sperm count $>1 \times 10^6$. TMSC is calculated by multiplying the concentration of sperm/mL × volume (mL) × motility (A + B) divided by 100[27] **(Flowchart 2)**.

(ART: assisted reproductive technology; ASRM: American Society for Reproductive Medicine; IVF: in vitro fertilization; HIV: human immunodeficiency virus; NICE: National Institute for Health and Clinical Excellence; TMSC: total motile sperm count)

Flowchart 2: Deciding for an IUI.

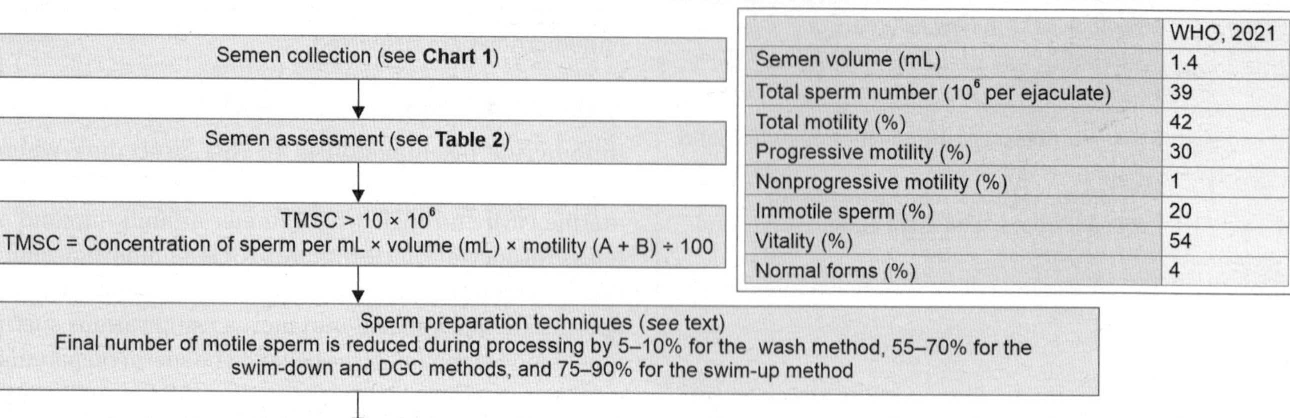

	WHO, 2021
Semen volume (mL)	1.4
Total sperm number (10^6 per ejaculate)	39
Total motility (%)	42
Progressive motility (%)	30
Nonprogressive motility (%)	1
Immotile sperm (%)	20
Vitality (%)	54
Normal forms (%)	4

(DGC: density gradient centrifugation; IUI: intrauterine insemination; TMSC: total motile sperm count; WHO: World Health Organization)

- Ejaculatory dysfunction or disability (physical or psychological) preventing vaginal sexual intercourse
- Mild male factor subfertility
- Unexplained subfertility
- Mild endometriosis
- Polycystic ovarian syndrome (PCOS) as second-line therapy
- Immunological subfertility
- Donor insemination in single mother/lesbian couples/ severe male factor or azoospermia or genetic issues where ICSI with partner sperms cannot be done
- HIV discordant couples.

Sperm Preparation Techniques for IUI

Many techniques are available for sperm preparation to be used in IUI or IVF/ICSI.[21] The most common employed techniques are:

- *Swim-up technique:* It involves the self-migration of sperms from the bottom fraction along with the centrifugation to remove debris and dead spermatozoa.
- *Density gradient technique:* It is based on the principle that motile spermatozoa have a higher density than dear and immotile spermatozoa. Density gradient centrifugation (DGC) involves passage of spermatozoa through different concentration of fractions of suspended silica particles (like 80% and 40%) with centrifugation.
- *Standard sperm wash:* It involves centrifugation only to remove seminal plasma.
- *Swim-down techniques*: It involves the reverse (top to bottom) migration of motile spermatozoa between differently concentrated human serum albumin. (Interested readers can read the sperm preparation techniques from WHO Manual or Cambridge Manual of IUI[22]).

ADVANCED ASSISTED REPRODUCTIVE TECHNIQUES BEYOND IUI

- In vitro fertilization
- Intracytoplasmic sperm injection
- Ovulation induction
- Oocyte retrieval/pick up (OCR or OPU)
- Embryo transfer
- Cryopreservation
- Advanced sperm selection techniques
- Assisted hatching
- Third party reproduction (donor egg/sperm program and surrogacy)
- Surgical sperm retrieval techniques

In Vitro Fertilization (IVF), Intracytoplasmic Sperm Injection (ICSI), and Ovulation Induction (OI)

Since the birth of the first IVF baby Louis brown in 1978, IVF has grown in leaps and bounds and has been a boon to millions of the infertile couples. As the knowledge and understanding of IVF grew, so were its success rates.[28] The basic procedural difference between IVF and ICSI is the final process of embryo formation. IVF is a conventional process where oocytes after retrieval are inseminated in Petri dish with spermatozoa in a culture media and kept in incubators overnight for the natural fertilization to happen. The fertilization is checked next day and embryos grown further to a cleavage stage (termed as day 2/3) or blastocyst stage (day 5). Good embryos are transferred back into the uterine cavity where implantation occurs. ICSI was invented in 1992 to overcome the issues related with IVF mainly being the couples with severe male factor problem or cases with fertilization failure with IVF or where sperms are retrieved surgically.[29] ICSI involves the process of retrieving oocytes, denuding them and then injecting a "selected" spermatozoa into the oocyte using a micromanipulator. Since ICSI involves a 1:1 interaction of sperm and oocyte and therefore in cases where either number of spermatozoa or oocyte or both are less, ICSI is generally recommended. In modern ART, however, ICSI has superseded IVF, whereby most centers nowadays employ ICSI universally for all indications **(Box 3)**.

Ovulation Induction

Ovulation induction in IVF/ICSI is termed COH or COS (controlled ovarian stimulation). The aim of any stimulation in IVF/ICSI is to create a balance to generate "optimal" number of oocytes which are "sufficient enough" to bear the procedural, technical, and physiological loss of IVF process to generate "optimal" embryos as well as "less enough" to cause no harm to the women [like ovarian hyperstimulation syndrome (OHSS)].[31] Although no magical number exists and it is difficult to control stimulation to achieve a desired number of so called "optimal" oocytes, it is assumed that generation of 6–15 oocytes is sufficient to achieve a desirable conception rate in fresh transfer cycle.[32] Above 15 oocytes, generally the embryos are cryopreserved and put later in the next cycles termed as frozen embryo transfer (read under the section "Cryopreservation").

BOX 3: Indications of IVF.[30]

- *Tubal pathology:* Traditionally, women with tubal issues are being treated by IVF and more so in modern era when IVF success rates have improved a lot thereby gaining advantage over tubal surgeries. However, women with proximal tubal obstruction offering hysteroscopic cannulation or selective salpingography with tubal catheterization provide a good chance of conception. When tubal surgery is not a realistic option, IVF is method of choice. Similarity patients with hydrosalpinx(s), IVF after salpingectomy is a better option rather tubal repair as it provides better chances of conception.
- *Unexplained infertility (idiopathic):* IVF is indicated if the duration is 2 years or longer.
- *Ovulatory disorders:* In WHO class 2, PCOS, IVF is advised as third line when other measures like ovulation induction or IUI have failed. In WHO class 3, again IVF with donor eggs program generally is a chosen line of treatment
- *Male infertility:* When the total motile sperm count (TMSC) <1 million, ICSI is the treatment of choice. IVF can be offered when TMSC >1 and <10 million
- *Endometriosis:* In case of severe endometriosis IVF is advised
- *Advanced age of the female:* In general, ovarian reserve starts diminishing after 35–37 years of the women's age hence speeding up the conception is generally the accepted norm among fertility specialists.

Indications of ICSI:[30]

- *Severe male factor and azoospermia:* When the total motile sperm count (TMSC) <1 million or <4% normal morphology and TMSC <5 million ICSI is advisable. Also, in cases of azoospermia where sperms are retrieved surgically via PESA/TESA/TESE or micro-TESE
- No or poor fertilization in the previous IVF cycle
- Cryopreserved oocytes

(ICSI: intracytoplasmic sperm injection; IVF: in vitro fertilization; PCOS: polycystic ovarian syndrome; PESA: percutaneous epididymal sperm aspiration; TESA: testicular sperm aspiration; TESE: testicular sperm extraction)

The COH involves:[33]

- Administering gonadotropins to stimulate the ovaries to generate more follicles
- Monitoring via ultrasound to track the growth of follicles
- Correlating with the hormones [LH, estradiol E2 and progesterone (P4) mainly]
- Blocking the premature LH surge by agonist or GnRH antagonist
- Providing the "Trigger" injection for a final maturation and retrieving oocytes after 34–36 hours

Two major protocols which are used in IVF are: Long protocol and antagonist protocol **(Table 4)**. Other less commonly employed protocols in IVF are: GnRH

TABLE 4: Comparison of long agonist protocol versus antagonist protocol.

Agonist (long protocol)	*Antagonist (fixed/flexible protocol)*
In this protocol, agonist starts from the day 21 of previous cycle (e.g., leuprolide 0.2–0.5 mg/day), continues during menstruation. *Patient is called on day 2–3 and* hormonal assessment *along with TVS is done and* gonadotropins *(150–225–300 IU) are started. The follicular growth is tracked and when DF reaches around 17 mm, an hCG trigger is given for a final maturation*	In this protocol, patient is directly called on the day 2/3 of her periods and after assessment, gonadotropins (150–225–300 IU) are started. The follicular growth is tracked and when DF reaches around 14 mm, GnRH antagonist is started (fixed protocol) and both gonadotropins and antagonist are continued till DF reaches 17 mm size when hCG/agonist trigger is given for a final maturation
Disadvantages of long protocol are: (1) More *days of injections hence less patient friendly,* (2) dose of gonadotropins *needed is more because of suppression,* (3) only *hCG trigger can be given which puts the patient at risk of OHSS*	Major advantage of antagonist protocol is that it allows the physician to administer GnRH agonist trigger for final maturation thereby avoiding OHSS cases. Other benefits of this protocol are a smaller number of injections, more patient friendly, and dose of gonadotropins used is less compared to long agonist protocol
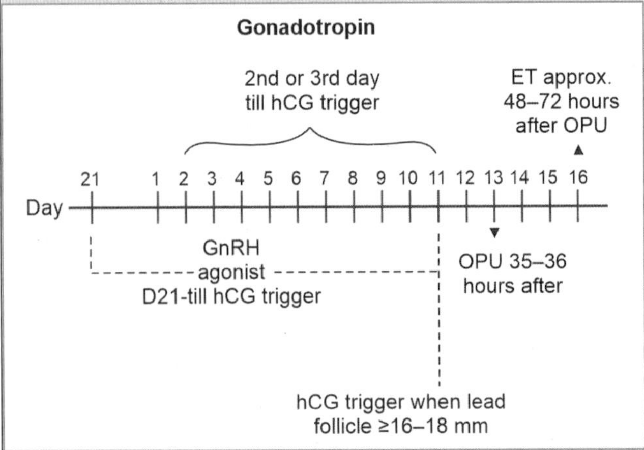	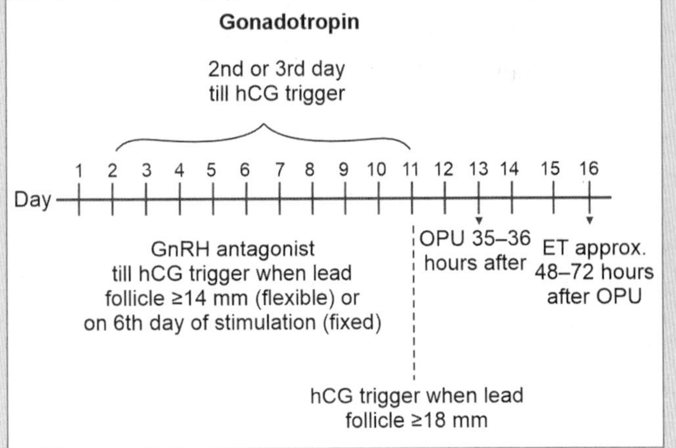

(DF: dominant follicle; ET: embryo transfer; GnRH: gonadotropin-releasing hormone; hCG: human chorionic gonadotropin; OHSS: ovarian hyperstimulation syndrome; OPU: oocyte pick up; TVS: transvaginal ultrasound)

flare/micro flare protocol, mixed agonist/antagonist protocol/natural cycle/modified natural cycle/minimum stimulation protocols.[34]

Oocyte Retrieval/Pick Up

Oocyte pick up is a surgical invasive procedure whereby oocytes are collected from the stimulated ovary using a transvaginal ultrasound. It involves a setup involving a suction unit (especially crafted for OPU procedures), a pickup needle which attaches to the suction unit, a transvaginal probe guide through which the needle is inserted under ultrasound guidance to aspirate the follicular fluid. The procedure is carried out under conscious sedation or general anesthesia.[35] The collected oocytes cumulus complex is assessed in the IVF laboratory and inseminated with prepared semen overnight. In cases of ICSI, the cumulus complex is denuded to remove the cumulus cells. The maturity of the oocyte is assessed and mature oocytes (metaphase II oocyte) are injected with a spermatozoa using a micromanipulator.

The OPU procedure is generally considered a safe procedure. Most common complications related to OPU procedure are: Pain persisting for few days, hemorrhage, trauma to pelvic organs, pelvic infection, or TO abscess. Anesthetic and other complications related to OHSS have also been encountered. Few case reports of injury to ureter, intestine, bladder hematoma, appendicitis, blood vessels, and even occasional deaths have been reported in literature.[36-38]

Embryo Transfer

Embryo transfer is the final and one of the most important steps of IVF process. It involves a team effort of the clinician, embryologist with the staff/colleague guiding and helping through ultrasound. The patient is positioned in lithotomy and the perineal area is cleaned with normal saline. The Cusco's speculum is inserted into vagina to retract the vaginal walls and cervix is appreciated. The outer part of the embryo transfer catheter is negotiated through cervical canal to position at just above the inner os. The embryologist loads the inner part of the embryo transfer catheter with the embryos and slides into the outer catheter to deposit the embryos at place with maximum implantation potential. In difficult transfers, there can be a problem in negotiating the cervix. An experienced specialist can overcome it by minor manipulation of the catheter along with the adjustment of the speculum. The cervix sometimes needs to be handled by Vulsellum or Allis forceps. A traumatic embryo transfer procedure can be detrimental to the ultimate success of the IVF procedure. Embryo transfer guidelines have been established by various IVF bodies around the world.[39]

Cryopreservation[40]

Cryopreservation is a technique whereby either embryos or gametes (oocytes and spermatozoa) are preserved by cooling them to a very low temperature. With the advancement in cryopreservation techniques, the slow time consuming conventional freezing techniques have been replaced by newer "vitrification" where embryos or gametes undergo rapid cooling which do not let ice crystals to develop.[41] The success and safety of cryopreserved embryos can be made out by the fact that it has generated a considerable debate among IVF clinicians whether universal "all freeze" strategy can be adopted or not.[42] Clearly, the data worldwide states a considerable increase in frozen embryo transfer procedures in last decade.[43] The common indications of cryopreservation are listed in **Table 5**.

TABLE 5: Indications of cryopreservation in IVF.[44-47]

Embryos	• Avoidance of OHSS • Elevated progesterone on the day of starting luteal phase • Fluid in the endometrial cavity just before transfer • Preimplantation Genetic screening • Single embryo transfer with "spare" embryos • Personal reasons
Spermatozoa	• Husband not available on the day of OPU • As a routine protocol during IVF cycle as it puts lot os stress on male on the final day of oocyte pick up and he might to be able to provide the semen sample • Diagnostic surgical sperm retrieval procedures • Patients undergoing treatments which may harm fertility like cancer chemo-/radiotherapy • Sudden death where sperms can be retrieved by SSR
Oocytes	• As a social freezing to delay childbearing • Patients undergoing treatments which may harm fertility like cancer chemo-/radiotherapy • In cases where SSR was done on the day of OPU and sperms were not retrieved

(IVF: in vitro fertilization; OHSS: ovarian hyperstimulation syndrome; OPU: oocyte pick up; SSR: surgical sperm retrieval)

Advanced Sperm Selection Techniques[48]

The sole purpose of these techniques is to select a better sperm out of the million that are ejaculated. The sperms selected by any of these techniques is presumed to be a better out of the whole lot, which is mature, structurally intact, and carries an integral DNA so that a better fertilization occurs. These techniques are mainly employed during ICSI and involve selecting sperms based on ultra-high magnification (IMSI-intracytoplasmic morphologically selected sperms), capacity to bind with hyaluronic acid (PICSI-physiological ICSI), separation based on surface antigens (AV-MACS-Annexin V-Magnetic Activated Cell Sorting), sperm selection based on sperm motility (Microfluidics).[49,50] Overall, a limited evidence exists in literature supporting these techniques when it comes to improving a success rate of an IVF cycle. More studies are needed especially randomized control trials (RCTs) before recommending these into a clinical practice.[50]

Preimplantation Genetic Testing Screening/Diagnosis

Embryo selection is an important part of the IVF process. It entirely depends upon the morphology of the embryo which is assessed on day 2/3 or day 5 of its development along with the time lapse of stages of its development. Based on morphological criteria, the best embryo is selected for the transfer.[51,52] The problem with the morphological assessment of the embryo is that it might not reveal the chromosomal status of the embryo which becomes an important reason for the nonimplantation of embryos or early miscarriages or genetic defects in the fetus. It is estimated that regardless of maternal age, only 44% of the best embryos, selected according to morphology, were found chromosomally normal which further decreased with increasing age.[53,54] Preimplantation genetic testing is a spectrum of assays where biopsy of the developing embryos is carried out to check the chromosomal status and only embryos with normal chromosomal status are transferred thereby increasing the success rate. Preimplantation genetic testing-monogenic is targeted to single gene disorders and is generally advised in couples with preexisting known genetic issues. Preimplantation genetic testing-aneuploidy assay screens all the 23 pairs of chromosomes and is advised in various scenarios like advanced maternal age, recurrent implantation failure, and recurrent miscarriages. The limitations of pre-invasive genetic testings (PGTs) especially in preinvasive genetic testing for aneuploidy (PGT-A) are occurrence of false positives and false negatives, mosaicism, failure to recognize microdeletions and microduplications along with de novo variants. PGT-A also has failed to gain acceptance as true evidence is lacking and the areas where its application can be more productive is lacking.[55]

Assisted Hatching

Assisted hatching is a technique whereby either artificial thinning or a frank break is given to of zona pellucida of the embryo.[56] It is proposed that it leads to a better hatching of the blastocyst thereby improving implantation rates. The assisted hatching of the embryo can be done with mechanical methods, chemical methods, laser application, or piezo technology. Whereas mechanical and chemical methods are technically difficult and require a lot of expertise, laser methods are easy to apply but need an expensive laser equipment. Various proposed indications of assisted hatching are:[57,58]

- *Advanced age of the female:* With age advancement, zona hardening is expected due to hormonal changes or lack of lysins enzymes
- *Frozen embryos:* Cryopreservation alters the glycoprotein matrix which hardens the zona
- *Recurrent implantation failure:* Nonhatching of the embryos is one of the proposed reasons for the multiple failure especially when other aspects of IVF procedure were up to the mark
- Thick zona pellucida of >15 μm

The problems associated with the assisted hatching are—one there is still no consensus whether their application in the above stated subset of population increases the success rate os an IVF/ICSI procedure. Second creating a breach into the embryos protective coating prematurely predisposes the embryo to detrimental uterine factors like toxins and microbiomes. Third, the possibility of monozygotic twinning or even death of the embryo may occur due to premature release of the inner cell mass.[56-59]

Third Party Reproduction (Donor Egg/Sperm Program and Surrogacy)

Third party reproduction refers to involvement of another person (different from the intending parents) to reproduce by either using his/her donated sperms or donated eggs or donated embryos or womb to carry pregnancy, later referred as surrogacy. Third party reproduction involves a lot of social, ethical, and psychological issues and is legally complex.[60,61] Recently, India too has updated its laws to streamline the IVF process with a particular focus on donor program and surrogacy.

Proposed indications for egg donation are:[62]

- Postmenopausal women
- Women with severely diminished ovarian reserve
- Premature ovarian failure
- Multiple IVF failures
- Advanced age of the female
- Surgically removed ovaries
- Inherited genetic conditions
- Poor responders
- Women with treatments like chemotherapy or radiotherapy which has an impact on ovarian reserve.

Proposed indications for sperm donation are:[62]

- *Male with severely deranged semen parameters or azoospermia:* Although most of these cases can be treated by ICSI, in those scenarios, where ICSI is not a chosen option or where surgical sperm retrieval method fails to extract sperms, donor sperm program is recommended.
- Inherited genetic conditions
- Single woman or lesbian couples

Proposed indications for gestational carrier (surrogacy):[62]

- Women with congenital absence of uterus or Mullerian agenesis
- Women with history of surgical removal of uterus [e.g., postpartum hemorrhage (PPH) in previous pregnancy]

- Women with recurrent miscarriages or recurrent implantation failure
- Women with severe endometrial issues like Asherman syndrome
- Women with adenomyosis or multiple fibroids when all other treatments have failed
- Women with medical conditions that can put her life in danger during pregnancy or where pregnancy is contraindicated like severe heart or renal issues, systemic lupus erythematosus (SLE), breast cancer, etc.

Azoospermia and Surgical Sperm Retrieval Techniques

Azoospermia is defined as the complete absence of spermatozoa in the semen. It is mandatory to examine centrifuge pallet of two separate semen samples before concluding the report.[63] Azoospermia occurs in 1% of the male population and is seen in 10–15% of the infertile men. It is categorized into three parts: (1) Pretesticular, (2) testicular, or (3) post-testicular azoospermia **(Box 4)**. However, clinically it is more practical to categorize azoospermia into obstructive and nonobstructive based on which the future management can be decided. In general, a detailed history, with basic hormones (follicle-stimulating hormone and total testosterone) along with clinical examination is generally sufficient to categorize azoospermia (*see* **Flowchart 3**). Patient with nonobstructive azoospermia (hypergonadotropic hypogonadism) should be offered karyotype and Y-chromosome polymerase chain reaction (PCR) to identify chromosomal problems which are found in 10–15% of azoospermic out of which 60–70% are found to carry a sex chromosomal abnormality (Klinefelter syndrome).[64-68] Men with hypogonadotropic hypogonadism should be enquired about external usage of steroids or androgens and be referred for a detailed MRI of brain to exclude pituitary lesions. In obstructive azoospermia, the involved pathology (obstruction) can be traced out by transrectal ultrasound along with a clinical examination and assessing semen parameters like low volume and absence of fructose. Obstructive azoospermia with suspected ejaculatory duct obstruction (EDO) can be

BOX 4: Categorization of azoospermia.

- *Pretesticular:*
  - Hypogonadotropic hypogonadism
  - Exogenous androgens or steroids
- *Testicular:* Primary testicular failure, genetic issues like Klinefelter syndrome, varicocele
- *Post-testicular:* Epididymal obstruction, ejaculatory duct obstruction, ejaculatory dysfunction, vasectomy

Flowchart 3: Diagnosis and management of azoospermia.

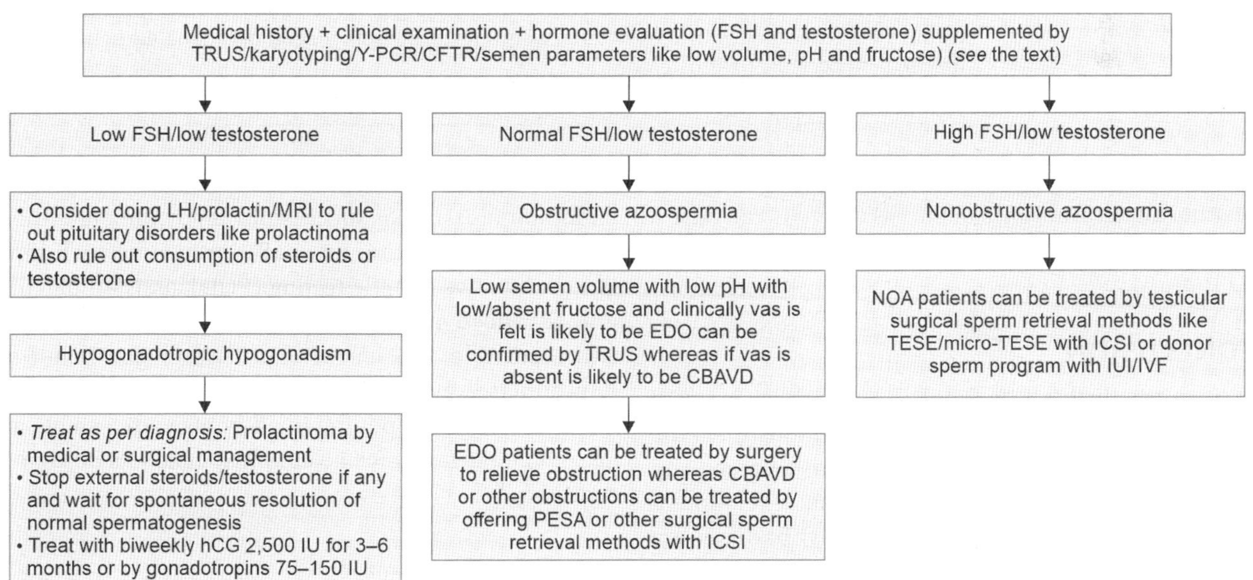

(CFTR: cystic fibrosis transmembrane conductance regulator; CBAVD: congenital bilateral absence of the vas deferens; EDO: ejaculatory duct obstruction; FSH: follicle-stimulating hormone; ICSI: intracytoplasmic sperm injection; LH: luteinizing hormone; NOA: nonobstructive azoospermia; PCR: polymerase chain reaction; PESA: percutaneous epididymal sperm aspiration; TESE: testicular sperm extraction; TRUS: trans rectal ultrasound)

TABLE 6: Surgical sperm retrieval methods.[69-72]

PESA (Percutaneous epididymal sperm aspiration)	It is a percutaneous method done mainly for obstructive azoospermia where a needle is inserted into epididymis to extract sperms. It is a simple, fast, and low-cost procedure that can be done with or without local anesthesia. It carries a minimal morbidity and can be repeated *Disadvantages:* Except for cases of CBAVD, where plenty of sperms can be retrieved for usage and cryopreservation. In other obstructive cases, only a few sperm are generally retrieved and hence cryopreservation is limited. It can lead to fibrosis and obstruction at aspiration site with a risk of hematoma/spermatocele
MESA (Microsurgical epididymal sperm aspiration)	It is an open technique applied in cases of obstructive azoospermia where epididymis is exposed to retrieve sperms. A large number of sperms can be retrieved and hence can be cryopreserved for a future use. It carries a lesser chance of hematoma formation *Disadvantages:* Surgical exploration is required which increases the cost and time and also needs an expert in microsurgery
TESA (Testicular sperm aspiration)	This is a percutaneous technique used in cases of nonobstructive azoospermia where a needle is inserted into testicular tissue to extract the pieces of seminiferous tubules which harbor sperms. Like PESA, it is fast and low-cost procedure where no microsurgical expertise is required and carries a mild postoperative discomfort *Disadvantages:* Relatively a blind procedure with a low success rate in NOA. only a few sperms are retrieved and hence scope of cryopreservation is limited. It carries a risk of hematoma and testicular atrophy
TESE [Testicular sperm extraction (single or multiple biopsies)]	Basically, it is a cut-open biopsy of testis to extract the seminiferous tubules which harbor sperms. It is a fast and repeatable procedure but has a low success rates in NOA
Micro-TESE (Microsurgical testicular sperm extraction)	This is a microsurgery which involves extensive exploration of testicular tissue where the testis is completely opened up, bivalved and under operative microscope the healthy seminiferous tubules are picked up to find sperms. It has a higher success rates compared to TESA/TESE. A larger number of sperm can be retrieved *Disadvantages:* The exploration might take a long time depending upon the pathology and needs a huge expertise and patience of the surgeon and attending embryologist. It is a costly procedure and carries a risk associated with anesthesia as well as testicular surgery complications like hematoma, fibrosis, and decreased testosterone

(CBAVD: congenital bilateral absence of the vas deferens; NOA: nonobstructive azoospermia)

offered a corrective surgery whereas other pathology like congenital bilateral absence of the vas deferens can have a good success with surgical sperm retrieval procedure and ICSI **(Table 6)**.[67,68]

CONCLUSION

Ever since the birth of the first IVF baby, the science of reproductive medicine has grown tremendously. As the understanding developed, newer treatments emerged and contributed to improved success rates. The development of ICSI helped couples with severely bad semen parameters to father their own child which in earlier days was difficult to achieve with conventional IVF. The surgical sperm retrieval techniques further aided males with azoospermia to use their own sperms for a pregnancy. Third party reproduction techniques allowed males and females with genetic issues, ovarian or testicular failure or endometrial issues to successfully enjoy the parenthood. With the introduction of some innovative techniques like preinvasive genetic testing (PGD), assisted hatching, and advanced sperm selection techniques, the patients with multiple unsuccessful attempts got benefited. Reproductive medicine is an ever-evolving science with lots of research happening in all fronts with a single aim, i.e., "to help every infertile couple to reach its destination successfully".

REFERENCES

1. Jain M, Singh M. (2022). Assisted Reproductive Technology (ART) Techniques. [online] Available from https://www.ncbi.nlm.nih.gov/books/NBK576409/. [Last accessed March, 2024].

2. Calhaz-Jorge C, De Geyter CH, Kupka MS, Wyns C, Mocanu E, Motrenko T, et al. Survey on ART and IUI: legislation, regulation, funding and registries in European countries: The European IVF-monitoring Consortium (EIM) for the European Society of Human Reproduction and Embryology (ESHRE). Hum Reprod Open. 2020;2020(1): hoz044.
3. Practice Committee of the American Society for Reproductive Medicine. Definitions of infertility and recurrent pregnancy loss: a committee opinion. Fertil Steril. 2013;99(1):63.
4. te Velde ER, Eijkemans R, Habbema HDF. Variation in couple fecundity and time to pregnancy, an essential concept in human reproduction. Lancet. 2000;355:1928-9.
5. Kamel RM. Management of the infertile couple: an evidence-based protocol. Reprod Biol Endocrinol. 2010;8:21.
6. Goverde AJ, McDonnell J, Vermeiden JP, Schats R, Rutten FF, Schoemaker J. Intrauterine insemination or in-vitro fertilisation in idiopathic subfertility and male subfertility: a randomised trial and cost-effectiveness analysis. Lancet. 2000;355:13-18.
7. Practice Committee of the American Society for Reproductive Medicine. Electronic address: asrm@asrm.org; Practice Committee of the American Society for Reproductive Medicine. Fertility evaluation of infertile women: a committee opinion. Fertil Steril. 2021;116(5):1255-65.
8. Inhorn MC, Patrizio P. Infertility around the globe: new thinking on gender, reproductive technologies and global movements in the 21st century. Hum Reprod Update. 2015;21(4):411-26.
9. Wang C, Mbizvo M, Festin MP, Björndahl L, Toskin I; other Editorial Board Members of the WHO Laboratory Manual for the Examination and Processing of Human Semen. Evolution of the WHO "Semen" processing manual from the first (1980) to the sixth edition (2021). Fertil Steril. 2022;117(2):237-45.
10. Boitrelle F, Shah R, Saleh R, Henkel R, Kandil H, Chung E, et al. The Sixth Edition of the WHO Manual for Human Semen Analysis: A Critical Review and SWOT Analysis. Life (Basel). 2021;11(12):1368.
11. The ESHRE Capri Workshop. Guidelines to the prevalence, diagnosis, treatment and management of infertility, 1996. Hum Reprod. 1996;11:1775-807.
12. Bauman JE. Basal body temperature: unreliable method of ovulation detection. Fertil Steril. 1981;36:729-33.
13. Rowe PJ, Comhaire FH, Hargreave TB, Mellows HJ. WHO Manual for the Standardized Investigation and Diagnosis of the Infertile Couple. Cambridge: Cambridge University Press; 1997.
14. Patil M. Assessing tubal damage. J Hum Reprod Sci. 2009;2(1):2-11.
15. Collins JA. Diagnostic Assessment of the Infertile Female Partner. Curr Probl Obstet Gynecol Fertil. 1988;11:6-42.
16. Boomsma CM, Heineman MJ, Cohlen BJ, Farquhar C. Semen preparation techniques for intrauterine insemination. Cochrane Database Syst Rev. 2007;(4):CD004507.
17. Sanocka D, Kurpisz M. Reactive oxygen species and sperm cells. Reprod Biol Endocrinol. 2004;2:12.
18. Ombelet W, Van der Auwera I, Bijnens H, Onofre J, Kremer C, Bruckers L, et al. Improving IUI success by performing modified slow-release insemination and a patient-centred approach in an insemination programme with partner semen: a prospective cohort study. Facts Views Vis Obgyn. 2021;13(4):359-67.
19. Cohlen B, Bijkerk A, Van der Poel S, Ombelet W. IUI: review and systematic assessment of the evidence that supports global recommendations. Hum Reprod Update. 2018;24(3):300-19.
20. Kim YJ, Park CW, Ku SY. Indications of intrauterine insemination for male and non-male factor infertility. Semin Reprod Med. 2014;32(4):306-12.
21. Vaughan DA, Sakkas D. Sperm selection methods in the 21st century. Biol Reprod. 2019 ;101(6):1076-82.
22. Dickey R, Brinsden P, Pyrzak R. Manual of Intrauterine Insemination and Ovulation Induction. Cambridge: Cambridge University Press; 2009.
23. Allahbadia GN. Intrauterine Insemination: Fundamentals Revisited. J Obstet Gynaecol India. 2017;67(6):385-92.
24. Yavas Y, Selub MR. Intrauterine insemination (IUI) pregnancy outcome is enhanced by shorter intervals from semen collection to sperm wash, from sperm wash to IUI time, and from semen collection to IUI time. Fertil Steril. 2004;82(6):1638-47.
25. Subak LL, Adamson GD, Boltz NL. Therapeutic donor insemination: a prospective randomized trial of fresh versus frozen sperm. Am J Obstet Gynecol. 1992;166(6 Pt 1):1597-604; discussion 1604-6.
26. Practice Committee of the American Society for Reproductive Medicine. Electronic address: asrm@asrm.org; Practice Committee of the American Society for Reproductive Medicine. Evidence-based treatments for couples with unexplained infertility: a guideline. Fertil Steril. 2020;113(2):305-22.
27. Starosta A, Gordon CE, Hornstein MD. Predictive factors for intrauterine insemination outcomes: a review. Fertil Res Pract. 2020;6(1):23.
28. Kamel RM. Assisted reproductive technology after the birth of Louise brown. J Reprod Infertil. 2013;14(3):96-109.
29. Rosenwaks Z, Pereira N. The pioneering of intracytoplasmic sperm injection: historical perspectives. Reproduction. 2017;154(6):F71-7.
30. Gardner DK, Weissman A, Howles CM, Shoham Z. Textbook of Assisted Reproductive Techniques, two volume set, 4th edition. Boca Raton, FL: CRC Press; 2012.
31. Gallos ID, Eapen A, Price MJ, Sunkara SK, Macklon NS, Bhattacharya S, et al. Controlled ovarian stimulation

protocols for assisted reproduction: a network meta-analysis. Cochrane Database Syst Rev. 2017;(3):CD012586.
32. Magnusson Å, Wennerholm UB, Källén K, Petzold M, Thurin-Kjellberg A, Bergh C. The association between the number of oocytes retrieved for IVF, perinatal outcome and obstetric complications. Hum Reprod. 2018;33(10):1939-47.
33. Drakopoulos P, Blockeel C, Stoop D, Camus M, de Vos M, Tournaye H, et al. Conventional ovarian stimulation and single embryo transfer for IVF/ICSI. How many oocytes do we need to maximize cumulative live birth rates after utilization of all fresh and frozen embryos? Hum Reprod. 2016;31(2):370-6.
34. Ubaldi F, Vaiarelli A, D'Anna R, Rienzi L. Management of poor responders in IVF: is there anything new? Biomed Res Int. 2014;2014:352098.
35. ESHRE Working Group on Ultrasound in ART, D'Angelo A, Panayotidis C, Amso N, Marci R, Matorras R, et al. Recommendations for good practice in ultrasound: oocyte pick up†. Hum Reprod Open. 2019;2019(4):hoz025.
36. Souza MDCB, Souza MM, Antunes RA, Tamm MA, Silva JBD, Mancebo ACA. Bladder hematoma: a complication from an oocyte retrieval procedure. JBRA Assist Reprod. 2019;23(1):75-8.
37. Huang H, Takai Y, Samejima K, Gomi Y, Narita T, Ichinose S, et al. Severe hemoperitoneum resulting from restart of letrozole after oocyte retrieval procedure: a case report. J Med Case Rep. 2021;15(1):327.
38. Levi-Setti PE, Cirillo F, Scolaro V, Morenghi E, Heilbron F, Girardello D, et al. Appraisal of clinical complications after 23,827 oocyte retrievals in a large assisted reproductive technology program. Fertil Steril. 2018;109(6):1038-1043.e1.
39. Practice Committee of the American Society for Reproductive Medicine. Electronic address: ASRM@asrm.org; Practice Committee of the American Society for Reproductive Medicine. Performing the embryo transfer: a guideline. Fertil Steril. 2017;107(4):882-96.
40. Mackens S, Santos-Ribeiro S, van de Vijver A, Racca A, Van Landuyt L, Tournaye H, et al. Frozen embryo transfer: a review on the optimal endometrial preparation and timing. Hum Reprod. 2017;32(11):2234-42.
41. Coello A, Pellicer A, Cobo A. Vitrification of human oocytes. Minerva Ginecol. 2018;70(4):415-23.
42. Zhu Q, Chen Q, Wang L, Lu X, Lyu Q, Wang Y, et al. REPLY: The 'Big Freeze': freeze-all should not be used for everyone. Hum Reprod. 2018;33(8):1579-80.
43. Kushnir VA, Barad DH, Albertini DF, Darmon SK, Gleicher N. Systematic review of worldwide trends in assisted reproductive technology 2004-2013. Reprod Biol Endocrinol. 2017;15(1):6.
44. Quaas AM, Hansen KR. Time to "cool off"? Examining indications for "elective deferred frozen embryo transfer". J Assist Reprod Genet. 2016;33(12):1551-2.
45. Cutting R. Single embryo transfer for all. Best Pract Res Clin Obstet Gynaecol. 2018;53:30-7.
46. Munné S, Kaplan B, Frattarelli JL, Child T, Nakhuda G, Shamma FN, et al.; STAR Study Group. Preimplantation genetic testing for aneuploidy versus morphology as selection criteria for single frozen-thawed embryo transfer in good-prognosis patients: a multicenter randomized clinical trial. Fertil Steril. 2019;112(6):1071-9.e7.
47. Labarta E, Mariani G, Paolelli S, Rodriguez-Varela C, Vidal C, Giles J, et al. Impact of low serum progesterone levels on the day of embryo transfer on pregnancy outcome: a prospective cohort study in artificial cycles with vaginal progesterone. Hum Reprod. 2021;36(3):683-92.
48. McDowell S, Kroon B, Ford E, Hook Y, Glujovsky D, Yazdani A. Advanced sperm selection techniques for assisted reproduction. Cochrane Database Syst Rev. 2014;(10):CD010461.
49. Lepine S, McDowell S, Searle LM, Kroon B, Glujovsky D, Yazdani A. Advanced sperm selection techniques for assisted reproduction. Cochrane Database Syst Rev. 2019;7(7):CD010461.
50. Baldini D, Ferri D, Baldini GM, Lot D, Catino A, Vizziello D, et al. Sperm Selection for ICSI: Do We Have a Winner? Cells. 2021;10(12):3566.
51. Gardner DK, Balaban B. Assessment of human embryo development using morphological criteria in an era of time-lapse, algorithms and 'OMICS': is looking good still important? Mol Hum Reprod. 2016;22(10):704-18.
52. Abeyta M, Behr B. Morphological assessment of embryo viability. Semin Reprod Med. 2014;32(2):114-26.
53. Lee A, Kiessling AA. Early human embryos are naturally aneuploid-can that be corrected? J Assist Reprod Genet. 2017;34(1):15-21.
54. Kotdawala AP, Mir P, Herrero J, Khajuria R, Lalitkumar PGL, Banker MR. Embryonic aneuploidy after preimplantation genetic screening: Age- and indication-matched comparative study between Indian and Spanish population. J Hum Reprod Sci. 2019;12(2):141-9.
55. Cooper AR, Jungheim ES. Preimplantation genetic testing: indications and controversies. Clin Lab Med. 2010;30(3):519-31.
56. Hammadeh ME, Fischer-Hammadeh C, Ali KR. Assisted hatching in assisted reproduction: a state of the art. J Assist Reprod Genet. 2011;28(2):119-28.
57. Practice Committee of the American Society for Reproductive Medicine; Practice Committee of the Society for Assisted Reproductive Technology. Role of assisted hatching in in vitro fertilization: a guideline. Fertil Steril. 2014;102(2):348-51.
58. Balaban B, Urman B, Alatas C, Mercan R, Mumcu A, Isiklar A. A comparison of four different techniques of assisted hatching. Hum Reprod. 2002;17(5):1239-43.
59. Lacey L, Hassan S, Franik S, Seif MW, Akhtar MA. Assisted hatching on assisted conception (in vitro fertilisation (IVF) and intracytoplasmic sperm injection (ICSI)). Cochrane Database Syst Rev. 2021;3(3):CD001894.

60. Benward J. Mandatory counseling for gamete donation recipients: ethical dilemmas. Fertil Steril. 2015;104(3):507-12.
61. Behjati Ardakani Z, Navabakhsh M, Tremayne S, Akhondi MM, Ranjbar F, Mohseni Tabrizi A. The Impact of Third Party Reproduction on Family and Kinship. J Reprod Infertil. 2021;22(1):3-15.
62. FOGSI. (2012). Surrogacy and Third Party Options (FOGSI MATTER PAGE 32-34). [online] Available from https://www.fogsi.org/wp-content/uploads/fogsi-focus/SURROGACY+AND+THIRD+PARTY+OPTIONS.pdf. [Last accessed March, 2024].
63. Jarow J, Sigman M, Kolettis PN, et al. The evaluation of the azoospermic male. Am Urol Assoc - Best Pract Statements. 2011.
64. Practice Committee of the American Society for Reproductive Medicine. Diagnostic evaluation of the infertile male: a committee opinion. Fertil Steril. 2015;103(3):e18-25.
65. Schlegel PN, Sigman M, Collura B, De Jonge CJ, Eisenberg ML, Lamb DJ, et al. Diagnosis and treatment of infertility in men: AUA/ASRM guideline part I. Fertil Steril. 2021; 115(1):54-61.
66. Schlegel PN, Sigman M, Collura B, De Jonge CJ, Eisenberg ML, Lamb DJ, et al. Diagnosis and Treatment of Infertility in Men: AUA/ASRM Guideline PART II. J Urol. 2021;205(1):44-51.
67. Practice Committee of the American Society for Reproductive Medicine in collaboration with the Society for Male Reproduction and Urology. Electronic address: asrm@asrm.org. The management of obstructive azoospermia: a committee opinion. Fertil Steril. 2019;111(5):873-80.
68. Practice Committee of the American Society for Reproductive Medicine. Electronic address: asrm@asrm.org. Management of nonobstructive azoospermia: a committee opinion. Fertil Steril. 2018;110(7):1239-45.
69. Shin DH, Turek PJ. Sperm retrieval techniques. Nat Rev Urol. 2013;10(12):723-30.
70. Esteves SC, Miyaoka R, Agarwal A. Sperm retrieval techniques for assisted reproduction. Int Braz J Urol. 2011;37(5): 570-83.
71. Esteves SC, Miyaoka R, Orosz JE, Agarwal A. An update on sperm retrieval techniques for azoospermic males. Clinics (Sao Paulo). 2013;68(Suppl 1):99-110.
72. Akhvlediani ND, Reva IA, Chernushenko AS, Pushkar DY. Sperm retrieval techniques in patients with non-obstructive azoospermia. Urologiia. 2021;(4):106-13.

Chapter 32

Fertility Preservation

Yogita Dogra

INTRODUCTION

Fertility preservation refers to the preservation of embryos, eggs, sperm, and reproductive tissues. This opens the door to having a child in the future. Adults and even some minors of both sexes have this option. A typical indication for fertility preservation is when a person's fertility is impaired by an illness or condition (medically indicated preservation) or when they want to postpone having children for personal issues (elective preservation).[1,2]

The likelihood of survival for many cancers has been rising consistently and is currently at about 80%. Following treatment, a lot of women are beginning to place more emphasis on life quality, with the potential of having children being a top concern. Invasive cancer affects about 2% of women in reproductive age, and after undergoing radiotherapy and chemotherapy for sterilization, they are at risk of ovarian failure.[1,3] The ovary, unlike the testis, has a set amount of oocytes and no germ stem cells, so there is no way for the oocyte pool to be replenished.[1]

INDICATIONS OF FERTILITY PRESERVATION[4]

- Patients receiving gonadotoxic medications for cancer
- Those with benign disorders who are receiving gonadotoxic treatments, including surgery, or who are at risk for premature ovarian insufficiency (POI), such as Turner syndrome
- People who are transgender (assigned females at birth)
- Female patients seeking oocyte cryopreservation due to age-related infertility
- Individual reasons for delaying parenthood may include waiting until one has found the ideal spouse or partner or until one is established in one's job.

FERTILITY PRESERVATION TECHNIQUES[5]

Fertility preservation techniques are shown in **Flowcharts 1 to 3**.

Flowchart 1: Ovarian protection.

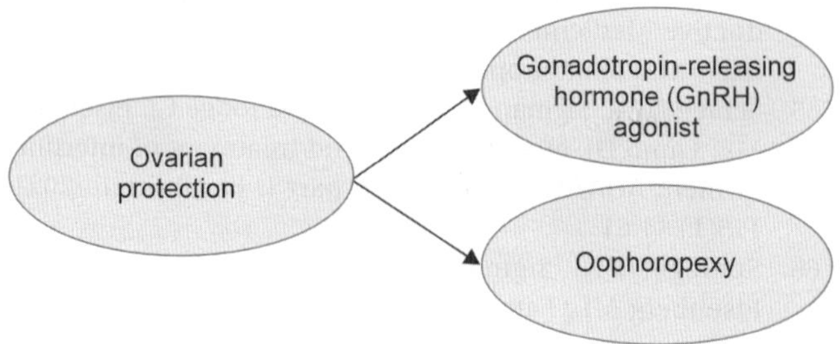

The challenges of fertility preservation techniques are mentioned in **Flowchart 4**.

PATIENT SELECTION AND PRE-FERTILITY PRESERVATION ASSESSMENT[4]

The factors to be kept in mind are given as follows:

Intrinsic factors:

- *Patient's health status:* Surgical/anesthetic risk, including thrombosis, infection, etc.; malignant contamination of ovary
- It is essential to obtain informed consent (patient/parent).
- Age (upper and lower limits for safety and efficacy)
- Evaluation of ovarian reserve

Extrinsic factors:

- *Predicted treatment's nature*: High/medium/low/uncertain risk of POI or infertility
    - Other pregnancy-related hazards, such as cardiac toxicity
    - Uterine radiotherapy
- Time, availability of local resources, expertise, and local criteria/funding

Factors that increase the risk of gonadotoxicity are:

- Diagnosis occurring at a later age

Flowchart 2: ART techniques.

ART techniques
Oocyte retrieval
IVM-GV stage oocytes
Ovarian tissue retrieval
Oocyte cryopreservation
IVF embryo cryopreservation
Tissue cryopreservation
IVM-GV staged oocytes in situ from ovary or after mild stimulation
Germinal vesicle
Mature oocyte
Ovarian transplantation
Cryopreservation of oocytes or embryos (after IVF)
Live births
Mature oocyte cryopreservation

(ART: assisted reproductive technique; GV: germinal vesicle; IVF: in vitro fertilization; IVM: in vitro maturation)

Flowchart 3: Estrogen sensitive tumours.

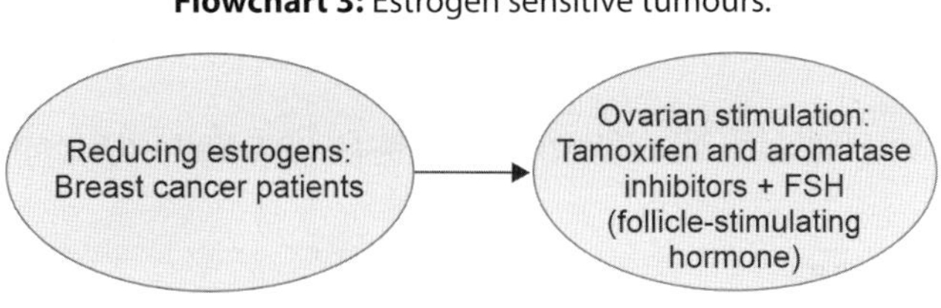

Flowchart 4: Fertility preservation techniques challenges.

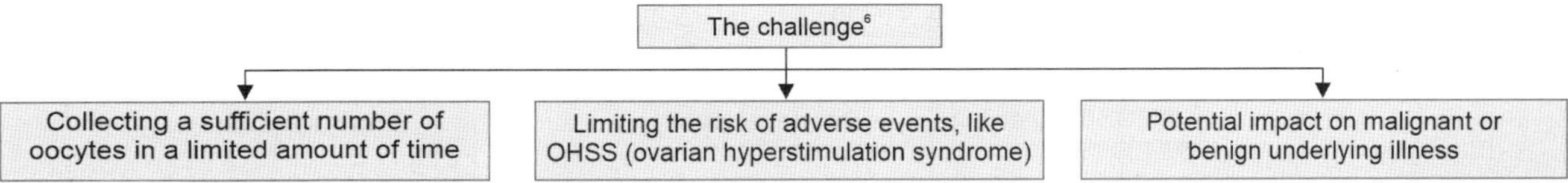

- Reduced anti-Mullerian hormone (AMH) levels prior to treatment
- Method of treatment and dosage (chemotherapy and radiotherapy).

OVARIAN STIMULATION PROTOCOLS[4]

GnRH Antagonist Protocol

The advantages are:

- Short duration of stimulation
- Gonadotropin-releasing hormone (GnRH) agonist as trigger, decreases OHSS (ovarian hyperstimulation syndrome) risk.

ESHRE guidelines 2020[4]	For ovarian stimulation in women seeking FP for medical reasons, the GnRH antagonist protocol is recommended for its feasibility in urgent situations, short time, and safety reasons
	For patients requiring ovarian stimulation where there is lack of urgency, the use of long protocol may be appropriate

(ESHRE: European Society of Human Reproduction and Embryology; FP: fertility preservation; GnRH: gonadotropin-releasing hormone)

Random Start Protocol[4,7]

Start ovarian stimulation (OS) on the day patient presents—any day of the cycle.

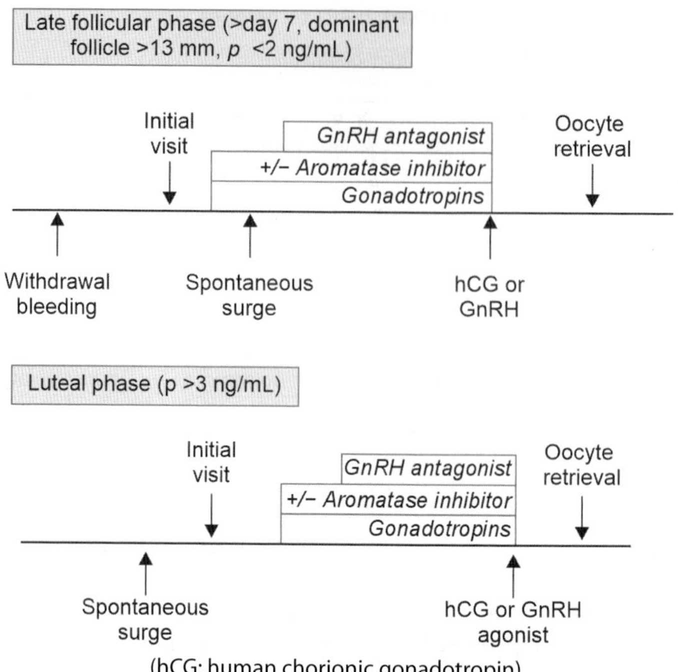

(hCG: human chorionic gonadotropin)

ESHRE guidelines 2020[4]	In urgent fertility preservation cycles, random start ovarian stimulation is an option.

Double Stimulation Protocol[4,8,9]

- This is also known as dual stimulation/DuoStim/Shanghai protocol.
- For poor responder patients or urgent fertility preservation.

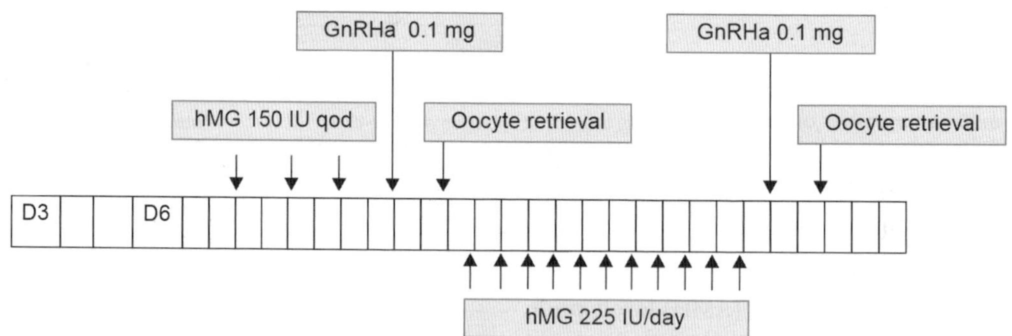

(GnRHa: gonadotropin-releasing hormone agonist; hMG: human menopausal gonadotropin)

- Clomiphene 25 mg from day 3 to day 13
- Letrozole 2.5 mg from day 3 to day 6 and day 17 to day 25
- Ibuprofen 0.6 g on day 14, 15, 28, 29
- Medroxyprogesterone acetate (MPA) 10 mg from day 25 to day 27
- Two stimulation protocols within the same menstrual cycle
- Two oocyte pickups performed approximately 2 weeks apart

ESHRE guidelines 2020[4]	Double stimulation can be considered for urgent fertility preservation cycles.

Luteal Halt Protocol/Crash Protocol[10,11]

- Decrease the days of luteal phase, by breaking down of the corpus luteum.
- *GnRH antagonists induces corpus luteum regression:* Single 3 mg dose or 2–3 consecutive daily 250 µg doses
- Serum progesterone levels decreased and menses starts in 2–4 days followed by OS.

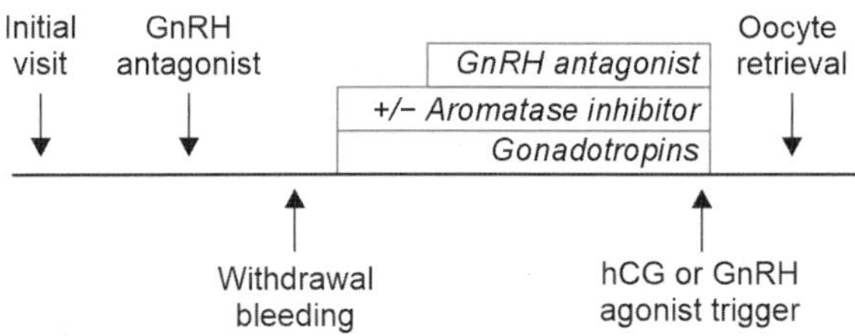

Estrogen-sensitive Tumors[12]

- Tamoxifen stimulation appears to produce a greater number of embryos and may offer a safe way of in vitro fertilization (IVF) and fertility preservation in breast cancer patients.[13]
- Combining low-dose FSH with tamoxifen (TamFSH-IVF) or letrozole (Letrozole-IVF) increases embryo yield compared to tamoxifen alone (Tam-IVF).[14]
- Although the letrozole regimen produces lower peak E2 levels, recurrence rates do not appear to be raised.[14]
- In cases of estrogen-sensitive tumors, the inclusion of letrozole had no negative effects on controlled ovarian stimulation (COS) results.[15]

*Letrozole-containing COS protocol **(Flowchart 5)**:*[16]

- Peak estradiol concentrations higher than natural cycle IVF
- Oocyte yield similar between aromatase inhibitor protocols and conventional stimulation
- Oocytes retrieved were lower with tamoxifen than with letrozole

ESHRE guidelines 2020[4]	In ovarian stimulation for fertility preservation in estrogen-sensitive disease the concomitant use of antiestrogen therapy, such as letrozole is probably recommended

TECHNIQUES OF FERTILITY PRESERVATION

Oocyte Cryopreservation

In order to achieve a live birth, around 20 vitrified oocytes are necessary, as the live birth rate per vitrified oocyte (oocyte donation) for the most experienced teams in the world is 5.7%.[17]

Flowchart 6 shows the eligible candidates for oocyte cryopreservation.

- *Choosing between oocyte and embryo cryopreservation:* Understand "embryos belong to the couple"
- *In case of separation/no consent from partner:* Embryos cannot be used.

The outcomes of oocyte cryopreservation are shown in **Flowchart 7**.

- Women with malignant disease should anticipate fewer oocyte retrievals for fertility preservation following controlled ovarian hyperstimulation (COH).[21]

Oocyte Cryopreservation: ESHRE Guidelines 2020[4]

ESHRE guidelines 2020[4]	Oocyte cryopreservation should be offered as an established option for fertility preservation.
	Women with a partner should be offered the option to cryopreserve unfertilized oocytes or to split the oocytes to attempt both embryo and oocyte cryopreservation.
	Women should be informed of accurate, center-specific expertise and live birth rates. They should also be informed that success rates after cryopreservation of oocytes at the time of a cancer diagnosis may be lower than in women without cancer.

Flowchart 5: Letrozole-containing controlled ovarian stimulation protocol.

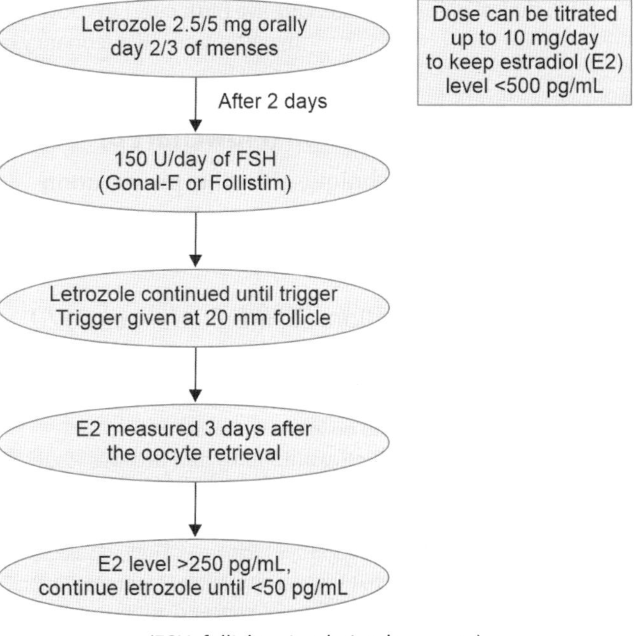

(FSH: follicle-stimulating hormone)

Flowchart 6: Eligible candidates for oocyte cryopreservation.

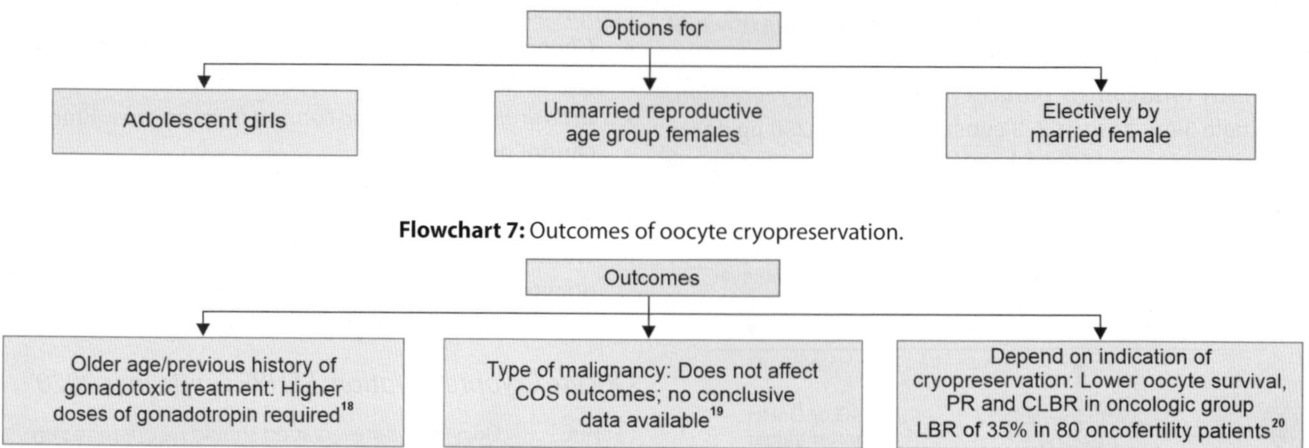

Flowchart 7: Outcomes of oocyte cryopreservation.

(CLBR: cumulative live birth rate; COS: controlled ovarian stimulation; LBR: live birth rate; PR: pregnancy rate)

Flowchart 8: Embryo cryopreservation.

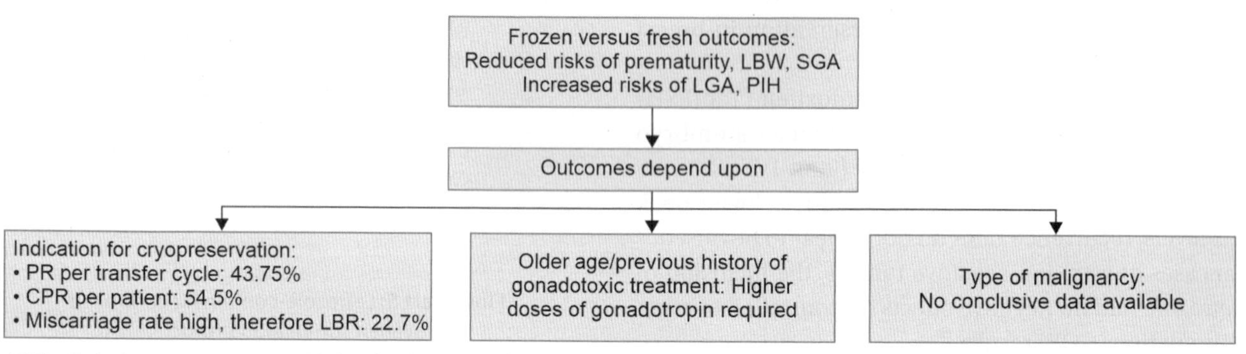

(CPR: clinical pregnancy rate; LBR: live birth rate; LBW: low birth weight; LGA: large for gestational age; PIH: pregnancy-induced hypertension; PR: pregnancy rate; SGA: small for gestational age)

Embryo Cryopreservation (Flowchart 8)[22,23]

- Most established technique of fertility preservation
- *Criteria:* Postpubertal females with a committed male partner or using donor sperm (with consent).

Embryo Cryopreservation: ESHRE Guidelines 2020[4]

ESHRE guidelines 2020[4]	Embryo cryopreservation is an established option for fertility.
	Women should be informed about the risk of losing reproductive autonomy and possible issues with ownership of stored embryos.
	Women should be informed of accurate, center-specific expertise and live birth rates. They should also be informed that success rates after cryopreservation of embryos at the time of cancer diagnosis may be lower than in women without cancer.

Ovarian Tissue Cryopreservation

Ovarian tissue transplantation (OTT) of cryopreserved ovarian tissue restores fertility.

After being frozen, the ovarian cortex can be kept for years in liquid nitrogen, giving the patient time to recover. After the patient has recovered, some of the cryopreserved tissue can be re-transplanted to those who have gone through menopause, and the ovarian grafts can reestablish a cyclic endocrine hormone milieu, including the ideal circumstances for conception, gestation, and parturition, possibly through IVF embryo transfer.[24-26]

The Edinburgh criteria for selection of ovarian tissue cryopreservation (OTC) is given in **Box 1**.

BOX 1: Edinburgh criteria for selection of ovarian tissue cryopreservation (OTC).[27]

- Age <30 years
- No previous chemotherapy or radiotherapy
- Realistic chance of long-term survival
- High risk of treatment-induced immediate ovarian failure (estimated at >50%)
- Informed consent from patient or (in the case of and incompetent child) from parents
- Negative human immunodeficiency virus (HIV) and hepatitis serology
- No existing children

Flowchart 9: Ovarian tissue transplantation.

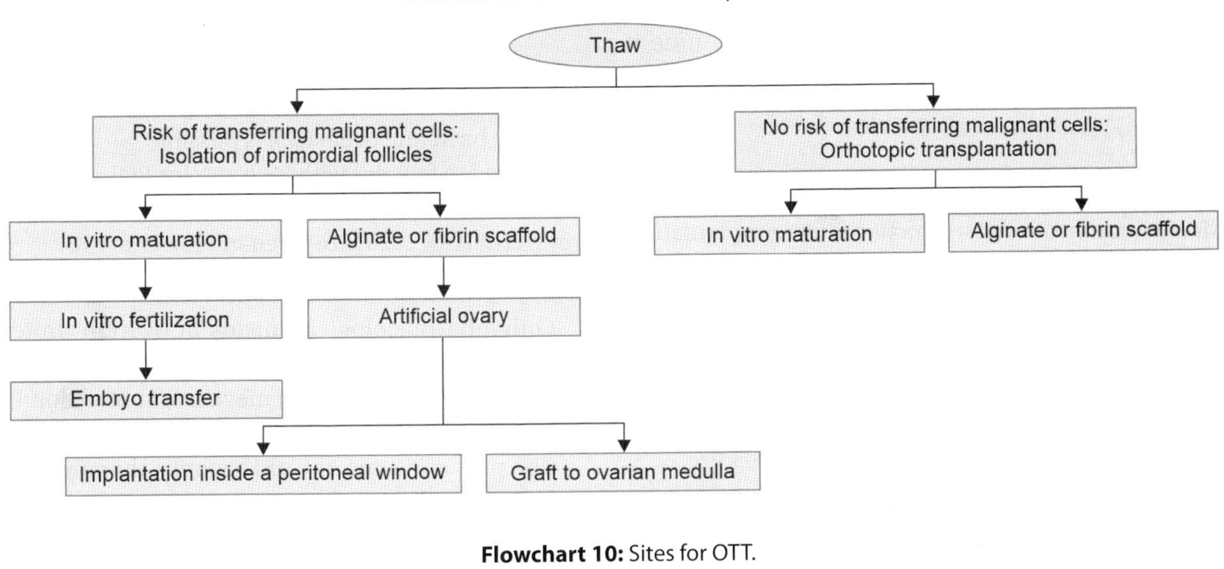

Flowchart 10: Sites for OTT.

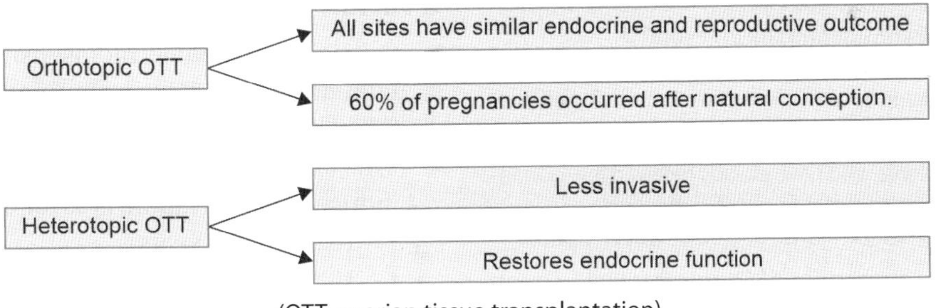

(OTT: ovarian tissue transplantation)

TECHNICAL ASPECTS OF OVARIAN TISSUE CRYOPRESERVATION

The Danish Protocol[1,25,28]

The majority of Danish patients have an entire ovary surgically removed, frequently during a laparoscopy. The benefits of removing the entire ovary are the reduced risk of problems during surgery and the potential to extend the window of potential fertility through repeated transplantations.

The cortex of the ovary is separated and treated before it is taken to the cryopreservation facility. Before being frozen, ovarian tissues can be transported for up to 5 hours on ice. This enables medical facilities lacking competence in cryopreservation to treat cancer patients locally while sending the ovarian tissue to a facility with that expertise. The thawing process consists of three steps, each of which takes 10 minutes. The time between the transplant and the tissue's revascularization appears to be crucial for follicle survival.

Ovarian Tissue Transplantation (Flowchart 9, 10)[1,29]

Heterotopic transplantation:

- Forearm subcutaneous
- Abdominal wall subcutaneous

ESHRE guidelines 2020[4]	For OTT, a one-step laparoscopy procedure should be performed, as it is considered safe without causing additional surgical risk.
	OTT at the orthotopic site is recommended to restore fertility.

Restoration of Ovarian Activity and Outcomes (Flowchart 11, 12)

Following grafting, ovarian activity typically restores within 3.5 and 6.5 months, coinciding with the time it takes follicles to develop from the primordial to the antral stage.[25,26,30]

After OTT:

- Restores ovarian activity in 95% patients
- Mean duration of ovarian function after reimplantation is 4–5 years

OTC/OTT Guidelines—ESHRE 2020[4]

ESHRE guidelines 2020[4]	It is recommended to offer OTC in patients undergoing moderate-/high-risk gonadotoxic treatment where oocyte/embryo cryopreservation is not feasible, or at patient preference.
	OTC should probably not be offered to patients with low ovarian reserve (AMH <0.5 ng/mL and AFC <5) or advanced age considering the unfavorable risk/benefit. Current evidence suggests that the efficiency of TC procedure is questionable above 36 years of age.
	Patients who have already received low gonadotoxic treatment or a previous course of chemotherapy, can be offered OTC as FP option.

(AMH: anti-Müllerian hormone; AFC: antral follicle count; FP: fertility preservation; OTC: ovarian tissue cryopreservation; TC: tissue cryopreservation)

COMBINATION TREATMENT[31]

Ovarian Tissue Cryopreservation ± Ovarian Stimulation

ESHRE guidelines 2020[4]	Ovarian stimulation can be performed immediately after OTC.
	OTC at the time of oocyte pick-up after ovarian stimulation should not be performed unless in a research context.
	Ovarian transposition can be performed at the same time as OTC in patients who will receive pelvic irradiation.

- A live birth rate of 50–60% is possible
- Offered to females with high risk for POI.

The OTC/OTT recommendations for specific patient groups (ESHRE 2020) are given in **Flowchart 13.**

Ovarian protection: GnRH agonist therapy:[32]

- Only pharmacological means of ovarian protection during chemotherapy **(Table 1)**.

Flowchart 14 shows the mechanism of action GnRH agonist therapy.

The pros and cons of GnRH agonist therapy are given in **Box 2** and **Flowchart 15**, respectively.

Ovarian Transposition[37-41]

- Ovaries are extremely radiosensitive, with an estimated fatal dosage of 2 Gy to eliminate 50% of nongrowing follicles present in the ovary.
- The age-dependent threshold dose for radiation-induced ovarian failure is as follows:
  - 18.4 Gy at 10 years
  - 16.5 Gy at 20 years
  - 14.3 Gy at 30 years

Flowchart 11: Restoration of ovarian activity and outcomes.

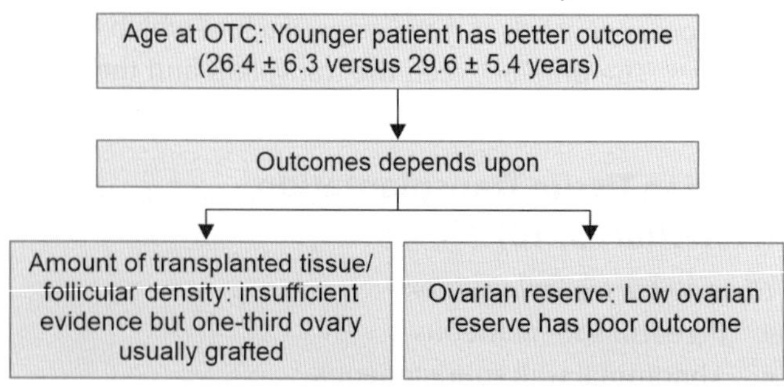

(OTC: ovarian tissue cryopreservation)

Flowchart 12: OTC/OTT safety concerns.

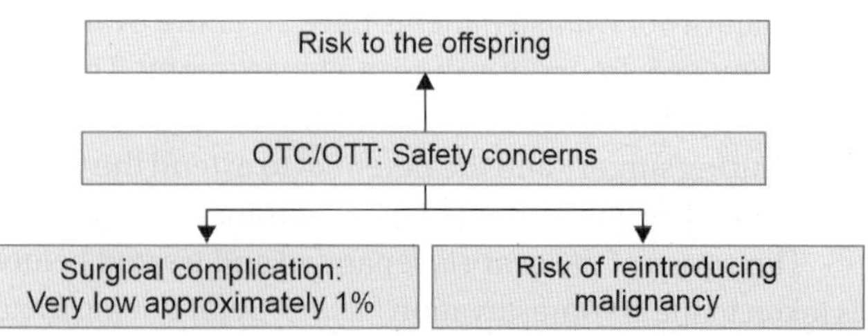

(OTC: ovarian tissue cryopreservation; OTT: ovarian tissue transplantation)

Flowchart 13: OTC/OTT recommendations for specific patient groups (ESHRE 2020).[4]

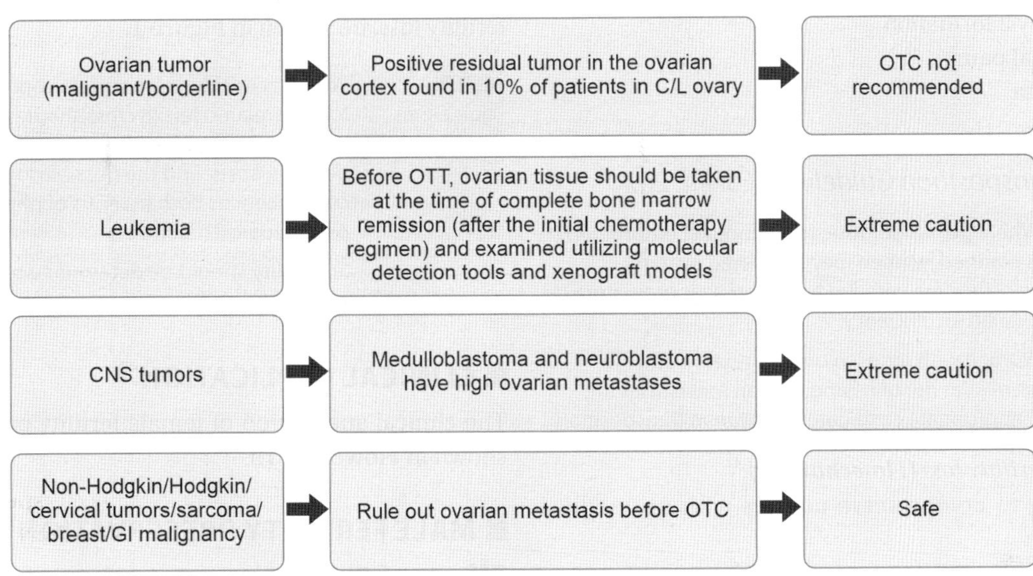

(C/L: contralateral; CNS: central nervous system; ESHRE: European Society of Human Reproduction and Embryology; GI: gastrointestinal; OTC: ovarian tissue cryopreservation; OTT: ovarian tissue transplantation)

TABLE 1: GnRH agonist dose and duration for ovarian protection.

Drug[36]	*Dose*	*Duration*
Inj. Leuprolide	3.75 mg IM	1 week before the start of chemotherapy and continue 4 weekly within 2 weeks before or after the final CT
Inj. Goserelin	3.6 mg SC	1 week before the start of chemotherapy and continue 4 weekly within 2 weeks before or after the final CT

(CT: computed tomography; IM: intramuscular; SC: subcutaneous)

Flowchart 14: Mechanism of action of gonadotropin-releasing hormone (GnRH) agonist therapy.

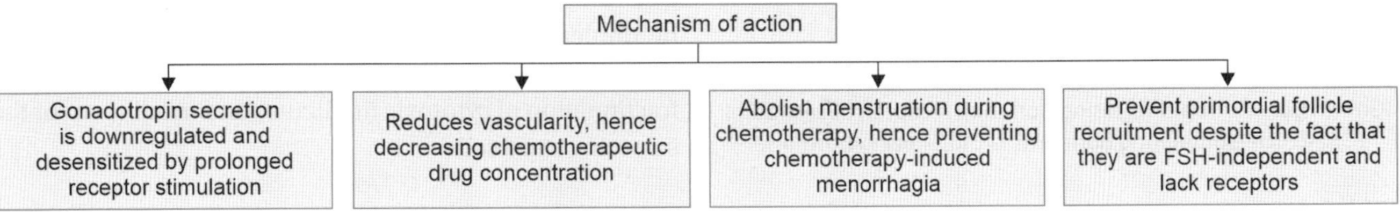

(FSH: follicle-stimulating hormone)
*Efficiency remains controversial

BOX 2: Pros of gonadotropin-releasing hormone (GnRH) agonist therapy.[33,34]

- There is some proof that GnRH agonists reduce FSH secretion, so interfering with "follicular burnout"
- Granulosa cells' cyclophosphamide-induced and BAX-mediated apoptotic pathway is inhibited by GnRHa
- Reduction of severe vaginal bleeding during chemotherapy
- Psychological benefit
- Can be coupled with other fertility preservation techniques without negative side effects

(BAX: B-cell lymphoma protein 2(Bcl-2)-associated protein X (Bax); FSH: follicle-stimulating hormone; GnRH: gonadotropin-releasing hormone; GnRHa: GnRH agonist)

Radiation treatment is recommended for:

- *Gynecological malignancies:* Cervix, vagina, uterus, and ovarian dysgerminoma
- *Non gynecological cancers*: Osteosarcoma, rhabdomyosarcoma, Hodgkins lymphoma, anorectal carcinomas

To protect ovaries from the direct ill effects of radiation, ovarian transposition prior to beginning radiotherapy has been suggested **(Flowchart 16, 17)**.

Common locations for ovarian transposition:

- Intra-abdominal paracolic gutters
- Anterior to psoas muscle
- Lower paracolic gutters

Complications of ovarian transposition: Occur in 12.8%

- Ovarian cyst formation
- Abdominal pain
- Hematoma
- Ischemia

Ovarian Transposition Guidelines: ESHRE 2020[4]

ESHRE guidelines 2020[4]	Where pelvic radiotherapy without chemotherapy is planned, women may be offered ovarian transposition with the aim to prevent premature ovarian insufficiency.
	Women with reduced ovarian reserve and women at risk of having ovarian metastases are inappropriate candidates for ovarian transposition.

Age-related fertility loss **(*Flowchart 18*)**:[42]

- Choosing to cryopreserve oocytes with no medical indication
- Also known as social/nonmedical/elective egg freezing.

Flowchart 15: Cons of gonadotropin-releasing hormone (GnRH) agonist therapy.[35]

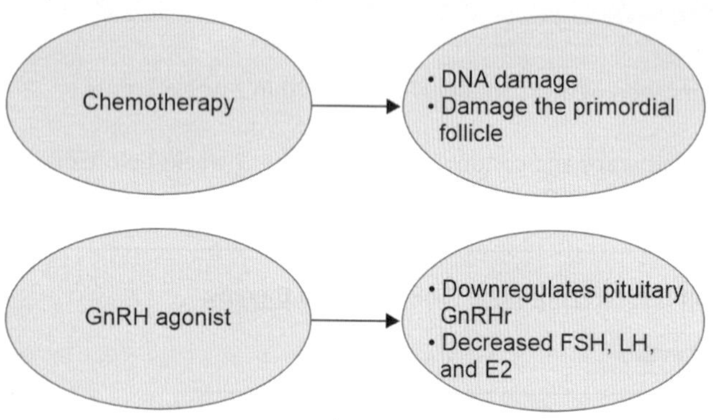

(GnRHr: gonadotropin-releasing hormone receptor; E2: estradiol; FSH: follicle-stimulating hormone; LH: luteinizing hormone)

Flowchart 16: Ovarian transposition.

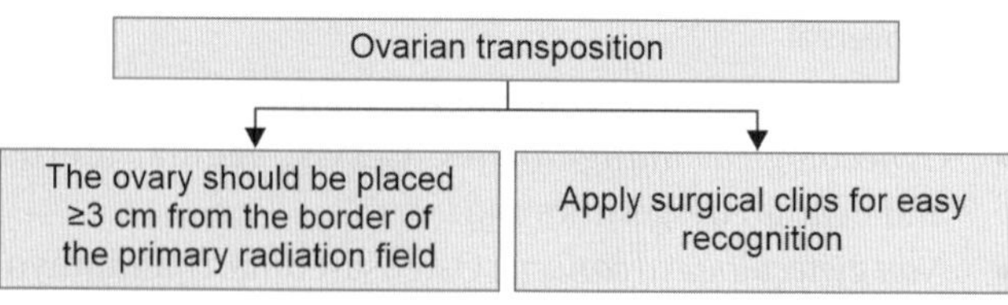

The issues with oocyte cryopreservation for age-related fertility loss are shown in **Figure 1**.

ESHRE guidelines 2020[4]	Women considering oocyte cryopreservation for age-related fertility loss should be fully informed regarding the success rates, risks, benefits, costs, and the possible long-term consequences both in terms of physical and psychological health.
	Suitability should be determined on a case-by-case basis.

CLINICAL APPLICATION[43]

The clinical application of female fertility preservation is shown in **Flowchart 19**.

MALE FERTILITY PRESERVATION[1,27,44-54]

Effects of Chemotherapy and Radiotherapy on the Testis

At all stages of life, the testis has been demonstrated to be very vulnerable to the harmful effects of irradiation and chemotherapy. The effect of combination chemotherapy on the spermatogenic epithelium depends on the drug type and dose. The threshold dosage of cyclophosphamide for infertility has been calculated to be between 7.5 and 9 g/m^2 and 10 g/m^2 in postpubertal males. Following radiation exposure, germline stem cells are acutely and dose-dependently reduced in the prepubescent testis. Over 6 Gy, the spermatogonial stem cell (SSC) pool might be depleted, leading to lifelong infertility. Recovery of spermatogenesis is dependent on the kind, dosage, and fractionation of cytotoxic medicines and irradiation for the surviving stem cells.

Current Options for the Preservation of Male Fertility

The current options for the preservation of male fertility are shown in **Flowchart 20**.

The options for ejaculated sperm collection are shown in **Box 3**.

Flowchart 17: Surgical techniques for ovarian transposition.

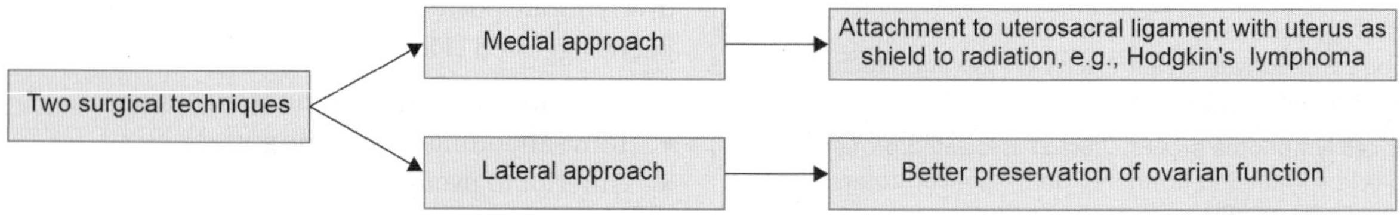

Flowchart 18: Pros and Cons of age related fertility preservation.

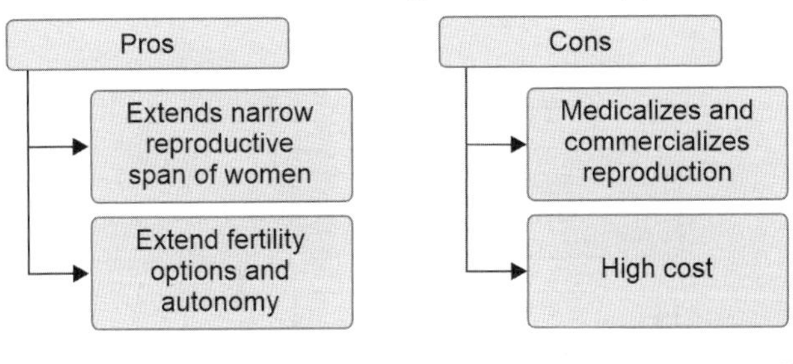

Male Fertility Preservation (Flowchart 21, 22, 23)[55]

Testicular biopsy does not appear to affect testicular growth but may result in intratesticular scarring in 6% of patients.

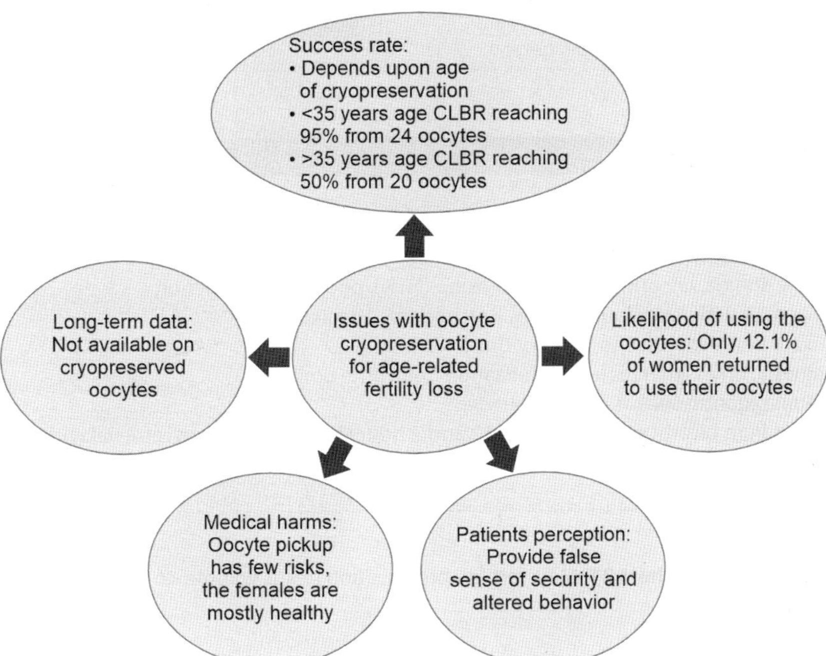

Fig. 1: Issues with oocyte cryopreservation for age-related fertility loss. (CLBR: cumulative live birth rate)

Flowchart 19: Clinical application of female fertility preservation.

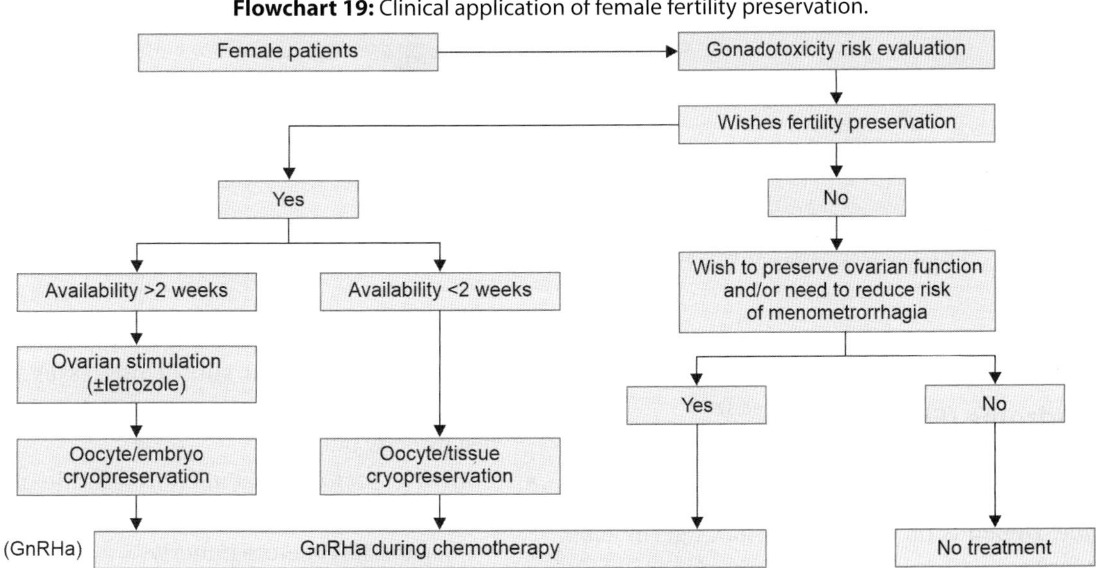

(GnRHa: gonadotropin-releasing hormone agonist)

Flowchart 20: Current options for the preservation of male fertility.

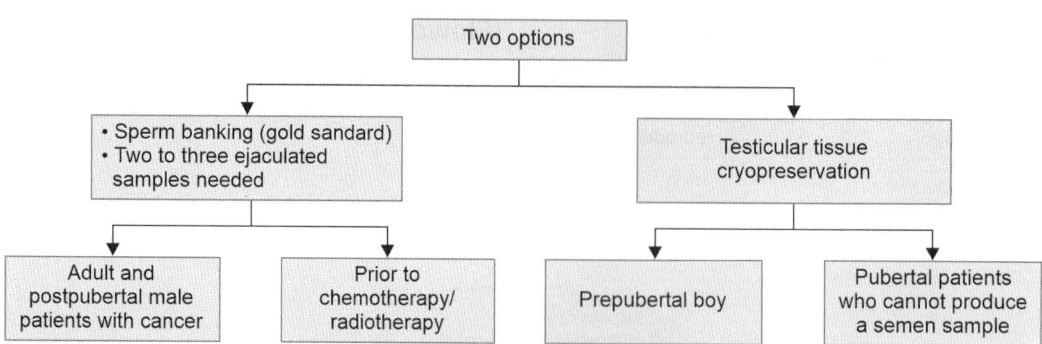

BOX 3: Options for ejaculated sperm collection.

- Masturbation
- Use of phosphodiesterase type 5 (PDE-5) inhibitors: 50 mg dose (25–100 mg)
  Side effects: Most common headache, flushing, dyspepsia
  Contraindication: Unstable angina, uncontrolled hypertension (HTN), poor cardiac function
- Vibratory stimulation: For men with neurologic injuries Psychogenic anejaculation
- Electroejaculation: Who are nonresponsive to penile vibratory stimulation
- Conducted under anesthesia, unless there is loss of sensation below the umbilicus
- Retrograde ejaculation: Alpha agonists such as pseudoephedrine
- Urinary alkalization with sodium bicarbonate 500 mg at 12 and 2 hours before the sperm retrieval with instillation of buffered sperm-friendly media into the bladder
- Surgical extraction of sperm

Flowchart 21: Male fertility preservation (testis volume <6 mL).

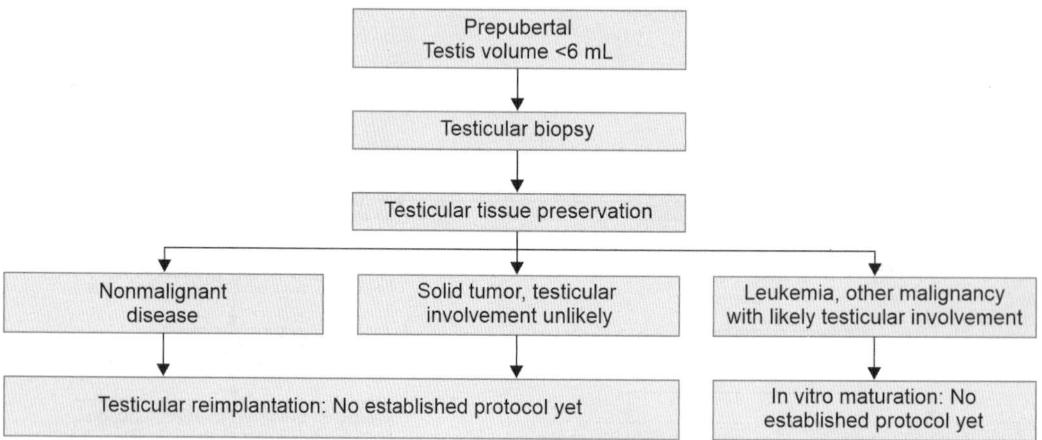

SPECIAL SITUATIONS

Turner Syndrome (Flowchart 24)[56,57]

- 45X or mosaic 45X/46XX
- *Likelihood of functional ovarian tissue in Turner syndrome:* Depends on presence of 46,XX germ cells in the ovaries
- *Opportunity:* A small subset possesses residual of ovarian follicles at birth/early childhood and will be lost by early puberty.
- *Limiting factor:* Absence of reliable marker for diagnosis before puberty
  - Serum AMH before puberty
  - Serum FSH after puberty

Flowchart 22: Male fertility preservation (testis volume 6–10 mL).

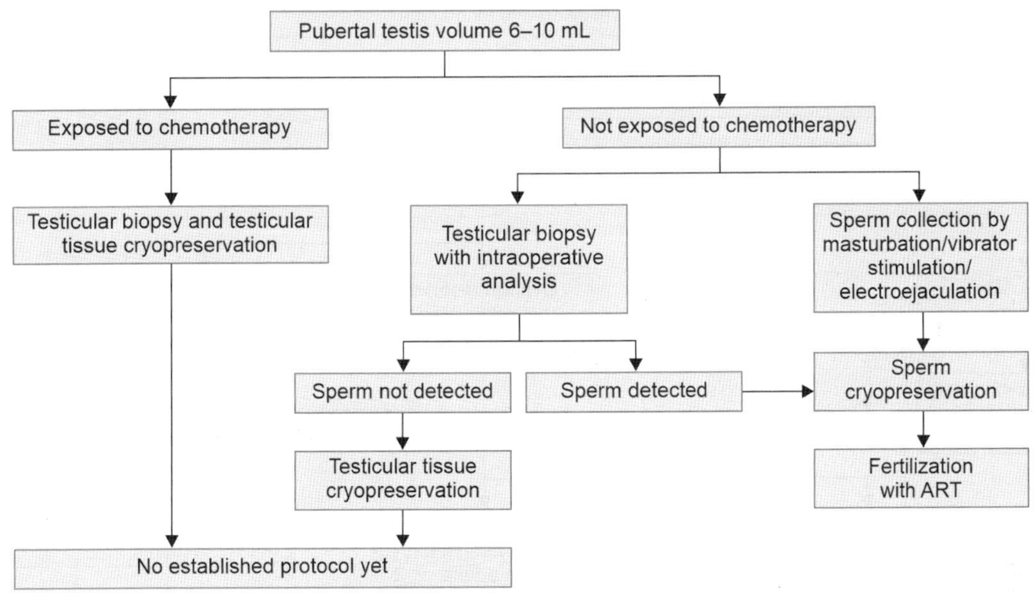

(ART: assisted reproductive technique)

Flowchart 23: Male fertility preservation (testis volume >10 mL).

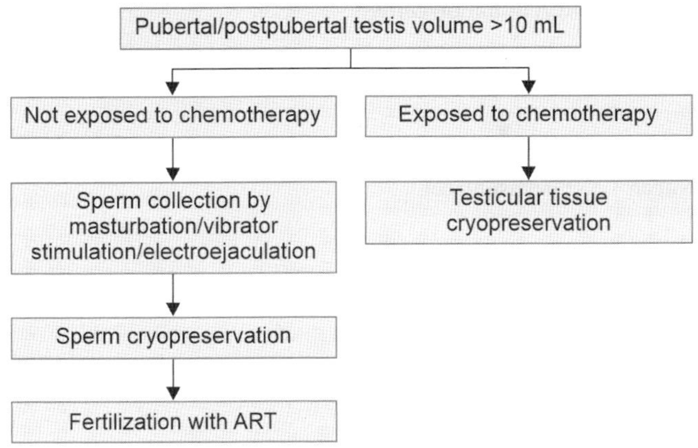

(ART: assisted reproductive technique)

- Remove as much tissue as possible, typically an entire ovary.

ESHRE Guidelines 2020[4]	OTC/OTT can be considered in patients with POI-associated genetic and chromosomal disorders but requires genetic counseling and should be performed within a research protocol.

BRCA Mutation Carriers (Flowchart 25)[58]

Prophylactic surgery:

- 35–40 years in BRCA1-mutated patients
- 45–50 years in women with BRCA2 mutation
- *Current recommendations:* Complete family planning by age 35–40 years old followed by prophylactic salpingo-oophorectomy
- BRCA1 carriers have lower serum AMH levels compared with women without BRCA mutations.
- BRCA1 mutations are associated with poor oocyte yield and low response to OS.
- Preimplantation genetic test for BRCA mutation carriers required, as risk of transmission is 50%.

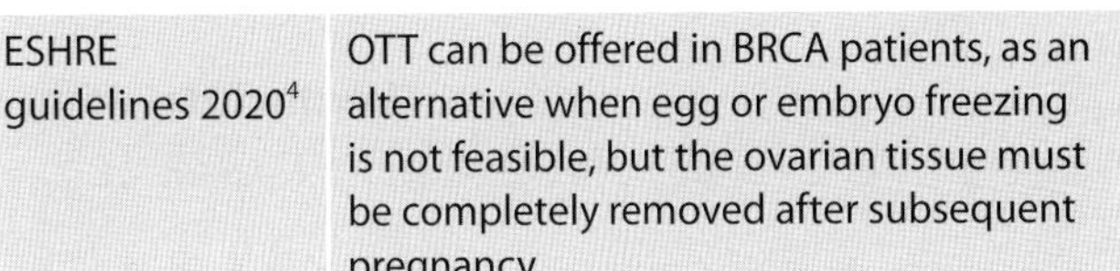

ESHRE guidelines 2020[4]	OTT can be offered in BRCA patients, as an alternative when egg or embryo freezing is not feasible, but the ovarian tissue must be completely removed after subsequent pregnancy

TRANSGENDER POPULATION[59]

Fertility preservation in Tans-female and Trans-male is depicted in **Flowchart 26 and 27)**.

CONCLUSION

- Fertility preservation should be given to all patients following gonadotoxic therapy.
- Referral process between an oncologist and an infertility expert
- *Oocyte cryopreservation:* Postpubescent unmarried females

Flowchart 24: Fertility preservation in turner syndrome.

- Fertility preservation counselling and evaluation
  - Ovarian reserve assessment, AMH, FSH, AFC, LH, Estradiol, inhibin-B
    - Prepubertal
      - Serum AMH >2 ng/mL → AMH drops <2 ng/mL in two consecutive measurements → Ovarian tissue freezing
      - Serum AMH <2 ng/mL → Ovarian tissue freezing
    - Postpubertal
      - Ovarian reserve not exhausted (46,XX/45,X)
        - Oocyte cryopreservation
        - Ovarian tissue cryopreservation
        - Embryo cryopreservation in adults
      - Ovarian reserve exhausted (45,X)
        - Embryo or oocyte donation
        - Adoption

Flowchart 25: Fertility preservation options in BRCA mutation carriers.

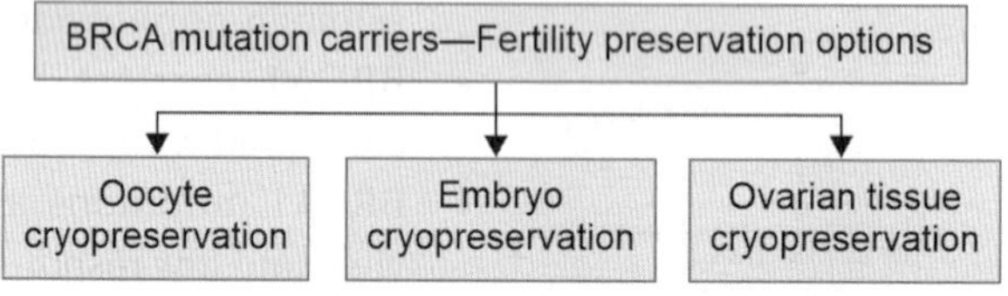

Flowchart 26: Fertility preservation in Trans-female.

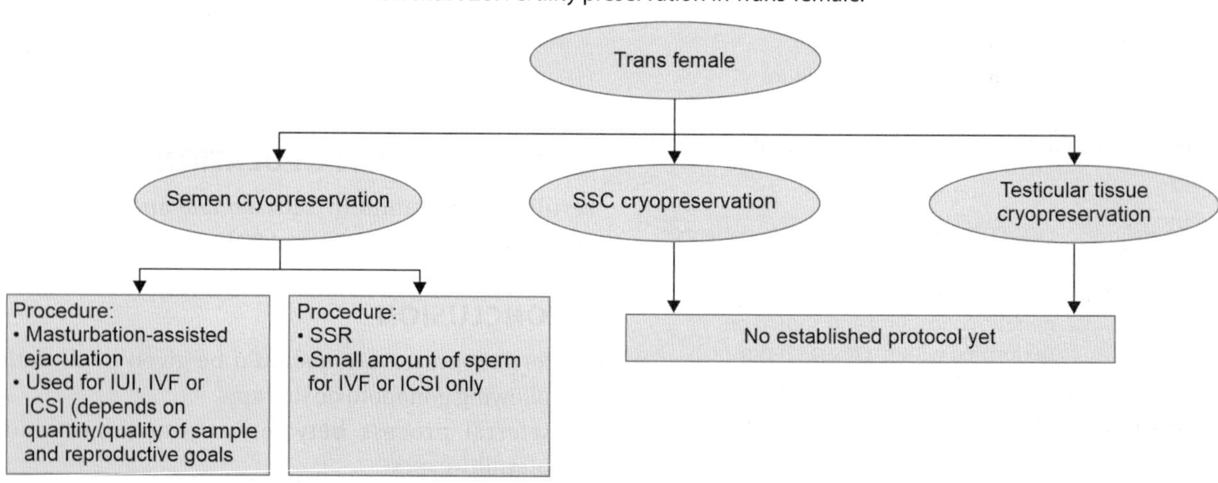

(ICSI: intracytoplasmic sperm injection; IUI: intrauterine insemination; IVF: in vitro fertilization; SSC: spermatogonial stem cell; SSR: surgical sperm retrieval)

Flowchart 27: Fertility preservation in Trans-male.

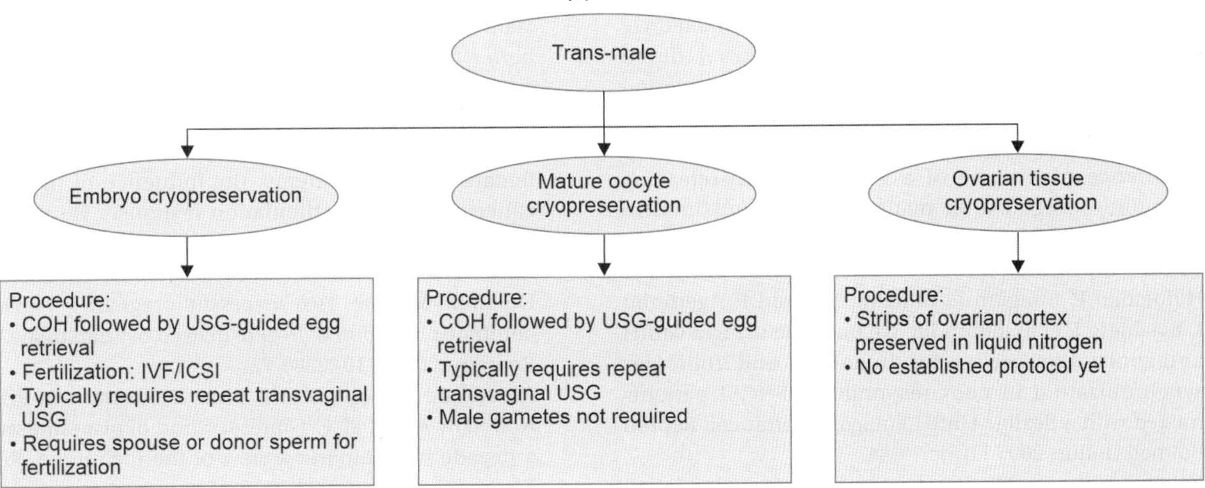

(COH: controlled ovarian hyperstimulation; ICSI: intracytoplasmic sperm injection; IVF: in vitro fertilization; USG: ultrasonography)

- *Embryo cryopreservation:* Requires partner or donor semen
- *OTC:* The only procedure that restores both reproductive and endocrine ovarian function.
- *In vitro maturation (IVM)*: An innovative technology
- *GnRH agonist*: The sole pharmacological therapy for ovarian protection
- *Ovarian transposition:* Provided to those at risk of pelvic radiation
- Young women and men can now utilize both proven and investigational medicines to address infertility caused by gonadotoxic therapy.
- OTC is becoming a well-established technology for fertility preservation on a global scale, and about 100 healthy children have been born using this method so far.
- The number of transplants is growing as women who have survived a disease return to have their fertility restored.
- In males, cryopreservation of sperm is the gold standard treatment for maintaining fertility, and young boys and men can have their testicular tissue cryopreserved with good outcomes; nevertheless, transplantation strategies to restore fertility must yet be developed.
- Most importantly, if there is a risk of gonadal damage and fertility loss, patients should be referred to an infertility specialist by hematologists and oncologists prior to the initiation of gonadotoxic treatment in order to receive the appropriate counseling on the available fertility preservation strategies.

REFERENCES

1. Kristensen SG, Andersen CY. Fertility preservation strategies. In: Gardner DK, Weissman A, Howles CM, Shoham Z (Eds). Textbook of Assisted Reproductive Techniques Vol 2: Clinical Perspectives, 5th edition. Boca Raton: CRC Press; 2018. pp. 784-97.
2. Fertility Preservation. Available from: https://my.clevelandclinic.org/health/treatments/17000-fertility-preservation [Last accessed April 2024].
3. Jemal A, Siegel R, Ward E, Murray T, Xu J, Thun MJ. Cancer statistics, 2007. CA Cancer J Clin. 2007;57(1):43-66.
4. European Society of Human Reproduction and Embryology. (2020). ESHRE Guideline: Female fertility preservation. Guideline of the European Society of Human Reproduction and Embryology. ESHRE Female Fertility Preservation Guideline Development Group 2020. Available from: https://www.eshre.eu/Guidelines-and-Legal/Guidelines/Female-fertility-preservation [Last accessed April, 2024].
5. Fisch B, Abir R. Female fertility preservation: past, present and future. Reproduction. 2018;156(1):F11-27.
6. Pinelli S, Basile S. Fertility preservation: current and future perspectives for oncologic patients at risk for iatrogenic premature ovarian insufficiency. BioMed Res Int. 2018;2018:6465903.
7. Talwar P, Arora PR, Mahajan N. Ovarian stimulation protocols in fertility preservation. In: Ghumman S (Ed). Principles and Practice of Controlled Ovarian Stimulation in ART. Springer: New Delhi; 2015. pp. 369-74.
8. Kuang Y, Chen Q, Hong Q, Lyu Q, Ai A, Fu Y, Shoham Z. Double stimulations during the follicular and luteal phases of poor responders in IVF/ICSI programmes (Shanghai protocol). Reprod Biomed Online. 2014;29(6): 684-91.

9. Vaiarelli A, Cimadomo D, Trabucco E, Vallefuoco R, Buffo L, Dusi L, et al. Double stimulation in the same ovarian cycle (DuoStim) to maximize the number of oocytes retrieved from poor prognosis patients: a multicenter experience and SWOT analysis. Front Endocrinol (Lausanne). 2018;9:317.
10. Anderson RA, Kinniburgh D, Baird DT. Preliminary experience of the use of a gonadotrophin-releasing hormone antagonist in ovulation induction/in-vitro fertilization prior to cancer treatment. Hum Reprod. 1999;14(10):2665-8.
11. Humaidan P, Bungum L, Bungum M, Hald F, Agerholm I, Blaabjerg J, et al. Reproductive outcome using a GnRH antagonist (cetrorelix) for luteolysis and follicular synchronization in poor responder IVF/ICSI patients treated with a flexible GnRH antagonist protocol. Reprod Biomed Online. 2005;11(6):679-84.
12. Cakmak H, Rosen MP. Random-start ovarian stimulation in patients with cancer. Curr Opin Obstet Gynecol. 2015;27(3):215-21.
13. Oktay K, Buyuk E, Davis O, Yermakova I, Veeck L, Rosenwaks Z. Fertility preservation in breast cancer patients: IVF and embryo cryopreservation after ovarian stimulation with tamoxifen. Hum Reprod. 2003;18(1):90-5.
14. Oktay K, Buyuk E, Libertella N, Akar M, Rosenwaks Z. Fertility preservation in breast cancer patients: a prospective controlled comparison of ovarian stimulation with tamoxifen and letrozole for embryo cryopreservation. J Clin Oncol. 2005;23(19):4347-53.
15. Bonardi B, Massarotti C, Bruzzone M, Goldrat O, Mangili G, Anserini P, et al. Efficacy and safety of controlled ovarian stimulation with or without letrozole co-administration for fertility preservation: a systematic review and meta-analysis. Front Oncol. 2020;10:574669.
16. Oktay K, Hourvitz A, Sahin G, Oktem O, Safro B, Cil A, et al. Letrozole reduces estrogen and gonadotropin exposure in women with breast cancer undergoing ovarian stimulation before chemotherapy. J Clin Endocrinol Metab. 2006;91(10):3885-90.
17. Donnez J, Dolmans MM. Ovarian tissue freezing: Current status. Curr Opin Obstet Gynecol. 2015;27:222-30.
18. Rodriguez-Wallberg KA, Waterstone M, Anastácio A. Ice age: Cryopreservation in assisted reproduction–An update. Reprod Biol. 2019;19(2):119-26.
19. Lefebvre T, Mirallié S, Leperlier F, Reignier A, Barrière P, Fréour T. Ovarian reserve and response to stimulation in women undergoing fertility preservation according to malignancy type. Reprod Biomed Online. 2018;37(2):201-7.
20. Cobo A, García-Velasco J, Domingo J, Pellicer A, Remohí J. Elective and onco-fertility preservation: factors related to IVF outcomes. Hum Reprod. 2018;33(12):2222-31.
21. Friedler S, Koc O, Gidoni Y, Raziel A, Ron-El R. Ovarian response to stimulation for fertility preservation in women with malignant disease: a systematic review and meta-analysis. Fertil Steril. 2012;97(1):125-33.
22. Maheshwari A, Pandey S, Amalraj Raja E, Shetty A, Hamilton M, Bhattacharya S. Is frozen embryo transfer better for mothers and babies? Can cumulative meta-analysis provide a definitive answer? Hum Reprod Update. 2018;24(1):35-58.
23. Alvarez RM, Ramanathan P. Fertility preservation in female oncology patients: the influence of the type of cancer on ovarian stimulation response. Hum Reprod. 2018;33(11):2051-9.
24. Andersen CY, Rosendahl M, Byskov AG, Loft A, Ottosen C, Dueholm M, et al. Two successful pregnancies following autotransplantation of frozen/thawed ovarian tissue. Hum Reprod. 2008;23(10):2266-72.
25. Rosendahl M, Schmidt KT, Ernst E, Rasmussen PE, Loft A, Byskov AG, et al. Cryopreservation of ovarian tissue for a decade in Denmark: a view of the technique. Reprod Biomed Online. 2011;22(2):162-71.
26. Donnez J, Dolmans MM, Pellicer A, Diaz-Garcia C, Serrano MS, Schmidt KT, et al. Restoration of ovarian activity and pregnancy after transplantation of cryopreserved ovarian tissue: a review of 60 cases of reimplantation. Fertil Steril. 2013;99(6):1503-13.
27. Wallace WH, Anderson RA, Irvine DS. Fertility preservation for young patients with cancer: who is at risk and what can be offered? Lancet Oncol. 2005;6(4):209-18.
28. Schmidt KL, Ernst E, Byskov AG, Nyboe Andersen A, Yding Andersen C. Survival of primordial follicles following prolonged transportation of ovarian tissue prior to cryopreservation. Hum Reprod. 2003;18:2654-9.
29. Gellert SE, Pors SE, Kristensen SG, Bay-Bjørn AM, Ernst E, Yding Andersen C. Transplantation of frozen-thawed ovarian tissue: an update on worldwide activity published in peer-reviewed papers and on the Danish cohort. J Assist Reprod Genet. 2018;35(4):561-70.
30. Schmidt KL, Andersen CY, Loft A, Byskov AG, Ernst E, Andersen AN. Follow-up of ovarian function post-chemotherapy following ovarian cryopreservation and transplantation. Hum Reprod. 2005;20(12):3539-46.
31. Dolmans MM, Marotta ML, Pirard C, Donnez J, Donnez O. Ovarian tissue cryopreservation followed by controlled ovarian stimulation and pick-up of mature oocytes does not impair the number or quality of retrieved oocytes. J Ovarian Res. 2014;7(1):1-5.
32. Clowse ME, Behera MA, Anders CK, Copland S, Coffman CJ, Leppert PC, et al. Ovarian preservation by GnRH agonists during chemotherapy: a meta-analysis. J Womens Health (Larchmt). 2009;18(3):311-9.
33. Scaruffi P, Stigliani S, Cardinali B, Massarotti C, Lambertini M, Sozzi F, et al. Gonadotropin releasing hormone agonists have an anti-apoptotic effect on cumulus cells. Int J Mol Sci. 2019;20(23):6045.
34. Meirow D, Rabinovici J, Katz D, Or R, Shufaro Y, Ben-Yehuda D. Prevention of severe menorrhagia in oncology patients with treatment-induced thrombocytopenia by

luteinizing hormone-releasing hormone agonist and depo-medroxyprogesterone acetate. Cancer. 2006;107(7):1634-41.

35. Dolmans MM, Taylor HS, Rodriguez-Wallberg KA, Blumenfeld Z, Lambertini M, von Wolff M, et al. Utility of gonadotropin-releasing hormone agonists for fertility preservation in women receiving chemotherapy: pros and cons. Fertil Steril. 2020;114(4):725-38.
36. Practice Committee of the American Society for Reproductive Medicine. Fertility preservation in patients undergoing gonadotoxic therapy or gonadectomy: a committee opinion. Fertil Steril. 2019;112(6):1022-33.
37. Wallace WH, Thomson AB, Saran F, Kelsey TW. Predicting age of ovarian failure after radiation to a field that includes the ovaries. Int J Radiat Oncol Biol Phys. 2005;62(3):738-44.
38. Moawad NS, Santamaria E, Rhoton-Vlasak A, Lightsey JL. Laparoscopic ovarian transposition before pelvic cancer treatment: ovarian function and fertility preservation. J Minim Invasive Gynecol. 2017;24(1):28-35.
39. Sella T, Mironov S, Hricak H. Imaging of transposed ovaries in patients with cervical carcinoma. AJR Am J Roentgenol. 2005;184(5):1602-10.
40. Clough KB, Goffinet F, Labib A, Renolleau C, Campana F, de la Rochefordiere A, et al. Laparoscopic unilateral ovarian transposition prior to irradiation: prospective study of 20 cases. Cancer. 1996;77(12):2638-45.
41. Grabenbauer GG, Girke P, Wildt L, Müller RG, Herbst M, Sauer R. Ovariopexy and the treatment of Hodgkin's disease. Strahlenther Onkol. 1991;167(5):273-6.
42. ESHRE Task Force on Ethics and Law; Dondorp W, de Wert G, Pennings G, Shenfield F, Devroey P, Tarlatzis B, Barri P, Diedrich K. Oocyte cryopreservation for age-related fertility loss. Hum Reprod. 2012;27(5):123-7.
43. Lambertini M, Peccatori FA, Demeestere I, Amant F, Wyns C, Stukenborg JB, et al; ESMO Guidelines Committee. Fertility preservation and post-treatment pregnancies in post-pubertal cancer patients: ESMO Clinical Practice Guidelines. Ann Oncol. 2020;31(12):1664-78.
44. Jahnukainen K, Ehmcke J, Hou M, Schlatt S. Testicular function and fertility preservation in male cancer patients. Best Pract Res Clin Endocrinol Metab. 2011;252:287-302.
45. Aubier F, Flamant F, Brauner R, Caillaud JM, Chaussain JM, Lemerle J. Male gonadal function after chemotherapy for solid tumors in childhood. J Clin Oncol. 1989;7(3):304-9.
46. Siimes MA, Rautonen J. Small testicles with impaired production of sperm in adult male survivors of childhood malignancies. Cancer 1990;65:1303-6.
47. Ridola V, Fawaz O, Aubier F, Bergeron C, de Vathaire F, Pichon F, et al. Testicular function of survivors of childhood cancer: a comparative study between ifosfamide-and cyclophosphamide-based regimens. Eur J Cancer. 2009;45(5):814-8.
48. Meistrich ML, Wilson G, Brown BW, Da Cunha MF, Lipshultz LI. Impact of cyclophosphamide on long-term reduction in sperm count in men treated with combination chemotherapy for Ewing and soft tissue sarcomas. Cancer. 1992;70(11):2703-12.
49. Rivkees SA, Crawford JD. The relationship of gonadal activity and chemotherapy-induced gonadal damage. JAMA. 1988;259(14):2123-5.
50. Jahnukainen K, Ehmcke J, Schlatt S. Testicular xenografts: A novel approach to study cytotoxic damage in juvenile primate testis. Cancer Res. 2006;66:3813-18.
51. Jahnukainen K, Ehmcke J, Nurmio M, Schlatt S. Irradiation causes acute and long-term spermatogonial depletion in cultured and xenotransplanted testicular tissue from juvenile nonhuman primates. Endocrinology. 2007;148(11):5541-8.
52. Rowley MJ, Leach DR, Warner GA, Heller CG. Effect of graded doses of ionizing radiation on the human testis. Radiat Res. 1974;59:665-78.
53. Centola GM, Keller JW, Henzler M, Rubin P. Effect of low-dose testicular irradiation on sperm count and fertility in patients with testicular seminoma. J Androl. 1994;15:608-13.
54. van Alphen MM, van den Kant HJG, de Rooij DG. Repopulation of the seminiferous epithelium of the rhesus monkey after X irradiation. Radiat Res. 1988;113:487-500.
55. Uijldert M, Meißner A, de Melker AA, van Pelt AMM, van de Wetering MD, van Rijn RR, et al. Development of the testis in pre-pubertal boys with cancer after biopsy for fertility preservation. Hum Reprod. 2017;32(12):2366-72.
56. Oktay K, Bedoschi G, Berkowitz K, Bronson R, Kashani B, McGovern P, et al. Fertility preservation in women with Turner syndrome: a comprehensive review and practical guidelines. J Pediatr Adolesc Gynecol. 2016;29(5):409-16.
57. Grynberg M, Bidet M, Benard J, Poulain M, Sonigo C, Cédrin-Durnerin I, et al. Fertility preservation in Turner syndrome. Fertil Steril. 2016;105(1):13-9.
58. Samuel JC, Ollila DW. Prophylaxis and screening options: Recommendations for young women with BRCA mutations. Breast Dis. 2006;23(1):31-5.
59. Mattawanon N, Spencer JB, Schirmer DA 3rd, Tangpricha V. Fertility preservation options in transgender people: a review. Rev Endocr Metab Disord. 2018;19(3):231-42.

Section

4

Urogynecology

Chapter 33

Anatomy of the Urinary Tract

Karishma Thariani, Ananya Aggarwal

INTRODUCTION

The urinary tract consists of the following:

- Upper urinary tract:
  - Kidneys
  - Ureters
- Lower urinary tract:
  - Urinary bladder
  - Urethra.

All these structures work in sync with each other in the process of urine formation, transportation, storage, and excretion. As students and practitioners of obstetrics and gynecology it is of utmost importance for us to understand the anatomical and functional considerate of the urinary tract, more so of the lower urinary tract. This chapter will, therefore, give you a quick overview of the often difficult to understand anatomy of the urinary tract and its clinical implications.

UPPER URINARY TRACT

Kidneys: The kidney filters the blood of solutes and then secretes, concentrates, and excretes urine into the lower urinary tract via the ureters.

Ureters: The ureters carry urine approximately 22–30 cm from the kidneys to the bladder for excretion. Gravity and peristaltic waves within the ureters propel urine from one end to the other. It is very important for every pelvic surgeon to know the anatomy of the ureter. In the majority of the patients, the course of the ureter is easily demarcated from the level of the pelvic brim **(Fig. 1)**.[1]

- At the level of the pelvic brim the ureter crosses the bifurcation of the common iliac artery and enters the pelvis.
- It is seen crossing the external iliac vessels from lateral to medial at the base of the infundibulopelvic ligaments.

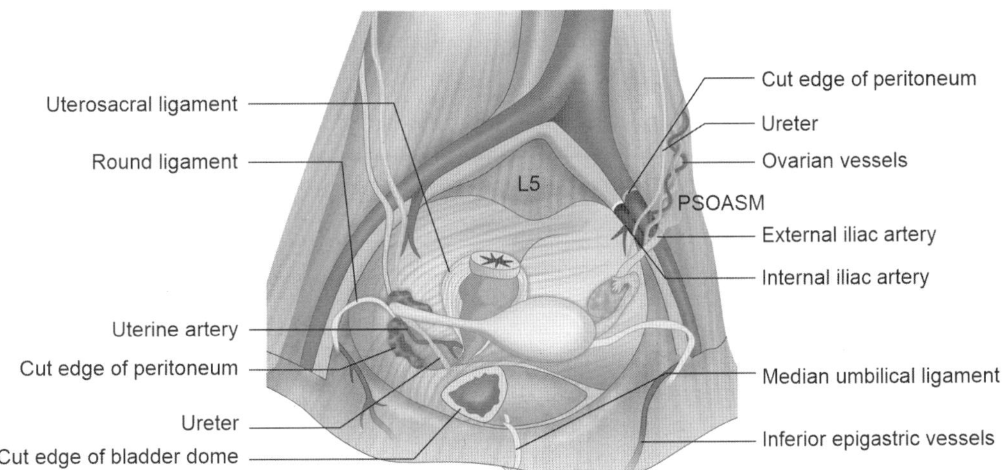

Fig. 1: Anatomy of the pelvic ureter. (PSOASM: PSOAS muscle)

- It further traverses down along the lateral pelvic wall anterior and medial to the internal iliac artery.
- It then courses medially and can be seen disappearing in the ureteric tunnel 1-1.5 cm lateral to the uterosacral insertion.
- Anatomically the ureter is 1-1.5 cm away from the vaginal fornices before it enters the bladder trigone.

Common sites of ureteric injury during gynecological procedures:

- Pelvic brim area while clamping the infundibulopelvic ligament
- The isthmic region while ligating the uterine vessels
- The vaginal apex while clamping or suturing the vaginal cuff

The prevalence of ureteric injuries in benign hysterectomies is 1.8%.

LOWER URINARY TRACT

The lower urinary tract consists of the bladder which is a storage organ and the urethra which is a conduit for the expulsion of urine. The bladder has two functions—urinary storage and expulsion. This is called the *micturition cycle* **(Fig. 2)**.

URINARY BLADDER

The urinary bladder is a hollow muscular organ, which acts as a reservoir for urine. When it is empty it lies entirely in the true pelvis, and when it is starts to fill it rises anterosuperiorly toward the abdomen.

The bladder wall consists of three histological layers **(Fig. 3)**[2]:

- *Inner layer:* It comprises multilayered transitional epithelium and is continuous with the ureters above and urethra below. It is impermeable to fluids and solutes functioning as an elastic barrier against urine.
- *Middle muscular layer (the detrusor muscle):* It is composed of smooth muscle fibers accounting for most of the thickness of the bladder wall. These muscle fibers allow for both the relaxation of the bladder during filling and contraction during emptying.
- *Outer adventitial layer:* It is composed of connective tissue which functions by holding the bladder in the

Bladder filling:
- Accommodation of volumes of urine at a low pressure and with normal sensation
- A bladder outlet that is closed at rest and stress
- Absence of involuntary bladder contractions (detrusor over activity)

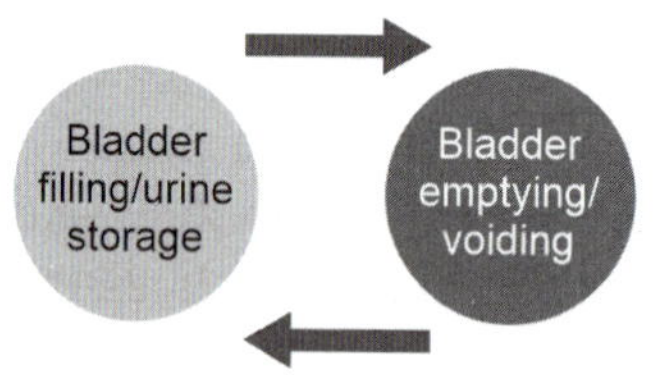

Bladder emptying:
- A coordinated contraction of bladder smooth muscle
- A concomitant lowering of resistance at the level bladder outlet
- Absence of anatomic obstruction

Fig. 2: The micturition cycle.

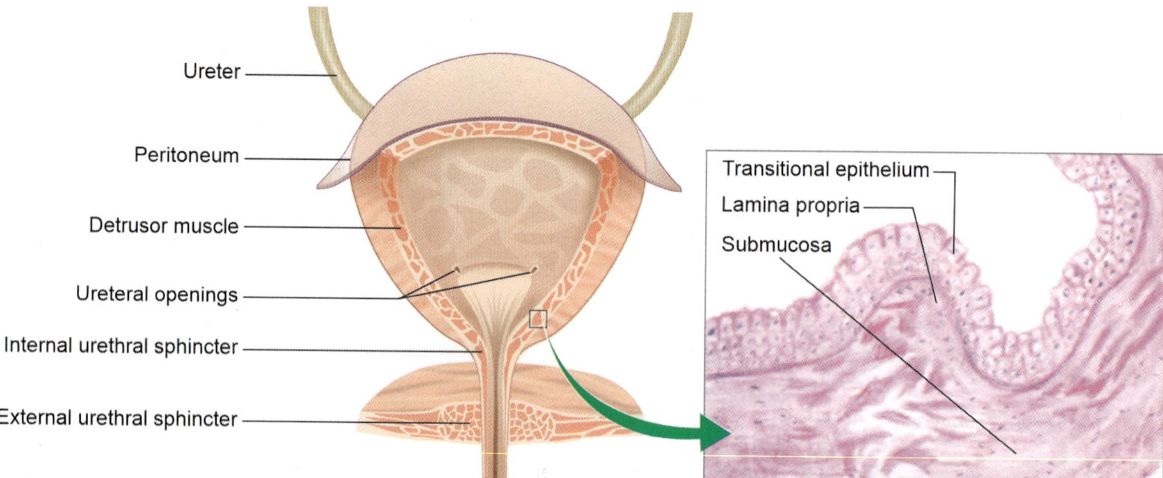

Fig. 3: The layers of the bladder wall.

pelvis via its various fascial condensations and is covered superiorly and posteriorly by peritoneum.

The base of the bladder is located posteroinferiorly and is triangular in shape. Each ureter forms the superolateral angle of the triangular base, while the internal urethral orifice forms the anteroinferior angle. This triangle, within the base of the bladder, is referred to as the trigone and can be seen on cystoscopy. Its smooth appearance is attributed to the lack of trabeculations, seen in the bladder mucosa elsewhere.

URETHRA AND SPHINCTER MECHANISMS

Urethra: The adult female urethra extends a distance of 3–4 cm from the internal urethral meatus of the bladder to the external urethral meatus. From beginning to end it is approximately 6 mm in diameter. Throughout its entire length it is embedded in the adventitia of the anterior vagina. Except during the passage of urine, the anterior and posterior walls of the urethra are in apposition, and the epithelium forms extensive longitudinal folds. The wall of the urethra consists of **(Fig. 4)**[3]:

- Epithelium (thrown into folds)
- Lamina propria (numerous veins)
- Muscular layer.

Urethral Sphincters

Internal Urethral Sphincter (IUS)

At the level of the bladder neck, the internal urethral sphincter (IUS) surrounds the proximal urethra and is seen as a continuation of the detrusor smooth muscle, therefore favoring proximal urethral closure by constricting its lumen. This sphincter is not under voluntary control and remains closed during bladder filling.

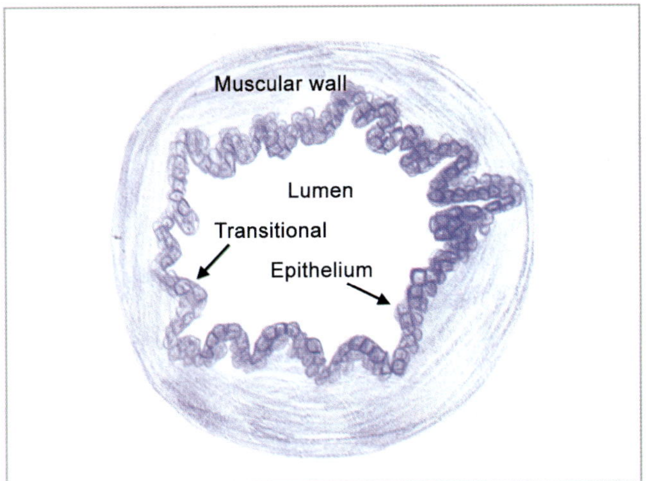

Fig. 4: The lumen of the urethra.

External Urethral Sphincter

It is a skeletal muscle, derived from the inner fibers of the levator ani muscle, surrounds the urethra as it traverses the deep perineal pouch therefore forming the external urethral sphincter (EUS). In females, the EUS begins at the inferior end of the bladder and includes:

- Sphincter urethrae muscle
- Compressor urethrae muscle, and
- Urethrovaginal sphincter.

The EUS is innervated by nerves originating from the Onuf's nucleus in the S2–S4 spinal segments (mainly via the pudendal nerve).

PELVIC FLOOR AND SUPPORTING STRUCTURES

In addition to the structures of the lower urinary tract, the pelvic floor musculature and surrounding structures also contribute to urinary continence. The pelvic viscera, levator ani muscles, pelvic ligaments, as well as overlying peritoneum all act in concert to assist the sphincters in maintaining continence. The pubourethral ligaments are the fascial support of the midurethra to the inner surface of the inferior pubis. Pubourethral ligament laxity is often attributed as a cause of stress incontinence, whereby there is rotational descent of the urethra and hence distraction with opening of the bladder neck/proximal urethra during times of stress such as coughing.

INNERVATION OF LOWER URINARY TRACT (TABLE 1)

Afferent (Sensory) Pathways

Sensation of bladder filling and fullness arises from stretch receptors in the detrusor muscle as well as the urothelium and is conveyed via the parasympathetic afferent nerves to the sacral spinal cord. There, they enter ascending spinal pathways to the brain stem (the pontine micturition center) and cerebral cortex.[4]

Efferent (Motor) Pathways

- Parasympathetic efferents innervate the detrusor muscle. These nerves originate from the sacral spinal cord (S2–S4) and causes detrusor muscle contraction via muscarinic receptors.

- Sympathetic efferents originate from T10 to L2 spinal segments and innervate smooth muscle of the bladder base, bladder neck, and proximal urethra. These fibers have an inhibitory function during bladder filling by causing relaxation of the detrusor via beta-3 adrenergic receptors.
- The somatic system consists of the pudendal and pelvic nerves which originate at cord segments S2–S4 (Onuf's nucleus) and innervate the urethral sphincter.

Pontine Micturition Center

The pontine micturition center (PMC) is located in the brain stem. It is responsible for coordinated micturition, at an appropriate time and place, by integrating afferent sensory feedback from the lower urinary tract and cortical input to assess the appropriateness of voiding under a particular circumstance.

TABLE 1: Innervation of lower urinary tract.

	Detrusor	*Urethral muscle*	*External sphincter*
Sympathetic (T10–L2)	Inhibit (β)	Stimulates alpha receptors	—
Parasympathetic (S2–4)	Stimulates muscarinic receptors	Inhibit	—
Somatic (S2–4)	—	—	Stimulates nicotinic receptors

HOW THE ANATOMY CORRELATES TO THE PHYSIOLOGY OF URINARY CONTINENCE?

The understanding of the anatomy is incomplete if its correlation with the physiology is not understood. The various anatomical components of the lower urinary tract work in sync with each other in a well-orchestrated manner to maintain urinary continence and unobstructed and complete voiding.[5]

Filling and Storage Phase

The bladder normally fills with urine by a series of peristaltic contractions of the ureters at a rate of between 0.5 and 5 mL/minute depending on the persons hydration levels. Under these conditions the bladder pressure rises only minimally, this property of the bladder is called *compliance*. Urethral closure meanwhile is maintained by the combined passive and active effects of its smooth and striated muscle components, its elastic content, and its blood supply. During the early stages of bladder filling, proprioceptive afferent impulses from stretch receptors within the bladder wall pass via the pelvic nerves to sacral dorsal. The descending impulses from the subcortical micturition centers subconsciously inhibit a detrusor motor response. As the bladder fills further the cortical inhibition followed by voluntary contraction of pelvic floor happens to maintain the urethral closure **(Flowchart 1)**.

Emptying phase: The emptying of urine is divided into two phases; the initiation phase and the voiding phase **(Flowchart 2)**.

Flowchart 1: The storage phase.

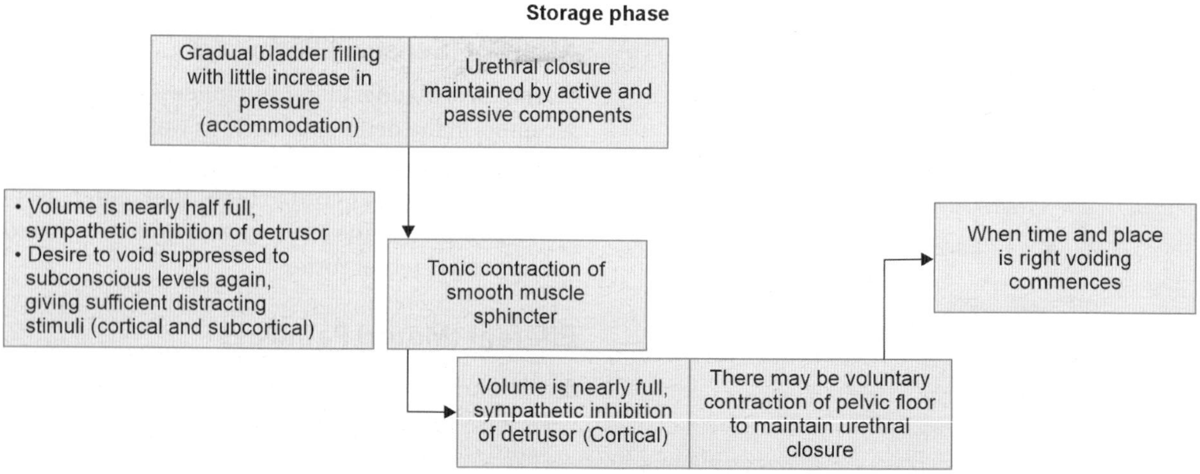

Storage phase

Gradual bladder filling with little increase in pressure (accommodation)

Urethral closure maintained by active and passive components

- Volume is nearly half full, sympathetic inhibition of detrusor
- Desire to void suppressed to subconscious levels again, giving sufficient distracting stimuli (cortical and subcortical)

Tonic contraction of smooth muscle sphincter

Volume is nearly full, sympathetic inhibition of detrusor (Cortical)

There may be voluntary contraction of pelvic floor to maintain urethral closure

When time and place is right voiding commences

Source: Center for Urogynecology and Pelvic Health.

Flowchart 2: The emptying phase.

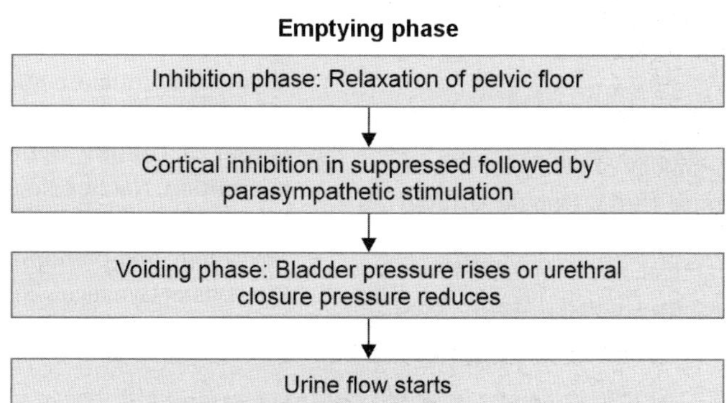

Source: Center for Urogynecology and Pelvic Health.

Flowchart 3: The pathophysiology of incontinence.

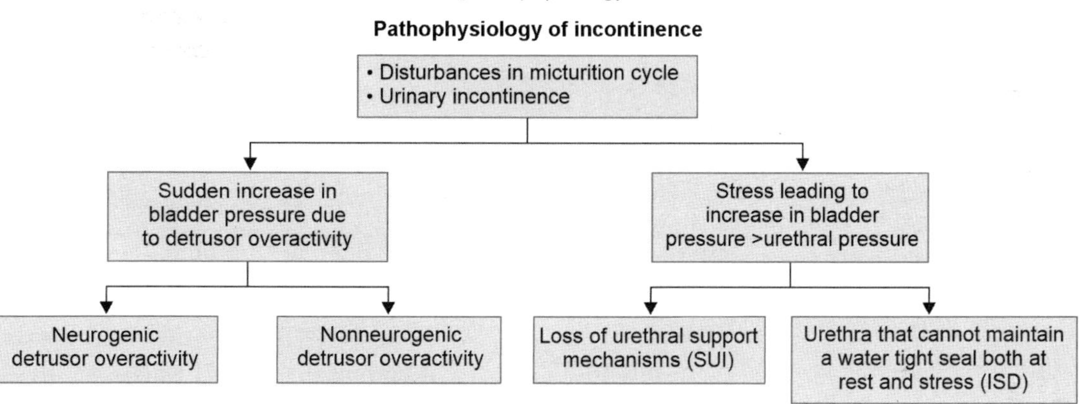

Source: Center for Urogynecology and Pelvic Health.

The initiation phase: Relaxation of the pelvic floor occurs early in the process of emptying and it is likely that simultaneous relaxation of the intrinsic striated muscle also occurs, leading to a marked fall in intra-urethral pressure. This is followed by a rise in intravesical pressure.

The voiding phase: A few seconds later the descending inhibitory influences from the cerebral cortex are suppressed, allowing the stimulation of the efferent parasympathetic impulses via the pelvic nerves to cause contraction of the detrusor. Depending on the relationship between the force of detrusor contraction and the residual urethral resistance, the intravesical pressure may rise to a variable extent (usually <60 cm H_2O), when the urethral pressure falls below the and increasing intravesical pressure, the urine flow begins.

HOW DOES INCONTINENCE OCCUR?

The pathophysiology of incontinence may be due to several disturbances of the micturition cycle described above. Assuming an intact lower urinary tract, urine flow occurs only when the intravesical pressure exceeds the maximum urethral pressure, or when the maximum urethral closure pressure becomes zero, any change or disturbance in this well synchronized system can lead to urinary incontinence **(Flowchart 3)**.

CONCLUSION

The understanding of the anatomy and physiology of the lower urinary tract is mandatory for all gynecologist and pelvic surgeons. It not only helps in understanding the basis of most urogynecological issues but also helps improving the outcomes of management. Surgical procedures can be simplified, complications avoided and even when they happen, recognized and managed appropriately with a sound knowledge of anatomy.

REFERENCES

1. Chancellor MB, Blaivas J. Physiology of the lower urinary tract. In: Kursh ED, McGuire EJ (Eds). Female Urology. Philadelphia: JB Lippincott Company; 1994. p. 39.
2. Kwok D, Tse V, Chan L. Practical anatomy and physiology of urinary continence. In: Chan L, Tse V (Eds). Multidisciplinary Care of Urinary Incontinence. Springer: London; 2013.
3. Brubaker LT, Saclarides TJ (Eds). The female pelvic floor: disorders of function and support. Philadelphia: FA Davis Company; 1996.
4. Fowler CJ, Griffiths D, de Groat WC. The neural control of micturition. Nat Rev Neurosci. 2008;9:453-66.
5. Keane DP, O'Sullivan S. Urinary incontinence: anatomy, physiology and pathophysiology. Baillieres Best Pract Res Clin Obstet Gynaecol. 2000;14(2):207-26.

Chapter

34

Female Stress Urinary Incontinence

Monika Gupta

INTRODUCTION

- Stress urinary incontinence (SUI) is the involuntary leakage of urine on effort or exertion, or on sneezing or coughing.
- Affects 4–35% of women.
- Stress urinary incontinence surgical management has seen revolutionary changes in last one and half decade.

INITIAL EVALUATION

The initial evaluation of patients desiring to undergo surgical intervention should include the following components.

History

The purpose of history taking is to determine the type of urinary incontinence (UI) that is most bothersome to the patient. It should include:

- Number of episodes of leaking in a day
- Usage of pads
- Obstetric histories
- Medical conditions such as diabetes mellitus neurological conditions, urinary tract infection (UTI)
- Intake of bladder irritants (tea, coffee, and alcohol) or any medication (diuretics, alpha agonist)
- *Questionnaires:* Validated tools for the assessment of severity of urinary incontinence and evaluation of quality of life. Examples of validated urinary incontinence questionnaires with good test-retest reliability are:
  - Urogenital distress inventory (UDI)
  - Incontinence impact questionnaire (IIQ)
  - Questionnaire for urinary incontinence diagnosis (QUID)
  - Incontinence-Quality of Life Questionnaire (I-QoL)
  - Incontinence severity index (ISI)
  - International Consultation on incontinence questionnaire (ICIQ).

Physical Examination

- Body mass index (BMI)
- Edema of extremities
- Abdominal examination for any mass or enlarged bladder
- Pelvic examination to rule out pelvic organ prolapse, cystocele, and vaginal atrophy
- Oxford scoring for grading pelvic floor muscle strength
- Neurological examination (anal wink reflex, bulbo-cavernosus reflex, and perineal sensation).

Demonstration of Stress Urinary Incontinence

Stress urinary incontinence should be objectively demonstrated before any anti-incontinence surgery is performed.

- *Cough stress test:* Visualization of fluid loss from urethra simultaneous with a cough
- *Postvoid residual (PVR) assessment:* Postvoid residual volume is then assessed within a few minutes by catheterization or ultrasound examination
- *Urinalysis:* To rule out urinary tract infection, hematuria, dehydration, or excessive fluid intake.

Urodynamics

Invasive urodynamics testing should not be performed before initiating noninvasive treatment. It may also be omitted before surgery in women with uncomplicated stress UI.

Urodynamics should be considered:

- When the diagnosis remains uncertain after history and physical examination.

- When the symptoms do not correlate with physical findings.
- After failed previous treatment.
- Early consideration in patients with neurogenic voiding dysfunction, prior history of radical pelvic surgery and pelvic radiation, and those at risk of upper urinary tract deterioration.

ADDITIONAL EVALUATION

Additional evaluation *should* be performed in the following scenarios:

- Lack of definitive diagnosis
- Inability to demonstrate SUI
- Known/suspected neurogenic lower urinary tract dysfunction
- Abnormal urinalysis
- Urgency-predominant mixed urinary incontinence
- Elevated postvoid residual volume
- High-grade pelvic organ prolapse (POP) (if SUI not demonstrated with POP reduction)
- Evidence of significant voiding dysfunction

Additional evaluation *may* be performed in the following scenarios:

- Concomitant overactive bladder symptoms
- Failure of prior anti-incontinence surgery
- Prior POP surgery.

TREATMENT

In patients who wish to undergo treatment, physicians should counsel regarding the availability of observation, pelvic floor muscle training, other non-surgical options, and surgical interventions. Physicians should also counsel patients on potential complications specific to the treatment options.

Lifestyle Advice

- Weight loss for women with increased weight
- Review of dietary and fluid intake
- Smoking cessation
- Voiding habits review and bladder training
- Review of medication.

Nonsurgical Management

- Pelvic floor muscle training by trained specialist or physiotherapist
- Biofeedback—as an adjunct to pelvic floor muscle training.

Medical Therapy

- Estrogen therapy if postmenopausal
- Tricyclic antidepressants—duloxetine (if women prefer pharmacological to surgical treatment or not suitable for surgical treatment).

Pessaries and Devices

- For women with stress or mixed UI with stress predominance, and those who are not ideal candiates for surgery, pessaries can reduce episodes of UI. Fittings for pessaries must be done under clinician's supervision **(Figs. 1A and B)**.

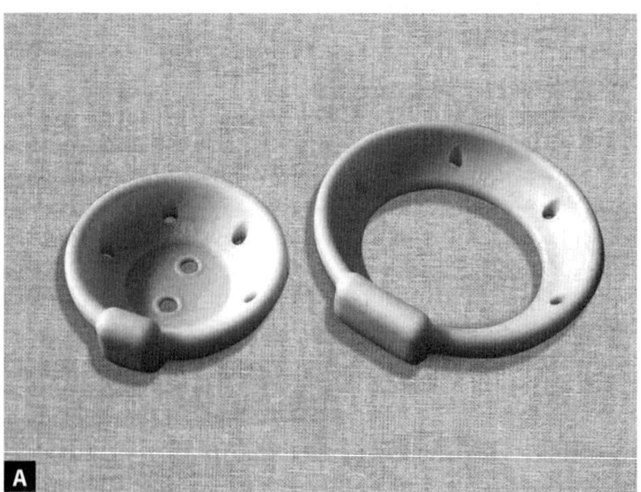

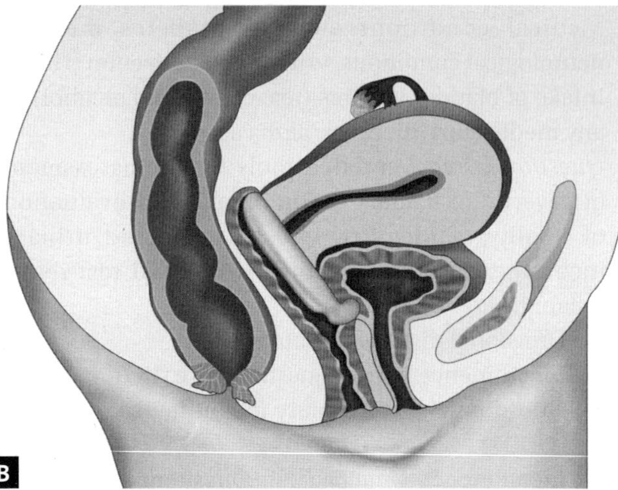

Figs. 1A and B: Continence pessary.

- Vaginal inserts (compress the bladder neck and urethra), urethral plugs (inserted into the urethra to prevent urine loss during activities) **(Figs. 2A to C)**

Surgical Management

Surgery considered only if above options have not been able to relieve the symptoms. Document patient refusal to conservative management if any and discuss individual patient condition needs before offering surgical options.

Midurethral Sling (Synthetic)

If a midurethral sling (MUS) surgery is selected, either the retropubic or transobturator midurethral sling may be offered. Specific risks and benefits of mesh as well as alternatives to a mesh sling should properly be discussed with the patient before the surgery.

For uncomplicated SUI, TOT (transobturator) is preferred over tension-free vaginal tape (TVT) (retropubic) due to lesser postop voiding dysfunction and lesser bladder perforation risk **(Fig. 3)**.

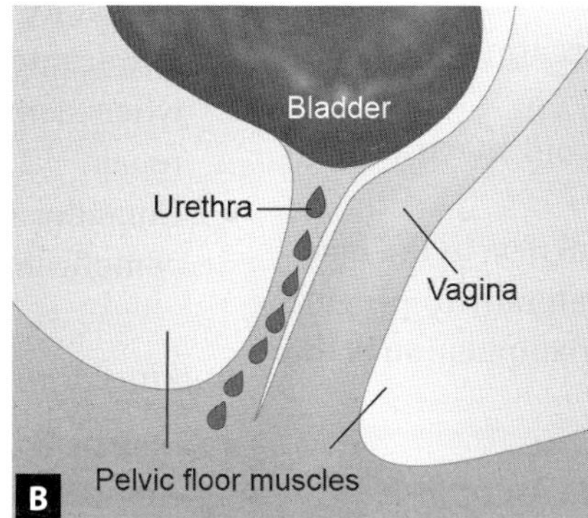

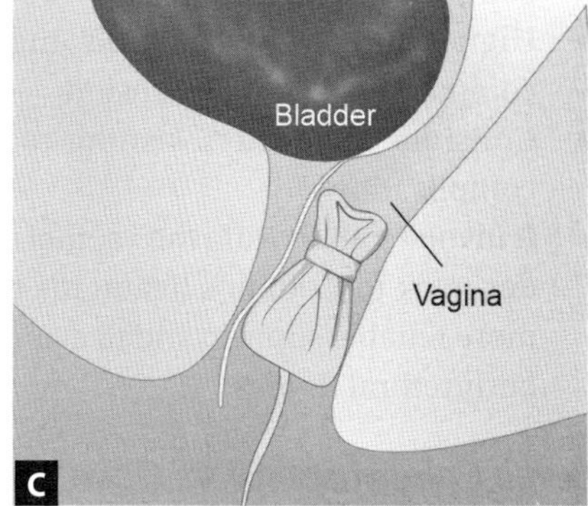

Figs. 2A to C: Vaginal inserts.

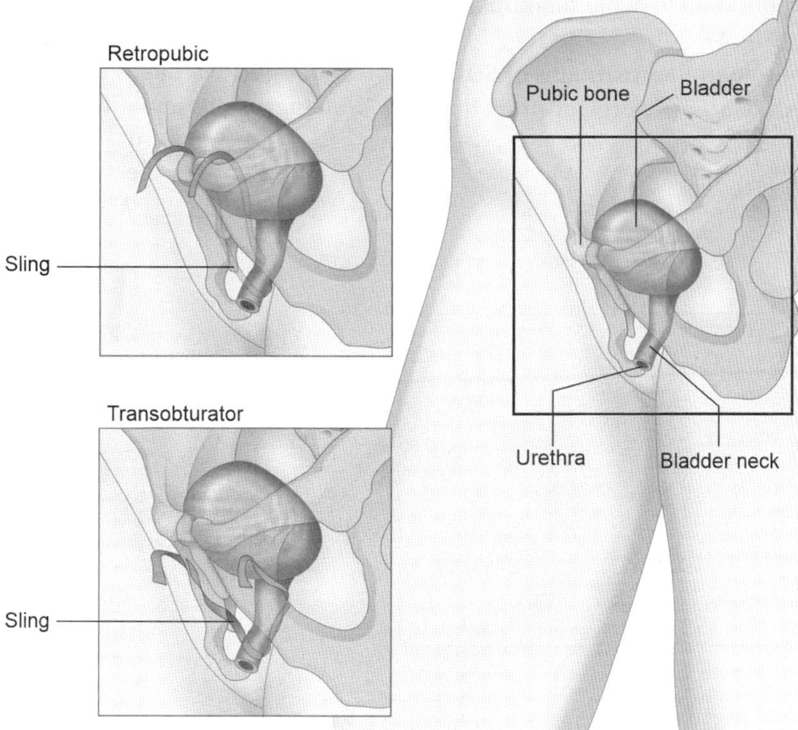

Fig. 3: Mid-urethral slings: Retropubic and transobtruator.
Source: Mayo Foundation for Medical Education and Research.

Synthetic slings have generally low rate of complications, high efficacy, and safety but small possibility of irreversible tape-related adverse events.

Autologous Fascia Pubovaginal Sling

- Similar success rates compared to MUS with longer operating time and possibly higher voiding dysfunction
- Lower rates of chronic pelvic pain, no risk of erosion, or extrusion
- Higher rates of postoperative morbidity
- Lower rate of bladder perforation during surgery compared to MUS
- Higher patient satisfaction and treatment success compared to colposuspension
- Consider in women wishing to avoid mesh-related complications
- It involves anchoring the vaginal wall next to the bladder neck to Cooper's ligaments bilaterally to provide more reliable support and to avoid manipulation of the periurethral tissue.

Burch Colposuspension: Open, Laparoscopic

This procedure involves anchoring the vaginal wall next to the bladder neck to Cooper's ligaments bilaterally to provide more reliable support and to avoid manipulation of the periurethral tissue **(Fig. 4)**.

It plays an important role in the treatment of SUI. It should be contemplated as an apt surgical procedure for woman with SUL, particularly if mesh insertion is a contraindication.

Mainly useful for SUI due to urethral hypermobility and is less effective in incontinence because of intrinsic sphincter deficiency (ISD).

- Inferior outcomes to pubovaginal slings for primary repair, possibly with less voiding dysfunction
- Outcomes similar or slightly less than synthetic MUS however longer operating time and recovery
- Laparoscopic approach when performed same technique as open has similar success rate with less morbidity than open approach.
- Lower rates of success, with higher retreatment rates, when compared to pubovaginal slings for primary repair
- Consider in women wishing to avoid mesh-related complications.

Bulking Agents

- May be a useful option for recurrent SUI with a well-supported urethra.
- Consider in women wishing to avoid mesh-related complications.
- For example, silicone, carbon-coated zirconium beads or hyaluronic acid/dextran copolymer **(Fig. 5)**.
- Major disadvantages are multiple sessions and poor long-term efficacy.

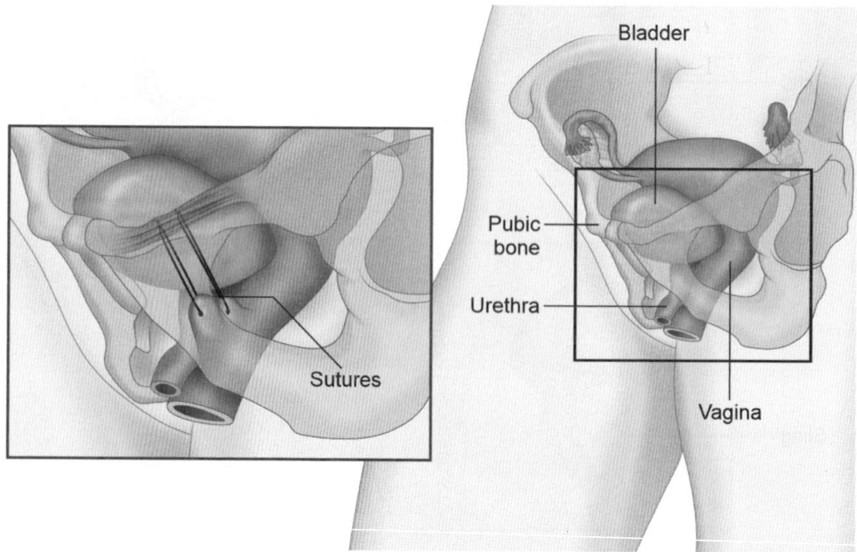

Fig. 4: Burch Colposuspention (retropubic).
Source: Mayo Foundation for Medical Education and Research.

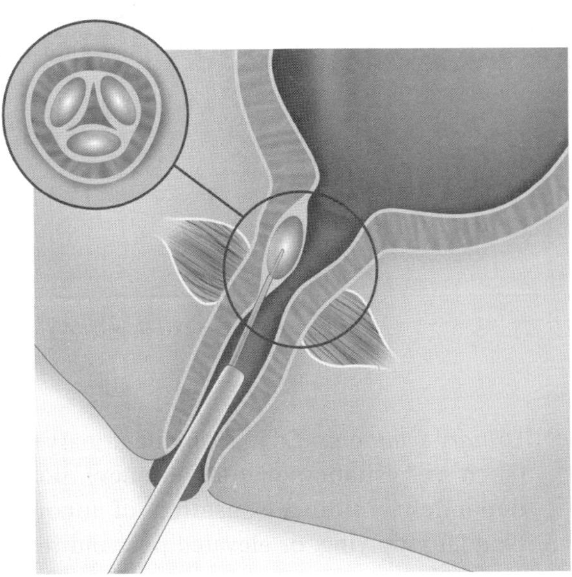

Fig. 5: Bulking agent supporting the urethera.

SPECIAL CASES

Mixed stress and urge incontinence: Manage according to the dominant problem, i.e., stress or urge.

Fixed immobile urethra:

- Pubovaginal sling
- Retropubic midurethral sling
- Urethral bulking agents

Concomitant surgery for pelvic organ prolapse repair and stress urinary incontinence: Any incontinence procedure

Concomitant neurogenic lower urinary tract dysfunction: Surgical treatment following appropriate evaluation and counseling

Childbearing, diabetes mellitus, obesity, geriatric: Surgical treatment following appropriate evaluation and counseling.

FOLLOW-UP

- First follow-up offered within the early postoperative period to assess any voiding problem, pain. Asymptomatic patient should be examined within 6 months postoperatively.
- Patients should be asked about residual incontinence
- Ease of voiding/force of stream
- Recent urinary tract infection
- Pain/sexual function
- New onset or worsened overactive bladder symptom
- Physical examination of all surgical incision sites, to evaluate healing, tenderness, mesh extrusion
- Postvoid residual should be obtained.

CONCLUSION

In most cases of uncomplicated female stress urinary incontinence (SUI), the initial management involves a variety of noninvasive measures, including behavioral modification, pelvic floor exercises with or without biofeedback and pharmacological therapy with Duloxitine.

Surgical correction of SUI aims at stabilizing the urethra and bladder neck in a well-supported retropubic position or augmenting the urethral resistance provided by the intrinsic sphincter unit.

No single procedure or intervention is optimal for all patients. Having a variety of treatment options offers the possibility of tailoring therapy to the desires and needs of the individual patient. The key to an optimal therapeutic outcome is an accurate diagnosis combined with the selection of an appropriate intervention that is acceptable to the patient after balancing multiple factors.

REFERENCES

1. Haylen BT, de Ridder D, Freeman RM, Swift SE, Berghmans B, Lee J, et al. An International Urogynecological Association (IUGA)/International Continence Society (ICS) joint report on the terminology for female pelvic floor dysfunction. International Urogynecological Association. International Continence Society. Neurourol Urodyn. 2010;29:4-20.
2. Sinha S, Agarwal MM, Vasudeva P, Khattar N, Madduri VKS, Yande S, et al. The Urological Society of India Guidelines for the Evaluation and Management of Nonneurogenic Urinary Incontinence in Adults (Executive Summary). Indian J Urol. 2019;35(3):185-8.
3. NICE Guidance - Urinary incontinence and pelvic organ prolapse in women: management: BJU Int. 2019; 123(5):777-803.
4. ACOG Practice Bulletin No. 63. Urinary incontinence in women. Obstet Gynecol. 2005;105:1533-45.
5. NICE clinical guideline 40: The management of urinary incontinence in women, developed by the National Collaborating Centre for Women's and Children's Health. 2006.

Chapter 35

Management of Pelvic Organ Prolapse

Achla Batra, Aakriti Batra

INTRODUCTION

Pelvic organ prolapse (POP) is the descent of pelvic structures into the vagina due to ligament or muscular weakness. POP is subcategorized according to the compartment of descent. Cystocele characterizes anterior wall herniation, rectocele refers to the posterior vaginal wall descent, and vaginal vault prolapse characterizes descent of the uterus, cervix, or apex of the vagina.[1] They can occur either singly or in combination.

EVALUATION

- *History:* Comprehensive obstetric history, chronic cough, constipation, weight lifting, injury or disease of pelvis, sexual history, effect of prolapse on quality of life and lifestyle.
- *Symptoms:* Asymptomatic, feeling of something coming out, bulge of prolapsed part outside introitus, limitation in physical activity, urinary, bowel, or sexual dysfunction.
   - *Urinary symptoms:* Changes in the urinary stream (such as hesitancy or intermittency), position-dependent voiding, sensation of incomplete bladder emptying, or elevated postvoid residual; patients may report pressing on the prolapse vaginally to aid with bladdery emptying; this is known as splinting.
   - *Bowel symptoms:* Sensation of incomplete bowel evacuation due to accumulation of stool into the prolapse, a sensation of anorectal blockage, and the need for vaginal or perineal splinting to help complete evacuation.
   - *Sexual dysfunction:* Dyspareunia, vaginal dryness, obstructed intercourse.

EXAMINATION

The examination of pelvic wall prolapse is shown in **Figures 1A to F**.

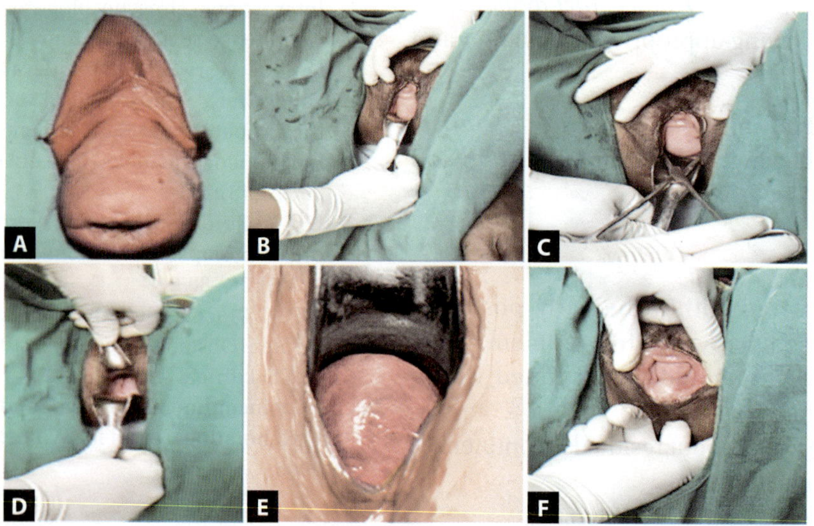

Figs. 1A to F: (A) Visual inspection on straining; (B) anterior compartment evaluation by retraction of posterior vaginal wall; (C) anterior vaginal central defect evaluation by lateral retraction of anterior vaginal wall; (D) apical compartment evaluation by retraction of anterior and posterior vaginal wall; (E) posterior compartment evaluation by retraction of anterior vaginal wall; and (F) posterior compartment evaluation by per rectal examination.

CLASSIFICATION OF PROLAPSE BY PELVIC ORGAN PROLAPSE QUANTIFICATION

The classification of prolapse by POP quantification is given in **Figure 2**.

SHAW'S CLASSIFICATION OF PELVIC ORGAN PROLAPSE

Figure 3 shows the Shaw's classification of POP.

Anterior vaginal wall:

- Upper two-thirds—Cystocele
- Lower one-third—Urethrocele

Posterior vaginal wall:

- Upper one-third—Enterocele
- Lower two-thirds—Rectocele

Uterine descent:

- Descent of cervix into vagina—First degree
- Descent of cervix up to introitus—Second degree

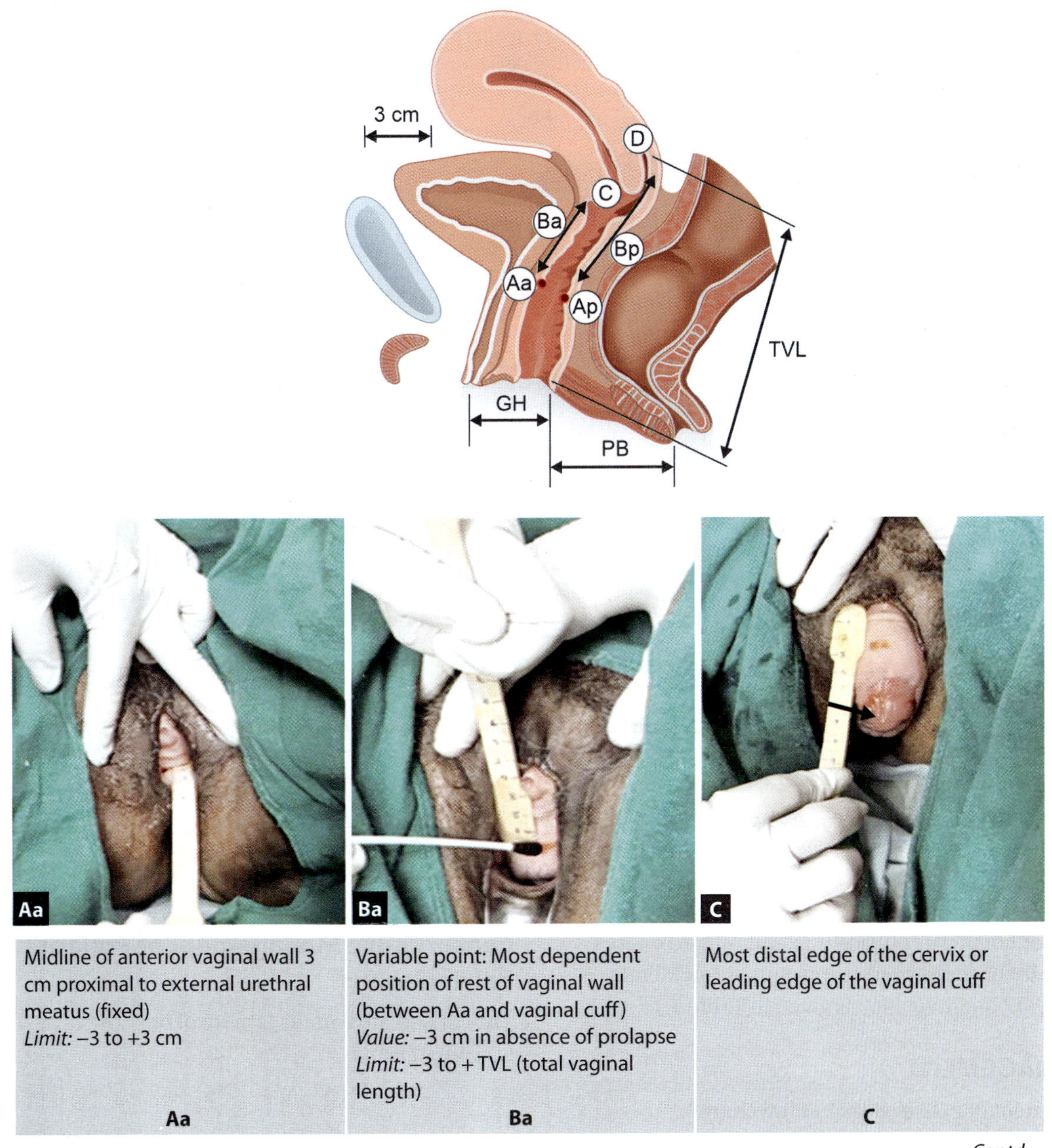

Aa	Ba	C
Midline of anterior vaginal wall 3 cm proximal to external urethral meatus (fixed) *Limit:* –3 to +3 cm	Variable point: Most dependent position of rest of vaginal wall (between Aa and vaginal cuff) *Value:* –3 cm in absence of prolapse *Limit:* –3 to + TVL (total vaginal length)	Most distal edge of the cervix or leading edge of the vaginal cuff

Contd...

Contd...

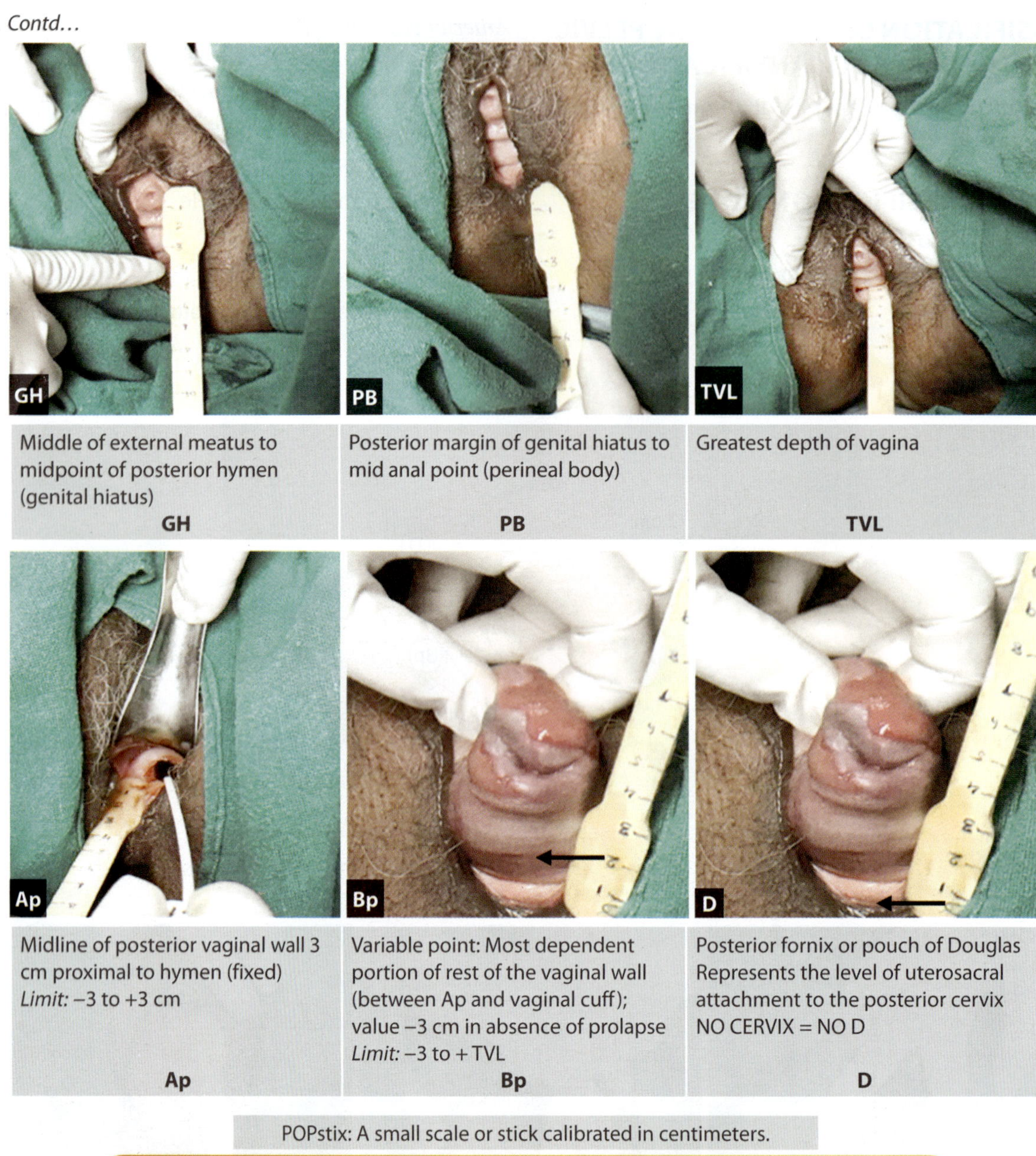

Fig. 2: Classification of prolapse by pelvic organ prolapse quantification.
Note: All examinations except TVL are done while maximal straining. TVL is done without straining.

- Descent of uterus outside introitus—Third degree
- Descent of cervix outside introitus—Procidentia

The POP quantification is given in **Table 1**.

MANAGEMENT

Management of POP is given in **Flowchart 1**.

Conservative Management of Prolapse of Pelvic Organs

The conservative management of prolapse of pelvic organs is given in **Box 1**.

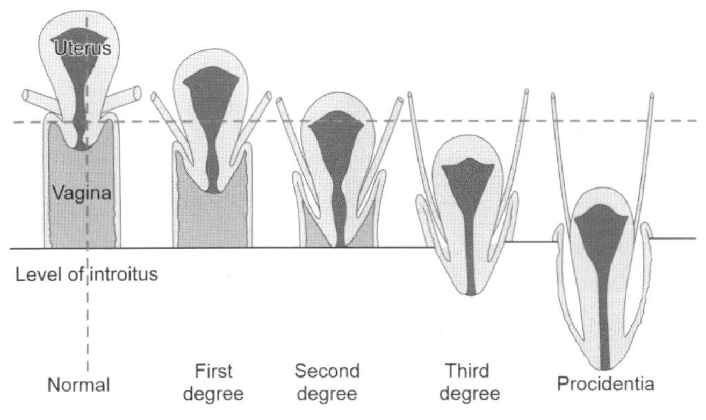

Fig. 3: Shaw's classification of pelvic organ prolapse.

TABLE 1: Pelvic organ prolapse quantification (POPQ) stages.

Stage 0	No prolapse is demonstrated Points Aa, Ap, Ba, Bp all are at –3 cm, and point C or D is between total vaginal length (TVL) and (TVL – 2 cm)
Stage 1	The most distal portion of prolapse is >1 cm above the level of hymen
Stage 2	The most distal portion of prolapse is ≤1 cm proximal or distal to plane of hymen
Stage 3	The most distal portion of prolapse >1 cm distal to plane of hymen but no farther than 2 cm less than the TVL
Stage 4	Complete to nearly complete eversion of the vagina. The most distal portion of prolapse protrudes to ≥ (TVL – 2 cm)

Flowchart 1: Management of pelvic organ prolapse (POP).

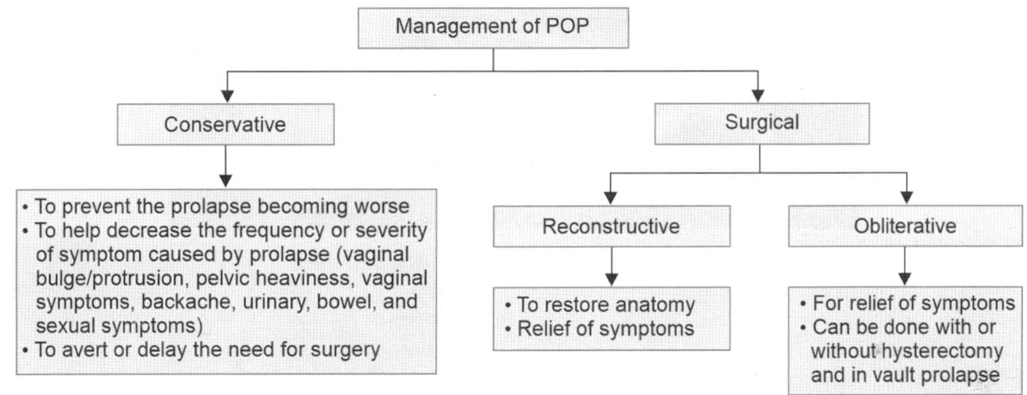

BOX 1: Conservative management of prolapse of pelvic organs.

- Asymptomatic women should be offered lifestyle modification
- Stage1 and stage 2 prolapse should be offered pelvic floor muscle training[2] (PFMT)
- Pessary can be effective in 85% cases

Life style modification:
- Avoid constipation
- Weight Loss
- Avoid heavy weight lifting
- Avoid high impact exercise

PFMT:
- Increases the muscle mass and causing elevation of the levator plate to a higher position in the pelvis
- Improves muscle coordination and timing to provide pelvic organ support at time of rise in intra-abdominal pressure

Pessary:
- Patient preference for avoidance of repeat surgery in recurrent prolapse
- To treat vaginal ulcerations caused by severe POP
- Presence of severe medical comorbidities that make the patient a poor surgical candidate
- Current pregnancy, to manage POP and cervical insufficiency
- Desire for future childbearing; the benefit of surgical repair of POP may be nullified by subsequent pregnancy and childbirth; short vaginal length
- Wide vaginal opening, or a history of hysterectomy can often have a fitting failure

(PFMT: pelvic floor muscle training; POP: pelvic organ prolapse)

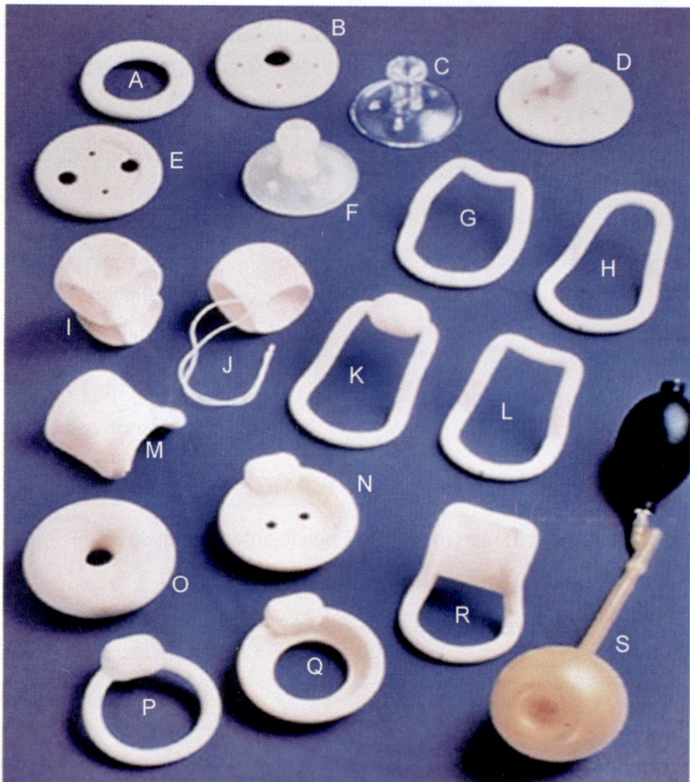

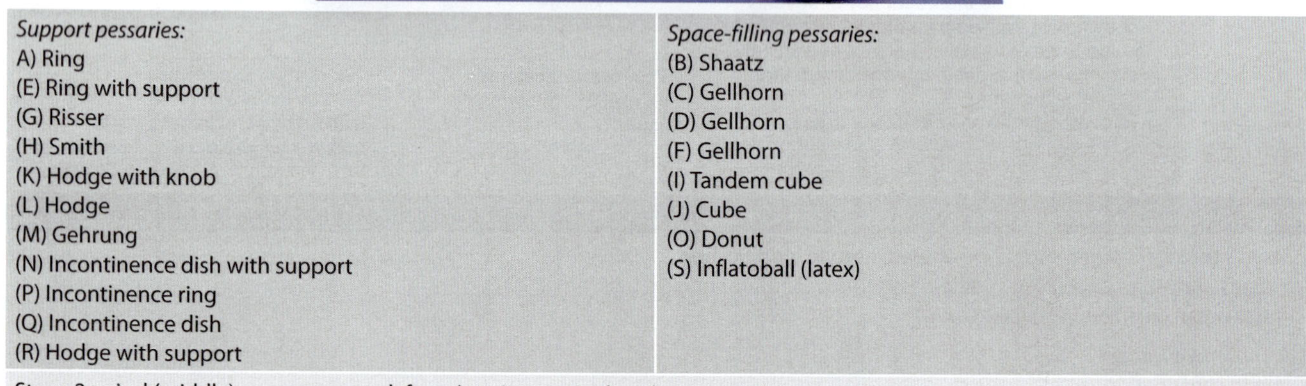

Support pessaries:	*Space-filling pessaries:*
A) Ring	(B) Shaatz
(E) Ring with support	(C) Gellhorn
(G) Risser	(D) Gellhorn
(H) Smith	(F) Gellhorn
(K) Hodge with knob	(I) Tandem cube
(L) Hodge	(J) Cube
(M) Gehrung	(O) Donut
(N) Incontinence dish with support	(S) Inflatoball (latex)
(P) Incontinence ring	
(Q) Incontinence dish	
(R) Hodge with support	

Stage 2 apical (middle) compartment defects (uterine, vaginal vault, enterocele)—ring pessaries
Stage 3 or 4 uterine or vaginal vault prolapse—Gellhorn or donut pessaries
Rectocele—Donut pessary
Wide genital hiatus—space-filling pessary

Fig. 4: Pessary types.

Pessary Types

The pessary types are shown in **Figure 4**.

Flowchart 2 shows reconstructive surgery for POP.

Table 2 shows the principles of vaginal surgeries.

The principles of abdominal/minimally invasive surgeries are given in **Table 3**.

CHOICE OF SURGERY

Factors affecting choice of surgery:

- Age, desire for fertility, desire for uterine preservation, etiology, symptoms
- Anatomical changes—supravaginal elongation of cervix
- Fitness of patient

Flowchart 2: Reconstructive surgery.

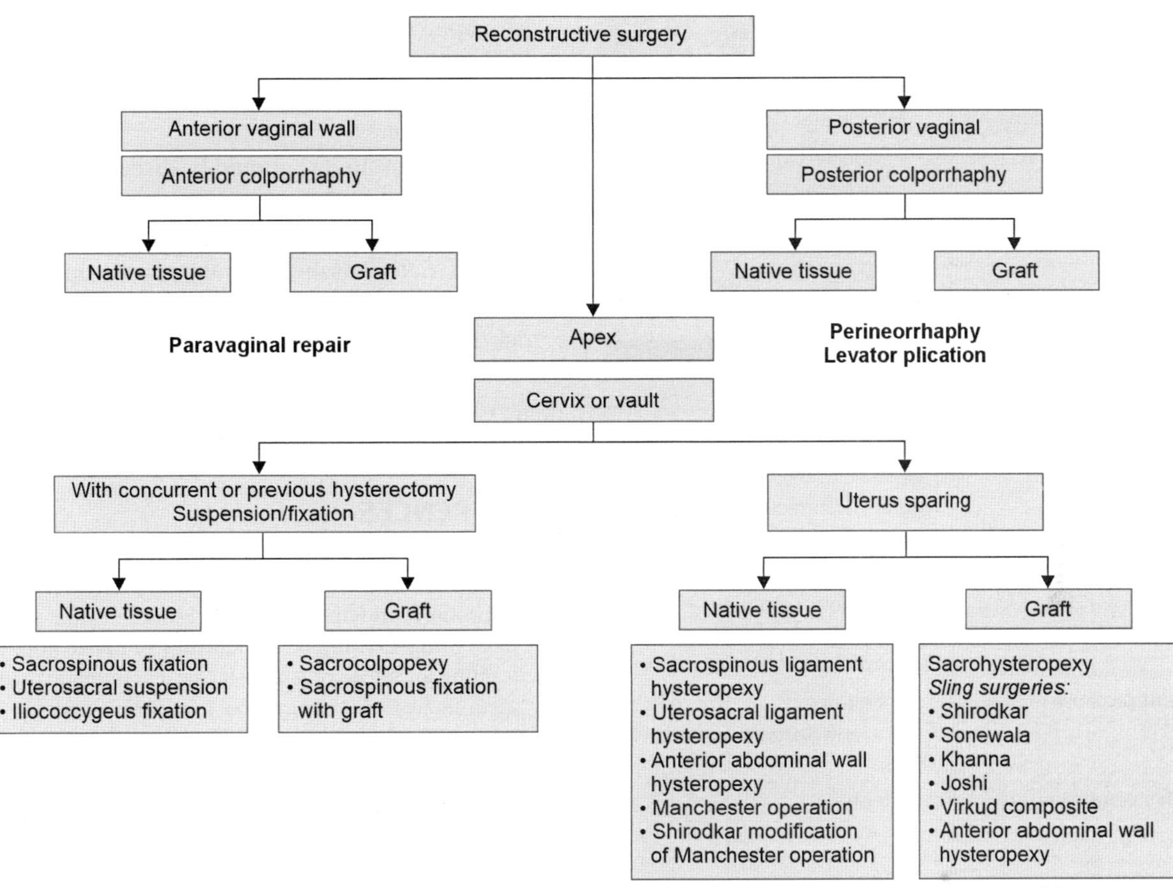

TABLE 2: Principles of vaginal surgeries.

Sacrospinous fixation	Vaginal apex or cervix fixed to sacrospinous ligament with suture or mesh
Uterosacral suspension	Vaginal apex or cervix suspended by suture to higher part of uterosacral ligament
Manchester	Amputation of cervix and shortening of Mackenrodt ligaments
Shirodkar modification of Manchester operation	Shortening and anterior transposition of uterosacral ligaments

TABLE 3: Principles of abdominal/minimally invasive surgeries.

Sacrocolpopexy/sacrohysteropexy	Vault/cervix attached to anterior longitudinal ligament (ALL) at sacral promontory (SP)
High uterosacral suspension/ hysteropexy	Vaginal apex or cervix suspended by suture to higher part of uterosacral
Anterior abdominal wall hysteropexy	Cervix attached by autologous rectus sheath sling or mesh to rectus sheath
Lateral pectopexy	Anterior cervix attached by mesh to pectineal ligament
Sling operations	*Shirodkar:* Mersilene tape or mesh from posterior cervix to ALL at level of SP *Sonewala:* From posterior cervix to ALL unilaterally on right side only *Khanna:* From cervix to anterior superior iliac spine *Virkud composite:* On right side from posterior cervix to SP and on left side from anterior cervix to anterior abdominal wall (rectus sheath)

TABLE 4: Reproductive age-group.

≥2nd degree apical prolapse	Surgery
Nulliparous prolapse	Sling surgery
With supravaginal elongation	Manchester Operation
Without supravaginal elongation	• Modified Manchester • Sacrospinous hysteropexy • Uterosacral hysteropexy • Abdominal wall fixation • Lateral pectopexy

TABLE 5: Postmenopausal.

Apical prolapse	Surgery
Second or third degree	Vaginal hysterectomy with McCaull suspension of apex
Fourth degree (procidentia)	• Vaginal hysterectomy with: – Sacrospinous fixation – High uterosacral suspension
Vault	• Sacrospinous colpopexy • Uterosacral colpopexy • Sacrocolpopexy
Recurrent prolapse	• Sacrocolpopexy • Colpocleisis if sexual activity not desired
Medically compromised	Colpocleisis

The choice of surgery in reproductive and postmenopausal age-group is given in **Tables 4 and 5**.

CONCLUSION

- Management of POP depends on stage of severity of symptoms, age of patient, desire for childbearing, desire for retaining uterus, and general condition of patient.[3,4]
- In patients with up to stage 1 prolapse, lifestyle modification and Kegel exercise help in alleviating symptoms.
- They should be offered to all women with prolapse as an adjuvant to any management option chosen.
- Pessary are the first-line treatment for all stages meeting the criteria for pessary insertion.
- Surgical treatment can be reconstructive with uterus-sparing apical fixation or hysterectomy with apical suspension. These can be done vaginal or abdominally with open route or laparoscopically.[5-7]
- In very old or medically poor risk patients, colpocleisis with or without removal of uterus can be done and is very effective.
- The choice of surgery should be discussed with patient and is done according to her choice. Success rate of most reconstructive surgeries are almost similar and in general, native-tissue vaginal prolapse operations are well tolerated, with good long-term efficacy, and low risk of major complications.[8,9]
- Mesh augmented apical suspension specially sacrocolpopexy/sacrohysteropexy has lower failure rate, but there is associated risk of mesh erosion.

REFERENCES

1. Madhu C, Swift S, Moloney-Geany S, Drake MJ. How to use the Pelvic Organ Prolapse Quantification (POP-Q) system? Neurourol Urodyn. 2018;37:S39-S43.
2. Li C, Gong Y, Wang B. The efficacy of pelvic floor muscle training for pelvic organ prolapse: a systematic review and meta-analysis. Int Urogynecol J. 2016;27(7):981-92.
3. Hagen S, Stark D. Conservative prevention and management of pelvic organ prolapse in women. Cochrane Database Syst Rev. 2011;(12):CD003882.
4. Bugge C, Adams EJ, Gopinath D, Stewart F, Dembinsky M, Sobiesuo P, et al. Pessaries (mechanical devices) for managing pelvic organ prolapse in women. Cochrane Database Syst Rev. 2020;11(11):CD004010.
5. Maher C, Feiner B, Baessler K, Christmann-Schmid C, Haya N, Brown J. Surgery for women with anterior compartment prolapse. Cochrane Database Syst Rev. 2016;11(11):CD004014.
6. Mowat A, Maher D, Baessler K, Christmann-Schmid C, Haya N, Maher C. Surgery for women with posterior compartment prolapse. Cochrane Database Syst Rev. 2018;3(3):CD012975.
7. Betschart C, Cervigni M, Contreras Ortiz O, Doumouchtsis SK, Koyama M, Medina C, et al. Management of apical compartment prolapse (uterine and vault prolapse): A FIGO Working Group report. Neurourol Urodyn. 2017;36(2):507-13.
8. Virkud A. Conservative Operations in Genital Prolapse. J Obstet Gynaecol India. 2016;66(3):144-8.
9. Deffieux X, Thubert T, Donon L, Hermieu JF, Le Normand L, Trichot C. Colpocleisis: guidelines for clinical practice. Prog Urol. 2016;26 Suppl 1:S61-72.

Chapter 36

Urinary Tract Injuries: Diagnosis and Management

Venus Bansal, Ananta Kanwar

INTRODUCTION

Urinary tract injuries have been considered one of the most potential injuries during the gynecological surgeries. In view of increasing medicolegal implications, there has been innovation and restructuring of steps of surgery and techniques to prevent urinary tract injuries, which has significantly decreased urological morbidities. Surgeon's vigilance and familiarity with anatomy and identification of injuries timely followed by an effective repair is the key to satisfactory postoperative recovery.

Ureteral injuries may occur in about 0.2%–1% of all pelvic surgeries and in 30% of radical hysterectomies. Obstetric and gynecological surgeries account for about 50% of all these injuries. They are quite rare during cesarean deliveries, but during cesarean hysterectomies the incidence of ureteral injuries increases to 3%. It is approximately seven times as compared to nonobstetric hysterectomies.[1]

The probability of bladder injury caused by obstetric and gynecological surgery is about 0.2%–20.0%. The foreign literature shows that the total incidence of urinary tract injury caused by laparoscopic surgery is about 0.05%–8.3%, of which ureteral injury accounts for 0.5%–3% and bladder injury accounts for 0.02%–8.3%.[2]

To understand injury and to plan the repair of the injury, getting thorough with the anatomy is of utmost importance.

SURGICAL ANATOMY OF URETER

In normal adults, the ureter is between 25 and 30 cm in length from the renal pelvis to the trigone of the bladder. The pelvic brim divides it into the abdominal and pelvic parts of ureter. Each of these components is approximately 12–15 cm in length.

In cross section, ureter is divided into distinct layers:

- The lumen lined by transitional epithelium
- Mucosal lining
- The muscular layer consisting of three types of muscular fibers: longitudinal, circular, and spiral smooth muscle fibers
- The adventitia which consists of the intercommunicating blood vessels
- The peritoneum lies over the ureter, hence making it a retroperitoneal organ.

Abdominal Ureter

Abdominal ureter runs on the anterior surface of psoas muscle and posterior to ovarian vessels till the level of pelvic brim. The right ureter is slightly lateral to inferior vena cava and enters pelvis over the bifurcation of right common iliac artery. The left ureter runs lateral to aorta and posterior to inferior mesenteric artery, ovarian vessels, and colon further entering the pelvis over left common iliac artery bifurcation.[3]

Pelvic Ureter

In the pelvis, it lies in a special connective tissue sheath that is attached to the peritoneum of the lateral pelvic wall and the "medial leaf" of the broad ligament. This explains why the ureter remains adhered to the peritoneum and moves with it when the retroperitoneal space is entered. It descends into the pelvis lateral to the sacrum and immediately ventral to the internal iliac (hypogastric) artery. The ureters then course medial to the internal iliac artery and its anterior branches. The ureters subsequently pass beneath the uterine artery; this relationship is often referred to as water under the bridge. It lies approximately 1–2 cm lateral to the cervix and courses within a "ureteral tunnel" surrounded by loose areolar tissue that allows for its peristalsis. Within this tunnel, the ureter roughly separates the anterior fibers of the cardinal ligament from the posterior fibers of the uterosacral ligament. Once through this tunnel, the ureter travels medially and anteromedially over the vaginal fornix to enter the trigone of the bladder.[4] Radiologic studies have revealed that, in 10% of cases involving cervical pathology, the ureter may be only 5 mm from the cervix.[5] Finally, the ureter courses obliquely within the bladder wall for approximately 1.5 cm to terminate at the ureteric orifices.

Above the pelvic brim, the blood supply of the ureter is derived from medial vessels. Below the pelvic brim, the blood supply originates laterally. Therefore, for abdominal ureter dissection and mobilization of the ureter should be approached from its lateral aspect. The pelvic ureter receives blood supply from the vessels it passes: the common iliac, internal iliac, uterine, and superior vesical vessels. The ureter's course runs medial to these vessels, and thus its blood supply reaches the ureter from lateral sources. This is important during ureteral isolation to avoid its devascularization injury.

SURGICAL ANATOMY OF BLADDER

The bladder is made of three layers—the mucosa made of specialized transitional epithelium with underlying lamina propria, a smooth muscle layer (detrusor muscle), and a fibrofatty serosal layer. An empty bladder is a pelvic organ and with increase in distension, it becomes an abdominal organ while occupying a preperitoneal position. Visceral peritoneum covers the confluence of the uterus, upper cervix, and posterior bladder, forming the vesicouterine pouch. The base of the bladder rests on the anterior vagina, supported by the vaginal attachments to the arcus tendineus fascia pelvis from the ischial spine to the pubic ramus. While the bladder floor and anterior vaginal wall are separable, the fusion plane between these structures can sometimes be densely adherent. Anteriorly, the bladder rests against the inner surface of the pubic bones and then, as it fills, also against the anterior abdominal wall.

The bladder is divided into a body (dome) and a fundus (base) approximately at the level of the ureteric orifices. The bladder dome is thin-walled and also distensible, whereas the bladder base is thicker and undergoes less distention during filling. The bladder base consists of the vesicle trigone and the detrusor loops. These loops are two U-shaped bands of fibers found at the vesical neck, where the urethra enters the bladder wall. Ureteric orifices lie within the trigone and empty here into the bladder.

The vascular supply to the bladder is derived from the internal iliac (hypogastric) artery via the superior and middle vesical arteries. Arborization of these vessels allows for redundancy and low risk of devascularization with pedicle ligation.

MANAGEMENT OF URETER INJURY

Prevention is better than cure, hence it becomes of utmost importance to identify the risk factors of iatrogenic ureter injuries, so that adequate precautions and steps can be taken to prevent it. The risk factors for iatrogenic ureter injury include congenital abnormalities [duplication of ureter, ectopic ureter (EU)], pelvic endometriosis, pelvic inflammatory disease or pelvic infection, previous pelvic surgery, previous pelvic irradiation, large pelvic tumors distorting the anatomy, and retroperitoneal fibrosis.

Types of Iatrogenic Injuries[6]

- *Avulsion:* It occurs when ureter is retracted forcefully in tissues weakened by necrosis or infection.
- *Transection:* It is sharp injury by scissors or scalpel. It mostly occurs when the ureter is engulfed within a tumor or adhesions.
- *Ligation:* When ureters are ligated by mistakenly identifying them as bleeding vessels; it commonly occurs during vaginal hysterectomy.
- *Crushing:* This injury mostly occurs when a surgeon is trying to desperately clamp a bleeder to control a hemorrhage. It is most commonly seen in radical hysterectomies. Crush injuries lead to necrosis, which further lead to fistula formation or stricture.
- *Devascularization:* While isolating the ureter, if extensive dissection is done, it leads to devascularization of the ureter which leads to ischemia, which later results in fistula or stricture.
- *Fulgration:* In cases of endometriosis, while using laparoscopic diathermy or laser for treatment of lesions, thermal injury to ureters is very common.
- *Kinking:* While taking a stitch during prolapse repairs, if the suture goes through the periureteric tissue or ureteric wall, the ureter may kink and lead to a decrease in the lumen of the ureter.

The best results of ureter repair are achieved when the iatrogenic injury is identified peroperatively. Diagnostic delays may vary from 2 days to 4 weeks and rarely may be later.

Postoperative presentation of ureter injury includes abdominal distension or excessive abdominal drain output, persistent fever, abdominal or flank pain, ileus, wound, or vaginal leakage. Flank pain is attributed to hydronephrosis, peritoneal irritation, or peritonitis.

PREVENTION OF URETERIC INJURY

After a clinical assessment of the patient, the decision to do additional investigations or procedures should be taken so that the risks of ureter injury can be minimized. Key steps to ensure this include:

- Appropriate operative approach and adequate exposure of the surgical field.
- Visualize, identify, and mobilize ureter whenever in doubt. Preferably identify ureters above the level of pathology.
- Seek early urosurgical assistance when appropriate.
- Preserve the para-ureteric fascia during dissection and mobilization of ureter.
- During pelvic lymph node dissection, the ureter is pushed medially, while during sigmoid colon dissection, the ureter may preferentially be reflected with descending colon.
- Plan for preoperative intravenous (IV) pyelography or computed tomography (CT)-IV urogram to understand any aberrant or distorted anatomy.
- Plan for preoperative ureteric stenting in case of expected difficult surgery.
- Avoid blind cauterization of bleeders near ureter and avoid taking broad pedicle for clamping artery or bleeders.
- Near the ureter, apply diathermy if required for minimum time, on minimum surface area and with minimum voltage.
- Before touching the ureter, the hot instruments should be cooled off.

PATIENT POPULATION AT HIGH RISK FOR UROLOGICAL INJURIES

- Multiparous females—in view of multiple cesarean deliveries and development of pelvic adhesions
- Obesity
- Intraoperative bleeding
- History of abdominal and pelvic surgeries
- Uterine size >12 weeks size
- Endometriosis
- Pelvic inflammatory disease
- Broad ligament fibroids (true and false), cervical fibroids
- Oncosurgery for pelvic pathology
- Congenital mullerian anomalies (mullerian anomalies are usually associated with urological anomaly)
- Large adnexal masses.

PEROPERATIVE APPROACHES FOR URETERAL DISSECTION

Medial Approach

Dissection of ureter is performed early before the pelvic peritoneum becomes edematous and before there is loss of ureteral peristalsis induced by surgical stress or Trendelenburg position. If no significant cul-de-sac or adnexal abnormality is there, uterus is anteverted with uterine manipulator, the ureters can be easily seen on pelvic side wall (posterior leaf of broad ligament). The peritoneum immediately above the ureter to be incised to create a peritoneal "window" to make division of the infundibulopelvic ligament or adnexal pedicle safer. Scissors are used to divide the peritoneum overlying the ureter, and are inserted into the incision created and then it can be spread. Thereafter, one blade of the scissors is placed on top of the ureter, with its blade visualized through the peritoneum, and the peritoneum divided. The ureter and its surrounding longitudinal endopelvic fascia sheath are dissected together away from the peritoneum without compromising its blood supply. This dissection is continued into the deep pelvis where the uterine vessels cross the ureter, lateral to the cardinal ligament insertion into the cervix.[7]

Superior Approach

The superior approach starts with dissection of rectosigmoid on left and cecum on right from the pelvic brim. Colon is freed from infundibulopelvic ligament vessels from roof of broad ligament to identify ureter. The ureter is reflected off the broad ligament and traced into the pelvis.

Lateral Approach

After displacing the uterus to the contralateral side, a pelvic side wall triangle is identified formed by the round ligament, lateral border by the external iliac artery, and medial border by the infundibulopelvic ligament. The peritoneum in the middle of the triangle is incised with scissors and the broad ligament opened by bluntly separating the extraperitoneal areolar tissues. The infundibulopelvic ligament is pulled medially with grasping forceps to expose the ureter at the pelvic brim where it crosses the common or external iliac artery. The operator then searches for the ureter distal to the pelvic brim and lateral to the infundibulopelvic ligament. The dissection is carried bluntly underneath and caudad to the round ligament, until the obliterated hypogastric artery is identified extraperitoneally. After the obliterated hypogastric artery has been identified extraperitoneally, it is an easy matter to develop the paravesical space by bluntly separating the areolar tissue on either side of the artery. The pararectal space is opened by blunt dissection proximal and medial to the uterine vessels, which lie on top of the cardinal ligament. After the pararectal space has been opened, the ureter is identified easily on the medial leaf of the broad ligament (really posterior leaf), which forms the medial border of the pararectal space.

PEROPERATIVE APPROACH FOR BLADDER DISSECTION

The round ligaments are divided at their midportion. Scissors or the same electrode are used to divide the vesicouterine peritoneal fold, starting at the left side and continuing across the midline to the right round ligament. The upper junction of the vesicouterine fold is identified as a white line firmly attached to the uterus, with 2–3 cm between it and the bladder dome. Incision is made below the white line while lifting the peritoneum covering the bladder. Bladder is mobilized off the uterus and upper vagina using scissors, or bluntly.

A bladder when densely adherent to the lower segment poses difficulty in pushing it down. Safe approach is lateral window technique. In the lateral area, the two leaves of broad ligament remains free and allows easy possibility for entry to dissect either vaginally or abdominally. In laparoscopic approach after cutting the cornual pedicles, the broad ligament is dissected down till the uterine bundle is identified. Once the uterine vascular bundle is identified, the space can be dissected just above these vessels to reach the lateral margins of cervix. Any fatty tissue should be moved with the bladder. Once the bladder is completely dissected and lifted off from the cervix below, midline adhesions of the bladder and pillars can be gradually separated using sharp dissection or harmonic ultracision staying near to cervix.[8]

DIAGNOSIS OF URETERIC INJURY DURING SURGERY

Visualizing the peristalsis of the ureter (Kelly's sign) is inadequate to demonstrate the intact ureter. So additional methods include:

- After complete retroperitoneal exposure of the ureter, look for hematoma, contusion, discoloration, and any obvious injury to the ureter.
- Anterograde IV methylene blue combined followed by cystoscopy (to avoid toxicity <2 mg/kg methylene blue should be used)
  - If methylene blue is seen coming through both ureters with peristalsis—ureters are safe.
  - If no methylene blue is seen or blood or air bubbles mixed with blood are seen on cystoscopy—ureters are injured.
  - Also, the appearance of methylene blue in the operative field is a sign of ureter injury.

DIAGNOSIS OF URETERIC INJURY POSTOPERATIVELY

With the increasing use of cautery devices in pelvic surgery, ureter injuries may not be noted until the postoperative period. Hence, their use has to be extremely careful and judicious as diffusion of thermal energy can cause occult ureteral injury resulting in delayed stricture or urinary leakage.

Imaging studies are the mainstay investigation for the diagnosis of delayed ureter injury. IV contrast-enhanced CT scan along with delayed image acquisition helps to know ureteral obstruction as well as urine extravasation. Renal ultrasound is also a helpful modality to identify obstruction (hydroureteronephrosis) and urinoma; however, it cannot detect urinary leaks.

Retrograde ureteropyelography may also help in diagnosis and initial therapy for ureteral injury. Opacification of the ureter can identify the site and severity of leakage or obstruction and helps to stent the ureter simultaneously. It helps in the spontaneous healing of the ureter. If stenting is not possible, a nephrostomy tube can be placed to protect the kidney. A percutaneous drain may be needed for large urinoma drainage. Surgical intervention should be done within 3–7 days postoperatively. Beyond this duration, the repair is done after 6–8 weeks after the inflammation has resolved.

Additional tests that can be done include testing vaginal fluid collected (if present) and assessing for creatinine levels. Creatinine content >10 mg/dL typically suggests urine. Also, a tampon test can be done in which the bladder is filled retrograde with methylene blue diluted saline and the soakage of the tampon is seen. If there is blue staining, it means the injury is in the bladder, and if colorless, then it is most likely a ureteric injury.

TECHNIQUES OF URETERAL REPAIR

General principles for repair of ureteral injuries:

- *In the case of ureteral ligation:* Assess the viability of ureter and stent placement can be tried.
- *Partial transection of ureter:* Primary repair over the ureteral stent should be done.
- *Total transection:*
  - Uncomplicated upper and middle thirds—ureteroureterostomy over ureteral stent or Boari flap for the proximal aspect of the mid ureter
  - Complicated upper and middle thirds—ureteroileal interposition
  - Lower third—ureteroneocystostomy with psoas hitch over the ureteral stent
  - Thermal injury—resection with management as per transaction

Ureteroureterostomy

Ureteroureterostomy is a suitable option for an acute injury in the midportion of the ureter (below the ureteropelvic junction and above the pelvic brim). Proximal and distal ureter is meticulously dissected from surrounding tissue, while taking care of conserving the blood supply of the ureter. Healthy edges of the ureter are obtained after debridement especially if it is after crush, ligation, or thermal injury. Proximal and vital ureter are spatulated for 1 cm to allow overlapping tension-free anastomosis. After approximating the posterior part or ureter, a Double-J ureter stent is placed with the help of a guiding wire. The ureter is then anastomosed using 4-0 or 5-0 absorbable sutures (running or interrupted). A portion or omentum can be freed and wrapped around the ureter to minimize leak and adhesions. A drain should be ideally placed in the pelvis for 2–3 days. The bladder is drained for 7–10 days and stent for at least 14 days.

Ureteroneocystostomy with Psoas Hitch

Ipsilateral psoas muscle is exposed after opening the peritoneum. Three 2-0 nonabsorbable sutures are placed parallel to psoas muscle and set aside. Bladder is distended through catheter with normal saline till near full capacity. Bladder is incised diagonally across the anterior dome. Full thickness stay sutures are taken at the midpoint of incision to provide traction. The suture should not perforate bladder epithelium and are tied in sequence to anchor the bladder. Bladder closure is started perpendicular to the original incision. If possible, the bladder closure can include the ureteral implantation without the need of a separate neocystostomy. Before implantation, the ureter edge is spatulated for 1 cm followed by tension-free anastomosis. The remainder bladder wall is closed using 2-0 or 3-0 absorbable sutures in one or two layers after placing Double-J ureteral stent. Perivesical fat and peritoneum can be used to provide additional coverage at the anastomosis. A drain should be ideally placed in the pelvis for 2–3 days. The bladder is drained for 7–10 days and stent for at least 14 days.

Boari Flap

The procedure should be done after ensuring adequate bladder capacity and compliance after a cystogram or cystometrogram as the bladder capacity decreases after the procedure. The gap between the ureter and distended bladder is measured. A 4–5 cm wide-based flap is outlined along a diagonal toward the contralateral side originating from the cephalad margin of the bladder. Flap is extended to the spatulated ureter in the form of a bladder tube followed by end-to-end anastomosis with the spatulated end. The repair is stented prior to completion. The bladder should be catheterized for 7–10 days and the stent should be removed after 2 weeks.

Ureteral–Ileal Interposition

In case of injuries to the proximal ureter, the bridging gap between the bladder and ureter is longer than what can be repaired with a psoas hitch or Boari flap. In such cases, a segment of the ileum can be interposed between the ureter and bladder. Ileal segment 12–15 cm proximal to the ileocecal valve is used to avoid malabsorption while preserving the mesentery of that segment. Psoas hitch can be utilized to minimize the length of the ileal segment to be resected to reduce metabolic derangement. Ureter is brought through the posterior peritoneum and anastomosed to the ileum segment (pro peristaltic orientation) with fine absorbable sutures. It is an end-to-side anastomosis which is completed over a ureteral stent. Foley catheter is kept for 7–10 days, and stent can be removed after 2 weeks. A cystogram is done before the removal of Foley catheter. Side effects of this procedure include extended postoperative recovery, effects on ipsilateral renal function, and mucous in urine.

SPECIAL CLINICAL CONDITIONS

Special consideration in certain clinical situations to avoid ureteric injuries may be required in cases when there is known or expected malignant pathology or benign pathology complicating and distorting the normal anatomy, while doing open or laparoscopic surgeries, while using electrocautery, with aberrant ureter anatomy.

Malignant versus Benign

In cases of gynecological malignancies, ureter can be directly involved due to spread of disease or indirectly involved due to mass effects. Therefore, ureters may get dilated and hence reflect as hydronephrosis in ultrasonography or radiological findings. Stent placement is more commonly required in malignant cases as compared to benign conditions. Caregiver has to weigh risk–benefit ratio for placing ureteric stent, considering significant decrease in quality-of-life post stent placement. Moreover, in patients with stenosis associated with radiotherapy, stent withdrawal after the improvement of hydronephrosis is difficult. In many cases with lymphadenopathy, we cannot expect stent withdrawal to be possible due to the poor prognosis of the primary disease.[9]

Open versus Laparoscopy Hysterectomy

Unlike open surgery where hemostasis is accomplished by pressure and careful application of fine clamps and ligatures, laparoscopic surgery relies on electrosurgery for achieving hemostasis. There are various methods and techniques for the ligation of the uterine vessels, with bipolar instruments being the most popular method. The higher power of bipolar coagulation leads to wider thermal damage and repeated electrocoagulation leads to heat accumulation which leads to unavoidable thermal damage.[10] In addition, larger vessel causes increased thermal injury as the current usage will be for longer duration.[11] Bipolar coagulation can result in ureter damage at the site of uterine artery coagulation. Such thermal damages usually present postoperatively.[12]

Advantages of clipping uterine artery include avoiding thermal spread with electrical device, withstanding higher arterial pressure compared to other instruments. Its disadvantages include technical difficulty in applying them. A complete dissection of vessels is done before applying the clip. In cases with aberrant anatomy, there may be vessel tearing, which mandates the need to use power source to control the bleed. Appropriately applied clips decrease the chances of ureter injury and rebleeding.

If cautery is being used for artery coagulation, it must be kept in mind that the lateral spread of heat is up to 3–4 mm. The power of bipolar cautery should be kept within 30 W. It should be ensured that there is no coupling of instruments, and the insulation of the instrument is intact and not touching any other organ. The energy should be used briefly as if the current is on for a longer duration, and the chances of remote site electrical injury is increased.

Aberrant Anatomy of Ureter

The incidence of duplication of ureter in adults is approximately 0.7%–4% with incomplete duplication more common than complete duplication.[13] EU with an incidence of 1:2,000 and 1:4,000 in autopsies is defined as any ureter, single or duplex, that opens at places other than the bladder trigone.[14]

RECENT ADVANCES

Ureteral injuries were associated with all methods of hysterectomy, irrespective of the method (vaginal, laparoscopic, or laparotomic). Intraoperative ureteral injury, a serious complication of abdominopelvic surgeries, can be avoided through ureter visualization. Near-infrared fluorescence (NIR-F) imaging offers real-time anatomical visualization of ureters during surgery. Pudexacianinium (ASP5354) chloride is an indocyanine green (ICG) derivative under investigation for intraoperative ureter visualization during colorectal or gynecologic surgery in adult and pediatric patients.

Image-guided surgery using NIR-F imaging technology offers real-time visual information for surgeons and potentially increases patient safety during operations.[15] Methylene blue and ICG are commercially available fluorescent dyes, but they are not approved by the US Food and Drug Administration (FDA) for ureter visualization. In addition, the excitation wavelength required by methylene blue limits penetration depth for identifying ureters, and ICG is minimally excreted into urine and thus is not suitable for ureteral imaging after IV administration.[16,17] Administration of the novel fluorescent imaging agent pudexacianinium, at doses of 1.0 mg and 3.0 mg, provides ureteral visualization for the full duration of minimally invasive, laparoscopic colorectal procedures.

MEDICOLEGAL CONSIDERATIONS

The most common speciality named in ureteral injury malpractice cases was gynecology followed by urology. The most common claim was intraoperative negligence. Intraoperative surgical consultation and ureteral identification should be considered if there is concern for ureteral involvement in the surgical field. Ureteric injury may not constitute negligence if it is demonstrated that the surgeon provided reasonable care that would be expected during the perioperative phases. Meticulous surgical technique and attention to pelvic anatomy is paramount to the effort of decreasing ureteral injury. Ureteric injury, in itself, does not constitute negligence, if it is demonstrated that the surgeon provided reasonable care, which would be expected of him/her during the preoperative, intraoperative, and postoperative phase. Delayed diagnosis resulted in more severe clinical sequelae and unfavorable legal outcomes for the surgeon. Surgeons should maintain a high degree of vigilance for ureteric injury in these patients, especially if patients re-present to hospital after initial discharge. Early imaging to rule out a visceral organ injury is recommended. Moreover, even if ureteric injury is repaired, a proper follow-up of kidney function is required for long duration, which may include a simple test like kidney function test (KFT) to as advanced as nuclear medicine. One must be mindful that patients often litigate not because an incident occurred, but because of a perceived or real lack of empathy and miscommunication between the team of health care providers and the patient and family.

CONCLUSION

To conclude, urinary tract injuries represent a significant potential risk during gynecologic surgeries. The prime focus should be, to predict the risk factors before surgery and take pre-emptive steps required beforehand, identify them during surgery, and treat the same if inadvertent injury has occurred. Prompt repair of urinary tract injuries decreases chances of severe renal complications, genitourinary fistula, sepsis, and death. Surgeons should have a low threshold for suspecting a urinary tract injury with the vague post operative symptoms of flank pain, abdominal pain, abnormal urine, fever or leakage of urine from abdominal incision or vaginal vault.

Intraoperative cystoscopy may help to evaluate bladder, urethra as well as ureters by assessing bilateral urine efflux. However its not mandatory in all cases and decision to perform cystoscopy is completely surgeon and case-dependent. With the increasing prevalence of robotic surgeries where the surgeon rely on visual clues only and lacks tactile stimulus we will be needing specific modality to decrease ureteric injuries.

The advent of new surgical techniques and careful restructuring of procedural steps has dramatically minimized these risks. Understanding the anatomical details is the key to precise and accurate surgical intervention and in best patient's interest.

REFERENCES

1. Ade-Ojo IP, Tijani O. A Review on the Etiology, Prevention, and Management of Ureteral Injuries During Obstetric and Gynecologic Surgeries. Int J Womens Health. 2021; 13:895-902.
2. Shi A, Ma H, Ma Y. Judgment and Prevention of Urinary Tract Injury in Gynecological Surgery Based on Data Mining. J Healthc Eng. 2022;2022:1270580.
3. Handa VL, Le LV. Te Linde's Operative Gynecology, 12th edition. Philadelphia: Wolters Kluver; 2020.
4. Hoffman BL, Schorge JO, Halvorson LM, Hamid CA, Corton MM, Schaffer JI. Williams Gynecology, 4th edition. New York: McGraw-Hill Education; 2020.
5. Gemer O, Simonocsky A, Huerta M, Kapustian V, Anteby E, Linov L. A radiological study on the anatomical proximity of the ureters and the cervix. Int Urogynecol J Pelvic Floor Dysfunct. 2007;18:991-5.
6. Cardozo L, Staskin D. Textbook of Female Urology and Urogynecology, 4th edition, volume 1. Boca Raton, Florida: CRC Press; 2017.
7. Donnez J. Atlas of Operative Laparoscopy and Hysteroscopy, 3rd edition. London: Informa Healthcare; 2007.
8. Seth S, Nagrath A. Preventing Bladder Injury at Hysterectomy in Post-Cesareans. J Gynecol Women's Health. 2017;3(2).
9. Kurata S, Tobu S, Udo K, Noguchi M. Outcomes of Ureteral Stent Placement for Hydronephrosis in Patients with Gynecological Malignancies. Curr Urol. 2017;10(3):126-31.
10. Liang J, Xing H, Chang Y. Thermal damage width and hemostatic effect of bipolar electrocoagulation, LigaSure, and Ultracision techniques on goat mesenteric vessels and optimal power for bipolar electrocoagulation. BMC Surg. 2019;19:147.
11. Carbonell AM, Joels CS, Kercher KW, Matthews BD, Sing RF, Heniford BT. A comparison of laparoscopic bipolar vessel sealing devices in the hemostasis of small-, medium-, and large-sized arteries. J Laparoendosc Adv Surg Tech A. 2003;13:377-80.

12. Alkatout I, Schollmeyer T, Hawaldar NA, Sharma N, Mettler L. Principles and safety measures of electrosurgery in laparoscopy. JSLS. 2012;16:130-9.
13. Privett JT, Jeans WD, Roylance J. The incidence and importance of renal duplication. Clin Radiol. 1976;27(4):521-30.
14. Jain P, Sarkar D, Maiti K, Gupta S, Pal DK. Rare cases of ectopic ureter: Analysis from a single centre with review of the literature. Turk J Urol. 2019;45(Supp. 1):S92-S97.
15. Slooter MD, Janssen A, Bemelman WA, Tanis PJ, Hompes R. Currently available and experimental dyes for intraoperative near-infrared fluorescence imaging of the ureters: a systematic review. Tech Coloproctol. 2019;23:305-13.
16. Verbeek FP, van der Vorst JR, Schaafsma BE, Swijnenburg RJ, Gaarenstroom KN, Elzevier HW, et al. Intraoperative near infrared fluorescence guided identification of the ureters using low dose methylene blue: a first in human experience. J Urol. 2013;190:574-9.
17. Friedman-Levi Y, Larush L, Diana M, Marchegiani F, Marescaux J, Goder N, et al. Optimization of liposomal indocyanine green for imaging of the urinary pathways and a proof of concept in a pig model. Surg Endosc. 2018;32:963-70.

Section

5 Aspects of Operative Gynecology

Chapter

37

Surgical Anatomy of the Female Pelvis

Tapasya Dhar

INTRODUCTION

The knowledge of the female pelvis is a prerequisite for a gynecologist to practice gynecological surgery. It helps us to safely achieve access, have an adequate exposure of tissues, ensure good hemostasis, and avoid any untoward injury to the viscera, nerves, and major blood vessels. Minimal invasive surgery provides a clear and magnified image of the pelvis and is the current novel medium of understanding normal as well as abnormal anatomy.

GENITAL STRUCTURES

Vagina

- The vagina is an elastic and canal-shaped organ with its attachment to pelvic wall.
- Both the anterior and posterior walls are in close contact with each other and lower vagina gets narrowed on it, with passage through the urogenital hiatus in the levator ani.
- The upper and lower vagina are at an angle of 120° maintained by the levator ani traction.

Relations of Vagina

- In lower one-third—with urethra and with perineal body and laterally to levator ani.
- In middle one-third—the bladder neck and trigone, rectum and levator laterally.
- Upper one-third—bladder and ureter, pouch of Douglas, and laterally to Mackenrodt's ligaments.

Endopelvic fascia: This fascia surrounds the muscularis of the pelvic organs, which attach it to the pelvic sidewalls and fuse to form ligaments.[1]

Uterus

- Uterus is a fibromuscular organ, and the dimensions vary according to the genetic makeup and previous pregnancies.
- In a woman of reproductive age, the body is double than the cervix, but before menarche and after menopause their sizes are identical.
- The endometrial cavity is surrounded by a muscular layer and the portion of the uterus lying above the insertion of the fallopian tubes is called fundus of the uterus and below it lies the body of the uterus.
- The muscle fibers of the uterus are arranged uniquely in a crisscross manner reflecting the origin from paired paramesonephric ducts.
- The uterus is lined by a specialized lining called the endometrium. The columnar epithelium further forms glands and stroma.
- The superficial portions of the endometrium subject to hormonal changes and this process is called menstruation.[2]

Cervix

- The cervix has two parts—the portio vaginalis, which is the part in the vagina and portio supravaginalis lying above the vagina and below the uterus.
- The cervix is made of thick fibrous connective tissue and smooth muscles. The smooth muscles and the connective fibrous tissue are reflected off the cervix during intrafascial hysterectomy.
- The upper part of cervical canal is the internal os, where the cervix joins the endometrium close to the isthmus. Below lies the external os, which contains the transition from squamous epithelium to columnar epithelium.[5]

Adnexa and Broad Ligament

Fallopian Tubes

The tubes are situated bilateral having a length of 7–12 cm and has four distinct parts, namely interstitial, isthmic, ampulla, and fimbrial end.

Ovary

- Ovary is attached to pelvic sidewalls by the infundibulopelvic ligaments and the contents are the ovarian artery and vein.
- On the medial side, the ovary is attached to the uterus via utero-ovarian ligament.
- The size of the ovary is 3 × 3 × 1.5 cm.
- The ovary has a cover of cuboidal/columnar epithelium along with a cortex and medulla.
- The cortex is made of a special stroma, which contains follicles, corpus luteum, and corpus albicans, and the medial portion has blood vessels and connective tissue.[5]

Round Ligament

- Round ligament is a homolog of gubernaculum testis.
- It arises from the anterolateral aspect of the uterus and enters the retroperitoneal tissue and deep inferior epigastric vessels lie medial to it and after that it enters the internal inguinal canal.
- Then it comes out of the external inguinal ring and attaches to the subcutaneous tissue in the labia majora.
- The ovaries, tubes, and the round ligaments are covered by a double-layered peritoneum called the *broad ligament*.
- The cardinal ligament and the uterosacrals form the lower border of the broad ligament **(Fig. 1)**.[5]

Sacral Promontory

Sacral promontory is the most pointed structure in the pelvis and is used as a point of reference in surface marking and innumerable surgeries. The sacral promontory forms the basis of orientation of the pelvis.

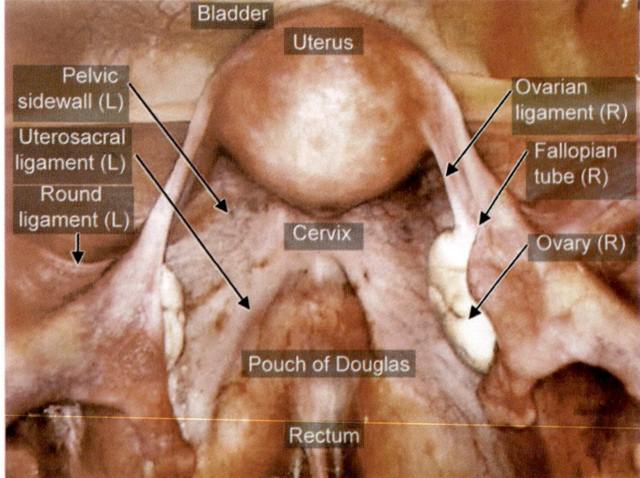

Fig. 1: Laparoscopic view of the pelvis including all the ligaments.

Importance

- Common iliac arteries and veins divide into internal and external iliac at this point.
- The *ureter* crosses from the lateral to medial side at the sacral promontory above the division of the iliac vessels.
- The superior hypogastric nerve divides to form left and right hypogastric nerves at this area.
- It forms the beginning point of *para-aortic lymph node dissection*. A skillful dissection done lateral to this point leads to the pararectal retroperitoneal space.
- The small bowel mesentery crosses this point to its way to right sacroiliac joint and delineates pelvic organs from abdominal organs.[1]

BLOOD VESSELS

External iliac vessels:

- These travel along iliopsoas muscle and supply the lower limbs.
- Ilio-obturator lymph node dissection can be done safely as these vessels have no branches in the antero-medial area.
- The anterior aspect of the external iliac gives off the inferior epigastric artery, which travels along the transversalis fascia below the anterior abdominal wall. During port placement in laparoscopy, extreme caution is necessary to prevent injury to this vessel.
- During the dissection of the pararectal spaces using the lateral approach, the external iliac vessels are the first structure that are visualized following the dissection of the peritoneum.[1]

Internal iliac vessels:

- After common iliac divides, the internal iliac gives off the anterior and posterior divisions.
- The anterior division mainly supplies the pelvic organs, and the posterior division pierces presacral fascia and supplies the gluteal area.
- The anterior division gives off the first branch, the uterine artery, following which it gives off the superior vesical artery at a level higher than the uterine artery and then continues as the obliterated hypogastric artery.[1]

Clinical Significance

- Uterine artery that crosses over the ureters is usually referred to as bridge over waters and is of clinical significance.
- Since there is a length of 5 cm of the internal iliac artery, which is devoid of any branches because the uterine artery arises further off, and this can be used for *internal iliac artery ligation* in severe postpartum hemorrhage.
- Running parallel to the arteries are the internal iliac veins therefore caution should be exercised that only the artery is ligated.
- This can be done by following the artery in a retrograde way around 3 cm from the origin of uterine artery.
- The internal iliac artery gives off the lateral branch, which is the obturator artery, and which is one of the contents of the lateral paravesical space.
- The obturator vessels lie below the obturator nerve, which is the lower limit of obturator lymph node dissection and any further dissection beyond this point leads to severe hemorrhage.[5]

UTERINE ARTERY

The anterior division of the internal iliac artery gives off the first branch which is the uterine artery. It crosses the ureter and then bifurcates into the descending and the ascending cervical branches. The uterine artery is visualized in the pararectal space.[3]

Clinical Significance

- As the ascending cervical branch is formed a few millimeters away before its entry into the uterus, it allows easy accessibility to clamp and coagulate the vessel in intrafascial hysterectomy.
- The uterine artery is anterior to the ureter and the uterine vein lies posterior to it. Hence, the ureters lies in the fork between these two, in case of an arterial bleed, lifting the ureter to close the opening of the uterine artery stops the bleeding. With venous bleeding, lifting of the ureters will not stop the bleeding but will give access to the vein **(Fig. 2)**.

NERVES

- The uterus, vagina, rectum, and bladder are supplied by the autonomic nervous system. They are innervated by motor and the sensory sympathetic and parasympathetic nerves.
- The inferior hypogastric nerve (T10–L2) gives the sympathetic nerves whereas the pelvic splanchnic nerve (S2–S4) gives the parasympathetic fibers.
- Both of these merge to form the inferior hypogastric plexus supplying the uterus and the urinary bladder.[1]

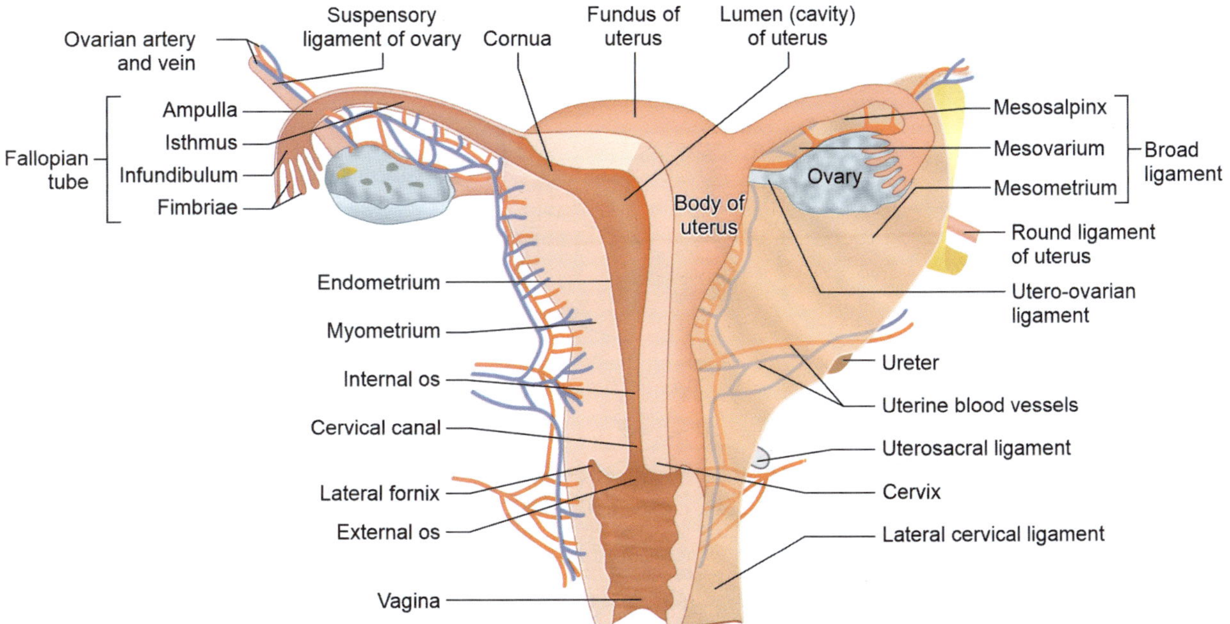

Fig. 2: Anatomy of the uterus, tube's and ovaries.

Clinical Significance

- Nowadays, nerve-sparing radical hysterectomy is the preferred hysterectomy, and there is a need for identifying the hypogastric plexus to preserving the bladder branches from the plexus.
- The bladder function post surgery is well-preserved by the nerve-sparing radical hysterectomy.

SPACES

The retroperitoneal spaces can be divided as follows:

- Pararectal space
- Paravesical space
- Prevesical space
- Rectovaginal space
- Presacral space.

These spaces of the pelvis are basically peritoneal depressions made by the reflection of the peritoneum lying between the pelvic organs and the pelvic walls. Further, these are exclusive spaces, which are created or developed as a result of dissection to gain an access to the important structures lying underneath or retract organs at the time of dissection.[2]

Pararectal Space

The retroperitoneum lying laterally from the rectum and seen bilaterally constitutes pararectal space. Pararectal space follows the sacral curve.

- Roof—posterior leaf of broad ligament
- Medial border—rectum
- Lateral border—internal iliac artery
- Cranially—uterine artery
- Floor—levator ani.

The pararectal space is divided into the medial and lateral pararectal spaces by the ureter.

The medial pararectal space is called *Okabayashi space.*

The lateral is called *Latzko space* **(Fig. 3)**.

Okabayashi space:

- *Contents:* Nerve fibers of the superior hypogastric plexus
- A plexus innervating the bladder
- The clinical significance is for nerve-sparing radical hysterectomy and in endometriosis nerve-sparing fertility surgery.[5]

The pararectal space is opened by two ways described as follows:

1. *Endometriotic or medial approach:*
   a. The incision here is carried out medial to infundibulopelvic ligaments. The first thing encountered is the ureter.
   b. The ureter is separated from the internal iliac artery and further dissected to reach the apex of the uterosacral ligaments.
   c. During the dissection, the uterine artery is seen.
   d. If the ureter is kept medial, the lateral pararectal space can be opened and further dissection of ureters is possible.
   e. By keeping the ureter lateral, the Okabayashi space is opened and parasympathetic nerves are visualized which are saved in *nerve-sparing radical hysterectomy.*[5]

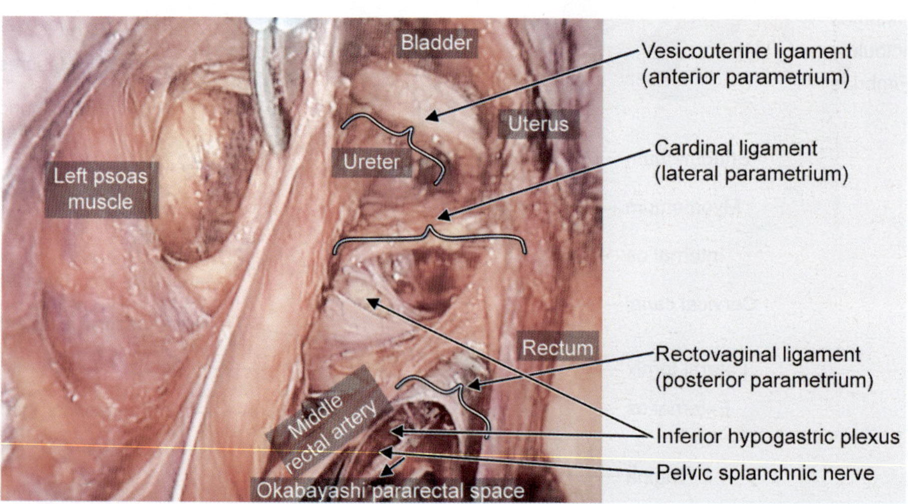

Fig. 3: Pararectal and latzko space.

2. *Oncological approach:*
   a. The incision on the peritoneum is made lateral to infundibulopelvic ligaments and exposes structures like the external iliac artery and vein.
   b. This space is used in oncology surgery for lymph node dissection **(Fig. 4)**.[4]

Paravesical Space

Paravesical space is a space in the retroperitoneal area lying lateral to the bladder. The relation to other structures are as follows:

- Laterally—pelvic sidewall
- Medial border—bladder
- Inferiorly—uterine artery **(Fig. 5)**
  - The paravesical, prevesical, and pararectal spaces communicate with each other. The dissection of these spaces is useful in radical hysterectomy.

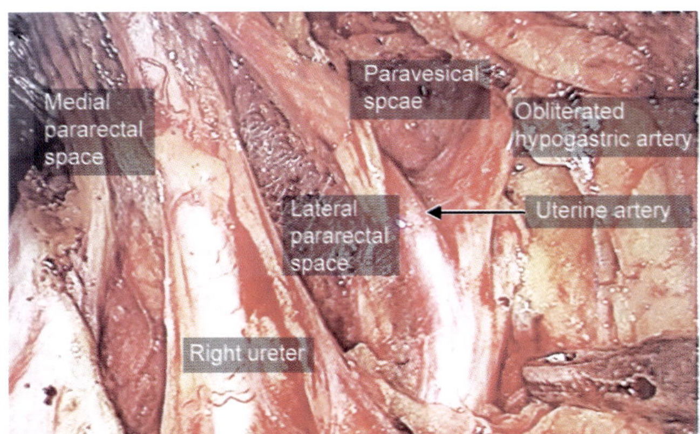

Fig. 4: Okabayashi space.

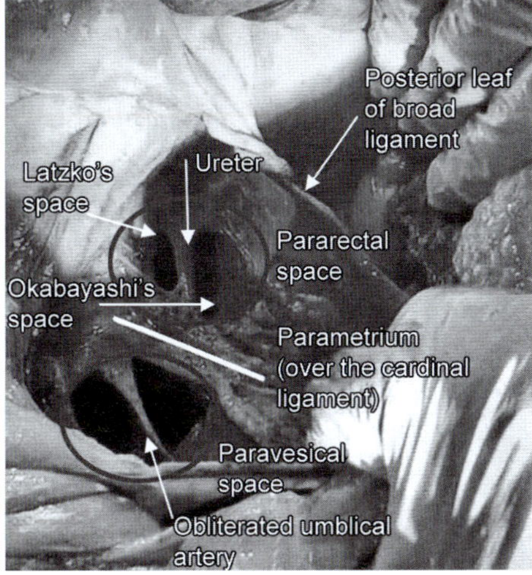

Fig. 5: Top view of pelvic spaces.

  - The obliterated hypogastric artery or lateral umbilical ligament divides it into medial and lateral areas.
  - To dissect the obturator and pelvic lymph nodes during radical hysterectomy, entry to this space is used.
  - The limit of dissection for medial paravesical space is levator ani muscle and the limit for the lateral dissection is obturator nerve **(Fig. 6)**.

Prevesical Space

Midline retroperitoneal space lies between the urinary bladder and the anterior abdominal wall.

This space is used especially in stress incontinence and sling surgeries where exposure to the bladder neck and urethra is required. It is useful in anterior exenterations surgeries in recurrent cervical cancer.[1]

Rectovaginal Space

- The space seen retroperitoneally is enclosed anteriorly by the uterus and posterior vaginal wall, posteriorly by the rectum, and laterally by the Mackenrodt's ligaments and the uterosacrals.
- The floor is by the levator ani and the roof is by peritoneum of pouch of Douglas.
- The Denonvilliers' fascia lies between vagina and the rectum and this space is created by dividing between two layers to reach the rectovaginal space.
- The end point of rectovaginal space dissection is done until levator ani is reached and this facilitates taking a large cuff of vagina in radical hysterectomy.
- Rectovaginal space dissection is useful in posterior sling surgery.[4]

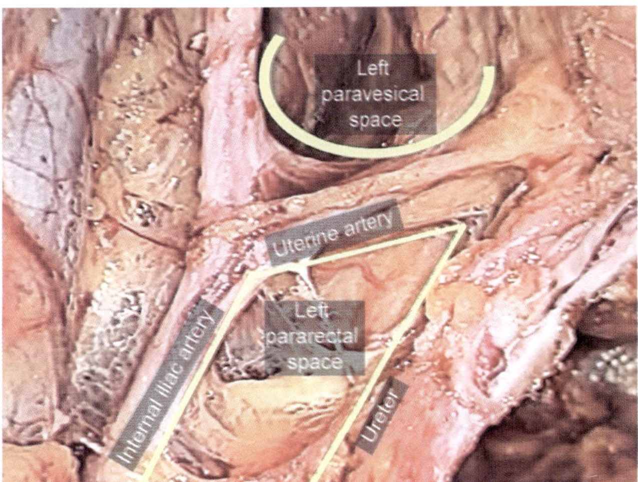

Fig. 6: Pararectal space.

Presacral Space

- The retroperitoneal space lying posterior to the rectum is covered anteriorly by mesorectum and posteriorly by Waldeyer's fascia.
- The presacral veins lie posterior to the Waldeyer's fascia.
- Clinical significance in posterior exenteration, deep infiltrating endometriosis surgery, sacrocolpopexy, and presacral neurectomy.[1]

Yabuki Space

- Also called the fourth space, it is a midline retroperitoneal space confined between the anterior aspect of uterus and ureter inserting into the bladder.
- The parasympathetic nerves innervating the bladder are present here and careful dissection in this area is required during *nerve-sparing radical hysterectomy.*[1]

REFERENCES

1. Puntambekar S, Manchanda R. Surgical pelvic anatomy in gynecologic oncology. Int J Gynaecol Obstet. 2018;143 (Suppl 2):86-92.
2. Cosma S, Ferraioli D, Mitidieri M, Ceccaroni M, Zola P, Micheletti L, et al. A simplified fascial model of pelvic anatomical surgery: going beyond parametrium-centered surgical anatomy. Anat Sci Int. 2021;96:20-9.
3. Brill AI, Rogers RM Jr, Gomel V. Anatomy and surgical dissection of the female pelvis: For the gynecologic surgeon. In: Munro MG, Gomel V (Eds). Reconstructive and Reproductive Surgery in Gynecology, 2nd edition. United States: CRC Press; 2018.
4. Muavha DA, Ras L, Jeffery S. Laparoscopic surgical anatomy for pelvic floor surgery. Best Pract Res Clin Obstet Gynaecol. 2019;54:89-102.
5. DeLancy JOL. Principals of anatomy and perioperative considerations. In: Jones HW III, Rock JA (Eds). Te Linde's Operative Gynecology, 11th edition. Philadelphia: Lippincott Williams & Wilkins; 2015.

38

Chapter

Preoperative Evaluation

Rajni Singh

INTRODUCTION

Preoperative evaluation starts with patient's counseling regarding failure of conservative management/medical management or need of surgical management.

Once the decision of surgical management is made, doctors proceed with proper history and physical examination to know the risk factors, thereby preparing team accordingly.

Evaluation should be based on patient-related risk factors and characteristics of the surgical procedure.

PATIENTS-RELATED RISK FACTORS

- *Age:* Age is the independent risk factor for perioperative complications and is associated with various medical disorders including diabetes mellitus, chronic obstructive pulmonary disease (COPD), organ damages, decreased functional capacity, arthritis, etc.
- *Diabetes:* Diabetic women have a high risk of perioperative complications including hypoglycemia, acute renal failure, neurological complications, stroke, and acute myocardial infarction.
- Diabetics with poor control are at greater risk of surgical site infection (SSI), silent coronary heart disease, and postoperative cardiovascular morbidity.
- *Hypertension:* Hypertensive patients should continue their oral antihypertensive medications up to the day of surgery. In patients taking diuretics, potassium levels should be closely monitored, since hypokalemia can potentiate the effects of muscle relaxants used during induction of anesthesia. Hypokalemia can increase the risk of cardiac arrhythmia and paralytic ileus.
- *Cardiac disease:* Patient having history of cardiac diseases should be evaluated by cardiology team and risk and benefit of surgery to be assessed.
- *Obesity:* Obesity is associated with various comorbid medical conditions [including hypertension, coronary artery disease (CAD), obstructive sleep apnea (OSA), diabetes mellitus].[1] It is associated with metabolic syndrome and thereby associated with higher risk of intraoperative and postoperative complications such as pneumonia, postoperative hypoxemia, unplanned reintubation, SSIs, wound complications, and venous thromboembolism.[2]
- Obesity is associated with an increase in SSIs, wound dehiscence, and incisional hernia possibly related to poor nutritional status, decreased antibiotic penetration, and decreased tissue oxygenation.[3]
- *OSA:* This is the most common type of sleep-disordered breathing. Patients with OSA are at higher risk for respiratory complications, postoperative cardiac events, and need for intensive care unit (ICU) care.[4]
- The STOP-BANG questionnaire is a validated screening tool for OSA **(Table 1)**.[5] The score can be used to predict increased risk of postoperative pulmonary and cardiac complications.
- Women with known OSA are advised to continue on continuous positive airway pressure (CPAP) therapy up to the day of surgery.
- *Smokers:* Smoking increases risk of SSI.

PREOPERATIVE TESTING

Routine testing is not recommended in low-risk cases and repeat testing is not required if reports are available within 4 months. Repeat testing to be done in case of deranged blood profile.[6]

Testing includes:

- Complete blood count (CBC)
- Thyroid-stimulating hormone (TSH)
- Coagulation profile
- Liver function test

TABLE 1: STOP-BANG questionnaire.

Snoring	Do you snore loudly (enough to be heard through closed doors, or your bed partner elbows you for snoring at night)?	Yes	No
Tired	Do you often feel tired, fatigued, or sleepy during the daytime?	Yes	No
Observed	Has anyone observed you stop breathing or choking/gasping during your sleep?	Yes	No
Pressure	Do you have or are you being treated for high blood pressure?	Yes	No
BMI	Is your body mass index (BMI) >35 kg/m^2?	Yes	No
Age	Older than 50 years of age?	Yes	No
Neck	• *For male:* Is your shirt collar 17 inches or larger? (measure around Adam's apple) • *For female:* Is your shirt collar 16 inches or larger?	Yes	No
Gender	Are you a male?	Yes	No

Note:
- *Score is 0–2:* Low risk
- *Score is 3–4:* Intermediate risks
- *Score is 5–8:* High risk of OSA (obstructive sleep apnea)

- Kidney function test
- Sugar profile
- Infection profile including hepatitis B surface antigen (HBsAg), anti-hepatitis C virus (anti-HCV), rapid plasma reagin (RPR) antigen, human immunodeficiency virus (HIV)
- ABO and Rh
- Pregnancy test (reproductive age group women)
- Electrocardiogram (ECG)
- Chest X-ray
- Ultrasound, specially transvaginal ultrasound (TVS)
- Magnetic resonance imaging (MRI) (specially for oncological cases)
- Computed tomography (CT) scan (do)
- Pap smear
  (**Table 2** specifies test for high-risk cases)

PREOPERATIVE STRATEGIES

- *Antibiotic prophylaxis:* SSI is defined as infection related to surgery/wound infection within 30 days of surgery. It consists of superficial incisional infections (bacterial)of the skin and subcutaneous tissue.[7]
- Antibiotic prophylaxis is best administered within 60 minutes of surgical incision.[13]

Table 3 shows the antibiotic prophylaxis regimens by procedure.

- *Thromboprophylaxis:* The American College of Chest Physicians (ACCP) provides guidelines for perioperative thromboprophylaxis based on risk categories (Caprini risk assessment score, **Table 4**).

The recommendation based on Caprini score is given in **Table 5**.

TABLE 2: Specific test for high-risk cases.

Cardiac disease[8]	• Baseline ECG • Stress testing • Echocardiogram
Diabetes mellitus[9]	• Baseline ECG • Serum electrolytes and creatinine • HbA1c within the last 4–6 weeks • Additional noninvasive cardiac testing if indicated
Hypertension	• Serum electrolytes and creatinine • Baseline ECG
Obesity[10]	• Screen for obstructive sleep apnea (STOP-BANG questionnaire) • Assessment for diabetes, cardiac disease
Obstructive sleep apnea[11,12]	• Echocardiogram • Serum electrolytes
Pulmonary disease (asthma, COPD)	• Pulmonary function tests • CXR
Liver disease	• CBC • Coagulation testing • Liver function tests

(CBC: complete blood count; COPD: chronic obstructive pulmonary disease CXR: chest X-ray; ECG: electrocardiogram; HbA1c: glycated hemoglobin)

MANAGING MEDICATION IN PERIOPERATIVE PERIOD

- *Warfarin:* Stop medication 5 days prior to surgery.
- Switch to low-molecular-weight heparin (LMWH)/unfractionated heparin.
- Check international normalized ratio (INR) prior to surgery.

TABLE 3: Antimicrobial prophylactic regimens by procedure.

Hysterectomy and urogynecology procedures, including those involving mesh	Cefazolin 2 g IV within 1 hour before procedure (3 g IV for weight >120 kg)
Alternatives	• Cefotetan, cefoxitin, cefuroxime Or • Ampicillin–sulbactam
History of immediate hypersensitivity to penicillin	*Clindamycin 900 mg IV q6 hours + one of these:* • Gentamicin 5 mg/kg IV • Quinolone 400 mg IV • Aztreonam 2 mg IV q4 hours
Alternatives	*Metronidazole 500 mg IV + one of these:* • Gentamicin 5 mg/kg IV • Quinolone 400 mg IV
Induced abortion or dilation and evacuation	• Doxycycline 200 mg 1 hour before procedure or • Metronidazole 500 mg BID for 5 days
Hysterosalpingogram or chromotubation with history of PID or dilated/abnormal fallopian tubes	Doxycycline 100 mg orally daily for 5 days

(BID: twice a day; IV: intravenous; PID: pelvic inflammatory disease)

TABLE 4: Caprini score.

Point 1	*Point 2*	*Point 3*	*Point 5*
• Age 41–60 years • Minor surgery • BMI >25 kg/m^2 • Swollen legs • Varicose veins • Pregnancy or postpartum • History of unexplained or recurrent spontaneous abortion • OCP/HRT • Sepsis (<1 month) • Serious lung disease including pneumonia (<1 month) • Acute myocardial infarction • Congestive heart failure (<1 month) • History of inflammatory bowel disease • Medical patient at bed rest	• Age 61–74 years • Arthroscopic surgery • Major open surgery (>45 minutes) • Laparoscopic surgery (>45 minutes) • Malignancy • Confined to bed (>72 hours) • Plaster cast • Central venous access	• Age ≥75 years • History of VTE • Family history of VTE • Factor V Leiden • Prothrombin 20210A • Lupus anticoagulant • Anticardiolipin antibodies • Elevated serum homocysteine • Heparin-induced thrombocytopenia • Other congenital or acquired thrombophilia	• Stroke (<1 month) • Elective arthroplasty • Hip, pelvis, or leg fracture • Acute spinal cord injury (<1 month)

(BMI: body mass index; HRT: hormone replacement therapy; OCP; oral contraceptive pills; VTE: venous thromboembolism)

TABLE 5: Recommendation based on Caprini score.

0	*Very low risk*	*Early mobilization*
1–2	Low risk	Mechanical prophylaxis with intermittent pneumatic compression
3–4	Moderate risk	Use LMWH, LDUH, or mechanical prophylaxis (preferably with intermittent pneumatic compression devices)
>5	High risk	Combined pharmacologic and mechanical methods (4 weeks)

(LDUH: low-dose unfractionated heparin; LMWH: low-molecular-weight heparin)

- *Corticosteroids:* Stress dose steroid is recommended, if patient has taken chronic steroid therapy >3 weeks/high dose (equivalent to >5 mg/day prednisolone).
- *Beta blockers: Continue till the day of surgery.*
- *Alpha 2 agonists:* Continue in the perioperative period due to risk of withdrawal and rebound hypertension.
- *Asthma medications:* No need to discontinue these medications for surgery.

- *Calcium channel blockers:* No need to discontinue these medications for surgery.
- *Diuretics:* Morning dose to be withheld in risk of hypokalemia.
- *Hypolipidemic nonstatin drugs:* To be discontinued day before surgery.
- *Statins:* No need to discontinue.
- *Oral hypoglycemic agents:* Switch to insulin prior to surgery.

CONCLUSION

Preoperative risk assesment decreases the intra and postoperative complications associated with surgery and thereby improves patient complainces and postop recovery.

REFERENCES

1. Committee on Practice Bulletins—Gynecology, American College of Obstetricians and Gynecologists. ACOG Practice Bulletin No. 84: Prevention of deep vein thrombosis and pulmonary embolism. Obstet Gynecol. 2007;110(2 Pt 1):429-40.
2. ACOG Practice Bulletin No. 195: Prevention of Infection after Gynecologic Procedures. Obstet Gynecol. 2018; 131(6):e172-e189.
3. Committee Opinion No. 619: Gynecologic surgery in the obese woman. Obstet Gynecol. 2015;125(1):274-8.
4. Bettelli G. Preoperative evaluation in geriatric surgery: comorbidity, functional status, and pharmacological history. Minerva Anestesiol. 2011;77(6):637-46.
5. Chung F, Abdullah HR, Liao P. STOP-Bang Questionnaire: A practical approach to screen for obstructive sleep apnea. Chest. 2016;149(3):631-8.
6. Douketis JD, Spyropoulos AC, Spencer FA, Mayr M, Jaffer AK, Eckman MH, et al. Perioperative management of antithrombotic therapy: Antithrombotic therapy and prevention of thrombosis, 9th ed: American College of Chest Physicians Evidence-Based Clinical Practice Guidelines. Chest. 2012;141(2 Suppl):e326S-e350S.
7. Espin-Basany E, Sanches-Garcia JL, Lopez-Cano M, Lozoya-Trujillo R, Medarde-Ferrer M, Armadans-Gil L, et al. Prospective, randomized study on antibiotic prophylaxis in colorectal surgery. Is it really necessary to use oral antibiotics? Int J Colorectal Dis. 2005;20(6):542-6.
8. Fleshier LA, Fleischmann KE, Auerbach AD, Barnason SA, Beckman JA, Bozkurt B, et al. 2014 ACC/AHA guideline on perioperative cardiovascular evaluation and management of patients undergoing noncardiac surgery: executive summary: a report of the American College of Cardiology/American Heart Association Task Force on Practice Guidelines. Circulation. 2014;130(24):2215-45.
9. Hamrahian AH, Roman S, Milan S. (2024). The management of the surgical patient taking glucocorticoids. [online] Available from https://medilib.ir/uptodate/show/146 [Last accessed March, 2024].
10. Johnson BE, Porter J. Preoperative evaluation of the gynecologic patient: considerations for improved outcomes. Obstet Gynecol. 2008;111(5):1183-94.
11. Lockhart EM, Willingham MD, Abdallah AB, Helsten DL, Bedair BA, Thomas J, et al. Obstructive sleep apnea screening and postoperative mortality in a large surgical cohort. Sleep Med. 2013;14(5):407-15.
12. Nagappa M, Patra J, Wong J, Subramani Y, Singh M, Ho G, et al. Association of STOP-bang questionnaire as a screening tool for sleep apnea and postoperative complications: a systematic review and bayesian meta-analysis of prospective and retrospective cohort studies. Anesth Analg. 2017;125(4):1301-8.
13. Rahn DD, Mamik MM, Sanses TVD, Matteson KA, Aschkenazi SO, Washington BB, et al. Society of Gynecologic Surgeons Systematic Review Group. Venous thromboembolism prophylaxis in gynecologic surgery: a systematic review. Obstet Gynecol. 2011;118(5):1111-25.

Chapter

39

Postoperative Management

Vandana R Bhuriya, Kirti Saini

INTRODUCTION

The outcome of a patient following a surgical procedure is best determined by the postoperative complications especially within 72 hours of surgery.

The key to minimize postoperative morbidity is appropriate preoperative assessment.

OBJECTIVES OF POSTOPERATIVE CARE

- Early mobilization and resumption of normal activities
- Reduction in postoperative complications
- Less hospital stay
- More patient satisfaction
- Decrease mental trauma to patient after surgery
- Less readmission rate

All the objectives can be achieved by Enhanced Recovery After Surgery (ERAS) program.

- ERAS is a paradigm shift in conventional perioperative care.
- It focuses mainly on minimizing the stress response of the operation, maintaining homeostasis, preventing catabolism with consequent loss of protein and muscle strength, and minimizes cell dysfunction.[1]

The program is based on perioperative optimization[2] which includes:

- Preoperative counseling
- Presurgical fasting and carbohydrate intake
- Prophylaxis for thromboembolism
- Standard anesthetic protocol
- Pain relief
- Optimized fluid administration
- Recovery of normal gastrointestinal function
- Early mobilization

Preoperative Counseling

Preoperative counseling reduces anxiety, increases patient satisfaction further improving postoperative recovery, and facilitates early discharge.

Aims:

- To recognize patients' expectations about surgical and anesthetic procedures and relieve their stress by explaining them about the route of surgery, operative time, etc.
- To make the patients aware about the care plan for the postoperative period, which includes postoperative pain control, length of hospital stay, and benefits of early mobilization and feeding.
- Advice regarding cessation of tobacco and alcohol and maintenance of nutrition prior to surgery.

Presurgical Fasting and Carbohydrate Drink

Before any surgery that requires sedation or general/regional anesthesia, light meal up to 6 hours before and ingestion of clear liquids up to 2 hours before the elective procedure is allowed.[2]

Oral carbohydrates have shown following beneficial effects in randomized control trials but are still not recommended **(Box 1)**.[3,4]

BOX 1: Beneficial effects of oral carbohydrates.

- Better preoperative well-being
- Reduction of postoperative insulin resistance
- Decreased protein breakdown
- Maintenance of lean body mass and muscle strength
- Provision of beneficial cardiac effects

Oral fluids including oral carbohydrates may not be safely administered in:

- Patients undergoing emergency surgery
- Patients with delayed gastric emptying
- Gastrointestinal motility disorders
- Obese and diabetic patients.

Prophylaxis for Venous Thromboembolism

- This should commence prior to induction of anesthesia.
- Mechanical prophylaxis is equally effective to that of heparin alone in reduction of venous thromboembolism (VTE).[1]
- Mechanical prophylaxis plus heparin = greater reduction in risk of VTE.[1]

Preoperatively, gynecologic surgical patients have been classified into four risk categories. This classification is used to determine the most appropriate thromboprophylaxis regime for specific patients. This is based on procedure type and duration, age, and presence of additional risk factors **(Box 2 and Table 1)**.[6]

BOX 2: Risk factors for venous thromboembolism (VTE).[5]

- Presence of malignancy
- Obesity (moderate obesity 75–90 kg or >20% of ideal body weight; morbid obesity 115 kg or >30% of ideal body weight)
- Advanced age
- Pelvic surgery
- Extrapelvic disease
- Previous history of VTE
- Use of preoperative corticosteroids
- Chemotherapy
- Immobility
- Hypercoagulable state
- Diabetes

Standard Anesthetic Protocol

The ERAS intraoperative recommendations are mentioned in **Box 3**.

Pain Relief

- Preemptive analgesia (analgesics block the activation of pain receptors before they are activated by the presence of noxious stimuli)

Superior pain control → Decreased need for analgesics.
- A multimodal approach including the preoperative use of gabapentin, oral or intravenous (IV) cyclooxygenase-2 (COX-2) inhibitors (celecoxib or parecoxib), and oral or IV paracetamol has been associated with reduced postoperative opioid use and it is included in ERAS protocol.[1]

Optimized Fluid Administration

Fluid overload can lead to:

- Electrolyte abnormalities
- Peripheral soft tissue edema that impairs mobility
- Small bowel edema leading to the delay in the return of bowel function
- Pulmonary congestion that leads to increased pulmonary morbidity.

Hypovolemia results in:

- Decrease in cardiac output impairing the supply of oxygen to tissues with consequent damage to organ systems.

One of the principles of ERAS protocol is to maintain intraoperative euvolemia.

TABLE 1: Risk categories of preoperative patients.

Low-risk level	*Moderate-risk level*	*High-risk level*	*Highest-risk level*
• Surgery time <30 minutes • Age <40 years • Risk factors—no risk factor ↓ Early ambulation	• Surgery time <30 minutes • Age <40 years • Risk factors Or • Surgery time <30 minutes • Age 40–60 years • No risk factors ↓ Low-molecular-weight heparin (LMWH) subcutaneous 12 hours before surgery + once a day until discharge postoperatively	• Surgery time <30 minutes • Age >60 years • Risk factor Or • Surgery time >30 minutes • Age >40 years • Risk factors ↓ LMWH + intermittent pneumatic compression (before surgery to patient became full ambulatory)	• Surgery time >30 minutes • Age >60 years • Prior venous thromboembolism • Hypercoagulable state ↓ LMWH + intermittent pneumatic compression (extend to 4 weeks)

BOX 3: Enhanced Recovery After Surgery (ERAS) intraoperative recommendations.[1]

Short-acting anesthetics:
- Continuous infusion of propofol (rapid onset, favorable antiemetic profile, rapid recovery)
- Regional anesthesia with or without concomitant general anesthesia (rapid awakening and less need of systemic opioids)
- Incisional infiltration of liposomal bupivacaine or bupivacaine

Continuous monitoring of body temperature prevents hypothermia as well as hyperthermia

Use of thermal blankets and warm intravenous fluids may be helpful

Prevention of postoperative nausea and vomiting by using two different classes of antiemetic agents 5HT3 antagonists (ondansetron), NK-1 antagonists (aprepitant), corticosteroids (dexamethasone), antihistamines (dimenhydrinate), anticholinergics (scopolamine), butyrophenones (haloperidol), and phenothiazines (chlorpromazine)

To achieve this, a goal-directed therapy is followed in which the hemodynamic parameters such as stroke volume, cardiac output, and peripheral vascular resistance are used to guide the use of IV fluids and inotropic therapy.[7]

To avoid intraoperative fluid overload, minimizing crystalloids and increasing the use of colloids is recommended. If a patient is hypotensive but euvolemic, the use of vasopressors is encouraged in preference to liberal administration of crystalloids.[1]

Recovery of Normal Gastrointestinal System

Early feeding and coffee intake prevents paralytic ileus.

Early Mobilization

- Another vital component of the ERAS protocol
- Protects against muscle and physical conditioning loss
- Reduce pulmonary and venous thromboembolic complications
- Improves insulin resistance
- Reduces hospital stay
- Contributes to the return of bowel function
- Decreases postoperative ileus rates.

COMPONENTS OF POSTOPERATIVE CARE

The components of postoperative care can be divided in terms of duration into early and late **(Table 2 and Flowchart 1)**.

Different types of fluids are described in **Table 3**.

TABLE 2: Postoperative care.

Early (<24 hours)	*Late (>24 hours)*
• Intravenous fluid management • Analgesic care • Postoperative blood glucose management • Electrolyte balance • Early thromboembolism prophylaxis	• Early oral feeding • Care of bowel • Care of bladder • Care of surgical site infection • Extended postoperative thromboembolism prophylaxis • Mental health

Flowchart 1: Postoperative fluid therapy (main goal—tissue perfusion).[8]

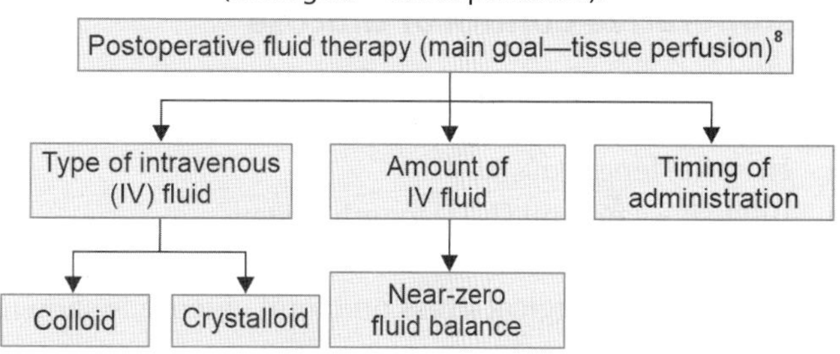

TABLE 3: Type of fluids.

Crystalloids	*Colloids*
Example: Normal saline Ringer lactate, dextrose normal saline • *Advantages:* – Contain small-sized molecule – Easy to metabolize • *Disadvantage:* – Can freely pass through double barrier of endothelium ↓ In large amount causes: - Pulmonary edema - Hyperchloremic acidosis	*Example:* Albumin solution Fresh frozen plasma, gelatins, dextrans, hydroxyethyl starch • *Advantages:* – Minor anti-inflammatory effect – Increase intravascular volume • *Disadvantage:* – High allergic risk

Amount of Fluid

- *Calculated by:* 2.5 mL/kg + 1 mmol/kg Na + 1 mmol/kg K^+
- Input output balance should be near to zero.
- *Patient should be examined carefully:*
    - Sign of fluid depletion:
        - Dry mucous membrane
        - Reduced skin turgor pressure
        - Decrease urine output

- Orthostatic hypotension
- Increase capillary refill time
- Tachycardia
- Low blood pressure (BP)
- Sign of fluid overloading:
  - Increase jugular venous pressure (JVP)
  - Sacral edema
  - Pulmonary edema.

Analgesia Care (Table 4)

Table 4 shows the analgesics in postoperative period.

TABLE 4: Analgesics in postoperative period.[9,10]

Analgesia	*Route of administration*	*Others*
Preemptive analgesia	IV, oral	
Paracetamol infusion	• Oral—2 g • IV—1 g 8 hourly • Rectal	• Well-tolerated • Produce rapid peak plasma level
NSAIDs	• Oral • Intramuscular • Rectal	Can affect gastric mucosa sensitive in asthmatic
Opioids		Prophylactic antiemetic to be given *Adverse effects:* • Respiratory depression • Hypercapnia • Increase risk of paralytic ileus
Thoracic epidural analgesia (TEA)	Epidural	• For open abdominal surgeries • Gynecologic cancer surgeries • Reduces time to return of gut function
Continuous wound infiltration (CWI)	Local	• Improve recovery and can be considered as alternative to TEA • Evidence of benefit is lacking

(IV: intravenous; NSAIDs: nonsteroidal anti-inflammatory drugs)

Postoperative Control of Glucose

Level of glucose affect:

- Increase perioperative mortality
- Hospital length of stay
- Intensive care unit (ICU) length of stay
- Postoperative infection.

The effect of surgical stress on blood glucose is shown in **Flowchart 2**.

Although the optimal glycemic target remains unclear, a reasonable goal in the majority of postoperative patients is to maintain blood glucose in the range of 80–130 mg/dL with the intent of decreasing postoperative infections and promote healing. **Flowchart 3** shows management of Hyperglycaemia in postoperative period.

- Blood sugar charting 2–4 hourly is mandatory is postoperative patients and it should be 1–2 hourly in critically ill patients.

Flowchart 2: Effect of surgical stress on blood glucose.

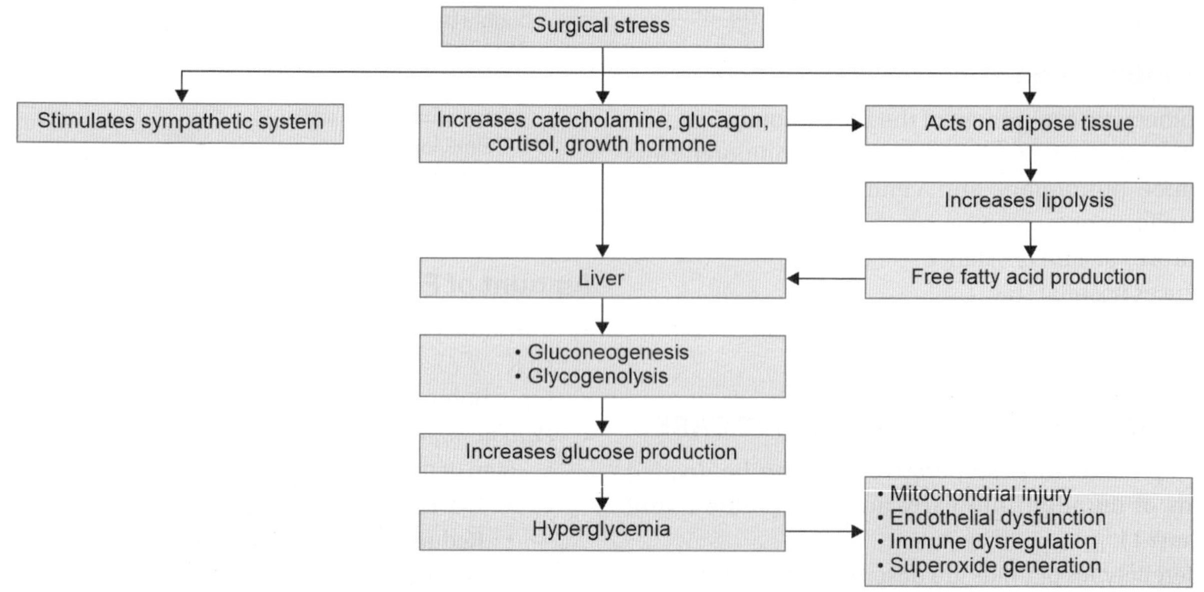

Flowchart 3: Management of hyperglycaemia in postoperative period.[11]

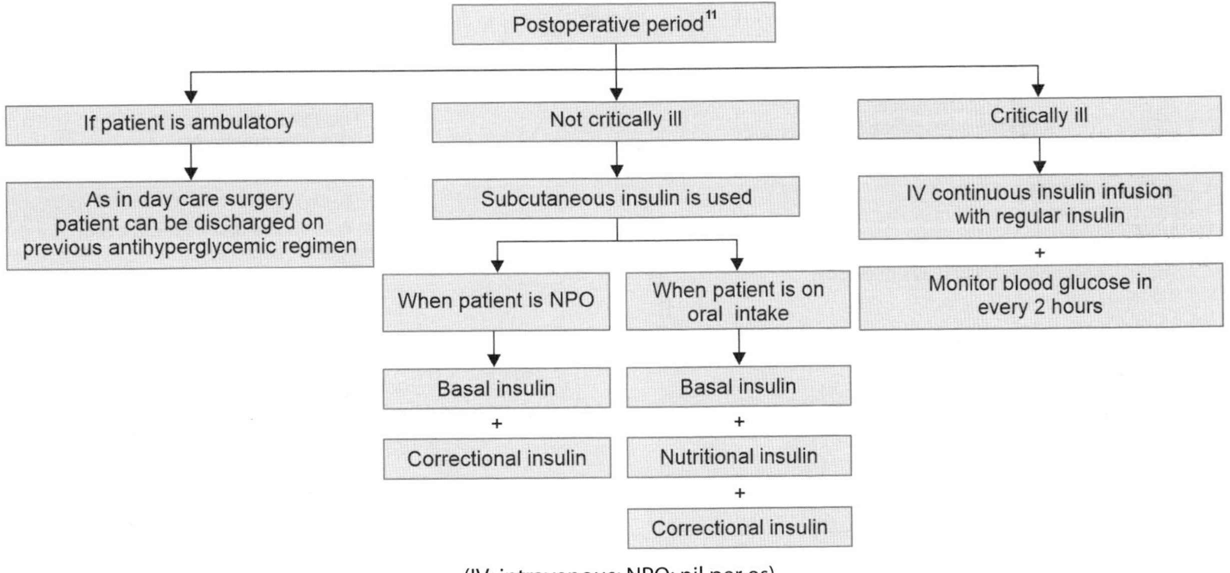

(IV: intravenous; NPO: nil per os)

TABLE 5: Components of Insulin regime.[11]

Basal insulin	***Long-acting insulin***
Correctional insulin	Either rapid- or short-acting insulin
Nutritional insulin	Either rapid- or short-acting

- Frequency of blood sugar charting is 1–2 hourly in critically ill patients.

Early Oral Intake

Recommendations as per ERAS protocol:

- Resumption of oral intake of liquids and solids within 24 hours after surgery
- Encourage fluid intake when the patient recovers from anesthesia
- Consider high-protein diets

Benefits:

- Enhanced recovery and accelerated return of bowel
- Have no effect on prevention of paralytic ileus which is prevented by:
  - Consumption of coffee reduces the risk of paralytic ileus from 30% to 10%.
  - Laxatives
  - Early ambulation and modal analgesia—reduces the paralytic ileus by two to five times
  - Gum chewing
  - Avoid use of Ryle's tube
  - Avoid excess fluids

Electrolyte Abnormalities

The serum electrolytes need to monitored in the postoperative period especially if the patient has diarrhoea, vomiting, paralytic ileus etc. The management of electrolyte (Sodium and Potassium) abnormalities is shown in **Flowchart 4 and 5.**[12]

Catheter Care

Table 6 gives the advantages and disadvantages of urinary catheter. The pathophysiology of catheter-induced urinary tract infection is given in **Flowchart 7.**

The points to remember in catheter care are shown in **Flowchart 8.**

Care of Bowel

Uncomplicated surgery: Regular diet on same operative day if no distension, bowel sounds are present, and patient is not nauseated from anesthesia.

Prolonged extensive surgery: Auscultate the bowel sounds and allow orally accordingly.

If postoperative abdominal distension present, differentiate between paralytic ileus and obstruction **(Table 7).**

Wound Care

Surgical site dressing:

- Remain in place for 48 hours after surgery, if not soaked
- Perform hand hygiene before dressing.

Flowchart 4: Electrolyte abnormalities.[12]

Sodium

Hyponatremia
<135 mEq—hyponatremia
<125 mEq—produce symptoms

Causes:
- Aggressive IV fluid use
- Venous absorption of large volume of distending media during hysteroscopy
- Due to pain
- SIADH

Sign and symptoms:
- Metabolic encephalopathy
- Respiratory distress

Correction:
Correction does not exceed 0.5 mEq/L/h
With a goal of Na^+—130 mEq/L
↓
Overaggressive correction cause central pontine myelinolysis

How to correct:
- No sign and symptoms careful use of isotonic fluid and if hypervolemia give Lasix
- Acute sign and symptoms present
  ↓
  3% NS in 100 mL infusion over 30 minutes (can be repeated two times if needed)

Hypernatremia
>145 mEq—hypernatremia

Causes:
Loss of hypotonic fluid such a diarrhea, sweating, and gastric secretion

Mechanism:
Hypernatremia
↓
Causes brain cell shrinkage
↓
Vascular bleeds and permanent neurologic damage

Compensatory mechanism:
To restore brain cell volume
↓
Brain secretes idiogenic osmoles
↓
Pull water back into cells

Correction:
First use isotonic or colloid fluid
↓
Use hypotonic solution

(IV: intravenous; NS: normal saline; SIADH: syndrome of inappropriate antidiuretic hormone secretion)

Flowchart 5: Potassium abnormalities.[12]

Potassium abnormalities

Hypokalemia <3.5 mEq/L

Causes:
Diarrhea

Serum K^+ level
- <2.5 mEq/L Cause muscle Necrosis
- <2.0 mEq/L Cause ascending Paralysis

Correction of K^+:
- Correction done at 20 mEq/h (measure Mg^{2+} level too as Mg^{2+} depletion causes hypokalemia)
- Can be given IV or orally depending on level
  IV infusion of KCl over 5–6 hours @ 20 mEq/h
  Syp Potklor

Hyperkalemia >5.5 mEq/L

Causes:
Inappropriate use of:
ACE Inhibitors
K^+ sparing drugs
NSAIDs

Sign and symptoms:
Mainly ECG changes
PR interval increases
P wave disappears
QRS widens

Correction:
10 U insulin with 50 mL of 50% dextrose

Calcium gluconate IV 10 mL 10% in 2–3 minutes
Beta-2 agonists
K^+ binder

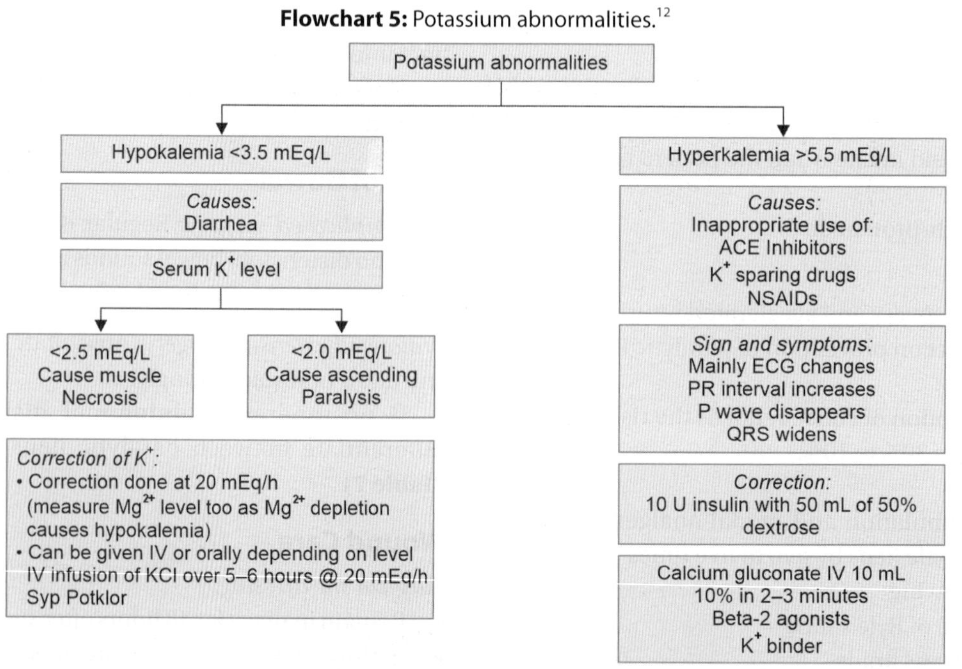

(IV: intravenous; NSAIDs: nonsteroidal anti-inflammatory drugs)

Flowchart 6: Route of surgery and return of Gut motility.[12]

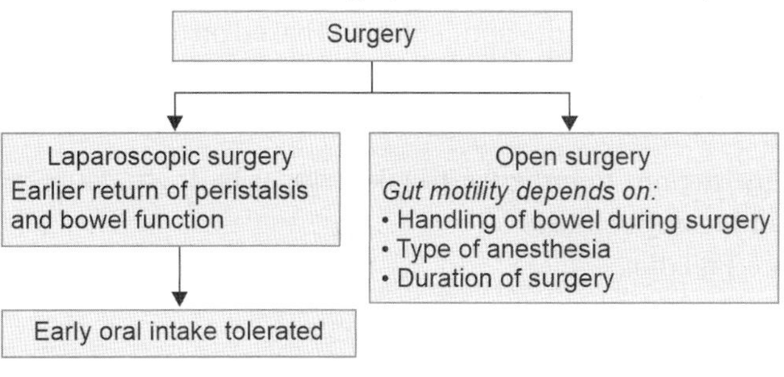

Flowchart 7: Pathophysiology of catheter-induced urinary tract infection (UTI).

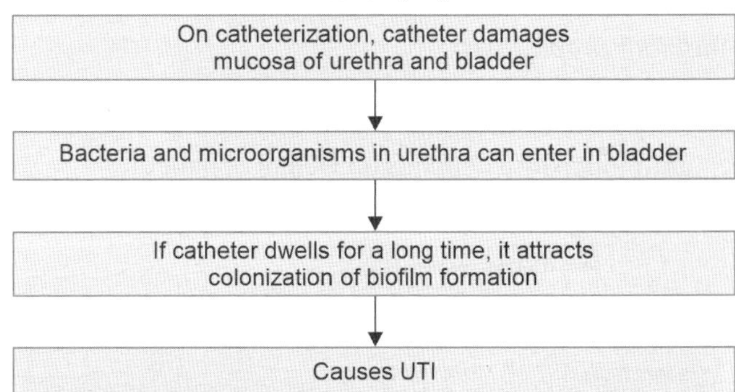

TABLE 6: Urinary catheter.

Advantages	*Disadvantages*
• Provides bladder decompression • Measures urine output • Helps in management of postoperative urinary retention • Prevents bladder injury	• Catheter-induced urinary retention • Causes discomfort to patient • Increases length of hospitalization • Urinary tract infection (UTI)

Flowchart 8: Points to remember in catheter care.

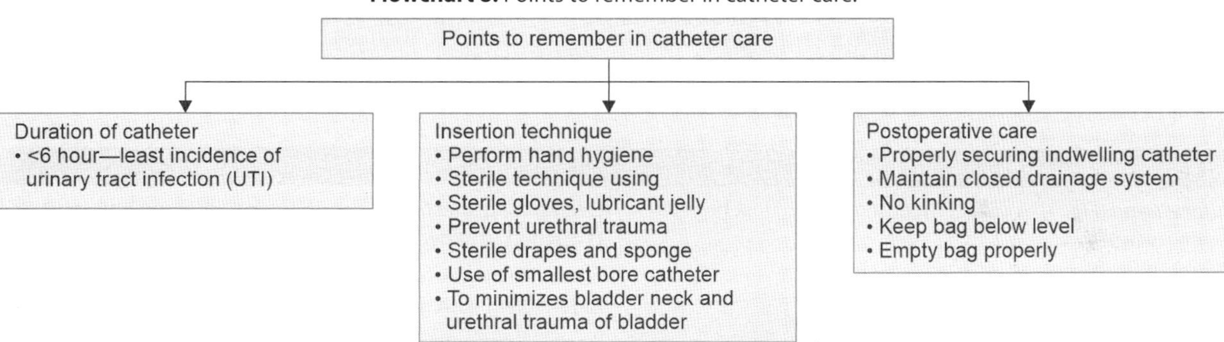

TABLE 7: Differentiating features between postoperative ileus and obstruction.

Clinical feature	*Paralytic ileus*	*Obstruction*
Onset of distension	Within 48–72 hours of surgery	Delayed up to 5–7 days
Bowel sounds	Absent or hypoactive	Peristaltic rushes Borborygmi
Abdominal pain	Mild discomfort	Progressive, severe cramping pain
Postoperative nausea and vomiting (PONV)	Present	Present
Abdominal X-ray	Distended loops of small and large bowel; gas usually present in colon	Single or multiple loops of distended bowel (usually small bowel) with air–fluid levels
Management	Conservative with nasogastric suction, enema	• Nasogastric decompression • Surgical exploration

Surgical Site Infection

The risk factors for surgical site infection (SSI) are shown in **Flowchart 9**.

Flowchart 10 shows the best practices to prevent surgical site infection.

Hand hygiene:

- Preoperative surgical scrubbing for 2–6 minutes is effective.
- Use antimicrobial or alcohol-based scrubs as it causes denaturation of proteins. Hence, effective in

gram-positive and gram-negative microorganisms both.

Antibiotic prophylaxis:

- Administered 30 minutes prior to incision
- Cefazolin is first-line recommended prophylactic antibiotic; dose depends on weight of patient.

Flowchart 9: Risk Factors for surgical site infection (SSI).

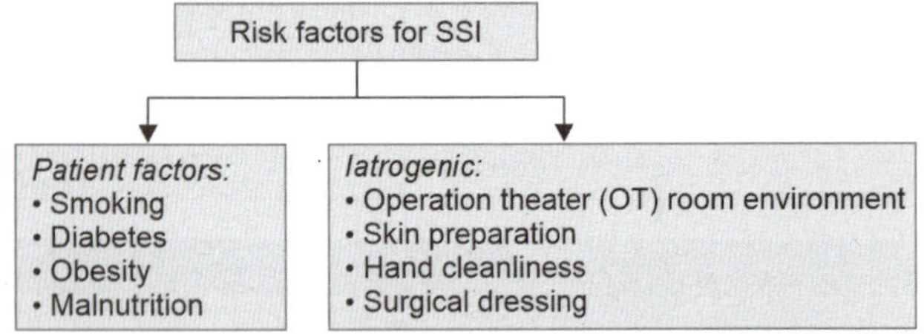

Surgical Site Preparation

Alcohol-based antiseptic solutions based on chlorhexidine gluconate for surgical site skin preparation in patients undergoing surgical procedures is recommended.

Instructions to patients: To take regular bath after surgery and change clothes

- Full diet
- Ambulation

Table 8 shows the postoperative wound infections.

CONCLUSION

The Goal of management in the postoperative period is primarily to prevent infections, to promote healing of

Flowchart 10: Best practices to prevent surgical site infection.[13]

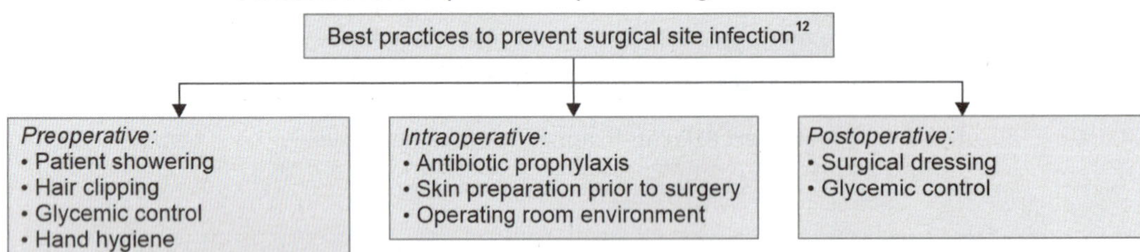

TABLE 8: Postoperative wound infections.

*Wound seroma **(Fig. 1)**:* A collection of serous fluid just beneath the skin surface Small amount managed by antiseptic dressing Larger amount—extensive incision and drainage	*Deep wound infection:* When purulent discharge is noted, wound is opened to allow drainage and removal of necrotic tissue *Fascia intact:* Mechanical debridement followed by loose wet to dry packing with gauze moistened with saline Antibiotic effective against anaerobes must be used *If infection involves the fascia—burst abdomen **(Fig. 2)**:* Surgical emergency as evisceration of bowel may occur Managed by identification of defect, freshening of fascial edges followed by reinforced closure of the fascia *Infected wound:* Daily dressing; can be left for delayed closure or closure of wound can be done after wound becomes healthy
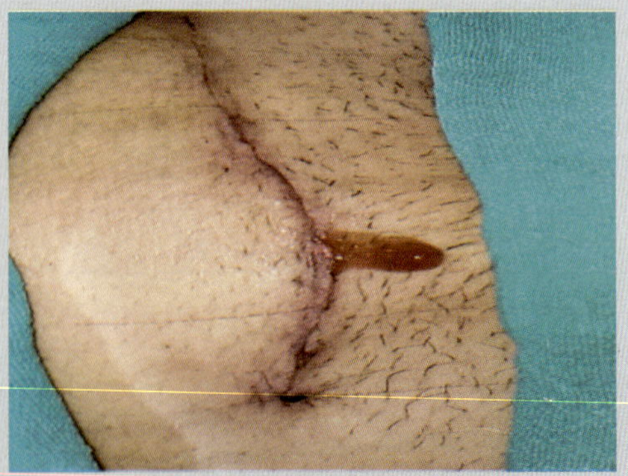 **Fig. 1:** Wound seroma.	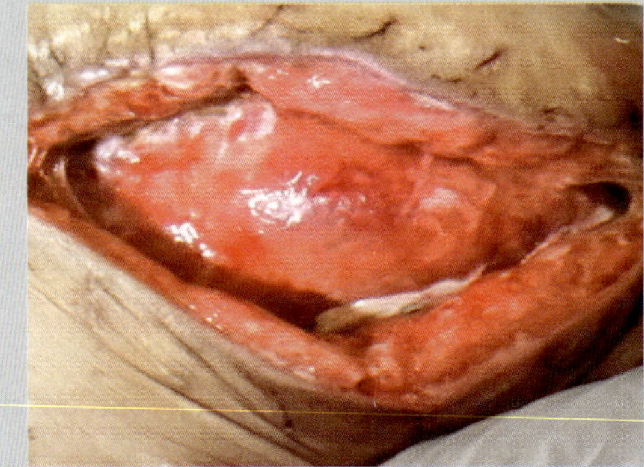 **Fig. 2:** Burst abdomen.

surgical wounds and to enable patient return to a state of optimal health. Appropriate fluid management,adequate pain relief, thromboembolism prophylaxis, surgical incision dressing, catheter care and early ambulation are critical components of postoperative management to ensure early recovery.

REFERENCES

1. Nelson G, Bakkum-Gamez J, Kalogera E, Glaser G, Altman A, Meyer LA, et al. Guidelines for perioperative care in gynecologic/oncology: Enhanced Recovery After Surgery (ERAS) Society recommendations-2019 update. Int J Gynecol Cancer. 2019;29(4):651-68.
2. Kalogera E, Dowdy SC. Enhanced recovery pathway in gynecologic surgery: improving outcomes through evidence-based medicine. Obstet Gynecol Clin North Am. 2016;43(3):551-73.
3. Nygren J, Thorell A, Ljungqvist O. Preoperative oral carbohydrate therapy. Curr Opin Anaesthesiol. 2015; 28(3):364-9.
4. Azagury DE, Ris F, Pichard C, Volonté F, Karsegard L, Huber O. Does perioperative nutrition and oral carbohydrate load sustainably preserve muscle mass after bariatric surgery? A randomized control trial. Surg Obes Relat Dis. 2015;11(4):920-6.
5. Lyman GH, Khorana AA, Kuderer NM, Lee AY, Arcelus JI, Balaban EP, et al.; American Society of Clinical Oncology Clinical Practice. Venous thromboembolism prophylaxis and treatment in patients with cancer: American Society of Clinical Oncology clinical practice guideline update. J Clin Oncol. 2013;31(17):2189-204.
6. Committee on Practice Bulletins—Gynecology, American College of Obstetricians and Gynecologists. ACOG Practice Bulletin No. 84: Prevention of deep vein thrombosis and pulmonary embolism. Obstet Gynecol. 2007;110(2 Pt 1): 429-40.
7. Gómez-Izquierdo JC, Feldman LS, Carli F, Baldini G. Meta-analysis of the effect of goal-directed therapy on bowel function after abdominal surgery. Br J Surg. 2015;102(6):577-89.
8. Walsh SR, Cook EJ, Bentley R, Farooq N, Gardner-Thorpe J, Tang T, et al. Perioperative fluid management: prospective audit. Int J Clin Pract. 2008;62:492-7.
9. Catro-Alves LJ, De Azevedo VL, De Freitas Braga TF, Gonclaves AC, DeOliveira GS Jr. The effect of neuraxial versus general anesthesia techniques on postoperative quality of recovery and analgesia after abdominal hysterectomy: a prospective, randomized, controlled trial. Anesth Analg. 2011;113(6):1480-6.
10. Ong CK, Seymour RA, Lirk P, Merry AF. Combining paracetamol (acetaminophen) with nonsteroidal antiinflammatory drugs: a qualitative systematic review of analgesic efficacy for acute postoperative pain. Anesth Analg. 2010;110(4):1170-9.
11. Dogra P, Anastasopoulou C, Jialal I. Diabetic perioperative management. In: StatPearls [Internet]. Treasure Island (FL): StatPearls Publishing; 2024.
12. World Health Organization. Global Guidelines for the Prevention of Surgical Site Infection, 2nd edition. Geneva: World Health Organization; 2018.

Chapter 40

Anesthesia for Gynecological Surgery

Rajesh Kumar Verma, Vivek Kaushal, Rohini Rao

INTRODUCTION

Gynecological surgeries have seen significant advances in the surgical technique to minimally invasive procedures which have been associated with comparable results when compared with open surgery. Various gynecological surgeries are listed in **Table 1**.

ANESTHETIC CONSIDERATION

NPO

- Solids—6–8 hours
- Pulp containing liquids/milk—4 hours
- Pulpless juices and clear water—2 hours

Also, preoperative benzodiazepines (alprazolam—0.5 mg) and H-2 blockers (ranitidine—150 mg) are given on the day of surgery with a sip of water as antianxiety and anti-aspiration prophylaxis.

TABLE 1: Types of gynecological surgical procedures.

Diagnostic procedures	• Basic laparoscopy • Hysteroscopy
External procedures	• Marsupialization of Bartholin gland cyst • Simple vulvectomy and ablation of condylomata • Labioplasty • De-epithelialization technique
Cervical surgical procedures	• Dilation and curettage
Ovarian, fallopian tube, and uterine surgical procedures	• Oophorectomy • Total salpingectomy • Endometrial ablation • Myomectomy (abdominal approach) • Total abdominal/vulval/laparoscopic vaginal/robotic hysterectomy • Uterine radiation seeding • Radical hysterectomy
	• Pelvic exenteration • Anterior (cystocele) and posterior (rectocele) colporrhaphy

Preoperative

Preanesthetic checkup should follow general guidelines, attention paid to general risk factors (obesity, increased age, and smoking) and quantification of metastasis. Cardiac function assessment and assessment of pulmonary complications are guided by the history and examination. Preoperative dyspnea in patients with advanced ovarian cancer can be difficult to investigate and can be a sequela of ascites. Consideration to change from a laparoscopic to an open surgical procedure in patients with concurrent diseases such as advanced ischemic heart disease and moderate-to-severe valvular lesion, chronic kidney disease can be made on a case-to-case basis **(Tables 2 and 3)**.

MULTIMODAL STRATEGIES FOR POSTOPERATIVE PAIN CONTROL[2]

Pregabalin/Gabapentin

Pregabalin in dose range of 100–300 mg OD to TID was found to reduce the pain scores and incidence of postoperative nausea vomiting (PONV). Gabapentin 1–2 hours prior to surgery in a single dose of 1,200 mg or with smaller doses of 100–400 mg given every 6–8 hours was also found to reduce the opioid consumption in the postoperative period. Adverse effects of these drugs were sedation and visual disturbances.

Cyclooxygenase Inhibitors

One of the ideal medications with opioid sparing effects hence reducing incidence of PONV. Both ketorolac and diclofenac intravenous (IV) or intramuscular (IM) can be used in therapeutic doses.

TABLE 2: Preoperative evaluation.

Hematology	Preoperative *hemogram* and prothrombin time (*PT)/international normalized ratio (INR)* assessed which are important before planning any neuraxial procedure and preoperative *blood grouping and cross matching* requiring intraoperative blood transfusion and with patients on antiplatelet therapy. Hb value 9–10 g/dL in an ASA-1 patient is considered safe for induction of anesthesia with valid consent for transfusion.
CVS (cardiovascular system)	*ECG:* Preoperatively can be ordered to identify any rhythm disorder and in coronary artery disease (CAD) patients to determine baseline. *Echocardiogram* can be ordered based on the case and any concurrent diseases [CAD, valvular heart disease (VHD)] and history of chemotherapy.
Respiratory system	*Chest X-ray* can be ordered to rule out any pulmonary metastasis in cancer (CA) patients. Presence of any pneumatoceles and bullae can be ruled out before giving general anesthesia. Pulmonary function test can be sought in chronic obstructive pulmonary disease (COPD) and asthmatic patients.
LFT (liver function test)	Induction agents (thiopentone) skeletal muscle relaxants (succinylcholine, vecuronium, and rocuronium) opioids (morphine) volatile anesthetics (halothane) are directly or indirectly metabolized or excreted by liver and their pharmacokinetics can be altered with hepatic diseases.
RFT (renal function test)	Serum urea and creatinine to calculate clearance for dosing or avoidance of nephrotoxic drugs, e.g., antimicrobials (amikacin, gentamycin, amphotericin), contrast agents, non-steroidal anti-inflammatory drugs (NSAIDS). Laparoscopic surgery causes intraabdominal hypertension and oliguria. Preoperative dialysis in end-stage renal disease (ESRD) patients.

TABLE 3: Anesthesia techniques.

Spinal anesthesia	*Drugs:* Bupivacaine 0.5% (H), ropivacaine (H)	• Most commonly used for open hysterectomies and prolapse surgeries • Cost effective and enhanced immediate postoperative recovery • Reduces postoperative opioid consumption due to addition of intrathecal adjuvants, e.g., morphine, fentanyl, and dexmedetomidine
Epidural anesthesia	*Drugs:* Bupivacaine 0.5/0.25/0.125% plain, ropivacaine with fentanyl and morphine	• No effect on gastric motility (avoid morphine) • Better postoperative pain control • Survival benefits for women with gynecological malignancies[1]
General anesthesia (GA)	• *Opioids:* Fentanyl, morphine (avoided in liver failure, reduce dose in renal failure), pethidine. • *Induction agents: Propofol* (avoid in known hypersensitivity and fat metabolism disorders), *ketamine* (used in acute shock), *etomidate* (mainly used in patients with coronary artery disease, transient adrenocortical suppression, myoclonus). • *Muscle relaxants: Suxamethonium* (faster onset, causes bradycardia, transient hyperkalemia, bradycardia, increase in IOP, ICP, myalgia). *Atracurium* (hypotension at higher doses due to histamine release), *vecuronium* (metabolites accumulate in renal failure), *rocuronium.* • *Reversal agents: Neostigmine* (anticholinesterase) with glycopyrrolate as they cause profound vagal stimulation, *sugammadex* (for failed intubation with aminosteroid relaxants like rocuronium)	One of the most commonly used modalities. Can be used in any case

Contd...

Contd...

General plus epidural anesthesia		• Epidural acts as adjuvant to general anesthesia in postoperative pain management in hysterectomies laparotomies for advanced gynecologic cancer surgeries • Reduces pulmonary complications post GA • Early mobilization • Reduces DVT (deep vein thrombosis) and pulmonary embolism • Reduced endocrine and metabolic response
General anesthesia adjuncts	Transversus abdominis plane (TAP) block	• Provides analgesia in Pfannenstiel and midline incisions • Reduction of postoperative pain, opioid and non-opioid consumption, and institutional length of stay
	Wound infiltration	Pelvic organ prolapse surgeries can be done under local anesthetic infiltration
	Intraperitoneal local anesthetics	• Intraperitoneal lidocaine is better than intravenous (IV) lidocaine • Opioid sparing effect
	Local anesthetic wound infusion (subcutaneous catheter)	• Provides adequate postoperative analgesia • Reducing general pain scores • Minimizes opioid requirement

Paracetamol

Equally efficacious as non-steroidal anti-inflammatory drugs (NSAIDS) at dose of 1–2 g OD or BD with opioid sparing effects and reduction in PONV.

Lidocaine/Ketamine Infusion

Intravenous (IV) lidocaine used during intraoperative period reduces postoperative pain and anti-analgesic consumption, led to faster return of gut function, and reduced hospital length of stay in abdominal surgeries. Ketamine when used in a limited subset of patients, such as those with chronic pain state who are on long-term opioids with continuous infusion with dose range of 50–600 μ/kg/minute was found to be useful in reducing pain scores and narcotic consumption.

POSTOPERATIVE NAUSEA VOMITING

A little >50% women (risk factor) undergoing gynecological procedures (risk factor) suffer from PONV. Strategies to reduce PONV include.

- Use of regional (spinal/epidural) rather than general anesthesia with local anesthetic infusion with minimal use of intrathecal opioids.
- Using TIVA (total intravenous anesthesia) with propofol in place of inhalational agents and nitrous oxide.
- Multimodal anesthesia with limited use of intraoperative and postoperative opioids.
- Liberal use of hydration in perioperative period.

Enhanced Recovery after Anesthesia[3]

- *Preoperative components:*
  - Dedicated preoperative counseling
  - Recommend tobacco (ideally 1 month before surgery) and alcohol cessation
  - Recognition and correction of anemia
  - Complex carbohydrates 2 hour prior to surgery
  - Elimination of oral mechanical bowel preparation
- *Intraoperative components:*
  - Before OR-CELECOXIB 400 orally, PCM 1,000 mg orally, gabapentin 600 mg orally.
  - Regional anesthesia
  - Multimodal pain management
  - Decrease fluid administration/increase colloid if needed
  - Sequential compression devices
  - Consider heparin or low molecular weight heparin (LMWH) for high-risk patients.

    - Antibiotics 60 minutes before skin incision
    - Alcohol-based cleansing unless contraindicated. Vaginal cleansing with 4% chlorhexidine gluconate
    - Hair clipping rather than shaving
    - Avoid drains and vaginal packs.
    - Maintain intraoperative normothermia.
- *Postoperative consideration:*
    - No NG tube
    - Start regular feed and chewing gum 4-hour post procedure.
    - Osmotic diuretics—senna and docusate sodium, magnesium oxide or hydroxide as needed
    - Maintain blood glucose between 100 and 200 mg/dL.
    - Multimodal analgesia with opioids, hydromorphone for postoperative pain.
    - IV and PCA regimens for continued pain despite titration of oral regimen.
    - Remove urinary catheter within 24 hours. Assess for requirement and removal of drains and vaginal packs.
    - Defined discharge protocols (e.g., oral analgesia, mobilization, ability to tolerate solids without nausea and vomiting)

CONCLUSION

A thorough medical history and preoperative evaluation is mandatory for planning of anesthesia techniques for gynecological surgery. Safe anesthesia leads to successful surgical outcomes.

REFERENCES

1. Lacassie HJ, Cartagena J, Branes J, Assel M, Echevarría GC. The relationship between neuraxial anesthesia and advanced ovarian cancer-related outcomes in the Chilean population. Anesth Analg. 2013;117:653-60.
2. Bauchat JR, Habib AS. Evidence-based anesthesia for major gynecologic surgery. Anesthesiol Clin. 2015;33(1):173-207.
3. Nelson G, Bakkum-Gamez J, Kalogera E, Glaser G, Altman A, Meyer LA, et al. Guidelines for perioperative care in gynecologic/oncology: Enhanced Recovery After Surgery (ERAS) Society recommendations—2019 update. Int J Gynecol Cancer. 2019;29:651-68.

Chapter

41 Perioperative Shock in Gynecology

Smiti Nanda, Nisha Malik

INTRODUCTION

Shock is defined as an acute clinical syndrome characterized by hypoperfusion of vital organs leading to cell hypoxia.[1] It is one of the life-threatening conditions that may be encountered in a gynecological patient.

CLASSIFICATION OF SHOCK

Flowchart 1 illustrates the types of shock based on their etiology. This chapter will focus on the two most commonly encountered shock states in gynecology during perioperative period: hemorrhagic and septic shock.

HEMORRHAGIC SHOCK

Massive bleeding, either intraoperative or postoperative, leading to hemorrhagic shock may be encountered in gynecological patients.

Risk Factors

The factors associated with increased risk of hemorrhage in gynecology are outlined in **Box 1**.

Pathophysiology

Flowchart 2 depicts the pathophysiological changes in hemorrhagic shock.

Clinical Presentation

The clinical signs and symptoms of hemorrhagic shock will vary depending on the volume and rate of blood loss and are depicted in **Box 2**.

Classification

A classification of hemorrhagic shock depending upon the volume of blood lost is as outlined in **Table 1**.

Flowchart 1: Classification of shock based on the etiology.[1]

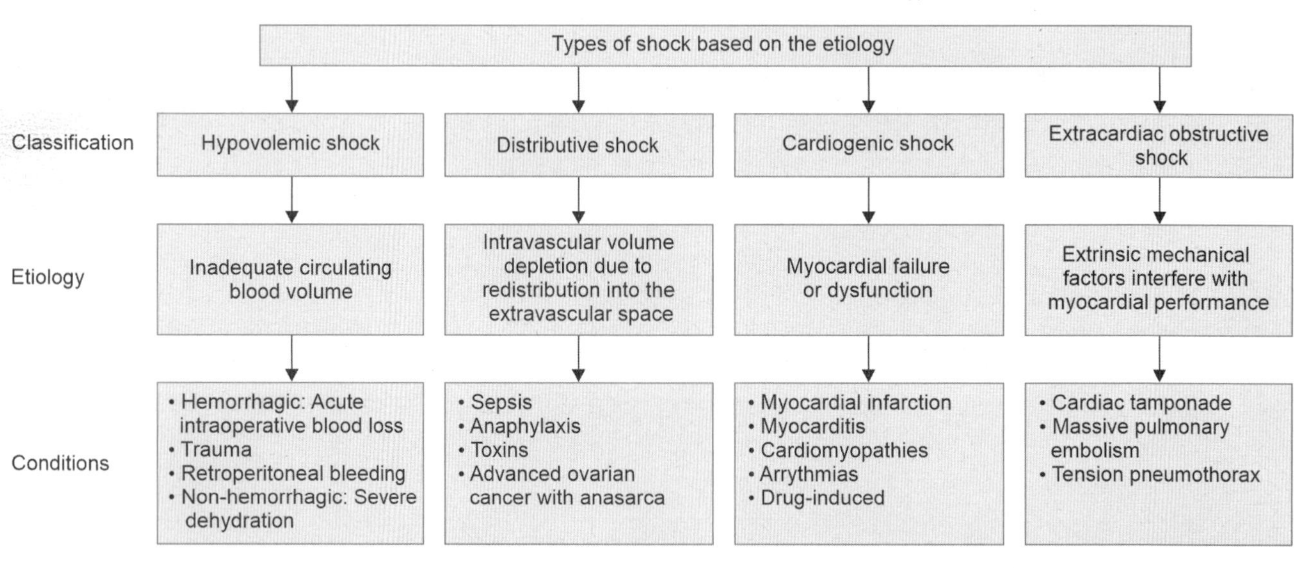

BOX 1: Risk factors for perioperative hemorrhagic shock in a gynecological patient.[2]

Preoperative factors:
- Anemia
- Personal history of coagulopathy
- Family history of coagulation disorders
- On anticoagulation therapy

Intraoperative factors:
- Distorted pelvic anatomy
- Dense adhesions
- Myomectomy
- Endometriosis
- Surgery for neoplasms

Flowchart 2: Pathophysiology of hemorrhagic shock.[1]

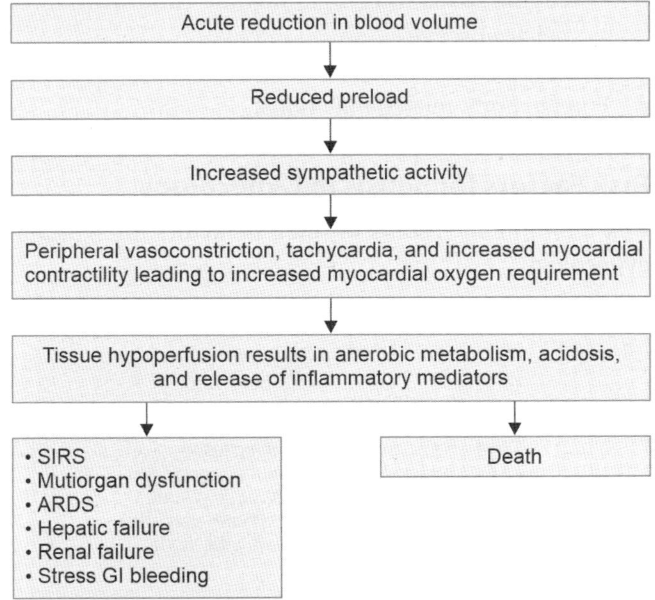

(ARDS: adult respiratory distress syndrome; GI: gastrointestinal; SIRS: systemic inflammatory sepsis syndrome)
Source: Adapted and modified from Smith.[1]

Prevention

The preoperative and intraoperative preventive strategies for hemorrhagic shock are as outlined in **Box 3**.

Management

Early recognition and immediate replacement of losses and prompt arrest of bleeding is vital for a favorable outcome. The aim of treatment is to maintain adequate tissue oxygenation and perfusion to prevent irreversible cellular damage. A useful mnemonic to prioritize the steps in management of hemorrhagic shock is "ORDER" as illustrated in **Flowchart 3**. **Flowchart 4** illustrates the blood components

BOX 2: Clinical manifestations of hemorrhagic shock.[1]

Early shock:
- CNS: Altered mental status
- CVS: Tachycardia, orthostatic hypotension
- Renal: Oliguria
- Respiratory: Tachypnea
- Hepatic: No change
- GIT: No change
- Hematological: Anemia
- Metabolic: None

Late shock:
- CNS: Obtunded
- CVS: Cardiac failure, arrthymias, hypotension
- Renal: Anuria
- Respiratory: Tachypnea, respiratory failure
- Hepatic: Liver failure
- GIT: Mucosal bleeding
- Hematological: Coagulopathy
- Metabolic: Acidosis, hypocalcemia, hypomagnesemia

(CNS: central nervous system; CVS: cardiovascular system; GIT: gastrointestinal system)
Source: Adapted and modified from Smith.[1]

TABLE 1: Classification of hemorrhagic shock.[1,2]

	Compensated	*Mild*	*Moderate*	*Severe*
Blood loss (mL)	≤1,000	1,000–1,500	1,500–2,000	>2,000
Volume loss (%)	<15	15–30	30–40	>40
Heart rate (bpm)	<100	>100	>120	>140
Blood pressure	Normal	Orthostatic change	Marked fall	Profound fall
Capillary refill	Normal	May be delayed	Usually delayed	Always delayed
Respiration	Normal	Mild increase	Moderate tachypnea	Marked tachypnea/respiratory collapse
Urinary output (mL/h)	>30	20–30	5–20	Anuria
Mental status	Normal or agitated	Agitated	Confused	Lethargic; obtunded

Source: Adapted from Martel.[2]

and pharmacological therapy for management of hemorrhagic shock.[1,2,4]

Treatment Goals

Box 4 depicts the treatment goals for managing shock.

Transfusion Complications

The complications associated with transfusion are as shown in **Flowchart 5**.[4]

BOX 3: Measures for prevention of hemorrhagic shock.[3]

Identification of patients at risk:
- Complete medical history
- Family history of bleeding diathesis
- Medication history
- Consultation with hematologist if coagulopathy
- Correction of anemia and coagulation profile

Intraoperative measures:
- Careful surgical technique
- Good exposure
- Experienced surgeon in complicated and radical surgeries
- Insertion of trocars away from large vessels during laparoscopy
- Secured ligation of vascular pedicles during hysterectomy
- Avoiding traction to vascular pedicles
- Proper surgical suturing of vaginal cuff

SEPTIC SHOCK

Septic shock, i.e., shock associated with infection, is a life-threatening condition with a case fatality rate exceeding 40%.[5] **Flowchart 6** illustrates the spectrum of infection syndromes.[1]

Gynecological patients undergoing any surgical procedure are particularly at risk of sepsis. The most common postoperative gynecological infections include the following:[6,7]

- *Surgical site infections or SSIs:*
  - Superficial incisional includes vaginal cuff cellulitis
  - Deep incisional includes pelvic cellulitis
  - Organ/space infections include tubo-ovarian or pelvic abscesses
- Urinary tract infections
- Foreign body-related infections.

Risk Factors for Sepsis in Gynecology

Box 5 enumerates some of the risk factors for sepsis in gynecology.[8]

Flowchart 3: Treatment algorithm for hemorrhagic shock ("ORDER")[1,2]

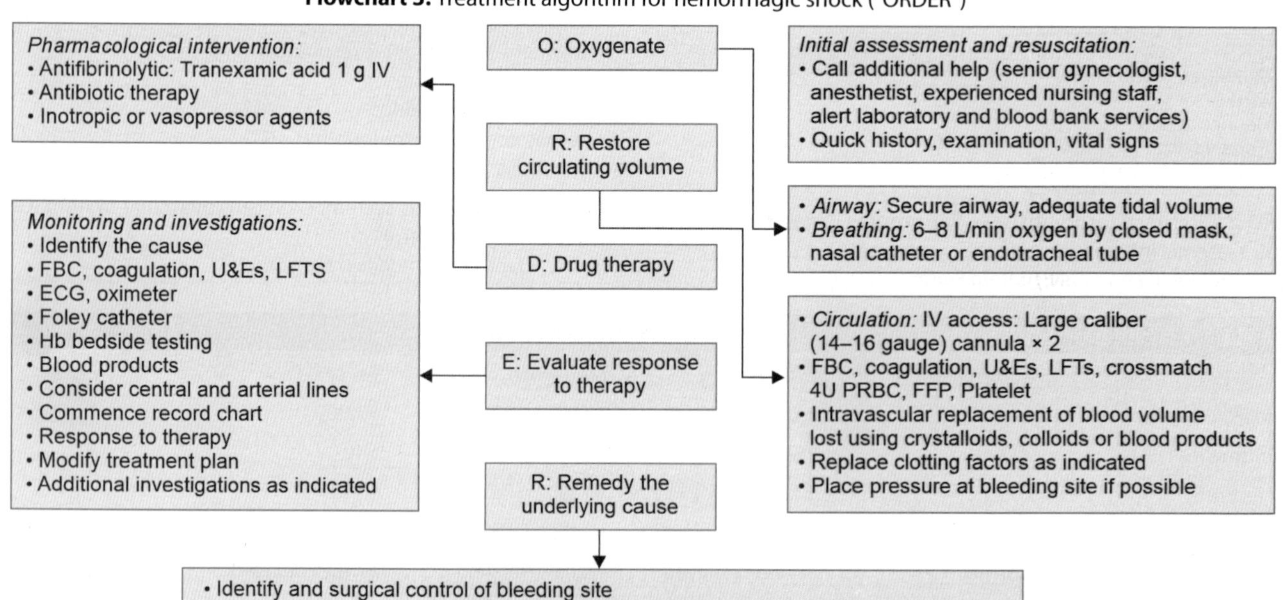

(ECG: electrocardiogram; FBC: full blood count; FFP: fresh frozen plasma; Hb: hemoglobin; IV: intravenous; LFTs: liver function tests; PRBC: packed red blood cell; U&E: urea and electrolytes;)
Source: Adapted and modified from Smith.[1]

Flowchart 4: Blood components and drug therapy for hemodynamic and cardiovascular support in hemorrhagic shock.[1,2,4]

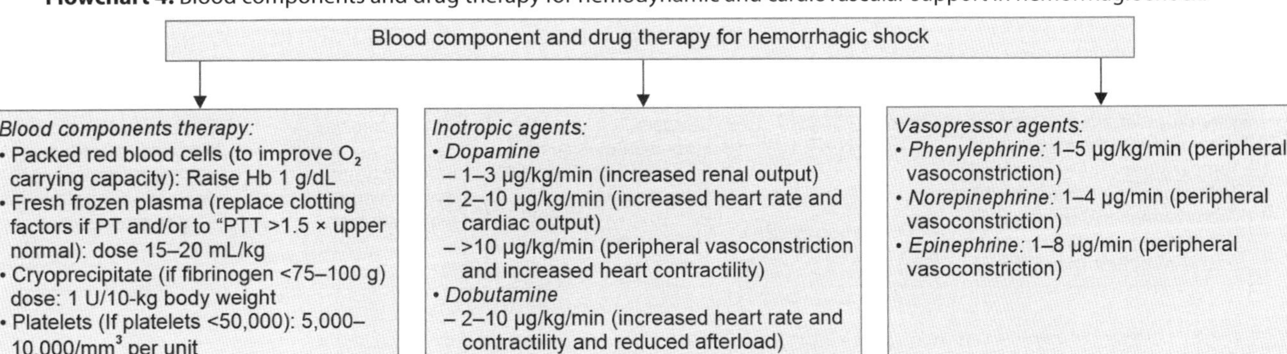

Source: Adapted and modified from: Martel.[2] Santoso et al.[4]

BOX 4: Treatment goals for hemorrhagic shock management.[1]

Cardiovascular/hemodynamic:

- *BP:* Maintain systolic at least 90 mm Hg; mean arterial pressure at least 60 mm Hg
- *Pulmonary capillary wedge pressure:* 14–18 mm Hg
- *Central venous pressure:* 12–15 cm H_2O
- *Oxygen saturation:* Arterial oxyhemoglobin saturation at least 92%

Pulmonary:
Blood gases: Maintain PaO_2 80–100 mm Hg; $PaCO_2$ 30–35 mm Hg; pH >7.35

Renal:
Maintain urine output of 20–30 mL/h (0.5 mL/kg/h); normalize BUN, creatinine

Hepatic:
Bilirubin <3 mg/dL

Mental status:
Restore

Treat the cause of hemorrhage:
Investigations:
- Hemoglobin >8 g/dL; platelets 100,000/mm^3
- INR <1.5; fibrinogen >75–100 mg/dL
- Serum lactate levels should be 2.2 mmol/L; suspect tissue hypoperfusion when elevated
- Avoid hypothermia, hypocalcemia, metabolic alkalosis, and hyperkalemia

(BP: blood pressure; BUN: blood urea nitrogen; INR: international normalized ratio; PaO_2: partial pressure of oxygen in arterial blood; $PaCO_2$: partial pressure of carbon dioxide in arterial blood)
Source: Adapted and modified from Smith.[1]

Flowchart 5: Transfusion-associated complications.[4]

Transfusion complications

Infectious:
- Hepatitis A, B, C
- HIV
- HTLV
- Parasitic diseases (malaria, Chagas disease, babesiosis)
- Creutzfeldt–Jakob disease

Immunologic:
- Acute hemolytic reactions
- Delayed hemolytic reactions
- Febrile illness
- Allergic anaphylactic reactions
- Transfusion-related acute lung injury (TRALI)

Miscellaneous:
- Citrate toxicity
- Hypocalcemia
- Hyperkalemia
- Hypothermia
- Volume overload

(HIV: human immunodeficiency virus; HTLV: human T-lymphotropic virus-1)

Flowchart 6: Spectrum of infection syndromes.[1]

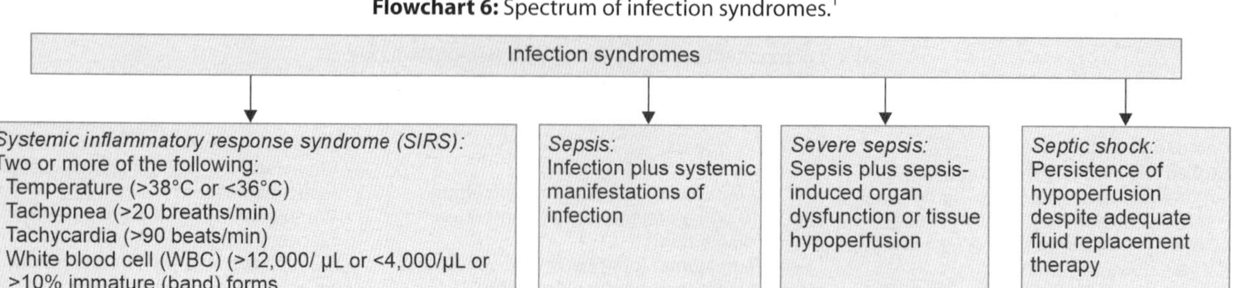

BOX 5: Risk factors for sepsis in gynecology.[8]

Host risk factors:
- Anemia
- Diabetes
- Malignancy
- Immunosuppression, e.g., steroid use, Alcohol
- Obesity
- Extremes of age
- Poor nutritional status
- History of radiation
- Colonization or infection with other organisms

Preoperative factors:
- Skin preparation
- Hair removal
- Poor glycemic control
- American Society of Anesthesiology (ASA) score of a minimum of 3
- Prolonged preoperative hospitalization

Intraoperative and postoperative factors:
- Prolonged surgery
- Route and mode of surgery
- Presence of indwelling catheters and surgical drains
- No use of antibiotic prophylaxis
- Hematoma
- Blood transfusion
- Increased intraoperative blood loss
- Prolonged hospitalization

Source: Adapted and modified from Ekanem.[8]

BOX 6: Microorganisms responsible for sepsis in gynecology.[8]

Gram-positive aerobes:
- *Staphylococcus aureus*
- *Enterococcus sp.*
- Group B hemolytic streptococcus
- *Staphylococcus pyogenes*
- *Staphylococcus epidermidis*
- Methicillin-resistant *staphylococcus aureus*

Gram-negative aerobes:
- *Klebsiella* sp.
- *Escherichia coli*
- *Pseudomonas aeruginosa*
- *Proteus* sp.

Anaerobes:
- *Clostridium* sp.
- Gardnerella vaginalis
- *Fusobacterium* sp.
- Bacteroides fragilis
- *Peptostreptococcus* sp.
- *Prevotella* sp.

Source: Adapted from Ekanem.[8]

Microbiology

The common causative pathogens involved in sepsis in gynecological patients are as shown in **Box 6.**[8]

Pathophysiology of Septic Shock

The pathophysiology of septic shock is as shown in **Flowchart 7.**

Clinical and Laboratory Diagnosis

Flowchart 8 illustrates the clinical and laboratory findings in patients with sepsis.[9] The workup in patients with septic shock is summarized in **Flowchart 9.**

Management of Septic Shock

Early goal-directed therapy is the key for successful management. The management of septic shock is as illustrated in **Table 2.**[11]

Flowchart 7: Pathophysiology of septic shock.[1]

- Endotoxins / Endogenous toxins
  - *Peripheral and pulmonary vascular effects:*
    - Vasoconstriction
    - Vasodilation
    - Maldistribution of blood flow
    - Endothelial cell destruction
    - Pulmonary edema/ARDS
    - Pulmonary hypertension
  - *Direct myocardial effects:*
    - Decreased LV ejection fraction
    - LV Dilatation
- *Cardiovascular insufficiency:*
  - Maldistribution of blood flow
  - Lactic acidemia
  - Increased mixed venous O_2
- Outcomes of cardiovascular insufficiency:
  - Reduced SVR → Unresponsive hypotension
  - *Severe myocardial depression:*
    - Decreased CO
    - Decreased LVEDV
    - → Unresponsive hypotension → Death
  - Multi-organ system failure → Death
  - Recovery

(ARDS: adult respiratory distress syndrome; CO: cardiac output; LV: left ventricle; LVEDV: left ventricular end diastolic volume; SVR: systemic vascular resistance)
Source: Adapted and modified from Smith.[1]

Flowchart 8: Clinical and laboratory diagnosis of sepsis.[9]

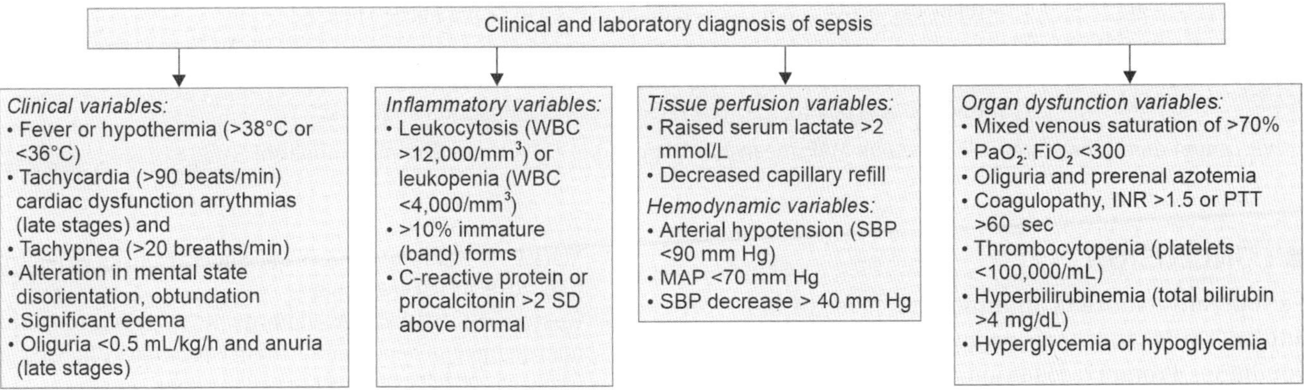

(FiO_2: fraction of inspired oxygen; INR: international normalized ratio; MAP: mean arterial pressure; PaO_2: partial pressure of oxygen in arterial blood; PTT: partial thromboplastin time; SBP: systolic blood pressure; SD: standard deviation; WBC: white blood count)
Source: Adapted and modified from Royal college of obstetricians and Gynaecologists.[9]

Flowchart 9: Workup in gynecological patient with signs and symptoms of septic shock.[10]

Workup

- **Full history and physical examination**
- *Blood investigations:*
  - FBC
  - U&Es
  - Glucose
  - LFTS
  - Acute-phase proteins: CRP, albumin, PCT
  - Coagulation studies: PT, APTT, INR, Platelet count, Fibrinogen, D-dimer, FDPs
  - ABGs, lactate
- *Septic screen as appropriate:*
  - Blood culture
  - Vaginal or cervical discharge
  - Urine
  - Sputum
  - Line tips
  - Wound swabs
  - Throat swab
  - Drain fluid
  - Stool
  - Cerebrospinal fluid
- *Imaging studies as appropriate:*
  - Ultrasound
  - Magnetic resonance imaging (MRI)
  - Computed tomographic (CT) scanning of the pelvis and abdomen
  - Chest X-ray (CXR) or CT of the chest
  - Echocardiography (Echo)

(ABG: arterial blood gas; APTT: activated partial thromboplastin time; CRP: C-reactive protein; FBC: full blood count; FDPs: fibrin degradation products; LFTs: liver function tests; INR: international normalized ratio; PCT: pro-calcitonin; PT: prothrombin time; U&E: urea and electrolytes)

TABLE 2: Management of severe sepsis and septic shock.[11]

Aim	• Correct the underlying cause • Optimize organ perfusion and oxygen delivery • Immediate resuscitation and treatment
Infection source control	• Obtain blood culture prior to antibiotic administration • Initial empiric broad-spectrum antibiotics to be stalled *within 1 hour of diagnosis* • Rapid identification of infection source and removal of infected/necrotic tissue
Shock management	• Measures most effective if achieved *within the first 6 hours of diagnosis* • Measure serum lactate • Restore CVP to 8–12 mm Hg • Restore MAP >65 mm Hg • Restore central venous O_2 saturation >70% or mixed venous saturation ≥65% • Fluid resuscitation with at least 30 mL/kg of IV crystalloid within the first 3 hours • Mechanical ventilation to reduce metabolic demand • First line vasoactive agents (nor epinephrine) when septic shock is fluid refractory
Additional therapies	• IV corticosteroids indicated in vasoactive-refractory shock • Addition of vasopressin indicated in vasoactive-refractory shock • Stress ulcer and pharmacologic VTE prophylaxis unless contraindicated

(CVP: central venous pressure; IV: intravenous; MAP: mean arterial pressure; VTE: venous thromboembolism)

CONCLUSION

Perioperative shock is an important cause of morbidity and mortality that may occur in many gynecological conditions. Prompt recognition, appropriate management, and multidisciplinary involvement are vital for improving the outcomes.

REFERENCES

1. Smith HO. Shock in the gynecologic patient editors. In: Rock JA, Jones HW (Eds). Te Linde's Operative Gynecology, 10th edition. Philadelphia: Lippincott-Williams & Wilkins; 2008. pp. 203-25.
2. Martel MJ. No. 115-Hemorrhagic Shock. J Obstet Gynaecol Can. 2018;40(12):e874-e882.
3. Gostout BS, Cliby WA, Podratz KC. Prevention and management of acute intraoperative bleeding. Clin Obstet Gynecol. 2002;45(2):481-91.
4. Santoso JT, Saunders BA, Grosshart K. Massive blood loss and transfusion in obstetrics and gynecology. Obstet Gynecol Surv. 2005;60(12):827-37.
5. Angus DC, Linde-Zwirble WT, Lidicker J, Clermont G, Carcillo J, Pinsky MR. Epidemiology of severe sepsis in the United States: analysis of incidence, outcome, and associated costs of care. Crit Care Med. 2001;29: 1303-10.

6. Byun JM, Jeong DH. Antibiotic prophylaxis for gynecologic cancer surgery. Taiwan J Obstet Gynecol. 2020;59:514-9.
7. Steiner HL, Strand EA. Surgical-site infection in gynecologic surgery: pathophysiology and prevention. Am J Obstet Gynecol. 2017;217:121-8.
8. Ekanem E, Oniya O, Saleh H, Konje JC. Surgical site infection in obstetrics and gynaecology: prevention and management. TOG. 2021;23(2):124-37.
9. Royal College of Obstetricians and Gynaecologists. Bacterial sepsis following pregnancy. Green Top Guideline No.64b. Available from: https://www.rcog.org.uk/media/bfnkzznd/gtg_64b.pdf [Last accessed April, 2024].
10. Fischerova D. Urgent care in gynaecology: resuscitation and management of sepsis and acute blood loss. Best Pract Res Clin Obstet Gynaecol. 2009;23(5):679-90.
11. Rhodes A, Evans LE, Alhazzani W, Levy MM, Antonelli M, Ferrer R, et al. Surviving Sepsis Campaign: International Guidelines for Management of Sepsis and Septic Shock: 2016. Crit Care Med. 2017;45(3):486-552.

Section

6

Complications of Early Pregnancy

Chapter

42

Early Pregnancy Loss

Ruchika Garg, Soniya Dhiman, Soni Bharti

INTRODUCTION

A pregnancy loss (miscarriage) is defined as the spontaneous demise of a pregnancy before 24 weeks of gestation or before the fetus reaches viability.[1] The presence of a nonviable intrauterine pregnancy with either an empty gestational sac or gestational sac containing an embryo or fetus without fetal heart activity within the first 12 + 6 weeks of gestation is defined as *early pregnancy loss*.[2]

INCIDENCE

Worldwide the pooled risk of miscarriage is 15.3%,[3] whereas the incidence of pregnancy loss in the Indian population is 6.9–8.5%.[4] Around 80% of miscarriages constitute early pregnancy losses. The rate increases with increase in age.

ETIOLOGY

Fetal and maternal factors contributed to early pregnancy loss. These factors are discussed in the following text in brief.

Fetal Factors

Almost half early pregnancy losses are euploid abortions while chromosomal abnormalities constitute another 50%.[5] Most common chromosomal abnormalities are trisomies of chromosomes 13, 16, 18, 21, and 22 (50–60%), monosomy X (9–13%), and triploidies (11–12%).[6]

Maternal Factors

- *Maternal age:* Advanced maternal age is a known risk factor for female subfertility, aneuploid abortions, stillbirths, and various other obstetric complications. European Society of Human Reproduction and Embryology (ESHRE) 2022 recommends informing women that the risk of pregnancy loss is lowest in women aged 20–35 years and rapidly increases after the age of 40 years.[1]
- *Medical disorders:* Endocrinopathies such as diabetes mellitus, obesity, thyroid disease, metabolic disorders, maternal obesity, maternal infections such as untreated syphilis, and viral infections such as cytomegalovirus (CMV), parvovirus, and zika virus are associated with early pregnancy losses.
- *Cancer:* Therapeutic doses of radiation and chemotherapy are abortifacient.
- *Antiphospholipid syndrome:* The antiphospholipid antibodies inhibit trophoblastic function and differentiation and activate the complement pathways at the maternal–fetal interface resulting in a local inflammatory response, resulting in recurrent miscarriage.
- *Inherited thrombophilia:* Factor V Leiden mutation, deficiencies of protein C, protein S, antithrombin III, and prothrombin gene mutation causes placental thrombosis, leading to recurrent pregnancy losses.
- *Environmental factors:* Exposure to ionizing radiation, lead, arsenic, and environmental pollutants can cause pregnancy loss by causing cell death, altering the growth of normal tissues, or interfering with normal cellular differentiation.
- *Behavioral factors:* Smoking, caffeine, alcohol consumption, substance abuse such as cannabis increases the risk of pregnancy loss.
- *Congenital uterine malformations:* Congenital uterine malformations such as septate uterus is associated with early pregnancy losses.

Paternal Factors

Increasing paternal age is associated with chromosomal abnormality of spermatozoa and thus early pregnancy loss.[7]

DIAGNOSIS

- *Clinical features:* A history of amenorrhea followed by vaginal discharge, bleeding, and abdominal cramps are suggestive features of pregnancy loss. Ectopic pregnancy is a significant differential diagnosis and needs to be ruled out.
- *Serum beta-hCG levels:* Serial falling levels of serum beta-hCG (beta human chorionic gonadotropin) levels are suggestive of failed pregnancy.
- *Ultrasonographic features:* Transvaginal ultrasound (TVS) with M mode is preferred over transabdominal ultrasound. Doppler should not be used due to theoretical temperature exposure. Usually, fetal cardiac activity can be detected at 6–6.5 weeks on TVS with crown-rump length (CRL) of 1–5 mm and mean sac diameter (MSD) of 13–18 mm.[8]

  For diagnosing early pregnancy loss, the Society of Radiologists (2013) and the American College of Radiology documented guidelines with the following sonographic findings:[9]
  - Embryo with a CRL of >7 mm with no heartbeat
  - MSD of ≥25 mm with no embryo
  - Absence of an embryo with a heartbeat if >2 weeks has elapsed following a scan that depicted a gestational sac without a yolk sac
  - Absence of an embryo with a heartbeat >11 days after a scan that depicted a gestational sac and yolk sac.

There are some suggestive findings on ultrasound for early pregnancy loss, but are not definitive. Such pregnancies should be reevaluated after 7–10 days for confirming the diagnosis.[10] Those findings are:

- Embryo with CRL of <7 mm and no heartbeat
- MSD of 16–24 mm and no embryo
- Absence of embryo with heartbeat 7–13 days after an ultrasound scan that showed a gestational sac without a yolk sac
- Absence of embryo with heartbeat 7–10 days after an ultrasound scan that showed a gestational sac with a yolk sac
- Absence of embryo for 6 weeks or longer after last menstrual period.

If fetal heart rate (FHR) is <100 and subchorionic hemorrhage is present, evaluate again after 7–10 days. When intrauterine pregnancy is not seen on ultrasonography (USG), serial beta-hCG monitoring should be done to rule out ectopic pregnancy.

MANAGEMENT

- *Expectant management:* Expectant management can be successful in achieving complete expulsion in 80% of women up to 8 weeks of the period of gestation, but they should be counseled regarding the need for surgical evacuation in case of incomplete abortion. Complete expulsion is defined as the absence of a gestational sac and an endometrial thickness of <30 mm.[11]
- *Medical management:* National Institute for Health and Care Excellence (NICE) guidelines outline the management of early pregnancy losses as follows:[12]
  - *Missed abortion:* A single dose of misoprostol 800 μg vaginally is recommended. One repeat dose can be given, if needed, after 3 hours after the first dose and within 7 days if there is no response to the first dose. Vaginal and sublingual are better than oral but sublingual causes more diarrhea. Addition of mifepristone 24 hours before misoprost reduces the need for surgical treatment.
  - *Incomplete abortion:* A single dose of 600/800 μg of misoprostol vaginally is recommended.
  - Medications for pain relief and antiemetics are also advised if needed.

    Women who are Rh(D) negative and unsensitized should receive 300 μg Rh(D)-immune globulin within 72 hours of the first misoprostol administration.
  - Follow-up USG to document the complete expulsion of POCs (products of conception) or serial beta-hCG monitoring should be done *after 7–14 days of expulsion.*
- *Surgical management:* Women with medical disorders such as severe anemia, bleeding disorders, or cardiovascular disease, or those who present with hemorrhage, hemodynamic instability, or signs of infection should be treated with surgical uterine evacuation. Many women prefer surgical evacuation to expectant or medical treatment because it provides more immediate completion of the process with less follow-up. Suction curettage with an electric vacuum source or manual vacuum aspirator, under local anesthesia or with sedation is performed. A single preoperative dose of doxycycline 200 mg is recommended to prevent infection after surgical management of early pregnancy loss.

- Hormonal contraception or IUDs can be used immediately after the completion of expulsion. Abstinence for 7–10 days is recommended to reduce the risk of infection.
- Psychological support and pre- and postprocedure counseling should be provided to reduce the anxiety and fear associated with miscarriage.

PREVENTION

- NICE recommends the use of 400 mg micronized vaginal progesterone twice daily till 16 weeks as prevention of miscarriage in women who have experienced vaginal bleeding in early pregnancy and have at least one previous miscarriage, though evidence does not suggest absolute efficacy in each pregnancy.[12]
- Bed rest is not recommended for prevention of early pregnancy loss. Prophylactic progesterone does not prevent early pregnancy (Cochrane 2008), but in women with three or more losses progesterone may help in prevention of pregnancy loss [American Society for Reproductive Medicine (ASRM) 2012].
- Preconception counseling to investigate genetic factors and inheritance of familial disorders.
- Preconception control of endocrinopathies and optimization of cardiovascular disorders.
- Folic acid supplementation of 0.4 mg daily is to be started in the preconception period as folate deficiency is linked to early miscarriage.[13]
- The use of low-dose aspirin and low-molecular-weight heparin helps in the prevention of miscarriages in thrombophilias.
- Avoiding known risk factors such as smoking, drinking alcohol, and illicit drug use and limit caffeine intake.
- Surgical correction of uterine anomalies if indicated.
- Seeking regular prenatal and antenatal care.

CONCLUSION

Empathy and effective communication are the key to manage early pregnancy loss each case should be given individualized treatment.

REFERENCES

1. ESHRE Guideline Group on RPL; Bender Atik R, Christiansen OB, Elson J, Kolte AM, Lewis S, Middeldorp S, et al. ESHRE guideline: recurrent pregnancy loss: an update in 2022. Hum Reprod Open. 2023;2023(1):hoad002.
2. National Institute for Health and Clinical Excellence. (2012). Ectopic pregnancy and miscarriage: diagnosis and initial management. Clinical Guideline [CG154]. Available from http://www.nice.org.uk/guidance/cg154/resources/guidance-ectopic-pregnancy and miscarriage-pdf [Last accessed March, 2024].
3. Quenby S, Gallos ID, Dhillon-Smith RK, Podesek M, Stephenson MD, Fisher J, et al. Miscarriage matters: the epidemiological, physical, psychological, and economic costs of early pregnancy loss. Lancet. 2021;397(10285):1658-67.
4. Kuppusamy P, Prusty RK, Chaaithanya IK, Gajbhiye RK, Sachdeva G. Pregnancy outcomes among Indian women: increased prevalence of miscarriage and stillbirth during 2015–2021. BMC Pregnancy Childbirth. 2023;23:150.
5. Alijotas-Reig J, Garrido-Gimenez C. Current concepts and new trends in the diagnosis and management of recurrent miscarriage. Obstet Gynecol Surv. 2013;68:445-66.
6. Jenderny J. Chromosome aberrations in a large series of spontaneous miscarriages in the German population and review of the literature. Mol Cytogenet. 2014;7:38.
7. Sartorius GA, Nieschlag E. Paternal age and reproduction. Hum Reprod Update. 2010;16(1):65-79.
8. Goldstein RB. Endovaginal sonography in very early pregnancy: new observations. Radiology. 1990;176(1):7-8.
9. Preisler J, Kopeika J, Ismail L, Vathanan V, Farren J, Abdallah Y, et al. Defining safe criteria to diagnose miscarriage: Prospective observational multicentre study. BMJ. 2015;351:h4579.
10. Doubilet PM, Benson CB, Bourne T, Blaivas M; Society of Radiologists in Ultrasound Multispecialty Panel on Early First Trimester Diagnosis of Miscarriage and Exclusion of a Viable Intrauterine Pregnancy. Diagnostic criteria for nonviable pregnancy early in the first trimester. Ultrasound Q. 2014;30(1):3-9.
11. Zhang J, Gilles JM, Barnhart K, Creinin MD, Westhoff C, Frederick MM; National Institute of Child Health Human Development (NICHD) Management of Early Pregnancy Failure Trial. A comparison of medical management with misoprostol and surgical management for early pregnancy failure. N Engl J Med. 2005;353:761-9.
12. National Institute for Health and Care Excellence. (2024). Ectopic pregnancy and miscarriage: diagnosis and initial management. NICE guideline [NG126]. [online] Available from https://www.nice.org.uk/guidance/ng126 [Last accessed March, 2024].
13. Gaskins AJ, Rich-Edwards JW, Hauser R, Williams PL, Gillman MW, Ginsburg ES, et al. Maternal prepregnancy folate intake and risk of spontaneous abortion and stillbirth. Obstet Gynecol. 2014;124(1):23-31.

Chapter 43

Ectopic Pregnancy

Parneet Kaur, Bhupinder Deep Kaur

INTRODUCTION

An ectopic pregnancy is defined as the implantation of fertilized ovum outside the endometrial lining of the uterine cavity.

The incidence of ectopic pregnancy is 1–2%.[1]

The incidence has risen in recent times in view of increased diagnostic test sensitivity, higher incidence of sexually transmitted diseases, and tubal-related infertility followed by assisted reproductive techniques.

Ectopic pregnancy is responsible for 6% of maternal deaths during early pregnancy.[2]

Fallopian tube remains the most common site of ectopic pregnancy, accounting for >90% of cases.

CLASSIFICATION (FLOWCHART 1)

Flowchart 1: Classification of ectopic pregnancy.[3,4]

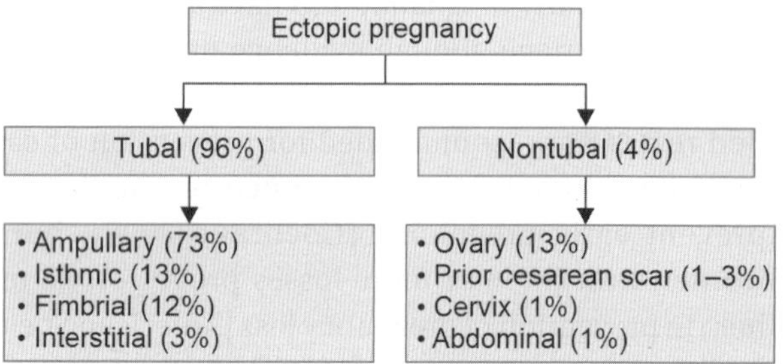

BOX 1: Risk factors of ectopic pregnancy.[5,6]

- *Risk factors:* Advancing age (>35 years)
- *Prior ectopic pregnancy:*
  - With previous one ectopic pregnancy—risk is 10%
  - With previous two or more ectopic pregnancies—risk of repeat ectopic increases to 25%
- *Prior tubal surgery:*
  - Salpingostomy
  - Tubectomy
  - Tubal reanastomosis
  - Lysis of adhesions
- H/o pelvic inflammatory disease (especially *Chlamydia trachomatis*) including pelvic tuberculosis
- Prior STD leading to distorted tubal anatomy
- Peritubal adhesion subsequent to salpingitis, appendicitis, endometriosis
- Congenital fallopian tube anomalies
- H/o infertility
- Use of assisted reproductive technique
- Smoking
- *Contraceptive method failure:*
  - Tubal sterilization (depends on sterilization technique)
  - Intrauterine contraceptive device (more incidence with LNG-IUS as compared to copper IUD)
- Previous pelvic or abdominal surgery
- In utero exposure to diethylstilbestrol

(H/o: history of; IUD: intrauterine device; LNG-IUS: levonorgestrel-releasing intrauterine system; STD: sexually transmitted disease)

**Heterotopic pregnancy:* Here, an ectopic pregnancy occurs along with intrauterine pregnancy (IUP).

The risk in naturally occurring pregnancies is 1:4,000.

Risk in women undergoing in vitro fertilization is 1:100 [ACOG (The American College of Obstetricians and Gynecologists)].[4]

TUBAL PREGNANCY

Risk Factors

In 50% of women diagnosed with ectopic pregnancy, no risk factor is identified,[5] but women who conceive with known risk factors should be evaluated to rule out ectopic pregnancy even in absence of symptoms **(Box 1)**.

Clinical Features

Ectopic pregnancy should be suspected in any women of reproductive age presenting with amenorrhea, abdominal pain, and bleeding per vaginum, irrespective of contraceptive use. It is a condition with very wide spectrum of clinical presentation, from absolutely normal parameters to a patient in shock requiring urgent resuscitation and surgical management.

Classical Triad (Fig. 1)

In one Indian study, classic triad was seen in 40.3% of cases.[7]

Presently, many cases are diagnosed with intact ectopic pregnancy in early pregnancy dating scans.

A high degree of suspicion is required to diagnose ectopic pregnancy **(Flowchart 2)**.

Signs

- *General physical examination (GPE):* Tachycardia, hypotension, pallor, and shock
- *Systemic examination:* Per abdomen and per vaginum assessment **(Table 1)**.[8]

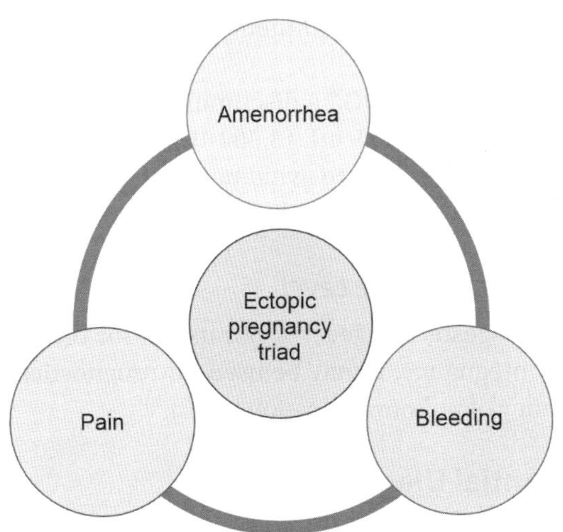

Fig. 1: Classical triad of ectopic pregnancy.

Diagnosis

Investigations specific for making diagnosis:

- Urine for pregnancy test
- Serum beta-hCG (beta-human chorionic gonadotropin) quantitative levels
- Transvaginal sonography (TVS)
- Serum progesterone levels.

Urine Pregnancy Test

Advantage: It is a simple test, easily available, and gives instant result. It should be performed in all cases irrespective of their sterilization status.[7]

Disadvantage: It does not differentiate between intrauterine and extrauterine (ectopic) pregnancy.

Transvaginal Sonography

Transvaginal sonography is imaging of choice [RCOG (Royal College of Obstetricians and Gynaecologists)][9] **(Table 2)**.[10]

Transabdominal sonography is only done if TVS is not available.

Findings on TVS in tubal ectopic pregnancy ***(Table 3)****:*[9,11,12] Most authors consider that a definitive diagnosis of tubal ectopic pregnancy can only be made when an extrauterine gestational sac (G-sac) containing a yolk sac or embryonic pole is visualized. A live extrauterine embryo is diagnostic of an ectopic pregnancy.[10]

Serum Beta-hCG Levels[4,8]

- With these assays, lower limit of detection being ≤5 mIU/mL.
- Serum beta-hCG level is useful for planning management of TVS-visualized ectopic pregnancy.

Flowchart 2: Clinical features of ectopic pregnancy.[8]

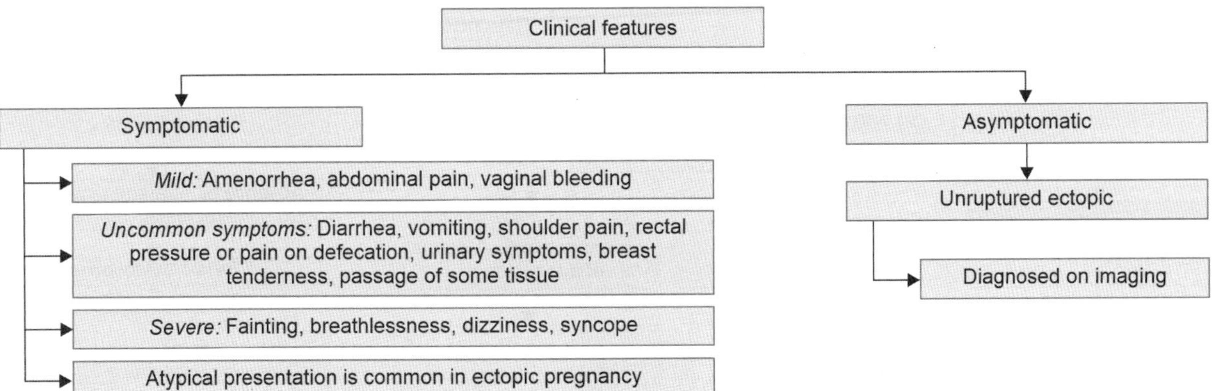

TABLE 1: Signs on per abdomen and per vaginum assessment.[8]

Per abdomen	*Per vaginum*
Muscle guarding	Uterus normal size or slightly bulky, may seem soft
Muscle rigidity	Cervical motion tenderness
Rebound tenderness	Adnexal mass
Abdominal distention	Fullness or tenderness in adnexa

TABLE 2: Findings on transvaginal sonography (TVS) if there is developing intrauterine pregnancy.[10]

Weeks	*Structure visible*
5^{+0} weeks	Empty G-sac (MSD 10 mm)
5^{+4} weeks	G-sac with yolk sac visible
6^{+0} weeks	G-sac and yolk sac with adjacent heartbeat and small embryo

(G-sac: gestational sac; MSD: mean sac diameter)

TABLE 3: Findings on transvaginal sonography (TVS) in tubal ectopic pregnancy.[9,11,12]

Finding	*Percentage*
Empty uterus	
Simple adnexal cyst	
Inhomogeneous/noncystic adnexal mass	50–60%
Adnexal mass—with gestational sac (G-sac) having yolk sac and fetal pole, with or without cardiac activity	20%
Identification of a hyperechoic ring called "bagel" or "tubal" sign representing empty G-sac	20%
Free fluid in pelvis	10–25%
Corpus luteum seen on color Doppler as ring of fire on ipsilateral side	70–75%
A suspicious mass moving separately from ovary called "blob" sign	60%

- Serial values need to be obtained, rather than a single value.
- Serial beta-hCG level assays are done to identify patterns that indicate either a growing or failing intrauterine pregnancy (IUP) or ectopic pregnancy.

Discriminatory levels: This is a predictable linked progression of laboratory and sonographic findings. Most of the studies consider beta-hCG levels of 1,500 mIU/mL as discriminatory levels at which intrauterine G-sac should be visible on TVS.[10,13-15] If the initial beta-hCG level exceeds this discriminatory level and no evidence IUP is there on TVS, then ectopic pregnancy remains a possibility **(Flowchart 3)**.[10]

When we are left with the dilemma regarding diagnosis, a repeat value of beta-hCG is done after 48 hours. And minimum expected rise of beta-hCG increase for a viable IUP is 53% (which was earlier taken as 66%). When beta-hCG rises more slowly than this, plateaus, or falls, the differential diagnosis is limited to failing IUP or ectopic pregnancy.

According to ACOG,[4] a discriminatory beta-hCG level as high as 3,500 mIU/mL (3,500 IU/L) should be used when a woman wishes to avoid unnecessary intervention in a potentially viable IUP, which she wishes to continue.

Serum Progesterone Level

Serum progesterone level is not useful for diagnosis of ectopic pregnancy. It may be used for prognostication of pregnancy of unknown location (PUL).[9,16]

Differential Diagnosis[6,10,13]

The differential diagnosis of ectopic pregnancy is given in **Table 4**.

Flowchart 3: Serum beta-hCG (beta-human chorionic gonadotropin) levels and transvaginal sonography (TVS) findings.[10]

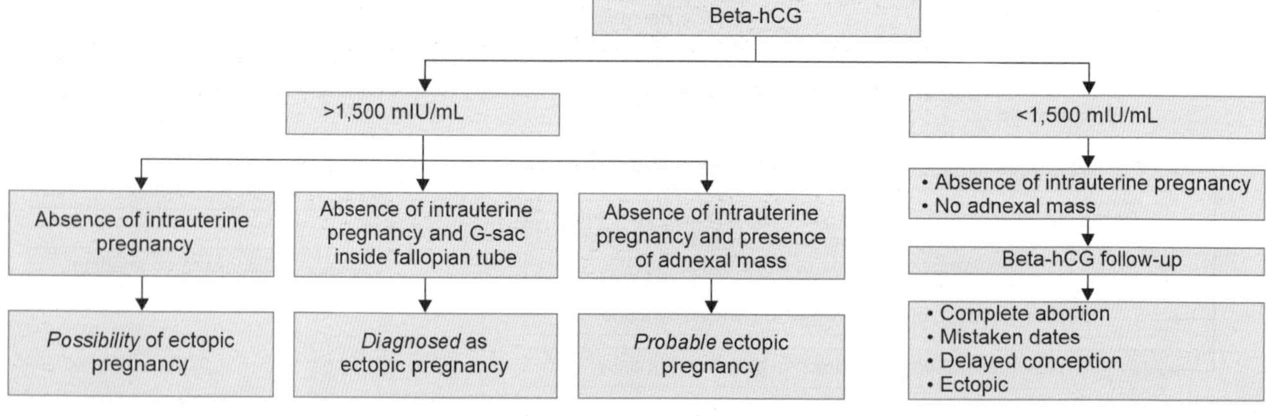

(G-sac: gestational sac)

TABLE 4: Differential diagnosis of ectopic pregnancy.

Condition	*Clinical assessment*	*Confirmatory investigation*
Threatened abortion	• *Symptoms:* Amenorrhea, vaginal bleed (spotting fresh) • *Signs:* P/V—os closed, uterus enlarged/soft, no adnexal mass or tenderness. UPT +ve	TVS (shows a developing IUP)
Incomplete abortion/inevitable abortion	• *Symptoms:* Amenorrhea, vaginal bleed (fresh, excessive amount, clots+) UPT +ve • *Signs:* P/V—os open, uterus enlarged/soft, no adnexal mass or tenderness, products of conception felt in cervical canal	TVS
Septic abortion	• *Symptoms:* Amenorrhea, vaginal bleeding, pain, fever, H/o D and C • *Signs:* Muscle guarding + • *P/V:* Cervical motion tenderness may be there, and adnexal tenderness is present • UPT ±	Clinical diagnosis: • CBC • TVS
Missed abortion	• *Symptoms:* Amenorrhea, vaginal bleed (dark altered colored blood) • *Signs:* P/V—os closed, uterus enlarged/soft, no adnexal mass or tenderness	TVS (shows IUP with no cardiac activity)
Acute PID	• *Symptoms:* No amenorrhea, pain lower abdomen, discharge P/V • *Signs:* P/S—cervix inflamed, hyperemic; mucopurulent discharge coming from external os • P/V—os closed, uterus normal size/firm, tenderness in the fornices, cervical motion tenderness ± • UPT –ve	*Clinical diagnosis:* Beta-hCG—nonpregnant levels
Corpus luteum rupture	*Symptoms:* Pain abdomen, amenorrhea ±, UPT –ve *Signs:* Tachycardia, hypotension, pallor	*Clinical:* TVS
Gestational trophoblastic disease	*Symptoms:* Amenorrhea, bleeding P/V, UPT +ve	TVS (snowstorm appearance)
Subchorionic hemorrhage	*Symptoms:* Amenorrhea, bleeding P/V, UPT +ve	TVS
Acute appendicitis	• *Symptoms:* No amenorrhea, pain lower abdomen (around umbilicus), fever, nausea/vomiting, UPT –ve • *Signs:* – P/A: McBurney point tenderness, rebound tenderness – P/V: os closed, uterus normal size/firm, no adnexal mass or tenderness	TAS/CT
Acute gastroenteritis	• *Symptoms:* No amenorrhea, pain lower abdomen, nausea/vomiting, UPT –ve • *Signs:* – P/A—NAD – P/V—os closed, uterus normal size/firm, no adnexal mass or tenderness	Clinical diagnosis
Torsion ovarian	Pain lower abdomen, nausea/vomiting, UPT –ve	TAS/TVS
Renal calculi	Colicky pain, nausea/vomiting, UPT –ve	Clinical diagnosis: USG for KUB

(Beta-hCG: beta-human chorionic gonadotropin; CBC: complete blood count; CT: computed tomography; H/o: history of; D and C: dilatation and curettage; KUB: kidneys, ureters, bladder; IUP: intrauterine pregnancy; NAD: no abnormality detected; P/A: per abdomen; PID: pelvic inflammatory disease; P/S: per speculum; P/V: per vaginum; TAS: transabdominal sonography; TVS: transvaginal sonography; UPT: urine pregnancy test)

Management

Flowchart 4 illustrates the management of ectopic pregnancy.

Expectant Management

The aim of expectant management is to select those patients where their ectopic pregnancy results in

Flowchart 4: Management of ectopic pregnancy.

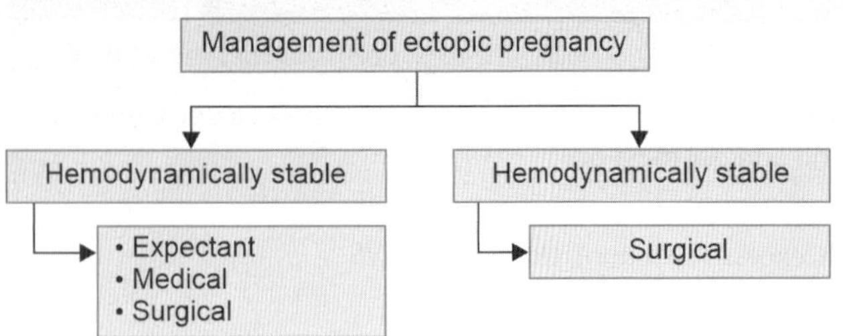

Flowchart 5: Expectant management of ectopic pregnancy.

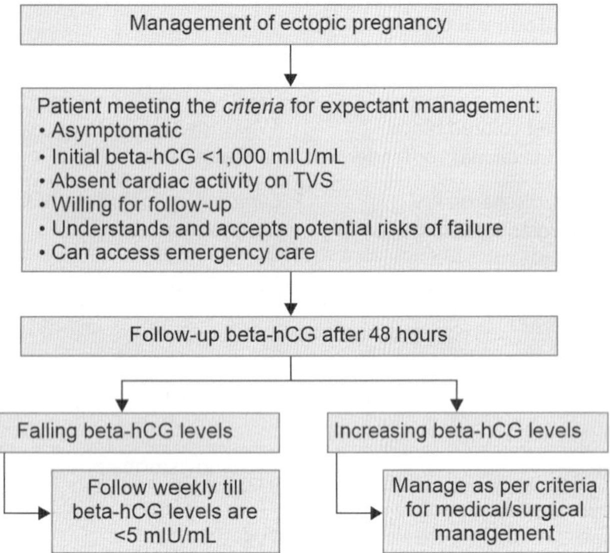

(Beta-hCG: beta-human chorionic gonadotropin; TVS: transvaginal sonography)

resolution rather than rupture. The evidence suggests that best predictor of successful expectant management is the initial beta-hCG levels. If initial levels are higher, the probability of spontaneous resolution decreases. In carefully selected patients, expectant management does not expose the patient to potentially serious side effects associated with methotrexate (MTX) **(Flowchart 5)**.[8,17,18]

The mean time to resolution is reported to be from 6 weeks to a maximum of 67 days. Success rate ranges from 47% to 100% depending on variable selection criteria.[18]

Also, in some cases expectant management need to be abandoned **(Box 2)**.

Previously, these patients were managed as inpatient, which is not cost-effective. Patients actually need close follow-up and frequent outpatient visits and counseling regarding when to seek urgent medical care.

BOX 2: Expectant management abandonment criteria.

Expectant management should be abandoned when:
- Significantly increased pain
- Increasing/plateau in beta-hCG levels
- Increase in adnexal mass on TVS
- Tubal rupture with hemoperitoneum
- Failure of beta-hCG levels to decrease

(Beta-hCG: beta-human chorionic gonadotropin; TVS: transvaginal sonography)

TABLE 5: Methotrexate (MTX) selection criteria.[8,17]

Criteria	*Consider MTX*	*Use MTX with caution*
Vitals	Normal	Normal
Intrauterine pregnancy	Absent	Absent
Abdominal pain	Not present	Mild
Adnexal mass	<3.5 cm	<3.5 cm
Appearance in adnexa	Only G-sac	G-sac with yolk sac or fetal node but no cardiac activity
Beta-hCG	<1,500 IU/L	1,500–5,000 IU/L
Patients' willingness to follow-up	Agreeable	Willing to follow-up

(Beta-hCG: beta-human chorionic gonadotropin; G-sac: gestational sac)

BOX 3: Contraindications to methotrexate (MTX).[9]

- Hemodynamic instability
- Presence of intrauterine pregnancy
- Breast feeding
- Noncompliance to follow-up
- Known sensitivity to MTX
- Chronic liver disease
- Preexisting blood dyscrasias
- Active pulmonary disease
- Immunodeficiency
- Peptic ulcer disease
- Chronic renal disease

Medical Management

- For unruptured tubal ectopic pregnancy—drug of choice—Inj. MTX.[4,9]
- For the medical management, there is a *methotrexate selection criteria* **(Table 5)**.[8,17]

Contraindications to MTX:[9] These are given in **Box 3**.

Dosage ***(Table 6)****:*[15,17] According to ACOG,[4] a single dose may be appropriate for patients with low initial beta-hCG

TABLE 6: Injection methotrexate (MTX) regimen.[15,17]

Single dose	*Two doses*	*Multiple doses*
Single IM dose of Inj. MTX 50 mg/m² on day 1	Single IM dose of Inj. MTX 50 mg/m² on day 1	*Dose:* Inj MTX 1 mg/kg IM on day 1, 3, 5, 7, alternate daily with folinic acid
Repeat beta-hCG on posttreatment day 4–7	Repeat 50 mg/m² IM on day 4	Inj. folinic acid 0.1 mg/kg IM on day 2, 4, 6, 8
There should be a ↓ in beta-hCG >15%.	Measure beta-hCG on posttreatment day 4–7	Repeat beta-hCG on Inj MTX days until >15% decrease from previous value
If fall in beta-hCG is >15%, patient is kept on follow-up with weekly beta-hCG levels till they reach nonpregnant levels (according to reference laboratory)	If fall in beta-hCG is >15%, patient is kept on follow-up with weekly beta-hCG levels till they reach nonpregnant levels (according to reference laboratory)	If ↓ in beta-hCG >15%, then stop Inj. MTX
If ↓ in beta-hCG <15%, give MTX 50 mg/m² and repeat beta-hCG on day 4 and 7 after second dose	If ↓ in beta-hCG <15%, administer MTX 50 mg/m² on day 7–11 and measure beta-hCG levels	Follow up patient with weekly beta-hCG levels till they reach nonpregnant level
If fall in beta-hCG is >15%, patient is kept on follow-up with weekly beta-hCG levels	If fall in beta-hCG is >15%, patient is kept on follow-up with weekly beta-hCG levels till they reach nonpregnant levels (according to reference laboratory)	
If, during follow-up, levels plateau or increase, MTX can be repeated	If ↓ in beta-hCG <15%, consider surgical management	

(Beta-hCG: beta-human chorionic gonadotropin; IM: intramuscular)

levels and two-dose regimen should be considered in women with initial high beta-hCG levels. Indian guidelines by FOGSI (The Federation of Obstetric and Gynecological Societies of India) recommend single-dose regimen. Overall, the success rate of single-dose MTX for tubal ectopic pregnancy ranges from 65–95% with 3–27% of patients requiring a second dose.[9]

Side effects of MTX:[4] The side effects of MTX are given in **Box 4**.

Cautions with MTX:[4]

- Women treated with MTX should be advised not to use alcohol.
- Women should also be advised to avoid folic acid supplements, foods that contain folic acid as they may decrease efficacy of MTX.
- Women should avoid nonsteroidal anti-inflammatory drugs (NSAIDs) as these may increase MTX toxicity.
- Sunlight exposure should also be avoided.
- Women should be advised to avoid vigorous activity and sexual intercourse.
- Women are advised against pregnancy till complete resolution.

BOX 4: Side effects of methotrexate.[4]

- Nausea/vomiting
- Stomatitis
- Abdominal pain
- Elevated liver enzymes
- Alopecia
- Pneumonitis

- The beta-hCG levels may rise initially to levels higher than pretreatment levels, but then should progressively decrease to nonpregnant level.
- Women treated with MTX may experience abdominal pain 2–3 days after administration of drug, which is due to cytotoxic effect of drug on trophoblastic tissue and tubal abortion. Pain can be managed expectantly if no signs of overt tubal rupture and significant hemoperitoneum are there. We can monitor patients with hemoglobin level and assessment of intraperitoneal hemorrhage by TVS **(Table 7)**.[19,20]
- Anechoic fluid is physiological but low-level echoes are typical of blood.

Follow-up of medical management—MTX:[5,9] This is described in **Box 5**.

TABLE 7: Assessment of hemoperitoneum on transvaginal sonography (TVS).[19,20]

Categorization	*Criteria*	*According to relation to posterior aspect of uterus*
Mild	When there is only echogenic fluid present in the pouch of Douglas	The fluid remains below isthmus
Moderate	When there are visible blood clots	• Fluid reaches body of uterus • Fluid is above fundus of uterus
Severe	• When there were blood clots and echogenic fluid present both in the pouch of Douglas and in the uterovesical space • Presence of blood in the Morrison's pouch (indicating severe intraperitoneal bleeding)	If fluid is present anterior to uterus, go for transabdominal sonography (TAS) to evaluate Morrison's pouch (if present, it means at least 700 mL of blood in peritoneal cavity)

BOX 5: Follow-up after medical management.[9]

- On average, the beta-hCG normalizes in 2–3 weeks. It can take up to 8 weeks in patients with high starting levels of beta-hCG
- Women are advised that there is no effect of MTX on ovarian reserve
- Pregnancy needs to be avoided for 3 months following MTX treatment.

(Beta-hCG: beta-human chorionic gonadotropin; MTX: methotrexate)

BOX 6: Criteria for switching from medical to surgical management.

Switch from medical to surgical management if:
- Beta-hCG levels—rising or plateauing
- Increase in adnexal mass on TVS
- Signs of internal hemorrhage

(Beta-hCG: beta-human chorionic gonadotropin; TVS: transvaginal sonography)

When to stop medical management and go for surgical management? **Box 6** gives the criteria for switching from medical to surgical management.

Surgical Management

Criteria:
- Ruptured tubal pregnancy, as per clinical assessment
- Failure of medical management
- Noncompliance to medical management follow-up
- Hemodynamic instability

Approach:[16] Laparoscopic surgery is a mainstay of treatment for both ruptured and unruptured ectopic pregnancies.[20]

- Laparoscopy is preferable to open approach in hemodynamically stable patients.[9]
- *Laparotomy:*
  - How to decide:
    - Clinical condition of patient
    - Availability of laparoscopy
    - Skill of the surgeon
  - Laparotomy is preferred in clinical setting where surgical management is a lifesaving procedure.

Surgical options for tubal pregnancy **(*Table 8*)**:[1,8]
- Always send tissue for histopathology
- Give anti-D prophylaxis.

INTERSTITIAL PREGNANCY[9,16,17]

Interstitial pregnancy occurs when the ectopic pregnancy implants in the interstitial part of the fallopian tube. This part of the fallopian tube is about 1–2 cm in length and traverses the muscular myometrium of the uterine wall, opening via the tubal ostium into the uterine cavity.

Ultrasound Criteria for Diagnosis

- Empty uterine cavity.
- Products of conception/G-sac located laterally in the interstitial (intramural) part of the tube and surrounded by <5 mm of myometrium in all imaging planes.
- Presence of the "interstitial line sign", which is a thin echogenic line extending from the central uterine cavity echo to the periphery of the interstitial sac.

3D ultrasound or magnetic resonance imaging (MRI) can be used as an adjunct to distinguish between angular and early IUP.

Management

- Management may be medical/surgical or conservative **(Flowchart 6)**.
- A single beta-hCG should be carried out at diagnosis. A repeat value after 48 hours can help decide further management.

TABLE 8: Salpingotomy versus salpingectomy.[1,8]

Salpingotomy	*Salpingectomy*
Linear incision along the length of the tube	• Complete or partial tube removal • It is preferable to remove affected tube as complete as possible to avoid risk of distal recurrence[21]
Preferred in unruptured cases	Preferred in ruptured cases
Prerequisite: • Damaged contralateral tube • Previous ectopic pregnancy • Previous abdominal surgery • Previous PID	*Prerequisite:* The contralateral tube is normal or fertility not desired
Follow-up with serum beta-hCG level is required to assess persistent trophoblastic tissue	Salpingectomy does not decrease anti-Müllerian hormone levels postoperatively but long-term effects on ovarian reserve are not known[22]
Persistent trophoblastic tissue, requires systemic methotrexate	As salpingectomy reduces the risk of epithelial ovarian cancer so proper counseling before surgery may be done according to desire of patient regarding future fertility[23]
Patients electing for this surgical management of ectopic pregnancy should be counseled regarding the residual risk (1:5) of requiring medical treatment after surgery	Studies have shown if contralateral tube is normal then there is similar intrauterine pregnancy rates and ectopic pregnancy rates in patients undergoing salpingectomy or salpingostomy[24]

(beta-hCG: beta-human chorionic gonadotropin; PID: pelvic inflammatory disease)

Flowchart 6: Management of interstitial pregnancy.

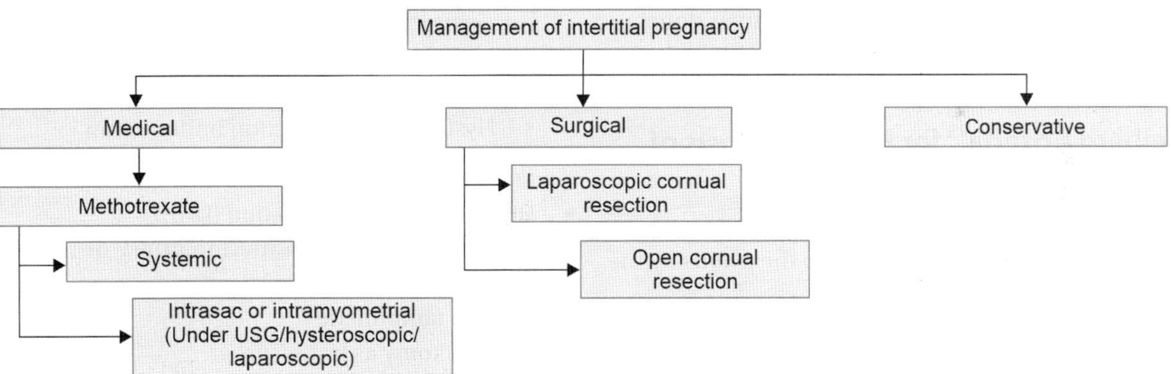

- First line of management is medical/conservative.
- Additional measures may be required for control of bleeding in form of diluted local vasopressin infiltration or uterine artery ligation.
- There are increased chance of uterine rupture or uterine scar dehiscence in future pregnancies and patient should be counseled accordingly.

OVARIAN PREGNANCY

Diagnosis

The ultrasound criteria suggestive of ovarian ectopic pregnancy are:[9,19]

- Empty uterine cavity.
- Cystic structure with wide echogenic ring on or within the ovary, generally seen surrounded by ovarian cortex and separate from corpus luteum.
- A yolk sac or embryo is seen rarely.
- Negative sliding organ sign—not possible to separate cystic structure from ovary on gentle probe palpation.
- Corpus luteum should be identified separate from the ovary (in the majority of cases, ovarian pregnancy is ipsilateral to the corpus luteum).
- A complex echogenic mass with free fluid in pouch of Douglas may represent ruptured ovarian ectopic pregnancy.

Differential Diagnosis[9]

- Corpus luteal cyst
- Tubal ectopic pregnancy stuck to ovary
- A second corpus luteum
- Ovarian germ cell tumors.

Management[5,16]

- A single-serum beta-hCG should be carried out at diagnosis to help with management.
- In some cases, a repeat serum beta-hCG in 48 hours may be useful in deciding further management.
- Management is mainly surgical though successful management with MTX has been published in literature.
- Laparoscopy is required for diagnosis and surgical management.
- Laparoscopic surgery is the mainstay of treatment in hemodynamically stable patients and open laparotomy is done if patient is bleeding or facilities or expertise is not available.
- Ovarian wedge resection is done in patients requiring future fertility, attempting to remove as little normal ovarian tissue as possible or oophorectomy is done.

Diagnosis is confirmed surgically and histologically.

Spiegelberg Criteria[25] for Final Diagnosis of Ovarian Ectopic Pregnancy

- The G-sac is located in the region of the ovary.
- The ectopic pregnancy is attached to the uterus by the ovarian ligament.
- Ovarian tissue in the wall of the G-sac is proved histologically.
- The tube on the involved side is intact.

CERVICAL PREGNANCY

Diagnosis

There are defined ultrasound criteria for diagnosis of cervical pregnancy, which are as follows:[9]

- Empty uterine cavity
- A barrel-shaped cervix
- A G-sac below the level of internal cervical os
- The absence of *sliding sign*
- Presence of blood flow around G-sac on color Doppler.

Sliding sign: This sign helps cervical pregnancies to be differentiated from incomplete abortions within the cervical canal. When probe pressure is applied on TVS, in miscarriages the G-sac slides against the cervical canal but it does not in an implanted cervical pregnancy.

Types of Cervical Pregnancy[19]

Cervical pregnancy may be:

- Partial if there is communication with cervical canal.
- Complete if there is no communication with cervical canal.
- All pregnancies implanted into posterior cervix should be described as cervical ectopic pregnancies regardless of her history of prior cesarean section.
- In cases of anterior implantation in women's past history of cesarean section, the differential diagnosis is to be made from cesarean scar ectopic pregnancy.

Management[16]

- A single beta-hCG should be carried out at diagnosis.
- *First line of treatment:* Medical/conservative
- *Indications for surgery:* Failure of medical management or life-threatening bleeding
- Surgical management should be carried out at tertiary care center.
- The treatment modalities are:
  - Suction and evacuation
  - Hysteroscopic resection under vision

Additional measures to reduce bleeding:

- Uterine artery embolization (UAE)
- Laparoscopic or open uterine artery or internal iliac artery (anterior division) ligation
- Balloon tamponade
- Hysterectomy as last resort.

CESAREAN SCAR PREGNANCY

Cesarean scar pregnancy (CSP) is a rare form of an ectopic pregnancy wherein the trophoblast invades a defective and thinned out myometrium in a scar which is a remnant of a previous cesarean.[26] The pregnancy is, within the uterine cavity and, unlike true ectopic pregnancies, may result in a live born infant. However, it may lead to severe maternal morbidity that is complicated by challenges in securing a timely diagnosis and optimal management.

Diagnosis

- Primarily by TVS
- MRI can be used as second-line investigation.

- *Diagnostic criteria for diagnosing CSP on TVS:*[9]
    - Empty uterine cavity
    - G-sac or solid mass of trophoblast located anteriorly at the level of internal os embedded at the site of the previous lower uterine cesarean section scar
    - Thin or absent layer of myometrium between G-sac and bladder
    - Evidence of prominent trophoblastic/placental circulation on Doppler examination
    - Empty endocervical canal.

Types of Cesarean Scar Pregnancies[17]

Table 9 lists the types of CSPs.

Management[16]

There are very limited data on the management of CSP (primarily case reports), with no consensus on the optimal approach. Following are the management options are described in literature:

- Medical therapy (MTX is the drug used)
- Surgical therapy (suction evacuation under hysteroscopy guidance)
- A combination of the two, with or without adjunct treatments

Adjunctive treatments are in form of:

- Cervical cerclage
- UAE
- Laparoscopic or open ligation of uterine artery or anterior division of internal iliac artery
- Balloon tamponade.

A serum beta-hCG level may be useful as a baseline prior to monitoring if conservative treatment is contemplated, but it does not have a role in the diagnosis of CSP.

According to RCOG FOGSI GCPR (Good Clinical Practice Recommendation) guidelines surgical approach is preferred over medical management. It is recommended that women with a CSP should be advised of the risks of another pregnancy and counseled regarding effective contraceptive methods, including LARC (long-acting reversible contraception) and permanent contraception.[27]

TABLE 9: Type I and type II cesarean scar pregnancy (CSP).[17]

Type I	*Type II*
• Endogenous-type cesarean scar pregnancy • Characterized by progression of the pregnancy to the cervico-isthmic space or uterine cavity • In select cases and with proper counseling, it can be treated expectantly	• Exogenous-type cesarean scar pregnancy • Characterized by deep invasion of the cesarean scar defect and progression toward the bladder and abdominal cavity • Can get complicated by uterine rupture and bleeding early in pregnancy • Many CSPs are believed to progress to placental accreta spectrum (PAS) disorders
Management: Type I cesarean scar pregnancy can be treated by expectant management or hysteroscopy	*Management:* Managed by laparoscopy or laparotomy, which allows for concurrent repair of the cesarean scar defect and removal of the pregnancy

CORNUAL PREGNANCY (RUDIMENTARY HORN PREGNANCY)

In the literature, the terms interstitial pregnancy and cornual pregnancy have been used synonymously. However, by strict definition, they are different entities. An interstitial pregnancy occurs in the interstitial or proximal portion of the fallopian tube within the muscular wall of the uterus lateral to the round ligament whereas, a cornual pregnancy implants in rudimentary horn of a unicornuate uterus.[19] RCOG classifies this pregnancy as cornual pregnancy.[9]

Diagnosis

- *Ultrasonography criteria for diagnosis of cornual pregnancy:*[11]
    - A single interstitial portion of the fallopian tube in the main uterine body
    - Products of conception/G-sac mobile and separate from the uterus surrounded by the myometrium
    - Vascular pedicle joining the G-sac to the unicornuate uterus
- *3D ultrasound* or *MRI* are helpful in diagnosis.
- A single beta-hCG level done at time of diagnosis is helpful in management and may be repeated after 48 hours if medical management is contemplated.

Management

- May be *medical/surgical* or *expectant*
- *Surgical approach is preferred:* Laparoscopic or open excision of rudimentary horn; during surgery, care should be taken for associated urinary tract anomalies.[16]

ABDOMINAL PREGNANCY

Abdominal pregnancies carry a nearly eightfold increased risk of maternal death compared with tubal pregnancies.

Diagnosis

The following ultrasound criteria have been suggested as being diagnostic of an early abdominal pregnancy.[28]

- Absence of an intrauterine G-sac
- Absence of both an evident dilated tube and a complex adnexal mass
- A gestational cavity surrounded by loops of bowel and separated from them by peritoneum
- A wide mobility similar to fluctuation of the sac, particularly evident with gentle pressure of the transvaginal probe toward the posterior cul-de-sac.

Management[17]

- Some early abdominal pregnancies are likely managed medically with MTX as a PUL with no establishment of a definitive pregnancy site.
- However, due to the risk of hemorrhage, abdominal pregnancies are typically managed surgically, either with laparoscopy or laparotomy.
- At the time of surgery, managing the placenta requires an individualized approach; life-threatening hemorrhage from the placental bed can occur if proper control of the vascular supply cannot be achieved. Successful tying off of the umbilical cord close to the placental insertion, followed by expectant management (with or without MTX), has been described in the literature.
- Preoperative embolization of feeding vessels has also been proposed to mitigate the risk of excessive bleeding.
- The risks of significant bleeding with attempted placental removal must be balanced against the complications associated with leaving the placenta in situ, including the risk of severe sepsis, delayed hemorrhage, and the need for repeat surgery.
- A multidisciplinary approach involving other surgical specialties (e.g., vascular surgery) may be the best approach.

ANGULAR PREGNANCY

The term angular pregnancy was initially defined as "implantation of the embryo in the lateral angle of the uterine cavity, just medial to the uterotubal junction".

- In a recent prospective study authors concluded that angular pregnancy is a variation of a normally implanted IUP rather than an ectopic pregnancy.[29]
- According to ESHRE (European Society of Human Reproduction and Embryology) recommendations the term angular pregnancy should be abandoned.[19]

INTRAMURAL PREGNANCY[19]

- The term intramural pregnancy is used to describe a pregnancy which is located within the uterus, but breaches the endometrial–myometrial junction and invades the myometrium of the uterine corpus above the internal os.
- Intramural pregnancies are relatively rare.
- They are located above the level of internal os which differentiates them from cervical and cesarean section scar pregnancies.
- They typically occur after previous myomectomy, but they can also be caused by scarring following uterine perforation or after classical cesarean section.
- They could be located anywhere within the uterine corpus, which makes the diagnosis difficult as compared to cervical and lower transverse cesarean section scar pregnancies.
- The defining feature is extension beyond the endometrial–myometrial junction above the level of the internal os.
- They can also be classified as:
  - Complete or
  - Partial.

HETEROTOPIC PREGNANCY[11,17]

Heterotopic pregnancy is the simultaneous presence of an intrauterine and extrauterine pregnancy. It occurs infrequently, with a reported rate of 1 in 30,000 spontaneous pregnancies and there is increased incidence in pregnancies conceived with assisted reproductive technology.

The fallopian tube is the most common site of ectopic pregnancy.

MANAGEMENT

- It should be according to clinical presentation and patients' preference
- Tubal heterotopic pregnancies are most frequently managed surgically with salpingectomy.

Flowchart 7: Pregnancy of unknown location (PUL) and beta-hCG (beta-human chorionic gonadotropin) follow-up.[8]

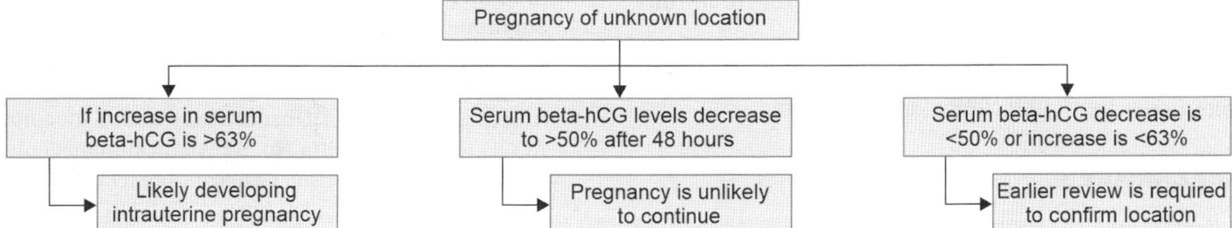

- Systemic MTX is contraindicated in the presence of a viable and desired IUP
- Targeted treatment of the ectopic pregnancy with intracardiac injection of potassium chloride has been described.
- Close follow-up of the IUP is recommended because there is an increased risk of loss of the IUP with the ectopic pregnancy.

PREGNANCY OF UNKNOWN LOCATION[17,30]

Pregnancy of unknown location is a transient state in which beta-hCG is positive but a transvaginal ultrasound fails to find either an intrauterine or extrauterine pregnancy. PUL is not a final diagnosis, but a classification that indicates a patient is at risk for ectopic pregnancy.

Possible Pregnancy Outcomes for Women Initially Classified as Having a PUL

- *IUP (34–40%):* It may be viable or nonviable.
- Failed PUL (44–69%) occurs when serum beta-hCG levels fall without intervention and the location of the pregnancy is never confirmed.
- Persistent PUL (2%) is defined as a sustained elevation in serial serum beta-hCG with no pregnancy visualized on TVS. This is not a final diagnosis but rather a classification of PUL.
- Ectopic pregnancy (8–14%).

Follow-up (Flowchart 7)[8]

- Inform her about need of meticulous follow-up as she may have ectopic pregnancy until location is determined.
- Pregnancy of unknown location, place more importance on clinical symptoms than serum beta-hCG levels and inform her where to report in case of emergency.
- Repeat beta-hCG after 48 hours (no role of serum progesterone levels) [NICE (National Institute for Health and Care Excellence)].[8]

Pregnancy of unknown location needs the most meticulous workup as we may disturb a developing IUP or miss an ectopic pregnancy. A large study showed the slowest or minimal rise for a normal viable IUP to be 24% at 1 day and 53% at 2 days.[31]

We can also look at accessary ultrasound findings in these cases. According to a study, in patients with PUL, high mean endometrial thickness (17 mm) predicted developing IUP and no IUP occurred in patients with endometrial thickness below 8 mm.[32]

When beta-hCG levels are not rising normally and ultrasonography cannot confirm pregnancy location, a dilatation and curettage or manual vacuum aspiration may be helpful. If evacuation of the uterus yields chorionic villi, then a failed IUP is diagnosed and treatment for an ectopic pregnancy may be avoided along with harmful effects of so.[13,15]

CHRONIC ECTOPIC PREGNANCY

A study showed that >95% tubal ectopic pregnancies became undetectable on ultrasound after 3 months.[33]

According to ESHRE[19] guidelines the term "chronic ectopic" should not be used in clinical practice. The term "residual ectopic pregnancy" should be used for an ectopic pregnancy, which presents as a discrete mass on ultrasound in a woman with negative urine pregnancy test.

CONCLUSION

- Ectopic pregnancy is a condition with varied clinical presentations and in spite of best available diagnostic modalities; many times, we are in dilemma.
- According to NICE guidelines clinicians should not be encouraged to rely on biochemistry results, rather should take clinical picture as the main priority.

- Plan best medical or surgical treatment for the patient keeping her past history and desire for future fertility in mind.
- Flexibility is the key. Never be afraid of changing the planned treatment if clinical picture alters.

REFERENCES

1. Brady PC. New Evidence to Guide Ectopic Pregnancy Diagnosis and Management. Obstet Gynecol Surv. 2017;72(10):618-25.
2. Crochet JR, Bastian LA, Chireau MV. Does this woman have an ectopic pregnancy?: the rational clinical examination systematic review. JAMA. 2013;309(16):1722-9.
3. Kumar V, Gupta J. Tubal ectopic pregnancy. BMJ Clin Evid. 2015;2015:1406.
4. American College of Obstetricians and Gynecologists' Committee on Practice Bulletins—Gynecology. ACOG Practice Bulletin No. 193: Tubal Ectopic Pregnancy. American College of Obstetricians and Gynecologists. Obstet Gynecol 2018;131:e91-103. [published correction appears in Obstet Gynecol. 2019;133(5):1059].
5. Panelli DM, Phillips CH, Brady PC. Incidence, diagnosis and management of tubal and nontubal ectopic pregnancies: a review. Fertil Res Pract. 2015;1:15.
6. Hendriks E, Rosenberg R, Prine L. Ectopic Pregnancy: Diagnosis and Management. Am Fam Physician. 2020; 101(10):599-606.
7. Tahmina S, Daniel M, Solomon P. Clinical Analysis of Ectopic Pregnancies in a Tertiary Care Centre in Southern India: A Six-Year Retrospective Study. J Clin Diagn Res. 2016;10(10):QC13-QC16.
8. National Institute for Health and Care Excellence. (2023). Ectopic pregnancy and miscarriage: diagnosis and initial management. NICE guideline [NG126]. Available from www.nice.org.uk/guidance/ng126 [Last accessed March, 2024].
9. Elson CJ, Salim R, Potdar N, Chetty M, Ross JA, Kirk EJ; on behalf of the Royal College of Obstetricians and Gynaecologists. Diagnosis and management of ectopic pregnancy (Green-top Guideline No. 21). BJOG. 2016;123:e15-e55.
10. Morin L, Van den Hof MC; Diagnostic Imaging Committee, Society of Obstetricians and Gynaecologists of Canada. RETIRED: Ultrasound evaluation of first trimester pregnancy complications. J Obstet Gynaecol Can. 2005; 27(6):581-91.
11. Kirk E, Bottomley C, Bourne T. Diagnosing ectopic pregnancy and current concepts in the management of pregnancy of unknown location. Hum Reprod Update. 2014;20(2):250-61.
12. Winder S, Reid S, Condous G. Ultrasound diagnosis of ectopic pregnancy. Australas J Ultrasound Med. 2011; 14(2):29-33.
13. Deutchman M, Tubay AT, Turok D. First trimester bleeding. Am Fam Physician. 2009;79(11):985-94.
14. Surampudi K, Gundabattula SR. The Role of Serum Beta hCG in Early Diagnosis and Management Strategy of Ectopic Pregnancy. J Clin Diagn Res. 2016;10(7):QC08-10.
15. American College of Obstetricians and Gynecologists. ACOG Practice Bulletin No. 94: Medical management of ectopic pregnancy. Obstet Gynecol. 2008;111:1479-85.
16. Gandhi A, Pandit S, Tank P. (2020). FOGSI GCPR on Ectopic Pregnancy. [online] Available from https://www.fogsi.org/wp-content/uploads/gcpr/Ectopic-Pregnancy.pdf [Last accessed March, 2024].
17. Po L, Thomas J, Mills K, Zakhari A, Tulandi T, Shuman M, et al. Guideline No. 414: Management of Pregnancy of Unknown Location and Tubal and Nontubal Ectopic Pregnancies. J Obstet Gynaecol Can. 2021;43(5):614-30.e1.
18. Craig LB, Khan S. Expectant management of ectopic pregnancy. Clin Obstet Gynecol. 2012;55(2):461-70.
19. ESHRE working group on Ectopic Pregnancy, Kirk E, Ankum P, Jakab A, Le Clef N, Ludwin A, Small R, et al. Terminology for describing normally sited and ectopic pregnancies on ultrasound: ESHRE recommendations for good practice. Hum Reprod Open. 2020;2020(4):hoaa055.
20. Ledger W, Clark J. Recent Advances in Obstetrics & Gynaecology: 25. United Kingdom: JP Medical Limited; 2014. pp. 29-40.
21. Ismail SIMF. Recurrent ectopic pregnancy after partial salpingectomy, distal and proximal, therapeutic and prophylactic. Gynecol Surg. 2009;6:281-2.
22. Mohamed AA, Yosef AH, James C, Al-Hussaini TK, Bedaiwy MA, Amer SAKS. Ovarian reserve after salpingectomy: a systematic review and meta-analysis. Acta Obstet Gynecol Scand. 2017;96(7):795-803.
23. Falconer H, Yin L, Grönberg H, Altman D. Ovarian cancer risk after salpingectomy: a nationwide population-based study. J Natl Cancer Inst. 2015;107(2):dju410.
24. Mol F, van Mello NM, Strandell A, Jurkovic D, Ross JA, Yalcinkaya TM, et al.; European Surgery in Ectopic Pregnancy (ESEP) study group. Salpingotomy versus salpingectomy in women with tubal pregnancy (ESEP study): an open-label, multicentre, randomised controlled trial. Lancet (London, England). 2014;383(9927):1483-9.
25. Birge O, Erkan MM, Ozbey EG, Arslan D. Medical management of an ovarian ectopic pregnancy: a case report. J Med Case Rep. 2015;9:290.
26. Hegde CV. Cesarean Scar Pregnancy: Some Management Options. J Obstet Gynaecol India. 2017;67(3):153-6.
27. Society for Maternal-Fetal Medicine (SMFM); Electronic address: pubs@smfm.org; Miller R, Timor-Tritsch IE, Gyamfi-Bannerman C. Society for Maternal-Fetal Medicine (SMFM) Consult Series #49: Cesarean scar pregnancy. Am J Obstet Gynecol. 2020;222(5):B2-B14.
28. Gerli S, Rossetti D, Baiocchi G, Clerici G, Unfer V, Di Renzo GC. Early ultrasonographic diagnosis and laparoscopic

treatment of abdominal pregnancy. Eur J Obstet Gynecol Reprod Biol. 2004;113(1):103-5.
29. Bollig KJ, Schust DJ. Refining Angular Pregnancy Diagnosis in the First Trimester: A Case Series of Expectant Management. Obstet Gynecol. 2020;135(1):175-84.
30. Reid S, Condous G. Is there a need to definitively diagnose the location of a pregnancy of unknown location? The case for "no". Fertil Steril. 2012;98:1085-90.
31. Barnhart KT, Sammel MD, Rinaudo PF, Zhou L, Hummel AC, Guo W. Symptomatic patients with an early viable intrauterine pregnancy: HCG curves redefined. Obstet Gynecol. 2004;104(1):50-5.
32. Moschos E, Twickler DM. Endometrial thickness predicts intrauterine pregnancy in patients with pregnancy of unknown location. Ultrasound Obstet Gynecol. 2008;32(7):929-34.
33. Dooley W, De Braud L, Memtsa M, Thanatsis N, Jauniaux E, Jurkovic D. Physical resolution of tubal ectopic pregnancy on ultrasound imaging following successful expectant management. Reprod Biomed Online. 2020;40(6):880-6.

44

Chapter

Recurrent Pregnancy Loss

Geetika Gupta Syal, Shalini Sharma

INTRODUCTION

- A *pregnancy loss* (miscarriage) is defined as the spontaneous loss of a pregnancy before the fetus reaches viability.[1]
- *Recurrent pregnancy loss* (RPL) is defined as three or more consecutive pregnancy losses at <20 weeks gestation or with a fetal weight <500 g.[2]
- Different definitions have been used in different guidelines **(Table 1)**.
- Affects approximately 1% of fertile couples.[2]

TYPES OF RECURRENT PREGNANCY LOSS

- *Primary RPL*: Multiple losses in a woman who has never delivered a liveborn.[2]
- *Secondary RPL*: Multiple pregnancy losses in a patient with a prior live birth.[2]
  - Successful pregnancy chances are >50% even after five miscarriages.[2]
  - Clinical investigations of pregnancy loss should be done after two-consecutive spontaneous miscarriages mainly in the presence of the following[3]:

TABLE 1: Definitions of RPL used in different guidelines.

Criteria	*WHO[4] (1977)*	*RCOG[5] (2023)*	*ASRM[6] (2012)*	*ESHRE[7] (2022)*	*DGGG/ÖEGGG/SGGG[8] (2018)*
Number of RPLs (n)	3	3	2	2	3
Sequence of loss	Not specified	Consecutive and non-consecutive	Consecutive	Consecutive and non-consecutive	Consecutive and non-consecutive
Type of pregnancy	All pregnancy losses, types are not specified	All pregnancy losses, types are not specified	Clinical pregnancies	Non-visualized pregnancy losses (biochemical pregnancy losses and/or resolved and treated pregnancies of unknown location), clinical pregnancies	All pregnancy losses, types are not specified
Evidence of loss	Not specified	Not specified	USG or histopathologic examination	Serum or urine β–hCG; USG	Not specified
Gestational age at pregnancy loss (weeks)	20 (22)	24	The first trimester of pregnancy, gestational weeks not specified	24	24
Fetal weight (grams)	<500	Not specified	Not specified	Not specified	<500

(ASRM: American Society of Reproductive Medicine; ESHRE: European Society of Human Reproduction and Embryology; hCG: human chorionic gonadotropin; RCOG: Royal College of Obstetricians and Gynecologists; RPL: recurrent pregnancy loss; USG: ultrasonography; WHO: World Health Organization)

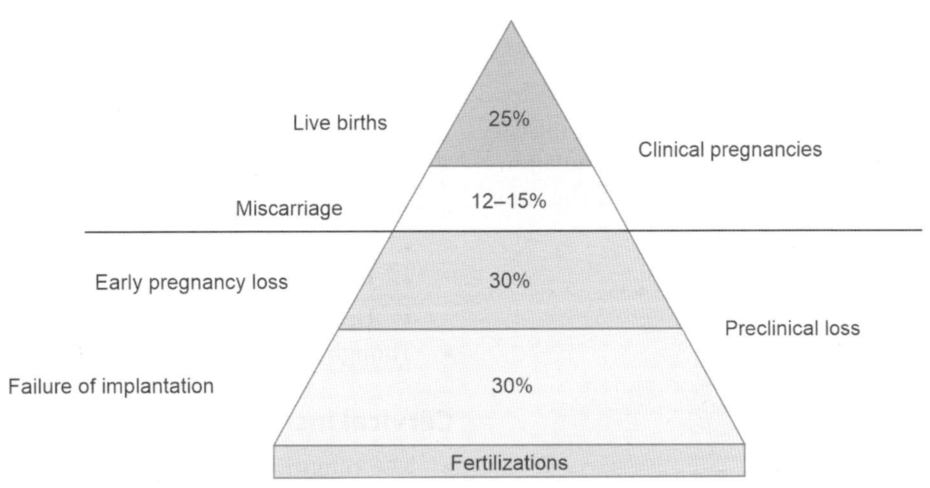

Fig. 1: Epidemiology of pregnancy loss.
Source: Taylor HS, Pal L, Sell E. Speroff's Clinical Gynecologic Endocrinology and Infertility. Philadelphia: Wolters Kluwer; 2019.

- Embryonic heart activity observed before any earlier pregnancy loss
- Normal karyotype on products of conception from an earlier loss
- Age of the female partner over 35 years
- Infertility

EPIDEMIOLOGY OF PREGNANCY LOSS (FIG. 1 AND TABLE 2)

- All chromosomally abnormal conceptions spontaneously abort, most before 10 weeks gestation.[3]
- Over 90% of conceptions having a normal karyotype continue.[3]
- 12–15% of clinically recognized pregnancies end in spontaneous miscarriage between 4 and 20 weeks of gestation.[3]

TABLE 2: Epidemiological factors influencing recurrent pregnancy loss (RPL).

Factor	*Effect on pregnancy loss*
Alcohol	No direct cause, but avoid excess
Smoking	No direct cause, but avoid excess
Caffeine	No direct cause, but avoid excess
Exercise	High-intensity exercise may pose risk
Family history	Though further studies needed, miscarriages may run in families[9]
Maternal age	Directly proportional with increasing age
Occupational/ environmental exposures	No evidence of direct effect
Paternal age	Advancing age associated with increased risk but maternal age is more significant
Previous pregnancy loss	Directly related with number of previous losses[10]
Stress	Not related with RPL
Weight/BMI	BMI >25 is associated with increased risk

ETIOLOGY

Recurrent pregnancy loss has been attributed to **(Fig. 2)**:

Genetic

Chromosomal Abnormalities

- Most common cause of RPL.[1]
- Associated with 70% of early pregnancy losses[11]
- Numerical chromosomal abnormalities are most common[12], include:
    - Trisomy (52%)
    - Polyploidy (21%)
    - Monosomy (13%)
- Maternal age >35 years is associated with increased chances of chromosomal abnormalities in the fetus.[11]

Parental Chromosomal Abnormalities

- In 2–6% of patients with RPL.[12]
- Balanced reciprocals or Robertsonian translocations are the most common abnormalities.[13]
- Though there are high chances of a future healthy live birth.

Anatomic

Congenital Uterine Anomalies (Fig. 3)

- Prevalence of 13.3% in those with a history of miscarriage compared with 5.5% in an unselected population.[1]
- In high-risk populations, septate uterus is the most common cause.[1]

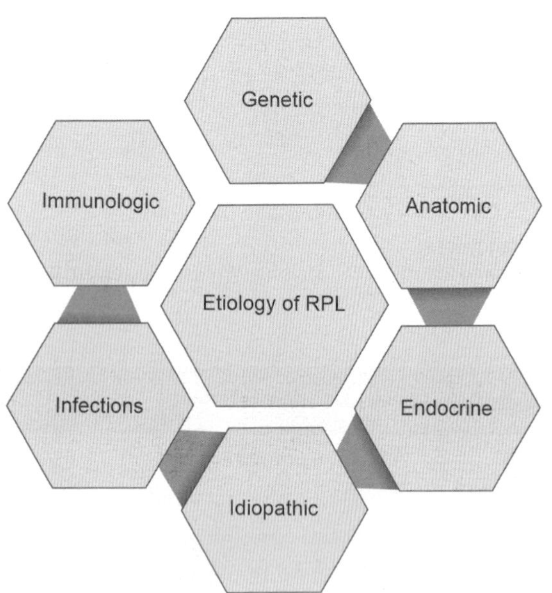

Fig. 2: Etiology of recurrent pregnancy loss.

- Most people with congenital uterine anomaly (CUA) experience a normal reproductive outcome.
- Responsible for second trimester losses.

Acquired Uterine Anomalies

- Acquired uterine malformations such as:
  - Leiomyoma
  - Polyps
  - Uterine adhesions
- Though evidence of direct association is lacking.[1]

Cervical Insufficiency

- The retentive power of the cervix [ostium internum uteri (internal os)] may be impaired functionally and/or anatomically due to the following conditions[15]:
  - Congenital uterine anomalies
  - Acquired (iatrogenic)—common, following:
    - Procedures involving dilatation of internal os [e.g., dilation and curettage (D&C), dilation and evacuation (D&E)]
    - Vaginal operative delivery through an undilated cervix and
    - Amputation of the cervix or cone biopsy or trachelectomy
  - Others—multiple gestations, prior preterm birth.

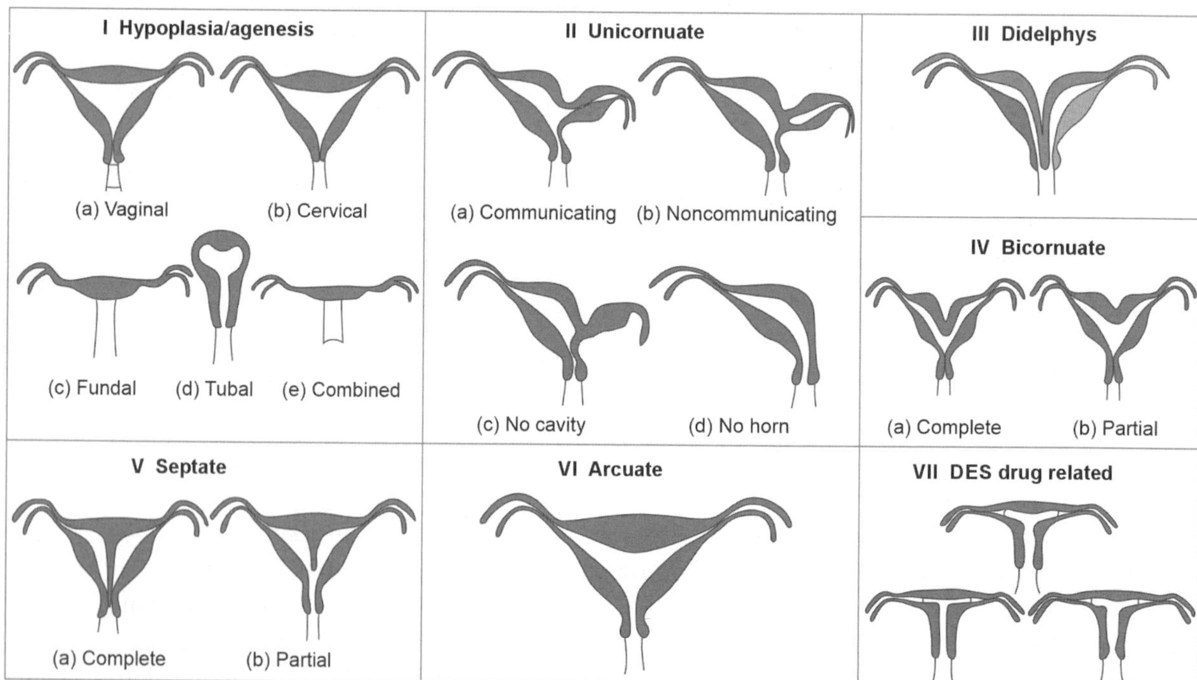

Fig. 3: Congenital uterine anomalies.[14]

- Commonly associated with the second trimester RPL[1]
- There is no recommendation of additional investigations in nonpregnant state.[1]

Infections

Hydrosalpinx

- Hydrosalpinx increases the risk of RPL
- Correcting it can reduce the risk among the women, especially in those undergoing in vitro fertilization (IVF).[1]

Chronic Endometritis

Evidence for direct association with RPL

Thrombophilia

Hereditary Thrombophilia

- These include:
  - Factor V Leiden gene mutations
  - Prothrombin gene mutations
  - Deficiencies in antithrombin, protein C, and protein S.
- Although they may be associated with adverse pregnancy outcomes, the evidence is weak for any association with RPL.[7]

Acquired Thrombophilia

- Antiphospholipid syndrome (APS) is the most common acquired thrombophilia.
- Prevalent in 5–20% in RPL patients and 2% in those with low-risk obstetric history.[6,7]
- It is the most important correctable cause of RPL.[5]

Endocrine

Implicated in approximately 17–20% of RPL.

Thyroid Disorders

- Overt hypothyroidism is a known important cause of RPL—should be treated.[1]
- Evidence finding a link between hyperthyroidism and RPL is lacking.[7]
- Subclinical hypothyroidism [elevated thyroid-stimulating hormone (TSH) and normal free thyroxine] and thyroid autoimmunity may also be associated but evidence is still needed.
- Newer guidelines suggest a new upper limit of normal for thyroid-stimulating hormone (TSH) of 4.0 mIU/L rather than the previous 2.5 mIU/L.[1]

Diabetes Mellitus

First trimester uncontrolled diabetes mellitus [high hemoglobin A1C (HbA1C)] increases the risk of fetal malformations, which may cause RPL.[5]

Polycystic Ovary Syndrome

- Some association between polycystic ovary syndrome (PCOS) and pregnancy loss risk.
- Hyperinsulinemia and hyperandrogenaemia, may be related, but evidence suggesting PCOS predisposes to RPL is lacking.[1]

Luteal Insufficiency

No evidence of its direct association with RPL due to non-uniform diagnostic criteria.[1]

Ovarian Reserve

- Some studies show an association between diminished ovarian reserves and RPL.
- Prognostic value of doing anti-Müllerian hormone (AMH) or antral follicle counts (AFC) has not been ascertained.[1]

Male Factors

- Higher sperm DNA fragmentation index (DFI) is associated with RPL.
- Though studies on treatment and prognosis are not ascertained.[1]

Idiopathic Recurrent Pregnancy Loss

- Some RPL causes are unexplained.
- According to most studies, greater than two-thirds of patients with RPL will have a live birth in a subsequent pregnancy.
- Treatment for these patients does not appear to offer any additional benefit.[1]

MANAGEMENT

- Multifactorial
- *Include:*
  - Lifestyle modification
  - Psychological support
  - Specific treatment of any identified cause.
- Couples should be counseled and treated together.
- The first visit should include[1]:
  - Complete history (medical, obstetric, family and lifestyle)

BOX 1: Useful investigative measures for evaluation of recurrent pregnancy loss.[13]

History:
- Detailed history pertaining to the pattern, trimester, and characteristics of prior miscarriages
- History of subfertility or infertility
- Menstrual history
- History of any gynecologic or obstetric infections
- History of personal or familial thrombotic disease, autoimmune disorders, medications for any ailment, environmental exposures, illicit drug use (e.g., caffeine, alcohol, smoking, marijuana, and *in utero diethylstilbestrol* exposure)
- Family history of RPL to be enquired.
- Signs or symptoms of thyroid, prolactin, glucose tolerance, and hyperandrogenic disorders (including polycystic ovarian syndrome)
- Genetic history and counseling of reproductive partners
- Previous diagnostic tests and treatments, including, if available, chromosome testing on products of conception

Physical examination:
- General physical examination with attention to:
  - Obesity
  - Hirsutism/acanthosis
  - Thyroid examination
  - Breast examination/galactorrhea
  - *Pelvic examination:*
    - Anatomy
    - Infection
    - Trauma
    - Estrogenization
    - Masculinization

- Lifestyle modifications should be advised such as:
  - Smoking
  - Excessive alcohol consumption
  - Overweight or underweight.

INVESTIGATIONS

Associations recommend different investigation guidelines [Royal College of Obstetricians and Gynecologists (RCOG)[5], American Society of Reproductive Medicine (ASRM)[6], and European Society of Human Reproduction and Embryology (ESHRE)[7]]. These are shown in **Table 3**.

Chromosomal Analysis (Box 2)

- Second pregnancy loss may warrant this test, which may lead to the need of further investigations.
- Along with psychological benefits, this may provide a reason for RPL **(Flowchart 1)**.

Ultrasound

- Transvaginal ultrasound scans remain the primary investigation for uterine anomalies.
- Hysterosalpingography is also a good screening test, though, 3D pelvic ultrasound will diagnose and classify CUAs accurately.[1]
- For complex CUAs or acquired uterine anomalies:
  - Magnetic resonance imaging (MRI)
  - Laparoscopy, and
  - Hysteroscopy can all be considered on an individual basis.

Thrombophilia Screening

Investigations for hereditary thrombophilias should only be carried out in the presence of additional risk factors, including a strong personal or family history of thrombosis or thrombophilia.[6,7]

Diagnosis of Antiphospholipid Syndrome Requires A Clinical Manifestation (Box 3)

Vascular thrombosis or pregnancy morbidity and the presence of at least one antiphospholipid antibody (aPL) on two or more occasions at least 12 weeks apart.

- The commonly screened aPLs are:
  - Lupus anticoagulant (LA)
  - Anticardiolipin antibodies [anti-cardiolipin antibody (ACA), immunoglobulin G (IgG) and immunoglobulin M (IgM)]
  - Anti beta-2 glycoprotein I antibodies (ab2GPI, IgG and IgM).

TABLE 3: Summary of recurrent pregnancy loss investigations used in different guidelines.

Disorder	*RCOG guideline (2023)*[5]	*ASRM guideline (2012)*[6]	*ESHRE guideline (2022)*[7]
General	–	Medical history relating to RPL, lifestyle	Medical, obstetric, and family history
Genetics	Cytogenetic analysis on POC on 3 and subsequent pregnancy loss, parental karyotyping if unbalanced structural chromosome abnormality, or there is unsuccessful or no pregnancy tissue available for testing	• Parental karyotyping • Karyotyping of POC may be useful	• Parental karyotyping not on a routine basis • Analysis of pregnancy tissue for explanatory purposes, using aCGH
Hereditary thrombophilia	Only in women with 2nd trimester loss	No screening unless in context of research or in case of additional risk factors for thrombophilia	No screening unless in context of research or in case of additional risk factors for thrombophilia
APS	LA and ACA on 2 occasions 12 weeks apart	LA, ACA, and ab2GPI	LA, ACA, and ab2GPI
Endocrine	TSH and TPO-Ab	TSH/HbA1C	TSH and TPO-Ab, if abnormal then do T4
Uterine anomaly	TVS/assess further with hysteroscopy/3D-US if abnormal	SHG/HSG/hysteroscopy	3D US, 2D ultrasound to rule out adenomyosis
Male factor	–	Sperm DFI not recommended	Sperm DFI can be considered for diagnostic purpose
Immunology	(NK cell, HLA typing) not recommended outside a research context.		

(ab2GPI: b2 glycoprotein I antibodies; ACA: anticardiolipin antibodies; aCGH: array-based comparative genomic hybridization; DFI: DNA fragmentation index; HLA: human leukocyte antigen; HSG: hysterosalpingogram; LA: lupus anticoagulant; NK: natural killer; POC: products of conception; SHG: sonohysterogram; TPO-Ab: thyroid peroxidase antibodies; TSH: thyroid-stimulating hormone; TVS: transvaginal scan; US: ultrasound)

Endocrinological and Metabolic Investigations

- Thyroid-stimulating hormone to rule out hypothyroidism.
- Thyroxine (T4) assessment and correction if TSH levels are >4.0 mIU/L.
- Hemoglobin A1C test to rule out and manage diabetes mellitus especially if there are other risk factors.
- Prolactin levels, if there is evidence of its abnormality (oligomenorrhea/amenorrhea)

Immunology Investigations

- Antinuclear antibodies (ANAs), as higher levels are associated with RPL
- Their presence has detrimental effects on further pregnancies.
- No uniform recommendation for therapy in these patients
- Other immunological markers such as:
  - Human leukocyte antigen (HLA)
  - Anti-histocompatibility (anti-HY) antibodies
  - Cytokines have a weak positive relation with RPL and their levels have no prognostic recommendations.[1]

Infection Investigations

- Spontaneous pregnancy losses are directly related to infections, but their role in causing RPL is still under investigation.
- Clinical signs and symptoms should lead to infection investigations.
- Routine screening for chronic endometritis has no evidence base, but samples taken for research are reasonable.[1]
- Rubella immunity screening may be offered to RPL patients **(Table 4)**.

Flowchart 1: Proposed evaluation for recurrent pregnancy loss.[16]

After 2nd miscarriage
For 1st trimester loss—work-up

a. History of alcohol intake and smoking
b. Screen for weight extremes
c. 3D US
d. APS screening (lupus, anti-beta-2 glycoprotein-1, anti-cardiolipin IgG and IgM)
e. Endocrine testing (Prolactin, TSH and TPO, and HbA1C)

→Manage abnormalities in a, b, c, d, and e accordingly
→Prophylactic vitamin D supplementation for all
→Psychological support

f. Cytogenetic analysis of POC

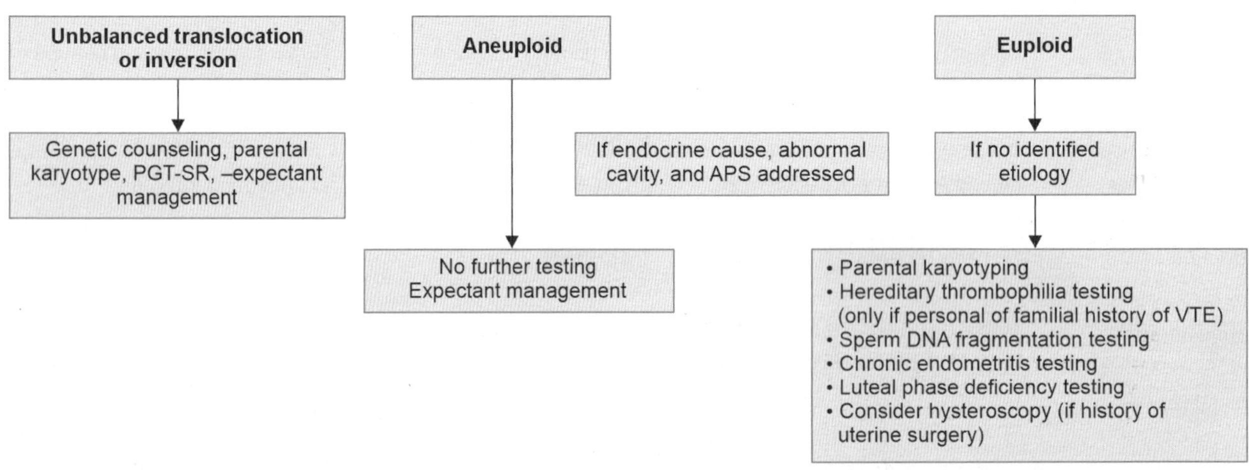

After 3rd miscarriage:
- If above algorithm has been followed in previous pregnancy loss—repeat POC chromosomal testing and complete full work-up if not already done
- If no testing—follow above algorithm
- If all possible causes have been addressed and POC aneuploidy is repetitive, consider expectant management versus PGT-A

(APS: antiphospholipid syndrome; HbA1C: hemoglobin A1C; IgG: immunoglobulin G; IgM: immunoglobulin M; PGT-A: preimplantation genetic testing for aneuploidy; PGT-SR: pre-implantation genetic testing for chromosomal structural rearrangements; POC: products of conception; TPO: thyroid peroxidase; TSH: thyroid-stimulating hormone; VTE: venous thromboembolism)

BOX 2: Genetic techniques to analyze products of conception (POC).

- Karyotyping
- Fluorescence in situ hybridization (FISH)
- Array-based comparative genomic hybridization (aCGH)
- Next generation sequencing (NGS)

Cervical Incompetence

- *History:*
  - Repeated spontaneous second trimester losses
  - History of preterm births
- *Internal examination*: May reveal presence of unilateral or bilateral cervical tear.[17]

Investigations

- *Currently there are no tests available to identify women with cervical insufficiency in the nonpregnant state.*[17]
- During pregnancy: Transvaginal ultrasound performed beyond 14 weeks of gestation, may show following features suggestive of cervical incompetence[17]:
  - Cervical length less than or equal to 25 mm
  - Funneling
  - Bulging of membranes into dilated internal os.

Male Factors

Poor quality sperms and increased sperm DNA fragmentation index (DFI)—are associated with RPL in most studies,

BOX 3: International Consensus Classification criteria for the antiphospholipid syndrome (APS).[6]

Antiphospholipid syndrome is present if one of the following clinical criteria and one of the laboratory criteria are met.

Clinical criteria:
- Vascular thrombosis
- *Pregnancy morbidity:*
  - One or more unexplained deaths of morphologically normal fetuses after the 10th week of gestation by ultrasound or direct examination of the fetus.
  - One or more premature births of a morphologically normal neonate before the 34th week of gestation because of eclampsia or severe preeclampsia or recognized features of placental insufficiency.
  - Three or more unexplained consecutive spontaneous abortions before the 10th week of gestation with maternal anatomic or hormonal abnormalities and paternal and maternal chromosomal causes excluded.

Laboratory criteria:
- Lupus anticoagulant presents in plasma on two or more occasions at least 12 weeks apart, or
- Anticardiolipin antibody of IgG or IgM isotype in serum or plasma present in medium or high titer [>40 GPL (IgG phospholipid) or MPL (IgM phospholipid) or >99th percentile], on two or more occasions at least 12 weeks apart, or
  Anti-beta-2-glycoprotein-I antibody of IgG and/or IgM isotype in serum or plasma (in titer greater than the 99th percentile), present on two or more occasions at least 12 weeks apart.

Source: Practice Committee of the American Society for Reproductive Medicine. Evaluation and treatment of recurrent pregnancy loss: a committee opinion. Fertil Steril. 2012;98(5):1103-11.

TABLE 4: Investigations for recurrent pregnancy loss (RPL).

	Recommendation	*To be assessed individually*	*Not recommended*
Infections	Rubella screening	Endometrial biopsy, HVS	TORCH screen
Anatomical abnormality	2D/3D pelvic scan	MRI, laparoscopy, hysteroscopy	–
Thrombophilia	Acquired thrombophilia work up	Hereditary thrombophilia	–
Endocrinological	HbA1C and thyroid stimulating hormone	T4, Prolactin	Testosterone level, SHBG, anti TPO-Ab
Immunological	–	ANA	• HLA • cytokines • NK cells
Chromosomal abnormality	Pregnancy specimen	Parental karyotyping	
Male-related Sperm DFI			

(ANA: antinuclear antibodies; anti TPO-Ab: anti-thyroid peroxidase antibodies; DFI: DNA fragmentation index; HLA: human leukocyte antigen; HVS: high vaginal swab; SHBG: sex hormone-binding globulin; TORCH: toxoplasmosis, rubella, cytomegalovirus, herpesm simplex)

but treatment and prognostic recommendations are still lacking.[1]

Treatments

Different treatment recommendations are published in different guidelines **(Table 5)**.

Chromosomal Disorders

- Genetic counseling.
- Reduced pregnancy losses have been seen in patients with preimplantation genetic testing (PGT), but there is no difference in the live birth rate.[1]

Anatomical Disorders

- There is a reduction in the pregnancy losses after hysteroscopic septal resection.
- Treating fibroids which distort the endometrial cavity has been associated with improved live births, however, treatment of other uterine malformations (polyps and endometrial adhesions) has not been associated with improved outcomes.[1]

Thrombophilia

Acquired

- Using UFH or LMWH with low-dose aspirin improves live birth.

TABLE 5: Different treatment guidelines.

	RCOG guideline[5] (2023)	ASRM guideline[6] (2012)	ESHRE guideline[7] (2022)
Chromosomal disorders		Genetic counseling recommended in all guidelines	
APS	LDA + heparin	LDA + UFH	• Prophylactic LMWH or UFH (start after positive UPT) • LDA (to be started preconceptionally)
Thrombophilia	Insufficient evidence	–	Insufficient evidence
Endocrinology	Insufficient evidence	Treatment of thyroid disorders, diabetes mellitus, and hyperprolactinemia	Treat maternal endocrine disorders
Uterine anatomy abnormality	• Insufficient evidence • Septum correction	Consider surgical correction	No benefit of septum correction
Idiopathic	Supportive measures	TLC, counseling, support	Supportive measures
Other treatment options	Progestogen supplementation in women with RPL who present with bleeding in early pregnancy	–	• Vaginal progesterone may improve live birth rate in women with three or more pregnancy losses • Repeated and high doses of IvIg very early in pregnancy may benefit women with four or more unexplained RPL

(IvIg: intravenous immunoglobulin; LDA: low-dose aspirin; LMWH: low-molecular-weight heparin; TLC: tender-loving care; UFH: unfractionated heparin; UPT: urine pregnancy test)

- LMWH has a better safety profile and reduced number of daily doses than UFH.
- After confirmation of pregnancy, APS patients should be started on prophylactic dose of LMWH and low-dose aspirin **(Table 6)**.

Inherited

- Anticoagulation recommendations are based on individualized risk of venous thromboembolic events in pregnancy.
- Venous thromboembolic event during the index pregnancy, posthospitalization management requires therapeutic anticoagulation.
- *Unfractionated heparin (UFH):* (Monitoring to keep aPTT 1.5 to 2.5 times normal OR LMWH: Consider monitoring through factor Xa levels in the third trimester).
- Thrombotic risk is greatest during the postpartum period.
- Anticoagulation should be reinitiated after delivery in doses reflecting predelivery treatment regimens **(Box 4)**.
- Women may continue injectable therapy or transition to oral anticoagulants (e.g., coumarin).
- Use of heparin or/of coumarin derivatives does not prohibit breastfeeding.

Immunological Treatments

Immunotherapy has not been associated with improved pregnancy outcomes.

Treatment of Endocrine Disorders

- Preconceptional advice for uncontrolled endocrine disorders is recommended.
- In hypothyroid patients, target level of TSH should be <4.0 mIU/L.[1]

Cervical Incompetence

Treatment Modalities[15]

- *Nonsurgical modalities:*
  - Decreased activity
  - Bed rest and pelvic rest—not effective
  - Vaginal pessary can be used in selected high-risk patients, but the evidence is not sufficient.
- *Surgical modalities*: *Cervical cerclage* **(Figs. 4A to C)**:
  - *Transvaginal cerclage:*
    - McDonald
    - Shirodkar
  - *Transabdominal cerclage*

TABLE 6: Thromboprophylaxis guidelines for inherited thrombophilia (ACCP guidelines).[18]

History of VTE	***Thrombophilia***	***Treatment during pregnancy***	***Treatment after delivery***
Previous single episode of VTE	All inherited thrombophilia	Prophylactic dose of UFH/ prophylactic or intermediate dose of LMWH	Prophylactic or intermediate dose of LMWH for 6 weeks
Previous multiple episodes of VTE (already receiving long-term anticoagulants)	All inherited thrombophilia	Adjusted dose LMWH or UFH, 75% of adjusted dose LMWH or intermediate dose LMWH	Prophylactic or intermediate dose of LMWH for 6 weeks
Previous multiple episodes of VTE (not receiving long-term anticoagulants)	All inherited thrombophilia	Prophylactic or intermediate or adjusted dose LMWH, intermediate or adjusted dose UFH	Prophylactic or intermediate dose of LMWH for 6 weeks
Asymptomatic patients with family history of VTE	Factor V Leiden mutation	Prophylactic or intermediate dose of LMWH	Vitamin K antagonists/prophylactic or intermediate dose of LMWH for 6 weeks
	Prothrombin gene mutation	Prophylactic or intermediate dose LMWH	Vitamin K antagonists/prophylactic or intermediate dose of LMWH for 6 weeks
	Protein C or S deficiency	Thromboprophylaxis not indicated	Prophylactic or intermediate dose of LMWH for 6 weeks
	Other	Thromboprophylaxis not indicated	Vitamin K antagonists/prophylactic or intermediate dose of LMWH for 6 weeks
Asymptomatic patients without family history of VTE	All inherited thrombophilia	Thromboprophylaxis not indicated	Thromboprophylaxis not indicated
Patients with high risk of preeclampsia	Irrespective of thrombophilia history	Start LDA from second trimester	Thromboprophylaxis not indicated

(ACCP: American College of Chest Physicians; LDA: low dose aspirin; LMWH: low molecular weight heparin; VTE: venous thromboembolism)

BOX 4: Anticoagulation regimens.

Low-molecular-weight heparin (LMWH)—prophylactic
- Enoxaparin, 40 mg SC once daily
- Dalteparin, 5,000 units SC once daily
- Tinzaparin, 4,500 units SC once daily
- Nadroparin, 2,850 units SC once daily

LMWH—intermediate dose
- Enoxaparin, 40 mg SC every 12 hours
- Dalteparin, 5,000 units SC every 12 hours

LMWH—adjusted dose (therapeutic)
- Enoxaparin, 1 mg/kg SC every 12 hours
- Dalteparin, 200 units/kg SC once daily
- Tinzaparin, 175 units/kg SC once daily
- Dalteparin, 100 units/kg every 12 hours

Note: anti-Xa level
- 0.6–1.0 units/mL 4 hours after last injection for q12 hour regimen
- Slightly higher doses for daily regimen

Unfractionated heparin (UFH)—prophylactic
- UFH, 5,000–7,500 units SC every 12 hours in first trimester
- UFH, 7,500–10,000 units SC every 12 hours in the second trimester
- UFH, 10,000 units SC every 12 hours in the third trimester, unless the aPTT is elevated.

UFH—adjusted dose (therapeutic)

UFH, 10,000 units or more SC every 12 hours in doses adjusted to target aPTT in the therapeutic range (1.5–2.5×) 6 hours after the injection

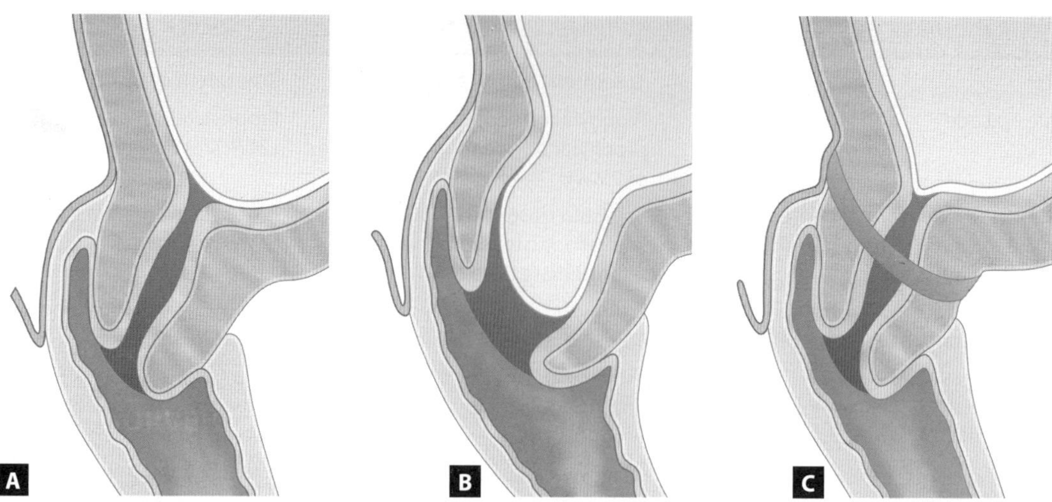

Figs. 4A to C: *Principles of cerclage operation:* (A) Competent cervix; (B) Incompetent cervix with herniation of the membranes; (C) Competency restored after cerclage operation.

TABLE 7: Treatment recommendations.

	Recommendation	***To be assessed individually***	***Not recommended***
Chromosomal disorders	PGT for structural rearrangements	PGT-A	–
Anatomical disorders	Surgical treatment of fibroids	Surgery of endometrial polyps, adhesions, and uterine septum	Surgical treatment to other Müllerian abnormalities
Thrombophilia	Low-dose aspirin for APS, UFH, or LMWH for other thrombophilias	–	–
Infection	Antibiotics in patient with active infection	–	Antibiotic prophylaxis
Immunological	–	–	Ivlg and steroids
Endocrine disorders	Treatment of hyperprolactinemia, diabetes mellitus, and overt thyroid disorders	–	Anti-TPO antibody positivity, subclinical hypothyroidism
Male factor	Lifestyle modification	Antioxidants	–
Idiopathic	Micronized progesterone	–	–
General measures	Smoking cessation Achieve normal BMI Supplement vitamin D Reassurance	Caffeine ≤3 cups a day	–

(APS: antiphospholipid antibody syndrome; BMI: body mass index; Ivlg: intravenous immunoglobulin; LMWH: low-molecular-weight heparin; PGT: preimplantation genetic testing; PGT-A: PGT for aneuploidy; RPL: recurrent pregnancy loss; UFH: unfractionated heparin)

Idiopathic Pregnancy Loss

- Use of progesterone in patients with RPL may decrease the pregnancy loss rate in subsequent pregnancies.
- Progesterone is beneficial in patients with one or more pregnancy losses and bleeding.[1]
- *Doses*[1]*:*
  - Patients with one or more pregnancy losses and bleeding
    - 400 mg BD up to 16 weeks of pregnancy
  - Patients with three or more pregnancy losses:

- 200 mg BD up to 12 weeks of pregnancy, in case of vaginal bleeding, increase dose up to 400 mg BD and continue till 16 weeks of pregnancy.

Male Factors

- No definitive treatment
- Lifestyle modification is indicated.

General Management (Table 7)

- Lifestyle modifications such as cessation of smoking, curtailing alcohol and caffeine intake, and achieving optimal body mass index.
- High-dose folic acid is indicated in patients with BMI ≥30
- Vitamin D supplementation
- Reassurance.

CONCLUSION

Recurrent pregnancy loss (RPL) is a complex and distressing condition affecting numerous women worldwide. Despite advances in medical technology and research, the underlying causes of RPL remain poorly understood, and effective treatments are limited. However, it is clear that RPL is a multifactorial condition. Early detection and management of underlying causes are crucial for improving pregnancy outcomes. Additionally, lifestyle modifications can also contribute to successful pregnancies. Further research is needed to unravel the mysteries of RPL, exploring novel biomarkers, therapeutic approaches, and personalized medicine strategies. By combining medical and emotional support, healthcare providers can empower women experiencing RPL to navigate their journey, fostering hope and resilience.

REFERENCES

1. Chester MR, Tirlapur A, Jayaprakasan K. Current management of recurrent pregnancy loss. The Obstetrician & Gynaecologist. 2022:24(4):260-71.
2. Cunningham FG, Leveno KJ, Dashe JS, Hoffman BL, Spong CY, Casey BM. Williams Obstetrics. 26th edition [New York]: McGraw Hill Medical; 2022.
3. Taylor HS, Pal L, Sell E. Speroff's Clinical Gynecologic Endocrinology and Infertility. Philadelphia: Wolters Kluwer; 2019.
4. World Health Organization. Recommended definitions, terminology, and format for statistical tables related to the perinatal period and use of a new certificate for cause of perinatal deaths. Modifications recommended by FIGO as amended 14 October 1976. Acta Obstet. Gynecol. Scand. 1977; 56:247-53.
5. Regan L, Rai R, Saravelos SH, Li TC. Recurrent miscarriage green-top Guideline No. 17. 2023.
6. Evaluation and treatment of recurrent pregnancy loss: a committee opinion. Fertility and Sterility. 2012; 98(5):1103-11.
7. Bender Atik R, Christiansen OB, Elson J, Kolte AM, Lewis S, Middeldorp S, et al. ESHRE guideline: recurrent pregnancy loss: an update in 2022. Human Reproduction Open. 2022; 2023(1).
8. Toth B, Würfel W, Bohlmann M, Zschocke J, Rudnik-Schöneborn S, Nawroth F, et al. Recurrent Miscarriage: Diagnostic and Therapeutic Procedures. Guideline of the DGGG, OEGGG and SGGG (S2k-Level, AWMF Registry Number 015/050). Geburtshilfe und Frauenheilkunde. 2018;78(4):364-81.
9. Woolner AMF, Nagdeve P, Raja EA, Bhattacharya S, Bhattacharya S. Family history and risk of miscarriage: a systematic review and meta-analysis of observational studies. Acta Obstet Gynecol Scand. 2020;99:1584-94.
10. Magnus MC, Wilcox AJ, Morken NH, Weinberg CR, Haberg SE. Role of maternal age and pregnancy history in risk of miscarriage: prospective register based study. BMJ 2019;364:l869.
11. Hogge WA, Byrnes AL, Lanasa MC, Surti U. The clinical use of karyotyping spontaneous abortions. Am J Obstet Gynecol. 2003;189:397-400.
12. Goddijn M, Leschot NJ. Genetic aspects of miscarriage. Baillieres Best Pract Res Clin Obstet Gynaecol. 2000;14:855-65.
13. Berek JS, Berek DL. Berek & Novak's gynecology. Philadelphia: Lippincott Williams & Wilkins; 2020.
14. Abd elsalam SM, Abd elmegeed NE, Said AHM, Sayed MA elghafar. Role of three-dimensional transvaginal sonography compared with magnetic resonance imaging in diagnosis of Mullerian duct anomalies. Egyptian Journal of Radiology and Nuclear Medicine. 2020;51(1).
15. Dc Dutta, Dutta DC. DC Dutta's textbook of obstetrics. New Delhi: Jaypee Brothers Medical Publishers; 2015.
16. Papas RS, Kutteh WH. A new algorithm for the evaluation of recurrent pregnancy loss redefining unexplained miscarriage: review of current guidelines. Current Opinion in Obstetrics & Gynecology. 2020;32(5):371-9.
17. Thakur M, Mahajan K. Cervical Insufficiency. 2022. In: StatPearls. Treasure Island (FL): StatPearls Publishing; 2024.
18. Bates SM, Greer IA, Middeldorp S, Veenstra DL, Prabulos AM, Vandvik PO. VTE, thrombophilia, antithrombotic therapy, and pregnancy. Chest. 2012;141(2):e691S-e736S.

Chapter 45

Hydatidiform Mole

Suparna Grover

INTRODUCTION

Gestational trophoblastic disease (GTD) is an umbrella term for a group of histologically distinct tumors originating from placenta with variable capacity for local invasion and spread.

They are classified as shown in **Flowchart 1**.[1]

This chapter focusses on hydatidiform moles and gestational trophoblastic neoplasia (GTN), which are discussed elsewhere.

HYDATIDIFORM MOLE

Hydatidiform mole is the result of an abnormal fertilization and is characterized by:

- *Abnormal trophoblastic proliferation and degeneration:* Hyperplasia of both cytotrophoblastic elements and syncytiotrophoblastic elements
- Stromal edema
- Villous structure is maintained.

Genetically, histologically and clinically, molar pregnancies are classified into two major types: (1) Complete and (2) partial mole.

EPIDEMIOLOGY

About 0.5–2.5 per 1,000 pregnancies with wide variation between different geographic regions and ethnicities;[2] rates of partial mole are higher than complete mole.

Risk Factors

- Complete mole is more common on both extremes of maternal age, i.e., teenage pregnancies and maternal age >35 years:
  - Risk with maternal age <15 years: Sixfold[3]
  - Risk with maternal age >45 years: Five to tenfold[4]
- *Geographical and ethnic variation:* Prevalence reported in various studies point to apparently twofold risk in Asian countries, even more so in Japan, Taiwan, and Philippines.[2]
- *History of molar pregnancy in past:*10 times the risk[2]
  - After one molar pregnancy, risk of recurrence 1%[5]
  - After two molar pregnancies, risk of recurrence >20%[6]
- *History of previous abortion*: Doubles the risk[7]
- Dietary deficiency of vitamin A/carotene

Flowchart 1: Classification of gestational trophoblastic disease.

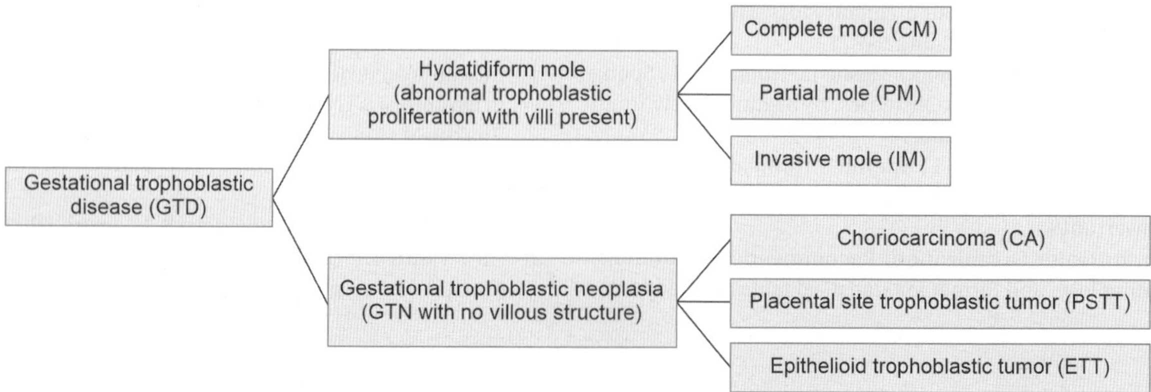

- Family clustering:
  - Associated with (a/w) autosomal recessive gene mutation: NLRP7 on chromosome 19q (48–80%) or KHDC3L on chromosome 6 (10–14%).[8,9]

PATHOPHYSIOLOGY

- Complete mole results from fertilization of a 23,X haploid ovum with inactivated genes by a single spermatozoa followed by duplication or, by two distinct sperms thus resulting in a diploid cell with both sets of nuclear chromosomes of paternal origin in both situations, as depicted in **Flowchart 2**.
- Recurrent familial molar pregnancies do not follow this pattern and are diploid, biparental.[10]
- Partial mole on the other hand is a result of "diandric triploidy". It may be formed if two sperms, either 23, X- or 23,Y-bearing, both fertilize a 23,X-containing haploid egg, whose genes have *not been* inactivated, thus resulting in a fertilized egg that is triploid.

This basic genetic difference leads to differences in clinical presentation and pathological changes observed in complete and partial mole as outlined in **Table 1**.

CLINICAL PRESENTATION

Patient may present in any of the following manners:

- Amenorrhea, asymptomatic with USG suggestive of (s/o) molar pregnancy
- Amenorrhea followed by complaint of bleeding per vaginum (P/V) with or without history of passage of grape-like vesicles with size of uterus larger in proportion to duration of pregnancy
- Amenorrhea followed by complaint of spotting P/V with size of uterus smaller than amenorrhea mimicking missed abortion
- Rarely, amenorrhea followed by any of the complications such as hyperemesis gravidarum, early-onset preeclampsia, eclampsia, abdominal distension due to theca lutein cysts or hemoptysis, respiratory distress, or neurological features like seizures secondary to metastasis.[14]

COMPLICATIONS

- Immediate and short-term complications of hydatidiform moles are described in **Table 2**.
- Long-term complication is post-molar GTN which usually presents within first 6 months of the molar pregnancy, hence justifying the post-molar surveillance

DIAGNOSIS

- *Clinical*: Molar pregnancy should always be one of the differential diagnoses in women with early pregnancy (8–16 weeks menorrhea) reporting with bleeding P/V.
  - History of vaginal bleeding with passage of grape-like vesicles.
  - Abdominal pain depending on the extent of uterine overdistension.
  - On general examination, pallor is often present and may be out of proportion to the visible blood loss.
  - Pulse rate may reveal tachycardia with hypertension or hypotension.

Flowchart 2: Pathogenesis of complete mole.

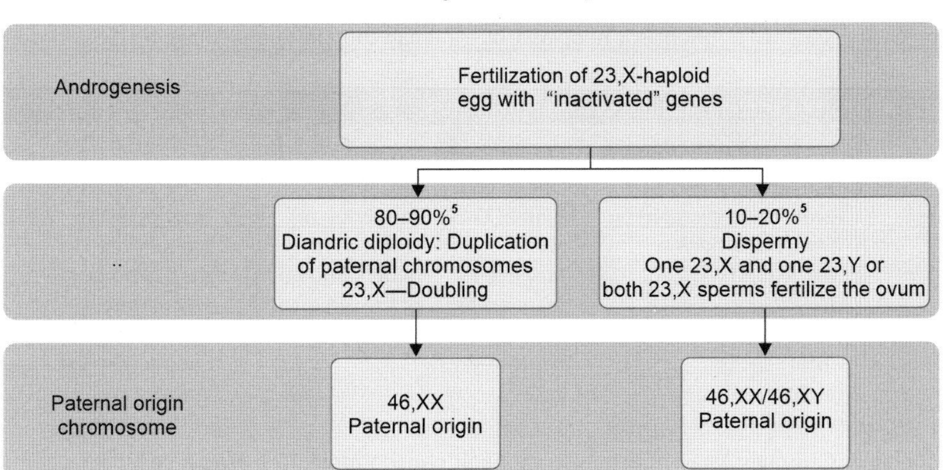

TABLE 1: Comparison of complete and partial mole.

	Complete mole	***Partial mole***
Genetic composition	Diandric diploid 46,XX or 46,XY chromosomes of paternal origin only (biandrogenic or dispermic mechanism)	• Diandric triploid 69, XXX or 69,XXY • One maternal set and two paternal sets of chromosomes
Gross pathology	• No fetal tissue or amnion • Clusters of vesicles with variable dimensions with a "bunch of grapes" appearance • Completely fills the endometrial cavity	• Fetus may be present but usually anomalous or just fetal membranes • Grossly hydropic changes in placental tissue may or may not be visible
Histological findings	Enlarged, edematous villi with central cistern, devoid of blood vessels and abnormal trophoblastic proliferation that diffusely involve the entire placenta	• The degree and extent of trophoblastic proliferation and villous edema is slight and focal only • Fetus, fetal membranes, and fetal vessels may be present
Immunostaining	p57KIP2 protein is absent.[11]	p57KIP2 protein is present.
P57(KIP2) gene is strongly paternally imprinted and maternally expressed, hence absent from purely androgenic complete moles.[11] Not useful in differentiating partial mole from hydropic degeneration of placenta. For latter, ploidy studies are required.		
Typical clinical presentation	Vaginal bleeding with history of passage of grape-like vesicles ± anemia, shock*	Signs and symptoms of an incomplete or missed abortion
*With common use of routine sonography in first trimester, many complete moles are now diagnosed before 12 weeks without any symptoms.		
Uterine size	Uterine enlargement in excess of gestational age is common. (25–50%)	Uncommon
USG findings: Uterus	Snow storm appearance: a complex, echogenic, intrauterine mass containing many small cystic spaces. Fetal tissues and amnionic sac are absent	• A growth-restricted fetus with thickened, hydropic placenta • Ratio of transverse diameter to AP diameter of >1.5[12]
USG findings: Ovaries	Bilateral theca lutein cysts due to prolonged exposure to excess of hCG (25–30%)	Less common
hCG levels	Depending on gestational age, can be very high	Always <100,000 mIU/mL
Additional features (due to high hCG levels)	Hyperemesis gravidarum (25%), preeclampsia (25%), hyperthyroidism (<10%) and theca lutein cysts developed in approximately one quarter of women, usually those presenting in second trimester	Rarely seen
Fetal prognosis	No fetus	Triploid, nonviable, and typically has multiple malformations with abnormal growth
Risk of post-molar GTN	15–20%[13] (mostly invasive mole, choriocarcinoma in one quarter to one third of post-molar GTN)	<5%[13]

(AP: anteroposterior; GTN: gestational trophoblastic neoplasia; hCG: human chorionic gonadotropin; USG: ultrasonography)

TABLE 2: Immediate and short-term complications of molar pregnancy.

Hemorrhage	***Hyperemesis gravidarum***	***Preeclampsia***	***Hyperthyroidism***	***Embolism***	***Metastasis***
• Overdistended uterine cavity due to hypertrophic villi • May lead to *anemia* and *shock* • Postevacuation uterotonics required to prevent uterine atony	• Due to very high hCG levels being secreted by hypertrophic villi • May lead to dehydration, electrocyte disturbances, hepatorenal dysfunction	• Due to very high hCG levels • May even manifest as eclampsia in neglected cases	hCG mimics the action of TSH leading to hyperstimulation of thyroid gland	The risk of trophoblastic embolization to the pulmonary vasculature, especially if uterine contractions are induced with an unripe cervix	• Lung is the most common metastatic site identified by pulmonary nodule on CXR or CT • Vaginal metastasis

(CT: computed tomography; CXR: chest X-ray; hCG: human chorionic gonadotropin; TSH: thyroid-stimulating hormone)

- On abdominal examination:
  - Uterine height more than period of amenorrhea
  - Doughy feel
  - No felt parts palpable, no external ballottement, no fetal heart sound (FHS) audible
- *Biochemical:*
  - Urine pregnancy test: To confirm pregnancy, if not already done
    - Beware of hook effect: Urine pregnancy test may be false negative [excessive beta-hCG (human chorionic gonadotropin) oversaturates urinary assay's targeting antibody][15]
  - Quantitative estimation of beta-hCG reveals levels higher than 1,00,000 mIU/mL in complete moles.
- *Radiological*:
  - USG is usually diagnostic for complete moles with typical "snow storm appearance"; echogenic uterine mass with multiple anechoic spaces of variable sizes.
  - Partial mole may be difficult to differentiate from missed abortion or incomplete abortion with hydropic degeneration of placenta.

MANAGEMENT OF MOLAR PREGNANCY

Principles of Management

These have been outlined in **Figure 1**.

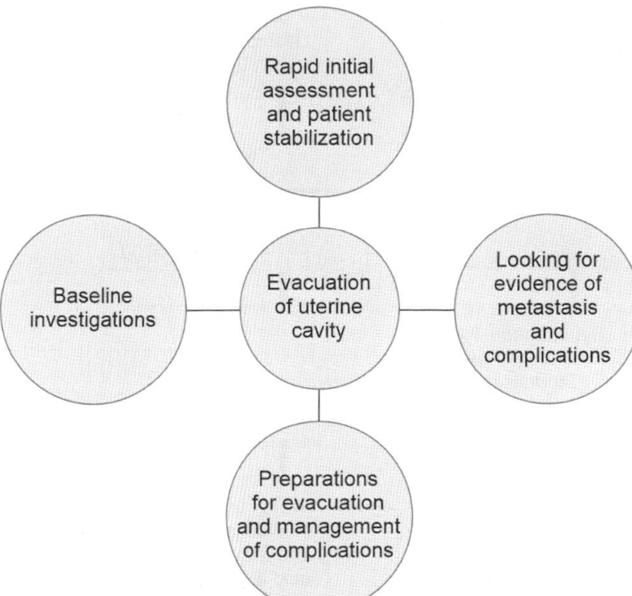

Fig. 1: Principles of management of molar pregnancy.

Rapid Initial Assessment

Hemodynamically unstable patient:

- If any evidence of respiratory distress, secure airway with assisted ventilation if required, consider possibility of embolism or lung metastases.
- If patient is in shock, resuscitation with intravenous crystalloids and arrange blood units for transfusion.
- If preeclampsia, control blood pressure with antihypertensive medications and see for features of imminent eclampsia.
  - Magnesium sulfate may be required.
- If patient has tachycardia, tremors etc., which is not explained by hemorrhage, and features s/o thyroid storm, consider antithyroid drugs and beta blockers.
- Preanesthetic evaluation for evacuation under anesthesia.
- Counseling regarding serious condition, complications, and plan of management.
- Urgent evacuation under anesthesia.

Hemodynamically stable patient:

- Send pre-evacuation investigations
- Secure intravenous line and send blood samples for crossmatching
- Preanesthetic evaluation for evacuation under anesthesia
- Counseling regarding condition, possible complications, plan of management, and need for long-term follow-up
- Evacuation can be done in a planned way by a senior gynecologist with anesthesia team but should not be delayed as far as possible.

Investigations to be Advised Pre-evacuation

- Quantitative estimation of beta-hCG reveals levels higher than 1,00,000 mIU/mL in complete moles
- Complete blood count to assess for anemia
- Blood grouping
- Thyroid function test may reveal hyperthyroidism
- Liver and renal function tests and coagulation profile may be altered in cases complicated by preeclampsia
- Chest X-ray for metastasis.

Preparation for Evacuation

- Secure two wide-bore intravenous lines.
- A central line may be placed for intraoperative central venous pressure (CVP) monitoring and rapid administration of fluids and blood transfusion.

- Arrange sufficient number of blood units (at least two) depending on general condition, amount of bleeding and degree of anemia
- Keep emergency drugs and equipment ready
- Keep uterotonic drugs ready
- No pre-evacuation prostaglandins or oxytocin infusion[14]
  - Any medication that leads to uterine contractions when cervical os is closed or unripe, has theoretical risk of increased chances of embolism of trophoblastic cells through uterine sinuses
- Pre-evacuation intravenous antibiotics
- Detailed counseling of patient and her family regarding the diagnosis, its potential complications, treatment plan, and need for postevacuation follow-up
- Written informed consent
- Check availability of suction apparatus, dilator set and 10- to 12-mm plastic suction cannula in operation theater (OT).

Uterine Evacuation in a Case of Molar Pregnancy

Vacuum aspiration is the method of choice.[5,14,16] Steps are detailed in **Flowchart 3**.

Important Clinical Points

- Procedure can usually be safely performed under general anesthesia although some prefer regional and local anesthesia.
- Bladder should be empty, but unnecessary catheterization should be avoided.
- Anterior lip of cervix can be very soft and so traumatic instruments such as vulsellum and tenaculum should NOT be used to grasp it.
- Dilator and cannula should not go beyond internal cervical os because overdistended uterine walls are very soft and prone to perforation even with gentle pressure.
- Start suction only when the cannula is safely placed just above internal os.
- Attempts to use narrower suction cannula will only lead to slowing down of evacuation, leading to more hemorrhage and incomplete emptying.
- Uterotonics should be started once evacuation is underway, to prevent uterine atony.[5,16]
- In case of uterine atony, successive uterotonics (oxytocin, prostaglandins), bimanual massage, and balloon tamponade may all be attempted.[5,16]
- Curettage should only be done once uterine tone is good and sustained.
- BE GENTLE, BE QUICK, and FOLLOW ASEPSIS.

Flowchart 3: Steps of dilatation and vacuum aspiration in cases of molar pregnancy.

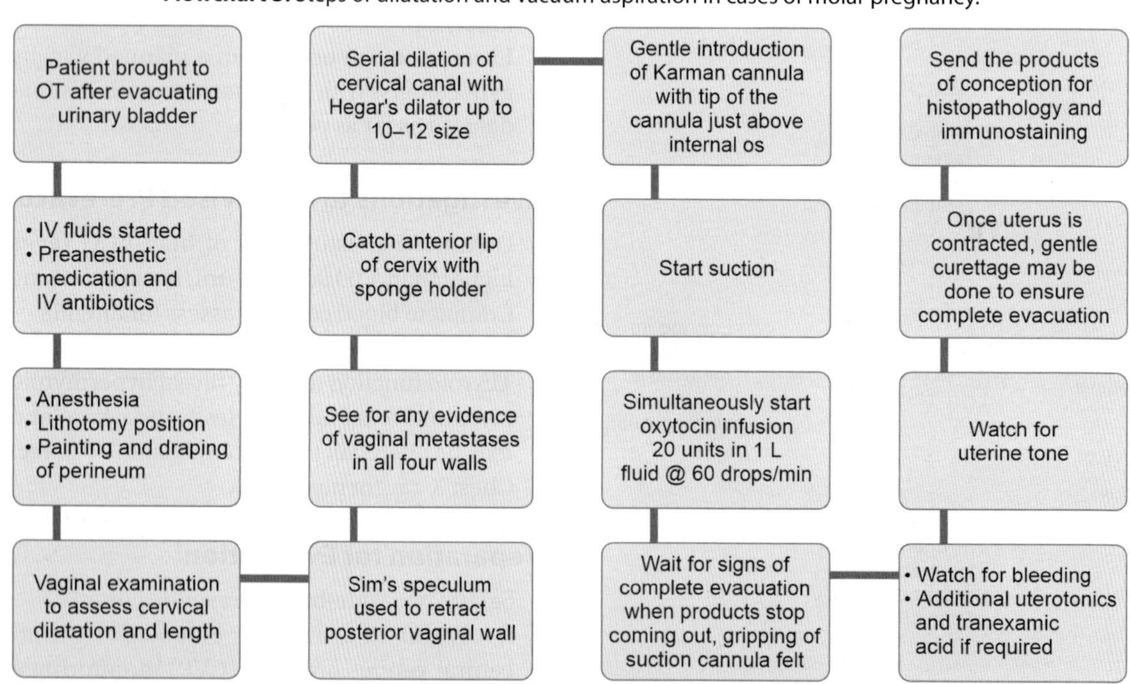

(IV: intravenous; OT: operation theater)

In Case of Perforation

- Discontinue suction immediately
- Increase oxytocin transfusion rate
- If patient bleeding profusely/hemodynamically unstable, immediate laparotomy should be performed.
  - Hysterotomy can be done to complete evacuation along with repair of uterine perforation.
  - Hysterectomy if repair not possible
  - If perforation is through a focus of invasive mole and fertility preservation is desired, segmental resection and repair may be attempted.
  - Gut injury must be ruled out.
- If patient is stable, laparoscopy may be done to rule out gut trauma or serious uterine injury and if hemostasis is adequate, uterine evacuation can be completed under laparoscopic guidance.

Respiratory Distress in Cases of Molar Pregnancy

- *Potential causes:*
  - Trophoblastic embolism
  - Severe anemia due to hemorrhage leading to cardiac failure
  - Preeclampsia leading to pulmonary edema/ARDS (acute respiratory distress syndrome)
  - Thyroid storm leading to high output failure
  - Iatrogenic fluid overload
- Preoperative Chest X-ray may help diagnose pulmonary metastasis.
- Invasive CVP monitoring may help in prevention and timely management.

POSTEVACUATION MONITORING

- Watch for general condition, vaginal bleeding, vitals, and intake output charting.
- Oxytocin infusion may be continued for 24 hours to prevent postoperative hemorrhage due to uterine atony.
- Baseline beta-hCG sent within 48 hours after evacuation
- Anti-D is to be given in all cases when maternal blood is Rh negative.[16]
  - Theoretically, it is not indicated in complete molar pregnancy because there are no fetal red blood cells (RBCs) that can lead to isoimmunization but since histopathological confirmation of distinction between complete and partial mole is not immediately available, it is safer to administer the immunoglobulin.

ROLE OF HYSTERECTOMY

- Usually not a first-line management
- Hysterectomy with mole in situ can be a better treatment option if maternal age > 40 years and family is completed.[17]
- Significant reduction in risk of post-molar GTN compared to uterine evacuation, odds ratio (OR) = 0.19.[17]
- Hysterectomy does not eliminate the need for post-molar monitoring for malignant sequelae because occult metastases at the time of diagnosis may have been missed.

ROLE OF REPEAT CURETTAGE

Currently, this is not recommended unless clinical features and USG are suggestive of retained products.

Second uterine curettage as an initial treatment for low-risk nonmetastatic GTN has been found to cure 40% cases without the morbidity posed by chemotherapy, which is the usual first-line treatment in a phase 2 trial.[18]

COUNSELING BEFORE DISCHARGE

- *Danger signs:* Bleeding, fever, and cardiorespiratory discomfort
- *Contraception:* Couple has to be advised the need to avoid pregnancy till the time the risk of persistent trophoblastic disease (PTD) can be safely ruled out.
  - Barrier contraceptives are safe but not a reliable method
  - Intrauterine contraceptive device (IUCD) is contraindicated.
  - Oral contraceptive (OC) pills are a safe and reliable option[5,14,16]
- *Follow-up plan:* Patient is asked to return for first follow-up 1 week after baseline postevacuation beta-hCG for repeat testing.

ROLE OF PROPHYLACTIC CHEMOTHERAPY

- Use of methotrexate or actinomycin D in limited course to prevent PTD
- Not recommended as a routine except cases with high risk of post-molar GTN who are not likely to be compliant to postevacuation surveillance[5,14]

MANAGEMENT OF THECA LUTEIN CYSTS

- Present in a third to a quarter of cases with molar pregnancy, usually bilateral 5–6 cm, may be larger.
- Presence of theca lutein cysts is considered a risk factor for PTD and malignant transformation.
- The natural history of theca lutein is spontaneous regression in 8 weeks following evacuation with falling hCG levels.
- Theca lutein cysts do not need any surgical management unless there is torsion or hemorrhage.

POST-MOLAR SURVEILLANCE

- Serial measurement of hCG is the primary means of surveillance for post-molar GTN as described in **Flowchart 4**.
- The duration of surveillance varies among different guidelines.[14,16]
    - In cases if complete mole, FIGO (International Federation of Gynaecology and Obstetrics) recommends monthly surveillance after normalization of hCG up to at least 6 months whereas UK guidelines recommend:
        - Surveillance till 6 months after evacuation if hCG normalizes in <56 days
        - Surveillance till 6 months after normalization if hCG normalizes in >56 days
- In cases of partial mole, surveillance may be stopped once hCG levels become undetectable for two consecutive monthly tests.[5]

Risk factors for post-molar persistent trophoblastic tumor:[4]

- Age > 40 years
- Uterine size beyond that appropriate for gestational age
- Theca lutein cysts
- hCG at diagnosis >1,00,000 IU/L

Flowchart 4: Postevacuation follow-up of molar pregnancy.

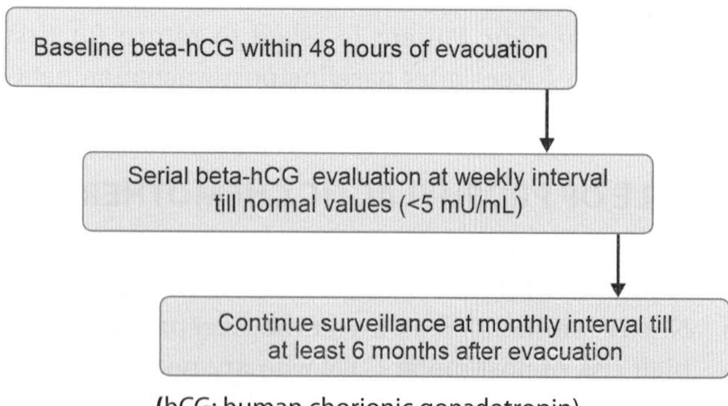

(hCG: human chorionic gonadotropin)

FIGO/WHO (World Health Organization) diagnostic criteria for post-molar GTN (2018):[16]

- Plateauing of hCG for 4 consecutive weekly values over 3 weeks
- An increase in hCG >10% for 3 consecutive weekly values over 2 weeks
- Histological diagnosis of choriocarcinoma or clinical/radiological evidence of metastases.

RARE SCENARIOS

Molar Ectopic Gestation

Management follows the general principles of ectopic gestation but post-molar surveillance remains the same as intrauterine molar pregnancy.

Molar Pregnancy with a Viable Twin[14]

- Counseling should describe the implications of continuing the pregnancy including possibility of spontaneous abortions (40%), perinatal outcome, and GTN
- If pregnancy is continued:
    - Invasive testing should be done for karyotyping the viable fetus.
    - Risk of preterm birth is 36%.
    - Being watchful for preeclampsia and growth restriction is important.
- If termination of pregnancy is chosen or is medically indicated, then depending on fetal size, medical methods of termination may be required.

INVASIVE MOLE

- Pathologic features similar to complete mole (edematous villi with trophoblastic proliferation) confined to uterus + invades myometrium.
- Locally aggressive, may extend to broad ligament, parametrium, vaginal wall metastasis to lungs
- Frequently metastases to lower genital tract or lungs through venous penetration
- *Clinical presentation:*
    - Similar to complete mole but after evacuation, bleeding and subinvolution should raise the suspicion.
    - On following the postevacuation beta-hCG levels, post-molar GTN is diagnosed.

- Confirmation of diagnosis is only possible if hysterectomy is done, which is rarely resorted to because the response to chemotherapy is usually good.

CONCLUSION

Molar pregnancy is a high risk condition but with timely diagnosis and evidence based management and follow up, complications and morbidity, both short term and long term can be minimised.

REFERENCES

1. le-Ming S, Mazur MT, Kurman RJ. Gestational trophoblastic disease and related tumor-like lesions. In: Kurman RJ, Ellenson LH, Ronnett BM (Eds). Blaustein's Pathology of the Female Genital Tract, 6th edition. New York: Springer; 2011. p. 1076.
2. Hammond C, Soper J. Gestational Trophoblastic Diseases. Glob Libr Women Med. 2008;1756-2228.
3. Bagshawe KD, Dent J, Webb J. Hydatidiform mole in England and Wales 1973-83. Lancet. 1986;2(8508): 673-7.
4. Horowitz NS, Eskander RN, Adelman MR, Burke W. Epidemiology, diagnosis, and treatment of gestational trophoblastic disease: A Society of Gynecologic Oncology evidenced-based review and recommendation. Gynecol Oncol. 2021;163(3):605-13.
5. Soper JT. Gestational Trophoblastic Disease: Current Evaluation and Management. Obstet Gynecol. 2021;137(2):355-70. Erratum in: Obstet Gynecol. 2022; 139(1):149.
6. Berkowitz RS, Im SS, Bernstein MR, Goldstein DP. Gestational trophoblastic disease. Subsequent pregnancy outcome, including repeat molar pregnancy. J Reprod Med. 1998;43(1):81-6.
7. Acaia B, Parazzini F, La Vecchia C, Ricciardiello O, Fedele L, Battista Candiani G. Increased frequency of complete hydatidiform mole in women with repeated abortion. Gynecol Oncol. 1988;31(2):310-4.
8. Murdoch S, Djuric U, Mazhar B, Seoud M, Khan R, Kuick R, et al. Mutations in NALP7 cause recurrent hydatidiform moles and reproductive wastage in humans. Nat Genet. 2006;38(3):300-2.
9. Slim R, Wallace EP. NLRP7 and the Genetics of Hydatidiform Moles: Recent Advances and New Challenges. Front Immunol. 2013;4:242.
10. Kalogiannidis I, Kalinderi K, Kalinderis M, Miliaras D, Tarlatzis B, Athanasiadis A. Recurrent complete hydatidiform mole: where we are, is there a safe gestational horizon? Opinion and mini-review. J Assist Reprod Genet. 2018;35(6):967-73.
11. Merchant SH, Amin MB, Viswanatha DS, Malhotra RK, Moehlenkamp C, Joste NE. p57KIP2 immunohistochemistry in early molar pregnancies: emphasis on its complementary role in the differential diagnosis of hydropic abortuses. Hum Pathol. 2005;36(2):180-6.
12. Fine C, Bundy AL, Berkowitz RS, Boswell SB, Berezin AF, Doubilet PM. Sonographic diagnosis of partial hydatidiform mole. Obstet Gynecol. 1989;73(3 Pt 1):414-8.
13. Albright BB, Shorter JM, Mastroyannis SA, Ko EM, Schreiber CA, Sonalkar S. Gestational Trophoblastic Neoplasia After Human Chorionic Gonadotropin Normalization Following Molar Pregnancy: A Systematic Review and Meta-analysis. Obstet Gynecol. 2020;135(1):12-23.
14. Tidy J, Seckl M, Hancock BW; on behalf of the Royal College of Obstetricians and Gynaecologists. Management of Gestational Trophoblastic Disease. Green-top Guideline No. 38 - June 2020. BJOG. 2021;128:e1-e27.
15. Al-Mahdili HA, Jones GR. High-dose hook effect in six automated human chorionic gonadotrophin assays. Ann Clin Biochem. 2010;47(Pt 4):383-5.
16. Ngan HY, Kohorn EI, Cole LA, Kurman RJ, Kim SJ, Lurain JR, et al. Trophoblastic disease. Int J Gynaecol Obstet. 2012;119 Suppl 2:S130-6. Erratum in: Int J Gynaecol Obstet. 2014;125(1):93.
17. Zhao P, Lu Y, Huang W, Tong B, Lu W. Total hysterectomy versus uterine evacuation for preventing post-molar gestational trophoblastic neoplasia in patients who are at least 40 years old: a systematic review and meta-analysis. BMC Cancer. 2019;19(1):13.
18. Osborne RJ, Filiaci VL, Schink JC, Mannel RS, Behbakht K, Hoffman JS, et al. Second Curettage for Low-Risk Nonmetastatic Gestational Trophoblastic Neoplasia. Obstet Gynecol. 2016;128(3):535-42.

Section

7

Menopause

46

Chapter

Menopause Transition and Climacteric

Sadhana Gupta, Amrita S Jaipuriar

INTRODUCTION

Menopause transition and climacteric can also be termed as perimenopause. This is the period of change in ovarian function. It is a journey from being fertile to becoming infertile and affects all women. Ovarian function is not completely abolished at perimenopause. In fact, the ovaries of many women are able to produce hormones for 1–2 years after menopause, although the production is decreased in comparison to premenopausal state.[1] Gradual changes occur before and after menopause, while ovarian function declines slowly toward its termination.

This transitional phase is defined as the climacteric. Climacteric is derived from "klimax" which is a Greek word for ladder. This accurately expresses the characteristic endocrine changes in women around the time of menopause. The menopausal transition is known to play a major role in the etiology of symptoms such as hot flashes, night sweats, uterine bleeding problems, and vulvovaginal atrophy. Mood changes, sleep disturbances, and sexual dysfunctions are also commonly reported and may be attributed to the hormonal aberrations experienced during the transition.

The hormonal environment in perimenopause is tumultuous, unlike menopause which is serene and steady. Perimenopausal symptoms may be compounded by factors such as adrenal fatigue, anovulatory cycles, and poor dietary habits.

Menopause is a state of decline in reproductive hormones. It indicates the termination of ovarian function which follows the aging. Menopause is confirmed when there is amenorrhea of 12 months since last menstruation. It gives rise to a variety of symptoms because the degenerative changes due to aging are combined with a decrease of estrogen production due to failing ovaries. Menopause can be natural or induced. A diagnosis of natural menopause is usually made retrospectively when spontaneous cessation of periods occurs for 1 year between the age of 45 and 55 years.

In India, the average age of menopause is 46.5 yrs. Worldwide the average age of menopause is 51 years. Women who smoke and are underweight tend to have earlier menopause, while women who are overweight often have a late menopause.

Menopause can also happen for reasons other than natural such as:

- *Premature menopause:* It is a condition when ovarian functions decline before the age of 40 years.
- This is called as premature ovarian failure or premature ovarian in sufficiency, reasons being smoking, radiation exposure, chemotherapeutic drugs, and surgery affecting the ovarian blood supply.
- *Surgical menopause:* Surgical menopause may follow the removal of one or both ovaries or radiation of the pelvis including the ovaries. In premenopausal women, abrupt menopause occurs in this condition and menopausal symptoms are more severe in comparison to women with natural menopause.

STAGING OF MENOPAUSE

Modified Anklesaria's Stages of Menopause (Table 1)

Stage I has its own limitations. It is retrospective and cannot be used for premature ovarian insufficiency, infertility, and also for hormone-based research studies.

Members of Stages of Reproductive Aging Workshop considered few factors for staging.

For example, menstrual cycle, hormonal factors, fertility, signs, and symptoms in other organ systems and uterine and ovarian anatomy.

TABLE 1: Modified Anklesaria's Indian Menopause Society (IMS) consensus group staging.

	Menopause		
Stage I	***Stage IIA***	***Stage IIB***	***Stage III***
• Roughly 2 years before menopause • Early (premenopausal symptoms): – IA—Vasomotor instability – IB—Early psychosomatic symptoms • Menstrual problems	• One year after last period • Atrophic changes • Genitourinary • Vasomotor • Weight gain • Osteopenia	• Up to 5 years after menopause • Intermediate (postmenopausal symptoms): – Late psychosomatic and genital symptoms – Sexual disorders – Residual changes from stage IIA osteopenia or osteoporosis	• From 5 years post menopause till • Late (postmenopausal) complications: – Residual changes from stage II – Ischemic heart changes – Other late complications, for example, Alzheimer's disease, osteoporosis
Prevent and treat	Treat	Treat	Treat and palliate

Source: Indian Menopause Society consensus statement, 2008.

TABLE 2: Five stages prior to final menstrual period (FMP).

						FMP 0		
Stages:	**–5**	**–4**	**–3**	**–2**	**–1**	**+1**		**+2**
Terminology:	Reproductive			Menopausal transition		Post menopause		
	Early	Peak	Late	Early	Late*	Early*		Late
				Perimenopause				
Duration of stage:	Variable			Variable		(a) 1 year	(b) 4 years	Until demise
Menstrual cycles:	Variable to regular	Regular		Variable cycle length (>7 days different from normal)	≥2 skipped cycles and an interval of amenorrhea (≥60 days)	Amenorrhea × 12 months	None	
Endocrine:	Normal FSH	↑FSH		↑FSH		↑FSH		

*Stages most likely to be characterized by vasomotor symptoms.
Stages/nomenclature of normal reproductive aging in women (FSH: follicle-stimulating hormone)
Source: Soules MR, Sherman S, Parrott E, Rebar R, Santoro N, Utian W, et al. Executive summary: Stages of reproductive aging workshop (STRAW). Parkcity, Utah, Menopause. 2001;8(6):402-7.

Main point of staging is the final menopausal period (FMP).

There are five stages prior to FMP and duration of the stages and age range, varies amongst women **(Table 2)**.

CHANGES AT PERIMENOPAUSE

Adrenal Fatigue

In middle age there is lot of stress of women because of the demand of work, family, children, aging parents, and society at large plays undue stress on many women. The result is a perpetual fight and fright cycle. Adrenal gland functions constantly releasing norepinephrine and epinephrine from adrenal medulla and corticosteroids, mineralocorticoids, and androgens from adrenal cortex. Long-term release of stress levels of mineralocorticoids can cause potassium deficiency and a magnesium imbalance, as well as chronic water retention and high blood pressure. The resultant magnesium insufficiency can affect many of the enzyme-driven metabolic pathways in the body.

Adrenal cortex makes all the sex hormones in small amounts and dehydroepiandrosterone (DHEA) in large amounts. DHEA is important for growth and repair of protein tissues and is a precursor to androstenedione, testosterone, and estrogens.

It is not a precursor to progesterone, aldosterone, pregnenolone, and cortisol.

Alterations in these adrenal hormones can predispose to multiple symptoms of perimenopause, aggression and anger from too much testosterone, oversensitivity, passivity, mental confusion, and agitation from too much estrogen in early phase of climacteric, and later depression from too little estrogen and progesterone.

Anovulatory Cycles (Fig. 1)

At climacteric due to the increasing follicle-stimulating hormone (FSH), the follicular phase is extended and estrogen levels remain sustained for longer periods.

Hormonal changes: Hormonal changes are the hallmark of the menopausal transition.

But can be highly variable and not always correlative with clinical symptoms.

The earliest detectable event in the menopausal transition is a decrease in inhibin.

The physiologic role of inhibin (which exists in two forms, A and B) is to reduce pituitary FSH secretion via negative feedback inhibition.

In the early transition, when the follicle cohort decreases, inhibin B declines, leading to a loss of pituitary inhibition, and FSH secretion is increased, although intermittently. Thus, the rise in FSH in the early transition is not due to decreased estradiol.

Falls in estradiol are more characteristic of the late menopausal transition, as the number of anovulatory cycles increase.

Increases in FSH and decreases in inhibin B generally represent early changes and are followed by decreases in both estradiol and inhibin A in the late stage of menopausal transition but there is significant variability.

Assessing FSH levels is not recommended as a test to determine the perimenopausal status.

The menstrual cycle begins to lengthen about 5 years before menopause.

However, menopause occurs without prior lengthening of the menstrual cycle in about 10% of women.

The levels of various hormones vary during the climacteric:

- First, there is decrease in the blood levels of inhibin, Followed by the increase in level of FSH, rising progressively from 4 to 5 years before menopause.
- Slightly increased FSH levels do not affect either the menstrual cycle or blood estradiol levels but reflect the decline of reproductive function due to aging.
- Elevated FSH levels are suggestive of poor quality of oocytes. Under such conditions, fertility can be considered to show a substantial decline.
- After elevation of FSH levels, LH levels also begin to increase from several years before menopause. An increase in LH levels is less marked than that of FSH **(Fig. 2)**.
- Although the function of the ovaries (where estradiol is produced) begins to decline before menopause, its production is maintained because of an increase in FSH levels.
- Estradiol levels are maintained, or are slightly increased in some cases, until about 1 year before

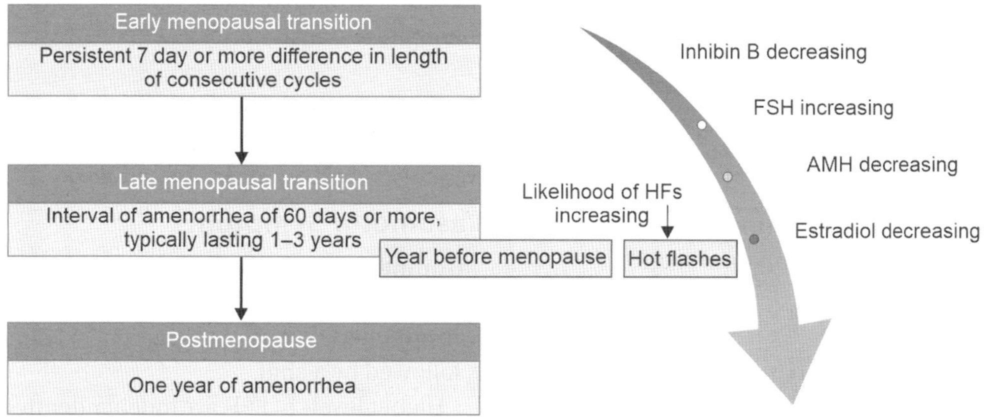

Fig. 1: Anovulatory cycles. (AMH: anti-Müllerian hormone; FSH: follicle-stimulating hormone)

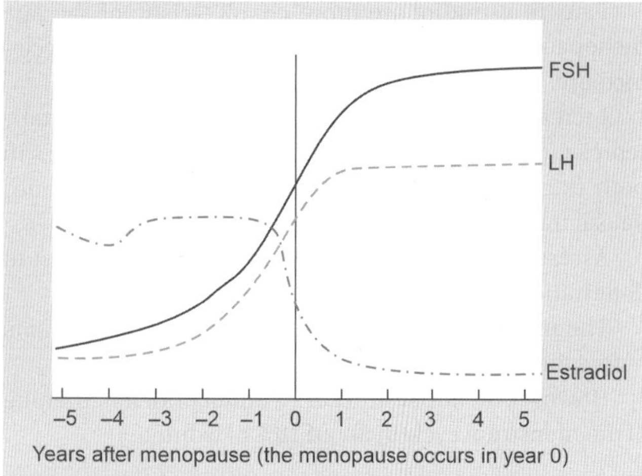

Fig. 2: Changes of hormone levels during the climacteric. (FSH: follicle-stimulating hormone; LH: luteinizing hormone)

menopause. At the time of menopause, both FSH and LH levels increase rapidly, but estradiol levels continue to decrease for a few years until it approaches the level similar to castration.

- Unlike estrogens, testosterone (a typical androgen) only decreases slightly after menopause. Androgen production by the adrenal glands decreases gradually with age but persists for at least 10 years after menopause.
- In postmenopausal women, low estrogen levels are mainly maintained by conversion of androgens in the peripheral adipose tissue.
- The more subcutaneous fat women have, the higher estrogen levels are.

Definitions and characteristics of the stages of reproductive aging, simplified from the STRAW (stages of reproductive aging workshop) definitions. Progression through the perimenopause is marked by increasing menstrual cycle variability that serves to define these stages. As will be discussed, this correlates with trends in hormonal levels. In addition to bleeding pattern changes, symptoms are also significant for increasingly prevalent VMS (vasomotor symptoms).

These typically begin either during the menopausal transition or early in the postmenopausal stage.

Anti-Müllerian Hormone

Anti-Müllerian hormone (AMH), such as inhibin, is a TGF-β (transforming growth factor-β) family peptide that is produced by the small, growing follicles of the ovary and is a marker of ovarian reserve. AMH has been found to decrease with age. While it remains a promising marker, AMH is not *currently* recommended as a routine diagnostic test for perimenopause.

Other Markers

Longitudinal studies of menstrual cycle hormones have revealed a progressive decline in urinary Progesterone metabolites across the menopausal transition, but it is not helpful for diagnostic purposes.

Testosterone secretion drops with menopause, a concomitant decline in sex hormone binding globulin (SHBG) leads to either maintenance or even an increase in the free androgen index (FAI) (Total testosterone × 100/ SHBG).

Adrenal androgens decline throughout adulthood and continue this decline throughout the transition. Androgen deficiency is not a common feature of menopause transition.

FACTORS INFLUENCING HORMONE LEVELS

Body mass index (BMI) and race impact hormones. Women with obesity have a lower rise in FSH and a less steep decline in estradiol as they traverse the FMP.

In obese women, the occurrence of climacteric symptoms is less frequent, and they have a relatively higher bone mineral density than in slender women, partly because of the production of estrogen by the adipose tissue.

Women who experience the cycle irregularity of the early menopause transition at a younger age than the median of 47 are more likely to have a longer, and more symptomatic transition than those whose onset is later in life.

Women who have more variable hormone levels over time are also more likely to have a symptomatic menopause transition.

SYMPTOMS

Climacteric symptoms can be classified as shown in **Box 1**.

Vasomotor Symptoms

Vasomotor symptoms are also known as hot flashes or hot flushes. There is sudden onset, intense feelings of warmth often involving the face, neck, or chest, at times it is followed by a chill.

Higher levels of FSH are independently associated with increased risk of VMS.

BOX 1: Symptoms of climacteric syndrome.

- *Symptoms of autonomic dystonia:*
  - *Vasomotor symptoms:* Feeling of warmth (facial hot flushes, and hot flushes), sweating, and sensitivity to cold
  - Sleep disorders
  - *Others:* Palpitations, headache, dizziness, and tinnitus
- *Mental disorders:*
  - Depression, mental lability, hypobulia, anxiety, and hypomnesia
- *Other symptoms:*
  - *Locomotor symptoms:* Shoulder stiffness, arthralgia, low back pain, and myalgia
  - *Gastrointestinal symptoms:* Abdominal pain, anorexia, nausea, vomiting, and diarrhea
  - *Others:* Fatigability, skin symptoms (eczema and pruritis), and dry mouth

Source: Sakmoto M, Mizuno M, Taketani Y. Principle of Obstetrics and Gynecology 1. Revised edition. Tokyo: Medical View Company Ltd.; 2001. p. 690.

Other risk factors are BMI, cigarette smoking, depressive symptoms, and alcohol intake.

In 70–80% women VMS are associated with the decrease of estrogen production, In 10–25% of the female population, hot flushes and sweating appear before menopause. At first, hot flushes develop during the night, and then in the daytime also. The onset is induced by stress. Hot flushes are frequently associated with sweating and palpitations. Their frequent occurrence during the night results in insomnia. One large study noted that women having frequent VMS had an increased risk of developing CVD (cardiovascular disease) over a 14-year period.[2,3]

Vaginal and Sexual Symptoms

Vaginal dryness, atrophy and associated dyspareunia and urinary symptoms, are common in the menopausal women. It is termed as genitourinary syndrome of menopause (GSM) and is associated with decreased vaginal collagen and elastin content, thinned epithelium, and increased vaginal pH related to decreasing estrogen.

Cross-sectional studies have suggested that increased pain during sexual intercourse may occur earlier in the transition, while decreases in desire may occur later in a woman's transition. The SWAN cohort noted a decrease in reported sexual functioning starting at around 20 months prior to the FMP, with the greatest rate of decline in the late transition and a slower decline in post menopause.

The decrease of estrogen levels gives rise to atrophic changes of the genitourinary tract.

These changes appear after the manifestations of climacteric symptoms. In other words, vaginitis, vulvar pruritis, vaginal dryness, dyspareunia, urinary tract infection, and frequency occur several years after menopause.

The skin also becomes thin and dry, which is considered to be caused by a decrease of cutaneous collagen fibers due to estrogen deprivation.

- *Sleep and mood changes:* Estrogens are known to maintain and improve brain function. Sleep is affected in women in menopause transition.[4]

  There is no conclusive evidence that estrogen can improve memory or are effective for preventing Alzheimer's disease and suppressing its progression.

  Women in the menopause transition have trouble falling asleep, early awakening, and interrupted sleep. All are the matter of concern and affect the health of these women. Sleep disturbances are strongly associated with vasomotor and psychological symptoms.

  Stress and lifestyle factors also affect the sleep. Poor sleep prior to the transition is predictive of worsening sleep symptoms throughout menopause. Sleep symptoms get stable after menopause.

  Three robust longitudinal studies—SWAN, the Penn Ovarian Aging Study, and the Harvard Study of Moods and cycles all demonstrated an increased risk for depressed mood during the menopausal transition,[5-7] including increased risk of new-onset major depressive disorder and anxiety in women with no prior history of a psychologic diagnosis.

  Major depressive episodes are defined by as having five or more symptoms, weight changes or change in appetite, insomnia or hypersomnia, psychomotor agitation or retardation, fatigue or loss of energy or interest, feelings of worthlessness or guilt, decreased concentration, recurrent thoughts of death one of which must include either depressed mood or loss of interest, causing significant distress or impairment in social, occupational, or other areas of functioning.[8]
- *Bleeding:* Abnormal menstrual bleeding (AUB) is quite a challenge to understand in the women in their menopause transition.

  Abnormal uterine bleeding presents a unique challenge in perimenopause because physiologic hormonal changes may mimic, or mask pathologic diagnoses.

Evaluation should be done to rule out serious conditions such as hyperplasia or cancer, and to find treatment options.

The PALM-COEIN classification of abnormal uterine bleeding should be utilized to establish a differential diagnosis of bleeding and for evaluation and management.[9]

- *Bone health:* Bone mineral density decreases rapidly after menopause. Menopause hormone therapy (MHT) or estrogen replacement therapy can prevent this decrease and reduces the risk of fracture due to osteoporosis.[10,11]

Cardiovascular Health

Estrogens affect lipid metabolism. With the decrease of estrogen levels there is decrease in high density lipoprotein (HDL) cholesterol, as well as an elevation of low-density lipoprotein (LDL) cholesterol and total cholesterol.

Most of the perimenopausal symptoms such as VMS and mood tend to improve with time. Ongoing targeted treatment is required for improvement in vaginal and urinary symptoms.

EVALUATION AND MANAGEMENT

About 4–6 years before cessation of menses, middle-aged women experience a progressive change in ovarian activity and a physiologic deterioration of hypothalamic-pituitary-ovarian axis function associated with fluctuating hormone levels of sexual hormones, estradiol, progesterone, and also androgens.[12]

During this transition, women face symptoms related to menopause (such as hot flushes, sleep disturbance, mood changes, memory complaints, and vaginal dryness). Neurological symptoms such as sleep disturbance, "brain fog" and mood changes are a major complaint of women transitioning menopause, with a significant impact on their quality of life, productivity, and physical health **(Flowchart 1)**.

Prior to attributing sleep problems to estrogen deficiency, sleep habits and presence of problems should be routinely assessed in clinic visits.

Depression, which can cause early morning awakening, restless leg syndrome, and sleep apnea should be ruled out.

The impact of the menopausal transition on mood is strongly related to quality of life.

At the time assessment prior history of major depression, medication use, VMS, and very upsetting life events (as loss of the partner or close family member) should be taken.

- *Cognitive function and cognitive disorders:* Despite the deep connection between estrogen and cognition, data regarding the relationship between hormone replacement therapy and the neuroprotective outcomes still remain conflicting. Studies have not shown any cognitive benefits of estrogen or combined estrogen-progestin therapy in women over the age of 65 years without underlying dementia.
- Young menopausal women without contraindication to menopausal hormone therapy (MHT) and with

Flowchart 1: Approach to evaluation and management of menopausal symptoms.

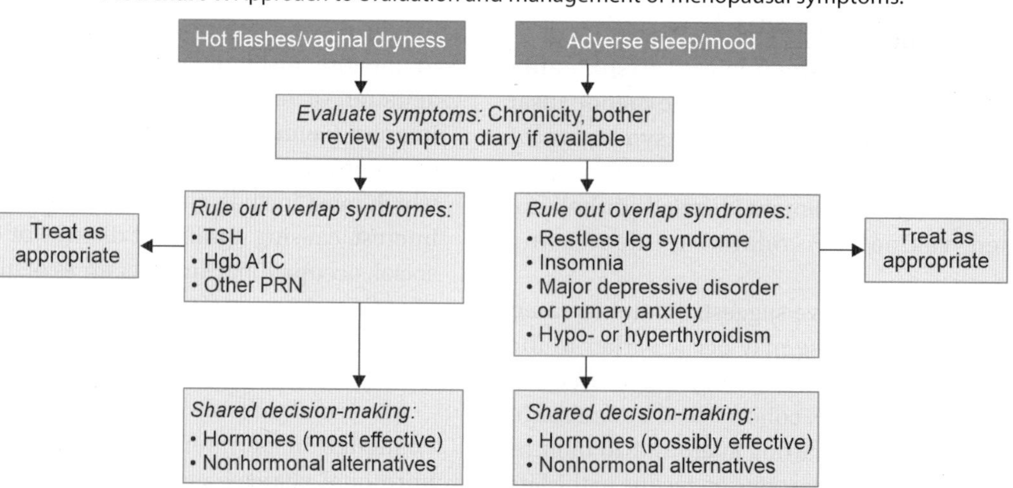

(Hgb A1C: hemoglobin A1C; PRN: pro re nata; TSH: thyroid-stimulating hormone)

impaired quality of life because of night sweats, VMS, or disrupted sleep can benefit from MHT.

- *Sleep disturbances:* MHT can improve sleep, reducing night sweats. The GABAergic sedating effects of progesterone can be considered in women with sleep issues.[13]
- *Mood and depressive symptoms:* Some data support a potential beneficial effect of MHT on mood. Estrogens can be considered in menopausal women with VMS as they may increase the response to anti-depressants.

Nonhormonal Management

Tender loving care, counseling, and psychological support are the key to management.

- *Selective serotonin receptor inhibitors (SSRIs):* The SSRIs are the class of antidepressants used most commonly for treating hot flashes.

  Paroxetine (Brisdelle) is an SSRI and is the only nonhormonal therapy that is specifically approved. Mirtazapine (Remeron) can be given to anxious and depressed patients. Common side-effects are drowsiness and weight gain. Clonazepam or etizolam can be given for sleep disturbance.
- *Gabapentin (Neurontin):* Gabapentin is approved to treat seizures, but it has also been shown to help reduce hot flashes. This drug is useful for women who cannot use estrogen therapy for health reasons and for those who also have migraines.
- *Fezolinetant (Veozah):* This medicine is a hormone-free option for treating menopause hot flashes. It works by blocking a pathway in the brain that helps regulate body temperature (central neural circuit).

Symptoms are grouped into two main categories, from among the four symptom sets that are most likely related to the menopausal transition.[11] Few symptoms mimic VMS, but because of their prevalence, ruling out thyroid disease and type 2 diabetes is prudent.

Vaginal dryness can sometimes be due to conditions other than menopause, e.g., lichen sclerosus et atrophicus. A physical examination is recommended.

Restless leg syndrome and chronic insomnia often overlap with menopause-related sleep concerns, and treatment will be most effective if both problems are addressed.

Primary mood disorders that are long-standing and severe are less likely to respond to hormone therapy (HT) alone. A thorough history will help the clinician determine whether single-agent therapy with hormones is indicated, or if co-management with a mental health professional may be more effective.

Hormones are likely the most efficient path to symptom control for women with VMS or vaginal dryness. In vaginal dryness topical hormones may be sufficient.

Women with mood and sleep issues may benefit from a trial of hormones. If they do not sufficiently improve, additional, non-hormonal options or co-management should be considered.[14]

Bleeding Problems

Bleeding problems can be diagnosed by history and physical examination followed by appropriate laboratory and imaging tests **(Flowchart 2)**.

A thorough menstrual, medication, surgical, and family history should be taken.

Examination should include BMI and thyroid gland. For uterus and cervix both speculum and bimanual assessment is mandatory.

Laboratory testing should include complete blood count, iron studies, thyroid assessment, liver function test, and evaluation for coagulation disorders.

Transvaginal ultrasonography is the imaging modality of choice to identify uterine and endometrial structural abnormalities, although further assessment with saline infusion sonohysterography (SIS) or office hysteroscopy may be needed for evaluation of intracavitary lesions.[10]

Endometrial biopsy is indicated for persistent abnormal uterine bleeding, in perimenopausal women with risk factors, including chronic anovulation and obesity.

Flowchart 2: International Federation of Gynecology and Obstetrics (FIGO) classification of abnormal uterine bleeding.

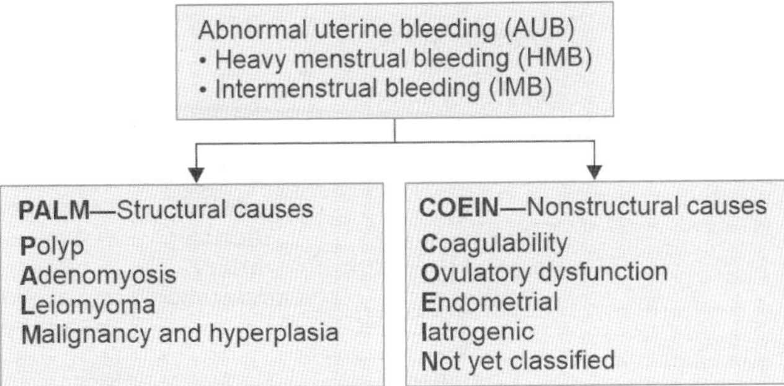

Source: Adapted from Munro MG, Critchley HO, Broder MS, Fraser IS; FIGO Working Group on Menstrual Disorders. FIGO classification system (PALM-COEIN) for causes of abnormal uterine bleeding in nongravid women of reproductive age. Int J Gynaecol Obstet. 2011;113(1):3-13.

Endometrial biopsy is very sensitive for detection of global endometrial pathology.

Hysteroscopy-guided endometrial biopsy is the gold standard.

For management of abnormal uterine bleeding **(Box 2)** in perimenopausal patients, appropriate evaluation, need for contraception, and management options available should be considered.

As shown in **Flowchart 3** management should be targeted toward underlying pathology when encountered. Otherwise, decisions should incorporate a patient-centered approach after discussion of the patient's goals. It is important to note that, when workup has ruled out pathology necessitating medical intervention, reassurance can be provided to the patient and it is reasonable to pursue expectant management. Additionally, bleeding abnormalities may be multifactorial; anovulatory bleeding concurrent with structural abnormalities is less likely to resolve after addressing the structural component alone. When symptoms are bothersome to the patient, treatment can range from medical to surgical management, and

BOX 2: Management of abnormal uterine bleeding.

- Expectant
- Non-hormonal medical treatments
  - Non-steroidal anti-inflammatory drugs
  - Anti-fibrinolytic agents
- Hormonal medical treatments
  - Cyclical or long-acting progestogens
  - Combined oral contraceptives
  - Gonadotropin-releasing hormone (GnRH) analogs (fibroid-associated)
  - Selective progesterone receptor modulators (fibroid-associated)
- Levonorgestrel intrauterine system (LNG-IUS)
- Endometrial ablation
- Uterine artery embolization (fibroid-associated)
- Hysterectomy

Flowchart 3: Approach to management of abnormal uterine bleeding in perimenopausal.

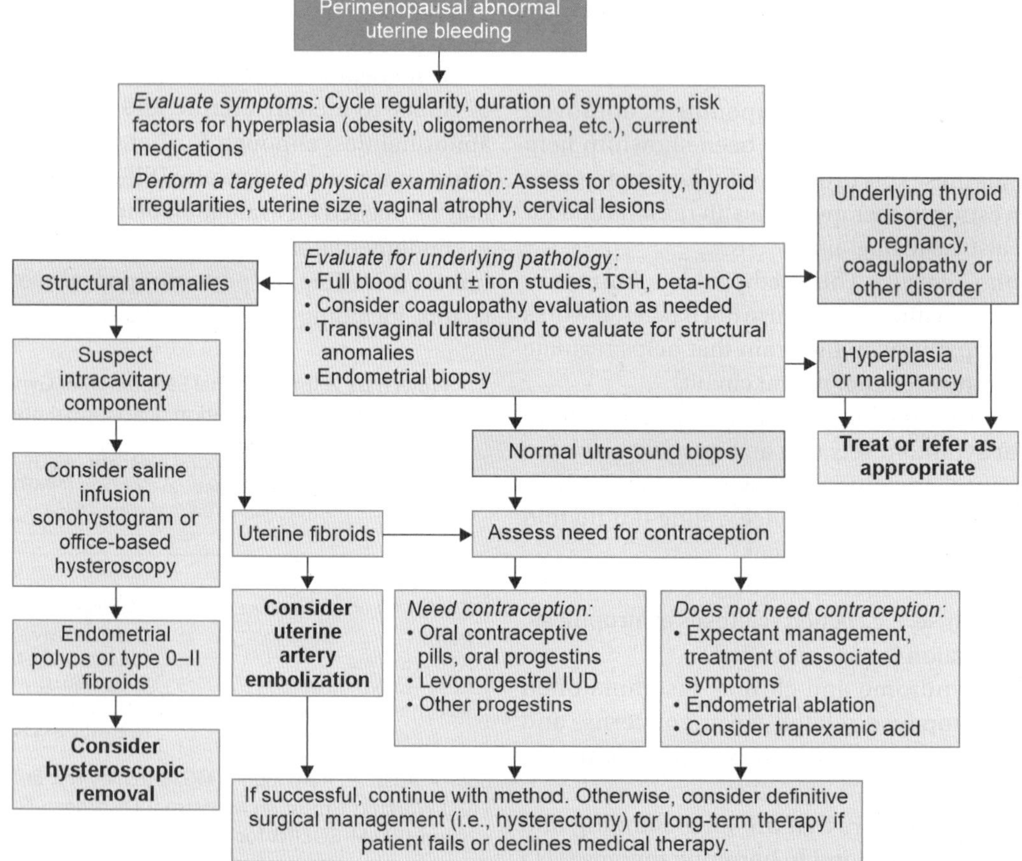

(hCG: human chorionic gonadotropin; IUD: intrauterine device; TSH: thyroid-stimulating hormone) definitive surgical management (i.e., a hysterectomy) can be considered for patients who have failed or declined other methods.

TREATMENT OF PERIMENOPAUSAL BLEEDING

Therapy must always be directed by proper diagnosis. Pregnancy and malignancy should be excluded. Treatment goals for patients with AUB include regulation of menstrual cycles, minimization of blood loss, and improvement in quality of life. For some, the knowledge that there is no serious problem and, especially for patients with dysfunctional anovulatory bleeding, is a normal part of the transition from the reproductive to the non-reproductive years.

Medications

- Hormonal contraceptives (oral, patch, or rings) can be given to perimenopausal women for irregular menstruation.
- Women at high risk of cardiovascular disease as myocardial infarction or stroke should be suggested progestin containing intrauterine device.
- Tranexamic acid offers a non-hormonal mechanism.
- Evaluation and treatment of symptoms should be personalized. Treatment should be based on type and severity of symptoms and also contraceptive need at the time of menopause transition should be considered.

The most commonly selected first-line choice for AUB treatment is combined oral contraceptives (COCs). The levonorgestrel intrauterine system (LNG-IUS) is the next most frequently selected option. If there is "breakthrough bleeding" with *COCs* (combined oral contraceptives), adjustment of the dose can be done or pill type can be changed. If persistent bleeding is there further diagnostic evaluation for coexisting organic pathology should be looked for. Those women with a contraindication to estrogen therapy, progesterone therapy or the *LNG-IUS* can be an alternative. Levonorgestrel causes thinning of endometrium so, the duration and amount of menstrual bleeding is reduced. This effect may begin during the first menstrual cycle after the placement of the device, and bleeding becomes progressively less over time.

Prostaglandins have a role in aberrant neovascularization leading to dysfunctional uterine bleeding. Non-steroidal anti-inflammatory drugs *(NSAIDs)* reduce prostaglandin synthesis.

Tranexamic acid reversibly blocks lysine-binding sites on plasminogen, preventing plasmin and fibrin polymer interaction and resulting in fibrin degradation, the stabilization of clots, and reduction of bleeding. It is contraindicated in women with increased risk for thromboembolism.

- *Non-medical approaches to abnormal uterine bleeding endometrial ablation:* Endometrial ablation is a minimally invasive alternative for the treatment of HMB.

  It is usually reserved for patient's refractory to medical treatment and who wish to avoid definitive treatment with hysterectomy. Although endometrial ablation is a less invasive surgical alternative to hysterectomy, it does not eliminate surgical risk and is followed by further surgery within 4 years in up to 38% of the women.
- *Radiological treatment: uterine artery embolization:* Uterine artery embolization (UAE), performed by an appropriately trained interventional radiologist, is a minimally invasive treatment option for uterine fibroids in perimenopausal women.

  Hysterectomy is the established surgical treatment for women who no longer plan to give birth and have completed their family.

  Myomectomy is a surgical procedure to remove uterine fibroids and reconstruct the uterus. It is used as a fertility-sparing option who want further pregnancy.

Endometrial Polyps

Endometrial polyps are usually removed as their malignant potential is uncertain and they may cause irregular bleeding. Removal can be done under a local anesthetic with hysteroscopic guidance.

Endometrial Hyperplasia of Low Malignant Potential

Endometrial hyperplasia of low malignant potential, (simple or complex hyperplasia without atypia), may be handled with conservative management. Endometrial resampling is done to check regression.

Progestogens, either orally or by LNG-IUS, are the treatment of choice.

Atypical Hyperplasia

Cases of atypical hyperplasia are usually treated with total hysterectomy and bilateral salpingo-oophorectomy.

When fertility-sparing required high-dose progestogens or LNG-IUS can be tried.

Gynecologic Malignancies Resulting in Abnormal Uterine Bleeding

This would need a detailed assessment of the patients and help of a gynecologist.

GENERAL MANAGEMENT

Menopause significantly increases the risk of cardiometabolic diseases, such as obesity, type 2 diabetes, cardiovascular diseases, non-alcoholic fatty liver disease (NAFLD)/metabolic associated fatty liver disease (MFFLD), and metabolic syndrome (MetS).

Management strategies for perimenopausal women include lifestyle modifications (physical activity and calorie-controlled diet), pharmacotherapy, bariatric surgery, and traditional health practices (yoga and meditation).

Diet

Traditional healthy diets, such as more intake of seafood, seaweeds, grains, dairy products, fresh vegetables, and fruits, and less consumption of fast foods, animal fats, sweets, and fried foods are associated with a reduced risk of MetS.

An energy-restricted diet combined with physical activity uses the energy stored in adipose tissue and results in gradual decreases in body weight.

Absence of high-energy refined grains was associated with increased probability of normal fasting blood glucose. And low-fat diet was related to reduced diastolic blood pressure and increased HDL-C levels.

These results suggest that diet is important in prevention of MetS in postmenopausal women.

In addition, nutrition, and supplementation, such as foods rich in vitamin D, Omega-3 fatty acids, antioxidants, phytochemicals, and probiotics are recommended for MetS, but further research is needed.

Medicines

Four medications approved for long-term use by the Food and Drug Administration (FDA) for the treatment of obesity—orlistat, phentermine/topiramate, naltrexone/bupropion, and liraglutide.

All in all climacteric or menopause transition is the mirror image of adolescence, which is the coming on to the reproductive years while perimenopause is the coming off from the reproductive years. It should be dealt with cautiously and delicately.

CONCLUSION

Perimenopause/menopause transition/climacteric is a period of change in ovarian functions leading a woman from fertile to infertile state. The fluctuations in the female hormones which occur during this period are responsible for the VMS and abnormal uterine bleeding. Mood changes, sleep disturbances, and sexual dysfunctions are also witnessed in this period.

Period of menopause transition can be divided into early or late. In early perimenopause there is variable cycle length and in late perimenopause the period of amenorrhea is >60 days. Obese women have less frequent climacteric symptoms. There is lower rise of FSH and less steep decline of estradiol. Hormone (Estrogen) therapy can be considered in women with VMS. Some data support a beneficial effect of MHT in mood changes.

Counseling and psychological support also have a good impact on the perimenopausal women with VMS, mood changes, and sleep disorders. Meditation and yoga help the women in a big way.

Bleeding disorders should be dealt carefully. History, physical examination, laboratory tests, and imaging should be done. Screening for breast cancer and cervical cancer is mandatory and in doubtful cases biopsy is necessary before putting the patient on MHT.

The perimenopausal women need special care as this period brings lot of physical, emotional, social, and psychological changes. They should be evaluated meticulously and treated carefully and tenderly.

REFERENCES

1. Delamater L, Santoro N. Management of the Perimenopause. Clin Obstet Gynecol. 2018;61(3):419-32.
2. van den Berg MJ, Herber-Gast GC, van der Schouw YT. Is an unfavourable cardiovascular risk profile a risk factor for vasomotor menopausal symptoms? Results of a population-based cohort study. BJOG. 2015;122:1252-8.
3. Szmuilowicz ED, Manson JE. Menopausal vasomotor symptoms and cardiovascular disease. Menopause. 2011;18:345-7.
4. Kravitz HM, Joffe H. Sleep during the perimenopause: a SWAN story. Obstet Gynecol Clin North Am. 2011;38: 567-86.
5. Bromberger JT, Kravitz HM. Mood and menopause: findings from the Study of Women's Health Across the

Nation (SWAN) over 10 years. Obstet Gynecol Clin North Am. 2011;38:609-25.
6. Cohen LS, Soares CN, Vitonis AF, Otto MW, Harlow BL. Risk for new onset of depression during the menopausal transition: the Harvard study of moods and cycles. Arch Gen Psychiatry. 2006;63:385-90.
7. Freeman EW, Sammel MD, Lin H, Nelson DB. Associations of hormones and menopausal status with depressed mood in women with no history of depression. Arch Gen Psychiatry. 2006;63:375-82.
8. First MB. Diagnostic and statistical manual of mental disorders, 5th edition, and clinical utility. J Nerv Ment Dis. 2013;201:727-9.
9. Munro MG, Critchley HO, Broder MS, Fraser IS, FIGO Working Group on Menstrual Disorders. FIGO classification system (PALM-COEIN) for causes of abnormal uterine bleeding in nongravid women of reproductive age. Int J Gynaecol Obstet. 2011;113:3-13.
10. Goldstein SR, Lumsden MA. Abnormal uterine bleeding in perimenopause. Climacteric. 2017;20:414-20.
11. Sherman S. Defining the menopausal transition. Am J Med. 2005;118(Suppl 12B):3-7.
12. Goldstein SR, Lumsden MA. Abnormal uterine bleeding in perimenopause. Climacteric. 2017;20(5):414-20.
13. Gava G, Orsili I, Alvisi S, Mancini I, Seracchioli R, Meriggiola MC. Cognition, mood and sleep in menopausal transition: the role of menopause hormone therapy. Medicina (Kaunas). 2019;55(10):668.
14. Jeong HG, Park H. Metabolic disorders in menopause. Metabolites. 2022;12(10):954.

Chapter 47

Premature Ovarian Failure

Rimmy Singla

INTRODUCTION

Premature ovarian failure (POF) is defined as a primary ovarian defect characterized by absent menarche (primary amenorrhea) or premature depletion of ovarian follicles/ arrested folliculogenesis before the age of 40 years (secondary amenorrhea).

PREMATURE OVARIAN FAILURE OR PRIMARY OVARIAN INSUFFICIENCY[1]

Premature ovarian failure is characterized by:
Amenorrhea >4 months

- Two consecutive follicle-stimulating hormone (FSH) level >30 mIU/mL in an interval of 1 month
- Age <40 years
- Usually permanent but ovarian activity can resume, and fertility reversed in 5–10% of cases.

Premature ovarian affects approximately 1% of women. POF accounts for 10–28% of women with primary amenorrhea and 4–18% of women with secondary amenorrhea. This heterogeneous disorder occurs from a variety of etiologies and it is diagnosed on the basis of elevated levels of serum FSH of generally >40 IU/L, which is detected on two occasions at least 4 weeks apart. It has been reported that 5–10% of women with POF have spontaneous ovulation, which leads to pregnancy. Four possible causes of premature menopause had been published by De Monaes Ruehsen and Tones in 1967:

1. Reduced number of germ cells at the time of birth
2. Acceleration of the normally constant follicular atresia processes
3. Postnatal destruction of germ cells in the ovaries
4. Functional ovarian failure.

The familial form of POF is rare; however, 4–31% of all cases are POF. In present scenario, in one case, a woman depicted with secondary amenorrhea and her progesterone withdrawal test was negative. She was getting her withdrawal bleed for the last 1 year only after taking a combination of estrogens and progesterone for 21 days. She was diagnosed to have hypothyroidism. Her serum antibodies for thyroid peroxidase was positive. Her FSH and luteinizing hormone (LH) levels were repeated on two occasions. They were found to be high which shows primary ovarian failure. When we tested for other hormones, they were within normal range and along with this her anti-Mullerian hormone (AMH) levels were also checked to prognosticate treatment result. Considering these factors, we advised the couple to go on with dual stimulation with self-gametes. The couple was well-informed about the donor oocyte program as well but they were eager on self-gametes and accepted treatment.

In these patients, in vitro fertilization (IVF) with donor egg was achieved to treatment option. Advanced techniques of cryopreserved ovarian tissue in transplantation as well as oocytes in vitro maturation, which were obtained from stem cells, made it feasible for few patients with primary ovarian insufficiency (POI) to utilize their oocyte for IVF.

Epidemiology

Premature ovarian failure affects approximately:

- One in 10,000 women by age 20 years
- One in 1,000 women by age 30 years
- One in 100 women by age 40 years.

The familial form of POF is rare, representing 4–31% of all cases of POF. The cause of POF still remains undetermined in majority of the cases (idiopathic). This idiopathic form of POF can show sporadic and familial forms.

Causes (Fig. 1)

- Induced (iatrogenic) POF/POI
- *Spontaneous POF/POI:* Cause usually unknown

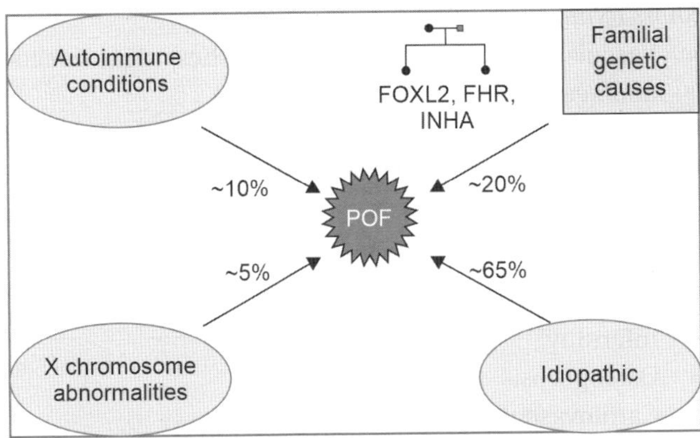

Fig. 1: Causes of premature ovarian failure (POF).

Two mechanisms:

1. *Follicular depletion:* Decreased germ cell migration, accelerated atretic process, acquired ovarian disease
2. Follicular dysfunction

The different causes of POF are given as follows:

- *Genetic POF:*
  - X chromosome disorders: Turner's
  - Genetic and cytogenetic causes: Various microdeletions
  - Fragile mental retardation 1 (FMRI) gene premutation: POF seen in 10–20% of carriers
  - LH (LHR-recessive)/FSH receptor mutations
  - POF in 46, XX females—gonadal dysgenesis and multiple malformation syndromes
- *Mutation in various enzymes involved in steroidogenesis:*
  - Steroidogenic enzyme defects-17a-hydroxylase, aromatase, 17,20-desmolase enzymes deficiency
  - Gonadotropin receptor antibodies
  - *Signal defects:* Defects in the gonadotropin structure and effect; secretion of biologically inactive gonadotropins; alpha- and beta chain defects; gonadotropin and postreceptor defects; circulating FSH-binding inhibitors
  - Abnormal gonadotropin receptor
  - Abnormality in the G-protein signaling pathway
- *Possible autoimmune causes:*
  - Acquired hemolytic and pernicious anemia
  - Asthma, chronic active hepatitis
  - Crohn's disease
  - Diabetes mellitus
  - Glomerulonephritis
  - Addison's disease
  - Hypoparathyroidism
  - Hypophysitis
  - Idiopathic thrombocytopenic purpura
  - Juvenile rheumatoid arthritis
  - Keratoconjunctivitis and Sjögren's syndrome
  - Malabsorption syndrome
  - Myasthenia gravis polyendocrinopathy
  - Primary biliary cirrhosis
  - Rheumatoid arthritis
  - Systemic lupus erythematosus (SLE)
  - Thyroid disorders, Grave's disease, and thyroiditis
  - Vitiligo
  - Alopecia
  - Perrault's syndrome
  - FSH receptor mutations/LH receptor mutations
- Idiopathic (resistant ovary syndrome)
- Infections (e.g., herpes zoster, cytomegalovirus): Lymphocytic oophoritis
- Iatrogenic—radiotherapy, chemotherapy—effect is dependent on age and dose, smoking, and environmental toxins.
- Approximately 20% of adults with idiopathic POI will experience hypothyroidism, most commonly Hashimoto thyroiditis. Following initial diagnosis of POI, it is appropriate to test thyrotropin levels for the presence of thyroid peroxidase antibodies.
- Although no recommendations for routine thyroid screening exist in this population, given the high prevalence of this disorder in patients with POI, it is acceptable to test for thyroid disease every 1–2 years. Patients with POI also have a 50% chance of developing adrenal insufficiency if they have adrenal autoimmunity. Patients should be tested for adrenal antibodies and if results are positive, they should undergo yearly corticotropin stimulation testing.
- Data are lacking on the follow-up of patients with negative test results. Diabetes mellitus, pernicious anemia, myasthenia gravis, rheumatoid arthritis, SLE, and dry eye syndrome also have been associated with POI, and testing should be based on symptomatology.
- The evaluation of other peptide factors of ovarian origin, such as inhibin B and AMH, may be useful to determine the follicular reserve when POF is suspected. Low levels of inhibin B may predict follicular depletion before the large FSH rise in amenorrhea.
- Screening for 21OH-Ab [or alternatively adrenocortical antibodies (ACA)] should be considered in women with POI of unknown cause or if an immune disorder is suspected.

- Refer POI patients with a positive 21OH-Ab/ACA test to an endocrinologist for testing of adrenal function and to rule out Addison's disease.
- Screening for thyroid (TPO-Ab) antibodies should be performed in women with POI of unknown cause or if an immune disorder is suspected.
- In patients with a positive TPO-Ab test, thyroid-stimulating hormone (TSH) should be measured every year. If 21OH-Ab/ACA and TPO-Ab are negative in women with POI, there is no indication for retesting later in life, unless signs or symptoms of these endocrine diseases develop.

Symptomatology and History-taking (Fig. 2)

- Last spontaneous menstrual cycle
- Prior pelvic surgeries, irradiation, or chemotherapy
- Symptoms of adrenal insufficiency
- Orthostatic hypotension
- Skin hyperpigmentation
- Unexplained weakness
- Salt craving
- Abdominal pain
- Anorexia
- Symptoms of hypothyroidism
- Family history of POF, male mental retardation, autoimmune disorders
- Symptoms of estrogen deprivation

Short-term:
- Vascular symptoms such as hot flushes, night sweats, and headaches
- Vaginal dryness, dyspareunia
- Urgency and stress urinary incontinence
- Irritability, forgetfulness
- Poor concentration, insomnia

Long-term:
- Infertility
- Osteoporosis
- Cardiovascular disease stroke
- *Psychological impact:* Depression.

Diagnosis of POF (Flowchart 1)

- Due to fluctuating ovarian function, symptoms may intermittently disappear, and it does not exclude a diagnosis of POF because some may experience sudden severe symptoms upon cessation of the contraceptive pill or some women with POF may not experience any symptoms. Young women with primary amenorrhea have less symptoms.
- Turner's syndrome is most common cause of POF due to this karyotype is always advised to rule out a Y cell line. When diagnosis is confirmed and Y chromosome identified, then patient should be counseled regarding gonadal removal because these patients have an increased potential for malignancy in the gonads.

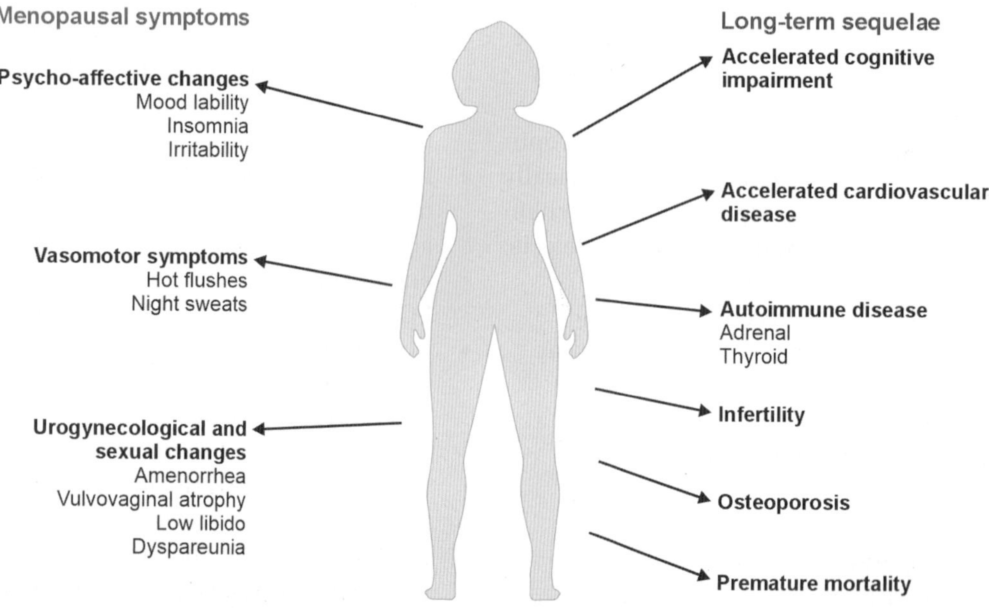

Fig. 2: Symptoms of premature ovarian failure (POF).

Flowchart 1: Diagnosis of premature ovarian failure (POF).

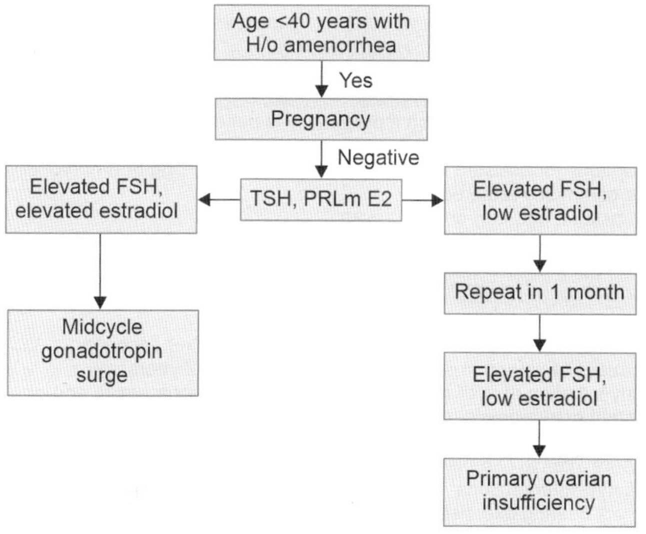

Individuals who had history of similar onset of early menopause and dementia in the family, FMR1 premutations also need to be checked. To exclude secondary causes of amenorrhea, it is advisable to perform a complete history and physical examination. POI and POF also confirm by pelvic ultrasonography with small uterus and ovaries.

Mandatory investigations:

- Chromosomal karyotype analysis
- Screening for thyroid antibody
- Pelvic ultrasound
- Routine screening for diabetes
- AMH level for fertility management.

Desirable investigations:

- *FMR1*: Permutation testing
- Screening for infections
- Bone mineral density (BMD)
- Autoimmune tests
- *Screening for 21:* Hydroxylase antibodies.

Follow-up: POF patients are always advised to be followed with 6–12-month intervals with periodic testing of TSH and calcium. If anti-adrenal antibodies are present, then ACTH stimulation test is suggested to assess adrenal reserve or referral to an endocrinologist.

Management[2]

As POF has cumulative negative effects over time, it is important for clinicians to make a timely diagnosis and begin appropriate strategies for:

- Education, counseling, and psychological support
- Prevention and treatment of estrogen deficiency with 17beta estradiol plus oral dydrogesterone
- *Maintain bone health:* Diet, exercise, calcium, vitamin D
- *Contraception:* Oral contraceptives have an occasional chance of causing fertility, though the chances are very low; hormonal replacement is an added advantage.
- *Fertility preservation:* Ovarian stimulation with oocyte retrieval followed by oocyte or embryo cryopreservation, ovarian tissue cryopreservation (experimental)
- Ovarian suppression with gonadotropin-releasing hormone (GnRH) prior to chemotherapy
- IVF with donor oocytes.

Hormonal Therapy[3]

Hormone replacement therapy (HRT) is the approved management for POI patients. Estrogens replacement is advised to decrease the risk of osteoporosis and cardiovascular disease.

- *Local estrogen:*
  - Estrogen therapy: 17-beta estradiol preferred to be conjugated equine estrogens (CEE) or ethinylestradiol (EE)
  - Combined therapy: 17-beta estradiol plus oral dydrogesterone/progesterone/levonorgestrel IUS (intrauterine system)
- Estrogen till natural age of menopause for bone health, local symptoms, and cardioprotection (no breaks)
- If intact uterus, progesterone for 12–14 days a month
- Alternatively, continuous combined preparation (better for endometrial protection)
- 17beta estradiol preferred to estrogens (CEE) or EE dye to lessen liver load
- Estrogen does not increase the risk of breast cancer before menopause.

Diet

- Always educate POF patients regarding their diet, calcium intake, vitamin D intake, and activity to improve their condition.
- According to North American Menopause Society, per day intake of 1,200 mg of elemental calcium maintenance of adequate vitamin D status.

Yoga/Meditation Activity

- Females with POI are always motivated to do different kinds of exercises.

Flowchart 2: Premature ovarian insufficiency (POI) management guidelines.[4,5]

- Inform in a sensitive and supportive manner
- Offer psychological support, refer to patient support groups, or educational resources Hormone replacement therapy (*see* below)
- Address symptoms related to menopause
- Treat sexual dysfunction
- Management of long-term consequences (*see* adjacent)

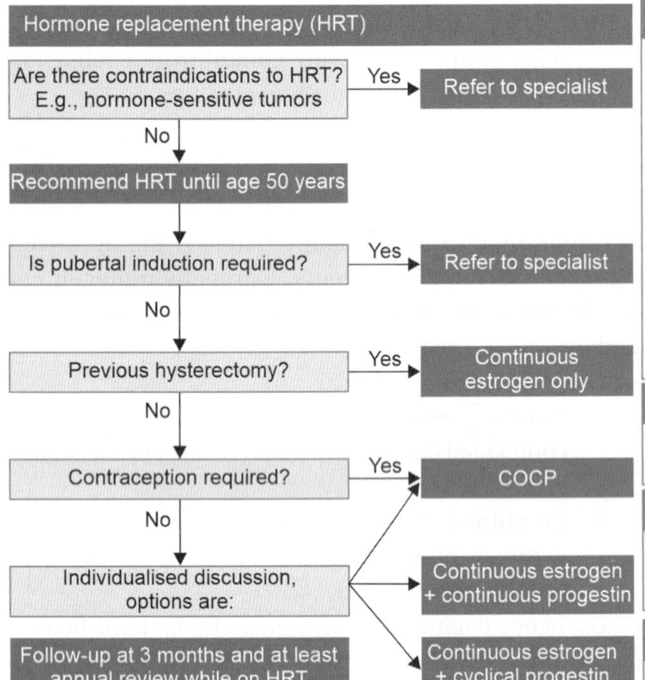

Management of long-term health consequences

1. Infertility
- Recommend contraception if pregnancy is not desired
- Referral to fertility specialist for counseling regarding:
  - Options for assisted reproduction therapy (donor egg or embryo is commonly required)
  - Obstetric and non-obstetric risks
  - Fitness for pregnancy

2. Osteoporosis
- Screen for osteoporosis with DXA scan
  - If initial DXA scan abnormal: Screen and treat other secondary causes of osteoporosis, repeat DXA scan every 2 years
- Early initiation of HRT recommended to maintain bone health and prevent osteoporosis (commonly require higher doses of HRT to maintain bone)
- Ensure vitamin D repletes
- Lifestyle interventions: ≥3 serves of dairy daily, encourage weight-bearing exercises, smoking cessation, minimize alcohol intake
- Refer to specialist if BMD declining despite adequate HRT
  - Initiation of osteoporosis specific treatment (e.g., bisphosphonates) requires specialist review in this young adult population
  - Caution with the use of osteoporosis specific agents in women desiring pregnancy

3. Cardiovascular disease
- Screen and treat cardiovascular risk factors
- Early initiation of HRT may reduce future risk of cardiovascular disease

4. Autoimmune conditions
- Annual TSH
- If baseline anti-adrenal antibodies or celiac serology abnormal→refer to specialist for further evaluation and management

5. Address specific causes of POI

E.g., tumer syndrome, breast cancer

(COCP: combined oral contraceptive pill; DXA: dual energy X-ray absorptiometry; HRT: hormone replacement therapy; TSH: thyroid-stimulating hormone)

- Weight-bearing exercises for 30 minutes per day, at least 3 days per week
- Outdoor sports.

CONCLUSION

Premature ovarian failure is a challenging issue as women are delaying having families, and this emotionally distressing problem must be dealt, on both the physical and psychological platform. POF's incidence rate is increasing secondary to improved survival of children with cancer, radiotherapy, endocrine disrupting chemicals, and many more. Variety of drug interventions such as different doses of steroids, estrogens, clomiphene, different dosage of gonadotropin, recombinant FSH (rFSH), and danazol are suggested to increase ovulation in females with POI. Only some random controlled trials are present which failed to define any development in rates of ovulation and pregnancy. Latest dual stimulation has been proved benefit for poor responders. Some other open corridors for these women are donor oocyte IVF, stem cell treatment, ovarian tissue cryopreservation, and mitochondrial replacement **(Flowchart 2)** guidelines.

REFERENCES

1. Nguyen HH, Milat F, Vincent A. Premature ovarian insufficiency in general practice: Meeting the needs of women. Available from: https://www.racgp.org.au/afp/2017/june/premature-ovarian-insufficiency-in-general-practice-meeting-the-needs-of-women/ [Last accessed April, 2024].
2. European Society for Human Reproduction and Embryology (ESHRE) Guideline Group on POI; Webber L, Davies M, Anderson R, Bartlett J, Braat D, Cartwright B, et al.

ESHRE guideline: Management of women with premature ovarian insufficiency. Hum Reprod. 2016;31(5):926-37.
3. Torrealday S, Kodaman P, Pal L. Premature Ovarian Insufficiency - an update on recent advances in understanding and management. F1000Res. 2017;6:2069.
4. European Society of Human Reproduction and Embryology. (2015). Management of women with premature ovarian insufficiency. Guideline of the European Society of Human Reproduction and Embryology. Available from: https://www.eshre.eu/Guidelines-and-Legal/Guidelines/Management-of-premature-ovarian-insufficiency.aspx [Last accessed April, 2024].
5. The American College of Obstetricians and Gynecologists. (2014). Primary Ovarian Insufficiency in Adolescents and Young Women. Available from: https://www.acog.org/clinical/clinical-guidance/committee-opinion/articles/2014/07/primary-ovarian-insufficiency-in-adolescents-and-young-women [Last accessed April, 2024].

Chapter 48

Menopause: Postmenopausal Bleeding

Yashodhara Pradeep, Ayush C Srivastava

INRODUCTION

Menopause is a retrospective diagnosis following 12 months of amenorrhea. For Indian women, the mean age at menopause reported in different studies appears to be rather young, between 41.9 and 49.4 years[1] and the average age of menopause is 46.6 years.[1]

DEFINITION

Postmenopausal bleeding (PMB) refers to any uterine bleeding in a woman who attained menopause.

- PMB is a cardinal sign for endometrial cancer.
- All menopausal women with PMB; after 6 months of amenorrhea should be evaluated for early detection of endometrial carcinoma.
- Any abnormal uterine bleeding (AUB) during perimenopausal transition should be evaluated to rule out endometrial hyperplasia (EH)/polyp/cancer.

Incidence of PMB: PMB accounts for 5% among all women attending gynecology outpatient clinic. It occurs in 4–11% of postmenopausal women (PMW); the incidence of PMB is higher after first year of menopause versus after 3 years of menopause.[2,3]

ETIOLOGY

The differential diagnosis of bleeding in postmenopausal patients is less broad than that for abnormal bleeding in premenopausal patients since the various causes of anovulation are not relevant.[4]

Based on the endometrial pathology, the prevalence is given in **Box 1**.

Structural abnormalities for PMB are mentioned in **Box 2**.

Other causes of PMB are listed in **Box 3**.

Box 4 lists the medications and other causes for PMB.

BOX 1: Etiology based on endometrial pathology.

• Polyp	37%
• Hypotrophy atrophy	30.8%
• Proliferative and secretory	14.5%
• Carcinoma	6.6%
• Fibroid	6.2%
• Hyperplasia without atypia	2%
• Hyperplasia with atypia	0.2%

BOX 2: Structural abnormalities for postmenopausal bleeding.

- Polyp benign/malignant
- Atrophy—hypoestrogenic state
- Leiomyoma uterus—more common in premenopausal women, after menopause the size of fibroid decreases. Although rare the diagnosis of uterine sarcoma should be considered in the event of size of fibroid start increasing and the symptoms of (pelvic pressure, pain, and bleeding)
- Adenomyosis—confirmed with haptoglobin (H/P)
- Endometrial hyperplasia **(Table 1)**
- Endometrial carcinoma **(Box 5)**
- Choriocarcinoma

BOX 3: Other causes of postmenopausal bleeding.

- Cervical pathology benign/malignant
- Forgotten IUCDs
- *Upper genital tract lesions:* PID, fallopian tube cancer, ovarian cancer
- *Disease of adjacent organs:* Diverticulitis
- Post-radiation therapy
- *Infection:* Endometritis

PID: Less common in postmenopausal women (PMW)

- Trauma
- *Systemic disease:* Renal, lymphoma, acute leukemia

(IUCDs: intrauterine contraceptive devices; PID: pelvic inflammatory disease)

ENDOMETRIAL HYPERPLASIA

Endometrial hyperplasia is a proliferation of endometrial glands. The EH is more common in perimenopausal and early postmenopausal patients.

There are different terminologies used from time to time. The most recent is WHO (World Health Organization) classification 2014.[5]

Endometrial hyperplasia without atypia or benign endometrial hyperplasia (BEH) in EIN (endometrial intraepithelial neoplasia) 2000. EH with atypia or EIN is a premalignant lesion **(Table 1)**.

Normal Endometrium

During menstrual cycle, there are dynamic changes in endometrium under the influence of estrogen, follicular growth, (proliferative endometrium gland to stroma ratio <2:1 ovulation followed by pregestational and luteal phase (secretary phase gland to stroma ratio is >2:1).

BOX 4: Medications and other causes for postmenopausal bleeding (PMB).

- Postmenopausal hormone therapy
- Benign/malignant
- Anticoagulant
- Antiplatelet agents/NSAIDs
- Tamoxifen
- Corticosteroids
- Chemotherapy
- Antibiotics
- Antipsychotic
- SSRIs
- Alcohol
- Vitamin E
- Garlic
- *Ginkgo biloba*

(NSAIDs: nonsteroidal anti-inflammatory drugs; SSRIs: selective serotonin reuptake inhibitors)

Types of Endometrial Hyperplasia

Endometrial hyperplasia without atypia or BEH: The gland to stroma ratio is increased >2:1, glands are mildly crowed dilated luminal outpouching, the feature of nuclear atypia is not present. The incidence is highest in early postmenopausal patients 142–213 per 100,000-woman years.

Endometrial hyperplasia with atypia (EIN): There is increase in gland to stromal ratio, disorganization of glands with luminal outpouching, cellular mitosis, and features of nuclear atypia is present. The incidence is highest in patients with >10 years of menopause 56/100,000-woman years.

The EIN 2000[6] classification is based on D-Scoring, computerized morphology to reduce the confusion present in terminology used in 1994—the simple hyperplasia; complex hyperplasia; simple hyperplasia with or without atypia; and complex EH with or without atypia.

The BEH has replaced other pathological features, which are used to describe cystic glandular hyperplasia, mild hyperplasia, and simple hyperplasia.

Risk factors for EH are similar to risk factors for endometrial carcinoma **(Box 5)**.

Clinical presentation: EH typically presents with AUB and is most common in perimenopausal or early postmenopausal patients. Rarely, patients will present with abnormal findings on cervical cytology.

Natural [endometrial biopsy (EB)] history: Coexistent endometrial carcinoma may be present in up to 40% of patients who have EH with atypia (EIN).

Diagnosis: EH is a histologic diagnosis based on characteristic findings on EB, dilatation and curettage (D and C), or hysterectomy specimen.

Subsequent evaluation: This was done in patients with a diagnosis of EH with atypia made by EB.

TABLE 1: Endometrial hyperplasia: Terminology and classification.

Classification system	***Histologic category***					
2014 WHO	Normal endometrium	Hyperplasia without atypia (nonneoplastic)		Atypical hyperplasia (neoplastic)		Endometrial carcinoma
2000 EIN		Benign endometrial hyperplasia		Endometrial intraepithelial neoplasia		
1994 WHO		Simple hyperplasia without atypia	Complex hyperplasia without atypia	Complex hyperplasia without atypia	Simple hyperplasia with atypia	

(EIN: endometrial intraepithelial neoplasia; WHO: World Health Organization)

BOX 5: Risk factors for endometrial cancer.

- Increasing age
- Unopposed estrogen therapy
- Tamoxifen therapy
- Early menarche
- Late menopause (after age 55 years)
- Nulliparity
- Polycystic ovary syndrome (chronic anovulation)
- Obesity
- Diabetes mellitus
- Estrogen-secreting tumor
- Lynch syndrome (hereditary nonpolyposis colorectal cancer)
- Cowden syndrome
- Family history of endometrial, ovarian, breast, or colon cancer

EVALUATION OF ENDOMETRIUM

- Transvaginal ultrasound[7,8] (sensitivity—96%, specificity—53%)
- Saline sonography to rule out focal hyperplasia
- Office EB is choice of procedure for endometrial sampling
- D and C (cervical stenosis, patient anxiety or pain, insufficient tissue on office EB, persistent PMB)
- Hysteroscopy and D and C (high suspicion of endometrial cancer and focal lesion).

Endometrial Sampling

Indications:

- All postmenopausal patients presenting with uterine PMB irrespective of amount of bleeding, spotting
- All PMW without PMB endometrial thickness ≥11 mm
- All PMW on hormone replacement therapy (HRT) endometrium is >8 mm
- Repeat EB in PMW with persistent bleeding, despite previous report is consistent with benign (preferable hysteroscopic guided)
- All women age ≥45 presenting with heavy menstrual bleeding (HMB) (AUB) cycle ≤21 days; menstrual flow ≥8 days and intermenstrual bleeding
- All women ≤45 years with body mass index (BMI) ≥30 kg/m^2; presenting with persistent AUB; history of (H/o) polycystic ovarian disease (PCOD); not responding to medical management and high risk for endometrial cancers (Lynch or Cowden syndrome)
- Any premenopausal patient with H/o prolonged amenorrhea with bleeding; history suggestive of PCOD
- Any abnormal glandular cervical cytology (AGC)
- Follow-up of patients with EH under treatment
- High risk for Lynch syndrome

Contraindications:

Absolute: Viable intrauterine pregnancy

Relative contraindications:

- Bleeding diathesis
- Woman on anticoagulants and her coagulation profile is with in normal limits; in case of uncontrolled coagulation parameters patient should be referred to physician to control diathesis.
- *Acute vaginal, cervical, and pelvic infections*: EB is deferred and done after treating infections.

Preprocedure and procedure task:

- Anesthesia/analgesia depends upon patients' choice and physician comfort
- Aseptic precaution
- Prophylactic antibiotic generally not required[9]
- Follow standard procedure for EB.

Endometrial suction (ES) devices used for ES: These include Pipelle[10,11]/endo cell—disposable plastic low-pressure device, 3 mm popularly used Karman cannula[12]—4–6 mm commonly used in public sector, high-pressure device; Vabra—plastic or stainless steel device 4 mm, and endometrial brush, like a brush used for endocervical sampling.

Side effect and complications:

- Uterine cramping
- Vasovagal reactions
- Rarely uterine perforation
- Excessive bleeding in a patient with undiagnosed coagulopathy
- *Postprocedure care:* Patient can resume usual activity including coitus when she feels comfortable.
- *Warning signs:* Patient should report immediately if unbearable pain, fever >37°C, foul-smelling vaginal discharge, or excessive bleeding.

Newer noninvasive techniques tumor cell-free DNA from peripheral blood and uterine lavage[13] for molecular testing are in experimental stage.

MANAGEMENT OF ENDOMETRIAL HYPERPLASIA[14-17]

Progestin Therapy

The choice of progestin therapy for treatment of endometrial hyperplasia is given in **Table 2**.

TABLE 2: Choice of progestin therapy for treatment of endometrial hyperplasia.

Progestins	*Doses—3–6 months, till repeat ES*	*Patient selection who can use it*
Medroxyprogesterone acetate (MPA)	10–20 mg orally daily	Can be used for pre- and postmenopausal patients with EH (any type)
Norethindrone acetate (NETA, also known as norethisterone acetate)	5–15 mg orally, daily	Can be used for pre- and postmenopausal patients with EH (any type)
Megestrol acetate	40–160 mg orally daily	Can be used for pre- and postmenopausal patients with EH (any type)
LNG IUD 52 mg preferred	Releases 20 µg/day initially	• The LNG 52 mg IUD is the preferred • Progestin therapy for pre- and postmenopausal patients with EH (any type)
Micronized progesterone (oral)	200–300 mg orally daily	Use only for patients with EH without atypia
Norethindrone (progestin-only contraceptive pill)	0.35 mg orally twice or three times daily	Use only for premenopausal patients with EH without atypia who need contraception
Combined estrogen–progestin oral contraceptive (COC)	Variable combinations are available	Use only for premenopausal patients with EH without atypia who need contraception
Depot medroxyprogesterone acetate (DMPA)	150 mg intramuscularly every 3 months	Use only for premenopausal patients with EH without atypia who need contraception

(EH: endometrial hyperplasia; ES: endometrial suction; LNG IUD: levonorgestrel-releasing intrauterine system)

MANAGEMENT OF PMB

Diagnosis

- History
- Examination
- Investigations

History

In case of a woman presenting with PMB, it is pertinent to ask about:

- Since when she is bleeding?
  - Is it spontaneous or precipitating factor trauma, intercourse, constipation?
  - The amount of bleeding, color (red or brown), associated with clots
  - *Associated symptoms:* Pain, fever, change in bladder or bowel habits
  - Is she on HRT?
  - H/o noncommunicable disease (NCD) such as diabetes mellites (DM), hypertension, dyslipidemia, cardiovascular disease (CVD), and coronary artery disease (CAD)
  - *H/o other medications:* Anticoagulants, aspirin, herbal, Ayush, or dietary supplements **(Table 1)**.
  - Family H/o breast, colon, or endometrial cancer
  - Whether she had any past H/o diagnosis of pathology associated with cervix, uterine, endometrial, or tubo-ovarian and related treatment.

Examination

- Physical examination
- *General examination:* Vitals [pulse, blood pressure (BP), respiration, temperature], check for anemia, lymphadenopathy
- *Height and weight:* BMI
- Lymphadenopathy
- Breast examination
- Vulval (local examination) to rule out any lesion, urethral caruncle, cystocele and rectocele
- *Pelvic examination:* To rule out lesions of vagina and cervix
- *Bimanual per vaginum (P/V) examination:* To know size, contour, surface smoothness and irregularity direction, and check for tenderness.

Investigations

- *Routine blood tests:*
  - Complete blood count (CBC)
  - Platelet count
  - *Blood sugar:* Fasting and postprandial

- Glucose challenge test
- Glycated hemoglobin (HbA1C)
- Lipid profile
- Thyroid-stimulating hormone (TSH)
- Renal function test (RFT)
- Liver function test (LFT)
- Urine examination routine and microscopic
- *Urine culture:* To rule out urine infection in PMW

- *Specific test:*
  - Cervical cytology LBC (liquid-based cytology); LBC:HPV (human papillomavirus)
  - *Ultrasound:* Transabdominal (TA) and transvaginal sonography (TVS) to rule out endometrial, cavity, myometrial and adnexal lesions to rule out structural lesions **(Box 1)**
  - *Ultrasonography (USG):* Doppler; 3D and 4D wherever indicated
  - Saline hysterography to rule out cavitary lesions polyp, submucous fibroid, etc.
  - *Computed tomography (CT)/magnetic resonance imaging (MRI):* Optional indication to be justified
  - *Office hysteroscopy/hysteroscopy:* In case of recurrent PMB
  - *EB:* Pipelle
  - D and C.

CHOICE OF TREATMENT FOR PMB

- The treatment depends upon the cause.
- Surgical total hysterectomy (TH); total hysterectomy with salpingo-oophorectomy (THBSO), treatment for structural abnormalities (polyp, leiomyoma, adenomyosis) of PMB **(Box 2)**
- For treatment of EH see **Table 2**.
- Appropriate management for other causes of PMB **(Boxes 3 and 4)**
- In case of medication as cause of PMB, withdraw or change the medication
- Treatment of endometrial cancer is beyond the scope of chapter.

CONCLUSION

- The choice of treatment is either observation, progestin therapy, or hysterectomy, which depends upon several factors: EH (with or without atypia) menopausal status, desire for fertility, contraceptive needs, and presence of risk factors for endometrial cancer.
- For most patients with EH without atypia, the recommendation is progestin therapy with endometrial sampling rather than hysterectomy **(Grade 2C)**.
- For patients with EH with atypia who are postmenopausal or premenopausal and have completed childbearing, the recommendation is hysterectomy rather than progestin therapy with endometrial sampling **(Grade 1B)**.
- For premenopausal patients with EH with atypia who wish to preserve fertility, the recommendations are progestin therapy rather than hysterectomy **(Grade 2B)**.
- For patients undergoing hysterectomy as treatment of choice for EH with atypia, the recommendation is hysterectomy without oophorectomy but bilateral salpingectomy **(Grade 2C)**.

REFERENCES

1. Prasad JB, Tyagi NK, Verma P. Age at Menopause in India: a systematic review. Diabetes Metab Syndr. 2021;15(1):373-7.
2. Astrup K, Olivarius Nde F. Frequency of spontaneously occurring postmenopausal bleeding in the general population. Acta Obstet Gynecol Scand. 2004;83(2):203-7.
3. Sharon Sung, Karen Carlson, Aaron Abramovitz. Postmenopausal Bleeding. NCBI Bookshelf. A service of the National Library of Medicine, National Institutes of Health. StatPearls. Treasure Island: StatPearls Publishing; September 18 2022.
4. T. Vanden Bosch, L Ameye, D Van Schoubroeck, T Bourne, D Timmerman. Intra-cavitary uterine pathology in women with abnormal uterine bleeding: a prospective study of 1220 women. Facts Views Vis Obgyn. 2015;7(1):17-24.
5. Emons G, Beckmann MW, Schmidt D, Mallmann P, and for the Uterus commission of the Gynecological Oncology Working Group (AGO). New WHO Classification of Endometrial Hyperplasias. Geburtshilfe Frauenheilkd. 2015;75:135-6.
6. Baak JP, Ørbo A, van Diest PJ, Jiwa M, de Bruin P, Broeckaert M, et al. Prospective multicenter evaluation of the morphometric D-score for prediction of the outcome of endometrial hyperplasias. Am J Surg Pathol. 2001;25(7):930-5.
7. Davidson KG, Dubinskey TJ. Ultrasonographic evaluation of the endometrium in postmenopausal vaginal bleeding. Radiol Clin North Am. 2003;41(4)769-80.
8. Goldstein SR. The role of transvaginal ultrasound or endometrial biopsy in the evaluation of the menopausal endometrium. Am J Obstet Gynecol. 2009;201(1):5-11.
9. ACOG Practice Bulletin. Clinical Management Guidelines for Obstetrician-Gynecologist. Antibiotic prophylaxis for gynecologic procedures. Obstetrics & Gynecology 108(1): p 225, July 2006. No.74;195.

10. Wagaarachchi PT, Firisena J. Efficiency of Pipelle device in sampling endometrium. Acta Obstet Gynecol Scand. 2000;79:793.
11. Sanam M, Majid MM. Comparison the Diagnostic Value of Dilatation and Curettage Versus Endometrial Biopsy by Pipelle–a Clinical Trial. Asian Pac J Cancer Prev. 2015;16(12):4971-5.
12. Suarez RA, Grimes DA, Majmudar B, Benigno BB. Diagnostic endometrial aspiration with the Karman cannula. J Reprod Med. 1983;28(1):41-4.
13. Nair N, Camacho-Vanegas O, Rykunov D, Dashkoff M, Camacho SC, Schumacher CA, et al. Genomic Analysis of Uterine Lavage Fluid Detects Early Endometrial Cancers and Reveals a Prevalent Landscape of Driver Mutations in Women without Histopathologic Evidence of Cancer: A Prospective Cross-Sectional Study. PLoS Med. 2016; 13(12):e1002206.
14. Abu Hashim H, Ghayaty E, El Rakhawy M. Levonorgestrel-releasing intrauterine system vs oral progestins for non-atypical endometrial hyperplasia: a systematic review and metaanalysis of randomized trials. Am J Obstet Gynecol. 2015;213:469-78.
15. Reed SD, Voigt LF, Newton KM, Garcia RH, Allison HK, Epplein M, et al. Progestin therapy of complex endometrial hyperplasia with and without atypia. Obstet Gynecol. 2009;113:655-62.
16. Gal D, Edman CD, Vellios F, Forney JP. Long-term effect of megestrol acetate in the treatment of endometrial hyperplasia. Am J Obstet Gynecol. 1983;146: 316-22.
17. Wildemeersch D, Dhont M. Treatment of nonatypical and atypical endometrial hyperplasia with a levonorgestrel-releasing intrauterine system. Am J Obstet Gynecol. 2003;188:1297-8.

Chapter 49

Menopausal Hormonal Therapy

Suvarna Satish Khadilkar, Madhva Prasad S

INTRODUCTION

Estrogen deficiency is associated with many to symptoms and pathological states which present after menopause. With many diseases receiving better medical care and subsequent increase in longevity, the average years spent after menopause is constantly on the rise. In other words, some women may spend about one-third of their years in the post-menopausal state. This has brought the concept of hormone replacement therapy (HRT) into prominence. HRT has evolved over time, with an initial phase of selective usage, followed by universal usage, followed by sudden stoppage due to side-effects followed by a selective gradual re-introduction. In its current form, the HRT is better referred to as menopausal hormone therapy (MHT), the term that would be preferred here. There is a wide range of benefits that may be accrued with the use of MHT and a reasonably large list of risks that can happen **(Fig. 1)**. The pattern of side effects and risks is quite complex based on the duration of therapy, type of medication and route, time (in relation to menopause occurrence) of initiation of therapy and prior medical condition of the woman.

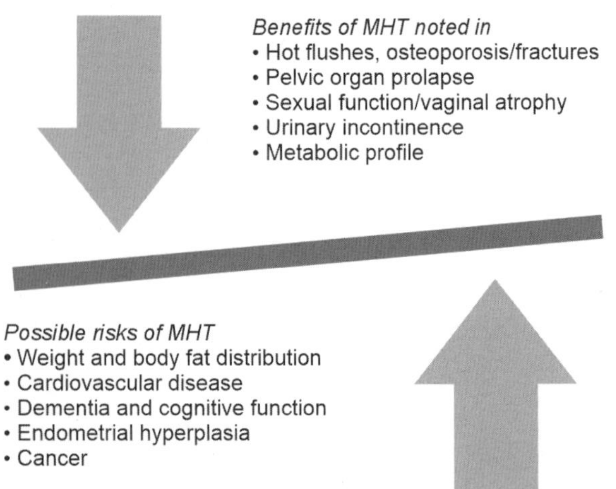

Fig. 1: Menopausal hormonal therapy (MHT)—benefits and risks.

The possible conditions which may get benefits and possible risk which one may accrue upon usage of MHT are listed in **Figure 1**. However, it should be noted that MHT has been shown to have an extremely complex pattern of risks and benefits. The type of medication, duration of treatment, age at the time of initiation, time since occurrence of menopause and prior medical status are all influencing factors on the benefits and risks.

Most of the information available regarding the benefits and risks consist of a composite understanding of the results of multiple studies that have already been performed, and a brief description of a few important ones are presented. Many studies are in progress on this topic,

- The *women's health initiative (WHI)* was one of the pioneer studies in this topic. Thus multicenter, multi-country study involved around 27,000 women with two main parts.
  - In one part, the estrogen only group (active treatment duration of around 5.9 years) was compared to placebo. Results of 7 years follow-up and (re-recruitment) 13 years follow-up are available.
  - The other part of the WHI was combined (estrogen plus progestin) in the intervention arm. Initial results were after 5.6 years of active therapy (intervention phase) and 2.4 years of follow-up after completion of the intervention. A further follow-up of around 13 years was done and the cohort is still being followed up.[1]
- The *Heart and Estrogen/progestin Replacement Study (HERS)* was a study which attempted to analyze benefits of estrogen-progestin supplementation focused on reduction of cardiovascular problems. However, since it showed no measurable benefits and a paradoxical

TABLE 1: Risks associated with continuous combined hormonal therapy.

Event (follow-up duration)	*Risk in general population (per 1,000)*	*Risk in treatment group (per 1,000)*	*Relative risk*
Fracture (5.6 years)	111	87	0.78
Breast cancer (5.6 years)	19	*24*	1.27
Stroke (3 years)	6	*8*	1.46
Gallbladder disease (6 years)	16	*27*	1.64
Death from lung cancer (8 years)	5	*9*	1.74
Coronary events (1 year)	2	*4*	1.89
VTE (1 year)	2	*7*	*4.28*

increase in the risk of cardiovascular morbidity, it was stopped without long term follow-up.[2]

- The *Effects of Postmenopausal Hormone Therapy (EPHT) trial* (MHT vs. placebo) was held in Eastern Europe on approximately 1,700 patients with follow-up data of around.
- *Women's International Study of Long-Duration Oestrogen After Menopause (WISDOM)* was an ambitious long duration multicentric trial wherein MHT as intervention was given only for 1 year. Due to poor funding, the trial was stopped early.
- The *Million Women Study* (1990s) prospectively recruited women with the main intention to look at cancer development in a cohort of women, of which 1/3rd were on MHT.[3]
- *Many small randomized controlled trials (RCTs)* with small patient numbers have been conducted. A systematic review by the Cochrane group[4] summarizes the findings.

Table 1 presents the risks associated with combined hormonal therapy. There is a reduction in the usage of continuous combined hormonal therapy reduces the risk of fracture but increases the risk of various adverse events. Though the absolute risk of occurrence of any of these events is very small, the occurrence of venous thromboembolism appears significant.

As can be seen in **Table 2**, the usage of estrogen-only hormonal therapy reduces the risk of fracture but increases the risk of various adverse events. Though the absolute risk of occurrence of any of these events is very small, the risk

TABLE 2: Risks associated with estrogen-only hormonal therapy.

Event (Follow-up of 7.1 years for all)	*Risk in general population (per 1,000)*	*Risk in treatment group (per 1,000)*	*Relative risk*
Fracture	141	103	0.73
Breast cancer	25	20	0.79
Coronary events	41	38	0.94
VTE	16	*21*	1.32
Stroke	24	*32*	1.33
Gallbladder disease	27	*47*	1.78

is somewhat comparable to the risks accrued upon taking combined estrogen-progestin combination hormonal therapy.

MENOPAUSAL HEALTH CONCERNS

Vasomotor Symptoms

The predominant reason for menopausal women to seek therapy is hot flushes. These vasomotor symptoms appear to occur in almost 75% perimenopausal women, for approximately 1–2 years. The most efficacious treatment for vasomotor symptoms is systemic estrogen therapy. Progesterone only options such as medroxyprogesterone acetate or megestrol can be considered when systemic estrogen is contraindicated. Recently, the selective serotonin reuptake inhibitors (SSRI)/serotonin and norepinephrine reuptake inhibitors (SNRI) group of drugs with ability to modify central neurotransmitter pathways are being utilized for this symptom with excellent results. Paroxetine is a commonly used option. Once initiated, MHT dosage should be reduced over many months. Other methods such as stellate ganglion block have not been proven to be useful.[5-8]

Urogenital Atrophy

The symptom complex of dyspareunia, dysuria, pruritus, vaginal dryness, and incontinence are now referred to as "genitourinary syndrome of menopause", for which lubricants are tried many times as the first option. These symptoms respond well to estrogen oral therapy. Topical estrogen cream, once to thrice a week or estradiol vaginal tablet twice a week or estrogen-vaginal ring are options. Topical dehydroepiandrosterone (DHEA) has been demonstrated to be useful.[9]

Osteoporosis

This condition of bone-fragility is an important menopausal concern, the occurrence of which is dictated by a combination of non-modifiable risk factors (Asian/white lineage, older age, smaller body frame, family history, oophorectomy, and early occurrence of menopause) and modifiable risk factors (vitamin D intake, calcium intake, sedentary lifestyle, and smoking). Chronic renal disease, hyperparathyroidism, and long-term steroid usage are also risk factors.

The objective assessment of osteoporosis is based on bone mineral density (BMD)—which can reveal normalcy, osteoporosis, or osteopenia.

A combination of history, clinical risk factors, and the BMD are combined to achieve a FRAX Score (fracture risk assessment).[10]

- *FRAX low risk patients* need lifestyle advice and sufficient oral calcium intake (1,200 mg) along with vitamin D (800 IU).
- *FRAX moderate risk patients* need lifestyle advice, sufficient calcium, and vitamin D intake. Pharmacological therapy is considered when any other specific risk factors exist (listed above).
- *FRAX high risk patients* need lifestyle advice, sufficient calcium, and vitamin D intake along with pharmacological management. Details are presented in **Table 3**. Usage of MHT definitely reduces hip fractures among women. However, in the background of availability of other non-hormonal medications for the prevention of osteoporosis, hormonal therapy is not initiated purely for the prevention of fractures.

TABLE 3: Pharmacological management options for osteoporosis.

Preparation and mechanism	***Drug, dosage, and comments***
Menopausal hormone therapy (MHT)	MHT should not be considered solely for the prevention of osteoporosis. Other agents should be utilized
Bisphosphonates: • Bone resorption reduction by hydroxyapatite crystal- binding • Hydroxyapatite breakdown inhibition • Calcification inhibition in areas of active bone turnover • Osteoblast and osteoclast apoptosis inhibition • The exact contribution of each component is variable	*Alendronate:* • 5 or 10 mg daily/35 or 70 mg weekly • First drug to be used clinically
	Risedronate: • 5 mg daily/35 mg weekly/150 mg monthly • Four times more potent than alendronate
	Zoledronate: • 5 mg once a year intravenous • Twenty times more potent than alendronate • For consideration in patients who may be non-compliant • Osteonecrosis of jaw is a complication
	Ibandronate: Has proven effect only on vertebral fractures, and not on non-vertebral fractures
Calcitonin (enhancement of bone matrix deposition)	• Intranasal spray is approved for treatment • Less efficacious than most alternatives • To be considered only if other agents are unacceptable
Denosumab: Binding to receptor-activator-of-nuclear-factor-kappa-ligand (RANKL) and resultant osteoclast inhibition reduction and bone resorption reduction	• It is a monoclonal antibody • 60 mg subcutaneous twice a year • It has been considered in other orthopedic issues also • Very little experience in India[11]
Raloxifene: Selective estrogen receptor modulated with positive effect in the bone	• 60 mg oral once a day for not more than 5 years • Long term use can increase risk of thromboembolism and stroke
Teriparatide-recombinant PTH: Osteoanabolic agent which causes improvement of architecture of bone trabeculae	• 20 µg subcutaneous daily • Cost is a prohibitive factor • Should not be combined with bisphosphonates • Should not be given for >2 years • Newer formulation abaloparatide shows promise[12]

Menopausal Hormone Therapy and Cardiovascular Disease

For a few years and probably almost a decade it was believed that "cardiovascular disease is worsened by menopausal hormonal therapy". However, it is now clear that this relationship is extremely complex. A detailed analysis of the WHI trial showed that the occurrence of cardiovascular disease post MHT relates to variables such as woman's age, time of utilization of hormonal therapy in relation to establishment of menopause, and duration of hormonal therapy. When MHT was initiated at a time very close to menopause attainment, the risk of cardiovascular adverse events was quite less. However, for those in early menopause and no specific contraindications, benefits of MHT slightly outweigh the cardiovascular risks.[13] Also, one clear principle is that CVD prevention is not an indication for starting menopausal hormonal therapy.

Menopausal Hormone Therapy and Stroke

The absolute risk of stroke after usage of MHT increases by around two additional events per 10,000 person-years (mainly ischemic subtype), with no specific impact of age at which MHT was initiated or time since menopause. Also, quite-interestingly, estrogen-progestin combinations appear to have similar propensity for stroke occurrence. Raloxifene and tibolone, when used in above-60 years women have slightly higher chance of stroke. Usage of transdermal hormones does not appear to increase risk of stroke.[14-17]

Menopausal Hormone Therapy and Breast Cancer

D'Alonzo et al. in their review have analyzed the complex relationship between MHT and breast cancer.[18] An initial analysis of the WHI showing increasing breast cancer led to a reduction in the utilization of MHT. Later, it became clear that the estrogen-progesterone combination—CEE + MPA had a high risk of breast cancer. The risk remained long after cessation of therapy. The risk was higher with continuous therapy as against intermittent therapy. The actual absolute risk of breast cancer after MHT is quite less—8 per 1,000 increases when combination MHT is utilized for 5 or more years. Initiation of MHT 1 year or after menopause showed higher breast cancer risk. Those currently utilizing MHT had higher risk than users who had discontinued. One drug with excellent safety profile (negligible risk of breast cancer development) is bazedoxifene.

Menopausal Hormone Therapy and Endometrial Cancer

Among patients receiving MHT and have uterus-in-situ, usage of only-estrogen regimens, usage of tibolone, usage of sequential combined therapy appears to increase endometrial cancer risk if used for <5 years. Continuous estrogen-progestin therapy appears to reduce the risk of endometrial cancer, with slightly higher benefit among obese women.[18]

Menopausal Hormone Therapy and Ovarian Cancer

A small but real risk of ovarian cancer is noted. A 50-year-old who utilizes MHT for 5 years, the risk of developing an ovarian cancer is around 1 in 1,000 users, with a mortality risk of around 1 in 1,700.[19]

Menopausal Hormone Therapy and Neurological problems

Alzheimer's disease and Parkinson's disease[20] are higher among the geriatric population. Though some data suggests that cognitive decline may be higher among those with hormonal therapy, a clear causation or explanation is not clear. On the other hand, studies using objective analysis suggest that there was an improvement in mood and mental state among those with HRT. A recent meta-analysis[21] suggests that estrogen replacement shows benefit in both Alzheimer's disease and Parkinson's disease.

Menopausal Hormone Therapy and All-cause Mortality[22]

In a cumulative 18-year follow-up, there was no increase in the increase in the all-cause mortality, cardiovascular mortality or cancer-related mortality among those who used MHT. Usage of combination of CEE + MPA does not look.

There is *no change in all-cause mortality, cardiovascular mortality, or cancer mortality* over a cumulative 18-year follow-up, with usage of CEE plus MPA (5.6 years) or CEE alone (7.2 years).

BIOCHEMICAL CHANGES IN MENOPAUSAL TRANSITION

There is a 10-fold increase in FSH value during the menopausal transition. Simultaneously, there is a decrease of estradiol up to 10–25 pg/mL, and of testosterone up to 290 ng/mL. Estradiol, which is the usual dominant estrogen, is replaced by estrone in the menopause. The ovary, the main site of estrogen production, is replaced by adrenals, adipose, and muscles (androstenedione aromatization). In the background of such complex changes, the main goal of MHT is to bring the physiological level of estrogen to 45–65 pg/mL, which can prevent any major menopausal symptoms. Dosage of MHT is the lowest reasonably achievable dose, with maximum acceptability and minimum side effects and least duration.

Synthetic Estrogens

Estradiol valerate, an artificially synthesized natural estrogen, appears to have good bioavailability, and is tolerated orally, with lesser metabolic burden on the liver and is almost completely converted to estradiol (by hydrolysis). However, when used orally, it can also get converted to estrone and circulate in the body in this form, reducing the therapeutic efficacy. Hence, non-oral form of estrogen administration, owing to preservation of estradiol/estrone ratio appears to be a more efficacious route.[23,24]

Conjugated equine estrogens: This is a chemical mixture consisting of a variety of estrogens sourced from menopausal mare's urine, containing equilin sulfate, estrone sulphate and few other compounds, in varying proportions. Sold under a common brand name of "Premarin", it is available as 0.375 mg, 0.625 mg (most commonly used dosage), and 1.25 mg.[25]

CONTRAINDICATIONS TO MENOPAUSAL HORMONE THERAPY

The contraindications to MHT are presented in **Table 4**.

Some pointers toward choice of hormone replacement preparation **(Box 1)**:

- Oral preparations can lead to lipid problems, coagulation abnormalities and liver enzyme elevation. Hence, those with health concerns such as hypertension, hypertriglyceridemia, thromboembolic disease, and anticonvulsant therapy should be considered for non-oral preparations of MHT.
- Predominance of urogenital symptoms should be pointed toward use of local vaginal preparations.
- Those with intact uterus should not be given estrogen-only preparations.

TABLE 4: Absolute and relative contraindications to menopausal hormone therapy (MHT).

Absolute contraindications	*Relative contraindications*
• Active endometrial, gynecological hormone-dependent cancers	• Migraine headaches
• Active breast cancer: immaterial of receptor status	• Superficial thrombophlebitis
• Known or suspected pregnancy	• Strong family history of breast cancer
• Undiagnosed, abnormal vaginal bleeding	• Uterine fibroids
• Liver disease with impaired/abnormal liver function	• Endometriosis
• Increased risk of thromboembolism	• Gallbladder disease
	• Impaired glucose tolerance
	• Hypertriglyceridemia

BOX 1: Menopausal hormonal therapy options.

- *Estrogen-only HTs*
  - *Conjugated equine estrogen (CEE):* 0.625 mg or 1.25 mg daily (most commonly studied)—was used in WHI (one arm)
  - *Estradiol (17-B estradiol):* 1 mg orally
  - *Estradiol valerate:* 2 mg
  - *Transdermal estradiol skin patches:* 0.014 mg, 0.025 mg, 0.05 mg, or 0.075 mg daily
  - *Intranasal puff of 17-B estradiol:* 0.15 mg or 0.3 mg daily
- *Continuous combined estrogen progestin replacement therapy (CCEPRT)*
  - *CEE 0.625 mg with MPA 2.5 mg daily:* Trials in which this was used are EPHT, HERS, PEPI, WAVE, WHI (combined arm), and WISDOM
  - EE 2.5 mg with MPA 10 mg daily
  - Estradiol 2 mg with 1 mg norethisterone daily
- *Sequential estrogen progestin replacement therapy (SEPRT)*
  - Estradiol 1 mg daily with MPA 5 mg for 12 days once a year
  - Estradiol 1 mg daily for 4 days, estradiol 1 mg plus 0.35 mg norethindrone daily for 3 days each week
  - Estradiol 2 mg days 1–22, 1 mg days 22–28, with norethisterone 1 mg days 13–22
  - CEE 0.625 with MPA 10 mg days 1–12 (PEPI 1995)
  - CEE 0.625 mg with micronized progesterone 200 mg days 1–12

Estrogenic Side Effects

Possible side effects of estrogen are breast tenderness, leg pain, headache, nausea, bloating, and vaginal discharge. Side effects may be experienced by around 14% of women on MHT.

Progestins

Medroxyprogesterone acetate is the most widely used progestin. Acne, backache, bloating, breast tenderness, dizziness, headache, greasy skin, weight gain, fatigue, and abdominal cramps are noted by around 20% of the women using this preparation for MHT.

For any MHT preparation, troublesome side effects which are either severe or prolonged, should prompt dosage reduction, alternative route of administration or alternative preparation.

Types of preparations available:

- Oral preparation
- Topical creams
- Transdermal patches
- Biodegradable implants
- Vaginal rings
- Vaginal pessaries
- Metered dose sprays.

Androgen Preparations

Dehydroepiandrosterone and DHEA-sulphate (which are interconvertible) are good sources of local estrogen in the menopausal transition. Topical application of intravaginal DHEA and testosterone patches are beneficial in genitourinary syndrome of menopause (GSM) and hypoactive sexual desire disorder. Other features of menopause may not benefit with this topical application.[26]

Tibolone

Tibolone, a steroidal molecule is a progestin, also has weak selective estrogenic activity and minimal androgenic activity. It effectively suppresses vasomotor symptoms, and also plays a role in improvement of libido and mood. At a recommended dose is 2.5 mg/day, there is improvement in GSM also no endometrial problems. Its main positive impact is on bone mass. However, long-term therapy appears to increase risk of stroke and breast cancer. For women who experience unacceptable side effects with first-line hormonal therapy, tibolone is apt. It is also used in those with premature ovarian failure or those who need add back therapy after gonadotropin hormone-releasing hormone (GnRH) usage.[15,27]

Raloxifene

Raloxifene, a second-generation SERM, has an estrogen-agonist effect on bone, and improves bone mineral density, and has established itself as a vital option in postmenopausal osteoporosis. Antagonist effect on the breast and uterus helps in prevention of breast cancer and reducing pelvic organ prolapse. A definite but small increase in the risk of venous thromboembolism has been noted. The preparation has been shown to be useful in a few milder types of schizophrenia. Raloxifene is administered as 60 mg tablets once a day, for a duration of <5 years.[28]

Bazedoxifene

Bazedoxifene, a reasonably new molecule, has interesting differential activity on various estrogen receptors. A combination of bazedoxifene 20 mg and conjugated estrogen 0.45 mg has been named TSEC (tissue selective estrogen complex),[29] which appears to have many ideal properties and is a good future candidate for first-line MHT.

- Reduces vasomotor symptoms significantly
- Bone-protective
- Improves sleep quality and overall quality of life
- No stimulatory effect on breast or endometrial tissue
- Favorable impact on GSM
- No worsening of any cardiovascular risk.

Ospemifene

Ospemifene, a recently introduced SERM administered orally (60 mg tablets) has been approved for management of GSM (vaginal dryness and dyspareunia—most important). Favorable impact on bone and breast have also been noted. Hot flashes (a common complaint), vaginal discharge, and hyperhidrosis are notable side-effects. Overall safety and acceptance are now well established in few studies and needs to be popularized in India.[30]

Decision Making Regarding Initiation of Menopausal Hormone Therapy

A detail **Flowchart 1** is given here. The patient must be explained about the complex patterns of risks and benefits of MHT for menopausal symptoms. Lowest possible dosage, for least possible duration, is considered. Electronic decision-making algorithms are now available.[31]

Deciding When to Stop Menopausal Hormone Therapy

Administration of MHT for >5 years is not recommended. Abrupt discontinuation is to be avoided. It may present similar to the original condition. To prevent this, weaning of medication is to be considered.[32]

Nutritional Supplementation

Isoflavones (daidzin, genistein, and daidzein) are phytoestrogens or plant estrogens, which are useful for

Flowchart 1: Approach to initiation of menopausal hormone therapy (MHT).

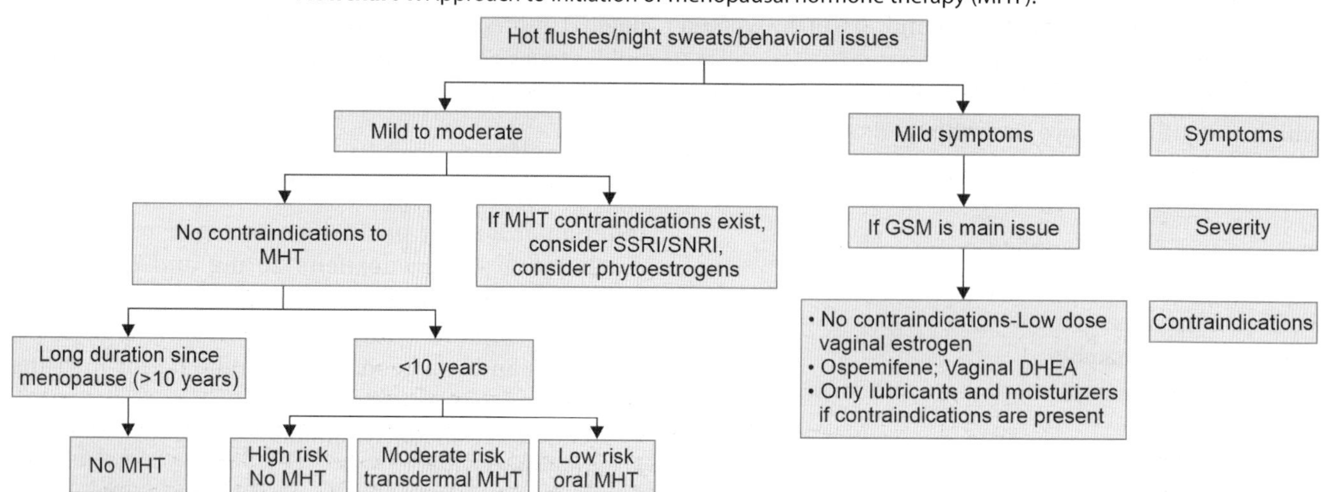

(DHEA: dehydroepiandrosterone; GSM: genitourinary syndrome of menopause; SSRI: selective serotonin reuptake inhibitors; SNRI: serotonin and norepinephrine reuptake inhibitors)

management of menopausal symptoms, especially the vasomotor. It has very less impact of GSM. Most patients experience at least a partial symptom reduction. For patients who are skeptical about formal hormonal therapy, this is an excellent option.[33]

CONCLUSION

A range of options for the administration of menopausal hormonal therapy is now available. The benefits obtained in the prevention/management of menopausal symptoms must be weighed against possible risks of therapy. Decision making is based on predominant symptom type, symptom severity, menopausal age, time since menopause, individual gynecological/medical factors which dictate the choice of dosage, route, and duration of therapy.

REFERENCES

1. Manson JE, Chlebowski RT, Stefanick ML, Aragaki AK, Rossouw JE, Prentice RL, et al. Menopausal hormone therapy and health outcomes during the intervention and extended poststopping phases of the Women's Health Initiative randomized trials. JAMA. 2013;310(13):1353-68.
2. Grady D, Herrington D, Bittner V, Blumenthal R, Davidson M, Hlatky M, et al. Cardiovascular disease outcomes during 6.8 years of hormone therapy: Heart and Estrogen/progestin Replacement Study follow-up (HERS II). JAMA. 2002;288(1):49-57.
3. The Million Women Study Collaborative Group. The Million Women Study: design and characteristics of the study population. Breast Cancer Res. 1999;1(1):73-80.
4. Marjoribanks J, Farquhar C, Roberts H, Lethaby A, Lee J. Long-term hormone therapy for perimenopausal and postmenopausal women. Cochrane Database Syst Rev. 2017;1(1):CD004143.
5. Prior JC, Nielsen JD, Hitchcock CL, Williams LA, Vigna YM, Dean CB. Medroxyprogesterone and conjugated oestrogen are equivalent for hot flushes: a 1-year randomized double-blind trial following premenopausal ovariectomy. Clin Sci (Lond). 2007;112(10):517-25.
6. Walega DR, Rubin LH, Banuvar S, Shulman LP, Maki PM. Effects of stellate ganglion block on vasomotor symptoms: findings from a randomized controlled clinical trial in postmenopausal women. Menopause. 2014;21(8):807-14.
7. Stubbs C, Mattingly L, Crawford SA, Wickersham EA, Brockhaus JL, McCarthy LH. Do SSRIs and SNRIs reduce the frequency and/or severity of hot flashes in menopausal women. J Okla State Med Assoc. 2017;110(5):272-74.
8. Goodwin JW, Green SJ, Moinpour CM, Bearden JD 3rd, Giguere JK, Jiang CS, et al. Phase III randomized placebo-controlled trial of two doses of megestrol acetate as treatment for menopausal symptoms in women with breast cancer. J Clin Oncol. 2008;26(10):1650-6.
9. Lethaby A, Ayeleke RO, Roberts H. Local oestrogen for vaginal atrophy in postmenopausal women. Cochrane Database Syst Rev. 2016;2016(8):CD001500.
10. Tella SH, Gallagher JC. Prevention and treatment of postmenopausal osteoporosis. J Steroid Biochem Mol Biol. 2014;142:155-70.
11. Yanbeiy ZA, Hansen KE. Denosumab in the treatment of glucocorticoid-induced osteoporosis: a systematic review and meta-analysis. Drug Des Devel Ther. 2019;13:2843-52.
12. Leder BZ, O'Dea LS, Zanchetta JR, Kumar P, Banks K, McKay K, et al. Effects of abaloparatide, a human parathyroid

hormone-related peptide analog, on bone mineral density in postmenopausal women with osteoporosis. J Clin Endocrinol Metab. 2015;100(2):697-706.
13. Chester RC, Kling JM, Manson JE. What the Women's Health Initiative has taught us about menopausal hormone therapy. Clin Cardiol. 2018;41(2):247-52.
14. Henderson VW, Lobo RA. Hormone therapy and the risk of stroke: perspectives 10 years after the Women's Health Initiative trials. Climacteric. 2012;15(3):229-34.
15. Formoso G, Perrone E, Maltoni S, Balduzzi S, Wilkinson J, Basevi V, et al. Short-term and long-term effects of tibolone in postmenopausal women. Cochrane Database Syst Rev. 2016;10(10):CD008536.
16. Canonico M, Carcaillon L, Plu-Bureau G, Oger E, Singh-Manoux A, Tubert-Bitter P, Postmenopausal hormone therapy and risk of stroke: impact of the route of estrogen administration and type of progestogen. Stroke. 2016; 47(7):1734-41.
17. Ott SM. In postmenopausal women, oral but not transdermal estrogen was associated with increased risk for stroke. Ann Intern Med. 2016;165(8):JC46.
18. D'Alonzo M, Bounous VE, Villa M, Biglia N. Current evidence of the oncological benefit-risk profile of hormone replacement therapy. Medicina (Kaunas). 2019;55(9):573.
19. Sjögren LL, Mørch LS, Løkkegaard E. Hormone replacement therapy and the risk of endometrial cancer: A systematic review. Maturitas. 2016; 91:25-35.
20. Gleason CE, Dowling NM, Wharton W, Manson JE, Miller VM, Atwood CS, et al. Effects of hormone therapy on cognition and mood in recently postmenopausal women: findings from the randomized, controlled KEEPS-Cognitive and Affective Study. PLoS Med. 2015;12(6):e1001833.
21. Song YJ, Li SR, Li XW, Chen X, Wei ZX, Liu QS, et al. The effect of estrogen replacement therapy on Alzheimer's disease and Parkinson's disease in postmenopausal women: a meta-analysis. Front Neurosci. 2020;14:157.
22. Manson JE, Aragaki AK, Rossouw JE, Anderson GL, Prentice RL, LaCroix AZ, et al. Menopausal hormone therapy and long-term all-cause and cause-specific mortality: the women's health initiative randomized trials. JAMA. 2017;318(10):927-38.
23. Järvinen A, Kainulainen P, Nissilä M, Nikkanen H, Kela M. Pharmacokinetics of estradiol valerate and medroxyprogesterone acetate in different age groups of postmenopausal women. Maturitas. 2004;47(3):209-17.
24. Ansbacher R. The pharmacokinetics and efficacy of different estrogens are not equivalent. Am J Obstet Gynecol. 2001;184(3):255-63.
25. Archer DF. The effect of the duration of progestogen Use on the occurrence of endometrial cancer in post-menopausal women. Menopause. 2001;8:245-51.
26. Portman DJ, Goldstein SR, Kagan R. Treatment of moderate to severe dyspareunia with intravaginal prasterone therapy: a review. Climacteric. 2019;22(1):65-72.
27. Huang G, Wharton W, Travison TG, Ho MH, Gleason C, Asthana S, et al. Effects of testosterone administration on cognitive function in hysterectomized women with low testosterone levels: a dose-response randomized trial. J Endocrinol Invest. 2015;38(4):455-61.
28. Khorsand I, Kashef R, Ghazanfarpour M, Mansouri E, Dashti S, Khadivzadeh T. The Beneficial and Adverse Effects of Raloxifene in Menopausal Women: A Mini Review. J Menopausal Med. 2018;24(3):183-7.
29. Lello S, Capozzi A, Scambia G. The Tissue-Selective Estrogen Complex (Bazedoxifene/Conjugated Estrogens) for the Treatment of Menopause. Int J Endocrinol. 2017:5064725.
30. Kagan R, Kellogg-Spadt S, Parish SJ. Practical Treatment Considerations in the Management of Genitourinary Syndrome of Menopause. Drugs Aging. 2019;36(10): 897-908.
31. Manson JE, Ames JM, Shapiro M, Gass ML, Shifren JL, Stuenkel CA, et al. Algorithm and mobile app for menopausal symptom management and hormonal/non-hormonal therapy decision making: a clinical decision-support tool from The North American Menopause Society. Menopause. 2015;22(3):247-53.
32. Hill DA, Crider M, Hill SR. Hormone therapy and other treatments for symptoms of menopause. Am Fam Physician. 2016;94(11):884-9.
33. Chen LR, Ko NY, Chen KH. Isoflavone supplements for menopausal women: a systematic review. Nutrients. 2019;11(11):2649.

Section

8 Gynecology Oncology

Chapter 50

Screening Strategies in Gynecologic Cancers

Priya Ganesh Kumar, Urvashi Miglani

INTRODUCTION

Screening protocols enable us to diagnose disease in asymptomatic patients. Mostly they allow us to clinch the diagnosis at an early stage so that the chances of cure are improved. For this, awareness about all the screening protocols is mandatory.

CERVICAL CANCER

Cervical cancer is preventable and curable, yet it has claimed the lives of more than 300,000 women in 2018. So, in May, 2018, the WHO (World Health Organization) Director-General announced a global call for action to eliminate cervical cancer, and in August, 2020 the World Health Assembly adopted the Global Strategy for Cervical Cancer Elimination.

To eliminate cervical cancer, all countries must reach and maintain an incidence rate of below 4 per 100,000 women. To achieve that goal, the following three issues need to be targeted:

1. *Vaccination:* 90% of girls fully vaccinated with the HPV (human papillomavirus) vaccine by the age of 15 years.
2. *Screening:* 70% of women screened using a high-performance test by the age of 35 years, and again by the age of 45 years.
3. *Treatment:* 90% of women with precancer treated and 90% of women with invasive cancer managed.

Each country should meet the 90-70-90 targets by 2030 to eliminate cervical cancer. The screening protocols are available from several organizations such as ASCCP (American Society for Colposcopy and Cervical Pathology), FOGSI (The Federation of Obstetric & Gynaecological Societies of India), and WHO. All these will be discussed in this chapter.

FOGSI GCPR

The FOGSI GCPR (Good Clinical Practice Recommendation) divides the health care systems into two categories: settings with good resources and settings with limited resources. On the basis of resources available, one can choose the screening and treatment modality. The following algorithms elucidate the screening strategies and further management of abnormal results as per FOGSI GCPR.[1]

Screening: General Principles (Flowcharts 1 to 11)

- *Age to stop:* 65 years with consistent negative results in last 15 years—women with no prior screening should undergo tests once at 65 years and, if negative, they should exit screening.
- Screening should be done for 20 years following abnormal reports ≥CIN2+ (cervical intraepithelial neoplasia 2+), irrespective of method of treatment.
- Following hysterectomy in which cervix was removed for benign causes: no need for screening, unless there is history of previous CIN
- Absence of cervix needs to be confirmed by clinical records or examination in a hysterectomized patient.
- Screening to be done at clinician's discretion if records are inadequate.
- In women with histopathologically confirmed CIN in hysterectomy specimen: Screening with HPV at 6 months and 18 months

Flowchart 1: Screening with primary HPV (human papillomavirus) testing in women aged >30 years.

HPV testing
HPV negative
HPV positive
VIA
Cytology
HPV 16/18 genotyping
Negative
Positive
≥ASCUS
NILM
Negative
Positive
Repeat HPV test at 1 year
Repeat HPV test at 1 year
HPV negative
HPV positive
HPV negative
HPV positive
Colposcopy ± biopsy
Return to routine 5 yearly screening protocol
Colposcopy ± biopsy
Return to routine 5 yearly screening protocol
Normal
Abnormal
Normal
Abnormal
Repeat HPV test at 1 year
Follow FOGSI CIN management GCPR
Repeat HPV test at 1 year
Follow FOGSI CIN management GCPR

Available HPV tests
The digene HPV test (Hybrid capture 2, Qiagen)
Cobas (Roche)

Affordable HPV test
careHPV (Qiagen)

(ASCUS: atypical squamous cells of undetermined significance; CIN: cervical intraepithelial neoplasia; FOGSI: The Federation of Obstetric & Gynaecological Societies of India; GCPR: Good Clinical Practice Recommendation; NILM: negative for intraepithelial lesion or malignancy)

Flowchart 2: Screening with Co-testing (HPV test with cytology) in women aged >30 years.

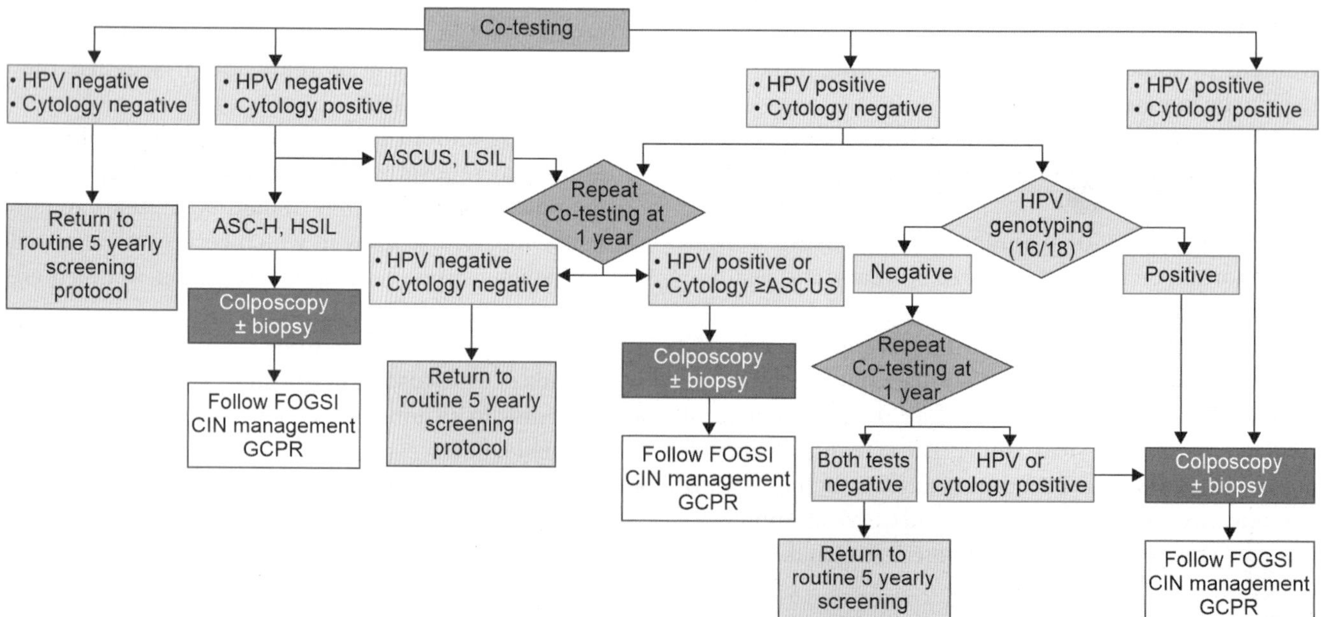

(ASC–H: atypical squamous cells cannot exclude HSI; HSIL: high-grade squamous intraepithelial lesion; LSIL: low-grade squamous intraepithelial lesion)

Flowchart 3: Screening with cytology: Management of ASCUS in women aged 30–64 years.

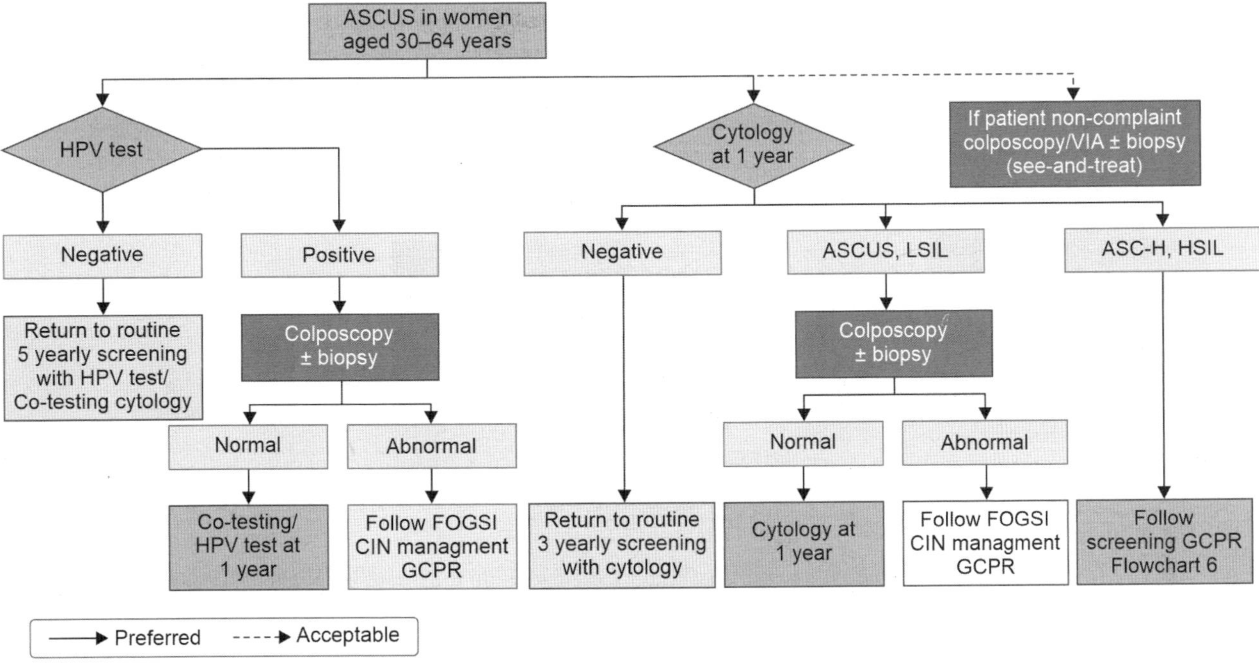

(ASC–H: atypical squamous cells cannot exclude HSI HSIL: high-grade squamous intraepithelial lesion; LSIL: low-grade squamous intraepithelial lesion; VIA: visual inspection with acetic acid)

Flowchart 4: Screening with cytology: Management of LSIL in women aged 30–64 years.

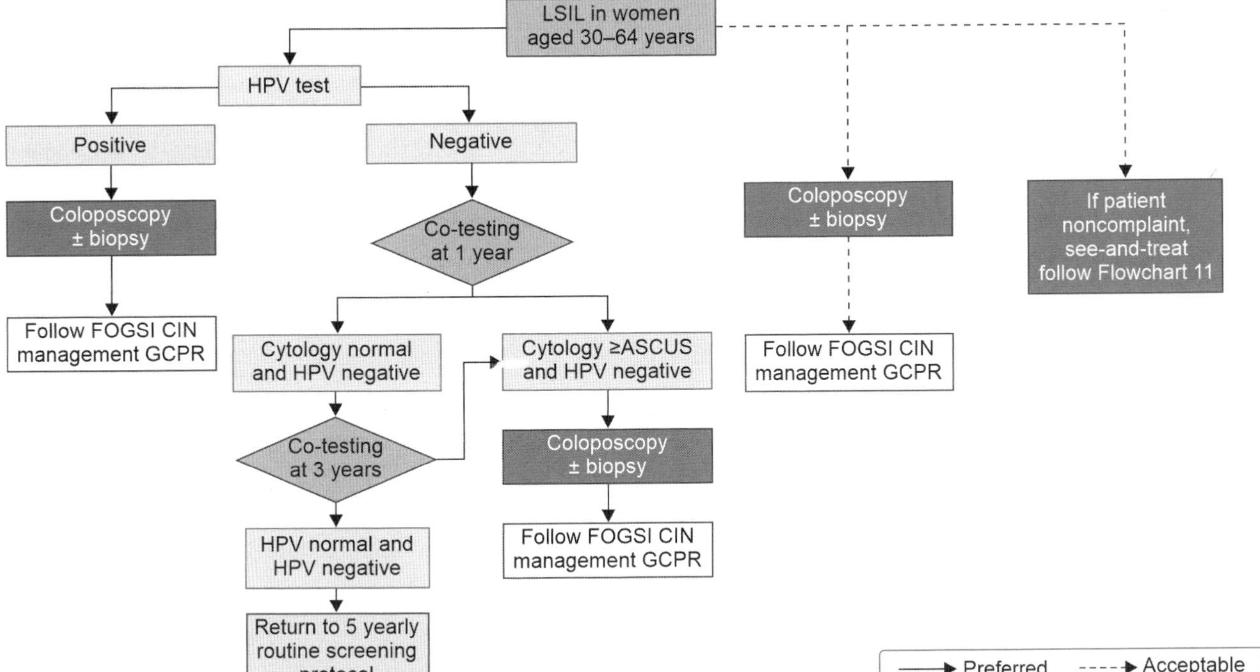

(HSIL: high-grade squamous intraepithelial lesion; LSIL: low-grade squamous intraepithelial lesion)

Flowchart 5: Screening with cytology: Management of ASCUS/LSIL in women aged 25–29 years.

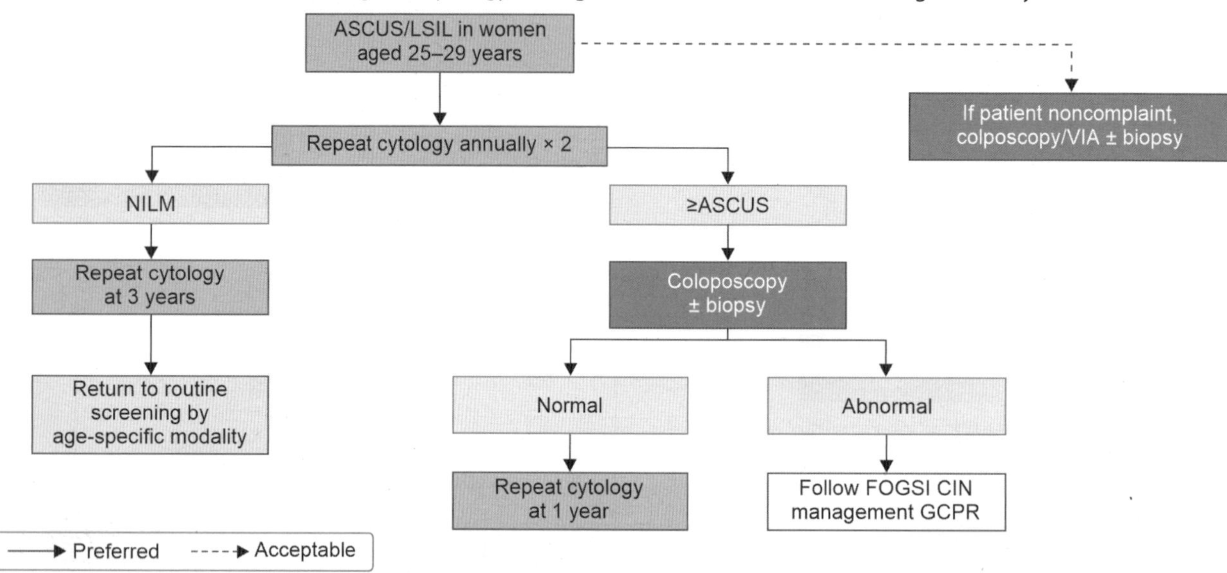

(HSIL: high-grade squamous intraepithelial lesion; LSIL: low-grade squamous intraepithelial lesion; NILM: negative for intraepithelial lesion or malignancy; VIA: visual inspection with acetic acid)

Flowchart 6: Screening with cytology: Management of ASC-H/HSIL in women aged >years.

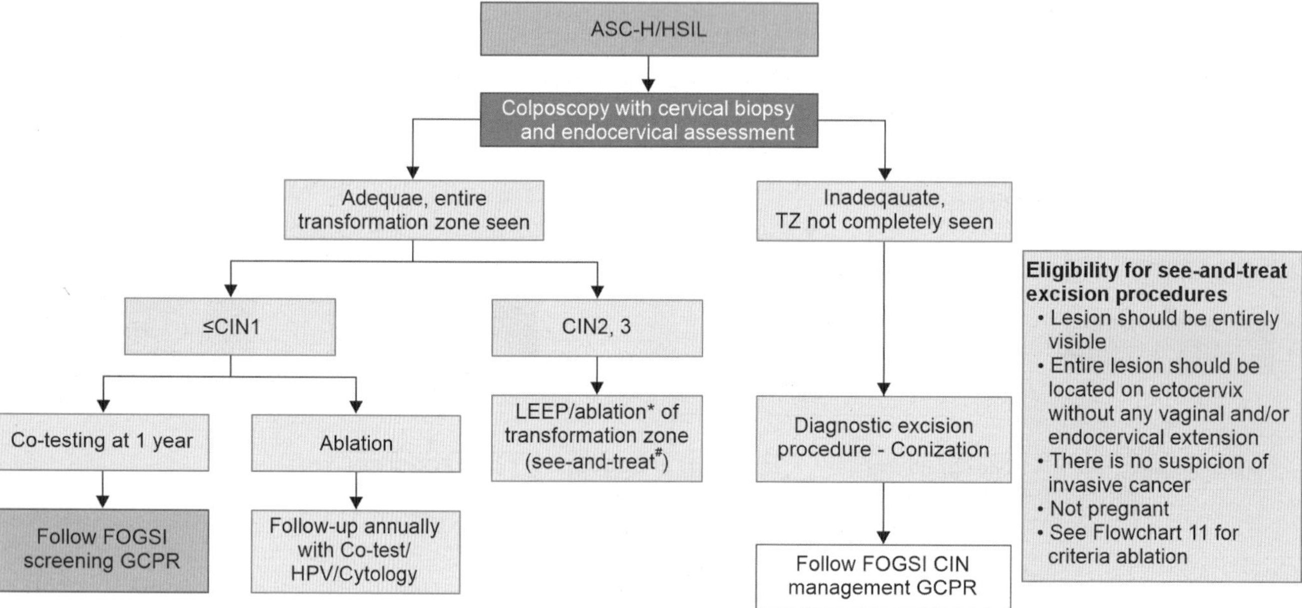

*Ablation can be considered in selected small lesions of CIN 2+

(HSIL: high-grade squamous intraepithelial lesion; LEEP: loop electrosurgical excision procedure; LSIL: low-grade squamous intraepithelial lesion)

Flowchart 7: Screening with cytology: Management of ASC-H/HSIL in women aged <30 years and desirous of pregnancy.

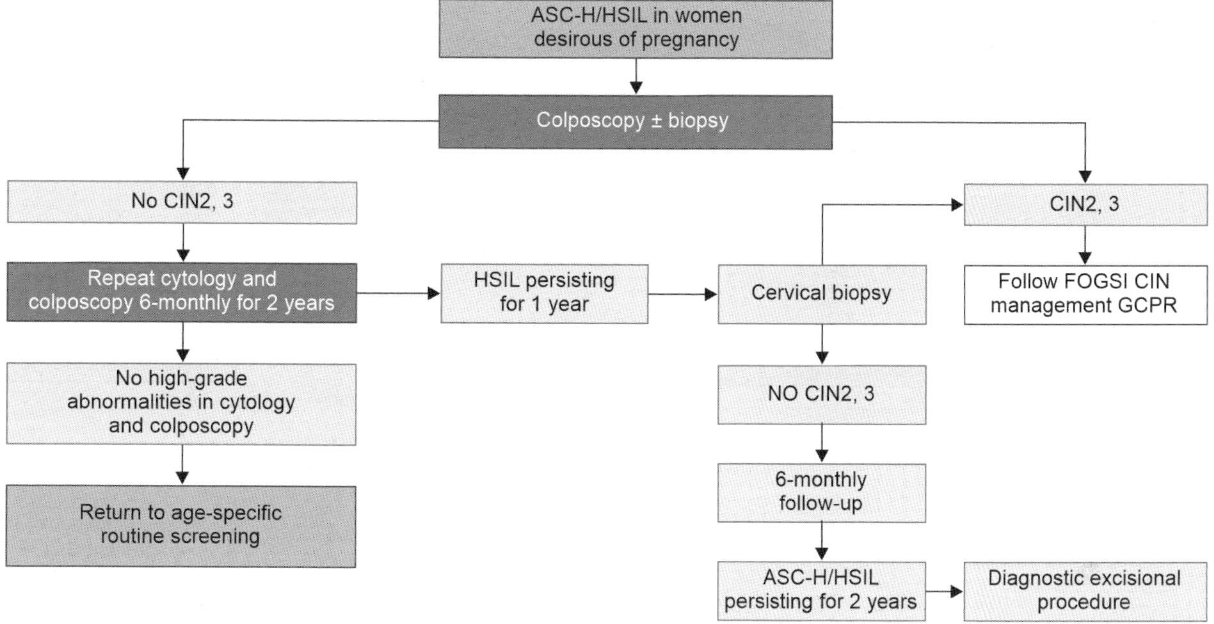

(HSIL: high-grade squamous intraepithelial lesion)

Flowchart 8: Screening with cytology: Management of abnormal glandular cells: Atypical endometrial cells.

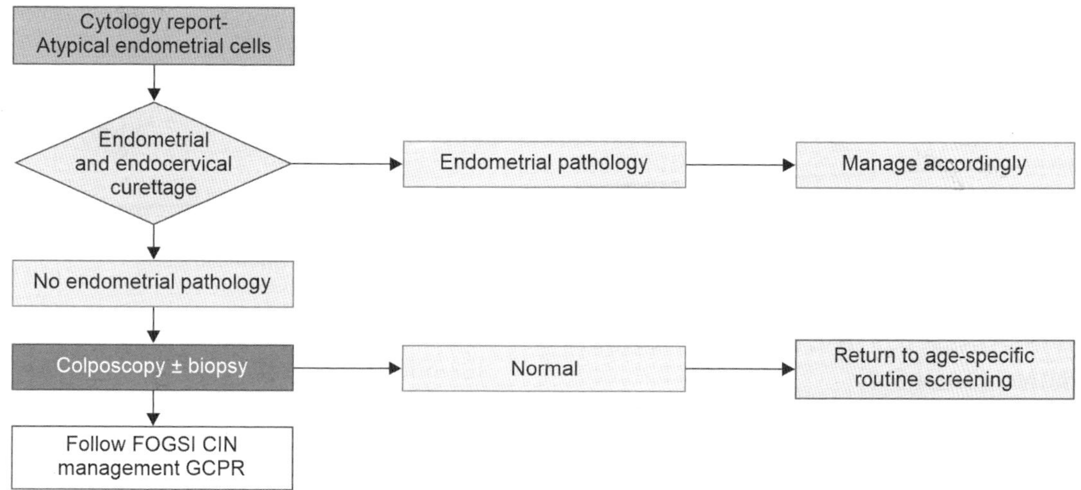

Flowchart 9: Screening with cytology: Management of abnormal glandular cells: AGC-NOS/atypical endocervical cells.

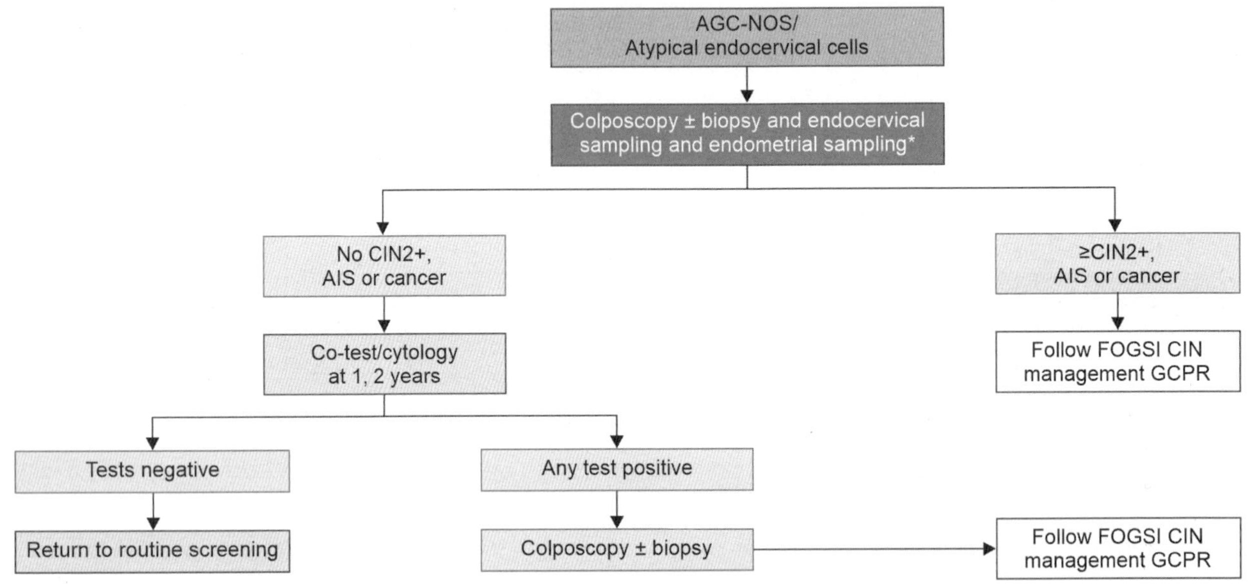

*Age >35 years, high risk factors, e.g., obesity, chronic anovulation
(AGC-NOS: atypical glandular cells – not otherwise specified; AIS: adenocarcinoma in situ)

Flowchart 10: Screening with cytology: Management of abnormal glandular cells: AGC-FN/AIS.

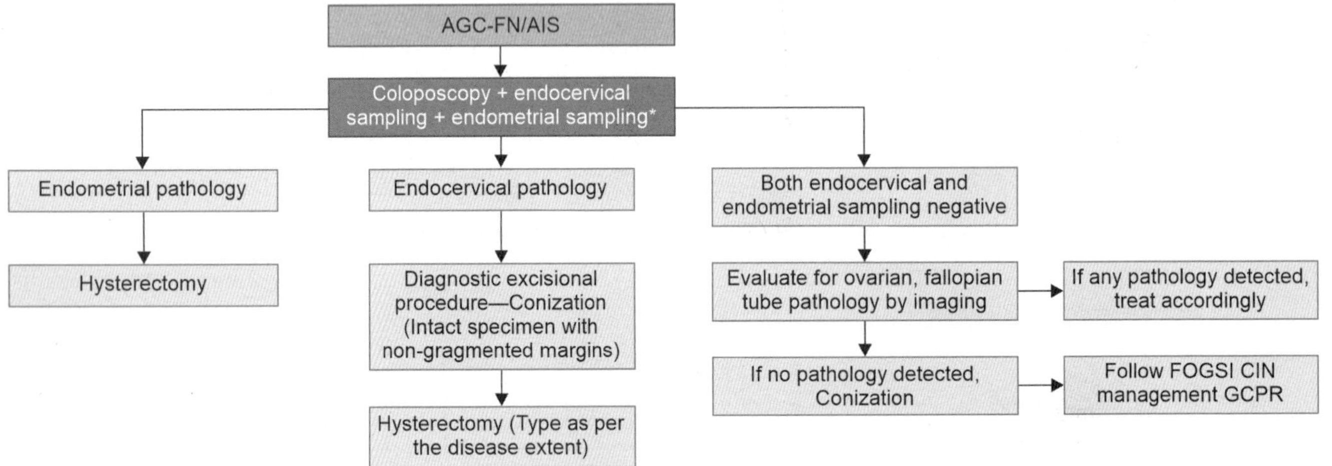

*Age >35 years, high risk factors, e.g., obesity, chronic anovulation
(AGC-FN: atypical glandular cells favoring neoplasia; AIS: adenocarcinoma in situ)

NEWER MODALITIES IN CYTOLOGY BASED SCREENING (TABLE 1)

Liquid-based cytology:

- Results in less inadequate samples (1.9%) and also allows reflex HPV and automated testing
- Sensitivity is comparable to conventional pap smears though is more sensitive for low-grade squamous intraepithelial lesion (LSIL) + lesions (11% more)
- It is more expensive.

P16INK 4A,Ki67:

- Identifies transforming HPV infection so is helpful in triaging low-grade smears
- Useful for both cytology and histology slides
- *Sensitivity*: 64–92%
- *Specificity*: 41–96%
- False-positive rates are high and it needs expertise.

Flowchart 11: Screening with VIA (visual inspection with acetic acid) in women aged 30–65 years.

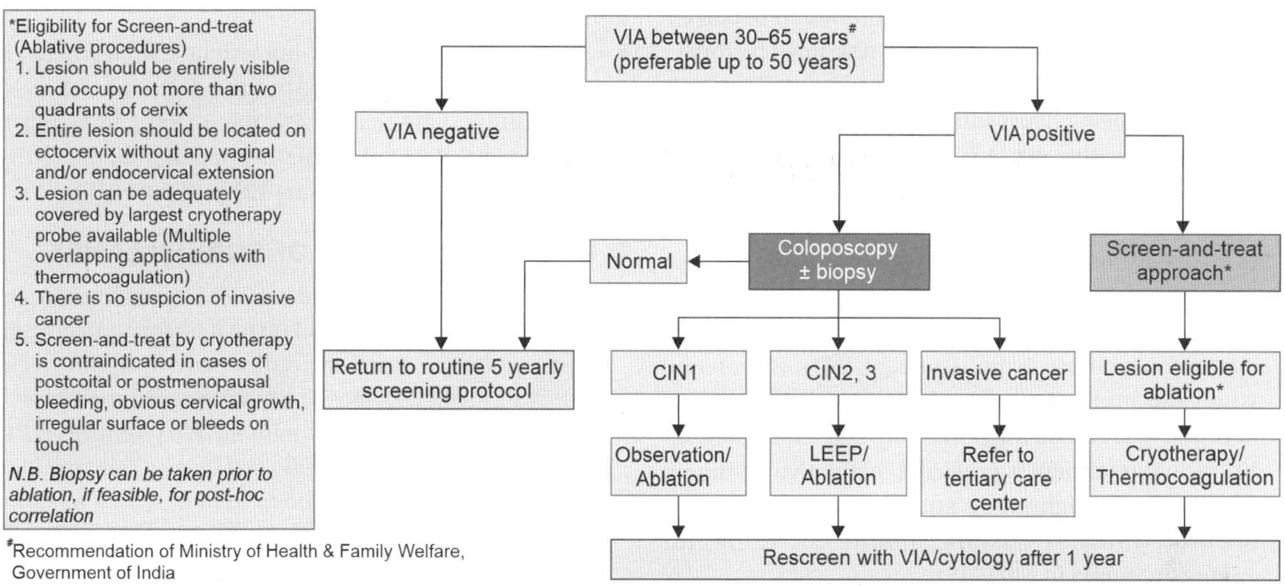

(LEEP: loop electrosurgical excision procedure)

TABLE 1: Newer modalities: Human papillomavirus (HPV)-based screening.

	Vaginal self-collection	*E6 E7 MRNA*	*DNA methylation (DNAme)*	*Markers of aberrant S-phase induction*
Options	Delphi device, Evalyn sampler	• APTIMA HPV GT (Hologic) • Oncotect (Incell Dx)	QIAsure Methylation Test (Qiagen)	Topoisomerase IIA (TOP2A) and MCM2 (BD)
Advantage	Acceptable eliminates cost of visiting clinician	• Effective triage for low-grade smears • Reduce colposcopy referral by 68% compared to 30% by HPV DNA testing	• Alternative triage for hrHPV automated, objective test • Run on the same sample as the HPV assay	Identify transforming HPV infection
Efficacy	• Sensitivity variable: 60–90% • Overall 3.4 (95% CI = 2.4–4.9) times more CIN2+ detected by self-collected HPV samples than by routine cytology	• Sensitivity: 90–95% Specificity: 42–61% • PPV: 67%	• CADM1-m18 combined with MAL-m1 methylation Sensitivity: 60.5–100% • Specificity: 22.7–83.3% (95% CI 78.4–87.4)	• Sensitivity: 67–99% • Specificity: 61–85%
Limitations	Inadequate sample possible, woman-dependent, reluctance to use medical devices	Expensive	Expensive, wide variation in reported sensitivity	Research settings

(CI: confidence interval; CIN: cervical intraepithelial neoplasia; hrHPV: high-risk HPV; PPV: positive predictive value)

GUIDELINES BY ASCCP (2019)

The key difference between 2019 guidelines and previous versions is the change from primarily test results-based algorithms to "risk-based" guidelines.[2] The following are the key changes from the prior guidelines:

- *Recommendations of these guidelines are based on risk and not on results*: Recommendations of colposcopy, treatment or surveillance will depend on a patient's risk of CIN3+ determined by a combination of current results and past history (including unknown history).

- *Deferral of colposcopy may be considered for certain patients*: Repeat HPV testing or co-testing at 1 year for HPV-positive, low-grade cytologic abnormalities after a documented negative screening HPV test or co-test.
- Guidance for expedited treatment is expanded (i.e., treatment without colposcopic biopsy). For nonpregnant patients 25 years or older, expedited treatment, defined as treatment without preceding colposcopic biopsy demonstrating CIN2+, is preferred when the immediate risk of CIN3+ is ≥60%, and is acceptable for those with risks between 25% and 60%.
- Excisional treatment is preferred to ablative treatment for histologic high-grade squamous intraepithelial lesion (HSIL). Excision should be done for adenocarcinoma in situ (AIS).
- Observation is preferred to treatment for CIN1.
- All positive primary HPV screening tests, regardless of genotype, should have additional reflex triage testing performed from the same laboratory specimen (e.g., reflex cytology).
- Continued surveillance with HPV testing or co-testing at 3-year intervals for at least 25 years is recommended after treatment and initial posttreatment management of histologic HSIL, CIN2, CIN3, or AIS. Surveillance may be continued beyond that if the subject's health allows.
- Surveillance with cytology alone is acceptable only if testing with HPV or co-testing is not feasible. Cytology is less sensitive than HPV testing so it is recommended at 6-month intervals when HPV testing or co-testing is recommended annually. Cytology is recommended annually when 3-year intervals are recommended for HPV or co-testing.
- Food and Drug Administration (FDA)-approved HPV assays should be used for management.
- For all indications, HPV mRNA and non-FDA approved HPV DNA tests should only be used as a co-test with cytology, till further data is available.

Clinical action thresholds: Clinical action thresholds determine management recommendations. CIN3 is the chosen end point for risk estimates in these guidelines because as compared to CIN2 this diagnosis is more pathologically reproducible and the HPV type distribution is closer to invasive cervical cancers.

For given current results and history combination, the immediate CIN3+ risk is examined.

- If this risk is 4% or greater, immediate management via colposcopy or treatment is indicated.
- If the immediate risk is <4%, the 5-year CIN3+ risk is examined to determine whether patients should return in 1, 3, or 5 years **(Flowchart 12)** (ASCCP 2019).

RECOMMENDATION OF SURVEILLANCE

- *Negative HPV test and co-test,* i.e., estimated 5-year CIN3+ risk of <0.15% based on past history and current test results: a return to routine screening at 5-year intervals using HPV-based testing is recommended.
- *Negative cytology results without HPV testing, HPV-positive ASC-US or LSIL followed by a colposcopy with results of less than CIN2, followed in turn by a negative follow-up HPV test or co-test, HPV-negative ASC-US,* i.e., estimated 5-year CIN3+ risk of 0.15–0.55% based on history and current test results, repeat testing in 3 years with HPV-based testing is recommended.
- *HPV-positive/NILM (negative for intraepithelial lesion or malignancy) or HPV-negative/LSIL with unknown previous screening history, histologic LSIL (CIN1) or less preceded by a low-grade co-test result (defined as HPV-positive LSIL, HPV-positive ASC-US, or repeated HPV-positive NILM), new HPV-positive ASC-US or LSIL results after a documented negative HPV test or co-test within an appropriate screening interval (approximately 5 years) or colposcopic examination less than CIN2 within the past year:* Estimated risk of CIN3+ based on history and current results that is below the threshold for immediate colposcopy (4.0% immediate risk) and above the 3-year follow-up threshold (≥0.55% at 5 years), repeat testing in 1 year with HPV-based testing is recommended.

RECOMMENDATION OF COLPOSCOPY

When patients have an estimated immediate risk of diagnosis of CIN3+ of 4.0% or greater based on history and current results, referral to colposcopy is recommended.

The HPV-positive ASC-US or HPV-positive LSIL cytology with unknown previous screening history and HPV-positive NILM cytology occurring at 2 consecutive annual visits): Colposcopy

The HPV-positive ASC-US or LSIL in patients whose preceding screening result was a negative HPV test or co-test within a routine screening interval (approximately 5 years). This additional information reduces the immediate CIN3+

Flowchart 12: Evaluation of patient risk.

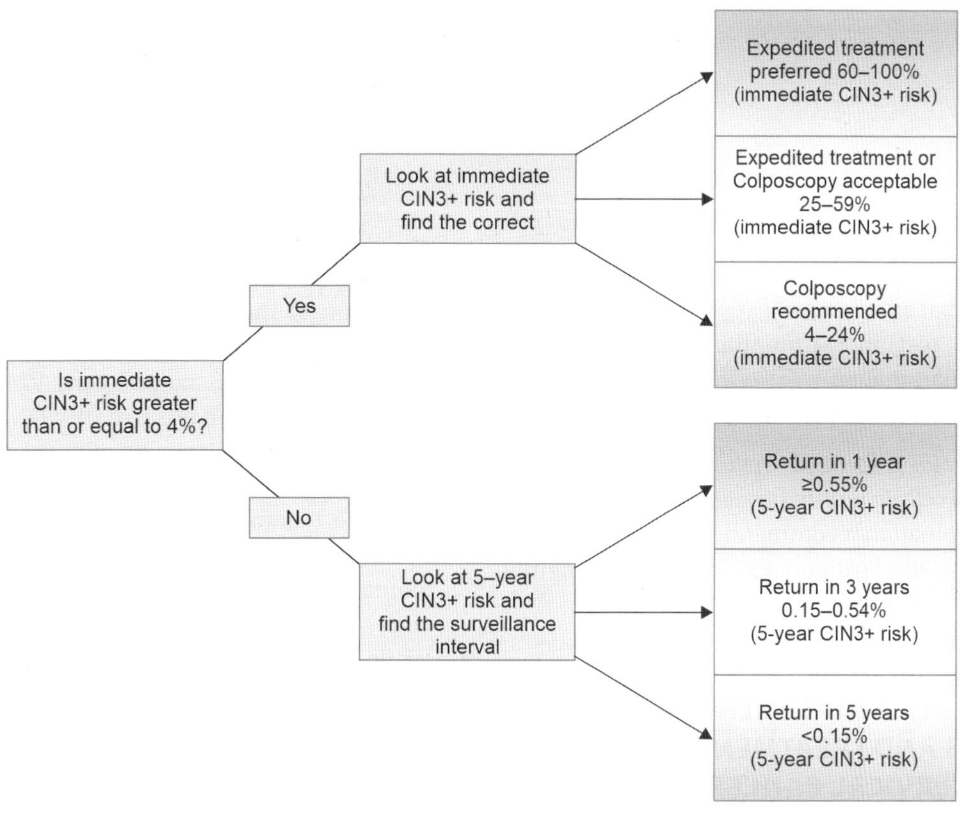

risk to approximately 2%, so 1-year surveillance may be done instead of immediate colposcopy.

EXCEPTIONS TO COLPOSCOPY THRESHOLD

- In *Patients with atypical squamous cells cannot exclude HSI cytology*, colposcopy is recommended regardless of HPV result because the immediate cancer risk for ASC-H is disproportionately high compared with the CIN3+ risk.
- *Patients with HPV 18-positive NILM*: HPV 18-positive NILM had a disproportionately high cancer risk compared with other results: 0.2% immediately and 0.56% at 5 years. This suggests that HPV 18-related CIN3 or AIS may be difficult to diagnose and/or more apt to rapidly progress from precancer to cancer.
- After two consecutive unsatisfactory screening tests

RECOMMENDATIONS OF TREATMENT (WOMEN AGED >25 YEARS)

High-grade squamous intraepithelial lesion cytology that is HPV 16-positive, HPV-positive HSIL cytology (regardless of genotype) in women who were underscreened (generally defined as no screening in >5 years):

- Estimated immediate risk of CIN3+ of 60% or greater based on history and current results
- Treatment using an excisional procedure without previous biopsy confirmation (preferred) or
- Colposcopy with biopsy (acceptable)

The HPV-negative HSIL cytology, HPV-positive ASC-H cytology, HPV-positive AGC, HPV-positive HSIL cytology:

- Estimated immediate risk of CIN3+ 25% or greater and <60% based on history and current results
- Treatment using an excisional procedure without previous biopsy confirmation or histologic evaluation with colposcopy and biopsy.

Primary HPV screening:

- *Primary HPV screening is used:* Do additional reflex triage test
- *Primary HPV screening test HPV 16 or HPV 18 positive and reflex triage testing not feasible:* Colposcopy with collection of triage test at colposcopy

Cytology that is NILM, but with absent transformation zone or endocervical cells **(Flowchart 13)**.

Flowchart 13: Repeat cytology screening after every 3 year.

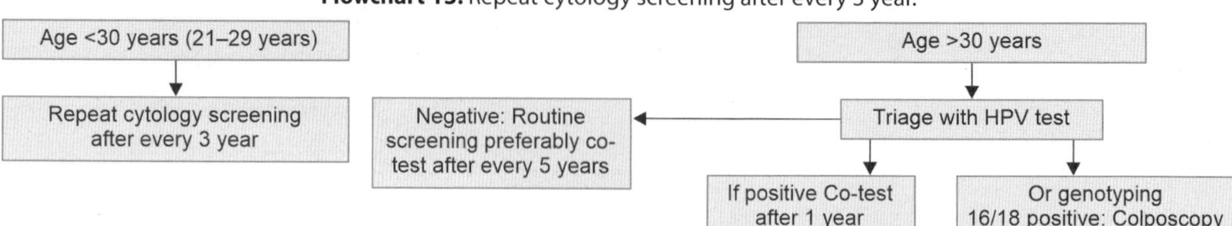

Unsatisfactory cytology:
- Repeat age-based screening (cytology, co-test, or primary HPV test) in 2–4 months after treatment of atrophy or inflammation.
- Triage using HPV testing is not recommended.

Benign endometrial cells in premenopausal patients or benign glandular cells in post-hysterectomy patients:
- No further evaluation recommended.
- For postmenopausal patients with benign endometrial cells, endometrial assessment is recommended.

WHO RECOMMENDATIONS AND GOOD PRACTICE STATEMENTS ON SCREENING AND TREATMENT TO PREVENT CERVICAL CANCER[4]

World Health Organization stresses on the use of two approaches for cervical screening:
1. *Screen-and-treat approach:* Decision to treat is based on a positive primary screening test only.
2. *Screen, triage and treat approach:* Decision to treat is based on a positive primary screening test followed by a positive second test (a "triage" test), with or without histologically confirmed diagnosis.

Recommendations and good practice statements: General population of women
- Use HPV DNA detection (samples taken by a health care provider or self-collected samples) as the primary screening test (with or without triage) rather than visual inspection with acetic acid (VIA) or cytology in screening.
- Start regular cervical cancer screening at the age of 30 years among the general population of women with screening at 5- to 10-year intervals (with HPV) and 3-year intervals (with VIA or cytology).
- In a screen-and-treat approach using HPV DNA detection as the primary screening test, WHO suggests treating women who test positive for HPV DNA among the general population of women.
- In a screen, triage and treat approach using HPV DNA detection as the primary screening test among the general population of women, WHO suggests using partial genotyping, colposcopy, VIA or cytology to triage women after a positive HPV DNA test.
- After the age of 50 years, screening may be stopped after two consecutive negative screening results consistent with the recommended regular screening intervals among both the general population of women and women living with HIV (human immunodeficiency virus).
- While transitioning to a program with a recommended regular screening interval, screening even just twice in a lifetime is beneficial.
- In the general population of women who have screened positive on an HPV DNA primary screening test and then negative on a triage test are retested with HPV DNA testing at 24 months and, if negative, move to the recommended regular screening interval.
- Women from the general population and women living with HIV who have screened positive on a cytology primary screening test and then have normal results on colposcopy are retested with HPV DNA testing at 12 months and, if negative, move to the recommended regular screening interval.
- Women from the general population who have been treated for histologically confirmed CIN2/3 or AIS, or treated as a result of a positive screening test are retested at 12 months with HPV DNA testing when available, rather than with cytology or VIA or co-testing and, if negative, move to the recommended regular screening interval.

RECOMMENDATIONS AND GOOD PRACTICE STATEMENTS: WOMEN LIVING WITH HIV

- HPV DNA detection in a screen, triage and treat approach starting at the age of 25 years with regular screening every 3–5 years.

- Women living with HIV who have screened positive on an HPV DNA primary screening test and then negative on a triage test, are retested with HPV DNA testing at 12 months and, if negative, move to the recommended regular screening interval.
- Women living with HIV who have been treated for histologically confirmed CIN2/3 or AIS, or treated as a result of a positive screening test are retested at 12 months with HPV DNA testing when available, rather than with cytology or VIA or co-testing, and, if negative, are retested again at 12 months and, if negative again, move to the recommended regular screening interval.

Screen-and-treat approaches:

- VIA as the primary screening test, followed by treatment
- HPV DNA (self- or clinician-collected) as the primary screening test, followed by treatment

Screen, triage and treat approaches:

- Cytology as the primary screening test, followed by colposcopy triage, followed by treatment
- HPV DNA as the primary screening test, followed by HPV16/18 triage (when already part of the HPV test), followed by treatment, and using VIA triage for those who screen negative for HPV16/18 5. HPV DNA as the primary screening test, followed by VIA triage, followed by treatment
- High-risk HPV DNA as the primary screening test, followed by colposcopy triage, followed by treatment.
- HPV DNA as the primary screening test, followed by cytology triage, followed by colposcopy and treatment.

The seven algorithms of WHO outline both the approaches **(Flowcharts 14 to 20)**.

ENDOMETRIAL CANCER

For the purpose of screening, the population may be divided into three groups:

1. At average risk for endometrial cancer
2. At increased risk for endometrial cancer
3. At high risk for endometrial cancer

At average risk for endometrial cancer: Endometrial cancer screening in the general population does not decrease mortality, so *not recommended.*[5,6]

At increased risk for endometrial cancer: History of unopposed estrogen therapy, late menopause, tamoxifen therapy, nulliparity, infertility or failure to ovulate, obesity, diabetes or hypertension

Flowchart 14: Primary VIA (visual inspection with acetic acid) screening (screen-and-treat) approach: For both the general population of women and women living with HIV (human immunodeficiency virus).

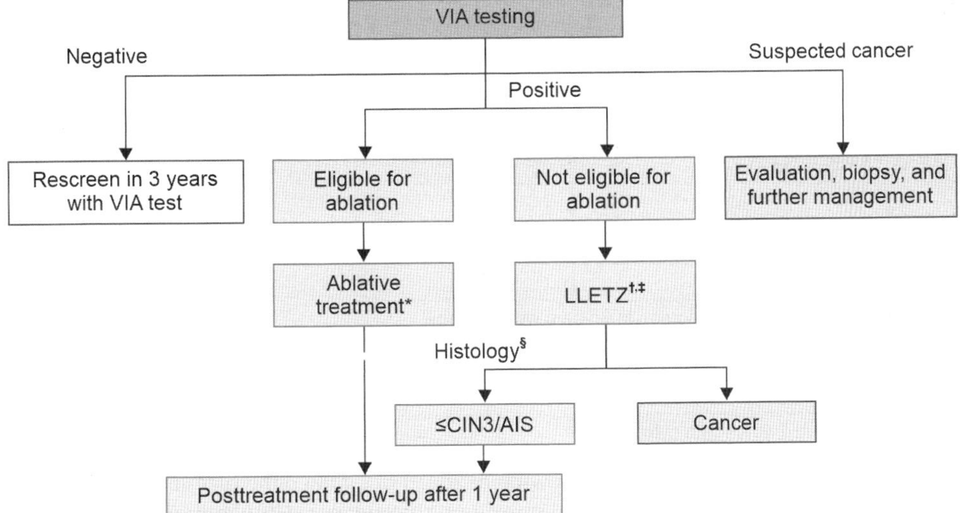

*Ablative treatment includes cryotherapy and thermal ablation.
†Cold knife conization (CKC) if LLETZ not available.
‡LLETZ and LEEP (loop electrosurgical excision procedure) indicate the same procedure.
§Histology may not be available in certain settings; women should be advised to attend follow-up after 1 year or to report earlier, if they have any of the symptoms of cervical cancer.
(AIS: adenocarcinoma in situ; LLETZ: large-loop excision of the transformation zone)

Flowchart 15: Primary HPV DNA test screening (screen-and-treat) approach: For the general population of women.

HPV DNA testing (self-sampled or collected by clinician)

Negative → Rescreen in 5 to 10 years with HPV DNA test

Positive → Determine eligibility for ablative treatment (after application of 3–5% acetic acid with or without magnification)

- Eligible for ablation → Ablative treatment* → Posttreatment follow-up after 1 year
- Not eligible for ablation → LLETZ†‡ → Histology§ → ≤CIN3/AIS; Cancer
- Suspected cancer → Evaluation, biopsy and further management

*Ablative treatment includes cryotherapy and thermal ablation.
†Cold knife conization (CKC) if LLETZ not available.
‡LLETZ and LEEP (loop electrosurgical excision procedure) indicate the same procedure.
§Histology may not be available in certain settings; women should be advised to attend follow-up after 1 year or to report earlier, if they have any of the symptoms of cervical cancer.
(AIS: adenocarcinoma in situ; LLETZ: large-loop excision of the transformation zone)

Flowchart 16: Primary cytology screening and colposcopy triage (screen, triage and treat approach): For both the general population of women and women living with HIV (human immunodeficiency virus).

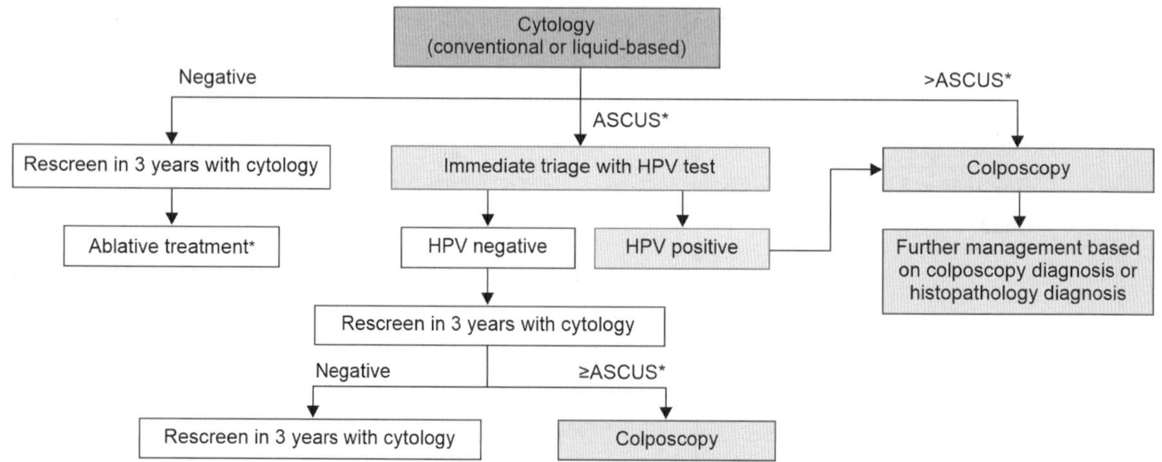

*Some programs prefer to use LSIL (low-grade squamous intraepithelial lesion) threshold.
(ASCUS: atypical squamous cells of undetermined significance; HPV: human papillomavirus)

- Routine screening not required in asymptomatic women but they should be informed about the risks and symptoms of endometrial cancer and advised to report any unexpected bleeding.
- Routine screening for endometrial cancer in asymptomatic tamoxifen users is not recommended but they should be informed about symptoms of cancer.

Flowchart 17: HPV DNA screening and HPV16/18 triage (screen, triage and treat approach): For both the general population of women and women living with HIV (human immunodeficiency virus).

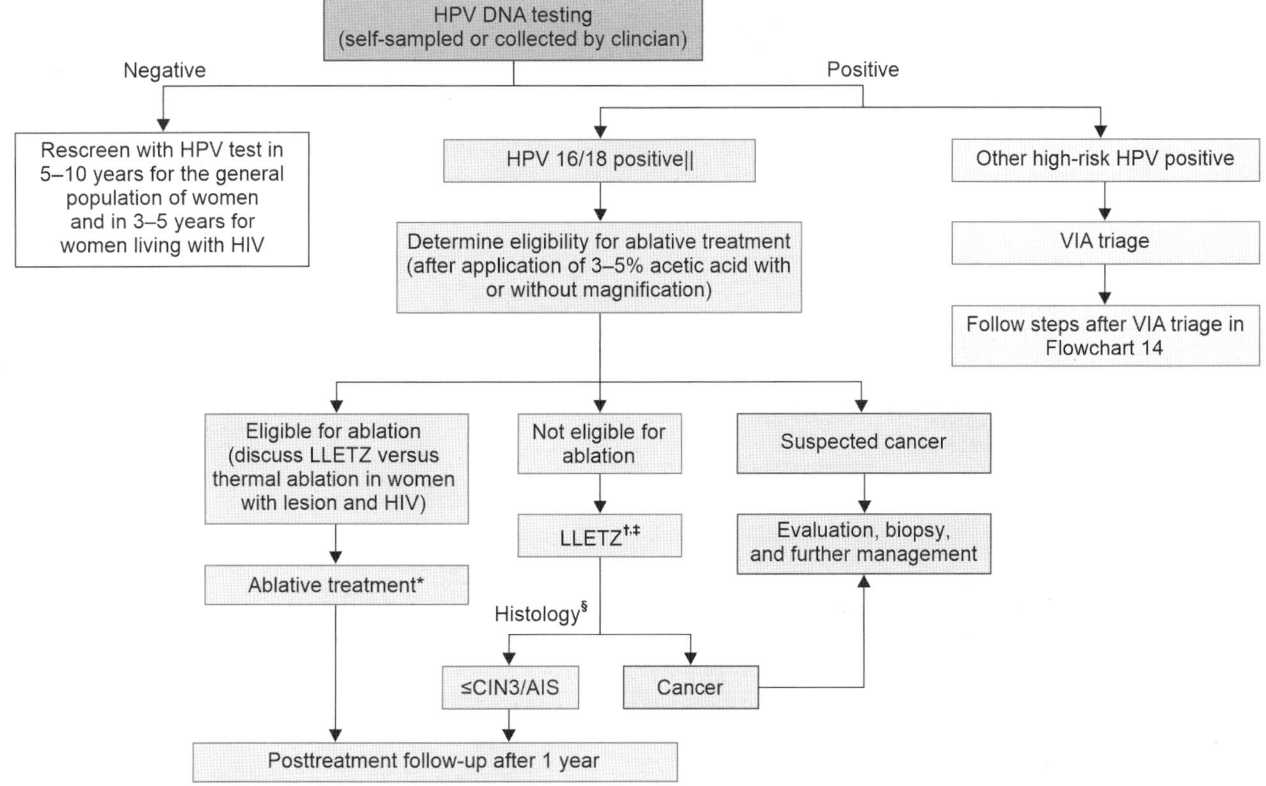

*Ablative treatment includes cryotherapy and thermal ablation.
†Cold knife conization (CKC) if LLETZ not available
‡LLETZ and LEEP (loop electrosurgical excision procedure) indicate the same procedure.
§Histology may not be available in certain settings; women should be advised to attend follow-up after 1 year or to report earlier, if they have any of the symptoms of cervical cancer.
||May or may not be positive for HPV 45.
(AIS: adenocarcinoma in situ; LLETZ: large-loop excision of the transformation zone; VIA: visual inspection with acetic acid)

- In women with adult granulosa cell tumor, if hysterectomy has not been performed, endometrial sampling is recommended. If this shows no evidence of (pre) malignancy, no further screening for endometrial malignancies is required.
- In patients with epithelial ovarian cancer undergoing fertility-sparing treatment, endometrial sampling is recommended at the time of diagnosis.[5]

At high risk for endometrial cancer: Lynch mutation carrier or those with strong family history

- Surveillance of the endometrium by gynecological examination, transvaginal ultrasound, and aspiration biopsy starting from the age of 35 years (annually until hysterectomy) should be offered to all Lynch syndrome mutation carrier.
- Prophylactic hysterectomy and bilateral salpingo-oophorectomy to be considered for patients at of Lynch syndrome at 40 years or completion of family.[5]

OVARIAN CANCER

Average risk women aged 45 years or older: Screening with transvaginal sonography alone, or CA 125 alone or both is *not recommended* as it does not decrease cancer-related mortality.[7]

- Evidence shows that markers such as HE4, mesothelin, spondin 2 do not increase early enough to be useful for screening.[7]
- *Prediction algorithms* like OVA 1, risk of malignancy index to be used only in preoperative assessment of patients with an adnexal mass and *not to be used for screening*.[7]

Flowchart 18: Primary HPV DNA screening and VIA triage (screen, triage and treat approach): For both the general population of women and women living with HIV (human immunodeficiency virus).

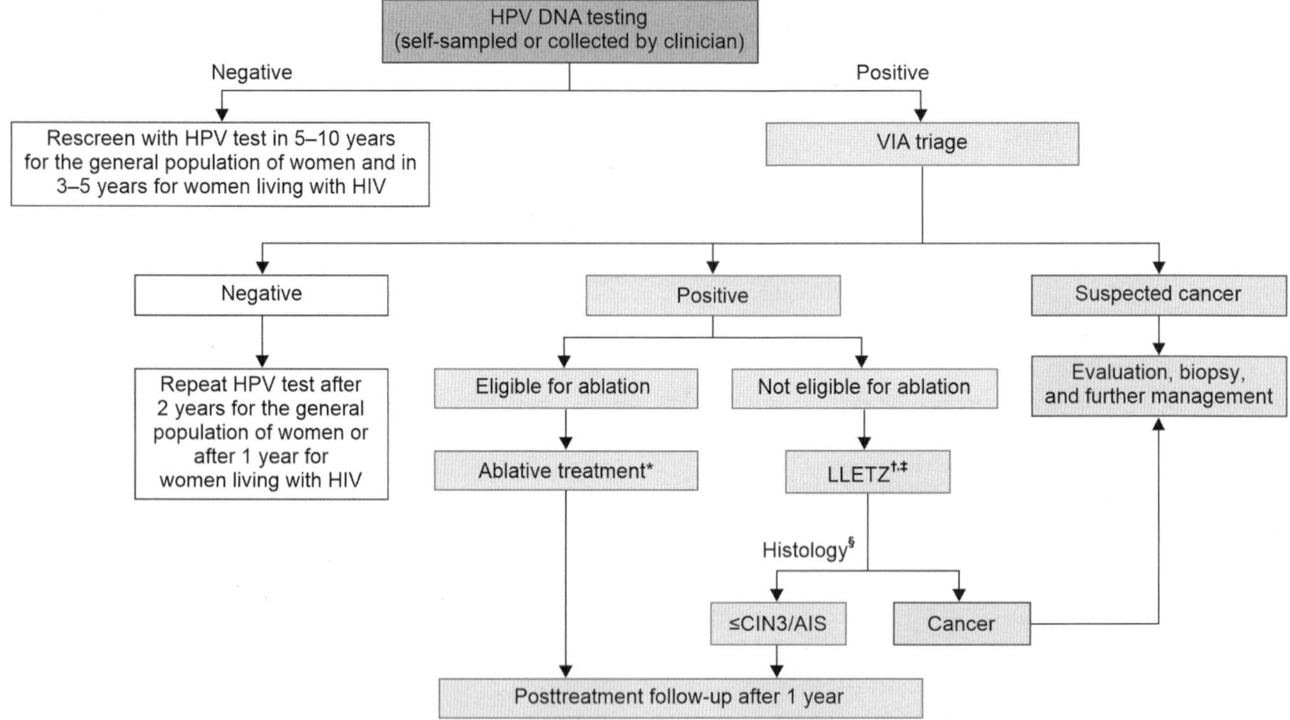

*Ablative treatment includes cryotherapy and thermal ablation.
†Cold knife conization (CKC) if LLETZ not available.
‡LLETZ and LEEP (loop electrosurgical excision procedure) indicate the same procedure.
§Histology may not be available in certain settings; women should be advised to attend follow-up after 1 year or to report earlier, if they have any of the symptoms of cervical cancer.
(AIS: adenocarcinoma in situ; LLETZ: large-loop excision of the transformation zone; VIA: visual inspection with acetic acid)

Flowchart 19: Primary HPV DNA screening and colposcopy triage (screen, triage and treat approach): For both the general population of women and women living with HIV (human immunodeficiency virus).

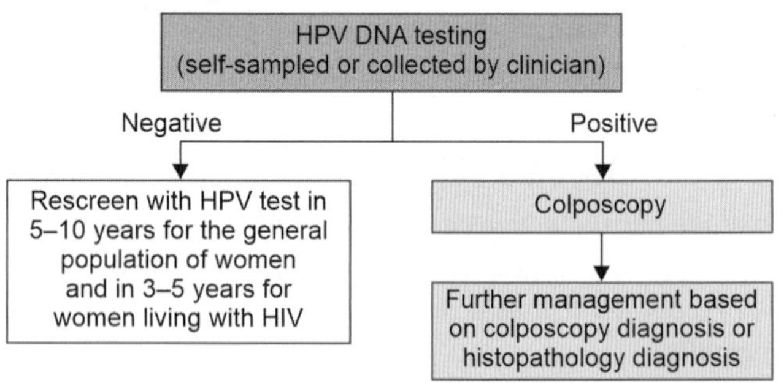

High-risk women (BRCA mutation carriers, family history of breast or ovarian cancer):[7]

- Risk-reducing salpingo-oophorectomy (RRSO) is preferred over screening (35–40 years for BRCA 1 and 40–45 years for BRCA 2).
- For women who decline RRSO, monitoring with CA 125 or sonography though of uncertain benefit may be considered at clinician's discretion starting at age 30–35 years. Evidence shows that it does not improve cancer-related mortality but improves the likelihood of diagnosis at early stage.[8,9]

VULVAL CANCER

- No evidence for specific screening for vulvar cancer.
- Encourage self-examination in women with lichen sclerosus
- Evaluation of any patient with symptoms (e.g., chronic vulvar pruritus) or signs (e.g., pigmented lesions, irregular ulcers) as they may require skin biopsy
- Regular inspection of the vulva should be done for women harboring SIL of the cervix, vagina, or anus.[10]

Flowchart 20: Primary HPV screening and cytology triage followed by colposcopy (screen, triage and treat approach): For both the general population of women and women living with HIV (human immunodeficiency virus).

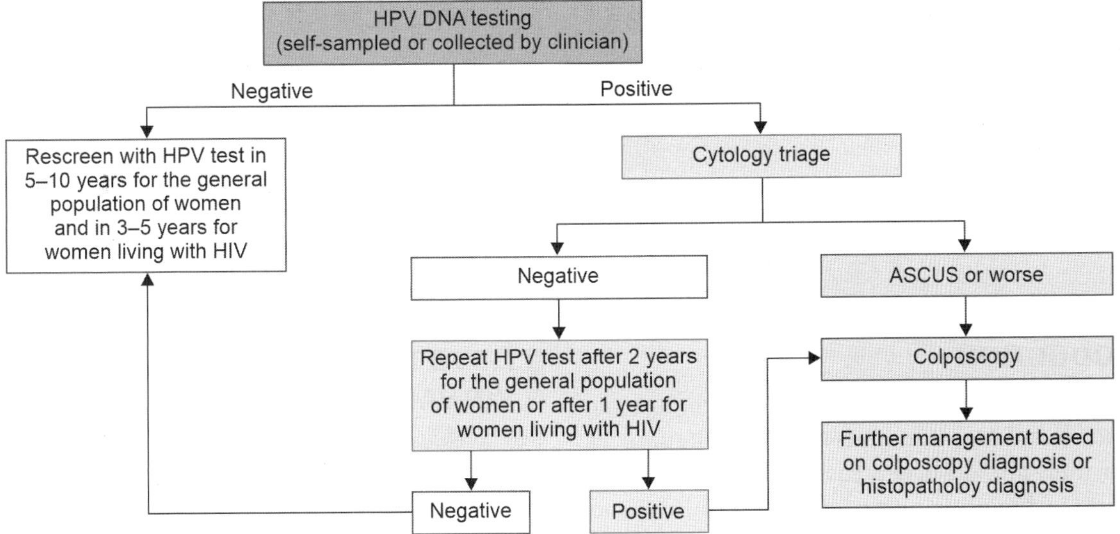

VAGINAL CANCER

- *Hysterectomy for benign disease*: No evidence to indicate routine screening for vaginal cancer
- *Hysterectomy for persistent HSIL after repeated excisional procedures of the cervix*: Regular vault smears and colposcopy.[11]

REFERENCES

1. World Health Organization. Global strategy to accelerate the elimination of cervical cancer as a public health problem. 17 November 2020. Global strategy.
2. FOGSI Gynaecologic Oncology Committee. (2018). FOGSI GCPR on Screening and Treatment of Preinvasive Lesions of Cervix and HPV vaccination. [online] Available from https://www.fogsi.org/wp-content/uploads/2018/03/FOGSI-GCPR-March-2018-final.pdf [Last accessed April, 2024].
3. Perkins RB, Guido RS, Castle PE, Chelmow D, Einstein MH, Garcia F, et al.; 2019 ASCCP Risk-Based Management Consensus Guidelines Committee. 2019 ASCCP risk-based management consensus guidelines for abnormal cervical cancer screening tests and cancer precursors. J Low Genit Tract Dis. 2020;24(2):102-31.
4. World Health Organization. WHO Guideline for Screening and Treatment of Cervical Pre-cancer Lesions for Cervical Cancer Prevention Geneva: World Health Organization; 2021. pp. 24-38.
5. Colombo N, Creutzberg C, Amant F, Bosse T, González-Martín A, Ledermann J, et al.; ESMO-ESGO-ESTRO Consensus Conference Working Group. ESMO-ESGO-ESTRO Consensus Conference on Endometrial Cancer: diagnosis, treatment and follow-up. Ann Oncol. 2016;27(1):16-41.
6. Koskas M, Amant F, Mirza MR, Creutzberg CL. Cancer of the corpus uteri: 2021 update. Int J Gynaecol Obstet. 2021;155:45-60.
7. National Comprehensive Cancer Network. Ovarian Cancer/Fallopian Tube Cancer/Primary Peritoneal Cancer. [online] Available from https://www.nccn.org/guidelines/nccn-guidelines/guidelines-detail?category=1&id=1453 [Last accessed April, 2024].
8. Daly MB, Pal T, Berry MP, Buys SS, Dickson P, Domchek SM, et al . Genetic/familial high-risk assessment: breast, ovarian, and pancreatic, version 2.2021, NCCN clinical practice guidelines in oncology. J Natl Compr Canc Netw. 2021;19(1):77-102.
9. Genetic/Familial High-Risk Assessment: Breast, Ovarian, and Pancreatic. [online] Available from https://www.nccn.org/guidelines/nccn-guidelines/guidelines-detail?category=2&id=1503 [Last accessed April, 2024].
10. Olawaiye AB, Cuello MA, Rogers LJ. Cancer of the vulva: 2021 update. Int J Gynecol Obstet. 2021;155:7-18.
11. Adams TS, Rogers LJ, Cuello MA. Cancer of the vagina: 2021 update. FIGO cancer report Int J Gynaecol Obstet. 2021;155:19-27.

Chapter 51

Tumors of Vulva and Vagina

Seema Singhal, Aarthi S Jayraj

INTRODUCTION

Vulvar cancer represents 4–5% of all female genital tract malignancies and for <1% of all female malignancies. Every year, there are 45,240 new cases of vulvar carcinoma in the world, accounting for an age-adjusted incidence rate of 0.85 per million women and a cumulative risk of 0.25% till age 74 years. Squamous cell carcinoma of the vulva accounts for 85–90% of all vulvar cancers followed by melanoma of the vulva. Other histological subtypes include basal cell carcinoma, Bartholin gland adenocarcinoma, sarcoma, etc.

RISK FACTORS

- Human papillomavirus (HPV) infection
- Multiple sexual partners/early age at onset of sexual activity
- Immunocompromised state [PLHA (people living with HIV/AIDS), inherited immunodeficiency syndromes]
- Smoking
- Vulval dermatoses such as lichen sclerosus and squamous hyperplasia
- Prior history of cervical cancer or precancer, vaginal precancer or cancer.

VULVAR PRECURSOR LESIONS

Premalignant lesions of the vulva are collectively known as vulvar squamous intraepithelial lesions (SILs) or vulvar intraepithelial neoplasia (VIN).

There are two types of VINs and they represent two pathways of development of vulvar carcinoma:

1. *uVIN or usual VIN:* Occurs due to HPV infection of vulvar skin.
2. *dVIN or differentiated VIN:* Associated with vulvar dermatoses.

ISSVD TERMINOLOGY FOR VULVAR SILs (2015)

	Entities	***Comments***
LSIL	Vulvar LSIL, flat condyloma, or human papillomavirus (HPV) effect	Vulvar intraepithelial neoplasia 1 (VIN 1)
HSIL	Vulvar HSIL, VIN usual type (uVIN)	Previously referred to as VIN 2 and VIN 3
dVIN	Lesions associated with vulvar dermatoses, mainly lichen sclerosus	Previously referred to as VIN simplex type

(dVIN: differentiated VIN; HSIL: high-grade squamous intraepithelial lesion; LSIL: low-grade squamous intraepithelial lesion)

DIFFERENTIATION BETWEEN UVIN AND DVIN

	uVIN	***dVIN***
Incidence	2.86/100,000	0.08/100,000
Proportion of vulvar carcinoma cases attributed to VIN	20%	80%
Risk of progression to vulvar carcinoma	9.7%	50%
Etiology	HPV infection	Vulvar dermatoses
Site of lesion	Multifocal	Unifocal
Age of patient	Younger	Older (>55 years)
Histopathology	Dyskaryotic cells in groups, koilocytes present	Isolated enlarged atypical cells, absence of koilocytes
Molecular characteristics	Positive for HPV DNA, p16-INK4a, MIB1	Positive for p53

(dVIN: differentiated vulvar intraepithelial neoplasia; HPV: human papillomavirus; uVIN: usual vulvar intraepithelial neoplasia)

FIGO STAGING OF VULVAR CANCER (REVISED IN 2021)

Stage	Description
Stage I	Tumor confined to the vulva
Stage IA	Tumor size ≤2 cm and stromal invasion ≤1 mm*
Stage IB	Tumor size >2 cm or stromal invasion >1 mm*
Stage II	Tumor of any size with extension to lower one-third of the urethra, lower one-third of the vagina, lower one-third of the anus with negative nodes
Stage III	Tumor of any size with extension to upper part of adjacent perineal structures, or with any number of nonfixed, nonulcerated lymph node
Stage IIIA	Tumor of any size with disease extension to upper two-thirds of the urethra, upper two-thirds of the vagina, bladder mucosa, rectal mucosa, or regional lymph node metastases ≤5 mm
Stage IIIB	Regional† lymph node metastases >5 mm
Stage IIIC	Regional† lymph node metastases with extracapsular spread
Stage IV	Tumor of any size fixed to bone, or fixed, ulcerated lymph node metastases, or distant metastases
Stage IVA	Disease fixed to pelvic bone, or fixed or ulcerated regional lymph node metastases
Stage IVB	Distant metastases

*Depth of invasion is measured from the basement membrane of the deepest, adjacent, dysplastic, tumor-free rete ridge (or nearest dysplastic rete peg) to the deepest point of invasion.
†Regional refers to inguinal and femoral lymph nodes.

Clinical presentation:
- Vulvar mass
- Pain in perineum
- Pruritis
- Bleeding from vulvar wound
- Enlarged inguinofemoral lymph nodes

Diagnosis:
- By punch biopsy of the vulvar lesion with a Keyes punch biopsy forceps
- Colposcopy for subclinical lesions.

HISTOLOGICAL SUBTYPES

Epithelial tumors:	
Squamous cell tumors and precursors:	
Squamous cell carcinoma:	8,070/3
Keratinizing	8,071/3
Nonkeratinizing	8,072/3
Basaloid	8,083/3
Warty	8,051/3
Verrucous	8,051/3
Basal cell carcinoma	8,090/3
Glandular tumors:	
Paget disease	8,542/3
Tumors arising from Bartholin and other specialized anogenital glands:	
Bartholin gland carcinomas:	
Adenocarcinoma	8,140/3
Squamous cell carcinoma	8,070/3
Adenosquamous carcinoma	8,560/3
Adenoid cystic carcinoma	8,200/3
Transitional cell carcinoma	8,120/3
Adenocarcinoma of mammary gland type	8,500/3
Adenocarcinoma of Skene gland origin	8,140/3
Phyllodes tumor, malignant	9,020/3
Adenocarcinomas of other types:	
Adenocarcinoma of sweat gland type	8,140/3
Adenocarcinoma of intestinal type	8,140/3
Neuroendocrine tumors:	
High-grade neuroendocrine carcinoma:	
Small-cell neuroendocrine carcinoma	8,041/3
Large-cell neuroendocrine carcinoma	8,013/3
Merkel cell tumor	8,247/3
Neuroectodermal tumors:	
Ewing sarcoma	9,364/3
Soft tissue tumors:	
Rhabdomyosarcoma:	
Embryonal	8,910/3
Alveolar	8,920/3
Leiomyosarcoma	8,890/3
Epithelioid sarcoma	8,804/3
Alveolar soft part sarcoma	9,581/3
Other sarcomas:	
Liposarcoma	8,850/3
Malignant peripheral nerve sheath tumor	9,540/3
Kaposi sarcoma	9,140/3
Fibrosarcoma	8,810/3
Dermatofibrosarcoma protuberans	8,832/1Δ
Melanocytic tumors:	
Malignant melanoma	8,720/3
Germ cell tumors:	
Yolk sac tumor	9,071/3
Lymphoid and myeloid tumors:	
Lymphomas	
Myeloid neoplasms	
Secondary tumors	

TREATMENT OF VULVAR CARCINOMA

Early-stage Vulvar Carcinoma

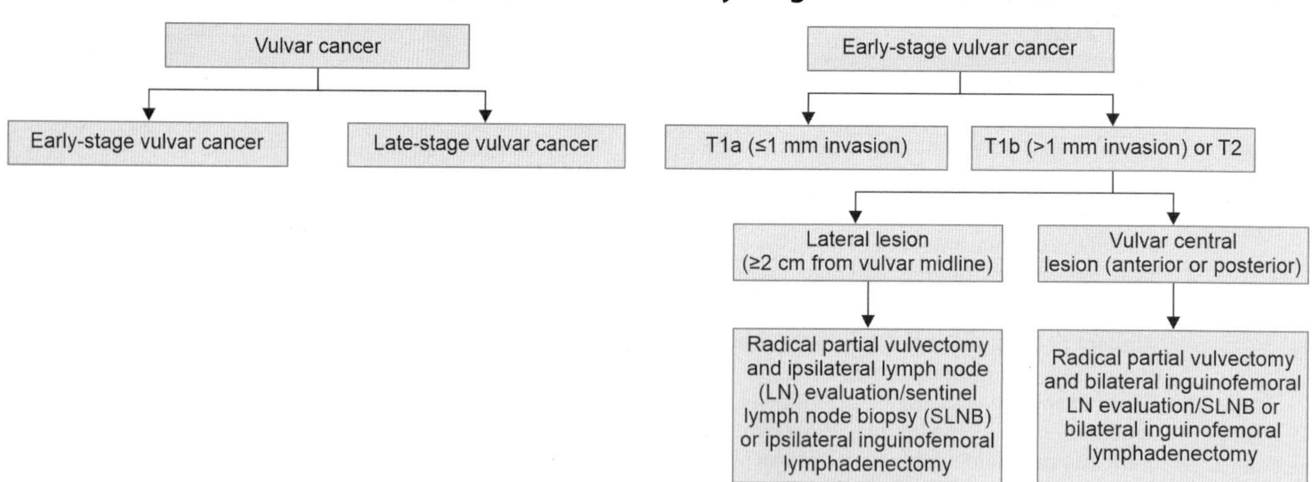

Lymphadenectomy in Vulvar Cancer

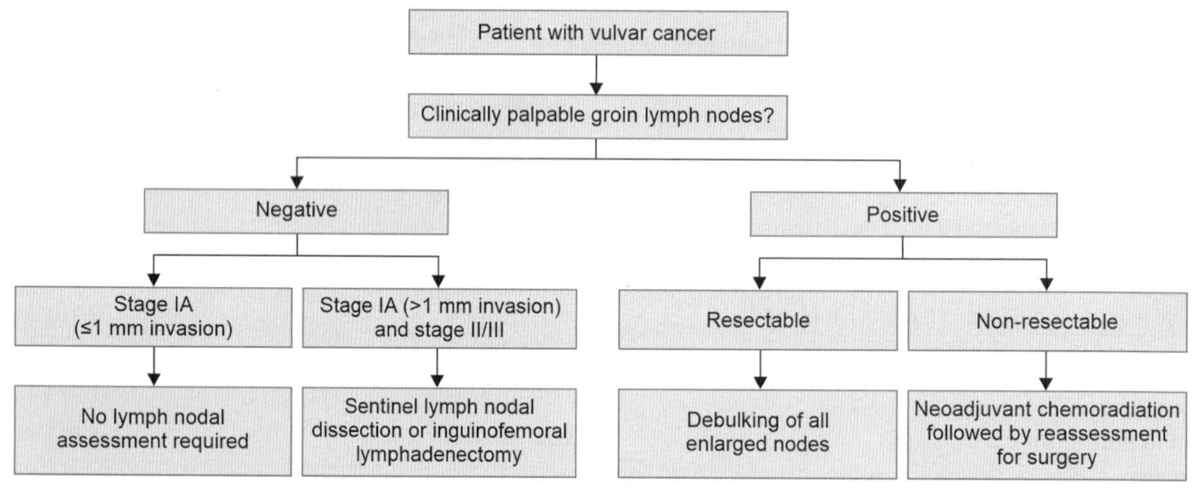

Sentinel Lymph Node Assessment

Margin Status Evaluation

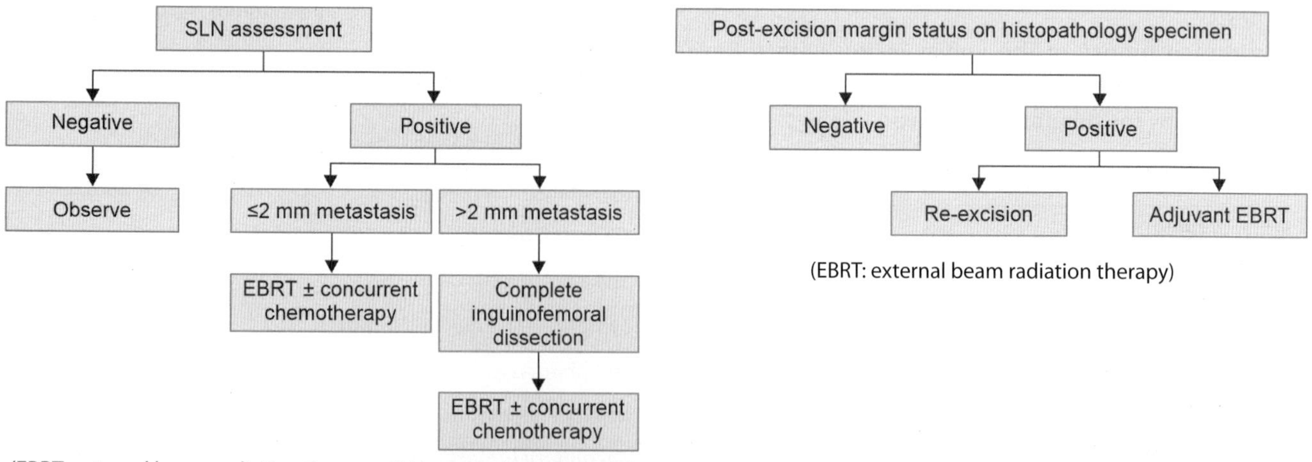

(EBRT: external beam radiation therapy)

(EBRT: external beam radiation therapy; SLN: sentinel lymph node)

ADVANCED VULVAR CARCINOMA

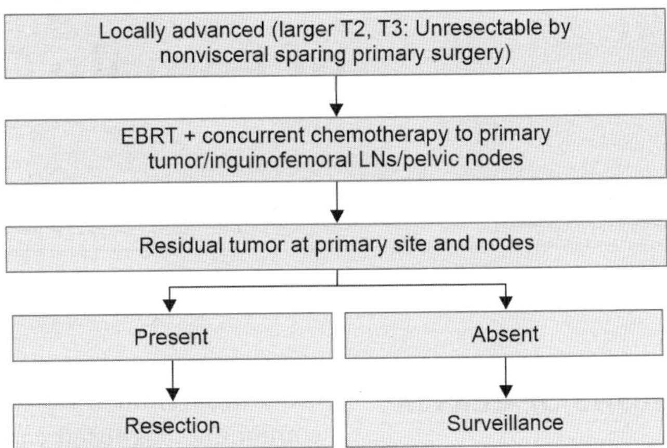

(EBRT: external beam radiation therapy; LNs: lymph nodes)

52

Chapter

Premalignant Lesions of Female Genital Tract

Rohini Rao, Megha Agarwal

PREMALIGNANT LESIONS OF CERVIX

INTRODUCTION

Cervical cancer is the fourth most common cancer among women worldwide and is a leading cause of mortality among women globally. It is almost entirely attributable to infection with the human papillomavirus (HPV) **(Fig. 1)**.

In 2020, an estimated 604,127 women were diagnosed with cervical cancer worldwide and about 341,831 women died from the disease. There was a global incidence of 13.3 cases per 100,000 women and mortality rate of 7.3 deaths per 100,000 women.[1]

Rates of cervical cancer have been estimated to be at least fourfold higher in countries with a low-ranking Human Development Index compared to those with a "very high" Human Development Index.[2] It is the most commonly diagnosed cancer in 23 countries with the leading cause of cancer death in 36 countries. The vast majority of these countries are in Sub-Saharan Africa, Melanesia, South America, and South-Eastern Asia.[1] In Indian scenario, it is the second most common cancer in women after breast cancer and about 123,907 women were diagnosed with cervical cancer in 2020 (18.3% of all cancers). Cervical cancer is a leading cause of cancer deaths in India, with approximately 77,348 deaths reported annually (11.4% of all cancers deaths).[1]

Human papillomavirus is the most common viral infection of the reproductive tract and is the cause of precancerous lesions that may progress to cancer. It is found to be the primary cause of cervical cancer, associated with nearly all cervical malignancies. Over 100 serotypes of

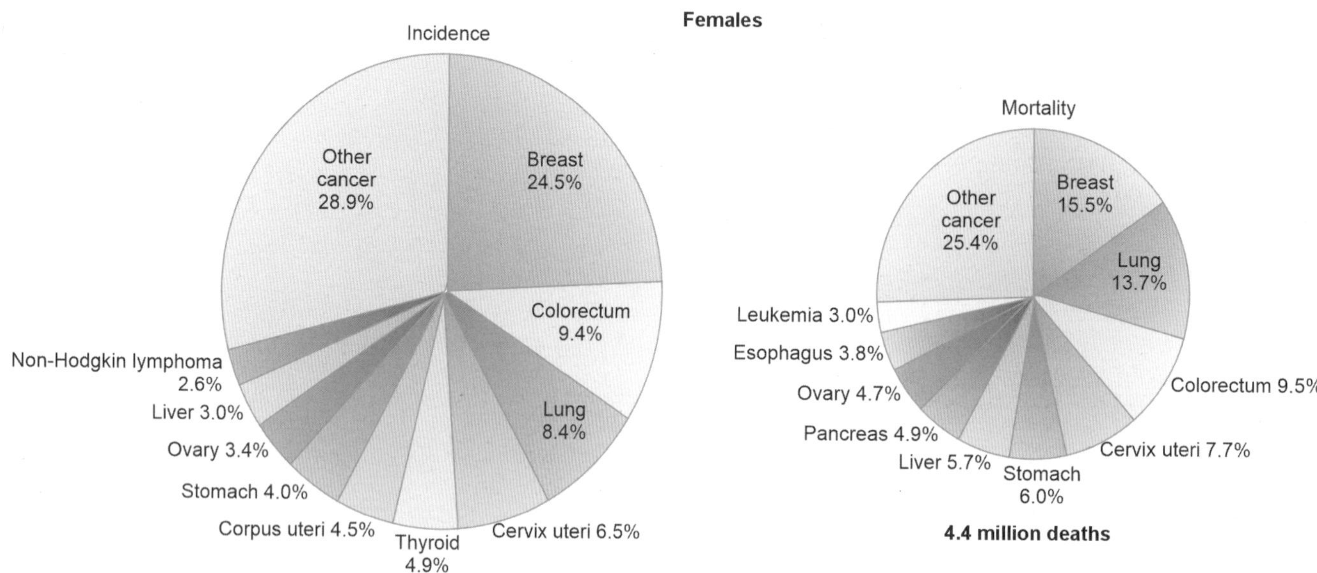

Fig. 1: Incidence of carcinoma cervix.
Source: GLOBOCAN 2020.

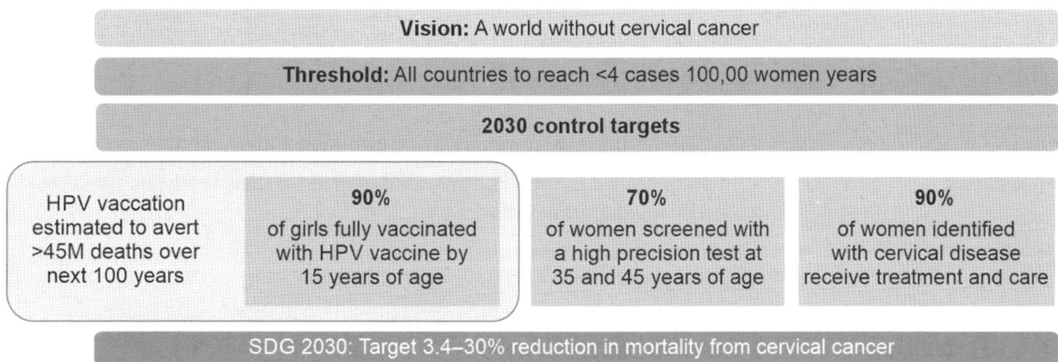

Fig. 2: A global initiative to eliminate cervical cancer by WHO. (SDG: Sustainable Development Goal; WHO: World Health Organization)

HPV have been discovered, of which 15–20 are oncogenic. The lag period between the oncogenic HPV infection and the invasive cervical cancer is 15–20 years. Based on the association with cervical cancer, genital HPVs are further grouped into high-risk types, probable high-risk types and low-risk types.[3] Worldwide, high-risk type HPV-16 and 18 contribute over 70% of all cervical cancer cases (the most prevalent being HPV-16 in at least 50–60% and HPV-18 in at least 10–12%). Similarly, in Indian women, the most common prevalent genotypes are HPV-16 and 18.[4] Other important cofactors include some sexually transmittable infections (STIs) [human immunodeficiency virus (HIV) and *Chlamydia trachomatis*], smoking, a higher number of childbirths, and long-term use of oral contraceptives.[5]

In cervical screening the primary goal is to prevent cervical cancer. This has to be achieved by the detection, treatment, and follow-up of preinvasive cervical lesions.[6,7] Modern understanding that almost all cervical cancers are caused by persistent infection with approximately 15 types of HPV. Cervical cancer is considered nearly completely preventable because of the highly effective primary (HPV vaccine) and secondary (screening) HPV-based screening measures.[8-10]

However, these measures have not been equitably implemented across and within countries. As of May, 2020, only 44% of women in low- and middle-income countries (LMICs) have ever been screened for cervical cancer, with the lowest prevalence among women in sub-Saharan compared with >60% in high-income countries.[11,12]

WHO LAUNCHED: A GLOBAL INITIATIVE TO ELIMINATE CERVICAL CANCER[13]

WHO launched a global initiative to eliminate cervical cancer by 2030 and it aims that 90% of girl should be fully vaccinated with HPV vaccine by 15 years of age as shown in **Figure 2**.

Tables 1 and 2 detail on the HPV vaccination.

TABLE 1: Human papillomavirus (HPV) vaccination.

Types of vaccines	• Bivalent • Quadrivalent • 9-Valent
License to use	9–45 years
Target age group (preferred)	9–14 years
Number of doses for girls aged <15 years and not immunocompromised or human immunodeficiency virus (HIV)-infected	Two doses
Number of doses for girls aged ≥15 years and are either immunocompromised and/or HIV-infected	Three doses
Interval	• *Two doses:* 0 and 6 months (second dose can be given between 5 and 15 months) • *Three doses:* Bivalent: 0, 1, 6 months • *Quadrivalent*: 0, 2, 6 months • 9 valent (0, 1–2, 6 months)
Catch-up vaccination (15–26 years)	*Three doses:* Counseling should be done in girls/women who have been sexually active regarding reduced efficacy and importance of screening from the age of 25–30 years
Older age groups (>26 years)	*Three doses:* Counseling should be done in women aged >26 years who have been sexually active regarding reduced efficacy in older age group and the importance of screening. Women who are in older age groups (>26 years) should first be screened in limited-resource settings

TABLE 2: HPV vaccination (special situations).

HIV-positive girls or immunocompromised girls	• Same age recommendation • Three doses
Interrupted doses	• As per age-based recommendation Continue with the remaining doses.
Pregnancy and lactation	• In pregnancy it is not recommended (if given inadvertently, no need for MTP) • Can be given during lactation
Sexual abuse victims	• Same age recommendation • Three doses • Preferably initiate at the time of examination at health care facility
Women with history of abnormal screening reports	• Same age recommendation • Counseling of women should be done regarding reduced efficacy in older age group and explained about the importance of regular follow-up

(HIV: human immunodeficiency virus; MTP: medical termination of pregnancy)

TABLE 3: Single-visit criteria approach strategies.

See and treat	***Screen and treat***
• In colposcopy clinics • In referred patient with abnormal cytology report • Colposcopy scoring indicating a high-grade lesion • Simultaneous treatment done by excision or ablation • Probability of overtreatment is very low because of high specificity of cytology	• In public health programs • Abnormal lesion detected by VIA (visual inspection with acetic acid) • Ablation criteria should be fulfilled • Treat immediately, with or without biopsy • In high prevalence areas, there is lower probability of overtreatment

Flowchart 1: Selecting the most appropriate screening test for practice.

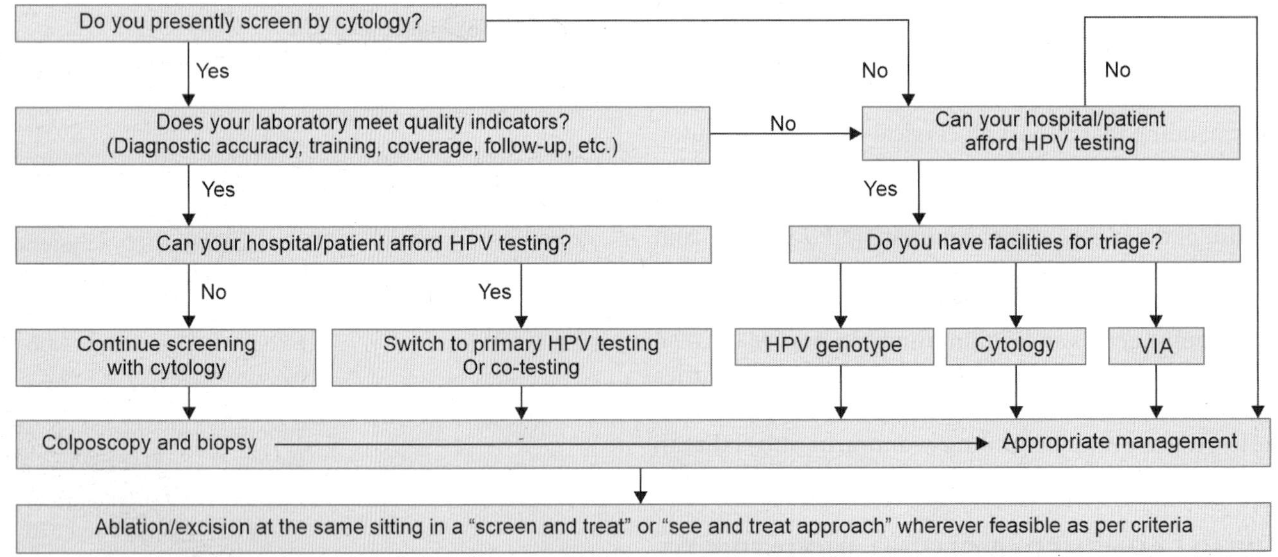

(VIA: visual inspection with acetic acid)

SCREENING OF PREINVASIVE LESIONS OF CERVIX

Single-visit criteria approach strategies are mentioned in **Table 3**.

SELECTING THE MOST APPROPRIATE SCREENING TEST FOR PRACTICE

Flowchart 1 shows selecting the most appropriate screening test for practice.

Table 4 mentions the resource-based strategies for cervical cancer screening and management of CIN (cervical intraepithelial neoplasia).

Tables 5 mention resource-based cervical cancer screening recommendation.

TABLE 4: Resource-based strategies for cervical cancer screening and management of cervical intraepithelial neoplasia (CIN).

Setting screening	*Tools triage*	*Tools management*	*Options single*	*Visit approach*
Good resource settings	Primary HPV test or Co-testing (HPV test + cytology) or cytology or VIA	• Cytology ± newer modalities* HPV test • HPV genotyping—16/18 • Colposcopy and biopsy† • VIA and biopsy†	• LEEP • Conization • Cryotherapy thermal coagulation	See and treat approach
Limited resource settings‡	VIA HPV	Colposcopy, if available biopsy†	Cryotherapy, LEEP, conization, thermal coagulation	Screen and treat or screen, see and treat approach

*Newer modalities (p16, Ki 67 testing, mRNA testing, E6, E7 protein testing)
†Biopsy from any abnormal area(s)
‡Affordable HPV test (if available), including self-sampling, can be used
(LEEP: loop electrosurgical excision procedure; VIA: visual inspection with acetic acid)

TABLE 5: Resource-based cervical cancer screening recommendation.

	Good resource settings	*Limited resource settings*
Modalities	HPV testing **(Flowchart 2)** • Primary HPV testing • Co-testing (HPV and cytology) **(Flowchart 3)** • Cytology • VIA (visual inspection with acetic acid)	VIA (Affordable HPV testing may be introduced if feasible)
Target age group (years)	25–65	30–65 (N.B.: In postmenopausal women, screening with VIA may not be as effective) **(Flowchart 4)**
Age to start (years)	• Cytology at 25 • Primary HPV testing /co-testing at 30	VIA at 30
Frequency	• Primary HPV testing or co-testing—every 5 years • Cytology—every 3 years	Every 5 years (at least one to three times in a lifetime)
Age to stop	• 65 years with consistent negative results in last 15 years • Women with no prior screening should undergo tests once at 65 years and, if negative, they should exit screening	
Follow-up method after treatment; interval	HPV testing (preferred) or cytology 12 months	VIA 12 months
Screening following abnormal reports ≥CIN2+, irrespective of method of treatment	20 years	
Screening in hysterectomized women	• Following hysterectomy in which cervix was removed for benign causes: No need for screening, unless there is history of previous cervical intraepithelial neoplasia (CIN) • Absence of cervix must be confirmed by clinical records or examination • If indications for hysterectomy are unclear, screening may be performed at clinician's discretion	

Contd...

Contd...

	Good resource settings	*Limited resource settings*
Follow-up in women with CIN in hysterectomy HPE report	Need to be screened with HPV at 6 months and 18 months	
Screening of immunocompromised women	Start within 1 year of initiation of sexual activity • HPV testing/co-testing/Cytology/VIA • Every 2–3 years	• VIA/affordable HPV testing if available • Every 3 years (at least twice as often as general population)

(HPE: histopathological examination; VIA: visual inspection with acetic acid)

TABLE 6: Terminology and three-tiered CIN system in cytologic and histologic diagnoses.

	Histology			*Cytology*	
Natural history model	*Dysplasia nomenclature*	*CIN nomenclature*	*LAST nomenclature*	*Papanicolaou Classification*	*The bethesda system*
Infection	Negative	Negative		I	NILM
	Squamous atypia	Squamous atypia		II	ASCUS
Precancer	Mild dysplasia	CIN1	LSIL	III	LSIL
	Moderate dysplasia	CIN3			
	Severe dysplasia	CIN3			HSIL
	Carcinoma *in situ*			IV	
Cancer	Carcinoma	CIN1	HSIL	V	Carcinoma

(ASCUS: atypical squamous cells of undetermined significance; CIN: cervical intraepithelial neoplasia; HSIL: high-grade squamous intraepithelial lesion; LSIL: low-grade squamous intraepithelial lesion; NILM: negative for intraepithelial lesion or malignancy)

TABLE 7: Epithelial cell abnormalities.

Abbreviation	*Terminology*
NILM	Negative for intraepithelial lesion or malignancy
ASCUS	Atypical squamous cell of undetermined significance **(Flowchart 5)**
ASC-H	Atypical squamous cell: Cannot exclude high-grade squamous intraepithelial lesion
LSIL	Low-grade squamous intraepithelial lesion **(Flowchart 6)**
HSIL	High-grade squamous intraepithelial lesion
SCC	Squamous cell carcinoma

TABLE 8: Glandular cell abnormalities

Abbreviation	*Terminology*
AGC	Atypical glandular cells (specify endocervical, endometrial or NOS, i.e., not otherwise significant)
AGC-FN	Atypical glandular cells: Favor neoplastic **(Flowcharts 7 to 9)**
AIS	Endocervical carcinoma in situ **(Flowchart 10)**
Endometrial cells	In a woman >40 years of age
Adenocarcinoma	
Others	

CLASSIFICATION OF PREINVASIVE CERVICAL DISEASE

Table 6 shows terminology and three-tiered CIN system in cytologic and histologic diagnoses.

Bethesda System for Reporting Cytology

Table 7 shows the abbreviation and terminology regarding epithelial cell abnormalities.

Table 8 shows the abbreviation and terminology regarding glandular cell abnormalities.

SCREENING WITH CYTOLOGY

The management of ASC-H/HSIL **(Flowchart 11)** (atypical squamous cell: cannot exclude high-grade squamous intraepithelial lesion/high-grade squamous intraepithelial lesion) in women aged >30 years is shown in **Flowchart 12** and women aged <30 years us shown in **Flowchart 13**.

Flowchart 2: Screening with primary human papillomavirus (HPV) testing in women aged >30 years.

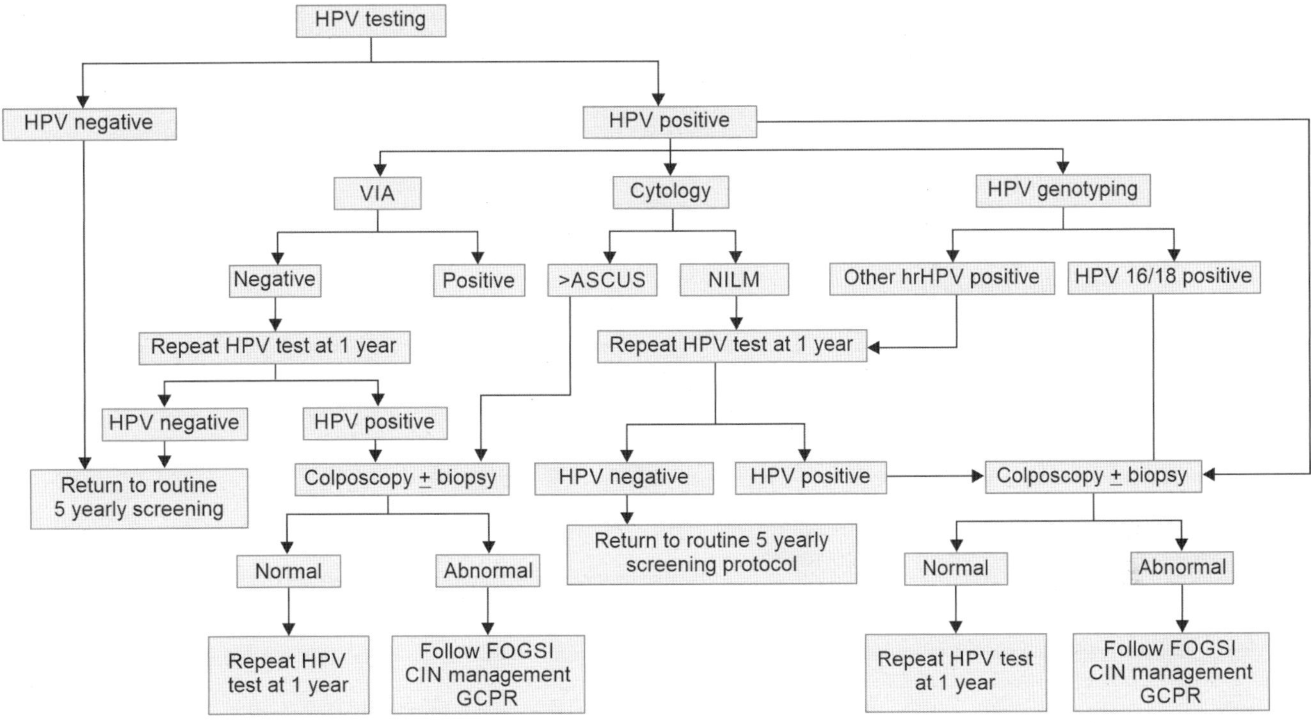

(ASCUS: atypical squamous cell of undetermined significance; CIN: cervical intraepithelial neoplasia; FOGSI: The Federation of Obstetric and Gynecological Societies of India; GCPR: good clinical practice recommendations; hrHPV: high-risk HPV; NILM: negative for intraepithelial lesion or malignancy; VIA: visual inspection with acetic acid)

Flowchart 3: Screening with Co-testing (HPV test with cytology) in women aged >30 years.

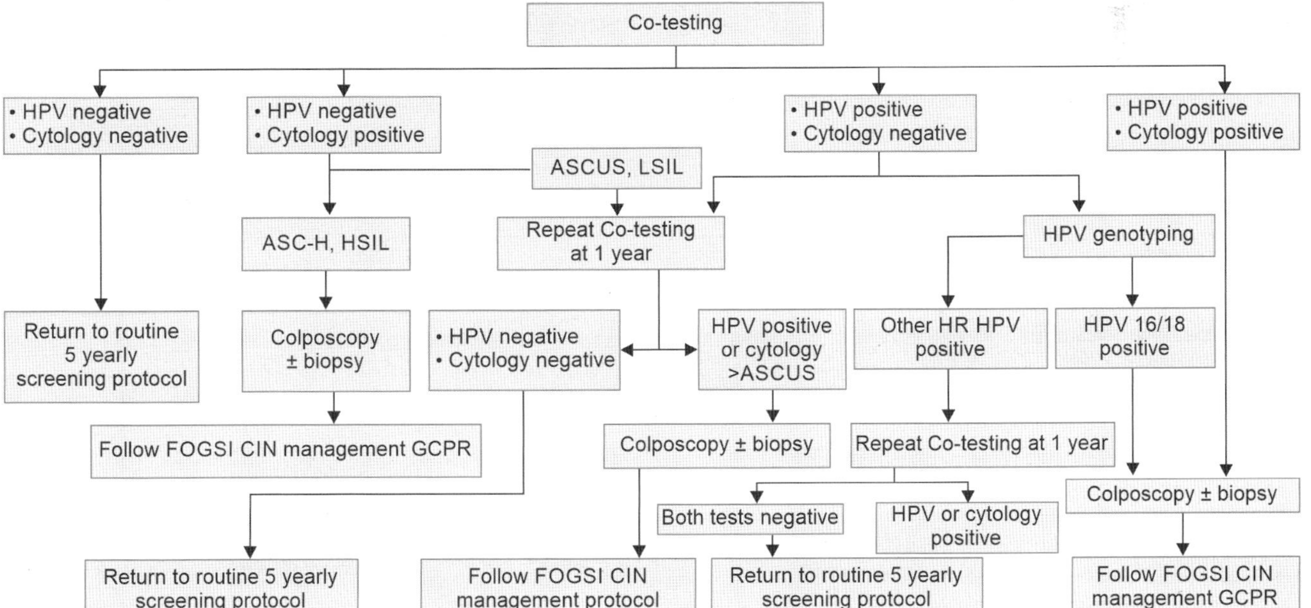

(ASCUS: atypical squamous cell of undetermined significance; ASC-H: atypical squamous cell: cannot exclude high-grade squamous intraepithelial lesion; CIN: cervical intraepithelial neoplasia; FOGSI: The Federation of Obstetric and Gynecological Societies of India; GCPR: good clinical practice recommendations; hrHPV: high-risk HPV; HSIL: high-grade squamous intraepithelial lesion; LSIL: low-grade squamous intraepithelial lesion)

Flowchart 4: Screening with VIA in women aged 30–65 years.

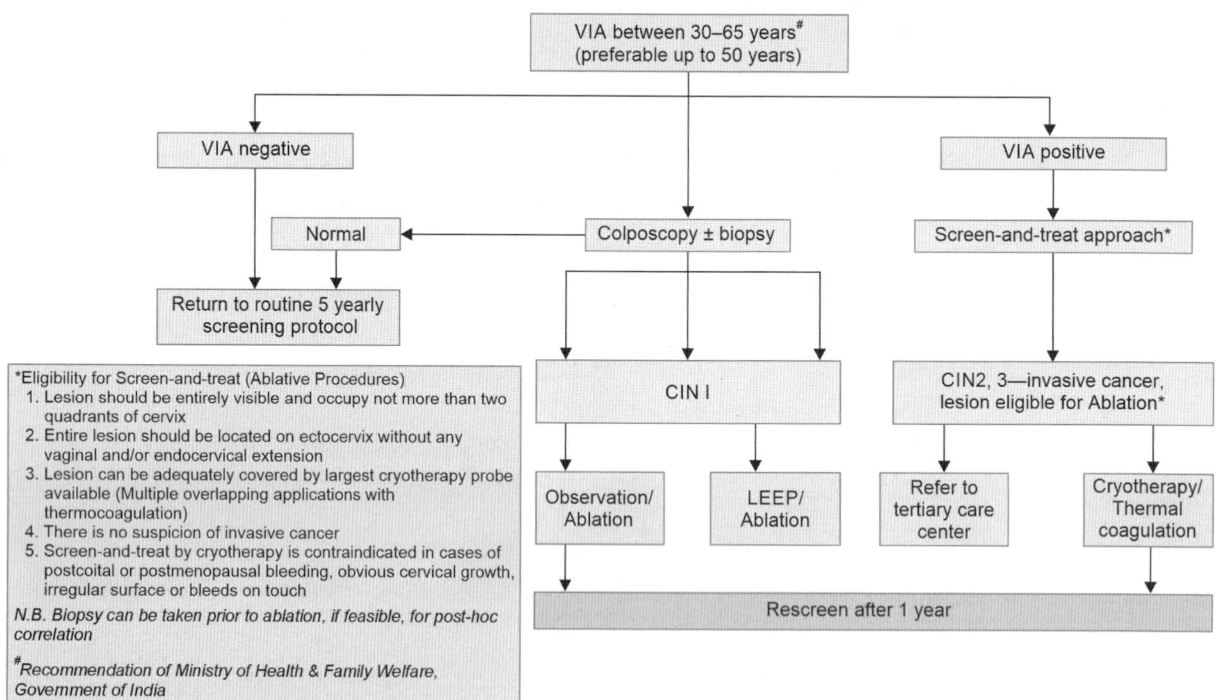

(CIN: cervical intraepithelial neoplasia; LEEP: loop electrosurgical excision procedure; VIA: visual inspection with acetic acid)

Flowchart 5: Screening with cytology management of atypical squamous cell of undetermined significance (ASCUS) in women aged ≥30 years.

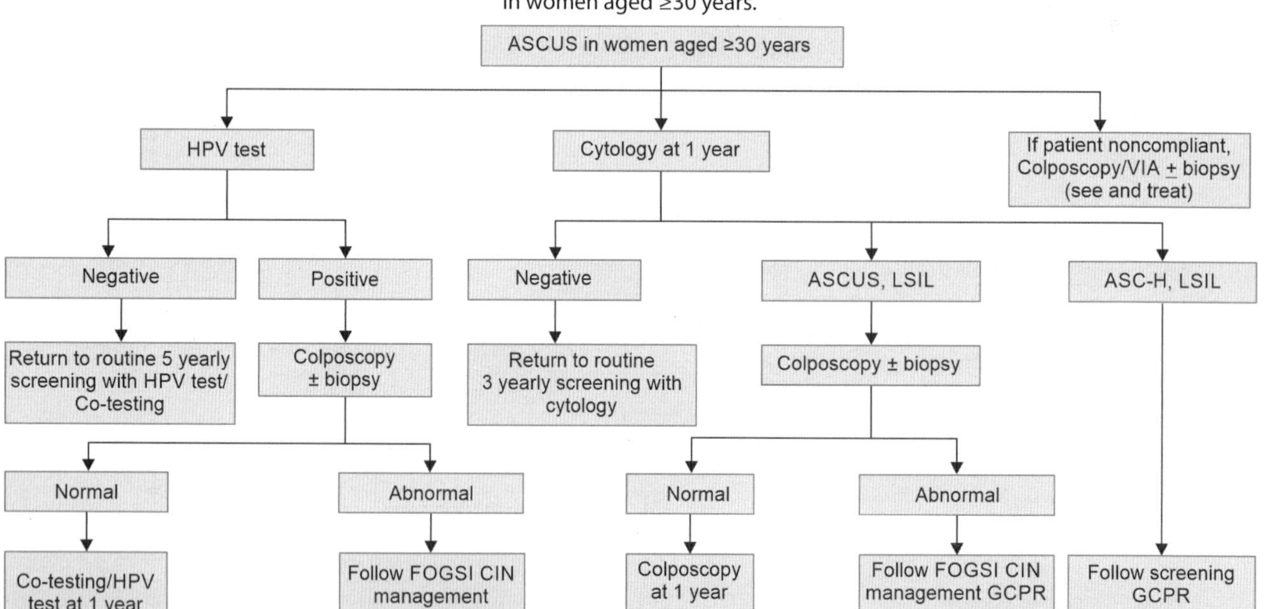

(ASCUS: atypical squamous cell of undetermined significance; ASC-H: atypical squamous cell: cannot exclude high-grade squamous intraepithelial lesion; CIN: cervical intraepithelial neoplasia; FOGSI: The Federation of Obstetric and Gynecological Societies of India; GCPR: good clinical practice recommendations; HSIL: high-grade squamous intraepithelial lesion; LSIL: low-grade squamous intraepithelial lesion; VIA: visual inspection with acetic acid)

Flowchart 6: Screening with cytology management of low-grade squamous intraepithelial lesion (LSIL) in women aged ≥30 years.

LSIL in women aged ≥30 years
HPV test
Positive
Negative
Colposcopy ± biopsy
If patient noncompliant, see and treat Followchart 12
Colposcopy ± biopsy
Co-testing at 1 year
Follow FOGSI CIN management GCPR
Cytology normal and HPV negative
Cytology ≥ASCUS or HPV positive
Follow FOGSI CIN management GPCR
Co-testing at 3 years
Colposcopy ± biopsy
Cotology normal and HPV negative
Follow FOGSI CIN management GPCR
Return to 5 yearly routine screening protocol

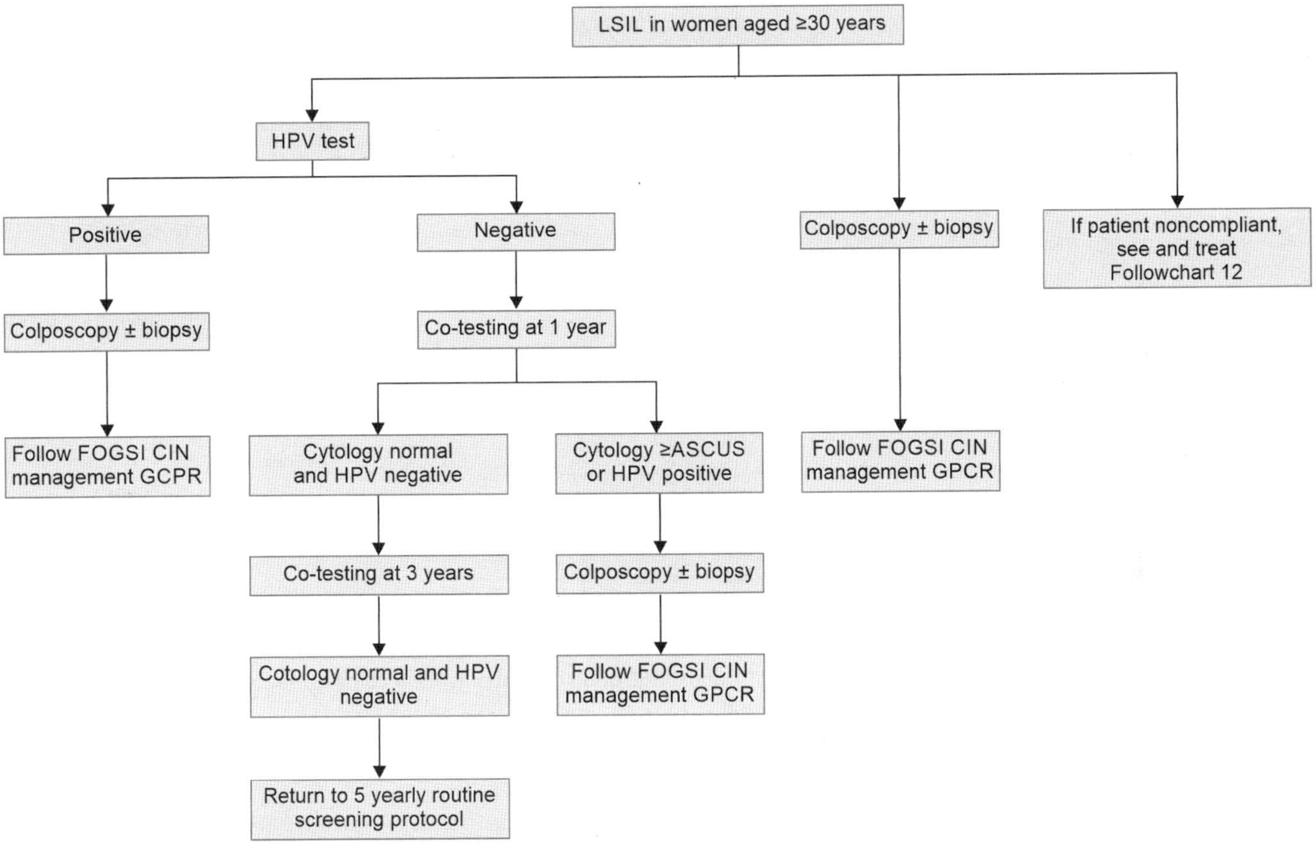

(ASCUS: atypical squamous cell of undetermined significance; CIN: cervical intraepithelial neoplasia; FOGSI: The Federation of Obstetric and Gynecological Societies of India; GCPR: good clinical practice recommendations)

Flowchart 7: Screening with cytology management of abnormal glandular cells: Atypical endometrial cells.

Cytology report—Atypical endometrial cells
Endometrial and endocervical curettage
Endometrial pathology
Manage accordingly
No endometrial pathology
Colposcopy ± biopsy
Normal
Return to age-specific routine screening
Follow FOGSI CIN management GCPR

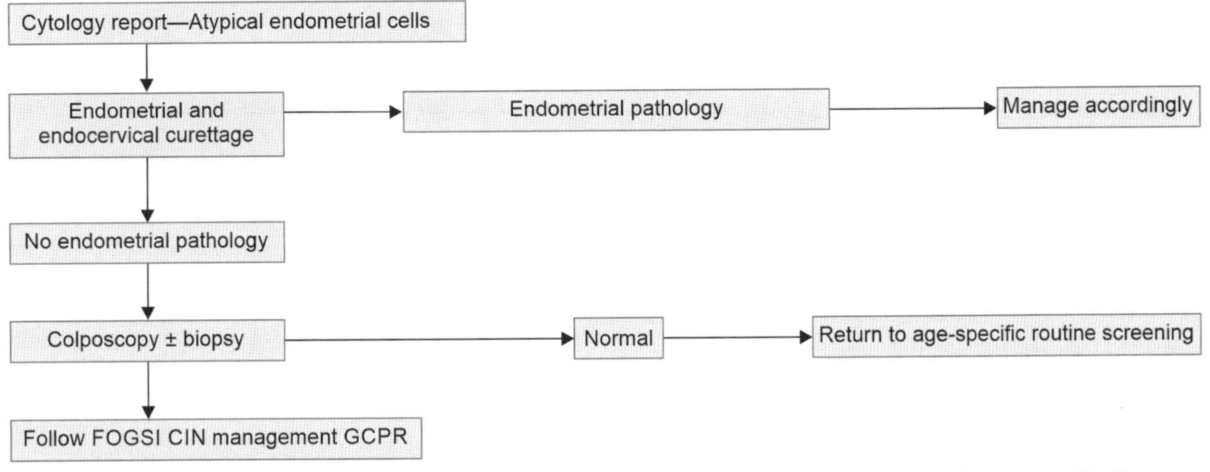

(CIN: cervical intraepithelial neoplasia; FOGSI: The Federation of Obstetric and Gynecological Societies of India; GCPR: good clinical practice recommendations)

Flowchart 8: Screening with cytology management of abnormal glandular cells: AGC-NOS/atypical endocervical cells.

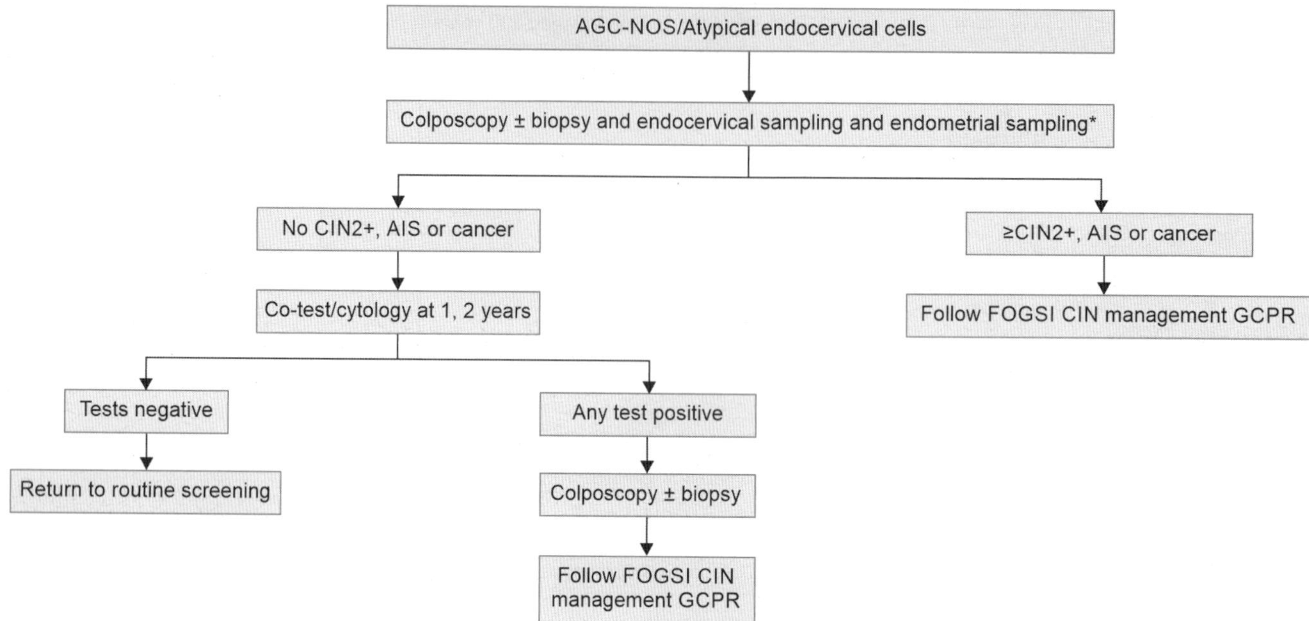

(AGC-NOS: atypical glandular cells-not otherwise specified; AIS: adenocarcinoma in situ; CIN: cervical intraepithelial neoplasia; FOGSI: The Federation of Obstetric and Gynecological Societies of India; GCPR: good clinical practice recommendations)
*Age >35 years, high risk factors, obesity, chronic anovulation

Flowchart 9: Screening with cytology management of abnormal glandular cells: AGC-FN/AIS.

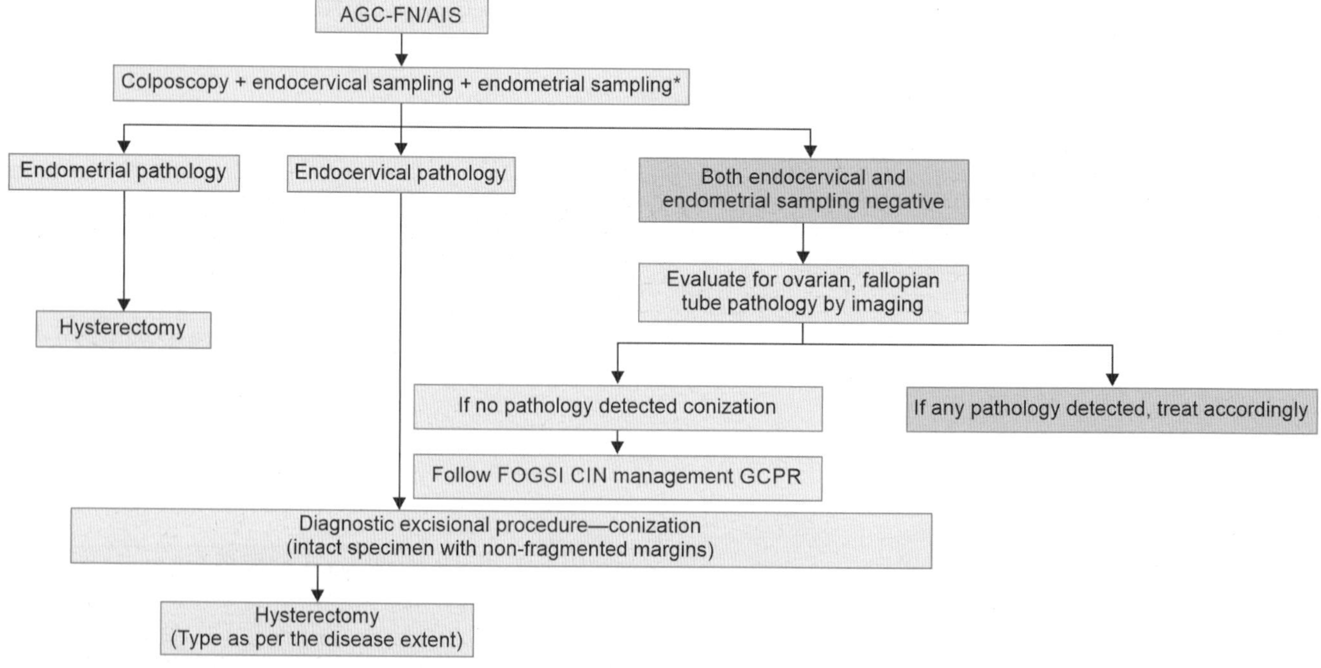

(AGC-FN: atypical glandular cells: favor neoplastic; AIS: adenocarcinoma in situ; CIN: cervical intraepithelial neoplasia; FOGSI: The Federation of Obstetric and Gynecological Societies of India; GCPR: good clinical practice recommendations)

Flowchart 10: Management of women with adenocarcinoma in situ (AIS).

AIS on biopsy or endocervical curettage (ECC)

Conization to rule out invasive disease

Hysterectomy

In young women desirous of fertility

Margins/ECC positive

Margins negative

Re-excise

Co-testing/Cytology colposcopy and ECC at 6 months

Follow-up

Flowchart 11: Screening with cytology management of ASCUS/LSIL in women aged <30 years.

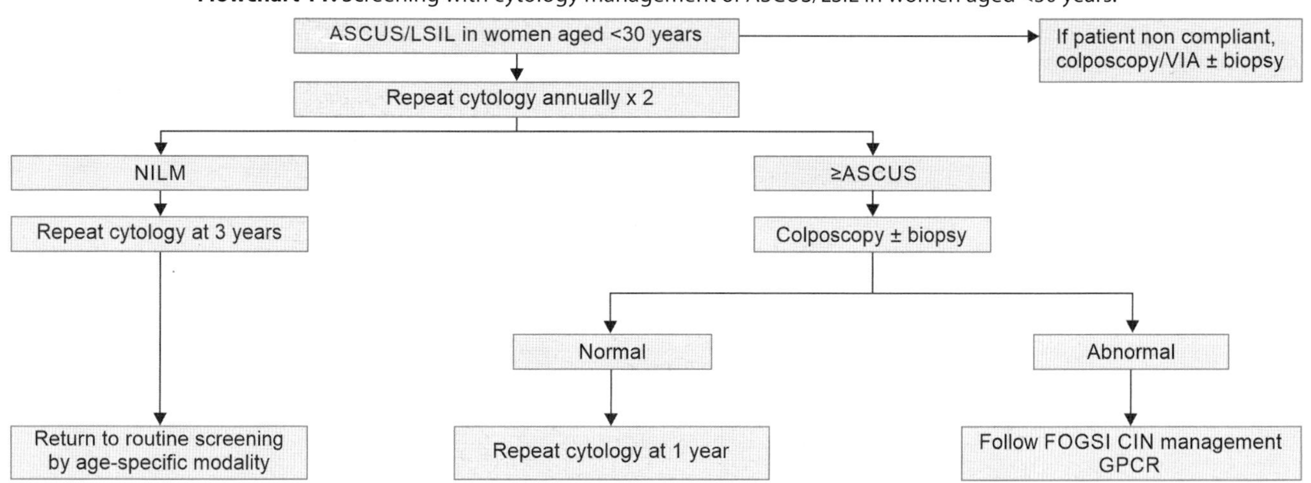

(ASCUS/LSIL: atypical squamous cell of undetermined significance/low-grade squamous intraepithelial lesion; CIN: cervical intraepithelial neoplasia; FOGSI: The Federation of Obstetric and Gynecological Societies of India; GCPR: good clinical practice recommendations; VIA: visual inspection with acetic acid)

Flowchart 12: Management of ASC-H/HSIL (atypical squamous cell: cannot exclude high-grade squamous intraepithelial lesion/high-grade squamous intraepithelial lesion) in women aged >30 years.

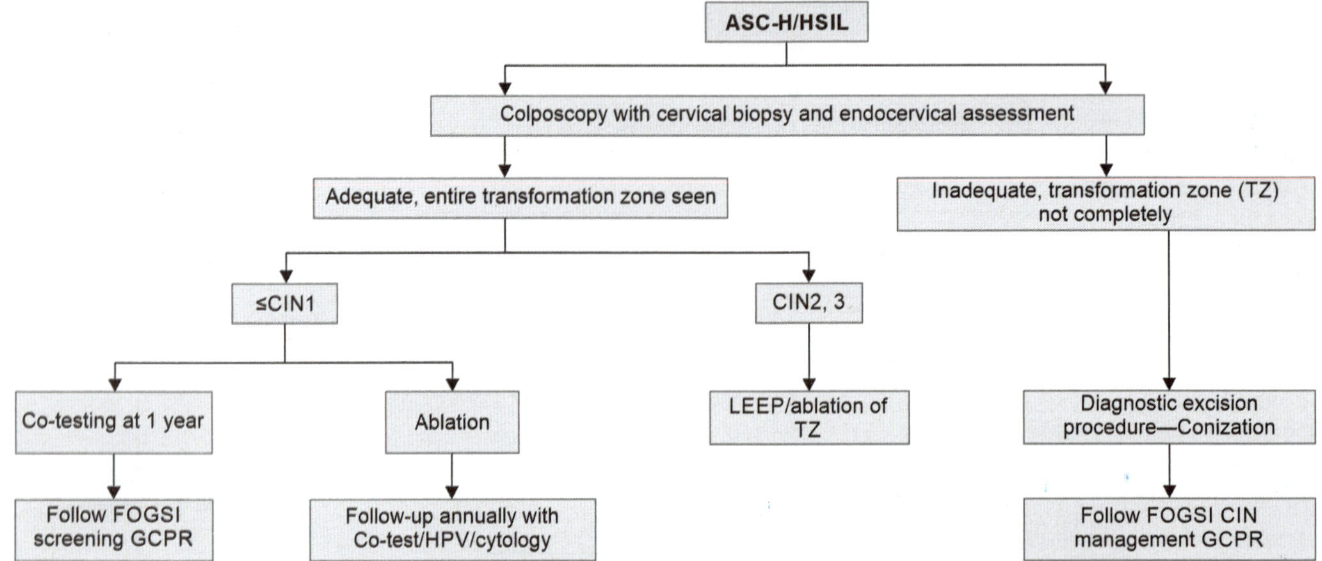

(CIN: cervical intraepithelial neoplasia; FOGSI: The Federation of Obstetric and Gynecological Societies of India; GCPR: good clinical practice recommendations; LEEP: loop electrosurgical excision procedure)

Flowchart 13: Screening with cytology management of ASC-H/HSIL in women aged <30 years and desirous of pregnancy.

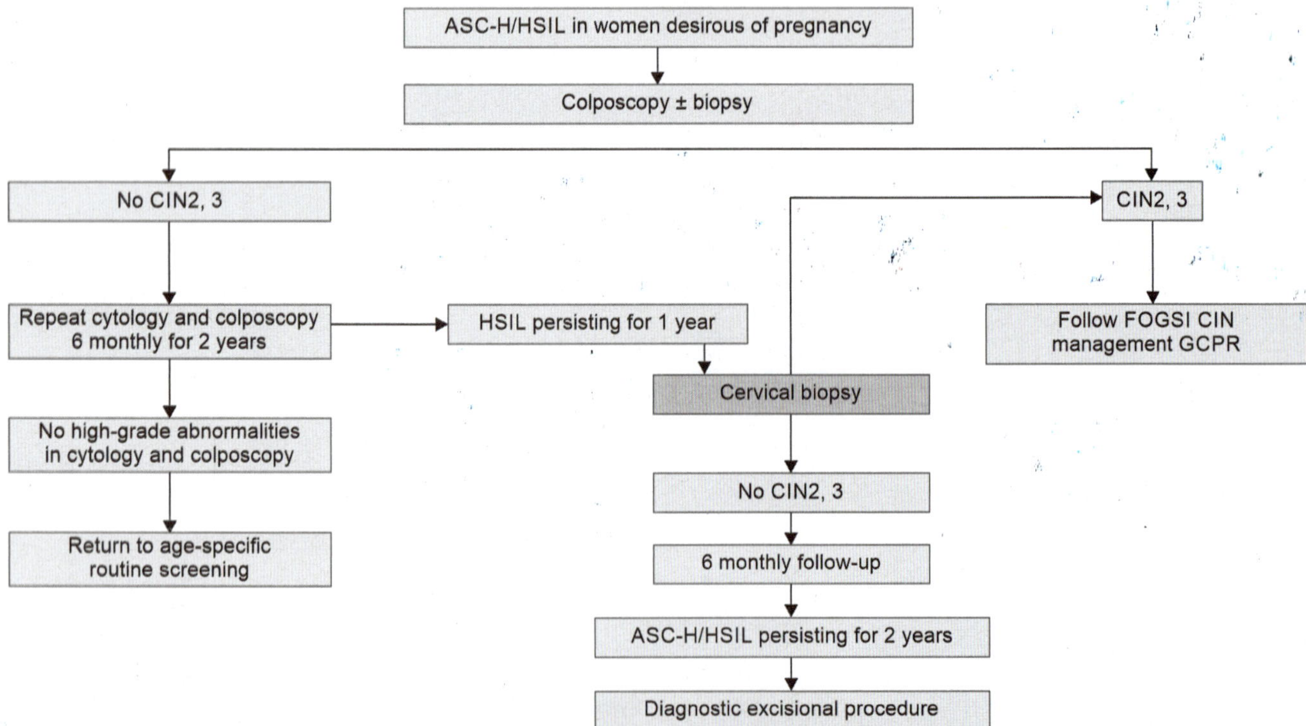

(ASC-H: atypical squamous cell: cannot exclude high-grade squamous intraepithelial lesion; CIN: cervical intraepithelial neoplasia; FOGSI: The Federation of Obstetric and Gynecological Societies of India; GCPR: good clinical practice recommendations; HSIL: high-grade squamous intraepithelial lesion)

Figure 3 shows lining of cervical epithelium with full-thickness atypia.

Figure 4 shows frequent atypical mitosis in superficial layers (HSIL).

NEWER MODALITIES: CYTOLOGY-BASED SCREENING

The newer modalities in cytology-based screening are shown in **Table 9**.

NEWER MODALITIES: HPV-BASED SCREENING

The newer modalities in HPV-based screening are shown in **Table 10**.

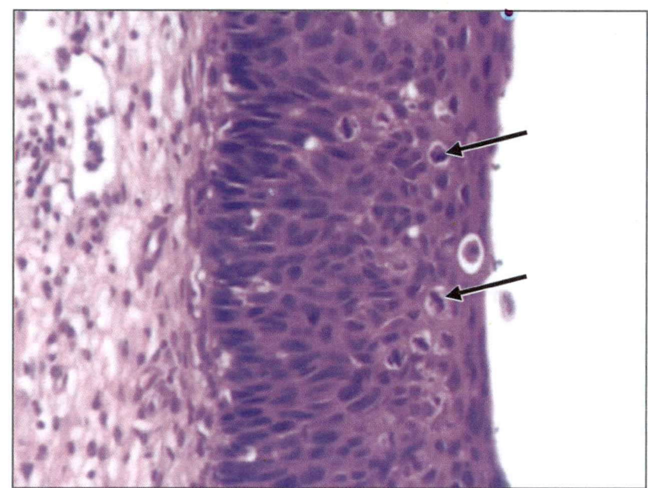

Fig. 3: Lining of cervical epithelium shows full-thickness atypia [high-grade squamous intraepithelial lesion (HSIL)] [hematoxylin and eosin (H&E) 40×].

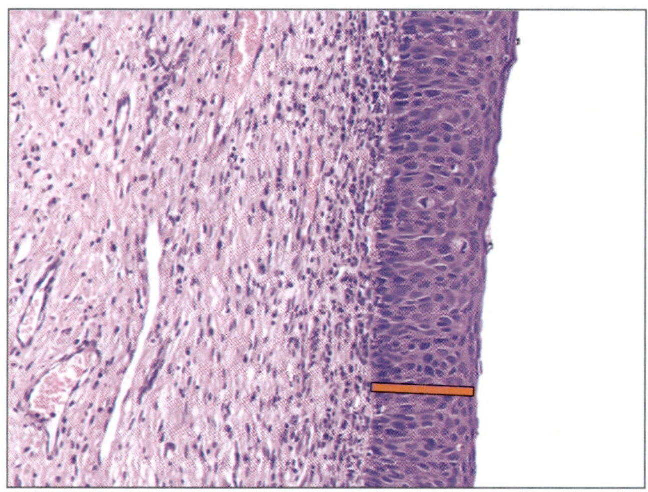

Fig. 4: Frequent atypical mitosis in superficial layers [high-grade squamous intraepithelial lesion (HSIL)] [hematoxylin and eosin (H&E) 400×].

TABLE 9: Newer modalities: Cytology-based screening.

	Liquid-based cytology (LBC)	***Biomarkers***
Options	ThinPrep (Hologic), SurePath (BD)	p16INK4a: CINtec (Roche), CINtec PLUS (p16INK4a + Ki-67)
Advantage	• Few inadequate Pap samples (1.9%) • Reflex HPV testing possible • Automated reading possible	Identify transforming HPV infection; detect CIN2 + disease; triage low-grade smears **(Flowchart 14)**; can be done with histology and cytology slides; reduce colposcopy referrals
Efficacy	Sensitivity not better than conventional (RR: 1.1); 11% more sensitivity for LSIL + lesions	Sensitivity: 64–92%, Specificity: 41–96% for low-grade smears
Limitations	More expensive, not cost-effective, automated not effective	Wide variation in reported sensitivity, lack of standardized reporting, need for substantial expertise, false positive rates high

(CIN: cervical intraepithelial neoplasia; LSIL: low-grade squamous intraepithelial lesion; RR: relative risk)

TABLE 10: Newer modalities: HPV-based screening.

	Vaginal self-collected	*E6 E7 mRNA*	*DNA methylation (DNAme)*	*Markers of aberrant S-phase induction*
Options	Delphi device, Evalyn sampler	• APTIMA, APTIMA HPV GT (Hologic) • Oncotect (Incell Dx)	QIAsure Methylation Test (Qiagen)	Topoisomerase IIA (TOP2A) and MCM2 (BD)
Advantage	Acceptable, eliminates cost of visiting a clinician	Effective triage for low-grade smears; reduces colposcopy referral by 68% compared to 30% by HPV DNA testing	Alternative triage for hrHPV; automated, objective test; run on the same sample as the HPV assay	Identify transforming HPV infection
Efficacy	Sensitivity variable: 60–90%; overall 3.4 (95% CI = 2.4-4.9) times more CIN2+ detected by self-collected HPV samples than by routine cytology	Sensitivity: 90–95%, specificity: 42–61%, PPV: 67%	CADM1-m18 combined with MAL-m1 methylation: Sensitivity: 60.5–100%, Specificity: 22.7–83.3% (95% CI: 78.4–87.4)	Sensitivity: 67–99%; specificity: 61–85%
Limitations	Inadequate sample possible, woman-dependent, reluctance to use medical devices	Expensive	Expensive, wide variation in reported sensitivity	Research settings

(CI: confidence interval; CIN: cervical intraepithelial neoplasia; hrHPV: high-risk human papillomavirus; PPV: positive predictive value)

Flowchart 14: Unsatisfactory smear.

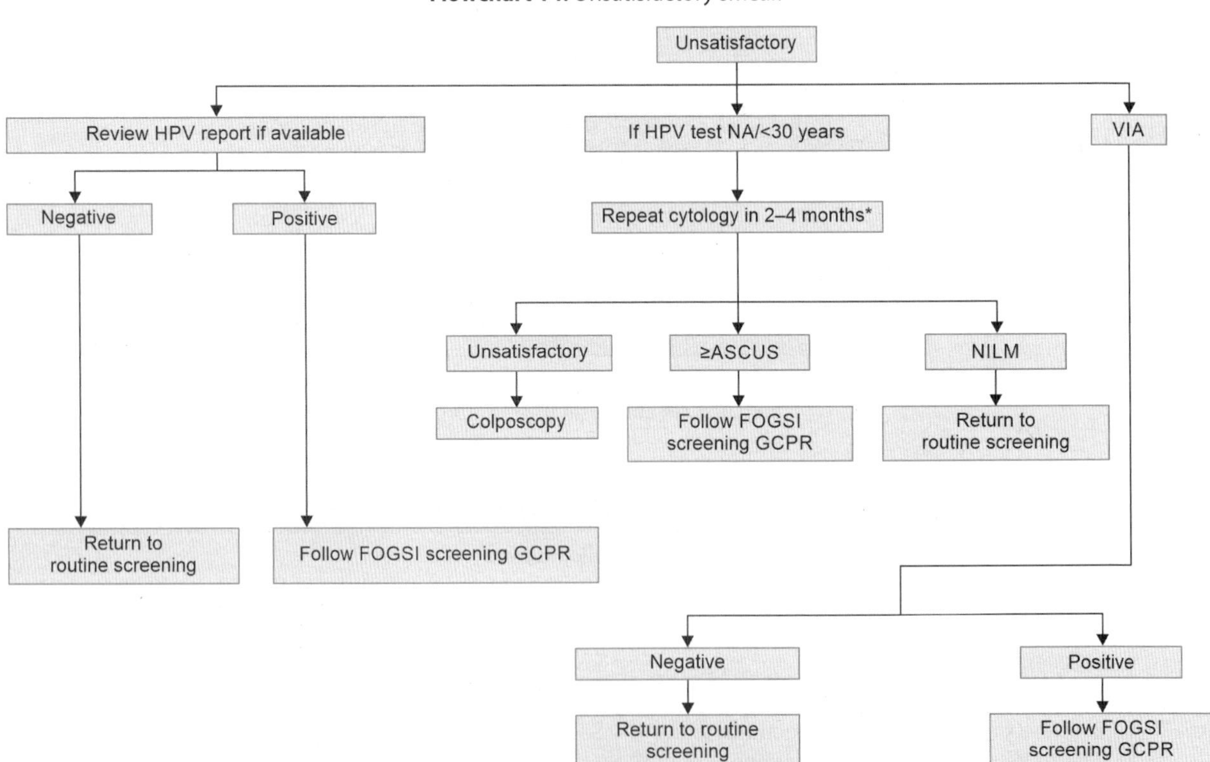

*Consider LBC if facilities available, in postmenopausal females local Estrogen × 3 weeks, and stop 1 week prior to testing.

(ASCUS: atypical squamous cell of undetermined significance; FOGSI: The Federation of Obstetric and Gynecological Societies of India; GCPR: good clinical practice recommendations; LBC: liquid-based cytology; NILM: negative for intraepithelial lesion or malignancy; VIA: visual inspection with acetic acid)

PREMALIGNANT LESIONS OF ENDOMETRIAL CANCER

INTRODUCTION

Endometrial cancer is the most common gynecological malignancy in the western world and sixth most common worldwide. Endometrial hyperplasia (EH) is the precursor of endometrial cancer **(Fig. 5)**.[1]

Endometrial hyperplasia is characterized by hyperplastic changes in endometrial glandular and stromal structures lining the uterine cavity.[14] EH is defined as irregular proliferation of the endometrial glands with an increase in the gland to stroma ratio when compared with proliferative endometrium.[15]

Premalignant lesions of the endometrium are a characteristic sequence of atypical EH or intraepithelial neoplasia that may give rise to endometrial carcinoma. Atypical hyperplasia/endometrioid intraepithelial neoplasia (AH/EIN) is the premalignant phase for endometrioid adenocarcinoma or type 1 endometrial carcinoma. Endometrial intraepithelial carcinoma (EIC) is thought to be the precursor lesion for the type 2 or serous carcinoma and a few of the endometrial clear cell carcinomas.

WHO classification (revised in 2014):
- Hyperplasia without atypia
- Hyperplasia with atypia

Risk of progression to endometrial cancer:
- Hyperplasia without atypia: <5% over 20 years (majority cases regress during follow-up)
- Hyperplasia with atypia: 29%

RISK FACTORS

Most common EH outturns from high levels of estrogens, combined with insufficient levels of progesterone.[16,17.]

Risk factors for the advancement of endometrial cancer include obesity, unbalanced estrogen therapy, PCOS, tamoxifen treatment, and nulliparity.[18]

Endometrial hyperplasia is common in women aged 50–54 years with body mass index (BMI) over 30.[19] The average age for EH is 52 years, which is 9 years lower than the average age for endometrial cancer. The increased risk of endometrial cancer among overweight (BMI > 25) and obese persons appears to be greater in postmenopausal than in younger women **(Table 11)**.[20]

PATHOPHYSIOLOGY

Accurate diagnosis of premalignant endometrial lesions and mitigation of risk factors can prevent progression to invasive carcinoma. During the cyclical hormonal endometrial remodeling, the endometrium responds to the estrogenic stimulation in a uniform programmed manner. The proliferative response is regulated by progestogenic effects. However, unopposed estrogenic stimulation of the endometrium can lead to the hyperproliferative glandular

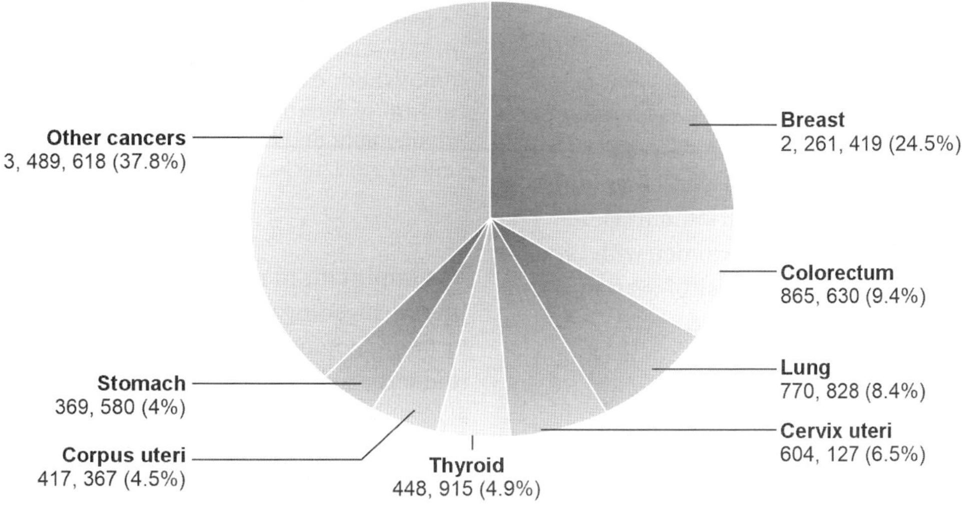

Fig. 5: Incidence of carcinoma endometrium.
Source: GLOBOCAN 2020.

TABLE 11: Characteristic and relative risk of endometrial cancer.

Characteristic	*Relative risk*
Nulliparity	2–3
Late menopause	2–4
Obesity	3–10
Diabetes mellitus	2.8
Unopposed estrogen therapy	4–8
Tamoxifen therapy	2–3
Lynch II syndrome	20

epithelium. This may lead to glandular remodeling and crowding of glands impinging on stromal spaces. Additionally, metaplastic changes may occur in glandular epithelium. If sporadic mutations accumulate in these hyperproliferative glands, prolonged or unopposed estrogen exposures act as a selection factor. The positively selected hyperproliferative clones then crowd out the normal glands **(Figs. 6A and B)**.

ENDOMETRIAL CANCER SCREENING

Most endometrial carcinomas are sporadic, but 10% of cases have a hereditary basis. Two genetic models in the development of familial endometrial cancer: Lynch II syndrome and patients with a predisposition for endometrial cancer alone. Both of which are inherited in an autosomal dominant fashion. The majority of studies focused on the increased incidence of endometrial cancer associated with Lynch II syndrome, a highly penetrant disorder (80–85%).[21] Lynch II syndrome is caused by an inherited mutation i of mismatch repair genes: *hMSH2, hMLH1, PMS1, PMS2,* or *hMSH6.*[22] This is characterized by early age (average age younger than 45 years) at onset of neoplastic lesions in a variety of tissues, including the colon, uterus, stomach, ureters, ovaries, and skin.[23] The lifetime risk of endometrial cancer in women with Lynch II syndrome is 32–60% and the lifetime risk of ovarian cancer is 10–12%.[24,25]

Screening for endometrial cancer for certain high-risk women, include families with hereditary nonpolyposis colorectal cancer. Most patients with endometrial cancer present with abnormal perimenopausal or postmenopausal uterine bleeding when the tumor is still confined to the uterus. Application of an appropriate and accurate diagnostic test in this situation usually results in early diagnosis and a high cure rate with timely treatment.

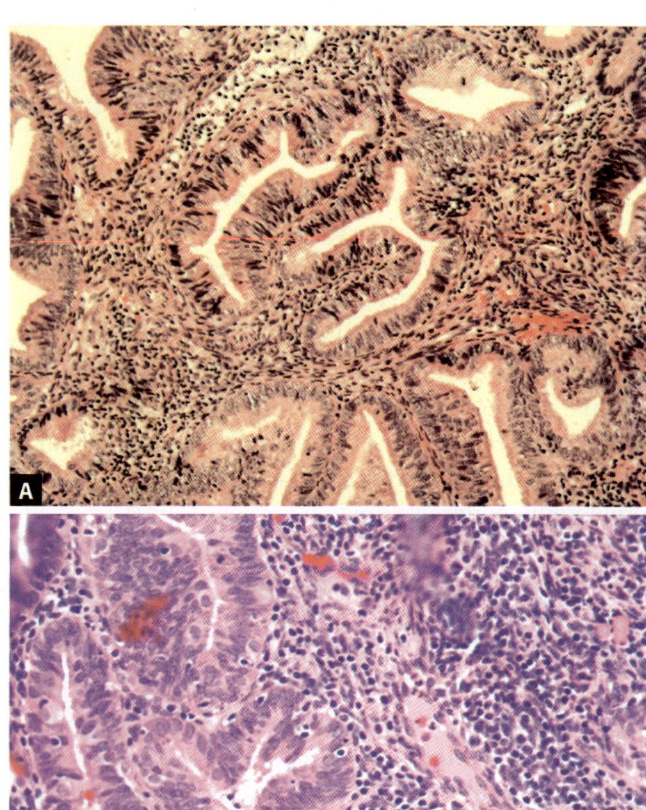

Figs. 6A and B: Atypical hyperplasia (complex hyperplasia with severe nuclear atypia) of endometrium. (A) The proliferative endometrial glands reveal considerable crowding and papillary infoldings. The endometrial stroma, although markedly diminished, can still be recognized between the glands. (B) Higher magnification demonstrates disorderly nuclear arrangement and nuclear enlargement and irregularity. Some contain small nucleoli.

It is important to recognize that the workup of abnormal uterine bleeding should include endometrial biopsy even in premenopausal patients, as 5% of cases are in women under the age of 40 years.

CLINICAL PRESENTATION

Patients with EH mostly present with abnormal uterine bleeding. This includes heavy menstrual bleeding, intermenstrual bleeding, irregular bleeding, unscheduled bleeding on hormone replacement therapy (HRT) and postmenopausal bleeding.[21] Occasionally patients may present with vaginal discharge.[26]

DIAGNOSIS

Diagnosis of endometrial cancer is given in **Flowchart 15**.

MANAGEMENT

Management of endometrial cancer is given in **Flowchart 16**.

Medical Management

Progestogens:

- Oral continues progestogens: Medroxyprogesterone 10–20 mg/day or norethisterone 10–15 mg/day
- Local intrauterine: Levonorgestrel-releasing intrauterine system (LNG-IUS) should be the first-line medical management due to higher disease regression rate

Duration of therapy: Minimum of 6 months, if fertility is not desired patient should be advised to retain LNG-IUS for 5 years as it reduces the risk of prolapse

Endometrial surveillance at a minimum of 6 monthly interval

Two negative biopsies 6 months apart should be obtained for confirmation of reversal of hyperplasia.

In high-risk patients annual follow-up should be advised after regression of disease.

INDICATION OF HYSTERECTOMY

Hysterectomy is indicated in women not wanting to preserve their fertility with EH without atypia:

- Progression to atypical hyperplasia occurs during follow-up
- No histological regression of hyperplasia despite 12 months of treatment
- Relapse of EH after completing progestogen treatment

Flowchart 15: Diagnosis of endometrial cancer.

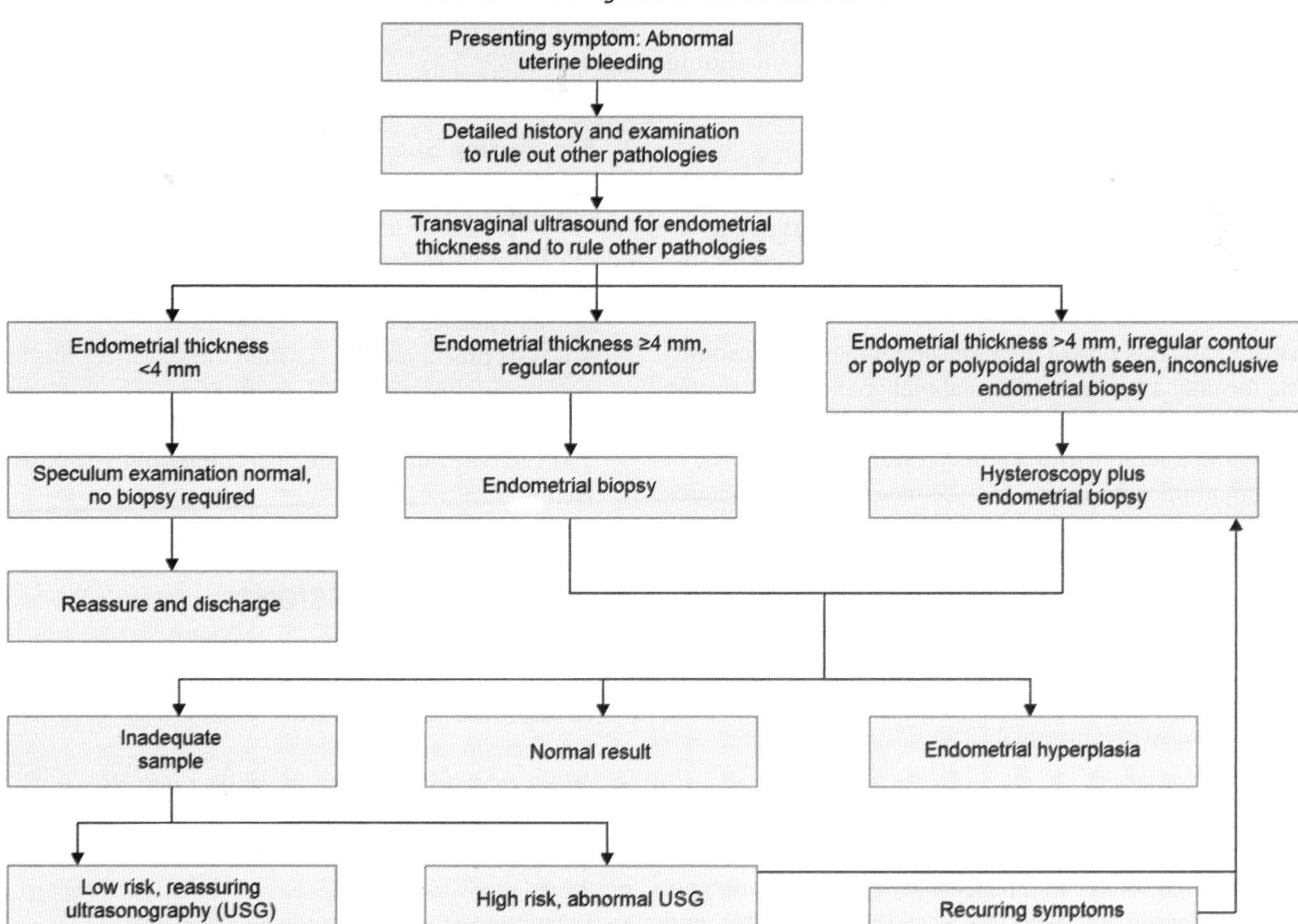

Flowchart 16: Management of endometrial cancer.

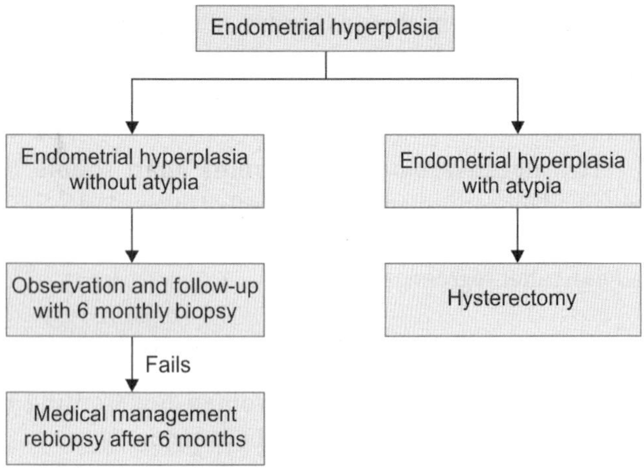

- Persistence of bleeding symptoms
- Woman declines to undergo endometrial surveillance or comply with medical treatment.

Patient wants to preserve fertility in EH without atypia: Medical management is with progestogens (LNG-IUS as first line) with 3 monthly biopsy and clinical review.

Once fertility is no longer required, patient should be offered hysterectomy in view of high risk of disease progression.

PREMALIGNANT LESIONS OF VULVA

INTRODUCTION

Vulvar cancer is one of the rare carcinomas and accounting for only 4% of gynecological malignancies. The most common subtype is squamous cell carcinoma (SCC) of the vulva and traditionally it has been regarded as a disease of postmenopausal women, but the mean age of incidence has fallen in recent years owing to the increase in HPV infections worldwide.[27,28] A preinvasive vulvar lesion is an area of skin or mucosa that has not yet progressed to invasive malignancy (cancer) but has the potential to do so. "Preinvasive vulvar lesions deserve specific attention because, beside the oncological risk, they affect not only functionality and body image but also psychosexual factors."[29]

PREVENTION

Primary Prevention (Vaccination)

Persistent HPV infection, particularly by HPV 16 subtype, has been associated with the development of HSIL and SCC of the vulva. The introduction of HPV vaccination as a primary prevention strategy in cervical cancer has been shown to also reduce the prevalence of noncervical premalignant lesions among vaccinated women.[30]

Secondary Prevention (Screening)

There is no specific screening for vulvar cancer. Self-examination in women should be encouraged in conditions like lichen sclerosus, which can develop into vulvar cancer. In addition, there should be early evaluation of any patient with symptoms (e.g., chronic vulvar pruritus) or signs (e.g., pigmented lesions, irregular ulcers), who may be a candidate for skin biopsy. Women with squamous intraepithelial lesion (SIL) of the cervix, vagina, or anus should undergo inspection of the vulva as part of their follow-up colposcopy visits.[30]

Tertiary Prevention (Management of Premalignant Lesions)

The approach to reduce vulvar cancer incidence is to the propitious treatment of predisposing preneoplastic lesions associated with vulvar cancer development.[30]

The two main pathological pathways that lead to vulvar SCC are:[31]

1. Keratinizing SCC usually occurs in older women and is often associated with lichen sclerosus and/or differentiated vulvar intraepithelial neoplasia (dVIN).
2. Warty/basaloid SCC by and large transpires in younger women, is caused by infection with oncogenic strains of HPV (particularly HPV 16, 18, 31, and 33), and has SIL as its precursor lesion. Often multifocal lesions are associated along with SIL in other parts of the lower genital tract (e.g., cervix, vagina, anus). HIV infection and cigarette smoking are also common predisposing factors.[30]

CLINICAL PROFILE OF PREMALIGNANT LESIONS

- More frequent in the age group 30–50 years
- Related to STI [herpes simplex virus 2 (HSV2), condyloma accuminata, gonorrhea, syphilis, *Gardnerella vaginalis*]
- *HPV-associated VIN seen in younger age*: HPV 6, 11—vulval condylomas while HPV 16, 18, 31, 33, 35 are associated with VIN (80%) and invasive cancer (40%).
- Associated CIN 10–80%
- VIN progression to cancer takes 20–30 years.

SYMPTOMS

- Mostly asymptomatic
- *Others:*
  - Itching
  - Burning sensation in vulva
  - Dysuria
  - Discharge/bleeding
  - Vulval ulcer
  - Dyspareunia
  - Warty growth/lump

MAJOR CLASSIFICATION SCHEMES PROPOSED FOR SQUAMOUS PRECURSOR LESIONS OF THE VULVA OVERTIME

The major classification schemes proposed for squamous precursor lesions of the vulva overtime are given in **Table 12**.

- *Vulvar high-grade squamous intraepithelial lesions (VHSIL)* are more common in women under 65 years and can transform into invasive cancer. "Approximately 60% of patients report itching and/or irritation, pain, or bleeding along with visible vulvar lesions." In others, lesions are diagnosed during a routine gynecological examination. The lesions differ widely from woman to woman in number, size, shape, color, and thickness.

  The several options for treating VHSIL: Surgical removal, ablation (removal with a laser or radiofrequency), or topical medications. If ablation or medication is used, the patient must have several biopsies beforehand to make sure no invasive cancer is present.
- *dVIN* is much rarer than VHSIL, as it is usually seen in older women. "Neoplasia" is the growth of abnormal cells or abnormal amounts of cells, which can start out benign but become cancerous. dVIN is more likely than VHSIL to transform into invasive cancer—sometimes quickly. It should always be completely removed surgically, according to the new consensus.
- *Vulvar Paget's disease in situ* is a glandular skin cancer of the vulva that may give rise to invasive cancer. Traditionally it was always treated surgically, but the consensus is now that imiquimod skin cream is effective and has minimal adverse effects. Radiation therapy can be used in some cases.
- *Vulvar melanoma in situ*—Melanoma is the most serious form of skin cancer, but vulvar melanoma in situ represents a noninvasive type (stage Ia). The recommended treatment is surgery that removes the lesion and an area of skin around it.

2020 WHO TERMINOLOGY

Human papillomavirus-associated squamous intraepithelial lesions include low-grade squamous intraepithelial lesion of the vulva (LSIL); HSIL HPV-independent VIN:

TABLE 12: Major classification schemes proposed for squamous precursor lesions of the vulva overtime.

1986 ISSVD	*2004 ISSVD* *2003 WHO*	*2005 Bethesda like*	*2012 LAST* *2014 WHO* *2015 ISSVD*
VIN1	VIN1 (only by WHO)	LG-VIL Condyloma VIN1	LSIL VIN1 Condyloma Mild Dysplasia Koilocytic Atypia
VIN2	uVIN VIN2 VIN3	HG-VIL VIN2,3 dVIN	HSIL VIN2–3 Moderate /severe dysplasia Bowen disease Bowenoid Dysplasia CIS
VIN3, severe atypia or VIN3, CIS			
VIN3, differintated type	dVIN		dVIN (not included in LAST guidelines)

(CIS: carcinoma in situ; dVIN: differentiated vulvar intraepithelial neoplasia; LG-VIL: low-grade vulvar intraepithelial lesion; HSIL: high-grade squamous intraepithelial lesion; ISSVD: The International Society for the Vulvovaginal Disease; LAST: Lower Anogenital Squamous Terminology; LSIL: low-grade squamous intraepithelial lesion; uVIN: usual-type VIN; WHO: World Health Organization)

Flowchart 17: 2020 World Health Organization (WHO) terminology.

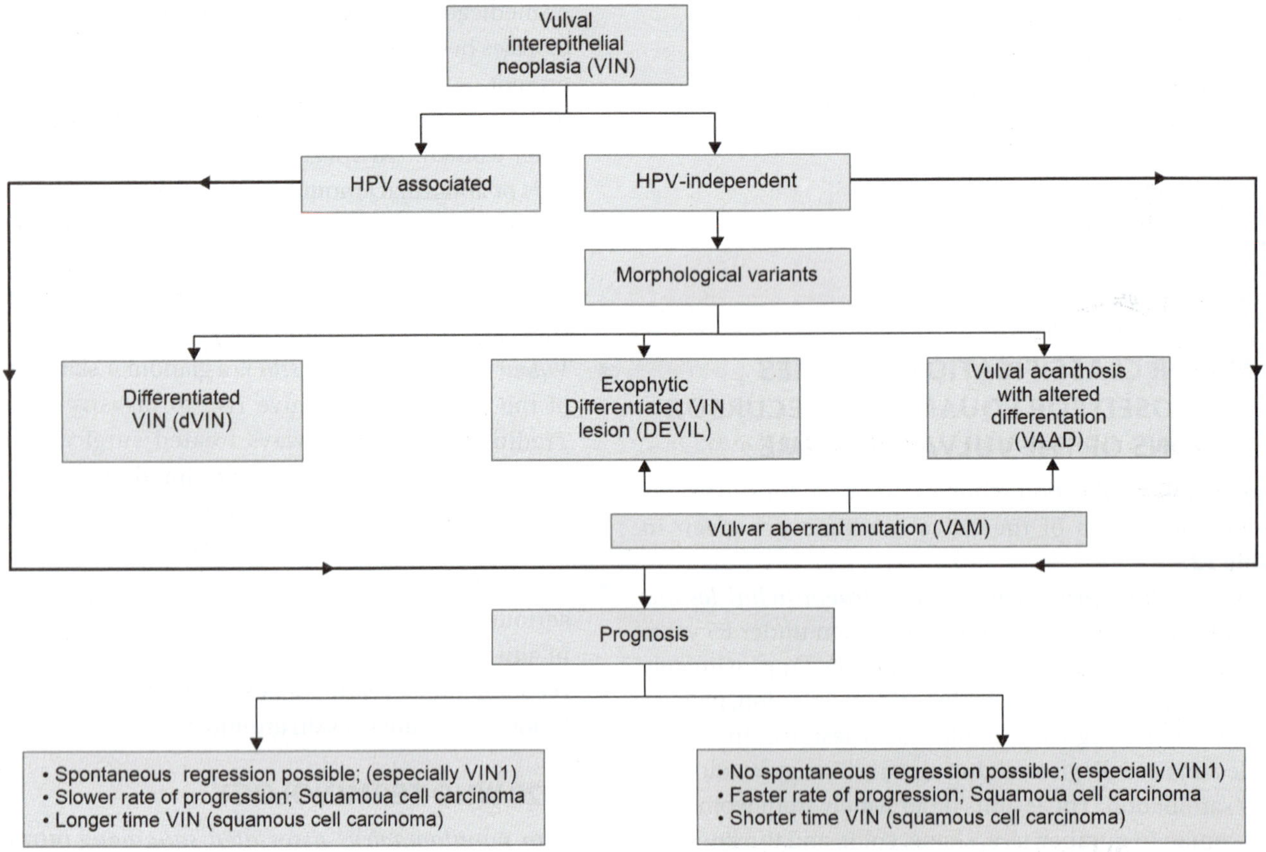

Vulval interepithelial neoplasia (VIN)

HPV associated

HPV-independent

Morphological variants

Differentiated VIN (dVIN)

Exophytic Differentiated VIN lesion (DEVIL)

Vulval acanthosis with altered differentiation (VAAD)

Vulvar aberrant mutation (VAM)

Prognosis

- Spontaneous regression possible; (especially VIN1)
- Slower rate of progression; Squamoua cell carcinoma
- Longer time VIN (squamous cell carcinoma)

- No spontaneous regression possible; (especially VIN1)
- Faster rate of progression; Squamoua cell carcinoma
- Shorter time VIN (squamous cell carcinoma)

dVIN; differentiated exophytic vulvar intraepithelial lesion (DEVIL); and vulvar acanthosis with altered differentiation (VAAD) **(Flowchart 17)**.

CLINICOPATHOLOGICAL CHARACTERISTICS OF THE 2020 WHO CLASSIFICATION OF VULVAR PRECANCEROUS LESIONS

Gross and histology:

- Lesions on vulva often are white, gray, pink, or dull red in color.
- Lesions look rough, raised from the surface and are often multifocal **(Fig. 7)**.

On histology:

- Cellular immaturity, nuclear abnormalities, and mitotic activity vary depending on the grade of VIN.
- Koilocytes may be present.
- Rete ridges are large and elongated.

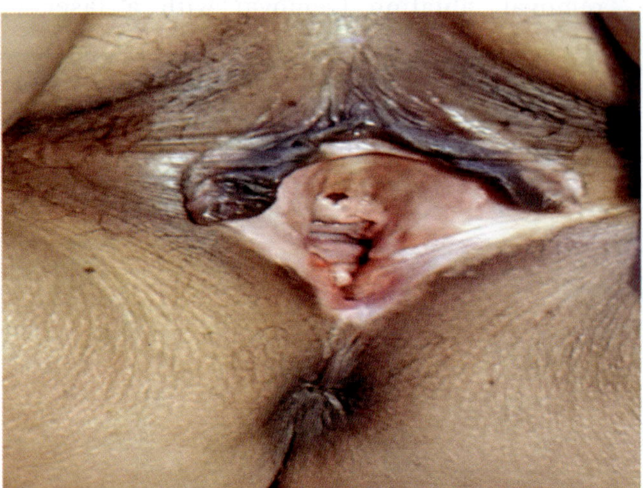

Fig. 7: White dull lesions [on histopathological examination (HPE) lichen sclerosis].

- The basement membrane is intact.
- Dermis is not involved.

DIFFERENTIAL DIAGNOSIS

Due to the variation in the clinical features of vulvar SILs, these lesions can mimic different diseases: lichen simplex chronicus, lichen sclerosus, lichen planus, psoriasis, and contact dermatitis.

TREATMENT OF VIN

Treatment is based on histologic findings, not on the presence or absence of HPV.

VIN1 (LSIL) is not treated: It is followed after 6 months, if regress, no further follow-up and if become VIN2 or VIN3 should be treated.

TREATMENT OF HSIL (VIN2, 3 OR CIS)

Laser Ablation

Advantages:
- Less disfiguring
- Depth of destruction can be adjusted
- Less bleeding

Disadvantages:
- Prolonged painful healing
- No tissue specimen
- Reoccurrence rate 20–25%.

Cavitational Ultrasonic Surgical Aspiration

Cavitational ultrasonic surgical aspiration (CUSA) is performed on non-hair-bearing vulval skin.

Ultrasound is used to cause cavitation and disruption of affected tissue.

Advantages:
- Less scaring
- Less pain
- Provides tissue for specimen

Disadvantages:
- Recurrence rate 35%
- Tissue specimen is fragmented.

Simple Vulvectomy

- Done in diffuse type in postmenopausal women
- Risk of recurrence is high (40–70%)
- Long-term follow-up is needed
- Patient may need skin grafting if there is disfigurement

Topical Medical Therapy

Topical IMIQUIMOD cream: Aldara
- Lesser toxicity
- Higher efficacy
- 77% cure rate
- 20% recurrence rate

Others:
- 5-FU (5-fluorouracil)
- Cidofovir
- PDT (photodynamic therapy) with 5 ALA (amino-levulinic acid).

FOLLOW-UP

- *VIN1:* Observation and follow-up after 6 months, if negative, then no need to follow up after 6 months; if becomes VIN2 or VIN3 then treated as VIN2 or VIN3
- *Follow-up of VIN2 or VIN3:* This is at 6 months, then at 1 year, and then annually.

CONCLUSION

Application of an appropriate and accurate screening and diagnostic tests usually result in early diagnosis and a high cure rate with timely treatment. The chart mentioned in the chapter outlines the strongly recommended screening and diagnostic protocol. It is important to recognize that the screening of CIN **(Flowcharts 18 to 22)** and workup of abnormal uterine bleeding is of utmost importance and can prevent further morbidities and mortalities.

Flowchart 18: Management of women with CIN1 on histology.

CIN1 on histology

Check preceding HPV/cytology/VIA report

HPV positive or ASCUS/LSIL or VIA positive

ASC-H, HSIL

Co-test/HPV/cytology/VIA at 1, 2 years

Co-testing at 1 and 2 years or cytology at 6 months

Test negative

Any test positive

Return to routine screening

Colposcopy ± biopsy

CIN1

CIN2/3

Repeat Co-test/HPV/cytology/VIA after 1 year

Diagnostic excisional procedure—LEEP

Test positive

Assess adequacy of excision

Colposcopy ± biopsy

Follow-up

CIN1 persisting for 2 years

Ablative procedure*, excisional procedure; if lesion not amenable for ablation or persisting CIN1 after ablation consider excision procedure

*Eligibility for ablative procedures
1. Lesion should be entirely visible and occupy not more than two quadrants of cervix
2. Entire lesion should be located on ectocervix without any vaginal and/or endocervical extension
3. Lesion can be adequately covered by largest cryotherapy probe available
4. There is no suspicion of invasive cancer

(ASCUS: atypical squamous cell of undetermined significance; CIN: cervical intraepithelial neoplasia; HSIL: high-grade squamous intraepithelial lesion; LEEP: loop electrosurgical excision procedure; LSIL: low-grade squamous intraepithelial lesion; VIA: visual inspection with acetic acid)

Flowchart 19: Management of women desirous of pregnancy with CIN1 on histology.

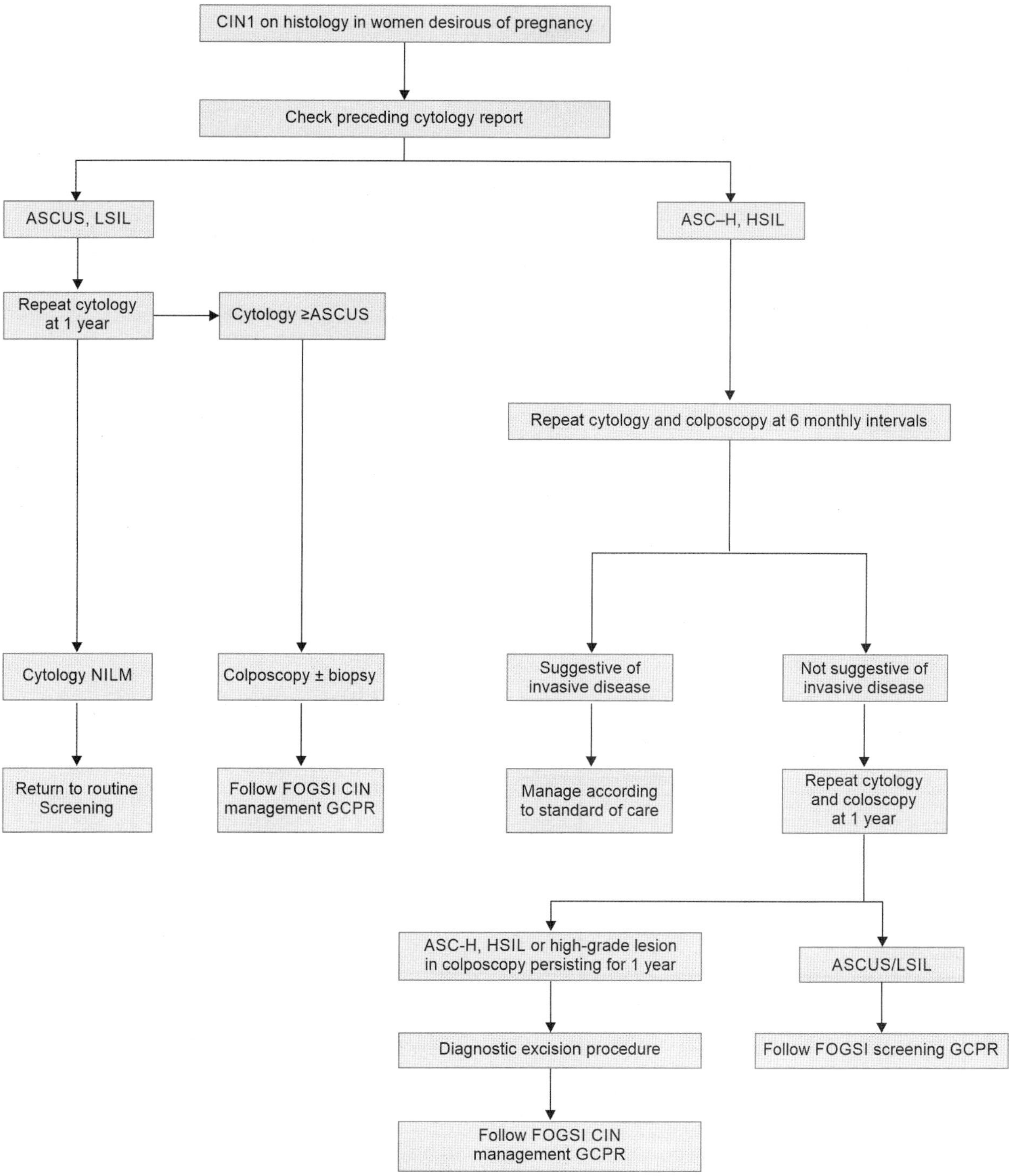

(ASCUS: atypical squamous cell of undetermined significance; ASC-H: atypical squamous cell: cannot exclude high-grade squamous intraepithelial lesion; CIN: cervical intraepithelial neoplasia; HSIL: high-grade squamous intraepithelial lesion; LEEP: loop electrosurgical excision procedure; LSIL: low-grade squamous intraepithelial lesion; VIA: visual inspection with acetic acid)

Flowchart 20: Management of women with CIN2, 3 on histology.

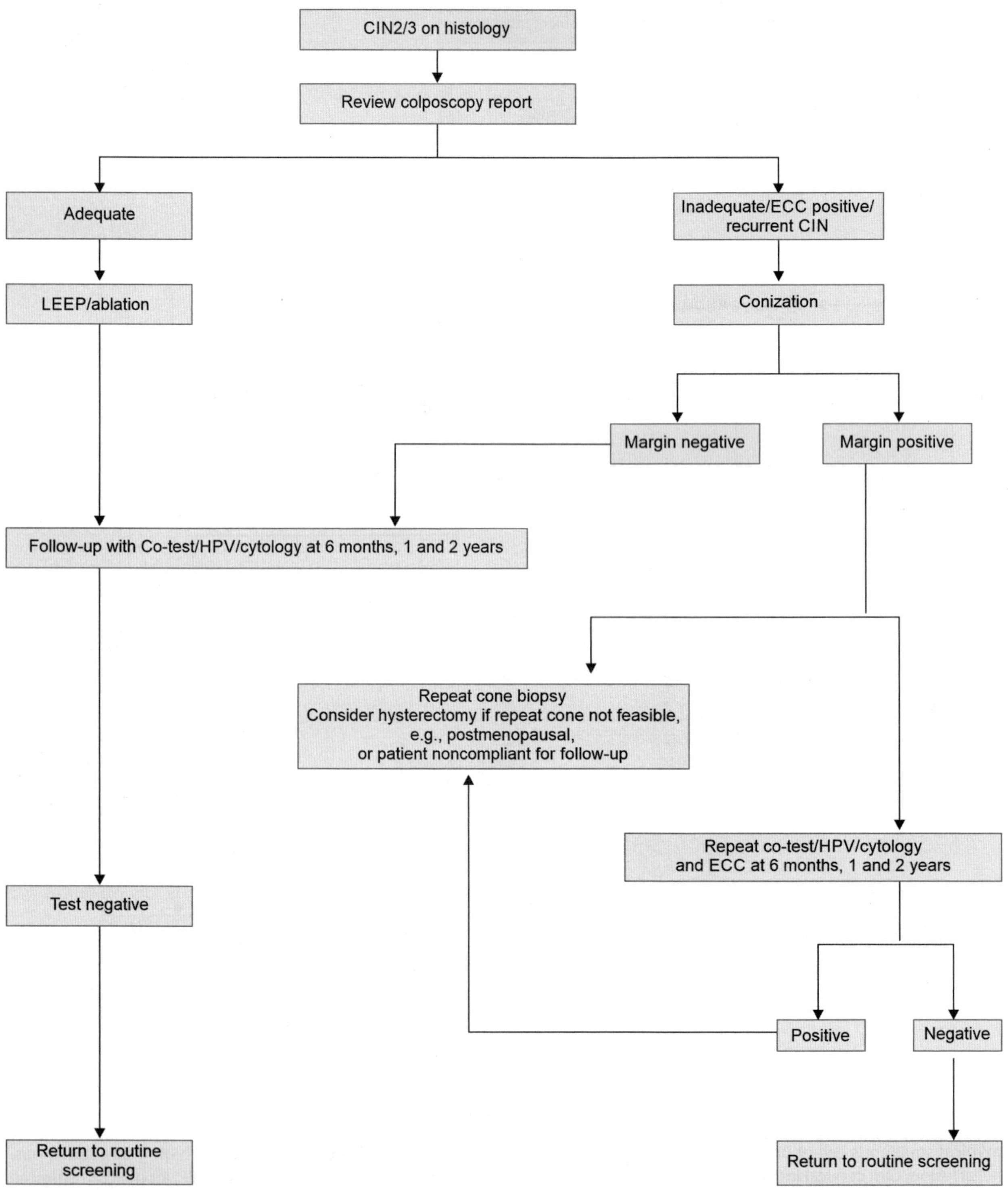

(CIN: cervical intraepithelial neoplasia; ECC: endocervical curettage; LEEP: loop electrosurgical excision procedure)

Flowchart 21: Management of women with CIN2, 3 on histology, desirous of pregnancy.

CIN2/3 young women desirous of pregnancy

Cytology and colposcopy at 6 months

CIN2 lesions, triage with p16, Ki-67

CIN3 with inadequate colposcopy

Lesson persists

Negative

Positive

Excisional procedure

Repeat cytology and colposcopy at 1 year

Adequate colposcopy

Inadequate colposcopy

Cytology negative/improving and normal colposcopy

High-grade cytology/colposcopy worsens at 1 year

Repeat Co-testing/cytology after 1 year

Treat using excision procedure or ablation if eligible

Any abnormal reports

Colposcopy ± biopsy

Follow FOGSI CIN management GCPR

Normal reports—follow-up with Co-test/cytology at 3 years

If normal, return to routine 5 yearly screening

(CIN: cervical intraepithelial neoplasia; FOGSI: The Federation of Obstetric and Gynecological Societies of India; GCPR: good clinical practice recommendations)

Flowchart 22: Management of women with cervical intraepithelial neoplasia (CIN) in pregnancy.

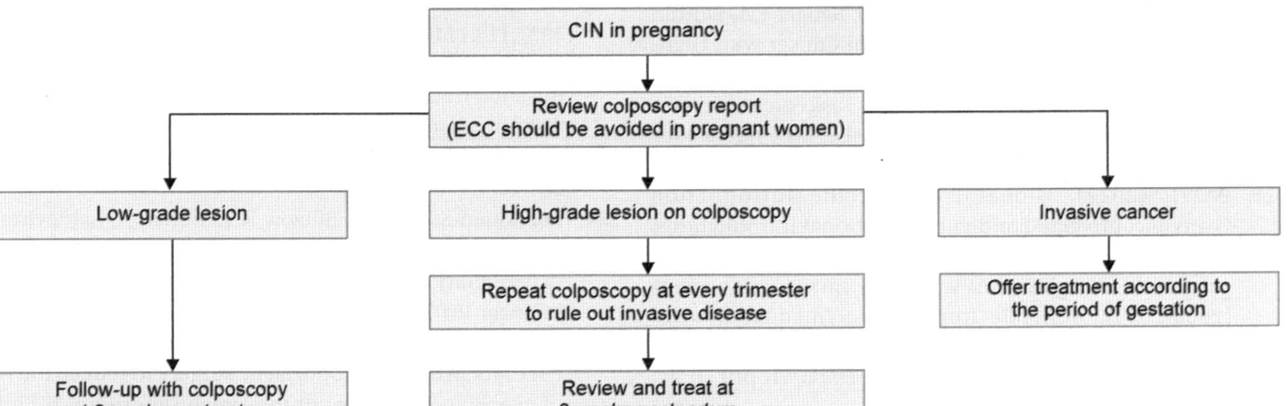

(ECC: endocervical curettage)

REFERENCES

1. Sung H, Ferlay J, Siegel RL, Laversanne M, Soerjomataram I, Jemal A, et al. Global cancer statistics 2020: GLOBOCAN estimates of incidence and mortality worldwide for 36 cancers in 185 countries. CA Cancer J Clin. 2021:71:209-49.
2. Bray F, Ferlay J, Soerjomataram I, Siegel RL, Torre LA, Jemal A. Global Cancer Statistics 2018: GLOBOCAN estimates of incidence and mortality worldwide for 36 cancers in 185 countries. CA Cancer J Clin. 2018;68(6):394-424.
3. Muñoz N, Bosch FX, de Sanjosé S, Herrero R, Castellsagué X, Shah KV, et al; International Agency for Research on Cancer Multicenter Cervical Cancer Study Group. Epidemiologic classification of human papillomavirus types associated with cervical cancer. N Eng J Med. 2003;348:518-27.
4. Schiffman M, Castle PE, Jeronimo J, Rodrigue AC, Wacholder S. Human papillomavirus and cervical cancer. Lancet 2007;370:890-907.
5. Herrero R, Murillo R. Cervical cancer. In: Thun M, Linet MS, Cerhan JR, Haiman CA, Schottenfeld D (Eds). Cancer Epidemiology and Prevention, 4th edition. New York: Oxford University Press; 2018. pp. 925-46.
6. Barron BA, Richart RM. Screening protocols for cervical neoplastic disease. Gynecol Oncol. 1981;12(2 Pt 2): S156-S167.
7. IARC. IARC Handbooks of Cancer Prevention Volume 10: Cervix Cancer Screening. Strategies IWGotEoCP, editor. Lyon: IARC Press; 2005.
8. Arbyn M, Xu L, Simoens C, Martin-Hirsch PP. Prophylactic vaccination against human papillomaviruses to prevent cervical cancer and its precursors. Cochrane Database Syst Rev. 2018;5(5):CD009069.
9. Hall MT, Simms KT, Lew JB, Smith MA, Brotherton JM, Saville M, et al. The projected timeframe until cervical cancer elimination in Australia: a modelling study. Lancet Public Health. 2019;4(1):e19-e27.
10. Melnikow J, Henderson JT, Burda BU, Senger CA, Durbin S, Soulsby MA. Screening for Cervical Cancer with High-Risk Human Papillomavirus Testing: A Systematic Evidence Review for the U.S. Preventive Services Task Force. Evidence Synthesis, No. 158. [Internet]. Rockville, MD: Agency for Healthcare Research and Quality (US); 2018.
11. PATH. (2022). Global HPV Vaccine Introduction Overview: projected and current national introductions, demonstration/pilot projects, gender-neutral vaccination programs, and global HPV vaccine introduction maps (2006-2023). Available from: https://media.path.org/documents/Global_Vaccine_Intro_Overview_Slides_Final_PATHwebsite_MAR_2022_qT92Wwh.pdf [Last accessed April, 2024].
12. Lemp JM, De Neve JW, Bussmann H, Chen S, Manne-Goehler J, Theilmann M, et al. Lifetime prevalence of cervical cancer screening in 55 low- and middle-income countries. JAMA. 2020;324:1532-42.
13. World Health Organization. Global strategy to accelerate the elimination of cervical cancer as a public health problem. Geneva: World Health Organization; 2020.
14. Tavassoli FA, Devilee P. World Health Organization Classification of Tumours. Pathology and Genetics. Tumours of the Breast and Female Genital Organs. Lyon: IARC Press; 2003. pp. 217-28.
15. Kurman RJ, Carcangiu ML, Herrington CS, Young RH, editors. WHO Classification of Tumours of Female Reproductive Organs, 4th edition. Lyon: IARC; 2014.
16. Ricci E, Moroni S. Parazzini F, Surace M, Benzi G, Salerio B, et al. Risk factors for endometrial hyperplasia: results from a case-control study. Int Gynecol Cancer. 2002;12:257-60.
17. Reed SD, Newton KM, Clinton WL, Epplein M, Garcia R, Allison K, et al. Incidence of endometrial hyperplasia. Am J Obstet Gynecol. 2009;200:678.e1-6.
18. Armstrong AJ, Hurd WW, Elguero S, Barker NM, Zanotti KM. Diagnosis and management of endometrial hyperplasia. J Minim Invasive Gynecol. 2012;19:562-71.
19. Lacey JV Jr, Chia VM. Endometrial hyperplasia and the risk of progression to carcinoma. Maturitas. 2009;63:39-44.
20. Reeves GK, Pirie K, Beral V, Green J, Spencer E, Bull D; Million Women Study Collaboration. Cancer incidence and mortality in relation to body mass index in the Million Women Study: cohort study. BMJ. 2007;335:1134.
21. Vasen HFA, Watson P, Mecklin JP, Lynch HT. New clinical criteria for hereditary nonpolyposis colorectal cancer (HNPCC, Lynch syndrome) proposed by the International Collaborative Group on HNPCC. Gastroenterology. 1999;116:1453-6.
22. Peltomäki P, Vasen HF. The International Collaborative Group on Hereditary Nonpolyposis Colorectal Cancer. Mutations predisposing to hereditary nonpolyposis colorectal cancer: Database and results of a collaborative study. The International Collaborative Group on Hereditary Nonpolyposis Colorectal Cancer. Gastroenterology. 1997;113:1146-58.
23. Umar A, Boland CR, Terdiman JP, Syngal S, de la Chapelle A, Rüschoff J, et al. Revised Bethesda guidelines for hereditary nonpolyposis colorectal cancer (Lynch Syndrome) and microsatellite instability. J Natl Cancer Inst. 2004;96: 261-8.
24. Renkonen-Sinisalo L, Bützow R, Leminen A, Lehtovirta P, Mecklin JP, Järvinen HJ. Surveillance for endometrial cancer in hereditary nonpolyposis colorectal cancer syndrome. Int J Cancer. 2007;120:821-4.
25. Schmeler KM, Lynch HT, Chen LM, Munsell MF, Soliman PT, Clark MB, et al. Prophylactic surgery to reduce the risk of gynecologic cancers in the Lynch syndrome. N Engl J Med. 2006;354:261-9.
26. Kurman RJ, Kaminski PF, Norris HJ. The behavior of endometrial hyperplasia. A long-term study of "untreated" hyperplasia in 170 patients. Cancer. 1985;56(2):403-12.

27. Barlow EL, Kang YJ, Hacker NF, Canfell K. Changing trends in vulvar cancer incidence and mortality rates in Australia since 1982. Int J Gynecol Cancer. 2015;25: 1683-9.
28. Kang YJ, Smith M, Barlow E, Coffey K, Hacker N, Canfell K. Vulvar cancer in high-income countries: increasing burden of disease. Int J Cancer. 2017;141:2174-86.
29. Wolters Kluwer Health: Lippincott. (2022). New Guidelines Address Treatment of Pre-Invasive Lesions of the Vulva. Available from: https://www.newswise.com/articles/new-guidelines-address-treatment-of-pre-invasive-lesions-of-the-vulva [Last accessed April, 2024].
30. Olawaiye AB, Cuello MA, Rogers LJ. Cancer of the vulva: 2021 update. Int J Gynaecol Obstet. 2021;155 Suppl 1 (Suppl 1):7-18.
31. Rakislova N, Clavero O, Alemany L, Saco A, Quirós B, Lloveras B, et al; VVAP study group. Histological characteristics of HPV-associated and -independent squamous cell carcinomas of the vulva: a study of 1,594 cases. Int J Cancer. 2017;141:2517-27.

Chapter

53

Carcinoma Cervix

Saritha Shamsunder, Versha Dhama

INTRODUCTION

Carcinoma cervix is the second most common cancer of females after breast in India and fourth most common worldwide according to Globacon 2020.[1] The overall worldwide mortality from cervical cancer is approximately 342,000 per annum.[1] In India, cervical cancer accounted for 9.4% of all cancers and 18.3% (123,907) of new cases in 2020.[1] In developing countries where health resources are constraint cervical cancer is the second most common cause of death in women.

RISK FACTORS

There are many risk factors proven to be involved in the development of cervical cancer, mentioned as follows:

- Early age of first intercourse
- Multiple sexual partners
- Cigarette-smoking
- High parity
- Low socioeconomic status
- Race[2] (black and Hispanic women)
- Chronic immune suppression
- Oral contraceptive pills (OCPs) usages (some studies found the use of OCPs have found to increase incidence of cervical cancer while others refute this hypothesis).[3,4]

All these risk factors are linked to sexual activity and exposure to sexually transmitted infections. Many of these sexually transmitted infections are considered to be the cofactors[5,6] in the causation of cervical cancer along with the causative organism human papillomavirus (HPV).

PATHOGENESIS OF CARCINOMA CERVIX

The key etiologic agent of cervical cancer is persistent infection with high-risk oncogenic HPV types. There are 15 high-risk HPV subtypes; two of which, HPV 16 and 18, are found in approximately 70% of cervical cancer cases. In India, the prevalence of women infected with HPV 16 and 18 is >90%.[7] The majority of immunocompetent individuals infected with HPV are able to clear the viral infection and remain asymptomatic. In a minority of patients, HPV infection persists and causes clinically detectable lesions that can progress to invasive cancer over a long time period.

Human papilloma virus has a predilection for epithelial cells of the cervix. Traumatic microabrasions, occuring during sexual intercourse, expose the naïve basal layer cells to HPV. HPV replication is dependent upon and utilizes the normal replicative machinery of the cervical cells, which is subverted by two viral proteins, E1 and E2. The mechanism by which HPV affects cellular growth and differentiation is through interaction of viral E6 and E7 proteins with tumor suppressor genes p53 and Rb, respectively. Also, E6 and E7 also modulate molecules involved in the innate and adaptive immune responses, potentially enabling HPV to evade immune surveillance, especially in the early stages of replication.

CLINICAL PRESENTATION AND EVALUATION

The most common presenting symptom of a patient with cervical cancer is vaginal bleeding in various forms as postcoital bleeding, postmenopausal bleeding, or irregular vaginal spotting/bleeding. Other symptoms may be weight loss, loss of appetite, foul-smelling vaginal discharge, and features of obstructive uropathy. Examination of such patients include general physical examination identifying involvement of cervical, inguinal, and supraclavicular lymph nodes. Pelvic examination done in two parts: inspection of cervix is done by using speculum examination followed by palpation by digital examination of cervix. Parametrium involvement is assessed by rectovaginal examination and presence of nodularity beyond cervix. If a visible tumor growth is encountered during examination, a punch biopsy from the lesion is enough to establish the diagnosis whereas for microscopic disease, colposcopic-directed biopsy is required. Once histopathological diagnosis is established, staging is done clinically supplemented by radiological and pathological findings as described below. The International Federation of Gynecology and Obstetrics (FIGO) 2018 staging[8] permits the use of any of the imaging modalities according to available resources, i.e., ultrasound, computed tomography (CT), magnetic resonance imaging (MRI), positron emission tomography (PET), to provide additional information on tumor size, nodal status, and local or systemic spread.

HISTOPATHOLOGY—WHO CLASSIFICATION OF FEMALE GENITAL TUMOURS[9]

It is essential that all cancers must be confirmed by microscopic examination. Cases are classified as carcinomas of the cervix if the primary growth is in the cervix. All histologic types must be included. The histopathologic types as described in the WHO Classification of Female Genital Tumours[9] are enumerated in **Box 1**.

BOX 1: World Health Organization (WHO) Classification of Female Genital Tumours.

1. *Squamous epithelial tumors:*
   - Squamous cell carcinoma, HPV-associated
   - Squamous cell carcinoma, HPV-independent
   - Squamous cell carcinoma NOS
2. *Glandular tumors:*
   - Adenocarcinoma NOS
   - Adenocarcinoma, HPV-associated
   - Adenocarcinoma, HPV-independent, gastric type
   - Adenocarcinoma, HPV-independent, clear cell type
   - Adenocarcinoma, HPV-independent, mesonephric type
   - Adenocarcinoma, HPV-independent, NOS
   - Endometrioid adenocarcinoma NOS
   - Carcinosarcoma NOS
   - Adenosquamous carcinoma
   - Mucoepidermoid carcinoma
   - Adenoid basal carcinoma
3. *Mixed epithelial and mesenchymal tumors:*
   - Adenosarcoma
4. *Germ cell tumors:*
   - Endodermal sinus tumor
   - Yolk sac tumor NOS
   - Choriocarcinoma NOS

(HPV: human papillomavirus; NOS: not otherwise specified)

Flowchart 1 mentions the workup of a case of suspected cancer cervix.

STAGING

Table 1 shows the FIGO Staging of Cancer Cervix Uteri (2018).[8]

TREATMENT OPTIONS

- *Surgical:*
  - Cone biopsy of cervix
  - Extrafascial hysterectomy
  - *Radical trachelectomy:* It involves removal of cervix, upper part of vagina and supporting tissue adjacent to these. This procedure is done where a woman has strongly desire to preserve childbearing.
  - Radical hysterectomy **(Table 2)** along with pelvic and/or pelvic lymphadenectomy.

Lymph-vascular space invasion (LVSI) is defined as the presence of carcinoma cells within the lymphatic and/or blood vessels. The presence of LVSI is a predictor of lymph node metastasis and associated with decreased survival of women with early staged carcinoma cervix. Hence, more extensive surgery is required in such cases.

However, recently published LACC trial phase 3 results indicated an increased risk of death and progression of

Flowchart 1: Workup of a case of suspected cancer cervix.

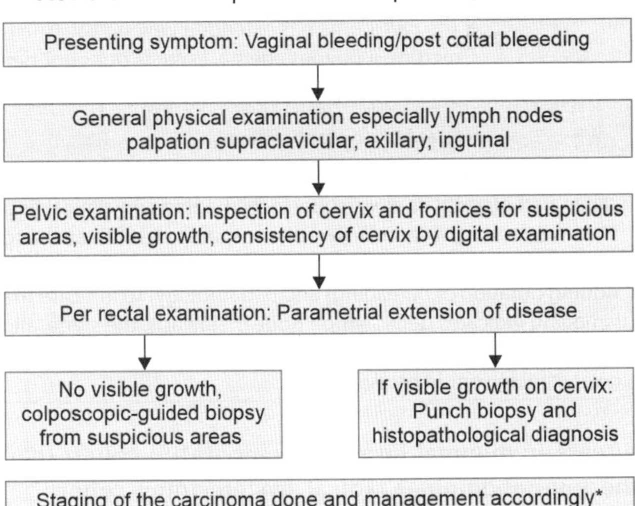

*The International Federation of Gynecology and Obstetrics (FIGO) 2018 staging permits the use of any of the imaging modalities according to available resources, i.e., ultrasound, computed tomography (CT), magnetic resonance imaging (MRI), positron emission tomography (PET), to provide additional information on tumor size, nodal status, and local or systemic spread. MRI is the best method of radiologic assessment of primary tumors >10 mm.

TABLE 1: The International Federation of Gynecology and Obstetrics (FIGO) Staging of Cancer Cervix Uteri (2018).

Stage	Description
I	The carcinoma is strictly confined to the cervix (extension to the uterine corpus should be disregarded)
IA	Invasive carcinoma that can be diagnosed only by microscopy, with maximum depth of invasion ≤5 mm*
IA1	Measured stromal invasion ≤3 mm in depth
IA2	Measured stromal invasion >3 and ≤5 mm in depth
IB	Invasive carcinoma with measured deepest invasion >5 mm (greater than stage IA); visible lesion limited to the cervix uteri with size measured by maximum tumor diameter†
IB1	Invasive carcinoma >5 mm depth of stromal invasion and ≤2 cm in greatest dimension
IB2	Invasive carcinoma >2 and ≤4 cm in greatest dimension
IB3	Invasive carcinoma >4 cm in greatest dimension
II	The carcinoma invades beyond the uterus but has not extended onto the lower third of the vagina or to the pelvic wall
IIA	Involvement limited to the upper two-thirds of the vagina without parametrial involvement
IIA1	Invasive carcinoma ≤4 cm in greatest dimension
IIA2	Invasive carcinoma >4 cm in greatest dimension
IIB	With parametrial involvement but not up to the pelvic wall
III	The carcinoma involves the lower third of the vagina and/or extends to the pelvic wall and/or causes hydronephrosis or nonfunctioning kidney and/or involves pelvic and/or para-aortic lymph nodes
IIIA	The carcinoma involves the lower third of the vagina, with no extension to the pelvic wall
IIIB	Extension to the pelvic wall and/or hydronephrosis or nonfunctioning kidney (unless known to be due to another cause)
IIIC	Involvement of pelvic and/or para-aortic lymph nodes (including micrometastases)‡, irrespective of tumor size and extent (with *r* and *p* notations)§
IIIC1	Pelvic lymph node metastasis only
IIIC2	Para-aortic lymph node metastasis
IV	The carcinoma has extended beyond the true pelvis or has involved (biopsy proven) the mucosa of the bladder or rectum. A bullous edema, as such, does not permit a case to be allotted to stage IV
IVA	Spread of the growth to adjacent pelvic organs
IVB	Spread to distant organs

*Imaging and pathology can be used, where available, to supplement clinical findings with respect to tumor size and extent, in all stages. Pathological findings supersede imaging and clinical findings.
†The involvement of vascular/lymphatic spaces should not change the staging. The lateral extent of the lesion is no longer considered.
‡Isolated tumor cells do not change the stage, but their presence should be recorded.
§Adding notation of *r* (imaging) and *p* (pathology) to indicate the findings that are used to allocate the case to stage IIIC. For example, if imaging indicates pelvic lymph node metastasis, the stage, allocation would be stage IIIC1r; if confirmed by pathological findings, it would be stage IIIC1p. The type of imaging modality or pathology technique used should always be documented. When in doubt, the lower staging should be assigned.

disease in minimal invasive surgery (MIS) cohort than the total abdominal radical hysterectomy cohort.[10-16] Therefore, surgeons should proceed with caution with the robotic-assisted radical hysterectomy approach and counsel their patients with cervical cancer on the results of the LACC trial. This will allow women to make the most informed decision about surgical approach to treat their cervical cancer.

- *Radiation therapy:*
  - External beam radiotherapy (EBRT)
  - Brachytherapy
  - Intensity modulated radiotherapy
- *Concurrent chemoradiation:* The use of chemotherapy to sensitize cells to radiation therapy improved local-regional control.

TABLE 2: Classification of radical hysterectomy.

	Simple extrafascial hysterectomy/type I/type A	***Modified radical hysterectomy/ type II/type B***	***Radical hysterectomy/ type III/type C***
Indication	Stage IA1	Stage IA1 with LVSI (lymph-vascular space invasion); IA2	Stage IB1 and IB2, selected stage IIA
Uterus and cervix	Removed	Removed	Removed
Ovaries	Optional removal	Optional removal	Optional removal
Vaginal margin	None	1–2 cm	Upper one quarter to one-third
Ureters	Not mobilized	Tunnel through broad ligament	Tunnel through broad ligament
Cardinal Ligaments	Divided at uterine and cervical border	Divided where ureter transits broad ligaments	Divided at pelvic side wall
Uterosacral ligaments	Divided at cervical border	Partially removed	Divided near sacral origin
Urinary bladder	Mobilized to base of bladder	Mobilized to upper vagina	Mobilized to middle vagina
Rectum	Not mobilized	Mobilized below cervix	Mobilized below cervix
Surgical approach	Laparotomy or laparoscopy or robotic surgery	Laparotomy or laparoscopy or robotic surgery	Laparotomy or laparoscopy or robotic surgery

STAGE-BASED MANAGEMENT OF CERVICAL CANCER

Table 3 shows the stage-based management of cervical cancer.

Microinvasive Cancer

Stage IA1: Cervical conization completes the treatment of microinvasive cancer unless surgical margins are involved. Incidence of pelvic lymph node metastasis in this stage is 1%. Therapeutic conization with or without pelvic lymphadenectomy, if LVSI is present or absent respectively, will suffice if childbearing is desired. Presence of LVSI does not change the stage but changes the management approach of the patient. If the woman is elderly or does not desire childbearing, extrafascial hysterectomy is the desired treatment along with or without pelvic lymphadenectomy.

Stage IA2: Pelvic lymphadenectomy or sentinel lymph node mapping is necessary as there is 3–8% incidence of pelvic lymph node metastasis in these lesions. Modified radical hysterectomy or radical trachelectomy (if requires childbearing) is the desired treatment of primary lesion. *If intermediate or high-risk factors are present in surgical specimen, adjuvant radiation/chemoradiation is advised.*

Invasive Cancer

Stage IB1, IB2, IIA1: These lesions are managed with radical trachelectomy or radical hysterectomy along

TABLE 3: Stage-based management of cervical cancer.

Stage	***Management***
IA1	
• Without LVSI	Conization or extrafascial hysterectomy
• With LVSI	Modified radical trachelectomy or Modified radical hysterectomy + Pelvic LND (lymphadenectomy) or SLN (sentinel lymph node)
IA2	Modified radical trachelectomy or modified radical hysterectomy + Pelvic LND or SLN
IB1	Modified radical trachelectomy or modified radical hysterectomy + Pelvic LND or SLN
IB2	Radical hysterectomy + Pelvic LND
IB3	Chemoradiation, pelvic field
IIA1	Radical hysterectomy+ Pelvic LND or chemoradiation
IIA2	Chemoradiation, pelvic field
IIB	Chemoradiation, pelvic field
IIIA	Chemoradiation, pelvic field
IIIB	Chemoradiation, pelvic + extended field
IIIC1	Chemoradiation, pelvic + extended field
IIIC2	Chemoradiation, pelvic + extended field + systemic chemotherapy
IVA	Chemoradiation, pelvic + extended field or pelvic exenteration
IVB	Systemic chemotherapy + radiation, pelvic or modified field

with pelvic lymphadenectomy and para-aortic lymph nodes evaluation. Following adjuvant therapy is recommended if:

- *Intermediate risk factors:* Radiation therapy
- *High-risk factors:* Chemoradiation

Stage IB3, IIA2: Concurrent chemoradiation is strongly recommended as primary treatment option for such lesions. Patients desiring surgical treatment can undergo radical hysterectomy with pelvic and para-aortic lymphadenectomy followed by chemoradiation depending on presence of intermediate or high-risk factors.

Stages IB3 to IIA2: Concurrent platinum-based chemoradiation (CCRT) is the preferred treatment option for these lesions. In areas where radiotherapy facilities are scarce, neoadjuvant chemotherapy (NACT) has been used followed by radical hysterectomy with pelvic lymphadenectomy.

FIGO Stages IIB–IVA

Concurrent platinum-based chemoradiation is considered the standard treatment for patients with locally advanced cervical cancer. A once-weekly infusion of cisplatin (40 mg/m^2 weekly with appropriate hydration) for 5–6 cycles during external beam therapy is a commonly used concurrent chemotherapy regimen. The primary goal of EBRT is to sterilize local disease and to shrink the tumor to facilitate subsequent intracavitary radiotherapy (ICRT). The combination of EBRT and ICRT maximizes the likelihood of locoregional control while minimizing the risk of treatment complications. Standard EBRT should deliver a dose of 45–50 Gy to the whole pelvis by the 2 or 4 field box technique encompassing the uterus, cervix, adnexal structures, parametria, and pelvic lymph nodes. The total combined dose with EBRT and ICRT should be in the range of 80–90 Gy **(Table 4 and Fig. 1)**

Table 5 shows the field design for the pelvic radiotherapy (EBRT).

FOLLOW-UP AFTER TREATMENT

Routine follow-up visits are recommended every 3–4 months for the first 2–3 years, then every 6 months until 5 years, and then annually for life. At each visit, history-taking and clinical examination are carried out to detect treatment complications and psychosexual morbidity as well as assess for recurrent disease.

TABLE 4: Classical 2D brachytherapy dose.

Point A	2 cm superior to the external cervical os and 2 cm superior to the external canal	Point representing the crossover of the ureter and uterine artery	7,000–8,000 cGy
Point B	3 cm lateral to point A	Represents pelvic lymph nodes	6,000 cGy

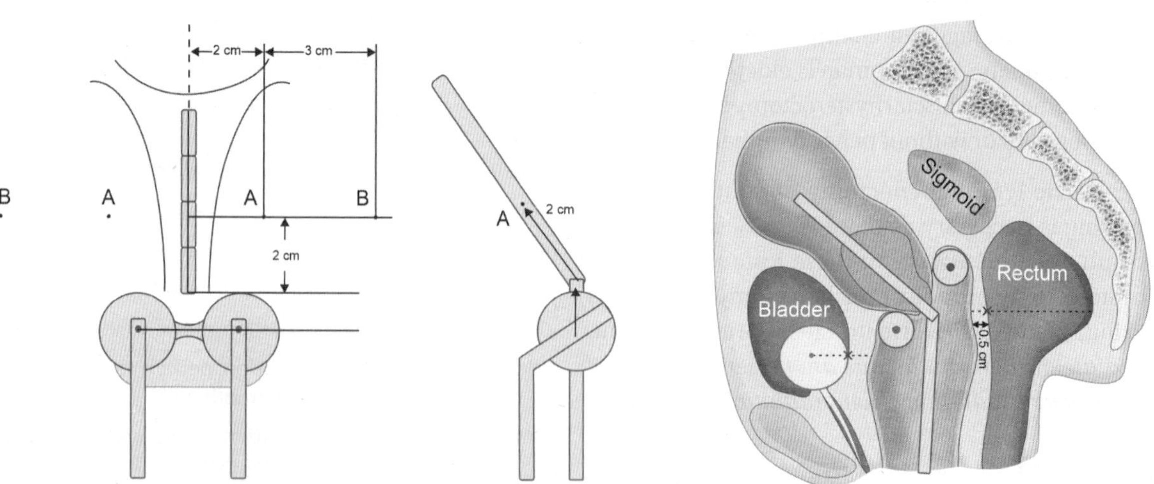

Fig. 1: Point A and point B of brachytherapy.[17]

TABLE 5: Field design for the pelvic radiotherapy [EBRT (external beam radiotherapy)].

Field	Border	Landmark
AP-PA fields	Superior	L4–5 vertebral interspace
	Inferior	2 cm below the obturator foramen or 3 cm inferior to distal disease, whichever is lower
	Lateral	1.5–2 cm lateral to the pelvic brim
Lateral fields	Superior inferior	Same as above
	Anterior	Anterior to the pubic symphysis
	Posterior	0.5 cm posterior to the anterior border of the S2/3 vertebral junction; may include the entire sacrum to cover the disease extent

CANCER CERVIX IN PREGNANCY

Cervical cancer is one of the most common malignancies in pregnancy, with an estimated incidence of 0.8–1.5 cases per 10,000 births. Most patients are diagnosed at an early stage of disease.[18-21] The management principle of cervical cancer in pregnancy is same as in nonpregnant women and broadly divided into following:

- *<20 weeks:* Treatment should be started without delay. The mode of therapy is decided according to the stage of disease. Spontaneous abortion occurs during radiation therapy.
- *>20 weeks:* Definitive treatment can be delayed in stages IA2 and IB1 and 1B2 if the woman wishes to continue the pregnancy.

Delivery: It should be conducted at a tertiary care center by classical cesarean followed by radical hysterectomy at or before 34 weeks of gestation.

WHO GLOBAL STRATEGY TO ELIMINATE CERVICAL CANCER

To eliminate cervical cancer, all countries must reach and maintain an incidence rate of below four per 100,000 women. Achieving that goal rests on three key pillars and their corresponding targets:

1. *Vaccination:* 90% of girls fully vaccinated with the HPV vaccine by the age of 15 years
2. *Screening:* 70% of women screened using a high-performance test by the age of 35 years, and again by the age of 45 years
3. *Treatment:* 90% of women with precancer treated and 90% of women with invasive cancer managed.

Each country should meet the 90-70-90 targets by 2030 to get on the path to eliminate cervical cancer within the next century.[22]

CONCLUSION

Carcinoma cervix ranks as the second most common cancer among females in India and the fourth most common globally. Several risk factors contribute to the development of cervical cancer, primarily linked to sexual activity and exposure to sexually transmitted infections. Persistent infection with high-risk oncogenic HPV types, notably HPV 16 and 18, is the primary etiological factor behind cervical cancer. Vaginal bleeding is the most common presenting symptom, along with other indicators such as weight loss, loss of appetite, and foul-smelling vaginal discharge. Clinical evaluation involves physical examination, including pelvic examination and biopsies for histopathological diagnosis, supplemented by staging using FIGO 2018 criteria. Treatment modalities include surgical interventions such as cone biopsy, hysterectomy, and lymphadenectomy, as well as radiation therapy and concurrent chemoradiation. Management strategies are tailored according to the stage of the disease, with options ranging from surgical interventions for early-stage disease to concurrent chemoradiation for locally advanced disease. Treatment decisions consider factors such as lymph node involvement, tumor size, and histopathological findings.

REFERENCES

1. Sung H, Ferlay J, Siegel RL, Laversanne M, Soerjomataram I, Jemal A, et al. Global cancer statistics 2020: GLOBOCAN estimates of incidence and mortality worldwide for 36 cancers in 185 countries. CA Cancer J Clin. 2021;71:209-49.
2. Pratte MA, Griffin A, Ogazi C, Yurasevecz S, Blanks CA, McCooey L, et al. Racial/Ethnic Disparities in Cervical Cancer Screening Services Among Contractors of the Connecticut Breast and Cervical Cancer Early Detection Program. Health Equity. 2018;2(1):30-6.
3. International Collaboration of Epidemiological Studies of Cervical Cancer. Comparison of risk factors for invasive squamous cell carcinoma and adenocarcinoma of the cervix: collaborative reanalysis of individual data on 8,097 women with squamous cell carcinoma and 1,374 women with adenocarcinoma from 12 epidemiological studies. Int J Cancer. 2007;120(4):885-91. Erratum in: Int J Cancer. 2007;120(11):2525. Berrington de González, Amy [removed]; Green, Jane [removed].

4. Ursin G, Peters RK, Henderson BE, d'Ablaing G 3rd, Monroe KR, Pike MC. Oral contraceptive use and adenocarcinoma of cervix. Lancet. 1994;344(8934):1390-4.
5. Abebe M, Eshetie S, Tessema B. Prevalence of sexually transmitted infections among cervical cancer suspected women at University of Gondar Comprehensive Specialized Hospital, North-west Ethiopia. BMC Infect Dis. 2021;21:378.
6. Madeleine MM, Anttila T, Schwartz SM, Saikku P, Leinonen M, Carter JJ, et al. Risk of cervical cancer associated with Chlamydia trachomatis antibodies by histology, HPV type and HPV cofactors. Int J Cancer. 2007;120(3):650-5.
7. Senapati R, Nayak B, Kar SK, Dwibedi B. HPV Genotypes distribution in Indian women with and without cervical carcinoma: Implication for HPV vaccination program in Odisha, Eastern India. BMC Infect Dis. 2017;17:30.
8. Bhatla N, Aoki D, Sharma DN, Sankaranarayanan R. Cancer of the cervix uteri: 2021 update. Int J Gynaecol Obstet. 2021;155 Suppl 1(Suppl 1):28-44.
9. International Agency for Research on Cancer (IARC). WHO Classification of Female Genital Tumours, 5th edition. Philippines: International Agency for Research on Cancer; 2020.
10. Ramirez PT, Adams S, Boggess JF, Burke WM, Frumovitz MM, Gardner GJ, et al. Robotic-assisted surgery in gynecologic oncology: a Society of Gynecologic Oncology consensus statement. Developed by the Society of Gynecologic Oncology's Clinical Practice Robotics Task Force. Gynecol Oncol. 2012;124:180-4.
11. Varghese A, Doglioli M, Fader AN. Updates and Controversies of Robotic-Assisted Surgery in Gynecologic Surgery. Clin Obstet Gynecol. 2019;62(4):733-48.
12. Fader AN. Surgery in cervical cancer. N Engl J Med. 2018;379:1955-7.
13. Sedlis A, Bundy BN, Rotman MZ, Lentz SS, Muderspach LI, Zaino RJ. A randomized trial of pelvic radiation therapy versus no further therapy in selected patients with stage IB carcinoma of the cervix after radical hysterectomy and pelvic lymphadenectomy: a Gynecologic Oncology Group Study. Gynecol Oncol. 1999;73:177-83.
14. Peters WA 3rd, Liu PY, Barrett RJ 2nd, Stock RJ, Monk BJ, Berek JS, et al. Concurrent chemotherapy and pelvic radiation therapy compared with pelvic radiation therapy alone as adjuvant therapy after radical surgery in high-risk early-stage cancer of the cervix. J Clin Oncol. 2000;18:1606-13.
15. Delgado G, Bundy B, Zaino R, Sevin BU, Creasman WT, Major F. Prospective surgical-pathological study of disease-free interval in patients with stage IB squamous cell carcinoma of the cervix: a Gynecologic Oncology Group study. Gynecol Oncol. 1990;38:352-7.
16. Pennington KP, Urban RR, Gray HJ. Revisiting minimally invasive surgery in the management of early-stage cervical cancer. J Natl Compr Cancer Netw. 2019;17:86-90.
17. Harkenrider MM, Alite F, Silva SR, Small W Jr. Image-Based Brachytherapy for the Treatment of Cervical Cancer. Int J Radiat Oncol Biol Phys. 2015;92(4):921-34.
18. Smith LH, Dalrymple JL, Leiserowitz GS, Danielsen B, Gilbert WM. Obstetrical deliveries associated with maternal malignancy in California, 1992 through 1997. Am J Obstet Gynecol 2001;184:1504.
19. Smith LH, Danielsen B, Allen ME, Cress R. Cancer associated with obstetric delivery: results of linkage with the California cancer registry. Am J Obstet Gynecol. 2003;189:1128-35.
20. Demeter A, Sziller I, Csapó Z, Szánthó A, Papp Z. Outcome of pregnancies after cold-knife conization of the uterine cervix during pregnancy. Eur J Gynaecol Oncol. 2002;23:207-10.
21. Duggan B, Muderspach LI, Roman LD, Curtin JP, d'Ablaing G 3rd, Morrow CP. Cervical cancer in pregnancy: reporting on planned delay in therapy. Obstet Gynecol. 1993;82 (4 Pt 1):598-602.
22. World Health Organization. (2024). Cervical Cancer Elimination Initiative. Available from: https://www.who.int/initiatives/cervical-cancer-elimination-initiative [Last accessed April, 2024].

Chapter 54

Ovarian Mass

Sharda Patra, Kanika Chopra

INTRODUCTION

Ovarian masses are commonly seen in premenopausal (8–35%) and postmenopausal women (3–17%).[1,2] These may present with symptoms rarely but in majority they present as an incidental finding during physical examination or imaging done for other reasons. Less commonly ovarian masses may present with acute pain in the abdomen requiring emergency management. The major concern is to differentiate between benign and malignant nature of the mass in order to plan further management accordingly. This differentiation is difficult in premenopausal women with no tests or algorithm available being clearly superior in terms of accuracy. Few exceptions are germ cell tumors with elevation of specific tumor markers such as alpha-fetoprotein and human chorionic gonadotropins.

CLINICAL PRESENTATION

Thorough history taking and examination is the key to arrive at the presumptive diagnosis. The various symptoms and signs are mentioned below.

Women with an ovarian mass may present with following symptoms:

- History of amenorrhea—possibility of ovarian ectopic to be kept in mind.
- Pelvic pain or pressure—pain can be unilateral or bilateral depending on the etiology. The pain can vary in its nature and severity and its association with menstrual cycles or act of coitus (present in cases of endometriomas). The pain can be more generalized in case of rupture or torsion of ovarian cyst.
- Abdominal fullness or pressure.
- Gastrointestinal discomfort (nausea, vomiting, diarrhea or constipation, decreased appetite).
- Urinary frequency or urgency.
- Weight loss.
- Abnormal uterine bleeding in case of hormone-producing tumor (sex chord stromal tumor)
- Fever, foul smelling vaginal discharge, vomiting and pain lower abdomen—possibility of tubo-ovarian abscess.

Signs

General physical and systematic examination should be done in detail with vitals.

- Presence of pallor, icterus, and generalized lymphadenopathy may suggest possibility of malignancy.
- Signs of virilization may also be there which may point toward hormone-producing tumor.
- Respiratory examination should be done in detail to rule out pleural effusion and consolidation.
- Breast and axillary examination to rule out malignant lump.

Per-abdominal and per-vaginal examination with per-rectal examination is done in cases of suspected ovarian mass to assess the size, consistency, mobility of the mass if palpable and nodularity in posterior cul-de-sac. Also, presence of free fluid, omental caking or organomegaly in abdominal examination may point toward malignant nature of the mass.

Clinical examination has poor sensitivity in detection of ovarian masses ranging from 15 to 51%.[3]

DIAGNOSTIC MODALITIES

Role of Ultrasonography

- Transvaginal scan is considered the first and the best diagnostic modality as it helps in characterization of the ovarian mass by subjective pattern recognition by an expert sonographer.
- The ultrasound examination should assess the size, composition of the mass (cystic, solid, or mixed);

laterality and the presence or absence of septation, mural nodules, papillary excrescences, or free fluid in the pelvis. Spectral color Doppler is useful to evaluate the vascularity of the pelvic lesions.

- In case of presence of pain, transvaginal scan can help in detecting probe tenderness as well and this may help in arriving at the diagnosis.
- Transabdominal scan is done in patients where masses extend beyond the pelvis and where TVS cannot be done as in prepubertal adolescents, virgin girls.
- Risk of malignancy can be characterized by the sonographic appearance of the ovarian mass and clinical characteristics of the patient.
- International ovarian tumor analysis (IOTA) is used in premenopausal women. It allows the sonographers with varying degree of expertise to use uniform terminology and classify ovarian masses as likely benign or malignant **(Tables 1 and 2)**. Using these rules the reported sensitivity of detection of ovarian malignancy is 95% and specificity 91%.[4]
- Newer classification system by American College of Radiology is developed to standardize the ultrasonography findings and stratify the risk of malignancy known as ORADS (ovarian and adnexal mass reporting and data system) **(Table 3)**.[5]

TABLE 1: Types of ovarian masses.

Benign	Benign	Malignant
Benign: • Functional cyst • Corpus luteal cyst • Lutein cyst (pregnancy related) • Hyper stimulated cyst	*Epithelial:* • Serous cystadenoma • Mucinous cystadenoma • Brenner tumor *Germ cell tumors* Mature cystic teratoma	*Epithelial ovarian cancer:* • Germ cell tumor – Dysgerminoma – Immature teratoma – Yolk sac tumor – Choriocarcinoma *Sex cord stromal tumor* Granulosa cell tumor *Secondary Metastasis*

TABLE 2: International Ovarian Tumor Analysis (IOTA) group ultrasound rules to classify masses as benign or malignant.

B-rules	*M-rules*
• Unilocular cysts • Presence of solid components (largest solid component <7 mm) • Presence of acoustic shadowing • Smooth multilocular tumor with a largest diameter <100 mm • No blood flow	• Irregular solid tumor • Ascites • At least four papillary structures • Irregular multilocular solid tumors with largest diameter ≥100 mm • Very strong blood flows

Role of Advanced Radiological Imaging

In case of suboptimal ultrasonography findings or very high suspicion of malignancy, magnetic resonance imaging can be planned. Computed tomography (CT) is not a primary modality for evaluation of adnexal masses but is useful in assessing the extent of metastasis when malignancy is suspected. CT scan can detect ascites, peritoneal implants, omental metastases, peritoneal implants, omental metastasis, pelvic or para-aortic lymph nodes, hepatic metastases and primary cancer site including pancreas or colon.

LABORATORY TESTS

- Serum beta-human chorionic gonadotropin (hCG) in case of suspected pregnancy.
- Complete blood count with total leukocyte count is case of suspected ovarian cyst rupture or tubo-ovarian abscess.
- Urine analysis
- Stool for occult blood

Role of Tumor Markers

- *CA-125:* It is a glycoprotein antigen and is most widely used tumor marker in assessment of ovarian masses but is highly nonspecific. It is known to be raised in many benign conditions such as endometriosis, fibroids, adenomyosis, and pelvic infections and is nonspecific in premenopausal women. In these women, single high value is of no significance and repeat testing after 4–6 weeks showing rising value is considered to be significant. In case of premenopausal women with CA-125 levels >200 U/L, referral to gynecologic oncologist is done.[6] CA-125 holds more significance in postmenopausal women as values >35 U/mL has sensitivity of 69–97% and specificity of 81–93% for diagnosing epithelial ovarian cancer.[7] Nearly 50% of early-stage ovarian cancers and 80% of advanced stage ovarian cancers show elevated CA-125 levels.[6] A postmenopausal woman with a pelvic mass and a raised CA-125 levels should be referred to a gynecologic oncologist.
- *HE-4:* It is human epididymis protein-4. It has sensitivity similar to CA-125 in detecting ovarian cancer. It is used in association with CA-125 in risk of

TABLE 3: Ovarian and adnexal mass reporting and data system (ORADS) risk stratification and management system.

ORADS score	*Description*	*Follow-up (premenopausal)*	*Follow-up (postmenopausal women)*
1.	• Normal ovary • Follicle simple cyst <3 cm • Corpus luteal cyst <3 cm	NA	
2. Almost certainly benign <1%	• Simple cyst <3 cm >3–5 cm >5 cm but <10 cm • Classic benign lesions • Non-simple unilocular cysts with smooth inner margins <3 cm, between 3–10 cm	NA In 8–12 weeks NA In 8–12 weeks	NA In 1 year In 8–12 weeks In 1 year In 8–12 weeks
3. Low-risk malignancy 1% to <10%	• Unilocular cyst ≥10 cm (simple or non-simple) • Typical dermoid cysts, endometriomas, hemorrhagic cysts ≥10 cm • Unilocular cyst, any size with irregular inner wall <3 mm height • Multilocular cyst <10 cm, smooth inner wall, CS (color score) = 1 to 3 • Solid smooth, any size, CS = 1	• Specialist ultrasound/MRI • Management by gynecologist	
4. Intermediate risk 10% to <50%	• Multilocular cyst, no solid component – ≥10 cm, smooth inner wall, CS = 1 to 3 – Any size, smooth inner wall, CS = 4 – Any size, irregular inner wall and/or irregular septation, any color score • Unilocular cyst with solid component • Any size, 0–3 papillary projections, CS = any • Multilocular cyst with solid component • Any size, CS = 1 to 2 • Solid Smooth, any size, CS = 2 to 3	• US specialist or MRI • Management by gynecologist with gynecologic oncologist consultation	
5. High risk >50%	• Unilocular cyst, any size, ≥4 papillary projections, CS = any • Multilocular cyst with solid component, any size, CS = 3 to 4 • Solid smooth, any size, CS = 4 • Solid irregular, any size, CS = any • Ascites and/or peritoneal nodules	Management by gynecologic oncologist	

malignancy algorithm. It is more useful in monitoring response to treatment or recurrence of ovarian cancer.

- Alfa-fetoprotein, hCG, and lactate dehydrogenase levels are recommended in women <40 years; in case of complex mass on ultrasound suggestive of germ cell tumors.
- *CEA* (Carcino embryonic antigen): These markers are useful in patients with bilateral ovarian masses and in those women who have gastrointestinal symptoms such as blood in stool or unexplained anemia. CA-125/CEA ratio of >25 suggests malignancy of ovary whereas ratio of <25 indicate extragenital malignancy with ovarian metastasis (Krukenberg's tumor).
- CA 19.9 is another nonspecific marker raised in diseases of pancreas, biliary tract, stomach, colorectum, and lung. It is also raised in mucinous cyst adenocarcinoma

MULTIMODAL TESTS

Risk of malignancy index I (RMI-I) combines three presurgical features which include:

1. Serum CA-125 (CA-125)
2. Menopausal status (M)
3. Ultrasound score (U)

Risk of malignancy index I (RMI) = U × M × CA-125

The ultrasound result is scored 1 point for each of the following characteristics—multilocular cysts, solid areas, metastases, ascites, and bilateral lesions. U = 0 (for an ultrasound score of 0), U = 1 (for an ultrasound score of 1), U = 3 (for an ultrasound score of 2–5). Menopausal status is 1 for premenopausal and 3 for postmenopausal women (achieved menopause or >50 years with history of hysterectomy). Serum CA-125 is measures in units per milliliter (U/mL).

The sensitivity of RMI-I score of >200 in detection of ovarian malignancies is 78% and specificity is 87%.[7]

Management

Surgery is performed in following cases:

- Patients with any menopausal status and ORADS-5 on imaging or any findings suggestive of metastatic disease (ascites, peritoneal masses).
- Postmenopausal patients with adnexal mass suspicious for malignancy and an elevated tumor marker (CA-125 >35 U/mL); a mass that is ≥10 cm in diameter or both an intermediate risk mass (ORADS 4) on imaging plus symptoms or risk factors for ovarian cancer or mass of 5–10 cm with symptoms suggestive of ovarian cancer.
- Premenopausal patients with an intermediate risk mass (ORADS 4) on imaging and a serum CA-125 level that is very elevated (>200 U/mL) or if a germ cell or sex cord-stromal tumor is suspected.

Ovarian cystectomy is performed in the case of premenopausal patient with preoperative and intraoperative low suspicion of malignancy. Minimally invasive route is preferred as compared to laparotomy.

Oophorectomy is preferred in postmenopausal patients and in one with high suspicion of malignancy. If frozen section of the mass suggests malignancy, then comprehensive surgical staging is done which includes following steps as given in **Box 1**.

Mean survival time for women with ovarian malignancy is significantly improved when managed by gynecologic oncologist. So, timely referral is the key.

BOX 1: Steps of staging laparotomy for ovarian malignancy.

- Vertical midline incision can be extended supra umbilically for large masses
- Peritoneal wash/collection of ascitic fluid for cytology 100 cc
- Assessment of whole of the abdomen in a clockwise manner starting from right iliac fossa, right paracolic gutter, right upper abdomen (liver, right hemidiaphragm), stomach, omentum, left upper abdomen (spleen, left hemi-diaphragm), left para-colic gutter, left iliac fossa, pelvis
- Random peritoneal biopsies are taken from bladder peritoneum, anteriorly, from the peritoneum of rectovaginal space, peritoneum from both iliac fossa from bilateral para-colic gutters, subdiaphragmatic peritoneum
- Total hysterectomy and bilateral salpingo-oophorectomy.
- Bilateral pelvic lymph node dissection atleast 6 from each side
- Para-aortic lymph node sampling till inferior mesenteric vein atleast 4 lymph nodes

SURVEILLANCE

Surveillance is performed for patients with low to intermediate risk masses on ultrasound and no indications for surgery. The risk of malignancy in these patients is considered to be low. Indications and frequency of follow-up are:

- *Patient of any menopausal status, ORADS 3 on imaging*: Repeat ultrasound in 3 and then 6 months.
- *Postmenopausal patients, ORADS 4 on imaging*: Repeat transvaginal ultrasound and CA 125 levels in 6 and 12 weeks, and then every 3–6 months for 1 year. A final ultrasound and CA 125 levels are done at 1 year.
- *Premenopausal women, ORADS 4 on imaging*: Repeat transvaginal ultrasound in 6 weeks, 3 months, and then 6 months. A final ultrasound is done at 1 year.

Surveillance is stopped in case the mass resolves or decreases in size.

Flowchart 1 depicts the algorithmic approach to an ovarian mass.[8,9]

CONCLUSION

Ovarian masses are commonly seen in premenopausal and postmenopausal women. Most important step in making a probable diagnosis of these masses is proper history taking and examination.

Transvaginal scan is considered the first and the best diagnostic modality as it helps in characterization of the ovarian mass by subjective pattern recognition by expert sonographers. International ovarian tumor analysis (IOTA) is used in premenopausal women. Another classification system used in ORADS (ovarian and adnexal mass reporting and data system) stratify the risk of malignancy by ultrasonography. Various markers like CA-125, HE-4, CEA,CA 19.9, LDH, Beta HcG, AFP can be done to know the type of ovarian malignancy. Management of the ovarian masses have to be individualized depending on the age of the patient, menopausal status, symptoms and ORADS scoring. Hence a careful and complete evaluation should be done for the type of ovarian mass and its management subsequently.

Flowchart 1: Algorithmic approach to an ovarian mass.

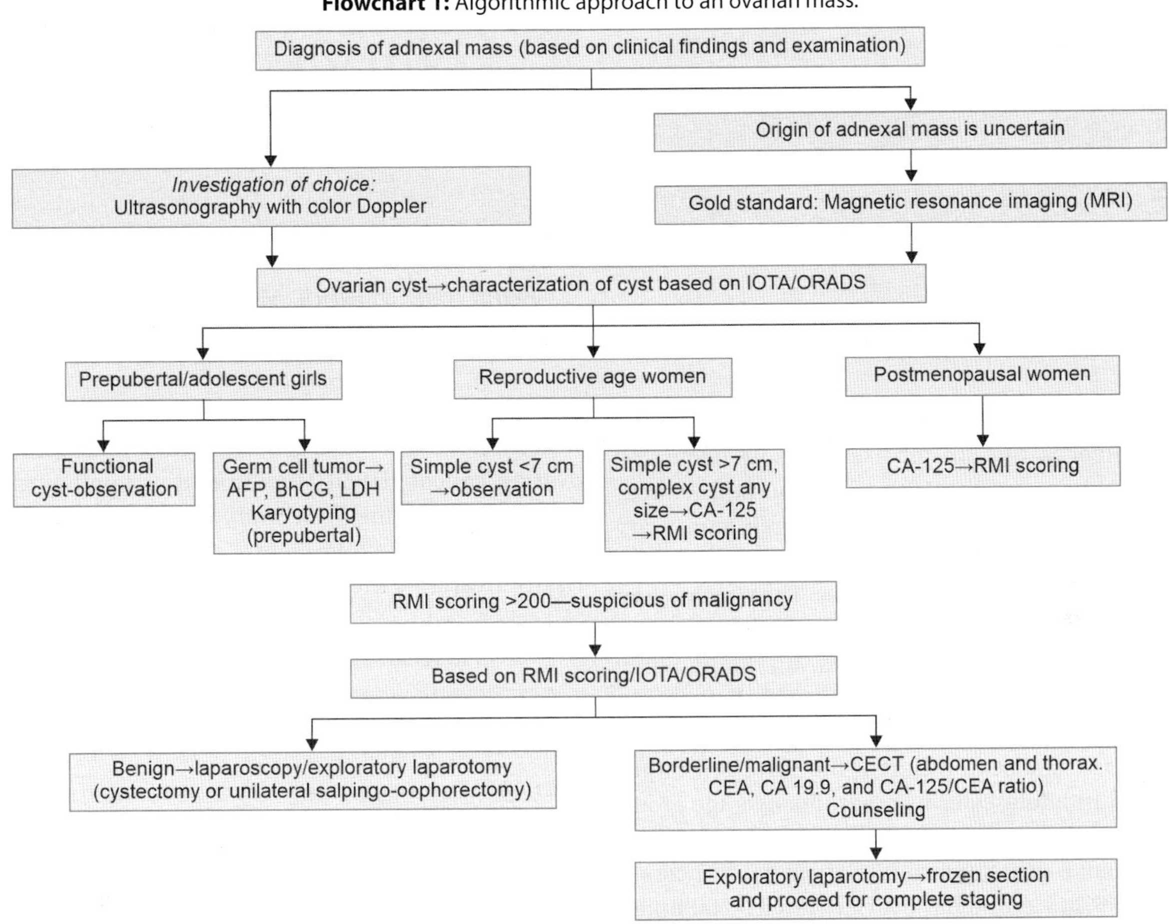

(AFP: alpha-fetoprotein; BhCG: beta-human chorionic gonadotropin; CEA: cancer embryonic antigen; CECT: contrast-enhanced computed tomography; IOTA: International ovarian tumor analysis; LDH: lactate dehydrogenase; ORADS: ovarian and adnexal mass reporting and data system; RMI: risk of malignancy index)

REFERENCES

1. Borgteldt C, Andolf E. Transvaginal sonographic ovarian findings in a random sample of women 24-40 years old. Ultrsound Obstet Gynaecol 1999;13:345.
2. Pavlik EJ, Ueland FR, Miller RW, Ubellacker JM, DeSimone CP, Elder J, et al. Frequency and disposition of ovarian abnormalities followed with serial transvaginal ultrasonography. Obstet Gynecol. 2013;22:210.
3. Ueland FR, Depriest PD, Desimone CP, Pavlik EJ, Lele SM, Kryscio RJ, et al. The accuracy of estimation under anaesthesia and transvaginal sonography in evaluating ovarian size. Gynecol Oncol. 2005;99:400-3.
4. Timmerman D, Ameye L, Fischerova D, Epstein E, Melis GB, Guerriero S, et al. Simple ultrasound rules to distinguish between benign and malignant adnexal masses before surgery: perspective validation by IOTA group. BMJ. 2010;341:C6839.
5. Andreotti RF, Timmerman D, Strachowski LM, Froyman W, Benacerraf BR, Bennett G, et al. O-RADS US risk stratification and management system: a consensus guidelines from the ACR-ovarian-Adnexal reporting and data trial. Radiology. 2020;294(1):168-85.
6. Myers ER, Bastian LA, Havrrilesky LJ, Kulasingam SL, Terplan MS, Cline KE, et al. Management of adnexal mass. Evidence Rep Technol Assess. 2006;130:1-145.
7. Anderson ES, Knudsen A, Rix P, Johansen B. Risk of malignancy index is the preoperative evaluation of patient with adnexal masses. Gynecol Oncol. 2003;90: 109-12.
8. RCOG. (2011). Management of Suspected Ovarian Masses in Premenopausal Women. Guideline no. 62. Available from: https://www.rcog.org.uk/media/yhujmdvr/gtg_62-1.pdf [Last accessed April, 2024].
9. RCOG. (2016). Ovarian cysts in postmenopausal women. Guideline no. 34. Available from: https://www.rcog.org.uk/guidance/browse-all-guidance/green-top-guidelines/ovarian-cysts-in-postmenopausal-women-green-top-guideline-no-34/ [Last accessed April, 2024].

Chapter 55

Ovarian Carcinoma

Kanica Batra Modi, Megha Mittal, Avneet Kaur

INTRODUCTION

Ovarian cancer is the 8th most commonly occurring cancer in women and the 18th most common cancer overall. According to Indian statistics, carcinoma ovary accounts for 45,701 new cases per year that accounts for 6.7% of all new malignancies in females. There were more than 313,000 new cases of ovarian cancer in 2020.

The age of presentation of ovarian malignancy depends on the type of histology. The epithelial ovarian cancer is an age-related disease and is considered mainly a postmenopausal disease whereas the germ cell tumor is diagnosed in adolescent and reproductive age group females.

The median 5-year survival rate of ovarian cancers is usually 49–56% in epithelial ovarian cancers/fallopian tube cancers, >90% in germ cell tumors of ovary.[1]

RISK FACTORS

Reproductive Risk Factors

Nulliparity, pelvic inflammatory disease, postmenopausal hormone replacement therapy (HRT), and ovarian stimulation/in vitro fertilization (IVF) are significantly associated with ovarian malignancy **(Flowchart 1)**.[2-5]

Obesity and Lifestyle

Obesity and smoking are associated with an increased risk of ovarian malignancy. Smoking especially is associated with mucinous carcinomas but a decreased risk for clear cell carcinomas.[6-11] Environmental factors have been investigated associated with the development of this neoplasm.

Hereditary Syndromes

About 23% of ovarian carcinomas have been related to hereditary conditions. Patients having two or more first-degree relatives with ovarian cancer including linkage with BRCA1 and BRCA2 genotypes [hereditary breast and ovarian cancer (HBOC) syndrome] or families affected by Lynch syndrome [hereditary nonpolyposis colorectal cancer (HNPCC) syndrome] are associated with increased risk of ovarian cancer, particularly early-onset disease.

Flowchart 1: Suspected ovarian malignancy.

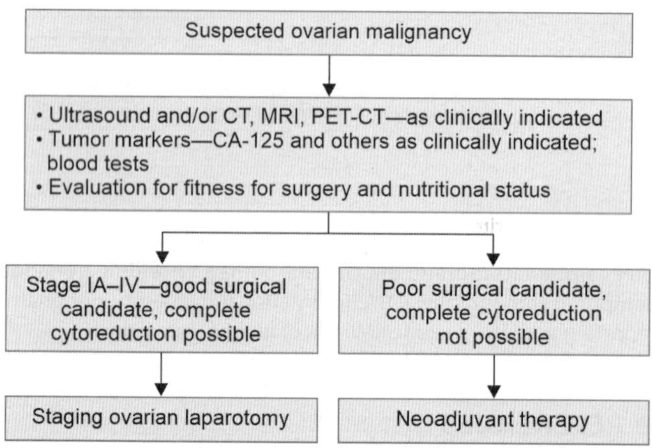

(CT: computed tomography; MRI: magnetic resonance imaging; PET: positron emission tomography)

TABLE 1: Classification of ovarian cancer.

Type 1	*Type 2*
Low-grade serous carcinomas, borderline serous tumors, low-grade endometrioid, mucinous and clear-cell carcinomas	High-grade serous carcinomas, carcinosarcomas, and undifferentiated cancers
Genetically stable	Aggressive, exhibit high genomic instability
KRAS, BRAF, ERBB2, PTEN, PIK3CA, B-Catenin gene (*CTNNB1*), *ARID1A*, and *PPP2R1A*	TP53, BRCA1/2, and mostly arise from STIC cells

(STIC: serous tubal intraepithelial carcinoma)

Other mutations associated are MLH1, MSH2, MSH6, and PMS2 germline mutations.[12-18]

According to clinical behavior and the molecular genetic abnormalities, ovarian cancer can be classified into two different types: type I and type II **(Table 1)**.

STAGING

FIGO Staging

Stage	Description
I	Tumor limited to ovaries (one or both) or fallopian tube(s)
IA	Tumor limited to one ovary (capsule intact) or fallopian tube, no tumor on ovarian or fallopian tube surface; no malignant cells in ascites or peritoneal washings
IB	Tumor limited to both ovaries; (capsules intact) or fallopian tubes; no tumor on ovarian or fallopian tube surface; no malignant cells in ascites or peritoneal washings
IC	Tumor limited to one or both ovaries or fallopian tubes, with any of the following:
IC1	Surgical spill
IC2	Capsule ruptured before surgery or tumor on ovarian or fallopian tube surface
IC3	Malignant cells in ascites or peritoneal washings
II	Tumor involves one or both ovaries or fallopian tubes with pelvic extension below pelvic brim or primary peritoneal cancer
IIA	Extension and/or implants on the uterus and/or fallopian tube(s) and/or ovaries
IIB	Extension to and/or implants on other pelvic tissues
III	Tumor involves one or both ovaries or fallopian tubes, or primary peritoneal cancer, with microscopically confirmed peritoneal metastasis outside the pelvis and/or metastasis to the retroperitoneal (pelvic and/or para-aortic) lymph nodes
IIIA1	Positive retroperitoneal lymph nodes only (histologically confirmed)
IIIAl(i)	Metastasis up to and including 10 mm in greatest dimension
IIIAl(ii)	Metastasis >10 mm in greatest dimension
IIIA2	Microscopic extrapelvic (above the pelvic brim) peritoneal involvement with or without positive retroperitoneal lymph nodes
IIIB	Macroscopic peritoneal metastasis beyond pelvis 2 cm or less in greatest dimension with or without metastasis to the retroperitoneal lymph nodes
IIIC	Macroscopic peritoneal metastasis beyond the pelvis >2 cm in greatest dimension with or without metastasis to the retroperitoneal lymph nodes (includes extension of tumor to capsule of liver and spleen without parenchymal involvement of either organ)
IV	Distant metastasis, including pleural effusion with positive cytology; liver or splenic parenchymal metastasis; metastasis to extra-abdominal organs (including inguinal lymph nodes and lymph nodes outside the abdominal cavity); and transmural involvement of intestine
IVA	Pleural effusion with positive cytology M1b
IVB	Liver or splenic parenchymal metastases; metastases to extra-abdominal organs (including inguinal lymph nodes and lymph nodes outside the abdominal cavity); transmural involvement of intestine

WHO HISTOLOGICAL CLASSIFICATION OF OVARIAN CANCER

The histological classification of ovarian tumors by the World Health Organization (WHO) is based on histogenetic principles, and this classification categorizes ovarian tumors with regard to their derivation from celomic surface epithelial cells, germ cells, and mesenchyme (the stroma and the sex cord).[19]

Epithelial ovarian cancers:
- *Serous (low/borderline/high grade)*: Predominantly found in stage 3 and 4
- Mucinous
- Endometrioid } → Associated with endometriosis
- Clear cell }
- Transitional cell tumors (Brenner tumors)
- Carcinosarcoma
- Mixed epithelial tumor
- Seromucinous
- Others

Germ cell tumors:
- Teratoma
- Dysgerminoma
- Yolk sac tumor
- Embryonal
- Choriocarcinoma

Sex cord stromal tumors:
- Fibroma
- Thecoma

- Pure sex cord stromal tumors
- Mixed sex cord stromal tumors

Mesenchymal tumors:

- Endometrial stromal sarcoma
- Leiomyoma
- Leiomyosarcoma

Miscellaneous tumors:

- Rete cystadenoma, adenoma and adenocarcinoma
- Wolffian tumor
- Solid pseudopapillary tumor
- Small cell carcinoma of the ovary, hypercalcemic type
- Wilms tumor

CLINICAL PRESENTATION

New onset abdominal/pelvic pain, distension, bloating, flatulence, and urinary symptoms such as urgency or frequency are the usual presenting complaints of patient with ovarian cancer. Abnormal uterine bleeding and irregular bowel habits are other unusual symptoms.

The location and the biology of ovarian tumors usually leads to delay in diagnosis. The current literature has enabled us to identify the at-risk population and detect the early-stage ovarian cancers. Because of the location of the ovaries and the biology of most epithelial cancers, it has been difficult to diagnose ovarian cancer at an earlier, more curable stage.[20-25]

SCREENING OF OVARIAN CANCER

Though literature does not recommend routine screening of general population. Large randomized studies and metanalysis have found that CA-125 and transvaginal ultrasound (US) have increased the likelihood of diagnosis at an early stage of ovarian cancer but have no effect on the overall ovarian cancer mortality rate.[26-28] Women with high-risk factors like BRCA mutations and family history of breast or ovarian cancer, risk-reducing salpingo-oophorectomy (RRSO) is generally preferred as it has known to prevent the primary peritoneal cancers and breast cancer.[29-33]

There are various prediction algorithms for assessing the likelihood of malignancy preoperatively. For example, the OVA1 test is a multivariate index assay (MIA) that uses five markers (including transthyretin, apolipoprotein A1, transferrin, beta-2 microglobulin, and CA-125).[34-38] However, the Society of Gynecologic Oncology (SGO) and the Food and Drug Administration (FDA) have stated that the OVA1 test should not be used as a screening tool to detect ovarian cancer in patients without any other signs of cancer, or as a stand-alone diagnostic tool.[39,40]

DIAGNOSIS AND WORKUP

Usually patients with ovarian cancer present to clinician with abdominal mass, ascites, and vague symptoms such as bloating and flatulence. The overall incidence of a symptomatic ovarian cyst in a premenopausal female being malignant is approximately 1:1000 increasing to 3:1000 till the age of 50 years.[41]

A careful physical examination along with TAS/TVS (transabdominal sonography/transvaginal sonography) and tumor markers [CA-125, alpha fetoprotein (AFP), lactate dehydrogenase (LDH), beta-human chorionic gonadotropin (beta-hCG)] are the first-line investigations in patients with ovarian masses **(Flowchart 2)**.[42,43] Further workup includes detailed family history, genetic testing, mammography, detailed gastrointestinal (GI) surveillance, and routine biochemical investigations (hemogram, liver, and kidney function test).

An initial evaluation with CA-125 and TVS pelvis can help calculate risk of malignancy index (RMI) I score. A recent systematic review36 showed the pooled sensitivities and specificities of an RMI I score of 200 in the detection of ovarian malignancies to be: RMI I sensitivity 78% [95% confidence interval (CI): 71–85%] and specificity 87% (95% CI: 83–91%).[42-50]

Calculation of the RMI I

The RMI I combines three presurgical features:

1. Serum CA-125 (CA-125)
2. Menopausal status (M)
3. Ultrasound score (U)

Flowchart 2: Initial evaluation of ovarian masses.

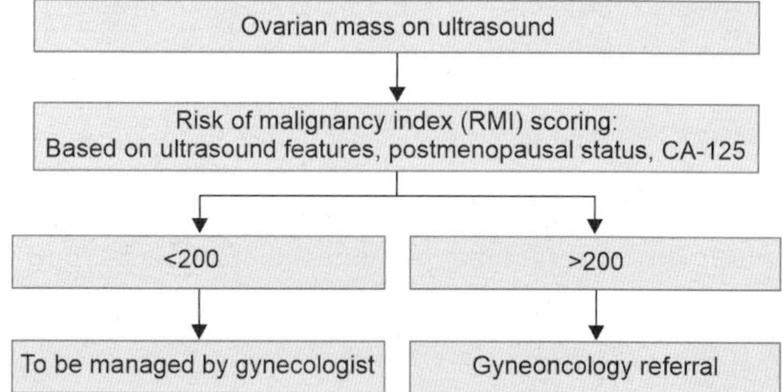

The RMI is a product of the ultrasound scan score, the menopausal status, and the serum CA-125 level (IU/mL) as follows: RMI = U × M × CA-125.

- The US result is scored 1 point for each of the following characteristics: multilocular cysts, solid areas, metastases, ascites, and bilateral lesions. U = 0 (for U of 0), U = 1 (for U of 1), and U = 3 (for U of 2–5).
- The menopausal status is scored as 1 = premenopausal and 3 = postmenopausal.
- Serum CA-125 is measured in IU/mL and can vary between zero to hundreds or even thousands of units.

Role of Imaging

Ultrasonography is typically used for initial evaluation and triaging into benign and malignant categories.[42-44] Abdominal and pelvic magnetic resonance imaging (MRI) or fludeoxyglucose-18–positron emission tomography (FDG-PET)/CT (computed tomography) scan can be used if results of USG are indeterminate. The NCCN Panel recommends PET/CT or MRI for indeterminate lesions if they will alter management.

For assessment of abdominopelvic metastasis for preoperative staging, estimation of resectability, and surgical planning, abdominal/pelvic CT or MRI are generally more useful than US.[51-53] Although CT is preferred by few oncologists, MRI has been shown to provide equivalent accuracy for staging and comparable accuracy for predicting peritoneal tumor volume, and can be useful if CT results are inconclusive. For assessing advanced disease, FDG-PET/CT may also be useful if CT results are indeterminate, and has been shown to have higher accuracy than CT for detection of metastases.[54-59]

MANAGEMENT

After the detailed workup and diagnosis, management comprises primarily surgery and chemotherapy **(Fig. 1)**. However, the tumor burden, operability, and desire of fertility guide the patient management.

Primary Therapy

- *Desirous of fertility:*
  - The prerequisites are:
    - *Stage 1A/1B:* Grade I/II ovarian tumors
    - XX karyotype
    - Compliance
    - Informed prognosis
  - Patients can either undergo unilateral or bilateral salpingo-oophorectomy with comprehensive surgical staging.
- Not desirous of fertility with good performance status and resectable tumor mass
  - Comprehensive surgical staging with total abdominal hysterectomy with bilateral salpingo-oophorectomy with debulking if needed.

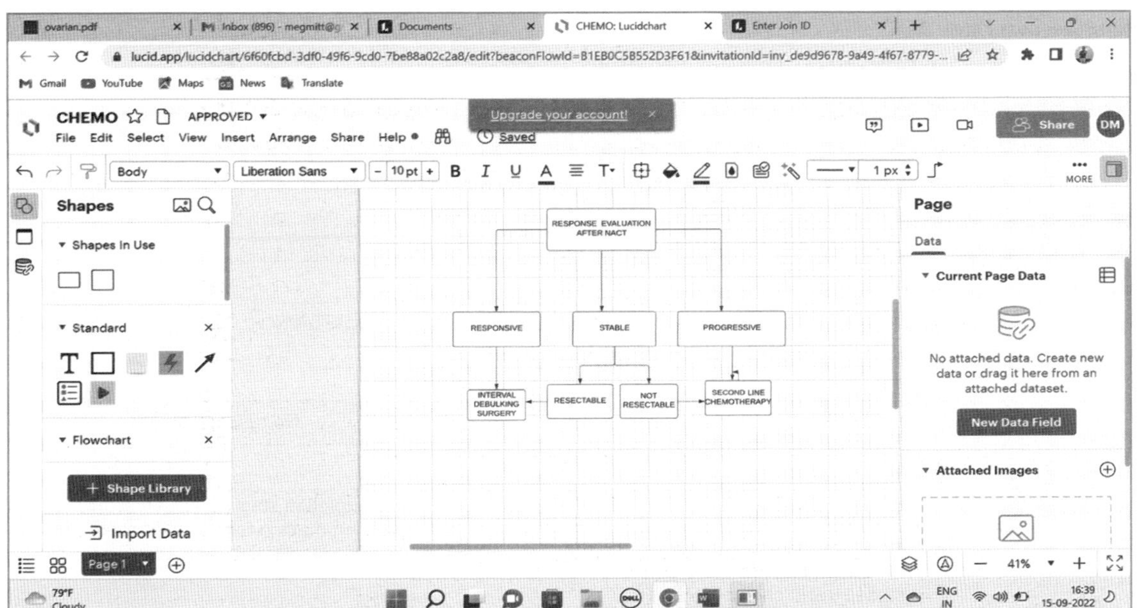

Fig. 1: Response evaluation after chemotherapy.

- Comprehensive surgical staging allows to rule out occult metastasis. It has been seen 30% of patients were upstaged after the primary surgery.

Box 1 shows the surgical principles for staging in ovarian cancer.

Box 2 shows the Surveillance of ovarian cancer patients.

- *Bulky tumor stage II–1V or poor surgical candidate:*
  - Patient undergoes neoadjuvant chemotherapy at least four cycles followed by interval debulking surgery depending on the tumor response.

BOX 1: Surgical principles for staging in ovarian cancer.

- Midline vertical incision
- Multiple cytological washings
- Intact tumor removal
- Complete abdominal exploration
- Total abdominal hysterectomy, with remaining tubes and ovaries
- Omentectomy
- Lymphadectomy/sampling
- Random peritoneal biopsies

BOX 2: Surveillance of ovarian cancer patients.

- Visit 2–4 monthly till 2 years and then 3–6 monthly till 3 years followed by annual checkups for pelvic examinations
- If CA-125 was elevated earlier it can be used in follow-up to detect recurrence
- Genetic surveillance if not done earlier
- CBC and chemistry panel
- Chest/abdominal/pelvic CT/MRI or PET as clinically indicated
- Long-term wellness care

(CBC: complete blood count; CT: computed tomography; MRI: magnetic resonance imaging; PET: positron emission tomography)

- *Diagnosed case of ovarian cancer by previous recent surgery **(Flowchart 3 and 4)**:*
  - Detailed history review, institutional pathologic review, metastatic workup such as chest X-ray, CT, or MRI, tumor markers, and genetic review.

The interval debulking procedure is similar to primary debulking procedure with aim of optimal cytoreduction.

Primary Adjuvant Therapy

Primary adjuvant therapy is guided by pathologic grading, staging, and histological type of the tumors.

- Less common ovarian histology such as carcinosarcoma, germ cell tumors, or borderline tumors.
- *Epithelial invasive tumors:*
  - Intravenous (IV) therapy—cisplatin/paclitaxel three to six cycles
  - Intraperitoneal (IP) therapy—in optimally debulked disease <1 cm in stage II/III patients.

Flowchart 3: Ovarian cancer by previous surgery.

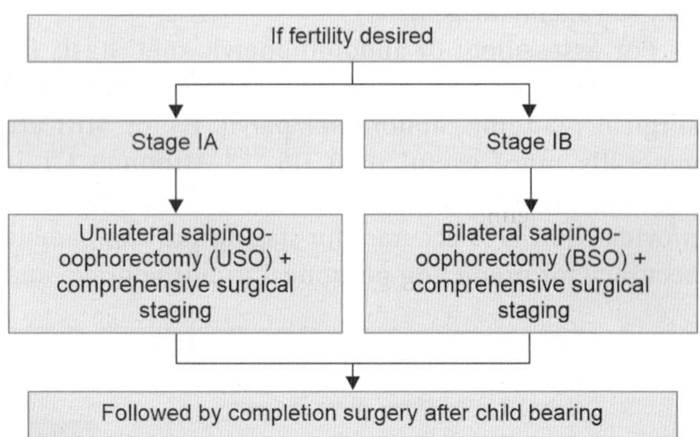

Flowchart 4: Ovarian cancer by previous surgery.

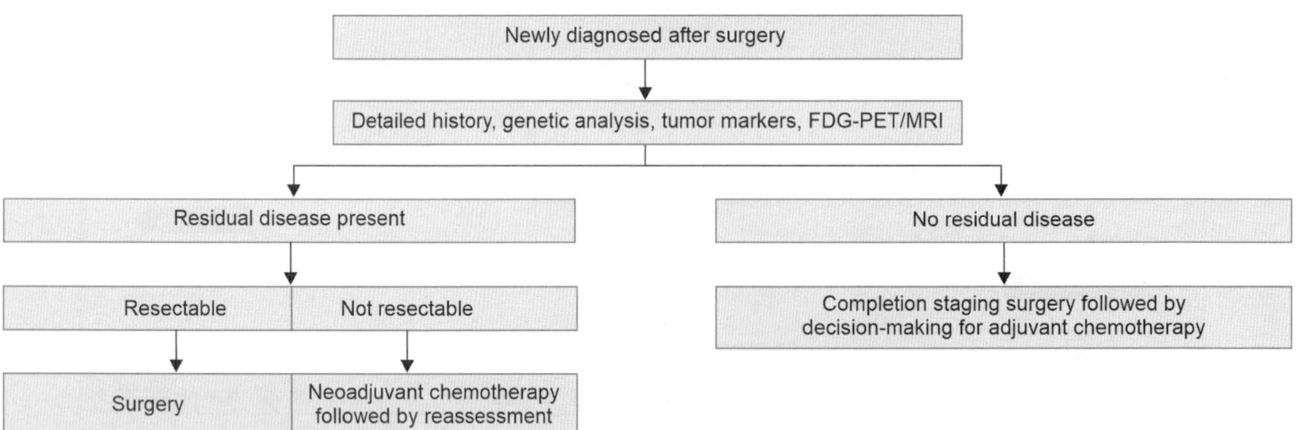

(FDG-PET: fludeoxyglucose-18–positron emission tomography; MRI: magnetic resonance imaging)

MONITORING

After the primary therapy, all the patients undergo periodic checkups to detect recurrences early.

Though few clinicians still debate early management of these recurrences as it has no effect on overall survival.[60]

RECURRENCE/PERSISTENT DISEASE

- The recurrent disease can be either clinical relapse or rising CA-125 levels, but management depends on whether adjuvant therapy was given or not.
- The progression of disease is measured by Response Evaluation Criteria in Solid Tumors (RECIST) criteria. The prognosis is poor if the disease progressed after two cycles of chemotherapy (platinum refractory) and for those who present with relapse within 6 months of primary adjuvant therapy (platinum-resistant). Usually such diseases are managed with second-line chemotherapy agents (such as doxorubicin, gemcitabine, and docetaxel).

REFERENCES

1. World Health Organization. Global Health Estimates 2020: Deaths by Cause, Age, Sex, by Country and by Region, 2000–2019. Geneva: World Health Organization; 2020.
2. Shen CC, Hu LY, Yang AC, Chiang YY, Hung JH, Tsai SJ. Risk of uterine, ovarian and breast cancer following pelvic inflammatory disease: a nationwide population-based retrospective cohort study. BMC Cancer. 2016;16:839.
3. Olsen CM, Nagle CM, Whiteman DC, Ness R, Pearce CL, Pike MC, et al; Ovarian Cancer Association Consortium. Obesity and risk of ovarian cancer subtypes: evidence from the Ovarian Cancer Association Consortium. Endocr Relat Cancer. 2013;20:251-62.
4. Bethea TN, Palmer JR, Adams-Campbell LL, Rosenberg L. A prospective study of reproductive factors and exogenous hormone use in relation to ovarian cancer risk among Black women. Cancer Causes Control. 2017;28:385-91.
5. Parazzini F, La Vecchia C, Negri E, Moroni S, dal Pino D, Fedele L. Pelvic inflammatory disease and risk of ovarian cancer. Cancer Epidemiol Biomarkers Prev. 1996;5:667-9.
6. Wentzensen N, Poole EM, Trabert B, White E, Arslan AA, Patel AV, et al. Ovarian cancer risk factors by histologic subtype: an analysis from the Ovarian Cancer Cohort Consortium. J Clin Oncol. 2016;34:2888-98.
7. Faber MT, Kjær SK, Dehlendorff C, Chang-Claude J, Andersen KK, Høgdall E, et al; Ovarian Cancer Association Consortium. Cigarette smoking and risk of ovarian cancer: a pooled analysis of 21 case-control studies. Cancer Causes Control. 2013;24:989-1004.
8. Collaborative Group on Epidemiological Studies of Ovarian Cancer; Beral V, Gaitskell K, Hermon C, Moser K, Reeves G, Peto R. Ovarian cancer and smoking: individual participant meta-analysis including 28,114 women with ovarian cancer from 51 epidemiological studies. Lancet Oncol. 2012;13:946-56.
9. Gram IT, Lukanova A, Brill I, Braaten T, Lund E, Lundin E, et al. Cigarette smoking and risk of histological subtypes of epithelial ovarian cancer in the EPIC cohort study. Int J Cancer. 2012;130:2204-10.
10. Tworoger SS, Gertig DM, Gates MA, Hecht JL, Hankinson SE. Caffeine, alcohol, smoking, and the risk of incident epithelial ovarian cancer. Cancer. 2008;112:1169-77.
11. Jordan SJ, Whiteman DC, Purdie DM, Green AC, Webb PM. Does smoking increase risk of ovarian cancer? A systematic review. Gynecol Oncol. 2006;103:1122-9.
12. Bonadona V, Bonaïti B, Olschwang S, Grandjouan S, Huiart L, Longy M, et al; French Cancer Genetics Network. Cancer risks associated with germline mutations in MLH1, MSH2, and MSH6 genes in Lynch syndrome. JAMA. 2011;305:2304-10.
13. Watson P, Vasen HFA, Mecklin JP, Bernstein I, Aarnio M, Järvinen HJ, et al. The risk of extra-colonic, extra-endometrial cancer in the Lynch syndrome. Int J Cancer. 2008;123:444-9.
14. Aarnio M, Sankila R, Pukkala E, Salovaara R, Aaltonen LA, de la Chapelle A, et al. Cancer risk in mutation carriers of DNA-mismatch-repair genes. Int J Cancer. 1999;81:214-8.
15. Suszynska M, Klonowska K, Jasinska AJ, Kozlowski P. Large-scale meta-analysis of mutations identified in panels of breast/ovarian cancer-related genes - Providing evidence of cancer predisposition genes. Gynecol Oncol. 2019;153:452-62.
16. Dominguez-Valentin M, Sampson JR, Seppälä TT, Ten Broeke SW, Plazzer JP, Nakken S, et al. Cancer risks by gene, age, and gender in 6350 carriers of pathogenic mismatch repair variants: findings from the Prospective Lynch Syndrome Database. Genet Med. 2020;22:15-25.
17. Karimi M, von Salomé J, Aravidis C, Silander G, Askmalm MS, Henriksson I, et al. A retrospective study of extracolonic, non-endometrial cancer in Swedish Lynch syndrome families. Hered Cancer Clin Pract. 2018;16:16.
18. Lu HM, Li S, Black MH, Lee S, Hoiness R, Wu S, et al. Association of Breast and Ovarian Cancers With Predisposition Genes Identified by Large-Scale Sequencing. JAMA Oncol. 2019;5:51-7.
19. Adhikari L, Hassell LA. WHO classification. [online] Available from https://www.pathologyoutlines.com/topic/ovarytumorwhoclassif.html [Last accessed April, 2024].
20. Committee on the State of the Science in Ovarian Cancer Research; Board on Health Care Services; Institute of Medicine; National Academies of Sciences, Engineering, and Medicine. Ovarian Cancers: Evolving Paradigms in Research and Care. Washington (DC): National Academies Press (US); 2016.
21. National Institutes of Health. (2024). A Study to Compare Two Surgical Procedures in Individuals With BRCA1

Mutations to Assess Reduced Risk of Ovarian Cancer. [online] Available from https://clinicaltrials.gov/ct2/show/NCT04251052 [Last accessed April, 2024].
22. Fotopoulou C, Hall M, Cruickshank D, Gabra H, Ganesan R, Hughes C, et al. British Gynaecological Cancer Society (BGCS) epithelial ovarian/fallopian tube/primary peritoneal cancer guidelines: recommendations for practice. Eur J Obstet Gynecol Reprod Biol. 2017;213:123-39.
23. Ledermann JA, Raja FA, Fotopoulou C, Gonzalez-Martin A, Colombo N, Sessa C; ESMO Guidelines Working Group. Newly diagnosed and relapsed epithelial ovarian carcinoma: ESMO Clinical Practice Guidelines for diagnosis, treatment and follow-up. Ann Oncol. 2013;24 Suppl 6:vi24-32.
24. National Institute for Health and Care Excellence (NICE). Ovarian Cancer: the recognition and initial management of ovarian cancer. Vol. Clinical guideline [CG122]. Cardiff, Wales: National Collaborating Centre for Cancer; 2011.
25. Goff BA, Mandel LS, Drescher CW, Urban N, Gough S, Schurman KM, et al. Development of an ovarian cancer symptom index: possibilities for earlier detection. Cancer. 2007;109:221-7.
26. Pinsky PF, Zhu C, Skates SJ, Black A, Partridge E, Buys SS, et al. Potential effect of the risk of ovarian cancer algorithm (ROCA) on the mortality outcome of the Prostate, Lung, Colorectal and Ovarian (PLCO) trial. Int J Cancer. 2013;132:2127-33.
27. Marchetti C, De Felice F, Perniola G, Lecce F, Vertechy L, Monti M, et al. Screening program in ovarian cancer: A logical step in clinical management? A meta-analysis. Curr Probl Cancer. 2018;42:235-40.
28. Jacobs IJ, Skates SJ, MacDonald N, Menon U, Rosenthal AN, Davies AP, et al. Screening for ovarian cancer: a pilot randomised controlled trial. Lancet. 1999;353:1207-10.
29. Eleje GU, Eke AC, Ezebialu IU, Ikechebelu JI, Ugwu EO, Okonkwo OO. Risk-reducing bilateral salpingo-oophorectomy in women with BRCA1 or BRCA2 mutations. Cochrane Database Syst Rev. 2018;8:CD012464.
30. Xiao YL, Wang K, Liu Q, Li J, Zhang X, Li HY. Risk Reduction and Survival Benefit of Risk-Reducing Salpingo-oophorectomy in Hereditary Breast Cancer: Meta-analysis and Systematic Review. Clin Breast Cancer. 2019;19:e48-e65.
31. Marchetti C, De Felice F, Palaia I, Perniola G, Musella A, Musio D, et al. Risk-reducing salpingooophorectomy: a meta-analysis on impact on ovarian cancer risk and all cause mortality in BRCA 1 and BRCA 2 mutation carriers. BMC Womens Health. 2014;14:150.
32. Reitsma W, de Bock GH, Oosterwijk JC, Bart J, Hollema H, Mourits MJ. Support of the 'fallopian tube hypothesis' in a prospective series of risk-reducing salpingo-oophorectomy specimens. Eur J Cancer. 2013;49:132-41.
33. Rebbeck TR, Kauff ND, Domchek SM. Meta-analysis of risk reduction estimates associated with risk-reducing salpingooophorectomy in BRCA1 or BRCA2 mutation carriers. J Natl Cancer Inst. 2009;101:80-7.
34. Ueland FR, Desimone CP, Seamon LG, Miller RA, Goodrich S, Podzielinski I, et al. Effectiveness of a multivariate index assay in the preoperative assessment of ovarian tumors. Obstet Gynecol. 2011;117:1289-97.
35. Bristow RE, Smith A, Zhang Z, Chan DW, Crutcher G, Fung ET, et al. Ovarian malignancy risk stratification of the adnexal mass using a multivariate index assay. Gynecol Oncol. 2013;128:252-9.
36. Longoria TC, Ueland FR, Zhang Z, Chan DW, Smith A, Fung ET, et al. Clinical performance of a multivariate index assay for detecting early-stage ovarian cancer. Am J Obstet Gynecol. 2014;210:78 e1-9.
37. Bristow RE, Hodeib M, Smith A, Chan DW, Zhang Z, Fung ET, et al. Impact of a multivariate index assay on referral patterns for surgical management of an adnexal mass. Am J Obstet Gynecol. 2013;209:581.e1-8.
38. Goodrich ST, Bristow RE, Santoso JT, Miller RW, Smith A, Zhang Z, et al. The effect of ovarian imaging on the clinical interpretation of a multivariate index assay. Am J Obstet Gynecol. 2014;211:65.e61-65.e11.
39. Fotopoulou C, Hall M, Cruickshank D, Gabra H, Ganesan R, Hughes C, et al. British Gynaecological Cancer Society (BGCS) epithelial ovarian/fallopian tube/primary peritoneal cancer guidelines: recommendations for practice. Eur J Obstet Gynecol Reprod Biol. 2017;213:123-39.
40. Asif N, Sattar A, Dawood MM, Rafi T, Aamir M, Anwar M. Pre-operative evaluation of ovarian mass: risk of malignancy index. JCPSP. 2004;14(3):128-31.
41. Oliver A, Overton C. Detecting ovarian disorders in primary care. Practitioner. 2014;258(1769):15-9, 2.
42. Expert Panel on Women's Imaging, Atri M, Alabousi A, et al. ACR Appropriateness Criteria((R)) Clinically Suspected Adnexal Mass, No Acute Symptoms. J Am Coll Radiol. 2019;16:S77-S93.
43. Glanc P, Benacerraf B, Bourne T, Brown D, Coleman BG, Crum C, et al. First International Consensus Report on Adnexal Masses: Management Recommendations. J Ultrasound Med. 2017;36:849-63.
44. Sadowski EA, Rockall AG, Maturen KE, Robbins JB, Thomassin-Naggara I. Adnexal lesions: Imaging strategies for ultrasound and MR imaging. Diagn Interv Imaging. 2019;100(10):635-46.
45. Morgante G, la Marca A, Ditto A, De Leo V. Comparison of two malignancy risk indices based on serum CA125, ultrasound score and menopausal status in the diagnosis of ovarian masses. Br J Obstet Gynaecol. 1999;106:524-7.
46. Aslam N, Tailor A, Lawton F, Carr J, Savvas M, Jurkovic D. Prospective evaluation of three different models for the preoperative diagnosis of ovarian cancer. BJOG. 2000;107:1347-53.
47. Mol BW, Boll D, De Kanter M, Heintz AP, Sijmons EA, Oei SG, et al. Distinguishing the benign and malignant adnexal mass: an external validation of prognostic models. Gynecol Oncol. 2001;80:162-7.

48. Manjunath AP, Pratapkumar, Sujatha K, Vani R. Comparison of three risk of malignancy indices in evaluation of pelvic masses. Gynecol Oncol. 2001;81:225-9.
49. Ma S, Shen K, Lang J. A risk of malignancy index in preoperative diagnosis of ovarian cancer. Chin Med J (Engl). 2003;116:396-9.
50. Andersen ES, Knudsen A, Rix P, Johansen B. Risk of malignancy index in the preoperative evaluation of patients with adnexal masses. Gynecol Oncol. 2003;90:109-12.
51. Van Holsbeke C, Van Calster B, Valentin L, Testa AC, Ferrazzi E, Dimou I, et al. External validation of mathematical models to distinguish between benign and malignant adnexal tumors: a multicenter study by the International Ovarian Tumor Analysis Group. Clin Cancer Res. 2007;13:4440-7.
52. Expert Panel on Women's Imaging; Kang SK, Reinhold C, Atri M, Benson CB, Bhosale PR, Jhingran A, et al. Appropriateness Criteria® Staging and Follow-Up of Ovarian Cancer. J Am Coll Radiol. 2018;15:S198-S207.
53. Dodge JE, Covens AL, Lacchetti C, Elit LM, Le T, Devries-Aboud M, et al; Gynecology Cancer Disease Site Group. Management of a suspicious adnexal mass: a clinical practice guideline. Curr Oncol. 2012;19:e244-57.
54. Foti PV, Attinà G, Spadola S, Caltabiano R, Farina R, Palmucci S, et al. MR imaging of ovarian masses: classification and differential diagnosis. Insights Imaging. 2016;7:21-41.
55. Castellucci P, Perrone AM, Picchio M, Ghi T, Farsad M, Nanni C, et al. Diagnostic accuracy of 18F-FDG PET/CT in characterizing ovarian lesions and staging ovarian cancer: correlation with transvaginal ultrasonography, computed tomography, and histology. Nucl Med Commun. 2007;28:589-95.
56. Nam EJ, Yun MJ, Oh YT, Kim JW, Kim JH, Kim S, et al. Diagnosis and staging of primary ovarian cancer: correlation between PET/CT, Doppler US, and CT or MRI. Gynecol Oncol. 2010;116:389-94.
57. Kitajima K, Murakami K, Yamasaki E, Kaji Y, Fukasawa I, Inaba N, et al. Diagnostic accuracy of integrated FDG-PET/contrast-enhanced CT in staging ovarian cancer: comparison with enhanced CT. Eur J Nucl Med Mol Imaging. 2008;35:1912-20.
58. Pfannenberg C, Königsrainer I, Aschoff P, Oksüz MO, Zieker D, Beckert S, et al. (18)F-FDG-PET/CT to select patients with peritoneal carcinomatosis for cytoreductive surgery and hyperthermic intraperitoneal chemotherapy. Ann Surg Oncol 2009;16:1295-303.
59. Yoshida Y, Kurokawa T, Kawahara K, Tsuchida T, Okazawa H, Fujibayashi Y, et al. Incremental benefits of FDG positron emission tomography over CT alone for the preoperative staging of ovarian cancer. AJR Am J Roentgenol. 2004;182:227-33.
60. Armstrong DK, Alvarez RD, Bakkum-Gamez JN, Barroilhet L, Behbakht K, et al. Ovarian Cancer, Version 2.2020, NCCN Clinical Practice Guidelines in Oncology. J Natl Compr Canc Netw. 2021;19(2):191-226.

Chapter 56

Radiation Therapy in Gynecological Cancers

Swarupa Mitra, David Simson, Soumitra Barik

INTRODUCTION

Gynecological malignancies are one of the commonest cancers seen in women, especially in developing countries. They result in significant morbidity and mortality. Radiation therapy constitutes an essential treatment modality in the management of most gynecological cancers. It offers radical treatment in certain malignancies arising in the female reproductive tract depending on their site and stage. While in others, radiation forms an important component of the adjuvant post operative treatment often along with concomitant chemotherapy. Going back to history, more than a century ago.

BASIC UNDERSTANDING OF RADIATION THERAPY

Radiotherapy is used as a treatment modality in approximately half of all cancer patients. Radiotherapy can be used as a curative intent or as a means to palliate a patient's symptoms. Radiotherapy was used as early as 1930 for treating gynecological malignancies such as radium sources used in brachytherapy for treating carcinoma cervix. Brachytherapy remains an effective mode of treatment in carcinoma cervix.

Mode of Action[1]

The mode of action of ionizing radiation is described in **Table 1**.

Types of Radiotherapy

The types of radiotherapy are shown in **Table 2**.

ROLE OF RADIOTHERAPY IN THE TREATMENT OF GYNECOLOGICAL CANCERS

Radical Radiotherapy

Box 1 shows types of radical radiotherapy.

TABLE 1: Mode of action.

Ionizing radiation acts by causing DNA damage through the tissue it passes to produce a biological effect.	
Types of ionizing radiation:	
Directly ionizing	Charged particles such as electrons or protons produce direct ionization of atoms of substance or tissue through which they pass
Indirectly ionizing	X-ray or gamma rays and neutrons produce ions by ejecting outer orbit electrons of atoms which interact in substance or matter through which it passes
Mechanism of action of DNA damage:	
Direct action	Heavy charged particles such as proton or alpha rays produce direct breakage of DNA strands
Indirect action	X-ray or gamma rays produce free electrons in matter by ejecting outer orbit electron from an atom. This free electron interacts with water molecules to produce hydroxy-free radicals, free radicals, in turn, cause DNA damage (single-stranded or double-stranded breaks)
Oxygen fixation hypothesis	The cells can repair the DNA damages produced by free radicals, but in presence of oxygen, the DNA-free radical complex interacts with oxygen to produce inorganic peroxide which is irreversible. In other words, oxygen fixes the DNA damage caused by free radicals, and cells cannot repair these damages
Biological action	When a cell with damaged DNA tries to divide, it will not be able to replicate DNA and subsequently die (called mitotic death). Biological action can take days, months to years to manifest in different tissues

TABLE 2: Types of radiotherapy.

External beam radiotherapy (EBRT)	Radiotherapy is delivered with external beam sources such as a cobalt-60 machine or a linear accelerator
Types of EBRT sources:	
Cobalt-60 teletherapy	Uses cobalt-60 as radioactive sources to produce gamma rays
Linear accelerator	Uses high-speed electron to hit the target to produce high-energy X-rays; the high-energy electrons also can be used to treat superficial tumors in the body
Energy of X-rays or gamma rays used:	
Cobalt-60 teletherapy	4 MV gamma rays
Linear accelerator	4–10 MV X-rays, 6–18 MeV electrons for clinical use
Technique used:	
Conventional 2D technique	To treat with direct or parallel opposed fields using bony landmarks with help of orthogonal planning X-rays
3D conformal therapy	To treat conformally to target volume with help of 3D CT anatomy
IMRT (intensity-modulated radiotherapy)	To treat with multiple numbers of beams (usually 5–9) to conformally target a volume and simultaneously spare a nearby organ at risk
IGRT (image-guided radiotherapy)	Delivering EBRT with help of image guidance (X-ray radiograph, CT imaging, or MR imaging)
Brachytherapy	Radiotherapy is delivered by placing radioactive sources in or close to the tumor in the body
Types of brachytherapy:	
Intracavitary brachytherapy	Radioactive sources are placed in the body cavity like the uterine cavity through the cervical os as treated in carcinoma cervix
Interstitial brachytherapy	Placement of radioactive sources directly in tissue with help of catheters
Mold brachytherapy	Placement of brachytherapy sources on skin surfaces with help of custom-made molds
Dose rate:	
Low-dose-rate (LDR) brachytherapy	Delivers 0.4–2 Gy/h
Medium-dose-rate (MDR) brachytherapy	Delivers 2–12 Gy/h
High-dose-rate (HDR) brachytherapy	Delivers >12 Gy/h
Ultra-low-dose-rate (ULDR) brachytherapy	Delivers 0.01–0.3 Gy/h
Pulse-dose-rate (PDR) brachytherapy	Delivers up to 3 Gy/h, which is utilized (pulsed) every hour, 24 pulses per day; a typical pulse lasts 10 minutes per hour[2]
Loading of sources:	
Manual after loading	Radioactive sources manually loaded to applicator placed in the body
Remote after loading	Radioactive sources remotely loaded to applicator placed in the body with help of remote after loader machine

(CT: computed tomography; MR: magnetic resonance)

Palliative Radiotherapy

When radiotherapy is not given to cure the patient rather than to decrease symptoms such as pain and bleeding, e.g., hemostatic radiotherapy is given in locally advanced carcinoma cervix.

CLASSIFICATION OF RADIATION THERAPY

There are two main types of radiation therapies:

1. External beam radiotherapy (EBRT) (teletherapy)
2. Internal radiation therapy (brachytherapy).

BOX 1: Types of Radical radiotherapy.

• Definitive radiotherapy	• Radiotherapy is used as the sole treatment modality with the primary intent to cure the patient, e.g., brachytherapy used in early stage (FIGO IA) cervical cancer
• Adjuvant radiotherapy	• Radiotherapy is used after surgery to decrease the risk of recurrence of disease, e.g., adjuvant radiotherapy used in cervical cancer in case of a large tumor (>4 cm), deep one-third stromal invasion or lymphovascular space invasion (any two factors present)
• Neoadjuvant Radiotherapy	• Radiotherapy is given before surgery to decrease the size of the tumor or to produce a pathological response, e.g., neoadjuvant radiotherapy is given along with chemotherapy in carcinoma esophagus
• Concomitant	• When radiation is delivered along with small doses of chemotherapy

(FIGO: The International Federation of Gynecology and Obstetrics)

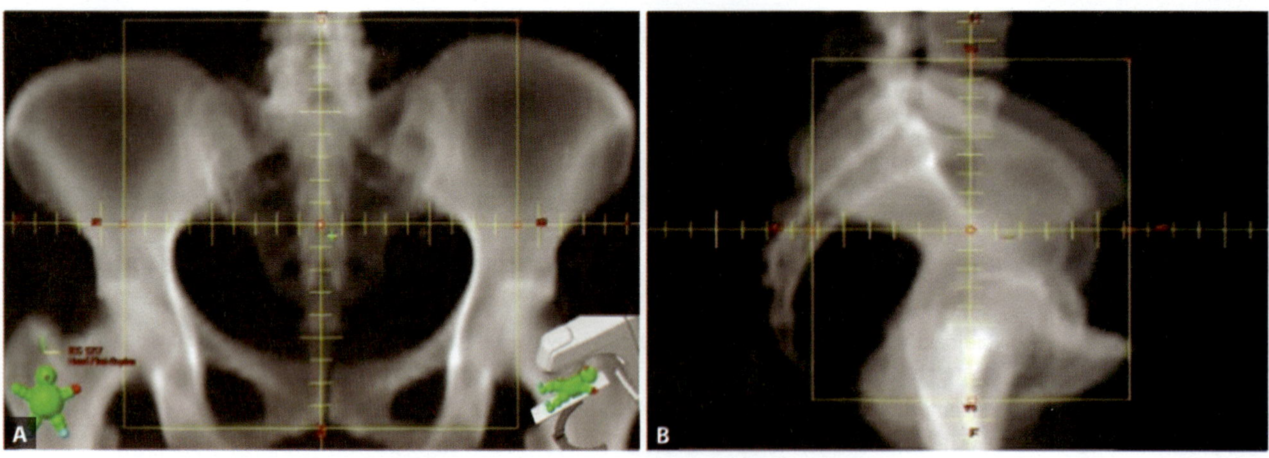

Figs. 1A and B: 2D conventional radiotherapy plan.

External Beam Radiotherapy/Teletherapy

Radiation is delivered to the patient from a source outside the patient. EBRT machines, *cobalt or linear accelerators*, produce ionizing radiation by decay of a radioactive nuclide such as cobalt-60 or produced electronically by the acceleration of electrons or other charged particles, like protons.

Linear accelerator: Electrons are accelerated to high energy and then exit the machine as an electron beam or they strike a target that produces X-rays (photons) These machines can thus generate either photon or electron beams of various energies, and their output is managed with sophisticated computer controls.

Today, linear accelerators are fast replacing the older cobalt-60 machines in most parts of the world.

Difference between photons and electrons: Photons are the more commonly used radiation mode since they can penetrate deep to reach internal organs.

Electrons are used for superficial targets such as the skin and breast, where the goal is to minimize radiation to deeper tissues and organs. Sometimes the two are combined to deliver the best possible tumor and normal tissue dose distribution.

Techniques of External Beam Radiotherapy

Two-dimensional radiation therapy: This is the older "square" style of radiotherapy planning; orthogonal radiographs were used to delineate the furthest extent of the tumor in relation to the bony anatomy of the patient. Generally, no attempt to shape the target could be possible and this led to irradiation of more normal tissue to the full target dose **(Figs. 1A and B)**.

During the 1990s, advances in imaging and in radiotherapy technology, especially computer technology, introduced the potential of 3D conformal radiation planning.

Three-dimensional conformal radiotherapy (3DCRT): The radiation beam can be conformed to the shape of the tumor by multileaf collimators **(Fig. 2)**. Collimators are a series of metal leaves in the path of the radiation beam which can shape the beam. One cone of beam deposits a large amount

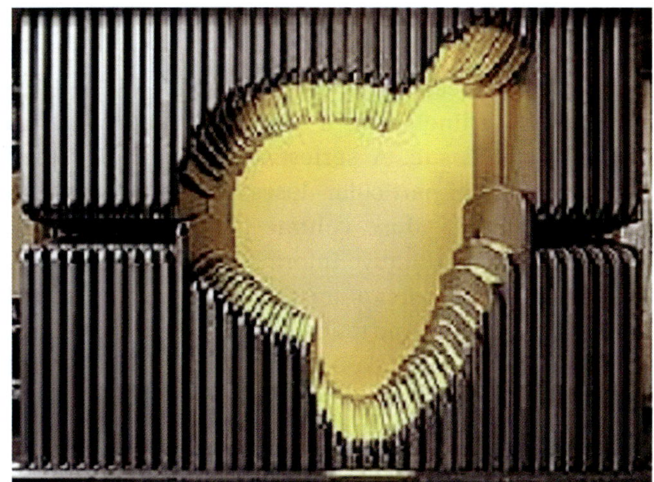

Fig. 2: Collimators to shape radiation beams.

of radiation to the healthy tissue in its path. But breaking it up into many lower energy beams to target the tumor from different angles helps better sparing of healthy tissues in the beam's path, resulting in lesser side effects.

Intensity-modulated radiotherapy (IMRT): One subset of 3DCRT splits each beam into many little beamlets of different intensities, customizing the dose and shape of the radiation to the tumor even more precisely.

Image-guided radiotherapy (IGRT): Imaging with an online scan is often used in conjunction with IMRT to ensure the radiation is delivered to the target area. IGRT allows for adjustments during treatment of tumors in organs that are moving, like lungs, and tumors located close to critical organs and tissues.

Stereotactic radiosurgery (SRS): This technique uses 3D imaging to determine the exact coordinates of the target and highly focused ablative doses of gamma rays or X-rays to converge on the tumor.

The Gamma Knife® is a treatment option that uses radioactive cobalt sources to focus multiple beams of radiation on a small area.

Linear accelerators can also be used to deliver stereotactic radiation therapy.

Stereotactic body radiation therapy (SBRT): When extracranial parts of the body are treated with SRS, it is called stereotactic body radiation therapy. Emerging areas for using SBRT include lung, liver, and bone.

Proton beam therapy: Protons are charged particles that deposit most of their energy at a given depth, minimizing risk to tissues beyond that point.

Proton beam therapy allows for highly specific targeting of tumors located near critical structures.

It is most commonly used the in treatment of pediatric, central nervous system (CNS), and intraocular malignancies.

Brachytherapy

Brachytherapy is a method of delivering radiation to tumors by placing radioactive sources either within or close to tumor tissue. Hence, a large radiation dose can be delivered directly while minimally affecting normal tissue. Brachytherapy is of two types: low dose rate (LDR) and high dose rate (HDR), depending on the length of time the radioactive sources remain in place.

Brachytherapy can be used alone or in combination with other therapies such as surgery, EBRT and chemotherapy.

Techniques of Brachytherapy in Gynecological Cancers

- *Intracavity brachytherapy*, in which the radiation source is placed within a body cavity, or a cavity created by surgery. For example, radiation can be placed in the vagina to treat cervical or endometrial cancer.
- *Intraluminal brachytherapy:* Where the source is placed in a lumen, for example the vagina after a total abdominal hysterectomy.
- *Interstitial brachytherapy:* Here the radiation source is placed within the tumor. This technique is used for instance in a vaginal vault recurrence.

TREATMENT PLANNING—PLANNING, DOSE, OARS, AND INDICATIONS

Treatment planning is the task to make sure a given radiation prescription is delivered in an optimized way. The aim is to deliver maximum dose to tumor with minimum dose to surrounding normal structures, thus achieving a higher cure rate with acceptable morbidity.

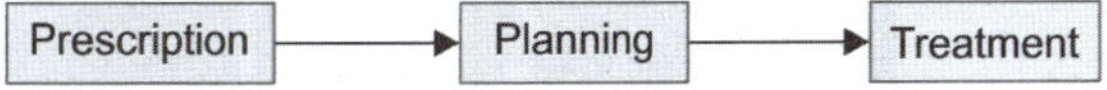

Defining the Tumor and Target Volumes for Radiotherapy (Fig. 3)

Gross tumor volume (GTV) is essentially the gross demonstrable location and extent of tumor as seen or palpated or imaged.

Clinical target volume (CTV) contains the demonstrable GTV plus a margin for subclinical disease spread.

This volume should be adequately treated if cure is to be achieved.

Planning target volume (PTV) is a geometric concept designed to ensure that the radiotherapy prescription dose is actually delivery to the CTV. It covers the uncertainties that may appear during treatment delivery **(Figs. 4A to C)**.

Defining the Organ at Risk

Organs at risk (OARs) are normal tissues whose radiation sensitivity may significantly influence treatment planning and/or prescribed dose.

After these volumes of interest are defined, the physicist, on his computerized planning system, creates radiation beams, with appropriate blocking to protect maximum normal tissue around. It is essential to maintain a tumor coverage with adequate dose with minimal dose to the OARs nearby.

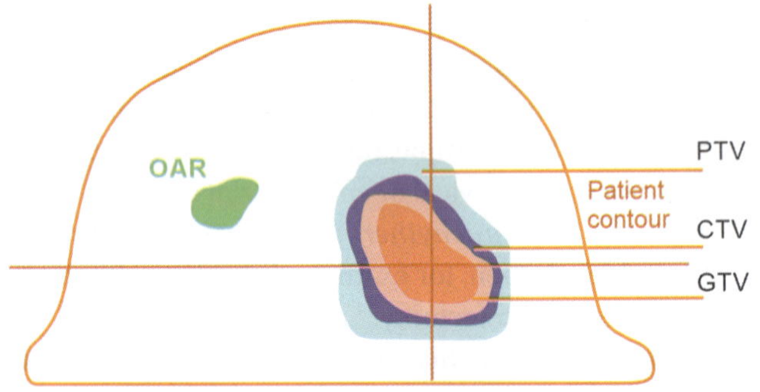

Fig. 3: Defining the tumor and target volumes for radiotherapy. (CTV: clinical target volume; GTV: gross target volume; PTV: planning target volume)

The dose distribution is evaluated by the physician with the help of isodose lines in different projections within a patient recreated.

An isodose line shows a specific dose level within a volume of tissue. A series of isodose lines thus depict regions of particular doses. This information is supplemented by dose-volume histograms (DVHs). They provide quantitative information about the doses delivered, as a function of fractional volume, by each of the defined volumes and the by critical structures around.

This planning may take a day or two before the patient is called for implementing the desired plan.

WORKFLOW IN THE RADIATION DEPARTMENT

External Beam Radiotherapy

- *Step 1: Counseling:* The radiation oncologist will counsel the patient regarding the intent of radiotherapy and possible side effects. After informed consent, the patient will be taken up for radiation planning.
- *Step 2: Bladder protocol:* After emptying the urinary bladder completely, she will be asked to drink 500–750 mL of water and wait for 45 minutes to 1 hour or until she feels her bladder full. The precise amount of water drank, and the time taken to bladder filling is noted, and this protocol is carried out every day before treatment to ensure reasonable reproducibility. Bladder protocol helps the radiosensitive small bowel to be pushed superiorly away from the radiation portals.

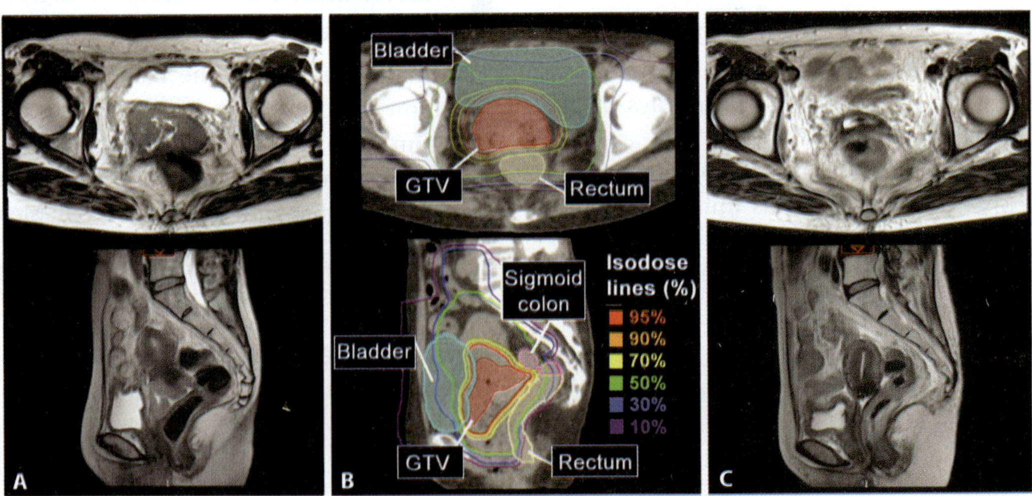

Figs. 4A to C: A treatment plan for a case of carcinoma cervix by intensity-modulated radiotherapy (IMRT) showing the gross target volume (GTV), organs at risk (OARs), and the isodose lines (%). (A) Before treatment; (B) Dose distribution of C-ion RT; and (C) 6 months after treatment. (RT: radiotherapy)

- *Step 3: Patient preparation:*
    - If the patient has constipation, give an enema before the radiation planning procedure, as the patient's rectum should be empty during radiation planning and daily treatment sessions.
    - An intravaginal lead marker will be kept at the lowest portion of the disease. It helps the radiation oncologist mark the radiation portal's lower extent.
    - In some cases, gastrografin oral contrast may be given well in advance to visualize the small bowel clearly on computed tomography (CT) images.
- *Step 4: Cast fabrication:* In this step, the patient will be taken to a simulator room, where a thermoplastic immobilization cast is made. This is made for her in the treatment position. The cast snugly fits the lower part of the abdomen, pelvis, and upper half of the thighs. The purpose of this cast is to immobilize the patient in the treatment position and reproduce the same every day during treatment **(Fig. 5)**.
- *Step 5: CT scan*: With the cast over the body, in the treatment position, a CT scan from L2 to mid-thigh will be acquired along with contrast.
- *Step 6: Contouring:* On the CT scan acquired during the simulation, the radiation oncologist will mark radiation targets, that is, cancer, and its lymph nodal drainage region. Also, the doctor will draw the normal structures in and around these targets. For example, in a typical pelvic case, the standard normal structures include urinary bladder, rectum, sigmoid colon, small bowel, and femoral heads. Remember, the contouring has to be done on each CT slice separately.

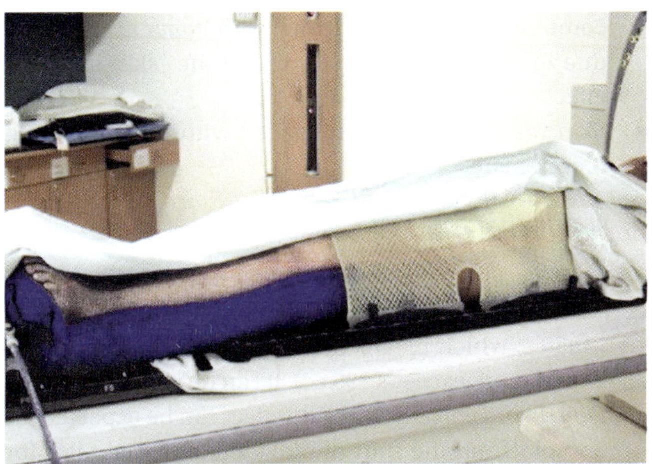

Fig. 5: Pelvic cast for immobilization.

- *Step 7: Physics planning:* After marking targets and the normal structures, the CT images will be transferred to the medical physics department for planning. The physics team carefully plans radiation beams with the help of software. This is according to the radiation techniques mentioned in the previous section, which select the beam angles around the patient's body, and decide each beam's weightage in a process called "optimization". This process helps us deliver the dose required to be given to the targets while bypassing or minimizing the doses to the normal structures. Both contouring and planning usually take a few days to be completed.
- Step *8: Plan evaluation:* The medical physics team might have optimized a few plans. Now, the radiation oncologist will evaluate each plan individually and must select the one which suits the patient the most. The plan selection could depend on the dose coverage of targets and dose limits achieved for various normal structures.
- *Step 9: Patient-specific quality assurance (QA):* Once the radiation oncologist selects the appropriate plan, the plan will be executed on the linear accelerator on a dummy machine. This ensures that the computer-generated plan is delivered to the patient safely without any glitches.
- *Step 10: Treatment delivery:* This is usually delivered by the radiation technologist. Before delivering the radiation plan, a CT scan of the patient will be taken to see the accuracy of the treatment position, ascertain the bladder filling, and ensure that the rectum is empty. After this, the plan will be executed on the patient.

Workflow in Brachytherapy

- There are many brachytherapy procedures done in gynecological cancers, as mentioned above. However, the workflow of the most common procedure—intracavitary brachytherapy or ICRT—is explained here. This is done in carcinoma cervix patients after finishing the prescribed course of EBRT.
- After informed consent, part preparation, and bowel preparation, the patient would be taken to the operation theater to place applicators into the uterus and vagina. This is usually done under spinal anesthesia. The central tube, also called the central tandem, is placed first in the uterus, followed by two ovoids in the right and left fornices of the vagina on either side of the central tandem. The central tandem and ovoids are then locked together with the help of a screw, and

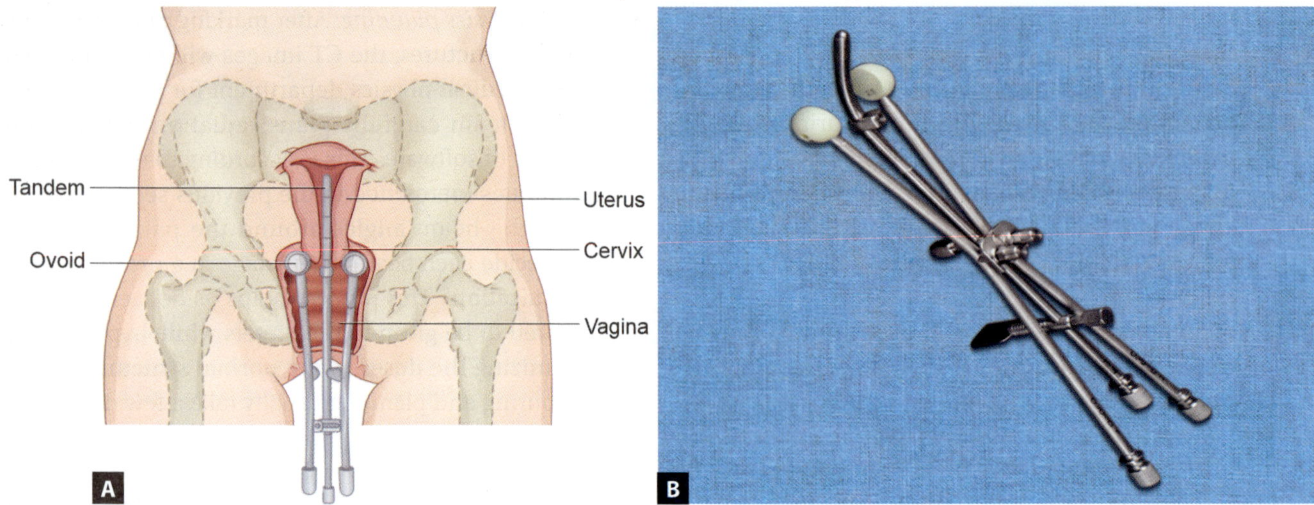

Figs. 6A and B: Applicators in intracavitary radiotherapy in carcinoma cervix.

the whole apparatus is held in position with the help of a "T" bandage **(Figs. 6A and B)**.

- After placement of the applicator inside the patient, a CT scan will be done. From here on, the workflow is similar to EBRT—contouring, physics planning, and plan evaluation. Once the plan has been approved, the patient will be shifted to the brachytherapy suite.
- The applicator will be connected to the brachytherapy machine with the help of connecting tubes. The radioactive material will now travel through these tubes and reside in the applicator at a designated position for a fixed time as planned by the medical physics team and the radiation oncologist.
- Once the treatment is finished, the radioactive seed will retract back into the brachytherapy machine. After the applicator removal, the patient is usually discharged on the same day.
- The workflow mentioned above is for ICRT. For the rest of the brachytherapy procedures, the general schema remains the same—application in OT, planning, and delivery of radiotherapy in the brachytherapy suite.

SIDE EFFECTS OF RADIOTHERAPY AND THEIR MANAGEMENT

The adverse effects of radiation can be divided into two.

1. *Acute:* Within 90 days from initiation of treatment[1]
2. *Late:* Side effects occurring after months or years after completion of radiotherapy.

Acute Side Effects

Radiation Dermatitis

Radiation dermatitis refers to skin toxicity. This toxicity is solely due to EBRT as the external beam enters the body through the skin. It manifests as:

- *Skin discoloration:* At the beginning of radiation, the skin might have a reddish hue, which later turns dark. Mostly this discoloration remains temporary.
- Skin desquamation and exfoliation in the region of radiation
- Skin ulcers might occur very rarely, which on very rare occasions heal with hypopigmented patches.
- Skin reactions are more commonly seen when treated which cobalt-60 machines. These reactions are less commonly seen when treated with linear accelerators due to a physical property called the "skin-sparing" effect.
- Skin reactions can be managed with topical steroid ointments.

Radiation Enteritis

Enteritis can be expected when the EBRT dose exceeds certain constraint levels laid down in the guidelines (V45 Gy >195 cc, which means the volume of bowel receiving 45 Gy and more exceeds 195 cc). It can be manifested as:

- *Diarrhea:* Typically, this side effect can be expected for 2 weeks from the start of radiation and might last for 2 weeks after completion of radiotherapy. The patient

might experience watery stools. This can be managed with the following:
- Probiotics
- Antisecretory agents like racecadotril
- Antimotility agents such as loperamide and lomotil (diphenoxylate and atropine)
- Antispasmodic agents like hyoscine in case of abdominal cramps
- In severe and uncontrolled or severe diarrhea, steroids (dexamethasone) can be given.
- If the patient has a fever along with loose stools, the patient might have an infective cause. In that case, an antibiotic might be required.

Late Side Effects

Proctitis

Proctitis is usually seen in the anterior wall of the rectum, as it often comes in the radiation field due to its close proximity to the uterus and cervix. Usually, it occurs when the total dose to the rectum exceeds 70–75 Gy with combined EBRT and brachytherapy. However, these days the incidence of proctitis is lower with highly sophisticated radiation planning algorithms and treatment delivery techniques. Proctitis manifests as:
- Tenesmus
- Occult blood in stools with consequent anemia
- Hematochezia
- Rectal fistula (extremely rare).

In cases of symptomatic radiation proctitis, mesalazine and or steroid enema can be used with very high success rates. If bleeding persists, sigmoidoscopy and argon plasma coagulation (APC) can be tried on the bleeding spots.

Cystitis

The bladder is much more resilient to radiation than the rectum; hence, cystitis is uncommon in the modern era of radiation practice. Minor bleeding will be stopped with continuous bladder irrigations. In case of uncontrolled cases, the following measures can be tried.
- Formalin instillation into the bladder—even though it is an effective technique to arrest bleeding, it causes severe pelvic discomfort to the patient and also causes constriction of the bladder. Due to these complications, this technique is not practiced routinely.
- Hyperbaric oxygen therapy in refractory cases is successful in 60–92% of cases.[2]
- APC of bleeding spots with the help of cystoscopy
  Note: Do not take a biopsy if one suspects radiation cystitis, as it could cause a fistula.
- If bleeding does not stop there, selective transarterial embolization (TAE) or surgical ligation of internal iliac arteries can be tried.[3]
- The last and final resort is cystectomy.

Sexual Dysfunction

Sexual dysfunction could be due to three main factors.
1. The disease on the cervix and vagina, when gets eradicated, can cicatrize the tissue causing shortening of vaginal length and stenosis.
2. Radiation by itself could cause tissue fibrosis in the vagina, causing stenosis. Radiation in the pelvic area will also affect the function of the ovaries and could cause permanent sterility.
3. Surgical techniques, in which the diseased part of the female genital tract is removed, will add to the problem.[4]

Engaging in regular coitus after the radiotherapy procedure is completed will prevent fibrosis from setting in and could result in better sexual function (vaginal stenosis and adhesions can start setting in after 3–6 months after completion of radiotherapy). Those patients who have severe dyspareunia or stenosis can use vaginal dilators. In extreme cases, vaginoplasty must be done.

Subacute Intestinal Obstruction

Subacute intestinal obstruction is a rare complication due to the adhesion of bowel loops. Most cases recover after conservative management, which includes "nil per oral" and continuous Ryle's tube aspiration. However, in rare instances, surgical adhesiolysis is required.[5]

FOLLOW-UP

The follow-up of these patients is done every:
- 3 monthly for the first 2 years
- Then, 6 monthly for a total of 5 years
- Annually after that.

During follow-up, a thorough history and physical examination specific to disease recurrence and side effects should be made. Imaging studies are mainly done based on the initial stage of cancer, histology, the time elapsed since the completion of treatment, and clinical findings during that follow-up.[6]

CONCLUSION

Gynecologic cancers were among the first malignancies to be treated with ionizing radiation. Today Radiotherapy has evolved immensely, to be able to deliver high doses of radiation to the tumour, and at the same time protecting all the normal organs around. As a result, curability and quality of life have improved grossly.

REFERENCES

1. Hall EJ, Giaccia AJ. Radiobiology for the Radiologist, 8th edition. Philadelphia: Wolters Kluwer; 2019.
2. Skowronek J, Piotrowski T. Pulsed dose rate brachytherapy: a method description and review of clinical application. Przegl Lek. 2002;59(1):31-6.
3. Okonogi N, Wakatsuki M, Kato S, Murata H, Kiyohara H, Karasawa K, et al; Working Group of Gynecological Tumors. Significance of concurrent use of weekly cisplatin in carbon-ion radiotherapy for locally advanced adenocarcinoma of the uterine cervix: A propensity score-matched analysis. Cancer Med. 2020;9:1400-8.
4. Radojevic MZ, Tomasevic A, Karapandzic VP, Milosavljevic N, Jankovic S, Folic M. Acute chemoradiotherapy toxicity in cervical cancer patients. Open Med (Wars). 2020;15(1):822-32.
5. Smit SG, Heyns CF. Management of radiation cystitis. Nat Rev Urol. 2010;7(4):206-14.
6. Korkmaz M, Şanal B, Aras B, Bozkaya H, Çınar C, Güneyli S, et al. The short- and long-term effectiveness of transcatheter arterial embolization in patients with intractable hematuria. Diagn Interv Imaging. 2016;97(2):197-201.

57

Chapter

Principles of Chemotherapy in Gynecological Malignancies

Sajjan Rajpurohit

INTRODUCTION

Gynecological malignancies must be treated using a multimodal approach that includes multidisciplinary research and conversation. Depending on the type of cancer, the clinical presentation, and the patient's preferences, the goals of systemic therapeutic delivery can vary.[1] The concepts governing systemic therapy choices that are a component of the typical care of this category of tumors are explored in this chapter.

BIOLOGY OF CANCER GROWTH

Chemotherapeutic medications can theoretically cure cancer, while sparing healthy cells by taking advantage of some intrinsic variations in their unique growth patterns.[2] The fact that the same chemotherapy regimen is not equally successful for all types of gynecological cancers is due to the unique properties of each tumor type. Cellular kinetics and biochemistry must be well understood in order to choose the right medications and prevent toxicity.[3]

Cell Cycle

All dividing cells follow the same basic pattern for replication. The amount of time needed to finish all five stages of the cell cycle is known as the cell generation time. Many biological processes, including protein synthesis, RNA synthesis, and DNA repair, take place during the G1 phase (G = Gap). The G0 phase, which is the resting phase, is thought to be prolonged in the cell. After a period of quiescence, G1 cells can either resume the cell cycle or terminally develop into the G0 phase. New DNA is created during the S phase. Cells have twice as much DNA during the G2 (premitotic) phase as they get ready for cell division. The M phase is when true mitosis and chromosome division actually occurs.

Cancer Cell Growth

Tumors are characterized by a Gompertzian growth pattern. Essentially, as a tumor mass grows, it takes progressively longer timeframes for it to double in size. When a cancer is microscopic and nonpalpable, growth is exponential. But as a tumor grows, fewer of its cells are replicating because the blood supply is constrained and interstitial pressure rises. Because more cells are in the active phase of the cell cycle when a tumor is in the exponential phase of Gompertzian growth, it should be more responsive to chemotherapy. Thus, metastases should be more sensitive to chemotherapy than the primary tumor.

This information can be used to our advantage, because advanced ovarian cancer is typically treated with surgery to remove the primary tumor, debulk big masses, and leave just microscopic residual disease for the adjuvant chemotherapy to address.[4] Also, it is assumed that more cells will enter the active phase of the cell cycle to speed up growth when a tumor mass shrank in response to treatment. This higher proportion of cells that may divide should also make a tumor more susceptible to chemotherapy.

The term "doubling time" refers to the amount of time required for a tumor to grow by twofold.

"Doubling period" describes the expansion of a whole heterogeneous tumor mass as opposed to the general term "cell cycle", which describes the activity of specific tumor cells. Various tumors can double at varying rates in people. The proportion of cells that are actively dividing, or "growth fraction", is thought to play a significant role in controlling the rate at which tumors expand and double in size. Typically, only a small percentage of the tumor will have cells that are rapidly proliferating. The remaining cells are dormant in the G0 stage.[5] In general, tumors that are cured by chemotherapy are those with a high growth fraction, such as gestational trophoblastic neoplasia.[6]

The surviving tumor cells should presumably be driven from the G0 phase into the more vulnerable phases of the cell cycle when the tumor volume is decreased by surgery or chemotherapy, making them susceptible to chemotherapy.[7]

Cell Kinetics

Chemotherapeutic agents typically work by first-order kinetics to kill a constant fraction of cells rather than a constant number. For instance, a single dose of a cytotoxic medication may kill between 10^2 and 10^4 logs of cells. This, however, is not curative since tumor burden may be 10^{12} cells or more. In order to achieve the level of cell death required to completely remove a tumor, chemotherapy treatments are frequently spaced out. In general, the quantity of live tumor cells present at the start of chemotherapy is inversely correlated with a cancer's prognosis. Some drugs achieve cell kill at several phases of the cell cycle.

These agents work throughout the entire replication cycle, from the G0 to the M phase. Cell cyclespecific agents act only on cells that are in a specific phase. Drugs that act at several stages of the cell cycle can be combined to increase the overall cell death.[8]

SYSTEMIC THERAPY

Systemic therapy describes the use of medications to treat cancer cells found throughout the body by travelling through the bloodstream. Endocrine therapy, chemotherapy, and targeted medications are some of these treatments. Systemic therapy can be used as a palliative treatment to help control the disease and symptoms for as long as feasible or as a curative treatment, frequently as part of a multimodality strategy.[9]

CHEMOTHERAPY

Cell division that is accelerated and uncontrolled is a hallmark of malignant tumors. Chemotherapy works by preventing cell division, which results in cell death, to generate its therapeutic benefits.

The following are some of the most often prescribed chemotherapeutic drugs for gynecological malignancies:

- *Carboplatin and cisplatin:* These are platinum agents that create intra- and interstrand DNA crosslinks that prevent DNA synthesis and cell division, resulting in death.
- *Paclitaxel:* This is a taxane, which stabilizes microtubules, stopping mitosis and leading to cell death.[10]
- *Liposomal doxorubicin:* An anthracycline that binds to DNA and prevents the synthesis of nucleic acids, causing cell death. A liposome tiny vesicle used to deliver doxorubicin enters the aberrant tumor vasculature and delivers the medication there.
- *Gemcitabine:* An antimetabolite that causes apoptosis by acting as a nucleoside analog to replace cytidine during DNA replication

The bulk of treatment-related side effects are brought on by the medications' impacts on the normal tissue cells' division processes. Acute side effects can happen right away during treatment, while late side effects can happen months or even years later.

Chemotherapy has the biggest effects on healthy cells with a high cell turnover rate, like bone marrow, gastrointestinal (GI) tract mucosal lining cells, and hair follicles. In between doses of treatment, the normal cells are typically able to repair the damage, reducing the likelihood of long-term damage. Each patient will suffer the adverse effects to a different extent, because the negative effects of the various chemotherapy medications vary.[11]

Candidates for chemotherapy should be healthy enough to withstand its effects. They should ideally have a World Health Organization (WHO) performance status (PS) of 0–2 and have normal renal, hepatic, and hematological function. Therapy offered to those who do not fit these criteria may increase morbidity or mortality.[12] Patients who report a new diagnosis of an advanced, quickly developing, chemosensitive malignancy are the only exception to this rule. In these circumstances, administering a less toxic chemotherapeutic drug, like carboplatin, to a patient with a poor PS may paradoxically relieve their symptoms and improve their PS. An expert oncologist should make the decision to administer chemotherapy to each patient because it is extremely individualized.

Cycles of intravenous (IV) chemotherapy are administered. Depending on the medicine combination utilized, the cycle's length varies. Deliveries of standard carboplatin and paclitaxel begin on the first day of a 21-day cycle. By administering the medication in this cyclical pattern, the normal cells have time to recuperate before the subsequent cycle, and the cancer cells can advance through the cell cycle to a more vulnerable stage, increasing the likelihood that the cancer cells will be killed.[13]

TARGETED THERAPY

Targeted medications are a more recent category of systemic therapy that are being utilized more frequently to treat cancer.[14] They have an effect on particular molecules that are important in the development or spread of cancer cells. When opposed to chemotherapy, the adverse effects may be less, because these targets are not abundant in normal cells.[15]

Two groups of targeted agents can be distinguished:

1. Monoclonal antibodies, such as the anti-angiogenic antibody bevacizumab, target specific antigens on the cell surface.
2. Small molecules, like poly-ADP ribose polymerase (PARP) inhibitors, can pass through the cell membrane and operate on intracellular targets.

Monoclonal Antibodies

A humanized monoclonal antibody called bevacizumab suppresses angiogenesis by specifically targeting the vascular endothelial growth factor (VEGF) receptor. In the following circumstances, it is approved for the treatment of gynecological malignancy:

- Advanced (stage 3 or 4) ovarian cancer as a first-line therapy with paclitaxel and carboplatin
- Platinum-resistant ovarian cancer in conjunction with paclitaxel, liposomal doxorubicin, or topotecan, or first platinum-sensitive ovarian cancer relapse in combination with carboplatin and gemcitabine
- Cervical cancer that is persistent, recurring, or metastatic in addition to paclitaxel plus topotecan or paclitaxel and cisplatin

Gastrointestinal perforation (0.9%), hemorrhage (24%), severe hypertension (18%), nephrotic syndrome (7%), infusion responses (3%), fistula (2%), thromboembolic events (2.6%), and nonhealing wounds are some of the side effects of bevacizumab. Thus, it needs to be used carefully in women with potential serosal involvement of the bowel, and must not be avoided within 4 weeks of planned surgical treatment.[16]

Poly-ADP Ribose Polymerase Inhibitors

Cancers in people with BRCA 1 or 2 mutations are more dependent on the cellular protein PARP for DNA repair.[17] A cell dies if the genetic damage cannot be repaired. High response rates to PARP inhibitors like olaparib or rucaparib are seen in these patients.[18] In advanced ovarian cancer patients with the BRCA mutation, the SOLO1 phase III clinical trial examined the efficacy of maintenance olaparib.[19] This study demonstrated that adding the oral PARP inhibitor olaparib for 2 years after receiving first-line platinum-based therapy dramatically increased the 3-year disease-free progression rate from 27% on placebo to 60%. Another PARP inhibitor, niraparib, is available for patients with platinum-sensitive recurrent ovarian high-grade serous carcinoma after second-line or later treatment, provided they have not previously received a PARP inhibitor. Patients must have a partial or complete response to platinum-based chemotherapy. Based on the NOVA trial, which revealed an improved progression-free survival in both groups, although the responses were more substantial in the BRCA-mutated group, it is available for both BRCA-mutated and wild-type patients. The most common mild-to-severe side effects of PARP inhibitors include fatigue (63%), nausea (77%), diarrhea (34%), and anemia (39%).

ENDOCRINE THERAPY

The goal of hormonal or endocrine therapy is to lessen estrogen's impact on the development of cancer cells. Every day, the treatment is often given as oral tablets. These are typically a well-tolerated class of medications, free of the serious side effects sometimes connected with chemotherapy treatment.[20] The selective estrogen receptor (ER) modulator tamoxifen reduces estrogen binding to the ER by competitive inhibition. Depending on the target tissue, it exhibits both agonist and mixed antagonist activity.[21]

A nonsteroidal aromatase inhibitor called letrozole prevents the transformation of androgens into estrogens in peripheral tissues.[22]

It is not clear how synthetic progestins like megestrol acetate and medroxyprogesterone acetate combat cancer. Theories include a pituitary-level action that lowers the production of gonadotropins and, consequently, estrogen, as well as local tissue effects at the ERs/progesterone receptors (PRs).[23]

Endocrine therapy may be beneficial for treating women with recurrent epithelial ovarian cancer who are asymptomatic. Letrozole has a 9% response rate, with 42% of patients experiencing stable illness, according to studies. Tamoxifen has a 10% response rate, and

32% of patients have stable disease. Megestrol acetate, medroxyprogesterone acetate, and tamoxifen have all shown effectiveness in studies for women with recurrent endometrial cancer, with reported response rates of up to 15–30%, mostly in lower-grade tumors.[23,24]

PHARMACOLOGIC PRINCIPLES

Dosing and Dose Intensity

Overall, treatment effectiveness depends on drug concentration and duration of exposure to critical tumor sites. Chemotherapeutic drugs may have a small therapeutic window or range. To get an optimal impact over a crucial threshold while preventing excessive toxicity, doses must be carefully determined. The body surface area (BSA) of the patient is used to calculate chemotherapy doses, which are then given in milligrams per meter squared (mg/m^2).

Body surface area is a better indicator of metabolic mass than body weight because it is less affected by abnormal adipose mass. This algorithm makes sure that each patient receives an equivalent amount of medication. Patient weights are measured before the start of each therapy session even though height is a constant variable.

Seldom is it necessary to account for tissue edema or ascites, because doses should be based on weight without this accompanying fluid.

Alternatively, the dosing of some drugs is more specific.

For instance, the reticuloendothelial system is responsible for the metabolism and elimination of the monoclonal antibody bevacizumab.

It is dosed only by patient weight (mg/kg).

Dosing for medications that are excreted through the kidneys, such carboplatin, may be determined using a calculation of the glomerular filtration rate (Calvert formula). The amount of drug administered over time is known as the dose intensity (or density). Its principal significance is in highly responsive tumors, which can be cured by chemotherapy.

However, it might not be possible to raise the dose to a level high enough to demonstrate a benefit without causing dose-limiting toxicity in other, less-sensitive tumor types. Yet, lowering the dose intensity to lessen toxicity can lead to less effective therapeutic outcomes.

Dose-dense chemotherapy: A regimen that administers treatment more frequently than every 3 weeks is known as dose-dense chemotherapy. It typically carries a higher hazardous level. Clinical trials are now evaluating its effectiveness in treating ovarian cancer using a regimen that combines weekly administration of paclitaxel and carboplatin every 3 weeks. Studies conducted to far have indicated some advantages in progression-free survival and overall survival, particularly in patients who have undergone surgery and still have >1 cm of residual disease. Although more research is necessary, patients with clear-cell or mucinous carcinomas or those on bevacizumab have not been found to benefit from dose-dense treatment.

Route of Administration

Chemotherapy may be administered systemically or regionally. Systemic therapy options include oral, IV, subcutaneous (SC), and intramuscular (IM) methods. Regional chemotherapy is aimed at delivering drugs directly into the cavity in which the tumor is located. Several agents clear the bodily cavity more slowly than they do through systemic circulation. As a result, higher concentrations of active medicines are exposed to cancer cells for a longer period of time. This technique has been most extensively studied in ovarian cancer, in which tumors are usually confined to the intraperitoneal (IP) space. Clinical studies have uniformly demonstrated a pharmacologic advantage favoring administration into the IP compartment. However, the existence of intra-abdominal adhesions, poor fluid circulation, fibrotic tumor encapsulation, and other factors, frequently prevent passive diffusion from penetrating into peritoneal tumor nodules and coexisting ascites. IP chemotherapy is often given to women with low residual illness due to these restrictions on medication penetration. During IV administration, several drugs known as vesicants require special care. If these extravasate into the SC tissue, it may cause significant discomfort and necrosis. Some medications require gradual infusion, either by a central venous catheter, which is preferred, or a quickly flowing peripheral IV catheter. If extravasation is suspected, the infusion is immediately stopped, the affected arm elevated, and ice packs applied. Plastic surgery should be consulted in serious circumstances. Agent activity and toxicity is influenced dramatically by drug inactivation, elimination, or excretion. Most of the time, the liver or kidneys are the main organs involved in this. As a result, drug activity may be diminished and toxicity exacerbated when normal hepatic or renal function is impaired. Moreover, medication toxicity frequently manifests itself more clearly in the elderly or undernourished. For instance, in cachectic women, a low serum creatinine level could not correctly reflect underlying renal function. Calculating a carboplatin dose using this artificially low figure could result in an excessive dosage and significant morbidity. Instead, a preset creatinine level may need to be selected (0.8 or 1.0 mg/dL) to aid safer dosing in some patients.

Drug Interactions and Reactions

Most women who receive chemotherapy are prescribed medication for other noncancerous conditions, such as hypertension. In addition, analgesics, antiemetics, and antibiotics are frequently given to women during chemotherapy. The majority of drug interactions are insignificant, but a few can significantly change a drug's toxicity. Drugs that are metabolized in the liver are particularly at risk for such interactions. For instance, using methotrexate while taking warfarin (Coumadin) will typically increase the anticoagulant effect and need lowering the amount of the warfarin. Despite a study of the patient's medical history and the prescription of preventative drugs, anaphylactic, allergic, or hypersensitive reactions might arise during or after chemotherapy administration. In order to handle these unforeseen but frequent problems, a treatment center needs to have a skilled nursing team and the necessary resources. A woman is encouraged to report any symptoms, such as flushing, pruritus, dyspnea, tachycardia, hoarseness, or dizziness, that could indicate an anaphylactic reaction before taking a medication. There must be immediate access to emergency supplies such extra oxygen, a ventilatory face mask and bag, or intubation equipment. Use of IV diphenhydramine (Benadryl) and/or corticosteroids may be sufficient for a regional hypersensitivity reaction. However, if a patient experiences generalized hypersensitivity or anaphylaxis, chemotherapy should be stopped right away, the emergency team should be informed, and emergency medications such as epinephrine (0.1–0.5 mg of a 1:10,000 solution) should be delivered.

DRUG RESISTANCE

In principle, larger tumor masses have a greater proportion of cells that have already developed resistance. Resistance may arise to one medicine or to several different agents, and it may be innate or learned. Intrinsic drug resistance is seen if tumors are first exposed to an agent and fail to respond. In contrast, tumors that were once responsive to a treatment no longer respond to it when they have acquired drug resistance. Sometimes a certain medicine causes this to happen.[25] More often, however, acquired resistance is "pleiotropic," meaning that a cancer is resistant to multiple chemotherapy agents. The P-glycoprotein or multidrug resistance pump frequently mediates this. Advanced ovarian cancer is a good example. Using platinum-based chemotherapy, the majority of patients will initially experience remission, but 80% will eventually relapse with tumors that are resistant to all cytotoxic drugs.

CURRENT MANAGEMENT OF GYNECOLOGICAL CANCERS

Ovarian, Tubal, and Primary Peritoneal Tumors

20% of all occurrences of ovarian or tubal cancer are early (stage I and II).[26] The primary form of therapy is upfront surgery. Patients at high risk, such as those with stage IC or II illness of any grade, clear-cell or high-grade tumors at any stage, receive adjuvant chemotherapy treatment. Meta-analyses, such as the 2009 Cochrane Review, which demonstrated a survival advantage for patients with high-risk characteristics or postoperative residual disease, provide support for adjuvant treatment. The recommended course of therapy is six cycles of chemotherapy with carboplatin, either alone or in combination with paclitaxel.[27,28]

Advanced ovarian cancer (stage III and IV) is treated with primary surgery, followed by adjuvant carboplatin and paclitaxel chemotherapy, if the disease may be optimally debulked at that time (typically stage IIIa and b). When extensive stage IIIc or IV disease is involved, neoadjuvant carboplatin and paclitaxel are administered for three cycles, followed by interval cytoreductive surgery, and then another three cycles of carboplatin and paclitaxel, if it is determined that optimal surgery is not possible or appropriate in the upfront setting. Moreover, bevacizumab might be taken into account, continuing for the whole year as a maintenance therapy. Following the initial round of chemotherapy, the PARP inhibitor olaparib has just been licensed for use as the first-line maintenance therapy for women with platinum-sensitive illness with the BRCA mutation.[29,30] This has the ability to not only delay the start of relapse but even to cure the illness. Regrettably, recurrence occurs in the majority of people with advanced ovarian cancer. Surgery can be beneficial for a small percentage of carefully chosen people. Although the primary goals of treatment for relapse management are often disease and symptom control and chemotherapy. A combination of carboplatin with liposomal doxorubicin or gemcitabine may be utilized if the patient has experienced a platinum-sensitive relapse (>6 months have passed since the patient's prior platinum-based treatment). Liposomal doxorubicin or weekly paclitaxel are frequently used as treatments when a patient has platinum resistance (<6 months have passed since the last platinum treatment). When the disease is fast developing on a computed tomography (CT) scan or generating symptoms, treatment is then given intermittently for as long as feasible with the goal of preserving quality of life in addition to prolonging life. The use of hormonal therapies is also possible in this situation. Regardless of BRCA status, certain women with platinum-sensitive illness after two courses of chemotherapy may also be treated with PARP inhibitors.

Cervical Tumors

Examination under anesthesia (EUA) is frequently a component of clinical staging, with the exception of early stages with tiny tumors, where correct clinical staging of cervical tumors is required for treatment planning. The conventional surgical procedure for early-stage tumors that are node-negative and are beyond stage 1A1 and up to stage 1B1 (limited to the cervix and 4 cm) would be a radical hysterectomy with or without bilateral salpingo-oophorectomy. The ovaries are often preserved in younger individuals, and ovarian metastases are uncommon. The purpose of pelvic lymphadenectomy is to facilitate surgical staging and determine whether postoperative adjuvant therapy is necessary.

When a radical trachelectomy is an option, in some circumstances fertility-sparing surgery may be chosen depending on the tumor's features as well as the patient's age or preferences. For high-risk individuals with recurrence-risk factors, adjuvant chemoradiotherapy is advised. They include parametrial invasion, unanticipated lymph node (LN) involvement, and positive resection margin. Adjuvant chemoradiation has a positive effect on survival in this population. In order to lower the chance of pelvic recurrence, (chemo-) radiotherapy may also be considered for patients with intermediate risk factors, such as tumors larger than 4 cm, close vaginal margins, profound stromal invasion, or the presence of lymphovascular space invasion (LVSI). The recommended course of treatment for tumors with stages >IB2 or when LNs are involved is primary chemoradiation, which can ensure high rates of local control even with stage IVA disease.

In order to reduce the radiation exposure to the organs at risk, particularly the bladder and colon, intensity-modulated radiotherapy (IMRT) techniques are being used more frequently to give treatment. The 5-week treatment regimen includes weekly IV infusions of cisplatin chemotherapy, which is often well tolerated but does raise the possibility of nausea and vomiting.[31] Due to the other pelvic structures' radiation sensitivity, the dose of radiation needed to treat cervical cancer effectively is more than what can be safely administered using only external beam procedures. Hence, after their initial external beam treatment, patients on a radical treatment track thereafter get brachytherapy.

This makes it possible to administer a tumoricidal dose to the tumor or cervix while causing only a tolerable amount of normal tissue damage. In order to put a radioactive source inside the applicators and deliver a high radiation dose to the target, brachytherapy necessitates inserting the applicators directly into the cervix or uterus. This can be done in a variety of methods, but in our facility, it entails an inpatient admission with three or four fractions administered over a few days.[32] The planning process for radiation distribution, which includes a team of expert physicists, radiographers, and clinical oncologists, can be complicated and frequently calls for both magnetic resonance imaging (MRI) and CT imaging. Patients with metastatic disease or those who are not well enough for aggressive therapy may be provided palliative sessions of radiation to control local disease, typically without chemotherapy. Poorly controlled pelvic disease can be uncomfortable and challenging to manage, thus radiation control of pelvic disease may still be advised even in cases of metastatic disease. Palliative chemotherapy may potentially be an option for metastatic cervical cancer.[33] Combination chemotherapy using platinum is advised. Bevacizumab increases the slight survival advantage.[34]

ENDOMETRIAL CANCER

Surgery is still the basis of treatment for endometrial cancer and often entails a hysterectomy and bilateral salpingo-oophrectomy with or without staging of the nearby LNs.[27] In some circumstances, it may be necessary to do omental biopsies as well as biopsies of any questionable pelvic or intra-abdominal lesions to help with surgical staging. Even though cytoreductive surgery is unlikely to be curative (stages III–IVA), it may occasionally be provided to patients with advanced disease since it may have a survival benefit and provide palliation for individuals with limited gross residual disease. Surgery may or may not be necessary when extrapelvic metastatic illness is present, depending on the specific symptoms, since treatment is palliative. Chemotherapy may only be suitable in these situations for palliation. Early-stage disease (IA) patients without major risk factors do not need adjuvant therapy, but those with intermediate- or high-risk factors should be given the option of adjuvant radiotherapy and/or chemotherapy. Adjuvant vaginal vault brachytherapy may be advantageous for those with a moderate risk of recurrence. Only at the vault does this lower local recurrence; it has no effect on survival. Adjuvant pelvic radiation may be beneficial for those who have higher risk factors, such as higher tumor stages and/or grades and the presence of LVSI, in order to lower the likelihood of a pelvic recurrence. Although it has not been demonstrated that pelvic radiation increases overall survival, it is more hazardous than brachytherapy alone. When a patient is thought to be at a high risk of recurrence, adjuvant chemotherapy with carboplatin and paclitaxel in combination is frequently administered before pelvic radiation. According to the PORTEC-3 research, patients who had chemotherapy in combination to radiotherapy had a 5% higher 5-year overall survival rate. In patients with stage 3 illness or serous histology, the advantages were greatest. In metastatic disease, the same platinum doublet may be administered for a typical six cycles to assist alleviate symptoms. Hormonal therapy, like Megace, may produce a response in patients with more sluggish illness, lower-grade tumors, and those reporting positive ERs or PRs results. A tiny portion of patients might experience an extended remission.

When a patient has an early-stage, low-grade condition and is deemed unfit for treatment or still needs fertility treatment, the Mirena intrauterine system (IUS) may be used alone.

VULVAL TUMORS

Squamous cell carcinomas, which are the most common type of vulval cancer, are best treated surgically if detected early. The procedure comprises either a radical modified vulvectomy or a radical local excision, and determining whether LN are involved is crucial for the prognosis. The LN must be evaluated by sentinel node biopsy or node dissection in patients with stage IB disease or higher and no palpable nodes. While radiotherapy can be an alternative to lower the risk of recurrence, it should also be taken into consideration in cases with close margins, large tumors >4 cm, and those with node-positive disease. Positive margins should be re-executed whenever possible. When a patient cannot have surgery or when the illness has progressed locally to the point where clear margins are impossible, primary chemoradiation may be an alternative. If the patient is healthy enough, concomitant chemotherapy with cisplatin would be administered together with a radiation dose of at least 60–65 Gy. The conventional strategy would be to administer 45 Gy to the pelvis over the course of 5 weeks, then enhance the remaining gross tumor.[28] If accessible, IMRT may be a different strategy that enables the delivery of a larger concurrent boost dosage to the primary tumor while administering pelvic radiation. uterine tumors Vascular tumors are uncommon. EUA plays a crucial role in planning. Similar to cervical tumors, vaginal tumors inside the proximal (upper) two-thirds of the vagina are often treated with radical chemoradiation. Treatment for those in the distal portion of the vagina usually resembles that for vulval malignancies.[35]

CONCLUSION

Effectiveness of chemotherapy is a dynamic process where a treating physician regularly balances the toxicity against the tumor response. Numerous factors influence toxicity and include the patient's baseline nutrition, overall health, extent of disease, and prior therapy. A clinician needs to have objective criteria for response when advising women to continue their therapy or transition to a different regimen. The complete response rate is the most crucial metric. These would include normal CA125 values (typically 35 U/mL), physical examination findings, and imaging test outcomes for ovarian cancer. Ultimately, women who have any possibility of cure are those who first achieve a complete response.[36,37] However, if chemotherapy results in a partial response, many women still view this as advantageous compared with supportive care, even if overall survival is uncertain.

REFERENCES

1. Aapro M, Jelkmann W, Constantinescu SN, Leyland-Jones B. Effects of erythropoietin receptors and erythropoiesis-stimulating agents on disease progression in cancer. Br J Cancer. 2012;106(7):1249-58.
2. Alazzam M, Tidy J, Osborne R, Coleman R, Hancock BW, Lawrie TA. Chemotherapy for resistant or recurrent

gestational trophoblastic neoplasia. Cochrane Database Syst Rev. 2012;12:CD008891.
3. Alvarez EA, Brady WE, Walker JL, Rotmensch J, Zhou XC, Kendrick JE, et al. Phase II trial of combination bevacizumab and temsirolimus in the treatment of recurrent or persistent endometrial carcinoma: a Gynecologic Oncology Group study. Gynecol Oncol. 2013;129(1):22-7.
4. Baumann KH, du Bois A, Meier W, Rau J, Wimberger P, Sehouli J, et al. A phase II trial (AGO 2.11) in platinum-resistant ovarian cancer: a randomized multicenter trial with sunitinib (SU11248) to evaluate dosage, schedule, tolerability, toxicity and effectiveness of a multitargeted receptor tyrosine kinase inhibitor monotherapy. Ann Oncol. 2012;23(9):2265-71.
5. Bohlius J, Wilson J, Seidenfeld J, Piper M, Schwarzer G, Sandercock J, et al. Recombinant human erythropoietins and cancer patients: updated meta-analysis of 57 studies including 9353 patients. J Natl Cancer Inst. 2006;98(10):708-14.
6. Bower M, Newlands ES, Holden L, Short D, Brock C, Rustin GJ, et al. EMA/CO for high-risk gestational trophoblastic tumors: results from a cohort of 272 patients. J Clin Oncol. 1997;15(7):2636-43. Erratum in: J Clin Oncol 1997;15(9):3168.
7. Burger RA, Brady MF, Bookman MA, Monk BJ, Walker JL, Homesley HD, et al. Risk factors for GI adverse events in a phase III randomized trial of bevacizumab in first-line therapy of advanced ovarian cancer: A Gynecologic Oncology Group Study. J Clin Oncol. 2014;32(12):1210-7.
8. Burstein HJ, Mangu PB, Somerfield MR, Schrag D, Samson D, Holt L, et al.; American Society of Clinical Oncology. et al. American Society of Clinical Oncology clinical practice guideline update on the use of chemotherapy sensitivity and resistance assays. J Clin Oncol. 2011;29(24):3328-30.
9. Cannistra SA, Matulonis UA, Penson RT, Hambleton J, Dupont J, Mackey H, et al. Phase II study of bevacizumab in patients with platinum-resistant ovarian cancer or peritoneal serous cancer. J Clin Oncol. 2007;25(33):5180-6. Erratum in: J Clin Oncol. 2008;26(10):1773.
10. Cantù MG, Buda A, Parma G, Rossi R, Floriani I, Bonazzi C, et al. Randomized controlled trial of single-agent paclitaxel versus cyclophosphamide, doxorubicin, and cisplatin in patients with recurrent ovarian cancer who responded to first-line platinum-based regimens. J Clin Oncol. 2002;20(5):1232-7.
11. Case AS, Rocconi RP, Kilgore LC, Barnes MN. Effectiveness of darbepoetin alfa versus epoetin alfa for the treatment of chemotherapy induced anemia in patients with gynecologic malignancies. Gynecol Oncol. 2006;101(3):499-502.
12. Chen H, Li J, Cui T, Hu L. Adjuvant gonadotropin-releasing hormone analogues for the prevention of chemotherapy induced premature ovarian failure in premenopausal women. Cochrane Database Syst Rev. 2011;(11):CD008018. Update in: Cochrane Database Syst Rev. 2019;3:CD008018.
13. Chura JC, Van Iseghem K, Downs LS Jr, Carson LF, Judson PL. Bevacizumab plus cyclophosphamide in heavily pretreated patients with recurrent ovarian cancer. Gynecol Oncol. 2007;107(2):326-30.
14. Coleman RL, Sill MW, Lankes HA, Fader AN, Finkler NJ, Hoffman JS, et al. A phase II evaluation of aflibercept in the treatment of recurrent or persistent endometrial cancer: a Gynecologic Oncology Group study. Gynecol Oncol. 2012;127(3):538-43. Erratum in: Gynecol Oncol. 2013;130(1):252-3.
15. Eisenhauer EA, Therasse P, Bogaerts J, Schwartz LH, Sargent D, Ford R, et al. New response evaluation criteria in solid tumours: revised RECIST guideline (version 1.1). Eur J Cancer. 2009;45(2):228-47.
16. Elgindy EA, El-Haieg DO, Khorshid OM, Ismail EI, Abdelgawad M, Sallam HN, et al. Gonadatrophin suppression to prevent chemotherapy-induced ovarian damage: a randomized controlled trial. Obstet Gynecol. 2013;121(1):78-86.
17. Fiorica JV, Brunetto VL, Hanjani P, Lentz SS, Mannel R, Andersen W, et al.; Gynecologic Oncology Group study. Phase II trial of alternating courses of megestrol acetate and tamoxifen in advanced endometrial carcinoma: a Gynecologic Oncology Group study. Gynecol Oncol. 2004;92(1):10-4.
18. Fleming GF, Filiaci VL, Marzullo B, Zaino RJ, Davidson SA, Pearl M, et al. Temsirolimus with or without megestrol acetate and tamoxifen for endometrial cancer: a gynecologic oncology group study. Gynecol Oncol. 2014;132(3): 585-92.
19. Gordon AN, Tonda M, Sun S, Rackoff W; Doxil Study 30-49 Investigators. Long-term survival advantage for women treated with pegylated liposomal doxorubicin compared with topotecan in a phase 3 randomized study of recurrent and refractory epithelial ovarian cancer. Gynecol Oncol. 2004;95(1):1-8.
20. Gotlieb WH, Amant F, Advani S, Goswami C, Hirte H, Provencher D, et al. Intravenous aflibercept for treatment of recurrent symptomatic malignant ascites in patients with advanced ovarian cancer: a phase 2, randomised, double-blind, placebo-controlled study. Lancet Oncol. 2012;13(2):154-62.
21. Tewari KS, Sill MW, Long HJ 3rd, Penson RT, Huang H, Ramondetta LM, et al. Improved survival with bevacizumab in advanced cervical cancer. N Engl J Med. 2014;370(8):734-43.
22. Tinker AV, Gebski V, Fitzharris B, Buck M, Stuart-Harris R, Beale P, et al. Phase II trial of weekly docetaxel for patients with relapsed ovarian cancer who have previously received paclitaxel—ANZGOG 02-01. Gynecol Oncol. 2007;104(3):647-53.
23. Hensley ML, Sill MW, Scribner DR Jr, Brown J, Debernardo RL, Hartenbach EM, et al. Sunitinib malate in the treatment of recurrent or persistent uterine leiomyosarcoma:

a Gynecologic Oncology Group phase II study. Gynecol Oncol. 2009;115(3):460-5.

24. Hurteau JA, Brady MF, Darcy KM, McGuire WP, Edmonds P, Pearl ML, et al. Randomized phase III trial of tamoxifen versus thalidomide in women with biochemical-recurrent-only epithelial ovarian, fallopian tube or primary peritoneal carcinoma after a complete response to first-line platinum/taxane chemotherapy with an evaluation of serum vascular endothelial growth factor (VEGF): A Gynecologic Oncology Group Study. Gynecol Oncol. 2010;119(3):444-50.
25. Long HJ 3rd, Bundy BN, Grendys EC Jr, Benda JA, McMeekin DS, Sorosky J, et al. Randomized phase III trial of cisplatin with or without topotecan in carcinoma of the uterine cervix: a Gynecologic Oncology Group Study. J Clin Oncol. 2005;23(21):4626-33.
26. Mackay HJ, Buckanovich RJ, Hirte H, Correa R, Hoskins P, Biagi J, et al. A phase II study single agent of aflibercept (VEGF Trap) in patients with recurrent or metastatic gynecologic carcinosarcomas and uterine leiomyosarcoma. A trial of the Princess Margaret Hospital, Chicago and California Cancer Phase II Consortia. Gynecol Oncol. 2012;125(1):136-40.
27. Mancari R, Signorelli M, Gadducci A, Carinelli S, De Ponti E, Sesana S, et al. Adjuvant chemotherapy in stage I-II uterine leiomyosarcoma: a multicentric retrospective study of 140 patients. Gynecol Oncol. 2014;133(3):531-6.
28. Gynecologic Oncology Group; Markman M, Blessing J, Rubin SC, Connor J, Hanjani P, Waggoner S. Phase II trial of weekly paclitaxel (80 mg/m2) in platinum and paclitaxel-resistant ovarian and primary peritoneal cancers: a Gynecologic Oncology Group study. Gynecol Oncol. 2006;101(3):436-40.
29. Kaufman B, Shapira-Frommer R, Schmutzler RK, Audeh MW, Friedlander M, Balmaña J, et al. Olaparib monotherapy in patients with advanced cancer and a germline BRCA1/2 mutation. J Clin Oncol. 2015;33(3):244-50.
30. Ledermann J, Harter P, Gourley C, Friedlander M, Vergote I, Rustin G, et al. Olaparib maintenance therapy in patients with platinum-sensitive relapsed serous ovarian cancer: a preplanned retrospective analysis of outcomes by BRCA status in a randomised phase 2 trial. Lancet Oncol. 2014;15(8):852-61. Erratum in: Lancet Oncol. 2015;16(4):e158.
31. Hesketh PJ. Chemotherapy-induced nausea and vomiting. N Engl J Med. 2008;358(23):2482-94.
32. Katsumata N, Yasuda M, Takahashi F, Isonishi S, Jobo T, Aoki D, Tsuda H, et al. Dose-dense paclitaxel once a week in combination with carboplatin every 3 weeks for advanced ovarian cancer: a phase 3, open-label, randomised controlled trial. Lancet. 2009;374(9698):1331-8.
33. Vogelzang NJ. Treatment options in metastatic renal carcinoma: an embarrassment of riches. J Clin Oncol. 2006;24(1):1-3.
34. Liu JF, Barry WT, Birrer M, Lee JM, Buckanovich RJ, Fleming GF, et al. Combination cediranib and olaparib versus olaparib alone for women with recurrent platinum-sensitive ovarian cancer: a randomised phase 2 study. Lancet Oncol. 2014;15(11):1207-14.
35. Matulonis UA, Berlin S, Ivy P, Tyburski K, Krasner C, Zarwan C, et al. Cediranib, an oral inhibitor of vascular endothelial growth factor receptor kinases, is an active drug in recurrent epithelial ovarian, fallopian tube, and peritoneal cancer. J Clin Oncol. 2009;27(33):5601-6.
36. Park JY, Jin KL, Kim DY, Kim JH, Kim YM, Kim KR, et al. Surgical staging and adjuvant chemotherapy in the management of patients with adult granulosa cell tumors of the ovary. Gynecol Oncol. 2012;125(1):80-6.
37. Peng C, Fan NC, Ligier M, Väänänen J, Leung PC. Expression and regulation of gonadotropin-releasing hormone (GnRH) and GnRH receptor messenger ribonucleic acids in human granulosa-luteal cells. Endocrinology. 1994;135(5):1740-6.

Chapter 58

Endometrial Cancer

Bindiya Gupta, Namita Batra

INTRODUCTION

Endometrial cancer is the most common cancer in developed countries and according to GLOBOCAN 2020, it is overall sixth most common cancer in women worldwide with 4,17,367 new cases diagnosed annually with an incidence of 1.05%. High-income countries have a higher incidence of 5.9% compared to low-resource countries, 4%.[1]

RISK FACTORS

The risk of endometrial cancer is directly proportional to age and body mass index (BMI) and has one of the strongest links with obesity. Obesity is a hyperestrogenic state due to aromatization of adrenal androgens into estrogen in the adipose tissues. It also is a proinflammatory state with high levels of C-reactive protein (CRP), tumor necrosis factor alpha (TNF-α), and interleukin 6 (IL-6). Every 5 kg/m^2 increase in BMI is associated with 54% higher risk of cancer.[2,3] Tamoxifen is a selective estrogen receptor modulator (SERM) with a proliferative effect on endometrium and long-term use for prevention or treatment of hormone sensitive breast cancer increases fourfold risk of endometrial cancer.[4] Smoking is associated with decreased risk of type I endometrial cancer and an increased risk of type II endometrial cancer.[5] Genetic predisposition is seen with Lynch II syndrome, Cowden syndrome, and BRCA 1 and BRCA 2 genes.[6] 3% of all endometrial cancers and 10% with mismatch repair deficiency will have Lynch syndrome **(Table 1)**.[7]

TABLE 1: Risk factors and relative risk of endometrial cancer.

Risk factor	*Relative risk*
Nulliparity	2–3
Late menopause (after 52 years)	2.4
Early menarche	1.5–2
Obesity: • Overweight • Obese	 1.5 >2.5
Diabetes	2.8
Tamoxifen	2–3
Unopposed estrogen therapy	4–8
Atypical endometrial hyperplasia	25–33(29%)
Lynch II (MSH2, MLH1, MSH6 mutations)	20 (13–49% lifetime risk)
BRCA	
Family history positive	
Insulin resistance	
Polycystic ovary syndrome (PCOS)	
Metabolic syndrome	
Hypertension	
Hyperthyroidism	
Cowden syndrome	

STAGING

The FIGO 2023 staging for cancer of the corpus uteri (surgical staging) is given in **Table 2A and B**.

CLASSIFICATION OF ENDOMETRIAL CANCER

Histopathology (WHO 2020) Classification[8]

The World Health Organization (WHO) 2020 classification of endometrial cancer is shown in **Box 1**.

TABLE 2A: 2023 FIGO staging of cancer of the endometrium.[a]

Stage	***Description***
I	Confined to the uterine corpus and ovary
IA	Disease limited to the endometrium OR non-aggressive histological type, i.e. low-grade endometroid, with invasion of less than half of myometrium with no or focal lymphovascular space involvement (LVSI) OR good prognosis disease
IA1	Non-aggressive histological type limited to an endometrial polyp OR confined to the endometrium
IA2	Non-aggressive histological types involving less than half of the myometrium with no or focal LVSI
IA3	Low-grade endometrioid carcinomas limited to the uterus and ovary[b]
IB	Non-aggressive histological types with invasion of half or more of the myometrium, and with no or focal LVSI[c]
IC	Aggressive histological types limited to a polyp or confined to the endometrium
II	Invasion of cervical stroma without extrauterine extension OR with substantial LVSI OR aggressive histological types with myometrial invasion
IIA	Invasion of the cervical stroma of non-aggressive histological types
IIB	Substantial LVSI [c]of non-aggressive histological types
IIC	Aggressive histological types with any myometrial involvement
III	Local and/or regional spread of the tumor of any histological subtype
IIIA	Invasion of uterine serosa, adnexa, or both by direct extension or metastasis
IIIA1	Spread to ovary or fallopian tube (except when meeting stage IA3 criteria)*b*
IIIA2	Involvement of uterine subserosa or spread through the uterine serosa
IIIB	Metastasis or direct spread to the vagina and/or to the parametria or pelvic peritoneum
IIIB1	Metastasis or direct spread to the vagina and/or the parametria
IIIB2	Metastasis to the pelvic peritoneum
IIIC	Metastasis to the pelvic or para-aortic lymph nodes or both[d]
IIIC1	Metastasis to the pelvic lymph nodes Micrometastasis
IIIC1i	
IIIC1ii	Macrometastasis
IIIC2	Metastasis to para-aortic lymph nodes up to the renal vessels, with or without metastasis to the pelvic lymph nodes
IIIC2i	Micrometastasis
IIIC2ii	Macrometastasis
IV	Spread to the bladder mucosa and/or intestinal mucosa and/or distance metastasis
IVA	Invasion of the bladder mucosa and/or the intestinal/bowel mucosa
IVB	Abdominal peritoneal metastasis beyond the pelvis
IVC	Distant metastasis, including metastasis to any extra- or intra-abdominal lymph nodes above the renal vessels, lungs, liver, brain, or bone

[a]Endometrial cancer is surgically staged and pathologically examined. In all stages, the grade of the lesion, the histological type and LVSI must be recorded. If available and feasible, molecular classification testing (*POLEmut*, MMRd, NSMP, p53abn) is encouraged in all patients with endometrial cancer for prognostic risk-group stratification and as factors that might influence adjuvant and systemic treatment decisions. Molecular subtype assignment can be done on a biopsy, in which case it need not be repeated on the hysterectomy specimen. When performed, these molecular classifications should be recorded in all stages.

[b]Low-grade EECs involving both the endometrium and the ovary are considered to have a good prognosis, and no adjuvant treatment is recommended if all the below criteria are met. Disease limited to low-grade endometrioid carcinomas involving the endometrium and ovaries (Stage IA3) must be distinguished from extensive spread of the endometrial carcinoma to the ovary (Stage IIIA1), by the following criteria: (1) no more than superficial myometrial invasion is present (<50%); (2) absence of extensive/substantial LVSI; (3) absence of additional metastases; and (4) the ovarian tumor is unilateral, limited to the ovary, without capsule invasion/rupture (equivalent to pT1a).

[c]LVSI as defined in WHO 2021: extensive/substantial, ≥5 vessels involved.

[d]Micrometastases are considered to be metastatic involvement (pN1 (mi)). The prognostic significance of isolated tumor cells (ITCs) is unclear. The presence of ITCs should be documented and is regarded as pN0(i+). According to TNM8, macrometastases are >2 mm in size, micrometastases are 0.2–2 mm and/or >200 cells, and isolated tumor cells are ≥0.2 mm and ≤200 cells.

(EEC: endometrioid carcinoma; LVSI: lymphovascular space involvement)

TABLE 2B: FIGO endometrial cancer stage with molecular classification.[a]

Stage designation	*Molecular findings in patients with early endometrial cancer (Stages I and II after surgical staging)*
Stage $IAm_{POLEmut}$	*POLEmut* endometrial carcinoma, confined to the uterine corpus or with cervical extension, regardless of the degree of LVSI or histological type
Stage $IICm_{p53abn}$	p53abn endometrial carcinoma confined to the uterine corpus with any myometrial invasion, with or without cervical invasion, and regardless of the degree of LVSI or histological type

[a]When the molecular classification is known:

- FIGO Stages I and II are based on surgical/anatomical and histological findings. In case the molecular classification reveals *POLEmut* or p53abn status, the FIGO stage is modified in the early stage of the disease. This is depicted in the FIGO stage by the addition of "m" for molecular classification, and a subscript is added to denote *POLEmut* or p53abn status, as shown below. MMRd or NSMP status do not modify early FIGO stages; however, these molecular classifications should be recorded for the purpose of data collection. When molecular classification reveals MMRd or NSMP, it should be recorded as Stage Im_{MMRd} or Stage Im_{NSMP} and Stage IIm_{MMRd} or Stage IIm_{NSMP}.
- FIGO Stages III and IV are based on surgical/anatomical findings. The stage category is not modified by molecular classification; however, the molecular classification should be recorded if known. When the molecular classification is known, it should be recorded as Stage IIIm or Stage IVm with the appropriate subscript for the purpose of data collection. For example, when molecular classification reveals p53abn, it should be recorded as Stage $IIIm_{p53abn}$ or Stage IVm_{p53abn}.

(LVSI: lymphovascular space involvement)

BOX 1: Histopathology [World Health Organization (WHO) 2020] classification.

- *Endometrioid carcinoma*: Adenocarcinoma; adenocarcinoma variants (with squamous differentiation; secretory variant; villoglandular variant; and ciliated cell variant)
  - Mucinous adenocarcinoma
  - Serous adenocarcinoma
  - Clear cell adenocarcinoma
  - Undifferentiated adenocarcinoma
  - Neuroendocrine tumors
  - Mixed carcinomas (carcinoma composed of more than one type with at least 10% of each component)
- Apart from the classification of endometrial carcinoma, carcinoma of the endometrium comprises mixed epithelial and mesenchymal tumors including: Adenomyoma, atypical polypoid adenomyoma, adenofibroma, and adenosarcoma
- *Carcinosarcoma*: Currently carcinosarcomas, in which both epithelial and mesenchymal components are malignant and aggressive tumors, are considered metaplastic carcinomas and are treated as aggressive carcinomas.

Note: The presence of unusual nuclear atypia in an architecturally low-grade tumor should prompt the evaluation of p53 and consideration of serous carcinoma. Adenocarcinomas with squamous differentiation are graded according to the microscopic features of the glandular component.

Histopathology grades:

1	GX: Grade cannot be assessed
2	G1: Well-differentiated
3	G2: Moderately differentiated*
4	G3: Poorly or undifferentiated

*Presence of nuclear atypia (pleomorphism and prominent nucleoli) raises the grade of a Grade 1 or Grade 2 tumor by 1.

Based on degree of differentiation:

G1	<5% of a nonsquamous or nonmorular solid growth pattern
G2	6–50% of a nonsquamous or nonmorular solid growth pattern
G3	>50% of a nonsquamous or nonmorular solid growth pattern

Traditionally Endometrial cancers were classified as type I or type II depending on the histological subtype. The type I cancers are more common, well differentiated and are as a result of unopposed estrogen stimulation and have a better prognosis. Type II or non-Endometrioid cancers include clear cell, serous, mucinous, squamous and adenosquamous which are estrogen independent, less differentiated, and associated with a poorer prognosis.[3] In 2023, FIGO staging has classified them into aggressive and non aggressive histology **(Box 2)**.

Molecular Classification[10]

The Cancer Genome Atlas (TCGA) did the molecular profiling of endometrial cancer which points toward a paradigm shift from morphological to molecular classification. The classification is based immunohistochemical markers for p53 and the mismatch repair proteins (MLH1, PMS2, MSH2, MSH6), and mutation analysis to detect pathogenic *POLE* mutations. 3% of endometrial cancers have so-called multiple classifying features, show more than one positive combination like p53-POLE, etc., and their clinicopathological outcomes depend on the better prognostic type **(Flowchart 1)**.

BOX 2: Aggressive and nonaggressive histologies (FIGO 2023 classification).

Nonaggressive histological types: low-grade (grade 1 and 2) EECs.
- Aggressive histological types
- High-grade EECs (grade 3), serous, clear cell, undifferentiated, mixed, mesonephric-like, gastrointestinal mucinous type carcinomas, and carcinosarcomas.

Note: It should be noted that high-grade EECs (grade 3) are a prognostically, clinically, and molecularly heterogenous disease, and the tumor type that benefits most from applying molecular classification for improved prognostication and for treatment decision-making. For practical purposes and to avoid undertreatment of patients, if the molecular classification is unknown, high-grade EECs were grouped together with the aggressive histological types in the actual FIGO classification.

Flowchart 1: The proactive molecular risk classifier for endometrial cancer (ProMiseE) algorithm molecular.

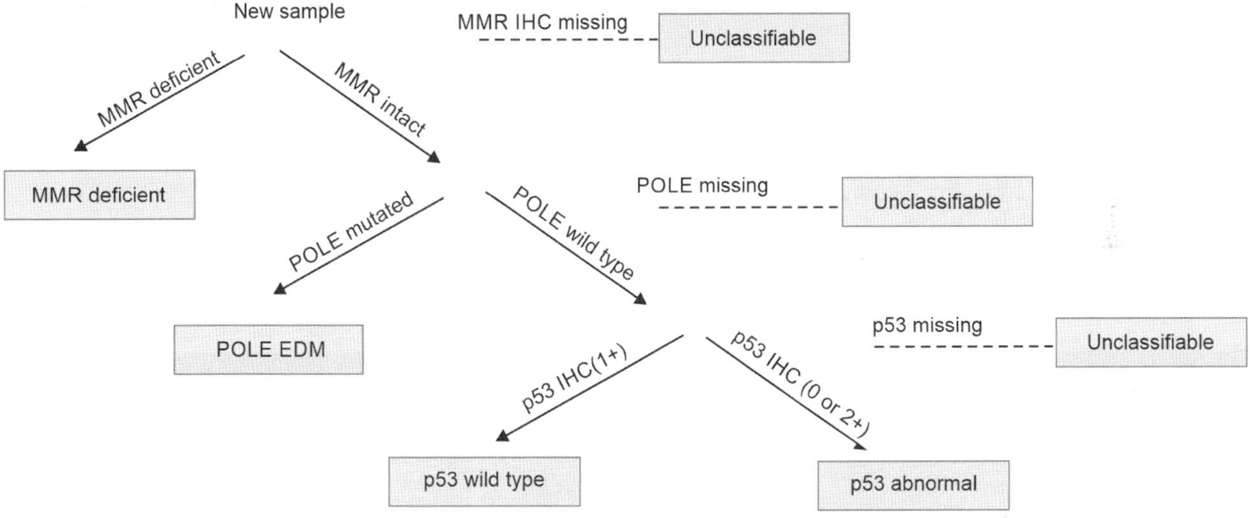

(IHC: immunohistochemistry; MMR: mismatch repair; POLE EDM: polymerase epsilon exonuclease domain mutated)

Four molecular groups (TGCA pragmatic molecular classifier nomenclature) is given in **Box 3**.

Prognostic Risk Groups According to ESGO–ESTRO Classification[11,12]

Table 3 shows the prognostic risk groups.

CLINICAL PRESENTATION

Postmenopausal bleeding and abnormal uterine bleeding including menstrual irregularities and intermenstrual bleeding are the most common symptoms of carcinoma endometrium. Mostly due to these symptoms diagnosis is made at an early stage. Patients with advanced disease may present with symptoms such as abdominal or pelvic pain, abdominal distension, early satiety and change in bowel bladder habits and metastatic symptoms.[13,14]

DIAGNOSIS[15]

Diagnosis of endometrial cancer is shown in **Flowchart 2**.

Endometrial aspiration should be the preferred for biopsy over the conventional dilation and curettage. In both postmenopausal and premenopausal women, endometrial aspiration biopsy with the Pipelle is the best method, with high detection rates of 99. 6% and 91%, respectively.[16]

Additional modalities:
- Serum CA-125 in advanced disease and serous carcinoma
- *Chest X-ray*: Diagnosis of pulmonary metastasis
- *Magnetic resonance imaging (MRI) pelvis*: Accurate means of assessing depth of myometrial invasion, cervical stromal involvement, parametrial spread, lymph node metastasis, and adnexal involvement
- *Contrast-enhanced computed tomography (CECT) abdomen and pelvis*: For advanced stage endometrial cancer and type 2 tumors; CT and MRI equivalent for nodal metastasis

BOX 3: Four molecular groups: The Cancer Genome Atlas (TGCA) pragmatic molecular classifier nomenclature of endometrial cancer (EC).

1. *Ultramutated POLE mutant endometrial carcinoma (4–12% EC):*
   a. Pathogenic POLE mutations
   b. Relatively younger women
   c. Good clinical outcomes
   d. Low reoccurrence rate

<1% disease specific mortality

2. *Hypermutated [mismatch repair deficient (MMRd)] endometrial carcinoma (23–36% EC):*
   a. Microsatellite instability
   b. Ten times higher mutations than MMR-stable patients
   c. 10% occur with Lynch syndrome with MMR germline defects
   d. Rest due to MLH1 mutations
   e. Intermediate stage-dependent prognosis
3. *Copy number high (p53 abnormal) (8–24% EC):*
   a. P53 mutations
   b. 20% of all endometrial carcinomas
   c. 50–70% disease-specific mortality
4. *Copy number low [no specific molecular profile (NSMP)] endometrial carcinoma (30–60%):*
   a. Tp53 wild-type and POLE wild-type MMR-proficient tumors with low somatic copy number alterations
   b. Diagnosis of exclusion
   c. Mostly endometroid carcinomas
   d. Intermediate to excellent stage-dependent and histopathology dependent prognosis

TABLE 3: Prognostic risk groups.

Risk group	*Molecular classification known*	*Molecular classification unknown*
Low	Stage IA endometrioid + low grade* + LVSI negative or focal	• Stage I–II POLEmut endometrial carcinoma, no residual disease • Stage IA MMRd/NSMP endometroid carcinoma + low grade* + LVSI negative or focal
Intermediate	• Stage IB endometroid + low grade* + LVSI negative or focal • Stage IA endometroid + high grade* + LVSI negative or focal • Stage IA nonendometrioid (serous, clear cell, undifferentiated) without myometrial invasion	• Stage IB MMRd/NSMP endometroid + low grade* + LVSI negative or focal • Stage IA MMRd/NSMP endometroid + high- grade* + LVSI negative or focal • Stage IA p53 abnormal and/or nonendometroid (serous, clear cell, undifferentiated) without myometrial invasion
High-to-intermediate	• Stage I endometroid + substantial LVSI regardless grade and depth of invasion • Stage IB endometroid high grade* regardless of LVSI status • Stage II	• Stage I MMRd/NSMP endometroid + substantial LVSI regardless grade and depth of invasion • Stage IB MMRd/NSMP endometroid high grade* regardless of LVSI status • Stage II MMRd/NSMP endometroid
High	• Stage III-IVA with no residual disease • Stage I-IVA nonendometroid (serous, clear cell, undifferentiated) with myometrial invasion and no residual disease	• Stage III-IVA MMRd/NSMP with no residual disease • Stage I–IVA p53 abnormal nonendometrial carcinoma with myometrial invasion and no residual disease • Stage I–IVA MMRd/NSMP nonendometrial carcinoma with myometrial invasion and no residual disease
Advanced metastatic	• Stage III–IVA with residual disease • Stage IVB	• Stage III–IVA with residual disease of any molecular type • Stage IVB of any molecular type

*Grades 1 and 2 are low-risk carcinomas and grade 3 are high-grade carcinomas
(LVSI: lymphovascular space invasion; MMRd: mismatch repair deficient; NSMP: nonspecific molecular profile; POLEmut: polymerase mutated)

Flowchart 2: Diagnosis of endometrial cancer.

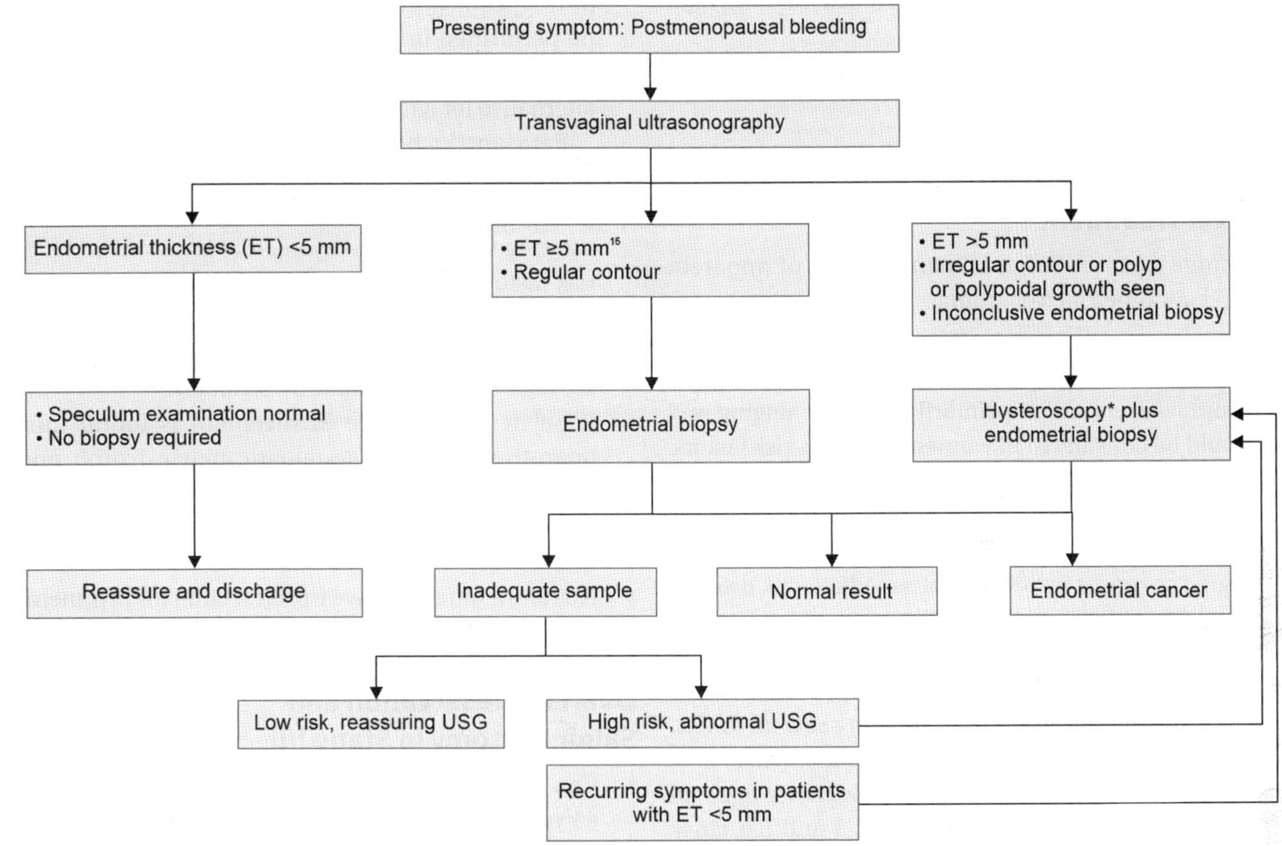

Presenting symptom: Postmenopausal bleeding
Transvaginal ultrasonography
Endometrial thickness (ET) <5 mm
• ET ≥5 mm[15]
• Regular contour
• ET >5 mm
• Irregular contour or polyp or polypoidal growth seen
• Inconclusive endometrial biopsy
• Speculum examination normal
• No biopsy required
Endometrial biopsy
Hysteroscopy* plus endometrial biopsy
Reassure and discharge
Inadequate sample
Normal result
Endometrial cancer
Low risk, reassuring USG
High risk, abnormal USG
Recurring symptoms in patients with ET <5 mm

* Hysteroscopy should be done at low pressure to prevent tumor dissipation through fallopian tubes.

Presenting symptom:
Abnormal uterine bleeding (AUB)
Transvaginal ultrasonography
Normal
ET postmenstrual on USG >12 mm[15]
Suspected structural anomaly
Endometrial tissue sampling:
• In women >40 years
• In women <40 years with high risk for carcinoma endometrium: Irregular bleeding, obesity associated with hypertension, PCOS, diabetes, ET >12 mm, family H/o carcinoma breast/ovary/endometrium/colon, tamoxifen use for HRT/carcinoma breast, HNPCC, late menopause

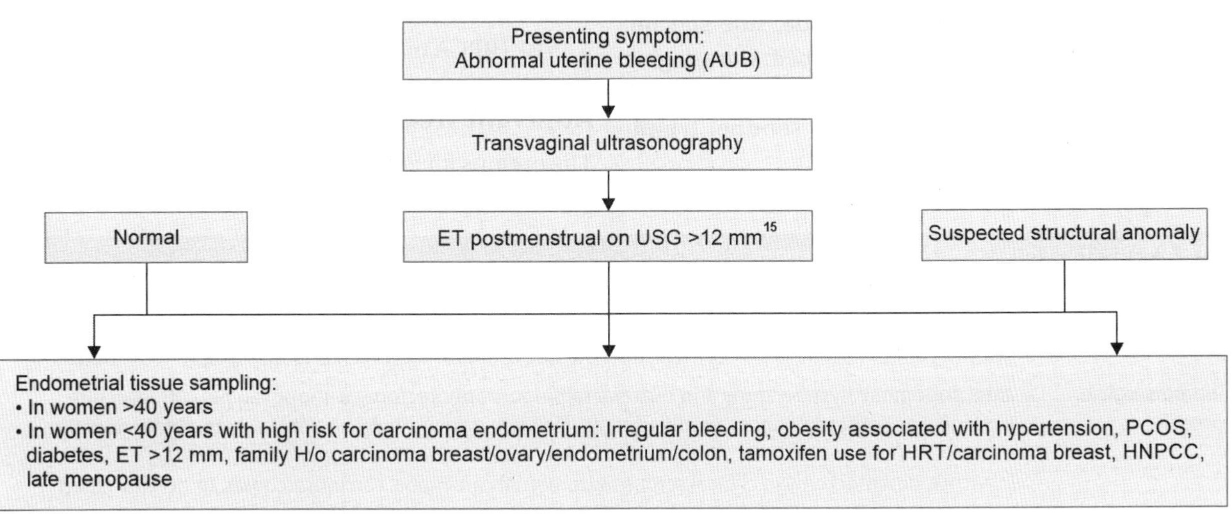

(H/o: history of; HNPCC: hereditary nonpolyposis colorectal cancer; HRT: hormone replacement therapy; PCOS: polycystic ovary syndrome; USG: ultrasonography)

- *Positron emission tomography-computed tomography (PET-CT)*: In high risk or advanced disease for lymph node and distant metastasis
- *Cystoscopy/proctoscopy*: If extension to bladder or rectum suspected

TREATMENT STAGE 1/2[17]

Surgical Treatment

Early-stage disease: Surgical management of apparent stage I/II endometrial carcinomas is given in **Table 4**.

Lymph Node Evaluation

- Lymphadenectomy is primarily used for staging and should be considered in women with high-risk factors. The therapeutic role of complete lymphadenectomy is controversial
- Sentinel lymph node biopsy can be considered for staging purposes in low and intermediate-risk disease. It can be omitted in cases without myometrial invasion. Systemic lymphadenectomy is not recommended in this group. Sentinel lymph node biopsy should be done by a surgeon who has experience in sentinel nodes. Dyes used are indocyanine green and methylene blue in combination with radioisotope which are injected superficially and deep in cervix at 3 and 9'o clock positions. The entire retroperitoneal spaces must be evaluated and suspicious lymph nodes should be removed regardless of mapping. Frozen section analysis may influence the decision to perform para-aortic lymphadenectomy in some cases. In case it is sentinel node negative on one side, side-specific lymphadenectomy should be performed. Ultrastaging of sentinel nodes should be done by pathologist. Overall detection rate is 96% with sentinel lymph nodes and 73% are bilateral.[18]
- Surgical lymph node staging (systematic pelvic and para-aortic (infrarenal) lymphadenectomy) should be performed in high-to-intermediate and high-risk disease. Sentinel lymph node biopsy is an acceptable alternative for lymph node staging in stage I/II.

  Para-aortic lymphadenectomy is indicated in suspicious aortic or common iliac nodes, grossly positive adnexa, grossly positive pelvic nodes, high-grade tumors showing full-thickness myometrial invasion, and high-risk histology including clear cell, papillary serous, or carcinosarcoma histologic subtypes.[19]
- Both macrometastasis and micrometastasis (<2 mm) are regarded as metastatic involvement.
- If pelvic lymph node involvement is found intra-operatively, further systematic pelvic lymph node dissection should be omitted. However, debulking of enlarged lymph nodes and para-aortic staging can be considered.
- No role of intraoperative frozen section for myometrial invasion.

Ovarian Preservation and Salpingectomy in Stage I/II

- Can be considered in premenopausal patients aged 45 years with low-grade endometroid carcinoma with <50% myometrial invasion and no obvious ovarian/extrauterine disease
- In cases of ovarian preservation, prophylactic salpingectomy is done.
- Ovarian preservation is not recommended in patients with positive family history involving ovarian cancer risk (BRCA mutation, Lynch syndrome)

Adjuvant Treatment[20-23]

The options for adjuvant therapy are given in **Table 5**.

TABLE 4: Early-stage disease surgical management.

Minimally invasive surgery (MIS)	• Preferred approach including patients with high-risk endometrial carcinomas • Avoid intraperitoneal tumor spillage including tumor rupture/morcellation even in endobag • *Relative contraindications of MIS*: Metastasis outside uterus and cervix (excluding lymph node metastasis)
Standard surgical approach	• Total abdominal hysterectomy and bilateral salpingo-oophorectomy without vaginal cuff resection • In stage II endometrial cancer, modified radical hysterectomy should be done only to obtain negative margins • Infracolic omentectomy in clinical serous endometrial carcinoma, carcinosarcomas, undifferentiated carcinoma; can be omitted in endometroid and clear cell stage I disease • Surgical restaging can be considered in previously incompletely staged patients with high-to-intermediate and high-risk disease if the outcome may imply for adjuvant treatment

TABLE 5: Options for adjuvant therapy.

Radiation therapy	• *External beam radiation therapy (EBRT):* Dose 45–50.4 Gy/25–28 mo/5–5.3 weeks, cobalt-60 used, target: pelvic lymph nodes and upper two-thirds of vagina • *Brachytherapy:* Vaginal cuff brachytherapy dose with EBRT: 6 Gy per cycle/2 Fx VCB/1 week apart, dose without EBRT: 7 Gy per cycle/3Fx VCB/1 week apart, iridium-192 used
Systemic chemotherapy	• *Single agent:* Paclitaxel, cisplatin, carboplatin, 5 FU, doxorubicin • *Combination:* Paclitaxel + carboplatin, paclitaxel + cisplatin + doxorubicin
Hormonal therapy	• High-dose progestins • Medroxyprogesterone acetate 100 mg TDS • Desogestrel acetate 80 mg BD • MPA + tamoxifen 20 mg OD or aromatase inhibitors (letrozole)
Biological therapy	• *Pembrolizumab:* In PD-L1+, MMR deficient tumors • *Serous cancers HER2-neu positive:* Trastuzumab • *Serous tumors HRD deficient:* Role of PARP inhibition investigated

(HRD: homologous recombination deficiency; PARP: Poly (adenosine diphosphate-ribose) polymerase)

TABLE 6: Adjuvant treatment recommendations in stage I/II disease.

Low risk	***No adjuvant treatment is recommended***
Intermediate risk	• Adjuvant brachytherapy (BT) to reduce vaginal reoccurrence can be considered • Omission of BT can be considered specially for patients 60 years of age
High-to-intermediate risk At least two of the factors: age >60 years, deep myometrial invasion, grade 3, serous or clear cell histology, and LVSI	• Adjuvant BT to reduce vaginal reoccurrence (BT) *preferred* • EBRT can be considered for substantial LVSI and for stage II • Adjuvant chemotherapy can be considered for substantial LVSI and/or high grade • Addition of chemotherapy to EBRT shows benefit in PFS and OS[20]
High risk	• Pelvic EBRT with concurrent and/or adjuvant chemotherapy • Addition of chemotherapy to EBRT shows benefit in PFS and OS especially in serous cancers and advanced stage[21] • Sequential chemotherapy and radiotherapy recommended • Chemotherapy alone is an alternative option

(EBRT: external beam radiation therapy; LVSI: lymphovascular space invasion; OS: overall survival; PFS: progression-free survival)

The adjuvant treatment recommendations in stage I/II disease are given in **Table 6**.

ADVANCED DISEASE: STAGE III/IV[22]

Advanced[d] disease—stage III/IV is shown in **Flowchart 3**.

RECURRENT DISEASE[24,25]

Flowchart 4 shows recurrent disease.

FERTILITY PRESERVATION, MANAGEMENT, AND FOLLOW-UP[26-31]

- Fertility preservation can be considered only in patients with strong desire to preserve fertility and with atypical endometrial hyperplasia (AEH)/endometrial carcinoma in situ [endometrioid intraepithelial neoplasia (EIN)] or grade 1 endometrioid endometrial carcinoma without myometrial invasion and without genetic risk factors.
- Patient must be willing for close follow-up and need for future hysterectomy after family completion or failure of treatment.
- Radiologic imaging (MRI pelvis) must be done to access the extension of disease.
- The management and follow-up for fertility preservation are shown in **Flowchart 5**.

FOLLOW-UP

Follow-up is given in **Table 7**.

- On each visit, patient history to be taken to rule out any symptoms of vaginal bleeding or brown discharge, change in bowel or bladder habit, development of

Flowchart 3: Advanced disease—stage III/IV.

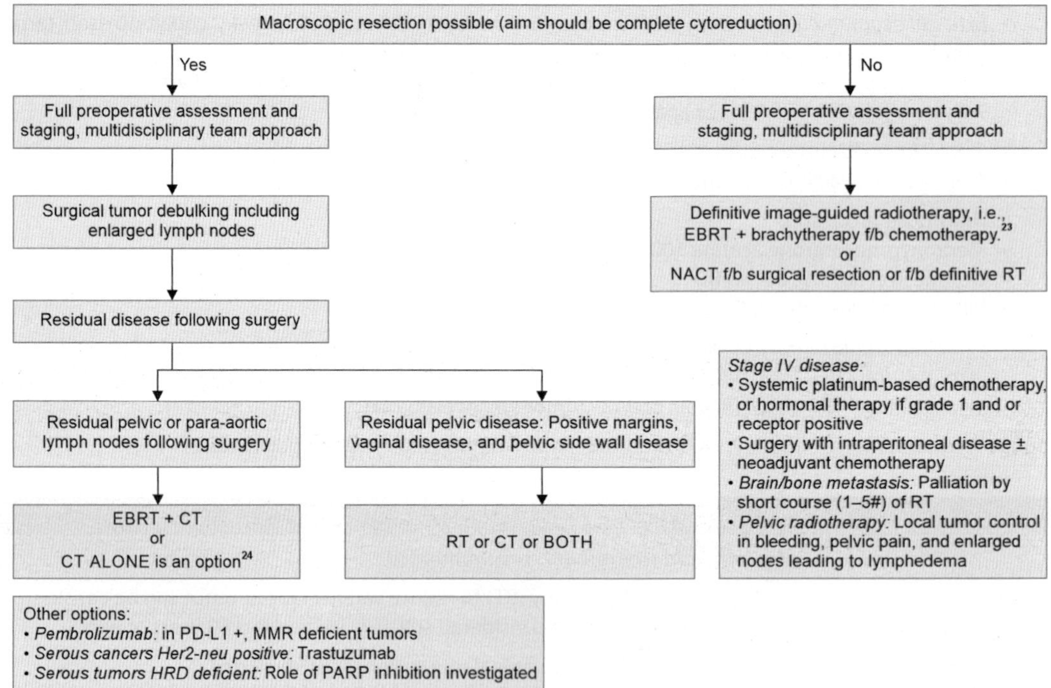

(CT: computed tomography; EBRT: external beam radiation therapy; f/b: followed by; RT: radiotherapy; PARP: Poly(ADP-Ribose) polymerase ; RT: radiotherapy)

Flowchart 4: Recurrent disease.

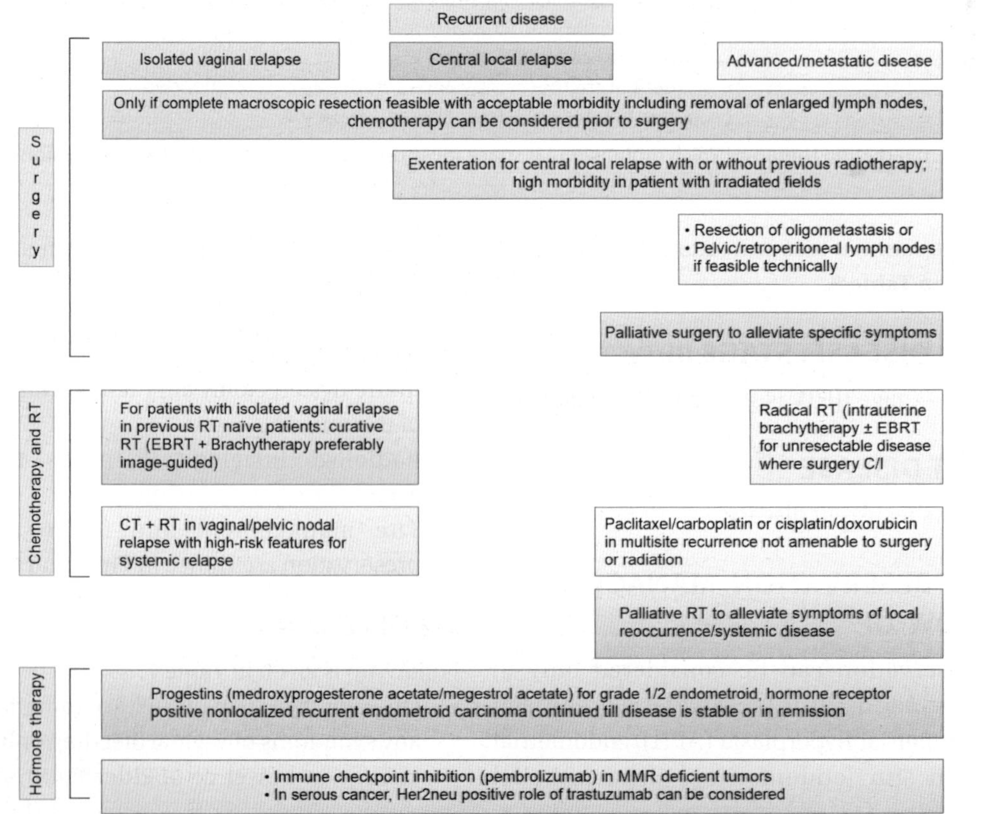

(C/I: Contraindicated; EBRT: external beam radiation therapy; MMR: mismatch repair; RT: radiotherapy)

Flowchart 5: Management and follow-up for fertility preservation.

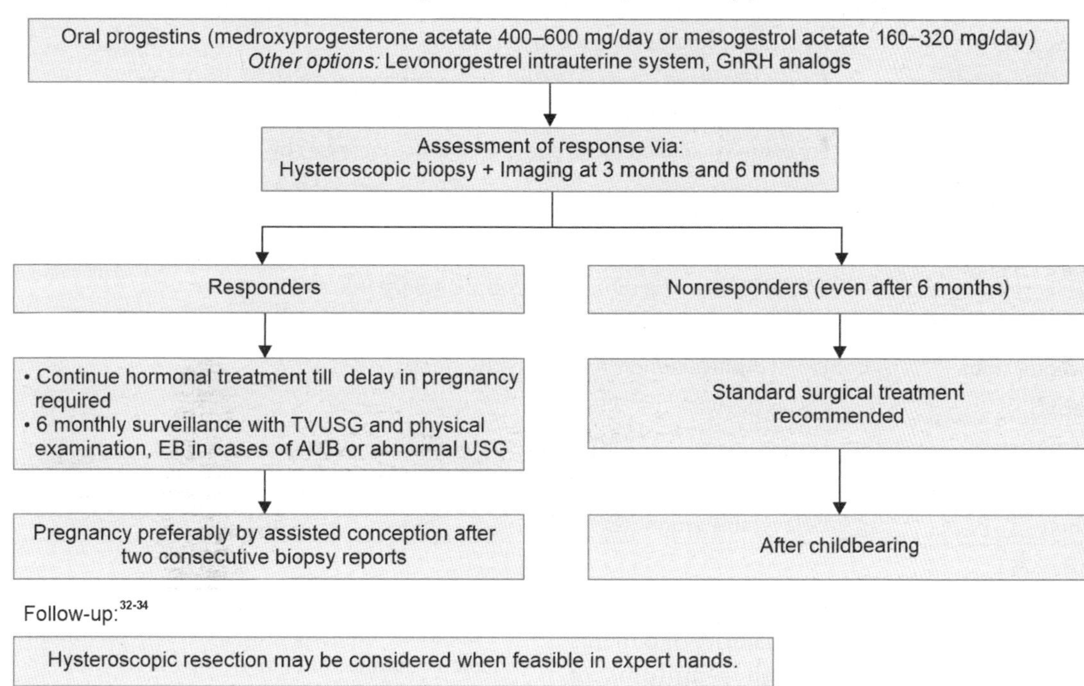

(AUB: abnormal uterine bleeding; EB: endometrial biopsy; GnRH: gonadotropin-releasing hormone; TVUSG: transvaginal ultrasonography)

TABLE 7: Follow-up.

Physical examination including: • Detailed history • General physical examination • Per abdomen examination • Per speculum examination • Bimanual examination • Rectovaginal examination	• *First 2–3 years*: Every 3–6 monthly • *The next 2–3 years (up to 5 years)*: Every 6 monthly • After 5 years: Annually • Stage 1A and 1B endometrioid adenocarcinoma (low risk) can be followed 6 monthly for 5 years and then can be discharged from follow-up • Patients with stage 1A confined to the endometrium/no myometrial invasion to be discharged from postoperation visit with no need for regular follow-up. • All other endometrial cancer patients (advanced stages, high-grade histology/high risk) to be followed by a gynecologic oncologist every 3 months for years 1 and 2, then every 6 months for a total of 5 years and then be discharged if no relapse[32]
Other surveillance tests	• Imaging tests are ordered if needed based on symptoms or suspicion of reoccurrence on examinations: CECT abdomen, pelvis, chest/PET-CT whole body/MRI abdomen, and pelvis • If CA-125 was high before treatment, then may be measured as a part of surveillance testing[33]

(CECT: contrast-enhanced computed tomography; MRI: magnetic resonance imaging; PET: positron emission tomography)

abdominal distension, persistent cough or shortness of breath, weight loss, lethargy, pain, or headaches.

- Follow-up examination includes focused history (for above symptoms), physical examination, including pelvic examination (speculum and bimanual and/or rectovaginal exam. This alone will detect relapse >80% of the time.[34]
- Imaging or blood investigations are not necessary unless symptoms suggestive of relapse are present.
- Pap smear is not indicated as it offers no benefit in detecting relapse.[3]

PROGNOSIS

Table 8 shows the prognostic factors.[35-38]

- The 5-year survival rates exceed 95% for stage 1 carcinoma endometrium and approaches 83% overall. The relapse rates for patients with early-stage disease range from 2% to 15%.[37,38]

TABLE 8: Prognostic factors.

Stage of the disease	*Most important prognostic factor*
Histopathological grade and type	*Nonendometrioid subtypes:* Poorer prognosis and survival: 33–58% only *Endometrioid subtypes:* 92% survival rates
Age	For every 1 year increase in age reoccurrence increases by 7%
Myometrial invasion	Distance of tumor myometrial junction from serosa is the most sensitive indicator of effect of myometrial invasion on survival *Deep myometrial invasion:* Higher risk of reoccurrence
Lymphovascular space invasion	Proportional to the tumor grade, reoccurrence risk, and mortality
Tumor size	<2 cm: Better 5-year survival[36]
Hormone receptor status	ER, PR positive tumors have better prognosis
Her2-neu status	Lesser survival
Lymph node metastasis	Lesser survival rates and more reoccurrences
Intraperitoneal metastasis	Lesser survival and more chances of reoccurrence
K-RAS mutation	Unfavorable prognosis

- The vast majority of patients who relapse (70–100%) do so within the first 3 years. The majority of patients (>75%) who relapse present with symptoms of recurrence between scheduled follow-up visits. 25% of patients with reoccurrence remain asymptomatic, but there is no overall survival difference with symptomatic reoccurrences, except in the rare case of localized recurrence.[39] Additional molecular markers associated with poor survical includes L1CAM, p53 positivity.

CONCLUSION

Endometrial cancer has undergone major changes due to the advent of molecular classification and revised staging. This has not only provided stronger prognostic significance but has also helped to guide adjuvant treatment.

REFERENCES

1. Sung H, Ferlay J, Siegel RL, Laversanne M, Soerjomataram I, Jemal A, et al. Global Cancer Statistics 2020: GLOBOCAN Estimates of Incidence and Mortality Worldwide for 36 Cancers in 185 Countries. CA Cancer J Clin. 2021;71(3):209-49.
2. Renehan AG, Tyson M, Egger M, Heller RF, Zwahlen M. Body-mass index and incidence of cancer: a systematic review and meta-analysis of prospective observational studies. Lancet. 2008;371:569-78.
3. Ellulu MS, Patimah I, Khaza'ai H, Rahmat A, Abed Y. Obesity and inflammation: the linking mechanism and the complications. Arch Med Sci. 2017;13(4):851-63.
4. An CK. Selective estrogen receptor modulators. Asian Spine J. 2016;4:787-91.
5. Zhou B, Yang L, Sun Q, Cong R, Gu H, Tang N, et al. Cigarette smoking and the risk of endometrial cancer: a meta-analysis. Am J Med. 2008;121:501-8.
6. Gammon A, Jasperson K, Champine M. Genetic basis of Cowden syndrome and its implications for clinical practice and risk management. Appl Clin Genet. 2016;9:83-92.
7. Post CCB, Stelloo E, Smit VTHBM, Ruano D, Tops CM, Vermij L, et al. Prevalence and prognosis of lynch syndrome and sporadic mismatch repair deficiency in endometrial cancer. J Natl Cancer Inst. 2021;113(9):1212-20.
8. Berek JS, Matias-Guiu X, Creutzberg C, et al; Endometrial Cancer Staging Subcommittee, FIGO Women's Cancer Committee. FIGO staging of endometrial cancer: 2023. Int J Gynaecol Obstet. 2023;162(2):383-394. Erratum in: Int J Gynaecol Obstet. 2023 6.
9. Watkins JC, Downing MJ, Crous-Bou M, Busch EL, Chen M, De Vivo I, et al. Endometrial tumor classification by histomorphology and biomarkers in the Nurses' Health Study. J Cancer Epidemiol. 2021;2021:8884364.
10. Alexa M, Hasenburg A, Battista MJ. The TCGA Molecular Classification of Endometrial Cancer and Its Possible Impact on Adjuvant Treatment Decisions. Cancers (Basel). 2021;13(6):1478.
11. Imboden S, Nastic D, Ghaderi M, Rydberg F, Siegenthaler F, Mueller MD, et al. Implementation of the 2021 molecular ESGO/ESTRO/ESP risk groups in endometrial cancer. Gynecol Oncol. 2021;162:394-400.
12. Coll-de la Rubia E, Martinez-Garcia E, Dittmar G, Gil-Moreno A, Cabrera S, Colas E. Prognostic biomarkers in endometrial cancer: A systematic review and meta-analysis. J Clin Med. 2020;9:1900.
13. Sung S, Carlson K, Abramovitz A. Postmenopausal Bleeding. In: StatPearls [Internet]. Treasure Island (FL): StatPearls Publishing; 2024.

14. Khafaga A, Goldstein SR. Abnormal uterine bleeding. Obstet Gynecol Clin North Am. 2019;46:595-605.
15. Chopra K, Agrawal S. (2019). AUB: Evaluation and Management. AOGD Bulletin 15. [online] Available from https://aogd.org/AOGD-Bulletin-September-2019.pdf [Last accessed April, 2024].
16. Abdelazim IA, Abdelrazak KM, Elbiaa AA, Al-Kadi M, Yehia AH. Accuracy of endometrial sampling compared to conventional dilatation and curettage in women with abnormal uterine bleeding. Arch Gynecol Obstet. 2015;291(5):1121-6.
17. Concin N, Matias-Guiu X, Vergote I, Cibula D, Mirza MR, Marnitz S, et al. ESGO/ESTRO/ESP guidelines for the management of patients with endometrial carcinoma. Int J Gynecol Cancer. 2021;31(1):12-39.
18. Wang L, Liu F. Meta-analysis of laparoscopy sentinel lymph node mapping in endometrial cancer. Arch Gynecol Obstet. 2018;298(3):505-10.
19. Bhatla N, Denny L. FIGO cancer report 2018. Int J Gynecol Obstet. 2018;143:2-3.
20. Hogberg T, Signorelli M, de Oliveira CF, Fossati R, Lissoni AA, Sorbe B, et al. Sequential adjuvant chemotherapy and radiotherapy in endometrial cancer–results from two randomised studies. Eur J Cancer. 2010;46(13):2422-31.
21. León-Castillo A, de Boer SM, Powell ME, Mileshkin LR, Mackay HJ, Leary A, et al; TransPORTEC consortium. Molecular classification of the PORTEC-3 trial for high-risk endometrial cancer: Impact on prognosis and benefit from adjuvant therapy. J Clin Oncol. 2020;38(29):3388.
22. van den Heerik ASVM, Horeweg N, de Boer SM, Bosse T, Creutzberg CL. Adjuvant therapy for endometrial cancer in the era of molecular classification: radiotherapy, chemoradiation and novel targets for therapy. Int J Gynecol Cancer. 2021;31(4):594-604.
23. Emons G, Tempfer C, Battista MJ, Mustea A, Vordermark D. Statement of the Uterus Committee of the Gynaecological Oncology Working Group (AGO) on the PORTEC-3 study. Geburtshilfe Frauenheilkd. 2018;78(10):923-6.
24. Rütten H, Verhoef C, van Weelden WJ, Smits A, Dhanis J, Ottevanger N, et al. Recurrent endometrial cancer: local and systemic treatment options. Cancers (Basel). 2021;13(24):6275.
25. Connor EV, Rose PG. Management strategies for recurrent endometrial cancer. Expert Rev Anticancer Ther. 2018;18:873-85.
26. Gunderson CC, Fader AN, Carson KA, Bristow RE. Oncologic and reproductive outcomes with progestin therapy in women with endometrial hyperplasia and grade 1 adenocarcinoma: a systematic review. Gynecol Oncol. 2012;125:477-82.
27. Britton H, Huang L, Lum A, Leung S, Shum K, Kale M, et al. Molecular classification defines outcomes and opportunities in young women with endometrial carcinoma. Gynecol Oncol. 2019;153:487-95.
28. Derbyshire AE, Ryan N, Crosbie EJ. Biomarkers needed to predict progestin response in endometrial cancer. BJOG. 2017;124:1584.
29. Barr CE, Ryan NAJ, Derbyshire AE, Wan YL, MacKintosh ML, McVey RJ, et al. Weight loss during intrauterine progestin treatment for obesity-associated atypical hyperplasia and early-stage cancer of the endometrium. Cancer Prev Res (Phila). 2021;14:1041-50.
30. Janda M, Robledo KP, Gebski V, Armes JE, Alizart M, Cummings M, et al. Complete pathological response following levonorgestrel intrauterine device in clinically stage 1 endometrial adenocarcinoma: Results of a randomized clinical trial. Gynecol Oncol. 2021;161: 143-51.
31. Kim MJ, Choe SA, Kim MK, Yun BS, Seong SJ, Kim YS. Outcomes of in vitro fertilization cycles following fertility-sparing treatment in stage IA endometrial cancer. Arch Gynecol Obstet. 2019;300:975-80.
32. Febbraro T, Robison K, Wilbur JS, Laprise J, Bregar A, Lopes V, et al. Adherence patterns to National Comprehensive Cancer Network (NCCN) guidelines for referral to cancer genetic professionals. Gynecol Oncol. 2015;138:109-14.
33. Salani R, Khanna N, Frimer M, Bristow RE, Chen LM. An update on post-treatment surveillance and diagnosis of relapse in women with gynecologic malignancies: Society of Gynecologic Oncology (SGO) recommendations. Gynecol Oncol. 2017;146:3-10.
34. Salani R, Pasinetti B, Carrara L, Gambino A, Odicino F, Pecorelli S. Pattern of failure and value of follow-up procedures in endometrial and cervical cancer patients. Gynecol Oncol. 2007;107:S241-7.
35. Amant F, Moerman P, Neven P, Timmerman D, Van Limbergen E, Vergote I. Endometrial cancer. Lancet. 2005;366(9484):491-505.
36. Lucic N, Draganovic D, Sibincic S, Ecim-Zlojutro V, Milicevic S. Myometrium invasion, tumour size and lymphovascular invasion as a prognostic factor in dissemination of pelvic lymphatics at endometrial carcinoma. Med Arch. 2017;71:325-9.
37. Bokhman JV. Two pathogenetic types of endometrial carcinoma. Gynecol Oncol. 1983;15(1):10-7.
38. Uharcek P. Prognostic factors in endometrial carcinoma. J Obstet Gynaecol Res. 2008;34:776-83.
39. Salani R, Nagel CI, Drennen E, Bristow RE. Recurrence pattern and surveillance for patients with early stage endometrial cancer. Gynecol Oncol. 2011;123:205-7.

Chapter

59

Gestational Trophoblastic Disease

Meenakshi B Chauhan, Aastha Chauhan

INTRODUCTION

Gestational trophoblastic disease is a heterogeneous group of interrelated lesions that arise from abnormal proliferation of placental trophoblasts during pregnancy.

CLASSIFICATION[1]

The classification of gestational trophoblastic disease is given in **Flowchart 1**.

EPIDEMIOLOGY

- Incidence is higher in Asian countries. It is 0.57–2 per 1,000 pregnancies.[2]
- Incidence is high in women in extremes of age.[3,4]
- History of previous hydatidiform mole.[4]

Risk factors for partial mole:[5,6]

- Use of oral contraceptive
- History of irregular menstruation

PATHOLOGY

The pathology of gestational trophoblastic disease is given in **Box 1**.

Recent development:

- Flow cytometry (diploidy or triploidy)
- Biomarkers—p57.

The difference between partial mole and complete mole is given in **Table 1**.

Flowchart 1: Classification of gestational trophoblastic disease (GTD).

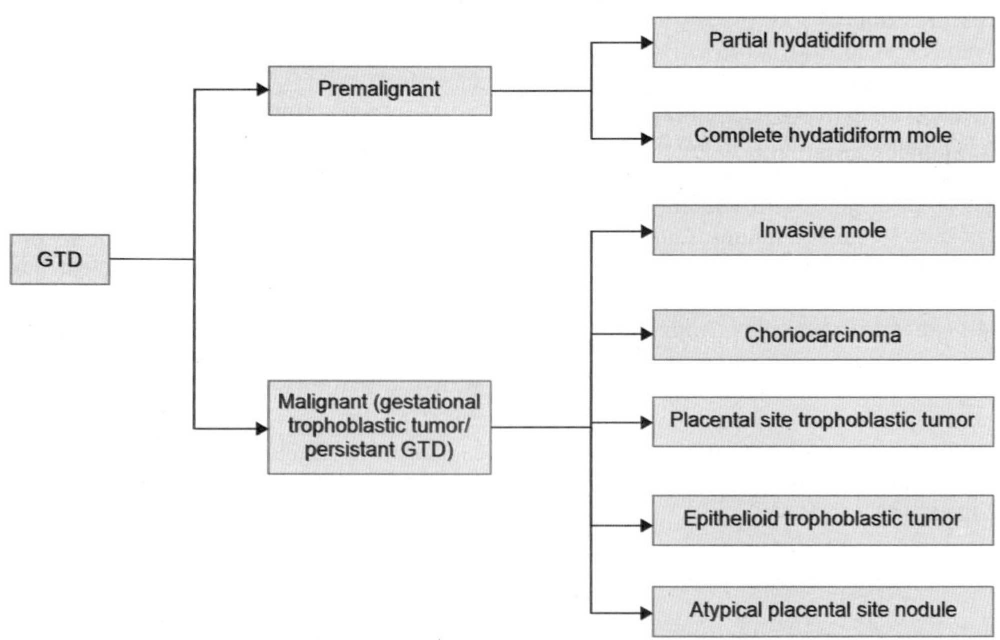

CLINICAL FEATURES

Common signs and symptoms are given in **Box 2.**

The less common clinical features are given in **Box 3.**[9]

INVESTIGATIONS

The investigations regarding gestational trophoblastic disease are given in **Box 4.**

MANAGEMENT OF GESTATIONAL TROPHOBLASTIC DISEASE

The management of gestational trophoblastic disease is shown in **Flowchart 2.**

BOX 1: Pathology of gestational trophoblastic disease.

- *Abnormal pregnancies:*
  - Benign neoplasia with malignant potential
  - Associated with an absent or abnormal fetus
- *Characterized histologically by:*
  - Trophoblastic proliferation (both syncytiotrophoblast and cytotrophoblast)
  - Edema of the villous stroma (hydropic change)

COMPLICATIONS OF GESTATIONAL TROPHOBLASTIC DISEASE

The complications of gestational trophoblastic disease are given in **Box 5.**

CHEMOTHERAPY IN GESTATIONAL TROPHOBLASTIC DISEASE[10]

Controversial role as prophylaxis.

Associated with reduction of post-molar gestational trophoblastic neoplasia (GTN) by 3–8%.

However, drugs used are toxic and may be avoided.[11]

Indications:

- When patient is not compliant in follow-up
- When risk of development of post-molar GTN is high.

The risk scoring system for the prediction of GTN in women with molar pregnancy is shown in **Table 2.**[11]

GESTATIONAL TROPHOBLASTIC NEOPLASMS/PERSISTENT GTD

Nonmetastatic diseases:

- Invasive mole
- Placental site trophoblastic tumor

TABLE 1: Difference between partial mole and complete mole.

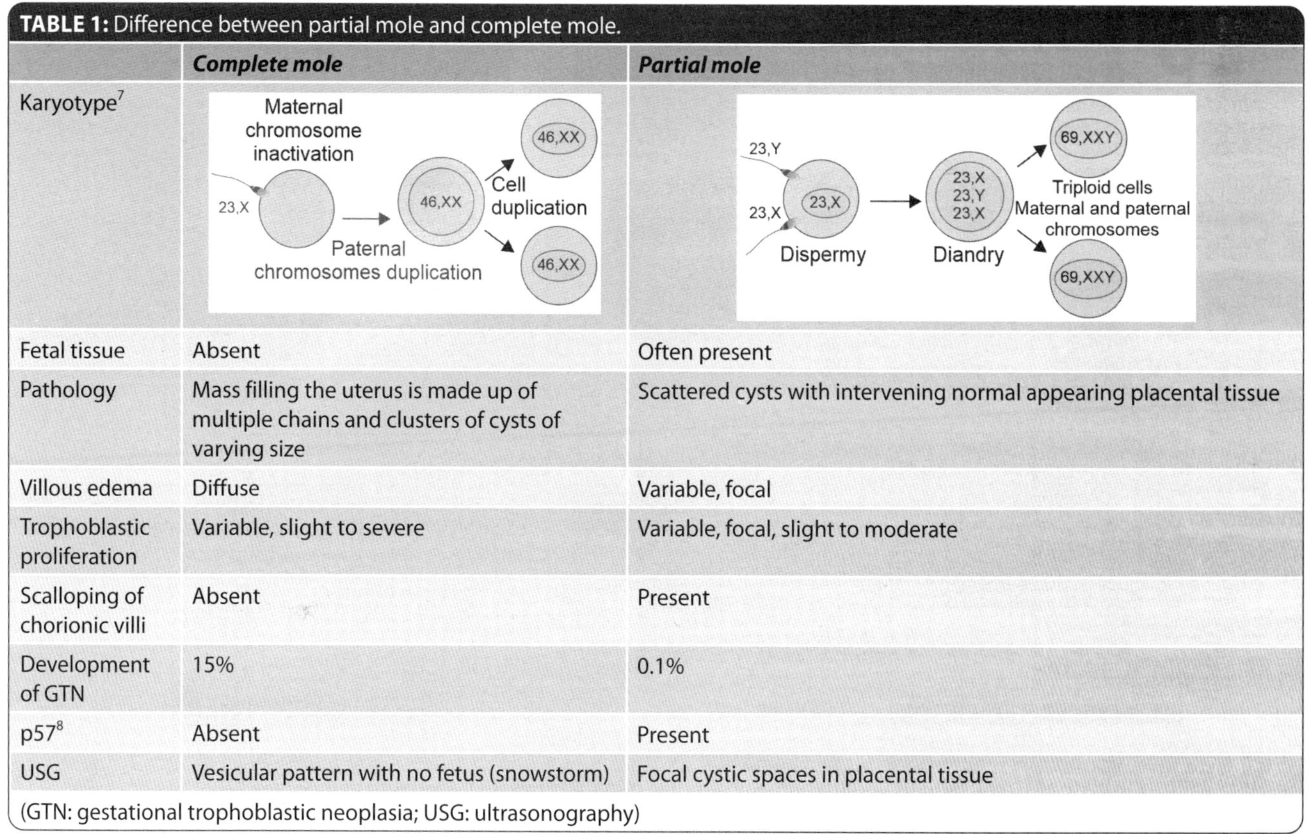

	Complete mole	*Partial mole*
Karyotype[7]		
Fetal tissue	Absent	Often present
Pathology	Mass filling the uterus is made up of multiple chains and clusters of cysts of varying size	Scattered cysts with intervening normal appearing placental tissue
Villous edema	Diffuse	Variable, focal
Trophoblastic proliferation	Variable, slight to severe	Variable, focal, slight to moderate
Scalloping of chorionic villi	Absent	Present
Development of GTN	15%	0.1%
p57[8]	Absent	Present
USG	Vesicular pattern with no fetus (snowstorm)	Focal cystic spaces in placental tissue

(GTN: gestational trophoblastic neoplasia; USG: ultrasonography)

BOX 2: Common signs and symptoms.

Signs:

- *General physical examination:*
  - Pallor
  - Raised blood pressure
- *Per abdomen:*
  - Uterus size more than gestational age
  - Consistency—doughy
  - Absence of fetal parts on palpation

Symptoms:

- Amenorrhea
- Vaginal bleeding—most common:
  - Spotting to profuse bleeding
  - Passage of vesicles per vaginum
  - Irregular bleeding
- Pain in abdomen

BOX 3: Other clinical features.

- Hyperemesis
- Hyperthyroidism due to thyrotropin like effect of hCG (human chorionic gonadotropin)
- Early-onset preeclampsia
- Abdominal distension (theca lutein cyst)

BOX 4: Investigations of gestational trophoblastic disease.

- Complete hemogram
- Blood group and Rh typing
- Thyroid function test
- Beta-hCG (beta-human chorionic gonadotropin)
- *Ultrasound:*
  - Complete mole: Multiple echoes(holes) within the placental mass and no fetus, characteristic snowstorm appearance
  - Partial mole: Focal cystic spaces within placenta and an increase in transverse diameter of gestational sac
  - Theca leutin cysts in the ovaries

Flowchart 2: Management of gestational trophoblastic disease.

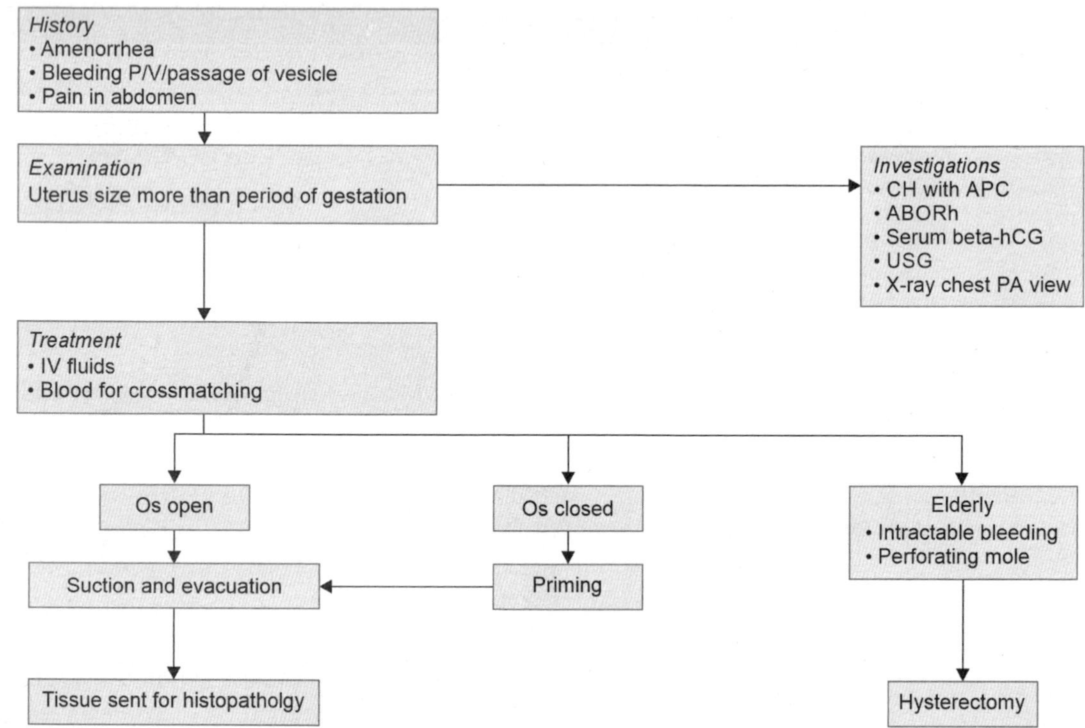

(APC: absolute platelet count; Beta-hCG: beta-human chorionic gonadotropin; CH: complete hemogram; IV: intravenous; PA: posteroanterior; P/V: per vaginum; USG: ultrasonography)

BOX 5: Complications of gestational trophoblastic disease.

Early:
- Excessive bleeding and shock
- Preeclampsia/eclampsia
- Embolization of the trophoblastic cells
- Uterine perforation may occur during evacuation
- Infection
- Disseminated intravascular coagulation (DIC)
- Thyroid storm

Late:
Malignant transformation

TABLE 2: Risk scoring system for the prediction of gestational trophoblastic neoplasia (GTN) in women with molar pregnancy.

Prognostic factor	*Score*			
	0	*1*	*2*	*3*
USG diagnosis	Partial	Complete	Recurrent	
Uterine size for GA (months)	Not >1	>1	>2	>3
hCG level (mIU/mL)	<50,000	>50,000 to <100,000	>100,000 to <1,000,000	>1,000,000
Diameter of theca lutein cysts (cm)	–	<6	>6 to <10	>10
Patient age (years)	–	<20	≥40	>50
Medical complications*	–	≥1	–	–

Low risk is score of <4; high risk is score ≥4
*Hyperthyroidism, preeclampsia, disseminated intravascular coagulation, hyperemesis, and trophoblastic embolization
(GA: gestational age; hCG: human chorionic gonadotropin; USG: ultrasonography)

- Epithelioid trophoblastic tumor
- Atypical placental site nodule

Metastatic diseases: Choriocarcinoma

Table 3 shows the risk of developing persistent gestational trophoblastic disease after pregnancy event.

Persistent GTD

Persistent gestational trophoblastic disease is shown **Box 6**.

GTN—Variants

Gestational trophoblastic neoplasia variant is shown **Table 4**.

TABLE 3: Risk of developing persistent gestational trophoblastic disease (GTD) after pregnancy event.

Pregnancy event	*Risk (%)*
Complete hydatidiform mole	15–20
Partial mole	1–4
Abortion	0.1%
Full-term pregnancy	0.01%

BOX 6: Persistent GTD.

- Plateau of human chorionic gonadotropin (hCG) lasting for 4 measurements over 3 weeks, i.e., days 1, 7, 14, and 21
- Increase in hCG of ≥10% for at least 3 values over 2 weeks, i.e., days 1, 7, and 14
- hCG level elevated for ≥6 months after molar evacuation
- Histologic diagnosis of choriocarcinoma

Note: Exclusion of a new pregnancy essential when evaluating patients with hCG rise.

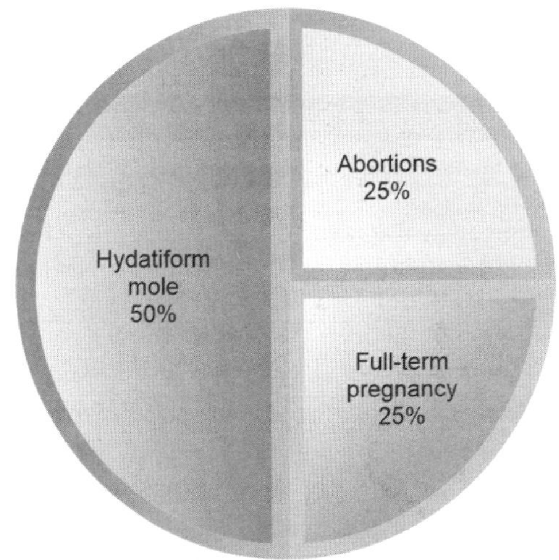

Fig. 1: Development following pregnancy event.

CHORIOCARCINOMA

Development following pregnancy event is shown in **Fig. 1**.

Clinical Features

The clinical features of choriocarcinoma are shown in **Box 7**.

The International Federation of Gynecology and Obstetrics (FIGO) staging of GTN is shown in **Box 8**.[1]

Table 5 shows World Health Organization scoring system based on prognostic factors modified as FIGO score.[1]

TABLE 4: Gestational trophoblastic neoplasia variant.

Type	*Gross appearance and spread*	*Histopathology*	*Treatment*
Choriocarcinoma	• Dark red or purple • Involves endometrium and may extend to myometrium and serosa	• Absence of Villi • Proliferation of syncytiotrophoblast and cytotrophoblast • Hemorrhage and necrosis	Chemotherapy
Placental site trophoblastic tumor (PSTT)	• Arises from placental implantation site • White-tan to yellow nodular masses	• Mainly cytotrophoblastic elements • Arising from intermediate trophoblast of placental bed	Hysterectomy
Epithelioid trophoblastic tumor (ETT)	Arise from intermediate trophoblastic cells of chorionic tissue	Discrete, nodular lesion with solid, cystic and hemorrhagic components; there is presence of nests of trophoblastic cells surrounded by extensive necrosis and a hyaline-like matrix	–
Invasive mole (chorioadenoma destruens)	• Locally invasive • May penetrate through myometrium and involve serora, parametrium, broad ligaments, and vaginal vault	Chorionic villi are maintained Trophoblast invades the myometrium	Chemotherapy/ hysterectomy may be needed

BOX 7: Clinical features of choriocarcinoma.

- Irregular vaginal bleeding after a pregnancy event—most common
- Asymptomatic—diagnosed by persistently raised human chorionic gonadotropin (hCG) levels after a molar pregnancy
- Due to metastatic disease—hemoptysis, seizures

BOX 8: International Federation of Gynecology and Obstetrics (FIGO) staging of gestational trophoblastic neoplasia.

- Gestational trophoblastic tumors strictly confined to the uterine corpus
- Gestational trophoblastic tumors extending to the adnexa or to the vagina, but limited to the genital structures
- Gestational trophoblastic tumors extending to the lungs, with or without genital tract involvement
- All other metastatic sites

TABLE 5: World Health Organization scoring system based on prognostic factors modified as FIGO score.

FIGO score	*0*	*1*	*2*	*4*
Age	<40	>40	–	–
Antecedent pregnancy	Mole	Abortion	Term	–
Interval from index pregnancy (months)	<4	4–6	7–12	>12
Pretreatment hCG	$<10^3$	$>10^3$–10^4	–	–
Largest tumor size including uterus	–	3–4	>5	–
Site of metastasis including uterus	Lung	Spleen, kidney	Gastrointestinal tract	Brain, liver
Number of metastasis identified	–	1–4	5–8	>8
Previous failed chemotherapy	–	–	Single drug	Two or more drugs

Note: Score of ≤6 indicates low risk and >6 indicates high-risk gestational trophoblastic neoplasia (GTN).
(FIGO: International Federation of Gynecology and Obstetrics; hCG: human chorionic gonadotropin)

Diagnosis of Gestational Trophoblastic Neoplasia

Diagnosis of GTN is given in **Flowchart 3**.

Surgery

Surgery of choriocarcinoma is shown in **Box 9**.

Salvage Therapies[1]

The salvage therapies for choriocarcinoma are shown in **Box 10**.

MANAGEMENT OF GESTATIONAL TROPHOBLASTIC NEOPLASM

The summary of management of GTN is shown in **Flowchart 4**.

FOLLOW-UP

Follow-up is given in **Flowchart 5**.

CONTRACEPTION[13]

- During entire interval of hCG follow-up
- Barrier methods and long-acting reversible contraception (LARC) methods can be used except intrauterine contraceptive device (IUCD)
- IUCD contraindicated till hCG is normal.
- Oral contraceptive pills can be safely used after molar evacuation.[14,15]

RECENT RECOMMENDATIONS[9]

- Anti-D prophylaxis is recommended following removal of a molar pregnancy.

Flowchart 3: Diagnosis of gestational trophoblastic neoplasia (GTN).

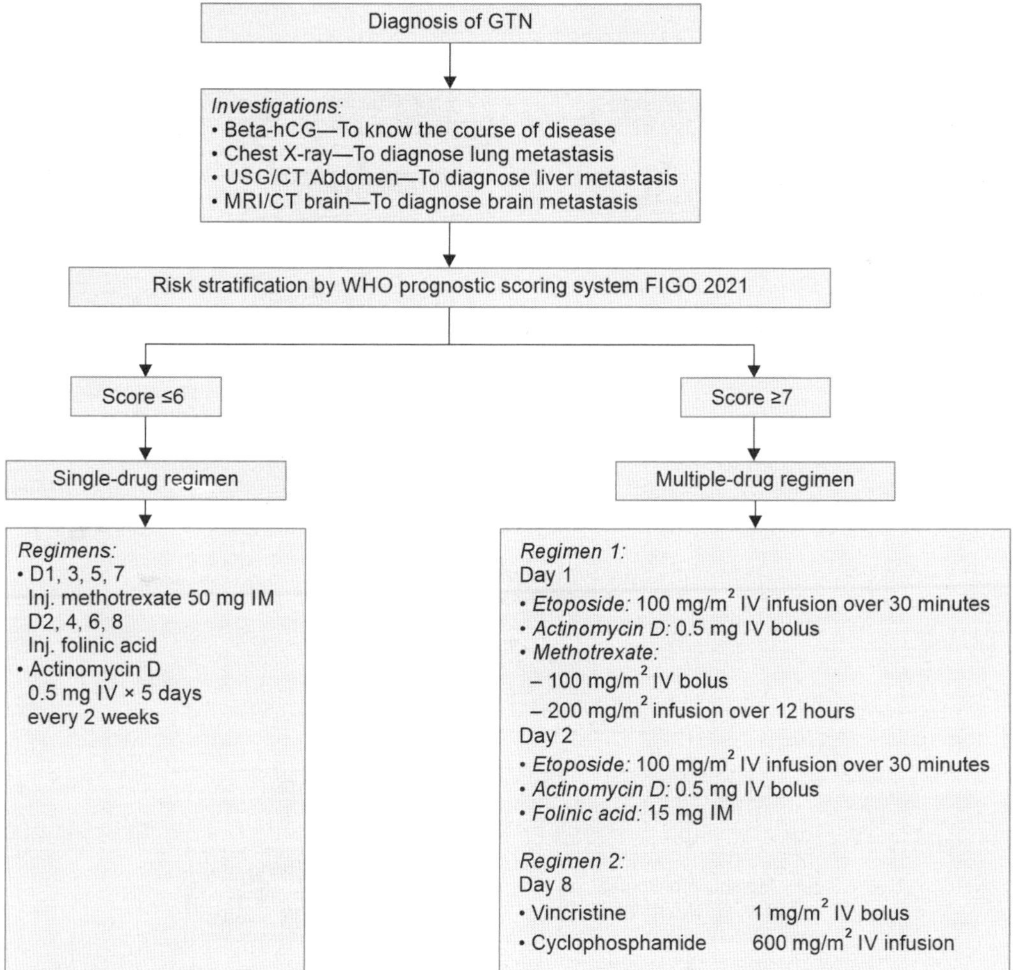

(FIGO: International Federation of Gynecology and Obstetrics; IM: intramuscular; IV: intravenous; WHO: World Health Organization)

BOX 9: Surgery of choriocarcinoma.

- *Hysterectomy:*
  - PSTT
  - ETT
  - Choriocarcinoma with adjuvant chemotherapy
    - If fertility not needed
- Thoracotomy
- Craniotomy
- Hepatic insection

(ETT: epithelioid trophoblastic tumor; PSTT: placental site trophoblastic tumor)

BOX 10: Salvage therapies for choriocarcinoma.

- EP-EMA (etoposide, cisplatin, etoposide, methotrexate and actinomycin-D)
- TP/TE (paclitaxel, cisplatin/paclitaxel, etoposide)
- MBE (methotrexate, bleomycin, etoposide)
- VIP or ICE (etoposide, ifosfamide, and cisplatin or carboplatin)
- BEP (bleomycin, etoposide, cisplatin)
- FA (5-fluorouracil, actinomycin-D)
- FAEV (floxuridine, actinomycin-D, etoposide, vincristine)
- High-dose chemotherapy with autologous bone marrow or stem cell transplant
- Immunotherapy with pembrolizumab[12]

Flowchart 4: Summary of management of gestational trophoblastic neoplasm (GTN).

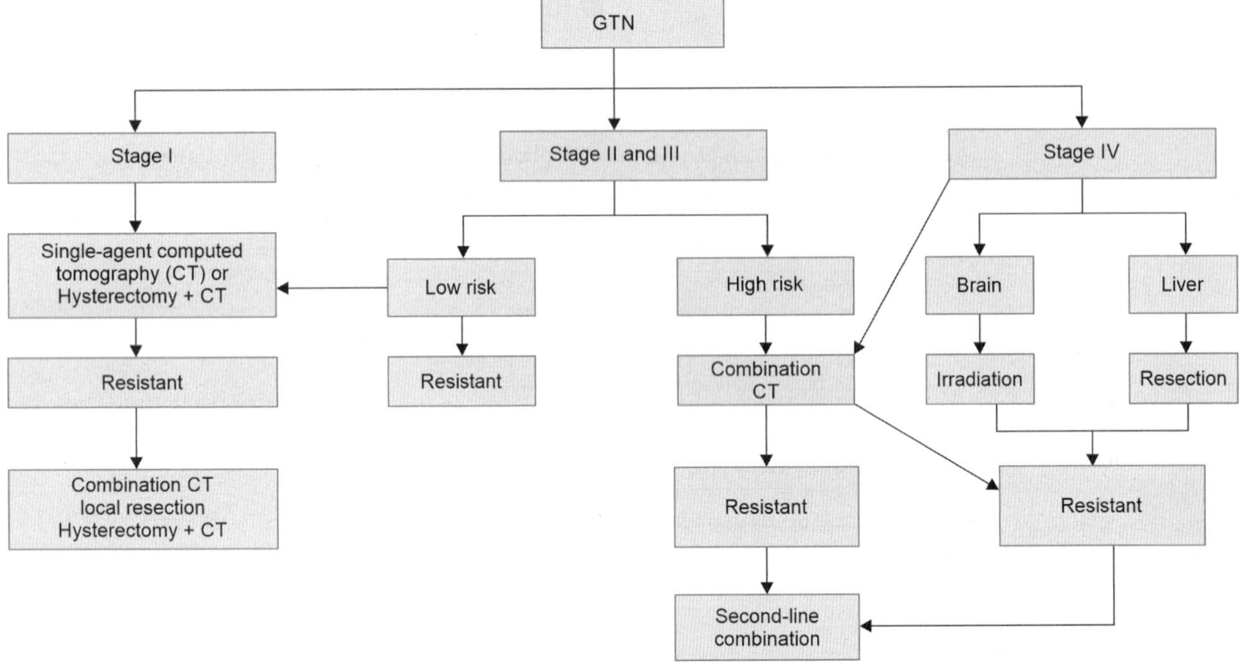

Flowchart 5: Follow-up.

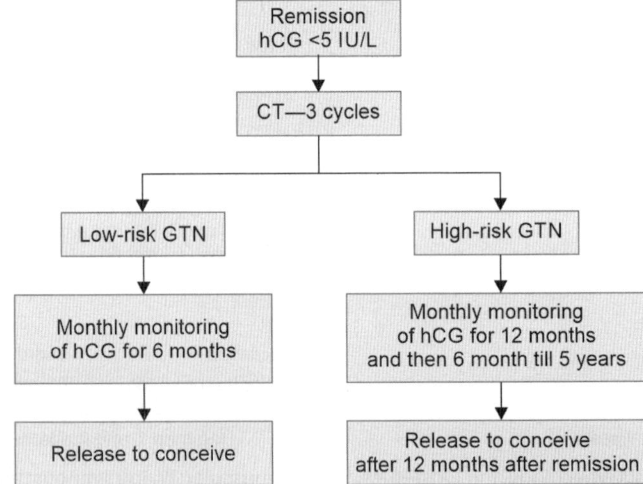

(CT: computed tomography; GTN: gestational trophoblastic neoplasia; hCG: human chorionic gonadotropin)

- Oxytocic infusion during evacuation is not recommended, but if the woman is experiencing significant hemorrhage prior to or during removal, use of oxytocin has to be weighed up against the risk of embolization.
- Women who undergo medical abortion or have a miscarriage should perform a urinary pregnancy test 3 weeks later.
- A urine hCG test should be performed in all cases of persistent or irregular vaginal bleeding continuing for >8 weeks after delivery or abortion.

CONCLUSION

Gestational trophoblastic disease is a group of lesions arising from abnormal proliferation of placental trophoblast. It can be premalignant or malignant. Mainstay of diagnosis is serum beta HCG and sonography. Chemotherapy is the treatment of choice for metastatic disease and also in selected cases of GTD. Follow up is essential in all cases and is done by Serum beta hCG estimation.

REFERENCES

1. Ngan HYS, Seckl MJ, Berkowitz RS, Xiang Y, Golfier F, Sekharan PK, et al. Diagnosis and management of gestational trophoblastic disease: 2021 update. Int J Gynecol Obstet. 2021;155 Suppl 1(Suppl 1):86-93.
2. Lurain JR. Gestational trophoblastic disease I: epidemiology, pathology, clinical presentation and diagnosis of gestational trophoblastic disease, and management of hydatidiform mole. Am J Obstet Gynecol. 2010;203:531-9.
3. Berkowitz RS, Cramer DW, Bernstein MR, Cassells S, Driscoll SG, Goldstein DP. Risk factors for complete molar pregnancy from a case-control study. Am J Obstet Gynecol. 1985;152(8):1016-20.
4. Lurain JR. Hydatidiform mole: recognition and management. Contemp Ob Gyn. 2019;64(03).
5. Sebire NJ, Foskett M, Fisher RA, Rees H, Seckl M, Newlands E. Risk of partial and complete hydatidiform molar pregnancy in relation to maternal age. BJOG. 2002;109;99-102.
6. Berkowitz RS, Bernstein MR, Harlow BL, Rice LW, Lage JM, Goldstein DP, et al. Case-control study of risk factors for partial molar pregnancy. Am J Obstet Gynecol. 1995;173(3 Pt 1):788-94.
7. Pattillo RA, Sasaki S, Katayama KP, Roesler M, Mattingly RF. Genesis of 46,XY hydatidiform mole. Am J Obstet Gynecol. 1981;141;104-5.
8. Atabaki Pasdar F, Khooei A, Fazel A, Rastin M, Tabasi N, Peirouvi T, et al. DNA flow cytometric analysis in variable types of hydropic placentas. Iran J Reprod Med. 2015;13(5):269-74.
9. Royal College of Obstetricians and Gynaecologists; Tidy J, Seckl M, Hancock BW. Management of Gestational Trophoblastic Disease. Royal College of Obstetricians and Gynaecologists. BJOG. 2021;128(3):e1-e27.
10. Uberti EM, Fajardo Mdo C, da Cunha AG, Rosa MW, Ayub AC, Graudenz Mda S, et al. Prevention of postmolar gestational trophoblastic neoplasia using prophylactic single bolus dose of actinomycin D in high-risk hydatidiform mole: a simple, effective, secure and low-cost approach without adverse effects on compliance to general follow-up or subsequent treatment. Gynecol Oncol. 2009;114; 299-305.
11. Wang Q, Fu J, Hu L, Fang F, Xie L, Chen H, et al. Prophylactic chemotherapy for hydatidiform mole to prevent gestational trophoblastic neoplasia. Cochrane Database Syst Rev. 2017;9:CD007289.
12. Ghorani E, Kaur B, Fisher RA, Short D, Joneborg U, Carlson JW, et al. Pembrolizumab is effective for drug-resistant gestational trophoblastic neoplasia. Lancet. 2017;390:2343-5.
13. Faculty of Sexual & Reproductive Healthcare. FSRH Guideline Executive Summary. Contraception After Pregnancy. London: FSRH; 2017.
14. Curry SL, Schlaerth JB, Kohorn EI, Boyce JB, Gore H, Twiggs LB, et al. Hormonal contraception and trophoblastic sequelae after hydatidiform mole (a Gyneclogic Oncology Group study). Am J Obstet Gynecol. 1989;160;805-9.
15. Braga A, Maestá I, Short D, Savage P, Harvey R, Seckl MJ. Hormonal contraception use before hCG remission does not increase the risk of gestational trophoblastic neoplasia following complete hydatidiform mole: a historical database review. BJOG. 2016;123;1330-5.

Section

9 Endoscopy/Laparoscopy

Chapter 60

Diagnostic Hysterolaparoscopy

Subash Mallya, Sonalica Suresh

INTRODUCTION

Hysterolaparoscopy is an excellent diagnostic modality to detect pathology in patients without overt clinical manifestations. It is useful in the diagnosis and treatment of infertility. Laparoscopy can reveal the presence of uterine malformations, tubal and ovarian pathology, peritubal adhesions, periadnexal adhesions and superficial peritoneal endometrial lesions. Diagnostic hysteroscopy is an equally important modality to detect uterine anomalies and other intrauterine pathologies. Diagnostic hysterolaparoscopy is a misnomer because when any pathology is detected during the surgery, appropriate corrective measures are to be taken.

INDICATIONS

- Infertility or subfertility
- Abnormal hysterosalpingogram with cornual or distal tubal block
- Suspicion of endometriosis
- Chronic pelvic pain
- Uterine anomalies
- Recurrent implantation failure or recurrent pregnancy loss
- Evaluation of pelvic mass.

CONTRAINDICATIONS

- Suspected intrauterine pregnancy
- Active cervicovaginal or pelvic infection
- Active uterine bleeding.

PREREQUISITES

- Clinical history and complete physical examination
- 2D/3D Ultrasound
- Done in the early follicular phase just after menstruation
- Risk assessment for high risk factors such as previous surgeries, abnormal body mass index (BMI).

PREOPERATIVE PREPARATION

- Informed consent to be taken including the risks, benefits and alternatives of the procedure and consent for possible treatment of identified pathology such as endometriosis which include excision or ablation of endometriosis, adhesiolysis and peritoneal biopsy. The possibility of conversion to laparotomy if procedure cannot be completed laparoscopically or if a complication occurs should also be discussed with the patient.
- Mechanical bowel preparation—two tablets of bisacodyl given on the previous day of surgery.
- Cervical preparation with misoprostol 200–400 µg 2 hours before surgery if entry difficulty is suspected.
- Preanesthetic check up
- Prophylactic antibiotics if needed.
- Preoperative check for the instruments, ergonomics. and team.

HYSTEROSCOPY

Hysteroscopy was first performed by Pantaleoni in 1869, and now it has become the minimally invasive choice for treatment of intrauterine pathologies.[1] From basic to super specialty unit wide range of instruments are available. The types of equipment used for hysteroscopy are:

Telescope and Sheath

- *Telescope:* The commonly used is the rigid rod lens telescopes **(Fig. 1A)** with different viewing angles. It has three parts—the eye piece, the barrel, and the objective lens. The diameter of the scope can be 1.9 mm/2.9 mm/4 mm. It can be with either centered lens (0°) or offset lens (12°, 30°) with angle of view opposite the light source. Most commonly used is the 2.9 mm 30° scope and for office hysteroscopy 1.9 mm is the best.[2]

- *Telescope sheath:* The telescope must be fitted to a sheath through which the distending medium can be infused into the uterine cavity. It is different for diagnostic and operative hysteroscopy. The diagnostic sheath is 4–5 mm in diameter, depending on the outer diameter of the telescope. It consists of an inner sheath **(Fig. 1B)** for influx of distension media and an operative channel to use operative instruments. The outer sheath is for the efflux of distension media **(Fig. 1)**.

An assembled hysteroscope is shown in **Figure 2**. In operative hysteroscopy a resectoscope **(Figs. 3A to F)** is used which has a continuous flow outer sheath, inner sheath with insulated ceramic tip and working element with electrode (loop/ball/Collins knife) attached to it.

Endo Vision Unit

- *Light source and cable* **(Fig. 4)**: Good image quality depends on the type of light source used. The available light sources are the xenon and the LED. LED has the advantage that it is not warm and hence the risk of burns is less. Light is usually transmitted through fiber optic cables with a diameter of 2.5–4.5 mm and length

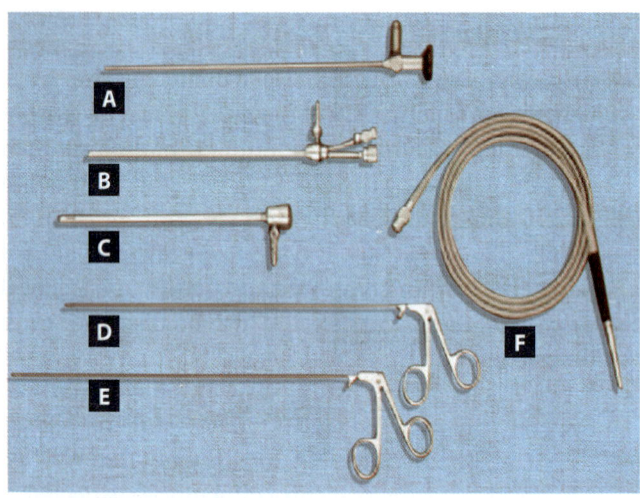

Figs. 1A to F: Diagnostic hysteroscope. (A) Telescope; (B) Inflow with operating channel; (C) Outer sheath; (D) and (E) 5F instruments; and (F) Light cable.

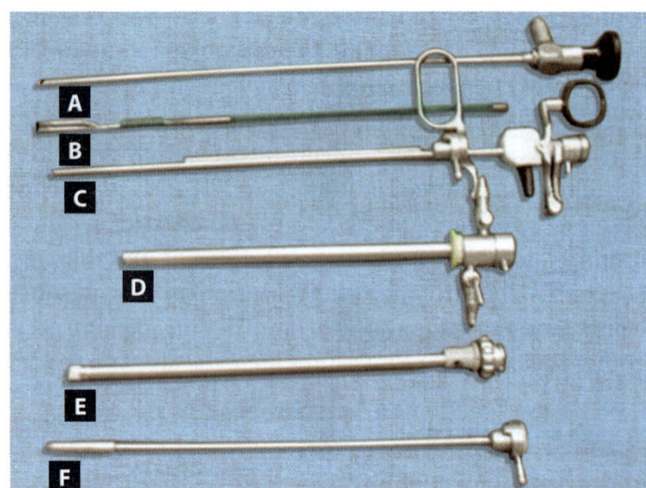

Figs. 3A to F: Resectoscope. (A) Telescope; (B) Electrode; (C) Working element; (D) Outer sheath; (E) Inner sheath; and (F) Obturator.

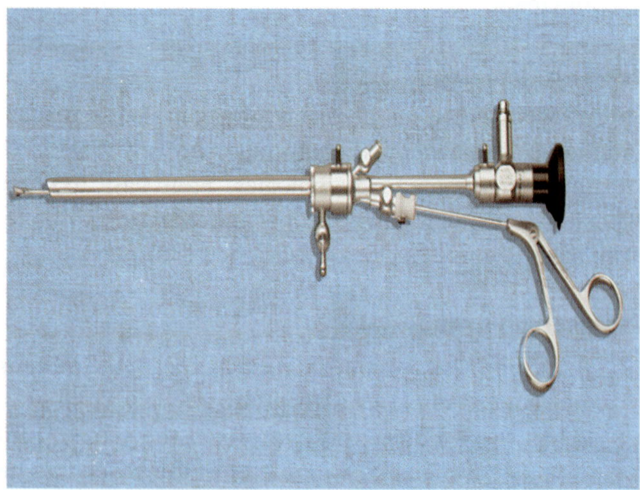

Fig. 2: Hysteroscope assembled.

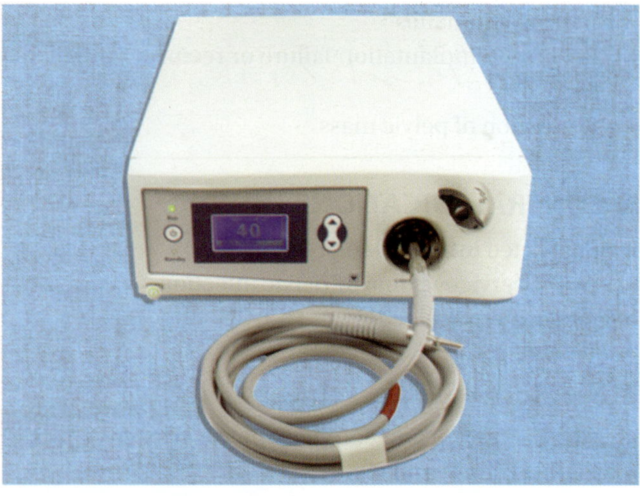

Fig. 4: Light source and cable.

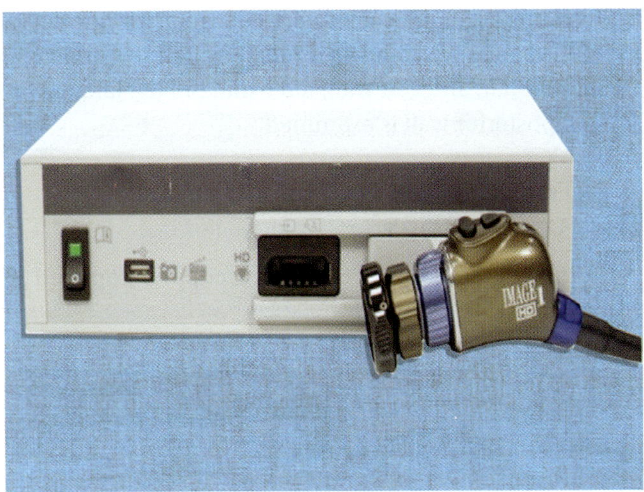

Fig. 5: Camera.

of 180–300 cm. The light source operating at higher power levels will produce more thermal energy and can cause inadvertent burns and hence the intensity of light has to be kept at a minimum.

- *Camera and monitor:* The camera **(Fig. 5)** is attached to the eye piece of the endoscope, where it captures the image and is processed and is sent to a monitor. Most monitors have the potential to display about 800 horizontal lines of resolution while high definition (HD) systems generally possess 1,080 lines. The high-definition cameras offer very high resolution and almost natural color inside the uterine cavity. The system usually includes facilities for storing and printing still images and video data.

Instruments

Mechanical instruments used in hysteroscopy are of 5F diameter and include various types of graspers and scissors **(Fig. 6)**.

Fluid Delivery System

A fluid delivery system is needed to control the flow and pressure needed to distend the uterine cavity. Distension media can be delivered either by:

- *Hysteromat:* Electronic suction irrigation pump **(Fig. 7)**—this is the ideal fluid delivery system where flow and pressure are electronically controlled. Settings are kept at 90–110 mm Hg inflow, 0.25 bar suction and flow rate of 200 mL/minute.
- *Simple gravity:* Hanging a 3-liter bottle of fluid 1.5 m above uterus can achieve a pressure of 70 mm Hg.

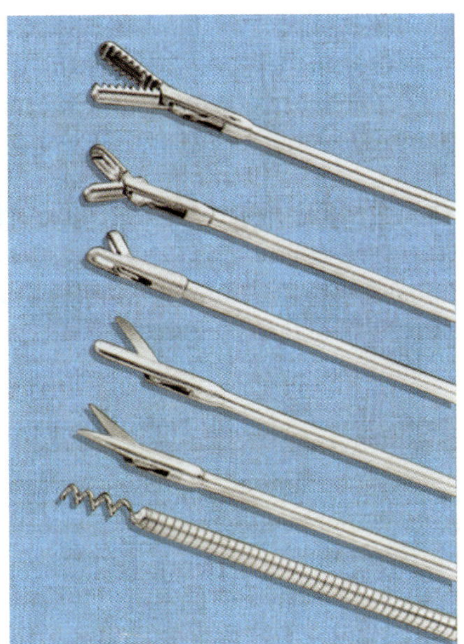

Fig. 6: Graspers and scissors.

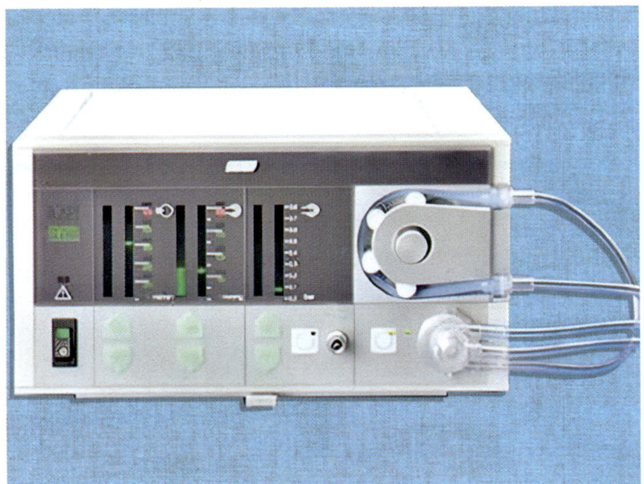

Fig. 7: Hysteromat.

- Using blood pressure cuff around the fluid bag helps to achieve a pressure of 80–120 mm Hg.

Distension Media

It is mainly of two types—gaseous and liquid media. CO_2 is the only gaseous distension media used in hysteroscopy. But because of its disadvantage of distorted view in the presence of mucus or blood and the need for hysteroflator for safe delivery of gas, it is presently not used. Most commonly used liquid distension media in diagnostic hysteroscopy is the low viscosity isotonic fluids such as normal saline 0.9% and Ringer lactate. In operative hysteroscopy the choice of the fluid used depends on the energy used for the procedure. With bipolar current normal saline is used as the distension media and with monopolar current hypotonic solution such as 1.5% glycine is used.

Procedure for Diagnostic Hysteroscopy

- Bladder must be empty.
- Patient positioning—lithotomy with buttocks placed beyond the edge of table, legs are flexed.
- Vaginal examination is done under anesthesia to assess uterine size and position.
- Assemble the hysteroscope, camera is white balanced, and inflow and outflow are attached and fluid is let in.
- Have a trained person for fluid monitoring.
- Make sure there are no air bubbles in the tubings.
- Vaginoscopic approach—vagina is distended, cervix, and external os located and hysteroscope is carefully inserted.
- Otherwise use speculum and vulsellum at 12 o'clock position. The uterine sound is passed and hysteroscope assembly is gently inserted through the external os.
- Follow the cervical crypts and visualize the internal os and negotiate the os to enter the uterine cavity.
- Examine systematically—cervical canal, uterine cavity for size, shape, contour-fundus, right ostia and right lateral wall, anterior wall, left ostia and left lateral wall and posterior wall is examined.

PATHOLOGY DETECTED DURING HYSTEROSCOPY

Pathology detected during hysteroscopy is depicted in **Table 1**, **Figures 8A to F**.

TABLE 1: Pathology detected during hysteroscopy.

	Pathology	*Corrective measures*
Cervix	• Cervical stenosis • Hypertrophic crypt	• Use of microscissors to relieve the stenosis • Resection of crypt
Uterine cavity	• Septum • Polyp • Fibroid • Intrauterine adhesions • Tubal block • Dysmorphic uterus	• Septal resection • Polypectomy • Myomectomy • Adhesiolysis • Tubal cannulation • Metroplasty

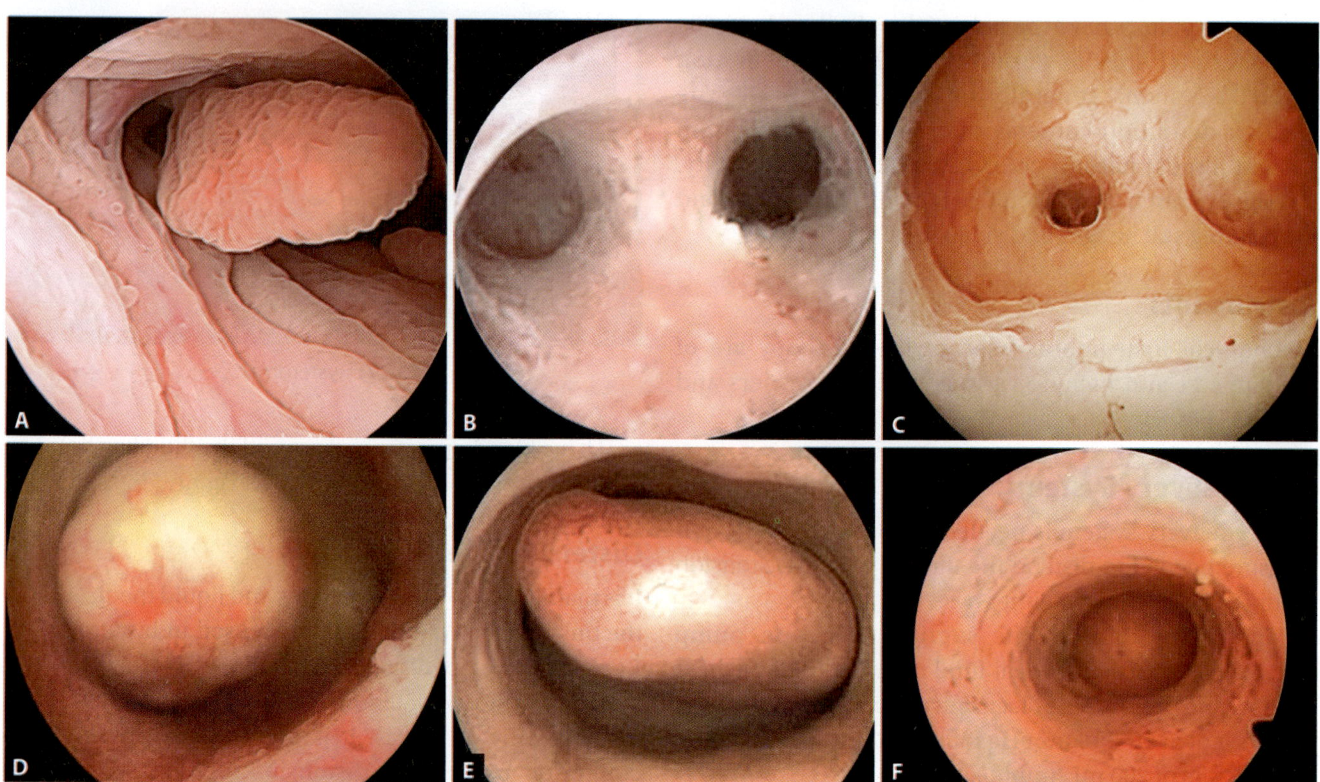

Figs. 8A to F: (A) Cervical polyp; (B) Uterine septum; (C) Intrauterine adhesions; (D) Submucous fibroid; (E) Endometrial polyp; and (F) Tubular uterus.

COMPLICATIONS OF HYSTEROSCOPY

Hysteroscopy—complications and prevention depicted in **Table 2**.

LAPAROSCOPY

Diagnostic laparoscopy is an option for thorough evaluation of peritoneal cavity and pelvic and abdominal organs. Laparoscopic surgery has several advantages in terms of reduced pain, better cosmetic results, decreased adhesion formation, faster recovery, and lower rate of infection. But also holds the disadvantages such as longer learning curve for surgeon, higher cost, and complications such as major vascular, gastrointestinal, and urinary tract injuries.

TABLE 2: Hysteroscopy—complications and prevention.

Complications	*Prevention*
Nerve injuries due to patient positioning	• Ideal lithotomy-moderate flexion, limited abduction, and external rotation • Use of silicon pads over stirrups • Avoid pressing on injury prone areas • Avoid leaning on thigh
• Distension media complications – Fluid overload—defined as fluid deficit >2500 mL for NS and >1,000 mL for glycine – Normal saline—pulmonary edema – Glycine—hyponatremia	• Use appropriate distension media and delivery system. • Trained person for monitoring fluid deficit • Work at lowest effective pressure • Keep operating time to a minimum
• Air embolism – First sign is sudden decrease in end-tidal CO_2	• Avoid Trendelenburg position. • Make sure there are no air bubbles in the tubings. • Monitor end tidal CO_2
• Perforation – Risk factors for perforation • Retroverted, acutely anteverted, or retroflexed uterus • Scarred uterus (previous uterine surgery) • Postmenopausal uterus • Infection • Recent pregnancy in last 6 months • Intrauterine adhesions • Uterine anomalies	• Assessment of risk factors • Visualize the internal os and enter into cavity, not to enter blindly
Hemorrhage	Inflated 12F Foleys is used for controlling hemorrhage

LAPAROSCOPIC EQUIPMENT

- *Telescope:* Depending on the viewing angle mainly two types of telescopes **(Fig. 9)** are used—(1) 0° straight forward telescope and (2) 30° forward oblique telescope. The diameter of the scope may vary from 5 to 10 mm.
- *Trocar and cannula:* They allow the insertion of laparoscopic instruments into the peritoneal cavity while maintaining the pressure created by the distending gas. The cannula is fitted with a Leur-type port for the attachment of insufflator tubing. Most frequently used is the 10 mm primary trocar and 5 mm secondary trocars. The tip of the trocar can be pyramidal, conical, bladed, or blunt tipped. Pyramidal tips are easier to insert. To visualize the layers of the abdominal wall during placement optical access trocars are also available. Both reusable metal trocars **(Fig. 10)** and disposable trocars **(Fig. 11)** are available.
- *Endo vision unit:* It consists of light source, light cable, camera, and monitor and is the same as that in hysteroscopy.
- *Insufflator:* CO_2 is the choice of gas for insufflation. The advantage of CO_2 is that it is rapidly absorbed by blood and converted to carbonic acid which immediately dissociates to bicarbonate and hydrogen ions which

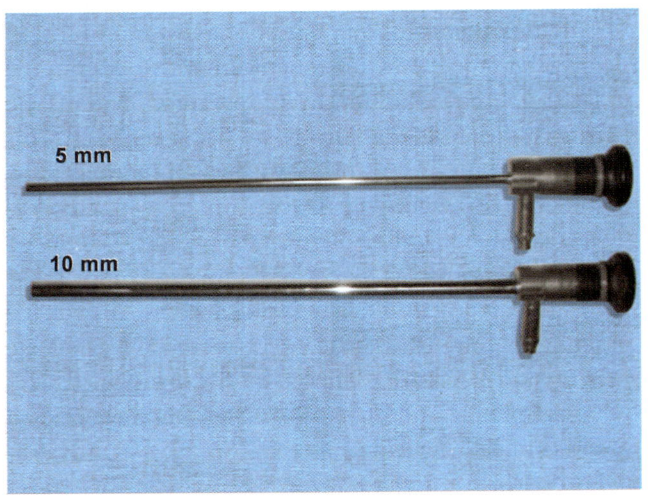

Fig. 9: Telescope.

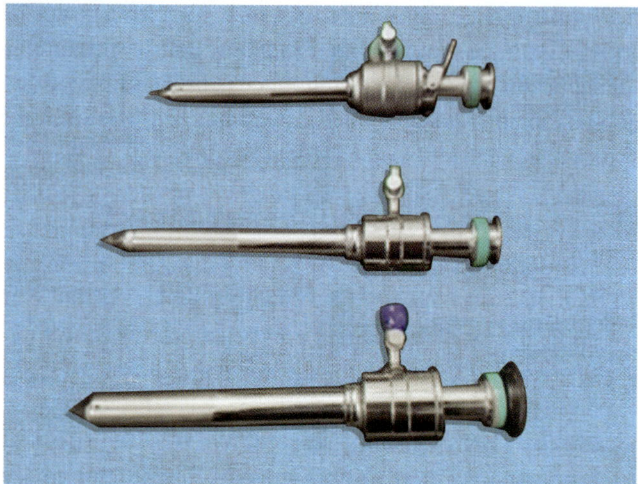

Fig. 10: Metal trocar.

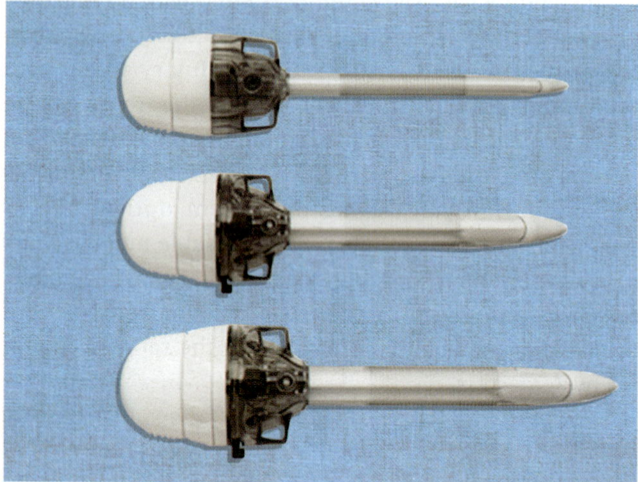

Fig. 11: Disposable trocar.

prevent gas embolism. Electronic CO_2 insufflator provides a continuous flow of gas to maintain an intraabdominal pressure ranging from 12 to 15 mm Hg. Before the introduction of primary trocar, the intraabdominal pressure is set at 20 mm Hg. This facilitates the insertion of trocar as it increases the distance between the trocar and retroperitoneal vessels. After insertion of trocar the pressure is reduced to 12–15 mm Hg as higher pressure can compromise the diaphragmatic movement and decrease the venous return by compressing the inferior vena cava. Flow rate is initially kept at 1 L/minute during insufflation with Veress needle and higher flow rates when connected to the trocar.

- *Operating instruments:* Laparoscopic hand instruments vary in length from 18 to 45 cm. Most commonly used is the 36 cm instrument in adult. For better ergonomics half of the instrument should be inside the abdomen and half outside, so that it acts as a class 1 lever, stabilizing the port and making surgery convenient. Most of the hand instruments have three parts—handle, insulated outer tube, and insert which makes the tip of instruments. The tip can be grasper, scissors, or forceps. Suction irrigation, electrocautery instruments—monopolar and bipolar, harmonic or vessel sealing device are the other instruments to be set in laparoscopy trolley for diagnostic laparoscopy.
- *Electrosurgical unit:* The electrosurgical unit includes a generator, two electrodes and the patient. Electrosurgery is the application of high frequency alternating current (AC) in surgery to achieve various thermal tissue effects including cutting and coagulation. Electrosurgery can be delivered through monopolar or bipolar instruments.

PROCEDURE FOR DIAGNOSTIC LAPAROSCOPY

Patient Positioning

The patient is placed in low lithotomy with buttock slightly over the edge of the table to allow use of an intrauterine manipulator. The legs are positioned in such a way that the patient's thigh should be in the same plane as the abdomen so that the laparoscopic instruments can be moved freely during surgery. The legs must be properly padded. The patient's abdomen should be below the waist of the surgeon. Make sure the patient cannot slide cephalad, and the arms are tucked. Vagina and abdomen are prepared and bladder is drained. If longer procedure is anticipated Foley catheter is put for continuous drainage.

Uterine manipulator placement—uterine manipulator can be placed to move the uterus during evaluation of pelvis.

Entry Technique

Abdominal access is attained by various techniques, which include open Hasson/Veress/direct trocar/Palmer's point entry. Umbilicus is usually chosen as the site for abdominal entry.

Veress Needle (Closed Entry Technique)

- Veress needle **(Fig. 12)** is used for creating pneumoperitoneum to separate the abdominal wall

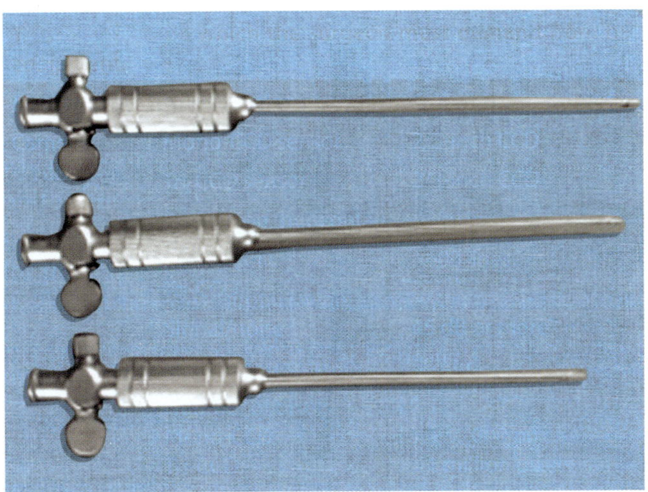

Fig. 12: Veress needle.

from internal organs so that primary trocar can be inserted safely.

- Veress needle varies from 8 to 12 cm in length with an external diameter of 2 mm. It has a spring-loaded tip that retracts as it pierces the abdominal wall allowing the blunt tip to engage on entry into the peritoneal cavity thus avoiding injury the abdominal organs.
- The operating table should be horizontal at the start of the procedure.
- Incision is made at the base of the umbilicus where the abdominal wall is thinnest and usually avascular.[3]
- The needle should be sharp and spring action and patency to be tested before insertion.
- The abdomen should be palpated to check for any masses and for the position of the aorta before insertion of Veress
- The lower abdominal wall is stabilized and Veress needle is held like a dart and inserted at right angles to the skin and should be pushed in just sufficiently to penetrate the fascia and the peritoneum.[4]
- *Tests of correct placement*
  - *Double click*: Two click sound to be heard, first when it pierces the rectus sheath and second when it pierces the peritoneum.
  - *Aspiration test (Palmer test):* 10 mL of saline is instilled into the peritoneal cavity through the needle and is aspirated back. If needle is in the peritoneal cavity, the saline cannot be aspirated back, and it falls into the dependent portion of peritoneal cavity.
  - *Hanging drop test:* A drop of saline instilled into the needle will be sucked in if the needle is in the peritoneal cavity, whereas it will remain in the needle if it is extraperitoneal.
  - Initial insufflation pressure of <8 mm Hg.
- Excessive lateral movement of the needle should be avoided.
- The needle is then connected to an insufflator with the flow set at 1 L/minute and intraabdominal pressure at 15–20 mm Hg.
- Loss of liver dullness over the lateral aspect of the right costal margin is the earliest reassuring sign and there will be symmetric distension of the abdomen when needle is intraperitoneal.
- Maximum of only two attempts to be made.
- Insufflate up to 20 mm Hg and insert the primary trocar.

Direct Trocar Entry

Direct trocar entry is a safe method with only one blind step for trocar insertion. Here trocar is inserted with the palm of right hand and index finger positioned 3 cm away from the trocar tip to guard against sudden uncontrolled entry into the abdomen. Trocar is inserted at 90 degree angle till the surgeon feels the resistance offered by the fascial layer. Abdominal wall is then lifted and trocar is advanced in a controlled fashion into the peritoneal cavity with a twisting motion with direction of trocar pointing towards pelvis. Laparoscope is then introduced and proper intraperitoneal placement is ensured and then pneumoperitoneum is created.

Open Hasson Technique

Make a small incision, either subumbilical or supraumbilical and directly incise the layers of abdominal wall, cut the peritoneum, and enter the abdomen. Since gas can escape around the incision, an olive is placed over the end of the trocar to occlude the incision and sutures are placed on the abdominal fascia and attached to the cannula.

It is useful in patients with previous abdominal surgery especially midline, pregnant women, very thin patients where little space exists between the abdominal wall and the spine.

Advantage

- Prevention of bowel injury caused by blind puncture with a needle and subsequent trocar.
- Avoid preperitoneal insufflation.
- Low incidence of vascular injuries.

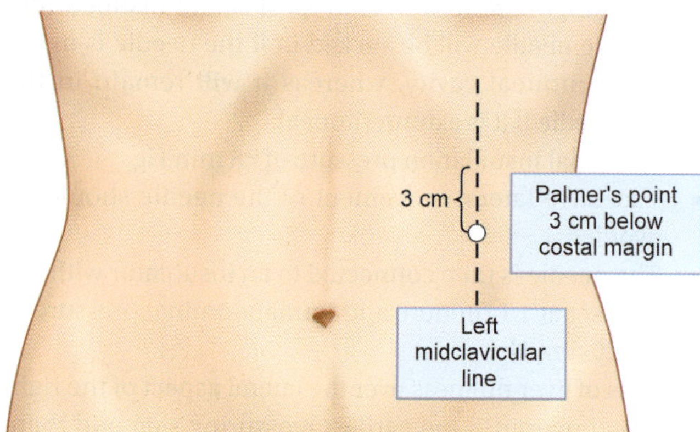

Fig. 13: Palmer's point.

Disadvantage

- More time needed
- Difficulty in maintenance of pneumoperitoneum.

Palmer's Point Entry

This is the safest initial entry site in high-risk patients. The point **(Fig. 13)** is 3 cm below the left costal margin in the mid clavicular line.[5] Splenomegaly has to be ruled out and make sure Ryle's tube is in place before entry at this point.

In high-risk cases:

- *Previous laparotomy:* Avoid scars, consider Hasson technique or Palmer's point
- *Obese:* Open technique or optical access trocars
- *Very thin patients:* Open Hasson or Palmer's point.

Procedure

- After insertion of primary trocar, reduce pressure to 12–15 mm Hg
- Area directly below the primary trocar entry site is evaluated for bleeding or signs of injury.
- Visual check is done by rotating the scope through 360° for a panoramic view of abdominal cavity and pelvis. Liver surface, gallbladder, falciform ligament, stomach, omentum, right and left hemidiaphragms, ascending colon, descending colon and transverse colon is viewed.
- Head down tilt is then given—Trendelenburg position
- After Trendelenburg positioning, bowel and omentum fall toward the upper abdomen to expose the pelvic structures.
- Secondary trocars are placed under direct vision perpendicular to the skin.

Evaluation of Pelvis

Examine:

- Uterus—shape, size, surface, and relation with tubes
- Fallopian tube—size, shape, surface, kinking, dilatation, stricture, and hydrosalpinx
- Ovary—size, shape, and presence of cyst
- Pelvic peritoneum for endometriotic implants, peritoneal defects or windows, fibrosis, or studding
- Pouch of Douglas
- Both ureters are visualized coursing from pelvic brim and peristalsis is noted.
- Chromopertubation—where methylene blue dye is injected into the uterine cavity using Rubins cannula and the fallopian tubes are checked for patency. The tubes are patent if the spillage of dye is seen from both tubes.

Pathology Detected During Laparoscopy

During visual assessment we may come across pathologies such as hydrosalpinx, tubal block, peritubal and periadnexal lesions, fibroids, endometriotic lesions, Müllerian anomalies, ovarian pathology, or tuberculous lesions. Indicated corrective measures can be performed accordingly.

Port Removal and Incision Closure

Gas tubing is disconnected from the trocar and ports on all cannulas are opened to deflate the abdominal cavity. All secondary cannulas are removed using laparoscopic visualization. Fascial closure using delayed absorbable suture is recommended for trocars measuring >10 mm to prevent incisional hernia. Skin incision is closed with subcuticular stitch 4-0 delayed absorbable suture.

COMPLICATIONS

The overall risk of serious complication from diagnostic laparoscopy is approximately two in 1,000, and include:

- Vascular injury
- Bowel injury—15% of bowel injuries may not be diagnosed at the time of laparoscopy.
- Genitourinary injuries
- Failure to gain entry into abdominal cavity and to complete the intended procedure.
- Incisional hernias
- Thromboembolic complications
- Pneumoperitoneum-related complications such as gas embolism and cardiac arrythmia and extraperitoneal insufflation

Frequent risks include shoulder tip pain, wound gaping, and infection.

PREVENTION OF LAPAROSCOPIC ENTRY COMPLICATIONS

- During entry patient should be supine and in flat position.
- Use entry technique with which you are more comfortable.
- Abdomen should be palpated to check for any masses and for position of the aorta.
- Avoid previous scars when choosing entry point.
- Make an adequate skin incision to avoid the need for excessive pressure to pass the trocar through the skin but avoid going to deep.[6,7]

CONCLUSION

Hysterolaparoscopy is an effective tool to assess infertility and simultaneously diagnose and treat uterine and pelvic lesions. It has a significant advantage over other modalities in view of low complication rates, minimal time requirement, and additional benefit of therapeutic interventions in few conditions.

REFERENCES

1. Stocker L, Umranikar A, Moors A, Umranikar S. An overview of hysteroscopy and hysteroscopic surgery. Obstet Gynaecol Reprod Med. 2013;23(5):146-53.
2. Cooper NAM, Clark TJ. Ambulatory hysteroscopy. Obstet Gynaecol. 2013;15:159-66.
3. CRoy GM, Bazzurini L, Solima E, Luciano AA. Safe technique for laparoscopic entry into the abdominal cavity. J Am Assoc Gynecol Laparosc. 2001;8(4):519-28.
4. Vilos GA, Vilos AG. Safe laparoscopic entry guided by Veress needle CO_2 insufflation pressure. J Am Assoc Gynecol Laparosc. 2003;10(3):415-20.
5. Frappell, J. Laparoscopic entry after previous surgery. Obstet Gynecol. 2012;14:207-9.
6. Pickett SD, Rodewald KJ, Billow MR, Giannios NM, Hurd WW. Avoiding major vessel injury during laparoscopic instrument insertion. Obstet Gynecol Clin North Am. 2010;37(3):387-97.
7. Brierley G, Arshad I, Shakir F, Visvathanan D, Arambage K. Vascular injury during laparoscopic gynaecological surgery: a methodological approach for prevention and management. Obstet Gynaecol. 2020;22(3):191-8.

61

Chapter

Endoscopy in Infertility

Divya Mahindru, Sarabjeet Singh, Jyotsna Gupta

INTRODUCTION

Infertility is defined as inability to achieve pregnancy after 1 year of regular and unprotected intercourse.[1] Approximately 80–85% of healthy young couples conceive within 1 year and an additional 10% would do so if they continue to try for another year.[2]

Infertility is a common phenomenon, affecting around 10–15% of couples with a great socioeconomic impact on their lives. With the increasing awareness and advanced facilities, society has become more acceptable of infertility and the couples have become more comfortable in seeking evaluation and treatment for the same.[3]

Infertility has been classified as primary or secondary infertility. Primary infertility is defined as the failure to achieve clinical pregnancy after 12 months of unprotected sexual exposure in a couple where the woman is in reproductive age group and is non-lactating. Secondary infertility refers to inability of the couple to conceive after a previous pregnancy.[4]

Identifying the cause of infertility has always been a challenge. Female factors contribute to about 40–45% cases of infertility whereas in about 25–40% cases male factor predominates. Anovulation, tubal factors, uterine factors, cervical factors, endometriosis, immunological factors, infections such as pelvic inflammatory disease (PID), adhesions, etc., account for most common causes of female infertility while abnormal spermatogenesis, infections, trauma to the genitals, endocrine disorders, sexual dysfunction generally cause male infertility. Subsequently, in about 20–30% couples routine infertility evaluation does not reveal any abnormality and are eventually diagnosed with unexplained infertility.[5]

The standard diagnostic tests for infertility evaluation are assessment of ovulation, tubal patency, hormonal evaluation, and semen analysis.[6] In the modern era of fertility management with tremendous improvement in success rates of assisted reproductive technology, the role of endoscopy (both laparo-hysteroscopy), especially diagnostic part, has been debated and understated.[7] Nevertheless, there are plenty of clinical scenarios where laparo-hysteroscopy is clinically recommended and justified. The primary advantage of offering laparoscopy along with hysteroscopy is that it not only helps the clinician to reach a more accurate diagnosis, but also offers therapeutic benefits.

Laparoscopy is a minimally invasive technique scheduled in the early to midfollicular phase of the cycle. It has the advantage of magnification, early recovery, better cosmetic benefit and is cost effective as patient is discharged early.[8] Infertile patients who undergo laparoscopy usually have failed conservative management and diagnostic laparoscopy aids in identification of pelvic adhesions, endometriosis, fibroids, adnexal masses. Laparoscopy usually requires an imaging system, an insufflation system, and special instruments, lasers, CO_2, uterine manipulator.

Clinical skill and patient selection are critical to the success of laparoscopic surgery. Most of the surgeries are without problems but serious complications can occur in a few cases.

Laparoscopic injuries frequently occur during the blind insertion of needles, trocars, and cannula through the abdominal wall. Women must be informed of the risks and potential complications such as bowel and vascular injury associated with laparoscopy.[9]

The most effective way to reduce complications of laparoscopic entry is to optimize insertion of the primary trocar and cannula.

ENTRY TECHNIQUES IN LAPAROSCOPY[10-14]

Closed

The primary incision for laparoscopy should be vertical from the base of the umbilicus. One should not incise so deeply as to enter the peritoneal cavity. A sharp disposable

Veress needle with a good spring action is recommended. The operating table should be horizontal when starting the procedure. The abdomen should be palpated to check for any before insertion of the Veress needle. After stabilizing the abdominal wall, a stab incision is made. The Veress needle is held like a dart and gently inserted perpendicular to the abdominal wall. There will be a sensation of initial resistance, followed by a given, at two points. The first point marks the needle traversing the fascia and the second as it penetrates the peritoneum. Excessive lateral movement of the needle should be avoided, as this may convert a small needlepoint injury in the wall of the bowel or vessel into a more complex tear.[14]

Open Technique with Hasson Cannula

The open cannula provides the surgeon with a safe method using a blunt ender trocar under direct vision to enter the abdomen, especially in a patient who has previously undergone intra-abdominal procedures. This technique avoids the use of sharp instruments after the initial skin incision.[14]

Direct trocar insertion is an acceptable alternative trocar insertion method. It is the most rapid method of entry and can be safely used in carefully selected cases.

A pneumoperitoneum is created before inserting the primary trocar. The primary trocar is inserted in a controlled manner perpendicular to the skin, through the incision at the thinnest part of the abdominal wall, in the base of the umbilicus. Insertion should be stopped immediately once the trocar is inside the abdominal cavity. Once the laparoscope has been introduced, the abdomen should be closely inspected for any adhesions, any evidence of damage, bleeding, or injury.

INDICATIONS OF LAPAROSCOPY IN INFERTILITY (SEE BOX 1)

BOX 1: Various proposed indications of laparoscopy and hysteroscopy in infertility management.

Laparoscopy:
- Endometriosis[15]
- Myomectomy[19]
- Adhesiolysis[16]
- Diagnostic laparoscopy[20]
- Hydrosalpinx (salpingectomy or clipping)
- Neo-salpingostomy[17,18]
- Fimbrioplasy[17,18]
- Sterilization reversal[17,18]
- Ovarian drilling[21]
  - Ovarian cystectomy[22]

Hysteroscopy:
- Diagnostic hysteroscopy[20]
- Endometrial polyps[23]
- Myomectomy[24,25]
- Adhesiolysis[26]
- Foreign body[27]
- Uterine septa[28]
- Hysteroscopic cannulation[17]

Endometriosis

Endometriosis is one of the prime reasons of offering laparoscopy in infertility management. Endometriosis affects fertility either directly by forming adhesions or by interfering with ovulation, tubal function, and implantation. Laparoscopy helps in removing endometrioma (cystectomy or drainage-former is better), removing adhesions, and fulgurating the endometriotic areas thereby restoring pelvic anatomy.[15]

Adhesiolysis

Pelvic adhesions either due to pelvic inflammation disease, tuberculosis, endometriosis, or previous surgery have been identified as a cause of infertility. In these patients, adhesions are scored and classified. Pregnancy was more likely to occur when >50% of the ovary was visible and when adhesions were filmy and avascular, rather than dense and vascular. Laparoscopic adhesiolysis in these cases improves the chances of pregnancy.[16]

Hydrosalpinx/Tubal Disease

Although, in the modern era of assisted reproductive technology, the role of tubal corrective surgeries has been debated, the later can be of use in various clinical scenarios to improve the pregnancy outcomes.[17] The patients with proximal tubal block(s) can be offered a hysteroscopic cannulation (+/– laparoscopic guidance) to clear the block following which the couple can try for a natural conception. Similarly, tubal corrective surgeries such as laparoscopic neo-salpingostomy, fimbrioplasty, microsurgical re-anastomosis can be offered to the patients with mid to distal blocks caused by hydrosalpinges, fimbrial phimosis, and peritubal adhesions usually PID or previous surgery or peritonitis (tuberculosis) and sterilization surgery.[17] The hydrosalpinx fluid impairs implantation and surgical intervention such as salpingectomy and clipping would remove the communication to the uterus and create a better noninfective helping increase pregnancy rates.[18]

Fibroids

Fibroids are a known but debatable cause of subfertility and pregnancy loss. A large fibroid can cause tubal occlusion at the cornua. A submucous fibroid can prevent successful nidation and implantation. They may alter the normal anatomical axis of the uterus thereby causing infertility. Laparoscopic myomectomy has been the preferred approach due to less intraoperative blood loss and postoperative pain with shorter hospital stay and early recovery.[19]

Ovarian Drilling in Polycystic Ovary Syndrome

Ovarian drilling or gonadotropins have been suggested as a second-line management of polycystic ovary syndrome (PCOS) in patients where the oral ovulogens have either failed to make the women pregnant despite documented ovulation for 6–12 months (clomiphene failure) or there is a resistance (clomiphene resistance) where the ovulogens were not sufficient to make women ovulate. Ovarian drilling compared to gonadotropin administration takes advantage of being one time procedure and is recommended for patients who come from remote areas where frequent follicle monitoring needed after gonadotropin administration is difficult. Also, it can be useful in patients who need laparoscopy for other indications such as suspected adhesions or tubal issues.[29]

HYSTEROSCOPY[30-32]

Hysteroscopy allows direct visual examination of the uterine cavity and cervical canal with miniaturized endoscopic equipment. It is an important diagnostic tool in infertility evaluation and helps in accurate localization of intrauterine pathology such as polyps, submucosal myomas, adhesions, foreign body, and many more.

For outpatient hysteroscopy or office hysteroscopy, patient selection has to be very meticulous. The patients in whom cervical stenosis is expected should not undergo office hysteroscopy.

Outpatient hysteroscopy should be conducted in an appropriately sized, equipped, and staffed treatment room. The surgeon should have the necessary skills and expertise to carry out hysteroscopy. Written patient information should be provided and consent for the procedure should be taken.[30-32]

Women should be given adequate analgesia with non-steroidal anti-inflammatory agents (NSAIDs) around 1 hour before their scheduled procedure. Routine cervical preparation before outpatient hysteroscopy is usually not advisable. Miniature hysteroscopes (2.7 mm with a 3–3.5 mm sheath) used for diagnostic outpatient hysteroscopy significantly reduce the discomfort experienced by the woman.[33]

Indications of Hysteroscopy in Infertility

Diagnostic Hysteroscopy[20]

Diagnostic hysteroscopy can be performed under local anesthesia as an office procedure or in an operating room if some surgical procedure is needed. It is usually performed in cases such as recurrent implantation failures after in vitro fertilization (IVF) or recurrent pregnancy loss or in cases with unexplained infertility.[20]

Hysteroscopic Polypectomy

There are functional and non-functional polyps. Hysteroscopy helps in identifying, localizing, and removing these polyps.[23]

Hysteroscopic Myomectomy

It is done for submucosal myomas **(Fig. 1)**. The International Federation of Gynecology and Obstetrics (FIGO) classifies submucosal myomas as type 0, 1, and 2. Type 0 is intracavitary, type 1 has <50% intramural component while type 2 has >50% intramural component. Hysteroscopic resection is done for type 0 and type 1 myomas while for type 2 laparoscopy is considered a better approach. For patients with recurrent miscarriage and intracavitary fibroids, surgery increases rates of viable pregnancy outcomes.[24,25]

Uterine Malformations

Congenital uterine malformations especially uterine septa contribute to pregnancy wastage and hysteroscopic resection of septa can improve the pregnancy outcomes.[28]

Intrauterine Adhesions

The manifestations of intrauterine adhesions include menstrual issues, impaired implantation, and infertility. Hysteroscopy not only confirms the presence of adhesions, but also helps incision of these adhesions.[26]

Tubal Cannulation

Hysteroscopy along with laparoscopy helps treat fallopian tube obstruction.[17]

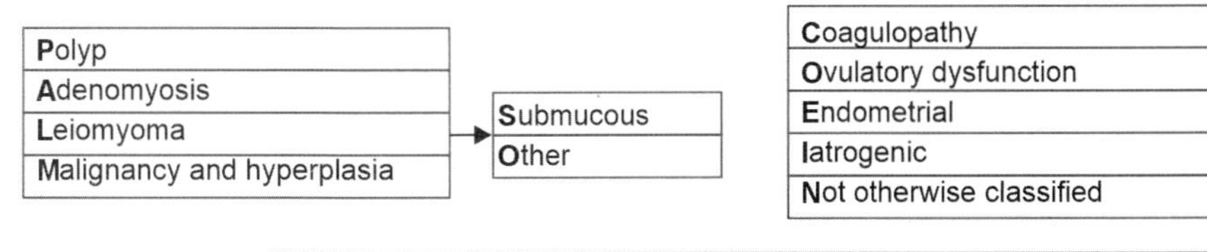

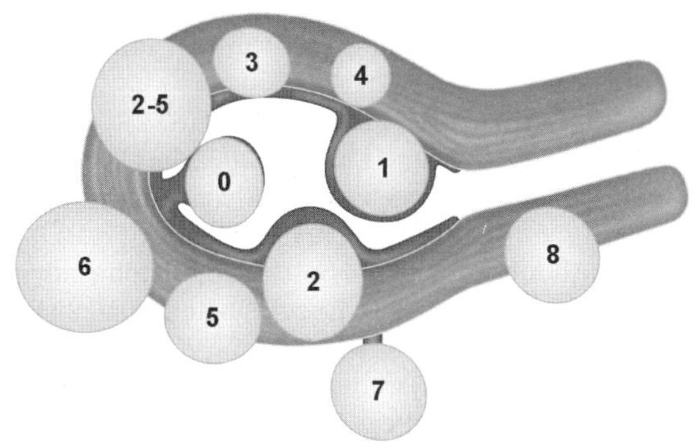

SM-Submucous	0	Pedunculated intracavitary
	1	<50% intramural
	2	≥50% intramural
	3	Contacts endometrium; 100% intramural
O-Other	4	Intramural
	5	Subserous ≥50% intramural
	6	Subserous <50% intramural
	7	Subserous pedunculated
	8	Other (specify e.g., cervical, parasitic)
Hybrid		Two numbers are listed separated by a hyphen. By convention, the first refers to the relationship with the endometrium while the second refers to the relationship to the serosa. One example is below
(contact both the endometrium and the serosal layer)	2-5	Submucous and subserous, each with less than half the diameter in the endometrial and peritoneal cavities, respectively.

Fig. 1: International Federation of Gynecology and Obstetrics (FIGO) leiomyoma subclassification system.
Source: Malcolm G Munro, MD.

TABLE 1: Risks and complications of laparo-hysteroscopy.[34,35]

Complications of laparoscopy[35]	*Complications of hysteroscopy*[34]
Injury to bowel, bladder, intestine or major blood vessel which may subsequently need laparotomy for a repair. Up to 15% injuries may go unnoticed during the surgery	Damage to uterus or nearby structures after perforation
Hernia at the site of entry	Infection
Failure to enter and complete the procedure	Bleeding
Wound bruising/wound infection/ wound gaping	Failure to enter and complete the procedure
Shoulder tip pain	Death (3–8 women in 1 lakh)
Blood transfusion	Anesthesia complications
Death (3–8 women in 1 lakh)	
Anesthesia complications	

COMPLICATIONS

The overall risk of serious complications during diagnostic laparoscopy or hysteroscopy is two in 1,000 women. The risk with laparoscopy increases in obese patients and patients with a previous history of abdominopelvic surgery. Laparoscopic procedures in difficult circumstances such as severe endometriosis and large myomas carry a more risk.[14] The serious and frequent risks associated with laparo-hysteroscopy are mentioned in **Table 1**.

CONCLUSION

Gynecological endoscopy remains the gold standard for evaluation of infertility. Because of the potential diagnostic and therapeutic benefits, endoscopic surgery has proved to be a boon in managing infertile patients. Laparo-hysteroscopy adds to accuracy and precision in evaluation of the patients and has revolutionized the field of infertility.

REFERENCES

1. Vander Borght M, Wyns C. Fertility and infertility: definition and epidemiology. Clin Biochem. 2018;62:2-10.
2. Fields E, Chard J, James D, Treasure T; Guideline Development Group. Fertility (update): summary of NICE guidance. BMJ. 2013;346:f650.
3. Pedro J, Brandão T, Schmidt L, Costa ME, Martins MV. What do people know about fertility? A systematic review on fertility awareness and its associated factors. Ups J Med Sci. 2018;123(2):71-81.
4. Larsen U. Primary and secondary infertility in sub-Saharan Africa. Int J Epidemiol. 2000;29(2):285-91.
5. Carson SA, Kallen AN. Diagnosis and management of infertility: a review. JAMA. 2021;326(1):65-76.
6. Hammond MG. Evaluation of the infertile couple. Obstet Gynecol Clin North Am. 1987;14(4):821-30.
7. Mahran A, Abdelraheim AR, Eissa A, Gadelrab M. Does laparoscopy still has a role in modern fertility practice? Int J Reprod Biomed. 2017 Dec;15(12):787-94.
8. Bouré L. General principles of laparoscopy. Vet Clin North Am Food Anim Pract. 2005;21(1):227-49.
9. Ahmad G, Baker J, Finnerty J, Phillips K, Watson A. Laparoscopic entry techniques. Cochrane Database Syst Rev. 2019;1(1):CD006583.
10. Raimondo D, Raffone A, Travaglino A, Ferla S, Maletta M, Rovero G, et al. Laparoscopic entry techniques: which should you prefer? Int J Gynaecol Obstet. 2023;160(3):742-50.
11. Varma R, Gupta JK. Laparoscopic entry techniques: clinical guideline, national survey, and medicolegal ramifications. Surg Endosc. 2008;22(12):2686-97.
12. Sutton CJG. Guideline on preventing entry-related gynaecological laparoscopic injuries: post-publication reflections of the senior author. Gynecol Surg. 2009;6:301-10.
13. Jamil M, Niaz K, Tahir F. Closed vs. open method of pneumoperitoneum at infra-umbilical site in laparoscopic surgery - a comparative study. J Pak Med Assoc. 2018; 68(10):1478-82.
14. RCOG. Green-top guideline: preventing entry-related gynaecological laparoscopic injuries. London: Royal College of Obstetricians and Gynaecologists; 2008. p. 49
15. Becker CM, Bokor A, Heikinheimo O, Horne A, Jansen F, Kiesel L, et al. ESHRE guideline: endometriosis. Hum Reprod Open. 2022;2022(2):hoac009.
16. Kavic SM, Kavic SM. Adhesions and adhesiolysis: the role of laparoscopy. JSLS. 2002;6(2):99-109.
17. Practice Committee of the American Society for Reproductive Medicine. Electronic address: ASRM@asrm.org. Role of tubal surgery in the era of assisted reproductive technology: a committee opinion. Fertil Steril. 2021;115(5):1143-50.
18. Ng KYB, Cheong Y. Hydrosalpinx - salpingostomy, salpingectomy or tubal occlusion. Best Pract Res Clin Obstet Gynaecol. 2019;59:41-7.
19. Pritts EA, Parker WH, Olive DL. Fibroids and infertility: an updated systematic review of the evidence. Fertil Steril. 2009;91(4):1215-23.
20. Kamath MS, Rikken JFW, Bosteels J. Does laparoscopy and hysteroscopy have a place in the diagnosis of unexplained infertility? Semin Reprod Med. 2020;38(1):29-35.
21. Mercorio A, Della Corte L, De Angelis MC, Buonfantino C, Ronsini C, Bifulco G, et al. Ovarian drilling: back to the future. Medicina (Kaunas). 2022;58(8):1002.
22. Hernandez-Nieto C, Lee JA, Gonzalez K, Mukherjee T, Copperman AB, Sandler B. Conservative treatment versus surgical excision of ovarian dermoid cysts: Impact on ovarian stimulation and IVF cycle success. Int J Gynaecol Obstet. 2020;148(3):392-98.
23. Vitale SG, Haimovich S, Laganà AS, Alonso L, Di Spiezio Sardo A, Carugno J et al. Endometrial polyps. An evidence-based diagnosis and management guide. Eur J Obstet Gynecol Reprod Biol. 2021;260:70-77.
24. Lasmar RB, Lasmar BP, Moawad NS. HYSTEROSCOPIC MYOMECTOMY. Medicina (Kaunas). 2022;58(11):1627.
25. Moawad NS, Palin H. Hysteroscopic Myomectomy. Obstet Gynecol Clin North Am. 2022 Jun;49(2):329-53.
26. Sanad AS, Aboulfotouh ME. Hysteroscopic adhesiolysis: efficacy and safety. Arch Gynecol Obstet. 2016;294(2): 411-6.
27. Moore JF, Carugno J. Hysteroscopy. 2023. In: StatPearls. Treasure Island (FL): StatPearls Publishing; 2023.
28. Practice Committee of the American Society for Reproductive Medicine. Electronic address: ASRM@asrm.org; Practice Committee of the American Society for Reproductive Medicine. Uterine septum: a guideline. Fertil Steril. 2016;106(3):530-40.
29. Thessaloniki ESHRE/ASRM-Sponsored PCOS Consensus Workshop Group. Consensus on infertility treatment related to polycystic ovary syndrome. Fertil Steril. 2008;89(3):505-22.
30. Stamenov GS, Vitale SG, Della Corte L, Vilos GA, Parvanov DA, Nikolova DN, et al. Hysteroscopy and female infertility: a fresh look to a busy corner. Hum Fertil (Camb). 2022;25(3):430-46.
31. Salazar CA, Isaacson KB. Office operative hysteroscopy: an update. J Minim Invasive Gynecol. 2018;25(2): 199-208.
32. Merviel P, Mergui JL, Sananes S, Antoine JM, Salat-Baroux J, Uzan S. Place de l'hystéroscopie dans le diagnostic et le traitement de l'infertilité [Role of hysteroscopy in the diagnosis and treatment of infertility]. Presse Med. 2000;29(23):1302-10.
33. Mazzon I, Favilli A, Grasso M, Horvath S, Bini V, Di Renzo GC, et al. Pain in diagnostic hysteroscopy: a multivariate analysis after a randomized, controlled trial. Fertil Steril. 2014;102(5):1398-403.
34. RCOG. Green-top guideline: Consent advice number 1: Diagnostic Hysteroscopy under general anaesthesia. London: Royal College of Obstetricians and Gynaecologists; 2008. p. 1-4.
35. RCOG. Green-top guideline: Consent advice number 1: Diagnostic Laparoscopy. London: Royal College of Obstetricians and Gynaecologists; 2008. p. 1-4.

62

Chapter

Total Laparoscopic Hysterectomy

Sushila Saini

INTRODUCTION

Hysterectomy is the commonest surgical procedure done by a gynecologist. According to the International Journal of Reproductive Medicine Volume 2014 (2014), the most common surgical approach was abdominal (74.7%), followed by vaginal (17.8%), and laparoscopic (6.6%) hysterectomy. Over the years, techniques have evolved from abdominal to total laparoscopic hysterectomy (TLH), first introduced in 1989 by Reich et al. (1989).[1] A traditional open hysterectomy requires an abdominal incision of 4–8 inches but in TLH only three to four small keyhole abdominal incisions ≤1 cm in size are required. Even a large uterus can be removed laparoscopically using this technique.

Total laparoscopic hysterectomy is defined by the removal of the uterus along with the cervix entirely by laparoscopic through vaginally or abdominally, along with laparoscopic closure of the vaginal cuff.

The selection of the route for hysterectomy is decided by:

- The size of the uterus,
- The other associated pathologies,
- The need for concurrent procedures,
- Surgeon's skills and training,
- Technological environment and support team availability in the operating room,
- Scheduled or emergent nature of the particular case and
- Finally, the patient's preference.

BENEFITS

Evidence-based studies have shown that laparoscopic hysterectomy (LH) is a better alternative to abdominal hysterectomy (AH) in terms of:

- Less postoperative morbidity,
- Early return to work,
- Better cosmetic results,
- Less requirement of antibiotics, blood, and IV fluid,
- Early discharge from hospital and
- Improved quality of health care.

INDICATIONS

The indications are mainly, (ACOG, 2021)[2]

- Uterine leiomyomas (51.4%)
- Abnormal uterine bleeding (41.7%)
- Endometriosis (30%), and
- Prolapse (18.2%).

CONTRAINDICATIONS

There are no absolute contraindications to a TLH for benign gynecologic diseases. Large size uterus can be removed by the laparoscopic route, it depends on the surgeon's expertise.

PREOPERATIVE MEASURES

- *Consent:* As per the ethical guidelines laid by Indian regulatory authorities, risks associated with anesthesia and surgery should be properly explained to patients and attendees with proper documentation of the consent process.

  In case of a history of previous surgery, tuberculous (TB), endometriosis, and pelvic inflammatory disease (PID), the possibility of bladder, or bowel injuries should be explained well before the surgery. In cases of big uterus or fibroids, grade 4 endometriosis or large masses there may chance of laparotomy, which should also be explained.
- *Bowel preparation:* We use two tablets of Dulcolax and four tablets of charcoal (Gasex™, Himalayas).
- *Thromboprophylaxis and pneumatic compression:* As per need.

INTRAOPERATIVE MEASURES

Positioning of Team, Patient, and Laparoscopy Tower

- The patient is positioned in dorsal decubitus, under general anesthesia, with oral tracheal intubation.
- The legs are positioned in 30° flexion, in lithotomy.
- The arms along the side of the body, to prevent brachial plexus injuries and buttocks extend slightly over the edge of the surgical table to facilitate the movement of the uterine manipulator.
- The bladder is catheterized if urine is not passed just before the operation to prevent any bladder injuries during entry into the abdomen.
- The laparoscopic tower is usually placed near the patient's right foot or in the middle of the legs, which allows the surgeon and the first assistant a clear vision. In cases of multiple monitors, they must be placed in front of each member of the surgical team.
- When the surgeon is right-handed, he or she will normally work from the left side of the patient. Then, the first assistant will be placed on the right side and the second assistant will be placed between the legs. Some right-handed surgeons prefer to be to the patient's right to allow ease of suturing cross table with the right hand.

Instruments

Laparoscopic instrument's trolley (**Fig. 1**):

- One 10 mm trocar and three 5 mm trocars
- 10 mm 30° telescope
- Graspers
- Bipolar shearer and bipolar
- Suction
- Monopolar hook
- Tooth and non-tooth needle holder.

Vaginal trolley (**Fig. 2**): Uterine manipulator, sims speculum, vulsellum, and vaginal dilators.

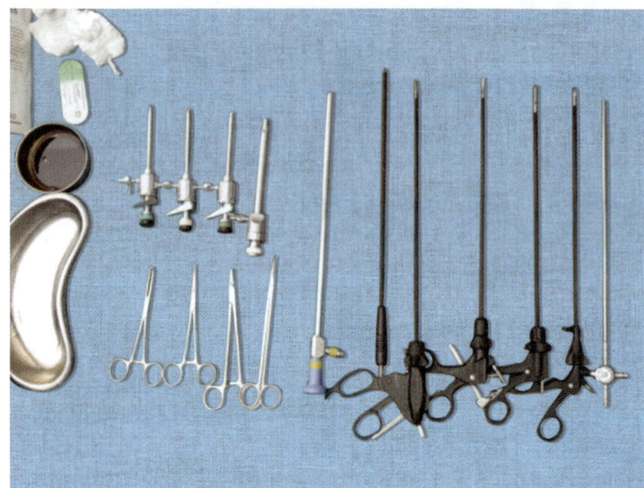

Fig. 1: Lap instrument's trolley.

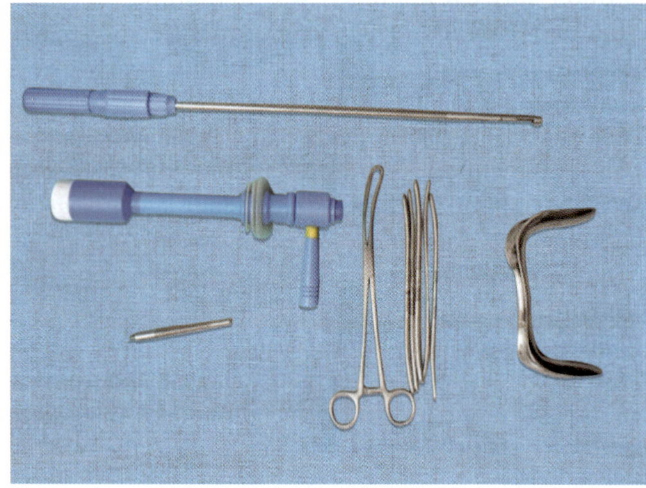

Fig. 2: Vaginal trolley

Surgical Steps

Abdominal Access: Creation of Pneumoperitoneum

Check before pneumoperitoneum creation:

- General anesthesia preferred for laparoscopy.
- The table must be placed in a neutral position.
- The pneumoperitoneum circuit must be checked.

Achieving intraperitoneal access using a recognized method: We can use any of the available methods to gain intraperitoneal access in which you had trained and confident.

At our center Jaipur Doorbeen Hospital (JDH), we follow direct entry with safety trocars which are actually like a 10-mm Veress as the trocar has retractable spring-like action as in Veress in which blades come out only when pressed against the firm surface such as rectus sheath and as soon as resistance is lost, blades also move inside and the blunt end came out so minimal chance of visceral injuries.

Direct trocar entry seemed to be far more dangerous than other methods, but in reality, it is considered a relatively safe alternative for closed laparoscopic access, when performed properly. The aim of this technique is to minimize complications related to the insertion of the Veress needle such as gas embolism. In this technique

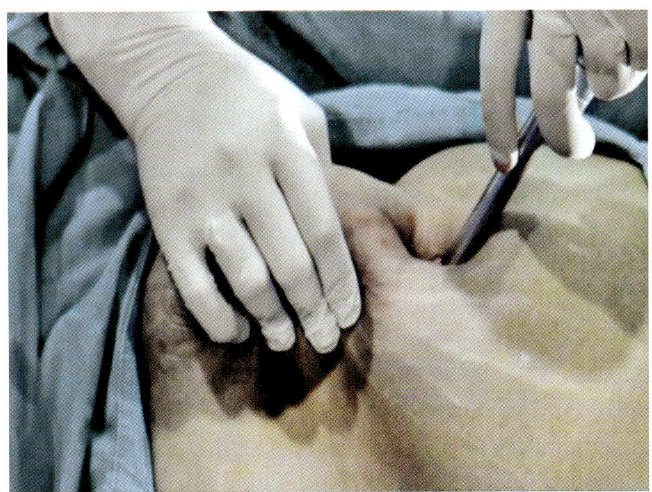

Fig. 3: Direct trocar entry.

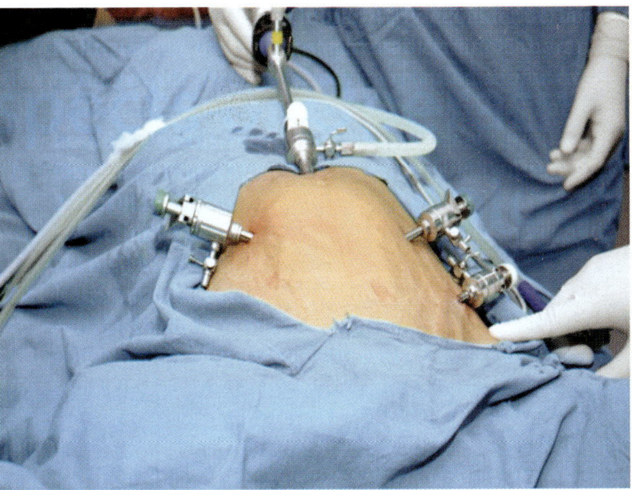

Fig. 4: Position of secondary trocars.

only one blind step (trocar) instead of three steps as in closed insufflated technique (Veress needle, insufflation, and trocar) so it is a faster method.

- A small skin incision, most often two fingers above the umbilicus is given, lifted the abdominal wall with a hand or clamps, insert 10 mm trocar by screwing hand movement **(Fig. 3)**.
- The rate of serious injury (serious vascular or bowel injury) is estimated to be 0.4/1,000, which is equivalent to that reported for the classical closed entry technique; *according to Cochrane Database of Systematic Reviews 2012, Direct trocar entry versus Closed entry technique.*

Inspection of the Abdominal Cavity

- Complete 360° inspection must be performed just after entry into the abdomen.
- Now Trendelenburg position should be requested as it allows the visualization of the promontory and the right ureter crossing the common and external iliac artery.

Secondary Trocars Insertion

We are ipsilateral surgeons, so we use two left and one right side 5 mm ports for most procedures. Secondary trocars, as a rule always inserted under vision.

First trocar inserted at 2 cm above and medial to anterior superior iliac spine (ASIS), second at 8 cm above and medial to ASIS or roughly in between the primary port and first 5 mm port on left side following *diamond baseball concept* **(Fig. 4)**. While placing this trocar, we should check the location of the inferior epigastric artery in the anterior abdominal wall. If it is not localized, always put the lateral trocar four fingers lateral to the midline or lateral to the rectus muscle.

Third trocar will be inserted on the right side, and it is the mirror image of second left port.

Inspection of the pelvis

- Good pelvic exposure is achieved by mobilization of the small bowel and sigmoid. The small bowel with its mesentery should be flipped from the pelvis to the abdomen following its mesenteric attachment.
- Perform adhesiolysis, if necessary and look for a transparent window between the omentum and peritoneum to separate adhesion bands.
- Now look for the size and anatomy of the uterus and adnexa.

Placement of Uterine Manipulator by the Second Assistant

For placement of uterine manipulator, hold the cervix with a vulsellum, measure the uterocervical length and select an adequate length size of attachment of manipulator, which will be set into the uterus and slowly screwed the manipulator into the uterus to fix it and till it reaches the fundus. Now the cervical ring is inserted after applying some lubricating gel to prevent any vaginal injury. If the cervical hook is also there in the manipulator, hold the anterior lip of the cervix by the hook before inserting the ring **(Figs. 5 and 6)**.

- The soft silicone rings are to be inserted while cutting the vault.

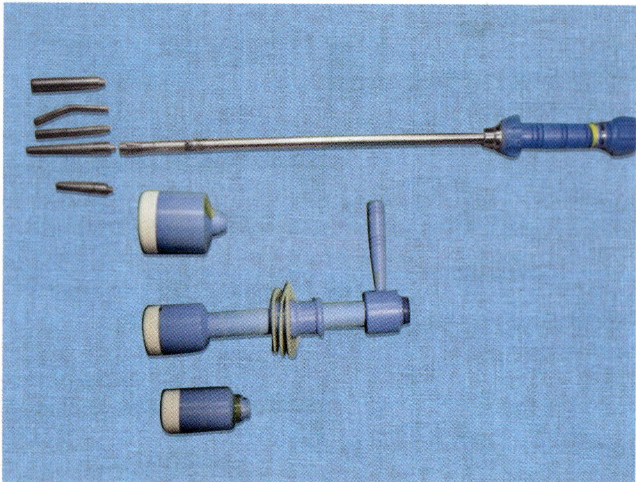

Fig. 5: Vaginal manipulator.

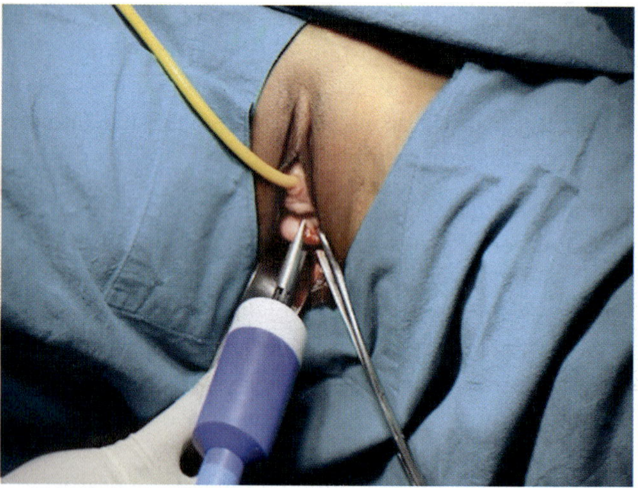

Fig. 6: Placement of vaginal manipulator.

- Now antevert the uterus entirely and check Pouch of Douglas (POD) and ovarian fossa for any pathology.
- Mobilize the sigmoid colon by dividing congenital adhesions between the sigmoid colon and the lateral pelvic wall to facilitate the exposure of the left ureter and left infundibulopelvic ligament and get good exposure into the pelvis by keeping the sigmoid colon out of the field.

STEPS OF TOTAL LAPAROSCOPIC HYSTERECTOMY

- Salpingectomy
- Tackling ovarian ligament
- Round ligament separation
- Opening of the broad ligament
- Opening of uterovesical (UV) fold of peritoneum and bladder separation
- Opening the posterior peritoneum
- Tacking B/L uterine arteries
- Opening the vault
- Delivery of uterus vaginally
- Vault closure.

Salpingectomy

We routinely perform salpingectomy as the first step of hysterectomy by laparoscopy. We see the pelvis from the cephalic side and the fundus of the uterus first by an endoscope inserted from the umbilical side. Hence, as the topmost structure of uterine adnexa, tubes are the first to be seen. So, we start with the salpingectomy first and then proceed to the ovarian pedicle.

During salpingectomy, hold the fimbrial end gently by non-tooth grasper in one hand and give mild traction and energy in the other hand. Now start coagulating and cutting the mesosalpinx very near the tube so we can preserve the mesosalpingeal cascade of vessels to maintain the blood supply of ovaries after hysterectomy. Do not go near the cornual end of the uterus, cut the tubes 1–2 cm away to prevent bleeding from cornual vessels **(Figs. 7A and B)**.

Take out both the tubes one by one. We routinely use bipolar shearer nowadays as it is a reusable instrument and cost-effective. It coagulates and cuts in the same sitting. This instrument cuts precisely at the site of coagulation done.

Tackling Ovarian Ligament

If oophorectomy is planned, then tacking infundibulopelvic (IP) ligament may be the first step or can proceed as usual and take out the ovaries at the end of the procedure.

While taking the IP ligament, always locate the ureter and then start coagulating the IP ligament. In cases of endometriosis, the ureter is adhered to the pelvic peritoneum and lifted off the base and attached to the ovarian base or IP ligament. It may cut along with the IP ligament. So always first locate and separate the ureter.

Hold the ovary and pull it away from the pelvic wall, give an excellent stretch to IP, and then start coagulating it, do not overcook, and then cut in small bits. Proceed in small bits toward the uterine cornual end and cut the utero-ovarian pedicle 1–2 cm away from cornua.

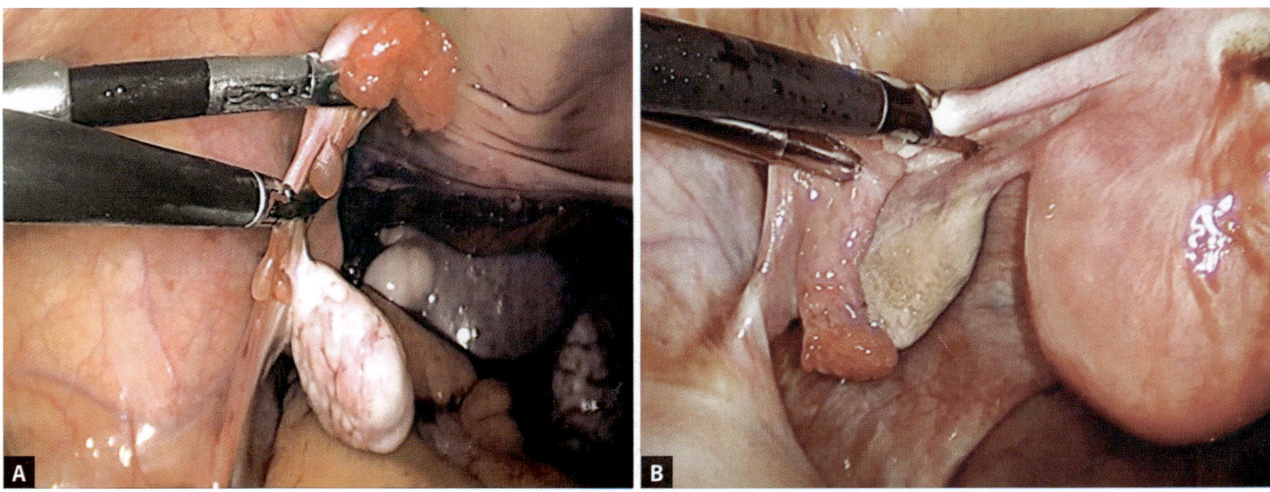

Figs. 7A and B: Grasping the fimbrial end by non-tooth grasper in one hand and bipolar shearer on the other hand.

The thumb rule is always hugging the ovaries, by this, we can prevent cornual bleeding and ureteric injury.

When oophorectomy is not planned, then directly takes utero-ovarian ligament 1-2 cm away from cornu and coagulates utero-ovarian anastomosis to separate the ovaries from the uterus **(Fig. 8)**, take them as a whole bunch in one go to coagulate them as they may bleed when cut partially if you take these in small bites.

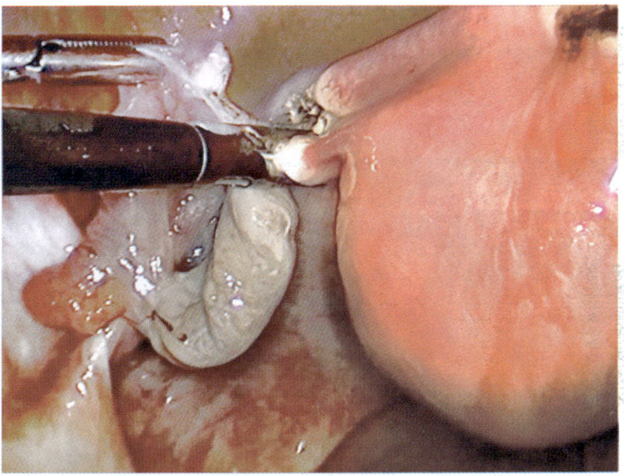

Fig. 8: Ovarian ligament is taken slight away from cornu.

Round Ligament Separation

Take the round ligament 2–3 cm away from the cornual end of the uterus and prevent retrograde bleeding from the cornual vessels that usually retract into the uterus **(Fig. 9)**.

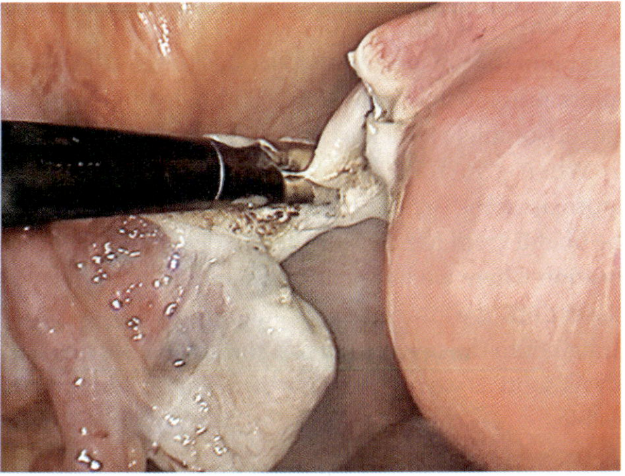

Fig. 9: Coagulation and cutting of the round ligament.

Opening of Both the Folds of the Broad Ligament

After cutting and separating the round ligament, we enter the broad ligament directly. At this step, we open the anterior and posterior fold of the broad ligament by inserting the grasper tip or holding the anterior and posterior leaf and letting the CO_2 enter to open them. We cut them until the base, separately, and uterine arteries are exposed at the bottom of broad ligament leaves **(Fig. 10)**.

The uterine manipulator is pushed cephalad, lateral to the opposite side of the dissection in a neutral position.

Opening of Uterovesical Fold of Peritoneum and Bladder Separation

Anterior broad ligament leaf dissection extended further by curving it anteriorly by lifting the UV fold by the

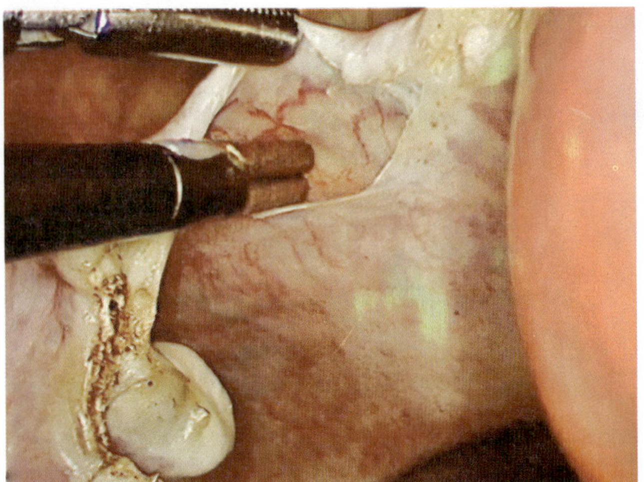

Fig. 10: Forming a plane between anterior and posterior leaf.

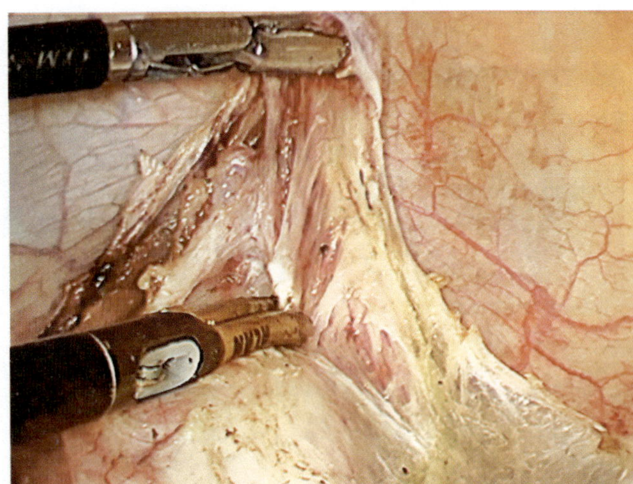

Fig. 12: Holding the bladder by nontraumatic grasper and separating it from the cervix.

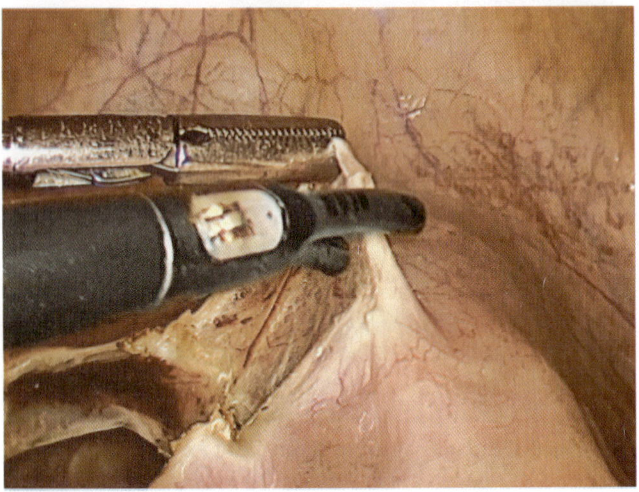

Fig. 11: Holding the UV fold by non-tooth grasper and opening it.

nontraumatic grasper. As soon as we cut the UV fold, CO_2 enters into loose areolar tissue and helps in making the space. Complete it till the contralateral side **(Fig. 11)**.

When you focus on the UV fold, you will find approximately 1–2 cm clear transparent space between the bladder and uterine attachment of the fold is the exact location where we have to cut the fold. At the time of this dissection, the uterine manipulator should be pushed straight into the pelvis to stretch the UV fold.

As soon as we open the UV fold, the bladder becomes visible.

For bladder dissection: Keep the uterus pushed straight into the pelvis by the uterine manipulator, lift the bladder with grasper vertically and start dissecting at the base, i.e., begin separating the bladder fibers from the cervix by sharp and blunt dissection. Sharp dissection should be preferred over blunt as this will lead to less bleeding and give good tissue planes to work further.

Proper bladder tissue should be held in the grasper, not the peritoneum only because if you hold only the peritoneum, the required pull-on bladder fiber will not transfer to the base and lead to loss of plane during bladder dissection. This technique also helps dissection of the adhered bladder of previous lower segment cesarean section (LSCS) or PID when we proceed from the central approach **(Fig. 12)**.

Bladder Separation in Previous Lower (Uterine) Segment Cesarean Section Cases

Two approaches—central and lateral approach.

The central approach, also known as a *fundal or direct* technique, where we directly attack the adhered area of the bladder by lifting the whole bladder tissue and start cutting the bladder fibers from the center of the cervix, keeping the instrument flushed with cervical tissue and continue cutting the thick white fibrous bands till we enter the standard vesicovaginal space as the adhesions are found in the fundal region only.

The lateral approach is also known as the *indirect technique*. We start separating from the lateral side of the bladder, looking for standard vesicovaginal space from the lateral side as there are no adhesions or fibrous tissue laterally **(Fig. 13)**.

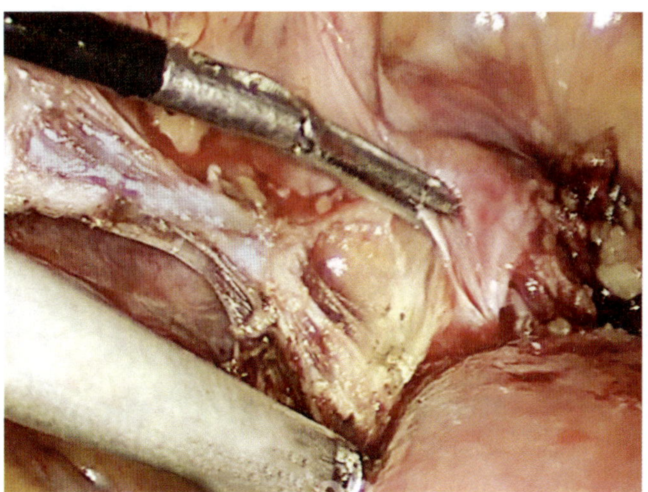

Fig. 13: Lateral approach for separating bladder in previous lower segment cesarean section (LSCS).

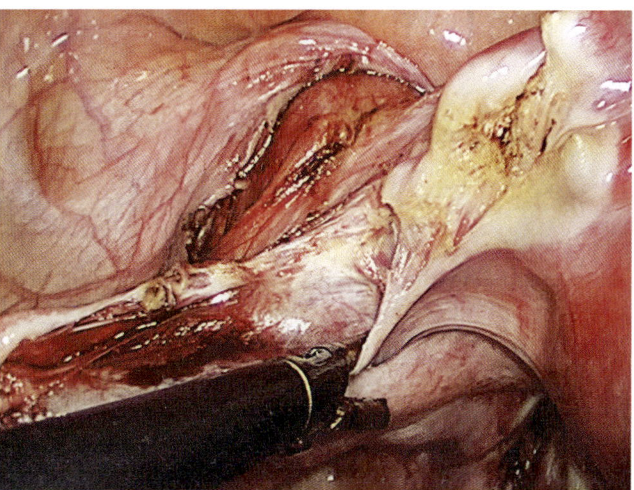

Fig. 14: Opening the posterior peritoneum above the UCL.

We start just above uterine vessels and bladder and progress into space below the adhered fundus, making a window by dissecting the bladder and vagina. Once enough space is created correctly, start cutting the adhered fundus of the bladder by lifting the whole bladder from the lateral window and separating it from the cervix.

But when comparing both the techniques, the central approach is more convenient as we routinely perform this only in every case. If cystotomy occurs at the bladder fundus, will be much easier to repair and this is entirely away from the ureters and uterine vessels.

While in the lateral approach, dissect directly over the uterine vessels, so there may be high chances of bleeding from uterine arteries and coagulating them without separation of the bladder, which may lead to a UV fistula. Secondly, if cystotomy occurs on the lateral side, it will be quite near to ureters and repair of it is also problematic.

Once the bladder is separated sufficiently down, a silicon cup of the vaginal manipulator is pushed into the fornices to make the vaginal wall prominent. This will lead to the fall of the bladder fundus caudally, now we can proceed with uterine vessel coagulation and cutting. We should never touch uterine arteries before separating the bladder.

Opening of the Posterior Peritoneum

The uterine manipulator should be pushed anteriorly and cranially and slightly lateral to the opposite side where the dissection will start **(Fig. 14)**.

The posterior peritoneum should be dissected toward the uterosacral ligaments (USL). Coagulation and the cutting of the USL are not always necessary but may be helpful in the case of a larger uterus for mobilization.

There is no need to cut the vaginal attachment of USL in normal-sized uteri because this will support the vault and help to prevent future vault prolapse.

Taking Uterine Vessels

Following essential steps should be followed while taking uterine vessels:

- It is performed at the level of ascending portion of the uterine artery.
- Uterine vessels should be adequately exposed and skeletonized **(Fig. 15)**.
- The posterior peritoneum must be separated towards the USL.
- The bladder should be pushed sufficiently down.
- The ureter should be identified, especially in the cases of distorted anatomy.
- Uterine vessels should be approached at perpendicular i.e., the tip of bipolar or sealer should be at 90° to the vessel to be taken, so the least diameter of the vessel comes into the jaw of the instrument for proper sealing of the vessel **(Fig. 16)**.
- Energy source should seal the vessel at the level of the intima to prevent postoperative bleeding if blood pressure (BP) rises while extubating or in the postoperative period.

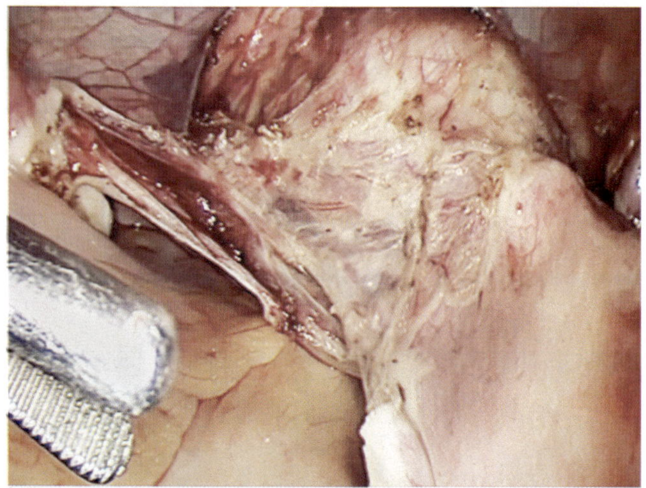

Fig. 15: Skeletonization of uterine vessel.

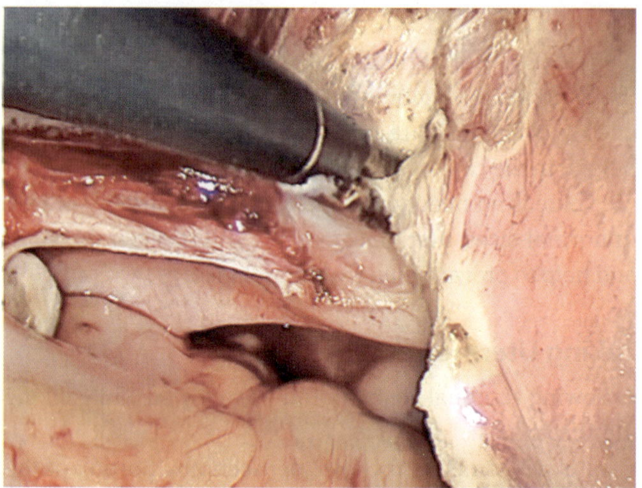

Fig. 16: Coagulation of uterine vessel at 90°.

- Do not overcook and no blackening or charring of tissue should be done as it will become sticky, and this charred tissue will be avulsed out with the jaw of the instrument, which may lead to profuse bleeding.

Once the uterine vessels are cut apart, the next clamp is just medial to the vascular pedicle, to separate the vascular pedicle from the uterine side and to cut the Mackenrodt's ligament. Mackenrodt's ligament is seen as white fibrous tissue just adjacent to the uterine vessel. When we cut this ligament, we should stop when the pink tissue becomes visible as it notifies the starting of vaginal tissue. Further going down may lead to ureteric injury **(Fig. 17)**.

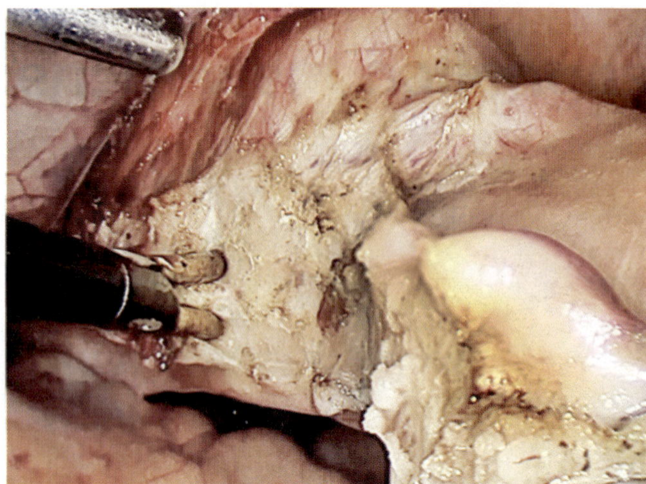

Fig. 17: Taking the Mackenrodt's ligament.

Colpotomy (Opening the Vault)

Before starting the cutting of the vagina and opening the vault, soft silicone rings should be placed into the vagina to prevent the gas leak after opening the vault. The uterine manipulator should be pushed cranially.

Also, check on both sides of the cervix that all attachments have already been dissected or not after pushing the colpotomizer vaginal ring sufficiently high up into the fornices and then moving the uterus right and left side.

We prefer to use an L hook attached with monopolar energy set on a 40-watt pure cutting current and starting from the right-anterior side and moving toward the left and then posterior moving slowly and circumferentially, taking care of small bleeders at angles with coagulation current delivered with heal of hook **(Figs. 18 and 19)**.

We should avoid excessive coagulation in the vaginal vault to prevent vault dehiscence and ensure good healing.

Uterus Retrieval

Specimens are retrieved vaginally, with or without morcellation, as per the size of the specimen and then the vagina is occluded to the prevention of gas leaks and to restore pneumoperitoneum.

We prefer to use two medium-sized gauze pads placed into a glove and placed into the middle of the vagina.

Closure of the Vault

The vaginal vault can be sutured with interrupted or continuous sutures, in one or two layers with delayed

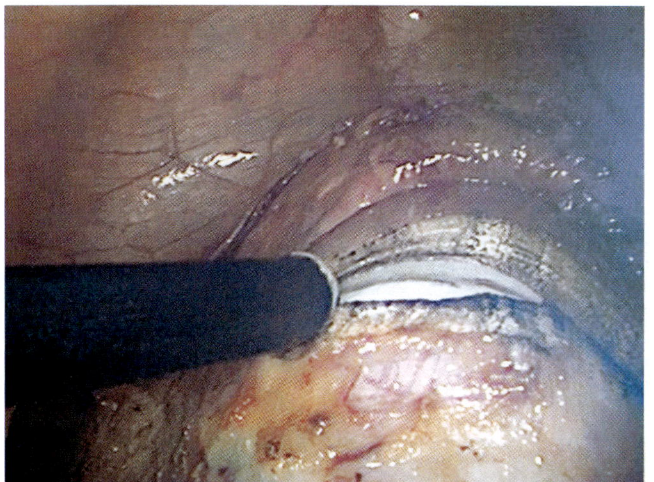

Fig. 18: Vault opened by monopolar hook.

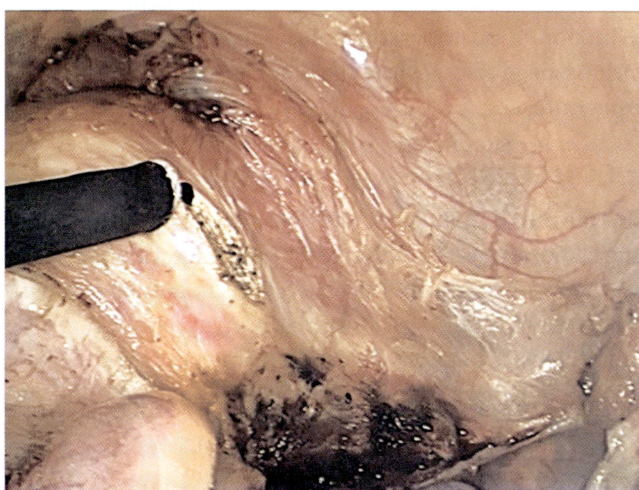

Fig. 19: Vault cut on the cup of the uterine manipulator an angle.

absorbable sutures by laparoscopically. We can also use barbed sutures to close the vault.

While taking a suture through the vagina, the entire thickness of the vagina should be included in the suture and passed through the sufficient width at the edges, including vaginal mucosa, fascia and the peritoneum from normal-looking healthy tissue, not through the energy-treated white dead tissue as they will slough off finally and both USL should be included as they also have the blood supply through the uterosacral vessel, and it may lead to oozing at the vault, which may be a cause of vault hematoma and this inclusion into the vault suture will help in the suspension of the vault.

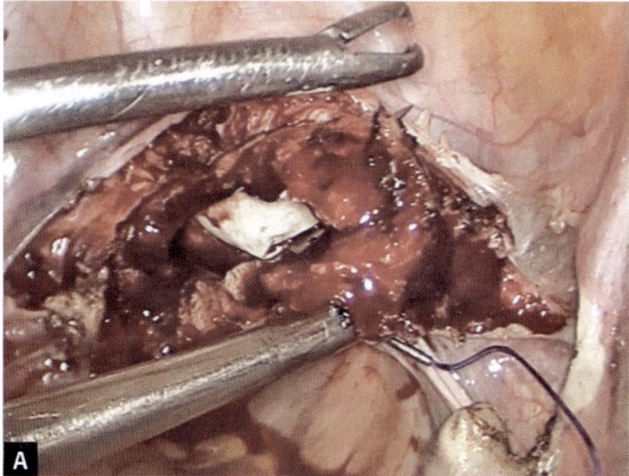

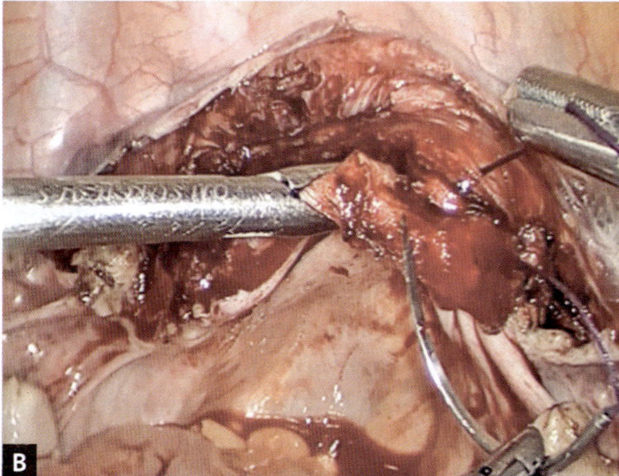

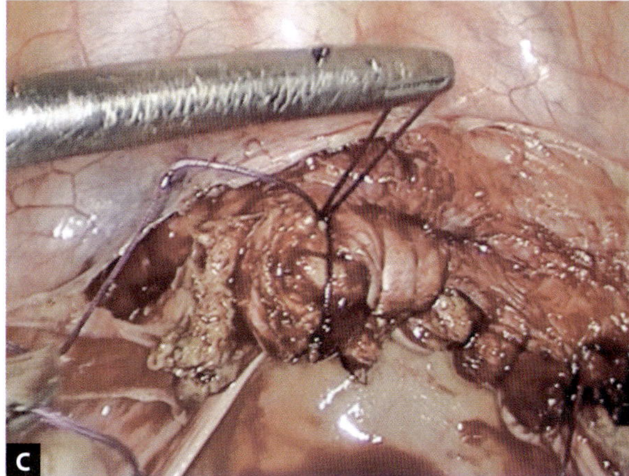

Figs. 20A to C: Vault suture in a continuous manner taking uterosacral ligament along with it.

Start from one side of the vaginal angle and tie it properly. Then we can start from another side, take the angle, continue suturing continuously until the angle, and secure it with the previously taken angle suture **(Figs. 20A to C)**.

We can take the second layer of sutures, including pubocervical and rectovaginal fascia, as they are detached from the peri cervical ring and with each other. This disruption of the peri cervical ring may lead to vault prolapse. This suture line is like an imbricating suture as we take the second layer of the suture line into LSCS.

We can do uterosacral shortening in cases of prolapse. At this time, we should be careful not to take the ureter into the bite or the ureter may be pulled inside the POD and got kinked so we can give a releasing incision into the peritoneum between the ureter and uterosacral ligament to prevent any kinking or inclusion of ureter into the suture.

At last, we can take purse-string Moskowitz suture into the POD to close it and prevent or treat enterocoel in cases of associated prolapse.

CHECKING FOR HEMOSTASIS AND ANY POSSIBLE INJURY

The pelvis should be washed with normal saline and aspirated after intra-abdominal pressure is reduced to 10 mm Hg. Vascular pedicles, vaginal vault, bowel, bladder, and ureters should be checked for no damage.

Cystoscopy is performed to see the bladder and ureteral injury if suspected.

If the ureteric injury is suspected, we recommend doing a complete dissection of the ureter. Alternatives include performing a retrograde ureterogram with the assistance of a urologist.

If you find long pedicle hypermobile ovaries, they can be fixed to round ligaments to prevent future torsion and prevent them from being stuck at the vault, which may be because of vault pain in the postoperative period.

All the secondary trocars should be removed under direct vision and examined for any bleeding. Evacuate the gas entirely before taking out the central 10 mm trocar. This trocar should be taken out after putting the endoscope inside to prevent omentum or bowel extrusion into the port. Fascia of ≥10 mm size should be sutured **(Fig. 21)**.

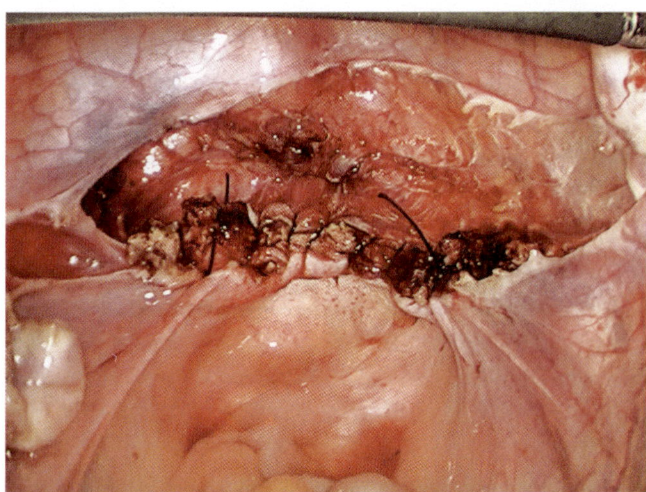

Fig. 21: Final view after vault suturing.

We may inject local subcutaneous anesthesia for the reduction of postoperative pain. Close the skin incisions with tissue glue or a simple nylon suture at the end of the procedure.

CONCLUSION

The current evidence shows that laparoscopic hysterectomy should be the preferred alternative to abdominal hysterectomy. Laparoscopic surgery is associated with the proven benefits of minimally invasive surgery. We should always perform the set steps in every case, making a protocol or step of surgery to master the technique, and minimize the complication.

REFERENCES

1. Reich H. New techniques in advanced laparoscopic surgery. Baillieres Clin Obstet Gynaecol. 1989;3:655-81.
2. Matteson KA, Butts SF. Choosing the Route of Hysterectomy for Benign Disease. The American college of obstetrics and gynecologists women's health care physicians. Committee Opinion: Number 701;2017.

Chapter

63

Complications of Endoscopic Surgery

Dilpreet Kaur Pandher, Shail Kaur

INTRODUCTION

Gyne endoscopy or minimal invasive surgery encompasses two types of procedures, laparoscopy and hysteroscopy. Laparoscopy is entry into the peritoneal cavity through a small incision and gas is insufflated into the abdominal cavity for better visualization of the contents, and energy devices are used for the surgical interventions. Minimal invasive surgery has many advantages but the procedure is also associated with the few complications specific to it's procedural aspects.

COMPLICATIONS OF LAPAROSCOPY

Overall rate of complications in laparoscopic surgeries is very low, with an incidence of around 4.64/1,000 procedures.[1]

Complications related to abdominal access can occur at two points:

1. Primary entry—for insufflation and camera placement.
2. Secondary ports—for putting instruments.

Half of the total complications occur at the time of primary entry for abdominal access.[2] 75% injuries are bowel or vascular in origin, **Table 1** describes the proportion of various complications.[3]

There are various risk factors that corelates with increased incidence of complications in laparoscopic surgery as mentioned in **Box 1**.

A total of 15% of conversions to open surgery are found to be due to access-related complications, most important ones are bleeding and bowel injury. Access-related vascular and gastrointestinal injuries are also the leading cause of mortality in laparoscopic surgeries.[4] The mortality rate associated with the laparoscopic surgeries is reported to be 3.33/1,00,000.[1]

Vascular Injuries

The incidence is 0.1–6.4/1,000 laparoscopy cases. The second common cause of procedure-related deaths, first being anesthesia-related causes.[5] Mortality rate among the patients suffering a major vascular injury during laparoscopy is reported to be 15%.[6] **Table 2** differentiates various characteristic features associated with major and minor vascular injuries, their prevention and management.

TABLE 1: Proportion of various complications.

Type of injury during abdominal access	*Proportion (%)*
Small bowel	25
Iliac artery	19
Colon	12
Iliac or other retroperitoneal vein	9
Secondary branches of a mesenteric vessel	7
Aorta	6
Inferior vena cava	4
Abdominal wall vessels	4
Bladder	3
Liver	2
Other	<2

BOX 1: High risk factors for laparoscopy.

- Prior surgery (pelvic or abdominal)
- Huge mass
- Poor cardiopulmonary reserve of the patient
- Distended bowel loops
- Tissue inflammation
- Surgeon skill of the procedure

TABLE 2: Characteristic feature of major and minor vessel injuries.

	Major vascular injuries ***Incidence: 0.1–1%***[7,8]	***Minor vascular injuries***
Vessels involved	• Aorta (lies at 2 cm distance from anterior abdominal wall in a thin patient) • Inferior vena cava • Right common Iliac artery (crossing midline)	• Omental and Mesenteric—during primary port entry • Abdominal wall—superior and inferior epigastric artery lacerations—during lateral trocar placement
Effect	Massive air embolism—if punctured by Veress needle	• Blood transfusion • Conversion to open surgery • Repeat surgery
Presentation	• Rapid exsanguination • Death • Free blood • Retroperitoneal bleed • Bleeding into mesentery	• May bleed sometime after surgery when tamponade effect of pneumoperitoneum is gone[9] • Hematoma 2–3 days postoperative • Abdominal wall pain • Bruising/external bleeding
Precautions		• Using smooth conical tip trocars[10] • Prior transillumination of abdominal wall
Management	• Reverse Trendelenburg position to minimize hypotension • Fluid resuscitation • Alert anesthesia team • Call vascular or trauma surgeon • *Immediate laparotomy:* Direct pressure and meticulous repair of vessel and blood transfusion	• Mostly conservative • *Indications of intervention:* Rapidly expanding hematoma, infected hematoma, hemodynamic instability • Attempt to control bleeding laparoscopically—compression (direct with gauze soaked in epinephrine), omentum, surgical/gelfoam, glue, clip/suture/cautery[11]
Convert to open surgery	Depends on: • Amount and rate of bleeding • Patient status—hemodynamic, advanced age, comorbidity • Inability of find source of bleeding • Surgeon comfort	

TABLE 3: Management of inferior epigastric artery bleed.

Tamponade	Placing a Foley catheter through the trocar site, inflating the balloon, traction and putting a clamp over it at the abdominal wall to maintain the pull[12]
Electrocoagulation	With bipolar forceps introduced through the contralateral port
Suturing	Enlarging the trocar site to visualize, clamp, and ligating the vessel

Management of Inferior Epigastric Artery Bleed

Inferior epigastric artery bleed is a form of minor vessel injury and may be encountered in clinical practice if simple practices like direct visualization and transcutaneous visualization through telescopic light are not followed before trocar insertion. **Table 3** represents management aspects in such cases.

Bowel Injury

This is the third leading cause of death following laparoscopic procedures.[5] It's incidence is 0.03–0.18%[13] with high mortality rate of 3.6%. incidence varies with the mode of entry; 0.4% with veress needle blind entry versus 1.1% with the open method. This contrary finding is possibly due to the selection of open technique for complicated cases and with anticipated adhesions.[13] Open entry technique carries less incidence of failed entry but there is no difference in visceral or vascular injury.[14] timing, site, type of injury along with it's management are described in **Table 4**.

Urinary Tract Injuries

Salient features of urinary bladder and ureteric injuries with prevention and management are described in **Table 5**.

TABLE 4: Bowel injury.

Timing of injury[1,15]	*Abdominal access (1/3rd)*	*Primary port (50%)*	*Veress needle (33%)*	*Secondary port (17%)*
Causes	Tissue dissection			
	Tissue grasping			
	Electrocoagulation			
Site[16]	Small bowel followed by stomach, liver and colon			
Small bowel injury management	Small veress puncture without tearing		Conservative	
	Trocar puncture		Primary suturing: one or two layers	
	Full thickness Injury ≥5 mm		Repaired in 2 layers	
	Laceration exceeding half of the luminal diameter		Segmental resection is recommended	
Stomach injury	Prevention	For supraumbilical ports: Decompression of stomach by nasogastric tube suctioning		
	Management	Repair in 2 layers, irrigation and suction of abdomen		
Colon	Puncture but no tearing	Antibiotic, copious irrigation and suction		
	Small rent with minimal soilage	• Defect closed in two layers with copious irrigation • Colostomy: rarely needed		
	Detection of injury/ leak	Filling the posterior cul-de-sac with normal saline, injecting air into the rectum through a rigid proctoscope or catheter-tipped bulb syringe, occluding the sigmoid proximally with bowel graspers, and looking for air leakage manifested as bubbles[12]		

TABLE 5: Bladder and ureter injuries.

Timing	*Mobilization off the lower uterine segment and cervix*[12]	
Incidence	Rare, 0.05– 0.66%	
Risk factors	Prior pelvic surgery, endometriosis distorting pelvic anatomy Midline, suprapubic trocar insertion with full bladder	
Prevention	Void immediately before procedure	
	One time drainage after induction of anesthesia (safer option)	
Detection	Defining bladder margins by filling with 300 mL saline	
	Continuous drainage with Foley catheter (helps in early recognition of injury)	Bloody urine Gaseous distention of urobag[17] Instillation of methylene blue dye into the bladder in suspected bladder injury cases
Management: Urology consultation	Puncture with veress needle	Repair generally not needed
	Small puncture <5–10 mm	Continuous bladder drainage (decompression) for 7–10 days
	• Large or irregular defects • Rents involving urethra or ureter	Watertight, multilayered repair using absorbable suture laparoscopic or open Foley catheter for 4–10 days depending upon size and location of tear[12]
Ureter Injury	Incidence	0.02–0.4%[12]
	Timing	Ligating and cutting ovarian and uterine vessels
	Detection	Cystoscopy to rule out ureteric injury
	Management	Stent placement, reimplantation, end to end anastomosis

TABLE 6: Port-site hernia.

Incidence[18]	0.21% (as per AAGL)	18% occurred in spite of fascial closure
Factors affecting[19]	• Trocar diameter >10 or 12 mm • Old age • More operative time • Postoperative surgical site infection	• Access technique • High body mass index • Excess tissue manipulation • Retrieving large specimen
Prevention[20-22]	• Using less diameter trocar • Bladeless trocars	• Radially dilating trocars • Repairing fascial defects
Management	Repair hernia to prevent complications	
Extraction site[23]	Rate	7.2%
Sites	Maximum at midline (13%)	Least at Pfannenstiel location (0.9%)
Prevention	Keeping extraction site off the midline or Pfannenstiel	

TABLE 7: Pneumoperitoneum related complications.

Type of complication	*Cause*
Emphysema: Subcutaneous, mediastinal	Improper positioning of pneumoperitoneum needle
Pneumothorax, cardiac arrhythmia, carbon dioxide retention	Physiological effect of insufflation
Postoperative pain: Abdominal or shoulder	Retained intra-abdominal gas causing diaphragm irritation
Air embolism	Venous injury

Hernia

Port site hernia is an important complication associated with laparoscopic surgery seen at port entry site, highest incidence at the specimen extraction site. Features associated with port site hernia are described in **Table 6**.

Pneumoperitoneum-related Complications

Laparoscopic surgery is associated with varied complications due to gaseous insufflation into the peritoneal cavity 24 as depicted in **Table 7**.

Electrosurgical Injuries

Electrocautery is an integral part of operative laparoscopy which too creates complications as enumerated in **Table 8**.

Nerve Injury

Laparoscopic surgery can cause neurological complications if position of the patient is inappropriate or for prolonged duration. Nerve injuries can cause persistent postoperative pain, paresthesia or numbness in the area of supply as detailed in **Table 9**. Lower limb neuropathies are commonly due to nerve compression or contusion and have spontaneous resolution usually within 3 months.[12]

General measures for neurological injury:
- Always sought for neurology consultation and physiotherapy in case of suspected nerve injury.
- 50% of patients with common peroneal and femoral neuropathies regain complete motor function by 12 months.
- *Sciatic nerve injuries:* Protracted and incomplete motor function recovery.
- Complete division may cause permanent disability.

Rare Complications

Laparoscopy is associated with some rare complications as mentioned below:
- Port site metastasis[29]
- Vulvar edema.[30,31]

HYSTEROSCOPY

It is a form of minimal invasive surgery for visualization and intervention within the uterine cavity by introducing a telescope through vagina and cervix into the endometrial cavity. Isotonic fluid media is used for distending the cavity. This procedure avoids many hysterectomies and has numerous advantages being a day care surgery with minimal intervention but is not absolutely complication

TABLE 8: Thermal injuries.

Bowel injury	Incidence	2/3rd bowel injuries occur due to dissection, electrocoagulation, tissue grasping (and 1/3rd during abdominal access)	
	Symptoms	• Manifest within 12–36 hours • May be delayed up to 5–7 days	
	Diagnosis[25]	• Free intra-abdominal air normally presents up to 1 week • Increasing amount of intra-abdominal air suggests ruptured viscera	
	Management	• Tissue is inverted and oversewn to healthy tissue at the margins • Resected with 1–2 cm margin and sutured • Resection and anastomosis for large injuries	
Bladder injury	In gynecological and urological procedures		
	Incidence	<0.5%[26,27]	
	Management	Ranges from catheterization to laparotomy	
Ureteral injuries	Incidence	0.08%[27]	<2% of pelvic procedures such as endometriosis[28]
	Cause	Excessive use of energy close to the field of ureter	
	Prevention	• Preoperative ureteral stenting in anticipated difficult cases • Identification of ureter during procedure • Confirm ureter integrity before closure	

TABLE 9: Nerve injuries.

Nerve	*Cause*	*Prevention*
Femoral neuropathy	Excessive hip flexion, abduction	• Thigh flexion <90° • Abduction <45° • Less operating time
Obturator neuropathy	• Direct injury during radical pelvic surgery or lymphadenectomy • Retropubic dissection • Excessive hip flexion	Careful dissection and positioning
Sciatic neuropathy	• Nerve stretching • Increased hip flexion and knee joint straightened or externally rotated	Stirrups should support the ankle and calf
• Brachial plexus injuries (incidence 0.16%) • Weakness of the upper arm and Erb palsy	• Steep Trendelenburg (body sliding relative to outstretched arms) • Shoulder braces positioned too medially	

free. Overall incidence of the complications associated with hysteroscopic procedure are about 0.22%–0.28%.[32] Diagnostic hysteroscopy carries lesser complication rate (0.13 %) as compared to the operative procedures (0.95%). Specific procedure related complications[33] with their incidence are enumerated in **Box 2**.

BOX 2: Complications of hysteroscopy.

Uterine perforation	(0.12%)
Fluid overload	(0.06%)
Intraoperative bleeding	(0.03%)
Bowel and bladder injuries	(0.02%)
Infection	(0.01%)

Uterine Perforation

Is a one of the commonly reported complications, can occur at various stages of the procedure as described in **Table 10**.

Fluid Overload

It is a serious and life-threatening condition but is preventable by taking simple measure as discussed in **Table 11**.

TABLE 10: Uterine perforation.

Causes	*Diagnosis*	*Prevention*	*Management*
Dilatation of cervix prior to introducing hysteroscope	• Endometrial cavity depressurizes and collapses • Omentum, bowel, or peritoneal structures are visualized • Sudden increase in the fluid deficit	Cervix and internal os negotiated under direct visualization	Extended observation postoperatively
Surgeries such as septum resection, myomectomy, and lysis of adhesions	Unexplained falling BP and medium leakage—may be due to injury to iliac vessels	Not activating the energy device during a thrusting or forward movement	Laparoscopy or laparotomy particularly if perforation is with energy device

TABLE 11: Fluid overload.

Prevention[34,35]		
Maintain low intrauterine fluid pressure	70–80 mm Hg	
Limit surgical operating time	1 hour or less	
Termination procedure if fluid deficit	With isotonic fluid	2,500 mL
	With hypo-osmolar fluids	750 mL in elderly patients or has comorbid conditions 1,000 mL for all other patients
Treatment		
Send stat sodium level	Administer loop diuretic (furosemide)	Severe hyponatremia—admit to an intensive care unit

BOX 3: Measures to control bleeding.

- Electrocoagulation
- Intrauterine balloon initially inflated to 2 to 5 mL (upto 30 mL) for 6–12 hours, avoid overdistension to prevent uterine rupture
- Emergency hysterectomy or uterine artery embolization

Bleeding

In case of excessive intra-operative bleeding, instant life saving measures are described in **Box 3**.

Gas Embolus

It is a grave condition, endangering the life of the patient. Incidence is quite less with the use of fluid distension media but still encountered in clinical practice. Various causes, clinical suspicion, diagnosis and resuscitative measures are mentioned in **Table 12**.

There are few other rare complications of the hysteroscopy as discussed below:

Infection: It is a rare complication. Avoid procedure if there is gross cervical, uterine, or tubal infection.[37]

TABLE 12: Gas embolism.

Findings	Decreased endtidal CO_2, millwheel murmur, cardiovascular collapse
Monitoring method	Transesophageal ECHO
More common with use of	CO_2 distension media[36]
With fluid distension	Entry of ambient air into uterine cavity can occur by repeated passes of the hysteroscope or the vapors created by use of electrosurgery
If suspicion	Stop procedure immediately
Prevention	Avoid Trendelenburg position

Sepsis: It may occur from unrecognized thermal bowel injury; urinary ascites can occur from an unrecognized bladder injury.

Technical issues: Poor visibility, equipment failure.[38]

CONCLUSION

Endoscopy is now routinely performed procedure in field of gynecology. Surgeons are well trained and equipped but they should also be well versed with the common complications for the patient safety. This knowledge will help them from getting caught in difficult intra-operative situations as well as medicolegal problems.

REFERENCES

1. Chapron C, Querleu D, Bruhat MA, Madelenat P, Fernandez H, Pierre F, Dubuisson JB. Surgical complications of diagnostic and operative gynaecological laparoscopy: a series of 29,966 cases. Hum Reprod. 1998;13(4):867-72.
2. Magrina JF. Complications of laparoscopic surgery. Clin Obstet Gynecol. 2002;45(2):469-80.
3. Chandler JG, Corson SL, Way LW. Three spectra of laparoscopic entry access injuries. J Am Coll Surg. 2001;192(4):478-90; discussion 490-1.
4. Fuller J, Ashar BS, Carey-Corrado J. Trocar-associated injuries and fatalities: an analysis of 1399 reports to the FDA. J Minim Invasive Gynecol. 2005;12:302.
5. Deziel DJ, Millikan KW, Economou SG, Doolas A, Ko ST, Airan MC. Complications of laparoscopic cholecystectomy: a national survey of 4,292 hospitals and an analysis of 77,604 cases. Am J Surg. 1993;165(1):9-14.
6. Shirk GJ, Johns A, Redwine DB. Complications of laparoscopic surgery: How to avoid them and how to repair them. J Minim Invasive Gynecol. 2006;13:352.
7. Sandadi S, Johannigman JA, Wong VL, Blebea J, Altose MD, Hurd WW. Recognition and management of major vessel injury during laparoscopy. J Minim Invasive Gynecol. 2010;17(6):692-702.
8. Geers J, Holden C. Major vascular injury as a complication of laparoscopic surgery: a report of three cases and review of the literature. Am Surg. 1996;62(5):377-9.
9. Fernández EM, Malagón AM, Arteaga I, Díaz H, Carrillo A. Conservative treatment of a huge abdominal wall hematoma after laparoscopic appendectomy. J Laparoendosc Adv Surg Tech A. 2005;15(6):634-7.
10. Tews G, Arzt W, Bohaumilitzky T, Füreder R, Frölich H. Significant reduction of operational risk in laparoscopy through the use of a new blunt trocar. Surg Gynecol Obstet. 1991;173(1):67-8.
11. Kuster GG, Fischer B. Pharmacologic hemostasis in laparoscopy: topical epinephrine facilitates cholecystectomy. Am Surg. 1993;59(5):281-4.
12. Handa, V. and Van Le, L., 2020. Te Linde's operative gynecology. 12th ed. Lippincott Williams & Wilkins, pp. 303-306.
13. Molloy D, Kaloo PD, Cooper M, Nguyen TV. Laparoscopic entry: a literature review and analysis of techniques and complications of primary port entry. Aust N Z J Obstet Gynaecol. 2002;42(3):246-54.
14. Ahmad G, O'Flynn H, Duffy JM, Phillips K, Watson A. Laparoscopic entry techniques. Cochrane Database Syst Rev. 2012;(2):CD006583.
15. Chapron C, Pierre F, Harchaoui Y, Lacroix S, Béguin S, Querleu D, Lansac J, Dubuisson JB. Gastrointestinal injuries during gynaecological laparoscopy. Hum Reprod. 1999;14(2):333-7.
16. Philosophe R. Avoiding complications of laparoscopic surgery. Fertil Steril. 2003;80(Suppl 4):30-9; quiz 54-6.
17. Sia-Kho E, Kelly RE. Urinary drainage bag distention: an indication of bladder injury during laparoscopy. J Clin Anesth. 1992;4(4):346-7.
18. Montz FJ, Holschneider CH, Munro MG. Incisional hernia following laparoscopy: a survey of the American Association of Gynecologic Laparoscopists. Obstet Gynecol. 1994;84(5):881-4.
19. Swank HA, Mulder IM, la Chapelle CF, Reitsma JB, Lange JF, Bemelman WA. Systematic review of trocar-site hernia. Br J Surg. 2012;99(3):315-23.
20. Johnson WH, Fecher AM, McMahon RL, Grant JP, Pryor AD. VersaStep trocar hernia rate in unclosed fascial defects in bariatric patients. Surg Endosc. 2006;20(10):1584-6.
21. Boike GM, Miller CE, Spirtos NM, Mercer LJ, Fowler JM, Summitt R, Orr JW Jr. Incisional bowel herniations after operative laparoscopy: a series of nineteen cases and review of the literature. Am J Obstet Gynecol. 1995;172(6):1726-31; discussion 1731-3.
22. Kadar N, Reich H, Liu CY, Manko GF, Gimpelson R. Incisional hernias after major laparoscopic gynecologic procedures. Am J Obstet Gynecol. 1993;168(5):1493-5.
23. Benlice C, Stocchi L, Costedio MM, Gorgun E, Kessler H. Impact of the Specific Extraction-Site Location on the Risk of Incisional Hernia After Laparoscopic Colorectal Resection. Dis Colon Rectum. 2016;59(8):743-50.
24. Aurora Pryor, William J Mann, Andrew T Bates. Complications of gynae laparoscopy. In: UpToDate, Jeffrey Marks, Tommaso Falcone, Wenliang Chen. July 2022.
25. Farooqui MO, Bazzoli JM. Significance of radiologic evidence of free air following laparoscopy. J Reprod Med. 1976;16(3):119-25.
26. Levy BF, De Guara J, Willson PD, Soon Y, Kent A, Rockall TA. Bladder injuries in emergency/expedited laparoscopic surgery in the absence of previous surgery: a case series. Ann R Coll Surg Engl. 2012;94(3): e118-20.
27. Wong JMK, Bortoletto P, Tolentino J, Jung MJ, Milad MP. Urinary tract injury in gynecologic laparoscopy for benign indication: a systematic review. Obstet Gynecol. 2018;131(1):100-108.
28. Gomel V, James C. Intraoperative management of ureteral injury during operative laparoscopy. Fertil Steril. 1991;55(2):416-9.
29. Ramirez PT, Wolf JK, Levenback C. Laparoscopic port-site metastases: etiology and prevention. Gynecol Oncol. 2003;91(1):179-89.
30. Pados G, Vavilis D, Pantazis K, Agorastos T, Bontis JN. Unilateral vulvar edema after operative laparoscopy: a case report and literature review. Fertil Steril. 2005;83(2):471-3.

31. Guven S, Guven ES, Ayhan A. Vulvar edema as a rare complication of laparoscopy. J Am Assoc Gynecol Laparosc. 2004;11(3):429-32.
32. Jansen FW, Vredevoogd CB, van Ulzen K, Hermans J, Trimbos JB, Trimbos-Kemper TC. Complications of hysteroscopy: a prospective, multicenter study. Obstet Gynecol. 2000;96(2):266-70.
33. Aydeniz B, Gruber IV, Schauf B, Kurek R, Meyer A, Wallwiener D. A multicenter survey of complications associated with 21,676 operative hysteroscopies. Eur J Obstet Gynecol Reprod Biol. 2002;104(2):160-4.
34. ACOG Committee opinion No. 800: The use of hysteroscopy for the diagnosis and treatment of intrauterine pathology. Obstet Gynecol. 2020 March.
35. AAGL Advancing Minimally Invasive Gynecology Worldwide; Munro MG, Storz K, Abbott JA, Falcone T, Jacobs VR, Muzii L, Tulandi T, Indman P, Istre O, Jacobs VR, Loffer FD, Nezhat CH, Tulandi T. AAGL Practice Report: Practice Guidelines for the Management of Hysteroscopic Distending Media: (Replaces Hysteroscopic Fluid Monitoring Guidelines. J Am Assoc Gynecol Laparosc. 2000;7:167-168.). J Minim Invasive Gynecol. 2013;20(2):137-48.
36. Brandner P, Neis KJ, Ehmer C. The etiology, frequency, and prevention of gas embolism during CO(2) hysteroscopy. J Am Assoc Gynecol Laparosc. 1999;6(4):421-8.
37. ACOG Practice Bulletin No. 195: Prevention of Infection After Gynecologic Procedures. Obstet Gynecol. 2018; 131(6):e172-e189.
38. Courdier S, Garbin O, Hummel M, Thoma V, Ball E, Favre R, Wattiez A. Equipment failure: causes and consequences in endoscopic gynecologic surgery. J Minim Invasive Gynecol. 2009;16(1):28-33.

Chapter 64

Laparoscopic Operating Theater

Raman Singla

INTRODUCTION

The success enjoyed by laparoscopic techniques in all fields imposes a change in thinking as concerns the organization of operating theaters.

Laparoscopic techniques have specific features:

- The technical equipment occupies more space and has a more important role in the quality of the surgery.
- More people are actively involved during the operation.
- The work is performed indirectly on a screen, which requires ergonomic positioning of the operating team.
- The specific aspect of not opening the stomach, which changes the operating height and imposes specific positioning on a suitable table.
- The surgeon is surrounded by specific wires and cables that need to be understood to achieve optimum results.

All of these factors should create further thought as concerns organizing an operating theater specifically for laparoscopic surgery.

SIZE OF THE OPERATING THEATER

The theater should be vast and well-lit. If the size of the theater may be easily explained by the presence of extra equipment, the lighting is essential for monitoring the unconscious patient. Monitoring the skin is in fact one of the first signs to be observed in the case of hemodynamic and respiratory problems, which can occur during these operations. It is, therefore, dangerous to perform laparoscopy in theaters which have poor lighting conditions due to windows being closed off.

The operating spaces are enlarged due to the fact that the members of the operating teams need to move around frequently. Therefore, when operating on the colon, the best position for the surgeon will be on the left for the left colon, on the right for the right colon and between the legs for the transverse.

OPERATING TABLE

The current trend adopted by the manufacturers of operating tables is aimed at versatility **(Fig. 1)**. Laparoscopy imposes certain criteria on the table. The laparoscopic operating field is higher than the usual field. On average, the instruments used measure 43 cm, half of this length being situated outside of the cavity. The operating field is, therefore, raised by around 30 cm, which in consequence means that the table has to be lowered by the same amount. This is further exaggerated when performing pelvic surgery, where the Trendelenburg position raises the operating field even higher. The laparoscopic table thus needs to be around 20 cm lower than traditional tables. In the absence of such tables, the surgeon may stand on a small stage. However, this must be wide enough for the surgeon, as well as for all of the pedals that he has to operate.

In laparoscopy, the organs are often moved aside by movement of the table. This movement must therefore be easy to perform. The optimum solution is electric controls.

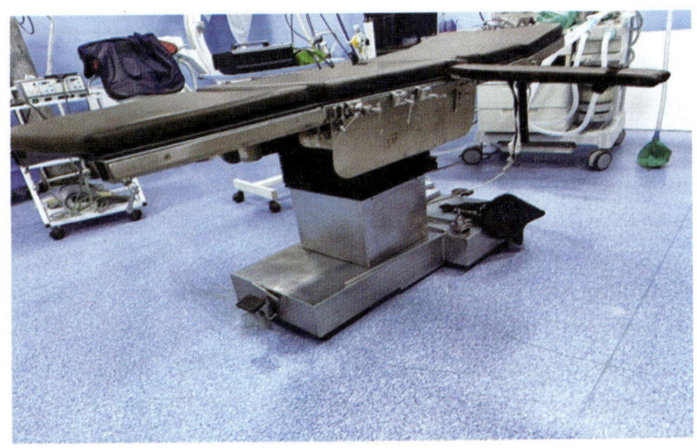

Fig. 1: Operating table.

POSITIONING OF THE PATIENTS

Ideally, the patient is laid flat, with her legs spread. Three spaces are thus created—the right, the left, and between the legs.

Traditionally, gynecologists worked on the left of the patient with an assistant on their left. The surgeon held the camera and worked with the left hand. The patient was then laid flat, legs spread and her arm on the side where the surgeon and assistant were positioned was protected alongside the body. The other arm was stretched out and supported the venous tract. In this way, all risks to the plexus brachial were eliminated. The positioning of the patients may therefore only be determined by close cooperation between surgeons and anesthetists. A sling should be placed on the relief of the acromion whenever the Trendelenburg position is required.

POSITIONING OF THE OPERATING TEAM

The positioning of the operating team depends on the indications and habits. However, the general principles remain unchanged. For reasons of clarity, we will use the example of pelvic surgery for the rest of this explanation.

Modern laparoscopic surgery (with a few exceptions) requires the use of both hands. Surgeons are therefore facing up to this requirement. To work comfortably for long periods, the surgeon must be in a position where the spinal column is straight, and with the elbows as close to the body as possible. His line of vision should be aligned with the axis of the work being performed and be situated between his hands. As the laparoscopy is viewed on a screen, the screen must be positioned so that this rule is respected. Thus, in our example, the surgeon's screen is best situated at the right foot of the patient. The visual fatigue from watching the screen means that it should be placed at a distance of between 1.6 and 6 times the length of the diagonal of the screen. If, in general, this compromise is easy to achieve for the surgeon, the same cannot be said for the assistants. The first assistant, in our example positioned opposite the surgeon, holds the camera arid works with an instrument with the right hand. His visual comfort is therefore equally important. If the theater only has one screen, the assistant will try to move back to improve his visual comfort. It is in this situation that, if the arm is not protected, there is a high risk of lesion of the right plexus brachial. The best situation is for the assistant to have a viewing screen, which in our example would be positioned at the left foot of the patient. If another assistant is required between the legs of the patient, he will look either at the screen on the left or the right.

The instrument table and instrument nurse are situated logically behind and to the left of the surgeon.

INSTRUMENT TROLLEY

Most laparoscopic trolleys have an insufflator, light, camera, and sometimes the lavage aspiration system stacked on it **(Fig. 2)**. The screen is generally at the top of this stack. The patient is connected to the trolley by a sort of umbilical cord which includes—the CO_2 cable, the light cable, the camera cable, and sometimes a lavage and an aspiration cable. In our example, this cord comes from the trolley which supports the surgeon's screen. However, the length of the cord is limited by the length of the optics cable, which should be as short as possible to avoid loss of light. These cables are usually around 2 m long.

Every time that the surgeon moves, the trolley and its attachments also need to be moved. These movements are difficult and create risks of disconnection and loss of asepsia. Furthermore, they are harmful to the equipment, making it more fragile. In these conditions, it is preferable to dissociate the trolley into two separate elements. On side, the screen and on the other, the rest of the equipment. The screen is in fact linked to the camera controls by a cable.

BASIC EQUIPMENT AND TECHNIQUES OF LAPAROSCOPIC SURGERY

Installation and Protection of the Equipment

Optimally, the equipment is installed on a mobile trolley, so that it may be used in a multidiscipline environment.

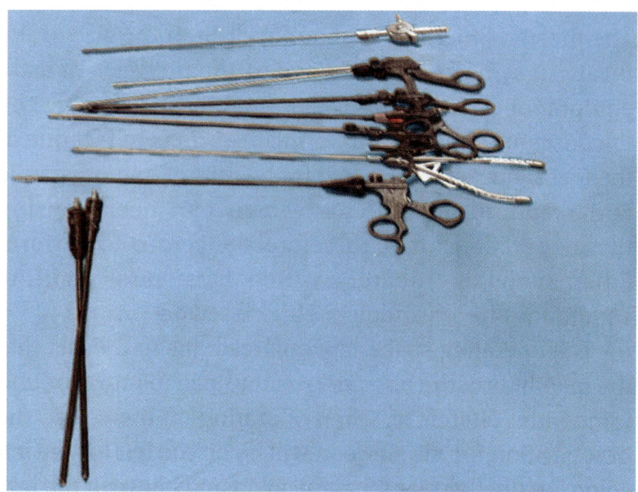

Fig. 2: Instrument trolley.

The pieces of equipment are positioned logically. The screen is positioned at the top (which, in certain cases, may be on a lateral arm). The various apparatus with the same length of cable is grouped together. Priority is given to the insufflator, which must be positioned so that the surgeon may read the display. It should be noted that the electrical power supply to this trolley must be separate from that of the electrical generator (which in principle, should not be on the trolley).

We will look at the various pieces of equipment successively.

Insufflator

The pneumoperitoneum allows viewing and to ensure that the organs are moved aside. The quality of the pneumoperitoneum is essential to how the endoscopic surgery is conducted, which at present is tending to increase in duration. The carbon dioxide (CO_2) used for the insufflations is a semi-inert gas and one which is cheap. Its diffusion in the peritoneum does not create any risk of emboly, thanks to its systematic diffusion regulated by the ventilation system. These qualities make it a gas of little danger.

It is essential to have an electronic insufflator, capable of monitoring the flow rate with regard to the intra-abdominal pressure **(Fig. 3)**. The intra-abdominal pressure is set to 15 mm Hg at maximum by the surgeon (P_{max}); depending on this pressure, the apparatus will display the flow rates varying between 0 and the maximum flow rate provided by the apparatus.

Modern insufflators have flow rates which are becoming higher and higher. In theory, these high flow rates do not present any harmful effects as long as the pressure remains below P_{imax}.

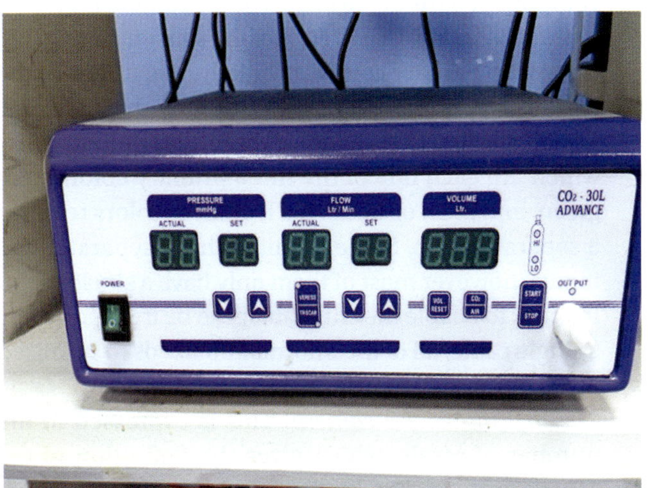

Fig. 3: Insufflator.

OPTICS SYSTEM

Considered as a gadget in 1984, the camera has become a true surgical instrument. Surgeons are generally not familiar with video techniques. However, it is the choice of the camera and light source, along with the knowledge of the basic principles of electronic viewing systems that will dictate the quality of the view available during surgery and thus the quality of the operation.

One of the advantages of the endoscopic approach is that of obtaining a virtually microsurgical view compared to that obtained by laparotomy. Whereas in fact at present, this view has become optoelectronic. In this unit, the quality of the image obtained in the end only depends on the quantity of light available at each step of this optical and electronic system.

This system may be artificially divided into three main parts:

- The production of the light (light source)
- The transmission of the light (endoscope and cable)
- The image acquisition (camera).

It is obvious that the final quality will be that of the worst element in the system. Therefore, both in the choice of the camera and the light source, all of the other elements in the system must also be taken into account.

Light Source (Fig. 4)

Different Types of Light Source

Several types of sources are currently available on the market. The two main types are halogen and xenon. The main difference between them is related to the difference of the color temperatures (5,000–5,600 K for halogen and 6,000–6,400 K for xenon). This difference may be remarked in the colors obtained, the xenon having a slightly bluish tint.

Power Supply

We have already said that the determining factor was the quantity of light available at each step of the system. The power supply of the source is therefore the first element to be taken into account. However, increasing the power poses a real problem as concerns the heat created. This light is defined as a cool light. There must be no ambiguity as concerns this definition. For 100% of energy consumed,

Fig. 4: Light source.

a normal light source (a light bulb) uses approximately 2% in light and 98% in heat. A cool light source lowers this ratio by creating more light but does not reduce the heat produced to zero. This implies a significant dissipation of heat, which increases as the power rating increases. The sources are protected against transmitting too much heat at present. The heat is essentially dissipated in transport, along the cable, in the connection with the endoscope and along the endoscope.

Regulation

The light sources have adjustments—a manual adjustment, which allows the light source to be adjusted to a power level by the surgeon. However, when video cameras are used, close-up viewing is hampered when there is too much light, whereas more distant views may be too dark. To counter this problem, the luminosity of most of the current light sources are adjustable.

This system is based on the following principle—the video camera transforms the signal into an electronic signal. This electronic signal is coded in order to be transported. The coding dissociates the luminance and chrominance of the image. The luminance is the quantity of light of the signal (black and white) which dictates the quality of the final image. When there is too much light in the image. (when the endoscope is near to the tissue) the luminance signal on the oscilloscope increases. On the other hand, when the luminosity is low, (distant view or red surroundings), the luminance is low and the electronic signal much weaker.

Fig. 5: Video camera.

Video Camera (Fig. 5)

Definitions

The first laparoscopic cameras really first appeared between 1984 and 1986. At the time they were not very sensitive, of around 20 lux and with modest definition, of the order of 1,50,000 pixel. Since then, the cameras have undergone considerable development. Simplifying, a camera is made up of several elements.

- *The nature of the sensor:* All of the cameras currently available are fitted with CCD (charged couple device) sensors. These sensors are electronic systems which transform the real image (photons) into electronic images which may be read on a screen. These sensors are not capable of analyzing colors. This is why the image has to be separated into primary colors by means of a prism or colored filters. There may be a single sensor (mono-CCD) which processes all of the component parts of the image, or the camera may have three sensors (tri-CCD), each of the sensors processing one of the primary colors separately. Regardless of the nature of the camera (mono or tri-CCD), the electronic signal produced is made of three component parts which are each one of the three primary colors (red, green, and blue, or RGB). Each of these colors requires a separate cable. Some monitors have separate RGB inputs, however most of them only have a single input. In order to enable the entire image to be transported by a single cable, all of the information needs to be mixed in a "composite" signal.
- *The definition of a camera:* This is expressed in the number of pixels, which gives the definition of the sensor and is given by the number of points which make up the image. The greater the number of pixels

that an image has, the better the definition of the image. Thus, the first cameras which were of low definition of 1,50,000 pixels have been replaced by very high-definition cameras over the last few years, of the order of 4,00,000–4,50,000 pixels for the mono-CCDs and over 1 million pixels for the tri-CCDs.

The definition of the cameras may also be expressed by the number of horizontal lines on the screen. A quality mono-CCD camera should have a resolution of over 300 lines, and a tri-CCD camera may have over 600.

If the resolution (number of lines) is the method used to judge the quality of a camera, it should be considered that a camera with around 350 lines is of satisfactory quality. However, if a comparison is made with the resolution of a 35-mm slide (2,000 lines), it may be observed that we are still some ways from optimum quality.

- *The signal/noise ratio:* The video image produced by the cameras may contain what is known as "noise". This noise appears in the form of a grain on the image and is more pronounced in dark or red areas, which is unfortunately often the case when used in laparoscopy. The amount of noise of the camera may be measured by the signal/noise ratio, which is expressed in decibels (S/N ratio). The higher the ratio, the less noise there will be in the image.
- *Regulation:* Some cameras are equipped with a system which makes viewing possible in poor lighting conditions. These systems operate by automatically increasing the gain of the camera and therefore at the cost of the quality of the image. It is preferable to have a powerful light source than a camera with automatic gain.

Finally, the cameras currently available are equipped with an automatic shutter, which is to say that the cameras are capable of adjusting their shutter speed to suit the lighting conditions. These shutters can usually be adjusted from 30 to over 10,000.

Lens (Fig. 6)

- Most cameras are supplied with lenses with focal lengths of 20–40 mm. Usually, with an optic of 110°, a 35-mm lens allows a full-screen image to be obtained.
- Some cameras have a zoom. If the zoom is not essential, it nevertheless offers a certain number of advantages:
  - Greater enlargement of the image
  - Full screen image, even when using small diameter or narrow angle optics. If a 35-mm lens is usually sufficient to obtain a full screen image, some endoscopes maintain a black circle on the screen. It is in these conditions that the zoom may be of use to restore the complete image.

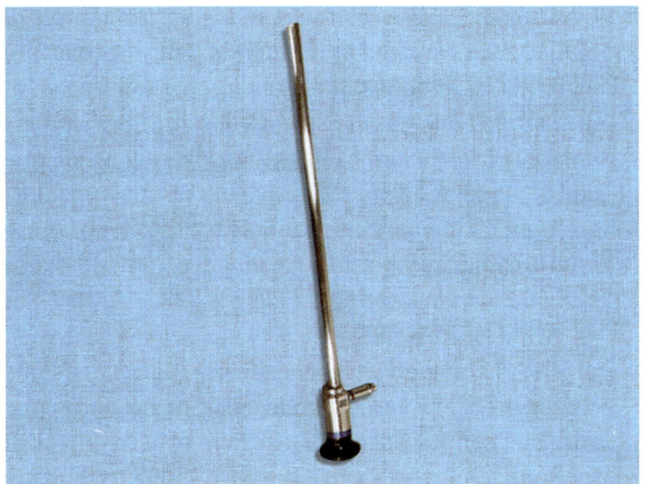

Fig. 6: Lens.

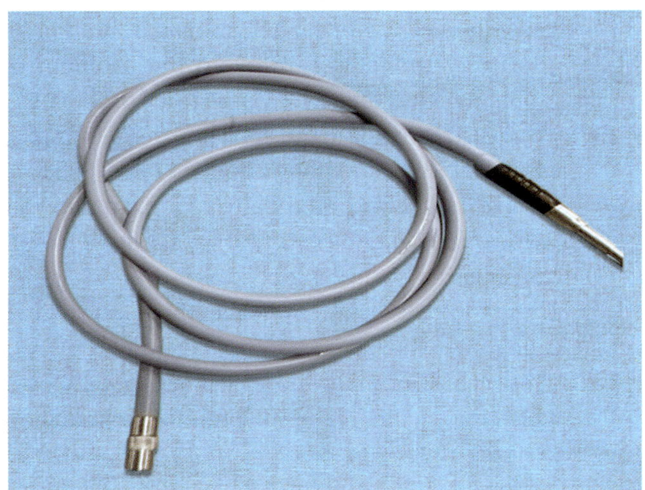

Fig. 7: Cables.

Optics and Cables

Cables (Fig. 7)

The cables which connect the endoscope to the light source are an element in the system, as when the light is transmitted, the cable is responsible for attenuating the light to a lesser or greater extent, depending on its type and especially its condition. There are two types of cable currently available on the market.

- *Optic cables:* These cables are made up of a bundle of optical fibers swaged at both ends. They have a very high quality of optical transmission but are fragile. In fact, progressively as they are used, some optical fibers break. The loss of optical fibers may be seen when one end of the cable is viewed in daylight. The broken fibers are seen as block spots. To increase the life of these optical cables, respect the following recommendations:
  - Handle them carefully.
  - Avoid twisting them.

  After the operation has been completed, the cable should preferably be disconnected from the endoscope and then connected to the light source. In fact, most of the sources currently available have a plug for holding the cable until it cools down.
- *Gel cables:* These cables are made up of a sheath that is filled with a clear optical gel, swaged at both ends by quartz. Manufactured to avoid the problem of alteration of optical fibers, they pose three problems:
  - The quartz swaging at the ends is extremely fragile, especially when the cable is hot.

  The slightest shock, on a bench for example, can cause the quartz end to crack and thus cause a loss in the transmission of the light:
  - These cables transmit more heat than optical fiber cables.
  - These cables are made more rigid by a metal sheath which makes them more difficult to maintain and to store.

Optics

The endoscope permits the light to be taken inside the cavity by means of the fibers and brings back the image by a set of lenses. As all optical systems, it has the capacity of attenuating the light.

This attenuation of the light is inversely proportional to the square of the lens diameter. This means that the smaller the diameter of the endoscope, the more light it will consume. The use of small diameter endoscopes must therefore be reserved for special indications, particularly in children.

This set of lenses also poses the problem of the accumulation of optical aberrations due to the stacking up of lenses, which leads to distortion of the image in the edges. These aberrations are even more visible when the endoscope is of poor quality. This is why that with the current explosion of endoscopy, some distributors offer poor quality endoscopes at low prices, and surgeons should be wary of them.

Monitor

The monitor is an important element in the viewing system. There are many selection criteria, but amongst them the essential is to have a monitor capable of restoring all of the resolution qualities of the camera used. This means that the number of horizontal lines of the monitor must be at least equal to the number of lines provided by the camera.

Elements for Selection

To conclude this technical section, we will look at the criteria which govern the selection of the camera and the light source. These criteria are numerous but certainly are essential. They are:

- *Quality of the image:* It is essential to use the best video equipment, as the camera is the surgeon's view. It is therefore important at the time of purchase for the surgeon to read the technical data sheet of the equipment and to take interest in the points described in the table here.
- *Price:* Obviously, quality has a price, and consequently the budget available remains a limiting factor. However, the importance of the viewing system must be emphasized in laparoscopy. It is preferable to delay the purchase of a laser or another expensive piece of equipment to the advantage of the quality of the basic equipment in the short term.
- *Coherency:* We have already seen earlier on in this section the importance of each of the elements in the viewing system. Optimal quality is required for each element, while avoiding any redundancy (e.g., adjustment on the camera and on the light source).

Table 1 summarizes the criteria which the surgeon must demand from his equipment (this table has been compiled from the experience of the authors, and is provided for information only):

LAVAGE-ASPIRATION SYSTEM

The lavage-aspiration system is an essential element in laparoscopic surgery. The water carries out several roles, as follows:

- Viewing
- Hydrodissection
- Protection of tissues
- Hemostasis (45°C)

TABLE 1: Criteria which the surgeon must demand from his equipment.

Equipment	*Specification*	*Criterion demanded*
Camera	Mono-CCD sensor	2/3-inch CCD
	Tri-CCD sensor	2/3-inch CCD
	Resolution (horizontal)	>300 lines
	Mono-CCD Tri-CCD	>600 lines
	Signal/noise	45 dB or more
	Minimum sensitivity	5–7 lux
	Number of pixels Mono-CCD Tri-CCD	 >3,50,000 pixels >1 million pixels
	Balance of white	Between 2,000 and 10,000
Lens		30 or 35 mm
Zoom		25–40 mm
Light source	Halogen xenon	250 watts 175–200 watts
Optic	Straight Angled	0°, 30°, or 45° 110° panoramic 10 mm
Cable	Optical fiber	

- Prevention of adhesions
- Resuscitation

The essential criteria for the apparatus which supplies and sucks up the water must satisfy these functions, which are as follows:

- High injection pressure, of around 1 bar
- High aspiration (0.4–0.6 bar)
- Optional heating
- Reserve.

Furthermore, advantage should be taken of the cannula effect. In fact, this sometimes requires a filter (for aspiration in the intestines) and other times not (aspiration of clots). The size of the injection channel diameter allows the pressure to be increased or reduced (pressure gun effect). Finally, the diameter of the aspiration channel must either be constant or increase, but never reduce. Different diameters are available.

TROCARS (FIG. 8)

The trocars are extremely important. It is via the trocars that the instruments pass through the wall. They are responsible for a large number of accidents and must be carefully selected.

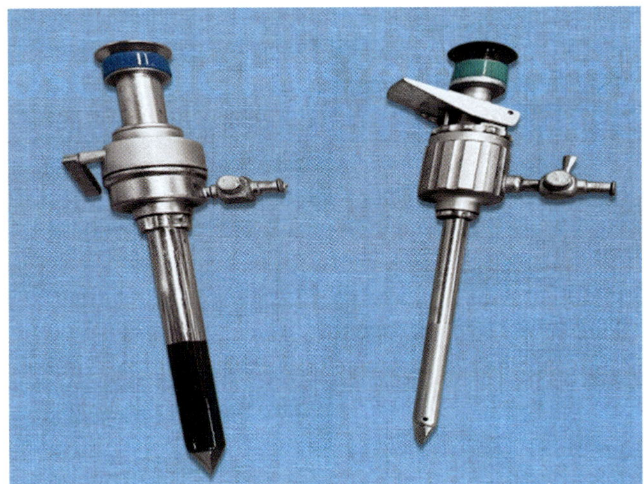

Fig. 8: Trocars.

Their point has either a conical or pyramid shape. The traumatism of the conical points is opposed to the reduction in the penetration force of the pyramid points. In fact, regardless of trends, the valves must have a valve for easy, atraumatic passage of the instruments. It must be possible to dismantle them completely to facilitate cleaning.

The size of the trocars is governed by the diameter of the instruments used. The easy solution would be to use the largest trocar for all requirements, the reduction systems making it possible to use all instruments with them. The choice would therefore be the 10–12 mm trocars. However, this shill to increasing the diameter must not be adopted.

INSTRUMENTS

The instruments must satisfy the essential criteria, making them robust, reliable, precise, ergonomic, and easy to clean.

Furthermore, this equipment must perform the various functions which are of use to the surgeon:

- Palpation, gripping
- Section, dissection
- Suturing
- Destruction
- Hemostasis.

Laparoscopic Scissors (Fig. 9)

These are fragile, microsurgical systems. A lot of scissors can take a monopolar connection. It should

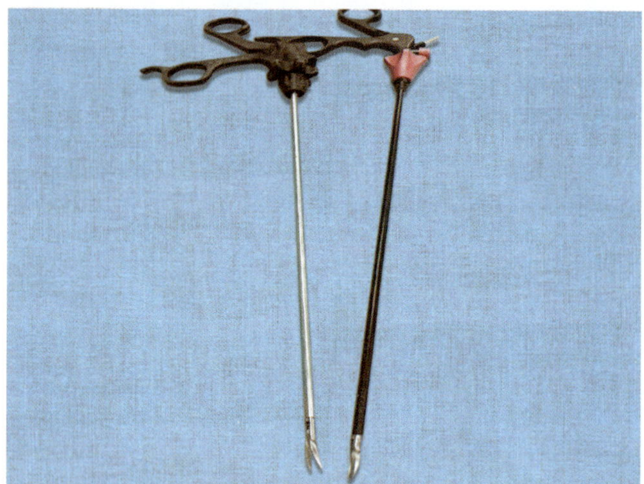

Fig. 9: Laparoscopic scissors.

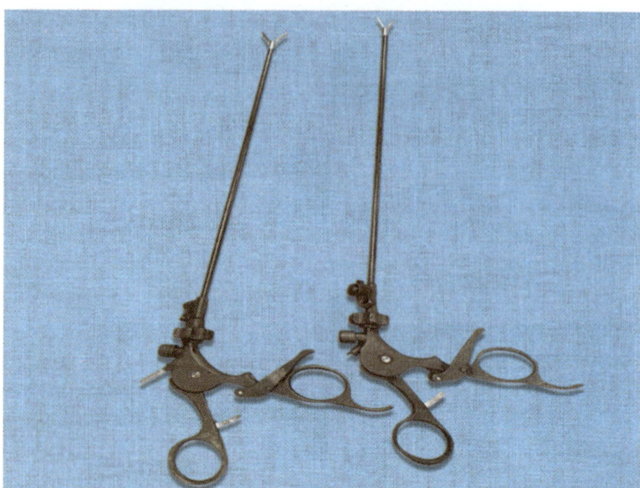

Fig. 10: Forceps.

be remembered that coagulation raises the seasons to a very high temperature, and this is responsible for them becoming blunt much quicker.

There are several different shapes of scissors available:

- *Straight:* These are for dissecting—both jaws may be active. These models are more efficient for dissecting. However, there are models available with one fixed jaw, which permits finer dissections to be carried out, especially when one of the dissected structures is vulnerable.
- *Curved:* This is the type currently preferred by surgeons. Their curve almost always makes it possible to have an approach angle of 90°, thus overcoming the single viewing angle of laparoscopy.
- *Hooked:* These are scissors suited to cutting sutures and associated pedicles. They are unsuitable for dissection, however.

Forceps (Fig. 10)

The forceps are used to grasp, present, dissect, and sometimes coagulate tissues. Most often, they are atraumatic, however the following forceps should be highlighted:

- *Flat:* Derived from microsurgery, they are only slightly traumatic but do not grasp the tissues very well. These are the most suited for dissection.
- *Grasping:* Specially designed for endoscopic gynecological surgery, they are considered to be atraumatic as they have a fine but strong grip, thus avoiding traumatisms of repeated holds.
- *Perforated:* Specially designed for moving intestines, they may be used in all operations, including with needles.
- *Biopsy:* These have been gradually replaced by the other types of forceps
- *Extraction:* Traumatic 5- or 10-mm forceps. Specially designed for trans-parietal extraction of excised tissues.
- *Babcock:* The solution proposed for abdominoperineal resections, for moving intestines.
- *Clip:* They may be re-usable or disposable. The clips used most often are made of titanium, but there are also forceps for resorbable clips.
- *Mechanical sutures:* They are rotary with a pistol handle, linear Endo Gia (Merlin), ELC (Ethicon). The active ends may be of different lengths. The sine most commonly used is 30 mm.

Dissectors (Fig. 11)

These are increasingly used in laparoscopy. They allow the vessels to be completely dissected on all of their faces.

Needle Holders (Fig. 12)

These are very similar to classic needle holders. They are of varying diameters and the active end is either straight or curved. They may have a closing system which is passive, spring operated or active by means of gearing. Some more recent models have a handle and are hand-held, which improves the grip and makes it easier to open and close the instrument.

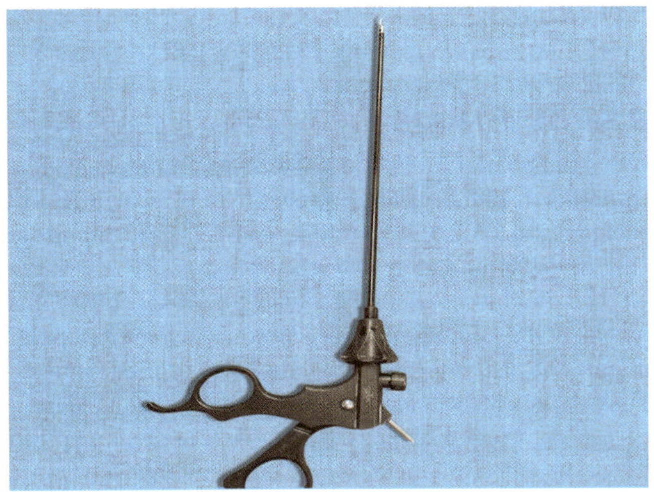

Fig. 11: Dissectors.

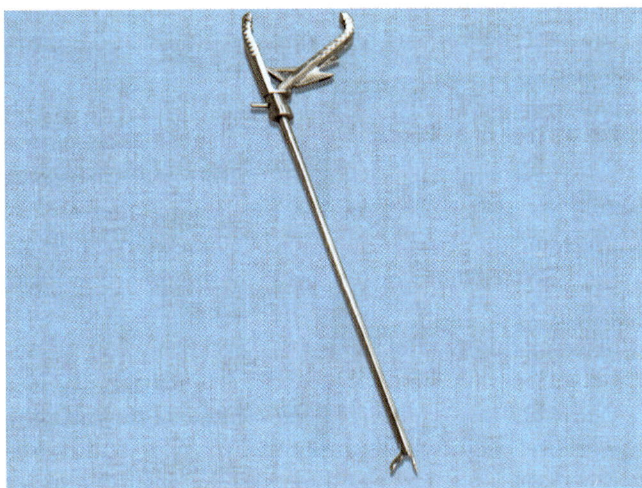

Fig. 12: Needle holders.

Multipurpose Instruments

It was Manhes who perfected the triton system. This original instrument allows simultaneous section-coagulation of the tissues by means of a retractable micro-point, pressurized lavage and aspiration.

At present, there are a number of manufacturers who produce multi-purpose instruments, based around the lavage, aspiration, and electro-surgical functions.

In fact, the number of surgical instruments designed for the endoscopic surgeon increases daily. To try to make an exhaustive description would be somewhat time-consuming. However, it should be reiterated that an experienced surgeon often only uses a few instruments. 90% endoscopic gynecological surgery can be performed with just five instruments—curved scissors, two pairs of grasping forceps, a bipolar hook, and a lavage/aspiration systems.

Fig. 13: Monopolar mode.

MONOPOLAR MODE (FIG. 13)

This mode as a single terminal and imposes a feedback electrode to be installed in the correct position. While any conductive instrument may be used, it must, however, be correctly insulated up to its active part.

In the monopolar mode, the surgeon can control six parameters:

Power

Too many surgeons ignore the power applied. It is true that most generators have settings in steps numbered from 1 to 10. In these cases, it is difficult to know what power rating a particular setting is exactly equal to (it is therefore necessary to check in the instruction manual and to display a table of correspondences in the operating theater).

The power is an important factor for the safety of the patient. A fixed power level can obviously not be proposed, as this is related to the size of the electrode—the finer the electrode, the lower the power level required to obtain the desired effect (concentration of energy).

Nature of the Electrical Wave

To obtain a section or coagulation, the generator supplies electrical waves which differ in their frequency and voltage. The higher the voltage, the higher the risk of electrical arcs and consequently the greater the danger for the patients. If the different voltages are compared, it may be seen that sections use low voltages, of around 2,000 and 3,000 Volts, and coagulation uses voltages of between 3,000 and 9,000 Volts.

Shape of the Electrode

Surgeons are very familiar with this parameter, as they know that the effect on the tissue varies depending on whether a monopolar "point" is used or a spatula. The finer the electrode, the more the power is concentrated. Certain forms of electrical wave allow the shape of the electrode to be varied—this is the case for mixed currents. With these currents, the finer the electrode the stronger the section, the wider the electrode, the better the coagulation element. Some electrodes are therefore of interest—this is the case of the spatula. Curved (disposable) scissors are of even more interest, as they combine the electrical effects (the function of the part of the instrument used—slice or face) with the mechanical effects (section and dissection).

Time that the Electrical Current is Applied

The length of time that the current is applied determines the extent of the effect. The longer the duration, the greater the effect on the tissues. In this way, a section may be too long and during a coagulation, the peripheral tissue damage may be too great.

Nature of the Tissue

Depending on the nature of the tissue, the surgeon will modify the electrical parameters used. Obviously, a sectioning effect is more efficient on the skin or on muscle (tissue with an aqueous content and therefore a good conductor) than on fat (tissue with low conductivity). This phenomenon is well known during parietal incisions in laparotomy.

Method of Applying the Energy

With a given current, the surgeon can privilege a particular element of the electrical effect by modifying the method of applying the energy.

This is an obvious check when using mixed currents:

- By selecting a mixed current and performing the actions in different sequences, a different effect is obtained. For example:
- *For a section:*
  - Tension the tissues
  - Activate the current
  - Touch the tissue
- *On the contrary, to obtain a coagulation with the same current:*
  - Release the tissues
  - Touch the tissue
  - Activate the current.

BIPOLAR MODE

When the bipolar mode is used, fewer parameters may be controlled by the surgeon. However, if there are two risks of accidents, the risk of surrounding burns is indeed present and requires this energy to be controlled.

The bipolar forceps are an instrument specific to this type of energy. In other words, contrary to the monopolar mode, not just any conductive instrument may be used. The two jaws of the forceps must be suitably insulated from each other up to their active ends. Closing the forceps on the tissue will close the electrical circuit. If the forceps are closed without any tissue between the jaws, the activation of the pedal will create a short circuit and cause the end of the forces to hear up.

Several factors may be controlled, however. They are as follows:

Size of the Electrode

There are several shapes of ends for the forceps. The best known are the usual Kleppinger type, where the two jaws are serrated and moved by an elastic force, and the single forceps, where the two jaws are smooth and are brought together by closing the handle. There are two usual widths for this type of forceps—1 and 3 mm. The finer the electrode, the more precise the coagulation, the less the surrounding tissue will be damaged and the lower the power level required. On the contrary, a wide end causes major tissue damage, which may be the desired effect if the surgeon is coagulating a large vessel. It should be remembered that there is a major increase in temperature, 1 cm around the forceps.

Power

Due to the absence of electrical risk, a high-power level does not present any major danger. However, it should be realized that the higher the power level, the more impedance there will be between the jaws. Too high a power level will create an electrical arc and charring. These two effects increase the risk of surrounding burns and lead to the tissue adhering to the forceps.

Time that the Electrical Current is Applied

To the same way, the length of time that the current is applied determines the final effect. Its duration is too short, this will create insufficient desiccation, and too long a duration will cause the tissue to adhere to the forceps and surrounding damage. This is why many generators are fitted with a system for automatic cut-out or an automatic drop in power. These systems do not guarantee that all of the tissue have been coagulated but indicate the point where the impedance having become too high, continuing the coagulation would only lead to a dangerous increase in temperature of the surrounding tissues. In fact, the technique for using the bipolar mode must be strict.

Tissue Concerned

The bipolar coagulation which causes major heating of the tissues between the two jaws of the forceps, certain tissues (the veins for example) explode easily, producing the opposite effect to that desired. In these cases, coagulate with a lower power level, placing a relatively large quantity of tissue between the forceps jaws.

CONCLUSION

The technical environment of laparoscopy has become a primordial factor for therapeutic success. Familiarity with both simple instruments and the basic energies will lead to a reduction in the risks for our patients while ensuring that these new techniques enjoy the therapeutic success. Which will no longer be reserved to certain elites but will become completely repeatable.

Section

10

Surgery for Obstetrical Complications

Chapter 65

Surgeries for Obstetrical Hemorrhage

Reeti Mehra, Radhika Gupta

INTRODUCTION

Hemorrhage still contributes to more than a quarter of maternal deaths all over the world [27% as per UNICEF (United Nations Children's Fund) data 2017] **(Fig. 1)**.

The blood flow to the uterus increases from 100 mL/min in nonpregnant state to about 700 mL near term. The increased flow and vascularity can lead to enormous blood loss, which is many times preventable.

DEFINITION

- Obstetrical hemorrhage is bleeding in pregnancy that occurs before, during, or after childbirth. Bleeding can be vaginal and less commonly intra-abdominal.

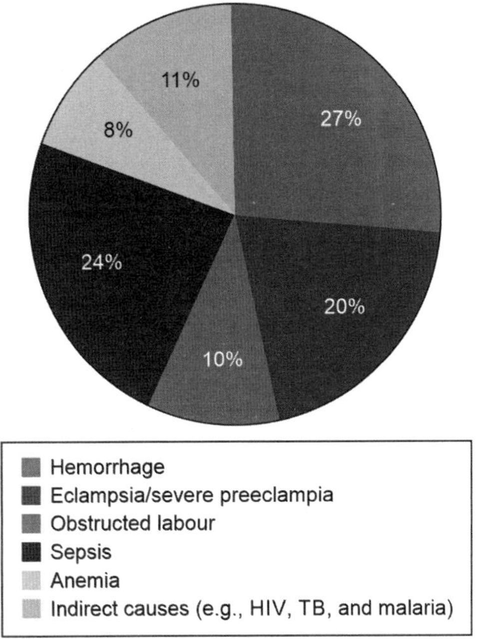

Fig. 1: Major causes of Maternal Mortality according to UNICEF 2017. (HIV: human immunodeficiency virus; TB: tuberculosis)

- Obstetrical hemorrhage may be due to different reasons at different periods of gestation, involving multiple risk factors for each cause.
  - Before 24 weeks it is called *early pregnancy bleeding*.
  - After 24 weeks and before childbirth, it is called as *antepartum hemorrhage*.
  - After childbirth it is called *postpartum hemorrhage (PPH)*.

CAUSES OF OBSTETRIC BLEEDING

The causes of obstetric bleeding are given in **Box 1**.

ECTOPIC PREGNANCY (FIG. 2)

- Implantation and development of blastocyst at any site other than normal endometrial lining of uterine cavity.
- Incidence of ectopic pregnancy is *1–2% worldwide* and is increasing over past years due to increased sexually transmitted diseases and artificial reproductive techniques and better diagnostic facilities.

BOX 1: Causes of obstetric bleeding.

Early pregnancy:
- Abortions
- Ectopic pregnancy
- Molar pregnancy

Antepartum hemorrhage:
- Placental abruption
- Placenta previa
- Morbidly adherent placenta

Intrapartum hemorrhage:
- Primary rupture uterus
- Secondary rupture uterus

Third-stage bleeding: Uterine inversion

Postpartum hemorrhage (4Ts):
- Atonic uterus (Tone)
- Retained products (Tissue)
- Genital injuries (Trauma)
- Coagulopathy (Thrombin)

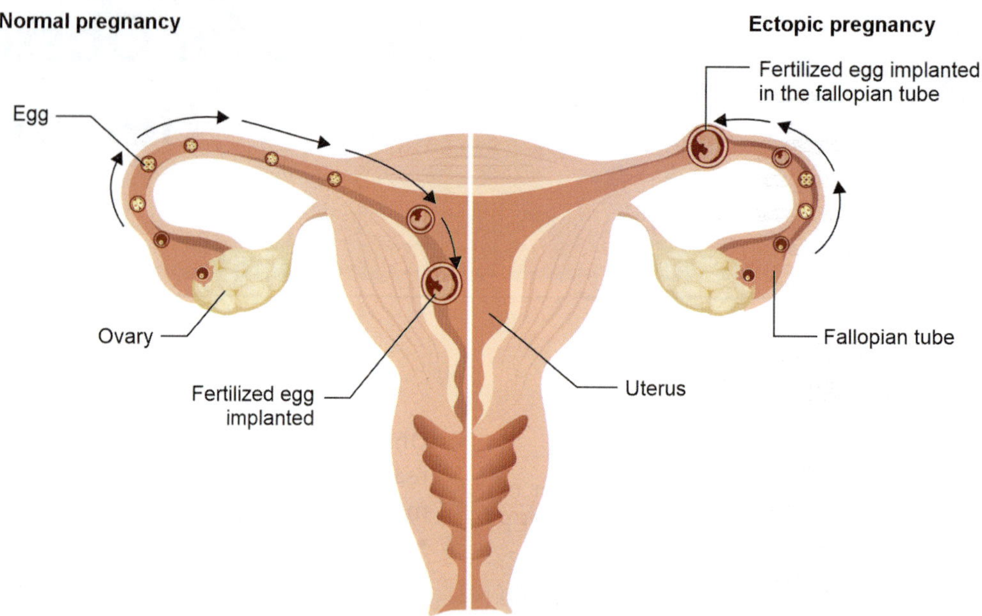

Fig. 2: Figure showing Ectopic gestation vs. normal intrauterine implantation.

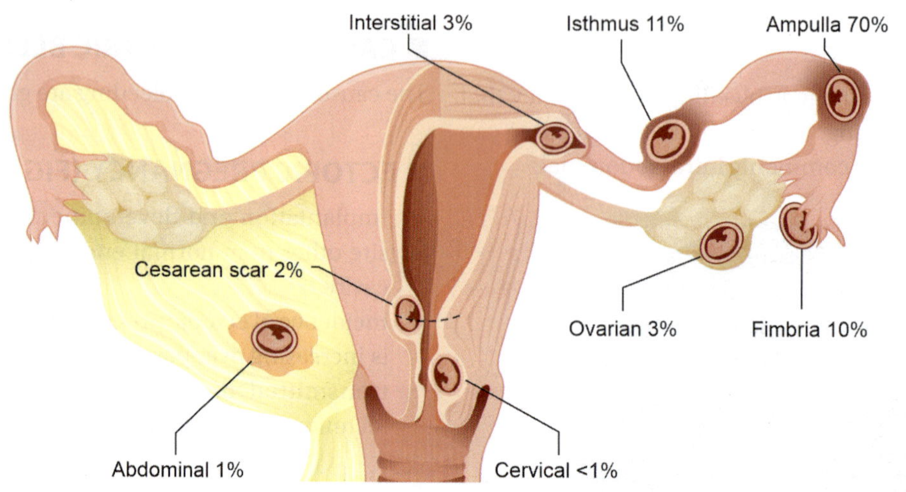

Fig. 3: Sites of ectopic pregnancy.

Sites of Ectopic Pregnancy (Fig. 3)

Figure 3 shows the sites of ectopic pregnancy.

Fallopian tube	95–97%
Ampullary	55%
Isthmic	20–25%
Infundibulum	17–20%
Interstitial (corneal)	2–4%
Ovarian	0.5%
Cervical	0.1%
Abdominal	<0.1%

- Patients mostly present in early pregnancy at from 6th to 10th weeks of gestation except interstitial pregnancy, where presentation can be in the second trimester.
- Most *common cause* of tubal ectopic is tubal damage postinflammation pelvic inflammatory disease leading to scarred tubes.

Other causes being intrauterine contraceptive devices, tuberculosis, prior tubal surgery, pregnancy after tubal ligation assisted reproduction technology (ART), developmental anomalies, and cigarette smoking.

Symptoms include pain, amenorrhea, giddiness, spotting, and some patients with ruptured ectopic may reach emergency room in shock with unstable vitals.

Diagnosis is made by symptoms, transvaginal ultrasound (TVS), and beta-human chorionic gonadotropin (hCG) assays and titers. TVS may show adnexal mass, gestational sac (1,500 discriminatory value for hCG) or in case of ruptured ectopic-free fluid in the pouch of Douglas.

TREATMENT (NATIONAL INSTITUTE FOR HEALTH AND CARE EXCELLENCE 2019)

- Expectant management is offered in patients with tubal ectopic pregnancy and:
  - Stable vitals and no pain
  - Small mass <3–5 cm and no cardiac activity
  - Beta-hCG <1,000 IU/mL (can be offered up to 1,500)
  - Are able to return for follow-up

Human chorionic gonadotropin follow-up is done on days 0, 4, and 7 and if a fall of 15% or more is there, then weekly follow-up till nonpregnant value.

- Medical management is offered with systemic methotrexate (single dose 50 mg/m^2) to women:
  - Who are hemodynamically stable, with and small mass, <3–5 cm and no cardiac activity
  - With beta-hCG <15,000 IU/mL
  - Who are able to return for follow-up

  Beta-hCG follow-up is done on days 4 and 7 and if there is a fall of 15% or more, then weekly beta-hCG is adviced, till it reaches nonpregnant value. Methotrexate may be repeated if the desirable fall in value of beta-hCG is not seen on day 4 **(Fig. 4)**.

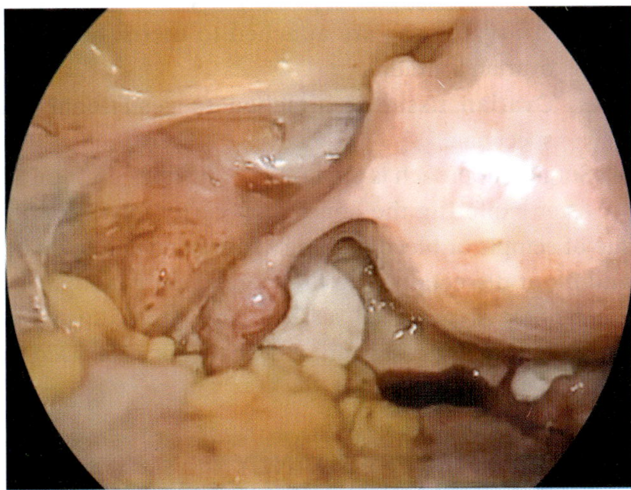

Fig. 4: Laparoscopic view of tubal ectopic gestation.

- Surgical management is offered:
  - If a woman is not able to come for follow-up
  - In significant pain
  - In ectopic pregnancy with a larger than 3–5 cm adnexal mass
  - In an ectopic pregnancy with visible cardiac activity on scan
  - In an ectopic pregnancy with serum hCG of <10–15,000 IU/mL

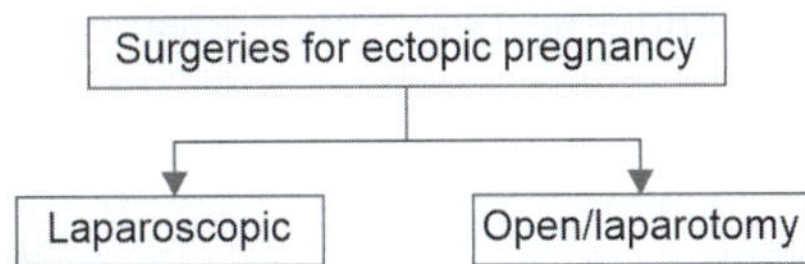

- Tubal ectopic
- Laparoscopic surgery or laparotomy

Conservative salpingostomy can be done for conserving the tube. In this procedure, after inserting required ports if laparoscopic:

- Tube is held with blunt forceps.
- Inject vasopressin (diluted in saline) in the mesosalpinx to decrease vascularity.
- Incision is given over antimesenteric border of the involved oviduct at ectopic site.
- Trophoblastic tissue is scooped out using pressure, forceps, suction or hydrodissection.
- Tubal lumen is irrigated to remove all trophoblastic tissue and removed in endobag preventing its spillage in peritoneal cavity, and then the tube is left to heal by secondary intention without suturing or can be sutured taking care not to include tubal mucosa, i.e., only the serosa and muscle are sutured (salpingotomy). Salpingostomy has shown better results of recanalization postprocedure versus salpingotomy.

  It should not be done in case of repeat ectopic in same tube, or if tube is severely damaged or patient's family is complete.

SALPINGOSTOMY DRAWBACKS

- Due to bleeding may require conversion to salpingectomy in 20% of cases
- Beta-hCG follow-up is necessary.
- Also, some patients may have persistent trophoblastic tissue requiring methotrexate administration.
- It should not be attempted in patients with ruptured tubes and unstable vitals.

Salpingectomy is the preferred surgery, specifically if other tube seems absolutely healthy.

Laparotomy: The tube is clamped cut and ligated with delayed absorbable suture. A distal suture to support the first one can be taken. Bilateral ovaries and contralateral tube examined carefully and hemostasis is checked.

CESAREAN SCAR ECTOPIC PREGNANCY

Cesarean scar ectopic pregnancy is a rare form of ectopic pregnancy where pregnancy gets implanted in previous pregnancy scar sites and there is thin myometrial tissue left between bladder and sac. This if advanced may lead to adherent placenta.

Surgical approaches can be:

- Abdominal
- *Laparoscopic or laparotomy:*
  - Here the abdominal cavity is reached followed by dissection of uterovesical fold.
  - If desired uterine arteries can be ligated bilaterally, scar site identified and incised, pregnancy tissue removed, scar excised or freshened till healthy myometrium is reached.
  - The scar site is sutured like hysterotomy incision by absorbable sutures.
- *Vaginal:*
  - Anterior colpotomy is made and then scar excision is done vaginally.
  - This is done only by surgeon experienced well with vaginal surgeries. Chances of failure and complications reported are less.
- *Hysteroscopic removal:*
  - Thickness between bladder and pregnancy should be >3 mm.
  - Scar pregnancy is identified and contents removed, scar site coagulated with bipolar cautery and then hysteroscope reinserted to confirm clearance of trophoblastic tissue.

INTERSTITIAL PREGNANCY

- When blastocyst implants in the interstitial portion of the tube it is called interstitial pregnancy.
- It can be dangerous and life-threatening if not diagnosed well in time leading to even rupture.
- The term cornual pregnancy is often used interchangeably with it, but here the implantation occurs in lateral and upper part of uterus. It should ideally be used for pregnancy in horn of uni- or bicornuate uterus. The term angular pregnancy should be used for pregnancy near tubal end of uterus, and here implantation occurs medial to round ligament.
- Interstitial pregnancy ruptures late as compared to other tubal ectopics because of thick myometrium surrounding
- *Ultrasound findings:*
  - Empty uterine cavity
  - Gestation sac 1 cm lateral to uterine end of tube
  - Interstitial line sign that is endometrial lining extending till cornua adjacent to tube
- Surgery may be laparoscopic or open depending on surgeon's expertise and patient's condition.

Cornuostomy: In this procedure incision is given in the most prominent part of ectopic after injecting vasopressin and then products of conception are removed, cavity is irrigated and then myometrium is sutured with absorbable sutures. Ipsilateral salpingectomy is done along.

Corneal resection: Here the size of ectopic is bigger. Vasopressin is injected and sometimes uterine artery is ligated to decrease bleeding then a purse-string suture is placed at base and then myometrium is incised as "v" shape, products of conception removed and cavity irrigated, bed bleeding controlled and then myometrium is sutured. Ipsilateral tube is removed.

Transcervical approach has also been attempted but failure and persistence is more.

OVARIAN ECTOPIC PREGNANCY

- Diagnosed mostly intraoperatively and confirmed postoperatively by *Spielberg criteria:*[1]
  - The tube and fimbria are intact and separate from ovary.
  - Sac occupies normal position of ovary.
  - Ovary and sac are attached to uterus by utero-ovarian ligament.
  - Ovarian tissue is attached to gestational sac specimen.
- Surgical approach is preferred.

After exploring the abdominal cavity for any other site of ectopic or bleeding, ovary is grasped and by sharp and blunt dissection. The pregnancy tissue is removed and hemostasis achieved. If size is large or bleeding is not controlled, wedge resection of ovary is done followed by suturing to achieve hemostasis.

ABDOMINAL PREGNANCY

- Implantation occurs outside reproductive system.

- If diagnosed early, removal can be done either laparoscopically or by laparotomy and placenta can also be removed with less chances of major hemorrhage.
- However, if diagnosed late then laparotomy is the only approach and placenta is usually not amenable to removal and should be kept in situ with minimal handling. Methotrexate has been given along with antibiotics.

HETEROTROPIC PREGNANCY

- In this type of pregnancy an ectopic pregnancy mostly tubal coexists with an intrauterine pregnancy.
- Incidence is 1:4,000 to 1:30,000 and is about 1:100 in in vitro fertilization (IVF) conception.
- Management includes surgical removal of ectopic pregnancy by salpingectomy. Conservative approach is not practiced due to inability to follow hCG as the results are affected by intrauterine pregnancy. However, surgical risk to intrauterine pregnancy is always there.[2]

POSTPARTUM HEMORRHAGE

- Postpartum hemorrhage is defined as blood loss of >500 mL after vaginal delivery or 1,000 mL after cesarian.
- A blood loss causing fall in hematocrit of >10% or any blood loss that threatens the hemodynamic stability of the patient has also been used to define PPH as different patients may have individual capacity to cope with blood loss. An anemic patient may get decompensated with even 300 mL of blood loss.
- The diagnosis of PPH is reserved for pregnancies beyond 20 weeks before which it will be called *postabortal hemorrhage*.

Uterine musculature acts as living ligatures followed by clotting and coagulation, which leads to normal hemostasis postdelivery. However, inability to gain tone due to multiple risk factors can cause excessive blood loss and can even prove fetal.[3]

Risk factors for uterine atony:
- Overdistended uterus
- Large fetus
- Polyhydramnios
- Multiple gestation
- Retained clots
- Induction of labor
- Prolonged labor
- Anesthesia agents
- Precipitate labor
- Chorioamnionitis
- Multiparity
- Placenta previa
- Previous history of PPH
- Morbidly adherent placenta.

Active management of third stage of labor has reduced the incidence of PPH.

Management

Initial management remains securing airway, breathing, circulation (ABC) of the patients.
- Insert two wide-bore cannulas and send samples for investigations and for blood cross match. Appropriate blood products are to be arranged.
- Intravenous fluid resuscitation should begin with warm fluids and colloids till blood is available, head end lowered and strict watch for vitals and urine output should be done.

Chandraharan and Arulkumaran proposed management algorithm called HEMOSTASIS to aid systemic management of PPH.

H	Ask for Help and Hands on uterus for massage
E	Ensure airway, breathing, circulation and ecbolics (drugs for uterine contraction), establish etiology
M	Massage uterus
O	Oxytocin infusion/prostaglandins
S	Shift to theater-aortic pressure or antishock garment
T	Tamponade or packing after excluding tissue and trauma, tranexamic acid 1 g IV
A	Apply compression sutures
S	Systemic pelvic devascularization—uterine, ovarian, internal iliac
I	Interventional radiology for uterine artery embolization
S	Subtotal/total abdominal hysterectomy

Uterus preserving surgeries: If medical and mechanical methods fail to control PPH then surgical method should be initiated.

Laparotomy is performed and further procedures are initiated. However, if placenta accreta is identified, peripartum hysterectomy should be performed without delay.
- *Stepwise uterine devascularization:*
  - Uterine artery ligation (unilateral followed by bilateral) is also called *O'Leary stitch* **(Fig. 5)**.

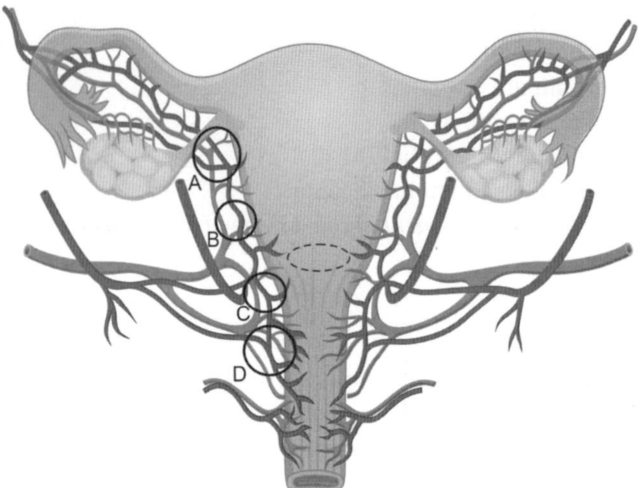

Fig. 5: Sites of devascularization of uterus for treatment of obstetric hemorrhage.

Uterine artery ascending branch is ligated above bladder reflection at the level of isthmus. Aim is to decrease the pulse pressure of the uterus. It is done by exteriorizing the uterus and holding broad ligament between thumb and fingers at base and suture (vicryl) is passed from broad ligament through transparent area from anterior to posterior and then brought back anteriorly through myometrium about 2 cm medial to uterine artery and tied.

- If not controlled by bilateral ligation, low uterine artery ligation is done at 3–5 cm below the above sutures (bladder flap is mobilized for this suture). In about 90% of cases uterine artery ligation is effective.
- If bleeding not controlled, ovarian vessel ligation can be done at infundibulopelvic ligament after identifying ovarian artery and ureter and avoiding ovarian vein.
- Internal iliac artery ligation is usually successful in 50% of the patients as it requires high surgical skills and clear knowledge of pelvic anatomy.

  Peritoneum over external iliac vessels is dissected making 4–5 cm incision from round ligament to infundibulopelvic ligament. Ureter is identified and left attached to peritoneum medially.

  Now identify common iliac artery at its bifurcation and internal iliac artery at medial border of psoas muscle.

  If possible, identify anterior and posterior branches of internal iliac artery and then anterior division is ligated with nonabsorbable suture by passing right angle from lateral to medial side of artery.

  Ligation is done at two places 2 cm distal to origin to avoid recanalization and vessel is not cut.

  External iliac artery is checked for pulsations and this method decreases pulse pressure by 85%.

- *Compression sutures:*
  - B-Lynch sutures: These are done if bleeding is not controlled in a cesarean section. Uterus is exteriorized and bimanually compressed to assess the advantage of applying compression sutures.

    If effective, vicryl is used on a large Mayo needle and uterus punctured 3 cm below uterine incision on right and then needle exited 3 cm above upper part of incision and 4 cm from lateral border then suture taken downward posteriorly and then at level of uterine incision, uterine cavity is entered posteriorly and then suture passed laterally and horizontally to left side and then needle exited from posterior uterine wall, suture taken upward and anteriorly above fundus then above incision on left side 3 cm above and 3–4 cm medial to lateral wall of uterus and finally needle is taken out of uterine cavity 3 cm below incision on left side; whole of suture is tightened to provide compression and then tied anteriorly **(Figs. 6A and B)**.

    Effective in 80% of cases of uterine atony and if bleeding controlled uterine incision sutured in two layers.
  - *Heyman sutures* are simple sutures and are applied if laparotomy is needed in postvaginal delivery because in these sutures hysterotomy is not needed.

    Two sutures are placed on both sides anteroposterior above bladder reflection and tied at fundus **(Fig. 7)**.
  - *Isthmic–cervical sutures:* These are placed at lower segment through anterior and posterior wall one on each side, right and left after dissecting bladder peritoneum downward. Cervical canal patency is ensured before tying knots. These sutures are helpful to control lower segment bleeding in placenta previa.
  - *Cho sutures:* Multiple square sutures are placed compressing anterior and posterior walls of uterus. Uterus is pierced at multiple points but this is an easy technique **(Figs. 8A and B)**.

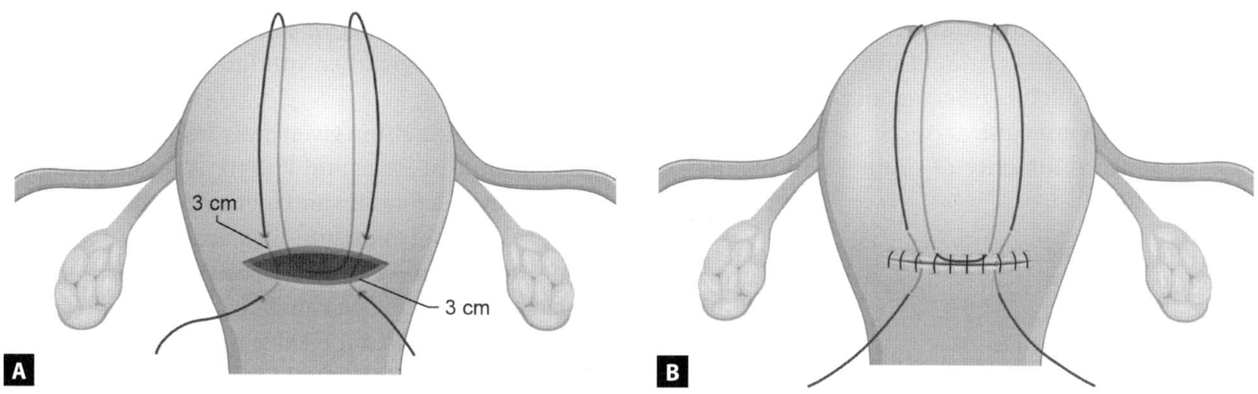

Figs. 6A and B: Technique of B-Lynch sutures.

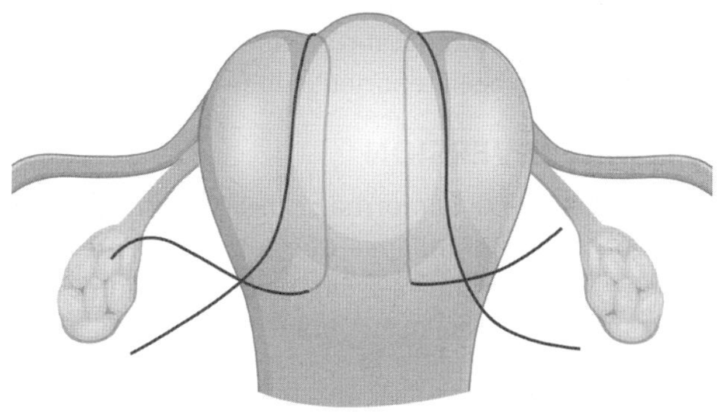

Fig. 7: Heyman sutures.

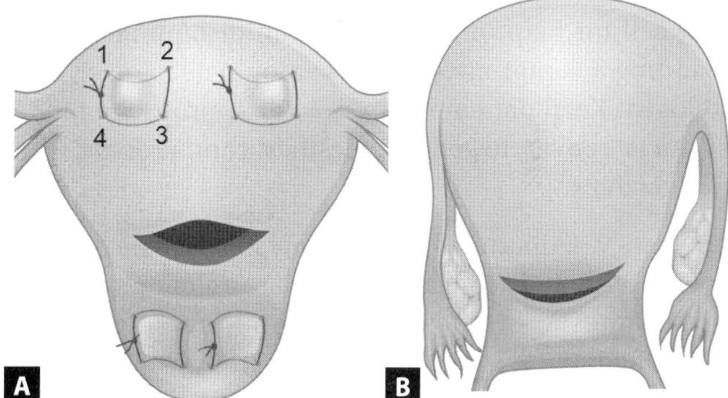

Figs. 8A and B: Cho's square sutures.

PERIPARTUM HYSTERECTOMY: INDICATIONS

- Most common are placenta accreta and atonic uterus
- Uterine rupture, lacerations or extension of hysterotomy
- Atonicity due to leiomyoma
- Undiagnosed malignancy

Hysterectomy for suspected placenta accreta: Placenta accreta can be diagnosed antepartum, during delivery or after birth of child.

When suspected, team is prepared and arrangements for blood, intensive care unit (ICU), and urologist are made. Ureteral stent placement before surgery reduces chances of ureteral injuries from 7% to 0%.

Steps in procedure are as follows:

1. Preoperative planning and preparation
2. Delivery scheduled at 34–36 weeks, the timing is individualized
3. Preparation for embolization or intravascular catheter placement if major bleeding is suspected
4. Patient is placed in lithotomy position for vaginal manipulation
5. Blood products arranged
6. Cystoscopy with placement of ureteral stents done if needed
7. Laparotomy and hysterotomy performed away from placental site (classical cesarean usually) and hysterotomy site closed without disturbing placenta
8. Retroperitoneal space is opened along infundibulopelvic ligament and utero-ovarian ligaments are tied preserving the ovaries.
9. Peritoneal dissection (cardinal ligament) is done up to uterine arteries and uterovesical fold is dissected, bladder mobilized away from uterus and uterine arteries, which are then ligated.
10. Dissection is continued below placental bed and hysterectomy completed taking out the uterus and cervix. Subtotal hysterectomy is faster but in such cases total hysterectomy needs to be done to achieve hemostasis.

11. The vaginal cuff is closed and hemostasis achieved.
12. In case of placenta percreta urinary bladder may be involved, which is resected and repaired by the assistance of a urologist.
13. Before closure integrity of ureters and bladder is ensured.

Hysterectomy for unexpected placenta accreta: As per American College of Obstetricians and Gynecologists (ACOG) if placenta percreta is found intraoperatively and if patient is not bleeding the management depends on the facilities available. Laparotomy incision may be closed and patient referred to a higher center without performing the cesarian. However, if accreta is found after the cesarian has been done and patient is bleeding, then call for help of the whole multidisciplinary team including a senior anesthetist, senior gynecologist, and urologist. Keep patient warm, ask for appropriate blood products, convert to general anesthesia and proceed with hysterectomy.

Emergency peripartum hysterectomy needs to be done in atonic uterus and accreta. The steps are as follows:
- Anesthesia, surgical teams, and nursing teams notified
- Vascular access achieved and blood products readied
- Surgical instruments for hysterectomy are opened
- Source of bleeding assessed and if atonic other options considered but if it is due to accreta then hysterectomy should be started
- The steps are then just as described previously under peripartum hysterectomy.
- After uterine arteries are ligated, patient stabilization can be done as bleeding will be decreased.
- Subtotal hysterectomy should be performed as it is faster with less blood loss. The vaginal angles are highly edematous, vascular, and friable and attempts to do a total hysterectomy will result in more hemorrhage in an already compromised patient and also prolongs the surgical time, further decompensating the patient
- Uterus amputated and cervix is closed

Complications of peripartum hysterectomy:
- Hemorrhage
- Bladder or bowel injury
- Ureteric injury
- Vascular injury
- ICU admission
- Surgical site infection
- Massive transfusion-related complications

Conservative treatment of placenta accreta: If diagnosed preoperatively, then uterine preservation can be attempted.

Hysterotomy incision given away from uterus and cord cut and ligated at base, and if there is no bleeding then placenta can be left in situ and uterine incision closed.

In presence of bleeding, hysterectomy is the only option. Uterine artery embolization or methotrexate can be given for enhancing placental resorption which usually takes 6 months placental resorption which usually takes 6 months. Consent for possibility of secondary hemorrhage and infection necessitating hysterectomy has to be taken.

In highly selected women triple-P technique involving en bloc removal of placenta and uterine wall by myometrial resection has also been reported in literature.

UTERINE INVERSION

Uterine inversion is a hemorrhagic disaster and can cause sudden postpartum collapse due to neurogenic shock even before hypovolemia sets in. Incidence is 1 in 20,000, uterine inversion can occur before or after placental separation.

Risk factors:
- Fundal placental attachment
- Uterine atony
- Mismanaged third stage of labor
- Morbidly adherent placenta

Management:
- Early detection and management increase the chances of reversion of uterus.
- Secure two wide-bore cannulas and arrange blood.
- *Johnsons maneuver* is promptly used to reposit uterus manually **(Figs. 9A and B)**.
  - All uterotonics are stopped and tocolytic or inhalational relaxants are given to prevent uterus from contracting and a hand is inserted in vagina pushing uterus in direction of umbilicus with upward pressure.
  - The repositioning of the uterus is started at the periphery and the fundus is the last to be reposited.
  - Hydrostatic pressure with saline can also be used with high success rate to reposit the uterus and has a high success rate.
  - If not initiated promptly there is high chance of failure because of formation of constriction ring at level of cervix.

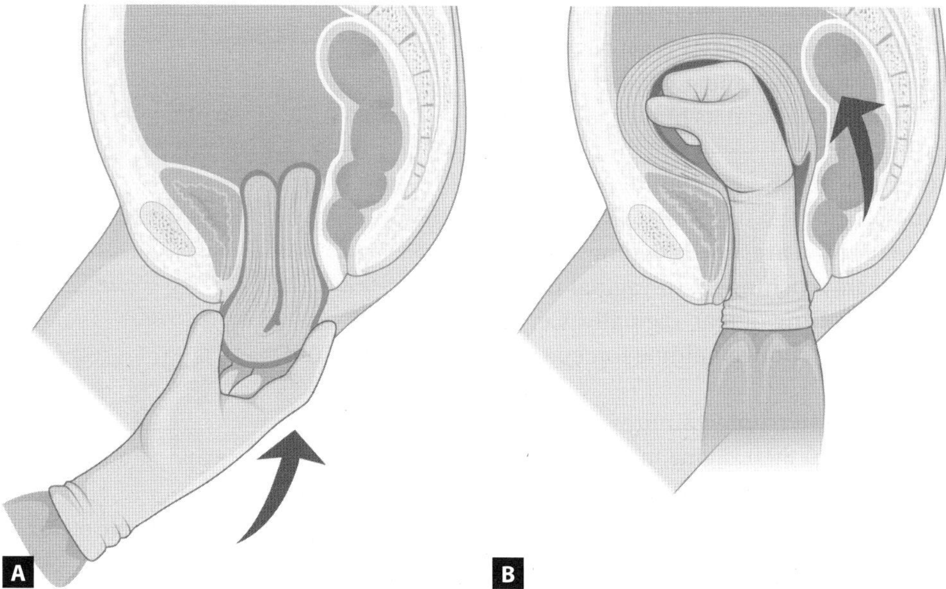

Figs. 9A and B: Johnson's maneuver for Uterine Inversion.

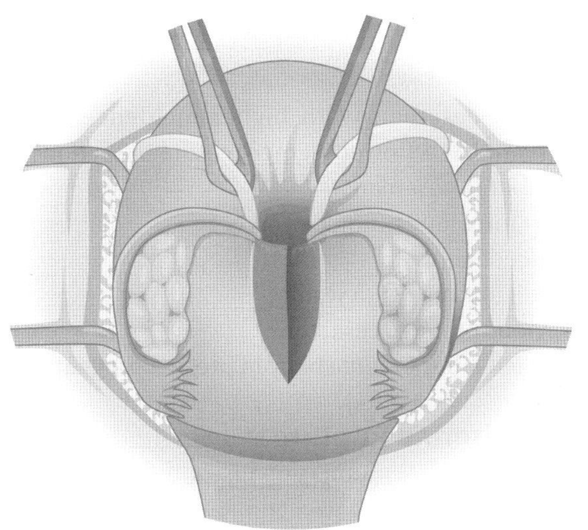

Fig. 10: *Haultains's incision* for uterine inversion management.

- Laparotomy is done if vaginal method is unsuccessful.

Huntington procedure is done where fundus is held with a suture or Allis forceps and upward pressure is applied or bilateral round ligaments are held with atraumatic forceps and traction applied but both these methods are technically difficult.

If constriction ring still prevents repositioning of uterus, then *Haultain's incision* **(Fig. 10)**, is given over myometrium posteriorly and then uterus reverted and held in position till it regains its tone and then uterotonics are started and incision is repaired.

Some surgeons prefer applying compression sutures to prevent recurrent inversion. There is a risk of rupture of this posterior hysterotomy incision in future pregnancy.

INJURIES TO THE BIRTH CANAL

Vulvovaginal Lacerations

According to ACOG (2016), 80% of women suffer some type of vaginal laceration during delivery.

Lacerations in lower one-third are easily detected and sutured. Those involving middle and upper third of vagina are comorbid with injuries of perineum and cervix.

Along with vaginal exploration, uterine exploration should also be done especially in vaginal birth after cesarians.

Cervical Lacerations

Cervical lacerations are not problematic unless they cause bleeding or extend into vagina.

They can be seen in precipitate labor or instrumental vaginal deliveries.

In all cases of PPH, cervix should be explored and sutured with continuous interlocking 1-0 catgut or vicryl.

Rarely cervix is avulsed from any vaginal wall called as colporrhexis.

Another rare injury is annular or circular detachment where the whole vaginal portion of cervix is avulsed.

Sometimes cervical tears extend to lower uterine segment involving uterine artery and its branches causing hemoperitoneum.

Occasionally, they may extend into peritoneal cavity. These need laparotomy.

Small cervical tears that are not bleeding do not need suturing. Large tears reaching up to vaginal fornix have good results after suturing of tear vaginally. Cervix is exposed and examined by sponge holding forceps and sutured under vision.

Continuous locking or interrupted sutures are taken from 1 cm above the apex and then vagina may be packed for pressure hemostasis.

Pelvic Hematomas

- Can be vulvar, vulvovaginal, paravaginal, or retroperitoneal hematomas
- If a pelvic vessel gets torn above pelvic fascia supralevator hematoma develops. Sometimes hematoma can be spontaneously formed by shearing off vessel without any laceration, usually with coagulopathy or after application of forceps.
- Pelvic hematoma may present as pain, pelvic pressure and inability to void.
- Sometimes supralevator or retroperitoneal hematoma may present as hypovolemia.

Management: It depends on size, location, duration, and expansion.

- Small painless hematomas not increasing in size can be managed conservatively.
- But if hematoma causes pain or retention of urine or enlarges in size then it needs to be drained.
- Blood loss is usually always underestimated.
- Hematoma drainage is done by incision at point of maximum distension and then evacuating clots followed by obliteration of cavity by absorbable sutures. Usually, no bleeding vessel is found. After obliteration the vagina is packed for 12–24 hours.
- Uterine artery embolization can be done for hematomas inaccessible surgically or where hemostasis has failed.

RUPTURE UTERUS (FIG. 11)

Rupture uterus can be classified as:

- Primary (in unscarred uterus) and secondary (in scarred uterus) or as

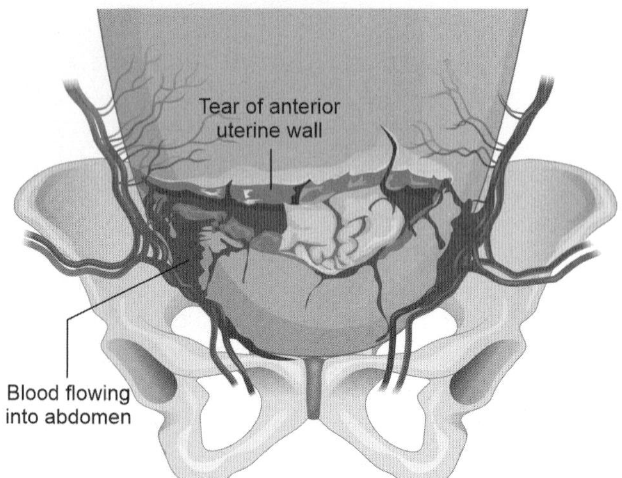

Fig. 11: Rupture uterus.

- Incomplete (scar dehiscence) and complete rupture (full thickness); most common type is partial or complete dehiscence of previous uterine scar.

Predisposing factors:

- Type of cesarean [with higher incidence in upper segment LSCS (lower segment cesarean section)]
- Number of cesareans
- Short interconception period
- Induced labor
- Injudicious use of oxytocin
- Previous preterm cesarean
- Pronged labor or large baby in present pregnancy.

Signs of Uterine Rupture

- Abnormal fetal heart rate [cardiotocography (CTG)]
- Scar tenderness
- Loss of uterine contour
- Hematuria
- Receding of station on per vaginum examination
- In second stage shortening of cord is pathognomonic of scar rupture.

Management depends on type of rupture, general condition of patient, desire of future childbearing, and extent of rupture.

In a patient with previous scar rupture which is clean and not bleeding much, repair of uterus can be done after freshening the margins of scar and if the patient does not desire fertility in future then along with repair tubal ligation should be done.

In a patient with rupture which is beyond the scope of repair, hysterectomy should be started and internal iliac artery ligation may be needed sometimes in case retroperitoneal hematoma is there.

CONCLUSION

Obstetric haemorrhage which contributes as a major cause of maternal mortality in India if managed in time can prevent lots of mortality as well as morbidity. Depending on the cause prompt action should be taken and principle remains the same first of all to maintain airway, breathing, circulation, arrange blood and blood products, apply medical methods whereever applicable and decide well in time when patient requires referral when in peripheral centre and when to proceed with surgery in a equipped centre.

REFERENCES

1. Rock J, Jones HW III. TeLinde's Operative Gynecology, 11th edition. Philadelphia: Wolters Kluver; 2015. pp. 1112, 1118, 1196, 1206.
2. Cunningham FG, Lenovo KJ, Bloom SL, Dashe JS, Hoffman BL, Casey BM, et al. Williams Obstetrics, 25th edition. United States: McGraw Hill; 2018. pp. 1686, 1688, 1690, 1692.
3. Walfish M, Neuman A, Wlody D. Maternal haemorrhage. Br J Anaesth. 2009;103:i47-i56.

Chapter 66

Episiotomy Repair and Management of Complex Perineal Lacerations

Kanan Kotdawala, Parul Kotdawala

INTRODUCTION

The term "episiotomy" is derived from ancient Greek words: EPISION—pubic region and OTOMY—surgical incision. Literature dates back to 18th century, when Sir Fielding Ould first performed surgical incision on pubic region to widen the birth canal.[1] With time this became most frequently performed procedure in mid-20th century. In the delivery of primiparous women, the episiotomy rate reached up to 90%.[2] Also it was very conveniently used in all instrumental deliveries.[3] But with time, the complications associated with it are getting noticed. And with reassessing the evidences, guidelines all over the world are now promoting a restricted use of episiotomy.[4]

CURRENT STATUS

Prevalence

The frequency of episiotomy is overall around 52% for all vaginal deliveries. It is more common in 1st time vaginal delivery (93%), as compared multipara with previous vaginal birth (30%). Episiotomy was a widely practiced intervention in USA till early 2000s. In 2006, ACOG (American College of Obstetricians and Gynecologists) recommended against its routine use. However, the selective use based on clinical judgment and maternal or fetal indications is still a well-accepted method. Till today, there are many countries including India; where episiotomy is very routinely performed in almost all primiparas.[5]

EPISIOTOMY

The technique of episiotomy was originally invented to curb the occurrence of irregular lacerations of perineum during labor and perineal tears of third and fourth degree. The general consensus is to make a well-directed and controlled incision over perineum during the second stage of labor to enlarge the vagina from its posterior facet to facilitate the delivery of the baby. Ideally, the episiotomy would relieve pressure on the perineum during child birth and leave an easily repairable incision in comparison to uncontrolled vaginal tears and trauma.[6] Regarding the indication of episiotomy; majority were given due to rigid perineum and due to maternal exhaustion. Some percentages of episiotomy were due to a routine practice of episiotomy in vaginal delivery of all primigravids and in all who had a history of perineal injury during previous child birth.[1]

However, there is evidence that episiotomy rate of >30% is not acceptable and episiotomy should be done on selective basis than done as a routine. There is evidence that episiotomy rate of >30% is not required and episiotomy should be employed on selective basis rather than as a routine procedure.[7]

Why Episiotomy?

- To provide extra space for delivery of the oncoming fetus
- To prevent excessive compression of head
- To prevent ragged deep tears.

Case Selection

- All instrumental deliveries (e.g., forceps; vacuum)
- All breech deliveries
- All abnormal presentations such as occipitoposterior; face to pubis
- Most primes with rigid perineum
- Preterm births.

Timing: When Episiotomy is Performed?

The episiotomy is generally performed when crowning of the presenting part occurs. The perineum is fully stretched and blanched, and the widest presenting diameter (BPD in case of cephalic presentation) has crossed the bony pelvis, and it is only the soft parts of perineum which is holding back the delivery.

Method of Episiotomy

- Generally, it is performed under local perineal infiltration or pudendal block or Saddle block.
- For a right-handed operator, an episiotomy on the right side is easier to suture and for left-handed one, an episiotomy on the left side of the patient is easier to suture.
- It is performed either with specially designed curved/angled scissor with blunt inner tip to avoid inadvertent trauma to rectum. Some surgeons prefer to perform episiotomy with knife.
- Cut on the inner side (vaginal mucosa) is generally bigger in comparison to outer side (vulvar skin).
- Types/direction of episiotomy—various directions have been employed with different intentions for episiotomy.
- Suturing techniques.

Angle Change with Stretching of Perineum

Due to stretching of the perineum, the angle at which the episiotomy cut is directed, changes once the delivery is complete, and the stretching effect is abolished. Hence a 45° angle while cutting becomes 25°–30° at the time of suturing **(Figs. 1A and B)**. This should be kept in mind, as sometimes the mediolateral suture line may almost become a midline!

SUTURING TECHNIQUES

The mucosa is always sutured with absorbable material either Catgut (biological material) or monofilament/rapid synthetic absorbable materials. Continuous locking stitches are employed for approximation of vaginal mucosa.

The muscles/perineal body is also sutured with same material. They can be approximated with continuous non-locking stitches or interrupted stitches.

The vulvar skin may be sutured with subcuticular absorbable material or interrupted mattress stitches with monofilament.

POSTOPERATIVE CARE

Pain relief along with anti-inflammatory medications are generally required in form of oral analgesics for 3–5 days. Seitz bath twice a day for 20 minutes each in warm water with antiseptic solution will clean the perineum and also improve the circulation for healing.

Antibiotics are generally not required and are avoided.

TYPES OF EPISIOTOMY

As such there are no international suggestions or standards for preferring the side (right or left) of the perineum for an episiotomy incision. But the episiotomy on the right side of the patient is easy to suture for a right-handed person, and conversely an episiotomy on the left side is easy to close for a left-handed person.

Although there are total seven types of episiotomy incisions, in regular practice only two of them are

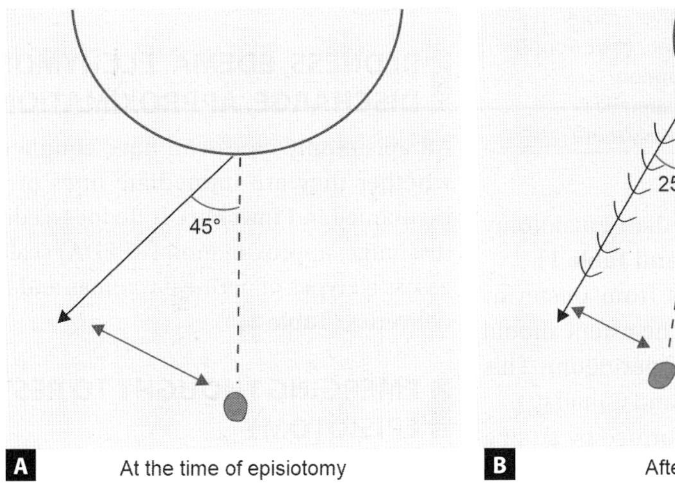

Figs. 1A and B: The evolution in distance of episiotomy scar from the anal sphincter.

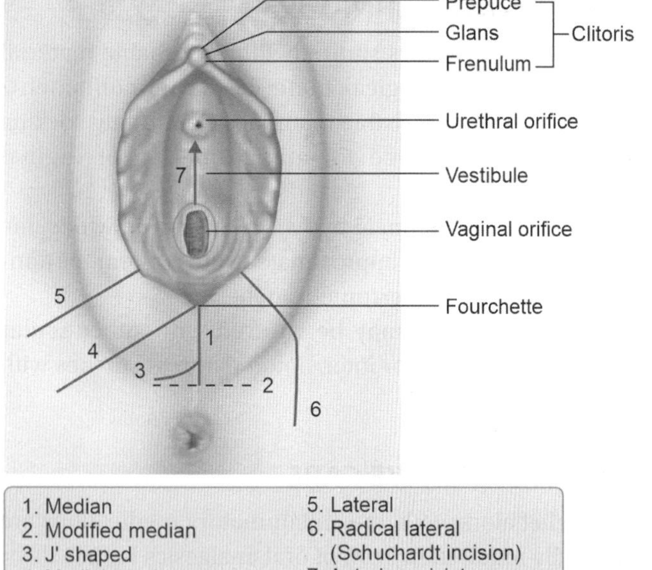

Fig. 2: Types of episiotomy.[8]

TABLE 1: Comparison—pros and cons of each type.[9]

	Median	*Mediolateral*
Merits	• Muscles are spared • Less blood loss • Easy repair • Least postoperative discomfort • Better healing • Dyspareunia least	• Safer for rectal involvement from extension • When needed, can extend the incision
Demerits	If extension happens, may involve the rectum	• Not a good apposition • More blood loss • More postoperative discomfort • Relatively more incidences of wound disruption • Dyspareunia is relatively more

employed in routine practice—the median episiotomy and the mediolateral episiotomy **(Fig. 2 and Table 1)**.

- *Median:* Median episiotomy starts from posterior fourchette and extends in midline. The extent should be approximately half of the size of perineum. This technique is commonly used in USA and Canada.
- *Modified median:* Episiotomy is performed by adding two transverse incisions in opposite directions perpendicular to the midline. There are data suggestive of increase in vaginal outlet diameter by 83% compared to median episiotomy.
- *"J" shaped:* This episiotomy starts with a midline incision and is then curved laterally to avoid the anus opening. It is combining the benefits of both median and mediolateral and at the same time avoiding risks of both.
- *Mediolateral:* An incision starting from midline and directed laterally-downward away from rectum.
- *Lateral:* It begins at vaginal introitus 1–2 cm lateral to the midline and is directed laterally and downward toward ischial tuberosity.
- *Radical (Schuchardt's incision):* Often considered a nonobstetrical incision, which carries deep into vaginal sulcus and is curved downward and laterally part way around the rectum. Very occasionally used to facilitate childbirth in complicated deliveries such as large head, difficult breech, or in correction for shoulder dystocia.
- *Anterior:* This incision may not be called an episiotomy. But this may rarely be required, especially in case of old scars and fibrosis blocking the passage of the baby.

Flowchart 1: Complications.

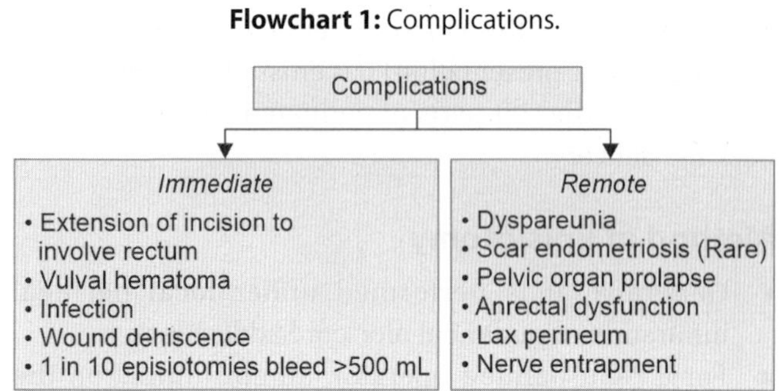

REDNESS, EDEMA, ECCHYMOSIS, DISCHARGE, APPROXIMATION SCALE

All episiotomy types can have couple of complications whether they are immediate ones or Remote ones as mentioned in **Flowchart 1**. Redness, edema, ecchymosis, discharge, approximation (REEDA) scale is developed to assess all types of perineal traumas and healing in vaginal deliveries **(Table 2)**.[10]

EMERGING THOUGHT TO RESTRICT EPISIOTOMY

Large data in published papers and prospective trials have shown that in most cases the assumed benefits of

TABLE 2: Redness, edema, ecchymosis, discharge, approximation (REEDA) scale.

REEDA					
Score	R: Redness	E: Edema	E: Ecchymosis	D: Discharge	A: Approximation
0	None	None	None	None	None
1	<0.25 cm of incision bilaterally	Perineal, <1 cm from incision	<0.25 cm B/L or 0.5 cm U/L of incision	Serous	Skin separation <3 mm
2	<0.5 cm of incision bilaterally	1–2 cm from incision	<0.25–1 cm B/L or 0.5–2 cm U/L of incision	Serosanguinous	Skin and subcutaneous fat separation
3	>0.5 cm of incision bilaterally	>2 cm from incision OR affects vulval area	>1 cm B/L or 2 cm U/L of incision	Bloody, purulent	Skin, subcutaneous fat, and fascial separation

Total score

Healed: 0

Moderately healed: 1–5

Mildly healed: 6–10

Not healed: 11–15

episiotomy are not confirmed and hence there has been a rethink about need of episiotomy in most cases. World Health Organization (WHO) and many national guidelines (ACOG in 2006, RCOG in 2007)[11] have proposed to restrict the frequency of episiotomy and a general consensus has emerged that ideal rates of episiotomy should be between 25 and 35% of all vaginal deliveries.[12]

American College of Obstetricians and Gynecologists (ACOG) guidelines[13] state that "the best available data do not support the liberal or routine use of episiotomy. However, there is a role for episiotomy for maternal or fetal indications such as avoiding severe maternal lacerations or to facilitate difficult births."

The WHO recommends an episiotomy rate of 10% as "a good goal to pursue," based on the data obtained in the English randomized controlled trial published in 1984.[14]

In the Cochrane systematic review, on "What would actually be the indications of episiotomy?" All the indications such as preterm delivery, breech delivery, macrosomia, shoulder dystocia, instrumental deliveries (forceps or vacuum extraction), nonreassuring fetal heart rate and rigid perineum or threat of severe perineal rupture have been questioned. The review further states "While it is clear that routine episiotomy should be avoided, there is no solid evidence corroborating ANY indication of episiotomy. The benefits of the use of episiotomy in these selected conditions remain controversial."[15-17]

Recently, it has been suggested that episiotomy should never be performed! Rather than protecting the pelvis, it has been shown to increase postpartum perineal pain, dyspareunia, blood loss, anal sphincter laceration, rectal damage, and anal.

MANAGEMENT OF COMPLEX PERINEAL LACERATIONS

Background

Around 60%–80% of women delivering vaginally develop perineal lacerations. From these, 3%–4% develop third-degree laceration and only 1% have fourth-degree laceration with five times more incidence in primiparae than multiparae.[18] Together, the third- and fourth-degree lacerations are termed as obstetric anal sphincter injury (OASIS).

Some risk factors for OASIS are:
- Operative deliveries, such as forceps/vacuum
- Midline episiotomy
- Increased birth weight/macrosomic baby
- History of previous OASIS
- Occipito-posterior presentation.

Classification of Perineal Lacerations

Perineal lacerations are classified into four basic categories:[19]

1. *First degree:* It is a superficial trauma to the vaginal mucosa and it may involve the perineal skin.
2. *Second degree:* In addition to first degree, lacerations involving the vaginal mucosa and perineal body
3. *Third degree:* In addition to second degree, the tear involves the anal sphincter. This is again classified into three subcategories:
   a. <50% tear of the anal sphincter
   b. >50% tear of the anal sphincter
   c. Tear extends to both; external and internal anal sphincters
4. *Fourth degree:* In addition to third degree, the tear involves the rectal mucosa. Most dreaded and severe perineal injury, disrupting perineal body (anchor for bulbospongiosus and transversus perinei), external anal sphincter, and internal anal sphincter, which leads to disruption of the rectal mucosa.

Sulcus and vaginal wall tearing: Sulcus tears refer to tears in upper vaginal wall in absence of major (third and fourth degree) perineal tears.[20]

Degrees of Perineal Lacerations

Figure 3 illustrates the basic anatomy of perineal and vaginal musculature trauma.

The degrees of tear in perineal lacerations are shown in **Figure 4**.

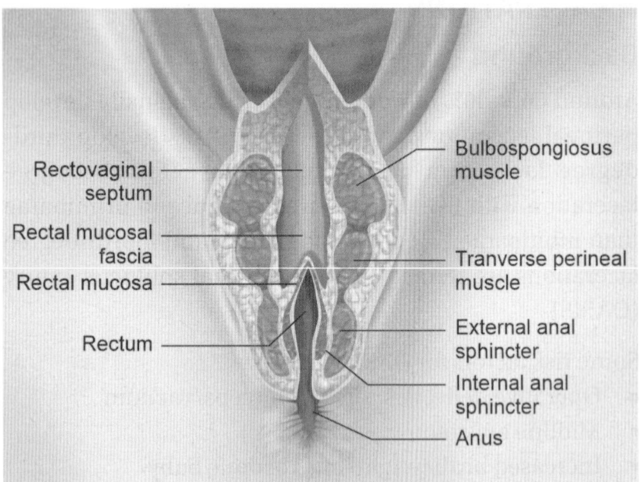

Fig. 3: Basic anatomy of perineal and vaginal musculature trauma. *Courtesy:* Dave Klemm.

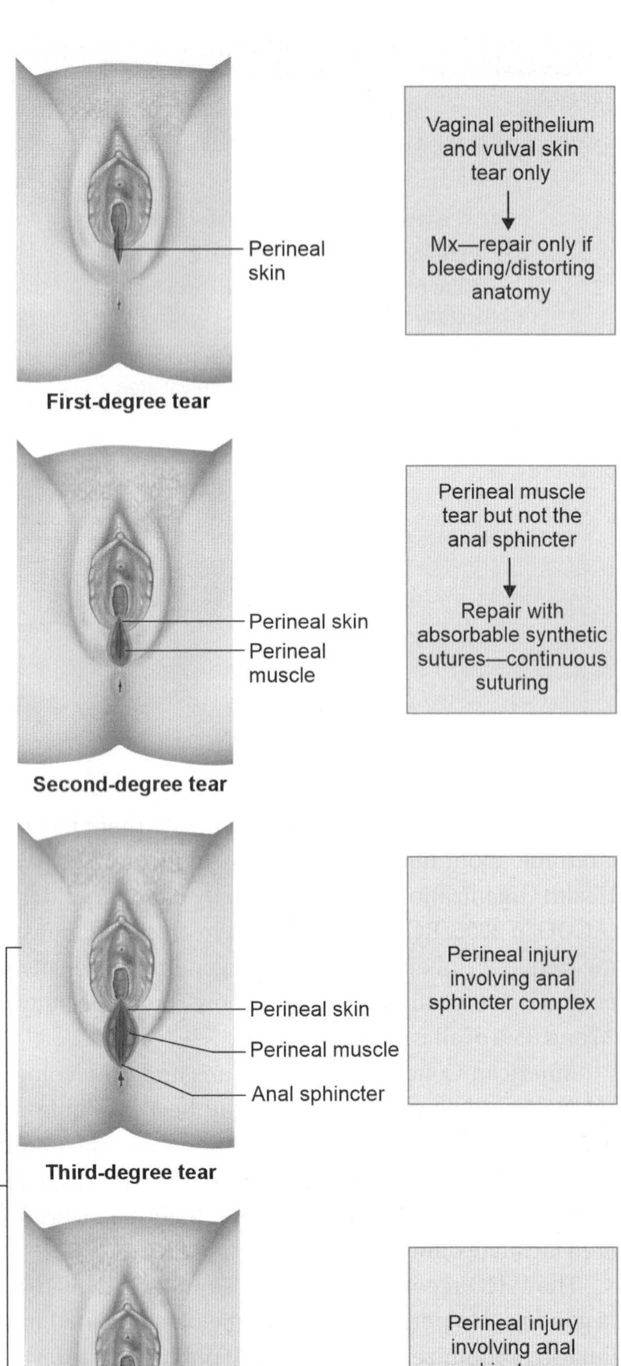

Fig. 4: Degrees of tear in perineal lacerations.

Management of Complex Perineal Lacerations[21]

- To reduce the risk:
  - Perineal massage
  - Warm compresses (during second stage of labor)
- Possibly reduces risk:
  - Manual perineal support during labor
  - Lateral birthing procedure
- Does not reduce the risk:
  - Delayed pushing
  - Perineal massage.

Perineal massage: Perineal massage seems to decrease the incidence of perineal tears needing suturing. Studies have shown that the incidence of third- and fourth-degree lacerations also decrease when massage was performed during the second stage of labor. Massage promotes perineal relaxation, rise in perineal blood flow, and better stretching of the vaginal tissue during labor. This may decrease the need for an episiotomy. Massage should be started after 34 weeks and is performed daily for 15–20 minutes till delivery. During the pushing stage of labor, place two fingers of a lubricated gloved hand just inside the vagina and move them from side to side. It puts mild, downward pressure on that area and can help to stretch it. Doing perineal massage at home, at the end of the third trimester also may help the tissue stretch more easily during labor.[22]

Keep the perineum warm: The area between the opening of the vagina and the anus is called the perineum. Placing a warm, damp cloth on this area during the pushing stage of labor may help the tissue to stretch more easily.

Control pushing: Aim for controlled, consistent pushing. Patient is advised to try not to push down hard all at once while the baby's head is coming through the vaginal opening. Pushing gently and slowly can give the tissue time to stretch.

Manual perineal protection/"hands-on" techniques:[23]

- Left hand supporting and slowing the head delivery
- Right hand protecting the perineum
- Proper communication with mother-not to push when head is crowning
- Consider episiotomy (risk groups and correct angles).

Repair of the Lacerations

First-degree lacerations are those which are not bleeding and have not altered the anatomy do not require surgical repair. If required, this can be sutured or approximated with adhesive skin glue. A trial compared the skin adhesive versus suturing for first-degree lacerations showed that the repair time was shorter in adhesive glue group. The overall patient pain scores were also slightly lower in the adhesive group. The surgeon may choose between suturing or adhesives for hemostatic first-degree lacerations.

While repairing second-degree lacerations, a continuous suture is preferred over interrupted suturing to minimize postoperative pain and the need for suture removal. Suturing is like an episiotomy repair; approximation of vaginal mucosa till introitus. Then the repair of the perineal body and the perineal tissue gaps and torn muscles. The skin is closed by absorbable subcuticular or interrupted sutures.

Third- and fourth-degree lacerations are repaired in a stepwise manner. The rectal mucosa is approximated by 3 "0" material starting above the apex of the laceration in a fourth-degree laceration. A layer of submucosal stitches will put the rectal/anal mucosa at rest. The retracted anal sphincter is identified and pulled up. This perhaps is the most vital step to avoid long-term incontinence of flatus and feces. The sphincter muscle with its fascial sheath is approximated by two sets of stitches perpendicular to each other, in a bid to avert a cut through phenomenon. When possible and identifiable, the internal anal sphincter should be repaired separately from the external anal sphincter **(Figs. 5 and 6)**. Once the rectal part of repair is completed, the perineal portion is closed in a manner similar to a second-degree tear.

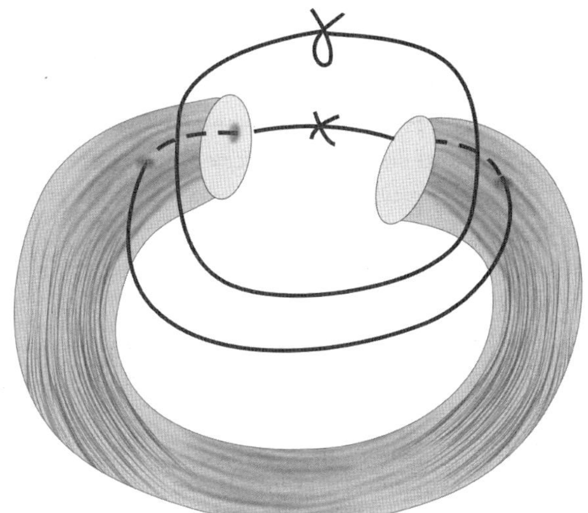

Fig. 5: End-to-end repair of anal sphincter in complete perineal tear.

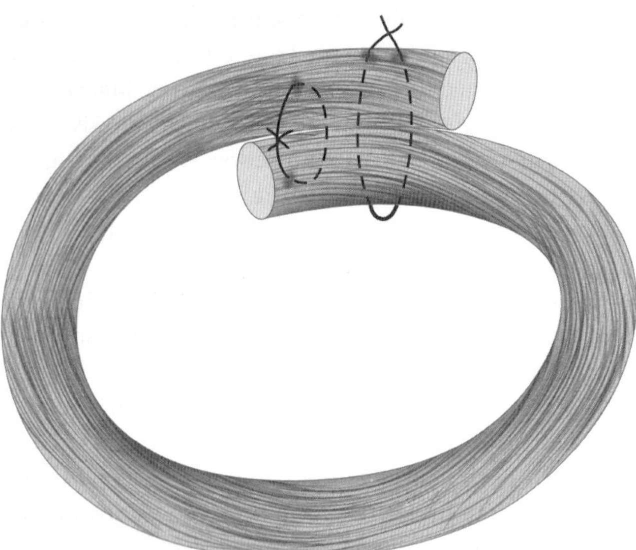

Fig. 6: Overlapping anal sphincter repair.

It is advisable to repair the "internal anal sphincter" separately with interrupted or mattress sutures. One should avoid overlapping of this sphincter muscle plain. As for the "external anal sphincter" we can use any technique, overlapping or end-to-end. For a partial tear the end-to-end method is preferred. For suturing 3-0 or 2-0 polyglactin (PG) or polydioxanone (PDS) are used.

A single dose of an antibiotic (second-generation cephalosporin) may be given following an OASIS repair. If there is contamination with rectal contents, a course of broad-spectrum antibiotics should be given for 5–7 days to avert the risk of infection and resulting wound breakdown. To avoid pressure on the sutures, the patient is advised to avoid straining at stool, is given stool softeners and mild laxatives, and may be put on liquids only for a day or two, and light diet for another 2–3 days. Reflex spasm of urinary sphincter is a possibility whenever perineal repair is carried out and should be watched for. Some form of physiotherapy afterwards is necessary to regain a better tone of the sphincter.

A counselling about the mode of delivery in next childbirth in such women who have undergone a repair of third- or fourth-degree laceration is vital. An informed consent should be clearly noted. The role of prophylactic episiotomy is not yet confirmed, and yet a mediolateral (more lateral) episiotomy is logical.

Although effect of warm compress on perineum during second stage of labor as well as the effect of perineal massage during later trimester and in second stage of labor are confirmed by many studies, these data are still inconclusive to recommend them for routine practice.

Complications: Early and Longer Term

Bleeding: Excessive bleeding is a most common complication of perineal lacerations. This is generally controlled quickly by pressure and prompt suturing of the wound. Occasional hematoma formation, when a deeper extension has happened, can lead to large amount of blood loss very quickly.

Pain: Pain and the time taken to repair the tears lead to a delay in mother–child bonding.

Infection: Although very uncommon, there always is a risk of infection and wound breakdown with any vaginal repair. Generally, the infection is a result of contamination of the suture site with feces or some long-standing perineal subclinical infection.

Long-term complications include pain, urinary or anal incontinence, and delayed return to sexual intercourse due to dyspareunia. These symptoms are worse in a case of episiotomy repair compared to those where natural tears were allowed. Flatus incontinence can persist for many years after an OASIS.

About 25% women who suffer from an OASIS injury may develop wound dehiscence within puerperium (first 6 weeks postpartum) and wound infection may occur in 20%. If the OASIS trauma is not identified clearly or the healing is poor, a rectovaginal fistula or a perineal fistula may develop. The time taken to regain normal sexual function following a perineal trauma depends upon the severity of the laceration.

Rarely, one of the branches of pudendal nerve gets entangled in the repair area and that may cause great difficulty and pain while sitting. Patient may require a ring to sit on. After identifying the problem, by locating the pain trigger, a releasing incision by cutting the fibrosis relieves this problem.

CONCLUSION

Episiotomy is an important tool in an obstetrician's armamentarium. If used judiciously, episiotomy can help easy and efficient delivery with minimum side-effects. But this time-tested procedure which was used regularly in past has its own limitations and new thoughts are emerging to restrict its use to gain maximum benefits in properly indicated scenarios.

REFERENCES

1. Ghulmiyyah L, Sinno S, Mirza F, Finianos E, Nassar AH. Episiotomy: history, present and future – a review. J Matern Fetal Neonatal Med. 2022;35(7):1386-91.
2. Glowm.com. (2011). Episiotomy. [online] Available from http://www.glowm.com/section-view/heading/Episiotomy/item/128 [Last accessed April, 2024].
3. NHS.UK. (2020). Episiotomy and perineal tears. [online] Available from https://www.nhs.uk/pregnancy/labour-and-birth/what-happens/episiotomy-and-perineal-tears/ [Last accessed April, 2024].
4. Deyaso ZF, Chekole TT, Bedada RG, Molla W, Uddo EB, Mamo TT. Prevalence of episiotomy practice and factors associated with it in Ethiopia, systematic review and meta-analysis. Womens Health. 2022;18:17455057221091660.
5. Kozhimannil KB, Karaca-Mandic P, Blauer-Peterson CJ, Shah NT, Snowden JM. Uptake and utilization of practice guidelines in hospitals in the United States: the case of routine episiotomy. Jt Comm J Qual Patient Saf. 2017;43(1):41-8.
6. Barjon K, Mahdy H. Episiotomy. In: StatPearls. Treasure Island (FL): StatPearls Publishing; 2022.
7. Marai W. A two years retrospective review of episiotomy at Jimma Teaching Hospital, southwestern Ethiopia. Ethiop Med J. 2002;40(2):141-8.
8. Kalis V, Laine K, de Leeuw J, Ismail K, Tincello D. Classification of episiotomy: towards a standardisation of terminology. BJOG Int J Obstet Gynaecol. 2012;119(5):522-6.
9. Garner DK, Patel AB, Hung J, Castro M, Segev TG, Plochocki JH, et al. Midline and mediolateral episiotomy: risk assessment based on clinical anatomy. Diagnostics. 2021;11(2):221.
10. Osmosis.org. (2022). REEDA: Episiotomy Healing Assessment Acronym. [online] Available from https://www.osmosis.org/answers/reeds-episiotomy-healing-assessment-acronym [Last accessed April, 2024].
11. Thiagamoorthy G, Johnson A, Thakar R, Sultan AH. National survey of perineal trauma and its subsequent management in the United Kingdom. Int Urogynecology J. 2014;25(12):1621-7.
12. Medscape. (2010). Changes in Episiotomy Practice: Evidence-based Medicine in Action. [online] Available from https://www.medscape.com/viewarticle/721538 [Last accessed April, 2024].
13. Google.com. (2006). ACOG 2006 guidelines episiotomy. [online] Available from https://www.google.com/search?q=acog+2006+guidelines+episiotomy&oq=acog+2006+guidelines+episiotomy&aqs=chrome..69i57j33i160l2.6551j0j7&sourceid=chrome&ie=UTF-8 [Last accessed April, 2024].
14. World Health Organization. Care in normal birth: a practical guide: report of a technical working group. Maternal and Newborn Health/Safe Motherhood Unit, Family and Reproductive Health, World Health Organization, 1996.
15. Melo I, Katz L, Coutinho I, Amorim MM. Selective episiotomy vs. implementation of a non episiotomy protocol: a randomized clinical trial. Reprod Health. 2014;11:66.
16. Helewa ME. Episiotomy and severe perineal trauma. Of science and fiction. CMAJ. 1997;156(6):811-3
17. Carroli G, Mignini L. Episiotomy for vaginal birth. Cochrane Database Syst Rev. 2009;(1):CD000081.
18. Ugwu EO, Iferikigwe ES, Obi SN, Eleje GU, Ozumba BC. Effectiveness of antenatal perineal massage in reducing perineal trauma and post-partum morbidities: A randomized controlled trial. J Obstet Gynaecol Res. 2018;44(7):1252-8.
19. Kalis V, Laine K, de Leeuw JW, Ismail KM, Tincello DG. Classification of episiotomy: towards a standardisation of terminology. BJOG. 2012;119(5):522-6.
20. Vieira F, Guimarães JV, Souza MCS, Sousa PML, Santos RF, Cavalcante AMRZ. Scientific evidence on perineal trauma during labor: Integrative review. Eur J Obstet Gynecol Reprod Biol. 2018;223:18-25.
21. Committee on Practice Bulletins-Obstetrics. ACOG Practice Bulletin No. 198: Prevention and Management of Obstetric Lacerations at Vaginal Delivery. Obstet Gynecol. 2018;132(3):e87-e102.
22. Demirel G, Golbasi Z. Effect of perineal massage on the rate of episiotomy and perineal tearing. Int J Gynaecol Obstet. 2015;131(2):183-6.
23. Bulchandani S, Watts E, Sucharitha A, Yates D, Ismail KM. Manual perineal support at the time of childbirth: a systematic review and meta-analysis. BJOG. 2015;122(9):1157-65.

Section

11

Miscellaneous

Chapter 67

Principles of Robotic Surgery

Anupama Bahadur, Rajlaxmi Mundhra, Gayatri Suresh

INTRODUCTION

A minimally invasive approach has revolutionized the field of surgery in the last couple of decades. Robot-assisted keyhole surgeries form the most advanced technology providing the surgeon the comfort of good ergonomics, and precision apart from the benefits of lesser trauma and quicker recovery to the patient. The technology is increasingly being explored in the field of gynecological, gastrointestinal, urological, and even cardiothoracic surgeries.

THE SURGICAL ROBOT

The robot is a computer-manipulated device designed to copy the complex surgical steps of the surgeon on the patient. The system merely reproduces the actions of the surgeon at the console with added advantages of extreme precision and are therefore referred to as *master-slave manipulators*. The initial robotic systems to achieve approval by the US Food and Drug Administration (FDA) include:[1]

- *Da Vinci* Surgical System (Intuitive Surgical, California)[2]
- The *ZEUS* system (Computer Motion, California).[3]

BASIC PARTS OF THE SURGICAL UNIT[4]

The surgical system comprises the following parts **(Flowchart 1)**:

- *Patient side cart:* The patient side cart bears the four robotic arms or manipulators which is docked to trocars inserted on the patient's body. These arms can accommodate either the endoscopic camera or surgical instruments such as scissors, grasper, needle driver etc. The computer-based software system is designed to transmit the visual images from the endoscope in the surgical field to the surgeon's console and in return transmits the surgeons hand movements to the "slave manipulators" mounted in the surgical field. The software is also capable of modifying the

Flowchart 1: Parts of surgical robot.[4]

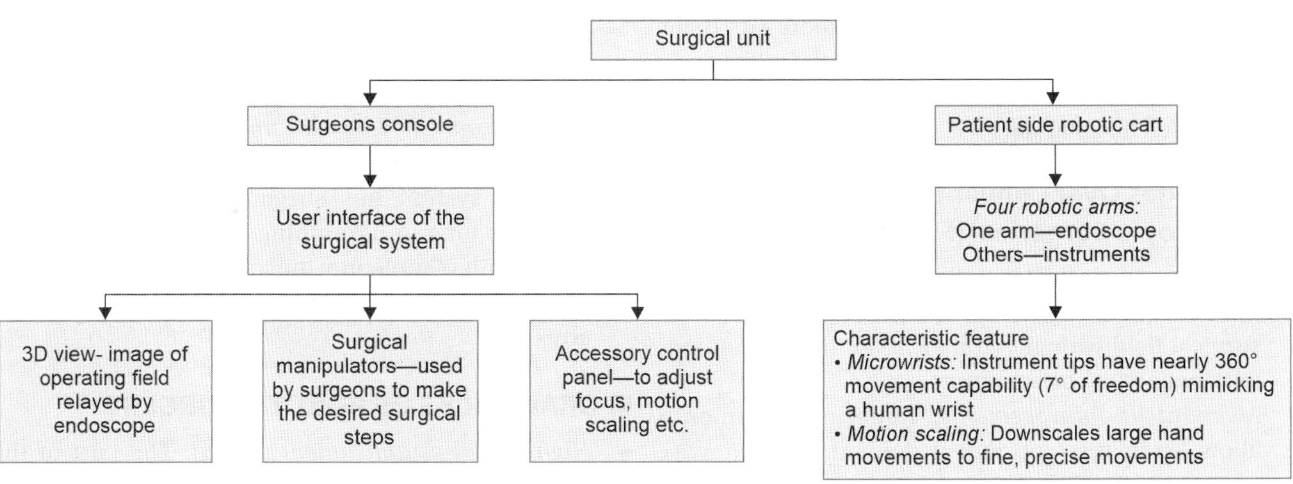

hand motions of the surgeon, for example, negate tremors and thereby enhance precision.

- *Surgeon console:* The console is the main user interface which has the high-resolution viewer where 3D images are transmitted from the endoscope can be visualized via a binocular device. They also have the surgical manipulators or the "master hand controls" and a control panel that facilitates adjusting focus and motion scaling. Hand movements at the level of the "masters" are what get transmitted to the surgical instruments at the robotic end. The ratio of movements at the level of the hand control and at the patient end can be adjusted which is known as motion scaling. For instance, a ratio of 3:1 allows a 1 cm movement of the robotic instrument for every 3 cm movement of the surgeon's hand control device. These hand controls are also equipped to filter out tremors thereby enhancing precision. The console is typically placed a couple of meters away from the operating table from where the surgeon is able to perform the surgery.

The four pillars on which the robotic system is built includes safety, intuitive control over the surgical instruments from the control, higher degree of freedom at the level of instrument tip and lastly, three-dimensional view of the surgical field. These principles allow robotic surgery to enjoy the advantages of minimally invasive surgery, while ensuring certain benefits of open surgery such as 3D vision, improved dexterity, and precision.

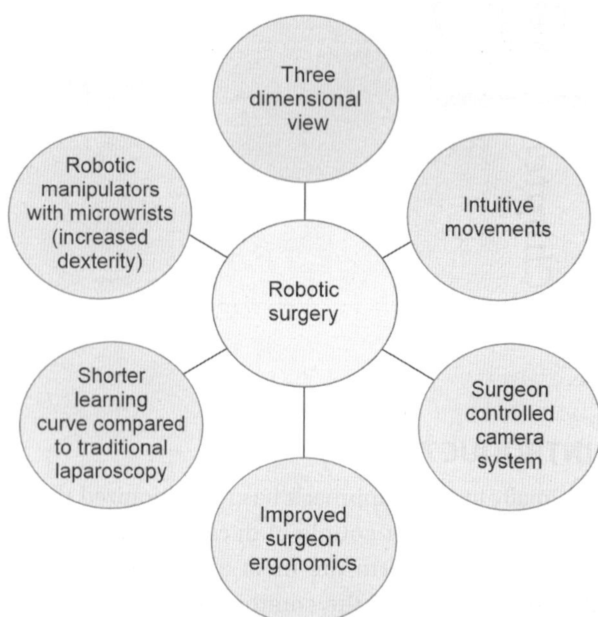

Fig. 1: Advantages of robotic surgery.[4,5]

ADVANTAGES OF ROBOTIC SURGERY

General Advantages of Minimally Invasive Approach[4,5]

- Minimal trauma to patient—smaller access incisions lead to decreased tissue handling, reduced tissue trauma and thereby lesser morbidity to the patient.
- Faster recovery.
- Decreased hospital stay.

Advantages Over Conventional Laparoscopic Surgeries[5] (Fig. 1)

- *Superior visualization:* Robotic surgical system provides the luxury of a three-dimensional view of the surgical field with added benefits of rapid zooming, automatic white balancing etc.
- *Mechanical improvements:* Robotic instruments allow "endowrist" movement similar to human wrist supporting 7° of freedom in contrast to 4° of freedom in conventional laparoscopy. This helps in easier suturing. Additionally, robotic trocars are either 8 mm or 11 mm in diameter docked onto robotic arms. They are able to better counter the torque produced by the abdominal wall on the instruments, particularly in obese patients and reduces the chances of fracture of instruments.
- *Stabilization of instruments:* With the help of the software, small movements and tremors by the surgeon are negated. This is in contrast to conventional laparoscopy where, tremors, and subtle movements are exaggerated.
- *Improved surgeon ergonomics:* Surgeon can perform the surgery while being comfortable seated in the console.
- *Surgeon controlled camera and intuitive movements:* In contrast to conventional laparoscopy, the robotic surgical system has surgeon-controlled camera allowing for rapid zooming in. Further, the hand movements of surgeon at the console and automatically transmitted to the instruments docked to the robotic arms.

DRAWBACKS OF ROBOTIC SURGERY

- *High initial establishment capital:* The cost of the robotic system ranges between 2 and 3 million USD

while cost of each robotic instrument ranges between 2,200 and 3,500 USD. On an average there is an expense of about 2,000 USD per case after establishment of the system.
- *Increased operating time:* In comparison to conventional laparoscopy, robotic surgeries take up extra time to allow for docking and doffing the robotic system.
- *Cost for procuring instruments:* Major limitation (limited number of uses of each instrument). Each robotic instrument needs to be replaced after ten uses.
- Surgical training and training of technical staff.
- Lack of tactile feedback during surgery—possibility of vascular avulsion and tearing of tissues.

AVENUES OF CURRENT CLINICAL APPLICATION

Robotic surgery is being explored in the fields of urology, surgical oncology, gynecology, ENT, and thoracic surgery. Robotic surgeries are particularly popular in deep pelvic surgeries, retroperitoneal surgeries where precision is of prime importance. With respect to gynecological surgeries, robotic surgery is gaining popularity in endometrial cancer surgeries and also for certain benign indications such as endometriosis, fibroid uterus **(Box 1)**.

BOX 1: Avenues for application of Robotic surgery.[4]

- Urology
- Surgical oncology
- Surgical gastroenterology
- Gynecology
- Cardiothoracic surgery
- ENT
- Neurosurgery, pediatric surgery

BOX 2: General principles of robotic surgery.[6]

- The robotic cart should be on the side of the target organ. Farther the target organ from the robotic cart and endoscope, greater are the chances of collision between instruments
- Interport distance should be >8 cm
- Endoscope port should be in line with the robotic cart
- Wide angles between the robotic arms to prevent collision between instruments
- Patient positioning should be carefully done. Cannot be altered after docking
- Surgery is technically easier when target organ is on a straight line from the camera port and robotic cart
- Robotic arms do not transmit tactile sensation to the surgeon and tensile characteristics of the tissues need to assessed visually by surgeon

OPERATIVE PROCEDURE

- Peritoneal access—in gynecological surgeries:
  - First port (8 mm or 12 mm) is inserted in midline, 20 cm from the pubic symphysis or 8–10 cm above the fundus of the uterus.
  - *Robotic accessory ports:* These are inserted at a distance of 8 mm from the midline port bilateral at the level of midclavicular line, lateral to the rectus muscle.
  - *Conventional accessory port:* This port may be inserted on either right or left, 8 mm away from the ipsilateral robotic accessory port. This port helps in specimen removal, suctioning, and suture introduction.
- After port placement, docking of ports to the robotic arm is done.
- Side-docking, where the equipment tower bearing the robotic arms is placed by the side of the patient is preferred in gynecological surgeries, so that vaginal manipulation is undisturbed.
- Robotic arms are connected to the surgical instruments via the robotic instrument adapter.
- Surgeon seated at the console visualizes the surgical field and performs surgery by making movements in the "master hand controls" which are translated to the surgical instruments inserted in the patient. Foot pedals are provided at the console to control camera, the "masters" and the energy devices.

CONCLUSION

Robotic surgery is a cutting-edge invention in the field of minimal access surgery that has far reaching implications. With its advantages of high degree of precision and dexterity, many complicated surgeries can now be performed via the minimal access route without compromising on patient safety. The only factor limiting incorporation of this technology is the cost concerns. Therefore, the challenges in expanding the field of robotic surgery would be to strike a balance between cost and clinical effectiveness.

REFERENCES

1. Hashizume M, Tsugawa K. Robotic surgery and cancer: the present state, problems and future vision. Jpn J Clin Oncol. 2004;34(5):227-37.
2. Intusurg.com. (2022). Surgical System, Intuitive Surgical. Available from: http://www.intusurg.com/html/davinci.html [Last accessed April, 2024].
3. Marescaux J, Rubino F. The ZEUS robotic system: experimental and clinical applications. Surg Clin North Am. 2003;83(6):1305-15, vii–viii.
4. Morris B. Robotic surgery: applications, limitations, and impact on surgical education. MedGenMed Medscape Gen Med. 2005;7(3):72.
5. Moorthy K, Munz Y, Dosis A, Hernandez J, Martin S, Bello F, et al. Dexterity enhancement with robotic surgery. Surg Endosc. 2004;18(5):790-5.
6. Prasad A. Robotic Surgery in India. J Young Med Res. 2014;12(2).

Chapter 68

Cosmetic Gynecology

Manish Subhash Mahajan

INTRODUCTION

Society is gradually overcoming the need of basic health care and is definitely interested in quality of life (QoL) improving services. To look beautiful and attractive is an innate desire for self-confidence which is an evolutionary attribute for sexual partner selection and species propagation.

The significant footfall at cosmetic and plastic surgery centers for various beautification procedures such as *nose reshaping, botox*, and *fillers* in facial aesthetics, breast reshaping, abdominoplasties, liposuction, body reshaping, and even more is quite explanatory. Cosmetic gynecology is simply extension of it.[1-4]

Cosmetic gynecology is newly evolving subspecialty of gynecology becoming popular day by day among patients and physicians. It is a broad term which aims to improve appearance as well as functionality of female genitalia.[5]

It includes use of various energy-based devices (EBDs), genital cosmetic surgical procedures and procedures under regenerative medicine.

Women undergo various physiological changes such as pregnancy, childbirth, gaining or losing weight, menopause, aging as well as certain pathological hormonal conditions such as hypothyroidism and insulin resistance which greatly affects the overall appearance including significant changes in appearance and functionality of vulva and vagina.[6]

Girls and women with polycystic ovary disease (PCOD) have body appearance concerns and altered body function can impact their psychological health of women with PCOS, which can also influence their relationships.[7-9]

Polycystic ovary syndrome individuals also have higher risk of female sexual dysfunctions[10,11] and such concerns should be provided with medical help.

Even when there is ongoing improvement in effective implementation of quality ante natal and intrapartum care, often there is ignorance and neglect about providing solutions for post-pregnancy body image changes. Such body image dissatisfaction significantly could affect the woman's dignity and confidence in the role of a wife or working woman apart from being a mother.[12,13]

With the improved life expectancy, for women expected to live more than one-third of their lives in postmenopausal period.[14]

Many such QoL-related concerns such as body image dissatisfaction, postpartum, and postmenopausal genital health alteration such as subjective feeling of vaginal laxity without demonstrable findings, and poor sexual satisfaction are reported in gynecology clinics but mostly remain unanswered because of lack of treatment options today and finding hopes with newer procedures under cosmetic gynecology.

These procedures have a great role in improving and/or restoring gynecological and sexual functions.

Apart from aesthetic enhancements or genital beautification, cosmetic gynecology procedures aim not only treating female sexual dysfunction but also for enhancing sexual pleasure when there are very few options available for women unlike for men in this apparent gender equal society.

OVERVIEW OF COSMETIC GYNECOLOGICAL PROCEDURES

Procedures under cosmetic gynecology are grossly divided into three broad categories:

1. EBDs
2. Minimally invasive procedure
3. Surgical procedures.

Energy-based Devices

Use of various machines that work on various biophysics principles which involves application of controlled thermal or nonthermal energy to the vulval or vaginal tissue stimulating collagen regeneration and strengthening, contracture of elastin fibers with tightening or firming effect, and improving blood flow at sight of application as well as inducing neovascularization.

These include lasers with different wavelengths (microablative CO_2 laser—10,600 nm, Erbium Yag laser 2,940 nm, nonablative 1,470 nm diode laser), radiofrequency devices, and high-intensity focused ultrasound (HIFU).

Energy-based devices are gaining fast popularity because of their noninvasive nature and good effectivity based on initial results **(Figs. 1A to C)**.[15]

The advantages of EBDs are listed in **Box 1**.

Although these technologies are used widely in dermatology application for face and skin rejuvenation and tightening since long time with good safety, their use for vulvovaginal tissue strengthening needs judicious and cautious approach until there is sufficient data on long-term safety or complications.[5,6,16-19]

Evaluation and consent for the use of EBDs are given in **Box 2**.

The probable indications for EBD use are listed in **Box 3**.

The contraindications for use of EBDs are given **Box 4**.

Minimally Invasive Procedures

Box 5 lists the types of minimally invasive procedures.

Platelet-rich Plasma

Platelet-rich plasma (PRP) is obtained by single-spin or double-spin centrifugation of whole blood and extracting buffy coat and clear plasma for injecting as therapeutic agent.

This autologous highly bioactive preparation has been proving tremendously effective as therapeutic application in various treatment modalities in orthopedics, dentistry, wound care, facial aesthetics, and hair regrowth treatment.

In gynecology, it has been used for thin endometrium, ovarian rejuvenation in cases of premature ovarian failure. It is now finding the application in various treatments for genital rejuvenation procedure under cosmetic gynecology and being used with variable success in treating female sexual dysfunctions, enhancing sexual sensation and pleasure, (G-spot and clitoral augmentation with PRP injections) vaginal dryness, and other genitourinary syndrome of menopause, early-grade stress urinary incontinence (SUI), lichen sclerosis, post-pelvic radiation,

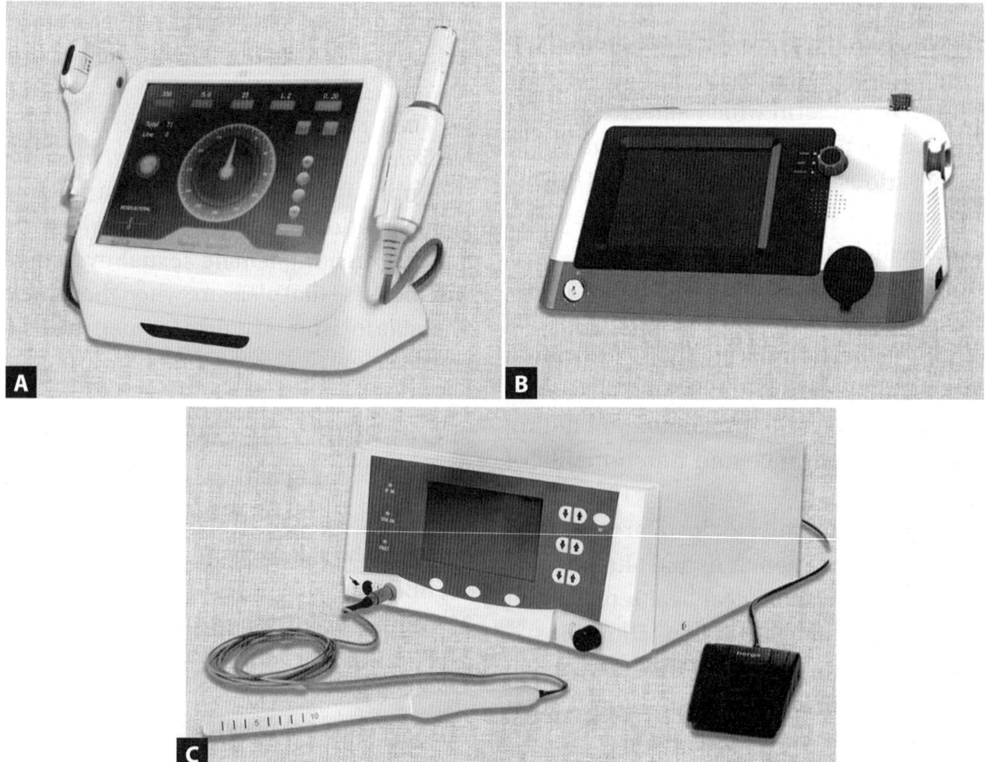

Figs. 1A to C: Energy-based devices.

BOX 1: Advantages of energy-based devices (EBDs).

- Good short-term safety profile
- Outpatient department (OPD) procedures
- Less time-consuming—"coffee break procedures"
- Noninvasive/minimally invasive
- Well-tolerated
- Minimal or no complications
- Bypasses risk of surgery—Anesthesia
- No or minimal downtime to resume routine life
- Faster recovery

BOX 2: Evaluation and consent for the use of energy-based devices (EBDs)

- Patient selection: Define indication—GSM, laxity, SUI
- Detailed medical history
- *History of previous such treatments:*
  - Examination
  - *SUI:* Grade, laxity, POP-Q
  - Rule out local infection, pregnancy, malignancy/premalignant lesions
  - Counseling and detailed discussion
  - Explain mechanism of action of EBD device to be used
  - Desired and expected effects
  - Explain it is a new modality of treatment, used for specific complaints, it is not yet included as standard treatment modality as compared to the gynecological surgical procedures
  - Variability of outcome including suboptimal OR no result
  - Possible risks—complications
  - Recurrence of symptoms and need of repeat session
  - Alternative treatment options
- *Documentation:* Written informed CONSENT

(GSM: genitourinary syndrome of menopause; POP-Q: pelvic organ prolapse quantification; SUI: stress urinary incontinence)

BOX 3: Probable indications for energy-based devices (EBDs) use.

- Atrophic vaginitis
- Vaginal laxity (postpartum/postmenopausal)
- Vulval laxity
- Chronic vaginal infections
- Mild–moderate stress urinary incontinence
- Sexual problems related to postpartum/postmenopausal laxity
- Genitourinary syndrome of menopause (GSM) (vaginal burning, itching, dryness, dyspareunia, dysuria)

prolapse, vesicovaginal fistulae, vulvovaginal rejuvenation, and contour restoration.[20-26]

The basis of therapeutic use of PRP is the high concentrations of cytokines, chemokines, growth factors, angiogenesis factors, differentiation factors in the alpha and delta granules of the tiny platelets. These bioactive molecules trigger the paracrine communication in target cells to promote wound healing as well as augmentation of normal tissues **(Box 6)**.

Vulvovaginal rejuvenation by PRP can be used to improve negative image of genitalia in laxity, scars, SUI, fistulae, and hypoestrogenic status. No wonder it is labeled as liquid gold or even elixir for the benefits it confers.

The PRP mechanisms for healing and regeneration are given in **Box 7**.

BOX 4: Contraindications for use of EBDs.

- Pregnancy
- Premalignant conditions and malignancy
- Active infection—urinary/vaginal/vulval
- Uncontrolled diabetes
- Chronic steroid therapy
- Patients under medications which cause photosensitivity
- Known collagen diseases
- Herpes infection
- Poor response in smokers

BOX 5: Minimally invasive procedures.

- Use of platelet-rich plasma (PRP)
- Regenerative medicine—use of stem cells
- Fillers—autologous fat fillers, hyaluronic acid
- Botox
- Threads
- Threads

BOX 6: Important bioactive molecules in platelet granules.

Growth factors:
- PDGF
- Insulin-like growth factor
- VEGF
- Platelet-derived angiogenic factor
- TGF-beta
- Fibroblast growth factor
- Epidermal growth factor
- Connective tissue growth factor
- Interleukin-8
- *Cytokines:* Fibronectin, vitronectin, sphingosine 1-phosphate

(PDGF: platelet-derived growth factor; TGF-beta: transforming growth factor-beta; VEGF: vascular endothelial growth factor)

BOX 7: Platelet-rich plasma (PRP) mechanisms for healing and regeneration.

- Tissue necrosis resolution
- Chemotaxis
- Cell regeneration
- Cell proliferation and migration
- Extracellular matrix synthesis
- Remodeling
- Angiogenesis
- Epithelialization

Application of Stem Cells and Regenerative Medicine in Gynecology

With progresses in understanding molecular biology, there has been a promising development in the therapeutic applications of autologous stem cells. Adult stem cells commonly derived from adipose tissues [adipose-derived stem cells (ADSCs)] and bone marrow [mesenchymal stem cells (MSCs)] serve as next level to PRP rejuvenation as evident by the growing number of publications and clinical trials exploring their use.

Fillers in Cosmetic Gynecology

Fillers, also known as injectable implants, are popularly used as lip, facial, or wrinkle fillers to create a smoother and/or fuller appearance in the face, including nasolabial folds (the lines extending from the sides of the nose to the edges of the mouth), cheeks, chin, lips, and back of the hands. As an extension to this, it finds application to achieve contouring of female genitalia in case of fat loss or displacement after childbirth and menopause. The flaccid or slack labia and less moist and dry labia are prone to repeated infection as well as lack youthful appearance.

Autologous fillers like fat fillers, PRP gel, or extrinsic fillers like hyaluronic acid apart from contour restoration and aesthetic enhancement of labia majora and mons, also helps for hydrating the skin and mucosa, restoring elasticity, and has a cushioning effect.

The result is not permanent and its longevity depends on nature of filler used and underlying tissue structure.

With improved techniques and improved safety in fat harvesting, the fat fillers are regaining popularity for such purpose.[27,28]

Botox

Therapeutic use of botulinum toxin (heat labile neurotoxin derived from *Clostridium botulinum*) is common in facial aesthetics for abolishing wrinkles to have younger appearance. Although in gynecology it has no cosmetic application, it is used for treating severe grade of vaginismus refractory to first line management. Botulinum toxin acts as a neuromodulator when injected in therapeutic doses and achieves temporary flaccid paralysis of muscles surrounding outer third of vagina by blocking presynaptic release of the acetylcholine at the neuromuscular junction.[29]

Carboxytherapy

Carboxytherapy, also known as carbon dioxide therapy, has therapeutic effect in wound healing and tissue rejuvenation

The mechanism of action is whenever CO_2 gas is injected in tissues carbonic acid is formed as injected gas immediately dissolves in the water, resulting in acidic pH at local tissue. Reduced pH results in Bohr effect which results in improved microcirculation, angiogenesis, vasodilation, weak oxyhemoglobin bond and, as a consequence, better tissue oxygenation.[30]

Carboxytherapy is an effective, acceptable, and safe treatment option for genitourinary syndrome of menopause due to specific effects of CO_2 and the absence of a proliferative influence on the endometrium and mammary glands. This implies the possibility of using carboxytherapy as an alternative treatment modality in patients who are unable to undergo hormone replacement therapy and other treatments.[31] It has also been used for cellulite treatment in women for contouring.[32]

Thread Lifts

Popular in facial aesthetic for fine lines, wrinkles, and facelift, use of biodegradable PDO (polydioxanone) threads is nonsurgical minimally invasive procedure that immediately lifts and tightens the skin while stimulating fibroblast activation and collagen production. Threads could be mono, screw, and cog threads and each serve a different purpose such as stimulating collagen production or lifting or to bring back volume in sagging portion.

Using thread lift sutures offers either as an alternative to or in conjunction with filler treatment, neuromodulator treatment, or surgical procedures.[33]

By placing barbed sutures along specific trajectories provides lifting effect to sunken and wrinkled skin of labia majora or mons pubis.

Effects are temporary as these threads dissolve in 6 months although effect can be seen for 12–24 months.

Implementing absorbable sutures have fewer adverse events, although the long-term efficacy of these procedures remains unclear.[34,35]

Few ongoing research are conducted for the use of such threads longitudinally in submucosal planes in lax vagina to induce collagen synthesis and firming effect thereby.

Surgical Procedures

Cosmetic gynecology surgical procedures are given in **Box 8**.

BOX 8: Cosmetic gynecology surgical procedures.

- Labia minoraplasty
- Hymenoplasty
- Clitoral hood reduction (hoodoplasty)
- Labia majoraplasty
- Monsplasty (pubic lift)
- Vaginoplasty
- Perineoplasty

Recently, there is surge in female genital cosmetic surgery (FGCS) with steady rise every year. Although it can provide positive and healthy results in needy patients, the majority of the rise in these surgeries is due to the effect of pornography and internet sensationalization by various commercial promotion by the medical industry. Scholars have criticized the tendency of depicting a fixed (normal) image of female genital aesthetic and to drive the need to alter, modify, or enhance the female genitals through surgery to fit with this image.[36]

Labia minoraplasty, first described as an aesthetic motive in the plastic surgery literature by Hodgkinson and Hait, is one of the most common FGCS and various techniques, such as deepithelization, linear resection, wedge resection, z-plasty or w-plasty incisions, which have been described for the excessively protruding or asymmetry of labia minora.[5]

While subjecting a woman for labiaplasty, principles of plastic microsurgery with preservation of blood supply must be followed meticulously.

It must be noted that hairless vulvas that have no protruding labia minora have been increasingly emphasized in Western culture and media[37] and not a necessarily universally acceptable beauty standard.

It is claimed that this procedure is demanded for undesired appearance, difficulty in wearing tight clothes, and obstructive tissue causing pain or discomfort during sexual intercourse. Labiaplasty questioned in view of function of labia minora for avoiding introital dryness with its rich sebaceous glands content as well as sensation during sexual stimulus.[38]

Labia minora reduction can result in protruding clitoris deformity (mini-penis deformity) which is corrected by simultaneous clitoral hoodoplasty surgery.[39]

Clitoral hood reduction can also be performed in isolation for improving sexual arousal by revealing more of a clitoris.[40]

Apart from fillers for labia majora augmentation, surgical lift can be performed for sagging labia majora.

Hymenoplasty is a popularly demanded procedure for reconstruction of torn hymen. The purpose is majorly sociocultural with a virginity taboo and surgery itself questioned on ethical grounds.

Monsplasty (pubic lift) is performed for sagging mons pubis and it is combined with surgical abdominoplasty surgery. Liposuction at mons and surrounding area also can serve the purpose in few cases.

Vaginoplasty and perineoplasty surgeries are performed for achieving tightening effect and has unclear boundary with the surgical pelvic floor repair procedures.

To provide recent update, American College of Obstetricians and Gynecologists (ACOG) Committee Opinion, Number 795, January 2020 (replacing committee opinion 378 September, 2007) provides clear alert regarding performing such surgeries.[41]

It has been mentioned that it must be informed to the patients that "such surgeries are not medically indicated, pose substantial risk, and their safety and effectiveness have not been established". Woman should be informed about the lack of high-quality data that supports the effectiveness of FCGS and counseled about their potential complications including pain, bleeding, infection, scarring, adhesions, altered sensations, dyspareunia and need for reoperation.

It is highly recommended to have detailed psychological assessment of a woman about firmness of decision of FCGS, explaining pros and cons, to decide that decision is not taken on the spur of the moment or at request of someone else and she must be made aware of the wide range of normal variations in the appearance of female genitalia.[42-44]

CONCLUSION

Cosmetic gynecology, a field which has garnered substantial attention over recent years, lacks a universally accepted nomenclature.[45]

Although it is becoming popular among society in patients as well as physicians, following facts must be taken into account:

- There are no set standards in treatment protocols in either modality.
- No uniformity in protocols
- There is currently no formal guidelines for the use of cosmetic/aesthetic genital alterations.
- No adequate randomized controlled trial (RCT) conducted
- Limited data on long-term safety or complications

- Currently there is no consensus of the optimal treatment duration and subsequent sessions.
- Many terms such as "vaginal laxity", "vaginal rejuvenation" are undefined terms and condition and more of a marketing terms rather than medical nomenclature.

Hence even when there are many promising results for improving many QoL concerns related to female genital health and sexual function, the quality data for long-term safety and/or complication is lacking and these treatments should be executed judiciously, defining the indications clearly, with adequate discussion and counseling with patients.

More studies and RCTs are required to have set protocols aiming at safety of this noble intent treatments.

REFERENCES

1. Tadinac M. Why do we all want to be young and beautiful (and women especially)? From the evolutionary psychological perspective. Acta Clin Croat. 2010;49(4):501-8.
2. Jonason PK, Raulston T, Rotolo A. More than just a pretty face and a hot body: multiple cues in mate-choice. J Soc Psychol. 2012;152(2):174-84.
3. Rhodes G. The evolutionary psychology of facial beauty. Annu Rev Psychol. 2006;57:199-226.
4. Singh D. Universal allure of the hourglass figure: an evolutionary theory of female physical attractiveness. Clin Plast Surg. 2006;33(3):359-70.
5. Güneş A, Alinsod RM. A mini-review of aesthetic gynecology and leading gynecology associations' approaches to this issue. Turk J Obstet Gynecol. 2018;15(2):105-111.
6. Mishra PC, Mihai D, Khurana N, Jabbar M. Cosmetic Gynecology: Present and Future Perspectives. Int J Innov Res Med Sci. 2021;6:845-53.
7. Bazarganipour F, Ziaei S, Montazeri A, Foroozanfard F, Kazemnejad A, Faghihzadeh S. Body image satisfaction and self-esteem status among the patients with polycystic ovary syndrome. Iran J Reprod Med. 2013;11(10):829-36.
8. Alur-Gupta S, Chemerinski A, Liu C, Lipson J, Allison K, Sammel MD, et al. Body-image distress is increased in women with polycystic ovary syndrome and mediates depression and anxiety. Fertil Steril. 2019;112(5):930-38.e1.
9. Scaruffi E, Franzoi IG, Civilotti C, Guglielmucci F, La Marca L, Tomelini M, et al. Body image, personality profiles and alexithymia in patients with polycystic ovary syndrome (PCOS). J Psychosom Obstet Gynaecol. 2019;40(4):294-303.
10. Loh HH, Yee A, Loh HS, Kanagasundram S, Francis B, Lim LL. Sexual dysfunction in polycystic ovary syndrome: a systematic review and meta-analysis. Hormones (Athens). 2020;19(3):413-23.
11. Fliegner M, Richter-Appelt H, Krupp K, Brunner F. Sexual Function and Socio-Sexual Difficulties in Women with Polycystic Ovary Syndrome (PCOS). Geburtshilfe Frauenheilkd. 2019;79(5):498-509.
12. Hodgkinson EL, Smith DM, Wittkowski A. Women's experiences of their pregnancy and postpartum body image: a systematic review and meta-synthesis. BMC Pregnancy Childbirth. 2014;14:330.
13. Fischman SH, Rankin EA, Soeken KL, Lenz ER. Changes in sexual relationships in postpartum couples. J Obstet Gynecol Neonatal Nurs. 1986;15(1):58-63.
14. Lobo RA. Menopause and aging. In: Strauss JF, Barbieri RL (Eds). Yen and Jaffe's Reproductive Endocrinology, 8th edition. Physiology, Pathophysiology, and Clinical Management. Netherlands: Elsevier; 2019. pp. 322-56.e9.
15. Juhász MLW, Korta DZ, Mesinkovska NA. Vaginal Rejuvenation: A Retrospective Review of Lasers and Radiofrequency Devices. Dermatol Surg. 2021;47(4):489-94.
16. Jorge Alberto Elías; Miriam Galich; Gabriela Corin; Pablo Naranjo Garcia; Victor Sivo; Devesa Nestor; Laureano Nuñez Management of Vaginal Atrophy, Vaginal Hyperlaxity and Stress Urinary Incontinence with Intravaginal High-Intensity Focused Ultrasound (HIFU). IJOGR. 2019;6(2):735-65.
17. Arroyo C. Fractional CO_2 laser treatment for vulvovaginal atrophy symptoms and vaginal rejuvenation in perimenopausal women. Int J Womens Health. 2017;9:591-5.
18. US Food and Drug Administration. (2018). Statement from FDA Commissioner Scott Gottlieb, M.D., on efforts to safeguard women's health from deceptive health claims and significant risks related to devices marketed for use in medical procedures for "vaginal rejuvenation". Available from: https://www.fda.gov/news-events/press-announcements/statement-fda-commissioner-scott-gottlieb-md-efforts-safeguard-womens-health-deceptive-health-claims [Last accessed April, 2024].
19. Ahluwalia J, Avram MM, Ortiz AE. Lasers and energy-based devices marketed for vaginal rejuvenation: A cross-sectional analysis of the MAUDE database. Lasers Surg Med. 2019;51(8):671-7.
20. Behnia-Willison F, Pour NR, Mohamadi B, Willison N, Rock M, Holten IW, et al. Use of Platelet-rich Plasma for Vulvovaginal Autoimmune Conditions Like Lichen Sclerosus. Plast Reconstr Surg Glob Open. 2016;4(11):e1124.
21. Gorlero F, Glorio M, Lorenzi P, Bruno-Franco M, Mazzei C. New approach in vaginal prolapse repair: mini-invasive surgery associated with application of platelet-rich fibrin. Int Urogynecol J. 2012; 23(6):715-22.
22. Morelli M, Rocca ML, Venturella R, Di Cello A, Del Negro S, Condorelli M, et al. Adjuvant use of platelet gel for

wound breakdown prevention in advanced vulvar cancer surgery: a retrospective study. Int J Gynecol Cancer. 2013;23(8):1490-4.
23. Shirvan MK, Alamdari DH, Ghoreifi A. A novel method for iatrogenic vesicovaginal fistula treatment: autologous platelet rich plasma injection and platelet rich fibrin glue interposition. J Urol. 2013;189(6):2125-9.
24. Samaie Nouroozi A, Alyasin A, Malek Mohammadi A, Mehrdad N, Mousavi SA, Vaezi M, et al. Autologous Platelet-Released Growth Factor and Sexual Dysfunction Amendment: A Pilot Clinical Trial of Successful Improvement Sexual Dysfunction after Pelvic Irradiation. Asian Pac J Cancer Prev. 2019;20(3):817-23.
25. Aguilar P, Hersant B, SidAhmed-Mezi M, Bosc R, Vidal L, Meningaud JP. Novel technique of vulvo-vaginal rejuvenation by lipofilling and injection of combined platelet-rich-plasma and hyaluronic acid: a case-report. Springerplus. 2016;5(1):1184.
26. Kim SH, Park ES, Kim TH. Rejuvenation Using Platelet-rich Plasma and Lipofilling for Vaginal Atrophy and Lichen Sclerosus. J Menopausal Med. 2017;23(1):63-8.
27. Salgado CJ, Tang JC, Desrosiers AE 3rd. Use of dermal fat graft for augmentation of the labia majora. J Plast Reconstr Aesthet Surg. 2012;65(2):267-70.
28. Fasola E, Gazzola R. Labia Majora Augmentation with Hyaluronic Acid Filler: Technique and Results. Aesthet Surg J. 2016;36(10):1155-63.
29. Pacik PT. Vaginismus: review of current concepts and treatment using botox injections, bupivacaine injections, and progressive dilation with the patient under anesthesia. Aesthetic Plast Surg. 2011;35(6):1160-4.
30. Sinozić T, Kovacević J. [Carboxytherapy—supportive therapy in chronic wound treatment]. Acta Med Croatica. 2013;67 Suppl 1:137-41.
31. Saidova AS, Sencha AN, Apolikhina IA. Carboxytherapy of genitourinary syndrome of menopause. Obstet Gynecol. 2020;5:113-21.
32. Pianez LR, Custódio FS, Guidi RM, de Freitas JN, Sant'Ana E. Effectiveness of carboxytherapy in the treatment of cellulite in healthy women: a pilot study. Clin Cosmet Investig Dermatol. 2016;9:183-90.
33. Halepas S, Chen XJ, Ferneini EM. Thread-Lift Sutures: Anatomy, Technique, and Review of Current Literature. J Oral Maxillofac Surg. 2020;78(5):813-20.
34. Tong LX, Rieder EA. Thread-Lifts: A Double-Edged Suture? A Comprehensive Review of the Literature. Dermatol Surg. 2019;45(7):931-40.
35. Agarwal B, Mishra P. Thread lift in aesthetic and regenerative gynecology. In: Jindal P, Malhotra N, Joshi S. (Eds). Aesthetic and Regenerative Gynecology. Singapore: Springer Nature Singapore; 2022.
36. Ashong AC, Batta HE. Sensationalising the female pudenda: an examination of public communication of aesthetic genital surgery. Glob J Health Sci. 2012;5(2):153-65.
37. Brotto LA, Bryce M, Todd N. Psychological Aspects and Approaches. Chapter 13: Female Genital Cosmetic Surgery. Available from: https://plasticsurgerykey.com/chapter-13-female-genital-cosmetic-surgery/ [Last accessed April, 2024].
38. Schober J, Aardsma N, Mayoglou L, Pfaff D, Martín-Alguacil N. Terminal innervation of female genitalia, cutaneous sensory receptors of the epithelium of the labia minora. Clin Anat. 2015;28(3):392-8.
39. Hamori CA. Postoperative clitoral hood deformity after labiaplasty. Aesthet Surg J. 2013;33(7):1030-6.
40. Goodman MP. Female cosmetic genital surgery. Obstet Gynecol. 2009;113(1):154-9.
41. Elective Female Genital Cosmetic Surgery: ACOG Committee Opinion Summary, Number 795. Obstet Gynecol. 2020;135(1):249-50.
42. Iglesia CB, Yurteri-Kaplan L, Alinsod R. Female genital cosmetic surgery: a review of techniques and outcomes. Int Urogynecol J. 2013;24(12):1997-2009.
43. Ostrzenski A. Cosmetic gynecology in the view of evidence-based medicine and ACOG recommendations: a review. Arch Gynecol Obstet. 2011;284(3):617-30.
44. Goodman MP. Female genital cosmetic and plastic surgery: A review. J Sex Med. 2011;8:1813-25.
45. Garcia B, Scheib S, Hallner B, Thompson N, Schiavo J, Peacock L. Cosmetic gynecology—a systematic review and call for standardized outcome measures. Int Urogynecol J. 2020;31(10):1979-95.

Chapter 69

Violence Against Women and Sexual Assault

Minakshi Rohilla, Supriya Kumari, Girija Shankar Mohanty

INTRODUCTION

As per World Health Organization (WHO) "Health is a state of complete physical, mental, and social well-being and not merely the absence of disease or infirmity". Violence against women is defined by UN as "any act of gender-based violence that results in, or is likely to result in, physical, sexual, or mental harm or suffering to women, including threats of such acts, coercion or arbitrary deprivation of liberty, whether occurring in public or in private life". Women are always considered as weaker, and their rights are restricted in this male-dominated society. Every woman has the right to explore her potentials naturally gifted and should have access to personal freedoms. However, gender-based discrimination in the world results in curtailing the care they need and their lifestyles are compromised by violence including in her own home. Workplace violence and new cyber-crimes are on high which poses threat to the security of women. As women constitutes approximately 50% of the whole world and each nations development depends upon women empowerment and its impossible if their rights are ignored and tortured. In 1980, the United Nations Convention on the Eliminations of all forms of Discrimination Against Women (CEDAW) emphasized that all women should have equal opportunities in public life and politically.

TYPES OF VIOLENCE

- *Physical violence:* This kind of violence is defined when one person controls another person's action by using their body or an object.
- *Sexual violence:* This kind of violence is defined as sexual activity without consent.
- *Emotional violence:* This kind of violence is defined when a person is made to feel stupid/worthless by harsh words or some actions.
- *Psychological violence:* This kind of violence is defined when individual is threatening to gain control.
- *Spiritual violence:* This kind of violence is defined when someone's spiritual beliefs are manipulated, dominated, or controlled by others.
- *Cultural violence:* This kind of violence is defined when the practices of her or his culture, religion, or tradition harms someone.
- *Verbal abuse:* This kind of violence is defined when some is harmed due to abusive language spoken or written.
- *Financial abuse:* This kind of violence is defined when financial resources are misutilized without person's consent.
- *Neglect:* This kind of violence is defined when someone does not provide care or assistance which is supposed to be done.
- *Sexual violence against women: Sexual violence, other than husband or intimate partner has been reported by 6% women.*[1] *The real prevalence is much higher owing to the stigma associated with this type of violence.*
- *Intimate partner violence:* Violence by an intimate partner or ex-partner affecting physical, psychological, sexual, or controlling behaviors of a woman.
- *Domestic violence:* Despite various laws, crime against women is increasing year on year basis. Incidence of such activity is so high that it is happening once in every 3 minutes, a dowry death occurs every 77 minutes caused by husband, or relatives and a woman is raped every 29 minutes.

PREVALENCE OF VIOLENCE AGAINST WOMEN AND GIRLS

As per a 2021 report by WHO, globally, approximately 736 million of women have been physically or sexually course that successfully intervene safety of women, reduce unequal gender power associations and address multiple risk factors that trigger violence.

The primary health care support is the entry point for referring women to the appropriate next level of support. However, most of the best practices are supportive of women victims after survival of violence and only very few societies discuss preventive practices. Fish bone analysis for the root cause of crime against women will definitely help to plan and take actions against such causes. Some preventive measures such as 24×7 women helpline number 181 need to be reachable, women only transport system during day or night will help to reduce outdone crimes. It is important for women to avoid any secluded travel routes or ways of stretched areas. Building of shops or parks on such routes should be preferred. Anonymous complaint boxes in every society building with regular supervision and actions can be maintained. Action taken and implementation should be displayed to all public notice boards, social media, and newspapers etc., for public awareness. All women organization should unite under one umbrella for protection and empowerment of women of the country.

CONCLUSION

Violence against women is preventable. While we continue to see women as less equal than men, disrespect and violence against women will continue to happen. If we want to stop this, we need to change our mind. Tender loving care (TLC) from either partner will help in better understanding of each other and reduce the VAW. The health sectors have a crucial role to play, the victims need to be provided with comprehensive health care. To achieve lasting change and to eliminate the root cause of violence, it is important to modify and impose regulations, develop, and implement strategies that indorse gender equality, assign assets to prevention, comeback, and devote in women's rights organizations.

REFERENCES

1. World Health Organization, on behalf of the United Nations Inter-Agency Working Group on Violence Against Women Estimation and Data. (2021). Violence against women prevalence estimates, 2018. Global, regional and national prevalence estimates for intimate partner violence against women and global and regional prevalence estimates for non-partner sexual violence against women. Available from: https://www.who.int/publications/i/item/9789240022256 [Last accessed April, 2024].
2. United Nations Office on Drugs and Crime. (2021). Killings of women and girls by their intimate partner or other family members Global estimates 2020. Available from: https://www.unodc.org/documents/data-and-analysis/statistics/crime/UN_BriefFem_251121.pdf [Last accessed April, 2024].
3. First Post. (2022). Crimes against women rose 15.3% in 2021, Delhi most unsafe: Key takeaways of NCRB report. Available from: https://www.firstpost.com/explainers/explained-crimes-against-women-rose-15-3-in-2021-delhi-most-unsafe-key-takeaways-of-ncrb-report-11144691.html [Last accessed April, 2024].
4. The Wire. (2022). Nearly 20% Increase in Rapes Across India in 2021, Rajasthan Had Highest Cases: NCRB. Available from: https://thewire.in/women/crimes-against-women-rape-cases-india-2021-ncrb-data#:~:text=Rajasthan%20reported%20the%20highest%20number,and%202%2C845%20in%20Uttar%20Pradesh [Last accessed April, 2024].
5. ICRW.org. (2011). International Men and Gender Equality Survey (IMAGES). Available from: https://www.icrw.org/publications/international-men-and-gender-equality-survey-images/ [Last accessed April, 2024].
6. UN Women and UNDP. (2021). COVID-19 Global Gender Response Tracker. Available from: https://wrd.unwomen.org/practice/resources/covid-19-global-gender-response-tracker [Last accessed April, 2024].
7. UN Women. Intensification of efforts to eliminate all forms of violence against women: Report of the Secretary-General; 2020. p. 4.
8. World Health Organization. (2019). RESPECT women: Preventing violence against women. Available from: https://www.who.int/publications/i/item/WHO-RHR-18.19 [Last accessed April, 2024].
9. United Nations, Department of Economic and Social Affairs, Statistics Division. (2020). Achieve gender equality and empower all women and girls. Available from: https://unstats.un.org/sdgs/report/2020/goal-05/ [Last accessed April, 2024].
10. United Nations Economic and Social Affairs. The World's Women 2015, Trends and Statistics 2015. p. 159.

Chapter

70 Blood Transfusion Safety Checklist and Standard Operating Procedures

Alpesh Gandhi

SECTION 1: BLOOD TRANSFUSION SAFETY CHECKLIST

- Indication must be present for blood transfusion. It should never be ordered unless it is worth the risk.
- It is always better to keep crossmatched blood ready for use but do not use without requirement.
- Group and screen samples used for provision of blood in pregnancy should be of <3 days old. Fresh sample is ideal.
- Except emergency, only stored and screened blood should be used.
- Blood of 1st relative/siblings should not be transfused unless in an emergency.
- Whole blood has no indication unless blood components are not available.
- Transfusion of single unit of blood has not much role.
- Patient's past history of (H/O) related to blood transfusion (BT), indication, reaction to or any complication because of BT should be evaluated, if any.
- Blood bag must not be issued in advance and has to be issued only when requested and required.
- All planned blood transfusions have to be performed during daytime before 6 PM. unless it is an emergency.
- Cryoprecipitate is transported with ice packs.
- Platelets should be gently shaking and should not be kept still during transport.
- Blood received from blood bank can be stored in fridge but not in deep freezer.
- Red cells received should be started within 60 minutes of leaving controlled storage and completed in maximum 4 hours.
- Visual inspection of the blood pack is done. Check for any leak, clots, discoloration turbidity, or hemolysis. If any check fails, return the blood to blood bank.
- Informed and written consent for transfusion of blood/blood components to be taken. When transfusion of all or specific blood components is refused by patient or relatives, this should be documented in the patient's clinical records.
- The blood bag should be verified by the attending doctor/nurse. This should include full name of patient, blood group and Rh of both the patient and the donor, name of the component, date of tapping, date of expiry, donor reference number, patient's reference number, date and time of issue, volume etc.
- ABO, rhesus D (RhD), and K (Kell) compatible red cell units, platelets, fresh frozen plasma (FFP), and cryoprecipitate should be transfused.
- Fresh frozen plasma and cryoprecipitate should ideally be of the same group as the recipient. If unavailable and emergency, FFP of a different ABO group is acceptable provided it does not have a high titer of anti-A or anti-B activity.
- Platelet concentrates should ideally be of the same ABO group as the recipient. When platelet concentrates are in short supply, administration of ABO-nonidentical platelets is an acceptable practice. If RhD-positive platelets are transfused to an RhD-negative woman of childbearing potential, anti-D immunoglobulin should be administered.
- Patient's identification has to be verified.
- All aseptic precautions have to be taken. Hands have to be washed and sterile gloves should be worn.
- No. 18 or 20 G intravenous (IV) line on forearm/central line should be taken.
- Blood is administered with special BT administration set with micron filter.
- Air should not be introduced into the administration set or the blood/blood components bag.
- Blood warming is usually not required, but it is good to keep the patient warm. If blood warming is required, use authenticated licensed blood warmer for it.

- Rate and units to be decided as per indication. Blood has to be started at the slow rate of 1 mL/minute.
- Monitor the vitals of the patient on blood/components transfusion. General condition, pulse rate, temperature, blood pressure (BP) and respiratory rate should be monitor at the beginning, after 15 minutes of starting of BT, then regularly at every 30 minutes, at the end of BT and after 1 hour of completion of blood transfusion. It is good to monitor oxygen saturation level, if pulse oximeter is available at the beginning, after 15 minutes of starting of BT and at every 30 minutes.
- Proper hydration is to be maintained when a patient is on BT. Urine output to be monitored.
- Monitoring chart has to be filled as per the measurement of monitoring parameters.
- No drug has to be added to the blood or blood product. Medication has to be given from another access, if required.
- Injection calcium gluconate is required when three or more units are given in a day.
- Signs for any blood transfusion reactions or complications should be watched for.
- Preserve the blood bag with label and BT set for few hours.
- Blood bag label to be stuck in nurses' note.
- Dispose the blood bag and BT set as per Hospital Biomedical Waste Management Guidelines
- When the blood group is unknown, in an extreme situation, red cells of group "O-Ve" can be given (although they may be incompatible for patients with irregular antibodies). In major obstetric hemorrhage, the provision of emergency blood with immediate issue of group O, Rh negative and K negative units, with a switch to group-specific blood as soon as feasible.
- In the case of simple, urticarial-type reactions with no other symptoms or signs, the patient has to be given antihistaminic and the transfusion may be continued at a slower rate may be continued at a slower rate with proper monitoring.
- If patient has an unexpected transfusion reaction, stop transfusion immediately. Call for the help. Check and monitor vital signs and oxygen saturation. Maintain IV access (do not flush existing line, change the IV set, maintain IV access with NS and use new IV line if required). Patient has to be catheterized.
- Check that the right pack has been given to the right patient. Administer therapy appropriate to the adverse event. Inform the responsible blood bank and senior.

Source: RCOG and WHO recommendations.

SECTION 2: INDICATIONS OF BLOOD TRANSFUSION IN OBSTETRICS PATIENTS/PATIENT SELECTION

Decision of blood transfusion has to be taken by a senior doctor/obstetrician.

Antepartum Period

Pregnancy <34 Weeks

- Hb <5 g% even without clinical signs of cardiac failure or hypoxia.
- In 5–7 g% in presence of signs and symptoms (S/S) impending heart failure.

Pregnancy >34 Weeks

- Hb <7 g% even without clinical signs of cardiac failure or hypoxia.
- Patient with severe anemia who is decompensated.

Anemia not due to hematinic deficiency (e.g., hemoglobinopathy or bone marrow failure syndromes). Hematologist should always be consulted.

Acute Hemorrhage

Always indicated if Hb <7 g% or if patient becomes hemodynamically unstable during hemorrhage.

Intrapartum Period

Hb <7 g% (in labor); decision of BT depends on medical history or symptoms.

Postpartum Period

Hb <7 g% (postpartum); decision of BT depends on medical history or symptoms.

Postoperative Period

In hemodynamically stable postoperative surgical patients, transfusion should be considered if Hb ≤8 g/dL or in presence of symptoms of inadequate oxygen delivery.

Patients in the Intensive Care Unit and with Sepsis

In critically ill, normovolemic patients, transfusion is considered with a target of Hb% 9 g/dL.

During the early resuscitative phase of severe sepsis if there is evidence of inadequate oxygen delivery to the tissues, blood transfusion is considered to achieve a target Hb% of 9 g/dL.

Blood transfusion should not be used to assist weaning from mechanical ventilation if the Hb is >7 g/dL.

SECTION 3: MASSIVE TRANSFUSION PROTOCOL

- Massive blood loss may be defined as the loss of 1 blood volume within a 24-hour period or 50% blood volume loss within 3 hours or a rate of loss of 150 mL/minute.
- In patients likely to need massive transfusion, massive hemorrhage and improper resuscitation can result in refractory coagulopathy. This patient needs massive transfusion protocol (MTP).
- Basic principle of MTP is *to treat aggressively, to prevent refractory coagulopathy and begin resuscitation with blood products as soon as possible.* The use of MTPs facilitates rapid availability of components.
- *Rule of four (4):* When there is massive hemorrhage or 1st possible sign of disseminated intra-vascular coagulation (DIC), administer blood products in a ratio of *four unit PRBC; four unit FFP; four unit PLT; four unit cryoprecipitate.*
- FFP should be used for volume expansion so that replacement of clotting factor may be started early. Until FFPs are made available volume expander should be used. Volume expander/FFP is to be administered rapidly over a period of 3–4 hours.
- The replacement therapy is guided by laboratory assessment.
- Platelet count should be maintained above 50,000/u. one unit of platelet (PRC) raises the platelet count by 6,000-10,000/u.
- Fibrinogen level should be maintained above 150 mg/dL.
- 4–8 units of cryoprecipitate generally is sufficient to achieve this goal.
- Maintain Hb above 9.0 g/dL by transfusing red cells.
- Injection tranexamic acid is very useful. It is to be given earlier in time during the bleeding. It can save 10–15% blood loss.

SECTION 4: BLOOD TRANSFUSION PROTOCOL

Blood Sample Collection

- Check that blood has been sent for crossmatching in high-risk obstetric patients and for patients planned for cesarean section (CS).
- Blood for crossmatch has to be sent well in advance, prior to the surgery, if planned.
- Sample collection has to be done in ethylenediaminetetraacetic acid (EDTA) and plain bulb.
- Fresh sample is ideal.

Details have to be Entered on Forms and Sample

- Blood requisition form has to be filled in and signed by a responsible doctor.
- Full name has to be written in readable letters and clearly on the blood sample with medical record depatment (MRD) number.
- Old report of blood group, Rh type, and cross match with diagnosis has to be mentioned in the form.
- Hb% and reason for transfusion or keeping cross-matched samples ready has to be written.
- The date and time when blood is required has to be mentioned.

Collection of Blood or Blood Products from Blood Bank

- Blood bag must not be issued in advance and has to be issued only when required.
- Patient's relative has to be informed on how the blood or blood products has to be brought.
- *Cryoprecipitate:* Has to be stored at 4–5°C till actual use and should be transported in ice pack.
- Cryoprecipitate is transported with ice packs.
- BT consent form has to be attached.
- Platelets should be agitated and should not be kept still during transport.

Procedure Prior to Blood Transfusion

- All planned blood transfusions have to be performed during daytime before 6 PM unless it is an emergency.
- Patient's identification has to be verified.
- Take informed written consent for blood/blood product transfusion.
- When transfusion of all or specific blood components is refused by patient or relatives, this should be documented in the patient's clinical records. Base line vital parameters—measure and record pulse rate, BP, temperature, respiratory rate.
- Check different types of labels on blood bag for different blood group, but do not rely only on color codes.

Blood group	*Label color*
Blood group A	Yellow
Blood group B	Pink
Blood group AB	White
Blood group O	Blue

- Check and match the details written on the label, on the blood bags and transfusion form issued by blood bank.
  - Patients full name
  - Unit number
  - Indoor patient department (IPD) number
  - Registration number
  - Date of collection, date of issue, and date of expiry
  - Blood group and Rh-typing
  - Human immunodeficiency virus (HIV) testing
- All previous reports of blood group have to be checked and confirmed that they are same.
- The blood bag has to be checked for any clotting, leak, frothing, discoloration, and/or fungus.

Protocol for Starting the Transfusion

- BT has to be started by standard protocol.
- Blood transfusion has to be started within half an hour of issue.
- Warming of blood bag is not done in case of a slow transfusion.
- Blood warming with blood warmer may be indicated for rapid transfusions of large volume.
- A vein on forearm has to be selected.
- All aseptic and antiseptic precautions have to be taken.
- Hands have to be washed.
- Sterile gloves should be worn.
- A 170–260 μm blood filter set/available BT set has to be used for transfusion.
- Blood filter blood set has to be inserted aseptically with no touch technique.
- All air has to be removed from the IV line.
- Methylated spirit has to be applied at the site of cannulation.
- An 18 Gauge IV cannula has to be used and fixed to skin.
- Air must not be introduced into the administration set or the blood/blood components bag.
- Ensure that blood component/blood product which is being transfused should be mixed by gently shaking the product.
- No drug has to be added to the blood or blood product.
- Blood has to be started at the rate of 1 mL/minute.
- Vital parameters have to be checked after 15 minutes.
- If no blood transfusion reaction present, the transfusion rate has to be increased by 4 mL/minute or at the rate decided by the transfusing doctor.
- Total blood/blood product has to be transfused within 4 hours of issuing it.

Notes to be Written on Indoor Case Paper

- Name and designation of the person starting the transfusion.
- Date and time have to be written on the case paper.
- The information written on bag slip/label has to be recorded on transfusion notes.
- Fresh instructions (orders) need to be written for the nurse.
- Monitoring chart has to be filled as per the measurement of monitoring parameters.

Monitoring of Blood Transfusion

- Symptoms of transfusion reaction such as breathlessness, itching, loin tenderness, etc., have to be watched for.
- The following parameters have to be checked after 15 minutes, and if normal, every 30 minutes till 1 hour after completion of blood transfusion.
  - General condition
  - Temperature
  - Pulse rate
  - Blood pressure
  - Respiratory rate
  - Urine output
  - Oxygen saturation is advisable if pulse oximeter is available.
  - Skin rashes and bleeding from any site has to be noted.
  - Color of urine changes to pink in case of transfusion reaction
  - Oozing from wound, surgical site, and IV sites should be checked.

Completion of Blood Transfusion

- Following completion of the transfusion, post-transfusion information has to be documented on the patient's chart.
- Blood bag label has to be removed and stuck to the blood transfusion notes after transfusion completion and not initially.

Discarding Empty Bags

- It has to be discarded as per the protocol and act for hospital biomedical waste.

- Bag has to be discarded in red bag.
- IV cannula has to be discarded after cutting in sharps (nonpuncture containers).

Concurrent Fluids and Medications

- Medication has to be given from another access if required.
- The blood bags should not be punctured with needles, except for insertion of the blood set.

Change of Intravenous Transfusion Set

- Transfusion set has to be changed after four consecutive units of red blood cell transfusion.
- Confirm that set has not been occluded.
- IV set has to be changed at least once every 24 hours.

When Blood is not Used?

- Blood bags have to be issued only when required immediately for transfusion.
- Unused blood bags must not be left in department's refrigerators.
- Do not waste blood and blood products.
- Blood bag out of temperature-controlled storage for >30 minutes should not be used.
- Unused blood and blood product bag(s) should be discarded in yellow bags (blood bank never takes bags back).

Blood Transfusion Reaction

- Transfusion has to be stopped immediately.
- CALL FOR HELP.
- Urgent/hasty action has to be taken due to possible life-threatening nature of acute transfusion reactions.
- Senior doctor on duty has to be informed whenever reaction occurs.
- Any adverse reaction to the transfusion of blood/blood components has to be reported to blood bank personnel immediately by the unit concerned/responsible doctor.
- Blood unit and administration set has to be removed.
- The intravenous cannula has to be left in place and a new administration set should be put.
- Intravenous saline infusion has to be started, so that urine output is maintained at 100 mL/hour.
- Injection chlorpheniramine 10–20 mg and injection hydrocortisone 100 mg have to be given intravenously.
- Injection adrenaline 0.5 mL 1:1,000 dilution has to be given intramuscularly in case of anaphylactic shock.
- It has to be repeated after 15 minutes if there is no significant response. (Intramuscular adrenaline is the first line treatment for anaphylaxis, with intravenous adrenaline reserved for unresponsive anaphylaxis or circulatory collapse).
- Give injection frusemide 20 mg IV stat if patient develops circulatory over loading or transfusion related acute lung injury (TRALI).
- In the case of simple, urticarial-type reactions with no other symptoms or signs, the patient has to be given antihistaminic and the transfusion may be continued at a slower rate.
- Diagnostic symptom(s) and sign(s) should be recorded.

Blood Transfusion Reaction Information

- Transfusion reaction form has to be obtained from the blood bank.
- All columns in the transfusion reaction form have to be filled in.
- Blood unit and administration set should be removed and sent to the blood bank.
- Blood samples should be taken from the other arm and sent to blood bank and other respective laboratory for the following tests:
    - Repeat group/cross match/Direct coomb's test (DCT) (goes to blood bank) coagulation studies, biochemistry—renal and liver function tests and first voided urine.

Checklist after Transfusion Reaction

- Senior has to be informed immediately.
- Patient has to be catheterized.
- Blood sample and urine sample has to be obtained.
- Vitals such as pulse rate, blood pressure, and SpO_2 saturation should be checked.
- Respective blood bank has to be informed with all required procedures.

Monitoring Parameters

- Dangerous if values rose as below. The following parameters has to be checked:
    - Temperature 21.5°C above baseline,
    - Hypotension/shock or hypertension, tachycardia, tachypnoea, wheeze, stridor, rigors or chills,
    - Nausea, vomiting or pain (local, chest, back) bleeding, and skin rash.

Blood Transfusion Notes

- Date
- Time
- Name
- MRD number
- Blood/component transfusion number
- Patient's blood group and Rh-typing
- Bag details
- PRBC/WB/ FFP/RDP/SDP/CRYOPRECIPITATE
- Donor ID number
- Date of testing
- Blood/component is crossmatched and compatibility is confirmed by blood transfusion officer (BTO).

Donor Blood Details

- Donor is checked for HIV, hepatitis B virus (HBV), hepatitis C virus (HCV), malaria parasite, syphilis, and cytomegalovirus (CMV) and found to be negative by BTO.
- MRD number
- Blood slip
- Patient's admission paper
- Details on bag details on slip
- Blood group
- Bag no
- DOC
- DOE

Checked and transfusion started by Dr.........

Transfusion started Time:

Transfusion completed Time:

Name of the Doctor:

Signature of Doctor:

SECTION 5: CONSENT FOR TRANSFUSION OF BLOOD AND BLOOD COMPONENTS BY DR ALPESH GANDHI

Patient Identification

Diagnosis: *Blood Group:*

Name of patient : ....

Sex: .... Age: .... Registration No.: ....

Attending doctor: ....

My/the patient's doctor .... has advised me/the patient for the transfusion of blood/blood products/red cells/platelets/fresh frozen plasma/cryoprecipitate and/or .... (blood products). The doctor has explained that I/the patient have following medical condition for which I/the patient need a transfusion of.. ....

The doctor has explained the benefits that are expected from my/the patient being transfused and as well, the risks associated with transfusion.

I/we understand that there is a small possibility that transfusion of blood/blood products can be incompatible with my/ the patient's body and can result in transfusion reactions such as fever, rashes, shortness of breath, shock, others, and on rare occasion death.

I/we understand that allergic reactions to blood/blood products like itching and fever are common but can be treated and may not always require stopping the transfusion.

I/we understand that in spite of careful screening according to national regulations there could be rare instances of acquiring life-threatening infections such as HIV I and II, Hepatitis B, Hepatitis C and other viruses/diseases which are yet unknown for which screening tests do not exist and which may not be recognized as an infection for many months or years.

The doctor has explained to me about the risks of not having the blood and/or blood products. I have had an opportunity to ask questions regarding transfusion of blood/blood products for myself/for the patient.

With my signature I give consent to administer blood/blood products for myself or for the patient. I agree this informed consent may serve for consent to give additional necessary blood/blood products till the end of this hospitalization or for the complete course of this illness if have been advised the future need for transfusion of blood/blood products.

I/we have read the above writing/have been explained in the language .... which I understand, about the need of the transfusion, its benefits, costs, and risks associated and other alternative managements and I/we hereby give my/our full valid consent for the transfusion.

I have understood the aforesaid and I am giving my consent willingly with a sound mental state without any coercion, undue influence, fraud, misrepresentation, or mistake of facts.

Date: .... Patient's signature: ....

Date: .... Relative's signature: ....

Consent of the Patient Surrogate

The patient is unable to consent because .... and hence I, ... (name and relationship with the patient) therefore give my consent on behalf of the patient after discussion with the doctor for transfusion.

In Case of Life-threatening Condition

Because of life-threatening medical condition, I have not provided the patient/relative with information sufficient to be considered Informed Consent and I have proceeded with ordering blood/blood products to be administered in sufficient quantity to alter, improve, or reserve a life-threatening medical condition.

Date: .... Doctor's signature: ....

Date: .... Relative's signature: ....

	Name	*Signature*	*Date*	*Time*
Patient/patient surrogate				

SECTION 6: MONITORING CHART DURING TRANSFUSION OF BLOOD OR COMPONENTS

Full name of the patient:

Age:

Indication:

Blood/type of components: Whole blood/PCV/FFP/platelets/cryoprecipitate

Unit number:

Patient reference number:

Donor's reference number:

Blood group and Rh (BGRh) A+Ve/B+Ve/AB+Ve/O+Ve/A-Ve/B-Ve/AB-Ve/O-Ve

Date:

Time:

Starting time of transfusion:	9.00 AM (for an example) (Please write exact time)		
Monitoring time	*Monitoring parameters*	*Findings/values*	Any complaints
9.00 AM			
	General condition		
	Temperature		
	Pulse rate		
	Blood pressure		
	Respiratory rate		
	Oxygen saturation		
	Urine output and color Any other C/O or reaction S/S		
9.15 AM			
	General condition		
	Temperature		
	Pulse rate		
	Blood pressure		
	Respiratory rate		
	Oxygen saturation		
	Urine output and color Any other C/O or reaction S/S		
9.45 AM and at every 30 minutes			
	General condition		
	Temperature		
	Pulse rate		
	Blood pressure		
	Respiratory rate		
	Oxygen saturation		
	Urine output and color Any other C/O or reaction S/S		
At the end of BT Or beginning of the new BT			
	General condition		
	Temperature		
	Pulse rate		
	Blood pressure		
	Respiratory rate		
	Oxygen saturation		
	Urine output and color Any other C/O or reaction S/S		
After 15 minutes of starting of 2nd BT			
	Same monitoring		

After 1 hour of completion of BT	
	General condition
	Temperature
	Pulse rate
	Blood pressure
	Respiratory rate
	Oxygen saturation
	Urine output and color Any other C/O or reaction S/S
Transfusion completion time:	
Comments at the completion of BT	
Name of nurse:	
Signature:	
Date:	
Time:	
Name of doctor:	
Signature:	
Date:	
Time:	

Index

Page numbers followed by *b* refer to box, *f* refer to figure, *fc* refer to flowchart, and *t* refer to table.

B

C

D

E

H

L

M

N

O

Q

R

S

T

U

V

W

X

Y

Z